ADVANCED ORGANIC CHEMISTRY

ADVANCED ORGANIC CHEMISTRY

REACTIONS, MECHANISMS, AND STRUCTURE

FOURTH EDITION

Jerry March

Professor of Chemistry
Adelphi University

A Wiley-Interscience Publication

JOHN WILEY & SONS

New York • Chichester • Brisbane • Toronto • Singapore

In recognition of the importance of preserving what has been written, it is a policy of John Wiley & Sons, Inc., to have books of enduring value published in the United States printed on acid-free paper, and we exert our best efforts to that end.

Library of Congress Cataloging in Publication Data:
March, Jerry, 1929–
    Advanced organic chemistry : reactions, mechanisms, and structure
/ Jerry March. — 4th ed.
        p.   cm.

    "A Wiley-Interscience publication."
    Includes bibliographical references and indexes.
    ISBN 0-471-60180-2 (alk. paper)
    1. Chemistry. Organic.   I. Title.

    QD251.2.M37   1992
    547—dc20   92-728
                                                                     CIP

Printed in the United States of America

10   9   8   7   6   5   4   3

This book is dedicated to the nearly 20,000 scientists whose names are listed in the Author Index, and to my wife, Beverly, and our children, Gale, David, and June

PREFACE

Knowledge of organic chemistry continues to move ahead on many fronts. New journals continue to appear and older ones increase in frequency of publication and/or in number of papers published. This fourth edition of *Advanced Organic Chemistry* has been thoroughly revised to reflect this growth. Every topic retained from the third edition has been brought up to date. Changes, ranging from minor to extensive, have been made on virtually every page of the third edition. More than 5000 references have been added. However, no changes were made in the organization: The structure of the fourth edition is essentially the same as that of the second and the third. Like the first three editions, the fourth is intended to be a textbook for a course in advanced organic chemistry taken by students who have had the standard undergraduate organic and physical chemistry courses.

I have attempted to give equal weight to the three fundamental aspects of the study of organic chemistry: reactions, mechanisms, and structure. A student who has completed a course based on this book should be able to approach the literature directly, with a sound knowledge of modern basic organic chemistry. I have treated lightly or not at all the major special areas of organic chemistry: terpenes, carbohydrates, proteins, polymerization and electrochemical reactions, steroids, etc. It is my opinion that these topics are best approached after the first year of graduate study, when the fundamentals have been mastered, either in advanced courses, or directly, by consulting the many excellent books and review articles available on these subjects.

The organization is based on reaction types, so the student can be shown that despite the large number of organic reactions, a relatively few principles suffice to explain nearly all of them. Accordingly, the reactions–mechanisms section of this book (Part 2) is divided into 10 chapters, each concerned with a different type of reaction. In the first part of each chapter the appropriate basic mechanisms are discussed along with considerations of reactivity and orientation, while the second part consists of numbered sections devoted to individual reactions, where the scope and the mechanism of each reaction are discussed. I have used numbered sections for the reactions, because I have found that students learn better when they are presented with clear outlines (for a further discussion of the arrangement of Part 2, see pp. 287–288). Since the methods for the preparation of individual classes of compounds (e.g., ketones, nitriles, etc.) are not treated all in one place, an index has been provided (Appendix B) by use of which all methods for the preparation of a given type of compound will be found. For each reaction, a list of *Organic Syntheses* references is given. Thus for most reactions the student can consult actual examples in *Organic Syntheses*.

The structure of organic compounds is discussed in the first five chapters of Part 1. This section provides a necessary background for understanding mechanisms and is also important in its own right. The discussion begins with chemical bonding and includes a chapter on stereochemistry. There follow two chapters on reaction mechanisms in general, one for ordinary reactions and the other for photochemical reactions. Part 1 concludes with two more chapters that give further background to the study of mechanisms.

In addition to reactions, mechanisms, and structure, the student should have some familiarity with the literature of organic chemistry. A chapter devoted to this topic has been placed in Appendix A, though many teachers may wish to cover this material at the beginning of the course.

In the third edition I included the new IUPAC names for organic transformations. Since then the rules have been broadened to cover additional cases; hence more such names are given in this edition. Furthermore, IUPAC has now published a new system for designating reaction mechanisms (see p. 290), and I now include some of the simpler of these new designations.

In treating a subject as broad as the basic structures, reactions, and mechanisms of organic chemistry, it is obviously not possible to cover each topic in great depth. Nor would this be desirable even if possible. Nevertheless, students will often wish to pursue individual topics further. An effort has therefore been made to guide the reader to pertinent review articles and books published since about 1965. In this respect, this book is intended to be a guide to the secondary literature (since about 1965) of the areas it covers. Furthermore, in a graduate course, students should be encouraged to consult primary sources. To this end, more than 15,000 references to original papers have been included.

Although basically designed for a one-year course on the graduate level, this book can also be used in advanced undergraduate courses as long as they are preceded by one-year courses in organic and physical chemistry. It can also be adapted, by the omission of a large part of its contents, to a one-semester course. Indeed, even for a one-year course, more is included than can be conveniently covered. Many individual sections can be easily omitted without disturbing continuity.

The reader will observe that this text contains much material that is included in first-year organic and physical chemistry courses, though in most cases it goes more deeply into each subject and, of course, provides references, which first-year texts do not. It has been my experience that students who have completed the first-year courses often have a hazy recollection of the material and greatly profit from a re-presentation of the material if it is organized in a different way. It is hoped that the organization of the material on reactions and mechanisms will greatly aid the memory and the understanding. In any given course the teacher may want to omit some chapters because the students already have an adequate knowledge of the material, or because there are other graduate courses that cover the areas more thoroughly. Chapters 1, 4, and 7 especially may fall into one of these categories.

Although this is a textbook, it has been designed to have reference value also. Students preparing for qualifying examinations and practicing organic chemists will find that Part 2 contains a survey of what is known about the mechanism and scope of about 580 reactions, arranged in an orderly manner based on reaction type and on which bonds are broken and formed. Also valuable for reference purposes are the previously mentioned lists of reactions classified by type of compound prepared (Appendix B) and of all of the *Organic Syntheses* references to each reaction.

Anyone who writes a book such as this is faced with the question of which units to use, in cases where international rules mandate one system, but published papers use another. Two instances are the units for energies and for bond distances. For energies, IUPAC mandates joules, and many journals do use this unit exclusively. However, organic chemists who publish in United States journals overwhelmingly use calories and this situation shows no signs of changing in the near future. Since previous editions of this book have been used extensively both in this country and abroad, I have now adopted the practice of giving virtually all energy values in both calories and joules. The question of units for bond distances is easier to answer. Although IUPAC does not recommend Ångstrom units, nearly all bond distances published in the literature anywhere in the world, whether in organic or in crystallographic journals, are in these units, though a few papers do use picometers. Therefore, I continue to use only Ångstrom units.

I am happy to acknowledge the assistance of chemists who have been kind enough to read portions of the manuscript of one or more of the editions and to send me their exceedingly helpful comments. I wish to thank Professors J. F. Bunnett, A. W. Burgstahler, D. J. Cram, P. de Mayo, E. L. Eliel, R. W. Griffin, Jr., G. S. Hammond, M. Kreevoy, J. Landesberg, S. Moon, G. A. Olah, G. C. Pimentel, W. H. Saunders, Jr., C. G. Swain, R. W. Taft, Jr., W. S. Trahanovsky, N. J. Turro, C. Walling, and R. Wistar, each of whom read one or more chapters of either the first or second editions; B. B. Jarvis and C. A. Bunton, who read the entire manuscript of the second edition; M. P. Doyle, who read the entire manuscript of the third edition; K. B. Wiberg, who offered valuable help in the preparation of the third edition; and S. Biali, R. D. Guthrie, E. A. Halevi, M. M. Kreevoy, T. M. Krygowski, A. G. Pinkus, and L. G. Wade, Jr., who sent comments that were exceedingly helpful. In addition, I wish to thank many of my colleagues at Adelphi University who have rendered assistance in various ways, among them F. Bettelheim, D. Davis, S. Z. Goldberg, J. Landesberg, S. Milstein, S. Moon, D. Opalecky, C. Shopsis, and S. Windwer. Dr. Goldberg rendered exceptionally valuable assistance in the preparation of the indexes. Special thanks are due to the Interscience division of John Wiley & Sons, Dr. Ted Hoffman, and the other editors at Wiley for their fine work in turning the manuscript into the finished book. I am also grateful to those readers who wrote to tell me about errors they discovered in the preceding editions or to make other comments. Such letters are always welcome.

Jerry March
Garden City, New York
November 1991

CONTENTS

BIBLIOGRAPHICAL NOTE

In this book the practices in citing references are slightly different from those prevailing elsewhere. The reader should note:

1. Author's initials are omitted in references. They will be found, however, in the author index.

2. For review articles, both the first and last page numbers are given, so that the reader may form an idea of the length of the article. If reference is made to only a portion of the article, these page numbers are also given.

3. When a journal is available both in Russian and in English, the page numbers of each article are, of course, different. The language of the journal title indicates whether the page number cited is to be found in the Russian or in the English version. For articles which have appeared in *Angewandte Chemie, International Edition in English,* both the English and German page numbers are given.

The following abbreviations are used for three common solvents:

DMF	Dimethylformamide	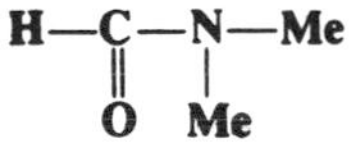 H—C—N—Me with O (double bond) and Me
THF	Tetrahydrofuran	 (ring structure with O)
HMPA	Hexamethylphosphoric triamide	$(Me_2N)_3P{=}O$

PART ONE

This book contains 19 chapters. Chapters 10 to 19, which make up Part 2, are directly concerned with organic reactions and their mechanisms. Chapters 1 to 9 may be thought of as an introduction to Part 2. The first five chapters deal with the structure of organic compounds. In these chapters are discussed the kinds of bonding important in organic chemistry, the three-dimensional structure of organic molecules, and the structure of species in which the valence of carbon is less than 4. Chapters 6 to 9 are concerned with other topics that help to form a background to Part 2: acids and bases, photochemistry, the relationship between structure and reactivity, and a general discussion of mechanisms and the means by which they are determined.

1
LOCALIZED
CHEMICAL BONDING

Localized chemical bonding may be defined as bonding in which the electrons are shared by two and only two nuclei. In Chapter 2 we shall consider *delocalized bonding*, in which electrons are shared by more than two nuclei.

Covalent Bonding[1]

Wave mechanics is based on the fundamental principle that electrons behave as waves (e.g., they can be diffracted) and that consequently a wave equation can be written for them, in the same sense that light waves, sound waves, etc. can be described by wave equations. The equation that serves as a mathematical model for electrons is known as the *Schrödinger equation*, which for a one-electron system is

$$\frac{\partial^2\psi}{\partial x^2} + \frac{\partial^2\psi}{\partial y^2} + \frac{\partial^2\psi}{\partial z^2} + \frac{8\pi^2 m}{h^2}(E - V)\,\psi = 0$$

where m is the mass of the electron, E is its total energy, V is its potential energy, and h is Planck's constant. In physical terms, the function ψ expresses the square root of the probability of finding the electron at any position defined by the coordinates x, y, and z, where the origin is at the nucleus. For systems containing more than one electron the equation is similar but more complicated.

The Schrödinger equation is a differential equation, which means that solutions of it are themselves equations. The solutions, however, are not differential equations, but simple equations for which graphs can be drawn. Such graphs, which are three-dimensional pictures that show the electron density, are called *orbitals* or electron clouds. Most students are familiar with the shapes of the s and p atomic orbitals (Figure 1.1). Note that each p orbital has a *node*—a region in space where the probability of finding the electron is extremely small.[2] Also note that in Figure 1.1 some lobes of the orbitals are labeled + and others −.

[1]The treatment of orbitals given here is necessarily simplified. For much fuller treatments of orbital theory as applied to organic chemistry, see Matthews *Quantum Chemistry of Atoms and Molecules*; Cambridge University Press: Cambridge, 1986; Clark *A Handbook of Computational Chemistry*; Wiley: New York, 1985; Albright; Burdett; Whangbo *Orbital Interactions in Chemistry*; Wiley: New York, 1985; McWeeny *Coulson's Valence*; Oxford University Press: Oxford, 1980; Murrell; Kettle; Tedder, *The Chemical Bond*; Wiley: New York, 1978; Dewar; Dougherty *The PMO Theory of Organic Chemistry*; Plenum: New York, 1975; Zimmerman *Quantum Mechanics for Organic Chemists*; Academic Press: New York, 1975; Borden *Modern Molecular Orbital Theory for Organic Chemists*; Prentice-Hall: Englewood Cliffs, NJ, 1975; Dewar *The Molecular Orbital Theory of Organic Chemistry*; McGraw-Hill: New York, 1969; Liberles *Introduction to Molecular Orbital Theory*; Holt, Rinehart, and Winston: New York, 1966.

[2]When wave-mechanical calculations are made according to the Schrödinger equation, the probability of finding the electron in a node is zero, but this treatment ignores relativistic considerations. When such considerations are applied, Dirac has shown that nodes do have a very small electron density: Powell *J. Chem. Educ.* **1968**, *45*, 558. See also Ellison and Hollingsworth *J. Chem. Educ.* **1976**, *53*, 767; McKelvey *J. Chem. Educ.* **1983**, *60*, 112; Nelson, *J. Chem. Educ.* **1990**, *67*, 643. For a review of relativistic effects on chemical structures in general, see Pyykkö *Chem. Rev.* **1988**, *88*, 563-594.

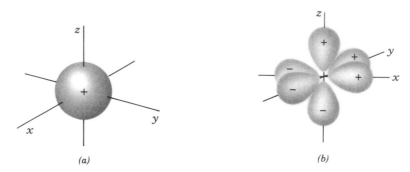

(a)   *(b)*

FIGURE 1.1   (a) the 1s orbital. (b) The three 2p orbitals.

These signs do not refer to positive or negative *charges*, since both lobes of an electron cloud must be negatively charged. They are the signs of the wave function ψ. When two parts of an orbital are separated by a node, ψ always has opposite signs on the two sides of the node. According to the Pauli exclusion principle, no more than two electrons can be present in any orbital, and they must have opposite spins.

Unfortunately, the Schrödinger equation can be solved exactly only for one-electron systems such as the hydrogen atom. If it could be solved exactly for molecules containing two or more electrons,[3] we would have a precise picture of the shape of the orbitals available to each electron (especially for the important ground state) and the energy for each orbital. Since exact solutions are not available, drastic approximations must be made. There are two chief general methods of approximation: the *molecular-orbital* method and the *valence-bond* method.

In the molecular-orbital method, bonding is considered to arise from the overlap of atomic orbitals. When any number of atomic orbitals overlap, they combine to form an equal number of new orbitals, called *molecular orbitals*. Molecular orbitals differ from atomic orbitals in that they are clouds that surround the nuclei of two or more atoms, rather than just one atom. In localized bonding the number of atomic orbitals that overlap is two (each containing one electron), so that two molecular orbitals are generated. One of these, called a *bonding orbital*, has a lower energy than the original atomic orbitals (otherwise a bond would not form), and the other, called an *antibonding orbital*, has a higher energy. Orbitals of lower energy fill first. Since the two original atomic orbitals each held one electron, both of these electrons can now go into the new molecular *bonding* orbital, since any orbital can hold two electrons. The antibonding orbital remains empty in the ground state. The greater the overlap, the stronger the bond, although total overlap is prevented by repulsion between the nuclei. Figure 1.2 shows the bonding and antibonding orbitals that arise by the overlap of two 1s electrons. Note that since the antibonding orbital has a node between the nuclei, there is practically no electron density in that area, so that this orbital cannot be expected to bond very well. Molecular orbitals formed by the overlap of two atomic orbitals when the centers of electron density are on the axis common to the two nuclei are called σ *(sigma)* orbitals, and the bonds are called σ bonds. Corresponding antibonding orbitals are designated σ^*. Sigma orbitals are formed not only by the overlap of two s orbitals, but also by the

---

[3]For a number of simple systems containing two or more electrons, such as the H_2 molecule or the He atom, approximate solutions are available that are so accurate that for practical purposes they are as good as exact solutions. See, for example, Roothaan; Weiss *Rev. Mod. Phys.* **1960**, *32*, 194; Kolos; Roothaan *Rev. Mod. Phys.* **1960**, *32*, 219. For a review, see Clark; Stewart *Q. Rev., Chem. Soc.* **1970**, *24*, 95-118.

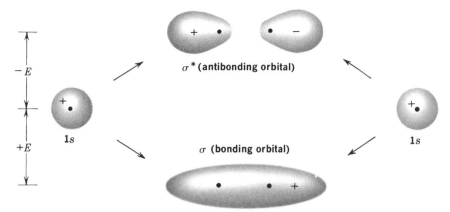

FIGURE 1.2   Overlap of two 1s orbitals gives rise to a σ and a σ* orbital.

overlap of any of the kinds of atomic orbital (s, p, d, or f) whether the same or different, but the two lobes that overlap must have the same sign: a positive s orbital can form a bond only by overlapping with another positive s orbital or with a-positive lobe of a p, d, or f orbital. Any σ orbital, no matter what kind of atomic orbitals it has arisen from, may be represented as approximately ellipsoidal in shape.

Orbitals are frequently designated by their symmetry properties. The σ orbital of hydrogen is often written Ψ_g. The g stands for *gerade*. A gerade orbital is one in which the sign on the orbital does not change when it is inverted through its center of symmetry. The σ* orbital is *ungerade* (designated Ψ_u). An ungerade orbital changes sign when inverted through its center of symmetry.

In molecular-orbital calculations, a wave function is formulated that is a linear combination of the atomic orbitals that have overlapped (this method is often called the *linear combination of atomic orbitals*, or LCAO). Addition of the atomic orbitals gives the bonding molecular orbital:

$$\Psi = c_A\psi_A + c_B\psi_B \tag{1}$$

The functions ψ_A and ψ_B are the functions for the atomic orbitals of atoms A and B, respectively, and c_A and c_B represent weighting factors. Subtraction is also a linear combination:

$$\Psi = c_A\psi_A - c_B\psi_B \tag{2}$$

This gives rise to the antibonding molecular orbital.

In the valence-bond method, a wave equation is written for each of various possible electronic structures that a molecule may have (each of these is called a *canonical form*), and the total Ψ is obtained by summation of as many of these as seem plausible, each with its weighting factor:

$$\Psi = c_1\psi_1 + c_2\psi_2 + \cdots \tag{3}$$

This resembles Eq. (1), but here each ψ represents a wave equation for an imaginary canonical form and each c is the amount contributed to the total picture by that form. For example, a wave function can be written for each of the following canonical forms of the hydrogen molecule:[4]

$$\text{H---H} \quad ^{\ominus}\overline{\text{H}} \ \text{H}^{\oplus} \quad ^{\oplus}\text{H} \ \overline{\text{H}}^{\ominus}$$

Values for c in each method are obtained by solving the equation for various values of each c and choosing the solution of lowest energy. In practice, both methods give similar solutions for molecules that contain only localized electrons, and these are in agreement with the Lewis structures long familiar to the organic chemist. Delocalized systems are considered in Chapter 2.

Multiple Valence

A univalent atom has only one orbital available for bonding. But atoms with a valence of 2 or more must form bonds by using at least two orbitals. An oxygen atom has two half-filled orbitals, giving it a valence of 2. It forms single bonds by the overlap of these with the orbitals of two other atoms. According to the principle of maximum overlap, the other two nuclei should form an angle of 90° with the oxygen nucleus, since the two available orbitals on oxygen are p orbitals, which are perpendicular. Similarly, we should expect that nitrogen, which has three mutually perpendicular p orbitals, would have bond angles of 90° when it forms three single bonds. However, these are not the observed bond angles. The bond angles are,[5] in water, 104°27′, and in ammonia, 106°46′. For alcohols and ethers the angles are even larger (see p. 22). A discussion of this will be deferred to p. 22, but it is important to note that covalent compounds do have definite bond angles. Although the atoms are continuously vibrating, the mean position is the same for each molecule of a given compound.

Hybridization

Consider the case of mercury. Its electronic structure is

$$[\textbf{Xe} \text{ core}]4f^{14}5d^{10}6s^2$$

Although it has no half-filled orbitals, it has a valence of 2 and forms two covalent bonds. We can explain this by imagining that one of the $6s$ electrons is promoted to a vacant $6p$ orbital to give the excited configuration

$$[\textbf{Xe} \text{ core}]4f^{14}5d^{10}6s^16p^1$$

In this state the atom has two half-filled orbitals, but they are not equivalent. If bonding were to occur by the overlap of these orbitals with the orbitals of external atoms, the two bonds would not be equivalent. The bond formed from the $6p$ orbital would be more stable than the one formed from the $6s$ orbital, since a larger amount of overlap is possible with the former. A more stable situation is achieved when, in the course of bond formation, the $6s$ and $6p$ orbitals combine to form two new orbitals that *are* equivalent; these are shown in Figure 1.3.

Since these new orbitals are a mixture of the two orginal orbitals, they are called *hybrid orbitals*. Each is called an *sp* orbital, since a merger of an s and a p orbital was required to

---

[4]In this book a pair of electrons, whether in a bond or unshared, is represented by a straight line.
[5]Bent *Chem. Rev.* **1961**, *61*, 275-311, p. 277.

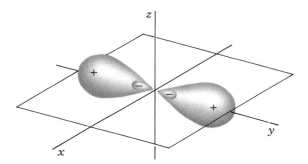

FIGURE 1.3 The two *sp* orbitals formed by mercury.

form it. The *sp* orbitals, each of which consists of a large lobe and a very small one, are atomic orbitals, although they arise only in the bonding process and do not represent a possible structure for the free atom. A mercury atom forms its two bonds by overlapping each of the large lobes shown in Figure 1.3 with an orbital from an external atom. This external orbital may be any of the atomic orbitals previously considered (*s*, *p*, *d*, or *f*) or it may be another hybrid orbital, although only lobes of the same sign can overlap. In any of these cases the molecular orbital that arises is called a σ orbital since it fits our previous definition of a σ orbital.

In general, because of mutual repulsion, equivalent orbitals lie as far away from each other as possible, so the two *sp* orbitals form an angle of 180°. This means that HgCl$_2$, for example, should be a linear molecule (in contrast to H$_2$O), and it is. This kind of hybridization is called *digonal hybridization*. An *sp* hybrid orbital forms a stronger covalent bond than either an *s* or a *p* orbital because it extends out in space in the direction of the other atom's orbital farther than the *s* or the *p* and permits greater overlap. Although it would require energy to promote a 6*s* electron to the 6*p* state, the extra bond energy more than makes up the difference.

Many other kinds of hybridization are possible. Consider boron, which has the electronic configuration

$$1s^2 2s^2 2p^1$$

yet has a valence of 3. Once again we may imagine promotion and hybridization:

$$1s^2 2s^2 2p^1 \xrightarrow{\text{promotion}} 1s^2 2s^1 2p_x^{\ 1} 2p_y^{\ 1} \xrightarrow{\text{hybridization}} 1s^2 (sp^2)^3$$

In this case there are three equivalent hybrid orbitals, each called *sp*^{2} (*trigonal hybridization*). This method of designating hybrid orbitals is perhaps unfortunate since nonhybrid orbitals are designated by single letters, but it must be kept in mind that *each* of the three orbitals is called *sp*^{2}. These orbitals are shown in Figure 1.4. The three axes are all in one plane and point to the corners of an equilateral triangle. This accords with the known structure of BF$_3$, a planar molecule with angles of 120°.

The case of carbon (in forming four single bonds) may be represented as

$$1s^2 2s^2 2p_x^{\ 1} 2p_y^{\ 1} \xrightarrow{\text{promotion}} 1s^2 2s^1 2p_x^{\ 1} 2p_y^{\ 1} 2p_z^{\ 1} \xrightarrow{\text{hybridization}} 1s^2 (sp^3)^4$$

There are four equivalent orbitals, each called *sp*^{3}, which point to the corners of a regular tetrahedron (Figure 1.4). The bond angles of methane would thus be expected to be 109°28′, which is the angle for a regular tetrahedron.

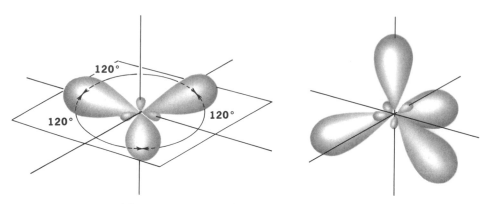

FIGURE 1.4 The three sp^2 and the four sp^3 orbitals.

Although the hybrid orbitals discussed in this section satisfactorily account for most of the physical and chemical properties of the molecules involved, it is necessary to point out that the sp^3 orbitals, for example, stem from only one possible approximate solution of the Schrödinger equation. The s and the three p atomic orbitals can also be combined in many other equally valid ways. As we shall see on p. 12, the four C—H bonds of methane do not always behave as if they are equivalent.

Multiple Bonds

If we consider the ethylene molecule in terms of the molecular-orbital concepts discussed so far, we have each carbon using sp^2 orbitals to form bonds with the three atoms to which it is connected. These sp^2 orbitals arise from hybridization of the $2s^1$, $2p_x{}^1$, and $2p_y{}^1$ electrons of the promoted state shown on p. 7. We may consider that any carbon atom that is bonded to only three different atoms uses sp^2 orbitals for this bonding. Each carbon of ethylene is thus bonded by three σ bonds: one to each hydrogen and one to the other carbon. Each carbon therefore has another electron in the $2p_z$ orbital that is perpendicular to the plane of the sp^2 orbitals. The two parallel $2p_z$ orbitals can overlap sideways to generate two new orbitals, a bonding and an antibonding orbital (Figure 1.5). Of course, in the ground state, both electrons go into the bonding orbital and the antibonding orbital remains vacant. Molecular orbitals formed by the overlap of atomic orbitals whose axes are parallel are called π orbitals if they are bonding and π* if they are antibonding.

In this picture of ethylene, the two orbitals that make up the double bond are not equivalent.[6] The σ orbital is ellipsoidal and symmetrical about the C—C axis. The π orbital is in the shape of two ellipsoids, one above the plane and one below. The plane itself represents a node for the π orbital. In order for the p orbitals to maintain maximum overlap, they must be parallel. This means that free rotation is not possible about the double bond, since the two p orbitals would have to reduce their overlap to allow one H—C—H plane to rotate with respect to the other. The six atoms of a double bond are therefore in a plane

---

[6]The double bond can also be pictured as consisting of two equivalent orbitals, where the centers of electron density point away from the C—C axis. This is the *bent-bond* or *banana-bond* picture. Support for this view is found in Pauling *Theoretical Organic Chemistry, The Kekulé Symposium*; Butterworth: London, 1959, pp. 2-5; Palke *J. Am. Chem. Soc.* **1986**, *108*, 6543. However, most of the literature of organic chemistry is written in terms of the σ–π picture, and in this book we will use it.

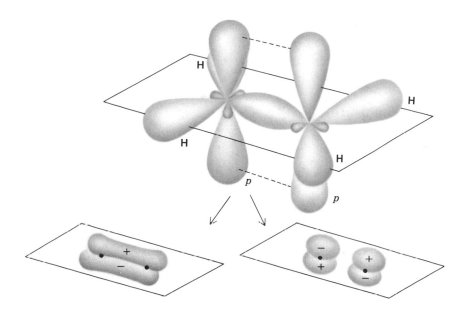

FIGURE 1.5 Overlapping p orbitals form a π and a π^* orbital. The σ orbitals are shown in the upper figure. They are still there in the states represented by the diagrams below, but have been removed from the picture for clarity.

with angles that should be about 120°. Double bonds are shorter than the corresponding single bonds because maximum stability is obtained when the p orbitals overlap as much as possible. Double bonds between carbon and oxygen or nitrogen are similarly represented: they consist of one σ and one π orbital.

In triple-bond compounds, carbon is connected to only two other atoms and hence uses sp hybridization, which means that the four atoms are in a straight line (Figure 1.6).[7] Each carbon has two p orbitals remaining, with one electron in each. These orbitals are perpendicular to each other and to the C—C axis. They overlap in the manner shown in Figure 1.7 to form two π orbitals. A triple bond is thus composed of one σ and two π orbitals. Triple bonds between carbon and nitrogen can be represented in a similar manner.

Double and triple bonds are important only for the first-row elements carbon, nitrogen, and oxygen.[8] For second-row elements multiple bonds are rare and compounds containing

FIGURE 1.6 The σ electrons of acetylene.

[7]For reviews of triple bonds, see Simonetta; Gavezzotti, in Patai *The Chemistry of the Carbon–Carbon Triple Bond*; Wiley: New York, 1978, pp. 1-56; Dale, in Viehe *Acetylenes*; Marcel Dekker: New York, 1969, pp. 3-96.
[8]This statement applies to the representative elements. Multiple bonding is also important for some transition elements. For a review of metal–metal multiple bonds, see Cotton *J. Chem. Educ.* **1983,** *60,* 713-720.

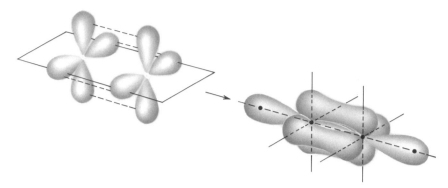

FIGURE 1.7 Overlap of p orbitals in a triple bond. For clarity, the σ orbitals have been removed from the drawing on the left, though they are shown on the right.

them are generally less stable[9] because these elements tend to form weaker π bonds than do the first-row elements.[10] The only ones of any importance at all are C=S bonds, and C=S compounds are generally much less stable than the corresponding C=O compounds (however, see $p\pi-d\pi$ bonding, p. 38). Stable compounds with Si=C and Si=Si bonds are rare, but examples have been reported,[11] including a pair of cis and trans Si=Si isomers.[12]

Photoelectron Spectroscopy

Although the four bonds of methane are equivalent according to most physical and chemical methods of detection (for example, neither the nmr nor the ir spectrum of methane contains peaks that can be attributed to different kinds of C—H bonds), there is one physical technique that shows that the eight valence electrons of methane can be differentiated. In this tech-

[9]For a review of double bonds between carbon and elements other than C, N, S, or O, see Jutzi *Angew. Chem. Int. Ed. Engl.* **1975**, *14*, 232-245 [*Angew. Chem. 87*, 269-283]. For reviews of multiple bonds involving silicon and germanium, see Barrau; Escudié; Satgé *Chem. Rev.* **1990**, *90*, 283-319 (Ge only); Raabe; Michl, in Patai and Rappoport *The Chemistry of Organic Silicon Compounds*, part 2, Wiley: New York, 1989, pp. 1015-1142; *Chem. Rev.* **1985**, *85*, 419-509 (Si only); Wiberg *J. Organomet. Chem.* **1984**, *273*, 141-177 (Si only); Gusel'nikov; Nametkin *Chem. Rev.* **1979**, *79*, 529-577 (Si only). For reviews of C=P and C≡P bonds, see Regitz *Chem. Rev.* **1990**, *90*, 191-213; Appel; Knoll *Adv. Inorg. Chem.* **1989**, *33*, 259-361; Markovski; Romanenko *Tetrahedron* **1989**, *45*, 6019-6090; Regitz; Binger *Angew. Chem. Int. Ed. Engl.* **1988**, *27*, 1484-1508 [*Angew. Chem. 100*, 1541-1565]; Appel; Knoll; Ruppert *Angew. Chem. Int. Ed. Engl.* **1981**, *20*, 731-744 [*Angew. Chem. 93*, 771-784]. For reviews of other second-row double bonds, see West *Angew. Chem. Int. Ed. Engl.* **1987**, *26*, 1201-1211 [*Angew. Chem. 99*, 1231-1241] (Si=Si bonds); Brook; Baines *Adv. Organometal. Chem.* **1986**, *25*, 1-44 (Si=C bonds); Kutney; Turnbull *Chem. Rev.* **1982**, *82*, 333-357 (S=S bonds). For reviews of multiple bonds between heavier elements, see Cowley; Norman *Prog. Inorg. Chem.* **1986**, *34*, 1-63; Cowley *Polyhedron* **1984**, *3*, 389-432; *Acc. Chem. Res.* **1984**, *17*, 386-392. For a theoretical study of multiple bonds to silicon, see Gordon *Mol. Struct. Energ.* **1986**, *1*, 101-148.

[10]For discussions, see Schmidt; Truong; Gordon *J. Am. Chem. Soc.* **1987**, *109*, 5217; Schleyer; Kost *J. Am. Chem. Soc.* **1988**, *110*, 2105.

[11]For Si=C bonds, see Brook; Nyburg; Abdesaken; Gutekunst; Gutekunst; Kallury; Poon; Chang; Wong-Ng *J. Am. Chem. Soc.* **1982**, *104*, 5667; Schaefer *Acc. Chem. Res.* **1982**, *15*, 283; Wiberg; Wagner; Riede; Müller *Organometallics* **1987**, *6*, 32. For Si=Si bonds, see West; Fink; Michl *Science* **1981**, *214*, 1343; Boudjouk; Han; Anderson *J. Am. Chem. Soc.* **1982**, *104*, 4992; Zilm; Grant; Michl; Fink; West *Organometallics* **1983**, *2*, 193; Fink; DeYoung; West; Michl *J. Am. Chem. Soc.* **1983**, *105*, 1070; Fink; Michalczyk; Haller; West; Michl *Organometallics* **1984**, *3*, 793; West *Pure Appl. Chem.* **1984**, *56*, 163-173; Masamune; Eriyama; Kawase *Angew. Chem. Int. Ed. Engl.* **1987**, *26*, 584 [*Angew. Chem. 99*, 601]; Shepherd; Campana; West *Heteroat. Chem.* **1990**, *1*, 1. For an Si=N bond, see Wiberg; Schurz; Reber; Müller *J. Chem. Soc., Chem. Commun.* **1986**, 591.

[12]Michalczyk; West; Michl *J. Am. Chem. Soc.* **1984**, *106*, 821, *Organometallics* **1985**, *4*, 826.

nique, called *photoelectron spectroscopy*,[13] a molecule or free atom is bombarded with vacuum uv radiation, causing an electron to be ejected. The energy of the ejected electron can be measured, and the difference between the energy of the radiation used and that of the ejected electron is the *ionization potential* of that electron. A molecule that contains several electrons of differing energies can lose any one of them as long as its ionization potential is less than the energy of the radiation used (a single molecule loses only one electron; the loss of two electrons by any individual molecule almost never occurs). A photoelectron spectrum therefore consists of a series of bands, each corresponding to an orbital of a different energy. The spectrum gives a direct experimental picture of all the orbitals present, in order of their energies, provided that radiation of sufficiently high energy is used.[14] Broad bands usually correspond to strongly bonding electrons and narrow bands to weakly bonding or nonbonding electrons. A typical spectrum is that of N_2, shown in Figure 1.8.[15] The N_2 molecule has the electronic structure shown in Figure 1.9. The two $2s$ orbitals of the nitrogen atoms combine to give the two orbitals marked 1 (bonding) and 2 (antibonding), while the six $2p$ orbitals combine to give six orbitals, three of which (marked 3, 4, and 5) are bonding. The three antibonding orbitals (not indicated in Figure 1.9) are unoccupied. Electrons ejected from orbital 1 are not found in Figure 1.8 because the ionization potential of these electrons is greater than the energy of the light used (they can be seen when higher-energy light is used). The broad band in Figure 1.8 (the individual peaks within this band are caused by different vibrational levels; see Chapter 7) corresponds to the four electrons in the degenerate orbitals 3 and 4. The triple bond of N_2 is therefore composed of these two orbitals and orbital 1. The bands corresponding to orbitals 2 and 5 are narrow; hence these orbitals contribute little to the bonding and may be regarded as the two unshared pairs of $\overline{N}{\equiv}\overline{N}$. Note that this result is contrary to that expected from a

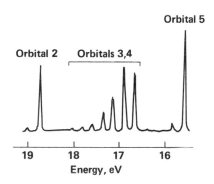

FIGURE 1.8 Photoelectron spectrum of N_2.[15]

[13]Only the briefest description of this subject is given here. For monographs, see Ballard *Photoelectron Spectroscopy and Molecular Orbital Theory*; Wiley: New York, 1978; Rabalais, *Principles of Ultraviolet Photoelectron Spectroscopy*; Wiley: New York, 1977; Baker; Betteridge *Photoelectron Spectroscopy*; Pergamon: Elmsford, NY, 1972; Turner; Baker; Baker; Brundle *High Resolution Molecular Photoelectron Spectroscopy*; Wiley: New York, 1970. For reviews, see Westwood *Chem. Soc. Rev.* **1989**, *18*, 317-345; Carlson *Annu. Rev. Phys. Chem.* **1975**, *26*, 211-233; Baker; Brundle; Thompson *Chem. Soc. Rev.* **1972**, *1*, 355-380; Bock; Mollère *J. Chem. Educ.* **1974**, *51*, 506-514; Bock; Ramsey *Angew. Chem. Int. Ed. Engl.* **1973**, *12*, 734-752 [*Angew. Chem. 85*, 773-792]; Turner *Adv. Phys. Org. Chem.* **1966**, *4*, 31-71. For the IUPAC descriptive classification of the electron spectroscopies, see Porter; Turner *Pure Appl. Chem.* **1987**, *59*, 1343-1406.
   [14]The correlation is not perfect, but the limitations do not seriously detract from the usefulness of the method. The technique is not limited to vacuum uv radiation. Higher energy radiation can also be used.
   [15]From Brundle; Robin, in Nachod; Zuckerman *Determination of Organic Structures by Physical Methods*, Vol. 3; Academic Press: New York, 1971, p. 18.

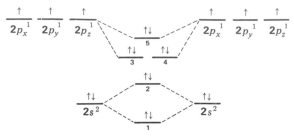

Nitrogen atom Nitrogen molecule Nitrogen atom
$\bar{N} \equiv \bar{N}$

FIGURE 1.9 Electronic structure of N_2 (inner-shell electrons omitted).[15]

naive consideration of orbital overlaps, where it would be expected that the two unshared pairs would be those of orbitals 1 and 2, resulting from the overlap of the filled $2s$ orbitals, and that the triple bond would be composed of orbitals 3, 4, and 5, resulting from overlap of the p orbitals. This example is one illustration of the value of photoelectron spectroscopy.

The photoelectron spectrum of methane[16] shows two bands,[17] at about 23 and 14 eV, and not the single band we would expect from the equivalency of the four C—H bonds. The reason is that ordinary sp^3 hybridization is not adequate to explain phenomena involving ionized molecules (such as the $CH_4^{\bullet+}$ radical ion, which is left behind when an electron is ejected from methane). For these phenomena it is necessary to use other combinations of atomic orbitals (see p. 8). The band at 23 eV comes from two electrons in a low-energy level (called the a_1 level), which can be regarded as arising from a combination of the $2s$ orbital of carbon with an appropriate combination of hydrogen $1s$ orbitals. The band at 14 eV comes from six electrons in a triply degenerate level (the t_2 level), arising from a combination of the three $2p$ orbitals of carbon with other combinations of $1s$ hydrogen orbitals. As was mentioned above, most physical and chemical processes cannot distinguish these levels, but photoelectron spectroscopy can.

Electronic Structures of Molecules

For each molecule, ion, or free radical that has only localized electrons, it is possible to draw an electronic formula, called a *Lewis structure*, that shows the location of these electrons. Only the valence electrons are shown. Valence electrons may be found in covalent bonds connecting two atoms or they may be unshared.[18] The student must be able to draw these structures correctly, since the position of electrons changes in the course of a reaction, and it is necessary to know where the electrons are initially before one can follow where they are going. To this end, the following rules operate:

1. The total number of valence electrons in the molecule (or ion or free radical) must be the sum of all outer-shell electrons "contributed" to the molecule by each atom plus the

[16]Brundle; Robin; Basch *J. Chem. Phys.* **1970,** *53*, 2196; Baker; Betteridge; Kemp; Kirby *J. Mol. Struct.* **1971,** *8*, 75; Potts; Price *Proc. R. Soc. London, Ser A* **1972,** *326*, 165.

[17]A third band, at 290 eV, caused by the $1s$ electrons of carbon, can also found if radiation of sufficiently high energy is used.

[18]It has been argued that although the Lewis picture of two electrons making up a covalent bond may work well for organic compounds, it cannot be successfully applied to the majority of inorganic compounds: Jørgensen *Top. Curr. Chem.* **1984,** *124*, 1-31.

negative charge or minus the positive charge, for the case of ions. Thus, for H_2SO_4, there are 2 (one for each hydrogen) + 6 (for the sulfur) + 24 (6 for each oxygen) = 32; while for SO_4^{2-}, the number is also 32, since each atom "contributes" 6 plus 2 for the negative charge.

2. Once the number of valence electrons has been ascertained, it is necessary to determine which of them are found in covalent bonds and which are unshared. Unshared electrons (either a single electron or a pair) form part of the outer shell of just one atom, but electrons in a covalent bond are part of the outer shell of both atoms of the bond. *First-row atoms* (B, C, N, O, F) *can have a maximum of eight valence electrons*, and usually have this number, although some cases are known where a first-row atom has only six or seven. Where there is a choice between a structure that has six or seven electrons around a first-row atom and one in which all such atoms have an octet, it is the latter that generally has the lower energy and that consequently exists. For example, ethylene is

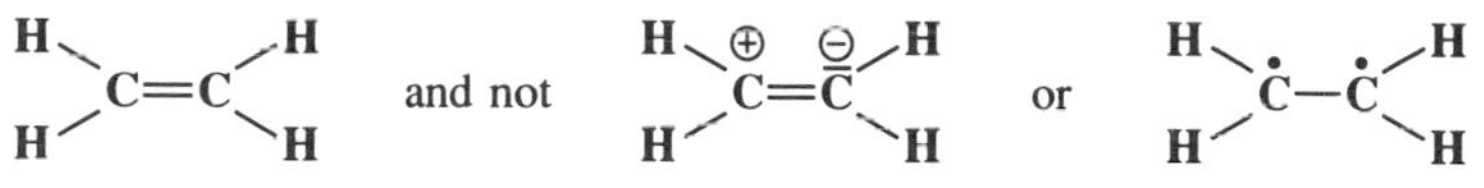

$$H_2C=CH_2 \quad \text{and not} \quad {}^{\oplus}H_2C=CH_2{}^{\ominus} \quad \text{or} \quad H_2\dot{C}-\dot{C}H_2$$

There are a few exceptions. In the case of the molecule O_2, the structure $|\dot{\underline{O}}-\dot{\underline{O}}|$ has a lower energy than $|O{=}O|$. Although first-row atoms are limited to 8 valence electrons, this is not so for second-row atoms, which can accommodate 10 or even 12 because they can use their empty d orbitals for this purpose.[19] For example, PCl_5 and SF_6 are stable compounds. In SF_6, one s and one p electron from the ground state $3s^2 3p^4$ of the sulfur are promoted to empty d orbitals, and the six orbitals hybridize to give six $sp^3 d^2$ orbitals, which point to the corners of a regular octahedron.

3. It is customary to show the formal charge on each atom. For this purpose an atom is considered to "own" all unshared electrons, but only *one-half of the electrons in covalent bonds*. The sum of electrons that thus "belong" to an atom is compared with the number "contributed" by the atom. An excess belonging to the atom results in a negative charge, and a deficiency results in a positive charge. The total of the formal charges on all atoms equals the charge on the whole molecule or ion. It should be noted that the counting procedure is not the same for determining formal charge as for determining the number of valence electrons. For both purposes an atom "owns" all unshared electrons, but for outer-shell purposes it "owns" both the electrons of the covalent bond, while for formal-charge purposes it "owns" only one-half of these electrons.

Examples of electronic structures are (as mentioned in footnote 4, in this book an electron pair, whether unshared or in a bond, is represented by a straight line):

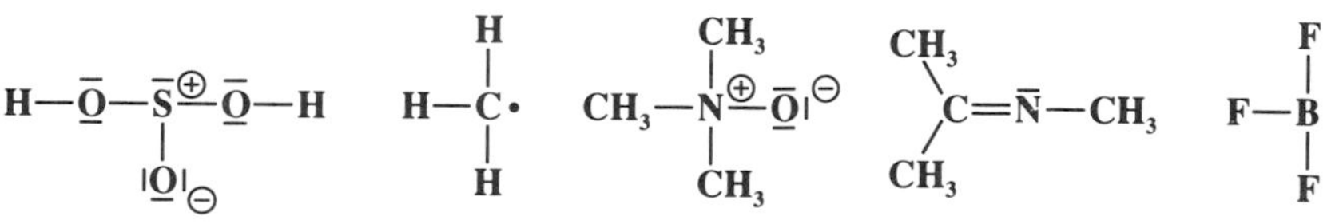

A coordinate-covalent bond, represented by an arrow, is one in which both electrons come from the same atom; i.e., the bond can be regarded as being formed by the overlap

[19]For a review concerning sulfur compounds with a valence shell larger than eight, see Salmond *Q. Rev., Chem. Soc.* **1968**, *22*, 235-275.

of an orbital containing two electrons with an empty one. Thus trimethylamine oxide would be represented

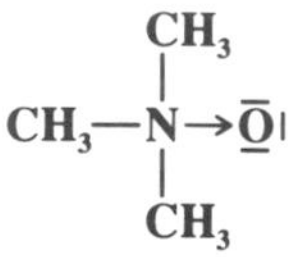

$$
\begin{array}{c}
\text{CH}_3 \\
| \\
\text{CH}_3\text{—N}\rightarrow\underline{\overline{\text{O}}}| \\
| \\
\text{CH}_3
\end{array}
$$

For a coordinate-covalent bond the rule concerning formal charge is amended, so that both electrons count for the donor and neither for the recipient. Thus the nitrogen and oxygen atoms of trimethylamine oxide bear no formal charges. However, it is apparent that the electronic picture is exactly the same as the picture of trimethylamine oxide given just above, and we have our choice of drawing an arrowhead or a charge separation. Some compounds, e.g., amine oxides, must be drawn one way or the other. It seems simpler to use charge separation, since this spares us from having to consider as a "different" method of bonding a way that is really the same as ordinary covalent bonding once the bond has formed.

Electronegativity

The electron cloud that bonds two atoms is not symmetrical (with respect to the plane that is the perpendicular bisector of the bond) except when the two atoms are the same and have the same substituents. The cloud is necessarily distorted toward one side of the bond or the other, depending on which atom (nucleus plus electrons) maintains the greater attraction for the cloud. This attraction is called *electronegativity*;[20] it is greatest for atoms in the upper-right corner of the periodic table and lowest for atoms in the lower-left corner. Thus a bond between fluorine and chlorine is distorted so that there is a higher probability of finding the electrons near the fluorine than near the chlorine. This gives the fluorine a partial negative charge and the chlorine a partial positive charge.

A number of attempts have been made to set up quantitative tables of electronegativity that indicate the direction and extent of electron-cloud distortion for a bond between any pair of atoms. The most popular of these scales, devised by Pauling, is based on bond energies (see p. 23) of diatomic molecules. The reasoning here is that if in a molecule A—B the electron distribution were symmetrical, the bond energy would be the mean of the energies of A—A and B—B, since in these cases the cloud must be undistorted. If the actual bond energy of A—B is higher than this (and it usually is), it is the result of the partial charges, since the charges attract each other and make a stronger bond, which requires more energy to break. It is necessary to assign a value to one element arbitrarily (F = 4.0). Then the electronegativity of another is obtained from the difference between the actual energy of A—B and the mean of A—A and B—B (this difference is called Δ) by the formula

$$
x_A - x_B = \sqrt{\frac{\Delta}{23.06}}
$$

where x_A and x_B are the electronegativities of the known and unknown atoms and 23.06 is an arbitrary constant. Part of the scale derived from this treatment is shown in Table 1.1.

[20]For a collection of articles on this topic, see Sen; Jørgensen *Electronegativity* (Vol. 6 of *Structure and Bonding*); Springer: New York, 1987. For a review, see Batsanov *Russ. Chem. Rev.* **1968**, *37*, 332-351.

TABLE 1.1 Electronegativities of some atoms on the Pauling[21] and Sanderson[25] scales

Element	Pauling	Sanderson	Element	Pauling	Sanderson
F	4.0	4.000	H	2.1	2.592
O	3.5	3.654	P	2.1	2.515
Cl	3.0	3.475	B	2.0	2.275
N	3.0	3.194	Si	1.8	2.138
Br	2.8	3.219	Mg	1.2	1.318
S	2.5	2.957	Na	0.9	0.835
I	2.5	2.778	Cs	0.7	0.220
C	2.5	2.746			

Other treatments[22] have led to scales that are based on different principles, e.g., the average of the ionization potential and the electron affinity,[23] the average one-electron energy of valence-shell electrons in ground-state free atoms,[24] or the "compactness" of an atom's electron cloud.[25] In some of these treatments electronegativities can be calculated for different valence states, for different hybridizations (e.g., sp carbon atoms are more electronegative than sp^2, which are still more electronegative than sp^3),[26] and even differently for primary, secondary, and tertiary carbon atoms. Also, electronegativities can be calculated for groups rather than atoms (Table 1.2).[27]

Electronegativity information can be obtained from nmr spectra. In the absence of a magnetically anisotropic group[28] the chemical shift of a ^1H or a ^{13}C nucleus is approximately proportional to the electron density around it and hence to the electronegativity of the atom or group to which it is attached. The greater the electronegativity of the atom or group, the lower the electron density around the proton and the further downfield the chemical shift. An example of the use of this correlation is found in the variation of chemical shift of the *ring* protons in the series toluene, ethylbenzene, isopropylbenzene, *t*-butylbenzene (there is a magnetically anisotropic group here, but its effect should be constant throughout the

TABLE 1.2 Some group electronegativities relative to H = 2.176[27]

CH₃	2.472	**CCl₃**	2.666
CH₃CH₂	2.482	**C₆H₅**	2.717
CH₂Cl	2.538	**CF₃**	2.985
CBr₃	2.561	**CN**	3.208
CHCl₂	2.602	**NO₂**	3.421

[21]Taken from Pauling *The Nature of the Chemical Bond*, 3rd ed.; Cornell University Press: Ithaca, NY, p. 93, except for the value for Na, which is from Ref. 25.

[22]For several sets of electronegativity values, see Huheey *Inorganic Chemistry*, 3rd ed., Harper and Row: New York, 1983, pp. 146-148; Mullay, in Sen and Jørgensen, Ref. 20, p. 9.

[23]Mulliken *J. Chem. Phys.* **1934**, *2*, 782; Iczkowski; Margrave *J. Am. Chem. Soc.* **1961**, *83*, 3547; Hinze; Jaffé *J. Am. Chem. Soc.* **1962**, *84*, 540.

[24]Allen *J. Am. Chem. Soc.* **1989**, *111*, 9003.

[25]See Sanderson *J. Am. Chem. Soc.* **1983**, *105*, 2259; *J. Chem. Educ.* **1988**, *65*, 112, 223.

[26]Walsh *Discuss. Faraday Soc.* **1947**, *2*, 18; Bergmann; Hinze, in Sen; Jørgensen, Ref. 20, pp. 146-190.

[27]Inamoto; Masuda *Chem. Lett.* **1982**, 1003. For a review of group electronegativities, see Wells *Prog. Phys. Org. Chem.* **1968**, *6*, 111-145. See also Bratsch *J. Chem. Educ.* **1988**, *65*, 223; Mullay *J. Am. Chem. Soc.* **1985**, *107*, 7271; Zefirov; Kirpichenok; Izmailov; Trofimov *Dokl. Chem.* **1987**, *296*, 440; Boyd; Edgecombe *J. Am. Chem. Soc.* **1988**, *110*, 4182.

[28]A magnetically anisotropic group is one that is not equally magnetized along all three axes. The most common such groups are benzene rings (see p. 41) and triple bonds.

series). It is found that the electron density surrounding the ring protons decreases[29] in the order given.[30] However, this type of correlation is by no means perfect, since all the measurements are being made in a powerful field, which itself may affect the electron density distribution. Coupling constants between the two protons of a system —CH—CH—X have also been found to depend on the electronegativity of X.[31]

When the difference in electronegativities is great, the orbital may be so far over to one side that it barely covers the other nucleus. This is an *ionic bond*, which is seen to arise naturally out of the previous discussion, leaving us with basically only one type of bond in organic molecules. Most bonds can be considered intermediate between ionic and covalent. We speak of percent ionic character of a bond, which indicates the extent of electron-cloud distortion. There is a continuous gradation from ionic to covalent bonds.

Dipole Moment

The *dipole moment* is a property of the molecule that results from charge separations like those discussed above. However, it is not possible to measure the dipole moment of an individual bond within a molecule; we can measure only the total moment of the molecule, which is the vectorial sum of the individual bond moments.[32] These individual moments are roughly the same from molecule to molecule,[33] but this constancy is by no means universal. Thus, from the dipole moments of toluene and nitrobenzene (Figure 1.10)[34] we should expect the moment of *p*-nitrotoluene to be about 4.36 D. The actual value 4.39 D is reasonable. However, the moment of *p*-cresol (1.57 D) is quite far from the predicted value of 1.11 D. In some cases, molecules may have substantial individual bond moments but no total moments at all because the individual moments are canceled out by the overall symmetry of the molecule. Some examples are CCl_4, *trans*-1,2-dibromoethene, and *p*-dinitrobenzene.

Because of the small difference between the electronegativities of carbon and hydrogen, alkanes have very small dipole moments, so small that they are difficult to measure. For

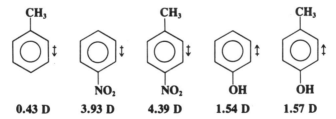

FIGURE 1.10 Some dipole moments, in debye units, measured in benzene. The arrow points to the negative part of the molecule.[34]

[29]This order is opposite to that expected from the field effect (p. 17). It is an example of the Baker–Nathan order (p.68).

[30]Moodie; Connor; Stewart *Can. J. Chem.* **1960**, *38*, 626.

[31]Williamson *J. Am. Chem. Soc.* **1963**, *85*, 516; Laszlo; Schleyer *J. Am. Chem. Soc.* **1963**, *85*, 2709; Niwa *Bull. Chem. Soc. Jpn.* **1967**, *40*, 2192.

[32]For methods of determining dipole moments and discussions of their applications, see Exner *Dipole Moments in Organic Chemistry*; Georg Thieme Publishers: Stuttgart, 1975. For tables of dipole moments, see McClellan *Tables of Experimental Dipole Moments*, Vol. 1; W.H. Freeman: San Francisco, 1963; Vol. 2, Rahara Enterprises: El Cerrito, CA, 1974.

[33]For example, see Koudelka; Exner *Collect. Czech. Chem. Commun.* **1985**, *50*, 188, 200.

[34]The values for toluene, nitrobenzene, and *p*-nitrotoluene are from McClellan, Ref. 32. The values for phenol and *p*-cresol were determined by Goode; Ibbitson *J. Chem. Soc.* **1960**, 4265.

example, the dipole moment of isobutane is 0.132 D[35] and that of propane is 0.085 D.[36] Of course, methane and ethane, because of their symmetry, have no dipole moments.[37] Few organic molecules have dipole moments greater than 7 D.

Inductive and Field Effects

The C—C bond in ethane has no polarity because it connects two equivalent atoms. However, the C—C bond in chlorocthane is polarized by the presence of the electronegative chlorine atom. This polarization is actually the sum of two effects. In the first of these, the C-1 atom, having been deprived of some of its electron density by the greater electronegativity of Cl,

$$\underset{\underset{2}{H_3C}}{\overset{\delta\delta+}{\longrightarrow}}\underset{\underset{1}{CH_2}}{\overset{\delta+}{\longrightarrow}}\overset{\delta-}{Cl}$$

is partially compensated by drawing the C—C electrons closer to itself, resulting in a polarization of this bond and a slightly positive charge on the C-2 atom. This polarization of one bond caused by the polarization of an adjacent bond is called the *inductive effect*. The effect is greatest for adjacent bonds but may also be felt farther away; thus the polarization of the C—C bond causes a (slight) polarization of the three methyl C—H bonds. The other effect operates not through bonds, but directly through space or solvent molecules, and is called the *field effect*.[38] It is often very difficult to separate the two kinds of effect, but it has been done in a number of cases, generally by taking advantage of the fact that the field effect depends on the geometry of the molecule but the inductive effect depends only on the nature of the bonds. For example, in isomers **1** and **2**[39] the inductive effect of the chlorine atoms on the position of the electrons in the COOH group (and hence on the acidity, see

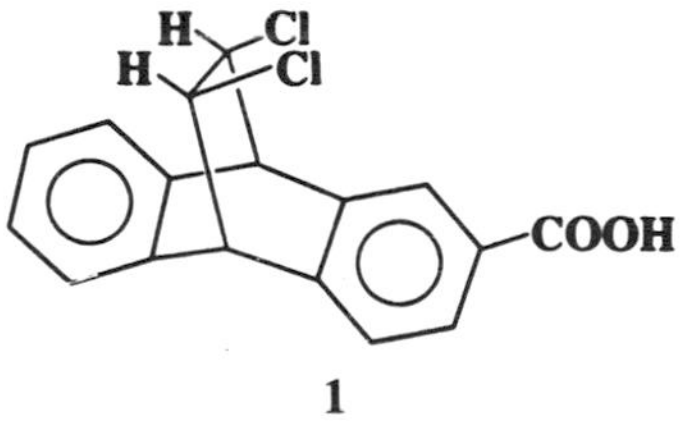

1	**2**
pK_a = 6.07	pK_a = 5.67

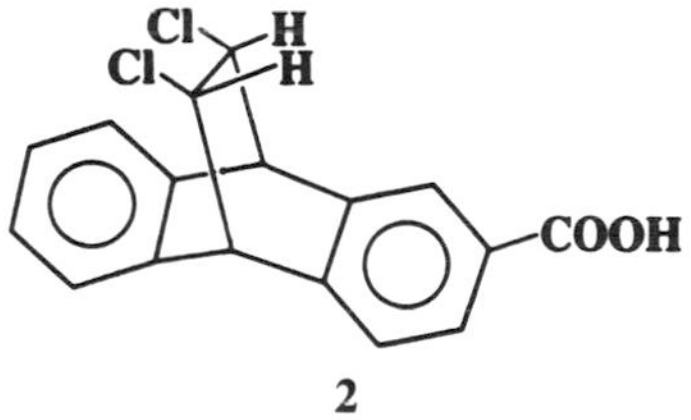

Chapter 8) should be the same since the same bonds intervene; but the field effect is different because the chlorines are closer in space to the COOH in **1** than they are in **2**. Thus a comparison of the acidity of **1** and **2** should reveal whether a field effect is truly operating. The evidence obtained from such experiments is overwhelming that field effects are much

[35]Maryott; Birnbaum *J. Chem. Phys.* **1956**, 24, 1022; Lide; Mann *J. Chem. Phys.* **1958**, 29, 914.
[36]Muenter; Laurie *J. Chem. Phys.* **1966**, 45, 855.
[37]Actually, symmetrical tetrahedral molecules like methane do have extremely small dipole moments, caused by centrifugal distortion effects; these moments are so small that they can be ignored for all practical purposes. For CH$_4$ μ is about 5.4 x 10^{-6} D: Ozier *Phys. Rev. Lett.* **1971**, 27, 1329; Rosenberg; Ozier; Kudian *J. Chem. Phys.* **1972**, 57, 568.
[38]Roberts; Moreland *J. Am. Chem. Soc.* **1953**, 75, 2167.
[39]This example is from Grubbs; Fitzgerald; Phillips; Petty *Tetrahedron* **1971**, 27, 935.

more important than inductive effects.[40] In most cases the two types of effect are considered together; in this book we will not attempt to separate them but will use the name *field effect* to refer to their combined action.[41]

Functional groups can be classified as electron-withdrawing ($-I$) or electron-donating ($+I$) groups relative to hydrogen. This means, for example, that NO_2, a $-I$ group, will draw electrons to itself more than a hydrogen atom would if it occupied the same position in the molecule.

$$O_2N \twoheadleftarrow CH_2 \leftarrow Ph$$

$$H \longrightarrow CH_2 \longrightarrow Ph$$

Thus, in α-nitrotoluene, the electrons in the N—C bond are farther away from the carbon atom than are the electrons in the H—C bond of toluene. Similarly, the electrons of the C—Ph bond are farther away from the ring in α-nitrotoluene than they are in toluene. Field effects are always comparison effects. We compare the $-I$ or $+I$ effect of one group with another (usually hydrogen). It is commonly said that, compared with hydrogen, the NO_2 group is electron-withdrawing and the O^- group electron-donating or electron-releasing. However, there is no actual donation or withdrawal of electrons, though these terms are convenient to use; there is merely a difference in the position of electrons due to the difference in electronegativity between H and NO_2 or between H and O^-.

Table 1.3 lists a number of the most common $-I$ and $+I$ groups.[42] It can be seen that compared with hydrogen, most groups are electron-withdrawing. The only electron-donating groups are groups with a formal negative charge (but not even all these), atoms of low

TABLE 1.3 Field effects of various groups relative to hydrogen
The groups are listed approximately in order of decreasing strength for both $-I$ and $+I$ groups

$+I$	$-I$		
O^-	NR_3^+	COOH	OR
COO^-	SR_2^+	F	COR
CR_3	NH_3^+	Cl	SH
CHR_2	NO_2	Br	SR
CH_2R	SO_2R	I	OH
CH_3	CN	OAr	C≡Cr
D	SO_2Ar	COOR	Ar
			CH=CR$_2$

[40]For example, see Dewar; Grisdale *J. Am. Chem. Soc.* **1962**, *84*, 3548; Stock *J. Chem. Educ.* **1972**, *49*, 400; Golden; Stock *J. Am. Chem. Soc.* **1972**, *94*, 3080; Liotta; Fisher; Greene; Joyner *J. Am. Chem. Soc.* **1972**, *94*, 4891; Wilcox; Leung *J. Am. Chem. Soc.* **1968**, *90*, 336; Butler *J. Chem. Soc. B* **1970**, 867, Adcock; Bettess; Rizvi *Aust. J. Chem.* **1970**, *23*, 1921; Rees; Ridd; Ricci *J. Chem. Soc., Perkin Trans. 2*, **1976**, 294; Topsom *Prog. Phys. Org. Chem.* **1976**, *12*, 1-20; *J. Am. Chem. Soc.* **1981**, *103*, 39; Grob; Kaiser; Schweizer *Helv. Chim. Acta* **1977**, *60*, 391; Reynolds *J. Chem. Soc., Perkin Trans. 2*, **1980**, 985, *Prog. Phys. Org. Chem.* **1983**, *14*, 165-203; Adcock; Butt; Kok; Marriott; Topsom *J. Org. Chem.* **1985**, *50*, 2551; Schneider; Becker *J. Phys. Org. Chem.* **1989**, *2*, 214; Bowden; Ghadir *J. Chem. Soc., Perkin Trans. 2* **1990**, 1333. Inductive effects may be important in certain systems. See, for example, Exner; Fiedler *Collect. Czech. Chem. Commun.* **1980**, *45*, 1251; Li; Schuster *J. Org. Chem.* **1987**, *52*, 3975.

[41]There has been some question as to whether it is even meaningful to maintain the distinction between the two types of effect: see Grob *Helv. Chim. Acta* **1985**, *68*, 882; Lenoir; Frank *Chem. Ber.* **1985**, *118*, 753; Sacher *Tetrahedron Lett.* **1986**, *27*, 4683.

[42]See also Ceppi; Eckhardt; Grob *Tetrahedron Lett.* **1973**, 3627.

electronegativity, such as Si,[43] Mg, etc., and perhaps alkyl groups. Alkyl groups[44] were formerly regarded as electron-donating, but many examples of behavior have been found that can be interpreted only by the conclusion that alkyl groups are electron-withdrawing compared with hydrogen.[45] In accord with this is the value of 2.472 for the group electronegativity of CH_3 (Table 1.2) compared with 2.176 for H. We shall see that when an alkyl group is attached to an unsaturated or trivalent carbon (or other atom), its behavior is best explained by assuming it is $+I$ (see, for example, pp. 168, 176, 270, 511), but when it is connected to a saturated atom, the results are not as clear, and alkyl groups seem to be $+I$ in some cases and $-I$ in others[46] (see also p. 271). Similarly, it is clear that the field-effect order of alkyl groups attached to unsaturated systems is tertiary > secondary > primary > CH_3, but this order is not always maintained when the groups are attached to saturated systems. Deuterium is electron-donating with respect to hydrogen.[47] Other things being equal, atoms with sp bonding generally have a greater electron-withdrawing power than those with sp^2 bonding, which in turn have more electron-withdrawing power than those with sp^3 bonding.[48] This accounts for the fact that aryl, vinylic, and alkynyl groups are $-I$. Field effects always decrease with increasing distance, and in most cases (except when a very powerful $+I$ or $-I$ group is involved), cause very little difference in a bond four bonds away or more. There is evidence that field effects can be affected by the solvent.[49]

For discussions of field effects on acid and base strength and on reactivity, see Chapters 8 and 9, respectively.

Bond Distances[50]

The distances between atoms in a molecule are characteristic properties of the molecule and can give us information if we compare the same bond in different molecules. The chief methods of determining bond distances and angles are x-ray diffraction (only for solids), electron diffraction (only for gases), and spectroscopic methods, especially microwave spectroscopy. The distance between the atoms of a bond is not constant, since the molecule is always vibrating; the measurements obtained are therefore average values, so that different methods give different results.[51] However, this must be taken into account only when fine distinctions are made.

Measurements vary in accuracy, but indications are that similar bonds have fairly constant lengths from one molecule to the next, though exceptions are known.[52] The variation is generally less than 1%. Table 1.4 shows distances for single bonds between two sp^3 carbons.

[43]For a review of field and other effects of silicon-containing groups, see Bassindale; Taylor, in Patai and Rappoport, Ref. 9, pp. 893-963.

[44]For a review of the field effects of alkyl groups, see Levitt; Widing *Prog. Phys. Org. Chem.* **1976,** *12,* 119-157.

[45]See Sebastian *J. Chem. Educ.* **1971,** *48,* 97.

[46]See, for example, Schleyer; Woodworth *J. Am. Chem. Soc.* **1968,** *90,* 6528; Wahl; Peterson *J. Am. Chem. Soc.* **1970,** *92,* 7238. The situation may be even more complicated. See, for example, Minot; Eisenstein; Hiberty; Anh *Bull. Soc. Chim. Fr.* **1980,** II-119.

[47]Streitwieser; Klein *J. Am. Chem. Soc.* **1963,** *85,* 2759.

[48]Bent *Chem. Rev.* **1961,** *61,* 275-311, p. 281.

[49]See Laurence; Berthelot; Lucon; Helbert; Morris; Gal *J. Chem. Soc., Perkin Trans. 2* **1984,** 705.

[50]For tables of bond distances and angles, see Allen; Kennard; Watson; Brammer; Orpen; Taylor *J. Chem. Soc., Perkin Trans. 2* **1987,** S1–S19; Tables of Interatomic Distances and Configurations in Molecules and Ions *Chem. Soc. Spec. Publ.* No. 11, 1958; Interatomic Distances Supplement *Chem. Soc. Spec. Publ.* No. 18, 1965; Harmony; Laurie; Kuczkowski; Schwendeman; Ramsay; Lovas; Lafferty; Maki *J. Phys.Chem. Ref. Data* **1979,** *8,* 619-721. For a review of molecular shapes and energies for many small organic molecules, radicals, and cations calculated by molecular orbital methods, see Lathan; Curtiss; Hehre; Lisle; Pople *Prog. Phys. Org. Chem.* **1974,** *11,* 175-261. For a discussion of substituent effects on bond distances, see Topsom *Prog. Phys. Org. Chem.* **1987,** *16,* 85-124.

[51]Burkert; Allinger *Molecular Mechanics*; ACS Monograph 177, American Chemical Society: Washington, 1982, pp. 6-9; Whiffen *Chem. Br.* **1971,** *7,* 57-61; Stals *Rev. Pure Appl. Chem.* **1970,** *20,* 1-22, pp. 2-5.

[52]Schleyer; Bremer *Angew. Chem. Int. Ed. Engl.* **1989,** *28,* 1226 [*Angew. Chem. 101,* 1264].

TABLE 1.4 Bond lengths between sp^3 carbons in some compounds

C—C bond in	Bond length, Å	C—C bond in	Bond length, Å
Diamond[53]	1.544	Cyclohexane[57]	1.540 ± 0.015
C_2H_6[54]	1.5324 ± 0.0011	t-Butyl chloride[58]	1.532
C_2H_5Cl[55]	1.5495 ± 0.0005	n-Butane to n-heptane[59]	1.531–1.534
C_3H_8[56]	1.532 ± 0.003	Isobutane[60]	1.535 ± 0.001

However, an analysis of C—OR bond distances in more than 2000 ethers and carboxylic esters (all with sp^3 carbon) shows that this distance increases with increasing electron withdrawal in the R group and as the C changes from primary to secondary to tertiary.[61] For these compounds, mean bond lengths of the various types ranged from 1.418 to 1.475 Å.

Bond distances for some important bond types are given in Table 1.5.[62] As can be seen in this table, carbon bonds are shortened by increasing s character. This is most often explained by the fact that, as the percentage of s character in a hybrid orbital increases, the orbital becomes more like an s orbital and hence is held more tightly by the nucleus than an orbital with less s character. However, other explanations have also been offered (see p. 31), and the matter is not completely settled.

Indications are that a C—D bond is slightly shorter than a corresponding C—H bond. Thus, electron-diffraction measurements of C_2H_6 and C_2D_6 showed a C—H bond distance of 1.1122 ± 0.0012 Å and a C—D distance of 1.1071 ± 0.0012 Å.[54]

Bond Angles

It might be expected that the bond angles of sp^3 carbon would always be the tetrahedral angle 109°28′, but this is so only where the four groups are identical, as in methane, neopentane, or carbon tetrachloride. In most cases the angles deviate a little from the pure tetrahedral value. For example, the C—C—Br angle in 2-bromopropane is 114.2°.[70] Similarly, slight variations are generally found from the ideal values of 120 and 180° for sp^2 and sp carbon, respectively. These deviations occur because of slightly different hybridizations, that is, a carbon bonded to four other atoms hybridizes one s and three p orbitals, but the four hybrid orbitals thus formed are generally not exactly equivalent, nor does each contain exactly 25% s and 75% p character. Because the four atoms have (in the most general case) different electronegativities, each makes its own demand for electrons from the carbon atom.[71] The carbon atom supplies more p character when it is bonded to more electronegative atoms, so that in chloromethane, for example, the bond to chlorine has somewhat more

[53]Lonsdale *Phil. Trans. R. Soc. London* **1947**, *A240*, 219.
[54]Bartell; Higginbotham *J. Chem. Phys.* **1965**, *42*, 851.
[55]Wagner; Dailey *J. Chem. Phys.* **1957**, *26*, 1588.
[56]Iijima *Bull. Chem. Soc. Jpn.* **1972**, *45*, 1291.
[57]Tables of Interatomic Distances, Ref. 50.
[58]Momany; Bonham; Druelinger *J. Am. Chem. Soc.* **1963**, *85*, 3075; also see Lide; Jen *J. Chem. Phys.* **1963**, *38*, 1504.
[59]Bonham; Bartell; Kohl *J. Am. Chem. Soc.* **1959**, *81*, 4765.
[60]Hilderbrandt; Wieser *J. Mol. Struct.* **1973**, *15*, 27.
[61]Allen; Kirby *J. Am. Chem. Soc.* **1984**, *106*, 6197; Jones; Kirby *J. Am. Chem. Soc.* **1984**, *106*, 6207.
[62]Except where noted, values are from Allen et al., Ref. 50. In this source, values are given to three significant figures.
[63]Costain; Stoicheff *J. Chem. Phys.* **1959**, *30*, 777.
[64]For a full discussion of alkyne bond distances, see Simonetta; Gavezzotti, Ref. 7.

TABLE 1.5   Bond distances
The values given are average lengths and do not necessarily apply exactly to the compounds mentioned[62]

Bond type	Length, Å	Typical compounds
C—C		
sp^3—sp^3	1.53	
sp^3—sp^2	1.51	Acetaldehyde, toluene, propene
sp^3—sp	1.47	Acetonitrile, propyne
sp^2—sp^2	1.48	Butadiene, glyoxal, biphenyl
sp^2—sp	1.43	Acrylonitrile, vinylacetylene
sp—sp	1.38	Cyanoacetylene, butadiyne
C=C		
sp^2—sp^2	1.32	Ethylene
sp^2—sp	1.31	Ketene, allenes
sp—sp^{63}	1.28	Butatriene, carbon suboxide
C≡C[64]		
sp—sp	1.18	Acetylene
C—H[65]		
sp^3—**H**	1.09	Methane
sp^2—**H**	1.08	Benzene, ethylene
sp—**H**[66]	1.08	**HCN**, acetylene
C—O		
sp^3—**O**	1.43	Dimethyl ether, ethanol
sp^2—**O**	1.34	Formic acid
C=O		
sp^2—**O**	1.21	Formaldehyde, formic acid
sp—**O**[57]	1.16	**CO₂**
C—N		
sp^3—**N**	1.47	Methylamine
sp^2—**N**	1.38	Formamide
C=N		
sp^2—**N**	1.28	Oximes, imines
C≡N		
sp—**N**	1.14	**HCN**
C—S		
sp^3—**S**	1.82	Methanethiol
sp^2—**S**	1.75	Diphenyl sulfide
sp—**S**	1.68	CH_3SCN
C=S		
sp—**S**	1.67	**CS₂**

C—halogen[67]	F	Cl	Br	I
sp^3—halogen	1.40	1.79	1.97	2.16
sp^2—halogen	1.34	1.73	1.88	2.10
sp—halogen	1.27[68]	1.63	1.79[69]	1.99[69]

[65]For an accurate method of C—H bond distance determination, see Henry *Acc. Chem. Res.* **1987,** *20,* 429-435.
[66]Bartell; Roth; Hollowell; Kuchitsu; Young *J. Chem. Phys.* **1965,** *42,* 2683.
[67]For reviews of carbon–halogen bonds, see Trotter, in Patai *The Chemistry of the Carbon–Halogen Bond,* pt. 1; Wiley: New York, 1973, pp. 49-62; Mikhailov *Russ. Chem. Rev.* **1971,** *40,* 983-997.
[68]Lide, *Tetrahedron* **1962,** *17,* 125.
[69]Rajput; Chandra *Bull. Chem. Soc. Jpn.* **1966,** *39,* 1854.
[70]Schwendeman; Tobiason *J. Chem. Phys.* **1965,** *43,* 201.
[71]For a review of this concept, see Bingel; Lüttke *Angew. Chem. Int. Ed. Engl.* **1981, 20,** 899-910 [*Angew. Chem.* *93,* 944-956].

than 75% p character, which of course requires that the other three bonds have somewhat less, since there are only three p orbitals (and one s) to be divided among the four hybrid orbitals.[72] Of course, in strained molecules, the bond angles may be greatly distorted from the ideal values (see p. 150).

For oxygen and nitrogen, angles of 90° are predicted from p^2 bonding. However, as we have seen (p. 6), the angles of water and ammonia are much larger than this, as are the angles of other oxygen and nitrogen compounds (Table 1.6); in fact, they are much closer to the tetrahedral angle of 109°28' than to 90°. These facts have led to the suggestion that in these compounds oxygen and nitrogen use sp^3 bonding, i.e., instead of forming bonds by the overlap of two (or three) p orbitals with $1s$ orbitals of the hydrogen atoms, they hybridize their $2s$ and $2p$ orbitals to form four sp^3 orbitals and then use only two (or three) of these for bonding with hydrogen, the others remaining occupied by unshared pairs (also called *lone pairs*). If this description is valid, and it is generally accepted by most chemists today,[78] it becomes necessary to explain why the angles of these two compounds are in fact not 109°28' but a few degrees smaller. One explanation that has been offered is that the unshared pair actually has a greater steric requirement than a pair in a bond, since there is no second nucleus to draw away some of the electron density and the bonds are thus crowded together. However, most evidence is that unshared pairs have smaller steric requirements than bonds[79] and the explanation most commonly accepted is that the hybridization is not pure sp^3. As we have seen above, an atom supplies more p character when it is bonded to more elec-

TABLE 1.6 Oxygen, sulfur, and nitrogen bond angles in some compounds

Angle	Value	Compound	Ref.
H—O—H	104°27'	Water	5
C—O—H	107–109°	Methanol	57
C—O—C	111°43'	Dimethyl ether	73
C—O—C	124 ± 5°	Diphenyl ether	74
H—S—H	92.1°	H_2S	74
C—S—H	99.4°	Methanethiol	74
C—S—C	99.1°	Dimethyl sulfide	75
H—N—H	106°46'	Ammonia	5
H—N—H	106°	Methylamine	76
C—N—H	112°	Methylamine	76
C—N—C	108.7°	Trimethylamine	77

[72]This assumption has been challenged; see Pomerantz; Liebman *Tetrahedron Lett.* **1975**, 2385.
[73]Blukis; Kasai; Myers *J. Chem. Phys.* **1963**, *38*, 2753.
[74]Abrahams *Q. Rev., Chem. Soc.* **1956**, *10*, 407-436.
[75]Iijima; Tsuchiya; Kimura *Bull. Chem. Soc. Jpn.* **1977**, *50*, 2564.
[76]Lide *J. Chem. Phys.* **1957**, *27*, 343.
[77]Lide; Mann *J. Chem. Phys.* **1958**, *28*, 572.
[78]An older theory holds that the bonding is indeed p^2, and that the increased angles come from repulsion of the hydrogen or carbon atoms. See Laing, *J. Chem. Educ.* **1987**, *64*, 124.
[79]See, for example, Pumphrey; Robinson *Chem. Ind. (London)* **1963**, 1903; Allinger; Carpenter; Karkowski *Tetrahedron Lett.* **1964**, 3345; Eliel; Knoeber *J. Am. Chem. Soc.* **1966**, *88*; 5347; **1968**, *90*; 3444; Jones; Katritzky; Richards; Wyatt; Bishop; Sutton *J. Chem. Soc. B* **1970**, 127; Blackburne; Katritzky; Takeuchi *J. Am. Chem. Soc.* **1974**, *96*, 682; *Acc. Chem. Res.* **1975**, *8*, 300-306; Aaron; Ferguson *J. Am. Chem. Soc.* **1976**, *98*, 7013; Anet; Yavari *J. Am. Chem. Soc.* **1977**, *99*, 2794; Vierhapper; Eliel *J. Org. Chem.* **1979**, *44*, 1081; Gust; Fagan *J. Org. Chem.* **1980**, *45*, 2511. For other views, see Lambert; Featherman *Chem. Rev.* **1975**, *75*, 611-626; Crowley; Morris; Robinson *Tetrahedron Lett.* **1976**, 3575; Breuker; Kos; van der Plas; van Veldhuizen *J. Org. Chem.* **1982**, *47*, 963.

tronegative atoms. An unshared pair may be considered to be an "atom" of the lowest possible electronegativity, since there is no attracting power at all. Consequently, the unshared pairs have more s and the bonds more p character than pure sp^3 orbitals, making the bonds somewhat more like p^2 bonds and reducing the angle. As seen in Table 1.6, oxygen, nitrogen, and sulfur angles generally increase with decreasing electronegativity of the substituents. Note that the explanation given above cannot explain why some of these angles are *greater* than the tetrahedral angle.

Bond Energies[80]

There are two kinds of bond energy. The energy necessary to cleave a bond to give the constituent radicals is called the *dissociation energy D*. For example, D for $H_2O \rightarrow HO + H$ is 118 kcal/mol (494 kJ/mol). However, this is not taken as the energy of the O—H bond in water, since D for $H-O \rightarrow H + O$ is 100 kcal/mol (418kJ/mol). The average of these two values, 109 kcal/mol (456 kJ/mol), is taken as the *bond energy E*. In diatomic molecules, of course, $D = E$.

D values may be easy or difficult to measure, but there is no question as to what they mean. With E values the matter is not so simple. For methane, the total energy of conversion from CH_4 to $C + 4H$ (at 0 K) is 393 kcal/mol (1644 kJ/mol).[81] Consequently, E for the C—H bond in methane is 98 kcal/mol (411 kJ/mol) at 0 K. The more usual practice, though, is not to measure the heat of atomization (i.e., the energy necessary to convert a compound to its atoms) directly but to calculate it from the heat of combustion. Such a calculation is shown in Figure 1.11.

Heats of combustion are very accurately known for hydrocarbons.[82] For methane the value at 25°C is 212.8 kcal/mol (890.4 kJ/mol), which leads to a heat of atomization of 398.0 kcal/mol (1665 kJ/mol) or a value of E for the C—H bond at 25°C of 99.5 kcal/mol

			kcal	kJ
$C_2H_{6\,(gas)} + 3\frac{1}{2}O_{2\,(gas)}$	$= 2CO_{2\,(gas)}$	$+ 3H_2O_{(liq.)}$	$+372.9$	$+1560$
$2CO_{2\,(gas)}$	$= 2C_{(graphite)}$	$+ 2O_{2\,(gas)}$	-188.2	-787
$3H_2O_{(liq.)}$	$= 3H_{2\,(gas)}$	$+ 1\frac{1}{2}O_{2\,(gas)}$	-204.9	-857
$3H_{2\,(gas)}$	$= 6H_{(gas)}$		-312.5	-1308
$2C_{(graphite)}$	$= 2C_{(gas)}$		-343.4	-1437
$C_2H_{6\,(gas)}$	$= 6H_{(gas)}$	$+ 2C_{(gas)}$	-676.1 kcal	-2829 kJ

FIGURE 1.11   Calculation of the heat of atomization of ethane at 25°C.

[80]For reviews including methods of determination, see Wayner; Griller *Adv. Free Radical Chem. (Greenwich, Conn.)* **1990,** *1,* 159-192; Kerr; *Chem. Rev.* **1966,** *66,* 465-500; Benson *J. Chem. Educ.* **1965,** *42,* 520-518; Wiberg, in Nachod; Zuckerman *Determination of Organic Structures by Physical Methods,* Vol. 3; Academic Press: New York, 1971, pp. 207-245.

[81]For the four steps, D values are 101 to 102, 88, 124, and 80 kcal/mol (423-427, 368, 519, and 335 kJ/mol), respectively, though the middle values are much less reliable than the other two: Knox; Palmer *Chem. Rev.* **1961,** *61,* 247-255; Brewer; Kester *J. Chem. Phys.* **1964,** *40,* 812; Linevsky *J. Chem. Phys.* **1967,** *47,* 3485.

[82]For values of heats of combustion of large numbers of organic compounds: hydrocarbons and others, see Cox; Pilcher, *Thermochemistry of Organic and Organometallic Compounds*; Academic Press: New York, 1970; Domalski *J. Phys. Chem. Ref. Data* **1972,** *1,* 221-277. For large numbers of heats-of-formation values (from which heats of combustion are easily calculated) see Stull; Westrum; Sinke *The Chemical Thermodynamics of Organic Compounds,* Wiley: New York, 1969.

(416 kJ/mol). This method is fine for molecules like methane in which all the bonds are equivalent, but for more complicated molecules assumptions must be made. Thus for ethane, the heat of atomization at 25°C is 676.1 kcal/mol or 2829 kJ/mol (Figure 1.11), and we must decide how much of this energy is due to the C—C bond and how much to the six C—H bonds. Any assumption must be artificial, since there is no way of actually obtaining this information, and indeed the question has no real meaning. If we make the assumption that E for each of the C—H bonds is the same as E for the C—H bond in methane (99.5 kcal/mol or 416 kJ/mol), then 6 × 99.5 (or 416) = 597.0 (or 2498), leaving 79.1 kcal/mol (331 kJ/mol) for the C—C bond. However, a similar calculation for propane gives a value of 80.3 (or 336) for the C—C bond, and for isobutane, the value is 81.6 (or 341). A consideration of heats of atomization of isomers also illustrates the difficulty. E values for the C—C bonds in pentane, isopentane, and neopentane, calculated from heats of atomization in the same way, are (at 25°C) 81.1, 81.8, and 82.4 kcal/mol (339, 342, 345 kJ/mol), respectively, even though all of them have twelve C—H bonds and four C—C bonds.

These differences have been attributed to various factors caused by the introduction of new structural features. Thus isopentane has a tertiary carbon whose C—H bond does not have exactly the same amount of s character as the C—H bond in pentane, which for that matter contains secondary carbons not possessed by methane. It is known that D values, which *can* be measured, are not the same for primary, secondary, and tertiary C—H bonds (see Table 5.3). There is also the steric factor. Hence it is certainly not correct to use the value of 99.5 kcal/mol (416 kJ/mol) from methane as the E value for all C—H bonds. Several empirical equations have been devised that account for these factors; the total energy can be computed[83] if the proper set of parameters (one for each structural feature) is inserted. Of course these parameters are originally calculated from the known total energies of some molecules which contain the structural feature.

Table 1.7 gives E values for various bonds. The values given are averaged over a large series of compounds. The literature contains charts that take account of hybridization (thus an sp^3 C—H bond does not have the same energy as an sp^2 C—H bond).[87]

Certain generalizations can be derived from the data in Table 1.7.

TABLE 1.7   Bond energy E values at 25°C for some important bond types[84]
E *values are arranged within each group in order of decreasing strength. The values are averaged over a large series of compounds.*

Bond	kcal/mol	kJ/mol	Bond	kcal/mol	kJ/mole
O—H	110–111	460–464	**C—S**[86]	61	255
C—H	96–99	400–415	**C—I**	52	220
N—H	93	390			
S—H	82	340	**C≡C**	199–200	835
			C=C	146–151	610–630
C—F	⋯	⋯	**C—C**	83–85	345–355
C—H	96–99	400–415			
C—O	85–91	355–380	**C≡N**	204	854
C—C	83–85	345–355	**C=O**	173–81	724–757
C—Cl	79	330			
C—N[85]	69–75	290–315	**C=N**[85]	143	598
C—Br	66	275			

[83]For a review, see Cox; Pilcher, Ref. 82, pp. 531–597. See also Gasteiger; Jacob; Strauss *Tetrahedron* **1979**, *35*, 139.
[84]These values, except where noted, are from Lovering; Laidler *Can. J. Chem.* **1960**, *38*, 2367; Levi; Balandin *Bull. Acad. Sci. USSR, Div. Chem. Sci.* **1960**, 149.
[85]Bedford; Edmondson; Mortimer *J. Chem. Soc.* **1962**, 2927.
[86]Grelbig; Pötter; Seppelt *Chem. Ber.* **1987**, *120*, 815.
[87]Ref. 83; Cox *Tetrahedron* **1962**, *18*, 1337.

1. There is a correlation of bond strengths with bond distances. A comparison of Tables 1.5 and 1.7 shows that, in general, *shorter bonds are stronger bonds*. Since we have already seen that increasing *s* character shortens bonds (p. 20), it follows that bond strengths increase with increasing *s* character.

2. Bonds become weaker as we move down the periodic table. Compare C—O and C—S or the four carbon–halogen bonds. This is a consequence of the first generalization, since bond distances must increase as we go down the periodic table because the number of inner electrons increases.

3. Double bonds are both shorter and stronger than the corresponding single bonds, but not twice as strong, because π overlap is less than σ overlap. This means that a σ bond is stronger than a π bond. The difference in energy between a single bond, say C—C, and the corresponding double bond is the amount of energy necessary to cause rotation around the double bond.[88]

---

[88]For a discussion of the different magnitudes of the bond energies of the two bonds of the double bond, see Miller *J. Chem. Educ.* **1978**, *55*, 778.

2
DELOCALIZED
CHEMICAL BONDING

Although the bonding of many compounds can be adequately described by a single Lewis structure (page 12), this is not sufficient for many other compounds. These compounds contain one or more bonding orbitals that are not restricted to two atoms, but that are spread out over three or more. Such bonding is said to be *delocalized*.[1] In this chapter we shall see which types of compounds must be represented in this way.

The two chief general methods of approximately solving the wave equation, discussed in Chapter 1, are also used for compounds containing delocalized bonds.[2] In the valence-bond method, several possible Lewis structures (called *canonical forms*) are drawn and the molecule is taken to be a weighted average of them. Each ψ in Eq. (3), Chapter 1,

$$\Psi = c_1\psi_1 + c_2\psi_2 + \cdots$$

represents one of these structures. This representation of a real structure as a weighted average of two or more canonical forms is called *resonance*. For benzene the canonical forms are **1** and **2**. Double-headed arrows are used to indicate resonance. When the wave equation is solved, it is found that the energy value obtained by considering that **1** and **2** participate equally is lower than that for **1** or **2** alone. If **3**, **4**, and **5** (called *Dewar structures*)

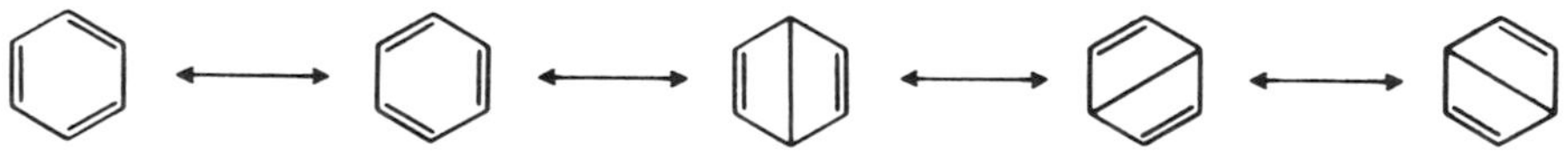

are also considered, the value is lower still. According to this method, **1** and **2** each contribute 39% to the actual molecule and the others 7.3% each.[3] The carbon–carbon bond order is 1.463 (not 1.5, which would be the case if only **1** and **2** contributed). In the valence-bond method, the *bond order* of a particular bond is the sum of the weights of those canonical forms in which the bonds is double plus 1 for the single bond that is present in all of them.[4] Thus, according to this picture, each C—C bond is not halfway between a single and a double bond but somewhat less. The energy of the actual molecule is obviously less than that of any one Lewis structure, since otherwise it would have one of those structures. The difference in energy between the actual molecule and the Lewis structure of lowest energy is call the *resonance energy*. Of course, the Lewis structures are not real, and their energies can only be estimated.

[1]The classic work on delocalized bonding is Wheland *Resonance in Organic Chemistry*; Wiley: New York, 1955.
[2]There are other methods. For a discussion of the free-electron method, see Streitwieser *Molecular Orbital Theory for Organic Chemists*; Wiley: New York, 1961, pp. 27-29. For the nonpairing method, in which benzene is represented as having three electrons between adjacent carbons, see Hirst; Linnett *J. Chem. Soc.* **1962**, 1035; Firestone *J. Org. Chem.* **1969**, *34*, 2621.
[3]Pullman; Pullman *Prog. Org. Chem.* **1958**, *4*, 31-71, p. 33.
[4]For a more precise method of calculating valence-bond orders, see Clarkson; Coulson; Goodwin *Tetrahedron* **1963**, *19*, 2153. See also Herndon; Párkányi *J. Chem. Educ.* **1976**, *53*, 689.

Qualitatively, the resonance picture is often used to describe the structure of molecules, but quantitative valence-bond calculations become much more difficult as the structures become more complicated (e.g., naphthalene, pyridine, etc.). Therefore the molecular-orbital method is used much more often for the solution of wave equations.[5] If we look at benzene by this method (qualitatively), we see that each carbon atom, being connected to three other atoms, uses sp^2 orbitals to form σ bonds, so that all 12 atoms are in one plane. Each carbon has a p orbital (containing one electron) remaining and each of these can overlap equally with the two adjacent p orbitals. This overlap of six orbitals (see Figure 2.1) produces six new orbitals, three of which (shown) are bonding. These three (called π orbitals) all occupy approximately the same space.[6] One of the three is of lower energy than

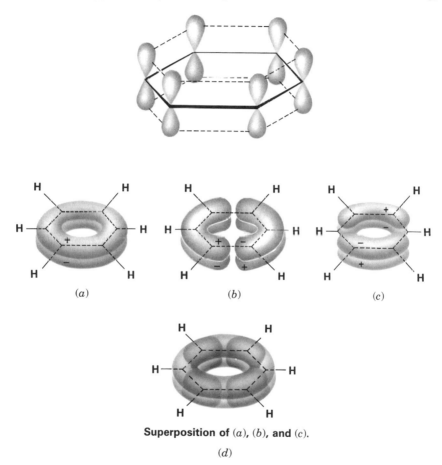

Superposition of (a), (b), and (c).

(d)

FIGURE 2.1   The six p orbitals of benzene overlap to form three bonding orbitals, (a), (b), and (c). The three orbitals superimposed are shown in (d).

[5]For a review of how mo theory explains localized and delocalized bonding, see Dewar *Mol. Struct. Energ.* **1988,** *5,* 1-61.

[6]According to the explanation given here, the symmetrical hexagonal structure of benzene is caused by both the σ bonds and the π orbitals. It has been contended, based on mo calculations, that this symmetry is caused by the σ framework alone, and that the π system would favor three localized double bonds: Shaik; Hiberty; Lefour; Ohanessian *J. Am. Chem. Soc.* **1987,** *109,* 363; Stanger; Vollhardt *J. Org. Chem.* **1988,** *53,* 4889. See also Cooper; Wright; Gerratt; Raimondi *J. Chem. Soc., Perkin Trans. 2* **1989,** 255, 263; Jug; Köster *J. Am. Chem. Soc.* **1990,** *112,* 6772; Aihara *Bull. Chem. Soc. Jpn.* **1990,** *63,* 1956.

the other two, which are degenerate. They each have the plane of the ring as a node and so are in two parts, one above and one below the plane. The two orbitals of higher energy (Figure 2.1b and c) also have another node. The six electrons that occupy this torus-shaped cloud are called the *aromatic sextet*. The carbon–carbon bond order for benzene, calculated by the molecular-orbital method, is 1.667.[7]

For planar unsaturated and aromatic molecules, many molecular-orbital calculations (*mo calculations*) have been made by treating the σ and π electrons separately. It is assumed that the σ orbitals can be treated as localized bonds and the calculations involve only the π electrons. The first such calculations were made by Hückel; such calculations are often called *Hückel molecular-orbital* (HMO) *calculations*.[8] Because electron–electron repulsions are either neglected or averaged out in the HMO method, another approach, the *self-consistent field* (SCF), or *Hartree–Fock*, method, was devised.[9] Although these methods give many useful results for planar unsaturated and aromatic molecules, they are often unsuccessful for other molecules; it would obviously be better if all electrons, both σ and π, could be included in the calculations. The development of modern computers has now made this possible.[10] Many such calculations have been made[11] using a number of methods, among them an extension of the Hückel method (EHMO)[12] and the application of the SCF method to all valence electrons.[13]

One type of mo calculation that includes all electrons is called *ab initio*.[14] Despite the name (which means "from first principles") this type does involve assumptions, though not very many. It requires a large amount of computer time, especially for molecules that contain more than about five or six atoms other than hydrogen. Treatments that use certain simplifying assumptions (but still include all electrons) are called *semi-empirical* methods.[15] One of the first of these was called CNDO (Complete Neglect of Differential Overlap),[16] but as computers have become more powerful, this has been superceded by more modern methods, including MINDO/3 (Modified Intermediate Neglect of Differential Overlap),[17] MNDO (Modified Neglect of Diatomic Overlap),[17] and AM1 (Austin Model 1), all of which were introduced by M. J. Dewar and co-workers.[18] Semi-empirical calculations are generally regarded as less accurate than ab initio methods,[19] but are much faster and cheaper. Indeed,

---

[7]The molecular-orbital method of calculating bond order is more complicated than the valence-bond method. See Ref. 3, p. 36; Clarkson; Coulson; Goodwin, Ref. 4.

[8]See Yates *Hückel Molecular Orbital Theory*; Academic Press: New York, 1978; Coulson; O'Leary; Mallion *Hückel Theory for Organic Chemists*; Academic Press: New York, 1978; Lowry; Richardson *Mechanism and Theory in Organic Chemistry*, 3rd ed., Harper and Row: New York, 1987, pp. 100-121.

[9]Roothaan *Rev. Mod. Phys.* **1951**, *23*, 69; Pariser; Parr *J. Chem. Phys.* **1952**, *21*, 466, 767; Pople *Trans. Faraday Soc.* **1953**, *49*, 1375, *J. Phys. Chem.* **1975**, *61*, 6; Dewar *The Molecular Orbital Theory of Organic Chemistry*; Mc-Graw-Hill: New York, 1969; Dewar, in *Aromaticity*, *Chem. Soc. Spec. Pub.* no. 21, 1967, pp. 177-215.

[10]For discussions of the progress made in quantum chemistry calculations, see Ramsden *Chem. Br.* **1978**, *14*, 396-403; Hall *Chem. Soc. Rev.* **1973**, *2*, 21-28.

[11]For a review of molecular-orbital calculatons on *saturated* organic compounds, see Herndon, *Prog. Phys. Org. Chem.* **1972**, *9*, 99-177.

[12]Hoffmann *J. Chem. Phys.* **1963**, *39*, 1397. See Yates, Ref. 8, pp. 190-201

[13]Dewar *The Molecular Orbital Theory of Chemistry*, Ref. 9; Jaffé *Acc. Chem. Res.* **1969**, *2*, 136-143; Kutzelnigg; Del Re; Berthier *Fortschr. Chem. Forsch.* **1971**, *22*, 1-222.

[14]Hehre; Radom; Schleyer; Pople *Ab Initio Molecular Orbital Theory*; Wiley: New York, 1986; Clark *A Handbook of Computational Chemistry*; Wiley: New York, 1985, pp. 233-317; Richards; Cooper *Ab Initio Molecular Orbital Calculations for Chemists*, 2nd ed., Oxford University Press: Oxford, 1983.

[15]For a review, see Thiel, *Tetrahedron* **1988**, *44*, 7393-7408.

[16]Pople; Santry; Segal *J. Chem. Phys.* **1965**, *43*, S129; Pople; Segal *J. Chem. Phys.* **1965**, *43*, S136; **1966**, *44*, 3289; Pople; Beveridge *Approximate Molecular Orbital Theory*; McGraw-Hill: New York, 1970.

[17]For a discussion of MNDO and MINDO/3, and a list of systems for which these methods have been used, with references, see Clark, Ref. 14, pp. 93-232. For a review of MINDO/3, see Lewis, *Chem. Rev.* **1986**, *86*, 1111-1123.

[18]First publications are, MINDO/3: Bingham; Dewar; Lo *J. Am. Chem. Soc.* **1975**, *97*, 1285; MNDO: Dewar; Thiel *J. Am. Chem. Soc.* **1977**, *99*, 4899; AM1: Dewar; Zoebisch; Healy; Stewart *J. Am. Chem. Soc.* **1985**, *107*, 3902.

[19]See however, Dewar; Storch *J. Am. Chem. Soc.* **1985**, *107*, 3898.

calculations for some very large molecules are possible only with the semi-empirical methods.[20]

Molecular orbital calculations, whether by ab initio or semi-empirical methods, can be used to obtain structures (bond distances and angles), energies (such as heats of formation), dipole moments, ionization energies, and other properties of molecules, ions, and radicals—not only of stable ones, but also of those so unstable that these properties cannot be obtained from experimental measurements.[21] Many of these calculations have been performed on transition states (p. 210); this is the only way to get this information, since transition states are not, in general, directly observable. Of course, it is not possible to check data obtained for unstable molecules and transition states against any experimental values, so that the reliability of the various mo methods for these cases is always a question. However, our confidence in them does increase when (1) different mo methods give similar results, and (2) a particular mo method works well for cases that can be checked against experimental methods.

Both the valence-bond and molecular-orbital methods show that there is delocalization in benzene. For example, each predicts that the six carbon–carbon bonds should have equal lengths, which is true. Since each method is useful for certain purposes, we shall use one or the other as appropriate.

Bond Energies and Distances in Compounds Containing Delocalized Bonds

If we add the energies of all the bonds in benzene, taking the values from a source like Table 1.7, the value for the heat of atomization turns out to be less than that actually found in benzene (Figure 2.2). The actual value is 1323 kcal/mol (5535 kJ/mol). If we use E values for a C=C double bond obtained from cyclohexene (148.8 kcal/mol; 622.6 kJ/mol), a C—C single bond from cyclohexane (81.8 kcal/mol, 342 kJ/mol), and C—H bonds from methane (99.5 kcal/mol, 416 kJ/mol), we get a total of 1289 kcal/mol (5390 kJ/mol) for structure **1** or **2**. By this calculation the resonance energy is 34 kcal/mol (145 kJ/mol). Of course, this is an arbitrary calculation since, in addition to the fact that we are calculating a heat of atomization for a nonexistent structure (**1**), we are forced to use E values that themselves do not have a firm basis in reality. The resonance energy can never be measured, only estimated, since we can measure the heat of atomization of the real molecule but can only make an intelligent guess at that of the Lewis structure of lowest energy. Another method frequently used for estimation of resonance energy involves measurements of heats of hy-

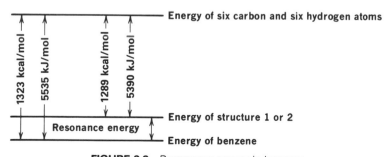

FIGURE 2.2   Resonance energy in benzene.

[20]Clark, Ref. 14, p. 141.
[21]Another method of calculating such properies is molecular mechanics (p. 149).

drogenation.[22] Thus, the heat of hydrogenation of cyclohexene is 28.6 kcal/mol (120 kJ/mol), so we might expect a hypothetical **1** or **2** with three double bonds to have a heat of hydrogenation of about 85.8 kcal/mol (360 kJ/mol). The real benzene has a heat of hydrogenation of 49.8 kcal/mol (208 kJ/mol), which gives a resonance energy of 36 kcal/mol (152 kJ/mol). By any calculation the real molecule is more stable than a hypothetical **1** or **2**.

The energies of the six benzene orbitals can be calculated from HMO theory in terms of two quantities, α and β. α is the amount of energy possessed by an isolated $2p$ orbital before overlap, while β (called the *resonance integral*) is an energy unit expressing the degree of stabilization resulting from π-orbital overlap. A negative value of β corresponds to stabilization, and the energies of the six orbitals are (lowest to highest): $\alpha + 2\beta$, $\alpha + \beta$, $\alpha + \beta$, $\alpha - \beta$, $\alpha - \beta$, and $\alpha - 2\beta$.[23] The total energy of the three occupied orbitals is $6\alpha + 8\beta$, since there are two electrons in each orbital. The energy of an ordinary double bond is $\alpha + \beta$, so that structure **1** or **2** has an energy of $6\alpha + 6\beta$. The resonance energy of benzene is therefore 2β. Unfortunately, there is no convenient way to calculate the value of β from molecular-orbital theory. It is often given for benzene as about 18 kcal/mol (76 kJ/mol); this number being half of the resonance energy calculated from heats of combustion or hydrogenation.

We might expect that bond distances in compounds exhibiting delocalization would lie between the values gives in Table 1.5. This is certainly the case for benzene, since the carbon–carbon bond distance is 1.40 Å,[24] which is between the 1.48 Å for an sp^2–sp^2 C—C single bond and the 1.32 Å of the sp^2–sp^2 C=C double bond.[25]

Kinds of Molecules That Have Delocalized Bonds

There are three main types of structure that exhibit delocalization:

1. *Double (or triple) bonds in conjugation.*[26] Benzene is, of course, an example, but the simplest is butadiene. In the molecular orbital picture (Figure 2.3), the overlap of four orbitals gives two bonding orbitals that contain the four electrons and two vacant antibonding orbitals. It can be seen that each orbital has one more node than the one of next lower energy. The energies of the four orbitals are (lowest to highest): $\alpha + 1.618\beta$, $\alpha + 0.618\beta$, $\alpha - 0.618\beta$, and $\alpha - 1.618\beta$; hence the total energy of the two occupied orbitals is $4\alpha + 4.472\beta$. Since the energy of two isolated double bonds is $4\alpha + 4\beta$, the resonance energy by this calculation is 0.472β.

In the resonance picture, these structures are considered to contribute:

$$CH_2=CH-CH=CH_2 \longleftrightarrow \overset{\oplus}{C}H_2-CH=CH-\overset{\ominus}{C}H_2 \longleftrightarrow \overset{\ominus}{C}H_2-CH=CH-\overset{\oplus}{C}H_2$$

$$\mathbf{6} \qquad\qquad\qquad \mathbf{7} \qquad\qquad\qquad \mathbf{8}$$

[22]For a review of heats of hydrogenation, with tables of values, see Jensen *Prog. Phys. Org. Chem.* **1976**, *12*, 189-228.

[23]For the method for calculating these and similar results given in this chapter, see Higasi; Baba; Rembaum *Quantum Organic Chemistry*; Interscience: New York, 1965. For values of calculated orbital energies and bond orders for many conjugated molecules, see Coulson; Streitwieser *Dictionary of π Electron Calculations*; W.H. Freeman: San Francisco, 1965.

[24]Bastiansen; Fernholt; Seip; Kambara; Kuchitsu *J. Mol. Struct.* **1973**, *18*, 163; Tamagawa; Iijima; Kimura *J. Mol. Struct.* **1976**, *30*, 243.

[25]The average C—C bond distance in aromatic rings is 1.38 Å: Allen; Kennard; Watson; Brammer; Orpen; Taylor *J. Chem. Soc., Perkin Trans. 2* **1987**, p. S8.

[26]For reviews of conjugation in open-chain hydrocarbons, see Simmons *Prog. Phys. Org. Chem.* **1970**, *7*, 1-50; Popov; Kogan *Russ. Chem. Rev.* **1968**, *37*, 119-141.

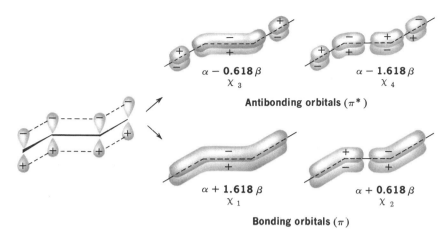

$$\alpha - 0.618\,\beta$$
$$\chi_3$$

$$\alpha - 1.618\,\beta$$
$$\chi_4$$

Antibonding orbitals (π^*)

$$\alpha + 1.618\,\beta$$
$$\chi_1$$

$$\alpha + 0.618\,\beta$$
$$\chi_2$$

Bonding orbitals (π)

FIGURE 2.3 The four π orbitals of butadiene, formed by overlap of four p orbitals.

In either picture the bond order of the central bond should be higher than 1 and that of the other carbon–carbon bonds less than 2, although neither predicts that the three bonds have equal electron density. Molecular-orbital bond orders of 1.894 and 1.447 have been calculated.[27]

Since about 1959 doubt has been cast on the reality of delocalization in butadiene and similar molecules. Thus, the bond lengths in butadiene are 1.34 Å for the double bonds and 1.48 Å for the single bond.[28] Since the typical single-bond distance of a bond that is not adjacent to an unsaturated group is 1.53 Å (p. 20), it has been argued that the shorter single bond in butadiene provides evidence for resonance. However, this shortening can also be explained by hybridization changes (see p. 20); and other explanations have also been offered.[29] Resonance energies for butadienes, calculated from heats of combustion or hydrogenation, are only about 4 kcal/mol (17 kJ/mol), and these values may not be entirely attributable to resonance. Thus, a calculation from heat of atomization data gives a resonance energy of 4.6 kcal/mol (19 kJ/mol) for *cis*-1,3-pentadiene, and −0.2 kcal/mol (−0.8 kJ/mol), for 1,4-pentadiene. These two compounds, each of which possesses two double bonds, two C—C single bonds, and eight C—H bonds, would seem to offer as similar a comparison as we could make of a conjugated with a nonconjugated compound, but they are nevertheless not strictly comparable. The former has three sp^3 C—H and five sp^2 C—H bonds, while the latter has two and six, respectively. Also, the two single C—C bonds of the 1,4-diene are both sp^2–sp^3 bonds, while in the 1,3-diene, one is sp^2–sp^3 and the other sp^2–sp^2. Therefore, it may be that some of the already small value of 4 kcal/mol (17 kJ/mol) is not resonance energy but arises from differing energies of bonds of different hybridization.[30]

[27]Coulson *Proc. R. Soc. London, Ser. A* **1939**, *169*, 413.
[28]Marais; Sheppard; Stoicheff *Tetrahedron* **1962**, 17, 163.
[29]Bartell *J. Am. Chem. Soc.* **1959**, *81*, 3497, *Tetrahedron* **1962**, *17*, 177, **1978**, *34*, 2891, *J. Chem. Educ.* **1968**, *45*, 754-767; Wilson *Tetrahedron*, **1962**, *17*, 191; Hughes *Tetrahedron* **1968**, *24*, 6423; Politzer; Harris *Tetrahedron* **1971**, 27, 1567.
[30]For negative views on delocalization in butadiene and similar molecules, see Dewar; Gleicher *J. Am. Chem. Soc.* **1965**, *87*, 692; Dewar; Schmeising *Tetrahedron* **1959**, *5*, 166, **1960**, *11*, 96; Brown *Trans. Faraday Soc.* **1959**, *55*, 694; Somayajulu *J. Chem. Phys.* **1959**, *31*, 919; Mikhailov *Bull. Acad. Sci. USSR, Div. Chem. Sci.* **1960**, 1284; *J. Gen. Chem. USSR* **1966**, *36*, 379. For positive views, see Miyazaki; Shigetani; Shinoda *Bull. Chem. Soc. Jpn.* **1971**, *44*, 1491; Berry *J. Chem. Phys.* **1962**, *30*, 936; Kogan; Popov *Bull. Acad. Sci. USSR, Div. Chem. Sci.* **1964**, 1306; Altmann; Reynolds *J. Mol. Struct.* **1977**, *36*, 149. In general, the negative argument is that resonance involving excited structures, such as **7** and **8**, is unimportant. See rule 6 on p. 35. An excellent discussion of the controversy is found in Popov; Kogan Ref. 26, pp. 119-124.

Although bond distances fail to show it and the resonance energy is low, the fact that butadiene is planar[31] shows that there is some delocalization, even if not as much as previously thought. Similar delocalization is found in other conjugated systems (e.g., $C=C-C=O$[32] and $C=C-C=N$), in longer systems with three or more multiple bonds in conjugation, and where double or triple bonds are conjugated with aromatic rings.

2. *Double (or triple) bonds in conjugation with a p orbital on an adjacent atom.* Where a *p* orbital is on an atom adjacent to a double bond, there are three parallel *p* orbitals that overlap. As previously noted, it is a general rule that the overlap of *n* atomic orbitals creates *n* molecular orbitals, so overlap of a *p* orbital with an adjacent double bond gives rise to three new orbitals, as shown in Figure 2.4. The middle orbital is a *nonbonding orbital* of zero bonding energy. The central carbon atom does not participate in the nonbonding orbital.

There are three cases: the original *p* orbital may have contained two, one, or no electrons. Since the original double bond contributes two electrons, the total number of electrons accommodated by the new orbitals is four, three, or two. A typical example of the first situation is vinyl chloride $CH_2=CH-Cl$. Although the *p* orbital of the chlorine atom is filled, it still overlaps with the double bond. The four electrons occupy the two molecular orbitals of lowest energies. This is our first example of resonance involving overlap between unfilled orbitals and a *filled* orbital. Canonical forms for vinyl chloride are:

$$CH_2=CH-\overline{\underline{Cl}} \longleftrightarrow \overset{\ominus}{CH_2}-CH=\overset{\oplus}{\underline{Cl}}$$

Any system containing an atom that has an unshared pair and that is directly attached to a multiple-bond atom can show this type of delocalization. Another example is the carbonate ion:

The bonding in allylic carbanions, e.g., $CH_2=CH-\overline{CH_2}{}^{\ominus}$, is similar.

The other two cases, where the original *p* orbital contains only one or no electron, are generally found only in free radicals and cations, respectively. Allylic free radicals have one electron in the nonbonding orbital. In allylic cations this orbital is vacant and only the bonding orbital is occupied. The orbital structures of the allylic carbanion, free radical, and cation differ from each other, therefore, only in that the nonbonding orbital is filled, half-filled, or empty. Since this is an orbital of zero bonding energy, it follows that the bonding π energies of the three species relative to electrons in the $2p$ orbitals of free atoms are the same. The electrons in the nonbonding orbital do not contribute to the bonding energy, positively or negatively.[33]

---

[31]Ref. 28; Fisher; Michl *J. Am. Chem. Soc.* **1987,** *109,* 1056; Wiberg; Rosenberg; Rablen *J. Am. Chem. Soc.* **1991,** *113,* 2890.

[32]For a treatise on $C=C-C=O$ systems, see Patai; Rappoport *The Chemistry of Enones,* two parts; Wiley: New York, 1989.

[33]It has been contended that here too, as with the benzene ring (Ref. 6), the geometry is forced upon allylic systems by the σ framework, and not the π system: Shaik; Hiberty; Ohanessian; Lefour *Nouv. J. Chim.* **1985,** *9,* 385. It has also been suggested, on the basis of ab initio calculations, that while the allyl cation has significant resonance stabilization, the allyl anion has little stabilization: Wiberg; Breneman; LePage *J. Am. Chem. Soc.* **1990,** *112,* 61.

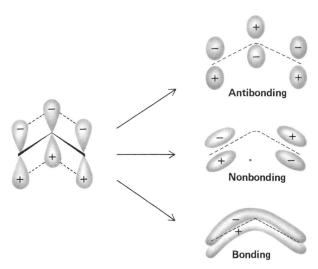

FIGURE 2.4 The three orbitals of an allylic system, formed by overlap of three *p* orbitals.

By the resonance picture, the three species may be described as having double bonds in conjugation with, respectively, an unshared pair, an unpaired electron, and an empty orbital (see Chapter 5):

$$CH_2\!=\!CH\;\;{}^{-}\overset{\ominus}{C}H_2 \longleftrightarrow \overset{\ominus}{C}H_2\!-\!CH\!=\!CH_2$$

$$CH_2\!=\!CH\!-\!\overset{\bullet}{C}H_2 \longleftrightarrow \overset{\bullet}{C}H_2\!-\!CH\!=\!CH_2$$

$$CH_2\!=\!CH\;\;\overset{\oplus}{C}H_2 \longleftrightarrow \overset{\oplus}{C}H_2\!-\!CH\!=\!CH_2$$

3. *Hyperconjugation.*   The third type of delocalization, called *hyperconjugation*, is discussed on p. 68.

We shall find examples of delocalization which cannot be strictly classified as belonging to any of these types.

Cross Conjugation[34]

In a cross-conjugated compound, three groups are present, two of which are not conjugated with each other, although each is conjugated with the third. Some examples[35] are

$$\underset{\underset{O}{\parallel}}{Ph\!-\!C\!-\!Ph} \qquad \overset{1}{C}H_2\!=\!\overset{2}{C}H\!-\!\underset{\underset{6}{CH_2}}{\overset{3}{C}}\!-\!\overset{4}{C}H\!=\!\overset{5}{C}H_2 \qquad CH_2\!=\!CH\!-\!O\!-\!CH\!=\!CH_2$$

9

---

[34]For a discussion, see Phelan; Orchin *J. Chem. Educ.* **1968**, *45*, 633-637.

[35]**9** is the simplest of a family of cross-conjugated alkenes, called *dendralenes*. For a review of these compounds, see Hopf *Angew. Chem. Int. Ed. Engl.* **1984**, *23*, 948-960 [*Angew. Chem. 96*, 947-958].

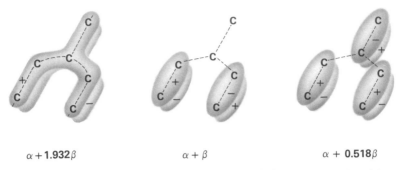

$$\alpha + 1.932\beta \qquad\qquad \alpha + \beta \qquad\qquad \alpha + 0.518\beta$$

FIGURE 2.5 The three bonding orbitals of 3-methylene-1,4-pentadiene (9).

Using the molecular-orbital method, we find that the overlap of six p orbitals in **9** gives six molecular orbitals, of which the three bonding orbitals are shown in Figure 2.5, along with their energies. Note that two of the carbon atoms do not participate in the $\alpha + \beta$ orbital. The total energy of the three occupied orbitals is $6\alpha + 6.900\beta$, so the resonance energy is 0.900β. Molecular-orbital bond orders are 1.930 for the C-1,C-2 bond, 1.859 for the C-3,C-6 bond and 1.363 for the C-2,C-3 bond.[34] Comparing these values with those for butadiene (p. 31), we see that the C-1,C-2 bond contains more and the C-3,C-6 bond less double-bond character than the double bonds in butadiene. The resonance picture supports this conclusion, since each C-1,C-2 bond is double in three of the five canonical forms, while the

$$\underset{\overset{|}{\text{CH}_2}}{\text{CH}_2\text{=CH—C—CH=CH}_2} \longleftrightarrow \underset{\overset{|}{\underset{\oplus \text{ or } \ominus}{\text{CH}_2}}}{\text{CH}_2\text{=CH—C=CH—}\overset{\ominus \text{ or } \oplus}{\text{CH}_2}} \longleftrightarrow \underset{\overset{|}{\underset{\oplus \text{ or } \ominus}{\text{CH}_2}}}{\overset{\ominus \text{ or } \oplus}{\text{CH}_2}\text{—CH=C—CH=CH}_2}$$

C-3,C-6 bond is double in only one. In most cases it is easier to treat cross-conjugated molecules by the molecular-orbital method than by the valence-bond method.

The Rules of Resonance

We have seen that one way of expressing the actual structure of a molecule containing delocalized bonds is to draw several possible structures and to assume that the actual molecule is a hybrid of them. These canonical forms have no existence except in our imaginations. The molecule does *not* rapidly shift between them. It is *not* the case that some molecules have one canonical form and some another. All the molecules of the substance have the same structure. That structure is always the same all the time and is a weighted average of all the canonical forms. In drawing canonical forms and deriving the true structures from them, we are guided by certain rules, among them the following:

1. All the canonical forms must be bona fide Lewis structures (see p. 12). For instance, none of them may have a carbon with five bonds.
2. The positions of the nuclei must be the same in all the structures. This means that all we are doing when we draw the various canonical forms is putting the *electrons* in in different

ways. For this reason, shorthand ways of representing resonance are easy to devise:

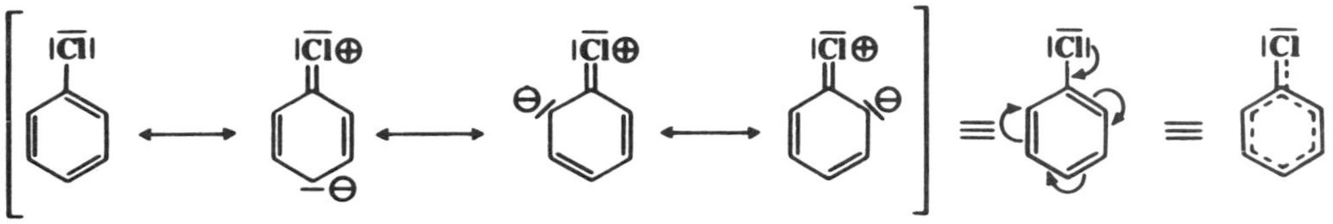

10   **11**

The resonance interaction of chlorine with the benzene ring can be represented as shown in **10** or **11** and both of these representations have been used in the literature to save space. However, we shall not use the curved-arrow method of **10** since arrows will be used in this book to express the actual movement of electrons in reactions. We will use representations like **11** or else write out the canonical forms. The convention used in dashed-line formulas like **11** is that bonds that are present in all canonical forms are drawn as solid lines while bonds that are not present in all forms are drawn as dashed lines. In most resonance, σ bonds are not involved, and only the π or unshared electrons are put in in different ways. This means that if we write one canonical form for a molecule, we can then write the others by merely moving π and unshared electrons.

3. All atoms taking part in the resonance, i.e., covered by delocalized electrons, must lie in a plane or nearly so (see p. 36). This, of course, does not apply to atoms that have the same bonding in all the canonical forms. The reason for planarity is maximum overlap of the p orbitals.

4. All canonical forms must have the same number of unpaired electrons. Thus $\overset{\bullet}{C}H_2$—CH=CH—$\overset{\bullet}{C}H_2$ is not a valid canonical form for butadiene.

5. The energy of the actual molecule is lower than that of any form, obviously. Therefore, delocalization is a stabilizing phenomenon.[36]

6. All canonical forms do not contribute equally to the true molecule. Each form contributes in proportion to its stability, the most stable form contributing most. Thus, for ethylene, the form $\overset{-}{C}H_2$—$\overset{+}{C}H_2$ has such a high energy compared to $CH_2{=}CH_2$ that it essentially does not contribute at all. We have seen the argument that such structures do not contribute even in such cases as butadiene.[30] Equivalent canonical forms, such as **1** and **2**, contribute equally. The greater the number of significant structures that can be written and the more nearly equal they are, the greater the resonance energy, other things being equal.

It is not always easy to decide relative stabilities of imaginary structures; the chemist is often guided by intuition.[37] However, the following rules may be helpful:

a. Structures with more covalent bonds are ordinarily more stable than those with fewer (compare **6** and **7**).

b. Stability is decreased by an increase in charge separation. Structures with formal charges are less stable than uncharged structures. Structures with more than two formal charges usually contribute very little. An especially unfavorable type of structure is one with two like charges on adjacent atoms.

---

[36]It has been argued that resonance is not a stabilizing phenomenon in all systems, especially in acyclic ions: Wiberg *Chemtracts: Org. Chem.* **1989**, *2*, 85. See also Ref. 120 in Chapter 8.

[37]A quantitative method for weighting canonical forms has been proposed by Gasteiger; Saller *Angew. Chem. Int. Ed. Engl.* **1985**, *24*, 687 [*Angew. Chem. 97*, 699].

c. Structures that carry a negative charge on a more electronegative atom are more stable than those in which the charge is on a less electronegative atom. Thus, **13** is more stable than **12**. Similarly, positive charges are best carried on atoms of low electronegativity.

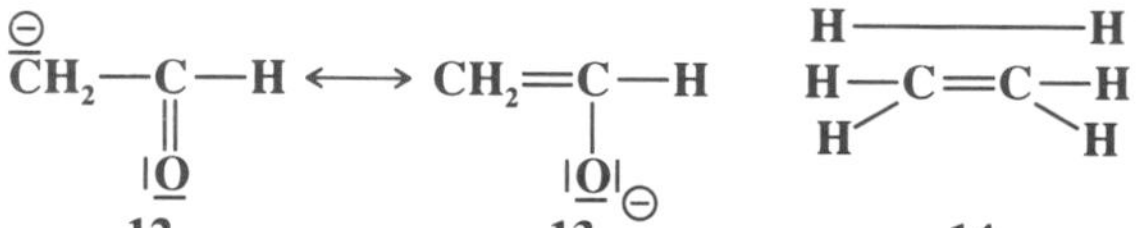

d. Structures with distorted bond angles or lengths are unstable, e.g., the structure **14** for ethane.

The Resonance Effect

Resonance always results in a different distribution of electron density than would be the case if there were no resonance. For example, if **15** were the actual structure of aniline, the

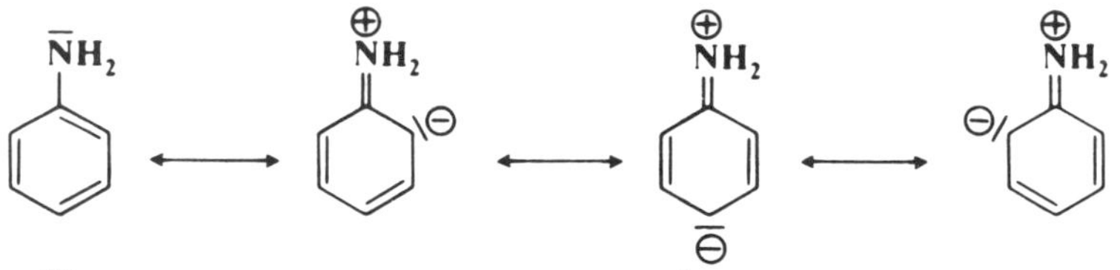

two unshared electrons of the nitrogen would reside entirely on that atom. Since the real structure is not **15** but a hybrid that includes contributions from the other canonical forms shown, the electron density of the unshared pair does not reside entirely on the nitrogen, but is spread over the ring. This decrease in electron density at one position (and corresponding increase elsewhere) is called the *resonance* or *mesomeric effect*. We loosely say that the NH_2 contributes or donates electrons to the ring by a resonance effect, although no actual contribution takes place. The "effect" is caused by the fact that the electrons are in a different place from that we would expect if there were no resonance. In ammonia, where resonance is absent, the unshared pair *is* located on the nitrogen atom. As with the field effect (p. 18), we think of a certain molecule (in this case ammonia) as a substrate and then see what happens to the electron density when we make a substitution. When one of the hydrogen atoms of the ammonia molecule is replaced by a benzene ring, the electrons are "withdrawn" by the resonance effect, just as when a methyl group replaces a hydrogen of benzene, electrons are "donated" by the field effect of the methyl. The idea of donation or withdrawal merely arises from the comparison of a compound with a closely related one or a real compound with a canonical form.

Steric Inhibition of Resonance

Rule 3 states that all the atoms covered by delocalized electrons must lie in a plane or nearly so. Many examples are known where resonance is reduced or prevented because the atoms are sterically forced out of planarity.

Bond lengths for the *o*- and *p*-nitro groups in picryl iodide are quite different.[38] Distance *a* is 1.45 Å, whereas *b* is 1.35 Å. The obvious explanation is that the oxygens of the *p*-nitro group are in the plane of the ring and thus in resonance with it, so that *b* has partial

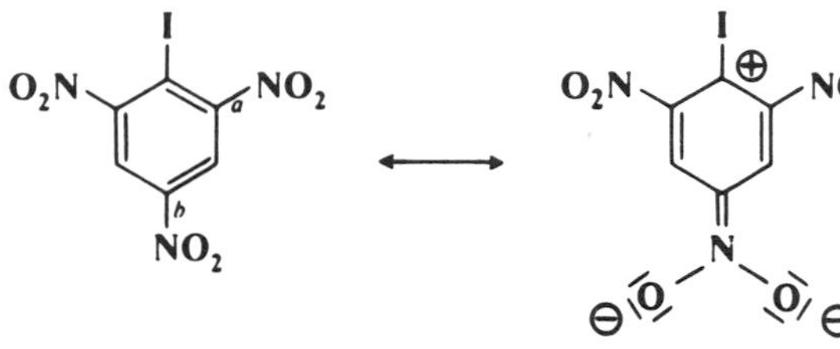

double-bond character, while the oxygens of the *o*-nitro groups are forced out of the plane by the large iodine atom.

The Dewar-type structure for the central ring of the anthracene system in **16** is possible only because the 9,10 substituents prevent the system from being planar.[39] **16** is the actual structure of the molecule and is not in resonance with forms like **17**, although in anthracene

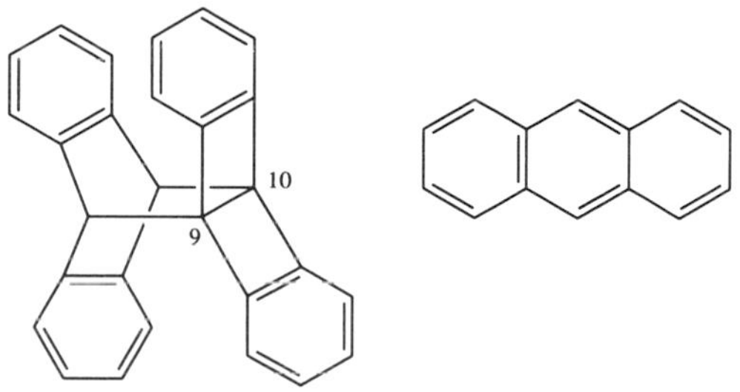

16   **17**

itself, Dewar structures and structures like **17** both contribute. This is a consequence of rule 2 (p. 34). In order for a **17**-like structure to contribute to resonance in **16**, the nuclei would have to be in the same positions in both forms.

Even the benzene ring can be forced out of planarity.[40] In [5]paracyclophane[41] (**18**) the presence of a short bridge (this is the shortest para bridge known for a benzene ring) forces the benzene ring to become boat-shaped. The parent **18** has so far not proven stable enough

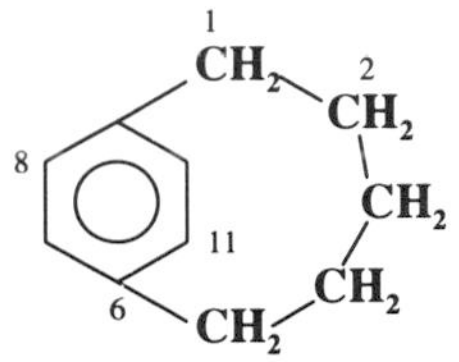

18

[38]Wepster, *Prog. Stereochem.* **1958**, *2*, 99-156, p. 125. For another example of this type of steric inhibition of resonance, see Exner; Folli; Marcaccioli; Vivarelli *J. Chem. Soc., Perkin Trans. 2* **1983**, 757.

[39]Applequist; Searle *J. Am. Chem. Soc.* **1964**, *86*, 1389.

[40]For a review of planarity in aromatic systems, see Ferguson; Robertson *Adv. Phys. Org. Chem.* **1963**, *1*, 203-281.

[41]For a monograph, see Keehn; Rosenfeld *Cyclophanes*, 2 vols.; Academic Press: New York, 1983. For reviews, see Bickelhaupt, *Pure Appl. Chem.* **1990**, *62*, 373-382; Vögtle; Hohner *Top. Curr. Chem.* **1978**, *74*, 1-29; Cram; Cram *Acc. Chem. Res.* **1971**, *4*, 204-213; Vögtle; Neumann *Top. Curr. Chem.* **1974**, *48*, 67-129; and reviews in *Top. Curr. Chem.* **1983**, *113*, 1-185; *115*, 1-163.

for isolation, but a uv spectrum was obtained and showed that the benzene ring was still aromatic, despite the distorted ring.[42] The 8,11-dichloro analog of **18** is a stable solid, and x-ray diffraction showed that the benzene ring is boat-shaped, with one end of the boat bending about 27° out of the plane, and the other about 12°.[43] This compound too is aromatic, as shown by uv and nmr spectra. [6]Paracyclophanes are also bent,[44] but in [7]paracyclophanes the bridge is long enough so that the ring is only moderately distorted. Similarly, [n,m]paracyclophanes (**19**), where n and m are both 3 or less (the smallest yet prepared is [2.2]paracyclophane), have bent (boat-shaped) benzene rings. All these compounds have properties that depart significantly from those of ordinary benzene compounds.

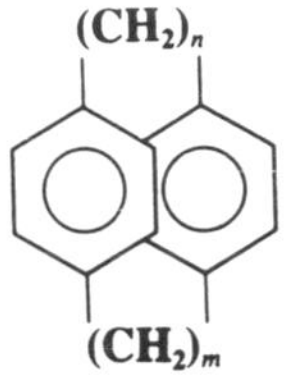

19 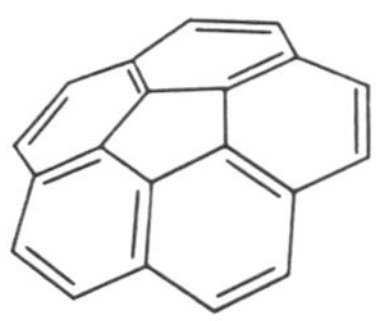   20 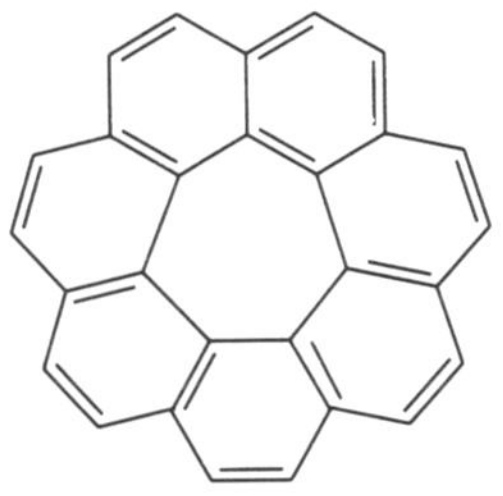   21

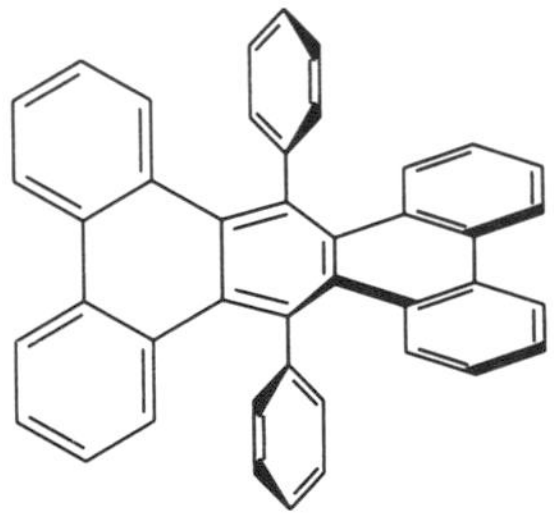

22 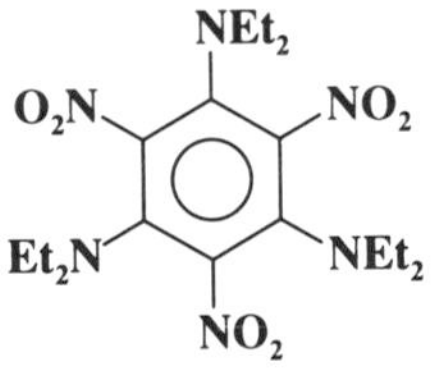   23

Other molecules in which benzene rings are forced out of planarity are corannulene (**20**),[45] (also called 5-circulene), 7-circulene (**21**),[46] **22**,[47] and **23**[48] (see also p. 161).

pπ–dπ Bonding. Ylides

We have mentioned (p. 9) that, in general, atoms of the second row of the periodic table do not form stable double bonds of the type discussed in Chapter 1 (π bonds formed by

[42]Jenneskens; de Kanter; Kraakman; Turkenburg; Koolhaas; de Wolf; Bickelhaupt; Tobe; Kakiuchi; Odaira *J. Am. Chem. Soc.* **1985**, *107*, 3716. See also Tobe; Kaneda; Kakiuchi; Odaira *Chem. Lett.* **1985**, 1301; Kostermans; de Wolf; Bickelhaupt *Tetrahedron Lett.* **1986**, *27*, 1095; van Zijl; Jenneskens; Bastiaan; MacLean; de Wolf; Bickelhaupt *J. Am. Chem. Soc.* **1986**, *108*, 1415; Rice; Lee; Remington; Allen; Clabo; Schaefer *J. Am. Chem. Soc.* **1987**, *109*, 2902.

[43]Jenneskens; Klamer; de Boer; de Wolf; Bickelhaupt; Stam *Angew. Chem. Int. Ed. Engl.* **1984**, *23*, 238 [*Angew. Chem. 96*, 236].

[44]See, for example, Liebe; Wolff; Krieger; Weiss; Tochtermann *Chem. Ber.* **1985**, *118*, 4144; Tobe; Ueda; Kakiuchi; Odaira; Kai; Kasai *Tetrahedron* **1986**, *42*, 1851.

[45]Barth; Lawton *J. Am. Chem. Soc.* **1971**, *93*, 1730; Scott; Hashemi; Meyer; Warren *J. Am. Chem. Soc.* **1991**, *113*, 7082.

[46]Yamamoto; Harada; Okamoto; Chikamatsu; Nakazaki; Kai; Nakao; Tanaka; Harada; Kasai *J. Am. Chem. Soc.* **1988**, *110*, 3578.

[47]Pascal; McMillan; Van Engen; Eason *J. Am. Chem. Soc.* **1987**, *109*, 4660.

[48]Chance; Kahr; Buda; Siegel *J. Am. Chem. Soc.* **1989**, *111*, 5940.

overlap of parallel p orbitals). However, there is another type of double bond that is particularly common for the second row atoms, sulfur and phosphorus. For example, such a double bond is found in the compound H_2SO_3, as written on the left. Like an ordinary

$$H\!-\!\overline{\underline{O}}\!-\!\overline{\underline{S}}\!-\!\overline{\underline{O}}\!-\!H \longleftrightarrow H\!-\!\overline{\underline{O}}\!-\!\overset{\oplus}{\underline{S}}\!-\!\overline{\underline{O}}\!-\!H$$
$$\underset{|\underline{O}}{\|} \qquad\qquad \underset{|\underline{O}|_\ominus}{|}$$

double bond, this double bond contains one σ orbital, but the second orbital is not a π orbital formed by overlap of half-filled p orbitals; instead it is formed by overlap of a filled p orbital from the oxygen with an empty d orbital from the sulfur. It is called a $p\pi$–$d\pi$ *orbital*.[49] Note that we can represent this molecule by two canonical forms but the bond is nevertheless localized, despite the resonance. Some other examples of $p\pi$–$d\pi$ bonding are

$$R_3P\!=\!\overline{O}| \longleftrightarrow \overset{\oplus}{R_3P}\!-\!\overline{O}|^{\ominus}$$

Phosphine oxides

$$R\!-\!\underset{\underset{|\underline{O}}{\|}}{\overset{|\overline{O}}{\overset{\|}{S}}}\!-\!R \longleftrightarrow R\!-\!\underset{\underset{|\underline{O}|_\ominus}{|}}{\overset{|\overline{O}|^\ominus}{\overset{\oplus}{S}}}\!-\!R$$

Sulfones

$$H\!-\!\underset{\overset{|}{O}H}{\overset{\overset{H}{|}}{P}}\!=\!\overline{O}| \longleftrightarrow H\!-\!\underset{\overset{|}{O}H}{\overset{\overset{H}{|}}{\overset{\oplus}{P}}}\!-\!\overline{\underline{O}}|^{\ominus}$$

Hypophosphorous acid

$$R\!-\!\underset{\underset{|\underline{O}}{\|}}{\overline{S}}\!-\!R \longleftrightarrow R\!-\!\underset{\underset{|\underline{O}|_\ominus}{|}}{\overset{\oplus}{\overline{S}}}\!-\!R$$

Sulfoxides

Nitrogen analogs are known for some of these phosphorus compounds, but they are less stable because the resonance is lacking. For example, amine oxides, analogs of phosphine oxides, can only be written $R_3\overset{\oplus}{N}\!-\!\overset{\ominus}{O}$. The $p\pi$–$d\pi$ canonical form is impossible since nitrogen is limited to eight outer-shell electrons.

In all the examples given above the atom that donates the electron pair is oxygen and, indeed, oxygen is the most common such atom. But in another important class of compounds, called *ylides*, this atom is carbon.[50] There are three main types of ylides—phosphorus,[51]

$$R_3P\!=\!CR_2 \longleftrightarrow \overset{\oplus}{R_3P}\!-\!\overline{C}R_2^{\ominus}$$

Phosphorus ylides

$$R_2S\!=\!CR_2 \longleftrightarrow \overset{\oplus}{R_2S}\!-\!\overline{C}R_2^{\ominus} \qquad \overset{\oplus}{R_3N}\!-\!\overline{C}R_2^{\ominus}$$

Sulfur ylides   Nitrogen ylides

[49]For a monograph, see Kwart; King *d-Orbitals in the Chemistry of Silicon, Phosphorus, and Sulfur*; Springer: New York, 1977.

[50]For a monograph, see Johnson *Ylid Chemistry*; Academic Press: New York, 1966. For reviews, see Morris, *Surv. Prog. Chem.* **1983**, *10*, 189-257; Hudson *Chem. Br.* **1971**, *7*, 287-294; Lowe *Chem. Ind.* (*London*) **1970**, 1070-1079. For a review on the formation of ylides from the reaction of carbenes and carbenoids with heteroatom lone pairs, see Padwa; Hornbuckle *Chem. Rev.* **1991**, *91*, 263-309.

[51]Although the phosphorus ylide shown has three R groups on the phosphorus atom, other phosphorus ylides are known where other atoms, e.g., oxygen, replace one or more of these R groups. When the three groups are all alkyl or aryl, the phosphorus ylide is also called a *phosphorane*.

nitrogen,[52] and sulfur ylides,[53] although arsenic,[54] selenium, etc., ylides are also known. Ylides may be defined as compounds in which a positively charged atom from group 15 or 16 of the periodic table is connected to a carbon atom carrying an unshared pair of electrons. Because of $p\pi$–$d\pi$ bonding, two canonical forms can be written for phosphorus and sulfur, but there is only one for nitrogen ylides. Phosphorus ylides are much more stable than nitrogen ylides (see also p. 957). Sulfur ylides also have a low stability.

In almost all compounds that have $p\pi$–$d\pi$ bonds, the central atom is connected to four atoms or three atoms and an unshared pair and the bonding is approximately tetrahedral. The $p\pi$–$d\pi$ bond, therefore, does not greatly change the geometry of the molecule in contrast to the normal π bond, which changes an atom from tetrahedral to trigonal.

AROMATICITY

In the nineteenth century it was recognized that aromatic compounds[55] differ greatly from unsaturated aliphatic compounds,[56] but for many years chemists were hard pressed to arrive at a mutually satisfactory definition of aromatic character [57] Qualitatively, there has never been real disagreement. Definitions have taken the form that aromatic compounds are characterized by a special stability and that they undergo substitution reactions more easily than addition reactions. The difficulty arises because these definitions are vague and not easy to apply in borderline cases. In 1925 Armit and Robinson[58] recognized that the aromatic properties of the benzene ring are related to the presence of a closed loop of electrons, the *aromatic sextet* (aromatic compounds are thus the arch examples of delocalized bonding), but it still was not easy to determine whether rings other than the benzene ring possessed such a loop. With the advent of magnetic techniques, most notably nmr, it is possible to determine experimentally whether or not a compound has a closed ring of electrons; aromaticity can now be defined as the *ability to sustain an induced ring current*. A compound with this ability is called *diatropic*. Although this definition also has its flaws,[59] it is the one most commonly accepted today. There are several methods of determining whether a compound can sustain a ring current, but the most important one is based on nmr chemical

[52]For a review of nitrogen ylides, see Musker *Fortschr. Chem. Forsch.* **1970**, *14*, 295-365.

[53]For a monograph on sulfur ylides, see Trost; Melvin *Sulfur Ylides*; Academic Press: New York, 1975. For reviews, see Fava in Bernardi; Csizmadia; Mangini *Organic Sulfur Chemistry;* Elsevier: New York, 1985, pp. 299-354; Belkin; Polezhaeva *Russ. Chem. Rev.* **1981**, *50*, 481-497; Block, in Stirling *The Chemistry of the Sulphonium Group*, part 2, Wiley: New York, 1981, pp. 680-702; Block *Reactions of Organosulfur Compounds*; Academic Press: New York, 1978, pp. 91-127.

[54]For reviews of arsenic ylides, see Lloyd; Gosney; Ormiston *Chem. Soc. Rev.* **1987**, *16*, 45-74; Yaozeng; Yanchang *Adv. Organomet. Chem.* **1982**, *20*, 115-157.

[55]For books on aromaticity, see Lloyd *The Chemistry of Conjugated Cyclic Compounds*; Wiley: New York, 1989; *Non-Benzenoid Conjugated Carbocyclic Compounds*; Elsevier: New York, 1984; Garratt *Aromaticity*; Wiley: New York, 1986; Balaban; Banciu; Ciorba *Annulenes, Benzo-, Hetero-, Homo-Derivatives and their Valence Isomers*, 3 vols.; CRC Press: Boca Raton, FL, 1987; Badger *Aromatic Character and Aromaticity*; Cambridge University Press: Cambridge, 1969; Snyder *Nonbenzenoid Aromatics*, 2 vols.; Academic Press: New York, 1969-1971; Bergmann; Pullman *Aromaticity, Pseudo-Aromaticity, and Anti-Aromaticity*; Israel Academy of Sciences and Humanities: Jerusalem, 1971; *Aromaticity*; *Chem. Soc. Spec. Pub.* no. 21, 1967. For reviews, see Gorelik *Russ. Chem. Rev.* **1990**, *59*, 116-133; Stevenson *Mol. Struct. Energ.* **1986**, *3*, 57-83; Sondheimer *Chimia* **1974**, *28*, 163-172; Cresp; Sargent *Essays Chem.* **1972**, *4*, 91-114; Figeys *Top. Carbocyclic Chem.* **1969**, *1*, 269-359; Garratt; Sargent *Adv. Org. Chem.* **1969**, *6*, 1-108; and papers in *Top. Curr. Chem.* **1990**, *153* and *Pure Appl. Chem.* **1980**, *52*, 1397-1667.

[56]For an account of the early history of aromaticity, see Snyder, in Snyder, Ref. 55, vol. 1, pp. 1-31. See also Balaban *Pure Appl. Chem.* **1980**, *52*, 1409.

[57]For a review of the criteria used to define aromatic character, see Jones *Rev. Pure Appl. Chem.* **1968**, *18*, 253-280. For methods of assigning aromaticity, see Jug; Köster *J. Phys. Org. Chem.* **1991**, *4*, 163; Zhou; Parr *J. Am. Chem. Soc.* **1989**, *111*, 7371; Katritzky; Barczynski; Musumarra; Pisano; Szafran *J. Am. Chem. Soc.* **1989**, *111*, 7; Schaad; Hess *J. Am. Chem. Soc.* **1972**, *94*, 3068, *J. Chem. Educ.* **1974**, *51*, 640. See also Ref. 85.

[58]Armit; Robinson *J. Chem. Soc.* **1925**, *127*, 1604.

[59]Jones, Ref. 57, pp. 266-274; Mallion *Pure Appl. Chem.* **1980**, *52*, 1541.

shifts.[60] In order to understand this, it is necessary to remember that, as a general rule, the value of the chemical shift of a proton in an nmr spectrum depends on the electron density of its bond; the greater the density of the electron cloud surrounding or partially surrounding a proton, the more upfield is its chemical shift (a lower value of δ). However, this rule has several exceptions; one is for protons in the vicinity of an aromatic ring. When an external magnetic field is imposed upon an aromatic ring (as in an nmr instrument), the closed loop of aromatic electrons circulates in a diamagnetic ring current, which sends out a field of its

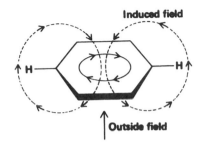

own. As can be seen in the diagram, this induced field curves around and in the area of the proton is parallel to the external field, so the field "seen" by the aromatic protons is greater than it would have been in the absence of the diamagnetic ring current. The protons are moved downfield (to higher δ) compared to where they would be if electron density were the only factor. Thus ordinary olefinic hydrogens are found at approximately 5 to 6 δ, while the hydrogens of benzene rings are located at about 7 to 8 δ. However, if there were protons located above or within the ring, they would be subjected to a *decreased* field and should appear at lower δ values than normal CH_2 groups (normal δ for CH_2 is approximately 1 to 2). The nmr spectrum of [10]paracyclophane (**24**) showed that this was indeed the case[61]

24

and that the CH_2 peaks were shifted to lower δ the closer they were to the middle of the chain.

It follows that aromaticity can be determined from an nmr spectrum. If the protons attached to the ring are shifted downfield from the normal olefinic region, we can conclude that the molecule is diatropic and hence aromatic. In addition, if the compound has protons above or within the ring (we shall see an example of the latter on p. 60), then if the compound is diatropic, these will be shifted upfield. One drawback to this method is that it cannot be applied to compounds that have no protons in either category, e.g., the dianion of squaric acid (p. 66). Unfortunately, ^{13}C nmr is of no help here, since these spectra do not show ring currents.[62]

---

[60]For a review of nmr and other magnetic properties with respect to aromaticity, see Haddon; Haddon; Jackman *Fortschr. Chem. Forsch.* **1971**, *16*, 103-220. For an example of a magentic method other than nmr, see Dauben; Wilson; Laity, in Snyder, Ref. 55, vol. 2, pp. 167-206.
[61]Waugh; Fessenden *J. Am. Chem. Soc.* **1957**, *79*, 846. See also Shapiro; Gattuso; Sullivan *Tetrahedron Lett.* **1971**, 223; Pascal; Winans; Van Engen *J. Am. Chem. Soc.* **1989**, *111*, 3007.
[62]For a review of ^{13}C spectra of aromatic compounds, see Günther; Schmickler *Pure Appl. Chem.* **1975**, *44*, 807-828.

It should be emphasized that the old and new definitions of aromaticity are not necessarily parallel. If a compound is diatropic and therefore aromatic under the new definition, it is more stable than the canonical form of lowest energy, but this does not mean that it will be stable to air, light, or common reagents, since *this* stability is determined not by the resonance energy but by the difference in free energy between the molecule and the transition states for the reactions involved; and these differences may be quite small, even if the resonance energy is large. A unified theory has been developed that relates ring currents, resonance energies, and aromatic character.[63]

The vast majority of aromatic compounds have a closed loop of six electrons in a ring (the aromatic sextet), and we consider these compounds first.[64]

Six-Membered Rings

Not only is the benzene ring aromatic, but so are many heterocyclic analogs in which one or more hetero atoms replace carbon in the ring.[65] When nitrogen is the hetero atom, little difference is made in the sextet and the unshared pair of the nitrogen does not participate in the aromaticity. Therefore, derivatives such as N-oxides or pyridinium ions are still aromatic. However, for nitrogen heterocycles there are more significant canonical forms (e.g., **25**) than for benzene. Where oxygen or sulfur is the hetero atom, it must be present

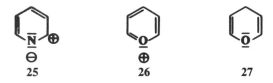

in its ionic form (**26**) in order to possess the valence of 3 that participation in such a system demands. Thus, pyran (**27**) is not aromatic, but the pyrylium ion (**26**) is.[66]

In systems of fused six-membered aromatic rings,[67] the principal canonical forms are usually not all equivalent. **28** has a central double bond and is thus different from the other two canonical forms of naphthalene, which are equivalent to each other.[68] For naphthalene,

[63]Haddon *J. Am. Chem. Soc.* **1979**, *101*, 1722; Haddon; Fukunaga *Tetrahedron Lett.* **1980**, *21*, 1191.

[64]Values of molecular-orbital energies for many aromatic systems, calculated by the HMO method, are given in Coulson; Streitwieser, Ref. 23. Values calculated by a variation of the SCF method are given by Dewar; Trinajstić *Collect. Czech. Chem. Commun.* **1970**, *35*, 3136, 3484.

[65]For reviews of aromaticity of heterocycles, see Katritzky; Karelson; Malhotra *Heterocycles* **1991**, *32*, 127-161; Cook; Katritzky; Linda *Adv. Heterocycl. Chem.* **1974**, *17*, 255-356.

[66]For a review of pyrylium salts, see Balaban; Schroth; Fischer *Adv. Heterocycl. Chem.* **1969**, *10*, 241-326.

[67]For books on this subject, see Gutman; Cyvin *Introduction to the Theory of Benzenoid Hydrocarbons*; Springer: New York, 1989; Dias *Handbook of Polycyclic Hydrocarbons—Part A: Benzenoid Hydrocarbons*; Elsevier: New York, 1987; Clar *Polycyclic Hydrocarbons*, 2 vols.; Academic Press: New York, 1964. For a "periodic table" that systematizes fused aromatic hydrocarbons, see Dias *Acc. Chem. Res.* **1985**, *18*, 241-248; *Top. Curr. Chem.* **1990**, *253*, 123-143, *J. Phys. Org. Chem.* **1990**, *3*, 765.

[68]As the size of a given fused ring system increases, it becomes more difficult to draw all the canonical forms. For discussions of methods for doing this, see Herndon *J. Chem. Educ.* **1974**, *51*, 10-15; Cyvin; Cyvin; Brunvoll; Chen *Monatsh. Chem.* **1989**, *120*, 833; Fuji; Xiaofeng; Rongsi *Top. Curr. Chem.* **1990**, *153*, 181; Wenchen; Wenjie *Top. Curr. Chem.* **1990**, *153*, 195; Sheng *Top. Curr. Chem.* **1990**, *153*, 211; Rongsi; Cyvin; Cyvin; Brunvoll; Klein *Top. Curr. Chem.* **1990**, *153*, 227, and references cited in these papers. For a monograph, see Cyvin; Gutman *Kekulé Structures in Benzenoid Hydrocarbons*; Springer, New York, 1988.

these are the only forms that can be drawn without consideration of Dewar forms or those with charge separation.[69] If we assume that the three forms contribute equally, the 1,2 bond has more double-bond character than the 2,3 bond. Molecular-orbital calculations show bond orders of 1.724 and 1.603, respectively (compare benzene, 1.667). In agreement with these predictions, the 1,2 and 2,3 bond distances are 1.36 and 1.415 Å, respectively,[70] and ozone preferentially attacks the 1,2 bond.[71] This nonequivalency of bonds, called *partial bond fixation*,[72] is found in nearly all fused aromatic systems. In phenanthrene, where the 9,10 bond is a single bond in only one of five forms, bond fixation becomes extreme and this bond is readily attacked by many reagents:[73]

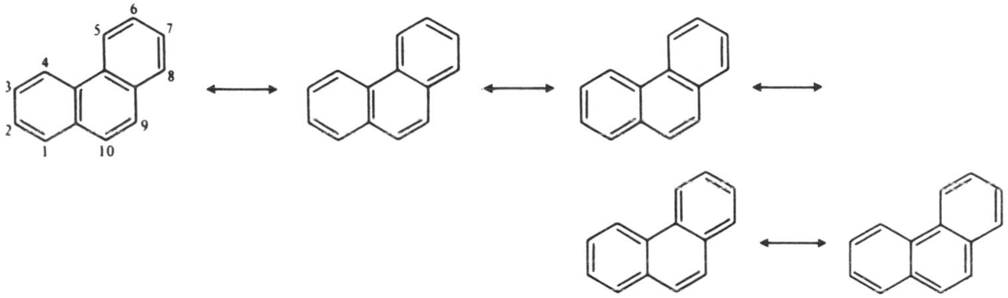

In general there is a good correlation between bond distances in fused aromatic compounds and bond orders. Another experimental quantity that correlates well with the bond order of a given bond in an aromatic system is the nmr coupling constant for coupling between the hydrogens on the two carbons of the bond.[74]

The resonance energies of fused systems increase as the number of principal canonical forms increases, as predicted by rule 6 (p. 35).[75] Thus, for benzene, naphthalene, anthracene, and phenanthrene, for which we can draw, respectively, two, three, four, and five principal canonical forms, the resonance energies are, respectively, 36, 61, 84, and 92 kcal/mol (152, 255, 351, and 385 kJ/mol), calculated from heat-of-combustion data.[76] Note that when phenanthrene, which has a total resonance energy of 92 kcal/mol (385 kJ/mol), loses the 9,10 bond by attack of a reagent such as ozone or bromine, two complete benzene rings remain, each with 36 kcal/mol (152 kJ/mol) that would be lost if benzene was similarly attacked. The fact that anthracene undergoes many reactions across the 9,10 positions can

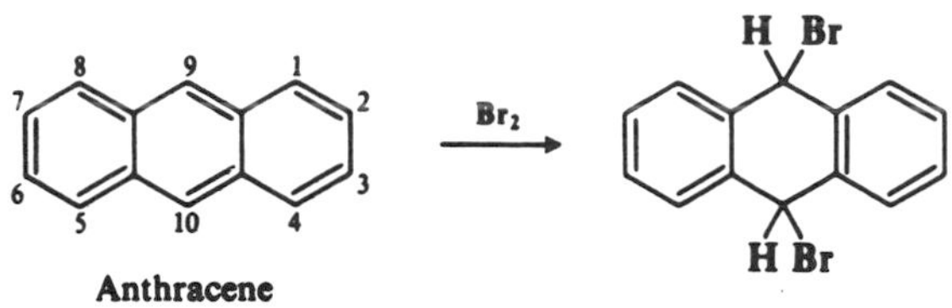

Anthracene

[69]For a modern valence bond description of naphthalene, see Sironi; Cooper; Gerratt; Raimondi *J. Chem. Soc., Chem. Commun.* **1989**, 675.
[70]Cruickshank *Tetrahedron* **1962**, *17*, 155.
[71]Kooyman *Recl. Trav. Chim. Pays-Bas* **1947**, *66*, 201.
[72]For a review, see Efros *Russ. Chem. Rev.* **1960**, *29*, 66-78.
[73]See also Lai *J. Am. Chem. Soc.* **1985**, *107*, 6678.
[74]Jonathan; Gordon; Dailey *J. Chem. Phys.* **1962**, *36*, 2443; Cooper; Manatt *J. Am. Chem. Soc.* **1969**, *91*, 6325.
[75]See Herndon *J. Am. Chem. Soc.* **1973**, *95*, 2404; Herndon; Ellzey *J. Am. Chem. Soc.* **1974**, *96*, 6631.
[76]Ref. 1, p. 98.

be explained in a similar manner. Resonance energies for fused systems can be estimated by counting canonical forms.[77]

Not all fused systems can be fully aromatic. Thus for phenalene (**29**) there is no way double bonds can be distributed so that each carbon has one single and one double bond.[78]

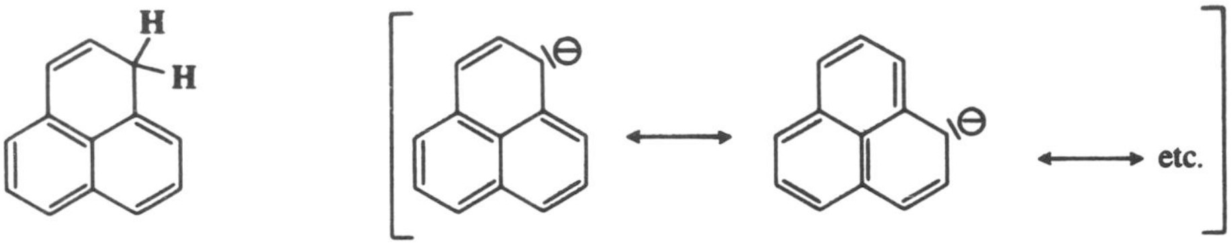

29   **30**

However, phenalene is acidic and reacts with potassium methoxide to give the corresponding anion (**30**), which is completely aromatic. So are the corresponding radical and cation, in which the resonance energies are the same (see p. 50)[79]

In a fused system there are not six electrons for each ring.[80] In naphthalene, if one ring is to have six, the other must have only four. One way to explain the greater reactivity of the ring system of naphthalene compared with benzene is to regard one of the naphthalene rings as aromatic and the other as a butadiene system.[81] This effect can become extreme, as in the case of triphenylene.[82] For this compound, there are eight canonical forms like **A**,

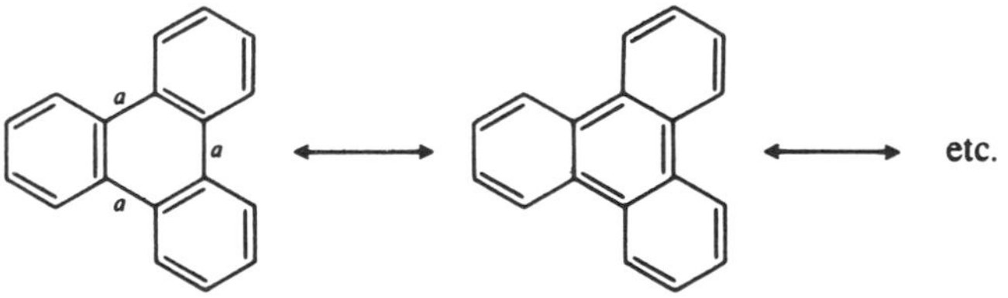

A   **B**

in which none of the three bonds marked *a* is a double bond and only one form (**B**) in which at least one of them is double. Thus the molecule behaves as if the 18 electrons were distributed so as to give each of the outer rings a sextet, while the middle ring is "empty." Since none of the outer rings need share any electrons with an adjacent ring, they are as stable as benzene; triphenylene, unlike most fused aromatic hydrocarbons, does not dissolve in concentrated sulfuric acid and has a low reactivity.[83] This phenomenon, whereby some rings in fused systems give up part of their aromaticity to adjacent rings, is called *annellation* and can be demonstrated by uv spectra[67] as well as reactivities.

In this book we will use a circle to represent single aromatic rings (as, for example, in **24**), but will show one canonical form for fused ring compounds (e.g., **28**). It would be misleading to use two circles for naphthalene, for example, because that would imply 12 aromatic electrons, although naphthalene has only ten.[84]

[77]Swinborne-Sheldrake; Herndon *Tetrahedron Lett.* **1975**, 755.

[78]For reviews of phenalenes, see Murata *Top. Nonbenzenoid Aromat. Chem.* **1973**, *1*, 159-190; Reid *Q. Rev., Chem. Soc.* **1965**, *19*, 274-302.

[79]Pettit *J. Am. Chem. Soc.* **1960**, *82*, 1972.

[80]For discussions of how the electrons in fused aromatic systems interact to form 4*n* + 2 systems, see Glidewell; Lloyd *Tetrahedron* **1984**, *40*, 4455, *J. Chem. Educ.* **1986**, *63*, 306; Hosoya *Top. Curr. Chem.* **1990**, *153*, 255.

[81]Meredith; Wright *Can. J. Chem.* **1960**, *38*, 1177.

[82]For a review of triphenylenes, see Buess; Lawson *Chem. Rev.* **1960**, *60*, 313-330.

[83]Clar; Zander *J. Chem. Soc.* **1958**, 1861.

[84]See Belloli *J. Chem. Educ.* **1983**, *60*, 190.

Five, Seven, and Eight-Membered Rings

Aromatic sextets can also be present in five- and seven-membered rings. If a five-membered ring has two double bonds and the fifth atom possesses an unshared pair of electrons, the ring has five p orbitals that can overlap to create five new orbitals—three bonding and two antibonding (Figure 2.6). There are six electrons for these orbitals: the four p orbitals of the double bonds each contribute one and the filled orbital contributes the other two. The six electrons occupy the bonding orbitals and constitute an aromatic sextet. The heterocyclic compounds pyrrole, thiophene, and furan are the most important examples of this kind of aromaticity, although furan has a lower degree of aromaticity than the other two.[85] Reso-

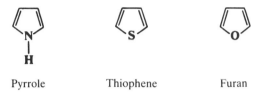

Pyrrole   Thiophene   Furan

nance energies for these three compounds are, respectively, 21, 29, and 16 kcal/mol (88, 121, and 67 kJ/mol).[86] The aromaticity can also be shown by canonical forms, e.g., for pyrrole:

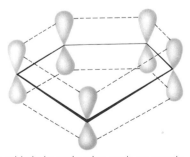

In contrast to pyridine, the unshared pair in canonical structure **A** in pyrrole is needed for the aromatic sextet. This is why pyrrole is a much weaker base than pyridine.

The fifth atom may be carbon if it has an unshared pair. Cyclopentadiene has unexpected acidic properties (p$K_a \approx 16$) since on loss of a proton, the resulting carbanion is greatly

FIGURE 2.6 Overlap of five p orbitals in molecules such as pyrrole, thiophene, and the cyclopentadienide ion.

[85]The order of aromaticity of these compounds is benzene > thiophene > pyrrole > furan, as calculated by an aromaticity index based on bond distance measurements. This index has been calculated for 5- and 6-membered monocyclic and bicyclic heterocycles: Bird *Tetrahedron* **1985**, *41*, 1409; **1986**, *42*, 89; **1987**, *43*, 4725.
[86]Ref. 1, p 99. See also Calderbank; Calvert; Lukins; Ritchie *Aust. J. Chem.* **1981**, *34*, 1835.

stabilized by resonance although it is quite reactive. The cyclopentadienide ion is usually represented as in **31**. Resonance in this ion is greater than in pyrrole, thiophene, and furan,

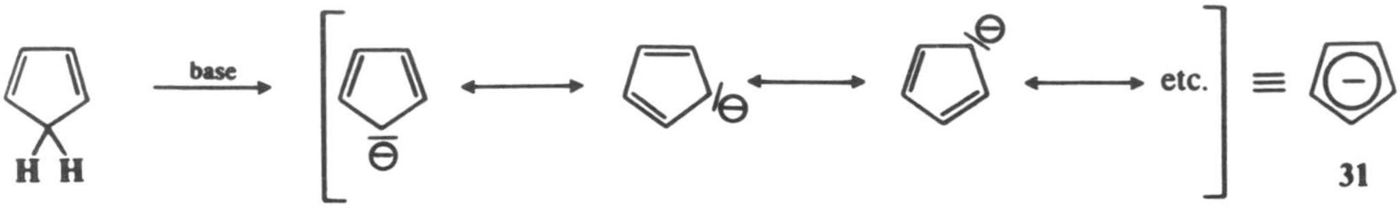

since all five forms are equivalent. The resonance energy for **31** has been estimated to be 24-27 kcal/mol (100-113 kJ/mol).[87] That all five carbons are equivalent has been demonstrated by labeling the starting compound with ^{14}C and finding all positions equally labeled when cyclopentadiene was regenerated.[88] As expected for an aromatic system, the cyclopentadienide ion is diatropic[89] and aromatic substitutions on it have been successfully carried out.[90] Indene and fluorene are also acidic ($pK_a \approx 20$ and 23, respectively) but less so than cyclopentadiene, since annellation causes the electrons to be less

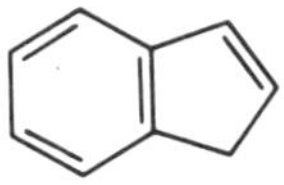

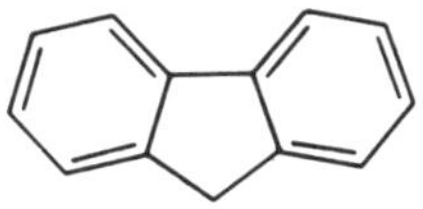

Fluorene 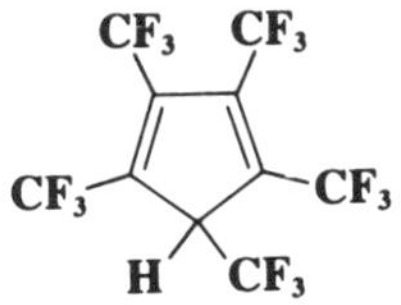   32

available to the five-membered ring. On the other hand, the acidity of 1,2,3,4,5-pentakis(trifluoromethyl)cyclopentadiene (**32**) is greater than that of nitric acid,[91] because of the electron-withdrawing effects of the trifluoromethyl groups (see p. 264).

In sharp contrast to cyclopentadiene is cycloheptatriene (**33**), which has no unusual acidity. This would be hard to explain without the aromatic sextet theory, since, on the

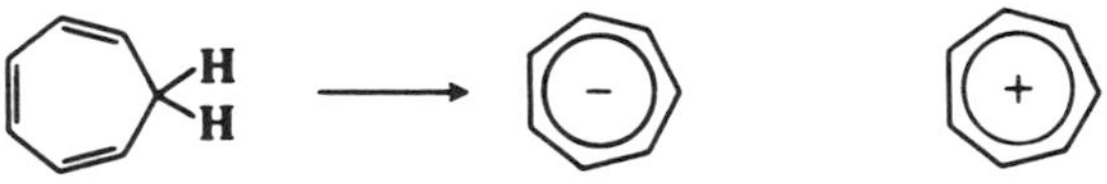

33   34   35

basis of resonance forms or a simple consideration of orbital overlaps, **34** should be as stable as the cyclopentadienyl anion (**31**). While **34** has been prepared in solution,[92] it is less stable than **31** and far less stable than **35**, in which **33** has lost not a proton but a hydride ion. The six double-bond electrons of **35** overlap with the empty orbital on the seventh carbon and there is a sextet of electrons covering seven carbon atoms. **35**, known as the *tropylium ion*, is quite stable.[93] Tropylium bromide, which could be completely covalent if the electrons of the bromine were sufficiently attracted to the ring, is actually an ionic compound:[94]

[87]Bordwell; Drucker; Fried *J. Org. Chem.* **1981**, *46*, 632.
[88]Tkachuk; Lee *Can. J. Chem.* **1959**, *37*, 1644.
[89]Bradamante; Marchesini; Pagani *Tetrahedron Lett.* **1971**, 4621.
[90]Webster *J. Org. Chem.* **1967**, *32*, 39; Rybinskaya; Korneva *Russ. Chem. Rev.* **1971**, *40*, 247-255.
[91]Laganis; Lemal *J. Am. Chem. Soc.* **1980**, *102*, 6633.
[92]Dauben; Rifi *J. Am. Chem. Soc.* **1963**, *85*, 3041; also see Breslow; Chang *J. Am Chem. Soc.* **1965**, *87*, 2200.
[93]For reviews, see Pietra *Chem. Rev.* **1973**, *73*, 293-364; Bertelli *Top. Nonbenzenoid Aromat. Chem.* **1973**, *1*, 29-46; Kolomnikova; Parnes *Russ. Chem. Rev.* **1967**, *36*, 735-753; Harmon, in Olah; Schleyer, *Carbonium Ions*, vol. 4; Wiley: New York, 1973, pp. 1579-1641.
[94]Doering; Knox *J. Am. Chem. Soc.* **1954**, *76*, 3203.

Just as with **31**, the equivalence of the carbons in **35** has been demonstrated by isotopic labeling.[95]

Another seven-membered ring that shows some aromatic character is tropone (**36**). This molecule would have an aromatic sextet if the two C=O electrons stayed away from the ring and resided near the electronegative oxygen atom. In fact, tropones are stable com-

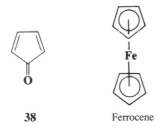

pounds, and tropolones (**37**) are found in nature.[96] However, analyses of dipole moments, nmr spectra, and x-ray diffraction measurements show that tropones and tropolones display appreciable bond alternations.[97] These molecules must be regarded as essentially nonaromatic, although with some aromatic character. Tropolones readily undergo aromatic substitution, emphasizing that the old and the new definitions of aromaticity are not always parallel. In sharp contrast to **36,** cyclopentadienone (**38**) has been isolated only in an argon

matrix below 38 K.[98] Above this temperature it dimerizes. Many earlier attempts to prepare it were unsuccessful.[99] As in **36**, the electronegative oxygen atom draws electron to itself, but in this case it leaves only four electrons and the molecule is unstable. Some derivatives of **38** have been prepared.[99]

Another type of five-membered aromatic compound is the *metallocenes* (also called *sandwich compounds*), in which two cyclopentadienide rings form a sandwich around a metallic ion. The best known of these is ferrocene, although others have been prepared

[95]Vol'pin; Kursanov; Shemyakin; Maimind; Neiman *J. Gen. Chem. USSR* **1959**, *29*, 3667.
[96]For reviews of tropones and tropolones, see Pietra *Acc. Chem. Res.* **1979**, *12*, 132-138; Nozoe *Pure Appl. Chem.* **1971**, *28*, 239-280.
[97]Bertelli; Andrews *J. Am. Chem. Soc.* **1969**, *91*, 5280; Bertelli; Andrews; Crews *J. Am. Chem. Soc.* **1969**, *91*, 5286; Schaefer; Reed *J. Am. Chem. Soc.* **1971**, *93*, 3902; Watkin; Hamor *J. Chem. Soc. B* **1971**, 2167; Barrow; Mills; Filippini *J. Chem. Soc., Chem. Commun.* **1973**, 66.
[98]Maier; Franz; Hartan; Lanz; Reisenauer *Chem. Ber.* **1985**, *118*, 3196.
[99]For a review of cyclopentadienone derivatives and of attempts to prepare the parent compound, see Ogliaruso; Romanelli; Becker *Chem. Rev.* **1965**, *65*, 261-367.

with Co, Ni, Cr, Ti, V, and many other metals.[100] Ferrocene is quite stable, subliming above 100°C and unchanged at 400°C. The two rings rotate freely.[101] Many aromatic substitutions have been carried out on metallocenes.[102] Metallocenes containing two metal atoms and three cyclopentadienyl rings have also been prepared and are known as *triple-decker sandwiches*.[103] Even tetradecker, pentadecker, and hexadecker sandwiches have been reported.[104]

The bonding in ferrocene may be looked upon in simplified molecular-orbital terms as follows.[105] Each of the cyclopentadienide rings has five molecular orbitals—three filled bonding and two empty antibonding orbitals (p. 45). The outer shell of the Fe atom possesses nine atomic orbitals, i.e., one $4s$, three $4p$, and five $3d$ orbitals. The six filled orbitals of the two cyclopentadienide rings overlap with the s, three p, and two of the d orbitals of the Fe to form twelve new orbitals, six of which are bonding. These six orbitals make up two ring-to-metal triple bonds. In addition further bonding results from the overlap of the empty antibonding orbitals of the rings with additional filled d orbitals of the iron. All told, there are eighteen electrons (ten of which may be considered to come from the rings and eight from iron in the zero oxidation state) in nine orbitals; six of these are strongly bonding and three weakly bonding or nonbonding.

The tropylium ion has an aromatic sextet spread over seven carbon atoms. An analogous ion, with the sextet spread over eight carbon atoms, is 1,3,5,7-tetramethylcyclooctatetraene

39

dictation (**39**). This ion, which is stable in solution at $-50°C$, is diatropic and approximately planar. **39** is not stable above about $-30°C$.[106]

Other Systems Containing Aromatic Sextets

Simple resonance theory predicts that pentalene (**40**), azulene (**41**), and heptalene (**42**) should be aromatic, although no nonionic canonical form can have a double bond at the

---

[100]For a monograph on metallocenes, see Rosenblum *Chemistry of the Iron Group Metallocenes*; Wiley: New York, 1965. For reviews, see Lukehart *Fundamental Transition Metal Organometallic Chemistry*; Brooks/Cole: Monterey, CA, 1985, pp. 85-118; Lemenovskii; Fedin *Russ. Chem. Rev.* **1986**, *55*, 127-142; Sikora; Macomber; Rausch *Adv. Organomet. Chem.* **1986**, *25*, 317-379; Pauson, *Pure Appl. Chem.* **1977**, *49*, 839-855; Nesmeyanov; Kochetkova *Russ. Chem. Rev.* **1974**, *43*, 710-715; Shul'pin; Rybinskaya *Russ. Chem. Rev.* **1974**, *43*, 716-732; Perevalova; Nikitina *Organomet. React.* **1972**, *4*, 163-419; Bublitz; Rinehart *Org. React.* **1969**, *17*, 1-154; Leonova; Kochetkova *Russ. Chem. Rev.* **1973**, *42*, 278-292; Rausch *Pure Appl. Chem.* **1972**, *30*, 523-538. For a bibliography of reviews on metallocenes, see Bruce *Adv. Organomet. Chem.* **1972**, *10*, 273-346, pp. 322-325.

[101]For a discussion of the molecular structure, see Haaland *Acc. Chem. Res.* **1979**, *12*, 415-422.

[102]For a review on aromatic substitution on ferrocenes, see Plesske *Angew. Chem. Int. Ed. Engl.* **1962**, *1*, 312-327, 394-399 [*Angew. Chem. 74*, 301-316, 347-352].

[103]For a review, see Werner *Angew. Chem. Int. Ed. Engl.* **1977**, *16*, 1-9 [*Angew. Chem. 89*, 1-10].

[104]See, for example, Siebert *Angew. Chem. Int. Ed. Engl.* **1985**, *24*, 943-958 [*Angew. Chem. 97*, 924-939].

[105]Rosenblum, Ref. 100, pp. 13-28; Coates; Green; Wade *Organometallic Compounds*, 3rd ed., vol. 2; Methuen: London, 1968, pp. 97-104; Grebenik; Grinter; Perutz *Chem. Soc. Rev.* **1988**, *17*, 453-490; pp. 460-464.

[106]This and related ions were prepared by Olah; Staral; Liang; Paquette; Melega; Carmody *J. Am. Chem. Soc.* **1977**, *99*, 3349. See also Radom; Schaefer *J. Am. Chem. Soc.* **1977**, *99*, 7522; Olah; Liang *J. Am. Chem. Soc.* **1976**, *98*, 3033; Willner; Rabinovitz *Nouv. J. Chim.* **1982**, *6*, 129.

ring junction. Molecular-orbital calculations show that azulene should be stable but not the other two, and this is borne out by experiment. Heptalene has been prepared[107] but reacts

40   **41**   **42**

readily with oxygen, acids, and bromine, is easily hydrogenated, and polymerizes on standing. Analysis of its nmr spectrum shows that it is not planar.[108] The 3,8-dibromo and 3,8-dicarbomethoxy derivatives of **42** are stable in air at room temperature but are not diatropic.[109] A number of methylated heptalenes and dimethyl 1,2-heptalenedicarboxylates have also been prepared and are stable nonaromatic compounds.[110] Pentalene has not been prepared,[111] but the hexaphenyl[112] and 1,3,5-tri-*t*-butyl derivatives[113] are known. The former is air-sensitive in solution. The latter is stable, but x-ray diffraction and photoelectron spectral data show bond alternation.[114] Pentalene and its methyl and dimethyl derivatives have been formed in solution, but they dimerize before they can be isolated.[115] Many other attempts to prepare these two systems have failed.

In sharp contrast to **40** and **42**, azulene, a blue solid, is quite stable and many of its derivatives are known.[116] Azulene readily undergoes aromatic substitution. Azulene may be regarded as a combination of **31** and **35** and, indeed, possesses a dipole moment of 0.8

43

D.[117] Interestingly, if two electrons are added to pentalene, a stable dianion (**43**) results.[118] It can be concluded that an aromatic system of electrons will be spread over two rings only if 10 electrons (not 8 or 12) are available for aromaticity.

[107]Dauben; Bertelli *J. Am. Chem. Soc.* **1961**, *83*, 4659; Vogel; Königshofen; Wassen; Müllen; Oth *Angew. Chem. Int. Ed. Engl.* **1974**, *13*, 732 [*Angew. Chem. 86*, 777]; Paquette; Browne; Chamot *Angew. Chem. Int. Ed. Engl.* **1979**, *18*, 546 [*Angew. Chem. 91*, 581]. For a review of heptalenes, see Paquette *Isr. J. Chem.* **1980**, *20*, 233-239.
[108]Bertelli, in Bergmann; Pullman, Ref. 55, p. 326. See also Stegemann; Lindner *Tetrahedron Lett.* **1977**, 2515.
[109]Vogel; Ippen *Angew. Chem. Int. Ed. Engl.* **1974**, *13*, 734 [*Angew. Chem. 86*, 778]; Vogel; Hogrefe *Angew. Chem. Int. Ed. Engl.* **1974**, *13*, 735 [*Angew. Chem. 86*, 779].
[110]Hafner; Knaup; Lindner *Bull. Soc. Chem. Jpn.* **1988**, *61*, 155.
[111]Metal complexes of pentalene have been prepared: Knox; Stone *Acc. Chem. Res.* **1974**, *7*, 321-328.
[112]LeGoff *J. Am. Chem. Soc.* **1962**, *84*, 3975. See also Hafner; Bangert; Orfanos *Angew. Chem. Int. Ed. Engl.* **1967**, *6*, 451 [*Angew. Chem. 79*, 414]; Hartke; Matusch *Angew. Chem. Int. Ed. Engl.* **1972**, *11*, 50 [*Angew. Chem. 84*, 61].
[113]Hafner; Süss *Angew. Chem. Int. Ed. Engl.* **1973**, *12*, 575 [*Angew. Chem. 85*, 626]. See also Hafner; Suda *Angew. Chem. Int. Ed. Engl.* **1976**, *15*, 314 [*Angew. Chem. 88*, 341].
[114]Kitschke; Lindner *Tetrahedron Lett.* **1977**, 2511; Bischof; Gleiter; Hafner; Knauer; Spanget-Larsen; Süss *Chem. Ber.* **1978**, *111*, 932.
[115]Bloch; Marty; de Mayo *J. Am. Chem. Soc.* **1971**, *93*, 3071; *Bull. Soc. Chim. Fr.* **1972**, 2031; Hafner; Dönges; Goedecke; Kaiser *Angew. Chem. Int. Ed. Engl.* **1973**, *12*, 337 [*Angew. Chem. 85*, 362].
[116]For a review on azulene, see Mochalin; Porshnev *Russ. Chem. Rev.* **1977**, *46*, 530-547.
[117]Tobler; Bauder; Günthard *J. Mol. Spectrosc.* **1965**, *18*, 239.
[118]Katz; Rosenberger; O'Hara *J. Am. Chem. Soc.* **1964**, *86*, 249. See also Willner; Becker; Rabinovitz *J. Am. Chem. Soc.* **1979**, *101*, 395.

Alternant and Nonalternant Hydrocarbons[119]

Aromatic hydrocarbons can be divided into two types. In alternant hydrocarbons, the conjugated carbon atoms can be divided into two sets such that no two atoms of the same set are directly linked. For convenience one set may be starred. Naphthalene is an alternant and azulene a nonalternant hydrocarbon:

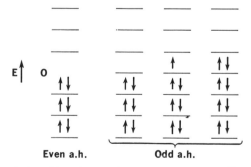

In alternant hydrocarbons, the bonding and antibonding orbitals occur in pairs; i.e., for every bonding orbital with an energy $-E$ there is an antibonding one with energy $+E$ (Figure 2.7[120]). Even-alternant hydrocarbons are those with an even number of conjugated atoms, i.e., an equal number of starred and unstarred atoms. For these hydrocarbons all the bonding orbitals are filled and the π electrons are uniformly spread over the unsaturated atoms.

As with the allylic system, odd-alternant hydrocarbons (which must be carbocations, carbanions, or radicals) in addition to equal and opposite bonding and antibonding orbitals also have a nonbonding orbital of zero energy. When an odd number of orbitals overlap, an odd number is created. Since orbitals of alternant hydrocarbons occur in $-E$ and $+E$ pairs, one orbital can have no partner and must therefore have zero bonding energy. For example, in the benzylic system the cation has an unoccupied nonbonding orbital, the free radical has one electron there and the carbanion two (Figure 2.8). As with the allylic system,

all three species have the same bonding energy. The charge distribution (or unpaired-electron distribution) over the entire molecule is also the same for the three species and can be calculated by a relatively simple process.[119]

FIGURE 2.7 Energy levels in odd- and even-alternant hydrocarbons.[120] The arrows represent electrons. The orbitals are shown as having different energies, but some may be degenerate.

[119]For discussions, see Jones *Physical and Mechanistic Organic Chemistry*, 2nd ed.; Cambridge University Press: Cambridge, 1984, pp. 122-129; Dewar *Prog. Org. Chem.* **1953**, *2*, 1-28.
[120]Taken from Dewar, Ref. 119, p. 8.

Energy

$\alpha - 2.101\beta$

$\alpha - 1.259\beta$

$\alpha - \beta$

α

$\alpha + \beta$

$\alpha + 1.259\beta$

$\alpha + 2.101\beta$

FIGURE 2.8   Energy levels for the benzyl cation, free radical, and carbanion. Since α is the energy of a p orbital (p. 30), the nonbonding orbital has no bonding energy.

For nonalternant hydrocarbons the energies of the bonding and antibonding orbitals are not equal and opposite and charge distributions are not the same in cations, anions, and radicals. Calculations are much more difficult but have been carried out.[121]

Aromatic Systems with Electron Numbers Other than Six

Ever since the special stability of benzene was recognized, chemists have been thinking about homologous molecules and wondering whether this stability is also associated with rings that are similar but of different sizes, such as cyclobutadiene (**44**), cyclooctatetraene (**45**), cyclodecapentaene[122] (**46**), etc. The general name *annulene* is given to these compounds,

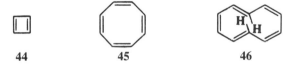

44   45   46

benzene being [6]annulene, and **44** to **46** being called, respectively, [4], [8], and [10]annulene. By a naïve consideration of resonance forms, these annulenes and higher ones should be as aromatic as benzene. Yet they proved remarkably elusive. The ubiquitous benzene ring is found in thousands of natural products, in coal and petroleum, and is formed by strong treatment of many noncyclic compounds. None of the other annulene ring systems has ever been found in nature and, except for cyclooctatetraene, their synthesis is not simple. Obviously, there is something special about the number six in a cyclic system of electrons.

Hückel's rule, based on molecular-orbital calculations,[123] predicts that electron rings will constitute an aromatic system only if the number of electrons in the ring is of the form

[121]Peters *J. Chem. Soc.* **1958,** 1023, 1028, 1039; Brown; Burden; Williams *Aust. J. Chem.* **1968,** *21*, 1939. For reviews, see Zahradnik, in Snyder, Ref. 55, vol. 2, pp. 1-80; Zahradnik *Angew. Chem. Int. Ed. Engl.* **1965,** *4*, 1039-1050 [*Angew. Chem. 77,* 1097-1109].

[122]The cyclodecapentaene shown here is the cis-trans-cis-cis-trans form. For other stereoisomers, see p. 58.

[123]For reviews of molecular-orbital calculations of nonbenzenoid cyclic conjugated hydrocarbons, see Nakajima *Pure Appl. Chem.* **1971,** *28*, 219-238; *Fortschr. Chem. Forsch.* **1972,** *32*, 1-42.

$4n + 2$, where n is zero or any position integer. Systems that contain 4n electrons are predicted to be nonaromatic. The rule predicts that rings of 2, 6, 10, 14, etc., electrons will be aromatic, while rings of 4, 8, 12, etc., will not be. This is actually a consequence of Hund's rule. The first pair of electrons in an annulene goes into the π orbital of lowest energy. After that the bonding orbitals are degenerate and occur in pairs of equal energy.

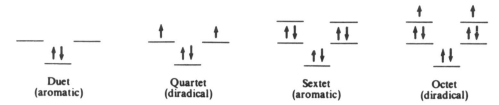

| Duet | Quartet | Sextet | Octet |
| (aromatic) | (diradical) | (aromatic) | (diradical) |

When there is a total of four electrons, Hund's rule predicts that two will be in the lowest orbital but the other two will be unpaired, so that the system will exist as a diradical rather than as two pairs. The degeneracy can be removed if the molecule is distorted from maximum molecular symmetry to a structure of lesser symmetry. For example, if **44** assumes a rectangular rather than a square shape, one of the previously degenerate orbitals has a lower energy than the other and will be occupied by two electrons. In this case, of course, the double bonds are essentially separate and the molecule is still not aromatic. Distortions of symmetry can also occur when one or more carbons are replaced by hetero atoms or in other ways.[124]

In the following sections systems with various numbers of electrons are discussed. When we look for aromaticity we look for: (1) the presence of a diamagnetic ring current; (2) equal or approximately equal bond distances, except when the symmetry of the system is disturbed by a hetero atom or in some other way; (3) planarity; (4) chemical stability; (5) the ability to undergo aromatic substitution.

Systems of Two Electrons[125]

Obviously, there can be no ring of two carbon atoms though a double bond may be regarded as a degenerate case. However, in analogy to the tropylium ion, a three-membered ring with a double bond and a positive charge on the third atom (the *cyclopropenyl cation*) is a $4n + 2$ system and hence is expected to show aromaticity. The unsubstituted **47** has been prepared,[126] as well as several derivatives, e.g., the trichloro, diphenyl, and dipropyl derivatives, and these are stable despite the angles of only 60°. In fact, the tripropylcyclopropenyl[127]

$$\left[\nabla\!\!\!\!\!\underset{\oplus}{} \longleftrightarrow {}^{\oplus}\nabla \longleftrightarrow \nabla^{\oplus} \right] \equiv \nabla\!\!\!\!\!\oplus$$

47

[124]For a discussion, see Hoffmann *Chem. Commun.* **1969**, 240.

[125]For reviews, see Billups; Moorehead, in Rappoport *The Chemistry of the Cyclopropyl Group*, pt. 2; Wiley: New York, 1987, pp. 1533-1574; Potts; Baum *Chem. Rev.* **1974**, 74, 189-213; Yoshida *Top. Curr. Chem.* **1973**, 40, 47-72; D'yakonov; Kostikov *Russ. Chem. Rev.* **1967**, 36, 557-563; Closs *Adv. Alicyclic Chem.* **1966**, 1, 53-127, pp. 102-126; Krebs *Angew. Chem. Int. Ed. Engl.* **1965**, 4, 10-22 [*Angew. Chem. 77,* 10-22].

[126]Breslow; Groves; Ryan *J. Am. Chem. Soc.* **1967**, 89, 5048; Farnum, Mehta; Silberman *J. Am. Chem. Soc.* **1967**, 89, 5048; Breslow; Groves *J. Am. Chem. Soc.* **1970**, 92, 984.

[127]Breslow; Höver; Chang *J. Am. Chem. Soc.* **1962**, 84, 3168.

and tricyclopropylcyclopropenyl[128] cations are among the most stable carbocations known, being stable even in water solution. The tri-t-butylcyclopropenyl cation is also very stable.[129] In addition, cyclopropenone and several of its derivatives are stable compounds,[130] in accord

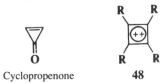

Cyclopropenone   **48**

with the corresponding stability of the tropones.[131] The ring system **47** is nonalternant and the corresponding radical and anion (which do not have an aromatic duet) have electrons in antibonding orbitals, so that their energies are much higher. As with **31** and **35**, the equivalence of the three carbon atoms in the triphenylcyclopropenyl cation has been demonstrated by ^{14}C labeling experiments.[132] The interesting dications **48** (R = Me or Ph) have been prepared,[133] and they too should represent aromatic systems of two electrons.[134]

Systems of Four Electrons. Antiaromaticity

The most obvious compound in which to look for a closed loop of four electrons is cyclobutadiene (**44**).[135] Hückel's rule predicts no aromatic character here, since 4 is not a number of the form $4n + 2$. There is a long history of attempts to prepare this compound and its simple derivatives, and, as we shall see, the evidence fully bears out Hückel's prediction—cyclobutadienes display none of the characteristics that would lead us to call them aromatic. More surprisingly, there is evidence that a closed loop of four electrons is actually *antiaromatic*.[136] If such compounds simply lacked aromaticity, we would expect them to be about as stable as similar nonaromatic compounds, but both theory and experiment show that they are *much less stable*.[137] An antiaromatic compound may be defined as a compound that is destabilized by a closed loop of electrons.

After years of attempts to prepare cyclobutadiene, the goal was finally reached by Pettit and co-workers.[138] It is now clear that **44** and its simple derivatives are extremely unstable

[128]Komatsu; Tomioka; Okamoto *Tetrahedron Lett.* **1980**, *21*, 947; Moss; Shen; Krogh-Jespersen; Potenza; Shugar; Munjal *J. Am. Chem. Soc.* **1986**, *108*, 134.

[129]Ciabattoni; Nathan *J. Am. Chem. Soc.* **1968**, *90*, 4495.

[130]See, for example, Kursanov; Vol'pin; Koreshkov *J. Gen. Chem. USSR* **1960**, *30*, 2855; Breslow; Oda *J. Am. Chem. Soc.* **1972**, *94*, 4787; Yoshida; Konishi; Tawara; Ogoshi *J. Am. Chem. Soc.* **1973**, *95*, 3043; Ref. 129.

[131]For a reveiw of cyclopropenones, see Eicher; Weber *Top. Curr. Chem. Soc.* **1975**, *57*, 1-109. For discussions of cyclopropenone structure, see Shäfer; Schweig; Maier; Sayrac; Crandall *Tetrahedron Lett.* **1974**, 1213; Tobey, in Bergmann; Pullman, Ref. 55, pp. 351-362; Greenberg; Tomkins; Dobrovolny; Liebman *J. Am. Chem. Soc.* **1983**, *105*, 6855.

[132]D'yakonov; Kostikov; Molchanov *J. Org. Chem. USSR* **1969**, *5*, 171; **1970**, *6*, 304.

[133]Freedman; Young *J. Am. Chem. Soc.* **1964**, *86*, 734; Olah; Bollinger; White *J. Am. Chem. Soc.* **1969**, *91*, 3667; Olah; Mateescu *J. Am. Chem. Soc.* **1970**, *92*, 1430; Olah; Staral *J. Am. Chem. Soc.* **1976**, *98*, 6290. See also Lambert; Holcomb *J. Am. Chem. Soc.* **1971**, *93*, 2994; Seitz; Schmiedel; Mann *Synthesis* **1974**, 578.

[134]See Pittman; Kress; Kispert *J. Org. Chem.* **1974**, *39*, 378. See, however, Krogh-Jespersen; Schleyer; Pople; Cremer *J. Am. Chem. Soc.* **1978**, *100*, 4301.

[135]For a monograph, see Cava; Mitchell *Cyclobutadiene and Related Compounds*; Academic Press: New York, 1967. For reviews, see Maier *Angew. Chem. Int. Ed. Engl.* **1988**, *27*, 309-332 [*Angew. Chem. 100*, 317-341]; **1974**, *13*, 425-438 [*Angew. Chem. 86*, 491-505]; Bally; Masamune *Tetrahedron* **1980**, *36*, 343-370; Vollhardt *Top. Curr. Chem.* **1975**, *59*, 113-136.

[136]For reviews of antiaromaticity, see Glukhovtsev; Simkin; Minkin; *Russ. Chem. Rev.* **1985**, *54*, 54-75; Breslow *Pure Appl. Chem.* **1971**, *28*, 111-130; *Acc. Chem. Res.* **1973**, *6*, 393-398; *Chem. Br.* **1968**, *4*, 100; *Angew. Chem. Int. Ed. Engl.* **1968**, *7*, 565-570 [*Angew. Chem. 80*, 573-578].

[137]For a discussion, see Bauld; Welsher; Cessac; Holloway *J. Am. Chem. Soc.* **1978**, *100*, 6920.

[138]Watts; Fitzpatrick; Pettit *J. Am. Chem. Soc.* **1965**, *87*, 3253, **1966**, *88*, 623. See also Cookson; Jones *J. Chem. Soc.* **1965**, 1881.

compounds with very short lifetimes (they dimerize by a Diels–Alder reaction; see **5-47**) unless they are stabilized in some fashion, either at ordinary temperatures embedded in the cavity of a hemicarcerand[138a] (see the structure of a carcerand, **30** on p. 89), or in matrices at very low temperatures (generally under 35 K). In either of these cases, the cyclobutadiene molecules are forced to remain apart from each other, and other molecules cannot get in. The structures of **44** and some of its derivatives have been studied a number of times using the low-temperature matrix technique.[139] The ground-state structure of **44** is a rectangular diene (not a diradical) as shown by the infrared (ir) spectra of **44** and deuterated **44** trapped in matrices[140] as well as by a photoelectron spectrum.[141] Molecular-orbital calculations agree.[142] The same conclusion was also reached in an elegant experiment in which 1,2-dideuterocyclobutadiene was generated. If **44** is a rectangular diene, the dideutero compound should exist as two isomers:

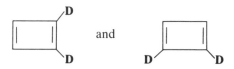

The compound was generated (as an intermediate that was not isolated) and two isomers were indeed found.[143] The cyclobutadiene molecule is not static, even in the matrices. There are two forms (**44a** and **44b**) which rapidly interconvert.[144]

47a   **47b**

There are some simple cyclobutadienes that are stable at room temperature for varying periods of time. These either have bulky substituents or carry certain other stabilizing substituents. Examples of the first type are tri-*t*-butylcyclobutadiene (**49**)[145] and the dithia

t-**Bu**   *t*-**Bu**

t-**Bu**   **H**

49   **50**

[138a]Cram; Tanner; Thomas *Angew. Chem. Int. Ed. Engl.* **1991**, *30*, 1024 [*Angew. Chem. 103*, 1048].

[139]See, for example, Lin; Krantz *J. Chem. Soc., Chem. Commun.* **1972**, 1111; Chapman; McIntosh; Pacansky *J. Am. Chem. Soc.* **1973**, *95*, 614; Maier; Mende *Tetrahedron Lett.* **1969**, 3155. For a review, see Sheridan *Org. Photochem.* **1987**, *8*, 159-248; pp. 167-181.

[140]Masamune; Souto-Bachiller; Machiguchi; Bertie *J. Am. Chem. Soc.* **1978**, *100*, 4889.

[141]Kreile; Münzel; Schweig; Specht *Chem. Phys. Lett.* **1986**, *124*, 140.

[142]See, for example, Borden; Davidson; Hart *J. Am. Chem. Soc.* **1978**, *100*, 388; Kollmar; Staemmler *J. Am. Chem. Soc.* **1978**, *100*, 4304; Jafri; Newton *J. Am. Chem. Soc.* **1978**, *100*, 5012; Ermer; Heilbronner *Angew. Chem. Int. Ed. Engl.* **1983**, *22*, 402 [*Angew. Chem. 95*, 414; Voter; Goddard *J. Am. Chem. Soc.* **1986**, *108*, 2830.

[143]Whitman; Carpenter *J. Am. Chem. Soc.* **1980**, *102*, 4272. See also Whitman; Carpenter *J. Am. Chem. Soc.* **1982**, *104*, 6473.

[144]Carpenter *J. Am. Chem. Soc.* **1983**, *105*, 1700; Huang; Wolfsberg *J. Am. Chem. Soc.* **1984**, *106*, 4039; Dewar; Merz; Stewart *J. Am. Chem. Soc.* **1984**, *106*, 4040; Orendt; Arnold; Radziszewski; Facelli; Malsch; Strub; Grant; Michl *J. Am. Chem. Soc.* **1988**, *110*, 2648. See, however, Arnold; Radziszewski; Campion; Perry; Michl *J. Am. Chem. Soc.* **1991**, *113*, 692.

[145]Masamune; Nakamura; Suda; Ona *J. Am. Chem. Soc.* **1973**, *95*, 8481; Maier; Alzérreca *Angew. Chem. Int. Ed. Engl.* **1973**, *12*, 1015 [*Angew. Chem. 85*, 1056]. For a discussion, see Masamune *Pure Appl. Chem.* **1975**, *44*, 861-884.

compound **50**.[146] These compounds are relatively stable because dimerization is sterically hindered. Examination of the nmr spectrum of **49** showed that the ring proton ($\delta = 5.38$) was shifted *upfield*, compared with the position expected for a nonaromatic proton, e.g., cyclopentadiene. As we shall see on p. 64, this indicates that the compound is antiaromatic. A similar investigation cannot be made for **50** because it has no ring proton, but x-ray crystallography showed that the central ring is a rectangular diene (as shown) with single and double-bond lengths of 1.59–1.60 and 1.34 Å, respectively.[147] The unusually long single-bond distance may be due to repulsion between the methyl groups. Photoelectron spectroscopy showed that **50** is not a diradical.[148]

The other type of stable cyclobutadiene has two electron-donating and two electron-withdrawing groups,[149] and is stable in the absence of water.[150] An example is **51**. The stability of these compounds is generally attributed to the resonance shown, a type of

51

resonance stabilization called the *push–pull* or *captodative effect*,[151] although it has been concluded from a photoelectron spectroscopy study that second order bond fixation is more important.[152] An x-ray crystallographic study of **51** has shown[153] the ring to be a distorted square with bond lengths of 1.46 Å and angles of 87° and 93°.

It is clear that simple cyclobutadienes, which could easily adopt a square planar shape if that would result in aromatic stabilization, do not in fact do so and are not aromatic. The high reactivity of these compounds is not caused merely by steric strain, since the strain should be no greater than that of simple cyclopropenes, which are known compounds. It is probably caused by antiaromaticity.[154]

The unfused cyclobutadiene system is stable in complexes with metals[155] (see Chapter 3), but in these cases electron density is withdrawn from the ring by the metal and there is

[146]Krebs; Kimling; Kemper *Liebigs Ann. Chem.* **1978**, 431.

[147]Irngartinger; Nixdorf; Riegler; Krebs; Kimling; Pocklington; Maier; Malsch; Schneider *Chem. Ber.* **1988**, *121*, 673. This paper also includes an x-ray structure of tetra-*t*-butylcyclobutadiene. See also Irngartinger; Nixdorf *Chem. Ber.* **1988**, *121*, 679; Dunitz; Krüger; Irngartinger; Maverick; Wang; Nixdorf *Angew. Chem. Int. Ed. Engl.* **1988**, *27*, 387 [*Angew. Chem. 100*, 415].

[148]Lauer; Müller; Schulte; Schweig; Krebs *Angew. Chem. Int. Ed. Engl.* **1974**, *13*, 544 [*Angew. Chem. 86*, 597]. See also Brown; Masamune *Can. J. Chem.* **1975**, *53*, 972; Lauer; Müller; Schulte; Schweig; Maier; Alzérreca *Angew. Chem. Int. Ed. Engl.* **1975**, *14*, 172 [*Angew. Chem. 87*, 194]; Irngartinger; Hase; Schulte; Schweig *Angew. Chem. Int. Ed. Engl.* **1977**, *16*, 187 [*Angew. Chem. 89*, 194].

[149]The presence of electron-donating and withdrawing groups on the same ring stabilizes 4n systems and destabilizes 4n + 2 systems. For a review of this concept, see Gompper; Wagner *Angew. Chem. Int. Ed. Engl.* **1988**, *27*, 1437-1455 [*Angew. Chem. 100*, 1492-1511].

[150]Gompper; Seybold *Angew. Chem. Int. Ed. Engl.* **1968**, *7*, 824 [*Angew. Chem. 80*, 804]; Neuenschwander; Niederhauser *Chimia* **1968**, *22*, 491, *Helv. Chim. Acta* **1970**, *53*, 519; Gompper; Mensch; Seybold *Angew. Chem. Int. Ed. Engl.* **1975**, *14*, 704 [*Angew. Chem. 87*, 711]; Gompper; Kroner; Seybold; Wagner *Tetrahedron* **1976**, *32*, 629.

[151]Manatt; Roberts *J. Org. Chem.* **1959**, *24*, 1336; Breslow; Kivelevich; Mitchell; Fabian; Wendel *J. Am. Chem. Soc.* **1965**, *87*, 5132; Hess; Schaad *J. Org. Chem.* **1976**, *41*, 3058.

[152]Gompper; Holsboer; Schmidt; Seybold *J. Am. Chem. Soc.* **1973**, *95*, 8479.

[153]Lindner; Gross *Chem. Ber.* **1974**, *107*, 598.

[154]For evidence, see Breslow; Murayama; Murahashi; Grubbs *J. Am. Chem. Soc.* **1973**, *95*, 6688; Herr *Tetrahedron* **1976**, *32*, 2835.

[155]For reviews, see Efraty *Chem. Rev.* **1977**, *77*, 691-744; Pettit *Pure Appl. Chem.* **1968**, *17*, 253-272; Maitlis *Adv. Organomet. Chem.* **1966**, *4*, 95-143; Maitlis; Eberius, in Snyder, Ref. 55, vol. 2. pp. 359-409.

no aromatic quartet. In fact, these cyclobutadiene–metal complexes can be looked upon as systems containing an aromatic duet. The ring is square planar,[156] the compounds undergo

R
□ ┤─Fe(CO)$_3$

aromatic substitution,[157] and nmr spectra of monosubstituted derivatives show that the C-2 and C-4 protons are equivalent.[157]

Two other systems that have been studied as possible aromatic or antiaromatic four-electron systems are **52** and **53**.[158] In these cases also the evidence supports antiaromaticity, not aromaticity. With respect to **52**, HMO theory predicts that an unconjugated

54   **52**   **53**

54 (i.e., a single canonical form) is more stable than a conjugated **52**,[159] so that **54** would actually lose stability by forming a closed loop of four electrons. The HMO theory is supported by experiment. Among other evidence, it has been shown that **55** (R = COPh) loses its proton in hydrogen-exchange reactions about 6000 times more slowly than **56**

Ph
├╲   H
│ ╲───R
Ph
55

Ph
├╲   H
│ ╲──R
Ph
56

(R = COPh).[160] Where R = CN, the ratio is about 10,000.[161] This indicates that **55** are much more reluctant to form carbanions (which would have to be cyclopropenyl carbanions) than **56**, which form ordinary carbanions. Thus the carbanions of **55** are less stable than corresponding ordinary carbanions. Although derivatives of cyclopropenyl anion have been prepared as fleeting intermediates (as in the exchange reactions mentioned above), all attempts to prepare the ion or any of its derivatives as relatively stable species have so far met with failure.[162]

In the case of **53**, the ion has been prepared and has been shown to be a diradical in the ground state,[163] as predicted by the discussion on p. 52.[164] Evidence that **53** is not only

[156]Dodge; Schomaker *Acta Crystallogr.* **1965,** *18,* 614; *Nature* **1960,** *186,* 798; Dunitz; Mez; Mills; Shearer *Helv. Chim. Acta* **1962,** *45* 647; Yannoni; Ceasar; Dailey *J. Am. Chem. Soc.* **1967,** *89,* 2833.

[157]Fitzpatrick; Watts; Emerson; Pettit *J. Am. Chem. Soc.* **1965,** *87,* 3255. For a discussion, see Pettit *J. Organomet. Chem.* **1975,** *100,* 205-217.

[158]For a review of cyclopentadienyl cations, see Breslow *Top. Nonbenzenoid Aromat. Chem.* **1973,** *1,* 81-94.

[159]Clark *Chem. Commun.* **1969,** 637; Ref. 136.

[160]Breslow; Brown; Gajewski *J. Am. Chem. Soc.* **1967,** *89,* 4383.

[161]Breslow; Douek *J. Am. Chem. Soc.* **1968,** *90,* 2698.

[162]See, for example, Breslow; Cortés; Juan; Mitchell *Tetrahedron Lett.* **1982,** *23,* 795. A triphenylcyclopropyl anion has been prepared in the gas phase, with a lifetime of 1-2 seconds: Bartmess; Kester; Borden; Köser *Tetrahedron Lett.* **1986,** *27,* 5931.

[163]Saunders; Berger; Jaffe; McBride; O'Neill; Breslow; Hoffman; Perchonock; Wasserman; Hutton; Kuck *J. Am. Chem. Soc.* **1973,** *95,* 3017.

[164]Derivatives of **53** show similar behavior; Breslow; Chang; Yager *J. Am. Chem. Soc.* **1963,** *85,* 2033; Volz *Tetrahedron Lett.* **1864,** 1899; Breslow; Hill; Wasserman *J. Am. Chem. Soc.* **1964,** *86,* 5349; Breslow; Chang; Hill; Wasserman *J. Am. Chem. Soc.* **1967,** *89,* 1112; Gompper; Glöckner *Angew. Chem. Int. Ed. Engl.* **1984,** *23,* 53 [*Angew. Chem.* **96,** 48].

nonaromatic but also antiaromatic comes from studies on **57** and **59**.[165] When **57** is treated with silver perchlorate in propionic acid, the molecule is rapidly solvolyzed (a reaction in

57	**58**	**59**	**53**

which the intermediate **58** is formed; see Chapter 5). Under the same conditions, **59** undergoes no solvolysis at all; i.e., **53** does not form. If **53** were merely nonaromatic, it should be about as stable as **58** (which of course has no resonance stabilization at all). The fact that it is so much more reluctant to form indicates that **53** is much less stable than **58**.

It is strong evidence for Hückel's rule that **52** and **53** are not aromatic while the cyclopropenyl cation (**47**) and the cyclopentadienyl anion (**31**) are, since simple resonance theory predicts no difference between **52** and **47** or **53** and **31** (the same number of equivalent canonical forms can be drawn for **52** as for **47** and for **53** as for **31**).

In compounds in which overlapping parallel p orbitals form a closed loop of $4n + 2$ electrons, the molecule is stabilized by resonance and the ring is aromatic. But the evidence given above (and additional evidence discussed below) indicates that when the closed loop contains $4n$ electrons, the molecule is *destabilized* by resonance. In summary, **44**, **52**, and **53** and their simple derivatives are certainly not aromatic and are very likely antiaromatic.

Systems of Eight Electrons

Cyclooctatetraene[166] (**45**) is not planar but tub-shaped.[167] Therefore we would expect that it is neither aromatic nor antiaromatic, since both these conditions require overlap of parallel p orbitals. The reason for the lack of planarity is that a regular octagon has angles of 135°,

45

while sp^2 angles are most stable at 120°. To avoid the strain, the molecule assumes a nonplanar shape, in which orbital overlap is greatly diminished.[168] Single- and double-bond distances in **45** are, respectively, 1.46 and 1.33 Å, which is expected for a compound made up of four individual double bonds.[167] The reactivity is also what would be expected for a linear polyene. However, the cyclooctadiendiynes **60** and **61** are planar conjugated eight-electron systems (the four extra triple-bond electrons do not participate), which nmr evidence show to be

[165]Breslow; Mazur *J. Am. Chem. Soc.* **1973**, *95*, 584; Breslow; Hoffman *J. Am. Chem.Soc.* **1972**, *94*, 2110. For further evidence, see Lossing; Treager *J. Am. Chem. Soc.* **1975**, *97*, 1579. See also Breslow; Canary *J. Am. Chem. Soc.* **1991**, *113*, 3950.

[166]For a monograph, see Fray; Saxton *The Chemistry of Cyclo-octatetraene and its Derivatives*; Cambridge University Press: Cambridge, 1978. For a review, see Paquette *Tetrahedron* **1975**, *31*, 2855-2883. For reviews of heterocyclic 8π systems, see Kaim *Rev. Chem. Intermed.* **1987**, *8*, 247-286; Schmidt *Angew. Chem. Int. Ed. Engl.* **1975**, *14*, 581-591 [*Angew. Chem. 87*, 603-613].

[167]Bastiansen; Hedberg; Hedberg *J. Chem. Phys.* **1957**, *27*, 1311.

[168]The compound perfluorotetracyclobutacyclooctatetraene has been found to have a planar cyclooctatetraene ring, although the corresponding tetracyclopenta analog is nonplanar: Einstein; Willis; Cullen; Soulen *J. Chem. Soc., Chem. Commun.* **1981**, 526. See also Paquette; Wang; Cottrell *J. Am. Chem. Soc.* **1987**, *109*, 3730.

antiaromatic.[169] There is evidence that part of the reason for the lack of planarity in **45** itself is that a planar molecular would have to be antiaromatic.[170] The cycloheptatrienyl anion (**34**) also has eight electrons but does not behave like an aromatic system.[92] The nmr spectrum

60   61   62

of the benzocycloheptatrienyl anion (**62**) shows that, like **49, 60,** and **61,** this compound is antiaromatic.[171]

Systems of Ten Electrons[172]

There are three geometrically possible isomers of [10]annulene—the all-cis (**63**), the mono-trans (**64**), and the cis–trans–cis–cis–trans (**46**). If Hückel's rule applies, they should

46   63   64

be planar. But it is far from obvious that the molecules would adopt a planar shape, since they must overcome considerable strain to do so. For a regular decagon (**63**) the angles would have to be 144°, considerably larger than the 120° required for sp^2 angles. Some of this strain would also be present in **64** but this kind of strain is eliminated in **46** since all the angles are 120°. However, it was pointed out by Mislow[173] that the hydrogens in the 1 and 6 positions should interfere with each other and force the molecule out of planarity.

Compounds **63** and **64** have been prepared[174] as crystalline solids at −80°C. Nmr spectra show that all the hydrogens lie in the olefinic region and neither compound is aromatic. From ^{13}C and proton nmr spectra it has been deduced that neither is planar. However, that the angle strain is not insurmountable has been demonstrated by the preparation of several compounds that have large angles but that are definitely planar 10-electron aromatic systems.

[169]For a review, see Huang; Sondheimer *Acc. Chem. Res.* **1982,** *15,* 96-102. See also Dürr; Klauck; Peters; von Schnering *Angew. Chem. Int. Ed. Engl.* **1983,** *22,* 332 [*Angew. Chem. 95,* 321]; Chan; Mak; Poon; Wong; Jia; Wang *Tetrahedron* **1986,** *42,* 655.

[170]Figeys; Dralants *Tetrahedron Lett.* **1971,** 3901; Buchanan *Tetrahedron Lett.* **1972,** 665.

[171]Staley; Orvedal *J. Am. Chem. Soc.* **1973,** *95,* 3382.

[172]For reviews, see Kemp-Jones; Masamune *Top. Nonbenzenoid Aromat. Chem.* **1973,** *1,* 121-157; Masamune; Darby *Acc. Chem. Res.* **1972,** *5,* 272-281; Burkoth; van Tamelen, in Snyder, Ref. 55, vol. 1, pp. 63-116; Vogel, in *Aromaticity,* Ref. 55, pp. 113-147.

[173]Mislow *J. Chem. Phys.* **1952,** *20,* 1489.

[174]Masamune; Hojo; Hojo; Bigam; Rabenstein *J. Am. Chem. Soc.* **1971,** *93,* 4966. [10]Annulenes had previously been prepared, but it was not known which ones: van Tamelen; Burkoth *J. Am. Chem. Soc.* **1967,** *89,* 151; van Tamelen; Greeley *Chem. Commun.* **1971,** 601; van Tamelen; Burkoth; Greeley *J. Am. Chem. Soc.* **1971,** *93,* 6120.

Among these are the dianion **65**, the anions **66** and **67**, and the azonine **68**.[175] **65**[176] has angles of about 135°, while **66**[177] and **67**[178] have angles of about 140°, which are not very far

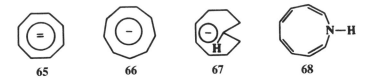

| 65 | 66 | 67 | 68 |

from 144°. The inner proton in **67**[179] (which is the mono-trans isomer of the all-cis **66**) is found far upfield in the nmr ($-3.5\ \delta$). For **63** and **64**, the cost in strain energy to achieve planarity apparently outweighs the extra stability that would come from an aromatic ring. To emphasize the delicate balance between these factors, we may mention that the oxygen analog of **68** (oxonin) and the N-carbethoxy derivative of **68** are nonaromatic and nonplanar, while **68** itself is aromatic and planar.[180]

So far **46** has not been prepared, despite many attempts. However, there are various ways of avoiding the interference between the two inner protons. The approach that has been most successful involves bridging the 1 and 6 positions.[181] Thus, 1,6-methano[10]annulene (**69**)[182] and its oxygen and nitrogen analogs **70**[183] and **71**[184] have been prepared and are stable compounds that undergo aromatic substitution and are diatropic.[185] For example, the perimeter protons of **69** are found at 6.9 to 7.3 δ, while the

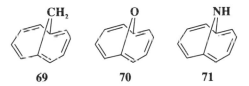

| 69 | 70 | 71 |

[175]For reviews of **68** and other nine-membered rings containing four double bonds and a hetero atom (*heteronins*), see Anastassiou *Acc. Chem. Res.* **1972**, *5*, 281-288, *Top. Nonbenzenoid Aromat. Chem.* **1973**, *1*, 1-27, *Pure Appl. Chem.* **1975**, *44*, 691-749. For a review of heteroannulenes in general, see Anastassiou; Kasmai *Adv. Heterocycl. Chem.* **1978**, *23*, 55-102.

[176]Katz *J. Am. Chem. Soc.* **1960**, *82*, 3784, 3785; Goldstein; Wenzel *J. Chem. Soc., Chem. Commun.* **1984**, 1654; Garkusha; Garbuzova; Lokshin; Todres *J. Organomet. Chem.* **1989**, *371*, 279. See also Noordik; van den Hark; Mooij; Klaassen *Acta Crystallogr. Sect. B.* **1974**, *30*, 833; Goldberg; Raymond; Harmon; Templeton *J. Am. Chem. Soc.* **1974**, *96*, 1348; Evans; Wink; Wayda; Little *J. Org. Chem.* **1981**, *46*, 3925; Heinz; Langensee; Müllen *J. Chem. Soc., Chem.Commun.* **1986**, 947.

[177]Katz; Garratt *J. Am. Chem. Soc.* **1964**, *86*, 5194; LaLancette; Benson *J. Am. Chem. Soc.* **1965**, *87*, 1941; Simmons; Chesnut; LaLancette *J. Am. Chem. Soc.* **1965**, *87*, 982; Paquette; Ley; Meisinger; Russell; Oku *J. Am. Chem. Soc.* **1974**, *96*, 5806; Radlick; Rosen *J. Am. Chem. Soc.* **1966**, *88*, 3461.

[178]Anastassiou; Gebrian *Tetrahedron Lett.* **1970**, 825.

[179]Boche; Weber; Martens; Bieberbach *Chem. Ber.* **1978**, *111*, 2480. See also Anastassiou; Reichmanis *Angew. Chem. Int. Ed. Engl.* **1974**, *13*, 728 [*Angew. Chem.* **86**, 784]; Boche; Bieberbach *Tetrahedron Lett.* **1976**, 1021.

[180]Anastassiou; Cellura *Chem. Commun.* **1969**, 903; Anastassiou; Gebrian *J. Am. Chem. Soc.* **1969**, *91*, 4011; Anastassiou; Cellura; Gebrian *Chem. Commun.* **1970**, 375; Anastassiou; Yamamoto *J. Chem. Soc., Chem. Commun.* **1972**, 286; Chiang; Paul; Anastassiou; Eachus *J. Am. Chem. Soc.* **1974**, *96*, 1636.

[181]For reviews of bridged [10]-, [14]-, and [18]annulenes, see Vogel *Pure Appl. Chem.* **1982**, *54*, 1015-1039; *Isr. J. Chem.* **1980**, *20*, 215-224; *Chimia* **1968**, *22*, 21-32; Vogel; Günther *Angew. Chem. Int. Ed. Engl.* **1967**, *6*, 385-401 [*Angew. Chem.* **79**, 429-446].

[182]Vogel; Roth *Angew. Chem. Int. Ed. Engl.* **1964**, *3*, 228 [*Angew. Chem.* **76**, 145]; Vogel; Böll *Angew. Int. Ed. Engl.* **1964**, *3*, 642 [*Angew. Chem.* **76**, 784]; Vogel; Böll; Biskup *Tetrahedron Lett.* **1966**, 1569.

[183]Vogel; Biskup; Pretzer; Böll *Angew. Chem. Int. Ed. Engl.* **1964**, *3*, 642 [*Angew. Chem.* **76**, 785]; Sondheimer; Shani *J. Am. Chem. Soc.* **1964**, *84*, 3168; Shani; Sondheimer *J. Am. Chem. Soc.* **1967**, *89*, 6310; Bailey; Mason *Chem. Commun.* **1967**, 1039.

[184]Vogel; Pretzer; Böll *Tetrahedron Lett.* **1965**, 3613. See also the first paper of Ref. 183.

[185]For another type of bridged diatropic [10]annulene, see Lidert; Rees *J. Chem. Soc., Chem. Commun.* **1982**, 499; Gilchrist; Rees; Tuddenham *J. Chem. Soc., Perkin Trans. 1* **1983**, 83; McCague; Moody; Rees *J. Chem. Soc., Perkin Trans. 1* **1984**, 165, 175; Gibbard; Moody; Rees *J. Chem. Soc., Perkin Trans. 1* **1985**, 731, 735.

bridge protons are at -0.5 δ. The crystal structure of **69** shows that the perimeter is nonplanar, but the bond distances are in the range 1.37 to 1.42 Å.[186] It has therefore been amply demonstrated that a closed loop of 10 electrons is an aromatic system, although some molecules that could conceivably have such a system are too distorted from planarity to be aromatic. A small distortion from planarity (as in **69**) does not prevent aromaticity, at least in part because the σ orbitals so distort themselves as to maximize the favorable (parallel) overlap of p orbitals to form the aromatic 10-electron loop.[187]

In **72**, where **69** is fused to two benzene rings in such a way that no canonical form can be written in which both benzene rings have six electrons, the aromaticity is reduced by annellation, as shown by the fact that the molecule rapidly converts to the more stable **73**,

in which both benzene rings can be fully aromatic[188] (this is similar to the cycloheptatriene–norcaradiene conversions discussed on p. 1135).

Systems of More than Ten Electrons: $4n + 2$ Electrons[189]

Extrapolating from the discussion of [10]annulene, we expect larger $4n + 2$ systems to be aromatic if they are planar. Mislow[173] predicted that [14]annulene (**74**) would possess the same type of interference as **46**, although in lesser degree. This is borne out by experiment. **74** is aromatic (it is diatropic; inner protons at 0.00 δ, outer protons at 7.6 δ),[190] but is completely destroyed by light and air in one day. X-ray analysis shows that although there are no alternating single and double bonds, the molecule is not planar.[191]

74

[186]Bianchi; Pilati; Simonetta *Acta Crystallogr., Sect. B* **1980**, *36*, 3146. See also Dobler; Dunitz *Helv. Chim Acta* **1965**, *48*, 1429.

[187]For a discussion, see Haddon *Acc. Chem. Res.* **1988**, *21*, 243-249.

[188]Hill; Giberson; Silverton *J. Am. Chem. Soc.* **1988**, *110*, 497. See also McCague; Moody; Rees; Williams *J. Chem. Soc., Perkin Trans. 1* **1984**, 909.

[189]For reviews of annulenes, with particular attention to their nmr spectra, see Sondheimer *Acc. Chem. Res.* **1972**, *5*, 81-91, *Pure Appl. Chem.* **1971**, *28*, 331-353, *Proc. R. Soc. London. Ser. A* **1967**, *297*, 173-204; Sondheimer; Calder; Elix; Gaoni; Garratt; Rowland; Sondheimer *Tetrahedron* **1971**, *27*, 3157; Oth; Müllen; Königshofen; Mann; Sakata; Vogel *Angew. Chem. Int. Ed. Engl.* **1974**, *13*, 284 [*Angew. Chem. 86*, 232]. For some other 14-electron aromatic systems, see Anastassiou, Elliott; Reichmanis *J. Am. Chem. Soc.* **1974**, *96*, 7823; Wife; Sondheimer *J. Am. Chem. Soc.* **1975**, *97*, 640; Ogawa; Kubo; Saikachi *Tetrahedron Lett.* **1971**, 4859; Oth; Müllen; Königshofen; Wassen; Vogel *Helv. Chim. Acta* **1974**, *57*, 2387; Willner; Gutman; Rabinovitz *J. Am. Chem. Soc.* **1977**, *99*, 4167; Röttele; Schröder *Chem. Ber.* **1982**, *115*, 248.

However, a number of stable bridged [14]annulenes have been prepared,[192] e.g., *trans*-15,16-dimethyldihydropyrene (**75**),[193] *syn*-1,6:8,13-diimino[14]annulene (**76**),[194] and *syn*- and *anti*-1,6:8,13-bismethano[14]annulene (**77** and **78**).[195] The dihydropyrene **75** (and its diethyl and dipropyl homologs) is undoubtedly aromatic: the π perimeter is approximately

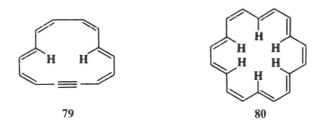

| 75 | 76 | 77 | 78 |

planar,[196] the bond distances are all 1.39 to 1.40 Å, and the molecule undergoes aromatic substitution[193] and is diatropic.[197] The outer protons are found at 8.14 to 8.67 δ, while the CH_3 protons are at -4.25 δ. **76** and **77** are also diatropic,[198] although x-ray crystallography indicates that the π periphery in at least **76** is not quite planar.[199] However, **78**, in which the geometry of the molecule greatly reduces the overlap of the p orbitals at the bridgehead positions with adjacent p orbitals, is definitely not aromatic,[200] as shown by nmr spectra[195] and x-ray crystallography, from which bond distances of 1.33 to 1.36 Å for the double bonds and 1.44 to 1.49 Å for the single bonds have been obtained.[201] In contrast, all the bond distances in **76** are ~1.38 to 1.40 Å.[199]

Another way of eliminating the hydrogen interferences of [14]annulene is to introduce one or more triple bonds into the system, as in dehydro[14]annulene (**79**).[202] All five known

| 79 | 80 |

[192]For a review, see Vogel *Pure Appl. Chem.* **1971**, *28*, 355-377.

[193]Boekelheide; Phillips *J. Am. Chem. Soc.* **1967**, *89*, 1695; Boekelheide; Miyasaka *J. Am. Chem. Soc.* **1967**, *89*, 1709. For reviews of dihydropyrenes, see Mitchell *Adv. Theor. Interesting Mol.* **1989**, *1*, 135-199; Boekelheide *Top. Nonbenzoid Arom. Chem.* **1973**, *1*, 47-79, *Pure Appl. Chem.* **1975**, *44*, 807-828.

[194]Vogel; Kuebart; Marco; Andree; Günther; Aydin *J. Am. Chem. Soc.* **1983**, *105*, 6982; Destro; Pilati; Simonetta; Vogel *J. Am. Chem. Soc.* **1985**, *107*, 3185, 3192. For the di —O— analog of **76**, see Vogel; Biskup; Vogel; Günther *Angew. Chem. Int. Ed. Engl.* **1966**, *5*, 734 [*Angew. Chem. 78*, 755].

[195]Vogel; Haberland; Günther *Angew. Chem. Int. Ed. Engl.* **1970**, *9*, 513 [*Angew. Chem. 82*, 510]; Vogel; Sombrock; Wagemann *Angew. Chem. Int. Ed. Engl.* **1975**, *14*, 564 [*Angew. Chem. 87*, 591].

[196]Hanson *Acta Crystallogr.* **1965**, *18*, 599, **1967**, *23*, 476.

[197]A number of annellated derivatives of **75** are less diatropic, as would be expected from the discussion on p. 44: Mitchell; Williams; Mahadevan; Lai; Dingle *J. Am. Chem. Soc.* **1982**, *104*, 2571 and other papers in this series.

[198]As are several other similarly bridged [14]annulenes; see, for example, Vogel; Reel *J. Am. Chem. Soc.* **1972**, *94*, 4388; Flitsch; Peeters *Chem. Ber.* **1973**, *106*, 1731; Huber; Lex; Meul; Müllen *Angew. Chem. Int. Ed. Engl.* **1981**, *20*, 391 [*Angew. Chem. 93*, 401]; Vogel; Nitsche; Krieg *Angew. Chem. Int. Ed. Engl.* **1981**, *20*, 811 [*Angew. Chem. 93*, 818]; Mitchell; Anker *Tetrahedron Lett.* **1981**, *22*, 5139; Vogel; Wieland; Schmalstieg; Lex *Angew. Chem. Int. Ed. Engl.* **1984**, *23*, 717 [*Angew. Chem. 96*, 717]; Neumann; Müllen *J. Am. Chem. Soc.* **1986**, *108*, 4105.

[199]Ganis; Dunitz *Helv. Chim. Acta* **1967**, *50*, 2369.

[200]For another such pair of molecules, see Vogel; Nitsche; Krieg, Ref. 198. See also Vogel; Schieb; Schulz; Schmidt; Schmickler; Lex *Angew. Chem. Int. Ed. Engl.* **1986**, *25*, 723 [*Angew. Chem. 98*, 729].

[201]Gramaccioli; Mimun; Mugnoli; Simonetta *Chem. Commun.* **1971**, 796. See also Destro; Simonetta *Tetrahedron* **1982**, *38*, 1443.

[202]For a review of dehydroannulenes, see Nakagawa *Top. Nonbenzenoid Aromat. Chem.* **1973**, *1*, 191-219.

dehydro[14]annulenes are diatropic. **79** can be nitrated or sulfonated.[203] The extra electrons of the triple bond do not form part of the aromatic system but simply exist as a localized bond. [18]Annulene (**80**) is diatropic:[204] the 12 outer protons are found at about $\delta = 9$ and the 6 inner protons at about $\delta = -3$. X-ray crystallography[205] shows that it is nearly planar, so that interference of the inner hydrogens is not important in annulenes this large. **80** is reasonably stable, being distillable at reduced pressures, and undergoes aromatic substitutions.[206] The C—C bond distances are not equal, but they do not alternate. There are 12 inner bonds of about 1.38 Å and 6 outer bonds of about 1.42 Å.[205] **80** has been estimated to have a resonance energy of about 37 kcal/mol (155 kJ/mol), similar to that of benzene.[207]

The known bridged [18]annulenes are also diatropic[208] as are most of the known dehydro[18]annulenes.[209] The dianions of open and bridged [16]annulenes[210] are also 18-electron aromatic systems.[211]

[22]Annulene[212] and dehydro[22]annulene[213] are also diatropic. In the latter compound there are 13 outer protons at 6.25 to 8.45 δ and 7 inner protons at 0.70 to 3.45 δ. Some aromatic bridged [22]annulenes are also known.[214] [26]Annulene has not yet been prepared, but several dehydro[26]annulenes are aromatic.[215] Furthermore, the dianion of 1,3,7,9,13,15,19,21-octadehydro[24]annulene is another 26-electron system that is aromatic.[216] Ojima and co-workers have prepared bridged dehydro derivatives of [26], [30], and [34] annulenes.[217] All of these are diatropic. The same workers prepared a bridged tetradehydro[38]annulene,[217] which showed no ring current. On the other hand, the dianion of the cyclophane **81** also has 38 perimeter electrons, and this species is diatropic.[218]

There is now no doubt that $4n + 2$ systems are aromatic if they can be planar, although **63** and **78** among others, demonstrate that not all such systems are in fact planar enough

[203]Gaoni; Sondheimer *J. Am. Chem. Soc.* **1964**, *86*, 521.

[204]Jackman; Sondheimer; Amiel; Ben-Efraim; Gaoni; Wolovsky; Bothner-By *J. Am. Chem. Soc.* **1962**, *84*, 4307; Gilles; Oth; Sondheimer; Woo *J. Chem. Soc. B* **1971**, 2177. For a thorough discussion, see Baumann; Oth *Helv. Chim. Acta* **1982**, *65*, 1885.

[205]Bregman; Hirshfeld; Rabinovich; Schmidt *Acta Crystallogr.* **1965**, *19*, 227; Hirshfeld; Rabinovich *Acta Crystallogr.* **1965**, *19*, 235.

[206]Calder; Garratt; Longuet-Higgins; Sondheimer; Wolovsky *J. Chem. Soc. C* **1967**, 1041; Woo; Sondheimer *Tetrahedron* **1970**, *26*, 3933.

[207]Oth; Bünzli; de Julien de Zélicourt *Helv. Chim. Acta* **1974**, *57*, 2276.

[208]For some examples, see DuVernet; Wennerström; Lawson; Otsubo; Boekelheide *J. Am. Chem. Soc.* **1978**, *100*, 2457; Ogawa; Sadakari; Imoto; Miyamoto; Kato; Taniguchi *Angew. Chem. Int. Ed. Engl.* **1983**, *22*, 417 [*Angew. Chem. 95*, 412]; Vogel; Sicken; Röhrig; Schmickler; Lex; Ermer *Angew. Chem. Int. Ed. Engl.* **1988**, *27*, 411 [*Angew. Chem. 100*, 450].

[209]Okamura; Sondheimer *J. Am. Chem. Soc.* **1967**, *89*, 5991; Ojima; Ejiri; Kato; Nakamura; Kuroda; Hirooka; Shibutani *J. Chem. Soc., Perkin Trans. 1* **1987**, 831; Sondheimer, Ref. 189. For two that are not, see Endo; Sakata; Misumi *Bull. Chem. Soc. Jpn.* **1971**, *44*, 2465.

[210]For a review of this type of polycyclic ion, see Rabinovitz; Willner; Minsky *Acc. Chem. Res.* **1983**, *16*, 298-304.

[211]Oth; Anthoine; Gilles *Tetrahedron Lett.* **1968**, 6265; Mitchell; Boekelheide *Chem. Commun.* **1970**, 1557; Oth; Baumann; Gilles; Schröder *J. Am. Chem. Soc.* **1972**, *94*, 3948. See also Brown; Sondheimer *Angew. Chem. Int. Ed. Engl.* **1974**, *13*, 337 [*Angew. Chem. 86*, 346]; Cresp; Sargent *J. Chem. Soc., Chem. Commun.* **1974**, 101; Schröder; Plinke; Smith; Oth *Angew. Chem. Int. Ed. Engl.* **1973**, *12*, 325 [*Angew. Chem. 85*, 350]; Rabinovitz; Minsky *Pure Appl. Chem.* **1982**, *54*, 1005-1014.

[212]McQuilkin; Metcalf; Sondheimer *Chem. Commun.* **1971**, 338.

[213]McQuilkin; Sondheimer *J. Am. Chem. Soc.* **1970**, *92*, 6341; Iyoda; Nakagawa *J. Chem. Soc., Chem. Commun.* **1972**, 1003. See also Kabuto; Kitahara; Iyoda; Nakagawa *Tetrahedron Lett.* **1976**, 2787; Akiyama; Nomoto; Iyoda; Nakagawa *Bull. Chem. Soc. Jpn.* **1976**, *49*, 2579.

[214]For example see Broadhurst; Grigg; Johnson *J. Chem. Soc., Perkin Trans. 1* **1972**, 2111; Ojima et al., Ref. 209; Yamamoto; Kuroda; Shibutani; Yoneyama; Ojima; Fujita; Ejiri; Yanagihara *J. Chem. Soc., Perkin Trans. 1* **1988**, 395.

[215]Metcalf; Sondheimer *J. Am. Chem. Soc.* **1971**, *93*, 5271; Iyoda; Nakagawa *Tetrahedron Lett.* **1972**, 4253; Ojima; Fujita; Matsumoto; Ejiri; Kato; Kuroda; Nozawa; Hirooka; Yoneyama; Tatemitsu *J. Chem. Soc., Perkin Trans. 1* **1988**, 385.

[216]McQuilkin; Garratt; Sondheimer *J. Am. Chem. Soc.* **1970**, *92*, 6682. See also Huber; Müllen; Wennerström *Angew. Chem. Int. Ed. Engl.* **1980**, *19*, 624 [*Angew. Chem. 92*, 636].

[217]Ojima et al., Ref. 215.

[218]Müllen; Unterberg; Huber; Wennerström; Norinder; Tanner; Thulin *J. Am. Chem. Soc.* **1984**, *106*, 7514.

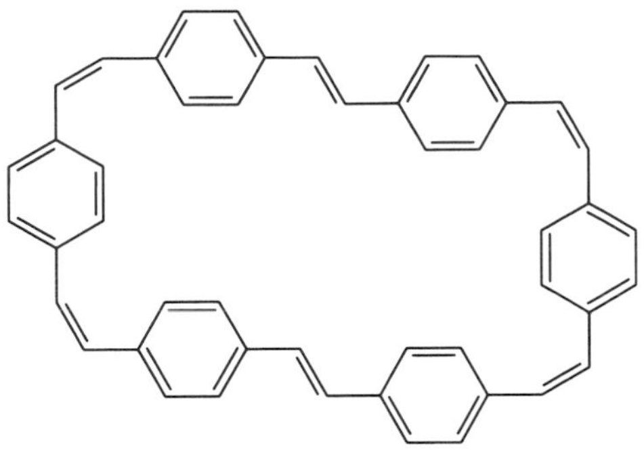

81

for aromaticity. The cases of **74** and **76** prove that absolute planarity is not required for aromaticity, but that aromaticity decreases with decreasing planarity.

The proton nmr spectrum of **82** (called kekulene) showed that in a case where electrons can form either aromatic sextets or larger systems, the sextets are preferred.[219] The 48 π electrons of **82** might, in theory, prefer structure **82a**, where each ring is a fused benzene

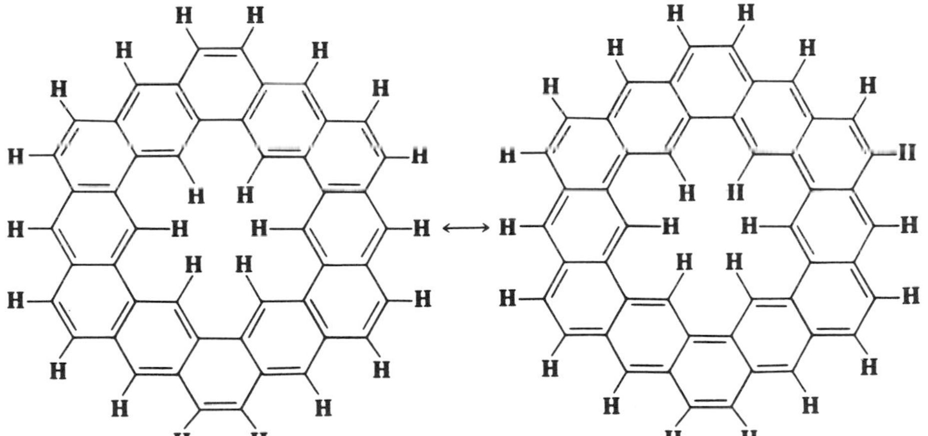

82a   **82b**

ring, or **82b**, which has a [30]annulene on the outside and an [18]annulene on the inside. The proton nmr spectrum of this compound shows three peaks at δ = 7.94, 8.37, and 10.45 in a ratio of 2:1:1. It is seen from the structure that **82** contains three groups of protons. The peak at 7.94 δ is attributed to the 12 ortho protons and the peak at 8.37 δ to the six external para protons. The remaining peak comes from the six inner protons. If the molecule preferred **82b**, we would expect to find this peak upfield, probably with a negative δ, as in the case of **80**. The fact that this peak is far downfield indicates that the electrons prefer to be in benzenoid rings. Note that in the case of the dianion of **81,** we have the opposite situation. In this ion, the 38-electron system is preferred even though 24 of these must come from the six benzene rings, which therefore cannot have aromatic sextets.

[219]Staab; Diederich *Chem. Ber.* **1983,** *116,* 3487; Staab; Diederich; Krieger; Schweitzer *Chem. Ber.* **1983,** *116,* 3504. For a similar molecule with 10 instead of 12 rings, see Funhoff; Staab *Angew. Chem. Int. Ed. Engl.* **1986,** *25,* 742 [*Angew. Chem. 98,* 757].

Systems of More than Ten Electrons: 4n Electrons[189]

As we have seen (p. 53), these systems are expected to be not only nonaromatic but actually antiaromatic. The chief criterion for antiaromaticity in annulenes is the presence of a *paramagnetic* ring current,[220] which causes protons on the outside of the ring to be shifted *upfield* while any inner protons are shifted *downfield*, in sharp contrast to a diamagnetic ring current, which causes shifts in the opposite directions. Compounds that sustain a paramagnetic ring current are called *paratropic*; we have already seen such behavior in certain four- and eight-electron systems. As with aromaticity, we expect that antiaromaticity will be at a maximum when the molecule is planar and when bond distances are equal.

The [12]annulene **83** has been prepared.[221] In solution this molecule undergoes rapid conformational mobility (as do many other annulenes),[222] so that above a certain temper-

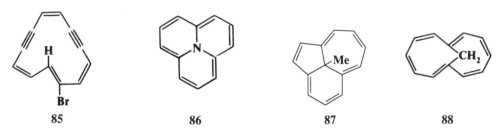

83   **84**

ature, in this case −150°C, all protons are magnetically equivalent. However, at −170°C the mobility is greatly slowed and the three inner protons are found at about 8 δ while the nine outer protons are at about 6 δ. **83** suffers from hydrogen interference and is certainly not planar. It is very unstable and above −50°C rearranges to **84**. Several bridged and dehydro[12]annulenes are known, e.g., 5-bromo-1,9-didehydro[12]annulene (**85**),[223] cycl[3.3.3]azine (**86**),[224] 9b-methyl-9bH-benzo[cd]azulene (**87**),[225] and 1,7-methano[12]-annulene (**88**).[226] In these compounds both hydrogen interference and conformational mo-

85   **86**   **87**   **88**

bility are prevented. In **86, 87,** and **88,** the bridge prevents conformational changes, while in **85** the bromine atom is too large to be found inside the ring. Nmr spectra show that all four compounds are paratropic, the inner proton of **85** being found at 16.4 δ. The dication of **77**[227] and the dianion of **69**[228] are also 12-electron paratropic species.

[220]Pople; Untch *J. Am. Chem. Soc.* **1966**, *88*, 4811; Longuet-Higgins, in *Aromaticity*, Ref. 55. pp. 109-111.

[221]Oth; Röttele; Schröder *Tetrahedron Lett.* **1970**, 61; Oth; Gilles; Schröder *Tetrahedron Lett.* **1970**, 67.

[222]For a review of conformational mobility in annulenes, see Oth *Pure Appl. Chem.* **1971**, *25*, 573-622.

[223]Untch; Wysocki *J. Am. Chem. Soc.* **1967**, *89*, 6386.

[224]Farquhar; Leaver *Chem. Commun.* **1969**, 24. For a review, see Matsuda; Gotou *Heterocycles* **1987**, *26*, 2757-2772.

[225]Hafner; Kühn *Angew. Chem. Int. Ed. Engl.* **1986**, *25*, 632 [*Angew. Chem. 98*, 648]. For a similar system, see Kohnz; Düll; Müllen *Angew. Chem. Int. Ed. Engl.* **1989**, *28*, 1343 [*Angew. Chem. 101*, 1375].

[226]Vogel; Königshofen; Müllen; Oth *Angew. Chem. Int. Ed. Engl.* **1974**, *13*, 281 [*Angew. Chem. 86*, 229]. See also Mugnoli; Simonetta *J. Chem. Soc., Perkin Trans. 2* **1976**, 822; Scott; Kirms; Günther; von Puttkamer *J. Am. Chem. Soc.* **1983**, *105*, 1372; Destro; Ortoleva; Simonetta; Todeschini *J. Chem. Soc., Perkin Trans. 2* **1983**, 1227.

[227]Müllen; Meul; Schade; Schmickler; Vogel *J. Am. Chem. Soc.* **1987**, *109*, 4992. This paper also reports a number of other bridged paratropic 12-, 16-, and 20-electron dianions and dications. See also Hafner; Thiele *Tetrahedron Lett.* **1984**, *25*, 1445.

[228]Schmalz; Günther *Angew. Chem. Int. Ed. Engl.* **1988**, *27*, 1692 [*Angew. Chem. 100*, 1754].

The results for [16]annulene are similar. The compound was synthesized in two different ways,[229] both of which gave **89**, which in solution is in equilibrium with **90**. Above $-50°C$ there is conformational mobility, resulting in the magnetic equivalence of all protons, but

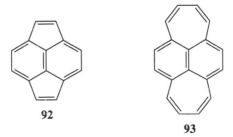

89   **90**   **91**

at $-130°C$ the compound is clearly paratropic: there are four protons at 10.56 δ and twelve at 5.35 δ. In the solid state, where the compound exists entirely as **89**, x-ray crystallography[230] shows that the molecules are nonplanar with almost complete bond alternation: the single bonds are 1.44 to 1.47 Å and the double bonds 1.31 to 1.35 Å. A number of dehydro and bridged [16]annulenes are also paratropic,[231] as are [20]annulene,[232] [24]annulene,[233] and **91**, a 28-electron system that is the tetraanion of [2₄]paracyclophanetetraene.[234] However, a bridged tetradehydro[32]annulene was atropic.[?17]

Both peracyclene (**92**)[235] (which because of strain is stable only in solution) and dipleiadiene (**93**)[236] are paratropic, as shown by nmr spectra. These molecules might have been expected to behave like naphthalenes with outer bridges, but instead, the outer π frameworks (12 and 16 electrons, respectively) constitute antiaromatic systems with an extra central double bond.

92

93

[229]Schröder; Oth *Tetrahedron Lett.* **1966**, 4083; Sondheimer; Gaoni *J. Am. Chem. Soc.* **1961**, *83*, 4863; Oth; Gilles *Tetrahedron Lett.* **1968**, 6259; Calder; Gaoni; Sondheimer *J. Am. Chem. Soc.* **1968**, *90*, 4946. For monosubstituted [16]annulenes, see Schröder; Kirsch; Oth *Chem. Ber.* **1974**, *107*, 460.
[230]Johnson; Paul; King *J. Chem. Soc. B* **1970**, 643.
[231]For example, see Calder; Garratt; Sondheimer *J. Am. Chem. Soc.* **1968**, *90*, 4954; Murata; Okazaki; Nakazawa *Angew. Chem. Int. Ed. Engl.* **1971**, *10*, 576 [*Angew. Chem. 83*, 623]; Ogawa; Kubo; Tabushi *Tetrahedron Lett.* **1973**, 361; Nakatsuji; Morigaki; Akiyama; Nakagawa *Tetrahedron Lett.* **1975**, 1233; Elix *Aust. J. Chem.* **1969**, *22*, 1951; Vogel; Kürshner; Schmickler; Lex; Wennerström; Tanner; Norinder; Krüger *Tetrahedron Lett.* **1985**, *26*, 3087.
[232]Metcalf; Sondheimer *J. Am. Chem. Soc.* **1971**, *93*, 6675. See also Oth; Woo; Sondheimer *J. Am. Chem. Soc.* **1973**, *95*, 7337; Nakatsuji; Nakagawa *Tetrahedron Lett.* **1975**, 3927; Wilcox; Farley **1984**, *106*, 7195.
[233]Calder; Sondheimer *Chem. Commun.* **1966**, 904. See also Stöckel; Sondheimer *J. Chem. Soc., Perkin Trans.1* **1972**, 355; Nakatsuji; Akiyama; Nakagawa *Tetrahedron Lett.* **1976**, 2623; Yamamoto et al., Ref. 214.
[234]Huber; Müllen; Wennerström, Ref. 216.
[235]Trost; Bright; Frihart; Brittelli *J. Am. Chem. Soc.* **1971**, *93*, 737; Trost; Herdle *J. Am. Chem. Soc.* **1976**, *98*, 4080.
[236]Vogel; Neumann; Klug; Schmickler; Lex *Angew. Chem. Int. Ed. Engl.* **1985**, *24*, 1046 [*Angew. Chem. 97*, 1044].

The fact that many $4n$ systems are paratropic even though they may be nonplanar and have unequal bond distances indicates that if planarity were enforced, the ring currents might be even greater. That this is true is dramatically illustrated by the nmr spectrum of the dianion of **75**[237] (and its diethyl and dipropyl homologs).[238] We may recall that in **75**, the outer protons were found at 8.14 to 8.67 δ with the methyl protons at −4.25 δ. For the dianion, however, which is forced to have approximately the same planar geometry but now has 16 electrons, the outer protons are shifted to about −3 δ while the methyl protons are found at about 21 δ, a shift of about 25 δ! We have already seen where the converse shift was made, when [16]annulenes that were antiaromatic were converted to 18-electron dianions that were aromatic.[211] In these cases, the changes in nmr chemical shifts were almost as dramatic. Heat-of-combustion measures also show that [16]annulene is much less stable than its dianion.[239]

We can therefore conclude that in $4n$ systems antiaromaticity will be at a maximum where a molecule is constrained to be planar (as in **52** or the dianion of **75**) but, where possible, the molecule will distort itself from planarity and avoid equal bond distances in order to reduce antiaromaticity. In some cases, such as cyclooctatraene, the distortion and bond alternation are great enough for antiaromaticity to be completely avoided. In other cases, e.g., **83** or **89**, it is apparently not possible for the molecules to avoid at least some p-orbital overlap. Such molecules show paramagnetic ring currents and other evidence of antiaromaticity, although the degree of antiaromaticity is not as great as in molecules such as **52** or the dianion of **75**.

Other Aromatic Compounds

We shall briefly mention three other types of aromatic compounds.

1. *Mesoionic compounds*[240] cannot be satisfactorily represented by Lewis structures not involving charge separation. Most of them contain five-membered rings. The most common

Sydnone

are the *sydnones*, stable aromatic compounds that undergo aromatic substitution when R' is hydrogen.

2. *The dianion of squaric acid.*[241] The stability of this system is illustrated by the fact that the pK_1 of squaric acid[242] is about 1.5 and the pK_2 is about 3.5,[243] which means that

---

[237]For a review of polycyclic dianions, see Rabinovitz; Cohen *Tetrahedron* **1988**, *44* 6957-6994.

[238]Mitchell; Klopfenstein; Boekelheide *J. Am. Chem. Soc.* **1969**, *91*, 4931. For another example, see Deger; Müllen; Vogel *Angew. Chem. Int. Ed. Engl.* **1978**, *17*, 957 [*Angew. Chem. 90*, 990].

[239]Stevenson; Forch *J. Am. Chem. Soc.* **1980**, *102*, 5985.

[240]For reviews, see Newton; Ramsden *Tetrahedron* **1982**, *38*, 2965-3011; Ollis; Ramsden *Adv. Heterocycl. Chem.* **1976**, *19*, 1-122; Ramsden *Tetrahedron* **1977**, *33*, 3203-3232; Yashunskii; Kholodov *Russ. Chem. Rev.* **1980**, *49*, 28-45; Ohta; Kato, in Snyder, Ref. 55, vol. 1, pp. 117-248.

[241]West; Powell *J. Am. Chem. Soc.* **1963**, *85*, 2577; Ito; West *J. Am. Chem. Soc.* **1963**, *85*, 2580.

[242]For a review of squaric acid and other nonbenzenoid quinones, see Wong; Chan; Luh, in Patai; Rappoport *The Chemistry of the Quinonoid Compounds*, vol. 2, pt. 2; Wiley: New York, 1988, pp. 1501-1563.

[243]Ireland; Walton *J. Phys. Chem.* **1967**, *71*, 751; MacDonald *J. Org. Chem.* **1968**, *33*, 4559.

even the second proton is given up much more readily than the proton of acetic acid, for example.[244] The analogous three-,[245] five-, and six-membered ring compounds are also known.[246]

3. *Homoaromatic compounds.* When cyclooctatetraene is dissolved in concentrated H_2SO_4, a proton adds to one of the double bonds to form the homotropylium ion **94**.[247] In this species an aromatic sextet is spread over seven carbons, as in the tropylium ion. The

94

eighth carbon is an sp^3 carbon and so cannot take part in the aromaticity. Nmr spectra show the presence of a diatropic ring current: H_b is found at $\delta = -0.3$; H_a at 5.1 δ; H_1 and H_7 at 6.4 δ; H_2–H_6 at 8.5 δ. This ion is an example of a *homoaromatic* compound, which may be defined as a compound that contains one or more[248] sp^3-hybridized carbon atoms in an otherwise conjugated cycle.[249] In order for the orbitals to overlap most effectively so as to close a loop, the sp^3 atoms are forced to lie almost vertically above the plane of the aromatic atoms.[250] In **94**, H_b is directly above the aromatic sextet and so is shifted far upfield in the nmr. All homoaromatic compounds so far discovered are ions, and it is questionable[251] as to whether homoaromatic character can exist in uncharged systems.[252] Homoaromatic ions of two and ten electrons are also known.

[244]There has been a controversy as to whether this dianion is in fact aromatic. See Aihara *J. Am. Chem. Soc.* **1981**, *103*, 1633.

[245]Eggerding; West *J. Am. Chem. Soc.* **1976**, *98*, 3641; Pericás; Serratosa *Tetrahedron Lett.* **1977**, 4437; Semmingsen; Groth *J. Am. Chem. Soc.* **1987**, *109*, 7238.

[246]For a monograph, see West *Oxocarbons*; Academic Press: New York, 1980. For reviews, see Serratosa *Acc. Chem. Res.* **1983**, *16*, 170-176; Schmidt *Synthesis* **1980**, 961-994; West *Isr. J. Chem.* **1980**, *20*, 300-307; West; Niu in Snyder, Ref. 55, vol. 1, pp. 311-345, and in Zabicky *The Chemistry of the Carbonyl Group*, vol. 2; Wiley: New York, 1970, pp. 241-275; Maahs; Hegenberg *Angew. Chem. Int. Ed. Engl.* **1966**, *5*, 888-893 [*Angew. Chem. 78*, 927-931].

[247]Rosenberg; Mahler; Pettit *J. Am. Chem. Soc.* **1962**, *84*, 2842; Keller; Pettit *J. Am. Chem.Soc.* **1966**, *88*, 604, 606; Winstein; Kaesz; Kreiter; Friedrich *J. Am. Chem. Soc.* **1965**, *87*, 3267; Winstein; Kreiter; Brauman *J. Am. Chem. Soc.* **1966**, *88*, 2047; Haddon *J. Am. Chem. Soc.* **1988**, *110*, 1108. See also Childs; Mulholland; Varadarajan; Yeroushalmi *J. Org. Chem.* **1983**, *48*, 1431.

[248]If a compound contains two such atoms it is *bishomoaromatic*; if three, *trishomoaromatic*, etc. For examples see Paquette, Ref. 249.

[249]For reviews, see Childs *Acc. Chem. Res.* **1984**, *17*, 347-352; Paquette *Angew. Chem. Int. Ed. Engl.* **1978**, *17*, 106-117 [*Angew. Chem. 90*, 114-125]; Winstein *Q. Rev., Chem. Soc.* **1969**, *23*, 141-176; *Aromaticity*, Ref. 55, pp. 5-45; and in Olah; Schleyer, *Carbonium Ions*; Wiley: New York, vol. 3, 1972, the reviews by Story; Clark, 1007-1098, pp. 1073-1093; Winstein 965-1005. (The latter is a reprint of the *Q. Rev., Chem. Soc.* review mentioned above.)

[250]Calculations show that only about 60% of the chemical shift difference between H_a and H_b is the result of the aromatic ring current, and that even H_a is shielded; it would appear at $\delta = \sim 5.5$ without the ring current: Childs; McGlinchey; Varadarajan *J. Am. Chem. Soc.* **1984**, *106*, 5974.

[251]Houk; Gandour; Strozier; Rondan; Paquette *J. Am. Chem. Soc.* **1979**, *101*, 6797; Paquette; Snow; Muthard; Cynkowski *J. Am. Chem. Soc.* **1979**, *101*, 6991. See however, Liebman; Paquette; Peterson; Rogers *J. Am. Chem. Soc.* **1986**, *108*, 8267.

[252]Examples of uncharged homo*anti*aromatic compounds have been claimed: Wilcox; Blain; Clardy; Van Duyne; Gleiter; Eckert-Maksić *J. Am. Chem. Soc.* **1986**, *108*, 7693; Scott; Cooney; Rogers; Dejroongruang *J. Am. Chem. Soc.* **1988**, *110*, 7244.

HYPERCONJUGATION

All of the delocalization discussed so far involves π electrons. Another type, called *hyperconjugation*, involves σ electrons.[253] When a carbon attached to at least one hydrogen is attached to an unsaturated atom or one with an unshared orbital, canonical forms such as **95** can be drawn. In such canonical forms there is no bond at all between the carbon and

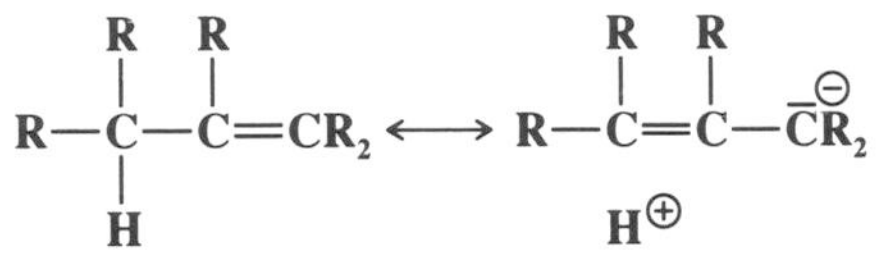

$$
\underset{\substack{| \\ \text{H}}}{\overset{\substack{\text{R} \quad \text{R} \\ | \quad |}}{\text{R}-\text{C}-\text{C}}}=\text{CR}_2 \longleftrightarrow \underset{\substack{| \\ \text{H}^{\oplus}}}{\overset{\substack{\text{R} \quad \text{R} \\ | \quad |}}{\text{R}-\text{C}=\text{C}}}-\overset{\ominus}{\text{CR}}_2
$$

95

hydrogen. The effect of **95** on the actual molecule is that the electrons in the C—H bond are closer to the carbon than they would be if **95** did not contribute at all.

Hyperconjugation in the above case may be regarded as an overlap of the σ orbital of the C—H bond and the π orbital of the C—C bond, analogous to the π–π–orbital overlap previously considered. As might be expected, those who reject the idea of resonance in butadiene (p. 31) believe it even less likely when it involves no-bond structures.

The concept of hyperconjugation arose from the discovery of apparently anomalous electron-release patterns for alkyl groups. By the field effect alone, the order of electron release for simple alkyl groups connected to an unsaturated system is *t*-butyl > isopropyl > ethyl > methyl, and this order is observed in many phenomena. Thus, the dipole moments in the gas phase of $PhCH_3$, PhC_2H_5, $PhCH(CH_3)_2$, and $PhC(CH_3)_3$ are, respectively, 0.37, 0.58, 0.65 and 0.70 D.[254]

However, Baker and Nathan observed that the rates of reaction with pyridine of *p*-substituted benzyl bromides (see reaction **0-43**) were about opposite that expected from

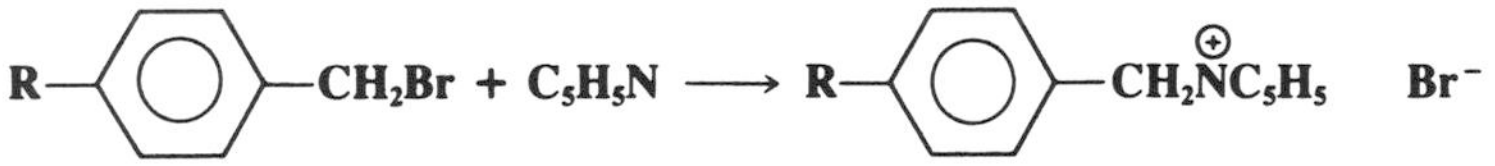

$$
\text{R}-\langle\bigcirc\rangle-\text{CH}_2\text{Br} + \text{C}_5\text{H}_5\text{N} \longrightarrow \text{R}-\langle\bigcirc\rangle-\text{CH}_2\overset{\oplus}{\text{N}}\text{C}_5\text{H}_5 \quad \text{Br}^-
$$

electron release by the field effect.[255] That is, the methyl-substituted compound reacted fastest and the *t*-butyl-substituted compounded reacted slowest.

This came to be called the *Baker–Nathan effect* and has since been found in many processes. Baker and Nathan explained it by considering that hyperconjugative forms contribute to the actual structure of toluene:

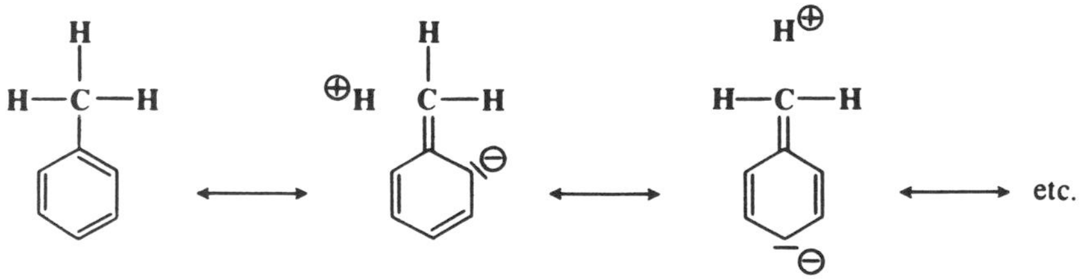

---

[253]For monographs, see Baker *Hyperconjugation*; Oxford University Press: Oxford, 1952; Dewar *Hyperconjugation*; Ronald Press: New York, 1962. For a review, see de la Mare *Pure Appl. Chem.* **1984**, *56*, 1755-1766.

[254]Baker; Groves *J. Chem. Soc.* **1939**, 1144.

[255]Baker; Nathan *J. Chem. Soc.* **1935**, 1840, 1844.

For the other alkyl groups, hyperconjugation is diminished because the number of C—H bonds is diminished and in *t*-butyl there are none; hence, with respect to this effect, methyl is the strongest electron donor and *t*-butyl the weakest.

However, the Baker–Nathan effect has now been shown not to be caused by hyperconjugation, but by differential solvation.[256] This was demonstrated by the finding that in certain instances where the Baker–Nathan effect was found to apply in solution, the order was completely reversed in the gas phase.[257] Since the molecular structures are unchanged in going from the gas phase into solution, it is evident that the Baker–Nathan order in these cases is not caused by a structural feature (hyperconjugation) but by the solvent. That is, each alkyl group is solvated to a different extent.[258]

At present the evidence is against hyperconjugation in the ground states of neutral molecules.[259] However, for carbocations and free radicals[260] and for excited states of molecules,[261] there is evidence that hyperconjugation is important. In hyperconjugation in the ground state of neutral molecules, which Muller and Mulliken call *sacrificial hyperconjugation*,[262] the canonical forms involve not only no-bond resonance but also a charge separation not possessed by the main form. In free radicals and carbocations, the canonical forms display no more charge separation than the main form. Muller and Mulliken call this *isovalent hyperconjugation*:

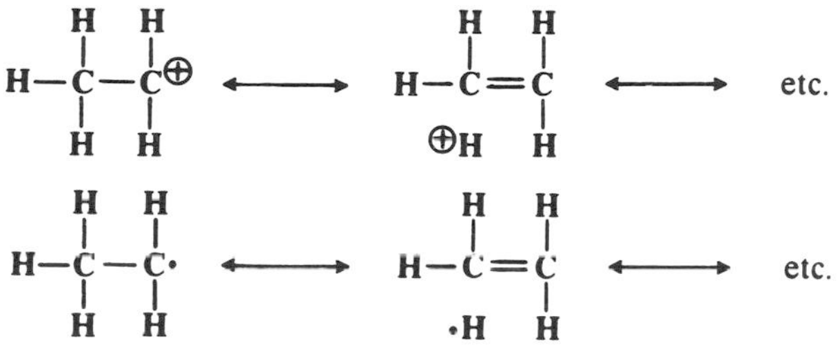

Even here the main form contributes more to the hybrid than the others.

TAUTOMERISM

There remains one topic to be discussed in our survey of chemical bonding in organic compounds. For most compounds all the molecules have the same structure, whether or not this structure can be satisfactorily represented by a Lewis formula. But for many other compounds there is a mixture of two or more structurally distinct compounds that are in rapid equilibrium. When this phenomenon, called *tautomerism*,[263] exists, there is a rapid shift back and forth among the molecules. In most cases, it is a proton that shifts from one atom of a molecule to another.

[256]This idea was first suggested by Schubert; Sweeney *J. Org. Chem.* **1956**, *21*, 119.

[257]Hehre; McIver; Pople; Schleyer *J. Am. Chem. Soc.* **1974**, *96*, 7162; Arnett; Abboud *J. Am. Chem. Soc.* **1975**, *97*, 3865; Glyde; Taylor *J. Chem. Soc., Perkin Trans. 2* **1977**, 678. See also Taylor *J. Chem. Res. (S)* **1985**, 318.

[258]For an opposing view, see Cooney; Happer *Aust. J. Chem.* **1987**, *40*, 1537.

[259]For some evidence in favor, see Laube; Ha *J. Am. Chem. Soc.* **1988**, *110*, 5511.

[260]Symons *Tetrahedron* **1962**, *18*, 333.

[261]Rao; Goldman; Balasubramanian *Can. J. Chem.* **1960**, *38*, 2508.

[262]Muller; Mulliken *J. Am. Chem. Soc.* **1958**, *80*, 3489.

[263]For reviews, see Toullec *Adv. Phys. Org. Chem.* **1982**, *18*, 1-77; Kol'tsov; Kheifets *Russ. Chem. Rev.* **1971**, *40*, 773-788, **1972**, *41*, 452-467; Forsén; Nilsson in Zabicky, Ref. 246, vol. 2, pp. 157-240.

Keto-Enol Tautomerism[264]

A very common form of tautomerism is that between a carbonyl compound containing an α hydrogen and its enol form:[264a]

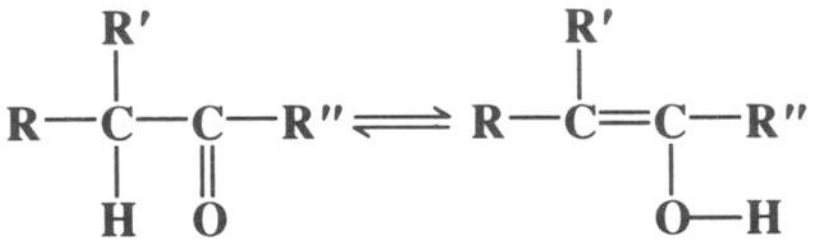

$$\underset{\text{Keto form}}{R-\overset{\overset{\textstyle R'}{|}}{\underset{\underset{\textstyle H}{|}}{C}}-\overset{\overset{\textstyle }{\|}}{\underset{\underset{\textstyle O}{}}{C}}-R''} \rightleftharpoons \underset{\text{Enol form}}{R-\overset{\overset{\textstyle R'}{|}}{C}=\overset{}{\underset{\underset{\textstyle O-H}{|}}{C}}-R''}$$

In simple cases (R" = H, alkyl, OR, etc.) the equilibrium lies well to the left (Table 2.1). The reason can be seen by examining the bond energies in Table 1.7. The keto form differs from the enol form in possessing a C—H, a C—C, and a C=O bond where the enol has a C=C, a C—O, and an O—H bond. The approximate sum of the first three is 359 kcal/mol (1500 kJ/mol) and of the second three is 347 kcal/mol (1452 kJ/mol). The keto form is therefore thermodynamically more stable by about 12 kcal/mol (48 kJ/mol) and enol forms cannot normally be isolated.[272a] In certain cases, however, a larger amount of the enol form

TABLE 2.1 The enol content of some carbonyl compounds

Compound	Enol content, %	Ref.
Acetone	6×10^{-7}	265
PhCOCH₃	1.1×10^{-6}	266
Cyclopentanone	1×10^{-6}	267
CH₃CHO	6×10^{-5}	268
Cyclohexanone	4×10^{-5}	267
Butanal	5.5×10^{-4}	269
(CH₃)₂CHCHO	1.4×10^{-2}	270
Ph₂CHCHO	9.1	271
CH₃COOEt	No enol found[a]	267
CH₃COCH₂COOEt	8.4	272
CH₃COCH₂COCH₃	80	272
PhCOCH₂COCH₃	89.2	267
EtOOCCH₂COOEt	7.7×10^{-3}	267
NCCH₂COOEt	2.5×10^{-1}	267

[a]Less than 1 part in 10 million.

[264]The mechanism for conversion of one tautomer to another is discussed in Chapter 12 (reaction **2-3**).

[264a]For a treatise, see Rappoport *The Chemistry of Enols*; Wiley: New York, 1990.

[265]Tapuhi; Jencks *J. Am. Chem. Soc.* **1982**, *104*, 5758; Chiang; Kresge; Tang; Wirz *J. Am. Chem. Soc.* **1984**, *106*, 460. See also Hine; Arata *Bull. Chem. Soc. Jpn.* **1976**, *49*, 3089; Guthrie *Can. J. Chem.* **1979**, *57*, 797, 1177; Dubois; El-Alaoui; Toullec *J. Am. Chem. Soc.* **1981**, *103*, 5393; Toullec *Tetrahedron Lett.* **1984**, *25*, 4401; Chiang; Kresge; Schepp *J. Am. Chem. Soc.* **1989**, *111*, 3977.

[266]Keeffe; Kresge; Toullec *Can. J. Chem.* **1986**, *64*, 1224.

[267]Gero *J. Org. Chem.* **1954**, *19*, 469, 1960; Keeffe, Kresge; Schepp *J. Am. Chem. Soc.* **1990**, *112*, 4862. See these papers for values for other simple compounds.

[268]Chiang; Hojatti; Keeffe; Kresge; Schepp; Wirz *J. Am. Chem. Soc.* **1987**, *109*, 4000.

[269]Bohne; MacDonald; Dunford *J. Am. Chem. Soc.* **1986**, *108*, 7867.

[270]Chiang; Kresge; Walsh *J. Am. Chem. Soc.* **1986**, *108*, 6314; Ref. 269.

[271]Chiang; Kresge; Krogh *J. Am. Chem. Soc.* **1988**, *110*, 2600.

[272]Moriyasu; Kato; Hashimoto *J. Chem. Soc., Perkin Trans. 2* **1986**, 515.

[272a]For reviews on the generation of unstable enols, see Kresge *Pure Appl. Chem.* **1991**, *63*, 213-221; Capon, in Rappoport, Ref. 264a, pp. 307-322.

is present, and it can even be the predominant form.[273] There are three main types of the more stable enols:[274]

1. Molecules in which the enolic double bond is in conjugation with another double bond. Some of these are shown in Table 2.1. As the table shows, carboxylic esters have a much smaller enolic content than ketones. In molecules like acetoacetic ester, the enol is also stabilized by internal hydrogen bonding, which is unavailable to the keto form:

$$CH_3-C\underset{O\cdots H-O}{\overset{CH}{\underset{\|}{C}}}C-OEt$$

2. Molecules that contain two or three bulky aryl groups.[275] An example is 2,2-dimesitylethenol (**96**). In this case the keto content at equilibrium is only 5%.[276] In cases

$$\underset{Ar}{\overset{Ar}{>}}C=C-H \rightleftharpoons \underset{Ar}{\overset{Ar}{>}}CH-C-H \qquad Ar = Me-\underset{Me}{\overset{Me}{\bigcirc}}-$$
$$\underset{96}{\overset{OH}{}} \qquad \underset{97}{\overset{O}{}}$$

such as this steric hindrance (p. 161) destabilizes the keto form. In **96** the two aryl groups are about 120° apart, but in **97** they must move closer together (~109.5°). Such compounds are often called *Fuson-type enols*.[277]

3. Highly fluorinated enols, an example being **98**.[278]

$$CF_2=\underset{OH}{\overset{}{C}}-CF_3 \xrightarrow[3\ h]{200°} CF_2H-\underset{O}{\overset{}{C}}-CF_3$$
$$\underset{98}{} \qquad \underset{99}{}$$

In this case the enol form is not more stable than the keto form (it is less stable, and converts to the keto form upon prolonged heating). It can however be kept at room temperature for long periods of time because the tautomerization reaction (**2-3**) is very slow, owing to the electron-withdrawing power of the fluorines.

Frequently, when the enol content is high, both forms can be isolated. The pure keto form of acetoacetic ester melts at $-39°C$, while the enol is a liquid even at $-78°C$. Each can be kept at room temperature for days if catalysts such as acids or bases are rigorously excluded.[279] Even the simplest enol, vinyl alcohol CH_2=CHOH, has been prepared in the

---

[273]For reviews of stable enols, see Kresge *Acc. Chem. Res.* **1990**, *23*, 43-48, *CHEMTECH*, **1986**, 250-254; Hart; Rappoport; Biali, in Rappoport, Ref. 264a, pp. 481-589; Hart, *Chem. Rev.* **1979**, *79*, 515-528; Hart; Sasaoka *J. Chem. Educ.* **1980**, *57*, 685-688.

[274]For some examples of other types, see Pratt; Hopkins *J. Am. Chem. Soc.* **1987**, *109*, 5553; Nadler; Rappoport; Arad; Apeloig *J. Am. Chem. Soc.* **1987**, *109*, 7873.

[275]For a review, see Rappoport; Biali *Acc. Chem. Res.* **1988**, *21*, 442-449. For a discussion of their structures, see Kaftory; Nugiel; Biali; Rappoport *J. Am. Chem. Soc.* **1989**, *111*, 8181.

[276]Biali; Rappoport *J. Am. Chem. Soc.* **1985**, *107*, 1007. See also Kaftory; Biali; Rappoport *J. Am. Chem. Soc.* **1985**, *107*, 1701; Nugiel; Rappoport *J. Am. Chem. Soc.* **1985**, *107*, 3669; Nadler; Rappoport *J. Am. Chem. Soc.* **1987**, *109*, 2112; O'Neill; Hegarty *J. Chem. Soc., Chem. Commun.* **1987**, 744; Becker; Andersson *Tetrahedron Lett.* **1987**, *28*, 1323.

[277]First synthesized by Fuson; see for example Fuson; Southwick; Rowland *J. Am. Chem. Soc.* **1944**, *66*, 1109.

[278]For a review, see Bekker; Knunyants *Sov. Sci. Rev. Sect. B* **1984**, *5*, 145-182.

[279]For an example of particularly stable enol and keto forms, which could be kept in the solid state for more than a year without significant interconversion, see Schulenberg *J. Am. Chem. Soc.* **1968**, *90*, 7008.

gas phase at room temperature, where it has a half-life of about 30 min.[280] The enol $Me_2C\!=\!CCHOH$ is indefinitely stable in the solid state at $-78°C$ and has a half-life of about 24 hours in the liquid state at $25°C$.[281]

The extent of enolization[281a] is greatly affected by solvent,[282] concentration, and temperature. Thus, acetoacetic ester has an enol content of 0.4% in water and 19.8% in toluene.[283] In this case, water reduces the enol concentration by hydrogen bonding with the carbonyl, making this group less available for internal hydrogen bonding. As an example of the effect of temperature, the enol content of pentan-2,4-dione $CH_3COCH_2COCH_3$ was found to be 95, 68, and 44%, respectively, at 22, 180, and $275°C$.[284]

When a strong base is present, both the enol and the keto form can lose a proton. The resulting anion (the *enolate ion*) is the same in both cases. Since **100** and **101** differ only in

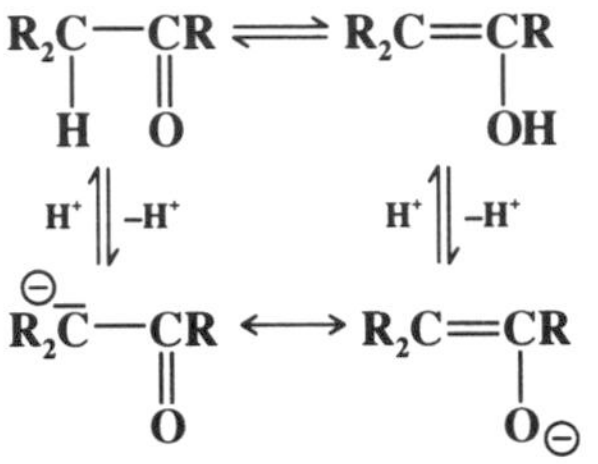

$$R_2C\!-\!CR \rightleftharpoons R_2C\!=\!CR$$
$$\underset{H}{|}\quad\underset{O}{||}\qquad\qquad\underset{OH}{|}$$

$$H^+\big\Uparrow\big\Downarrow{-H^+}\qquad\qquad H^+\big\Uparrow\big\Downarrow{-H^+}$$

$$\overset{\ominus}{R_2\bar{C}}\!-\!CR \longleftrightarrow R_2C\!=\!CR$$
$$\underset{O}{||}\qquad\qquad\underset{O\ominus}{|}$$

100   **101**

placement of electrons, *they* are not tautomers but canonical forms. The true structure of the enolate ion is a hybrid of **100** and **101** although **101** contributes more, since in this form the negative charge is on the more electronegative atom.

Other Proton-Shift Tautomerism

In all such cases, the anion resulting from removal of a proton from either tautomer is the same because of resonance. Some examples are:[285]

1. Phenol-keto tautomerism.[286]

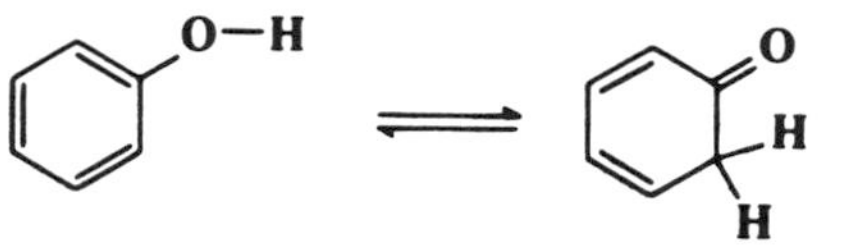

Phenol   Cyclohexadienone

[280]Saito *Chem. Phys. Lett.* **1976,** *42,* 399. See also Capon; Rycroft; Watson; Zucco *J. Am. Chem. Soc.* **1981,** *103,* 1761; Holmes; Lossing *J. Am. Chem. Soc.* **1982,** *104,* 2648; McGarritty; Cretton; Pinkerton; Schwarzenbach; Flack *Angew. Chem. Int. Ed. Engl.* **1983,** *22,* 405 [*Angew. Chem. 95,* 426]; Rodler; Blom; Bauder *J. Am. Chem. Soc.* **1984,** *106,* 4029; Capon; Guo; Kwok; Siddhanta; Zucco *Acc. Chem. Res.* **1988,** *21,* 135-140.
[281]Chin; Lee; Park; Kim *J. Am. Chem. Soc.* **1988,** *110,* 8244.
[281a]For a review of keto-enol equilibrium constants, see Toullec, in Rappoport, Ref. 264a, pp. 323-398.
[282]For an extensive study, see Mills; Beak *J. Org. Chem.* **1985,** *50,* 1216.
[283]Meyer *Leibigs Ann. Chem.* **1911,** *380,* 212. See also Ref. 272.
[284]Hush; Livett; Peel; Willett *Aust. J. Chem.* **1987,** *40,* 599.
[285]For a review of the use of x-ray crystallography to determine tautomeric forms, see Furmanova *Russ. Chem. Rev.* **1981,** *50,* 775-791.
[286]For reviews, see Ershov; Nikiforov *Russ. Chem. Rev.* **1966,** *35,* 817-833; Forsén; Nilsson, Ref. 263, pp. 168-198.

For most simple phenols this equilibrium lies well to the side of the phenol, since only on that side is there aromaticity. For phenol itself there is no evidence for the existence of the keto form.[287] However, the keto form becomes important and may predominate: (1) where certain groups, such as a second OH group or an N=O group, are present;[288] (2) in systems of fused aromatic rings;[289] (3) in heterocyclic systems. In many heterocyclic compounds in the liquid phase or in solution, the keto form is more stable,[290] although in vapor phase the positions of many of these equilibria are reversed.[291] For example, in the equilibrium between 4-pyridone (**102**) and 4-hydroxypyridine (**103**), **102** is the only form detectable in ethanolic solution, while **103** predominates in the vapor phase.[291]

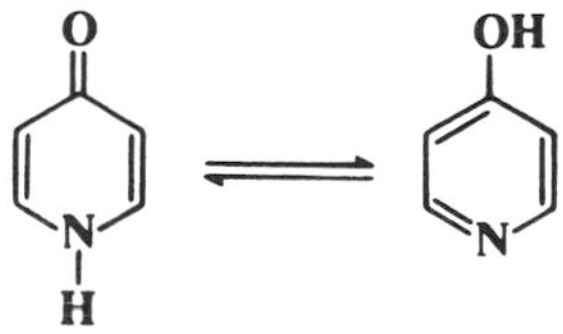

$$\textbf{102} \qquad\qquad \textbf{103}$$

2. Nitroso-oxime tautomerism.

$$R_2CH\!-\!N\!=\!O \;\rightleftharpoons\; R_2C\!=\!N\!-\!OH$$

Nitroso   Oxime

This equilibrium lies far to the right, and as a rule nitroso compounds are stable only when there is no α hydrogen.

3. Aliphatic nitro compounds are in equilibrium with aci forms.

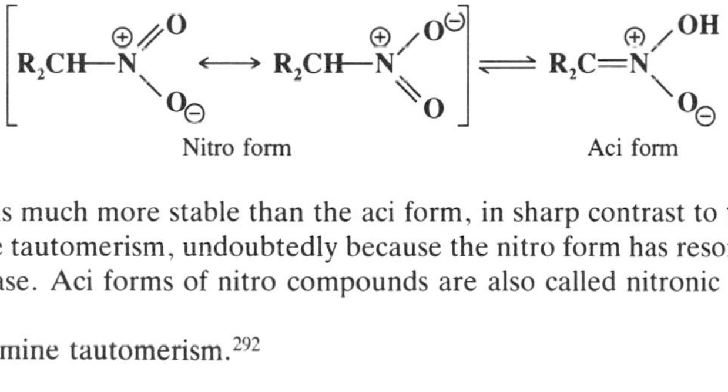

Nitro form   Aci form

The nitro form is much more stable than the aci form, in sharp contrast to the parallel case of nitroso-oxime tautomerism, undoubtedly because the nitro form has resonance not found in the nitroso case. Aci forms of nitro compounds are also called nitronic acids and azinic acids.

4. Imine-enamine tautomerism.[292]

$$R_2CH\!-\!CR\!=\!NR \;\rightleftharpoons\; R_2C\!=\!CR\!-\!NHR$$

Imine   Enamine

---

[287]Keto forms of phenol and some simple derivatives have been generated as intermediates with very short lives, but long enough for spectra to be taken at 77 K. Lasne; Ripoll; Denis *Tetrahedron Lett.* **1980**, *21*, 463. See also Capponi; Gut; Wirz *Angew. Chem. Int. Ed. Engl.* **1986**, *25*, 344 [*Angew. Chem.* **98**, 358].

[288]Ershov; Nikiforov, Ref. 286. See also Highet; Chou *J. Am. Chem. Soc.* **1977**, *99*, 3538.

[289]See, for example, Majerski; Trinajstić *Bull. Chem. Soc. Jpn.* **1970**, *43*, 2648.

[290]For a monograph on tautomerism in heterocyclic compounds, see Elguero; Marzin; Katrizky; Linda *The Tautomerism of Heterocycles*; Academic Press: New York, 1976. For reviews, see Katritzky; Karelson; Harris *Heterocycles* **1991**, *32*, 329-369; Beak *Acc. Chem. Res.* **1977**, *10*, 186-192; Katritzky *Chimia* **1970**, *24*, 134-146.

[291]Beak; Fry; Lee; Steele *J. Am. Chem. Soc.* **1976**, *98*, 171.

[292]For reviews, see Shainyan; Mirskova *Russ. Chem. Rev.* **1979**, *48*, 107-117; Mamaev; Lapachev *Sov. Sci. Rev. Sect. B.* **1985**, *7*, 1-49. The second review also includes other closely related types of tautomerization.

Enamines are normally stable only when there is no hydrogen on the nitrogen (R_2C=CR—NR_2). Otherwise, the imine form predominates.[293]

Ring-chain tautomerism[294] (as in sugars) consists largely of cyclic analogs of the previous examples. There are many other highly specialized cases of proton-shift tautomerism.

Valence Tautomerism

This type of tautomerism is discussed on p. 1134.

[293]For examples of the isolation of primary and secondary enamines, see Shin; Masaki; Ohta *Bull. Chem. Soc. Jpn.* **1971**, *44*, 1657; de Jeso; Pommier *J. Chem. Soc., Chem. Commun.* **1977**, 565.

[294]For a monograph, see Valters; Flitsch *Ring-Chain Tautomerism*; Plenum: New York, 1985. For reviews, see Valters *Russ. Chem. Rev.* **1973**, *42*, 464-476, **1974**, *43*, 665-678; Escale; Verducci *Bull. Soc. Chim. Fr.* **1974**, 1203-1206.

3
BONDING WEAKER
THAN COVALENT

In the first two chapters we discussed the structure of molecules each of which is an aggregate of atoms in a distinct three-dimensional arrangement held together by bonds with energies on the order of 50 to 100 kcal/mol (200 to 400 kJ/mol). There are also very weak attractive forces *between* molecules, on the order of a few tenths of a kilocalorie per mole. These forces, called van der Waals forces, are caused by electrostatic attractions such as those between dipole and dipole, induced dipole and induced dipole, etc, and are responsible for liquefaction of gases at sufficiently low temperatures. The bonding discussed in this chapter has energies of the order of 2 to 10 kcal/mol (9 to 40 kJ/mol), intermediate between the two extremes, and produces clusters of molecules. We will also discuss compounds in which portions of molecules are held together without any attractive forces at all.

HYDROGEN BONDING

A *hydrogen bond* is a bond between a functional group A—H and an atom or group of atoms B in the same or a different molecule.[1] With exceptions to be noted later, hydrogen bonds are formed only when A is oxygen, nitrogen, or fluorine and when B is oxygen, nitrogen, or fluorine. The oxygen may be singly or doubly bonded and the nitrogen singly, doubly, or triply bonded. The bonds are usually represented by dotted lines, as shown in the following examples:

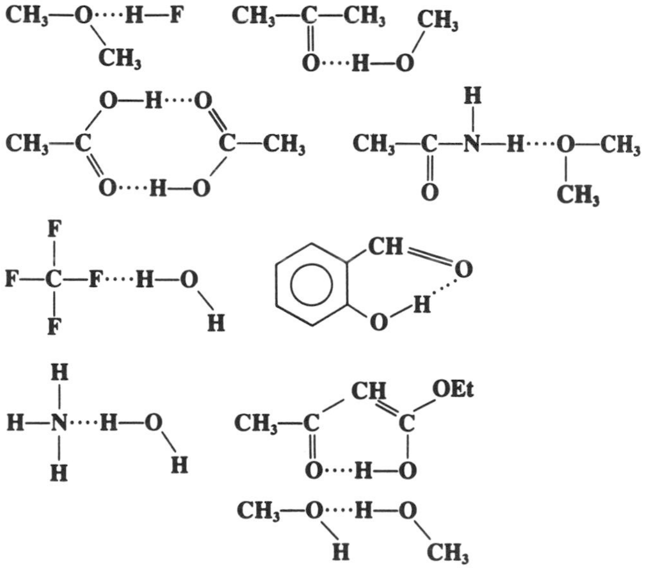

[1]For a treatise, see Schuster; Zundel; Sandorfy *The Hydrogen Bond*; 3 vols., North Holland Publishing Co.: Amsterdam, 1976. For a monograph, see Joesten; Schaad *Hydrogen Bonding*; Marcel Dekker: New York, 1974. For

Hydrogen bonds can exist in the solid and liquid phases and in solution. Even in the gas phase, compounds that form particularly strong hydrogen bonds may still be associated.[2] Acetic acid, for example, exists in the gas phase as a dimer, as shown above, except at very low pressures.[3] In solution and in the liquid phase, hydrogen bonds rapidly form and break. The mean lifetime of the $NH_3 \cdots H_2O$ bond is 2×10^{-12} sec.[4] Except for a few very strong hydrogen bonds,[5] such as the $FH \cdots F^-$ bond (which has an energy of about 50 kcal/mol or 210 kJ/mol), the strongest hydrogen bonds are the $FH \cdots F$ bond and the bonds connecting one carboxylic acid with another. The energies of these bonds are in the range of 6 to 8 kcal/mol or 25 to 30 kJ/mol (for carboxylic acids, this refers to the energy of each bond). Other $OH \cdots O$ and $NH \cdots N$ bonds have energies of 3 to 6 kcal/mol (12 to 25 kJ/mol). To a first approximation, the strength of hydrogen bonds increases with increasing acidity of A—H and basicity of B, but the parallel is far from exact.[6] A quantitative measure of the strengths of hydrogen bonds has been established, involving the use of an α scale to represent hydrogen-bond donor acidities and a β scale for hydrogen-bond acceptor basicities.[7] The use of the β scale, along with another parameter, ξ, allows hydrogen bond basicities to be related to proton transfer basicities (pK values).[8]

When two compounds whose molecules form hydrogen bonds with each other are both dissolved in water, the hydrogen bond between the two molecules is usually greatly weakened or completely removed,[9] because the molecules generally form hydrogen bonds with the water molecules rather than with each other, especially since the water molecules are present in such great numbers.

Many studies have been made of the geometry of hydrogen bonds,[10] and the evidence shows that in most (though not all) cases the hydrogen is on or near the straight line formed by A and B.[11] This is true both in the solid state (where x-ray crystallography and neutron diffraction have been used to determine structures),[12] and in solution.[13] It is significant that the vast majority of intramolecular hydrogen bonding occurs where *six-membered rings* (counting the hydrogen as one of the six) can be formed, in which linearity of the hydrogen bond is geometrically favorable, while five-membered rings, where linearity is usually not

---

reviews, see Meot-Ner *Mol. Struct. Energ.* **1987**, *4*, 71-103; Deakyne *Mol. Struct. Energ.* **1987**, *4*, 105-141; Joesten *J. Chem. Educ.* **1982**, *59*, 362-366; Gur'yanova; Gol'dshtein; Perepelkova *Russ. Chem. Rev.* **1976**, *45*, 792-806; Pimentel; McClellan *Annu. Rev. Phys. Chem.* **1971**, *22*, 347-385; Kollman; Allen *Chem. Rev.* **1972**, *72*, 283-303; Huggins *Angew. Chem. Int. Ed. Engl.* **1971**, *10*, 147-151 [*Angew. Chem. 83*, 163-168]; Rochester, in Patai *The Chemistry of the Hydroxyl Group*, pt. 1; Wiley: New York, 1971, 327-392, pp. 328-369. See also Hamilton; Ibers *Hydrogen Bonding in Solids*; W.A. Benjamin: New York, 1968.

[2]For a review of energies of hydrogen bonds in the gas phase, see Curtiss; Blander *Chem. Rev.* **1988**, *88*, 827-841.

[3]For a review of hydrogen bonding in carboxylic acids and acid derivatives, see Hadži; Detoni, in Patai *The Chemistry of Acid Derivatives*, pt. 1; Wiley: New York, 1979, pp. 213-266.

[4]Emerson; Grunwald; Kaplan; Kromhout *J. Am. Chem. Soc.* **1960**, *82*, 6307.

[5]For a review of very strong hydrogen bonding, see Emsley *Chem. Soc. Rev.* **1980**, *9*, 91-124.

[6]For reviews of the relationship between hydrogen bond strength and acid–base properties, see Pogorelyi; Vishnyakova *Russ. Chem. Rev.* **1984**, *53*, 1154-1167; Epshtein *Russ. Chem. Rev.* **1979**, *48*, 854-867.

[7]For reviews, see Abraham; Doherty; Kamlet; Taft *Chem. Br.* **1986**, 551-554; Kamlet; Abboud; Taft *Prog. Phys. Org. Chem.* **1981**, *13*, 485-630. For a comprehensive table and α and β values, see Kamlet; Abboud; Abraham; Taft *J. Org. Chem.* **1983**, *48*, 2877. For a criticism of the β scale, see Laurence; Nicolet; Helbert *J. Chem. Soc., Perkin Trans. 2* **1986**, 1081. See also Nicolet; Laurence; Luçon *J. Chem. Soc., Perkin Trans. 2* **1987**, 483; Abboud; Roussel; Gentric; Sraidi; Lauransan; Guihéneuf; Kamlet; Taft *J. Org. Chem.* **1988**, *53*, 1545; Abraham; Grellier; Prior; Morris; Taylor *J. Chem. Soc., Perkin Trans. 2* **1990**, 521.

[8]Kamlet; Gal; Maria; Taft *J. Chem. Soc., Perkin Trans. 2* **1985**, 1583.

[9]Stahl; Jencks *J. Am. Chem. Soc.* **1986**, *108*, 4196.

[10]For reviews, see Etter *Acc. Chem. Res.* **1990**, *23*, 120-126; Taylor; Kennard *Acc. Chem. Res.* **1984**, *17*, 320-326.

[11]See Stewart *The Proton: Applications to Organic Chemistry*; Academic Press: New York, 1985, pp. 148-153.

[12]A statisical analysis of x-ray crystallographic data has shown that most hydrogen bonds in crystals are nonlinear by about 10 to 15°: Kroon; Kanters; van Duijneveldt-van de Rijdt; van Duijneveldt; Vliegenthart *J. Mol. Struct.* **1975**, *24*, 109. See also Ceccarelli; Jeffrey; Taylor *J. Mol. Struct.* **1981**, *70*, 255; Taylor; Kennard; Versichel *J. Am. Chem. Soc.* **1983**, *105*, 5761; **1984**, *106*, 244.

[13]For reviews of a different aspect of hydrogen bond geometry: the angle between A—H $\cdots$ B and the rest of the molecule, see Legon; Millen *Chem. Soc. Rev.* **1987**, *16*, 467-498, *Acc. Chem. Res.* **1987**, *20*, 39-46.

favored (though it is known), are much rarer. Except for the special case of $FH\cdots F^-$ bonds (see p. 78), the hydrogen is not equidistant between A and B. For example, in ice the O—H distance is 0.97 Å, while the H$\cdots$O distance is 1.79 Å.[14]

In certain cases x-ray crystallography has shown that a single H—A can form simultaneous hydrogen bonds with two B atoms (*bifurcated* or *three-center hydrogen bonds*). An example is an adduct (**1**) formed from pentane-2,4-dione (in its enol form) and diethylamine, in

1   **2**   **3**

which the O—H hydrogen simultaneously bonds[15] to an O and an N (the N—H hydrogen forms a hydrogen bond with the O of another pentane-2,4-dione molecule).[16] On the other hand, in the adduct (**2**) formed from 1,8-biphenylenediol and HMPA, the B atom (in this case oxygen) forms simultaneous hydrogen bonds with two A—H hydrogens.[17] Another such case is found in methyl hydrazine carboxylate **3**.[18]

Hydrogen bonding has been detected in many ways, including measurements of dipole moments, solubility behavior, freezing-point lowering, and heats of mixing, but the most important way is by the effect of the hydrogen bond on ir[19] and other spectra. The ir frequencies of groups such as O—H or C=O are shifted when the group is hydrogen bonded. Hydrogen bonding always moves the peak toward lower frequencies, for both the A—H and the B groups, though the shift is greater for the former. For example, a free OH group of an alcohol or phenol absorbs at about 3590 to 3650 cm^{-1}, while a hydrogen-bonded OH group is found about 50 to 100 cm^{-1} lower.[20] In many cases, in dilute solution, there is partial hydrogen bonding, that is, some OH groups are free and some are hydrogen bonded. In such cases two peaks appear. Infrared spectroscopy can also distinguish between inter- and intramolecular hydrogen bonding, since intermolecular peaks are intensified by an increase in concentration while intramolecular peaks are unaffected. Other types of spectra that have been used for the detection of hydrogen bonding include Raman, electronic,[21] and nmr.[22] Since hydrogen bonding involves a rapid movement of protons from one atom to another, nmr records an average value. Hydrogen bonding can be detected because it usually produces a chemical shift to a lower field. Hydrogen bonding changes with temper-

[14]Pimentel; McClellan *The Hydrogen Bond*; W.H. Freeman: San Francisco, 1960, p. 260.

[15]Emsley; Freeman; Parker; Dawes; Hurthouse *J. Chem. Soc., Perkin Trans. 1* **1986**, 471.

[16]For some other three-center hydrogen bonds, see Taylor; Kennard; Versichel *J. Am. Chem. Soc.* **1984**, *106*, 244; Jeffrey; Mitra *J. Am. Chem. Soc.* **1984**, *106*, 5546; Staab; Elbl; Krieger *Tetrahedron Lett.* **1986**, *27*, 5719.

[17]Hine; Ahn; Gallucci; Linden *J. Am. Chem. Soc.* **1984**, *106*, 7980; Hine; Hahn; Miles *J. Org. Chem.* **1986**, *51*, 577.

[18]Caminati; Fantoni; Schäfer; Siam; Van Alsenoy *J. Am. Chem. Soc.* **1986**, *108*, 4364.

[19]For reviews of the use of ir spectra to detect hydrogen bonding, see Symons *Chem. Soc. Rev.* **1983**, *12*, 1-34; Egorochkin; Skobeleva *Russ. Chem. Rev.* **1979**, *48*, 1198-1211; Tichý *Adv. Org. Chem.* **1965**, *5*, 115-298; Ratajczak; Orville-Thomas *J. Mol. Struct.* **1968**, *1*, 449. For a review of studies by ir of the shapes of intramolecular hydrogen-bonded compounds, see Aaron *Top. Stereochem.* **1979**, *11*, 1-52. For a review of the use of rotational spectra to study hydrogen bonding, see Legon *Chem. Soc. Rev.* **1990**, *19*, 197-237.

[20]Tichý, Ref. 19, contains a lengthy table of free and intramolecularly hydrogen-bonding peaks.

[21]For a discussion of the effect of hydrogen bonding on electronic spectra, see Lees; Burawoy *Tetrahedron* **1963**, *19*, 419.

[22]For a review of the use of nmr to detect hydrogen bonding, see Davis; Deb *Adv. Magn. Reson.* **1970**, *4*, 201-270.

ature and concentration, and comparison of spectra taken under different conditions also serves to detect and measure it. As with infrared spectra, intramolecular hydrogen bonding can be distinguished from intermolecular by its constancy when the concentration is varied.

Hydrogen bonds are important because of the effects they have on the properties of compounds, among them:

1. Intermolecular hydrogen bonding raises boiling points and frequently melting points.

2. If hydrogen bonding is possible between solute and solvent, this greatly increases solubility and often results in large or even infinite solubility where none would otherwise be expected. It is interesting to speculate what the effect on the human race would be if ethanol had the same solubility in water as ethane or chloroethane.

3. Hydrogen bonding causes lack of ideality in gas and solution laws.

4. As previously mentioned, hydrogen bonding changes spectral absorption positions.

5. Hydrogen bonding, especially the intramolecular variety, changes many chemical properties. For example, it is responsible for the large amount of enol present in certain tautomeric equilibria (see p. 71). Also, by influencing the conformation of molecules (see Chapter 4), it often plays a significant role in determining reaction rates.[23] Hydrogen bonding is also important in maintaining the three-dimensional structures of protein and nucleic acid molecules.

Besides oxygen, nitrogen, and fluorine, there is evidence that weaker hydrogen bonding exists in other systems.[24] Although many searches have been made for hydrogen bonding where A is carbon,[25] only three types of C—H bonds have been found that are acidic enough to form weak hydrogen bonds. These are found in terminal alkynes, $RC \equiv CH$,[26] chloroform and some other halogenated alkanes, and HCN. Weak hydrogen bonds are formed by compounds containing S—H bonds.[27] There has been much speculation regarding other possibilities for B. There is evidence that Cl can form weak hydrogen bonds,[28] but Br and I form very weak bonds if at all.[29] However, the *ions* Cl^-, Br^-, and I^- form hydrogen bonds that are much stronger than those of the covalently bonded atoms.[30] As we have already seen, the $FH \cdots F^-$ bond is especially strong. In this case the hydrogen is equidistant from the fluorines.[31] Similarly, a sulfur atom[27] can be the B component in weak hydrogen bonds,[32] but the SH^- ion forms much stronger bonds.[33] Hydrogen bonding has been directly observed (by nmr and ir) between a negatively charged carbon (see Carbanions, Chapter 5) and an

[23]For reviews of the effect of hydrogen bonding on reactivity, see Hibbert; Emsley *Adv. Phys. Org. Chem.* **1990**, *26*, 255-379; Sadekov; Minkin; Lutskii *Russ. Chem. Rev.* **1970**, *39*, 179-195.

[24]For a review, see Pogorelyi *Russ. Chem. Rev.* **1977**, *46*, 316-336.

[25]For a monograph on this subject, see Green *Hydrogen Bonding by C—H Groups*; Wiley: New York, 1974. See also Taylor; Kennard *J. Am. Chem. Soc.* **1982**, *104*, 5063; Harlow; Li; Sammes *J. Chem. Soc., Perkin Trans. 1* **1984**, 547; Nakai; Inoue; Yamamoto; Ōki *Bull. Chem. Soc. Jpn.* **1989**, *62*, 2923; Seiler; Dunitz *Helv. Chim. Acta* **1989**, *72*, 1125.

[26]For a review, see Hopkinson, in Patai *The Chemistry of the Carbon-Carbon Triple Bond*, pt. 1, Wiley: New York, 1978, pp. 75-136. See also DeLaat; Ault *J. Am. Chem. Soc.* **1987**, *109*, 4232.

[27]For reviews of hydrogen bonding in sulfur-containing compounds, see Zuika; Bankovskii *Russ. Chem. Rev.* **1973**, *42*, 22-36; Crampton, in Patai *The Chemistry of the Thiol Group*, pt. 1; Wiley: New York, 1974, pp. 379-396; Ref. 24.

[28]For a review of hydrogen bonding to halogens, see Smith, in Patai *The Chemistry of the Carbon–Halogen Bond*, pt. 1; Wiley: New York, 1973, pp. 265-300. See also Bastiansen; Fernholt; Hedberg; Seip *J. Am. Chem. Soc.* **1985**, *107*, 7836.

[29]West; Powell; Whatley; Lee; Schleyer *J. Am. Chem. Soc.* **1962**, *84*, 3221; Fujimoto; Takeoka; Kozima *Bull. Chem. Soc. Jpn.* **1970**, *43*, 991; Azrak; Wilson *J. Chem. Phys.* **1970**, *52*, 5299.

[30]Allerhand; Schleyer *J. Am. Chem. Soc.* **1963**, *85*, 1233; McDaniel; Valleé *Inorg. Chem.* **1963**, *2*, 996; Fujiwara; Martin *J. Am. Chem. Soc.* **1974**, *96*, 7625; French; Ikuta; Kebarle *Can. J. Chem.* **1982**, *60*, 1907.

[31]A few exceptions have been found, where the presence of an unsymmetrical cation causes the hydrogen to be closer to one fluorine than to the other: Williams; Schneemeyer *J. Am. Chem. Soc.* **1973**, *95*, 5780.

[32]Vogel; Drago *J. Am. Chem. Soc.* **1970**, *92*, 5347; Mukherjee; Palit; De *J. Phys. Chem.* **1970**, *74*, 1389; Schaefer; McKinnon; Sebastian; Peeling; Penner; Veregin *Can. J. Chem.* **1987**, *65*, 908; Marstokk; Møllendal; Uggerrud *Acta Chem. Scand.* **1989**, *43*, 26.

[33]McDaniel; Evans *Inorg. Chem.* **1966**, *5*, 2180; Sabin *J. Chem. Phys.* **1971**, *54*, 4675.

OH group in the same molecule.[34] Another type of molecule in which carbon is the B component are isocyanides, R—$\overset{\oplus}{N}\equiv\overset{\ominus}{C}$, which form rather strong hydrogen bonds.[35] There is evidence that double and triple bonds, aromatic rings,[36] and even cyclopropane rings[37] may be the B component of hydrogen bonds, but these bonds are very weak. An interesting case is that of the *in*-bicyclo[4.4.4]-1-tetradecyl cation **4** (see out–in isomerism, p. 133). Nmr

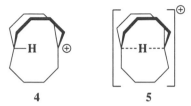

4   **5**

and ir spectra show that the actual structure of this ion is **5**, in which both the A and the B component of the hydrogen bond is a carbon.[38]

Deuterium also forms hydrogen bonds; in some systems these seem to be stronger than the corresponding hydrogen bonds; in others, weaker.[39]

ADDITION COMPOUNDS

When the reaction of two compounds results in a product that contains all the mass of the two compounds, the product is called an *addition compound*. There are several kinds. In the rest of this chapter we will discuss addition compounds in which the molecules of the starting materials remain more or less intact and weak bonds hold two or more molecules together. We can divide them into four broad classes: electron donor-acceptor complexes, complexes formed by crown ethers and similar compounds, inclusion compounds, and catenanes.

Electron Donor–Acceptor (EDA) Complexes[40]

In *EDA complexes*,[41] there is always a donor molecule and an acceptor. The donor may donate an unshared pair (an *n* donor) or a pair of electrons in a π orbital of a double bond or aromatic system (a π donor). One test for the presence of an EDA complex is the electronic spectrum. These complexes generally exhibit a spectrum (called a *charge-transfer*

[34]Ahlberg; Davidsson; Johnsson; McEwen; Rönnqvist *Bull. Soc. Chim. Fr.* **1988**, 177.

[35]Ferstandig *J. Am. Chem. Soc.* **1962**, *84*, 3553; Allerhand; Schleyer *J. Am. Chem. Soc.* **1963**, *85*, 866.

[36]For example, see Bakke; Chadwick *Acta Chem. Scand., Ser. B* **1988**, *42*, 223: Atwood; Hamada; Robinson; Orr; Vincent *Nature* **1991**, *349*, 683.

[37]Joris; Schleyer; Gleiter *J. Am. Chem. Soc.* **1968**, *90*, 327; Yoshida; Ishibe; Kusumoto *J. Am. Chem. Soc.* **1969**, *91*, 2279.

[38]McMurry; Lectka; Hodge *J. Am. Chem. Soc.* **1989**, *111*, 8867. See also Sorensen; Whitworth *J. Am. Chem. Soc.* **1990**, *112*, 8135.

[39]Dahlgren; Long *J. Am. Chem. Soc.* **1960**, *82*, 1303; Creswell; Allred *J. Am. Chem. Soc.* **1962**, *84*, 3966; Singh; Rao *Can. J. Chem.* **1966**, *44*, 2611; Cummings; Wood *J. Mol. Struct.* **1974**, *23*, 103.

[40]For monographs, see Foster *Organic Charge-Transfer Complexes*; Academic Press: New York, 1969; Mulliken; Person, *Molecular Complexes*; Wiley: New York, 1969; Rose, *Molecular Complexes*; Pergamon: Elmsford, NY, 1967. For reviews, see Poleshchuk; Maksyutin, *Russ. Chem. Rev.* **1976**, *45*, 1077-1090; Banthorpe *Chem. Rev.* **1970**, *70*, 295-322; Kosower *Prog. Phys. Org. Chem.* **1965**, *3*, 81-163; Foster *Chem. Br.* **1976**, *12*, 18-23.

[41]These have often been called *charge-transfer complexes*, but this term implies that the bonding involves charge transfer, which is not always the case, so that the more neutral name EDA complex is preferable. See Ref. 59.

spectrum) that is not the same as the sum of the spectra of the two individual molecules.[42] Because the first excited state of the complex is relatively close in energy to the ground state, there is usually a peak in the visible or near-uv region and EDA complexes are often colored. Many EDA complexes are unstable and exist only in solutions in equilibrium with their components, but others are stable solids. In most EDA complexes the donor and acceptor molecules are present in an integral ratio, most often 1:1, but complexes with nonintegral ratios are also known. There are several types of acceptor; we will discuss complexes formed by two of them.

1. *Complexes in which the acceptor is a metal ion and the donor an olefin or an aromatic ring* (*n* donors do not give EDA complexes with metal ions but form covalent bonds instead).[43] Many metal ions form complexes, which are often stable solids, with olefins, dienes (usually conjugated, but not always), alkynes, and aromatic rings. The generally accepted picture of the bonding in these complexes,[44] first proposed by Dewar,[45] can be illustrated for the complex in which silver ion is bonded to an olefin. There are two bonds between the metal ion and the olefin. One is a σ bond formed by overlap of the filled π orbital of the olefin with the empty $5s$ orbital of the silver ion, and the other a π bond

$$\begin{array}{c} R_2C \\ \parallel \;\; Ag^+ \\ R_2C \end{array}$$

formed by overlap of a filled $4d$ orbital of the silver ion and an empty antibonding π^* orbital of the olefin. The bond is not from the silver ion to one atom but to the whole π center. The net result is that some electron density is transferred from the olefin to the metal ion.[46]

Among the compounds that form complexes with silver and other metals are benzene[47] (represented as in **6**) and cyclooctatetraene. When the metal involved has a coordination number greater than 1, more than one donor molecule participates. In many cases, this extra electron density comes from CO groups, which in these complexes are called carbonyl groups. Thus, benzenechromium tricarbonyl (**7**) is a stable compound.[48] Three arrows are

[42]For examples of EDA complexes that do not show charge-transfer spectra, see Dewar; Thompson *Tetrahedron Suppl.* **1966,** *7,* 97; Bentley; Dewar *Tetrahedron Lett.* **1967,** 5043.

[43]For monographs, see Collman; Hegedus; Norton; Finke *Principles and Applications of Organotransition Metal Chemistry,* 2nd ed; University Science Books: Mill Valley, CA, 1987; Alper *Transition Metal Organometallics in Organic Synthesis,* 2 vols.; Academic Press: New York, 1976, 1978; King *Transition-Metal Organic Chemistry;* Academic Press: New York, 1969; Green, *Organometallic Compounds,* vol. 2; Methuen: London, 1968; For general reviews, see Churchill; Mason *Adv. Organomet. Chem.* **1967,** *5,* 93-135; Cais, in Patai *The Chemistry of Alkenes,* vol. 1; Wiley: New York, 1964, pp. 335-385. Among the many reviews limited to certain classes of complexes are: transition metals–dienes, Nakamura *J. Organomet. Chem.* **1990,** *400,* 35-48; metals–cycloalkynes and arynes, Bennett; Schwemlein *Angew. Chem. Int. Ed. Engl.* **1989,** *28,* 1296-1320 [*Angew. Chem. 101,* 1349-1373]; metals–pentadienyl ions, Powell *Adv. Organomet. Chem.* **1986,** *26,* 125-164; complexes of main-group metals, Jutzi; Noltes *Organomet. Chem.* **1986,** *26,* 217-295; intramolecular complexes, Omae *Angew. Chem. Int. Ed. Engl.* **1982,** *21,* 889-902 [*Angew. Chem. 94,* 902-915]; transition metals–olefins and acetylenes, Pettit; Barnes *Fortschr. Chem. Forsch.* **1972,** *28,* 85-139; Quinn; Tsai *Adv. Inorg. Chem. Radiochem.* **1969,** *12,* 217-373; Pt- and Pd-olefins and acetylenes, Hartley *Chem. Rev.* **1969,** *69,* 799-844; silver ion–olefins and aromatics, Beverwijk; van der Kerk; Leusink; Noltes *Organomet. Chem. Rev., Sect. A* **1970,** *5,* 215-280; metals–substituted olefins, Jones *Chem. Rev.* **1968,** *68,* 785-806; transition metals–allylic compounds, Clarke *J. Organomet. Chem.* **1974,** *80,* 155-173; transition metals–arenes, Silverthorn *Adv. Organomet. Chem.* **1976,** *14,* 47-137; metals–organosilicon compounds, Haiduc; Popa *Adv. Organomet. Chem.* **1977,** *15,* 113-146; metals–carbocations, Pettit; Haynes, in Olah; Schleyer *Carbonium Ions,* vol. 5; Wiley: New York, 1976, pp. 2263-2302; metals–seven- and eight-membered rings, Bennett *Adv. Organomet. Chem.* **1966,** *4,* 353-387. For a list of review articles on this subject, see Bruce *Adv. Organomet. Chem.* **1972,** *10,* 273-346, pp. 317-321.

[44]For reviews, see Pearson *Metallo-organic Chemistry;* Wiley: New York, 1985; Ittel; Ibers *Adv. Organomet. Chem.* **1976,** *14,* 33-61; Hartley *Chem. Rev.* **1973,** *73,* 163-190; *Angew. Chem. Int. Ed. Engl.* **1972,** *11,* 596-606 [*Angew. Chem. 84,* 657-667].

[45]Dewar *Bull. Soc. Chim. Fr.* **1951,** *18,* C79.

[46]For a discussion of how the nature of the metal ion affects the stability of the complex, see p. 263.

[47]For a monograph, see Zeiss; Wheatley; Winkler *Benzenoid-Metal Complexes;* Ronald Press: New York, 1966.

[48]Nicholls; Whiting *J. Chem. Soc.* **1959,** 551. For reviews of arene–transition-metal complexes, see Uemura *Adv. Met.-Org. Chem.* **1991,** *2,* 195-245; Silverthorn *Adv. Organomet. Chem.* **1975,** *13,* 47-137.

shown, since all three aromatic bonding orbitals contribute some electron density to the metal. Metallocenes (p. 47) may be considered a special case of this type of complex, although the bonding in metallocenes is much stronger.

| **6** | **7** | **8** |

In a number of cases olefins that are too unstable for isolation have been isolated in the form of metal complexes. As example is norbornadienone, which was isolated in the form of its iron–tricarbonyl complex (**8**).[49] The free dienone spontaneously decomposes to carbon monoxide and benzene (see **7-36**).

The donor (or ligand) molecules in these complexes are classified by the prefix *hapto*[50] and/or the descriptor η^n (the Greek letter eta), where n indicates how many atoms the ligand uses to bond with the metal.[51] Thus, ethene is a dihapto or η^2 ligand, and benzene hexahapto or η^6. Ferrocene can be called bis(η^5-cyclopentadienyl)iron(II).This system can be extended to compounds in which only a single σ bond connects the organic group to the metal, e.g., C_6H_5—Li (a monohapto or η^1 ligand), and to complexes in which the organic group is an ion, e.g., π-allyl complexes such as **9**, in which the allyl ligand is trihapto or η^3.

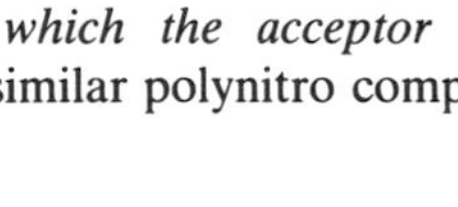

| **9** | allyllithium |

Note that in a compound such as allyllithium, where a σ bond connects the carbon to the metal, the allyl group is referred to as monohapto or η^1.

2. *Complexes in which the acceptor is an organic molecule.* Picric acid, 1,3,5-trinitrobenzene, and similar polynitro compounds are the most important of these.[52] Picric

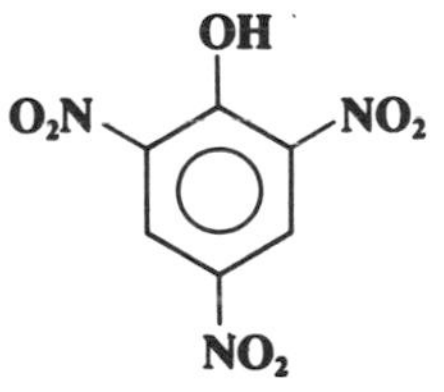

Picric acid

acid forms addition compounds with many aromatic hydrocarbons, aromatic amines, aliphatic amines, olefins, and other compounds. These addition compounds are usually solids with definite melting points and are often used as derivatives of the compounds in question. They are called picrates, though they are not salts of picric acid but addition compounds.

[49]Landesberg; Sieczkowski *J. Am. Chem. Soc.* **1971**, *93*, 972.
[50]For a discussion of how this system originated, see Cotton *J. Organomet. Chem.* **1975**, *100*, 29.
[51]Another prefix used for complexes is μ (mu), which indicates that the ligand bridges two metal atoms.
[52]For a review, see Parini *Russ. Chem. Rev.* **1962**, *31*, 408-417; for a review of complexes in which the acceptor is an organic cation, see Kampar *Russ. Chem. Rev.* **1982**, *51*, 107-118; also see Ref. 40.

Unfortunately, salts of picric acid are also called picrates. Similar complexes are formed between phenols and quinones (quinhydrones).[53] Olefins that contain electron-withdrawing substituents also act as acceptor molecules as do carbon tetrahalides[54] and certain anhydrides.[55] A particularly strong olefin acceptor is tetracyanoethylene.[56]

The bonding in these cases is more difficult to explain than in the previous case, and indeed no really satisfactory explanation is available.[57] The difficulty is that although the donor has a pair of electrons to contribute (both n donors and π donors are found here), the acceptor does not have a vacant orbital. Simple attraction of the dipole-induced-dipole type accounts for some of the bonding[58] but is too weak to explain the bonding in all cases;[59] e.g., nitromethane, with about the same dipole moment as nitrobenzene, forms much weaker complexes. Some other type of bonding clearly must also be present in many EDA complexes. The exact nature of this bonding, called *charge-transfer bonding*, is not well understood, but it presumably involves some kind of donor–acceptor interaction.

Crown Ether Complexes and Cryptates[60]

Crown ethers are large-ring compounds containing several oxygen atoms, usually in a regular pattern. Examples are 12-crown-4 (**10**),[61] dicyclohexano-18-crown-6 (**11**), and 15-crown-5 (**12**). These compounds have the property[62] of forming complexes with positive ions, gen-

[53]For a review of quinone complexes, see Foster; Foreman, in Patai *The Chemistry of the Quinonoid Compounds*, pt. 1, Wiley: New York, 1974, pp. 257-333.

[54]See Blackstock; Lorand; Kochi *J. Org. Chem.* **1987**, *52*, 1451.

[55]For a review of anhydrides as acceptors, see Foster, in Patai, Ref. 3, pp. 175-212.

[56]For a review of complexes formed by tetracyanoethylene and other polycyano compounds, see Melby, in Rappoport *The Chemistry of the Cyano Group*, Wiley: New York, 1970, pp. 639-669. See also Fatiadi *Synthesis* **1987**, 959-978.

[57]For reviews, see Bender *Chem. Soc. Rev.* **1986**, *15*, 475-502; Kampar; Neilands *Russ. Chem. Rev.* **1986**, *55*, 334-342; Bent *Chem. Rev.* **1968**, *68*, 587-648.

[58]See, for example, Le Fèvre; Radford; Stiles *J. Chem. Soc. B* **1968**, 1297.

[59]Mulliken; Person *J. Am. Chem. Soc.* **1969**, *91*, 3409.

[60]For a treatise, see Atwood; Davies; MacNicol *Inclusion Compounds*, 3 vols.; Academic Press: New York, 1984. For monographs, see Weber et al. *Crown Ethers and Analogs*; Wiley: New York, 1989; Vögtle *Host Guest Complex Chemistry I, II, and III (Top. Curr. Chem. 98, 101, 121)*; Springer: Berlin, 1981, 1982, 1984; Vögtle; Weber *Host Guest Complex Chemistry/Macrocycles*; Springer: Berlin, 1985 [this book contains nine articles from the *Top. Curr. Chem.* vols. just mentioned]; Hiraoka *Crown Compounds*; Elsevier: New York, 1982; De Jong; Reinhoudt *Stability and Reactivity of Crown-Ether Complexes*; Academic Press: New York, 1981; Izatt; Christensen *Synthetic Multidentate Macrocyclic Compounds*; Academic Press: New York, 1978. For reviews, see McDaniel; Bradshaw; Izatt *Heterocycles* **1990**, *30*, 665-706; Sutherland, *Chem. Soc. Rev.* **1986**, *15*, 63-91; Sutherland in Takeuchi; Marchand *Applications of NMR Spectroscopy to Problems in Stereochemistry and Conformational Analysis*; VCH: New York, 1986; Franke; Vögtle *Top. Curr. Chem.* **1986**, *132*, 135-170; Cram *Angew. Chem. Int. Ed. Engl.* **1986**, *25*, [*Angew. Chem. 98*, 1041-1060]; Vögtle; Löhr; Franke; Worsch *Angew. Chem. Int. Ed. Engl.* **1985**, *24*, 727-742 [*Angew. Chem. 97*, 721]; Gutsche *Acc. Chem. Res.* **1983**, *16*, 161-170; Tabushi; Yamamura *Top. Curr. Chem.* **1983**, *113*, 145-182; Stoddart *Prog. Macrocyclic Chem.* **1981**, *2*, 173-250; De Jong; Reinhoudt *Adv. Phys. Org. Chem.* **1980**, *17*, 279-433; Vögtle; Weber, in Patai *The Chemistry of Functional Groups, Supplement E*; Wiley: New York, 1980, pp. 59-156; Poonia *Prog. Macrocyclic Chem.* **1979**, *1*, 115-155; Reinhoudt; De Jong *Prog. Macrocyclic Chem.* **1979**, *1*, 157-217; Cram; Cram *Acc. Chem. Res.* **1978**, *11*, 8-14; *Science*, **1974**, *183*, 803-809; Knipe *J. Chem. Educ.* **1976**, *53*, 618-622; Gokel; Durst *Synthesis* **1976**, 168-184, *Aldrichimica Acta* **1976**, *9*, 3-12; Lehn *Struct. Bonding (Berlin)* **1973**, *16*, 1-69; Christensen; Eatough; Izatt *Chem. Rev.* **1974**, *74*, 351-384; Pedersen; Frensdorff *Angew. Chem. Int. Ed. Engl.* **1972**, *11*, 16-25 [*Angew. Chem. 84*, 16-26]. For a monograph on the synthesis of crown ethers, see Gokel; Korzeniowski *Macrocyclic Polyether Synthesis*; Springer: New York, 1982. For reviews, see Krakowiak; Bradshaw; Zamecka-Krakowiak *Chem. Rev.* **1989**, *89*, 929-972; Jurczak; Pietraszkiewicz *Top. Curr. Chem.* **1986**, *130*, 183-204; Gokel; Dishong; Schultz; Gatto *Synthesis* **1982**, 997-1012; Bradshaw; Stott *Tetrahedron* **1980**, *36*, 461-510; Laidler; Stoddart, in Patai *The Chemistry of Functional Groups, Supplement E*; Wiley: New York, 1980, pp. 3-42. For reviews of acyclic molecules with similar properties, see Vögtle *Chimia* **1979**, *33*, 239-251; Vögtle; Weber *Angew. Chem. Int. Ed. Engl.* **1979**, *18*, 753-776 [*Angew. Chem. 91*, 813-837]. For a review of cryptands that hold two positive ions, see Lehn *Pure Appl. Chem.* **1980**, *52*, 2441-2459. The 1987 Nobel Prize in Chemistry was awarded to Charles J. Pedersen, Donald J. Cram, and Jean-Marie Lehn for their work in this area. The three Nobel lectures were published in two journals (respectively, CJP, DJC, J-ML): *Angew. Chem. Int. Ed. Engl.* **1988**, *27* [*Angew. Chem. 100*] pp. 1021-1027 [1053-1059], 1009-1020 [1041-1052], 89-112 [91-116]; and *Chem. Scr.* **1988**, *28*, pp. 229-235, 263-274, 237-262. See also the series *Advances in Supramolecular Chemistry*.

[61]Cook; Caruso; Byrne; Bowers; Speck; Liotta *Tetrahedron Lett.* **1974**, 4029.

[62]Discovered by Pedersen *J. Am. Chem. Soc.* **1967**, *89*, 2495, 7017. For an account of the discovery, see Schroeder; Petersen *Pure Appl. Chem.* **1988**, *60*, 445.

erally metallic ions (though not usually ions of transition metals) or ammonium and sub-stituted ammonium ions.[62a] The crown ether is called the *host* and the ion is the *guest*. In most cases the ions are held tightly in the center of the cavity.[63] Each crown ether binds

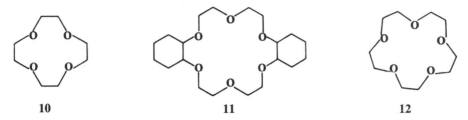

10   11   12

different ions, depending on the size of the cavity. For example, **10** binds Li^{+}[64] but not K^{+},[65] while **11** binds K^{+} but not Li^{+}.[66] Similarly, **11** binds Hg^{2+} but not Cd^{2+} or Zn^{2+}, and Sr^{2+} but not Ca^{2+}.[67] The complexes can frequently be prepared as well-defined sharp-melting solids.

Apart from their obvious utility in separating mixtures of cations,[68] crown ethers have found much use in organic synthesis (see the discussion on p. 363). Chiral crown ethers have been used for the resolution of racemic mixtures (p. 121). Although crown ethers are most frequently used to complex cations, amines, phenols, and other neutral molecules have also been complexed[69] (see p. 133 for the complexing of anions).[70]

Macrocycles containing nitrogen or sulfur atoms,[71] e.g., **13** and **14**,[72] have similar prop-erties, as do those containing more than one kind of hetero atom, e.g., **15**,[73] **16**,[74] or **17**.[75] Bicyclic molecules like **16** can surround the enclosed ion in three dimensions, binding it even more tightly than the monocyclic crown ethers. Bicyclics and cycles of higher order[76] are called *cryptands* and the complexes formed are called *cryptates* (monocylics are also sometimes called cryptands). The tricyclic cryptand **17** has ten binding sites and a spherical cavity.[75] Another molecule with a spherical cavity (though not a cryptand) is **18,** which

[62a]For a monograph, see Inoue; Gokel *Cation Binding by Macrocycles*; Marcel Dekker: New York, 1990.

[63]For reviews of thermodynamic and kinetic data for this type of interaction, see Izatt; Bradshaw; Nielsen; Lamb; Christensen; Sen *Chem. Rev.* **1985**, *85*, 271-339; Parsonage; Staveley, in Atwood; Davies; MacNicol, Ref. 60, vol. 3, pp. 1-36.

[64]Anet; Krane; Dale; Daasvatn; Kristiansen *Acta Chem. Scand.* **1973**, *27*, 3395.

[65]Certain derivatives of 14-crown-4 and 12-crown-3 show very high selectivity for Li^{+} compared to the other alkali metal ions. See Bartsch; Czech; Kang; Stewart; Walkowiak; Charewicz; Heo; Son *J. Am. Chem. Soc.* **1985**, *107*, 4997; Dale; Eggestad; Fredriksen; Groth *J. Chem. Soc., Chem. Commun.* **1987**, 1391; Dale; Fredriksen *Pure Appl. Chem.* **1989**, *61*, 1587.

[66]Izatt; Nelson; Rytting; Haymore; Christensen *J. Am. Chem. Soc.* **1971**, *93*, 1619.

[67]Kimura; Iwashima; Ishimori; Hamaguchi *Chem. Lett.* **1977**, 563.

[68]Crown ethers have been used to separate isotopes of cations, e.g., ^{44}Ca from ^{40}Ca. For a review, see Heumann *Top. Curr. Chem.* **1985**, *127*, 77-132.

[69]For reviews, see Vögtle; Müller; Watson *Top. Curr. Chem.* **1984**, *125*, 131-164; Weber, *Prog. Macrocycl. Chem.* **1987**, *3*, 337-419; Diederich *Angew. Chem. Int. Ed. Engl.* **1988**, *27*, 362-386 [*Angew. Chem. 100*, 372-396].

[70]A neutral molecule (e.g., urea) and a metal ion (e.g., Li^{+}) were made to be joint guests in a macrocyclic host, with the metal ion acting as a bridge that induces a partial charge on the urea nitrogens: van Staveren; van Eerden; van Veggel; Harkema; Reinhoudt *J. Am. Chem. Soc.* **1988**, *110*, 4994. See also Rodrigue; Bovenkamp; Murchie; Buchanan; Fortier *Can. J. Chem.* **1987**, *65*, 2551; Fraser; Fortier; Markiewicz; Rodrigue; Bovenkamp *Can. J. Chem.* **1987**, *65*, 2558.

[71]For reviews of sulfur-containing macroheterocycles, see Voronkov; Knutov *Sulfur Rep.* **1986**, *6*, 137-256, *Russ. Chem. Rev.* **1982**, *51*, 856-871. For a review of those containing S and N, see Reid; Schröder *Chem. Soc. Rev.* **1990**, *19*, 239-269.

[72]For a review of **14** and its derivatives, see Chaudhuri; Wieghardt *Prog. Inorg. Chem.* **1987**, *35*, 329-436.

[73]Dietrich; Lehn; Sauvage *Chem. Commun.* **1970**, 1055.

[74]Newcomb; Gokel; Cram *J. Am. Chem. Soc.* **1974**, *96*, 6810.

[75]Graf; Lehn *J. Am. Chem. Soc.* **1975**, *97*, 5022.

[76]For reviews, see Potvin; Lehn *Prog. Macrocycl. Chem.* **1987**, *3*, 167-239; Kiggen; Vögtle *Prog. Macrocycl. Chem.* **1987**, *3*, 309-336; Dietrich, in Atwood; Davies; MacNicol, Ref. 60, vol. 2, pp. 337-405; Parker *Adv. Inorg. Radiochem.* **1983**, *27*, 1-26; Lehn *Acc. Chem. Res.* **1978**, *11*, 49-57, *Pure Appl. Chem.* **1977**, *49*, 857-870.

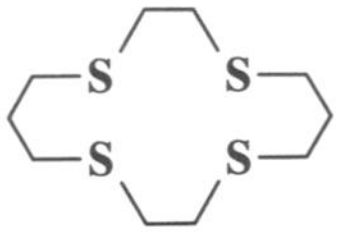

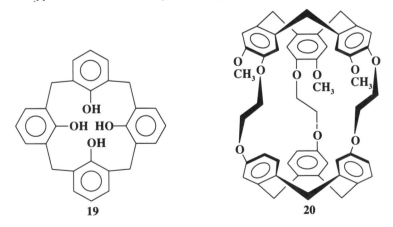

13 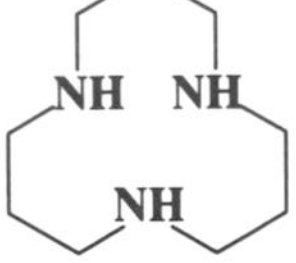   **14** 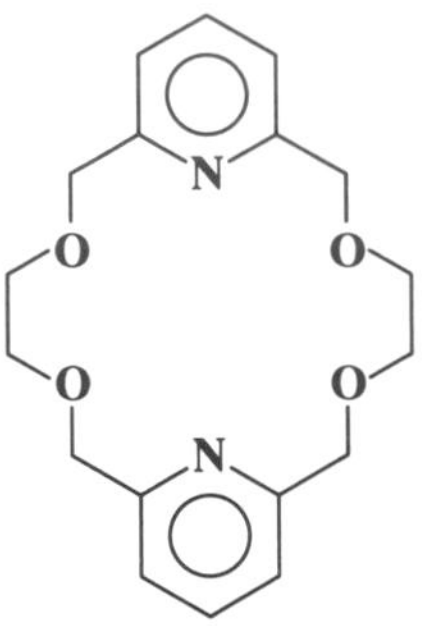   **15** 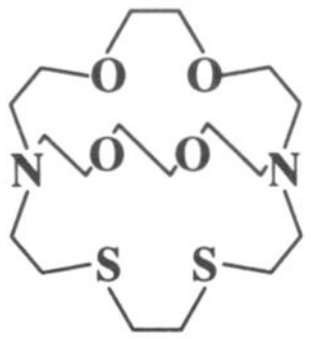   **16**

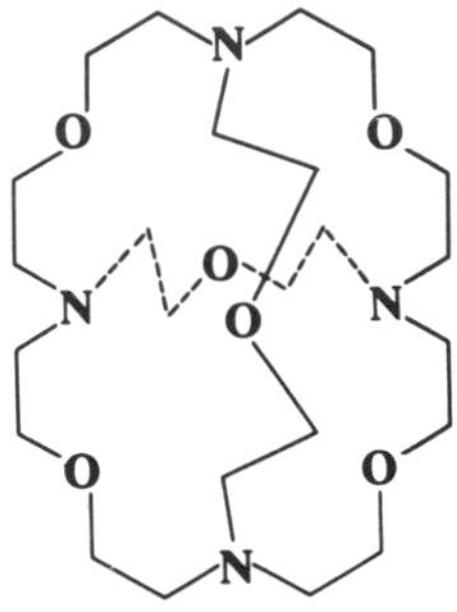

17 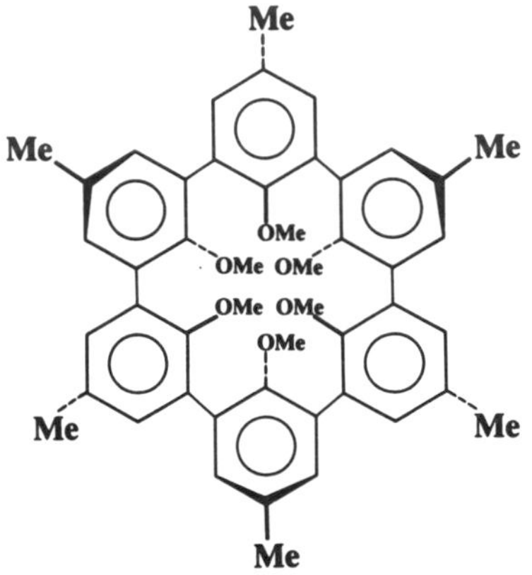   **18**

complexes Li$^+$ and Na$^+$ (preferentially Na$^+$), but not K$^+$, Mg^{2+}, or Ca^{2+}.[77] Molecules such as these, whose cavities can be occupied only by spherical entities, have been called *spherands*.[77] Other types are *calixarenes*, e.g., **19**,[78] *cryptophanes*, e.g., **20**,[79] *hemispherands* (an

19   **20**

[77]Cram; Dicker *J. Chem. Soc., Chem. Commun.* **1982**, 1219; Cram; Doxsee *J. Org. Chem.* **1986**, *51*, 5068; Cram *CHEMTECH* **1987**, 120, *Chemtracts: Org. Chem.* **1988**, *1*, 89; Paek; Knobler; Maverick; Cram *J. Am. Chem. Soc.* **1989**, *111*, 8662; Bryant; Ho; Knobler; Cram *J. Am. Chem. Soc.* **1990**, *112*, 5837.

[78]For monographs, see Vicens; Böhmer *Calixarenes: A Versatile Class of Macrocyclic Compounds*; Kluwer: Dordrecht, 1991; Gutsche *Calixarenes*; Royal Society of Chemistry: Cambridge, 1989. For reviews, see Gusche *Prog. Macrocycl. Chem.* **1987**, *3*, 93-165, *Top. Curr. Chem.* **1984**, *123*, 1-47.

[79]For reviews, see Collet *Tetrahedron* **1987**, *43*, 5725-5759, in Atwood; Davies; MacNicol, Ref. 60, Vol. 1, pp. 97-121.

example is **21**[80]), and *podands*.[81] The last-named are host compounds in which two or more arms come out of a central structure. Examples are **22**[82] and **23**.[83] **23**, also called an *octopus molecule*, binds simple cations such as Na⁺, K⁺, and Ca²⁺. *Lariat ethers* are compounds

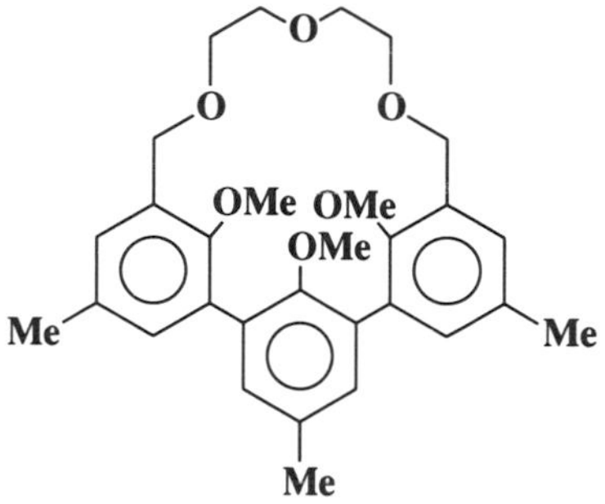

21

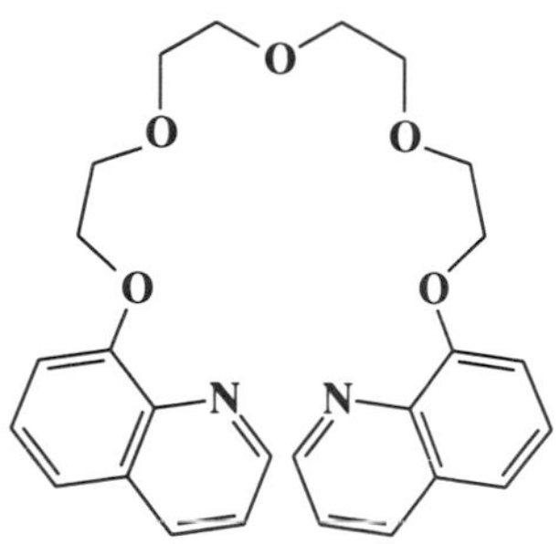

22

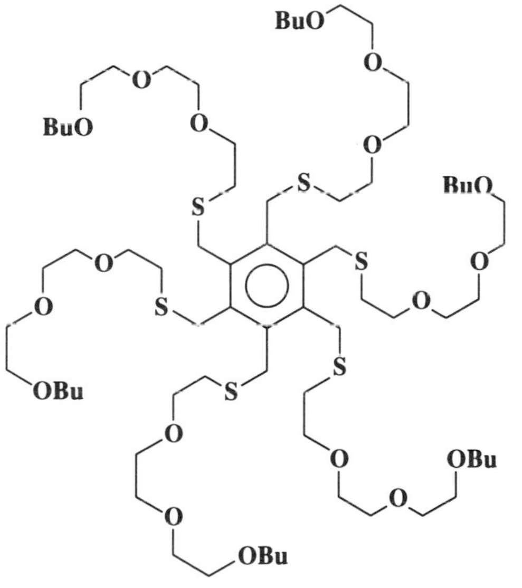

23

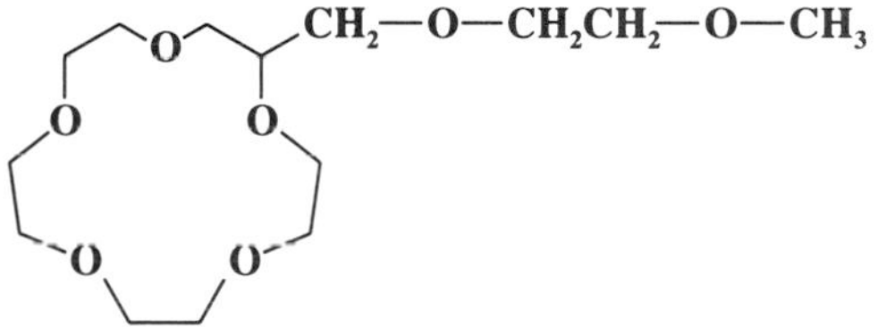

24

containing a crown ether ring with one or more side chains that can also serve as ligands.[84] An example is **24**.

The bonding in these complexes is the result of ion-dipole attractions between the hetero atoms and the positive ions.

As we have implied, the ability of these host molecules to bind guests is often very specific, enabling the host to pull just one molecule or ion out of a mixture. This is called

[80]Lein; Cram *J. Am. Chem. Soc.* **1985**, *107*, 448.

[81]For reviews, see Kron; Tsvetkov *Russ. Chem. Rev.* **1990**, *59*, 283-298; Menger *Top. Curr. Chem.* **1986**, *136*, 1-15.

[82]Tümmler; Maass; Weber; Wehner; Vögtle *J. Am. Chem. Soc.* **1977**, *99*, 4683.

[83]Vögtle; Weber *Angew. Chem. Int. Ed. Engl.* **1974**, *13*, 814 [*Angew. Chem. 13*, 896].

[84]See Gatto; Dishong; Diamond *J. Chem. Soc., Chem. Commun.* **1980**, 1053; Gatto; Gokel *J. Am. Chem. Soc.* **1984**, *106*, 8240; Nakatsuji; Nakamura; Yonetani; Yuya; Okahara *J. Am. Chem. Soc.* **1988**, *110*, 531.

molecular recognition.[85] In general, cryptands, with their well-defined three-dimensional cavities, are better for this than monocyclic crown ethers or ether derivatives. An example is the host **25,** which selectively binds the dication **26** (*n* = 5) rather than **26** (*n* = 4), and **26** (*n* = 6) rather than **26** (*n* = 7).[86] The host **27,** which is water-soluble, forms 1:1 complexes

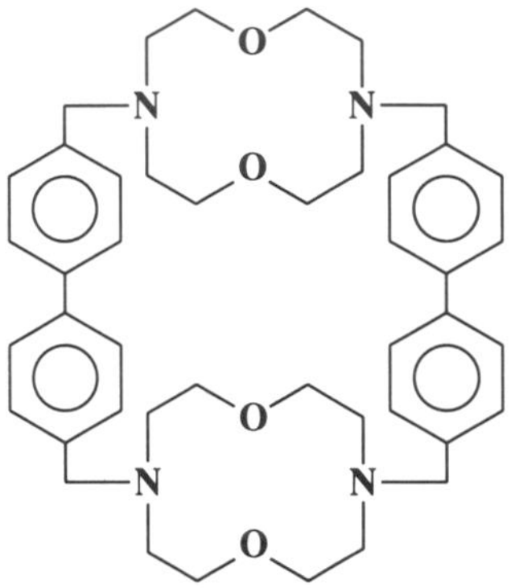

$$H_3\overset{+}{N}-(CH_2)_n-\overset{+}{N}H_3$$

26

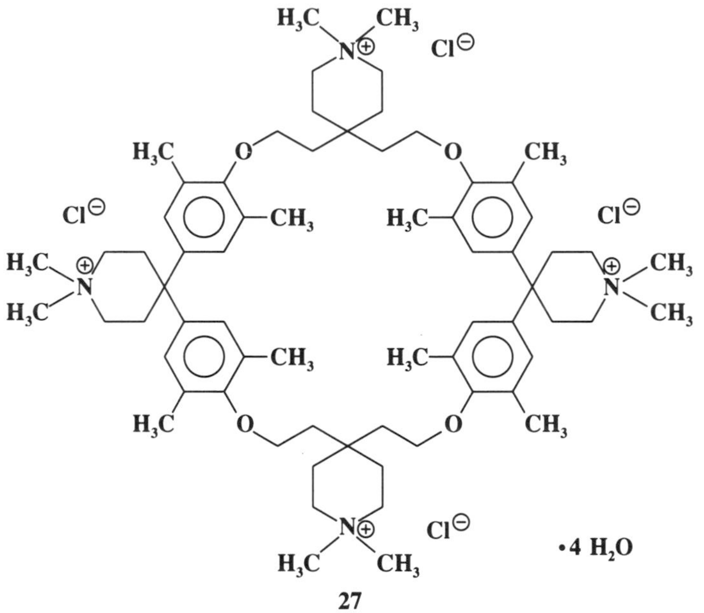

25

27

$\cdot 4 \ H_2O$

with neutral aromatic hydrocarbons such as pyrene and fluoranthene, and even (though more weakly) with biphenyl and naphthalene, and is able to transport them through an aqueous phase.[87]

---

[85]For reviews, see Rebek *Angew. Chem. Int. Ed. Engl.* **1990,** *29*, 245-255 [*Angew. Chem. 102*, 261-272], *Acc. Chem. Res.* **1990,** *23*, 399-404, *Top. Curr. Chem.* **1988,** *149*, 189-210, *Mol. Struct. Energ.* **1988,** *10*, 219-250; Diederich *J. Chem. Educ.* **1990,** *67*, 813-820; Hamilton *J. Chem. Educ.* **1990,** *67*, 821-828; Raevskii *Russ. Chem. Rev.* **1990,** *59*, 219-233.
[86]Mageswaran; Mageswaran; Sutherland *J. Chem. Soc., Chem. Commun.* **1979,** 722.
[87]Diederich; Dick *J. Am. Chem. Soc.* **1984,** *106*, 8024; Diederich; Griebel *J. Am. Chem. Soc.* **1984,** *106*, 8037. See also Vögtle; Müller; Werner; Losensky *Angew. Chem. Int. Ed. Engl.* **1987,** *26*, 901 [*Angew. Chem. 99*, 930].

Of course, it has long been known that molecular recognition is very important in biochemistry. The action of enzymes and various other biological molecules is extremely specific because these molecules also have host cavities that are able to recognize only one or a few particular types of guest molecules. It is only in recent years that organic chemists have been able to synthesize non-natural hosts that can also perform crude (compared to biological molecules) molecular recognition. The macrocycle **28** has been used as a catalyst, for the hydrolysis of acetyl phosphate and the synthesis of pyrophosphate.[88]

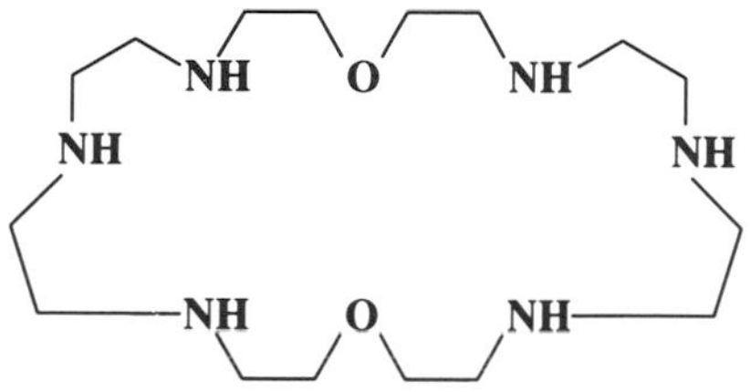

28

No matter what type of host, the strongest attractions occur when combination with the guest causes the smallest amount of distortion of the host.[89] That is, a fully preorganized host will bind better than a host whose molecular shape must change in order to accomodate the guest.

Inclusion Compounds

This type of addition compound is different from either the EDA complexes or the crown ether type of complexes previously discussed. Here the host forms a crystal lattice which has spaces large enough for the guest to fit into. There is no bonding between the host and the guest except van der Waals forces. There are two main types, depending on the shape of the space.[90] The spaces in *inclusion compounds* are in the shape of long tunnels or channels, while the other type, often called *clathrate*,[91] or *cage compounds* have spaces that are completely enclosed. In both types the guest molecule must fit into the space and potential guests that are too large or too small will not go into the lattice, so that the addition compound will not form.

One important host molecule among the inclusion compounds is urea.[92] Ordinary crystalline urea is tetragonal, but when a guest is present, urea crystallizes in a hexagonal lattice, containing the guest in long channels (Figure 3.1).[93] The hexagonal type of lattice can form only when a guest molecule is present, showing that van der Waals forces between the host and the guest, while small, are essential to the stability of the structure. The diameter of the channel is about 5 Å, and which molecules can be guests is dependent only on their

---

[88]Hosseini; Lehn *J. Am. Chem. Soc.* **1987**, *109*, 7047. For a discussion, see Mertes; Mertes *Acc. Chem. Res.* **1990**, *23*, 413-418.

[89]See Cram, *Angew. Chem. Int. Ed. Engl.* **1986**, *25*, 1039-1057 [*Angew. Chem. 98*, 1041-1060].

[90]For a treatise that includes both types, see Atwood; Davies; MacNicol, Ref. 60. For reviews, see Weber *Top. Curr. Chem.* **1987**, *140*, 1-20; Gerdil *Top. Curr. Chem.* **1987**, *140*, 71-105; Mak; Wong *Top. Curr. Chem.* **1987**, *140*, 141-164. For a review of channels with helical shapes, see Bishop; Dance *Top. Curr. Chem.* **1988**, *149*, 137-188.

[91]For reviews, see Goldberg *Top. Curr. Chem.* **1988**, *149*, 1-44; Weber; Czugler *Top. Curr. Chem.* **1988**, *149*, 45-135; MacNicol; McKendrick; Wilson *Chem. Soc. Rev.* **1978**, *7*, 65-87.

[92]For a review of urea and thiourea inclusion compounds, see Takemoto; Sonoda, in Atwood; Davies; MacNicol, Ref. 60, vol. 2, pp. 47-67.

[93]This picture is taken from a paper by Montel *Bull. Soc. Chim. Fr.* **1955**, 1013.

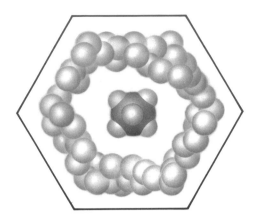

FIGURE 3.1 Guest molecule in a urea lattice.[93]

shapes and sizes and not on any electronic or chemical effects. For example, octane and 1-bromooctane are suitable guests for urea, but 2-bromooctane, 2-methylheptane, and 2-methyloctane are not. Also both dibutyl maleate and dibutyl fumarate are guests; neither diethyl maleate or diethyl fumarate is a guest, but dipropyl fumarate is a guest and dipropyl maleate is not.[94] In these complexes, there is usually no integral molar ratio (though by chance there may be). For example, the octane–urea ratio is 1:6.73.[95] A deuterium quadrupole echo spectroscopy study of a urea complex showed that the urea molecules do not remain rigid, but undergo 180° flips about the C=O axis at the rate of more than 10^6 per second at 30°C.[96]

The complexes are solids but are not useful as derivatives, since they melt, with decomposition of the complex, at the melting point of urea. They are useful, however, in separating isomers that would be quite difficult to separate otherwise. Thiourea also forms inclusion compounds though with channels of larger diameter, so that n-alkanes cannot be guests but, for example, 2-bromooctane, cyclohexane, and chloroform readily fit.

The most important host for clathrates is hydroquinone.[97] Three molecules, held together by hydrogen bonding, make a cage in which fits one molecule of guest. Typical guests are methanol (but not ethanol), SO_2, CO_2, and argon (but not neon). In contrast to the inclusion compounds, the crystal lattices here can exist partially empty. Another host is water. Usually six molecules of water form the cage and many guest molecules, among them Cl_2, propane, and methyl iodide, can fit. The water clathrates, which are solids, can normally be kept only

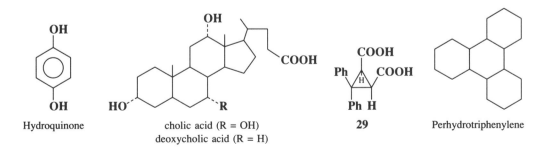

Hydroquinone cholic acid (R = OH) **29** Perhydrotriphenylene
deoxycholic acid (R = H)

[94]Radell; Connolly; Cosgrove *J. Org. Chem.* **1961,** *26,* 2960.
[95]Redlich; Gable; Dunlop; Millar *J. Am. Chem. Soc.* **1950,** *72,* 4153.
[96]Heaton; Vold; Vold *J. Am. Chem. Soc.* **1989,** *111,* 3211.
[97]For a review, see MacNicol, in Atwood; Davies; MacNicol, Ref. 60, vol. 2, pp. 1-45.

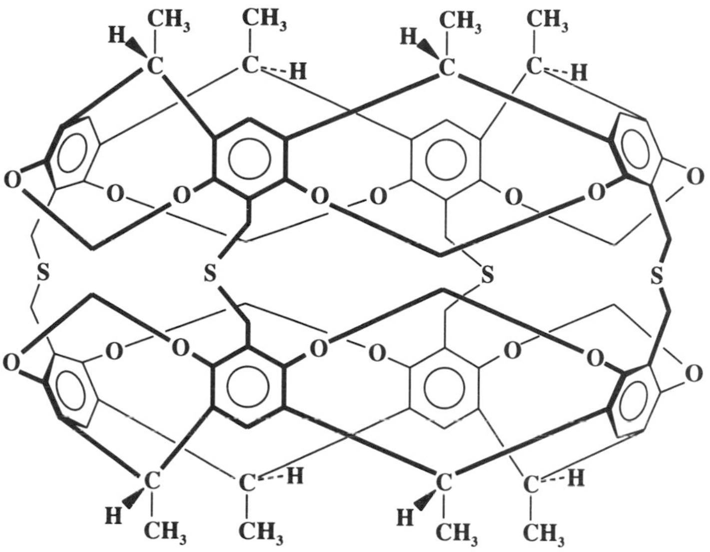

30

at low temperatures; at room temperature, they decompose.[98] Another inorganic host is sodium chloride (and some other alkali halides), which can encapsulate organic molecules such as benzene, naphthalene, and diphenylmethane.[99]

Among other hosts[100] for inclusion and/or clathrate compounds are deoxycholic acid,[101] cholic acid,[102] small ring compounds such as **29**,[103] perhydrotriphenylene,[104] and the compound **30**, which has been called a *carcerand*.[105]

Cyclodextrins

There is one type of host that can form both channel and cage complexes. This type is called *cyclodextrins* or *cycloamyloses*.[106] The host molecules are made up of six, seven, or eight glucose units connected in a large ring, called, respectively, α-, β-, or γ-cyclodextrin (Figure 3.2 shows the β or seven-membered ring compound). The three molecules are in the shape of hollow truncated cones (Figure 3.3) with primary OH groups projecting from the narrow

---

[98]For a monograph on water clathrates, see Berecz; Balla-Achs *Gas Hydrates*; Elsevier: New York, 1983. For reviews, see Jeffrey, in Atwood; Davies, MacNicol, Ref. 60, vol. 1, pp. 135-190; Cady *J. Chem. Educ.* **1983**, *60*, 915-918; Byk; Fomina *Russ. Chem. Rev.* **1968**, *37*, 469-491.

[99]Kirkor; Gebicki; Phillips; Michl *J. Am. Chem. Soc.* **1986**, *108*, 7106.

[100]See also Toda *Pure App. Chem.* **1990**, *62*, 417-422, *Top. Curr. Chem.* **1988**, *149*, 211-238, **1987**, *140*, 43-69; Davies; Finocchiaro; Herbstein, in Atwood; Davies; MacNicol, Ref. 60, vol. 2, pp. 407-453.

[101]For a review, see Giglio, in Atwood; Davies; MacNicol, Ref. 60, vol. 2, pp. 207-229.

[102]See Miki; Masui; Kasei; Miyata; Shibakami; Takemoto *J. Am. Chem. Soc.* **1988**, *110*, 6594.

[103]Weber; Hecker; Csöregh; Czugler *J. Am. Chem. Soc.* **1989**, *111*, 7866.

[104]For a review, see Farina, in Atwood; Davies; MacNicol, Ref. 60, vol. 2, pp. 69-95.

[105]Cram; Karbach; Kim; Baczynskyj; Marti; Sampson; Kalleymeyn *J. Am. Chem. Soc.* **1988**, *110*, 2554; Sherman; Knobler; Cram *J. Am. Chem. Soc.* **1991**, *113*, 2194.

[106]For a monograph, see Bender; Komiyama *Cyclodextrin Chemistry*; Springer: New York, 1978. For reviews, see, in Atwood; Davies; MacNicol, Ref. 60, the reviews, by Saenger, vol. 2, 231-259, Bergeron, vol. 3, 391-443, Tabushi, Vol. 3, 445-471, Breslow, vol. 3, 473-508; Croft; Bartsch *Tetrahedron* **1983**, *39*, 1417-1474; Tabushi; Kuroda *Adv. Catal.* **1983**, *32*, 417-466; Tabushi *Acc. Chem. Res.* **1982**, *15*, 66-72; Saenger *Angew. Chem. Int. Ed. Engl.* **1980**, *19*, 344-362 [*Angew. Chem. 92*, 343-361]; Bergeron *J. Chem. Ed.* **1977**, *54*, 204-207; Griffiths; Bender *Adv. Catal.* **1973**, *23*, 209-261.

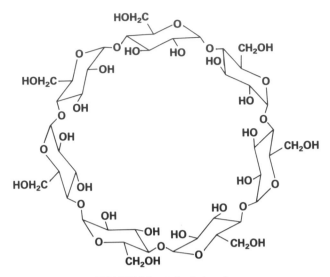

FIGURE 3.2 β-Cyclodextrin.

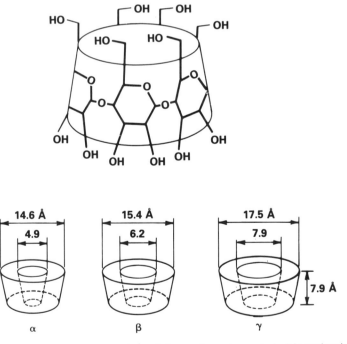

FIGURE 3.3 Shape and dimensions of the α-, β-, and γ-cyclodextrin molecules.[107]

[107]Szejtle, Ref. 109, p. 332; Nickon; Silversmith *The Name Game*; Pergamon: Elmsford, NY, p. 235.

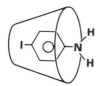

FIGURE 3.4 Schematic drawing of the complex of α-cyclodextrin and p-iodoaniline.[108]

side of the cones and secondary OH group from the wide side. As expected for carbohydrate molecules, all of them are soluble in water and the cavities normally fill with water molecules held in place by hydrogen bonds (6, 12, and 17 H_2O molecules for the α, β, and γ forms, respectively), but the insides of the cones are less polar than the outsides, so that nonpolar organic molecules readily displace the water. Thus the cyclodextrins form 1:1 cage complexes with many guests, ranging in size from the noble gases to large organic molecules. A guest molecule must not be too large or it will not fit, though many stable complexes are known in which one end of the guest molecule protrudes from the cavity (Figure 3.4). On the other hand, if the guest is too small, it may go through the bottom hole (though some small polar molecules such as methanol do form complexes in which the cavity also contains some water molecules). Since the cavities of the three cyclodextrins are of different sizes (Figure 3.3), a large variety of guests can be accommodated. Since cyclodextrins are nontoxic (they are actually small starch molecules), they are now used industrially to encapsulate foods and drugs.[109]

The cyclodextrins also form channel-type complexes, in which the host molecules are stacked on top of each other, like coins in a row.[110] For example, α-cyclodextrin (cyclohexaamylose) forms cage complexes with acetic, propionic, and butyric acids, but channel complexes with valeric and higher acids.

Catenanes and Rotaxanes[111]

These compounds contain two or more independent portions that are not bonded to each other by any valence forces but nevertheless must remain linked. *Catenanes* are made up of two or more rings held together as links in a chain, while in *rotaxanes* a linear portion

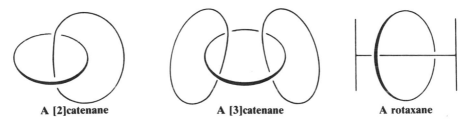

A [2]catenane   A [3]catenane   A rotaxane

is threaded through a ring and cannot get away because of bulky end groups. Catenanes and rotaxanes can be prepared by statistical methods or directed syntheses.[112] An example of a statistical synthesis of a rotaxane is a reaction where a compound **A** is bonded at two

[108]Modified from Saenger; Beyer; Manor *Acta Crystallogr. Sect. B* **1976**, *32*, 120.

[109]For reviews, see Pagington *Chem. Br.* **1987**, *23*, 455-458; Szejtli, in Atwood; Davies; MacNicol, Ref. 60, vol. 3, 331-390.

[110]See Saenger, Ref. 106.

[111]For a monograph, see Schill *Catenanes, Rotaxanes, and Knots*; Academic Press: New York, 1971. For a review, see Schill, in Chiurdoglu *Conformational Analysis*; Academic Press: New York, 1971, pp. 229-239.

[112]For discussions, see Ref. 111; Walba *Tetrahedron* **1985**, *41*, 3161-3212.

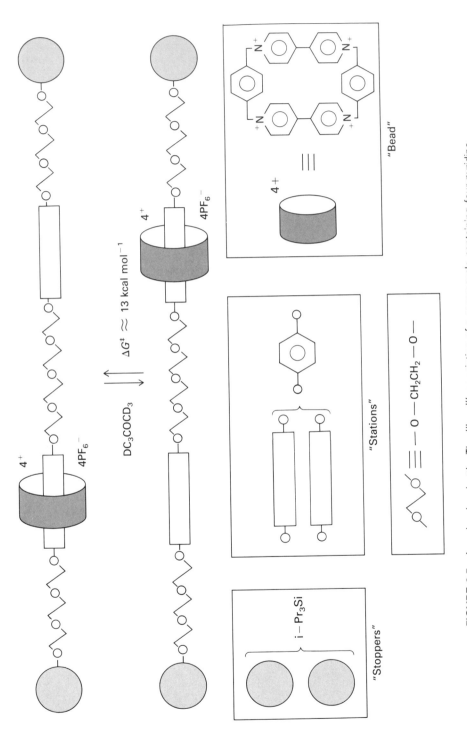

FIGURE 3.5 A molecular shuttle. The "bead", consisting of a macrocycle containing four pyridine rings and two benzene rings, moves back and forth between the two "stations", which are benzene rings. The "stoppers" prevent the bead from falling off the chain.[115]

positions to another compound **B** in the presence of a large ring **C**. It is hoped that some **A** molecules would by chance be threaded through **C** before combining with the two **B** molecules, so that some rotaxane (**D**) would be formed along with the normal product **E**.[113]

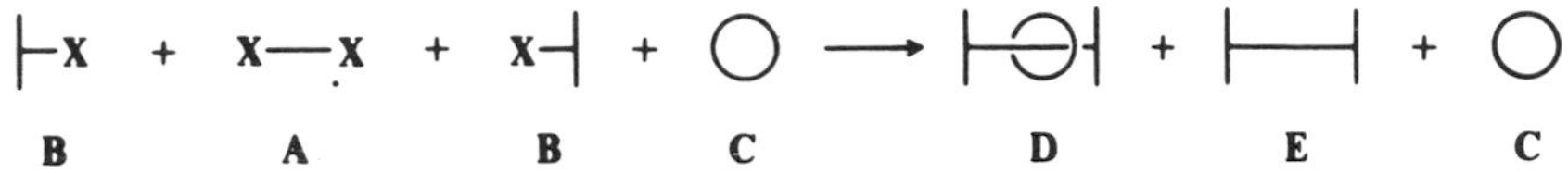

In a directed synthesis, the separate parts of the molecule are held together by other bonds that are later cleaved. See **9-65** for statistical and directed syntheses of catenanes.[114]

A rotaxane that is also an inclusion compound is shown schematically in Figure 3.5.[115] In this molecule the bulky end groups (or "stoppers") are triisopropylsilyl groups (i-Pr$_3$Si—) and the chain consists of a series of —O—CH$_2$CH$_2$—O— groups, but also contains two benzene rings. The ring (or "bead") around the chain is a macrocycle containing two benzene rings and four pyridine rings, and is preferentially attracted to one of the benzene rings in the chain. (The benzene moiety serves as a "station" for the "bead".) However, in this particular compound the symmetry of the chain makes the two "stations" equivalent, so that the "bead" is equally attracted to them, and the "bead" actually moves back and forth rapidly between the two "stations", as shown by the temperature dependence of the nmr spectrum.[116] This molecule has been called a *molecular shuttle*.

[113]Schemes of this type were carried out by Harrison and Harrison *J. Am. Chem. Soc.* **1967**, *89*, 5723; Ogino *J. Am. Chem. Soc.* **1981**, *103*, 1303. For a different kind of statistical syntheszis of a rotaxane, see Harrison *J. Chem. Soc., Chem. Commun.* **1972**, 231; *J. Chem. Soc., Perkin Trans. 1* **1974**, 301; Schill; Beckmann; Schweikert; Fritz *Chem. Ber.* **1986,** *119*, 2647. See also Agam; Graiver; Zilkha *J. Am. Chem. Soc.* **1976**, *98*, 5206.

[114]For a directed synthesis of a rotaxane, see Schill; Züllenkopf *Liebigs Ann. Chem.* **1969**, *721*, 53; Schill; Zürcher; Vetter *Chem. Ber.* **1973**, *106*, 228.

[115]Adapted from a diagram in Anelli; Spencer; Stoddart *J. Am. Chem. Soc.* **1991**, *113*, 5131.

[116]Ref. 115. For a review of the synthesis and properties of molecules of this type, see Philp; Stoddart *Synlett* **1991**, 445-458.

4

STEREOCHEMISTRY

In the previous chapters we discussed electron distribution in organic molecules. In this chapter we discuss the three-dimensional structure of organic compounds.[1] The structure may be such that *stereoisomerism*[2] is possible. Stereoisomers are compounds made up of the same atoms bonded by the same sequence of bonds but having different three-dimensional structures which are not interchangeable. These three-dimensional structures are called *configurations*.

OPTICAL ACTIVITY AND CHIRALITY

Any material that rotates the plane of polarized light is said to be *optically active*. If a pure compound is optically active, the molecule is nonsuperimposable on its mirror image. If a molecule is superimposable on its mirror image, the compound does not rotate the plane of polarized light; it is *optically inactive*. The property of nonsuperimposability of an object on its mirror image is called *chirality*. If a molecule is not superimposable on its mirror image, it is *chiral*. If it is superimposable on its mirror image, it is *achiral*. The relationship between optical activity and chirality is absolute. No exceptions are known, and many thousands of cases have been found in accord with it (however, see p. 98). The ultimate criterion, then, for optical activity is chirality (nonsuperimposability on the mirror image). This is both a necessary and a sufficient condition.[3] This fact has been used as evidence for the structure determination of many compounds, and historically the tetrahedral nature of carbon was deduced from the hypothesis that the relationship might be true.

If a molecule is nonsuperimposable on its mirror image, the mirror image must be a different molecule, since superimposability is the same as identity. In each case of optical activity of a pure compound there are two and only two isomers, called *enantiomers* (sometimes *enantiomorphs*), which differ in structure only in the left- and right-handedness of

---

[1]For books on this subject, see Sokolov *Introduction to Theoretical Stereochemistry*; Gordon and Breach: New York, 1991; Bassindale *The Third Dimension in Organic Chemistry*; Wiley: New York, 1984; Nógrádi, *Stereochemistry*; Pergamon: Elmsford, NY, 1981; Kagan *Organic Stereochemistry*; Wiley: New York, 1979; Testa *Principles of Organic Stereochemistry*; Marcel Dekker: New York, 1979; Izumi; Tai *Stereo-Differentiating Reactions;* Academic Press: New York, Kodansha Ltd.: Tokyo, 1977; Natta; Farina *Stereochemistry*; Harper and Row: New York, 1972; Eliel *Elements of Stereochemistry*; Wiley: New York, 1969; Mislow *Introduction to Stereochemistry*; W. A. Benjamin: New York, 1965. Three excellent treatments of stereochemistry that, though not recent, contain much that is valid and useful, are Eliel *Stereochemistry of Carbon Compounds*; McGraw-Hill: New York, 1962; Wheland *Advanced Organic Chemistry*, 3rd ed.; Wiley: New York, 1960, pp. 195-514; Shriner; Adams; Marvel, in Gilman *Advanced Organic Chemistry*; vol. 1, 2nd ed.; Wiley: New York, 1943, pp. 214-488. For a historical treatment, see Ramsay *Stereochemistry*; Heyden & Son, Ltd.: London, 1981.

[2]The IUPAC 1974 Recommendations, Section E, Fundamental Stereochemistry, give definitions for most of the terms used in this chapter, as well as rules for naming the various kinds of stereoisomers. They can be found in *Pure Appl. Chem.* **1976**, *45*, 13-30 and in *Nomenclature of Organic Chemistry*; Pergamon: Elmsford, NY, 1979 (the "Blue Book").

[3]For a discussion of the conditions for optical activity in liquids and crystals, see O'Loane *Chem. Rev.* **1980**, *80*, 41-61. For a discussion of chirality as applied to molecules, see Quack *Angew. Chem. Int. Ed. Engl.* **1989**, *28*, 571-586 [*Angew. Chem. 101*, 588-604].

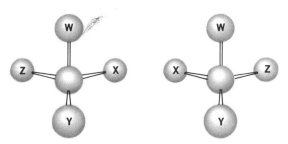

FIGURE 4.1 Enantiomers.

their orientations (Figure 4.1). Enantiomers have identical[4] physical and chemical properties except in two important respects:

1. They rotate the plane of polarized light in opposite directions, though in equal amounts. The isomer that rotates the plane to the left (counterclockwise) is called the *levo isomer* and is designated (−), while the one that rotates the plane to the right (clockwise) is called the *dextro isomer* and is designated (+). Because they differ in this property they are often called *optical antipodes*.

2. They react at different rates with other chiral compounds. These rates may be so close together that the distinction is practically useless, or they may be so far apart that one enantiomer undergoes the reaction at a convenient rate while the other does not react at all. This is the reason that many compounds are biologically active while their enantiomers are not. Enantiomers react at the same rate with achiral compounds.[4a]

In general, it may be said that enantiomers have identical properties in a symmetrical environment, but their properties may differ in an unsymmetrical environment.[5] Besides the important differences previously noted, enantiomers may react at different rates with achiral molecules if an optically active *catalyst* is present; they may have different solubilities in an optically active *solvent*; they may have different indexes of refraction or absorption spectra *when examined with circularly polarized light*, etc. In most cases these differences are too small to be useful and are often too small to be measured.

Although pure compounds are always optically active if they are composed of chiral molecules, mixtures of equal amounts of enantiomers are optically inactive since the equal and opposite rotations cancel. Such mixtures are called *racemic mixtures*[6] or *racemates*.[7] Their properties are not always the same as those of the individual enantiomers. The properties in the gaseous or liquid state or in solution usually are the same, since such a mixture is nearly ideal, but properties involving the solid state,[8] such as melting points, solubilities, and heats of fusion, are often different. Thus racemic tartaric acid has a melting point of 204-206°C and a solubility in water at 20°C of 206 g/liter, while for the (+) or the (−)

[4]Interactions between electrons, nucleons, and certain components of nucleons (e.g., bosons), called *weak interactions*, violate parity; that is, mirror image interactions do not have the same energy. It has been contended that interactions of this sort cause one of a pair of enantiomers to be (slightly) more stable than the other. See Tranter *J. Chem. Soc., Chem. Commun.* **1986**, 60, and references cited therein. See also Ref. 13.

[4a]For a reported exception, see Hata *Chem. Lett.* **1991**, 155.

[5]For a review of discriminating interactions between chiral molecules, see Craig; Mellor *Top. Curr. Chem.* **1976**, *63*, 1-48.

[6]Strictly speaking, the term *racemic mixture* applies only when the mixture of molecules is present as separate solid phases, but in this book we shall use this expression to refer to any equimolar mixture of enantiomeric molecules, liquid, solid, gaseous, or in solution.

[7]For a monograph on the properties of racemates and their resolution, see Jacques; Collet; Wilen *Enantiomers, Racemates, and Resolutions*; Wiley: New York, 1981.

[8]For a discussion, see Wynberg; Lorand *J. Org. Chem.* **1981**, *46*, 2538 and references cited therein.

enantiomer, the corresponding figures are 170°C and 1390 g/liter. The separation of a racemic mixture into its two optically active components is called *resolution*. The presence of optical activity always proves that a given compound is chiral, but its absence does not prove that the compound is achiral. A compound that is optically inactive may be achiral, or it may be a racemic mixture (see also p. 98).

Dependence of Rotation on Conditions of Measurement

The *amount* of rotation α is not a constant for a given enantiomer; it depends on the length of the sample vessel, the temperature, the solvent[9] and concentration (for solutions), the pressure (for gases), and the wavelength of light.[10] Of course, rotations determined for the same compound under the same conditions are identical. The length of the vessel and the concentration or pressure determine the number of molecules in the path of the beam and α is linear with this. Therefore, a number is defined, called the *specific rotation* $[\alpha]$, which is

$$[\alpha] = \frac{\alpha}{lc} \text{ for solutions} \qquad [\alpha] = \frac{\alpha}{ld} \text{ for pure compounds}$$

where α is the observed rotation, l is the cell length in decimeters, c is the concentration in grams per milliliter, and d is the density in the same units. The specific rotation is usually given along with the temperature and wavelength, in this manner: $[\alpha]_{546}^{25}$. These conditions must be duplicated for comparison of rotations, since there is no way to put them into a simple formula. The expression $[\alpha]_D$ means that the rotation was measured with sodium D light; i.e., $\lambda = 589$ nm. The molar rotation $[M]_\lambda^t$ is the specific rotation times the molecular weight divided by 100.

It must be emphasized that although the value of α changes with conditions, the molecular structure is unchanged. This is true even when the changes in conditions are sufficient to change not only the amount of rotation but even the direction. Thus one of the enantiomers of aspartic acid, when dissolved in water, has $[\alpha]_D$ equal to $+4.36°$ at 20°C and $-1.86°$ at 90°C, though the molecular structure is unchanged. A consequence of such cases is that there is a temperature at which there is *no* rotation (in this case 75°C). Of course, the other enantiomer exhibits opposite behavior. Other cases are known in which the direction of rotation is reversed by changes in wavelength, solvent, and even concentration.[11] In theory, there should be no change in $[\alpha]$ with concentration, since this is taken into account in the formula, but associations, dissociations, and solute–solvent interactions often cause nonlinear behavior. For example, $[\alpha]_D^{24}$ for $(-)$-2-ethyl-2-methylsuccinic acid in $CHCl_3$ is $-5.0°$ at $c = 16.5$, $-0.7°$ at $c = 10.6$, $+1.7°$ at $c = 8.5$, and $+18.9°$ at $c = 2.2$.[12]

What Kinds of Molecules Display Optical Activity?

Although the ultimate criterion is, of course, nonsuperimposability on the mirror image (chirality), other tests may be used that are simpler to apply but not always accurate. One such test is the presence of a *plane of symmetry*.[13] A plane of symmetry[14] (also called a

[9]A good example is found in Kumata; Furukawa; Fueno *Bull. Chem. Soc. Jpn.* **1970**, *43*, 3920.

[10]For a review of polarimetry, see Lyle; Lyle, in Morrison, Ref. 88, vol. 1, pp. 13-27.

[11]For examples, see Shriner; Adams; Marvel, Ref. 1, pp. 291-301.

[12]Krow; Hill *Chem. Commun.* **1968**, 430.

[13]For a theoretical discussion of the relationship between symmetry and chirality, including parity violation (Ref. 4), see Barron *Chem. Soc. Rev.* **1986**, *15*, 189-223.

[14]The definitions of plane, center, and alternating axis of symmetry are taken from Eliel *Elements of Stereochemistry*, Ref. 1, pp. 6,7. See also Lemière; Alderweireldt *J. Org. Chem.* **1980**, *45*, 4175.

mirror plane) is a plane passing through an object such that the part on one side of the plane is the exact reflection of the part on the other side (the plane acting as a mirror). *Compounds possessing such a plane are always optically inactive*, but there are a few cases known in which compounds lack a plane of symmetry and are nevertheless inactive. Such compounds possess a *center of symmetry*, such as in α-truxillic acid, or an *alternating axis of symmetry* as in **1**.[15] A center of symmetry[14] is a point within an object such that a straight

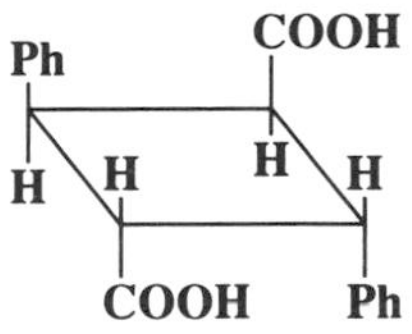

α-truxillic acid 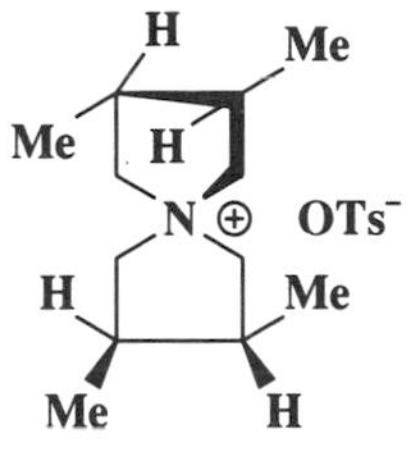   **1**

line drawn from any part or element of the object to the center and extended an equal distance on the other side encounters an equal part or element. An alternating axis of symmetry[14] of order *n* is an axis such that when an object containing such an axis is rotated by $360°/n$ about the axis and then reflection is effected across a plane at right angles to the axis, a new object is obtained that is indistinguishable from the original one. Compounds that lack an alternating axis of symmetry are always chiral.

A molecule that contains just one *chiral carbon atom* (defined as a carbon atom connected to four different groups; also called an *asymmetric carbon atom*) is always chiral and hence optically active.[16] As seen in Figure 4.1, such a molecule cannot have a plane of symmetry, whatever the identity of W, X, Y, and Z, as long as they are all different. However, the presence of a chiral carbon is neither a necessary nor a sufficient condition for optical activity, since optical activity may be present in molecules with no chiral atom[17] and since some molecules with two or more chiral carbon atoms are superimposable on their mirror images and hence inactive. Examples of such compounds will be discussed subsequently.

Optically active compounds may be classified into several categories.

1. *Compounds with a chiral carbon atom.* If there is only one such atom, the molecule must be optically active. This is so no matter how slight the differences are among the four groups. For example, optical activity is present in

$$BrCH_2CH_2CH_2CH_2CH_2CH_2-\underset{\underset{CH_3}{|}}{CH}-CH_2CH_2CH_2CH_2CH_2Br$$

Optical activity has been detected even in cases[18] such as 1-butanol-1-*d*, where one group is hydrogen and another deuterium.[19]

$$CH_3CH_2CH_2-\underset{\underset{D}{|}}{\overset{\overset{H}{|}}{C}}-OH$$

[15]McCasland; Proskow *J. Am. Chem. Soc.* **1955**, *77*, 4688.
[16]For discussions of the relationship between a chiral carbon and chirality, see Mislow; Siegel *J. Am. Chem. Soc.* **1984**, *106*, 3319; Brand; Fisher *J. Chem. Educ.* **1987**, *64*, 1035.
[17]For a review of such molecules, see Nakazaki *Top. Stereochem.* **1984**, *15*, 199-251.
[18]For reviews of compounds where chirality is due to the presence of deuterium or tritium, see Barth; Djerassi *Tetrahedron* **1981**, *24*, 4123-4142; Arigoni; Eliel *Top. Stereochem.* **1969**, *4*, 127-243; Verbit *Prog. Phys. Org. Chem.* **1970**, *7*, 51-127. For a review of compounds containing chiral methyl groups, see Floss; Tsai; Woodard *Top. Stereochem.* **1984**, *15*, 253-321.
[19]Streitwieser; Schaeffer *J. Am. Chem. Soc.* **1956**, *78*, 5597.

However, the amount of rotation is greatly dependent on the nature of the four groups, in general increasing with increasing differences in polarizabilities among the groups. Alkyl groups have very similar polarizabilities[20] and the optical activity of 5-ethyl-5-propylundecane is too low to be measureable at any wavelength between 280 and 580 nm.[21]

2. *Compounds with other quadrivalent chiral atoms.*[22]   Any molecule containing an atom that has four bonds pointing to the corners of a tetrahedron will be optically active if the four groups are different. Among atoms in this category are Si,[23] Ge, Sn,[24] and N (in quaternary salts or N-oxides).[25] In sulfones the sulfur bonds tetrahedrally, but since two of the groups are always oxygen, no chirality normally results. However, the preparation[26] of

an optically active sulfone (**2**) in which one oxygen is ^{16}O and the other ^{18}O illustrates the point that slight differences in groups are all that is necessary. This has been taken even further with the preparation of the ester **3**, both enantiomers of which have been prepared.[27] Optically active chiral phosphates **4** have similarly been made.[28]

3. *Compounds with tervalent chiral atoms.*   Atoms with pyramidal bonding[29] might be expected to give rise to optical activity if the atom is connected to three different groups, since the unshared pair of electrons is analogous to a fourth group, necessarily different from the others. For example, a secondary or tertiary amine where X, Y, and Z are different

would be expected to be chiral and thus resolvable. Many attempts have been made to resolve such compounds, but until 1968 all of them failed because of *pyramidal inversion*, which is a rapid oscillation of the unshared pair from one side of the XYZ plane to the other, thus converting the molecule into its enantiomer.[30] For ammonia there are 2×10^{11}

[20]For a discussion of optical activity in paraffins, see Brewster *Tetrahedron* **1974**, *30*, 1807.

[21]Wynberg; Hekkert; Houbiers; Bosch *J. Am. Chem. Soc.* **1965**, *87*, 2635; Wynberg and Hulshof *Tetrahedron* **1974**, *30*, 1775; Ten Hoeve; Wynberg *J. Org. Chem.* **1980**, *45*, 2754.

[22]For reviews of compounds with asymmetric atoms other than carbon, see Aylett *Prog. Stereochem.* **1969**, *4*, 213-217; Belloli *J. Chem. Educ.* **1969**, *46*, 640-644; Sokolov; Reutov *Russ. Chem. Rev.* **1965**, *34*, 1-12.

[23]For reviews of stereochemistry of silicon, see Corriu; Guérin; Moreau, in Patai; Rappoport *The Chemistry of Organic Silicon Compounds*; pt. 1: Wiley: New York, 1989, pp. 305-370, *Top. Stereochem.* **1984**, *15*, 43-198; Maryanoff; Maryanoff, in Morrison, Ref. 88, vol. 4, pp. 355-374.

[24]For reviews of the stereochemistry of Sn and Ge compounds, see Gielen *Top. Curr. Chem.* **1982**, *104*, 57-105; *Top. Stereochem.* **1981**, *12*, 217-251.

[25]For a review, see Davis; Jenkins, in Morrison, Ref. 88, vol. 4, pp. 313-353. The first resolution of a quaternary ammonium salt of this type was done by Pope; Peachey *J. Chem. Soc.* **1899**, *75*, 1127.

[26]Stirling *J. Chem. Soc.* **1963**, 5741; Sabol; Andersen *J. Am. Chem. Soc.* **1969**, *91*, 3603; Annunziata; Cinquini; Colonna *J. Chem. Soc., Perkin Trans. 1* **1972**, 2057.

[27]Lowe; Salamone *J. Chem. Soc., Chem. Commun.* **1984**, 466; Lowe; Parratt *J. Chem. Soc., Chem. Commun.* **1985**, 1075.

[28]Abbott; Jones; Weinman; Knowles *J. Am. Chem. Soc.* **1978**, *100*, 2558; Cullis; Lowe *J. Chem. Soc., Chem. Commun.* **1978**, 512. For a review, see Lowe *Acc. Chem. Res.* **1983**, *16*, 244-251.

[29]For a review of the stereochemistry at trivalent nitrogen, see Raban and Greenblatt, in Patai *The Chemistry of Functional Groups, Supplement F*, pt. 1, Wiley: New York, 1982, pp. 53-83.

[30]For reviews of the mechanism of, and the effect of structure on, pyramidal inversion, see Lambert *Top. Stereochem.* **1971**, *6*, 19-105; Rauk; Allen; Mislow *Angew. Chem. Int. Ed. Engl.* **1970**, *9*, 400-414 [*Angew. Chem. 82*, 453-468]; Lehn *Fortschr. Chem. Forsch.* **1970**, *15*, 311-377; Mislow *Pure Appl. Chem.* **1968**, *25*, 549-562.

inversions every second. The inversion is less rapid in substituted ammonias[31] (amines, amides, etc.). Two types of nitrogen atom invert particularly slowly, namely, a nitrogen atom in a three-membered ring and a nitrogen atom connected to another atom bearing an unshared pair. Even in such compounds, however, for many years pyramidal inversion proved too rapid to permit isolation of separate isomers. This goal was accomplished[25] only when compounds were synthesized in which both features are combined: a nitrogen atom in a three-membered ring connected to an atom containing an unshared pair. For example, the two isomers of 1-chloro-2-methylaziridine (**5** and **6**) were separated and do not interconvert at room temperature.[32] In suitable cases this barrier to inversion can result in compounds that are optically active solely because of a chiral tervalent nitrogen atom. For

example, **7** has been resolved into its separate enantiomers.[32a] Note that in this case too, the nitrogen is connected to an atom with an unshared pair. Conformational stability has also been demonstrated for oxaziridines,[33] diaziridines, e.g., **8**,[34] triaziridines, e.g. **9**,[35] and

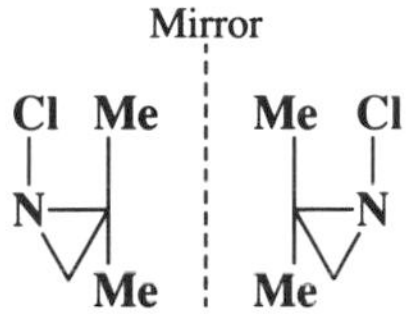

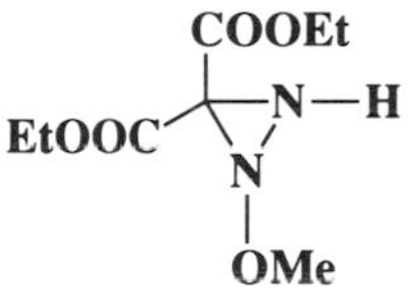

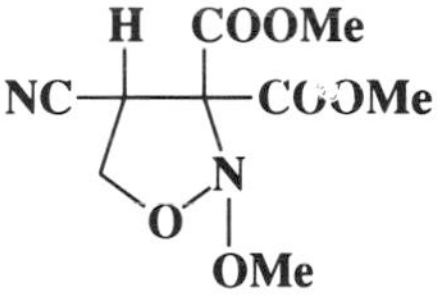

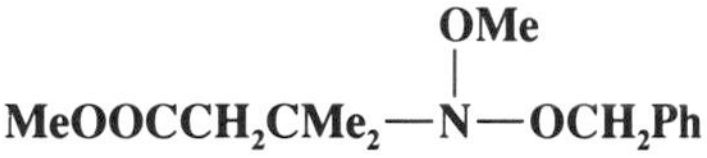

---

[31]For example, see Andose; Lehn; Mislow; Wagner *J. Am. Chem. Soc.* **1970**, *92*, 4050; Stackhouse; Baechler; Mislow *Tetrahedron Lett.* **1971**, 3437, 3441.

[32]Brois *J. Am. Chem. Soc.* **1968**, *90*, 506, 508. See also Shustov; Kadorkina; Kostyanovsky; Rauk *J. Am. Chem. Soc.* **1988**, *110*, 1719; Lehn; Wagner *Chem. Commun.* **1968**, 148; Felix; Eschenmoser *Angew. Chem. Int. Ed. Engl.* **1968**, *7*, 224 [*Angew. Chem. 80*, 197]; Kostyanovsky; Samoilova; Chervin *Bull. Acad. Sci. USSR, Div. Chem. Sci.* **1968**, 2705, *Tetrahedron Lett.* **1969**, 719. For a review, see Brois *Trans. N.Y. Acad. Sci.* **1969**, *31*, 931-951.

[32a]Schurig; Leyrer *Tetrahedron: Asymmetry* **1990**, *1*, 865.

[33]Boyd *Tetrahedron Lett.* **1968**, 4561; Boyd; Spratt; Jerina *J. Chem. Soc. C* **1969**, 2650; Montanari; Moretti; Torre *Chem. Commun.* **1968**, 1694, **1969**, 1086; Bucciarelli; Forni; Moretti; Torre; Prosyanik; Kostyanovsky *J. Chem. Soc., Chem. Commun.* **1985**, 998; Bucciarelli; Forni; Moretti; Torre; Brückner; Malpezzi *J. Chem. Soc., Perkin Trans. 2* **1988**, 1595. See also Mannschreck; Linss; Seitz *Liebigs Ann. Chem.* **1969**, 727, 224; Forni; Moretti; Torre; Brückner; Malpezzi; Di Silvestro *J. Chem. Soc., Perkin Trans. 2* **1984**, 791. For a review of oxaziridines, see Schmitz *Adv. Heterocycl. Chem.* **1979**, *24*, 63-107.

[34]Rudchenko; D'yachenko; Zolotoi; Atovmyan; Chervin; Kostyanovsky *Tetrahedron* **1982**, *38*, 961; Shustov; Denisenko; Chervin; Asfandiarov; Kostyanovsky *Tetrahedron* **1985**, *41*, 5719 and references cited in these papers. See also Mannschreck; Radeglia; Gründemann; Ohme *Chem. Ber.* **1967**, *100*, 1778.

[35]Hilpert; Hoesch; Dreiding *Helv. Chim. Acta* **1985**, *68*, 1691, **1987**, *70*, 381.

1,2-oxazolidines, e.g., **10,**[36] even though in this case the ring is five-membered. However, note that the nitrogen atom in **10** is connected to two oxygen atoms.

Another compound in which nitrogen is connected to two oxygens is **11**. In this case there is no ring at all, but it has been resolved into (+) and (−) enantiomers ($[\alpha]_D^{20} \approx \pm 3°$).[37] This compound and several similar ones reported in the same paper are the first examples of compounds whose optical activity is solely due to an acyclic tervalent chiral nitrogen atom. However, **11** is not optically stable and racemizes at 20°C with a half-life of 1.22 hr. A similar compound (**11,** with OCH$_2$Ph replaced by OEt) has a longer half-life— 37.5 hr at 20°C.

In molecules in which the nitrogen atom is at a bridgehead, pyramidal inversion is of course prevented. Such molecules, if chiral, can be resolved even without the presence of the two structural features noted above. For example, optically active **12** (Tröger's base)

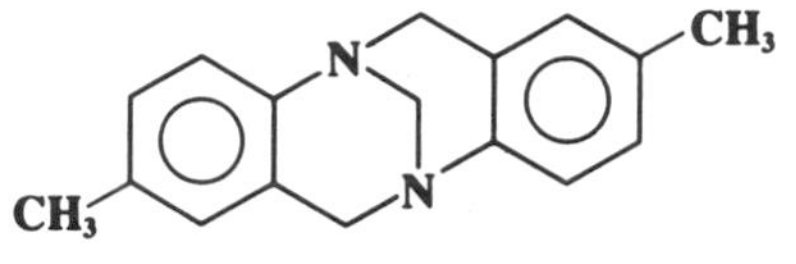

12    13

has been prepared.[38] Phosphorus inverts more slowly and arsenic still more slowly.[39] Non-bridgehead phosphorus,[40] arsenic, and antimony compounds have also been resolved, e.g., **13**.[41] Sulfur exhibits pyramidal bonding in sulfoxides, sulfinic esters, sulfonium salts, and sulfites. Examples of each of these have been resolved.[42] An interesting example is

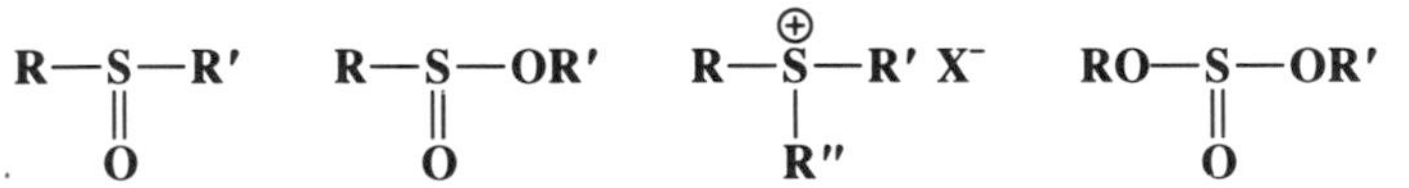

(+)-Ph^{12}CH$_2$SO^{13}CH$_2$Ph, a sulfoxide in which the two alkyl groups differ only in ^{12}C versus ^{13}C but which has $[\alpha]_{280} = +0.71°$.[43]

4. Suitably substituted adamantanes. Adamantanes bearing four different substituents at the bridgehead positions are chiral and optically active and **14,** for example, has been

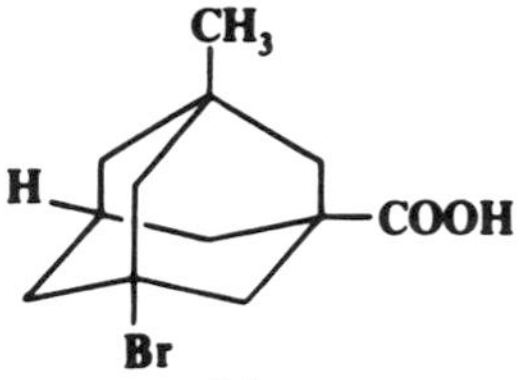

14

[36]Müller; Eschenmoser *Helv. Chim. Acta* **1969,** *52,* 1823; Dobler; Dunitz; Hawley *Helv. Chim. Acta* **1969,** *52,* 1831.

[37]Kostyanovsky; Rudchenko; Shtamburg; Chervin; Nasibov *Tetrahedron* **1981,** *37,* 4245; Kostyanovsky; Rudchenko *Doklad. Chem.* **1982,** *263,* 121. See also Rudchenko; Ignatov; Chervin; Kostyanovsky *Tetrahedron* **1988,** *44,* 2233.

[38]Prelog; Wieland *Helv. Chim. Acta* **1944,** *27,* 1127.

[39]For reviews, see Yambushev; Savin *Russ. Chem. Rev.* **1979,** *48,* 582-595; Gallagher; Jenkins *Top. Stereochem.* **1968,** *3,* 1-96; Kamai; Usacheva *Russ. Chem. Rev.* **1966,** *35,* 601-613.

[40]For a review of chiral phosphorus compounds, see Valentine, in Morrison, Ref. 88, vol. 4, pp. 263-312.

[41]Horner; Fuchs *Tetrahedron Lett.* **1962,** 203.

[42]For reviews of chiral organosulfur compounds, see Andersen, in Patai; Rappoport; Stirling *The Chemistry of Sulphones and Sulphoxides*; Wiley: New York, 1988, pp. 55-94; and in Stirling *The Chemistry of the Sulphonium Group*, pt. 1, Wiley: New York, 1981, pp. 229-312; Barbachyn; Johnson, in Morrison, Ref. 88, vol. 4, pp. 227-261; Cinquini; Cozzi; Montanari, in Bernardi; Csizmadia; Mangini *Organic Sulfur Chemistry*; Elsevier: New York, 1985, pp. 355-407; Mikołajczyk; Drabowicz *Top. Stereochem.* **1982,** *13,* 333-468.

[43]Andersen; Colonna; and Stirling, *J. Chem. Soc., Chem. Commun.* **1973,** 645.

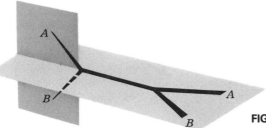

FIGURE 4.2 Perpendicular disymmetric planes.

resolved.[44] This type of molecule is a kind of expanded tetrahedron and has the same symmetry properties as any other tetrahedron.

5. *Restricted rotation giving rise to perpendicular disymmetric planes.* Certain compounds that do not contain asymmetric atoms are nevertheless chiral because they contain a structure that can be schematically represented as in Figure 4.2. For these compounds we can draw two perpendicular planes neither of which can be bisected by a plane of symmetry. If either plane could be so bisected, the molecule would be superimposable on its mirror image, since such a plane would be a plane of symmetry. These points will be illustrated by examples.

Biphenyls containing four large groups in the ortho positions cannot freely rotate about the central bond because of steric hindrance.[45] In such compounds the two rings are in perpendicular planes. If either ring is symmetrically substituted, the molecule has a plane of symmetry. For example, consider:

Mirror

Ring B is symmetrically substituted. A plane drawn perpendicular to ring B contains all the atoms and groups in ring A; hence it is a plane of symmetry and the compound is achiral. On the other hand, consider:

Mirror

There is no plane of symmetry and the molecule is chiral; many such compounds have been resolved. Note that groups in the para position cannot cause lack of symmetry. Isomers that

---

[44]Hamill; McKervey *Chem. Commun.* **1969,** 864; Applequist; Rivers; Applequist *J. Am. Chem. Soc.* **1969,** *91,* 5705.

[45]When the two rings of a biphenyl are connected by a bridge, rotation is of course impossible. For a review of such compounds, see Hall *Prog. Stereochem.* **1969,** *4,* 1-42.

can be separated only because rotation about single bonds is prevented or greatly slowed are called *atropisomers*.[46]

It is not always necessary for four large ortho groups to be present in order for rotation to be prevented. Compounds with three and even two groups, if large enough, can have hindered rotation and, if suitably substituted, can be resolved. An example is biphenyl-2,2'-bissulfonic acid.[47] In some cases, the groups may be large enough to slow rotation greatly but not to prevent it completely. In such cases, optically active compounds can be prepared that slowly racemize on standing. Thus, **15** loses its optical activity with a half-life

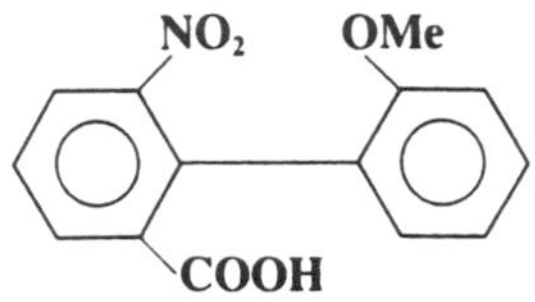

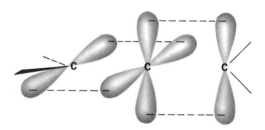

NO₂   OMe

COOH

15

of 9.4 min in ethanol at 25°C.[48] Compounds with greater rotational stability can often be racemized if higher temperatures are used to supply the energy necessary to force the groups past each other.[49] Many analogous cases are known, where optical activity arises from hindered rotation of other types of aromatic ring, e.g., binaphthyls, bipyrryls, etc.

In allenes the central carbon is *sp*-bonded. The remaining two *p* orbitals are perpendicular to each other and each overlaps with the *p* orbital of one adjacent carbon atom, forcing the

two remaining bonds of each carbon into perpendicular planes. Thus allenes fall into the category represented by Figure 4.2:

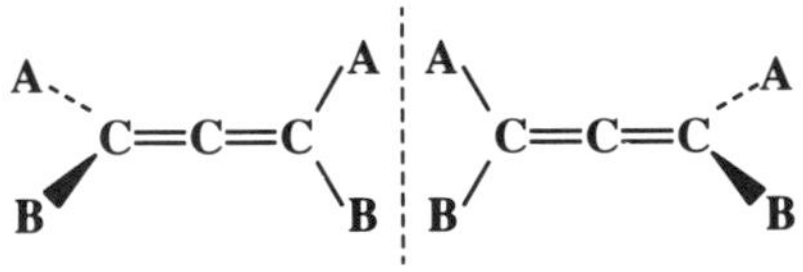

$$A\!\cdot\!\cdot\!\!\diagdown \quad \diagup A \quad A\diagdown \quad \quad \cdot\!\cdot\!A$$
C=C=C   C=C=C
B   B   B   B

Mirror

[46]For a review, see Ōki *Top. Stereochem.* **1983,** *14,* 1-81.
[47]Patterson; Adams *J. Am. Chem. Soc.* **1935,** *57,* 762.
[48]Stoughton; Adams *J. Am. Chem. Soc.* **1932,** *54,* 4426.
[49]For a monograph on the detection and measurement of restricted rotations, see Ōki *Applications of Dynamic NMR Spectroscopy to Organic Chemistry*; VCH: New York, 1985.

Like biphenyls, allenes are chiral only if both sides are unsymmetrically substituted.[50] For example,

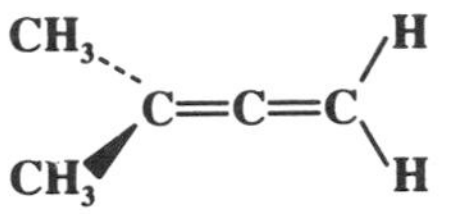

Inactive 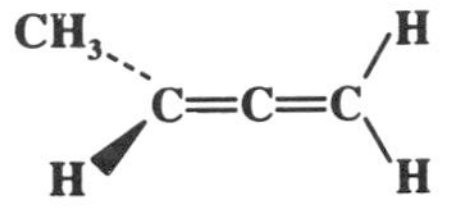   Inactive 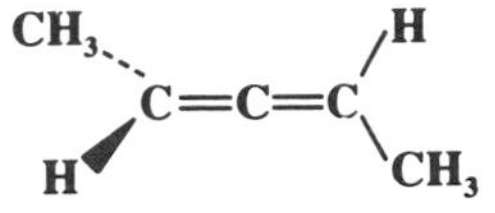   Active

These cases are completely different from the cis-trans isomerism of compounds with one double bond (p. 127). In the latter cases the four groups are all in one plane, the isomers are not enantiomers, and neither is chiral, while in allenes the groups are in two perpendicular planes and the isomers are a pair of optically active enantiomers.

When three, five, or any *odd* number of cumulative double bonds exist, orbital overlap causes the four groups to occupy one plane and cis–trans isomerism is observed. When four, six, or any *even* number of cumulative double bonds exist, the situation is analogous to that in the allenes and optical activity is possible. **16** has been resolved.[51]

Among other types of compounds that contain the system illustrated in Figure 4.2 and

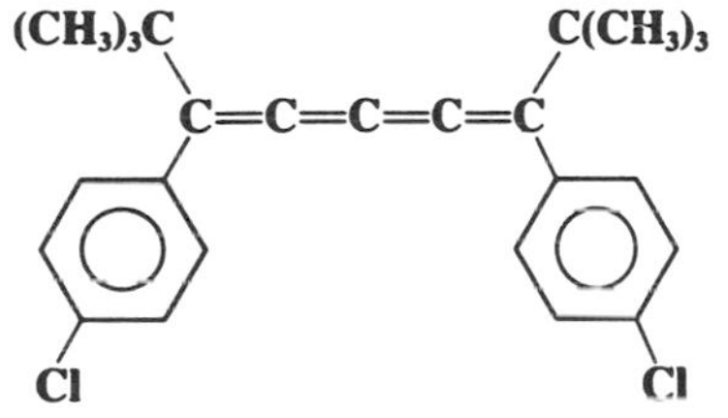

16

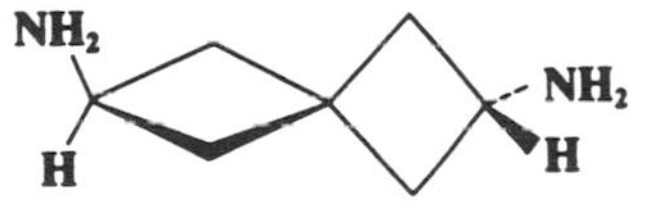

17 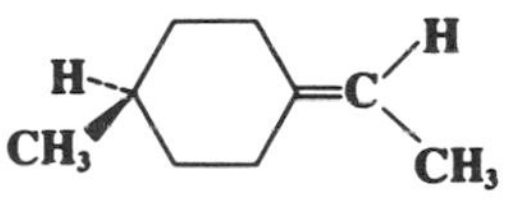   **18**

that are similarly chiral if both sides are dissymmetric are spiranes, e.g., **17**, and compounds with exocyclic double bonds, e.g., **18**.

6. Chirality due to a helical shape.[52]   Several compounds have been prepared that are chiral because they have a shape that is actually helical and can therefore be left- or right-handed in orientation. The entire molecule is usually less than one full turn of the helix, but this does not alter the possibility of left- and right-handedness. An example is hexahelicene,[53] in which one side of the molecule must lie above the other because of

[50]For reviews of allene chirality, see Runge, in Landor *The Chemistry of the Allenes*, vol. 3; Academic Press: New York, 1982, pp. 579-678, and in Patai *The Chemistry of Ketenes, Allenes, and Related Compounds*, pt. 1; Wiley: New York, 1980, pp. 99-154; Rossi; Diversi *Synthesis* **1973**, 25-36.
[51]Nakagawa; Shingū; Naemura *Tetrahedron Lett.* **1961**, 802.
[52]For a review, see Meurer; Vögtle *Top. Curr. Chem.* **1985**, *127*, 1-76. See also Ref. 54.
[53]Newman; Lednicer *J. Am. Chem. Soc.* **1956**, *78*, 4765. Optically active heptahelicene has also been prepared, as have higher helicenes: Flammang-Barbieux; Nasielski; Martin *Tetrahedron Lett.* **1967**, 743; Martin; Baes *Tetrahedron* **1975**, *31*, 2135; Bernstein; Calvin; Buchardt, *J. Am. Chem. Soc.* **1972**, *94*, 494, **1973**, *95*, 527; Defay; Martin *Bull. Soc. Chim. Belg.* **1984**, *93*, 313. Even pentahelicene is crowded enough to be chiral: Goedicke; Stegemeyer *Tetrahedron Lett.* **1970**, 937; Bestmann; Roth *Chem. Ber.* **1974**, *107*, 2923.

crowding.[54] Others are *trans*-cyclooctene (see also p. 128), in which the carbon chain must lie above the double bond on one side and below it on the other,[55] and suitably substituted

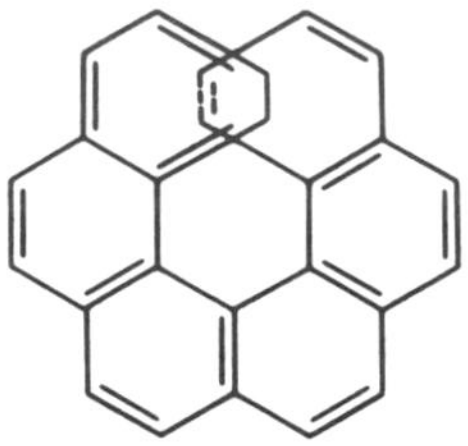

hexahelicene 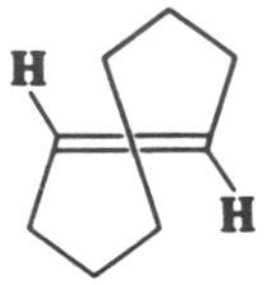   *trans*-cyclooctene 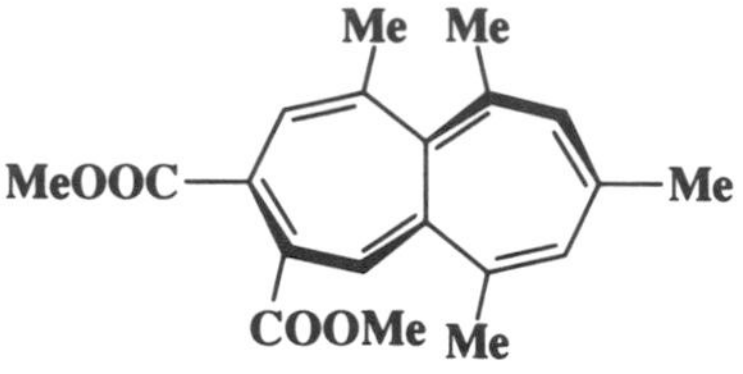   **19**

heptalenes. Heptalene itself is not planar (p. 49), and its twisted structure makes it chiral, but the enantiomers rapidly interconvert. However, bulky substituents can hinder the interconversion and several such compounds, including **19**, have been resolved.[56]

   7. *Optical activity caused by restricted rotation of other types.* Substituted paracyclophanes may be optically active and **20**, for example, has been resolved.[57] In this case chirality

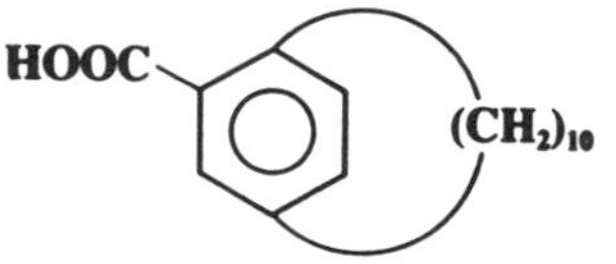

20

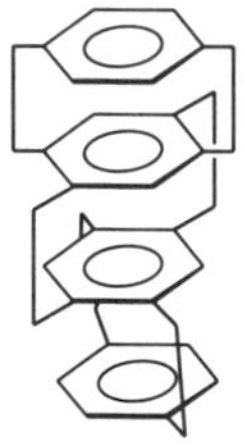

21

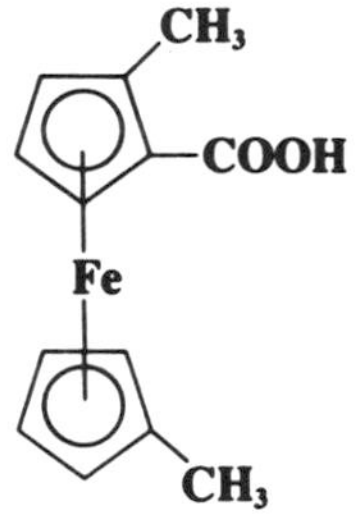

22 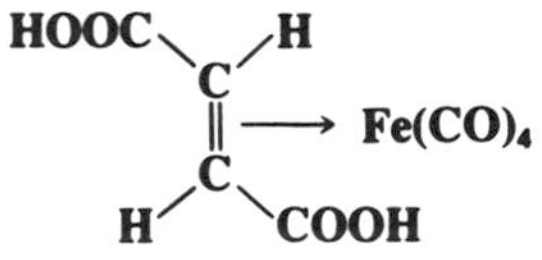   **23**

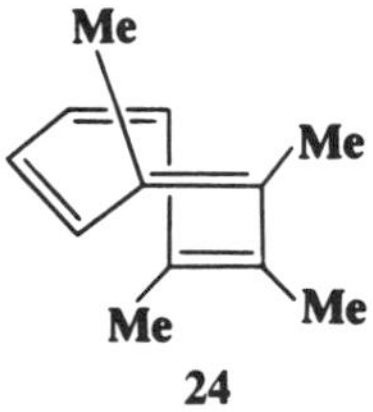

24

[54]For reviews of the helicenes, see Laarhoven; Prinsen *Top. Curr. Chem.* **1984**, *125*, 63-130; Martin *Angew. Chem. Int. Ed. Engl.* **1974**, *13*, 649-660 [*Angew. Chem. 86*, 727-738].
[55]Cope; Ganellin; Johnson; Van Auken; Winkler *J. Am. Chem. Soc.* **1963**, *85*, 3276. Also see Levin; Hoffmann *J. Am. Chem. Soc.* **1972**, *94*, 3446.
[56]Hafner; Knaup; Lindner; Flöter *Angew. Chem. Int. Ed. Engl.* **1985**, *24*, 212 [*Angew. Chem. 97*, 209]; Bernhard; Brügger; Daly; Schönholzer; Weber; Hansen *Helv. Chim. Acta* **1985**, *68*, 415.
[57]Blomquist; Stahl; Meinwald; Smith *J. Org. Chem.* **1961**, *26*, 1687. For a review of chiral cyclophanes and related molecules, see Schlögl *Top. Curr. Chem.* **1984**, *125*, 27-62.

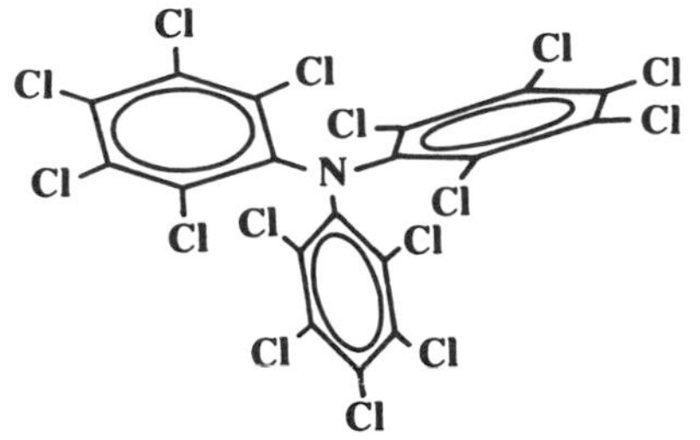

perchlorotriphenylamine

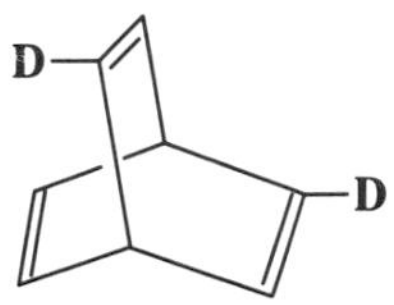

25

results because the benzene ring cannot rotate in such a way that the carboxyl group goes through the alicyclic ring. Many chiral layered cyclophanes, e.g. **21,** have been prepared.[58] Metallocenes substituted with at least two different groups on one ring are also chiral.[59] Several hundred such compounds have been resolved, one example being **22.** Chirality is also found in other metallic complexes of suitable geometry.[60] For example, fumaric acid–iron tetracarbonyl (**23**) has been resolved.[61] 1,2,3,4-Tetramethylcyclooctatetraene (**24**) is also chiral.[62] This molecule, which exists in the tub form (p. 57), has neither a plane nor an alternating axis of symmetry. Another compound that is chiral solely because of hindered rotation is the propellor-shaped perchlorotriphenylamine, which has been resolved.[63] The 2,5-dideuterio derivative (**25**) of barrelene is chiral, though the parent hydrocarbon and the monodeuterio derivative are not. **25** has been prepared in optically active form[64] and is another case where chirality is due to isotopic substitution.

The main molecular chain in compound **26** has the form of a Möbius strip (see Figure 15.8).[65] This molecule has no chiral carbons, nor does it have a rigid shape, but it too has

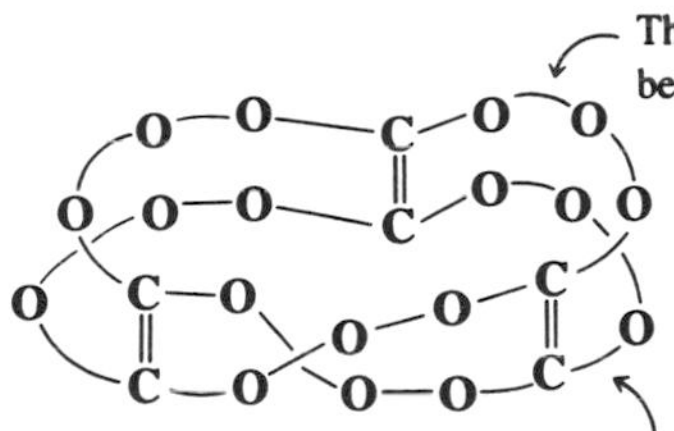

There is a CH_2—CH_2 group between each of the two oxygens

26

There is a CH_2 group between each C and O

neither a plane nor an alternating axis of symmetry. **26** has been synthesized and has, in fact, been shown to be chiral.[66] Another interesting type of chirality has been proposed,

[58]Nakazaki; Yamamoto; Tanaka; Kametani *J. Org. Chem.* **1977,** *42,* 287.

[59]For reviews on the stereochemistry of metallocenes, see Schlögl *J. Organomet. Chem.* **1986,** *300,* 219-248, *Top. Stereochem.* **1967,** *1,* 39-91, *Pure Appl. Chem.* **1970,** *23,* 413-432.

[60]For reviews of such complexes, see Paiaro *Organomet. Chem. Rev., Sect. A* **1970,** *6,* 319-335.

[61]Paiaro; Palumbo; Musco; Panunzi *Tetrahedron Lett.* **1965,** 1067; also see Paiaro; Panunzi *J. Am. Chem. Soc.* **1964,** *86,* 5148.

[62]Paquette; Gardlik; Johnson; McCullough *J. Am. Chem. Soc.* **1980,** *102,* 5026.

[63]Hayes; Nagumo; Blount; Mislow *J. Am. Chem. Soc.* **1980,** *102,* 2773; Okamoto; Yashima; Hatada; Mislow *J. Org. Chem.* **1984,** *49,* 557.

[64]Lightner; Paquette; Chayangkoon; Lin; Peterson *J. Org. Chem.* **1988,** *53,* 1969.

[65]For a review of chirality in Möbius-strip molecules catenanes, and knots, see Walba *Tetrahedron* **1985,** *41,* 3161-3212.

[66]Walba; Richards; Haltiwanger *J. Am. Chem. Soc.* **1982,** *104,* 3219.

though no example is yet known.[67] Rings containing 50 or more members should be able to exist as knots (**27**). Such a knot would be nonsuperimposable on its mirror image. A compound of this type has been synthesized (by the copper ion method discussed in **9-65**),

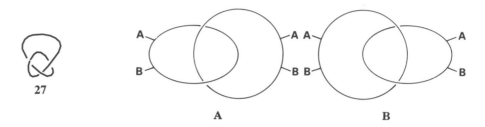

though not yet resolved.[68] Catenanes and rotaxanes (see p. 91) can also be chiral if suitably substituted.[69] For example, **A** and **B** are nonsuperimposable mirror images.

Creation of a Chiral Center

Any structural feature of a molecule that gives rise to optical activity may be called a *chiral center*. In many reactions a new chiral center is created, e.g.,

$$CH_3CH_2COOH + Br_2 \xrightarrow{P} CH_3CHBrCOOH$$

If the reagents and reaction conditions are all symmetrical, the product must be a racemic mixture. No optically active material can be created if all starting materials and conditions are optically inactive.[70] This statement also holds when one begins with a racemic mixture. Thus racemic 2-butanol, treated with HBr, must give racemic 2-bromobutane.

The Fischer Projection

For a thorough understanding of stereochemistry it is useful to examine molecular models (like those depicted in Figure 4.1). However, this is not feasible when writing on paper or a blackboard. In 1891 Emil Fischer greatly served the interests of chemistry by inventing the Fischer projection, a method of representing tetrahedral carbons on paper. By this convention, the model is held so that the two bonds in front of the paper are horizontal and those behind the paper are vertical.

[67]Frisch; Wasserman *J. Am. Chem. Soc.* **1961,** *83,* 3789.

[68]Dietrich-Buchecker; Guilhem; Pascard; Sauvage *Angew. Chem. Int. Ed. Engl.* **1990,** *29,* 1154 [*Angew. Chem. 102,* 1202].

[69]For a discussion of the stereochemistry of these compounds, see Schill *Catenanes, Rotaxanes, and Knots*; Academic Press: New York, 1971, pp. 11-18.

[70]There is one exception to this statement. In a very few cases racemic mixtures may crystalize from solution in such a way that all the (+) molecules go into one crystal and the (−) molecules into another. If one of the crystals crystallizes before the other, a rapid filtration results in optically active material. For a discussion, see Pincock; Wilson *J. Chem. Educ.* **1973,** *50,* 455.

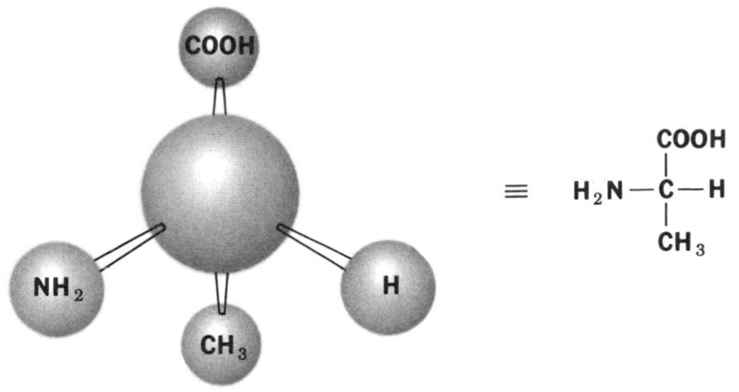

$$\equiv\quad H_2N-\underset{\underset{CH_3}{|}}{\overset{\overset{COOH}{|}}{C}}-H$$

In order to obtain proper results with these formulas, it should be remembered that they are projections and must be treated differently from the models in testing for superimposability. Every plane is superimposable on its mirror image; hence with these formulas there must be added the restriction that they may not be taken out of the plane of the blackboard or paper. Also they may not be rotated 90°, though 180° rotation is permissible:

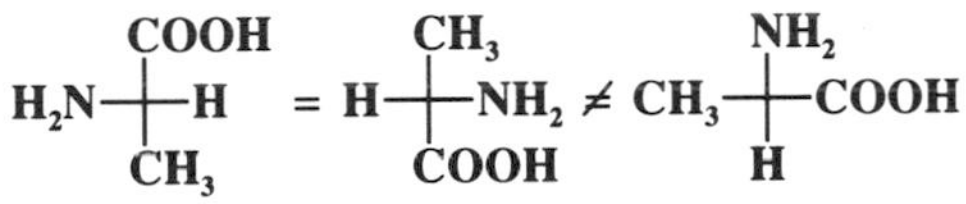

$$H_2N \overset{COOH}{\underset{CH_3}{+}} H \;=\; H \overset{CH_3}{\underset{COOH}{+}} NH_2 \;\ne\; CH_3 \overset{NH_2}{\underset{H}{+}} COOH$$

It is also permissible to keep any one group fixed and to rotate the other three clockwise or counterclockwise (because this can be done with models):

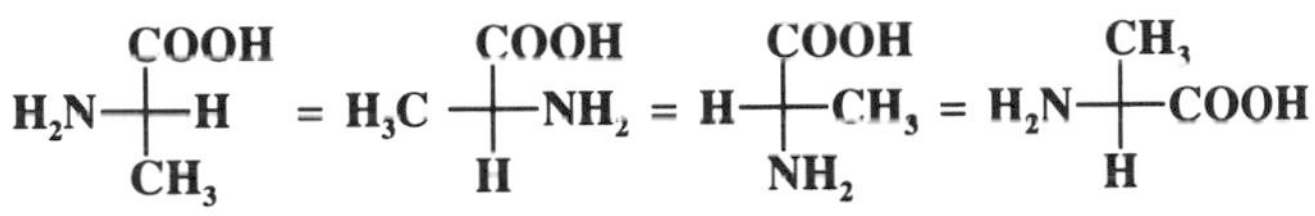

$$H_2N \overset{COOH}{\underset{CH_3}{+}} H \;=\; H_3C \overset{COOH}{\underset{H}{+}} NH_2 \;=\; H \overset{COOH}{\underset{NH_2}{+}} CH_3 \;=\; H_2N \overset{CH_3}{\underset{H}{+}} COOH$$

However, the *interchange* of any two groups results in the conversion of an enantiomer into its mirror image (this applies to models as well as to the Fischer projections).

With these restrictions Fischer projections may be used instead of models to test whether a molecule containing asymmetric carbons is superimposable on its mirror image. However, there are no such conventions for molecules whose chirality arises from anything other than chiral atoms; when such molecules are examined on paper, three-dimensional pictures must be used. With models or three-dimensional pictures there are no restrictions about the plane of the paper.

Absolute Configuration

Suppose we have two test tubes, one containing ($-$)-lactic acid and the other the ($+$) enantiomer. One test tube contains **28** and the other **29**. How do we know which is which? Chemists in the early part of this century pondered this problem and decided that they could

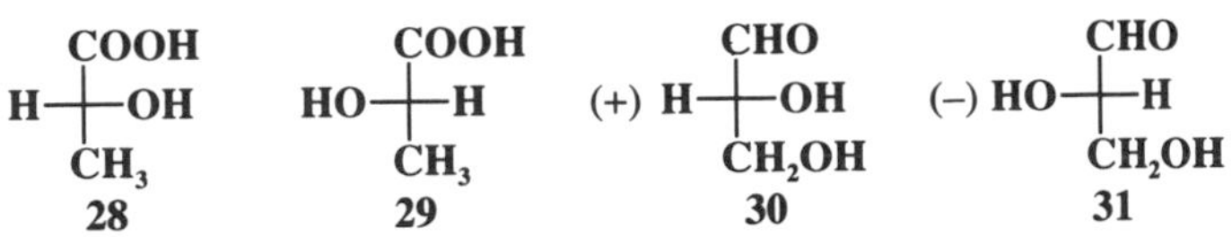

$$H \overset{COOH}{\underset{CH_3}{+}} OH \qquad HO \overset{COOH}{\underset{CH_3}{+}} H \qquad (+)\; H \overset{CHO}{\underset{CH_2OH}{+}} OH \qquad (-)\; HO \overset{CHO}{\underset{CH_2OH}{+}} H$$

　　　　28　　　　　　　**29**　　　　　　　**30**　　　　　　　**31**

not know—for lactic acid or any other compound. Therefore Rosanoff proposed that one compound be chosen as a standard and a configuration be arbitrarily assigned to it. The compound chosen was glyceraldehyde because of its relationship to the sugars. The (+) isomer was assigned the configuration shown in **30** and given the label D. The (−) isomer, designated to be **31,** was given the label L. Once a standard was chosen, other compounds could then be related to it. For example, (+)-glyceraldehyde, oxidized with mercuric oxide, gives (−)-glyceric acid:

$$(+)\;\; H\!\!-\!\!\underset{\underset{\displaystyle CH_2OH}{|}}{\overset{\overset{\displaystyle CHO}{|}}{|}}\!\!-\!\!OH \;\;\xrightarrow{\;HgO\;}\;\; (-)\;\; H\!\!-\!\!\underset{\underset{\displaystyle CH_2OH}{|}}{\overset{\overset{\displaystyle COOH}{|}}{|}}\!\!-\!\!OH$$

Since it is highly improbable that the configuration at the central carbon changed, it can be concluded that (−)-glyceric acid has the same configuration as (+)-glyceraldehyde and therefore (−)-glyceric acid is also called D. This example emphasizes that molecules with the same configuration need not rotate the plane of polarized light in the same direction. This fact should not surprise us when we remember that the same compound can rotate the plane in opposite directions under different conditions.

Once the configuration of the glyceric acids was known (in relation to the glyceraldehydes), it was then possible to relate other compounds to either of these, and each time a new compound was related, others could be related to *it*. In this way many thousands of compounds were related, indirectly, to D- or L-glyceraldehyde, and it was determined that **28,** which has the D configuration, is the isomer that rotates the plane of polarized light to the left. Even compounds without asymmetric atoms, such as biphenyls and allenes, have been placed in the D or L series.[71] When a compound has been placed in the D or L series, its *absolute configuration* is said to be known.[72]

In 1951 it became possible to determine whether Rosanoff's guess was right. Ordinary x-ray crystallography cannot distinguish between a D and a L isomer, but by use of a special technique, Bijvoet was able to examine sodium rubidium tartrate and found that Rosanoff had made the correct choice.[73] It was perhaps historically fitting that the first true absolute configuration should have been determined on a salt of tartaric acid, since Pasteur made his great discoveries on another salt of this acid.

In spite of the former widespread use of D and L to denote absolute configuration, the method is not without faults. The designation of a particular enantiomer as D or L can depend on the compounds to which it is related. Examples are known where an enantiomer can, by five or six steps, be related to a known D compound, and by five or six other steps, be related to the L enantiomer of the same compound. In a case of this sort, an arbitrary choice of D or L must be used. Because of this and other flaws, the DL system is no longer used, except for certain groups of compounds such as carbohydrates and amino acids.

---

[71]The use of small *d* and *l* is now discouraged, since some authors used it for rotation, and some for configuration. However, a racemic mixture is still a *dl* mixture, since there is no ambiguity here.

[72]For lists of absolute configurations of thousands of compounds, with references, mostly expressed as (*R*) or (*S*) rather than D or L, see Klyne; Buckingham *Atlas of Stereochemistry*, 2nd ed., 2 vols.; Oxford University Press: Oxford, 1978; Jacques; Gros; Bourcier; Brienne; Toullec *Absolute Configurations* (vol. 4 of Kagan *Stereochemistry*), Georg Thieme Publishers: Stuttgart, 1977.

[73]Bijvoet; Peerdeman; van Bommel *Nature* **1951,** *168,* 271. For a list of organic structures whose absolute configurations have been determined by this method, see Allen; Rogers *Chem. Commun.* **1966,** 838; Allen; Neidle; Rogers *Chem. Commun.* **1968,** 308, **1969,** 452; Neidle; Rogers; Allen *J. Chem. Soc. C* **1970,** 2340.

The Cahn–Ingold–Prelog System

The system that has replaced the DL system is the *Cahn–Ingold–Prelog* system, in which the four groups on an asymmetric carbon are ranked according to a set of sequence rules.[74] For our purposes we confine ourselves to only a few of these rules, which are sufficient to deal with the vast majority of chiral compounds.

1. Substituents are listed in order of decreasing atomic number of the atom directly joined to the carbon.

2. Where two or more of the atoms connected to the asymmetric carbon are the same, the atomic number of the second atom determines the order. For example, in the molecule Me_2CH—$CHBr$—CH_2OH, the CH_2OH group takes precedence over the Me_2CH group because oxygen has a higher atomic number than carbon. Note that this is so even though there are two carbons in Me_2CH and only one oxygen in CH_2OH. If two or more atoms connected to the second atom are the same, the third atom determines the precedence, etc.

3. All atoms except hydrogen are formally given a valence of 4. Where the actual valence is less (as in nitrogen, oxygen, or a carbanion), phantom atoms (designated by a subscript $_0$) are used to bring the valence up to four. These phantom atoms are assigned an atomic number of zero and necessarily rank lowest. Thus the ligand —$\overset{\oplus}{N}HMe_2$ ranks higher than —NMe_2.

4. A tritium atom takes precedence over deuterium, which in turn takes precedence over ordinary hydrogen. Similarly, any higher isotope (such as ^{14}C) takes precedence over any lower one.

5. Double and triple bonds are counted as if they were split into two or three single bonds, respectively, as in the examples in Table 4.1 (note the treatment of the phenyl group).

TABLE 4.1 How four common groups are treated in the Cahn–Ingold–Prelog system

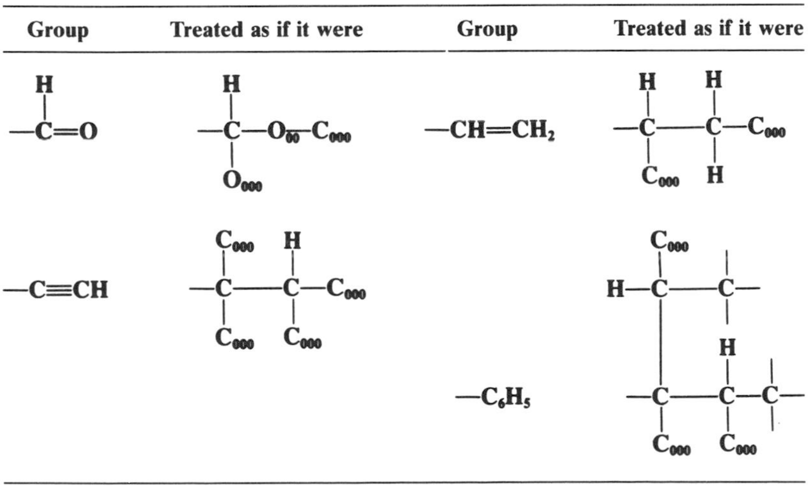

[74]For descriptions of the system and sets of sequence rules, see Ref. 2; Cahn; Ingold; Prelog *Angew. Chem. Int. Ed. Engl.* **1966**, *5*, 385-415 [*Angew. Chem. 78*, 413-447]; Cahn *J. Chem. Educ.* **1964**, *41*, 116; Fernelius; Loening; Adams *J. Chem. Educ.* **1974**, *51*, 735. See also Prelog and Helmchen *Angew. Chem., Int. Ed. Engl.* **1982**, *21*, 567-583 [*Angew. Chem. 94*, 614-631].

Note that in a C=C double bond, the two carbon atoms are *each* regarded as being connected to two carbon atoms and that one of the latter is counted as having three phantom substituents.

As an exercise, we shall compare the four groups in Table 4.1. The first atoms are connected, respectively, to (H, O, O), (H, C, C), (C, C, C), and (C, C, C). That is enough to establish that —CHO ranks first and —CH=CH₂ last, since even one oxygen outranks three carbons and three carbons outrank two carbons and a hydrogen. To classify the remaining two groups we must proceed further along the chains. We note that —C₆H₅ has two of its (C, C, C) carbons connected to (C, C, H), while the third is (₀₀₀) and is thus preferred to —C≡CH, which has only one (C, C, H) and two (₀₀₀)s.

By application of the above rules, some groups in descending order of precedence are COOH, COPh, COMe, CHO, CH(OH)₂, *o*-tolyl, *m*-tolyl, *p*-tolyl, phenyl, C≡CH, *t*-butyl, cyclohexyl, vinyl, isopropyl, benzyl, neopentyl, allyl, *n*-pentyl, ethyl, methyl, deuterium, and hydrogen. Thus the four groups of glyceraldehyde are arranged in the sequence: OH, CHO, CH₂OH, H.

Once the order is determined, the molecule is held so that the lowest group in the sequence is pointed away from the viewer. Then if the other groups, in the order listed, are oriented clockwise, the molecule is designated *R*, and if counterclockwise, *S*. For glyceraldehyde, the (+) enantiomer is *R*:

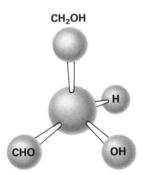

Note that when a compound is written in the Fischer projection, the configuration can easily be determined without constructing the model.[75] If the lowest-ranking group is either at the top or the bottom (because these are the two positions pointing away from the viewer), the *R* configuration is present if the other three groups in descending order are clockwise, e.g.,

$$\begin{array}{cccc}
\text{H} & \text{OH} & \text{H} & \text{OH} \\
\text{HCO}\!-\!\!\!\overset{|}{\underset{|}{\text{C}}}\!\!\!-\!\text{CH}_2\text{OH} & \text{HOCH}_2\!-\!\!\!\overset{|}{\underset{|}{\text{C}}}\!\!\!-\!\text{CHO} & \text{HOCH}_2\!-\!\!\!\overset{|}{\underset{|}{\text{C}}}\!\!\!-\!\text{CHO} & \text{HCO}\!-\!\!\!\overset{|}{\underset{|}{\text{C}}}\!\!\!-\!\text{CH}_2\text{OH} \\
\text{OH} & \text{H} & \text{OH} & \text{H}
\end{array}$$

(*R*)-Glyceraldehyde   (*S*)-Glyceraldehyde

[75]For a discussion of how to determine *R* or *S* from other types of formula, see Eliel *J. Chem. Educ.* **1985,** *62,* 223.

If the lowest-ranking group is not at the top or bottom, one can simply interchange it with the top or bottom group, bearing in mind that in so doing, one is inverting the configuration, e.g.:

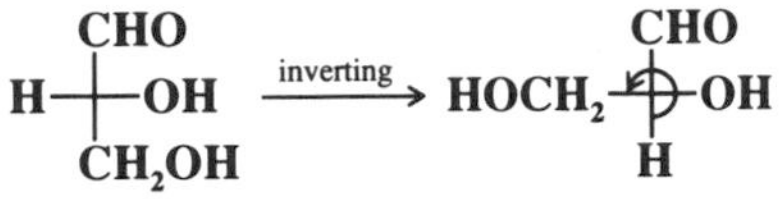

$$
\begin{array}{ccc}
\text{CHO} & & \text{CHO} \\
\text{H}\!-\!\!\!|\!-\!\text{OH} & \xrightarrow{\text{inverting}} & \text{HOCH}_2\!-\!\!\!\langle\!\!|\!\!\rangle\!-\!\text{OH} \\
\text{CH}_2\text{OH} & & \text{H}
\end{array}
$$

(*S*)-Glyceraldehyde

Therefore the original compound was (*R*)-glyceraldehyde.

The Cahn–Ingold–Prelog system is unambiguous and easily applicable in most cases. Whether to call an enantiomer *R* or *S* does not depend on correlations, but the configuration must be known before the system can be applied and this does depend on correlations. The Cahn–Ingold–Prelog system has also been extended to chiral compounds that do not contain chiral atoms.[76]

Methods of Determining Configuration[77]

In all the methods,[78] it is necessary to relate the compound of unknown configuration to another whose configuration is known. The most important methods of doing this are:

1. Conversion of the unknown to, or formation of the unknown from, a compound of known configuration without disturbing the chiral center. See the glyceraldehyde–glyceric acid example above (p. 108). Since the chiral center was not disturbed, the unknown obviously has the same configuration as the known. This does not necessarily mean that if the known is *R*, the unknown is also *R*. This will be so if the sequence is not disturbed but not otherwise. For example, when (*R*)-1-bromo-2-butanol is reduced to 2-butanol without dis-

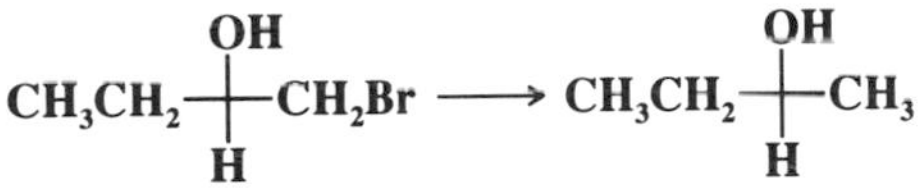

$$
\begin{array}{ccc}
\text{OH} & & \text{OH} \\
\text{CH}_3\text{CH}_2\!-\!\!\!|\!-\!\text{CH}_2\text{Br} & \longrightarrow & \text{CH}_3\text{CH}_2\!-\!\!\!|\!-\!\text{CH}_3 \\
\text{H} & & \text{H} \\
R & & S
\end{array}
$$

turbing the chiral center, the product is the *S* isomer, even though the configuration is unchanged, because CH_3CH_2 ranks lower than $BrCH_2$ but higher than CH_3.

2. Conversion at the chiral center if the mechanism is known. Thus, the SN2 mechanism proceeds with inversion of configuration at an asymmetric carbon (see p. 294). It was by a series of such transformations that lactic acid was related to alanine:

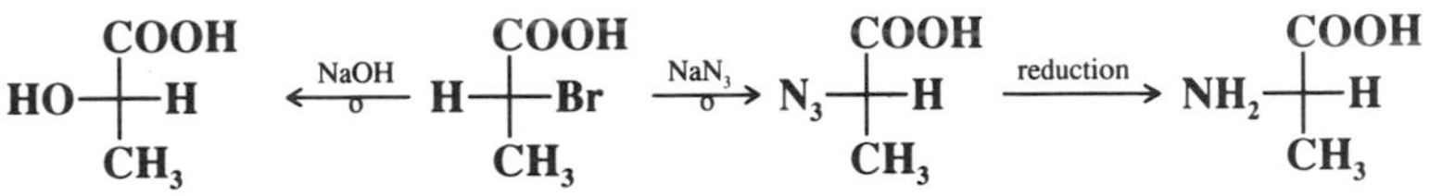

$$
\begin{array}{cccc}
\text{COOH} & \text{COOH} & \text{COOH} & \text{COOH} \\
\text{HO}\!-\!\!|\!-\!\text{H} & \text{H}\!-\!\!|\!-\!\text{Br} & \text{N}_3\!-\!\!|\!-\!\text{H} & \text{NH}_2\!-\!\!|\!-\!\text{H} \\
\text{CH}_3 & \text{CH}_3 & \text{CH}_3 & \text{CH}_3
\end{array}
$$

(*S*)-(+)-Lactic acid $\xleftarrow{\text{NaOH}}$ ... R $\xrightarrow{\text{NaN}_3}$ S $\xrightarrow{\text{reduction}}$ (*S*)-(+)-Alanine

See also the discussion on p. 295.

---

[76]For a discussion of these rules, as well as for a review of methods for establishing configurations of chiral compounds not containing chiral atoms, see Krow *Top. Stereochem.* **1970**, *5*, 31-68.

[77]For a monograph, see Kagan *Determination of Configuration by Chemical Methods* (vol. 3 of Kagan *Stereochemistry*); Georg Thieme Publishers: Stuttgart, 1977. For reviews, see Brewster, in Bentley; Kirby *Elucidation of Organic Structures by Physical and Chemical Methods*, 2nd ed. (vol. 4 of Weissberger *Techniques of Chemistry*), pt. 3; Wiley: New York, 1972, pp. 1-249; Klyne; Scopes *Prog. Stereochem.* **1969**, *4*, 97-166; Schlenk *Angew. Chem. Int. Ed. Engl.* **1965**, *4*, 139-145 [*Angew. Chem. 77*, 161-168]. For a review of absolute configuration of molecules in the crystalline state, see Addadi; Berkovitch-Yellin; Weissbuch; Lahav; Leiserowitz *Top. Stereochem.* **1986**, *16*, 1-85.

[78]Except the x-ray method of Bijvoet.

3. Biochemical methods. In a series of similar compounds, such as amino acids or certain types of steroids, a given enzyme will usually attack only molecules with one kind of configuration. If the enzyme attacks only the L form of eight amino acids, say, then attack on the unknown ninth amino acid will also be on the L form.

4. Optical comparison. It is sometimes possible to use the sign and extent of rotation to determine which isomer has which configuration. In a homologous series, the rotation usually changes gradually and in one direction. If the configurations of enough members of the series are known, the configurations of the missing ones can be determined by extrapolation. Also certain groups contribute more or less fixed amounts to the rotation of the parent molecule, especially when the parent is a rigid system such as a steroid.

5. The special x-ray method of Bijvoet gives direct answers and has been used in a number of cases.[73]

Other methods have also been used, including optical rotatory dispersion,[79] circular dichroism,[79] nmr, and asymmetric synthesis (see p. 118).

The Cause of Optical Activity

The question may be asked: Just why does a chiral molecule rotate the plane of polarized light? Theoretically, the answer to this question is known and in a greatly simplified form may be explained as follows.[80]

Whenever any light hits any molecule in a transparent material, the light is slowed because of interaction with the molecule. This phenomenon on a gross scale is responsible for the refraction of light and the decrease in velocity is proportional to the refractive index of the material. The extent of interaction depends on the polarizability of the molecule. Plane-polarized light may be regarded as being made up of two kinds of circularly polarized light. Circularly polarized light has the appearance (or would have, if one could see the wave) of a helix propagating around the axis of light motion, and one kind is a left-handed and the other a right-handed helix. As long as the plane-polarized light is passing through a symmetrical region, the two circularly polarized components travel at the same speed. However, a chiral molecule has a different polarizability depending on whether it is approached from the left or the right. One circularly polarized component approaches the molecule, so to speak, from the left and sees a different polarizability (hence on a gross scale, a different refractive index) than the other and is slowed to a different extent. This would seem to mean that the left- and right-handed circularly polarized components travel at different velocities, since each has been slowed to a different extent. However, it is not possible for two components of the same light to be traveling at different velocities. What actually takes place, therefore, is that the faster component "pulls" the other towards it, resulting in rotation of the plane. Empirical methods for the prediction of the sign and amount of rotation based on bond refractions and polarizabilities of groups in a molecule have been devised,[81] and have given fairly good results in many cases.

In liquids and gases the molecules are randomly oriented. A molecule that is optically inactive because it has a plane of symmetry will very seldom be oriented so that the plane

[79]See Ref. 191 for books and reviews on optical rotatory dispersion.

[80]For longer, nontheoretical discussions, see Eliel, *Stereochemistry of Carbon Compounds*, Ref. 1, pp. 398-412; Wheland, Ref. 1, pp. 204-211. For theoretical discussions, see Caldwell; Eyring *The Theory of Optical Activity* Wiley: New York, 1971; Buckingham; Stiles *Acc. Chem. Res.* **1974,** *7*, 258-264; Mason *Q. Rev., Chem. Soc.* **1963,** *17*, 20-66.

[81]Brewster *Top. Stereochem.* **1967,** *2*, 1-72, *J. Am. Chem. Soc.* **1959,** *81*, 5475, 5483, 5493; Davis; Jensen *J. Org. Chem.* **1970,** *35*, 3410; Jullien; Requin; Stahl-Larivière *Nouv. J. Chim.* **1979,** *3*, 91; Sathyanarayana; Stevens *J. Org. Chem.* **1987,** *52*, 3170; Wroblewski; Applequist; Takaya; Honzatko; Kim; Jacobson; Reitsma; Yeung; Verkade *J. Am. Chem. Soc.* **1988,** *110*, 4144.

of the polarized light coincides with the plane of symmetry. When it is so oriented, that particular molecule does not rotate the plane but all others not oriented in that manner do rotate the plane, even though the molecules are achiral. There is no net rotation because, the molecules being present in large number and randomly oriented, there will always be another molecule later on in the path of the light that is oriented exactly opposite and will rotate the plane back again. Even though nearly all molecules rotate the plane individually, the total rotation is zero. For chiral molecules, however (if there is no racemic mixture), no opposite orientation is present and there is a net rotation.

Molecules with More than One Chiral Center

When a molecule has two chiral centers, each has its own configuration and can be classified R or S by the Cahn–Ingold–Prelog method. There are a total of four isomers, since the first center may be R or S and so may the second. Since a molecule can have only one mirror image, only one of the other three can be the enantiomer of **A**. This is **B** (the mirror image

$$
\begin{array}{cccc}
& V & V & V & V \\
R\ U{-}\!\!\!\!+\!\!\!\!{-}W & S\ W{-}\!\!\!\!+\!\!\!\!{-}U & R\ U{-}\!\!\!\!+\!\!\!\!{-}W & S\ W{-}\!\!\!\!+\!\!\!\!{-}U \\
R\ X{-}\!\!\!\!+\!\!\!\!{-}Z & S\ Z{-}\!\!\!\!+\!\!\!\!{-}X & S\ Z{-}\!\!\!\!+\!\!\!\!{-}X & R\ X{-}\!\!\!\!+\!\!\!\!{-}Z \\
& Y & Y & Y & Y \\
& \mathbf{A} & \mathbf{B} & \mathbf{C} & \mathbf{D}
\end{array}
$$

of an R center is *always* an S center). **C** and **D** are a second pair of enantiomers and the relationship of **C** and **D** to **A** and **B** is designated by the term *diastereomer*. Diastereomers may be defined as *stereoisomers that are not enantiomers*. **C** and **D** being enantiomers, must have identical properties, except as noted on p. 95; the same is true for **A** and **B**. However, the properties of **A** and **B** are not identical with those of **C** and **D**. They have different melting points, boiling points, solubilities, reactivity, and all other physical, chemical, and spectral properties. The properties are usually *similar* but not *identical*. In particular, diastereomers have different specific rotations; indeed one diastereomer may be chiral and rotate the plane of polarized light while another may be achiral and not rotate at all (an example is presented below).

It is now possible to see why, as mentioned on p. 95, enantiomers react at different rates with other chiral molecules but at the same rate with achiral molecules. In the latter case, the activated complex formed from the R enantiomer and the other molecule is the mirror image of the activated complex formed from the S enantiomer and the other molecule. Since the two activated complexes are enantiomeric, their energies are the same and the rates of the reactions in which they are formed must be the same (see Chapter 6). However, when an R enantiomer reacts with a chiral molecule that has, say, the R configuration, the activated complex has two chiral centers with configurations R and R, while the activated complex formed from the S enantiomer has the configurations S and R. The two activated complexes are diastereomeric, do not have the same energies, and consequently are formed at different rates.

Although four is the maximum possible number of isomers when the compound has two chiral centers (chiral compounds without a chiral carbon, or with one chiral carbon and another type of chiral center, also follow the rules described here), some compounds have fewer. When the three groups on one chiral atom are the same as those on the other, one of the isomers (called a *meso* form) has a plane of symmetry and hence is optically inactive, even though it has two chiral carbons. Tartaric acid is a typical case. There are only three isomers of tartaric acid: a pair of enantiomers and an inactive meso form. For compounds

$$
\begin{array}{ccc}
\text{COOH} & \text{COOH} & \text{COOH} \\
\text{H}-\!\!\!-\text{OH} & \text{HO}-\!\!\!-\text{H} & \text{H}-\!\!\!-\text{OH} \\
\text{HO}-\!\!\!-\text{H} & \text{H}-\!\!\!-\text{OH} & \text{H}-\!\!\!-\text{OH} \\
\text{COOH} & \text{COOH} & \text{COOH} \\
\end{array}
$$

dl pair   meso
the three stereoisomers of tartaric acid

that have two chiral atoms, meso forms are found only where the four groups on one of the chiral atoms are the same as those on the other chiral atom.

In most cases with more than two chiral centers, the number of isomers can be calculated from the formula 2^n, where n is the number of chiral centers, although in some cases the actual number is less than this, owing to meso forms.[82] An interesting case is that of 2,3,4-pentanetriol (or any similar molecule). The middle carbon is not asymmetric when the 2- and 4-carbons are both R (or both S) but is asymmetric when one of them is R and the

$$
\begin{array}{cccc}
\text{CH}_3 & \text{CH}_3 & \text{CH}_3 & \text{CH}_3 \\
S \;\; \text{H}-\!\!\!-\text{OH} & S \;\; \text{H}-\!\!\!-\text{OH} & S \;\; \text{H}-\!\!\!-\text{OH} & R \;\; \text{HO}-\!\!\!-\text{H} \\
\text{H}-\!\!\!-\text{OH} & \text{HO}-\!\!\!-\text{H} & \text{HO}-\!\!\!-\text{H} & \text{H}-\!\!\!-\text{OH} \\
R \;\; \text{H}-\!\!\!-\text{OH} & R \;\; \text{H}-\!\!\!-\text{OH} & S \;\; \text{HO}-\!\!\!-\text{H} & R \;\; \text{H}-\!\!\!-\text{OH} \\
\text{CH}_3 & \text{CH}_3 & \text{CH}_3 & \text{CH}_3 \\
\end{array}
$$

meso   meso   dl pair

other S. Such a carbon is called a *pseudoasymmetric* carbon. In these cases there are four isomers: two meso forms and one *dl* pair. The student should satisfy himself or herself, remembering the rules governing the use of the Fischer projections, that these isomers are different, that the meso forms are superimposable on their mirror images, and that there are no other stereoisomers. Two diastereomers that have a different configuration at only one chiral center are called *epimers*.

In compounds with two or more chiral centers, the absolute configuration must be separately determined for each center. The usual procedure is to determine the configuration at one center by the methods discussed on pp. 111–112 and then to relate the configuration at that center to the others in the molecule. One method is x-ray crystallography, which, as previously noted, cannot be used to determine the absolute configuration at any chiral center but which does give relative configurations of all the chiral centers in a molecule and hence the absolute configurations of all once the first is independently determined. Other physical and chemical methods have also been used for this purpose.

The problem arises how to name the different stereoisomers of a compound when there are more than two.[2] Enantiomers are virtually always called by the same name, being distinguished by R and S or D and L or $(+)$ and $(-)$. In the early days of organic chemistry, it was customary to give each pair of enantiomers a different name or at least a different prefix (such as *epi-*, *peri-*, etc.). Thus the aldohexoses are called glucose, mannose, idose,

[82]For a method of generating all stereoisomers consistent with a given empirical formula, suitable for computer use, see Nourse; Carhart; Smith; Djerassi *J. Am. Chem. Soc.* **1979**, *101*, 1216; **1980**, *102*, 6289.

etc., although they are all 2,3,4,5,6-pentahydroxyhexanal (in their open-chain forms). This practice was partially due to lack of knowledge about which isomers had which configurations. Today it is customary to describe *each chiral position* separately as either *R* or *S* or, in special fields, to use other symbols. Thus, in the case of steroids, groups above the "plane" of the ring system are designated β, and those below it α. Solid lines are often used to depict β groups and dashed lines for α groups. An example is

1α-Chloro-5-cholesten-3β-ol

For many open-chain compounds prefixes are used that are derived from the names of the corresponding sugars and that describe the whole system rather than each chiral center separately. Two such common prefixes are *erythro*- and *threo*-, which are applied to systems

Erythro *dl* pair   Threo *dl* pair

containing two asymmetric carbons when two of the groups are the same and the third is different.[83] The erythro pair has the identical groups on the same side when drawn in the Fischer convention, and if Y were changed to Z, it would be meso. The threo pair has them on opposite sides, and if Y were changed to Z, it would still be a *dl* pair. Another system[84] for designating stereoisomers[85] uses the terms *syn* and *anti*. The "main chain" of the molecule is drawn in the common zig-zag manner. Then if two nonhydrogen substituents are on the same side of the plane defined by the main chain, the designation is syn; otherwise anti.

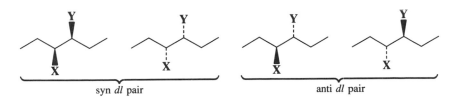

syn *dl* pair   anti *dl* pair

[83]For more general methods of designating diastereomers, see Carey; Kuehne *J. Org. Chem.* **1982,** *47,* 3811; Boguslavskaya *J. Org. Chem. USSR* **1986,** *22,* 1412; Seebach; Prelog *Angew. Chem. Int. Ed. Engl.* **1982,** *21,* 654-660 [*Angew. Chem. 94,* 696-702]; Brewster *J. Org. Chem.* **1986,** *51,* 4751. See also Tavernier *J. Chem. Educ.* **1986,** *63,* 511; Brook *J. Chem. Educ.* **1987,** *64,* 218.
[84]For still another system, see Seebach; Prelog, Ref. 83.
[85]Masamune; Kaiho; Garvey *J. Am. Chem. Soc.* **1982,** *104,* 5521.

Asymmetric Synthesis

Organic chemists often wish to synthesize a chiral compound in the form of a single enantiomer or diastereomer, rather than as a mixture of stereoisomers. There are two basic ways in which this can be done.[86] The first way, which is more common, is to begin with a single stereoisomer, and to use a synthesis that does not affect the chiral center (or centers), as in the glyceraldehyde–glyceric acid example on p. 108. The optically active starting compound can be obtained by a previous synthesis, or by resolution of a racemic mixture (p. 120), but it is often more convenient to obtain it from nature, since many compounds, such as amino acids, sugars, and steroids are present in nature in the form of a single enantiomer or diastereomer. These compounds are regarded as a *chiral pool*; that is, readily available compounds that can be used as starting materials.[87]

The other basic method is called *asymmetric synthesis*,[88] or *stereoselective synthesis*. As was mentioned before, optically active materials cannot be created from inactive starting materials and conditions; hence true asymmetric synthesis is impossible, except in the manner previously noted.[70] However, when a new chiral center is created, the two possible configurations need not be formed in equal amounts if anything is present that is not symmetric. We discuss asymmetric synthesis under four headings:

1. *Active substrate.* If a new chiral center is created in a molecule that is already optically active, the two diastereomers are not (except fortuitously) formed in equal amounts. The reason is that the direction of attack by the reagent is determined by the groups already there. For certain additions to the carbon–oxygen double bond of ketones containing an

---

[86]For a monograph that covers both ways, including a list of commercially available optically active starting compounds, see Morrison; Scott *Asymmetric Synthesis*, vol. 4; Academic Press: New York, 1984. For a monograph covering a more limited area, see Williams *Synthesis of Optically Active α-Amino Acids*; Pergamon: Elmsford, NY, 1989. For reviews on both ways, see Crosby *Tetrahedron* **1991**, *47*, 4789-4846; Mori *Tetrahedron* **1989**, *45*, 3233-3298.

[87]For books on the synthesis of optically active compounds starting from natural products, see Coppola; Schuster *Asymmetric Synthesis*; Wiley: New York, 1987 (amino acids as starting compounds); Hanessian *Total Synthesis of Natural Products: The Chiron Approach*; Pergamon: Elmsford, NY, 1983 (mostly carbohydrates as starting compounds). For reviews, see Jurczak; Pikul; Bauer *Tetrahedron* **1986**, *42*, 447-488; Hanessian *Aldrichimica Acta* **1989**, *22*, 3-15; Jurczak; Gotębiowski *Chem. Rev.* **1989**, *89*, 149-164.

[88]For a treatise on this subject, see Morrison *Asymmetric Synthesis*, 5 vols. [vol. 4 co-edited by Scott]; Academic Press: New York, 1983-1985. For books, see Nógrádi *Stereoselective Synthesis*; VCH: New York, 1986; Eliel; Otsuka *Asymmetric Reactions and Processes in Chemistry*; American Chemical Society: Washington, 1982; Morrison; Mosher *Asymmetric Organic Reactions*; Prentice-Hall: Englewood Cliffs, NJ, 1971, paperback reprint, American Chemical Society: Washington, 1976; Izumi; Tai, Ref. 1. For reviews, see Ward *Chem. Soc. Rev.* **1990**, *19*, 1-19; Whitesell *Chem. Rev.* **1989**, *89*, 1581-1590; Fujita; Nagao *Adv. Heterocycl. Chem.* **1989**, *45*, 1-36; Kochetkov; Belikov *Russ. Chem. Rev.* **1987**, *56*, 1045-1067; Oppolzer *Tetrahedron* **1987**, *43*, 1969-2004; Seebach; Imwinkelried; Weber *Mod. Synth. Methods* **1986**, *4*, 125-259; ApSimon; Collier *Tetrahedron* **1986**, *42*, 5157-5254; Mukaiyama; Asami *Top. Curr. Chem.* **1985**, *127*, 133-167; Martens *Top. Curr. Chem.* **1984**, *125*, 165-246; Duhamel; Duhamel; Launay; Plaquevent *Bull. Soc.Chim. Fr.* **1984**, II-421-II-430; Mosher; Morrison *Science* **1983**, *221*, 1013-1019; Schöllkopf *Top. Curr. Chem.*

asymmetric α carbon, *Cram's rule* predicts which diastereomer will predominate.[89] If the molecule is observed along its axis, it may be represented as in **34** (see p. 139), where S, M, and L stand for small, medium, and large, respectively. The oxygen of the carbonyl

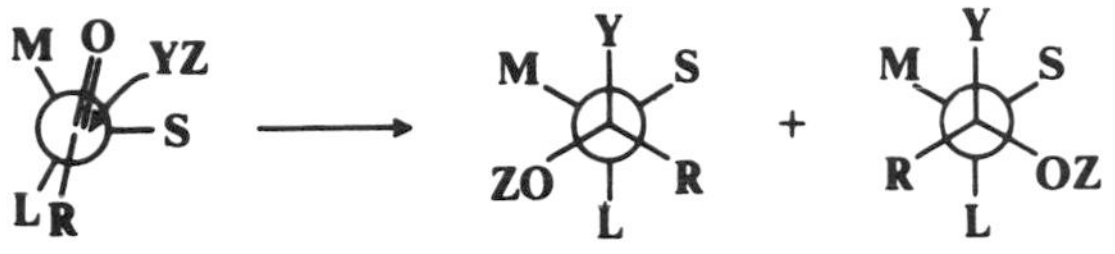

Major product   Minor product

orients itself between the small- and the medium-sized groups. The rule is that the incoming group preferentially attacks on the side of the plane containing the small group. By this rule, it can be predicted that **33** will be formed in larger amounts than **32**.

Many reactions of this type are known, in some of which the extent of favoritism approaches 100% (for an example see **2-11**).[90] The farther away the reaction site is from the chiral center, the less influence the latter has and the more equal the amounts of diastereomers formed.

In a special case of this type of asymmetric synthesis, a compound (**35**) with achiral molecules, but whose crystals are chiral, was converted by ultraviolet light to a single enantiomer of a chiral product (**36**).[91]

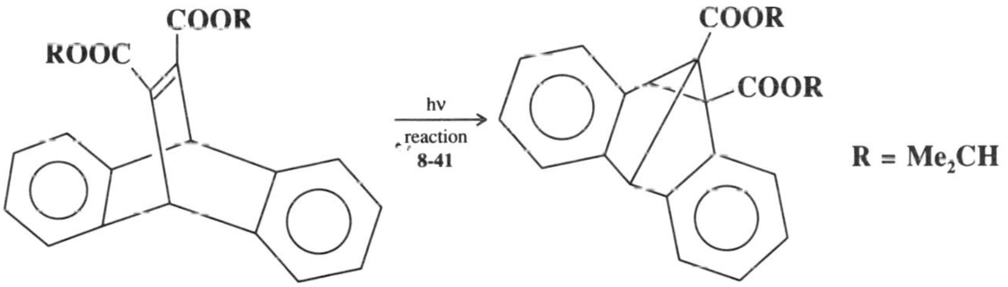

35   **36**
not chiral   optically active

1983, *109*, 65-84; Quinkert; Stark *Angew. Chem. Int. Ed. Engl.* **1983**, *22*, 637 655 [*Angew. Chem. 95*, 651-669]; Tramontini *Synthesis* **1982**, 605-644; Drauz; Kleeman; Martens *Angew. Chem. Int. Ed. Engl.* **1982**, *21*, 584-608 [*Angew. Chem. 94*, 590-613]; Wynberg *Recl. Trav. Chim. Pays-Bas* **1981**, *100*, 393-399; Bartlett *Tetrahedron* **1980**, *36*, 2-72; ApSimon; Seguin *Tetrahedron* **1979**, *35*, 2797-2842; Valentine; Scott *Synthesis* **1978**, 329-356; Scott; Valentine *Science* **1974**, *184*, 943-952; Kagan; Fiaud *Top. Stereochem.* **1978**, *10*, 175-285; ApSimon, in Bentley; Kirby, Ref. 77, pp. 251-408; Boyd; McKervey *Q. Rev., Chem. Soc* **1968**, *22*, 95-122; Goldberg *Sel. Org. Transform.* **1970**, *1*, 363-394; Klabunovskii; Levitina *Russ. Chem. Rev.* **1970**, *39*, 1035-1049; Inch *Synthesis* **1970**, 466-473; Mathieu; Weill-Raynal *Bull. Soc. Chim. Fr.* **1968**, 1211-1244; Amariglio; Amariglio; Duval *Ann. Chim. (Paris)* [14] **1968**, *3*, 5-25; Pracejus *Fortschr. Chem. Forsch.* **1967**, *8*, 493-553; Velluz; Valls; Mathieu *Angew. Chem. Int. Ed. Engl.* **1967**, *6*, 778-789 [*Angew. Chem. 79*, 774-785].

[89]Cram; Elhafez *J. Am. Chem. Soc.* **1952**, *74*, 5828; Cram; Kopecky *J. Am. Chem. Soc.* **1959**, *81*, 2748; Leitereg; Cram *J. Am. Chem. Soc.* **1968**, *90*, 4019. For reviews, see Ref. 5 in Chapter 16. For discussions, see Salem *J. Am. Chem. Soc.* **1973**, *95*, 94; Anh *Top. Curr. Chem* **1980**, *88*, 145-162, pp. 151-161; Eliel, in Morrison, Ref. 88, vol. 2, pp. 125-155.

[90]For other examples and references to earlier work, see Eliel, Ref. 89; Eliel; Koskimies; Lohri *J. Am. Chem. Soc.* **1978**, *100*, 1614; Still; McDonald *Tetrahedron Lett.* **1980**, *21*, 1031; Still; Schneider *Tetrahedron Lett.* **1980**, *21*, 1035.

[91]Evans; Garcia-Garibay; Omkaram; Scheffer; Trotter; Wireko *J. Am. Chem. Soc.* **1986**, *108*, 5648; Garcia-Garibay; Scheffer; Trotter; Wireko *Tetrahedron Lett.* **1987**, *28*, 4789. For an earlier example, see Penzien; Schmidt *Angew. Chem. Int. Ed. Engl.* **1969**, *8*, 608 [*Angew. Chem. 81*, 628].

It is often possible to convert an achiral compound to a chiral compound by (1) addition of a chiral group; (2) running an asymmetric synthesis, and (3) cleavage of the original chiral group. An example is conversion of the achiral 2-pentanone to the chiral 4-methyl-3-heptanone **39**.[92] In this case more than 99% of the product was the (S) enantiomer.

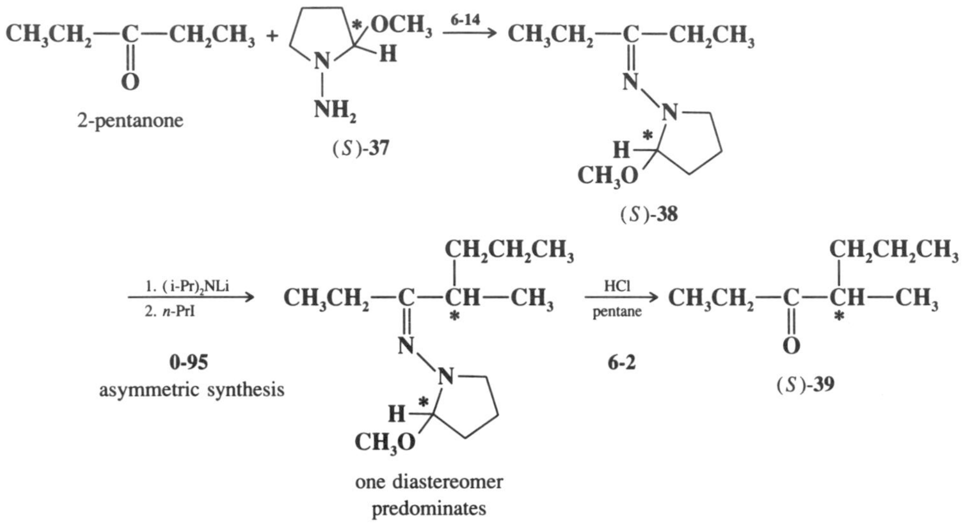

The compound **37** is called a *chiral auxiliary* because it is used to induce asymmetry and then removed.

2. *Active reagent.* A pair of enantiomers can be separated by an active reagent that reacts faster with one of them than it does with the other (this is also a method of resolution). If the absolute configuration of the reagent is known, the configuration of the enantiomers can often be determined by a knowledge of the mechanism and by seeing which diastereomer is preferentially formed.[93] Creation of a new chiral center in an inactive molecule can also be accomplished with an active reagent, though it is rare for 100% selectivity to be observed. An example[94] is the reduction of methyl benzoylformate with optically active N-benzyl-3-(hydroxymethyl)-4-methyl-1,4-dihydropyridine (**40**) to produce mandelic acid that contained about 97.5% of the S-(+) isomer and 2.5% of the R-(−) isomer (for another

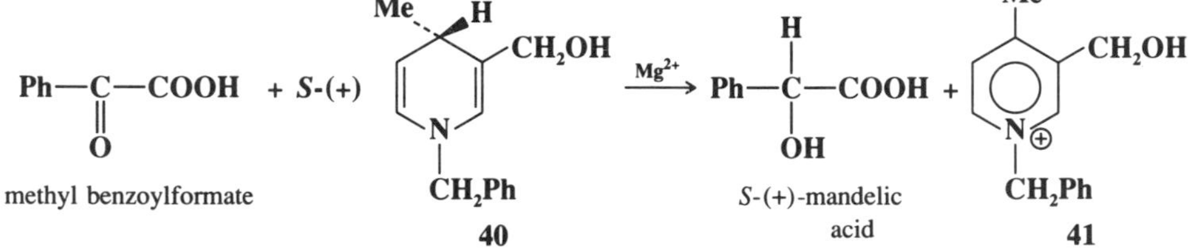

[92]Enders; Eichenauer; Baus; Schubert; Kremer *Tetrahedron* **1984**, *40*, 1345.

[93]See, for example, Horeau *Tetrahedron Lett.* **1961**, 506; Marquet; Horeau *Bull. Soc. Chim. Fr.* **1967**, 124; Brockmann; Risch *Angew. Chem. Int. Ed. Engl.* **1974**, *13*, 664 [*Angew. Chem. 86*, 707]; Potapov; Gracheva; Okulova *J. Org. Chem. USSR* **1989**, *25*, 311.

[94]Meyers; Oppenlaender *J. Am. Chem. Soc.* **1986**, *108*, 1989. For reviews of asymmetric reduction, see Morrison *Surv. Prog. Chem.* **1966**, *3*, 147-182; Yamada; Koga *Sel. Org. Transform.* **1970**, *1*, 1-33; Ref. 232 in Chapter 15. See also Morrison, Ref. 88, vol. 2.

example, see p. 786). Note that the other product, **41,** is not chiral. Reactions like this, in which one reagent (in this case **40**) gives up its chirality to another, are called *self-immolative.* In this intramolecular example:

$$\text{83 \% overall yield}$$

chirality is transferred from one atom to another in the same molecule.[95]

A reaction in which an inactive substrate is converted selectively to one of two enantiomers is called an *enantioselective* reaction, and the process is called *asymmetric induction.* These terms apply to reactions in this category and in categories 3 and 4.

When an optically active substrate reacts with an optically active reagent to form two new chiral centers, it is possible for both centers to be created in the desired sense. This type of process is called *double asymmetric synthesis*[96] (for an example, see p. 942).

3. *Active catalyst or solvent.*[97] Many such examples are present in the literature, among them reduction of ketones and substituted alkenes to optically active (though not optically pure) secondary alcohols and substituted alkanes by treatment with hydrogen and a chiral homogeneous hydrogenation catalyst (reactions **6-25** and **5-9**),[98] the treatment of aldehydes or ketones with organometallic compounds in the presence of a chiral catalyst (see **6-29**), and the conversion of alkenes to optically active epoxides by treatment with a hydroperoxide and a chiral catalyst (see **5-36**). In some instances, notably in the homogeneous catalytic hydrogenation of alkenes (**5-9**), the ratio of enantiomers prepared in this way is as high as 98:2.[99] Other examples of the use of a chiral catalyst or solvent are the reaction between secondary alkyl Grignard reagents and vinylic halides (**0-87**) in the presence of chiral transition-metal complexes:[100]

$$\text{Ph—CH—MgCl} + \text{CH}_2{=}\text{CHBr} \xrightarrow[\text{complex}]{\text{chiral}} \text{Ph—CH—CH}{=}\text{CH}_2$$

with Me groups below the first CH and the product CH.

the conversion of chlorofumaric acid (in the form of its diion) to the $(-)$-*threo* isomer of the diion of chloromalic acid by treatment with H_2O and the enzyme fumarase,[101]

[95]Goering; Kantner; Tseng *J. Org. Chem.* **1983,** *48,* 715.

[96]For a review, see Masamune; Choy; Petersen; Sita *Angew. Chem. Int. Ed. Engl.* **1985,** *24,* 1-30 [*Angew. Chem. 97,* 1-31].

[97]For a monograph, see Morrison *Asymmetric Synthesis,* vol. 5; Academic Press: New York, 1985. For reviews, see Tomioka *Synthesis* **1990,** 541-549; Consiglio; Waymouth *Chem. Rev.* **1989,** *89,* 257-276; Brunner, in Hartley *The Chemistry of the Metal–Carbon Bond,* vol. 5; Wiley: New York, 1989, pp. 109-146; Noyori; Kitamura *Mod. Synth. Methods* **1989,** *5,* 115-198; Pfaltz *Mod. Synth. Methods* **1989,** *5,* 199-248; Kagan *Bull. Soc. Chim. Fr.* **1988,** 846-853; Brunner *Synthesis* **1988,** 645-654; Wynberg *Top. Stereochem.* **1986,** *16,* 87-129. See also papers in *Tetrahedron: Asymmetry* **1991,** *2,* 481-732.

[98]For reviews of these and related topics, see Zief; Crane *Chromatographic Separations;* Marcel Dekker: New York, 1988; Brunner *J. Organomet. Chem.* **1986,** *300,* 39-56; Bosnich; Fryzuk *Top. Stereochem.* **1981,** *12,* 119-154.

[99]See Vineyard; Knowles; Sabacky; Bachman; Weinkauff *J. Am. Chem. Soc.* **1977,** *99,* 5946; Fryzuk; Bosnich *J. Am. Chem. Soc.* **1978,** *100,* 5491.

[100]For reviews of chiral transition metal complex catalysts, see Brunner *Top. Stereochem.* **1988,** *18,* 129-247; Hayashi; Kumada *Acc. Chem. Res.* **1982,** *15,* 395-401.

[101]Findeis; Whitesides *J. Org. Chem.* **1987,** *52,* 2838. For a monograph on enzymes as chiral catalysts, see Rétey; Robinson *Stereospecificity in Organic Chemistry and Enzymology;* Verlag Chemie: Deerfield Beach, FL, 1982. For reviews, see Klibanov *Acc. Chem. Res.* **1990,** *23,* 114-120; Jones *Tetrahedron* **1986,** *42,* 3351-3403; Jones, in Morrison, Ref. 88, vol. 5, pp. 309-344; Švedas; Galaev *Russ. Chem. Rev.* **1983,** *52,* 1184-1202. See also Simon; Bader; Günther; Neumann; Thanos *Angew. Chem. Int. Ed. Engl.* **1985,** *24,* 539-553 [*Angew. Chem. 97,* 541-555].

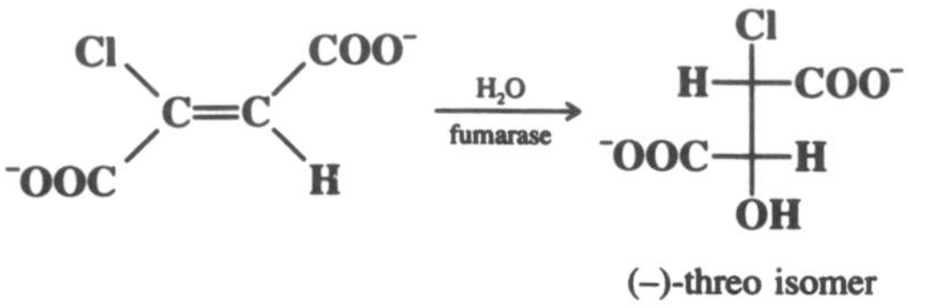

(−)-threo isomer

and the preparation of optically active alcohols by the treatment of Grignard reagents with aldehydes in optically active ether solvents.[102]

4. *Reactions in the presence of circularly polarized light.*[103]   If the light used to initiate a photochemical reaction (Chapter 7) of achiral reagents is circularly polarized, then, in theory, a chiral product richer in one enantiomer might be obtained. However, such experiments have not proved fruitful. In certain instances, the use of left and right circularly polarized light *has* given products with opposite rotations[104] (showing that the principle is valid), but up to now the extent of favoritism has always been less than 1%.

Methods of Resolution[105]

A pair of enantiomers can be separated in several ways, of which conversion to diastereomers and separation of these by fractional crystallization is the most often used. In this method and in some of the others, both isomers can be recovered, but in some methods it is necessary to destroy one.

1. *Conversion to diastereomers.*   If the racemic mixture to be resolved contains a carboxyl group (and no strongly basic group), it is possible to form a salt with an optically active base. Since the base used is, say, the S form, there will be a mixture of two salts

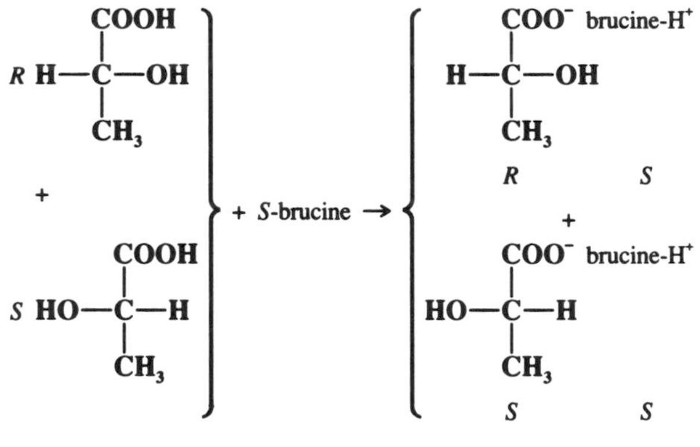

---

[102]See, for example, Blomberg; Coops *Recl. Trav. Chim. Pays-Bas* **1964**, *83*, 1083; Inch; Lewis; Sainsbury; Sellers *Tetrahedron Lett.* **1969**, 3657; Jalander; Strandberg *Acta Chem. Scand., Ser. B* **1983**, *37*, 15. See also Seebach; Kalinowski; Langer; Crass; Wilka *Org. Synth. VII*, 41.

[103]For a review, See Buchardt *Angew. Chem. Int. Ed. Engl.* **1974**, *13*, 179-185 [*Angew. Chem. 86*, 222]. For a discussion, see Barron *J. Am. Chem. Soc.* **1986**, *108*, 5539.

[104]See, for example, Moradopour; Nicoud; Balavoine; Kagan; Tsoucaris *J. Am. Chem. Soc.* **1971**, *93*, 2353; Bernstein; Calvin; Buchardt *J. Am. Chem. Soc.* **1972**, *94*, 494, **1973**, *95*, 527, *Tetrahedron Lett.* **1972**, 2195; Nicoud; Kagan *Isr. J. Chem.* **1977**, *15*, 78. See also Zandomeneghi; Cavazza; Pietra *J. Am. Chem. Soc.* **1984**, *106*, 7261.

[105]For a monograph, see Ref. 7. For reviews, see Wilen; Collet; Jacques *Tetrahedron* **1977**, *33*, 2725-2736; Wilen *Top. Stereochem.* **1971**, *6*, 107-176; Boyle *Q. Rev., Chem. Soc.* **1971**, *25*, 323-341; Buss; Vermeulen *Ind. Eng. Chem.* **1968**, *60* (8), 12-28.

produced having the configurations *SS* and *RS*. Although the acids are enantiomers, the salts are diastereomers and have different properties. The property most often used for separation is differential solubility. The mixture of diastereomeric salts is allowed to crystallize from a suitable solvent. Since the solubilities are different, the initial crystals formed will be richer in one diastereomer. Filtration at this point will already have achieved a partial resolution. Unfortunately, the difference in solubilities is rarely if ever great enough to effect total separation with one crystallization. Usually fractional crystallizations must be used and the process is long and tedious. Fortunately, naturally occurring optically active bases (mostly alkaloids) are readily available. Among the most commonly used are brucine, ephedrine, strychnine, and morphine. Once the two diastereomers have been separated, it is easy to convert the salts back to the free acids and the recovered base can be used again.

Most resolution is done on carboxylic acids and often, when a molecule does not contain a carboxyl group, it is converted to a carboxylic acid before resolution is attempted. However, the principle of conversion to diastereomers is not confined to carboxylic acids, and other groups[106] may serve as handles to be coupled to an optically active reagent.[107] Racemic bases can be converted to diastereomeric salts with active acids. Alcohols[108] can be converted to diastereomeric esters, aldehydes to diastereomeric hydrazones, etc. Even hydrocarbons can be converted to diastereomeric inclusion compounds,[109] with urea. Urea is not chiral, but the cage structure is.[110] Chiral crown ethers have been used to separate mixtures of enantiomeric alkyl- and arylammonium ions, by the formation of diastereomeric complexes[111] (see also category 3, below). *trans*-Cyclooctene (p. 104) was resolved by conversion to a platinum complex containing an optically active amine.[112]

Although fractional crystallization has always been the most common method for the separation of diastereomers, its tediousness and the fact that it is limited to solids prompted a search for other methods. Fractional distillation has given only limited separation, but gas chromatography[113] and preparative liquid chromatography[114] have proved more useful and, in many cases, have supplanted fractional crystallization, especially where the quantities to be resolved are small.[115]

[106]For summaries of methods used to resolve particular types of compounds, see Boyle, Ref. 105; Eliel *Stereochemistry of Carbon Compounds*, Ref. 1, pp. 49-63.

[107]For an extensive list of reagents that have been used for this purpose and of compounds resolved, see Wilen *Tables of Resolving Agents and Optical Resolutions*; University of Notre Dame Press: Notre Dame, IN, 1972.

[108]For a review of resolution of alcohols, see Klyashchitskii; Shvets *Russ. Chem. Rev.* **1972**, *41*, 592-602.

[109]For reviews of chiral inclusion compounds, including their use for resolution, see Prelog; Kovačević; Egli *Angew. Chem. Int. Ed. Engl.* **1989**, *28*, 1147-1152 [*Angew. Chem. 101*, 1173-1178]; Worsch; Vögtle *Top. Curr. Chem.* **1987**, *140*, 21-41; Toda *Top. Curr. Chem.* **1987**, *140*, 43-69; Stoddart *Top. Stereochem.* **1987**, *17*, 207-288; Sirlin *Bull. Soc. Chim. Fr.* **1984**, II-5-II-40; Arad-Yellin; Green; Knossow; Tsoucaris, in Atwood; Davies; MacNicol *Inclusion Compounds*, vol. 3; Academic Press: New York, 1984, pp. 263-295; Stoddart *Prog. Macrocyclic Chem.* **1981**, *2*, 173-250; Cram et al., Pure Appl. Chem. **1975**, *43*, 327-349; Cram; Cram *Science* **1974**, *183*, 803-809.

[110]See Schlenk, *Liebigs Ann. Chem.* **1973**, 1145, 1156, 1179, 1195. Inclusion complexes of tri-*o*-thymotide can be used in a similar manner: see Arad-Yellin; Green; Knossow; Tsoucaris *J. Am. Chem. Soc.* **1983**, *105*, 4561.

[111]See, for example, Kyba; Koga; Sousa; Siegel; Cram *J. Am. Chem. Soc.* **1973**, *95*, 2692; Sogah; Cram *J. Am. Chem. Soc.* **1979**, *101*, 3035; Lingenfelter; Helgeson; Cram *J. Org. Chem.* **1981**, *46*, 393; Pearson; Leigh; Sutherland *J. Chem. Soc., Perkin Trans. 1* **1979**, 3113; Bussman; Lehn; Oesch; Plumeré; Simon *Helv. Chim. Acta* **1981**, *64*, 657; Davidson; Bradshaw; Jones; Dalley; Christensen; Izatt; Morin; Grant *J. Org. Chem.* **1984**, *49*, 353. See also Toda; Tanaka; Omata; Nakamura; Ōshima *J. Am. Chem. Soc.* **1983**, *105*, 5151.

[112]Ref. 55. For a review, see Tsuji *Adv. Org. Chem.* **1969**, *6*, 109-255, pp. 220-227.

[113]See, for example, Casanova; Corey *Chem. Ind. (London)* **1961**, 1664; Gil-Av; Nurok *Proc. Chem. Soc.* **1962**, 146; Gault; Felkin *Bull. Soc. Chim. Fr.* **1965**, 742; Vitt; Saporovskaya; Gudkova; Belikov *Tetrahedron Lett.* **1965**, 2575; Westley; Halpern; Karger *Anal. Chem.* **1968**, *40*, 2046; Kawa; Yamaguchi; Ishikawa *Chem. Lett.* **1982**, 745.

[114]For example, See Pirkle; Hoekstra *J. Org. Chem.* **1974**, *39*, 3904; Pirkle; Hauske *J. Org. Chem.* **1977**, *42*, 1839; Helmchen; Nill *Angew. Chem. Int. Ed. Engl.* **1979**, *18*, 65 [*Angew. Chem. 91*, 66]; Meyers; Slade; Smith; Mihelich; Hershenson; Liang *J. Org. Chem.* **1979**, *44*, 2247; Goldman; Kustanovich; Weinstein; Tishbee; Gil-Av *J. Am. Chem. Soc.* **1982**, *104*, 1093.

[115]For monographs on the use of liquid chromatography to effect resolutions, see Lough *Chiral Liquid Chromatography*; Blackie and Sons: London, 1989; Krstulović *Chiral Separations by HPLC*; Ellis Horwood: Chichester, 1989; Zief; Crane, Ref. 98. For a review, see Karger *Anal. Chem.* **1967**, *39* (8), 24A-50A.

2. *Differential absorption.* When a racemic mixture is placed on a chromatographic column, if the column consists of chiral substances, then in principle the enantiomers should move along the column at different rates and should be separable without having to be converted to diastereomers.[115] This has been successfully accomplished with paper, column, thin-layer,[116] and gas and liquid chromatography.[117] For example, racemic mandelic acid has been almost completely resolved by column chromatography on starch.[118] Many workers have achieved separations with gas and liquid chromatography by the use of columns packed with chiral absorbents.[119] Columns packed with chiral materials are now commercially available and are capable of separating the enantiomers of certain types of compounds.[120]

3. *Chiral recognition.* The use of chiral hosts to form diastereomeric inclusion compounds was mentioned above. But in some cases it is possible for a host to form an inclusion compound with one enantiomer of a racemic guest, but not the other. This is called *chiral recognition*. One enantiomer fits into the chiral host cavity, the other does not. More often, both diastereomers are formed, but one forms more rapidly than the other, so that if the guest is removed it is already partially resolved (this is a form of kinetic resolution, see category 6). An example is use of the chiral crown ether **42** partially to resolve the racemic amine salt **43**.[121] When an aqueous solution of **43** was mixed with a solution of optically active **42** in chloroform, and the layers separated, the chloroform layer contained about

42   **43**

twice as much of the complex between **42** and (*R*)-**43** as of the diastereomeric complex. Many other chiral crown ethers and cryptands have been used, as have been cyclodextrins,[122]

[116]Weinstein *Tetrahedron Lett.* **1984**, *25*, 985.

[117]For monographs, see Allenmark *Chromatographic Enantioseparation*; Ellis Horwood: Chichester, 1988; König *The Practice of Enantiomer Separation by Capillary Gas Chromatography*; Hüthig: Heidelberg, 1987. For reviews, see Schurig; Nowotny *Angew. Chem. Int. Ed. Engl.* **1990**, *29*, 939-957 [*Angew. Chem. 102*, 969-986]; Pirkle; Pochapsky *Chem. Rev.* **1989**, *89*, 347-362, *Adv. Chromatogr.* **1987**, *27*, 73-127; Okamoto *CHEMTECH* **1987**, 176-181; Schurig *Angew. Chem. Int. Ed. Engl.* **1984**, *23*, 747-765 [*Angew. Chem. 96*, 733-752]; Blaschke *Angew. Chem. Int. Ed. Engl.* **1980**, *19*, 13-24 [*Angew. Chem. 92*, 14-25]; Rogozhin; Davankov *Russ. Chem. Rev.* **1968**, *37*, 565-575. See also many articles in the journal *Chirality.*

[118]Ohara; Fujita; Kwan *Bull. Chem. Soc. Jpn.* **1962**, *35*, 2049; Ohara; Ohta; Kwan *Bull. Chem. Soc. Jpn.* **1964**, *37*, 76. See also Blaschke; Donow *Chem. Ber.* **1975**, *108*, 2792; Hess; Burger; Musso *Angew. Chem. Int. Ed. Engl.* **1978**, *17*, 612 [*Angew. Chem. 90*, 645].

[119]See, for example, Gil-Av; Feibush; Charles-Sigler *Tetrahedron Lett.* **1966**, 1009; Gil-Av; Tishbee; Hare *J. Am. Chem. Soc.* **1980**, *102*, 5115; Hesse; Hagel *Liebigs Ann. Chem.* **1976**, 996; Schlögl; Widhalm *Chem. Ber.* **1982**, *115*, 3042; Koppenhoefer; Allmendinger; Nicholson *Angew. Chem. Int. Ed. Engl.* **1985**, *24*, 48 [*Angew. Chem. 97*, 46]; Dobashi; Hara *J. Am. Chem. Soc.* **1985**, *107*, 3406, *Tetrahedron Lett.* **1985**, *26*, 4217, *J. Org. Chem.* **1987**, *52*, 2490; Konrad; Musso *Liebigs Ann. Chem.* **1986**, 1956; Pirkle; Pochapsky; Mahler; Corey; Reno; Alessi *J. Org. Chem.* **1986**, *51*, 4991; Okamoto; Aburatani; Kaida; Hatada *Chem. Lett.* **1988**, 1125; Ehlers; König; Lutz; Wenz; tom Dieck *Angew. Chem. Int. Ed. Engl.* **1988**, *27*, 1556 [*Angew. Chem. 100*, 1614]; Hyun; Park; Baik *Tetrahedron Lett.* **1988**, *29*, 4735; Schurig; Nowotny; Schmalzing *Angew. Chem. Int. Ed. Engl.* **1989**, *28*, 736 [*Angew. Chem. 101*, 785]; Ôi; Shijo; Miyano *Chem. Lett.* **1990**, 59; Erlandsson; Marle; Hansson; Isaksson; Pettersson; Pettersson *J. Am. Chem. Soc.* **1990**, *112*, 4573.

[120]See, for example, Pirkle and Welch *J. Org. Chem.* **1984**, *49*, 138.

[121]Cram; Cram, Ref. 109. See also Yamamoto; Fukushima; Okamoto; Hatada; Nakazaki *J. Chem. Soc., Chem. Commun.* **1984**, 1111; Kanoh; Hongoh; Katoh; Motoi; Suda *J. Chem. Soc., Chem. Commun.* **1988**, 405; Bradshaw; Huszthy; McDaniel; Zhu; Dalley; Izatt; Lifson *J. Org. Chem.* **1990**, *55*, 3129.

[122]See, for example, Hamilton; Chen *J. Am. Chem. Soc.* **1988**, *110*, 5833.

cholic acid,[123] and other kinds of hosts.[109] Of course, enzymes are generally very good at chiral recognition, and much of the work in this area has been an attempt to mimic the action of enzymes.

4. *Biochemical processes.*[124]   The chiral compound that reacts at different rates with the two enantiomers may be present in a living organism. For instance, a certain bacterium may digest one enantiomer but not the other. This method is limited, since it is necessary to find the proper organism and since one of the enantiomers is destroyed in the process. However, when the proper organism is found, the method leads to a high extent of resolution since biological processes are usually very stereoselective.

5. *Mechanical separation.*[125]   This is the method by which Pasteur proved that racemic acid was actually a mixture of (+)- and (−)-tartaric acids.[126] In the case of racemic sodium ammonium tartrate the enantiomers crystallize separately—all the (+) molecules going into one crystal and all the (−) into another. Since the crystals too are nonsuperimposable, their appearance is not identical and a trained crystallographer can separate them with tweezers.[127] However, this is seldom a practical method, since few compounds crystallize in this manner. Even sodium ammonium tartrate does so only when it is crystallized below 27°C. A more useful variation of the method, though still not very common, is the seeding of a racemic solution with something that will cause only one enantiomer to crystallize.[128] An interesting example of the mechanical separation technique was reported in the isolation of heptahelicene (p. 103). One enantiomer of this compound, which incidentally has the extremely high rotation of $[\alpha]_D^{20} = +6200°$, spontaneously crystallizes from benzene.[129] In the case of 1,1'-binaphthyl, optically active crystals can be formed simply by heating polycrystalline racemic samples of the compound at 76–150°. A phase change from one crystal form to another takes place.[130] It may be noted that 1,1'-binaphthyl is one of the few compounds that can be resolved by the Pasteur tweezer method. In some cases resolution can be achieved by enantioselective crystallization in the presence of a chiral additive.[131] Spontaneous resolution has also been achieved by sublimation. In the case of the norborneol derivative **44**,

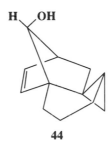

44

[123]See Miyata; Shibakana; Takemoto *J. Chem. Soc., Chem. Commun.* **1988**, 655.

[124]For a review, see Sih; Wu *Top. Stereochem.* **1989**, *19*, 63-125.

[125]For reviews, see Collet; Brienne; Jacques *Chem. Rev.* **1980**, *80*, 215-230; *Bull. Soc. Chim. Fr.* **1972**, 127-142, **1977**, 494-498. For a discussion, see Curtin; Paul *Chem. Rev.* **1981**, *81*, 525-541, pp. 535-536.

[126]Besides discovering this method of resolution, Pasteur also discovered the method of conversion to diastereomers and separation by fractional crystallization and the method of biochemical separation (and, by extension, kinetic resolution).

[127]This is a case of optically active materials arising from inactive materials. However, it may be argued that an optically active investigator is required to use the tweezers. Perhaps a hypothetical human being constructed entirely of inactive molecules would be unable to tell the difference between left- and right-handed crystals.

[128]For a review of the seeding method, see Secor *Chem. Rev.* **1963**, *63*, 297.

[129]Martin et al., Ref. 53. See also Wynberg; Groen *J. Am. Chem. Soc.* **1968**, *90*, 5339. For a discussion of other cases, see McBride; Carter *Angew. Chem. Int. Ed. Engl.* **1991**, *30*, 293 [*Angew. Chem. 103*, 298].

[130]Wilson; Pincock *J. Am. Chem. Soc.* **1975**, *97*, 1474; Kress; Duesler; Etter; Paul; Curtin *J. Am. Chem. Soc.* **1980**, *102*, 7709. See also Lu; Pincock *J. Org. Chem.* **1978**, *43*, 601; Gottarelli; Spada *J. Org. Chem.* **1991**, *56*, 2096. For a discussion and other examples, see Agranat; Perlmutter-Hayman; Tapuhi *Nouv. J. Chem.* **1978**, *2*, 183.

[131]Addadi; Weinstein; Gati; Weissbuch; Lahav *J. Am. Chem. Soc.* **1982**, *104*, 4610. See also Weissbuch; Addadi; Berkovitch-Yellin; Gati; Weinstein; Lahav; Leiserowitz *J. Am. Chem. Soc.* **1983**, *105*, 6615.

when the racemic solid is subjected to sublimation, the ($+$) molecules condense into one crystal and the ($-$) molecules into another.[132] In this case the crystals are superimposable, unlike the situation with sodium ammonium tartrate, but the investigators were able to remove a single crystal, which proved optically active.

6. *Kinetic resolution.*[133] Since enantiomers react with chiral compounds at different rates, it is sometimes possible to effect a partial separation by stopping the reaction before completion. This method is very similar to the asymmetric syntheses discussed on p. 102. An important application of this method is the resolution of racemic alkenes by treatment with optically active diisopinocampheylborane,[134] since alkenes do not easily lend themselves to conversion to diastereomers if no other functional groups are present. Another example is the resolution of allylic alcohols such as **45** with one enantiomer of a chiral epoxidizing agent (see **5-36**).[135] In the case of **45** the discrimination was extreme. One enantiomer was converted to the epoxide and the other was not, the rate ratio (hence the selectivity factor)

| racemic | (R)-enantiomer | (S)-enantiomer |

45

being more than 100. Of course, in this method only one of the enantiomers of the original racemic mixture is obtained, but there are at least two possible ways of getting the other: (1) use of the other enantiomer of the chiral reagent; (2) conversion of the product to the starting compound by a reaction that preserves the stereochemistry.

Reactions catalyzed by enzymes can be utilized for this kind of resolution.[136]

7. *Deracemization.* In this type of process, one enantiomer is converted to the other, so that a racemic mixture is converted to a pure enantiomer, or to a mixture enriched in one enantiomer. This is not quite the same as the methods of resolution previously mentioned, though an outside optically active substance is required. For example, the racemic thioester **46** was placed in contact with a certain optically active amide. After 28 days the solution contained 89% of one enantiomer and 11% of the other.[137] To effect the deracem-

46

[132]Paquette; Lau *J. Org. Chem.* **1987**, *52*, 1634.

[133]For reviews, see Kagan; Fiaud *Top. Stereochem.* **1988**, *18*, 249-330; Brown *Chem. Ind. (London)* **1988**, 612-617.

[134]Brown; Ayyangar; Zweifel *J. Am. Chem. Soc.* **1964**, *86*, 397.

[135]Martin; Woodard; Katsuki; Yamada; Ikeda; Sharpless *J. Am. Chem. Soc.* **1981**, *103*, 6237. See also Kobayashi; Kusakabe; Kitano; Sato *J. Org. Chem.* **1988**, *53*, 1586; Kitano; Matsumoto; Sato *Tetrahedron* **1988**, *44*, 4073; Carlier; Mungall; Schröder; Sharpless *J. Am. Chem. Soc.* **1988**, *110*, 2978; Discordia; Dittmer *J. Org. Chem.* **1990**, *55*, 1414. For other examples, see Miyano; Lu; Viti; Sharpless *J. Org. Chem.* **1985**, *50*, 4350; Paquette; DeRussy; Cottrell *J. Am. Chem. Soc.* **1988**, *110*, 890; Weidert; Geyer; Horner *Liebigs Ann. Chem.* **1989**, 533; Katamura; Ohkuma; Tokunaga; Noyori *Tetrahedron: Assymetry* **1990**, *1*, 1; Hayashi; Miwata; Oguni *J. Chem. Soc., Perkin Trans. 1* **1991**, 1167.

[136]For example, see Schwartz; Madan; Whitesell; Lawrence *Org. Synth.* **69**, 1; Guibé-Jampel; Rousseau; Salaün *J. Chem. Soc., Chem. Commun.* **1987**, 1080; Francalanci; Cesti; Cabri; Bianchi; Martinengo; Foá *J. Org. Chem.* **1987**, *52*, 5079; Mohr; Rösslein; Tamm *Tetrahedron Lett.* **1989**, *30*, 2513; Kazlauskas *J. Am. Chem. Soc.* **1989**, *111*, 4953.

[137]Pirkle; Reno *J. Am. Chem. Soc.* **1987**, *109*, 7189. For another example, see Reider; Davis; Hughes; Grabowski *J. Org. Chem.* **1987**, *52*, 955.

ization two conditions are necessary: (1) the enantiomers must complex differently with the optically active substance; (2) they must interconvert under the conditions of the experiment. In this case the presence of a base (Et_3N) was necessary for the interconversion to take place.

Optical Purity[138]

Suppose we have just attempted to resolve a racemic mixture by one of the methods described in the previous section. How do we know that the two enantiomers we have obtained are pure? For example, how do we know that the (+) isomer is not contaminated by, say, 20% of the (−) isomer and vice versa? If we knew the value of [α] for the pure material ($[α]_{max}$), we could easily determine the purity of our sample by measuring its rotation. For example, if $[α]_{max}$ is + 80° and our (+) enantiomer contains 20% of the (−) isomer, [α] for the sample will be + 48°.[139] We define *optical purity* as

$$\text{Percent optical purity} = \frac{[α]_{obs}}{[α]_{max}} \times 100$$

Assuming a linear relationship between [α] and concentration, which is true for most cases, the optical purity is equal to the percent excess of one enantiomer over the other:

$$\text{Optical purity} = \text{percent enantiomeric excess} = \frac{[R] - [S]}{[R] + [S]} \times 100 = \% R - \% S$$

But how do we determine the value of $[α]_{max}$? It is plain that we have two related problems here, namely, what are the optical purities of our two samples and what is the value of $[α]_{max}$. If we solve one, the other is also solved. Several methods for solving these problems are known.

One of these methods involves the use of nmr.[140] Suppose we have a nonracemic mixture of two enantiomers and wish to know the proportions. We convert the mixture into a mixture of diastereomers with an optically pure reagent and look at the nmr spectrum of the resulting mixture, e.g.,

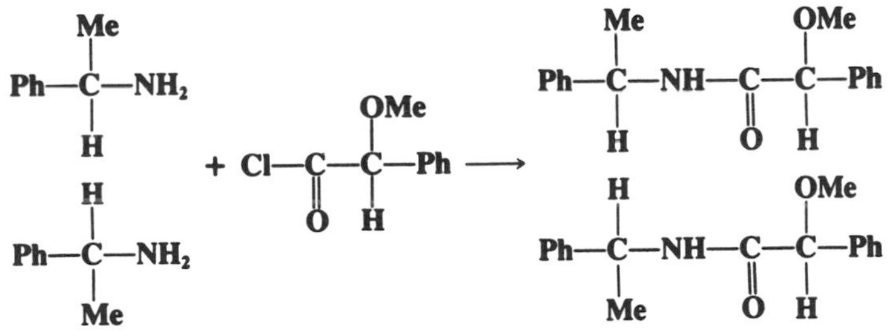

If we examined the nmr spectrum of the starting mixture, we would find only one peak (split into a doublet by the C—H) for the Me protons, since enantiomers give identical nmr

[138]For a review, see Raban; Mislow *Top. Stereochem.* **1967**, *2*, 199-230.

[139]If a sample contains 80% (+) and 20% (−) isomer, the (−) isomer cancels an equal amount of (+) isomer and the mixture behaves as if 60% of it were (+) and the other 40% inactive. Therefore the rotation is 60% of 80° or 48°. This type of calculation, however, is not valid for cases in which [α] is dependent on concentration (p. 96); see Horeau *Tetrahedron Lett.* **1969**, 3121.

[140]Raban; Mislow *Tetrahedron Lett.* **1965**, 4249, **1966**, 3961; Jacobus; Raban *J. Chem. Educ.* **1969**, *46*, 351; Tokles; Snyder *Tetrahedron Lett.* **1988**, *29*, 6063. For a review, see Yamaguchi, in Morrison, Ref. 88, vol. 1, pp. 125-152. See also Ref. 138.

spectra.[141] But the two amides are not enantiomers and each Me gives its own doublet. From the intensity of the two peaks, the relative proportions of the two diastereomers (and hence of the original enantiomers) can be determined. Alternatively, the unsplit OMe peaks could have been used. This method was satisfactorily used to determine the optical purity of a sample of 1-phenylethylamine (the case shown above),[142] as well as other cases, but it is obvious that sometimes corresponding groups in diastereomeric molecules will give nmr signals that are too close together for resolution. In such cases one may resort to the use of a different optically pure reagent. ^{13}C nmr can be used in a similar manner.[143] It is also possible to use these spectra to determine the absolute configuration of the original enantiomers by comparing the spectra of the diastereomers with those of the original enantiomers.[144] From a series of experiments with related compounds of known configurations it can be determined in which direction one or more of the ^{1}H or ^{13}C nmr peaks are shifted by formation of the diastereomer. It is then assumed that the peaks of the enantiomers of unknown configuration will be shifted the same way.

A closely related method does not require conversion of enantiomers to diastereomers but relies on the fact that (in principle, at least) enantiomers have different nmr spectra *in a chiral solvent*, or when mixed with a chiral molecule (in which case transient diastereomeric species may form). In such cases the peaks may be separated enough to permit the proportions of enantiomers to be determined from their intensities.[145] Another variation, which gives better results in many cases, is to use an achiral solvent but with the addition of a *chiral lanthanide shift reagent* such as tris[3-trifluoroacetyl-*d*-camphorato]europium-(III).[146] Lanthanide shift reagents have the property of spreading nmr peaks of compounds with which they can form coordination compounds, e.g., alcohols, carbonyl compounds, amines, etc. Chiral lanthanide shift reagents shift the peaks of the two enantiomers of many such compounds to different extents.

Another method, involving gas chromatography,[147] is similar in principle to the nmr method. A mixture of enantiomers whose purity is to be determined is converted by means of an optically pure reagent into a mixture of two diastereomers. These diastereomers are then separated by gas chromatography (p. 121) and the ratios determined from the peak areas. Once again, the ratio of diastereomers is the same as that of the original enantiomers. High-pressure liquid chromatography has been used in a similar manner and has wider applicability.[148] The direct separation of enantiomers by gas or liquid chromatography on a chiral column has also been used to determine optical purity.[149]

---

[141]Though enantiomers give identical nmr spectra, the spectrum of a single enantiomer may be different from that of the racemic mixture, even in solution. See Williams; Pitcher; Bommer; Gutzwiller; Uskoković *J. Am. Chem. Soc.* **1969**, *91*, 1871.

[142]Ref. 138, pp. 216-218.

[143]For a method that relies on diastereomer formation without a chiral reagent, see Feringa; Smaardijk; Wynberg *J. Am. Chem. Soc.* **1985**, *107*, 4798; Feringa; Strijtveen; Kellogg *J. Org. Chem.* **1986**, *51*, 5484. See also Pasquier; Marty *Angew. Chem. Int. Ed. Engl.* **1985**, *24*, 315 [*Angew. Chem. 97*, 328]; Luchinat; Roelens *J. Am. Chem. Soc.* **1986**, *108*, 4873.

[144]See Dale; Mosher *J. Am. Chem. Soc.* **1973**, *95*, 512; Rinaldi *Prog. NMR Spectrosc.* **1982**, *15*, 291-352; Faghih; Fontaine; Horibe; Imamura; Lukacs; Olesker; Seo *J. Org. Chem.* **1985**, *50*, 4918; Trost et al. *J. Org. Chem.* **1986**, *51*, 2370.

[145]For reviews of nmr chiral solvating agents, see Weisman, in Morrison, Ref. 88, vol. 1, pp. 153-171; Pirkle; Hoover *Top. Stereochem.* **1982**, *13*, 263-331. For literature references, see Sweeting; Anet *Org. Magn. Reson.* **1984**, *22*, 539. See also Pirkle; Tsipouras *Tetrahedron Lett.* **1985**, *26*, 2989; Parker; Taylor *Tetrahedron* **1987**, *43*, 5451.

[146]Whitesides; Lewis *J. Am. Chem. Soc.* **1970**, *92*, 6979, **1971**, *93*, 5914; Sweeting; Crans; Whitesides *J. Org. Chem.* **1987**, *52*, 2273. For a monograph on chiral lanthanide shift reagents, see Morrill *Lanthanide Shift Reagents in Stereochemical Analysis*; VCH: New York, 1986. For reviews, see Fraser, in Morrison, Ref. 88, vol. 1, pp. 173-196; Sullivan *Top. Stereochem.* **1978**, *10*, 287-329.

[147]Charles; Fischer; Gil-Av *Isr. J. Chem.* **1963**, *1*, 234; Halpern; Westley *Chem. Commun.* **1965**, 246; Vitt; Saporovskaya; Gudkova; Belikov *Tetrahedron Lett.* **1965**, 2575; Guetté; Horeau *Tetrahedron Lett.* **1965**, 3049; Westley; Halpern *J. Org. Chem.* **1968**, *33*, 3978.

[148]For a review, see Pirkle; Finn, in Morrison, Ref. 88, vol. 1, pp. 87-124.

[149]For reviews, see in Morrison, Ref. 88, vol. 1, the articles by Schurig, pp. 59-86 and Pirkle; Finn, pp. 87-124.

Other methods[150] involve isotopic dilution,[151] kinetic resolution,[152] ^{13}C nmr relaxation rates of diastereomeric complexes,[153] and circular polarization of luminescence.[154]

CIS–TRANS ISOMERISM

Compounds in which rotation is restricted may exhibit cis–trans isomerism.[155] These compounds do not rotate the plane of polarized light (unless they also happen to be chiral), and the properties of the isomers are not identical. The two most important types are isomerism resulting from double bonds and that resulting from rings.

Cis–Trans Isomerism Resulting from Double Bonds

It has been mentioned (p. 9) that the two carbon atoms of a $C=C$ double bond and the four atoms directly attached to them are all in the same plane and that rotation around the double bond is prevented. This means that in the case of a molecule $WXC=CYZ$, stereoisomerism exists when $W \neq X$ and $Y \neq Z$. There are two and only two isomers (**E** and **F**), each superimposable on its mirror image unless one of the groups happens to carry a

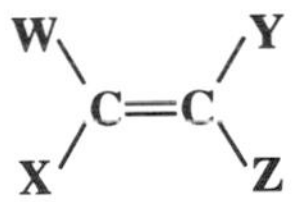

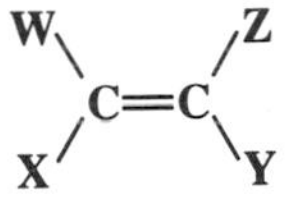

chiral center. Note that **E** and **F** are diastereomers, by the definition given on p. 113. There are two ways to name such isomers. In the older method, one isomer is called *cis* and the other *trans*. When $W = Y$, **E** is the cis and **F** the trans isomer. Unfortunately, there is no easy way to apply this method when the four groups are different. The newer method, which can be applied to all cases, is based on the Cahn–Ingold–Prelog system (p. 109). The two groups at each carbon are ranked by the sequence rules. Then that isomer with the two higher ranking groups on the same side of the double bond is called Z (for the German word *zusammen* meaning *together*); the other is E (for *entgegen* meaning *opposite*).[156] A few examples are shown. Note that the Z isomer is not necessarily the one that would be called cis under the older system (e.g., **47**, **48**). Like cis and trans, E and Z are used as prefixes; e.g., **48** is called (E)-1-bromo-1,2-dichloroethene.

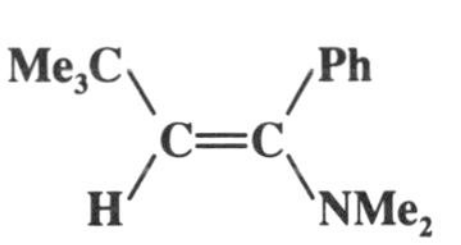

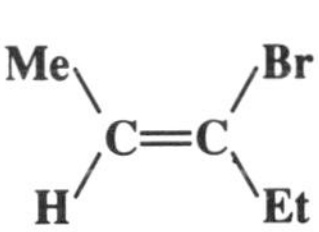

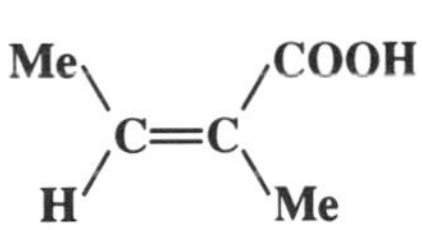

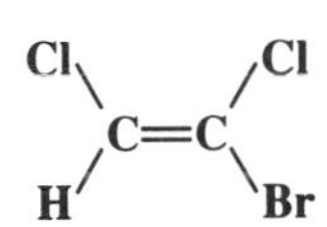

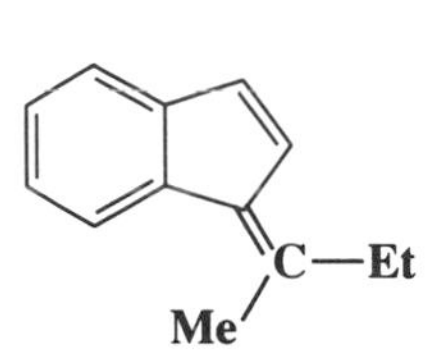

[150]See also Leitich *Tetrahedron Lett.* **1978**, 3589; Hill; Zens; Jacobus *J. Am. Chem. Soc.* **1979**, *101*, 7090; Matsumoto; Yajima; Endo *Bull. Chem. Soc. Jpn.* **1987**, *60*, 4139.

[151]Berson; Ben-Efraim *J. Am. Chem. Soc.* **1959**, *81*, 4083. For a review, see Andersen; Gash; Robertson, in Morrison, Ref. 88, vol. 1, pp. 45-57.

[152]Horeau *J. Am. Chem. Soc.* **1964**, *86*, 3171, *Bull. Soc. Chim. Fr.* **1964**, 2673; Horeau; Guetté; Weidmann *Bull. Soc. Chim. Fr.* **1966**, 3513. For a review, see Schoofs; Guetté, in Morrison, Ref. 88, vol. 1, pp. 29-44.

[153]Hofer; Keuper *Tetrahedron Lett.* **1984**, *25*, 5631.

[154]Eaton, *Chem. Phys. Lett.* **1971**, *8*, 251; Schippers; Dekkers *Tetrahedron* **1982**, *38*, 2089.

[155]Cis-trans isomerism was formerly called *geometrical isomerism*.

[156]For a complete description of the system, see Ref. 2.

This type of isomerism is also possible with other double bonds, such as C=N,[157] N=N, or even C=S,[158] though in these cases only two or three groups are connected to the double-bond atoms. In the case of imines, oximes, and other C=N compounds, if W = Y **49** may be called syn and **50** anti, though E and Z are often used here too. In azo compounds there is no ambiguity. **51** is always syn or Z regardless of the nature of W and Y.

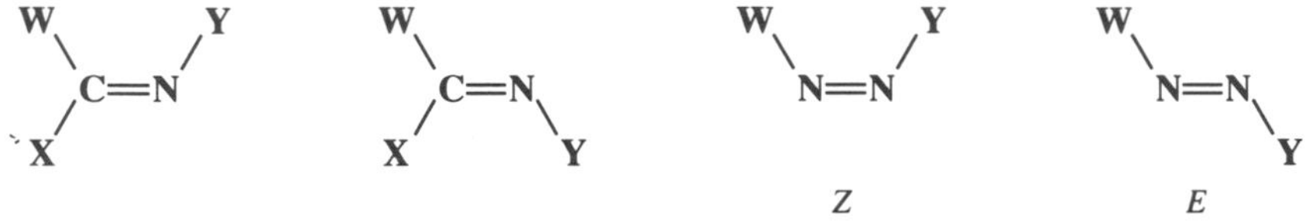

		Z	E
49	**50**	**51**	

If there is more than one double bond[159] in a molecule and if W ≠ X and Y ≠ Z for each, the number of isomers in the most general case is 2^n, although this number may be decreased if some of the substituents are the same, as in

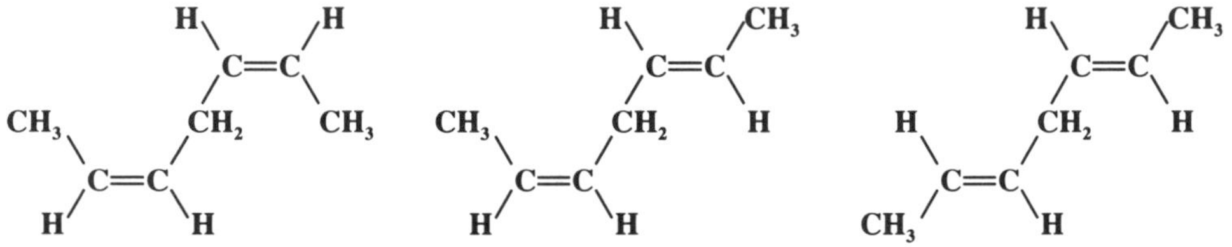

cis-cis or   cis-trans or   trans-trans or
Z, Z   Z, E   E, E

When a molecule contains a double bond and an asymmetric carbon, there are four isomers, a cis pair of enantiomers and a trans pair:

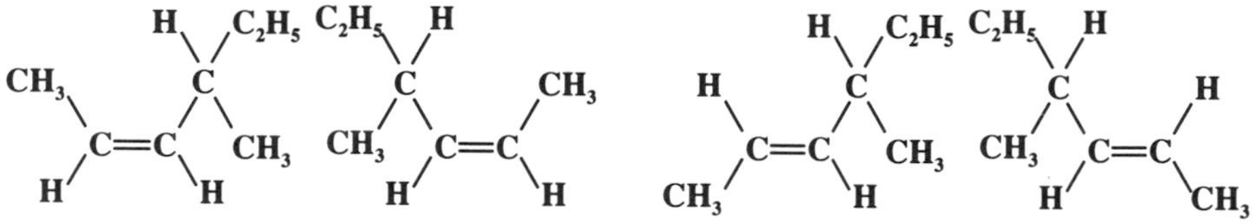

Z or cis *dl* pair   E or trans *dl* pair

Double bonds in small rings are so constrained that they must be cis. From cyclopropene (a known system) to cycloheptene, double bonds in a stable ring cannot be trans. However, the cyclooctene ring is large enough to permit trans double bonds to exist (see p. 104), and for rings larger than 10- or 11-membered, trans isomers are more stable[160] (see also p. 158).

In a few cases, single-bond rotation is so slowed that cis and trans isomers can be isolated even where no double bond exists[161] (see also p. 162). One example is

[157]For reviews of isomerizations about C=N bonds, see, in Patai *The Chemistry of the Carbon-Nitrogen Double Bond*; Wiley: New York, 1970, the articles by McCarty, 363-464 (pp. 364-408), and Wettermark, 565-596 (pp. 574-582).
[158]King; Durst *Can. J. Chem.* **1966**, *44*, 819.
[159]This rule does not apply to allenes, which do not show cis-trans isomerism at all (see p. 103).
[160]Cope; Moore; Moore *J. Am. Chem. Soc.* **1959**, *81*, 3153.
[161]For a review, see Ref. 49, pp. 41-71.

N-methyl-N-benzylthiomesitylide (**52** and **53**),[162] the isomers of which are stable in the crystalline state but interconvert with a half-life of about 25 hr in CDCl₃ at 50°C.[163] This

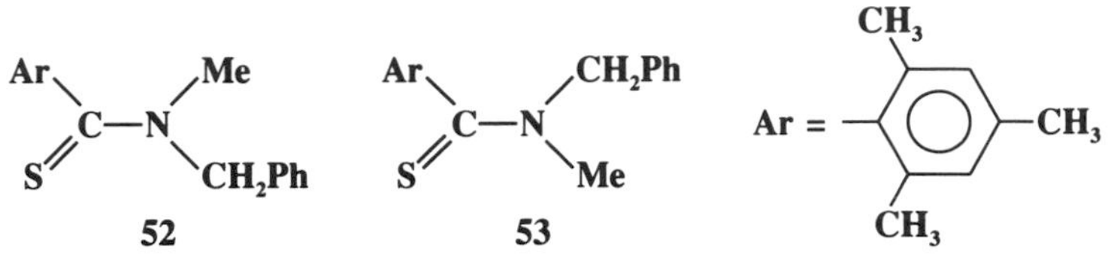

type of isomerism is rare; it is found chiefly in certain amides and thioamides, because resonance gives the single bond some double-bond character and slows rotation.[46] (For other examples of restricted rotation about single bonds, see pp. 161–163).

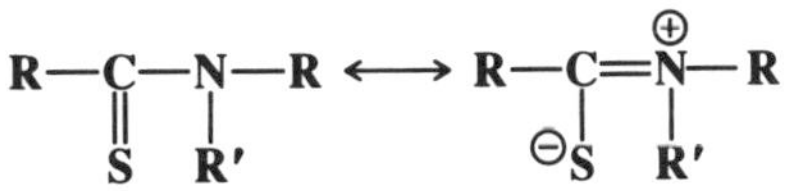

Conversely, there are compounds in which nearly free rotation is possible around what are formally C=C double bonds. These compounds, called *push–pull* or *captodative* ethylenes, have two electron-withdrawing groups on one carbon and two electron-donating groups on the other (**54**).[164] The contribution of diionic canonical forms such as the one

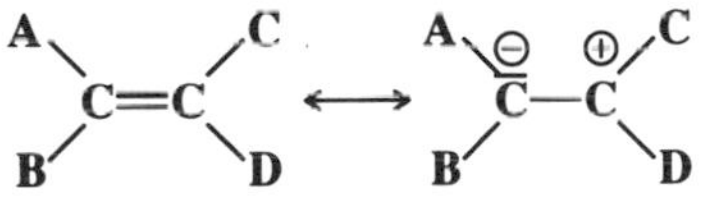

54

A, B = electron-withdrawing
C, D = electron-donating

shown decreases the double-bond character and allows easier rotation. For example, the compound **55** has a barrier to rotation of 13 kcal/mol (55 kJ/mol)[165], compared to a typical value of about 62-65 kcal/mol (260-270 kJ/mol) for simple alkenes.

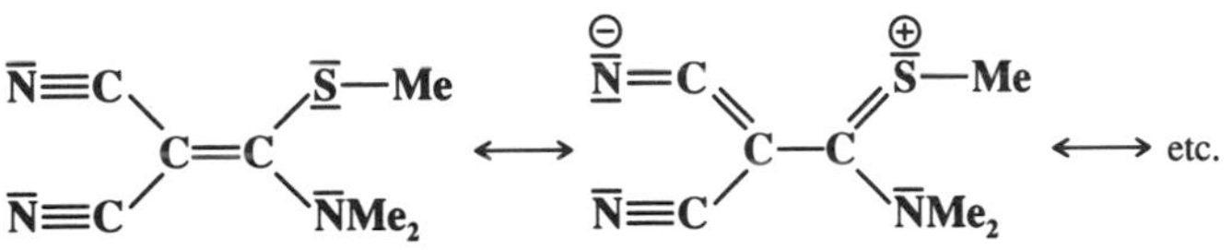

55

Since they are diastereomers, cis–trans isomers always differ in properties; the differences may range from very slight to considerable. The properties of maleic acid are so different from those of fumaric acid (Table 4.2) that it is not surprising that they have different names.

[162]Mannschreck *Angew. Chem. Int. Ed. Engl.* **1965**, *4*, 985 [*Angew. Chem.* 77, 1032]. See also Toldy; Radics *Tetrahedron Lett.* **1966**, 4753; Völter; Helmchen *Tetrahedron Lett.* **1978**, 1251; Walter; Hühnerfuss *Tetrahedron Lett.* **1981**, *22*, 2147.
[163]This is another example of atropisomerism (p. 102).
[164]For reviews, see Sandström *Top. Stereochem.* **1983**, *14*, 83-181; Ref. 49, pp. 111-125.
[165]Sandström; Wennerbeck *Acta Chem. Scand., Ser. B* **1978**, *32*, 421.

TABLE 4.2 Some properties of maleic and fumaric acids

Property	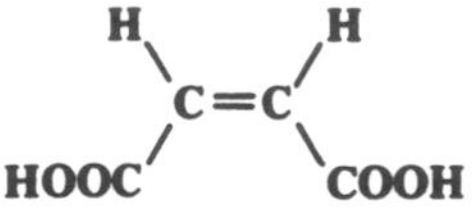 Maleic acid	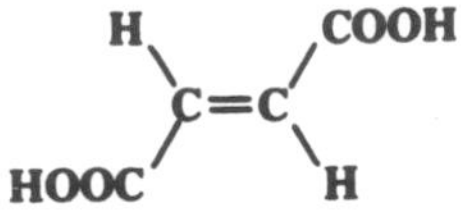 Fumaric acid
Melting point, °C	130	286
Solubility in water at 25°C, g/liter	788	7
K_1 (at 25°C)	1.5×10^{-2}	1×10^{-3}
K_2 (at 25°C)	2.6×10^{-7}	3×10^{-5}

Since they generally have more symmetry than cis isomers, trans isomers in most cases have higher melting points and lower solubilities in inert solvents. The cis isomer usually has a higher heat of combustion, which indicates a lower thermochemical stability. Other noticeably different properties are densities, acid strengths, boiling points, and various types of spectra, but the differences are too involved to be discussed here.

Cis–Trans Isomerism of Monocyclic Compounds

Although rings of four carbons and larger are not generally planar (see p. 148), they will be treated as such in this section, since the correct number of isomers can be determined when this is done[166] and the principles are easier to visualize (see p. 145).

The presence of a ring, like that of a double bond, prevents rotation. Cis and trans isomers are possible whenever there are two carbons on a ring, each of which is substituted by two different groups. The two carbons need not be adjacent. Examples are

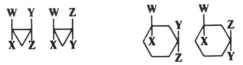

As with double bonds, W may equal Y and X may equal Z, but W may not equal X and Y may not equal Z if cis and trans isomers are to be possible. There is an important difference from the double-bond case: The substituted carbons are chiral carbons. This means that there are not *only* two isomers. In the most general case, where W, X, Y, and Z are all different, there are four isomers since neither the cis nor the trans isomer is superimposable on its mirror image. This is true regardless of ring size or which carbons are involved, except that in rings of even-numbered size when W, X, Y, and Z are at opposite corners, no chirality is present, e.g., **56**. In this case the substituted carbons are *not* chiral carbons. Note also that

56

---

[166]For a discussion of why this is so, see Leonard; Hammond; Simmons *J. Am. Chem. Soc.* **1975**, *97*, 5052.

a plane of symmetry exists in such compounds. When W = Y and X = Z, the cis isomer is always superimposable on its mirror image and hence is a meso compound, while the trans isomer consists of a *dl* pair, except in the case noted above. Again, the cis isomer has a plane of symmetry while the trans does not.

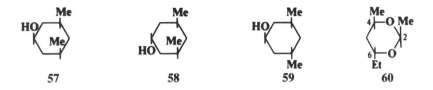

cis meso   trans *dl* pair

Rings with more than two differently substituted carbons can be dealt with on similar principles. In some cases it is not easy to tell the number of isomers by inspection.[82] The best method for the student is to count the number *n* of differently substituted carbons (these will usually be asymmetric, but not always, e.g., in **56**) and then to draw 2^n structures, crossing out those that can be superimposed on others (usually the easiest method is to look for a plane of symmetry). By this means it can be determined that for 1,2,3-cyclohexanetriol there are two meso compounds and a *dl* pair; and for 1,2,3,4,5,6-hexachlorocyclohexane there are seven meso compounds and a *dl* pair. The drawing of these structures is left as an exercise for the student.

Similar principles apply to heterocyclic rings as long as there are carbons (or other ring atoms) containing two different groups.

Cyclic stereoisomers containing only two differently substituted carbons are named either cis or trans, as previously indicated. The *Z, E* system is not used for cyclic compounds. However, cis–trans nomenclature will not suffice for compounds with more than two differently substituted atoms. For these compounds, a system is used in which the configuration of each group is given with respect to a reference group, which is chosen as the group attached to the lowest-numbered ring member bearing a substituent giving rise to cis–trans isomerism. The reference group is indicated by the symbol *r*. Three stereoisomers named according to this system are *c*-3,*c*-5-dimethylcyclohexan-*r*-1-ol (**57**), *t*-3,*t*-5-dimethylcyclohexan-*r*-1-ol (**58**), and *c*-3,*t*-5-dimethylcyclohexan-*r*-1-ol (**59**). The last

57	**58**	**59**	**60**

example demonstrates the rule that when there are two otherwise equivalent ways of going around the ring, one chooses the path that gives the cis designation to the first substituent after the reference. Another example is *r*-2,*c*-4-dimethyl-*t*-6-ethyl-1,3-dioxane (**60**).

Cis–Trans Isomerism of Fused and Bridged Ring Systems

Fused bicyclic systems are those in which two rings share two and only two atoms. In such systems there is no new principle. The fusion may be cis or trans, as illustrated by *cis*- and *trans*-decalin. However, when the rings are small enough, the trans configuration is impossible and the junction must be cis. The smallest trans junction that has been prepared when one ring is four-membered is a four-five junction; *trans*-bicyclo[3.2.0]heptane (**61**) is known.[167] For the bicyclo[2.2.0] system (a four-four fusion), only cis compounds have been

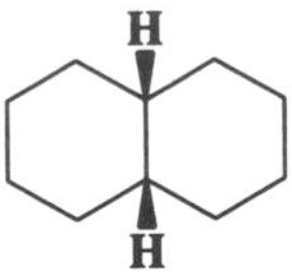

cis-Decalin 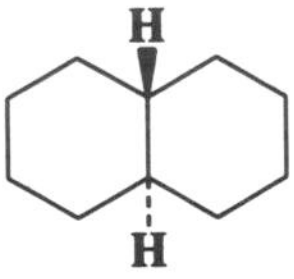   *trans*-Decalin 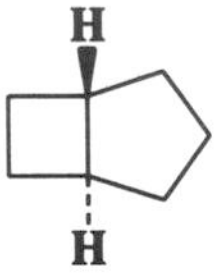   **61** 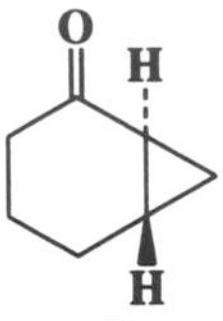   **62**

made. The smallest known trans junction when one ring is three-membered is a six-three junction (a bicyclo[4.1.0] system). An example is **62**.[168] When one ring is three-membered and the other eight-membered (an eight-three junction), the trans-fused isomer is more stable than the corresponding cis-fused isomer.[169]

In *bridged* bicyclic ring systems, two rings share more than two atoms. In these cases there may be fewer than 2^n isomers because of the structure of the system. For example, there are only two isomers of camphor (a pair of enantiomers), although it has two chiral

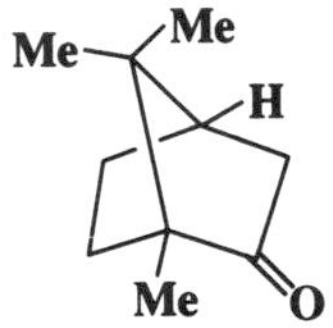

camphor

carbons. In both isomers the methyl and hydrogen are cis. The trans pair of enantiomers is impossible in this case, since the bridge *must* be cis. The smallest bridged system so far prepared in which the bridge is trans is the [4.3.1] system; the trans ketone **63** has been

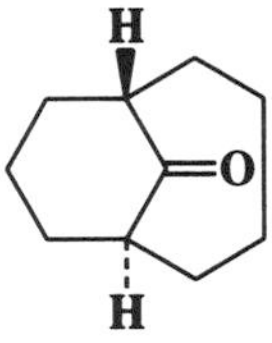

63

prepared.[170] In this case there are four isomers, since both the trans and the cis (which has also been prepared) are pairs of enantiomers.

When one of the bridges contains a substituent, the question arises as to how to name the isomers involved. When the two bridges that do *not* contain the substituent are of unequal length, the rule generally followed is that the prefix *endo-* is used when the substituent is

[167]Meinwald; Tufariello; Hurst *J. Org. Chem.* **1964**, *29*, 2914.
[168]Paukstelis; Kao *J. Am. Chem. Soc.* **1972**, *94*, 4783. For references to other examples, see Gassman; Bonser *J. Am. Chem. Soc.* **1988**, *110*, 2309.
[169]Corbally; Perkins; Carson; Laye; Steele *J. Chem. Soc., Chem. Commun.* **1978**, 778.
[170]Winkler; Hey; Williard *Tetrahedron Lett.* **1988**, *29*, 4691.

closer to the longer of the two unsubstituted bridges; the prefix *exo-* is used when the substituent is closer to the shorter bridge; e.g.,

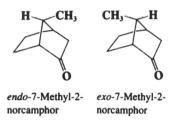

exo-2-Norborneol   endo-2-Norborneol

When the two bridges not containing the substituent are of equal length, this convention cannot be applied, but in some cases a decision can still be made; e.g., if one of the two bridges contains a functional group, the endo isomer is the one in which the substituent is closer to the functional group:

endo-7-Methyl-2-   exo-7-Methyl-2-
norcamphor   norcamphor

Out–In Isomerism

Another type of stereoisomerism, called *out–in isomerism*,[171] is found in salts of tricyclic diamines with nitrogen at the bridgeheads. In cases where k, l, and $m > 6$, the N—H bonds can be inside the molecular cavity or outside, giving rise to three isomers, as shown. Simmons and Park[172] isolated several such isomers with k, l, and m varying from 6 to 10. In the 9,9,9

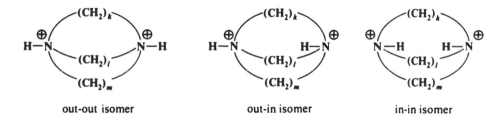

out-out isomer   out-in isomer   in-in isomer

compound, the cavity of the in–in isomer is large enough to encapsulate a chloride ion that is hydrogen bonded to the two N—H groups. The species thus formed is a cryptate, but differs from the cryptates discussed at p. 83 in that there is a negative rather than a positive ion enclosed.[173] Even smaller ones (e.g., the 4,4,4 compound) have been shown to form

[171]For reviews, see Alder *Tetrahedron* **1990**, *46*, 683-713, *Acc. Chem. Res.* **1983**, *16*, 321-327.
[172]Simmons; Park *J. Am. Chem. Soc.* **1968**, *90*, 2428; Park; Simmons *J. Am. Chem. Soc.* **1968**, *90*, 2429, 2431; Simmons; Park; Uyeda; Habibi *Trans. N.Y. Acad. Sci.* **1970**, *32*, 521. See also Dietrich; Lehn; Sauvage *Tetrahedron Lett.* **1969**, 2885, 2889, *Tetrahedron* **1973**, *29*, 1647; Dietrich; Lehn; Sauvage; Blanzat *Tetrahedron* **1973**, *29*, 1629.
[173]For reviews, see Schmidtchen; Gleich; Schummer *Pure. Appl. Chem.* **1989**, *61*, 1535-1546; Pierre; Baret *Bull. Soc. Chim. Fr.* **1983**, II-367-II-380. See also Hosseini; Lehn *Helv. Chim. Acta* **1988**, *71*, 749.

mono-inside-protonated ions.[174] Out–in and in–in isomers have also been prepared in analogous all-carbon tricyclic systems.[175]

In the compound **64**, which has four quaternary nitrogens, a halide ion has been encapsulated without a hydrogen being present on a nitrogen.[176] This ion does not display out–in isomerism.

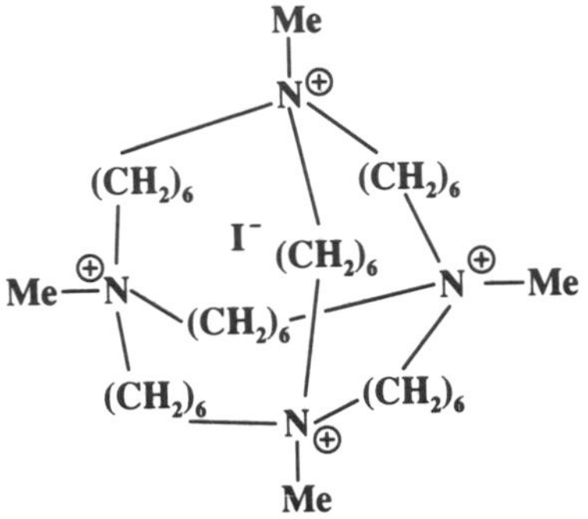

64

Enantiotopic and Diastereotopic Atoms, Groups, and Faces[177]

Many molecules contain atoms or groups which appear to be equivalent but which a close inspection will show to be actually different. We can test whether two atoms are equivalent by replacing each of them in turn with some other atom or group. If the new molecules created by this process are identical, the original atoms are equivalent; otherwise not. We can distinguish three cases.

1. In the case of malonic acid $CH_2(COOH)_2$, propane CH_2Me_2, or any other molecule of the form CH_2Y_2,[178] if we replace either of the CH_2 hydrogens by a group Z, the identical compound results. The two hydrogens are thus equivalent. Equivalent atoms and groups need not, of course, be located on the same carbon atom. For example, all the chlorine atoms of hexachlorobenzene are equivalent as are the two bromine atoms of 1,3-dibromopropane.

2. In the case of ethanol CH_2MeOH, if we replace one of the CH_2 hydrogens by a group Z, we get one enantiomer of the compound ZCHMeOH (**65**), while replacement of the other hydrogen gives the *other* enantiomer (**66**). Since the two compounds that result upon

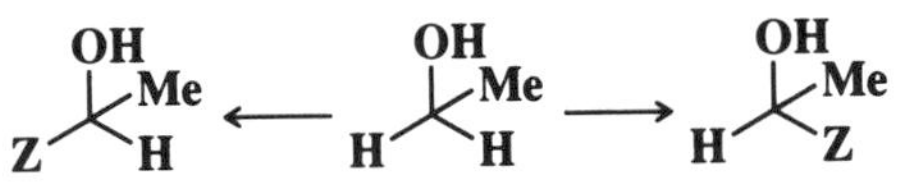

65   **66**

[174]Alder; Moss; Sessions *J. Chem. Soc., Chem. Commun.* **1983**, 997, 1000; Alder; Orpen; Sessions *J. Chem. Soc., Chem. Commun.* **1983**, 999; Dietrich; Lehn; Guilhem; Pascard *Tetrahedron Lett.* **1989**, *30*, 4125; Wallon; Peter-Katalinić; Werner; Müller; Vögtle *Chem. Ber.* **1990**, *123*, 375.

[175]Park; Simmons *J. Am. Chem. Soc.* **1972**, *94*, 7184; Gassman; Thummel *J. Am. Chem. Soc.* **1972**, *94*, 7183; Gassman; Hoye *J. Am. Chem. Soc.* **1981**, *103*, 215; McMurry; Hodge *J. Am. Chem. Soc.* **1984**, *106*, 6450; Winkler; Hey; Williard *J. Am. Chem. Soc.* **1986**, *108*, 6425.

[176]Schmidtchen; Müller *J. Chem. Soc., Chem. Commun.* **1984**, 1115. See also Schmidtchen *J. Am. Chem. Soc.* **1986**, *108*, 8249, *Top. Curr. Chem.* **1986**, *132*, 101-133.

[177]These terms were coined by Mislow. For lengthy discussions of this subject, see Eliel *Top. Curr. Chem.* **1982**, *105*, 1-76, *J. Chem. Educ.* **1980**, *57*, 52; Mislow; Raban *Top. Stereochem.* **1967**, *1*, 1-38. See also Ault *J. Chem. Educ.* **1974**, *51*, 729; Kaloustian; Kaloustian *J. Chem. Educ.* **1975**, *52*, 56; Jennings *Chem. Rev.* **1975**, *75*, 307-322.

[178]In the case where Y is itself a chiral group, this statement is only true when the two Y groups have the same configuration.

replacement of H by Z (**65** and **66**) are not identical but enantiomeric, the hydrogens are *not* equivalent. We define as *enantiotopic* two atoms or groups that upon replacement with a third group give enantiomers. In any symmetrical environment the two hydrogens behave as equivalent, but in a dissymmetrical environment they may behave differently. For example, in a reaction with a chiral reagent they may be attacked at different rates. This has its most important consequences in enzymatic reactions,[179] since enzymes are capable of much greater discrimination than ordinary chiral reagents. An example is found in the Krebs cycle, in biological organisms, where oxaloacetic acid (**67**) is converted to α-oxoglutaric acid (**69**) by a sequence that includes citric acid (**68**) as an intermediate. When **67** is labeled with ^{14}C at the 4 position, the label is found only at C-1 of **69**, despite the fact that **68** is not

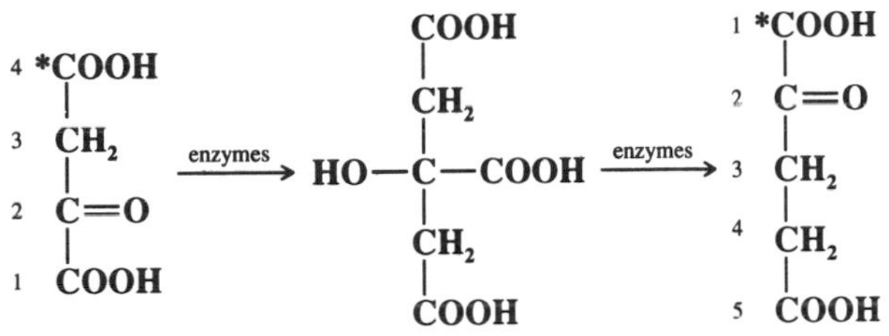

chiral. The two CH_2COOH groups of **68** are enantiotopic and the enzyme easily discriminates between them.[180] Note that the X atoms or groups of any molecule of the form CX_2WY are always enantiotopic if neither W nor Y is chiral, though enantiotopic atoms and groups may also be found in other molecules, e.g., the hydrogen atoms in 3-fluoro-3-chlorocyclopropene (**70**). In this case, substitution of an H by a group Z makes the C-3

70

atom asymmetric and substitution at C-1 gives the opposite enantiomer from substitution at C-2.

The term *prochiral*[181] is used for a compound or group that has two enantiotopic atoms or groups, e.g., CX_2WY. That atom or group X that would lead to an R compound if preferred to the other is called *pro-R*. The other is *pro-S*; e.g.,

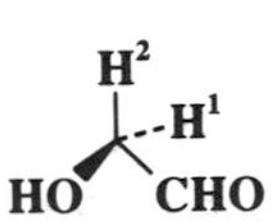

[179]For a review, see Benner; Glasfeld; Piccirilli *Top. Stereochem.* **1989**, *19*, 127-207. For a nonenzymatic example, see Job; Bruice *J. Am. Chem. Soc.* **1974**, *96*, 809.

[180]The experiments were carried out by Evans; Slotin *J. Biol. Chem.* **1941**, *141*, 439; Wood; Werkman; Hemingway; Nier *J. Biol. Chem.* **1942**, *142*, 31. The correct interpretation was given by Ogston *Nature* **1948**, *162*, 963. For discussion, see Hirschmann, in Florkin; Stotz *Comprehensive Biochemistry*, vol. 12, pp. 236-260, Elsevier: New York, 1964; Cornforth *Tetrahedron* **1974**, *30*, 1515; Vennesland *Top. Curr. Chem.* **1974**, *48*, 39-65; Eliel *Top. Curr. Chem.*, Ref. 177, pp. 5-7, 45-70.

[181]Hanson *J. Am. Chem. Soc.* **1966**, *88*, 2731; Hirschmann; Hanson *Tetrahedron* **1974**, *30*, 3649.

3. Where two atoms or groups in a molecule are in such positions that replacing each of them in turn by a group Z gives rise to diastereomers, the atoms or groups are called *diastereotopic*. Some examples are the CH_2 groups of 2-chlorobutane (**71**), vinyl chloride (**72**), and chlorocyclopropane (**73**) and the two olefinic hydrogens of **74**. Diastereotopic atoms and groups are different in any environment, chiral or achiral. These hydrogens react

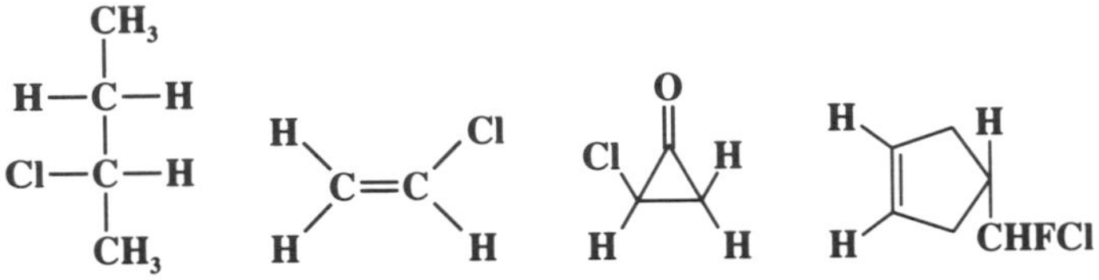

at different rates with achiral reagents, but an even more important consequence is that in nmr spectra, diastereotopic hydrogens theoretically give different peaks and split each other. This is in sharp contrast to equivalent or enantiotopic hydrogens, which are indistinguishable in the nmr, except when chiral solvents are used, in which case enantiotopic (but not equivalent) protons give different peaks.[182] The term *isochronous* is used for hydrogens that are indistinguishable in the nmr.[183] In practice, the nmr signals from diastereotopic protons are often found to be indistinguishable, but this is merely because they are very close together. Theoretically they are distinct, and they have been resolved in many cases. When they appear together, it is sometimes possible to resolve them by the use of lanthanide shift reagents (p. 126) or by changing the solvent or concentration. Note that X atoms or groups CX_2WY are diastereotopic if either W or Y is chiral.

Just as there are enantiotopic and diastereotopic atoms and groups, so we may distinguish *enantiotopic and diastereotopic faces* in trigonal molecules. Again we have three cases: (1) In formaldehyde or acetone (**G**), attack by an achiral reagent A from either face of the molecule gives rise to the same transition state and product; the two faces are thus equivalent. (2) In butanone or acetaldehyde (**H**), attack by an achiral A at one face gives a transition

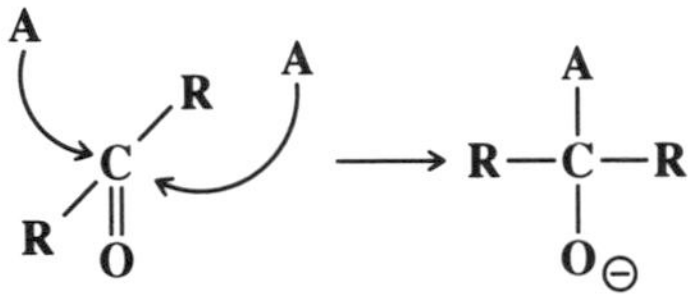

[182]Pirkle *J. Am. Chem. Soc.* **1966**, *88*, 1837; Burlingame; Pirkle *J. Am. Chem. Soc.* **1966**, *88*, 4294; Pirkle; Burlingame *Tetrahedron Lett.* **1967**, 4039.
[183]For a review of isochronous and nonisochronous nuclei in the nmr, see van Gorkom; Hall *Q. Rev., Chem. Soc.* **1968**, *22*, 14-29. For a discussion, see Silverstein; LaLonde *J. Chem. Educ.* **1980**, *57*, 343.

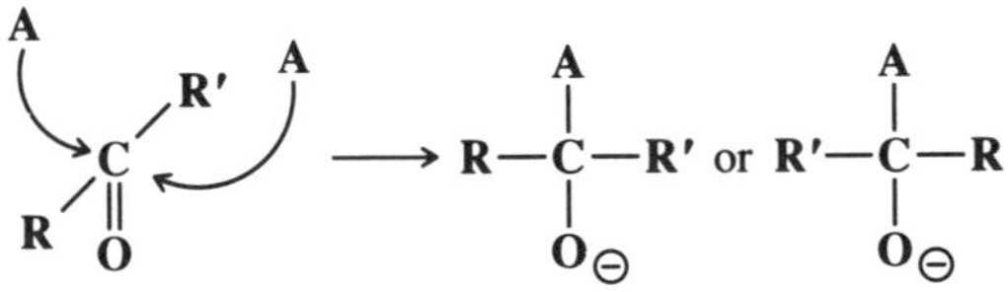

state and product that are the enantiomers of those arising from attack at the other face. Such faces are enantiotopic. As we have already seen (p. 106), a racemic mixture must result in this situation. However, attack at an enantiotopic face by a chiral reagent gives diastereomers, which are not formed in equal amounts. (3) In a case like **75**, the two faces are

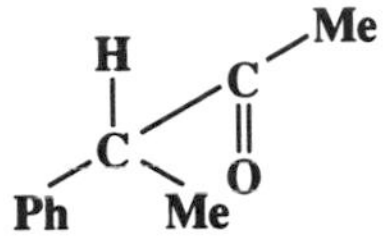

75

obviously not equivalent and are called diastereotopic. Enantiotopic and diastereotopic faces can be named by an extension of the Cahn–Ingold–Prelog system.[181] If the three groups as arranged by the sequence rules have the order X > Y > Z, that face in which the groups in this sequence are clockwise (as in **J**) is the *Re* face (from Latin *rectus*) whereas **K** shows the *Si* face (from Latin *sinister*).

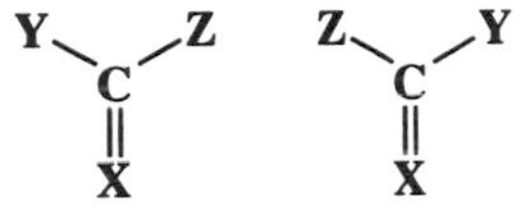

J   **K**

Stereospecific and Stereoselective Syntheses

Any reaction in which only one of a set of stereoisomers is formed exclusively or predominantly is called a *stereoselective* synthesis.[184] The same term is used when a mixture of two or more stereoisomers is exclusively or predominantly formed at the expense of other stereoisomers. In a *stereospecific* reaction, a given isomer leads to one product while another stereoisomer leads to the opposite product. All stereospecific reactions are necessarily stereoselective, but the converse is not true. These terms are best illustrated by examples. Thus, if maleic acid treated with bromine gives the *dl* pair of 2,3-dibromosuccinic acid while fumaric acid gives the meso isomer (this is the case), the reaction is stereospecific as well as stereoselective because two opposite isomers give two opposite isomers:

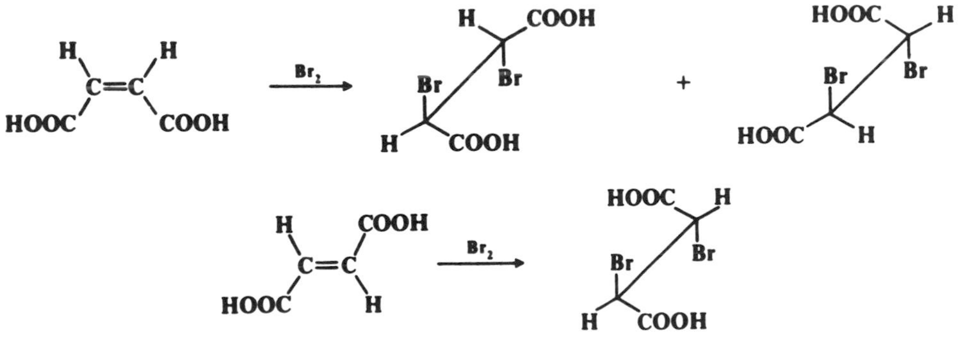

---

[184]For a further discussion of these terms and of stereoselective reactions in general, see Eliel *Stereochemisty of Carbon Compounds*, Ref. 1, pp. 434-446. For a review of how certain reactions can be run with stereocontrol, see Bartlett *Tetrahedron* **1980**, *36*, 2-72.

However, if both maleic and fumaric acid gave the *dl* pair or a mixture in which the *dl* pair predominated, the reaction would be stereoselective but not stereospecific. If more or less equal amounts of *dl* and meso forms were produced in each case, the reaction would be nonstereoselective. A consequence of these definitions is that if a reaction is carried out on a compound that has no stereoisomers, it cannot be stereospecific, but at most stereoselective. For example, addition of bromine to methylacetylene could (and does) result in preferential formation of *trans*-1,2-dibromopropene, but this can be only a stereoselective, not a stereospecific reaction.

CONFORMATIONAL ANALYSIS

If two different three-dimensional arrangements in space of the atoms in a molecule are interconvertible merely by free rotation about bonds, they are called *conformations*; if not, *configurations*.[185] Configurations represent *isomers* that can be separated, as previously discussed in this chapter. Conformations represent *conformers*, which are rapidly interconvertible and thus nonseparable. The terms "conformational isomer" and "rotamer" are sometimes used instead of "conformer." A number of methods have been used to determine conformations.[186] These include x-ray and electron diffraction, ir, Raman, uv, nmr,[187] and microwave spectra,[188] photoelectron spectroscopy,[189] supersonic molecular jet spectroscopy,[190] and optical rotatory dispersion and circular dichroism measurements.[191] Some of these methods are useful only for solids. It must be kept in mind that the conformation of a molecule in the solid state is not necessarily the same as in solution.[192] Conformations can be *calculated* by a method called molecular mechanics (p. 149).

[185]For books on conformational analysis, see Dale *Stereochemistry and Conformational Analysis*; Verlag Chemie: Deerfield Beach, FL, 1978; Chiurdoglu *Conformational Analysis*; Academic Press: New York, 1971; Eliel; Allinger; Angyal; Morrison *Conformational Analysis*; Wiley: New York, 1965; Hanack *Conformation Theory*; Academic Press: New York, 1965. For reviews, see Dale *Top. Stereochem.* **1976**, *9*, 199-270; Truax; Wieser *Chem. Soc. Rev.* **1976**, *5*, 411-429; Eliel *J. Chem. Educ.* **1975**, *52*, 762-767; Bastiansen; Seip; Boggs *Perspect. Struct. Chem.* **1971**, *4*, 60-165; Bushweller; Gianni, in Patai *The Chemistry of Functional Groups, Supplement E*; Wiley: New York, 1980, pp. 215-278.
[186]For a review, see Eliel; Allinger; Angyal; Morrison, Ref. 185, pp. 129-188.
[187]For monographs on the use of nmr to study conformational questions, see Ōki, Ref. 49; Marshall *Carbon–Carbon and Carbon–Proton NMR Couplings*; VCH: New York, 1983. For reviews, see Anet; Anet, in Nachod; Zuckerman *Determination of Organic Structures by Physical Methods*, vol. 3; Academic Press: New York, 1971, pp. 343-420; Kessler *Angew. Chem. Int. Ed. Engl.* **1970**, *9*, 219-235 [*Angew. Chem. 82*, 237-253]; Ivanova; Kugatova-Shemyakina *Russ. Chem. Rev.* **1970**, *39*, 510-528; Anderson *Q. Rev. Chem. Soc.* **1965**, *19*, 426-439; Franklin; Feltkamp *Angew. Chem. Int. Ed. Engl.* **1965**, *4*, 774-783 [*Angew. Chem. 77*, 798-807]; Johnson *Adv. Magn. Reson.* **1965**, *1*, 33-102. See also Whitesell; Minton *Stereochemical Analysis of Alicyclic Compounds by C-13 NMR Spectroscopy*; Chapman and Hall: New York, 1987.
[188]For a review see Wilson *Chem. Soc. Rev.* **1972**, *1*, 293-318.
[189]For a review, see Klessinger; Rademacher *Angew. Chem. Int. Ed. Engl.* **1979**, *18*, 826-837 [*Angew. Chem. 91*, 885-896].
[190]Breen; Warren; Bernstein; Seeman *J. Am. Chem. Soc.* **1987**, *109*, 3453.
[191]For monographs, see Kagan *Determination of Configurations by Dipole Moments, CD, or ORD* (vol. 2 of Kagan, *Stereochemistry*); Georg Thieme Publishers: Stuttgart, 1977; Crabbé *ORD and CD in Chemistry and Biochemistry*; Academic Press: New York, 1972, *Optical Rotatory Dispersion and Circular Dichroism in Organic Chemistry*; Holden-Day: San Francisco, 1965; Snatzke *Optical Rotatory Dispersion and Circular Dichroism in Organic Chemistry*; Sadtler Research Laboratories: Philadelphia, 1967; Velluz; Legrand; Grosjean *Optical Circular Dichroism*; Academic Press: New York, 1965. For reviews, see Smith, *Chem. Rev.* **1983**, *83*, 359-377; Håkansson, in Patai *The Chemistry of Acid Derivatives*, pt. 1; Wiley: New York, 1979, pp. 67-120; Hudec; Kirk *Tetrahedron* **1976**, *32*, 2475-2506; Schellman *Chem. Rev.* **1975**, *75*, 323-331; Velluz; Legrand *Bull. Soc. Chim. Fr.* **1970**, 1785-1795; Barrett, in Bentley; Kirby, Ref. 77, pt. 1, 1972, pp. 515-610; Snatzke *Angew. Chem. Int. Ed. Engl.* **1968**, *7*, 14-25 [*Angew. Chem. 18*, 15-26]; Crabbé in Nachod; Zuckerman, Ref. 187, vol. 3, pp. 133-205; Crabbé; Klyne *Tetrahedron* **1967**, *23*, 3449; Crabbé *Top. Stereochem.* **1967**, *1*, 93-198; Eyring; Liu; Caldwell *Chem. Rev.* **1968**, *68*, 525-540.
[192]See Kessler; Zimmermann; Förster; Engel; Oepen; Sheldrick *Angew. Chem. Int. Ed. Engl.* **1981**, *20*, 1053 [*Angew. Chem. 93*, 1085].

Conformation in Open-Chain Systems[193]

For any open-chain single bond that connects two sp^3 carbon atoms, an infinite number of conformations are possible, each of which has a certain energy associated with it. For ethane there are two extremes, a conformation of highest and one of lowest potential energy, depicted in two ways as:

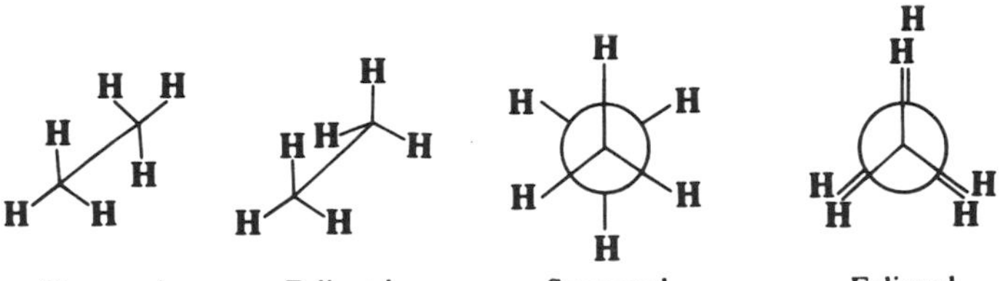

Staggered   Eclipsed   Staggered   Eclipsed

In *Newman projection formulas* (the two figures on the right) the observer looks at the C—C bond head on. The three lines emanating from the center of the circle represent the bonds coming from the front carbon, with respect to the observer.

The staggered conformation is the conformation of lowest potential energy for ethane. As the bond rotates, the energy gradually increases until the eclipsed conformation is reached, when the energy is at a maximum. Further rotation decreases the energy again. Figure 4.3 illustrates this. The *angle of torsion*, which is a dihedral angle, is the angle between the XCC and the CCY planes, as shown:

For ethane the difference in energy is about 2.9 kcal/mol (12 kJ/mol).[194] This difference is called the *energy barrier*, since in free rotation about a single bond there must be enough rotational energy present to cross the barrier every time two hydrogen atoms are opposite each other. There has been much speculation about the cause of the barriers and many explanations have been suggested.[195] It has been concluded from molecular-orbital calculations that the barrier is caused by repulsion between overlapping filled molecular orbitals.[196] That is, the ethane molecule has its lowest energy in the staggered conformation because in this conformation the orbitals of the C—H bonds have the least amount of overlap with the C—H orbitals of the adjacent carbon.

At ordinary temperatures enough rotational energy is present for the ethane molecule rapidly to rotate, though it still spends most of its time at or near the energy minimum. Groups larger than hydrogen cause larger barriers. When the barriers are large enough, as

---

[193]For a review, see Berg; Sandström *Adv. Phys. Org. Chem.* **1989,** *25,* 1-97.
[194]Lide *J. Chem. Phys.* **1958, 29,** 1426; Weiss; Leroi *J. Chem. Phys.* **1968,** *48,* 962; Hirota; Saito; Endo *J. Chem. Phys.* **1979,** *71,* 1183.
[195]For a review of methods of measuring barriers, of attempts to explain barriers, and of values of barriers, see Lowe *Prog. Phys. Org. Chem.* **1968,** *6,* 1-80. For other reviews of this subject, see Oosterhoff *Pure Appl. Chem.* **1971,** *25,* 563-571; Wyn-Jones; Pethrick *Top. Stereochem.* **1970,** *5,* 205-274; Pethrick; Wyn-Jones *Q. Rev., Chem. Soc.* **1969,** *23,* 301-324; Brier *J. Mol. Struct.* **1970,** *6,* 23-36; Lowe *Science* **1973,** *179,* 527-533.
[196]See Pitzer *Acc. Chem. Res.* **1983,** *16,* 207-210. See, however, Bader, Cheeseman; Laidig; Wiberg; Breneman *J. Am. Chem. Soc.* **1990,** *112,* 6350.

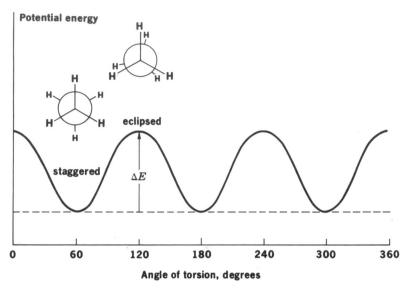

FIGURE 4.3 Conformational energy diagram for ethane.

in the case of suitably substituted biphenyls (p. 101) or the diadamantyl compound mentioned on p. 142, rotation at room temperature is completely prevented and we speak of configurations, not conformations. Even for compounds with small barriers, cooling to low temperatures may remove enough rotational energy for what would otherwise be conformational isomers to become configurational isomers.

A slightly more complicated case than ethane is that of a 1,2-disubstituted ethane (YCH_2—CH_2Y or YCH_2—CH_2X),[197] such as *n*-butane, for which there are four extremes: a fully staggered conformation, called *anti*, *trans*, or *antiperiplanar*; another staggered con-

anti, trans, or
antiperiplanar

anticlinal

gauche or
synclinal
L

synperiplanar

formation, called *gauche* or *synclinal*; and two types of eclipsed conformations, called *synperiplanar* and *anticlinal*. An energy diagram for this system is given in Figure 4.4. Although there is constant rotation about the central bond, it is possible to estimate what percentage of the molecules are in each conformation at a given time. For example, it was concluded from a consideration of dipole moment and polarizability measurements that for 1,2-dichloroethane in CCl_4 solution at 25°C about 70% of the molecules are in the anti and

[197]For discussions of the conformational analysis of such systems, see Kingsbury *J. Chem. Educ.* **1979**, *56*, 431-437; Wiberg; Murcko *J. Am. Chem. Soc.* **1988**, *110*, 8029; Allinger; Grev; Yates; Schaefer **1990**, *112*, 114.

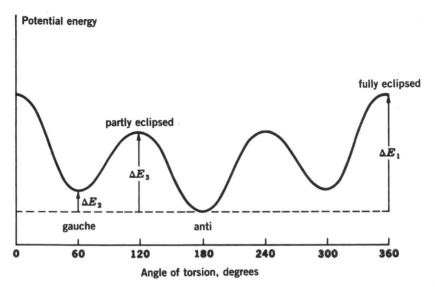

FIGURE 4.4   Conformational energy for YCH_2—CH_2Y or YCH_2—CH_2X. For *n*-butane, ΔE_1 = 4 to 6, ΔE_2 = 0.9, and ΔE_3 = 3.4 kcal/mol (17–25, 3.8, 14 kJ/mol, respectively).

about 30% in the gauche conformation.[198] The corresponding figures for 1,2-dibromoethane are 89% anti and 11% gauche.[199] The eclipsed conformations are unpopulated and serve only as pathways from one staggered conformation to another. Solids normally consist of a single conformer.

It may be observed that the gauche conformation of butane (**L**) or any other similar molecule is chiral. The lack of optical activity in such compounds arises from the fact that **L** and its mirror image are always present in equal amounts and interconvert too rapidly for separation.

For butane and for most other molecules of the forms YCH_2—CH_2Y and YCH_2—CH_2X, the anti conformer is the most stable, but exceptions are known. One group of exceptions consists of molecules containing small electronegative atoms, especially fluorine and oxygen. Thus 2-fluoroethanol,[200] 1,2-difluoroethane,[201] and 2-fluoroethyl trichloroacetate $(FCH_2CH_2OCOCCl_3)$[202] exist predominantly in the gauche form and compounds such as 2-chloroethanol and 2-bromoethanol[200] also prefer the gauche form. There is as yet no generally accepted explanation for this behavior.[203] It was believed that the favorable gauche conformation of 2-fluoroethanol was the result of intramolecular hydrogen bonding,

[198]Aroney; Izsak; Le Fèvre *J. Chem. Soc.* **1962**, 1407; Le Fèvre; Orr *Aust. J. Chem.* **1964**, *17*, 1098.

[199]The anti form of butane itself is also more stable than the gauche form: Schrumpf *Angew. Chem. Int. Ed. Engl.* **1982**, *21*, 146 [*Angew. Chem. 94*, 152].

[200]Wyn-Jones; Orville-Thomas *J. Mol. Struct.* **1967**, *1*, 79; Buckley; Giguère; Yamamoto *Can. J. Chem.* **1968**, *46*, 2917; Davenport; Schwartz *J. Mol. Struct.* **1978**, *50*, 259; Huang; Hedberg *J. Am. Chem. Soc.* **1989**, *111*, 6909.

[201]Klaboe; Nielsen *J. Chem. Phys.* **1960**, *33*, 1764; Abraham; Kemp *J. Chem. Soc. B* **1971**, 1240; Bulthuis; van den Berg; Maclean *J. Mol. Struct.* **1973**, *16*, 11; van Schaick; Geise; Mijlhoff; Renes *J. Mol. Struct.* **1973**, *16*, 23; Friesen; Hedberg *J. Am. Chem. Soc.* **1980**, *102*, 3987; Fernholt; Kveseth *Acta Chem. Scand., Ser. A* **1980**, *34*, 163.

[202]Abraham; Monasterios *Org. Magn. Reson.* **1973**, *5*, 305.

[203]It has been proposed that the preference for the gauche conformation in these molecules is an example of a more general phoenomenon, known as the *gauche effect*, i.e., a tendency to adopt that structure that has the maximum number of gauche interactions between adjacent electron pairs or polar bonds. This effect is ascribed to nuclear electron attactive forces between the groups or unshared pairs: Wolfe; Rauk; Tel; Csizmadia *J. Chem. Soc. B* **1971**, 136; Wolfe *Acc. Chem. Res.* **1972**, *5*, 102-111. See also Phillips; Wray *J. Chem. Soc., Chem. Commun.* **1973**, 90; Radom; Hehre; Pople *J. Am. Chem. Soc.* **1972**, *94*, 2371; Zefirov *J. Org. Chem. USSR* **1974**, *10*, 1147; Juaristi *J. Chem. Educ.* **1979**, *56*, 438.

but this explanation does not do for molecules like 2-fluoroethyl trichloroacetate and has in fact been ruled out for 2-fluoroethanol as well.[204] Other exceptions are known, where small electronegative atoms are absent. For example 1,1,2,2-tetrachloroethane and 1,1,2,2-tetrabromoethane both prefer the gauche conformation,[205] even though 1,1,2,2-tetrafluoroethane prefers the anti.[206] Also, both 2,3-dimethylpentane and 3,4-dimethylhexane prefer the gauche conformation,[207] and 2,3-dimethylbutane shows no preference for either.[208] Furthermore, the solvent can exert a powerful effect. For example, the compound 2,3-dinitro-2,3-dimethylbutane exists entirely in the gauche conformation in the solid state, but in benzene, the gauche–anti ratio is 79:21; while in CCl_4 the anti form is actually favored (gauche–anti ratio 42:58).[209]

In one case two conformational isomers of a single aliphatic hydrocarbon, 3,4-di(1-adamantyl)-2,2,5,5-tetramethylhexane, have proven stable enough for isolation at room temperature.[210] The two isomers **M** and **N** were separately crystallized, and the struc-

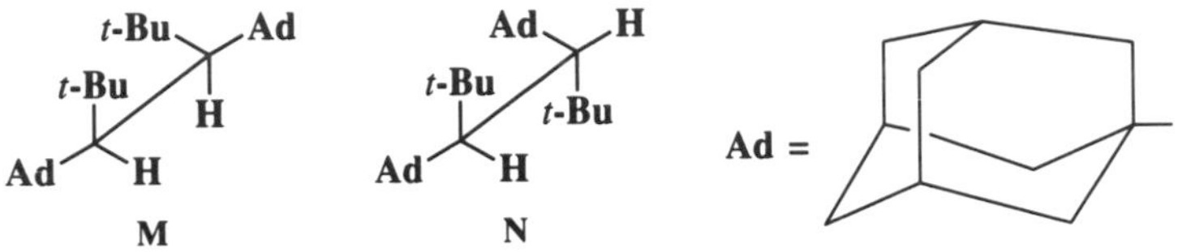

tures proved by x-ray crystallography. (The actual dihedral angles are distorted from the 60° angles shown in the drawings, owing to steric hindrance between the large groups.)

All the conformations so far discussed have involved rotation about sp^3–sp^3 bonds. Many studies have also been made of compounds with sp^3–sp^2 bonds.[211] For example, propanal (or any similar molecule) has four extreme conformations, two of which are called *eclipsing* and the other two *bisecting*. For propanal the eclipsing conformations have lower energy than the other two, with **P** favored over **Q** by about 1 kcal/mol (4 kJ/mol).[212] As has already

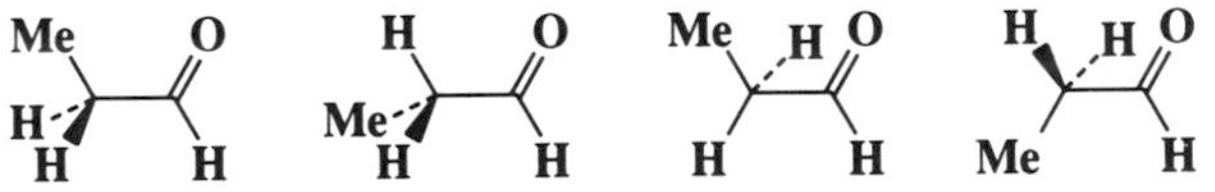

been pointed out (p. 128), for a few of these compounds, rotation is slow enough to permit cis–trans isomerism, though for simple compounds rotation is rapid. For example, acetaldehyde has a lower rotational barrier (about 1 kcal/mol or 4 kJ/mol) than ethane.[213]

[204]Griffith; Roberts *Tetrahedron Lett.* **1974**, 3499.

[205]Kagarise *J. Chem. Phys.* **1956**, *24*, 300.

[206]Brown; Beagley *J. Mol. Struct.* **1977**, *38*, 167.

[207]Ritter; Hull; Cantow *Tetrahedron Lett.* **1978**, 3093.

[208]Lunazzi; Macciantelli; Bernardi; Ingold *J. Am. Chem. Soc.* **1977**, *99*, 4573.

[209]Tan; Chia; Huang; Kuok; Tang *J. Chem. Soc., Perkin Trans. 2* **1984**, 1407.

[210]Flamm-ter Meer; Beckhaus; Peters; von Schnering; Fritz; Rüchardt *Chem. Ber.* **1986**, *119*, 1492; Rüchardt; Beckhaus *Angew. Chem. Int. Ed. Engl.* **1985**, *24*, 529-538 [*Angew. Chem.* 97, 531-540].

[211]For reviews, see Sinegovskaya; Keiko; Trofimov *Sulfur Rep.* **1987**, *7*, 337-378 (for enol ethers and thioethers); Karabatsos; Fenoglio *Top. Stereochem.* **1970**, *5*, 167-203; Jones; Owen *J. Mol. Struct.* **1973**, *18*, 1-32 (for carboxylic esters). See also Schweizer; Dunitz *Helv. Chim. Acta* **1982**, *65*, 1547; Chakrabarti; Dunitz *Helv. Chim. Acta* **1982**, *65*, 1555; Cossé-Barbi; Massat; Dubois *Bull. Soc. Chim. Belg.* **1985**, *94*, 919; Dorigo; Pratt; Houk *J. Am. Chem. Soc.* **1987**, *109*, 6591.

[212]Butcher; Wilson *J. Chem. Phys.* **1964**, *40*, 1671; Allinger; Hickey *J. Mol. Struct.* **1973**, *17*, 233; Gupta *Can. J. Chem.* **1985**, *63*, 984.

[213]Davidson; Allen *J. Chem. Phys.* **1971**, *54*, 2828.

Conformation in Six-membered Rings[214]

For cyclohexane there are two extreme conformations in which all the angles are tetra-hedral.[215] These are called the *boat* and the *chair* conformations and in each the ring is said to be *puckered*. The chair conformation is a rigid structure, but the boat form is flexible[216]

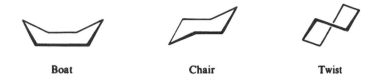

Boat   **Chair**   **Twist**

and can easily pass over to a somewhat more stable form known as the *twist* conformation. The twist form is about 1.5 kcal/mol (6.3 kJ/mol) more stable than the boat because it has less eclipsing interaction (see p. 156).[217] The chair form is more stable than the twist form by about 5 kcal/mol (21 kJ/mol).[218] In the vast majority of compounds containing a cyclo-hexane ring, the molecules exist almost entirely in the chair form. Yet it is known that the boat or twist form exists transiently. An inspection of the chair form shows that six of its bonds are directed differently from the other six:

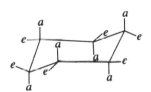

On each carbon, one bond is directed up or down and the other more or less in the "plane" of the ring. The up or down bonds are called *axial* and the others *equatorial*. The axial bonds point alternately up and down. If a molecule were frozen into a chair form, there would be isomerism in monosubstituted cyclohexanes. For example, there would be an equatorial methylcyclohexane and an axial isomer. However, it has never been possible to isolate isomers of this type at room temperature.[219] This proves the transient existence of the boat or twist form, since in order for the two types of methylcyclohexane to be non-separable, there must be rapid interconversion of one chair form to another (in which all axial bonds become equatorial and vice versa) and this is possible only through a boat or twist conformation. Conversion of one chair form to another requires an activation energy of about 10 kcal/mol (42 kJ/mol)[220] and is very rapid at room temperature.[221] However, by

[214]For reviews, see Jensen; Bushweller *Adv. Alicyclic Chem.* **1971**, *3*, 139-194; Robinson; Theobald *Q. Rev., Chem. Soc.* **1967**, *21*, 314-330; Eliel *Angew. Chem. Int. Ed. Engl.* **1965**, *4*, 761-774 [*Angew. Chem. 77*, 784-797].

[215]The C—C—C angles in cyclohexane are actually 111.5° [Davis; Hassel *Acta Chem. Scand.* **1963**, *17*,1181; Geise; Buys; Mijlhoff *J. Mol. Struct.* **1971**, *9*, 447; Bastiansen; Fernholt; Seip; Kambara; Kuchitsu *J. Mol. Struct.* **1973**, *18*, 163], but this is within the normal tetrahedral range (see p. 20).

[216]See Dunitz *J. Chem. Educ.* **1970**, *47*, 488.

[217]For a review of nonchair forms, see Kellie; Riddell *Top. Stereochem.* **1974**, *8*, 225-269.

[218]Margrave; Frisch; Bautista; Clarke; Johnson *J. Am. Chem. Soc.* **1963**, *85*, 546; Squillacote; Sheridan; Chapman; Anet *J. Am. Chem. Soc.* **1975**, *97*, 3244.

[219]Wehle; Fitjer *Tetrahedron Lett.* **1986**, *27*, 5843, have succeeded in producing two conformers that are indefinitely stable in solution at room temperature. However, the other five positions of the cyclohexane ring in this case are all spirosubstituted with cyclobutane rings, greatly increasing the barrier to chair–chair interconversion.

[220]Jensen; Noyce; Sederholm; Berlin *J. Am. Chem. Soc.* **1962**, *84*, 386; Anet; Ahmad; Hall *Proc. Chem. Soc.* **1964**, 145; Bovey; Hood; Anderson; Kornegay *J. Chem. Phys.* **1964**, *41*, 2041; Anet; Bourn *J. Am. Chem. Soc.* **1967**, *89*, 760. See also Strauss *J. Chem. Educ.* **1971**, *48*, 221.

[221]For reviews of chair–chair interconversions, see Ōki, Ref. 49, pp. 287-307; Anderson *Top. Curr. Chem.* **1974**, *45*, 139-167.

working at low temperatures, Jensen and Bushweller were able to obtain the pure equatorial conformers of chlorocyclohexane and trideuteriomethoxycyclohexane as solids and in solution.[222] Equatorial chlorocyclohexane has a half-life of 22 years in solution at −160°C.

In some molecules the twist conformation is actually preferred. An example is **76**, in which hydrogen bonding stabilizes the otherwise high-energy form.[223] Of course, in certain

76	norbornane	twistane

bicyclic compounds, the six-membered ring is forced to maintain a boat or twist conformation, as in norbornane or twistane.

In monosubstituted cyclohexanes, the substituent normally prefers the equatorial position because in the axial position there is interaction between the substituent and the axial hydrogens in the 3 and 5 positions, but the extent of this preference depends greatly on the nature of the group. Alkyl groups have a greater preference than polar groups and for alkyl groups the preference increases with size. For polar groups, size seems to be unimportant. Both the large HgBr[224] and HgCl[225] groups and the small F group have been reported to have little or no conformational preference (the HgCl group actually shows a slight preference for the axial position). Table 4.3 gives approximate values of the free energy required for various groups to go from the equatorial position to the axial (these are called *A* values),[226] though it must be kept in mind that they vary somewhat with physical state, temperature, and solvent.[227]

In disubstituted compounds, the rule for alkyl groups is that the conformation is such that as many groups as possible adopt the equatorial position. How far it is possible depends on the configuration. In a *cis*-1,2-disubstituted cyclohexane, one substituent must be axial and the other equatorial. In a *trans*-1,2 compound both may be equatorial or both axial. This is also true for 1,4-disubstituted cyclohexanes, but the reverse holds for 1,3 compounds: the trans isomer must have the *ae* conformation and the cis isomer may be *aa* or *ee*. For alkyl groups, the *ee* conformation predominates over the *aa* but for other groups this is not necessarily so. For example, both *trans*-1,4-dibromocyclohexane and the corresponding dichloro compound have the *ee* and *aa* conformations about equally populated[228] and most *trans*-1,2-dihalocyclohexanes exist predominantly in the *aa* conformation.[229] Note that in the

[222]Jensen; Bushweller *J. Am. Chem. Soc.* **1966**, *88*, 4279; **1969**, *91*, 3223.

[223]Stolow *J. Am. Chem. Soc.* **1961**, *83*, 2592, **1964**, *86*, 2170; Stolow; McDonagh; Bonaventura *J. Am. Chem. Soc.* **1964**, *86*, 2165. For some other examples, see Camps; Iglesias *Tetrahedron Lett.* **1985**, *26*, 5463; Fitjer; Scheuermann; Klages; Wehle; Stephenson; Binsch *Chem. Ber.* **1986**, *119*, 1144.

[224]Jensen; Gale *J. Am. Chem. Soc.* **1959**, *81*, 6337.

[225]Anet; Krane; Dodderel; Praeger *Tetrahedron Lett.* **1974**, 3255.

[226]Except where otherwise indicated, these values are from Jensen; Bushweller, Ref. 214. See also Ref. 238.

[227]See, for example, Ford; Allinger *J. Org. Chem.* **1970**, *35*, 3178. For a critical review of the methods used to obtain these values, see Jensen; Bushweller, Ref. 214.

[228]Atkinson; Hassel *Acta Chem. Scand.* **1959**, *13*, 1737; Abraham; Rossetti *Tetrahedron Lett.* **1972**, 4965, *J. Chem. Soc., Perkin Trans. 2* **1973**, 582. See also Hammarström; Berg; Liljefors *Tetrahedron Lett.* **1987**, *28*, 4883.

[229]Hageman; Havinga *Recl. Trav. Chim. Pays-Bas* **1969**, *88*, 97; Klaeboe *Acta Chem. Scand.* **1971**, *25*, 695; Abraham; Xodo; Cook; Cruz *J. Chem. Soc., Perkin Trans. 2* **1982**, 1503; Samoshin; Svyatkin; Zefirov *J. Org. Chem. USSR* **1988**, *24*, 1080, and references cited in these papers. *trans*-1,2-Difluorocyclohexane exists predominantly in the *ee* conformation: see Zefirov; Samoshin; Subbotin; Sergeev *J. Org. Chem. USSR* **1981**, *17*, 1301.

TABLE 4.3   Free-energy differences between equatorial and axial substituents on a cyclohexane ring (*A* values)[226]

Group	Approximate $-\Delta G°$, kcal/mole	kJ/mole	Group	Approximate $-\Delta G°$, kcal/mole	kJ/mole
HgCl[225]	-0.25	-1.0	NO$_2$	1.1	4.6
HgBr	0	0	COOEt	1.1–1.2	4.6–5.0
D[237]	0.008	0.03	COOMe	1.27–1.31	5.3–5.5
CN	0.15–0.25	0.6–1.0	COOH	1.36–1.46	5.7–6.1
F	0.25	1.0	NH$_2$[230]	1.4	5.9
C≡CH	0.41	1.7	CH=CH$_2$[231]	1.7	7.1
I	0.46	1.9	CH$_3$[232]	1.74	7.28
Br	0.48–0.62	2.0–2.6	C$_2$H$_5$	~1.75	~7.3
OTs	0.515	2.15	iso-Pr	~2.15	~9.0
Cl	0.52	2.2	C$_6$H$_{11}$[233]	2.15	9.0
OAc	0.71	3.0	SiMe$_3$[234]	2.4–2.6	10–11
OMe[238]	0.75	3.1	C$_6$H$_5$[235]	2.7	11
OH	0.92–0.97	3.8–4.1	*t*-Bu[236]	4.9	21

latter case the two halogen atoms are anti in the *aa* conformation but gauche in the *ee* conformation.[239]

Since compounds with alkyl equatorial substituents are generally more stable, *trans*-1,2 compounds, which can adopt the *ee* conformation, are thermodynamically more stable than their *cis*-1,2 isomers, which must exist in the *ae* conformation. For the 1,2-dimethylcyclohexanes, the difference in stability is about 2 kcal/mol (8 kJ/mol). Similarly, *trans*-1,4 and *cis*-1,3 compounds are more stable than their stereoisomers.

An interesting anomaly is *all-trans*-1,2,3,4,5,6-hexaisopropylcyclohexane, in which the six isopropyl groups prefer the axial position, although the six ethyl groups of the corresponding hexaethyl compound prefer the equatorial position.[240] The alkyl groups of these compounds can of course only be all axial or all equatorial, and it is likely that the molecule prefers the all-axial conformation because of unavoidable strain in the other conformation.

Incidentally, we can now see, in one case, why the correct number of stereoisomers could be predicted by assuming planar rings, even though they are not planar (p. 130). In the

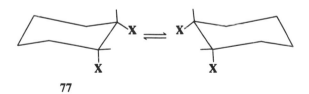

77

[230]Buchanan; Webb *Tetrahedron Lett.* **1983**, *24*, 4519.
[231]Eliel; Manoharan *J. Org. Chem.* **1981**, *46*, 1959.
[232]Booth; Everett *J. Chem. Soc., Chem. Commun.* **1976**, 278.
[233]Hirsch *Top. Stereochem.* **1967**, *1*, 199-222.
[234]Kitching; Olszowy; Drew; Adcock *J. Org. Chem.* **1982**, *47*, 5153.
[235]Squillacote; Neth *J. Am. Chem. Soc.* **1987**, *109*, 198.
[236]Manoharan; Eliel *Tetrahedron Lett.* **1984**, *25*, 3267.
[237]Anet; O'Leary *Tetrahedron Lett.* **1989**, *30*, 1059.
[238]Schneider; Hoppen *Tetrahedron Lett.* **1974**, 579.
[239]For a case of a preferential diaxial conformation in 1,3 isomers, see Ochiai; Iwaki; Ukita; Matsuura; Shiro; Nagao *J. Am. Chem. Soc.* **1988**, *110*, 4606.
[240]Golan; Goren; Biali *J. Am. Chem. Soc.* **1990**, *112*, 9300.

case of both a *cis*-1,2-XX-disubstituted and a *cis*-1,2-XY-disubstituted cyclohexane, the molecule is nonsuperimposable on its mirror image; neither has a plane of symmetry. However, in the former case (**77**) conversion of one chair form to the other (which of course happens rapidly) turns the molecule into its mirror image, while in the latter case (**78**) rapid interconversion does not give the mirror image but merely the conformer in which the original axial and equatorial substituents exchange places. Thus the optical inactivity of **77**

is not due to a plane of symmetry but to a rapid interconversion of the molecule and its mirror image. A similar situation holds for *cis*-1,3 compounds. However, for *cis*-1,4 isomers (both XX and XY) optical inactivity arises from a plane of symmetry in both conformations. All *trans*-1,2- and *trans*-1,3-disubstituted cyclohexanes are chiral (whether XX or XY), while *trans*-1,4 compounds (both XX and XY) are achiral, since all conformations have a plane of symmetry.

The conformation of a group can be frozen into a desired position by putting into the ring a large alkyl group (most often *t*-butyl), which greatly favors the equatorial position.[241]

The principles involved in the conformational analysis of six-membered rings containing one or two trigonal atoms, e.g., cyclohexanone and cyclohexene, are similar.[242]

Conformation in Six-Membered Rings Containing Hetero Atoms

In six-membered rings containing hetero atoms,[243] the basic principles are the same; i.e., there are chair, twist, and boat forms, axial and equatorial groups, etc., but in certain compounds a number of new factors enter the picture. We deal with only two of these.[244]

1. In 5-alkyl-substituted 1,3-dioxanes, the 5-substituent has a much smaller preference for the equatorial position than in cyclohexane derivatives;[245] the *A* values are much lower.

[241]This idea was suggested by Winstein; Holness *J. Am. Chem. Soc.* **1955**, *77*, 5561. There are a few known compounds in which a *t*-butyl group is axial. See, for example, Vierhapper *Tetrahedron Lett.* **1981**, *22*, 5161.

[242]For a monograph, see Rabideau *The Conformational Analysis of Cyclohexenes, Cyclohexadienes, and Related Hydroaromatic Compounds*; VCH: New York, 1989. For reviews, see Vereshchagin *Russ. Chem. Rev.* **1983**, *52*, 1081-1095; Johnson *Chem. Rev.* **1968**, *68*, 375-413. See also Lambert; Clikeman; Taba; Marko; Bosch; Xue *Acc. Chem. Res.* **1987**, *20*, 454-458; Ref. 185; Ref. 214.

[243]For monographs, see Glass *Conformational Analysis of Medium-Sized Heterocycles*; VCH: New York, 1988; Riddell *The Conformational Analysis of Heterocyclic Compounds*; Academic Press: New York, 1980. For reviews, see Juaristi *Acc. Chem. Res.* **1989**, *22*, 357-364; Crabb; Katritzky *Adv. Heterocycl. Chem.* **1984**, *36*, 1-173; Eliel *Angew. Chem. Int. Ed. Engl.* **1972**, *11*, 739-750 [*Angew. Chem. 84*, 779-791], *Pure Appl. Chem.* **1971**, *25*, 509-525, *Acc. Chem. Res.* **1970**, *3*, 1-8; Lambert *Acc. Chem. Res.* **1971**, *4*, 87-94; Romers; Altona; Buys; Havinga *Top. Stereochem.* **1969**, *4*, 39-97; Bushweller; Gianni, Ref. 185, pp. 232-274.

[244]These factors are discussed by Eliel, Ref. 243.

[245]Riddell; Robinson *Tetrahedron* **1967**, *23*, 3417; Eliel; Knoeber *J. Am. Chem. Soc.* **1968**, *90*, 3444. See also Abraham; Banks; Eliel; Hofer; Kaloustian *J. Am. Chem. Soc.* **1972**, *94*, 1913; Eliel; Alcudia *J. Am. Chem. Soc.* **1974**, *96*, 1939.

This indicates that the lone pairs on the oxygens have a smaller steric requirement than the C—H bonds in the corresponding cyclohexane derivatives. Similar behavior is found in the

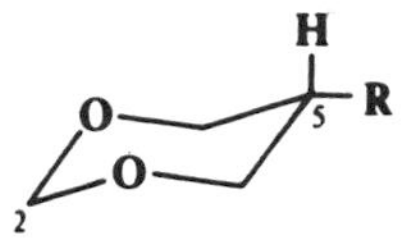

1,3-dithianes.[246] With certain nonalkyl substituents (e.g., F, NO_2, SOMe, NMe_3^+) the axial position is actually preferred.[247]

2. An alkyl group located on a carbon α to a hetero atom prefers the equatorial position, which is of course the normally expected behavior, but a *polar* group in such a location prefers the *axial* position. An example of this phenomenon, known as the *anomeric effect*,[248] is the greater stability of a α-glucosides over β-glucosides. A number of explanations have

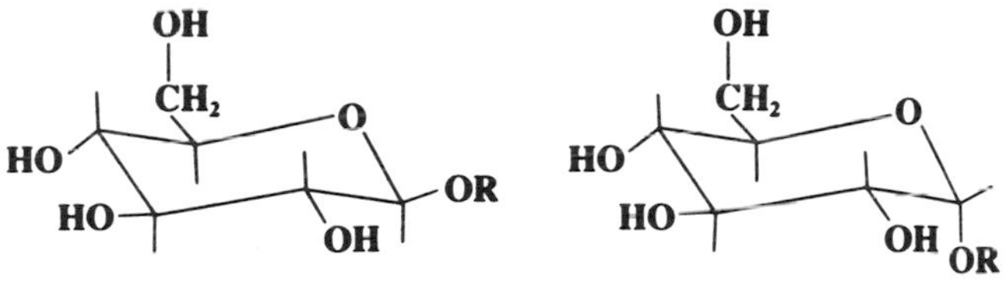

A β-glucoside   An α-glucoside
79   **80**

been offered for the anomeric effect. The one[249] that has received the most acceptance[250] is that one of the lone pairs of the polar atom connected to the carbon (an oxygen atom in the case of **80**) can be stabilized by overlapping with an antibonding orbital of the bond between the carbon and the other polar atom:

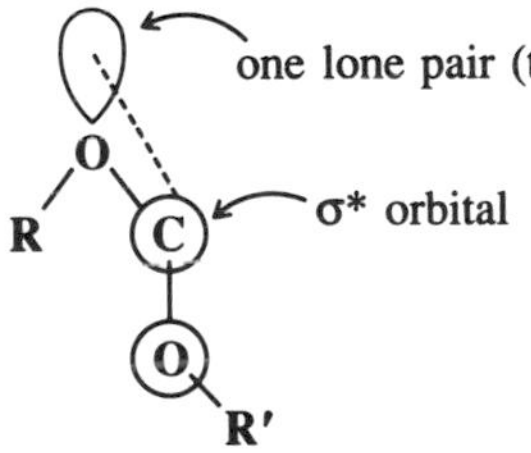

one lone pair (the other not shown)

σ* orbital

[246]Hutchins; Eliel *J. Am. Chem. Soc.* **1969**, *91*, 2703.
[247]Kaloustian; Dennis; Mager; Evans; Alcudia; Eliel *J. Am. Chem. Soc.* **1976**, *98*, 956. See also Eliel; Kandasamy; Sechrest *J. Org. Chem.* **1977**, *42*, 1533.
[248]For books on this subject, see Kirby *The Anomeric Effect and Related Stereoelectronic Effects at Oxygen*; Springer: New York, 1983; Szarek; Horton *Anomeric Effect*; American Chemical Society: Washington, 1979. For reviews see Deslongchamps *Stereoelectronic Effects in Organic Chemistry*; Pergamon: Elmsford, NY, 1983, pp. 4-26; Zefirov *Tetrahedron* **1977**, *33*, 3193-3202; Zefirov; Shekhtman *Russ. Chem. Rev.* **1971**, *40*, 315-329; Lemieux *Pure Appl. Chem.* **1971**, *27*, 527-547; Angyal *Angew. Chem. Int. Ed. Engl.* **1969**, *8*, 157-166 [*Angew. Chem. 81*, 172-182]; Martin *Ann. Chim. (Paris)* [14] **1971**, *6*, 205-218.
[249]See Romers; Altona; Buys; Havinga *Top. Stereochem.* **1969**, *4*, 39-97, pp. 73-77; Wolfe; Whangbo; Mitchell *Carbohydr. Res.* **1979**, *69*, 1.
[250]For some evidence for this explanation, see Fuchs; Ellencweig; Tartakovsky; Aped *Angew. Chem. Int. Ed. Engl.* **1986**, *25*, 287 [*Angew. Chem. 98*, 289]; Praly; Lemieux *Can. J. Chem.* **1987**, *65*, 213; Booth; Khedhair; Readshaw *Tetrahedron* **1987**, *43*, 4699. For evidence against it, see Box *Heterocycles* **1990**, *31*, 1157-1181.

This can happen only if the two orbitals are in the positions shown. The situation can also be represented by this type of hyperconjugation (called "negative hyperconjugation"):

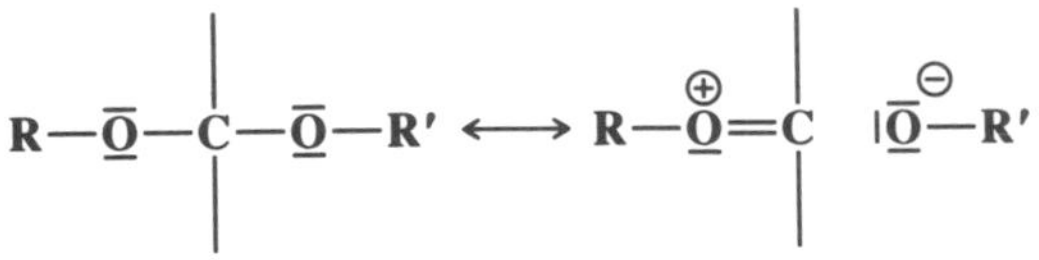

$$R-\overline{\underline{O}}-\overset{|}{\underset{|}{C}}-\overline{\underline{O}}-R' \longleftrightarrow R-\overset{\oplus}{\underline{O}}=\overset{|}{C} \quad |\overset{\ominus}{\underline{O}}-R'$$

It is possible that simple repulsion between parallel dipoles in **79** also plays a part in the greater stablity of **80**.

Conformation in Other Rings

Three-membered rings must be planar, but they seem to be the only saturated rings that generally are. Cyclobutane[251] is not planar but exists as in **81**, with an angle between the planes of about 35°.[252] The deviation from planarity is presumably caused by eclipsing in the planar form (see p. 156). Oxetane, in which eclipsing is less, is closer to planarity, with

$$\bigvee \qquad \begin{matrix} CH_2-O \\ | \qquad | \\ CH_2-CH_2 \end{matrix}$$

81 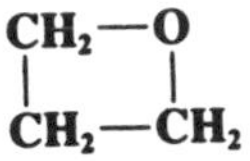   Oxetane

an angle between the planes of about 10°.[253] Cyclopentane might be expected to be planar, since the angles of a regular pentagon are 108°, but it is not so, also because of eclipsing effects.[254] There are two puckered conformations, the *envelope* and the *half-chair*. There is little energy difference between these two forms and many five-membered ring systems have conformations somewhere in between them.[255] Although in the envelope conformation one

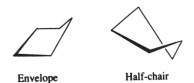

Envelope   Half-chair

carbon is shown above the others, ring motions cause each of the carbons in rapid succession to assume this position. The puckering rotates around the ring in what may be called a *pseudorotation*.[256] In substituted cyclopentanes and five-membered rings in which at least

[251]For reviews of the stereochemistry of four-membered rings, see Legon *Chem. Rev.* **1980**, *80*, 231-262; Moriarty *Top. Stereochem.* **1974**, *8*, 271-421; Cotton; Frenz *Tetrahedron* **1974**, *30*, 1587-1594.

[252]Dows; Rich *J. Chem. Phys.* **1967**, *47*, 333; Stone; Mills *Mol. Phys.* **1970**, *18*, 631; Miller; Capwell *Spectrochim. Acta, Part A* **1971**, *27*, 947; Miller; Capwell; Lord; Rea *Spectrochim. Acta, Part A* **1972**, *28*, 603. However, some cyclobutane derivatives are planar, at least in the solid state: for example, see Margulis; Fischer *J. Am. Chem. Soc.* **1967**, *89*, 223; Margulis *Chem. Commun.* **1969**, 215; *J. Am. Chem. Soc.* **1971**, *93*, 2193.

[253]Luger; Buschmann *J. Am. Chem. Soc.* **1984**, *106*, 7118.

[254]For reviews of the conformational analysis of five-membered rings, see Fuchs *Top. Stereochem.* **1978**, *10*, 1-94; Legon, Ref. 251.

[255]Willy; Binsch; Eliel *J. Am. Chem. Soc.* **1970**, *92*, 5394; Lipnick *J. Mol. Struct.* **1974**, *21*, 423.

[256]Kilpatrick; Pitzer; Spitzer *J. Am. Chem. Soc.* **1947**, *69*, 2438; Pitzer; Donath *J. Am. Chem. Soc.* **1959**, *81*, 3213; Durig; Wertz *J. Chem. Phys.* **1968**, *49*, 2118; Lipnick *J. Mol. Struct.* **1974**, *21*, 411; Poupko; Luz; Zimmermann *J. Am. Chem. Soc.* **1982**, *104*, 5307.

one atom does not contain two substituents (such as tetrahydrofuran, cyclopentanone, etc.), one conformer may be more stable than the others. The barrier to planarity in cyclopentane has been reported to be 5.2 kcal/mol (22 kJ/mol).[257]

Rings larger than six-membered are always puckered[258] unless they contain a large number of sp^2 atoms (see the section on strain in medium rings, p. 155). It should be noted that axial and equatorial hydrogens are found only in the chair conformations of six-membered rings. In rings of other sizes the hydrogens protrude at angles that generally do not lend themselves to classification in this way,[259] though in some cases the terms "pseudo-axial" and "pseudo-equatorial" have been used to classify hydrogens in rings of other sizes.[260]

Molecular Mechanics

Molecular mechanics (also known as *force field calculations*)[261] is a method for the calculation of conformational geometries.[262] It is used to calculate bond angles and distances, as well as total potential energies, for each conformation of a molecule.[263] Molecular orbital calculations (p. 28) can also give such information, but molecular mechanics is generally easier, cheaper (requires less computer time), and/or more accurate. In mo calculations, positions of the nuclei of the atoms are assumed, and the wave equations take account only of the electrons. Molecular mechanics calculations ignore the electrons, and study only the positions of the nuclei. Another important difference is that in an mo calculation each molecule is treated individually, but in molecular mechanics, parameters are obtained for small, simple molecules and then used in the calculations for larger or more complicated ones.

Molecular mechanics calculations use an empirically devised set of equations for the potential energy of molecules. These include terms for vibrational bond stretching, bond angle bending, and other interactions between atoms in a molecule. All these are summed up:

$$V = \Sigma V_{stretch} + \Sigma V_{bend} + \Sigma V_{torsion} + \Sigma V_{VDW}$$

V_{VDW} sums up the interactions (van der Waals) between atoms of a molecule that are not bonded to each other. The set of functions, called the force field, contains adjustable pa-

[257]Carreira; Jiang; Person; Willis *J. Chem. Phys.* **1972,** *56,* 1440.

[258]For reviews of conformations in larger rings, see Arshinova *Russ. Chem. Rev.* **1988,** *57,* 1142-1161; Ounsworth; Weiler *J. Chem. Educ.* **1987,** *64,* 568-572; Ōki, Ref. 49, pp. 307-321; Casanova; Waegell *Bull. Soc. Chim. Fr.* **1975,** 911-921; Anet *Top. Curr. Chem.* **1974,** *45,* 169-220; Dunitz *Pure Appl. Chem.* **1971,** *25,* 495-508, *Perspect. Struct. Chem.* **1968,** *2,* 1-70; Tochtermann *Fortchr. Chem. Forsch.* **1970,** *15,* 378-444; Dale *Angew. Chem. Int. Ed. Engl.* **1966,** *5,* 1000-1021 [*Angew. Chem. 78,* 1070-1093]. For a monograph, see Glass *Conformational Analysis of Medium-Sized Heterocycles*; VCH: New York, 1988. Also see the monographs by Hanack and Eliel; Allinger; Angyal; Morrison, Ref. 185.

[259]For definitions of axial, equatorial, and related terms for rings of any size, see Anet *Tetrahedron Lett.* **1990,** *31,* 2125.

[260]For a discussion of the angles of the ring positions, see Cremer *Isr. J. Chem.* **1980,** *20,* 12.

[261]Sometimes called the *Westheimer method*, because of the pioneering work of F. H. Westheimer: Westheimer; Mayer *J. Chem. Phys.* **1946,** *14,* 733; Westheimer *J. Chem. Phys.* **1947,** *15,* 252; Rieger; Westheimer *J. Am. Chem. Soc.* **1950,** *72,* 19.

[262]For a monograph, see Burkert; Allinger *Molecular Mechanics*; American Chemical Society: Washington, 1982. For reviews, see Osawa; Musso *Angew. Chem. Int. Ed. Engl.* **1983,** *22,* 1-12 [*Angew. Chem. 95,* 1-12], *Top. Stereochem.* **1982,** *13,* 117-193; Boyd; Lipkowitz *J. Chem. Educ.* **1982,** *59,* 269-274; Cox *J. Chem. Educ.* **1982,** *59,* 275-278; Ermer *Struct. Bonding (Berlin)* **1976,** *27,* 161-211; Allinger *Adv. Phys. Org. Chem.* **1976,** *13,* 1-82; Altona; Faber *Top. Curr. Chem.* **1974,** *45,* 1-38. For worked out calculations, using the MMP2 program, see Clark *A Handbook of Computational Chemistry*; Wiley: New York, 1985. See also the series *Advances in Molecular Modeling*.

[263]For an alternative approach, that gives geometries based on electrostatic forces, see Kirpichenok; Zefirov *J. Org. Chem. USSR* **1987,** *23* 607, 623; Zefirov; Samoshin; Svyatkin; Mursakulov *J. Org. Chem. USSR* **1987,** *23,* 634.

rameters that are optimized to get the best fit of known properties of the molecules. The assumption is made that corresponding parameters and force constants can be transferred from one molecule to another. Molecular mechanics is therefore based on experimental data.

In a typical molecular mechanics calculation for a molecule[264] a trial geometrical structure is assumed (bond distances, angles, torsion angles, etc.). Hydrogen atoms are generally not explicitly considered (their positions are calculated later, from standard geometric parameters). The computer searches the trial structure and constructs a list of interaction terms: bond distances, atoms attached to a common atom (bond angles), atoms attached to adjacent atoms (torsion angles), and nonbonded interactions, and then chooses the force field parameters for these interactions from a list stored in the program. It then calculates the potential energy of the trial structure, using the V equation given above. The computer next goes through an energy minimization process by plotting small changes in geometrical coordinates against energy, looking for places in the curve where the first derivatives of V are equal to zero, which means that the total energy V is at a minimum. This must be done separately for each stable conformation, since there is no known method for finding the lowest V for a molecule (e.g., the anti conformation of *n*-butane). If appropriate trial structures are entered, the computer will find the lowest V for the anti and gauche conformations, separately. This can be a handicap for large molecules (e.g., 2,3-dimethylundecane) which may have many stable conformations (that is, energy minima). The computer will find only those minima recognized by the investigator.[265] Molecular mechanics can also be used to study energy maxima (barriers), but in much less detail.[266] A number of force field computer programs are available, among them the Allinger MM2, MMP2, and MM3[267] force fields, and the Bartell MUB-2 force field.[268]

A molecular mechanics calculation gives the total potential energy of each conformation. If the mole fractions of all the conformations are known, or can be calculated, the enthalpy of formation of the compound can be obtained.[269]

Even though molecular mechanics has given satisfactory results (that is, results that agree with experimental measurements) for many molecules, it is still not totally reliable, since it does fail in certain cases. A further limitation is that it can be used only in cases for which transferable parameters can be obtained from simple molecules. Molecular orbital calculations do not have this limitation.

STRAIN

Steric strain[270] exists in a molecule when bonds are forced to make abnormal angles. This results in a higher energy than would be the case in the absence of angle distortions. There

---

[264]This description is from Burkert; Allinger, Ref. 262, pp. 63-65.

[265]For methods of dealing with this difficulty, see Li; Scheraga *Proc. Natl. Acad. Sci. USA* **1987**, *84*, 6611; Saunders *J. Am. Chem. Soc.* **1987**, *109*, 3150; Wilson; Cui; Moskowitz; Schmidt *Tetrahedron Lett.* **1988**, *29*, 4373; Billeter; Howard; Kuntz; Kollman *J. Am. Chem. Soc.* **1988**, *110*, 8385.

[266]See Burkert; Allinger, Ref. 262, pp. 72-76.

[267]See Allinger; Yuh; Lii *J. Am. Chem. Soc.* **1989**, *111*, 8551; Allinger; Chen; Rahman; Pathiaseril *J. Am. Chem. Soc.* **1991**, *113*, 4505.

[268]For a list of programs and sources, see Burkert; Allinger, Ref. 262, pp. 317-319. See also Clark, Ref. 262, p. 10. Improved MM2 parameters for aldehydes and ketones are reported by Bowen; Pathiaseril; Profeta; Allinger *J. Org. Chem.* **1987**, *52*, 5162. For extensions of MM2 to other systems, see Bowen; Reddy; Patterson; Allinger *J. Org. Chem.* **1988**, *53*, 5471; Frierson; Imam; Zalkow; Allinger *J. Org. Chem.* **1988**, *53*, 5248; Tai; Allinger *J. Am. Chem. Soc.* **1988**, *110*, 2050; Podlogar; Raber *J. Org. Chem.* **1989**, *54*, 5032.

[269]See Clark, Ref. 262, pp. 173-184. See also DeTar *J. Org. Chem.* **1987**, *52*, 1851.

[270]For a monograph, see Greenberg; Liebman *Strained Organic Molecules*; Academic Press: New York, 1978. For reviews, see Wiberg *Angew. Chem. Int. Ed. Engl.* **1986**, *25*, 312-322 [*Angew. Chem. 98*, 312-322]; Greenberg; Stevenson *Mol. Struct. Energ.* **1986**, *3*, 193-266; Liebman; Greenberg *Chem. Rev.* **1976**, *76*, 311-365. For a review of the concept of strain, see Cremer; Kraka *Mol. Struct. Energ.* **1988**, *7*, 65-138.

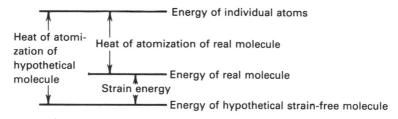

FIGURE 4.5   Strain energy calculation.

are, in general, two kinds of structural features that result in sterically caused abnormal bond angles. One of these is found in small-ring compounds, where the angles must be less than those resulting from normal orbital overlap. Such strain is called *small-angle strain*. The other arises when nonbonded atoms are forced into close proximity by the geometry of the molecule. These are called *nonbonded interactions*.

Strained molecules possess *strain energy*. That is, their potential energies are higher than they would be if strain were absent.[271] The strain energy for a particular molecule can be estimated from heat of atomization or heat of combustion data. A strained molecule has a lower heat of atomization than it would have if it were strain-free (Figure 4.5). As in the similar case of resonance energies (p. 29), strain energies can not be known exactly, because the energy of a real molecule can be measured, but not the energy of a hypothetical un-strained model. It is also possible to calculate strain energies by molecular mechanics, not only for real molecules, but also for those that cannot be made.[272]

Strain in Small Rings

Three-membered rings have a great deal of angle strain, since 60° angles represent a large departure from the tetrahedral angles. In sharp contrast to other ethers, ethylene oxide is quite reactive, the ring being opened by many reagents (see p. 353). Ring opening, of course, relieves the strain.[273] Cyclopropane,[274] which is even more strained[275] than ethylene oxide, is also cleaved more easily than would be expected for an alkane.[276] Thus, pyrolysis at 450 to 500°C converts it to propene, bromination gives 1,3-dibromopropane,[277] and it can be hydrogenated to propane (though at high pressure).[278] Other three-membered rings are similarly reactive.[279]

There is much evidence, chiefly derived from nmr coupling constants, that the bonding in cyclopropanes is not the same as in compounds that lack small-angle strain.[280] For a

[271]For discussions, see Wiberg; Bader; Lau *J. Am. Chem. Soc.* **1987**, *109*, 985, 1001.

[272]For a review, see Rüchardt; Beckhaus, Ref. 210. See also Burkert; Allinger, Ref. 262, pp. 169-194; Allinger, Ref. 262, pp. 45-47.

[273]For reviews of reactions of cyclopropanes and cyclobutanes, see Trost *Top. Curr. Chem.* **1986**, *133*, 3-82; Wong; Lau; Tam *Top. Curr. Chem.* **1986**, *133*, 83-157.

[274]For a treatise, see Rappoport *The Chemistry of the Cyclopropyl Group*, 2 pts.; Wiley: New York, 1987.

[275]For reviews of strain in cyclopropanes, see, in Ref. 274, the papers by Wiberg, pt. 1., pp. 1-26; Liebman; Greenberg, pt. 2, pp. 1083-1119; Liebman; Greenberg *Chem. Rev.* **1989**, *89*, 1225-1246.

[276]For reviews of ring-opening reactions of cyclopropanes, see Wong; Hon; Tse; Yip; Tanko; Hudlicky *Chem. Rev.* **1989**, *89*, 165-198; Reissig, in Ref. 274, pt. 1, pp. 375-443.

[277]Ogg; Priest *J. Am. Chem. Soc.* **1938**, *60*, 217.

[278]Shortridge; Craig; Greenlee; Derfer; Boord *J. Am. Chem. Soc.* **1948**, *70*, 946.

[279]For a review of the pyrolysis of three- and four-membered rings, see Frey *Adv. Phys. Org. Chem.* **1966**, *4*, 147-193.

[280]For discussions of bonding in cyclopropanes, see Bernett *J. Chem. Educ.* **1967**, *44*, 17-24; de Meijere *Angew. Chem. Int. Ed. Engl.* **1979**, *18*, 809-826 [*Angew. Chem. 91*, 867-884]; Honegger; Heilbronner; Schmelzer *Nouv. J. Chem.* **1982**, *6*, 519; Cremer; Kraka *J. Am. Chem. Soc.* **1985**, *107*, 3800, 3811; Slee *Mol. Struct. Energ.* **1988**, *5*, 63-114; Ref. 284.

normal carbon atom, one *s* and three *p* orbitals are hybridized to give four approximately equivalent sp^3 orbitals, each containing about 25% *s* character. But for a cyclopropane carbon atom, the four hybrid orbitals are far from equivalent. The two orbitals directed to the outside bonds have more *s* character than a normal sp^3 orbital, while the two orbitals involved in ring bonding have less, because the more *p*-like they are the more they resemble ordinary *p* orbitals, whose preferred bond angle is 90° rather than 109.5°. Since the small-angle strain in cyclopropanes is the difference between the preferred angle and the real angle of 60°, this additional *p* character relieves some of the strain. The external orbitals have about 33% *s* character, so that they are approximately sp^2 orbitals, while the internal orbitals have about 17% *s* character, so that they may be called approximately sp^5 orbitals.[281] Each of the three carbon–carbon bonds of cyclopropane is therefore formed by overlap of two sp^5 orbitals. Molecular-orbital calculations show that such bonds are not completely σ in character. In normal C—C bonds, sp^3 orbitals overlap in such a way that the straight line connecting the nuclei becomes an axis about which the electron density is symmetrical. But in cyclopropane, the electron density is directed *away from* the ring. Figure 4.6 shows the direction of orbital overlap.[282] For cylopropane, the angle (marked θ) is 21°. Cyclobutane exhibits the same phenomenon but to a lesser extent, θ being 7°.[282] Molecular orbital calculations also show that the maximum electron densities of the C—C σ orbitals are bent away from the ring, with θ = 9.4° for cyclopropane and 3.4° for cyclobutane.[283] The bonds in cyclopropane are called *bent bonds*, and are intermediate in character between σ and π, so that cyclopropanes behave in some respects like double-bond compounds.[284] For one thing, there is much evidence, chiefly from uv spectra,[285] that a cyclopropane ring is conjugated with an adjacent double bond and that this conjugation is greatest for the conformation shown in *a* in Figure 4.7 and least or absent for the conformation shown in *b*, since overlap of the double-bond π orbital with two of the *p*-like orbitals of the cyclopropane ring is greatest in conformation *a*. However, the conjugation between a cyclopropane ring and a double bond is less than that between two double bonds.[286] For other examples of the similarities in behavior of a cyclopropane ring and a double bond, see p. 755.

Four-membered rings also exhibit angle strain, but much less, and are less easily opened.

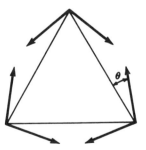

FIGURE 4.6 Orbital overlap in cyclopropane. The arrows point toward the center of electron density.

[281]Randić; Maksić *Theor. Chim. Acta* **1965**, *3*, 59; Foote *Tetrahedron Lett.* **1963**, 579; Weigert; Roberts *J. Am. Chem. Soc.* **1967**, *89*, 5962.

[282]Coulson; Moffitt *Philos. Mag.* **1949**, *40*, 1; Coulson; Goodwin *J. Chem. Soc.* **1962**, 2851, **1963**, 3161; Peters *Tetrahedron* **1963**, *19*, 1539; Hoffmann; Davidson *J. Am. Chem. Soc.* **1971**, *93*, 5699.

[283]Wiberg; Bader; Lau, Ref. 271; Cremer; Kraka, Ref. 280.

[284]For reviews, see Tidwell, in Ref. 274, pt. 1, pp. 565-632; Charton, in Zabicky *The Chemistry of Alkenes*, vol. 2, pp. 511-610, Wiley: New York, 1970.

[285]See, for example, Cromwell; Hudson *J. Am. Chem. Soc.* **1953**, *75*, 872; Kosower; Ito *Proc. Chem. Soc.* **1962**, 25; Dauben; Berezin *J. Am. Chem. Soc.* **1967**, *89*, 3449; Jorgenson; Leung *J. Am. Chem. Soc.* **1968**, *90*, 3769; Heathcock; Poulter *J. Am. Chem. Soc.* **1968**, *90*, 3766; Tsuji; Shibata; Hienuki; Nishida *J. Am. Chem. Soc.* **1978**, *100*, 1806; Drumright; Mas; Merola; Tanko *J. Org. Chem.* **1990**, *55*, 4098.

[286]Staley *J. Am. Chem. Soc.* **1967**, *89*, 1532; Pews; Ojha *J. Am. Chem. Soc.* **1969**, *91*, 5769. See, however, Noe; Young *J. Am. Chem. Soc.* **1982**, *104*, 6218.

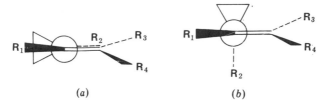

FIGURE 4.7 Conformations of α-cyclopropylalkenes. Conformation a leads to maximum conjugation and conformation b to minimum conjugation.

Cyclobutane is more resistant than cyclopropane to bromination, and though it can be hydrogenated to butane, more strenuous conditions are required. Nevertheless, pyrolysis at 420°C gives two molecules of ethylene. As mentioned earlier (p. 148), cyclobutane is not planar.

Many highly strained compounds containing small rings in fused systems have been prepared,[287] showing that organic molecules can exhibit much more strain than simple cyclopropanes or cyclobutanes.[288] Table 4.4 shows a few of these compounds.[289] Perhaps the most interesting are cubane, prismane, and the substituted tetrahedrane, since preparation of these ring systems had been the object of much endeavor. Prismane has the structure that Ladenburg proposed as a possible structure for benzene. The bicyclobutane molecule is bent, with the angle θ between the planes equal to $126 \pm 3°$.[290] The rehybridization effect,

described above for cyclopropane, is even more extreme in this molecule. Calculations have shown that the central bond is essentially formed by overlap of two p orbitals with little or no s character.[291] *Propellanes* are compounds in which two carbons, directly connected, are also connected by three other bridges. The one in the table is the smallest possible propellane,[292] and is in fact more stable than the larger [2.1.1]propellane and [2.2.1]propellane, which have been isolated only in solid matrixes at low temperature.[293]

In certain small-ring systems, including small propellanes, the geometry of one or more carbon atoms is so constrained that all four of their valences are directed to the same side of a plane ("inverted tetrahedron"), as in **81**.[294] An example is 1,3-dehydroadamantane, **82**

[287]For reviews discussing the properties of some of these as well as related compounds, see the reviews in *Chem. Rev.* **1989**, *89*, 975-1270, and the following: Jefford *J. Chem. Educ.* **1976**, *53*, 477-482; Seebach *Angew. Chem. Int. Ed. Engl.* **1965**, *4*, 121-131 [*Angew. Chem.* **77**, 119-129]; Greenberg; Liebman, Ref. 270, pp. 210-220. For a review of bicyclo[*n.m.*0]alkanes, see Wiberg *Adv. Alicyclic Chem.* **1968**, *2*, 185-254.
[288]For a useful classification of strained polycyclic systems, see Gund; Gund *J. Am. Chem. Soc.* **1981**, *103*, 4458.
[289]For a computer program that generates IUPAC names for complex bridged systems, see Rücker; Rücker *Chimia* **1990**, *44*, 116.
[290]Haller; Srinivasan *J. Chem. Phys.* **1964**, *41*, 2745.
[291]Schulman; Fisanick *J. Am. Chem. Soc.* **1970**, *92*, 6653; Newton; Schulman *J. Am. Chem. Soc.* **1972**, *94*, 767.
[292]Wiberg; Walker *J. Am. Chem. Soc.* **1982**, *104*, 5239; Wiberg; Waddell *J. Am. Chem. Soc.* **1990**, *112*, 2194; Seiler *Helv. Chim. Acta* **1990**, *73*, 1574; Bothe; Schlüter *Chem. Ber.* **1991**, *124*, 587. For reviews of small-ring propellanes, see Wiberg *Chem. Rev.* **1989**, *89*, 975-983; Ginsburg, in Ref. 274, pt. 2, pp. 1193-1221. For a discussion of the formation of propellanes, see Ginsburg *Top. Curr. Chem.* **1987**, *137*, 1-17.
[293]Walker; Wiberg; Michl *J. Am. Chem. Soc.* **1982**, *104*, 2056; Wiberg; Walker; Pratt; Michl *J. Am. Chem. Soc.* **1983**, *105*, 3638.
[294]For a review, see Wiberg *Acc. Chem. Res.* **1984**, *17*, 379-386.

Structural formula of compound prepared	Systematic name of ring system	Common name if any	Ref.
	Bicyclo[1.1.0]butane	Bicyclobutane	303
	$\Delta^{1,4}$-Bicyclo[2.2.0]hexene		304
	Tricyclo[1.1.0.0^{2,4}]butane	Tetrahedrane	305
	Pentacyclo[5.1.0.0^{2,4}.0^{3,5}.0^{6,8}]octane	Octabisvalene	306
	Tricyclo[1.1.1.0^{1,3}]pentane	A [1.1.1]propellane	292
	Tetracyclo[2.2.0.0^{2,6}.0^{3,5}]hexane	Prismane	295
	Pentacyclo[4.2.0.0^{2,5}.0^{3,8}.0^{4,7}]octane	Cubane	296
	Pentacyclo[5.4.1.0^{3,1}.0^{5,9}.0^{8,11}]dodecane	4[Peristylane]	297
	Hexacyclo[5.3.0.0^{2,6}.0^{3,10}.0^{4,9}.0^{5,8}]decane	Pentaprismane	298
	Tricyclo[3.1.1.1^{2,4}]octane	Diasterane	299
	Hexacyclo[4.4.0.0^{2,4}.0^{3,9}.0^{5,8}.0^{7,10}]decane		300
	Nonacyclo[10.8.0.0^{2,11}.0^{4,9}.0^{4,19}.0^{6,17}.0^{7,16}.0^{9,14}.0^{14,19}]eicosane	A double tetraasterane	301
	Undecacyclo[9.9.0.0^{1,5}.0^{2,12}.0^{2,18}.0^{3,7}.0^{6,10}.0^{8,12}.0^{11,15}.0^{13,17}.0^{16,20}]eicosane	Pagodane	302

(which is also a propellane).[307] X-ray crystallography of the 5-cyano derivative of **82** shows that the four carbon valences at C-1 and C-3 are all directed "into" the molecule and none

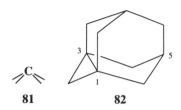

81   **82**

point outside.[308] **82** is quite reactive; it is unstable in air, readily adds hydrogen, water, bromine, or acetic acid to the C_1—C_3 bond, and is easily polymerized. When two such atoms are connected by a bond (as in **82**), the bond is very long (the C_1—C_3 bond length in the 5-cyano derivative of **82** is 1.64 Å), as the atoms try to compensate in this way for their enforced angles. The high reactivity of the C_1—C_3 bond of **82** is not only caused by strain, but also by the fact that reagents find it easy to approach these atoms since there are no bonds (e.g., C—H bonds on C-1 or C-3) to get in the way.

Strain in Medium Rings[309]

In rings larger than four-membered, there is no small-angle strain, but there are three other kinds of strain. In the chair form of cyclohexane, which does not exhibit any of the three kinds of strain, all six carbon–carbon bonds have the two attached carbons in the gauche conformation. However, in five-membered rings and in rings containing from 7 to 13 carbons any conformation in which all the ring bonds are gauche contains transannular interactions,

[295]Katz; Acton *J. Am. Chem. Soc.* **1973**, *95*, 2738. See also Viehe; Merényi; Oth; Senders; Valange *Angew. Chem. Int. Ed. Engl.* **1964**, *3*, 755 [*Angew. Chem. 76*, 923]; Wilzbach; Kaplan *J. Am. Chem. Soc.* **1965**, *87*, 4004.

[296]Eaton; Cole *J. Am. Chem. Soc.* **1964**, *86*, 3157; Barborak; Watts; Pettit *J. Am. Chem. Soc.* **1966**, *88*, 1328; Hedberg; Hedberg; Eaton; Nodari; Robiette *J. Am. Chem. Soc.* **1991**, *113*, 1514. For a review of cubanes, see Griffin; Marchand *Chem. Rev.* **1989**, *89*, 997-1010.

[297]Paquette; Fischer; Browne; Doecke *J. Am. Chem. Soc.* **1985**, *105*, 686.

[298]Eaton; Or; Branca; Shankar *Tetrahedron* **1986**, *42*, 1621. See also Dauben; Cunningham *J. Org. Chem.* **1983**, *48*, 2842.

[299]Otterbach; Musso *Angew. Chem. Int. Ed. Engl.* **1987**, *26*, 554 [*Angew. Chem. 99*, 588].

[300]Allred; Beck *J. Am. Chem. Soc.* **1973**, *95*, 2393.

[301]Hoffmann; Musso *Angew. Chem. Int. Ed. Engl.* **1987**, *26*, 1006 [*Angew. Chem. 99*, 1036].

[302]Rihs *Tetrahedron Lett.* **1983**, *24*, 5857.

[303]Lemal; Menger; Clark *J. Am. Chem. Soc.* **1963**, *85*, 2529; Wiberg; Lampman *Tetrahedron Lett.* **1963**, 2173. For reviews of preparations and reactions of this system, see Hoz, in Ref. 274, pt. 2, pp. 1121-1192; Wiberg; Lampman; Ciula; Connor; Schertler; Lavanish *Tetrahedron* **1965**, *21*, 2749-2769; Wiberg *Rec. Chem. Prog.* **1965**, *26*, 143-154; Wiberg, Ref. 287. For a review of [*n*.1.1] systems, see Meinwald; Meinwald *Adv. Alicyclic Chem.* **1966**, *1*, 1-51.

[304]Casanova; Bragin; Cottrell *J. Am. Chem. Soc.* **1978**, *100*, 2264.

[305]Maier; Pfriem; Schäfer; Malsch; Matusch *Chem. Ber.* **1981**, *114*, 3965; Maier; Pfriem; Malsch; Kalinowski; Dehnicke *Chem. Ber.* **1981**, *114*, 3988; Irngartinger; Goldmann; Jahn; Nixdorf; Rodewald; Maier; Malsch; Emrich *Angew. Chem. Int. Ed. Engl.* **1984**, *23*, 993 [*Angew. Chem. 96*, 967]; Maier; Fleischer *Tetrahedron Lett.* **1991**, *32*, 57. For reviews of attempts to synthesize tetrahedrane, see Maier *Angew. Chem. Int. Ed. Engl.* **1988**, *27*, 309-332 [*Angew. Chem. 100*, 317-341]; Zefirov; Koz'min; Abramenkov *Russ. Chem. Rev.* **1978**, *47*, 163-171. For a review of tetrahedranes and other cage molecules stabilized by steric hindrance, see Maier; Rang; Born, in Olah *Cage Hydrocarbons*; Wiley: New York, 1990, pp. 219-259. See also Maier; Born *Angew. Chem. Int. Ed. Engl.* **1989**, *28*, 1050 [*Angew. Chem. 101*, 1085].

[306]Rücker; Trupp *J. Am. Chem. Soc.* **1988**, *110*, 4828.

[307]Pincock and Torupka *J. Am. Chem. Soc.* **1969**, *91*, 4593; Pincock; Schmidt; Scott; Torupka *Can. J. Chem.* **1972**, *50*, 3958; Scott; Pincock *J. Am. Chem. Soc.* **1973**, *95*, 2040.

[308]Gibbons; Trotter *Can. J. Chem.* **1973**, *51*, 87.

[309]For reviews, see Gol'dfarb; Belen'kii *Russ. Chem. Rev.* **1960**, *29*, 214-235; Raphael *Proc. Chem. Soc.* **1962**, 97-105; Sicher *Prog. Stereochem.* **1962**, *3*, 202-264.

i.e., interactions between the substituents on C-1 and C-3 or C-1 and C-4, etc. These interactions occur because the internal space is not large enough for all the quasi-axial hydrogen atoms to fit without coming into conflict. The molecule can adopt other conformations in which this *transannular strain* is reduced, but then some of the carbon–carbon bonds must adopt eclipsed or partially eclipsed conformations. The strain resulting from eclipsed conformations is called *Pitzer strain.* For saturated rings from 3- to 13-membered (except for the chair form of cyclohexane) there is no escape from at least one of these two types of strain. In practice each ring adopts conformations that minimize both sorts of strain as much as possible. For cyclopentane, as we have seen (p. 148), this means that the molecule is not planar. In rings larger than 9-membered, Pitzer strain seems to disappear, but transannular strain is still present.[310] For 9- and 10-membered rings, some of the transannular and Pitzer strain may be relieved by the adoption of a third type of strain, *large-angle strain.* Thus, C—C—C angles of 115 to 120° have been found in x-ray diffraction of cyclononylamine hydrobromide and 1,6-diaminocyclodecane dihydrochloride.[311]

The amount of strain in cycloalkanes is shown in Table 4.5,[312] which lists heats of combustion per CH_2 group. As can be seen, cycloalkanes larger than 13-membered are as strain-free as cyclohexane.

Transannular interactions can exist across rings from 8- to 11-membered and even larger.[313] Such interactions can be detected by dipole and spectral measurements. For example, that the carbonyl group in **83a** is affected by the nitrogen (**83b** is probably another canonical form) has been demonstrated by photoelectron spectroscopy, which shows that

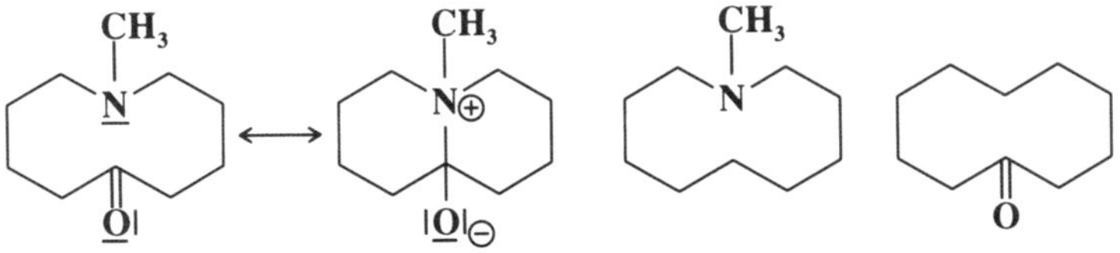

the ionization potentials of the nitrogen n and C=O π orbitals in **83** differ from those of the two comparison molecules **84** and **85**,[314] It is significant that when **83** accepts a proton,

TABLE 4.5 Heats of combustion in the gas phase for cycloalkanes, per CH_2 group[312]

Size of ring	$-\Delta H_c$, (g)		Size of ring	$-\Delta H_c$, (g)	
	kcal/mol	kJ/mol		kcal/mol	kJ/mol
3	166.3	695.8	10	158.6	663.6
4	163.9	685.8	11	158.4	662.7
5	158.7	664.0	12	157.8	660.2
6	157.4	658.6	13	157.7	659.8
7	158.3	662.3	14	157.4	658.6
8	158.6	663.6	15	157.5	659.0
9	158.8	664.4	16	157.5	659.0

[310]Huber-Buser; Dunitz *Helv. Chim. Acta* **1960**, *43*, 760.
[311]Bryan; Dunitz *Helv. Chim. Acta* **1960**, *43*, 1; Dunitz; Venkatesan *Helv. Chim. Acta* **1961**, *44*, 2033.
[312]Gol'dfarb; Belen'kii, Ref. 309, p. 218.
[313]For a review, see Cope; Martin; McKervey *Q. Rev., Chem. Soc.* **1966**, *20*, 119-152.
[314]Spanka; Rademacher *J. Org. Chem.* **1986**, *51*, 592. See also Spanka; Rademacher; Duddeck *J. Chem. Soc., Perkin Trans. 2* **1988**, 2119; Leonard; Fox; Ōki *J. Am. Chem. Soc.* **1954**, *76*, 5708.

it goes to the oxygen rather than to the nitrogen. Many examples of transannular reactions are known. A few are:

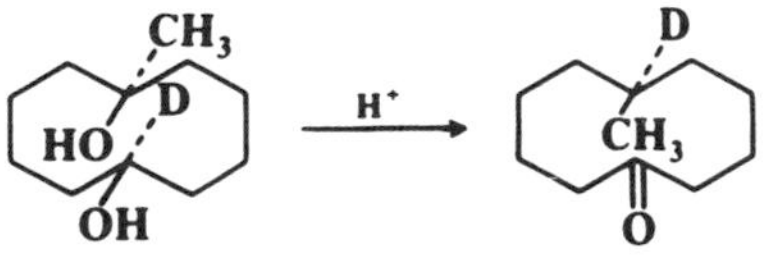

Ref. 315

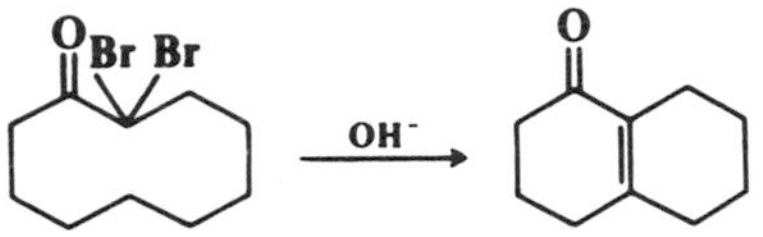

Ref. 316

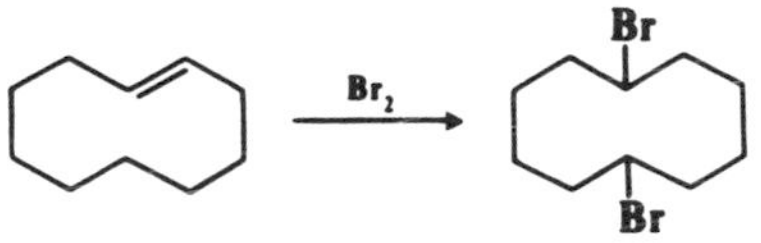

Ref. 317

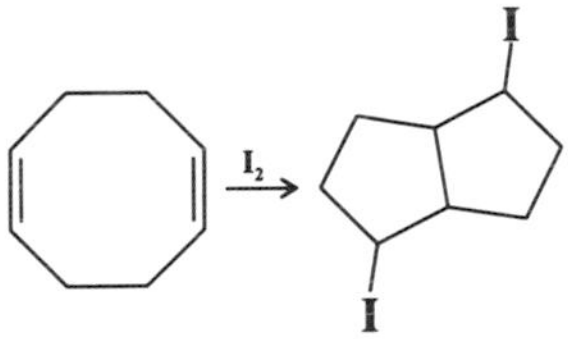

Ref. 318

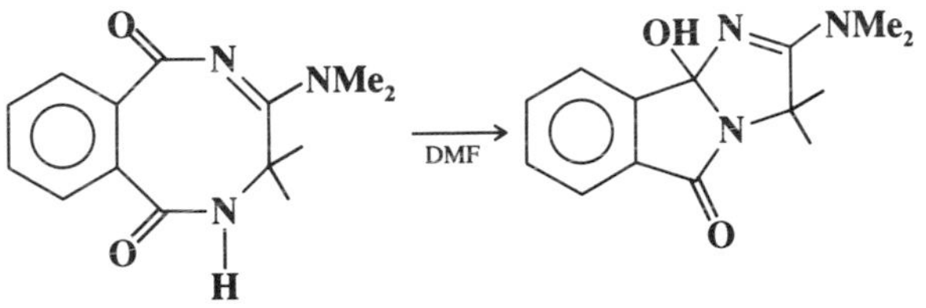

Ref. 319

In summary, we can divide saturated rings into four groups, of which the first and third are more strained than the other two.[320]

1. *Small rings* (3- and 4-membered). Small-angle strain predominates.

2. *Common rings* (5-, 6-, and 7-membered). Largely unstrained. The strain that is present is mostly Pitzer strain.

3. *Medium rings* (8- to 11-membered). Considerable strain; Pitzer, transannular, and large-angle strain.

4. *Large rings* (12-membered and larger). Little or no strain.

[315]Prelog; Küng *Helv. Chim. Acta* **1956**, *39*, 1394.
[316]Schenker; Prelog *Helv. Chim. Acta* **1953**, *36*, 896.
[317]Sicher; Závada; Svoboda *Collect. Czech. Chem. Commun.* **1962**, *27*, 1927.
[318]Uemura; Fukuzawa; Toshimitsu; Okano; Tezuka; Sawada *J. Org. Chem.* **1983**, *48*, 270.
[319]Schläpfer-Dähler; Prewo; Bieri; Germain; Heimgartner *Chimia* **1988**, *42*, 25.
[320]For a review on the influence of ring size on the properties of cyclic systems, see Granik *Russ. Chem. Rev.* **1982**, *51*, 119-134.

Unsaturated Rings[321]

Double bonds can exist in rings of any size. As expected, the most highly strained are the three-membered rings. Small-angle strain, which is so important in cyclopropane, is even greater in cyclopropene[322] because the ideal angle is greater. In cyclopropane, the bond angle is forced to be 60°, about 50° smaller than the tetrahedral angle; but in cyclopropene, the angle, also about 60°, is now about 60° smaller than the ideal angle of 120°. Thus, the angle is cyclopropene is about 10° more strained than in cyclopropane. However, this additional strain is offset by a decrease in strain arising from another factor. Cyclopropene, lacking two hydrogens, has none of the eclipsing strain present in cyclopropane. Cyclopropene has been prepared[323] and is stable at liquid–nitrogen temperatures, though on warming even to −80°C it rapidly polymerizes. Many other cyclopropenes are stable at room temperature and above.[322] The highly strained benzocyclopropene,[324] in which the cyclopropene ring is fused to a benzene ring, has been prepared[325] and is stable for weeks at room temperature, though it decomposes on distillation at atmospheric pressure.

benzocyclopropene

As previously mentioned, double bonds in relatively small rings must be cis. A stable trans double bond[326] first appears in an eight-membered ring (*trans*-cyclooctene, p. 104), though the transient existence of *trans*-cyclohexene and cycloheptene has been demonstrated.[327] Above about 11 members, the trans isomer is more stable than the cis.[160] It has proved possible to prepare compounds in which a trans double bond is shared by two cycloalkene rings (e.g., **86**). Such compounds have been called [*m. n*]*betweenanenes*, and

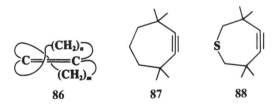

86   **87**   **88**

[321]For a review of strained double bonds, see Zefirov; Sokolov *Russ. Chem. Rev.* **1967**, *36*, 87-100. For a review of double and triple bonds in rings, see Johnson *Mol. Struct. Energ.* **1986**, *3*, 85-140.

[322]For reviews of cyclopropenes, see Baird *Top. Curr. Chem.* **1988**, *144*, 137-209; Halton; Banwell, in Ref. 274, pt. 2, pp. 1223-1339; Closs *Adv. Alicyclic Chem.* **1966**, *1*, 53-127; For a discussion of the bonding and hybridization, see Allen *Tetrahedron* **1982**, *38*, 645.

[323]Dem'yanov; Doyarenko *Bull. Acad. Sci. Russ.* **1922**, *16*, 297, *Ber.* **1923**, *56*, 2200; Schlatter *J. Am. Chem. Soc.* **1941**, *63*, 1733; Wiberg; Bartley *J. Am. Chem. Soc.* **1960**, *82*, 6375; Stigliani; Laurie; Li *J. Chem. Phys.* **1975**, *62*, 1890.

[324]For reviews of cycloproparenes, see Halton *Chem. Rev.* **1989**, *89*, 1161-1185, **1973**, *73*, 113-126; Billups; Rodin; Haley *Tetrahedron* **1988**, *44*, 1305-1338; Halton; Stang *Acc. Chem. Res.* **1987**, *20*, 443-448; Billups *Acc. Chem. Res.* **1978**, *11*, 245-251.

[325]Vogel; Grimme; Korte *Tetrahedron Lett.* **1965**, 3625. Also see Anet; Anet *J. Am. Chem. Soc.* **1964**, *86*, 526; Müller; Bernardinelli; Thi *Chimia* **1988**, *42*, 261; Neidlein; Christen; Poignée; Boese; Bläser; Gieren; Ruiz-Pérez; Hübner *Angew. Chem. Int. Ed. Engl.* **1988**, *27*, 294 [*Angew. Chem. 100*, 292].

[326]For reviews of trans cycloalkenes, see Nakazaki; Yamamoto; Naemura *Top. Curr. Chem.* **1984**, *125*, 1-25; Marshall *Acc. Chem. Res.* **1980**, *13*, 213-218.

[327]Bonneau; Joussot-Dubien; Salem; Yarwood *J. Am. Chem. Soc.* **1979**, *98*, 4329; Wallraff; Michl *J. Org. Chem.* **1986**, *51*, 1794; Squillacote; Bergman; De Felippis *Tetrahedron Lett.* **1989**, *30*, 6805.

several have been prepared with m and n values from 8 to 26.[328] The double bonds of the smaller betweenanenes, as might be expected from the fact that they are deeply buried within the bridges, are much less reactive than those of the corresponding cis–cis isomers. The smallest unstrained cyclic triple bond is found in cyclononyne.[329] Cyclooctyne has been isolated,[330] but its heat of hydrogenation shows that it is considerably strained. There have been a few compounds isolated with triple bonds in seven-membered rings. 3,3,7,7-Tetramethylcycloheptyne (**87**) dimerizes within an hour at room temperature,[331] but the thia derivative **88,** in which the C—S bonds are longer than the corresponding C—C bonds in **87**, is indefinitely stable even at 140°C.[332] Cycloheptyne itself has not been isolated, though its transient existence has been shown.[333] Cyclohexyne[334] and its 3,3,6,6-tetramethyl derivative[335] have been trapped at 77 K, and in an argon matrix at 12 K, respectively, and ir spectra have been obtained. Transient six- and even five-membered rings containing triple bonds have also been demonstrated.[336] A derivative of cyclopentyne has been trapped in a matrix.[337] Although cycloheptyne and cyclohexyne have not been isolated at room temperatures, Pt(0) complexes of these compounds have been prepared and are stable.[338] The smallest cyclic allene[339] so far isolated is 1-*t*-butyl-1,2-cyclooctadiene **89**.[340] The parent 1,2-cyclooctadiene has not been isolated. It has been shown to exist transiently, but rapidly

89

dimerizes.[341] The presence of the *t*-butyl group apparently prevents this. The transient existence of 1,2-cycloheptadiene has also been shown,[342] and both 1,2-cyclooctadiene and 1,2-cycloheptadiene have been isolated in platinum complexes.[343] 1,2-Cyclohexadiene has been trapped at low temperatures, and its structure has been proved by spectral studies.[344]

[328]Marshall; Lewellyn *J. Am. Chem. Soc.* **1977**, *99*, 3508; Nakazaki; Yamamoto; Yanagi *J. Chem. Soc., Chem. Commun.* **1977**, 346; *J. Am. Chem. Soc.* **1979**, *101*, 147; Ceré; Paolucci; Pollicino; Sandri; Fava *J. Chem. Soc., Chem. Commun.* **1980**, 755; Marshall; Flynn *J. Am. Chem. Soc.* **1983**, *105*, 3360. For reviews, see Ref. 326. For a review of these and similar compounds, see Borden *Chem. Rev.* **1989**, *89*, 1095-1109.
[329]For reviews of triple bonds in rings, see Meier *Adv. Strain Org. Chem.* **1991**, *1*, 215-272; Krebs; Wilke *Top. Curr. Chem.* **1983**, *109*, 189-233; Nakagawa in Patai *The Chemistry of the C—C Triple Bond*, pt. 2; Wiley: New York, 1978, pp. 635-712; Krebs, in Viehe *Acetylenes*; Marcel Dekker: New York, 1969, pp. 987-1062. For a list of strained cycloalkynes that also have double bonds, see Meier; Hanold; Molz; Bissinger; Kolshorn; Zountsas *Tetrahedron* **1986**, *42*, 1711.
[330]Blomquist; Liu *J. Am. Chem. Soc.* **1953**, *75*, 2153. See also Bühl; Gugel; Kolshorn; Meier *Synthesis* **1978**, 536.
[331]Krebs; Kimling *Angew. Chem. Int. Ed. Engl.* **1971**, *10*, 509 [*Angew. Chem. 83*, 540]; Schmidt; Schweig; Krebs *Tetrahedron Lett.* **1974**, 1471.
[332]Krebs; Kimling *Tetrahedron Lett.* **1970**, 761.
[333]Wittig; Meske-Schüller *Liebigs Ann. Chem.* **1968**, *711*, 65; Krebs; Kimling, Ref. 331; Bottini; Frost; Anderson; Dev *Tetrahedron* **1973**, *29*, 1975.
[334]Wentrup; Blanch; Briehl; Gross *J. Am. Chem. Soc.* **1988**, *110*, 1874.
[335]See Sander; Chapman *Angew. Chem. Int. Ed. Engl.* **1988**, *27*, 398 [*Angew. Chem. 100*, 402]; Krebs; Colcha; Müller; Eicher; Pielartzik; Schnöckel *Tetrahedron Lett.* **1984**, *25*, 5027.
[336]See, for example, Wittig; Mayer *Chem. Ber.* **1963**, *96*, 329, 342; Wittig; Weinlich *Chem. Ber.* **1965**, *98*, 471; Bolster; Kellogg *J. Am. Chem. Soc.* **1981**, *103*, 2868; Gilbert; Baze *J. Am. Chem. Soc.* **1983**, *105*, 664.
[337]Chapman; Gano; West; Regitz; Maas *J. Am. Chem. Soc.* **1981**, *103*, 7033.
[338]Bennett; Robertson; Whimp; Yoshida *J. Am. Chem. Soc.* **1971**, *93*, 3797.
[339]For reviews of cyclic allenes, see Johnson *Adv. Theor. Interesting Mol.* **1989**, *1*, 401-436, *Chem. Rev.* **1989**, *89*, 1111-1124; Thies *Isr. J. Chem.* **1985**, *26*, 191-195; Schuster; Coppola *Allenes in Organic Synthesis*; Wiley: New York, 1984, pp. 38-56.
[340]Price; Johnson *Tetrahedron Lett.* **1986**, *27*, 4679.
[341]See Marquis; Gardner *Tetrahedron Lett.* **1966**, 2793.
[342]Wittig; Dorsch; Meske-Schüller *Liebigs Ann. Chem.* **1968**, *711*, 55.
[343]Visser; Ramakers *J. Chem. Soc., Chem. Commun.* **1972**, 178.
[344]Wentrup; Gross; Maquestiau; Flammang *Angew. Chem. Int. Ed. Engl.* **1983**, *22*, 542 [*Angew. Chem. 95*, 551]. 1,2,3-Cyclohexatriene has also been trapped: Shakespeare; Johnson *J. Am. Chem. Soc.* **1990**, *112*, 8578.

Cyclic allenes in general are less strained than their acetylenic isomers.[345] The cyclic cumulene 1,2,3-cyclononatriene has also been synthesized and is reasonably stable in solution at room temperature in the absence of air.[346]

In bridged bicyclic compounds double bonds at the bridgehead are impossible in small systems. This is the basis of *Bredt's rule*,[347] which states that elimination to give a double bond in a bridged bicyclic system (e.g., **90**) always leads away from the bridgehead. This rule no longer applies when the rings are large enough. In determining whether a bicyclic

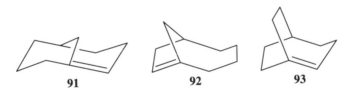

90

system is large enough to accommodate a bridgehead double bond, the most reliable criterion is the size of the ring in which the double bond is located.[348] Bicyclo[3.3.1]non-1-ene[349] (**91**) and bicyclo[4.2.1]non-1(8)ene[350] (**92**) are stable compounds. Both can be looked upon as

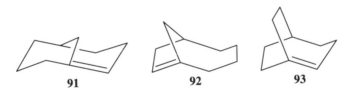

91   **92**   **93**

derivatives of *trans*-cyclooctene, which is of course a known compound. **91** has been shown to have a strain energy of the same order of magnitude as that of *trans*-cyclooctene.[351] On the other hand, in bicyclo[3.2.2]non-1-ene (**93**), the largest ring that contains the double bond is *trans*-cycloheptene, which is as yet unknown. **93** has been prepared, but dimerized before it could be isolated.[352] Even smaller systems ([3.2.1] and [2.2.2]), but with imine double bonds (**94-96**), have been obtained in matrixes at low temperatures.[353] These com-

[345]Moore; Ward *J. Am. Chem. Soc.* **1963**, *85*, 86.

[346]Angus; Johnson *J. Org. Chem.* **1984**, *49*, 2880.

[347]For reviews, see Shea *Tetrahedron* **1980**, *36*, 1683-1715; Buchanan *Chem. Soc. Rev.* **1974**, *3*, 41-63; Köbrich *Angew. Chem. Int. Ed. Engl.* **1973**, *12*, 464-473 [*Angew. Chem. 85*, 494-503]. For reviews of bridgehead olefins, see Billups; Haley; Lee *Chem. Rev.* **1989**, *89*, 1147-1159; Warner *Chem. Rev.* **1989**, *89*, 1067-1093; Szeimies *React. Intermed.* (*Plenum*) **1983**, *3*, 299-366; Keese *Angew. Chem. Int. Ed. Engl.* **1975**, *14*, 528-538 [*Angew. Chem. 87*, 568-578].

[348]For a discussion and predictions of stability in such compounds, see Maier; Schleyer *J. Am. Chem. Soc.* **1981**, *103*, 1891.

[349]Marshall; Faubl *J. Am. Chem. Soc.* **1967**, *89*, 5965, **1970**, *92*, 948; Wiseman *J. Am. Chem. Soc.* **1967**, *89*, 5966; Wiseman; Pletcher *J. Am. Chem. Soc.* **1970**, *92*, 956; Kim; White *J. Am. Chem. Soc.* **1975**, *97*, 451; Becker *Helv. Chim. Acta* **1977**, *60*, 81. For the preparation of optically active **91**, see Nakazaki; Naemura; Nakahara *J. Org. Chem.* **1979**, *44*, 2438.

[350]Wiseman; Chan; Ahola *J. Am. Chem. Soc.* **1969**, *91*, 2812; Carruthers; Qureshi *Chem. Commun.* **1969**, 832; Becker *Tetrahedron Lett.* **1975**, 2207.

[351]Lesko; Turner *J. Am. Chem. Soc.* **1968**, *90*, 6888; Burkert *Chem. Ber.* **1977**, *110*, 773.

[352]Wiseman; Chong *J. Am. Chem. Soc.* **1969**, *91*, 7775.

[353]Sheridan; Ganzer *J. Am. Chem. Soc.* **1983**, *105*, 6158; Ref. 354.

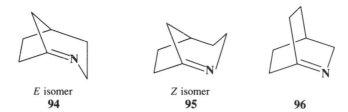

E isomer Z isomer

94   **95**   **96**

pounds are destroyed on warming. **94** and **95** are the first reported example of E-Z isomerism at a strained bridgehead double bond.[354]

Strain Due to Unavoidable Crowding[355]

In some molecules, large groups are so close to each other that they cannot fit into the available space in such a way that normal bond angles are maintained. It has proved possible to prepare compounds with a high degree of this type of strain. For example, success has been achieved in synthesizing benzene rings containing ortho t-butyl groups. The 1,2,3-tri-t-butyl compounds **97**[356] (see p. 873), **98**,[357] and **99**[358] have been prepared, as well as the 1,2,3,4-tetra-t-butyl compound **100**.[359] That these molecules are strained is demon-

F   **F**   MeOOC   **COOMe**   **COOMe**

F   COO-t-Bu   **COOMe**

97   **98**   **99**   **100**

strated by uv and ir spectra, which show that the ring is not planar in 1,2,4-tri-t-butylbenzene, and by a comparison of the heats of reaction of this compound and its 1,3,5 isomer, which show that the 1,2,4 compound possesses about 22 kcal/mol (92 kJ/mol) more strain energy than its isomer[360] (see also p. 1117). X-ray diffraction of **98** shows a nonplanar, boat conformation for the ring.[357] SiMe$_3$ groups are larger than CMe$_3$ groups, and it has proven possible to prepare C$_6$(SiMe$_3$)$_6$. This compound has a chair-shaped ring in the solid state, and a mixture of chair and boat forms in solution.[361] Even smaller groups can sterically interfere in ortho positions. In hexaisopropylbenzene, the six isopropyl groups are so crowded that they cannot rotate but are lined up around the benzene ring, all pointed in

[354]Radziszewski; Downing; Wentrup; Kaszynski; Jawdosiuk; Kovacic; Michl *J. Am. Chem. Soc.* **1985**, *107*, 2799.
[355]For reviews, see Tidwell *Tetrahedron* **1978**, *34*, 1855-1868; Voronenkov; Osokin *Russ. Chem. Rev.* **1972**, *41*, 616-629. For a review of early studies, see Mosher; Tidwell *J. Chem. Educ.* **1990**, *67*, 9-14. For a review of van der Waals radii, see Zefirov; Zorkii *Russ. Chem. Rev.* **1989**, *58*, 421-440.
[356]Viehe; Merényi; Oth; Valange *Angew. Chem. Int. Ed. Engl.* **1964**, *3*, 746 [*Angew. Chem. 76*, 890].
[357]Maas; Fink; Wingert; Blatter; Regitz *Chem. Ber.* **1987**, *120*, 819.
[358]Arnett; Bollinger *Tetrahedron Lett.* **1964**, 3803.
[359]Maier; Schneider *Angew. Chem. Int. Ed. Engl.* **1980**, *19*, 1022 [*Angew. Chem. 92*, 1056]. For another example, see Krebs; Franken; Müller *Tetrahedron Lett.* **1981**, *22*, 1675.
[360]Arnett; Sanda; Bollinger; Barber *J. Am. Chem. Soc.* **1967**, *89*, 5389; Krüerke; Hoogzand; Hübel *Chem. Ber.* **1961**, *94*, 2817; Dale *Chem. Ber.* **1961**, *94*, 2821. See also Barclay; Brownstein; Gabe; Lee *Can. J. Chem.* **1984**, *62*, 1358.
[361]Sakurai; Ebata; Kabuto; Sekiguchi *J. Am. Chem. Soc.* **1990**, *112*, 1799.

the same direction.[362] This compound is an example of a *geared molecule*.[363] The isopropyl groups fit into each other in the same manner as interlocked gears. Another example is **101** (which is a stable enol).[364] In this case each ring can rotate about its C—aryl bond only by forcing the other to rotate as well. In the case of triptycene derivatives such as **102**, a

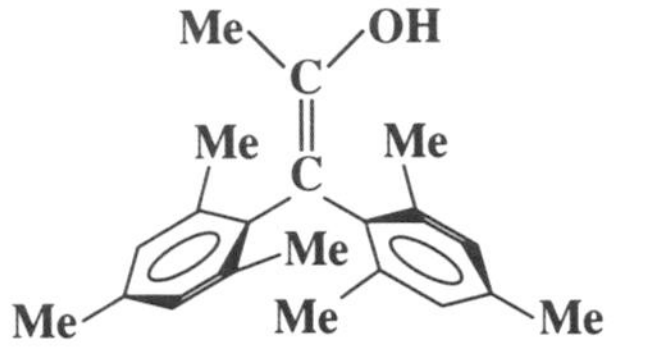

101 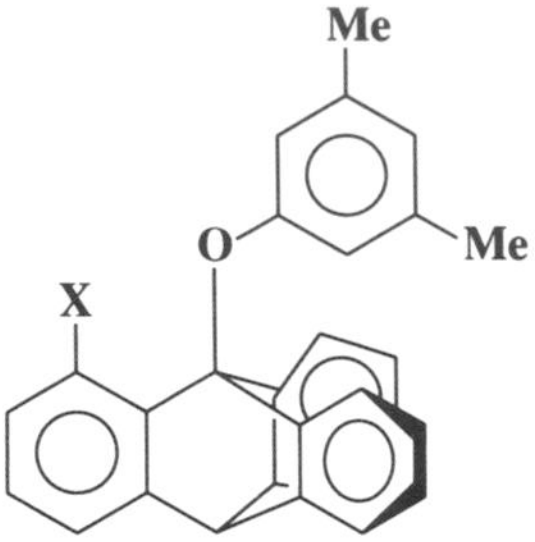   **102**

complete 360° rotation of the aryl group around the O—aryl bond requires the aryl group to pass over three rotational barriers; one of which is the C—X bond and other two the "top" C—H bonds of the other two rings. As expected, the C—X barrier is the highest, ranging from 10.3 kcal/mol (43.1 kJ/mol) for X = F to 17.6 kcal/mole (73.6 kJ/mol) for X = *t*-butyl.[365] In another instance, it has proved possible to prepare cis and trans isomers of 5-amino-2,4,6-triiodo-N,N,N',N'-tetramethylisophthalamide because there is no room for the CONMe$_2$ groups to rotate, caught as they are between two bulky iodine atoms.[366]

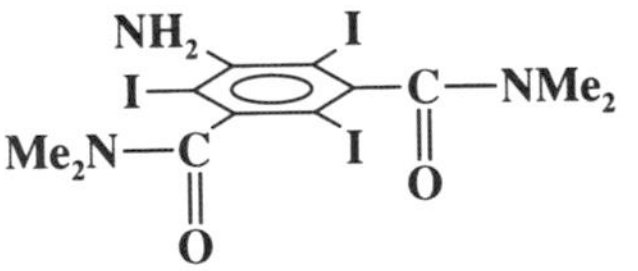

cis 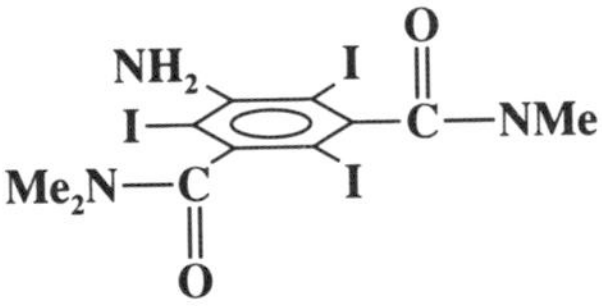   trans

The trans isomer is chiral and has been resolved, while the cis isomer is a meso form. Another example of cis–trans isomerism resulting from restricted rotation about single bonds[367] is found in 1,8-di-*o*-tolylnapthalene[368] (see also p. 128).

[362]Arnett; Bollinger *J. Am. Chem. Soc.* **1964**, *86*, 4730; Hopff; Gati *Helv. Chim. Acta* **1965**, *48*, 509; Siegel; Gutiérrez; Schweizer; Ermer; Mislow *J. Am. Chem. Soc.* **1986**, *108*, 1569. For the similar structure of hexakis(dichloromethyl)benzene, see Kahr; Biali; Schaefer; Buda; Mislow *J. Org. Chem.* **1987**, *52*, 3713.

[363]For reviews, see Iwamura; Mislow *Acc. Chem. Res.* **1988**, *21*, 175-182; Mislow *Chemtracts: Org. Chem.* **1989**, *2*, 151-174, *Chimia* **1986**, *40*, 395-402; Berg; Liljefors; Roussel; Sandström *Acc. Chem. Res.* **1985**, *18*, 80-86.

[364]Nugiel; Biali; Rappoport *J. Am. Chem. Soc.* **1984**, *106*, 3357.

[365]Yamamoto; Ōki *Bull. Chem. Soc. Jpn.* **1986**, *59*, 3597. For reviews of similar cases, see Yamamoto *Pure Appl. Chem.* **1990**, *62*, 569-574; Ōki, Ref. 49, pp. 269-284.

[366]Ackerman; Laidlaw; Snyder *Tetrahedron Lett.* **1969**, 3879; Ackerman; Laidlaw *Tetrahedron Lett.* **1969**, 4487. See also Cuyegkeng; Mannschreck *Chem. Ber.* **1987**, *120*, 803.

[367]For a monograph on restricted rotation about single bonds, see Ōki, Ref. 49. For reviews, see Förster; Vögtle *Angew. Chem. Int. Ed. Engl.* **1977**, *16*, 429-441 [*Angew. Chem. 89*, 443-455]; Ōki *Angew. Chem. Int. Ed. Engl.* **1976**, *15*, 87-93 [*Angew. Chem. 88*, 67-74].

[368]Clough; Roberts *J. Am. Chem. Soc.* **1976**, *98*, 1018. For a study of rotational barriers in this system, see Cosmo; Sternhell *Aust. J. Chem.* **1987**, *40*, 1107.

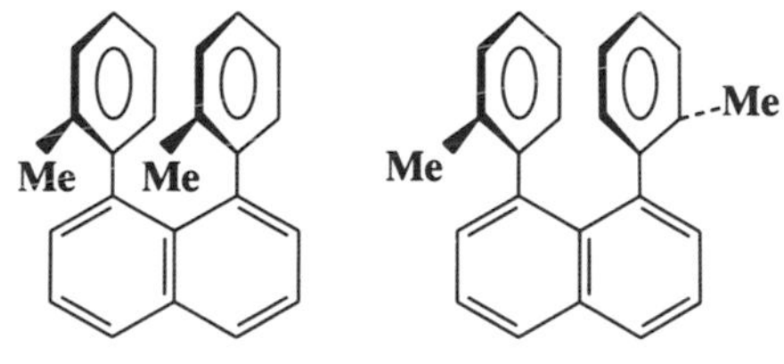

cis   trans

There are many other cases of intramolecular crowding that result in the distortion of bond angles. We have already mentioned hexahelicene (p. 103) and bent benzene rings (p. 37). The compounds tri-*t*-butylamine and tetra-*t*-butylmethane are as yet unknown. In the latter, there is no way for the strain to be relieved and it is questionable whether this compound can ever be made. In tri-*t*-butylamine the crowding can be eased somewhat if the three bulky groups assume a planar instead of the normal pyramidal configuration. In tri-*t*-butylcarbinol, coplanarity of the three *t*-butyl groups is prevented by the presence of the OH group, and yet this compound has been prepared.[369] Tri-*t*-butylamine should have less steric strain than tri-*t*-butylcarbinol and it should be possible to prepare it.[370] The tetra-*t*-butylphosphonium cation $(t\text{-Bu})_4P^+$ has been prepared.[371] Although steric effects are nonadditive in crowded molecules, a quantitative measure has been proposed by D. F. DeTar, based on molecular mechanics calculations. This is called *formal steric enthalpy* (FSE), and values have been calculated for alkanes, alkenes, alcohols, ethers, and methyl esters.[372] For example, some FSE values for alkanes are: butane 0.00; 2,2,3,3-tetramethylbutane 7.27; 2,2,4,4,5-pentamethylhexane 11.30; and tri-*t*-butylmethane 38.53.

The two carbon atoms of a C=C double bond and the four groups attached to them are normally in a plane, but if the groups are large enough, significant deviation from planarity can result.[373] The compound tetra-*t*-butylethene(**103**) has not been prepared,[374] but the tetraaldehyde **104**, which should have about the same amount of strain, has been made. X-ray crystallography shows that **104** is twisted out of a planar shape by an angle of 28.6°.[375]

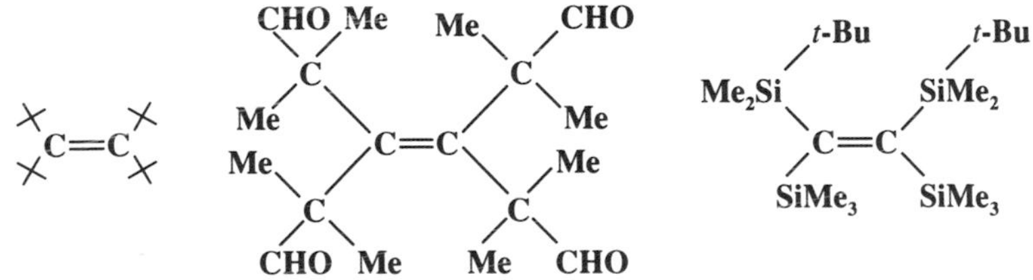

103   104   105

[369]Bartlett; Lefferts *J. Am. Chem. Soc.* **1955,** *77,* 2804; Bartlett; Tidwell *J. Am. Chem. Soc.* **1968,** *90,* 4421.

[370]For attempts to prepare tri-*t*-butylamine, see Back; Barton *J. Chem. Soc., Perkin Trans 1* **1977,** 924. For the preparation of di-*t*-butylmethylamine and other sterically hindered amines, see Kopka; Fataftah; Rathke *J. Org. Chem.* **1980,** *45,* 4616; Audeh; Fuller; Hutchinson; Lindsay Smith *J. Chem. Res. (S)* **1979,** 270.

[371]Schmidbaur; Blaschke; Zimmer-Gasser; Schubert *Chem. Ber.* **1980,** *113,* 1612.

[372]DeTar; Binzet; Darba *J. Org. Chem.* **1985,** *50,* 2826, 5298, 5304.

[373]For reviews, see Luef; Keese *Top. Stereochem.* **1991,** *20,* 231-318; Sandström, Ref. 164, pp. 160-169.

[374]For a list of crowded alkenes that have been made, see Drake; Rabjohn; Tempesta; Taylor *J. Org. Chem.* **1988,** *53,* 4555. See also Garratt; Payne; Tocher *J. Org. Chem.* **1990,** *55,* 1909.

[375]Krebs; Nickel; Tikwe; Kopf *Tetrahedron Lett.* **1985,** *26,* 1639.

Also, the C=C double bond distance is 1.357 Å, significantly longer than a normal C=C bond of 1.32 Å (Table 1.5). Z-1,2-Bis(t-butyldimethylsilyl)-1,2-bis(trimethylsilyl)ethene (**105**) has an even greater twist, but could not be made to undergo conversion to the E isomer, probably because the groups are too large to slide past each other.[376] A different kind of double bond strain is found in tricyclo[4.2.2.2^{2,5}]dodeca-1,5-diene (**106**),[377] cubene (**107**),[378] and homocub-4(5)-ene (**108**).[379] In these molecules, the four groups on the double bond are all forced to be on one side of the double-bond plane.[380] In **106** the angle between

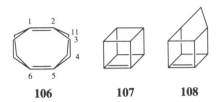

106　　　　**107**　　**108**

the line C$_1$–C$_2$ (extended) and the plane defined by C$_2$, C$_3$, and C$_{11}$ is 27°. An additional source of strain in this molecule is the fact that the two double bonds are pushed into close proximity by the four bridges. In an effort to alleviate this sort of strain the bridge bond distances (C$_3$–C$_4$) are 1.595 Å, which is considerably longer than the 1.53 Å expected for a normal sp^3–sp^3 C—C bond (Table 1.5). **107** and **108** have not been isolated, but have been generated as intermediates that were trapped by reaction with other compounds.[378,379]

[376]Sakurai; Ebata; Kabuto; Nakadaira *Chem. Lett.* **1987**, 301.

[377]Wiberg; Matturo; Okarma; Jason *J. Am. Chem. Soc.* **1984**, *106*, 2194; Wiberg; Adams; Okarma; Matturo; Segmuller *J. Am. Chem. Soc.* **1984**, *106*, 2200.

[378]Eaton; Maggini *J. Am. Chem. Soc.* **1988**, *110*, 7230.

[379]Hrovat; Borden *J. Am. Chem. Soc.* **1988**, *110*, 7229.

[380]For a review of such molecules, see Borden *Chem. Rev.* **1989**, *89*, 1095-1109. See also Hrovat; Borden *J. Am. Chem. Soc.* **1988**, *110*, 4710.

5

CARBOCATIONS, CARBANIONS, FREE RADICALS, CARBENES, AND NITRENES

There are four types of organic species in which a carbon atom has a valence of only 2 or 3.[1] They are usually very short-lived, and most exist only as intermediates that are quickly converted to more stable molecules. However, some are more stable than others and fairly stable examples have been prepared of three of the four types. The four types of species are *carbocations* (**A**), *free radicals* (**B**), *carbanions* (**C**), and *carbenes* (**D**). Of the four, only carbanions have a complete octet around the carbon. There are many other organic ions

$$
\underset{\textstyle \underset{\textstyle R}{|}}{\overset{\textstyle \overset{\textstyle R}{|}}{R-C^{\oplus}}} \qquad
\underset{\textstyle \underset{\textstyle R}{|}}{\overset{\textstyle \overset{\textstyle R}{|}}{R-C\cdot}} \qquad
\underset{\textstyle \underset{\textstyle R}{|}}{\overset{\textstyle \overset{\textstyle R}{|}}{R-\underset{}{C}|^{\ominus}}} \qquad
\overset{\textstyle \overset{\textstyle R}{|}}{R-\underline{C}} \qquad
R-\overline{N}|
$$

A   B   C   D   E

and radicals with charges and unpaired electrons on atoms other than carbon, but we will discuss only *nitrenes* (**E**), the nitrogen analogs of carbenes. Each of the five types is discussed in a separate section, which in each case includes brief summaries of the ways in which the species form and react. These summaries are short and schematic. The generation and fate of the five types are more fully treated in appropriate places in Part 2 of this book.

CARBOCATIONS[2]

Nomenclature

First we must say a word about the naming of **A**. For many years these species were called "carbonium ions," though it was suggested[3] as long ago as 1902 that this was inappropriate

[1]For general references, see Isaacs *Reactive Intermediates in Organic Chemistry*; Wiley: New York, 1974; McManus *Organic Reactive Intermediates*; Academic Press: New York, 1973. Two serial publications devoted to review articles on this subject are *Reactive Intermediates* (Wiley) and *Reactive Intermediates (Plenum)*.

[2]For a treatise, see Olah; Schleyer *Carbonium Ions*, 5 vols.; Wiley: New York, 1968-1976. For monographs, see Vogel *Carbocation Chemistry*; Elsevier: New York, 1985; Bethell; Gold *Carbonium Ions*; Academic Press: New York, 1967. For reviews, see Saunders; Jiménez-Vázquez *Chem. Rev.* **1991**, *91*, 375-397; Arnett; Hofelich; Schriver *React. Intermed.* (*Wiley*) **1987**, *3*, 189-226; Bethell; Whittaker *React. Intermed.* (*Wiley*) **1981**, *2*, 211-250; Bethell *React. Intermed.* (*Wiley*) **1978**, *1*, 117-161; Olah *Chem. Scr.* **1981**, *18*, 97-125, *Top. Curr. Chem.* **1979**, *80*, 19-88, *Angew. Chem. Int. Ed. Engl.* **1973**, *12*, 173-212 [*Angew. Chem. 85*, 183-225] (this review has been reprinted as Olah *Carbocations and Electrophilic Reactions*; Wiley: New York, 1974); Isaacs, Ref. 1, pp. 92-199; McManus; Pittman, in McManus, Ref. 1, pp. 193-335; Buss; Schleyer; Allen *Top. Stereochem.* **1973**, *7*, 253-293; Olah; Pittman *Adv. Phys. Org. Chem.* **1966**, *4*, 305-347. For reviews of dicarbocations, see Lammertsma; Schleyer; Schwarz *Angew. Chem. Int. Ed. Engl.* **1989**, *28*, 1321-1341 [*Angew. Chem. 101*, 1313-1335]; Lammertsma *Rev. Chem. Int.* **1988**, *9*, 141-169; Pagni *Tetrahedron* **1984**, *40*, 4161-4215; Prakash; Rawdah; Olah *Angew. Chem. Int. Ed. Engl.* **1983**, *22*, 390-401 [*Angew. Chem. 95*, 356-367]. See also the series *Advances in Carbocation Chemistry*.

[3]Gomberg *Ber.* **1902**, *35*, 2397.

because "-onium" usually refers to a covalency higher than that of the neutral atom. Nevertheless, the name "carbonium ion" was well established and created few problems[4] until some years ago, when George Olah and his co-workers found evidence for another type of intermediate in which there is a positive charge at a carbon atom, but in which the formal covalency of the carbon atom is five rather than three. The simplest example is the methanonium ion CH_5^+ (see p. 580). Olah proposed[5] that the name "carbonium ion" be henceforth reserved for pentacoordinated positive ions, and that **A** be called "carbenium ions." He also proposed the term "carbocation" to encompass both types. IUPAC has accepted these definitions.[6] Although some authors still refer to **A** as carbonium ions and others call them carbenium ions, the general tendency is to refer to them simply as *carbocations*, and we will follow this practice. The pentavalent species are much rarer than **A**, and the use of the term "carbocation" for **A** causes little or no confusion.

Stability and Structure

Carbocations are intermediates in several kinds of reactions. The more stable ones have been prepared in solution and in some cases even as solid salts. In solution the carbocation may be free (this is more likely in polar solvents, in which it is solvated) or it may exist as an ion pair,[7] which means that it is closely associated with a negative ion, called a *counterion* or *gegenion*. Ion pairs are more likely in nonpolar solvents.

Among simple alkyl carbocations[8] the order of stability is tertiary > secondary > primary. Many examples are known of rearrangements of primary or secondary carbocations to tertiary, both in solution and in the gas phase. Since simple alkyl cations are not stable in ordinary strong-acid solutions, e.g., H_2SO_4, the study of these species was greatly facilitated by the discovery that many of them could be kept indefinitely in stable solutions in mixtures of fluorosulfuric acid and antimony pentafluoride. Such mixtures, usually dissolved in SO_2 or SO_2ClF, are among the strongest acidic solutions known and are often called *super acids*.[9] The original experiments involved the addition of alkyl fluorides to SbF_5.[10]

$$RF + SbF_5 \longrightarrow R^+ SbF_6^-$$

Subsequently it was found that the same cations could also be generated from alcohols in super acid–SO_2 at $-60°C$[11] and from alkenes by the addition of a proton from super acid or $HF–SbF_5$ in SO_2 or SO_2ClF at low temperatures.[12] Even alkanes give carbocations in super acid by loss of H^-. For example,[13] isobutane gives the *t*-butyl cation

$$Me_3CH \xrightarrow{\text{FSO}_3\text{H–SbF}_5} Me_3C^+ \ SbF_5FSO_3^- + H_2$$

[4]For a history of the term "carbonium ion", see Traynham *J. Chem. Educ.* **1986**, *63*, 930.

[5]Olah *CHEMTECH* **1971**, *1*, 566, *J. Am. Chem. Soc.* **1972**, *94*, 808.

[6]Gold; Loening; McNaught; Sehmi *Compendium of Chemical Terminology: IUPAC Recommendations*; Blackwell Scientific Publications: Oxford, 1987.

[7]For a treatise, see Szwarc *Ions and Ion Pairs in Organic Reactions*, 2 vols.; Wiley: New York, 1972-1974.

[8]For a review, see Olah; Olah, in Olah; Schleyer, Ref. 2, vol. 2, pp. 715-782.

[9]For a review of carbocations in super acid solutions, see Olah; Prakash; Sommer, in *Superacids*; Wiley: New York, 1985, pp. 65-175.

[10]Olah; Baker; Evans; Tolgyesi; McIntyre; Bastien *J. Am. Chem. Soc.* **1964**, *86*, 1360; Brouwer; Mackor *Proc. Chem. Soc.* **1964**, 147; Kramer *J. Am. Chem. Soc.* **1969**, *91*, 4819.

[11]Olah; Comisarow; Cupas; Pittman *J. Am. Chem. Soc.* **1965**, *87*, 2997; Olah; Sommer; Namanworth *J. Am. Chem. Soc.* **1967**, *89*, 3576.

[12]Olah; Halpern *J. Org. Chem.* **1971**, *36*, 2354. See also Herlem *Pure Appl. Chem.* **1977**, *49*, 107.

[13]Olah; Lukas *J. Am. Chem. Soc.* **1967**, *89*, 4739.

No matter how they are generated, study of the simple alkyl cations has provided dramatic evidence for the stability order. Both propyl fluorides gave the isopropyl cation; all four butyl fluorides[14] gave the *t*-butyl cation, and all seven of the pentyl fluorides tried gave the *t*-pentyl cation. *n*-Butane, in super acid, gave only the *t*-butyl cation. To date no primary cation has survived long enough for detection. Neither methyl nor ethyl fluoride gave the corresponding cations when treated with SbF_5. At low temperatures, methyl fluoride gave chiefly the methylated sulfur dioxide salt $(CH_3OSO)^+$ SbF_6^- [15] while ethyl fluoride rapidly formed the *t*-butyl and *t*-hexyl cations by addition of the initially formed ethyl cation to ethylene molecules also formed.[16] At room temperature, methyl fluoride also gave the *t*-butyl cation.[17] In accord with the stability order, hydride ion is abstracted from alkanes by super acid most readily from tertiary and least readily from primary positions.

The stability order can be explained by hyperconjugation and by the field effect. In the hyperconjugation explanation,[18] we compare a primary carbocation with a tertiary. It is seen that many more canonical forms are possible for the latter:

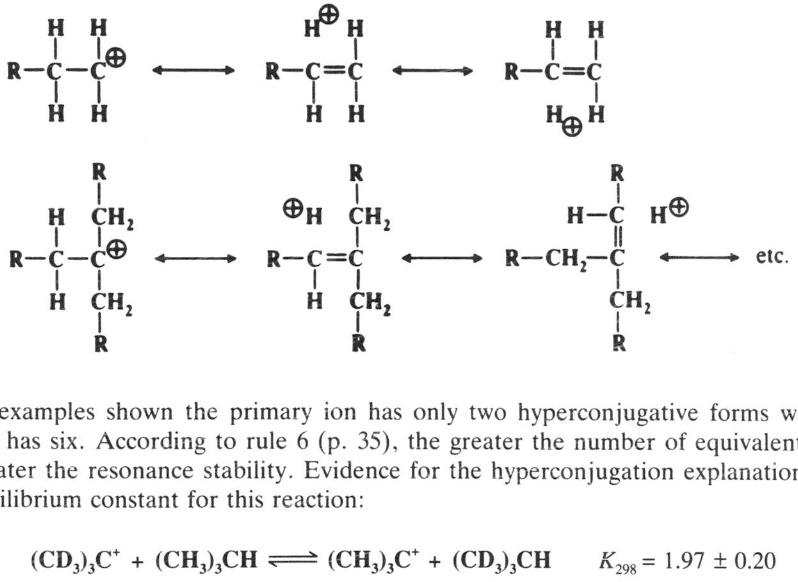

In the examples shown the primary ion has only two hyperconjugative forms while the tertiary has six. According to rule 6 (p. 35), the greater the number of equivalent forms, the greater the resonance stability. Evidence for the hyperconjugation explanation is that the equilibrium constant for this reaction:

$$(CD_3)_3C^+ + (CH_3)_3CH \rightleftharpoons (CH_3)_3C^+ + (CD_3)_3CH \qquad K_{298} = 1.97 \pm 0.20$$
$$\mathbf{1} \qquad\qquad\qquad\qquad \mathbf{2}$$

is 1.97, showing that **2** is more stable than **1**.[19] This is a β secondary isotope effect; there is less hyperconjugation in **1** than in **2** (see p. 228).[20]

[14]The *sec*-butyl cation has been prepared by slow addition of *sec*-butyl chloride to SbF_5-SO_2ClF solution at $-110°C$ [Saunders; Hagen; Rosenfeld *J. Am. Chem. Soc.* **1968**, *90*, 6882] and by allowing molecular beams of the reagents to impinge on a very cold surface [Saunders; Cox; Ohlmstead *J. Am. Chem. Soc.* **1973**, *95*, 3018; Saunders; Cox; Lloyd *J. Am. Chem. Soc.* **1979**, *101*, 6656; Myhre; Yannoni *J. Am. Chem. Soc.* **1981**, *103*, 230].

[15]Peterson; Brockington; Vidrine *J. Am. Chem. Soc.* **1976**, *98*, 2660; Olah; Donovan; Lin *J. Am. Chem. Soc.* **1976**, *98*, 2661; Calves; Gillespie *J. Chem. Soc., Chem. Commun.* **1976**, 506; Olah; Donovan *J. Am. Chem. Soc.* **1978**, *100*, 5163.

[16]Ref. 8, p. 722.

[17]Olah; DeMember; Schlosberg *J. Am. Chem. Soc.* **1969**, *91*, 2112; Bacon; Gillespie *J. Am. Chem. Soc.* **1971**, *91*, 6914.

[18]For a review of molecular-orbital theory as applied to carbocations, see Radom; Poppinger; Haddon, in Olah; Schleyer, Ref. 2, vol. 5, pp. 2303-2426.

[19]Meot-Ner *J. Am. Chem. Soc.* **1987**, *109*, 7947.

[20]If only the field effect were operating, **1** would be more stable than **2**, since deuterium is electron-donating with respect to hydrogen (p. 19), assuming that the field effect of deuterium could be felt two bonds away.

The field effect explanation is that the electron-donating effect of alkyl groups increases the electron density at the charge-bearing carbon, reducing the net charge on the carbon, and in effect spreading the charge over the α carbons. It is a general rule that the more concentrated any charge is, the less stable the species bearing it will be.

The most stable of all alkyl cations is the *t*-butyl cation. Even the relatively stable *t*-pentyl and *t*-hexyl cations fragment at higher temperatures to produce the *t*-butyl cation, as do all other alkyl cations with four or more carbons so far studied.[21] Methane,[22] ethane, and propane, treated with super acid, also yield *t*-butyl cations as the main product (see **2-18**). Even paraffin wax and polyethylene give *t*-butyl cation. Solid salts of *t*-butyl and *t*-pentyl cations, e.g., Me_3C^+ SbF_6^-, have been prepared from super-acid solutions and are stable below $-20°C$.[23]

Where the positive carbon is in conjugation with a double bond the stability is greater because of increased delocalization due to resonance and because the positive charge is

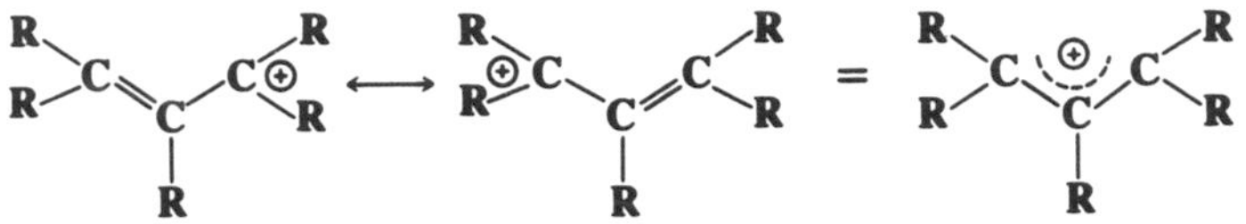

spread over two atoms instead of being concentrated on one (see the molecular-orbital picture of this species on p. 33). Each of the two atoms has a charge of about $\frac{1}{2}$ (the charge is exactly $\frac{1}{2}$ if all of the R groups are the same). Stable allylic-type cations[24] have been prepared by the solution of conjugated dienes in concentrated sulfuric acid, e.g.,[25]

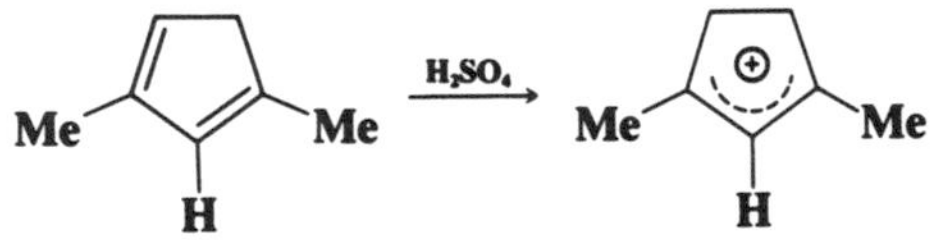

Both cyclic and acyclic allylic cations have been produced in this way. Stable allylic cations have also been obtained by the reaction between alkyl halides, alcohols, or olefins (by hydride extraction) and SbF_5 in SO_2 or SO_2ClF.[26] Divinylmethyl cations[27] are more stable than the simple allylic type, and some of these have been prepared in concentrated sulfuric acid.[28] Arenium ions (p. 502) are important examples of this type. Propargyl cations ($RC{\equiv}CCR_2^+$) have also been prepared.[29]

[21]Ref. 13; Ref. 8, pp. 750-764.
[22]Olah; Klopman; Schlosberg *J. Am. Chem. Soc.* **1969,** *91,* 3261. See also Hogeveen; Gaasbeek *Recl. Trav. Chim. Pays-Bas* **1968,** *87,* 319.
[23]Olah; Svoboda; Ku *Synthesis* **1973,** 492; Ref. 13.
[24]For reviews, see Deno, in Olah; Schleyer, Ref. 2, vol. 2, pp. 783-806; Richey, in Zabicky *The Chemistry of Alkenes,* vol. 2, Wiley: New York, 1970, pp. 39-114.
[25]Deno; Richey; Hodge; Wisotsky *J. Am. Chem. Soc.* **1962,** *84,* 1498; Deno; Richey; Friedman; Hodge; Houser; Pittman *J. Am. Chem. Soc.* **1963,** *85,* 2991.
[26]Olah; Comisarow *J. Am. Chem. Soc.* **1964,** *86,* 5682; Olah; Clifford; Halpern; Johanson *J. Am. Chem. Soc.* **1971,** *93,* 4219; Olah; Liang *J. Am. Chem. Soc.* **1972,** *94,* 6434; Olah; Spear *J. Am. Chem. Soc.* **1975,** *97,* 1539.
[27]For a review of divinylmethyl and trivinylmethyl cations, see Sorensen, in Olah; Schleyer, Ref. 2, vol. 2, pp. 807-835.
[28]Deno; Pittman *J. Am. Chem. Soc.* **1964,** *86,* 1871.
[29]Pittman; Olah *J. Am. Chem. Soc.* **1965,** *87,* 5632; Olah; Spear; Westerman; Denis *J. Am. Chem. Soc.* **1974,** *96,* 5855.

Canonical forms can be drawn for benzylic cations,[30] similar to those shown above for allylic cations, e.g.,

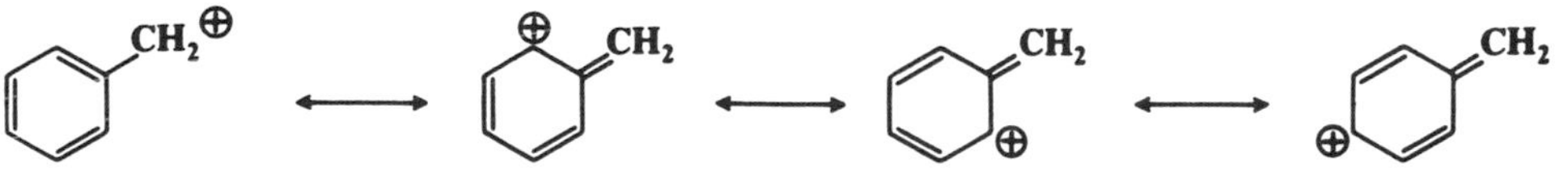

A number of benzylic cations have been obtained in solution as SbF_6^- salts.[31] Diarylmethyl and triarylmethyl cations are still more stable. Triphenylchloromethane ionizes in polar solvents that do not, like water, react with the ion. In SO_2, the equilibrium

$$Ph_3CCl \rightleftharpoons Ph_3C^+ + Cl^-$$

has been known for many years. Both triphenylmethyl and diphenylmethyl cations have been isolated as solid salts[32] and, in fact, Ph_3C^+ BF_4^- and related salts are available commercially. Arylmethyl cations are further stabilized if they have electron-donating substituents in ortho or para positions.[33]

Cyclopropylmethyl cations[34] are even more stable than the benzyl type. **5** has been prepared by solution of the corresponding alcohol in 96% sulfuric acid,[35] and **3, 4,** and similar ions by solution of the alcohols in $FSO_3H–SO_2–SbF_5$.[36] This special stability, which

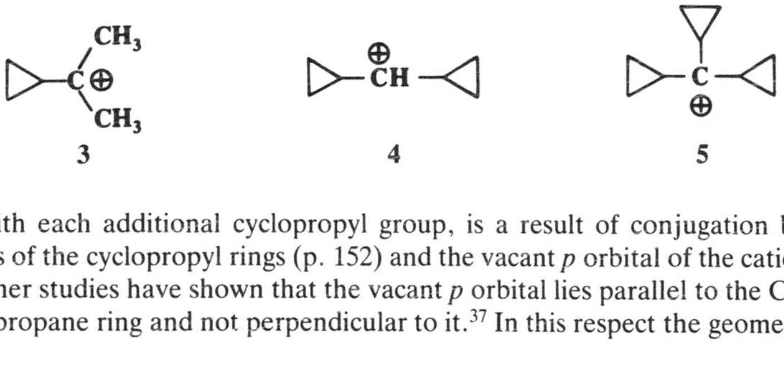

increases with each additional cyclopropyl group, is a result of conjugation between the bent orbitals of the cyclopropyl rings (p. 152) and the vacant p orbital of the cationic carbon. Nmr and other studies have shown that the vacant p orbital lies parallel to the C-2,C-3 bond of the cyclopropane ring and not perpendicular to it.[37] In this respect the geometry is similar

[30]For a review of benzylic, diarylmethyl, and triarymethyl cations, see Freedman, in Olah; Schleyer, Ref. 2, vol. 4, pp. 1501-1578.

[31]Bollinger; Comisarow; Cupas; Olah *J. Am. Chem. Soc.* **1967**, *89*, 5687; Olah; Porter; Jeuell; White *J. Am. Chem. Soc.* **1972**, *94*, 2044.

[32]Volz *Angew. Chem. Int. Ed. Engl.* **1963**, *2*, 622 [*Angew. Chem. 75*, 921]; Volz; Schnell *Angew. Chem. Int. Ed. Engl.* **1965**, *4*, 873 [*Angew. Chem. 77*, 864].

[33]Goldacre; Phillips *J. Chem. Soc.* **1949**, 1724; Deno; Schriesheim *J. Am. Chem. Soc.* **1955**, *77*, 3051.

[34]For reviews, see in Olah; Schleyer, Ref. 2, vol. 3: Richey, pp. 1201-1294; Wiberg; Hess; Ashe, pp. 1295-1345.

[35]Deno; Richey; Liu; Hodge; Houser; Wisotsky *J. Am. Chem. Soc.* **1962**, *84*, 2016.

[36]Pittman; Olah *J. Am. Chem. Soc.* **1965**, *87*, 2998; Deno; Liu; Turner; Lincoln; Fruit *J. Am. Chem. Soc.* **1965**, *87*, 3000.

[37]For example, see Ree; Martin *J. Am. Chem. Soc.* **1970**, *92*, 1660; Kabakoff; Namanworth *J. Am. Chem. Soc.* **1970**, *92*, 3234; Buss; Gleiter; Schleyer *J. Am. Chem. Soc.* **1971**, *93*, 3927; Poulter; Spillner *J. Am. Chem. Soc.* **1974**, *96*, 7591; Childs; Kostyk; Lock; Mahendran *J. Am. Chem. Soc.* **1990**, *112*, 8912; Ref. 35.

to that of a cyclopropane ring conjugated with a double bond (p. 152). Cyclopropylmethyl cations are further discussed on pp. 323–324. The stabilizing effect just discussed is unique to cyclopropyl groups. Cyclobutyl and larger cyclic groups are about as effective at stabilizing a carbocation as ordinary alkyl groups.[38]

Another structural feature that increases carbocation stability is the presence, adjacent to the cationic center, of a hetero atom bearing an unshared pair,[39] e.g., oxygen,[40] nitrogen,[41] or halogen.[42] Such ions are stabilized by resonance:

$$
\overset{\overset{\textstyle R}{\textstyle |}}{R-\underset{\oplus}{C}-\bar{O}-Me} \longleftrightarrow \overset{\overset{\textstyle R}{\textstyle |}}{R-C=\underset{\oplus}{\bar{O}}-Me}
$$

The methoxymethyl cation can be obtained as a stable solid, $MeOCH_2^+$ SbF_6^-.[43] Carbocations containing either a β or γ silicon atom are also stabilized,[44] relative to similar ions without the silicon atom.

Simple acyl cations RCO^+ have been prepared[45] in solution and the solid state.[46] The acetyl cation CH_3CO^+ is about as stable as the t-butyl cation (see, for example, Table 5.1). The 2,4,6-trimethylbenzoyl and 2,3,4,5,6-pentamethylbenzoyl cations are especially stable (for steric reasons) and are easily formed in 96% H_2SO_4.[47] These ions are stabilized by a canonical form containing a triple bond (**G**), though the positive charge is principally located on the carbon,[48] so that **F** contributes more than **G**.

$$
\underset{\oplus}{R-C=\bar{O}} \longleftrightarrow R-C\equiv\underset{\oplus}{\bar{O}}
$$

$$
\textbf{F} \qquad\qquad \textbf{G}
$$

The stabilities of most other stable carbocations can also be attributed to resonance. Among these are the tropylium, cyclopropenium, and other aromatic cations discussed in Chapter 2. Where resonance stability is completely lacking, as in the phenyl ($C_6H_5^+$) or

[38]Sorensen; Miller; Ranganayakulu *Aust. J. Chem.* **1973**, *26*, 311.

[39]For a review, see Hevesi *Bull. Soc. Chim. Fr.* **1990**, 697-703. For examples of stable solutions of such ions, see Kabuss *Angew. Chem. Int. Ed. Engl.* **1966**, *5*, 675 [*Angew. Chem. 78*, 714]; Dimroth; Heinrich *Angew. Chem. Int. Ed. Engl.* **1966**, *5*, 676 [*Angew. Chem. 78*, 715]; Tomalia; Hart *Tetrahedron Lett.* **1966**, 3389; Ramsey; Taft *J. Am. Chem. Soc.* **1966**, *88*, 3058; Olah; Liang; Mo *J. Org. Chem.* **1974**, *39*, 2394; Borch *J. Am. Chem. Soc.* **1968**, *90*, 5303; Rabinovitz; Bruck *Tetrahedron Lett.* **1971**, 245.

[40]For a review of ions of the form $R_2C—OR'$, see Rakhmankulov; Akhmatdinov; Kantor *Russ. Chem. Rev.* **1984**, *53*, 888-899. For a review of ions of the form $R'\overset{+}{C}(OR)_2$ and $\overset{+}{C}(OR)_3$, see Pindur; Müller; Flo; Witzel *Chem. Soc. Rev.* **1987**, *16*, 75-87.

[41]For a review of such ions where nitrogen is the hetero atom, see Scott; Butler, in Olah; Schleyer, Ref. 2, vol. 4, pp. 1643-1696.

[42]For reviews of such ions where the hetero atom is halogen, see Allen; Tidwell *Adv. Carbocation Chem.* **1989**, *1*, 1-44; Olah; Mo, in Olah; Schleyer, Ref. 2, vol. 5, pp. 2135-2262. *Adv. Fluorine Chem.* **1973**, *7*, 69-112. For the preparation, in superacid solution, of the ions CX_3^+ (X = Cl, Br, I), see Olah; Heiliger; Prakash *J. Am. Chem. Soc.* **1989**, *111*, 8020.

[43]Olah; Svoboda *Synthesis* **1973**, 52.

[44]For a review and discussion of the causes, see Lambert *Tetrahedron* **1990**, *46*, 2677-2689. See also Lambert; Chelius *J. Am. Chem. Soc.* **1990**, *112*, 8120.

[45]For reviews of acyl cations, see Al-Talib; Tashtoush *Org. Prep. Proced. Int.* **1990**, *22*, 1-36; Olah; Germain; White; in Olah; Schleyer, Ref. 2, vol. 5, pp. 2049-2133. For a review of the preparation of acyl cations from acyl halides and Lewis acids, see Lindner *Angew. Chem. Int. Ed. Engl.* **1970**, *9*, 114-123 [*Angew. Chem. 82*, 143-153].

[46]See, for example, Olah; Kuhn; Tolgyesi; Baker *J. Am. Chem. Soc.* **1962**, *84*, 2733; Deno; Pittman; Wisotsky *J. Am. Chem. Soc.* **1964**, *86*, 4370; Olah; Dunne; Mo; Szilagyi *J. Am. Chem. Soc.* **1972**, *94*, 4200; Olah; Svoboda *Synthesis* **1972**, 306.

[47]Hammett; Deyrup *J. Am. Chem. Soc.* **1933**, *55*, 1900; Newman; Deno *J. Am. Chem. Soc.* **1951**, *73*, 3651.

[48]Boer *J. Am. Chem. Soc.* **1968**, *90*, 6706; Le Carpentier; Weiss *Acta Crystallogr. Sect. B* **1972**, 1430. See also Olah; Westerman *J. Am. Chem. Soc.* **1973**, *95*, 3706.

vinyl cations, the ion, if formed at all, is usually very short-lived.[49] Neither vinyl[50] nor phenyl cation has as yet been prepared as a stable species in solution.[51]

Various quantitative methods have been developed to express the relative stabilities of carbocations.[52] One of the most common of these, though useful only for relatively stable cations that are formed by ionization of alcohols in acidic solutions, is based on the equation[53]

$$H_R = pK_{R^+} - \log \frac{C_{R^+}}{C_{ROH}}$$

pK_{R^+} is the pK value for the reaction $R^+ + 2H_2O \rightleftharpoons ROH + H_3O^+$ and is a measure of the stability of the carbocation. H_R is an easily obtainable measurement of the acidity of a solvent (see p. 256) and approaches pH at low concentrations of acid. In order to obtain pK_{R^+} for a cation R^+, one dissolves the alcohol ROH in an acidic solution of known H_R. Then the concentrations of R^+ and ROH are obtained, generally from spectra, and pK_{R^+} is easily calculated.[54] A measure of carbocation stability that applies to less-stable ions is the dissociation energy $D(R^+{-}H^-)$ for the cleavage reaction $R{-}H \rightarrow R^+ + H^-$, which can be obtained from photoelectron spectroscopy and other measurements. Some values of $D(R^+{-}H^-)$ are shown in Table 5.1.[57] Within a given class of ion, e.g. primary, secondary, allylic, aryl, etc., $D(R^+{-}H^-)$ has been shown to be a linear function of the logarithm of the number of atoms in R^+, with larger ions being more stable.[56]

TABLE 5.1 Heterolytic $R{-}H \rightarrow R^+ + H^-$ dissociation energies in the gas phase

| | $D(R^+{-}H^-)$ | | |
Ion	kcal/mol	kJ/mol	Ref.
CH_3^+	314.6	1316	55
$C_2H_5^+$	276.7	1158	55
$(CH_3)_2CH^+$	249.2	1043	55
$(CH_3)_3C^+$	231.9	970.3	55
$C_6H_5^+$	294	1230	56
$H_2C{=}CH^+$	287	1200	56
$H_2C{=}CH{-}CH_2^+$	256	1070	56
cyclopentyl	246	1030	56
$C_6H_5CH_2^+$	238	996	56
CH_3CO^+	230	962	56

[49]For a review of destabilized carbocations, see Tidwell *Angew. Chem. Int. Ed. Engl.* **1984,** *23,* 20-32 [*Angew. Chem. 96,* 16-28].

[50]Solutions of aryl-substituted vinyl cations have been reported to be stable for at least a short time at low temperatures. Nmr spectra have been obtained: Abram; Watts *J. Chem. Soc., Chem. Commun.* **1974,** 857; Siehl; Carnahan; Eckes; Hanack *Angew. Chem. Int. Ed. Engl.* **1974,** *13,* 675 [*Angew. Chem. 86,* 677]. The 1-cyclobutenyl cation has been reported to be stable in the gas phase: Franke; Schwarz; Stahl *J. Org. Chem.* **1980,** *45,* 3493. See also Siehl; Koch *J. Org. Chem.* **1984,** *49,* 575.

[51]For a monograph, see Stang; Rappoport; Hanack; Subramanian *Vinyl Cations,* Academic Press: New York, 1979. For reviews of aryl and/or vinyl cations, see Hanack *Pure Appl. Chem.* **1984,** *56,* 1819-1830, *Angew. Chem. Int. Ed. Engl.* **1978,** *17,* 333-341 [*Angew. Chem. 90,* 346-359], *Acc. Chem. Res.* **1976,** *9,* 364-371; Rappoport *Reactiv. Intermed. (Plenum)* **1983,** *3,* 427-615; Ambroz; Kemp *Chem. Soc. Rev.* **1979,** *8,* 353-365; Richey; Richey, in Olah; Schleyer, Ref. 2, vol. 2, pp. 899-957; Richey, Ref. 24, pp. 42-49; Modena; Tonellato *Adv. Phys. Org. Chem.* **1971,** *9,* 185-280; Stang *Prog. Phys. Org. Chem.* **1973,** *10,* 205-325. See also Charton *Mol. Struct. Energ.* **1987,** *4,* 271-316.

[52]For reviews, see Bagno; Scorrano; More O'Ferrall *Rev.Chem. Intermed.* **1987,** *7,* 313-352; Bethell; Gold, Ref. 2, pp. 59-87.

[53]Deno; Jaruzelski; Schriesheim *J. Am. Chem. Soc.* **1955,** *77,* 3044; Deno; Schriesheim *J. Am. Chem. Soc.* **1955,** *77,* 3051; Deno; Berkheimer; Evans; Peterson *J. Am. Chem. Soc.* **1959,** *81,* 2344.

[54]For a list of stabilities of 39 typical carbocations, see Arnett; Hofelich *J. Am. Chem. Soc.* **1983,** *105,* 2889. See also Schade; Mayr; Arnett *J. Am. Chem. Soc.* **1988,** *110,* 567; Schade; Mayr *Tetrahedron* **1988,** *44,* 5761.

[55]Schultz; Houle; Beauchamp *J. Am. Chem. Soc.* **1984,** *106,* 3917.

[56]Lossing; Holmes *J. Am. Chem. Soc.* **1984,** *106,* 6917.

[57]Refs. 55, 56. See also Staley; Wieting; Beauchamp *J. Am. Chem. Soc.* **1977,** *99,* 5964; Arnett; Petro *J. Am. Chem. Soc.* **1978,** *100,* 5408; Arnett; Pienta *J. Am. Chem. Soc.* **1980,** *102,* 3329.

Since the central carbon of tricoordinated carbocations has only three bonds and no other valence electrons, the bonds are sp^2 and should be planar.[58] Raman, ir, and nmr spectroscopic data on simple alkyl cations show this to be so.[59] Other evidence is that carbocations are difficult or impossible to form at bridgehead atoms in [2.2.1] systems,[60] where they cannot be planar (see p. 301). However, larger bridgehead ions can exist. For example, the adamantyl cation (**6**) has been synthesized, as the SF_6^- salt.[61] Among other bridgehead cations that have been prepared in super-acid solution at $-78°C$ are the dodecahydryl cation (**7**)[62]

| **6** | **7** | **8** |

and the 1-trishomobarrelyl cation (**8**).[63] In the latter case the instability of the bridgehead position is balanced by the extra stability gained from the conjugation with the three cyclopropyl groups.

Triarylmethyl cations[64] are propeller-shaped, though the central carbon and the three ring carbons connected to it are in a plane:[65]

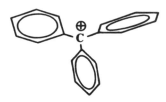

The three benzene rings cannot be all in the same plane because of steric hindrance, though increased resonance energy would be gained if they could.

An important tool for the investigation of carbocation structure is measurement of the ^{13}C nmr chemical shift of the carbon atom bearing the positive charge.[66] This shift approximately correlates with electron density on the carbon. ^{13}C chemical shifts for a number of ions are given in Table 5.2.[67] As shown in the table, the substitution of an ethyl for a methyl or a methyl for a hydrogen causes a downfield shift, indicating that the central carbon

[58]For discussions of the stereochemistry of carbocations, see Henderson *Chem. Soc. Rev.* **1973**, *2*, 397-413; Buss; Schleyer; Allen, Ref. 2; Schleyer in Chiurdoglu *Conformational Analysis*; Academic Press: New York, 1971, pp. 241-249; Hehre *Acc. Chem. Res.* **1975**, *8*, 369-376; Ref. 30, pp. 1561-1574.

[59]Olah; DeMember; Commeyras; Bribes *J. Am. Chem. Soc.* **1971**, *93*, 459; Olah et al., Ref. 10; Yannoni; Kendrick; Myhre; Bebout; Petersen *J. Am. Chem. Soc.* **1989**, *111*, 6440.

[60]For a review of bridgehead carbocations, see Fort, in Olah; Schleyer, Ref. 2, vol. 4, pp. 1783-1835.

[61]Schleyer; Fort; Watts; Comisarow; Olah *J. Am. Chem. Soc.* **1964**, *86*, 4195; Olah; Prakash; Shih; Krishnamurthy; Mateescu; Liang; Sipos; Buss; Gund; Schleyer *J. Am. Chem. Soc.* **1985**, *107*, 2764. See also Kruppa; Beauchamp *J. Am. Chem. Soc.* **1986**, *108*, 2162; Laube *Angew. Chem. Int. Ed. Engl.* **1986**, *25*, 349 [*Angew. Chem. 98*, 368].

[62]Olah; Prakash; Fessner; Kobayashi; Paquette *J. Am. Chem. Soc.* **1988**, *110*, 8599.

[63]de Meijere; Schallner *Angew. Chem. Int. Ed. Engl.* **1973**, *12*, 399 [*Angew. Chem. 85*, 400].

[64]For a review of crystal-structure determinations of triarylmethyl cations and other carbocations that can be isolated in stable solids, see Sundaralingam; Chwang, in Olah; Schleyer, Ref. 2, vol. 5, pp. 2427-2476.

[65]Sharp; Sheppard *J. Chem. Soc.* **1957**, 674; Gomes de Mesquita; MacGillavry; Eriks *Acta Crystallogr.* **1965**, *18*, 437; Schuster; Colter; Kurland *J. Am. Chem. Soc.* **1968**, *90*, 4679.

[66]For reviews of the nmr spectra of carbocations, see Young *Prog. Nucl. Magn. Reson. Spectrosc.* **1979**, *12*, 261-286; Farnum *Adv. Phys. Org. Chem.* **1975**, *11*, 123-175.

[67]Olah; White *J. Am. Chem. Soc.* **1968**, *90*, 1884, **1969**, *91*, 5801. For ^{13}C nmr data for additional ions, see Olah; Donovan *J. Am. Chem. Soc.* **1977**, *99*, 5026; Olah; Prakash; Liang *J. Org. Chem.* **1977**, *42*, 2666.

TABLE 5.2 ^{13}C chemical-shift values, in parts per million from $^{13}CS_2$, for the charged carbon atom of some carbocations in $SO_2ClF–SbF_5$, $SO_2–FSO_3H–SbF_5$, or $SO_2–SbF_5$[67]

Ion	Chemical shift	Temp., °C	Ion	Chemical shift	Temp., °C
Et$_2$MeC$^+$	-139.4	-20	C(OH)$_3^+$	$+28.0$	-50
Me$_2$EtC$^+$	-139.2	-60	PhMe$_2$C$^+$	-61.1	-60
Me$_3$C$^+$	-135.4	-20	PhMeCH$^+$	-40[68]	
Me$_2$CH$^+$	-125.0	-20	Ph$_2$CH$^+$	-5.6	-60
Me$_2$COH$^+$	-55.7	-50	Ph$_3$C$^+$	-18.1	-60
MeC(OH)$_2^+$	-1.6	-30	Me$_2$(cyclopropyl)C$^+$	-86.8	-60
HC(OH)$_2^+$	$+17.0$	-30			

becomes somewhat more positive. On the other hand, the presence of hydroxy or phenyl groups decreases the positive character of the central carbon. The ^{13}C chemical shifts are not always in exact order of carbocation stabilities as determined in other ways. Thus the chemical shift shows that the triphenylmethyl cation has a more positive central carbon than diphenylmethyl cation, though the former is more stable. Also, the 2-cyclopropylpropyl and 2-phenylpropyl cations have shifts of -86.8 and -61.1, respectively, though we have seen that according to other criteria a cyclopropyl group is better than a phenyl group at stabilizing a carbocation.[68] The reasons for this discrepancy are not fully understood.[66,69]

Nonclassical Carbocations

These are discussed at pp. 312–326.

The Generation and Fate of Carbocations

Carbocations, stable or unstable, are usually generated in one of two general ways:

1. A direct ionization, in which a group attached to a carbon atom leaves with its pair of electrons (see Chapters 10, 13, 17, 18):

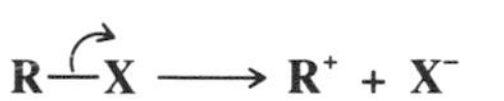

$$R–X \longrightarrow R^+ + X^- \quad \text{(may be reversible)}$$

2. A proton or other positive species adds to one atom of an unsaturated system, leaving the adjacent carbon atom with a positive charge (see Chapters 11, 15, 16).

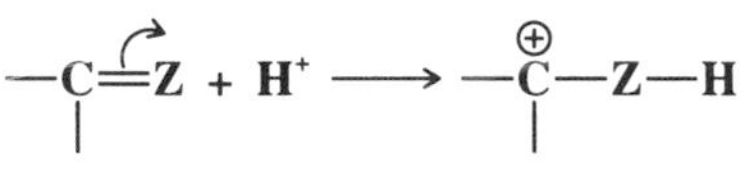

$$-\overset{|}{C}=Z + H^+ \longrightarrow -\overset{|}{\overset{\oplus}{C}}-Z-H$$

Formed by either process, carbocations are most often short-lived transient species and react further without being isolated.

[68]Olah; Porter; Kelly *J. Am. Chem. Soc.* **1971**, *93*, 464.
[69]For discussions, see Brown; Peters *J. Am. Chem. Soc.* **1973**, *95*, 2400, **1977**, *99*, 1712; Olah; Westerman; Nishimura *J. Am. Chem. Soc.* **1974**, *96*, 3548; Wolf; Harch; Taft; Hehre *J. Am. Chem. Soc.* **1975**, *97*, 2902; Fliszár *Can. J. Chem.* **1976**, *54*, 2839; Kitching; Adcock; Aldous *J. Org. Chem.* **1979**, *44*, 2652. See also Larsen; Bouis *J. Am. Chem. Soc.* **1975**, *97*, 4418; Volz; Shin; Streicher *Tetrahedron Lett.* **1975**, 1297; Larsen *J. Am. Chem. Soc.* **1978**, *100*, 330.

The two chief pathways by which carbocations react to give stable products are the reverse of the two pathways just described.

1. The carbocation may combine with a species possessing an electron pair (a Lewis acid–base reactions, see Chapter 8):

$$R^+ + \overline{Y}^- \longrightarrow R{-}Y$$

This species may be OH^-, halide ion, or any other negative ion, or it may be a neutral species with a pair to donate, in which case, of course, the immediate product must bear a positive charge (see Chapters 10, 13, 15, 16).

2. The carbocation may lose a proton (or much less often, another positive ion) from the adjacent atom (see Chapters 11, 17):

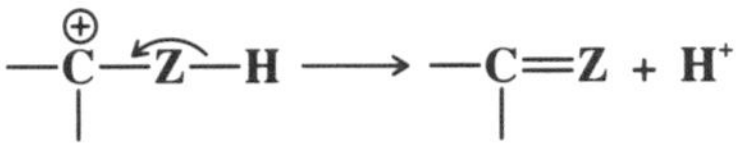

$$\overset{\oplus}{-\underset{|}{C}}{-}Z{-}H \longrightarrow -C{=}Z + H^+$$

Carbocations can also adopt two other pathways that lead not to stable products, but to other carbocations:

3. *Rearrangement.* An alkyl or aryl group or a hydrogen (sometimes another group) migrates with its electron pair to the positive center, leaving another positive charge behind (see Chapter 18):

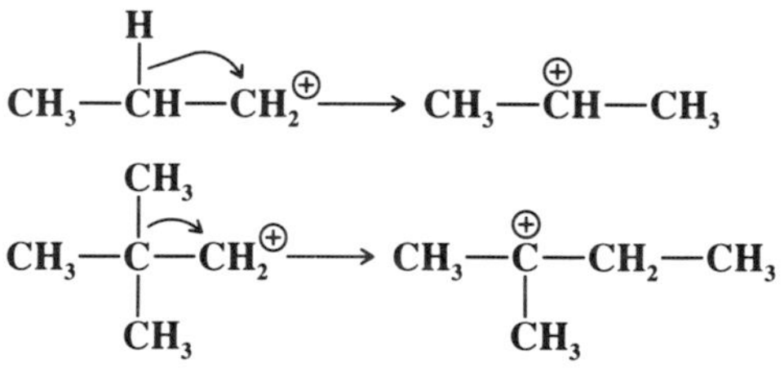

$$CH_3{-}\underset{\overset{|}{H}}{CH}{-}CH_2^{\oplus} \longrightarrow CH_3{-}\overset{\oplus}{CH}{-}CH_3$$

$$CH_3{-}\underset{\underset{CH_3}{|}}{\overset{\overset{CH_3}{|}}{C}}{-}CH_2^{\oplus} \longrightarrow CH_3{-}\underset{\underset{CH_3}{|}}{\overset{\oplus}{C}}{-}CH_2{-}CH_3$$

4. *Addition.* A carbocation may add to a double bond, generating a positive charge at a new position (see Chapters 11, 15):

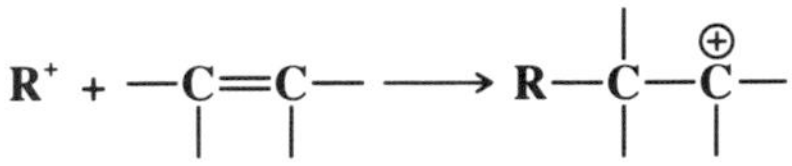

$$R^+ + -C{=}C- \longrightarrow R{-}\underset{|}{\overset{|}{C}}{-}\overset{\oplus}{\underset{|}{C}}-$$

$$\mathbf{9}$$

Whether formed by pathway 3 or 4, the new carbocation normally reacts further in an effort to stabilize itself, usually by pathway 1 or 2. However, **9** can add to another alkene molecule, and this product can add to still another, etc. This is one of the mechanisms for vinyl polymerization.

CARBANIONS

Stability and Structure[70]

An *organometallic compound* is a compound that contains a bond between a carbon atom and a metal atom. Many such compounds are known, and organometallic chemistry is a very large area, occupying a borderline region between organic and inorganic chemistry. Many carbon–metal bonds, e.g., carbon–mercury bonds, are undoubtedly covalent, but in bonds between carbon and the more active metals the electrons are closer to the carbon. Whether the position of the electrons in a given bond is close enough to the carbon to justify calling the bond ionic and the carbon moiety a carbanion depends on the metal, on the structure of the carbon moiety, and on the solvent and in some cases is a matter of speculation. In this section we discuss carbanions with little reference to the metal. In the next section we shall deal with the structures of organometallic compounds.

By definition, every carbanion possesses an unshared pair of electrons and is therefore a base. When a carbanion accepts a proton, it is converted to its conjugate acid (see Chapter 8). The stability of the carbanion is directly related to the strength of the conjugate acid. The weaker the acid, the greater the base strength and the lower the stability of the carbanion.[71] By stability here we mean stability toward a proton donor; the lower the stability, the more willing the carbanion is to accept a proton from any available source and hence to end its existence as a carbanion. Thus the determination of the order of stability of a series of carbanions is equivalent to a determination of the order of strengths of the conjugate acids, and one can obtain information about relative carbanion stability from a table of acid strengths like Table 8.1.

Unfortunately, it is not easy to measure acid strengths of very weak acids like the conjugate acids of simple unsubstituted carbanions. There is little doubt that these carbanions are very unstable in solution, and in contrast to the situation with carbocations, efforts to prepare solutions in which carbanions such as ethyl or isopropyl exist in a relatively free state have not yet been successful. Nor has it been possible to form these carbanions in the gas phase. Indeed, there is evidence that simple carbanions such as ethyl and isopropyl are unstable towards loss of an electron, which converts them to radicals.[72] Nevertheless, there have been several approaches to the problem. Applequist and O'Brien[73] studied the position of equilibrium for the reaction

$$RLi + R'I \rightleftharpoons RI + R'Li$$

in ether and ether–pentane. The reasoning in these experiments was that the R group that forms the more stable carbanion would be more likely to be bonded to lithium than to

[70]For monographs, see Buncel; Durst *Comprehensive Carbanion Chemistry*, pts. A, B, and C; Elsevier: New York, 1980, 1984, 1987; Bates; Ogle *Carbanion Chemistry*; Springer: New York, 1983; Stowell *Carbanions in Organic Synthesis*; Wiley: New York, 1979; Cram *Fundamentals of Carbanion Chemistry*; Academic Press: New York, 1965. For reviews, see Staley *React. Intermed. (Wiley)* **1985**, *3*, 19-43; Staley; Dustman *React. Intermed. (Wiley)* **1981**, *2*, 15-57; le Noble *React. Intermed. (Wiley)* **1978**, *1*, 27-67; Solov'yanov; Beletskaya *Russ. Chem. Rev.* **1978**, *47*, 425-439; Isaacs, Ref. 1, pp. 234-293; Kaiser; Slocum, in McManus, Ref. 1, pp. 337-422; Ebel *Fortchr. Chem. Forsch.* **1969**, *12*, 387-439; Cram *Surv. Prog. Chem.* **1968**, *4*, 45-68; Reutov; Beletskaya *Reaction Mechanisms of Organometallic Compounds*; North Holland Publishing Co.: Amsterdam, 1968, pp. 1-64; Streitwieser; Hammons *Prog. Phys. Org. Chem.* **1965**, *3*, 41-80. For reviews of nmr spectra of carbanions, see Young, Ref. 66; O'Brien, in *Comprehensive-Carbanion Chemistry*, pt. A, cited above, pp. 271-322. For a review of dicarbanions, see Thompson; Green *Tetrahedron* **1991**, *47*, 4223-4285.
[71]For a monograph on hydrocarbon acidity, see Reutov; Beletskaya; Butin *CH-Acids*; Pergamon: Elmsford, NY, 1978. For a review, see Fischer; Rewicki *Prog. Org. Chem.* **1968**, *7*, 116-161.
[72]See Graul; Squires *J. Am. Chem. Soc.* **1988**, *110*, 607; Schleyer; Spitznagel; Chandrasekhar *Tetrahedron Lett.* **1986**, *27*, 4411.
[73]Applequist; O'Brien *J. Am. Chem. Soc.* **1963**, *85*, 743.

iodine. Carbanion stability was found to be in this order: vinyl > phenyl > cyclopropyl > ethyl > n-propyl > isobutyl > neopentyl > cyclobutyl > cyclopentyl. In a somewhat similar approach, Dessy and co-workers[74] treated a number of alkylmagnesium compounds with a number of alkylmercury compounds in THF, setting up the equilibrium

$$R_2Mg + R'_2Hg \rightleftharpoons R_2Hg + R'_2Mg$$

where the group of greater carbanion stability is linked to magnesium. The carbanion stability determined this way was in the order phenyl > vinyl > cyclopropyl > methyl > ethyl > isopropyl. The two stability orders are in fairly good agreement, and they show that stability of simple carbanions decreases in the order methyl > primary > secondary. It was not possible by the experiments of Dessy and co-workers to determine the position of t-butyl, but there seems little doubt that it is still less stable. We can interpret this stability order solely as a consequence of the field effect since resonance is absent. The electron-donating alkyl groups of isopropyl result in a greater negative charge density at the central carbon atom (compared with methyl), thus decreasing its stability. The results of Applequist and O'Brien show that β branching also decreases carbanion stability. Cyclopropyl occupies an apparently anomalous position, but this is probably due to the large amount of s character in the carbanionic carbon (see p. 178).

A different approach to the problem of hydrocarbon acidity and hence carbanion stability is that of Shatenshtein and co-workers, who treated hydrocarbons with deuterated potassium amide and measured the rates of hydrogen exchange.[75] The experiments did not measure *thermodynamic* acidity, since rates were measured, not positions of equilibria. They measured *kinetic* acidity, i.e., which compounds gave up protons most rapidly (see p. 214 for the distinction between thermodynamic and kinetic control of product). Measurements of rates of hydrogen exchange enable one to compare acidities of a series of acids against a given base even where the positions of the equilibria cannot be measured because they lie too far to the side of the starting materials, i.e., where the acids are too weak to be converted to their conjugate bases in measurable amounts. Although the correlation between thermodynamic and kinetic acidity is far from perfect,[76] the results of the rate measurements, too, indicated that the order of carbanion stability is methyl > primary > secondary > tertiary.[75]

However, experiments in the gas phase gave different results. In reactions of OH⁻ with alkyltrimethylsilanes it is possible for either R or Me to cleave. Since the R or Me comes

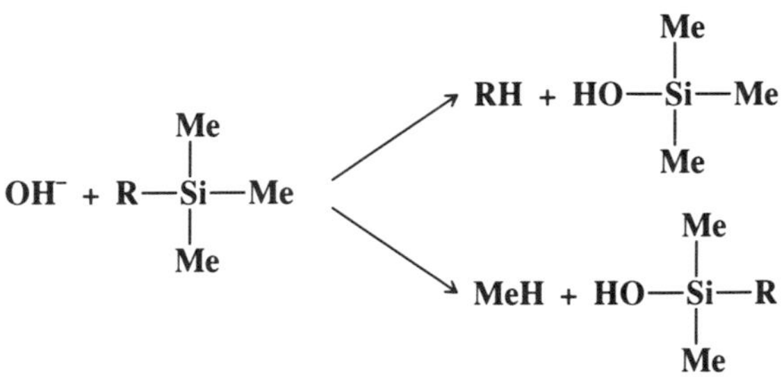

off as a carbanion or incipient carbanion, the product ratio RH:MeH can be used to establish the relative stabilities of various R groups. From these experiments a stability order of

[74]Dessy; Kitching; Psarras; Salinger; Chen; Chivers *J. Am. Chem. Soc.* **1966**, *88*, 460.

[75]For reviews, see Jones *Surv. Prog. Chem.* **1973**, *6*, 83-112; Shatenshtein; Shapiro *Russ. Chem. Rev.* **1968**, *37*, 845-854.

[76]For example, see Bordwell; Matthews; Vanier *J. Am. Chem. Soc.* **1975**, *97*, 442.

neopentyl > cyclopropyl > *t*-butyl > *n*-propyl > methyl > isopropyl > ethyl was found.[77] On the other hand, in a different kind of gas-phase experiment, Graul and Squires were able to observe CH_3^- ions, but not the ethyl, isopropyl, or *t*-butyl ions.[78]

Many carbanions are far more stable than the simple kind mentioned above. The increased stability is due to certain structural features:

1. *Conjugation of the unshared pair with an unsaturated bond:*

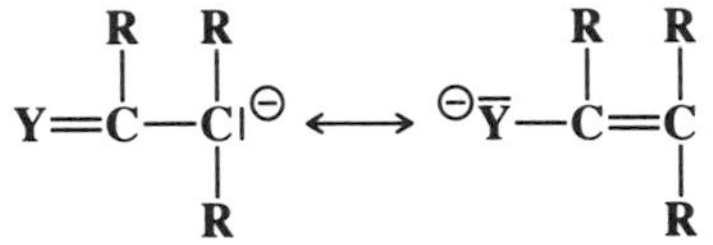

$$Y=\overset{\overset{\displaystyle R}{|}}{C}-\overset{\overset{\displaystyle R}{|}}{\underset{\underset{\displaystyle R}{|}}{C}}|^{\ominus} \longleftrightarrow {}^{\ominus}\overline{Y}-\overset{\overset{\displaystyle R}{|}}{C}=\overset{\overset{\displaystyle R}{|}}{\underset{\underset{\displaystyle R}{|}}{C}}$$

In cases where a double or triple bond is located α to the carbanionic carbon, the ion is stabilized by resonance in which the unshared pair overlaps with the π electrons of the double bond. This factor is responsible for the stability of the allylic[79] and benzylic[80] types of carbanions:

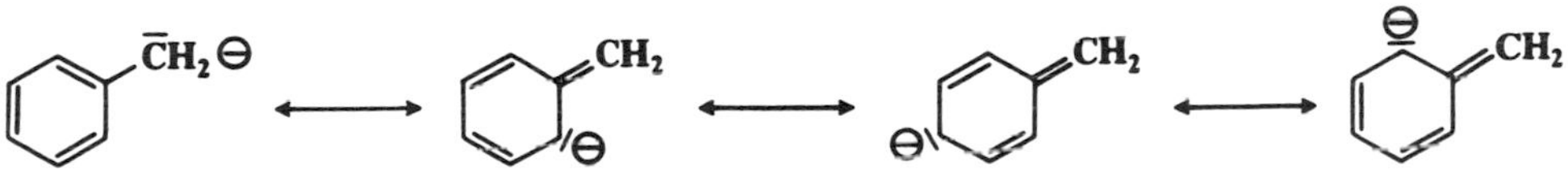

$$R-CH=CH-\overline{C}H_2{}^{\ominus} \longleftrightarrow R-\overset{\ominus}{\overline{C}H}-CH=CH_2$$

Diphenylmethyl and triphenylmethyl anions are still more stable and can be kept in solution indefinitely if water is rigidly excluded.[81] X-ray crystallographic structures have been obtained for Ph_2CH^- and Ph_3C^- enclosed in crown ethers.[82]

Where the carbanionic carbon is conjugated with a carbon–oxygen or carbon–nitrogen multiple bond (Y = O or N), the stability of the ion is greater than that of the triarylmethyl anions, since these electronegative atoms are better capable of bearing a negative charge than carbon. However, it is questionable whether ions of this type should be called carbanions at all, since in the case of enolate ions, for example, **K** contributes more to the hybrid than **J** though such ions react more often at the carbon than at the oxygen. Enolate ions can also be kept in stable solutions. A nitro group is particularly effective in stabilizing a negative

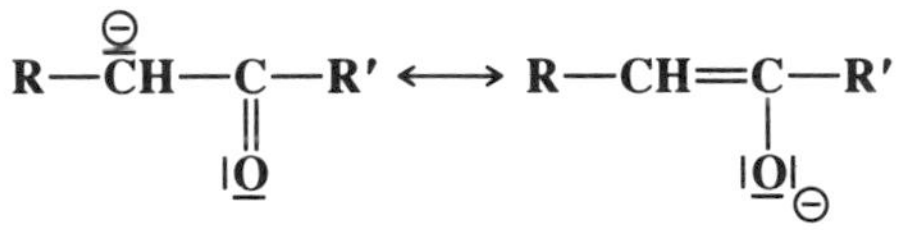

$$R-\overset{\ominus}{\overline{C}H}-\underset{\underset{\displaystyle |\underline{O}}{\|}}{C}-R' \longleftrightarrow R-CH=\underset{\underset{\displaystyle |\underline{O}|_{\ominus}}{|}}{C}-R'$$

$$\mathbf{J} \qquad\qquad\qquad \mathbf{K}$$

[77]DePuy; Gronert; Barlow; Bierbaum; Damrauer *J. Am. Chem. Soc.* **1989**, *111*, 1968. The same order (for *t*-Bu, Me, i-Pr, and Et) was found in gas-phase cleavages of alkoxides (**2-41**): Tumas; Foster; Brauman *J. Am. Chem. Soc.* **1984**, *106*, 4053.

[78]Graul; Squires, Ref. 72.

[79]For a review of allylic anions, see Richey, Ref. 24, pp. 67-77.

[80]Although benzylic carbanions are more stable than the simple alkyl type, they have not proved stable enough for isolation so far. The benzyl carbanion has been formed and studied in submicrosecond times; Bockrath; Dorfman *J. Am. Chem. Soc.* **1974**, *96*, 5708.

[81]For a review of spectrophotometric investigations of this type of carbanion, see Buncel; Menon, in Buncel; Durst, Ref. 70, pp. 97-124.

[82]Olmstead; Power *J. Am. Chem. Soc.* **1985**, *107*, 2174.

charge on an adjacent carbon, and the anions of simple nitro alkanes can exist in water. Thus pK_a for nitromethane is 10.2. Dinitromethane is even more acidic ($pK_a = 3.6$).

In contrast to the stability of cyclopropylmethyl cations (p. 169), the cyclopropyl group exerts only a weak stabilizing effect on an adjacent carbanionic carbon.[83]

By combining a very stable carbanion with a very stable carbocation, Okamoto and

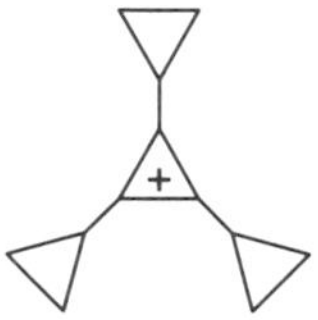

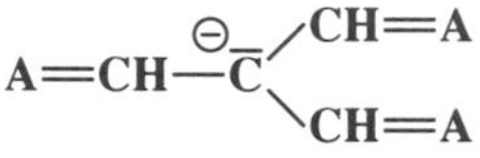

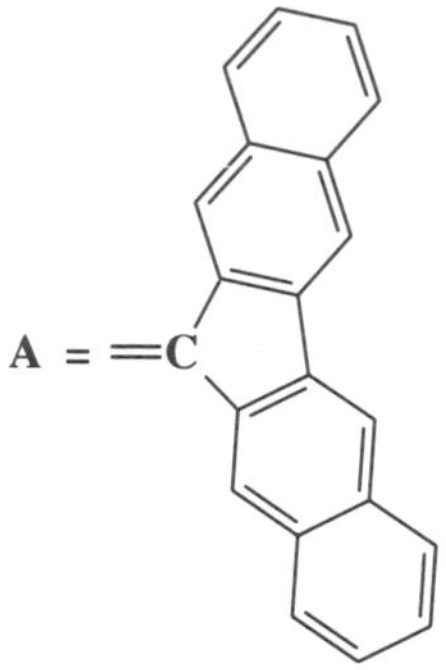

10

co-workers were able to isolate the salt **10,** as well as several similar salts, as stable solids. These are salts that consist entirely of carbon and hydrogen.[84]

2. *Carbanions increase in stability with an increase in the amount of s character at the carbanionic carbon.* Thus the order of stability is

$$RC\equiv C^- > R_2C=CH^- \approx Ar^- > R_3C-CH_2^-$$

Acetylene, where the carbon is sp-hybridized with 50% s character, is much more acidic than ethylene[85] (sp^2, 33% s), which in turn is more acidic than ethane, with 25% s character. Increased s character means that the electrons are closer to the nucleus and hence of lower energy. As previously mentioned, cyclopropyl carbanions are more stable than methyl, owing to the larger amount of s character as a result of strain (see p. 152).

3. *Stabilization by sulfur[86] or phosphorus.* Attachment to the carbanionic carbon of a sulfur or phosphorus atom causes an increase in carbanion stability, though the reasons for this are in dispute. One theory is that there is overlap of the unshared pair with an empty d orbital[87] ($p\pi$–$d\pi$ bonding, see p. 38). For example, a carbanion containing the SO_2R group would be written

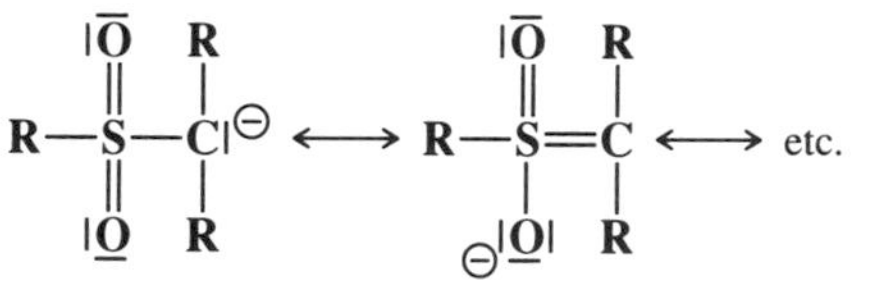

[83]Perkins; Ward *J. Chem. Soc., Perkin Trans. 1* **1974,** 667; Perkins; Peynircioglu *Tetrahedron* **1985,** *41,* 225.

[84]Okamoto; Kitagawa; Takeuchi; Komatsu; Kinoshita; Aonuma; Nagai; Miyabo *J. Org. Chem.* **1990,** *55,* 996. See also Okamoto; Kitagawa; Takeuchi; Komatsu; Miyabo *J. Chem. Soc., Chem. Commun.* **1988,** 923.

[85]For a review of vinylic anions, see Richey, Ref. 24, pp. 49-56.

[86]For reviews of sulfur-containing carbanions, see Oae; Uchida in Patai; Rappoport; Stirling *The Chemistry of Sulphones and Sulphoxides;* Wiley: New York, 1988, pp. 583-664; Wolfe, in Bernardi; Csizmadia; Mangini *Organic Sulfur Chemistry*; Elsevier, New York, 1985, pp. 133-190; Block *Reactions of Organosulfur Compounds*; Academic Press: New York, 1978, pp. 42-56; Durst; Viau *Intra-Sci. Chem. Rep.* **1973,** 7 (3), 63-74. For a review of selenium-stabilized carbanions, see Reich, in Liotta *Organoselenium Chemistry*; Wiley: New York, 1987, pp. 243-276.

[87]For support for this theory, see Wolfe; LaJohn; Bernardi; Mangini; Tonachini *Tetrahedron Lett.* **1983,** *24,* 3789; Wolfe; Stolow; LaJohn *Tetrahedron Lett.* **1983,** *24,* 4071.

However, there is evidence against *d*-orbital overlap; and the stabilizing effects have been attributed to other causes.[88] An α silicon atom also stabilizes carbanions.[89]

4. *Field effects.* Most of the groups that stabilize carbanions by resonance effects (either the kind discussed in paragraph 1 above or the kind discussed in paragraph 3) have electron-withdrawing field effects and thereby stabilize the carbanion further by spreading the negative charge, though it is difficult to separate the field effect from the resonance effect. However, in a nitrogen ylide $R_3\overset{\oplus}{N}-\overset{\ominus}{C}R_2$ (see p. 39), where a positive nitrogen is adjacent to the negatively charged carbon, only the field effect operates. Ylides are more stable than the corresponding simple carbanions. Carbanions are stabilized by a field effect if there is any hetero atom (O, N, or S) connected to the carbanionic carbon, provided that the hetero atom bears a positive charge in at least one important canonical form,[90] e.g.,

$$\underset{\underset{\text{Me}}{|}}{\overset{\underset{\parallel}{|\underline{O}}}{Ar-C-\bar{N}-CH_2^{\ominus}}} \longleftrightarrow \underset{\underset{\ominus\,|\underline{O}|\quad Me}{}}{Ar-C=\overset{\oplus}{N}-CH_2^{\ominus}}$$

5. Certain carbanions are stable because they are aromatic (see the cyclopentadienyl anion p. 46, and other aromatic anions in Chapter 2).

6. *Stabilization by a nonadjacent π bond.*[91]   In contrast to the situation with carbocations (see pp. 314-316), there have been fewer reports of carbanions stabilized by interaction with a nonadjacent π bond. One that may be mentioned is **13,** formed when optically active camphenilone (**11**) was treated with a strong base (potassium *t*-butoxide).[92] That **13** was

11   **12**   **13**

truly formed was shown by the following facts: (1) A proton was abstracted: ordinary CH_2 groups are not acidic enough for this base; (2) recovered **11** was racemized: **13** is symmetrical and can be attacked equally well from either side; (3) when the experiment was performed in deuterated solvent, the rate of deuterium uptake was equal to the rate of racemization; and (4) recovered **11** contained up to three atoms of deuterium per molecule, though if **12** were the only ion, no more than two could be taken up. Ions of this type, in which a

[88]Bernardi; Csizmadia; Mangini; Schlegel; Whangbo; Wolfe *J. Am. Chem. Soc.* **1975,** *97,* 2209; Epiotis; Yates; Bernardi; Wolfe *J. Am. Chem. Soc.* **1976,** *98,* 5435; Lehn; Wipff *J. Am. Chem. Soc.* **1976,** *98,* 7498; Borden; Davidson; Andersen; Denniston; Epiotis *J. Am. Chem. Soc.* **1978,** *100,* 1604; Bernardi; Bottoni; Venturini; Mangini *J. Am. Chem. Soc.* **1986,** *108,* 8171.
[89]Wetzel; Brauman *J. Am. Chem. Soc.* **1988,** *110,* 8333.
[90]For a review of such carbanions, see Beak; Reitz *Chem. Rev.* **1978,** *78,* 275-316. See also Rondan; Houk; Beak; Zajdel; Chandrasekhar; Schleyer *J. Org. Chem.* **1981,** *46,* 4108.
[91]For reviews, see Werstiuk *Tetrahedron* **1983,** *39,* 205-268; Hunter; Stothers; Warnhoff, in de Mayo *Rearrangements in Ground and Excited States,* vol. 1; Academic Press: New York, 1980, pp. 410-437.
[92]Nickon; Lambert *J. Am. Chem. Soc.* **1966,** *88,* 1905. Also see Brown; Occolowitz *Chem. Commun.* **1965,** 376; Grutzner; Winstein *J. Am. Chem. Soc.* **1968,** *90,* 6562; Staley; Reichard *J. Am. Chem. Soc.* **1969,** *91,* 3998; Hunter; Johnson; Stothers; Nickon; Lambert; Covey *J. Am. Chem. Soc.* **1972,** *94,* 8582; Miller *J. Am. Chem. Soc.* **1969,** *91,* 751; Werstiuk; Yeroushalmi; Timmins *Can. J. Chem.* **1983,** *61,* 1945; Lee; Squires *J. Am. Chem.Soc.* **1986,** *108,* 5078; Peiris; Ragauskas; Stothers *Can. J. Chem.* **1987,** *65,* 789; Shiner; Berks; Fisher *J. Am. Chem. Soc.* **1988,** *110,* 957.

negatively charged carbon is stabilized by a carbonyl group two carbons away, are called *homoenolate ions*.

Overall, functional groups in the α position stabilize carbanions in the following order: $NO_2 > RCO > COOR > SO_2 > CN \approx CONH_2 > Hal > H > R$.

It is unlikely that free carbanions exist in solution. Like carbocations, they are usually in ion pairs or else solvated.[93] Among experiments which demonstrated this was the treatment of $PhCO\bar{C}HMe^-$ M^+ with ethyl iodide, where M^+ was Li^+, Na^+, or K^+. The half-lives of the reaction were[94] for Li, 31×10^{-6}; Na, 0.39×10^{-6}; and K, 0.0045×10^{-6}, demonstrating that the species involved were not identical. Similar results[95] were obtained with Li, Na, and Cs triphenylmethides Ph_3C^- M^+.[96] Where ion pairs are unimportant, carbanions are solvated. Cram[70] has demonstrated solvation of carbanions in many solvents. There may be a difference in the structure of a carbanion depending on whether it is free (e.g., in the gas phase) or in solution. The negative charge may be more localized in solution in order to maximize the electrostatic attraction to the counterion.[97]

The structure of simple unsubstituted carbanions is not known with certainty since they have not been isolated, but it seems likely that the central carbon is sp^3-hybridized, with the unshared pair occupying one apex of the tetrahedron. Carbanions would thus have pyramidal structures similar to those of amines.

The methyl anion CH_3^- has been observed in the gas phase and reported to have a pyramidal structure.[98] If this is a general structure for carbanions, then any carbanion in which the three R groups are different should be chiral and reactions in which it is an intermediate should give retention of configuration. Attempts have been made to demonstrate this but without success.[99] A possible explanation is that pyramidal inversion takes place here, as in amines, so that the unshared pair and the central carbon rapidly oscillate from one side of the plane to the other. There is, however, other evidence for the sp^3 nature of the central carbon and for its tetrahedral structure. Carbons at bridgeheads, though extremely reluctant to undergo reactions in which they must be converted to carbocations, undergo with ease reactions in which they must be carbanions and stable bridgehead carbanions are known.[100]

[93]For reviews of carbanion pairs, see Hogen-Esch *Adv. Phys. Org. Chem.* **1977**, *15*, 153-266; Jackman; Lange *Tetrahedron* **1977**, *33*, 2737-2769. See also Ref 7.
[94]Zook; Gumby *J. Am. Chem. Soc.* **1960**, *82*, 1386.
[95]Solov'yanov; Karpyuk; Beletskaya; Reutov *J. Org. Chem. USSR* **1981**, *17*, 381. See also Solov'yanov; Beletskaya; Reutov *J. Org. Chem. USSR* **1983**, *19*, 1964.
[96]For other evidence for the existence of carbanionic pairs, see Hogen-Esch; Smid *J. Am. Chem. Soc.* **1966**, *88*, 307, 318; **1969**, *91*, 4580; Abatjoglou; Eliel; Kuyper *J. Am. Chem. Soc.* **1977**, *99*, 8262; Solov'yanov; Karpyuk; Beletskaya; Reutov *Doklad. Chem.* **1977**, *237*, 668; DePalma; Arnett *J. Am. Chem. Soc.* **1978**, *100*, 3514; Buncel; Menon *J. Org. Chem.* **1979**, *44*, 317; O'Brien; Russell; Hart *J. Am. Chem. Soc.* **1979**, *101*, 633; Streitwieser; Shen *Tetrahedron Lett.* **1979**, 327; Streitwieser *Acc. Chem. Res.* **1984**, *17*, 353.
[97]See Schade; Schleyer; Geissler; Weiss *Angew. Chem. Int. Ed. Engl.* **1986**, *21*, 902 [*Angew. Chem.* **98**, 922].
[98]Ellison; Engelking; Lineberger *J. Am. Chem. Soc.* **1978**, *100*, 2556.
[99]Retention of configuration has never been observed with simple carbanions. Cram has obtained retention with carbanions stabilized by resonance. However, these carbanions are known to be planar or nearly planar, and retention was caused by asymmetric solvation of the planar carbanions (see p. 574).
[100]For other evidence that carbanions are pyramidal, see Streitwieser; Young *J. Am. Chem. Soc.* **1969**, *91*, 529; Peoples; Grutzner *J. Am. Chem. Soc.* **1980**, *102*, 4709.

Also, reactions at vinylic carbons proceed with retention,[101] indicating that the intermediate **14** has sp^2 hybridization and not the sp hybridization that would be expected in the analogous carbocation. A cyclopropyl anion can also hold its configuration.[102]

$$\underset{R}{\overset{R}{>}}C=C\overset{R}{\underset{\ominus}{<}}$$

14

Carbanions in which the negative charge is stabilized by resonance involving overlap of the unshared-pair orbital with the π electrons of a multiple bond are essentially planar, as would be expected by the necessity for planarity in resonance, though unsymmetrical solvation or ion-pairing effects may cause the structure to deviate somewhat from true planarity.[103] Cram and co-workers have shown that where chiral carbanions possessing this type of resonance are generated, retention, inversion, or racemization can result, depending on the solvent (see p. 574). This result is explained by unsymmetrical solvation of planar or near-planar carbanions. However, some carbanions that are stabilized by adjacent sulfur or phosphorus, e.g.,

$$Ar-SO_2-\overset{R}{\underset{R'}{C}}|^{\ominus} \qquad Ar-\overset{R}{\underset{R'}{N}}-SO_2-\overset{R}{\underset{R'}{C}}|^{\ominus} \qquad Ar-\overset{|\overline{O}|}{\underset{_\ominus|\underline{O}|}{P}}-\overset{R}{\underset{R'}{C}}|^{\ominus}$$

$$K^+$$

are inherently chiral, since retention of configuration is observed where they are generated, even in solvents that cause racemization or inversion with other carbanions.[104] The configuration about the carbanionic carbon, at least for some of the α-sulfonyl carbanions, seems to be planar,[105] and the inherent chirality is caused by lack of rotation about the C—S bond.[106]

[101]Curtin; Harris *J. Am. Chem. Soc.* **1951**, *73*, 2716, 4519; Braude; Coles *J. Chem. Soc.* **1951**, 2078; Nesmeyanov; Borisov *Tetrahedron* **1957**, *1*, 158. Also see Miller; Lee *J. Am. Chem. Soc.* **1959**, *81*, 6313; Hunter; Cram *J. Am. Chem. Soc.* **1964**, *86*, 5478; Walborsky; Turner *J. Am. Chem. Soc.* **1972**, *94*, 2273; Arnett; Walborsky *J. Org. Chem.* **1972**, *37*, 3678; Feit; Melamed; Speer; Schmidt *J. Chem. Soc., Perkin Trans. 1* **1984**, 775; Chou; Kass *J. Am. Chem. Soc.* **1991**, *113*, 4357.
[102]Walborsky; Motes *J. Am. Chem. Soc.* **1970**, *92*, 2445; Motes; Walborsky *J. Am. Chem. Soc.* **1970**, *92*, 3697; Boche; Harms; Marsch *J. Am. Chem. Soc.* **1988**, *110*, 6925. For a monograph on cyclopropyl anions, cations, and radicals, see Boche; Walborsky *Cyclopropane Derived Reactive Intermediates*; Wiley: New York, 1990. For a review, see Boche; Walborsky, in Rappoport *The Chemistry of the Cyclopropyl Group*, pt. 1; Wiley: New York, 1987, pp. 701-808 (the monograph includes and updates the review).
[103]See the discussion in Cram *Fundamentals of Carbanion Chemistry*; Academic Press: New York, 1965, pp. 85-105.
[104]Cram; Nielsen; Rickborn *J. Am. Chem. Soc.* **1960**, *82*, 6415; Cram; Wingrove *J. Am. Chem. Soc.* **1962**, *84*, 1496; Corey; Kaiser *J. Am. Chem. Soc.* **1961**, *83*, 490; Goering; Towns; Dittmer *J. Org. Chem.* **1962**, *27*, 736; Corey; Lowry *Tetrahedron Lett.* **1965**, 803; Bordwell; Phillips; Williams *J. Am. Chem. Soc.* **1968**, *90*, 426; Annunziata; Cinquini; Colonna; Cozzi *J. Chem. Soc., Chem. Commun.* **1981**, 1005; Chassaing; Marquet; Corset; Froment *J. Organomet. Chem.* **1982**, *232*, 293. For a discussion, see Ref. 103, pp. 105-113.
[105]Boche; Marsch; Harms; Sheldrick *Angew. Chem. Int. Ed. Engl.* **1985**, *24*, 573 [*Angew. Chem. 97*, 577]; Gais; Vollhardt; Hellmann; Paulus; Lindner *Tetrahedron Lett.* **1988**, *29*, 1259; Gais; Müller; Vollhardt; Lindner *J. Am. Chem. Soc.* **1991**, *113*, 4002. For a contrary view, see Trost; Schmuff *J. Am. Chem. Soc.* **1985**, *107*, 396.
[106]Grossert; Hoyle; Cameron; Roe; Vincent *Can. J. Chem.* **1987**, *65*, 1407.

The Structure of Organometallic Compounds[107]

Whether a carbon–metal bond is ionic or polar-covalent is determined chiefly by the electronegativity of the metal and the structure of the organic part of the molecule. Ionic bonds become more likely as the negative charge on the metal-bearing carbon is decreased by resonance or field effects. Thus the sodium salt of acetoacetic ester has a more ionic carbon–sodium bond than methylsodium.

Most organometallic bonds are polar-covalent. Only the alkali metals have electronegativities low enough to form ionic bonds with carbon, and even here the behavior of lithium alkyls shows considerable covalent character. The simple alkyls and aryls of sodium, potassium, rubidium, and cesium[108] are nonvolatile solids[109] insoluble in benzene or other organic solvents, while alkyllithiums are soluble, although they too are generally nonvolatile solids. Alkyllithiums do not exist as monomeric species in hydrocarbon solvents or ether.[110] In benzene and cyclohexane, freezing-point-depression studies have shown that alkyllithiums are normally hexameric unless steric interactions favor tetrameric aggregates.[111] Nmr studies, especially measurements of $^{13}C-^6Li$ coupling, have also shown aggregation in hydrocarbon solvents.[112] Boiling-point-elevation studies have been performed in ether solutions, where alkyllithiums exist in two- to fivefold aggregates.[113] Even in the gas phase[114] and in the solid state,[115] alkyllithiums exist as aggregates. X-ray crystallography has shown that methyllithium has the same tetrahedral structure in the solid state as in ether solution.[115] However, *t*-butyllithium is monomeric in THF, though dimeric in ether and tetrameric in hydrocarbon solvents.[116] Neopentyllithium exists as a mixture of monomers and dimers in THF.[117]

The C—Mg bond in Grignard reagents is covalent and not ionic. The actual structure of Grignard reagents in solution has been a matter of much controversy over the years.[118] In 1929 it was discovered[119] that the addition of dioxane to an ethereal Grignard solution precipitates all the magnesium halide and leaves a solution of R_2Mg in ether; i.e., there can

[107]For a monograph, see Elschenbroich; Salzer *Organometallics*; VCH: New York, 1989. For reviews, see Oliver, in Hartley; Patai *The Chemistry of the Metal–Carbon Bond*, vol. 2; Wiley: New York, 1985, pp. 789-826; Coates; Green; Wade *Organometallic Compounds*, 3rd ed., vol. 1; Methuen: London, 1967. For a review of the structures of organodialkali compounds, see Grovenstein, in Buncel; Durst, Ref. 70, pt. C, pp. 175-221.

[108]For a review of x-ray crystallographic studies of organic compounds of the alkali metals, see Schade; Schleyer *Adv. Organomet. Chem.* **1987**, *27*, 169-278.

[109]X-ray crystallography of potassium, rubidium, and cesium methyls shows completely ionic crystal lattices: Weiss; Sauermann *Chem. Ber.* **1970**, *103*, 265; Weiss; Köster *Chem. Ber.* **1977**, *110*, 717.

[110]For reviews of the structure of alkyllithium compounds, see Setzer; Schleyer *Adv. Organomet. Chem.* **1985**, *24*, 353-451; Schleyer *Pure Appl. Chem.* **1984**, *56*, 151-162; Brown *Pure Appl. Chem.* **1970**, *23*, 447-462, *Adv. Organomet. Chem.* **1965**, *3*, 365-395; Kovrizhnykh; Shatenshtein *Russ. Chem. Rev.* **1969**, *38*, 840-849. For reviews of the structures of lithium enolates and related compounds, see Boche *Angew. Chem. Int. Ed. Engl.* **1989**, *28*, 277-297 [*Angew. Chem. 101*, 286-306]; Seebach *Angew. Chem. Int. Ed. Engl.* **1988**, *27*, 1624-1654 [*Angew. Chem. 100*, 1685-1715]. For a review of the use of nmr to study these structures, see Günther; Moskau; Bast; Schmalz *Angew. Chem. Int. Ed. Engl.* **1987**, *26*, 1212-1220 [*Angew. Chem. 99*, 1242-1250]. For monographs on organolithium compounds, see Wakefield *Organolithium Methods*; Academic Press: New York, 1988, *The Chemistry of Organolithium Compounds*; Pergamon: Elmsford, NY, 1974.

[111]Lewis; Brown *J. Am. Chem. Soc.* **1970**, *92*, 4664; Brown; Rogers *J. Am. Chem. Soc.* **1957**, *79*, 1859; Weiner; Vogel; West *Inorg. Chem.* **1962**, *1*, 654.

[112]Fraenkel; Henrichs; Hewitt; Su *J. Am. Chem. Soc.* **1984**, *106*, 255; Thomas; Jensen; Young *Organometallics* **1987**, *6*, 565. See also Kaufman; Gronert; Streitwieser *J. Am. Chem. Soc.* **1988**, *110*, 2829.

[113]Wittig; Meyer; Lange *Liebigs Ann. Chem.* **1951**, *571*, 167. See also McGarrity; Ogle *J. Am. Chem. Soc.* **1985**, *107*, 1805; Bates; Clarke; Thomas *J. Am. Chem. Soc.* **1988**, *110*, 5109.

[114]Berkowitz; Bafus; Brown *J. Phys. Chem.* **1961**, *65*, 1380; Brown; Dickerhoof; Bafus *J. Am. Chem. Soc.* **1962**, *84*, 1371; Chinn; Lagow *Organometallics* **1984**, *3*, 75; Plavšić; Srzić; Klasinc *J. Phys. Chem.* **1986**, *90*, 2075.

[115]Dietrich *Acta Crystallogr.* **1963**, *16*, 681; Weiss; Lucken *J. Organomet. Chem.* **1964**, *2*, 197; Weiss; Sauermann; Thirase *Chem. Ber.* **1983**, *116*, 74.

[116]Bauer; Winchester; Schleyer *Organometallics* **1987**, *6*, 2371.

[117]Fraenkel; Chow; Winchester *J. Am. Chem. Soc.* **1990**, *112*, 6190.

[118]For reviews, see Ashby *Bull. Soc. Chim. Fr.* **1972**, 2133-2142, *Q. Rev., Chem. Soc.* **1967**, *21*, 259-285; Wakefield *Organomet. Chem. Rev.* **1966**, *1*, 131-156; Bell *Educ. Chem.* **1973**, 143-145.

[119]Schlenk; Schlenk *Ber.* **1929**, *62B*, 920.

be no RMgX in the solution since there is no halide. The following equilibrium, now called the *Schlenk equilibrium*, was proposed as the composition of the Grignard solution:

$$2RMgX \rightleftharpoons R_2Mg + MgX_2 \rightleftharpoons R_2Mg \cdot MgX_2$$

15

in which **15** is a complex of some type. Much work has demonstrated that the Schlenk equilibrium actually exists and that the position of the equilibrium is dependent on the identity of R, X, the solvent, the concentration, and the temperature.[120] It has been known for many years that the magnesium in a Grignard solution, no matter whether it is RMgX, R_2Mg, or MgX_2, can coordinate with two molecules of ether in addition to the two covalent bonds:

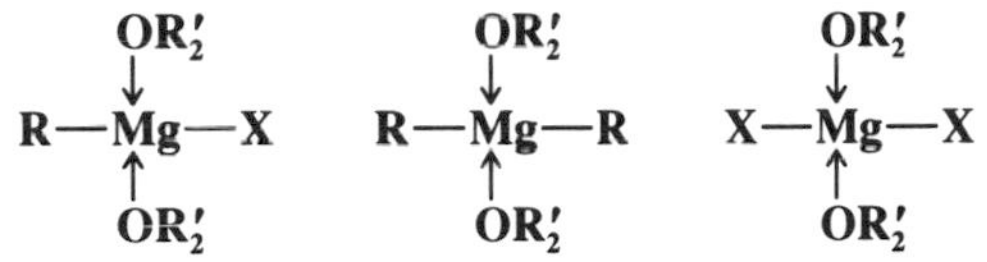

$$\begin{array}{ccc} OR_2' & OR_2' & OR_2' \\ \downarrow & \downarrow & \downarrow \\ R-Mg-X & R-Mg-R & X-Mg-X \\ \uparrow & \uparrow & \uparrow \\ OR_2' & OR_2' & OR_2' \end{array}$$

Rundle and co-workers performed x-ray-diffraction studies on solid phenylmagnesium bromide dietherate and on ethylmagnesium bromide dietherate, which they obtained by cooling ordinary ethereal Grignard solutions until the solids crystallized.[121] They found that the structures were monomeric:

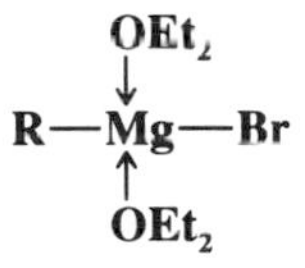

$$\begin{array}{c} OEt_2 \\ \downarrow \\ R-Mg-Br \qquad R = ethyl, phenyl \\ \uparrow \\ OEt_2 \end{array}$$

These solids still contained ether. When ordinary ethereal Grignard solutions prepared from bromomethane, chloromethane, bromoethane, and chloroethane were evaporated at about 100°C under vacuum so that the solid remaining contained no ether, x-ray diffraction showed *no* RMgX but a mixture of R_2Mg and MgX_2.[122] These results indicate that in the presence of ether $RMgX \cdot 2Et_2O$ is the preferred structure, while the loss of ether drives the Schlenk equilibrium to $R_2Mg + MgX_2$. However, conclusions drawn from a study of the solid materials do not necessarily apply to the structures in solution.

Boiling-point-elevation and freezing-point-depression measurements have demonstrated that in tetrahydrofuran at all concentrations and in ether at low concentrations (up to about 0.1 M) Grignard reagents prepared from alkyl bromides and iodides are monomeric, i.e., there are few or no molecules with two magnesium atoms.[123] Thus, part of the Schlenk equilibrium is operating

$$2RMgX \rightleftharpoons R_2Mg + MgX_2$$

[120]See Parris; Ashby *J. Am. Chem. Soc.* **1971**, *93*, 1206; Salinger; Mosher *J. Am. Chem. Soc.* **1964**, *86*, 1782; Kirrmann; Hamelin; Hayes *Bull. Soc. Chim. Fr.* **1963**, 1395.
[121]Guggenberger; Rundle *J. Am. Chem. Soc.* **1968**, *90*, 5375; Stucky; Rundle *J. Am. Chem. Soc.* **1964**, *86*, 4825.
[122]Weiss *Chem. Ber.* **1965**, *98*, 2805.
[123]Ashby; Becker *J. Am. Chem. Soc.* **1963**, *85*, 118; Ashby; Smith *J. Am. Chem. Soc.* **1964**, *86*, 4363; Vreugdenhil; Blomberg *Recl. Trav. Chim. Pays-Bas* **1963**, *82*, 453, 461.

but not the other part; i.e., **15** is not present in measurable amounts. This was substantiated by ^{25}Mg nmr spectra of the ethyl Grignard reagent in THF, which showed the presence of three peaks, corresponding to EtMgBr, Et$_2$Mg, and MgBr$_2$.[124] That the equilibrium between RMgX and R$_2$Mg lies far to the left for "ethylmagnesium bromide" in ether was shown by Smith and Becker, who mixed 0.1 M ethereal solutions of Et$_2$Mg and MgBr$_2$ and found that a reaction occurred with a heat evolution of 3.6 kcal/mol (15 kJ/mol) of Et$_2$Mg, and that the product was *monomeric* (by boiling-point-elevation measurements).[125] When either solution was added little by little to the other, there was a linear output of heat until almost a 1:1 molar ratio was reached. Addition of an excess of either reagent gave no further heat output. These results show that at least under some conditions the Grignard reagent is largely RMgX (coordinated with solvent) but that the equilibrium can be driven to R$_2$Mg by evaporation of all the ether or by addition of dioxane.

For some aryl Grignard reagents it has proved possible to distinguish separate nmr chemical shifts for ArMgX and Ar$_2$Mg.[126] From the area under the peaks it is possible to calculate the concentrations of the two species, and from them, equilibrium constants for the Schlenk equilibrium. These data show[126] that the position of the equilibrium depends very markedly on the aryl group and the solvent but that conventional aryl Grignard reagents in ether are largely ArMgX, while in THF the predominance of ArMgX is less, and with some aryl groups there is actually more Ar$_2$Mg present. Separate nmr chemical shifts have also been found for alkyl RMgBr and R$_2$Mg in HMPA[127] and in ether at low temperatures.[128] When Grignard reagents from alkyl bromides or chlorides are prepared in triethylamine the predominant species is RMgX.[129] Thus the most important factor determining the position of the Schlenk equilibrium is the solvent. For primary alkyl groups the equilibrium constant for the reaction as written above is lowest in Et$_3$N, higher in ether, and still higher in THF.[130]

However, Grignard reagents prepared from alkyl bromides or iodides in ether at higher concentrations (0.5 to 1 M) contain dimers, trimers, and higher polymers, and those prepared from alkyl chlorides in ether at all concentrations are dimeric,[131] so that **15** is in solution, probably in equilibrium with RMgX and R$_2$Mg; i.e., the complete Schlenk equilibrium seems to be present.

The Grignard reagent prepared from 1-chloro-3,3-dimethylpentane in ether undergoes rapid inversion of configuration at the magnesium-containing carbon (demonstrated by nmr; this compound is not chiral).[132] The mechanism of this inversion is not completely known.

It might be mentioned that matters are much simpler for organometallic compounds with less-polar bonds. Thus Et$_2$Hg and EtHgCl are both definite compounds, the former a liquid and the latter a solid.

The Generation and Fate of Carbanions

The two principal ways in which carbanions are generated are parallel with the ways of generating carbocations.

[124]Benn; Lehmkuhl; Mehler; Rufińska *Angew. Chem. Int. Ed. Engl.* **1984,** *23*, 534 [*Angew. Chem. 96*, 521].
[125]Smith; Becker *Tetrahedron* **1966,** *22*, 3027.
[126]Evans; Khan *J. Chem. Soc. A* **1967,** 1643; Evans; Fazakerley *Chem. Commun.* **1968,** 974.
[127]Ducom *Bull. Chem. Soc. Fr.* **1971,** 3518, 3523, 3529.
[128]Ashby; Parris; Walker *Chem. Commun.* **1969,** 1464; Parris; Ashby, Ref. 120.
[129]Ashby; Walker *J. Org. Chem.* **1968,** *33*, 3821.
[130]Parris; Ashby, Ref. 120.
[131]Ashby; Smith, Ref. 123.
[132]Whitesides; Witanowski; Roberts *J. Am. Chem. Soc.* **1965,** *87*, 2854; Whitesides; Roberts *J. Am. Chem. Soc.*
1965, *87*, 4878. Also see Witanowski; Roberts *J. Am. Chem. Soc.* **1966,** *88*, 737; Fraenkel; Cottrell; Dix *J. Am. Chem. Soc.* **1971,** *93*, 1704; Pechhold; Adams; Fraenkel *J. Org. Chem.* **1971,** *36*, 1368; Maercker; Geuss *Angew. Chem. Int. Ed. Engl.* **1971,** *10*, 270 [*Angew. Chem. 83*, 288].

1. A group attached to a carbon leaves without its electron pair:

$$R{-}H \longrightarrow \bar{R}^{\ominus}{+}\ H^+$$

The leaving group is most often a proton. This is a simple acid–base reaction, and a base is required to remove the proton.[133] However, other leaving groups are known (see Chapter 12):

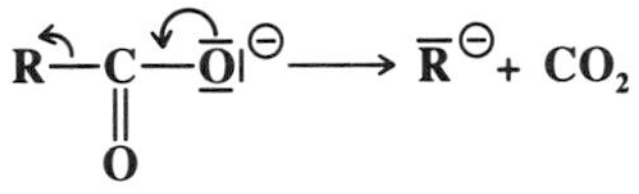

$$R{-}C{-}\underset{\underset{O}{\parallel}}{\ddot{O}}|^{\ominus} \longrightarrow \bar{R}^{\ominus}{+}\ CO_2$$

2. A negative ion adds to a carbon–carbon double or triple bond (see Chapter 15):

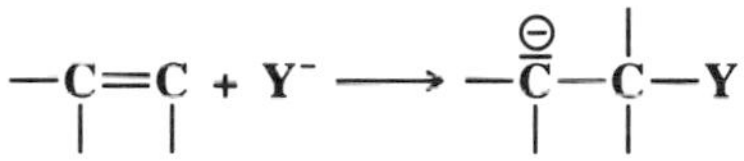

$$-C{=}C\ +\ Y^- \longrightarrow -\overset{\ominus}{C}{-}C{-}Y$$

The addition of a negative ion to a carbon–oxygen double bond does not give a carbanion, since the negative charge resides on the oxygen.

The most common reaction of carbanions is combination with a positive species, usually a proton, or with another species that has an empty orbital in its outer shell (a Lewis acid–base reaction):

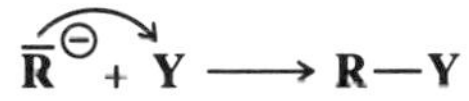

$$\bar{R}^{\ominus}\ +\ Y \longrightarrow R{-}Y$$

Carbanions may also form a bond with a carbon that already has four bonds, by displacing one of the four groups (S_N2 reaction, see Chapter 10):

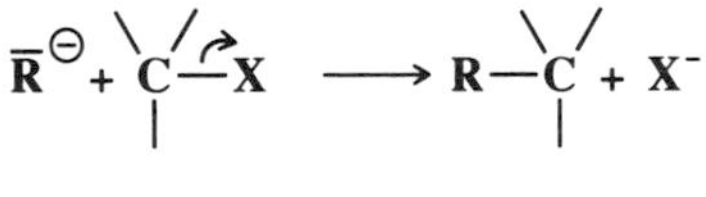

$$\bar{R}^{\ominus}+\ C{-}X \longrightarrow R{-}C\ +\ X^-$$

Like carbocations, carbanions can also react in ways in which they are converted to species that are still not neutral molecules. They can add to double bonds (usually $C{=}O$ double bonds; see Chapters 10 and 16),

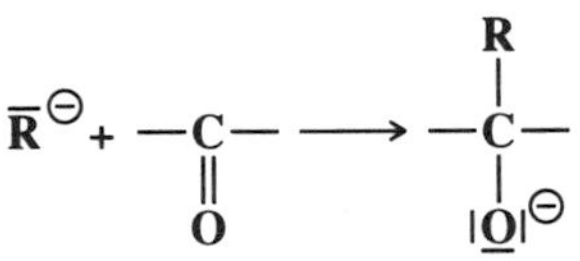

$$\bar{R}^{\ominus}+\ -\underset{\underset{O}{\parallel}}{C}{-} \longrightarrow -\overset{\overset{R}{|}}{\underset{\underset{|\underline{O}|^{\ominus}}{|}}{C}}{-}$$

or rearrange, though this is rare (see Chapter 18),

$$Ph_3C\bar{C}H_2{}^{\ominus} \longrightarrow Ph_2\bar{C}CH_2Ph$$

---

[133]For a review of such reactions, see Durst, in Buncel; Durst, Ref. 70, pt. B, pp. 239-291.

or be oxidized to free radicals.[134] A system in which a carbocation [Ph(p-Me$_2$NC$_6$H$_4$)$_2$C$^+$] oxidizes a carbanion [(p-NO$_2$C$_6$H$_4$)$_3$C$^-$] to give two free radicals, reversibly, so that all four species are present in equilibrium, has been demonstrated.[135]

Organometallic compounds that are not ionic but polar-covalent behave very much as if they were ionic and give similar reactions.

FREE RADICALS

Stability and Structure[136]

A *free radical* (often simply called a *radical*) may be defined as a species that contains one or more unpaired electrons. Note that this definition includes certain stable inorganic molecules such as NO and NO$_2$, as well as many individual atoms, such as Na and Cl. As with carbocations and carbanions, simple alkyl radicals are very reactive. Their lifetimes are extremely short in solution, but they can be kept for relatively long periods frozen within the crystal lattices of other molecules.[137] Many spectral[138] measurements have been made on radicals trapped in this manner. Even under these conditions the methyl radical decomposes with a half-life of 10 to 15 min in a methanol lattice at 77 K.[139] Since the lifetime of a radical depends not only on its inherent stability, but also on the conditions under which it is generated, the terms *persistent* and *stable* are usually used for the different senses. A stable radical is inherently stable; a persistent radical has a relatively long lifetime under the conditions at which it is generated, though it may not be very stable.

Associated with the spin of an electron is a magnetic moment, which can be expressed by a quantum number of $+\frac{1}{2}$ or $-\frac{1}{2}$. According to the Pauli principle, any two electrons occupying the same orbital must have opposite spins, so the total magnetic moment is zero for any species in which all the electrons are paired. In radicals, however, one or more electrons are unpaired, so there is a net magnetic moment and the species is paramagnetic. Radicals can therefore be detected by magnetic-susceptibility measurements, but for this technique a relatively high concentration of radicals is required. A much more important technique is *electron spin resonance* (esr), also called *electron paramagnetic resonance* (epr).[140] The principle of esr is similar to that of nmr, except that electron spin is involved

---

[134]For a review, see Guthrie, in Buncel; Durst, Ref. 70, pt. A, pp. 197-269.

[135]Arnett; Molter; Marchot; Donovan; Smith *J. Am. Chem. Soc.* **1987**, *109*, 3788. See also Ref. 84.

[136]For monographs, see Alfassi *Chemical Kinetics of Small Organic Radicals*, 4 vols.; CRC Press: Boca Raton, FL, 1988; Nonhebel; Tedder; Walton *Radicals*; Cambridge University Press: Cambridge, 1979; Nonhebel; Walton *Free-Radical Chemistry*; Cambridge University Press: Cambridge, 1974; Kochi *Free Radicals*, 2 vols.; Wiley: New York, 1973; Hay *Reactive Free Radicals*; Academic Press: New York, 1974; Pryor *Free Radicals*; McGraw-Hill: New York, 1966. For reviews, see Kaplan *React. Intermed. (Wiley)* **1985**, *3*, 227-303; **1981**, *2*, 251-314; **1978**, *1*, 163-196; Griller; Ingold *Acc. Chem. Res.* **1976**, *9*, 13-19; Huyser, in McManus, Ref. 1, pp. 1-59; Isaacs, Ref. 1, pp. 294-374.

[137]For a review of the use of matrices to study radicals and other unstable species, see Dunkin *Chem. Soc. Rev.* **1980**, *9*, 1-23; Jacox *Rev. Chem. Intermed.* **1978**, *2*, 1-36. For a review of the study of radicals at low temperatures, see Mile *Angew. Chem. Int. Ed. Engl.* **1968**, *7*, 507-519 [*Angew. Chem. 80*, 519-531].

[138]For a review of infrared spectra of radicals trapped in matrices, see Andrews *Annu. Rev. Phys. Chem.* **1971**, *22*, 109-132.

[139]Sullivan; Koski *J. Am. Chem. Soc.* **1963**, *85*, 384.

[140]For monographs, see Wertz; Bolton *Electron Spin Resonance*; McGraw-Hill: New York, 1972 [reprinted by Chapman and Hall: New York, and Methuen: London, 1986]; Assenheim *Introduction to Electron Spin Resonance*; Plenum: New York, 1967; Bersohn; Baird *An Introduction to Electron Paramagnetic Resonance*; W.A. Benjamin: New York, 1966. For reviews, see Bunce *J. Chem. Educ.* **1987**, *64*, 907-914; Hirota; Ohya-Nishiguchi, in Bernasconi *Investigation of Rates and Mechanisms of Reactions*, 4th ed., pt. 2; Wiley: New York, 1986, pp. 605-655; Griller; Ingold *Acc. Chem. Res.* **1980**, *13*, 193-200; Norman *Chem. Soc. Rev.* **1980**, *8*, 1-27; Fischer, in Kochi, Ref. 136, vol. 2, pp. 435-491; Russell, in Nachod; Zuckerman *Determination of Organic Structures by Physical Methods*, vol. 3; Academic Press: New York, 1971, pp. 293-341; Rassat *Pure Appl. Chem.* **1971**, *25*, 623-634; Kevan *Methods Free-Radical Chem.* **1969**, *1*, 1-33; Geske *Prog. Phys. Org. Chem.* **1967**, *4*, 125-211; Norman; Gilbert *Adv. Phys. Org. Chem.* **1967**, *5*, 53-119; Schneider; Möbius; Plato *Angew. Chem. Int. Ed. Engl.* **1965**, *4*, 856-867 [*Angew. Chem. 77*, 888-900]. For a review on the application of esr to photochemistry, see Wan *Adv. Photochem.* **1974**, *9*, 1-145. For a review of the related ENDOR method, see Kurreck; Kirste; Lubitz *Angew. Chem. Int. Ed. Engl.* **1984**, *23*, 173-194 [*Angew. Chem. 96*, 171-193]. See also Poole *Electron Spin Resonance. A Comprehensive Treatise on Experimental Techniques*, 2nd ed.; Wiley: New York, 1983.

rather than nuclear spin. The two electron spin states ($m_s = \frac{1}{2}$ and $m_s = -\frac{1}{2}$) are ordinarily of equal energy, but in a magnetic field the energies are different. As in nmr, a strong external field is applied and electrons are caused to flip from the lower state to the higher by the application of an appropriate radio-frequency signal. Inasmuch as two electrons paired in one orbital must have opposite spins which cancel, an esr spectrum arises only from species that have one or more unpaired electrons, i.e., free radicals.

Since only free radicals give an esr spectrum, the method can be used to detect the presence of radicals and to determine their concentration. Furthermore, information concerning the electron distribution (and hence the structure) of free radicals can be obtained from the splitting pattern of the esr spectrum (esr peaks are split by nearby protons).[141] Fortunately (for the existence of most free radicals is very short), it is not necessary for a radical to be persistent for an esr spectrum to be obtained. Esr spectra have been observed for radicals with lifetimes considerably less than 1 sec. Failure to observe an esr spectrum does not prove that radicals are not involved, since the concentration may be too low for direct observation. In such cases the *spin trapping* technique can be used.[142] In this technique a compound is added that is able to combine with very reactive radicals to produce more persistent radicals; the new radicals can be observed by esr. The most important spin-trapping compounds are nitroso compounds, which react with radicals to give fairly stable nitroxide radicals:[143] $RN=O + R'\cdot \rightarrow RR'N-O\cdot$.

Because there is an equal probability that a given unpaired electron will have a quantum number of $+\frac{1}{2}$ or $-\frac{1}{2}$, radicals cause two lines or groups of lines to appear on an electronic spectrum, and are sometimes referred to as *doublets*.

Another magnetic technique for the detection of free radicals uses an ordinary nmr instrument. It was discovered[144] that if an nmr spectrum is taken during the course of a reaction, certain signals may be enhanced, either in a positive or negative direction; others may be reduced. When this type of behavior, called *chemically induced dynamic nuclear polarization*[145] (CIDNP), is found in the nmr spectrum of the product of a reaction, it means that *at least a portion of that product was formed via the intermediacy of a free radical*.[146] For example, the question was raised whether radicals were intermediates in the exchange reaction between ethyl iodide and ethyllithium (**2-39**):

$$EtI + EtLi \rightleftharpoons EtLi + EtI$$

Curve *a* in Figure 5.1[147] shows an nmr spectrum taken during the course of the reaction. Curve *b* is a reference spectrum of ethyl iodide (CH$_3$ protons at $\delta = 1.85$; CH$_2$ protons at

[141]For reviews of the use of esr spectra to determine structures, see Walton *Rev. Chem. Intermed.* **1984**, *5*, 249-291; Kochi *Adv. Free-Radical Chem.* **1975**, *5*, 189-317. For esr spectra of a large number of free radicals, see Bielski; Gebicki *Atlas of Electron Spin Resonance Spectra*; Academic Press: New York, 1967.
[142]For reviews, see Janzen; Haire *Adv. Free Radical Chem. (Greenwich, Conn.)* **1990**, *1*, 253-295; Gasanov; Freidlina *Russ. Chem. Rev.* **1987**, *56*, 264-274; Perkins *Adv. Phys. Org. Chem.* **1980**, *17*, 1-64; Zubarev; Belevskii; Bugaenko *Russ. Chem. Rev.* **1979**, *48*, 729-745; Evans *Aldrichimica Acta* **1979**, *12*, 23-29; Janzen *Acc. Chem. Res.* **1971**, *4*, 31-40. See also the collection of papers on this subject in *Can. J. Chem.* **1982**, *60*, 1379-1636.
[143]For a series of papers on nitroxide radicals, see *Pure Appl. Chem.* **1990**, *62*, 177-316.
[144]Ward; Lawler *J. Am. Chem. Soc.* **1967**, *89*, 5518; Ward; Lawler; Cooper *J. Am. Chem. Soc.* **1969**, *91*, 746; Bargon; Fischer; Johnsen *Z. Naturforsch., Teil A* **1967**, *22*, 1551; Bargon; Fischer *Z. Naturforsch., Teil A* **1967**, *22*, 1556; Lepley *J. Am. Chem. Soc.* **1968**, *90*, 2710, **1969**, *91*, 749; Lepley; Landau *J. Am. Chem. Soc.* **1969**, *91*, 748.
[145]For a monograph on CIDNP, see Lepley; Closs *Chemically Induced Magnetic Polarization*; Wiley: New York, 1973. For reviews, see Adrian *Rev. Chem. Intermed.* **1986**, *7*, 173-194; Closs; Miller; Redwine *Acc. Chem. Res.* **1985**, *18*, 196-202; Lawler; Ward, in Nachod; Zuckerman, Ref. 140, vol. 5, 1973, pp. 99-150; Ward, in Kochi, Ref. 136, vol. 1, pp. 239-273; *Acc. Chem. Res.* **1972**, *5*, 18-24; Closs *Adv. Magn. Reson.* **1974**, *7*, 157-229; Lawler *Acc. Chem. Res.* **1972**, *5*, 25-32; Kaptein *Adv. Free-Radical Chem.* **1975**, *5*, 319-380; Bethell; Brinkman *Adv. Phys. Org. Chem.* **1973**, *10*, 53-128.
[146]A related technique is called chemically induced dynamic electron polarization (CIDEP). For a review, see Hore; Joslin; McLauchlan *Chem. Soc. Rev.* **1979**, *8*, 29-61.
[147]Ward; Lawler; Cooper, Ref. 144.

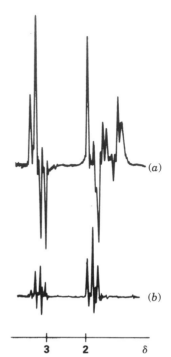

FIGURE 5.1[147]   (a) Nmr spectrum taken during reaction between EtI and EtLi in benzene (the region between 2.5 and 3.5 δ was scanned with an amplitude twice that of the remainder of the spectrum). The signals at 1.0 to 1.6 δ are due to butane, some of which is also formed in the reaction. (b) Reference spectrum of EtI.

$\delta = 3.2$). Note that in curve *a* some of the ethyl iodide signals are enhanced; others go below the base line (*negative enhancement*; also called *emission*). Thus the ethyl iodide formed in the exchange shows CIDNP and hence was formed via a free-radical intermediate. CIDNP results when protons in a reacting molecule become dynamically coupled to an unpaired electron while traversing the path from reactants to products. Although the presence of CIDNP almost always means that a free radical is involved,[148] its absence does not prove that a free-radical intermediate is necessarily absent, since reactions involving free-radical intermediates can also take place without observable CIDNP. Also, the presence of CIDNP does not prove that *all* of a product was formed via a free-radical intermediate, only that some of it was.

As with carbocations, the stability order of free radicals is tertiary > secondary > primary, explainable by hyperconjugation, analogous to that in carbocations (p. 167):

$$
\begin{array}{ccc}
\overset{\displaystyle H \quad H}{\underset{\displaystyle H \quad H}{R-C-C\cdot}} & \longleftrightarrow & \overset{\displaystyle \cdot H \quad H}{\underset{\displaystyle H \quad H}{R-C=C}} & \longleftrightarrow & \overset{\displaystyle H \quad H}{\underset{\displaystyle \cdot H \quad H}{R-C=C}}
\end{array}
$$

---

[148]It has been shown that CIDNP can also arise in cases where para hydrogen (H₂ in which the nuclear spins are opposite) is present: Eisenschmid; Kirss; Deutsch; Hommeltoft; Eisenberg; Bargon; Lawler; Balch *J. Am. Chem. Soc.* **1987,** *109,* 8089.

With resonance possibilities, the stability of free radicals increases;[149] some can be kept indefinitely.[150] Benzylic and allylic[151] radicals for which canonical forms can be drawn similar to those shown for the corresponding cations (pp. 168, 169) and anions (p. 177) are more stable than simple alkyl radicals but still have only a transient existence under ordinary conditions. However, the triphenylmethyl and similar radicals[152] are stable enough to exist in solution at room temperature, though in equilibrium with a dimeric form. The concen-

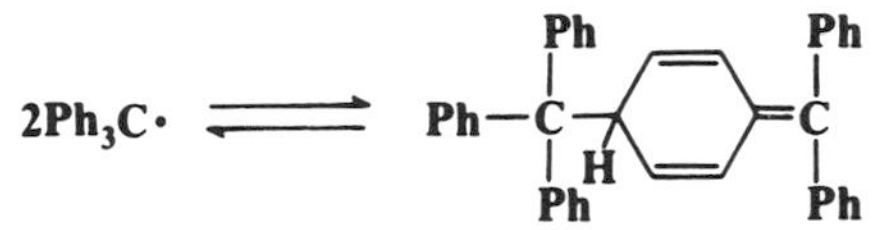

$$2Ph_3C\cdot \;\rightleftharpoons\; $$

16

tration of triphenylmethyl radical in benzene solution is about 2% at room temperature. For many years it was assumed that $Ph_3C\cdot$, the first stable free radical known,[153] dimerized to hexaphenylethane (Ph_3C—CPh_3),[154] but uv and nmr investigations have shown that the true structure is **16**.[155] Although triphenylmethyl-type radicals are stabilized by resonance:

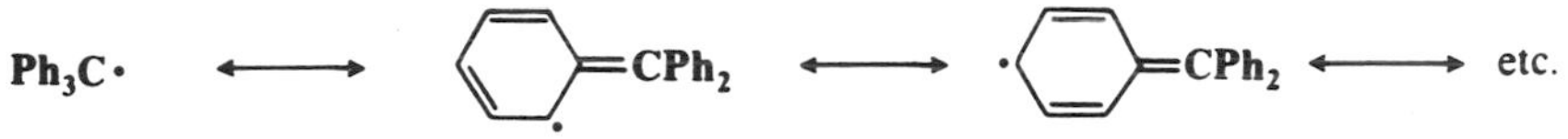

$$Ph_3C\cdot \;\longleftrightarrow\; \qquad \longleftrightarrow\; \qquad \longleftrightarrow\; \text{etc.}$$

it is steric hindrance to dimerization and not resonance that is the major cause of their stability.[156] This was demonstrated by the preparation of the radicals **17** and **18**.[157] These

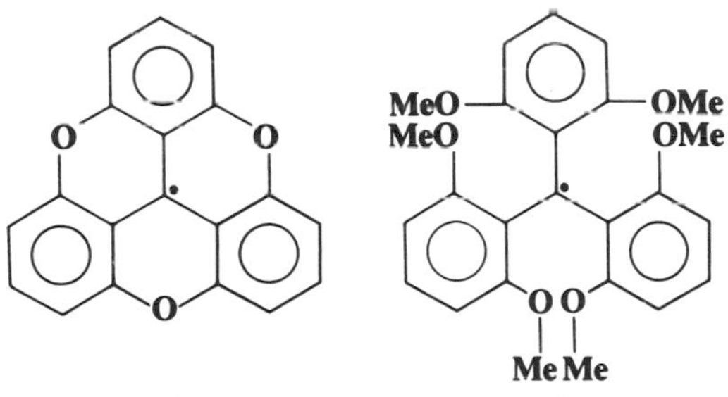

17   18

[149]For a discussion, see Robaugh; Stein *J. Am. Chem. Soc.* **1986,** *108,* 3224.

[150]For a monograph on stable radicals, including those in which the unpaired electron is not on a carbon atom, see Forrester; Hay; Thomson *Organic Chemistry of Stable Free Radicals*; Academic Press: New York, 1968.

[151]For an electron diffraction study of the allyl radical, see Vajda; Tremmel; Rozsondai; Hargittai; Maltsev; Kagramanov; Nefedov *J. Am. Chem. Soc.* **1986,** *108,* 4352.

[152]For a review, see Sholle; Rozantsev *Russ. Chem. Rev.* **1973,** *42,* 1011-1020.

[153]Gomberg *J. Am. Chem. Soc.* **1900,** *22,* 757, *Ber.* **1900,** *33,* 3150.

[154]Hexaphenylethane has still not been prepared, but substituted compounds [hexakis(3,5-di-*t*-butyl-4-biphenylyl)ethane and hexakis(3,5-di-*t*-butylphenyl)ethane] have been shown by x-ray crystallography to be nonbridged hexaarylethanes in the solid state: Stein; Winter; Rieker *Angew. Chem. Int. Ed. Engl.* **1978,** *17,* 692 [*Angew. Chem. 90,* 737]; Kahr; Van Engen; Mislow *J. Am. Chem. Soc.* **1986,** *108,* 8305; Yannoni; Kahr; Mislow *J. Am. Chem. Soc.* **1988,** *110,* 6670. In solution, both dissociate into free radicals.

[155]Lankamp; Nauta; MacLean *Tetrahedron Lett.* **1968,** 249; Staab; Brettschneider; Brunner *Chem. Ber.* **1970,** *103,* 1101; Volz; Lotsch; Schnell *Tetrahedron* **1970,** *26,* 5343; McBride *Tetrahedron* **1974,** *30,* 2009. See also Guthrie; Weisman *Chem. Commun.* **1969,** 1316; Takeuchi; Nagai; Tokura *Bull. Chem. Soc. Jpn.* **1971,** *44,* 753. For an example where a secondary benzilic radical undergoes this type of dimerization, see Peyman; Peters; von Schnering; Rüchardt *Chem. Ber.* **1990,** *123,* 1899.

[156]For a review of steric effects in free radical chemistry, see Rüchardt *Top. Curr. Chem.* **1980,** *88,* 1-32.

[157]Sabacky; Johnson; Smith; Gutowsky; Martin *J. Am. Chem. Soc.* **1967,** *89,* 2054.

radicals are electronically very similar, but **17,** being planar, has much less steric hindrance to dimerization than Ph$_3$C•, while **18,** with six groups in ortho positions, has much more. On the other hand, the planarity of **17** means that it has a maximum amount of resonance stabilization, while **18** must have much less, since its degree of planarity should be even less than Ph$_3$C•, which itself is propeller-shaped and not planar. Thus if resonance is the chief cause of the stability of Ph$_3$C•, **18** should dimerize and **17** should not, but if steric hindrance is the major cause, the reverse should happen. In the event, it was found[157] that **18** gave no evidence of dimerization, even in the solid state, while **17** existed primarily in the dimeric form, which is dissociated to only a small extent in solution,[158] indicating that steric hindrance to dimerization is the major cause for the stability of triarylmethyl radicals. A similar conclusion was reached in the case of (NC)$_3$C•, which dimerizes readily though considerably stabilized by resonance.[159] Nevertheless, that resonance is still an important contributing factor to the stability of radicals is shown by the facts that (1) the radical t-Bu(Ph)$_2$C• dimerizes more than Ph$_3$C•, while p-PhCOC$_6$H$_4$(Ph$_2$)C• dimerizes less.[160] The latter has more canonical forms than Ph$_3$C•, but steric hindrance should be about the same (for attack at one of the two rings). (2) A number of radicals (p-XC$_6$H$_4$)$_3$C•, with X = F, Cl, O$_2$N, CN, etc. do not dimerize, but are kinetically stable.[161] Completely chlorinated triarylmethyl radicals are more stable than the unsubstituted kind, probably for steric reasons, and many are quite inert in solution and in the solid state.[162]

It has been postulated that the stability of free radicals is enhanced by the presence at the radical center of both an electron-donating and an electron-withdrawing group.[163] This is called the *push–pull* or *captodative effect* (see also pp. 129). The effect arises from increased resonance, e.g.:

$$R-\overset{\bullet}{\underset{\underline{NR'_2}}{C}}-C\equiv\overline{N} \longleftrightarrow R-\overset{\ominus}{\underset{\underset{\oplus}{\bullet NR'_2}}{C}}-C\equiv\overline{N} \longleftrightarrow R-\underset{\underset{\oplus}{\bullet NR'_2}}{C}=C=\overset{\ominus}{\underline{N}} \longleftrightarrow$$

$$R-\underset{\underline{NR'_2}}{C}=C=\overset{\bullet}{N} \longleftrightarrow R-\underset{\underset{\oplus}{NR'_2}}{C}-\overset{\bullet}{C}=\overline{N}^{\ominus}$$

There is some evidence in favor[164] of the captodative effect, some of it from esr studies.[165] However, there is also experimental[166] and theoretical[167] evidence against it. There is evidence that while FCH$_2$• and F$_2$CH• are more stable than CH$_3$•, the radical CF$_3$• is less stable; that is, the presence of the third F destabilizes the radical.[168]

[158]Müller; Moosmayer; Rieker; Scheffler *Tetrahedron Lett.* **1967,** 3877. See also Neugebauer; Hellwinkel; Aulmich *Tetrahedron Lett.* **1978,** 4871.

[159]Kaba; Ingold *J. Am. Chem. Soc.* **1976,** 98, 523.

[160]Zarkadis; Neumann; Marx; Uzick *Chem. Ber.* **1985,** 118, 450; Zarkadis; Neumann; Uzick *Chem. Ber.* **1985,** 118, 1183.

[161]Dünnebacke; Neumann; Penenory; Stewen *Chem. Ber.* **1989,** 122, 533.

[162]For reviews, see Ballester *Adv. Phys. Org. Chem.* **1989,** 25, 267-445, pp. 354-405, *Acc. Chem. Res.* **1985,** 18, 380-387. See also Hegarty; O'Neill *Tetrahedron Lett.* **1987,** 28, 901.

[163]For reviews, see Sustmann; Korth *Adv. Phys. Org. Chem.* **1990,** 26, 131-178; Viehe; Janousek; Merényi; Stella *Acc. Chem. Res.* **1985,** 18, 148-154.

[164]For a summary of the evidence, see Pasto *J. Am. Chem. Soc.* **1988,** 110, 8164. See also Ref. 163.

[165]See, for example Korth; Lommes; Sustmann; Sylvander; Stella *New J. Chem.* **1987,** 11, 365; Sakurai; Kyushin; Nakadaira; Kira *J. Phys. Org. Chem.* **1988,** 1, 197; Rhodes; Roduner *Tetrahedron Lett.* **1988,** 29, 1437; Viehe; Merényi; Janousek *Pure Appl. Chem.* **1988,** 60, 1635; Creary; Sky; Mehrsheikh-Mohammadi *Tetrahedron Lett.* **1988,** 29, 6839; Bordwell; Lynch *J. Am. Chem. Soc.* **1989,** 111, 7558.

[166]See, for example Beckhaus; Rüchardt *Angew. Chem. Int. Ed. Engl.* **1987,** 26, 770 [*Angew. Chem.* 99, 807]; Neumann; Penenory; Stewen; Lehnig *J. Am. Chem. Soc.* **1989,** 111, 5845; Bordwell; Bausch; Cheng; Cripe; Lynch; Mueller *J. Org. Chem.* **1990,** 55, 58; Bordwell; Harrelson *Can. J. Chem.* **1990,** 68, 1714.

[167]See Pasto, Ref. 164.

[168]Jiang; Li; Wang *J. Org. Chem.* **1989,** 54, 5648.

Certain radicals with the unpaired electron not on a carbon are also very stable.[169] Diphenylpicrylhydrazyl is a solid that can be kept for years. We have already mentioned nitroxide radicals. **19** is a nitroxide radical so stable that reactions can be performed on it

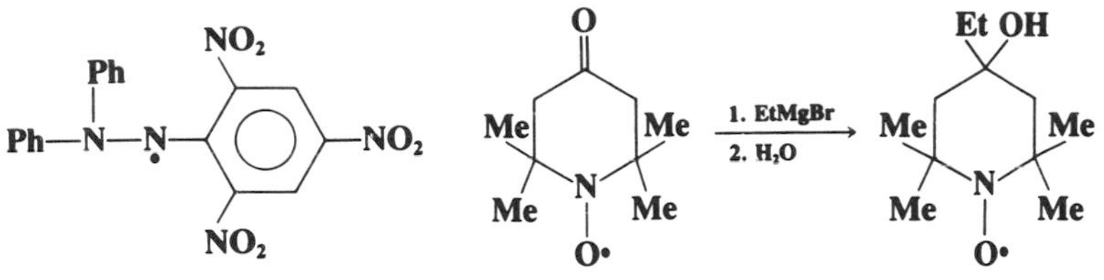

Diphenylpicrylhydrazyl   **19**

without affecting the unpaired electron[170] (the same is true for some of the chlorinated triarylmethyl radicals mentioned above[171]).

Dissociation energies (D values) of R—H bonds provide a measure of the relative inherent stability of free radicals R.[172] Table 5.3 lists such values.[173] The higher the D value, the less stable the radical.

TABLE 5.3 D_{298} values for some R—H bonds[173]
Free-radical stability is in the reverse order

R	D kcal/mol	kJ/mol
Ph·	111	464
CF_3·	107	446
CH_2=CH·	106	444
cyclopropyl[174]	106	444
Me·	105	438
Et·	100	419
Me_3CCH_2·	100	418
Pr·	100	417
Cl_3C·	96	401
Me_2CH·	96	401
Me_3C·[175]	95.8	401
cyclohexyl	95.5	400
$PhCH_2$·	88	368
HCO·	87	364
CH_2=CH—CH_2·	86	361

[169]For reviews of radicals with the unpaired electron on atoms other than carbon, see, in Kochi, Ref. 136, vol. 2, the reviews by Nelson, pp. 527-593 (N-centered); Bentrude, pp. 595-663 (P-centered); Kochi, pp. 665-710 (O-centered); Kice, pp. 711-740 (S-centered); Sakurai, pp. 741-807 (Si, Ge, Sn, and Pb-centered).

[170]Neiman; Rozantsev; Mamedova *Nature* **1963**, *200*, 256. For reviews of such radicals, see Aurich, in Patai *The Chemistry of Functional Groups, Supplement F*, pt. 1, Wiley: New York, 1982, pp. 565-622 [This review has been reprinted, and new material added, in Breuer; Aurich; Nielsen *Nitrones, Nitronates, and Nitroxides*; Wiley: New York, 1989, pp. 313-399]; Rozantsev; Sholle *Synthesis* **1971**, 190-202, 401-414.

[171]See Ballester; Veciana; Riera; Castañer; Armet; Rovira *Chem. Soc., Chem. Commun.* **1983**, 982.

[172]It has been claimed that relative D values do not provide such a measure: Nicholas; Arnold *Can. J. Chem.* **1984**, *62*, 1850, 1860.

[173]Except where noted, these values are from Kerr, in Weast *Handbook of Chemistry and Physics*, 69th ed.; CRC Press: Boca Raton, FL, 1988, p. F-183. For another list of D values, see McMillen; Golden *Annu. Rev. Phys. Chem.* **1982**, *33*, 493. See also Tsang *J. Am. Chem. Soc.* **1985**, *107*, 2872; Holmes; Lossing; Maccoll *J. Am. Chem. Soc.* **1988**, *110*, 7339; Holmes; Lossing *J. Am. Chem. Soc.* **1988**, *110*, 7343; Roginskii *J. Org. Chem. USSR* **1989**, *25*, 403.

[174]For a review of cyclopropyl radicals, see Walborsky *Tetrahedron* **1981**, *37*, 1625-1651. See also Boche; Walborsky, Ref. 102.

[175]This value is from Gutman *Acc. Chem. Res.* **1990**, *23*, 375-380.

There are two possible structures for simple alkyl radicals.[176] They might have sp^2 bonding, in which case the structure would be planar, with the odd electron in a p orbital, or the bonding might be sp^3, which would make the structure pyramidal and place the odd electron in an sp^3 orbital. Esr spectra of $CH_3\cdot$ and other simple alkyl radicals as well as other evidence indicate that these radicals have planar structures.[177] This is in accord with the known loss of optical activity when a free radical is generated at a chiral carbon.[178] In addition, electronic spectra of the CH_3 and CD_3 radicals (generated by flash photolysis) in the gas phase have definitely established that under these conditions the radicals are planar or near-planar.[179] Ir spectra of $CH_3\cdot$ trapped in solid argon led to a similar conclusion.[180]

Evidence from studies on bridgehead compounds shows that though a planar configuration is more stable, pyramidal structures are not impossible. In contrast to the situation with carbocations, free radicals have often been generated at bridgeheads, although studies have shown that bridgehead free radicals are less rapidly formed than the corresponding open-chain radicals.[181] In sum, the available evidence indicates that though simple alkyl free radicals prefer a planar, or near-planar shape, the energy difference between a planar and a pyramidal free radical is not great. However, free radicals in which the carbon is connected to atoms of high electronegativity, e.g., $CF_3\cdot$, prefer a pyramidal shape;[182] increasing the electronegativity increases the deviation from planarity.[183] Cyclopropyl radicals are also pyramidal.[184]

Free radicals with resonance are definitely planar, though triphenylmethyl-type radicals are propeller-shaped,[185] like the analogous carbocations (p. 172).

A number of diradicals (also called biradicals) are known.[186] When the unpaired electrons of a diradical are widely separated, e.g., as in $\cdot CH_2CH_2CH_2CH_2\cdot$, the species behaves spectrally like two doublets. When they are close enough for interaction or can interact through an unsaturated system (as in trimethylenemethane,[187] they can have total spin numbers of $+1$, 0, or -1, since each electron could be either $+\frac{1}{2}$ or $-\frac{1}{2}$. Spectroscopically

[176]For a review, see Kaplan, in Kochi, Ref. 136, vol. 2, pp. 361-434.

[177]See, for example, Cole; Pritchard; Davidson; McConnell *Mol. Phys.* **1958**, *1*, 406; Fessenden; Schuler *J. Chem. Phys.* **1963**, *39*, 2147; Symons *Nature* **1969**, *222*, 1123, *Tetrahedron Lett.* **1973**, 207; Bonazzola; Leray; Roncin *J. Am. Chem. Soc.* **1977**, *99*, 8348; Giese; Beckhaus *Angew. Chem. Int. Ed. Engl.* **1978**, *17*, 594 [*Angew. Chem. 90*, 635]; Ref. 98. See, however, Paddon-Row; Houk *J. Am. Chem. Soc.* **1981**, *103*, 5047.

[178]There are a few exceptions. See p. 682.

[179]Herzberg; Shoosmith *Can. J. Phys.* **1956**, *34*, 523; Herzberg *Proc. R. Soc. London, Ser. A* **1961**, *262*, 291. See also Tan; Winer; Pimentel *J. Chem. Phys.* **1972**, *57*, 4028; Yamada; Hirota; Kawaguchi *J. Chem. Phys.* **1981**, *75*, 5256.

[180]Andrews; Pimentel *J. Chem. Phys.* **1967**, *47*, 3637; Milligan; Jacox *J. Chem. Phys.* **1967**, *47*, 5146.

[181]Lorand; Chodroff; Wallace *J. Am. Chem. Soc.* **1968**, *90*, 5266; Fort; Franklin *J. Am. Chem. Soc.* **1968**, *90*, 5267; Humphrey; Hodgson; Pincock *Can. J. Chem.* **1968**, *46*, 3099; Oberlinner; Rüchardt *Tetrahedron Lett.* **1969**, 4685; Danen; Tipton; Saunders *J. Am. Chem. Soc.* **1971**, *93*, 5186; Fort; Hiti *J. Org. Chem.* **1977**, *42*, 3968; Lomas *J. Org. Chem.* **1987**, *52*, 2627.

[182]Fessenden; Schuler *J. Chem. Phys.* **1965**, *43*, 2704; Rogers; Kispert *J. Chem. Phys.* **1967**, *46*, 3193; Pauling *J. Chem. Phys.* **1969**, *51*, 2767.

[183]For example, 1,1-dichloroalkyl radicals are closer to planarity than the corresponding 1,1-difluoro radicals, though still not planar: Chen; Tang; Montgomery; Kochi *J. Am. Chem. Soc.* **1974**, *96*, 2201. For a discussion, see Krusic; Bingham *J. Am. Chem. Soc.* **1976**, *98*, 230.

[184]See Deycard; Hughes; Lusztyk; Ingold *J. Am. Chem. Soc.* **1987**, *109*, 4954.

[185]Adrian *J. Chem. Phys.* **1958**, *28*, 608; Andersen *Acta Chem. Scand.* **1965**, *19*, 629.

[186]For a monograph, see Borden *Diradicals*; Wiley: New York, 1982. For reviews, see Johnston; Scaiano *Chem. Rev.* **1989**, *89*, 521-547; Doubleday; Turro; Wang *Acc. Chem. Res.* **1989**, *22*, 199-205; Scheffer; Trotter *Rev. Chem. Intermed.* **1988**, *9*, 271-305; Wilson *Org. Photochem.* **1985**, *7*, 339-466; Borden *React. Intermed. (Wiley)* **1985**, *3*, 151-188, **1981**, *2*, 175-209; Borden; Davidson *Acc. Chem. Res.* **1981**, *14*, 69-76; Salem; Rowland *Angew. Chem. Int. Ed. Engl.* **1972**, *11*, 92-111 [*Angew. Chem. 84*, 86-106]; Salem *Pure Appl. Chem.* **1973**, *33*, 317-328; Jones *J. Chem. Educ.* **1974**, *51*, 175-181; Morozova; Dyatkina *Russ. Chem. Rev.* **1968**, *37*, 376-391. See also Döhnert; Koutecký *J. Am. Chem. Soc.* **1980**, *102*, 1789. For a series of papers on diradicals, see *Tetrahedron* **1982**, *38*, 735-867.

[187]For reviews of trimethylenemethane, see Borden; Davidson *Ann. Rev. Phys. Chem.* **1979**, *30*, 125-153; Bergman; in Kochi, Ref. 136, vol. 1, pp. 141-149.

they are called *triplets*,[188] since each of the three possibilities is represented among the molecules and gives rise to its own spectral peak. In triplet *molecules* the two unpaired

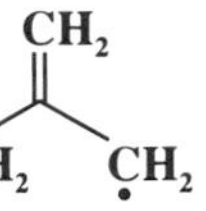

$$\text{CH}_2$$

$$\overset{\bullet}{\text{C}}\text{H}_2 \qquad \overset{\bullet}{\text{C}}\text{H}_2$$

Trimethylenemethane

electrons have the same spin. Radicals with both unpaired electrons on the same carbon are discussed under carbenes.

The Generation and Fate of Free Radicals[189]

Free radicals are formed from molecules by breaking a bond so that each fragment keeps one electron.[190] The energy necessary to break the bond is supplied in one of two ways.

1. *Thermal cleavage.* Subjection of any organic molecule to a high enough temperature in the gas phase results in the formation of free radicals. When the molecule contains bonds with D values or 20 to 40 kcal/mol (80 to 170 kJ/mol), cleavage can be caused in the liquid phase. Two common examples are cleavage of diacyl peroxides[192] and of azo compounds;[193]

$$\text{R}-\overset{\underset{\|}{\text{O}}}{\text{C}}-\text{O}-\text{O}-\overset{\underset{\|}{\text{O}}}{\text{C}}-\text{R} \xrightarrow{\Delta} 2\text{R}-\overset{\underset{\|}{\text{O}}}{\text{C}}-\text{O}\cdot$$

$$\text{R}-\text{N}{=}\text{N}-\text{R} \xrightarrow{\Delta} 2\text{R}\cdot + \text{N}_2$$

2. *Photochemical cleavage* (see p. 236). The energy of light of 600 to 300 nm is 48 to 96 kcal/mol (200 to 400 kJ/mol), which is of the order of magnitude of covalent-bond energies. Typical examples are photochemical cleavage of chlorine and of ketones;

$$\text{Cl}_2 \xrightarrow{hv} 2\text{Cl}\cdot$$

$$\text{R}-\overset{\underset{\|}{\text{O}}}{\text{C}}-\text{R} \xrightarrow[\text{vapor phase}]{hv} \text{R}-\overset{\underset{\|}{\text{O}}}{\text{C}}\cdot + \text{R}\cdot$$

[188]For discussions of the triplet state, see Wagner; Hammond *Adv. Photochem.* **1968**, *5*, 21-156; Turro *J. Chem. Educ.* **1969**, *46*, 2-6. For a discussion of esr spectra of triplet states, see Wasserman; Hutton *Acc. Chem. Res.* **1977**, *10*, 27-32.

[189]For a summary of methods of radical formation, see Giese *Radicals in Organic Synthesis: Formation of Carbon–Carbon Bonds*; Pergamon: Elmsford, NY, 1986, pp. 267-281. For a review on formation of free radicals by thermal cleavage, see Brown *Pyrolytic Methods in Organic Chemistry*; Academic Press: New York, 1980, pp. 44-61.

[190]It is also possible for free radicals to be formed by the collision of two nonradical species. For a review, see Harmony *Methods Free-Radical Chem.* **1974**, *5*, 101-176.

[191]For a review of homolytic cleavage of carbon–metal bonds, see Barker; Winter, in Hartley; Patai, Ref. 107, pp. 151-218.

[192]For a review of free radical mechanisms involving peroxides in solution, see Howard, in Patai *The Chemistry of Peroxides*; Wiley: New York, 1983, pp. 235-258. For a review of pyrolysis of peroxides in the gas phase, see Batt; Liu, in the same volume, pp. 685-710. See also Chateauneuf; Lusztyk; Ingold *J. Am. Chem. Soc.* **1988**, *110*, 2877, 2886.

[193]For a review of the cleavage of azoalkanes, see Engel *Chem. Rev.* **1980**, *80*, 99-150. For summaries of later work, see Adams; Burton; Andrews; Weisman; Engel *J. Am. Chem. Soc.* **1986**, *108*, 7935; Schmittel; Rüchardt *J. Am. Chem. Soc.* **1987**, *109*, 2750.

Radicals are also formed from other radicals, either by the reaction between a radical and a molecule (which *must* give another radical, since the total number of electrons is odd) or by cleavage of a radical to give another radical, e.g.,

$$Ph-\underset{\underset{O}{\|}}{C}-O\cdot \longrightarrow Ph\cdot + CO_2$$

Radicals can also be formed by oxidation or reduction, including electrolytic methods.

Reactions of free radicals either give stable products (termination reactions) or lead to other radicals, which themselves must usually react further (propagation reactions). The most common termination reactions are simple combinations of similar or different radicals:

$$R\cdot + R'\cdot \longrightarrow R-R'$$

Another termination process is disproportionation:[194]

$$2CH_3-CH_2\cdot \longrightarrow CH_3-CH_3 + CH_2=CH_2$$

There are four principal propagation reactions, of which the first two are most common:

1. *Abstraction of another atom or group, usually a hydrogen atom* (see Chapter 14):

$$R\cdot + R'-H \longrightarrow R-H + R'\cdot$$

2. *Addition to a multiple bond* (see Chapter 15):

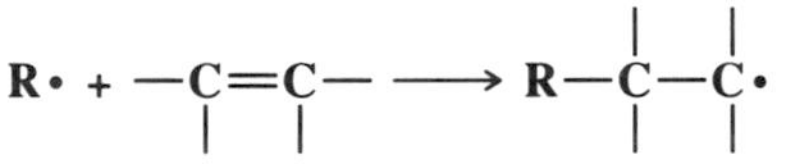

$$R\cdot + -\underset{|}{\overset{|}{C}}=\underset{|}{\overset{|}{C}}- \longrightarrow R-\underset{|}{\overset{|}{C}}-\underset{|}{\overset{|}{C}}\cdot$$

The radical formed here may add to another double bond, etc. This is one of the chief mechanisms for vinyl polymerization.

3. *Decomposition.*   This can be illustrated by the decomposition of the benzoxy radical (above).

4. *Rearrangement:*

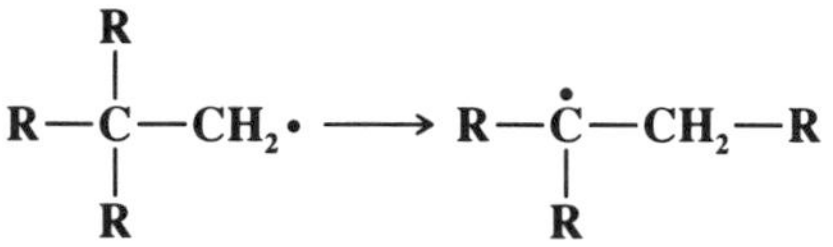

$$R-\underset{\underset{R}{|}}{\overset{\overset{R}{|}}{C}}-CH_2\cdot \longrightarrow R-\underset{\underset{R}{|}}{\overset{\cdot}{C}}-CH_2-R$$

This is less common than rearrangement of carbocations, but it does occur (though not when R = alkyl or hydrogen; see Chapter 18).

Besides these reactions, free radicals can be oxidized to carbocations or reduced to carbanions.[195]

[194]For reviews of termination reactions, see Pilling *Int. J. Chem. Kinet.* **1989**, *21*, 267-291; Khudyakov; Levin; Kuz'min *Russ. Chem. Rev.* **1980**, *49*, 982-1002; Gibian; Corley *Chem. Rev.* **1973**, *73*, 441-464.

[195]For a review of the oxidation and reduction of free radicals, see Khudyakov and Kuz'min *Russ. Chem. Rev.* **1978**, *47*, 22-42.

Radical Ions[196]

Several types of radical anions are known with the unpaired electron or the charge or both on atoms other than carbon. Important examples are semiquinones[197] (**20**) and ketyls[198]

$$\begin{array}{c}|\overline{O}\cdot \\ \\ \text{(ring)} \\ \\ |\underline{O}|_{\ominus} \\ \textbf{20}\end{array} \qquad \begin{array}{c}-\overset{\cdot}{\underset{|}{C}}-Ar \\ |\underline{O}|_{\ominus} \\ \textbf{21}\end{array}$$

(**21**). Reactions in which alkali metals are reducing agents often involve radical anion intermediates, e.g., reaction **5-10**:

$$\text{(ring)} + \overset{\cdot}{Na} \longrightarrow \text{(ring)} \overset{\ominus}{\underset{}{\;}} Na^+ \longrightarrow \text{products}$$

Several types of radical cation are also known.[199]

CARBENES

Stability and Structure[200]

Carbenes are highly reactive species, practically all having lifetimes considerably under 1 sec. With exceptions noted below (p. 200), carbenes have been isolated only by entrapment in matrices at low temperatures (77 K or less).[201] The parent species CH_2 is usually called

[196]For a monograph, see Kaiser; Kevan *Radical Ions*; Wiley: New York, 1968. For reviews, see Gerson; Huber *Acc. Chem. Res.* **1987**, *20*, 85-90; Todres *Tetrahedron* **1985**, *41*, 2771-2823; Russell; Norris, in McManus, Ref. 1, pp. 423-448; Holy; Marcum *Angew. Chem. Int. Ed. Engl.* **1971**, *10*, 115-124 [*Angew. Chem. 83*, 132-142]; Bilevich; Okhlobystin *Russ. Chem. Rev.***1968**, *37*, 954-968; Szwarc *Prog. Phys. Org. Chem.* **1968**, *6*, 322-438. For a related review, see Chanon; Rajzmann; Chanon *Tetrahedron* **1990**, *46*, 6193-6299. For a series of papers on this subject, see *Tetrahedron* **1986**, *42*, 6097-6349.

[197]For a review of semiquinones, see Depew; Wan, in Patai; Rappoport *The Chemistry of the Quinonoid Compounds*, vol. 2, pt. 2; Wiley: New York, 1988, pp. 963-1018.

[198]For a review of ketyls, see Russell, in Patai; Rappoport *The Chemistry of Enones*, pt. 1; Wiley: New York, 1989, pp. 471-512.

[199]For reviews, see Roth *Acc. Chem. Res.* **1987**, *20*, 343-350; Courtneidge; Davies *Acc. Chem. Res.* **1987**, *20*, 90-97; Hammerich; Parker *Adv. Phys. Org. Chem.* **1984**, *20*, 55-189; Symons *Chem. Soc. Rev.* **1984**, *13*, 393-439; Bard; Ledwith; Shine *Adv. Phys. Org. Chem.* **1976**, *13*, 155-278.

[200]For monographs, see Jones; Moss *Carbenes*, 2 vols.; Wiley: New York, 1973-1975; Kirmse *Carbene Chemistry*, 2nd ed.; Academic Press: New York, 1971; Rees; Gilchrist *Carbenes, Nitrenes, and Arynes*; Nelson: London, 1969. For reviews, see Minkin; Simkin; Glukhovtsev *Russ. Chem. Rev.* **1989**, *58*, 622-635; Moss; Jones *React. Intermed. (Wiley)* **1985**, *3*, 45-108, **1981**, *2*, 59-133, **1978**, *1*, 69-115; Isaacs, Ref. 1, pp. 375-407; Bethell *Adv. Phys. Org. Chem.* **1969**, *7*, 153-209; Bethell, in McManus, Ref. 1, pp. 61-126; Closs *Top. Stereochem.* **1968**, *3*, 193-235; Herold; Gaspar *Fortschr. Chem. Forsch.* **1966**, *5*, 89-146; Rozantsev; Fainzil'berg; Novikov *Russ. Chem. Rev.* **1965**, *34*, 69-88. For a theoretical study, see Liebman; Simons *Mol. Struct. Energ.* **1986**, *1*, 51-99.

[201]For example, see Murray; Trozzolo; Wasserman; Yager *J. Am. Chem. Soc.* **1962**, *84*, 3213; Brandon; Closs; Hutchison *J. Chem. Phys.* **1962**, *37*, 1878; Milligan; Mann; Jacox; Mitsch *J. Chem. Phys.* **1964**, *41*, 1199; Nefedov; Maltsev; Mikaelyan *Tetrahedron Lett.* **1971**, 4125; Wright *Tetrahedron* **1985**, *41*, 1517. For reviews, see Zuev; Nefedov *Russ. Chem. Rev.* **1989**, *58*, 636-643; Sheridan *Org. Photochem.* **1987**, *8*, 159-248, pp. 196-216; Trozzolo *Acc. Chem. Res.* **1968**, *1*, 329-335.

methylene, though derivatives are more often named by the carbene nomenclature. Thus CCl_2 is generally known as dichlorocarbene, though it can also be called dichloromethylene.

The two nonbonded electrons of a carbene can be either paired or unpaired. If they are paired, the species is spectrally a *singlet*, while, as we have seen (p. 193), two unpaired electrons appear as a *triplet*. An ingenious method of distinguishing between the two possibilities was developed by Skell,[202] based on the common reaction of addition of carbenes to double bonds to form cyclopropane derivatives (**5-50**). If the singlet species adds to *cis*-2-butene, the resulting cyclopropane should be the cis isomer since the movements of

$$
\begin{array}{ccc}
\underset{Me}{\overset{H}{\diagdown}}C=C\underset{Me}{\overset{H}{\diagup}} \overset{CH_2}{} & \longrightarrow & \underset{Me}{\overset{H}{\diagdown}}C-C\underset{Me}{\overset{H}{\diagup}}\overset{CH_2}{}
\end{array}
$$

the two pairs of electrons should occur either simultaneously or with one rapidly succeeding another. However, if the attack is by a triplet species, the two unpaired electrons cannot both go into a new covalent bond, since by Hund's rule they have parallel spins. So one of the unpaired electrons will form a bond with the electron from the double bond that has the opposite spin, leaving two unpaired electrons that have the same spin and therefore cannot form a bond at once but must wait until, by some collision process, one of the

$$
\underset{Me}{\overset{H}{\diagdown}}C-C\underset{Me}{\overset{H}{\diagup}}\overset{\uparrow CH_2}{} \longrightarrow \underset{Me}{\overset{H}{\diagdown}}C-C\underset{Me}{\overset{H}{\diagup}}\overset{CH_2\uparrow}{} \xrightarrow{\text{collision}}
$$

$$
\underset{Me}{\overset{H}{\diagdown}}C-\overset{\bullet}{C}HMe \overset{\uparrow CH_2}{} \longrightarrow \underset{Me}{\overset{H}{\diagdown}}C-\overset{\bullet}{C}HMe \overset{CH_2}{}
$$

electrons can reverse its spin. During this time, there is free rotation about the C—C bond and a mixture of *cis*- and *trans*-1,2-dimethylcyclopropanes should result.[203]

The results of this type of experiment show that CH_2 itself is usually formed as a singlet species, which can decay to the triplet state, which consequently has a lower energy (molecular-orbital calculations and experimental determinations show that the difference in energy between singlet and triplet CH_2 is about 8 to 10 kcal/mol or 33 to 42 kJ/mol[204]). However, it is possible to prepare triplet CH_2 directly by a photosensitized decomposition of diazomethane.[205] CH_2 is so reactive[206] that it generally reacts as the singlet before it has

[202]Skell; Woodworth *J. Am. Chem. Soc.* **1956**, *78*, 4496; Skell *Tetrahedron* **1985**, *41*, 1427.

[203]These conclusions are generally accepted though the reasoning given here may be oversimplified. For discussions, see Closs, Ref. 200, pp. 203-210; Bethell *Adv. Phys. Org.Chem.*, Ref. 200, pp. 194-200; Hoffmann *J. Am. Chem. Soc.* **1968**, *90*, 1475.

[204]See, for example, Hay; Hunt; Goddard *Chem. Phys. Lett.* **1972**, *13*, 30; Dewar; Haddon; Weiner *J. Am. Chem. Soc.* **1974**, *96*, 253; Frey; Kennedy *J. Chem. Soc., Chem. Commun.* **1975**, 233; Lucchese; Schaefer *J. Am. Chem. Soc.* **1977**, *99*, 6765; Roos; Siegbahn *J. Am. Chem. Soc.* **1977**, *99*, 7716; Lengel; Zare *J. Am. Chem. Soc.* **1978**, *100*, 7495; Borden; Davidson, Ref. 187, pp. 128, 134; Leopold; Murray; Lineberger *J. Chem. Phys.* **1984**, *81*, 1048.

[205]Kopecky; Hammond; Leermakers *J. Am. Chem. Soc.* **1961**, *83*, 2397, **1962**, *84*, 1015; Duncan; Cvetanović *J. Am. Chem. Soc.* **1962**, *84*, 3593.

[206]For a review of the kinetics of CH_2 reactions, see Laufer *Rev. Chem. Intermed.* **1981**, *4*, 225-257.

a chance to decay to the triplet state.[207] As to other carbenes, some react as triplets, some as singlets, and others as singlets or triplets, depending on how they are generated.

There is a limitation to the use of stereospecificity of addition as a diagnostic test for singlet or triplet carbenes.[208] When carbenes are generated by photolytic methods, they are often in a highly excited singlet state. When they add to the double bond, the addition is stereospecific; but the cyclopropane formed carries excess energy; i.e., it is in an excited state. It has been shown that under certain conditions (low pressures in the gas phase) the excited cyclopropane may undergo cis-trans isomerization *after* it is formed, so that triplet carbene may seem to be involved although in reality the singlet was present.[209]

The most common carbenes are CH_2 and CCl_2,[210] but many others have been reported, e.g.,[211]

$$\cdot \overset{\bullet}{C}Ph_2 \qquad R_2C=C=\overset{-}{C} \qquad H\overset{-}{C}-\underset{\underset{O}{\|}}{C}-R \qquad \triangleright|$$

Studies of the ir spectrum of CCl_2 trapped at low temperatures in solid argon indicate that the ground state for this species is the singlet.[212]

The geometrical structure of triplet methylene can be investigated by esr measurements,[213] since triplet species are diradicals. Such measurements made on triplet CH_2 trapped in matrices at very low temperatures (4 K) show that triplet CH_2 is a bent molecule, with an angle of about 136°.[214] Epr measurements cannot be made on singlet species, but from electronic spectra of CH_2 formed in flash photolysis of diazomethane it was concluded that singlet CH_2 is also bent, with an angle of about 103°.[215] Singlet CCl_2[212] and CBr_2[216] are also

| Triplet methylene | Singlet methylene |

bent, with angles of 100 and 114°, respectively. It has long been known that triplet aryl carbenes are bent.[217]

[207]Decay of singlet and triplet CH_2 has been detected in solution, as well as in the gas phase: Turro; Cha; Gould *J. Am. Chem. Soc.* **1987**, *109*, 2101.

[208]For other methods of distinguishing singlet from triplet carbenes, see Hendrick; Jones *Tetrahedron Lett.* **1978**, 4249, Creary *J. Am. Chem. Soc.* **1980**, *102*, 1611.

[209]Rabinovitch; Tschuikow-Roux; Schlag *J. Am. Chem. Soc.* **1959**, *81*, 1081; Frey *Proc. R. Soc. London, Ser. A* **1959**, *251*, 575. It has been reported that a singlet carbene (CBr_2) can add nonstereospecifically: Lambert; Larson; Bosch *Tetrahedron Lett.* **1983**, *24*, 3799.

[210]For reviews of halocarbenes, see Burton; Hahnfeld *Fluorine Chem. Rev.* **1977**, *8*, 119-188; Margrave; Sharp; Wilson *Fort. Chem. Forsch.* **1972**, *26*, 1-35, pp. 3-13.

[211]For reviews of unsaturated carbenes, see Stang *Acc. Chem. Res.* **1982**, *15*, 348-354; *Chem. Rev.* **1978**, *78*, 383-403. For a review of carbalkoxycarbenes, see Marchand; Brockway *Chem. Rev.* **1974**, *74*, 431-469. For a review of arylcarbenes, see Schuster *Adv. Phys. Org. Chem.* **1986**, *22*, 311-361. For a review of carbenes with neighboring hetero atoms, see Taylor *Tetrahedron* **1982**, *38*, 2751-2772.

[212]Andrews *J. Chem. Phys.* **1968**, *48*, 979.

[213]The technique of spin trapping (p. 187) has been applied to the detection of transient triplet carbenes: Forrester; Sadd *J. Chem. Soc., Perkin Trans. 2* **1982**, 1273.

[214]Wasserman; Kuck; Hutton; Yager *J. Am. Chem. Soc.* **1970**, *92*, 7491; Wasserman; Yager; Kuck *Chem. Phys. Lett.* **1970**, *7*, 409; Wasserman; Kuck; Hutton; Anderson; Yager *J. Chem. Phys.* **1971**, *54*, 4120; Bernheim; Bernard; Wang; Wood; Skell *J. Chem. Phys.* **1970**, *53*, 1280, **1971**, *54*, 3223.

[215]Herzberg; Shoosmith *Nature* **1959**, *183*, 1801; Herzberg *Proc. R. Soc. London, Ser. A* **1961**, *262*, 291; Herzberg; Johns *Proc. R. Soc. London, Ser. A* **1967**, *295*, 107, *J. Chem. Phys.* **1971**, *54*, 2276.

[216]Ivey; Schulze; Leggett; Kohl *J. Chem. Phys.* **1974**, *60*, 3174.

[217]Trozzolo; Wasserman; Yager *J. Am. Chem. Soc.* **1965**, *87*, 129; Senthilnathan; Platz *J. Am. Chem. Soc.* **1981**, *103*, 5503; Gilbert; Griller; Nazran *J. Org. Chem.* **1985**, *50*, 4738.

Flash photolysis of $CHBr_3$ produced the intermediate CBr[218]

$$CHBr_3 \xrightarrow[\text{photolysis}]{\text{flash}} \cdot \overline{C}-Br$$

This is a *carbyne*. The intermediates CF and CCl were generated similarly from $CHFBr_2$ and $CHClBr_2$, respectively.

The Generation and Fate of Carbenes[219]

Carbenes are chiefly formed in two ways, though other pathways are also known.

1. In α elimination, a carbon loses a group without its electron pair, usually a proton, and then a group with its pair, usually a halide ion:[220]

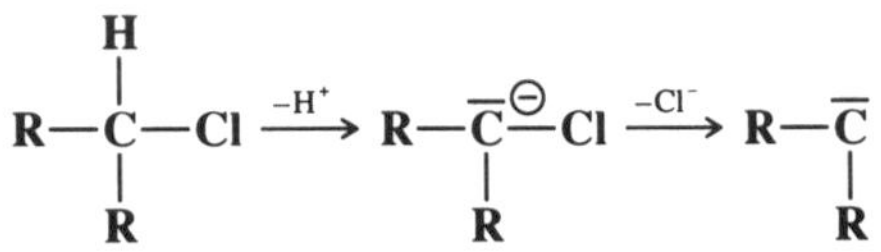

$$R-\underset{R}{\overset{H}{\underset{|}{\overset{|}{C}}}}-Cl \xrightarrow{-H^+} R-\underset{R}{\overset{|}{\underset{|}{\overline{C}}}}{\overset{\ominus}{}}Cl \xrightarrow{-Cl^-} R-\underset{R}{\overset{|}{\underset{|}{\overline{C}}}}$$

The most common example is formation of dichlorocarbene by treatment of chloroform with a base (see reaction **0-3**), but many other examples are known, a few of which are

$$CCl_3-COO^- \xrightarrow{\Delta} CCl_2 + CO_2 + Cl^- \qquad \text{Ref. 221}$$

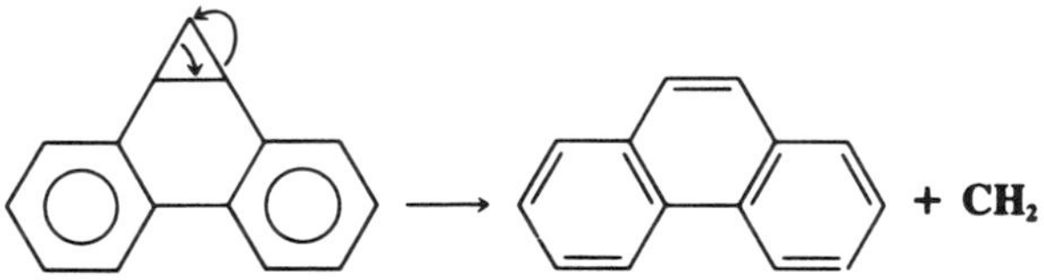

$\longrightarrow$ $+ CH_2$ 　　Ref. 222

$$Ph-Hg-\underset{F}{\overset{F}{\underset{|}{\overset{|}{C}}}}-F \xrightarrow{NaI} CF_2 + PhHgI + NaF \qquad \text{Ref. 223}$$

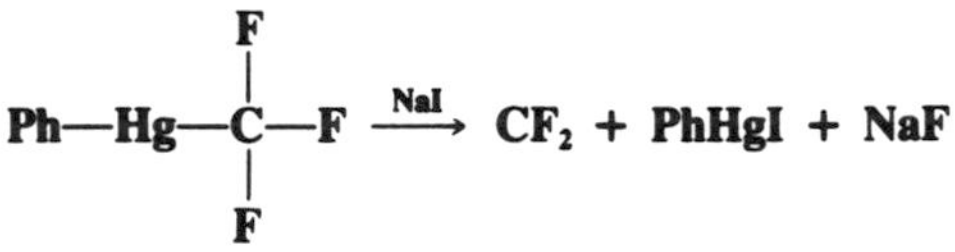

2. Disintegration of compounds containing certain types of double bonds:

$$R_2C{\overset{\frown}{\underset{\curvearrowright}{=}}}Z \longrightarrow R_2\overline{C} + \overline{Z}$$

[218]Ruzsicska; Jodhan; Choi; Strausz *J. Am. Chem. Soc.* **1983**, *105*, 2489.

[219]For reviews, see Jones *Acc. Chem. Res.* **1974**, *7*, 415-421; Kirmse, in Bamford; Tipper *Comprehensive Chemical Kinetics*, vol. 9; Elsevier: New York, 1973, pp. 373-415; Ref. 200. For a review of electrochemical methods of carbene generation, see Petrosyan; Niyazymbetov *Russ. Chem. Rev.* **1989**, *58*, 644-653.

[220]For a review of formation of carbenes in this manner, see Kirmse *Angew. Chem. Int. Ed. Engl.* **1965**, *4*, 1-10 [*Angew. Chem. 77*, 1-10].

[221]Wagner *Proc. Chem. Soc.* **1959**, 229.

[222]Richardson; Durrett; Martin; Putnam; Slaymaker; Dvoretzky *J. Am. Chem. Soc.* **1965**, *87*, 2763. For reviews of this type of reaction, see Hoffmann *Angew. Chem. Int. Ed. Engl.* **1971**, *10*, 529,537 [*Angew. Chem. 83*, 595-603]; Griffin *Angew. Chem. Int. Ed. Engl.* **1971**, *10*, 537-547 [*Angew. Chem. 83*, 604-613]. See also Hoffmann *Acc. Chem. Res.* **1985**, *18*, 248-253.

[223]Seyferth; Hopper; Darragh *J. Am. Chem. Soc.* **1969**, *91*, 6536; Seyferth *Acc. Chem. Res.* **1972**, *5*, 65-74.

The two most important ways of forming CH_2 are examples: the photolysis of ketene

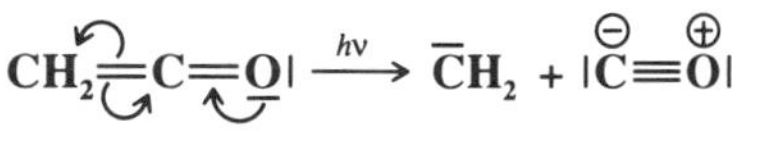

$$CH_2 = C = O| \xrightarrow{h\nu} \bar{C}H_2 + |\overset{\ominus}{C} \equiv \overset{\oplus}{O}|$$

and the isoelectronic decomposition of diazomethane.[224]

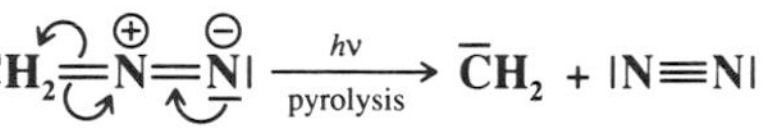

$$CH_2 = \overset{\oplus}{N} = \overset{\ominus}{N}| \xrightarrow[\text{pyrolysis}]{h\nu} \bar{C}H_2 + |N \equiv N|$$

Diazirines (isomeric with diazoalkanes) also give carbenes:[225]

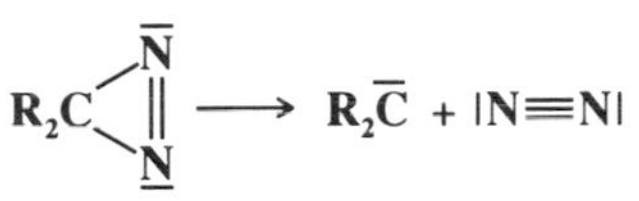

$$R_2C \overset{\bar{N}}{\underset{N}{\big\langle}} \Big\| \longrightarrow R_2\bar{C} + |N \equiv N|$$

Because most carbenes are so reactive, it is often difficult to prove that they are actually present in a given reaction. In many instances where a carbene is *apparently* produced by an α elimination or by disintegration of a double-bond compound there is evidence that no free carbene is actually involved. The neutral term *carbenoid* is used where it is known that a free carbene is not present or in cases where there is doubt. α-Halo organometallic compounds R_2CXM are often called carbenoids because they readily give α elimination reactions[226] (for example, see **2-39**).

The reactions of carbenes are more varied than those of the species previously discussed in this chapter.

1. Additions to carbon–carbon double bonds have already been mentioned. Carbenes also add to aromatic systems, but the immediate products rearrange, usually with ring enlargement (see **5-50**). Additions of carbenes to other double bonds, such as C=N (**6-61** and **6-62**), and to triple bonds have also been reported.

2. An unusual reaction of carbenes is that of insertion into C—H bonds (**2-20**). Thus CH_2 reacts with methane to give ethane and with propane to give *n*-butane and isobutane.

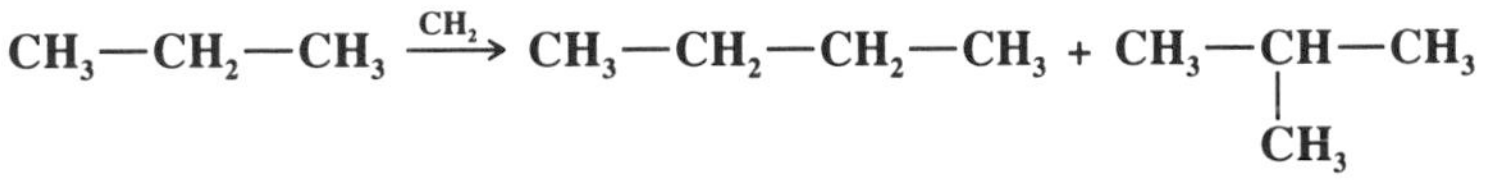

$$CH_3-CH_2-CH_3 \xrightarrow{CH_2} CH_3-CH_2-CH_2-CH_3 + CH_3-\underset{\underset{CH_3}{|}}{CH}-CH_3$$

This reaction is virtually useless for synthetic purposes but illustrates the extreme reactivity of carbene. Treatment in the liquid phase of an alkane such as pentane with carbene formed from the photolysis of diazomethane gives the three possible products in statistical ratios[227] demonstrating that carbene is displaying no selectivity. For many years, it was a generally accepted principle that the lower the selectivity the greater the reactivity; however, this

[224]For a review, see Regitz; Maas *Diazo Compounds*; Academic Press: New York, 1986, pp. 170-184.
[225]For a treatise, see Liu *Chemistry of Diazirines*, 2 vols.; CRC Press: Boca Raton, FL, 1987. For reviews, see Liu *Chem. Soc. Rev.* **1982**, *11*, 127-140; Frey *Adv. Photochem.* **1966**, *4*, 225-256.
[226]For a review, see Nefedov; D'yachenko; Prokof'ev *Russ.Chem. Rev.* **1977**, *46*, 941-966.
[227]Doering; Buttery; Laughlin; Chaudhuri *J. Am. Chem. Soc.* **1956**, *78*, 3224; Richardson; Simmons; Dvoretzky *J. Am. Chem. Soc.* **1961**, *83*, 1934; Halberstadt; McNesby *J. Am. Chem. Soc.* **1967**, *89*, 3417.

principle is no longer regarded as general because many exceptions have been found.[228] Singlet CH_2 generated by photolysis of diazomethane is probably the most reactive organic species known, but triplet CH_2 is somewhat less reactive, and other carbenes are still less reactive. The following series of carbenes of decreasing reactivity has been proposed on the basis of discrimination between insertion and addition reactions: CH_2 > HCCOOR > PhCH > BrCH ≈ ClCH.[229] Dihalocarbenes generally do not give insertion reactions at all. Insertion of carbenes into other bonds has also been demonstrated, though not insertion into C—C bonds.[229a]

Two carbenes that are stable at room temperature have been reported.[230] These are **22** and **23**. In the absence of oxygen and moisture **22** exists as stable crystals with a melting point of 240-241°C.[230a] Its structure was proved by x-ray crystallography. **23**, which is in resonance with an ylide form and with a form containing a P≡C bond, is a red oil that

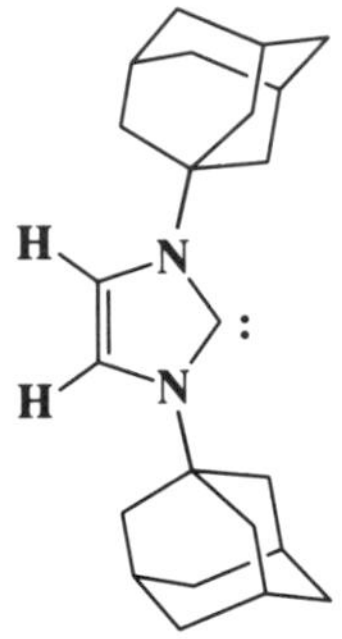

22

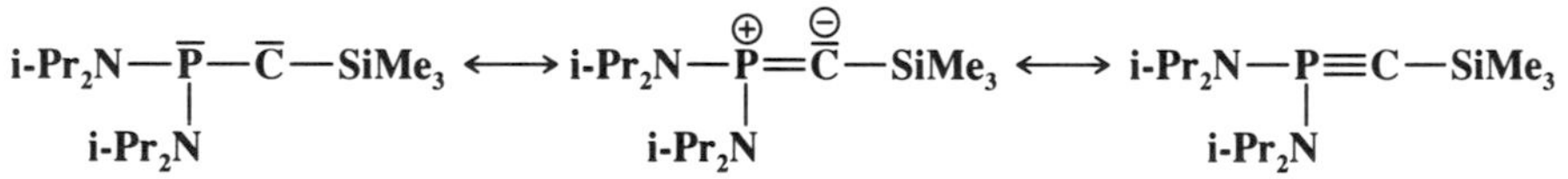

23

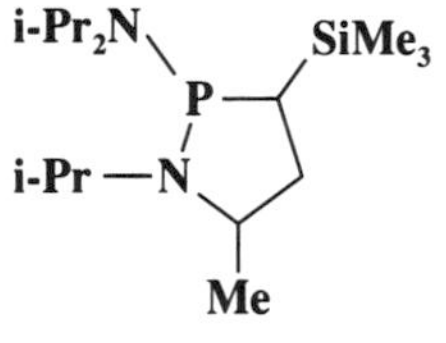

24

[228]For reviews of this question, see Buncel; Wilson *J. Chem. Educ.* **1987**, *64*, 475-480; Johnson *Tetrahedron* **1980**, *36*, 3461-3480, *Chem. Rev.* **1975**, *75*, 755-765; Giese *Angew. Chem. Int. Ed. Engl.* **1977**, *16*, 125-136 [*Angew. Chem.* **89**, 162-173]; Pross *Adv. Phys. Org. Chem.* **1977**, *14*, 69-132. See also Ritchie; Sawada *J. Am. Chem. Soc.* **1977**, *99*, 3754; Argile; Ruasse *Tetrahedron Lett.* **1980**, *21*, 1327; Godfrey *J. Chem. Soc., Perkin Trans. 2* **1981**, 645; Kurz; El-Nasr *J. Am. Chem. Soc.* **1982**, *104*, 5823; Srinivasan; Shunmugasundaram; Arumugam *J. Chem. Soc., Perkin Trans. 2* **1985**, 17; Bordwell; Branca; Cripe *Isr. J. Chem.* **1985**, *26*, 357; Formosinho *J. Chem. Soc., Perkin Trans. 2* **1988**, 839; Johnson; Stratton *J. Chem. Soc., Perkin Trans. 2* **1988**, 1903. For a group of papers on this subject, see *Isr. J. Chem.* **1985**, *26*, 303-428.

[229]Closs; Coyle *J. Am. Chem. Soc.* **1965**, *87*, 4270.

[229a]See, for example, Doering; Knox; Jones *J. Org. Chem.* **1959**, *24*, 136; Franzen *Liebigs Ann. Chem.* **1959**, *627*, 22; Bradley; Ledwith *J. Chem. Soc.* **1961**, 1495; Frey; Voisey *Chem. Commun.* **1966**, 454; Seyferth; Damrauer; Mui; Jula *J. Am. Chem. Soc.* **1968**, *90*, 2944; Tomioka; Ozaki; Izawa *Tetrahedron* **1985**, *41*, 4987; Frey; Walsh; Watts *J. Chem. Soc., Chem. Commun.* **1989**, 284.

[230]For a discussion, see Regitz *Angew. Chem. Int. Ed. Engl.* **1991**, *30*, 674 [*Angew. Chem. 103*, 691].

[230a]Arduengo; Harlow; Kline *J. Am. Chem. Soc.* **1991**, *113*, 361.

undergoes internal insertion (the carbene carbon inserts into one of the C—H bonds of an isopropyl group to give **24**) when heated to 300°C.[231]

3. It would seem that dimerization should be an important reaction of carbenes

$$R_2\overline{C} + R_2\overline{C} \longrightarrow R_2C=CR_2$$

but it is not, because the reactivity is so great that the carbene species do not have time to find each other and because the dimer generally has so much energy that it dissociates again. Apparent dimerizations have been observed, but it is likely that the products in many reported instances of "dimerization" do not arise from an actual dimerization of two carbenes but from attack by a carbene on a molecule of carbene precursor, e.g.,

$$R_2\overline{C} + R_2CN_2 \longrightarrow R_2C=CR_2 + N_2$$

4. Alkylcarbenes can undergo rearrangement, with migration of alkyl or hydrogen.[231a] Indeed these rearrangements are generally so rapid[232] that additions to multiple bonds and insertion reactions, which are so common for CH_2, are seldom encountered with alkyl or dialkyl carbenes. Unlike rearrangement of the species previously encountered in this chapter, most rearrangements of carbenes directly give stable molecules. Some examples are

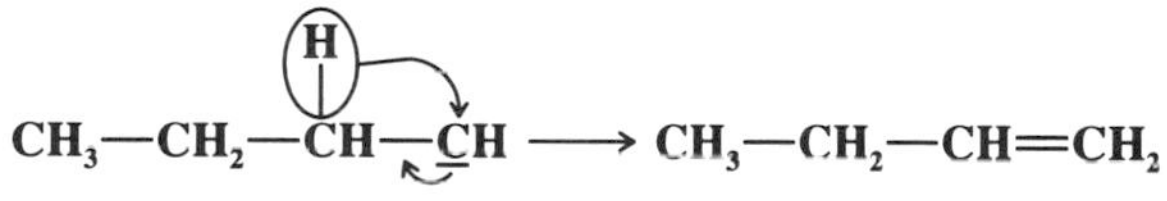

$$CH_3—CH_2—CH—\overset{..}{C}H \longrightarrow CH_3—CH_2—CH=CH_2 \qquad \text{Ref. 233}$$

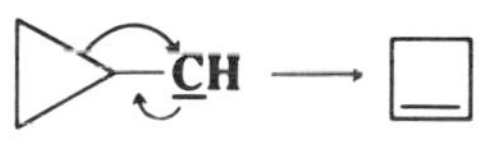

Ref. 234

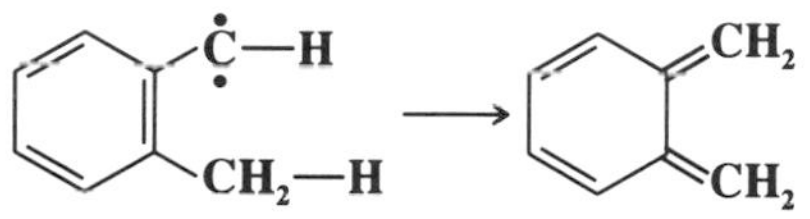

Ref. 235

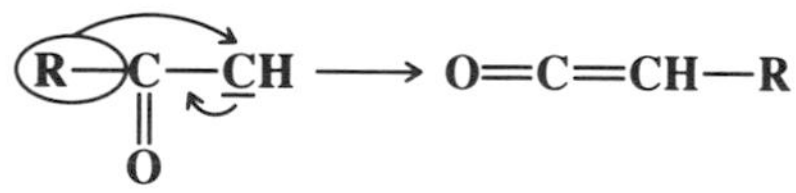

Ref. 236

The rearrangement of acylcarbenes to ketenes is called the Wolff rearrangement (**8-8**). A few rearrangements in which carbenes rearrange to other carbenes are also known.[237] Of course, the new carbene must stabilize itself in one of the ways we have mentioned.

[231]Igau; Grutzmacher; Baceiredo; Bertrand *J. Am. Chem. Soc.* **1991**, *113*, 6463; Igau; Baceiredo; Trinquier; Bertrand *Angew. Chem. Int. Ed. Engl.* **1989**, *28*, 621 [*Angew. Chem. 101*, 617]. See also Gillette; Baceiredo; Bertrand *Angew. Chem. Int. Ed. Engl.* **1990**, *29*, 1429 [*Angew. Chem. 102*, 1486].
[231a]For reviews of carbene and nitrene rearrangements, see Brown, Ref. 189, pp. 115-163; Wentrup *Adv. Heterocycl. Chem.* **1981**, *28*, 231-361, *React. Intermed. (Plenum)* **1980**, *1*, 263-319, *Top. Curr. Chem.* **1976**, *62*, 173-251; Jones, in de Mayo, Ref. 91, vol. 1, pp. 95-160; Schaefer *Acc. Chem. Res.* **1979**, *12*, 288-296; Kirmse, Ref. 200, pp. 457-496.
[232]The activation energy for the 1,2-hydrogen shift has been estimated at 1.1 kcal/mole (4.5 kJ/mol), an exceedingly low value: Stevens; Liu; Soundararajan; Paike *Tetrahedron Lett.* **1989**, *30*, 481.
[233]Kirmse; Doering *Tetrahedron* **1960**, *11*, 266. For kinetic studies of the rearrangement: Cl—C̄—CHR₂ → ClCH=CR₂, see Liu; Bonneau *J. Am. Chem. Soc.* **1989**, *111*, 6873; Jackson; Soundararajan; White; Liu; Bonneau; Platz *J. Am. Chem. Soc.* **1989**, *111*, 6874; Ho; Krogh-Jespersen; Moss; Shen; Sheridan; Subramanian *J. Am. Chem. Soc.* **1989**, *111*, 6875; LaVilla; Goodman *J. Am. Chem. Soc.* **1989**, *111*, 6877.
[234]Friedman; Shechter *J. Am. Chem. Soc.* **1960**, *82*, 1002.
[235]McMahon; Chapman *J. Am. Chem. Soc.* **1987**, *109*, 683.
[236]Friedman; Berger *J. Am. Chem. Soc.* **1961**, *83*, 492, 500.
[237]For a review, see Jones *Acc. Chem. Res.* **1977**, *10*, 353-359.

5. Triplet carbenes can abstract hydrogen or other atoms to give free radicals, e.g.,

$$\cdot \overset{..}{C}H_2 + CH_3CH_3 \longrightarrow CH_3\cdot + \cdot CH_2CH_3$$

This is not surprising, since triplet carbenes are free radicals. But singlet carbenes can also give this reaction, though in this case only halogen atoms are abstracted, not hydrogen.[238]

NITRENES

Nitrenes,[239] R—N, are the nitrogen analogs of carbenes, and most of what we have said about carbenes also applies to them. Nitrenes are too reactive for isolation under ordinary conditions. Alkyl nitrenes have been isolated by trapping in matrices at 4 K,[240] while aryl nitrenes, which are less reactive, can be trapped at 77 K.[241] The ground state of NH, and probably of most nitrenes,[242] is a triplet, though nitrenes can be generated in both triplet and singlet states. In additions of EtOOC—N to C=C double bonds two species are involved,

$$R—\overset{..}{\underset{..}{N}} \qquad R—\overset{..}{\underset{\cdot}{N}}\cdot$$

Singlet   Triplet

one of which adds stereospecifically and the other not. By analogy with Skell's proposal involving carbenes (p. 196) these are taken to be the singlet and triplet species, respectively.[243]

The two principal means of generating nitrenes are analogous to those used to form carbenes.

1. *Elimination.*   An example is

$$R—\underset{\overset{|}{H}}{N}—OSO_2Ar \overset{base}{\longrightarrow} R—N + B—H + ArSO_2O^-$$

2. *Breakdown of certain double-bond compounds.*   The most common method of forming nitrenes is photolytic or thermal decomposition of azides,[244]

$$R—\overset{..}{\underset{..}{N}}=\overset{\oplus}{N}=\overset{..}{\underset{..}{N}}{}^{\ominus} \overset{\Delta \; or \; h\nu}{\longrightarrow} R—N + N_2$$

[238]Roth *J. Am. Chem. Soc.* **1971,** *93,* 1527, 4935, *Acc. Chem. Res.* **1977,** *10,* 85-91.

[239]For monographs, see Scriven *Azides and Nitrenes;* Academic Press: New York, 1984; Lwowski *Nitrenes;* Wiley: New York, 1970. For reviews, see Scriven *React. Intermed. (Plenum)* **1982,** *2,* 1-54; Lwowski *React. Intermed. (Wiley)* **1985,** *3,* 305-332, **1981,** *2,* 315-334, **1978,** *1,* 197-227, *Angew. Chem. Int. Ed. Engl.* **1967,** *6,* 897-906 [*Angew. Chem.* **79,** 922-931]; Abramovitch, in McManus, Ref. 1, pp. 127-192; Hünig *Helv. Chim. Acta* **1971,** *54,* 1721-1747; Belloli *J. Chem. Educ.* **1971,** *48,* 422-426; Kuznetsov; Ioffe *Russ. Chem. Rev.* **1989,** *58,* 732-746 (N- and O-nitrenes); Meth-Cohn *Acc. Chem. Res.* **1987,** *20,* 18-27 (oxycarbonylnitrenes); Abramovitch; Sutherland *Fortsch. Chem. Forsch.* **1970,** *16,* 1-33 (sulfonyl nitrenes); Ioffe; Kuznetsov *Russ. Chem. Rev.* **1972,** *41,* 131-146 (N-nitrenes).

[240]Wasserman; Smolinsky; Yager *J. Am. Chem. Soc.* **1964,** *86,* 3166. For the structure of CH_3—N, as determined in the gas phase, see Carrick; Brazier; Bernath; Engelking *J. Am. Chem. Soc.* **1987,** *109,* 5100.

[241]Smolinsky; Wasserman; Yager *J. Am. Chem. Soc.* **1962,** *84,* 3220. For a review, see Sheridan, Ref. 201, pp. 159-248.

[242]A few nitrenes have been shown to have singlet ground states. See Sigman; Autrey; Schuster *J. Am. Chem. Soc.* **1988,** *110,* 4297.

[243]McConaghy; Lwowski *J. Am. Chem. Soc.* **1967,** *89,* 2357, 4450; Mishra; Rice; Lwowski *J. Org. Chem.* **1968,** *33,* 481.

[244]For reviews, see Dyall, in Patai; Rappoport *The Chemistry of Functional Groups, Supplement D,* pt. 1; Wiley: New York, 1983, pp. 287-320; Dürr; Kober *Top. Curr. Chem.* **1976,** *66,* 89-114; L'Abbé *Chem. Rev.* **1969,** *69,* 345-363.

The unsubstituted nitrene NH has been generated by photolysis of or electric discharge through NH_3, N_2H_4, or HN_3.

The reactions of nitrenes are also similar to those of carbenes.[245] As in that case, many reactions in which nitrene intermediates are suspected probably do not involve free nitrenes. It is often very difficult to obtain proof in any given case that a free nitrene is or is not an intermediate.

1. *Insertion* (see **2-12**).   Nitrenes, especially acyl nitrenes and sulfonyl nitrenes, can insert into C—H and certain other bonds, e.g.,

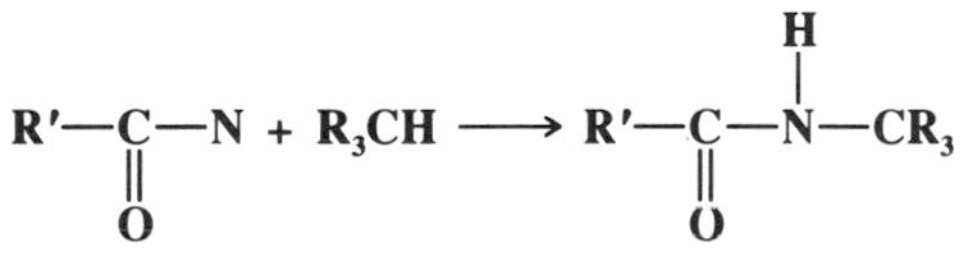

$$R'-\underset{\underset{O}{\|}}{C}-N\ +\ R_3CH\ \longrightarrow\ R'-\underset{\underset{O}{\|}}{C}-\overset{\overset{H}{|}}{N}-CR_3$$

2. *Addition to C=C bonds* (see **5-42**):

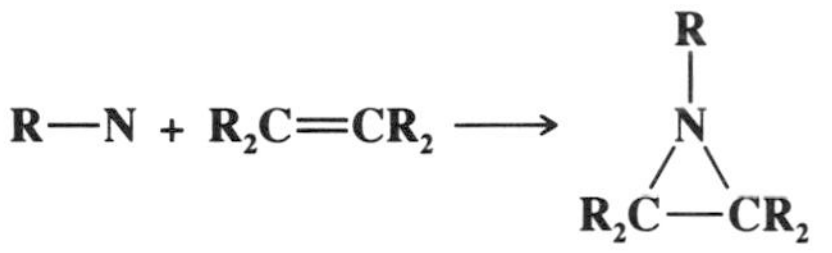

$$R-N\ +\ R_2C=CR_2\ \longrightarrow\ \overset{\overset{R}{|}}{\underset{R_2C-CR_2}{N}}$$

3. *Rearrangements.*[231a]   Alkyl nitrenes do not generally give either of the two preceding reactions because rearrangement is more rapid, e.g.,

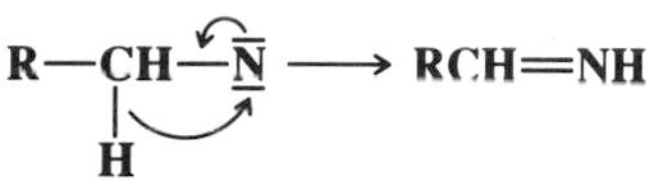

$$R-\underset{\underset{H}{|}}{CH}-\overset{\frown}{\underline{N}}\ \longrightarrow\ RCH=NH$$

Such rearrangements are so rapid that it is usually difficult to exclude the possibility that a free nitrene was never present at all, i.e., that migration takes place at the same time that the nitrene is formed[246] (see p. 1091).

4. *Abstraction*, e.g.,

$$R-N\ +\ R-H\ \longrightarrow\ R-\overset{\cdot}{\underset{}{N}}-H\ +\ R\cdot$$

5. *Dimerization.*   One of the principal reactions of NH is dimerization to diimide N_2H_2. Azobenzenes are often obtained in reactions where aryl nitrenes are implicated:[247]

$$2Ar-N\ \longrightarrow\ Ar-N=N-Ar$$

It would thus seem that dimerization is more important for nitrenes than it is for carbenes, but again it has not been proved that free nitrenes are actually involved.

---

[245]For a discussion of nitrene reactivity, see Subbaraj; Subba Rao; Lwowski *J. Org. Chem.* **1989**, *54*, 3945.

[246]For example, see Moriarty; Reardon *Tetrahedron* **1970**, *26*, 1379; Abramovitch; Kyba *J. Am. Chem. Soc.* **1971**, *93*, 1537.

[247]See, for example, Leyva Platz; Persy; Wirz *J. Am. Chem. Soc.* **1986**, *108*, 3783.

At least two types of *nitrenium ions*, the nitrogen analogs of carbocations, can exist as intermediates, though much less work has been done in this area than on carbocations. In one type (**25**) the nitrogen is bonded to two atoms and in the other (**26**) to only one atom.[248]

$$R-\overset{\oplus}{\underset{}{\underline{N}}}-R' \qquad R-\overset{}{\underset{\underset{R'}{|}}{C}}=\underline{N}^{\oplus}$$

$$\textbf{25} \qquad\qquad \textbf{26}$$

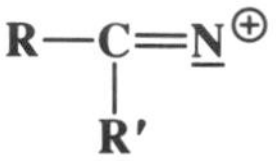

When R = H in **25** the species is a protonated nitrene. Like carbenes and nitrenes, nitrenium ions can exist in singlet or triplet states.[249]

[248]For reviews of **25**, see Abramovitch; Jeyaraman, in Scriven *Azides and Nitrenes*, Ref. 239, pp. 297-357; Gassman *Acc. Chem. Res.* **1970,** *3*, 26-33. For a review of **26**, see Lansbury, in Lwowski *Nitrenes*, Ref. 239, pp. 405-419.
[249]Gassman; Cryberg *J. Am. Chem. Soc.* **1969,** *91*, 5176.

6

MECHANISMS AND
METHODS OF
DETERMINING THEM

A mechanism is the actual process by which a reaction takes place—which bonds are broken, in what order, how many steps are involved, the relative rate of each step, etc. In order to state a mechanism completely, we should have to specify the positions of all atoms, including those in solvent molecules, and the energy of the system, at every point in the process. A proposed mechanism must fit all the facts available. It is always subject to change as new facts are discovered. The usual course is that the gross features of a mechanism are the first to be known and then increasing attention is paid to finer details. The tendency is always to probe more deeply, to get more detailed descriptions.

Although for most reactions gross mechanisms can be written today with a good degree of assurance, no mechanism is known completely. There is much about the fine details which is still puzzling, and for some reactions even the gross mechanism is not yet clear. The problems involved are difficult because there are so many variables. Many examples are known where reactions proceed by different mechanisms under different conditions. In some cases there are several proposed mechanisms, each of which completely explains all the data.

Types of Mechanism

In most reactions of organic compounds one or more covalent bonds are broken. We can divide organic mechanisms into three basic types, depending on how the bonds break.

1. If a bond breaks in such a way that both electrons remain with one fragment, the mechanism is called *heterolytic*. Such reactions do not necessarily involve ionic intermediates, though they usually do. The important thing is that the electrons are never unpaired. For most reactions it is convenient to call one reactant the *attacking reagent* and the other the *substrate*. In this book we shall always designate as the substrate that molecule that supplies carbon to the new bond. When carbon–carbon bonds are formed, it is necessary to be arbitrary about which is the substrate and which the attacking reagent. In heterolytic reactions the reagent generally brings a pair of electrons to the substrate or takes a pair of electrons from it. A reagent that brings an electron pair is called a *nucleophile* and the reaction is *nucleophilic*. A reagent that takes an electron pair is called an *electrophile* and the reaction is *electrophilic*. In a reaction in which the substrate molecule becomes cleaved, part of it (the part not containing the carbon) is usually called the *leaving group*. A leaving group that carries away an electron pair is called a *nucleofuge*. If it comes away without the electron pair, it is called an *electrofuge*.

2. If a bond breaks in such a way that each fragment gets one electron, free radicals are formed and such reactions are said to take place by *homolytic* or *free-radical mechanisms*.

3. It would seem that all bonds must break in one of the two ways previously noted. But there is a third type of mechanism in which electrons (usually six, but sometimes some other number) move in a closed ring. There are no intermediates, ions or free radicals, and it is impossible to say whether the electrons are paired or unpaired. Reactions with this type of mechanism are called *pericyclic*.[1]

Examples of all three types of mechanisms are given in the next section.

Types of Reaction

The number and range of organic reactions is so great as to seem bewildering, but actually almost all of them can be fitted into just six categories. In the description of the six types that follows, the immediate products are shown, though in many cases they then react with something else. All the species are shown without charges, since differently charged reactants can undergo analogous changes. The descriptions given here are purely formal and are for the purpose of classification and comparison. All are discussed in detail in Part 2 of this book.

1. *Substitutions.*   If heterolytic, these can be classified as nucleophilic or electrophilic depending on which reactant is designated as the substrate and which as the attacking reagent (very often Y must first be formed by a previous bond cleavage).
   a. Nucleophilic substitution (Chapters 10, 13).

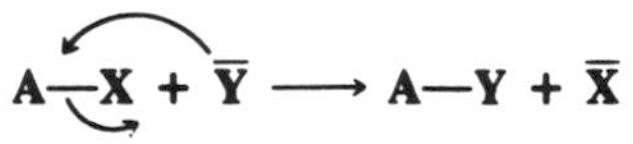

$$A{-}X + \overline{Y} \longrightarrow A{-}Y + \overline{X}$$

   b. Electrophilic substitution (Chapters 11, 12).

$$A{-}X + Y \longrightarrow A{-}Y + X$$

   c. Free-radical substitution (Chapter 14).

$$A{-}X + Y{\cdot} \longrightarrow A{-}Y + X{\cdot}$$

In free-radical substitution, Y· is usually produced by a previous free-radical cleavage, and X· goes on to react further.
   2. *Additions to double or triple bonds* (Chapters 15, 16).   These reactions can take place by all three of the mechanistic possibilities.
   a. Electrophilic addition (heterolytic).

$$A{=}B + Y{-}W \longrightarrow A{-}\overset{Y}{\underset{|}{B}} + \overline{W} \longrightarrow A{-}\overset{W\ \ Y}{\underset{|\ \ \ |}{B}}$$

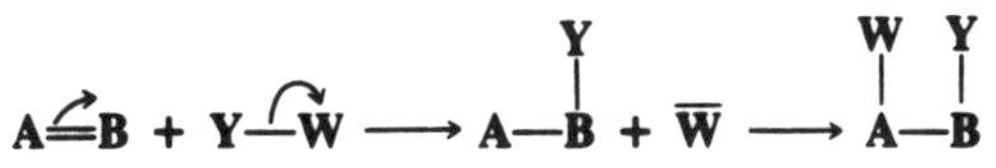

   b. Nucleophilic addition (heterolytic).

$$A{=}B + Y{-}W \longrightarrow \overline{A}{-}\overset{Y}{\underset{|}{B}} + W \longrightarrow A{-}\overset{W\ \ Y}{\underset{|\ \ \ |}{B}}$$

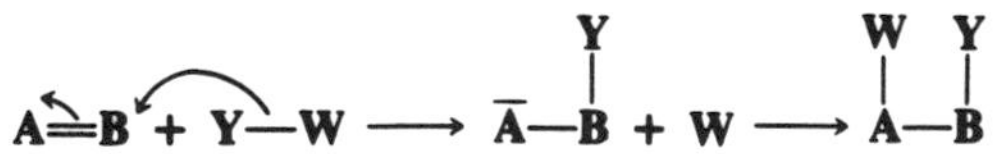

[1]For a classification of pericyclic reactions, see Hendrickson *Angew. Chem. Int. Ed. Engl.* **1974**, *13*, 47-76 [*Angew. Chem.* *86*, 71-100].

c. Free-radical addition (homolytic).

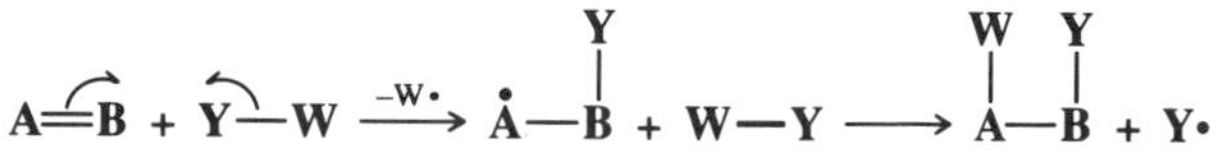

$$A\overset{\frown}{=}B + Y\overset{\frown}{-}W \xrightarrow{-W\cdot} \overset{\cdot}{A}-\overset{|}{\underset{}{B}}\overset{Y}{|} + W-Y \longrightarrow \overset{W}{\underset{|}{A}}-\overset{Y}{\underset{|}{B}} + Y\cdot$$

d. Simultaneous addition (pericyclic).

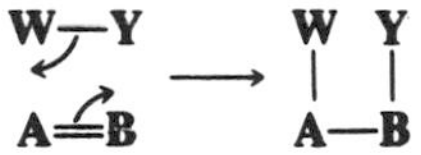

$$\overset{W-Y}{\underset{A\overset{}{=}B}{\overset{\swarrow}{\curvearrowleft}}} \longrightarrow \overset{W}{\underset{A}{\underset{|}{}}}\overset{Y}{\underset{B}{\underset{|}{}}}$$

The examples show Y and W coming from the same molecule, but very often (except in simultaneous addition) they come from different molecules. Also, the examples show the Y—W bond cleaving at the same time that Y is bonding to B, but often (again except for simultaneous addition) this cleavage takes place earlier.

3. β *Elimination* (Chapter 17).

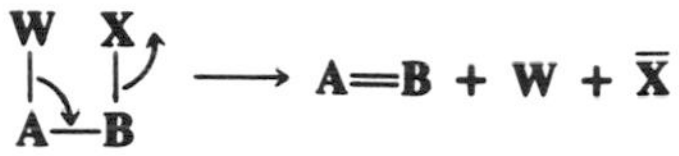

$$\overset{W}{\underset{A}{\underset{\curvearrowleft}{|}}}\overset{X}{\underset{B}{\underset{}{\nearrow}}} \longrightarrow A{=}B + W + \overline{X}$$

These reactions can take place by either heterolytic or pericyclic mechanisms. Examples of the latter are shown on p. 1006. Free-radical β eliminations are extremely rare. In heterolytic eliminations W and X may or may not leave simultaneously and may or may not combine.

4. *Rearrangement* (Chapter 18). Many rearrangements involve migration of an atom or group from one atom to another. There are three types, depending on how many electrons the migrating atom or group carries with it.

a. Migration with electron pair (nucleophilic).

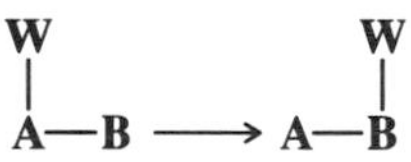

$$\overset{W}{\underset{A}{\underset{}{|}}}-B \longrightarrow A-\overset{W}{\underset{B}{\underset{}{|}}}$$

b. Migration with one electron (free-radical).

$$\overset{W}{\underset{A}{\underset{}{|}}}-\overset{\cdot}{B} \longrightarrow \overset{\cdot}{A}-\overset{W}{\underset{B}{\underset{}{|}}}$$

c. Migration without electrons (electrophilic; rare).

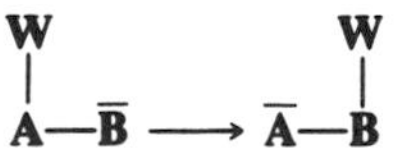

$$\overset{W}{\underset{A}{\underset{}{|}}}-\overline{B} \longrightarrow \overline{A}-\overset{W}{\underset{B}{\underset{}{|}}}$$

The illustrations show 1,2 rearrangements, in which the migrating group moves to the adjacent atom. These are the most common, although longer rearrangements are also possible. There are also some rearrangements that do not involve simple migration at all (see Chapter 18). Some of the latter involve pericyclic mechanisms.

5. *Oxidation and reduction* (Chapter 19).   Many oxidation and reduction reactions fall naturally into one of the four types mentioned above, but many others do not. For a description of oxidation–reduction mechanistic types, see p. 1159.

6. Combinations of the above.

Note that arrows are used to show movement of *electrons*. An arrow always follows the motion of electrons and never of a nucleus or anything else (it is understood that the rest of the molecule follows the electrons). Ordinary arrows (double-headed) follow electron pairs, while single-headed arrows follow unpaired electrons. Double-headed arrows are also used in pericyclic reactions for convenience, though in these reactions we do not really know how or in which direction the electrons are moving.

Thermodynamic Requirements for Reaction

In order for a reaction to take place spontaneously, the free energy of the products must be lower than the free energy of the reactants; i.e., ΔG must be negative. Reactions can go the other way, of course, but only if free energy is added. Like water on the surface of the earth, which only flows downhill and never uphill (though it can be carried or pumped uphill), molecules seek the lowest possible potential energy. Free energy is made up of two components, enthalpy H and entropy S. These quantities are related by the equation

$$\Delta G = \Delta H - T\Delta S$$

The enthalpy change in a reaction is essentially the difference in bond energies (including resonance, strain, and solvation energies) between the reactants and the products. The enthalpy change can be calculated by totaling the bond energies of all the bonds broken, subtracting from this the total of the bond energies of all the bonds formed, and adding any changes in resonance, strain, or solvation energies. Entropy changes are quite different, and refer to the disorder or randomness of the system. The less order in a system, the greater the entropy. The preferred conditions in nature are *low* enthalpy and *high* entropy, and in reacting systems, enthalpy spontaneously decreases while entropy spontaneously increases.

For many reactions entropy effects are small and it is the enthalpy that mainly determines whether the reaction can take place spontaneously. However, in certain types of reaction entropy is important and can dominate enthalpy. We shall discuss several examples.

1. In general, liquids have lower entropies than gases, since the molecules of gas have much more freedom and randomness. Solids, of course, have still lower entropies. Any reaction in which the reactants are all liquids and one or more of the products is a gas is therefore thermodynamically favored by the increased entropy; the equilibrium constant for that reaction will be higher than it would otherwise be. Similarly, the entropy of a gaseous substance is higher than that of the same substance dissolved in a solvent.

2. In a reaction in which the number of product molecules is equal to the number of reactant molecules, e.g., A + B → C + D, entropy effects are usually small, but if the number of molecules is increased, e.g., A → B + C, there is a large gain in entropy because more arrangements in space are possible when more molecules are present. Reactions in which a molecule is cleaved into two or more parts are therefore thermodynamically favored by the entropy factor. Conversely, reactions in which the number of product molecules is less than the number of reactant molecules show entropy decreases, and in such cases there must be a sizable decrease in enthalpy to overcome the unfavorable entropy change.

3. Although reactions in which molecules are cleaved into two or more pieces have favorable entropy effects, many potential cleavages do not take place because of large increases in enthalphy. An example is cleavage of ethane into two methyl radicals. In this case a bond of about 79 kcal/mol (330 kJ/mol) is broken, and no new bond is formed to compensate for this enthalpy increase. However, ethane can be cleaved at very high temperatures, which illustrates the principle that *entropy becomes more important as the temperature increases*, as is obvious from the equation $\Delta G = \Delta H - T\Delta S$. The enthalpy term is independent of temperature, while the entropy term is directly proportional to the absolute temperature.

4. An acyclic molecule has more entropy than a similar cyclic molecule because there are more conformations (compare hexane and cyclohexane). Ring opening therefore means a gain in entropy and ring closing a loss.

Kinetic Requirements for Reaction

Just because a reaction has a negative ΔG does not necessarily mean that it will take place in a reasonable period of time. A negative ΔG is a *necessary* but not a *sufficient* condition for a reaction to occur spontaneously. For example, the reaction between H_2 and O_2 to give H_2O has a large negative ΔG, but mixtures of H_2 and O_2 can be kept at room temperature for many centuries without reacting to any significant extent. In order for a reaction to take place, *free energy of activation* ΔG^+ must be added.[2] This situation is illustrated in Figure 6.1,[3] which is an energy profile for a one-step reaction without an intermediate. In this type of diagram the horizontal axis (called the *reaction coordinate*)[4] signifies the progression of the reaction. $\Delta G_f^{\ddagger}$ is the free energy of activation for the forward reaction. If the reaction shown in Figure 6.1 is reversible, $\Delta G_r^{\ddagger}$ must be greater than $\Delta G_f^{\ddagger}$, since it is the sum of ΔG and $\Delta G_f^{\ddagger}$.

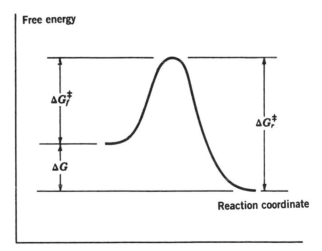

FIGURE 6.1   Free-energy profile of a reaction without an intermediate where the products have a lower free energy than the reactants.

---

[2]For mixtures of H_2 and O_2 this can be done by striking a match.
[3]Strictly speaking, this is an energy profile for a reaction of the type XY + Z → X + YZ. However, it may be applied, in an approximate way, to other reactions.
[4]For a review of reaction coordinates and structure–energy relationships, see Grunwald *Prog. Phys. Org. Chem.* **1990**, *17*, 55-105.

When a reaction between two or more molecules has progressed to the point corresponding to the top of the curve, the term *transition state* is applied to the positions of the nuclei and electrons. The transition state possesses a definite geometry and charge distribution but has no finite existence; the system passes through it. The system at this point is called an *activated complex*.[5]

In the *transition-state theory*[6] the starting materials and the activated complex are taken to be in equilibrium, the equilibrium constant being designated K^+. According to the theory, all activated complexes go on to product at the same rate (which, though at first sight surprising, is not unreasonable, when we consider that they are all "falling downhill") so that the rate constant (see p. 220) of the reaction depends only on the position of the equilibrium between the starting materials and the activated complex, i.e., on the value of K^+. ΔG^+ is related to K^+ by

$$\Delta G^+ = -2.3RT \log K^+$$

so that a higher value of ΔG^+ is associated with a smaller rate constant. The rates of nearly all reactions increase with increasing temperature because the additional energy thus supplied helps the molecules to overcome the activation energy barrier. Some reactions have no free energy of activation at all, meaning that K^+ is essentially infinite and that virtually all collisions lead to reaction. Such processes are said to be *diffusion-controlled*.[7]

Like ΔG, ΔG^+ is made up of enthalpy and entropy components

$$\Delta G^+ = \Delta H^+ - T\Delta S^+$$

ΔH^+, the *enthalpy of activation*, is the difference in bond energies, including strain, resonance, and solvation energies, between the starting compounds and the *transition state*. In many reactions bonds have been broken or partially broken by the time the transition state is reached; the energy necessary for this is ΔH^+. It is true that additional energy will be supplied by the formation of new bonds, but if this occurs after the transition state, it can affect only ΔH and not ΔH^+.

Entropy of activation ΔS^+, which is the difference in entropy between the starting compounds and the transition state, becomes important when two reacting molecules must approach each other in a specific orientation in order for the reaction to take place. For example, the reaction between a simple noncyclic alkyl chloride and hydroxide ion to give an alkene (**7-13**) takes place only if, in the transition state, the reactants are oriented as shown.

Not only must the OH$^-$ be near the hydrogen, but the hydrogen must be oriented anti to the chlorine atom.[8] When the two reacting molecules collide, if the OH$^-$ should be near

[5]For a discussion of transition states, see Laidler *J. Chem. Educ.* **1988**, *65*, 540.

[6]For fuller discussions, see Kreevoy; Truhlar, in Bernasconi, Ref. 25, pt. 1, pp. 13-95; Moore; Pearson *Kinetics and Mechanism*, 3rd ed.; Wiley: New York, 1981, pp. 137-181; Klumpp *Reactivity in Organic Chemistry*; Wiley: New York, 1982; pp. 227-378.

[7]For a monograph on diffusion-controlled reactions, see Rice, *Comprehensive Chemical Kinetics*, Vol. 25 (edited by Bamford; Tipper; Compton); Elsevier: New York, 1985.

[8]As we shall see in Chapter 17, with some molecules elimination is also possible if the hydrogen is oriented syn, instead of anti, to the chlorine atom. Of course, this orientation also requires a considerable loss of entropy.

the chlorine atom or near R^1 or R^2, no reaction can take place. In order for a reaction to occur, the molecules must surrender the freedom they normally have to assume many possible arrangements in space and adopt only that one that leads to reaction. Thus, a considerable loss in entropy is involved, i.e., ΔS^+ is negative.

Entropy of activation is also responsible for the difficulty in closing rings[9] larger then six-membered. Consider a ring-closing reaction in which the two groups that must interact are situated on the ends of a ten-carbon chain. In order for reaction to take place, the groups must encounter each other. But a ten-carbon chain has many conformations, and in only a few of these are the ends of the chain near each other. Thus, forming the transition state requires a great loss of entropy.[10] This factor is also present, though less so, in closing rings of six members or less (except three-membered rings), but with rings of this size the entropy loss is less than that of bringing two individual molecules together. For example, a reaction between an OH group and a COOH group in the same molecule to form a lactone with a five- or six-membered ring takes place much faster than the same reaction between a molecule containing an OH group and another containing a COOH group. Though ΔH^+ is about the

$$CH_2 \begin{matrix} CH_2-C-OH \\ \\ CH_2-OH \end{matrix} \quad \longrightarrow \quad CH_2 \begin{matrix} CH_2-C \\ \\ CH_2 \end{matrix} O$$

Faster

$$CH_3CH_2-C-OH + HO-CH_3 \longrightarrow CH_3CH_2-C-O-CH_3$$

Slower

same, ΔS^+ is much less for the cyclic case. However, if the ring to be closed has three or four members, small-angle strain is introduced and the favorable ΔS^+ may not be sufficient to overcome the unfavorable ΔH^+ change. Table 6.1 shows the relative rate constants for the closing of rings of 3 to 23 members all by the same reaction.[11] Reactions in which the transition state has more disorder than the starring compounds,e.g., the pyrolytic conversion of cyclopropane to propene, have positive ΔS^+ values and are thus favored by the entropy effect.

Reactions with intermediates are two-step (or more) processes. In these reactions there is an energy "well." There are two transition states, each with an energy higher than the intermediate (Figure 6.2). The deeper the well, the more stable the intermediate. In Figure 6.2a, the second peak is higher than the first. The opposite situation is shown in Figure 6.2b. Note that in reactions in which the second peak is higher than the first, the overall ΔG^+ is less than the sum of the ΔG^+ values for the two steps. Minima in free-energy-profile diagrams (*intermediates*) correspond to real species which have a finite though very short

[9]For discussions of the entropy and enthalpy of ring-closing reactions, see De Tar; Luthra *J. Am. Chem. Soc.* **1980**, *102*, 4505; Mandolini *Bull. Soc. Chim. Fr.* **1988**, 173. For a related discussion, see Menger *Acc. Chem. Res.* **1985**, *18*, 128-134.

[10]For reviews of the cyclization of acyclic molecules, see Nakagaki; Sakuragi; Mutai *J. Phys. Org. Chem.* **1989**, *2*, 187-204; Mandolini *Adv. Phys. Org. Chem.* **1986**, *22*, 1-111. For a review of the cyclization and conformation of hydrocarbon chains, see Winnik *Chem. Rev.* **1981**, *81*, 491-524. For a review of steric and electronic effects in heterolytic ring closures, see Valters *Russ. Chem. Rev.* **1982**, *51*, 788-801.

[11]The values for 4, 5, and 6 are from Mandolini *J. Am.Chem. Soc.* **1978**, *100*, 550; the others are from Galli; Illuminati; Mandolini; Tamborra *J. Am. Chem. Soc.* **1977**, *99*, 2591. See also Illuminati; Mandolini *Acc. Chem. Res.* **1981**, *14*, 95-102. See, however, van der Kerk; Verhoeven; Stirling *J. Chem. Soc., Perkin Trans. 2* **1985**, 1355; Benedetti; Stirling *J. Chem. Soc., Perkin Trans. 2* **1986**, 605.

TABLE 6.1   Relative rate constants at 50°C (Eight-membered ring = 1) for the reaction $Br(CH_2)_{n-2}CO_2^- \rightarrow (CH_2)_{n-2} \overset{\displaystyle \frown}{\underset{O}{|}} C{=}O$, where $n =$ the ring size[11]

Ring size	Relative rate
3	21.7
4	5.4×10^3
5	1.5×10^6
6	1.7×10^4
7	97.3
8	1.00
9	1.12
10	3.35
11	8.51
12	10.6
13	32.2
14	41.9
15	45.1
16	52.0
18	51.2
23	60.4

existence. These may be the carbocations, carbanions, free radicals, etc., discussed in Chapter 5 or molecules in which all the atoms have their normal valences. In either case, under the reaction conditions they do not live long (because $\Delta G_2^{\ddagger}$ is small) but rapidly go on to products. Maxima in these curves, however, do not correspond to actual species but only to transition states in which bond breaking and/or bond making have partially taken place. Transition states have only a transient existence with an essentially zero lifetime.[12]

The Baldwin Rules for Ring Closure

In previous sections, we discussed, in a general way, the kinetic and thermodynamic aspects of ring-closure reactions. J. E. Baldwin has supplied a more specific set of rules for certain closings of 3- to 7-membered rings.[13] These rules distinguish two types of ring closure, called

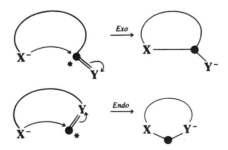

---

[12]Despite their transient existences, it is possible to study transition states of certain reactions in the gas phase with a technique called laser femtochemistry: Zewail; Bernstein *Chem. Eng. News* **1988**, *66*, No. 45 (Nov. 7), 24-43. For another method, see Collings; Polanyi; Smith; Stolow; Tarr *Phys. Rev. Lett.* **1987**, *59*, 2551.

[13]Baldwin *J. Chem. Soc., Chem. Commun.* **1976**, 734; Baldwin in *Further Perspectives in Organic Chemistry (Ciba Foundation Symposium 53)*; Elsevier North Holland: Amsterdam, 1979, pp. 85-99. See also Baldwin; Thomas; Kruse; Silberman *J. Org. Chem.* **1977**, *42*, 3846; Baldwin; Lusch *Tetrahedron* **1982**, *38*, 2939; Anselme *Tetrahedron Lett.* **1977**, 3615; Fountain; Gerhardt *Tetrahedron Lett.* **1978**, 3985.

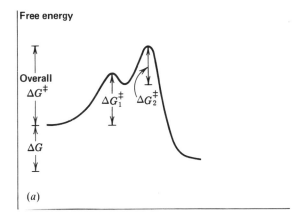

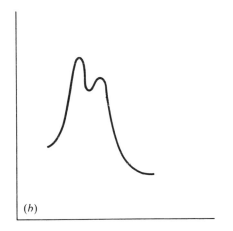

FIGURE 6.2 (*a*) Free-energy profile for a reaction with an intermediate. $\Delta G_1^{\ddagger}$ and $\Delta G_2^{\ddagger}$ are the free energy of activation for the first and second stages, respectively. (*b*) Free-energy profile for a reaction with an intermediate in which the first peak is higher than the second.

Exo and *Endo*, and three kinds of atoms at the starred positions: *Tet* for sp^3, *Trig* for sp^2, and *Dig* for *sp*. The following are Baldwin's rules for closing rings of 3 to 7 members.

Rule 1. Tetrahedral systems
(a) 3 to 7-*Exo-Tet* are all favored processes
(b) 5 to 6-*Endo-Tet* are disfavored

Rule 2. Trigonal systems
(a) 3 to 7-*Exo-Trig* are favored
(b) 3 to 5-*Endo-Trig* are disfavored[14]
(c) 6 to 7-*Endo-Trig* are favored

[14]For some exceptions to the rule in this case, see Trost; Bonk *J. Am. Chem. Soc.* **1985,** *107,* 1778; Auvray; Knochel; Normant *Tetrahedron Lett.* **1985,** *26,* 4455; Torres; Larson *Tetrahedron Lett.* **1986,** *27,* 2223.

Rule 3. Digonal systems
  (a) 3 to 4-*Exo-Dig* are disfavored
  (b) 5 to 7-*Exo-Dig* are favored
  (c) 3 to 7-*Endo-Dig* are favored

"Disfavored" does not mean it cannot be done—only that it is more difficult than the favored cases. These rules are empirical and have a stereochemical basis. The favored pathways are those in which the length and nature of the linking chain enables the terminal atoms to achieve the proper geometries for reaction. The disfavored cases require severe distortion of bond angles and distances. Many cases in the literature are in substantial accord with these rules.

Kinetic and Thermodynamic Control

There are many cases in which a compound under a given set of reaction conditions can undergo competing reactions to give different products:

$$A \underset{\longrightarrow C}{\overset{\longrightarrow B}{\rightleftharpoons}}$$

Figure 6.3 shows a free-energy profile for a reaction in which B is thermodynamically more stable than C (lower ΔG), but C is formed faster (lower ΔG^+). If neither reaction is reversible, C will be formed in larger amount because it is formed faster. The product is said to be *kinetically controlled*. However, if the reactions are reversible, this will not necessarily be the case. If such a process is stopped well before the equilibrium has been established, the reaction will be kinetically controlled since more of the faster-formed product will be present. However, if the reaction is permitted to approach equilibrium, the predominant or even exclusive product will be B. Under these conditions the C that is first formed reverts to A,

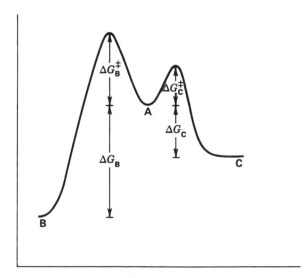

FIGURE 6.3   Free-energy profile illustrating kinetic versus thermodynamic control of product. The starting compound (**A**) can react to give either **B** or **C**.

while the more stable B does so much less. We say the product is *thermodynamically controlled*.[15] Of course, Figure 6.3 does not describe all reactions in which a compound A can give two different products. In many cases the more stable product is also the one that is formed faster. In such cases the product of kinetic control is also the product of thermodynamic control.

The Hammond Postulate

Since transition states have zero lifetimes, it is impossible to observe them directly and information about their geometries must be obtained from inference. In some cases our inferences can be very strong. For example, in the SN2 reaction (p. 294) between CH_3I and I^- (a reaction in which the product is identical to the starting compound), the transition state should be perfectly symmetrical. In most cases, however, we cannot reach such easy conclusions, and we are greatly aided by the *Hammond postulate*,[16] which states that for any single reaction step, *the geometry of the transition state for that step resembles the side to which it is closer in free energy*. Thus, for an exothermic reaction like that shown in Figure 6.1, the transition state resembles the reactants more than the products, though not much more because there is a substantial ΔG^+ on both sides. The postulate is most useful in dealing with reactions with intermediates. In the reaction illustrated in Figure 6.2a, the first transition state lies much closer in energy to the intermediate than to the reactants, and we can predict that the geometry of the transition state resembles that of the intermediate more than it does that of the reactants. Likewise, the second transition state also has a free energy much closer to that of the intermediate than to the products, so that both transition states resemble the intermediate more than they do the products or reactants. This is generally the case in reactions that involve very reactive intermediates. Since we usually know more about the structure of intermediates than of transition states, we often use our knowledge of intermediates to draw conclusions about the transition states (for examples, see pp. 340, 750).

Microscopic Reversibility

In the course of a reaction the nuclei and electrons assume positions that at each point correspond to the lowest free energies possible. If the reaction is reversible, these positions must be the same in the reverse process, too. This means that the forward and reverse reactions (run under the same conditions) must proceed by the same mechanism. This is called the *principle of microscopic reversibility*. For example, if in a reaction A → B there is an intermediate C, then C must also be an intermediate in the reaction B → A. This is a useful principle since it enables us to know the mechanism of reactions in which the equilibrium lies far over to one side. Reversible photochemical reactions are an exception, since a molecule that has been excited photochemically does not have to lose its energy in the same way (Chapter 7).

Marcus Theory

It is often useful to compare the reactivity of one compound with that of similar compounds. What we would like to do is to find out how a reaction coordinate (and in particular the

---

[15]For a discussion of thermodynamic vs. kinetic control, see Klumpp, Ref. 6, pp. 36-89.

[16]Hammond *J. Am. Chem. Soc.* **1955**, *77*, 334. For a discussion, see Fărcaşiu *J. Chem. Educ.* **1975**, *52*, 76-79.

transition state) changes when one reactant molecule is replaced by a similar molecule. Marcus theory is a method for doing this.[17]

In this theory the activation energy ΔG^+ is thought of as consisting of two parts.

1. An *intrinsic* free energy of activation, which would exist if the reactants and products had the same $\Delta G°$.[18] This is a kinetic part, called the *intrinsic barrier* ΔG_{int}^+.

2. A thermodynamic part, which arises from the $\Delta G°$ for the reaction.

The Marcus equation says that the overall ΔG^+ for a one-step reaction is[19]

$$\Delta G^+ = \Delta G_{int}^+ + \tfrac{1}{2}\Delta G^\Delta + \frac{(\Delta G^\Delta)^2}{16(\Delta G_{int}^+ - w^R)}$$

where the term ΔG^Δ stands for

$$\Delta G^\Delta = \Delta G° - w^R + w^P$$

w^R, a work term, is the free energy required to bring the reactants together and w^P is the work required to form the successor configuration from the products.

For a reaction of the type $AX + B \rightarrow BX + A$, the intrinsic barrier[20] ΔG_{int}^+ is taken to be the average ΔG^+ for the two symmetrical reactions

$$AX + A \longrightarrow AX + A \qquad \Delta G_{A,A}^+$$

$$BX + B \longrightarrow BX + B \qquad \Delta G_{B,B}^+$$

so that

$$\Delta G^+ = \tfrac{1}{2}(\Delta G_{A,A}^+ + \Delta G_{B,B}^+)$$

One type of process that can successfully be treated by the Marcus equation is the SN2 mechanism (p. 294)

$$R{-}X + Y \longrightarrow R{-}Y + X$$

When R is CH_3 the process is called *methyl transfer*.[21] For such reactions the work terms w^R and w^P are assumed to be very small compared to $\Delta G°$, and can be neglected, so that the Marcus equation simplifies to

$$\Delta G^+ = \Delta G_{int}^+ + \tfrac{1}{2}\Delta G° + \frac{(\Delta G°)^2}{16\Delta G_{int}^+}$$

The Marcus equation allows ΔG^+ for $RX + Y \rightarrow RY + X$ to be calculated from the barriers of the two symmetrical reactions $RX + X \rightarrow RX + X$ and $RY + Y \rightarrow RY + Y$. The results of such calculations are generally in agreement with the Hammond postulate.

Marcus theory can be applied to any single-step process where something is transferred

---

[17]For reviews, see Albery *Annu. Rev. Phys. Chem.* **1980**, *31*, 227-263; Kreevoy; Truhlar, in Bernasconi, Ref. 25, pt. 1, pp. 13-95.

[18]$\Delta G°$ is the standard free energy; that is, ΔG at atmospheric pressure.

[19]Albery; Kreevoy, Ref. 21, pp. 98-99.

[20]For discussions of intrinsic barriers, see Lee *J. Chem. Soc., Perkin Trans. 2* **1989**, 943, *Chem. Soc. Rev.* **1990**, *19*, 133-145.

[21]For a review of Marcus theory applied to methyl transfer, see Albery; Kreevoy *Adv. Phys. Org. Chem.* **1978**, *16*, 87-157. See also Ref. 20; Lewis; Kukes; Slater *J. Am. Chem. Soc.* **1980**, *102*, 1619; Lewis, Hu *J. Am. Chem. Soc.* **1984**, *106*, 3292; Lewis; McLaughlin; Douglas *J. Am.Chem. Soc.* **1985**, *107*, 6668; Lewis *Bull. Soc. Chim. Fr.* **1988**, 259.

from one particle to another. It was originally derived for electron transfers,[22] and then extended to transfers of H^+ (see p. 258), H^-,[23] and $H\cdot$[24] as well as methyl transfers.

METHODS OF DETERMINING MECHANISMS

There are a number of commonly used methods for determining mechanisms.[25] In most cases one method is not sufficient, and the problem is generally approached from several directions.

Identification of Products

Obviously any mechanism proposed for a reaction must account for all the products obtained and for their relative proportions, including products formed by side reactions. Incorrect mechanisms for the von Richter reaction (**3-25**) were accepted for many years because it was not realized that nitrogen was a major product. A proposed mechanism cannot be correct if it fails to predict the products in approximately the observed proportions. For example, any mechanism for the reaction

$$CH_4 + Cl_2 \xrightarrow{h\nu} CH_3Cl$$

that fails to account for the formation of a small amount of ethane cannot be correct (see **4-1**), and any mechanism proposed for the Hofmann rearrangement (**8-14**):

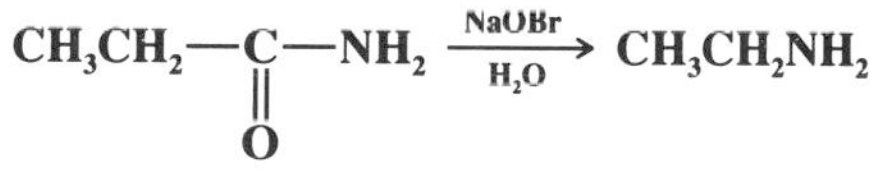

$$CH_3CH_2-\underset{\underset{O}{\|}}{C}-NH_2 \xrightarrow[H_2O]{NaOBr} CH_3CH_2NH_2$$

must account for the fact that the missing carbon appears as CO_2.

Determination of the Presence of an Intermediate

Intermediates are postulated in many mechanisms. There are several ways, none of them foolproof,[26] for attempting to learn whether or not an intermediate is present and, if so, its structure.

1. *Isolation of an intermediate.* It is sometimes possible to isolate an intermediate from a reaction mixture by stopping the reaction after a short time or by the use of very mild conditions. For example, in the Neber rearrangement (**8-13**)

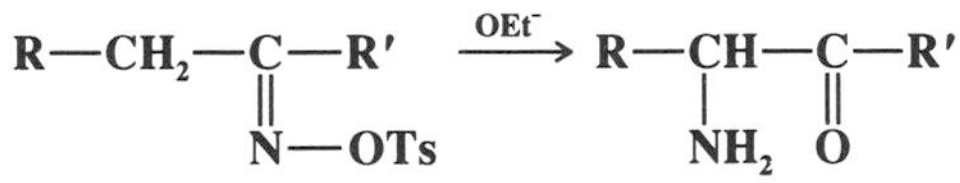

$$R-CH_2-\underset{\underset{N-OTs}{\|}}{C}-R' \xrightarrow{OEt^-} R-\underset{\underset{NH_2}{|}}{CH}-\underset{\underset{O}{\|}}{C}-R'$$

[22]Marcus *J. Phys. Chem.* **1963**, *67*, 853, *Annu. Rev. Phys. Chem.* **1964**, *15*, 155-196; Eberson *Electron Transfer Reactions in Organic Chemistry*; Springer: New York, 1987.

[23]Kreevoy; Lee *J. Am. Chem. Soc.* **1984**, *106*, 2550; Lee; Ostović; Kreevoy *J. Am. Chem. Soc.* **1988**, *110*, 3989; Kim; Lee; Kreevoy *J. Am. Chem. Soc.* **1990**, *112*, 1889.

[24]See for example Dneprovskii; Eliscenkov *J. Org. Chem. USSR* **1988**, *24*, 243.

[25]For a treatise on this subject, see Bernasconi *Investigation of Rates and Mechanisms of Reactions*, 4th ed. (vol. 6 of Weissberger *Techniques of Chemistry*), 2 pts.; Wiley: New York, 1986. For a monograph, see Carpenter *Determination of Organic Reaction Mechanisms*; Wiley: New York, 1984.

[26]For a discussion, see Martin *J. Chem. Educ.* **1985**, *62*, 789.

the intermediate **1** has been isolated. If it can be shown that the isolated compound gives the same product when subjected to the reaction conditions and at a rate no slower than

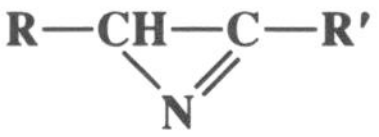

$$R—CH—C—R'$$

1

the starting compound, this constitutes strong evidence that the reaction involves that intermediate, though it is not conclusive, since the compound may arise by an alternate path and by coincidence give the same product.

   2. *Detection of an intermediate.*   In many cases an intermediate cannot be isolated but can be detected by ir, nmr, or other spectra.[27] The detection by Raman spectra of NO_2^+ was regarded as strong evidence that this is an intermediate in the nitration of benzene (see **1-2**). Free radical and triplet intermediates can often be detected by esr and by CIDNP (see Chapter 5). Free radicals (as well as radical ions and EDA complexes) can also be detected by a method that does not rely on spectra. In this method a double-bond compound is added to the reaction mixture, and its fate traced.[28] One possible result is cis-trans conversion. For example, *cis*–stilbene is isomerized to the trans isomer in the presence of RS· radicals, by this mechanism:

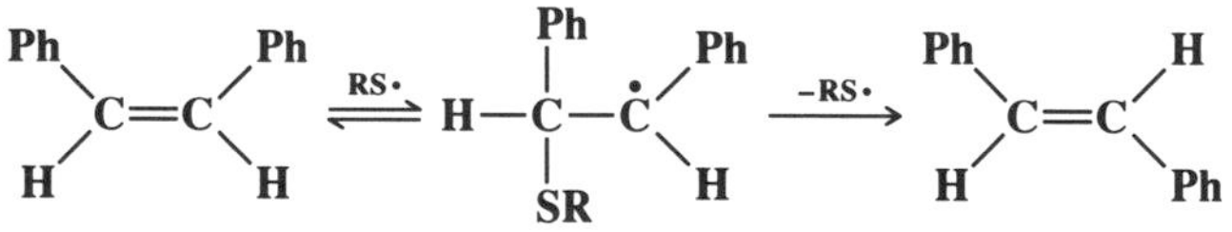

cis-stilbene   *trans*-stilbene

Since the trans isomer is more stable than the cis, the reaction does not go the other way, and the detection of the isomerized product is evidence for the presence of the RS· radicals.

   3. *Trapping of an intermediate.*   In some cases, the suspected intermediate is known to be one that reacts in a given way with a certain compound. The intermediate can then be trapped by running the reaction in the presence of that compound. For example, benzynes (p. 646) react with dienes in the Diels–Alder reaction (**5-47**). In any reaction where a benzyne is a suspected intermediate, the addition of a diene and the detection of the Diels–Alder adduct indicate that the benzyne was probably present.

   4. *Addition of a suspected intermediate.*   If a certain intermediate is suspected, and if it can be obtained by other means, then under the same reaction conditions it should give the same products. This kind of experiment can provide conclusive negative evidence: if the correct products are not obtained, the suspected compound is not an intermediate. However, if the correct products are obtained, this is not conclusive since they may arise by coincidence. The von Richter reaction (**3-25**) provides us with a good example here too. For many years it had been assumed that an aryl cyanide was an intermediate, since cyanides are easily hydrolyzed to carboxylic acids (**6-5**). In fact, in 1954, *p*-chlorobenzonitrile was shown to give *p*-chlorobenzoic acid under normal von Richter conditions.[29] However, when the experiment was repeated with 1-cyanonaphthalene, no 1-naphthoic acid was obtained, although

---

[27]For a review on the use of electrochemical methods to detect intermediates, see Parker *Adv. Phys. Org. Chem.* **1983**, *19*, 131-222. For a review of the study of intermediates trapped in matrixes, see Sheridan *Org. Photochem.* **1987**, *8*, 159-248.
[28]For a review, see Todres *Tetrahedron* **1987**, *43*, 3839-3861.
[29]Bunnett; Rauhut; Knutson; Bussell *J. Am. Chem. Soc.* **1954**, *76*, 5755.

2-nitronaphthalene gave 13% 1-naphthoic acid under the same conditions.[30] This proved that 2-nitronaphthalene must have been converted to 1-naphthoic acid by a route that does not involve 1-cyanonaphthalene. It also showed that even the conclusion that *p*-chlorobenzonitrile was an intermediate in the conversion of *m*-nitrochlorobenzene to *p*-chlorobenzoic acid must now be suspect, since it is not likely that the mechanism would substantially change in going from the naphthalene to the benzene system.

The Study of Catalysis[31]

Much information about the mechanism of a reaction can be obtained from a knowledge of which substances catalyze the reaction, which inhibit it, and which do neither. Of course, just as a mechanism must be compatible with the products, so must it be compatible with its catalysts. In general, catalysts perform their actions by providing an alternate pathway for the reaction in which $\Delta G^{\ddagger}$ is less than it would be without the catalyst. Catalysts do not change ΔG.

Isotopic Labeling[32]

Much useful information has been obtained by using molecules that have been isotopically labeled and tracing the path of the reaction in that way. For example, in the reaction

$$\text{RCOO}^- + \text{BrCN} \longrightarrow \text{RCN}$$

does the CN group in the product come from the CN in the BrCN? The use of ^{14}C supplied the answer, since $R^{14}CO_2^-$ gave *radioactive* RCN.[33] This surprising result saved a lot of labor, since it ruled out a mechanism involving the replacement of CO_2 by CN (see **6-59**). Other radioactive isotopes are also frequently used as tracers, but even stable isotopes can be used. An example is the hydrolysis of esters

$$\underset{\underset{O}{\|}}{R-C}-OR' + H_2O \longrightarrow \underset{\underset{O}{\|}}{R-C}-OH + R'OH$$

Which bond of the ester is broken, the acyl—O or the alkyl—O bond? The answer is found by the use of $H_2^{18}O$. If the acyl—O bond breaks, the labeled oxygen will appear in the acid; otherwise it will be in the alcohol (see **0-10**). Although neither compound is radioactive, the one that contains ^{18}O can be determined by submitting both to mass spectrometry. In a similar way, deuterium can be used as a label for hydrogen. In this case it is not necessary to use mass spectrometry, since ir and nmr spectra can be used to determine when deuterium has been substituted for hydrogen. ^{13}C is also nonradioactive; it can be detected by ^{13}C nmr.[34]

In the labeling technique, it is not generally necessary to use completely labeled compounds. Partially labeled material is usually sufficient.

[30]Bunnett; Rauhut *J. Org. Chem.* **1956**, *21*, 944.

[31]For treatises, see Jencks *Catalysis in Chemistry and Enzymology*; McGraw-Hill: New York, 1969; Bender *Mechanisms of Homogeneous Catalysis from Protons to Proteins*; Wiley: New York, 1971. For reviews, see Coenen *Recl. Trav. Chim. Pays-Bas* **1983**, *102*, 57-64; and in Bernasconi, Ref. 25, pt. 1, the articles by Keeffe; Kresge, pp. 747-790; Haller; Delgass, pp. 951-979.

[32]For reviews see Wentrup, in Bernasconi, Ref. 25, pt. 1, pp. 613-661; Collins *Adv. Phys. Org. Chem.* **1964**, *2*, 3-91. See also the series *Isotopes in Organic Chemistry*.

[33]Douglas; Eccles; Almond *Can. J. Chem.* **1953**, *31*, 1127; Douglas; Burditt *Can. J. Chem.* **1958**, *36*, 1256.

[34]For a review, see Hinton; Oka; Fry *Isot. Org. Chem.* **1977**, *3*, 41-104.

Stereochemical Evidence[35]

If the products of a reaction are capable of existing in more than one stereoisomeric form, the form that is obtained may give information about the mechanism. For example, (+)-malic acid was discovered by Walden[36] to give (−)-chlorosuccinic acid when treated with PCl_5 and the (+) enantiomer when treated with $SOCl_2$, showing that the mechanisms of these apparently similar conversions could not be the same (see pp. 295, 327). Much useful information has been obtained about nucleophilic substitution, elimination, rearrangement, and addition reactions from this type of experiment. The isomers involved need not be enantiomers. Thus, the fact that cis-2-butene treated with $KMnO_4$ gives meso-2,3-butanediol and not the racemic mixture is evidence that the two OH groups attack the double bond from the same side (see reaction **5-35**).

Kinetic Evidence[37]

The rate of a homogeneous reaction[38] is the rate of disappearance of a reactant or appearance of a product. The rate nearly always changes with time, since it is usually proportional to concentration and the concentration of reactants decreases with time. However, the rate is not always proportional to the concentration of all reactants. In some cases a change in the concentration of a reactant produces no change at all in the rate, while in other cases the rate may be proportional to the concentration of a substance (a catalyst) that does not even appear in the stoichiometric equation. A study of which reactants affect the rate often tells a good deal about the mechanism.

If the rate is proportional to the change in concentration of only one reactant (A), the rate law (the rate of change of concentration of A with time t) is

$$\text{Rate} = \frac{-d[\mathbf{A}]}{dt} = k[\mathbf{A}]$$

where k is the rate constant for the reaction. There is a minus sign because the concentration of A decreases with time. A reaction that follows such a rate law is called a first-order reaction. The units of k for a first-order reaction are sec^{-1}. The rate of a second-order reaction is proportional to the concentration of two reactants, or to the square of the concentration of one:

$$\frac{-d[\mathbf{A}]}{dt} = k[\mathbf{A}][\mathbf{B}] \qquad \text{or} \qquad \frac{-d[\mathbf{A}]}{dt} = k[\mathbf{A}]^2$$

For a second-order reaction the units are liters mol^{-1} sec^{-1} or some other units expressing the reciprocal of concentration or pressure per unit time interval.

Similar expressions can be written for third-order reactions. A reaction whose rate is proportional to [A] and to [B] is said to be first order in A and in B, second order overall.

[35]For lengthy treatments of the relationship between stereochemistry and mechanism, see Billups; Houk; Stevens, in Bernasconi, Ref. 25, pt. 1, pp. 663-746; Eliel *Stereochemistry of Carbon Compounds*; McGraw-Hill: New York, 1962; Newman *Steric Effects in Organic Chemistry*; Wiley, New York, 1956.

[36]Walden *Ber.* **1896**, *29*, 136, **1897**, *30*, 3149, **1899**, *32*, 1833.

[37]For the use of kinetics in determining mechanisms, see Connors *Chemical Kinetics*; VCH: New York, 1990; Zuman; Patel *Techniques in Organic Reaction Kinetics*; Wiley: New York, 1984; Drenth; Kwart *Kinetics Applied to Organic Reactions*; Marcel Dekker: New York, 1980; Hammett *Physical Organic Chemistry*, 2nd ed.; McGraw-Hill: New York, 1970, pp. 53-100; Gardiner *Rates and Mechanisms of Chemical Reactions*; W.A. Benjamin: New York, 1969; Leffler; Grunwald *Rates and Equilibria of Organic Reactions*; Wiley: New York, 1963; Jencks, Ref. 31, pp. 555-614; Refs. 6 and 25.

[38]A homogeneous reaction occurs in one phase. Heterogeneous kinetics have been studied much less.

A reaction rate can be measured in terms of any reactant or product, but the rates so determined are not necessarily the same. For example, if the stoichiometry of a reaction is $2A + B \rightarrow C + D$ then, on a molar basis, A must disappear twice as fast as B, so that $-d[A]/dt$ and $-d[B]/dt$ are not equal but the former is twice as large as the latter.

The rate law of a reaction is an experimentally determined fact. From this fact we attempt to learn the *molecularity*, which may be defined as the number of molecules that come together to form the activated complex. It is obvious that if we know how many (and which) molecules take part in the activated complex, we know a good deal about the mechanism. The experimentally determined rate order is not necessarily the same as the molecularity. Any reaction, no matter how many steps are involved, has only one rate law, but each step of the mechanism has its own molecularity. For reactions that take place in one step (reactions without an intermediate) the order is the same as the molecularity. A first-order, one-step reaction is always unimolecular; a one-step reaction that is second order in A always involves two molecules of A; if it is first order in A and in B, then a molecule of A reacts with one of B, etc. For reactions that take place in more than one step, the order *for each step* is the same as the molecularity *for that step*. This fact enables us to predict the rate law for any proposed mechanism, though the calculations may get lengthy at times.[39] If any one step of a mechanism is considerably slower than all the others (this is usually the case), the rate of the overall reaction is essentially the same as that of the slow step, which is consequently called the *rate-determining* step.[40]

For reactions that take place in two or more steps, two broad cases can be distinguished:

1. The first step is slower than any subsequent step and is consequently rate-determining. In such cases, the rate law simply includes the reactants that participate in the slow step. For example, if the reaction $A + 2B \rightarrow C$ has the mechanism

$$A + B \xrightarrow{\text{slow}} I$$

$$I + B \xrightarrow{\text{fast}} C$$

where I is an intermediate, the reaction is second order, with the rate law

$$\text{Rate} = \frac{-d[A]}{dt} = k[A][B]$$

2. When the first step is not rate-determining, determination of the rate law is usually much more complicated. For example, consider the mechanism

$$A + B \underset{k_{-1}}{\overset{k_1}{\rightleftharpoons}} I$$

$$I + B \xrightarrow{k_2} C$$

where the first step is a rapid attainment of equilibrium, followed by a slow reaction to give C. The rate of disappearance of A is

$$\frac{-d[A]}{dt} = k_1[A][B] - k_{-1}[I]$$

[39]For a discussion of how order is related to molecularity in many complex situations, see Szabó, in Bamford; Tipper *Comprehensive Chemical Kinetics*, vol. 2; Elsevier: New York, 1969, pp. 1-80.

[40]Many chemists prefer to use the term *rate-limiting step* or *rate-controlling step* for the slow step, rather than *rate-determining step*. See the definitions in Gold; Loening; McNaught; Sehmi *IUPAC Compendium of Chemical Terminology*; Blackwell Scientific Publications: Oxford, 1987, p. 337. For a discussion of rate-determining steps, see Laidler *J. Chem. Educ.* **1988,** *65*, 250.

Both terms must be included because A is being formed by the reverse reaction as well as being used up by the forward reaction. This equation is of very little help as it stands since we cannot measure the concentration of the intermediate. However, the combined rate law for the formation and disappearance of I is

$$\frac{d[\mathbf{I}]}{dt} = k_1[\mathbf{A}][\mathbf{B}] - k_{-1}[\mathbf{I}] - k_2[\mathbf{I}][\mathbf{B}]$$

At first glance we seem no better off with this equation, but we can make the assumption that *the concentration of I does not change with time*, since it is an intermediate that is used up (going either to A + B or to C) as fast as it is formed. This assumption, called the assumption of the *steady state*,[41] enables us to set $d[\mathbf{I}]/dt$ equal to zero and hence to solve for [I] in terms of the measurable quantities [A] and [B]:

$$[\mathbf{I}] = \frac{k_1[\mathbf{A}][\mathbf{B}]}{k_2[\mathbf{B}] + k_{-1}}$$

We now insert this value for [I] into the original rate expression to obtain

$$\frac{-d[\mathbf{A}]}{dt} = \frac{k_1 k_2 [\mathbf{A}][\mathbf{B}]^2}{k_2[\mathbf{B}] + k_{-1}}$$

Note that this rate law is valid whatever the values of k_1, k_{-1}, and k_2. However, our original hypothesis was that the first step was faster than the second, or that

$$k_1[\mathbf{A}][\mathbf{B}] \gg k_2[\mathbf{I}][\mathbf{B}]$$

Since the first step is an equilibrium

$$k_1[\mathbf{A}][\mathbf{B}] = k_{-1}[\mathbf{I}]$$

we have

$$k_{-1}[\mathbf{I}] \gg k_2[\mathbf{I}][\mathbf{B}]$$

Canceling [I], we get

$$k_{-1} \gg k_2[\mathbf{B}]$$

We may thus neglect $k_2[\text{B}]$ in comparison with k_{-1} and obtain

$$\frac{-d[\mathbf{A}]}{dt} = \frac{k_1 k_2}{k_{-1}} [\mathbf{A}][\mathbf{B}]^2$$

The overall rate is thus third order: first order in A and second order in B. Incidentally, if the first step is rate-determining (as was the case in the preceding paragraph), then

$$k_2[\mathbf{B}] \gg k_{-1} \quad \text{and} \quad \frac{-d[\mathbf{A}]}{dt} = k_1[\mathbf{A}][\mathbf{B}]$$

which is the same rate law we deduced from the rule that where the first step is rate-determining, the rate law includes the reactants that participate in that step.

   It is possible for a reaction to involve A and B in the rate-determining step, though only [A] appears in the rate law. This occurs when a large excess of B is present, say 100 times

[41]For a discussion, see Raines; Hansen *J. Chem. Educ.* **1988**, *65*, 757.

the molar quantity of A. In this case the complete reaction of A uses up only 1 mole of B, leaving 99 moles. It is not easy to measure the change in concentration of B with time in such a case, and it is seldom attempted, especially when B is also the solvent. Since [B], for practical purposes, does not change with time, the reaction appears to be first order in A though actually both A and B are involved in the rate-determining step. This is often referred to as a *pseudo-first-order* reaction. Pseudo-order reactions can also come about when one reactant is a catalyst whose concentration does not change with time because it is replenished as fast as it is used up and when a reaction is conducted in a medium that keeps the concentration of a reactant constant, e.g., in a buffer solution where H^+ or OH^- is a reactant. Pseudo-first-order conditions are frequently used in kinetic investigations for convenience in experimentation and calculations.

What is actually being measured is the change in concentration of a product or a reactant with time. Many methods have been used to make such measurements.[42] The choice of a method depends on its convenience and its applicability to the reaction being studied. Among the most common methods are:

1. *Periodic or continuous spectral readings.* In many cases the reaction can be carried out in the cell while it is in the instrument. Then all that is necessary is that the instrument be read, periodically or continuously. Among the methods used are ir and uv spectroscopy, polarimetry, nmr, and esr.[43]

2. *Quenching and analyzing.* A series of reactions can be set up and each stopped in some way (perhaps by suddenly lowering the temperature or adding an inhibitor) after a different amount of time has elapsed. The materials are then analyzed by spectral readings, titrations, chromatography, polarimetry, or any other method.

3. *Removal of aliquots at intervals.* Each aliquot is then analyzed as in method 2.

4. *Measurement of changes in total pressure, for gas-phase reactions.*[44]

5. *Calorimetric methods.* The output or absorption of heat can be measured at time intervals.

Special methods exist for kinetic measurements of very fast reactions.[45]

In any case what is usually obtained is a graph showing how a concentration varies with time. This must be interpreted[46] to obtain a rate law and a value of k. If a reaction obeys simple first- or second-order kinetics, the interpretation is generally not difficult. For example, if the concentration at the start is A_0, the first-order rate law

$$\frac{-d[\mathbf{A}]}{dt} = k[\mathbf{A}] \qquad \text{or} \qquad \frac{-d[\mathbf{A}]}{[\mathbf{A}]} = k\,dt$$

can be integrated between the limits $t = 0$ and $t = t$ to give

$$-\ln\frac{[\mathbf{A}]}{\mathbf{A}_0} = kt \qquad \text{or} \qquad \ln[\mathbf{A}] = -kt + \ln\mathbf{A}_0$$

[42]For a monograph on methods of interpreting kinetic data, see Zuman; Patel, Ref. 37. For a review of methods of obtaining kinetic data, see Batt, in Bamford; Tipper, Ref. 39, vol. 1, 1969, pp. 1-111.

[43]For a review of esr to measure kinetics, see Norman *Chem. Soc. Rev.* **1979**, *8*, 1-27.

[44]For a review of the kinetics of reactions in solution at high pressures, see le Noble *Prog. Phys. Org. Chem.* **1967**, *5*, 207-330. For reviews of synthetic reactions under high pressure, see Matsumoto; Sera; Uchida *Synthesis* **1985**, 1-26; Matsumoto; Sera *Synthesis* **1985**, 999-1027.

[45]For reviews, see Connors, Ref. 37, pp. 133-186; Zuman; Patel, Ref. 37, pp. 247-327; Krüger *Chem. Soc. Rev.* **1982**, *11*, 227-255; Hague, in Bamford; Tipper, Ref. 39, vol. 1, pp. 112-179, Elsevier, New York, 1969; Bernasconi, Ref. 25, pt. 2. See also Bamford; Tipper, Ref. 39, vol. 24, 1983.

[46]For discussions, much fuller than that given here, of methods for interpreting kinetic data, see Connors, Ref. 37, pp. 17-131; Ritchie *Physical Organic Chemistry*, 2nd ed.; Marcel Dekker: New York, 1990, pp. 1-35; Zuman; Patel, Ref. 37; Margerison, in Bamford; Tipper, Ref. 39, vol. 1, pp. 343-421, 1969; Moore; Pearson, Ref. 6, pp. 12-82; in Bernasconi, Ref. 25, pt. 1, the articles by Bunnett, pp. 251-372, Noyes, pp. 373-423, Bernasconi, pp. 425-485, Wiberg, pp. 981-1019.

Therefore, if a plot of $\ln[A]$ against t is linear, the reaction is first order and k can be obtained from the slope. For first-order reactions it is customary to express the rate not only by the rate constant k but also by the *half-life*, which is the time required for half of any given quantity of a reactant to be used up. Since the half-life $t_{1/2}$ is the time required for $[A]$ to reach $A_0/2$, we may say that

$$\ln \frac{A_0}{2} = kt_{1/2} + \ln A_0$$

so that

$$t_{1/2} = \frac{\ln\left(\dfrac{A_0}{A_0/2}\right)}{k} = \frac{\ln 2}{k} = \frac{0.693}{k}$$

For the general case of a reaction first order in A and first order in B, second order overall, integration is complicated, but it can be simplified if equimolar amounts of A and B are used, so that $A_0 = B_0$. In this case

$$\frac{-d[A]}{dt} = k[A][B]$$

is equivalent to

$$\frac{-d[A]}{dt} = k[A]^2 \quad \text{or} \quad \frac{-d[A]}{[A]^2} = k\, dt$$

Integrating as before gives

$$\frac{1}{[A]} - \frac{1}{A_0} = kt$$

Thus, under equimolar conditions, if a plot of $1/[A]$ against t is linear, the reaction is second order with a slope of k. It is obvious that the same will hold true for a reaction second order in A.[47]

Although many reaction-rate studies do give linear plots, which can therefore be easily interpreted, the results in many other studies are not so simple. In some cases a reaction may be first order at low concentrations but second order at higher concentrations. In other cases fractional orders are obtained, and even negative orders. The interpretation of complex kinetics often requires much skill and effort. Even where the kinetics are relatively simple, there is often a problem in interpreting the data because of the difficulty of obtaining precise enough measurements.[48]

Nmr spectra can be used to obtain kinetic information in a completely different manner from that mentioned on p. 223. This method, which involves the study of nmr line shapes,[49] depends on the fact that nmr spectra have an inherent time factor: if a proton changes its environment less rapidly than about 10^3 times per second, an nmr spectrum shows a separate peak for each position the proton assumes. For example, if the rate of rotation around the

[47]We have given the integrated equations for simple first- and second-order kinetics. For integrated equations for a large number of kinetic types, see Margerison, Ref. 46, p. 361.

[48]See Hammett, Ref. 37, pp. 62-70.

[49]For a monograph, see Oki *Applications of Dynamic NMR Spectroscopy to Organic Chemistry*; VCH: New York, 1985. For reviews, see Fraenkel, in Bernasconi, Ref. 25, pt. 2, pp. 547-604; Aganov; Klochkov; Samitov *Russ. Chem. Rev.* **1985**, *54*, 931-947; Roberts *Pure Appl. Chem.* **1979**, *51*, 1037-1047; Binsch *Top. Stereochem.* **1968**, *3*, 97-192; Johnson *Adv. Magn. Reson.* **1965**, *1*, 33-102.

C—N bond of N,N-dimethylacetamide is slower than 10^3 rotations per second, the two N-methyl groups each have separate chemical shifts since they are not equivalent, one being

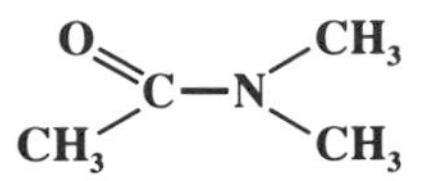

$$\underset{CH_3}{\overset{O}{\diagdown}}C-N\underset{CH_3}{\overset{CH_3}{\diagup}}$$

cis to the oxygen and the other trans. However, if the environmental change takes place more rapidly than about 10^3 times per second, only one line is found, at a chemical shift that is the weighted average of the two individual positions. In many cases, two or more lines are found at low temperatures, but as the temperature is increased, the lines coalesce because the interconversion rate increases with temperature and passes the 10^3 per second mark. From studies of the way line shapes change with temperature it is often possible to calculate rates of reactions and of conformational changes. This method is not limited to changes in proton line shapes but can also be used for other atoms that give nmr spectra and for esr spectra.

Several types of mechanistic information can be obtained from kinetic studies.

1. From the order of a reaction, information can be obtained about which molecules and how many take part in the rate-determining step. Such knowledge is very useful and often essential in elucidating a mechanism. For any mechanism that can be proposed for a given reaction, a corresponding rate law can be calculated by the methods discussed on pp. 221-223. If the experimentally obtained rate law fails to agree with this, the proposed mechanism is wrong. However, it is often difficult to relate the order of a reaction to the mechanism, especially when the order is fractional or negative. In addition, it is frequently the case that two or more proposed mechanisms for a reaction are kinetically indistinguishable, i.e., they predict the same rate law.

2. Probably the most useful data obtained kinetically are the rate constants themselves. They are important since they can tell us the effect on the rate of a reaction of changes in the structure of the reactants (see Chapter 9), the solvent, the ionic strength, the addition of catalysts, etc.

3. If the rate is measured at several temperatures, in most cases a plot of ln k against $1/T$ (T stands for absolute temperature) is nearly linear[50] with a negative slope, and fits the equation

$$\ln k = \frac{-E_a}{RT} + \ln A$$

where R is the gas constant and A a constant called the *frequency factor*. This permits the calculation of E_a, which is the Arrhenius activation energy of the reaction. ΔH^+ can then be obtained by

$$E_a = \Delta H^+ + RT$$

It is also possible to use these data to calculate ΔS^+ by the formula[51]

$$\frac{\Delta S^+}{4.576} = \log k - 10.753 - \log T + \frac{E_a}{4.576T}$$

---

[50]For a review of cases where such a plot is nonlinear, see Blandamer; Burgess; Robertson; Scott *Chem. Rev.* **1982**, *82*, 259-286.
[51]For a derivation of this equation, see Bunnett, in Bernasconi, Ref. 25, pt. 1, p. 287.

for energies in calorie units. For joule units the formula is

$$\frac{\Delta S^+}{19.15} = \log k - 10.753 - \log T + \frac{E_a}{19.15T}$$

One then obtains ΔG^+ from $\Delta G^+ = \Delta H^+ - T\Delta S^+$.

Isotope Effects

When a hydrogen in a reactant molecule is replaced by deuterium, there is often a change in the rate. Such changes are known as *deuterium isotope effects*[52] and are expressed by the ratio k_H/k_D. The ground-state vibrational energy (called the zero-point vibrational energy) of a bond depends on the mass of the atoms and is lower when the reduced mass is higher.[53] Therefore, D—C, D—O, D—N bonds, etc., have lower energies in the ground state than the corresponding H—C, H—O, H—N bonds, etc. Complete dissociation of a deuterium bond consequently requires more energy than that for a corresponding hydrogen bond in the same environment (Figure 6.4). If an H—C, H—O, or H—N bond is not broken at all in a reaction or is broken in a non-rate-determining step, substitution of deuterium for hydrogen causes no change in the rate (see below for an exception to this statement), but

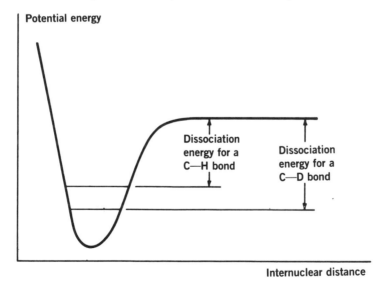

FIGURE 6.4 A C—D bond has a lower zero-point energy than does a corresponding C—H bond; thus the dissociation energy is higher.

[52]For a monograph, see Melander; Saunders *Reaction Rates of Isotopic Molecules*; Wiley: New York, 1980. For reviews, see Isaacs *Physical Organic Chemistry*; Longman Scientific and Technical: Essex, 1987, pp. 255-281; Lewis *Top. Curr. Chem.* **1978**, *74*, 31-44; Saunders, in Bernasconi, Ref. 25, pp. 565-611; Bell *The Proton in Chemistry*, 2nd ed.; Cornell University Press: Ithaca, NY, 1973, pp. 226-296, *Chem. Soc. Rev.* **1974**, *3*, 513-544; Bigeleisen; Lee; Mandel *Annu. Rev. Phys. Chem.* **1973**, 24, 407-440; Wolfsberg *Annu. Rev. Phys. Chem.* **1969**, *20*, 449-478; Saunders *Surv. Prog. Chem.* **1966**, *3*, 109-146; Simon; Palm *Angew. Chem. Int. Ed. Engl.* **1966**, *5*, 920-933 [*Angew. Chem. 78*, 993-1007]; Jencks, Ref. 31, pp. 243-281. For a review of temperature dependence of primary isotope effects as a mechanistic criterion, see Kwart *Acc. Chem. Res.* **1982**, 15, 401-408. For a review of the effect of pressure on isotope effects, see Isaacs, *Isot. Org. Chem.* **1984**, *6*, 67-105. For a review of isotope effects in the study of reactions in which there is branching from a common intermediate, see Thibblin; Ahlberg *Chem. Soc. Rev.* **1989**, *18*, 209-224. See also the series *Isotopes in Organic Chemistry*.
[53]The reduced mass μ of two atoms connected by a covalent bond is $\mu = m_1m_2/(m_1 + m_2)$.

if the bond is broken in the rate-determining step, the rate must be lowered by the substitution.

This provides a valuable diagnostic tool for determination of mechanism. For example, in the bromination of acetone (**2-4**)

$$CH_3COCH_3 + Br_2 \longrightarrow CH_3COCH_2Br$$

the fact that the rate is independent of the bromine concentration led to the postulate that the rate-determining step was prior tautomerization of the acetone:

$$CH_3COCH_3 \; \Longleftrightarrow \; CH_3\overset{\displaystyle OH}{\underset{|}{C}}{=}CH_2$$

In turn, the rate-determining step of the tautomerization involves cleavage of a C—H bond (see **2-3**). Thus there should be a substantial isotope effect if deuterated acetone is brominated. In fact, k_H/k_D was found to be about 7.[54] Deuterium isotope effects usually range from 1 (no isotope effect at all) to about 7 or 8, though in a few cases, larger[55] or smaller values have been reported.[56] Values of k_H/k_D smaller that 1 are called *inverse isotope effects*. Isotope effects are greatest when, in the transition state, the hydrogen is symmetrically bonded to the atoms between which it is being transferred.[57] Also, calculations show that isotope effects are at a maximum when the hydrogen in the transition state is on the straight line connecting the two atoms between which the hydrogen is being transferred and that for sufficiently nonlinear configurations they decrease to $k_H/k_D = 1$ to 2.[58] Of course, in open systems there is no reason for the transition state to be nonlinear, but this is not the case in many intramolecular mechanisms, e.g., in a 1,2 migration of a hydrogen

$$\overset{\displaystyle H}{\underset{|}{C}}{-}C \qquad \overset{\displaystyle H}{C{-}C} \longrightarrow \overset{\displaystyle H}{C}{-}\overset{\displaystyle H}{\underset{|}{C}}$$

Transition
state

To measure isotope effects it is not always necessary to prepare deuterium-enriched starting compounds. It can also be done by measuring the change in deuterium concentration at specific sites between a compound containing deuterium in natural abundance and the reaction product, using a high field nmr instrument.[59]

[54]Reitz; Kopp Z. Phys. Chem., Abt. A **1939**, *184*, 429.

[55]For an example of a reaction with a deuterium isotope effect of 24.2, see Lewis; Funderburk J. Am. Chem. Soc. **1967**, *89*, 2322. The high isotope effect in this case has been ascribed to *tunneling* of the proton: because it is so small a hydrogen atom can sometimes get through a thin potential barrier without going over the top, i.e., without obtaining the usually necessary activation energy. A deuterium, with a larger mass, is less able to do this. The phenomenon of tunneling is a consequence of the uncertainty principle. k_H/k_T for the same reaction is 79: Lewis; Robinson J. Am. Chem. Soc. **1968**, *90*, 4337. An even larger deuterium isotope effect (~50) has been reported for the oxidation of benzyl alcohol. This has also been ascribed to tunneling: Roecker; Meyer J. Am. Chem. Soc. **1987**, *109*, 746. For discussions of high isotope effects, see Kresge; Powell J. Am. Chem. Soc. **1981**, *103*, 201; Caldin; Mateo; Warrick J. Am. Chem. Soc. **1981**, *103*, 202. For arguments that high isotope effects can be caused by factors other than tunneling, see McLennan Aust. J. Chem. **1979**, *32*, 1883; Thibblin J. Phys. Org. Chem. **1988**, *1*, 161; Kresge; Powell J. Phys. Org. Chem. **1990**, *3*, 55.

[56]For a review of a method for calculating the magnitude of isotope effects, see Sims; Lewis Isot. Org. Chem. **1984**, *6*, 161-259.

[57]Kwart; Latimore J. Am. Chem. Soc. **1971**, *93*, 3770; Pryor; Kneipp J. Am. Chem Soc. **1971**, *93*, 5584; Bell; Cox J. Chem. Soc. B **1971**, 783; Bethell; Hare; Kearney J. Chem. Soc.. Perkin Trans. 2 **1981**, 684, and references cited in these papers. See, however, Motell; Boone; Fink Tetrahedron **1978**, *34*, 1619.

[58]More O'Ferrall J. Chem. Soc. B **1970**, 785, and references cited therein.

[59]Pascal; Baum; Wagner; Rodgers; Huang J. Am. Chem. Soc. **1986**, *108*, 6477.

The substitution of tritium for hydrogen gives isotope effects that are numerically larger. Isotope effects have also been observed with other elements, but they are much smaller, about 1.02 to 1.10. For example, k_{12C}/k_{13C} for

$$\text{Ph*CH}_2\text{Br} + \text{CH}_3\text{O}^- \xrightarrow{\text{CH}_3\text{OH}} \text{Ph*CH}_2\text{OCH}_3$$

is 1.053.[60] Although they are small, heavy-atom isotope effects can be measured quite accurately and are often very useful.[61]

Deuterium isotope effects have been found even where it is certain that the C—H bond does not break at all in the reaction. Such effects are called *secondary isotope effects*,[62] the term *primary isotope effect* being reserved for the type discussed previously. Secondary isotope effects can be divided into α and β effects. In a β secondary isotope effect, substitution of deuterium for hydrogen β to the position of bond breaking slows the reaction. An example is solvolysis of isopropyl bromide:

$$(\text{CH}_3)_2\text{CHBr} + \text{H}_2\text{O} \xrightarrow{k_H} (\text{CH}_3)_2\text{CHOH}$$

$$(\text{CD}_3)_2\text{CHBr} + \text{H}_2\text{O} \xrightarrow{k_D} (\text{CD}_3)_2\text{CHOH}$$

where k_H/k_D was found to be 1.34.[63] The cause of β isotope effects has been a matter of much controversy, but they are most likely due to hyperconjugation effects in the transition state. The effects are greatest when the transition state has considerable carbocation character.[64] Although the C—H bond in question is not broken in the transition state, the carbocation is stabilized by hyperconjugation involving this bond. Because of hyperconjugation, the difference in vibrational energy between the C—H bond and the C—D bond in the transition state is less than it is in the ground state, so the reaction is slowed by substitution of deuterium for hydrogen.

Support for hyperconjugation as the major cause of β isotope effects is the fact that the effect is greatest when D is anti to the leaving group[65] (because of the requirement that all atoms in a resonance system be coplanar, planarity of the D—C—C—X system would most greatly increase the hyperconjugation), and the fact that secondary isotope effects can be transmitted through unsaturated systems.[66] There is evidence that at least some β isotope effects are steric in origin[67] (e.g., a CD_3 group has a smaller steric requirement than a CH_3 group) and a field-effect explanation has also been suggested (CD_3 is apparently a better electron donor than CH_3[68]), but hyperconjugation is the most probable cause in most instances.[69] Part of the difficulty in attempting to explain these effects is their small size,

[60]Stothers; Bourns *Can. J. Chem.* **1962**, *40*, 2007. See also Ando; Yamataka; Tamura; Hanafusa *J. Am. Chem. Soc.* **1982**, *104*, 5493.

[61]For a review of carbon isotope effects, see Willi *Isot. Org. Chem.* **1977**, *3*, 237-283.

[62]For reviews, see Westaway *Isot. Org. Chem.* **1987**, *7*, 275-392; Sunko; Hehre *Prog. Phys. Org. Chem.* **1983**, *14*, 205-246; Shiner, in Collins; Bowman *Isotope Effects in Chemical Reactions*; Van Nostrand-Reinhold: Princeton, 1970, pp. 90-159; Laszlo; Welvart *Bull. Soc. Chim. Fr.* **1966**, 2412-2438; Halevi *Prog. Phys. Org. Chem.* **1963**, *1*, 109-221. For a review of model calculations of secondary isotope effects, see McLennan *Isot. Org. Chem.* **1987**, *7*, 393-480. See also Ref. 56.

[63]Leffek; Llewellyn; Robertson *Can. J. Chem.* **1960**, *38*, 2171.

[64]Bender; Feng *J. Am. Chem. Soc.* **1960**, *82*, 6318; Jones; Bender *J. Am. Chem. Soc.* **1960**, *82*, 6322.

[65]Shiner; Murr; Heinemann *J. Am. Chem. Soc.* **1963**, *85*, 2413; Shiner; Humphrey *J. Am. Chem. Soc.* **1963**, *85*, 2416; Shiner; Jewett *J. Am. Chem. Soc.* **1964**, *86*, 945; DeFrees; Hehre; Sunko *J. Am. Chem. Soc.* **1979**, *101*, 2323. See also Siehl; Walter *J. Chem. Soc., Chem. Commun.* **1985**, 76.

[66]Shiner; Kriz *J. Am. Chem. Soc.* **1964**, *86*, 2643.

[67]Bartell *J. Am. Chem. Soc.* **1961**, *83*, 3567; Brown; Azzaro; Koelling; McDonald *J. Am. Chem. Soc.* **1966**, *88*, 2520; Kaplan; Thornton *J. Am. Chem. Soc.* **1967**, *89*, 6644; Carter; Dahlgren *Acta Chem. Scand.* **1970**, *24*, 633; Leffek; Matheson *Can. J. Chem.* **1971**, *49*, 439; Sherrod; Boekelheide *J. Am. Chem. Soc.* **1972**, *94*, 5513.

[68]Halevi; Nussim; Ron *J. Chem. Soc.* **1963**, 866; Halevi; Nussim *J. Chem. Soc.* **1963**, 876.

[69]Karabatsos; Sonnichsen; Papaioannou; Scheppele; Shone *J. Am. Chem. Soc.* **1967**, *89*, 463; Kresge; Preto *J. Am. Chem. Soc.* **1967**, *89*, 5510; Jewett; Dunlap *J. Am. Chem. Soc.* **1968**, *90*, 809; Sunko; Szele; Hehre *J. Am. Chem. Soc.* **1977**, *99*, 5000; Kluger; Brandl *J. Org. Chem.* **1986**, *51*, 3964.

ranging only as high as about 1.5.[70] Another complicating factor is that they can change with temperature. In one case[71] k_H/k_D was 1.00 ± 0.01 at $0°C$, 0.90 ± 0.01 at $25°C$, and 1.15 ± 0.09 at $65°C$. Whatever the cause, there seems to be a good correlation between β secondary isotope effects and carbocation character in the transition state, and they are thus a useful tool for probing mechanisms.

The other type of secondary isotope effect results from a replacement of hydrogen by deuterium at the carbon containing the leaving group. These (called α *secondary isotope effects*) are varied, with values so far reported[72] ranging from 0.87 to 1.26.[73] These effects are also correlated with carbocation character. Nucleophilic substitutions that do not proceed through carbocation intermediates (SN2 reactions) have α isotope effects near unity.[74] Those that do involve carbocations (SN1 reactions) have higher α isotope effects, which depend on the nature of the leaving group.[75] The accepted explanation for α isotope effects is that one of the bending C—H vibrations is affected by the substitution of D for H more or less strongly in the transition state than in the ground state.[76] Depending on the nature of the transition state, this may increase or decrease the rate of the reaction. α isotope effects on SN2 reactions can vary with concentration,[77] an effect attributed to a change from a free nucleophile to one that is part of an ion pair[78] (see p. 350). This illustrates the use of secondary isotope effects as a means of studying transition state structure. γ secondary isotope effects have also been reported.[79]

Another kind of isotope effect is the *solvent isotope effect*.[80] Reaction rates often change when the solvent is changed from H_2O to D_2O or from ROH to ROD. These changes may be due to any of three factors or a combination of all of them.

1. The solvent may be a reactant. If an O—H bond of the solvent is broken in the rate-determining step, there will be a primary isotope effect. If the molecules involved are D_2O or D_3O^+ there may also be a secondary effect caused by the O—D bonds that are not breaking.

2. The substrate molecules may become labeled with deuterium by rapid hydrogen exchange, and then the newly labeled molecule may become cleaved in the rate-determining step.

3. The extent or nature of solvent-solute interactions may be different in the deuterated and nondeuterated solvents; this may change the energies of the transition state and hence the activation energy of the reaction. These are secondary isotope effects. Two physical models for this third factor have been constructed.[81]

[70]A value for k_{CH_3}/k_{CD_3} of 2.13 was reported for one case: Liu; Wu *Tetrahedron Lett.* **1986**, *27*, 3623.

[71]Halevi; Margolin *Proc. Chem. Soc.* **1964**, 174.

[72]A value of 2.0 has been reported in one case, for a cis-trans isomerization, rather than a nucleophilic substitution: Caldwell; Misawa; Healy; Dewar *J. Am. Chem. Soc.* **1987**, *109*, 6869.

[73]Shiner; Buddenbaum; Murr; Lamaty *J. Am. Chem. Soc.* **1968**, *90*, 418; Harris; Hall; Schleyer *J. Am. Chem. Soc.* **1971**, *93*, 2551.

[74]For reported exceptions, see Tanaka; Kaji; Hayami *Chem.Lett.* **1972**, 1223; Westaway *Tetrahedron Lett.* **1975**, 4229.

[75]Shiner; Dowd *J. Am. Chem. Soc.* **1971**, *93*, 1029; Shiner; Fisher *J. Am. Chem. Soc.* **1971**, *93*, 2553; Willi; Ho; Ghanbarpour *J. Org. Chem.* **1972**, *37*, 1185; Shiner; Neumann; Fisher *J. Am. Chem. Soc.* **1982**, *104*, 354; and references cited in these papers.

[76]Streitwieser; Jagow; Fahey; Suzuki *J. Am. Chem. Soc.* **1958**, *80*, 2326.

[77]Westaway; Waszczylo; Smith; Rangappa *Tetrahedron Lett.* **1985**, *26*, 25.

[78]Westaway; Lai *Can. J. Chem.* **1988**, *66*, 1263.

[79]Leffek; Llewellyn; Robertson *J. Am. Chem. Soc.* **1960**, *82*, 6315, *Chem. Ind. (London)* **1960**, 588; Werstiuk; Timmins; Cappelli *Can. J. Chem.* **1980**, *58*, 1738.

[80]For reviews, see Alvarez; Schowen *Isot. Org. Chem.* **1987**, *7*, 1-60; Kresge; More O'Ferrall, Powell *Isot. Org. Chem.* **1987**, *7*, 177-273; Schowen *Prog. Phys. Org. Chem.* **1972**, *9*, 275-332; Gold *Adv. Phys. Org. Chem.* **1969**, *7*, 259-331; Laughton; Robertson, in Coetzee; Ritchie *Solute–Solvent Interactions;* Marcel Dekker: New York, 1969, pp. 399-538. For a review of the effect of isotopic changes in the solvent on the properties of nonreacting solutes, see Arnett; McKelvey, in Coetzee; Ritchie, cited above, pp. 343-398.

[81]Swain; Bader *Tetrahedron* **1960**, *10*, 182; Bunton; Shiner *J. Am. Chem. Soc.* **1961**, *83*, 42, 3207, 3214; Swain; Thornton *J. Am. Chem. Soc.* **1961**, *83*, 3884, 3890. See also Mitton; Gresser; Schowen *J. Am. Chem. Soc.* **1969**, *91*, 2045.

It is obvious that in many cases the first and third factors at least, and often the second, are working simultaneously. Attempts have been made to separate them.[82]

The methods described in this chapter are not the only means of determining mechanisms. In an attempt to elucidate a mechanism, the investigator is limited only by his or her ingenuity.

[82]More O'Ferrall; Koeppl; Kresge *J. Am. Chem. Soc.* **1971**, *93*, 9.

7
PHOTOCHEMISTRY

Most reactions carried out in organic chemistry laboratories take place between molecules all of which are in their ground electronic states. In a *photochemical reaction*,[1] however, a reacting molecule has been previously promoted by absorption of light to an electronically excited state. A molecule in an excited state must lose its extra energy in some manner; it cannot remain in the excited condition for long. However, a chemical reaction is not the only possible means of relinquishing the extra energy. In this chapter we first discuss electronically excited states and the processes of promotion to these states. Then we examine the possible pathways open to the excited molecule, first the physical and then the chemical pathways. The subject of electronic spectra is closely related to photochemistry.

Excited States and the Ground State

Electrons can move from the ground-state energy level of a molecule to a higher level (i.e., an unoccupied orbital of higher energy) if outside energy is supplied. In a photochemical process this energy is in the form of light. Light of any wavelength has associated with it an energy value given by $E = h\nu$, where ν is the frequency of the light (ν = velocity of light c divided by the wavelength λ) and h is Planck's constant. Since the energy levels of a molecule are quantized, the amount of energy required to raise an electron in a given molecule from one level to a higher one is a fixed quantity. Only light with exactly the frequency corresponding to this amount of energy will cause the electron to move to the higher level. If light of another frequency (too high or too low) is sent through a sample, it will pass out without a loss in intensity, since the molecules will not absorb it. However, if light of the correct frequency is passed in, the energy will be used by the molecules for electron promotion and hence the light that leaves the sample will be diminished in intensity or altogether gone. A *spectrophotometer* is an instrument that allows light of a given frequency to pass through a sample and that detects (by means of a phototube) the amount of light that has been transmitted, i.e., not absorbed. A spectrophotometer compares the intensity of the transmitted light with that of the incident light. Automatic instruments gradually and continuously change the frequency, and an automatic recorder plots a graph of absorption vs. frequency or wavelength.

---

[1]There are many books on photochemistry. Some recent ones are Michl; Bonačić-Koutecký *Electronic Aspects of Organic Photochemistry*; Wiley: New York, 1990; Scaino *Handbook of Organic Photochemistry*, 2 vols.; CRC Press: Boca Raton, FL, 1989; Coxon; Halton *Organic Photochemistry*, 2nd ed.; Cambridge University Press: Cambridge, 1987; Coyle *Photochemistry in Organic Synthesis*; Royal Society of Chemistry: London, 1986, *Introduction to Organic Photochemistry*; Wiley: New York, 1986; Horspool *Synthetic Organic Photochemistry*; Plenum: New York, 1984; Margaretha *Preparative Organic Photochemistry*, *Top. Curr. Chem.* **1982**, *103*; Turro *Modern Molecular Photochemistry*; W.A. Benjamin: New York, 1978; Rohatgi-Mukherjee *Fundamentals of Photochemistry*; Wiley: New York, 1978; Barltrop; Coyle *Principles of Photochemistry*; Wiley: New York, 1978. For a comprehensive older treatise, see Calvert; Pitts *Photochemistry*; Wiley: New York, 1966. For a review of the photochemistry of radicals and carbenes, see Scaiano; Johnston *Org. Photochem.* **1989**, *10*, 309-355. For a history of photochemistry, see Roth *Angew. Chem. Int. Ed. Engl.* **1989**, *28*, 1193-1207 [*Angew. Chem. 101*, 1220-1234]. For a glossary of terms used in photochemistry, see Braslavsky; Houk *Pure Appl. Chem.* **1988**, *60*, 1055-1106. See also the series, *Advances in Photochemistry*, *Organic Photochemistry*, and *Excited States*.

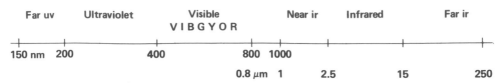

FIGURE 7.1   The uv, visible, and ir portions of the spectrum.

The energy of electronic transitions corresponds to light in the visible, uv, and far-uv regions of the spectrum (Figure 7.1). Absorption positions are normally expressed in wavelength units, usually nanometers (nm).[2] If a compound absorbs in the visible, it is colored, possessing a color complementary to that which is absorbed.[3] Thus a compound absorbing in the violet is yellow. The far-uv region is studied by organic chemists less often than the visible or ordinary uv regions because special vacuum instruments are required owing to the fact that oxygen and nitrogen absorb in these regions.

From these considerations it would seem that an electronic spectrum should consist of one or more sharp peaks, each corresponding to the transfer of an electron from one electronic level to another. Under ordinary conditions the peaks are seldom sharp. In order to understand why, it is necessary to realize that molecules are constantly vibrating and rotating and that these motions are also quantized. A molecule at any time is not only in a given electronic state but also in a given vibrational and rotational state. The difference between two adjacent vibrational levels is much smaller than the difference between adjacent electronic levels, and the difference between adjacent rotational levels is smaller still. A typical situation is shown in Figure 7.2. When an electron moves from one electronic level to another, it moves from a given vibrational and rotational level within that electronic level to some vibrational and rotational level at the next electronic level. A given sample contains a large number of molecules, and even if all of them are in the ground electronic state, they are still distributed among the vibrational and rotational states (though the ground vibrational state V_0 is most heavily populated). This means that not just one wavelength of light will be absorbed but a number of them close together, with the most probable transition causing the most intense peak. But in molecules containing more than a few atoms there are so many possible transitions and these are so close together that what is observed is a relatively broad band. The height of the peak depends on the number of molecules making the transition and is proportional to log ε, where ε is the *extinction coefficient*. The extinction coefficient can be expressed by $\varepsilon = E/cl$, where c is the concentration in moles per liter, l is the cell length in centimeters, and $E = \log I_0/I$, where I_0 is the intensity of the incident light and I of the transmitted light. The wavelength is usually reported as λ_{max}, meaning that this is the top of the peak. Purely vibrational transitions, such as between V_0 and V_1 of E_1, which require much less energy, are found in the ir region and are the basis of ir spectra. Purely rotational transitions are found in the far-ir and microwave (beyond the far-ir) regions.

A uv or visible absorption peak is caused by the promotion of an electron in one orbital (usually a ground-state orbital) to a higher orbital. Normally the amount of energy necessary to make this transition depends mostly on the nature of the two orbitals involved and much less on the rest of the molecule. Therefore, a simple functional group such as the C=C double bond always causes absorption in the same general area. A group that causes absorption is called a *chromophore*.

[2]Formerly, millimicrons (mμ) were frequently used; numerically they are the same as nanometers.
[3]For monographs, see Zollinger *Color Chemistry*; VCH: New York, 1987; Gordon; Gregory *Organic Chemistry in Colour*; Springer: New York, 1983; Griffiths *Colour and Constitution of Organic Molecules*; Academic Press: New York, 1976. See also Fabian; Zahradník *Angew. Chem. Int. Ed. Engl.* **1989**, *28*, 677-694 [*Angew. Chem. 101*, 693-710].

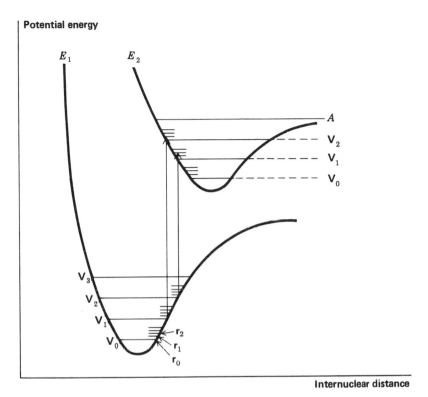

FIGURE 7.2   Energy curves for a diatomic molecule. Two possible transitions are shown. When an electron has been excited to the point marked *A*, the molecule may cleave (p. 236).

Singlet and Triplet States. "Forbidden" Transitions

In most organic molecules, all electrons in the ground state are paired, with each member of a pair possessing opposite spin as demanded by the Pauli principle. When one of a pair of electrons is promoted to an orbital of higher energy, the two electrons no longer share an orbital, and the promoted electron may, in principle, have the same spin as its former partner or the opposite spin. As we saw in Chapter 5, a molecule in which two unpaired electrons have the same spin is called a *triplet*, while one in which all spins are paired is a *singlet*. Thus, at least in principle, for every excited singlet state there is a corresponding triplet state. In most cases, the triplet state has a lower energy than the corresponding singlet because of Hund's rule. Therefore, a different amount of energy and hence a different wavelength is required to promote an electron from the ground state (which is almost always a singlet) to an excited singlet than to the corresponding triplet state.

It would thus seem that promotion of a given electron in a molecule could result either in a singlet or a triplet excited state depending on the amount of energy added. However, this is often not the case because transitions between energy levels are governed by selection rules, which state that certain transitions are "forbidden." There are several types of "forbidden" transitions, two of which are more important than the others.

1. *Spin-forbidden transitions.*   Transitions in which the spin of an electron changes are not allowed, because a change from one spin to the opposite involves a change in angular momentum and such a change would violate the law of conservation of angular momentum.

Therefore, singlet–triplet and triplet–singlet transitions are forbidden, whereas singlet–singlet and triplet–triplet transitions are allowed.

   2. *Symmetry-forbidden transitions.*   Among the transitions in this class are those in which a molecule has a center of symmetry. In such cases, a $g \rightarrow g$ or $u \rightarrow u$ transition (see p. 5) is "forbidden," while a $g \rightarrow u$ or $u \rightarrow g$ transition is allowed.

We have put the word "forbidden" into quotation marks because these transitions are not actually forbidden but only highly improbable. In most cases promotions from a singlet ground state to a triplet excited state are so improbable that they cannot be observed, and it is safe to state that in most molecules only singlet–singlet promotions take place. However, this rule does break down in certain cases, most often when a heavy atom (such as iodine) is present in the molecule, in which cases it can be shown from spectra that singlet–triplet promotions are occurring.[4] Symmetry-forbidden transitions can frequently be observed, though usually with low intensity.

Types of Excitation

When an electron in a molecule is promoted (normally only one electron in any molecule), it usually goes into the lowest available vacant orbital, though promotion to higher orbitals is also possible. For most organic molecules there are consequently four types of electronic excitation:

   1. $\sigma \rightarrow \sigma^*$. Alkanes, which have no n or π electrons, can be excited only in this way.[5]
   2. $n \rightarrow \sigma^*$. Alcohols, amines,[6] ethers, etc. can also be excited in this manner.
   3. $\pi \rightarrow \pi^*$. This pathway is open to alkenes as well as to aldehydes, carboxylic esters, etc.
   4. $n \rightarrow \pi^*$. Aldehydes, ketones, carboxylic esters, etc. can undergo this promotion as well as the other three.

The four excitation types above are listed in what is normally the order of decreasing energy. Thus light of the highest energy (in the far uv) is necessary for $\sigma \rightarrow \sigma^*$ excitation, while $n \rightarrow \pi^*$ promotions are caused by ordinary uv light. However, the order may sometimes be altered in some solvents.

   In 1,3-butadiene (and other compounds with two conjugated double bonds) there are two π and two π^* orbitals (p. 31). The energy difference between the higher π (χ_2) and the lower π^* (χ_3) orbital is less than the difference between the π and π^* orbitals of ethylene. Therefore 1,3-butadiene requires less energy than ethylene, and thus light of a higher wavelength, to promote an electron. This is a general phenomenon, and it may be stated that, in general, *the more conjugation in a molecule, the more the absorption is displaced toward higher wavelengths* (see Table 7.1).[7] When a chromophore absorbs at a certain wavelength and the substitution of one group for another causes absorption at a longer wavelength, a *bathochromic shift* is said to have occurred. The opposite kind of shift is called *hypsochromic*.

   Of the four excitation types listed above, the $\pi \rightarrow \pi^*$ and $n \rightarrow \pi^*$ are far more important in organic photochemistry than the other two. Compounds containing C=O groups can be excited in both ways, giving rise to at least two peaks in the uv.

[4]For a review of photochemical heavy-atom effects, see Koziar; Cowan *Acc. Chem. Res.* **1978,** *11*, 334-341.
[5]An *n* electron is one in an unshared pair.
[6]For a review of the photochemistry of amines, see Malkin; Kuz'min *Russ. Chem. Rev.* **1985,** *54*, 1041-1057.
[7]Bohlmann; Mannhardt *Chem. Ber.* **1956,** *89*, 1307.

TABLE 7.1 Ultraviolet absorption[7]
of CH_3—$(CH{=}CH)_n$—CH_3 for some
values of n

n	nm
2	227
3	263
6	352
9	413

As we have seen, a chromophore is a group that causes a molecule to absorb light. Examples of chromophores in the visible or uv are $C{=}O$, $N{=}N$,[8] Ph, and NO_2. Some chromophores in the far uv (beyond 200 nm) are $C{=}C$, $C{\equiv}C$, Cl, and OH. An *auxochrome* is a group that displaces (through resonance) and usually intensifies the absorption of a chromophore present in the same molecule. Groups such as Cl, OH, and NH_2 are generally regarded as auxochromes since they shift (usually bathochromically) the uv and visible bands of chromophores such as Ph or $C{=}O$ (see Table 7.2).[9] Since auxochromes are themselves chromophores (to be sure, generally in the far-uv), it is sometimes difficult to decide which group in a molecule is an auxochrome and which a chromophore. For example, in acetophenone (PhCOMe) is the chromophore Ph or $C{=}O$? In such cases the distinction becomes practically meaningless.

TABLE 7.2 Some uv peaks of substituted benzenes in water, or water with a
trace of methanol (for solubility)
Note how auxochromes shift and usually intensify the peaks[9]

	Primary band		Secondary band	
	λ_{max}, nm	ϵ_{max}	λ_{max}, nm	ϵ_{max}
PhH	203.5	7,400	254	204
PhCl	209.5	7,400	263.5	190
PhOH	210.5	6,200	270	1,450
PhOMe	217	6,400	269	1,480
PhCN	224	13,000	271	1,000
PhCOOH	230	11,600	273	970
PhNH$_2$	230	8,600	280	1,430
PhO$^-$	235	9,400	287	2,600
PhAc	245.5	9,800		
PhCHO	249.5	11,400		
PhNO$_2$	268.5	7,800		

[8]For a review of the azo group as a chromophore, see Rau *Angew. Chem. Int. Ed. Engl.* **1973**, 224-235 [*Angew. Chem.* **85**, 248-258].
[9]These values are from Jaffé; Orchin *Theory and Applications of Ultraviolet Spectroscopy*; Wiley: New York, 1962, p. 257.

Nomenclature and Properties of Excited States

An excited state of a molecule can be regarded as a distinct chemical species, different from the ground state of the same molecule and from other excited states. It is obvious that we need some method of naming excited states. Unfortunately, there are several methods in use, depending on whether one is primarily interested in photochemistry, spectroscopy, or molecular-orbital theory.[10] One of the most common methods simply designates the original and newly occupied orbitals, with or without a superscript to indicate singlet or triplet. Thus the singlet state arising from promotion of a π to a π^* orbital in ethylene would be the $^1(\pi,\pi^*)$ state or the π,π^* singlet state. Another very common method can be used even in cases where one is not certain which orbitals are involved. The lowest-energy excited state is called S_1, the next S_2, etc., and triplet states are similarly labeled T_1, T_2, T_3, etc. In this notation the ground state is S_0. Other notational systems exist, but in this book we shall confine ourselves to the two types just mentioned.

The properties of excited states are not easy to measure because of their generally short lifetimes and low concentrations, but enough work has been done for us to know that they often differ from the ground state in geometry, dipole moment and acid or base strength.[11] For example, acetylene, which is linear in the ground state, has a trans geometry

$$\underset{H}{\overset{}{\diagdown}}C\equiv C\overset{H}{\diagup}$$

with approximately sp^2 carbons in the $^1(\pi,\pi^*)$ state.[12] Similarly, the $^1(\pi,\pi^*)$ and the $^3(\pi,\pi^*)$ states of ethylene have a perpendicular and not a planar geometry,[13] and the $^1(n,\pi^*)$ and $^3(n,\pi^*)$ states of formaldehyde are both pyramidal.[14] Triplet species tend to stabilize themselves by distortion, which relieves interaction between the unpaired electrons. Obviously, if the geometry is different, the dipole moment will probably differ also and the change in geometry and electron distribution often results in a change in acid or base strength.[15] For example, the S_1 state of 2-naphthol is a much stronger acid (p$K = 3.1$) than the ground state (S_0) of the same molecule (p$K = 9.5$).[16]

Photolytic Cleavage

We have said that when a molecule absorbs a quantum of light, it is promoted to an excited state. Actually, that is not the only possible outcome. Because the energy of visible and uv light is of the same order of magnitude as that of covalent bonds (Table 7.3), another

---

[10]For discussions of excited-state notation and other terms in photochemistry, see Pitts; Wilkinson; Hammond *Adv. Photochem.* **1963**, *1*, 1-21; Porter; Balzani; Moggi *Adv. Photochem.* **1974**, *9*, 147-196. See also Braslavsky; Houk, Ref. 1.

[11]For reviews of the structures of excited states, see Zink; Shin *Adv. Photochem.* **1991**, *16*, 119-214; Innes *Excited States* **1975**, *2*, 1-32; Hirakawa; Masamichi *Vib. Spectra Struct.* **1983**, *12*, 145-204.

[12]Ingold; King *J. Chem. Soc.* **1953**, 2702, 2704, 2708, 2725, 2745. For a review of acetylene photochemistry, see Coyle *Org. Photochem.* **1985**, *7*, 1-73.

[13]Mercer; Mulliken *Chem. Rev.* **1969**, *69*, 639-656.

[14]Robinson; Di Giorgio *Can. J. Chem.* **1958**, *36*, 31; Buenker; Peyerimhoff *J. Chem. Phys.* **1970**, *53*, 1368; Garrison; Schaefer; Lester *J. Chem. Phys.* **1974**, *61*, 3039; Streitwieser; Kohler *J. Am. Chem. Soc.* **1988**, *110*, 3769. For reviews of excited states of formaldehyde, see Buck *Recl. Trav. Chim. Pays-Bas* **1982**, *101*, 193-198, 225-233; Moule; Walsh *Chem. Rev.* **1975**, *75*, 67-84.

[15]For a review of acid–base properties of excited states, see Ireland; Wyatt *Adv. Phys. Org. Chem.* **1976**, *12*, 131-221.

[16]Weller *Z. Phys. Chem. (Frankfurt am Main)* **1955**, *3*, 238, *Discuss. Faraday Soc.* **1959**, *27*, 28.

TABLE 7.3 Typical energies for some
covalent single bonds (see Table 1.7) and the
corresponding approximate wavelengths

Bond	E		
	kcal/mol	kJ/mol	nm
C—H	95	397	300
C—O	88	368	325
C—C	83	347	345
Cl—Cl	58	243	495
O—O	35	146	820

possibility is that the molecule may cleave into two parts, a process known as *photolysis*.
There are three situations that can lead to cleavage:

1. The promotion may bring the molecule to a vibrational level so high that it lies above
the right-hand portion of the E_2 curve (line A in Figure 7.2). In such a case the excited
molecule cleaves at its first vibration.

2. Even where the promotion is to a lower vibrational level, one which lies wholly within
the E_2 curve (such as V_1 or V_2), the molecule may still cleave. As Figure 7.2 shows,
equilibrium distances are greater in excited states than in the ground state. The *Franck-
Condon principle* states that promotion of an electron takes place much faster than a single
vibration (the promotion takes about 10^{-15} sec; a vibration about 10^{-12} sec). Therefore,
when an electron is suddenly promoted, even to a low vibrational level, the distance between
the atoms is essentially unchanged and the bond finds itself in a compressed condition like
a pressed-in spring; this condition may be relieved by an outward surge that is sufficient to
break the bond.

3. In some cases the excited state is entirely dissociative (Figure 7.3), i.e., there is no

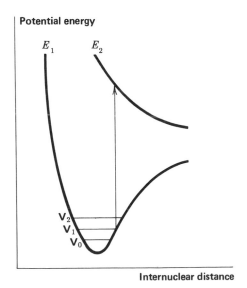

FIGURE 7.3 Promotion to a dissociative state results in bond cleavage.

distance where attraction outweighs repulsion, and the bond must cleave. An example is the hydrogen molecule, where a $\sigma \rightarrow \sigma^*$ promotion always results in cleavage.

A photolytic cleavage can break the molecule into two smaller molecules or into two free radicals (see p. 243). Cleavage into two ions, though known, is much rarer. Once free radicals are produced by a photolysis, they behave like free radicals produced in any other way (Chapter 5) except that they may be in excited states, and this can cause differences in behavior.

The Fate of the Excited Molecule: Physical Processes

When a molecule has been photochemically promoted to an excited state, it does not remain there for long. Most promotions are from the S_0 to the S_1 state. As we have seen, promotions from S_0 to triplet states are "forbidden." Promotions to S_2 and higher singlet states take place, but in liquids and solids these higher states usually drop very rapidly to the S_1 state (about 10^{-13} to 10^{-11} sec). The energy lost when an S_2 or S_3 molecule drops to S_1 is given up in small increments to the environment by collisions with neighboring molecules. Such a process is called an *energy cascade*. In a similar manner, the initial excitation and the decay from higher singlet states initially populate many of the vibrational levels of S_1, but these also cascade, down to the lowest vibrational level of S_1. Therefore, in most cases, the lowest vibrational level of the S_1 state is the only important excited singlet state.[17] This state can undergo various physical and chemical processes. In the following list, we describe the physical pathways open to molecules in the S_1 and excited triplet states. These pathways are also shown in a modified Jablonski diagram (Figure 7.4) and in Table 7.4.

1. A molecule in the S_1 state can cascade down through the vibrational levels of the S_0 state and thus return to the ground state by giving up its energy in small increments to the environment, but this is generally quite slow because the amount of energy is large. The process is called *internal conversion* (IC). Because it is slow, most molecules in the S_1 state adopt other pathways.[18]

2. A molecule in the S_1 state can drop to some low vibrational level of the S_0 state all at once by giving off the energy in the form of light. This process, which generally happens within 10^{-9} sec, is called *fluorescence*. This pathway is not very common either (because it is relatively slow), except for small molecules, e.g., diatomic, and rigid molecules, e.g., aromatic. For most other compounds fluorescence is very weak or undetectable. For compounds that do fluoresce, the fluorescence emission spectra are usually the approximate mirror images of the absorption spectra. This comes about because the fluorescing molecules all drop from the lowest vibrational level of the S_1 state to various vibrational levels of S_0, while excitation is from the lowest vibrational level of S_0 to various levels of S_1 (Figure 7.5). The only peak in common is the one (called the 0–0 peak) that results from transitions between the lowest vibrational levels of the two states. In solution, even the 0–0 peak may be noncoincidental because the two states are solvated differently. Fluorescence nearly always arises from a $S_1 \rightarrow S_0$ transition, though azulene (p. 49) and its simple derivatives are exceptions,[19] emitting fluorescence from $S_2 \rightarrow S_0$ transitions.

---

[17]For a review of physical and chemical processes undergone by higher states, see Turro; Ramamurthy; Cherry; Farneth *Chem. Rev.* **1978**, *78*, 125-145.

[18]For a monograph on radiationless transitions, see Lin *Radiationless Transitions*; Academic Press: New York, 1980. For reviews, see Kommandeur *Recl. Trav. Chim. Pays-Bas* **1983**, *102*, 421-428; Freed *Acc. Chem. Res.* **1978**, *11*, 74-80.

[19]For other exceptions, see Gregory; Hirayama; Lipsky *J. Chem. Phys.* **1973**, *58*, 4697; Sugihara; Wakabayashi; Murata; Jinguji; Nakazawa; Persy; Wirz *J. Am. Chem. Soc.* **1985**, *107*, 5894, and references cited in these papers. See also Ref. 17, pp. 126-129.

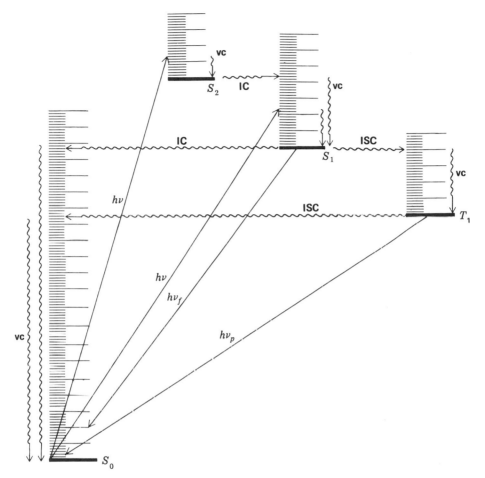

FIGURE 7.4 Modified Jablonski diagram showing transitions between excited states and the ground state. Radiative processes are shown by straight lines, radiationless processes by wavy lines. *IC* = internal conversion; *ISC* = intersystem crossing, *vc* = vibrational cascade; $h\nu_f$ = fluorescence; $h\nu_p$ = phosphorescence.

Because of the possibility of fluorescence, any chemical reactions of the S_1 state must take place very fast, or fluorescence will occur before they can happen.

3. Most molecules (though by no means all) in the S_1 state can undergo an *intersystem crossing* (ISC) to the lowest triplet state T_1.[20] An important example is benzophenone, of which approximately 100% of the molecules that are excited to the S_1 state cross over to the T_1.[21] Intersystem crossing from singlet to triplet is of course a "forbidden" pathway, since the angular-momentum problem (p. 233) must be taken care of, but this often takes place by compensations elsewhere in the system. Intersystem crossings take place without loss of energy. Since a singlet state usually has a higher energy than the corresponding

[20]Intersystem crossing from S_1 to T_2 and higher triplet states has also been reported in some aromatic molecules: Li; Lim *Chem. Phys.* **1972**, *57*, 605; Sharf; Silbey *Chem. Phys. Lett.* **1970**, *5*, 314. See also Schlag; Schneider; Fischer *Annu. Rev. Phys. Chem.* **1971**, *22*, 465-526, pp. 490-494. There is evidence that ISC can also occur from the S_2 state of some molecules: Samanta *J. Am. Chem. Soc.* **1991**, *113*, 7427.
[21]Moore; Hammond; Foss *J. Am. Chem. Soc.* **1961**, *83*, 2789.

TABLE 7.4   Physical processes undergone by
excited molecules
*The superscript v indicates
vibrationally excited state: excited states higher
than S_1 or T_1 are omitted*

$S_0 + h\nu \rightarrow S_1^v$	Excitation
$S_1^v \rightsquigarrow S_1 + \text{heat}$	Vibrational relaxation
$S_1 \rightarrow S_0 + h\nu$	Fluorescence
$S_1 \rightsquigarrow S_0 + \text{heat}$	Internal conversion
$S_1 \rightsquigarrow T_1^v$	Intersystem crossing
$T_1^v \rightsquigarrow T_1 + \text{heat}$	Vibrational relaxation
$T_1 \rightarrow S_0 + h\nu$	Phosphorescence
$T_1 \rightsquigarrow S_0 + \text{heat}$	Intersystem crossing
$S_1 + A_{(S_0)} \rightarrow S_0 + A_{(S_1)}$	Singlet–singlet transfer (photosensitization)
$T_1 + A_{(S_0)} \rightarrow S_0 + A_{(T_1)}$	Triplet–triplet transfer (photosensitization)

triplet, this means that energy must be given up. One way for this to happen is for the S_1 molecule to cross to a T_1 state at a high vibrational level and then for the T_1 to cascade down to its lowest vibrational level (see Figure 7.4). This cascade is very rapid (10^{-12} sec). When T_2 or higher states are populated, they too rapidly cascade to the lowest vibrational level of the T_1 state.

4. A molecule in the T_1 state may return to the S_0 state by giving up heat (intersystem crossing) or light (this is called *phosphorescence*).[22] Of course, the angular-momentum difficulty exists here, so that both intersystem crossing and phosphorescence are very slow ($\sim 10^{-3}$ to 10^1 sec). This means that T_1 states generally have much longer lifetimes than S_1 states. When they occur in the same molecule, phosphorescence is found at lower frequencies than fluorescence (because of the higher difference in energy between S_1 and S_0 than between T_1 and S_0) and is longer-lived (because of the longer lifetime of the T_1 state).

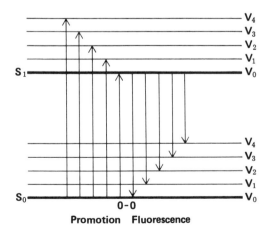

FIGURE 7.5   Promotion and fluorescence between S_1 and S_0 states.

[22]For a review of physical processes of triplet states, see Lower; El-Sayed *Chem. Rev.* **1966,** *66,* 199-241. For a review of physical and chemical processes of triplet states see Wagner; Hammond *Adv. Photochem.* **1968,** *5,* 21-156.

5. If nothing else happens to it first, a molecule in an excited state (S_1 or T_1) may transfer its excess energy all at once to another molecule in the environment, in a process called *photosensitization*.[23] The excited molecule (which we shall call D for donor) thus drops to S_0 while the other molecule (A for acceptor) becomes excited:

$$\mathbf{D^* + A \longrightarrow A^* + D}$$

Thus there are *two* ways for a molecule to reach an excited state—by absorption of a quantum of light or by transfer from a previously excited molecule.[24] The donor D is also called a *photosensitizer*. This energy transfer is subject to the *Wigner spin-conservation rule*, which is actually a special case of the law of conservation of momentum we encountered previously. According to the Wigner rule, the total electron spin does not change after the energy transfer. For example, when a triplet species interacts with a singlet these are some allowed possibilities:[25]

D*	A	D	A*	
($\uparrow\uparrow$)* +	$\uparrow\downarrow$ $\longrightarrow$	$\uparrow\downarrow$ +	($\uparrow\uparrow$)*	singlet and triplet
	$\longrightarrow$	$\uparrow\uparrow\downarrow$ +	$\uparrow$	doublet and doublet (two radicals)
	$\longrightarrow$	$\uparrow\uparrow$ +	$\downarrow$ + $\uparrow$	triplet and two doublets
	$\longrightarrow$	$\uparrow\downarrow$ +	$\uparrow$ + $\uparrow$	singlet and two doublets

In all these cases the products have three electrons spinning "up" and the fourth "down" (as do the starting molecules). However, formation of, say, two triplets ($\uparrow\uparrow$ + $\downarrow\downarrow$) or two singlets ($\uparrow\downarrow$ + $\uparrow\downarrow$), whether ground states or excited, would violate the rule.

In the two most important types of photosensitization, both of which are in accord with the Wigner rule, a triplet excited state generates another triplet and a singlet generates a singlet:

$$\mathbf{D}_{T_1} + \mathbf{A}_{S_0} \longrightarrow \mathbf{A}_{T_1} + \mathbf{D}_{S_0} \quad \text{triplet–triplet transfer}$$

$$\mathbf{D}_{S_1} + \mathbf{A}_{S_0} \longrightarrow \mathbf{A}_{S_1} + \mathbf{D}_{S_0} \quad \text{singlet–singlet transfer}$$

Singlet–singlet transfer can take place over relatively long distances, e.g., 40 Å, but triplet transfer normally requires a collision between the molecules.[26] Both types of photosensitization can be useful for creating excited states when they are difficult to achieve by direct irradiation. Photosensitization is therefore an important method for carrying out photochemical reactions when a molecule cannot be brought to the desired excited state by direct absorption of light. Triplet–triplet transfer is especially important because triplet states are usually much more difficult to prepare by direct irradiation than singlet states (often impossible) and because triplet states, having longer lifetimes, are much more likely than singlets to transfer energy by photosensitization. Photosensitization can also be accomplished by electron transfer.[27]

[23]For reviews, see Albini *Synthesis* **1981**, 249-264; Turro; Dalton; Weiss *Org. Photochem.* **1969**, *2*, 1-62.

[24]There is also a third way: in certain cases excited states can be produced directly in ordinary reactions. For a review, see White; Miano; Watkins; Breaux *Angew. Chem. Int. Ed. Engl.* **1974**, *13*, 229-243 [*Angew. Chem. 86*, 292-307].

[25]For another table of this kind, see Calvert; Pitts, Ref. 1, p. 89.

[26]Long-range triplet–triplet transfer has been observed in a few cases: Bennett; Schwenker; Kellogg *J. Chem. Phys.* **1964**, *41*, 3040; Ermolaev; Sveshnikova *Izv. Akad. Nauk SSSR, Ser. Fiz.* **1962**, *26*, 29 [*C. A.* **1962**, *57*, 1688], *Opt. Spectrosc. (USSR)* **1964**, *16*, 320.

[27]For a review, see Kavarnos; Turro *Chem. Rev.* **1986**, *86*, 401-449. See also Mariano, Ref. 35.

In choosing a photosensitizer one should avoid a compound that absorbs in the same region as the acceptor because the latter will then compete for the light.[28] For examples of the use of photosensitization to accomplish reactions, see **5-37, 5-49.**

The Fate of the Excited Molecule: Chemical Processes

Although both excited singlet and triplet species can undergo chemical reactions, they are much more common for triplets, simply because these generally have much longer lifetimes. Excited singlet species, in most cases, have a lifetime of less than 10^{-10} sec and undergo one of the physical processes already discussed before they have a chance to react chemically. Therefore, photochemistry is largely the chemistry of triplet states.[29] Table 7.5[30] lists many of the possible chemical pathways that can be taken by an excited molecule.[31] The first four of these are unimolecular reactions; the others are bimolecular. In the case of bimolecular reactions it is rare for two excited molecules to react with each other (because the concentration of excited molecules at any one time is generally low); reactions are between an excited molecule and an unexcited molecule of either the same or another species. The reactions listed in Table 7.5 are primary processes. Secondary reactions often follow, since the primary products are frequently radicals or carbenes; even if they are ordinary molecules, they are often in upper vibrational levels and so have excess energy. In almost all cases the primary products of photochemical reactions are in their ground states, though exceptions are known.[32] Of the reactions listed in Table 7.5, the most common are cleavage into radicals (1), decomposition into molecules (2), and (in the presence of a suitable acceptor molecule) photosensitization (7), which we have already discussed. The following are some specific examples of reaction categories (1) to (6). Other examples are discussed in Part 2 of this book.[33]

TABLE 7.5 Primary photochemical reactions of an excited molecule **A—B—C**[30]
 Examples are given in the text; the most common are (1), (2), *and, in the presence of a suitable acceptor molecule,* (7)

(A—B—C) $\longrightarrow$ A—B· + C·	Simple cleavage into radicals	(1)
(A—B—C) $\longrightarrow$ E + F	Decomposition into molecules	(2)
(A—B—C) $\longrightarrow$ A—C—B	Intramolecular rearrangement	(3)
(A—B—C) $\longrightarrow$ A—B—C'	Photoisomerization	(4)
(A—B—C) $\xrightarrow{RH}$ A—B—C—H + R·	Hydrogen-atom abstraction	(5)
(A—B—C) $\longrightarrow$ (ABC)$_2$	Photodimerization	(6)
(A—B—C) $\xrightarrow{A}$ ABC + A*	Photosensitization	(7)

[28]For a review of other complications that can take place in photosensitized reactions, see Engel; Monroe *Adv. Photochem.* **1971,** *8,* 245-313.

[29]For a review of the chemical reactions of triplet states, see Wagner; Hammond, Ref. 22. For other reviews of triplet states, see *Top. Curr. Chem.,* **1975,** vols. 54 and 55.

[30]Adapted from Calvert; Pitts, Ref. 1, p. 367.

[31]For a different kind of classification of photochemical reactions, see Dauben; Salem; Turro *Acc. Chem. Res.* **1975,** *8,* 41. For reviews of photochemical reactions where the molecules are geometrically constrained, see Ramamurthy *Tetrahedron* **1986,** *42,* 5753-5839; Ramamurthy; Eaton *Acc. Chem. Res.* **1988,** *21,* 300-306; Turro; Cox; Paczkowski *Top. Curr. Chem.* **1985,** *129,* 57-97.

[32]Turro; Lechtken; Lyons; Hautala; Carnahan; Katz *J. Am. Chem. Soc.* **1973,** *95,* 2035.

[33]For monographs on the use of photochemistry for synthesis, see Ninomiya; Naito *Photochemical Synthesis*; Academic Press: New York, 1989; Coyle *Photochemistry in Organic Synthesis*; Royal Society of Chemistry: London, 1986; Schönberg *Preparative Organic Photochemistry*; Springer: Berlin, 1968.

Category 1. *Simple cleavage into radicals.*[34]Aldehydes and ketones absorb in the 230 to 330 nm region. This is assumed to result from an $n \rightarrow \pi^*$ singlet–singlet transition. The excited aldehyde or ketone can then cleave.[35]

$$R'—\underset{\underset{O}{\|}}{C}—R \xrightarrow{h\nu} R'—\underset{\underset{O}{\|}}{C}\cdot + R\cdot$$

When applied to ketones, this is called *Norrish Type I cleavage* or often just *Type I cleavage*. In a secondary process, the acyl radical R'—CO• can then lose CO to give R'• radicals. Another example of a category 1 process is cleavage of Cl_2 to give two Cl atoms. Other bonds that are easily cleaved by photolysis are the O—O bonds of peroxy compounds and the C—N bonds of aliphatic azo compounds R—N=N—R.[36] The latter is an important source of radicals R•, since the other product is the very stable N_2.

Category 2. *Decomposition into molecules.* Aldehydes (though not generally ketones) can also cleave in this manner:

$$R—\underset{\underset{O}{\|}}{C}—H \xrightarrow{h\nu} R—H + CO$$

This is an extrusion reaction (see Chapter 17). In another example of a process in category 2, aldehydes and ketones with a γ hydrogen can cleave in still another way (a β elimination, see Chapter 17):

$$R_2CH—CR_2—\underset{\underset{O}{\|}}{C}—R' \xrightarrow{h\nu} R_2C{=}CR_2 + R_2CH—\underset{\underset{O}{\|}}{C}—R'$$

This reaction, called *Norrish Type II cleavage*,[37] involves intramolecular abstraction of the γ hydrogen followed by cleavage of the resulting diradical[38] (a secondary reaction) to give an enol that tautomerizes to the aldehyde or ketone product.[39]

[34]For reviews, see Jackson; Okabe *Adv. Photochem.* **1986**, *13*, 1-94; Kresin; Lester *Adv. Photochem.* **1986**, *13*, 95-163.

[35]For full discussions of aldehyde and ketone photochemistry, see Formosinho; Arnaut *Adv. Photochem.* **1991**, *16*, 67-117; Newton, in Coyle, Ref. 33, pp. 39-60; Lee; Lewis *Adv. Photochem.* **1980**, *12*, 1-96; Calvert; Pitts, Ref. 1, pp. 368-427; Coyle; Carless *Chem. Soc. Rev.* **1972**, *1*, 465-480; Pitts; Wan, in Patai *The Chemistry of the Carbonyl Group*; Wiley: New York, 1966, pp. 823-916; Dalton; Turro *Annu. Rev. Phys. Chem.* **1970**, *21*, 499-560; Bérces, in Bamford; Tipper *Comprehensive Chemical Kinetics*, vol. 5; Elsevier: New York, 1972, pp. 277-380; Turro; Dalton; Dawes; Farrington; Hautala; Morton; Niemczyk; Shore *Acc. Chem. Res.* **1972**, *5*, 92-101; Wagner *Top. Curr. Chem.* **1976**, *66*, 1-52; Wagner; Hammond, Ref. 22, pp. 87-129. For reviews of the photochemistry of cyclic ketones, see Weiss *Org. Photochem.* **1981**, *5*, 347-420; Chapman; Weiss *Org. Photochem.* **1973**, *3*, 197-288; Morton; Turro *Adv. Photochem.* **1974**, *9*, 197-309. For reviews of the photochemistry of α-diketones, see Rubin *Top. Curr. Chem.* **1985**, *129*, 1-56, **1969**, *13*, 251-306; Monroe *Adv. Photochem.* **1971**, *8*, 77-108. For a review of the photochemistry of protonated unsaturated carbonyl compounds, see Childs *Rev. Chem. Intermed.* **1980**, *3*, 285-314. For reviews of the photochemistry of C=S compounds, see Coyle *Tetrahedron* **1985**, *41*, 5393-5425; Ramamurthy *Org. Photochem.* **1985**, *7*, 231-338. For a review of the chemistry of C=N compounds, see Mariano *Org. Photochem.* **1987**, *9*, 1-128.

[36]For reviews of the photochemistry of azo compounds, see Adam; Oppenländer *Angew. Chem. Int. Ed. Engl.* **1986**, *25*, 661-672 [*Angew. Chem. 98*, 659-670]; Dürr; Ruge *Top. Curr. Chem.* **1976**, *66*, 53-87; Drewer, in Patai *The Chemistry of the Hydrazo, Azo, and Azoxy Groups*, pt. 2; Wiley: New York, 1975, pp. 935-1015.

[37]For thorough discussions of the mechanism, see Wagner, in de Mayo *Rearrangements in Ground and Excited States*, vol. 3; Academic Press: New York, 1980, pp. 381-444, *Acc. Chem. Res.* **1971**, *4*, 168-177; Dalton; Turro, Ref. 35, pp. 526-538.

[38]For reviews of the diradicals produced in this reaction, see Wilson *Org. Photochem.* **1985**, *7*, 339-466, pp. 349-373; Scaiano; Lissi; Encina *Rev. Chem. Intermed.* **1978**, *2*, 139-196. For a review of a similar process, where δ hydrogens are abstracted, see Wagner *Acc. Chem. Res.* **1989**, *22*, 83-91.

[39]This mechanism was proposed by Yang; Yang *J. Am. Chem. Soc.* **1958**, *80*, 2913. Among the evidence for this mechanism is the fact that the diradical intermediate has been trapped: Wagner; Zepp *J. Am. Chem. Soc.* **1972**, *94*, 287; Wagner; Kelso; Zepp *J. Am. Chem. Soc.* **1972**, *94*, 7480; Adam; Grabowski; Wilson *Chem. Ber.* **1989**, *122*, 561. See also Caldwell; Dhawan; Moore *J. Am. Chem. Soc.* **1985**, *107*, 5163.

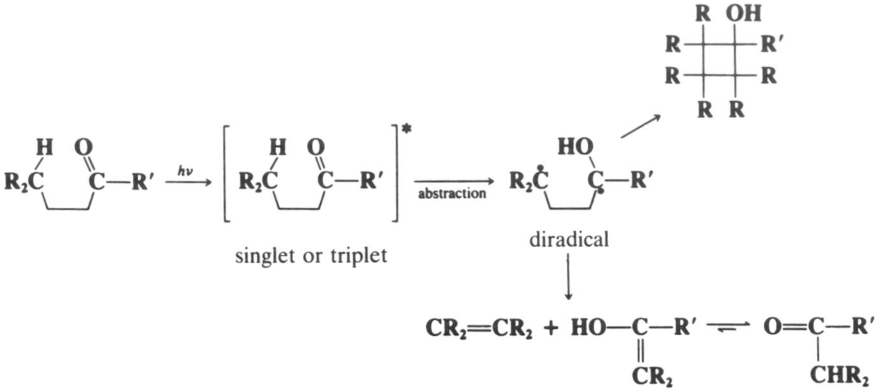

$$\underset{\substack{\text{singlet or triplet}}}{R_2\overset{H}{C}-\overset{O}{\overset{\|}{C}}-R'} \xrightarrow{h\nu} \left[\underset{}{R_2\overset{H}{C}-\overset{O}{\overset{\|}{C}}-R'}\right]^* \xrightarrow[\text{abstraction}]{} \underset{\substack{\text{diradical}}}{R_2\overset{HO}{\overset{|}{C}}\quad \overset{}{\overset{.}{C}}-R'} \nearrow \begin{array}{c} R\overset{}{-}OH \\ R\overset{}{-}\overset{}{-}R' \\ R\overset{}{-}\overset{}{-}R \\ R\quad R \end{array}$$

$$\downarrow$$

$$CR_2{=}CR_2 + \underset{\substack{\| \\ CR_2}}{HO{-}C{-}R'} \rightleftharpoons \underset{\substack{| \\ CHR_2}}{O{=}C{-}R'}$$

Both singlet and triplet n,π^* states undergo the reaction.[40] The intermediate diradical can also cyclize to a cyclobutanol, which is often a side product. Carboxylic esters, anhydrides, and other carbonyl compounds can also give this reaction.[41] The photolysis of ketene to CH_2 (p. 199) is still another example of a reaction in category 2. Both singlet and triplet CH_2 are generated, the latter in two ways:

$$CH_2{=}C{=}O \xrightarrow{h\nu} CH_2{=}C{=}O(S_1) \rightarrow \overline{C}H_2 + CO$$
$$\downarrow$$
$$CH_2{=}C{=}O(T_1) \rightarrow \cdot\overset{.}{C}H_2 + CO$$

Category 3. Intramolecular rearrangement. Two examples are the rearrangement of the trimesityl compound **1** to the enol ether **2**,[42] and irradiation of *o*-nitrobenzaldehydes **3** to give *o*-nitrosobenzoic acids **4**.[43]

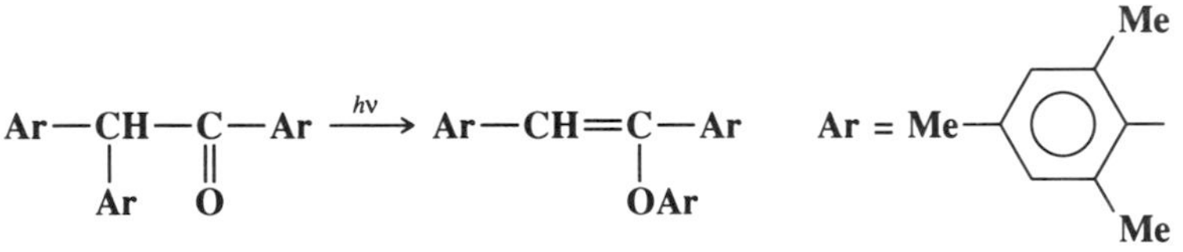

$$\underset{\mathbf{1}}{\underset{\substack{| \quad \| \\ Ar \quad O}}{Ar{-}CH{-}C{-}Ar}} \xrightarrow{h\nu} \underset{\mathbf{2}}{\underset{\substack{| \\ OAr}}{Ar{-}CH{=}C{-}Ar}} \qquad Ar = Me{-}\underset{}{\overset{Me}{\underset{Me}{\bigcirc}}}{-}$$

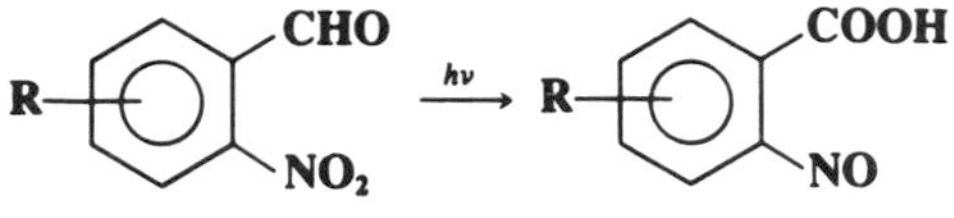

3   →   **4**

[40]Wagner; Hammond *J. Am. Chem. Soc.* **1965**, *87*, 4009; Dougherty *J. Am. Chem. Soc.* **1965**, *87*, 4011; Ausloos; Rebbert *J. Am. Chem. Soc.* **1964**, *86*, 4512; Casey; Boggs *J. Am. Chem. Soc.* **1972**, *94*, 6457.

[41]For a review of the photochemistry of carboxylic acids and acid derivatives, see Givens; Levi, in Patai *The Chemistry of Acid Derivatives*, pt. 1; Wiley: New York, 1979, pp. 641-753.

[42]Hart; Lin *Tetrahedron Lett.* **1985**, *26*, 575; Wagner; Zhou *J. Am. Chem. Soc.* **1988**, *110*, 611.

[43]For a review of this and closely related reactions, see Morrison, in Feuer *The Chemistry of the Nitro and Nitroso Groups*, pt. 1; Wiley: New York, 1969, pp. 165-213, 185-191. For a review of photochemical rearrangements of benzene derivatives, see Kaupp *Angew. Chem. Int. Ed. Engl.* **1980**, *19*, 243-275 [*Angew. Chem.* 92, 245-276]. See also Yip; Sharma *Res. Chem. Intermed.* **1989**, *11*, 109.

Category 4. *Photoisomerization.* The most common reaction in this category is photochemical cis–trans isomerization.[44] For example, *cis*-stilbene can be converted to the trans isomer:

$$\underset{H}{\overset{Ph}{\diagdown}}C=C\underset{H}{\overset{Ph}{\diagup}} \quad \xrightarrow{h\nu} \quad \underset{H}{\overset{Ph}{\diagdown}}C=C\underset{Ph}{\overset{H}{\diagup}}$$

The isomerization takes place because the excited states, both S_1 and T_1, of many olefins have a perpendicular instead of a planar geometry (p. 236), so cis–trans isomerism disappears upon excitation. When the excited molecule drops back to the S_0 state, either isomer can be formed. A useful example is the photochemical conversion of *cis*-cyclooctene to the much less stable trans isomer.[45] Another interesting example of this isomerization involves azo crown ethers. The crown ether **5**, in which the N=N bond is anti, preferentially binds NH_4^+, Li^+, and Na^+, but the syn isomer preferentially binds K^+ and Rb^+ (see p. 83). Thus,

5   **6**

ions can be selectively put in or taken out of solution merely by turning a light source on or off.[46]

In another example, the trans azo compound **6** is converted to its cis isomer when exposed to light. In this case[47] the cis isomer is a stronger acid than the trans. The trans isomer is dissolved in a system containing a base, wherein a liquid membrane separates two sides, one of which is illuminated, the other kept dark. On the illuminated side, the light converts the trans isomer to the cis. The cis isomer, being a stronger acid, donates its proton to the base, converting cis ArOH to cis ArO^-. This ion migrates to the dark side, where it rapidly reverts to the trans ion, which reacquires a proton. Because each cycle forms one H_3O^+ ion in the illuminated compartment and one OH^- ion in the dark compartment, the process

[44]For reviews of cis–trans isomerizations, see Sonnet *Tetrahedron* **1980**, *36*, 557-604; Schulte-Frohlinde; Görner *Pure Appl. Chem.* **1979**, *51*, 279-297; Saltiel; Charlton, in de Mayo, Ref. 37, pp. 25-89; Saltiel; Chang; Megarity; Rousseau; Shannon; Thomas; Uriarte *Pure Appl. Chem.* **1975**, *41*, 559-579; Saltiel; D'Agostino, Megarity, Metts; Neuberger; Wrighton; Zafiriou *Org. Photochem.* **1979**, *3*, 1-113. For reviews of the photochemistry of alkenes, see Leigh; Srinivasan *Acc. Chem. Res.* **1987**, *20*, 107-114; Steinmetz *Org. Photochem.* **1987**, *8*, 67-158; Adam; Oppenländer, Ref. 36; Mattes; Farid *Org. Photochem.* **1984**, *6*, 233-326; Kropp *Org. Photochem.* **1979**, *4*, 1-142; Morrison *Org. Photochem.* **1979**, *4*, 143-190; Kaupp *Angew. Chem. Int. Ed. Engl.* **1978**, *17*, 150-168 [*Angew. Chem. 90*, 161-179]. For a review of the photochemistry of allenes and cumulenes, see Johnson *Org. Photochem.* **1985**, *7*, 75-147.
[44a]For a review of the photoisomerization of stilbenes, see Waldeck *Chem. Rev.* **1991**, *91*, 415-436.
[45]Deyrup; Betkouski *J. Org. Chem.* **1972**, *37*, 3561.
[46]Shinkai; Nakaji; Nishida; Ogawa; Manabe *J. Am. Chem. Soc.* **1980**, *102*, 5860. See also Irie; Kato *J. Am. Chem. Soc.* **1985**, *107*, 1024; Shinkai; Miyazaki; Manabe *J. Chem. Soc., Perkin Trans. 1* **1987**, 449; Shinkai; Yoshida; Manabe; Fuchita *J. Chem. Soc., Perkin Trans. 1* **1988**, 1431; Akabori; Kumagai; Habata; Sato *J. Chem. Soc., Perkin Trans. 1* **1989**, 1497; Shinkai; Yoshioka; Nakayama; Manabe *J. Chem. Soc., Perkin Trans. 2* **1990**, 1905. For a review, see Shinkai; Manabe *Top. Curr. Chem.* **1984**, *121*, 67-104.
[47]Haberfield *J. Am. Chem. Soc.* **1987**, *109*, 6177.

reverses the normal reaction whereby these ions neutralize each other.[48] Thus the energy of light is used to do chemical work.[49] Two other examples of category 4 reactions are[44]

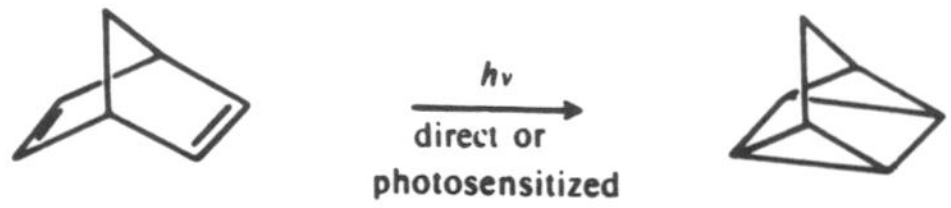

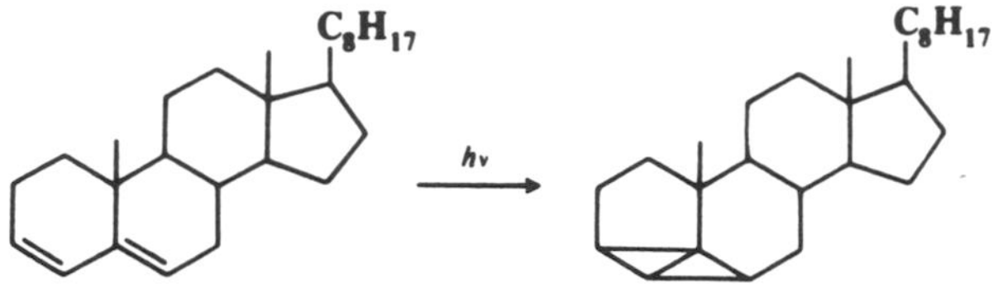

Cholesta-3, 5-diene

These examples illustrate that the use of photochemical reactions can make it very easy to obtain compounds that would be difficult to get in other ways. Reactions similar to these are discussed at **5-49.**

Category 5. Hydrogen atom abstraction. When benzophenone is irradiated in isopropyl alcohol, the initially formed S_1 state crosses to the T_1 state, which abstracts hydrogen from the solvent to give the radical **7. 7** then abstracts another hydrogen to give benzhydrol (**8**) or dimerizes to benzpinacol (**9**):

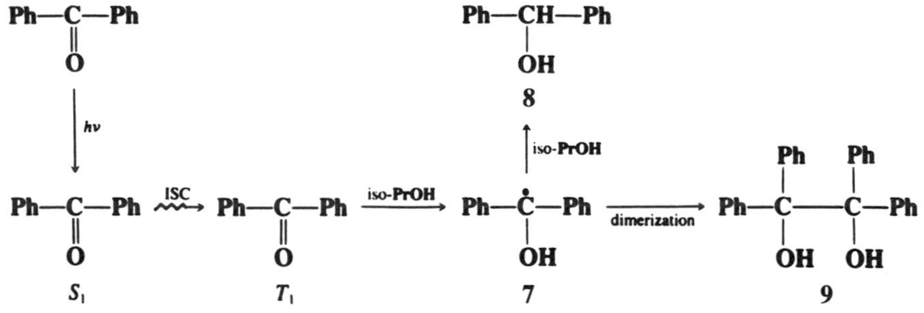

An example of intramolecular abstraction has already been given (p. 243).

Category 6. Photodimerization. An example is dimerization of cyclopentenone:[51]

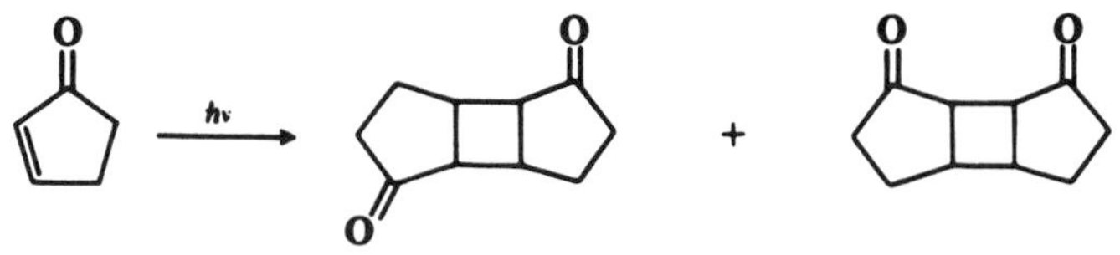

See **5-49** for a discussion of this and similar reactions.

[48]Haberfield *J. Am. Chem. Soc.* **1987,** *109,* 6178.

[49]For a review of instances where macrocycles change in response to changes in light, pH, temperature, etc., see Beer *Chem. Soc. Rev.* **1989,** *18,* 409-450. For an example not involving a macrocycle, see Feringa; Jager; de Lange; Meijer *J. Am. Chem. Soc.* **1991,** *113,* 5468.

[50]Hammond; Turro; Fischer *J. Am. Chem. Soc.* **1961,** *83,* 4674; Dauben; Cargill *Tetrahedron* **1961,** *15,* 197; Dauben; Wipke *Pure Appl. Chem.* **1964,** *9,* 539.

[51]Eaton *J. Am. Chem. Soc.* **1962,** *84,* 2344, 2454, *Acc. Chem. Res.* **1968,** *1,* 50. For a review of the photochemistry of α,β-unsaturated ketones, see Schuster, in Patai; Rappoport *The Chemistry of Enones,* pt. 2; Wiley: New York, 1989, pp. 623-756.

The Determination of Photochemical Mechanisms[52]

The methods used for the determination of photochemical mechanisms are largely the same as those used for organic mechanisms in general (Chapter 6): product identification, isotopic tracing, the detection and trapping of intermediates, and kinetics. There are, however, a few new factors: (1) there are generally many products in a photochemical reaction, as many as 10 or 15; (2) in measuring kinetics, there are more variables, since we can study the effect on the rate of the intensity or the wavelength of light; (3) in the detection of intermediates by spectra we can use the technique of *flash photolysis*, which can detect extremely short-lived intermediates.

In addition to these methods, there are two additional techniques.

1. The use of emission (fluorescence and phosphorescence) as well as absorption spectroscopy. From these spectra the presence of as well as the energy and lifetime of singlet and triplet excited states can often be calculated.

2. The study of quantum yields. The *quantum yield* is the fraction of absorbed light that goes to produce a particular result. There are several types. A *primary quantum yield* for a particular process is the fraction of molecules absorbing light that undergo that particular process. Thus, if 10% of all the molecules that are excited to the S_1 state cross over to the T_1 state, the primary quantum yield for that process is 0.10. However, primary quantum yields are often difficult to measure. A *product quantum yield* (usually designated Φ) for a product P that is formed from a photoreaction of an initially excited molecule A can be expressed as

$$\Phi = \frac{\text{number of molecules of } \mathbf{P} \text{ formed}}{\text{number of quanta absorbed by } \mathbf{A}}$$

Product quantum yields are much easier to measure. The number of quanta absorbed can be determined by an instrument called an *actinometer*, which is actually a standard photochemical system whose quantum yield is known. An example of the information that can be learned from quantum yields is the following. If the quantum yield of a product is finite and invariant with changes in experimental conditions, it is likely that the product is formed in a primary rate-determining process. Another example: in some reactions, the product quantum yields are found to be well over 1 (perhaps as high as 1000). Such a finding indicates a chain reaction (see p. 678 for a discussion of chain reactions).

---

[52]For a review, see Calvert; Pitts, Ref. 1, pp. 580-670.

8

ACIDS AND BASES

Two acid-base theories are used in organic chemistry today—the Brønsted theory and the Lewis theory.[1] These theories are quite compatible and are used for different purposes.[2]

Brønsted Theory

According to this theory, an acid is defined as a *proton donor*[3] and a base as a *proton acceptor* (a base must have a pair of electrons available to share with the proton; this is usually present as an unshared pair, but sometimes is in a π orbital). An acid-base reaction is simply the transfer of a proton from an acid to a base. (Protons do not exist free in solution but must be attached to an electron pair). When the acid gives up a proton, the species remaining still retains the electron pair to which the proton was formerly attached. Thus the new species, in theory at least, can reacquire a proton and is therefore a base. It is referred to as the *conjugate base* of the acid. All acids have a conjugate base, and all bases have a *conjugate acid*. All acid-base reactions fit the equation

$$\text{A—H} + \overline{\text{B}} \; \rightleftharpoons \; \overline{\text{A}} + \text{B—H}$$

$$\text{Acid}_1 \qquad \text{Base}_2 \qquad \text{Base}_1 \qquad \text{Acid}_2$$

No charges are shown in this equation, but an acid always has a charge one positive unit higher than that of its conjugate base.

Acid strength may be defined as the tendency to give up a proton and *base strength* as the tendency to accept a proton. Acid–base reactions occur because acids are not equally strong. If an acid, say HCl, is placed in contact with the conjugate base of a weaker acid, say acetate ion, the proton will be transferred because the HCl has a greater tendency to lose its proton than acetic acid. That is, the equilibrium

$$\text{HCl} + \text{CH}_3\text{COO}^- \; \rightleftharpoons \; \text{CH}_3\text{COOH} + \text{Cl}^-$$

lies well to the right. On the other hand, treatment of acetic acid with chloride ion gives essentially no reaction, since the weaker acid already has the proton.

This is always the case for any two acids, and by measuring the positions of the equilibrium the relative strengths of acids and bases can be determined.[4] Of course, if the two acids involved are close to each other in strength, a measurable reaction will occur from both sides, though the position of equilibrium will still be over to the side of the weaker acid

---

[1]For monographs on acids and bases, see Stewart *The Proton: Applications to Organic Chemistry*; Academic Press: New York, 1985; Bell *The Proton in Chemistry*, 2nd ed.; Cornell University Press: Ithaca, NY, 1973; Finston; Rychtman *A New View of Current Acid–Base Theories*; Wiley: New York, 1982.

[2]For discussion of the historical development of acid–base theory, see Bell *Q. Rev., Chem. Soc.* **1947**, *1*, 113-125; Bell *The Proton in Chemistry*, 1st ed.; Cornell University Press: Ithaca, NY, 1959, pp. 7-17.

[3]According to IUPAC terminology (Bunnett; Jones *Pure Appl. Chem.* **1988**, *60*, 1115), an acid is a *hydron* donor. IUPAC recommends that the term *proton* be restricted to the nucleus of the hydrogen isotope of mass 1, while the nucleus of the naturally occurring element (which contains about 0.015% deuterium) be called the *hydron* (the nucleus of mass 2 has always been known as the *deuteron*). This accords with the naturally-occurring negative ion, which has long been called the *hydride* ion. In this book, however, we will continue to use *proton* for the naturally occurring form, because most of the literature uses this term.

[4]Although equilibrium is reached in most acid–base reactions extremely rapidly (see p. 254), some are slow (especially those in which the proton is given up by a carbon) and in these cases time must be allowed for the system to come to equilibrium.

(unless the acidities are equal within experimental limits). In this manner it is possible to construct a table in which acids are listed in order of acid strength (Table 8.1).[5] Next to each acid in Table 8.1 is shown its conjugate base. It is obvious that if the acids in such a table are listed in *decreasing* order of acid strength, the bases must be listed in *increasing* order of base strength, since the stronger the acid, the weaker must be its conjugate base. The pK_a values in Table 8.1 are most accurate in the middle of the table. They are much harder to measure[6] for very strong and very weak acids, and these values must be regarded as approximate. Qualitatively, it can be determined that $HClO_4$ is a stronger acid than H_2SO_4, since a mixture of $HClO_4$ and H_2SO_4 in 4-methyl-2-pentanone can be titrated to an $HClO_4$ end point without interference by H_2SO_4.[7] Similarly, $HClO_4$ can be shown to be stronger than HNO_3 or HCl. However, this is not quantitative, and the value of -10 in the table is not much more than an educated guess. The values for RNO_2H^+, $ArNO_2H^+$, HI, $RCNH^+$ and RSH_2^+ must also be regarded as highly speculative.[8] A wide variety of pK_a values has been reported for the conjugate acids of even such simple bases as acetone[9] (-0.24 to -7.2), diethyl ether (-0.30 to -6.2), ethanol (-0.33 to -4.8), methanol (-0.34 to -4.9), and 2-propanol (-0.35 to -5.2), depending on the method used to measure them.[10] Very accurate values can be obtained only for acids weaker than hydronium ion and stronger than water.

The bottom portion of Table 8.1 consists of very weak acids[11] (pK_a above ~ 17). In most of these acids, the proton is lost from a carbon atom, and such acids are known as *carbon acids*. pK_a values for such weak acids are often difficult to measure and are known only approximately. The methods used to determine the relative positions of these acids are discussed in Chapter 5.[12] The acidity of carbon acids is proportional to the stability of the carbanions that are their conjugate bases (see p. 175).

The extremely strong acids at the top of the table are known as *super acids* (see p. 166).[13] The actual species present in the FSO_3H–SbF_5 mixture are probably $H[SbF_5(SO_3F)]$ and $H[SbF_2(SO_3F)_4]$.[14] The addition of SO_3 causes formation of the still stronger $H[SbF_4(SO_3F)_2]$, $H[SbF_3(SO_3F)_3]$, and $H[(SbF_5)_2(SO_3F)]$.[14]

By the use of tables such as Table 8.1, it is possible to determine whether a given acid will react with a given base. For tables in which acids are listed in order of decreasing strength, the rule is that *any acid will react with any base in the table that is below it but not with any above it*.[15] It must be emphasized that the order of acid strength in Table 8.1 applies

[5]Table 8.1 is a thermodynamic acidity scale and applies only to positions of equilibria. For the distinction between thermodynamic and kinetic acidity, see p. 176.

[6]For a review of methods of determining pK_a values, see Cookson *Chem. Rev.* **1974**, *74*, 5-28.

[7]Kolthoff; Bruckenstein, in Kolthoff; Elving *Treatise on Analytical Chemistry*, vol. 1, pt. 1; Wiley: New York, 1959, pp. 475-542, p. 479.

[8]For reviews of organic compounds protonated at O, N, or S, see Olah; White; O'Brien *Chem. Rev.* **1970**, *70*, 561-591; Olah; White; O'Brien, in Olah; Schleyer *Carbonium Ions*, vol. 4; Wiley: New York, 1973, pp. 1697-1781.

[9]For discussions of pK_a determinations for the conjugate acids of ketones, see Bagno; Lucchini; Scorrano *Bull. Soc. Chim. Fr.* **1987**, 563; Toullec *Tetrahedron Lett.* **1988**, *29*, 5541.

[10]Rochester *Acidity Functions*; Academic Press: New York, 1970. For discussion of the basicity of such compounds, see Liler *Reaction Mechanisms in Sulfuric Acid*; Academic Press: New York, 1971, pp. 118-139.

[11]For a monograph on very weak acids, see Reutov; Beletskaya; Butin *CH-Acids*; Pergamon: New York, 1978. For other discussions, see Cram *Fundamentals of Carbanion Chemistry*; Academic Press: New York, 1965, pp. 1-45; Streitwieser; Hammons *Prog. Phys. Org. Chem.* **1965**, *3*, 41-80.

[12]For reviews of methods used to measure the acidity of carbon acids, see Jones *Q. Rev., Chem. Soc.* **1971**, *25*, 365-378; Fischer; Rewicki *Prog. Org. Chem.* **1968**, *7*, 116-161; Reutov; Beletskaya; Butin, Ref. 11, Chapter 1 [an earlier version of this chapter appeared in *Russ. Chem. Rev.* **1974**, *43*, 17-31]; Ref. 6. For reviews on acidities of carbon acids, see Gau; Assadourian *Prog. Phys. Org. Chem.* **1987**, *16*, 237-285; in Buncel; Durst *Comprehensive Carbanion Chemistry*, pt. A; Elsevier: New York, 1980, the reviews by Pellerite; Brauman, pp. 55-96 (gas phase acidities); and Streitwieser; Juaristi; Nebenzahl, pp. 323-381.

[13]For a monograph, see Olah; Prakash; Sommer *Superacids*; Wiley: New York, 1985. For a review, see Gillespie; Peel *Adv. Phys. Org. Chem.* **1971**, *9*, 1-24. For a review of solid superacids, see Arata *Adv. Catal.* **1990**, *37*, 165-211. For a review of methods of measuring superacidity, see Jost; Sommer *Rev. Chem. Intermed.* **1988**, *9*, 171-199.

[14]Gillespie *Acc. Chem. Res.* **1968**, *1*, 202-209.

[15]These reactions are equilibria. What the rule actually says is that the position of equilibrium will be such that the weaker acid predominates. However, this needs to be taken into account only when the acid and base are close to each other in the table (within about 2 pK units).

TABLE 8.1 pK_a values for many types of acids

The values in boldface are exact values; the others are approximate, especially above 18 and below − 2[16]

Acid	Base	Approximate pK_a (relative to water)	Ref.
Super acids:			
$HF–SbF_5$	$SbF_6{}^-$		19
$FSO_3H–SbF_5–SO_3$			14
$FSO_3H–SbF_5$			14, 19
FSO_3H	$FSO_3{}^-$		14
RNO_2H^+	RNO_2	− 12	20
$ArNO_2H^+$	$ArNO_2$	− 11	20
$HClO_4$	$ClO_4{}^-$	− 10	21
HI	I^-	− 10	21
$RCNH^+$	RCN	− 10	22
$R{-}\overset{+}{\underset{OH}{C}}{-}H$	$R{-}\underset{O}{\overset{\parallel}{C}}{-}H$	− 10	23
H_2SO_4	$HSO_4{}^-$		
HBr	Br^-	− 9	21
$Ar{-}\overset{+}{\underset{OH}{C}}{-}OR^{17}$	$Ar{-}\underset{O}{\overset{\parallel}{C}}{-}OR$	− 7.4	20
HCl	Cl^-	− 7	21
$RSH_2{}^+$	RSH	− 7	20
$Ar{-}\overset{+}{\underset{OH}{C}}{-}OH^{17}$	$Ar{-}\underset{O}{\overset{\parallel}{C}}{-}OH$	− 7	24
$Ar{-}\overset{+}{\underset{OH}{C}}{-}H$	$Ar{-}\underset{O}{\overset{\parallel}{C}}{-}H$	− 7	25
$R{-}\overset{+}{\underset{OH}{C}}{-}R$	$R{-}\underset{O}{\overset{\parallel}{C}}{-}R$	− 7	9, 22, 26
$ArSO_3H$	$ArSO_3{}^-$	− 6.5	27
$R{-}\overset{+}{\underset{OH}{C}}{-}OR^{17}$	$R{-}\underset{O}{\overset{\parallel}{C}}{-}OR$	− 6.5	20
$ArOH_2{}^+$	ArOH	− 6.4	28
$R{-}\overset{+}{\underset{OH}{C}}{-}OH^{17}$	$R{-}\underset{O}{\overset{\parallel}{C}}{-}OH$	− 6	20
$Ar{-}\overset{+}{\underset{OH}{C}}{-}R$	$Ar{-}\underset{O}{\overset{\parallel}{C}}{-}R$	− 6	25, 29
$Ar{-}\overset{+}{\underset{H}{O}}{-}R$	$Ar{-}O{-}R$	− 6	28, 30
$CH(CN)_3$	$^-C(CN)_3$	− 5	31
Ar_3NH^+	Ar_3N	− 5	32
$H{-}\overset{+}{\underset{OH}{C}}{-}H$	$H{-}\underset{O}{\overset{\parallel}{C}}{-}H$	− 4	33
$R{-}\overset{+}{\underset{H}{O}}{-}R$	$R{-}O{-}R$	− 3.5	22, 30, 34
$R_3COH_2{}^+$	R_3COH	− 2	34
$R_2CHOH_2{}^+$	R_2CHOH	− 2	34, 35
$RCH_2OH_2{}^+$	RCH_2OH	− 2	22, 34, 35
H_3O^+	H_2O	**− 1.74**	36

TABLE 8.1 (Continued)

Acid	Base	Approximate pK_a (relative to water)	Ref.
Ar—C—NH$_2^{17}$ ‖ OH$^+$	Ar—C—NH$_2$ ‖ O	−1.5	37
HNO$_3$	NO$_3^-$	−1.4	21
R—C—NH$_2^{17}$ ‖ OH$^+$	R—C—NH$_2$ ‖ O	−0.5	37
Ar$_2$NH$_2^+$	Ar$_2$NH	1	32
HSO$_4^-$	SO$_4^{2-}$	**1.99**	38
HF	F$^-$	**3.17**	38
HONO	NO$_2^-$	**3.29**	38
ArNH$_3^+$	ArNH$_2$	3–5	39
ArNR$_2$II$^+$	ArNR$_2$	3–5	39
RCOOH	RCOO$^-$	4–5	39
HCOCH$_2$CHO	HCOC̄HCHO	5	40
H$_2$CO$_3^{18}$	HCO$_3^-$	**6.35**	38
H$_2$S	HS$^-$	**7.00**	38
ArSH	ArS$^-$	6–8	41
CH$_3$COCH$_2$COCH$_3$	CH$_3$COC̄HCOCH$_3$	9	40
HCN	CN$^-$	9.2	42
NH$_4^+$	NH$_3$	**9.24**	38
ArOH	ArO$^-$	8–11	43
RCH$_2$NO$_2$	RC̄HNO$_2$	10	44
R$_3$NH$^+$	R$_3$N	10–11	39
RNH$_3^+$	RNH$_2$	10–11	39
HCO$_3^-$	CO$_3^{2-}$	**10.33**	38
RSH	RS$^-$	10–11	41
R$_2$NH$_2^+$	R$_2$NH	11	39
NCCH$_2$CN	NCC̄HCN	11	40, 45
CH$_3$COCH$_2$COOR	CH$_3$COC̄HCOOR	11	40
CH$_3$SO$_2$CH$_2$SO$_2$CH$_3$	CH$_3$SO$_2$C̄HSO$_2$CH$_3$	12.5	46
EtOOCCH$_2$COOEt	EtOOCC̄HCOOEt	13	40
CH$_3$OH	CH$_3$O$^-$	15.2	47, 48
H$_2$O	OH$^-$	**15.74**	49
⬠ (cyclopentadiene)	⬠$^{\ominus}$ (cyclopentadienyl anion)	16	50
RCH$_2$OH	RCH$_2$O$^-$	16	47
RCH$_2$CHO	RC̄HCHO	16	51
R$_2$CHOH	R$_2$CHO$^-$	16.5	47
R$_3$COH	R$_3$CO$^-$	17	47
RCONH$_2$	RCONH$^-$	17	52
RCOCH$_2$R	RCOC̄HR	19–20	53
(indene)	(indenyl anion)$^{\ominus}$	20	54, 55
(fluorene)	(fluorenyl anion)$^{\ominus}$	23	54, 55
ROOCCH$_2$R	ROOCC̄HR	24.5	40

TABLE 8.1 (*Continued*)

Acid	Base	Approximate pK_a (relative to water)	Ref.
RCH$_2$CN	RC̄HCN	25	40, 56
HC≡CH	HC≡C$^-$	25	57
Ar$_3$CH	Ar$_3$C$^-$	31.5	54, 58
Ar$_2$CH$_2$	Ar$_2$CH$^-$	33.5	54, 58
H$_2$	H$^-$	35	59
NH$_3$	NH$_2{}^-$	38	60
PhCH$_3$	PhCH$_2{}^-$	40	61
CH$_2$=CHCH$_3$	[CH$_2$ ≏ CH ≏ CH$_2$]$^-$	43	62
PhH	Ph$^-$	43	63
CH$_2$=CH$_2$	CH$_2$=CH$^-$	44	64
cyclo-C$_3$H$_6$	cyclo-C$_3$H$_5{}^-$	46	65
CH$_4$	CH$_3{}^-$	48	66
C$_2$H$_6$	C$_2$H$_5{}^-$	50	67
(CH$_3$)$_2$CH$_2$	(CH$_3$)$_2$CH$^-$	51	67
(CH$_3$)$_3$CH	(CH$_3$)$_3$C$^-$	—	68

[16]In this table we do not give pK_a values for individual compounds (with a few exceptions), only average values for functional groups. Extensive tables of pK values for many carboxylic and other acids and amines are given in Ref. 39. Values for more than 5500 organic acids are given in Serjeant; Dempsey *Ionisation Constants of Organic Acids in Aqueous Solution*; Pergamon: Elmsford, NY, 1979; Kortüm; Vogel; Andrussow *Dissociation Constants of Organic Acids in Aqueous Solution*; Butterworth: London, 1961. The index in the 1979 volume covers both volumes. Kortüm; Vogel; Andrussow *Pure Appl. Chem.* **1960**, *1*, 190-536 give values for 631 carboxylic acids and 110 phenols. Ref. 20 gives hundreds of values for very strong acids (very weak bases). Perrin *Dissociation Constants of Organic Bases in Aqueous Solution*; Butterworth: London, 1965, and Supplement, 1972 list pK values for more than 7000 amines and other bases. Collumeau *Bull. Soc. Chim. Fr.* **1968**, 5087-5112 gives pK values for about 800 acids and bases. Bordwell *Acc. Chem. Res.* **1988**, *21*, 456-463 gives values for more than 300 acids in dimethyl sulfoxide. For inorganic acids and bases, see Perrin, Ref. 42, *Pure Appl. Chem.* **1969**, *20*, 133-236.

[17]Carboxylic acids, esters, and amides are shown in this table to be protonated on the carbonyl oxygen. There has been some controversy on this point, but the weight of evidence is in that direction. See, for example, Katritzky; Jones *Chem. Ind. (London)* **1961**, 722; Ottenheym; van Raayen; Smidt; Groenewege; Veerkamp *Recl. Trav. Chim. Pays-Bas* **1961**, *80*, 1211; Stewart; Muenster *Can. J. Chem.* **1961**, *39*, 401; Smith; Yates *Can. J. Chem.* **1972**, *50*, 771; Benedetti; Di Blasio; Baine *J. Chem. Soc. Perkin Trans. 2* **1980**, 500; Ref. 8; Homer; Johnson, in Zabicky *The Chemistry of Amides*; Wiley: New York, 1970, pp. 188-197. It has been shown that some amides protonate at nitrogen: see Perrin *Acc. Chem. Res.* **1989**, *22*, 268-275. For a review of alternative proton sites, see Liler *Adv. Phys. Org. Chem.* **1975**, *11*, 267-392.

[18]This value includes the CO$_2$ usually present. The value for H$_2$CO$_3$ alone is 3.9 (Ref. 21).

[19]Brouwer; van Doorn *Recl. Trav. Chim. Pays-Bas* **1972**, *91*, 895; Gold; Laali; Morris; Zdunek *J. Chem. Soc., Chem. Commun.* **1981**, 769; Sommer; Canivet; Schwartz; Rimmelin *Nouv. J. Chim.* **1981**, *5*, 45.

[20]Arnett *Prog. Phys. Org. Chem.* **1963**, *1*, 223-403, pp. 324-325.

[21]Bell, Ref. 1.

[22]Deno; Wisotsky *J. Am. Chem. Soc.* **1963**, *85*, 1735; Deno; Gaugler; Wisotsky *J. Org. Chem.* **1966**, *31*, 1967.

[23]Levy; Cargioli; Racela *J. Am. Chem. Soc.* **1970**, *92*, 6238. See, however, Brouwer; van Doorn *Recl. Trav. Chim. Pays-Bas* **1971**, *90*, 1010.

[24]Stewart; Granger *Can. J. Chem.* **1961**, *39*, 2508.

[25]Yates; Stewart *Can. J. Chem.* **1959**, *37*, 664; Stewart; Yates *J. Am. Chem. Soc.* **1958**, *80*, 6355.

[26]Lee *Can. J. Chem.* **1970**, *48*, 1919.

[27]Cerfontain; Koeberg-Telder; Kruk *Tetrahedron Lett.* **1975**, 3639.

[28]Arnett; Wu *J. Am. Chem. Soc.* **1960**, *82*, 5660; Koeberg-Telder; Lambrechts; Cerfontain *Recl. Trav. Chim. Pays-Bas* **1983**, *102*, 293.

[29]Fischer; Grigor; Packer; Vaughan *J. Am. Chem. Soc.* **1961**, *83*, 4208.

[30]Arnett; Wu *J. Am. Chem. Soc.* **1960**, *82*, 4999.

[31]Boyd *J. Phys. Chem.* **1963**, *67*, 737.

[32]Arnett; Quirk; Burke *J. Am. Chem. Soc.* **1970**, *92*, 1260.

[33]McTigue; Sime *Aust. J. Chem.* **1963**, *16*, 592.

[34]Deno; Turner *J. Org. Chem.* **1966**, *31*, 1969.

[35]Lee; Demchuk *Can. J. Chem.* **1987**, *65*, 1769; Chandler; Lee *Can. J. Chem.* **1990**, *68*, 1757.

[36]For a discussion, see Campbell; Waite *J. Chem. Educ.* **1990**, *67*, 386.

[37]Cox; Druet; Klausner; Modro; Wan; Yates *Can. J. Chem.* **1981**, *59*, 1568; Grant; McTigue; Ward *Aust. J. Chem.* **1983**, *36*, 2211.

when a given acid and base react without a solvent or, when possible, in water. In other solvents the order may be greatly different (see p. 272). In the gas phase, where solvation effects are completely or almost completely absent, acidity orders may also differ greatly.[69] For example, in the gas phase, toluene is a stronger acid than water and t-butoxide ion is a weaker base than methoxide ion[70] (see also pp. 270-272). It is also possible for the acidity order to change with temperature. For example, above 50°C the order of base strength is $BuOH > H_2O > Bu_2O$; from 1 to 50°C the order is $BuOH > Bu_2O > H_2O$; while below 1°C the order becomes $Bu_2O > BuOH > H_2O$.[71]

[38]Bruckenstein; Kolthoff; in Kolthoff; Elving *Treatise on Analytical Chemistry*, vol. 1, pt. 1; Wiley: New York, 1959, pp. 432-433.

[39]Brown; McDaniel; Häflinger, in Braude; Nachod *Determination of Organic Structures by Physical Methods*, vol. 1; Academic Press: New York, 1955, pp. 567-662.

[40]Pearson; Dillon *J. Am. Chem. Soc.* **1953**, *75*, 2439.

[41]Crampton, in Patai *The Chemistry of the Thiol Group*, pt. 1; Wiley: New York, 1974, pp. 396-410.

[42]Perrin *Ionisation Constants of Inorganic Acids and Bases in Aqueous Solution*, 2nd ed.; Pergamon: Elmsford, NY, 1982.

[43]Rochester, in Patai *The Chemistry of the Hydroxyl Group*, pt. 1; Wiley: New York, 1971, p. 374.

[44]Cram *Chem. Eng. News* **1963**, *41*(No. 33, Aug. 19), 94.

[45]Bowden; Stewart *Tetrahedron* **1965**, *21*, 261.

[46]Hine; Philips; Maxwell *J. Org. Chem.* **1970**, *35*, 3943. See also Ang; Lee *Aust. J. Chem.* **1977**, *30*, 521.

[47]Reeve; Erikson; Aluotto *Can. J. Chem.* **1979**, *57*, 2747.

[48]See also Mackay; Bohme *J. Am. Chem. Soc.* **1978**, *100*, 327; Olmstead; Margolin; Bordwell *J. Org. Chem.* **1980**, *45*, 3295.

[49]Harned; Robinson *Trans. Faraday Soc.* **1940**, *36*, 973.

[50]Streitwieser; Nebenzahl *J. Am. Chem. Soc.* **1976**, *98*, 2188.

[51]Guthrie; Cossar *Can. J. Chem.* **1986**, *64*, 2470.

[52]Homer; Johnson, Ref. 17, pp. 238-240.

[53]Tapuhi; Jencks *J. Am. Chem. Soc.* **1982**, *104*, 5758; Guthrie; Cossar; Klym *J. Am. Chem. Soc.* **1984**, *106*, 1351; Chiang; Kresge; Tang; Wirz *J. Am. Chem. Soc.* **1984**, *106*, 460.

[54]Streitwieser; Ciuffarin; Hammons *J. Am. Chem. Soc.* **1967**, *89*, 63.

[55]Streitwieser; Hollyhead; Pudjaatmaka; Owens; Kruger; Rubenstein; MacQuarrie; Brokaw; Chu; Niemeyer *J. Am. Chem. Soc.* **1971**, *93*, 5088.

[56]For a review of the acidity of cyano compounds, see Hibbert, in Patai; Rappoport *The Chemistry of Triple-bonded Functional Groups*, pt. 1; Wiley: New York, 1983, pp. 699-736.

[57]Cram, Ref. 11, p. 19. See also Dessy; Kitching; Psarras; Salinger; Chen; Chivers *J. Am. Chem. Soc.* **1966**, *88*, 460.

[58]Streitwieser; Hollyhead; Sonnichsen; Pudjaatmaka; Chang; Kruger *J. Am. Chem. Soc.* **1971**, *93*, 5096.

[59]Buncel; Menon *J. Am. Chem. Soc.* **1977**, *99*, 4457.

[60]Buncel; Menon *J. Organomet. Chem.* **1977**, *141*, 1.

[61]Streitwieser; Ni *Tetrahedron Lett.* **1985**, *26*, 6317; Albrecht; Schneider *Tetrahedron* **1986**, *42*, 4729.

[62]Boerth; Streitwieser *J. Am. Chem. Soc.* **1981**, *103*, 6443.

[63]Streitwieser; Scannon; Niemeyer *J. Am. Chem. Soc.* **1972**, *94*, 7936.

[64]Maskornick; Streitwieser *Tetrahedron Lett.* **1972**, 1625; Streitwieser; Boerth *J. Am. Chem. Soc.* **1978**, *100*, 755.

[65]This value is calculated from results given in Streitwiesser; Caldwell; Young *J. Am. Chem. Soc.* **1969**, *91*, 529. For a review of acidity and basicity of cyclopropanes, see Battiste; Coxon, in Rappoport *The Chemistry of the Cyclopropyl Group*, pt. 1; Wiley: New York, 1987, pp. 255-305.

[66]This value is calculated from results given in Streitwieser; Taylor *J. Chem. Soc. D* **1970**, 1248.

[67]These values are based on those given in Ref. 44 but are corrected to the newer scale of Streitwieser; Refs. 63 and 64.

[68]Breslow and co-workers report a value of 71 [Breslow; Goodin *J. Am. Chem. Soc.* **1976**, *98*, 6076; Breslow; Grant *J. Am. Chem. Soc.* **1977**, *99*, 7745], but this was obtained by a different method, and is not comparable to the other values in Table 8.1. A more comparable value is about 53. See also Juan; Schwarz; Breslow *J. Am. Chem. Soc.* **1980**, *102*, 5741.

[69]For a review of acidity and basicity scales in the gas phase and in solution, see Gal; Maria *Prog. Phys. Org. Chem.* **1990**, *17*, 159-238.

[70]Brauman; Blair *J. Am. Chem. Soc.* **1970**, *92*, 5986; Bohme; Lee-Ruff; Young *J. Am. Chem. Soc.* **1972**, *94*, 4608, 5153.

[71]Gerrard; Macklen *Chem. Rev.* **1959**, *59*, 1105-1123. For other examples, see Calder; Barton *J. Chem. Educ.* **1971**, *48*, 338; Hambly *Rev. Pure Appl. Chem.* **1965**, *15*, 87-100, p. 88.

The Mechanism of Proton Transfer Reactions

Proton transfers between oxygen and nitrogen acids and bases are usually extremely fast.[72] In the thermodynamically favored direction they are generally diffusion controlled.[73] In fact, a *normal acid* is defined[74] as one whose proton transfer reactions are completely diffusion controlled, except when the conjugate acid of the base to which the proton is transferred has a pK value very close (differs by < about 2 pK units) to that of the acid. The normal acid–base reaction mechanism consists of three steps:

1.   **HA + |B ⇌ AH⋯|B**

2.   **AH⋯|B ⇌ A|⋯HB**

3.   **A|⋯HB ⇌ A| + HB**

The actual proton transfer takes place in the second step—the first step is formation of a hydrogen-bonded complex. The product of the second step is another hydrogen-bonded complex, which dissociates in the third step.

However, not all such proton transfers are diffusion controlled. For example, if an internal hydrogen bond exists in a molecule, reaction with an external acid or base is often much slower.[75] In a case such as this:

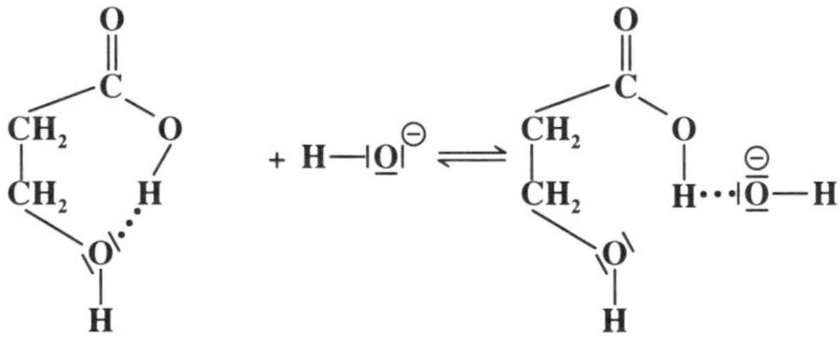

3 - Hydroxypropanoic
acid

the OH⁻ ion can form a hydrogen bond with the acidic hydrogen only if the internal hydrogen bond breaks. Therefore only some of the collisions between OH⁻ ions and 3-hydroxypropanoic acid molecules result in proton transfer. In many collisions the OH⁻ ions will come away empty-handed, resulting in a lower reaction rate. Note that this affects only the rate, not the equilibrium. Another factor that can create lower rates is a molecular structure in which the acidic proton is protected within a molecular cavity (e.g., the in–in and out–in isomers shown on p. 133). See also the proton sponges mentioned on p. 268. Proton transfers between an acidic and a basic group within the same molecule can also be slow, if the two groups are too far apart for hydrogen bonding. In such cases participation of solvent molecules may be necessary.

---

[72]For reviews of such proton transfers, see Hibbert *Adv. Phys. Org. Chem.* **1986,** *22,* 113-212; Crooks, in Bamford; Tipper *Chemical Kinetics,* vol. 8; Elsevier: New York, 1977, pp. 197-250.
[73]Kinetic studies of these very fast reactions were first carried out by Eigen. See Eigen *Angew. Chem. Int. Ed. Engl.* **1964,** *3,* 1-19 [*Angew. Chem.* **1963,** *75,* 489-509].
[74]See, for example, Hojatti; Kresge; Wang *J. Am. Chem. Soc.* **1987,** *109,* 4023.
[75]For an example of a slow proton transfer from F₃CCOOH to (PhCH₂)₃N, see Ritchie; Lu *J. Am. Chem. Soc.* **1989,** *111,* 8542.

Proton transfers to or from a carbon atom[76] in most cases are much slower than those strictly between oxygen or nitrogen atoms. At least three factors can be responsible for this,[77] not all of them applying in every case:

1. Hydrogen bonding is very weak or altogether absent for carbon (Chapter 3).

2. Many carbon acids, upon losing the proton, form carbanions that are stabilized by resonance. Structural reorganization (movement of atoms to different positions within the molecule) may accompany this. Chloroform, HCN, and 1-alkynes do not form resonance-stabilized carbanions, and these[78] behave kinetically as normal acids.[79]

3. There may be considerable reorganization of solvent molecules around the ion as compared to the neutral molecule.[80]

In connection with factors 2 and 3, it has been proposed[77] that any factor that stabilizes the product (e.g., by resonance or solvation) lowers the rate constant if it develops late on the reaction coordinate, but increases the rate constant if it develops early. This is called the Principle of Imperfect Synchronization.

Measurements of Solvent Acidity[81]

When a solute is added to an acidic solvent it may become protonated by the solvent. If the solvent is water and the concentration of solute is not very great, then the pH of the solution is a good measure of the proton-donating ability of the solvent. Unfortunately, this is no longer true in concentrated solutions because activity coefficients are no longer unity. A measurement of solvent acidity is needed which works in concentrated solutions and applies to mixed solvents as well. The Hammett acidity function[82] is a measurement that is used for acidic solvents of high dielectric constant.[83] For any solvent, including mixtures of solvents (but the proportions of the mixture must be specified), a value H_0 is defined as

$$H_0 = pK_{BH_w^+} - \log \frac{[BH^+]}{[B]}$$

H_0 is measured by using "indicators" that are weak bases (B) and so are partly converted, in these acidic solvents, to the conjugate acids BH^+. Typical indicators are o-nitroanilinium ion, with a pK in water of -0.29, and 2,4-dinitroanilinium ion, with a pK in water of -4.53. For a given solvent, $[BH^+]/[B]$ is measured for one indicator, usually by spectrophotometric means. Then, using the known pK in water ($pK_{BH_w^+}$) for that indicator, H_0 can be calculated for that solvent system. In practice, several indicators are used, so that an average H_0 is

[76]For reviews of proton transfers to and from carbon, see Hibbert, in Bamford; Tipper, Ref. 72, pp. 97-196; Kreevoy *Isot. Org. Chem.* **1976**, *2*, 1-31; Leffek *Isot. Org. Chem.* **1976**, *2*, 89-125.

[77]See Bernasconi *Tetrahedron* **1985**, *41*, 3219.

[78]Lin; Chiang; Dahlberg; Kresge *J. Am. Chem. Soc.* **1983**, *105*, 5380; Bednar; Jencks *J. Am. Chem. Soc.* **1985**, *107*, 7117, 7126, 7135; Kresge; Powell *J. Org. Chem.* **1986**, *51*, 822; Formosinho; Gal *J. Chem. Soc., Perkin Trans. 2* **1987**, 1655.

[79]Not all 1-alkynes behave as normal acids; see Aroella; Arrowsmith; Hojatti; Kresge; Powell; Tang; Wang *J. Am. Chem. Soc.* **1987**, *109*, 7198.

[80]See Bernasconi; Terrier *J. Am. Chem. Soc.* **1987**, *109*, 7115; Kurz *J. Am. Chem. Soc.* **1989**, *111*, 8631.

[81]For fuller treatments, see Hammett *Physical Organic Chemistry*, 2nd ed.; McGraw-Hill: New York, 1970, pp. 263-313; Jones *Physical and Mechanistic Organic Chemistry*, 2nd ed.; Cambridge University Press: Cambridge, 1984, pp. 83-93; Arnett; Scorrano *Adv. Phys. Org. Chem.* **1976**, *13*, 83-153.

[82]Hammett; Deyrup *J. Am. Chem. Soc.* **1932**, *54*, 2721.

[83]For a monograph on acidity functions, see Rochester, Ref. 10. For reviews, see Ref. 81; Cox; Yates *Can. J. Chem.* **1983**, *61*, 2225-2243; Boyd, in Coetzee; Ritchie *Solute–Solvent Interactions*; Marcel Dekker: New York, 1969, pp. 97-218; Vinnik *Russ. Chem. Rev.* **1966**, *35*, 802-817; Liler, Ref. 10, pp. 26-58.

taken. Once H_0 is known for a given solvent system, pK_a values in it can be calculated for any other acid–base pair.

The symbol h_0 is defined as

$$h_0 = \frac{a_{H^+} f_I}{f_{HI^+}}$$

where a_{H^+} is the activity of the proton and f_I and f_{HI^+} are the activity coefficients of the indicator and conjugate acid of the indicator,[84] respectively. H_0 is related to h_0 by

$$H_0 = -\log h_0$$

so that H_0 is analogous to pH and h_0 to $[H^+]$, and indeed in dilute aqueous solution $H_0 = $ pH.

H_0 reflects the ability of the solvent system to donate protons, but it can be applied only to acidic solutions of high dielectric constant, mostly mixtures of water with acids such as nitric, sulfuric, perchloric, etc. It is apparent that the H_0 treatment is valid only when f_I/f_{HI^+} is independent of the nature of the base (the indicator). Since this is so only when the bases are structurally similar, the treatment is limited. Even when similar bases are compared, many deviations are found.[85] Other acidity scales[86] have been set up, among them H_- for bases with a charge of -1, H_R for aryl carbinols,[87] H_C for bases that protonate on carbon,[88] and H_A for unsubstituted amides.[89] It is now clear that there is no single acidity scale that can be applied to a series of solvent mixtures, irrespective of the bases employed.[90]

Although most acidity functions have been applied only to acidic solutions, some work has also been done with strongly basic solutions.[91] The H_- function, which is used for highly acidic solutions when the base has a charge of -1, can also be used for strongly basic solvents, in which case it measures the ability of these solvents to abstract a proton from a neutral acid BH.[92] When a solvent becomes protonated, its conjugate acid is known as a *lyonium ion.*

Another approach to the acidity function problem was proposed by Bunnett and Olsen,[93] who derived the equation

$$\log \frac{[SH^+]}{[S]} + H_0 = \phi(H_0 + \log [H^+]) + pK_{SH^+}$$

[84]For a review of activity coefficient behavior of indicators in acid solutions, see Yates; McClelland *Prog. Phys. Org. Chem.* **1974**, *11*, 323-420.

[85]For example, see Kresge; Barry; Charles; Chiang *J. Am. Chem. Soc.* **1962**, *84*, 4343; Katritzky; Waring; Yates *Tetrahedron* **1963**, *19*, 465; Arnett; Mach *J. Am. Chem. Soc.* **1964**, *86*, 2671; Jorgenson; Hartter *J. Am. Chem. Soc.* **1963**, *85*, 878; Kreevoy; Baughman *J. Am. Chem. Soc.* **1973**, *95*, 8178; García; Leal; Herrero; Palacios *J. Chem. Soc., Perkin Trans. 2* **1988**, 1759; Ref. 32.

[86]For lengthy tables of many acidity scales, with references, see Cox; Yates, Ref. 83. For an equation that is said to combine the vast majority of acidity functions, see Zalewski; Sarkice; Geltz *J. Chem. Soc., Perkin Trans. 2* **1983**, 1059.

[87]Deno; Jaruzelski; Schriesheim *J. Am. Chem. Soc.* **1955**, *77*, 3044; Deno; Berkheimer; Evans; Peterson *J. Am. Chem. Soc.* **1959**, *81*, 2344.

[88]Reagan *J. Am. Chem. Soc.* **1969**, *91*, 5506.

[89]Yates; Stevens; Katritzky *Can. J. Chem.* **1964**, *42*, 1957; Yates; Riordan *Can. J. Chem.* **1965**, *43*, 2328; Edward; Wong *Can. J. Chem.* **1977**, *55*, 2492; Liler; Marković *J. Chem. Soc., Perkin Trans. 2* **1982**, 551.

[90]Hammett, Ref. 81, p. 278; Rochester, Ref. 10, p. 21.

[91]For another approach to solvent basicity scales, see Catalán; Gómez; Couto; Laynez *J. Am. Chem. Soc.* **1990**, *112*, 1678.

[92]For reviews, see Rochester *Q. Rev., Chem. Soc.* **1966**, *20*, 511-525; Rochester, Ref. 10, pp. 234-264; Bowden *Chem. Rev.* **1966**, *66*, 119-131 (the last review is reprinted in Coetzee and Ritchie, Ref. 83, pp. 186-215).

[93]Bunnett; Olsen *Can. J. Chem.* **1966**, *44*, 1899, 1917; Bunnett; McDonald; Olsen *J. Am. Chem. Soc.* **1974**, *96*, 2855.

where S is a base that is protonated by an acidic solvent. Thus the slope of a plot of log $([SH^+]/[S]) + H_0$ against $H_0 + \log[H^+]$ is the parameter ϕ, while the intercept is the pK_a of the lyonium ion SH^+ (referred to infinite dilution in water). The value of ϕ expresses the response of the equilibrium $S + H^+ \rightleftharpoons SH^+$ to changing acid concentration. A negative ϕ indicates that the log of the ionization ratio $[SH^+]/[S]$ increases, as the acid concentration increases, more rapidly than $-H_0$. A positive ϕ value indicates the reverse. The Bunnett–Olsen equation given above is a linear free-energy relationship (see p. 281) that pertains to acid-base equilibria. A corresponding equation that applies to kinetic data is

$$\log k_\psi + H_0 = \phi(H_0 + \log[\mathbf{H^+}]) + \log k_2^0$$

where k_ψ is the pseudo-first-order rate constant for a reaction of a weakly basic substrate taking place in an acidic solution and k_2^0 is the second-order rate constant at infinite dilution in water. In this case ϕ characterizes the response of the reaction rate to changing acid concentration of the solvent. The Bunnett–Olsen treatment has also been applied to basic media, where, in a group of nine reactions in concentrated NaOMe solutions, no correlation was found between reaction rates and either H_- or stoichiometric base concentration but where the rates were successfully correlated by a linear free-energy equation similar to those given above.[94]

A treatment partially based on the Bunnett–Olsen one is that of Bagno, Scorrano, and More O'Ferrall,[95] which formulates medium effects (changes in acidity of solvent) on acid–base equilibria. An appropriate equilibrium is chosen as reference, and the acidity dependence of other reactions compared with it, by use of the linear free-energy equation

$$\log \frac{K'}{K_0'} = m^* \log \frac{K}{K_0}$$

where the K values are the equilibrium constants for the following:

K for the reaction under study in any particular medium
K' for the reference reaction in the same medium
K_0 for the reaction under study in a reference solvent
K_0' for the reference reaction in the same reference solvent

and m^* is the slope of the relationship [corresponding to $(1 - \phi)$ of the Bunnett–Olsen treatment]. This equation has been shown to apply to many acid–base reactions.

Another type of classification system was devised by Bunnett[96] for reactions occurring in moderately concentrated acid solutions. Log $k_\psi + H_0$ is plotted against log a_{H_2O}, where K_ψ is the pseudo-first-order rate constant for the protonated species and a_{H_2O} is the activity of water. Most such plots are linear or nearly so. According to Bunnett, the slope of this plot w tells something about the mechanism. Where w is between -2.5 and 0, water is not involved in the rate-determining step; where w is between 1.2 and 3.3, water is a nucleophile in the rate-determining step; where w is between 3.3 and 7, water is a proton-transfer agent. These rules hold for acids in which the proton is attached to oxygen or nitrogen.

[94]More O'Ferrall *J. Chem. Soc., Perkin Trans. 2* **1972**, 976.
[95]Bagno; Scorrano; More O'Ferrall *Rev. Chem. Intermed.* **1987**, *7*, 313-352. See also Marziano; Cimino; Passerini *J. Chem. Soc., Perkin Trans. 2* **1973**, 1915; Lucchini; Modena; Scorrano; Cox; Yates *J. Am. Chem. Soc.* **1982**, *104*, 1958; Sampoli; De Santis; Marziano *J. Chem. Soc., Chem. Commun.* **1985**, 110; Cox *Acc. Chem. Res.* **1987**, *20*, 27-31.
[96]Bunnett *J. Am. Chem. Soc.* **1961**, *83*, 4956, 4968, 4973, 4978.

Acid and Base Catalysis[97]

Many reactions are catalyzed by acids, bases, or both. In such cases the catalyst is involved in a fundamental way in the mechanism. Nearly always the first step of such a reaction is a proton transfer between the catalyst and the substrate.

Reactions can be catalyzed by acid or base in two different ways, called *general* and *specific catalysis*. If the rate of an acid-catalyzed reaction run in a solvent S is proportional to $[SH^+]$, the reaction is said to be subject to *specific acid catalysis*, the acid being the lyonium ion SH^+. The acid that is put into the solvent may be stronger or weaker than SH^+, but the rate is proportional only to the $[SH^+]$ that is actually present in the solution (derived from $S + HA \rightleftharpoons SH^+ + A^-$). The identity of HA makes no difference except insofar as it determines the position of equilibrium and hence the $[SH^+]$. Most measurements have been made in water, where SH^+ is H_3O^+.

In *general acid catalysis*, the rate is increased not only by an increase in $[SH^+]$ but also by an increase in the concentration of other acids (e.g., in water by phenols or carboxylic acids). These other acids increase the rate even when $[SH^+]$ is held constant. In this type of catalysis the strongest acids catalyze best, so that, in the example given, an increase in the phenol concentration catalyzes the reaction much less than a similar increase in $[H_3O^+]$. This relationship between acid strength of the catalyst and its catalytic ability can be expressed by the *Brønsted catalysis equation*[98]

$$\log k = \alpha \log K_a + C$$

where k is the rate constant for a reaction catalyzed by an acid of ionization constant K_a. According to this equation, when $\log k$ is plotted against $\log K_a$ for catalysis of a given reaction by a series of acids, a straight line should be obtained with slope α and intercept C. Although straight lines are obtained in many cases, this is not always the case. The relationship usually fails when acids of different types are compared. For example, it is much more likely to hold for a group of substituted phenols than for a collection of acids that contains both phenols and carboxylic acids. The Brønsted equation is another linear free-energy relationship (see p. 281).

Analogously, there are *general* and *specific* (S^- from an acidic solvent SH) *base-catalyzed reactions*. The Brønsted law for bases is

$$\log k = \beta \log K_b + C$$

The Brønsted equations relate a rate constant k to an equilibrium constant K_a. In Chapter 6 we saw that the Marcus equation also relates a rate term (in that case $\Delta G^{\ddagger}$) to an equilibrium term $\Delta G°$. When the Marcus treatment is applied to proton transfers[99] between a carbon and an oxygen (or a nitrogen), the simplified[100] equation (p. 216)

$$\Delta G^{\ddagger} = \Delta G_{int}^{\ddagger} + \tfrac{1}{2}\Delta G° + \frac{(\Delta G°)^2}{16\,\Delta G_{int}^{\ddagger}}$$

[97]For reviews, see Stewart, Ref. 1, pp. 251-305; Hammett, Ref. 81, pp. 315-345; Willi, in Bamford; Tipper, Ref. 72, pp. 1-95; Jones, Ref. 81, pp. 72-82; Bell, Ref. 1, pp. 159-193; Jencks *Catalysis in Chemistry and Enzymology;* McGraw-Hill: New York, 1969, pp. 163-242; Bender *Mechanisms of Homogeneous Catalysis from Protons to Proteins;* Wiley: New York, 1971, pp. 19-144.

[98]For reviews, see Klumpp *Reactivity in Organic Chemistry;* Wiley: New York, 1982, pp. 167-179; Bell, in Chapman; Shorter *Correlation Analysis in Chemistry: Recent Advances;* Plenum Press: 1978, pp. 55-84; Kresge *Chem. Soc. Rev.* **1973,** *2,* 475-503.

[99]For applications of Marcus theory to proton transfers, see Marcus *J. Phys. Chem.* **1968,** *72,* 891; Kreevoy; Konasewich *Adv. Chem. Phys.* **1971,** *21,* 243; Kresge *Chem. Soc. Rev.* **1973,** *2,* 475-503.

[100]Omitting the work terms.

where

$$\Delta G_{int}^{\ddagger} = \tfrac{1}{2} (\Delta G_{O,O}^{\ddagger} + \Delta G_{C,C}^{\ddagger})$$

can be further simplified: Because proton transfers between oxygen and oxygen (or nitrogen and nitrogen) are much faster than those between carbon and carbon, $\Delta G_{O,O}^{\ddagger}$ is much smaller than $\Delta G_{C,C}^{\ddagger}$ and we can write[101]

$$\Delta G^{\ddagger} = \tfrac{1}{2} \Delta G_{C,C}^{\ddagger} + \tfrac{1}{2} \Delta G^{\circ} + \frac{(\Delta G^{\circ})^2}{8 \, \Delta G_{C,C}^{\ddagger}}$$

Thus, if the carbon part of the reaction is kept constant and only the A of HA is changed (where A is an oxygen or nitrogen moiety), then $\Delta G^{\ddagger}$ is dependent only on ΔG°. Differentiation of this equation yields the Brønsted α:

$$\frac{d\Delta G^{\ddagger}}{d\Delta G^{\circ}} = \alpha = \tfrac{1}{2} \left(1 + \frac{\Delta G^{\circ}}{2 \, \Delta G_{C,C}^{\ddagger}} \right)$$

The Brønsted law is therefore a special case of the Marcus equation.

A knowledge of whether a reaction is subject to general or specific acid catalysis supplies information about the mechanism. For any acid-catalyzed reaction we can write

Step 1 $A \overset{SH^+}{\rightleftharpoons} AH^+$

Step 2 $AH^+ \longrightarrow$ products

If the reaction is catalyzed only by the specific acid SH^+, it means that step 1 is rapid and step 2 is rate-controlling, since an equilibrium has been rapidly established between A and the strongest acid present in the solution, namely, SH^+ (since this is the strongest acid that can be present in S). On the other hand, if step 2 is faster, there is no time to establish equilibrium and the rate-determining step must be step 1. This step is affected by all the acids present, and the rate reflects the sum of the effects of each acid (general acid catalysis). General acid catalysis is also observed if the slow step is the reaction of a hydrogen-bond complex $A \cdots HB$, since each complex reacts with a base at a different rate. A comparable discussion can be used for general and specific base catalysis.[102] Further information can be obtained from the values α and β in the Brønsted catalysis equations, since these are approximate measures of the extent of proton transfer in the transition state. In most cases values of α and β are between 1 and 0. A value of α or β near 0 is generally taken to mean that the transition state resembles the reactants; i.e., the proton has been transferred very little when the transition state has been reached. A value of α or β near 1 is taken to mean the opposite; i.e., in the transition state the proton has been almost completely transferred. However, cases are known in which these generalizations are not followed,[103] and their theoretical basis has been challenged.[104] In general, the proton in the transition state lies closer to the weaker base.

[101]Albery *Annu. Rev. Phys. Chem.* **1980,** *31,* 227-263, p. 244.
[102]For discussions of when to expect general or specific acid or base catalysis, see Jencks *Acc. Chem. Res.* **1976,** *9,* 425-432; Stewart; Srinivasan *Acc. Chem. Res.* **1978,** *11,* 271-277; Guthrie *J. Am. Chem. Soc.* **1980,** *102,* 5286.
[103]See, for example, Bordwell; Boyle *J. Am. Chem. Soc.* **1972,** *94,* 3907; Davies *J. Chem. Soc., Perkin Trans. 2* **1974,** 1018; Agmon *J. Am. Chem. Soc.* **1980,** *102,* 2164; Murray; Jencks *J. Am. Chem. Soc.* **1988,** *110,* 7561.
[104]Pross; Shaik *New J. Chem.* **1989,** *13,* 427; Lewis, *J. Phys. Org. Chem.* **1990,** *3,* 1.

Lewis Acids and Bases. Hard and Soft Acids and Bases

At about the same time that Brønsted proposed his acid–base theory, Lewis put forth a broader theory. A base in the Lewis theory is the same as in the Brønsted one, namely, a compound with an available pair of electrons, either unshared or in a π orbital. A *Lewis acid*, however, is any species with a vacant orbital.[105] In a Lewis acid–base reaction the unshared pair of the base forms a covalent bond with the vacant orbital of the acid, as represented by the general equation

$$A + \overline{B} \longrightarrow A\text{---}B$$

in which charges are not shown, since they may differ. A specific example is

$$BF_3 + \overline{N}H_3 \longrightarrow F_3\overset{\ominus}{B}\text{---}\overset{\oplus}{N}H_3$$

In the Brønsted picture, the acid is a proton donor, but in the Lewis picture the proton itself is the acid since it has a vacant orbital. A Brønsted acid becomes, in the Lewis picture, the compound that gives up the actual acid. The advantage of the Lewis theory is that it correlates the behavior of many more processes. For example, $AlCl_3$ and BF_3 are Lewis acids because they have only six electrons in the outer shell and have room for eight. $SnCl_4$ and SO_3 have eight, but their central elements, not being in the first row of the periodic table, have room for ten or twelve. Other Lewis acids are simple cations, like Ag^+. The simple reaction $A + \overline{B} \rightarrow A\text{---}B$ is not very common in organic chemistry, but the scope of the Lewis picture is much larger because reactions of the types

$$A^1 + A^2\text{---}B \longrightarrow A^1\text{---}B + A^2$$
$$B^1 + A\text{---}B^2 \longrightarrow A\text{---}B^1 + B^2$$
$$A^1\text{---}B^1 + A^2\text{---}B^2 \longrightarrow A^1\text{---}B^2 + A^2\text{---}B^1$$

which are very common in organic chemistry, are also Lewis acid–base reactions. In fact, all reactions in which a covalent bond is formed through one species contributing a filled and the other a vacant orbital may be regarded as Lewis acid–base reactions.

When a Lewis acid combines with a base to give a negative ion in which the central atom has a higher-than-normal valence, the resulting salt is called an *ate complex*.[106] Examples are

$$Me_3B + LiMe \longrightarrow Me_4B^- \ Li^+$$
Ate complex

$$Ph_5Sb + LiPh \longrightarrow Ph_6Sb^- \ Li^+$$
Ate complex

Ate complexes are analogous to the onium salts formed when a Lewis base expands its valence, e.g.,

$$Me_3N + MeI \longrightarrow Me_4N^+ \ I^-$$
Onium salt

---

[105]For a monograph on Lewis acid–base theory, see Jensen *The Lewis Acid–Base Concept*; Wiley: New York, 1980. For a discussion of the definitions of Lewis acid and base, see Jensen *Chem. Rev.* **1978**, *78*, 1-22.
[106]For a review of ate complexes, see Wittig *Q. Rev., Chem. Soc.* **1966**, *20*, 191-210.

Far fewer quantitative measurements have been made of Lewis acid strength compared to that of Brønsted acids.[107] A simple table of Lewis acidities based on some quantitative measurement (such as that given for Brønsted acids in Table 8.1) is not feasible because Lewis acidity depends on the nature of the base. Qualitatively, the following approximate sequence of acidity of Lewis acids of the type MX_n has been suggested, where X is a halogen atom or an inorganic radical: $BX_3 > AlX_3 > FeX_3 > GaX_3 > SbX_5 > SnX_4 > AsX_5 > ZnX_2 > HgX_2$.

The facility with which an acid–base reaction takes place depends of course on the strengths of the acid and the base. But it also depends on quite another quality, called the *hardness* or *softness* of the acid or base.[108] Hard and soft acids and bases have these characteristics:

Soft bases. The donor atoms are of low electronegativity and high polarizability and are easy to oxidize. They hold their valence electrons loosely.

Hard bases. The donor atoms are of high electronegativity and low polarizability and are hard to oxidize. They hold their valence electrons tightly.

Soft acids. The acceptor atoms are large, have low positive charge, and contain unshared pairs of electrons (*p* or *d*) in their valence shells. They have high polarizability and low electronegativity.

Hard acids. The acceptor atoms are small, have high positive charge, and do not contain unshared pairs in their valence shells. They have low polarizability and high electronegativity.

A qualitative listing of the hardness of some acids and bases is given in Table 8.2.[109] The treatment has also been made quantitative,[110] with the following operational definition:

$$\eta = \frac{I - A}{2}$$

In this equation η, the *absolute hardness*, is half the difference between I, the ionization potential, and A, the electron affinity. The softness, σ, is the reciprocal of η. Values of η for some molecules and ions are given in Table 8.3.[111] Note that the proton, which is involved in all Brønsted acid–base reactions, is the hardest acid listed, with $\eta = \infty$ (it has no ionization potential). The above equation cannot be applied to anions, because electron affinities cannot be measured for them. Instead, the assumption is made that η for an anion X^- is the same as that for the radical $X\bullet$.[112] Other methods are also needed to apply the treatment to polyatomic cations.[112]

[107]For reviews of the quantitative aspects of Lewis acidity, see Satchell; Satchell *Q. Rev., Chem. Soc.* **1971**, *25*, 171-199, *Chem. Rev.* **1969**, *69*, 251-278. See also Maria; Gal *J. Phys. Chem.* **1985**, *89*, 1296; Larson; McMahon *J. Am. Chem. Soc.* **1985**, *107*, 766; Larson; Szulejko; McMahon *J. Am. Chem. Soc.* **1988**, *110*, 7604; Sandström; Persson; Persson *Acta Chem. Scand.* **1990**, *44*, 653; Laszlo; Teston-Henry *Tetrahedron Lett.* **1991**, *32*, 3837.

[108]Pearson *J. Am. Chem. Soc.* **1963**, *85*, 3533, *Science* **1966**, *151*, 172; Pearson; Songstad *J. Am. Chem. Soc.* **1967**, *89*, 1827. For a monograph on the concept, see Ho *Hard and Soft Acids and Bases Principle in Organic Chemistry*; Academic Press: New York, 1977. For reviews, see Pearson, *J. Chem. Educ.* **1987**, *64*, 561-567; Ho *Tetrahedron* **1985**, *41*, 1-86, *J. Chem. Educ.* **1978**, *55*, 355-360, *Chem. Rev.* **1975**, *75*, 1-20; Pearson, in Chapman; Shorter *Advances in Linear Free-Energy Relationships*; Plenum Press: New York, 1972, pp. 281-319; Pearson *Surv. Prog. Chem.* **1969**, *5*, 1-52 [portions of this article slightly modified also appear in Pearson *J. Chem. Educ.* **1968**, *45*, 581-587, 643-648]; Garnovskii; Osipov; Bulgarevich *Russ. Chem. Rev.* **1972**, *41*, 341-359; Seyden-Penne *Bull. Soc. Chim. Fr.* **1968**, 3871-3878. For a collection of papers, see Pearson *Hard and Soft Acids and Bases*; Dowden, Hutchinson, and Ross: Stroudsberg, PA, 1973.

[109]Taken from larger listings in Pearson, Ref. 108.

[110]Parr; Pearson *J. Am. Chem. Soc.* **1983**, *105*, 7512; Pearson *Inorg. Chem.* **1988**, *27*, 734. *J. Org. Chem.* **1989**, *54*, 1423. See also Orsky; Whitehead *Can. J. Chem.* **1987**, *65*, 1970.

[111]Note that there is not always a strict correlation between the values in Table 8.3 and the categories of Table 8.2.

[112]Pearson *J. Am. Chem. Soc.* **1988**, *110*, 7684.

TABLE 8.2 Hard and soft acids and bases[109]

Hard bases	Soft bases	Borderline bases
H_2O OH^- F^-	R_2S RSH RS^-	$ArNH_2$ C_5H_5N
AcO^- SO_4^{2-} Cl^-	I^- R_3P $(RO)_3P$	N_3^- Br^-
CO_3^{2-} NO_3^- ROH	CN^- RCN CO	NO_2^-
RO^- R_2O NH_3	C_2H_4 C_6H_6	
RNH_2	H^- R^-	

Hard acids	Soft acids	Borderline acids
H^+ Li^+ Na^+	Cu^+ Ag^+ Pd^{2+}	Fe^{2+} Co^{2+} Cu^{2+}
K^+ Mg^{2+} Ca^{2+}	Pt^{2+} Hg^{2+} BH_3	Zn^{2+} Sn^{2+} Sb^{3+}
Al^{3+} Cr^{2+} Fe^{3+}	$GaCl_3$ I_2 Br_2	Bi^{3+} BMe_3 SO_2
BF_3 $B(OR)_3$ $AlMe_3$	CH_2 carbenes	R_3C^+ NO^+ GaH_3
$AlCl_3$ AlH_3 SO_3		$C_6H_5^+$
RCO^+ CO_2		
HX (hydrogen-bonding molecules)		

TABLE 8.3 Some absolute hardness values in electron volts[110]

Cations		Molecules		Anions[b]	
Ion	η	Compound	η	Ion	η
H^+	∞	HF	11.0	F^-	7.0
Al^{3+}	45.8	CH_4	10.3	H^-	6.4
Li^+	35.1	BF_3	9.7	OH^-	5.7
Mg^{2+}	32.6	H_2O	9.5	NH_2^-	5.3
Na^+	21.1	NH_3	8.2	CN^-	5.1
Ca^{2+}	19.5	HCN	8.0	CH_3^-	4.9
K^+	13.6	$(CH_3)_2O$	8.0	Cl^-	4.7
Zn^{2+}	10.9	CO	7.9	$CH_3CH_2^-$	4.4
Cr^{3+}	9.1	C_2H_2	7.0	Br^-	4.2
Cu^{2+}	8.3	$(CH_3)_3N$	6.3	$C_6H_5^-$	4.1
Pt^{2+}	8.0	H_2S	6.2	SH^-	4.1
Sn^{2+}	7.9	C_2H_4	6.2	$(CH_3)_2CH^-$	4.0
Hg^{2+}	7.7	$(CH_3)_2S$	6.0	I^-	3.7
Fe^{2+}	7.2	$(CH_3)_3P$	5.9	$(CH_3)_3C^-$	3.6
Pd^{2+}	6.8	CH_3COCH_3	5.6		
Cu^+	6.3	C_6H_6	5.3		
		HI	5.3		
		C_5H_5N	5.0		
		C_6H_5OH	4.8		
		$CH_2{}^a$	4.7		
		C_6H_5SH	4.6		
		Cl_2	4.6		
		$C_6H_5NH_2$	4.4		
		Br_2	4.0		
		I_2	3.4		

^{a}For singlet state.
^{b}The same as for the corresponding radical.

Once acids and bases have been classified as hard or soft, a simple rule can be given: *hard acids prefer to bond to hard bases, and soft acids prefer to bond to soft bases (the HSAB principle)*.[112a] The rule has nothing to do with acid or base *strength* but merely says that the product A—B will have extra stability if both A and B are hard or if both are soft. Another rule is that a soft Lewis acid and a soft Lewis base tend to form a covalent bond, while a hard acid and a hard base tend to bond ionically.

One application of the first rule given above is found in complexes between alkenes or aromatic compounds and metal ions (p. 80). Alkenes and aromatic rings are soft bases and should prefer to complex with soft acids. Thus, Ag^+, Pt^{2+}, and Hg^{2+} complexes are common, but complexes of Na^+, Mg^{2+}, or Al^{3+} are rare. Chromium complexes are also common, but in such complexes the chromium is in a low or zero oxidation state (which softens it) or attached to other soft ligands. In another application, we may look at this reaction:

$$CH_3-\underset{\underset{O}{\|}}{C}-SR' + OR^- \rightleftharpoons CH_3-\underset{\underset{O}{\|}}{C}-OR + SR'^-$$

The HSAB principle predicts that the equilibrium should lie to the right, because the hard acid CH_3CO^+ should have a greater affinity for the hard base RO^- than for the soft base RS^-. Indeed, thiol esters are easily cleaved by OR^- or hydrolyzed by dilute base (OH^- is also a hard base).[113] Another application of the rule is discussed on p. 349.[114]

The Effects of Structure on the Strengths of Acids and Bases[115]

The structure of a molecule can affect its acidity or basicity in a number of ways. Unfortunately, in most molecules two or more of these effects (as well as solvent effects) are operating, and it is usually very difficult or impossible to say how much each effect contributes to the acid or base strength.[116] Small differences in acidity or basicity between similar molecules are particularly difficult to interpret. It is well to be cautious when attributing them to any particular effect.

1. *Field effects.* These were discussed on p. 17. As an example of the influence of field effects on acidity, we may compare the acidity of acetic acid and nitroacetic acid:

$$H-CH_2-\underset{\underset{O}{\|}}{C}-O-H \qquad O_2N-CH_2-\underset{\underset{O}{\|}}{C}-O-H$$

$$pK_a = 4.76 \qquad\qquad pK_a = 1.68$$

[112a]For proofs of this principle, see Chattaraj; Lee; Parr *J. Am. Chem. Soc.* **1991**, *113*, 1855.

[113]Wolman, in Patai *The Chemistry of the Thiol Group*, pt. 2; Wiley: New York, 1974, p. 677; Maskill *The Physical Basis of Organic Chemistry*; Oxford University Press: Oxford, 1985, p. 159.

[114]See also Bochkov *J. Org. Chem. USSR* **1986**, *22*, 1830, 1837.

[115]For a monograph, see Hine *Structural Effects on Equilibria in Organic Chemistry*; Wiley: New York, 1975. For reviews, see Taft *Prog. Phys. Org. Chem.* **1983**, *14*, 247-350; Petrov *Russ. Chem. Rev.* **1983**, *52*, 1144-1155 (NH acids); Bell, Ref. 1, pp. 86-110; Barlin; Perrin, in Bentley; Kirby *Elucidation of Organic Structures by Physical and Chemical Methods*, 2nd ed. (vol. 4 of Weissberger *Techniques of Chemistry*), pt. 1; Wiley: New York, 1972, pp. 611-676. For discussions, see Bolton; Hepler *Q. Rev., Chem. Soc.* **1971**, *25*, 521-532; Barlin; Perrin *Q. Rev., Chem. Soc.* **1966**, *20*, 75-101; Thirot *Bull. Soc. Chim. Fr.* **1967**, 3559; Liler, Ref. 10, pp. 59-144. For a monograph on methods of estimating pK values by analogy, extrapolation, etc., see Perrin; Dempsey; Serjeant *pK$_a$ Prediction for Organic Acids and Bases*; Chapman and Hall: New York, 1981.

[116]The varying degrees by which the different factors that affect gas-phase acidities of 25 acids has been calculated: Taft; Koppel; Topsom; Anvia *J. Am. Chem. Soc.* **1990**, *112*, 2047.

The only difference in the structure of these molecules is the substitution of NO_2 for H. Since NO_2 is a strongly electron-withdrawing group, it withdraws electron density from the negatively charged COO^- group in the anion of nitroacetic acid (compared with the anion of acetic acid) and, as the pK_a values indicate, nitroacetic acid is about 1000 times stronger than acetic acid.[117] Any effect that results in electron withdrawal from a negatively charged center is a stabilizing effect because it spreads the charge. Thus, $-I$ groups increase the acidity of uncharged acids such as acetic because they spread the negative charge of the anion. However, $-I$ groups also increase the acidity of any acid, no matter what the charge. For example, if the acid has a charge of $+1$ (and its conjugate base is therefore uncharged), a $-I$ group destabilizes the positive center (by increasing and concentrating the positive charge) of the acid, a destabilization that will be relieved when the proton is lost. In general we may say that *groups that withdraw electrons by the field effect increase acidity and decrease basicity, while electron-donating groups act in the opposite direction.* Another example is the molecule $(C_6F_5)_3CH$, which has three strongly electron-withdrawing C_6F_5 groups and a pK_a of 16,[118] compared with Ph_3CH, with a pK_a of 31.5 (Table 8.1), an acidity enhancement of about 10^{15}. Table 8.4 shows pK_a values for some acids. An approximate idea of field effects can be obtained from this table. In the case of the chlorobutyric acids note how the effect decreases with distance. It must be remembered, however, that field effects are not the sole cause of the acidity differences noted and that in fact solvation effects may be more important in many cases (see pp. 269-272).[119]

2. Resonance effects.   Resonance that stabilizes a base but not its conjugate acid results in the acid having a higher acidity than otherwise expected and vice versa. An example is found in the higher acidity of carboxylic acids compared with primary alcohols.

The $RCOO^-$ ion is stabilized by resonance not available to the RCH_2O^- ion (or to $RCOOH$).[120] Note that the $RCOO^-$ is stabilized not only by the fact that there are two equivalent canonical forms but also by the fact that the negative charge is spread over both oxygen atoms and is therefore less concentrated than in RCH_2O^-. The same effect is found in other compounds containing a $C=O$ or $C\equiv N$ group. Thus amides $RCONH_2$ are more acidic than amines RCH_2NH_2; esters RCH_2COOR' than ethers RCH_2CH_2OR'; and ketones RCH_2COR' than alkanes RCH_2CH_2R' (Table 8.1). The effect is enhanced when two carbonyl groups are attached to the same carbon (because of additional resonance and spreading

[117]For a review of the enhancement of acidity by NO_2, see Lewis, in Patai *The Chemistry of Functional Groups, Supplement F.* pt. 2; Wiley: New York, 1982, pp. 715-729.

[118]Filler; Wang *Chem. Commun.* **1968**, 287.

[119]For discussions, see Edward *J. Chem. Educ.* **1982**, *59*, 354; Schwartz *J. Chem. Educ.* **1981**, *58*, 778.

[120]It has been contended that resonance delocalization plays only a minor role in the increased strength of carboxylic acids compared to alcohols, and the ". . . higher acidity of acids arises principally because the electrostatic potential of the acidic hydrogens is more positive in the neutral acid molecule . . .": Siggel; Thomas *J. Am. Chem. Soc.* **1986**, *108*, 4360; Siggel; Streitwieser; Thomas *J. Am. Chem. Soc.* **1988**, *110*, 8022; Thomas; Carroll; Siggel *J. Org. Chem.* **1988**, *53*, 1812. For contrary views, see Exner *J. Org. Chem.* **1988**, *53*, 1810; Dewar; Krull *J. Chem. Soc., Chem. Commun.* **1990**, 333; Perrin *J. Am. Chem. Soc.* **1991**, *113*, 2865. See also Godfrey *Tetrahedron Lett.* **1990**, *31*, 5181.

TABLE 8.4 pK values for some acids[39]

Acid	pK	Acid	pK
HCOOH	3.77	ClCH₂COOH	2.86
CH₃COOH	4.76	Cl₂CHCOOH	1.29
CH₃CH₂COOH	4.88	Cl₃CCOOH	0.65
CH₃(CH₂)ₙCOOH	4.82–4.95		
(n = 2 to 7)		O₂NCH₂COOH	1.68
(CH₃)₂CHCOOH	4.86	(CH₃)₃N⁺CH₂COOH	1.83
(CH₃)₃CCOOH	5.05	HOOCCH₂COOH	2.83
		PhCH₂COOH	4.31
FCH₂COOH	2.66	⁻OOCCH₂COOH	5.69
ClCH₂COOH	2.86		
BrCH₂COOH	2.86		
ICH₂COOH	3.12	O₃SCH₂COOH	4.05
		HOCH₂COOH	3.83
ClCH₂CH₂CH₂COOH	4.52	H₂C=CHCH₂COOH	4.35
CH₃CHClCH₂COOH	4.06		
CH₃CH₂CHClCOOH	2.84		

of charge); for example, β-keto esters are more acidic than simple ketones or carboxylic esters (Table 8.1). Extreme examples of this effect are found in the molecules tricyano-

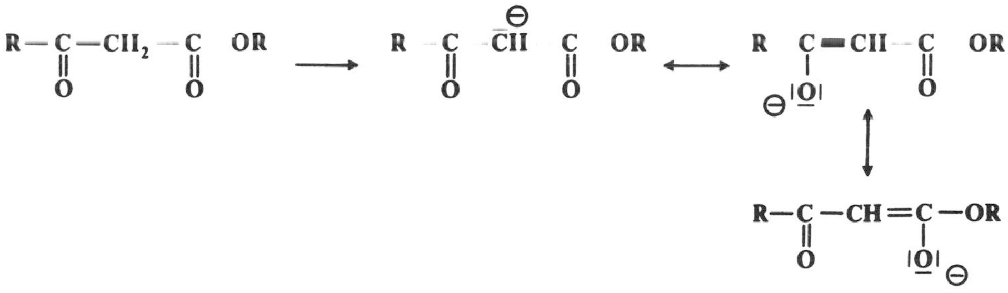

methane (NC)₃CH, with a pK_a of -5, and 2-(dicyanomethylene)-1,1,3,3-tetracyanopropene (NC)₂C=C[CH(CN)₂]₂, whose first pK_a is below -8.5 and whose second pK_a is -2.5.

Resonance effects are also important in aromatic amines. *m*-Nitroaniline is a weaker base than aniline, a fact that can be accounted for by the $-I$ effect of the nitro group. But

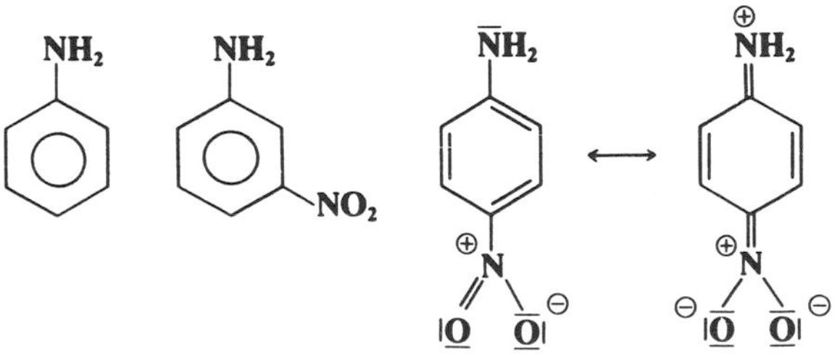

pK_a of conjugate[122]   4.60   2.47   1.11
              acid   **A**

p-nitroaniline is weaker still, though the $-I$ effect should be less because of the greater distance. We can explain this result by taking into account the canonical form **A**. Because **A** contributes to the resonance hybrid,[121] the electron density of the unshared pair is lower in *p*-nitroaniline than in *m*-nitroaniline, where a canonical form such as **A** is impossible. The basicity is lower in the para compound for two reasons, both caused by the same effect: (1) the unshared pair is less available for attack by a proton, and (2) when the conjugate acid is formed, the resonance stabilization afforded by **A** is no longer available because the previously unshared pair is now being shared by the proton. The acidity of phenols is affected by substituents in a similar manner.

In general, resonance effects lead to the same result as field effects. That is, here too, electron-withdrawing groups increase acidity and decrease basicity, and electron-donating groups act in the opposite manner. As a result of both resonance and field effects, charge dispersal leads to greater stability.

3. *Periodic table correlations.*   When comparing Brønsted acids and bases that differ in the position of an element in the periodic table:

a. Acidity increases and basicity decreases in going from left to right across a row of the periodic table. Thus acidity increases in the order $CH_4 < NH_3 < H_2O < HF$, and basicity decreases in the order $CH_3^- > NH_2^- > OH^- > F^-$. This behavior can be explained by the increase in electronegativity upon going from left to right across the table. It is this effect that is responsible for the great differences in acidity between carboxylic acids, amides, and ketones: $RCOOH \gg RCONH_2 \gg RCOCH_3$.

b. Acidity increases and basicity decreases in going down a column of the periodic table, despite the decrease in electronegativity. Thus acidity increases in the order $HF < HCl < HBr < HI$ and $H_2O < H_2S$, and basicity decreases in the order $NH_3 > PH_3 > AsH_3$. This behavior is related to the size of the species involved. Thus, for example, F^-, which is much smaller than I^-, attracts a proton much more readily because its negative charge occupies a smaller volume and is therefore more concentrated (note that F^- is also much harder than I^- and is thus more attracted to the hard proton; see p. 263). This rule does not always hold for positively charged acids. Thus, although the order of acidity for the group 16 hydrides is $H_2O < H_2S < H_2Se$, the acidity order for the positively charged ions is $H_3O^+ > H_3S^+ > H_3Se^+$.[123]

Lewis acidity is also affected by periodic table considerations. In comparing acid strengths of Lewis acids of the form MX_n:[107]

c. Acids that require only one electron pair to complete an outer shell are stronger than those that require two. Thus $GaCl_3$ is stronger than $ZnCl_2$. This results from the relatively smaller energy gain in adding an electron pair that does not complete an outer shell and from the buildup of negative charge if two pairs come in.

d. Other things being equal, the acidity of MX_n decreases in going down the periodic table because as the size of the molecule increases, the attraction between the positive nucleus and the incoming electron pair is weaker. Thus BCl_3 is a stronger acid than $AlCl_3$.[124]

4. *Statistical effects.*   In a symmetrical diprotic acid, the first dissociation constant is twice as large as expected since there are two equivalent ionizable hydrogens, while the second constant is only half as large as expected because the conjugate base can accept a proton at two equivalent sites. So K_1/K_2 should be 4, and approximately this value is found

---

[121]See, however, Lipkowitz *J. Am. Chem. Soc.* **1982**, *104*, 2647; Krygowski; Maurin *J. Chem. Soc., Perkin Trans. 2* **1989**, 695.

[122]Smith, in Patai *The Chemistry of the Amino Group*; Wiley: New York, 1968, pp. 161-204.

[123]Taft, Ref. 115, pp. 250-254.

[124]Note that Lewis acidity *decreases*, whereas Brønsted acidity *increases*, going down the table. There is no contradiction here when we remember that in the Lewis picture the actual acid in all Brønsted acids is the same, namely, the proton. In comparing, say, HI and HF, we are not comparing different Lewis acids but only how easily F^- and I^- give up the proton.

for dicarboxylic acids where the two groups are sufficiently far apart in the molecule that they do not influence each other. A similar argument holds for molecules with two equivalent basic groups.[125]

5. *Hydrogen bonding.* Internal hydrogen bonding can greatly influence acid or base strength. For example, the pK for *o*-hydroxybenzoic acid is 2.98, while the value for the para isomer is 4.58. Internal hydrogen bonding between the OH and COO$^-$ groups of the conjugate base of the ortho isomer stabilizes it and results in an increased acidity.

6. *Steric effects.* The proton itself is so small that direct steric hindrance is seldom encountered in proton transfers. Steric effects are much more common in Lewis acid–base reactions in which larger acids are used. Spectacular changes in the order of base strength have been demonstrated when the size of the acid was changed. Table 8.5 shows the order of base strength of simple amines when compared against acids of various size.[126] It can be seen that the usual order of basicity of amines (when the proton is the reference acid) can be completely inverted by using a large enough acid. The strain caused by formation of a covalent bond when the two atoms involved each have three large groups is called *face strain* or *F strain*.

Steric effects can indirectly affect acidity or basicity by affecting the resonance (see p. 37). For example, *o-t*-butylbenzoic acid is about 10 times as strong as the para isomer, because the carboxyl group is forced out of the plane by the *t*-butyl group. Indeed, virtually all ortho benzoic acids are stronger than the corresponding para isomers, regardless of whether the group on the ring is electron-donating or electron-withdrawing.

Steric effects can also be caused by other types of strain. 1,8-Bis(diethylamino)-2,7-dimethoxynaphthalene (**1**) is an extremely strong base for a tertiary amine (pK_a of the

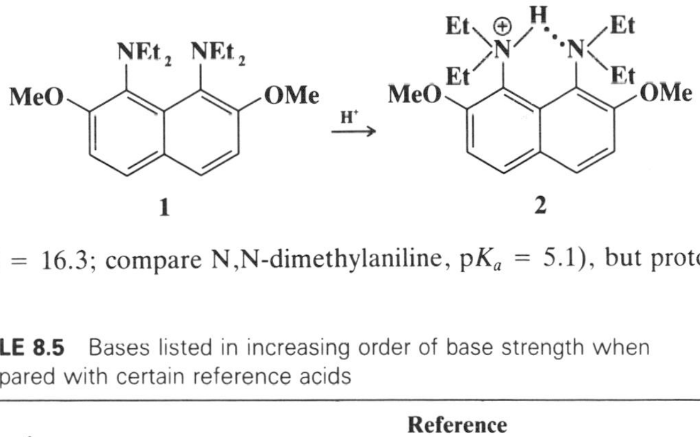

conjugate acid = 16.3; compare N,N-dimethylaniline, pK_a = 5.1), but proton transfers to

TABLE 8.5 Bases listed in increasing order of base strength when compared with certain reference acids

Increasing order of base strength[a]	Reference acid		
	H$^+$ or BMe$_3$	BMe$_3$	B(CMe$_3$)$_3$
	NH$_3$	Et$_3$N	Me$_3$N	Et$_3$N
	Me$_3$N	NH$_3$	Me$_2$NH	Et$_2$NH
	MeNH$_2$	Et$_2$NH	NH$_3$	EtNH$_2$
	Me$_2$NH	EtNH$_2$	MeNH$_2$	NH$_3$

[a]The order of basicity (when the reference acids were boranes) was determined by the measurement of dissociation pressures.

[125]The effect discussed here is an example of a symmetry factor. For an extended discussion, see Eberson, in Patai *The Chemistry of Carboxylic Acids and Esters*; Wiley: New York, 1969, pp. 211-293.
[126]Brown *J. Am. Chem. Soc.* **1945**, *67*, 378, 1452, *Boranes in Organic Chemistry*; Cornell University Press: Ithaca, NY, 1972, pp. 53-64. See also Brown; Krishnamurthy; Hubbard *J. Am. Chem. Soc.* **1978**, *100*, 3343.

and from the nitrogen are exceptionally slow; slow enough to be followed by a uv spectrophotometer.[127] **1** is severely strained because the two nitrogen lone pairs are forced to be near each other.[128] Protonation relieves the strain: one lone pair is now connected to a hydrogen, which forms a hydrogen bond to the other lone pair (shown in **2**). The same effects are found in 4,5-bis(dimethylamino)fluorene (**3**)[129] and 4,5-bis(dimethylamino)-

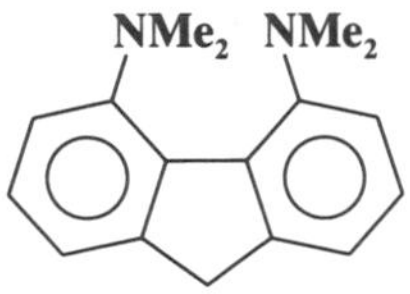

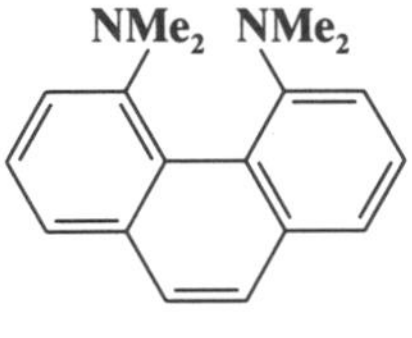

phenanthrene (**4**).[130] Compounds such as **1, 3,** and **4** are known as *proton sponges*.[131] Another type of proton sponge is quino[7,8-*h*]quinoline (**5**).[132] Protonation of this compound also gives a stable monoprotonated ion similar to **2**, but the steric hindrance found in **1, 3,** and

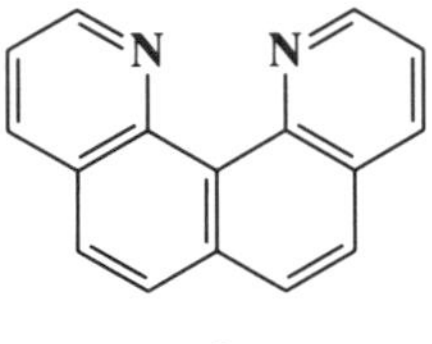

4 is absent. Therefore **5** is a much stronger base than quinoline (**6**) (pK_a values of the conjugate acids are 12.8 for **5** and 4.9 for **6**), but proton transfers are not abnormally slow.

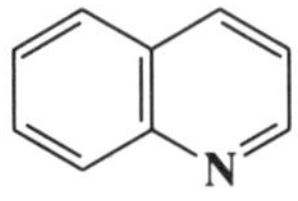

Another type of steric effect is the result of an entropy effect. The compound 2,6-di-*t*-butylpyridine is a weaker base than either pyridine or 2,6-dimethylpyridine.[133] The reason is that the conjugate acid (**7**) is less stable than the conjugate acids of non-sterically

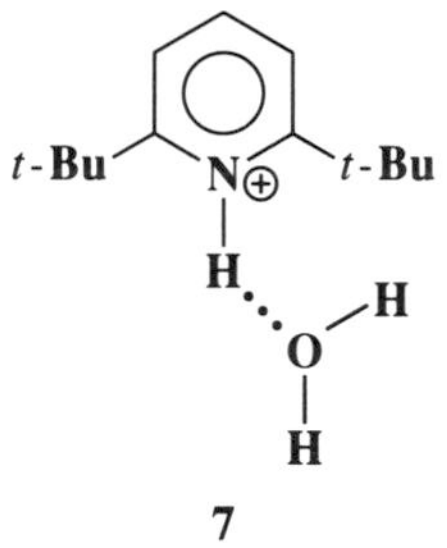

[127]Alder; Goode; Miller; Hibbert; Hunte; Robbins *J. Chem. Soc., Chem. Commun.* **1978,** 89; Hibbert; Hunte *J. Chem. Soc., Perkin Trans. 2* **1983,** 1895; Barnett; Hibbert *J. Am. Chem. Soc.* **1984,** *106,* 2080; Hibbert; Simpson *J. Chem. Soc., Perkin Trans. 2* **1987,** 243, 613.

[128]For a review of the effect of strain on amine basicities, see Alder *Chem. Rev.* **1989,** *89,* 1215-1223.

[129]Staab; Saupe; Krieger *Angew. Chem. Int. Ed. Engl.* **1983,** *22,* 731 [*Angew. Chem. 95,* 748].

[130]Saupe; Krieger; Staab *Angew. Chem. Int. Ed. Engl.* **1986,** *25,* 451 [*Angew. Chem. 98,* 460].

[131]For a review, see Staab; Saupe *Angew. Chem. Int. Ed. Engl.* **1988,** *27,* 865-879 [*Angew. Chem. 895-909*].

[132]Zirnstein; Staab *Angew. Chem. Int. Ed. Engl.* **1987,** *26,* 460 [*Angew. Chem. 99,* 460]; Krieger; Newsom; Zirnstein; Staab *Angew. Chem. Int. Ed. Engl.* **1989,** *28,* 84 [*Angew. Chem. 101,* 72]. See also Schwesinger; Missfeldt; Peters; Schnering *Angew. Chem.Int. Ed. Engl.* **1987,** *26,* 1165 [*Angew. Chem. 99,* 1210]; Alder; Eastment; Hext; Moss; Orpen; White *J. Chem. Soc., Chem. Commun.* **1988,** 1528; Staab; Zirnstein; Krieger *Angew. Chem. Int. Ed. Engl.* **1989,** *28,* 86 [*Angew. Chem. 101,* 73].

[133]Brown; Kanner *J. Am. Chem. Soc.* **1953,** *75,* 3865; **1966,** *88,* 986.

hindered pyridines. In all cases the conjugate acids are hydrogen-bonded to a water molecule, but in the case of **7** the bulky *t*-butyl groups restrict rotations in the water molecule, lowering the entropy.[134]

The conformation of a molecule can also affect its acidity. The following pK_a values were determined for these compounds:[135]

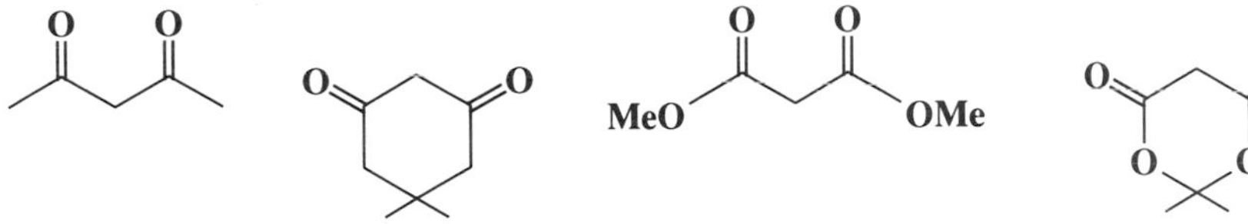

	8	9	10	11
pK_a	13.3	11.2	15.9	7.3

Since ketones are stronger acids than carboxylic esters (Table 8.1), we are not surprised that **8** is a stronger acid than **10**. But cyclization of **8** to **9** increases the acidity by only 2.1 pK units while cyclization of **10** to **11** increases it by 8.6 units. Indeed, it has long been known that **11** (called Meldrum's acid) is an unusually strong acid for a 1,3-diester. In order to account for this very large cyclization effect, molecular orbital calculations were carried out two conformations of methyl acetate and of its enolate ion by two groups.[136] Both found

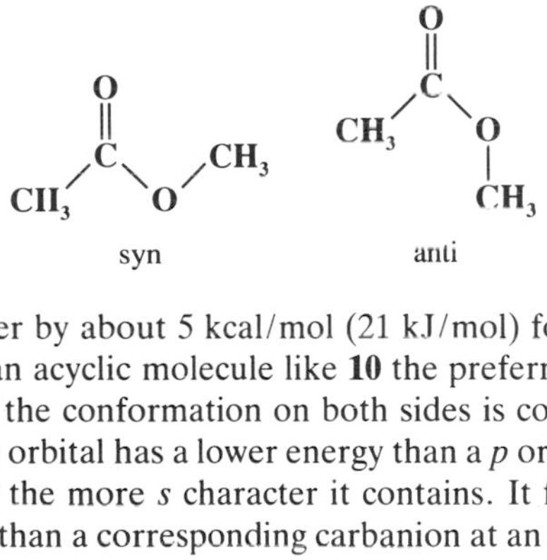

that loss of a proton is easier by about 5 kcal/mol (21 kJ/mol) for the syn than for the anti conformer of the ester. In an acyclic molecule like **10** the preferred conformations are anti, but in Meldrum's acid (**11**) the conformation on both sides is constrained to be syn.

7. Hybridization. An *s* orbital has a lower energy than a *p* orbital. Therefore the energy of a hybrid orbital is lower the more *s* character it contains. It follows that a carbanion at an *sp* carbon is more stable than a corresponding carbanion at an sp^2 carbon. Thus $HC{\equiv}C^-$, which has more *s* character in its unshared pair than $CH_2{=}CH^-$ or $CH_3CH_2^-$ (*sp* vs. sp^2 vs. sp^3, respectively), is a much weaker base. This explains the relatively high acidity of acetylenes and HCN. Another example is that alcohol and ether oxygens, where the unshared pair is sp^3, are more strongly basic than carbonyl oxygens, where the unshared pair is sp^2 (Table 8.1).

The Effects of the Medium on Acid and Base Strength

Structural features are not the only factors that affect acidity or basicity. The same compound can have its acidity or basicity changed when the conditions are changed. The effect of

[134]Meot-Ner; Sieck *J. Am. Chem. Soc.* **1983**, *105*, 2956; Hopkins; Jahagirdar; Moulik; Aue; Webb; Davidson; Pedley *J. Am. Chem. Soc.* **1984**, *106*, 4341; Meot-Ner; Smith *J. Am. Chem. Soc.* **1991**, *113*, 862, and references cited in these papers. See also Benoit; Fréchette; Lefebvre *Can. J. Chem.* **1988**, *66*, 1159.
[135]Arnett; Harrelson *J. Am. Chem. Soc.* **1987**, *109*, 809.
[136]Wang; Houk *J. Am. Chem. Soc.* **1988**, *110*, 1870; Wiberg; Laidig *J. Am. Chem. Soc.* **1988**, *110*, 1872.

temperature (p. 253) has already been mentioned. More important is the effect of the solvent, which can exert considerable influence on acid and base strengths by differential solvation.[137] If a base is more solvated than its conjugate acid, its stability is increased relative to the conjugate acid. For example, Table 8.5 shows that toward the proton, where steric effects are absent, methylamine is a stronger base than ammonia and dimethylamine is stronger still.[138] These results are easily explainable if one assumes that methyl groups are electron-donating. However, trimethylamine, which should be even stronger, is a weaker base than dimethylamine or methylamine. This apparently anomalous behavior can be explained by differential hydration.[139] Thus, NH_4^+ is much better hydrated (by hydrogen bonding to the water solvent) than NH_3 because of its positive charge.[140] It has been estimated that this effect contributes about 11 pK units to the base strength of ammonia.[141] When methyl groups replace hydrogen, this difference in hydration decreases[142] until, for trimethylamine, it contributes only about 6 pK units to the base strength.[141] Thus two effects act in opposite directions, the field effect increasing the basicity as the number of methyl groups increases and the hydration effect decreasing it. When the effects are added, the strongest base is dimethylamine and the weakest is ammonia. If alkyl groups are electron-donating, one would expect that in the gas phase,[143] where the solvation effect does not exist, the basicity order of amines toward the proton should be $R_3N > R_2NH > RNH_2 > NH_3$, and this has indeed been confirmed, for R = Me as well as R = Et and Pr.[144] Aniline too, in the gas phase, is a stronger base than NH_3,[145] so its much lower basicity in aqueous solution (pK_a of $PhNH_3^+$ 4.60 compared with 9.24 for aqueous NH_4^+) is caused by similar solvation effects and not by resonance and field electron-withdrawing effects of a phenyl group. Similarly, pyridine[146] and pyrrole[147] are both much less basic than NH_3 in aqueous solution (pyrrole[148] is neutral in aqueous solution) but *more* basic in the gas phase. These examples in particular

[137]For reviews of the effects of solvent, see Epshtein; Iogansen *Russ. Chem. Rev.* **1990**, *59*, 134-151; Dyumaev; Korolev *Russ. Chem. Rev.* **1980**, *49*, 1021-1032. For a review of the effects of the solvent dimethyl sufoxide, see Taft; Bordwell *Acc. Chem. Res.* **1988**, *21*, 463-469.

[138]For a review of the basicity of amines, see Ref. 122.

[139]Trotman-Dickenson *J. Chem. Soc.* **1949**, 1293; Pearson *J. Am. Chem. Soc.* **1948**, *70*, 204; Pearson; Williams *J. Am. Chem. Soc.* **1954**, *76*, 258; Hall *J. Am. Chem. Soc.* **1957**, *79*, 5441; Arnett; Jones; Taagepera; Henderson; Beauchamp; Holtz; Taft *J. Am. Chem. Soc.* **1972**, *94*, 4724; Aue; Webb; Bowers *J. Am. Chem. Soc.* **1972**, *94*, 4726, **1976**, *98*, 311, 318; Mucci; Domain; Benoit *Can. J. Chem.* **1980**, *58*, 953. See also Drago; Cundari; Ferris *J. Org. Chem.* **1989**, *54*, 1042.

[140]For discussions of the solvation of ammonia and amines, see Jones; Arnett *Prog. Phys. Org. Chem.* **1974**, *11*, 263-420; Grunwald; Ralph *Acc. Chem. Res.* **1971**, *4*, 107-113.

[141]Condon *J. Am. Chem. Soc.* **1965**, *87*, 4481, 4485.

[142]For two reasons: (1) the alkyl groups are poorly solvated by the water molecules, and (2) the strength of the hydrogen bonds of the BH$^+$ ions decreases as the basicity of B increases: Lau; Kebarle *Can. J. Chem.* **1981**, *59*, 151.

[143]For reviews of acidities and basicities in the gas phase, see Liebman *Mol. Struct. Energ.* **1987**, *4*, 49-70; Dixon; Lias *Mol. Struct. Energ.* **1987**, *2*, 269-314; Bohme, in Patai, Ref. 117, pp. 731-762; Bartmess; McIver, in Bowers *Gas Phase Ion Chemistry*, vol. 2; Academic Press: New York, 1979, pp. 88-121; Kabachnik *Russ. Chem. Rev.* **1979**, *48*, 814-827; Kebarle *Annu. Rev. Phys. Chem.* **1977**, *28*, 445-476; Arnett *Acc. Chem. Res.* **1973**, *6*, 404-409. For a comprehensive table of gas-phase basicities, see Lias; Liebman; Levin *J. Phys. Chem. Ref. Data* **1984**, *13*, 695-808. See also the tables of gas-phase acidities and basicities in Meot-Ner; Kafafi *J. Am. Chem. Soc.* **1988**, *110*, 6297; Headley *J. Am. Chem. Soc.* **1987**, *109*, 2347; McMahon; Kebarle *J. Am. Chem. Soc.* **1985**, *107*, 2612, **1977**, *99*, 2222, 3399; Wolf; Staley; Koppel; Taagepera; McIver; Beauchamp; Taft *J. Am. Chem. Soc.* **1977**, *99*, 5417; Cumming; Kebarle *J. Am. Chem. Soc.* **1977**, *99*, 5818, **1978**, *100*, 1835, *Can. J. Chem.* **1978**, *56*, 1; Bartmess; Scott; McIver *J. Am. Chem. Soc.* **1979**, *101*, 6046; Fujio; McIver; Taft *J. Am. Chem. Soc.* **1981**, *103*, 4017; Lau; Nishizawa; Tse; Brown; Kebarle *J. Am. Chem. Soc.* **1981**, *103*, 6291.

[144]Munson *J. Am. Chem. Soc.* **1965**, *87*, 2332; Brauman; Riveros; Blair *J. Am. Chem. Soc.* **1971**, *93*, 3914; Briggs; Yamdagni; Kebarle *J. Am. Chem. Soc.* **1972**, *94*, 5128; Aue; Webb; Bowers, Ref. 139.

[145]Briggs; Yamdagni; Kebarle, Ref. 144, Dzidic *J. Am. Chem. Soc.* **1972**, *94*, 8333; Ikuta; Kebarle *Can. J. Chem.* **1983**, *61*, 97.

[146]Taagepera; Henderson; Brownlee; Beauchamp; Holtz; Taft *J. Am. Chem. Soc.* **1972**, *94*, 1369; Taft; Taagepera; Summerhays; Mitsky *J. Am. Chem. Soc.* **1973**, *95*, 3811; Briggs; Yamdagni; Kebarle, Ref. 144.

[147]Yamdagni; Kebarle *J. Am. Chem. Soc.* **1973**, *95*, 3504.

[148]For a review of the basicity and acidity of pyrroles, see Catalan; Abboud; Elguero *Adv. Heterocycl. Chem.* **1987**, *41*, 187-274.

show how careful one must be in attributing relative acidities or basicities to any particular effect.

For simple alcohols the order of gas-phase *acidity* is completely reversed from that in aqueous solution. In solution the acidity is in the order H_2O > $MeCH_2OH$ > Me_2CHOH > Me_3COH, but in the gas phase the order is precisely the opposite.[149] Once again solvation effects can be invoked to explain the differences. Comparing the two extremes, H_2O and Me_3COH, we see that the OH^- ion is very well solvated by water while the bulky Me_3CO^- is much more poorly solvated because the water molecules cannot get as close to the oxygen. Thus in solution H_2O gives up its proton more readily. When solvent effects are absent, however, the intrinsic acidity is revealed and Me_3COH is a stronger acid than H_2O. This result demonstrates that simple alkyl groups cannot be simply regarded as electron-donating. If methyl is an electron-donating group, then Me_3COH should be an intrinsically weaker acid than H_2O, yet it is stronger. A similar pattern is found with carboxylic acids, where simple aliphatic acids such as propanoic are stronger than acetic acid in the gas phase,[150] though weaker in aqueous solution (Table 8.4). The evidence in these and other cases[151] is that alkyl groups can be electron-donating when connected to unsaturated systems but in other systems may have either no effect or may actually be electron-withdrawing. The explanation given for the intrinsic gas-phase acidity order of alcohols as well as the basicity order of amines is that alkyl groups, because of their polarizability, can spread both positive and negative charges.[152] It has been calculated that even in the case of alcohols the field effects of the alkyl groups are still operating normally, but are swamped by the greater polarizability effects.[153] Polarizability effects on anionic centers are a major factor in gas-phase acid–base reactions.[154]

It has been shown (by running reactions on ions that are solvated in the gas phase) that solvation by even one molecule of solvent can substantially affect the order of basicities.[155]

An important aspect of solvent effects is the effect on the orientation of solvent molecules when an acid or base is converted to its conjugate. For example, consider an acid RCOOH converted to $RCOO^-$ in aqueous solution. The solvent molecules, by hydrogen bonding, arrange themselves around the COO^- group in a much more orderly fashion than they had been arranged around the COOH group (because they are more strongly attracted to the negative charge). This represents a considerable loss of freedom and a decrease in entropy. Thermodynamic measurements show that for simple aliphatic and halogenated aliphatic acids in aqueous solution at room temperature, the entropy ($T\Delta S$) usually contributes much more to the total free-energy change ΔG than does the enthalpy ΔH.[156] Two examples are shown in Table 8.6.[157] Resonance and field effects of functional groups therefore affect the acidity of RCOOH in two distinct ways. They affect the enthalpy (electron-withdrawing

[149]Baird *Can. J. Chem.* **1969**, *47*, 2306; Brauman; Blair, Ref. 70; Arnett; Small; McIver; Miller *J. Am. Chem. Soc.* **1974**, *96*, 5638; Blair; Isolani; Riveros *J. Am. Chem. Soc.* **1973**, *95*, 1057; McIver; Scott; Riveros *J. Am. Chem. Soc.* **1973**, *95*, 2706. The alkylthiols behave similarly: gas-phase acidity increases with increasing group size while solution (aqueous) acidity decreases: Bartmess; McIver *J. Am. Chem. Soc.* **1977**, *99*, 4163.

[150]For a table of gas-phase acidities of 47 simple carboxylic acids, see Caldwell; Renneboog; Kebarle *Can. J. Chem.* **1989**, *67*, 611.

[151]Brauman; Blair *J. Am. Chem. Soc.* **1971**, *93*, 4315; Kwart; Takeshita *J. Am. Chem. Soc.* **1964**, *86*, 1161; Fort; Schleyer *J. Am. Chem. Soc.* **1964**, *86*, 4194; Holtz; Stock *J. Am. Chem. Soc.* **1965**, *87*, 2404; Laurie; Muenter *J. Am. Chem. Soc.* **1966**, *88*, 2883.

[152]Brauman; Blair, Ref. 70; Munson, Ref. 144; Brauman; Riveros; Blair, Ref. 144; Huheey *J. Org. Chem.* **1971**, *36*, 204; Radom *Aust. J. Chem.* **1975**, *28*, 1; Aitken; Bahl; Bomben; Gimzewski; Nolan; Thomas *J. Am. Chem. Soc.* **1980**, *102*, 4873.

[153]Taft; Taagepera; Abboud; Wolf; DeFrees; Hehre; Bartmess; McIver *J. Am. Chem. Soc.* **1978**, *100*, 7765. For a scale of polarizability parameters, see Hehre; Pau; Headley; Taft; Topsom *J. Am. Chem. Soc.* **1986**, *108*, 1711.

[154]Bartmess; Scott; McIver *J. Am. Chem. Soc.* **1979**, *101*, 6056.

[155]Bohme; Rakshit; Mackay *J. Am. Chem. Soc.* **1982**, *104*, 1100.

[156]Bolton; Hepler, Ref. 115; Ref. 71. See also Wilson; Georgiadis; Bartmess *J. Am. Chem. Soc.* **1991**, *113*, 1762.

[157]Bolton; Hepler, Ref. 115, p. 529; Hambly, Ref. 71, p. 92.

TABLE 8.6   Thermodynamic values for the ionizations of acetic and chloroacetic acids in H_2O at 25°C[157]

Acid	pK_a	ΔG		ΔH		$T\Delta S$	
		kcal/mol	kJ/mole	kcal/mole	kJ/mol	kcal/mol	kJ/mol
CH₃COOH	4.76	+6.5	+27	−0.1	−0.4	−6.6	−28
ClCH₂COOH	2.86	+3.9	+16	−1.1	−4.6	−5.0	−21
Cl₃CCOOH	0.65	+0.9	+3.8	+1.5	+6.3	+0.6	+2.5

groups increase acidity by stabilizing RCOO⁻ by charge dispersal), but they also affect the entropy (by lowering the charge on the COO⁻ group and by changing the electron-density distribution in the COOH group, electron-withdrawing groups alter the solvent orientation patterns around both the acid and the ion, and consequently change ΔS).

A change from a protic to an aprotic solvent can also affect the acidity or basicity, since there is a difference in solvation of anions by a protic solvent (which can form hydrogen bonds) and an aprotic one.[158] The effect can be extreme: in DMF, picric acid is stronger than HBr,[159] though in water HBr is far stronger. This particular result can be attributed to size. That is, the large ion $(O_2N)_3C_6H_2O^-$ is better solvated by DMF than the smaller ion Br⁻.[160] The ionic strength of the solvent also influences acidity or basicity, since it has an influence on activity coefficients.

In summary, solvation can have powerful effects on acidity and basicity. In the gas phase the effects discussed in the previous section, especially resonance and field effects, operate unhindered by solvent molecules. As we have seen, electron-withdrawing groups generally increase acidity (and decrease basicity); electron-donating groups act in the opposite way. In solution, especially aqueous solution, these effects still largely persist (which is why pK values in Table 8.4 do largely correlate with resonance and field effects), but in general are much weakened, and occasionally reversed.[119]

[158]For a review, see Parker *Q. Rev., Chem. Soc.* **1962,** *16,* 163-187.
[159]Sears; Wolford; Dawson *J. Electrochem. Soc.* **1956,** *103,* 633.
[160]Miller; Parker *J. Am. Chem. Soc.* **1961,** *83,* 117.

9
EFFECTS OF STRUCTURE ON REACTIVITY

When the equation for a reaction of, say, carboxylic acids, is written, it is customary to use the formula RCOOH, which implies that all carboxylic acids undergo the reaction. Since most compounds with a given functional group do give more or less the same reactions, the custom is useful, and the practice is used in this book. It allows a large number of individual reactions to be classified together and serves as an aid both for memory and understanding. Organic chemistry would be a huge morass of unconnected facts without the symbol R. Nevertheless, it must be borne in mind that a given functional group does not always react the same way, regardless of what molecule it is a part of. The reaction at the functional group is influenced by the rest of the molecule. This influence may be great enough to stop the reaction completely or to make it take an entirely different course. Even when two compounds with the same functional group undergo the same reaction, the rates and/or the positions of equilibrium are usually different, sometimes slightly, sometimes greatly, depending on the structures of the compounds. The greatest variations may be expected when additional functional groups are present.

The effects of structure on reactivity can be divided into three major types: field, resonance (or mesomeric), and steric.[1] In most cases two or all three of these are operating, and it is usually not easy to tell how much of the rate enhancement (or decrease) is caused by each of the three effects.

Resonance and Field Effects

It is often particularly difficult to separate resonance and field effects; they are frequently grouped together under the heading of *electrical effects*.[2] Field effects were discussed on pp. 17-19. Table 1.3 contains a list of some $+I$ and $-I$ groups. As for resonance effects, on p. 36 it was shown how the electron density distribution in aniline is not the same as it would be if there were no resonance interaction between the ring and the NH_2 group. Most groups that contain an unshared pair on an atom connected to an unsaturated system display a similar effect; i.e., the electron density on the group is less than expected, and the density on the unsaturated system is greater. Such groups are said to be electron-donating by the resonance effect ($+M$ groups). Alkyl groups, which do not have an unshared pair, are also $+M$ groups, presumably because of hyperconjugation.

On the other hand, groups that have a multiple-bonded electronegative atom directly connected to an unsaturated system are $-M$ groups. In such cases we can draw canonical

---

[1]For a monograph, see Klumpp *Reactivity in Organic Chemistry*; Wiley: New York, 1982. For a general theoretical approach to organic reactivity, see Pross *Adv. Phys. Org. Chem.* **1985**, *21*, 99-196.

[2]For reviews of the study of electrical effects by ab initio mo methods, see Topsom *Prog. Phys. Org. Chem.* **1987**, *16*, 125-191, *Mol. Struct. Energ.* **1987**, *4*, 235-269.

forms in which electrons have been taken from the unsaturated system into the group,
e.g.,

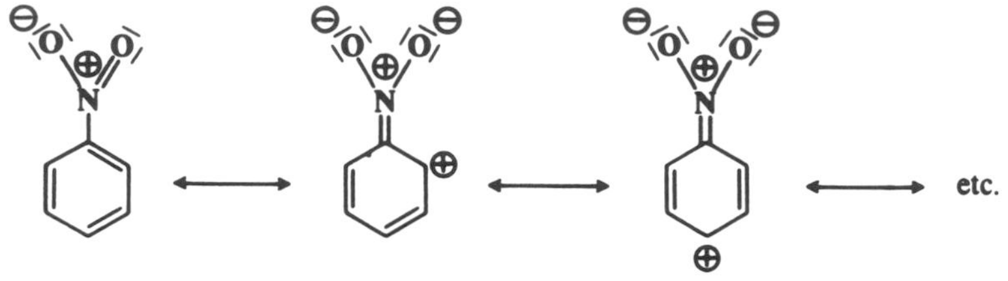

Table 9.1 contains a list of some $+M$ and $-M$ groups.

The resonance effect of a group, whether $+M$ or $-M$, operates only when the group is directly connected to an unsaturated system, so that, for example, in explaining the effect of the CH_3O group on the reactivity of the COOH in $CH_3OCH_2CH_2COOH$, only the field effect of the CH_3O need be considered. This is one way of separating the two effects. In p-methoxybenzoic acid both effects must be considered. The field effect operates through space, solvent molecules, or the σ bonds of a system, while the resonance effect operates through π electrons.

It must be emphasized once again that neither by the resonance nor by the field effect are any electrons actually being donated or withdrawn, though these terms are convenient (and we shall use them). As a result of both effects, the electron-density distribution is not the same as it would be without the effect (see pp. 18, 36). One thing that complicates the study of these effects on the reactivity of compounds is that a given group may have an effect in the transition state which is considerably more or less than it has in the unreacting molecule.

An example will show the nature of electrical effects (resonance and field) on reactivity. In the alkaline hydrolysis of aromatic amides (**0-11**), the rate-determining step is the attack of hydroxide ion at the carbonyl carbon:

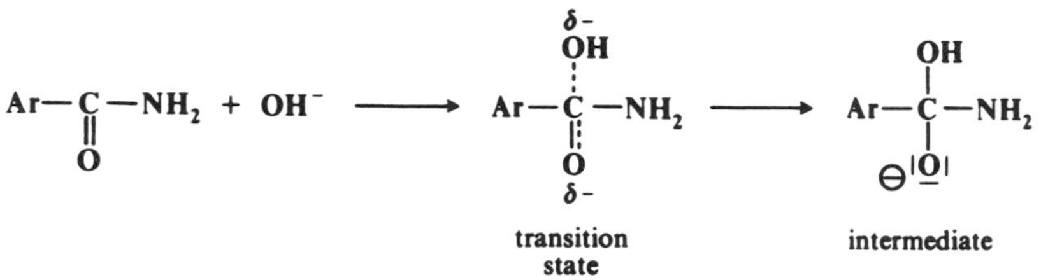

In the transition state, which has a structure somewhere between that of the starting amide (**1**) and the intermediate (**2**), the electron density on the carbonyl carbon is increased. Therefore, electron-withdrawing groups $(-I$ or $-M)$ on the aromatic ring will lower the free energy of the transition state (by spreading the negative charge). These groups have much less effect on the free energy of **1**. Since G is lowered for the transition state, but not substantially for **1**, $\Delta G^{\ddagger}$ is lowered and the reaction rate is increased (Chapter 6). Conversely, electron-donating groups $(+I$ or $+M)$ should decrease the rate of this reaction. Of course, many groups are $-I$ and $+M$, and for these it is not always possible to predict which effect will predominate.

TABLE 9.1 Some groups with $+M$ and $-M$ effects, not listed in order of strength of effect
Ar appears in both lists because it is capable of both kinds of effect

$+M$ groups		$-M$ groups	
O⁻	SR	NO₂	CHO
S⁻	SH	CN	COR
NR₂	Br	COOH	SO₂R
NHR	I	COOR	SO₂OR
NH₂	Cl	CONH₂	NO
NHCOR	F	CONHR	Ar
OR	R	CONR₂	
OH	Ar		
OCOR			

Steric Effects

It occasionally happens that a reaction proceeds much faster or much slower than expected on the basis of electrical effects alone. In these cases it can often be shown that steric effects are influencing the rate. For example, Table 9.2 lists relative rates for the Sₙ2 ethanolysis of certain alkyl halides (see p. 294).[3] All these compounds are primary bromides; the branching is on the second carbon, so that field-effect differences should be small. As Table 9.2 shows, the rate decreases with increasing β branching and reaches a very low value for neopentyl bromide. This reaction is known to involve an attack by the nucleophile from a position opposite to that of the bromine (see p. 294). The great decrease in rate can be attributed to *steric hindrance*, a sheer physical blockage to the attack of the nucleophile. Another example of steric hindrance is found in 2,6-disubstituted benzoic acids, which are difficult to esterify no matter what the resonance or field effects of the groups in the 2 or the 6 position. Similarly, once 2,6-disubstituted benzoic acids *are* esterified, the esters are difficult to hydrolyze.

Not all steric effects decrease reaction rates. In the hydrolysis of RCl by an Sₙ1 mechanism (see p. 298), the first step, which is rate-determining, involves ionization of the alkyl chloride to a carbocation:

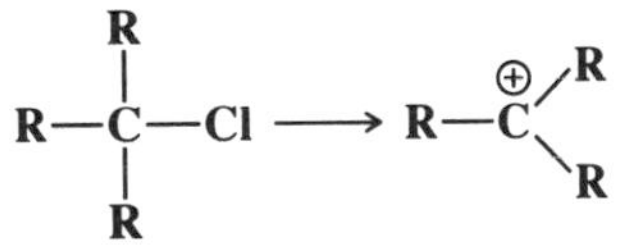

$$R\!-\!\overset{\displaystyle R}{\underset{\displaystyle R}{\overset{|}{\underset{|}{C}}}}\!-\!Cl \longrightarrow R\!-\!\overset{\oplus}{C}\overset{\displaystyle R}{\underset{\displaystyle R}{\diagdown}}$$

The central carbon in the alkyl chloride is sp^3-hybridized, with angles of about 109.5°, but

TABLE 9.2 Relative rates of reaction of RBr with ethanol[3]

R	Relative rate
CH₃	17.6
CH₃CH₂	1
CH₃CH₂CH₂	0.28
(CH₃)₂CHCH₂	0.030
(CH₃)₃CCH₂	4.2×10^{-6}

[3]Hughes *Q. Rev., Chem. Soc.* **1948**, *2*, 107-131.

when it is converted to the carbocation, the hybridization becomes sp^2 and the preferred angle is 120°. If the halide is tertiary and the three alkyl groups are large enough, they will be pushed together by the enforced tetrahedral angle, resulting in strain (see p. 163). This type of strain is called *B strain*[4] (for back strain), and it can be relieved by ionization to the carbocation.[5]

The rate of ionization (and hence the solvolysis rate) of a molecule in which there is B strain is therefore expected to be larger than in cases where B strain is not present. Table 9.3 shows that this is so.[6] Substitution of ethyl groups for the methyl groups of *t*-butyl chloride does not cause B strain; the increase in rate is relatively small, and the rate smoothly rises with the increasing number of ethyl groups. The rise is caused by normal field and resonance (hyperconjugation) effects. Substitution by one isopropyl group is not greatly different. But with the second isopropyl group the crowding is now great enough to cause B strain, and the rate is increased tenfold. Substitution of a third isopropyl group increases the rate still more. Another example where B strain increases solvolysis rates is found with the highly crowded molecules tri-*t*-butylcarbinol, di-*t*-butylneopentylcarbinol, *t*-butyldineopentylcarbinol, and trineopentylcarbinol, where rates of solvolysis of the *p*-nitrobenzoate esters are faster than that of *t*-butyl nitrobenzoate by factors of 13,000, 19,000, 68,000, and 560, respectively.[7]

Another type of strain, that can affect rates of cyclic compounds, is called *I strain* (internal strain).[8] This type of strain results from changes in ring strain in going from a tetrahedral to a trigonal carbon or vice versa. For example, as mentioned above, SN1 solvolysis of an alkyl halide involves a change in the bond angle of the central carbon from about 109.5° to about 120°. This change is highly favored in 1-chloro-1-methylcyclopentane because it relieves eclipsing strain (p. 156); thus this compound undergoes solvolysis in 80% ethanol at

	t-BuCl	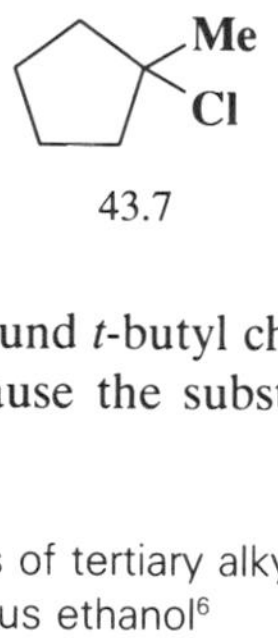 (cyclopentane Me, Cl)	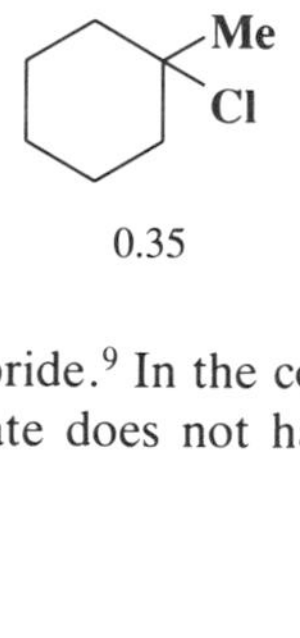 (cyclohexane Me, Cl)
Relative solvolysis rates	1.0	43.7	0.35

25°C 43.7 times faster than the reference compound *t*-butyl chloride.[9] In the corresponding cyclohexyl compound this factor is absent because the substrate does not have eclipsing

TABLE 9.3 Rates of hydrolysis of tertiary alkyl chlorides at 25°C in 80% aqueous ethanol[6]

Halide	Rate	Halide	Rate
Me₃CCl	0.033	Et₃CCl	0.099
Me₂EtCCl	0.055	Me₂(iso-Pr)CCl	0.029
MeEt₂CCl	0.086	Me(iso-Pr)₂CCl	0.45

[4]For a discussion, see Brown *Boranes in Organic Chemistry*; Cornell University Press: Ithaca, NY, 1972, pp. 114-121.
[5]For reviews of the effects of strain on reactivity, see Stirling *Tetrahedron* **1985**, *41*, 1613-1666, *Pure Appl. Chem.* **1984**, *56*, 1781-1796.
[6]Brown; Fletcher *J. Am. Chem. Soc.* **1949**, *71*, 1845.
[7]Bartlett; Tidwell *J. Am. Chem. Soc.* **1968**, *90*, 4421.
[8]For a discussion, see Ref. 4, pp. 105-107, 126-128.
[9]Brown; Borkowski *J. Am. Chem. Soc.* **1952**, *74*, 1894. See also Brown; Ravindranathan; Peters; Rao; Rho *J. Am. Chem. Soc.* **1977**, *99*, 5373.

strain (p. 156), and this compound undergoes the reaction at about one-third the rate of
t-butyl chloride. The reasons for this small decrease in rate are not clear. Corresponding
behavior is found in the other direction, in changes from a trigonal to a tetrahedral carbon.
Thus cyclohexanone undergoes addition reactions faster than cyclopentanone. Similar considerations apply to larger rings. Rings of 7 to 11 members exhibit eclipsing and transannular
strain; and in these systems reactions in which a tetrahedral carbon becomes trigonal generally proceed faster than in open-chain systems.[10]

Conformational effects on reactivity can be considered under the heading of steric effects,[11] though in these cases we are considering not the effect of a group X and that of
another group X' upon reactivity at a site Y but the effect of the conformation of the
molecule. Many reactions fail entirely unless the molecules are able to assume the proper
conformation. An example is the rearrangement of N-benzoylnorephedrine. The two dia-

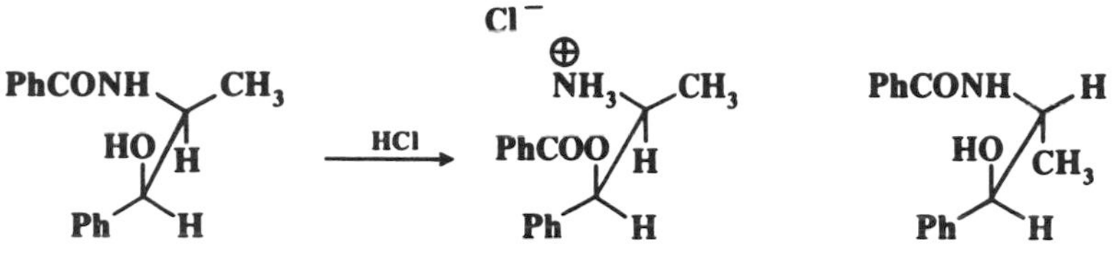

stereomers of this compound behave very differently when treated with alcoholic HCl. In
one of the isomers nitrogen-to-oxygen migration takes place, while the other does not react
at all.[12] In order for the migration to take place, the nitrogen must be near the oxygen
(gauche to it). When **3** assumes this conformation, the methyl and phenyl groups are anti
to each other, which is a favorable position, but when **4** has the nitrogen gauche to the
oxygen, the methyl must be gauche to the phenyl, which is so unfavorable that the reaction
does not occur. Other examples are electrophilic additions to C=C double bonds (see p.
735) and E2 elimination reactions (see p. 983). Also, many examples are known where axial
and equatorial groups behave differently.[13]

In steroids and other rigid systems, a functional group in one part of the molecule can
strongly affect the rate of a reaction taking place at a remote part of the same molecule by
altering the conformation of the whole skeleton. An example of this effect, called *conformational transmission*, is found in ergost-7-en-3-one (**5**) and cholest-6-en-3-one (**6**), where
6 condenses with benzaldehyde 15 times faster than **5**.[14] The reaction site in both cases is

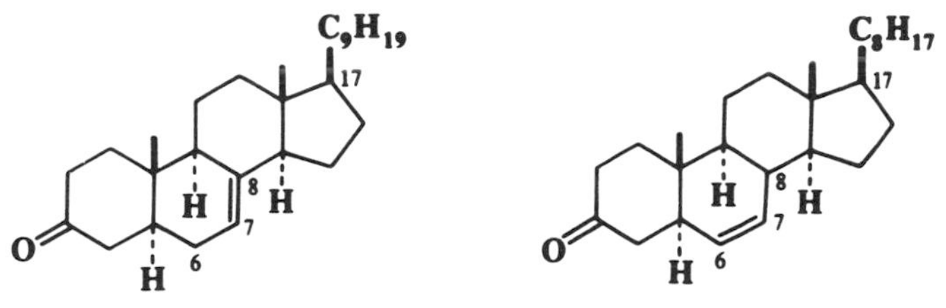

[10]See, for example. Schneider; Thomas *J. Am. Chem. Soc.* **1980**, *102*, 1424.
[11]For reviews of conformational effects, see Green; Arad-Yellin; Cohen *Top. Stereochem.* **1986**, *16*, 131-218; Ōki *Acc. Chem. Res.* **1984**, *17*, 154-159; Seeman *Chem. Rev.* **1983**, *83*, 83-134. See also Ōki; Tsukahara; Moriyama; Nakamura *Bull. Chem. Soc. Jpn.* **1987**, *60*, 223, and other papers in this series.
[12]Fodor; Bruckner; Kiss; Ōhegyi *J. Org. Chem.* **1949**, *14*, 337.
[13]For a discussion, see Eliel *Stereochemistry of Carbon Compounds*; McGraw-Hill: New York, 1962, pp. 219-234.
[14]Barton; McCapra; May; Thudium *J. Chem. Soc.* **1960**, 1297.

the carbonyl group, and the rate increases because moving the double bond from the 7 to the 6 position causes a change in conformationat the carbonyl group (the difference in the side chain at C-17 does not affect the rate).

Quantitative Treatments of the Effect of Structure on Reactivity[15]

Suppose a reaction is performed on a substrate molecule that can be represented as XGY, where Y is the site of the reaction, X a variable substituent, and G a skeleton group to which X and Y are attached, and we find that changing X from H to CH_3 results in a rate increase by a factor, say, 10. We would like to know just what part of the increase is due to each of the effects previously mentioned. The obvious way to approach such a problem is to try to find compounds in which one or two of the factors are absent or at least negligible. This is not easy to do acceptably because factors that seem negligible to one investigator do not always appear so to another. The first attempt to give numerical values was that of Hammett.[16] For the cases of m- and p-XC_6H_4Y, Hammett set up the equation

$$\log \frac{k}{k_0} = \sigma\rho$$

where k_0 is the rate constant or equilibrium constant for X = H, k is the constant for the group X, ρ is a constant for a given reaction under a given set of conditions, and σ is a constant characteristic of the group X. The equation is called the *Hammett equation*.

The value of ρ was set at 1.00 for ionization of XC_6H_4COOH in water at 25°C. σ_m and σ_p values were then calculated for each group (for a group X, σ is different for the meta and para positions). Once a set of σ values was obtained, ρ values could be obtained for other reactions from the rates of just two X-substituted compounds, if the σ values of the X groups were known (in practice, at least four well-spaced values are used to calculate ρ because of experimental error and because the treatment is not exact). With the ρ value thus calculated and the known σ values for other groups, rates can be predicted for reactions that have not yet been run.

The σ values are numbers that sum up the total electrical effects (resonance plus field) of a group X when attached to a benzene ring. The treatment usually fails for the ortho position. The Hammett treatment has been applied to many reactions and to many functional groups and correlates quite well an enormous amount of data. Jaffé's review article[16] lists ρ values for 204 reactions,[17] many of which have different ρ values for different conditions.

[15]For monographs, see Exner *Correlation Analysis of Chemical Data*; Plenum: New York, 1988; Johnson *The Hammett Equation*; Cambridge University Press: Cambridge, 1973; Shorter *Correlation Analysis of Organic Reactivity*; Wiley: New York, 1982, *Correlation Analysis in Organic Chemistry*; Clarendon Press: Oxford, 1973; Chapman; Shorter *Correlation Analysis in Chemistry: Recent Advances*; Plenum: New York, 1978, *Advances in Linear Free Energy Relationships*; Plenum: New York, 1972; Wells *Linear Free Energy Relationships*; Academic Press: New York, 1968. For reviews, see Connors *Chemical Kinetics*; VCH: New York, 1990, pp. 311-383; Lewis, in Bernasconi *Investigation of Rates and Mechanisms of Reactions* (vol. 6 of Weissberger *Techniques of Chemistry*), 4th ed.; Wiley: New York, 1986, pp. 871-901; Hammett, Ref. 2, pp. 347-390; Jones *Physical and Mechanistic Organic Chemistry*, 2nd ed.; Cambridge University Press: Cambridge, 1984, pp. 38-68; Charton, *CHEMTECH* **1974**, 502-511, **1975**, 245-255; Hine *Structural Effects in Organic Chemistry*; Wiley: New York, 1975, pp. 55-102; Afanas'ev *Russ. Chem. Rev.* **1971**, *40*, 216-232; Laurence; Wojtkowiak *Ann. Chim. (Paris)* **1970**, [14] *5*, 163-191. For a historical perspective, see Grunwald *CHEMTECH* **1984**, 698.

[16]For a review, see Jaffé *Chem. Rev.* **1953**, *53*, 191.

[17]Additional ρ values are given in Wells *Chem. Rev.* **1963**, *63*, 171-218 and van Bekkum; Verkade; Wepster *Recl. Trav. Chim. Pays-Bas* **1959**, *78*, 821-827.

Among them are reactions as disparate as the following:
Rate constants for

$$\text{ArCOOMe} + \text{OH}^- \longrightarrow \text{ArCOO}^-$$

$$\text{ArCH}_2\text{Cl} + \text{I}^- \longrightarrow \text{ArCH}_2\text{I}$$

$$\text{ArNH}_2 + \text{PhCOCl} \longrightarrow \text{ArNHCOPh}$$

$$\text{ArH} + \text{NO}_2^+ \longrightarrow \text{ArNO}_2$$

$$\text{ArCO}_2\text{OCMe}_3 \longrightarrow \text{decomposition (a free-radical process)}$$

Equilibrium constants for

$$\text{ArCOOH} + \text{H}_2\text{O} \rightleftharpoons \text{ArCOO}^- + \text{H}_3\text{O}^+$$

$$\text{ArCHO} + \text{HCN} \rightleftharpoons \text{ArCH(CN)OH}$$

The Hammett equation has also been shown to apply to many physical measurements, including ir frequencies and nmr chemical shifts.[18] The treatment is reasonably successful whether the substrates are attacked by electrophilic, nucleophilic, or free-radical reagents, the important thing being that the mechanism be the same *within* a given reaction series.

However, there are many reactions that do not fit the treatment. These are mostly reactions where the attack is directly on the ring and where the X group can enter into direct resonance interaction with the reaction site in the transition state (that is, the substrate is XY rather than XGY). For these cases, two new sets of σ values have been devised: σ⁺ values (proposed by H. C. Brown) for cases in which an electron-donating group interacts with a developing positive charge in the transition state (this includes the important case of electrophilic aromatic substitutions; see Chapter 11), and σ⁻ values, where electron-withdrawing groups interact with a developing negative charge. Table 9.4 gives σ, σ⁺, and σ⁻ values for some common X groups.[19] As shown in the table, σ is not very different from σ⁺ for most electron-withdrawing groups. σ_m^- values are not shown in the table, since they are essentially the same as the σ_m values.

A positive value of σ indicates an electron-withdrawing group and a negative value an electron-donating group. The constant ρ measures the susceptibility of the reaction to electrical effects.[20] Reactions with a positive ρ are helped by electron-withdrawing groups and vice versa. The following ρ values for the ionization of some carboxylic acids illustrate this:[21]

XC_6H_4—COOH	1.00	XC_6H_4—CH=CH—COOH	0.47
XC_6H_4—CH$_2$—COOH	0.49	XC_6H_4—CH$_2$CH$_2$—COOH	0.21

---

[18]For a review of Hammett treatment of nmr chemical shifts, see Ewing, in Chapman; Shorter *Correlation Analysis in Chemistry: Recent Advances*; Plenum, New York, 1978, pp. 357-396.

[19]Unless otherwise noted, σ values are from Exner, in Chapman; Shorter, Ref. 18, pp. 439-540, and σ⁺ values from Okamoto; Inukai; Brown *J. Am. Chem. Soc.* **1958,** *80,* 4969 and Brown; Okamoto *J. Am. Chem. Soc.* **1958,** *80,* 4979. σ⁻ values, except as noted, are from Jaffe, Ref. 16. Exner, pp. 439-540, has extensive tables giving values for more than 500 groups, as well as σ⁺, σ⁻, σ_I, σ_R°, and E_s values for many of these groups. Other large tables of the various sigma values are found in Hansch; Leo; Taft *Chem. Rev.* **1991,** *91,* 165-195. For tables of σ_p, σ_m, σ⁺, σ_I, and σ_R° values of many groups containing Si, Ge, Sn, and Pb atoms, see Egorochkin; Razuvaev *Russ. Chem. Rev.* **1987,** *56,* 846-858. For values for heteroaromatic groups, see Mamaev; Shkurko; Baram *Adv. Heterocycl. Chem.* **1987,** *42,* 1-82.

[20]For discussions of the precise significance of ρ, see Dubois; Ruasse; Argile *J. Am. Chem. Soc.* **1984,** *106,* 4840; Ruasse; Argile; Dubois *J. Am. Chem. Soc.* **1984,** *106,* 4846; Lee; Shim; Chung; Kim; Lee *J. Chem. Soc., Perkin Trans. 2* **1988,** 1919.

[21]Jones, Ref. 15, p. 42.

TABLE 9.4 σ, σ^+, and σ^- values for some common groups[19]

Group	σ_p	σ_m	σ_p^+	σ_m^+	σ_p^-
O^-	-0.81^{31}	-0.47^{31}	-4.27^{32}	-1.15^{32}	
NMe_2	-0.63	-0.10	-1.7		
NH_2	-0.57	-0.09	-1.3	-0.16	
OH	-0.38^{22}	0.13^{22}	-0.92^{23}		
OMe	-0.28^{22}	0.10	-0.78	0.05	
CMe_3	-0.15	-0.09	-0.26	-0.06	
Me	-0.14	-0.06	-0.31	-0.10^{24}	
H	0	0	0	0	0
Ph	0.05^{25}	0.05	-0.18	0^{25}	
COO^-	0.11^{31}	0.02^{31}	-0.41^{32}	-0.10^{32}	
F	0.15	0.34	-0.07	0.35	
Cl	0.24	0.37	0.11	0.40	
Br	0.26	0.37	0.15	0.41	
I	0.28^{25}	0.34	0.14	0.36	
$N{=}NPh^{26}$	0.34	0.28	0.17		
COOH	0.44	0.35	0.42	0.32	0.73
COOR	0.44	0.35	0.48	0.37	0.68
COMe	0.47	0.36			0.87
CF_3	0.53	0.46		0.57^{24}	
NH_3^+	0.60^{31}	0.86^{31}			
CN^{27}	0.70	0.62	0.66	0.56	1.00
SO_2Me	0.73	0.64			
NO_2	0.81	0.71	0.79	0.73^{24}	1.27
NMe_3^+	0.82^{28}	0.88^{28}	0.41	0.36	
N_2^+	1.93^{29}	1.65^{29}	1.88^{29}		3^{30}

This example shows that the insertion of a CH_2 or a $CH{=}CH$ group diminishes electrical effects to about the same extent, while a CH_2CH_2 group diminishes them much more. A ρ greater than 1 would mean that the reaction is more sensitive to electrical effects than is the ionization of XC_6H_4COOH ($\rho = 1.00$).

Similar calculations have been made for compounds with two groups X and X' on one ring, where the σ values are sometimes additive and sometimes not,[33] for other ring systems such as naphthalene[34] and heterocyclic rings,[35] and for ethylenic and acetylenic systems.[36]

[22]Matsui; Ko; Hepler *Can. J. Chem.* **1974**, *52*, 2906.

[23]de la Mare; Newman *Tetrahedron Lett.* **1982**, 1305 give this value as -1.6.

[24]Amin; Taylor *Tetrahedron Lett.* **1978**, 267.

[25]Sjöström; Wold *Chem. Scr.* **1976**, *9*, 200.

[26]Byrne; Happer; Hartshorn; Powell *J. Chem. Soc., Perkin Trans. 2* **1987**, 1649.

[26a]For a review of directing and activating effects of C=O, C=C, C≡N, and C=S groups, see Charton, in Patai *The Chemistry of Double-bonded Functional Groups*, vol. 2, pt. 1; Wiley: New York, 1989, pp. 239-298.

[27]For a review of directing and activating effects of CN and C≡C groups, see Charton, in Patai; Rappoport *The Chemistry of Functional Groups, Supplement C*, pt. 1; Wiley: New York, 1983, pp. 269-323.

[28]McDaniel; Brown *J. Org. Chem.* **1958**, *23*, 420.

[29]Ustynyuk; Subbotin; Buchneva; Gruzdneva; Kazitsyna *Doklad. Chem.* **1976**, *227*, 175.

[30]Lewis; Johnson *J. Am. Chem. Soc.* **1959**, *81*, 2070.

[31]Hine *J. Am. Chem. Soc.* **1960**, *82*, 4877.

[32]Binev; Kuzmanova; Kaneti; Juchnovski *J. Chem. Soc., Perkin Trans. 2* **1982**, 1533.

[33]Stone; Pearson *J. Org. Chem.* **1961**, 26, 257.

[34]Berliner; Winikov *J. Am. Chem. Soc.* **1959**, *81*, 1630; see also Wells; Ehrenson; Taft. Ref. 48.

[35]For reviews, see Charton, in Chapman; Shorter, Ref. 18, pp. 175-268; Tomasik; Johnson *Adv. Heterocycl. Chem.* **1976**, *20*, 1-64.

[36]For reviews of the application of the Hammett treatment to unsaturated systems, see Ford; Katritzky; Topsom, in Chapman; Shorter, Ref. 18, pp. 269-311; Charton *Prog. Phys. Org. Chem.* **1973**, *10*, 81-204.

The Hammett equation is a *linear free-energy relationship (LFER)*. This can be demonstrated as follows for the case of equilibrium constants (for rate constants a similar demonstration can be made with $\Delta G^{\ddagger}$ instead of ΔG). For each reaction, where X is any group,

$$\Delta G = -RT \ln K$$

For the unsubstituted case,

$$\Delta G_0 = -RT \ln K_0$$

The Hammett equation can be rewritten

$$\log K - \log K_0 = \sigma\rho$$

so that

$$\frac{-\Delta G}{2.3RT} + \frac{\Delta G_0}{2.3RT} = \sigma\rho$$

and

$$-\Delta G = \sigma\rho 2.3RT - \Delta G_0$$

For a given reaction under a given set of conditions, σ, R, T, and ΔG_0 are all constant, so that σ is linear with ΔG.

The Hammett equation is not the only LFER.[37] Some, like the Hammett equation, correlate structural changes in reactants, but the Grunwald–Winstein relationship (see p. 360) correlates changes in solvent and the Brønsted relation (see p. 258) relates acidity to catalysis. The Taft equation is a structure–reactivity equation that correlates only field effects.[38]

Taft, following Ingold,[39] assumed that for the hydrolysis of carboxylic esters, steric and resonance effects will be the same whether the hydrolysis is catalyzed by acid or base (see the discussion of ester–hydrolysis mechanisms, reaction **0-10**). Rate differences would therefore be caused only by the field effects of R and R' in RCOOR'. This is presumably a good system to use for this purpose because the transition state for acid-catalyzed hydrolysis (**7**) has a greater positive charge (and is hence destabilized by $-I$ and stabilized by $+I$ substituents) than the starting ester, while the transition state for base-catalyzed hydrolysis (**8**)

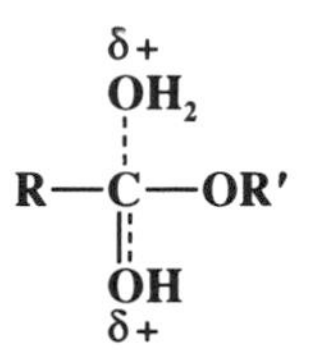

$$\begin{array}{cc}
\overset{\delta+}{\underset{}{OH_2}} & \overset{\delta-}{\underset{}{OH}} \\[4pt]
R\!-\!\overset{|}{\underset{\underset{\delta+}{OH}}{C}}\!-\!OR' & R\!-\!\overset{|}{\underset{\underset{\delta-}{O}}{C}}\!-\!OR' \\[16pt]
\mathbf{7} & \mathbf{8}
\end{array}$$

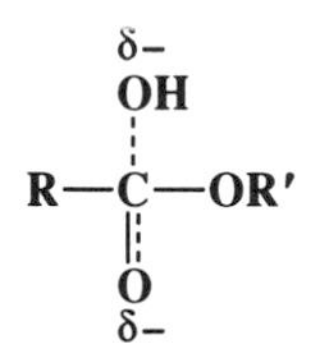

[37]For a discussion of physicochemical preconditions for LFERs, see Exner *Prog. Phys. Org. Chem.* **1990**, *18*, 129-161.
   [38]For reviews of the separation of resonance and field effects, see Charton *Prog. Phys. Org. Chem.* **1981**, *13*, 119-251; Shorter *Q. Rev., Chem. Soc.* **1970**, *24*, 433-453; *Chem. Br.* **1969**, *5*, 269-274. For a review of field and inductive effects, see Reynolds *Prog. Phys. Org. Chem.* **1983**, *14*, 165-203. For a review of field effects on reactivity, see Grob *Angew. Chem. Int. Ed. Engl.* **1976**, *15*, 569-575 [*Angew. Chem. 88*, 621-627].
   [39]Ingold *J. Chem. Soc.* **1930**, 1032.
   [40]For another set of field-effect constants, based on a different premise, see Draffehn; Ponsold *J. Prakt. Chem.* **1978**, *320*, 249.

has a greater negative charge than the starting ester. Field effects of substitutents X could therefore be determined by measuring the rates of acid- and base-catalyzed hydrolysis of a series XCH_2COOR', where R' is held constant.[35] From these rate constants, a value σ_I could be determined by the equation[41]

$$\sigma_I \equiv 0.181 \left[\log \left(\frac{k}{k_0} \right)_B - \log \left(\frac{k}{k_0} \right)_A \right]$$

In this equation $(k/k_0)_B$ is the rate constant for basic hydrolysis of XCH_2COOR' divided by the rate constant for basic hydrolysis of CH_3COOR', $(k/k_0)_A$ is the similar rate-constant ratio for acid catalysis, and 0.181 is an arbitrary constant. σ_I is a substituent constant for a group X, substituted at a saturated carbon, that reflects only field effects.[42] Once a set of σ_I values was obtained, it was found that the equation

$$\log \frac{k}{k_0} = \rho_I \sigma_I$$

holds for a number of reactions, among them:[43]

$$\textbf{RCH}_2\textbf{OH} \longrightarrow \textbf{RCH}_2\textbf{O}^-$$

$$\textbf{RCH}_2\textbf{Br} + \textbf{PhS}^- \longrightarrow \textbf{RCH}_2\textbf{SPh} + \textbf{Br}^-$$

$$\text{Acetone} + \textbf{I}_2, \text{ catalyzed by } \textbf{RCOOH} \longrightarrow$$

$$o\text{-Substituted-}\textbf{ArNH}_2 + \textbf{PhCOCl} \longrightarrow \textbf{ArNHCOPh}$$

As with the Hammett equation, σ_I is constant for a given reaction under a given set of conditions. For very large groups the relationship may fail because of the presence of steric effects, which are not constant. The equation also fails when X enters into resonance with the reaction center to different extents in the initial and transition states. A list of some σ_I values is given in Table 9.5.[44] The σ_I values are about what we would expect for pure field-effect values (see p. 18) and are additive, as field effects (but not resonance or steric effects) would be expected to be. Thus, in moving a group one carbon down the chain, there is a decrease by a factor of 2.8 ± 0.5 (compare the values of R and RCH_2 in Table 9.5 for R = Ph and CH_3CO). An inspection of Table 9.5 shows that σ_I values for most groups are fairly close to the σ_m values (Table 9.4) for the same groups. This is not surprising, since σ_m values would be expected to arise almost entirely from field effects, with little contribution from resonance.

Since σ_p values represent the sum of resonance and field effects, these values can be divided into resonance and field contributions if σ_I is taken to represent the field-effect

---

[41]The symbol σ_F is also used in the literature; sometimes in place of σ_I, and sometimes to indicate only the field (not the inductive) portion of the total effect (p. 17).

[42]There is another set of values (called σ^* values) that are also used to correlate field effects. These are related to σ_I values by = $\sigma_{I(X)} = 0.45\sigma^*_{(XCH_2)}$. We discuss only σ_I, and not σ^* values.

[43]Wells, Ref. 17, p. 196.

[44]These values are from Bromilow; Brownlee; Lopez; Taft, Ref. 52, except that the values for NHAc, OH, and I are from Wells; Ehrenson; Taft, Ref. 48, the values for Ph and NMe_3^+ are from Ref. 51 and Taft; Deno; Skell, Ref. 47, and the value for CMe_3 is from Seth-Paul; de Meyer-van Duyse; Tollenaere *J. Mol. Struct.* **1973**, *19*, 811. The values for the CH_2Ph and CH_2COCH_3 groups were calculated from σ^* values by the formula given in footnote 42. For much larger tables of σ_I and σ_R values, see Charton, Ref. 38. See also Ref. 19 and Taylor; Wait *J. Chem. Soc., Perkin Trans. 2* **1986**, 1765.

TABLE 9.5 σ_I and σ_R° values for some groups[44]

Group	σ_I	σ_R°	Group	σ_I	σ_R°
CMe$_3$	−0.07	−0.17	OMe	0.27	−0.42
Me	−0.05	−0.13	OH	0.27	−0.44
H	0	0	I	0.39	−0.12
PhCH$_2$	0.04		CF$_3$	0.42	0.08
NMe$_2$[45]	0.06	−0.55	Br	0.44	−0.16
Ph	0.10	−0.10	Cl	0.46	−0.18
CH$_3$COCH$_2$	0.10		F	0.50	−0.31
NH$_2$	0.12	−0.50	CN	0.56	0.08
CH$_3$CO	0.20	0.16	SO$_2$Me	0.60	0.12
COOEt	0.20	0.16	NO$_2$	0.65	0.15
NHAc	0.26	−0.22	NMe$_3$+[46]	0.86	

portion.[47] The resonance contribution σ_R[48] is defined as

$$\sigma_R = \sigma_p - \sigma_I$$

As it stands, however, this equation is not very useful because the σ_R value for a given group, which should be constant if the equation is to have any meaning, is actually not constant but depends on the nature of the reaction.[49] In this respect, the σ_I values are much better. Although they vary with solvent in some cases, σ_I values are essentially invariant throughout a wide variety of reaction series. However, it is possible to overcome[50] the problem of varying σ_R values by using a special set of σ_R values, called σ_R°,[51] that measure the ability to delocalize π electrons into or out of an unperturbed or "neutral" benzene ring. Several σ_R° scales have been reported; the most satisfactory values are obtained from ^{13}C chemical shifts of substituted benzenes.[52] Table 9.5 lists some values of σ_R°, most of which were obtained in this way.[53]

An equation such as

$$\log \frac{k}{k_0} = \rho_I \sigma_I + \rho_R \sigma_R^\circ$$

[45]For σ_R° values for some other NR$_2$ groups, see Korzhenevskaya; Titov; Chotii; Chekhuta *J. Org.Chem. USSR* **1987**, *28*, 1109.

[46]Although we give a σ_I value for NMe$_3$+, (and *F* values for three charged groups in Table 9.6), it has been shown that charged groups (called polar substituents) cannot be included with uncharged groups (dipolar substituents) in one general scale of electrical substituent effects: Marriott; Reynolds; Topsom *J. Org. Chem.* **1985**, *50*, 741.

[47]Roberts; Moreland *J. Am. Chem. Soc.* **1953**, *75*, 2167; Taft *J. Am. Chem. Soc.* **1957**, *79*, 1045, *J. Phys. Chem.* **1960**, *64*, 1805; Taft; Lewis *J. Am. Chem. Soc.* **1958**, *80*, 2436; Taft; Deno; Skell *Annu. Rev. Phys. Chem.* **1958**, *9*, 287-314, pp. 290-293.

[48]For reviews of the σ_I and σ_R concept as applied to benzenes and naphthalenes, respectively, see Ehrenson; Brownlee; Taft *Prog. Phys. Org. Chem.* **1973**, *10*, 1-80; Wells; Ehrenson; Taft *Prog. Phys. Org. Chem.* **1968**, *6*, 147-322. See also Taft; Topsom *Prog. Phys. Org. Chem.* **1987**, *16*, 1-83; Charton *Prog. Phys. Org. Chem.* **1987**, *16*, 287-315.

[49]Taft; Lewis *J. Am. Chem. Soc.* **1959**, *81*, 5343; Reynolds; Dais; MacIntyre; Topsom; Marriott; von Nagy-Felsobuki; Taft *J. Am. Chem. Soc.* **1983**, *105*, 378.

[50]For a different way of overcoming this problem, see Happer; Wright *J. Chem. Soc., Perkin Trans. 2* **1979**, 694.

[51]Taft; Ehrenson; Lewis; Glick *J. Am. Chem. Soc.* **1959**, *81*, 5352.

[52]Bromilow; Brownlee; Lopez; Taft *J. Org. Chem.* **1979**, *44*, 4766. See also Marriott; Topsom *J. Chem. Soc., Perkin Trans. 2* **1985**, 1045.

[53]For a set of σ_R values for use in XY$^+$ systems, see Charton *Mol. Struct. Energ.* **1987**, *4*, 271-317.

which treats resonance and field effects separately, is known as a *dual substituent parameter equation*.[54]

The only groups in Table 9.5 with negative values of σ_I are the alkyl groups methyl and *t*-butyl. There has been some controversy on this point.[55] One opinion is that σ_I values decrease in the series methyl, ethyl, isopropyl, *t*-butyl (respectively, -0.046, -0.057, -0.065, -0.074).[56] Other evidence, however, has led to the belief that all alkyl groups have approximately the same field effect and that the σ_I values are invalid as a measure of the intrinsic field effects of alkyl groups.[57]

Another attempt to divide σ values into resonance and field contributions[58] is that of Swain and Lupton, who have shown that the large number of sets of σ values (σ_m, σ_p, σ_p^-, σ_p^+, σ_I, σ_R°, etc., as well as others we have not mentioned) are not entirely independent and that linear combinations of two sets of new values F (which expresses the field-effect contribution) and R (the resonance contribution) satisfactorily express 43 sets of values.[59] Each set is expressed as

$$\sigma = fF + rR$$

where f and r are weighting factors. Some F and R values for common groups are given in Table 9.6.[60] From the calculated values of f and r, Swain and Lupton calculated that the

TABLE 9.6 F and R values for some groups[60]

Group	F	R	Group	F	R
COO⁻	-0.27	0.40	OMe	0.54	-1.68
Me₃C	-0.11	-0.29	CF₃	0.64	0.76
Et	-0.02	-0.44	I	0.65	-0.12
Me	-0.01	-0.41	Br	0.72	-0.18
H	0	0	Cl	0.72	-0.24
Ph	0.25	-0.37	F	0.74	-0.60
NH₂	0.38	-2.52	NHCOCH₃	0.77	-1.43
COOH	0.44	0.66	CN	0.90	0.71
OH	0.46	-1.89	NMe₃⁺	1.54	
COOEt	0.47	0.67	N₂⁺	2.36	2.81
COCH₃	0.50	0.90			

[54]There are also three-parameter equations. See, for example de Ligny and van Houwelingen *J. Chem. Soc., Perkin Trans. 2* **1987**, 559.

[55]For a discussion, see Shorter, in Chapman; Shorter *Advances in Linear Free Energy Relationships*, Ref. 15, pp. 98-103.

[56]For support for this point of view, see Levitt; Widing *Prog. Phys. Org. Chem.* **1976**, *12*, 119-157; Taft; Levitt *J. Org. Chem.* **1977**, *42*, 916; MacPhee; Dubois *Tetrahedron Lett.* **1978**, 2225; Screttas *J. Org. Chem.* **1979**, *44*, 3332; Hanson *J. Chem. Soc., Perkin Trans. 2* **1984**, 101.

[57]For support for this point of view, see, for example, Ritchie *J. Phys. Chem.* **1961**, *65*, 2091; Bordwell; Drucker; McCollum *J. Org. Chem.* **1976**, *41*, 2786; Bordwell; Fried *Tetrahedron Lett.* **1977**, 1121; Charton *J. Am. Chem. Soc.* **1977**, *99*, 5687, *J. Org. Chem.* **1979**, *44*, 903; Adcock; Khor *J. Org. Chem.* **1978**, *43*, 1272; DeTar *J. Org. Chem.* **1980**, *45*, 5166, *J. Am. Chem. Soc.* **1980**, *102*, 7988.

[58]Yukawa and Tsuno have still another approach, also involving dual parameters: Yukawa; Tsuno *Bull. Chem. Soc. Jpn.* **1959**, *32*, 971. For a review and critique of this method, see Shorter, in Chapman; Shorter, Ref. 18, pp. 119-173, pp. 126-144. This article also discusses the Swain-Lupton and Taft σ_I, σ_R approaches. For yet other approaches, see Afanas'ev *J. Org. Chem. USSR* **1981**, *17*, 373, *J. Chem. Soc., Perkin Trans. 2* **1984**, 1589; Ponec *Coll. Czech. Chem. Commun.* **1983**, *48*, 1564.

[59]Swain; Lupton *J. Am. Chem. Soc.* **1968**, *90*, 4328; Swain; Unger; Rosenquist; Swain *J. Am. Chem. Soc.* **1983**, *105*, 492.

[60]Taken from a much longer list in Swain; Unger; Rosenquist; Swain, Ref. 59. Long tables of R and F values are also given in Hansch; Leo; Taft, Ref. 19.

importance of resonance, % R, is 20% for σ_m, 38% for σ_p, and 62% for $\sigma_p{}^+$.[61] This is another dual substituent parameter approach.

Taft was also able to isolate steric effects.[62] For the acid-catalyzed hydrolysis of esters in aqueous acetone, log (k/k_0) was shown to be insensitive to polar effects.[63] In cases where resonance interaction was absent, this value was proportional only to steric effects (and any others[64] that are not field or resonance). The equation is

$$\log \frac{k}{k_0} = E_s$$

Some E_s values are given in Table 9.7,[65] where hydrogen is taken as standard, with a value of 0.[66] This treatment is more restricted than those previously discussed, since it requires more assumptions, but the E_s values are approximately in order of the size of the groups. Charton has shown that E_s values for substituents of types CH_2X, CHX_2, and CX_3 are linear functions of the van der Waals radii for these groups.[67]

Two other steric parameters are independent of any kinetic data. Charton's υ values are derived from van der Waals radii,[68] and Meyer's V^a values from the volume of the portion of the substituent that is within 0.3 nm of the reaction center.[69] The V^a values are obtained by molecular mechanics calculations based on the structure of the molecule. Table 9.7 gives υ and V^a values for some groups.[70] As can be seen in the table, there is a fair, but not

TABLE 9.7 E_s, υ, and V^a values for some groups[65]

Group	E_s	υ	$V^a \times 10^2$	Group	E_s	υ	$V^a \times 10^2$
H	0	0		Cyclohexyl	−2.03	0.87	6.25
F	−0.46	0.27	1.22	iso-**Bu**	−2.17	0.98	5.26
CN	−0.51			sec-**Bu**	−2.37	1.02	6.21
OH	−0.55			CF₃	−2.4	0.91	3.54
OMe	−0.55		3.39	*t*-**Bu**	−2.78	1.24	7.16
NH₂	−0.61			NMe₃⁺	−2.84		
Cl	−0.97	0.55	2.54	Neopentyl	−2.98	1.34	5.75
Me	−1.24	0.52	2.84	CCl₃	−3.3	1.38	6.43
Et	−1.31	0.56	4.31	CBr₃	−3.67	1.56	7.29
I	−1.4	0.78	4.08	(Me₃CCH₂)₂CH	−4.42	2.03	
Pr	−1.6	0.68	4.78	Et₃C	−5.04	2.38	
iso-**Pr**	−1.71	0.76	5.74	Ph₃C	−5.92	2.92	

[61]The Swain-Lupton treatment has been criticized by Reynolds; Topsom *J. Org. Chem.* **1984**, *49*, 1989; Hoefnagel; Oosterbeek; Wepster *J. Org. Chem.* **1984**, *49*, 1993; and Charton *J. Org. Chem.* **1984**, *49*, 1997. For a reply to these criticisms, see Swain *J. Org. Chem.* **1984**, *49*, 2005. A study of the rates of dediazoniation reactions (**3-23**) was more in accord with the Taft and Charton (Ref. 38) σ_I and σ_R values than with the Swain-Lupton F and R values: Nakazumi; Kitao; Zollinger *J. Org. Chem.* **1987**, *52*, 2825.

[62]For reviews of quantitative treatments of steric effects, see Gallo; Roussel; Berg *Adv. Heterocycl. Chem.* **1988**, *43*, 173-299; Gallo *Prog. Phys. Org. Chem.* **1983**, *14*, 115-163; Unger; Hansch *Prog. Phys. Org. Chem.* **1976**, *12*, 91-118.

[63]Another reaction used for the quantitative measurement of steric effects is the aminolysis of esters (**0-55**); De Tar; Delahunty *J. Am. Chem. Soc.* **1983**, *105*, 2734.

[64]It has been shown that E_s values include solvation effects: McClelland; Steenken *J. Am. Chem. Soc.* **1988**, *110*, 5860.

[65]E_s, υ, and V^a values are taken from longer tables in respectively, Ref. 62, Charton *J. Am. Chem. Soc.* **1975**, *97*, 1552, *J. Org. Chem.* **1976**, *41*, 2217; and Ref. 69.

[66]In Taft's original work, Me was given the value 0. The E_s values in Table 9.7 can be converted to the orginal values by adding 1.24.

[67]Charton *J. Am. Chem. Soc.* **1969**, *91*, 615.

[68]Charton, Ref. 65. See also Charton *J. Org. Chem.* **1978**, *43*, 3995; Idoux; Schreck *J. Org. Chem.* **1978**, *43*, 4002.

[69]Meyer *J. Chem. Soc., Perkin Trans. 2* **1986**, 1567.

[70]For a discussion of the various steric parameters, see DeTar, Ref. 57.

perfect, correlation among the E_s, v, and V^a values. Other sets of steric values, e.g., E'_s,[71] E^*_s,[72] Ω_s[73] and $\mathcal{S}_f$,[74] have also been proposed.[70]

Since the Hammett equation has been so successful in the treatment of the effects of groups in the meta and para positions, it is not surprising that attempts have been made to apply it to ortho positions also.[75] The effect on a reaction rate or equilibrium constant of a group in the ortho position is called the *ortho effect*.[76] Despite the many attempts made to quantify ortho effects, so far no set of values commands general agreement. However, the Hammett treatment is successful for ortho compounds when the group Y in $o\text{-}XC_6H_4Y$ is separated from the ring; e.g., ionization constants of $o\text{-}XC_6H_4OCH_2COOH$ can be successfully correlated.[77]

Linear free-energy relationships can have mechanistic implications. If $\log(k/k_0)$ is linear with the appropriate σ, it is likely that the same mechanism operates throughout the series. If not, a smooth curve usually indicates a gradual change in mechanism, while a pair of intersecting straight lines indicates an abrupt change,[78] though nonlinear plots can also be due to other causes, such as complications arising from side reactions. If a reaction series follows σ^+ or σ^- better than σ it generally means that there is extensive resonance interaction in the transition state.[79]

Information can also be obtained from the magnitude and sign of ρ. For example, a strongly negative ρ value indicates a large electron demand at the reaction center, from which it may be concluded that a highly electron-deficient center, perhaps an incipient carbocation, is involved. Conversely, a positive ρ value is associated with a developing negative charge in the transition state.[80] The $\sigma\rho$ relationship even applies to free-radical reactions, because free radicals can have some polar character (p. 679), though ρ values here are usually small (less than about 1.5) whether positive or negative. Reactions involving cyclic transition states (p. 206) also exhibit very small ρ values.

[71]MacPhee; Panaye; Dubois *Tetrahedron* **1978**, *34*, 3553, *J. Org. Chem.* **1980**, *45*, 1164; Dubois; MacPhee; Panaye *Tetrahedron Lett.* **1978**, 4099; *Tetrahedron* **1980**, *36*, 919. See also Datta; Sharma *J. Chem. Res. (S)* **1987**, 422.

[72]Fellous; Luft *J. Am. Chem. Soc.* **1973**, *95*, 5593.

[73]Komatsuzaki; Sakakibara; Hirota *Tetrahedron Lett.* **1989**, *30*, 3309, *Chem. Lett.* **1990**, 1913.

[74]Beckhaus *Angew. Chem. Int. Ed. Engl.* **1978**, *17*, 593 [*Angew. Chem. 90*, 633].

[75]For reviews, see Fujita; Nishioka *Prog. Phys. Org. Chem.* **1976**, *12*, 49-89; Charton *Prog. Phys. Org. Chem.* **1971**, *8*, 235-317; Shorter, Ref. 55, pp. 103-110. See also Segura *J. Org. Chem.* **1985**, *50*, 1045; Robinson; Horton; Fosheé; Jones; Hanissian; Slater *J. Org. Chem.* **1986**, *51*, 3535.

[76]This is not the same as the ortho effect discussed on p. 514.

[77]Charton *Can. J. Chem.* **1960**, *38*, 2493.

[78]For a discussion, see Schreck *J. Chem. Educ.* **1971**, *48*, 103-107.

[79]See, however, Gawley *J. Org. Chem.* **1981**, *46*, 4595.

[80]For another method of determining transition state charge, see Williams *Acc. Chem. Res.* **1984**, *17*, 425-430.

PART TWO

In Part 2 of this book we shall be directly concerned with organic reactions and their mechanisms. The reactions have been classified into 10 chapters, based primarily on reaction type: substitutions, additions to multiple bonds, eliminations, rearrangements, and oxidation–reduction reactions. Five chapters are devoted to substitutions; these are classified on the basis of mechanism as well as substrate. Chapters 10 and 13 include nucleophilic substitutions at aliphatic and aromatic substrates, respectively. Chapters 12 and 11 deal with electrophilic substitutions at aliphatic and aromatic substrates, respectively. All free-radical substitutions are discussed in Chapter 14. Additions to multiple bonds are classified not according to mechanism, but according to the type of multiple bond. Additions to carbon–carbon multiple bonds are dealt with in Chapter 15; additions to other multiple bonds in Chapter 16. One chapter is devoted to each of the three remaining reaction types: Chapter 17, eliminations; Chapter 18, rearrangements; Chapter 19, oxidation–reduction reactions. This last chapter covers only those oxidation–reduction reactions that could not be conveniently treated in any of the other categories (except for oxidative eliminations).

Each chapter in Part 2 consists of two main sections. The first section of each chapter (except Chapter 19) deals with mechanism and reactivity. For each reaction type the various mechanisms are discussed in turn, with particular attention given to the evidence for each mechanism and to the factors that cause one mechanism rather than another to prevail in a given reaction. Following this, each chapter contains a section on reactivity, including, where pertinent, a consideration of orientation and the factors affecting it.

The second main section of each chapter is a treatment of the reactions belonging to the category indicated by the title of the chapter. It is not possible to discuss in a book of this nature all or nearly all known reactions. However, an attempt has been made to include all the important reactions of standard organic chemistry which can be used to prepare relatively pure compounds in reasonable yields. In order to present a well-rounded picture and to include some reactions that are traditionally discussed in textbooks, a number of reactions that do not fit into the above category have been included. The scope of the coverage is apparent from the fact that more than 90% of the individual preparations given in *Organic Syntheses* are treated. However, certain special areas have been covered only lightly or not at all. Among these are electrochemical and polymerization reactions, and the preparation and reactions of heterocyclic compounds, carbohydrates, steroids, and compounds containing phosphorus, silicon, arsenic, boron, and mercury. The basic principles involved in these areas are of course no different from those in the areas more fully treated. Even with these omissions, however, some 580 reactions are treated in this book.

Each reaction is discussed in its own numbered section.[1] These are numbered consec-

---

[1]The classification of reactions into sections is, of course, to some degree arbitrary. Each individual reaction (for example, $CH_3Cl + CN^- \rightarrow CH_3CN$ and $C_2H_5Cl + CN^- \rightarrow C_2H_5CN$) is different, and custom generally decides how we group them together. Individual preferences also play a part. Some chemists would say that $C_6H_5N_2^+$ + CuCN $\rightarrow C_6H_5CN$ and $C_6H_5N_2^+$ + CuCl $\rightarrow C_6H_5Cl$ are examples of the "same" reaction. Others would say they are not, but that $C_6H_5N_2^+$ + CuCl $\rightarrow C_6H_5Cl$ and $C_6H_5N_2^+ \rightarrow CuBr + C_6H_5Br$ are examples of the "same" reaction. No claim is made that the classification system used in this book is more valid than any other. For another way of classifying reactions, see Fujita *J. Chem. Soc., Perkin Trans. 2* **1988,** 597.

utively within a chapter. The *first* digit in each number is the *second* digit of the chapter number. Thus, reaction **6-1** is the first reaction of Chapter 16 and reaction **3-21** is the twenty-first reaction of Chapter 13. The second part of the reaction number has no other significance. The order in which the reactions are presented is not arbitrary but is based on an orderly outline that depends on the type of reaction. The placement of each reaction in a separate numbered section serves as an aid to both memory and understanding by setting clear boundary lines between one reaction and another, even if these boundary lines must be arbitrary, and by clearly showing the relationship of each reaction to all the others. Within each section, the scope and utility of the reaction are discussed and references are given to review articles, if any. If there are features of the mechanism that especially pertain to that reaction, these are also discussed within the section rather than in the first part of the chapter where the discussion of mechanism is more general.

IUPAC Nomenclature for Transformations

There has long been a need for a method of naming reactions. As most students know well, many reactions are given the names of their discoverers or popularizers (e.g., Clemmensen, Diels–Alder, Prins, Wittig, Cope, Corey–Winter). This is useful as far as it goes, but each name must be individually memorized, and there are many reactions that do not have such names. The IUPAC Commission on Physical Organic Chemistry has produced a *system* for naming not reactions, but transformations (a reaction includes all reactants; a transformation shows only the substrate and product, omitting the reagents). The advantages of a systematic method are obvious. Once the system is known, no memorization is required; the name can be generated directly from the equation. The system includes rules for naming eight types of transformation: substitutions, additions, eliminations, attachments and detachments, simple rearrangements, coupling and uncoupling, insertions and extrusions, and ring opening and closing. We give here only the most basic rules for the first three of these types, which however will suffice for naming many transformations.[2] The complete rules give somewhat different names for speech-writing and indexing. In this book we give only the speech-writing names.

Substitutions. A name consists of the entering group, the syllable "de," and the leaving group. If the leaving group is hydrogen, it may be omitted (in all examples, the substrate is written on the left).

$$CH_3CH_2Br + CH_3O^- \longrightarrow CH_3CH_2-O-CH_3 \quad \textbf{Methoxy-de-bromination}$$

$$\text{C}_6\text{H}_6 + HNO_3 \xrightarrow{H_2SO_4} \text{C}_6\text{H}_5NO_2 \quad \begin{array}{l} \textbf{Nitro-de-hydrogenation} \\ \text{or } \textbf{Nitration} \end{array}$$

Multivalent substitutions are named by a modification of this system that includes suffixes such as "bisubstitution" and "tersubstitution."

$$CH_2Cl_2 + 2EtO^- \longrightarrow CH_2(OEt)_2 \quad \textbf{Diethoxy-de-dichloro-bisubstitution}$$

$$CH_3CHO + Ph_3P=CH_2 \longrightarrow CH_3CH=CH_2 \quad \textbf{Methylene-de-oxo-bisubstitution}$$

$$CH_3C\equiv N + H_2O \xrightarrow{H^+} CH_3-\underset{\underset{O}{\|}}{C}-OH \quad \textbf{Hydroxy, oxo-de-nitrilo-tersubstitution}$$

(Note: the nitrilo group is ≡N.)

[2]For the complete rules, as so far published, see Jones; Bunnett *Pure Appl. Chem.* **1989**, *61*, 725-768.

Additions. For simple 1,2-additions, the names of both addends are given followed by the suffix "addition." The addends are named in order of priority in the Cahn–Ingold–Prelog system (p. 109), the lower-ranking addend coming first. Multivalent addition is indicated by "biaddition," etc.

$$CH_3—CH{=}CH_2 + HBr \longrightarrow CH_3—CH_2—CH_2—Br \quad \textbf{Hydro-bromo-addition}$$

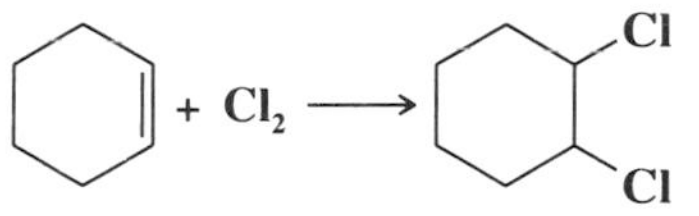

$$CH_3—\underset{\underset{O}{\|}}{C}—H + HCN \longrightarrow CH_3—\underset{\underset{OH}{|}}{CH}—CN \quad \textit{O}\text{-}\textbf{Hydro-}\textit{C}\text{-}\textbf{cyano-addition}$$

$$CH_3—C{\equiv}CH + H_2O \longrightarrow CH_3—\underset{\underset{O}{\|}}{C}—CH_3 \quad \textbf{Dihydro-oxo-biaddition}$$

Eliminations are named the same way as additions, except that "elimination" is used instead of "addition."

$$CH_3—\underset{\underset{Br}{|}}{CH}—\underset{\underset{Br}{|}}{C}(CH_3)_2 \xrightarrow{\ Zn\ } CH_3CH{=}C(CH_3)_2 \quad \textbf{Dibromo-elimination}$$

$$CH_3CH_2—\underset{\underset{SO_3^-\ Na^+}{|}}{CH}—OH \longrightarrow CH_3CH_2CHO \quad \textit{O}\text{-}\textbf{Hydro-}\textit{C}\text{-}\textbf{sulfonato-elimination}$$

$$CH_3CH_2—\underset{\underset{Br}{|}}{CH}—Br \xrightarrow{\ NH_2^-\ } CH_3C{\equiv}C \quad \textbf{Dihydro-dibromo-bielimination}$$

In the reaction sections of this book, we shall give IUPAC names for most transformations (these names will be printed in the same typeface used above), including examples of all eight types.[3] As will become apparent, some transformations require more rules than we have given here.[2] However, it is hoped that the simplicity of the system will also be apparent.

Two further notes: (1) Many transformations can be named using either of two reactants as the substrate. For example, the transformation **methylene-de-oxo-bisubstitution** above, can also be named **ethylidene-de-triphenylphosphoranediyl-bisubstitution.** In this book, unless otherwise noted, we will show only those names in which the substrate is considered to undergo the reactions indicated by the titles of the chapters. Thus the name we give to **1-12** (ArH + RCl → ArR) is **alkyl-de-hydrogenation,** not **aryl-de-chlorination,** though the latter name is also perfectly acceptable under the IUPAC system. (2) The IUPAC rules recognize that some transformations are too complex to be easily fitted into the system, so they also include a list of names for some complex transformations, which are IUPAC approved, but nonsystematic (for some examples, see reactions **2-44, 8-36, 9-63**).

---

[3]For some examples, see: attachments (**8-29, 9-28**), detachments (**9-48, 9-56**), simple rearrangements (**8-7, 8-31**), coupling (**0-86, 9-35**), uncoupling (**9-9, 9-61**), insertions (**2-20, 8-9**), extrusions (**7-47, 7-51**), ring opening (**0-18, 0-49**), ring closing (**0-13, 5-47**).

IUPAC System for Symbolic Representation of Mechanisms

In addition to providing a system for naming transformations, the IUPAC Commission on Physical Organic Chemistry has also produced one for representing mechanisms.[4] As we shall see in Part Two, many mechanisms (though by no means all) are commonly referred to by designations such as S_N2, $A_{AC}2$, E1cB, $S_{RN}1$, etc., many of them devised by C.K. Ingold and his co-workers. While these designations have been useful (and we shall continue to use them in this book), the sheer number of them can be confusing, especially since the symbols do not give a direct clue to what is happening. For example, there is no way to tell directly from the symbols how S_N2' is related to S_N2 (see p. 328). The IUPAC system is based on a very simple description of bond changes.[5] The letter A represents formation of a bond (association); D the breaking of a bond (dissociation). These are *primitive changes*. The basic description of a mechanism consists of these letters, with subscripts to indicate where the electrons are going. In any mechanism the *core atoms* are defined as (a) the two atoms in a multiple bond that undergoes addition, or (b) the two atoms that will be in a multiple bond after elimination, or (c) the single atom at which substitution takes place.

As an example of the system, this is how an E1cB mechanism (p. 991) would be represented:

Step 1 $\quad H-\overline{\underline{O}}|^{\ominus} \quad -\overset{|}{C}-\overset{|}{C}-Cl \longrightarrow H_2O + {}^{\ominus}\overline{C}-\overset{|}{C}-Cl \quad A_nD_E \text{ (or } A_{xh}D_H)$

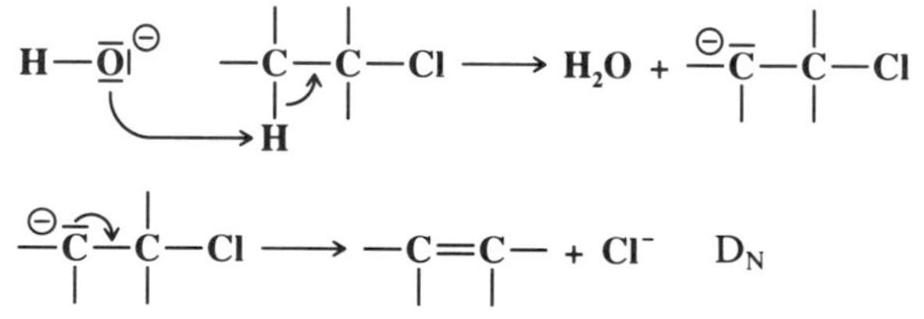

Step 2 $\quad {}^{\ominus}\overline{C}-\overset{|}{C}-Cl \longrightarrow -C=C- + Cl^- \quad D_N$

Overall designation: $A_nD_E + D_N$ (or $A_{xh}D_H + D_N$)

In this case the overall reaction is:

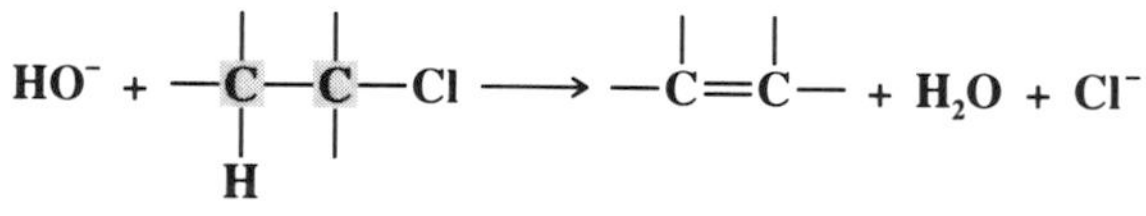

$$HO^- + -\overset{|}{\underset{H}{C}}-\overset{|}{C}-Cl \longrightarrow -C=C- + H_2O + Cl^-$$

and the core atoms are the two shaded carbons.

Step 1, First Symbol
A bond is being formed between O and H. Bond formation is represented by A. For this particular case the system gives two choices for subscript. In any process, the subscript is N if a core atom is forming a bond to a nucleophile (A_N) or breaking a bond to a nucleofuge (D_N). If a noncore atom is doing the same thing, lowercase n is used instead. Since H and O are non-core atoms, the lowercase n is used, and the formation of the O—H bond is designated by A_n. However, because involvement of H^+ is so common in organic mechanisms, the rules allow an alternative. The subscript H or h may replace N or n. The symbol xh denotes that the H^+ comes from or goes to an unspecified carrier atom X. Thus the

---

[4]Guthrie *Pure Appl. Chem.* **1989**, *61*, 23-56. For a briefer description, see Guthrie and Jencks *Acc. Chem. Res.* **1989**, *22*, 343-349.

[5]There are actually two IUPAC systems. The one we use in this book (Ref. 4) is intended for general use. A more detailed system, which describes every conceivable change happening in a system, and which is designed mostly for computer handling and storage, is given by Littler *Pure Appl. Chem.* **1989**, *61*, 57-81. The two systems are compatible; the Littler system uses the same symbols as the Guthrie system, but has additional symbols.

term A_{xh} means that a bond is being formed between H (moving without electrons) and an outside atom, in this case O. The same subscript, xh, would be used if the outside atom were any other nucleophilic atom, say, N or S.

Step 1, Second Symbol
A bond is being broken between C and H. The symbol is D. In any process, the subscript is E if a core atom is forming a bond to an electrophile (A_E) or breaking a bond to an electrofuge (D_E). Since C is a core atom, the symbol here is D_E. Alternatively, the symbol could be D_H. The rules allow A_H or D_H to replace A_E or D_E if the electrophile or electrofuge is H^+. Because a core atom is involved in this primitive change the H in the subscript is capitalized.

Step 1, Combined Symbols
In step 1 two bond changes take place simultaneously. In such cases they are written together, with no space or punctuation:

$$A_n D_E \text{ or } A_{xh} D_{II}$$

Step 2
Only one bond is broken in this step and no bonds are formed. (The movement of a pair of unshared electrons into the C—C bond, forming a double bond, is not designated by any symbol. In this system bond multiplicity changes are understood without being specified.) Thus the symbol is D. The broken bond is between a core atom (C) and a nucleofuge (Cl), so the designation is D_N.

Overall Designation
This can be either $A_n D_N + D_N$ or $A_{xh} D_H + D_N$. The + symbol shows that there are two separate steps. If desired, rate-limiting steps can be shown by the symbol ‡. In this case, if the first step is the slow step [old designation (E1cB)$_I$], the designation would be $A_n D_E^\ddagger + D_N$ or $A_{xh} D_H^\ddagger + D_N$.

For most mechanisms (other than rearrangements), there will be only two A or D terms with uppercase subscripts, and the nature of the reaction can be immediately recognized by looking at them. If both are A, the reaction is an addition; if both are D (as in $A_n D_E + D_N$) it is an elimination. If one is A and the other D, the reaction is a substitution.

We have given here only a brief description of the system. Other IUPAC designations will be shown in Part Two, where appropriate. For more details, further examples, and additional symbols, see Ref. 4.

Organic Syntheses References

At the end of each numbered section there is a list of *Organic Syntheses* references (abbreviated OS). With the exception of a few very common reactions (**2-3, 2-22, 2-24,** and **2-38**) the list includes *all* OS references for each reaction. The volumes of OS that have been covered are Collective Volumes **I** to **VII** and individual volumes **66** to **69**. Where no OS references are listed at the end of a section, the reaction has not been reported in OS through volume **69**. These listings thus constitute a kind of index to OS.[6] Certain ground

---

[6]Two indexes to *Organic Syntheses* have been published as part of the series. One of these, Liotta; Volmer *Organic Syntheses Reaction Guide*; Wiley: New York, 1991, which covers the series through volume 68, is described on p. 1257. The other, which covers the series through Collective Volume V, is Shriner; Shriner *Organic Syntheses Collective Volumes I, II, III, IV, V, Cumulative Indices*; Wiley: New York, 1976. For an older index to *Organic Syntheses* (through volume 45), see Sugasawa; Nakai *Reaction Index of Organic Syntheses*; Wiley: New York, 1967.

rules were followed in assembling these lists. A reaction in which two parts of a molecule independently undergo simultaneous reaction is listed under both reactions. Similarly, if two reactions happen (or might happen) rapidly in succession without the isolation of an intermediate, the reactions are listed in both places. For example, at OS **IV**, 266 is

$$\text{(cyclic ether)} \xrightarrow[\text{H}_2\text{SO}_4]{\text{POCl}_3} \text{Cl(CH}_2)_4\text{O(CH}_2)_4\text{Cl}$$

This reaction is treated as **0-68** followed by **0-16** and is listed in both places. However, certain reactions are not listed because they are trivial examples. An instance of this is the reaction found at OS **III**, 468:

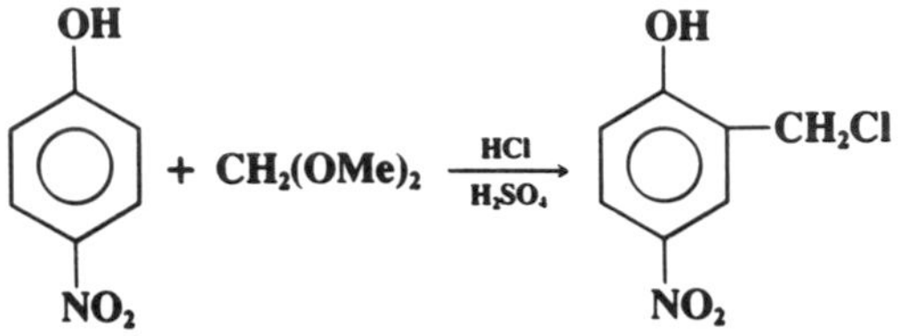

This is a chloromethylation reaction and is consequently listed at **1-24.** However, in the course of the reaction formaldehyde is generated from the acetal. This reaction is not listed at **0-6** (hydrolysis of acetals), because it is not really a preparation of formaldehyde.

10

ALIPHATIC NUCLEOPHILIC SUBSTITUTION

In nucleophilic substitution the attacking reagent (the nucleophile) brings an electron pair to the substrate, using this pair to form the new bond, and the leaving group (the nucleofuge) comes away with an electron pair:

$$R \overset{\frown}{-} \overline{X} + \overline{Y} \longrightarrow R-Y + \overline{X}$$

This equation says nothing about charges. Y may be neutral or negatively charged; RX may be neutral or positively charged; so there are four charge types, examples of which are

Type I	$R-I + OH^-$	$\longrightarrow R-OH + I^-$
Type II	$R-I + NMe_3$	$\longrightarrow R-\overset{\oplus}{N}Me_3 + I^-$
Type III	$R-\overset{\oplus}{N}Me_3 + OH^-$	$\longrightarrow R-OH + NMe_3$
Type IV	$R-\overset{\oplus}{N}Me_3 + H_2S$	$\longrightarrow R-\overset{\oplus}{S}H_2 + NMe_3$

In all cases, Y must have an unshared pair of electrons, so that all nucleophiles are Lewis bases. When Y is the solvent, the reaction is called *solvolysis*. Nucleophilic substitution at an aromatic carbon is considered in Chapter 13.

Nucleophilic substitution at an alkyl carbon is said to *alkylate* the nucleophile. For example, the above reaction between RI and NMe₃ is an *alkylation* of trimethylamine. Similarly, nucleophilic substitution at an acyl carbon is an *acylation* of the nucleophile.

MECHANISMS

Several distinct mechanisms are possible for aliphatic nucleophilic substitution reactions, depending on the substrate, nucleophile, leaving group, and reaction conditions. In all of them, however, the attacking reagent carries the electron pair with it, so that the similarities are greater than the differences. Mechanisms that occur at a saturated carbon atom are considered first.[1] By far the most common are the SN1 and SN2 mechanisms.

---

[1]For a monograph on this subject, see Hartshorn *Aliphatic Nucleophilic Substitution*; Cambridge University Press: Cambridge, 1973. For reviews, see Katritzky; Brycki *Chem. Soc. Rev.* **1990**, *19*, 83-105; Richard *Adv. Carbocation Chem.* **1989**, *1*, 121-169; Bazilevskii; Koldobskii; Tikhomirov *Russ. Chem. Rev.* **1986**, *55*, 948-965; de la Mare; Swedlund, in Patai *The Chemistry of the Carbon–Halogen Bond*, pt. 1; Wiley: New York, 1973, pp. 409-490. For some older books, see Thornton *Solvolysis Mechanisms*; Ronald Press: New York, 1964; Bunton *Nucleophilic Substitution at a Saturated Carbon Atom*; American Elsevier: New York, 1963; Streitwieser *Solvolytic Displacement Reactions*; McGraw-Hill: New York, 1962.

The S$_N$2 Mechanism

S$_N$2 stands for *substitution nucleophilic bimolecular*. The IUPAC designation (p. 290) is A$_N$D$_N$. In this mechanism there is *backside attack*: the nucleophile approaches the substrate from a position 180° away from the leaving group. The reaction is a one-step process with no intermediate (see, however, pp. 297-298 and 305). The C—Y bond is formed as the C—X bond is broken:

$$\overline{Y} + \overset{}{\underset{}{C}}{-}X \longrightarrow Y\cdots\overset{}{\underset{}{C}}\cdots X \longrightarrow Y{-}\overset{}{\underset{}{C}}{-} + \overline{X}$$

1

The energy necessary to break the C—X bond is supplied by simultaneous formation of the C—Y bond. The position of the atoms at the top of the curve of free energy of activation can be represented as **1**. Of course the reaction does not stop here: this is the transition state. The group X must leave as the group Y comes in, because at no time can the carbon have more than eight electrons in its outer shell. When the transition state is reached, the central carbon atom has gone from its initial sp^3 hybridization to an sp^2 state with an approximately perpendicular p orbital. One lobe of this p orbital overlaps with the nucleophile and the other with the leaving group. This is why a frontside S$_N$2 mechanism has never been observed. In a hypothetical frontside transition state, both the nucleophile and the leaving group would have to overlap with the same lobe of the p orbital. The backside mechanism involves the maximum amount of overlap throughout the course of the reaction. During the transition state the three nonreacting substituents and the central carbon are approximately coplanar. They will be exactly coplanar if both the entering and the leaving group are the same.

There is a large amount of evidence for the S$_N$2 mechanism. First there is the kinetic evidence. Since both the nucleophile and the substrate are involved in the rate-determining step (the only step, in this case), the reaction should be first order in each component, second order overall, and satisfy the rate expression

$$\text{Rate} = k[\mathbf{RX}][\mathbf{Y}] \tag{1}$$

This rate law has been found to apply. It has been noted that the 2 in S$_N$2 stands for bimolecular. It must be remembered that this is not always the same as second order (see p. 221). If a large excess of nucleophile is present—for example, if it is the solvent—the mechanism may still be bimolecular, though the experimentally determined kinetics will be first order:

$$\text{Rate} = k[\mathbf{RX}] \tag{2}$$

As previously mentioned (p. 223), such kinetics are called *pseudo-first order*.

The kinetic evidence is a necessary but not a sufficient condition; we will meet other mechanisms that are also consistent with these data. Much more convincing evidence is obtained from the fact that the mechanism predicts inversion of configuration when substitution occurs at a chiral carbon and this has been observed many times. This inversion of configuration (see p. 111) is called the *Walden inversion* and was observed long before the S$_N$2 mechanism was formulated by Hughes and Ingold.[2]

---

[2]Cowdrey; Hughes; Ingold; Masterman; Scott *J. Chem. Soc.* **1937,** 1252. The idea that the addition of one group and removal of the other are simultaneous was first suggested by Lewis in *Valence and the Structure of Atoms and Molecules;* Chemical Catalog Company: New York, 1923, p. 113. The idea that a one-step substitution leads to inversion was proposed by Olsen *J. Chem. Phys.* **1933,** *1*, 418.

At this point it is desirable for us to see just how it was originally proved that a given substitution reaction proceeds with inversion of configuration, even before the mechanism was known. Walden presented a number of examples[3] in which inversion *must* have taken place. For example, (+)-malic acid could be converted to (+)-chlorosuccinic acid by thionyl chloride and to (−)-chlorosuccinic acid by phosphorus pentachloride:

$$
\begin{array}{ccccc}
\text{COOH} & & \text{COOH} & & \text{COOH} \\
| & \xleftarrow{\text{SOCl}_2} & | & \xrightarrow{\text{PCl}_5} & | \\
(+)\text{CHCl} & & (+)\text{CHOH} & & (-)\text{CHCl} \\
| & & | & & | \\
\text{CH}_2\text{COOH} & & \text{CH}_2\text{COOH} & & \text{CH}_2\text{COOH}
\end{array}
$$

One of these must be an inversion and the other a retention of configuration, but the question is which is which? The signs of rotation are of no help in answering this question since, as we have seen (p. 108), rotation need not be related to configuration. Another example discovered by Walden is

$$
\begin{array}{ccccc}
\text{COOH} & & \text{COOH} & & \text{COOH} \\
| & \xleftarrow[\text{H}_2\text{O}]{\text{Ag}_2\text{O}} & | & \xrightarrow{\text{KOH}} & | \\
(+)\text{CHOH} & & (+)\text{CHCl} & & (-)\text{CHOH} \\
| & & | & & | \\
\text{CH}_2\text{COOH} & & \text{CH}_2\text{COOH} & & \text{CH}_2\text{COOH}
\end{array}
$$

Once again, one reaction and only one must be an inversion, but which?[4] It may also be noticed [illustrated by the use of thionyl chloride on (+)-malic acid and treatment of the product with KOH] that it is possible to convert an optically active compound into its enantiomer.[5]

A series of experiments designed to settle the matter of exactly where inversion takes place was performed by Phillips, Kenyon, and co-workers. In 1923, Phillips carried out the following cycle:[6]

$$
\begin{array}{ccccc}
\text{CH}_2\text{Ph} & & \text{CH}_2\text{Ph} & & \text{CH}_2\text{Ph} \\
| & \xrightarrow[A]{\text{TsCl}} & | & \xrightarrow[B]{\substack{\text{EtOH}\\ \text{K}_2\text{CO}_3}} & | \\
\text{Me}-\text{CH}-\text{OH} & & \text{Me}-\text{CH}-\text{OTs} & & \text{Me}-\text{CH}-\text{OEt} \\
\alpha = +33.0° & & \alpha = +31.1° & & \alpha = -19.9°
\end{array}
$$

$$\text{K}\Big\downarrow C$$

$$
\begin{array}{ccc}
\text{CH}_2\text{Ph} & & \text{CH}_2\text{Ph} \\
| & \xrightarrow[D]{\text{EtBr}} & | \\
\text{Me}-\text{CH}-\text{OK} & & \text{Me}-\text{CH}-\text{OEt} \\
& & \alpha = +23.5°
\end{array}
$$

In this cycle, (+)-1-phenyl-2-propanol is converted to its ethyl ether by two routes, path *AB* giving the (−) ether, and path *CD* giving the (+) ether. Therefore, at least one of the four steps must be an inversion. It is extremely unlikely that there is inversion in step *A*,

---

[3]Walden *Ber.* **1893**, *26*, 210, **1896**, *29*, 133, **1899**, *32*, 1855.

[4]For a discussion of these cycles, see Kryger; Rasmussen *Acta Chem. Scand.* **1972**, *26*, 2349.

[5]The student may wonder just what the mechanism is in cases where retention of configuration is involved since it certainly is not simple SN2. As we shall see later, the reaction between malic acid and thionyl chloride is an SNi process (p. 326), while a neighboring-group mechanism (p. 308) is involved in the treatment of chlorosuccinic acid with silver oxide.

[6]Phillips *J. Chem. Soc.* **1923**, *123*, 44. For analyses of such cycles and general descriptions of more complex ones, see Garwood; Cram *J. Am. Chem. Soc.* **1970**, *92*, 4575; Cram; Cram *Fortschr. Chem. Forsch.* **1972**, *31*, 1-43.

C, or D, since in all these steps the C—O bond is unbroken, and in none of them could the oxygen of the bond have come from the reagent. There is therefore a high probability that A, C, and D proceeded with retention, leaving B as the inversion. A number of other such cycles were carried out, always with nonconflicting results.[7] These experiments not only definitely showed that certain specific reactions proceed with inversion, but also established the configurations of many compounds.

Walden inversion has been found at a primary carbon atom by the use of a chiral substrate containing a deuterium and a hydrogen atom at the carbon bearing the leaving group.[8] Inversion of configuration has also been found for SN2 reactions proceeding in the gas phase.[9]

Another kind of evidence for the SN2 mechanism comes from compounds with potential leaving groups at bridgehead carbons. If the SN2 mechanism is correct, these compounds should not be able to react by this mechanism, since the nucleophile cannot approach from the rear. Among the many known examples of unsuccessful reaction attempts at bridgeheads

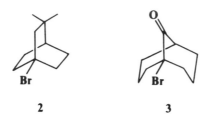

2   **3**

under SN2 conditions[10] are treatment of the [2.2.2] system **2** with ethoxide ion[11] and treatment of the [3.3.1] system **3** with sodium iodide in acetone.[12] In these cases, open-chain analogs underwent the reactions readily. As a final example of evidence for the SN2 mechanism, the reaction between optically active 2-octyl iodide and radioactive iodide ion may be mentioned:

$$C_6H_{13}CHMeI + *I^- \longrightarrow C_6H_{13}CHMe*I + I^-$$

We expect racemization in this reaction, since if we start with the pure R isomer, at first each exchange will produce an S isomer, but with increasing concentration of S isomer, *it* will begin to compete for I^- with the R isomer, until at the end a racemic mixture is left. The point investigated was a comparison of the rate of inversion with the rate of uptake of radioactive $*I^-$. It was found[13] that the rates were identical within experimental error:

Rate of inversion $2.88 \pm 0.03 \times 10^{-5}$

Rate of exchange $3.00 \pm 0.25 \times 10^{-5}$

[7]For example, see Kenyon; Phillips; Turley *J. Chem. Soc.* **1925**, *127*, 399; Kenyon; Phillips; Taylor *J. Chem. Soc.* **1933**, 173; Kenyon; Phillips; Shutt *J. Chem. Soc.* **1935**, 1663.

[8]Streitwieser *J. Am. Chem. Soc.* **1953**, *75*, 5014.

[9]Lieder; Brauman *J. Am. Chem. Soc.* **1974**, *96*, 4028; Speranza; Angelini *J. Am. Chem. Soc.* **1980**, *102*, 3115. For a review of nucleophilic displacements in the gas phase, see Riveros; José; Takashima *Adv. Phys. Org. Chem.* **1985**, *21*, 197-240.

[10]For a review of bridgehead reactivity in nucleophilic substitution reactions, see Müller; Mareda, in Olah *Cage Hydrocarbons*; Wiley: New York, 1990, pp. 189-217. For a review of reactions at bridgehead carbons, see Fort; Schleyer *Adv. Alicyclic Chem.* **1966**, *1*, 283-370.

[11]Doering; Levitz; Sayigh; Sprecher; Whelan *J. Am. Chem. Soc.* **1953**, *75*, 1008. Actually, a slow substitution was observed in this case, but not by an SN2 mechanism.

[12]Cope; Synerholm *J. Am. Chem. Soc.* **1950**, *72*, 5228.

[13]Hughes; Juliusburger; Masterman; Topley; Weiss *J. Chem. Soc.* **1935**, 1525.

What was actually measured was the rate of racemization, which is twice the rate of inversion, since each inversion creates, in effect, two racemic molecules. The significance of this result is that it shows that every act of exchange is an act of inversion.

Eschenmoser and co-workers have provided strong evidence that the transition state in an S_N2 reaction must be linear.[14] Base treatment of methyl α-tosyl-*o*-toluenesulfonate (**4**) gives the *o*-(l-tosylethyl)benzenesulfonate ion (**6**). The role of the base is to remove the α

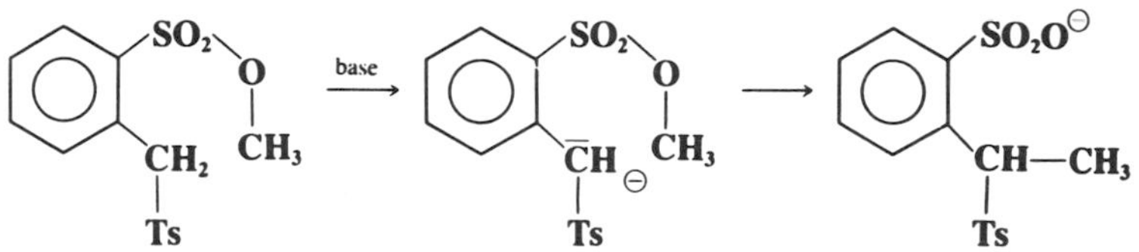

proton to give the ion **5**. It might be supposed that the negatively charged carbon of **5** attacks the methyl group in an internal S_N2 process:

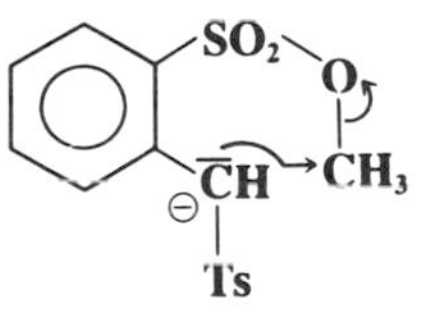

but this is not the case. Crossover experiments[14] (p. 555) have shown that the negatively charged carbon attacks the methyl group of another molecule rather than the nearby one in the same molecule, that is, the reaction is intermolecular and not intramolecular, despite the more favorable entropy of the latter pathway (p. 211). The obvious conclusion is that intramolecular attack does not take place because complete linearity cannot be attained. This behavior is in sharp contrast to that in cases in which the leaving group is not constrained (p. 309), where intramolecular S_N2 mechanisms operate freely.

There is evidence, both experimental and theoretical, that there are intermediates in at least some S_N2 reactions in the gas phase, in charge type I reactions, where a negative ion nucleophile attacks a neutral substrate. Two energy minima, one before and one after the transition state appear in the reaction coordinate (Figure 10.1).[15] These minima correspond to unsymmetrical ion–dipole complexes.[16] Theoretical calculations also show such minima in certain solvents, e.g., DMF, but not in water.[17]

For a list of some of the more important reactions that operate by the S_N2 mechanism, see Table 10.7.

[14]Tenud; Farooq; Seibl; Eschenmoser *Helv. Chim. Acta* **1970**, *53*, 2059. See also King; McGarrity *J. Chem. Soc., Chem. Commun.* **1979**, 1140.
[15]Taken from Chandrasekhar; Smith; Jorgensen, Ref. 16.
[16]Olmstead; Brauman *J. Am. Chem. Soc.* **1977**, *99*, 4219; Pellerite; Brauman *J. Am. Chem. Soc.* **1980**, *102*, 5993; Wolfe; Mitchell; Schlegel *J. Am. Chem. Soc.* **1981**, *103*, 7692; Chandrasekhar; Smith; Jorgensen *J. Am. Chem. Soc.* **1985**, *107*, 154; Evanseck; Blake; Jorgensen *J. Am. Chem. Soc.* **1987**, *109*, 2349; Kozaki; Morihashi; Kikuchi *J. Am. Chem. Soc.* **1989**, *111*, 1547; Jorgensen *Acc. Chem. Res.* **1989**, *22*, 184-189.
[17]Chandrasekhar; Jorgensen *J. Am. Chem. Soc.* **1985**, *107*, 2974.

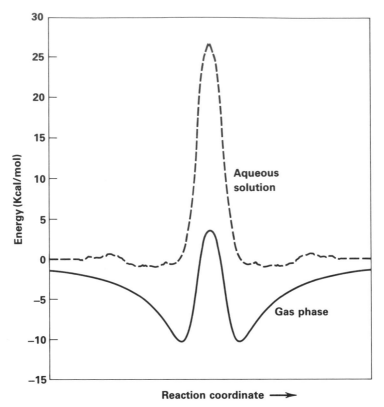

FIGURE 10.1 Free-energy profile for the gas phase (solid line) and aqueous solution (dashed line) SN2 reaction between CH_3Cl and Cl^-, from molecular orbital calculations.[15]

The SN1 Mechanism

The most ideal version of the SN1 mechanism (*substitutional nucleophilic unimolecular*) consists of two steps (once again, possible charges on the substrate and nucleophile are not shown):

Step 1 $R—X \overset{slow}{\rightleftharpoons} R^+ + X$

Step 2 $R^+ + Y \overset{fast}{\longrightarrow} R—Y$

The first step is a slow ionization of the substrate and is the rate-determining step. The second is a rapid reaction between the intermediate carbocation and the nucleophile. The ionization is always assisted by the solvent,[18] since the energy necessary to break the bond is largely recovered by solvation of R^+ and of X. For example the ionization of t-BuCl to t-Bu$^+$ and Cl$^-$ in the gas phase without a solvent requires 150 kcal/mol (630 kJ/mol). In the absence of a solvent such a process simply would not take place, except at very high temperatures. In water this ionization requires only 20 kcal/mol (84 kJ/mol). The difference

[18]For reviews of solvolysis, see Okamoto *Adv. Carbocation Chem.* **1989**, *1*, 171-218; Blandamer; Scott; Robertson *Prog. Phys. Org. Chem.* **1985**, *15*, 149-196; Robertson *Prog. Phys. Org. Chem.* **1967**, *4*, 213-280. For a review of the solvolytic cleavage of t-butyl substrates, see Dvorko; Ponomareva; Kulik *Russ. Chem. Rev.* **1984**, *53*, 547-560.

is solvation energy. In cases where the role of the solvent is solely to assist in departure of the leaving group from the frontside, that is, where there is a complete absence of backside (SN2) participation by solvent molecules, the mechanism is called *limiting* SN1. There is kinetic and other evidence[19] that in pulling X away from RX, two molecules of a protic solvent form weak hydrogen bonds with X

$$\mathbf{R-X}\begin{matrix} \cdot\cdot\mathbf{H-O-R} \\ \cdot\cdot\mathbf{H-O-R} \end{matrix} \longrightarrow \mathbf{R^+}$$

In the IUPAC system the SN1 mechanism is $D_N + A_N$ or $D_N^\ddagger + A_N$ (where $\ddagger$ denotes the rate-determining step). The IUPAC designations for the SN1 and SN2 mechanisms thus clearly show the essential differences between them: $A_N D_N$ indicates that bond breaking is concurrent with bond formation; $D_N + A_N$ shows that the former happens first.

In looking for evidence for the SN1 mechanism the first thought is that it should be a first-order reaction following the rate law

$$\text{Rate} = k[\mathbf{RX}] \tag{3}$$

Since the slow step involves only the substrate, the rate should be dependent only on the concentration of that. Although the solvent is necessary to assist in the process of ionization, it does not enter the rate expression, because it is present in large excess. However, the simple rate law given in Eq. (3) is not sufficient to account for all the data. Many cases are known where pure first-order kinetics are followed, but in many other cases more complicated kinetics are found. We can explain this by taking into account the reversibility of the first step. The X formed in this step competes with Y for the cation and the rate law must be modified as follows (see Chapter 6):

$$\mathbf{RX} \underset{k_{-1}}{\overset{k_1}{\rightleftharpoons}} \mathbf{R^+} + \mathbf{X}$$

$$\mathbf{R^+} + \mathbf{Y} \xrightarrow{k_2} \mathbf{RY}$$

$$\text{Rate} = \frac{k_1 k_2[\mathbf{RX}][\mathbf{Y}]}{k_{-1}[\mathbf{X}] + k_2[\mathbf{Y}]} \tag{4}$$

At the beginning of the reaction, when the concentration of X is very small, $k_{-1}[\mathbf{X}]$ is negligible compared with $k_2[\mathbf{Y}]$ and the rate law is reduced to Eq. (3). Indeed, SN1 reactions generally do display simple first-order kinetics in their initial stages. Most kinetic studies of SN1 reactions fall into this category. In the later stages of SN1 solvolyses, [X] becomes large and Eq. (4) predicts that the rate should decrease. This is found to be the case for diarylmethyl halides,[20] though not for *t*-butyl halides, which follow Eq. (3) for the entire reaction.[21] An explanation for this difference is that *t*-butyl cations are less selective than the relatively stable diarylmethyl type (p. 169). Although halide ion is a much more powerful nucleophile than water, there is much more water available since it is the solvent.[22] The selective diphenylmethyl cation survives many collisions with solvent molecules before combining with a reactive halide, but the less selective *t*-butyl ion cannot wait for a reactive but relatively rare halide ion and combines with the solvent.

[19]Blandamer; Burgess; Duce; Symons; Robertson; Scott *J. Chem. Res. (S)* **1982**, 130.
[20]Benfey; Hughes; Ingold *J. Chem. Soc.* **1952**, 2488.
[21]Bateman; Hughes; Ingold *J. Chem. Soc.* **1940**, 960.
[22]In the experiments mentioned, the solvent was actually "70%" or "80%" aqueous acetone. "80%" aqueous acetone consists of 4 vol of dry acetone and 1 vol of water.

If the X formed during the reaction can decrease the rate, at least in some cases, it should be possible to *add* X from the outside and further decrease the rate in that way. This retardation of rate by addition of X is called *common-ion effect* or the *mass-law effect*. Once again, addition of halide ions decreases the rate for diphenylmethyl but not for *t*-butyl halides.

One factor that complicates the kinetic picture is the *salt effect*. An increase in ionic strength of the solution usually increases the rate of an SN1 reaction (p. 359). But when the reaction is of charge type II, where both Y and RX are neutral, so that X is negatively charged (and most solvolyses are of this charge type), the ionic strength increases as the reaction proceeds and this increases the rate. This effect must be taken into account in studying the kinetics. Incidentally, the fact that the addition of outside ions *increases* the rate of most SN1 reactions makes especially impressive the *decrease* in rate caused by the common ion.

It may be noted that the pseudo-first-order rate law for an SN2 reaction in the presence of a large excess of Y [Eq. (2)] is the same as that for an ordinary SN1 reaction [Eq. (3)]. It is thus not possible to tell these cases apart by simple kinetic measurements. However, we can often distinguish between them by the common-ion effect mentioned above. Addition of a common ion will not markedly affect the rate of an SN2 reaction beyond the effect caused by other ions. Unfortunately, as we have seen, not all SN1 reactions show the common-ion effect, and this test fails for *t*-butyl and similar cases.

Kinetic studies also provide other evidence for the SN1 mechanism. If this mechanism operates essentially as shown on p. 298, the rate should be the same for a given substrate under a given set of conditions, *regardless of the identity of the nucleophile or its concentration*. In one experiment that demonstrates this, benzhydryl chloride (Ph_2CHCl) was treated in SO_2 with the nucleophiles fluoride ion, pyridine, and triethylamine at several concentrations of each nucleophile.[23] In each case the initial rate of the reaction was approximately the same when corrections were made for the salt effect. The same type of behavior has been shown in a number of other cases, even when the reagents are as different in their nucleophilicities (see p. 348) as H_2O and OH^-.

It is normally not possible to detect the carbocation intermediate of an SN1 reaction directly, because its lifetime is very short. However, in the case of 3,4'-dimethoxydiphenylmethyl acetate (**7**) and certain other substrates in polar solvents it was possible to initiate

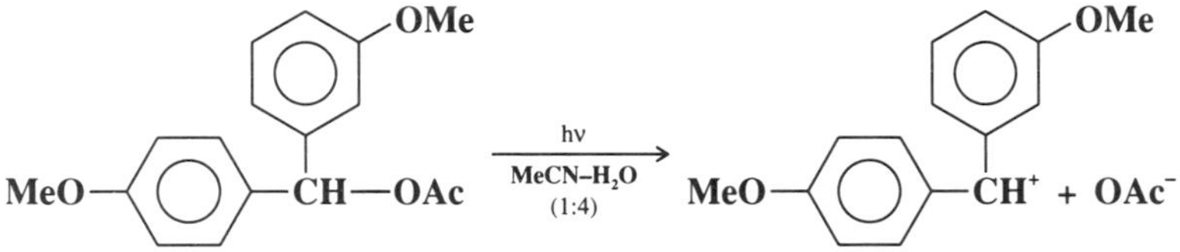

the reaction photolytically, and under these conditions the uv spectra of the intermediate carbocations could be obtained,[24] providing additional evidence for the SN1 mechanism.

Further evidence for the SN1 mechanism is that reactions run under SN1 conditions fail or proceed very slowly at the bridgehead positions[10] of [2.2.1] (norbornyl) systems[25] (e.g. 1-chloroapocamphane, **8**). If SN1 reactions require carbocations and if carbocations must

[23]Bateman; Hughes; Ingold *J. Chem. Soc.* **1940**, 1011.
[24]McClelland; Kanagasabapathy; Steenken *J. Am. Chem. Soc.* **1988**, *110*, 6913.
[25]For a review, see Fort, in Olah; Schleyer *Carbonium Ions*, vol. 4; Wiley: New York, 1973, pp. 1783-1835.

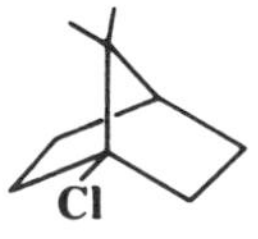

8    9 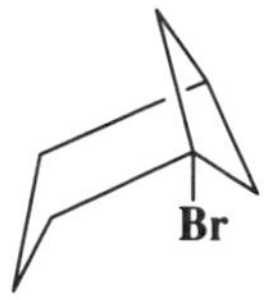   10    11

be planar or nearly planar, then it is no surprise that bridgehead 1-norbornyl carbon atoms, which cannot assume planarity, do not become the seat of carbocations. As an example, **8**, boiled 21 hr with 30% KOH in 80% ethanol or 48 hr with aqueous ethanolic silver nitrate, gave no reaction in either case,[26] though analogous open-chain systems reacted readily. According to this theory, SN1 reactions should be possible with larger rings, since near-planar carbocations might be expected there. This turns out to be the case. For example, [2.2.2] bicyclic systems undergo SN1 reactions much faster than smaller bicyclic systems, though the reaction is still slower than with open-chain systems.[27] Proceeding to a still larger system, the bridgehead [3.2.2] cation **9** is actually stable enough to be kept in solution in SbF_5-SO_2ClF at temperatures below $-50°C$[28] (see also p. 345). Other small bridgehead systems that undergo SN1 reactions are the [3.1.1] (e.g., **10**)[29] and the cubyl (e.g., **11**)[30] systems. Ab initio calculations show that the cubyl cation, though it cannot be planar, requires less energy to form than the 1-norbornyl cation.[31]

Certain nucleophilic substitution reactions that normally involve carbocations can take place at norbornyl bridgeheads[32] (though it is not certain that carbocations are actually involved in all cases) if the leaving group used is of the type that cannot function as a nucleophile (and thus come back) once it has gone, e.g.,

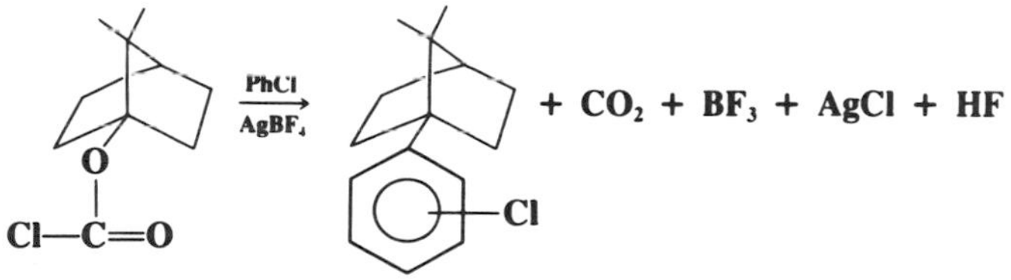

In this example,[33] chlorobenzene is the nucleophile (see **1-12**).

Additional evidence for the SN1 mechanism—in particular, for the intermediacy of carbocations—is that solvolysis rates of alkyl chlorides in ethanol parallel carbocation stabilities as determined by heats of ionization measured in superacid solutions (p. 166).[34]

[26]Bartlett; Knox *J. Am. Chem. Soc.* **1939**, *61*, 3184.

[27]For synthetic examples, see Kraus; Hon *J. Org. Chem.* **1985**, *50*, 4605.

[28]Olah; Liang; Wiseman; Chong *J. Am. Chem. Soc.* **1972**, *74*, 4927.

[29]Della; Pigou; Tsanaktsidis *J. Chem. Soc., Chem. Commun.* **1987**, 833.

[30]Eaton; Yang; Xiong *J. Am. Chem. Soc.* **1990**, *112*, 3225; Moriarty; Tuladhar; Penmasta; Awasthi *J. Am. Chem. Soc.* **1990**, *112*, 3228.

[31]Hrovat; Borden *J. Am. Chem. Soc.* **1990**, *112*, 3227.

[32]Ref. 26; Beak; Trancik *J. Am. Chem. Soc.* **1968**, *90*, 2714; Clive; Denyer *Chem. Commun.* **1971**, 1112; White; McGirk; Aufdermarsh; Tiwari; Todd *J. Am. Chem. Soc.* **1973**, *95*, 8107; Beak; Harris *J. Am. Chem. Soc.* **1974**, *96*, 6363.

[33]For a review of reactions with the OCOCl leaving group, see Beak *Acc. Chem. Res.* **1976**, *9*, 230-236.

[34]Arnett; Petro *J. Am. Chem. Soc.* **1978**, *100*, 5408; Arnett; Petro; Schleyer *J. Am. Chem. Soc.* **1979**, *101*, 522; Arnett; Pienta *J. Am. Chem. Soc.* **1980**, *102*, 3329; Arnett; Molter *Acc. Chem. Res.* **1985**, *18*, 339-346.

Ion Pairs in the SN1 Mechanism[35]

Like the kinetic evidence, the stereochemical evidence for the SN1 mechanism is less clear-cut than it is for the SN2 mechanism. If there is a free carbocation, it is planar (p. 172), and the nucleophile should attack with equal facility from either side of the plane, resulting in complete racemization. Although many first-order substitutions do give complete racemization, many others do not. Typically there is 5 to 20% inversion, though in a few cases, a small amount of retention of configuration has been found. These and other results have led to the conclusion that in many SN1 reactions at least some of the products are not formed from free carbocations but rather from *ion pairs*. According to this concept,[36] SN1 reactions proceed in this manner:

$$R—X \rightleftharpoons R^+X^- \rightleftharpoons R^+\|X^- \rightleftharpoons R^+ + X^-$$
$$\quad\quad\quad\quad \mathbf{12} \quad\quad\quad \mathbf{13} \quad\quad\quad \mathbf{14}$$

where **12** is an *intimate*, *contact*, or *tight* ion pair, **13** a *loose*, or *solvent-separated* ion pair, and **14** the dissociated ions (each surrounded by molecules of solvent).[37] The reaction in which the intimate ion pair recombines to give the original substrate is referred to as *internal return*. The reaction products can result from attack by the nucleophile at any stage. In the intimate ion pair **12**, R^+ does not behave like the free cation of **14**. There is probably significant bonding between R^+ and X^- and asymmetry may well be maintained.[38] X^- "solvates" the cation on the side from which it departed, while solvent molecules near **12** can only solvate it from the opposite side. Nucleophilic attack by a solvent molecule on **12** thus leads to inversion.

A complete picture of the possibilities for solvolysis reactions in a solvent SH (ignoring the possibilities of elimination or rearrangement—see Chapters 17 and 18) is the following,[39] though in any particular case it is unlikely that all these reactions occur:

$$
\begin{array}{ccc}
\text{SR} & \text{SR} & \delta\text{SR} + (1-\delta)\text{RS} \\
\uparrow & \uparrow & \uparrow \\
\text{SH}\big|_{(SN2)} & \text{SH}\big|_B & \text{SH}\big| \\
RX \rightleftharpoons & R^+X^- \rightleftharpoons & R^+\|X^- \searrow \\
& A\big\| & \big\| \quad\quad \text{SH} \rightarrow \tfrac{1}{2}\text{SR} + \tfrac{1}{2}\text{RS} \\
& & \quad\quad R^+ + X^- \\
XR \rightleftharpoons & X^-R^+ \rightleftharpoons & X^-\|R^+ \nearrow \\
\text{SH}\big|_{(SN2)} & \text{SH}\big| & \text{SH}\big| \\
\downarrow & \downarrow & \downarrow \\
\text{RS} & \text{RS} & \delta\text{RS} + (1-\delta)\text{SR}
\end{array}
$$

In this scheme RS and SR represent enantiomers, etc., and δ represents some fraction. The following are the possibilities: (1) Direct attack by SH on RX gives SR (complete inversion) in a straight SN2 process. (2) If the intimate ion pair $R^+ X^-$ is formed, the solvent can attack at this stage. This can lead to total inversion if reaction A does not take place or to a combination of inversion and racemization if there is competition between A and B. (3) If the solvent-separated ion pair is formed, SH can attack here. The stereochemistry is not

[35]For reviews of ion pairs in SN reactions, see Beletskaya *Russ. Chem. Rev.* **1975**, *44*, 1067-1090; Harris *Prog. Phys. Org. Chem.* **1974**, *11*, 89-173; Raber; Harris; Schleyer, in Szwarc *Ions and Ion Pairs in Organic Reactions*, vol. 2; Wiley: New York, 1974, pp. 247-374.

[36]Proposed by Winstein; Clippinger; Fainberg; Heck; Robinson *J. Am. Chem. Soc.* **1956**, *78*, 328.

[37]For a review of the energy factors involved in the recombination of ion pairs, see Kessler; Feigel *Acc. Chem. Res.* **1982**, *15*, 2-8.

[38]Fry; Lancelot; Lam; Harris; Bingham; Raber; Hall; Schleyer *J. Am. Chem. Soc.* **1970**, *92*, 2538.

[39]Shiner; Fisher *J. Am. Chem. Soc.* **1971**, *93*, 2553.

maintained as tightly and more racemization (perhaps total) is expected. (4) Finally, if free R^+ is formed, it is planar, and attack by SH gives complete racemization.

The ion-pair concept thus predicts that SN1 reactions can display either complete racemization or partial inversion. The fact that this behavior is generally found is evidence that ion pairs are involved in many SN1 reactions. There is much other evidence for the intervention of ion pairs:[40]

1. The compound 2-octyl brosylate was labeled at the sulfone oxygen with ^{18}O and solvolyzed. The unreacted brosylate recovered at various stages of solvolysis had the ^{18}O considerably, though not completely, scrambled:[41]

In an intimate ion pair, the three oxygens become equivalent:

Similar results were obtained with several other sulfonate esters.[42] The possibility must be considered that the scrambling resulted from ionization of one molecule of $ROSO_2Ar$ to R^+ and $ArSO_2O^-$ followed by attack by the $ArSO_2O^-$ ion on *another* carbocation or perhaps on a molecule of $ROSO_2Ar$ in an SN2 process. However, this was ruled out by solvolyzing unlabeled substrate in the presence of labeled $HOSO_2Ar$. These experiments showed that there was some intermolecular exchange (3 to 20%), but not nearly enough to account for the amount of scrambling found in the original experiments. Similar scrambling was found in solvolysis of labeled carboxylic esters R—^{18}O—COR', where the leaving group is $R'COO^-$.[43] In this case also, the external addition of $RCOO^-$ did not result in significant exchange. However, it has been proposed that the scrambling could result from a concerted process, not involving ion-pair intermediates, and there is some evidence for this view.[44]

2. The *special salt effect*. The addition of $LiClO_4$ or $LiBr$ in the acetolysis of certain tosylates produced an initial steep rate acceleration that then decreased to the normal linear acceleration (caused by the ordinary salt effect).[45] This is interpreted as follows: the ClO_4^-

[40]For further evidence beyond that given here, see Winstein; Baker; Smith *J. Am. Chem. Soc.* **1964**, *86*, 2072; Streitwieser; Walsh *J. Am. Chem. Soc.* **1965**, *87*, 3686; Sommer; Carey *J. Org. Chem.* **1967**, *32*, 800, 2473; Kwart; Irvine *J. Am. Chem. Soc.* **1969**, *91*, 5541; Harris; Becker; Fagan; Walden *J. Am. Chem. Soc.* **1974**, *96*, 4484; Bunton; Huang; Paik *J. Am. Chem. Soc.* **1975**, *97*, 6262; Humski; Sendijarević; Shiner *J. Am. Chem. Soc.* **1976**, *98*, 2865; Maskill; Thompson; Wilson *J. Chem. Soc., Chem. Commun.* **1981**, 1239; McManus; Safavy; Roberts *J. Org. Chem.* **1982**, *47*, 4388; Ref. 35; McLennan; Stein; Dobson *Can. J. Chem.* **1986**, *64*, 1201; Kinoshita; Komatsu; Ikai; Kashimura; Tanikawa; Hatanaka; Okamoto *J. Chem. Soc., Perkin Trans. 2* **1988**, 1875; Ronco; Petit; Guyon; Villa *Helv. Chim. Acta* **1988**, *71*, 648; Kevill; Kyong; Weitl *J. Org. Chem.* **1990**, *55*, 4304.
[41]Diaz; Lazdins; Winstein *J. Am. Chem. Soc.* **1968**, *90*, 1904.
[42]Goering; Thies *J. Am. Chem. Soc.* **1968**, *90*, 2967, 2968; Goering; Jones *J. Am. Chem. Soc.* **1980**, *102*, 1628; Yukawa; Morisaki; Tsuji; Kim; Ando *Tetrahedron Lett.* **1981**, *22*, 5187; Chang; le Noble *J. Am. Chem. Soc.* **1983**, *105*, 3708; Paradisi; Bunnett *J. Am. Chem. Soc.* **1985**, *107*, 8223; Fujio; Sanematsu; Tsuno; Sawada; Takai *Tetrahedron Lett.* **1988**, *29*, 93.
[43]Goering; Levy *J. Am. Chem. Soc.* **1962**, *84*, 3853, **1964**, *86*, 120; Goering; Hopf *J. Am. Chem. Soc.* **1971**, *93*, 1224.
[44]Dietze; Wojciechowski *J. Am. Chem. Soc.* **1990**, *112*, 5240.
[45]Ref. 36; Winstein; Klinedinst; Clippinger *J. Am. Chem. Soc.* **1961**, *83*, 4986; Cristol; Noreen; Nachtigall *J. Am. Chem. Soc.* **1972**, *94*, 2187.

(or Br⁻) traps the solvent-separated ion pair to give $R^+ \parallel ClO_4^-$ which, being unstable under these conditions, goes to product. Hence, the amount of solvent-separated ion pair that would have returned to the starting material is reduced, and the rate of the overall reaction is increased. The special salt effect has been directly observed by the use of picosecond absorption spectroscopy.[46]

3. We have previously discussed the possibilities of racemization or inversion of the *product* RS of a solvolysis reaction. However, the formation of an ion pair followed by internal return can also affect the stereochemistry of the *substrate* molecule RX. Cases have been found where internal return racemizes an original optically active RX, an example being solvolysis in aqueous acetone of α-*p*-anisylethyl *p*-nitrobenzoate,[47] while in other cases partial or complete retention is found, for example, solvolysis in aqueous acetone of *p*-chlorobenzhydryl *p*-nitrobenzoate.[48] Racemization of RX is presumably caused by the pathway: $RX \rightleftarrows R^+X^- \rightleftarrows X^-R^+ \rightleftarrows XR$. Evidence for ion pairs is that, in some cases where internal return involves racemization, it has been shown that such racemization is *faster* than solvolysis. For example, optically active *p*-chlorobenzhydryl chloride racemizes about 30 times faster than it solvolyzes in acetic acid.[49]

Molecular orbital calculations[50] made on *t*-BuCl show that the C Cl distance in the intimate ion pair is 2.9 Å and the onset of the solvent-separated ion pair takes place at about 5.5 Å (compare the C—Cl bond length of 1.8 Å).

In a few cases, SN1 reactions have been found to proceed with partial retention (20 to 50%) of configuration. Ion pairs have been invoked to explain some of these.[51] For example, it has been proposed that the phenolysis of optically active α-phenylethyl chloride, in which the ether of net retained configuration is obtained, involves a four-center mechanism:

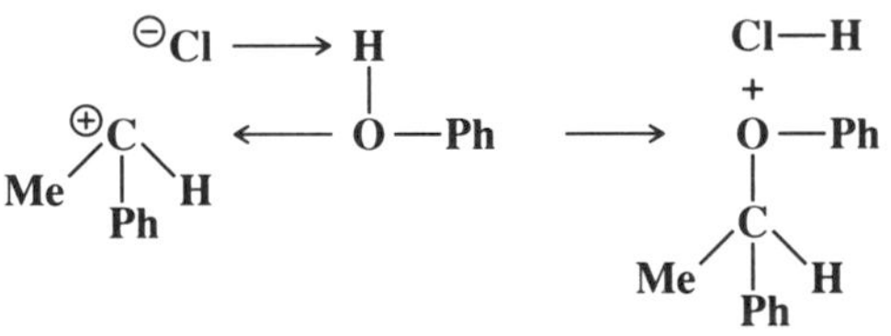

This conclusion is strengthened by the fact that partial retention was obtained in this system only with chloride or other neutral leaving groups; with leaving groups bearing a positive charge, which are much less likely to form hydrogen bonds with the solvent, no retention was found.[52] Partial retention can also arise when the ion pair is shielded at the backside by an additive such as acetonitrile, acetone, or aniline.[53]

The difference between the SN1 and SN2 mechanisms is in the timing of the steps. In the SN1 mechanism, first X leaves, then Y attacks. In the SN2 case, the two things happen simultaneously. One could imagine a third possibility: first the attack of Y and then the removal of X. This is not possible at a saturated carbon, since it would mean more than

[46]Simon; Peters *J. Am. Chem. Soc.* **1982**, *104*, 6142.

[47]Goering; Briody; Sandrock, *J. Am. Chem. Soc.* **1970**, *92*, 7401.

[48]Goering; Briody; Levy *J. Am. Chem. Soc.* **1963**, *85*, 3059.

[49]Winstein; Gall; Hojo; Smith *J. Am. Chem. Soc.* **1960**, *82*, 1010. See also Shiner; Hartshorn; Vogel *J. Org. Chem.* **1973**, *38*, 3604.

[50]Jorgensen; Buckner; Huston; Rossky *J. Am. Chem. Soc.* **1987**, *109*, 1891.

[51]Okamoto; Yamada; Nitta; Shingu *Bull. Chem. Soc. Jpn.* **1966**, *39*, 299; Okamoto; Takeuchi; Inoue *J. Chem. Soc., Perkin Trans. 2* **1980**, 842; Okamoto *Pure Appl. Chem.* **1984**, *56*, 1797-1808. For a similar mechanism with amine nucleophiles, see Lee; Kim; Kang; Lee *J. Org. Chem.* **1988**, *53*, 2678; Lee; Kim; Lee; Kim *J. Phys. Org. Chem.* **1989**, *2*, 35.

[52]Okamoto; Kinoshita; Shingu *Bull. Chem. Soc. Jpn.* **1970**, *43*, 1545.

[53]Okamoto; Nitta; Dohi; Shingu *Bull. Chem. Soc. Jpn.* **1971**, *44*, 3220; Kinoshita; Ueno; Ikai; Fujiwara; Okamoto *Bull. Chem. Soc. Jpn.* **1988**, *61*, 3273; Kinoshita et al., Ref. 40.

eight electrons in the outer shell of carbon. However, this type of mechanism is possible and indeed occurs at other types of substrate (p. 331; Chapter 13).

Mixed SN1 and SN2 Mechanisms

Some reactions of a given substrate under a given set of conditions display all the characteristics of SN2 mechanisms; other reactions seem to proceed by SN1 mechanisms, but cases are found that cannot be characterized so easily. There seems to be something in between, a mechanistic "borderline" region.[54] At least two broad theories have been devised to explain these phenomena. One theory holds that intermediate behavior is caused by a mechanism that is neither "pure" SN1 nor "pure" SN2, but some "in-between" type. According to the second theory, there is no intermediate mechanism at all, and borderline behavior is caused by simultaneous operation, in the same flask, of both the SN1 and SN2 mechanisms; that is, some molecules react by the SN1, while others react by the SN2 mechanism.

One formulation of the intermediate-mechanism theory is that of Sneen.[55] The formulation is in fact very broad and applies not only to borderline behavior but to all nucleophilic substitutions at a saturated carbon.[56] According to Sneen, all SN1 and SN2 reactions can be accommodated by one basic mechanism (the *ion-pair mechanism*). The substrate first ionizes to an intermediate ion pair which is then converted to products:

$$\mathbf{RX} \underset{}{\overset{k_1}{\rightleftharpoons}} \mathbf{R^+ \ X^-} \overset{k_2}{\rightleftharpoons} \text{products}$$

The difference between the SN1 and SN2 mechanisms is that in the former case the *formation* of the ion pair (k_1) is rate-determining, while in the SN2 mechanism its *destruction* (k_2) is rate determining. Borderline behavior is found where the rates of formation and destruction of the ion pair are of the same order of magnitude.[57] However, a number of investigators have asserted that these results could also be explained in other ways.[58]

There is evidence for the Sneen formulation where the leaving group has a positive charge. In this case there is a cation-molecule pair ($RX^+ \rightarrow R^+ \ X$)[59] instead of the ion pair that would be present if the leaving group were uncharged. Katritzky, le Noble, and co-workers found that when such a reaction was run at varying high pressures, there was a minimum in the plot of rate constant vs. pressure.[60] A minimum of this sort usually indicates a change in mechanism, and the interpretation in this case was that the normal SN2 mechanism operates at higher pressures and the cation–molecule mechanism at lower pressures.

---

[54]For an essay on borderline mechanisms in general, see Jencks *Chem. Soc. Rev.* **1982**, *10*, 345-375.

[55]Weiner; Sneen *J. Am. Chem. Soc.* **1965**, *87*, 292; Sneen; Larsen *J. Am. Chem. Soc.* **1969**, *91*, 362, 6031; Sneen; Felt; Dickason *J. Am. Chem. Soc.* **1973**, *95*, 638; Sneen *Acc. Chem. Res.* **1973**, *6*, 46-53.

[56]Including substitution at an allylic carbon; see Sneen; Bradley *J. Am. Chem. Soc.* **1972**, *94*, 6975; Sneen; Carter *J. Am. Chem. Soc.* **1972**, *94*, 6990; Bordwell; Mecca *J. Am. Chem. Soc.* **1975**, *97*, 123, 127; Bordwell; Wiley; Mecca *J. Am. Chem. Soc.* **1975**, *97*, 132; Kevill; Degenhardt *J. Am. Chem. Soc.* **1979**, *101*, 1465.

[57]For evidence for this point of view, see Ref. 55; Sneen; Carter; Kay *J. Am. Chem. Soc.* **1966**, *88*, 2594; Sneen; Robbins *J. Am. Chem. Soc.* **1972**, *94*, 7868; Graczyk; Taylor *J. Am. Chem. Soc.* **1974**, *96*, 3255; Peeters; Anteunis *J. Org. Chem.* **1975**, *40*, 312; Pross; Aronovitch; Koren *J. Chem. Soc., Perkin Trans. 2* **1978**, 197; Blandamer; Robertson; Scott; Vrielink *J. Am. Chem. Soc.* **1980**, *102*, 2585; Stein; Tencer; Moffatt; Dawe; Sweet *J. Org. Chem.* **1980**, *45*, 3539; Stein; Moffatt *Can. J. Chem.* **1985**, *63*, 3433; Stein *Can. J. Chem.* **1987**, *65*, 363.

[58]See, for example, Gregory; Kohnstam; Queen; Reid *Chem. Commun.* **1971**, 797; Kurz; Harris *J. Am. Chem. Soc.* **1970**, *92*, 4117; Raber; Harris; Hall; Schleyer *J. Am. Chem. Soc.* **1971**, *93*, 4821; McLennan *J. Chem. Soc., Perkin Trans. 2* **1972**, 1577, **1974**, 481; *Acc. Chem. Res.* **1976**, *9*, 281-287, *Tetrahedron Lett.* **1975**, 4689; McLennan; Martin *Tetrahedron Lett.* **1973**, 4215; Raaen; Juhlke; Brown; Collins *J. Am. Chem. Soc.* **1974**, *96*, 5928; Gregoriou *Tetrahedron Lett.* **1974**, 233, **1976**, 4605, 4767; Queen; Matts *Tetrahedron Lett.* **1975**, 1503; Stein *J. Org. Chem.* **1976**, *41*, 519; Stephan *Bull. Soc. Chim. Fr.* **1977**, 779; Katritzky; Musumarra; Sakizadeh *J. Org. Chem.* **1981**, *46*, 3831. For a reply to some of these objections, see Sneen; Robbins, Ref. 57. For a discussion, see Klumpp *Reactivity in Organic Chemistry*; Wiley: New York, 1982, pp. 442-450.

[59]For ion–molecule pairs in other solvolysis reactions, see Thibblin *J. Chem. Soc., Perkin Trans. 2* **1987**, 1629.

[60]Katritzky; Sakizadeh; Gabrielsen; le Noble *J. Am. Chem. Soc.* **1984**, *106*, 1879.

An alternative view that also favors an intermediate mechanism is that of Schleyer and co-workers,[61] who believe that the key to the problem is varying degrees of nucleophilic solvent assistance to ion-pair formation. They have proposed an SN2 (intermediate) mechanism.[62]

Among the experiments that have been cited for the viewpoint that borderline behavior results from simultaneous SN1 and SN2 mechanisms is the behavior of 4-methoxybenzyl chloride in 70% aqueous acetone.[63] In this solvent, hydrolysis (that is, conversion to 4-methoxybenzyl alcohol) occurs by an SN1 mechanism. When azide ions are added, the alcohol is still a product, but now 4-methoxybenzyl azide is another product. Addition of azide ions increases the rate of ionization (by the salt effect) but *decreases* the rate of hydrolysis. If more carbocations are produced but fewer go to the alcohol, then some azide must be formed by reaction with carbocations—an SN1 process. However, the rate of ionization is always *less* than the total rate of reaction, so some azide must also form by an SN2 mechanism.[63] Thus, the conclusion is that SN1 and SN2 mechanisms operate simultaneously.[64]

Some nucleophilic substitution reactions that seem to involve a "borderline" mechanism actually do not. Thus, one of the principal indications that a "borderline" mechanism is taking place has been the finding of partial racemization and partial inversion. However, Weiner and Sneen have demonstrated that this type of stereochemical behavior is quite consistent with a strictly SN2 process. These workers studied the reaction of optically active 2-octyl brosylate in 75% aqueous dioxane, under which conditions inverted 2-octanol was obtained in 77% optical purity.[65] When sodium azide was added, 2-octyl azide was obtained along with the 2-octanol, *but the latter was now 100% inverted*. It is apparent that, in the original case, 2-octanol was produced by two different processes: an SN2 reaction leading to inverted product, and another process in which some intermediate leads to racemization or retention. When azide ions were added, they scavenged this intermediate, so that the entire second process now went to produce azide, while the SN2 reaction, unaffected by addition of azide, still went on to give inverted 2-octanol. What is the nature of the intermediate in the second process? At first thought we might suppose that it is a carbocation, so that this would be another example of simultaneous SN1 and SN2 reactions. However, solvolysis of 2-octyl brosylate in pure methanol or of 2-octyl methanesulfonate in pure water, in the absence of azide ions, gave methyl 2-octyl ether or 2-octanol, respectively, *with 100% inversion of configuration*, indicating that the mechanism in these solvents was pure SN2. Since methanol and water are more polar than 75% aqueous dioxane and since an increase in polarity of solvent increases the rate of SN1 reactions at the expense of SN2 (p. 356), it is extremely unlikely that any SN1 process could occur in 75% aqueous dioxane. The intermediate in the second process is thus not a carbocation. What it is is suggested by the fact that, in the absence of azide ions, the amount of inverted 2-octanol decreased with an

[61]Bentley; Schleyer *J. Am. Chem. Soc.* **1976,** *98,* 7658; Bentley; Bowen; Morten; Schleyer *J. Am. Chem. Soc.* **1981,** *103,* 5466.

[62]For additional evidence for this view, see Laureillard; Casadevall; Casadevall *Tetrahedron* **1984,** *40,* 4921, *Helv. Chim. Acta* **1984,** *67,* 352; McLennan *J. Chem. Soc., Perkin Trans. 2* **1981,** 1316. For evidence against the SN2(intermediate) mechanism, see Allen; Kanagasabapathy; Tidwell *J. Am. Chem. Soc.* **1985,** *107,* 4513; Fărcaşiu; Jähme; Rüchardt *J. Am. Chem. Soc.* **1985,** *107,* 5717; Dietze; Jencks *J. Am. Chem. Soc.* **1986,** *108,* 4549; Dietze; Hariri; Khattak *J. Org. Chem.* **1989,** *54,* 3317; Coles; Maskill *J. Chem. Soc., Perkin Trans. 2* **1987,** 1083; Richard; Amyes; Vontor *J. Am. Chem. Soc.* **1991,** *113,* 5871.

[63]Kohnstam; Queen; Shillaker *Proc. Chem. Soc.* **1959,** 157; Amyes; Richard *J. Am. Chem. Soc.* **1990,** *112,* 9507. For other evidence supporting the concept of simultaneous mechanisms, see Pocker *J. Chem. Soc.* **1959,** 3939, 3944; Casapieri; Swart *J. Chem. Soc.* **1961,** 4342, **1963,** 1254; Ceccon; Papa; Fava *J. Am. Chem. Soc.* **1966,** *88,* 4643; Okamoto; Uchida; Saitô; Shingu *Bull. Chem. Soc. Jpn.* **1966,** *39,* 307; Guinot; Lamaty *Chem. Commun.* **1967,** 960; Queen *Can. J. Chem.* **1979,** *57,* 2646; Katritzky; Musumarra; Sakizadeh; El-Shafie; Jovanovic *Tetrahedron Lett.* **1980,** *21,* 2697; Richard; Rothenberg; Jencks *J. Am. Chem. Soc.* **1984,** *106,* 1361; Richard; Jencks *J. Am. Chem. Soc.* **1984,** *106,* 1373, 1383; Katritzky; Brycki *J. Phys. Org. Chem.* **1988,** *1,* 1; Stein *Can. J. Chem.* **1989,** *67,* 297.

[64]These data have also been explained as being in accord with the ion-pair mechanism: Sneen; Larsen *J. Am. Chem. Soc.* **1969,** *91,* 6031.

[65]Weiner; Sneen *J. Am. Chem. Soc.* **1965,** *87,* 287.

increasing percentage of dioxane in the solvent. Thus the intermediate is an oxonium ion formed by an S_N2 attack *by dioxane*. This ion is not a stable product but reacts with water in another S_N2 process to produce 2-octanol with retained configuration. The entire process can be shown as follows:

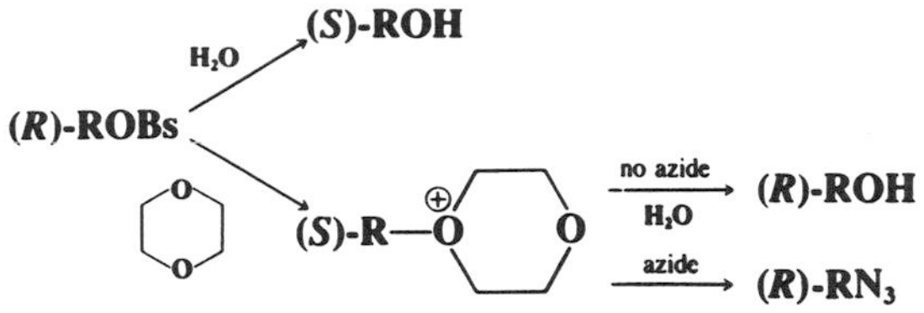

That part of the original reaction that resulted in retention of configuration[66] is thus seen to stem from two successive S_N2 reactions and not from any "borderline" behavior.[67]

SET Mechanisms

In certain reactions where nucleophilic substitutions would seem obviously indicated, there is evidence that radicals and/or radical ions are actually involved.[68] The first step in such a process is transfer of an electron from the nucleophile to the substrate to form a radical anion:

Step 1 $R—X + \overline{Y}^- \longrightarrow R—X^{\cdot} + Y\cdot$

Mechanisms that begin this way are called *SET (single electron transfer) mechanisms*.[69] Once formed, the radical ion cleaves:

Step 2 $R—X^{\underline{\cdot}} \longrightarrow R\cdot + \overline{X}^-$

The radicals formed in this way can go on to product by reacting with the Y· produced in Step 1 or with the original nucleophilic ion Y⁻, in which case an additional step is necessary:

Step 3 $R\cdot + Y\cdot \longrightarrow R—Y$

or

Step 3 $R\cdot + \overline{Y}^- \longrightarrow R—Y^{\underline{\cdot}}$

Step 4 $R—Y^{\underline{\cdot}} + R—X \longrightarrow R—Y + R—X^{\underline{\cdot}}$

In the latter case, the radical ion $R—X^{\underline{\cdot}}$ is formed by Step 4 as well as by Step 1, so that a chain reaction (p. 678) can take place.

[66]According to this scheme, the configuration of the isolated RN₃ should be retained. It was, however, largely inverted, owing to a competing S_N2 reaction where N₃⁻ directly attacks ROBs.

[67]For other examples, see Streitwieser; Walsh; Wolfe *J. Am. Chem. Soc.* **1965,** *87,* 3682; Streitwieser; Walsh *J. Am. Chem. Soc.* **1965,** *87,* 3686; Beronius; Nilsson; Holmgren *Acta Chem. Scand.* **1972,** *26,* 3173. See also Knier; Jencks *J. Am. Chem. Soc.* **1980,** *102,* 6789.

[68]Kerber; Urry; Kornblum *J. Am. Chem. Soc.* **1965,** *87,* 4520; Kornblum; Michel; Kerber *J. Am. Chem. Soc.* **1966,** *88,* 5660, 5662; Russell; Danen *J. Am. Chem. Soc.* **1966,** *88,* 5663; Bank; Noyd *J. Am. Chem. Soc.* **1973,** *95,* 8203; Ashby; Goel; Park *Tetrahedron Lett.* **1981,** *22,* 4209. For discussions of the relationship between S_N2 and SET mechanisms, see Lewis *J. Am. Chem. Soc.* **1989,** *111,* 7576; Shaik *Acta Chem. Scand.* **1990,** *44,* 205-221.

[69]For reviews, see Savéant *Adv. Phys. Org. Chem.* **1990,** *26,* 1-130; Rossi; Pierini; Palacios *J. Chem. Educ.* **1989,** *66,* 720; Ashby *Acc. Chem. Res.* **1988,** *21,* 414-421; Chanon; Tobe *Angew. Chem. Int. Ed. Engl.* **1982,** *21,* 1-23 [*Angew. Chem.* **94,** 27-49]. See also Pross *Acc. Chem. Res.* **1985,** *18,* 212-219; Chanon *Acc. Chem. Res.* **1987,** *20,* 214-221.

One type of evidence for an SET mechanism is the finding of some racemization. A totally free radical would of course result in a completely racemized product RY, but it has been suggested[70] that inversion can also take place in some SET processes. The suggestion is that in Step 1 the Y^- still approaches from the back side, even though an ordinary SN2 mechanism will not follow, and that the radical R•, once formed, remains in a solvent cage with Y• still opposite X^-, so that Steps 1, 2, and 3 can lead to inversion.

$$\overline{Y}^- + R—X \longrightarrow [Y•\ R—X^{\ddot{-}}]_{\text{solvent}\atop\text{cage}} \longrightarrow [Y•\ R•\ X^-]_{\text{solvent}\atop\text{cage}} \longrightarrow Y—R + X^-$$

Reactions with SET mechanisms typically show predominant, though not 100%, inversion.

Other evidence cited[71] for SET mechanisms has been detection of radical or radical ion intermediates by esr[72] or CIDNP; the finding that such reactions can take place at 1-norbornyl bridgeheads;[73] and the formation of cyclic side products when the substrate has a double bond in the 5,6 position (such substrates are called *radical probes*).

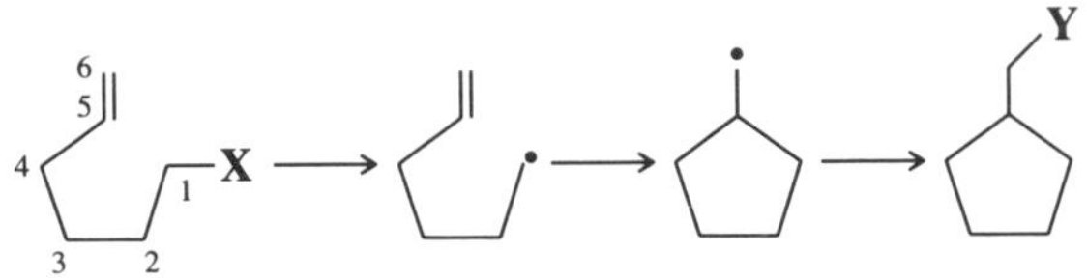

Free radicals with double bonds in this position are known to cyclize readily (p. 744).[74]

The SET mechanism is chiefly found where X = I or NO_2 (see **0-94**). A closely related mechanism, the SRN1, takes place with aromatic substrates (Chapter 13).[75] In that mechanism the initial attack is by an electron donor, rather than a nucleophile.

The mechanisms so far considered can, in theory at least, operate on any type of saturated (or for that matter unsaturated) substrate. There are other mechanisms that are more limited in scope.

The Neighboring-Group Mechanism[76]

It is occasionally found with certain substrates that (1) the rate of reaction is greater than expected, and (2) the configuration at a chiral carbon is *retained* and not inverted or racemized. In these cases there is usually a group with an unshared pair of electrons β to the leaving group (or sometimes farther away). The mechanism operating in such cases is called the *neighboring-group mechanism* and consists essentially of two SN2 substitutions, each

---

[70]Ashby; Pham *Tetrahedron Lett.* **1987**, *28*, 3183; Daasbjerg; Lund; Lund *Tetrahedron Lett.* **1989**, *30*, 493.

[71]See also Chanon; Tobe, Ref. 69; Fuhlendorff; Lund; Lund; Pedersen *Tetrahedron Lett.* **1987**, *28*, 5335.

[72]See, for example Russell; Pecoraro *J. Am. Chem. Soc.* **1979**, *101*, 3331.

[73]Santiago; Morris; Rossi *J. Chem. Soc., Chem. Commun.* **1988**, 220.

[74]For criticisms of this method for demonstrating SET mechanisms, see Newcomb; Kaplan *Tetrahedron Lett.* **1988**, *29*, 3449; Newcomb; Kaplan; Curran *Tetrahedron Lett.* **1988**, *29*, 3451; Newcomb; Curran *Acc. Chem. Res.* **1988**, *21*, 206-214; Newcomb *Acta Chem. Scand.* **1990**, *44*, 299. For replies to the criticism, see Ashby *Acc. Chem. Res.* **1988**, *21*, 414-421; Ashby; Pham; Amrollah-Madjdabadi *J. Org. Chem.* **1991**, *56*, 1596.

[75]In this book we make the above distinction between the SET and SRN1 mechanisms. However, many workers use the designation SET to refer to the SRN1, the chain version of the SET, or both.

[76]For a monograph, see Capon; McManus *Neighboring Group Participation*, vol. 1; Plenum: New York, 1976.

causing an inversion so the net result is retention of configuration.[77] In the first step of this reaction the neighboring group acts as a nucleophile, pushing out the leaving group but still retaining attachment to the molecule. In the second step the external nucleophile displaces the neighboring group by a backside attack:

Step 1

$$R-\overset{\overset{\displaystyle R}{|}}{\underset{\underset{\displaystyle R}{|}}{C}}\overset{\overset{\displaystyle Z|}{\diagdown}}{\underset{\underset{\displaystyle X}{\diagup}}{\quad}}\overset{\overset{\displaystyle R}{|}}{\underset{\underset{\displaystyle R}{|}}{C}}-R \longrightarrow R-\overset{\overset{\displaystyle Z^{\oplus}}{\diagup\diagdown}}{\underset{\underset{\displaystyle R}{|}}{C}}\overset{}{\underset{\underset{\displaystyle R}{|}}{C}}-R \;+\; X^{-}$$

Step 2

$$R-\overset{\overset{\displaystyle Z^{\oplus}}{\diagup\diagdown}}{\underset{\underset{\displaystyle R}{|}}{C}}\overset{}{\underset{\underset{\displaystyle R}{|}}{C}}-R + \overline{Y} \longrightarrow R-\overset{\overset{\displaystyle Z|}{|}}{\underset{\underset{\displaystyle R}{|}}{C}}\overset{\overset{\displaystyle R}{|}}{\underset{\underset{\displaystyle Y}{|}}{C}}-R$$

The reaction obviously must go faster than if Y were attacking directly, since if the latter process were faster, *it* would be happening. The neighboring group Z is said to be lending *anchimeric assistance.* The rate law followed in the neighboring-group mechanism is the first-order law shown in Eq. (2) or (3); that is, Y does not take part in the rate-determining step.

The reason attack by Z is faster than that by Y is that the group Z is more available. In order for Y to react, it must collide with the substrate, but Z is immediately available by virtue of its position. A reaction between the substrate and Y involves a large decrease in entropy of activation (ΔS^{*}), since the reactants are far less free in the transition state than before. Reaction of Z involves a much smaller loss of ΔS^{*} (see p. 211).[78]

It is not always easy to determine when a reaction rate has been increased by anchimeric assistance. In order to be certain, it is necessary to know what the rate would be without participation by the neighboring group. The obvious way to examine this question is to compare the rates of the reaction with and without the neighboring group, for example, $HOCH_2CH_2Br$ vs. CH_3CH_2Br. However, this will certainly not give an accurate determination of the extent of participation, since the steric and field effects of H and OH are not the same. Futhermore, no matter what the solvent, the shell of solvent molecules that surrounds the polar protic OH group must differ greatly from that which surrounds the nonpolar H. Because of these considerations, it is desirable to have a large increase in the rate, preferably more than fiftyfold, before a rate increase is attributed to neighboring-group participation.

The first important evidence for the existence of this mechanism was the demonstration that retention of configuration can occur if the substrate is suitable. It was shown that the threo *dl* pair of 3-bromo-2-butanol when treated with HBr gave *dl*-2,3-dibromobutane, while the erythro pair gave the meso isomer:[79]

---

[77]There is evidence that this kind of process can happen intermolecularly (e.g., RX + Z⁻ → RZ + Y⁻). In this case Z⁻ acts as a catalyst for the reaction RX + Y⁻ → RY: McCortney; Jacobson; Vreeke; Lewis *J. Am. Chem. Soc.* **1990**, *112*, 3554.

[78]For a review of the energetics of neighboring-group participation, see Page *Chem. Soc. Rev.* **1973**, *2*, 295-323.

[79]Winstein; Lucas *J. Am. Chem. Soc.* **1939**, *61*, 1576, 2845.

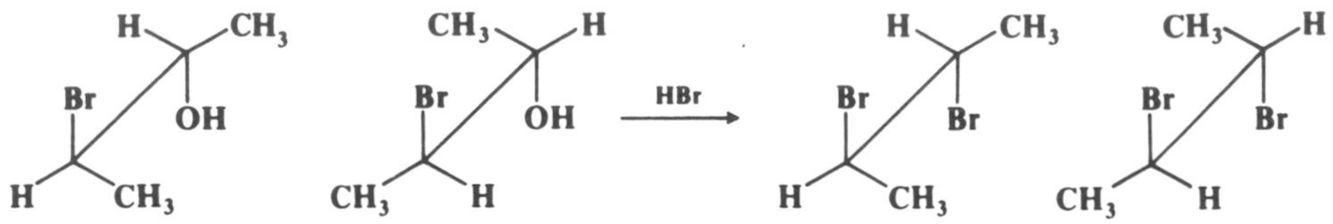

threo *dl* pair   dl pair

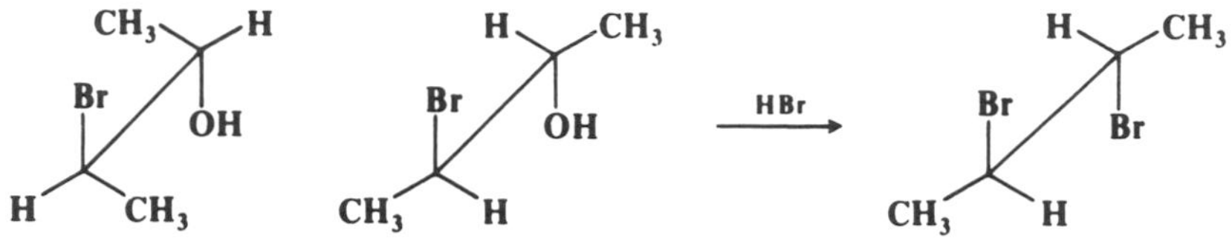

erythro *dl* pair   meso

This indicated that retention had taken place. Note that both products are optically inactive and so cannot be told apart by differences in rotation. The meso and *dl* dibromides have different boiling points and indexes of refraction and were identified by these properties. Even more convincing evidence was that either of the two threo isomers alone gave not just one of the enantiomeric dibromides, but the *dl* pair. The reason for this is that the intermediate present after the attack by the neighboring group (**15**) is symmetrical, so the external

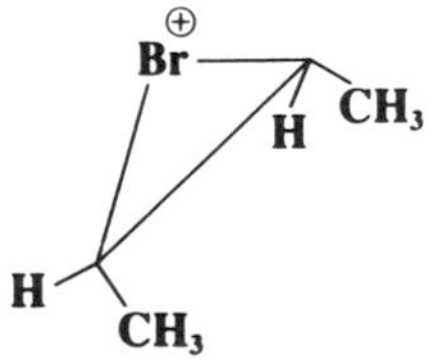

15

nucleophile Br⁻ can attack both carbon atoms equally well. **15** is a bromonium ion, the existence of which has been demonstrated in several types of reactions.

Although **15** is symmetrical, intermediates in most neighboring-group mechanisms are not, and it is therefore possible to get not a simple substitution product but a rearrangement. This will happen if Y attacks not the carbon atom from which X left, but the one to which Z was originally attached:

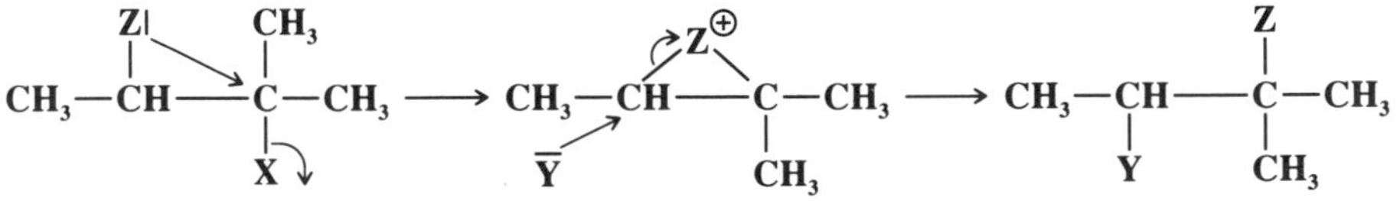

In such cases substitution and rearrangement products are often produced together. For a discussion of rearrangements, see Chapter 18.

Another possibility is that the intermediate may be stable or may find some other way to stabilize itself. In such cases, Y never attacks at all and the product is cyclic. These are simple internal S_N2 reactions. Two examples are formation of epoxides and lactones:

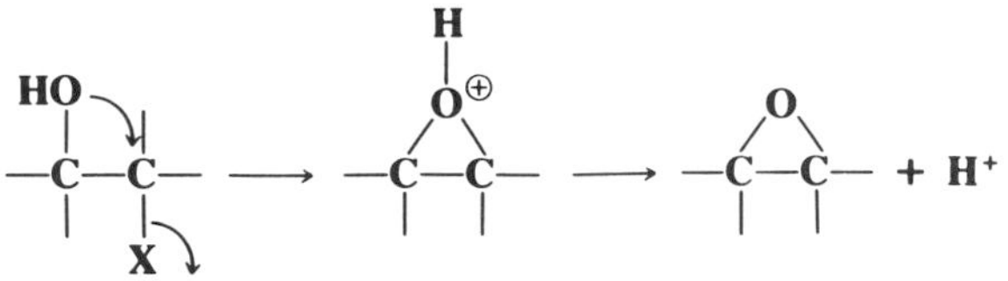

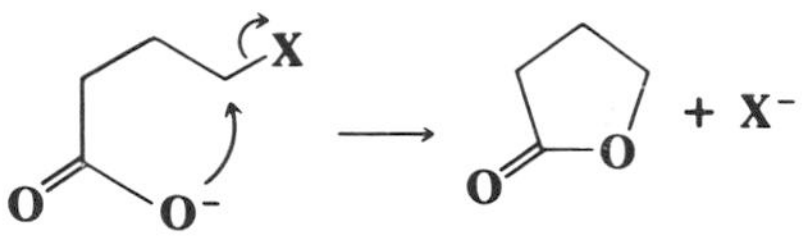

The fact that acetolysis of both 4-methoxy-1-pentyl brosylate (**16**) and 5-methoxy-2-pentyl brosylate (**17**) gave the same mixture of products is further evidence for participation by a

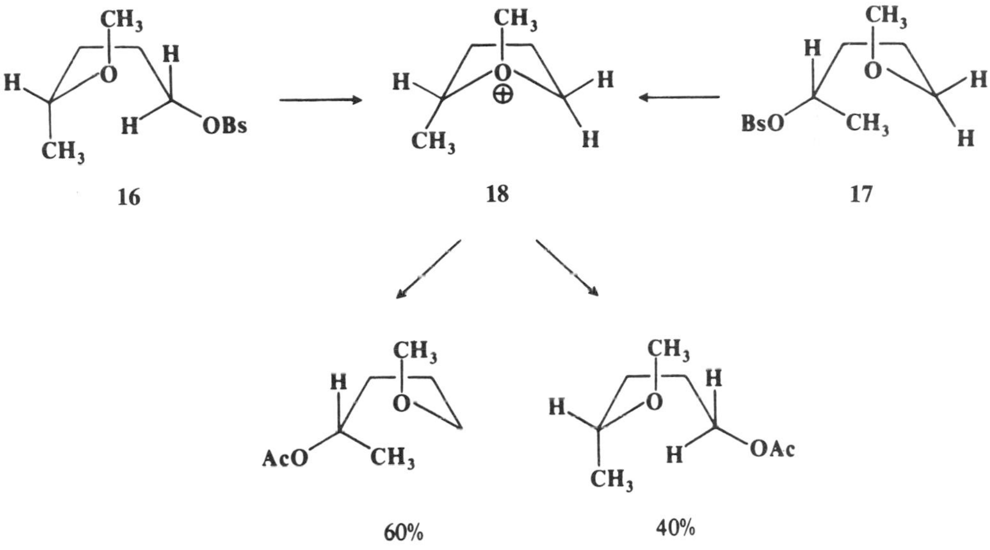

neighboring group.[80] In this case the intermediate **18** is common to both substrates.

The neighboring-group mechanism operates only when the ring size is right for a particular type of Z. For example, for $MeO(CH_2)_nOBs$, neighboring-group participation was important for $n = 4$ or 5 (corresponding to a five- or six-membered intermediate) but not for $n = 2$, 3, or 6.[81] However, optimum ring size is not the same for all reactions, even with a particular Z. In general, the most rapid reactions occur when the ring size is three, five, or six, depending on the reaction type. The likelihood of four-membered ring neighboring-group participation is increased when there are alkyl groups α or β to the neighboring group.[82]

The following are some of the more important neighboring groups: COO^- (but not COOH), COOR, COAr, OCOR,[83] OR, OH, O^-,[84] NH_2, NHR, NR_2, NHCOR, SH, SR,

[80]Allred; Winstein *J. Am. Chem. Soc.* **1967**, *89*, 3991, 3998.

[81]Winstein; Allred; Heck; Glick *Tetrahedron* **1958**, *3*, 1; Allred; Winstein *J. Am. Chem. Soc.* **1967**, *89*, 4012.

[82]Eliel; Clawson; Knox *J. Org. Chem.* **1985**, *50*, 2707; Eliel; Knox *J. Am. Chem. Soc.* **1985**, *107*, 2946.

[83]For an example of OCOR as a neighboring group where the ring size is seven-membered, see Wilen; Delguzzo; Saferstein *Tetrahedron* **1987**, *43*, 5089.

[84]For a review of oxygen functions as neighboring groups, see Perst *Oxonium Ions in Organic Chemistry*; Verlag Chemie: Deerfield Beach, FL, 1971, pp. 100-127. There is evidence that the oxygen in an epoxy group can also act as a neighboring group: Francl; Hansell; Patel; Swindell *J. Am. Chem. Soc.* **1990**, *112*, 3535.

S⁻,[85] I, Br, and Cl. The effectiveness of halogens as neighboring groups decreases in the order I > Br > Cl.[86] Cl is a very weak neighboring group and can be shown to act in this way only when the solvent does not interfere. For example, when 5-chloro-2-hexyl tosylate is solvolyzed in acetic acid, there is little participation by the Cl, but when the solvent is changed to trifluoroacetic acid, which is much less nucleophilic, neighboring-group participation by the Cl becomes the major reaction pathway.[87] Thus, Cl acts as a neighboring group *only when there is need for it* (for other examples of the *principle of increasing electron demand*, see below; p. 315).

A number of intermediates of halogen participation (halonium ions),[88] e.g., **19** and **20**, have been prepared as stable salts in SbF_5–SO_2 or SbF_5–SO_2ClF solutions.[89] Some have even

$$\overset{\oplus}{X}\ SbF_6^-$$

19

$$\overset{\oplus}{X}\ SbF_6^- \qquad X = Br,\ Cl,\ or\ I$$

20

been crystallized. Attempts to prepare four-membered homologs of **19** and **20** were not successful.[90] There is no evidence that F can act as a neighboring group.[86]

The principle that a neighboring group lends assistance in proportion to the need for such assistance also applies to differences in leaving-group ability. Thus, p-$NO_2C_6H_4SO_2O$ (the nosylate group) is a better leaving group than p-$MeC_6H_4SO_2O$ (the tosylate group). Experiments have shown that the OH group in *trans*-2-hydroxycyclopentyl arenesulfonates:

$$\text{OH}\quad\text{H}$$
$$\text{H}\quad\text{OSO}_2\text{Ar}$$

acts as a neighboring group when the leaving group is tosylate but not when it is nosylate, apparently because the nosylate group leaves so rapidly that it does not require assistance.[91]

Neighboring-Group Participation by π and σ Bonds. Nonclassical Carbocations[92]

For all the neighboring groups listed in the preceding section, the nucleophilic attack is made by an atom with an unshared pair of electrons. In this section we consider neighboring-

[85]For a review of sulfur-containing neighboring groups, see Block *Reactions of Organosulfur Compounds*; Academic Press: New York, 1978, pp. 141-145.

[86]Peterson *Acc. Chem. Res.* **1971**, *4*, 407-413, and references cited therein.

[87]Peterson; Bopp; Chevli; Curran; Dillard; Kamat *J. Am. Chem. Soc.* **1967**, *89*, 5902. See also Reich; Reich *J. Am. Chem. Soc.* **1974**, *96*, 2654.

[88]For a monograph, see Olah *Halonium Ions*; Wiley: New York, 1975. For a review, see Koster, in Patai; Rappoport *The Chemistry of Functional Groups, Supplement D*, pt. 2; Wiley: New York, 1983, pp. 1265-1351.

[89]See, for example Olah; Bollinger *J. Am. Chem. Soc.* **1967**, *89*, 4744, **1968**, *90*, 947; Olah; Peterson *J. Am. Chem. Soc.* **1968**, *90*, 4675; Peterson; Clifford; Slama *J. Am. Chem. Soc.* **1970**, *92*, 2840; Bonazza; Peterson *J. Org. Chem.* **1973**, *38*, 1015; Henrichs; Peterson *J. Am. Chem. Soc.* **1973**, *95*, 7449, *J. Org. Chem.* **1976**, 41, 362; Olah; Liang; Staral *J. Am. Chem. Soc.* **1974**, *96*, 8112; Vančik; Percač; Sunko *J. Chem. Soc., Chem. Commun.* **1991**, 807.

[90]Olah; Bollinger; Mo; Brinich *J. Am. Chem. Soc.* **1972**, *94*, 1164.

[91]Haupt; Smith *Tetrahedron Lett.* **1974**, 4141.

[92]For monographs, see Olah; Schleyer *Carbonium Ions*, vol. 3; Wiley: New York, 1972; Bartlett *Nonclassical Ions*; W.A. Benjamin: New York, 1965. For reviews, see Barkhash *Top. Curr. Chem.* **1984**, *116/117*, 1-265; Kirmse *Top. Curr. Chem.* **1979**, *80*, 125-311, pp. 196-288; McManus; Pittman, in McManus *Organic Reactive Intermediates*; Academic Press: New York, 1973, pp. 302-321; Bethell; Gold *Carbonium Ions*; Academic Press: New York, 1967; pp. 222-282. For a related review, see Prakash; Iyer *Rev. Chem. Intermed.* **1988**, *9*, 65-116.

group participation by C=C π bonds and C—C and C—H σ bonds. There has been a great deal of controversy over whether such bonds can act as neighboring groups and about the existence and structure of the intermediates involved. These intermediates are called *nonclassical* (or *bridged*) carbocations. In classical carbocations (Chapter 5) the positive charge is localized on one carbon atom or delocalized by resonance involving an unshared pair of electrons or a double or triple bond in the allylic position. In a nonclassical carbocation, the positive charge is delocalized by a double or triple bond that is not in the allylic position or by a single bond. Examples are the 7-norbornenyl cation (**21**), the norbornyl cation (**22**),

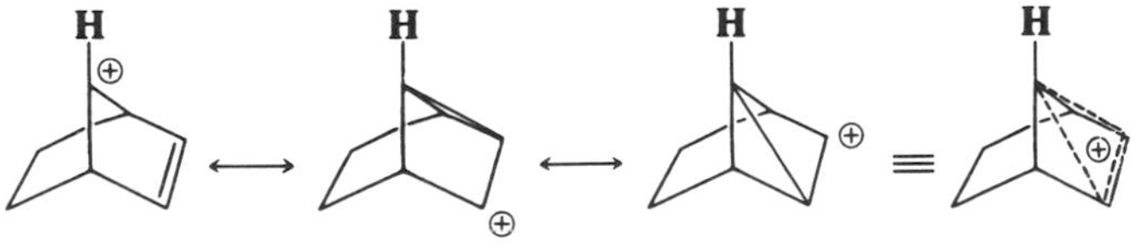

| 21a | 21b | 21c | 21d |

| 22a | 22b | 22c |

| 23a | 23b | 23c |

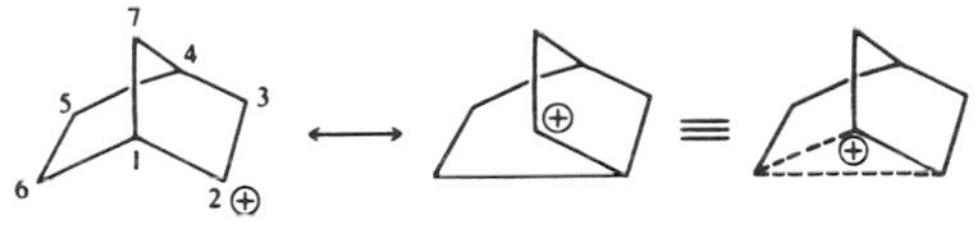

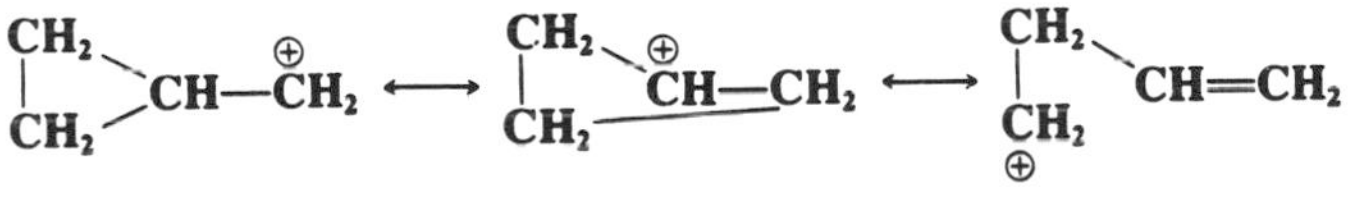

and the cyclopropylmethyl cation (**23**). **21** is called a *homoallylic* carbocation, because in **21a** there is one carbon atom between the positively charged carbon and the double bond. Many of these carbocations can be produced in more than one way if the proper substrates are chosen. For example, **22** can be generated by the departure of a leaving group from **24**

| 24 | 22 | 25 |

or from **25**.[93] The first of these pathways is called the σ route to a nonclassical carbocation, because participation of a σ bond is involved. The second is called the π route.[94] The argument against the existence of nonclassical carbocations is essentially that the structures **21a, 21b, 21c** (or **22a, 22b,** etc.) are not canonical forms but real structures and that there is rapid equilibration among them.

In discussing nonclassical carbocations we must be careful to make the distinction between neighboring-group participation and the existence of nonclassical carbocations.[95] If a nonclassical carbocation exists in any reaction, then an ion with electron delocalization, as shown

[93]Lawton *J. Am. Chem. Soc.* **1961**, *83*, 2399; Bartlett; Bank; Crawford; Schmid *J. Am. Chem. Soc.* **1965**, *88*, 1288.
[94]Winstein; Carter *J. Am. Chem. Soc.* **1961**, *83*, 4485.
[95]This was pointed out by Cram *J. Am. Chem. Soc.* **1964**, *86*, 3767.

in the above examples, is a discrete reaction intermediate. If a carbon-carbon double or single bond participates in the departure of the leaving group to form a carbocation, it may be that a nonclassical carbocation is involved, but there is no necessary relation. In any particular case either or both of these possibilities can be taking place.

In the following pages we consider some of the evidence bearing on the questions of the participation of π and σ bonds and on the existence of nonclassical carbocations,[96] though a thorough discussion is beyond the scope of this book.[78]

1. C=C *as a neighboring group.*[97] The most striking evidence that C=C can act as a neighboring group is that acetolysis of **26**-OTs is 10^{11} times faster than that of **27**-OTs and *proceeds with retention of configuration.*[98] The rate data alone do not necessarily prove that

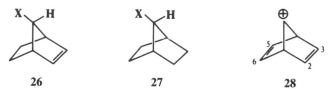

acetolysis of **26**-OTs involves a nonclassical intermediate (**21d**), but it is certainly strong evidence that the C=C group assists in the departure of the OTs. Evidence that **21** is indeed a nonclassical ion comes from an nmr study of the relatively stable norbornadienyl cation (**28**). The spectrum shows that the 2 and 3 protons are not equivalent to the 5 and 6 protons.[99] Thus there is interaction between the charged carbon and one double bond, which is evidence for the existence of **21d**.[100] In the case of **26** the double bond is geometrically fixed in an especially favorable position for backside attack on the carbon bearing the leaving group (hence the very large rate enhancement), but there is much evidence that other double bonds in the homoallylic position,[101] as well as in positions farther away,[102] can also lend anchimeric assistance, though generally with much lower rate ratios. One example of the latter is the compound β-(*syn*-7-norbornenyl)ethyl brosylate (**29**) which at 25°C undergoes

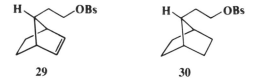

[96]The arguments against nonclassical ions are summed up in Brown *The Nonclassical Ion Problem*; Plenum: New York, 1977. This book also includes rebuttals by Schleyer. See also Brown *Pure Appl. Chem.* **1982**, *54*, 1783-1796.

[97]For reviews, see Story; Clark, in Olah; Schleyer, Ref. 92, vol. 3, 1972, pp. 1007-1060; Richey, in Zabicky *The Chemistry of Alkenes*, vol. 2; Wiley: New York, 1970, pp. 77-101.

[98]Winstein; Shatavsky *J. Am. Chem. Soc.* **1956**, *78*, 592.

[99]Story; Saunders *J. Am. Chem. Soc.* **1962**, *84*, 4876; Story; Snyder; Douglass; Anderson; Kornegay *J. Am. Chem. Soc.* **1963**, *85*, 3630. For a discussion, see Story; Clark, Ref. 97, pp. 1026-1041. See also Lustgarten; Brookhart; Winstein *J. Am. Chem. Soc.* **1972**, *94*, 2347.

[100]For further evidence for the nonclassical nature of **21**, see Winstein; Ordronneau *J. Am. Chem. Soc.* **1960**, *82*, 2084; Brookhart; Diaz; Winstein *J. Am. Chem. Soc.* **1966**, *88*, 3135; Richey; Lustgarten *J. Am. Chem. Soc.* **1966**, *88*, 3136; Gassman; Patton *J. Am. Chem. Soc.* **1969**, *91*, 2160; Richey; Nichols; Gassman; Fentiman; Winstein; Brookhart; Lustgarten *J. Am. Chem. Soc.* **1970**, *92*, 3783; Gassman; Doherty *J. Am. Chem. Soc.* **1982**, *104*, 3742; Laube *J. Am. Chem. Soc.* **1989**, *111*, 9224.

[101]For examples, see Shoppee *J. Chem. Soc.* **1946**, 1147; LeBel; Huber *J. Am. Chem. Soc.* **1963**, *85*, 3193; Closson; Kwiatkowski *Tetrahedron* **1965**, *21*, 2779; Cristol; Nachtigall *J. Am. Chem. Soc.* **1968**, *90*, 7132; Masamune; Takada; Nakatsuka; Vukov; Cain *J. Am. Chem. Soc.* **1969**, *91*, 4322; Hess *J. Am. Chem. Soc.* **1969**, *91*, 5657; Brown; Peters; Ravindranathan *J. Am. Chem. Soc.* **1975**, *97*, 7449; Lambert; Finzel *J. Am. Chem. Soc.* **1983**, *105*,1954; Schleyer; Bentley; Koch; Kos; Schwarz *J. Am. Chem. Soc.* **1987**, *109*, 6953.

[102]For examples, see LeNy *C. R. Acad. Sci.* **1960**, *251*, 1526; Goering; Closson *J. Am. Chem. Soc.* **1961**, *83*, 3511; Bartlett; Trahanovsky; Bolon; Schmid *J. Am. Chem. Soc.* **1965**, *87*, 1314; Bly; Swindell *J. Org. Chem.* **1965**, *30*, 10; Marvell; Sturmer; Knutson *J. Org. Chem.* **1968**, *33*, 2991; Cogdell *J. Org. Chem.* **1972**, *37*, 2541; Ferber; Gream *Aust. J. Chem.* **1981**, *34*, 1051; Kronja; Polla; Borčić *J. Chem. Soc., Chem. Commun.* **1983**, 1044; Orlović; Borčić; Humski; Kronja; Imper; Polla; Shiner *J. Org. Chem.* **1991**, *56*, 1874; Ref. 94.

acetolysis about 140,000 times faster than the saturated analog **30**.[103] Triple bonds[104] and allenes[105] can also act as neighboring groups.

We have already seen evidence that participation by a potential neighboring group can be reduced or eliminated if an outside nucleophile is present that is more effective than the neighboring group in attacking the central carbon (p. 312), or if a sufficiently good leaving group is present (p. 312). In another example of the principle of increasing electron demand, Gassman and co-workers have shown that neighboring-group participation can also be reduced if the stability of the potential carbocation is increased. They found that the presence of a *p*-anisyl group at the 7 position of **26** and **27** exerts a powerful leveling effect on the rate differences. Thus, solvolysis in acetone–water at 85°C of **31** was only about 2.5 times

31	**32**	**33** $Ar = p\text{-MeOC}_6\text{H}_4$

$$Ar' = p\text{-NO}_2\text{C}_6\text{H}_4$$

faster than that of the saturated compound **32**.[106] Furthermore, both **31** and its stereoisomer **33** gave the same mixture of solvolysis products, showing that the stereoselectivity in the solvolysis of **26** is not present here. The difference between **31** and **26** is that in the case of **31** the positive charge generated at the 7 position in the transition state is greatly stabilized by the *p*-anisyl group. Apparently the stabilization by the *p*-anisyl group is so great that further stabilization that would come from participation by the C=C bond is not needed.[107] The use of a phenyl instead of a *p*-anisyl group is not sufficient to stop participation by the double bond completely, though it does reduce it.[108] These results permit us to emphasize our previous conclusion that *a neighboring group lends anchimeric assistance only when there is sufficient demand for it.*[109]

The ability of C=C to serve as a neighboring group can depend on its electron density. When the strongly electron-withdrawing CF_3 group was attached to a double bond carbon of **34**, the solvolysis rate was lowered by a factor of about 10^6.[110] A second CF_3 group had

Relative Rates

$R^1 = R^2 = H$	1.4×10^{12}
$R^1 = H, R^2 = CF_3$	1.5×10^6
$R^1 = R^2 = CF_3$	1

$$Mos = MeO-\!\!\left\langle\bigcirc\right\rangle\!\!-SO_2-$$

34

[103]Bly; Bly; Bedenbaugh; Vail *J. Am. Chem. Soc.* **1967**, *89*, 880.

[104]See, for example, Closson; Roman *Tetrahedron Lett.* **1966**, 6015; Hanack; Herterich; Vött *Tetrahedron Lett.* **1967**, 3871; Lambert; Papay; Mark *J. Org. Chem.* **1975**, *40*, 633; Peterson; Vidrine *J. Org. Chem.* **1979**, *44*, 891. For a review of participation by triple bonds and allylic groups, see Rappoport *React. Intermed. (Plenum)* **1983**, *3*, 440-453.

[105]Jacobs; Macomber *Tetrahedron Lett.* **1967**, 4877; Bly; Koock *J. Am. Chem. Soc.* **1969**, *91*, 3292, 3299; Von Lehman; Macomber *J. Am. Chem. Soc.* **1975**, *97*, 1531.

[106]Gassman; Zeller; Lamb *Chem. Commun.* **1968**, 69.

[107]Nevertheless, there is evidence from ¹³C nmr spectra that some π participation is present, even in the cation derived from **31**: Olah; Berrier; Arvanaghi; Prakash *J. Am. Chem. Soc.* **1981**, *103*, 1122.

[108]Gassman; Fentiman *J. Am. Chem. Soc.* **1969**, *91*, 1545, **1970**, *92*, 2549.

[109]For a discussion of the use of the tool of increasing electron demand to probe neighboring-group activity by double bonds, sigma bonds, and aryl rings, see Lambert; Mark; Holcomb; Magyar *Acc. Chem. Res.* **1979**, *12*, 317-324.

[110]Gassman; Hall *J. Am. Chem. Soc.* **1984**, *106*, 4267.

an equally strong effect. In this case two CF_3 groups decrease the electron density of the C=C bond to the point that the solvolysis rate for **34** ($R^1 = R^2 = CF_3$) was about the same as (actually about 17 times slower than) the rate for the saturated substrate **27** (X = OMos). Thus, the two CF_3 groups completely remove the ability of the C=C bond to act as a neighboring group.

2. *Cyclopropyl*[111] *as a neighboring group.*[112]   On p. 152 we saw that the properties of a cyclopropane ring are in some ways similar to those of a double bond. Therefore it is not surprising that a suitably placed cyclopropyl ring can also be a neighboring group. Thus *endo-anti*-tricyclo-[3.2.1.0^{2,4}]octan-8-yl *p*-nitrobenzoate (**35**) solvolyzed about 10^{14} times

Ar = *p*-NO$_2$C$_6$H$_4$

35   **36**   **37**

faster that the *p*-nitrobenzoate of **27**-OH.[113] Obviously, a suitably placed cyclopropyl ring can be even more effective[114] as a neighboring group than a double bond.[115] The need for suitable placement is emphasized by the fact that **37** solvolyzed only about five times faster than **27**-OBs,[116] while **36** solvolyzed three times *slower* than **27**-OBs.[117] In the case of **35** and of all other cases known where cyclopropyl lends considerable anchimeric assistance, the developing *p* orbital of the carbocation is orthogonal to the participating bond of the cyclopropane ring.[118] An experiment designed to test whether a developing *p* orbital that would be parallel to the participating bond would be assisted by that bond showed no rate enhancement.[118] This is in contrast to the behavior of cyclopropane rings directly attached to positively charged carbons, where the *p* orbital is parallel to the plane of the ring (pp. 169, 324). Rate enhancements, though considerably smaller, have also been reported for suitably placed cyclobutyl rings.[119]

3. *Aromatic rings as neighboring groups.*[120]   There is a great deal of evidence that aromatic rings in the β position can function as neighboring groups. Stereochemical evidence

[111]In this section we consider systems in which at least one carbon separates the cyclopropyl ring from the carbon bearing the leaving group. For a discussion of systems in which the cyclopropyl group is directly attached to the leaving-group carbon, see p. 323.

[112]For a review, see Haywood-Farmer *Chem. Rev.* **1974**, *74*, 315-350.

[113]Tanida; Tsuji; Irie *J. Am. Chem. Soc.* **1967**, *89*, 1953; Battiste; Deyrup; Pincock; Haywood-Farmer *J. Am. Chem. Soc.* **1967**, *89*, 1954.

[114]For a competitive study of cyclopropyl vs. double-bond participation, see Lambert; Jovanovich; Hamersma; Koeng; Oliver *J. Am. Chem. Soc.* **1973**, *95*, 1570.

[115]For other evidence for anchimeric assistance by cyclopropyl, see Sargent; Lowry; Reich *J. Am. Chem. Soc.* **1967**, *89*, 5985; Battiste; Haywood-Farmer; Malkus; Seidl; Winstein *J. Am. Chem. Soc.* **1970**, *92*, 2144; Coates; Yano *Tetrahedron Lett.* **1972**, 2289; Masamune; Vukov; Bennett; Purdham *J. Am. Chem. Soc.* **1972**, *94*, 8239; Gassman; Creary *J. Am. Chem. Soc.* **1973**, *95*, 2729; Costanza; Geneste; Lamaty; Roque *Bull. Soc. Chim. Fr.* **1975**, 2358; Takakis; Rhodes *Tetrahedron Lett.* **1983**, *24*, 4959.

[116]Battiste; Deyrup; Pincock; Haywood-Farmer, Ref. 113; Haywood,-Farmer; Pincock *J. Am. Chem. Soc.* **1969**, *91*, 3020.

[117]Haywood-Farmer; Pincock; Wells *Tetrahedron* **1966**, *22*, 2007; Haywood-Farmer; Pincock, Ref. 116. For some other cases where there was little or no rate enhancement by cyclopropyl, see Wiberg; Wenzinger *J. Org. Chem.* **1965**, *30*, 2278; Sargent; Taylor; Demisch *Tetrahedron Lett.* **1968**, 2275; Rhodes; Takino *J. Am. Chem. Soc.* **1970**, *92*, 4469; Hanack; Krause *Liebigs Ann.* **1972**, *760*, 17.

[118]Gassman; Seter; Williams *J. Am. Chem. Soc.* **1971**, *93*, 1673. For a discussion, see Haywood-Farmer; Pincock, Ref. 116. See also Chenier; Jenson; Wulff *J. Org. Chem.* **1982**, *47*, 770.

[119]For example, see Sakai; Diaz; Winstein *J. Am. Chem. Soc.* **1970**, *92*, 4452; Battiste; Nebzydoski *J. Am. Chem. Soc.* **1970**, *92*, 4450; Schipper; Driessen; de Haan; Buck *J. Am. Chem. Soc.* **1974**, *96*, 4706; Ohkata; Doecke; Klein; Paquette *Tetrahedron Lett.* **1980**, *21*, 3253.

[120]For a review, see Lancelot; Cram; Schleyer, in Olah; Schleyer, Ref. 92, vol. 3, 1972, pp. 1347-1483.

was obtained by solvolysis of L-*threo*-3-phenyl-2-butyl tosylate (**38**) in acetic acid.[121] Of the acetate product 96% was the threo isomer and only about 4% was erythro. Moreover, both

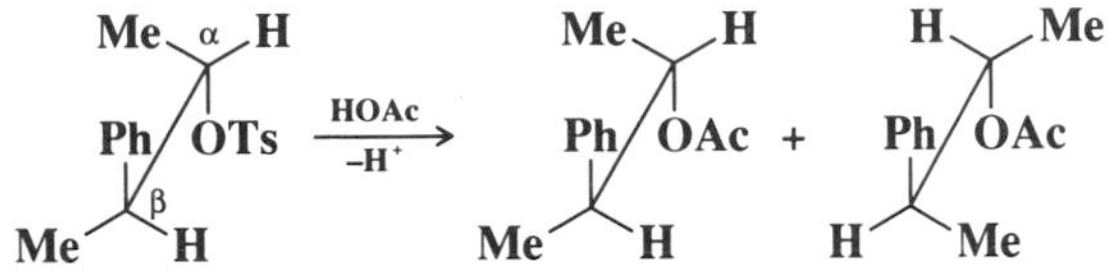

the (+) and (−) threo isomers (**39** and **40**) were produced in approximately equal amounts (a racemic mixture). When solvolysis was conducted in formic acid, even less erythro isomer was obtained. This result is similar to that found on reaction of 3-bromo-2-butanol with HBr (p. 309) and leads to the conclusion that configuration is retained because phenyl acts as a neighboring group. However, evidence from rate studies is not so simple. If β-aryl groups assist the departure of the leaving group, solvolysis rates should be enhanced. In general they are not. However, solvolysis rate studies in 2-arylethyl systems are complicated by the fact that, for primary and secondary systems, two pathways can exist.[122] In one of these (designated k_Δ), the aryl, behaving as a neighboring group, pushes out the leaving group to give a bridged ion, called a *phenonium ion* (**41**), and is in turn pushed out by the solvent

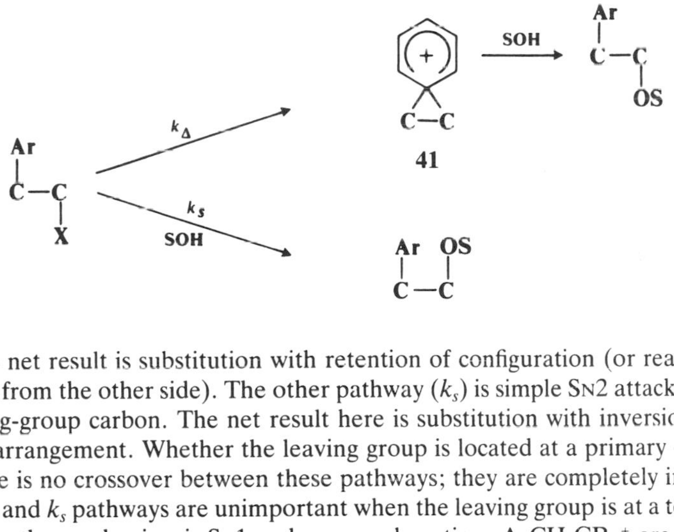

SOH, so the net result is substitution with retention of configuration (or rearrangement, if **41** is opened from the other side). The other pathway (k_s) is simple SN2 attack by the solvent at the leaving-group carbon. The net result here is substitution with inversion and no possibility of rearrangement. Whether the leaving group is located at a primary or a secondary carbon, there is no crossover between these pathways; they are completely independent.[123] (Both the k_Δ and k_s pathways are unimportant when the leaving group is at a tertiary carbon. In these cases the mechanism is SN1 and open carbocations $ArCH_2CR_2{}^+$ are intermediates. This pathway is designated k_c.) Which of the two pathways (k_s or k_Δ) predominates in any given case depends on the solvent and on the nature of the aryl group. As expected from the results we have seen for Cl as a neighboring group (p. 312), the k_Δ/k_s ratio is highest for solvents that are poor nucleophiles and so compete very poorly with the aryl group. For several common solvents the k_Δ/k_s ratio increases in the order EtOH < CH₃COOH <

[121]Cram *J. Am. Chem. Soc.* **1949**, *71*, 3863, **1952**, *74*, 2129.

[122]Winstein; Heck *J. Am. Chem. Soc.* **1956**, *78*, 4801; Brookhart; Anet; Cram; Winstein *J. Am. Chem. Soc.* **1966**, *88*, 5659; Lee; Unger; Vassie *Can. J. Chem.* **1972**, *50*, 1371.

[123]Harris; Schadt; Schleyer; Lancelot *J. Am. Chem. Soc.* **1969**, *91*, 7508; Brown; Kim; Lancelot; Schleyer *J. Am. Chem. Soc.* **1970**, *92*, 5244; Brown; Kim *J. Am. Chem. Soc.* **1971**, *93*, 5765.

HCOOH $<$ CF$_3$COOH.[124] In accord with this, the following percentages of retention were obtained in solvolysis of 1-phenyl-2-propyl tosylate at 50°C: solvolysis in EtOH 7%, CH$_3$COOH 35%, HCOOH 85%.[124] This indicates that k_s predominates in EtOH (phenyl participates very little), while k_Δ predominates in HCOOH. Trifluoroacetic acid is a solvent of particularly low nucleophilic power, and in this solvent the reaction proceeds entirely by k_Δ;[125] deuterium labeling showed 100% retention.[126] This case provides a clear example of neighboring-group rate enhancement by phenyl: the rate of solvolysis of PhCH$_2$CH$_2$OTs at 75°C in CF$_3$COOH is 3040 times the rate for CH$_3$CH$_2$OTs.[125]

With respect to the aromatic ring, the k_Δ pathway is electrophilic aromatic substitution (Chapter 11). We predict that groups on the ring which activate that reaction (p. 507) will increase, and deactivating groups will decrease, the rate of this pathway. This prediction has been borne out by several investigations. The *p*-nitro derivative of **38** solvolyzed in acetic acid 190 times slower than **38,** and there was much less retention of configuration; the acetate produced was only 7% threo and 93% erythro.[127] At 90°C, acetolysis of *p*-ZC$_6$H$_4$CH$_2$CH$_2$OTs gave the rate ratios shown in Table 10.1.[128] Throughout this series k_s is fairly constant, as it should be since it is affected only by the rather remote field effect of Z. It is k_Δ that changes substantially as Z is changed from activating to deactivating. The evidence is thus fairly clear that participation by aryl groups depends greatly on the nature of the group. For some groups, e.g., *p*-nitrophenyl, in some solvents, e.g., acetic acid, there is essentially no neighboring-group participation at all,[129] while for others, e.g., *p*-methoxy-phenyl, neighboring-group participation is substantial. The combined effect of solvent and structure is shown in Table 10.2, where the figures shown were derived by three different methods.[130] The decrease in neighboring-group effectiveness when aromatic rings are substituted by electron-withdrawing groups is reminiscent of the similar case of C=C bonds substituted by CF$_3$ groups (p. 315).

Several phenonium ions have been prepared as stable ions in solution where they can be studied by nmr, among them **42,**[131] **43,**[132] and the unsubstituted **41.**[133] These were

TABLE 10.1 Approximate k_Δ/k_s ratios for acetolysis of *p*-ZC$_6$H$_4$CH$_2$CH$_2$OTs at 90°C[128]

Z	k_Δ/k_s
MeO	30
Me	11
H	1.3
Cl	0.3

TABLE 10.2 Percent of product formed by the k_Δ pathway in solvolysis of *p*-ZC$_6$H$_4$CH$_2$CHMeOTs[130]

Z	Solvent	Percent by k_Δ
H	CH$_3$COOH	35–38
H	HCOOH	72–79
MeO	CH$_3$COOH	91–93
MeO	HCOOH	99

[124]Diaz; Lazdins; Winstein *J. Am. Chem. Soc.* **1968,** *90,* 6546; Diaz; Winstein *J. Am. Chem. Soc.* **1969,** *91,* 4300. See also Schadt; Lancelot; Schleyer *J. Am. Chem. Soc.* **1978,** *100,* 228.

[125]Nordlander; Deadman *J. Am. Chem. Soc.* **1968,** *90,* 1590; Nordlander; Kelly *J. Am. Chem. Soc.* **1969,** *91,* 996.

[126]Jablonski; Snyder *J. Am. Chem. Soc.* **1969,** *91,* 4445.

[127]Thompson; Cram *J. Am. Chem. Soc.* **1969,** *91,* 1778. See also Tanida; Tsuji; Ishitobi; Irie *J. Org. Chem.* **1969,** *34,* 1086; Kingsbury; Best *Bull. Chem. Soc. Jpn.* **1972,** *45,* 3440.

[128]Coke; McFarlane; Mourning; Jones *J. Am. Chem. Soc.* **1969,** *91,* 1154; Jones; Coke *J. Am. Chem. Soc.* **1969,** *91,* 4284. See also Harris; Schadt; Schleyer; Lancelot, Ref. 123.

[129]The k_Δ pathway is important for *p*-nitrophenyl in CF$_3$COOH: Ando; Shimizu; Kim; Tsuno; Yukawa *Tetrahedron Lett.* **1973,** 117.

[130]Lancelot; Schleyer *J. Am. Chem. Soc.* **1969,** *91,* 4291, 4296; Lancelot; Harper; Schleyer *J. Am. Chem. Soc.* **1969,** *91,* 4294; Schleyer; Lancelot *J. Am. Chem. Soc.* **1969,** *91,* 4297.

[131]Olah; Comisarow; Namanworth; Ramsey *J. Am. Chem. Soc.* **1967,** *89,* 5259; Ramsey; Cook; Manner *J. Org. Chem.* **1972,** *37,* 3310.

[132]Olah; Comisarow; Kim *J. Am. Chem. Soc.* **1969,** *91,* 1458. See, however, Ramsey; Cook; Manner, Ref. 131.

[133]Olah; Porter *J. Am. Chem. Soc.* **1971,** *93,* 6877; Olah; Spear; Forsyth *J. Am. Chem. Soc.* **1976,** *98,* 6284.

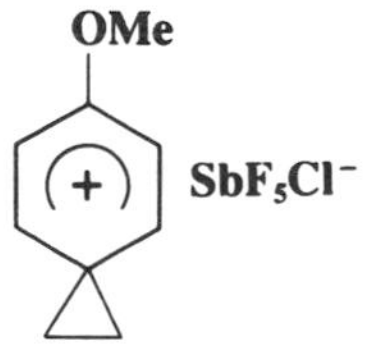

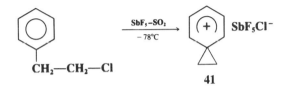

42 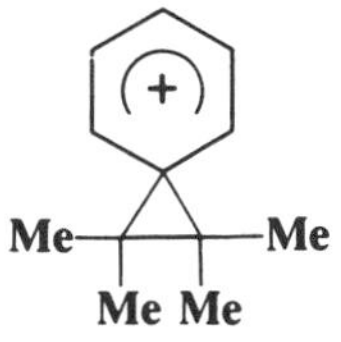   43

prepared[134] by the method shown for **41:** treatment of the corresponding β-arylethyl chloride with SbF_5–SO_2 at low temperatures. These conditions are even more extreme than the

Ph—CH_2—CH_2—Cl $\xrightarrow[-78°C]{SbF_5–SO_2}$

41

solvolysis in CF_3COOH mentioned earlier. The absence of any nucleophile at all eliminates not only the k_s pathways but also nucleophilic attack on **41.** Although **41** is not in equilibrium with the open-chain ion $PhCH_2CH_2{}^+$ (which is primary and hence unstable), **43** is in equilibrium with the open-chain tertiary ions $Ph\overset{\oplus}{C}Me_2CMe_2$ and $Ph\overset{\oplus}{C}MeCMe_3$, though only **43** is present in appreciable concentration. Proton and ^{13}C nmr show that **41, 42,** and **43** are classical carbocations where the only resonance is in the six-membered ring. The three-

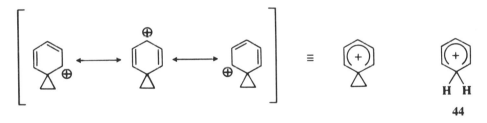

44

membered ring is a normal cyclopropane ring that is influenced only to a relatively small extent by the positive charge on the adjacent ring. Nmr spectra show that the six-membered rings have no aromatic character but are similar in structure to the arenium ions, e.g., **44,** that are intermediates in electrophilic aromatic substitution (Chapter 11). A number of phenonium ions, including **41,** have also been reported to be present in the gas phase, where their existence has been inferred from reaction products and from ^{13}C labeling.[135]

It is thus clear that β-aryl groups can function as neighboring groups.[136] Much less work

[134]For some others, see Olah; Singh; Liang *J. Org. Chem.* **1984,** *49,* 2922; Olah; Singh *J. Am. Chem.Soc.* **1984,** *106,* 3265.

[135]Fornarini; Sparapani; Speranza *J. Am. Chem. Soc.* **1988,** *110,* 34, 42; Fornarini; Muraglia *J. Am. Chem. Soc.* **1989,** *111,* 873; Mishima; Tsuno; Fujio *Chem. Lett.* **1990,** 2277.

[136]For additional evidence, see Tanida *Acc. Chem. Res.* **1968,** *1,* 239-245; Kingsbury; Best *Tetrahedron Lett.* **1967,** 1499; Braddon; Wiley; Dirlam; Winstein *J. Am. Chem. Soc.* **1968,** *90,* 1901; Tanida; Ishitobi; Irie *J. Am. Chem. Soc.* **1968,** *90,* 2688; Brown; Tritle *J. Am. Chem. Soc.* **1968,** *90,* 2689; Bentley; Dewar *J. Am. Chem. Soc.* **1970,** *92,* 3996; Raber; Harris; Schleyer *J. Am. Chem. Soc.* **1971,** *93,* 4829; Shiner; Seib *J. Am. Chem. Soc.* **1976,** *98,* 862; Faïn; Dubois *Tetrahedron Lett.* **1978,** 791; Yukawa; Ando; Token; Kawada; Matsuda; Kim; Yamataka *Bull. Chem. Soc. Jpn.* **1981,** *54,* 3536; Ferber; Gream *Aust. J. Chem.* **1981,** *34,* 2217; Fujio; Goto; Seki; Mishima; Tsuno; Sawada; Takai *Bull. Chem. Soc. Jpn.* **1987,** *60,* 1097. For a discussion of evidence obtained from isotope effects, see Scheppele *Chem. Rev.* **1972,** *72,* 511-532, pp. 522-525.

has been done on aryl groups located in positions farther away from the leaving group, but there is evidence that these too can lend anchimeric assistance.[137]

4. *The carbon-carbon single bond as a neighboring group.*[138]

a. *The 2-norbornyl system.* In the investigations to determine whether a C—C σ bond can act as a neighboring group, by far the greatest attention has been paid to the 2-norbornyl system.[139] Winstein and Trifan found that solvolysis in acetic acid of optically active *exo*-2-norbornyl brosylate (**45**) gave a racemic mixture of the two exo acetates; no endo isomers were formed:[140]

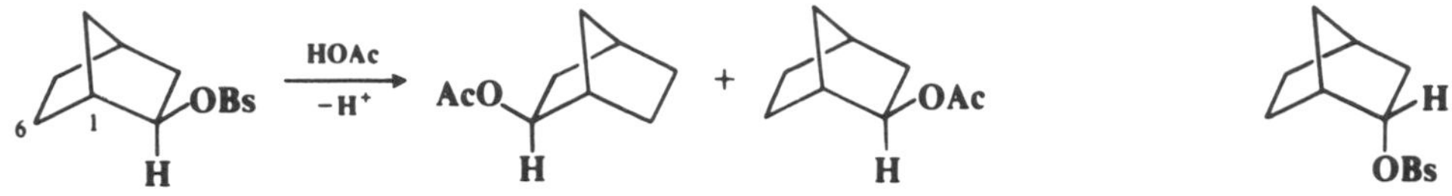

Futhermore, **45** solvolyzed about 350 times faster than its endo isomer **48**. Similar high exo/endo rate ratios have been found in many other [2.2.1] systems. These two results—(1) that solvolysis of an optically active exo isomer gave only racemic exo isomers and (2) the high exo/endo rate ratio—were interpreted by Winstein and Trifan as indicating that the 1,6 bond assists in the departure of the leaving group and that a nonclassical intermediate (**49**)

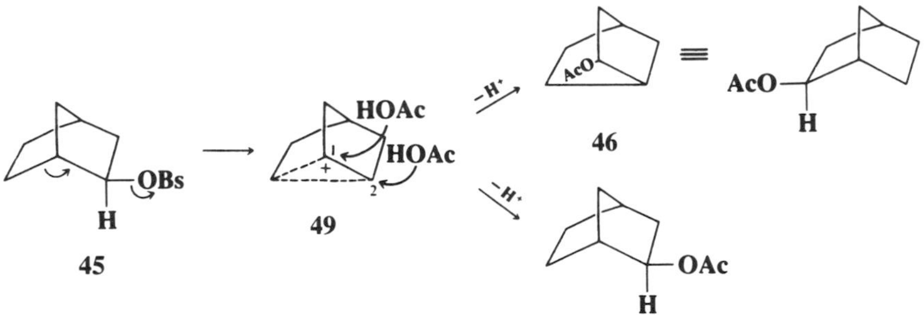

is involved. They reasoned that solvolysis of the endo isomer **48** is not assisted by the 1,6 bond because it is not in a favorable position for backside attack, and that consequently solvolysis of **48** takes place at a "normal" rate. Therefore the much faster rate for the solvolysis of **45** must be caused by anchimeric assistance. The stereochemistry of the product is also explained by the intermediacy of **49**, since in **49** the 1 and 2 positions are equivalent and would be attacked by the nucleophile with equal facility, but only from the exo direction in either case. Incidentally, acetolysis of **48** also leads exclusively to the exo acetates (**46**

[137]Heck; Winstein *J. Am. Chem. Soc.* **1957**, *79*, 3105; Muneyuki; Tanida *J. Am. Chem. Soc.* **1968**, *90*, 656; Ouellette; Papa; Attea; Levin *J. Am. Chem. Soc.* **1970**, *92*, 4893; Jackman; Haddon *J. Am. Chem. Soc.* **1974**, *96*, 5130; Gates; Frank; von Felten *J. Am. Chem. Soc.* **1974**, *96*, 5138; Ando; Yamawaki; Saito *Bull. Chem. Soc. Jpn.* **1978**, *51*, 219.

[138]For a review pertaining to studies of this topic at low temperatures, see Olah *Angew. Chem. Int. Ed. Engl.* **1973**, *12*, 173-212, pp. 192-198 [*Angew. Chem. 85*, 183-225].

[139]For reviews, see Olah; Prakash; Williams *Hypercarbon Chemistry*; Wiley: New York, 1987, pp. 157-170; Grob *Angew. Chem. Int. Ed. Engl.* **1982**, *21*, 87-96 [*Angew. Chem. 94*, 87-96]; Sargent, in Olah; Schleyer, Ref. 92, vol. 3, 1972, pp. 1099-1200; Sargent *Q. Rev., Chem. Soc.* **1966**, *20*, 301-371; Gream *Rev. Pure Appl. Chem.* **1966**, *16*, 25-60; Ref. 92. For a closely related review, see Kirmse *Acc. Chem. Res.* **1986**, *19*, 36-41. See also Ref. 143.

[140]Winstein; Trifan *J. Am. Chem. Soc.* **1952**, *74*, 1147, 1154; Winstein; Clippinger; Howe; Vogelfanger *J. Am. Chem. Soc.* **1965**, *87*, 376.

and **47**), so that in this case Winstein and Trifan postulated that a classical ion (**50**) is first formed and then converted to the more stable **49**. Evidence for this interpretation is that the product from solvolysis of **48** is not racemic but contains somewhat more **47** than **46** (corresponding to 3 to 13% inversion, depending on the solvent),[140] suggesting that when **50** is formed, some of it goes to give **47** before it can collapse to **49**.

The concepts of σ participation and the nonclassical ion **49** have been challenged by H. C. Brown,[96] who has suggested that the two results can also be explained by postulating that **45** solvolyzes without participation of the 1,6 bond to give the classical ion **50** which is in rapid equilibrium with **51**. This rapid interconversion has been likened to the action of

a windshield wiper.[141] Obviously, in going from **50** to **51** and back again, **49** must be present, but in Brown's view it is a transition state and not an intermediate. Brown's explanation for the stereochemical result is that exclusive exo attack is a property to be expected from any 2-norbornyl system, not only for the cation but even for reactions not involving cations, because of steric hindrance to attack from the endo side. There is a large body of data that shows that exo attack on norbornyl systems is fairly general in many reactions. As for the obtention of a racemic mixture, this will obviously happen if **50** and **51** are present in equal amounts, since they are equivalent and exo attack on **50** and **51** gives, respectively, **47** and **46**. Brown explains the high exo/endo rate ratios by contending that it is not the endo rate that is normal and the exo rate abnormally high, but the exo rate that is normal and the endo rate abnormally *low*, because of steric hindrance to removal of the leaving group in that direction.[142]

A vast amount of work has been done[143] on solvolysis of the 2-norbornyl system in an effort to determine whether the 1,6 bond participates and whether **49** is an intermediate. Most,[144] although not all,[145] chemists now accept the intermediacy of **49**.

Besides the work done on solvolysis of 2-norbornyl compounds, the 2-norbornyl cation

[141]Another view is somewhere in between: There are two interconverting ions, but each is asymmetrically bridged: Bielmann; Fuso; Grob *Helv. Chim. Acta* **1988**, *71*, 312; Flury; Grob; Wang; Lennartz; Roth *Helv. Chim. Acta* **1988**, *71*, 1017.

[142]For evidence against steric hindrance as the only cause of this effect, see Menger; Perinis; Jerkunica; Glass *J. Am. Chem. Soc.* **1978**, *100*, 1503.

[143]For thorough discussions, see Lenoir; Apeloig; Arad; Schleyer *J. Org. Chem.* **1988**, *53*, 661; Grob *Acc. Chem. Res.* **1983**, *16*, 426-431; Brown *Acc. Chem. Res.* **1983**, *16*, 432-440; Walling *Acc. Chem. Res.* **1983**, *16*, 448-454; Refs. 92, 96, and 139. For commentary on the controversy, see Arnett; Hofelich; Schriver *Rect. Intermed. (Wiley)* **1985**, *3*, 189-226, pp. 193-202.

[144]For some recent evidence in favor of a nonclassical **49**, see Arnett; Petro; Schleyer *J. Am. Chem. Soc.* **1979**, *101*, 522: Albano; Wold *J. Chem. Soc., Perkin Trans. 2* **1980**, 1447; Wilcox; Tuszynski *Tetrahedron Lett.* **1982**, *23*, 3119; Kirmse; Siegfried *J. Am. Chem. Soc.* **1983**, *105*, 950; Creary; Geiger *J. Am. Chem. Soc.* **1983**, *105*, 7123; Chang; le Noble *J. Am. Chem. Soc.* **1984**, *106*, 810; Kirmse; Brandt *Chem. Ber.* **1984**, *117*, 2510; Wilcox; Brungardt *Tetrahedron Lett.* **1984**, *25*, 3403; Lajunen *Acc. Chem. Res.* **1985**, *18*, 254-258; Sharma; Sen Sharma; Hiraoka; Kebarle *J. Am. Chem. Soc.* **1985**, *107*, 3747; Servis; Domenick; Forsyth; Pan *J. Am. Chem. Soc.* **1987**, *109*, 7263; Lenoir et al., Ref. 143.

[145]For some recent evidence against a nonclassical **49**, see Dewar; Haddon; Komornicki; Rzepa *J. Am. Chem. Soc.* **1977**, *99*, 377; Lambert; Mark *J. Am. Chem. Soc.* **1978**, *100*, 2501; Christol; Coste; Pietrasanta; Plénat; Renard *J. Chem. Soc., (S)* **1978**, 62; Brown; Ravindranathan; Rao; Chloupek; Rei *J. Org. Chem.* **1978**, *43*, 3667; Brown; Rao *J. Org. Chem.* **1979**, *44*, 133, 3536, **1980**, 45, 2113; Liu; Yen; Hwang *J. Chem. Res.(S)* **1980**, 152; Werstiuk; Dhanoa; Timmins *Can. J. Chem.* **1983**, *61*, 2403; Brown; Chloupek; Takeuchi *J. Org. Chem.* **1985**, *50*, 826; Brown; Ikegami; Vander Jagt *J. Org. Chem.* **1985**, *50*, 1165; Nickon; Swartz; Sainsbury; Toth *J. Org. Chem.* **1986**, *51*, 3736. See also Brown *Top. Curr. Chem.* **1979**, *80*, 1-18.

has also been extensively studied at low temperatures; there is much evidence that under these conditions the ion is definitely nonclassical. Olah and co-workers have prepared the 2-norbornyl cation in stable solutions at temperatures below $-150°C$ in $SbF_5–SO_2$ and $FSO_3H–SbF_5–SO_2$, where the structure is static and hydride shifts are absent.[146] Studies by proton and ^{13}C nmr, as well as by laser Raman spectra and x-ray electron spectroscopy, led to the conclusion[147] that under these conditions the ion is nonclassical.[148] A similar result has been reported for the 2-norbornyl cation in the solid state where at 77 K and even 5 K, ^{13}C nmr spectra gave no evidence of the freezing out of a single classical ion.[149]

Olah and co-workers represented the nonclassical structure as a corner-protonated nortricyclane (52); the symmetry is better seen when the ion is drawn as in 53. Almost all the

| Nortricyclane | 52 | 53 | 54 |

positive charge resides on C-1 and C-2 and very little on the bridging carbon C-6. Other evidence for the nonclassical nature of the 2-norbornyl cation in stable solutions comes from heat of reaction measurements that show that the 2-norbornyl cation is more stable (by about 6-10 kcal/mol or 25-40 kJ/mol) than would be expected without the bridging.[150] Studies of ir spectra of the 2-norbornyl cation in the gas phase also show the nonclassical structure.[151] Ab inito calculations show that the nonclassical structure corresponds to an energy minimum.[152]

The spectra of other norbornyl cations have also been investigated at low temperatures. Spectra of the tertiary 2-methyl- and 2-ethylnorbornyl cations show less delocalization,[153] and the 2-phenylnorbornyl cation (54) is essentially classical,[154] as are the 2-methoxy-[155] and 2-chloronorbornyl cations.[156] We may recall (p. 170) that methoxy and halo groups also

[146]The presence of hydride shifts (p. 1069) under solvolysis conditions has complicated the interpretation of the data.

[147]Olah; White; DeMember; Commeyras; Lui *J. Am. Chem. Soc.* **1970**, *92*, 4627; Olah *J. Am. Chem. Soc.***1972**, *94*, 808; *Acc. Chem. Res.* **1976**, *9*, 41-52; Olah; Liang; Mateescu; Riemenschneider *J. Am. Chem. Soc.* **1973**, *95*, 8698; Saunders; Kates *J. Am. Chem. Soc.* **1980**, *102*, 6867, **1983**, *105*, 3571; Olah; Prakash; Arvanaghi; Anet *J. Am. Chem. Soc.* **1982**, *104*, 7105; Olah; Prakash; Saunders *Acc. Chem. Res.* **1983**, *16*, 440-448. See also Schleyer; Lenoir; Mison; Liang; Prakash; Olah *J. Am. Chem. Soc.* **1980**, *102*, 683; Johnson; Clark *J. Am. Chem. Soc.* **1988**, *110*, 4112.

[148]This conclusion has been challenged: Fong *J. Am. Chem. Soc.* **1974**, *96*, 7638; Kramer *Adv. Phys. Org. Chem.* **1975**, *11*, 177-224; Brown; Periasamy; Kelly; Giansiracusa *J. Org. Chem.* **1982**, *47*, 2089; Kramer; Scouten *Adv. Carbocation Chem.* **1989**, *1*, 93-120. See, however, Olah; Prakash; Farnum; Clausen *J. Org. Chem.* **1983**, *48*, 2146.

[149]Yannoni; Macho; Myhre *J. Am. Chem. Soc.* **1982**, *104*, 907, 7380, *Bull. Soc. Chim. Belg.* **1982**, *91*, 422; Myhre; Webb; Yannoni *J. Am. Chem. Soc.* **1990**, *112*, 8991.

[150]For some examples, see Hogeveen; Gaasbeek *Recl. Trav. Chim. Pays-Bas* **1969**, *88*, 719; Hogeveen *Recl. Trav. Chim. Pays-Bas* **1970**, *89*, 74; Solomon; Field *J. Am. Chem. Soc.* **1976**, *98*, 1567; Staley; Wieting; Beauchamp *J. Am. Chem. Soc.* **1977**, *99*, 5964; Arnett; Petro *J. Am. Chem. Soc.* **1978**, *100*, 2563; Arnett; Pienta; Petro *J. Am. Chem. Soc.* **1980**, *102*, 398; Saluja; Kebarle *J. Am. Chem. Soc.* **1979**, *101*, 1084; Schleyer; Chandrasekhar *J. Org. Chem.* **1981**, *46*, 225; Lossing; Holmes *J. Am. Chem. Soc.* **1984**, *106*, 6917.

[151]Koch; Liu; DeFrees; Sunko; Vančik *Angew. Chem. Int. Ed. Engl.* **1990**, *29*, 183 [*Angew. Chem. 102*, 198].

[152]See, for example Koch; Liu; DeFrees *J. Am. Chem. Soc.* **1989**, *111*, 1527.

[153]Olah; DeMember; Lui; White *J. Am. Chem. Soc.* **1969**, *91*, 3958. See also Laube *Angew. Chem. Int. Ed. Engl.* **1987**, *26*, 560 [*Angew. Chem. 99*, 578]; Forsyth; Panyachotipun *J. Chem. Soc., Chem. Commun.* **1988**, 1564.

[154]Olah; Liang *J. Am. Chem. Soc.* **1974**, *96*, 195; Olah; White; DeMember; Commeyras; Lui, Ref. 147; Farnum; Mehta *J. Am. Chem. Soc.* **1969**, *91*, 3256; Ref. 153. See also Schleyer; Kleinfelter; Richey *J. Am. Chem. Soc.* **1963**, *85*, 479; Farnum; Wolf *J. Am. Chem. Soc.***1974**, *96*, 5166.

[155]Nickon; Lin *J. Am. Chem. Soc.* **1969**, *91*, 6861. See also Montgomery; Grendze; Huffman *J. Am. Chem. Soc.* **1987**, *109*, 4749.

[156]Fry; Farnham *J. Org. Chem.* **1969**, *34*, 2314.

stabilize a positive charge. ^{13}C nmr data show that electron-withdrawing groups on the benzene ring of **54** cause the ion to become less classical, while electron-donating groups enhance the classical nature of the ion.[157]

b. *The cyclopropylmethyl system.* Apart from the 2-norbornyl system, the greatest amount of effort in the search for C—C participation has been devoted to the cyclopropylmethyl system.[158] It has long been known that cyclopropylmethyl substrates solvolyze with abnormally high rates and that the products often include not only unrearranged cyclopropylmethyl but also cyclobutyl and homoallylic compounds. An example is[159]

$$\triangleright\!\!-CH_2Cl \xrightarrow{\text{EtOH–H}_2\text{O}} \triangleright\!\!-CH_2OH \; + \; \overset{OH}{\square} \; + \; CH_2\!\!=\!\!CHCH_2CH_2OH$$

$$\sim 48\% \qquad\qquad \sim 47\% \qquad\qquad \sim 5\%$$

Cyclobutyl substrates also solvolyze abnormally rapidly and give similar products. Furthermore, when the reactions are carried out with labeled substrates, considerable, though not complete, scrambling is observed. For these reasons it has been suggested that a common intermediate (some kind of nonclassical intermediate, e.g., **23**, p. 313) is present in these cases. This common intermediate could then be obtained by three routes:

$$\overset{3}{\underset{4}{\triangleright}}\!\!-\overset{1}{C}H_2\!\!-\!X \xrightarrow{\text{bent } \sigma \text{ route}} \textbf{23} \xleftarrow{\sigma \text{ route}} \overset{X}{\square}$$

$$\downarrow \pi \text{ route}$$

$$CH_2\!\!=\!\!CH\!\!-\!\!CH_2\!\!-\!\!CH_2\!\!-\!X$$

In recent years much work has been devoted to the study of these systems, and it is apparent that matters are not so simple. Though there is much that is still not completely understood, some conclusions can be drawn.

i. In solvolysis of simple primary cyclopropylmethyl systems the rate is enhanced because of participation by the σ bonds of the ring.[160] The ion that forms initially is an unrearranged cyclopropylmethyl cation[161] that is *symmetrically* stabilized, that is, both the 2,3 and 2,4 σ bonds help stabilize the positive charge. We have already seen (p. 169) that a cyclopropyl group stabilizes an adjacent positive charge even better than a phenyl group. One way of representing the structure of this cation is as shown in **55**. Among the evidence that **55** is a

$$\overset{3}{\underset{4}{\triangleright}}\!\!-\overset{1}{C}H_2\!\!\overset{\oplus}{} \longleftrightarrow \overset{\oplus}{\underset{}{\diagup}}\!\!=\longleftrightarrow \underset{\oplus}{\diagup}\!\!=$$

55

symmetrical ion is that substitution of one or more methyl groups in the 3 and 4 positions increases the rate of solvolysis of cyclopropylcarbinyl 3,5-dinitrobenzoates by approximately

[157]Olah; Prakash; Liang *J. Am. Chem. Soc.* **1977**, *99*, 5683; Farnum; Botto; Chambers; Lam *J. Am. Chem. Soc.* **1978**, *100*, 3847. See also Olah; Berrier; Prakash *J. Org. Chem.* **1982**, *47*, 3903.
[158]For reviews, see in Olah; Schleyer, Ref. 92, vol. 3, 1972, the articles by Richey, pp. 1201-1294, and by Wiberg; Hess; Ashe, pp. 1295-1345; Hanack; Schneider *Fortschr. Chem. Forsch.* **1967**, *8*, 554-607, *Angew. Chem. Int. Ed. Engl.* **1967**, *6*, 666-677 [*Angew. Chem. 79*, 709-720]; Sarel; Yovell; Sarel-Imber *Angew. Chem. Int. Ed. Engl.* **1968**, *7*, 577-588 [*Angew. Chem. 90*, 592-603].
[159]Roberts; Mazur *J. Am. Chem. Soc.* **1951**, *73*, 2509.
[160]See, for example, Roberts; Snyder *J. Org. Chem.* **1979**, *44*, 2860, and references cited therein.
[161]Wiberg; Ashe *J. Am. Chem. Soc.* **1968**, *90*, 63.

a factor of 10 for *each* methyl group.[162] If only one of the σ bonds (say, the 2,3 bond) stabilizes the cation, then methyl substitution at the 3 position should increase the rate, and a second methyl group at the 3 position should increase it still more, but a second methyl group at the 4 position should have little effect.[163]

ii. The most stable geometry of simple cyclopropylmethyl cations is the bisected one shown on p. 169. There is much evidence that in systems where this geometry cannot be obtained, solvolysis is greatly slowed.[164]

iii. Once a cyclopropylmethyl cation is formed, it can rearrange to two other cyclopropylmethyl cations:

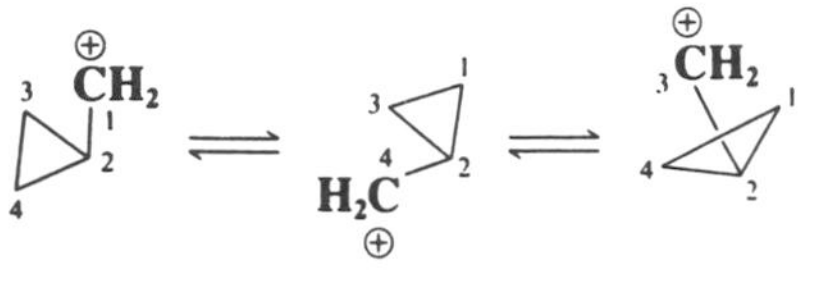

This rearrangement, which accounts for the scrambling, is completely stereospecific.[165] The rearrangements probably take place through a nonplanar cyclobutyl cation intermediate or transition state. The formation of cyclobutyl and homoallylic products from a cyclopropylmethyl cation is also completely stereospecific. These products may arise by direct attack of the nucleophile on **55** or on the cyclobutyl cation intermediate.[165] A planar cyclobutyl cation is ruled out in both cases because it would be symmetrical and the stereospecificity would be lost.

iv. The rate enhancement in the solvolysis of secondary cyclobutyl substrates is probably caused by participation by a bond leading directly to **55**, which accounts for the fact that solvolysis of cyclobutyl and of cyclopropylmethyl substrates often gives similar product

mixtures. There is no evidence that requires the cyclobutyl cations to be intermediates in most secondary cyclobutyl systems, though tertiary cyclobutyl cations can be solvolysis intermediates.

v. The unsubstituted cyclopropylmethyl cation has been generated in super-acid solutions at low temperatures, where ^{13}C nmr spectra have led to the conclusion that it consists of a mixture of the bicyclobutonium ion **23** and the bisected cyclopropylmethyl cation **55**, in equilibrium with **23**.[166] Molecular orbital calculations show that these two species are energy minima, and that both have nearly the same energy.[167]

[162]Schleyer; Van Dine *J. Am. Chem. Soc.* **1966,** *88,* 2321.

[163]For a summary of additional evidence for the symmetrical nature of cyclopropylmethyl cations, see Wiberg; Hess; Ashe, Ref. 158, pp. 1300-1303.

[164]For example, see Ree; Martin *J. Am. Chem. Soc.* **1970,** *92,* 1660; Rhodes; DiFate *J. Am. Chem. Soc.* **1972,** *94,* 7582. See, however, Brown; Peters *J. Am. Chem. Soc.* **1975,** *97,* 1927.

[165]Wiberg; Szeimies *J. Am. Chem. Soc.* **1968,** *90,* 4195, **1970,** *92,* 571; Majerski; Schleyer *J. Am. Chem. Soc.* **1971,** *93,* 665.

[166]Staral; Yavari; Roberts; Prakash; Donovan; Olah *J. Am. Chem. Soc.* **1978,** *100,* 8016. See also Olah; Jeuell; Kelly; Porter *J. Am. Chem. Soc.* **1972,** *94,* 146; Olah; Spear; Hiberty; Hehre *J. Am. Chem. Soc.* **1976,** *98,* 7470; Saunders; Siehl *J. Am. Chem. Soc.* **1980,** *102,* 6868; Brittain; Squillacote; Roberts *J. Am. Chem. Soc.* **1984,** *106,* 7280; Siehl; Koch *J. Chem. Soc., Chem. Commun.* **1985,** 496; Prakash; Arvanaghi; Olah *J. Am. Chem. Soc.* **1985,** *107,* 6017; Myhre; Webb; Yannoni *J. Am. Chem. Soc.* **1990,** *112,* 8992.

[167]Koch; Liu; DeFrees *J. Am. Chem. Soc.* **1988,** *110,* 7325; Saunders; Laidig; Wiberg; Schleyer *J. Am. Chem. Soc.* **1988,** *110,* 7652.

c. *Methyl as a neighboring group.* Both the 2-norbornyl and cyclopropylmethyl system contain a σ bond that is geometrically constrained to be in a particularly favorable position for participation as a neighboring group. However, there have been a number of investigations to determine whether a C—C bond can lend anchimeric assistance even in a simple open-chain compound such as neopentyl tosylate. On solvolysis, neopentyl systems undergo almost exclusive rearrangement and **56** must lie on the reaction path, but the two questions

Neopentyl tosylate   **56**

that have been asked are: (1) Is the departure of the leaving group concerted with the formation of the CH_3—C bond (that is, does the methyl participate)? (2) Is **56** an intermediate or only a transition state? With respect to the first question, there is evidence, chiefly from isotope effect studies, that indicates that the methyl group in the neopentyl system does indeed participate,[168] though it may not greatly enhance the rate. As to the second question, evidence that **56** is an intermediate is that small amounts of cyclopropanes (10 to 15%) can be isolated in these reactions.[169] **56** is a protonated cyclopropane and would give cyclopropane on loss of a proton.[170] In an effort to isolate a species that has structure **56,** the 2,3,3-trimethyl-2-butyl cation was prepared in super-acid solutions at low temperatures.[171] However, proton and ^{13}C nmr, as well as Raman spectra, showed this to be a pair of rapidly equilibrating open ions.

57

Of course, **57** must lie on the reaction path connecting the two open ions, but it is evidently a transition state and not an intermediate. However, evidence from x-ray photoelectron spectroscopy (ESCA) has shown that the 2-butyl cation is substantially methyl bridged.[172]

5. *Hydrogen as a neighboring group.* The questions relating to hydrogen are similar to those relating to methyl. There is no question that hydride can migrate, but the two questions are: (1) Does the hydrogen participate in the departure of the leaving group? (2) Is **58** an intermediate or only a transition state? There is some evidence that a β hydrogen can

[168]For example, see Dauben; Chitwood *J. Am. Chem. Soc.* **1968,** *90,* 6876; Ando; Morisaki *Tetrahedron Lett.* **1979,** 121; Shiner; Seib *Tetrahedron Lett.* **1979,** 123; Shiner; Tai *J. Am. Chem. Soc.* **1981,** *103,* 436; Yamataka; Ando *J. Am. Chem. Soc.* **1982,** *104,* 1808; Yamataka; Ando; Nagase; Hanamura; Morokuma *J. Org. Chem.* **1984,** *49,* 631. For an opposing view, see Zamashchikov; Rudakov; Bezbozhnaya; Matveev *J. Org. Chem. USSR* **1984,** *20,* 11.
[169]Skell; Starer *J. Am. Chem. Soc.* **1960,** 82, 2971; Silver *J. Am. Chem. Soc.* **1960,** *82,* 2971; Friedman; Bayless *J. Am. Chem. Soc.* **1969,** *91,* 1790; Friedman; Jurewicz *J. Am. Chem. Soc.* **1969,** *91,* 1800, 1803; Dupuy; Hudson; Karam *Tetrahedron Lett.* **1971,** 3193; Silver; Meek *Tetrahedron Lett.* **1971,** 3579; Dupuy; Hudson *J. Chem. Soc., Perkin Trans. 2* **1972,** 1715.
[170]For further discussions of protonated cyclopropanes, see pp. 757, 1056.
[171]Olah; White *J. Am. Chem. Soc.* **1969,** *91,* 5801; Olah; Comisarow; Kim *J. Am. Chem. Soc.* **1969,** *91,* 1458; Olah; DeMember; Commeyras; Bribes *J. Am. Chem. Soc.* **1971,** *93,* 459.
[172]Johnson; Clark, Ref. 147. See also Carneiro; Schleyer; Koch; Raghavachari *J. Am. Chem. Soc.* **1990,** *112,* 4064.

participate.[173] Evidence that **58** can be an intermediate in solvolysis reactions comes from a study of the solvolysis in trifluoroacetic acid of deuterated *sec*-butyl tosylate **59**. In this

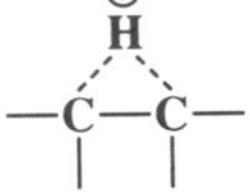

$$-\overset{|}{\underset{|}{C}}\overset{\overset{\oplus}{H}}{\cdots}\overset{|}{\underset{|}{C}}-$$

58

solvent of very low nucleophilic power, the products were an equimolar mixture of **60** and **61**,[174] but *no* **62** or **63** was found. If this reaction did not involve neighboring hydrogen at

$$\underset{\textbf{59}}{\overset{\overset{\text{OTs}}{|}}{CH_3CH_2CDCD_3}} \xrightarrow{\text{CF}_3\text{COOH}} \underset{\textbf{60}}{\overset{\overset{\text{OOCCF}_3}{|}}{CH_3CH_2CDCD_3}} + \underset{\textbf{61}}{\overset{\overset{\text{OOCCF}_3}{|}}{CH_3CHCDHCD_3}}$$

all (pure SN2 or SN1), the product would be only **60**. On the other hand, if hydrogen does migrate, but only open cations are involved, then there should be an equilibrium among

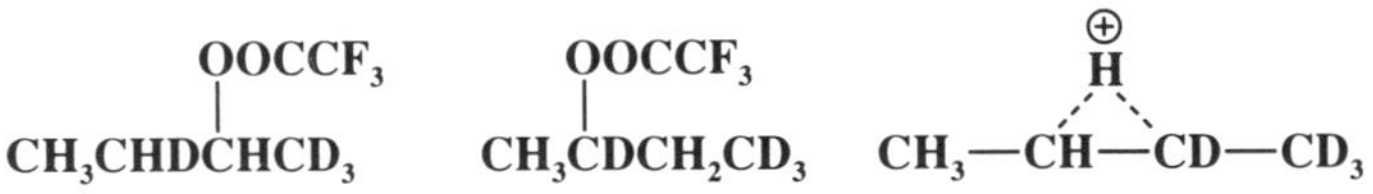

$$\underset{\textbf{62}}{\overset{\overset{\text{OOCCF}_3}{|}}{CH_3CHDCHCD_3}} \qquad \underset{\textbf{63}}{\overset{\overset{\text{OOCCF}_3}{|}}{CH_3CDCH_2CD_3}} \qquad \underset{\textbf{64}}{\overset{\overset{\oplus}{H}}{CH_3-CH\cdots CD-CD_3}}$$

these four cations:

$$CH_3CH_2\overset{\oplus}{C}DCD_3 \rightleftharpoons CH_3\overset{\oplus}{C}HCDHCD_3 \rightleftharpoons CH_3CDH\overset{\oplus}{C}HCD_3 \rightleftharpoons CH_3\overset{\oplus}{C}DCH_2CD_3$$

leading not only to **60** and **61**, but also to **62** and **63**. The results are most easily compatible with the intermediacy of the bridged ion **64** which can then be attacked by the solvent equally at the 2 and 3 positions. Attempts to prepare **58** as a stable ion in super-acid solutions at low temperatures have not been successful.[172]

The SNi Mechanism

In a few reactions, nucleophilic substitution proceeds with retention of configuration, even where there is no possibility of a neighboring-group effect. In the SNi mechanism (*substitution nucleophilic internal*) part of the leaving group must be able to attack the substrate, detaching

[173]See, for example, Shiner; Jewett *J. Am. Chem. Soc.* **1965**, *87*, 1382; Pánková; Sicher; Tichý; Whiting *J. Chem. Soc. B* **1968**, 365; Tichý; Hapala; Sicher *Tetrahedron Lett.* **1969**, 3739; Myhre; Evans *J. Am. Chem. Soc.* **1969**, *91*, 5641; Inomoto; Robertson; Sarkis *Can. J. Chem.* **1969**, *47*, 4599; Shiner; Stoffer *J. Am. Chem. Soc.* **1970**, *92*, 3191; Krapcho; Johanson *J. Org. Chem.* **1971**, *36*, 146; Chuit; Felkin; Le Ny; Lion; Prunier *Tetrahedron* **1972**, *28*, 4787; Stéhelin; Lhomme; Ourisson *J. Am. Chem. Soc.* **1971**, *93*, 1650; Stéhelin; Kanellias; Ourisson *J. Org. Chem.* **1973**, *38*, 847, 851; Hiršl-Starčević; Majerski; Sunko *J. Org. Chem.* **1980**, *45*, 3388; Buzek; Schleyer; Sieber; Koch; Carneiro; Vančik; Sunko *J. Chem. Soc., Chem. Commun.* **1991**, 671; Imhoff; Ragain; Moore; Shiner *J. Org. Chem.* **1991**, *56*, 3542.
[174]Dannenberg; Goldberg; Barton; Dill; Weinwurzel; Longas *J. Am. Chem. Soc.* **1981**, *103*, 7764. See also Dannenberg; Barton; Bunch; Goldberg; Kowalski *J. Org. Chem.* **1983**, *48*, 4524; Allen; Ambidge; Tidwell *J. Org. Chem.* **1983**, *48*, 4527.

itself from the rest of the leaving group in the process. The IUPAC designation is D_N + $A_N D_e$. The first step is the same as the very first step of the S_N1 mechanism—dissociation into an intimate ion pair.[175] But in the second step part of the leaving group attacks, necessarily from the front since it is unable to get to the rear. This results in retention of configuration:

Step 1 $R-OSOCl \longrightarrow R^+$

Step 2

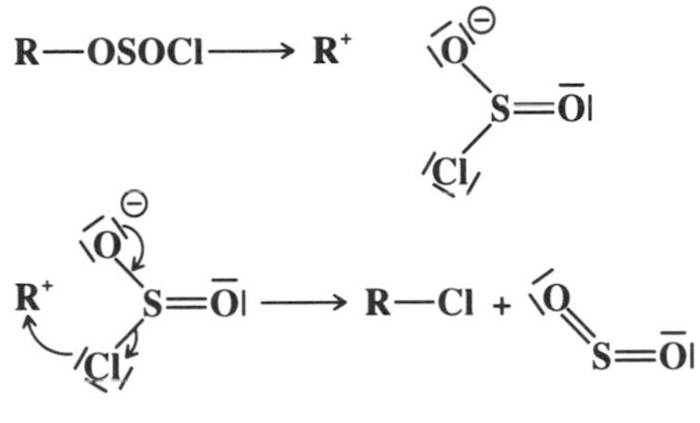

The example shown is the most important case of this mechanism yet discovered, since the reaction of alcohols with thionyl chloride to give alkyl halides usually proceeds in this way, with the first step in this case being $ROH + SOCl_2 \rightarrow ROSOCl$ (these alkyl chlorosulfites can be isolated).

Evidence for this mechanism is as follows: the addition of pyridine to the mixture of alcohol and thionyl chloride results in the formation of alkyl halide with *inverted* configuration. Inversion results because the pyridine reacts with ROSOCl to give $ROSO\overset{\oplus}{N}C_5H_5$ before anything further can take place. The Cl^- freed in this process now attacks from the rear. The reaction between alcohols and thionyl chloride is second order, which is predicted by this mechanism, but the decomposition by simple heating of ROSOCl is first order.[176]

The S_Ni mechanism is relatively rare. Another example is the decomposition of ROCOCl (alkyl chloroformates) into RCl and CO_2.[177]

Nucleophilic Substitution at an Allylic Carbon. Allylic Rearrangements

Allylic substrates undergo nucleophilic substitution reactions especially rapidly (see p. 341), but we discuss them in a separate section because they are usually accompanied by a certain kind of rearrangement known as an *allylic rearrangement*.[178] When allylic substrates are treated with nucleophiles under S_N1 conditions, two products are usually obtained: the normal one and a rearranged one.

$$R-CH=CH-CH_2X \xrightarrow{Y^-} R-CH=CH-CH_2Y + R-\underset{\underset{Y}{|}}{CH}-CH=CH_2$$

[175] Lee; Finlayson *Can. J. Chem.* **1961**, *39*, 260; Lee; Clayton; Lee; Finlayson *Tetrahedron* **1962**, *18*, 1395.
[176] Lewis; Boozer *J. Am. Chem. Soc.* **1952**, *74*, 308.
[177] Lewis; Herndon; Duffey *J. Am. Chem. Soc.* **1961**, *83*, 1959; Lewis; Witte *J. Chem. Soc. B* **1968**, 1198. For other examples, see Hart; Elia *J. Am. Chem. Soc.* **1961**, *83*, 985; Stevens; Dittmer; Kovacs *J. Am. Chem. Soc.* **1963**, *85*, 3394; Kice; Hanson *J. Org. Chem.* **1973**, *38*, 1410; Cohen; Solash *Tetrahedron Lett.* **1973**, 2513; Verrinder; Hourigan; Prokipcak *Can. J. Chem.* **1978**, *56*, 2582.
[178] For a review, see DeWolfe, in Bamford; Tipper *Comprehensive Chemical Kinetics*, vol. 9; Elsevier: New York, 1973, pp. 417-437. For comprehensive older reviews, see DeWolfe; Young *Chem. Rev.* **1956**, *56*, 753-901; in Patai *The Chemistry of Alkenes*; Wiley: New York, 1964, the sections by Mackenzie, pp. 436-453 and DeWolfe; Young, pp. 681-738.

Two products are formed because an allylic type of carbocation is a resonance hybrid

$$R-CH=CH-CH_2{}^{\oplus} \longleftrightarrow R-\overset{\oplus}{C}H-CH=CH_2$$

so that C-1 and C-3 each carry a partial positive charge and both are attacked by Y. Of course, an allylic rearrangement is undetectable in the case of symmetrical allylic cations, as in the case where R = H, unless isotopic labeling is used. This mechanism has been called the SN1' mechanism. The IUPAC designation is $1/D_N + 3/A_N$, the numbers 1 and 3 signifying the *relative* positions where the nucleophile attacks and from which the nucleofuge leaves.

As with other SN1 reactions, there is clear evidence that SN1' reactions can involve ion pairs. If the intermediate attacked by the nucleophile is a completely free carbocation, then, say,

$$CH_3CH=CHCH_2Cl \quad \text{and} \quad CH_3CHClCH=CH_2$$

65   **66**

should give the same mixture of alcohols when reacting with hydroxide ion, since the carbocation from each should be the same. When treated with 0.8 N aqueous NaOH at 25°C, **65** gave 60% $CH_3CH=CHCH_2OH$ and 40% $CH_3CHOHCH=CH_2$, while **66** gave the products in yields of 38 and 62%, respectively.[179] This phenomenon is called the *product spread*. In this case, and in most others, the product spread is in the direction of the starting compound. With increasing polarity of solvent, the product spread decreases and in some cases is entirely absent. It is evident that in such cases the high polarity of the solvent stabilizes completely free carbocations. There is other evidence for the intervention of ion pairs in many of these reactions. When $H_2C=CHCMe_2Cl$ was treated with acetic acid, both acetates were obtained, but also some $ClCH_2CH=CMe_2$,[180] and the isomerization was faster than the acetate formation. This could not have arisen from a completely free Cl^- returning to the carbon, since the rate of formation of the rearranged chloride was unaffected by the addition of external Cl^-. All these facts indicate that the first step in these reactions is the formation of an unsymmetrical intimate ion pair that undergoes a considerable amount of internal return and in which the counterion remains close to the carbon from which it departed. Thus, **65** and **66**, for example, give rise to two *different* intimate ion pairs. The field of the anion polarizes the allylic cation, making the nearby carbon atom more electrophilic, so that it has a greater chance of attracting the nucleophile.[181]

Nucleophilic substitution at an allylic carbon can also take place by an SN2 mechanism, in which case no allylic rearrangement usually takes place. However, allylic rearrangements can also take place under SN2 conditions, by the following mechanism, in which the nucleophile attacks at the γ carbon rather than the usual position:[182]

---

[179]DeWolfe; Young, *Chem. Rev.*, Ref. 178, give several dozen such examples.

[180]Young; Winstein; Goering *J. Am. Chem. Soc.* **1951**, *73*, 1958.

[181]For additional evidence for the involvement of ion pairs in SN1' reactions, see Goering; Linsay *J. Am. Chem. Soc.* **1969**, *91*, 7435; d'Incan; Viout *Bull. Soc. Chim. Fr.* **1971**, 3312; Astin; Whiting *J. Chem. Soc., Perkin Trans. 2* **1976**, 1157; Kantner; Humski; Goering *J. Am. Chem. Soc.* **1982**, *104*, 1693; Thibblin *J. Chem. Soc., Perkin Trans. 2* **1986**, 313; Ref. 56.

[182]For a review of the SN2' mechanism, see Magid *Tetrahedron* **1980**, *36*, 1901-1930, pp. 1901-1910.

The IUPAC designation is $3/1/A_ND_N$. This mechanism is a second-order allylic rearrangement; it usually comes about where SN2 conditions hold but where α substitution sterically retards the normal SN2 mechanism. There are thus few well-established cases of the SN2' mechanism on substrates of the type C=C—CH₂X, while compounds of the form C=C—CR₂X give the SN2' rearrangement almost exclusively when they give bimolecular reactions at all. Increasing the size of the nucleophile can also increase the extent of the SN2' reaction at the expense of the SN2.[183] In certain cases the leaving group can also have an affect on whether the rearrangment occurs. Thus PhCH=CHCH₂X, treated with LiAlH₄, gave 100% SN2 reaction (no rearrangement) when X = Br or Cl, but 100% SN2' when X = PPh₃⁺ Br⁻.[184]

The SN2' mechanism as shown above involves the simultaneous movement of three pairs of electrons. However, Bordwell has contended that there is no evidence that requires that this bond making and bond breaking be in fact concerted,[185] and that a true SN2' mechanism is a myth. There is evidence both for[186] and against[187] this proposal.

The stereochemistry of SN2' reactions has been investigated. It has been found that both syn[188] (the nucleophile enters on the side from which the leaving group departs) and anti[189]

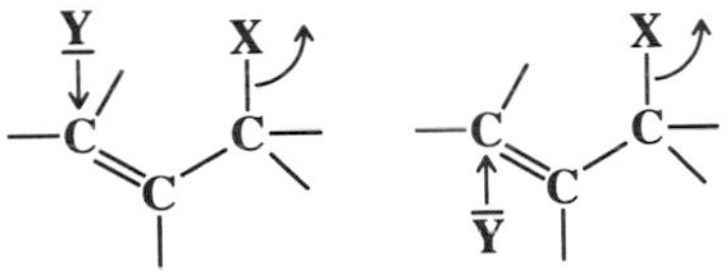

reactions can take place, depending on the nature of X and Y,[190] though the syn pathway predominates in most cases.

When a molecule has in an allylic position a nucleofuge capable of giving the SNi reaction, it is possible for the nucleophile to attack at the γ position instead of the α position. This is called the SNi' mechanism and has been demonstrated on 2-buten-1-ol and 3-buten-2-ol,

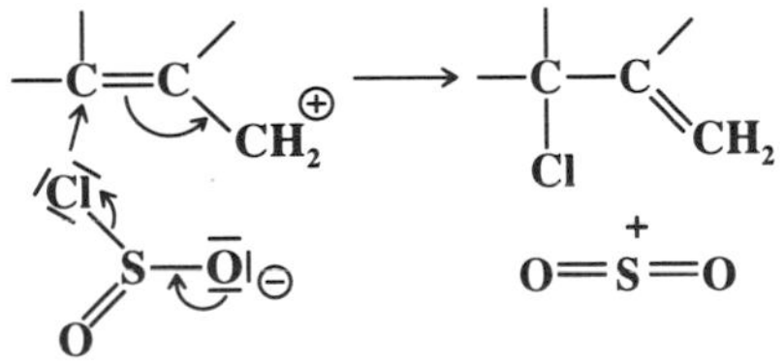

[183]Bordwell; Clemens; Cheng *J. Am. Chem. Soc.***1987,** *109,* 1773.
[184]Hirabe; Nojima; Kusabayashi *J. Org. Chem.* **1984,** *49,* 4084.
[185]Bordwell; Schexnayder *J. Org. Chem.* **1968,** *33,* 3240; Bordwell; Mecca *J. Am. Chem. Soc.* **1972,** *94,* 5829; Bordwell *Acc. Chem. Res.* **1970,** *3,* 281-290, pp. 282-285. See also de la Mare; Vernon *J. Chem. Soc. B* **1971,** 1699; Dewar *J. Am. Chem. Soc.* **1984,** *106,* 209.
[186]See Uebel; Milaszewski; Arlt *J. Org. Chem.* **1977,** *42,* 585.
[187]See Fry *Pure Appl. Chem.* **1964,** *8,* 409; Georgoulis; Ville *J. Chem. Res. (S)* **1978,** 248, *Bull. Soc. Chim. Fr.* **1985,** 485; Meislich; Jasne *J. Org. Chem.* **1982,** *47,* 2517.
[188]See, for example, Stork; White *J. Am. Chem. Soc.* **1956,** *78,* 4609; Jefford; Sweeney; Delay *Helv. Chim. Acta* **1972,** *55,* 2214; Kirmse; Scheidt; Vater *J. Am. Chem. Soc.* **1978,** *100,* 3945; Gallina; Ciattini *J. Am. Chem. Soc.* **1979,** *101,* 1035; Magid; Fruchey *J. Am. Chem. Soc.* **1979,** *101,* 2107; Bäckvall; Vågberg; Genêt *J. Chem. Soc., Chem. Commun.* **1987,** 159.
[189]See, for example, Borden; Corey *Tetrahedron Lett.* **1969,** 313; Takahashi; Satoh *Bull. Chem. Soc. Jpn.* **1975,** *48,* 69; Staroscik; Rickborn *J. Am. Chem. Soc.* **1971,** *93,* 3046; See also Liotta *Tetrahedron Lett.* **1975,** 523; Stork; Schoofs *J. Am. Chem. Soc.* **1979,** *101,* 5081.
[190]Stork; Kreft *J. Am. Chem. Soc.* **1977,** *99,* 3850, 3851; Oritani; Overton *J. Chem. Soc., Chem. Commun.* **1978,** 454; Bach; Wolber *J. Am. Chem. Soc.* **1985,** *107,* 1352. See also Chapleo; Finch; Roberts; Woolley; Newton; Selby *J. Chem. Soc., Perkin Trans. 1* **1980,** 1847; Stohrer *Angew. Chem. Int. Ed. Engl.* **1983,** *22,* 613 [*Angew. Chem. 95,* 642].

both of which gave 100% allylic rearrangement when treated with thionyl chloride in ether.[191] Ordinary allylic rearrangements (SN1') or SN2' mechanisms could not be expected to give 100% rearrangement in *both* cases. In the case shown, the nucleophile is only part of the leaving group, not the whole. But it is also possible to have reactions in which a simple leaving group, such as Cl, comes off to form an ion pair and then returns not to the position whence it came but to the allylic position:

$$R—CH=CH—CH_2Cl \longrightarrow R—CH=CH—CH_2{}^+\ Cl^- \longrightarrow R—\underset{\underset{Cl}{|}}{CH}—CH=CH_2$$

Most SNi' reactions are of this type.

Allylic rearrangements have also been demonstrated in propargyl systems, e.g.,[192]

$$PhC\equiv CCH_2OTs + MeMgBr \xrightarrow{\ CuBr\ } Ph—\underset{\underset{Me}{|}}{C}=C=CH_2 \quad \text{(Reaction 0-87)}$$

The product in this case is an allene,[193] but such shifts can also give triple-bond compounds or, if Y = OH, an enol will be obtained that tautomerizes to an α,β-unsaturated aldehyde or ketone.

$$R—C\equiv C—\underset{\underset{X}{|}}{CR_2} \xrightarrow{\ OH^-\ } R—\underset{\underset{OH}{|}}{C}=C=CR_2 \longrightarrow R—\underset{\underset{O}{\|}}{C}—C=CR_2$$

When X = OH, this conversion of acetylenic alcohols to unsaturated aldehydes or ketones is called the *Meyer–Schuster rearrangement*.[194] The propargyl rearrangement can also go the other way; that is, 1-haloalkenes, treated with organocopper compounds, give alkynes.[195]

Nucleophilic Substitution at an Aliphatic Trigonal Carbon. The Tetrahedral Mechanism

All the mechanisms so far discussed take place at a saturated carbon atom. Nucleophilic substitution is also important at trigonal carbons, especially when the carbon is double-bonded to an oxygen, a sulfur, or a nitrogen. Nucleophilic substitution at vinylic carbons is considered in the next section; at aromatic carbons in Chapter 13.

Substitution at a carbonyl group (or the corresponding nitrogen and sulfur analogs) most often proceeds by a second-order mechanism, which in this book is called the *tetrahedral*[196]

---

[191]Young, *J. Chem. Educ.* **1962**, *39*, 456. For other examples, see Pegolotti; Young *J. Am. Chem. Soc.* **1961**, *83*, 3251; Mark *Tetrahedron Lett.* **1962**, 281; Czernecki; Georgoulis; Labertrande; Prévost *Bull. Soc. Chim. Fr.* **1969**, 3568; Lewis; Witte, Ref. 177; Corey; Boaz *Tetrahedron Lett.* **1984**, *25*, 3055.

[192]Vermeer; Meijer; Brandsma *Recl. Trav. Chim.Pays-Bas* **1975**, *94*, 112.

[193]For reviews of such rearrangements, see Schuster; Coppola *Allenes in Organic Synthesis*; Wiley: New York, 1984, pp. 12-19, 26-30; Taylor *Chem. Rev.* **1967**, *67*, 317-359, pp. 324-328.

[194]For a review, see Swaminathan; Narayanan *Chem. Rev.* **1971**, *71*, 429-438. For discussions of the mechanism, see Edens; Boerner; Chase; Nass; Schiavelli *J. Org. Chem.* **1977**, *42*, 3403; Andres; Cardenas, Silla; Tapia *J. Am. Chem. Soc.* **1988**, *110*, 666.

[195]Corey; Boaz *Tetrahedron Lett.* **1984**, *25*, 3059, 3063.

[196]This mechanism has also been called the "addition–elimination mechanism," but in this book we limit this term to the type of mechanism shown on p. 335.

mechanism.[197] The IUPAC designation is $A_N + D_N$. S_N1 mechanisms, involving carbocations, are sometimes found with these substrates, especially with essentially ionic substrates such as $RCO^+ BF_4^-$; there is evidence that in certain cases simple S_N2 mechanisms can take place, especially with a very good leaving group such as Cl^-;[198] and an SET mechanism has also been reported.[199] However, the tetrahedral mechanism is by far the most prevalent. Although this mechanism displays second-order kinetics, it is not the same as the S_N2 mechanism previously discussed. In the tetrahedral mechanism, first Y attacks to give an intermediate containing both X and Y, and then X leaves. This sequence, impossible at a saturated carbon, is possible at an unsaturated one because the central carbon can release a pair of electrons to the oxygen and so preserve its octet:

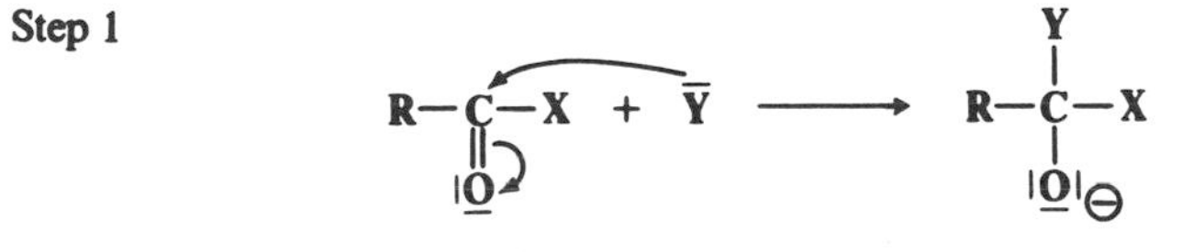

Step 1

$$R-\underset{\underset{|\underline{O}|}{\|}}{C}-X \ + \ \overline{Y} \longrightarrow R-\underset{\underset{|\underline{O}|\ominus}{|}}{\overset{\overset{Y}{|}}{C}}-X$$

67

Step 2

$$R-\underset{\underset{|\underline{O}|\ominus}{|}}{\overset{\overset{Y}{|}}{C}}-X \longrightarrow R-\underset{\underset{|\underline{O}}{\|}}{\overset{\overset{Y}{|}}{C}} \ + \ \overline{X}$$

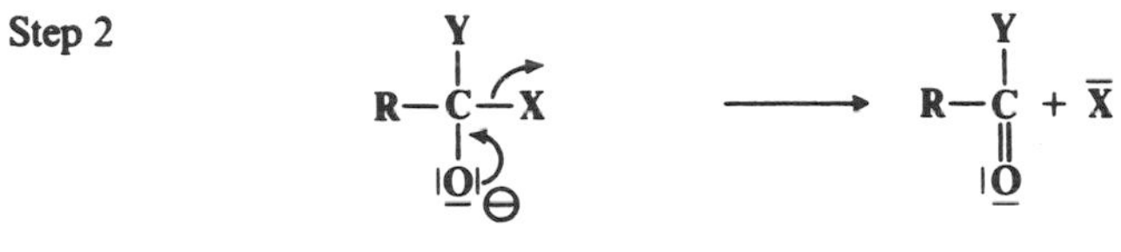

When reactions are carried out in acid solution, there may also be a preliminary and a final step:

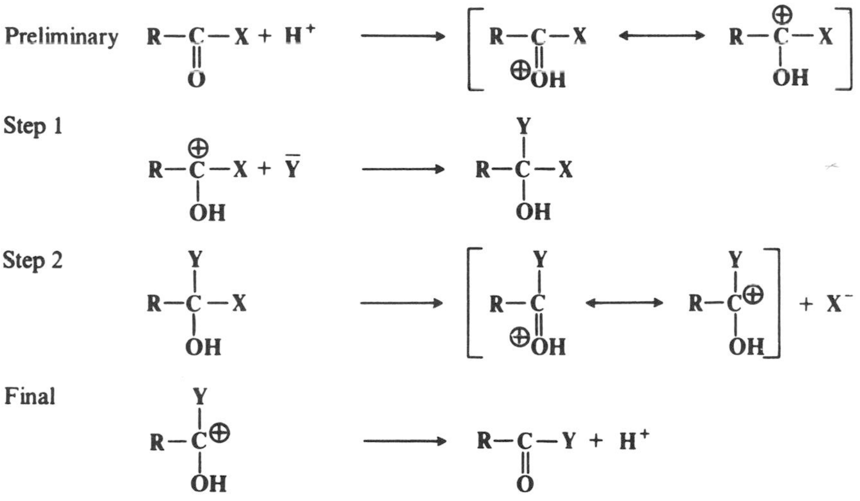

Preliminary

$$R-\underset{\underset{O}{\|}}{C}-X + H^+ \longrightarrow \left[R-\underset{\underset{\oplus OH}{\|}}{C}-X \longleftrightarrow R-\underset{\underset{OH}{|}}{\overset{\oplus}{C}}-X \right]$$

Step 1

$$R-\underset{\underset{OH}{|}}{\overset{\oplus}{C}}-X + \overline{Y} \longrightarrow R-\underset{\underset{OH}{|}}{\overset{\overset{Y}{|}}{C}}-X$$

Step 2

$$R-\underset{\underset{OH}{|}}{\overset{\overset{Y}{|}}{C}}-X \longrightarrow \left[R-\underset{\underset{\oplus OH}{\|}}{\overset{\overset{Y}{|}}{C}} \longleftrightarrow R-\underset{\underset{OH}{|}}{\overset{\overset{Y}{|}}{C}}^{\oplus} \right] + X^-$$

Final

$$R-\underset{\underset{OH}{|}}{\overset{\overset{Y}{|}}{C}}^{\oplus} \longrightarrow R-\underset{\underset{O}{\|}}{C}-Y + H^+$$

[197]For reviews of this mechanism, see Talbot, in Bamford; Tipper, Ref. 178, vol. 10, 1972, pp. 209-223; Jencks *Catalysis in Chemistry and Enzymology*; McGraw-Hill: New York, 1969, pp. 463-554; Satchell; Satchell, in Patai *The Chemistry of Carboxylic Acids and Esters*; Wiley: New York, 1969, pp. 375-452; Johnson *Adv. Phys. Org. Chem.* **1967**, *5*, 237-330.

[198]For a review, see Williams *Acc. Chem. Res.* **1989**, *22*, 387-392. For examples, see Kevill; Foss *J. Am. Chem. Soc.* **1969**, *91*, 5054; Haberfield; Trattner *Chem. Commun.* **1971**, 1481; Shpan'ko; Goncharov; Litvinenko *J. Org. Chem. USSR* **1979**, *15*, 1472, 1478; De Tar *J. Am. Chem. Soc.* **1982**, *104*, 7205; Bentley; Carter; Harris *J. Chem. Soc., Perkin Trans. 2* **1985**, 983; Shpan'ko; Goncharov *J. Org. Chem. USSR* **1987**, *23*, 2287; Guthrie; Pike *Can. J. Chem.* **1987**, *65*, 1951; Kevill; Kim *Bull. Soc. Chim. Fr.* **1988**, 383, *J. Chem. Soc., Perkin Trans. 2* **1988**, 1353; Bentley; Koo *J. Chem. Soc., Perkin Trans. 2* **1989**, 1385. See however, Buncel; Um; Hoz *J. Am. Chem. Soc.* **1989**, *111*, 971.

[199]Bacaloglu; Blaskó; Bunton; Ortega *J. Am. Chem. Soc.* **1990**, *112*, 9336.

The hydrogen ion is a catalyst. The reaction rate is increased because it is easier for the nucleophile to attack the carbon when the electron density of the latter has been decreased.[200]

Evidence for the existence of the tetrahedral mechanism is as follows:[201]

1. The kinetics are first order each in the substrate and in the nucleophile, as predicted by the mechanism.

2. There is other kinetic evidence in accord with a tetrahedral intermediate. For example, the rate "constant" for the reaction between acetamide and hydroxylamine is not constant but decreases with increasing hydroxylamine concentration.[202] This is not a smooth decrease; there is a break in the curve. A straight line is followed at low hydroxylamine concentration and another straight line at high concentration. This means that the identity of the rate-determining step is changing. Obviously, this cannot happen if there is only one step: there must be two steps and hence an intermediate. Similar kinetic behavior has been found in other cases as well,[203] in particular, plots of rate against pH are often bell-shaped.

3. Basic hydrolysis has been carried out on carboxylic esters labeled with ^{18}O in the carbonyl group.[204] If this reaction proceeded by the normal SN2 mechanism, all the ^{18}O would remain in the carbonyl group, even if, in an equilibrium process, some of the carboxylic acid formed went back to the starting material:

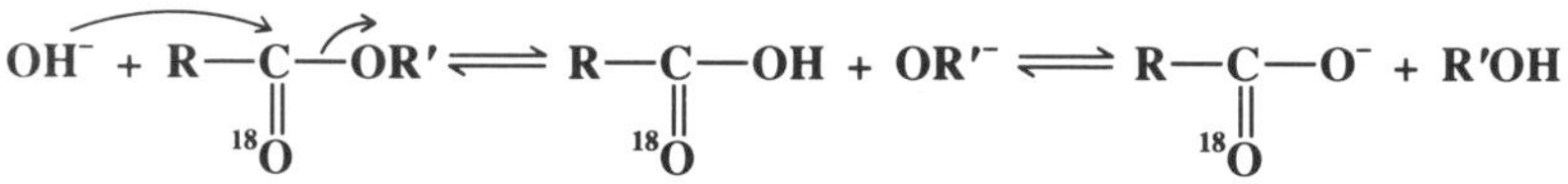

$$OH^- + R-C-OR' \rightleftharpoons R-C-OH + OR'^- \rightleftharpoons R-C-O^- + R'OH$$

On the other hand, if the tetrahedral mechanism operates

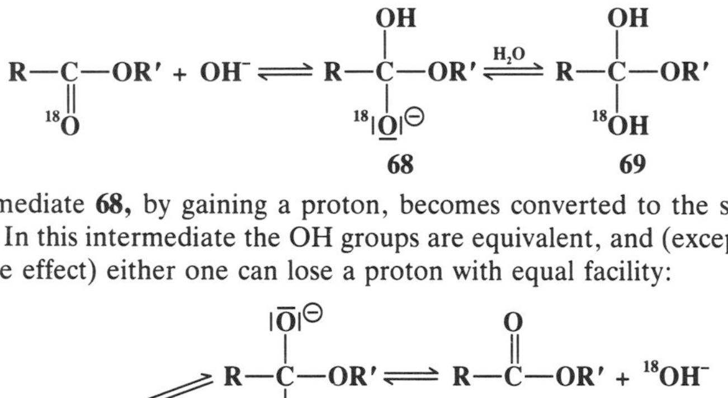

$$R-C-OR' + OH^- \rightleftharpoons R-C-OR' \xrightarrow{H_2O} R-C-OR'$$
$$\underset{\mathbf{68}}{} \qquad \underset{\mathbf{69}}{}$$

then the intermediate **68,** by gaining a proton, becomes converted to the symmetrical intermediate **69.** In this intermediate the OH groups are equivalent, and (except for the small $^{18}O/^{16}O$ isotope effect) either one can lose a proton with equal facility:

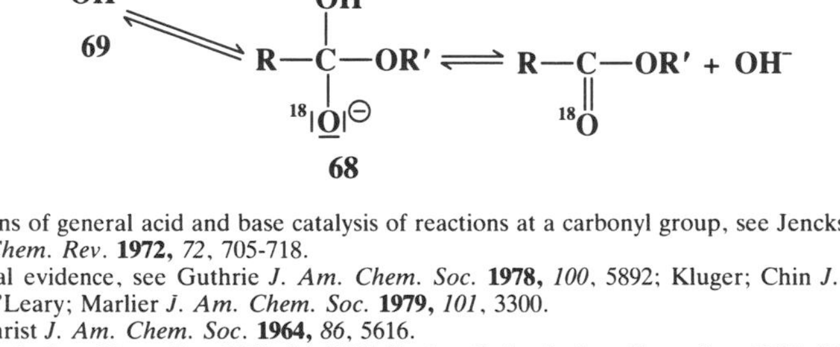

$$R-C-OR' \rightleftharpoons R-C-OR' + ^{18}OH^-$$
$$\underset{\mathbf{70}}{}$$

$$R-C-OR' \rightleftharpoons R-C-OR' + OH^-$$
$$\underset{\mathbf{68}}{}$$

[200]For discussions of general acid and base catalysis of reactions at a carbonyl group, see Jencks *Acc. Chem. Res.* **1976,** *9,* 425-432, *Chem. Rev.* **1972,** *72,* 705-718.

[201]For additional evidence, see Guthrie *J. Am. Chem. Soc.* **1978,** *100,* 5892; Kluger; Chin *J. Am. Chem. Soc.* **1978,** *100,* 7382; O'Leary; Marlier *J. Am. Chem. Soc.* **1979,** *101,* 3300.

[202]Jencks; Gilchrist *J. Am. Chem. Soc.* **1964,** *86,* 5616.

[203]Hand; Jencks *J. Am. Chem. Soc.* **1962,** *84,* 3505; Bruice; Fedor *J. Am. Chem. Soc.* **1964,** *86,* 4886; Johnson *J. Am. Chem. Soc.* **1964,** *86,* 3819; Fedor; Bruice *J. Am. Chem. Soc.* **1964,** *86,* 5697, **1965,** *87,* 4138; Kevill; Johnson *J. Am. Chem. Soc.* **1965,** *87,* 928; Leinhard; Jencks *J. Am. Chem. Soc.* **1965,** *87,* 3855; Schowen; Jayaraman; Kershner *J. Am. Chem. Soc.* **1966,** *88,* 3373.

[204]Bender *J. Am. Chem. Soc.* **1951,** *73,* 1626; Bender; Thomas *J. Am. Chem. Soc.* **1961,** *83,* 4183, 4189.

The intermediates **68** and **70** can now lose OR' to give the acid (not shown in the equations given), or they can lose OH to regenerate the carboxylic ester. If **68** goes back to ester, the ester will still be labeled, but if **70** reverts to ester, the ^{18}O will be lost. A test of the two possible mechanisms is to stop the reaction before completion and to analyze the recovered ester for ^{18}O. This is just what was done by Bender, who found that in alkaline hydrolysis of methyl, ethyl, and isopropyl benzoates, the esters had lost ^{18}O. A similar experiment carried out for acid-catalyzed hydrolysis of ethyl benzoate showed that here too the ester lost ^{18}O. However, alkaline hydrolysis of substituted benzyl benzoates showed *no* ^{18}O loss.[205] This result does not necessarily mean that no tetrahedral intermediate is involved in this case. If **68** and **70** do not revert to ester, but go entirely to acid, no ^{18}O loss will be found even with a tetrahedral intermediate. In the case of benzyl benzoates this may very well be happening, because formation of the acid relieves steric strain. Another possibility is that **68** loses OR' before it can become protonated to **69**.[206] Even the experiments that *do* show ^{18}O loss do not *prove* the existence of the tetrahedral intermediate, since it is possible that ^{18}O is lost by some independent process not leading to ester hydrolysis. To deal with this possibility, Bender and Heck[207] measured the rate of ^{18}O loss in the hydrolysis of ethyl trifluorothioloacetate-^{18}O:

$$F_3C-\overset{\underset{||}{18}}{\underset{O}{C}}-SEt + H_2O \underset{k_2}{\overset{k_1}{\rightleftharpoons}} \text{intermediate} \xrightarrow{k_3} F_3CCOOH + EtSH$$

This reaction had previously been shown[208] to involve an intermediate by the kinetic methods mentioned on p. 332. Bender and Heck showed that the rate of ^{18}O loss and the value of the partitioning ratio k_2/k_3 as determined by the oxygen exchange technique were exactly in accord with these values as previously determined by kinetic methods. Thus the original ^{18}O-exchange measurements showed that there is a tetrahedral species present, though not necessarily on the reaction path, while the kinetic experiments showed that there is some intermediate present, though not necessarily tetrahedral. Bender and Heck's results demonstrate that there is a tetrahedral intermediate and that it lies on the reaction pathway.

   4. In some cases, tetrahedral intermediates have been isolated[209] or detected spectrally.[210]

   Several studies have been made of the directionality of approach by the nucleophile.[211] Menger has proposed for reactions in general, and specifically for those that proceed by the tetrahedral mechanism, that there is no single definable preferred transition state, but rather a "cone" of trajectories. All approaches within this cone lead to reaction at comparable rates; it is only when the approach comes outside of the cone that the rate falls.

   Directionality has also been studied for the second step. Once the tetrahedral intermediate (**67**) is formed, it loses Y (giving the product) or X (reverting to the starting compound). Deslongchamps has proposed that one of the factors affecting this choice is the conformation of the intermediate; more specifically, the positions of the lone pairs. In this view, a leaving

[205]Bender; Matsui; Thomas; Tobey *J. Am. Chem. Soc.* **1961**, *83*, 4193. See also Shain; Kirsch *J. Am. Chem. Soc.* **1968**, *90*, 5848.
   [206]For evidence for this possibility, see McClelland *J. Am. Chem. Soc.* **1984**, *106*, 7579.
   [207]Bender; Heck *J. Am. Chem. Soc.* **1967**, *89*, 1211.
   [208]Fedor; Bruice *J. Am. Chem. Soc.* **1965**, *87*, 4138.
   [209]Rogers; Bruice *J. Am. Chem. Soc.* **1974**, *96*, 2481; Khouri; Kaloustian *J. Am. Chem. Soc.* **1986**, *108*, 6683.
   [210]For reviews, see Capon; Dosunmu; Sanchez *Adv. Phys. Org. Chem.* **1985**, *21*, 37-98; McClelland; Santry *Acc. Chem. Res.* **1983**, *16*, 394-399; Capon; Ghosh; Grieve *Acc. Chem. Res.* **1981**, *14*, 306-312. See also Lobo; Marques; Prabhakar; Rzepa *J. Chem. Soc., Chem. Commun.* **1985**, 1113; van der Wel; Nibbering *Recl. Trav. Chim. Pays-Bas* **1988**, *107*, 479, 491.
   [211]For discussions, see Menger *Tetrahedron* **1983**, *39*, 1013-1040; Liotta; Burgess; Eberhardt *J. Am. Chem. Soc.* **1984**, *106*, 4849.

group X or Y can depart only if the other two atoms on the carbon both have an orbital antiperiplanar to the C—X or C—Y bond. For example, consider an intermediate

$$
\begin{array}{c}
\text{OR} \\
| \\
\text{R}'\text{—C—X} \\
| \\
\underset{\_}{\text{O}}
\end{array}
$$

formed by attack of OR$^-$ on a substrate R'COX. Cleavage of the C—X bond with loss of X can take place from conformation **A,** because the two lone-pair orbitals marked * are

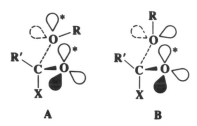

antiperiplanar to the C—X bond, but not from **B** because only the O$^-$ has such an orbital. If the intermediate is in conformation **B,** the OR may leave (if X has a lone-pair orbital in the proper position) rather than X. This factor is called *stereoelectronic control*.[212] Of course, there is free rotation in acyclic intermediates, and many conformations are possible, but some are preferred, and cleavage reactions may take place faster than rotation, so stereoelectronic control can be a factor in some situations. Much evidence has been presented for this concept.[213] More generally, the term *stereoelectronic effects* refers to any case in which orbital position requirements affect the course of a reaction. The backside attack in the SN2 mechanism is an example of a stereoelectronic effect.

Some nucleophilic substitutions at a carbonyl carbon are *catalyzed* by nucleophiles.[214] There occur, in effect, two tetrahedral mechanisms:

$$
\underset{\underset{\text{O}}{\|}}{\text{R—C—X}} + \underset{\text{Catalyst}}{\text{Z}} \longrightarrow \underset{\underset{\text{O}}{\|}}{\text{R—C—Z}} + \text{Y} \longrightarrow \underset{\underset{\text{O}}{\|}}{\text{R—C—Y}}
$$

(For an example, see **0-9.**) When this happens internally, we have an example of a neighboring-group mechanism at a carbonyl carbon.[215] For example, the hydrolysis of phthalamic

[212]It has also been called the "antiperiplanar lone pair hypothesis (ALPH)." For a reinterpretation of this factor in terms of the principle of least nuclear motion (see **5-10**), see Hosie; Marshall; Sinnott *J. Chem. Soc., Perkin Trans. 2* **1984,** 1121; Sinnott *Adv. Phys. Org. Chem.* **1988,** *24,* 113-204.

[213]For monographs, see Kirby *The Anomeric Effect and Related Stereoelectronic Effects at Oxygen*; Springer: New York, 1983; Deslongchamps *Stereoelectronic Effects in Organic Chemistry*; Pergamon: New York, 1983. For lengthy treatments, see Sinnott, Ref. 212; Gorenstein *Chem. Rev.* **1987,** *87,* 1047-1077; Deslongchamps *Heterocycles* **1977,** *7,* 1271-1317, *Tetrahedron* **1975,** *31,* 2463-2490. For additional evidence, see Deslongchamps; Barlet; Taillefer *Can. J. Chem.* **1980,** *58,* 2167; Perrin; Arrhenius *J. Am. Chem. Soc.* **1982,** *104,* 2839; Briggs; Evans; Glenn; Kirby *J. Chem. Soc., Perkin Trans. 2* **1983,** 1637; Deslongchamps; Guay; Chênevert *Can. J. Chem. 63,* **1985,** 2493; Ndibwami; Deslongchamps *Can. J. Chem.* **1986,** *64,* 1788; Hegarty; Mullane *J. Chem. Soc., Perkin Trans. 2* **1986,** 995. For evidence against the theory, see Perrin; Nuñez *J. Am. Chem. Soc.* **1986,** *108,* 5997, **1987,** *109,* 522.

[214]For reviews of nucleophilic catalysis, see Bender *Mechanisms of Homogeneous Catalysis from Protons to Proteins*; Wiley: New York, 1971, pp. 147-179; Jencks, Ref. 197, pp. 67-77; Johnson, Ref. 197, pp. 271-318. For a review where Z = a tertiary amine (the most common case), see Cherkasova; Bogatkov; Golovina *Russ. Chem. Rev.* **1977,** *46,* 246-263.

[215]For reviews, see Kirby; Fersht *Prog. Bioorg. Chem.* **1971,** *1,* 1-82; Capon *Essays Chem.* **1972,** *3,* 127-156.

acid (**71**) takes place as follows:

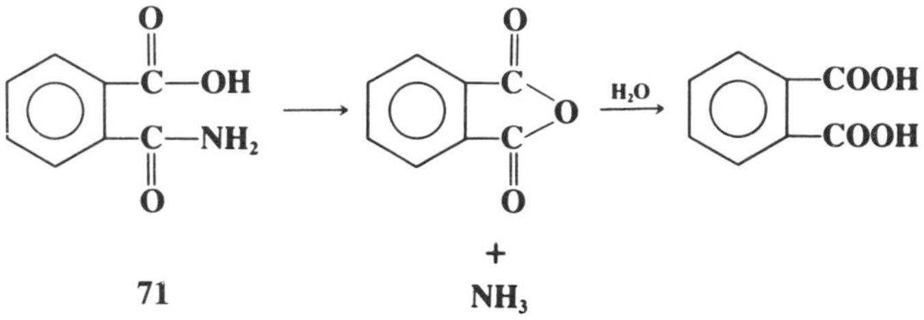

Evidence comes from comparative rate studies.[216] Thus **71** was hydrolyzed about 10^5 times faster than benzamide ($PhCONH_2$) at about the same concentration of hydrogen ions. That this enhancement of rate was not caused by the resonance or field effects of COOH (an electron-withdrawing group) was shown by the fact both *o*-nitrobenzamide and terephthalamic acid (the para isomer of **71**) were hydrolyzed more slowly than benzamide. Many other examples of neighboring-group participation at a carbonyl carbon have been reported.[217] It is likely that nucleophilic catalysis is involved in enzyme catalysis of ester hydrolysis.

The attack of a nucleophile on a carbonyl group can result in substitution or addition (Chapter 16), though the first step of each mechanism is the same. The main factor that determines the product is the identity of the group X in RCOX. When X is alkyl or hydrogen, addition usually takes place. When X is halogen, OH, OCOR, NH_2, etc., the usual reaction is substitution.

For a list of some of the more important reactions that operate by the tetrahedral mechanism, see Table 10.8.

Nucleophilic Substitution at a Vinylic Carbon

Nucleophilic substitution at a vinylic carbon[218] is difficult (see p. 341), but many examples are known. The most common mechanisms are the tetrahedral mechanism and the closely related *addition–elimination mechanism*. Both of these mechanisms are impossible at a saturated substrate. The addition-elimination mechanism has been demonstrated for the reaction between 1,1-dichloroethene (**72**) and ArS^-, catalyzed by EtO^-.[219] The product was

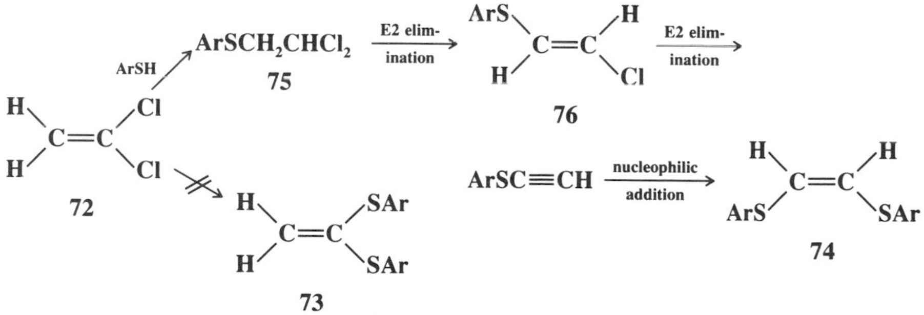

[216]Bender; Chow; Chloupek *J. Am. Chem. Soc.* **1958**, *80*, 5380.

[217]For examples, see Bruice; Pandit *J. Am. Chem. Soc.* **1960**, *82*, 5858; Zimmering; Westhead; Morawetz *Biochim. Biophys. Acta* **1957**, *25*, 376; Kirby; McDonald; Smith *J. Chem. Soc., Perkin Trans. 2* **1974**, 1495; Martin; Tan *J. Chem. Soc., Perkin Trans. 2* **1974**, 129; Kluger; Lam *J. Am. Chem. Soc.* **1978**, *100*, 2191; Page; Render; Bernáth *J. Chem. Soc., Perkin Trans. 2* **1986**, 867.

[218]For reviews, see Rappoport *Recl. Trav. Chim. Pays-Bas* **1986**, *104*, 309-349, *React. Intermed. (Plenum)* **1983**, *3*, 427-615, *Adv. Phys. Org. Chem.* **1969**, *7*, 1-114; Shainyan *Russ. Chem. Rev.* **1986**, *55*, 511-530; Modena *Acc. Chem. Res.* **1971**, *4*, 73-80

[219]Truce; Boudakian *J. Am. Chem. Soc.* **1956**, *78*, 2748.

not the 1,1-dithiophenoxy compound **73** but the "rearranged" compound **74.** Isolation of **75** and **76** showed that an addition–elimination mechanism had taken place. In the first step ArSH adds to the double bond (nucleophilic addition, p. 741) to give the saturated **75.** The second step is an E2 elimination reaction (p. 983) to give the alkene **76.** A second elimination and addition give **74.**

The tetrahedral mechanism, often also called addition–elimination (Ad_N-E), takes place with much less facility than with carbonyl groups, since the negative charge of the intermediate must be borne by a carbon, which is less electronegative than oxygen, sulfur, or nitrogen:

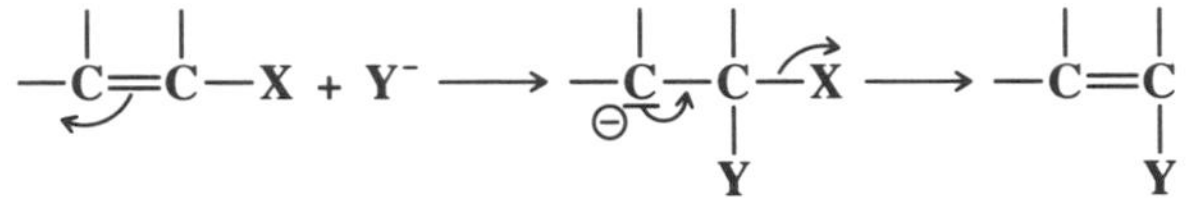

Such an intermediate can also stabilize itself by combining with a positive species. When it does, the reaction is nucleophilic addition to a C=C double bond (see Chapter 15). It is not surprising that with vinylic substrates addition and substitution often compete. For chloroquinones, where the charge is spread by resonance, tetrahedral intermediates have been isolated:[220]

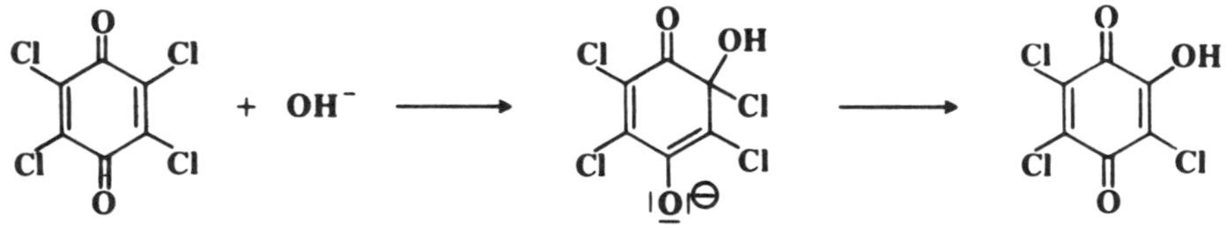

Isolated

In the case of Ph(MeO)C=C(NO$_2$)Ph + RS$^-$, the intermediate lived long enough to be detected by uv spectroscopy.[221]

Since both the tetrahedral and addition–elimination mechanisms begin the same way, it is usually difficult to tell them apart, and often no attempt is made to do so. The strongest kind of evidence for the addition–elimination sequence is the occurrence of a "rearrangement" (as in the conversion of **72** to **74**), but of course the mechanism could still take place even if no rearrangement is found. Evidence[222] that a tetrahedral or an addition–elimination mechanism takes place in certain cases (as opposed, for example, to an SN1 or SN2 mechanism) is that the reaction rate increases when the leaving group is changed from Br to Cl to F (this is called the *element effect*).[223] This clearly demonstrates that the carbon–halogen bond does not break in the rate-determining step (as it would in both the SN1 and SN2 mechanisms), because fluorine is by far the poorest leaving group among the halogens in both the SN1 and SN2 reactions (p. 352). The rate is faster with fluorides in the cases cited, because the superior electron-withdrawing character of the fluorine makes the carbon of the C—F bond more positive and hence more susceptible to nucleophilic attack.

Ordinary vinylic substrates react very poorly if at all by these mechanisms, but substitution is greatly enhanced in substrates of the type ZCH=CHX, where Z is an electron-withdrawing

[220]Hancock; Morrell; Rhom *Tetrahedron Lett.* **1962,** 987.

[221]Bernasconi; Fassberg; Killion; Rappoport *J. Am. Chem. Soc.* **1989,** *112,* 3169, *J. Org. Chem.* **1990,** *55,* 4568.

[222]Additional evidence comes from the pattern of catalysis by amines, similar to that discussed for aromatic substrates on p. 643. See Rappoport; Peled *J. Am. Chem. Soc.* **1979,** *101,* 2682, and references cited therein.

[223]Beltrame; Favini; Cattania; Guella *Gazz. Chim. Ital.* **1968,** *98,* 380. See also Rappoport; Rav-Acha *Tetrahedron Lett.* **1984,** *25,* 117; Solov'yanov; Shtern; Beletskaya; Reutov *J. Org. Chem. USSR* **1983,** *19,* 1945; Avramovitch; Weyerstahl; Rappoport *J. Am. Chem. Soc.* **1987,** *109,* 6687.

group such as HCO, RCO,[224] EtOOC, ArSO$_2$, NC, F, etc., since these β groups stabilize the carbanion:

$$ZCH=CHX \xrightarrow{\;Y^-\;} Z-\overset{\overset{\displaystyle Y}{|}}{\underset{\underset{\displaystyle X}{|}}{\overset{\ominus}{C}}}-\overset{}{\underset{\underset{\displaystyle H}{|}}{CH}} \longrightarrow ZCH=CHY$$

Many such examples are known. In most cases where the stereochemistry has been investigated, retention of configuration is observed,[225] but stereoconvergence (the same product mixture from an E or Z substrate) has also been observed,[226] especially where the carbanionic carbon bears two electron-withdrawing groups. It is not immediately apparent why the tetrahedral mechanism should lead to retention, but this behavior has been ascribed, on the basis of molecular orbital calculations, to hyperconjugation involving the carbanionic electron pair and the substituents on the adjacent carbon.[227]

Vinylic substrates are in general very reluctant to undergo SN1 reactions, but they can be made to do so in two ways:[228] (1) By the use of an α group that stabilizes the vinylic cation. For example, α-aryl vinylic halides ArCBr=CR$_2'$ have often been shown to give SN1 reactions.[229] SN1 reactions have also been demonstrated with other stabilizing groups: cyclopropyl,[230] vinylic,[231] alkynyl,[232] and an adjacent double bond (R$_2$C=C=CR'X).[233] (2) Even without α stabilization, by the use of a very good leaving group, e.g., OSO$_2$CF$_3$ (triflate).[234] The stereochemical outcome of SN1 reactions at a vinylic substrate is often randomization,[235] that is, either a cis or a trans substrate gives a 1:1 mixture of cis and trans products, indicating that vinylic cations are linear. Another indication that vinylic cations prefer to be linear is the fact that reactivity in cycloalkenyl systems decreases with decreasing ring size.[236] However, a linear vinylic cation need not give random products.[237] The empty p orbital lies in the plane of the double bond, so entry of the nucleophile can be and often

[224]For a review, see Rybinskaya; Nesmeyanov; Kochetkov *Russ. Chem. Rev.* **1969**, *38*, 433-456.

[225]Rappoport *Adv. Phys. Org. Chem.*, Ref. 218, pp. 31-62; Shainyan, Ref. 218, pp. 516-520. See also Rappoport; Gazit *J. Am. Chem. Soc.* **1987**, *109*, 6698.

[226]See Rappoport; Gazit *J. Org. Chem.* **1985**, *50*, 3184, *J. Am. Chem. Soc.* **1986**, *51*, 4112; Park; Ha *Bull. Chem. Soc. Jpn.* **1990**, *63*, 3006.

[227]Apeloig; Rappoport *J. Am. Chem. Soc.* **1979**, *101*, 5095.

[228]For reviews of the SN1 mechanism at a vinylic substrate, see Stang; Rappoport; Hanack; Subramanian *Vinyl Cations*, Chapter 5; Academic Press: New York, 1979; Stang *Acc. Chem. Res.* **1978**, *11*, 107-114, *Prog. Phys. Org. Chem.* **1973**, *10*, 205-325; Rappoport *Acc. Chem. Res.* **1976**, *9*, 265-273; Subramanian; Hanack *J. Chem. Educ.* **1975**, *52*, 80-86; Hanack *Acc. Chem. Res.* **1970**, *3*, 209-216; Modena; Tonellato *Adv. Phys. Org. Chem.* **1971**, *9*, 185-280, pp. 231-253; Grob *Chimia* **1971**, *25*, 87-91; Rappoport; Bässler; Hanack *J. Am. Chem. Soc.* **1970**, *92*, 4985-4987.

[229]For a review, see Stang; Rappoport; Hanack; Subramanian, Ref. 228, Chapter 6.

[230]Sherrod; Bergman *J. Am. Chem. Soc.* **1969**, *91*, 2115, **1971**, *93*, 1925; Kelsey; Bergman *J. Am. Chem. Soc.* **1970**, *92*, 238, **1971**, *93*, 1941; Hanack; Bässler *J. Am. Chem. Soc.* **1969**, *91*, 2117; Hanack; Bässler; Eymann; Heyd; Kopp *J. Am. Chem. Soc.* **1974**, *96*, 6686.

[231]Grob; Spaar *Tetrahedron Lett.* **1969**, 1439, *Helv. Chim. Acta* **1970**, *53*, 2119.

[232]Hassdenteufel; Hanack *Tetrahedron Lett.* **1980**, 503. See also Kobayashi; Nishi; Koyama; Taniguchi *J. Chem. Soc., Chem. Commun.* **1980**, 103.

[233]Schiavelli; Gilbert; Boynton; Boswell *J. Am. Chem. Soc.* **1972**, *94*, 5061.

[234]See, for example, Stang; Summerville *J. Am. Chem. Soc.* **1969**, *91*, 4600; Clarke; Bergman *J. Am. Chem. Soc.* **1972**, *94*, 3627, **1974**, *96*, 7934; Summerville; Schleyer *J. Am. Chem. Soc.* **1972**, *94*, 3629, **1974**, *96*, 1110; Eckes; Subramanian; Hanack *Tetrahedron Lett.* **1973**, 1967; Hanack; Märkl; Martinez *Chem. Ber.* **1982**, *115*, 772.

[235]Rappoport; Apeloig *J. Am. Chem. Soc.* **1969**, *91*, 6734; Kelsey; Bergman, Ref. 230.

[236]Pfeifer; Bahn; Schleyer; Bocher; Harding; Hummel; Hanack; Stang *J. Am. Chem. Soc.* **1971**, *93*, 1513.

[237]For examples of inversion, see Clarke; Bergman, Ref. 234; Summerville; Schleyer, Ref. 234.

is influenced by the relative size of R^1 and R^2.[238] It must be emphasized that even where vinylic substrates do give SN1 reactions, the rates are generally lower than those of the corresponding saturated compounds.

Alkynyl cations are so unstable that they cannot be generated even with very good leaving groups. However, one way in which they have been generated was by formation of a tritiated substrate.

$$R-C{\equiv}C-T \xrightarrow{\beta \text{ decay}} R-C{\equiv}C-{}^3He \xrightarrow[\text{fast}]{\text{very}} R-C{\equiv}C^+ + {}^3He$$

When the tritium (half-life 12.26 y) decays it is converted to the helium-3 isotope, which, of course, does not form covalent bonds, and so immediately departs, leaving behind the alkynyl cation. When this was done in the presence of benzene, $RC{\equiv}CC_6H_5$ was isolated.[239] The tritium-decay technique has also been used to generate vinylic and aryl cations.[240]

Besides the mechanisms already discussed, another mechanism, involving an *elimination–addition* sequence, has been observed in vinylic systems (a similar mechanism is known for aromatic substrates, p. 646). An example of a reaction involving this mechanism is the reaction of 1,2-dichloroethane with ArS^- and OEt^- to produce **74**. The mechanism may be formulated as:

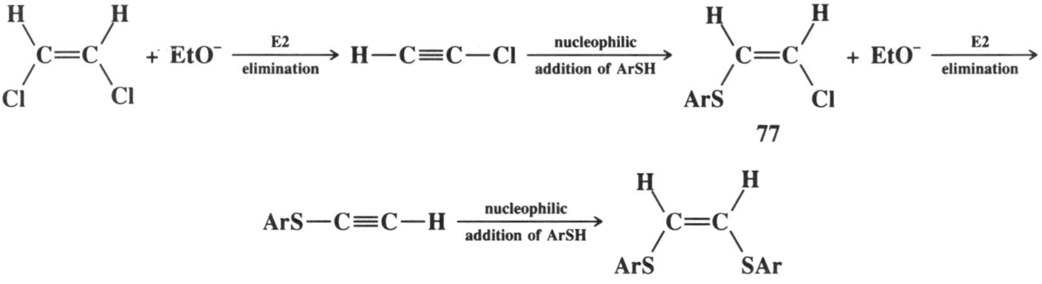

The steps are the same as in the addition–elimination mechanism, but in reverse order. Evidence for this sequence[241] is as follows: (1) The reaction does not proceed without ethoxide ion, and the rate is dependent on the concentration of this ion and not on that of ArS^-. (2) Under the same reaction conditions, chloroacetylene gave **77** and **74**. (3) **77**, treated with ArS^-, gave no reaction but, when EtO^- was added, **74** was obtained. It is interesting that the elimination–addition mechanism has even been shown to occur in five- and six-membered cyclic systems, where triple bonds are greatly strained.[242] Note that both the addition–elimination and elimination–addition sequences, as shown above, lead to overall retention of configuration, since in each case both addition and elimination are anti.

[238]Maroni; Melloni; Modena *J. Chem. Soc., Chem. Commun.* **1972**, 857.

[239]Angelini; Hanack; Vermehren; Speranza *J. Am. Chem. Soc.* **1988**, *110*, 1298.

[240]For a review, see Cacace *Adv. Phys. Org. Chem.* **1970**, *8*, 79-149. See also Angelini; Fornarini; Speranza *J. Am. Chem. Soc.* **1982**, *104*, 4773; Fornarini; Speranza *Tetrahedron Lett.* **1984**, *25*, 869, *J. Am. Chem. Soc.* **1985**, *107*, 5358.

[241]Truce; Boudakian; Heine; McManimie *J. Am. Chem. Soc.* **1956**, *78*, 2743; Flynn; Badiger; Truce *J. Org. Chem.* **1963**, *28*, 2298. See also Shainyan; Mirskova *J. Org. Chem. USSR* **1984**, *20*, 885, 1989, **1985**, *21*, 283.

[242]Montgomery; Scardiglia; Roberts *J. Am. Chem. Soc.* **1965**, *87*, 1917; Montgomery; Clouse; Crelier; Applegate *J. Am. Chem. Soc.* **1967**, *89*, 3453; Caubere; Brunet *Tetrahedron* **1971**, *27*, 3515; Bottini; Corson; Fitzgerald; Frost *Tetrahedron* **1972**, *28*, 4883.

The elimination–addition sequence has also been demonstrated for certain reactions of saturated substrates, e.g., $ArSO_2CH_2CH_2SO_2Ar$.[243] Treatment of this with ethoxide proceeds as follows:

$$ArSO_2CH_2CH_2SO_2Ar \xrightarrow[\text{E2 elimination}]{\text{EtO}^-} ArSO_2CH{=}CH_2 \xrightarrow[\text{addition}]{\text{EtO}^-} ArSO_2CH_2CH_2OEt$$

Mannich bases (see **6-16**) of the type $RCOCH_2CH_2NR_2$ similarly undergo nucleophilic substitution by the elimination–addition mechanism.[244] The nucleophile replaces the NR_2 group.

The simple SN2 mechanism has never been convincingly demonstrated for vinylic substrates.[245]

REACTIVITY

A large amount of work has been done on this subject. Though a great deal is known, much is still poorly understood, and many results are anomalous and hard to explain. In this section only approximate generalizations are attempted. The work discussed here, and the conclusions reached, pertain to reactions taking place in solution. Some investigations have also been caried out in the gas phase.[246]

The Effect of Substrate Structure

The effect on the reactivity of a change in substrate structure depends on the mechanism.

1. *Branching at the α and β carbons.* For the SN2 mechanism, branching at either the α or the β carbon decreases the rate. Tertiary systems seldom[247] react by the SN2 mechanism and neopentyl systems react so slowly as to make such reactions, in general, synthetically useless.[248] Table 10.3 shows average relative rates for some alkyl substrates.[249] The reason for these low rates is almost certainly steric.[250] The transition state **1** is more crowded when larger groups are close to the central carbon.

TABLE 10.3 Average relative SN2 rates for some alkyl substrates[249]

R	Relative rate	R	Relative rate
Methyl	30	Isobutyl	0.03
Ethyl	1	Neopentyl	10^{-5}
Propyl	0.4	Allyl	40
Butyl	0.4	Benzyl	120
Isopropyl	0.025		

[243]Kader; Stirling *J. Chem. Soc.* **1962**, 3686. For another example, see Popov; Piskunova; Matvienko *J. Org. Chem. USSR* **1986**, *22*, 1299.

[244]For an example, see Andrisano; Angeloni; De Maria; Tramontini *J. Chem. Soc. C* **1967**, 2307.

[245]For discussions, see Miller *Tetrahedron* **1977**, *33*, 1211; Texier; Henri-Rousseau; Bourgois *Bull. Soc. Chim. Fr.* **1979**, II-11,86; Rappoport *Acc. Chem. Res.* **1981**, *14*, 7-15; Rappoport; Avramovitch *J. Org. Chem.* **1982**, *47*, 1397.

[246]See, for example DePuy; Gronert; Mullin; Bierbaum *J. Am. Chem. Soc.* **1990**, *112*, 8650.

[247]For a reported example, see Edwards; Grieco *Can. J. Chem.* **1974**, *52*, 3561.

[248]SN2 reactions on neopentyl tosylates have been conveniently carried out in the solvents HMPA and Me$_2$SO: Lewis; Gustafson; Erman *Tetrahedron Lett.* **1967**, 401; Paquette; Philips *Tetrahedron Lett.* **1967**, 4645; Stephenson; Solladié; Mosher *J. Am. Chem. Soc.* **1972**, *94*, 4184; Anderson; Stephenson; Mosher *J. Am. Chem. Soc.* **1974**, *96*, 3171.

[249]This table is from Streitwieser, Ref. 1, p. 13. Also see Table 9.2.

[250]For evidence, see Caldwell; Magnera; Kebarle *J. Am. Chem. Soc.* **1984**, *106*, 959.

The tetrahedral mechanism for substitution at a carbonyl carbon is also slowed or blocked completely by α or β branching for similar reasons. For example, esters of the formula

$$Y\cdots\overset{\displaystyle \diagdown \diagup}{\underset{\displaystyle |}{C}}\cdots X$$

1

R_3CCOOR' cannot generally be hydrolyzed by the tetrahedral mechanism (see **0-10**), nor can acids R_3CCOOH be easily esterified.[251] Synthetic advantage can be taken of this fact, for example, when in a molecule containing two ester groups only the less hindered one is hydrolyzed.

For the SN1 mechanism, α branching increases the rate, as shown in Table 10.4.[252] We can explain this by the stability order of alkyl cations (tertiary > secondary > primary). Of course, the rates are not actually dependent on the stability of the ions, but on the difference in free energy between the starting compounds and the transition states. We use the Hammond postulate (p. 215) to make the assumption that the transition states resemble the cations and that anything (such as α branching) that lowers the free energy of the ions also lowers it for the transition states. For simple alkyl groups, the SN1 mechanism is important under all conditions only for tertiary substrates.[253] As previously indicated (p. 306), secondary substrates generally react by the SN2 mechanism,[254] except that the SN1 mechanism may become important at high solvent polarities. Table 10.4 shows that isopropyl bromide reacts less than twice as fast as ethyl bromide in the relatively nonpolar 60% ethanol (compare this with the 10^4 ratio for *t*-butyl bromide, where the mechanism is certainly SN1), but in the more polar water the rate ratio is 11.6. The 2-adamantyl system is an exception; it is a secondary system that reacts by the SN1 mechanism because backside attack is hindered for steric reasons.[255] Because there is no SN2 component, this system provides an opportunity for comparing the pure SN1 reactivity of secondary and tertiary substrates. It has been found that substitution of a methyl group for the α hydrogen of 2-adamantyl substrates (thus changing a secondary to a tertiary system) increases solvolysis rates by a factor of about 10^8.[256] Simple primary substrates react by the SN2 mechanism (or with participation by neighboring alkyl or hydrogen) but not by the SN1 mechanism, even when solvolyzed in

TABLE 10.4   Relative rates of solvolysis of RBr in two solvents[252]

RBr substrate	In 60% ethanol at 55°C	In water at 50°C
MeBr	2.08	1.05
EtBr	1.00	1.00
iso-**PrBr**	1.78	11.6
t-**BuBr**	2.41×10^4	1.2×10^6

[251]For a molecular mechanics study of this phenomenon, see DeTar; Binzet; Darba *J. Org. Chem.* **1987**, *52*, 2074.

[252]These values are from Streitwieser, Ref. 1, p. 43, where values are also given for other conditions. Methyl bromide reacts faster than ethyl bromide (and in the case of 60% ethanol, ispropyl bromide) because most of it (probably all) reacts by the SN2 mechanism.

[253]For a report of an SN1 mechanism at a primary carbon, see Zamashchikov; Bezbozhnaya; Chanysheva *J. Org. Chem. USSR* **1986**, *22*, 1029.

[254]See Raber; Harris *J. Chem. Educ.* **1972**, *49*, 60; Lambert; Putz; Mixan *J. Am. Chem. Soc.* **1972**, *94*, 5132; Nordlander; McCrary *J. Am. Chem. Soc.* **1972**, *94*, 5133; Ref. 38; Dietze; Jencks, Ref. 62; Dietze; Hariri; Khattak, Ref. 62.

[255]Fry; Harris; Bingham; Schleyer *J. Am. Chem. Soc.* **1970**, *92*, 2540; Schleyer; Fry; Lam; Lancelot *J. Am. Chem. Soc.* **1970**, *92*, 2542. See also Pritt; Whiting *J. Chem. Soc., Perkin Trans.* 2 **1975**, 1458. For an ab initio molecular orbital study of the 2-adamantyl cation, see Dutler; Rauk; Sorensen; Whitworth *J. Am. Chem. Soc.* **1989**, *111*, 9024.

[256]Fry; Engler; Schleyer *J. Am. Chem. Soc.* **1972**, *94*, 4628. See also Gassman; Pascone *J. Am. Chem. Soc.* **1973**, *95*, 7801.

solvents of very low nucleophilicity (e.g., trifluoroacetic acid or trifluoroethanol[257]), and even when very good leaving groups (e.g., OSO_2F) are present[258] (see, however, p. 359).

For some tertiary substrates, the rate of SN1 reactions is greatly increased by the relief of B strain in the formation of the carbocation (see p. 276). Except where B strain is involved, β branching has little effect on the SN1 mechanism, except that carbocations with β branching undergo rearrangements readily. Of course, isobutyl and neopentyl are primary substrates, and for this reason react very slowly by the SN1 mechanism, but not more slowly than the corresponding ethyl or propyl compounds.

To sum up, primary and secondary substrates generally react by the SN2 mechanism and tertiary by the SN1 mechanism. However, tertiary substrates seldom undergo nucleophilic substitution at all. Elimination is always a possible side reaction of nucleophilic substitutions (wherever a β hydrogen is present), and with tertiary substrates it usually predominates. With a few exceptions, nucleophilic substitutions at a tertiary carbon have little or no preparative value. However, tertiary substrates that can react by the SET mechanism (e.g., p-$NO_2C_6H_4CMe_2Cl$) give very good yields of substitution products when treated with a variety of nucleophiles.[259]

2. *Unsaturation at the α carbon.* Vinylic, acetylenic,[260] and aryl substrates are very unreactive toward nucleophilic substitutions. For these systems both the SN1 and SN2 mechanisms are greatly slowed or stopped altogether. One reason that has been suggested for this is that sp^2 (and even more, sp) carbons have a higher electronegativity than sp^3 carbons and thus a greater attraction for the electrons of the bond. As we have seen (p. 269), an sp—H bond has a higher acidity than an sp^3—H bond, with that of an sp^2—H bond in between. This is reasonable; the carbon retains the electrons when the proton is lost and an sp carbon, which has the greatest hold on the electrons, loses the proton most easily. But in nucleophilic substitution, the leaving group *carries off* the electron pair, so the situation is reversed and it is the sp^3 carbon that loses the leaving group and the electron pair most easily. It may be recalled (p. 20) that bond distances decrease with increasing s character. Thus the bond length for a vinylic or aryl C—Cl bond is 1.73 Å compared with 1.78 Å for a saturated C—Cl bond. Other things being equal, a shorter bond is a stronger bond.

Of course we have seen (p. 337) that SN1 reactions at vinylic substrates can be accelerated by α substituents that stabilize that cation, and that reactions by the tetrahedral mechanism can be accelerated by β substituents that stabilize the carbanion. Also, reactions at vinylic substrates can in certain cases proceed by addition–elimination or elimination–addition sequences (pp. 335, 338).

In contrast to such systems, substrates of the type RCOX are usually much *more* reactive than the corresponding RCH$_2$X. Of course, the mechanism here is almost always the tetrahedral one. Three reasons can be given for the enhanced reactivity of RCOX: (1) The carbonyl carbon has a sizable partial positive charge that makes it very attractive to nucleophiles. (2) In an SN2 reaction a σ bond must break in the rate-determining step, which requires more energy than the shift of a pair of π electrons, which is what happens in a tetrahedral mechanism. (3) A trigonal carbon offers less steric hindrance to a nucleophile than a tetrahedral carbon.

For reactivity in aryl systems, see Chapter 13.

3. *Unsaturation at the β carbon.* SN1 rates are increased when there is a double bond in the β position, so that allylic and benzylic substrates react rapidly (Table 10.5).[261] The

[257]Dafforn; Streitwieser *Tetrahedron Lett.* **1970**, 3159.

[258]Cafferata; Desvard; Sicre *J. Chem. Soc., Perkin Trans. 2* **1981**, 940.

[259]Kornblum et al. *J. Org. Chem.* **1987**, *52*, 196.

[260]For a discussion of SN reactions at acetylenic substrates, see Miller; Dickstein *Acc. Chem. Res.* **1976**, *9*, 358-363.

[261]Streitwieser, Ref. 1, p. 75. Actually, the figures for Ph$_2$CHOTs and Ph$_3$COTs are estimated from the general reactivity of these substrates.

TABLE 10.5   Relative rates for
the S$_N$1 reaction between ROTs
and ethanol at 25°C[261]

Group	Relative rate
Et	0.26
iso-**Pr**	0.69
CH$_2$=CHCH$_2$	8.6
PhCH$_2$	100
Ph$_2$CH	~10^5
Ph$_3$C	~10^{10}

reason is that allylic (p. 168) and benzylic (p. 169) cations are stabilized by resonance. As
shown in Table 10.5, a second and a third phenyl group increase the rate still more, because
these carbocations are more stable yet. It should be remembered that allylic rearrangements
are possible with allylic systems.

In general, S$_N$1 rates at an allylic substrate are increased by any substituent in the 1 or
3 position that can stabilize the carbocation by resonance or hyperconjugation.[262] Among
these are alkyl, aryl, and halo groups.

S$_N$2 rates for allylic and benzylic systems are also increased (see Table 10.3), probably
owing to resonance possibilities in the transition state. Evidence for this in benzylic systems
is that the rate of the reaction

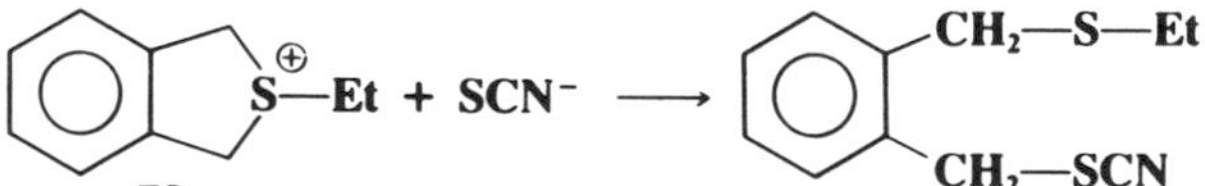

was 8000 times slower than the rate with $(PhCH_2)_2SEt^+$.[263] The cyclic **78** does not have the
proper geometry for conjugation in the transition state.

Triple bonds in the β position (in propargyl systems) have about the same effect as double
bonds.[264] Alkyl, aryl, halo, and cyano groups, among others, in the 3 position of allylic
substrates increase S$_N$2 rates, owing to increased resonance in the transition state, but alkyl
and halo groups in the 1 position decrease the rates because of steric hindrance.

4. α *substitution.*   Compounds of the formula ZCH$_2$X, where Z = RO, RS, or R$_2$N
undergo S$_N$1 reactions very rapidly,[265] because of the increased resonance in the carbocation.
These groups have an unshared pair on an atom directly attached to the positive carbon,
which stabilizes the carbocation (p. 170). The field effects of these groups would be expected
to decrease S$_N$1 rates (see Section 6, p. 344), so the resonance effect is far more important.

When Z in ZCH$_2$X is RCO,[266] HCO, ROCO, NH$_2$CO, NC, or F$_3$C,[267] S$_N$1 rates are
decreased compared to CH$_3$X, owing to the electron-withdrawing field effects of these

[262]For a discussion of the relative reactivities of different allylic substrates, see DeWolfe; Young, in Patai, Ref.
178, pp. 683-688, 695-697.
[263]King; Tsang; Abdel-Malik; Payne *J. Am. Chem. Soc.* **1985,** *107,* 3224.
[264]Hatch; Chiola *J. Am. Chem. Soc.* **1951,** *73,* 360; Jacobs; Brill *J. Am. Chem. Soc.* **1953,** *75,* 1314.
[265]For a review of the reactions of α-haloamines, sulfides, and ethers, see Gross; Höft *Angew. Chem. Int. Ed.
Engl.* **1967,** *6,* 335-355 [*Angew. Chem. 79,* 358-378].
[266]For a review of α-halo ketones, including reactivity, see Verhé; De Kimpe, in Patai; Rappoport, Ref. 88, pt.
1, pp. 813-931. This review has been reprinted, and new material added, in De Kimpe; Verhé *The Chemistry of
α-Haloketones, α-Haloaldehydes, and α-Haloimines*; Wiley: New York, 1988, pp. 225-368.
[267]Allen; Jansen; Koshy; Mangru; Tidwell *J. Am. Chem. Soc.* **1982,** *104,* 207; Liu; Kuo; Shu *J. Am. Chem. Soc.*
1982, *104,* 211; Gassman; Harrington *J. Org. Chem.* **1984,** *49,* 2258; Allen; Girdhar; Jansen; Mayo; Tidwell *J. Org.
Chem.* **1986,** *51,* 1324; Allen; Kanagasabapathy; Tidwell *J. Am. Chem. Soc.* **1986,** *108,* 3470; Richard *J. Am. Chem.
Soc.* **1989,** *111,* 1455.

groups. Furthermore, carbocations[268] with an α CO or CN group are greatly destabilized because of the partial positive charge on the adjacent carbon (**79**). S$_N$1 reactions have been carried out on such compounds,[269] but the rates are very low. For example, from a comparison of the solvolysis rates of **80** and **81**, a rate-retarding effect of $10^{7.3}$ was estimated for the

$$\underset{\textbf{79}}{\overset{\displaystyle R\overset{\oplus}{\underset{R}{-}}\overset{\delta+}{C}\overset{}{\underset{O\delta-}{\overset{\|}{-}}}R}{}}$$

80   **81**

C$=$O group.[270] However, when a different kind of comparison is made: RCOCR$_2'$X vs. HCR$_2'$X (where X = a leaving group), the RCO had only a small or negligible rate-retarding effect, indicating that resonance stabilization[271]

$$\underset{\textbf{C}}{R-\overset{R}{\underset{\oplus}{C}}-\overset{}{\underset{|\underline{O}}{\overset{\|}{C}}}-R} \longleftrightarrow \underset{\textbf{D}}{R-\overset{R}{C}=\overset{}{\underset{|\underline{O}_{\oplus}}{C}}-R}$$

may be offsetting the inductive destabilization for this group.[272] For a CN group also, the rate-retarding effect is reduced by this kind of resonance.[273] A carbocation with an α COR group has been isolated.[274]

When S$_N$2 reactions are carried out on these substrates, rates are greatly increased for certain nucleophiles (e.g., halide or halide-like ions), but decreased or essentially unaffected by others.[275] For example, α-chloroacetophenone (PhCOCH$_2$Cl) reacts with KI in acetone at 75° about 32,000 times faster than 1-chlorobutane,[276] but α-bromoacetophenone reacts with the nucleophile triethylamine 0.14 times as fast as iodomethane.[275] The reasons for this varying behavior are not clear, but those nucleophiles that form a "tight" transition state (one in which bond making and bond breaking have proceeded to about the same extent) are more likely to accelerate the reaction.[277]

[268]For reviews of such carbocations, see Bégué; Charpentier-Morize *Acc. Chem. Res.* **1980**, *13*, 207-212; Charpentier-Morize *Bull. Soc. Chim. Fr.* **1974**, 343-351.

[269]For reviews, see Creary *Acc. Chem. Res.* **1985**, *18*, 3-8; Creary; Hopkinson; Lee-Ruff *Adv. Carbocation Chem.* **1989**, *1*, 45-92; Charpentier-Morize; Bonnet-Delpon *Adv. Carbocation Chem.* **1989**, *1*, 219-253.

[270]Creary *J. Org. Chem.* **1979**, *44*, 3938.

[271]**D**, which has the positive charge on the more electronegative atom, is less stable than **C**, according to rule c on p. 36, but it nevertheless seems to be contributing in this case.

[272]Creary; Geiger *J. Am. Chem. Soc.* **1982**, *104*, 4151; Creary *J. Am. Chem. Soc.* **1984**, *106*, 5568. See however Takeuchi; Yoshida; Ohga; Tsugeno; Kitagawa *J. Org. Chem.* **1990**, *55*, 6063.

[273]Gassman; Saito; Talley *J. Am. Chem. Soc.* **1980**, *102*, 7613.

[274]Takeuchi; Kitagawa; Okamoto *J. Chem. Soc., Chem. Commun.* **1983**, 7. See also Dao; Maleki; Hopkinson; Lee-Ruff *J. Am. Chem. Soc.* **1986**, *108*, 5237.

[275]Halvorsen; Songstad *J. Chem. Soc., Chem. Commun.* **1978**, 327.

[276]Bordwell; Brannen *J. Am. Chem. Soc.* **1964**, *86*, 4645. For some other examples, see Conant; Kirner; Hussey *J. Am. Chem. Soc.* **1925**, *47*, 488; Sisti; Lowell *Can. J. Chem.* **1964**, *42*, 1896.

[277]For discussions of possible reasons, see McLennan; Pross *J. Chem. Soc., Perkin Trans. 2* **1984**, 981; Yousaf; Lewis *J. Am. Chem. Soc.* **1987**, *109*, 6137; Lee; Shim; Chung; Lee *J. Chem. Soc., Perkin Trans. 2* **1988**, 975; Yoh; Lee *Tetrahedron Lett.* **1988**, *29*, 4431.

When Z is SOR or SO_2R (e.g., α-halo sulfoxides and sulfones), nucleophilic substitution is retarded.[278] The SN1 mechanism is slowed by the electron-withdrawing effect of the SOR or SO_2R group,[279] and the SN2 mechanism presumably by the steric effect.

5. β *substitution.* For compounds of the type ZCH_2CH_2X, where Z is any of the groups listed in the previous section as well as halogen or phenyl, SN1 rates are lower than for unsubstituted systems, because the resonance effects mentioned in Section 4 are absent, but the field effects are still there, though smaller. These groups in the β position do not have much effect on SN2 rates unless they behave as neighboring groups and enhance the rate through anchimeric assistance,[280] or unless their size causes the rates to decrease for steric reasons.[281]

6. *The effect of electron-donating and electron-withdrawing groups.* If substitution rates of series of compounds $p\text{-}ZC_6H_4CH_2X$ are measured, it is possible to study the electronic effects of groups Z on the reaction. Steric effects of Z are minimized or eliminated, because Z is so far from the reaction site. For SN1 reactions electron-withdrawing Z decrease the rate and electron-donating Z increase it,[282] because the latter decrease the energy of the transition state (and of the carbocation) by spreading the positive charge, e.g.,

while electron-withdrawing groups concentrate the charge. The Hammett σρ relationship (p. 278) correlates fairly successfully the rates of many of these reactions (with $σ^+$ instead of σ). ρ values are generally about -4, which is expected for a reaction where a positive charge is created in the transition state.

For SN2 reactions no such simple correlations are found.[283] In this mechanism bond breaking is about as important as bond making in the rate-determining step, and substituents have an effect on both processes, often in opposite directions. The unsubstituted benzyl chloride and bromide solvolyze by the SN2 mechanism.[282]

For Z = alkyl, the Baker–Nathan order (p. 68) is usually observed both for SN1 and SN2 reactions.

In para-substituted benzyl systems, steric effects have been removed, but resonance and field effects are still present. However, Holtz and Stock studied a system that removes not only steric effects but also resonance effects. This is the 4-substituted bicyclo[2.2.2]octylmethyl tosylate system (**82**).[284] In this system steric effects are completely

$$Z\text{—}\langle\rangle\text{—}CH_2OTs$$

82

[278]Bordwell; Jarvis *J. Org. Chem.* **1968**, *33*, 1182; Loeppky; Chang *Tetrahedron Lett.* **1968**, 5414; Cinquini; Colonna; Landini; Maia *J. Chem. Soc., Perkin Trans. 2* **1976**, 996.
[279]See, for example Creary; Mehrsheikh-Mohammadi; Eggers *J. Am. Chem. Soc.* **1987**, *109*, 2435.
[280]For example, substrates of the type $RSCH_2CH_2X$ are so prone to the neighboring-group mechanism that ordinary SN2 reactions have only recently been observed: Sedaghat-Herati; McManus; Harris *J. Org. Chem.* **1988**, *53*, 2539.
[281]See, for example, Okamoto; Kita; Araki; Shingu *Bull. Chem.Soc. Jpn.* **1967**, *40*, 1913.
[282]Jorge; Kiyan; Miyata; Miller *J. Chem. Soc., Perkin Trans. 2* **1981**, 100; Vitullo; Grabowski; Sridharan *J. Chem. Soc., Chem. Commun.* **1981**, 737.
[283]See Sugden; Willis *J. Chem. Soc.* **1951**, 1360; Baker; Nathan *J. Chem. Soc.* **1935**, 1840; Hayami; Tanaka; Kurabayashi; Kotani; Kaji *Bull. Chem. Soc. Jpn.* **1971**, *44*, 3091; Westaway; Waszczylo *Can. J. Chem.* **1982**, *60*, 2500; Lee; Sohn; Oh; Lee *Tetrahedron* **1986**, *42*, 4713.
[284]Holtz; Stock *J. Am. Chem. Soc.* **1965**, *87*, 2404.

absent, owing to the rigidity of the molecules, and only field effects operate. By this means Holtz and Stock showed that electron-withdrawing groups increase the rate of SN2 reactions. This can be ascribed to stabilization of the transition state by withdrawal of some of the electron density.

For substrates that react by the tetrahedral mechanism, electron-withdrawing groups increase the rate and electron-donating groups decrease it.

7. *Cyclic substrates.* Cyclopropyl substrates are extremely resistant to nucleophilic attack.[285] For example, cyclopropyl tosylate solvolyzes about 10^6 times more slowly than cyclobutyl tosylate in acetic acid at 60°C.[286] When such attack does take place, the result is generally not normal substitution (though exceptions are known,[287] especially when an α stabilizing group such as aryl or alkoxy is present) but ring opening:[286]

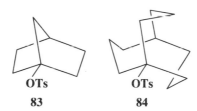

There is much evidence that the ring opening is usually concerted with the departure of the leaving group[288] (as in the similar case of cyclobutyl substrates, p. 324), from which we can conclude that if the 2,3 bond of the cyclopropane ring did not assist, the rates would be lower still. It has been estimated[289] that without this assistance the rates of these already slow reactions would be further reduced by a factor of perhaps 10^{12}. For a discussion of the stereochemistry of the ring opening, see p. 1119. For larger rings, we have seen (p. 276) that, because of I strain, cyclohexyl substrates solvolyze slower than analogous compounds in which the leaving group is attached to a ring of 5 or of from 7 to 11 members.

8. *Bridgeheads.*[10] The SN2 mechanism is impossible at bridgeheads (p. 296). SN1 reactions can take place if the rings are large enough (p. 301).[290] Solvolytic reactivity at bridgehead positions spans a wide range; e.g., from $k = 4 \times 10^{-17}$ s^{-1} for **83** (very slow)

OTs   OTs

83   **84**

to 3×10^6 s^{-1} for the [3.3.3] compound **84** (very fast);[291] a range of 22 orders of magnitude. Molecular mechanics calculations show that SN1 bridgehead reactivity is determined by strain changes between the substrate and the carbocation intermediate.[292]

[285]For reviews, see Friedrich, in Rappoport *The Chemistry of the Cyclopropyl Group*, pt. 1; Wiley: New York, 1987, pp. 633-700; Aksenov; Terent'eva; Savinykh *Russ. Chem. Rev.* **1980**, *49*, 549-557.

[286]Roberts; Chambers *J. Am. Chem. Soc.* **1951**, *73*, 5034.

[287]For example, see Kirmse; Schütte *J. Am. Chem. Soc.* **1967**, *89*, 1284; Landgrebe; Becker *J. Am. Chem. Soc.* **1967**, *89*, 2505; Howell; Jewett *J. Am. Chem. Soc.* **1971**, *93*, 798; van der Vecht; Steinberg; de Boer *Recl. Trav. Chim. Pays-Bas* **1978**, *96*, 313; Engbert; Kirmse *Liebigs Ann. Chem.* **1980**, 1689; Turkenburg; de Wolf; Bickelhaupt; Stam; Konijn *J. Am. Chem. Soc.* **1982**, *104*, 3471; Banert *Chem. Ber.* **1985**, *118*, 1564; Vilsmaier; Weber; Weidner *J. Org. Chem.* **1987**, *52*, 4921.

[288]For example, see Schleyer; Van Dine; Schöllkopf; Paust *J. Am. Chem. Soc.* **1966**, *88*, 2868; DePuy; Schnack; Hausser *J. Am. Chem. Soc.* **1966**, *88*, 3343; Jefford; Medary *Tetrahedron* **1967**, *23*, 4123; Jefford; Wojnarowski *Tetrahedron* **1969**, *25*, 2089; Hausser; Uchic *J. Org. Chem.* **1972**, *37*, 4087.

[289]Sliwinski; Su; Schleyer *J. Am. Chem. Soc.* **1972**, *94*, 133; Brown; Rao; Ravindranathan *J. Am. Chem. Soc.* **1978**, *100*, 7946.

[290]For a review of organic synthesis using bridgehead carbocations, see Kraus; Hon; Thomas; Laramay; Liras; Hanson *Chem. Rev.* **1989**, *89*, 1591-1598.

[291]Bentley; Roberts *J. Org. Chem.* **1988**, *50*, 5852.

[292]Gleicher; Schleyer *J. Am. Chem. Soc.* **1967**, *89*, 582; Bingham; Schleyer *J. Am. Chem. Soc.* **1971**, *93*, 3189; Müller; Blanc; Mareda *Chimia* **1987**, *41*, 399; Müller; Mareda *Helv. Chim. Acta* **1987**, *70*, 1017; Ref. 291.

TABLE 10.6 List of groups in approximately descending order of reactivity toward S$_N$1 and S$_N$2 reactions
Z is RCO, HCO, ROCO, NH$_2$CO, NC, or a similar group

S$_N$1 reactivity	S$_N$2 reactivity
Ar$_3$CX	Ar$_3$CX
Ar$_2$CHX	Ar$_2$CHX
ROCH$_2$X, RSCH$_2$X, R$_2$NCH$_2$X	ArCH$_2$X
R$_3$CX	ZCH$_2$X
ArCH$_2$X	
$-\overset{\mid}{C}=\overset{\mid}{C}CH_2X$	$-\overset{\mid}{C}=\overset{\mid}{C}CH_2X$
R$_2$CHX	RCH$_2$X ≈ RCHDX ≈ RCHDCH$_2$X
RCH$_2$X ≈ R$_3$CCH$_2$X	R$_2$CHX
RCHDX	R$_3$CX
RCHDCH$_2$X	ZCH$_2$CH$_2$X
$-\overset{\mid}{C}=\overset{\mid}{C}X$	R$_3$CCH$_2$X
ZCH$_2$X	
ZCH$_2$CH$_2$X	$-\overset{\mid}{C}=\overset{\mid}{C}X$
ArX	ArX
[2.2.1] Bridgehead-**X**	Bridgehead-**X**

TABLE 10.7 The more important synthetic reactions of Chapter 10 that take place by the S$_N$2 mechanism (R = primary, often secondary, alkyl). Catalysts are not shown[a]

0-1	RX + OH$^-$ → ROH
0-12	RX + OR'$^-$ → ROR'
0-13	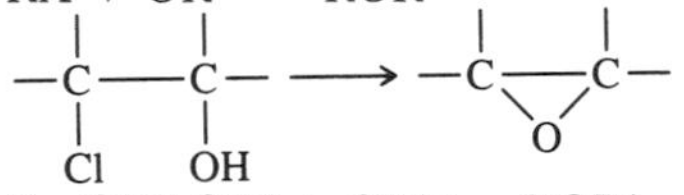 $-\overset{\mid}{\underset{\mid}{C}}-\overset{\mid}{\underset{OH}{C}}-$ ⟶ $-\overset{\mid}{C}\overset{}{\underset{O}{\diagdown\diagup}}\overset{\mid}{C}-$ $\quad$ (Cl)
0-14	R—OSO$_2$OR″ + OR'$^-$ → ROR'
0-16	2ROH → ROR
0-18	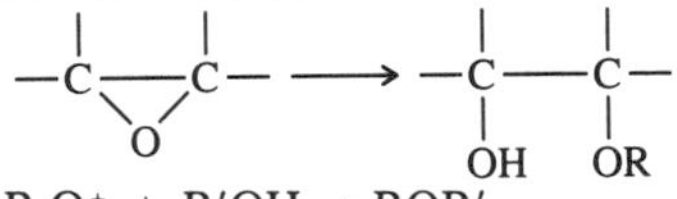 $-\overset{\mid}{C}\overset{}{\underset{O}{\diagdown\diagup}}\overset{\mid}{C}-$ ⟶ $-\overset{\mid}{\underset{OH}{C}}-\overset{\mid}{\underset{OR}{C}}-$
0-19	R$_3$O$^+$ + R'OH → ROR'
0-24	RX + R'COO$^-$ → R'COOR
0-31	RX + OOH$^-$ → ROOH
0-35	RX + SH$^-$ → RSH
0-36	RX + R'S$^-$ → RSR'
0-38	RX + S$_2$$^{2-}$ → RSSR
0-41	RX + SO$_3$$^{2-}$ → RSO$_2$O$^-$
0-42	RX + SCN$^-$ → RSCN
0-43	RX + R'$_2$NH → RR'$_2$N
0-43	RX + R'$_3$N → RR'$_3$N$^+$ X$^-$
0-44	RX + (CH$_2$)$_6$N$_4$ → N$_2$(CH$_2$)$_6$NR$^+$ X$^-$ $\xrightarrow{\text{H}^+}$ RNH$_2$
0-49	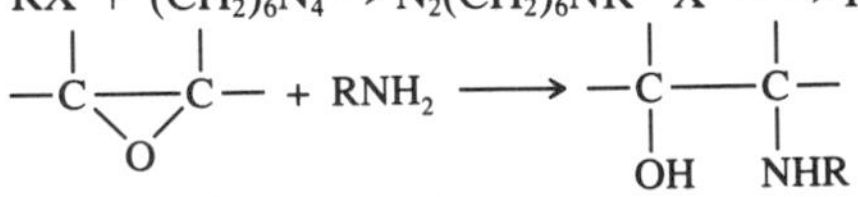 $-\overset{\mid}{C}\overset{}{\underset{O}{\diagdown\diagup}}\overset{\mid}{C}-$ + RNH$_2$ ⟶ $-\overset{\mid}{\underset{OH}{C}}-\overset{\mid}{\underset{NHR}{C}}-$
0-58	RX + R'CONH$^-$ → RNHCOR'

TABLE 10.7 *(Continued)*

0-60	RX + NO$_2^-$ → RNO$_2$ + RONO
0-61	RX + N$_3^-$ → RN$_3$
0-62	RX + NCO$^-$ → RNCO
0-62	RX + NCS$^-$ → RNCS

0-65	RX + X$'^-$ → RX$'$
0-66	R—OSO$_2$OR$'$ + X$^-$ → RX
0-67	ROH + PCl$_5$ → RCl
0-68	ROR$'$ + 2HI → RI + R$'$I
0-69	$-\overset{|}{C}\underset{\diagup O\diagdown}{-}\overset{|}{C}- + HX \longrightarrow -\overset{|}{C}-\overset{|}{C}-$ OH X
0-70	R—O—COR$'$ + LiI → RI + R$'$COO$^-$

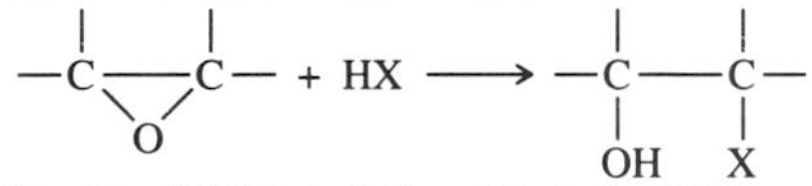

0-76	RX + LiAlH$_4$ → RH
0-77	R—OSO$_2$R$'$ + LiAlH$_4$ → RH
0-80	$-\overset{|}{C}\underset{\diagup O\diagdown}{-}\overset{|}{C}- + LiAlH_4 \longrightarrow -\overset{|}{C}-\overset{|}{C}-$ OH H

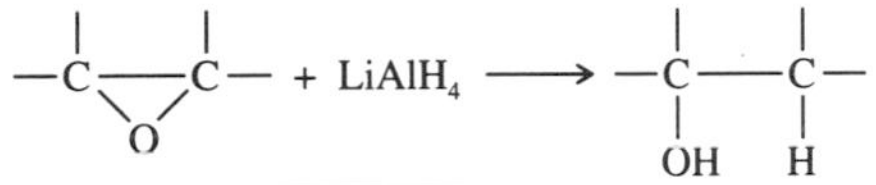

0-87	RX + R$'_2$CuLi → RR$'$
0-93	$-\overset{|}{C}\underset{\diagup O\diagdown}{-}\overset{|}{C}- + RMgX \longrightarrow -\overset{|}{C}-\overset{|}{C}-$ OII R
0-94	RX + H$\overline{C}$(CO$_2$R$'$)$_2$ → RCH(CO$_2$R$'$)$_2$
0-95	RX + R$''\overline{C}$II—COR$'$ → RCR$''$—COR$'$
0-96	RX + R$'\overline{C}$HCOO$^-$ → RR$'$CHCOO$^-$
0-97	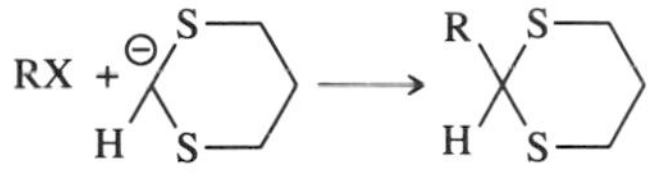 RX + (dithiane) → (substituted dithiane)
0-100	RX + R$'$C≡C$^\ominus$ → RC≡CR$'$
0-101	RX + CN$^-$ → RCN

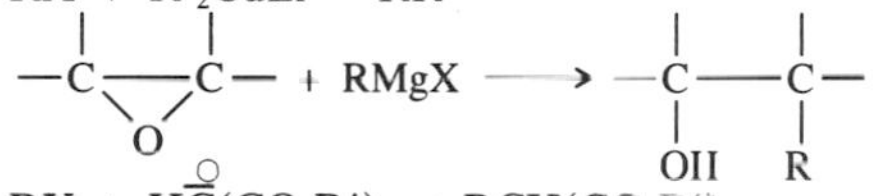

"This is schematic list only. Some of these reactions may also take place by other mechanisms and the scope may vary greatly. See the discussion of each reaction for details.

9. *Deuterium substitution.* α and β secondary isotope effects affect the rate in various ways (p. 228). The measurement of α secondary isotope effects provides a means of distinguishing between SN1 and SN2 mechanisms, since for SN2 reactions the values range from 0.95 to 1.06 per α D, while for SN1 reactions the values are higher.[293] This method is especially good because it provides the minimum of perturbation of the system under study; changing from α H to α D hardly affects the reaction, while other probes, such as changing a substituent or the polarity of the solvent, may have a much more complex effect.

Table 10.6 is an approximate listing of groups in order of SN1 and SN2 reactivity. Table 10.7 shows the main reactions that proceed by the SN2 mechanism (if R = primary or, often, secondary alkyl); Table 10.8 shows the main reactions that proceed by the tetrahedral mechanism.

[293]Ref. 39. For a review of secondary isotope effects in SN2 reactions, see Westaway *Isot. Org. Chem.* **1987,** 7, 275-392.

TABLE 10.8 The more important synthetic reactions of Chapter 10 that take place by the tetrahedral mechanism. Catalysts are not shown

0-8	$RCOX + H_2O \rightarrow RCOOH$
0-9	$RCOOCOR' + H_2O \rightarrow RCOOH + R'COOH$
0-10	$RCO_2R' + H_2O \rightarrow RCOOH + R'OH$
0-11	$RCONR_2' + H_2O \rightarrow RCOOH + R_2'NH$ $\qquad (R' = H, alkyl, aryl)$
0-20	$RCOX + R'OH \rightarrow RCO_2R'$
0-21	$RCOOCOR + R'OH \rightarrow RCO_2R'$
0-22	$RCOOH + R'OH \rightarrow RCO_2R'$
0-23	$RCO_2R' + R''OH \rightarrow RCO_2R'' + R'OH$
0-27	$RCOX + R'COO^- \rightarrow RCOOCOR'$
0-31	$RCOX + H_2O_2 \rightarrow RCO_3H$

0-37	$RCOX + R'SH \rightarrow RCOSR'$
0-52	$RCOX + NHR_2' \rightarrow RCONR_2'$ $\qquad (R' = H, alkyl, aryl)$
0-53	$RCOOCOR + NHR_2' \rightarrow RCONR_2'$ $\qquad (R' = H, alkyl, aryl)$
0-54	$RCOOH + NHR_2' \xrightarrow[\text{agent}]{\text{coupling}} RCONR_2'$ $\qquad (R' = H, alkyl, aryl)$
0-55	$RCO_2R' + NHR_2''$ $\qquad (R'' = H, alkyl, aryl)$
0-74	$RCOOH + SOCl_2 \rightarrow RCOCl$
0-83	$RCOX + LiAlH(O\text{-}t\text{-}Bu)_3 \rightarrow RCHO$
0-85	$RCONR_2' + LiAlH_4 \rightarrow RCHO$
0-104	$RCOX + R_2CuLi \rightarrow RCOR'$
0-108	$2RCH_2CO_2R' \rightarrow RCH_2COCHRCO_2R'$

The Effect of the Attacking Nucleophile[294]

Any species that has an unshared pair (i.e., any Lewis base) can be a nucleophile, whether it is neutral or has a negative charge. The rates of SN1 reactions are independent of the identity of the nucleophile, since it does not appear in the rate-determining step.[295] This may be illustrated by the effect of changing the nucleophile from H_2O to OH^- for a primary and a tertiary substrate. For methyl bromide, which reacts by an SN2 mechanism, the rate is multiplied more than 5000 by the change to the more powerful nucleophile OH^-, but for *t*-butyl bromide, which reacts by an SN1 mechanism, the rate is unaffected.[296] A change in nucleophile can, however, change the *product* of an SN1 reaction. Thus solvolysis of benzyl tosylate in methanol gives benzyl methyl ether (the nucleophile is the solvent methanol). If the more powerful nucleophile Br^- is added, the rate is unchanged, but the product is now benzyl bromide.

For SN2 reactions in solution there are four main principles that govern the effect of the nucleophile on the rate, though the nucleophilicity order is not invariant but depends on substrate, solvent, leaving group, etc.

1. A nucleophile with a negative charge is always a more powerful nucleophile than its conjugate acid (assuming the latter is also a nucleophile). Thus OH^- is more powerful than H_2O, NH_2^- more powerful than NH_3, etc.

---

[294]For a monograph, see Harris; McManus *Nucleophilicity*; American Chemical Society: Washington, 1987. For reviews, see Klumpp *Reactivity in Organic Chemistry*; Wiley: New York, 1982, pp. 145-167, 181-186; Hudson, in Klopman *Chemical Reactivity and Reaction Paths*; Wiley: New York, 1974, pp. 167-252.

[295]It is, however, possible to measure the rates of reaction of nucleophiles with fairly stable carbocations: see Ritchie *Acc. Chem. Res.* **1972**, *5*, 348-354; Ritchie; Minasz; Kamego; Sawada *J. Am. Chem. Soc.* **1977**, *99*, 3747; McClelland; Banait; Steenken *J. Am. Chem. Soc.* **1986**, *108*, 7023.

[296]Bateman; Cooper; Hughes; Ingold *J. Chem. Soc.* **1940**, 925.

2. In comparing nucleophiles whose attacking atom is in the same row of the periodic table, nucleophilicity is approximately in order of basicity, though basicity is thermodynamically controlled and nucleophilicity is kinetically controlled. So an approximate order of nucleophilicity is $NH_2^- > RO^- > OH^- > R_2NH > ArO^- > NH_3 > $ pyridine $> F^- > H_2O > ClO_4^-$, and another is $R_3C^- > R_2N^- > RO^- > F^-$ (see Table 8.1). This type of correlation works best when the structures of the nucleophiles being compared are similar, as with a set of substituted phenoxides. Within such a series, linear relationships can often be established between nucleophilic rates and pK values.[297]

3. Going down the periodic table, nucleophilicity increases, though basicity decreases. Thus the usual order of halide nucleophilicity is $I^- > Br^- > Cl^- > F^-$ (though as we shall see below, this order is solvent-dependent). Similarly, any sulfur nucleophile is more powerful than its oxygen analog, and the same is true for phosphorus vs. nitrogen. The main reason for this distinction between basicity and nucleophilic power is that the smaller negatively charged nucleophiles are more solvated by the usual polar protic solvents; that is, because the negative charge of Cl^- is more concentrated than the charge of I^-, the former is more tightly surrounded by a shell of solvent molecules that constitute a barrier between it and the substrate. This is most important for protic polar solvents in which the solvent may be hydrogen-bonded to small nucleophiles. Evidence for this is that many nucleophilic substitutions with small negatively charged nucleophiles are much more rapid in aprotic polar solvents than in protic ones[298] and that, in DMF, an aprotic solvent, the order of nucleophilicity was $Cl^- > Br^- > I^-$.[299] Another experiment was the use of $Bu_4N^+ X^-$ and LiX as nucleophiles in acetone, where X^- was a halide ion. The halide ion in the former salt is much less associated than in LiX. The relative rates with LiX were Cl^-, 1; Br^-, 5.7; I^-, 6.2, which is in the normal order, while with $Bu_4N^+ X^-$, where X^- is much freer, the relative rates were Cl^-, 68; Br^-, 18; I^-, 3.7.[300] In a further experiment halide ions were allowed to react with the molten salt $(n\text{-}C_5H_{11})_4N^+ X^-$ at 180°C in the absence of a solvent.[301] Under these conditions, where the ions are unsolvated and unassociated, the relative rates were Cl^-, 620; Br^-, 7.7; I^-, 1. In the gas phase, where no solvent is present, an approximate order of nucleophilicity was found to be $OH^- > F^- \sim MeO^- > MeS^- \gg Cl^- > CN^- > Br^-$,[302] providing further evidence that solvation is responsible for the effect in solution.

However, solvation is not the entire answer since, even for *uncharged* nucleophiles, nucleophilicity increases going down a column in the periodic table. These nucleophiles are not so greatly solvated and changes in solvent do not greatly affect their nucleophilicity.[303] To explain these cases we may use the principle of hard and soft acids and bases (p. 261).[304] The proton is a hard acid, but an alkyl substrate (which may be considered to act as a Lewis acid toward the nucleophile considered as a base) is a good deal softer. According to the principle given on p. 263, we may then expect the alkyl group to prefer softer nucleophiles than the proton does. Thus the larger, more polarizable (softer) nucleophiles have a greater (relative) attraction toward an alkyl carbon than toward a proton.

[297]See, for example, Jokinen; Luukkonen; Ruostesuo; Virtanen; Koskikallio *Acta Chem. Scand.* **1971**, *25*, 3367; Bordwell; Hughes *J. Org. Chem.* **1983**, *48*, 2206, *J. Am. Chem. Soc.* **1984**, *106*, 3234.

[298]Parker *J. Chem. Soc.* **1961**, 1328 has a list of about 20 such reactions.

[299]Weaver; Hutchison *J. Am. Chem. Soc.* **1964**, *86*, 261; See also Rodewald; Mahendran; Bear; Fuchs *J. Am. Chem. Soc.* **1968**, *90*, 6698; Fuchs; Mahendran *J. Org. Chem.* **1971**, *36*, 730; Müller; Siegfried *Helv. Chim. Acta* **1971**, *54*, 2675; Liotta; Grisdale; Hopkins *Tetrahedron Lett.* **1975**, 4205; Bordwell; Hughes *J. Org. Chem.* **1981**, *46*, 3570. For a contrary result in liquid SO_2, see Lichtin; Puar; Wasserman *J. Am. Chem. Soc.* **1967**, *89*, 6677.

[300]Winstein; Savedoff; Smith; Stevens; Gall *Tetrahedron Lett.* **1960**, no. 9, 24.

[301]Gordon; Varughese *Chem. Commun.* **1971**, 1160. See also Ford; Hauri; Smith *J. Am. Chem. Soc.* **1974**, *96*, 4316.

[302]Olmstead; Brauman *J. Am. Chem. Soc.* **1977**, *99*, 4219. See also Tanaka; Mackay; Payzant; Bohme *Can. J. Chem.* **1976**, *54*, 1643.

[303]Parker *J. Chem. Soc.* **1961**, 4398.

[304]Pearson *Surv. Prog. Chem.* **1969**, *5*, 1-52, pp. 21-38.

4. The freer the nucleophile, the greater the rate.[305] We have already seen one instance of this.[300] Another is that the rate of attack by $(EtOOC)_2CBu^- Na^+$ in benzene was increased by the addition of substances (for example, 1,2-dimethoxyethane, adipamide) that specifically solvated the Na^+ and thus left the anion freer.[306] In a nonpolar solvent such as benzene, salts such as $(EtOOC)_2CBu^- Na^+$ usually exist as ion-pair aggregations of large molecular weights.[307] Similarly, it was shown that the half-life of the reaction between $C_6H_5COCHEt^-$ and ethyl bromide depended on the positive ion: K^+, 4.5×10^{-3}; Na^+, 3.9×10^{-5}; Li^+, 3.1×10^{-7}.[308] Presumably, the potassium ion leaves the negative ion most free to attack most rapidly. Further evidence is that in the gas phase,[309] where nucleophilic ions are completely free, without solvent or counterion, reactions take place orders of magnitude faster than the same reactions in solution.[302] It has proven possible to measure the rates of reaction of OH^- with methyl bromide in the gas phase, with OH^- either unsolvated or solvated with one, two, or three molecules of water.[310] The rates were, with the number of water molecules in parentheses: (0) 1.0×10^{-9}; (1) 6.3×10^{-10}; (2) 2×10^{-12}; (3) 2×10^{-13} cm^3 molecule^{-1} s^{-1}. This provides graphic evidence that solvation of the nucleophile decreases the rate. The rate of this reaction in aqueous solution is 2.3×10^{-25} cm^3 molecule^{-1} s^{-1}. Similar results were found for other nucleophiles and other solvents.[311] In solution too, studies have been made of the effect of solvation of the nucleophile by a specific number of water molecules. When the salt $(n\text{-}C_6H_{13})_4N^+ F^-$ was allowed to react with n-octyl methanesulfonate, the relative rate fell from 822 for no water molecules to 96 for 1.5 water molecules to 1 for 6 water molecules.[312]

In Chapter 3 we saw that cryptands specifically solvate the alkali metal portion of salts like KF, KOAc, etc. Synthetic advantage can be taken of this fact to allow anions to be freer, thus increasing the rates of nucleophilic substitutions and other reactions (see p. 364).

However, the four rules given above do not always hold. One reason is that steric influences often play a part. For example, the t-butoxide ion Me_3CO^- is a stronger base than OH^- or OEt^-, but a much poorer nucleophile because its large bulk hinders it from closely approaching a substrate.

The following overall nucleophilicity order for SN2 mechanisms (in protic solvents) was given by Edwards and Pearson:[313] $RS^- > ArS^- > I^- > CN^- > OH^- > N_3^- > Br^- > ArO^- > Cl^- > $ pyridine $> AcO^- > H_2O$. A quantitative relationship[314] (the *Swain-Scott equation*) has been worked out similar to the linear free-energy equations considered in Chapter 9:[315]

$$\log \frac{k}{k_0} = sn$$

[305]For a review of the effect of nucleophile association on nucleophilicity, see Guibe; Bram *Bull. Soc. Chim. Fr.* **1975**, 933-948.

[306]Zaugg; Horrom; Borgwardt *J. Am. Chem. Soc.* **1960**, *82*, 2895; Zaugg; Leonard *J. Org. Chem.* **1972**, *37*, 2253. See also Solov'yanov; Dem'yanov; Beletskaya; Reutov *J. Org. Chem. USSR* **1976**, *12*, 714, 2215; Solov'yanov; Ahmed; Beletskaya; Reutov *J. Org. Chem. USSR* **1987**, *23*, 1243; Jackman; Lange *J. Am. Chem. Soc.* **1981**, *103*, 4494.

[307]See, for example Williard; Carpenter *J. Am. Chem. Soc.* **1986**, *108*, 462.

[308]Zook; Gumby *J. Am. Chem. Soc.* **1960**, *82*, 1386. See also Cacciapaglia; Mandolini *J. Org. Chem.* **1988**, *53*, 2579.

[309]For some other measurements of rates of SN2 reactions in the gas phase, see Barlow; Van Doren; Bierbaum *J. Am. Chem. Soc.* **1988**, *110*, 7240; Merkel; Havlas; Zahradna!ek *J. Am. Chem. Soc.* **1988**, *110*, 8355.

[310]Bohme; Mackay *J. Am. Chem. Soc.* **1981**, *103*, 978; Bohme; Raksit *J. Am. Chem. Soc.* **1984**, *106*, 3447. See also Hierl; Ahrens; Henchman; Viggiano; Paulson; Clary *J. Am. Chem. Soc.* **1986**, *108*, 3142.

[311]Bohme; Raksit *Can. J. Chem.* **1985**, *63*, 3007.

[312]Landini; Maia; Rampoldi *J. Org. Chem.* **1989**, *54*, 328.

[313]Edwards; Pearson *J. Am. Chem. Soc.* **1962**, *84*, 16.

[314]Swain; Scott *J. Am. Chem. Soc.* **1953**, *75*, 141.

[315]This is not the only equation that has been devised in an attempt to correlate nucleophilic reactivity. For reviews of attempts to express nucleophilic power quantitatively, see Ritchie *Pure Appl. Chem.* **1978**, *50*, 1281-1290; Duboc, in Chapman; Shorter *Correlation Analysis in Chemistry: Recent Advances*; Plenum: New York, 1978, pp. 313-355; Ibne-Rasa *J. Chem. Educ.* **1967**, *44*, 89-94. See also Hoz; Speizman *J. Org. Chem.* **1983**, *48*, 2904; Kawazoe; Ninomiya; Kohda; Kimoto *Tetrahedron Lett.* **1986**, *27*, 2897; Kevill; Fujimoto *J. Chem. Res. (S)* **1988**, 408.

where n is the nucleophilicity of a given group, s is the sensitivity of a substrate to nucleophilic attack, and k_0 is the rate for H_2O, which is taken as the standard and for which n is assigned a value of zero. s is defined as 1.0 for methyl bromide. Table 10.9 contains values of n for some common nucleophiles.[316] The order is similar to that of Edwards and Pearson. The Swain–Scott equation can be derived from Marcus theory.[317]

It is now evident that an absolute order of either nucleophilicity[318] or leaving-group ability, even in the gas phase where solvation is not a factor, does not exist, because they have an effect on each other. When the nucleophile and leaving group are both hard or both soft, the reaction rates are relatively high, but when one is hard and the other soft, rates are reduced.[319] Although this effect is smaller than the effects in paragraphs 1 and 4 above, it still prevents an absolute scale of either nucleophilicity or leaving-group ability. There has been controversy as to whether the selectivity of a reaction should increase with decreasing reactivity of a series of nucleophiles, or whether the opposite holds. There is evidence for both views.[320]

For substitution at a carbonyl carbon, the nucleophilicity order is not the same as it is at a saturated carbon, but follows the basicity order more closely. The reason is presumably that the carbonyl carbon, with its partial positive charge, resembles a proton more than does the carbon at a saturated center. That is, a carbonyl carbon is a much harder acid than a saturated carbon. The following nucleophilicity order for these substrates has been determined:[321] $Me_2C—NO^- > EtO^- > MeO^- > OH^- > OAr^- > N_3^- > F^- > H_2O > Br^- \sim I^-$. Soft bases are ineffective at a carbonyl carbon.[322] In a reaction carried out in the gas phase with alkoxide nucleophiles OR^- solvated by only one molecule of an alcohol $R'OH$, it was found that both RO^- and $R'O^-$ attacked the formate substrate ($HCOOR''$) about equally, though in the unsolvated case, the more basic alkoxide is the better nucleophile.[323] In this study, the product ion $R''O^-$ was also solvated by one molecule of ROH or $R'OH$.

If, adjacent to the attacking atom on the nucleophile, there is an atom containing one or more unshared pairs, the nucleophilicity is enhanced. Examples of such nucleophiles are HO_2^-, $Me_2C=NO^-$, NH_2NH_2, etc. This is called the *alpha effect*,[324] and the reasons for it

TABLE 10.9 Nucleophilicities of some common reagents[316]

Nucleophile	n	Nucleophile	n
SH⁻	5.1	**Br⁻**	3.5
CN⁻	5.1	**PhO⁻**	3.5
I⁻	5.0	**AcO⁻**	2.7
PhNH₂	4.5	**Cl⁻**	2.7
OH⁻	4.2	**F⁻**	2.0
N₃⁻	4.0	**NO₃⁻**	1.0
Pyridine	3.6	**H₂O**	0.0

[316]From Wells *Chem. Rev.* **1963**, *63*, 171-219, p. 212. See also Koskikallio *Acta Chem. Scand.* **1969**, *23*, 1477, 1490.

[317]Albery; Kreevoy *Adv. Phys. Org. Chem.* **1978**, *16*, 87-157, pp. 113-115.

[318]However, for a general model of intrinsic nucleophilicity in the gas phase, see Pellerite; Brauman *J. Am. Chem. Soc.* **1983**, *105*, 2672.

[319]Olmstead; Brauman, Ref. 302.

[320]For discussions, see Dietze; Jencks *J. Am. Chem. Soc.* **1989**, *111*, 5880.

[321]Hudson; Green *J. Chem. Soc.* **1962**, 1055; Bender; Glasson *J. Am. Chem. Soc.* **1959**, *81*, 1590; Jencks; Gilchrist *J. Am. Chem. Soc.* **1968**, *90*, 2622.

[322]For theoretical treatments of nucleophilicity at a carbonyl carbon, see Buncel; Shaik; Um; Wolfe *J. Am. Chem. Soc.* **1988**, *110*, 1275, and references cited therein.

[323]Baer; Stoutland; Brauman *J. Am. Chem. Soc.* **1989**, *111*, 4097.

[324]For reviews, see Grekov; Veselov *Russ. Chem. Rev.* **1978**, *47*, 631-648; Fina; Edwards *Int. J. Chem. Kinet.* **1973**, *5*, 1-26.

are not completely understood. Several possible explanations have been offered.[325] One is that the ground state of the nucleophile is destabilized by repulsion between the adjacent pairs of electrons;[326] another is that the transition state is stabilized by the extra pair of electrons;[327] a third is that the adjacent electron pair reduces solvation of the nucleophile.[328] Evidence supporting the third explanation is that there was no alpha effect in the reaction of HO_2^- with methyl formate in the gas phase,[329] though HO_2^- shows a strong alpha effect in solution. The alpha effect is substantial for substitution at a carbonyl or other unsaturated carbon, at some inorganic atoms,[330] and for reactions of a nucleophile with a carbocation,[331] but is generally smaller or absent entirely for substitution at a saturated carbon.[332]

The Effect of the Leaving Group

1. *At a saturated carbon.*   The leaving group comes off more easily the more stable it is as a free entity. This is usually inverse to its basicity, and the best leaving groups are the weakest bases. Thus iodide is the best leaving group among the halides and fluoride the poorest. Since XH is always a weaker base than X^-, nucleophilic substitution is always easier at a substrate RXH^+ than at RX. An example of this effect is that OH and OR are not leaving groups from ordinary alcohols and ethers but can come off when the groups are protonated, that is, converted to ROH_2^+ or $RORH^+$.[333] Reactions in which the leaving group does not come off until it has been protonated have been called SN1cA or SN2cA, depending on whether after protonation the reaction is an SN1 or SN2 process (these designations are often shortened to A1 and A2). The cA stands for conjugate acid, since the substitution takes place on the conjugate acid of the substrate. The IUPAC designations for these mechanisms are, respectively, $A_h + D_N + A_N$ and $A_h + A_N D_N$; that is, the same designations as SN1 and SN2, with A_h to show the preliminary step. When another electrophile assumes the role of the proton, the symbol A_e is used instead. The ions ROH_2^+ and $RORH^+$ can be observed as stable entities at low temperatures in super-acid solutions.[334] At higher temperatures they cleave to give carbocations.

It is obvious that the best nucleophiles (e.g., NH_2^-, OH^-) cannot take part in SN1cA or SN2cA processes, because they would be converted to their conjugate acids under the acidic conditions necessary to protonate the leaving groups.[335] Because SN1 reactions do not require powerful nucleophiles but do require good leaving groups, most of them take place under

[325]For discussions, see Wolfe; Mitchell; Schlegel; Minot; Eisenstein *Tetrahedron Lett.* **1982**, *23*, 615; Hoz; Buncel *Isr. J. Chem.* **1985**, *26*, 313.

[326]Buncel; Hoz *Tetrahedron Lett.* **1983**, *24*, 4777. For evidence that this is not the sole cause, see Oae; Kadoma *Can. J. Chem.* **1986**, *64*, 1184.

[327]See Hoz *J. Org. Chem.* **1982**, *47*, 3545; Laloi-Diard; Verchere; Gosselin; Terrier *Tetrahedron Lett.* **1984**, *25*, 1267.

[328]For other explanations, see Hudson; Hansell; Wolfe; Mitchell *J. Chem. Soc., Chem. Commun.* **1985**, 1406; Shustov *Doklad. Chem.* **1985**, *280*, 80. For a discussion, see Herschlag; Jencks *J. Am. Chem. Soc.* **1990**, *112*, 1951.

[329]DePuy; Della; Filley; Grabowski; Bierbaum *J. Am. Chem. Soc.* **1983**, *105*, 2481; Buncel; Um *J. Chem. Soc., Chem. Commun.* **1986**, 595; Terrier; Degorre; Kiffer; Laloi *Bull. Soc. Chim. Fr.* **1988**, 415. For some evidence against this explanation, see Moss; Swarup; Ganguli *J. Chem. Soc., Chem. Commun.* **1987**, 860.

[330]For example, see Kice; Legan *J. Am. Chem. Soc.* **1973**, *95*, 3912.

[331]Dixon; Bruice *J. Am. Chem. Soc.* **1971**, *93*, 3248, 6592.

[332]Gregory; Bruice *J. Am. Chem. Soc.* **1967**, *89*, 4400; Oae; Kadoma; Yano *Bull. Chem. Soc. Jpn.* **1969**, *42*, 1110; McIsaac; Subbaraman; Subbaraman; Mulhausen; Behrman *J. Org. Chem.* **1972**, *37*, 1037. See, however, Beale *J. Org. Chem.* **1972**, *37*, 3871; Buncel; Wilson; Chuaqui *J. Am. Chem. Soc.* **1982**, *104*, 4896, *Int. J. Chem. Kinet.* **1982**, *14*, 823.

[333]For a review of ORH^+ as a leaving group, see Staude; Patat, in Patai *The Chemistry of the Ether Linkage*; Wiley: New York, 1967, pp. 22-46.

[334]Olah; O'Brien *J. Am. Chem. Soc.* **1967**, *89*, 1725; Olah; Sommer; Namanworth *J. Am. Chem. Soc.* **1967**, *89*, 3576; Olah; Olah, in Olah; Schleyer, Ref. 92, vol. 2, 1970, pp. 743-747.

[335]Even in the gas phase, NH_3 takes a proton from $CH_3OH_2^+$ rather than acting as a nucleophile: Okada; Abe; Taniguchi; Yamabe *J. Chem. Soc., Chem. Commun.* **1989**, 610.

acidic conditions. In contrast, SN2 reactions, which do require powerful nucleophiles (which are generally strong bases), most often take place under basic or neutral conditions.

Another circumstance that increases leaving-group power is ring strain. Ordinary ethers do not cleave at all and protonated ethers only under strenuous conditions, but epoxides[336] are cleaved quite easily and protonated epoxides even more easily. Aziridines[337] and epi-

sulfides, three-membered rings containing, respectively, nitrogen and sulfur, are also easily cleaved (see p. 368).[338]

Although halides are common leaving groups in nucleophilic substitution for synthetic purposes, it is often more convenient to use alcohols. Since OH does not leave from ordinary alcohols, it must be converted to a group that does leave. One way is protonation, mentioned above. Another is conversion to a reactive ester, most commonly a sulfonic ester. The sulfonic ester groups *tosylate, brosylate, nosylate,* and *mesylate* are better leaving groups

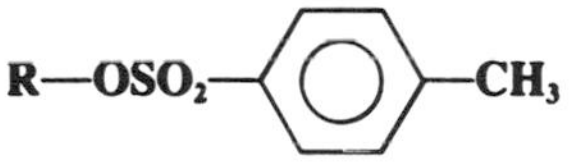

ROTs

p-Toluenesulfonates
Tosylates

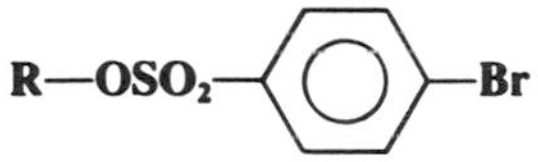

ROBs

p-Bromobenzenesulfonates
Brosylates

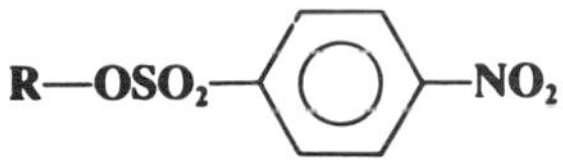

RONs

p-Nitrobenzenesulfonates
Nosylates

ROMs

Methanesulfonates
Mesylates

than halides and are frequently used. Other leaving groups are still better, and compounds containing these groups make powerful alkylating agents. Among them are oxonium ions (ROR_2^+),[339] alkyl perchlorates $(ROClO_3)$,[340] ammonioalkanesulfonate esters (*betylates*) $(ROSO_2(CH_2)_nNMe_3^+)$,[341] alkyl fluorosulfonates $(ROSO_2F)$,[342] and the fluorinated com-

[336]For a review of the reactions of epoxides, see Smith *Synthesis* **1984**, 629-656. For a review of their synthesis and reactions, see Bartók; Láng, in Patai *The Chemistry of Functional Groups, Supplement E*; Wiley: New York, 1980, pp. 609-681.

[337]For a review of aziridine cleavages in the synthesis of natural products, see Kametani; Honda *Adv. Heterocycl. Chem.* **1986**, *39*, 181-236.

[338]There is evidence that relief of ring strain is not the only factor responsible for the high rates of ring-opening of 3-membered rings: Di Vona; Illuminati; Lillocci *J. Chem. Soc., Perkin Trans. 2* **1985**, 1943; Bury; Earl; Stirling *J. Chem. Soc., Chem. Commun.* **1985**, 393.

[339]For a monograph, see Perst, Ref. 84. For reviews, see Perst, in Olah; Schleyer, Ref. 92, vol. 5, 1976, pp. 1961-2047; Granik; Pyatin; Glushkov *Russ. Chem. Rev.* **1971**, *40*, 747-759. For a discussion of their use, see Curphey *Org. Synth.* **VI**, 1021.

[340]Baum; Beard *J. Am. Chem. Soc.* **1974**, *96*, 3233. See also Kevill; Lin *Tetrahedron Lett.* **1978**, 949.

[341]King; Loosmore; Aslam; Lock; McGarrity *J. Am. Chem. Soc.* **1982**, *104*, 7108; King; Lee *Can. J. Chem.* **1981**, *59*, 356, 362; King; Skonieczny; Poole *Can. J. Chem.* **1983**, *61*, 235.

[342]Ahmed; Alder; James; Sinnott; Whiting *Chem. Commun.* **1968**, 1533; Ahmed; Alder *Chem. Commun.* **1969**, 1389; Alder *Chem. Ind. (London)* **1973**, 983. For a discussion of the hazards involved in the use of these and other alkylating agents, see Alder; Sinnott; Whiting; Evans *Chem. Br.* **1978**, 324.

pounds *triflates*[343] and *nonaflates*.[343] *Tresylates* are about 400 times less reactive than triflates, but still about 100 times more reactive than tosylates.[344] Halonium ions ($RClR^+$, $RBrR^+$,

R—OSO$_2$CF$_3$

ROTf

Trifluoromethanesulfonates

Triflates

R—OSO$_2$C$_4$F$_9$

Nonafluorobutanesulfonates

Nonaflates

R—OSO$_2$CH$_2$CF$_3$

2,2,2-Trifluoroethanesulfonates

Tresylates

RIR^+), which can be prepared in super-acid solutions (p. 312) and isolated as solid SbF_6^- salts, are also extremely reactive in nucleophilic substitution.[345] Of the above types of compound, the most important in organic synthesis are tosylates, mesylates, oxonium ions, and triflates. The others have been used mostly for mechanistic purposes.

NH_2, NHR, and NR_2 are extremely poor leaving groups,[346] but the leaving-group ability of NH_2 can be greatly improved by converting a primary amine RNH_2 to the ditosylate $RNTs_2$. The NTs_2 group has been successfully replaced by a number of nucleophiles.[347] Another way of converting NH_2 into a good leaving group has been extensively developed by Katritzky and co-workers.[348] In this method the amine is converted to a pyridinium compound (**86**) by treatment with a pyrylium salt (frequently a 2,4,6-triphenylpyrylium salt, **85**).[349] When the salt is heated, the counterion acts as a nucleophile. In some cases a

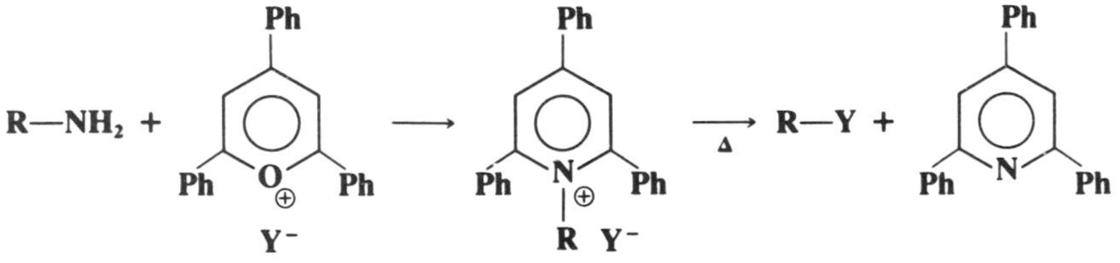

nonnucleophilic ion such as BF_4^- is used as the counterion for the conversion **85 → 86**, and then Y^- is added to **86**. Among the nucleophiles that have been used successfully in this reaction are I^-, Br^-, Cl^-, F^-, OAc^-, N_3^-, NHR_2, and H^-. Ordinary NR_2 groups are good leaving groups when the substrate is a Mannich base (these are compounds of the form $RCOCH_2CH_2NR_2$; see reaction **6-16**).[350] The elimination-addition mechanism applies in this case.

[343]For reviews of triflates, nonaflates, and other fluorinated ester leaving groups, see Stang; Hanack; Subramanian *Synthesis* **1982**, 85-126; Howells; Mc Cown *Chem. Rev.* **1977**, 77, 69-92, pp. 85-87.

[344]Crossland; Wells; Shiner *J. Am. Chem. Soc.* **1971**, 93, 4217.

[345]Peterson; Clifford; Slama, Ref. 89; Olah; DeMember; Schlosberg; Halpern *J. Am. Chem. Soc.* **1972**, 94, 156; Peterson; Waller *J. Am. Chem. Soc.* **1972**, 94, 5024; Olah; Svoboda *Synthesis* **1973**, 203; Olah; Mo *J. Am. Chem. Soc.* **1974**, 96, 3560.

[346]For a review of the deamination of amines, see Baumgarten; Curtis, in Patai *The Chemistry of Functional Groups, Supplement F*, pt. 2; Wiley: New York, 1982, pp. 929-997.

[347]For references, see Müller; Thi *Helv. Chim. Acta* **1980**, 63, 2168; Curtis; Knutson; Baumgarten *Tetrahedron Lett.* **1981**, 22, 199.

[348]For reviews, see Katritzky; Marson *Angew. Chem. Int. Ed. Engl.* **1984**, 23, 420-429 [*Angew. Chem.* 96, 403-413]; Katritzky *Tetrahedron* **1980**, 36, 679-699. For reviews of the use of such leaving groups to study mechanistic questions, see Katritzky; Sakizadeh; Musumarra *Heterocycles* **1985**, 23, 1765-1813; Katritzky; Musumarra *Chem. Soc. Rev.* **1984**, 13, 47-68.

[349]For discussions of the mechanism, see Katritzky; Brycki *J. Am. Chem. Soc.* **1986**, 108, 7295, and other papers in this series.

[350]For a review of Mannich bases, see Tramontini *Synthesis* **1973**, 703-775.

Probably the best leaving group is N_2 from the species RN_2^+, which can be generated in several ways,[351] of which the two most important are the treatment of primary amines with nitrous acid (see p. 635 for this reaction)

$$\textbf{RNH}_2 + \textbf{HONO} \longrightarrow \textbf{RN}_2{}^+$$

and the protonation of diazo compounds[352]

$$\textbf{R}_2\textbf{C}{=}\overset{\oplus}{\textbf{N}}{=}\overset{\ominus}{\textbf{N}} + \textbf{H}^+ \longrightarrow \textbf{R}_2\textbf{CHN}_2{}^+$$

No matter how produced, RN_2^+ are usually too unstable to be isolable,[353] reacting presumably by the SN1 or SN2 mechanism.[354] Actually, the exact mechanisms are in doubt because the rate laws, stereochemistry, and products have proved difficult to interpret.[355] If there are free carbocations they should give the same ratio of substitution to elimination to rearrangements, etc. as carbocations generated in other SN1 reactions, but they often do not. "Hot" carbocations (unsolvated and/or chemically activated) that can hold their configuration have been postulated,[356] as have ion pairs, in which OH^- (or OAc^-, etc., depending on how the diazonium ion is generated) is the counterion.[357] One class of aliphatic diazonium salts of which several members have been isolated as stable salts are the cyclopropeniumyldiazonium salts:[358]

$$\text{R = Me or i-Pr}$$
$$\text{X}^- = \text{BF}_4{}^- \text{ or SbCl}_6{}^-$$

Diazonium ions generated from ordinary aliphatic primary amines are usually useless for preparative purposes, since they lead to a mixture of products giving not only substitution by any nucleophile present, but also elimination and rearrangements if the substrate permits. For example, diazotization of n-butylamine gave 25% 1-butanol, 5.2% 1-chlorobutane, 13.2% 2-butanol, 36.5% butenes (consisting of 71% 1-butene, 20% trans-2-butene, and 9% cis-2-butene), and traces of butyl nitrites.[359]

[351]For reviews, see Kirmse Angew. Chem. Int. Ed. Engl. **1976**, 15, 251-261 [Angew. Chem. 88, 273-283]; Collins Acc. Chem. Res. **1971**, 4, 315-322; Moss Chem. Eng. News **1971**, 49, 28-36 (No. 48, Nov. 22).

[352]For a treatise, see Regitz; Maas Diazo Compounds; Academic Press: New York, 1986. For reviews of the reactions of aliphatic diazo compounds with acids, see Hegarty, in Patai The Chemistry of Diazonium and Diazo Groups, pt. 2; Wiley: New York, 1978, pp. 511-591, pp. 571-575; More O'Ferrall Adv. Phys. Org. Chem. **1967**, 5, 331-399. For review of the structures of these compounds, see Studzinskii; Korobitsyna Russ. Chem. Rev. **1970**, 39, 834-843.

[353]Aromatic diazonium salts can, of course, be isolated (see Chapter 13), but only a few aliphatic diazonium salts have been prepared (see also Ref. 358). For reviews see Laali; Olah Rev. Chem. Intermed. **1985**, 6, 237-253; Bott, in Patai; Rappoport The Chemistry of Functional Groups, Supplement C, pt. 1; Wiley: New York, 1983, pp. 671-697; Bott Angew. Chem. Int. Ed. Engl. **1979**, 18, 259-265 [Angew. Chem. 91, 279-285]. The simplest aliphatic diazonium ion $CH_3N_2^+$ has been prepared at $-120°$ in super-acid solution, where it lived long enough for an nmr spectrum to be taken: Berner; McGarrity J. Am. Chem. Soc. **1979**, 101, 3135.

[354]For an example of a diazonium ion reacting by an SN2 mechanism, see Mohrig; Keegstra; Maverick; Roberts; Wells J. Chem. Soc., Chem. Commun. **1974**, 780.

[355]For reviews of the mechanism, see Manuilov; Barkhash Russ. Chem. Rev. **1990**, 59, 179-192; Saunders; Cockerill Mechanisms of Elimination Reactions; Wiley: New York, 1973, pp. 280-317; in Olah; Schleyer, Ref. 92, vol. 2, **1970**, the articles by Keating; Skell, pp. 573-653; and by Friedman, pp. 655-713; White; Woodcock, in Patai The Chemistry of the Amino Group; Wiley: New York, 1968, pp. 440-483; Ref. 351.

[356]Semenow; Shih; Young J. Am. Chem. Soc. **1958**, 80, 5472. For a review of "hot" or "free" carbocations, see Keating; Skell, Ref. 355.

[357]Collins, Ref. 351; Collins; Benjamin J. Org. Chem. **1972**, 37, 4358; White; Field J. Am. Chem. Soc. **1975**, 97, 2148; Cohen; Daniewski; Solash J. Org. Chem. **1980**, 45, 2847; Maskill; Thompson; Wilson J. Chem. Soc., Perkin Trans. 2 **1984**, 1693; Connor; Maskill Bull. Soc. Chim. Fr. **1988**, 342.

[358]Weiss; Wagner; Priesner; Macheleid J. Am. Chem. Soc. **1985**, 107, 4491.

[359]Whitmore; Langlois J. Am. Chem. Soc. **1932**, 54, 3441; Streitwieser; Schaeffer J. Am. Chem. Soc. **1957**, 79, 2888.

In the SN1cA and SN2cA mechanisms (p. 352) there is a preliminary step, the addition of a proton, before the normal SN1 or SN2 process occurs. There are also reactions in which the substrate *loses* a proton in a preliminary step. In these reactions there is a carbene intermediate.

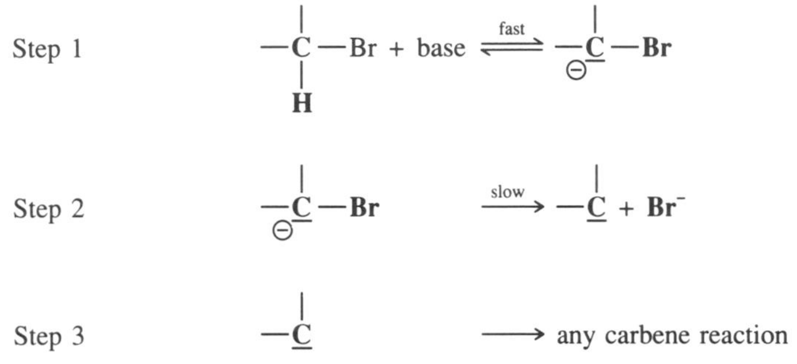

Step 1 $\qquad$ $-\overset{|}{\underset{\underset{\mathbf{H}}{|}}{\mathbf{C}}}-\mathbf{Br}$ + base $\overset{\text{fast}}{\rightleftharpoons}$ $-\overset{|}{\underset{\ominus}{\mathbf{C}}}-\mathbf{Br}$

Step 2 $\qquad$ $-\overset{|}{\underset{\ominus}{\mathbf{C}}}-\mathbf{Br}$ $\overset{\text{slow}}{\longrightarrow}$ $-\overset{|}{\underline{\mathbf{C}}} + \mathbf{Br}^-$

Step 3 $\qquad$ $-\overset{|}{\underline{\mathbf{C}}}$ $\qquad\longrightarrow$ any carbene reaction

Once formed by this process, the carbene may undergo any of the normal carbene reactions (see p. 199). When the net result is substitution, this mechanism has been called the SN1cB (for conjugate base) mechanism.[360] Though the slow step is an SN1 step, the reaction is second order; first order in substrate and first order in base.

Table 10.10 lists some leaving groups in approximate order of ability to leave. The order of leaving-group ability is about the same for SN1 and SN2 reactions.

2. *At a carbonyl carbon.* In both the SN1 and SN2 mechanisms the leaving group departs during the rate-determining step and so directly affects the rate. In the tetrahedral mechanism at a carbonyl carbon, the bond between the substrate and leaving group is still intact during the slow step. Nevertheless, the nature of the leaving group still affects the reactivity in two ways: (1) By altering the electron density at the carbonyl carbon, the rate of the reaction is affected. The greater the electron-withdrawing character of X, the greater the partial positive charge on C and the more rapid the attack by a nucleophile. (2) The nature of the leaving group affects the *position of equilibrium.* In the intermediate **67** (p. 331) there is competition between X and Y as to which group leaves. If X is a poorer leaving group than Y, then Y will preferentially leave and **67** will revert to the starting compounds. Thus there is a partitioning factor between **67** going on to product (loss of X) or back to starting compound (loss of Y). The sum of these two factors causes the sequence of reactivity to be RCOCl > RCOOCOR' > RCOOAr > RCOOR' > RCONH$_2$ > RCONR'$_2$ > RCOO$^-$.[361] Note that this order is approximately the order of decreasing stability of the leaving-group anion. If the leaving group is bulky, it may exert a steric effect and retard the rate for this reason.

[360]Pearson; Edgington *J. Am. Chem. Soc.* **1962**, *84*, 4607.

[361]RCOOH would belong in this sequence just after RCOOAr, but it fails to undergo many reactions for a special reason. Many nucleophiles, instead of attacking the C=O group, are basic enough to take a proton from the acid, converting it to the unreactive RCOO$^-$.

TABLE 10.10 Leaving groups listed in approximate order of decreasing ability to leave. Groups that are common leaving groups at saturated and carbonyl carbons are indicated

Substrate RX	Common leaving groups	
	At saturated carbon	At carbonyl carbon
RN_2^+	×	
$ROR_2'^+$		
$ROSO_2C_4F_9$		
$ROSO_2CF_3$	×	
$ROSO_2F$		
ROTs, etc.[a]	×	
RI	×	
RBr	×	
ROH_2^+	× (conjugate acid of alcohol)	
RCl	×	× (acyl halides)
$RORH^+$	× (conjugate acid of ether)	
$RONO_2$, etc.[a]		
$RSR_2'^{+363}$		
$RNR_3'^+$	×	
RF		
$ROCOR'^{364}$	×	× (anhydrides)
RNH_3^+		
$ROAr^{365}$		× (aryl esters)
ROH		× (carboxylic acids)
ROR		× (alkyl esters)
RH		
RNH_2		× (amides)
RAr		
RR		

[a]ROTs, etc., includes esters of sulfuric and sulfonic acids in general, for example, $ROSO_2OH$, $ROSO_2OR$, $ROSO_2R$, etc. $RONO_2$, etc., includes inorganic ester leaving groups, such as $ROPO(OH)_2$, $ROB(OH)_2$, etc.

The Effect of the Reaction Medium[362]

The effect of solvent polarity on the rate of SN1 reactions depends on whether the substrate is neutral or positively charged. For neutral substrates, which constitute the majority of cases, the more polar the solvent, the faster the reaction, since there is a greater charge in the transition state than in the starting compound (Table 10.11[366]) and the energy of an ionic transition state is reduced by polar solvents. However, when the substrate is positively charged, the charge is more spread out in the transition state than in the starting ion, and

[362]For a monograph, see Reichardt *Solvents and Solvent Effects in Organic Chemistry*, 2nd ed.; VCH: New York, 1988. For reviews, see Klumpp, Ref. 294, pp. 186-203; Bentley; Schleyer *Adv. Phys. Org.* Chem. **1977**, *14*, 1-67.

[363]For a review of the reactions of sulfonium salts, see Knipe, in Stirling *The Chemistry of the Sulphonium Group*, pt. 1; Wiley: New York, 1981, pp. 313-385. See also Badet; Julia; Lefebvre *Bull. Soc. Chim. Fr.* **1984**, II-431.

[364]For a review of SN2 reactions of carboxylic esters, where the leaving group is OCOR', see McMurry *Org. React.* **1976**, *24*, 187-224.

[365]Nitro substitution increases the leaving-group ability of ArO groups, and alkyl picrates [2,4,6-ROC₆H₂(NO₂)₃] react at rates comparable to tosylates: Sinnott; Whiting *J. Chem. Soc. B* **1971**, 965. See also Page; Pritt; Whiting *J. Chem. Soc., Perkin Trans. 2* **1972**, 906.

[366]This analysis is due to Ingold *Structure and Mechanism in Organic Chemistry*, 2d ed.; Cornell University Press: Ithaca, NY, 1969, pp. 457-463.

TABLE 10.11 Transition states for SN1 reactions of charged and uncharged substrates, and for SN2 reactions of the four charge types[366]

Reactants and transition states		Charge in the transition state relative to starting materials	How an increase in solvent polarity affects the rate
SN2	Type I $RX + Y^- \longrightarrow Y^{\delta-}\cdots R\cdots X^{\delta-}$	Dispersed	Small decrease
	Type II $RX + Y \longrightarrow Y^{\delta+}\cdots R\cdots X^{\delta-}$	Increased	Large increase
	Type III $RX^+ + Y^- \longrightarrow Y^{\delta-}\cdots R\cdots X^{\delta+}$	Decreased	Large decrease
	Type IV $RX^+ + Y \longrightarrow Y^{\delta+}\cdots R\cdots X^{\delta+}$	Dispersed	Small decrease
SN1	$RX \longrightarrow R^{\delta+}\cdots X^{\delta-}$	Increased	Large increase
	$RX^- \longrightarrow R^{\delta+}\cdots X^{\delta-}$	Dispersed	Small decrease

a greater solvent polarity slows the reaction. Even for solvents with about the same polarity, there is a difference between protic and aprotic solvents.[367] SN1 reactions of un-ionized substrates are more rapid in protic solvents, which can form hydrogen bonds with the leaving group. Examples of protic solvents are water, alcohols, and carboxylic acids, while some polar aprotic solvents are dimethylformamide (DMF), dimethyl sulfoxide,[368] acetonitrile, acetone, sulfur dioxide, and hexamethylphosphoramide [$(Me_2N)_3PO$], HMPA.[369]

For SN2 reactions, the effect of the solvent depends on which of the four charge types the reaction belongs to (p. 293). In types I and IV, an initial charge is dispersed in the transition state, so the reaction is hindered by polar solvents. In type III initial charges are *decreased* in the transition state, so that the reaction is even more hindered by polar solvents. Only type II, where the reactants are uncharged but the transition state has built up a charge, is aided by polar solvents. These effects are summarized in Table 10.11.[366] Westaway has proposed a "solvation rule" for SN2 reactions, which states that changing the solvent will not change the structure of the transition state for type I reactions, but will change it for type II reactions.[370] For SN2 reactions also, the difference between protic and aprotic solvents must be considered.[371] For reactions of types I and III the transition state is more solvated in polar aprotic solvents than in protic ones,[372] while (as we saw on p. 349) the original charged nucleophile is less solvated in aprotic solvents[373] (the second factor is generally much greater than the first[374]). So the change from, say, methanol to dimethyl sulfoxide should greatly increase the rate. As an example, the relative rates at 25°C for the reaction between methyl iodide and Cl^- were[298] in MeOH, 1; in $HCONH_2$ (still protic though a weaker acid), 12.5; in HCONHMe, 45.3; and $HCONMe_2$, 1.2×10^6. The change in rate in going from a protic to an aprotic solvent is also related to the *size* of the attacking anion. Small ions are solvated best in protic solvents, since hydrogen bonding is most important for them, while large anions are solvated best in aprotic solvents (protic solvents have highly developed structures held together by hydrogen bonds; aprotic solvents have much looser

[367]See, for example Ponomareva; Dvorko; Kulik; Evtushenko *Doklad. Chem.* **1983**, *272*, 291.

[368]For reviews of reactions in dimethyl sulfoxide, see Buncel; Wilson *Adv. Phys. Org. Chem.* **1977**, *14*, 133-202; Martin; Weise; Niclas *Angew. Chem. Int. Ed. Engl.* **1967**, *6*, 318-334 [*Angew. Chem. 79*, 340-357].

[369]For reviews of HMPA, see Normant *Russ. Chem. Rev.* **1970**, *39*, 457-484, *Bull. Soc. Chim. Fr.* **1968**, 791-826, *Angew. Chem. Int. Ed. Engl.* **1967**, *6*, 1046-1067 [*Angew. Chem. 79*, 1029-1050].

[370]Westaway *Can. J. Chem.* **1978**, *56*, 2691; Westaway; Lai *Can. J. Chem.* **1989**, *67*, 345.

[371]For reviews of the effects of protic and aprotic solvents, see Parker *Chem. Rev.* **1969**, *69*, 1-32, *Adv. Phys. Org. Chem.* **1967**, *5*, 173-235, *Adv. Org. Chem.* **1965**, *5*, 1-46; Madaule-Aubry *Bull. Soc. Chim. Fr.* **1966**, 1456.

[372]However, even in aprotic solvents, the transition state is less solvated than the charged nucleophile: Magnera; Caldwell; Sunner; Ikuta; Kebarle *J. Am. Chem. Soc.* **1984**, *106*, 6140.

[373]See, for example, Fuchs; Cole *J. Am. Chem. Soc.* **1973**, *95*, 3194.

[374]See, however, Haberfield; Clayman; Cooper *J. Am. Chem. Soc.* **1969**, *91*, 787.

structures, and it is easier for a large anion to be fitted in). So the rate of attack by small anions is most greatly increased by the change from a protic to an aprotic solvent. This may have preparative significance. The review articles in Ref. 371 have lists of several dozen reactions of charge types I and III in which yields are improved and reaction times reduced in polar aprotic solvents. Reaction types II and IV are much less susceptible to the difference between protic and aprotic solvents.

Since for most reactions SN1 rates go up and SN2 rates go down in solvents of increasing polarity, it is quite possible for the same reaction to go by the SN1 mechanism in one solvent and the SN2 in another. Table 10.12 is a list of solvents in order of ionizing power;[375] a solvent high on the list is a good solvent for SN1 reactions. Trifluoroacetic acid, which was not studied by Smith, Fainberg, and Winstein, has greater ionizing power than any solvent listed in Table 10.12.[376] Because it also has very low nucleophilicity, it is an excellent solvent for SN1 solvolyses. Other good solvents for this purpose are 1,1,1-trifluoroethanol CF_3CH_2OH, and 1,1,1,3,3,3-hexafluoro-2-propanol $(F_3C)_2CHOH$.[377]

We have seen how the polarity of the solvent influences the rates of SN1 and SN2 reactions. The ionic strength of the medium has similar effects. In general, the addition of an external salt affects the rates of SN1 and SN2 reactions in the same way as an increase in solvent polarity, though this is not quantitative; different salts have different effects.[378] However, there are exceptions: though the rates of SN1 reactions are usually increased by the addition of salts (this is called the *salt effect*), addition of the leaving-group ion often decreases the rate (the common-ion effect, p. 300). There is also the special salt effect of $LiClO_4$, mentioned on p. 303. In addition to these effects, SN1 rates are also greatly accelerated when there are ions present that specifically help in pulling off the leaving group.[379] Especially important are Ag^+, Hg^{2+}, and Hg_2^{2+}, but H^+ helps to pull off F (hydrogen bonding).[380] Even primary halides have been reported to undergo SN1 reactions when assisted by metal ions.[381] This does not mean, however, that reactions in the presence of metallic ions invariably proceed

TABLE 10.12 Relative rates of ionization of p-methoxyneophyl toulenesulfonate in various solvents[375]

Solvent	Relative rate	Solvent	Relative rate
HCOOH	153	**Ac₂O**	0.020
H₂O	39	Pyridine	0.013
80% **EtOH-H₂O**	1.85	Acetone	0.0051
AcOH	1.00	**EtOAc**	6.7×10^{-4}
MeOH	0.947	Tetrahydrofuran	5.0×10^{-4}
EtOH	0.370	**Et₂O**	3×10^{-5}
Me₂SO	0.108	**CHCl₃**	
Octanoic acid	0.043	Benzene	Lower still
MeCN	0.036	Alkanes	
HCONMe₂	0.029		

[375]Smith; Fainberg; Winstein *J. Am. Chem. Soc.* **1961**, *83*, 618.

[376]Refs. 87, 125; Streitwieser; Dafforn *Tetrahedron Lett.* **1969**, 1263.

[377]Schadt; Schleyer; Bentley *Tetrahedron Lett.* **1974**, 2335.

[378]See, for example, Duynstee; Grunwald; Kaplan *J. Am. Chem. Soc.* **1960**, *82*, 5654; Bunton; Robinson *J. Am. Chem. Soc.* **1968**, *90*, 5965.

[379]For a review, see Kevill, in Patai; Rappoport, Ref. 88, pt. 2, pp. 933-984.

[380]For a review of assistance by metallic ions, see Rudakov; Kozhevnikov; Zamashchikov *Russ. Chem. Rev.* **1974**, *43*, 305-316. For an example of assistance in removal of F by H^+, see Coverdale; Kohnstam *J. Chem. Soc.* **1960**, 3906.

[381]Zamashchikov; Rudakov; Litvinenko; Uzhik *Doklad. Chem.* **1981**, *258*, 186; Zamashchikov; Rudakov; Bezbozhnaya; Matveev *J. Org. Chem. USSR* **1984**, *20*, 424. See, however, Kevill; Fujimoto *J. Chem. Soc., Chem. Commun.* **1983**, 1149.

by the S_N1 mechanism. It has been shown that alkyl halides can react with $AgNO_2$ and $AgNO_3$ by the S_N1 or S_N2 mechanism, depending on the reaction conditions.[382]

The effect of solvent has been treated quantitatively (for S_N1 mechanisms, in which the solvent pulls off the leaving group) by a linear free-energy relationship[383]

$$\log \frac{k}{k_0} = mY$$

where m is characteristic of the substrate (defined as 1.00 for t-BuCl) and is usually near unity, Y is characteristic of the solvent and measures its "ionizing power," and k_0 is the rate in a standard solvent, 80% aqueous ethanol at 25°C. This is known as the Grunwald–Winstein equation, and its utility is at best limited. Y values can of course be measured for solvent *mixtures* too, and this is one of the principal advantages of the treatment, since it is not easy otherwise to assign a polarity arbitrarily to a given mixture of solvents.[384] The treatment is most satisfactory for different proportions of a given solvent pair. For wider comparisons the treatment is not so good quantitatively, although the Y values do give a reasonably good idea of solvolyzing power.[385] Table 10.13 contains a list of some Y values.[386]

Ideally, Y should measure only the ionizing power of the solvent, and should not reflect any backside attack by a solvent molecule in helping the nucleofuge to leave (nucleophilic assistance; k_s, p. 317). Actually, there is evidence that many solvents do lend some nucleophilic assistance,[387] even with tertiary substrates.[387a] It was proposed that a better measure of solvent "ionizing power" would be a relationship based on 2-adamantyl substrates, rather than t-BuCl, since the structure of this system completely prevents backside nucleophilic assistance (p. 340). Such a scale, called Y_{OTs}, was developed, with m defined as 1.00 for 2-adamantyl tosylate.[388] Some values of Y_{OTs} are given in Table 10.13. These values, which are actually based on both 1- and 2-adamantyl tosylates (both are equally impervious to nucleophilic assistance and show almost identical responses to solvent ionizing power[389]) are called Y_{OTs} because they apply only to tosylates. It has been found that solvent "ionizing power" depends on the leaving group, so separate scales[390] have been set up for OTf,[391] Cl,[392] Br,[392] I,[393] and other nucleofuges,[394] all based on the corresponding adamantyl compounds.

[382]Kornblum; Jones; Hardies *J. Am. Chem. Soc.* **1966**, *88*, 1704; Kornblum; Hardies *J. Am. Chem. Soc.* **1966**, *88*, 1707.

[383]Grunwald; Winstein *J. Am. Chem. Soc.* **1948**, *70*, 846.

[384]For reviews of polarity scales of solvent mixtures, see Reichardt, Ref. 362, pp. 339-405; Langhals *Angew. Chem. Int. Ed. Engl.* **1982**, *21*, 724-733 [*Angew. Chem. 94*, 739-749].

[385]For a criticism of the Y scale, see Abraham; Doherty; Kamlet; Harris; Taft *J. Chem. Soc., Perkin Trans. 2* **1987**, 1097.

[386]Y values are from Fainberg; Winstein *J. Am. Chem. Soc.* **1956**, *78*, 2770, except for the value for CF_3CH_2OH which is from Shiner; Dowd; Fisher; Hartshorn; Kessick; Milakofsky; Rapp *J. Am. Chem. Soc.* **1969**, *91*, 4838. Y_{OTs} values are from Bentley; Llewellyn, Ref. 390, pp. 143-144. Z values are from Ref. 396. $E_T(30)$ values are from Reichardt; Dimroth *Fortschr. Chem. Forsch.* **1969**, *11*, 1-73; Reichardt *Angew. Chem. Int. Ed. Engl.* **1979**, *18*, 98-110 [*Angew. Chem. 91*, 119-131]; Reichardt; Harbusch-Görnert *Liebigs Ann. Chem.* **1983**, 721-743; Laurence; Nicolet; Reichardt *Bull. Soc. Chim. Fr.* **1987**, 125; Laurence; Nicolet; Lucon; Reichardt *Bull. Soc. Chim. Fr.* **1987**, 1001; Reichardt; Eschner; Schäfer *Liebigs Ann. Chem.* **1990**, 57. Values for many additional solvents are given in the last five papers. Many values from all of these scales are given in Reichardt, Ref. 384.

[387]A scale of solvent nucleophilicity (as opposed to ionizing power), called the N_T scale, has been developed: Kevill; Anderson *J. Org. Chem.* **1991**, *56*, 1845.

[387a]For discussions, with references, see Kevill; Anderson *J. Am. Chem. Soc.* **1986**, *108*, 1579; McManus; Neamati-Mazreah; Karaman; Harris *J. Org. Chem.* **1986**, *51*, 4876; Abraham; Doherty; Kamlet; Harris; Taft *J. Chem. Soc., Perkin Trans. 2* **1987**, 913.

[388]Schadt; Bentley; Schleyer *J. Am. Chem. Soc.* **1976**, *98*, 7667.

[389]Bentley; Carter *J. Org. Chem.* **1983**, *48*, 579.

[390]For a review of these scales, see Bentley; Llewellyn *Prog. Phys. Org. Chem.* **1990**, *17*, 121-158.

[391]Kevill; Anderson *J. Org. Chem.* **1985**, *50*, 3330. See also Creary; McDonald *J. Org. Chem.* **1985**, *50*, 474.

[392]Bentley; Carter *J. Am. Chem. Soc.* **1982**, *104*, 5741. See also Liu; Sheu *J. Org. Chem.* **1991**, *56*, 3021.

[393]Bentley; Carter; Roberts *J. Org. Chem.* **1984**, *49*, 5183.

[394]See Kevill; Bahari; Anderson *J. Am. Chem. Soc.* **1984**, *106*, 2895; Bentley; Roberts *J. Org. Chem.* **1985**, *50*, 4821; Takeuchi; Ikai; Shibata; Tsugeno *J. Org. Chem.* **1988**, *53*, 2852; Kevill; Bahnke *Tetrahedron* **1988**, *44*, 7541; Hawkinson; Kevill *J. Org. Chem.* **1988**, *53*, 3857, **1989**, *54*, 154; Kevill; Hawkinson *J. Org. Chem.* **1990**, *55*, 5394.

TABLE 10.13 Y, Y_{OTs}, Z, and E_T (30) values for some solvents[386]

Solvent	Y	Y_{OTs}	Z	$E_T(30)$
CF₃COOH		4.57		
H₂O	3.5	4.1	94.6	63.1
(CF₃)₂CHOH		3.82		65.3
HCOOH	2.1	3.04		
H₂O—EtOH (1:1)	1.7	1.29	90	55.6
CF₃CH₂OH	1.0	1.77		59.8
HCONH₂	0.6		83.3	56.6
80% EtOH	0.0	0.0	84.8	53.7
MeOH	−1.1	−0.92	83.6	55.4
AcOH	−1.6	−0.9	79.2	51.7
EtOH	−2.0	−1.96	79.6	51.9
90% dioxane	−2.0	−2.41	76.7	46.7
iso-PrOH	−2.7	−2.83	76.3	48.4
95% acetone	−2.8	−2.95	72.9	48.3
t-BuOH	−3.3	−3.74	71.3	43.9
MeCN		−3.21	71.3	45.6
Me₂SO			71.1	45.1
HCONMe₂		−4.14	68.5	43.8
Acetone			65.7	42.2
HMPA				40.9
CH₂Cl₂				40.7
Pyridine			64.0	40.5
CHCl₃			63.2	39.1
PhCl				37.5
THF				37.4
Dioxane				36.0
Et₂O				34.5
C₆H₆			54	34.3
PhMe				33.9
CCl₄				32.4
n-Octane				31.1
n-Hexane				31.0
Cyclohexane				30.9

In order to include a wider range of solvents than those in which any of the Y values can be conveniently measured, other attempts have been made at correlating solvent polarities.[395] Kosower found that the position of the charge-transfer peak (see p. 79) in the uv spectrum of the complex (**87**) between iodide ion and 1-methyl- or 1-ethyl-4-carbometh-

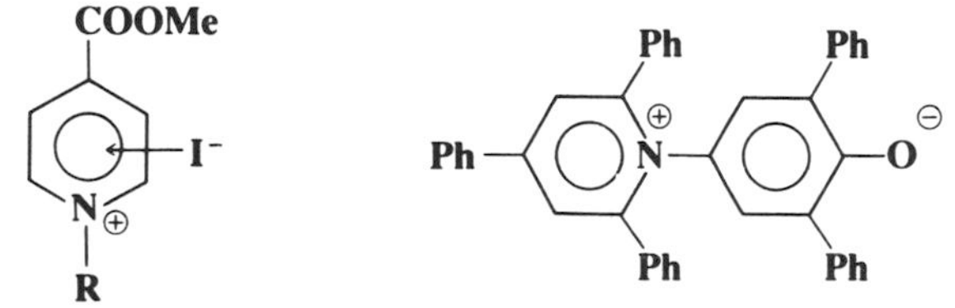

COOMe

87

R = Me or Et

88

[395]For reviews of solvent polarity scales, see Abraham; Grellier; Abboud; Doherty; Taft *Can. J. Chem.* **1988**, *66*, 2673-2686; Kamlet; Abboud; Taft *Prog. Phys. Org. Chem.* **1981**, *13*, 485-630; Shorter *Correlation Analysis of Organic Reactivity*; Wiley: New York, 1982, pp. 127-172; Reichardt, Ref. 386; Reichardt; Dimroth, Ref. 386; Abraham *Prog. Phys. Org. Chem.* **1974**, *11*, 1-87; Koppel; Palm, in Chapman; Shorter *Advances in Linear Free Energy Relationships*; Plenum: New York, 1972, pp. 203-280; Ref. 384. See also Chastrette; Carretto *Tetrahedron* **1982**, *38*, 1615; Chastrette; Rajzmann; Chanon; Purcell *J. Am. Chem. Soc.* **1985**, *107*, 1.

oxypyridinium ion was dependent on the polarity of the solvent.[396] From these peaks, which are very easy to measure, Kosower calculated transition energies that he called Z values. Z values are thus measures of solvent polarity analogous to Y values. Another scale is based on the position of electronic spectra peaks of the pyridinium-N-phenolbetaine **88** in various solvents.[397] Solvent polarity values on this scale are called $E_T(30)$[398] values. $E_T(30)$ values are related to Z values by the expression[399]

$$Z = 1.41E_T(30) + 6.92$$

Table 10.13 shows that Z and $E_T(30)$ values are generally in the same order as Y values. Other scales, the π^* scale,[400] the π^*_{azo} scale,[401] and the Py scale,[402] are also based on spectral data.[403]

The effect of solvent on nucleophilicity has already been discussed (pp. 349-350).

Phase Transfer Catalysis and Ultrasound

A difficulty that occasionally arises when carrying out nucleophilic substitution reactions is that the reactants do not mix. For a reaction to take place the reacting molecules must collide. In nucleophilic substitutions the substrate is usually insoluble in water and other polar solvents, while the nucleophile is often an anion, which is soluble in water but not in the substrate or other organic solvents. Consequently, when the two reactants are brought together, their concentrations in the same phase are too low for convenient reaction rates. One way to overcome this difficulty is to use a solvent that will dissolve both species. As we saw on p. 358, a dipolar aprotic solvent may serve this purpose. Another way, which is used very often, is *phase transfer catalysis*.[404]

In this method, a catalyst is used to carry the nucleophile from the aqueous into the organic phase. As an example, simply heating and stirring a two-phase mixture of 1-chlorooctane for several days with aqueous NaCN gives essentially no yield of 1-cyanooctane. But if a small amount of an appropriate quaternary ammonium salt is added, the product

[396]Kosower *J. Am. Chem. Soc.* **1958**, *80*, 3253, 3261, 3267; Kosower; Wu; Sorensen *J. Am. Chem. Soc.* **1961**, *83*, 3147. See also Larsen; Edwards; Dobi *J. Am. Chem. Soc.* **1980**, *102*, 6780.
[397]Dimroth; Reichardt; Siepmann; Bohlmann *Liebigs Ann. Chem.* **1963**, *661*, 1; Dimroth; Reichardt *Liebigs Ann. Chem.* **1969**, *727*, 93. See also Haak; Engberts *Recl. Trav. Chim. Pays-Bas* **1986**, *105*, 307.
[398]The symbol E_T comes from *energy, transition*. The (30) is used because the ion **88** bore this number in the first paper of Ref. 397. Values based on other ions have also been reported: See, for example Reichardt; Harbusch-Görnert; Schäfer *Liebigs Ann. Chem.* **1988**, 839.
[399]Reichardt; Dimroth, Ref. 386, p. 32.
[400]Kamlet; Abboud; Taft *J. Am. Chem. Soc.* **1977**, *99*, 6027; Doherty; Abraham; Harris; Taft; Kamlet *J. Org. Chem.* **1986**, *51*, 4872; Kamlet; Doherty; Abboud; Abraham; Taft *CHEMTECH* **1986**, 566-576, and other papers in this series. See also Doan; Drago *J. Am. Chem. Soc.* **1982**, *104*, 4524; Kamlet; Abboud; Taft, Ref. 395; Bekárek *J. Chem. Soc., Perkin Trans. 2* **1986**, 1425; Abe *Bull. Chem. Soc. Jpn.* **1990**, *63*, 2328.
[401]Buncel; Rajagopal *J. Org. Chem.* **1989**, *54*, 798.
[402]Dong; Winnik *Can. J. Chem.* **1984**, *62*, 2560.
[403]For a review of such scales, see Buncel; Rajagopal *Acc. Chem. Res.* **1990**, *23*, 226-231.
[404]For monographs, see Dehmlow; Dehmlow *Phase Transfer Catalysis*, 2nd ed.; Verlag Chemie: Deerfield Beach, FL, 1983; Starks; Liotta *Phase Transfer Catalysis*; Academic Press: New York, 1978; Weber; Gokel *Phase Transfer Catalysis in Organic Synthesis*; Springer: New York, 1977. For reviews, see Mąkosza; Fedoryński *Adv. Catal.* **1987**, *35*, 375-422; Gallo; Mąkosza; Dou; Hassanaly *Adv. Heterocycl. Chem.* **1984**, *36*, 175-234; Montanari; Landini; Rolla *Top. Curr. Chem.* **1982**, *101*, 147-200; Alper *Adv. Organomet. Chem.* **1981**, *19*, 183-211; Gallo; Dou; Hassanaly *Bull. Soc. Chim. Belg.* **1981**, *90*, 849-879; Dehmlow *Chimia* **1980**, *34*, 12-20, *Angew. Chem. Int. Ed. Engl.* **1977**, *16*, 493-505, **1974**, *13*, 170-174 [*Angew. Chem. 89*, 521-533; *86*, 187-196]; Mąkosza *Surv. Prog. Chem.* **1980**, *9*, 1-53; Starks, *CHEMTECH* **1980**, 110-117; Sjöberg *Aldrichimica Acta* **1980**, *13*, 55-58; McIntosh *J. Chem. Educ.* **1978**, *55*, 235-238; Gokel; Weber *J. Chem. Educ.* **1978**, *55*, 350-354; Weber; Gokel *J. Chem. Educ.* **1978**, *55*, 429-433; Liotta, in Izatt; Christensen *Synthetic Multidentate Macrocyclic Compounds*; Academic Press: New York, 1978, pp. 111-205; Brändström *Adv. Phys. Org. Chem.* **1977**, *15*, 267-330; Jones *Aldrichimica Acta* **1976**, *9*, 35-45; Dockx *Synthesis* **1973**, 441-456.

is quantitatively formed in about 2 hr.[405] There are two principal types of phase transfer catalyst. Though the action of the two types is somewhat different, the effects are the same. Both get the anion into the organic phase and allow it to be relatively free to react with the substrate.

1. *Quaternary ammonium or phosphonium salts.* In the above-mentioned case of NaCN, the uncatalyzed reaction does not take place because the CN^- ions cannot cross the interface between the two phases, except in very low concentration. The reason is that the Na^+ ions are solvated by the water, and this solvation energy would not be present in the organic phase. The CN^- ions cannot cross without the Na^+ ions because that would destroy the electrical neutrality of each phase. In contrast to Na^+ ions, quaternary ammonium (R_4N^+)[406] and phosphonium (R_4P^+) ions with sufficiently large R groups are poorly solvated in water and prefer organic solvents. If a small amount of such a salt is added, three equilibria are set up:

Organic phase $\quad Q^+CN^- + RCl \xrightarrow{\quad 4 \quad} RCN + Q^+Cl^-$

Aqueous phase $\quad Q^+CN^- + Na^+Cl^- \rightleftharpoons Na^+CN^- + Q^+Cl^- \qquad Q^+ = R_4N^+ \text{ or } R_4P^+$

The Na^+ ions remain in the aqueous phase; they cannot cross. The Q^+ ions do cross the interface and carry an anion with them. At the beginning of the reaction the chief anion present is CN^-. This gets carried into the organic phase (equilibrium 1) where it reacts with RCl to produce RCN and Cl^-. The Cl^- then gets carried into the aqueous phase (equilibrium 2). Equilibrium 3, taking place entirely in the aqueous phase, allows Q^+CN^- to be regenerated. All the equilibria are normally reached much faster than the actual reaction (4), so the latter is the rate-determining step.

In some cases, the Q^+ ions have such a low solubility in water that virtually all remain in the organic phase.[407] In such cases the exchange of ions (equilibrium 3) takes place across the interface. Still another mechanism (*the interfacial mechanism*) can operate where OH^- extracts a proton from an organic substrate.[408] In this mechanism, the OH^- ions remain in the aqueous phase and the substrate in the organic phase; the deprotonation takes place at the interface.[409]

2. *Crown ethers and other cryptands.*[410] We saw in Chapter 3 that certain cryptands are able to surround certain cations. In effect, a salt like KCN is converted by dicyclohexano-18-crown-6 into a new salt (**89**) whose anion is the same, but whose cation is now a much larger species with the positive charge spread over a large volume and hence much less

89   90   91

[405]Starks; Liotta, Ref. 404, p. 2.

[406]Bis-quaternary ammonium salts have also been used: Lissel; Feldman; Nir; Rabinovitz *Tetrahedron Lett.* **1989,** *30,* 1683.

[407]Landini; Maia; Montanari *J. Chem. Soc. Commun.* **1977,** 112, *J. Am. Chem. Soc.* **1978,** *100,* 2796.

[408]For a review, see Rabinovitz; Cohen; Halpern *Angew. Chem. Int. Ed. Engl.* **1986,** *25,* 960-970 [*Angew. Chem.* 98, 958-968].

[409]This mechanism was proposed by Mąkosza *Pure Appl. Chem.* **1975,** *43,* 439. See also Dehmlow; Thieser; Sasson; Pross *Tetrahedron* **1985,** *41,* 2927; Mason; Magdassi; Sasson *J. Org. Chem.* **1990,** *55,* 2714.

[410]For a review of this type of phase transfer catalysis, see Liotta, in Patai, Ref. 336, pp. 157-174.

concentrated. This larger cation is much less solubilized by water than K^+ and much more attracted to organic solvents. Though KCN is generally insoluble inorganic solvents, the cryptate salt is soluble in many of them. In these cases we do not need an aqueous phase at all but simply add the salt to the organic phase. Suitable cryptands have been used to increase greatly the rates of reactions where F^-, Br^-, I^-, OAc^-, and CN^- are nucleophiles.[411] Certain compounds that are not cryptands can act in a similar manner. One example is the podand tris(3,6-dioxaheptyl)amine (**90**), also called TDA-1.[412] Another, not related to the crown ethers, is the pyridyl sulfoxide **91**.[413]

Both of the above-mentioned catalyst types get the anions into the organic phase, but there is another factor as well. There is evidence that sodium and potassium salts of many anions, even if they could be dissolved in organic solvents, would undergo reactions very slowly (dipolar aprotic solvents are exceptions) because in these solvents the anions exist as ion pairs with Na^+ or K^+ and are not free to attack the substrate (p. 350). Fortunately, ion pairing is usually much less with the quaternary ions and with the positive cryptate ions, so the anions in these cases are quite free to attack. Such anions are sometimes referred to as "naked" anions.

Not all quaternary salts and cryptands work equally well in all situations. Some experimentation is often required to find the optimum catalyst.

Although phase transfer catalysis has been most often used for nucleophilic substitutions, it is not confined to these reactions. Any reaction that needs an insoluble anion dissolved in an organic solvent can be accelerated by an appropriate phase transfer catalyst. We shall see some examples in later chapters. In fact, in principle, the method is not even limited to anions, and a small amount of work has been done in transferring cations,[414] radicals, and molecules.[415] The reverse type of phase transfer catalysis has also been reported: transport into the aqueous phase of a reactant that is soluble in organic solvents.[416]

The catalysts mentioned above are soluble. Certain cross-linked polystyrene resins, as well as alumina[417] and silica gel, have been used as insoluble phase transfer catalysts. These, called *triphase catalysts*,[418] have the advantage of simplified product work-up and easy and quantitative catalyst recovery, since the catalyst can easily be separated from the product by filtration.

Another technique used to increase reaction rates is *ultrasound*.[419] In this technique the reaction mixture is subjected to high-energy sound waves, most often 20 KHz, but sometimes higher (a frequency of 20 KHz is about the upper limit of human hearing). When these

[411]See, for example, Liotta; Harris; McDermott; Gonzalez; Smith *Tetrahedron Lett.* **1974**, 2417; Sam; Simmons *J. Am. Chem. Soc.* **1974**, *96*, 2252; Durst *Tetrahedron Lett.* **1974**, 2421.

[412]Soula *J. Org. Chem.* **1985**, *50*, 3717.

[413]Furukawa; Ogawa; Kawai; Oae *J. Chem. Soc., Perkin Trans. 1* **1984**, 1833. See also Fujihara; Imaoka; Furukawa; Oae *J. Chem. Soc., Perkin Trans. 1* **1986**, 333.

[414]See Armstrong; Godat *J. Am. Chem. Soc.* **1979**, *101*, 2489; Iwamoto; Yoshimura; Sonoda; Kobayashi *Bull. Chem. Soc. Jpn.* **1983**, *56*, 796.

[415]See, for example, Dehmlow; Slopianka *Chem. Ber.* **1979**, *112*, 2765.

[416]Mathias; Vaidya *J. Am. Chem. Soc.* **1986**, *108*, 1093; Fife; Xin *J. Am. Chem. Soc.* **1987**, *109*, 1278.

[417]Quici; Regen *J. Org. Chem.* **1979**, *44*, 3436.

[418]For reviews, see Regen *Nouv. J. Chim.* **1982**, *6*, 629-637; *Angew. Chem. Int. Ed. Engl.* **1979**, *18*, 421-429 [*Angew. Chem. 91*, 464-472]. See also Molinari; Montanari; Quici; Tundo *J. Am. Chem. Soc.* **1979**, *101*, 3920; Bogatskii; Luk'yanenko; Pastushok; Parfenova *Doklad. Chem.* **1985**, *283*, 210; Pugia; Czech; Czech; Bartsch *J. Org. Chem.* **1986**, *51*, 2945.

[419]For monographs, see Ley; Low *Ultrasound in Synthesis*; Springer: New York, 1989; Mason; Lorimer *Sonochemistry*; Wiley: New York, 1988; Suslick *Ultrasound*; VCH: New York, 1988. For reviews, see Giguere *Org. Synth. Theory Appl.* **1989**, *1*, 103-172; Einhorn; Einhorn; Luche *Synthesis* **1989**, 787-813; Goldberg; Sturkovich; Lukevics *Heterocycles* **1989**, *29*, 597-627; Abdulla *Aldrichimica Acta* **1988**, *21*, 31-42; Moon *CHEMTECH* **1987**, 434-437; Lorimer; Mason *Chem. Soc. Rev.* **1987**, *16*, 275-311; Lindley; Mason *Chem. Soc. Rev.* **1987**, *16*, 239-274; Boudjouk *J. Chem. Educ.* **1986**, *63*, 427; Bremner *Chem. Br.* **1986**, 633-638; Suslick *Adv. Organomet. Chem.* **1986**, *25*, 73-119, *Mod. Synth. Methods* **1986**, *4*, 1-60. See also the series *Advances in Sonochemistry*.

waves are passed through a mixture, small bubbles form (*cavitation*). Collapse of these bubbles produces powerful shock waves that greatly increase the temperatures and pressures within these tiny regions, resulting in an increased reaction rate.[420] In the common instance where a metal, as a reactant or catalyst, is in contact with a liquid phase, a further effect is that the surface of the metal is cleaned and/or eroded by the ultrasound, allowing the liquid-phase molecules to come into closer contact with the metal atoms. Among the advantages of ultrasound is that it may increase yields, reduce side reactions, and permit the use of lower temperatures and/or pressures. It has been postulated that ultrasound has its best results with reactions that proceed, at least partially, through free-radical intermediates.[421]

Ambident Nucleophiles. Regioselectivity

Some nucleophiles have a pair of electrons on each of two or more atoms, or canonical forms can be drawn in which two or more atoms bear an unshared pair. In these cases the nucleophile may attack in two or more different ways to give different products. Such reagents are called *ambident nucleophiles*.[422] In most cases a nucleophile with two potentially attacking atoms can attack with either of them, depending on conditions, and mixtures are often obtained, though this is not always the case. For example, the nucleophile NCO^- usually gives only isocyanates RNCO and not the isomeric cyanates ROCN.[423] When a reaction can potentially give rise to two or more structural isomers (e.g., ROCN or RNCO) but actually produces only one, the reaction is said to be *regioselective*[424] (compare the definitions of stereoselective, p. 137 and enantioselective, p. 119). Some important ambident nucleophiles are:

1. *Ions of the type* —CO—$\overset{\ominus}{\text{CR}}$—CO—. These ions, which are derived by removal of a proton from malonic esters, β-keto esters, β-diketones, etc., are resonance hybrids:

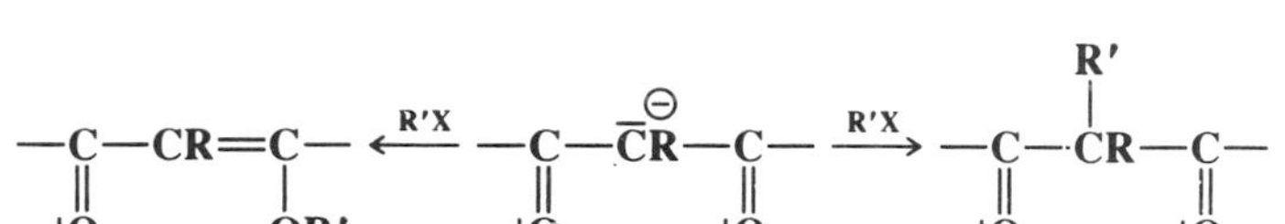

$$-\underset{|\underline{O}}{\overset{||}{C}}-\overset{\ominus}{CR}-\underset{|\underline{O}}{\overset{||}{C}}- \longleftrightarrow -\underset{|\underline{O}}{\overset{|}{C}}=CR-\underset{|\underline{O}_\ominus}{\overset{||}{C}}- \longleftrightarrow -\underset{|\underline{O}}{\overset{||}{C}}-CR=\underset{|\underline{O}_\ominus}{\overset{|}{C}}-$$

They can thus attack a saturated carbon with their carbon atoms (C-alkylation) or with their oxygen atoms (O-alkylation):

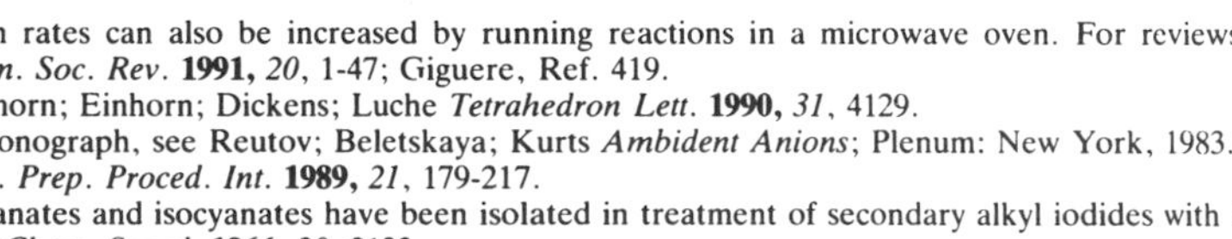

$$-\underset{|\underline{O}}{\overset{||}{C}}-CR=\underset{OR'}{\overset{|}{C}}- \overset{R'X}{\longleftarrow} -\underset{|\underline{O}}{\overset{||}{C}}-\overset{\ominus}{CR}-\underset{|\underline{O}}{\overset{||}{C}}- \overset{R'X}{\longrightarrow} -\underset{|\underline{O}}{\overset{||}{C}}-\overset{R'}{\underset{|}{CR}}-\underset{|\underline{O}}{\overset{||}{C}}-$$

With unsymmetrical ions, three products are possible, since either oxygen can attack. With a carbonyl substrate the ion can analogously undergo C-acylation or O-acylation.

[420]Reaction rates can also be increased by running reactions in a microwave oven. For reviews, see Mingos; Baghurst *Chem. Soc. Rev.* **1991**, *20*, 1-47; Giguere, Ref. 419.
[421]See Einhorn; Einhorn; Dickens; Luche *Tetrahedron Lett.* **1990**, *31*, 4129.
[422]For a monograph, see Reutov; Beletskaya; Kurts *Ambident Anions*; Plenum: New York, 1983. For a review, see Black *Org. Prep. Proceed. Int.* **1989**, *21*, 179-217.
[423]Both cyanates and isocyanates have been isolated in treatment of secondary alkyl iodides with NCO^-: Holm; Wentrup *Acta Chem. Scand.* **1966**, *20*, 2123.
[424]This term was introduced by Hassner *J. Org. Chem.* **1968**, *33*, 2684.

2. Compounds of the type $CH_3CO—CH_2—CO—$ can give up two protons, if treated with 2 moles of a strong enough base, to give dicarbanions:

$$CH_3—CO—CH_2—CO— \xrightarrow[\text{of base}]{2 \text{ moles}} {}^{\ominus}\overline{C}H_2—CO—\overset{\ominus}{\overline{C}}H—CO—$$
$$\textbf{92}$$

Such ions are ambident nucleophiles, since they have two possible attacking carbon atoms, aside from the possibility of attack by oxygen. In such cases, the attack is virtually always by the more basic carbon.[425] Since the hydrogen of a carbon bonded to two carbonyl groups is more acidic than that of a carbon bonded to just one (see Chapter 8), the CH group of **92** is less basic than the CH_2 group, so the latter attacks the substrate. This gives rise to a useful general principle: whenever we desire to remove a proton at a given position for use as a nucleophile but there is a stronger acidic group in the molecule, it may be possible to take off both protons; if it is, then attack is always by the desired position since it is the ion of the weaker acid. On the other hand, if it is desired to attack with the more acidic position, all that is necessary is to remove just one proton.[426] For example, ethyl acetoacetate can be alkylated at either the methyl or the methylene group (**0-94**):

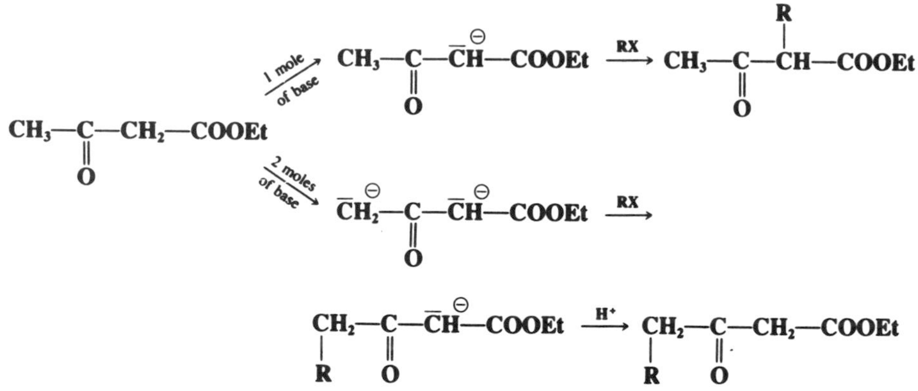

3. *The CN⁻ ion.* This nucleophile can give nitriles RCN (**0-101**) or isocyanides $RN{\equiv}C$.
4. *The nitrite ion.* This ion can give nitrite esters $R—O—N{=}O$ (**0-32**) or nitro compounds RNO_2 (**0-60**), which are not esters.
5. Phenoxide ions (which are analogous to enolate ions) can undergo C-alkylation or O-alkylation:

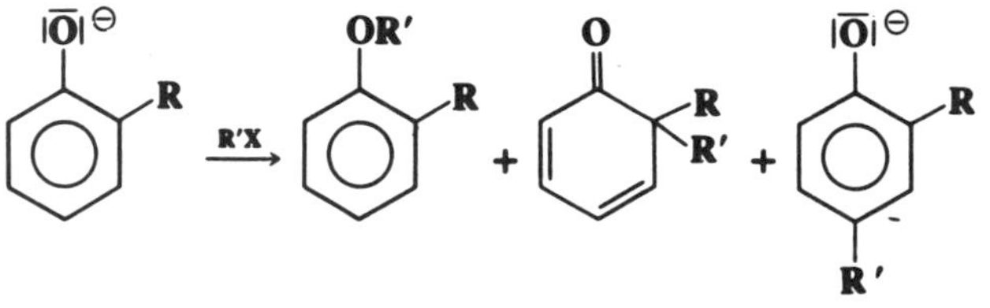

---

[425]For an exception, see Trimitsis; Hinkley; TenBrink; Faburada; Anderson; Poli; Christian; Gustafson; Erdman; Rop *J. Org. Chem.* **1983**, *48*, 2957.
[426]The use of this principle was first reported by Hauser; Harris *J. Am. Chem. Soc.* **1958**, *80*, 6360. It has since been applied many times. For reviews, see Thompson; Green *Tetrahedron* **1991**, *47*, 4223-4285; Kaiser; Petty; Knutson *Synthesis* **1977**, 509-550; Harris; Harris *Org. React.* **1969**, *17*, 155-211.

6. Removal of a proton from an aliphatic nitro compound gives a carbanion ($R_2\overset{\ominus}{C}$—NO_2) that can be alkylated at oxygen or carbon.[427] O-Alkylation gives nitronic esters, which are generally unstable to heat but break down to give an oxime and an aldehyde or ketone.

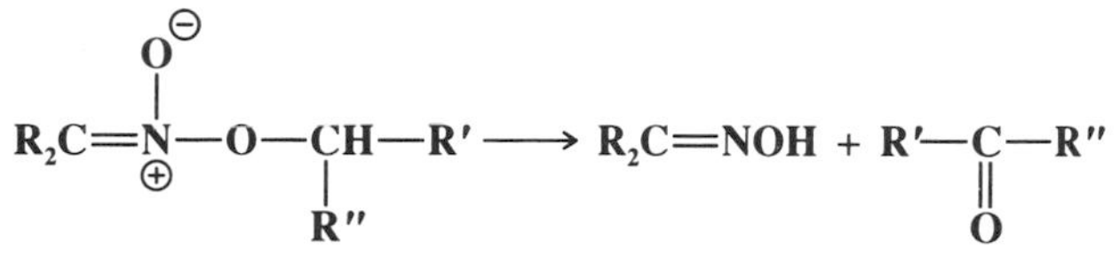

$$R_2C=\underset{\underset{\oplus}{|}}{\overset{\overset{\overset{\ominus}{O}}{|}}{N}}-O-\underset{\underset{R''}{|}}{CH}-R' \longrightarrow R_2C=NOH + R'-\underset{\underset{O}{\|}}{C}-R''$$

There are many other ambident nucleophiles.

It would be useful to have general rules as to which atom of an ambident nucleophile will attack a given substrate under a given set of conditions.[428] Unfortunately, the situation is complicated by the large number of variables. It might be expected that the more electronegative atom would always attack, but this is often not the case. Where the products are determined by thermodynamic control (p. 214), the principal product is usually the one in which the atom of higher basicity has attacked (i.e., $C > N > O > S$).[429] However, in most reactions, the products are kinetically controlled and matters are much less simple. Nevertheless, the following generalizations can be made, while recognizing that there are many exceptions and unexplained results. As in the discussion of nucleophilicity in general (p. 348), there are two major factors: the polarizability (hard-soft character) of the nucleophile and solvation effects.

1. The principle of hard and soft acids and bases states that hard acids prefer hard bases and soft acids prefer soft bases (p. 263). In an SN1 mechanism the nucleophile attacks a carbocation, which is a hard acid. In an SN2 mechanism the nucleophile attacks the carbon atom of a molecule, which is a softer acid. The more electronegative atom of an ambident nucleophile is a harder base than the less electronegative atom. We may thus make the statement: As the character of a given reaction changes from SN1-like to SN2-like, an ambident nucleophile becomes more likely to attack with its less electronegative atom.[430] Therefore, changing from SN1 to SN2 conditions should favor C attack by CN^-, N attack by NO_2^-, C attack by enolate or phenoxide ions, etc. As an example, primary alkyl halides are attacked (in protic solvents) by the carbon atom of the anion of CH_3COCH_2COOEt, while α-chloro ethers, which react by the SN1 mechanism, are attacked by the oxygen atom. However, this does not mean that attack is by the less electronegative atom in all SN2 reactions and by the more electronegative atom in all SN1 reactions. The position of attack also depends on the nature of the nucleophile, the solvent, the leaving group, and other conditions. The rule merely states that increasing the SN2 character of the transition state makes attack by the less electronegative atom more likely.

2. All negatively charged nucleophiles must of course have a positive counterion. If this ion is Ag^+ (or some other ion that specifically helps in removing the leaving group, p. 359), rather than the more usual Na^+ or K^+, then the transition state is more SN1-like. Therefore

[427]For a review, see Erashko; Shevelev; Fainzil'berg *Russ. Chem. Rev.* **1966**, *35*, 719-732.

[428]For reviews, see Jackman; Lange *Tetrahedron* **1977**, *33*, 2737-2769; Reutov; Kurts *Russ. Chem. Rev.* **1977**, *46*, 1040-1056; Gompper; Wagner *Angew. Chem. Int. Ed. Engl.* **1976**, *15*, 321-333 [*Angew. Chem. 88*, 389-401]; Shevelev *Russ. Chem. Rev.* **1970**, *39*, 844-858.

[429]For an example, see Bégué; Charpentier-Morize; Née *J. Chem. Soc., Chem. Commun.* **1989**, 83.

[430]This principle, sometimes called *Kornblum's rule*, was first stated by Kornblum; Smiley; Blackwood; Iffland *J. Am. Chem. Soc.* **1955**, *77*, 6269.

the use of Ag$^+$ promotes attack at the more electronegative atom. For example, alkyl halides treated with NaCN generally give mostly RCN, but the use of AgCN increases the yield of isocyanides RNC.[431]

3. In many cases the solvent influences the position of attack. The freer the nucleophile, the more likely it is to attack with its more electronegative atom, but the more this atom is encumbered by either solvent molecules or positive counterions, the more likely is attack by the less electronegative atom. In protic solvents, the more electronegative atom is better solvated by hydrogen bonds than the less electronegative atom. In polar aprotic solvents, neither atom of the nucleophile is greatly solvated, but these solvents are very effective in solvating cations. Thus in a polar aprotic solvent the more electronegative end of the nucleophile is freer from entanglement by both the solvent and the cation, so that a change from a protic to a polar aprotic solvent often increases the extent of attack by the more electronegative atom. An example is attack by sodium β-naphthoxide on benzyl bromide, which resulted in 95% O-alkylation in dimethyl sulfoxide and 85% C-alkylation in 2,2,2-trifluoroethanol.[432] Changing the cation from Li$^+$ to Na$^+$ to K$^+$ (in nonpolar solvents) also favors O- over C-alkylation[433] for similar reasons (K$^+$ leaves the nucleophile much freer than Li$^+$), as does the use of crown ethers, which are good at solvating cations (p. 82).[434] Alkylation of the enolate ion of cyclohexanone in the gas phase, where the nucleophile is completely free, showed only O-alkylation and no C-alkylation.[435]

4. In extreme cases, steric effects can govern the regioselectivity.[436]

Ambident Substrates

Some substrates (e.g., 1,3-dichlorobutane) can be attacked at two or more positions. We may call these *ambident substrates*. In the example given, there happen to be two leaving groups in the molecule, but there are two kinds of substrates that are inherently ambident (unless symmetrical). One of these, the allylic type, has already been discussed (p. 327). The other is the epoxy (or the similar aziridine or episulfide) substrate.[437]

[431]Actually, this reaction is more complicated than it seems on the surface; see Austad; Songstad; Stangeland *Acta Chem. Scand.* **1971**, *25*, 2327; Carretero; García Ruano *Tetrahedron Lett.* **1985**, *26*, 3381.

[432]Kornblum; Berrigan; le Noble *J. Chem. Soc.* **1963**, *85*, 1141; Kornblum; Seltzer; Haberfield *J. Am. Chem. Soc.* **1963**, *85*, 1148. For other examples, see le Noble; Puerta *Tetrahedron Lett.* **1966**, 1087; Brieger; Pelletier *Tetrahedron Lett.* **1965**, 3555; Heiszwolf; Kloosterziel *Recl. Trav. Chim. Pays-Bas* **1970**, *89*, 1153, 1217; Kurts; Masias; Beletskaya; Reutov *J. Org. Chem. USSR* **1971**, *7*, 2323; Schick; Schwarz; Finger; Schwarz *Tetrahedron* **1982**, *38*, 1279.

[433]Kornblum; Seltzer; Haberfield, Ref. 432; Kurts; Beletskaya; Masias; Reutov *Tetrahedron Lett.* **1968**, 3679. See, however, Sarthou; Bram; Guibe *Can. J. Chem.* **1980**, *58*, 786.

[434]Smith; Hanson *J. Org. Chem.* **1971**, *36*, 1931; Kurts; Dem'yanov; Beletskaya; Reutov *J. Org. Chem. USSR* **1973**, *9*, 1341; Cambillau; Sarthou; Bram *Tetrahedron Lett.* **1976**, 281; Akabori; Tuji *Bull. Chem. Soc. Jpn.* **1978**, *51*, 1197. See also Zook; Russo; Ferrand; Stotz *J. Org. Chem.* **1968**, *33*, 2222; le Noble; Palit *Tetrahedron Lett.* **1972**, 493.

[435]Jones; Kass; Filley; Barkley; Ellison *J. Am. Chem. Soc.* **1985**, *107*, 109.

[436]See, for example O'Neill; Hegarty *J. Org. Chem.* **1987**, *52*, 2113.

[437]For reviews of S$_N$ reactions at such substrates, see Rao; Paknikar; Kirtane *Tetrahedron* **1983**, *39*, 2323-2367; Behrens; Sharpless *Aldrichimica Acta* **1983**, *16*, 67-79; Enikolopiyan *Pure Appl. Chem.* **1976**, *48*, 317-328; Fokin; Kolomiets *Russ. Chem. Rev.* **1976**, *45*, 25-42; Wohl *Chimia* **1974**, *28*, 1-5; Kirk *Chem. Ind. (London)* **1973**, 109-116; Buchanan; Sable *Sel. Org. Transform.* **1972**, *2*, 1-95; Dermer; Ham *Ethylenimine and Other Aziridines*; Academic Press: New York, 1969, pp. 206-273; Akhrem; Moiseenkov; Dobrynin *Russ. Chem. Rev.* **1968**, *37*, 448-462; Gritter, in Patai, Ref. 333, pp. 390-400.

Substitution of the free epoxide, which generally occurs under basic or neutral conditions, usually involves an SN2 mechanism. Since primary substrates undergo SN2 attack more readily than secondary, unsymmetrical epoxides are attacked in neutral or basic solution at the less highly substituted carbon, and stereospecifically, with inversion at that carbon. Under acidic conditions, it is the protonated epoxide that undergoes the reaction. Under these conditions the mechanism can be either SN1 or SN2. In SN1 mechanisms, which favor tertiary carbons, we might expect that attack would be at the more highly substituted carbon, and this is indeed the case. However, even when protonated epoxides react by the SN2 mechanism, attack is usually at the more highly substituted position.[438] Thus, it is often possible to change the direction of ring opening by changing the conditions from basic to acidic or vice versa. In the ring opening of 2,3-epoxy alcohols, the presence of Ti(O-i-Pr)$_4$ increases both the rate and the regioselectivity, favoring attack at C-3 rather than C-2.[439] When an epoxide ring is fused to a cyclohexane ring, SN2 ring opening invariably gives diaxial rather than diequatorial ring opening.[440]

Cyclic sulfates (**93**), prepared from 1,2-diols, react in the same manner as epoxides, but usually more rapidly:[441]

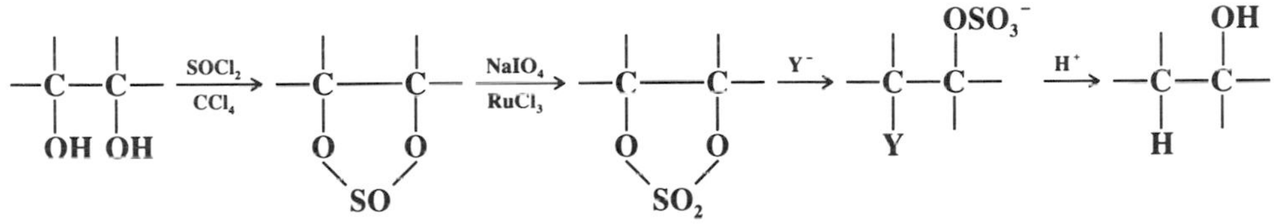

REACTIONS

The reactions in this chapter are classified according to the attacking atom of the nucleophile in the order O, S, N, halogen, H, C. For a given nucleophile, reactions are classified by the substrate and leaving group, with alkyl substrates usually considered before acyl ones. Nucleophilic substitutions at a sulfur atom are treated at the end.

Not all the reactions in this chapter are actually nucleophilic substitutions. In some cases the mechanisms are not known with enough certainty even to decide whether a nucleophile, an electrophile, or a free radical is attacking. In other cases (such as **0-76**), conversion of one compound to another can occur by two or even all three of these possibilities, depending on the reagent and the reaction conditions. However, one or more of the nucleophilic mechanisms previously discussed do hold for the overwhelming majority of the reactions in this chapter. For the alkylations, the SN2 is by far the most common mechanism, as long as R is primary or secondary alkyl. For the acylations, the tetrahedral mechanism is the most common.

[438]Addy; Parker *J. Chem. Soc.* **1963**, 915; Biggs; Chapman; Finch; Wray *J. Chem. Soc. B* **1971**, 55.
[439]Caron; Sharpless *J. Org. Chem.* **1985**, *50*, 1557. See also Chong; Sharpless *J. Org. Chem.* **1985**, *50*, 1560; Behrens; Sharpless *J. Org. Chem.* **1985**, *50*, 5696.
[440]Murphy; Alumbaugh; Rickborn *J. Am. Chem. Soc.* **1969**, *91*, 2649. For a method of overriding this preference, see McKittrick; Ganem *J. Org. Chem.* **1985**, *50*, 5897.
[441]Gao; Sharpless *J. Am. Chem. Soc.* **1988**, *110*, 7538; Kim; Sharpless *Tetrahedron Lett.* **1989**, *30*, 655.

Oxygen Nucleophiles

A. Attack by OH at an Alkyl Carbon

0-1 Hydrolysis of Alkyl Halides
Hydroxy-de-halogenation

$$RX + H_2O \longrightarrow ROH_2{}^+ \xrightarrow{-H^+} ROH + H^+$$
$$RX + OH^- \longrightarrow ROH$$

Alkyl halides can be hydrolyzed to alcohols. Hydroxide ion is usually required, except that especially active substrates such as allylic or benzylic types can be hydrolyzed by water. Ordinary halides can also be hydrolyzed by water,[442] if the solvent is HMPA or N-methyl-2-pyrrolidone.[443] In contrast to most nucleophilic substitutions at saturated carbons, this reaction can be performed on tertiary substrates without significant interference from elimination side reactions. The reaction is not frequently used for synthetic purposes, because alkyl halides are usually obtained from alcohols.

Vinylic halides are unreactive (p. 341), but they can be hydrolyzed to ketones at room temperature with mercuric trifluoroacetate, or with mercuric acetate in either trifluoroacetic

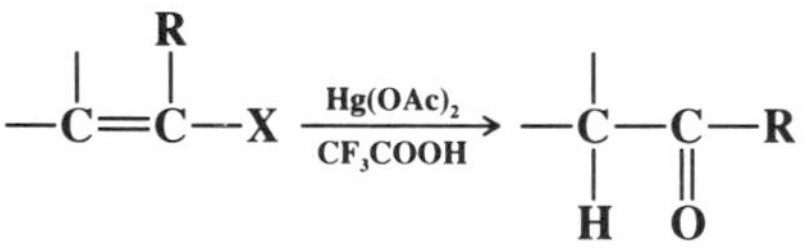

$$-\overset{\displaystyle |}{C}=\overset{\displaystyle \overset{R}{|}}{C}-X \xrightarrow[\text{CF}_3\text{COOH}]{\text{Hg(OAc)}_2} -\overset{\displaystyle \underset{H}{|}}{C}-\overset{\displaystyle \underset{O}{\|}}{C}-R$$

acid or acetic acid containing BF$_3$ etherate.[444] Primary bromides and iodides give alcohols when treated with bis(tributyltin)oxide Bu$_3$Sn—O—SnBu$_3$ in the presence of silver salts.[445]

OS **II**, 408; **III**, 434; **IV**, 128; **VI**, 142, 1037.

0-2 Hydrolysis of *gem*-Dihalides
Oxo-de-dihalo-bisubstitution

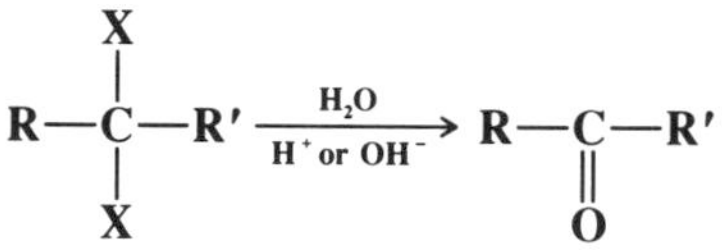

$$R-\overset{\displaystyle \overset{X}{|}}{\underset{\displaystyle \underset{X}{|}}{C}}-R' \xrightarrow[\text{H}^+ \text{ or OH}^-]{\text{H}_2\text{O}} R-\overset{\displaystyle \underset{O}{\|}}{C}-R'$$

[442]It has been proposed that the mechanism of the reaction of primary halides with water is not the ordinary SN2 mechanism, but that the rate-determining process involves a fluctuation of solvent configuration: Kurz; Kurz *Isr. J. Chem.* **1985**, *26*, 339; Kurz; Lee; Love; Rhodes *J. Am. Chem. Soc.* **1986**, *108*, 2960.
[443]Hutchins; Taffer *J. Org. Chem.* **1983**, *48*, 1360.
[444]Martin; Chou *Tetrahedron Lett.* **1978**, 1943; Yoshioka; Takasaki; Kobayashi; Matsumoto *Tetrahedron Lett.* **1979**, 3489.
[445]Gingras; Chan *Tetrahedron Lett.* **1989**, *30*, 279.

gem-Dihalides can be hydrolyzed with either acid or basic catalysis to give aldehydes or ketones.[446] Formally, the reaction may be regarded as giving R—C(OH)XR', which is unstable and loses HX to give the carbonyl compound. For aldehydes, strong bases cannot be used, because the product undergoes the aldol reaction (**6-39**) or the Cannizzaro reaction (**9-69**).

OS **I**, 95; **II**, 89, 133, 244, 549; **III**, 538, 788; **IV**, 110, 423, 807. Also see OS **III**, 737.

0-3 Hydrolysis of 1,1,1-Trihalides
Hydroxy,oxo-de-trihalo-tersubstitution

$$RCX_3 + H_2O \longrightarrow RCOOH$$

This reaction is similar to the previous one. The utility of the method is limited by the lack of availability of trihalides, though these compounds can be prepared by addition of CCl_4 and similar compounds to double bonds (**5-33**) and by the free-radical halogenation of methyl groups on aromatic rings (**4-1**). When the hydrolysis is carried out in the presence of an alcohol, a carboxylic ester can be obtained directly.[447] 1,1-Dichloroalkenes can also be hydrolyzed to carboxylic acids, by treatment with H_2SO_4. In general 1,1,1-trifluorides do not undergo this reaction,[448] though exceptions are known.[449]

Aryl 1,1,1-trihalomethanes can be converted to acyl halides by treatment with sulfur trioxide.[450]

$$ArCCl_3 + SO_3 \longrightarrow Ar-\overset{\displaystyle O}{\underset{\displaystyle \|}{C}}-Cl + (ClSO_2)_2O$$

Chloroform is more rapidly hydrolyzed with base than dichloromethane or carbon tetrachloride and gives not only formic acid but also carbon monoxide.[451] Hine[452] has shown that the mechanism of chloroform hydrolysis is quite different from that of dichloromethane or carbon tetrachloride, though superficially the three reactions appear similar. The first step is the loss of a proton to give CCl_3^- which then loses Cl^- to give dichlorocarbene CCl_2, which is hydrolyzed to formic acid or carbon monoxide.

$$HCCl_3 \underset{OH^-}{\overset{}{\rightleftharpoons}} CCl_3^- \xrightarrow{-Cl^-} \overline{C}Cl_2 \xrightarrow{H_2O} HCOOH \text{ or } CO$$

This is an example of an SN1cB mechanism (p. 356). The other two compounds react by the normal mechanisms. Carbon tetrachloride cannot give up a proton and dichloromethane is not acidic enough.

OS **III**, 270; **V**, 93. Also see OS **I**, 327.

[446]For a review, see Salomaa, in Patai *The Chemistry of the Carbonyl Group*, vol. 1; Wiley: New York, 1966, pp. 177-210.
[447]See, for example, Le Fave; Scheurer *J. Am. Chem. Soc.* **1950**, *72*, 2464.
[448]Sheppard; Sharts *Organic Fluorine Chemistry*; W.A. Benjamin: New York, 1969, pp. 410-411; Hudlický, *Chemistry of Organic Fluorine Compounds*, 2nd ed.; Ellis Horwood: Chichester, 1976, pp. 273-274.
[449]See, for example, Kobayashi; Kumadaki *Acc. Chem. Res.* **1978**, *11*, 197-204.
[450]Rondestvedt *J. Org. Chem.* **1976**, *41*, 3569, 3574, 3576. For another method, see Nakano; Ohkawa; Matsumoto; Nagai *J. Chem. Soc., Chem. Commun.* **1977**, 808.
[451]For a review, see Kirmse *Carbene Chemistry*, 2nd ed.; Academic Press: New York, 1971, pp. 129-141.
[452]Hine *J. Am. Chem. Soc.* **1950**, *72*, 2438. Also see le Noble *J. Am. Chem. Soc.* **1965**, *87*, 2434.

0-4 Hydrolysis of Alkyl Esters of Inorganic Acids
Hydroxy-de-sulfonyloxy-substitution, etc.

$$
\left.\begin{array}{l}
\text{R—OSO}_2\text{R'} \\
\text{R—OSO}_2\text{OH} \\
\text{R—OSO}_2\text{OR'} \\
\text{R—OSOR'} \\
\text{R—ONO}_2 \\
\text{R—ONO} \\
\text{R—OPO(OH)}_2 \\
\text{R—OPO(OR')}_2 \\
\text{R—OB(OH)}_2 \\
\text{and others}
\end{array}\right\} \xrightarrow{\text{H}_2\text{O}} \text{R—OH}
$$

Esters of inorganic acids, including those given above and others, can be hydrolyzed to alcohols. The reactions are most successful when the ester is that of a strong acid, but it can be done for esters of weaker acids by the use of hydroxide ion (a more powerful nucleophile) or acidic conditions (which make the leaving group come off more easily). When vinylic substrates are hydrolyzed, the products are aldehydes or ketones.

$$\text{R}_2\text{C}=\text{CH—X} \xrightarrow{\text{H}_2\text{O}} \text{R}_2\text{C}=\text{CH—OH} \longrightarrow \text{R}_2\text{CH—CHO}$$

These reactions are all considered at one place because they are formally similar, but though some of them involve R—O cleavage and are thus nucleophilic substitutions at a saturated carbon, others involve cleavage of the bond between the inorganic atom and oxygen and are thus nucleophilic substitutions at a sulfur, nitrogen, etc. It is even possible for the same ester to be cleaved at either position, depending on the conditions. Thus benzhydryl p-toluenesulfinate ($\text{Ph}_2\text{CHOSOC}_6\text{H}_4\text{CH}_3$) was found to undergo C—O cleavage in HClO_4 solutions and S—O cleavage in alkaline media.[453] In general, the weaker the corresponding acid, the less likely is C—O cleavage. Thus, sulfonic acid esters $\text{ROSO}_2\text{R'}$ generally give C—O cleavage,[454] while nitrous acid esters RONO usually give N—O cleavage.[455] Esters of sulfonic acids that are frequently hydrolyzed are mentioned on p. 353. For hydrolysis of sulfonic acid esters, see also **0-114.**

OS **VI**, 852. See also OS **67**, 13.

0-5 Hydrolysis of Diazo Ketones
Hydro,hydroxy-de-diazo-bisubstitution

$$\text{R—}\underset{\overset{\|}{\text{O}}}{\text{C}}\text{—CHN}_2 + \text{H}_2\text{O} \xrightarrow{\text{H}^+} \text{R—}\underset{\overset{\|}{\text{O}}}{\text{C}}\text{—CH}_2\text{OH}$$

Diazo ketones are relatively easy to prepare (see **0-112**). When treated with acid, they add a proton to give α-keto diazonium salts, which are hydrolyzed to the alcohols by the SN1 or SN2 mechanism.[456] Relatively good yields of α-hydroxy ketones can be prepared in this

[453]Bunton; Hendy J. Chem. Soc. **1963**, 627. For another example, see Batts J. Chem. Soc. B **1966**, 551.

[454]Barnard; Robertson Can. J. Chem. **1961**, 39, 881. See also Drabicky; Myhre; Reich; Schmittou J. Org. Chem. **1976**, 41, 1472.

[455]For a discussion of the mechanism of hydrolysis of alkyl nitrites, see Williams Nitrosation; Cambridge University Press: Cambridge, 1988, pp. 162-163.

[456]Dahn; Gold Helv. Chim. Acta **1963**, 46, 983; Thomas; Leveson Int. J. Chem. Kinet. **1983**, 15, 25. For a review of the acid-promoted decomposition of diazo ketones, see Smith; Dieter Tetrahedron **1981**, 37, 2407-2439.

way, since the diazonium ion is somewhat stabilized by the presence of the carbonyl group, which discourages N_2 from leaving because that would result in an unstable α-carbonyl carbocation.

0-6 Hydrolysis of Acetals, Enol Ethers, and Similar Compounds[457]

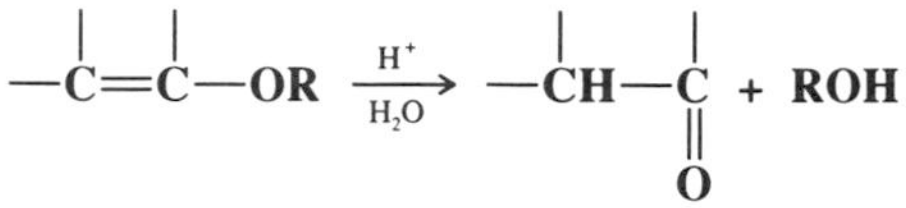

$$-\overset{|}{C}=\overset{|}{C}-OR \xrightarrow[H_2O]{H^+} -\overset{|}{C}H-\overset{|}{\underset{\overset{\|}{O}}{C}} + ROH \quad \textbf{3/Hydro-de-\textit{O}-alkylation}$$

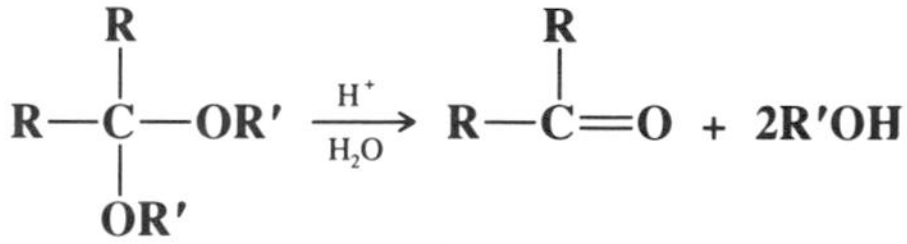

$$R-\overset{\overset{R}{|}}{\underset{\underset{OR'}{|}}{C}}-OR' \xrightarrow[H_2O]{H^+} R-\overset{R}{\underset{}{C}}=O + 2R'OH \quad \textbf{\textit{O}-Alkyl-\textit{C}-alkoxy-elimination}$$

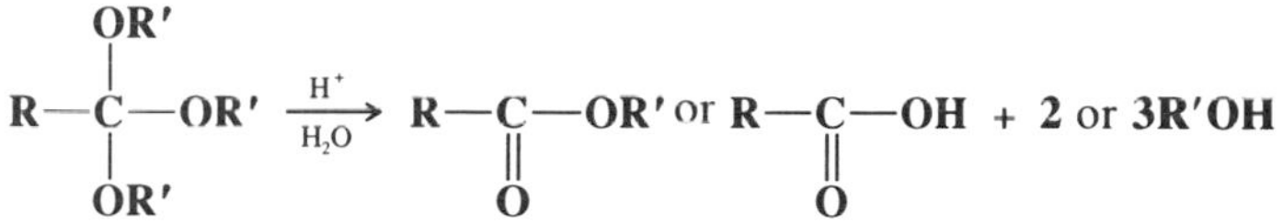

$$R-\overset{\overset{OR'}{|}}{\underset{\underset{OR'}{|}}{C}}-OR' \xrightarrow[H_2O]{H^+} R-\overset{}{\underset{\overset{\|}{O}}{C}}-OR' \text{ or } R-\overset{}{\underset{\overset{\|}{O}}{C}}-OH + 2 \text{ or } 3R'OH$$

The alkoxyl group OR is not a leaving group, so these compounds must be converted to the conjugate acids before they can be hydrolyzed. Although 100% sulfuric acid and other concentrated strong acids readily cleave simple ethers,[458] the only acids used preparatively for this purpose are HBr and HI (**0-68**). However, acetals, ketals, and ortho esters[459] are easily cleaved by dilute acids. These compounds are hydrolyzed with greater facility because carbocations of the type $RO-\overset{\oplus}{\underset{|}{C}}-$ are greatly stabilized by resonance (p. 170). The reactions therefore proceed by the S$_N$1 mechanism,[460] as shown for acetals:[461]

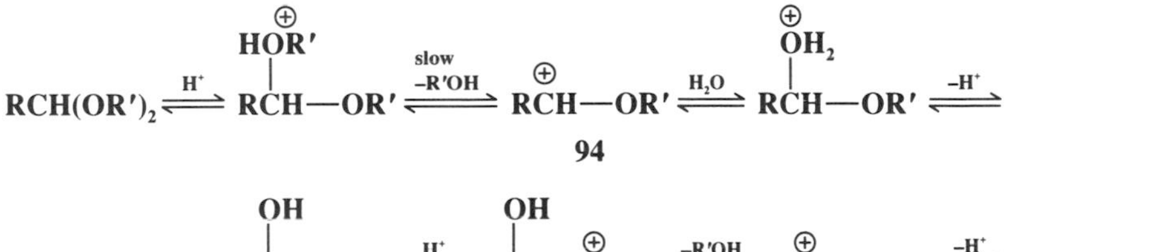

$$RCH(OR')_2 \underset{H^+}{\rightleftharpoons} RCH-OR' \underset{\overset{slow}{-R'OH}}{\rightleftharpoons} R\overset{\oplus}{C}H-OR' \underset{H_2O}{\rightleftharpoons} RCH-OR' \underset{-H^+}{\rightleftharpoons}$$

with $\overset{\oplus}{HOR'}$ above the first RCH—OR' and $\overset{\oplus}{OH_2}$ above the last, labeled **94**

$$RCH-OR' \rightleftharpoons R\overset{OH}{\underset{}{C}}H-\overset{\oplus}{OR'}H \underset{-R'OH}{\rightleftharpoons} R\overset{\oplus}{C}H-OH \underset{-H^+}{\rightleftharpoons} RCH=O$$

with OH above the first RCH—OR' and H⁺ below the first arrow

hemiacetal

This mechanism (which is an S$_N$1cA or A1 mechanism) is the reverse of that for acetal formation by reaction of an aldehyde and an alcohol (**6-6**). Among the facts supporting the

[457]For reviews, see Bergstrom, in Patai, Ref. 336, pp. 881-902; Cockerill; Harrison, in Patai *The Chemistry of Functional Groups, Supplement A*, pt. 1; Wiley: New York, 1977, pp. 149-329; Cordes; Bull *Chem. Rev.* **1974**, *74*, 581-603; Cordes *Prog. Phys. Org. Chem.* **1967**, *4*, 1-44; Salomaa, Ref. 446, pp. 184-198; Pindur; Müller; Flo; Witzel *Chem. Soc. Rev.* **1987**, *16*, 75-87 (ortho esters); Cordes, in Patai, Ref. 197, pp. 632-656 (ortho esters); DeWolfe *Carboxylic Ortho Acid Derivatives*; Academic Press: New York, 1970, pp. 134-146 (ortho esters); Rekasheva *Russ. Chem. Rev.* **1968**, *37*, 1009-1022 (enol ethers).

[458]Jaques; Leisten *J. Chem. Soc.* **1964**, 2683. See also Olah; O'Brien *J. Am. Chem. Soc.* **1967**, *89*, 1725.

[459]For a review of the reactions of ortho esters, see Pavlova; Davidovich; Rogozhin *Russ. Chem. Rev.* **1986**, *55*, 1026-1041.

[460]For a review of the mechanisms of hydrolysis of acetals and thioacetals, see Satchell; Satchell *Chem. Soc. Rev.* **1990**, *19*, 55-81.

[461]Kreevoy; Taft *J. Am. Chem. Soc.* **1955**, *77*, 3146, 5590.

mechanism are:[462] (1) The reaction proceeds with *specific* H_3O^+ catalysis (see p. 259). (2) It is faster in D_2O. (3) Optically active ROH are not racemized. (4) Even with *t*-butyl alcohol the R—O bond does not cleave, as shown by ^{18}O labeling.[463] (5) In the case of acetophenone ketals, the intermediate corresponding to **94** [$Ar\overset{\oplus}{C}Me(OR)_2$] could be trapped with sulfite ions (SO_3^{2-}).[464] (6) Trapping of this ion did not affect the hydrolysis rate,[464] so the rate-determining step must come earlier. (7) In the case of 1,1-dialkoxyalkanes, intermediates corresponding to **94** were isolated as stable ions in super-acid solution at $-75°C$, where their spectra could be studied.[465] (8) Hydrolysis rates greatly increase in the order $CH_2(OR')_2$ $< RCH(OR')_2 < R_2C(OR')_2 < RC(OR')_3$, as would be expected for a carbocation intermediate. Formation of **94** is usually the rate-determining step (as marked above), but there is evidence that at least in some cases this step is fast, and the rate-determining step is loss of R'OH from the protonated hemiacetal.[466] Rate-determining addition of water to **94** has also been reported.[467]

While the A1 mechanism shown above operates in most acetal hydrolyses, it has been shown that at least two other mechanisms can take place with suitable substrates.[468] In one of these mechanisms the second and third of the above steps are concerted, so that the mechanism is SN2cA (or A2). This has been shown, for example, in the hydrolysis of 1,1-diethoxyethane, by isotope effect studies:[469]

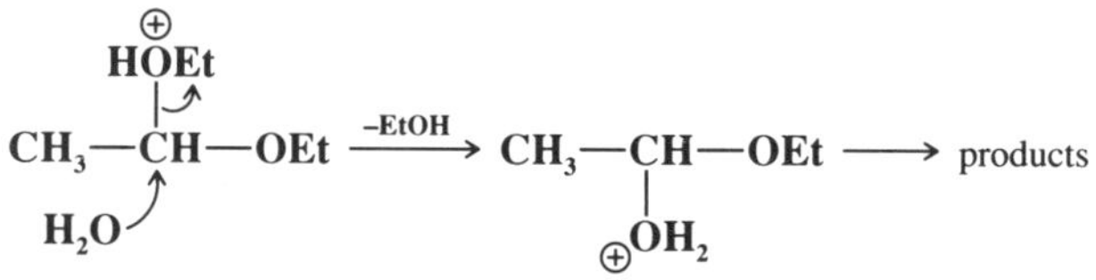

$$\overset{\overset{\oplus}{HOEt}}{\underset{\underset{H_2O}{\nearrow}}{CH_3-\underset{\nwarrow}{CH}-OEt}} \xrightarrow{-EtOH} CH_3-\underset{\underset{\oplus}{OH_2}}{CH}-OEt \longrightarrow products$$

In the second mechanism, the first and second steps are concerted. In the case of hydrolysis of 2-(*p*-nitrophenoxy)tetrahydropyran, *general* acid catalysis was shown[470] demonstrating that the substrate is protonated in the rate-determining step (p. 259). Reactions in which a substrate is protonated in the rate-determining step are called A-SE2 reactions.[471] However, if protonation of the substrate were all that happens in the slow step, then the proton in the transition state would be expected to lie closer to the weaker base (p. 259). Because the substrate is a much weaker base than water, the proton should be largely transferred. Since the Brønsted coefficient was found to be 0.5, the proton was actually transferred only

[462]For a discussion of these, and of other evidence, see Cordes *Prog. Phys. Org. Chem.*, Ref. 457.

[463]Cawley; Westheimer *Chem. Ind. (London)* **1960**, 656.

[464]Young; Jencks *J. Am. Chem. Soc.* **1977**, *99*, 8238. See also Jencks *Acc. Chem. Res.* **1980**, *13*, 161-169; McClelland; Ahmad *J. Am. Chem. Soc.* **1978**, *100*, 7027, 7031; Young; Bogseth; Rietz *J. Am. Chem. Soc.* **1980**, *102*, 6268. However, in the case of simple aliphatic acetals, **94** could not be trapped: Amyes; Jencks *J. Am. Chem. Soc.* **1988**, *110*, 3677.

[465]See White; Olah *J. Am. Chem. Soc.* **1969**, *91*, 2943; Akhmatdinov; Kantor; Imashev; Yasman; Rakhmankulov *J. Org. Chem. USSR* **1981**, *17*, 626.

[466]Jensen; Lenz *J. Am. Chem. Soc.* **1978**, *100*, 1291; Finley; Kubler; McClelland *J. Org. Chem.* **1980**, *45*, 644; Przystas; Fife *J. Am. Chem. Soc.* **1981**, *103*, 4884; Chiang; Kresge *J. Org. Chem.* **1985**, *50*, 5038; Fife; Natarajan *J. Am. Chem. Soc.* **1986**, *108*, 2425, 8050; McClelland; Sørensen *Acta Chem. Scand.* **1990**, *44*, 1082.

[467]Toullec; El-Alaoui *J. Org. Chem.* **1985**, *50*, 4928; Fife; Natarajan, Ref. 466.

[468]For a review, see Fife *Acc. Chem. Res.* **1972**, *5*, 264-272. For a discussion, see Wann; Kreevoy *J. Org. Chem.* **1981**, *46*, 419.

[469]Kresge; Weeks *J. Am. Chem. Soc.* **1984**, *106*, 7140. See also Fife *J. Am. Chem. Soc.* **1967**, *89*, 3228; Craze; Kirby; Osborne *J. Chem. Soc., Perkin Trans. 2* **1978**, 357; Amyes; Jencks *J. Am. Chem. Soc.* **1989**, *111*, 7888, 7900.

[470]Fife; Jao *J. Am. Chem. Soc.* **1968**, *90*, 4081; Fife; Brod *J. Am. Chem. Soc.* **1970**, *92*, 1681. For other examples, see Kankaanperä; Lahti *Acta Chem. Scand.* **1969**, *23*, 2465; Mori; Schaleger *J. Am. Chem. Soc.* **1972**, *94*, 5039; Capon; Nimmo *J. Chem. Soc., Perkin Trans. 2* **1975**, 1113; Eliason; Kreevoy *J. Am. Chem. Soc.* **1978**, *100*, 7037; Jensen; Herold; Lenz; Trusty; Sergi; Bell; Rogers *J. Am. Chem. Soc.* **1979**, *101*, 4672.

[471]For a review of A-SE2 reactions, see Williams; Kreevoy *Adv. Phys. Org. Chem.* **1968**, *6*, 63-101.

about halfway. This can be explained if the basicity of the substrate is increased by partial breaking of the C—O bond. The conclusion is thus drawn that steps 1 and 2 are concerted. The hydrolysis of ortho esters in most cases is also subject to general acid catalysis.[472]

The hydrolysis of acetals and ortho esters is governed by the stereoelectronic control factor previously discussed (see **A** and **B** on p. 334)[473] though the effect can generally be seen only in systems where conformational mobility is limited, especially in cyclic systems.

Particularly convenient reagents for acetals are wet silica gel[474] and Amberlyst-15 (a sulfonic acid-based polystyrene cation exchange resin).[475] Acetals and ketals can be converted to ketones under nonaqueous conditions by treatment with BF_3 etherate-I^- in $CHCl_3$ or MeCN,[476] with triphenylphosphine dibromide PPh_3Br_2,[477] with $SmCl_3$–Me_3SiCl,[478] or with Me_3SiI in CH_2Cl_2 or $CHCl_3$.[479] They can also be hydrolyzed with $LiBF_4$ in wet MeCN.[480]

Although acetals, ketals, and ortho esters are easily hydrolyzed by acids, they are extremely resistant to hydrolysis by bases. An aldehyde or ketone can therefore be protected from attack by a base by conversion to the acetal or ketal (**6-6**), and then can be cleaved with acid. Thioacetals, thioketals, *gem*-diamines, and other compounds that contain any two of the groups OR, OCOR, NR_2, NHCOR, SR, and halogen on the same carbon can also be hydrolyzed to aldehydes or ketones, in most cases, by acid treatment. Thioacetals $RCH(SR')_2$ and thioketals $R_2C(SR')_2$ are among those compounds generally resistant to acid hydrolysis. Because conversion to these compounds (**6-11**) serves as an important method for protection of aldehydes and ketones, many methods have been devised to cleave them to the parent carbonyl compounds. Among reagents[481] used for this purpose are $HgCl_2$,[482] H_2O_2–HCl,[483] *t*-BuBr–Me_2SO,[484] Me_2SO–HCl–dioxane,[485] $Cu(NO_3)_2$ on clay (clay-cop),[486] $CuSO_4$ on silica gel,[487] *m*-chloroperoxybenzoic acid and CF_3COOH in CH_2Cl_2,[488] $GaCl_3$–H_2O,[489] phenyl dichlorophosphate–DMF–NaI,[490] bis(triflouroacetoxy)iodobenzene $(CF_3CO_2)_2IPh$,[491] diphosphorus tetraiodide P_2I_4 in Ac_2O,[492] and benzeneseleninic anhydride $(PhSeO)_2O$.[493] Electrochemical methods have also been used.[494]

[472]See Bergstrom; Cashen; Chiang; Kresge *J. Org. Chem.* **1979**, *44*, 1639; Ahmad; Bergstrom; Cashen; Chiang; Kresge; McClelland; Powell *J. Am. Chem. Soc.* **1979**, *101*, 2669; Chiang; Kresge; Lahti; Weeks *J. Am. Chem. Soc.* **1983**, *105*, 6852; Santry; McClelland *J. Am. Chem. Soc.* **1983**, *105*, 6138; Fife; Przystas *J. Chem. Soc., Perkin Trans. 2* **1987**, 143.

[473]See, for example, Kirby *Acc. Chem. Res.* **1984**, *17*, 305-311; Bouab; Lamaty; Moreau *Can. J. Chem.* **1985**, *63*, 816. See, however, Ratcliffe; Mootoo; Andrews; Fraser-Reid *J. Am. Chem. Soc.* **1989**, *111*, 7661.

[474]Huet; Lechevallier; Pellet; Conia *Synthesis* **1978**, 63.

[475]Coppola *Synthesis* **1984**, 1021.

[476]Mandal; Shrotri; Ghogare *Synthesis* **1986**, 221.

[477]Wagner; Heitz; Mioskowski *J. Chem. Soc., Chem. Commun.* **1989**, 1619.

[478]Ukaji; Koumoto; Fujisawa *Chem. Lett.* **1989**, 1623.

[479]Jung; Andrus; Ornstein *Tetrahedron Lett.* **1977**, 4175. See also Balme; Goré *J. Org. Chem.* **1983**, *48*, 3336.

[480]Lipshutz; Harvey *Synth. Commun.* **1982**, *12*, 267.

[481]For references to other reagents, see Gröbel; Seebach *Synthesis* **1977**, 357-402, pp. 359-367; Cussans; Ley; Barton *J. Chem. Soc., Perkin Trans. 1* **1980**, 1654.

[482]Corey; Erickson *J. Org. Chem.* **1971**, *36*, 3553. For a mechanistic study, see Satchell; Satchell *J. Chem. Soc., Perkin Trans. 2* **1987**, 513.

[483]Olah; Narang; Salem *Synthesis* **1980**, 657, 659.

[484]Olah; Mehrotra; Narang *Synthesis* **1982**, 151.

[485]Prato; Quintily; Scorrano; Sturaro *Synthesis* **1982**, 679.

[486]Laszlo; Cornélis *Aldrichimica Acta* **1988**, *21*, 97-103, p. 101.

[487]Caballero; Gros *J. Chem. Res. (S)* **1989**, 320.

[488]Cossy *Synthesis* **1987**, 1113.

[489]Saigo; Hashimoto; Kihara; Umehara; Hasegawa *Chem. Lett.* **1990**, 831.

[490]Liu; Wiszniewski *Tetrahedron Lett.* **1988**, *29*, 5471.

[491]Stork; Zhao *Tetrahedron Lett.* **1989**, *30*, 287.

[492]Shigemasa; Ogawa; Sashiwa; Saimoto *Tetrahedron Lett.* **1989**, *30*, 1277.

[493]Cussans; Ley; Barton, Ref. 481.

[494]See Platen; Steckhan *Chem. Ber.* **1984**, *117*, 1679; Schulz-von Itter; Steckhan *Tetrahedron* **1987**, *43*, 2475.

Enol ethers are readily hydrolyzed by acids; the rate-determining step is protonation of the substrate. However, protonation does not take place at the oxygen but at the β carbon,[495] because that gives rise to the stable carbocation **95**.[496] After that the mechanism is similar to the A1 mechanism given above for the hydrolysis of acetals.

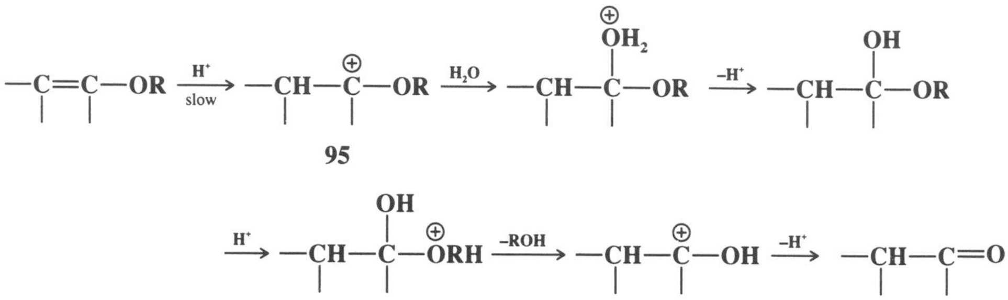

$$
-\overset{|}{\underset{|}{C}}=\overset{|}{\underset{|}{C}}-OR \xrightarrow[\text{slow}]{H^+} -\overset{|}{\underset{|}{C}}H-\overset{\oplus}{\underset{|}{C}}-OR \xrightarrow{H_2O} -\overset{|}{\underset{|}{C}}H-\overset{\overset{\oplus}{OH_2}}{\underset{|}{C}}-OR \xrightarrow{-H^+} -\overset{|}{\underset{|}{C}}H-\overset{OH}{\underset{|}{C}}-OR
$$

95

$$
\xrightarrow{H^+} -\overset{|}{\underset{|}{C}}H-\overset{\overset{OH}{|}}{\underset{|}{\overset{\oplus}{C}}}-ORH \xrightarrow{-ROH} -\overset{|}{\underset{|}{C}}H-\overset{\oplus}{\underset{|}{C}}-OH \xrightarrow{-H^+} -\overset{|}{\underset{|}{C}}H-\overset{|}{C}=O
$$

Among the facts supporting this mechanism (which is an A-SE2 mechanism because the substrate is protonated in the rate-determining step) are: (1) ^{18}O labeling shows that in $ROCH=CH_2$ it is the vinyl-oxygen bond and not the RO bond that cleaves;[497] (2) the reaction is subject to general acid catalysis;[498] (3) there is a solvent isotope effect when D_2O is used.[498] Enamines are also hydrolyzed by acids (see **6-2**); the mechanism is similar. Ketene dithioacetals $R_2C=C(SR')_2$ also hydrolyze by a similar mechanism, except that the initial protonation step is partially reversible.[499] Furans represent a special case of enol ethers that are cleaved by acid to give 1,4 diones. Thus

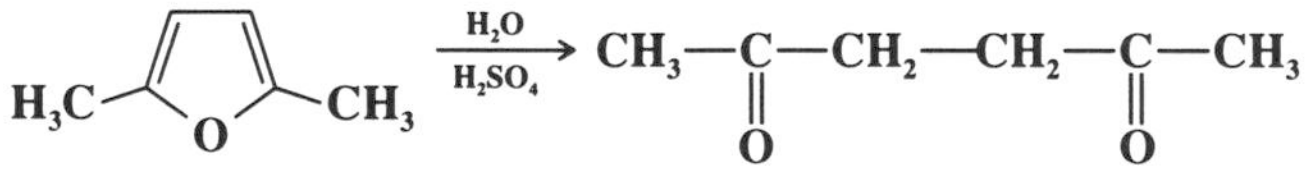

$$
\underset{H_3C}{\overbrace{}}\underset{O}{}\underset{CH_3}{} \xrightarrow[H_2SO_4]{H_2O} CH_3-\overset{O}{\underset{\|}{C}}-CH_2-CH_2-\overset{O}{\underset{\|}{C}}-CH_3
$$

Oxonium ions are cleaved by water to give an alcohol and an ether:

$$
R_3O^+ BF_4^- + H_2O \longrightarrow R_2O + ROH + HBF_4
$$

OS **I**, 67, 205; **II**, 302, 305, 323; **III**, 37, 127, 465, 470, 536, 541, 641, 701, 731, 800; **IV**, 302, 499, 660, 816, 903; **V**, 91, 292, 294, 703, 716, 937, 967, 1088; **VI**, 64, 109, 312, 316, 361, 448, 496, 683, 869, 893, 905, 996; **VII**, 12, 162, 241, 249, 251, 263, 271, 287, 381, 495; **68**, 25, 92; **69**, 31, 55, 148.

0-7 Hydrolysis of Epoxides
(3)OC-seco-hydroxy-de-alkoxy-substitution

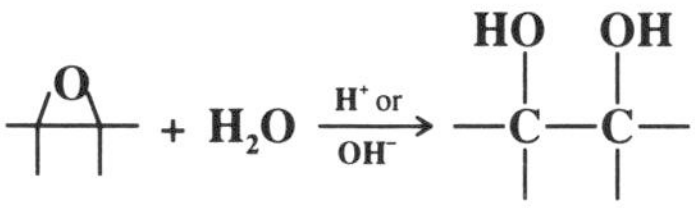

$$
\overset{O}{\overbrace{\triangle}} + H_2O \xrightarrow[OH^-]{H^+ \text{ or}} -\overset{HO}{\underset{|}{C}}-\overset{OH}{\underset{|}{C}}-
$$

[495]Jones; Wood *J. Chem. Soc.* **1964**, 5400; Okuyama; Fueno; Furukawa *Bull. Chem. Soc. Jpn.* **1970**, *43*, 3256; Kreevoy; Eliason *J. Phys. Chem.* **1969**, *72*, 1313; Lienhard; Wang *J. Am. Chem. Soc.* **1969**, *91*, 1146; Kresge; Chen *J. Am. Chem. Soc.* **1972**, *94*, 2818; Burt; Chiang; Kresge; Szilagyi *Can. J. Chem.* **1984**, *62*, 74.
[496]See Chwang; Kresge; Wiseman *J. Am. Chem. Soc.* **1979**, *101*, 6972.
[497]Kiprianova; Rekasheva *Dokl. Akad. Nauk SSSR* **1962**, *142*, 589.
[498]Fife *J. Am. Chem. Soc.* **1965**, *87*, 1084; Salomaa; Kankaanperä; Lajunen *Acta Chem. Scand.* **1966**, *20*, 1790; Kresge; Chiang *J. Chem. Soc. B* **1967**, *53*, 58; Kresge; Yin *Can. J. Chem.* **1987**, *65*, 1753.
[499]For a review, see Okuyama *Acc. Chem. Res.* **1986**, *19*, 370-376.

The hydrolysis of epoxides is a convenient method for the preparation of *vic*-diols. The reaction is catalyzed by acids or bases (see discussion of the mechanism on p. 369). Among acid catalysts the reagent of choice is perchloric acid, since side reactions are minimized with this reagent.[500] Dimethyl sulfoxide is a superior solvent for the alkaline hydrolysis of epoxides.[501]

OS **V**, 414.

B.   Attack by OH at an Acyl Carbon

0-8   Hydrolysis of Acyl Halides
Hydroxy-de-halogenation

$$\text{RCOCl} + \text{H}_2\text{O} \longrightarrow \text{RCOOH}$$

Acyl halides are so reactive that hydrolysis is easily carried out. In fact, most simple acyl halides must be stored under anhydrous conditions lest they react with water in the air. Consequently, water is usually a strong enough nucleophile for the reaction, though in difficult cases hydroxide ion may be required. The reaction is seldom synthetically useful, because acyl halides are normally prepared from acids. The reactivity order is F < Cl < Br < I.[502] If a carboxylic acid is used as the nucleophile, an exchange may take place (see **0-74**). The mechanism[502] of hydrolysis can be either SN1 or tetrahedral, the former occurring in highly polar solvents and in the absence of strong nucleophiles.[503] There is also evidence for the SN2 mechanism in some cases.[504]

Hydrolysis of acyl halides is not usually catalyzed by acids, except for acyl fluorides, where hydrogen bonding can assist in the removal of F.[505]

OS **II**, 74.

0-9   Hydrolysis of Anhydrides
Hydroxy-de-acyloxy-substitution

$$\underset{\overset{\|}{\text{O}}}{\text{R}-\text{C}}-\text{O}-\underset{\overset{\|}{\text{O}}}{\text{C}-\text{R}'} + \text{H}_2\text{O} \longrightarrow \underset{\overset{\|}{\text{O}}}{\text{R}-\text{C}}-\text{OH} + \underset{\overset{\|}{\text{O}}}{\text{R}'-\text{C}}-\text{OH}$$

Anhydrides are somewhat more difficult to hydrolyze than acyl halides, but here too water is usually a strong enough nucleophile. The mechanism is usually tetrahedral. Only under acid catalysis does the SN1 mechanism occur and seldom even then.[506] Anhydride hydrolysis can also be catalyzed by bases. Of course, OH⁻ attacks more readily than water, but other bases can also catalyze the reaction. This phenomenon, called *nucleophilic catalysis* (p. 334), is actually the result of two successive tetrahedral mechanisms. For example, pyridine catalyzes the hydrolysis of acetic anhydride in this manner.[507]

[500]Fieser; Fieser *Reagents for Organic Synthesis*, vol. 1; Wiley: New York, 1967, p. 796.

[501]Berti; Macchia; Macchia *Tetrahedron Lett.* **1965**, 3421.

[502]For a review, see Talbot, Ref. 197, pp. 226-257. For a review of the mechanisms of reactions of acyl halides with water, alcohols, and amines, see Kivinen, in Patai *The Chemistry of Acyl Halides*; Wiley: New York, 1972, pp. 177-230.

[503]Bender; Chen *J. Am. Chem. Soc.* **1963**, *85*, 30. See also Song; Jencks *J. Am. Chem. Soc.* **1989**, *111*, 8470; Bentley; Koo; Norman *J. Org. Chem.* **1991**, *56*, 1604.

[504]Bentley; Carter; Harris, Ref. 198; Guthrie; Pike, Ref. 198. See also Lee; Sung; Uhm; Ryu *J. Chem. Soc., Perkin Trans. 2* **1989**, 1697.

[505]Bevan; Hudson *J. Chem. Soc.* **1953**, 2187; Satchell *J. Chem. Soc.* **1963**, 555.

[506]Satchell *Q. Rev., Chem. Soc.* **1963**, *17*, 160-203, pp. 172-173. For a review of the mechanism, see Talbot, Ref. 197, pp. 280-287.

[507]Butler; Gold *J. Chem. Soc.* **1961**, 4362; Fersht; Jencks *J. Am. Chem. Soc.* **1970**, *92*, 5432, 5442; Deady; Finlayson *Aust. J. Chem.* **1983**, *36*, 1951.

$$CH_3-\overset{\underset{\displaystyle O}{\|}}{C}-O-\overset{\underset{\displaystyle O}{\|}}{C}-CH_3 + \underset{N}{\bigcirc} \longrightarrow CH_3-\overset{\underset{\displaystyle O}{\|}}{C}-\overset{\oplus}{N}\bigcirc + \overset{\ominus}{O}-\overset{\underset{\displaystyle O}{\|}}{C}-CH_3$$

$$CH_3-\overset{\underset{\displaystyle O}{\|}}{C}-\overset{\oplus}{N}\bigcirc + H_2O \longrightarrow CH_3-\overset{\underset{\displaystyle O}{\|}}{C}-\overset{\oplus}{O}H_2 + \underset{N}{\bigcirc}$$

Many other nucleophiles similarly catalyze the reaction.
OS **I**, 408; **II**, 140, 368, 382; **IV**, 766; **V**, 8, 813.

0-10 Hydrolysis of Carboxylic Esters
Hydroxy-de-alkoxylation

$$R-\overset{\underset{\displaystyle O}{\|}}{C}-OR' \quad \overset{\underset{H^+}{H_2O}}{\rightleftarrows} \quad RCOOH + R'OH$$
$$\overset{\underset{H_2O}{OH^-}}{\longrightarrow} RCOO^- + R'OH$$

Ester hydrolysis is usually catalyzed by acids or bases. Since OR is a much poorer leaving group than halide or OCOR, water alone does not hydrolyze most esters. When bases catalyze the reaction, the attacking species is the more powerful nucleophile OH$^-$. This reaction is called *saponification* and gives the salt of the acid. Acids catalyze the reaction by making the carbonyl carbon more positive and therefore more susceptible to attack by the nucleophile. Both reactions are equilibrium reactions, so they are practicable only when there is a way of shifting the equilibrium to the right. Since formation of the salt does just this, ester hydrolysis is almost always done for preparative purposes in basic solution, unless the compound is base-sensitive. Ester hydrolysis can also be catalyzed[508] by metal ions, by cyclodextrins,[509] by enzymes,[510] and by nucleophiles (see **0-9**).[197] Among other compounds used to cleave carboxylic esters have been methanesulfonic acid,[511] guanidine,[512] Dowex-50,[513] Me$_3$SiI,[514] MeSiCl$_3$–NaI,[515] and KOSiMe$_3$.[516] Phenolic esters can be similarly cleaved; in fact the reaction is usually faster for these compounds.[517] Lactones also undergo the reaction[518] (though if the lactone is five- or six-membered, the hydroxy acid often spontaneously relactonizes) and thiol esters (RCOSR') give thiols R'SH. Sterically hindered esters are hydrolyzed with difficulty (p. 340), though this can be accomplished at room temperature with "anhydrous hydroxide," generated via the reaction of 2 moles of *t*-BuOK with 1 mole

[508]For a list of catalysts and reagents that have been used to convert carboxylic esters to acids, with references, see Larock *Comprehensive Organic Transformations*; VCH: New York, 1989, pp. 981-985.
[509]See Bender; Komiyama *Cyclodextrin Chemistry*; Springer: New York, 1978, pp. 34-41. The mechanism is shown in Saenger *Angew. Chem. Int. Ed. Engl.* **1980**, *19*, 344-362 [*Angew. Chem. 92*, 343-361].
[510]For reviews of ester hydrolysis catalyzed by pig liver esterase, see Zhu; Tedford *Tetrahedron* **1990**, *46*, 6587-6611; Ohno; Otsuka *Org. React.* **1989**, *37*, 1-55. For reviews of enzymes as catalysts in synthetic organic chemistry, see Wong *Chemtracts: Org. Chem.* **1990**, *3*, 91-111, *Science* **1989**, *244*, 1145-1152; Whitesides; Wong *Angew. Chem. Int. Ed. Engl.* **1985**, *24*, 617-638 [*Angew. Chem. 97*, 617-638].
[511]Loev *Chem. Ind. (London)* **1964**, 193.
[512]Kunesch; Miet; Poisson *Tetrahedron Lett.* **1987**, *28*, 3569.
[513]Basu; Sarkar; Ranu *Synth. Commun.* **1989**, *19*, 627.
[514]Ho; Olah *Angew. Chem. Int. Ed. Engl.* **1976**, *15*, 774 [*Angew. Chem. 88*, 847]; Jung; Lyster *J. Am. Chem. Soc.* **1977**, *99*, 968. For a review of this reagent, see Olah; Narang *Tetrahedron* **1982**, *38*, 2225-2277.
[515]Olah; Husain; Singh; Mehrotra *J. Org. Chem.* **1983**, *48*, 3667.
[516]Laganis; Chenard *Tetrahedron Lett.* **1984**, *25*, 5831.
[517]For a method of hydrolyzing phenolic esters in the presence of other esters, see Blay; Cardona; Garcia; Pedro *Synthesis* **1989**, 438.
[518]For a review of the mechanisms of lactone hydrolysis, see Kaiser; Kézdy *Prog. Bioorg. Chem.* **1976**, *4*, 239-267, pp. 254-265.

of water.[519] Hindered esters can also be cleaved with *n*-propyllithium.[520] For esters insoluble in water the rate of two-phase ester saponification can be greatly increased by the application of ultrasound.[521] Phase-transfer techniques have also been applied.[522]

Ingold[523] has classified the acid- and base-catalyzed hydrolyses of esters (and the formation of esters, since these are reversible reactions and thus have the same mechanisms) into eight possible mechanisms (Table 10.14), depending on the following criteria: (1) acid- or base-catalyzed, (2) unimolecular or bimolecular, and (3) acyl cleavage or alkyl cleavage.[525] All eight of these are SN1, SN2, or tetrahedral mechanisms. The acid-catalyzed mechanisms are shown with reversible arrows. They are not only reversible but symmetrical; that is, the mechanisms for ester formation are exactly the same as for hydrolysis, except that H replaces R. Internal proton transfers, such as shown for **B** and **C,** may not actually be direct but may take place through the solvent. There is much physical evidence to show that esters are initially protonated on the carbonyl and not on the alkyl oxygen (Chapter 8, Ref. 17). We have nevertheless shown the $A_{AC}1$ mechanism as proceeding through the ether-protonated intermediate **A,** since it is difficult to envision OR' as a leaving group here. It is of course possible for a reaction to proceed through an intermediate even if only a tiny concentration is present. The designations $A_{AC}1$, etc., are those of Ingold. The $A_{AC}2$ and $A_{AC}1$ mechanisms are also called A2 and A1, respectively. It may be noted that the $A_{AC}1$ mechanism is actually the same as the SN1cA mechanism for this type of substrate and that $A_{AL}2$ is analogous to SN2cA. Some authors use A1 and A2 to refer to all types of nucleophilic substitution in which the leaving group first acquires a proton. The base-catalyzed reactions are not shown with reversible arrows, since they are reversible only in theory and not in practice. Hydrolyses taking place under neutral conditions are classified as B mechanisms.

Of the eight mechanisms, seven have actually been observed in hydrolysis of carboxylic esters. The one that has not been observed is the $B_{AC}1$ mechanism.[526] The most common mechanisms are the $B_{AC}2$ for basic catalysis and the $A_{AC}2$[527] for acid catalysis, that is, the two tetrahedral mechanisms. Both involve acyl-oxygen cleavage. The evidence is: (1) hydrolysis with $H_2^{18}O$ results in the ^{18}O appearing in the acid and not in the alcohol;[528] (2) esters with chiral R' groups give alcohols with *retention* of configuration;[529] (3) allylic R' gives no allylic rearrangement;[530] (4) neopentyl R' gives no rearrangement;[531] all these facts indicate that the O—R' bond is not broken. It has been concluded that two molecules of water are required in the $A_{AC}2$ mechanism.

$$R \overset{\oplus}{\underset{\underset{OH}{|}}{C}} OR' + I\overset{\frown}{O} - H \overset{\frown}{+} I\overset{..}{O}H_2 \longrightarrow R \underset{\underset{OH}{|}}{\overset{\overset{OH}{|}}{C}} OR' + H_3O^+$$

[519]Gassman; Schenk *J. Org. Chem.* **1977,** *42,* 918.
[520]Lion; Dubois; MacPhee; Bonzougou *Tetrahedron* **1979,** *35,* 2077.
[521]Moon; Duchin; Cooney *Tetrahedron Lett.* **1979,** 3917.
[522]Dehmlow; Naranjo *J. Chem. Res., (S)* **1979,** 238; Loupy; Pedoussaut; Sansoulet *J. Org. Chem.* **1986,** *51,* 740.
[523]Ingold, Ref. 366, pp. 1129-1131.
[524]As given here, the IUPAC designations for $B_{AC}1$ and $B_{AL}1$ are the same, but Rule A.2 adds further symbols so that they can be distinguished: Su-AL for $B_{AL}1$ and Su-AC for $B_{AC}1$. See the IUPAC rules: Guthrie *Pure Appl. Chem.* **1989,** *61,* 23-56, p. 49.
[525]For reviews of the mechanisms of ester hydrolysis and formation, see Kirby, in Bamford; Tipper, Ref. 178, vol. 10, 1972, pp. 57-207; Euranto, in Patai, Ref. 197, pp. 505-588.
[526]This is an SN1 mechanism with OR' as leaving group, which does not happen.
[527]For a discussion of this mechanism with specific attention to the proton transfers involved, see Zimmermann; Rudolph *Angew. Chem. Int. Ed. Engl.* **1965,** *4,* 40-49 [*Angew. Chem. 77,* 65-74].
[528]For one of several examples, see Polanyi; Szabo *Trans. Faraday Soc.* **1934,** *30,* 508.
[529]Holmberg *Ber.* **1912,** *45,* 2997.
[530]Ingold; Ingold *J. Chem. Soc.* **1932,** 758.
[531]Norton; Quayle *J. Am. Chem. Soc.* **1940,** *62,* 1170.

TABLE 10.14 Classification of the eight mechanisms for ester hydrolysis and formation[523]

	Name		Type	Mechanism
	Ingold	**IUPAC[524]**		
Acid catalysis — Acyl cleavage	$A_{AC}1$	$A_h + D_N + A_N + D_h$	S_N1	A
Acid catalysis — Acyl cleavage	$A_{AC}2$	$A_h + A_N + A_{xh}D_h + D_N + D_h$	Tetra-hedral	B
Acid catalysis — Alkyl cleavage	$A_{AL}1$	$A_h + D_N + A_N + D_h$	S_N1	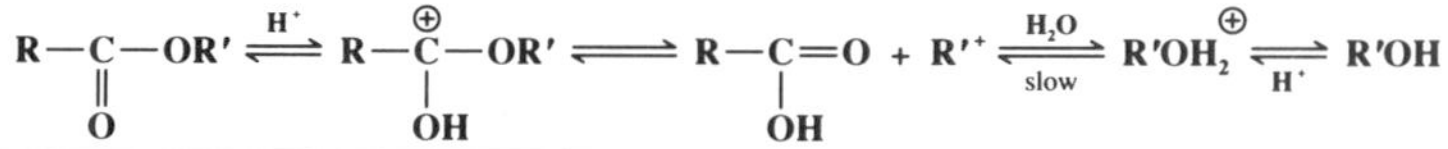 C
Acid catalysis — Alkyl cleavage	$A_{AL}2$	$A_h + A_N D_N + D_h$	S_N2	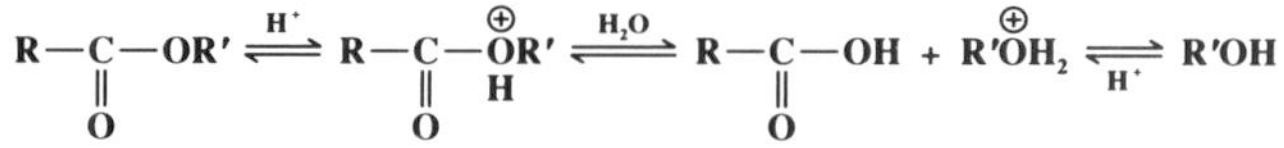
Basic catalysis — Acyl cleavage	$B_{AC}1$	$D_N + A_N + A_{xh}D_h$	S_N1	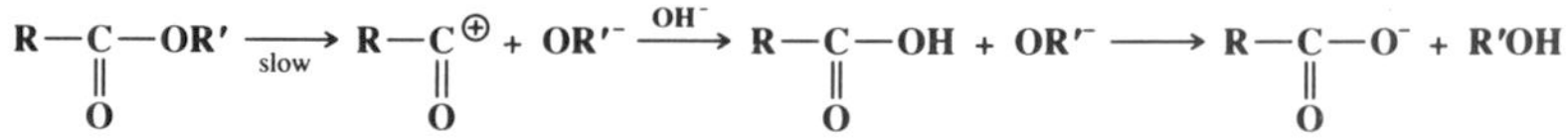
Basic catalysis — Acyl cleavage	$B_{AC}2$	$A_N + D_N + A_{xh}D_h$	Tetra-hedral	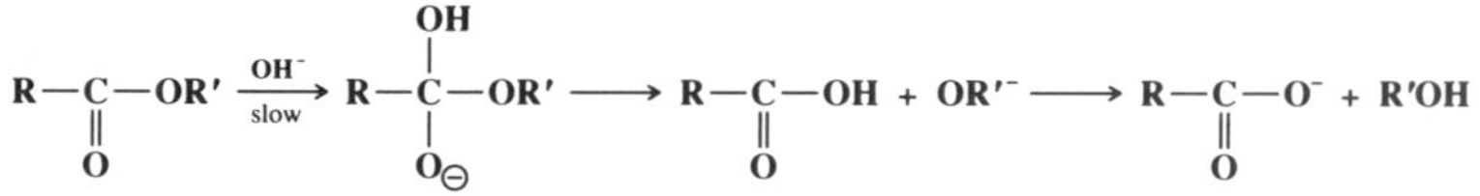
Basic catalysis — Alkyl cleavage	$B_{AL}1$	$D_N + A_N + A_{xh}D_h$	S_N1	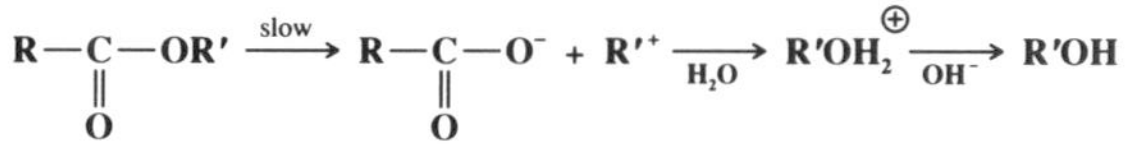
Basic catalysis — Alkyl cleavage	$B_{AL}2$	$A_N D_N$	S_N2	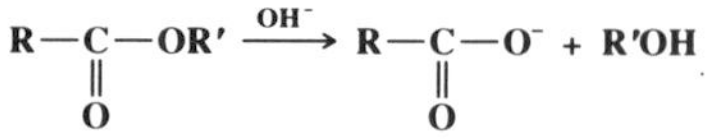

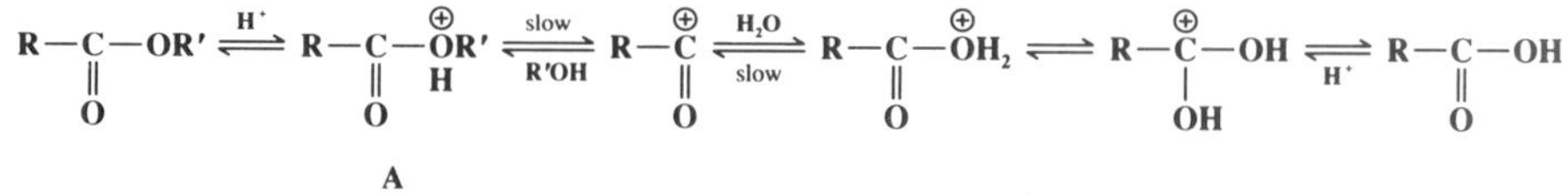

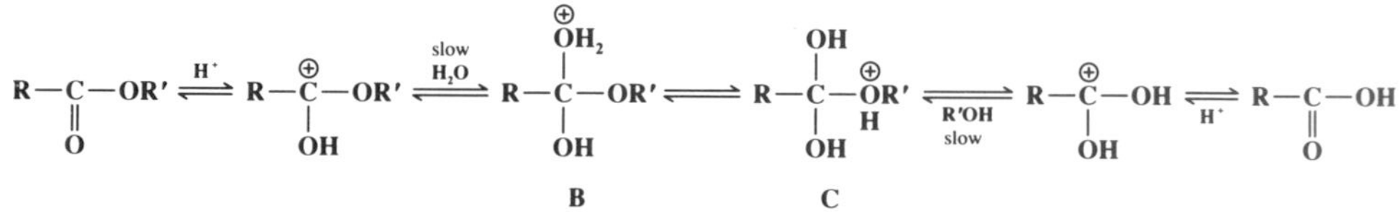

A:
$$R-\underset{O}{\overset{\|}{C}}-OR' \underset{H^+}{\rightleftharpoons} R-\underset{O}{\overset{\|}{C}}-\overset{\oplus}{OR'H} \xrightarrow[R'OH]{slow} R-\overset{\oplus}{\underset{O}{\overset{\|}{C}}} \underset{slow}{\overset{H_2O}{\rightleftharpoons}} R-\overset{\oplus}{C}(OH)-OH_2 \rightleftharpoons R-\underset{OH}{\overset{\oplus}{C}}-OH \underset{H^+}{\rightleftharpoons} R-\underset{O}{\overset{\|}{C}}-OH$$

B:
$$R-\underset{O}{\overset{\|}{C}}-OR' \underset{H^+}{\rightleftharpoons} R-\overset{\oplus}{\underset{OH}{C}}-OR' \underset{slow}{\overset{H_2O}{\rightleftharpoons}} R-\underset{OH}{\overset{\overset{\oplus}{OH_2}}{\underset{|}{C}}}-OR' \rightleftharpoons R-\underset{OH}{\overset{OH}{\underset{|}{C}}}-\overset{\oplus}{OR'H} \xrightarrow[slow]{R'OH} R-\overset{\oplus}{\underset{OH}{C}}-OH \rightleftharpoons R-\underset{O}{\overset{\|}{C}}-OH$$

C:
$$R-\underset{O}{\overset{\|}{C}}-OR' \underset{H^+}{\rightleftharpoons} R-\overset{\oplus}{\underset{OH}{C}}-OR' \rightleftharpoons R-\underset{O}{\overset{\|}{C}}=O + \overset{\oplus}{R'} \overset{H_2O}{\underset{slow}{\rightleftharpoons}} R-\underset{O}{\overset{\|}{C}}=O + R'\overset{\oplus}{OH_2} \underset{H^+}{\rightleftharpoons} R'OH$$

$$R-\underset{O}{\overset{\|}{C}}-OR' \underset{H^+}{\rightleftharpoons} R-\overset{\oplus}{\underset{OH}{C}}-OR' \overset{H_2O}{\longrightarrow} R-\underset{O}{\overset{\|}{C}}-OH + R'\overset{\oplus}{OH_2} \underset{H^+}{\rightleftharpoons} R'OH$$

$$R-\underset{O}{\overset{\|}{C}}-OR' \xrightarrow{slow} R-\overset{\oplus}{\underset{O}{\overset{\|}{C}}} + OR'^- \xrightarrow{OH^-} R-\underset{O}{\overset{\|}{C}}-OH + OR'^- \longrightarrow R-\underset{O}{\overset{\|}{C}}-O^- + R'OH$$

$$R-\underset{O}{\overset{\|}{C}}-OR' \xrightarrow[slow]{OH^-} R-\underset{O\ominus}{\overset{OH}{\underset{|}{C}}}-OR' \longrightarrow R-\underset{O}{\overset{\|}{C}}-OH + OR'^- \longrightarrow R-\underset{O}{\overset{\|}{C}}-O^- + R'OH$$

$$R-\underset{O}{\overset{\|}{C}}-OR' \xrightarrow{slow} R-\underset{O}{\overset{\|}{C}}-O^- + \overset{\oplus}{R'} \xrightarrow{H_2O} R'\overset{\oplus}{OH_2} \xrightarrow{OH^-} R'OH$$

$$R-\underset{O}{\overset{\|}{C}}-OR' \xrightarrow{OH^-} R-\underset{O}{\overset{\|}{C}}-O^- + R'OH$$

If this is so, the protonated derivatives **B** and **C** would not appear at all. This conclusion stems from a value of *w* (see p. 257) of about 5, indicating that water acts as a proton donor here as well as a nucleophile.[532] Termolecular processes are rare, but in this case the two water molecules are already connected by a hydrogen bond. (A similar mechanism, called B$_{AC}$3, also involving two molecules of water, has been found for esters that hydrolyze without a catalyst.[533] Such esters are mostly those containing halogen atoms in the R group.)

The other mechanism involving acyl cleavage is the A$_{AC}$1 mechanism. This is rare, being found only where R is very bulky, so that bimolecular attack is sterically hindered, and only in ionizing solvents. The mechanism has been demonstrated for esters of 2,4,6-trimethylbenzoic acid (mesitoic acid). This acid depresses the freezing point of sulfuric acid four times as much as would be predicted from its molecular weight, which is evidence for the equilibrium

$$\text{ArCOOH} + 2\text{H}_2\text{SO}_4 \rightleftharpoons \text{ArCO}^\oplus + \text{H}_3\text{O}^+ + 2\text{HSO}_4^-$$

In a comparable solution of benzoic acid the freezing point is depressed only twice the predicted amount, indicating only a normal acid–base reaction. Further, a sulfuric acid solution of methyl mesitoate when poured into water gave mesitoic acid, while a similar solution of methyl benzoate similarly treated did not.[534] The A$_{AC}$1 mechanism is also found when acetates of phenols or of primary alcohols are hydrolyzed in concentrated (more than 90%) H$_2$SO$_4$ (the mechanism under the more usual dilute acid conditions is the normal A$_{AC}$2).[535]

The mechanisms involving alkyl-oxygen cleavage are ordinary S$_N$1 and S$_N$2 mechanisms in which OCOR (an acyloxy group) or its conjugate acid is the leaving group. Two of the four mechanisms, the B$_{AL}$1 and A$_{AL}$1 mechanisms, occur most readily when R′ comes off as a stable carbocation, that is, when R′ is tertiary alkyl, allylic, benzylic, etc. For acid catalysis, most esters with this type of alkyl group (especially tertiary alkyl) cleave by this mechanism, but even for these substrates, the B$_{AL}$1 mechanism occurs only in neutral or weakly basic solution, where the rate of attack by OH$^-$ is so slowed that the normally slow (by comparison) unimolecular cleavage takes over. These two mechanisms have been established by kinetic studies, ^{18}O labeling, and isomerization of R′.[536] Secondary and benzylic acetates hydrolyze by the A$_{AC}$2 mechanism in dilute H$_2$SO$_4$, but in concentrated acid the mechanism changes to A$_{AL}$1.[535] Despite its designation, the B$_{AL}$1 mechanism is actually uncatalyzed (as is the unknown B$_{AC}$1 mechanism).

The two remaining mechanisms, B$_{AL}$2 and A$_{AL}$2, are very rare, the B$_{AL}$2 because it requires OH$^-$ to attack an alkyl carbon when an acyl carbon is also available, and the A$_{AL}$2 because it requires water to be a nucleophile in an S$_N$2 process. Both have been observed, however. The B$_{AL}$2 has been seen in the hydrolysis of β-lactones under neutral conditions[537] (because cleavage of the C—O bond in the transition state opens the four-membered ring and relieves strain), the alkaline hydrolysis of methyl 2,4,6-tri-*t*-butyl benzoate,[538] and in the unusual reaction[539]

$$\text{ArCOOMe} + \text{RO}^- \longrightarrow \text{ArCOO}^- + \text{ROMe}$$

[532]Martin *J. Am. Chem. Soc.* **1962,** *84,* 4130. See also Lane; Cheung; Dorsey *J. Am. Chem. Soc.* **1968,** *90,* 6492; Yates; McClelland *J. Am. Chem. Soc.* **1967,** *89,* 2686; Yates *Acc. Chem. Res.* **1971,** *6,* 136-144; Huskey; Warren; Hogg *J. Org. Chem.* **1981,** *46,* 59.

[533]Euranto; Kanerva; Cleve *J. Chem. Soc., Perkin Trans. 2* **1984,** 2085; Neuvonen *J. Chem. Soc., Perkin Trans. 2* **1986,** 1141; Euranto; Kanerva *Acta Chem. Scand., Ser. B* **1988,** *42* 717.

[534]Treffers; Hammett *J. Am. Chem. Soc.* **1937,** *59,* 1708. For other evidence for this mechanism, see Bender; Chen *J. Am. Chem. Soc.* **1963,** *85,* 37.

[535]Yates, Ref. 532; Al-Shalchi; Selwood; Tillett *J. Chem. Res. (S)* **1985,** 10.

[536]For discussions, see Kirby, Ref. 525, pp. 86-101; Ingold, Ref. 366, pp. 1137-1142, 1157-1163.

[537]Cowdrey; Hughes; Ingold; Masterman; Scott *J. Chem. Soc.* **1937,** 1264; Long; Purchase *J. Am. Chem. Soc.* **1950,** *73,* 3267.

[538]Barclay; Hall; Cooke *Can. J. Chem.* **1962,** *40,* 1981.

[539]Sneen; Rosenberg *J. Org. Chem.* **1961,** *26,* 2099. See also Müller; Siegfried *Helv. Chim. Acta* **1974,** *57,* 987.

When it does occur, the BAL2 mechanism is easy to detect, since it is the only one of the base-catalyzed mechanisms that requires inversion at R'. However, in the last example given, the mechanism is evident from the nature of the product, since the ether could have been formed in no other way. The AAL2 mechanism has been reported in the acid cleavage of γ-lactones.[539a]

To sum up the acid-catalysis mechanisms, AAC2 and AAL1 are the common mechanisms, the latter for R' that give stable carbocations, the former for practically all the rest. AAC1 is rare, being found mostly with strong acids and sterically hindered R. AAL2 is even rarer. For basic catalysis, BAC2 is almost universal; BAL1 occurs only with R' that give stable carbocations and then only in weakly basic or neutral solutions; BAL2 is very rare; and BAC1 has never been observed.

The above results pertain to reactions in solution. In the gas phase[540] reactions can take a different course, as illustrated by the reaction of carboxylic esters with MeO⁻, which in the gas phase was shown to take place only by the BAL2 mechanism,[541] even with aryl esters,[542] where this means that an SN2 mechanism takes place at an aryl substrate. However, when the gas-phase reaction of aryl esters was carried out with MeO⁻ ions, each of which was solvated with a single molecule of MeOH or H_2O, the BAC2 mechanism was observed.[542]

In the special case of alkaline hydrolysis of N-substituted aryl carbamates, there is another mechanism[543] involving elimination–addition:[544]

$$R-NH-\underset{\underset{O}{\|}}{C}-OAr \overset{OH^-}{\rightleftharpoons} R-\overset{\ominus}{N}-\underset{\underset{O}{\|}}{C}-OAr \longrightarrow$$

$$R-N=C=O \overset{H_2O}{\longrightarrow} R-NH-\underset{\underset{O}{\|}}{C}-OH \longrightarrow CO_2 + RNH_2$$
$$+$$
$$OAr^-$$

This mechanism does not apply to unsubstituted or N,N-disubstituted aryl carbamates, which hydrolyze by the normal mechanisms. Carboxylic esters substituted in the α position by an electron-withdrawing group (e.g., CN or COOEt) can also hydrolyze by a similar mechanism involving a ketene intermediate.[545] These elimination–addition mechanisms usually are referred to as E1cB mechanisms, because that is the name given to the elimination portion of the mechanism (p. 991).

The acid-catalyzed hydrolysis of enol esters RCOOCR'=CR''₂ can take place either by the normal AAC2 mechanism or by a mechanism involving initial protonation on the double-bond carbon, similar to the mechanism for the hydrolysis of enol ethers given in **0-6**,[546]

[539a]Moore; Schwab *Tetrahedron Lett.* **1991**, *32*, 2331.
[540]Takashima; José; do Amaral; Riveros *J. Chem. Soc. Chem. Commun.* **1983**, 1255.
[541]Comisarow *Can. J. Chem.* **1977**, *55*, 171.
[542]Fukuda; McIver *J. Am. Chem. Soc.* **1979**, *101*, 2498.
[543]For a review of elimination–addition mechanisms at a carbonyl carbon, see Williams; Douglas *Chem. Rev.* **1975**, *75*, 627-649.
[544]Bender; Homer *J. Org. Chem.* **1965**, *30*, 3975; Williams *J. Chem. Soc., Perkin Trans. 2* **1972**, 808, **1973**, 1244; Hegarty; Frost *J. Chem. Soc., Perkin Trans. 2* **1973**, 1719; Menger; Glass *J. Org. Chem.* **1974**, *39*, 2469; Sartoré; Bergon; Calmon *J. Chem. Soc., Perkin Trans. 2* **1977**, 650; Moravcová; Večeřa *Collect. Czech. Chem. Commun.* **1977**, *42*, 3048; Broxton; Chung *J. Org. Chem.* **1986**, *51*, 3112.
[545]Casanova; Werner; Kiefer *J. Am. Chem. Soc.* **1967**, *89*, 2411; Holmquist; Bruice *J. Am. Chem. Soc.* **1969**, *91*, 2993, 3003; Campbell; Lawrie *Chem. Commun.* **1971**, 355; Kirby; Lloyd *J. Chem. Soc., Perkin Trans. 2* **1976**, 1762; Broxton; Duddy *J. Org. Chem.* **1981**, *46*, 1186; Inoue; Bruice *J. Am. Chem. Soc.* **1982**, *104*, 1644, *J. Org. Chem.* **1983**, *48*, 3559, **1986**, *51*, 959; Alborz; Douglas *J. Chem. Soc., Perkin Trans. 2* **1982**, 331; Thea; Cevasco; Guanti; Kashefi-Naini; Williams *J. Org. Chem.* **1985**, *50*, 1867; Isaacs; Najem *Can. J. Chem.* **1986**, *64*, 1140, *J. Chem. Soc., Perkin Trans. 2* **1988**, 557.
[546]Alkynyl esters also hydrolyze by this mechanism; see Allen; Kitamura; Roberts; Stang; Tidwell *J. Am. Chem. Soc.* **1988**, *110*, 622.

depending on reaction conditions.[547] In either case, the products are the carboxylic acid RCOOH and the aldehyde or ketone $R_2''CHCOR'$.

OS **I**, 351, 360, 366, 379, 391, 418, 523; **II**, 1, 5, 53, 93, 194, 214, 258, 299, 416, 422, 474, 531, 549; **III**, 3, 33, 101, 209, 213, 234, 267, 272, 281, 300, 495, 510, 526, 531, 615, 637, 652, 705, 737, 774, 785, 809 (but see OS **V**, 1050), 833, 835; **IV**, 15, 55, 169, 317, 417, 444, 532, 549, 555, 582, 590, 608, 616, 628, 630, 633, 635, 804; **V**, 8, 445, 509, 687, 762, 887, 985, 1031; **VI**, 75, 121, 560, 690, 824, 913, 1024; **VII**, 4, 190, 210, 297, 319, 323, 356, 411; **65**, 203; **66**, 37, 87, 173; **67**, 76, 170; **68**, 175, 198; **69**, 1, 19. Ester hydrolyses with concomitant decarboxylation are listed at reaction **2-40**.

0-11 Hydrolysis of Amides
Hydroxy-de-amination

$$R-\underset{\underset{O}{\|}}{C}-NH_2 \quad \overset{\overset{OH^-}{H_2O}}{\underset{\underset{H_2O}{H^+}}{\rightleftarrows}} \quad \begin{array}{l} R-\underset{\underset{O}{\|}}{C}-O^- + NH_3 \\[2em] R-\underset{\underset{O}{\|}}{C}-OH + NH_4^+ \end{array}$$

Unsubstituted amides ($RCONH_2$) can be hydrolyzed with either acidic or basic catalysis, the products being, respectively, the free acid and the ammonium ion or the salt of the acid and ammonia. N-Substituted ($RCONHR'$) and N,N-disubstituted ($RCONR_2''$) amides can be hydrolyzed analogously, with the primary or secondary amine, respectively (or their salts), being obtained instead of ammonia. Lactams, imides, cyclic imides, hydrazides, etc., also undergo the reaction. Water alone is not sufficient to hydrolyze most amides, since NH_2 is even a poorer leaving group than OR.[548] Prolonged heating is often required, even with acidic or basic catalysts.[549] In difficult cases, nitrous acid, NOCl, N_2O_4,[550] or a similar compound can be used (unsubstituted amides only[551]).

$$R-\underset{\underset{O}{\|}}{C}-NH_2 + HONO \longrightarrow R-\underset{\underset{O}{\|}}{C}-OH + N_2$$

These reactions involve a diazonium ion (see **2-49**) and are much faster than ordinary hydrolysis; for benzamide the nitrous acid reaction took place 2.5×10^7 times faster than ordinary hydrolysis.[552] Another procedure for difficult cases involves treatment with aqueous sodium peroxide.[553] In still another method, the amide is treated with water and t-BuOK at room temperature.[554] The strong base removes the proton from **96**, thus preventing the reaction marked k_{-1}. Amide hydrolysis can also be catalyzed by nucleophiles (see p. 334).

---

[547]See, for example, Noyce; Pollack *J. Am. Chem. Soc.* **1969**, *91*, 119, 7158; Monthéard; Camps; Chatzopoulos; Benzaïd *Bull. Soc. Chim. Fr.* **1984**, II-109. For a discussion, see Euranto *Pure Appl. Chem.* **1977**, *49*, 1009-1020.

[548]The very low rate of amide hydrolysis by water alone has been measured: Kahne; Still *J. Am. Chem. Soc.* **1988**, *110*, 7529.

[549]For a list of catalysts and reagents that have been used to hydrolyze amides, with references, see Ref. 508, pp. 988-989.

[550]Kim; Kim; Park *Tetrahedron Lett.* **1990**, *31*, 3893.

[551]N-Substituted amides can be converted to N-nitrosoamides, which are more easily hydrolyzable than the original amide. For example, see Rull; Serratosa; Vilarrasa *Tetrahedron Lett.* **1977**, 4549. For another method of hydrolyzing N-substituted amides, see Flynn; Zelle; Grieco *J. Org. Chem.* **1983**, *48*, 2424.

[552]Ladenheim; Bender *J. Am. Chem. Soc.* **1960**, *82*, 1895.

[553]Vaughan; Robbins *J. Org. Chem.* **1975**, *40*, 1187.

[554]Gassman; Hodgson; Balchunis *J. Am. Chem. Soc.* **1976**, *98*, 1275.

The same framework of eight possible mechanisms that was discussed for ester hydrolysis can also be applied to amide hydrolysis.[555] Both the acid- and base-catalyzed hydrolyses are essentially irreversible, since salts are formed in both cases. For basic catalysis[556] the mechanism is BAC2.

$$R-\overset{\underset{\|}{O}}{C}-NR_2' + OH^- \underset{k_{-1}}{\overset{\underset{slow}{k_1}}{\rightleftharpoons}} R-\overset{\underset{\|}{O}}{\underset{\ominus}{\overset{\overset{OH}{|}}{C}}}-NR_2' \overset{k_2}{\longrightarrow}$$

96

$$R-\overset{\underset{\|}{O}}{C}-OH + NR_2'^- \longrightarrow R-\overset{\underset{\|}{O}}{C}-O^- + R_2'NH$$

There is much evidence for this mechanism, similar to that discussed for ester hydrolysis. In certain cases, kinetic studies have shown that the reaction is second order in OH^-, indicating that **96** can lose a proton to give **97**.[557] Depending on the nature of R', **97** can

$$R-\overset{\underset{\|}{O}}{\underset{\ominus}{\overset{\overset{OH}{|}}{C}}}-NR_2' \longrightarrow R-\overset{\underset{\|}{O}}{\underset{\ominus}{\overset{\overset{O^\ominus}{|}}{C}}}-NR_2' \quad \overset{a}{\underset{BH}{\nearrow}}_{\searrow} \quad \begin{matrix} R-\overset{\underset{\|}{O}}{C}-O^\ominus + NR_2'^\ominus \\ \\ R-\overset{\underset{\|}{O}}{C}-O^\ominus + HNR_2' \end{matrix}$$

96   **97**

cleave directly to give the two negative ions (path *a*) or become N-protonated prior to or during the act of cleavage (path *b*), in which case the products are obtained directly and a final proton transfer is not necessary.[558] Studies of the effect, on the rate of hydrolysis and on the ratio k_{-1}/k_2, of substituents on the aromatic rings in a series of amides $CH_3CONHAr$ led to the conclusion that path *a* is taken when Ar contains electron-withdrawing substituents and path *b* when electron-donating groups are present.[559] The presence of electron-withdrawing groups helps stabilize the negative charge on the nitrogen, so that $NR_2'^-$ can be a leaving group (path *a*). Otherwise, the C—N bond does not cleave until the nitrogen is protonated (either prior to or in the act of cleavage), so that the leaving group, *even in the base-catalyzed reaction*, is not $NR_2'^-$ but the conjugate NHR_2' (path *b*). Though we have shown formation of **96** as the rate-determining step in the BAC2 mechanism, this is true

[555]For reviews, see O'Connor *Q. Rev., Chem. Soc.* **1970**, *24*, 553-564; Talbot, Ref. 197, pp. 257-280; Challis; Challis, in Zabicky *The Chemistry of Amides*; Wiley: New York, 1970, pp. 731-857.

[556]For a comprehensive list of references, see DeWolfe; Newcomb *J. Org. Chem.* **1971**, *36*, 3870.

[557]Biechler; Taft *J. Am. Chem. Soc.* **1957**, *79*, 4927. For evidence that a similar intermediate can arise in base-catalyzed ester hydrolysis, see Khan; Olagbemiro *J. Org. Chem.* **1982**, *47*, 3695.

[558]Eriksson; Holst *Acta Chem. Scand.* **1966**, *20*, 1892; Eriksson *Acta Chem. Scand.* **1968**, *22*, 892, *Acta Pharm. Suec.* **1969**, *6*, 139-162.

[559]Bender; Thomas *J. Am. Chem. Soc.* **1961**, *83*, 4183; Pollack; Bender *J. Am. Chem. Soc.* **1970**, *92*, 7190; Kershner; Schowen *J. Am. Chem. Soc.* **1971**, *93*, 2014; Schowen; Hopper; Bazikian *J. Am. Chem. Soc.* **1972**, *94*, 3095. See also Ref. 556; Gani; Viout *Tetrahedron Lett.* **1972**, 5241; Menger; Donohue *J. Am. Chem. Soc.* **1973**, *95*, 432; Pollack; Dumsha *J. Am. Chem. Soc.* **1973**, *95*, 4463; Kijima; Sekiguchi *J. Chem. Soc., Perkin Trans. 2* **1987**, 1203.

only at high base concentrations. At lower concentrations of base, the cleavage of **96** or **97** becomes rate-determining.[560]

For acid catalysis, matters are less clear. The reaction is generally second order, and it is known that amides are primarily protonated on the oxygen (Chapter 8, Ref. 17). Because of these facts it has been generally agreed that most acid-catalyzed amide hydrolysis takes place by the $A_{AC}2$ mechanism.

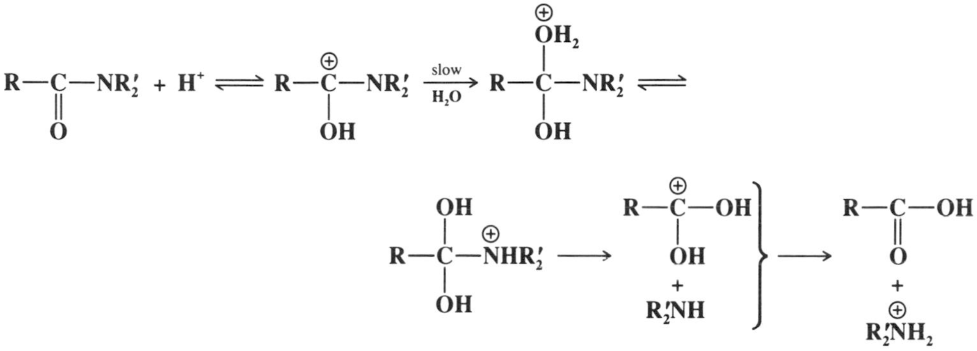

$$R-\underset{\underset{O}{\parallel}}{C}-NR'_2 + H^+ \rightleftharpoons R-\underset{\underset{OH}{|}}{\overset{\oplus}{C}}-NR'_2 \xrightarrow[\text{H}_2\text{O}]{\text{slow}} R-\underset{\underset{OH}{|}}{\overset{\overset{\oplus}{OH_2}}{C}}-NR'_2 \rightleftharpoons$$

$$R-\underset{\underset{OH}{|}}{\overset{\overset{OH}{|}}{C}}-\overset{\oplus}{NHR'_2} \longrightarrow \left. \begin{array}{c} R-\overset{\oplus}{C}-OH \\ | \\ OH \\ + \\ R'_2NH \end{array} \right\} \longrightarrow \begin{array}{c} R-C-OH \\ \parallel \\ O \\ + \\ \overset{\oplus}{R'_2NH_2} \end{array}$$

Further evidence for this mechanism is that a small but detectable amount of ^{18}O exchange (see p. 332) has been found in the acid-catalyzed hydrolysis of benzamide.[561] (^{18}O exchange has also been detected for the base-catalyzed process,[562] in accord with the $B_{AC}2$ mechanism). Kinetic data have shown that three molecules of water are involved in the rate-determining step,[563] suggesting that, as in the $A_{AC}2$ mechanism for ester hydrolysis (**0-10**), additional water molecules take part in a process such as

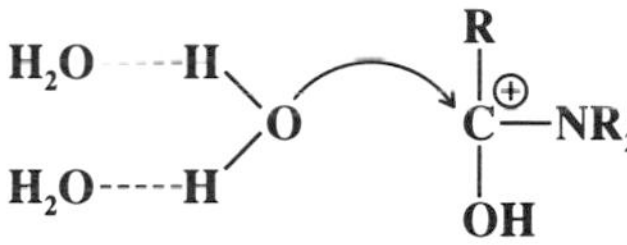

$$\begin{array}{c} H_2O \cdots H \\ \diagdown \\ \hspace{1cm} O \\ \diagup \\ H_2O \cdots H \end{array} \overset{\overset{R}{|}}{\underset{\underset{OH}{|}}{\overset{\oplus}{C}}}-NR'_2$$

The four mechanisms involving alkyl—N cleavage (the AL mechanisms) do not apply to this reaction. They are not possible for unsubstituted amides, since the only N—C bond is the acyl bond. They are possible for N-substituted and N,N-disubstituted amides, but in these cases they give entirely different products and are not amide hydrolyses at all.

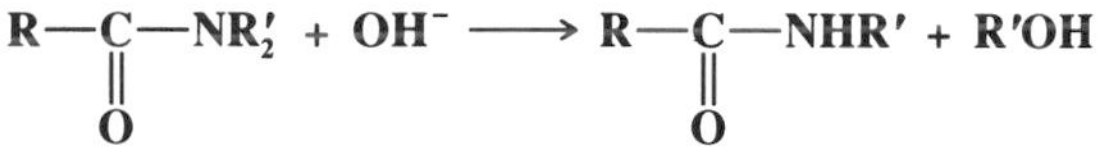

$$R-\underset{\underset{O}{\parallel}}{C}-NR'_2 + OH^- \longrightarrow R-\underset{\underset{O}{\parallel}}{C}-NHR' + R'OH$$

This reaction, while rare, has been observed for various N-t-butyl amides in 98% sulfuric acid, where the mechanism was the $A_{AL}1$ mechanism,[564] and for certain amides containing

[560]Schowen; Jayaraman; Kershner *J. Am. Chem. Soc.* **1966**, *88*, 3373. See also Gani; Viout *Tetrahedron* **1976**, *32*, 1669, 2883; Bowden; Bromley *J. Chem. Soc., Perkin Trans. 2* **1990**, 2103.

[561]McClelland *J. Am. Chem. Soc.* **1975**, *97*, 5281; Bennet; Ślebocka-Tilk; Brown; Guthrie; Jodhan *J. Am. Chem. Soc.* **1990**, *112*, 8497.

[562]Bender; Thomas, Ref. 559, Bunton; Nayak; O'Connor *J. Org. Chem.* **1968**, *33*, 572; Ślebocka-Tilk; Bennet; Hogg; Brown *J. Am. Chem. Soc.* **1991**, *113*, 1288; Ref. 561.

[563]Moodie; Wale; Whaite *J. Chem. Soc.*, **1963**, 4273; Yates; Stevens *Can. J. Chem.* **1965**, *43*, 529; Yates; Riordan *Can. J. Chem.* **1965**, *43*, 2328.

[564]Lacey *J. Chem. Soc.* **1960**, 1633; Druet; Yates *Can. J. Chem.* **1984**, *62*, 2401.

an azo group, where a BAL1 mechanism was postulated.[565] Of the two first-order acyl cleavage mechanisms, only the AAc1 has been observed, in concentrated sulfuric acid solutions.[566] Of course, the diazotization of unsubstituted amides might be expected to follow this mechanism, and there is evidence that this is true.[552]

OS **I**, 14, 111, 194, 201, 286; **II,** 19, 25, 28, 49, 76, 208, 330, 374, 384, 457, 462, 491, 503, 519, 612; **III,** 66, 88, 154, 256, 410, 456, 586, 591, 661, 735, 768, 813; **IV,** 39, 42, 55, 58, 420, 441, 496, 664; **V,** 27, 96, 341, 471, 612, 627; **VI,** 56, 252, 507, 951, 967; **VII,** 4, 287; **65,** 119, 173; **67,** 52; **68,** 83; **69,** 55.

The oxidation of aldehydes to carboxylic acids can proceed by a nucleophilic mechanism, but more often it does not. The reaction is considered in Chapter 14 (**4-6**). Basic cleavage of β-keto esters and the haloform reaction could be considered at this point, but they are also electrophilic substitutions and are treated in Chapter 12 (**2-43** and **2-44**).

C. Attack by OR at an Alkyl Carbon

0-12 Alkylation with Alkyl Halides. The Williamson Reaction
Alkoxy-de-halogenation

$$RX + OR'^- \longrightarrow ROR'$$

The *Williamson reaction*, discovered in 1850, is still the best general method for the preparation of unsymmetrical ethers or, for that matter, symmetrical ones.[567] The reaction can also be carried out with aromatic R', though C-alkylation is sometimes a side reaction (see p. 366).[568] The normal method involves treatment of the halide with alkoxide or aroxide ion prepared from an alcohol or phenol, but it is also possible to mix the halide and alcohol or phenol directly with solid KOH in Me$_2$SO[569] or with HgO and HBF$_4$ in CH$_2$Cl$_2$.[570] The reaction is not successful for tertiary R (because of elimination), and low yields are obtained with secondary R. Many other functional groups can be present in the molecule without interference. Ethers with one tertiary group *can* be prepared by treatment of an alkyl halide or sulfate ester (**0-14**) with a tertiary alkoxide R'O$^-$, which is prepared by removal of a proton from a tertiary alcohol with methylsulfinyl carbanion,[571] or with a copper(I) tertiary alkoxide.[572] Di-*t*-butyl ether was prepared in high yield by direct attack by *t*-BuOH on the *t*-butyl cation (at $-80°C$ in SO$_2$ClF).[573] Di-*t*-alkyl ethers in general have proved difficult to make, but they can be prepared in low-to-moderate yields by treatment of a tertiary halide with Ag$_2$CO$_3$ or Ag$_2$O.[574] Active halides such as Ar$_3$CX may react directly with the alcohol without the need for the more powerful nucleophile alkoxide ion.[575] Even tertiary halides have been converted to ethers in this way, with no elimination.[576] The mechanism is these cases is of course SN1. *t*-Butyl halides can be converted to aryl *t*-butyl ethers by treatment

[565]Stodola *J. Org. Chem.* **1972,** *37*, 178.
[566]Duffy; Leisten *J. Chem. Soc.* **1960,** 545, 853; Barnett; O'Connor *J. Chem. Soc., Chem. Commun.* **1972,** 525, *J. Chem. Soc., Perkin Trans. 2* **1972,** 2378.
[567]For a review, see Feuer; Hooz, in Patai, Ref. 333, pp. 446-450, 460-468.
[568]For a list of reagents used to convert alcohols and phenols to ethers, see Ref. 508, pp. 446-448.
[569]Benedict; Bianchi; Cate *Synthesis* **1979,** 428; Johnstone; Rose *Tetrahedron* **1979,** *35*, 2169. See also Loupy; Sansoulet; Vaziri-Zand *Bull. Soc. Chim. Fr.* **1987,** 1027.
[570]Barluenga; Alonso-Cires; Campos; Asensio *Synthesis* **1983,** 53.
[571]Sjöberg; Sjöberg *Acta Chem. Scand.* **1972,** *26*, 275.
[572]Whitesides; Sadowski; Lilburn *J. Am. Chem. Soc.* **1974,** *96*, 2829.
[573]Olah; Halpern; Lin *Synthesis* **1975,** 315. For another synthesis of di-*t*-butyl ether, see Masada; Yonemitsu; Hirota *Tetrahedron Lett.* **1979,** 1315.
[574]Masada; Sakajiri *Bull. Chem. Soc. Jpn.* **1978,** *51*, 866.
[575]For a review of reactions in which alcohols serve as nucleophiles, see Salomaa; Kankaanperä; Pihlaja, in Patai *The Chemistry of the Hydroxyl Group*, pt. 1; Wiley: New York, 1971, pp. 454-466.
[576]Biordi; Moelwyn-Hughes, *J. Chem. Soc.* **1962,** 4291.

with phenols and an amine such as pyridine.[577] Aryl alkyl ethers can be prepared from alkyl halides by treatment with an aryl acetate (instead of a phenol) in the presence of K_2CO_3 and a crown ether.[578]

gem-Dihalides react with alkoxides to give acetals, and 1,1,1-trihalides give ortho esters.[579] Both aryl alkyl and dialkyl ethers can be efficiently prepared with the use of phase transfer catalysis (p. 362)[580] and with micellar catalysis.[581]

Hydroxy groups can be protected[582] by reaction of their salts with chloromethyl methyl ether.

$$RO^- + CH_3OCH_2Cl \longrightarrow ROCH_2OCH_3$$

This protecting group is known as MOM (methoxymethyl) and such compounds are called MOM ethers. The resulting acetals are stable to bases and are easily cleaved with mild acid treatment (**0-6**). Another protecting group, the 2-methoxyethoxymethyl group (the MEM group), is formed in a similar manner: $RO^- + MeOCH_2CH_2OCH_2Cl \rightarrow ROCH_2OCH_2CH_2OMe$. Both MOM and MEM groups can be cleaved with dialkyl- and diarylboron halides such as Me_2BBr.[583] Phenacyl bromides ($ArCOCH_2Br$) have also been used to protect hydroxy groups.[584] The resulting ethers can easily be hydrolyzed with zinc and acetic acid.

Aryl cyanates[585] can be prepared by reaction of phenols with cyanogen halides in the presence of a base: $ArO^- + ClCN \rightarrow ArOCN + Cl^-$.[586] This reaction has also been applied to certain alkyl cyanates.[587]

Though most Williamson reactions proceed by the SN2 mechanism, there is evidence (see p. 308) that in some cases the SET mechanism can take place, especially with alkyl iodides.[588]

OS **I**, 75, 205, 258, 296, 435; **II**, 260; **III**, 127, 140, 209, 418, 432, 544; **IV**, 427, 457, 558, 590, 836; **V**, 251, 258, 266, 403, 424, 684; **VI**, 301, 361, 395, 683; **VII**, 34, 386, 435; **65**, 68, 173; **68**, 92; **69**, 148.

0-13 Epoxide Formation
(3)*OC-cyclo*-Alkoxy-de-halogenation

[577]Masada; Oishi *Chem. Lett.* **57**, 1978. For another method, see Camps; Coll; Moretó, *Synthesis* **1982**, 186.
[578]Banerjee; Gupta; Singh *J. Chem. Soc., Chem. Commun.* **1982**, 815.
[579]For a review of the formation of ortho esters by this method, see DeWolfe, Ref. 457, pp. 12-18.
[580]For reviews, see Starks; Liotta, Ref. 404, pp. 128-138; Weber; Gokel *Phase Transfer Catalysis in Organic Synthesis*, Ref. 404, pp. 73-84. For the use of phase transfer catalysis to convert, selectively, one OH group of a diol or triol to an ether, see de la Zerda; Barak; Sasson *Tetrahedron* **1989**, *45*, 1533.
[581]Jursić *Tetrahedron* **1988**, *44*, 6677.
[582]For other protecting groups for OH, see Greene, *Protective Groups in Organic Synthesis*; Wiley: New York, 1981, pp. 10-113; Corey; Gras; Ulrich *Tetrahedron Lett.* **1976**, 809 and references cited therein.
[583]Guindon; Yoakim; Morton *J. Org. Chem.* **1984**, *49*, 3912. For other methods, see Williams; Sakdarat *Tetrahedron Lett.* **1983**, *24*, 3965; Hanessian; Delorme; Dufresne *Tetrahedron Lett.* **1984**, *25*, 2515; Rigby; Wilson *Tetrahedron Lett.* **1984**, *25*, 1429.
[584]Hendrickson; Kandall *Tetrahedron Lett.* **1970**, 343.
[585]For reviews of alkyl and aryl cyanates, see Jensen; Holm in Patai *The Chemistry of Cyanates and Their Thio Derivatives*, pt. 1; Wiley: New York, 1977, pp. 569-618; Grigat; Pütter *Angew. Chem. Int. Ed. Engl.* **1967**, *6*, 206-218 [*Angew. Chem.* **79**, 219-231].
[586]Grigat; Pütter *Chem. Ber.* **1964**, *97*, 3012; Martin; Bauer *Org. Synth.* **VII**, 435.
[587]Kauer; Henderson *J. Am. Chem. Soc.* **1964**, *86*, 4732.
[588]Ashby; Bae; Park; Depriest; Su *Tetrahedron Lett.* **1984**, *25*, 5107.

This is a special case of **0-12.** The base removes the proton from the OH group and the epoxide then attacks in an internal S$_N$2 reaction.[589] Many epoxides have been made in this way.[590] The method can also be used to prepare larger cyclic ethers: five- and six-membered rings. Additional treatment with base yields the glycol (**0-7**).

OS **I**, 185, 233; **II**, 256; **III**, 835; **VI**, 560; **VII**, 164, 356; **66,** 160.

0-14 Alkylation with Inorganic Esters
Alkoxy-de-sulfonyloxy-substitution

$$R—OSO_2OR'' + R'O^- \longrightarrow ROR'$$

The reaction of alkyl sulfates with alkoxide ions is quite similar to **0-12** in mechanism and scope. Other inorganic esters can also be used. One of the most common usages of the reaction is the formation of methyl ethers of alcohols and phenols by treatment of alkoxides or aroxides with methyl sulfate. The alcohol or phenol can be methylated directly, by treatment with dimethyl sulfate and alumina in cyclohexane.[591] Carboxylic esters sometimes give ethers when treated with alkoxides (B$_{AL}$2 mechanism, p. 381) in a very similar process (see also **0-23**).

t-Butyl ethers can be prepared by treating the compound *t*-butyl 2,2,2-trichloroacetimidate with an alcohol or phenol in the presence of boron trifluoride etherate.[592]

$$Cl_3C—\overset{\displaystyle \|}{\underset{\displaystyle NH}{C}}—O\text{-}t\text{-}Bu + ROH \xrightarrow{BF_3\text{-}Et_2O} t\text{-}BuOR$$

OS **I**, 58, 537; **II**, 387, 619; **III**, 127, 564, 800; **IV**, 588; **VI**, 737, 859, **VII**, 41. Also see OS **V**, 431.

0-15 Alkylation with Diazo Compounds
Hydro,alkoxy-de-diazo-bisubstitution

$$CH_2N_2 + ROH \xrightarrow{HBF_4} CH_3OR$$

$$R_2CN_2 + ArOH \longrightarrow R_2CHOAr$$

Reaction with alcohols is general for diazo compounds, but it is most often performed with diazomethane to produce methyl ethers or with diazo ketones to produce α-keto ethers, since these kinds of diazo compounds are most readily available. With diazomethane[593] the method is expensive and requires great caution. It is used chiefly to methylate alcohols and phenols that are expensive or available in small amounts, since the conditions are mild and high yields are obtained. Hydroxy compounds react better as their acidity increases; ordinary alcohols do not react at all unless a catalyst such as HBF$_4$[594] or silica gel[595] is present. The more acidic phenols react very well in the absence of a catalyst. Oximes, and ketones that

[589]See, for example, Swain; Ketley; Bader *J. Am. Chem. Soc.* **1959,** *81,* 2353; Knipe *J. Chem. Soc., Perkin Trans. 2* **1973,** 589.
[590]For a review, see Berti *Top. Stereochem.* **1973,** 7, 93-251, pp. 187-209.
[591]Ogawa; Ichimura; Chihara; Teratani; Taya *Bull. Chem. Soc. Jpn.* **1986,** *59,* 2481.
[592]Armstrong; Brackenridge; Jackson; Kirk *Tetrahedron Lett.* **1988,** *29,* 2483.
[593]For a review of diazomethane, see Pizey *Synthetic Reagents,* vol. 2; Wiley: New York, 1974, pp. 65-142.
[594]Neeman; Caserio; Roberts; Johnson *Tetrahedron* **1959,** *6,* 36.
[595]Ohno; Nishiyama; Nagase *Tetrahedron Lett.* **1979,** 4405; Ogawa; Hagiwara; Chihara; Teratani; Taya *Bull. Chem. Soc. Jpn.* **1987,** *60,* 627.

have substantial enolic contributions, give O-alkylation to form, respectively, O-alkyl oximes and enol ethers. The mechanism[596] is as in **0-5**:

$$H_2C \overset{\oplus}{=} N \overset{\ominus}{=} N| + ROH \longrightarrow H_3C \overset{\oplus}{-} N \equiv N \xrightarrow[S_N1 \text{ or } S_N2]{RO^-} CH_3OR$$

Diazoalkanes can also be converted to ethers by thermal or photochemical cleavage in the presence of an alcohol. These are carbene or carbenoid reactions.[597] Similar intermediates are involved when diazoalkanes react with alcohols in the presence of *t*-BuOCl to give acetals.[598]

$$R_2CN_2 + 2R'OH \xrightarrow{t\text{-BuOCl}} R_2C(OR')_2$$

OS **V**, 245. Also see OS **V**, 1099.

0-16   Dehydration of Alcohols
    Alkoxy-de-hydroxylation

$$2ROH \xrightarrow{H_2SO_4} ROR + H_2O$$

The dehydration of alcohols to form ethers[599] is analogous to **0-12** and **0-14**, but the species from which the leaving group departs is ROH_2^+ or $ROSO_2OH$. The former is obtained directly on treatment of alcohols with sulfuric acid and may go, by an S$_N$1 or S$_N$2 pathway, directly to the ether if attacked by another molecule of alcohol. On the other hand, it may, again by either an S$_N$1 or S$_N$2 route, be attacked by the nucleophile HSO_4^-, in which case it is converted to $ROSO_2OH$, which in turn may be attacked by an alcohol molecule to give ROR. Elimination is always a side reaction and, in the case of tertiary alkyl substrates, completely predominates. Good yields of ethers were obtained by heating diarylcarbinols [ArAr'CHOH → (ArAr'CH)$_2$O] with TsOH in the solid state.[600]

The ether prepared is symmetrical. Mixed ethers can be prepared if one group is tertiary alkyl and the other primary or secondary, since the latter group is not likely to compete with the tertiary group in the formation of the carbocation, while a tertiary alcohol is a very poor nucleophile.[601] If one group is not tertiary, the reaction of a mixture of two alcohols leads to all three possible ethers. Diols can be converted to cyclic ethers,[602] though the reaction is most successful for five-membered rings. Thus, 1,6-hexanediol gives mostly 2-ethyltetrahydrofuran. However, 5-, 6-, and 7-membered rings have been prepared with AlPO$_4$–Al$_2$O$_3$,[603] with BuSnCl$_3$,[604] and with a Nafion-H acid catalyst[605] (the last-named reagent was also used to make an 8-membered ring). This reaction is also important in preparing furfural derivatives from aldoses, with concurrent elimination:

$$HOCH_2-CHOH-CHOH-CHOH-CHO \xrightarrow{H^+} \overset{}{\underset{O}{\text{⟨furan⟩}}}-CHO$$

[596]Kreevoy; Thomas *J. Org. Chem.* **1977,** 42, 3979. See also McGarrity; Smyth *J. Am. Chem. Soc.* **1980,** *102,* 7303.

[597]Bethell; Howard *J. Chem. Soc. B* **1969,** 745; Bethell; Newall; Whittaker *J. Chem. Soc. B* **1971,** 23; Noels; Demonceau; Petiniot; Hubert; Teyssié *Tetrahedron* **1982,** *38,* 2733.

[598]Baganz; May *Angew. Chem. Int. Ed. Engl.* **1966,** *5,* 420 [*Angew. Chem. 78,* 448].

[599]For a review, see Ref. 567, pp. 457-460, 468-470.

[600]Toda; Takumi; Akehi *J. Chem. Soc., Perkin Trans. 2* **1990,** 1270.

[601]See, for example, Jenner *Tetrahedron Lett.* **1988,** *29,* 2445.

[602]For a list of reagents, with references, see Ref. 508, pp. 449-450.

[603]Costa; Riego *Synth. Commun.* **1987,** *17,* 1373.

[604]Tagliavini; Marton; Furlani *Tetrahedron* **1989,** *45,* 1187.

[605]Olah; Fung; Malhotra *Synthesis* **1981,** 474.

Phenols and primary alcohols form ethers when heated with dicyclohexylcarbodiimide[606] (see **0-22**). 1,2-Diols can be converted to epoxides by treatment with dimethylformamide dimethyl acetal [$(MeO)_2CHNMe_2$],[607] with diethyl azodicarboxylate [$EtOOCN=NCOOEt$] and Ph_3P,[608] with a dialkoxytriphenylphosphorane,[609] or with $TsCl-NaOH-PhCH_2NEt_3^+ Cl^-$.[610]

OS **I**, 280; **II**, 126; **IV**, 25, 72, 266, 350, 393, 534; **V**, 539, 1024; **VI**, 887; **69**, 205. Also see OS **V**, 721.

0-17 Transetherification
Hydroxy-de-alkoxylation
Alkoxy-de-hydroxylation

$$ROR' + R''OH \longrightarrow ROR'' + R'OH$$

The exchange of one alkoxy group for another is very rare for *ethers*, though it has been accomplished with reactive R, for example, diphenylmethyl with *p*-toluenesulfonic acid as a catalyst,[611] and by treatment of alkyl aryl ethers with alkoxide ions: $ROAr + R'O^- \rightarrow ROR' + ArO^-$.[612] However, acetals and ortho esters undergo transetherification readily,[613] for example,[614]

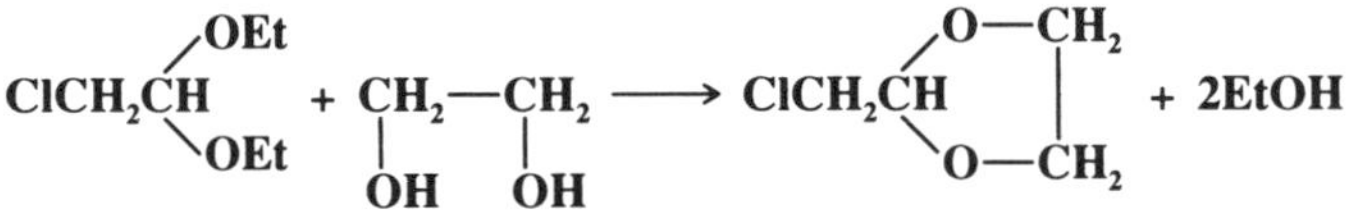

because, as we have seen (**0-6**), departure of the leaving group from an acetal gives a particularly stable carbocation. These are equilibrium reactions, and most often the equilibrium is shifted by removing the lower-boiling alcohol by distillation. Enol ethers can be prepared by treating an alcohol with an enol ester or a different enol ether, with mercuric acetate as a catalyst,[615] e.g.,

$$ROCH=CH_2 + R'OH \xrightarrow{Hg(OAc)_2} R'OCH=CH_2 + ROH$$

1,2-Diketones can be converted to α-keto enol ethers by treatment with an alkoxytrimethylsilane $ROSiMe_3$.[616]

OS **VI**, 298, 491, 584, 606, 869; **VII**, 334; **65**, 32; **68**, 92. Also see OS **V**, 1080, 1096.

[606]Vowinkel *Chem. Ber.* **1962**, *95*, 2997, **1963**, *96*, 1702, **1966**, *99*, 42.
[607]Neumann *Chimia* **1969**, *23*, 267.
[608]Guthrie; Jenkins; Yamasaki; Skelton; White *J. Chem. Soc., Perkin Trans. 1* **1981**, 2328 and references cited therein. For a review of diethyl azodicarboxylate–Ph_3P, see Mitsunobu *Synthesis* **1981**, 1-28.
[609]Robinson; Barry; Kelly; Evans *J. Am. Chem. Soc.* **1985**, *107*, 5210; Kelly; Evans *J. Org. Chem.* **1986**, *51*, 5490. See also Hendrickson; Hussoin *Synlett* **1990**, 423.
[610]Szeja *Synthesis* **1985**, 983.
[611]Pratt; Draper *J. Am. Chem. Soc.* **1949**, *71*, 2846.
[612]Zoltewicz; Sale *J. Org. Chem.* **1970**, *35*, 3462.
[613]For reviews, see Ref. 575, pp. 458-463; DeWolfe, Ref. 457, pp. 18-29, 146-148.
[614]McElvain; Curry *J. Am. Chem. Soc.* **1948**, *70*, 3781.
[615]Watanabe; Conlon *J. Am. Chem. Soc.* **1957**, *79*, 2828; Büchi; White *J. Am. Chem. Soc.* **1964**, *86*, 2884. For a review, see Shostakovskii; Trofimov; Atavin; Lavrov *Russ. Chem. Rev.* **1968**, *37*, 907-919. For a discussion of the mechanism, see Gareev *J. Org. Chem. USSR* **1982**, *18*, 36.
[616]Ponaras; Meah *Tetrahedron Lett.* **1986**, *27*, 4953.

0-18 Alcoholysis of Epoxides
(3)OC-seco-Alkoxy-de-alkoxylation

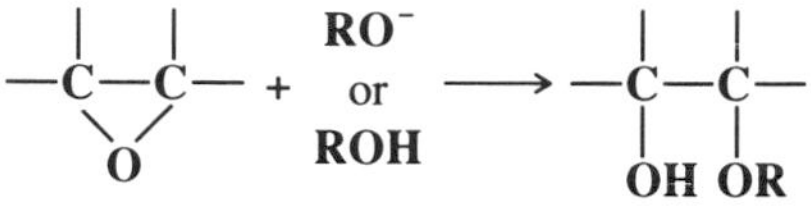

This reaction is analogous to **0-7**. It may be acid, base, or alumina[617] catalyzed, and may occur by either an SN1 or SN2 mechanism. Many of the β-hydroxy ethers produced in this way are valuable solvents, for example, diethylene glycol, Cellosolve, etc. Aziridines can similarly be converted to β-amino ethers.[618]

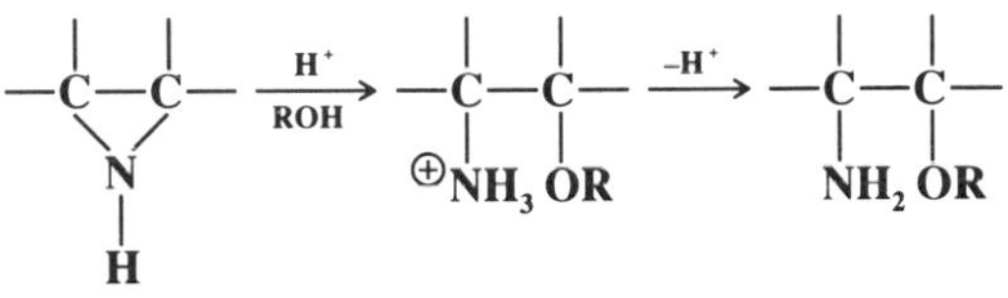

In the *Payne rearrangement*, a 2,3-epoxy alcohol is converted to an isomeric one, by treatment with aqueous base:[619]

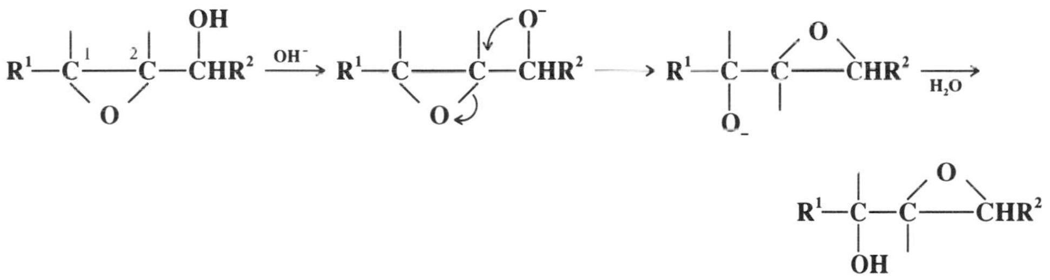

The reaction results in inverted configuration at C-2. Of course, the product can also revert to the starting material by the same pathway, so a mixture of epoxy alcohols is generally obtained.

0-19 Alkylation with Onium Salts
Alkoxy-de-hydroxylation

$$R_3O^+ + R'OH \longrightarrow ROR' + R_2O$$

Oxonium ions are excellent alkylating agents, and ethers can be conveniently prepared by treating them with alcohols or phenols.[620] Quaternary ammonium salts can sometimes also be used.[621]

OS **65,** 140; **66,** 29.

---

[617]See Posner; Rogers *J. Am. Chem. Soc.* **1977,** *99,* 8208, 8214.
[618]For a review, see Dermer; Ham, Ref. 437, pp. 224-227, 256-257.
[619]Payne *J. Org. Chem.* **1962,** *27,* 3819; Behrens; Ko; Sharpless; Walker *J. Org. Chem.* **1985,** *50,* 5687.
[620]Granik; Pyatin; Glushkov, Ref. 339, p. 749.
[621]For an example, see Vogel; Büchi *Org. Synth.* **66,** 29.

D. Attack by OR at an Acyl Carbon

0-20 Alcoholysis of Acyl Halides
Alkoxy-de-halogenation

$$R-\underset{\underset{O}{\|}}{C}-X + R'OH \longrightarrow R-\underset{\underset{O}{\|}}{C}-OR'$$

The reaction between acyl halides and alcohols or phenols is the best general method for the preparation of carboxylic esters. The reaction is of wide scope, and many functional groups do not interfere. A base is frequently added to combine with the HX formed. When aqueous alkali is used, this is called the *Schotten–Baumann procedure*, but pyridine is also frequently used. Both R and R' may be primary, secondary, or tertiary alkyl or aryl. Enolic esters can also be prepared by this method, though C-acylation competes in these cases. In difficult cases, especially with hindered acids or tertiary R', the alkoxide can be used instead of the alcohol.[622] Activated alumina has also been used as a catalyst, for tertiary R'.[623] Thallium salts of phenols give very high yields of phenolic esters.[624] Phase transfer catalysis has been used for hindered phenols.[625]

When phosgene is the acyl halide, haloformic esters or carbonates can be obtained.

$$Cl-\underset{\underset{O}{\|}}{C}-Cl \xrightarrow{ROH} RO-\underset{\underset{O}{\|}}{C}-Cl \xrightarrow{ROH} RO-\underset{\underset{O}{\|}}{C}-OR$$

An important example is the preparation of carbobenzoxy chloride ($PhCH_2OCOCl$) from phosgene and benzyl alcohol. This compound is widely used for protection of amino groups during peptide synthesis (see **0-52**).

As with **0-8,** the mechanism can be S_N1 or tetrahedral.[502] Pyridine catalyzes the reaction by the nucleophilic catalysis route (see **0-9**).

Acyl halides can also be converted to carboxylic acids by using ethers instead of alcohols, in MeCN in the presence of certain catalysts such as cobalt(II) chloride.[626]

$$R-\underset{\underset{O}{\|}}{C}-X + R'-O-R'' \xrightarrow[\text{MeCN}]{\text{CoCl}_2} R-\underset{\underset{O}{\|}}{C}-OR' + R''Cl$$

This is a method for the cleavage of ethers (see also **0-68**).

OS **I,** 12; **III,** 142, 144, 167, 187, 623, 714; **IV,** 84, 263, 478, 479, 608, 616, 788; **V,** 1, 166, 168, 171; **VI,** 199, 259, 312, 824. **VII,** 190; **65,** 203; **69,** 1.

0-21 Alcoholysis of Anhydrides
Alkoxy-de-acyloxy-substitution

$$R-\underset{\underset{O}{\|}}{C}-O-\underset{\underset{O}{\|}}{C}-R'' + R'OH \longrightarrow R-\underset{\underset{O}{\|}}{C}-OR' + R''COOH$$

---

[622]For an example, see Kaiser; Woodruff, *J. Org. Chem.* **1970,** *35,* 1198.
[623]Nagasawa; Yoshitake; Amiya; Ito *Synth. Commun.* **1990,** *20,* 2033.
[624]Taylor, McLay; McKillop *J. Am. Chem. Soc.* **1968,** *90,* 2422.
[625]Illi, *Tetrahedron Lett.* **1979,** 2431. For another method, see Nekhoroshev; Ivakhnenko; Okhlobystin *J. Org. Chem. USSR* **1977,** *13,* 608.
[626]See Ahmad; Iqbal *Chem. Lett.* **1987,** 953, and references cited therein.

The scope of this reaction is similar to that of **0-20.** Though anhydrides are somewhat less reactive than acyl halides, they are often used to prepare carboxylic esters. Acids, Lewis acids, and bases are often used as catalysts—most often, pyridine.[627] Catalysis by pyridine is of the nucleophilic type (see **0-9**). 4-(N,N-Dimethylamino)pyridine is a better catalyst than pyridine and can be used in cases where pyridine fails.[628] A nonbasic catalyst is cobalt(II) chloride.[629] Formic anhydride is not a stable compound but esters of formic acid can be prepared by treating alcohols[630] or phenols[631] with acetic–formic anhydride. Cyclic anhydrides give monoesterified dicarboxylic acids, for example,

$$\begin{array}{c}
CH_2-C \overset{O}{\diagup} \\
\qquad\qquad O \;+\; ROH \;\longrightarrow\; \begin{array}{l} CH_2-COOR \\ CH_2-COOH \end{array} \\
CH_2-C \underset{O}{\diagdown}
\end{array}$$

Alcohols can also be acylated by mixed organic-inorganic anhydrides, such as acetic-phosphoric anhydride $MeCOOPO(OH)_2$[632] (see **0-33**).

OS **I**, 285, 418; **II**, 69, 124; **III**, 11, 127, 141, 169, 237, 281, 428, 432, 690, 833; **IV**, 15, 242, 304; **V**, 8, 459, 591, 887; **VI**, 121, 245, 560, 692; **67**, 76; **69**, 19.

0-22 Esterification of Carboxylic Acids
Alkoxy-de-hydroxylation

$$RCOOH + R'OH \overset{H^+}{\rightleftharpoons} RCOOR' + H_2O$$

The esterification of carboxylic acids with alcohols[633] is the reverse of **0-11** and can be accomplished only if a means is available to drive the equilibrium to the right.[634] There are many ways of doing this, among which are: (1) addition of an excess of one of the reactants, usually the alcohol; (2) removal of the ester or the water by distillation; (3) removal of water by azeotropic distillation; and (4) removal of water by use of a dehydrating agent or a molecular sieve. When R' is methyl, the most common way of driving the equilibrium is by adding excess MeOH; when R' is ethyl, it is preferable to remove water by azeotropic distillation.[635] The most common catalysts are H_2SO_4 and TsOH, though some reactive acids (e.g., formic,[636] trifluoroacetic[637]) do not require a catalyst. Besides methyl and ethyl, R' may be other primary or secondary alkyl groups, but tertiary alcohols usually give carbocations and elimination. Phenols can sometimes be used to prepare phenolic esters, but yields are generally very low.

[627]For a list of catalysts, with references, see Ref. 508, pp. 980-981.
[628]For reviews, see Scriven *Chem. Soc. Rev.* **1983**, *12*, 129-161; Höfle; Steglich; Vorbrüggen *Angew. Chem. Int. Ed. Engl.* **1978,** *17*, 569-583 [*Angew. Chem. 90*, 602-615].
[629]Ahmad; Iqbal *J. Chem. Soc., Chem. Commun.* **1987**, 114.
[630]For example, see Stevens; van Es *Recl. Trav. Chim. Pays-Bas,* **1964,** *83*, 1287; van Es; Stevens *Recl. Trav. Chim. Pays-Bas* **1965**, *84*, 704.
[631]For example, see Stevens; van Es *Recl. Trav. Chim. Pays-Bas* **1964**, *83*, 1294; Sōfuku; Muramatsu; Hagitani *Bull. Chem. Soc. Jpn.* **1967**, *40*, 2942.
[632]Fatiadi *Carbohydr. Res.* **1968**, *6*, 237.
[633]For a review of some methods, see Haslam *Tetrahedron* **1980**, *36*, 2409-2433.
[634]For a list of reagents, with references, see Ref. 508, pp. 966-972.
[635]Newman *An Advanced Organic Laboratory Course*; Macmillan: New York, 1972, pp. 8-10.
[636]Formates can be prepared if diisopropyl ether is used to remove water by azeotropic distillation: Werner, *J. Chem. Res. (S)* **1980**, 196.
[637]Johnston; Knipe; Watts *Tetrahedron Lett.* **1979**, 4225.

γ- and δ-hydroxy acids are easily lactonized by treatment with acids, or often simply on standing, but larger and smaller lactone rings cannot be made in this manner, because

$$CH_3-CH-CH_2-CH_2-COOH \longrightarrow$$
$$\qquad\quad |$$
$$\qquad\quad OH$$

polyester formation occurs more readily.[638] Often the conversion of a group such as keto or halogen, γ or δ to a carboxyl group, to a hydroxyl group gives the lactone directly, since the hydroxy acid cyclizes too rapidly for isolation. β-Substituted β-hydroxy acids can be converted to β-lactones by treatment with benzenesulfonyl chloride in pyridine at 0 to 5°C.[639] ε-Lactones (seven-membered rings) have been made by cyclization of ε-hydroxy acids at high dilution.[640] Macrocyclic lactones[641] can be prepared indirectly in very good yields by conversion of the hydroxy acids to 2-pyridinethiol esters and adding these to refluxing xylene.[642]

$$HO-(CH_2)_n-COOH \longrightarrow HO-(CH_2)_n-C-S\quad \xrightarrow{\Delta}$$
$$\qquad\qquad\qquad\qquad\qquad\qquad\quad ||$$
$$\qquad\qquad\qquad\qquad\qquad\qquad\quad O$$

A closely related method, which often gives higher yields, involves treatment of the hydroxy acids with 1-methyl- or 1-phenyl-2-halopyridinium salts, especially 1-methyl-2-chloropyridinium iodide (*Mukaiyama's reagent*).[643] Another method uses organotin oxides.[644]

---

[638]For a review of the synthesis of lactones and lactams, see Wolfe; Ogliaruso, in Patai *The Chemistry of Acid Derivatives*, pt. 2; Wiley: New York, 1979, pp. 1062-1330. For a list of methods for converting hydroxy acids to lactones, with references, see Ref. 508, pp. 941-943.

[639]Adam; Baeza; Liu *J. Am. Chem. Soc.* **1972**, *94*, 2000. For other methods of converting β-hydroxy acids to β-lactones, see Merger *Chem. Ber.* **1968**, *101*, 2413; Blume *Tetrahedron Lett.* **1969**, 1047.

[640]Lardelli; Lamberti; Weller; de Jonge *Recl. Trav.Chim. Pays-Bas* **1967**, *86*, 481.

[641]For reviews on the synthesis of macrocyclic lactones, see Nicolaou *Tetrahedron* **1977**, *33*, 683-710; Back *Tetrahedron* **1977**, *33*, 3041-3059; Masamune; Bates; Corcoran *Angew. Chem. Int. Ed. Engl.* **1977**, *16*, 585-607 [*Angew. Chem.* **89**, 602-624].

[642]Corey; Nicolaou; Melvin *J. Am. Chem. Soc.* **1975**, *97*, 653, 655; Corey; Brunelle; Stork *Tetrahedron Lett.* **1976**, 3405; Corey; Brunelle; *Tetrahedron Lett.* **1976**, 3409; Wollenberg; Nimitz; Gokcek *Tetrahedron Lett.* **1980**, *21*, 2791; Thalmann; Oertle; Gerlach *Org. Synth. VII*, 470. See also Schmidt; Heermann *Angew. Chem. Int. Ed. Engl.* **1979**, *18*, 308 [*Angew. Chem.* **91**, 330].

[643]For a review of reactions with this and related methods, see Mukaiyama *Angew. Chem. Int. Ed. Engl.* **1979**, *18*, 707-721 [*Angew. Chem.* **91**, 798-812].

[644]Steliou; Szczygielska-Nowosielska; Favre; Poupart; Hanessian *J. Am. Chem. Soc.* **1980**, *102*, 7578; Steliou; Poupart *J. Am. Chem. Soc.* **1983**, *105*, 7130. For some other methods, see Masamune; Kamata; Schilling *J. Am. Chem. Soc.* **1975**, *97*, 3515; Scott; Naples *Synthesis* **1976**, 738; Kurihara; Nakajima; Mitsunobu *Tetrahedron Lett.* **1976**, 2455; Corey; Brunelle; Nicolaou *J. Am. Chem. Soc.* **1977**, *99*, 7359; Vorbrüggen; Krolikiewicz *Angew. Chem. Int. Ed. Engl.* **1977**, *16*, 876 [*Angew. Chem.* **89**, 914]; Nimitz; Wollenberg *Tetrahedron Lett.* **1978**, 3523; Inanaga; Hirata; Saeki; Katsuki; Yamaguchi *Bull. Chem. Soc. Jpn.* **1979**, *52*, 1989; Venkataraman; Wagle *Tetrahedron Lett.* **1980**, *21*, 1893; Schmidt; Dietsche *Angew. Chem. Int. Ed. Engl.* **1981**, *20*, 771 [*Angew. Chem.* **93**, 786]; Taniguchi; Kinoshita; Inomata; Kotake *Chem. Lett.* **1984**, 1347; Cossy; Pete *Bull. Soc. Chim. Fr.* **1988**, 989.

Esterification is catalyzed by acids (not bases) in ways that were discussed on p. 379.[525] The mechanisms are usually AAC2, but AAC1 and AAL1 have also been observed.[645] Certain acids, such as 2,6-di-ortho-substituted benzoic acids, cannot be esterified by the AAC2 mechanism because of steric hindrance (p. 340). In such cases, esterification can be accomplished by dissolving the acid in 100% H_2SO_4 (forming the ion RCO^+) and pouring the solution into the alcohol (AAC1 mechanism). The reluctance of hindered acids to undergo the normal AAC2 mechanism can sometimes be put to advantage when, in a molecule containing two COOH groups, only the less hindered one is esterified. The AAC1 pathway cannot be applied to unhindered carboxylic acids.

Another way to esterify a carboxylic acid is to treat it with an alcohol in the presence of a dehydrating agent.[634] One of these is dicyclohexylcarbodiimide (DCC), which is converted

DCC   DHU

in the process to dicyclohexylurea (DHU). The mechanism[646] has much in common with the nucleophilic catalysis mechanism; the acid is converted to a compound with a better leaving group. However, the conversion is not by a tetrahedral mechanism (as it is in nucleophilic catalysis), since the C—O bond remains intact during this step:

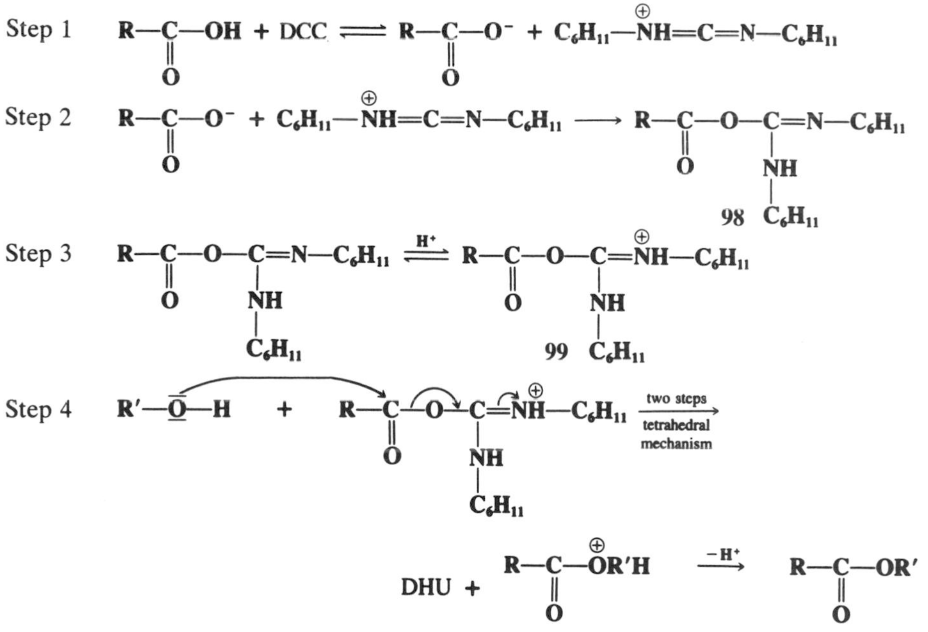

Evidence for this mechanism was the preparation of O-acylureas similar to **98** and the finding that when catalyzed by acids they react with alcohols to give esters.[647]

[645]For a review of aspects of the mechanism, see Ref. 575, pp. 466-481.
[646]Smith; Moffatt; Khorana *J. Am. Chem. Soc.* **1958,** *80,* 6204; Balcom; Petersen *J. Org. Chem.* **1989,** *54,* 1922.
[647]Doleschall; Lempert *Tetrahedron Lett.* **1963,** 1195.

However, there are limitations to the use of DCC; yields are variable and N-acylureas are side products. Many other dehydrating agents[648] have been used, including an alkyl chloroformate and Et_3N,[649] pyridinium salts–Bu_3N,[643] phenyl dichlorophosphate $PhOPOCl_2$,[650] DCC and an aminopyridine,[651] 2-chloro-1,3,5-trinitrobenzene and pyridine,[652] di-2-pyridyl carbonate,[653] polystyryl diphenylphosphine,[654] (trimethylsilyl)ethoxyacetylene,[655] 1,1'-carbonylbis(3-methylimidazolium) triflate (CBMIT),[656] Amberlyst-15,[657] diethyl azodicarboxylate $EtOOCN{=}NCOOEt$ and Ph_3P[658] (when these reagents are used the procedure is called the *Mitsunobu esterification reaction*[659]), chlorosulfonyl isocyanate $ClSO_2NCO$,[660] chlorosilanes,[661] $MeSO_2Cl$–Et_3N,[662] Ph_3P–CCl_4–

100   **101**

Et_3N,[663] and N,N'-carbonyldiimidazole (**100**).[664] In the latter case easily alcoholyzed imidazolides (**101**) are intermediates. BF_3 promotes the esterification by converting the acid to $RCO^+\ BF_3OH^-$, so the reaction proceeds by an $A_{AC}1$ type of mechanism. The use of BF_3–etherate is simple and gives high yields.[665] Carboxylic esters can also be prepared by treating carboxylic acids with *t*-butyl ethers and acid catalysts.[666]

$$RCOOH + t\text{-Bu-OR}' \longrightarrow RCOOR' + H_2C{=}CMe_2 + H_2O$$

Carboxylic acids can be converted to *t*-butyl esters by treatment with *t*-butyl 2,2,2-trichloroacetimidate (see **0-14**) and BF_3–Et_2O.[592]

OS **I**, 42, 138, 237, 241, 246, 254, 261, 451; **II**, 260, 264, 276, 292, 365, 414, 526; **III**, 46, 203, 237, 381, 413, 526, 531, 610; **IV**, 169, 178, 302, 329, 390, 398, 427, 506, 532, 635, 677; **V**, 80, 762, 946; **VI**, 471, 797; **VII**, 93, 99, 210, 319, 356, 386, 470; **66,** 22, 142; **67,** 76. Also see OS **III**, 536, 742.

[648]For a list of many of these with references, see Arrieta; García; Lago; Palomo *Synth. Commun.* **1983**, *13*, 471.
[649]Kim; Lee; Kim *J. Org. Chem.* **1985**, *50*, 560.
[650]Liu; Chan; Lee *Tetrahedron Lett.* **1978**, 4461. García; Arrieta; Palomo *Synth. Commun.* **1982**, *12*, 681. See also Ueda; Oikawa *J. Org. Chem.* **1985**, *50*, 760.
[651]Hassner; Alexanian *Tetrahedron Lett.* **1978**, 4475; Neises; Steglich *Angew. Chem. Int. Ed. Engl.* **1978**, *17*, 522 [*Angew. Chem. 90*, 556]; Boden; Keck *J. Org. Chem.* **1985**, *50*, 2394.
[652]Takimoto; Inanaga; Katsuki; Yamaguchi *Bull. Chem. Soc. Jpn.* **1981**, *54*, 1470. See also Kim; Yang *Synth. Commun.* **1981**, *11*, 121; Takimoto; Abe; Kodera; Ohta *Bull. Chem. Soc. Jpn.* **1983**, *56*, 639.
[653]Kim; Ko *Tetrahedron Lett.* **1984**, *25*, 4943. For a review of 2-pyridyl reagents, see Kim *Org. Prep. Proced. Int.* **1988**, *20*, 145-172.
[654]Caputo; Corrado; Ferreri; Palumbo *Synth. Commun.* **1986**, *16*, 1081.
[655]Kita; Akai; Yamamoto; Taniguchi; Tamura *Synthesis* **1989**, 334.
[656]Saha; Schultz; Rapoport *J. Am. Chem. Soc.* **1989**, *111*, 4856.
[657]Petrini; Ballini; Marcantoni; Rosini *Synth. Commun.* **1988**, *18*, 847.
[658]Mitsunobu; Yamada *Bull. Chem. Soc. Jpn.* **1967**, *40*, 2380; Camp; Jenkins *Aust. J. Chem.* **1988**, *41*, 1835.
[659]For discussions of the mechanism, see Varasi; Walker; Maddox *J. Org. Chem.* **1987**, *52*, 4235; Hughes; Reamer; Bergan; Grabowski *J. Am. Chem. Soc.* **1988**, *110*, 6487; Crich; Dyker; Harris *J. Org. Chem.* **1989**, *54*, 257; Camp; Jenkins *J. Org. Chem.* **1989**, *54*, 3045, 3049.
[660]Keshavamurthy; Vankar; Dhar *Synthesis* **1982**, 506. For a review of $ClSO_2NCO$, see Dhar; Murthy *Synthesis* **1988**, 437-450.
[661]Nakao; Oka; Fukumoto *Bull. Chem. Soc. Jpn.* **1981**, *54*, 1267; Brook; Chan *Synthesis* **1983**, 201.
[662]Chandrasekaran; Turner *Synth. Commun.* **1982**, *12*, 727.
[663]Hashimoto; Furukawa *Bull. Chem. Soc. Jpn.* **1981**, *54*, 2227; Ramaiah *J. Org. Chem.* **1985**, *50*, 4991.
[664]For a review, see Staab; Rohr *Newer Methods Prep. Org. Chem.* **1968**, *5*, 61-108. See also Morton; Mangroo; Gerber *Can. J. Chem.* **1988**, *66*, 1701.
[665]For examples, see Marshall; Erickson; Folsom *Tetrahedron Lett.* **1970**, 4011; Kadaba *Synthesis* **1972**, 628, *Synth. Commun.* **1974**, *4*, 167.
[666]Derevitskaya; Klimov; Kochetkov *Tetrahedron Lett.* **1970**, 4269. See also Mohacsi *Synth. Commun.* **1982**, *12*, 453.

0-23 Alcoholysis of Carboxylic Esters. Transesterification
Alkoxy-de-alkoxylation

$$R\overset{\text{H}^+}{\underset{\text{or OH}^-}{\rightleftharpoons}} \quad R\text{—}\underset{\underset{O}{\|}}{C}\text{—}OR' + R''OH \; \rightleftharpoons \; R\text{—}\underset{\underset{O}{\|}}{C}\text{—}OR'' + R'OH$$

Transesterification is catalyzed[667] by acids or bases.[668] It is an equilibrium reaction and must be shifted in the desired direction. In many cases low-boiling esters can be converted to higher-boiling ones by the distillation of the lower-boiling alcohol as fast as it is formed. This reaction has been used as a method for the acylation of a primary OH in the presence of a secondary OH: The diol is treated with ethyl acetate in the presence of Woelm neutral alumina.[669] Regioselectivity has also been accomplished by using enzymes (lipases) as catalysts.[670] Lactones are easily opened by treatment with alcohols to give open-chain hydroxy esters:

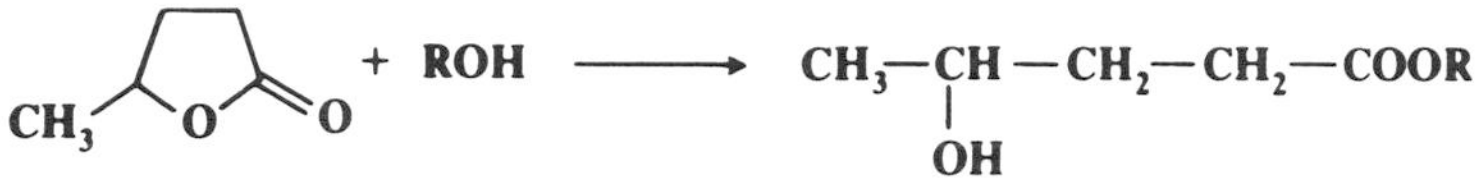

$$\underset{CH_3}{\text{(lactone)}} + ROH \longrightarrow CH_3\text{—}\underset{\underset{OH}{|}}{CH}\text{—}CH_2\text{—}CH_2\text{—}COOR$$

Transesterification has been caried out with phase-transfer catalysis, without an added solvent.[671] In another procedure, RCOOR' are converted to RCOOR'' by treatment of the ester and an alcohol R''OH with *n*-BuLi, which converts the R''OH to R''OLi.[672]

Transesterification occurs by mechanisms[673] that are identical with those of ester hydrolysis—cxccpt that ROII replaces IIOII—that is, by the acyl–oxygen fission mechanisms. When alkyl fission takes place, the products are the *acid* and the *ether*:

$$R\text{—}\underset{\underset{O}{\|}}{C}\text{—}O\text{—}R' + R''OH \longrightarrow R\text{—}\underset{\underset{O}{\|}}{C}\text{—}OH + ROR''$$

Therefore, transesterification reactions frequently fail when R' is tertiary, since this type of substrate most often reacts by alkyl–oxygen cleavage. In such cases, the reaction is of the Williamson type with OCOR as the leaving group (see **0-14**).

With enol esters, the free alcohol is the enol of a ketone, so such esters easily undergo the reaction

$$CH_2\!=\!\underset{\underset{CH_3}{|}}{C}\text{—}OCOR + R'OH \longrightarrow RCOOR' + CH_2\!=\!\underset{\underset{CH_3}{|}}{C}\text{—}OH \rightleftharpoons CH_3\text{—}\underset{\underset{CH_3}{|}}{C}\!=\!O$$

[667]For a list of catalysts, with references, see Ref. 508, pp. 985-987.
[668]For some methods of transesterification under neutral conditions, see Bittner, Barneis; Felix *Tetrahedron Lett.* **1975**, 3871; Hashimoto; Furukawa; Kuroda *Tetrahedron Lett.* **1980**, *21*, 2857; Olah; Narang; Salem; Gupta *Synthesis* **1981**, 142; Otera; Yano; Kawabata; Nozaki *Tetrahedron Lett.* **1986**, *27*, 2383; Imwinkelried; Schiess; Seebach *Org. Synth.* **65**, 230.
[669]Posner; Oda *Tetrahedron Lett.* **1981**, *22*, 5003; Rana; Barlow; Matta *Tetrahedron Lett.* **1981**, *22*, 5007. See also Costa; Riego *Can. J. Chem.* **1987**, *65*, 2327.
[670]Therisod; Klibanov *J. Am. Chem. Soc.* **1987**, *109*, 3977. See also Wang; Lalonde; Momongan; Bcrgbreiter; Wong *J. Am. Chem. Soc.* **1988**, *110*, 7200.
[671]Barry; Bram; Petit *Tetrahedron Lett.* **1988**, *29*, 4567. See also Nishiguchi; Taya *J. Chem. Soc., Perkin Trans. 1* **1990**, 172.
[672]Meth-Cohn *J. Chem. Soc., Chem. Commun.* **1986**, 695.
[673]For a review, see Koskikallio, in Patai, Ref. 197, pp. 103-136.

Hence, enol esters such as isopropenyl acetate are good acylating agents for alcohols.[674] Isopropenyl acetate can also be used to convert other ketones to the corresponding enol acetates in an exchange reaction:[675]

$$\underset{\underset{OAc}{|}}{Me-C}=CH_2 + \underset{\underset{O}{||}}{R-C}-CHR_2' \xrightarrow{H^*} \underset{\underset{OAc}{|}}{R-C}=CR_2' + Me_2CO$$

Enol esters can also be prepared in the opposite type of exchange reaction, catalyzed by mercuric acetate[676] or Pd(II) chloride,[677] e.g.,

$$RCOOH + R'COOCH=CH_2 \underset{\underset{}{\overset{Hg(OAc)_2}{\underset{H_2SO_4}{\rightleftharpoons}}}}{} RCOOCH=CH_2 + R'COOH$$

A closely related reaction is equilibration of a dicarboxylic acid and its diester to produce monoesters:

$$ROCO(CH_2)_nCOOR + HOOC(CH_2)_nCOOH \rightleftharpoons 2ROCO(CH_2)_nCOOH$$

OS **II**, 5, 122, 360; **III**, 123, 146, 165, 231, 281, 581, 605; **IV**, 10, 549, 630, 977; **V**, 155, 545, 863; **VI**, 278; **VII**, 4, 164, 411; **65**, 98, 230; **67**, 170; **68**, 77, 92, 155, 210. See also OS **VII**, 87; **66**, 108.

Alcoholysis of amides is possible but is seldom performed,[678] except for the imidazolide type of amide (**101**).

E.   Attack by OCOR at an Alkyl Carbon

0-24   Alkylation of Carboxylic Acid Salts
Acyloxy-de-halogenation

$$RX + R'COO^- \xrightarrow{HMPA} R'COOR$$

Sodium salts of carboxylic acids, including hindered acids such as mesitoic, rapidly react with primary and secondary bromides and iodides at room temperature in dipolar aprotic solvents, especially HMPA, to give high yields of carboxylic esters.[679] The mechanism is S_N2. Another method uses phase transfer catalysis.[680] With this method good yields of esters have been obtained from primary, secondary, benzylic, allylic, and phenacyl halides.[681] In another procedure, which is applicable to long-chain primary halides, the dry carboxylate salt and the halide, impregnated on alumina as a solid support, are subjected to irradiation by microwaves in a commercial microwave oven.[682] In still another method, carboxylic acids

---

[674]Jeffery; Satchell *J. Chem. Soc.* **1962**, 1906; Rothman; Hecht; Pfeffer; Silbert *J. Org. Chem.* **1972**, *37*, 3551.
[675]For examples, see Deghenghi; Engel *J. Am. Chem. Soc.* **1960**, *82*, 3201; House; Trost *J. Org. Chem.* **1965**, *30*, 2502.
[676]For example, see Hopff; Osman *Tetrahedron* **1968**, *24*, 2205, 3887; Mondal; van der Meer; German; Heikens *Tetrahedron* **1974**, *30*, 4205.
[677]Henry *J. Am. Chem. Soc.* **1971**, *93*, 3853, *Acc. Chem. Res.* **1973**, *6*, 16-24.
[678]For example, see Czarnik *Tetrahedron Lett.* **1984**, *25*, 4875. For a list of references, see Ref. 508, pp. 989-990.
[679]Parker, *Adv. Org. Chem.* **1965**, *5*, 1-46, p. 37; Alvarez; Watt *J. Org. Chem.* **1968**, *33*, 2143; Mehta *Synthesis* **1972**, 262; Shaw; Kunerth *J. Org. Chem.* **1974**, *39*, 1968; Larock *J. Org. Chem.* **1974**, *39*, 3721; Pfeffer; Silbert *J. Org. Chem.* **1976**, *41*, 1373.
[680]For reviews of phase transfer catalysis of this reaction, see Starks; Liotta, Ref. 404, pp. 140-155; Weber; Gokel *Phase Transfer Catalysis in Organic Synthesis*, Ref. 404, pp. 85-95.
[681]For an alternative method for phenacyl halides, see Clark; Miller *Tetrahedron Lett.* **1977**, 599.
[682]Bram; Loupy; Majdoub; Gutierrez; Ruiz-Hitzky *Tetrahedron* **1990**, *46*, 5167. See also Barry; Bram; Decodts; Loupy; Orange; Petit; Sansoulet *Synthesis* **1985**, 40; Arrad; Sasson *J. Am. Chem. Soc.* **1988**, *110*, 185; Dakka; Sasson; Khawaled; Bram; Loupy *J. Chem. Soc., Chem. Commun.* **1991**, 853.

have been esterified by treatment with primary or secondary halides in benzene in the presence of DBU (p. 1023).[683] In most cases good yields of esters can be obtained only with one of these methods. Without phase transfer catalysts and in protic solvents, the reaction is useful only for fairly active R, such as benzylic, allylic, etc. (S$_N$1 mechanism), but not for tertiary alkyl, since elimination occurs instead.[684] Sodium salts are often used, but potassium, silver, cesium,[685] and substituted ammonium salts have also been used. Lactones can be prepared from halo acids by treatment with base (see **0-22**). This has most often been accomplished with γ and δ lactones, but macrocyclic lactones (e.g., 11 to 17 members) have also been prepared in this way.[686]

Cooper(I) carboxylates give esters with primary (including neopentyl without rearrangement), secondary, and tertiary alkyl, allylic, and vinylic halides.[687] A simple S$_N$ mechanism is obviously precluded in this case. Vinylic halides can be converted to vinylic acetates by treatment with sodium acetate if palladium(II) chloride is present.[688]

A carboxylic acid (not the salt) can be the nucleophile if F$^-$ is present.[689] Dihalides have been converted to diesters by this method.[689] A COOH group can be conveniently protected by reaction of its ion with a phenacyl bromide (ArCOCH$_2$Br).[584] The resulting ester is easily cleaved when desired with zinc and acetic acid. Dialkyl carbonates can be prepared without phosgene (see **0-20**) by phase-transfer catalyzed treatment of primary akyl halides with dry KHCO$_3$ and K$_2$CO$_3$.[690]

Other leaving groups can also be replaced by OCOR. Alkyl chlorosulfites (ROSOCl) and other derivatives of sulfuric, sulfonic, and other inorganic acids can be treated with carboxylate ions to give the corresponding esters. The use of dimethyl sulfate[691] or trimethyl phosphate[692] allows sterically hindered COOH groups to be methylated. With certain substrates, carboxylic acids are strong enough nucleophiles for the reaction. Examples of such substrates are trialkyl phosphites P(OR)$_3$[693] and acetals of dimethylformamide.[694]

$$(RO)_2CHNMe_2 + R'COOH \longrightarrow R'COOR + ROH + HCONMe_2$$

This is an S$_N$2 process, since inversion is found at R. Another good leaving group is NTs$_2$; ditosylamines react quite well with acetate ion in dipolar aprotic solvents:[695] RNTs$_2$ + OAc$^-$ → ROAc. Ordinary primary amines have been converted to acetates and benzoates by the Katritzky pyrylium–pyridinium method (p. 354).[696] Quaternary ammonium salts can be cleaved by heating with AcO$^-$ in an aprotic solvent.[697] Oxonium ions can also be used as substrates:[698] R$_3$O$^+$ + R'COO$^-$ → R'COOR + R$_2$O.

[683]Ono; Yamada; Saito; Tanaka; Kaji *Bull. Chem. Soc. Jpn.* **1978**, *51*, 2401; Mal *Synth. Commun.* **1986**, *16*, 331.

[684]See, however, Moore; Foglia; McGahan *J. Org. Chem.* **1979**, *44*, 2425.

[685]See Kruizinga; Strijtveen; Kellogg *J. Org. Chem.* **1981**, *46*, 4321; Dijkstra; Kruizinga; Kellogg *J. Org. Chem.* **1987**, *52*, 4230.

[686]For example, see Galli; Mandolini *Org. Synth. VI*, 698; Kruizinga; Kellogg *J. Chem. Soc. Chem. Commun.* **1979**, 286; *J. Am. Chem. Soc.* **1981**, *103*, 5183; Regen; Kimura *J. Am. Chem. Soc.* **1982**, *104*, 2064; Kimura; Regen *J. Org. Chem.* **1983**, *48*, 1533.

[687]Lewin; Goldberg *Tetrahedron Lett.* **1972**, 491; Klumpp; Bos; Schakel; Schmitz; Vrielink *Tetrahedron Lett.* **1975**, 3429.

[688]Kohll; van Helden *Recl. Trav. Chim. Pays-Bas* **1968**, *87*, 481; Volger *Recl. Trav. Chim. Pays-Bas* **1968**, *87*, 501; Yamaji; Fujiwara; Asano; Teranishi *Bull. Chem. Soc. Jpn.* **1973**, *46*, 90.

[689]Clark; Emsley; Hoyte *J. Chem. Soc. Perkin Trans. 1* **1977**, 1091. See also Barluenga; Alonso-Cires; Campos; Asensio *Synthesis* **1983**, 649.

[690]Lissel; Dehmlow *Chem. Ber.* **1981**, *114*, 1210.

[691]Grundy; James; Pattenden *Tetrahedron Lett.* **1972**, 757.

[692]Harris; Patel *Chem. Ind. (London)* **1973**, 1002.

[693]Szmuszkovicz *Org. Prep. Proceed. Int.* **1972**, *4*, 51.

[694]Vorbrüggen *Angew. Chem. Int. Ed. Engl.* **1963**, *2*, 211 [*Angew. Chem. 75*, 296]; Brechbühler; Büchi; Hatz; Schreiber; Eschenmoser *Angew. Chem. Int. Ed. Engl.* **1963**, *2*, 212 [*Angew. Chem. 75*, 296].

[695]Andersen; Uh *Synth. Commun.* **1972**, *2*, 297; Curtis; Schwartz; Hartman; Pick; Kolar; Baumgarten *Tetrahedron Lett* **1977**, 1969.

[696]See Katritzky; Gruntz; Kenny; Rezende; Sheikh *J. Chem. Soc., Perkin Trans 1* **1979**, 430.

[697]Wilson; Joule *Tetrahedron* **1968**, *24*, 5493.

[698]Raber; Gariano; Brod; Gariano; Guida; Guida; Herbst *J. Org. Chem.* **1979**, *44*, 1149.

In a variation of this reaction, alkyl halides can be converted to carbamates, by treatment with a secondary amine and K_2CO_3 under phase transfer conditions.[699]

$$RX + R_2'NH + K_2CO_3 \xrightarrow{Bu_4NH^+ HSO_4^-} R-O-\underset{\underset{O}{\|}}{C}-NR_2'$$

OS **II**, 5; **III**, 650; **IV**, 582; **V**, 580; **VI**, 273, 576, 698.

0-25 Cleavage of Ethers with Acetic Anhydride
Acyloxy-de-alkoxylation

$$R-O-R' + Ac_2O \xrightarrow{FeCl_3} ROAc + R'OAc$$

Dialkyl ethers can be cleaved by treatment with anhydrous ferric chloride in acetic anhydride.[700] In this reaction both R groups are converted to acetates. Yields are moderate to high. Ethers can also be cleaved by the mixed anhydride acetyl tosylate:[701]

$$R_2O + CH_3-\underset{\underset{O}{\|}}{C}-OTs \longrightarrow RO-\underset{\underset{O}{\|}}{C}-CH_3 + ROTs$$

Epoxides give β-hydroxyalkyl carboxylates when treated with a carboxylic acid or a carboxylate ion and a suitable catalyst.[702]
OS **67**, 114.

0-26 Alkylation of Carboxylic Acids with Diazo Compounds
Hydro,acyloxy-de-diazo-bisubstitution

$$R_2CN_2 + R'COOH \longrightarrow R'COOCHR_2$$

Carboxylic acids can be converted to esters with diazo compounds in a reaction essentially the same as **0-15.** In contrast to alcohols, carboxylic acids undergo the reaction quite well at room temperature, since the reactivity of the reagent increases with acidity. The reaction is used where high yields are important or where the acid is sensitive to higher temperatures. Because of availability, the diazo compounds most often used are diazomethane[593] (for methyl esters)

$$CH_2N_2 + RCOOH \longrightarrow RCOOCH_3$$

and diazo ketones. The mechanism is as shown in **0-15.**
OS **V**, 797.

F. Attack by OCOR at an Acyl Carbon

0-27 Acylation of Carboxylic Acids with Acyl Halides
Acyloxy-de-halogenation

$$RCOCl + R'COO^- \longrightarrow RCOOCOR'$$

---

[699]Gómez-Parra; Sánchez; Torres *Synthesis* **1985**, 282, *J. Chem. Soc., Perkin Trans. 2* **1987**, 695. For another method, with lower yields, see Yoshida; Ishii; Yamashita *Chem. Lett.* **1984**, 1571.
[700]Ganem; Small *J. Org. Chem.* **1974**, *39*, 3728.
[701]Karger; Mazur *J. Am. Chem. Soc.* **1968**, *90*, 3878. See also Coffi-Nketsia; Kergomard; Tautou *Bull. Soc. Chim. Fr.* **1967**, 2788.
[702]See Otera; Matsuzaki *Synthesis* **1986,** 1019; Deardorff; Myles *Org. Synth. 67*, 114.

Unsymmetrical as well as symmetrical anhydrides are often prepared by the treatment of an acyl halide with a carboxylic acid salt. If a metallic salt is used, Na^+, K^+, or Ag^+ are the most common cations, but more often pyridine or another tertiary amine is added to the free acid and the salt thus formed is treated with the acyl halide. Mixed formic anhydrides are prepared from sodium formate and an aryl halide, by use of a solid-phase copolymer of pyridine-1-oxide.[703] Symmetrical anhydrides can be prepared by reaction of the acyl halide with aqueous NaOH or $NaHCO_3$ under phase transfer conditions.[704]

OS **III**, 28, 422, 488; **IV**, 285; **VI**, 8, 910; **66**, 132. See also OS **VI**, 418.

0-28   Acylation of Carboxylic Acids with Carboxylic Acids
     Acyloxy-de-hydroxylation

$$2RCOOH \xrightleftharpoons{P_2O_5} (RCO)_2O + H_2O$$

Anhydrides can be formed from two molecules of an ordinary carboxylic acid only if a dehydrating agent is present so that the equilibrium can be driven to the right. Common dehydrating agents[705] are acetic anhydride, trifluoroacetic anhydride, dicyclohexylcarbodi-imide,[706] methoxyacetylene,[707] and P_2O_5. Among other reagents used have been trimethyl-silylethoxyacetylene $Me_3SiC{\equiv}COEt$,[708] tetracyanoethylene and a base,[709] 1,1,1-trichloro-3,3,3-trifluoroacetone and pyridine,[710] diphenyl phosphorochloridate $(PhO)_2POCl$,[711] and phenyl N-phenylphosphoramidochloridate $(PhO)(PhNH)POCl$.[711] The method is very poor for the formation of mixed anhydrides, which in any case generally undergo disproportion-ation to the two simple anhydrides when they are heated. However, simple heating of dicarboxylic acids does give cyclic anhydrides, provided that the ring formed contains five, six, or seven members, e.g.,

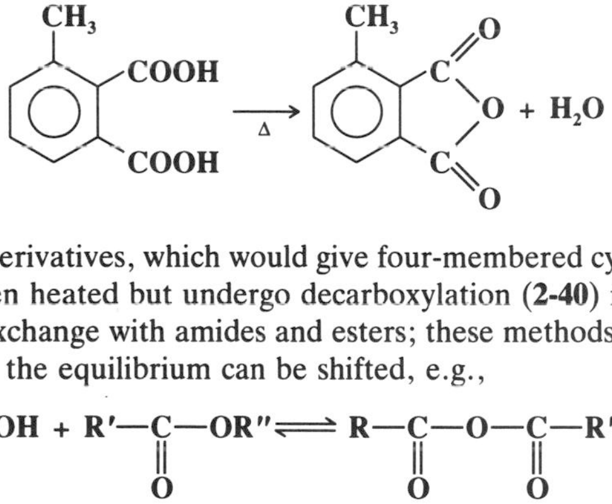

Malonic acid and its derivatives, which would give four-membered cyclic anhydrides, do not give this reaction when heated but undergo decarboxylation (**2-40**) instead.

    Carboxylic acids exchange with amides and esters; these methods are sometimes used to prepare anhydrides if the equilibrium can be shifted, e.g.,

[703]Fife; Zhang *J. Org. Chem.* **1986**, *51*, 3744. See also Fife; Zhang *Tetrahedron Lett.* **1986**, *27*, 4933, 4937. For a review of acetic formic anhydride see Strazzolini; Giumanini; Cauci *Tetrahedron* **1990**, *46* 1081-1118.

[704]Plusquellec; Roulleau; Lefeuvre; Brown *Tetrahedron* **1988**, *44*, 2471; Wang; Hu; Cui *J. Chem. Res. (S)* **1990**, 84.

[705]For lists of other dehydrating agents with references, see Ref. 508, pp. 965-966; Ogliaruso; Wolfe, in Patai, Ref. 638, pt. 1, pp. 437-438.

[706]For example, see Schüssler; Zahn *Chem. Ber.* **1962**, *95*, 1076; Rammler; Khorana *J. Am. Chem. Soc.* **1963**, *85*, 1997. See also Hata; Tajima; Mukaiyama *Bull. Chem. Soc. Jpn.* **1968**, *41*, 2746.

[707]See, for example, Eglinton; Jones; Shaw; Whiting *J. Chem. Soc.* **1954**, 1860; Arens; Doornbos *Recl. Trav. Chim. Pays-Bas* **1955**, *74*, 79.

[708]Kita; Akai; Yoshigi; Nakajima; Yasuda; Tamura *Tetrahedron Lett.* **1984**, *25*, 6027.

[709]Voisin; Gastambide *Tetrahedron Lett.* **1985**, *26*, 1503.

[710]Abdel-Baky; Giese *J. Org. Chem.* **1986**, *51*, 3390.

[711]Mestres; Palomo *Synthesis* **1981**, 218.

Enolic esters are especially good for this purpose, because the equilibrium is shifted by formation of the ketone.

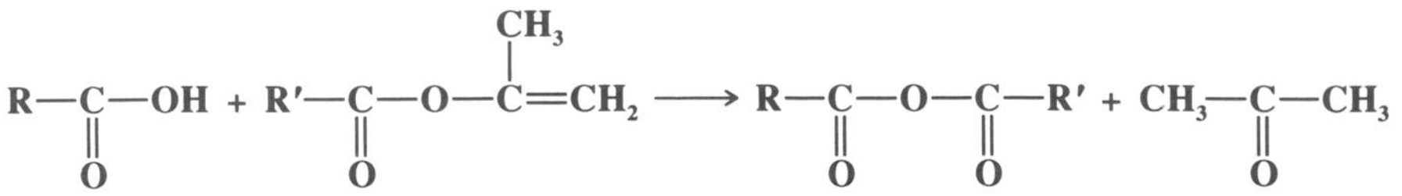

$$R—\underset{\underset{O}{\|}}{C}—OH + R'—\underset{\underset{O}{\|}}{C}—O—\underset{\overset{\overset{CH_3}{|}}{}}{C}{=}CH_2 \longrightarrow R—\underset{\underset{O}{\|}}{C}—O—\underset{\underset{O}{\|}}{C}—R' + CH_3—\underset{\underset{O}{\|}}{C}—CH_3$$

Carboxylic acids also exchange with anhydrides; indeed, this is how acetic anhydride acts as a dehydrating agent in this reaction.

Anhydrides can be formed from certain carboxylic acid salts; for example, by treatment of trimethylammonium carboxylates with phosgene:[712]

$$2RCOO^{\ominus} \overset{\oplus}{N}HEt_3 \xrightarrow{COCl_2} RCOOCOR + 2\overset{\oplus}{N}HEt_3 \ Cl^- + CO_2$$

or of thallium(I) carboxylates with thionyl chloride,[624] or of sodium carboxylates with CCl_4 and a catalyst such as CuCl or $FeCl_2$.[713]

OS **I**, 91, 410; **II**, 194, 368, 560; **III**, 164, 449; **IV**, 242, 630, 790; **V**, 8, 822. Also see OS **VI**, 757; **VII**, 506.

G. Other Oxygen Nucleophiles

0-29 Formation of Oxonium Salts

$$RX + R_2O \xrightarrow{AgBF_4} R_3O^+ \ BF_4^- + AgX \quad \textbf{Dialkyloxonio-de-halogenation}$$

$$RX + R_2'CO \xrightarrow{AgBF_4} R_2'C{=}\overset{\oplus}{O}—R \ BF_4^- + AgX$$

Alkyl halides can be alkylated by ethers or ketones to give oxonium salts, if a very weak, negatively charged nucleophile is present to serve as a counterion and a Lewis acid is present to combine with X^-.[714] A typical procedure consists of treating the halide with the ether or the ketone in the presence of $AgBF_4$ or $AgSbF_6$. The Ag^+ serves to remove X^- and the BF_4^- or SbF_6^- acts as the counterion. Another method involves treatment of the halide with a complex formed between the oxygen compound and a Lewis acid, e.g., $R_2O—BF_3 + RF \rightarrow R_3O^+ \ BF_4^-$, though this method is most satisfactory when the oxygen and halogen atoms are in the same molecule so that a cyclic oxonium ion is obtained. Ethers and oxonium ions also undergo exchange reactions:

$$2R_3O^+ \ BF_4^- + 3R_2'O \rightleftharpoons 2R_3'O^+ \ BF_4^- + 3R_2O$$

OS **V**, 1080, 1096, 1099; **VI**, 1019.

0-30 Reaction of Halides with Oxide Ion
Oxy-de-dihalo-*aggre*-substitution

$$2RX + O^{2-} \longrightarrow ROR + 2X^-$$

[712]Rinderknecht; Ma *Helv. Chim. Acta* **1964**, *47*, 152. See also Nangia; Chandrasekaran *J. Chem. Res., (S)* **1984**, 100.
[713]Weiss; Havelka; Nefedov *Bull. Acad. Sci. USSR, Div. Chem. Sci.* **1978**, *27*, 193.
[714]Meerwein; Hederich; Wunderlich *Arch. Pharm.* **1958**, *291/63*, 541. For a review, see Perst, Ref. 84, pp. 22-39.

Alkyl halides can be converted to symmetrical ethers by treatment with oxide ion generated in situ by a reaction between an organotin oxide and fluoride ion in the presence of a quaternary ammonium iodide or a crown ether.[715]

$$R_3'Sn-O-SnR_3' + 2F^- \longrightarrow 2R_3'SnF + O^{2-}$$

The procedure was used for R = primary alkyl and benzylic. Some unsymmetrical ethers ROR″ were also made, by using R″OSnR$_3'$ instead of R$_3'$SnOSnR$_3'$.

0-31   Preparation of Peroxides and Hydroperoxides
       Hydroperoxy-de-halogenation

$$RX + OOH^- \longrightarrow ROOH$$

Hydroperoxides can be prepared by treatment of alkyl halides, esters of sulfuric or sulfonic acids, or alcohols with hydrogen peroxide in basic solution, where it is actually HO_2^-.[716] Sodium peroxide is similarly used to prepare dialkyl peroxides ($2RX + Na_2O_2 \to ROOR$). Another method, which gives primary, secondary, or tertiary hydroperoxides and peroxides, involves treatment of the halide with H_2O_2 or a peroxide in the presence of silver trifluoroacetate.[717] Peroxides can also be prepared[718] by treatment of alkyl bromides or tosylates with potassium superoxide KO_2 in the presence of crown ethers (though alcohols may be side products[719]) and by the reaction between alkyl triflates and germanium or tin peroxide.[720]

Diacyl peroxides and acyl hydroperoxides can similarly be prepared[721] from acyl halides or anhydrides

$$\text{PhCCl} + \text{H}_2\text{O}_2 \xrightarrow{\text{OH}^-} \text{PhCOOCPh}$$
$$\overset{\|}{\text{O}} \qquad\qquad \overset{\|}{\text{O}}\ \overset{\|}{\text{O}}$$

$$(\text{CH}_3\text{C})_2\text{O} + \text{H}_2\text{O}_2 \xrightarrow{\text{H}_2\text{SO}_4} \text{CH}_3\text{COOH}$$
$$\quad\overset{\|}{\text{O}} \qquad\qquad\qquad \overset{\|}{\text{O}}$$

and from carboxylic acids.[722] Diacyl peroxides can also be prepared by the treatment of carboxylic acids with hydrogen peroxide in the presence of dicyclohexylcarbodiimide,[723] H_2SO_4, methanesulfonic acid, or some other dehydrating agent. Mixed alkyl–acyl peroxides (peresters) can be made from acyl halides and hydroperoxides.

$$\text{R}-\overset{\|}{\underset{\text{O}}{\text{C}}}-\text{X} + \text{R}'\text{OOH} \longrightarrow \text{R}-\overset{\|}{\underset{\text{O}}{\text{C}}}-\text{OOR}'$$

OS **III**, 619, 649; **V**, 805, 904; **VI**, 276.

[715]Harpp; Gingras *J. Am. Chem. Soc.* **1988**, *110*, 7737.
[716]For a review, see Hiatt, in Swern *Organic Peroxides*, vol. 2, Wiley: New York, 1971, pp. 1-151. For a review of hydrogen peroxide, see Pandiarajan, in Pizey, Ref. 593, vol. 6, 1985, pp. 60-155.
[717]Cookson; Davies; Roberts *J. Chem. Soc., Chem. Commun.* **1976**, 1022. For another preparation of unsymmetrical peroxides, see Bourgeois; Montaudon; Maillard *Synthesis* **1989**, 700.
[718]Johnson; Nidy; Merritt *J. Am. Chem. Soc.* **1978**, *100*, 7960.
[719]Alcohols have also been reported to be the main products: San Filippo; Chern; Valentine *J. Org. Chem.* **1975**, *40*, 1678; Corey; Nicolaou; Shibasaki; Machida; Shiner *Tetrahedron Lett.* **1975**, 3183.
[720]Salomon; Salomon *J. Am. Chem. Soc.* **1979**, *101*, 4290.
[721]For a review of the synthesis and reactions of acyl peroxides and peresters, see Bouillon; Lick; Schank, in Patai, *The Chemistry of Peroxides*; Wiley: New York, 1983, pp. 279-309. For a review of the synthesis of acyl peroxides, see Hiatt, Ref. 716, vol. 2, pp. 799-929.
[722]See Silbert; Siegel; Swern *J. Org. Chem.* **1962**, *27*, 1336.
[723]Greene; Kazan *J. Org. Chem.* **1963**, *28*, 2168.

0-32 Preparation of Inorganic Esters
Nitrosooxy-de-hydroxylation, etc.

$$ROH + HONO \xrightarrow{H^+} RONO$$

$$ROH + HONO_2 \xrightarrow{H^+} RONO_2$$

$$ROH + SOCl_2 \longrightarrow ROSOOR$$

$$ROH + POCl_3 \longrightarrow PO(OR)_3$$

$$ROH + SO_3 \longrightarrow ROSO_2OH$$

$$ROH + (CF_3SO_2)_2O \longrightarrow ROSO_2CF_3$$

The above transformations show a few of the many inorganic esters that can be prepared by attack of an inorganic acid or, better, its acid halide or anhydride, on an alcohol.[724] Although for convenience all these similar reactions are grouped together, these are not all nucleophilic substitutions at R. The other possible pathway is nucleophilic substitution at the inorganic central atom:[725]

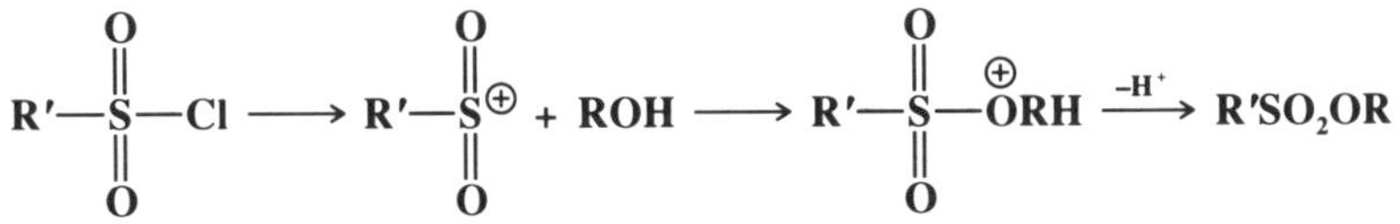

$$R'\!-\!\overset{\overset{\displaystyle O}{\|}}{\underset{\underset{\displaystyle O}{\|}}{S}}\!-\!Cl \longrightarrow R'\!-\!\overset{\overset{\displaystyle O}{\|}}{\underset{\underset{\displaystyle O}{\|}}{S}}{}^{\oplus} + ROH \longrightarrow R'\!-\!\overset{\overset{\displaystyle O}{\|}}{\underset{\underset{\displaystyle O}{\|}}{S}}\!-\!\overset{\oplus}{O}RH \xrightarrow{-H^+} R'SO_2OR$$

or a corresponding S$_N$2 type (see p. 496). In such cases there is no alkyl-O cleavage. Mono esters of sulfuric acid (alkylsulfuric acids), which are important industrially because their salts are used as detergents, can be prepared by treating alcohols with SO$_3$, H$_2$SO$_4$, Cl-SO$_2$OH, or SO$_3$ complexes.[726] Alcohols are often converted to silyl ethers, for protection and other synthetic purposes: ROH + Me$_3$CSiCl → ROSiMe$_3$.[727] Alkyl nitrites[728] can be conveniently prepared by an exchange reaction ROH + R'ONO → RONO + R'OH, where R = t-Bu.[729] Primary amines can be converted to alkyl nitrates (RNH$_2$ → RONO$_2$) by treatment with N$_2$O$_4$ at −78°C in the presence of an excess of amidine base.[730]

Alkyl halides are often used as substrates instead of alcohols. In such cases the *salt* of the inorganic acid is usually used and the mechanism is nucleophilic substitution at the carbon atom. An important example is the treatment of alkyl halides with silver nitrate to form alkyl nitrates. This is used as a test for alkyl halides. In some cases there is competition from the central atom. Thus nitrite ion is an ambident nucleophile that can give nitrites or nitro compounds (see **0-60**).[731] Dialkyl or aryl alkyl ethers can be cleaved with anhydrous sulfonic acids.[732]

$$ROR' + R''SO_2OH \longrightarrow ROSO_2R'' + R'OH$$

[724]For a review, see Ref. 575, pp. 481-497.
[725]For an example involving nitrite formation, see Aldred; Williams; Garley *J. Chem. Soc., Perkin Trans. 2* **1982,** 777.
[726]For a review, see Sandler; Karo, *Organic Functional Group Preparations*, 2d ed., vol. 3; Academic Press: New York, 1989, pp. 129-151.
[727]For a review, see Lalonde; Chan *Synthesis* **1985,** 817-845.
[728]For a review of alkyl nitrites, see Williams *Nitrosation*; Cambridge University Press: Cambridge, 1988, pp. 150-172.
[729]Doyle; Terpstra; Pickering; LePoire *J. Org. Chem.* **1983,** *48,* 3379. For a review of the nitrosation of alcohols, see Ref. 728, pp. 150-156.
[730]Barton; Narang *J. Chem. Soc., Perkin Trans.1* **1977,** 1114.
[731]For a review of formation of nitrates from alkyl halides, see Boguslavskaya; Chuvatkin; Kartashov *Russ. Chem. Rev.* **1988,** *57,* 760-775.
[732]Klamann; Weyerstahl *Chem. Ber.* **1965,** *98,* 2070.

R″ may be alkyl or aryl. For dialkyl ethers, the reaction does not end as indicated above, since R′OH is rapidly converted to R′OR′ by the sulfonic acid (reaction **0-16**), which in turn is further cleaved to R′OSO$_2$R″ so that the product is a mixture of the two sulfonates. For aryl alkyl ethers, cleavage always takes place to give the phenol, which is not converted to the aryl ether under these conditions. Ethers can also be cleaved in a similar manner by mixed anhydrides of sulfonic and carboxylic acids[733] (prepared as in **0-33**). β-Hydroxyalkyl perchlorates[734] and sulfonates can be obtained from epoxides.[735] Epoxides and oxetanes give dinitrates when treated with N$_2$O$_5$,[736] e.g.,

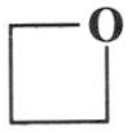

$$\boxed{}\!\!-\!\!O \quad + \; N_2O_5 \;\longrightarrow\; NO_2O\!-\!CH_2CH_2CH_2\!-\!ONO_2$$

oxetane

Aziridines and azetidines react similarly, giving nitramine nitrates; e.g., N-butylazetidine gave NO$_2$OCH$_2$CH$_2$CH$_2$N(Bu)NO$_2$.[736]
OS **II**, 106, 108, 109, 112, 204, 412; **III**, 148, 471; **IV**, 955; **V**, 839; **66**, 211; **67**, 1, 13. Also see OS **II**, 111.

0-33 Preparation of Mixed Organic–Inorganic Anhydrides
Nitrooxy-de-acyloxy-substitution

$$(RCO)_2O + HONO_2 \;\longrightarrow\; RCOONO_2$$

Mixed organic–inorganic anhydrides are seldom isolated, though they are often intermediates when acylation is carried out with acid derivatives catalyzed by inorganic acids. Sulfuric, perchloric, phosphoric, and other acids form similar anhydrides, most of which are unstable or not easily obtained because the equilibrium lies in the wrong direction. These intermediates are formed from amides, carboxylic acids, and esters, as well as anhydrides. Organic anhydrides of phosphoric acid are more stable than most others and, for example, RCOOPO(OH)$_2$ can be prepared in the form of its salts.[737] Mixed anhydrides of carboxylic and sulfonic acids (RCOOSO$_2$R′) are obtained in high yields by treatment of sulfonic acids with acyl halides or (less preferred) anhydrides.[738]
OS **I**, 495; **VI**, 207; **VII**, 81.

0-34 Alkylation of Oximes

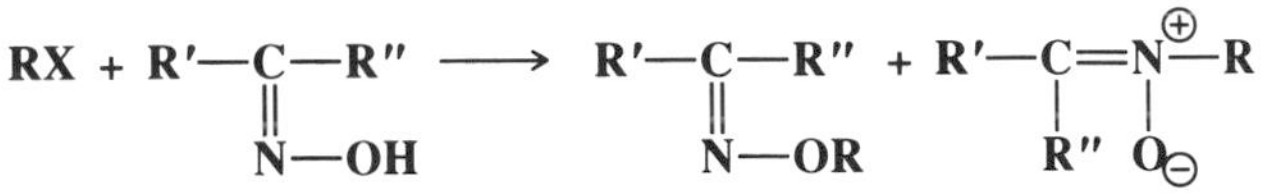

$$RX + R'\!-\!\underset{\underset{N-OH}{\|}}{C}\!-\!R'' \;\longrightarrow\; R'\!-\!\underset{\underset{N-OR}{\|}}{C}\!-\!R'' \;+\; R'\!-\!\underset{\underset{R''\;\;O_{\ominus}}{|}}{C}\!\!=\!\!\overset{\oplus}{N}\!-\!R$$

A nitrone

Oximes can be alkylated by alkyl halides or sulfates. N-Alkylation is a side reaction, yielding a nitrone.[739] The relative yield of oxime ether and nitrone depends on the nature of the

[733]Karger; Mazur *J. Org. Chem.* **1971**, *36*, 532, 540.
[734]For a review of the synthesis and reactions of organic perchlorates, see Zefirov; Zhdankin; Koz'min *Russ. Chem. Rev.* **1988**, *57*, 1041-1053.
[735]Zefirov; Kirin; Yur'eva; Zhdankin *J. Org. Chem. USSR* **1987**, *23*, 1264.
[736]Golding; Millar; Paul; Richards *Tetrahedron Lett.* **1988**, *29*, 2731, 2735.
[737]Avison *J. Chem. Soc.* **1955**, 732.
[738]Karger; Mazur *J. Org. Chem.* **1971**, *36*, 528.
[739]For a review of nitrones, see Torssell *Nitrile Oxides, Nitrones, and Nitronates in Organic Synthesis*; VCH: New York, 1988, pp. 75-93.

reagents, including the configuration of the oxime, and on the reaction conditions.[740] For example, *anti*-benzaldoximes give nitrones, while the syn isomers give oxime ethers.[741]

OS **III**, 172; **V**, 1031. Also see OS **V**, 269; **VI**, 199.

Sulfur Nucleophiles

Sulfur compounds[742] are better nucleophiles than their oxygen analogs (p. 349), so in most cases these reactions take place faster and more smoothly than the corresponding reactions with oxygen nucleophiles. There is evidence that some of these reactions take place by SET mechanisms.[743]

0-35 Attack by SH at an Alkyl Carbon. Formation of Thiols[744]
Mercapto-de-halogenation

$$RX + H_2S \longrightarrow RSH_2^+ \longrightarrow RSH + H^+$$
$$RX + HS^- \longrightarrow RSH$$

Sodium sulfhydride (NaSH) is a much better reagent for the formation of thiols (mercaptans) from alkyl halides than H_2S and is used much more often. It is easily prepared by bubbling H_2S into an alkaline solution. The reaction is most useful for primary halides. Secondary substrates give much lower yields, and the reaction fails completely for tertiary halides because elimination predominates. Sulfuric and sulfonic esters can be used instead of halides. Thioethers (RSR) are often side products.[745] The conversion can also be accomplished under neutral conditions by treatment of a primary halide with F^- and a tin sulfide such as $Ph_3SnSSnPh_3$.[746] An indirect method for the conversion of an alkyl halide to a thiol consists of treatment with thiourea to give an isothiuronium salt, which with alkali or a high-molecular-weight amine is cleaved to the thiol:

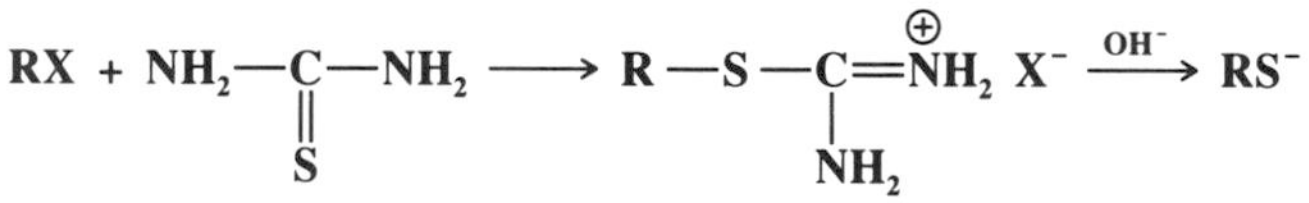

$$RX + NH_2{-}\overset{\displaystyle}{\underset{\displaystyle S}{C}}{-}NH_2 \longrightarrow R{-}S{-}\overset{\displaystyle \oplus}{\underset{\displaystyle NH_2}{C}}{=}NH_2 \; X^- \xrightarrow{OH^-} RS^-$$

Another indirect method is hydrolysis of Bunte salts (see **0-39**).

Thiols have also been prepared from alcohols. One method involves treatment with H_2S and a catalyst such as Al_2O_3,[747] but this is limited to primary alcohols. Another method involves treatment with Lawesson's reagent (see **6-11**).[748] Still another method, involving the use of a fluoropyridinium salt and sodium N,N-dimethylthiocarbamate, can be applied

[740]For a review, see Reutov; Beletskaya; Kurts, Ref. 422, pp. 262-272.

[741]Buehler *J. Org. Chem.* **1967**, *32*, 261.

[742]For monographs on sulfur compounds, see Bernardi; Csizmadia; Mangini *Organic Sulfur Chemistry*; Elsevier: New York, 1985; Oae *Organic Chemistry of Sulfur*; Plenum: New York, 1977. For monographs on selenium compounds, see Krief; Hevesi *Organoselenium Chemistry I*; Springer: New York, 1988; Liotta *Organoselenium Chemistry*; Wiley: New York, 1987.

[743]See Ashby; Park; Goel; Su *J. Org. Chem.* **1985**, *50*, 5184.

[744]For a review, see Wardell, in Patai *The Chemistry of the Thiol Group*, pt. 1; Wiley: New York, 1974, pp. 179-211.

[745]For a method of avoiding thioether formation, see Vasil'tsov; Trofimov; Amosova *J. Org. Chem. USSR* **1983**, *19*, 1197.

[746]Gingras; Harpp *Tetrahedron Lett.* **1990**, *31*, 1397.

[747]Lucien; Barrault; Guisnet; Maurel *Nouv. J. Chim.* **1979**, *3*, 15.

[748]Nishio *J. Chem. Soc., Chem. Commun.* **1989**, 205.

to primary, secondary, allylic, and benzylic alcohols.[749] When epoxides are substrates, the products are β-hydroxy thiols:[750]

$$-\overset{|}{\underset{O}{C}}\overset{|}{\underset{}{C}}- \ + \ SH^- \ \xrightarrow{\ H_2O\ } \ -\overset{|}{\underset{OH}{C}}-\overset{|}{\underset{SH}{C}}-$$

Tertiary nitro compounds give thiols ($RNO_2 \rightarrow RSH$) when treated with sulfur and sodium sulfide, followed by amalgamated aluminum.[751]

OS **III**, 363, 440; **IV**, 401, 491; **V**, 1046; **65**, 50. Also see OS **II**, 345, 411, 573; **IV**, 232; **V**, 223; **VI**, 620.

0-36 Attack by S at an Alkyl Carbon. Formation of Thioethers
Alkylthio-de-halogenation

$$RX \ + \ R'S^- \ \longrightarrow \ RSR'$$

Thioethers (sulfides) can be prepared by treatment of alkyl halides with salts of thiols (thiolate ions).[752] R' may be alkyl or aryl. As in **0-35**, RX cannot be a tertiary halide, and sulfuric and sulfonic esters can be used instead of halides. As in the Williamson reaction (**0-12**), yields are improved by phase-transfer catalysis.[753] Instead of RS^- ions, thiols themselves can be used, if the reaction is run in benzene in the presence of DBU (p. 1023).[754] Neopentyl bromide was converted to Me_3CCH_2SPh in good yield by treatment with PhS^- in liquid NH_3 at $-33°C$ under the influence of light.[755] This probably takes place by an $S_{RN}1$ mechanism (see p. 648). Vinylic sulfides can be prepared by treating vinylic bromides with PhS^- in the presence of a nickel complex,[756] and with R_3SnPh in the presence of $Pd(PPh_3)_4$.[757]

R can be tertiary if an alcohol is the substrate, e.g.,[758]

$$R_3COH \ + \ R_3'CSH \ \xrightarrow{\ H_2SO_4\ } \ R_3CSCR_3'$$

This reaction is analogous to **0-16.** Primary and secondary alcohols can be converted to alkyl aryl sulfides ($ROH \rightarrow RSAr$) in high yields by treatment with Bu_3P and an N-(arylthio)succinimide in benzene.[759] Thioethers RSR' can be prepared from an alcohol ROH and a halide R'Cl by treatment with tetramethylthiourea $Me_2NC(=S)NMe_2$ followed by NaH.[760]

Thiolate ions are also useful for the demethylation of certain ethers,[761] esters, amines, and quaternary ammonium salts. Aryl methyl ethers[762] can be cleaved by heating with EtS^-

[749]Hojo; Yoshino; Mukaiyama *Chem. Lett* **1977**, 133, 437. For another method, see Alper; Sibtain *J. Org. Chem.* **1988**, *53*, 3306.
[750]For a review, see Ref. 744, pp. 246-251.
[751]Kornblum; Widmer *J. Am. Chem. Soc.* **1978**, *100*, 7086.
[752]For a review, see Peach, in Patai, Ref. 744, pt. 2, pp. 721-735.
[753]For a review of the use of phase transfer catalysis to prepare sulfur-containing compounds, see Weber; Gokel *Phase Transfer Catalysis in Organic Synthesis*, Ref. 404, pp. 221-233.
[754]Ono; Miyake; Saito; Kaji *Synthesis* **1980**, 952. See also Ferreira; Comasseto; Braga *Synth. Commun.* **1982**, *12*, 595; Ando; Furuhata; Tsumaki; Sekiguchi *Synth. Commun.* **1982**, *12*, 627.
[755]Pierini; Peñéñory; Rossi *J. Org. Chem.* **1985**, *50*, 2739.
[756]Cristau; Chabaud; Labaudiniere; Christol *J. Org. Chem.* **1986**, *51*, 875.
[757]Carpita; Rossi; Scamuzzi *Tetrahedron Lett.* **1989**, *30*, 2699. For another method, see Ogawa; Hayami; Suzuki *Chem. Lett.* **1989**, 769.
[758]Fehnel; Carmack *J. Am. Chem. Soc.* **1949**, *71*, 84; Cain; Evans; Lee *J. Chem. Soc.* **1962**, 1694.
[759]Walker *Tetrahedron Lett.* **1977**, 4475. See the references in this paper for other methods of converting alcohols to sulfides. See also Cleary *Synth. Commun.* **1989**, *19*, 737.
[760]Fujisaki; Fujiwara; Norisue; Kajigaeshi *Bull. Chem. Soc. Jpn.* **1985**, *58*, 2429.
[761]For a review, see Evers *Chem. Scr.* **1986**, *26*, 585-597.
[762]Certain other sulfur-containing reagents also cleave methyl and other ethers: see Hanessian; Guindon *Tetrahedron Lett.* **1980**, *21*, 2305; Williard; Fryhle *Tetrahedron Lett.* **1980**, *21*, 3731; Node; Nishide; Fuji; Fujita *J. Org. Chem.* **1980**, *45*, 4275. For cleavage with selenium-containing reagents, see Evers; Christiaens *Tetrahedron Lett.* **1983**, *24*, 377. For a review of the cleavage of aryl alkyl ethers, see Tiecco *Synthesis* **1988**, 749-759.

in the dipolar aprotic solvent DMF: $ROAr + EtS^- \rightarrow ArO^- + EtSR$.[763] Carboxylic esters and lactones are cleaved (the lactones give ω-alkylthio carboxylic acids) with a thiol and $AlCl_3$ or $AlBr_3$.[764] Esters and lactones are similarly cleaved in high yield by phenyl selenide ion $PhSe^-$.[765] Allylic sulfides have been prepared by treating allylic carbonates ROCOOMe (R = an allylic group) with a thiol and a Pd(0) catalyst.[766] A good method for the de-methylation of quaternary ammonium salts consists of refluxing them with PhS^- in butan-one:[767]

$$R_3\overset{\oplus}{N}Me + PhS^- \xrightarrow{\text{MeCOEt}} R_3N + PhSMe$$

A methyl group is cleaved more readily than other simple alkyl groups (such as ethyl), though loss of these groups competes, but benzylic and allylic groups cleave even more easily, and this is a useful procedure for the cleavage of benzylic and allylic groups from quaternary ammonium salts, even if methyl groups are also present.[768]

Symmetrical thioethers can also be prepared by treatment of an alkyl halide with sodium sulfide,[769] in a reaction similar to **0-30**.

$$2RX + Na_2S \longrightarrow RSR$$

This reaction can be carried out internally, by treatment of sulfide ions with 1,4- or 1,5-dihalides, to prepare five- and six-membered sulfur-containing heterocyclic rings.

$$\text{R—CH—CH}_2\text{—CH}_2\text{—CH—R'} + S^{2-} \longrightarrow$$
$$\overset{|}{\text{Cl}} \qquad\qquad \overset{|}{\text{Cl}}$$

Certain larger rings have also been closed in this way.[770]

gem-Dihalides can be converted to thioacetals $RCH(SR')_2$,[771] and acetals have been converted to monothioacetals $R_2C(OR')(SR'')$,[772] and to thioacetals.[773]

Selenides and tellurides can be prepared similarly.[774] When epoxides are substrates, β-hydroxy sulfides are obtained in a manner analogous to that mentioned in **0-35**. Epoxides can also be directly converted to episulfides,[775] by treatment with a phosphine sulfide such as Ph_3PS[776] or with thiourea and titanium tetraisopropoxide.[777]

[763]Feutrill; Mirrington *Tetrahedron Lett.* **1970**, 1327, *Aust. J. Chem.* **1972**, 25, 1719, 1731.
[764]Node; Nishide; Ochiai; Fuji; Fujita *J. Org. Chem.* **1981**, 46, 5163.
[765]Scarborough; Smith *Tetrahedron Lett.* **1977**, 4361; Liotta; Santiesteban *Tetrahedron Lett.* **1977**, 4369; Liotta; Sunay; Santiesteban; Markiewicz *J. Org. Chem.* **1981**, 46, 2605; Kong; Chen; Zhou *Synth. Commun.* **1988**, 18, 801.
[766]Trost; Scanlan *Tetrahedron Lett.* **1986**, 27, 4141.
[767]Shamma; Deno; Remar *Tetrahedron Lett.* **1966**, 1375. For alternative procedures, see Hutchins; Dux *J. Org. Chem.* **1973**, 38, 1961; Posner; Ting *Synth. Commun.* **1974**, 4, 355.
[768]Kametani; Kigasawa; Hiiragi; Wagatsuma; Wakisaka *Tetrahedron Lett.* **1969**, 635.
[769]For another reagent, see Harpp; Gingras; Aida; Chan *Synthesis* **1987**, 1122.
[770]See Hammerschmidt; Bieber; Vögtle *Chem. Ber.* **1978**, 111, 2445; Singh; Mehrotra; Regen *Synth. Commun.* **1981**, 11, 409.
[771]See, for example Wähälä; Ojanperä; Häyri; Hase *Synth. Commun.* **1987**, 17, 137.
[772]Masaki; Serizawa; Kaji *Chem. Lett.* **1985**, 1933; Sato; Kobayashi; Gojo; Yoshida; Otera; Nozaki *Chem. Lett.* **1987**, 1661.
[773]Park; Kim *Chem. Lett.* **1989**, 629.
[774]Brandsma; Wijers *Recl. Trav. Chim. Pays-Bas* **1963**, 82, 68; Clarembeau; Krief *Tetrahedron Lett.* **1984**, 25, 3625. For a review of nucleophilic selenium, see Monahan; Brown; Waykole; Liotta, in Liotta, Ref. 742, pp. 207-241.
[775]For a review of episulfide information, see Fokin; Kolomiets *Russ. Chem. Rev.* **1975**, 44, 138-153.
[776]Chan; Finkenbine *J. Am. Chem. Soc.* **1972**, 94, 2880.
[777]Gao; Sharpless *J. Org. Chem.* **1988**, 53, 4114. For other methods, see Calō; Lopez; Marchese; Pesce *J. Chem. Soc., Chem. Commun.* **1975**, 621; Takido; Kobayashi; Itabashi *Synthesis* **1986**, 779; Bouda; Borredon; Delmas; Gaset *Synth. Commun.* **1987**, 17, 943, **1989**, 19, 491.

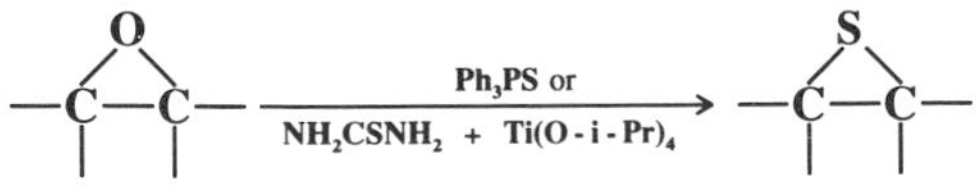

$$-\overset{\displaystyle O}{\underset{|}{C}}\overset{}{\underset{|}{C}}-\xrightarrow[\text{NH}_2\text{CSNH}_2 \; + \; \text{Ti(O-i-Pr)}_4]{\text{Ph}_3\text{PS or}}-\overset{\displaystyle S}{\underset{|}{C}}\overset{}{\underset{|}{C}}-$$

Alkyl halides, treated with thioethers, give sulfonium salts.[778]

$$RI + R_2'S \longrightarrow R_2'SR^+ \; I^-$$

Other leaving groups have also been used for this purpose.[779]
Alcohols, when treated with a thiol acid and zinc iodide, give thiol esters:[780]

$$ROH + R'-\overset{}{\underset{\displaystyle O}{C}}-SH \xrightarrow{\text{ZnI}_2} R-S-\overset{}{\underset{\displaystyle O}{C}}-R'$$

This method is an alternative to **0-37** as a way to prepare thiol esters.

OS **II**, 31, 345, 547, 576; **III**, 332, 751, 763; **IV**, 396, 667, 892, 967; **V**, 562, 780, 1046; **VI**, 5, 31, 268, 364, 403, 482, 556, 601, 683, 704, 737, 833, 859; **VII**, 453; **65**, 150. See also OS **VI**, 776.

0-37 Attack by SH or SR at an Acyl Carbon[781]

$$RCOCl + H_2S \longrightarrow R-\overset{}{\underset{\displaystyle O}{C}}-SH \qquad \textbf{Mercapto-de-halogenation}$$

$$RCOCl + R'SH \longrightarrow R-\overset{}{\underset{\displaystyle O}{C}}-SR' \qquad \textbf{Alkylthio-de-halogenation}$$

Thiol acids and thiol esters[782] can be prepared in this manner, which is analogous to **0-8** and **0-23**. Anhydrides[783] and aryl esters (RCOOAr)[784] are also used as substrates, but the reagents in these cases are usually SH⁻ and SR⁻. Thiol esters can also be prepared by treatment of carboxylic acids with trisalkylthioboranes B(SR)₃,[785] with P₄S₁₀–Ph₃SbO,[786] or with a thiol RSH and either polyphosphate ester or phenyl dichlorophosphate PhOPOCl₂.[787] Esters RCOOR' can be converted to thiol esters RCOSR" by treatment with trimethylsilyl sulfides Me₃SiSR" and AlCl₃.[788]

OS **III**, 116, 599; **IV**, 924, 928; **VII**, 81; **66**, 108.

[778]For a review of the synthesis of sulfonium salts, see Lowe, in Stirling, Ref. 363, pp. 267-312.
[779]See Badet; Jacob; Julia *Tetrahedron* **1981**, *37*, 887; Badet; Julia *Tetrahedron Lett.* **1979**, 1101, and references cited in the latter paper.
[780]Gauthier; Bourdon; Young *Tetrahedron Lett.* **1986**, *27*, 15.
[781]For a review, see Satchell *Q. Rev., Chem. Soc.* **1963**, *17*, 160-203, pp. 182-184.
[782]For a review of these compounds, see Scheithauer; Mayer *Top. Sulfur Chem.* **1979**, *4*, 1-373.
[783]Ahmad; Iqbal *Tetrahedron Lett.* **1986**, *27*, 3791.
[784]Hirabayashi; Mizuta; Mazume *Bull. Chem. Soc. Jpn.* **1965**, *38*, 320.
[785]Pelter; Levitt; Smith; Jones *J. Chem. Soc., Perkin Trans. 1* **1977**, 1672.
[786]Nomura; Miyazaki; Nakano; Matsuda *Chem. Ber.* **1990**, *123*, 2081.
[787]Imamoto; Kodera; Yokoyama *Synthesis* **1982**, 134; Liu; Sabesan *Can. J. Chem.* **1980**, *58*, 2645. For other methods of converting carboxylic acids to thiol esters, see the references given in these papers. See also Dellaria; Nordeen; Swett *Synth. Commun.* **1986**, *16*, 1043.
[788]Mukaiyama; Takeda; Atsumi *Chem. Lett.* **1974**, 187. See also Hatch; Weinreb *J. Org. Chem.* **1977**, *42*, 3960; Cohen; Gapinski *Tetrahedron.* **1978**, 4319.

0-38   Formation of Disulfides
Dithio-de-dihalo-*aggre*-substitution

$$2RX + S_2^{2-} \longrightarrow RSSR + 2X^-$$

Disulfides can be prepared by treatment of alkyl halides with disulfide ions and also indirectly by the reaction of Bunte salts (see **0-39**) with acid solutions of iodide, thiocyanate ion, or thiourea,[789] or by pyrolysis or treatment with hydrogen peroxide. Alkyl halides also give disulfides when refluxed with sulfur and NaOH,[790] and with piperidinium tetrathiotungstate or piperidinium tetrathiomolybdate.[791]

There are no OS references, but a similar preparation of a polysulfide may be found in OS **IV**, 295.

0-39   Formation of Bunte Salts
Sulfonatothio-de-halogenation

$$RX + S_2O_3^{2-} \longrightarrow R{-}S{-}SO_3^- + X^-$$

Primary and secondary but not tertiary alkyl halides are easily converted to Bunte salts ($RSSO_3^-$) by treatment with thiosulfate ion.[792] Bunte salts can be hydrolyzed with acids to give the corresponding thiols[793] or converted to disulfides, tetrasulfides, or pentasulfides.[794]

OS **VI**, 235.

0-40   Alkylation of Sulfinic Acid Salts
Alkylsulfonyl-de-halogenation

$$RX + R'SO_2^- \longrightarrow R{-}SO_2{-}R' + X^-$$

Alkyl halides or alkyl sulfates, treated with the salts of sulfinic acids, give sulfones.[795] Alkyl sulfinates R'SO—OR may be side products.[796] Sulfonic acids themselves can be used, if DBU (p. 1023) is present.[797] Sulfones have also been prepared by treatment of alkyl halides with tosylhydrazide.[798]

OS **IV**, 674. See also OS **VI**, 1016.

0-41   Attack by Sulfite Ion
Sulfonato-de-halogenation

$$RX + SO_3^{2-} \longrightarrow R{-}SO_2O^- + X^-$$

Salts of sulfonic acids can be prepared by treatment of primary or secondary alkyl halides with sulfite ion.[799] Even tertiary halides have been used, though the yields are low. Epoxides treated with bisulfite give β-hydroxy sulfonic acids.[800]

[789]Milligan; Swan *J. Chem. Soc.* **1962**, 2712.
[790]Chorbadjiev; Roumian; Markov *J. Prakt. Chem.* **1977**, *319*, 1036.
[791]Dhar; Chandrasekaran *J. Org. Chem.* **1989**, *54*, 2998.
[792]For a review of Bunte salts, see Distler *Angew. Chem. Int. Ed. Engl.* **1967**, *6*, 544-553 [*Angew. Chem. 79*, 520-529].
[793]Kice *J. Org. Chem.* **1963**, *28*, 957.
[794]Milligan; Saville; Swan *J. Chem. Soc.* **1963**, 3608.
[795]For a review, see Schank, in Patai; Rappoport; Stirling *The Chemistry of Sulphones and Sulphoxides*; Wiley: New York, 1988, pp. 165-231, pp. 177-188.
[796]See, for example Meek; Fowler *J. Org. Chem.* **1968**, *33*, 3422; Kiełbasiński; Żurawiński; Drabowicz; Mikołajczyk *Tetrahedron* **1988**, *44*, 6687.
[797]Biswas; Mal *J. Chem. Res. (S)* **1988**, 308.
[798]Ballini; Marcantoni; Petrini *Tetrahedron* **1989**, *45*, 6791.
[799]For a review, see Gilbert *Sulfonation and Related Reactions*; Wiley: New York, 1965, pp. 136-148, 161-163.
[800]For a discussion, see Yoneda; Griffin; Carlyle *J. Org. Chem.* **1975**, *40*, 375.

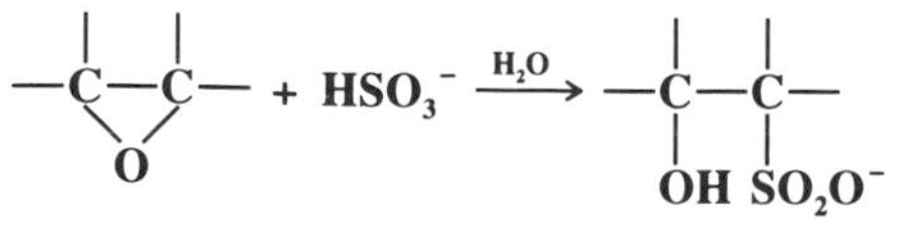

$$-\overset{\mid}{\underset{\diagdown_{O}\diagup}{C}}-\overset{\mid}{C}- + HSO_3^- \xrightarrow{H_2O} -\overset{\mid}{\underset{OH}{C}}-\overset{\mid}{\underset{SO_2O^-}{C}}-$$

OS **II,** 558, 564; **IV,** 529.

0-42　Formation of Alkyl Thiocyanates
Thiocyanato-de-halogenation

$$RX + SCN^- \longrightarrow RSCN + X^-$$

Alkyl halides or sulfuric or sulfonic esters can be heated with sodium or potassium thiocyanate to give alkyl thiocyanates,[801] though the attack by the analogous cyanate ion (**0-62**) gives exclusive N-alkylation. Primary amines can be converted to thiocyanates by the Katritzky pyrylium–pyridinium method (p. 354).[802]

　　OS **II,** 366.

Nitrogen Nucleophiles

A.　Attack by NH$_2$, NHR, or NR$_2$ at an Alkyl Carbon

0-43　Alkylation of Amines
Amino-de-halogenation

$$3RX + NH_3 \longrightarrow R_3N \quad + RX \longrightarrow R_4N^+ \quad X^-$$

$$2RX + R'NH_2 \longrightarrow R_2R'N \quad + RX \longrightarrow R_3R'N^+ \quad X^-$$

$$RX + R'R''NH \longrightarrow RR'R''N \quad + RX \longrightarrow R_2R'R''N^+ \quad X^-$$

$$RX + R'R''R'''N \longrightarrow RR'R''R'''N^+ \quad X^-$$

The reaction between alkyl halides and ammonia or primary amines is not usually a feasible method for the preparation of primary or secondary amines, since they are stronger bases than ammonia and preferentially attack the substrate. However, the reaction is very useful for the preparation of tertiary amines[803] and quaternary ammonium salts. If ammonia is the nucleophile,[804] the three or four alkyl groups on the nitrogen of the product must be identical. If a primary, secondary, or tertiary amine is used, then different alkyl groups can be placed on the same nitrogen atom. The conversion of tertiary amines to quaternary salts is called the *Menshutkin reaction*.[805] It is sometimes possible to use this method for the preparation of a primary amine by the use of a large excess of ammonia or a secondary amine by the use of a large excess of primary amine. However, the limitations of this approach can be seen in the reaction of a saturated solution of ammonia in 90% ethanol with ethyl bromide

[801]For a review of thiocyanates, see Guy, in Patai *The Chemistry of Cyanates and Their Thio Derivatives,* pt. 2; pp. 819-886, Wiley: New York, 1977, pp. 819-886.

[802]Katritzky; Gruntz; Mongelli; Rezende *J. Chem. Soc., Perkin Trans. 1* **1979,** 1953. For the conversion of primary alcohols to thiocyanates, see Tamura; Kawasaki; Adachi; Tanio; Kita *Tetrahedron Lett.* **1977,** 4417.

[803]For reviews of this reaction, see Gibson, in Patai, Ref. 355, pp. 45-55; Spialter; Pappalardo *The Acyclic Aliphatic Tertiary Amines*; Macmillan: New York, 1965, pp. 14-29.

[804]For a review of ammonia as a synthetic reagent, see Jeyaraman, in Pizey, Ref. 593, vol. 5, 1983, pp. 9-83.

[805]For a review of stereoselectivity in this reaction, especially where the tertiary nitrogen is included in a ring, see Bottini, *Sel. Org. Transform.* **1970,** *1,* 89-142. For a review of quaternization of heteroaromatic rings, see Zoltewicz; Deady *Adv. Heterocycl. Chem.* **1978,** *22,* 71-121.

in a 16:1 molar ratio, under which conditions the yield of primary amine was 34.2% (at a 1:1 ratio the yield was 11.3%).[806] One type of substrate that does give reasonable yields of primary amine (provided a large excess of NH_3 is used) are α-halo acids, which are converted to amino acids.

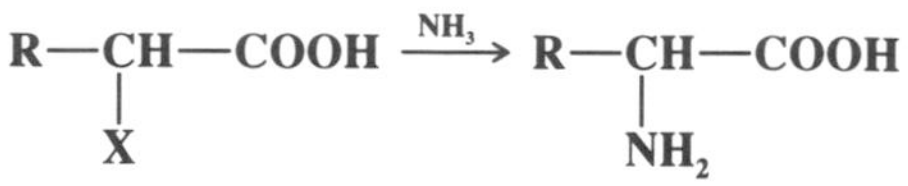

$$R-\underset{\underset{X}{|}}{CH}-COOH \xrightarrow{NH_3} R-\underset{\underset{NH_2}{|}}{CH}-COOH$$

Primary amines can be prepared from alkyl halides by **0-44,** by **0-63,** by **0-61** followed by reduction of the azide (**9-53**), or by the Gabriel synthesis (**0-58**).

The immediate product in any particular step is the protonated amine, which, however, rapidly loses a proton to another molecule of ammonia or amine in an equilibrium process, e.g.,

$$RX + R_2NH \longrightarrow R_3\overset{\oplus}{N}H + R_2NH \rightleftharpoons R_3N + R_2\overset{\oplus}{N}H_2$$

When it is desired to convert a primary or secondary amine directly to the quaternary salt (*exhaustive alkylation*), the rate can be increased by the addition of a nonnucleophilic strong base that serves to remove the proton from $RR'NH_2^+$ or $RR'R''NH^+$ and thus liberates the amine to attack another molecule of RX.[807]

The conjugate bases of ammonia and of primary and secondary amines (NH_2^-, RNH^-, R_2N^-) are sometimes used as nucleophiles,[808] but in most cases offer no advantage over ammonia or amines, since the latter are basic enough. This is in contrast to the analogous methods **0-1, 0-12, 0-35,** and **0-36.** Primary arylamines are easily alkylated, but diaryl- and triarylamines are very poor nucleophiles. However, the reaction has been carried out with diarylamines.[809] Sulfates or sulfonates can be used instead of halides. The reaction can be carried out intramolecularly to give cyclic amines, with three-, five-, and six-membered (but not four-membered) rings being easily prepared. Thus, 4-chloro-1-aminobutane treated with base gives pyrrolidine, and 2-chloroethylamine gives aziridine[810] (analogous to **0-13**):

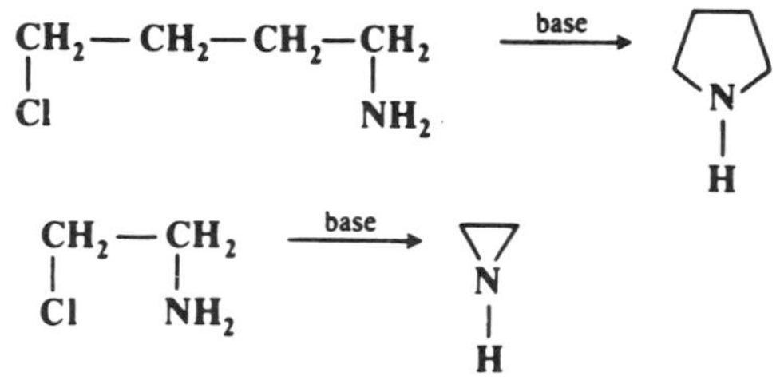

Four-membered cyclic amines (azetidines) have been prepared in a different way:[811]

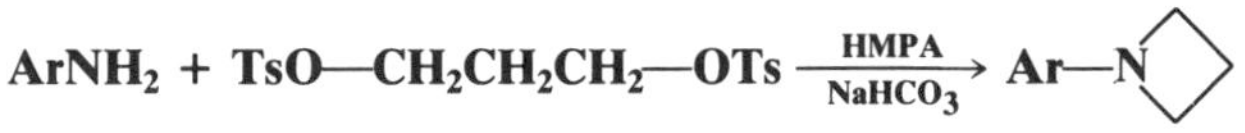

$$ArNH_2 + TsO-CH_2CH_2CH_2-OTs \xrightarrow[NaHCO_3]{HMPA} Ar-N\diamondsuit$$

This reaction was also used to close five-, six-, and seven-membered rings.

[806]Werner *J. Chem. Soc.* **1918,** *113*, 899.
[807]Sommer; Jackson *J. Org. Chem.* **1970,** *35*, 1558; Sommer; Lipp; Jackson *J. Org. Chem.* **1971,** *36*, 824.
[808]For a discussion of the mechanism of the reaction between a primary halide and Ph_2NLi, see DePue; Collum *J. Am. Chem. Soc.* **1988,** *110*, 5524.
[809]Patai; Weiss *J. Chem. Soc.* **1959,** 1035.
[810]For a review of aziridine formation by this method, see Dermer; Ham, Ref. 437, pp. 1-59.
[811]Juaristi; Madrigal *Tetrahedron* **1989,** *45*, 629.

As usual, tertiary substrates do not give the reaction at all but undergo preferential elimination. However, tertiary (but not primary or secondary) halides R_3CCl can be converted to primary amines R_3CNH_2 by treatment with NCl_3 and $AlCl_3$[812] in a reaction related to **0-50.**

Phosphines behave similarly, and compounds of the type R_3P and R_4P^+ X^- can be so prepared. The reaction between triphenylphosphine and quaternary salts of nitrogen heterocycles in an aprotic solvent is probably the best way of dealkylating the heterocycles, e.g.,[813]

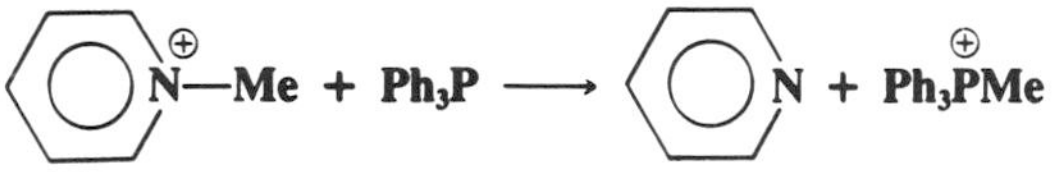

OS **I,** 23, 48, 102, 300, 488; **II,** 85, 183, 290, 328, 374, 397, 419, 563; **III,** 50, 148, 254, 256, 495, 504, 523, 705, 753, 774, 813, 848; **IV,** 84, 98, 383, 433, 466, 582, 585, 980; **V,** 88, 124, 306, 361, 434, 499, 541, 555, 608, 736, 751, 758, 769, 825, 883, 985, 989, 1018, 1085, 1145; **VI,** 56, 75, 104, 106, 175, 552, 652, 704, 818, 967; **67,** 105, 133; **68,** 188, 227. Also see OS **II,** 395; **IV,** 950.

0-44   Conversion of Alkyl Halides to Primary Amines with Hexamethylenetetramine
  Amino-de-halogenation (overall transformation)

$$RX + (CH_2)_6N_4 \longrightarrow N_3(CH_2)_6\overset{\oplus}{N}R\ X^- \xrightarrow[EtOH]{HCl} RNH_2$$

Primary amines can be prepared from alkyl halides by the use of hexamethylenetetramine[814] followed by cleavage of the resulting salt with ethanolic HCl. The method, called the *Delépine reaction,* is most successful for active halides such as allylic and benzylic halides and α-halo ketones, and for primary iodides.

OS **V,** 121.

0-45   Conversion of Alkyl Halides to Secondary Amines with Cyanamide
  Imino-de-dihalo-*aggre*-substitution (overall transformation)

$$2RX + {}^{2-}N{-}CN \longrightarrow R_2N{-}CN \xrightarrow[2.OH^-]{1.H_3O^+} R_2NH$$

A convenient way of obtaining secondary amines without contamination by primary or tertiary amines involves treatment of alkyl halides with the sodium or calcium salt of cyanamide $NH_2{-}CN$ to give disubstituted cyanamides, which are then hydrolyzed and decarboxylated to secondary amines. Good yields are obtained when the reaction is carried out under phase-transfer conditions.[815] R may be primary, secondary, allylic, or benzylic. 1,ω-Dihalides give cyclic secondary amines.

OS **I,** 203.

[812]Kovacic; Lowery *J. Org. Chem.* **1969,** *34,* 911; Strand; Kovacic *J. Am. Chem. Soc.* **1973,** *95,* 2977.
[813]For example, see Deady; Finlayson; Korytsky *Aust. J. Chem.* **1979,** *32,* 1735.
[814]For a review of the reactions of this reagent, see Blažević; Kolbah; Belin; Šunjić; Kajfež *Synthesis* **1979,** 161-176.
[815]Jończyk; Ochal; Mąkosza *Synthesis* **1978,** 882.

0-46 Replacement of a Hydroxy by an Amino Group
Amino-de-hydroxylation

$$
\underset{\underset{CN}{\overset{OH}{|}}}{R-C-R'} + NH_3 \longrightarrow \underset{\underset{CN}{\overset{NH_2}{|}}}{R-C-R'}
$$

Cyanohydrins can be converted to amines by treatment with ammonia. The use of primary or secondary amines instead of ammonia leads to secondary and tertiary cyanoamines, respectively. It is more common to perform the conversion of an aldehyde or ketone directly to the cyanoamine without isolation of the cyanohydrin (see **6-50**). α-Hydroxy ketones (acyloins and benzoins) behave similarly.[816] The conversion ROH → RNH$_2$ can be accomplished for primary and secondary alcohols by treatment with hydrazoic acid (HN$_3$), diisopropyl azodicarboxylate (i-Pr–OOCN=NCOO–i-Pr), and excess Ph$_3$P in THF, followed by water or aqueous acid.[817] This is a type of Mitsunobu reaction (see **0-22**). Other alcohol-to-amine Mitsunobu reactions have also been reported.[818] Primary and secondary alcohols ROH (but not methanol) can be converted to tertiary amines[819] R$_2'$NR by treatment with the secondary amine R$_2'$NH and (t-BuO)$_3$Al in the presence of Raney nickel.[820] The use of aniline gives secondary amines PhNHR. Allylic alcohols ROH react with primary (R'NH$_2$) or secondary (R$_2'$NH) amines in the presence of platinum or palladium complexes, to give secondary (RNHR') or tertiary (RNR$_2'$) allylic amines.[821]

β-Amino alcohols give aziridines when treated with triphenylphosphine dibromide in the presence of triethylamine:[822]

$$
\underset{\underset{OH}{\overset{}{|}} \quad \underset{NHR'}{\overset{}{|}}}{R_2CH-CH_2} + Ph_3PBr_2 \xrightarrow{Et_3N} R-\overset{\displaystyle R}{\underset{\underset{R'}{\overset{}{|}}}{\triangle_N}}
$$

The fact that inversion takes place at the OH carbon indicates that an S$_N$2 mechanism is involved, with OPPh$_3$ as the leaving group.

Alcohols can be converted to amines in an indirect manner.[823] The alcohols are converted to alkyloxyphosphonium perchlorates which in DMF successfully *monoalkylate* not only secondary but also primary amines.[824]

$$
ROH \xrightarrow[\text{2.NH}_4\text{ClO}_4]{\text{1.CCl}_4-\text{P(NMe}_2)_3} R\overset{\oplus}{O}P(NMe_2)_3 \quad CClO_4^- \xrightarrow[\text{R'R''NH}]{\text{DMF}} RR'R''N + OP(NMe_2)_3
$$

[816]For example, see Klemmensen; Schroll; Lawesson *Ark. Kemi* **1968**, *28*, 405.
[817]Fabiano; Golding; Sadeghi *Synthesis* **1987**, 190.
[818]See, for example, Henry; Marcin; McIntosh; Scola; Harris; Weinreb *Tetrahedron Lett.* **1989**, *30*, 5709; Edwards; Stemerick; McCarthy *Tetrahedron Lett.* **1990**, *31*, 3417.
[819]For other methods of converting certain alcohols to secondary and tertiary amines, see Murahashi; Kondo; Hakata *Tetrahedron Lett.* **1982**, *23*, 229; Baiker; Richarz *Tetrahedron Lett.* **1977**, *Helv. Chim. Acta* **1978**, *61*, 1169, *Synth. Commun.* **1978**, *8*, 27; Grigg; Mitchell; Sutthivaiyakit; Tongpenyai *J. Chem. Soc., Chem. Commun* **1981**, 611; Arcelli; Bui-The-Khai; Porzi *J. Organomet. Chem.* **1982**, *235*, 93; Kelly; Eskew; Evans *J. Org. Chem.* **1986**, *51*, 95; Huh; Tsuji; Kobayashi; Okuda; Watanabe *Chem. Lett.* **1988**, 449.
[820]Botta; De Angelis; Nicoletti *Synthesis* **1977**, 722.
[821]Atkins; Walker; Manyik *Tetrahedron Lett.* **1970**, 3821; Tsuji; Takeuchi; Ogawa; Watanabe *Chem. Lett.* **1986**, 293.
[822]Okada; Ichimura; Sudo *Bull. Chem. Soc. Jpn.* **1970**, *43*, 1185. See also Pfister *Synthesis* **1984**, 969; Suzuki; Tani *Chem. Lett.* **1984**, 2129; Marsella *J. Org. Chem.* **1987**, *52*, 467.
[823]For some other indirect methods, see White; Ellinger *J. Am. Chem. Soc.* **1965**, *87*, 5261; Burgess; Penton; Taylor *J. Am. Chem. Soc.* **1970**, *92*, 5224; Hendrickson; Joffee *J. Am. Chem. Soc.* **1973**, *95*, 4083; Trost; Keinan *J. Org. Chem.* **1979**, *44*, 3451; Ref 619 in Chapter 19.
[824]Castro; Selve *Bull. Soc. Chim. Fr.* **1971**, 4368. For a similar method, see Tanigawa; Murahashi; Moritani *Tetrahedron Lett.* **1975**, 471.

Thus by this means secondary as well as tertiary amines can be prepared in good yields.

A solution of the sodium salt of N-methylaniline in HMPA can be used to cleave the methyl group from aryl methyl ethers:[825] $ArOMe + PhNMe^- \rightarrow ArO^- + PhNMe_2$. This reagent also cleaves benzylic groups. In a similar reaction, methyl groups of aryl methyl ethers can be cleaved with lithium diphenylphosphide Ph_2PLi.[826] This reaction is specific for methyl ethers and can be carried out in the presence of ethyl ethers with high selectivity.

OS **II**, 29, 231; **IV**, 91, 283; **VI**, 567, 788; **VII**, 501. Also see OS **I**, 473; **III**, 272, 471.

0-47 Transamination
Alkylamino-de-amination

$$RNH_2 + R'NH^- \longrightarrow RR'NH + NH_2^-$$

Where the nucleophile is the conjugate base of a primary amine, NH_2 can be a leaving group. The method has been used to prepare secondary amines.[827] In another process, primary amines are converted to secondary amines in which both R groups are the same ($2RNH_2 \rightarrow R_2NH + NH_3$)[828] by refluxing in xylene in the presence of Raney nickel.[829] Quaternary salts can be dealkylated with ethanolamine.[830]

$$R_4N^+ + NH_2CH_2CH_2OH \longrightarrow R_3N + R\overset{\oplus}{N}H_2CH_2CH_2OH$$

In this reaction, methyl groups are cleaved in preference to other saturated alkyl groups. A similar reaction takes place between a Mannich base (see **6-16**) and a secondary amine, where the mechanism is elimination–addition (see p. 338). See also **9-5**.

OS **V**, 1018.

0-48 Alkylation of Amines with Diazo Compounds
Hydro,dialkylamino-de-diazo-bisubstitution

$$CR_2N_2 + R_2'NH \xrightarrow{BF_3} CHR_2NR_2'$$

The reaction of diazo compounds with amines is similar to **0-15**.[831] The acidity of amines is not great enough for the reaction to proceed without a catalyst, but BF_3, which converts the amine to the F_3B-NHR_2' complex, enables the reaction to take place. Cuprous cyanide can also be used as a catalyst.[832] The most common substrate is diazomethane,[593] in which case this is a method for the methylation of amines. Ammonia has been used as the amine but, as in the case of **0-43**, mixtures of primary, secondary, and tertiary amines are obtained. Primary aliphatic amines give mixtures of secondary and tertiary amines. Secondary amines give successful alkylation. Primary aromatic amines also give the reaction, but diaryl or arylalkylamines react very poorly.

[825]Loubinoux; Coudert; Guillaumet *Synthesis* **1980**, 638.
[826]Ireland; Walba *Org. Synth. VI*, 567.
[827]Baltzly; Blackman *J. Org. Chem.* **1963**, *28*, 1158.
[828]In a similar manner, a mixture of primary amines can be converted to a mixed secondary amine. For a review of the mechanism, see Geller *Russ. Chem. Rev.* **1978**, *47*, 297-306.
[829]De Angelis; Grgurina; Nicoletti *Synthesis* **1979**, 70; See also Ballantine; Purnell; Rayanakorn; Thomas; Williams *J. Chem. Soc., Chem. Commun.* **1981**, 9; Arcelli; Bui-The-Khai; Porzi *J. Organomet. Chem.* **1982**, *231*, C31; Jung; Fellmann; Garrou *Organometallics* **1983**, *2*, 1042; Tsuji; Shida; Takeuchi; Watanabe *Chem. Lett.* **1984**, 889; Bank; Jewett *Tetrahedron Lett.* **1991**, *32*, 303.
[830]Hünig; Baron *Chem. Ber.* **1957**, *90*, 395, 403.
[831]Müller; Huber-Emden; Rundel *Liebigs. Ann. Chem.* **1959**, *623*, 34.
[832]Saegusa; Ito; Kobayashi; Hirota; Shimizu *Tetrahedron Lett.* **1966**, 6131.

0-49 Amination of Epoxides
(3)*OC-seco*-Amino-de-alkoxylation

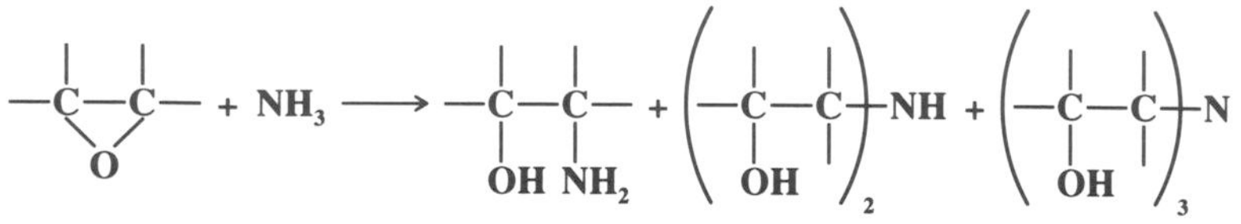

$$-\overset{|}{\underset{\underset{O}{\diagdown\diagup}}{C}}-\overset{|}{\underset{}{C}}- \; + \; NH_3 \; \longrightarrow \; -\overset{|}{\underset{OH}{C}}-\overset{|}{\underset{NH_2}{C}}- \; + \; \left(-\overset{|}{\underset{OH}{C}}-\overset{|}{\underset{}{C}}-\right)_{\!2}\!\!NH \; + \; \left(-\overset{|}{\underset{OH}{C}}-\overset{|}{\underset{}{C}}-\right)_{\!3}\!\!N$$

The reaction between epoxides and ammonia is a general and useful method for the preparation of β-hydroxyamines.[833] Ammonia gives largely the primary amine, but also some secondary and tertiary amines. The useful solvents, the ethanolamines, are prepared by this reaction. For another way of accomplishing this conversion, see **0-51**. Primary and secondary amines give, respectively, secondary and tertiary amines,[834] e.g.,

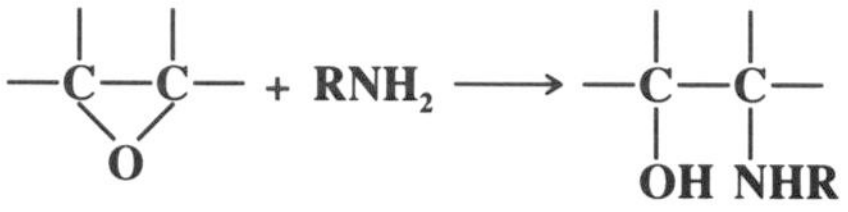

$$-\overset{|}{\underset{\underset{O}{\diagdown\diagup}}{C}}-\overset{|}{\underset{}{C}}- \; + \; RNH_2 \; \longrightarrow \; -\overset{|}{\underset{OH}{C}}-\overset{|}{\underset{NHR}{C}}-$$

Episulfides, which can be generated in situ in various ways, react similarly to give β-amino thiols,[835] and aziridines give 1,2-diamines.[836] Triphenylphosphine similarly reacts with epoxides to give an intermediate that undergoes elimination to give olefins (see the Wittig reaction, **6-47**).

There are no OS references, but see OS **VI**, 652 for a related reaction.

0-50 Amination of Alkanes
Amino-de-hydrogenation or **Amination**

$$R_3CH \; + \; NCl_3 \; \xrightarrow[\text{0-10°C}]{\text{AlCl}_3} \; R_3CNH_2$$

Alkanes, arylalkanes, and cycloalkanes can be aminated, at tertiary positions only, by treatment with trichloroamine and aluminum chloride at 0 to 10°C.[837] For example, *p*-MeC₆H₄CHMe₂ gives *p*-MeC₆H₄CMe₂NH₂, methylcyclopentane gives 1-amino-1-methyl-cyclopentane, and adamantane gives 1-aminoadamantane, all in good yields. This is a useful reaction, since there are not many other methods for the preparation of *t*-alkyl amines. The mechanism has been rationalized as an S_N1 process with H⁻ as the leaving group:[837]

$$NCl_3 \; + \; AlCl_3 \; \longrightarrow \; (Cl_2N\!-\!AlCl_3)^-\,Cl^+$$

$$R_3CH \; \xrightarrow{Cl^+} \; R_3C^+ \; \xrightarrow{NCl_2^-} \; R_3CNCl_2 \; \xrightarrow[2H^+]{-2Cl^+} \; R_3CNH_2$$

See also **2-11**.
OS **V**, 35.

[833]For an example, see McManus; Larson; Hearn *Synth. Commun.* **1973**, *3*, 177.
[834]For improved methods, see Carre; Houmounou; Caubere *Tetrahedron Lett.* **1985**, *26*, 3107; Fujiwara; Imada; Baba; Matsuda *Tetrahedron Lett.* **1989**, *30*, 739; Yamada; Yumoto; Yamamoto *Tetrahedron Lett.* **1989**, *30*, 4255; Chini; Crotti; Macchia *Tetrahedron Lett.* **1990**, *31*, 4661.
[835]Reynolds; Massad; Fields; Johnson *J. Org. Chem.* **1961**, *26*, 5109; Reynolds, Fields; Johnson *J. Org. Chem.* **1961**, *26*, 5111, 5116, 5119, 5125; Wineman; Gollis; James; Pomponi *J. Org. Chem.* **1962**, *27*, 4222.
[836]For a review, see Dermer; Ham, Ref. 437, pp. 262-268.
[837]Kovacic; Chaudhary *Tetrahedron* **1967**, *23*, 3563; Strand; Kovacic, Ref. 812; Wnuk; Chaudhary; Kovacic *J. Am. Chem. Soc.* **1976**, *98*, 5678, and references cited in these papers.

0-51 Formation of Isocyanides
Haloform–isocyanide transformation

$$CHCl_3 + RNH_2 \xrightarrow{OH^-} R\overset{\oplus}{-N}\equiv\overset{\ominus}{C}$$

Reaction with chloroform under basic conditions is a common test for primary amines, both aliphatic and aromatic, since isocyanides have very strong bad odors. The reaction probably proceeds by an SN1cB mechanism with dichlorocarbene as an intermediate:

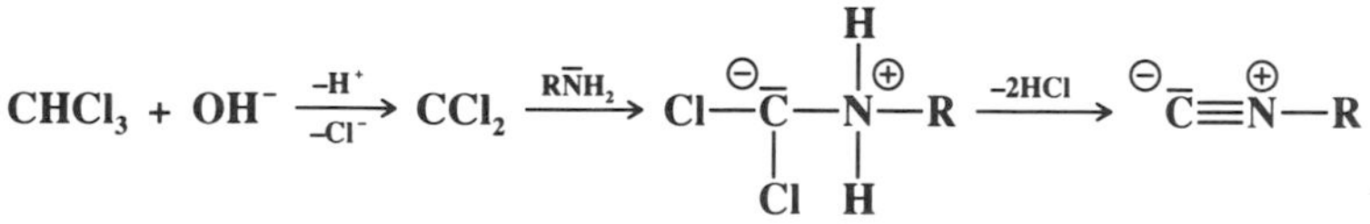

$$CHCl_3 + OH^- \xrightarrow[-Cl^-]{-H^+} CCl_2 \xrightarrow{R\bar{N}H_2} \underset{\underset{\displaystyle Cl\ \ H}{|\ \ |}}{Cl-\overset{\ominus}{\underset{}{C}}-\overset{H}{\overset{|}{\underset{}{N}}}\overset{\oplus}{-}R} \xrightarrow{-2HCl} \overset{\ominus}{C}\equiv\overset{\oplus}{N}-R$$

The reaction can also be used synthetically for the preparation of isocyanides, though yields are generally not high.[838] An improved procedure has been reported.[839] When secondary amines are involved, the adduct cannot lose two moles of HCl. Instead it is hydrolyzed to an N,N-disubstituted formamide:[840]

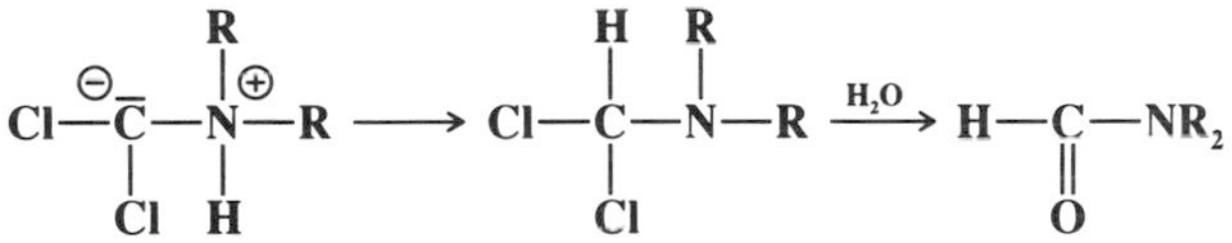

$$\underset{\underset{\displaystyle Cl\ \ H}{|\ \ |}}{Cl-\overset{\ominus}{\underset{}{C}}-\overset{R}{\overset{|}{\underset{}{N}}}\overset{\oplus}{-}R} \longrightarrow \underset{\underset{\displaystyle Cl}{|}}{Cl-\overset{H}{\overset{|}{\underset{}{C}}}-\overset{R}{\overset{|}{\underset{}{N}}}-R} \xrightarrow{H_2O} \underset{\underset{\displaystyle O}{||}}{H-\overset{}{\underset{}{C}}-NR_2}$$

A completely different way of preparing isocyanides involves the reaction of epoxides or oxetanes with trimethylsilyl cyanide and zinc iodide, e.g.,[841]

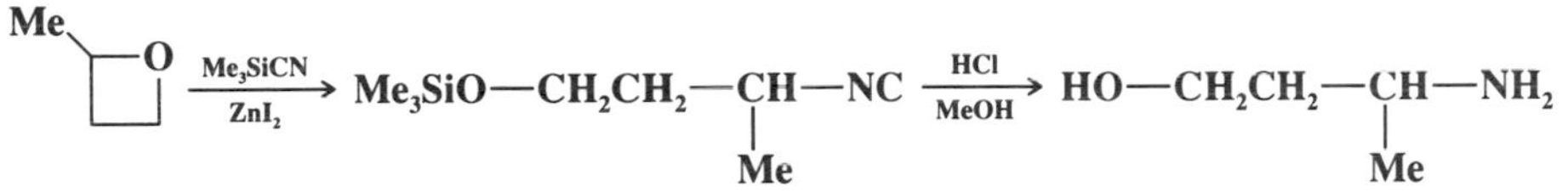

$$\overset{Me}{\underset{}{\boxed{}O}} \xrightarrow[ZnI_2]{Me_3SiCN} Me_3SiO-CH_2CH_2-\underset{\underset{\displaystyle Me}{|}}{CH}-NC \xrightarrow[MeOH]{HCl} HO-CH_2CH_2-\underset{\underset{\displaystyle Me}{|}}{CH}-NH_2$$

$$\textbf{102}$$

The products can be hydrolyzed to hydroxyamines, e.g., **102**.
OS **VI**, 232.

B. Attack by NH_2, NHR, or NR_2 at an Acyl Carbon[842]

0-52 Acylation of Amines by Acyl Halides
Amino-de-halogenation

$$RCOX + NH_3 \longrightarrow RCONH_2 + HX$$

The treatment of acyl halides with ammonia or amines is a very general reaction for the preparation of amides.[843] The reaction is highly exothermic and must be carefully controlled,

[838]For a review of isocyanides, see Periasamy; Walborsky *Org. Prep. Proced. Int.* **1979**, *11*, 293-311.
[839]Weber; Gokel *Tetrahedron Lett.* **1972**, 1637; Weber; Gokel; Ugi *Angew. Chem. Int. Ed. Engl.* **1972**, *11*, 530 [*Angew. Chem. 84*, 587].
[840]Saunders; Murray *Tetrahedron* **1959**, *6*, 88; Frankel; Feuer; Bank *Tetrahedron Lett.* **1959**, no. 7, 5.
[841]Gassman; Haberman *Tetrahedron Lett.* **1985**, *26*, 4971, and references cited therein.
[842]For a review, see Challis; Butler, in Patai, Ref. 355, pp. 279-290.
[843]For a review, see Beckwith, in Zabicky, Ref. 555, pp. 73-185.

usually by cooling or dilution. Ammonia gives unsubstituted amides, primary amines give N-substituted amides, and secondary amines give N,N-disubstituted amides. Arylamines can be similarly acylated. In some cases aqueous alkali is added to combine with the liberated HCl. This is called the *Schotten–Baumann procedure*, as in **0-20**.

Hydrazine and hydroxylamine also react with acyl halides to give, respectively, hydrazides RCONHNH$_2$[844] and hydroxamic acids RCONHOH,[845] and these compounds are often made in this way. When phosgene is the acyl halide, both aliphatic and aromatic primary amines give chloroformamides ClCONHR that lose HCl to give isocyanates RNCO.[846] This is one of the most common methods for the preparation of isocyanates.[847] Thiophosgene,[847a] sim-

$$Cl—\underset{\underset{O}{\|}}{C}—Cl + RNH_2 \longrightarrow Cl—\underset{\underset{O}{\|}}{C}—NHR \xrightarrow{-HCl} O=C=N—R$$

ilarly treated, gives isothiocyanates. A safer substitute for phosgene in this reaction is trichloromethyl chloroformate CCl$_3$OCOCl.[848] When chloroformates ROCOCl are treated with primary amines, carbamates ROCONHR' are obtained.[849] An example of this reaction is the use of benzyl chloroformate to protect the amino group of amino acids and peptides:

$$PhCH_2—O—\underset{\underset{O}{\|}}{C}—Cl + H_2NR \longrightarrow PhCH_2—O—\underset{\underset{O}{\|}}{C}—NHR$$

The PhCH$_2$OCO group is called the carbobenzoxy group, and is often abbreviated Cbz or Z. Another important group similarly used is the *t*-butoxycarbonyl group Me$_3$COCO, abbreviated as Boc. In this case, the chloride Me$_3$COCOCl is unstable, so the anhydride (Me$_3$COCO)$_2$O is used instead, in an example of **0-53**. Amino groups in general are often protected by conversion to amides. The treatment of acyl halides with lithium nitride gives N,N-diacyl amides (triacylamines):[850]

$$3RCOCl + Li_3N \longrightarrow (RCO)_3N$$

The reactions proceed by the tetrahedral mechanism.[851]

OS **I**, 99, 165; **II**, 76, 208, 278, 328, 453; **III**, 167, 375, 415, 488, 490, 613; **IV**, 339, 411, 521, 620, 780; **V**, 201, 336; **VI**, 382, 715; **VII**, 56, 287, 307; **67**, 187; **68**, 83. See also OS **VII**, 302.

0-53 Acylation of Amines by Anhydrides
Amino-de-acyloxy-substitution

$$R—\underset{\underset{O}{\|}}{C}—O—\underset{\underset{O}{\|}}{C}—R' + NH_3 \longrightarrow R—\underset{\underset{O}{\|}}{C}—NH_2 + R'COOH$$

[844]For a review of hydrazides, see Paulsen; Stoye, in Zabicky, Ref. 555, pp. 515-600.
[845]For an improved method, see Ando; Tsumaki *Synth. Commun.* **1983**, *13*, 1053.
[846]For reviews of the preparation of isocyanates and isothiocyanates, see, respectively, the articles by Richter; Ulrich, pp. 619-818, and Drobnica; Kristián; Augustín pp. 1003-1221, in Patai *The Chemistry of Cyanates and Their Thio Derivatives*, pt. 2; Wiley: New York, 1977.
[847]For examples, see Ozaki *Chem. Rev.* **1972**, *72*, 457-496, pp. 457-460. For a review of the industrial preparation of isocyanates by this reaction, see Twitchett *Chem. Soc. Rev.* **1974**, *3*, 209-230.
[847a]For a review of thiophosgene, see Sharma *Sulfur Rep.* **1986**, *5*, 1-100.
[848]Kurita; Iwakura *Org. Synth. VI*, 715.
[849]For an improved procedure, see Raucher; Jones *Synth. Commun.* **1985**, *15*, 1025.
[850]Baldwin; Blanchard; Koening *J. Org. Chem.* **1965**, *30*, 671.
[851]Kivinen, Ref. 502; Bender; Jones *J. Org. Chem.* **1962**, *27*, 3771. See also Song; Jencks *J. Am. Chem. Soc.* **1989**, *111*, 8479.

This reaction, similar in scope and mechanism[852] to **0-52,** can be carried out with ammonia or primary or secondary amines.[853] However, ammonia and primary amines can also give imides, in which two acyl groups are attached to the nitrogen. This is especially easy with cyclic anhydrides, which produce cyclic imides.[854]

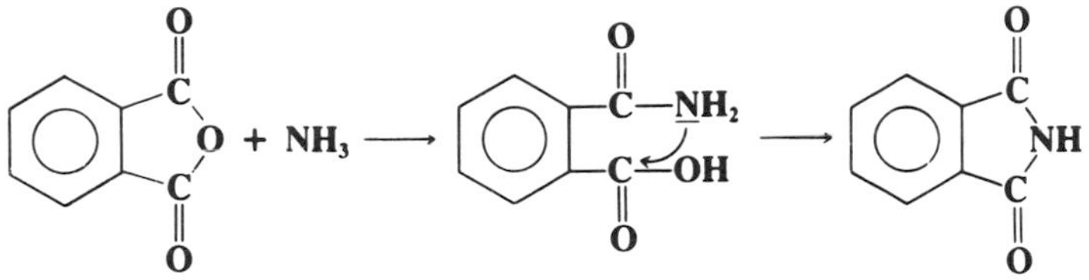

The second step in this case, which is much slower than the first, is the attack of the amide nitrogen on the carboxylic carbon. Unsubstituted and N-substituted amides have been used instead of ammonia. Since the other product of this reaction is RCOOH, this is a way of "hydrolyzing" such amides in the absence of water.[855]

Even though formic anhydride is not a stable compound (see p. 542), amines can be formylated with the mixed anhydride of acetic and formic acids HCOOCOMe[856] or with a mixture of formic acid and acetic anhydride. Acetamides are not formed with these reagents. Secondary amines can be acylated in the presence of a primary amine by conversion to their salts and addition of 18-crown-6.[857] The crown ether complexes the primary ammonium salt, preventing its acylation, while the secondary ammonium salts, which do not fit easily into the cavity, are free to be acylated.

OS **I,** 457; **II,** 11; **III,** 151, 456, 661, 813; **IV,** 5, 42, 106, 657; **V,** 27, 373, 650, 944, 973; **VI,** 1; **VII,** 4, 70; **66,** 132.

0-54   Acylation of Amines by Carboxylic Acids
   Amino-de-hydroxylation

$$\text{RCOOH} + \text{NH}_3 \longrightarrow \text{RCOO}^- \text{NH}_4^+ \xrightarrow{\text{pyrolysis}} \text{RCONH}_2$$

When carboxylic acids are treated with ammonia or amines, salts are obtained. The salts of ammonia or primary or secondary amines can be pyrolyzed to give amides,[858] but the method is less convenient than **0-52, 0-53,** and **0-55** and is seldom of preparative value.[859] Lactams are produced fairly easily from γ- or δ-amino acids,[860] e.g.,

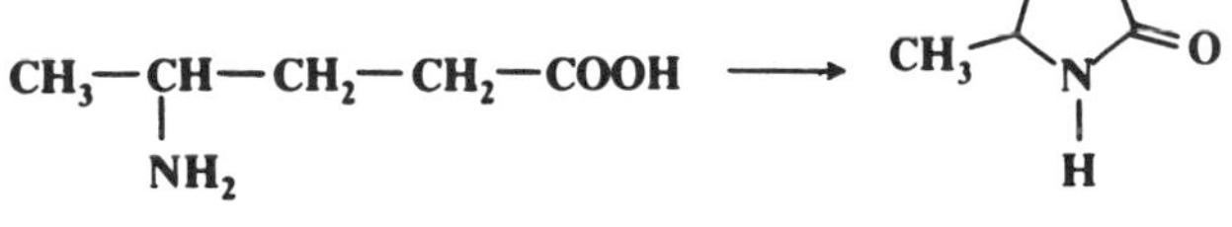

Although treatment of carboxylic acids with amines does not directly give amides, the reaction can be made to proceed in good yield at room temperature or slightly above by

[852]For a discussion of the mechanism, see Kluger; Hunt *J. Am. Chem. Soc.* **1989,** *111*, 3325.
[853]For a review, see Beckwith, in Zabicky, Ref. 555, pp. 86-96.
[854]For reviews of imides, see Wheeler; Rosado, in Zabicky, Ref. 555, pp. 335-381; Hargreaves; Pritchard; Dave *Chem. Rev.* **1970,** *70*, 439-469 (cyclic imides).
[855]Eaton; Rounds; Urbanowicz; Gribble *Tetrahedron Lett.* **1988,** *29*, 6553.
[856]For the formylation of amines with the mixed anhydride of formic and trimethylacetic acid, see Vlietstra; Zwikker; Nolte; Drenth *Recl. Trav. Chim. Pays-Bas* **1982,** *101*, 460.
[857]Barrett; Lana *J. Chem. Soc., Chem. Commun.* **1978,** 471.
[858]For example, see Mitchell; Reid *J. Am. Chem. Soc.* **1931,** *53*, 1879.
[859]For a review of amide formation from carboxylic acids, see Beckwith, in Zabicky, Ref. 555, pp. 105-109.
[860]See, for example, Bladé-Font *Tetrahedron Lett.* **1980,** *21*, 2443.

the use of coupling agents,[861] the most important of which is dicyclohexylcarbodiimide. This is very convenient and is used[862] a great deal in peptide synthesis.[863] The mechanism is probably the same as in **0-22** up to the formation of **99**. This intermediate is then attacked by another molecule of RCOO$^-$ to give the anhydride $(RCO)_2O$, which is the actual species that reacts with the amine:

$$R\!-\!\overset{\underset{\displaystyle O}{\|}}{C}\!-\!\overset{\oplus}{O}\!-\!\overset{\underset{\displaystyle NH}{|}}{C}\!=\!NH\!-\!C_6H_{11} + R\!-\!\overset{\underset{\displaystyle O}{\|}}{C}\!-\!\overset{\ominus}{O} \xrightarrow[\substack{\text{tetrahedral} \\ \text{mechanism}}]{\text{two steps}} R\!-\!\overset{\underset{\displaystyle O}{\|}}{C}\!-\!O\!-\!\overset{\underset{\displaystyle O}{\|}}{C}\!-\!R + DHU$$

99 C_6H_{11}

The anhydride has been isolated from the reaction mixture and then used to acylate an amine.[864] Other promoting agents[865] are N,N'-carbonyldiimidazole (**100**, p. 396),[664] which behaves as in reaction **0-22**, $POCl_3$,[866] $TiCl_4$,[867] sulfuryl chloride fluoride SO_2ClF,[868] benzotriazol-1-yl diethyl phosphate,[869] $Ti(OBu)_4$,[870] molecular sieves,[871] N,N,N',N'-tetramethyl(succinimido)uronium tetrafluoroborate,[872] CBMIT[656] (p. 396), Lawesson's reagent (p. 893),[873] chlorosulfonyl isocyanate,[660] P_2I_4,[874] pyridinium salts–Bu_3N,[875] and a mixture of Bu_3P and PhCNO.[876] Certain dicarboxylic acids form amides simply on treatment with primary aromatic amines. In these cases the cyclic anhydride is an intermediate and is the species actually attacked by the amine.[877] Carboxylic acids can also be converted to amides by heating with amides of carboxylic acids (exchange),[878] sulfonic acids, or phosphoric acids, e.g.,[879]

$$RCOOH + Ph_2PONH_2 \longrightarrow RCONH_2 + Ph_2POOH$$

or by treatment with trisalkylaminoboranes $[B(NHR')_3]$, with trisdialkylaminoboranes $[B(NR'_2)_3]$,[880]

$$RCOOH + B(NR'_2)_3 \longrightarrow RCONR'_2$$

or with bis(diorganoamino)magnesium reagents $(R_2N)_2Mg$.[881]

[861]For a review of peptide synthesis with dicyclohexylcarbodiimide and other coupling agents, see Klausner; Bodansky *Synthesis* **1972**, 453-463.

[862]It was first used this way by Sheehan; Hess *J. Am. Chem. Soc.* **1955**, 77, 1067.

[863]For a treatise on peptide synthesis, see Gross; Meienhofer *The Peptides*, 3 vols.; Academic Press: New York, 1979-1981. For a monograph, see Bodanszky; Bodanszky *The Practice of Peptide Synthesis*; Springer: New York, 1984.

[864]Schüssler; Zahn *Chem. Ber.* **1962**, 95, 1076; Rebek; Feitler *J. Am. Chem. Soc.* **1974**, 96, 1606. There is evidence that some of the **99** is converted to products by another mechanism. See Rebek; Feitler *J. Am. Chem. Soc.* **1973**, 95, 4052.

[865]For a list of reagents, with references, see Ref. 508, pp. 972-976.

[866]Klosa *J. Prakt. Chem.* **1963**, [4] 19, 45.

[867]Wilson; Weingarten *Can. J. Chem.* **1970**, 48, 983.

[868]Olah; Narang; Garcia-Luna *Synthesis* **1980**, 661.

[869]Kim; Chang; Ko *Tetrahedron Lett.* **1985**, 26, 1341.

[870]Shteinberg; Kondratov; Shein *J. Org. Chem. USSR* **1988**, 24, 1774.

[871]Cossy; Pale-Grosdemange *Tetrahedron Lett.* **1989**, 30, 2771.

[872]Bannwarth; Knorr *Tetrahedron Lett.* **1991**, 32, 1157.

[873]Thorsen; Andersen; Pedersen; Yde; Lawesson *Tetrahedron* **1985**, 41, 5633.

[874]Suzuki; Tsuji; Hiroi; Sato; Osuka *Chem. Lett.* **1983**, 449.

[875]Bald; Saigo; Mukaiyama *Chem. Lett.* **1975**, 1163. See also Mukaiyama; Aikawa; Kobayashi *Chem. Lett.* **1976**, 57.

[876]Grieco; Clark; Withers *J. Org. Chem.* **1979**, 44, 2945.

[877]Higuchi; Miki; Shah; Herd *J. Am. Chem. Soc.* **1963**, 85, 3655.

[878]For example, see Schindbauer *Monatsh. Chem.* **1968**, 99, 1799.

[879]Zhmurova; Voitsekhovskaya; Kirsanov *J. Gen. Chem. USSR* **1959**, 29, 2052. See also Kopecký; Šmejkal *Chem. Ind. (London)* **1966**, 1529; Liu; Chan; Lee *Synth. Commun.* **1979**, 9, 31.

[880]Pelter; Levitt; Nelson *Tetrahedron* **1970**, 26, 1539; Pelter; Levitt *Tetrahedron* **1970**, 26, 1545, 1899.

[881]Sanchez; Vest; Despres *Synth. Commun.* **1989**, 19, 2909.

An important technique, discovered by R. B. Merrifield in 1963[882] and since used for the synthesis of many peptides,[883] is called *solid phase synthesis* or *polymer-supported synthesis*.[884] The reactions used are the same as in ordinary synthesis, but one of the reactants is anchored onto a solid polymer. For example, if it is desired to couple two amino acids (to form a dipeptide), the polymer selected might be polystyrene with CH_2Cl side chains (Fig. 10.2, **103**). One of the amino acids, protected by a *t*-butoxycarbonyl group (Boc), would then be coupled to the side chains (step A). It is not necessary that all the side chains be converted, but a random selection will be. The Boc group is then removed by hydrolysis with trifluoroacetic acid in CH_2Cl_2 (step B) and the second amino acid is coupled to the first, using DCC or some other coupling agent (step C). The second Boc group is removed (step D), resulting in a dipeptide that is still anchored to the polymer. If this dipeptide is the desired product, it can be cleaved from the polymer by various methods,[885] one of which is treatment with HF (step E). If a longer peptide is wanted, additional amino acids can be added by repeating steps C and D.

The basic advantage of the polymer support techniques is that the polymer (including all chains attached to it) is easily separated from all other reagents, because it is insoluble in the solvents used. Excess reagents, other reaction products (such as DHU), side products, and the solvents themselves are quickly washed away. Purification of the polymeric species (such as **104**, **105**, and **106**) is rapid and complete. The process can even be automated,[886] to the extent that six or more amino acids can be added to a peptide chain in one day. Commercial automated peptide synthesizers are now available.[887]

Although the solid phase technique was first developed for the synthesis of peptide chains and has seen considerable use for this purpose, it has also been used to synthesize chains of polysaccharides and polynucleotides; in the latter case, solid phase synthesis has almost completely replaced synthesis in solution.[888] The technique has been applied less often to reactions in which only two molecules are brought together (nonrepetitive syntheses), but many examples have been reported.[889]

OS **I**, 3, 82, 111, 172, 327; **II**, 65, 562; **III**, 95, 328, 475, 590, 646, 656, 768; **IV**, 6, 62, 513; **V**, 670, 1070; **69**, 55. Also see OS **III**, 360; **VI**, 263; **67**, 69.

0-55 Acylation of Amines by Carboxylic Esters
Amino-de-alkoxylation

$$RCOOR' + NH_3 \longrightarrow RCONH_2 + R'OH$$

[882]Merrifield *J. Am. Chem. Soc.* **1963**, *85*, 2149.

[883]For a monograph on solid state peptide synthesis, see Birr *Aspects of the Merrifield Peptide Synthesis*; Springer: New York, 1978. For reviews, see Bayer *Angew. Chem. Int. Ed. Engl.* **1991**, *30*, 113-129 [*Angew. Chem. 103*, 117-133]; Kaiser *Acc. Chem. Res.* **1989**, *22*, 47-54; Jacquier *Bull. Soc. Chim. Fr.* **1989**, 220-236; Barany; Kneib-Cordonier; Mullen *Int. J. Pept. Protein Res.* **1987**, *30*, 705-739; Andreev; Samoilova; Davidovich; Rogozhin *Russ. Chem. Rev.* **1987**, *56*, 366-381; in vol. 2 of Ref. 863, the articles by Barany; Merrifield, pp. 1-184, Fridkin, pp. 333-363; Erickson; Merrifield, in Neurath; Hill; Boeder *The Proteins*, 3rd ed., vol. 2; Academic Press: New York, 1976, pp. 255-527. For R. B. Merrifield's Nobel Prize lecture, see Merrifield *Angew. Chem. Int. Ed. Engl.* **1985**, *24*, 799-810 [*Angew. Chem. 97*, 801-812], *Chem. Scr.* **1985**, *25*, 121-131.

[884]For monographs on solid phase synthesis in general, see Laszlo *Preparative Organic Chemistry Using Supported Reagents*; Academic Press: New York, 1987; Mathur; Narang; Williams *Polymers as Aids in Organic Chemistry*; Academic Press: New York 1980; Hodge; Sherrington *Polymer-supported Reactions in Organic Synthesis*; Wiley: New York, 1980. For reviews, see Sheppard, *Chem. Br.* **1983**, 402-414; Pillai; Mutter *Top. Curr. Chem.* **1982**, *106*, 119-175; Akelah; Sherrington *Chem. Rev.* **1981**, *81*, 557-587; Akelah *Synthesis* **1981**, 413-438; Rebek *Tetrahedron* **1979**, *35*, 723-731; McKillop; Young *Synthesis* **1979**, 401-422, 481-500; Neckers, *CHEMTECH* **1978** (Feb.), 108-116; Crowley; Rapoport *Acc. Chem. Res.* **1976**, *9*, 135-144; Patchornik; Kraus *Pure Appl. Chem.* **1975**, *43*, 503-526.

[885]For some of these methods, see Whitney; Tam; Merrifield *Tetrahedron* **1984**, *40*, 4237.

[886]This was first reported by Merrifield; Stewart; Jernberg *Anal. Chem.* **1966**, *38*, 1905.

[887]For a discussion of automated organic synthesis, see Frisbee; Nantz; Kramer; Fuchs *J. Am. Chem. Soc.* **1984**, *106*, 7143. For an improved method, see Schnorrenberg; Gerhardt *Tetrahedron* **1989**, *45*, 7759.

[888]For a review, see Bannwarth *Chimia* **1987**, *41*, 302-317.

[889]For reviews, see Fréchet *Tetrahedron* **1981**, *37*, 663-683; Fréchet, in Hodge; Sherrington, Ref. 884, pp. 293-342, Leznoff, *Acc. Chem. Res.* **1978**, *11*, 327-333, *Chem. Soc. Rev.* **1974**, *3*, 64-85.

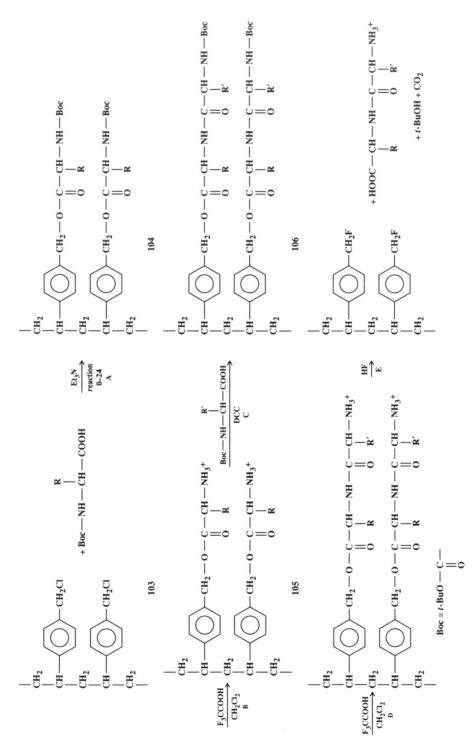

FIGURE 10.2 An outline of dipeptide synthesis by the solid phase technique.

The conversion of carboxylic esters to amides is a useful reaction, and unsubstituted, N-substituted, and N,N-disubstituted amides can be prepared this way from the appropriate amine.[890] Both R and R' can be alkyl or aryl. An especially good leaving group is *p*-nitrophenyl. Many simple esters (R = Me, Et, etc.) are not very reactive, and strongly basic catalysis has been used,[891] as well as catalysis by cyanide ion,[892] and high pressure.[893] β-Keto esters undergo the reaction especially easily.[894] In another procedure, esters are treated with dimethylaluminum amides Me_2AlNRR' to give good yields of amides under mild conditions.[895] The reagents are easily prepared from Me_3Al and NH_3 or a primary or secondary amine or their salts. The ester-to-amide conversion has also been accomplished electrochemically, by passing electric current in the cathodic compartment.[896]

As in **0-52** hydrazides and hydroxamic acids can be prepared from carboxylic esters, with hydrazine and hydroxylamine, respectively. Both hydrazine and hydroxylamine react more rapidly than ammonia or primary amines (the alpha effect, p. 351). Imidates $RC(=NH)OR'$ give amidines $RC(=NH)NH_2$. Lactones, when treated with ammonia or primary amines, give lactams. Lactams are also produced from γ- and δ-amino esters in an internal example of this reaction. Isopropenyl formate is a useful compound for the formylation of primary and secondary amines.[897]

$$R_2NH + HCOOCMe=CH_2 \longrightarrow R_2NCHO + CH_2=CMeOH \rightleftharpoons MeCOMe$$

Although more studies have been devoted to the mechanism of the acylation of amines with carboxylic esters than with other reagents, the mechanistic details are not yet entirely clear.[898] In its broad outlines, the mechanism appears to be essentially BAc2.[899] Under the normal basic conditions, the reaction is general base-catalyzed,[900] indicating that a proton is being transferred in the rate-determining step and that two molecules of amine are involved.[901]

[890]For a review, see Ref. 843, pp. 96-105. For a list of reagents, with references, see Ref. 508, pp. 987-988.
[891]For references, see Ref. 893.
[892]Högberg; Ström; Ebner; Rämsby *J. Org. Chem.* **1987**, *52*, 2033.
[893]Matsumoto; Hashimoto; Uchida; Okamoto; Otani *Chem. Ber.* **1989**, *122*, 1357.
[894]Labelle; Gravel *J. Chem. Soc., Chem. Commun.* **1985**, 105.
[895]Basha; Lipton; Weinreb *Tetrahedron Lett.* **1977**, 4171, *Org. Synth.* VI, 492; Levin; Turos; Weinreb *Synth. Commun.* **1982**, *12*, 989; Barrett; Dhanak *Tetrahedron Lett.* **1987**, *28*, 3327. For the extension of this method to the formation of hydrazides, see Benderly; Stavchansky *Tetrahedron Lett.* **1988**, *29*, 739.
[896]Arai; Shaw; Nozawa; Kawai; Nakajima *Tetrahedron Lett.* **1987**, *28*, 441.
[897]van Melick; Wolters *Synth. Commun.* **1972**, *2*, 83.
[898]For a discussion of the mechanism, see Satchell; Satchell, Ref. 197, pp. 410-431.
[899]Bunnett; Davis *J. Am. Chem. Soc.* **1960**, *82*, 665; Bruice; Donzel; Huffman; Butler *J. Am. Chem. Soc.* **1967**, *89*, 2106.
[900]Bunnett; Davis, Ref. 899, Jencks; Carriuolo *J. Am. Chem. Soc.* **1960**, *82*, 675; Bruice; Mayahi *J. Am. Chem. Soc.* **1960**, *82*, 3067.
[901]Blackburn; Jencks *J. Am. Chem. Soc.* **1968**, *90*, 2638; Bruice; Felton *J. Am. Chem. Soc.* **1969**, *91*, 2799; Felton; Bruice *J. Am. Chem. Soc.* **1969**, *91*, 6721; Nagy; Reuliaux; Bertrand; Van Der Mensbrugghe; Leseul; Nagy *Bull. Soc. Chim. Belg.* **1985**, *94*, 1055.

Alternatively, another base, such as H_2O or OH^-, can substitute for the second molecule of amine. With some substrates and under some conditions, especially at low pH, the breakdown of **107** can become rate-determining.[902] The reaction also takes place under acidic conditions and is general acid-catalyzed, so that breakdown of **107** is rate-determining and proceeds as follows:[903]

$$
\begin{array}{ccc}
\text{R'O} \!\!\downarrow \text{H}\!\!-\!\!\text{A} & & \\
\underset{\underset{\ominus}{\overset{|}{\text{C}}\!\!-\!\!\text{O}|}}{\text{R}\!\!-\!\!\overset{|}{\text{C}}\!\!-\!\!\text{NHR''}} & \xrightarrow{\text{slow}} & \text{R}\!\!-\!\!\underset{\text{O}}{\overset{\|}{\text{C}}}\!\!-\!\!\text{NHR''} + \text{R'OH} + \text{A}^-
\end{array}
$$

HA may be $R''NH_3^+$ or another acid. **107** may or may not be further protonated on the nitrogen. Even under basic conditions, a proton donor may be necessary to assist leaving-group removal. Evidence for this is that the rate is lower with NR_2^- in liquid ammonia than with NHR_2 in water, apparently owing to the lack of acids to protonate the leaving oxygen.[904]

In the special case of β-lactones, where small-angle strain is an important factor, alkyl-oxygen cleavage is observed (BAL2 mechanism, as in the similar case of hydrolysis of β-lactones, **0-10**), and the product is not an amide but a β-amino acid:

$$
\text{H}_3\overset{\frown}{\text{N}} \quad + \quad \underset{\underset{\text{O}\!-\!\!\text{C}=\text{O}}{\overset{|\qquad|}{}}}{\text{CH}_2\!\!-\!\!\text{CH}_2} \quad \longrightarrow \quad \overset{\oplus}{\text{NH}_3}\text{CH}_2\text{CH}_2\overset{\ominus}{\text{COO}}
$$
$$
\text{β-Alanine}
$$

A similar result has been found for certain sterically hindered esters.[905] This reaction is similar to **0-43,** with OCOR as the leaving group.

OS **I**, 153, 179; **II**, 67, 85; **III**, 10, 96, 108, 404, 440, 516, 536, 751, 765; **IV**, 80, 357, 441, 486, 532, 566, 819; **V**, 168, 301, 645; **VI**, 203, 492, 620, 936; **VII**, 4, 30, 41, 411; **65**, 173; **67**, 52; **68**, 77. Also see OS **I**, 5; **V**, 582; **VII**, 75.

0-56   Acylation of Amines by Amides
Alkylamino-de-amination

$$
\text{RCONH}_2 + \text{R'}\overset{\oplus}{\text{NH}}_3 \longrightarrow \text{RCONHR'} + \text{NH}_4^+
$$

This is an exchange reaction and is usually carried out with the salt of the amine.[906] The leaving group is usually NH_2 rather than NHR or NR_2 and primary amines (in the form of their salts) are the most common reagents. BF_3 can be added to complex with the leaving ammonia. The reaction is often used to convert urea to substituted ureas: $NH_2CONH_2 + RNH_3^+ \rightarrow NH_2CONHR + NH_4^+$.[907] N-R-Substituted amides are converted to N-R'-sub-stituted amides by treatment with N_2O_4 to give an N-nitroso compound, followed by treat-

[902]Hansen *Acta Chem. Scand.* **1963**, *17*, 1307; Satterthwait; Jencks *J. Am. Chem. Soc.* **1974**, *96*, 7018, 7031; Blackburn; Jencks, Ref. 901; Gresser; Jencks *J. Am. Chem. Soc.* **1977**, *99*, 6963, 6970. See also Yang; Jencks *J. Am. Chem. Soc.* **1988**, *110*, 2972.

[903]Blackburn; Jencks, Ref. 901.

[904]Bunnett; Davis, Ref. 899.

[905]Zaugg; Helgren; Schaefer *J. Org. Chem.* **1963**, *28*, 2617. See also Weintraub; Terrell *J. Org. Chem.* **1965**, *30*, 2470; Harada; Kinoshita *Bull. Chem. Soc. Jpn.* **1967**, *40*, 2706.

[906]For a list of procedures, with references, see Ref. 508, pp. 990-991.

[907]For a discussion of the mechanism, see Chimishkyan; Snagovskii; Gulyaev; Leonova; Kusakin *J. Org. Chem. USSR* **1985**, *21*, 1955.

ment of this with a primary amine $R'NH_2$.[908] Lactams can be converted to ring-expanded lactams if a side chain containing an amino group is present on the nitrogen. A strong base

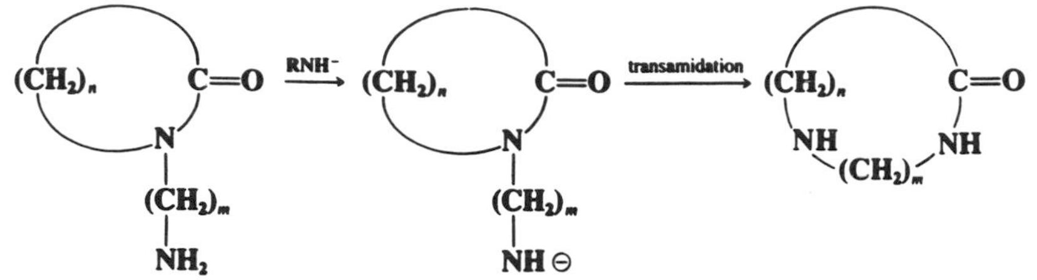

is used to convert the NH_2 to NH^-, which then acts as a nucleophile, expanding the ring by means of a transamidation.[909] The discoverers call it the Zip reaction, by analogy with the action of zippers.[910]

OS **I**, 302 (but see **V**, 589), 450, 453; **II**, 461; **III**, 151, 404; **IV**, 52, 361. See also OS **67**, 60.

0-57 Acylation of Amines by Other Acid Derivatives

Acid derivatives that can be converted to amides include thiol acids RCOSH, thiol esters RCOSR,[911] acyloxyboranes $RCOB(OR')_2$,[912] silicic esters $(RCOO)_4Si$, 1,1,1-trihalo ketones $RCOCX_3$,[913] α-keto nitriles, acyl azides, and nonenolizable ketones (see the Haller–Bauer reaction **2-33**).

OS **III**, 394; **IV**, 6, 569; **V**, 160, 166; **VI**, 1004.

C.   Attack by NHCOR

0-58   N-Alkylation of Amides and Imides
Acylamino-de-halogenation

$$RX + {}^{\ominus}NHCOR' \longrightarrow RNHCOR'$$

Amides are very weak bases, far too weak to attack alkyl halides, so they must first be converted to their conjugate bases. By this method, unsubstituted amides can be converted to N-substituted, or N-substituted to N,N-disubstituted, amides.[914] Esters of sulfuric or sulfonic acids can also be substrates. Tertiary substrates give elimination. O-Alkylation is at times a side reaction.[915] Both amides and sulfonamides have been alkylated under phase transfer conditions.[916]

---

[908]Garcia; Vilarrasa *Tetrahedron Lett.* **1982**, *23*, 1127.

[909]Kramer; Guggisberg; Hesse; Schmid *Angew. Chem. Int. Ed. Engl.* **1977**, *16*, 861 [*Angew. Chem. 89*, 899], *Helv. Chim. Acta* **1978**, *61*, 1342; Askitoğlu; Guggisberg; Hesse *Helv. Chim. Acta* **1985**, *68*, 750. For a carbon analog, see Nakashita; Hesse *Helv. Chim. Acta* **1983**, *66*, 845; Süsse; Hájiček; Hesse *Helv. Chim. Acta* **1985**, *68*, 1986.

[910]For a review of this reaction, and of other ring expansions to form macrocyclic rings, see Stach; Hesse *Tetrahedron* **1988**, *44*, 1573-1590.

[911]For a discussion of the mechanism, see Douglas *Acc. Chem. Res.* **1986**, *19*, 186-192.

[912]The best results are obtained when the acyloxyboranes are made from a carboxylic acid and catecholborane (p. 615): Collum; Chen; Ganem *J. Org. Chem.* **1978**, *43*, 4393.

[913]See, for example Salim; Nome; Rezende *Synth. Commun.* **1989**, *19*, 1181; Druzian; Zucco; Rezende; Nome *J. Org. Chem.* **1989**, *54*, 4767.

[914]For procedures, see Luh; Fung *Synth. Commun.* **1979**, *9*, 757; Koziara; Zawadzki; Zwierzak *Synthesis* **1979**, 527; Gajda; Koziara; Zawadzki; Zwierzak *Synthesis* **1979**, 549; Yamawaki; Ando; Hanafusa *Chem. Lett.* **1981**, 1143; Sukata *Bull. Chem. Soc. Jpn.* **1985**, *58*, 838.

[915]For a review of alkylation of amides, see Challis; Challis, Ref. 555, pp. 734-754.

[916]Gajda; Zwierzak *Synthesis* **1981**, 1005; Burke; Spillane *Synthesis* **1985**, 935.

The *Gabriel synthesis*[917] for converting halides to primary amines is based on this reaction. The halide is treated with potassium phthalimide and the product hydrolyzed (**0-11**):

$$RX + \overset{\ominus}{N}\underset{}{\big\langle}\text{(phthalimide)} \longrightarrow R-N\underset{}{\big\langle}\text{(phthalimide)} \overset{H^+}{\longrightarrow} RNH_3{}^+ + \underset{}{\big\langle}\text{benzene}\big\rangle\begin{array}{l}-COOH\\-COOH\end{array}$$

It is obvious that the primary amines formed in this reaction will be uncontaminated by secondary or tertiary amines (unlike **0-43**). The reaction is usually rather slow but can be conveniently speeded by the use of a dipolar aprotic solvent such as DMF[918] or with a crown ether.[919] Hydrolysis of the phthalimide, whether acid- or base-catalyzed (acid catalysis is used far more frequently), is also usually very slow, and better procedures are generally used. A common one is the Ing–Manske procedure,[920] in which the phthalimide is heated

$$R-N\underset{}{\big\langle}\text{(phthalimide)} + NH_2NH_2 \longrightarrow RNH_2 + \begin{array}{l}HN\\HN\end{array}\underset{}{\big\langle}\text{(phthalhydrazide)}$$

with hydrazine in an exchange reaction, but other methods have been introduced, using Na$_2$S in aqueous THF or acetone,[921] NaBH$_4$–2-propanol followed by acetic acid;[922] 40% aqueous methylamine,[923] and *n*-pentylamine.[924]

N-Alkyl amides or imides can also be prepared starting from alcohols by treatment of the latter with equimolar amounts of the amide or imide, Ph$_3$P, and diethyl azodicarboxylate (EtOOCN=NCOOEt) at room temperature (the Mitsunobu reaction, see p. 396).[925]

An alternative to the Gabriel synthesis, in which alkyl halides can be converted to primary amines in good yields, involves treatment of the halide with the strong base guanidine followed by alkaline hydrolysis.[926] In another alternative,[927] the sodium salt of diphenyl-

[917]For a review, see Gibson; Bradshaw *Angew. Chem. Int. Ed. Engl.* **1968**, *7*, 919-930 [*Angew. Chem. 80*, 986-996].

[918]For example, see Sheehan; Bolhofer *J. Am. Chem. Soc.* **1950**, *72*, 2786. See also Landini; Rolla *Synthesis* **1976**, 389.

[919]Soai; Ookawa; Kato *Bull. Chem. Soc. Jpn.* **1982**, *55*, 1671.

[920]Ing; Manske *J. Chem. Soc.* **1926**, 2348.

[921]Kukolja; Lammert *J. Am. Chem. Soc.* **1975**, *97*, 5582.

[922]Osby; Martin; Ganem *Tetrahedron Lett.* **1984**, *25*, 2093.

[923]Wolfe; Hasan *Can. J. Chem.* **1970**, *48*, 3572.

[924]Kasztreiner; Szilágyi; Kośáry; Huszti *Acta. Chim. Acad. Sci. Hung.* **1975**, *84*, 167 [*Chem. Abstr. 83*, 113084].

[925]Mitsunobu; Wada; Sano *J. Am. Chem. Soc.* **1972**, *94*, 679; Grunewald; Paradkar; Pazhenchevsky; Pleiss; Sall; Seibel; Reitz *J.Org. Chem.* **1983**, *48*, 2321; Ślusarska; Zwierzak *Liebigs Ann. Chem.* **1986**, 402; Kolasa; Miller *J. Org. Chem.* **1987**, *52*, 4978; Sammes; Thetford *J. Chem. Soc., Perkin Trans. 1* **1989**, 655.

[926]Hebrard; Olomucki *Bull. Soc. Chim. Fr.* **1970**, 1938.

[927]For other methods, see Mukaiyama; Taguchi; Nishi *Bull. Chem. Soc. Jpn.* **1971**, *44*, 2797; Hendrickson; Bergeron; Sternbach *Tetrahedron* **1975**, *31*, 2517; Hendrickson; Bergeron; Giga; Sternbach *J. Am. Chem. Soc.* **1973**, *95*, 3412; Clarke; Elliott; Jones *J. Chem. Soc., Perkin Trans 1* **1978**, 1088; Mukaiyama; Tsuji; Watanabe *Chem. Lett.* **1978**, 1057; Zwierzak; Pilichowska *Synthesis* **1982**, 922; Calverley *Synth. Commun.* **1983**, *13*, 601; Harland; Hodge; Maughan; Wildsmith *Synthesis* **1984**, 941; Grehn; Ragnarsson *Synthesis* **1987**, 275; Dalla Croce; La Rosa; Ritieni *J. Chem. Res. (S)* **1988**, 346; Yinglin; Hongwen *Synthesis* **1990**, 122.

phosphinamide Ph_2PONH_2 is alkylated with primary[928] or secondary[929] alkyl halides or with alcohols in the presence of $MeSO_2Cl$,[930] which converts ROH to $ROSO_2Me$. Hydrolysis of Ph_2PONHR with HCl gives the amine.

Amides can also be alkylated with diazo compounds, as in **0-48.** Salts of sulfonamides $(ArSO_2NH^-)$ can be used to attack alkyl halides to prepare N-alkyl sulfonamides (Ar-SO_2NHR) that can be further alkylated to $ArSO_2NRR'$. Hydrolysis of the latter is a good method for the preparation of secondary amines. Secondary amines can also be made by crown-ether assisted alkylation of $F_3CCONHR$ (R = alkyl or aryl) and hydrolysis of the resulting $F_3CCONRR'$.[931]

Internal N-alkylation has been used to prepare the highly strained compounds α-lactams.[932]

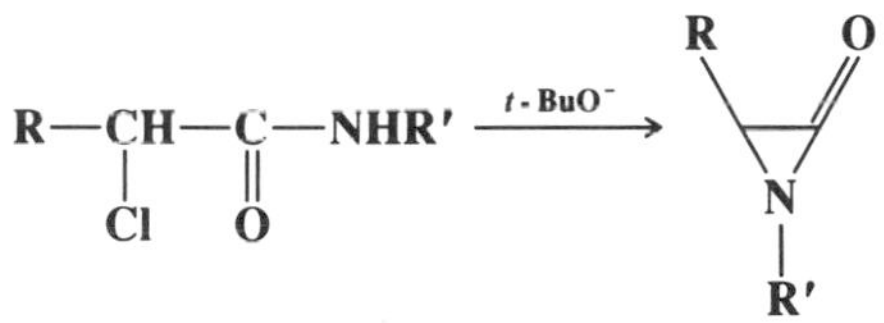

$$R{-}\underset{\underset{Cl}{|}}{C}H{-}\underset{\underset{O}{\|}}{C}{-}NHR' \xrightarrow{\ t\text{-}BuO^-\ } \underset{\underset{R'}{|}}{\overset{\overset{\displaystyle R\quad O}{\diagup\!\!\diagdown}}{N}}$$

OS **I**, 119, 203, 271; **II**, 25, 83, 208; **III**, 151; **IV**, 810; **V**, 1064; **VI**, 951; **VII**, 501.

0-59   N-Acylation of Amides and Imides
Acylamino-de-halogenation

$$\textbf{RCOCl + H_2NCOR' \longrightarrow RCONHCOR'}$$

Imides can be prepared by the attack of amides or their salts on acyl halides, anhydrides, and carboxylic acids or esters.[933] The best synthetic method for the preparation of acyclic imides is the reaction between an amide and an anhydride at 100°C catalyzed by H_2SO_4.[934] When acyl chlorides are treated with amides in a 2:1 molar ratio at low temperatures in the presence of pyridine, the products are N,N-diacylamides $(RCO)_3N$.[935]

This reaction is often used to prepare urea derivatives, an important example being the preparation of barbituric acid:[936]

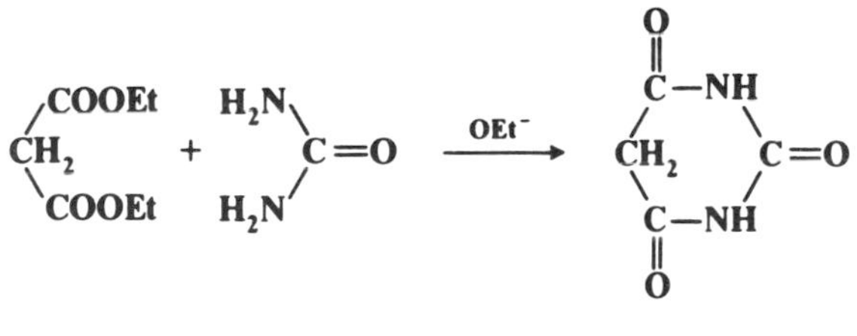

$$\underset{\underset{COOEt}{\diagdown}}{\overset{\overset{COOEt}{\diagup}}{CH_2}} + \underset{\underset{H_2N}{\diagup}}{\overset{\overset{H_2N}{\diagdown}}{C}}{=}O \xrightarrow{\ OEt^-\ } \overset{\displaystyle O}{\underset{\underset{\displaystyle O}{\overset{\|}{C}}}{\underset{\underset{C-NH}{\diagdown\ \diagup}}{\overset{\overset{C-NH}{\diagup\ \diagdown}}{CH_2}}}{C}{=}O$$

[928]Zwierzak; Podstawczyńska *Angew. Chem. Int. Ed. Engl.* **1977**, *16*, 702 [*Angew. Chem. 89*, 737].
[929]Ślusarska; Zwierzak *Synthesis* **1980**, 717.
[930]Ślusarska; Zwierzak *Synthesis* **1981**, 155.
[931]Nordlander; Catalane; Eberlein; Farkas; Howe; Stevens; Tripoulas *Tetrahedron Lett.* **1978**, 4987. For other methods, see Zwierzak; Brylikowska-Piotrowicz *Angew. Chem. Int. Ed. Engl.* **1977**, *16*, 107 [*Angew. Chem. 89*, 109]; Briggs; Brown; Jiricny; Meidine *Synthesis* **1980**, 295; Ref. 928.
[932]Baumgarten; Fuerholzer; Clark; Thompson *J. Am. Chem. Soc.* **1963**, *85*, 3303; Quast; Leybach *Chem. Ber.* **1991**, *124*, 849. For a review of α-lactams, see Lengyel; Sheehan *Angew. Chem. Int. Ed. Engl.* **1968**, *7*, 25-36 [*Angew. Chem. 80*, 27-37].
[933]For a review, see Challis; Challis, Ref. 555, pp. 759-773.
[934]Baburao; Costello; Petterson; Sander *J. Chem. Soc. C* **1968**, 2779; Davidson; Skovronek *J. Am. Chem. Soc.* **1958**, *80*, 376.
[935]For example, see LaLonde; Davis *J. Org. Chem.* **1970**, *35*, 771.
[936]For a review of barbituric acid, see Bojarski; Mokrosz; Bartoń; Paluchowska *Adv. Heterocycl. Chem.* **1985**, *38*, 229-297.

When the substrate is oxalyl chloride (ClCOCOCl) and the reagent an unsubstituted amide, an acyl isocyanate (RCONCO) is formed. The "normal" product (RCONHCOCOCl) does not form, or if it does, it rapidly loses CO and HCl.[937]

OS **II**, 60, 79, 422; **III**, 763; **IV**, 245, 247, 496, 566, 638, 662, 744; **V**, 204, 944.

D. Other Nitrogen Nucleophiles

0-60 Formation of Nitro Compounds[938]
Nitro-de-halogenation

$$RX + NO_2^- \longrightarrow RNO_2$$

Sodium nitrite can be used to form nitro compounds with primary or secondary alkyl bromides or iodides, though the method is of limited scope. Silver nitrite gives nitro compounds only when RX is a primary bromide or iodide. Nitrite esters are an important side product in all these cases (**0-32**) and become the major product (by an SN1 mechanism) when secondary or tertiary halides are treated with silver nitrite.

OS **I**, 410; **IV**, 368, 454, 724.

0-61 Formation of Azides
Azido-de-halogenation

$$RX + N_3^- \longrightarrow RN_3$$

$$RCOX + N_3^- \longrightarrow RCON_3$$

Alkyl azides can be prepared by treatment of the appropriate halide with azide ion.[939] Phase transfer catalysis[940] and ultrasound[941] have been used. Other leaving groups have also been used,[942] for example, OH,[943] OMs, OTs,[944] and OAc.[945] Epoxides react with NaN_3, with HN_3 in DMF,[946] or with HN_3–Et_3Al[947] to give β-azido alcohols; these are easily converted to aziridines,[948] e.g.,

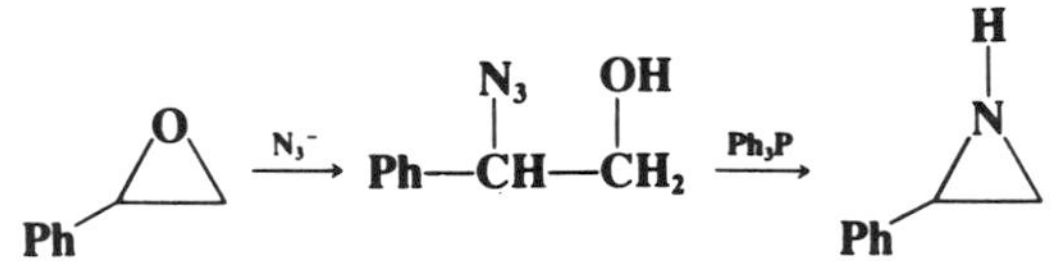

[937]Speziale; Smith *J. Org. Chem.* **1962**, *27*, 3742; Speziale; Smith; Fedder *J. Org. Chem.* **1965**, *30*, 4306.

[938]For reviews, see Larson, in Feuer *The Chemistry of the Nitro and Nitroso Groups*, pt. 1; Wiley: New York, 1969, pp. 325-339; Kornblum *Org. React.* **1962**, *12*, 101-156.

[939]For reviews, see Scriven; Turnbull *Chem. Rev.* **1988**, *88*, 297-368; Biffin; Miller; Paul, in Patai *The Chemistry of the Azido Group*; Wiley: New York, 1971, pp. 57-119.

[940]See Reeves; Bahr *Synthesis* **1979**, 823; Nakajima; Oda; Inouye *Tetrahedron Lett.* **1978**, 3107; Marti; Rico; Ader; de Savignac; Lattes *Tetrahedron Lett.* **1989**, *30*, 1245.

[941]Priebe *Acta Chem. Scand., Ser. B* **1984**, *38*, 895.

[942]See, for example, Svetlakov; Mikheev; Fedotov *J. Org. Chem. USSR* **1971**, *7*, 2304; Hojo; Kobayashi; Soai; Ikeda; Mukaiyama *Chem. Lett.* **1977**, 635; Murahashi; Tanigawa; Imada; Taniguchi *Tetrahedron Lett.* **1986**, *27*, 227.

[943]See, for example, Viaud; Rollin *Synthesis* **1990**, 130.

[944]Scriven; Turnbull, Ref. 939, p. 306.

[945]Murahashi; Taniguchi; Imada; Tanigawa *J. Org. Chem.* **1989**, *54*, 3292.

[946]Saito; Bunya; Inaba; Moriwake; Torii *Tetrahedron Lett.* **1985**, *26*, 5309.

[947]Mereyala; Frei *Helv. Chim. Acta* **1986**, *69*, 415.

[948]See, for example, Ittah; Sasson; Shahak; Tsaroom; Blum *J. Org. Chem.* **1978**, *43*, 4271. For the mechanism of the conversion to aziridines, see Pöchlauer; Müller; Peringer *Helv. Chim. Acta* **1984**, *67*, 1238.

This conversion has been used as a key step in the preparation of optically active aziridines from optically active 1,2-diols (prepared by **5-35**).[949] Even hydrogen can be the leaving group: Benzylic hydrogens have been replaced by N_3 by treatment with HN_3 in $CHCl_3$ in the presence of DDQ (p. 1163).[950]

Tertiary alkyl azides can be prepared by stirring tertiary alkyl chlorides with NaN_3 and $ZnCl_2$ in CS_2[951] or by treating tertiary alcohols with NaN_3 and CF_3COOH[952] or with HN_3 and $TiCl_4$[953] or BF_3.[954] Acyl azides, which can be used in the Curtius reaction (**8-15**), can be similarly prepared from acyl halides or anhydrides.[955]

OS **III**, 846; **IV**, 715; **V**, 273, 586; **VI**, 95, 207, 210, 910; **VII**, 433; **69**, 205. See also OS **VII**, 206.

0-62   Formation of Isocyanates and Isothiocyanates
Isocyanato-de-halogenation
Isothiocyanato-de-halogenation

$$\text{RX} + \text{NCO}^- \longrightarrow \text{RNCO}$$

$$\text{RX} + \text{NCS}^- \longrightarrow \text{RNCS}$$

When the reagent is the thiocyanate ion, S-alkylation is an important side reaction (**0-42**), but the cyanate ion practically always gives exclusive N-alkylation.[422] Primary alkyl halides have been converted to isocyanates by treatment with sodium nitrocyanamide $NaNCNNO_2$ and *m*-chloroperbenzoic acid, followed by heating of the initially produced $RN(NO_2)CN$.[956] When alkyl halides are treated with NCO^- in the presence of ethanol, carbamates can be prepared directly (see **6-8**).[957] Acyl halides give the corresponding acyl isocyanates and isothiocyanates.[958] For the formation of isocyanides, see **0-101**.

OS **III**, 735.

0-63   Formation of Bis(trimethylsilyl)amines
Bis(trimethylsilyl)amino-de-halogenation

$$\text{RX} + (\text{Me}_3\text{Si})_2\text{N}^-\ \text{Na}^+ \longrightarrow \text{RN(SiMe}_3)_2$$

Primary alkyl, allylic, and benzylic bromides, iodides, and tosylates react with sodium bis(trimethylsilyl)amide to give derivatives that are easily hydrolyzed to produce amine salts in high overall yields.[959]

$$\text{RN(SiMe}_3)_2 \xrightarrow[\text{H}^+]{\text{H}_2\text{O}} \text{RNH}_3^+ + (\text{Me}_3\text{Si})_2\text{O}$$

This is therefore an indirect way of converting halides to primary amines.

[949]Lohray; Gao; Sharpless *Tetrahedron Lett.* **1989**, *30*, 2623.
[950]Guy; Lemor; Doussot; Lemaire *Synthesis* **1988**, 900.
[951]Miller *Tetrahedron Lett.* **1975**, 2959. See also Koziara; Zwierzak *Tetrahedron Lett.* **1987**, *28*, 6513.
[952]Balderman; Kalir *Synthesis* **1978**, 24.
[953]Hassner; Fibiger; Andisik *J. Org. Chem.* **1984**, *49*, 4237.
[954]See, for example, Adam; Andrieux; Plat *Tetrahedron* **1985**, *41*, 399.
[955]For a review of acyl azides, see Lwowski, in Patai, Ref. 939, pp. 503-554.
[956]Manimaran; Wolford; Boyer *J. Chem. Res. (S)* **1989**, 331.
[957]Argabright; Rider; Sieck *J. Org. Chem.* **1965**, *30*, 3317; Effenberger; Drauz; Förster; Müller *Chem. Ber.* **1981**, *114*, 173.
[958]For reviews of acyl isocyanates, see Tsuge, in Patai, Ref. 585, pt. 1, pp. 445-506; Nuridzhanyan *Russ. Chem. Rev.* **1970**, *39*, 130-139; Lozinskii; Pel'kis *Russ. Chem. Rev.* **1968**, *37*, 363-375.
[959]Bestmann; Wölfel *Chem. Ber.* **1984**, *117*, 1250.

0-64 Formation of Azoxy Compounds
Alkyl-*NNO*-azoxy-de-halogenation

$$RX + R'N{=}N{-}O^- \longrightarrow R{-}\overset{\oplus}{\underset{O_\ominus}{N}}{=}N{-}R'$$
$$\textbf{108}$$

The reaction between alkyl halides and alkanediazotates (**108**) gives azoxyalkanes.[960] R and R' may be the same or different, but neither may be aryl or tertiary alkyl. The reaction is regioselective; only the isomer shown is obtained.

Halogen Nucleophiles[961]

A. Attack at an Alkyl Carbon

0-65 Halide Exchange
Halo-de-halogenation

$$RX + X'^- \rightleftharpoons RX' + X^-$$

Halide exchange, sometimes call the *Finkelstein reaction*, is an equilibrium process, but it is often possible to shift the equilibrium.[962] The reaction is most often applied to the preparation of iodides and fluorides. Iodides can be prepared from chlorides or bromides by taking advantage of the fact that sodium iodide, but not the bromide or chloride, is soluble in acetone. When an alkyl chloride or bromide is treated with a solution of sodium iodide in acetone, the equilibrium is shifted by the precipitation of sodium chloride or bromide. Since the mechanism is S_N2, the reaction is much more successful for primary halides than for secondary or tertiary halides; sodium iodide in acetone can be used as a test for primary bromides or chlorides. Tertiary chlorides can be converted to iodides by treatment with excess NaI in CS_2, with $ZnCl_2$ as catalyst.[963] Vinylic bromides give vinylic iodides with retention of configuration when treated with KI and a nickel bromide-zinc catalyst,[964] or with KI and CuI in hot HMPA.[965]

Fluorides[966] are prepared by treatment of other alkyl halides with any of a number of fluorinating agents, among them anhydrous HF (which is useful only for reactive substrates such as benzylic or allylic), AgF, KF, HgF_2, Bu_4N^+ HF_2^-,[967] BrF_3,[968] $Et_3N{\cdot}2HF$,[969] and, for polyhalo compounds (such as chloroform), HF plus SbF_3.[970] The equilibria in these cases

[960]For reviews, see Yandovskii; Gidaspov; Tselinskii *Russ. Chem. Rev.* **1980**, *49*, 237-248; Moss *Acc. Chem. Res.* **1974**, *7*, 421-427.

[961]For a review of the formation of carbon–halogen bonds, see Hudlicky; Hudlicky, in Patai; Rappoport, Ref. 88, pt. 2, pp. 1021-1172.

[962]For a list of reagents for alkyl halide interconversion, see Ref. 508, pp. 337-339.

[963]Miller; Nunn *J. Chem. Soc., Perkin Trans 1* **1976**, 416.

[964]Takagi; Hayama; Inokawa *Chem. Lett.* **1978**, 1435.

[965]Suzuki; Aihara; Yamamoto; Takamoto; Ogawa *Synthesis* **1988**, 236.

[966]For reviews of the introduction of fluorine into organic compounds, see Mann *Chem. Soc. Rev.* **1987**, *16*, 381-436; Rozen; Filler *Tetrahedron* **1985**, *41*, 1111-1153; Hudlický, Ref. 448, pp. 24-169; Sheppard; Sharts, Ref. 448, pp. 52-184, 409-430.

[967]Bosch; Camps; Chamorro; Gasol; Guerrero *Tetrahedron Lett.* **1987**, *28*, 4733. See also Cox; Terpinski; Lawrynowicz *J. Org. Chem.* **1984**, *49*, 3216.

[968]Kartashov; Chuvatkin; Kurskii; Boguslavskaya *J. Org. Chem. USSR* **1988**, *24*, 2279.

[969]Giudicelli; Picq; Veyron *Tetrahedron Lett.* **1990**, *31*, 6527.

[970]For reviews of the use of halogen exchange to prepare alkyl fluorides, see Sharts; Sheppard *Org. React.* **1974**, *21*, 125-406; Hudlický, Ref. 448, pp. 91-136.

are shifted because the alkyl fluoride once formed has little tendency to react, owing to the extremely poor leaving-group ability of fluorine. Phase transfer catalysis of the exchange reaction is a particularly effective way of preparing both fluorides and iodides.[971]

Primary alkyl chlorides can be converted to bromides with ethyl bromide, N-methyl-2-pyrrolidinone and a catalytic amount of NaBr,[972] with LiBr under phase-transfer conditions,[973] and with Bu_4N^+ Br^-.[974] For secondary and tertiary alkyl chlorides, treatment in CH_2Cl_2 with excess gaseous HBr and an anhydrous $FeBr_3$ catalyst has given high yields[975] (this procedure is also successful for chroride-to-iodide conversions). Alkyl chlorides or bromides can be prepared from iodides by treatment with HCl or HBr in the presence of HNO_3, making use of the fact that the leaving I^- is oxidized to I_2 by the HNO_3.[976] Primary iodides give the chlorides when treated with PCl_5 in $POCl_3$.[977] Alkyl fluorides and chlorides are converted to the bromides and iodides (and alkyl fluorides to the chlorides) by heating with the corresponding HX in excess amounts.[978]

OS **II**, 476; **IV**, 84, 525; **66**, 87.

0-66   Formation of Alkyl Halides from Esters of Sulfuric and Sulfonic Acids
Halo-de-sulfonyloxy-substitution, etc.

$$ROSO_2R' + X^- \longrightarrow RX$$

Alkyl sulfates, tosylates, and other esters of sulfuric and sulfonic acids can be converted to alkyl halides with any of the four halide ions.[979] Neopentyl tosylate reacts with Cl^-, Br^-, or I^- without rearrangement in HMPA.[980] Similarly, allylic tosylates can be converted to chlorides without allylic rearrangement by reaction with LiCl in the same solvent.[981] Inorganic esters are intermediates in the conversion of alcohols to alkyl halides with $SOCl_2$, PCl_5, PCl_3, etc. (**0-67**), but are seldom isolated.

OS **I**, 25; **II**, 111, 404; **IV**, 597, 753; **V**, 545.

0-67   Formation of Alkyl Halides from Alcohols
Halo-de-hydroxylation

$$ROH + HX \longrightarrow RX$$

$$ROH + SOCl_2 \longrightarrow RCl$$

Alcohols can be converted to alkyl halides with several reagents,[982] the most common of which are halogen acids HX and inorganic acid halides such as $SOCl_2$,[983] PCl_5, PCl_3, $POCl_3$, etc.[984] HBr is usually used for alkyl bromides and HI for alkyl iodides. These reagents are

[971]For reviews, see Starks; Liotta, Ref. 404, pp. 112-125; Weber; Gokel *Phase Transfer Catalysis in Organic Synthesis.* Ref. 404, pp. 117-124. See also Clark; Macquarrie *Tetrahedron Lett.* **1987,** *28,* 111; Bram; Loupy; Pigeon *Synth. Commun.* **1988,** *18,* 1661.
[972]Willy; McKean; Garcia *Bull. Chem. Soc. Jpn.* **1976,** *49,* 1989. See also Babler; Spina *Synth. Commun.* **1984,** *14,* 1313.
[973]Sasson; Weiss; Loupy; Bram; Pardo *J. Chem. Soc., Chem. Commun.* **1986,** 1250; Loupy; Pardo *Synth. Commun.* **1988,** *18,* 1275.
[974]Bidd; Whiting *Tetrahedron Lett.* **1984,** *25,* 5949.
[975]Yoon; Kochi *J. Org. Chem.* **1989,** *54,* 3028.
[976]Svetlakov; Moisak; Averko-Antonovich *J. Org. Chem. USSR* **1969,** *5,* 971.
[977]Bartley; Carman; Russell-Maynard *Aust. J. Chem.* **1985,** *38,* 1879.
[978]Namavari; Satyamurthy; Phelps; Barrio *Tetrahedron Lett.* **1990,** *31,* 4973.
[979]For a list of reagents, with references, see Ref. 508, pp. 360-362.
[980]Stephenson; Solladié; Mosher, Ref. 248.
[981]Stork; Grieco; Gregson *Tetrahedron Lett.* **1969,** 1393.
[982]For a list of reagents, with references, see Ref. 508, pp. 353-360.
[983]For a review of thionyl chloride $SOCl_2$, see Pizey, Ref. 593, vol. 1, 1974, pp. 321-357.
[984]For a review, see Brown, in Patai, Ref. 575, pt. 1, pp.595-622.

often generated in situ from the halide ion and an acid such as phosphoric or sulfuric. The use of HI sometimes results in reduction of the alkyl iodide to the alkane (**0-76**) and, if the substrate is unsaturated, can also reduce the double bond.[985] The reaction can be used to prepare primary, secondary, or tertiary halides, but alcohols of the isobutyl or neopentyl type often give large amounts of rearrangement products. Tertiary chlorides are easily made with concentrated HCl, but primary and secondary alcohols react with HCl so slowly that a catalyst, usually zinc chloride, is required.[986] Primary alcohols give good yields of chlorides upon treatment with HCl in HMPA.[987] The inorganic acid chlorides $SOCl_2$, PCl_3, etc., give primary, secondary, or tertiary alkyl chlorides with much less rearrangement than is observed with HCl.

Analogous bromides and iodides, especially PBr_3, have also been used, but they are more expensive and used less often than HBr or HI, though some of them may also be generated in situ (e.g., PBr_3 from phosphorous and bromine). Secondary alcohols always gives *some* rearranged bromides if another secondary position is available, even with PBr_3, PBr_5, or $SOBr_2$; thus 3-pentanol gives both 2- and 3-bromopentane. Such rearrangement can be avoided by converting the alcohol to a sulfonate and then using **0-66**,[988] or by the use of phase transfer catalysis.[989] HF does not generally convert alcohols to alkyl fluorides.[990] The most important reagent for this purpose is the commercially available diethylaminosulfur trifluoride Et_2NSF_3 (DAST),[991] which converts primary, secondary, tertiary, allylic, and benzylic alcohols to fluorides in high yields under mild conditions.[992] Fluorides have also been prepared from alcohols by treatment with SF_4,[993] SeF_4,[994] TsF,[995] and indirectly, by conversion to a sulfate or tosylate, etc. (**0-66**).

Primary, secondary, and tertiary alcohols can be converted to any of the four halides by treatment with the appropriate NaX, KX, or NH_4X in polyhydrogen fluoride–pyridine solution.[996] This method is even successful for neopentyl halides. Another reagent that converts neopentyl alcohol to neopentyl chloride, in 95% yield, is PPh_3–CCl_3CN.[997]

Other reagents[998] have also been used, for example, $(RO)_3PRX$[999] and R_3PX_2[1000] (made from R_3P and X_2), which give good yields for primary (including neopentyl), secondary,

[985]Jones; Pattison *J. Chem. Soc. C* **1969**, 1046.
[986]Phase-transfer catalysts have been used instead of $ZnCl_2$; Landini; Montanari; Rolla *Synthesis* **1974**, 37.
[987]Fuchs; Cole *Can. J. Chem.* **1975**, *53*, 3620.
[988]Cason; Correia *J. Org. Chem.* **1961**, *26*, 3645.
[989]Dakka; Sasson *Tetrahedron Lett.* **1987**, *28*, 1223.
[990]For an exception, see Hanack; Eggensperger; Hähnle *Liebigs Ann. Chem.* **1962**, *652*, 96; See also Politanskii; Ivanyk; Sarancha; Shevchuk *J. Org. Chem. USSR* **1974**, *10*, 697.
[991]For a review of this reagent, see Hudlický *Org. React.* **1988**, *35*, 513-637.
[992]Middleton *J. Org. Chem.* **1975**, *40*, 574.
[993]For reviews, see Wang *Org. React.* **1985**, *34*, 319-400; Kollonitsch *Isr. J. Chem.* **1978**, *17*, 53-59; Boswell; Ripka; Scribner; Tullock *Org. React.* **1974**, *21*, 1-124.
[994]Olah; Nojima; Kerekes *J. Am. Chem. Soc.* **1974**, *96*, 925.
[995]Shimizu; Nakahara; Yoshioka *Tetrahedron Lett.* **1985**, *26*, 4207. For another method, see Olah; Li *Synlett* **1990**, 267.
[996]Olah; Welch *Synthesis* **1974**, 653; Olah; Welch; Vankar; Nojima; Kerekes; Olah *J. Org. Chem.* **1979**, *44*, 3872; Alvernhe; Lacombe; Laurent; Rousset *J. Chem. Res., (S)* **1983**, 246.
[997]Matveeva; Yalovskaya; Cherepanov; Kurts; Bundel' *J. Org. Chem. USSR* **1989**, *25*, 587.
[998]For some other reagents, not listed here, see Echigo; Mukaiyama *Chem. Lett.* **1978**, 465; Barton; Stick; Subramanian *J. Chem. Soc., Perkin Trans. 1* **1976**, 2112; Savel'yanov; Nazarov; Savel'yanova; Suchkov *J. Org. Chem. USSR* **1977**, *13*, 604; Jung; Hatfield *Tetrahedron Lett.* **1978**, 4483; Sevrin; Krief *J. Chem. Soc., Chem. Commun.* **1980**, 656; Olah; Gupta; Malhotra; Narang *J. Org. Chem.* **1980**, *45*, 1638; Hanessian; Leblanc; Lavallée *Tetrahedron Lett.* **1982**, *23*, 4411; Cristol; Seapy *J. Org. Chem.* **1982**, *47*, 132; Richter; Tucker *J. Org. Chem.* **1983**, *48*, 2625; Imamoto; Matsumoto; Kusumoto; Yokoyama *Synthesis* **1983**, 460; Ref. 515; Toto; Doi *J. Org. Chem.* **1987**, *52*, 4999; Camps; Gasol; Guerrero *Synthesis* **1987**, 511; Schmidt; Brooks *Tetrahedron Lett.* **1987**, *28*, 767; Collingwood; Davies; Golding *Tetrahedron Lett.* **1987**, *28*, 4445; Kozikowski; Lee *Tetrahedron Lett.* **1988**, *29*, 3053; Classon; Liu; Samuelsson *J. Org. Chem.* **1988**, *53*, 6126; Munyemana; Frisque-Hesbain; Devos; Ghosez *Tetrahedron Lett.* **1989**, *30*, 3077; Ernst; Winkler *Tetrahedron Lett.* **1989**, *30*, 3081.
[999]Rydon *Org. Synth. VI*, 830.
[1000]Wiley; Hershkowitz; Rein; Chung *J. Am. Chem. Soc.* **1964**, *86*, 964; Wiley; Rein; Hershkowitz *Tetrahedron Lett.* **1964**, 2509; Schaefer; Weinberg *J. Org. Chem.* **1965**, *30*, 2635; Kaplan *J. Org. Chem.* **1966**, *31*, 3454; Weiss; Snyder *J. Org. Chem.* **1971**, *36*, 403; Garegg; Johansson; Samuelsson *Synthesis* **1984**, 168.

and tertiary halides without rearrangements,[1001] Me_2SBr_2[1002] (prepared from Me_2S and Br_2), $Me_3SiCl–SeO_2$,[1003] and a mixture of PPh_3 and CCl_4[1004] (or CBr_4[1005]).

$$ROH + Ph_3P + CCl_4 \longrightarrow RCl + Ph_3PO + HCCl_3$$

The last method converts allylic alcohols[1006] to the corresponding halides without allylic rearrangements.[1007] A simple method that is specific for benzylic and allylic alcohols (and does not give allylic rearrangement) involves reaction with N-chloro- or N-bromosuccinimide and methyl sulfide.[1008] The specificity of this method is illustrated by the conversion, in 87% yield, of (Z)-$HOCH_2CH_2CMe{=}CHCH_2OH$ to (Z)-$HOCH_2CH_2Me{=}CHCH_2Cl$. Only the allylic OH group was affected. Allylic and benzylic alcohols can also be converted to bromides or iodides with $NaX–BF_3$ etherate,[1009] and to iodides with AlI_3.[1010]

When the reagent is HX, the mechanism is SN1cA or SN2cA; i.e., the leaving group is not OH^-, but OH_2 (p. 352). The leaving group is not OH^- with the other reagents either, since in these cases the alcohol is first converted to an inorganic ester, e.g., ROSOCl with $SOCl_2$ (0-32). The leaving group is therefore $OSOCl^-$ or a similar group (0-66). These may react by the SN1 or SN2 mechanism and, in the case of ROSOCl, by the SNi mechanism (p. 326).

OS **I**, 25, 36, 131, 142, 144, 292, 294, 533; **II**, 91, 136, 159, 246, 308, 322, 358, 399, 476; **III**, 11, 227, 370, 446, 698, 793, 841; **IV**, 106, 169, 323, 333, 576, 681; **V**, 1, 249, 608; **VI**, 75, 628, 634, 638, 781, 830, 835; **VII**, 210, 319, 356; **65**, 119, 211. Also see OS **III**, 818; **IV**, 278, 383, 597.

0-68   Formation of Alkyl Halides from Ethers
Halo-de-alkoxylation

$$ROR' + HI \longrightarrow RI + R'OH$$

Ethers can be cleaved by heating with concentrated HI or HBr.[1011] HCl is seldom successful.[1012] HBr reacts more slowly than HI, but it is often a superior reagent, since it causes fewer side reactions. Phase transfer catalysis has also been used.[1013] Dialkyl ethers and alkyl aryl ethers can be cleaved. In the latter case the alkyl–oxygen bond is the one broken. As in **0-67** the actual leaving group is not OR'^-, but OHR'. Although alkyl aryl ethers always cleave so as to give an alkyl halide and a phenol, there is no general rule for dialkyl ethers. Often cleavage occurs from both sides, and a mixture of two alcohols and two alkyl halides is obtained. However, methyl ethers are usually cleaved so that methyl iodide or bromide is a product. An excess of HI or HBr converts the alcohol product into alkyl halide, so that dialkyl ethers (but not alkyl aryl ethers) are converted to 2 moles of alkyl halide. This

---

[1001]For reviews of reactions with these reagents, see Castro *Org. React.* **1983**, *29*, 1-162; Mackie, in Cadogan *Organophosphorus Reagents in Organic Synthesis*; Academic Press: New York, 1979; pp. 433-466.
[1002]Furukawa; Inoue; Aida; Oae *J. Chem. Soc., Chem. Commun.* **1973**, 212.
[1003]Lee; Kang *J. Org. Chem.* **1988**, *53*, 3634.
[1004]For a review, see Appel, *Angew. Chem. Int. Ed. Engl.* **1975**, *14*, 801-811 [*Angew. Chem. 87*, 863-874]. For a general review of this and related reagents, see Appel; Halstenberg, in Cadogan, Ref. 1001, pp. 387-431. For a discussion of the mechanism, see Slagle, Huang, Franzus *J. Org. Chem.* **1981**, *46*, 3526.
[1005]Katritzky; Nowak-Wydra; Marson *Chem. Scr.* **1987**, *27*, 477; Wagner; Heitz; Mioskowski *Tetrahedron Lett.* **1989**, *30*, 557.
[1006]For a review of the conversion of allylic alcohols to allylic halides, see Magid *Tetrahedron* **1980**, *36*, 1901-1930, pp. 1924-1926.
[1007]Snyder *J. Org. Chem.* **1972**, *37*, 1466; Axelrod; Milne; van Tamelen *J. Am. Chem. Soc.* **1973**, *92*, 2139.
[1008]Corey; Kim; Takeda *Tetrahedron Lett.* **1972**, 4339.
[1009]Vankar; Rao *Tetrahedron Lett.* **1985**, *26*, 2717; Mandal; Mahajan *Tetrahedron Lett.* **1985**, *26*, 3863.
[1010]Sarmah; Barua *Tetrahedron* **1989**, *45*, 3569.
[1011]For reviews of ether cleavage in general, see Bhatt; Kulkarni *Synthesis* **1983**, 249-282; Ref. 333. For a review of cleavage of aryl alkyl ethers, see Tiecco, Ref. 762.
[1012]Cleavage with HCl has been accomplished in the presence of surfactants: Juršić *J. Chem. Res. (S)* **1989**, 284.
[1013]Landini; Montanari; Rolla *Synthesis* **1978**, 771.

procedure is often carried out so that a mixture of only two products is obtained instead of four. Cyclic ethers (usually tetrahydrofuran derivatives) can be similarly cleaved (see **0-69** for epoxides). Ethers have also been cleaved with Lewis acids such as BF_3, BCl_3, Me_2BBr,[1014] BBr_3,[1015] or $AlCl_3$.[1016] In such cases, the departure of the OR is assisted by complex formation with the Lewis acid:

$$R\overset{\oplus}{-}\underset{|}{\overset{|}{O}}\overset{\ominus}{-}BF_3$$
$$\underset{R'}{|}$$

Lewis acids are also used in conjunction with acyl halides. The reagent $NaI–BF_3$ etherate selectively cleaves ethers in the order benzylic ethers > alkyl methyl ethers > aryl methyl ethers.[1017]

Dialkyl and alkyl aryl ethers can be cleaved with iodotrimethylsilane:[1017a] ROR' + $Me_3SiI \rightarrow RI + Me_3SiOR$.[1018] A more convenient and less expensive alternative, which gives the same products, is a mixture of chlorotrimethylsilane and NaI.[1019] A mixture of $SiCl_4$ and NaI has also been used,[1020] as has diiodosilane SiH_2I_2.[1021] Alkyl aryl ethers can also be cleaved with LiI to give alkyl iodides and salts of phenols[1022] in a reaction similar to **0-70**. Triphenyldibromophosphorane (Ph_3PBr_2) cleaves dialkyl ethers to give 2 moles of alkyl bromide.[1023]

A closely related reaction is cleavage of oxonium salts.

$$R_3O^+ \ X^- \longrightarrow RX + R_2O$$

For these substrates, HX is not required, and X can be any of the four halide ions.

t-Butyldimethylsilyl ethers $ROSiMe_2CMe_3$ can be converted to bromides RBr by treatment with Ph_3PBr_2,[1024] $Ph_3P–CBr_4$,[1025] or BBr_3.[1026] Alcohols are often protected by conversion to this kind of silyl ether.[1027]

OS **I**, 150; **II**, 571; **III**, 187, 432, 586, 692, 753, 774, 813; **IV**, 266, 321; **V**, 412; **VI**, 353. See also OS **65**, 68; **67**, 210.

0-69   Formation of Halohydrins from Epoxides
(3) *OC-seco*-**Halo-de-alkoxylation**

$$-\overset{|}{\underset{|}{C}}-\overset{|}{\underset{\diagdown \ \diagup}{C}}- \ + \ HX \longrightarrow -\overset{|}{\underset{|}{C}}-\overset{|}{\underset{|}{C}}-$$
$$\underset{O}{} \qquad\qquad \underset{OH \ \ X}{}$$

[1014]Guindon; Yoakim; Morton *Tetrahedron Lett.* **1983**, *24*, 2969; Guindon; Bernstein; Anderson *Tetrahedron Lett.* **1987**, *28*, 2225; Guindon; Therien; Girard; Yoakim *J. Org. Chem.* **1987**, *52*, 1680.
[1015]Manson; Musgrave *J. Chem. Soc.* **1963**, 1011; McOmie; Watts; West *Tetrahedron* **1968**, *24*, 2289; Egly; Pousse; Brini *Bull. Soc. Chim. Fr.* **1972**, 1357; Press *Synth. Commun.* **1979**, *9*, 407; Niwa; Hida; Yamada *Tetrahedron Lett.* **1981**, *22*, 4239.
[1016]For a review, see Johnson, in Olah *Friedel-Crafts and Related Reactions*, vol. 4; Wiley: New York, 1965, pp. 1-109.
[1017]Vankar; Rao *J. Chem. Res. (S)* **1985**, 232. See also Mandal; Soni; Ratnam *Synthesis* **1985**, 274.
[1017a]For a review of this reagent, see Olah; Prakash; Krishnamurti *Adv. Silicon Chem.* **1991**, *1*, 1-64.
[1018]Jung; Lyster *J. Org. Chem.* **1977**, *42*, 3761; *Org. Synth.* VI, 353.
[1019]Morita; Okamoto; Sakurai *J. Chem. Soc., Chem. Commun.* **1978**, 874; Olah; Narang; Gupta; Malhotra *J. Org. Chem.* **1979**, *44*, 1247; Amouroux; Jatczak; Chastrette *Bull. Soc. Chim. Fr.* **1987**, 505.
[1020]Bhatt; El-Morey *Synthesis* **1982**, 1048.
[1021]Keinan; Perez *J. Org. Chem.* **1987**, *52*, 4846.
[1022]Harrison *Chem. Commun.* **1969**, 616.
[1023]Anderson; Freenor *J. Org. Chem.* **1972**, *37*, 626.
[1024]Aizpurua; Cossío; Palomo *J. Org. Chem.* **1986**, *51*, 4941.
[1025]Mattes; Benezra *Tetrahedron Lett.* **1987**, *28*, 1697.
[1026]Kim; Park *J. Org. Chem.* **1988**, *53*, 3111.
[1027]See Corey; Venkateswarlu *J. Am. Chem. Soc.* **1972**, *94*, 6190.

This is a special case of **0-68** and is frequently used for the preparation of halohydrins. In contrast to the situation with open-chain ethers and with larger rings, many epoxides react with all four hydrohalic acids, though with HF[1028] the reaction is unsuccessful with simple aliphatic and cycloalkyl epoxides.[1029] HF does react with more rigid epoxides, such as those in steroid systems. The reaction can applied to simple epoxides[1030] if polyhydrogen fluoride–pyridine is the reagent. The epoxide-to-fluorohydrin conversion has also been carried out with SiF_4 and a tertiary amine.[1031] Chloro-, bromo-, and iodohydrins can also be prepared[1032] by treating epoxides with Ph_3P and X_2.[1033] Epoxides can be converted directly to 1,2-dichloro compounds by treatment with $SOCl_2$ and pyridine,[1034] with Ph_3P and CCl_4,[1035] or with Ph_3PCl_2.[1036] These are two-step reactions: a halohydrin is formed first and is then converted by the reagents to the dihalide (**0-67**). As expected, inversion is found at both carbons. Meso epoxides were cleaved enantioselectively with the chiral reagents B-halodiisopino-campheylboranes (see **5-12**), where the halogen was Cl, Br, or I.[1037]

Acyl chlorides react with ethylene oxide in the presence of NaI to give 2-iodoethyl esters.[1038]

$$R\!-\!\underset{\underset{O}{\|}}{C}\!-\!Cl + H_2C\!-\!CH_2 + NaI \xrightarrow{MeCN} R\!-\!\underset{\underset{O}{\|}}{C}\!-\!O\!-\!CH_2CH_2I$$

OS **I**, 117; **VI**, 424.

0-70   Cleavage of Carboxylic Esters with Lithium Iodide
Iodo-de-acyloxy-substitution

$$R'COOR + LiI \xrightarrow[\Delta]{pyridine} RI + R'COOLi$$

Carboxylic esters where R is methyl or ethyl can be cleaved by heating with lithium iodide in refluxing pyridine or a higher-boiling amine.[1039] The reaction is useful where a molecule is sensitive to acid and base (so that **0-10** cannot be used) or where it is desired to cleave selectively only one ester group in a molecule containing two or more. For example, refluxing O-acetyloleanolic acid methyl ester with LiI in s-collidine cleaved only the 17-carbomethoxy

[1028]For a review of reactions HF with epoxides, see Sharts; Sheppard, Ref. 966. For a related review, see Yoneda *Tetrahedron* **1991**, *47*, 5329-5365.

[1029]Shahak; Manor; Bergmann *J. Chem. Soc. C* **1968**, 2129.

[1030]Olah; Meidar *Isr. J. Chem.* **1978**, *17*, 148.

[1031]Shimizu; Yoshioka *Tetrahedron Lett.* **1988**, *29*, 4101. For other methods, see Muehlbacher; Poulter *J. Org. Chem.* **1988**, *53*, 1026; Ichihara; Hanafusa *J. Chem. Soc., Chem. Commun.* **1989**, 1848.

[1032]Einhorn; Luche *J. Chem. Soc., Chem. Commun.* **1986**, 1368; Ciaccio; Addess; Bell *Tetrahedron Lett.* **1986**, *27*, 3697; Spawn; Drtina; Wiemer *Synthesis* **1986**, 315.

[1033]Palumbo; Ferreri; Caputo *Tetrahedron Lett.* **1983**, *24*, 1307.

[1034]Campbell; Jones; Wolfe *Can. J. Chem.* **1966**, *44*, 2339.

[1035]Isaccs; Kirkpatrick *Tetrahedron Lett.* **1972**, 3869.

[1036]Sonnet; Oliver *J. Org. Chem.* **1976**, *41*, 3279; *Org. Synth. VI*, 424. This method also applies to Ph_3PBr_2. For another method, see Echigo; Watanabe; Mukaiyama *Chem. Lett.* **1977**, 1013.

[1037]Srebnik; Joshi; Brown *Isr. J. Chem.* **1989**, *29*, 229.

[1038]Belsner; Hoffmann *Synthesis* **1982**, 239. See also Roloff *Chimia* **1985**, *39*, 392; Iqbal; Khan; Srivastava *Tetrahedron Lett.* **1988**, *29*, 4985.

[1039]Taschner; Liberek *Rocz. Chem.* **1956**, *30*, 323 [*Chem. Abstr.* **1957**, *51*, 1039]. For a review, see Ref. 364.

group, not the 3-acetyl group.[1040] Esters RCOOR' and lactones can also be cleaved with a mixture of Me$_3$SiCl and NaI to give R'I and RCOOH.[1041]

0-71 Conversion of Diazo Ketones to α-Halo Ketones
Hydro,halo-de-diazo-bisubstitution

$$RCOCHN_2 + HBr \longrightarrow RCOCH_2Br$$

When diazo ketones are treated with HBr or HCl, they give the respective α-halo ketones. HI does not give the reaction, since it reduces the product to a methyl ketone (**0-82**). α-Fluoro ketones can be prepared by addition of the diazo ketone to polyhydrogen fluoride–pyridine.[1042] This method is also successful for diazoalkanes.

Diazotization of α-amino acids in the above solvent at room temperature gives α-fluoro carboxylic acids.[1043] If this reaction is run in the presence of excess KCl or KBr, the corresponding α-chloro or α-bromo acid is obtained instead.[1044]

OS **III**, 119.

0-72 Conversion of Amines to Halides
Halo-de-amination

$$RNH_2 \longrightarrow RNTs_2 \xrightarrow[\text{DMF}]{I^-} RI$$

Primary alkyl amines RNH$_2$ can be converted[1045] to alkyl halides by (1) conversion to RNTs$_2$ (p. 354) and treatment of this with I$^-$ or Br$^-$ in DMF,[347] (2) diazotization with *t*-butyl nitrite and a metal halide such as TiCl$_4$ in DMF,[1046] or (3) the Katritzky pyrylium–pyridinium method (p. 354).[1047] Alkyl groups can be cleaved from secondary and tertiary aromatic amines by concentrated HBr in a reaction similar to **0-68**, e.g.,[1048]

$$ArNR_2 + HBr \longrightarrow RBr + ArNHR$$

Tertiary aliphatic amines are also cleaved by HI, but useful products are seldom obtained. Tertiary amines can be cleaved by reaction with phenyl chloroformate:[1049] R$_3$N + ClCOOPh → RCl + R$_2$NCOOPh. α-Chloroethyl chloroformate behaves similarly.[1050] Alkyl halides may be formed when quaternary ammonium salts are heated: R$_4$N$^+$ X$^-$ → R$_3$N + RX.[1051]

OS **66**, 151. See also OS **I**, 428.

0-73 Conversion of Tertiary Amines to Cyanamides. The von Braun Reaction
Bromo-de-dialkylamino-substitution

$$R_3N + BrCN \longrightarrow R_2NCN + RBr$$

[1040]Elsinger; Schreiber; Eschenmoser *Helv. Chim. Acta* **1960**, *43*, 113.
[1041]Olah; Narang; Gupta; Malhotra, Ref. 1019. See also Kolb; Barth *Synth. Commun.* **1981**, *11*, 763.
[1042]Olah; Welch *Synthesis* **1974**, 896; Olah; Welch; Vankar; Nojima; Kerekes; Olah, Ref. 996.
[1043]Olah; Prakash; Chao *Helv. Chim. Acta* **1981**, *64*, 2528; Faustini; De Munary; Panzeri; Villa; Gandolfi *Tetrahedron Lett.* **1981**, *22*, 4533; Barber; Keck; Rétey *Tetrahedron Lett.* **1982**, *23*, 1549.
[1044]Olah; Shih; Prakash *Helv. Chim. Acta* **1983**, *66*, 1028.
[1045]For another method, see Lorenzo; Molina; Vilaplana *Synthesis* **1980**, 853.
[1046]Doyle; Bosch; Seites *J. Org. Chem.* **1978**, *43*, 4120.
[1047]Katritzky; Horvath; Plau *Synthesis* **1979**, 437; Katritzky; Chermprapai; Patel *J. Chem. Soc., Perkin Trans. 1* **1980**, 2901.
[1048]Chambers; Pearson *J. Org. Chem.* **1963**, *28*, 3144.
[1049]Hobson; McCluskey *J. Chem. Soc. C* **1967**, 2015. For a review, see Cooley; Evain *Synthesis* **1989**, 1-7.
[1050]Olofson; Martz; Senet; Piteau; Malfroot *J. Org. Chem.* **1984**, *49*, 2081; Olofson; Abbott *J. Org. Chem.* **1984**, *49*, 2795. See also Campbell; Pilipauskas; Khanna; Rhodes *Tetrahedron Lett.* **1987**, *28*, 2331.
[1051]For examples, see Ko; Leffek *Can. J. Chem.* **1970**, *48*, 1865, **1971**, *49*, 129; Deady; Korytsky *Tetrahedron Lett.* **1979**, 451.

The *von Braun reaction*, which involves the cleavage of tertiary amines by cyanogen bromide to give an alkyl bromide and a disubstituted cyanamide, has been applied to many tertiary amines.[1052] Usually, the R group that cleaves is the one that gives the most reactive halide (for example, benzyl or allyl). For simple alkyl groups, the smallest are the most readily cleaved. One or two of the groups on the amine may be aryl, but they do not cleave. Cyclic amines have been frequently cleaved by this reaction. Secondary amines also give the reaction, but the results are usually poor.[1053]

The mechanism consists of two successive nucleophilic substitutions, with the tertiary amine as the first nucleophile and the liberated bromide ion as the second:

Step 1 $\text{NC}\overset{\curvearrowright}{-}\text{Br} + \text{R}_3\overline{\text{N}}$ $\longrightarrow$ $\text{NC}-\overset{\oplus}{\text{N}}\text{R}_3 + \text{Br}^-$

Step 2 $\text{R}\overset{\oplus}{\underset{\curvearrowleft}{-}}\text{NR}_2\text{CN} + \text{Br}^- \longrightarrow \text{RBr} + \text{R}_2\text{NCN}$

The intermediate N-cyanoammonium bromide has been trapped, and its structure confirmed by chemical, analytical, and spectral data.[1054] The BrCN in this reaction has been called a *counterattack reagent*; that is, a reagent that accomplishes, in one flask, two transformations designed to give the product.[1055]

OS **III**, 608.

B. Attack at an Acyl Carbon

0-74 Formation of Acyl Halides from Carboxylic Acids
Halo-de-hydroxylation

$$\text{RCOOH} + \text{SOCl}_2 \longrightarrow \text{RCOCl}$$

The same inorganic acid halides that convert alcohols to alkyl halides (**0-67**) also convert carboxylic acids to acyl halides.[1056] The reaction is the best and the most common method for the preparation of acyl chlorides. Bromides and iodides[1057] are also made in this manner, but much less often. Thionyl chloride[983] is the best reagent, since the by-products are gases and the acyl halide is easily isolated, but PX_3 and PX_5 (X = Cl or Br) are also commonly used.[1058] Hydrogen halides do not give the reaction. A particularly mild procedure, similar to one mentioned in **0-67**, involves reaction of the acid with Ph_3P in CCl_4, whereupon acyl chlorides are produced without obtaining any acidic compound as a by-product.[1059] Acyl fluorides can be prepared by treatment of carboxylic acids with cyanuric fluoride.[1060] Acid salts are also sometimes used as substrates. Acyl halides are also used as reagents in an exchange reaction:

$$\text{RCOOH} + \text{R}'\text{COCl} \rightleftharpoons \text{RCOCl} + \text{R}'\text{COOH}$$

[1052]For a review, see Cooley; Evain, Ref. 1049.

[1053]For a detailed discussion of the scope of the reaction and of the ease of cleavage of different groups, see Hageman *Org. React.* **1953**, pp. 205-225.

[1054]Fodor; Abidi *Tetrahedron Lett.* **1971**, 1369; Fodor; Abidi; Carpenter *J. Org. Chem.* **1974**, *39*, 1507. See also Paukstelis; Kim *J. Org. Chem.* **1974**, *39*, 1494.

[1055]For a review of counterattack reagents, see Hwu; Gilbert *Tetrahedron* **1989**, *45*, 1233-1261.

[1056]For a review, see Ansell, in Patai, Ref. 502, pp. 35-68.

[1057]Carboxylic acids and some of their derivatives react with diiodosilane SiH_2I_2 to give good yields of acyl iodides: Keinan; Sahai *J. Org. Chem.* **1990**, *55*, 3922.

[1058]For a list of reagents, with references, see Ref. 508, pp. 963-964.

[1059]Lee *J. Am. Chem. Soc.* **1966**, *88*, 3440. For other methods of preparing acyl chlorides, see Venkataraman; Wagle *Tetrahedron Lett.* **1979**, 3037; Devos; Remion; Frisque-Hesbain; Colens; Ghosez *J. Chem. Soc., Chem. Commun.* **1979**, 1180.

[1060]Olah; Nojima; Kerekes *Synthesis* **1973**, 487. For other methods of preparing acyl fluorides, see Mukaiyama; Tanaka *Chem. Lett.* **1976**, 303; Ishikawa; Sasaki *Chem. Lett.* **1976**, 1407.

which probably involves an anhydride intermediate. This is an equilibrium reaction that must be driven to the desired side. Oxalyl chloride and bromide are frequently used as the acyl halide reagent, since oxalic acid decomposes to CO and CO_2, and the equilibrium is thus driven to the side of the other acyl halide.

OS **I**, 12, 147, 394; **II**, 74, 156, 169, 569; **III**, 169, 490, 547, 555, 613, 623, 712, 714; **IV**, 34, 88, 154, 263, 339, 348, 554, 608, 616, 620, 715, 739, 900; **V**, 171, 258, 887; **VI**, 95, 190, 549, 715; **VII**, 467; **66**, 87, 116, 121.

0-75   Formation of Acyl Halides from Acid Derivatives
Halo-de-acyloxy-substitution
Halo-de-halogenation

$$(RCO)_2O + HF \longrightarrow RCOF$$

$$RCOCl + HF \longrightarrow RCOF$$

These reactions are most important for the preparation of acyl fluorides.[1061] Acyl chlorides and anhydrides can be converted to acyl fluorides by treatment with polyhydrogen fluoride–pyridine solution[996] or with liquid HF at $-10°C$.[1062] Formyl fluoride, which is a stable compound, was prepared by the latter procedure from the mixed anhydride of formic and acetic acids.[1063] Acyl fluorides can also be obtained by reaction of acyl chlorides with KF in acetic acid[1064] or with diethylaminosulfur trifluoride (DAST).[1065] Carboxylic esters and anhydrides can be converted to acyl halides other than fluorides by the inorganic acid halides mentioned in **074**, as well as with Ph_3PX_2 (X = Cl or Br),[1066] but this is seldom done. Halide exchange can be carried out in a similar manner. When halide exchange is done, it is always acyl bromides and iodides that are made from chlorides, since chlorides are by far the most readily available.[1067]

OS **II**, 528; **III**, 422; **V**, 66, 1103. See also OS **IV**, 307.

Hydrogen as Nucleophile

The reactions in this section (**0-76** to **0-85**) are reductions and could have been considered in Chapter 19. They are treated here because they involve replacement of a leaving group by hydrogen, which frequently attacks as the nucleophile hydride ion. However, not all the reactions in this section are true nucleophilic substitutions and for some of them more than one kind of mechanism may be involved, depending on the reagents and on the conditions. When cleavage of a carbon-hetero atom bond is accomplished by catalytic hydrogenation, the reaction is called *hydrogenolysis*.

A.   Attack at an Alkyl Carbon

0-76   Reduction of Alkyl Halides
Hydro-de-halogenation or **Dehalogenation**

$$RX + LiAlH_4 \longrightarrow RH$$

[1061]For lists of reagents converting acid derivatives to acyl halides, see Ref. 508, pp. 977, 980, 985.
[1062]Olah; Kuhn *J. Org. Chem.* **1961**, *26*, 237.
[1063]Olah; Kuhn *J. Am. Chem. Soc.* **1960**, *82*, 2380.
[1064]Emsley; Gold; Hibbert; Szeto *J. Chem. Soc., Perkin Trans. 2* **1988**, 923.
[1065]Markovski; Pashinnik *Synthesis* **1975**, 801.
[1066]Burton; Koppes *J. Chem. Soc., Chem. Commun.* **1973**, 425, *J. Org. Chem.* **1975**, *40*, 3026; Anderson; Kono *Tetrahedron Lett.* **1973**, 5121.
[1067]For methods of converting acyl chlorides to bromides or iodides, see Schmidt; Russ; Grosse *Synthesis* **1981**, 216; Hoffmann; Haase *Synthesis* **1981**, 715.

This type of reduction can be accomplished with many reducing agents,[1068] the most common being lithium aluminum hydride.[1069] This reagent reduces almost all types of alkyl halide, including vinylic, bridgehead, and cyclopropyl halides.[1070] Reduction with lithium aluminum deuteride serves to introduce deuterium into organic compounds. An even more powerful reducing agent, reportedly the strongest S_N2 nucleophile known, is lithium triethylborohydride $LiEt_3BH$. This reagent rapidly reduces primary, secondary, allylic, benzylic, and neopentyl halides, but not tertiary (these give elimination) or aryl halides.[1071] Another powerful reagent, which reduces primary, secondary, tertiary, allylic, vinylic, aryl, and neopentyl halides, is a complex formed from lithium trimethoxyaluminum hydride $LiAlH(OMe)_3$ and CuI.[1072] A milder reducing agent is $NaBH_4$ in a dipolar aprotic solvent such as Me_2SO, DMF, or sulfolane,[1073] which at room temperature or above reduces primary, secondary, and some tertiary[1074] halides in good yield without affecting other functional groups that would be reduced by $LiAlH_4$, for example, COOH, COOR, CN.[1075] Other reducing agents[1076] are zinc (with acid or base), $SnCl_2$, chromium(II) ion,[1077] either in the form of simple chromous salts (for active substrates or *gem*-dihalides[1078]) or complexed with ethylenediamine or ethanolamine (for ordinary alkyl halides[1079]), tris(trimethylsilyl)silane $(Me_3Si)_3SiH–NaBH_4$,[1080] $SmI_2–THF–HMPA$,[1081] and Et_3SiH in the presence of $AlCl_3$.[1082] The last two methods are good for primary, secondary, and tertiary halides. Sodium arsenite and base, diethyl phosphonate Et_3N,[1083] phosphorus tris(dimethylamide) $(Me_2N)_3P$,[1084] a metal carbonyl such as $Fe(CO)_5$ and a hydrogen donor,[1085] or organotin hydrides R_nSnH_{4-n}[1086] (chiefly Bu_3SnH).[1087] can be used to reduce just one halogen of a *gem*-dihalide or a 1,1,1-trihalide.[1088] The organotin hydride $(MeOCH_2CH_2OCH_2CH_2CH_2)_3SnH$ reduces

[1068]For reviews, see Hudlický *Reductions in Organic Chemistry*; Ellis Horwood: Chichester, 1984, pp. 62-67, 181; Pinder *Synthesis* **1980**, 425-452. For a list of reagents, see Ref. 508, pp. 18-24.

[1069]For a review of $LiAlH_4$, see Pizey, Ref. 593, vol. 1, 1974, pp. 101-294. For monographs on complex metal hydrides, see Seyden-Penne *Reductions by the Alumino- and Borohydrides*; VCH: New York, 1991; Hajós *Complex Hydrides*; Elsevier: New York, 1979.

[1070]Jefford; Kirkpatrick; Delay *J. Am. Chem. Soc.* **1972**, *94*, 8905; Krishnamurthy; Brown *J. Org. Chem.* **1982**, *47*, 276.

[1071]Brown; Kim; Krishnamurthy *J. Org. Chem.* **1980**, *45*, 1; Krishnamurthy; Brown *J. Org. Chem.* **1980**, *45*, 849, **1983**, *48*, 3085.

[1072]Masamune; Rossy; Bates *J. Am. Chem. Soc.* **1973**, *95*, 6452; Masamune; Bates; Georghiou *J. Am. Chem. Soc.* **1974**, *96*, 3686.

[1073]Bell; Vanderslice; Spehar *J. Org. Chem.* **1969**, *34*, 3923; Hutchins; Hoke; Keogh; Koharski *Tetrahedron Lett.* **1969**, 3495; Vol'pin; Dvolaitzky; Levitin *Bull. Soc. Chim. Fr.* **1970**, 1526; Hutchins; Kandasamy; Dux; Maryanoff; Rotstein; Goldsmith; Burgoyne; Cistone; Dalessandro; Puglis *J. Org. Chem.* **1978**, *43*, 2259.

[1074]Hutchins; Bertsch; Hoke *J. Org. Chem.* **1971**, *36*, 1568.

[1075]For the use of $NaBH_4$ under phase transfer conditions, see Bergbreiter; Blanton *J. Org. Chem.* **1987**, *52*, 472.

[1076]For some other reducing agents, not mentioned here, see Akiba; Shimizu; Ohnari; and Ohkata *Tetrahedron Lett.* **1985**, *26*, 3211; Kim; Yi *Bull. Chem. Soc. Jpn.* **1985**, *58*, 789; Cole; Kirwan; Roberts; Willis *J. Chem. Soc., Perkin Trans. 1* **1991**, 103; and Ref. 1068.

[1077]For reviews, see Hanson *Synthesis* **1974**, 1-8, pp. 2-5; Hanson; Premuzic *Angew. Chem. Int. Ed. Engl.* **1968**, *7*, 247-252 [*Angew. Chem. 80*, 271-276]. For a review of the mechanisms of reduction of alkyl halides by metal complexes, see Kochi *Organometallic Mechanisms and Catalysis*; Academic Press: New York, 1978, pp. 138-177.

[1078]Castro; Kray *J. Am. Chem. Soc.* **1966**, *88*, 4447.

[1079]Kochi; Mocadlo *J. Am. Chem. Soc.* **1966**, *88*, 4094; Kochi; Powers *J. Am. Chem. Soc.* **1970**, *92*, 137.

[1080]Lesage; Chatgilialoglu; Griller *Tetrahedron Lett.* **1989**, *30*, 2733. See also Ballestri; Chatgilialoglu; Clark; Griller; Giese; Kopping *J. Org. Chem.* **1991**, *56*, 678.

[1081]Inanaga; Ishikawa; Yamaguchi *Chem. Lett.* **1987**, 1485. See also Molander; Hahn *J. Org. Chem.* **1986**, *51*, 1135. For reviews of SmI_2, see Soderquist *Aldrichimica Acta* **1991**, *24*, 15-23; Kagan *New J. Chem.* **1990**, *14*, 453-460.

[1082]Doyle; McOsker; West *J. Org. Chem.* **1976**, *41*, 1393; Parnes; Romanova; Vol'pin *J. Org. Chem. USSR* **1988**, *24*, 254.

[1083]Hirao; Kohno; Ohshiro; Agawa *Bull. Chem. Soc. Jpn.* **1983**, *56*, 1881.

[1084]Downie; Lee *Tetrahedron Lett.* **1968**, 4951.

[1085]For reviews, see Freidlina; Gasanov; Kuz'mina; Chukovskaya *Russ. Chem. Rev.* **1985**, *54*, 662-675; Chukovskaya; Freidlina; Kuz'mina *Synthesis* **1983**, 773-784.

[1086]Seyferth; Yamazaki; Alleston *J. Org. Chem.* **1963**, *28*, 703.

[1087]For reviews of organotin hydrides, see Neumann *Synthesis* **1987**, 665-683; Kuivila *Synthesis* **1970**, 499-509, *Acc. Chem. Res.* **1968**, *1*, 299-305.

[1088]See, for example Chukovskaya; Freidlina; Kuz'mina, Ref. 1085.

alkyl halides and is water soluble, unlike Bu_3SnH.[1089] Reduction, especially of bromides and iodides, can also be effected by catalytic hydrogenation,[1090] and electrochemically.[1091] A good reducing agent for the removal of all halogen atoms in a polyhalo compound (including vinylic, allylic, geminal, and even bridgehead halogens) is lithium[1092] or sodium[1093] and *t*-BuOH in THF. Propargylic halides can often be reduced with allylic rearrangement to give allenes.[1094]

$$\underset{\underset{X}{|}}{R_2C}-C{\equiv}CH \xrightarrow{\text{LiAlH}_4} R_2C{=}C{=}CH_2$$

The choice of a reducing agent usually depends on what other functional groups are present. Each reducing agent reduces certain groups and not others. This type of selectivity is called *chemoselectivity*. A chemoselective reagent is one that reacts with one functional group (e.g., halide) but not another (e.g., C=O). For example, there are several reagents that reduce only the halogen of α-halo ketones, leaving the carbonyl group intact.[1095] Among them are i-Pr_2NLi,[1096] CH_3SNa,[1097] aqueous $TiCl_3$,[1098] NaI in aqueous acid–THF,[1099] PI_3 or P_2I_4,[1100] nickel boride,[1101] sodium formaldehyde sulfoxylate,[1102] i-Bu_2AlH–$SnCl_2$,[1103] NaHS–$SnCl_2$,[1104] $AlCl_3$–EtSH,[1105] $MeSiCl_3$–NaI,[515] and sodium hydrosulfite $Na_2S_2O_4$.[1106] Both $NaBH_3CN$–$SnCl_2$[1107] and the *n*-butyllithium ate complex (p. 260) of B-*n*-butyl-9-BBN[1108] (see p. 785) reduce tertiary alkyl, benzylic, and allylic halides, but do not react with primary or secondary alkyl or aryl halides. Another highly selective reagent, in this case for primary and secondary iodo and bromo groups, is sodium cyanoborohydride $NaBH_3CN$ in HMPA.[1109] Most of the reducing agents mentioned reduce chlorides, bromides, and iodides, but organotin hydrides also reduce fluorides.[1110] See page 1206 for a discussion of selectivity in reduction reactions.

[1089]Light; Breslow *Tetrahedron Lett.* **1990,** *31*, 2957.

[1090]For a discussion, see Rylander *Hydrogenation Methods;* Academic Press: New York, 1985.

[1091]For reviews, see Fry *Synthetic Organic Electrochemistry,* 2nd ed.; Wiley: New York, 1989, pp. 136-151; Feoktistov, in Baizer; Lund *Organic Electrochemistry;* Marcel Dekker: New York, 1983, pp. 259-284.

[1092]For example, see Bruck; Thompson; Winstein *Chem. Ind.(London)* **1960,** 405; Gassman; Pape *J. Org. Chem.* **1964,** *29*, 160; Fieser; Sachs *J. Org. Chem.* **1964,** *29*, 1113; Nazer *J. Org. Chem.* **1965,** *30*, 1737; Berkowitz *Synthesis* **1990,** 649.

[1093]For example, see Gassman; Aue; Patton *J. Am. Chem. Soc.* **1968,** *90*, 7271; Gassman; Marshall *Org. Synth.* V, 424.

[1094]For examples, see Crandall; Keyton; Kohne *J. Org. Chem.* **1968,** *33*, 3655; Claesson; Olsson *J. Am. Chem. Soc.,* **1979,** *101*, 7302.

[1095]For a review of reductive dehalogenation of polyhalo ketones, see Noyori; Hayakawa *Org. React.* **1983,** *29*, 163-344.

[1096]Dubois; Lion; Dugast *Tetrahedron Lett.* **1983,** *24*, 4207.

[1097]Ōki; Funakoshi; Nakamura *Bull. Chem. Soc. Jpn.* **1971,** *44*, 828. See also Inoue; Hata; Imoto *Chem. Lett.* **1975,** 1241.

[1098]Ho; Wong *Synth. Commun.* **1973,** *3*, 237; McMurry *Acc. Chem. Res.* **1974,** *7*, 281-286, pp. 284-285; Pradhan; Patil *Tetrahedron Lett.* **1989,** *30*, 2999. See also Clerici; Porta *Tetrahedron Lett.* **1987,** *28*, 1541.

[1099]Gemal; Luche *Tetrahedron Lett.* **1980,** *21*, 3195. See also Olah; Arvanaghi; Vankar *J. Org. Chem.* **1980,** *45*, 3531; Ho *Synth. Commun.* **1981,** *11*, 101; Ono; Kamimura; Suzuki *Synthesis* **1987,** 406.

[1100]Denis; Krief *Tetrahedron Lett.* **1981,** *22*, 1431.

[1101]Sarma; Borbaruah; Sharma *Tetrahedron Lett.* **1985,** *26*, 4657.

[1102]Harris *Synth. Commun.* **1987,** *17*, 1587.

[1103]Oriyama; Mukaiyama *Chem. Lett.* **1984,** 2069.

[1104]Ono; Maruyama; Kamimura *Synthesis* **1987,** 1093.

[1105]Fuji; Node; Kawabata; Fujimoto *J. Chem. Soc., Perkin Trans. 1* **1987,** 1043.

[1106]Chung; Hu *Synth. Commun.* **1982,** *12*, 261.

[1107]Kim; Ko *Synth. Commun.* **1985,** *15*, 603.

[1108]Toi; Yamamoto; Sonoda; Murahashi *Tetrahedron* **1981,** *37*, 2261.

[1109]Hutchins; Kandasamy; Maryanoff; Masilamani; Maryanoff *J. Org. Chem.* **1977,** *42*, 82.

[1110]Fluorides can also be reduced by a solution of K and dicyclohexano-18-crown-6 in toluene or diglyme: Ohsawa; Takagaki; Haneda; Oishi *Tetrahedron Lett.* **1981,** *22*, 2583. See also Brandänge; Dahlman; Ölund *Acta Chem. Scand., Ser. B* **1983,** *37*, 141.

With lithium aluminum hydride and most other metallic hydrides, the mechanism usually consists of simple nucleophilic substitution with attack by hydride ion that may or may not be completely free. The mechanism is S_N2 rather than S_N1, since primary halides react better than secondary or tertiary (tertiary generally give alkenes or do not react at all) and since Walden inversion has been demonstrated. However, rearrangements found in the reduction of bicyclic tosylates with $LiAlH_4$ indicate that the S_N1 mechanism can take place.[1111] There is evidence that $LiAlH_4$ and other metal hydrides can also reduce halides by an SET mechanism,[1112] especially those, such as vinylic,[1113] cyclopropyl,[1114] or bridgehead halides, that are resistant to nucleophilic substitution. Reduction of halides by $NaBH_4$ in 80% aqueous diglyme[1115] and by BH_3 in nitromethane[1116] takes place by an S_N1 mechanism. $NaBH_4$ in sulfolane reduces tertiary halides possessing a β hydrogen by an elimination–addition mechanism.[1117]

With other reducing agents the mechanism is not always nucleophilic substitution. For example, reductions with organotin hydrides generally[1118] take place by free-radical mechanisms,[1119] as do those with $Fe(CO)_5$[1120] and $(Me_3Si)_3SiH–NaBH_4$.[1080] Alkyl halides, including fluorides and polyhalides, can be reduced with magnesium and a secondary or tertiary alcohol (most often 2-propanol).[1121] This is actually an example of the occurrence in one step of the sequence:

$$RX \longrightarrow RMgX \xrightarrow{H^+} RH$$

More often the process is carried out in two separate steps (**2-38** and **2-23**).

OS **I**, 357, 358, 548; **II**, 320, 393; **V**, 424; **VI**, 142, 376, 731; **68**, 32. See also OS **69**, 66.

0-77 Reduction of Tosylates and Similar Compounds
Hydro-de-sulfonyloxy-substitution

$$RCH_2OTs + LiAlH_4 \longrightarrow RCH_3$$

Tosylates and other sulfonates can be reduced[1122] with $LiAlH_4$,[1123] with $NaBH_4$ in a dipolar aprotic solvent,[1124] with $LiEt_3BH$, with i-Bu_2AlH (DIBALH),[1125] or with $Bu_3SnH–NaI$.[1126] The scope of the reaction seems to be similar to that of **0-76**. When the reagent is $LiAlH_4$, alkyl tosylates are reduced more rapidly than iodides or bromides if the solvent is Et_2O,

[1111]Appleton; Fairlie; McCrindle *Chem. Commun.* **1967**, 690; Kraus; Chassin *Tetrahedron Lett.* **1970**, 1443.

[1112]Ashby; DePriest; Goel *Tetrahedron Lett.* **1981**, *22*, 1763, 3729; Singh; Khurana; Nigam *Tetrahedron Lett.* **1981**, *22*, 2901; Srivastava; le Noble *Tetrahedron Lett.* **1984**, *25*, 4871; Ashby; Pham *J. Org. Chem.* **1986**, *51*, 3598; Hatem; Meslem; Waegell *Tetrahedron Lett.* **1986**, *27*, 3723; Ashby; Pham; Amrollah-Majdjabadi *J. Org. Chem.* **1991**, *56*, 1596. See however Hirabe; Takagi; Muraoka; Nojima; Kusabayashi *J. Org. Chem.* **1985**, *50*, 1797; Park; Chung; Newcomb *J. Org. Chem.* **1987**, *52*, 3275.

[1113]Chung *J. Org. Chem.* **1980**, *45*, 3513.

[1114]McKinney; Anderson; Keyes; Schmidt *Tetrahedron Lett.* **1982**, *23*, 3443; Hatem; Waegell *Tetrahedron* **1990**, *46*, 2789.

[1115]Bell; Brown *J. Am. Chem. Soc.* **1966**, *88*, 1473.

[1116]Matsumura; Tokura *Tetrahedron Lett.* **1969**, 363.

[1117]Jacobus *Chem. Commun.* **1970**, 338; Ref. 1074.

[1118]For an exception, see Carey; Tramper *Tetrahedron Lett.* **1969**, 1645.

[1119]Kuivila; Menapace *J. Org. Chem.* **1963**, *28*, 2165; Menapace; Kuivila *J. Am. Chem. Soc.* **1964**, *86*, 3047; Tanner; Singh *J. Org. Chem.* **1986**, *51*, 5182.

[1120]Nelson; Detre; Tanabe *Tetrahedron Lett.* **1973**, 447; Freidlina et al., Ref. 1085.

[1121]Bryce-Smith; Wakefield; Blues *Proc. Chem. Soc.* **1963**, 219.

[1122]For a list of substrate types and reagents, with references, see Ref. 508, pp. 28-31.

[1123]For examples, see Rapoport; Bonner *J. Am. Chem. Soc.* **1951**, *73*, 2872; Eschenmoser; Frey *Helv. Chim. Acta* **1952**, *35*, 1660; Dimitriadis; Massy-Westropp *Aust. J. Chem.* **1982**, *35*, 1895.

[1124]Hutchins; Hoke; Keogh; Koharski, Ref. 1073.

[1125]Janssen; Hendriks; Godefroi *Recl. Trav. Chim. Pays-Bas* **1984**, *103*, 220.

[1126]Ueno; Tanaka; Okawara *Chem. Lett.* **1983**, 795.

but the order is reversed in diglyme.[1127] The reactivity difference is great enough so that a tosylate function can be reduced in the presence of a halide and vice versa.

OS **VI**, 376, 762; **68,** 138. See also OS **VII,** 66.

0-78 Hydrogenolysis of Alcohols[1128]
Hydro-de-hydroxylation or **Dehydroxylation**

$$\text{ROH} + \text{H}_2 \xrightarrow{\text{catalyst}} \text{RH}$$

The hydroxyl groups of most alcohols can seldom be cleaved by catalytic hydrogenation and alcohols are often used as solvents for hydrogenation of other compounds. However, benzyl-type alcohols undergo the reaction readily and have often been reduced.[1129] Diaryl and triarylcarbinols are similarly easy to reduce and this has been accomplished with LiAlH_4–AlCl_3,[1130] with NaBH_4 in F_3CCOOH,[1131] and with iodine, water, and red phosphorus (OS **I,** 224). Other reagents have been used,[1132] among them Fe(CO)_5,[1133] $\text{Me}_3\text{SiCl–MeI–}$ MeCN,[1134] $\text{Et}_3\text{SiH–BF}_3$,[1135] $\text{SmI}_2\text{–THF–HMPA}$,[1136] $\text{NaBH}_4\text{–F}_3\text{CCOOH}$,[1137] P_2I_4,[1138] Me_2SiI_2,[1139] and tin and HCl. 1,3-Diols are especially susceptible to hydrogenolysis. Tertiary alcohols can be reduced by catalytic hydrogenolysis when the catalyst is Raney nickel.[1140] Allylic alcohols (and ethers and acetates) can be reduced (often with accompanying allylic rearrangement) with Zn amalgam and HCl, as well as with certain other reagents.[1141] α-Acetylenic alcohols are converted to alkynes by reduction of their cobalt carbonyl complexes with NaBH_4 and CF_3COOH.[1142] Reagents that reduce the OH group of α-hydroxy ketones without affecting the C=O group include lithium diphenylphosphide Ph_2PLi,[1143] red phosphorus–iodine,[1144] and Me_3SiI.[1145]

Alcohols can also be reduced indirectly by conversion to a sulfonate and reduction of that compound (**0-77**). The two reactions can be carried out without isolation of the sulfonate if the alcohol is treated with pyridine–SO_3 in THF, and LiAlH_4 then added.[1146] Another indirect reduction that can be done in one step involves treatment of the alcohol (primary, secondary, or benzylic) with NaI, Zn, and Me_3SiCl.[1147] In this case the alcohol is first converted to the iodide, which is reduced. For other indirect reductions of OH, see **0-81.**

[1127]Krishnamurthy *J. Org. Chem.* **1980,** *45,* 2550.
[1128]For a review, see Müller, in Patai *The Chemistry of Functional Groups, Supplement E*, pt. 1; Wiley: New York, 1980, pp. 515-522.
[1129]For reviews, see Rylander, Ref. 1090, pp. 157-163, *Catalytic Hydrogenation over Platinum Metals*; Academic Press: New York, 1967, pp. 449-468. For a review of the stereochemistry of hydrogenolysis, see Klabunovskii *Russ. Chem. Rev.* **1966,** *35,* 546-558.
[1130]Blackwell; Hickinbottom *J. Chem. Soc.* **1961,** 1405; Avendaño; de Diego; Elguero *Monatsh. Chem.* **1990,** *121,* 649.
[1131]For a review, see Gribble; Nutaitis *Org. Prep. Proced. Int.* **1985,** *17,* 317-384.
[1132]For a list of reagents, with references, see Ref. 508, pp. 27-28.
[1133]Alper; Sališová *Tetrahedron Lett.* **1980,** *21,* 801.
[1134]Sakai; Miyata; Utaka; Takeda *Tetrahedron Lett.* **1987,** *28,* 3817.
[1135]Orfanopoulos; Smonou *Synth. Commun.* **1988,** *18,* 833; Smonou; Orfanopoulos *Tetrahedron Lett.* **1988,** *29,* 5793.
[1136]Kusuda; Inanaga; Yamaguchi *Tetrahedron Lett.* **1989,** *30,* 2945.
[1137]Nutaitis; Bernardo *Synth. Commun.* **1990,** *20,* 487.
[1138]Suzuki; Tani; Kubota; Sato; Tsuji; Osuka *Chem. Lett* **1983,** 247.
[1139]Ando; Ikeno *Tetrahedron Lett.* **1979,** 4941; Wiggins *Synth. Commun.* **1988,** *18,* 741.
[1140]Krafft; Crooks *J. Org. Chem.* **1988,** *53,* 432. For another catalyst, see Parnes; Shaapuni; Kalinkin; Kursanov *Bull. Acad. Sci. USSR, Div. Chem. Sci.* **1974,** *23,* 1592.
[1141]For discussion, see Elphimoff-Felkin; Sarda *Org. Synth.* VI, 769; *Tetrahedron* **1977,** *33,* 511. For another reagent, see Lee; Alper *Tetrahedron Lett.* **1990,** *31,* 4101.
[1142]Nicholas; Siegel *J. Am. Chem. Soc.* **1985,** *107,* 4999.
[1143]Leone-Bay *J. Org. Chem.* **1986,** *51,* 2378.
[1144]Ho; Wong *Synthesis* **1975,** 161.
[1145]Ho *Synth. Commun.* **1979,** *9,* 665.
[1146]Corey; Achiwa *J. Org. Chem.* **1969,** *34,* 3667.
[1147]Morita; Okamoto; Sakurai *Synthesis* **1981,** 32.

The mechanisms of most alcohol reductions are obscure.[1148] Hydrogenolysis of benzyl alcohols can give inversion or retention of configuration, depending on the catalyst.[1149]
OS **I,** 224; **IV,** 25, 218, 482; **V,** 339; **VI,** 769.

0-79   Replacement of Alkoxyl by Hydrogen
Hydro-de-alkoxylation or **Dealkoxylation**

$$-\overset{|}{C}(OR)_2 \xrightarrow{\text{LiAlH}_4 - \text{AlCl}_3} -\overset{|}{C}HOR + ROH$$

$$-\overset{|}{C}(OR)_3 \xrightarrow{\text{LiAlH}_4} -\overset{|}{C}H(OR)_2 + ROH$$

Simple ethers are not normally cleaved by reducing agents, although such cleavage has sometimes been reported (for example, tetrahydrofuran treated with LiAlH_4–AlCl_3[1150] or with a mixture of $\text{LiAlH}(O\text{-}t\text{-Bu})_3$ and Et_3B[1151] gave 1-butanol; the latter reagent also cleaves methyl alkyl ethers).[1152] Certain types of ethers can be cleaved quite well by reducing agents.[1153] Among these are allyl aryl,[1154] vinyl aryl,[1155] and benzylic ethers[1129] (for epoxides, see **0-80**). Acetals and ketals are resistant to LiAlH_4 and similar hydrides, and carbonyl groups are often converted to acetals or ketals for protection. However, a combination of LiAlH_4 and AlCl_3[1156] does reduce acetals and ketals, removing one group, as shown above.[1157] The actual reducing agents in this case are primarily chloroaluminum hydride AlH_2Cl and dichloroaluminum hydride AlHCl_2, which are formed from the reagents.[1158] This conversion can also be accomplished with DIBALH,[1159] with Nafion-H,[1160] with monochloroborane–etherate BH_2Cl–Et_2O,[1161] as well as with other reagents.[1162] Ortho esters are easily reduced to acetals by LiAlH_4 alone, offering a route to aldehydes, which are easily prepared by hydrolysis of the acetals (**0-6**).
OS **III,** 693; **IV,** 798; **V,** 303. Also see OS **III,** 742; **VII,** 386.

0-80   Reduction of Epoxides
(3)*OC-seco*-Hydro-de-alkoxylation

$$-\overset{|}{\underset{\diagdown}{C}}\!\!-\!\!\overset{|}{\underset{\diagup}{C}}- + \text{LiAlH}_4 \longrightarrow -\overset{|}{C}H-\overset{|}{\underset{|}{C}}- $$
$$\underset{O}{} \qquad\qquad\qquad\qquad \underset{OH}{}$$

[1148]For discussions of the mechanisms of the hydrogenolysis of benzyl alcohols, see Khan; McQuillin; Jardine *Tetrahedron Lett.* **1966,** 2649, *J. Chem. Soc. C* **1967,** 136; Garbisch; Schreader; Frankel *J. Am. Chem. Soc.* **1967,** 89, 4233; Mitsui; Imaizumi; Esashi *Bull. Chem. Soc. Jpn.* **1970,** 43, 2143.
[1149]Mitsui; Kudo; Kobayashi *Tetrahedron* **1969,** 25, 1921; Mitsui; Imaizumi; Esashi, Ref. 1148.
[1150]Bailey; Marktscheffel *J. Org. Chem.* **1960,** 25, 1797.
[1151]Krishnamurthy; Brown *J. Org. Chem.* **1979,** 44, 3678.
[1152]For a review of ether reduction, see Müller, Ref. 1128, pp. 522-528.
[1153]For a list of reagents, with references, see Ref. 508, pp. 501-504.
[1154]Tweedie; Cuscurida *J. Am. Chem. Soc.* **1957,** 79, 5463.
[1155]Tweedie; Barron *J. Org. Chem.* **1960,** 25, 2023. See also Hutchins; Learn *J. Org. Chem.* **1982,** 47, 4380.
[1156]For a review of reductions by metal hydride–Lewis acid combinations, see Rerick, in Augustine *Reduction*; Marcel Dekker: New York, 1968, pp. 1-94.
[1157]Eliel; Badding; Rerick *J. Am. Chem. Soc.* **1962,** 84, 2371.
[1158]Ashby; Prather *J. Am. Chem. Soc.* **1966,** 88, 729; Diner; Davis; Brown *Can. J. Chem.* **1967,** 45, 207.
[1159]See, for example, Zakharkin; Khorlina *Bull. Acad. Sci. USSR, Div. Chem. Sci.* **1959,** 2156; Takano; Akiyama; Sato; Ogasawara *Chem. Lett.* **1983,** 1593.
[1160]Olah; Yamato; Iyer; Prakash *J. Org. Chem.* **1986,** 51, 2826.
[1161]Borders; Bryson *Chem. Lett.* **1984,** 9.
[1162]For lists of other reagents that accomplish this conversion, with references, see Tsunoda; Suzuki; Noyori *Tetrahedron Lett.* **1979,** 4679; Kotsuki; Ushio; Yoshimura; Ochi *J. Org. Chem.* **1987,** 52, 2594; Ref. 508, pp. 463-465.

Reduction of epoxides is a special case of **0-79** and is easily carried out.[1163] The most common reagent is $LiAlH_4$, which reacts by the S_N2 mechanism, giving inversion of configuration. An epoxide on a substituted cyclohexane ring cleaves in such a direction as to give an axial alcohol. As expected for an S_N2 mechanism, cleavage usually occurs so that a tertiary alcohol is formed if possible. If not, a secondary alcohol is preferred. However, for certain substrates, the epoxide ring can be opened the other way by reduction with $NaBH_3CN$–BF_3,[1164] with Me_3SiCl–Zn,[1165] with dicyclopentadienyltitanium chloride and 1,4-cyclohexadiene,[1166] or with BH_3 in tetrahydrofuran.[1167] The reaction has also been carried out with other reagents, for example, sodium amalgam in EtOH, Li in ethylenediamine,[1168] Bu_3SnH–NaI,[1169] and by catalytic hydrogenolysis.[1170] Chemoselective and regioselective ring opening (e.g., of allylic epoxides and of epoxy ketones and esters) has been achieved with NaHTe,[1171] SmI_2,[1172] sodium bis(2-methoxyethoxy)aluminum hydride (Red-Al),[1173] and H_2 and a Pd-phosphine catalyst.[1174] Highly hindered epoxides can be conveniently reduced, without rearrangement, with lithium triethylborohydride.[1175]

Epoxides can be reductively halogenated (the product is the alkyl bromide or iodide rather than the alcohol) with Me_3SiCl–NaX–$(Me_2SiH)_2O$ (1,1,3,3-tetramethyldisiloxane).[1176]

See **9-46** for another type of epoxide reduction.

0-81 Reductive Cleavage of Carboxylic Esters
Hydro-de-acyloxylation or **Deacyloxylation**

$$R\!-\!O\!-\!\underset{\underset{O}{\|}}{C}\!-\!R' \xrightarrow[\text{EtNH}_2]{\text{Li}} RH + R'COO^-$$

The alkyl group R of certain carboxylic esters can be reduced to RH[1177] by treatment with lithium in ethylamine.[1178] The reaction is successful when R is a tertiary or a sterically hindered secondary alkyl group. A free-radical mechanism is likely.[1179] Similar reduction, also by a free-radical mechanism, has been reported with sodium in HMPA–t-BuOH.[1180] In the latter case, tertiary R groups give high yields of RH, but primary and secondary R are converted to a mixture of RH and ROH. Both of these methods provide an indirect method

[1163]For a list of reagents, with references, see Ref. 508, pp. 505-508.
[1164]Hutchins; Taffer; Burgoyne *J. Org. Chem.* **1981**, *46*, 5214.
[1165]Vankar; Arya; Rao *Synth. Commun.* **1983**, *13*, 869. See also Vankar; Chaudhuri; Rao *Tetrahedron Lett.* **1987**, *28*, 551.
[1166]RajanBabu; Nugent; Beattie *J. Am. Chem. Soc.* **1990**, *112*, 6408.
[1167]For a review of epoxide reduction with BH_3, see Cragg, *Organoboranes in Organic Synthesis;* Marcel Dekker: New York, 1973, pp. 345-348. See also Yamamoto; Toi; Sonoda; Murahashi *J. Chem. Soc., Chem. Commun.* **1976**, 672.
[1168]Brown; Ikegami; Kawakami *J. Org. Chem.* **1970**, *35*, 3243.
[1169]Bonini; Di Fabio *Tetrahedron Lett.* **1988**, *29*, 819.
[1170]For a review, see Rylander, *Catalytic Hydrogenation over Platinum Metals*, Ref. 1129, pp. 478-485.
[1171]Osuka; Taka-Oka; Suzuki *Chem. Lett.* **1984**, 271.
[1172]Molander; La Belle; Hahn *J. Org. Chem.* **1986**, *51*, 5259; Otsubo; Inanaga; Yamaguchi *Tetrahedron Lett.* **1987**, *28*, 4437. See also Miyashita; Hoshino; Suzuki; Yoshikoshi *Chem. Lett.* **1988**, 507.
[1173]Gao; Sharpless *J. Org. Chem.* **1988**, *53*, 4081.
[1174]Oshima; Yamazaki; Shimizu; Nizar; Tsuji *J. Am. Chem. Soc.* **1989**, *111*, 6280.
[1175]Krishnamurthy; Schubert; Brown *J. Am. Chem. Soc.* **1973**, *95*, 8486.
[1176]Aizpurua; Palomo *Tetrahedron Lett.* **1984**, *25*, 3123.
[1177]For a review of some of the reactions in this section and some others, see Hartwig *Tetrahedron* **1983**, *39*, 2609-2645.
[1178]Barrett; Godfrey; Hollinshead; Prokopiou; Barton; Boar; Joukhadar; McGhie; Misra *J. Chem. Soc., Perkin Trans. 1* **1981**, 1501.
[1179]Barrett; Prokopiou; Barton; Boar; McGhie *J. Chem. Soc., Chem. Commun.* **1979**, 1173.
[1180]Deshayes; Pete *Can. J. Chem.* **1984**, *62*, 2063.

of accomplishing **0-78** for tertiary R.[1181] The same thing can be done for primary and secondary R by treating alkyl chloroformates ROCOCl with tri-*n*-propylsilane in the presence of *t*-butyl peroxide[1182] and by treating thiono ethers ROC(=S)W (where W can be OAr or other groups) with Ph$_2$SiH$_2$ and a free radical initiator.[1183] Allylic acetates can be reduced with NaBH$_4$ and a palladium complex,[1184] with *p*-bis(diphenylhydrosilyl)benzene,[1185] and with SmI$_2$–Pd(0).[1186] The last reagent converts propargylic acetates to allenes R^{1}C≡CR^{2}R^{3}OAc → R^{1}CH=C=CR^{2}R^3.[1186] For other carboxylic ester reductions, see **9-40, 9-42,** and **9-43.**

OS **VII,** 139.

0-82   Reduction of the C—N Bond
Hydro-de-amination or **Deamination**

$$\text{RNH}_2 + \text{NH}_2\text{OSO}_2\text{OH} \xrightarrow[\text{0°C}]{\text{OH}^-} \text{RH} + \text{N}_2 + \text{SO}_4{}^{2-}$$

Primary amines have been reduced to RH with hydroxylamine-O-sulfonic acid and aqueous NaOH.[1187] It is postulated that R—N=N—H is an intermediate that decomposes to the carbocation. The reaction has also been accomplished with difluoroamine HNF$_2$;[1188] the same intermediates are postulated in this case. An indirect means of achieving the same result is the conversion of the primary amine to the sulfonamide RNHSO$_2$R' (**0-116**) and treatment of this with NH$_2$OSO$_2$OH.[1189] Other indirect methods involve reduction of N,N-ditosylates (p. 354) with NaBH$_4$ in HMPA[1190] and modifications of the Katritzky pyrylium–pyridinium method.[1191] Allylic and benzylic amines[1129] can be reduced by catalytic hydrogenolysis. Enamines are cleaved to olefins with alane AlH$_3$,[1192] e.g.,

and with 9-BBN (p. 785) or borane methyl sulfide (BMS).[1193] Since enamines can be prepared from ketones (**6-14**), this is a way of converting ketones to alkenes. In the latter case BMS gives retention of configuration [an (*E*) isomer gives the (*E*) product] while 9-BBN gives the other isomer.[1193] Diazo ketones are reduced to methyl ketones by HI: RCOCHN$_2$ + HI → RCOCH$_3$.[1194]

[1181]For other methods, see Barton; Crich; Löbberding; Zard *J. Chem. Soc., Chem. Commun.* **1985,** 646; Barton; Crich *J. Chem. Soc., Perkin Trans. 1* **1986,** 1603.
[1182]Jackson; Malek *J. Chem. Soc., Perkin Trans. 1* **1980,** 1207.
[1183]See Barton; Jang; Jaszberenyi *Tetrahedron Lett.* **1990,** *31*, 4681 and references cited therein. For similar methods, see Nozaki; Oshima; Utimoto *Bull. Chem. Soc. Jpn.* **1990,** *63*, 2578; Kirwan; Roberts; Willis *Tetrahedron Lett.* **1990,** *31*, 5093.
[1184]Hutchins; Learn; Fulton *Tetrahedron Lett.* **1980,** *21*, 27. See also Ipaktschi *Chem. Ber.* **1984,** *117*, 3320.
[1185]Sano; Takeda; Migita *Chem. Lett.* **1988,** 119. See also Keinan; Greenspoon *Isr. J. Chem.* **1984,** *24*, 82.
[1186]Tabuchi; Inanaga; Yamaguchi *Tetrahedron Lett.* **1986,** *27*, 601, 5237. See also Ref. 1136.
[1187]Doldouras; Kollonitsch *J. Am. Chem. Soc.* **1978,** *100*, 341.
[1188]Bumgardner; Martin; Freeman *J. Am. Chem. Soc.* **1963,** *85*, 97.
[1189]Nickon; Hill *J. Am. Chem. Soc.* **1964,** *86*, 1152.
[1190]Hutchins; Cistone; Goldsmith; Heuman *J. Org. Chem.* **1975,** *40*, 2018.
[1191]See Katritzky; Bravo-Borja; El-Mowafy; Lopez-Rodriguez *J. Chem. Soc., Perkin Trans. 1* **1984,** 1671.
[1192]Coulter; Lewis; Lynch *Tetrahedron* **1968,** *24*, 4489.
[1193]Singaram; Goralski; Rangaishenvi; Brown *J. Am. Chem. Soc.* **1989,** *111*, 384.
[1194]For example, see Pojer; Ritchie; Taylor *Aust. J. Chem.* **1968,** *21*, 1375.

Quaternary ammonium salts can be cleaved with LiAlH$_4$

$$R_4N^+ + LiAlH_4 \longrightarrow R_3N + RH$$

as can quaternary phosphonium salts R$_4$P$^+$. Other reducing agents have also been used, for example, lithium triethylborohydride (which preferentially cleaves methyl groups)[1195] and sodium in liquid ammonia. When quaternary salts are reduced with sodium amalgam in water, the reaction is known as the *Emde reduction*. However, this reagent is not applicable to the cleavage of ammonium salts with four *saturated* alkyl groups. Of course, aziridines[1170] can be reduced in the same way as epoxides (0-80).

Nitro compounds RNO$_2$ can be reduced to RH[1196] by sodium methylmercaptide CH$_3$SNa in an aprotic solvent[1197] or by Bu$_3$SnH.[1198] Both reactions have free-radical mechanisms.[1199] Tertiary nitro compounds can be reduced to RH by NaHTe.[1200] Bu$_3$SnH also reduces iso-cyanides RNC (prepared from RNH$_2$ by formylation followed by 7-41) to RH,[1201] a reaction that can also be accomplished with Li or Na in liquid NH$_3$,[1202] or with K and a crown ether in toluene.[1203] α-Nitro ketones can be reduced to ketones with Na$_2$S$_2$O$_4$–Et$_3$SiH in HMPA–H$_2$O.[1204]

Hydrogenolysis with a Pt catalyst in the gas phase has been reported to reduce nitro compounds, as well as primary and secondary amines.[1205]

OS **III**, 148; **IV**, 508; **68**, 227.

For reduction of the C—S bond, see **4-36**.

B. Attack at an Acyl Carbon

0-83 Reduction of Acyl Halides
Hydro-de-halogenation or Dehalogenation

$$RCOCl \xrightarrow[-78°C]{LiAlH(O\text{-}t\text{-}Bu)_3} RCHO$$

Acyl halides can be reduced to aldehydes[1206] by treatment with lithium tri-t-butoxyaluminum hydride in diglyme at $-78°C$.[1207] R may be alkyl or aryl and may contain many types of substituents, including NO$_2$, CN, and EtOOC groups. The reaction stops at the aldehyde stage because steric hindrance prevents further reduction under these conditions. Acyl halides can also be reduced to aldehydes by hydrogenolysis with palladium-on-barium sulfate

[1195]Cooke; Parlman *J. Org. Chem.* **1975**, *40*, 531.
[1196]For a method of reducing allylic nitro groups, see Ono; Hamamoto; Kamimura; Kaji *J. Org. Chem.* **1986**, *51*, 3734.
[1197]Kornblum; Carlson; Smith *J. Am. Chem. Soc.* **1979**, *101*, 647; Kornblum; Widmer; Carlson *J. Am. Chem. Soc.* **1979**, *101*, 658.
[1198]For reviews, see Ono, in Feuer; Nielsen *Nitro Compounds; Recent Advances in Synthesis and Chemistry*; VCH: New York, 1990, pp. 1-135, pp. 1-45; Rosini; Ballini *Synthesis* **1988**, 833-847, pp. 835-837; Ono; Kaji *Synthesis* **1986**, 693-704. For discussions of the mechanism, see Korth; Sustmann; Dupuis; Geise *Chem. Ber.* **1987**, *120*, 1197; Kamimura; Ono *Bull. Chem. Soc. Jpn.* **1988**, *61*, 3629.
[1199]For a discussion of the mechanism with Bu$_3$SnH, see Tanner; Harrison; Chen; Kharrat; Wayner; Griller; McPhee *J. Org. Chem.* **1990**, *55*, 3321. If an α substituent is present, it may be reduced instead of the NO$_2$. For a mechanistic discussion, see Bowman; Crosby; Westlake *J. Chem. Soc., Perkin Trans. 2* **1991**, 73.
[1200]Suzuki; Takaoka; Osuka *Bull. Chem. Soc. Jpn.* **1985**, *58*, 1067.
[1201]Barton; Bringmann; Motherwell *Synthesis* **1980**, 68.
[1202]See Niznik; Walborsky *J. Org. Chem.* **1978**, *43*, 2396; Yadav; Reddy; Joshi *Tetrahedron Lett.* **1988**, *44*, 7243.
[1203]Ohsawa; Mitsuda; Nezu; Oishi *Tetrahedron Lett.* **1989**, *30*, 845.
[1204]Kamimura; Kurata; Ono *Tetrahedron Lett.* **1989**, *30*, 4819.
[1205]Guttieri; Maier *J. Org. Chem.* **1984**, *49*, 2875.
[1206]For a review of the formation of aldehydes from acid derivatives, see Fuson, in Patai, Ref. 446, pp. 211-232. For a review of the reduction of acyl halides, see Wheeler, in Patai, Ref. 502, pp. 231-251.
[1207]Brown; McFarlin *J. Am. Chem. Soc.* **1958**, *80*, 5372; Brown; Subba Rao *J. Am. Chem. Soc.* **1958**, *80*, 5377.

as catalyst. This is called the *Rosenmund reduction*.[1208] A more convenient hydrogenolysis procedure involves palladium-on-charcoal as the catalyst, with ethyldiisopropylamine as acceptor of the liberated HCl and acetone as the solvent.[1209] The reduction of acyl halides to aldehydes has also been carried out[1210] with Bu_3SnH,[1211] with $Bu_3GeH-Pd(PPh_3)_4$,[1212] with $NaBH_4$ in a mixture of DMF and THF,[1213] and with ions of the form $HM(CO)_4^-$ (M = Fe, Cr, W).[1214] In some of these cases, the mechanisms are free-radical. There are several indirect methods for the conversion of acyl halides to aldehydes, most of them involving prior conversion of the halides to certain types of amides (see **0-85**). There is also a method in which the COOH group is replaced by a completely different CHO group (**0-110**). Also see **9-45**.

OS **III**, 551, 627; **VI**, 529, 1007. Also see OS **III**, 818; **VI**, 312.

0-84 Reduction of Carboxylic Acids, Esters, and Anhydrides to Aldehydes[1215]
Hydro-de-hydroxylation or **Dehydroxylation** (overall transformation)

$$RCOOH \xrightarrow[\text{MeNH}_2]{\text{Li}} RCH{=}N{-}Me \xrightarrow{\text{H}_2\text{O}} RCHO$$

With most reducing agents, reduction of carboxylic acids generally gives the primary alcohol (**9-38**) and the isolation of aldehydes is not feasible. However, simple straight-chain carboxylic acids have been reduced to aldehydes[1216] by treatment with Li in $MeNH_2$ or NH_3 followed by hydrolysis of the resulting imine,[1217] with borane–Me_2S followed by pyridinium chlorochromate,[1218] with isobutylmagnesium bromide and a titanium-complex catalyst followed by hydrolysis,[1219] with thexylchloroborane–Me_2S[1220] or thexylbromoborane–Me_2S[1221] (see **5-12** for the thexyl group), with $LiAlH(O\text{-}t\text{-}Bu)_3$ and chloromethylene dimethylammonium chloride[1222] $Me_2N{=}CHCl^+$ Cl^- in pyridine,[1223] and with diaminoaluminum hydrides.[1224] Caproic and isovaleric acids have been reduced to aldehydes in 50% yields or better with DIBALH (i-Bu_2AlH) at −75 to −70°C.[1225]

[1208]For a review, see Ref. 1170, pp. 398-404. For a discussion of the Pt catalyst, see Maier; Chettle; Rai; Thomas *J. Am. Chem. Soc.* **1986**, *108*, 2608.

[1209]Peters; van Bekkum *Recl. Trav. Chim. Pays-Bas* **1971**, *90*, 1323, **1981**, *100*, 21. See also Burgstahler; Weigel; Shaefer *Synthesis* **1976**, 767.

[1210]For some other methods, see Wagenknecht *J. Org. Chem.* **1972**, *37*, 1513; Smith; Smith *J. Chem. Soc., Chem. Commun.* **1975**, 459; Leblanc; Moise; Tirouflet *J. Organomet. Chem.* **1985**, *292*, 225; Corriu; Lanneau; Perrot *Tetrahedron Lett.* **1988**, *29*, 1271. For a list of reagents, with references, see Ref. 508, pp. 620-621.

[1211]Kuivila *J. Org. Chem.* **1960**, *25*, 284; Walsh; Stoneberg; Yorke; Kuivila *J. Org. Chem.* **1969**, *34*, 1156; Four; Guibe *J. Org. Chem.* **1981**, *46*, 4439; Lusztyk; Lusztyk; Maillard; Ingold *J. Am. Chem. Soc.* **1984**, *106*, 2923.

[1212]Geng; Lu *J. Organomet. Chem.* **1989**, *376*, 41.

[1213]Babler; Invergo *Tetrahedron Lett.* **1981**, *22*, 11; Babler *Synth. Commun.* **1982**, *12*, 839. For the use of $NaBH_4$ and metal ions, see Entwistle; Boehm; Johnstone; Telford *J. Chem. Soc., Perkin Trans. 1* **1980**, 27.

[1214]Cainelli; Manescalchi; Umani-Ronchi *J. Organomet. Chem.* **1984**, *276*, 205; Kao; Gaus; Youngdahl; Darensbourg *Organometallics* **1984**, *3*, 1601.

[1215]For a review, see Cha *Org. Prep. Proced. Int.* **1989**, *21*, 451-477.

[1216]For other reagents, see Hubert; Eyman; Wiemer *J. Org. Chem.* **1984**, *49*, 2279; Corriu; Lanneau; Perrot *Tetrahedron Lett.* **1987**, *28*, 3941; Cha; Kim; Yoon; Kim *Tetrahedron Lett.* **1987**, *28*, 6231. See also the lists in Ref. 508, pp. 619-622.

[1217]Bedenbaugh; Bedenbaugh; Bergin; Adkins *J. Am. Chem. Soc.* **1970**, *92*, 5774; Burgstahler; Worden; Lewis *J. Org. Chem.* **1963**, *28*, 2918.

[1218]Brown; Rao; Kulkarni *Synthesis* **1979**, 704.

[1219]Sato; Jinbo; Sato *Synthesis* **1981**, 871.

[1220]Brown; Cha; Yoon; Nazer *J. Org. Chem.* **1987**, *52*, 5400.

[1221]Cha; Kim; Lee *J. Org. Chem.* **1987**, *52*, 5030.

[1222]For the preparation of this reagent, see Fujisawa; Sato *Org. Synth.* **66**, 121.

[1223]Fujisawa; Mori; Tsuge; Sato *Tetrahedron Lett.* **1983**, *24*, 1543.

[1224]Muraki; Mukaiyama *Chem. Lett.* **1974**, 1447, **1975**, 215.

[1225]Zakharkin; Khorlina *J. Gen. Chem. USSR* **1964**, *34*, 1021; Zakharkin; Sorokina *J. Gen. Chem. USSR* **1967**, *37*, 525.

Carboxylic esters have been reduced to aldehydes with DIBALH at $-70°C$, with di-aminoaluminum hydrides,[1224] with $LiAlH_4$–Et_2NH,[1226] and with $NaAlH_4$ at -65 to $-45°C$, and (for phenolic esters) with $LiAlH(O\text{-}t\text{-}Bu)_3$ at $0°C$.[1227] Aldehydes have also been prepared by reducing ethyl thiol esters RCOSEt with Et_3SiH and a Pd–C catalyst.[1228]

Anhydrides, both aliphatic and aromatic, as well as mixed anhydrides of carboxylic and carbonic acids, have been reduced to aldehydes in moderate yields with disodium tetracar-bonylferrate $Na_2Fe(CO)_4$.[1229]

Also see **9-40** and **9-42**.

OS **VI**, 312; **66**, 121; **69**, 55.

0-85 Reduction of Amides to Aldehydes
Hydro-de-dialkylamino-substitution

$$RCONR'_2 + LiAlH_4 \longrightarrow RCHO + NHR'_2$$

N,N-Disubstituted amides can be reduced to amines with $LiAlH_4$ (see **9-39**), but also to aldehydes.[1230] Keeping the amide in excess gives the aldehyde rather than the amine. Some-times it is not possible to prevent further reduction and primary alcohols are obtained instead. Other reagents[1231] that give good yields of aldehydes are DIBALH,[1232] $LiAlH(O\text{-}t\text{-}Bu)_3$, $LiAlH_4$–EtOH,[1233] $NaAlH_4$,[1234] and diaminoaluminum hydrides.[1235]

Aldehydes have been prepared from carboxylic acids or acyl halides by first converting them to certain types of amides that are easily reducible. The following are some examples:[1236]

1. *Reissert compounds*[1237] (**109**) are prepared from the acyl halide by treatment with quinoline and cyanide ion. Treatment of **109** with sulfuric acid gives the corresponding aldehyde.

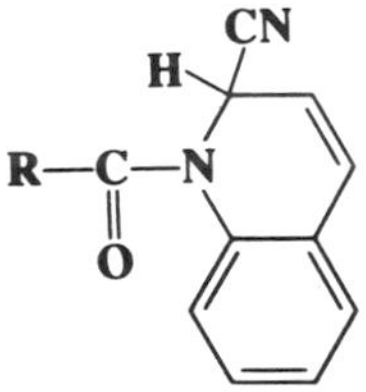

109 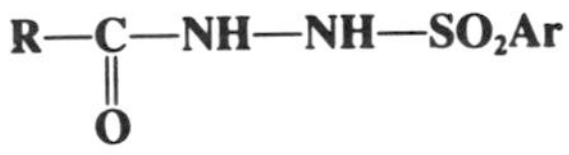 **110** 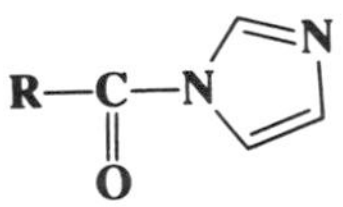 **101**

2. Acyl sulfonylhydrazides (**110**) are cleaved with base to give aldehydes. This is known as the *McFadyen–Stevens reduction* and is applicable only to aromatic aldehydes or aliphatic

[1226]Cha; Kwon *J. Org. Chem.* **1987**, *52*, 5486.

[1227]Zakharkin; Khorlina *Tetrahedron Lett.* **1962**, 619, *Bull. Acad. Sci. USSR, Div. Chem. Sci.* **1963**, 288, **1964**, 435; Zakharkin; Gavrilenko; Maslin; Khorlina *Tetrahedron Lett.* **1963**, 2087; Zakharkin; Gavrilenko; Maslin *Bull. Acad. Sci. USSR, Div. Chem. Sci.* **1964**, 867; Weissman; Brown *J. Org. Chem.* **1966**, *31*, 283.

[1228]Fukuyama; Lin; Li *J. Am. Chem. Soc.* **1990**, *112*, 7050.

[1229]Watanabe; Yamashita; Mitsudo; Igami; Takegami *Bull. Chem. Soc. Jpn.* **1975**, *48*, 2490; Watanabe; Yamashita; Mitsudo; Igami; Tomi; Takegami *Tetrahedron Lett.* **1975**, 1063.

[1230]For a review, see Fuson, in Patai, Ref. 446, pp. 220-225.

[1231]For a list of reagents, with references, see Ref. 508, pp. 623-624.

[1232]Zakharkin; Khorlina *Bull. Acad. Sci. USSR, Div. Chem. Sci* **1959**, 2046.

[1233]Brown; Tsukamoto *J. Am. Chem. Soc.* **1964**, *86*, 1089.

[1234]Zakharkin; Maslin; Gavrilenko *Tetrahedron* **1969**, *25*, 5555.

[1235]Muraki; Mukaiyama *Chem. Lett.* **1975**, 875.

[1236]For other examples, see Brown; Tsukamoto *J. Am. Chem. Soc.* **1961**, *83*, 4549; Doleschall *Tetrahedron* **1976**, *32*, 2549; Atta-ur-Rahman; Basha *J. Chem. Soc., Chem. Commun.* **1976**, 594; Izawa; Mukaiyama *Bull. Chem. Soc. Jpn.* **1979**, *52*, 555; Craig; Ekwuribe; Fu; Walker *Synthesis* **1981**, 303.

[1237]For reviews of Reissert compounds, see Popp; Uff *Heterocycles* **1985**, *23*, 731-740; Popp *Bull. Soc. Chim Belg* **1981**, *90*, 609-613, *Adv. Heterocycl. Chem.* **1979**, *24*, 187-214, **1968**, *9*, 1-25.

aldehydes with no α hydrogen.[1238] RCON=NH (see **0-82**) has been proposed as an intermediate in this reaction.[1239]

3. Imidazoles (**101**)[664] can be reduced to aldehydes with $LiAlH_4$.
4. See also the Sonn-Müller method (**6-28**).

OS **67**, 69. See OS **IV**, 641, **VI**, 115 for the preparation of Reissert compounds.

Carbon Nucleophiles

In any heterolytic reaction in which a new carbon–carbon bond is formed[1240] one carbon atoms attacks as a nucleophile and the other as an electrophile. The classification of a given reaction as nucleophilic or electrophilic is a matter of convention and is usually based on analogy. Although not discussed in this chapter, **1-12** to **1-28** and **2-15** to **2-20** are nucleophilic substitutions with respect to one reactant, though, following convention, we classify them with respect to the other. Similarly, all the reactions in this section (**0-86** to **0-113**) would be called electrophilic substitution (aromatic or aliphatic) if we were to consider the reagent as the substrate.

A. Attack at an Alkyl Carbon. In **0-86** to **0-93** the nucleophile is a "carbanion" part of an organometallic compound, often a Grignard reagent. There is much that is still not known about the mechanisms of these reactions and many of them are not nucleophilic substitutions at all. In those reactions that are nucleophilic substitutions, the attacking carbon brings a pair of electrons with it to the new C—C bond, whether or not free carbanions are actually involved. The connection of two alkyl or aryl groups is called *coupling*. Reactions **0-86** to **0-93** include both symmetrical and unsymmetrical coupling reactions. The latter are also called *cross-coupling reactions*. Other coupling reactions are considered in later chapters.

0-86   Coupling of Alkyl Halides. The Wurtz Reaction
De-halogen-coupling

$$2RX + Na \longrightarrow RR$$

The coupling of alkyl halides by treatment with sodium to give a symmetrical product is called the *Wurtz reaction*. Side reactions (elimination and rearrangement) are so common that the reaction is seldom used. Mixed Wurtz reactions of two alkyl halides are even less feasible because of the number of products obtained. A somewhat more useful reaction (though still not very good) takes place when a mixture of an alkyl and an aryl halide is treated with sodium to give an alkylated aromatic compound (the *Wurtz–Fittig reaction*).[1241] However, the coupling of two aryl halides with sodium is impractical (but see **3-16**). Other metals have also been used to effect Wurtz reactions,[1242] notably silver, zinc,[1243] iron,[1244] activated copper,[1245] and pyrophoric lead.[1246] Lithium, under the influence of ultrasound,

[1238]Babad; Herbert; Stiles *Tetrahedron Lett.* **1966**, 2927; Dudman; Grice; Reese *Tetrahedron Lett.* **1980**, *21*, 4645.

[1239]For discussions, see Cacchi; Paolucci *Gazz. Chem. Ital.* **1974**, *104*, 221; Matin; Craig; Chan *J. Org. Chem.* **1974**, *39*, 2285.

[1240]For a monograph that discusses most of the reactions in this section, see Stowell *Carbanions in Organic Synthesis*; Wiley: New York, 1979. For a review, see Noyori, in Alper *Transition Metal Organometallics in Organic Synthesis*, vol. 1; Academic Press: New York, 1976, pp. 83-187.

[1241]For an example, see Kwa; Boelhouwer *Tetrahedron* **1970**, *25*, 5771.

[1242]For a list of reagents, including metals and other reagents, with references, see Ref. 508, pp. 47-48.

[1243]See, for example, Nosek *Collect. Czech. Chem. Commun.* **1964**, *29*, 597.

[1244]Nozaki; Noyori *Tetrahederon* **1966**, *22*, 2163; Onsager *Acta Chem. Scand., Ser. B* **1978**, *32*, 15.

[1245]Ginah; Donovan; Suchan; Pfennig; Ebert *J. Org. Chem.* **1990**, *55*, 584.

[1246]Mészáros *Tetrahedron Lett.* **1967**, 4951; Azoo; Grimshaw *J. Chem. Soc. C* **1968**, 2403.

has been used to couple alkyl, aryl, and benzylic halides.[1247] Metallic nickel, prepared by the reduction of nickel halides with Li, dimerizes benzylic halides to give $ArCH_2CH_2Ar$.[1248] The coupling of alkyl halides has also been achieved electrochemically.[1249]

One type of Wurtz reaction that is quite useful is the closing of small rings, especially three-membered rings.[1250] For example, 1,3-dibromopropane can be converted to cyclopropane by Zn and NaI.[1251] Two highly strained molecules that have been prepared this way are bicyclobutane[1252] and tetracyclo[3.3.1.1^{3,7}.0^{1,3}]decane.[1253] Three- and four-membered

rings can also be closed in this manner with certain other reagents,[1254] including benzoyl peroxide,[1255] *t*-BuLi,[1256] (phenylsulfonyl)methylene dilithium $PhSO_2CHLi_2$[1257] and lithium amalgam,[1258] as well as electrochemically.[1259]

Vinylic halides can be coupled to give 1,3-butadienes by treatment with activated copper powder in a reaction analogous to the Ullmann reaction (**3-16**).[1260] This reaction is stereo-specific, with retention of configuration at both carbons. Vinylic halides can also be

coupled[1261] with CuCl,[1262] with Zn–NiCl$_2$,[1263] and with *n*-BuLi in ether in the presence of MnCl$_2$.[1264]

[1247]Han; Boudjouk *Tetrahedron Lett.* **1981**, *22*, 2757.

[1248]Inaba; Matsumoto; Rieke *J. Org. Chem.* **1984**, *49*, 2093. For some other reagents that accomplish this, see Sayles; Kharasch *J. Org. Chem.* **1961**, *26*, 4210; Cooper *J. Am. Chem. Soc.* **1973**, *95*, 4158; Ho; Olah *Synthesis* **1977**, 170; Ballatore; Crozet; Surzur *Tetrahedron Lett.* **1979**, 3073; Yamada; Momose *Chem. Lett.* **1981**, 1277; Iyoda; Sakaitani; Otsuka; Oda *Chem. Lett.* **1985**, 127.

[1249]Folest; Nedelec; Perichon *J. Chem. Res. (S)* **1989**, 394.

[1250]For a review, see Freidlina; Kamyshova; Chukovskaya *Russ. Chem. Rev.* **1982**, *51*, 368-376. For reviews of methods of synthesizing cyclopropane rings, see, in Rappoport *The Chemistry of the Cyclopropyl Group*, pt. 1; Wiley: New York, 1987, the reviews by Tsuji; Nishida, pp. 307-373, and Verhé; De Kimpe, pp. 445-564.

[1251]For a discussion of the mechanism, see Applequist; Pfohl *J. Org. Chem.* **1978**, *43*, 867.

[1252]Wiberg; Lampman *Tetrahedron Lett.* **1963**, 2173; Lampman; Aumiller *Org. Synth. VI*, 133.

[1253]Pincock; Schmidt; Scott; Torupka *Can. J. Chem.* **1972**, *50*, 3958.

[1254]For a list of reagents, with references, see Ref. 508, pp. 87-88.

[1255]Kaplan *J. Am. Chem. Soc.* **1967**, *89*, 1753, *J. Org. Chem.* **1967**, *32*, 4059.

[1256]Bailey; Gagnier *Tetrahedron Lett.* **1982**, *23*, 5123.

[1257]Eisch; Dua; Behrooz *J. Org. Chem.* **1985**, *50*, 3674.

[1258]Connor; Wilson *Tetrahedron Lett.* **1967**, 4925.

[1259]Rifi *J. Am. Chem. Soc.* **1967**, *89*, 4442, *Org. Synth. VI*, 153.

[1260]Cohen; Poeth *J. Am. Chem. Soc.* **1972**, *94*, 4363.

[1261]For some other methods, see Jones *J. Org. Chem.* **1967**, *32*, 1667; Semmelhack; Helquist; Gorzynski *J. Am. Chem. Soc.* **1972**, *94*, 9234; Wellmann; Steckhan *Synthesis* **1978**, 901; Miyahara; Shiraishi; Inazu; Yoshino *Bull. Chem. Soc. Jpn.* **1979**, *52*, 953; Grigg; Stevenson; Worakun *J. Chem. Soc., Chem. Commun.* **1985**, 971; Vanderesse; Fort; Becker; Caubere *Tetrahedron Lett.* **1986**, *27*, 3517.

[1262]Kauffmann; Sahm *Angew. Chem. Int. Ed. Engl.* **1967**, *6*, 85 [*Angew. Chem. 79*, 101]; Toda; Takehira *J. Chem. Soc., Chem. Commun.* **1975**, 174.

[1263]Takagi; Mimura; Inokawa *Bull. Chem. Soc. Jpn.* **1984**, *57*, 3517.

[1264]Cahiez; Bernard; Normant *J. Organomet. Chem.* **1976**, *113*, 99.

It seems likely that the mechanism of the Wurtz reaction consists of two basic steps. The first is halogen–metal exchange to give an organometallic compound (RX + M → RM), which in many cases can be isolated (2-38). Following this, the organometallic compound reacts with a second molecule of alkyl halide (RX + RM → RR). This reaction and its mechanism are considered in the next section (0-87).

OS **III**, 157; **V**, 328, 1058; **VI**, 133, 153.

0-87 The Reaction of Alkyl Halides with Organometallic Reagents[1265]
Alkyl-de-halogenation

$$RX + R_2'CuLi \longrightarrow R\!-\!R'$$

The reagents lithium dialkylcopper[1266] (also called *Gilman reagents*) react with alkyl bromides, chlorides, and iodides in ether or THF to give good yields of the cross-coupling products.[1267] The reaction is of wide scope.[1268] R may be primary alkyl, allylic, benzylic, aryl, vinylic, or allenic, and may contain keto, COOH, COOR, or $CONR_2$ groups. The reaction at a vinylic substrate occurs stereospecifically, with retention of configuration.[1269] When the reagent and substrate are both vinylic, yields are low, but the reaction can be made to go (to give 1,3-butadienes) stereospecifically in high yields by the use of $ZnBr_2$ and a Pd(0) complex.[1270] Many *gem*-dihalides do not react, but when the two halogens are on a carbon α to an aromatic ring[1271] or on a cyclopropane ring,[1272] both halogens can be replaced by R, e.g., $PhCHCl_2$ → $PhCHMe_2$. However, 1,2-dibromides give exclusive elimination[1273] (7-29). R' in $R_2'CuLi$ may be primary alkyl, vinylic, allylic, or aryl. Thus, in the reaction as so far described, neither R nor R' may be secondary or tertiary alkyl. However, secondary and tertiary alkyl coupling can be achieved (on primary RX) by the use of $R_2'CuLi \cdot PBu_3$[1274] (though this procedure introduces problems in the work-up) or by the use of $PhS(R')CuLi$,[1275] which selectively couples a secondary or tertiary R' with a primary iodide RI to give RR'.[1276] From the opposite standpoint, coupling to a secondary R can be achieved in high yield with the reagents $R_2'Cu(CN)Li_2$,[1277] where R' is primary alkyl or vinylic (but not aryl).[1278] The reagents $RCu(PPh_2)Li$, $RCu(NR_2')Li$, and $Cu(PR_2')Li$ (R' = cyclohexyl) are more stable than R_2CuLi and can be used at higher

---

[1265]For a review of the reactions in this section, see Naso; Marchese, in Patai; Rappoport, Ref. 88, pt. 2, pp. 1353-1449.

[1266]For the structure of Me_2CuLi (a cyclic dimer), see Pearson; Gregory *J. Am. Chem. Soc.* **1976**, *98*, 4098. See also Lipshutz; Kozlowski; Breneman *Tetrahedron Lett.* **1985**, *26*, 5911. For reviews of the structure and reactions of organocopper compounds, see Power *Prog. Inorg. Chem.* **1991**, *39*, 75-112; Collman; Hegedus; Norton; Finke *Principles and Applications of Organotransition Metal Chemistry*, 2nd ed.; University Science Books: Mill Valley, CA, 1987, pp. 682-698.

[1267]Corey; Posner *J. Am. Chem. Soc.* **1967**, *89*, 3911, **1968**, *90*, 5615; Whitesides; Fischer; San Filippo; Bashe; House *J. Am. Chem. Soc.* **1969**, *91*, 4871; Bergbreiter; Whitesides *J. Org. Chem.* **1975**, *40*, 779.

[1268]For a review of this reaction, see Posner *Org. React.* **1975**, *22*, 253-400. For a review of organocopper reagents, see Normant *Synthesis* **1972**, 63-80. For examples of the use of this reaction in this synthesis of natural products, see Posner *An Introduction to Synthesis Using Organocopper Reagents*; Wiley: New York, 1980, pp. 68-81. For lists of substrates and reagents, with references, see Ref. 508, pp. 206-210, 304-306, 788.

[1269]Corey; Posner, Ref. 1267; Klein; Levene *J. Am. Chem. Soc.* **1972**, *94*, 2520.

[1270]Jabri; Alexakis; Normant *Tetrahedron Lett.* **1981**, *22*, 959, **1982**, *23*, 1589, *Bull. Soc. Chim. Fr.* **1983**, II-321, II-332.

[1271]Posner; Brunelle *Tetrahedron Lett.* **1972**, 293.

[1272]See, for example, Kitatani; Hiyama; Nozaki *Bull. Chem. Soc. Jpn* **1977**, *50*, 1600.

[1273]Posner; Ting *Synth. Commun.* **1973**, *3*, 281.

[1274]Whitesides; Fischer; San Filippo; Bashe; House, Ref. 1267.

[1275]Prepared as in Ref. 1285 or treatment of PhSCu with RLi: Posner; Brunelle; Sinoway *Synthesis* **1974**, 662.

[1276]Posner; Whitten; Sterling *J. Am. Chem. Soc.* **1973**, *95*, 7788; Posner; Whitten *Tetrahedron Lett.* **1973**, 1815.

[1277]For reviews of these and other "higher order" organocuprates, see Lipshutz; Wilhelm; Kozlowski *Tetrahedron* **1984**, *40*, 5005-5038, Lipshutz *Synthesis* **1987**, 325-341, *Synlett* **1990**, 119-128. See also Bertz *J. Am. Chem. Soc.* **1990**, *112*, 4031; Lipshutz; Sharma; Ellsworth *J. Am. Chem. Soc.* **1990**, *112*, 4032.

[1278]Lipshutz; Wilhelm; Floyd *J. Am. Chem. Soc.* **1981**, *103*, 7672.

temperatures.[1279] With an allenic substrate, reaction with R(CN)CuLi can give ordinary displacement (with retention of configuration)[1280] or an SN2' reaction to produce an alkyne.[1281] In the latter case, a chiral allene gave a chiral alkyne. The fact that R_2'CuLi do not react with ketones provides a method for the alkylation of ketones[1282] (see also **0-95** and **0-99**), though halogen–metal exchange (**2-39**) is a side reaction and can become the main reaction.[1283]

$$R_2C\!-\!C\!-\!R'' + R_2'CuLi \longrightarrow R_2C\!-\!C\!-\!R''$$
$$\underset{Br}{|}\;\underset{O}{\|} \qquad\qquad\qquad \underset{R'}{|}\;\underset{O}{\|}$$

When α,α'-dibromo ketones (**111**) are treated with Me$_2$CuLi in ether at $-78°$C and the mixture quenched with methanol, *mono*methylation takes place[1284] (no dimethylation is observed). It has been suggested that the reaction involves cyclization (**0-86**) to a cyclopropanone followed by nucleophilic attack to give the enolate ion **112** which is protonated by

the methanol. If methyl iodide is added instead of methanol, an α,α'-dimethyl ketone is obtained, presumably from SN2 attack by **112** on methyl iodide (**0-95**). Only halides that are highly reactive to SN2 attack (e.g., methyl and benzylic halides) react successfully with **112**. Primary, secondary, *and tertiary* monoalkylation of **111** can be achieved if **111** is treated with a lithium *t*-butoxy(alkyl)copper reagent[1285] instead of Me$_2$CuLi. For example, 2,6-dibromocyclohexanone, treated with lithium *t*-butoxy(*t*-butyl)copper, gave 66% 2-*t*-butyl-cyclohexanone. This is one of the few methods for introducing a tertiary alkyl group α to a carbonyl group. When dialkylcopperzinc reagents R$_2$CuZnCl couple with allylic halides, almost complete allylic rearrangement occurs (SN2'), and the reaction is diastereoselective if the allylic halide contains a δ alkoxy group.[1286]

For the preparation of R_2'CuLi reagents, see **2-35**.

A much older reaction is the coupling of alkyl halides with Grignard reagents.[1287] Grignard reagents have the advantage that they are usually simpler to prepare than the corresponding R_2'CuLi, but the reaction is much narrower in scope. Grignard reagents couple only with active halides: allylic (though allylic rearrangements are common) and benzylic. They also couple with tertiary alkyl halides, but generally in low or moderate yields.[1288] Aryl Grignard

[1279]Bertz; Dabbagh; Villacorta *J. Am. Chem. Soc.* **1982**, *104*, 5824; Bertz; Dabbagh *J. Org. Chem.* **1984**, *49*, 1119.
[1280]Mooiweer; Elsevier; Wijkens; Vermeer *Tetrahedron Lett.* **1985**, *26*, 65.
[1281]Corey; Boaz *Tetrahedron Lett.* **1984**, *25*, 3059, 3063. For the reaction of these reagents with haloalkynes, see Yeh; Knochel *Tetrahedron Lett.* **1989**, *30*, 4799.
[1282]Dubois; Lion; Moulineau *Tetrahedron Lett.* **1971**, 177; Dubois; Fournier; Lion *Bull. Soc. Chim. Fr.* **1976**, 1871.
[1283]See Corey; Posner, Ref. 1267; Wakselman; Mondon *Tetrahedron Lett.* **1973**, 4285.
[1284]Posner; Sterling *J. Am. Chem. Soc.* **1973**, *95*, 3076. See also Posner; Sterling; Whitten; Lentz; Brunelle *J. Am. Chem. Soc.* **1975**, *97*, 107; Lion; Dubois *Tetrahedron* **1975**, *31*, 1223. Ph$_2$CuLi behaves similarly: see Lei; Doubleday; Turro *Tetrahedron Lett.* **1986**, *27*, 4671.
[1285]Prepared by treating CuI with *t*-BuOLi in THF at 0°C and adding RLi to this solution.
[1286]Nakamura; Sekiya; Arai; Aoki *J. Am. Chem. Soc.* **1989**, *111*, 3091.
[1287]For reviews, see Raston; Salem, in Hartley *The Chemistry of the Metal–Carbon Bond*, vol. 4; Wiley: New York, 1987, pp. 161-306, pp. 269-283; Kharasch; Reinmuth *Grignard Reactions of Nonmetallic Substances*; Prentice-Hall: Englewood Cliffs, NJ, 1954, pp. 1046-1165.
[1288]See, for example, Ohno; Shimizu; Ishizaki; Sasaki; Eguchi *J. Org. Chem.* **1988**, *53*, 729.

reagents usually give better yields in these reactions than alkyl Grignard reagents. Furthermore, because Grignard reagents react with the C=O group (**6-29, 6-32**), they cannot be used to couple with halides containing ketone, COOR, or amide functions. Though the coupling of Grignard reagents with ordinary alkyl halides is usually not useful for synthetic purposes, small amounts of symmetrical coupling product are commonly formed while Grignard reagents are being prepared. Grignard reagents can be made to couple with alkyl halides in good yields by the use of certain catalysts.[1289] Among these are Cu(I) salts, which permit the coupling of Grignard reagents with primary alkyl halides in good yield[1290] (organocopper salts are probably intermediates here), and iron(III)[1291] or palladium[1292] complexes, which allow the coupling of Grignard reagents and vinylic halides. Grignard reagents prepared from primary or secondary[1293] alkyl or aryl halides can be coupled with vinylic or aryl halides in high yields in the presence of a nickel(II) catalyst.[1294] When a chiral nickel(II) catalyst is used, optically active hydrocarbons can be prepared from achiral reagents.[1295] Neopentyl iodides also couple with aryl Grignard reagents in the presence of a nickel(II) catalyst.[1295a]

Other organometallic compounds[1296] have also been used to couple with alkyl halides.[1297] Organosodium and organopotassium compounds are more reactive than Grignard reagents and couple even with less reactive halides. The difficulty is in preparing and keeping them long enough for the alkyl halide to be added. Alkenes can be prepared by the coupling of vinylic lithium compounds with primary halides[1298] or of vinylic halides with alkyllithiums in the presence of a Pd or Ru catalyst.[1299] When treated with organocopper compounds and Lewis acids (e.g., n-BuCu·BF$_3$), allylic halides give substitution with almost complete allylic rearrangement, irrespective of the degree of substitution at the two ends of the allylic system.[1300]

Organoaluminum compounds couple very well with tertiary (to give products containing a quaternary carbon) and benzylic halides at $-78°C$.[1301] This reaction can also be applied to allylic, secondary, and some primary halides, but several days standing at room temperature is required (see also **0-90**). Products containing a quaternary carbon can also be

[1289]For reviews, see Erdik *Tetrahedron* **1984**, *40*, 641-657; Kochi, Ref. 1077, pp. 374-398.

[1290]Tamura; Kochi *J. Am. Chem. Soc.* **1971**, *93*, 1485, *Synthesis* **1971**, 303, *J. Organomet. Chem.* **1972**, *42*, 205; Onuma; Hashimoto *Bull. Chem. Soc. Jpn.* **1972**, *45*, 2582; Derguini-Boumechal; Linstrumelle *Tetrahedron Lett.* **1976**, 3225; Mirviss *J. Org. Chem.* **1989**, *54*, 1948.

[1291]Tamura; Kochi *Synthesis* **1971**, 303, *J. Am. Chem. Soc.* **1971**, *93*, 1487; Smith; Kochi *J. Org. Chem.* **1976**, *41*, 502; Walborsky; Banks *J. Org. Chem.* **1981**, *46*, 5074; Molander; Rahn; Shubert; Bonde *Tetrahedron Lett.* **1983**, *24*, 5449.

[1292]Dang; Linstrumelle *Tetrahedron Lett.* **1978**, 191; Ratovelomanana; Linstrumelle; Normant *Tetrahedron Lett.* **1985**, *26*, 2575; Rossi; Carpita *Tetrahedron Lett.* **1986**, *27*, 2529; Minato; Suzuki; Tamao *J. Am. Chem. Soc.* **1987**, *109*, 1257; Fiandanese; Marchese; Mascolo; Naso; Ronzini *Tetrahedron Lett.* **1988**, *29*, 3705. For other references, see Ref. 508, pp. 201-202.

[1293]Hayashi; Konishi; Kobori; Kumada; Higuchi; Hirotsu *J. Am. Chem. Soc.* **1984**, *106*, 158.

[1294]Corriu; Masse *J. Chem. Soc., Chem. Commun.* **1972**, 144; Tamao; Sumitani; Kumada *J. Am. Chem. Soc.* **1972**, *94*, 4374. For a review, see Kumada *Pure Appl. Chem.* **1980**, *52*, 669-679.

[1295]For a review, see Hayashi; Kumada, in Morrison *Asymmetric Synthesis*, vol. 5; Academic Press: New York, **1985**, pp. 147-169. See also Cross; Kellogg *J. Chem. Soc., Chem. Commun.* **1987**, 1746; Iida; Yamashita *Bull. Chem. Soc. Jpn.* **1988**, *61*, 2365.

[1295a]Yuan; Scott *Tetrahedron Lett.* **1991**, *32*, 189.

[1296]For lists of reagents and substrates, with references, see Ref. 508, pp. 57-67.

[1297]For a review of the coupling of organic halides with organotin, mercury, and copper compounds catalyzed by palladium complexes, see Beletskaya *J. Organomet. Chem.* **1983**, *250*, 551-564. For a review of palladium-assisted coupling, see Larock *Organomercury Compounds in Organic Synthesis*; Springer: New York, **1985**, pp. 249-262.

[1298]Linstrumelle *Tetrahedron Lett.* **1974**, 3809; Millon; Lorne; Linstrumelle *Synthesis* **1975**, 434; Duhamel; Poirier *J. Am. Chem. Soc.* **1977**, *99*, 8356.

[1299]Murahashi; Yamamura; Yanagisawa; Mita; Kondo *J. Org. Chem.* **1979**, *44*, 2408.

[1300]Yamamoto; Yamamoto; Yatagai; Maruyama *J. Am. Chem. Soc.* **1980**, *102*, 2318. See also Lipshutz; Ellsworth; Dimock *J. Am. Chem. Soc.* **1990**, *112*, 5869.

[1301]Miller *J. Org. Chem.* **1966**, *31*, 908; Kennedy *J. Org. Chem.* **1970**, *35*, 532. See also Kennedy; Sivaram *J. Org. Chem.* **1973**, *38*, 2262; Sato; Kodama; Sato *J. Organomet. Chem.* **1978**, *157*, C30.

obtained by treatment of tertiary halides with dialkyl or diaryl zinc reagents in CH_2Cl_2,[1302] with Me_4Si and $AlCl_3$,[1303] or with alkyltitanium reagents $RTiCl_3$ and R_2TiCl_2.[1304] The titanium method can also be used with secondary halides ($R_2CHCl \rightarrow R_2CHMe$), tertiary ethers ($R_3COR' \rightarrow R_3CMe$), and *gem*-dihalides ($R_2CCl_2 \rightarrow R_2CMe_2$).[1305] Vinylic aluminum compounds (in the presence of a suitable transition-metal catalyst) couple with allylic halides, acetates, and alcohol derivatives to give 1,4-dienes,[1306] and with vinylic and benzylic halides to give 1,3-dienes and allylic arenes, respectively.[1307] Arylpalladium salts "ArPdX" prepared from arylmercury compounds and lithium palladium chloride couple with allylic chlorides in moderate yields, though allylic rearrangements can occur.[1308] The advantage of this procedure is that the aryl group may contain nitro, ester, or aldehyde groups, etc., which cannot be present in a Grignard reagent. Allylic, benzylic, vinylic, and aryl halides couple with organotin reagents in a reaction catalyzed by palladium complexes.[1309] Such functional groups as COOR, CN, OH, and CHO may be present in either reagent, but the substrate may not bear a β hydrogen on an sp^3 carbon, because that results in elimination. Organosilanes $RSiMe_3$ or $RSiMe_2F$ (where R can be vinylic, allylic, or alkynyl) couple with vinylic, allylic, and aryl bromides and iodides R'X, in the presence of certain catalysts, to give RR' in good yields.[1310] Alkenylboranes ($R_2'C{=}CHBZ_2$; Z = various groups) couple in high yields with vinylic, alkynyl, aryl, benzylic, and allylic halides in the presence of tetrakis(triphenylphosphine)palladium $Pd(PPh_3)_4$ and a base to give $R_2'C{=}CHR$.[1311] 9-Alkyl-9-BBN compounds (p. 785) also couple with vinylic and aryl halides[1312] as well as with α-halo ketones, nitriles, and esters.[1313]

gem-Dichlorides have been prepared by coupling alkyl halides to $RCCl_3$ compounds electrochemically, in an undivided cell with a sacrificial anode:[1314]

$$RCCl_3 + R'X + 2e^- \longrightarrow RCCl_2R' + Cl^- + X^-$$

R' could also be Cl, in which case the product bears a CCl_3 group.[1315]

Much study has been devoted to the mechanisms of these reactions,[1316] but firm conclusions are still lacking, in part because the mechanisms vary depending on the metal, the R group, the catalyst, if any, and the reaction conditions. Two basic pathways can be envi-

[1302]Reetz; Wenderoth; Peter; Steinbach; Westermann *J. Chem. Soc., Chem. Commun.* **1980**, 1202. See also Klingstedt; Frejd *Organometallics* **1983**, *2*, 598.

[1303]Bolestova; Parnes; Latypova; Kursanov *J. Org. Chem. USSR* **1981**, *17*, 1203.

[1304]Reetz; Westermann; Steinbach *Angew. Chem. Int. Ed. Engl.* **1980**, *19*, 900, 901 [*Angew. Chem. 92*, 931, 933].

[1305]Reetz; Steinbach; Wenderoth *Synth. Commun.* **1982**, *11*, 261.

[1306]Lynd; Zweifel *Synthesis* **1974**, 658; Matsushita; Negishi *J. Am. Chem. Soc.* **1981**, *103*, 2882, *J. Chem. Soc., Chem. Commun.* **1982**, 160. For similar reactions with other metals, see Larock; Bernhardt; Driggs *J. Organomet. Chem.* **1978**, *156*, 45; Yoshida; Tamao; Takahashi; Kumada *Tetrahedron Lett.* **1978**, 2161; Brown; Campbell *J. Org. Chem.* **1980**, *45*, 550; Baeckström; Björkling; Högberg; Norin *Acta Chem. Scand., Ser. B* **1984**, *38*, 779.

[1307]Negishi *Acc. Chem. Res.* **1982**, *15*, 340-348; Negishi; Luo *J. Org. Chem.* **1983**, *48*, 1560; Negishi; Takahashi; Baba; Van Horn; Okukado *J. Am. Chem. Soc.* **1987**, *109*, 2393; Negishi; Takahashi; Baba *Org. Synth. 66*, 60.

[1308]Heck *J. Am. Chem. Soc.* **1968**, *90*, 5531. For a review of palladium-assisted coupling, see Heck *Palladium Reagents in Organic Syntheses*; Academic Press: New York, 1985, pp. 208-214, 242-249.

[1309]For a review, see Stille *Angew. Chem. Int. Ed. Engl.* **1986**, *25*, 508-524 [*Angew. Chem. 98*, 504-519]. See also Stille; Simpson *J. Am. Chem. Soc.* **1987**, *109*, 2138; Bumagin; Andryukhova; Beletskaya *Doklad. Chem.* **1989**, *307*, 211; Stork; Isaacs *J. Am. Chem. Soc.* **1990**, *112*, 7399; Laborde; Lesheski; Kiely *Tetrahedron Lett.* **1990**, *31*, 1837. For a review of the mechanism, see Bumagin; Beletskaya *Russ. Chem. Rev.* **1990**, *59*, 1174-1184.

[1310]Hatanaka; Hiyama *J. Org. Chem.* **1988**, *53*, 918, **1989**, *54*, 268.

[1311]Brown; Molander *J. Org. Chem.* **1981**, *46*, 645; Miyaura; Yamada; Suginome; Suzuki *J. Am. Chem. Soc.* **1985**, *107*, 972; Sato; Miyaura; Suzuki *Chem. Lett.* **1989**, 1405; Rivera; Soderquist *Tetrahedron Lett.* **1991**, *32*, 2311; and references cited in these papers. For a review, see Matteson *Tetrahedron* **1989**, *45*, 1859-1885.

[1312]Miyaura; Ishiyama; Sasaki; Ishikawa; Satoh; Suzuki *J. Am. Chem. Soc.* **1989**, *111*, 314. See also Soderquist; Santiago *Tetrahedron Lett.* **1990**, *31*, 5541.

[1313]Brown; Joshi; Pyun; Singaram *J. Am. Chem. Soc.* **1989**, *111*, 1754. For another such coupling, see Matteson; Tripathy; Sarkar; Sadhu *J. Am. Chem. Soc.* **1989**, *111*, 4399.

[1314]Nédélec; Aït Haddou Mouloud; Folest; Périchon *J. Am. Chem. Soc.* **1988**, *53*, 4720.

[1315]For the transformation RX → RCF₃, see Chen; Wu *J. Chem. Soc., Chem. Commun.* **1989**, 705.

[1316]For a review, see Beletskaya; Artamkina; Reutov *Russ. Chem. Rev.* **1976**, *45*, 330-347.

sioned: a nucleophilic substitution process (which might be S_N1 or S_N2) and a free-radical mechanism. This could be an SET pathway, or some other route that provides radicals. In either case the two radicals R• and R'• would be in a solvent cage:

$$\text{RX} + \text{R'M} \longrightarrow \begin{bmatrix} \text{R•}+ \text{R'•} \\ + \text{MX} \end{bmatrix} \longrightarrow \text{RR'}$$
Solvent cage

It is necessary to postulate the solvent cage because, if the radicals were completely free, the products would be about 50% RR', 25% RR, and 25% R'R'. This is generally not the case; in most of these reactions RR' is the predominant or exclusive product.[1317] An example where an S_N2 mechanism has been demonstrated (by the finding of inversion of configuration at R) is the reaction between allylic or benzylic lithium reagents with secondary halides.[1318] Similarly, inversion has been shown in the reaction of 2-bromobutane with Ph_2CuLi[1274] (though the same reaction with 2-iodobutane has been reported to proceed with racemization[1319]). The fact that in some of these cases the reaction can be successfully applied to aryl and vinylic substrates indicates that a simple S_N process cannot be the only mechanism. One possibility is that the reagents first undergo an exchange reaction: $ArX + RM \rightarrow RX + ArM$, and then a nucleophilic substitution takes place. On the other hand, there is much evidence that many coupling reactions involving organometallic reagents with simple alkyl groups occur by free-radical mechanisms. Among the evidence[1320] is the observation of CIDNP in reactions of alkyl halides with simple organolithium reagents[1321] (see p. 187), the detection of free radicals by esr spectroscopy[1322] (p. 186), and the formation of 2,3-dimethyl-2,3-diphenylbutane when the reaction was carried out in the presence of cumene[1323] (this product is formed when a free radical abstracts a hydrogen from cumene to give $Ph\overset{\bullet}{C}Me_2$, which dimerizes). Evidence for free-radical mechanisms has also been found for the coupling of alkyl halides with simple organosodium compounds (Wurtz),[1324] with Grignard reagents,[1325] and with lithium dialkylcopper reagents.[1326] Free radicals have also been implicated in the metal-ion-catalyzed coupling of alkyl and aryl halides with Grignard reagents.[1327]

For symmetrical coupling of organometallic reagents ($2RM \rightarrow RR$), see **4-33** to **4-35**.

OS **I**, 186; **III**, 121; **IV**, 748; **V**, 1092; **VI**, 407, 675; **VII**, 77, 172, 245, 326, 485; **66**, 60; **68**, 130, 162; **69**, 120.

[1317]When a symmetrical distribution of products *is* found, this is evidence for a free-radical mechanism: the solvent cage is not efficient and breaks down.

[1318]Sauer; Braig *Tetrahedron Lett.* **1969**, 4275; Sommer; Korte *J. Org. Chem.* **1970**, *35*, 22; Korte; Kinner; Kaska *Tetrahedron Lett.* **1970**, 603. See also Schlosser; Fouquet *Chem. Ber.* **1974**, *107*, 1162, 1171.

[1319]Lipshutz; Wilhelm *J. Am. Chem. Soc.* **1982**, *104*, 4696; Lipshutz; Wilhelm; Nugent; Little; Baizer *J. Org. Chem.* **1983**, *48*, 3306.

[1320]For other evidence, see Muraoka; Nojima; Kusabayashi; Nagase *J. Chem. Soc., Perkin Trans. 2* **1986**, 761.

[1321]Ward; Lawler; Cooper *J. Am. Chem. Soc.* **1969**, *91*, 746; Lepley; Landau *J. Am. Chem. Soc.* **1969**, *91*, 748; Podoplelov; Leshina; Sagdeev; Kamkha; Shein *J. Org. Chem. USSR* **1976**, *12*, 488. For a review, see Ward; Lawler; Cooper, in Lepley; Closs *Chemically Induced Magnetic Polarization*; Wiley: New York, 1973, pp. 281-322.

[1322]Russell; Lamson *J. Am. Chem. Soc.* **1969**, *91*, 3967.

[1323]Bryce-Smith *Bull. Soc. Chim. Fr.* **1963**, 1418.

[1324]Garst; Cox *J. Am. Chem. Soc.* **1970**, *92*, 6389; Kasukhin; Gragerov *J. Org. Chem. USSR* **1971**, *7*, 2087; Garst; Hart *J. Chem Soc., Chem. Commun.* **1975**, 215.

[1325]Gough; Dixon *J. Org. Chem.* **1968**, *33*, 2148; Ward; Lawler; Marzilli *Tetrahedron Lett.* **1970**, 521; Kasukhin; Ponomarchuk; Buteiko *J. Org. Chem. USSR* **1972**, *8*, 673; Singh; Tayal; Nigam *J. Organomet. Chem.* **1972**, *42*, C9.

[1326]Ashby; DePriest; Tuncay; Srivastava *Tetrahedron Lett.* **1982**, *23*, 5251; Ashby; Coleman *J. Org. Chem.* **1987**, *52*, 4554; Bertz; Dabbagh; Mujsce *J. Am. Chem. Soc.* **1991**, *113*, 631.

[1327]Norman; Waters *J. Chem. Soc.* **1957**, 950; Frey *J. Org. Chem.* **1961**, *26*, 5187; Slaugh *J. Am. Chem. Soc.* **1961**, *83*, 2734; Davies; Done; Hey *J. Chem. Soc. C* **1969**, 1392, 2021, 2056; Abraham; Hogarth *J. Organomet. Chem.* **1968**, *12*, 1, 497; Tamura; Kochi *J. Am. Chem. Soc.* **1971**, *93*, 1483, 1485, 1487, *J. Organomet. Chem.* **1971**, *31*, 289, **1972**, *42*, 205; Lehr; Lawler *J. Am. Chem. Soc.* **1986**, *106*, 4048.

0-88 Allylic and Propargylic Coupling with a Halide Substrate
De-halogen-coupling

$$2R-\overset{\overset{\textstyle R}{|}}{C}=\overset{\overset{\textstyle R}{|}}{C}-CH_2Br + Ni(CO)_4 \longrightarrow R-\overset{\overset{\textstyle R}{|}}{C}=\overset{\overset{\textstyle R}{|}}{C}-CH_2CH_2-\overset{\overset{\textstyle R}{|}}{C}=\overset{\overset{\textstyle R}{|}}{C}-R + NiBr_2 + 4CO$$

Because of the presence of the 1,5-diene moiety in many naturally occurring compounds, a great deal of effort has been expended in searching for methods to couple[1328] allylic groups.[1329] In one of these methods, allylic halides, tosylates, and acetates can be symmetrically coupled by treatment with nickel carbonyl[1330] at room temperature in a solvent such as THF or DMF to give 1,5-dienes.[1331] The order of halide reactivity is I > Br > Cl. With unsymmetrical allylic substrates, coupling nearly always takes place at the less-substituted end. The reaction can be performed intramolecularly; large (11- to 20-membered) rings can be made in good yields (60 to 80%) by the use of high dilution. An example[1332] is

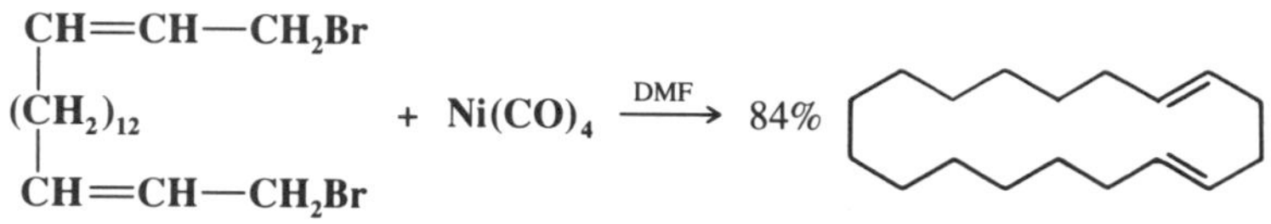

It is likely that the mechanism involves reaction of the allylic compound with Ni(CO)$_4$ to give one or more π-allyl complexes, one of which may be **113**, which can then lose CO to

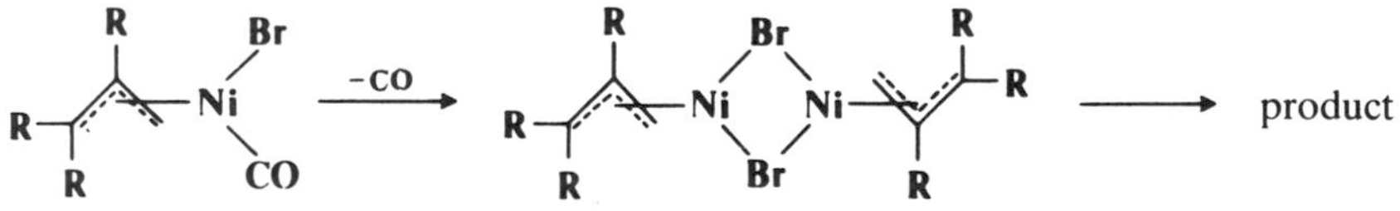

113 **114**

give a π-allylnickel bromide (**114**) which reacts further, perhaps with CO, to give the product. The complexes **114** can be isolated from the solution and crystallized as stable solids.

Unsymmetrical coupling can be achieved by treating an alkyl halide directly with **114**, in a polar aprotic solvent.[1333] In this case too, unsymmetrical allylic groups couple at the less

$$R'X + 114 \longrightarrow R'-CH_2-\overset{\overset{\textstyle R}{|}}{C}=\overset{\overset{\textstyle R}{|}}{C}-R$$

[1328]For a review of some allylic coupling reactions, see Magid *Tetrahedron* **1980**, *36*, 1901-1930, pp. 1910-1924.

[1329]In this section are discussed methods in which one molecule is a halide. For other allylic coupling reactions, see **0-87, 0-90,** and **0-91.**

[1330]For a review of the use of organonickel compounds in organic synthesis, see Tamao; Kumada, in Hartley, Ref. 1287, pp. 819-887.

[1331]For reviews, see Collman et al., Ref. 1266, pp. 739-748; Billington *Chem. Soc. Rev.* **1985**, *14*, 93-120; Kochi, Ref. 1077, pp. 398-408; Semmelhack *Org. React.* **1972**, *19*, 115-198, pp. 162-170; Baker *Chem. Rev.* **1973**, *73*, 487-530, pp. 512-517; Heimbach; Jolly; Wilke *Adv. Organomet. Chem.* **1970**, *8*, 29-86, pp. 30-39.

[1332]Corey; Wat *J. Am. Chem. Soc.* **1967**, *89*, 2757. See also Corey; Helquist *Tetrahedron Lett.* **1975**, 4091; Reijnders; Blankert; Buck *Recl. Trav. Chim. Pays-Bas* **1978**, *97*, 30.

[1333]Corey; Semmelhack *J. Am. Chem. Soc.* **1967**, *89*, 2755. For a review, see Semmelhack, Ref. 1331, pp. 147-162. For a discussion of the preparation and handling of π-allylnickel halides, see Semmelhack, Ref. 1331, pp. 144-146.

substituted end. The mechanism here cannot be simple nucleophilic substitution, since aryl and vinylic halides undergo the reaction as well as or better than simple primary bromides. There is evidence that free radicals are involved.[1334] Hydroxy or carbonyl groups in the alkyl halide do not interfere. When **114** reacts with an allylic halide, a mixture of three products is obtained because of halogen–metal interchange. For example, allyl bromide treated with **114** prepared from methallyl bromide gave an approximately statistical mixture of 1,5-hexadiene, 2-methyl-1,5-hexadiene, and 2,5-dimethyl-1,5-hexadiene.[1335]

The reaction between primary and secondary halides and allyltributylstannane provides another method for unsymmetrical coupling $RX + CH_2=CHCH_2SnBu_3 \rightarrow RCH_2CH=CH_2$.[1336]

Symmetrical coupling of allylic halides can also be accomplished by heating with magnesium in ether,[1337] with a cuprous iodide–dialkylamide complex,[1338] with $CrCl_3$–$LiAlH_4$,[1339] with Te^{2-} ions,[1340] with ion powder in DMF,[1341] or electrochemically.[1342] The coupling of two different allylic groups has been achieved by treatment of an allylic bromide with an allylic Grignard reagent in THF containing HMPA,[1343] or with an allylic tin reagent.[1344] This type of coupling can be achieved with almost no allylic rearrangement in the substrate (and almost complete allylic rearrangement in the reagent) by treatment of allylic halides with lithium allylic boron ate complexes ($RCH=CHCH_2\overset{\ominus}{B}R''_3\ Li^+$).[1345]

In another method for the coupling of two different allylic groups,[1346] a carbanion derived from a β,γ-unsaturated thioether couples with an allylic halide.[1347] The product contains an SPh group that must be removed (with Li in ethylamine) to give the 1,5-diene, but this

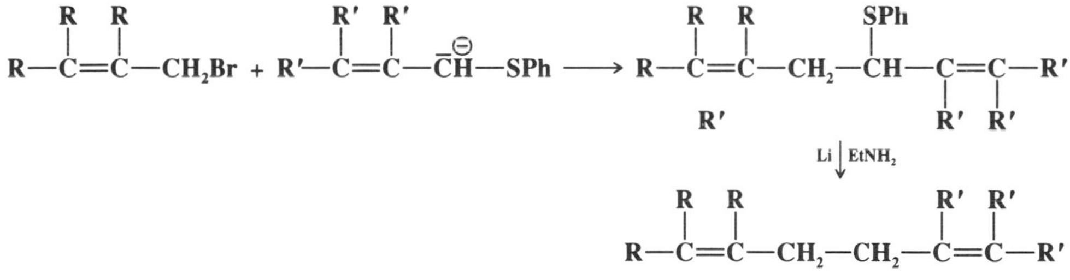

method has the advantage that, unlike most of the methods previously discussed, the coupling preserves the original positions and configurations of the two double bonds; no allylic rearrangements take place.

[1334]Hegedus; Thompson *J. Am. Chem. Soc.* **1985**, *107*, 5663.

[1335]Corey; Semmelhack; Hegedus *J. Am. Chem. Soc.* **1968**, *90*, 2416.

[1336]See Keck; Yates *J. Am. Chem. Soc.* **1982**, *104*, 5829; Migita; Nagai; Kosugi *Bull. Chem. Soc. Jpn* **1983**, *56*, 2480.

[1337]Turk; Chanan *Org. Synth. III*, 121.

[1338]Kitagawa; Oshima; Yamamoto; Nozaki *Tetrahedron Lett.* **1975**, 1859.

[1339]Okude; Hiyama; Nozaki *Tetrahedron Lett.* **1977**, 3829.

[1340]Clive; Anderson; Moss; Singh *J. Org. Chem.* **1982**, *47*, 1641.

[1341]Hall; Hurley *Can. J. Chem.* **1969**, *47*, 1238.

[1342]Tokuda; Endate; Suginome *Chem. Lett.* **1988**, 945.

[1343]Stork; Grieco; Gregson *Tetrahedron Lett.* **1969**, 1393; Grieco *J. Am. Chem. Soc.* **1969**, *91*, 5660.

[1344]Godschalx; Stille *Tetrahedron Lett.* **1980**, *21*, 2599; **1983**, *24*, 1905; Hosomi; Imai; Endo; Sakurai *J. Organomet. Chem.* **1985**, *285*, 95. See also Yanagisawa; Norikate; Yamamoto *Chem. Lett.* **1988**, 1899.

[1345]Yamamoto; Yatagai; Maruyama *J. Am. Chem. Soc.* **1981**, *103*, 1969.

[1346]For other procedures, see Axelrod; Milne; van Tamelen *J. Am. Chem Soc.* **1970**, *92*, 2139; Morizawa; Kanemoto; Oshima; Nozaki *Tetrahedron Lett.* **1982**, *23*, 2953.

[1347]Biellmann; Ducep *Tetrahedron Lett.* **1969**, 3707.

In a method for propargylating an alkyl halide without allylic rearrangement, the halide is treated with lithio-1-trimethylsilylpropyne (**115**) which is a lithium compound protected

$$RX + LiCH_2\overset{3}{}-C\equiv\overset{1}{C}-SiMe_3 \longrightarrow RCH_2-C\equiv C-SiMe \xrightarrow[2.CN^-]{1.Ag^+} R-CH_2-C\equiv CH$$

115

by an SiMe$_3$ group.[1348] Attack by the ambident nucleophile at its 1 position (which gives an allene) takes place only to a small extent, because of steric blockage by the large SiMe$_3$ group. The SiMe$_3$ group is easily removed by treatment with Ag$^+$ followed by CN$^-$. **115** is prepared by treating propynyllithium with Me$_3$SiCl to give MeC≡CSiMe$_3$ from which a proton is removed with BuLi. R may be primary or allylic.[1349] On the other hand, propargylic halides can be alkylated with essentially complete allylic rearrangement, to give allenes, by treatment with Grignard reagents and metallic salts,[1350] or with dialkylcuprates R$_2$Cu.[1351]

$$R-C\equiv C-CH_2X + R'MgX \xrightarrow{FeCl_3} R-\underset{R'}{\underset{|}{C}}=C=CH_2$$

OS **III**, 121; **IV**, 748; **VI**, 722.

0-89   Coupling of Organometallic Reagents with Esters of Sulfuric and Sulfonic Acids
Alkyl-de-sulfonyloxy-substitution, etc.

$$ROSO_2OR' + R''_2CuLi \longrightarrow RR''$$

Lithium dialkylcopper reagents couple with alkyl tosylates.[1352] High yields are obtained with primary tosylates; secondary tosylates give lower yields.[1353] Aryl tosylates do not react. Vinylic triflates[1354] couple very well to give alkenes.[1355] Vinylic triflates also couple with allylic cuprates, to give 1,4-dienes.[1356] Tosylates and other sulfonates and sulfates also couple with Grignard reagents,[1357] most often those prepared from aryl or benzylic halides.[1358] Alkyl sulfates and sulfonates generally make better substrates in reactions with Grignard reagents than the corresponding halides (**0-87**). The method is useful for primary and secondary R. Allylic tosylates can be symmetrically coupled with Ni(CO)$_4$ (see **0-88**). Propargylic tosylates couple with vinylic cuprates to give vinylic allenes.[1359] Vinylic triflates, in the presence of Pd(Ph$_3$P)$_4$ and LiCl, couple with organotin compounds R'SnMe$_3$, where R' can be alkyl,

[1348]Corey; Kirst; Katzenellenbogen *J. Am. Chem. Soc.* **1970**, *92*, 6314.
[1349]For an alternative procedure, see Ireland; Dawson; Lipinski *Tetrahedron Lett.* **1970**, 2247.
[1350]Pasto; Chou; Waterhouse; Shults; Hennion *J. Org. Chem.* **1978**, *43*, 1385; Jeffery-Luong; Linstrumelle *Tetrahedron Lett.* **1980**, *21*, 5019.
[1351]Pasto; Chou; Fritzen; Shults; Waterhouse; Hennion *J. Org. Chem.* **1978**, *43*, 1389. See also Tanigawa; Murahashi *J. Org. Chem.* **1980**, *45*, 4536.
[1352]Johnson; Dutra *J. Am. Chem. Soc.* **1973**, *95*, 7777, 7783. For examples, see Posner *An Introduction to Synthesis Using Organocopper Reagents*, Ref. 1268, pp. 85-90.
[1353]Secondary tosylates give higher yields when they contain an O or S atom: Hanessian; Thavonekham; DeHoff *J. Org. Chem.* **1989**, *54*, 5831.
[1354]For a review of coupling reactions of vinylic triflates, see Scott; McMurry *Acc. Chem. Res.* **1988**, *21*, 47-54.
[1355]McMurry; Scott *Tetrahedron Lett.* **1980**, *21*, 4313; Tsushima; Araki; Murai *Chem. Lett.* **1989**, 1313.
[1356]Lipshutz; Elworthy *J. Org. Chem.* **1990**, *55*, 1695.
[1357]For a review, see Kharasch; Reinmuth, Ref. 1287, pp. 1277-1286.
[1358]For an example involving an allylic rearrangement (conversion of a silylalkyne to a silylallene), see Danheiser; Tsai; Fink *Org. Synth.* 66, 1.
[1359]Baudouy; Goré *J. Chem. Res. (S)* **1981**, 278. See also Elsevier; Vermeer *J. Org. Chem.* **1989**, *54*, 3726.

allylic, vinylic, or alkynyl.[1360] The reaction has been performed intramolecularly, to prepare large-ring lactones.[1361]

OS **I**, 471; **II**, 47, 360; **VII**, 351; **66**, 1; **68**, 116.

0-90   Coupling Involving Alcohols
De-hydroxyl-coupling

$$2ROH \xrightarrow[-78°C]{MeLi-TiCl_3} RR$$

Allylic or benzylic alcohols can be symmetrically coupled[1362] by treatment with methyllithium and titanium trichloride at -78°C[1363] or by refluxing with $TiCl_3$ and $LiAlH_4$.[1364] When the substrate is an allylic alcohol, the reaction is not regiospecific, but a mixture of normal coupling and allylically rearranged products is found. A free-radical mechanism is involved.[1365] Another reagent that symmetrically couples allylic and benzylic alcohols is $NbCl_5$–$NaAlH_4$.[1366] The $TiCl_3$–$LiAlH_4$ reagent can also convert 1,3-diols to cyclopropanes, provided that at least one α phenyl is present,[1367] e.g.,

$$Ph-\underset{\underset{OH}{|}}{CH}-CH_2-\underset{\underset{OH}{|}}{CH}-Ph \xrightarrow[LiAlH_4]{TiCl_3} \quad \underset{Ph}{\triangle}Ph \quad \text{(cis and trans)}$$

Tertiary alcohols react with trimethylaluminum at 80 to 200°C to give methylation.[1368] The presence of side products from elimination and rearrangement, as well as the lack of

$$R_3COH + Me_3Al \xrightarrow{80-200°C} R_3CMe$$

stereospecificity,[1369] indicate an S_N1 mechanism. The reaction can also be applied to primary and secondary alcohols if these contain an aryl group in the α position. Higher trialkylaluminums are far less suitable, because reduction competes with alkylation (see also reactions of Me_3Al with ketones, **6-29**, and with carboxylic acids, **6-32**). Me_2TiCl_2 also reacts with tertiary alcohols in the same way.[1370] Allylic alcohols couple with a reagent prepared from MeLi, CuI, and R'Li in the presence of $(Ph_3PNMePh)^+ I^-$ to give alkenes that are products of allylic rearrangement.[1371] The reaction gives good yields with primary, secondary, and

[1360]Scott; Stille *J. Am. Chem. Soc.* **1986**, *108*, 3033; Kwon; McKee; Stille *J. Org. Chem.* **1990**, *55*, 3114. For discussions of the mechanism, see Stang; Kowalski; Schiavelli; Longford *J. Am. Chem. Soc.* **1989**, *111*, 3347; Stang; Kowalski *J. Am. Chem. Soc.* **1989**, *111*, 3356.

[1361]Stille; Tanaka *J. Am. Chem. Soc.* **1987**, *109*, 3785.

[1362]For a review, see Lai *Org. Prep. Proceed. Int.* **1980**, *12*, 363-391, pp. 377-388.

[1363]Sharpless; Hanzlik; van Tamelen *J. Am. Chem. Soc.* **1968**, *90*, 209.

[1364]McMurry; Silvestri; Fleming; Hoz; Grayston *J. Org. Chem.* **1978**, *43*, 3249. For another method, see Nakanishi; Shundo; Nishibuchi; Otsuji *Chem. Lett.* **1979**, 955.

[1365]van Tamelen; Åkermark; Sharpless *J. Am. Chem. Soc.* **1969**, *91*, 1552.

[1366]Sato; Oshima *Chem. Lett.* **1982**, 157. For a reagent that couples benzhydrols, see Pri-Bar; Buchman; Blum *Tetrahedron Lett.* **1977**, 1443.

[1367]Baumstark; McCloskey; Tolson; Syriopoulos *Tetrahedron Lett.* **1977**, 3003; Walborsky; Murati *J. Am. Chem. Soc.* **1980**, *102*, 426.

[1368]Meisters; Mole *J. Chem. Soc., Chem. Commun.* **1972**, 595; Harney; Meisters; Mole *Aust. J. Chem.* **1974**, *27*, 1639.

[1369]Salomon; Kochi *J. Org. Chem.* **1973**, *38*, 3715.

[1370]Reetz; Westermann; Steinbach *J. Chem. Soc., Chem. Commun.* **1981**, 237.

[1371]Tanigawa; Ohta; Sonoda; Murahashi *J. Am. Chem. Soc.* **1978**, *100*, 4610; Goering; Tseng *J. Org. Chem.* **1985**, *50*, 1597. For another procedure, see Yamamoto; Maruyama *J. Organomet. Chem.* **1978**, *156*, C9.

$$
\underset{\substack{|\\R}}{R}\!\!-\!\!\overset{\substack{R\\|}}{C}\!\!=\!\!\overset{\substack{R\\|}}{C}\!\!-\!\!\overset{\substack{R\\|}}{\underset{\substack{|\\R}}{C}}\!\!-\!\!OH \xrightarrow[\text{(Ph}_3\text{PNMePh)}^+ \text{ I}^-]{\text{MeLi}-\text{CuI}-\text{R'Li}} R\!\!-\!\!\overset{\substack{R\\|}}{\underset{\substack{|\\R'}}{C}}\!\!-\!\!\overset{\substack{R\\|}}{C}\!\!=\!\!\overset{\substack{R\\|}}{C}\!\!-\!\!R
$$

tertiary alcohols, and with alkyl and aryllithiums.[1372] Allylic alcohols also couple with certain Grignard reagents[1373] in the presence of a nickel complex to give both normal products and the products of allylic rearrangement.

0-91 Coupling of Organometallic Reagents with Carboxylic Esters
Alkyl-de-acyloxy-substitution

$$
R\!\!-\!\!\overset{\substack{R\\|}}{C}\!\!=\!\!\overset{\substack{R\\|}}{C}\!\!-\!\!\overset{\substack{R\\|}}{\underset{\substack{|\\R}}{C}}\!\!-\!\!OAc + R'_2\text{CuLi} \longrightarrow R\!\!-\!\!\overset{\substack{R\\|}}{C}\!\!=\!\!\overset{\substack{R\\|}}{C}\!\!-\!\!\overset{\substack{R\\|}}{\underset{\substack{|\\R}}{C}}\!\!-\!\!R' + R\!\!-\!\!\overset{\substack{R\\|}}{\underset{\substack{|\\R'}}{C}}\!\!-\!\!\overset{\substack{R\\|}}{C}\!\!=\!\!\overset{\substack{R\\|}}{C}\!\!-\!\!R
$$

Lithium dialkylcopper reagents couple with allylic acetates to give normal coupling products or those resulting from allylic rearrangement, depending on the substrate.[1374] A mechanism involving a σ-allylic copper(III) complex has been suggested.[1375] With propargyl substrates, the products are allenes.[1376] Allenes are also obtained when propargyl acetates are treated

$$
\text{RC}\!\!\equiv\!\!\text{C}\!\!-\!\!\text{CR}_2\!\!-\!\!\text{OAc} + \text{R}'_2\text{CuLi} \longrightarrow \text{RR}'\text{C}\!\!=\!\!\text{C}\!\!=\!\!\text{CR}_2
$$

with methylmagnesium iodide.[1377] Lithium dialkylcopper reagents also give normal coupling products with enol acetates of β-dicarbonyl compounds.[1378] It is also possible to carry out the coupling of allylic acetates with Grignard reagents, if catalytic amounts of cuprous salts are present.[1379] With this method yields are better and regioselectivity can be controlled by a choice of cuprous salts. Allylic, benzylic, and cyclopropylmethyl acetates couple with trialkylaluminums,[1380] and allylic acetates couple with aryl and vinylic tin reagents, in the presence of a palladium-complex catalyst.[1381] Allylic acetates can be symmetrically

[1372]For the allylation of benzylic alcohols, see Cella *J. Org. Chem.* **1982**, *47*, 2125.

[1373]Buckwalter; Burfitt; Felkin; Joly-Goudket; Naemura; Salomon; Wenkert; Wovkulich *J. Am. Chem. Soc.* **1978**, *100*, 6445; Felkin; Joly-Goudket; Davies *Tetrahedron Lett.* **1981**, *22*, 1157; Consiglio; Morandini; Piccolo *J. Am. Chem. Soc.* **1981**, *103*, 1846, and references cited in these papers. For a review, see Felkin; Swierczewski *Tetrahedron* **1975**, *31*, 2735-2748. For other procedures, see Mukaiyama; Imaoka; Izawa *Chem. Lett.* **1977**, 1257; Fujisawa; Iida; Yukizaki; Sato *Tetrahedron Lett.* **1983**, *24*, 5745.

[1374]Rona; Tökes; Tremble; Crabbé *Chem. Commun.* **1969**, 43; Anderson; Henrick; Siddall *J. Am. Chem. Soc.* **1970**, *92*, 735; Goering; Singleton *J. Am. Chem. Soc.* **1976**, *98*, 7854; Gallina; Ciattini *J. Am. Chem. Soc.* **1979**, *101*, 1035; Goering; Kantner *J. Org. Chem.* **1984**, *49*, 422. For examples of the use of this reaction with allylic and propargyl substrates, see Posner, Ref. 1352, pp. 91-104.

[1375]Goering; Kantner *J. Org. Chem.* **1983**, *48*, 721; Goering; Kantner; Seitz *J. Org. Chem.* **1985**, *50*, 5495.

[1376]Crabbé; Barreiro; Dollat; Luche *J. Chem. Soc., Chem. Commun.* **1976**, 183, and references cited therein.

[1377]Roumestant; Gore *Bull. Soc. Chim. Fr.* **1972**, 591, 598.

[1378]Casey; Marten *Synth. Commun.* **1973**, *3*, 321, *Tetrahedron Lett.* **1974**, 925. See also Posner; Brunelle *J. Chem. Soc., Chem. Commun.* **1973**, 907; Kobayashi; Takei; Mukaiyama *Chem. Lett.* **1973**, 1097.

[1379]Tseng; Paisley; Goering *J. Org. Chem.* **1986**, *51*, 2884; Tseng; Yen; Goering *J. Org. Chem.* **1986**, *51*, 2892; Underiner; Paisley; Schmitter; Lesheski; Goering *J. Org. Chem.* **1989**, *54*, 2369; Bäckvall; Sellén; Grant *J. Am. Chem. Soc.* **1990**, *112*, 6615. See also Hiyama; Wakasa *Tetrahedron Lett.* **1985**, *26*, 3259.

[1380]Itoh; Oshima; Sasaki; Yamamoto; Hiyama; Nozaki *Tetrahedron Lett.* **1979**, 4751; Gallina *Tetrahedron Lett.* **1985**, *26*, 519; Tolstikov; Dzhemilev *J. Organomet. Chem.* **1985**, *292*, 133.

[1381]Del Valle; Stille; Hegedus *J. Org. Chem.* **1990**, *55*, 3019. For another method, see Legros; Fiaud *Tetrahedron Lett.* **1990**, *31*, 7453.

coupled by treatment with $Ni(CO)_4$ (reaction **0-88**) or with Zn and a palladium-complex catalyst,[1382] or converted to unsymmetrical 1,5-dienes by treatment with an allylic stannane $R_2C=CHCH_2SnR_3$ in the presence of a palladium complex.[1383]

0-92 Coupling of Organometallic Reagents with Compounds Containing the Ether Linkage[1384]
Alkyl-de-alkoxy-substitution

$$R_2C(OR')_2 + R''MgX \longrightarrow R_2CR''(OR') + R'OMgX$$

$$RC(OR')_3 + R''MgX \longrightarrow RCR''(OR')_2 + R'OMgX$$

Acetals,[1385] ketals, and ortho esters[1386] react with Grignard reagents to give, respectively, ethers and acetals (or ketals). The latter can be hydrolyzed to aldehydes or ketones (**0-6**). This procedure is a way of converting a halide R''X (which may be alkyl, aryl, vinylic, or alkynyl) to an aldehyde R''CHO, increasing the length of the carbon chain by one carbon (see also **0-102**). The ketone synthesis generally gives lower yields. Acetals, including allylic acetals, also give this reaction with organocopper compounds and BF_3.[1387] Acetals also undergo substitution when treated with silyl enol ethers or allylic silanes, with a Lewis acid catalyst,[1388] e.g.,

$$RCH_2-\underset{\underset{OR''}{|}}{CH}-OR'' + CH_2=\underset{\underset{R'}{|}}{C}-OSiMe_3 \xrightarrow{TiCl_4} RCH_2-\underset{\underset{OR''}{|}}{CH}-CH_2-\underset{\underset{O}{||}}{C}-R'$$

Tertiary amines can be prepared by the reaction of amino ethers with Grignard reagents,[1389] $(R_2NCH_2-OR' + R''MgX \rightarrow R_2NCH_2-R'')$ or with lithium dialkylcopper reagents.[1390] This method, when followed by treatment of the amine with a chloroformate (see **0-72**) allows an alkyl halide RX to be converted to its homolog RCH_2X in only two laboratory steps[1391] (see also p. 476):

$$RX \xrightarrow[\textbf{2-38}]{Mg} RMgX \xrightarrow{R'OCH_2NR'_2} RCH_2NR'_2 \xrightarrow{ClCOOAr} RCH_2Cl + R'_2NCOOAr$$

Ordinary ethers are not cleaved by Grignard reagents (in fact, diethyl ether and THF are the most common solvents for Grignard reagents), though more active organometallic compounds often do cleave them.[1392] Allylic ethers can be cleaved by Grignard reagents in

[1382]Sasaoka; Yamamoto; Kinoshita; Inomata; Kotake *Chem. Lett.* **1985**, 315.

[1383]Trost; Keinan *Tetrahedron Lett.* **1980**, *21*, 2595.

[1384]For a review, see Trofimov; Korostova *Russ. Chem. Rev.* **1975**, *44*, 41-55.

[1385]For a review of coupling reactions of acetals, see Mukaiyama; Murakami *Synthesis* **1987**, 1043-1054. For a discussion of the mechanism, see Abell; Massy-Westropp *Aust. J. Chem.* **1985**, *38*, 1031. For a list of substrates and reagents, with references, see Ref. 508, pp. 404-405.

[1386]For a review of the reaction with ortho esters, see DeWolfe, Ref. 457, pp. 44-45, 224-230.

[1387]Normant; Alexakis; Ghribi; Mangeney *Tetrahedron* **1989**, *45*, 507; Alexakis; Mangeney; Ghribi; Marek; Sedrani; Guir; Normant *Pure Appl. Chem.* **1988**, *60*, 49-56.

[1388]See Mori; Ishihara; Flippen; Nozaki; Yamamoto; Bartlett; Heathcock *J. Org. Chem.* **1990**, *55*, 6107, and references cited therein.

[1389]For example, see Miginiac; Mauzé *Bull. Soc. Chim. Fr.* **1968**, 2544; Eisele; Simchen *Synthesis* **1978**, 757; Kapnang; Charles *Tetrahedron Lett.* **1983**, *24*, 1597; Morimoto; Takahashi; Sekiya *J. Chem. Soc., Chem. Commun.* **1984**, 794; Mesnard; Miginiac *J. Organomet. Chem.* **1989**, *373*, 1. See also Bourhis; Bosc; Golse *J. Organomet. Chem.* **1983**, *256*, 193.

[1390]Germon; Alexakis; Normant *Bull. Soc. Chim. Fr.* **1984**, II-377.

[1391]Yankep; Charles *Tetrahedron Lett.* **1987**, *28*, 427.

[1392]For a review of the reactions of ethers with Grignard reagents, see Kharasch; Reinmuth, Ref. 1287, pp. 1013-1045.

THF if CuBr is present.[1393] The reaction takes place either with or without allylic rearrangement.[1394] Propargylic ethers give allenes.[1395] Vinylic ethers can also be cleaved by Grignard reagents in the presence of a catalyst, in this case, a nickel complex.[1396] Silyl enol ethers $R_2C{=}CROSiMe_3$ behave similarly.[1397]

Certain acetals and ketals can be dimerized in a reaction similar to **0-86** by treatment with $TiCl_4$–$LiAlH_4$, e.g.,[1398]

$$\underset{\overset{|}{OEt}}{PhCH{-}OEt} \xrightarrow[\text{LiAlH}_4]{\text{TiCl}_4} 85\%\ \underset{\overset{|}{OEt}\ \overset{|}{OEt}}{PhCH{-}CHPh}$$

Also see **0-93**.

OS **II**, 323; **III**, 701. Also see OS **V**, 431.

0-93 The Reaction of Organometallic Reagents with Epoxides
3(*OC*)-*seco*-Alkyl-de-alkoxy-substitution

$$\underset{O}{-\overset{|}{C}{-}\overset{|}{C}{-}} + RMgX \longrightarrow R{-}\overset{|}{\underset{|}{C}}{-}\overset{|}{\underset{|}{C}}{-}OMgX$$

The reaction between Grignard reagents and epoxides is very valuable and is often used to increase the length of a carbon chain by two carbons.[1399] The Grignard reagent may be aromatic or aliphatic, though tertiary Grignard reagents give low yields. As expected for an S_N2 process, attack is at the less substituted carbon. Lithium dialkylcopper reagents also give the reaction,[1400] often producing higher yields, and have the additional advantage that they do not react with ester, ketone, or carboxyl groups so that the epoxide ring of epoxy esters, ketones, and carboxylic acids can be selectively attacked, often in a regioselective manner.[1401] The use of BF_3 increases the reactivity of R_2CuLi, enabling it to be used with thermally unstable epoxides.[1402] The reaction has also been performed with other organometallic compounds, e.g., of Li, Al, etc.[1403]

[1393]Commercon; Bourgain; Delaumeny; Normant; Villieras *Tetrahedron Lett.* **1975**, 3837; Claesson; Olsson *Chem. Soc., Chem. Commun.* **1987**, 621.

[1394]Normant; Commercon; Gendreau; Bourgain; Villieras *Bull. Soc. Chim. Fr.* **1979**, II-309; Gendreau; Normant *Tetrahedron* **1979**, *35*, 1517; Calo; Lopez; Pesce *J. Chem. Soc., Perkin Trans. 1* **1988**, 1301. See also Valverde; Bernabé; Garcia-Ochoa; Gómez *J. Org. Chem.* **1990**, *55*, 2294.

[1395]Alexakis; Marek; Mangeney; Normant *Tetrahedron Lett.* **1989**, *30*, 2387, *J. Am. Chem. Soc.* **1990**, *112*, 8042.

[1396]Wenkert; Michelotti; Swindell; Tingoli *J. Org. Chem.* **1984**, *49*, 4894; Kocieński; Dixon; Wadman *Tetrahedron Lett.* **1988**, *29*, 2353.

[1397]Hayashi; Katsuro; Kumada *Tetrahedron Lett.* **1980**, *21*, 3915.

[1398]Ishikawa; Mukaiyama *Bull Chem. Soc. Jpn.* **1978**, *51*, 2059.

[1399]For a review, see Kharasch; Reinmuth, Ref. 1287, pp. 961-1012. For a thorough discussion, see Schaap; Arens *Recl. Trav. Chim. Pays-Bas* **1968**, *87*, 1249. For improved procedures, see Huynh; Derguini-Boumechal; Linstrumelle *Tetrahedron Lett.* **1979**, 1503; Schrumpf; Grätz; Meinecke; Fellenberger *J. Chem. Res. (S)* **1982**, 162.

[1400]For examples of the use of this reactions, see Posner, Ref. 1352, pp. 103-113. See also Lipshutz; Kozlowski; Wilhelm *J. Am. Chem. Soc.* **1982**, *104*, 2305.

[1401]Johnson; Herr; Wieland *J. Org. Chem.* **1973**, *38*, 4263; Hartman; Livinghouse; Rickborn *J. Org. Chem.* **1973**, *38*, 4346; Hudrlik; Peterson; Rona *J. Org. Chem.* **1975**, *40*, 2263; Chong; Sharpless *Tetrahedron Lett.* **1985**, *26*, 4683; Chong; Cyr; Mar *Tetrahedron Lett.* **1987**, *28*, 5009; Larchevêque; Petit *Tetrahedron Lett.* **1987**, *28*, 1993.

[1402]See, for example, Alexakis; Jachiet; Normant *Tetrahedron* **1986**, *42*, 5607.

[1403]For lists of organometallic reagents that react with epoxides, see Wardell; Paterson, in Hartley; Patai *The Chemistry of the Metal–Carbon Bond*, vol. 2; Wiley: New York, 1985, pp. 307-310; Ref. 508, pp. 512-520.

When *gem*-disubstituted epoxides (**116**) are treated with Grignard reagents (and sometimes other epoxides), the product may be **117**, that is, the new alkyl group may appear on

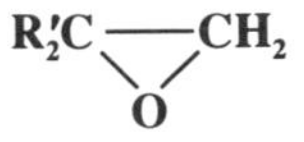

$$R_2'C \overset{\displaystyle O}{\overline{}} CH_2 \qquad R_2'CH \overset{\displaystyle |}{\underset{OH}{-}} CHR$$

116 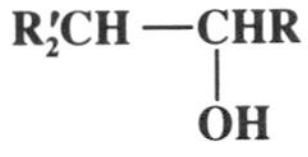   **117**

the same carbon as the OH. In such cases, the epoxide is isomerized to an aldehyde or a ketone before reacting with the Grignard reagent. Halohydrins are often side products.

When the substrate is a vinylic epoxide,[1404] Grignard reagents generally give a mixture of the normal product and the product of allylic rearrangement.[1405]

$$R-MgX + CH_2=CH-CH-CH_2 \longrightarrow RCH_2CH=CHCH_2OMgX$$

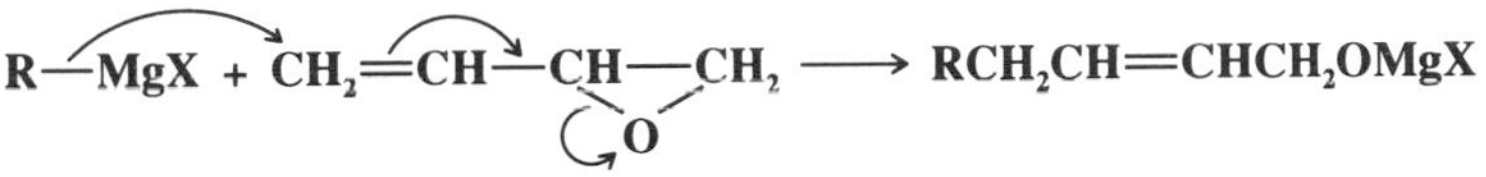

The latter often predominates. In the case of R_2CuLi,[1406] acyclic substrates give mostly allylic rearrangement.[1405] The double bond of the "vinylic" epoxide can be part of an enolate ion if the substrate is cyclic. In this case R_2CuLi give exclusive allylic rearrangement (S_N2'), while Grignard and organolithium reagents give normal substitution, e.g.,[1407]

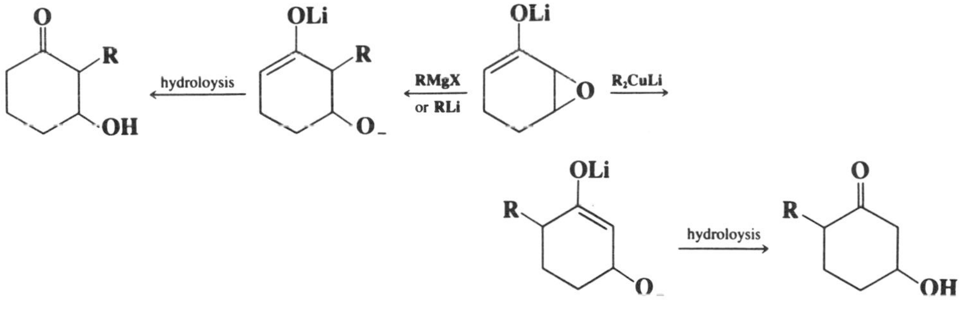

An organometallic equivalent that opens epoxides is a hydrosilane, e.g., Me_3SiH, and carbon monoxide, catalyzed by dicobalt octacarbonyl:[1408]

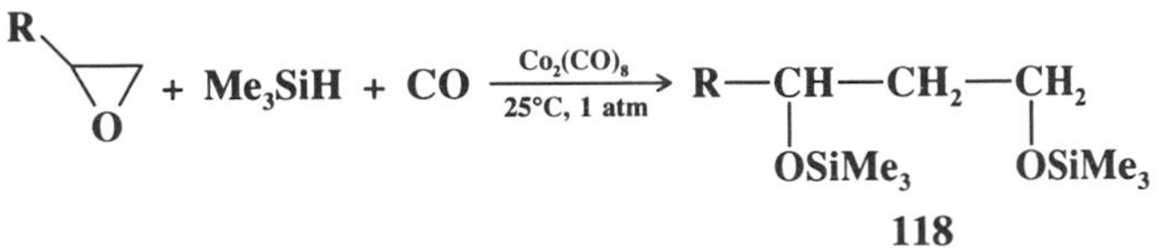

$$\underset{O}{\overset{R}{\triangle}} + Me_3SiH + CO \xrightarrow[25°C, 1\ atm]{Co_2(CO)_8} R-\underset{OSiMe_3}{\overset{|}{CH}}-CH_2-\underset{OSiMe_3}{\overset{|}{CH_2}}$$

118

[1404]For a list of organometallic reagents that react with vinylic epoxides, with references, see Ref. 508, pp. 123-124.

[1405]Anderson *J. Am. Chem. Soc.* **1970**, *92*, 4978; Johnson; Herr; Wieland, Ref. 1401; Marshall; Trometer; Blough; Crute *J. Org. Chem.* **1988**, *53*, 4274; Marshall; Trometer; Cleary *Tetrahedron* **1989**, *45*, 391.

[1406]For a review of the reactions of vinylic epoxides with organocopper reagents, see Marshall *Chem. Rev.* **1989**, *89*, 1503-1511.

[1407]Wender; Erhardt; Letendre *J. Am. Chem. Soc.* **1981**, *103*, 2114.

[1408]Murai; Kato; Murai; Toki; Suzuki; Sonoda *J. Am. Chem. Soc.* **1984**, *106*, 6093.

The 1,3-disilyl ether **118** can be hydrolyzed to a 1,3-diol.[1409]
Aziridines have been similarly opened, to give amines.[1410]
OS **I**, 306; **VII**, 501; **69**, 1, 80.

0-94 Alkylation at a Carbon Bearing an Active Hydrogen
Bis(ethoxycarbonyl)methyl-de-halogenation, etc.

$$RX + Z—\overset{\ominus}{C}H—Z' \longrightarrow Z—\underset{\underset{R}{|}}{C}H—Z'$$

Compounds that contain two (or three, but this is rare) strong electron-withdrawing groups on a carbon atom are more acidic than compounds without such groups (p. 264) and are easily converted to their corresponding enolate ions (p. 72). These enolate ions can attack alkyl halides, resulting in their alkylation.[1411] Z and Z' may be COOR', CHO, COR', CONR$_2'$, COO$^-$, CN,[1412] NO$_2$, SOR', SO$_2$R',[1413] SO$_2$OR', SO$_2$NR$_2'$ or similar groups.[1414] A carbon atom with any two of these (the same or different) will give up a proton (if it has one) to a suitable base. Some commonly used bases are sodium ethoxide and potassium *t*-butoxide, each in its respective alcohol as solvent. With particularly acidic compounds (e.g., β-diketones—Z, Z' = COR'), sodium hydroxide in water or aqueous alcohol or acetone, or even sodium carbonate,[1415] is a strong enough base for the reaction. If at least one Z group is COOR', saponification is a possible side reaction. In addition to the groups listed above, Z may also be phenyl, but if two phenyl groups are on the same carbon, the acidity is less than in the other cases and a stronger base must be used. However, the reaction can be successfully carried out with diphenylmethane with NaNH$_2$ as the base.[1416] The solvent used in the reaction must not be acidic enough to protonate either the enolate ion or the base, which in most cases rules out water. The use of polar aprotic solvents, e.g., DMF or Me$_2$SO, markedly increases the rate of alkylation[1417] but also increases the extent of alkylation at the oxygen rather than the carbon (p. 368). Phase transfer catalysis has also been used.[1418]

Usually the reaction is carried out on a CH$_2$ group connected to two Z groups. In such cases it is possible to alkylate twice, first removing the proton with a base, then alkylating with RX, then removing the proton from ZCHRZ', and finally alkylating the resulting enolate ion with the same or a different RX. The reaction is successful for primary and secondary alkyl, allylic (with allylic rearrangement possible), and benzylic RX, but fails for tertiary halides, since these undergo elimination under the reaction conditions (see, however,

[1409]For another method of converting epoxides to 1,3-diols, see Pelter; Bugden; Rosser *Tetrahedron Lett.* **1985**, 26, 5097.
[1410]See, for example Eis; Ganem *Tetrahedron Lett.* **1985**, 26, 1153; Onistschenko; Buchholz; Stamm *Tetrahedron* **1987**, 43, 565.
[1411]For dicussions of reactions **0-94** and **0-95**, see House *Modern Synthetic Reactions*, 2nd ed.; W. A. Benjamin: New York, 1972, pp. 492-570, 586-595; Carruthers *Some Modern Methods of Organic Synthesis*, 3rd ed.; Cambridge University Press: Cambridge, 1986, pp. 1-26.
[1412]For reviews of the reactions of malononitrile CH$_2$(CN)$_2$, see Fatiadi *Synthesis* **1978**, 165-204, 241-282; Freeman *Chem. Rev.* **1969**, 69, 591-624.
[1413]For a review of compounds with two SO$_2$R groups on the same carbon (*gem*-disulfones), see Neplyuev; Bazarova; Lozinskii *Russ. Chem. Rev.* **1986**, 55, 883-900.
[1414]For lists of examples, with references, see Ref. 508, pp. 764-772*ff*, 894-896.
[1415]See, for example, Fedoryński; Wojciechowski; Matacz; Mąkosza *J. Org. Chem.* **1978**, 43, 4682.
[1416]Murphy; Hamrick; Hauser *Org. Synth.* V. 523.
[1417]Zaugg; Horrom; Borgwardt, Ref. 306; Zaugg; Dunnigan; Michaels; Swett; Wang; Sommers; DeNet *J. Org. Chem.* **1961**, 26, 644; Johnstone; Tuli; Rose *J. Chem. Res. (S)* **1980**, 283.
[1418]See Sukhanov; Trappel'; Chetverikov; Yanovskaya *J. Org. Chem. USSR* **1985**, 21, 2288; Tundo; Venturello; Angeletti *J. Chem. Soc., Perkin Trans. 1* **1987**, 2159.

p. 466). Various functional groups may be present in RX as long as they are not sensitive to base. Side reactions that may cause problems are the above-mentioned competing O-alkylation, elimination (if the enolate ion is a strong enough base), and dialkylation.

An important example of this reaction is the *malonic ester synthesis*, in which both Z groups are COOEt. The product can be hydrolyzed and decarboxylated (**2-40**) to give a carboxylic acid. An illustration is the preparation of 2-ethylpentanoic acid from malonic ester:

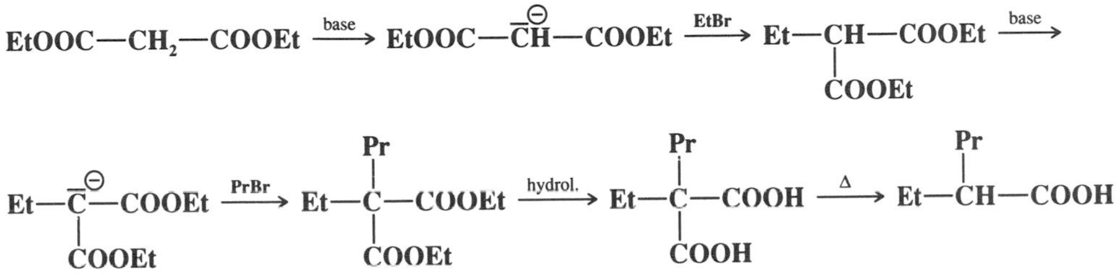

$$EtOOC—CH_2—COOEt \xrightarrow{base} EtOOC—\overset{\ominus}{C}H—COOEt \xrightarrow{EtBr} Et—CH—COOEt \xrightarrow{base}$$
$$\overset{|}{COOEt}$$

$$Et—\overset{\ominus}{\underset{COOEt}{C}}—COOEt \xrightarrow{PrBr} Et—\overset{Pr}{\underset{COOEt}{C}}—COOEt \xrightarrow{hydrol.} Et—\overset{Pr}{\underset{COOH}{C}}—COOH \xrightarrow{\Delta} Et—\overset{Pr}{C}H—COOH$$

It is obvious that many carboxylic acids of the formulas RCH_2COOH and $RR'CHCOOH$ can be synthesized by this method (for some other ways of preparing such acids, see **0-96**, **0-98**, and **0-99**). Another important example is the *acetoacetic ester synthesis*, in which Z is COOEt and Z' is $COCH_3$. In this case the product can be decarboxylated with acid or dilute base (**2-40**) to give a ketone or cleaved with concentrated base (**2-43**) to give a carboxylic ester and a salt of acetic acid:

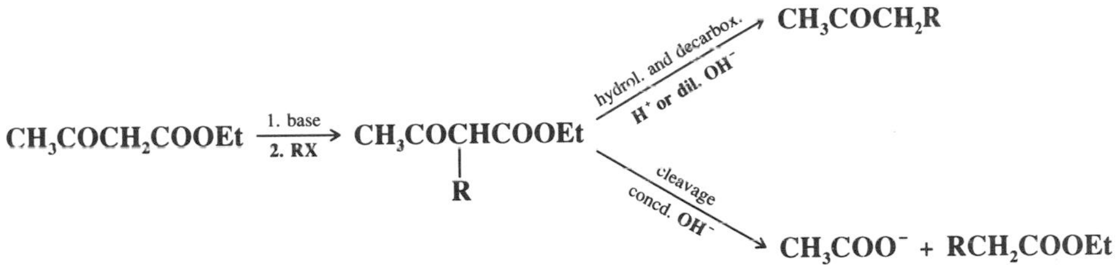

$$CH_3COCH_2COOEt \xrightarrow[2.\ RX]{1.\ base} CH_3COCHCOOEt$$
$$\overset{|}{R}$$

with branches:
$$\xrightarrow[H^+\ or\ dil.\ OH^-]{hydrol.\ and\ decarbox.} CH_3COCH_2R$$
$$\xrightarrow[concd.\ OH^-]{cleavage} CH_3COO^- + RCH_2COOEt$$

Another way of preparing ketones involves alkylation[1419] of β-keto sulfoxides[1420] or sulfones,[1421] e.g.,

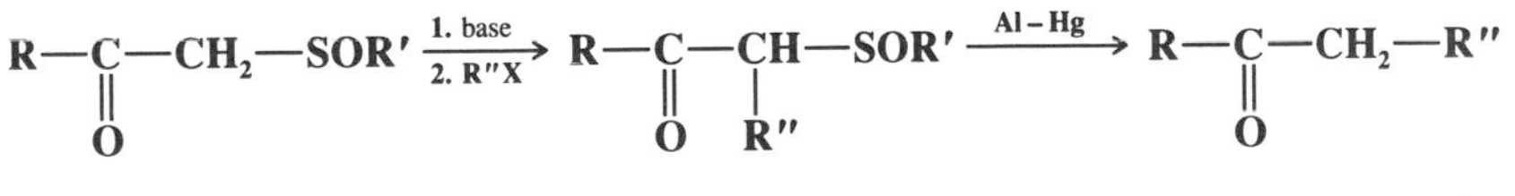

$$R—\overset{\underset{O}{||}}{C}—CH_2—SOR' \xrightarrow[2.\ R''X]{1.\ base} R—\overset{\underset{O}{||}}{C}—\overset{\underset{R''}{|}}{C}H—SOR' \xrightarrow{Al-Hg} R—\overset{\underset{O}{||}}{C}—CH_2—R''$$

since the product in this case is easily reduced to a ketone in high yields with aluminum amalgam or by electrolysis.[1422] The β-keto sulfoxides or sulfones are easily prepared (**0-109**). Other examples of the reaction are the *cyanoacetic ester synthesis*, in which Z is COOEt and Z' is CN (as in the malonic ester synthesis, the product here can be hydrolyzed and decarboxylated), and the *Sorensen* method of amino acid synthesis, in which the reaction is applied to N-acetylaminomalonic ester $(EtOOC)_2CHNHCOCH_3$. Hydrolysis and decarboxylation of the product in this case gives an α-amino acid. The amino group is also frequently protected by conversion to a phthalimido group.

[1419]For a review of the synthetic uses of β-keto sulfoxides, sulfones, and sulfides, see Trost *Chem. Rev.* **1978**, *78*, 363-382. For a review of asymmetric synthesis with chiral sulfoxides, see Solladié *Synthesis* **1981**, 185-196.
[1420]Gassman; Richmond *J. Org. Chem.* **1966**, *31*, 2355. Such sulfoxides can be alkylated on the other side of the C=O group by the use of two moles of base: Kuwajima; Iwasawa *Tetrahedron Lett.* **1974**, 107.
[1421]House; Larson *J. Org. Chem.* **1968**, *33*, 61; Kurth; O'Brien *J. Org. Chem.* **1985**, 3846.
[1422]Lamm; Samuelsson *Acta Chem. Scand.* **1969**, *23*, 691.

The reaction is not limited to Z—CH$_2$—Z' compounds. Other acidic CH hydrogens, which include, for example, the methyl hydrogens of α-aminopyridines, the methyl hydrogens of ynamines of the form CH$_3$C≡CNR$_2$[1423] (the product in this case can be hydrolyzed to an amide RCH$_2$CH$_2$CONR$_2$), the CH$_2$ hydrogens of cyclopentadiene and its derivatives (p. 46), hydrogens connected to a triple-bond carbon (**0-100**), and the hydrogen of HCN (**0-101**) can also be removed with a base and the resulting ion alkylated (see also **0-95** to **0-98**).

Alkylation takes place at the most acidic position of a reagent molecule; for example, acetoacetic ester (CH$_3$COCH$_2$COOEt) is alkylated at the methylene and not at the methyl group, because the former is more acidic than the latter and hence gives up its proton to the base. However, if 2 moles of base are used, then not only is the most acidic proton removed but also the second most acidic. Alkylation of this doubly charged anion then takes place at the less acidic position (see p. 366). This technique has been used to alkylate many compounds in the second most acidic position.[1424]

When ω,ω'-dihalides are used, ring closures can be effected:[1425]

$$
\begin{array}{c}
\text{CH}_2\text{X} \qquad \text{COOEt} \\
\diagup \qquad \diagup \\
(\text{CH}_2)_n \;+\; \text{CH}_2 \quad \xrightarrow{\text{base}} \quad (\text{CH}_2)_n \quad \text{C} \\
\diagdown \qquad \diagdown \\
\text{CH}_2\text{X} \qquad \text{COOEt}
\end{array}
\qquad
\begin{array}{c}
\text{CH}_2 \quad \text{COOEt} \\
\diagup \diagdown \\
\text{C} \\
\diagdown \diagup \\
\text{CH}_2 \quad \text{COOEt}
\end{array}
$$

This method has been used to close rings of from three ($n = 0$) to seven members, although five-membered ring closures proceed in highest yields. Another ring-closing method involves internal alkylation.[1426]

$$
\text{X}(\text{CH}_2)_n\text{CH}(\text{COOEt})_2 \;\xrightarrow{\text{base}}\; (\text{CH}_2)_n \quad \text{C}(\text{COOEt})_2
$$

This method has been shown to be applicable to medium rings (10 to 14 members) without the use of high-dilution techniques.[1427]

The mechanism of these reactions is usually S$_N$2 with inversion taking place at a chiral RX, though there is strong evidence that an SET[1428] mechanism is involved in certain cases,[1429] especially where the nucleophile is an α-nitro carbanion[1430] and/or the substrate contains a nitro or cyano[1431] group. Tertiary alkyl groups can be introduced by an S$_N$1 mechanism if the ZCH$_2$Z' compound (not the enolate ion) is treated with a tertiary carbocation generated in situ from an alcohol or alkyl halide and BF$_3$ or AlCl$_3$,[1432] or with a tertiary alkyl perchlorate.[1433]

[1423]Corey; Cane *J. Org. Chem.* **1970,** *35,* 3405.

[1424]For a list of references, see Ref. 508, pp. 772-773. See also Ref. 426.

[1425]Zefirov; Kuznetsova; Kozhushkov; Surmina; Rashchupkina *J. Org. Chem. USSR* **1983,** *19,* 474.

[1426]For example, see Knipe; Stirling *J. Chem. Soc. B* **1968,** 67; Gosselck; Winkler *Tetrahedron Lett.* **1970,** 2437; Walborsky; Murari *Can. J. Chem.* **1984,** *62,* 2464. For a review of this method as applied to the synthesis of β-lactams, see Bose; Manhas; Chatterjee; Abdulla *Synth. Commun.* **1971,** *1,* 51-73. For a list of examples, see Ref. 508, pp. 81, 83-84.

[1427]Deslongchamps; Lamothe; Lin *Can. J. Chem.* **1984,** *62,* 2395, **1987,** *65,* 1298; Brillon; Deslongchamps *Can. J. Chem.* **1987,** *65,* 43, 56.

[1428]These SET mechanisms are often called S$_{RN}$1 mechanisms. See also Ref. 75.

[1429]Kerber; Urry; Kornblum *J. Am. Chem. Soc.* **1965,** *87,* 4520; Kornblum; Michel; Kerber *J. Am. Chem. Soc.* **1966,** *88,* 5660, 5662; Russell; Danen *J. Am. Chem. Soc.* **1966,** *88,* 5663; Russell; Ros *J. Am. Chem. Soc.* **1985,** *107,* 2506; Ashby; Argyropoulos *J. Org. Chem.* **1985,** *50,* 3274; Bordwell; Wilson *J. Am. Chem. Soc.* **1987,** *109,* 5470; Bordwell; Harrelson *J. Am. Chem. Soc.* **1989,** *111,* 1052.

[1430]For a review of mechanisms with these nucleophiles, see Bowman *Chem. Soc. Rev.* **1988,** *17,* 283-316.

[1431]Kornblum; Fifolt *Tetrahedron* **1989,** *45,* 1311.

[1432]For example, see Boldt; Militzer *Tetrahedron Lett* **1966,** 3599; Crimmins; Hauser *J. Org. Chem.* **1967,** *32,* 2615; Boldt; Militzer; Thielecke; Schulz *Liebigs Ann. Chem.* **1968,** *718,* 101.

[1433]Boldt; Thielecke *Angew. Chem. Int. Ed. Engl.* **1966,** *5,* 1044 [*Angew. Chem. 78,* 1058]; Boldt; Ludwieg; Militzer *Chem. Ber.* **1970,** *103,* 1312.

Other leaving groups are sometimes used. Sulfates, sulfonates, and epoxides give the expected products. Acetals can behave as substrates, one OR group being replaced by ZCHZ' in a reaction similar to **0-92**.[1434] Ortho esters behave similarly, but the product loses R'OH to give an enol ether.[1435]

$$ \text{ZCH}_2\text{Z}' + \text{RC(OR')}_3 \xrightarrow{\text{Ac}_2\text{O}} \text{ZZ'C=CROR'} $$

The SO_2Ph group of allylic sulfones can be a leaving group if a palladium(0) complex is present.[1436] The NR_2 group from Mannich bases such as $RCOCH_2CH_2NR_2$ can also act as a leaving group in this reaction (elimination–addition mechanism, p. 338). A nitro group can be displaced[1437] from α-nitro esters, ketones, nitriles, and α,α-dinitro compounds,[1438] and even from simple tertiary nitro compounds of the form R_3CNO_2[1439] or ArR_2CNO_2[1440] by salts of nitroalkanes, e.g.,

$$ \underset{\overset{|}{\text{NO}_2}}{\text{Me}_2\text{C}} - \text{COOEt} + \text{Me}_2\overset{\ominus}{\text{C}}\text{NO}_2 \longrightarrow \underset{\overset{|}{\text{Me}_2\text{C}-\text{NO}_2}}{\text{Me}_2\text{C}} - \text{COOEt} $$

These reactions take place by SET mechanisms.[1441] However, with α-nitro sulfones it is the sulfone group that is displaced, rather than the nitro group.[1442] The SO_2R group of allylic sulfones can be replaced by CHZZ' ($C=CCH_2-SO_2R \rightarrow C=CCH_2-CHZZ'$) if an $Mo(CO)_6$ catalyst is used.[1443] Alkylation α to a nitro group can be achieved with the Katritzky pyrylium–pyridinium reagents.[1444] This reaction probably has a free-radical mechanism.[1445]

Palladium can be the leaving atom if the substrate is a π-allylpalladium complex (an η^3 complex). Ions of ZCHZ' compounds react with such complexes[1446] in the presence of triphenylphosphine,[1447] e.g.,

$$ \xrightarrow{\text{Ph}_3\text{P}} \quad \text{EtCH}=\text{CPrCH}_2\text{CH(COOEt)}_2 $$

E and *Z* isomers

[1434]Yufit; Krasnaya; Levchenko; Kucherov *Bull. Acad. Sci. USSR, Div. Chem. Sci.* **1967,** 123; Aleskerov; Yufit; Kucherov *Bull. Acad. Sci. USSR, Div. Chem. Sci.* **1972,** *21,* 2279.

[1435]For a review, see DeWolfe, Ref. 457, pp. 231-266.

[1436]Trost; Schmuff; Miller *J. Am. Chem. Soc.* **1980,** *102,* 5979.

[1437]For reviews, see Kornblum, in Patai, Ref. 346, pt. 1, pp. 361-393; Kornblum *Angew. Chem. Int. Ed. Engl.* **1975,** *14,* 734-745 [*Angew. Chem. 87,* 797-808]. For reviews of aliphatic S_N reactions in which NO_2 is a leaving group, see Tamura; Kamimura; Ono *Synthesis* **1991,** 423-434; Kornblum, in Feuer; Nielsen, Ref. 1198, pp. 46-85.

[1438]Kornblum; Kelly; Kestner *J. Org. Chem.* **1985,** *50,* 4720.

[1439]Kornblum; Erickson *J. Org. Chem.* **1981,** *46,* 1037.

[1440]Kornblum; Carlson; Widmer; Fifolt; Newton; Smith *J. Org. Chem.* **1978,** *43,* 1394.

[1441]For a review of the mechanism, see Beletskaya; Drozd *Russ. Chem. Rev.* **1979,** *48,* 431-448. See also Kornblum; Wade *J. Org. Chem.* **1987,** *52,* 5301; Ref. 1430; Ref. 1437.

[1442]Kornblum; Boyd; Ono *J. Am. Chem. Soc.* **1974,** *96,* 2580.

[1443]Trost; Merlic *J. Org. Chem.* **1990,** *55,* 1127.

[1444]Katritzky; de Ville; Patel *Tetrahedron* **1981,** *37,* Suppl. 1, 25; Katritzky; Kashmiri; Wittmann *Tetrahedron* **1984,** *40,* 1501.

[1445]Katritzky; Chen; Marson; Maia; Kashmiri *Tetrahedron* **1986,** *42,* 101.

[1446]For a review of the use of η^3-allylpalladium complexes to form C—C bonds, see Tsuji, in Hartley; Patai, Ref. 1403, vol. 3, 1985, pp. 163-199.

[1447]For reviews, see Trost *Angew. Chem. Int. Ed. Engl.* **1989,** *28,* 1173-1192 [*Angew. Chem. 101,* 1199-1219], *Chemtracts: Org. Chem.* **1988,** *1,* 415-435, *Aldrichimica Acta* **1981,** *14,* 43-50, *Acc. Chem. Res.* **1980,** *13,* 385-393, *Tetrahedron* **1977,** *33,* 2615-2649; Tsuji; Minami *Acc. Chem. Res.* **1987,** *20,* 140-145; Tsuji *Tetrahedron* **1986,** *42,* 4361-4401, *Organic Synthesis with Palladium Compounds*; Springer: Berlin, 1981, pp. 45-51, 125-132; Heck *Palladium Reagents in Organic Synthesis*; Academic Press: New York, 1985, pp. 130-166; Hegedus, in Buncel; Durst *Comprehensive Carbanion Chemistry,* vol. 5, pt. B; Elsevier: New York, 1984, pp. 30-44.

When the Pd bears chiral ligands, these reactions can be enantioselective.[1448] π-Allylmolybdenum compounds behave similarly.[1449] Because palladium compounds are expensive, a catalytic synthesis, which uses much smaller amounts of the complex, was developed. That is, a substrate such as an allylic acetate, alcohol, amine, or nitro compound[1450] is treated with the nucleophile, and a catalytic amount of a palladium salt is added. The π-allylpalladium complex is generated in situ. Alkene–palladium complexes (introducing the nucleophile at a vinylic rather than an allylic carbon) can also be used.[1451]

OS **I**, 248, 250; **II**, 262, 279, 384, 474; **III**, 213, 219, 397, 405, 495, 705; **IV**, 10, 55, 288, 291, 623, 641, 962; **V**, 76, 187, 514, 523, 559, 743, 767, 785, 848, 1013; **VI**, 223, 320, 361, 482, 503, 587, 781, 991; **VII**, 339, 411; **66**, 75; **68**, 56; **69**, 38. See also OS **68**, 210.

0-95 Alkylation of Ketones, Nitriles, and Carboxylic Esters
α-**Acylalkyl-de-halogenation,** etc.

$$\underset{\text{O}}{\text{RCH}_2-\overset{\|}{\text{C}}-\text{R}'} \xrightarrow{\text{base}} \underset{\text{O}}{\text{R}\overset{\ominus}{\text{CH}}-\overset{\|}{\text{C}}-\text{R}'} \xrightarrow{\text{R}''\text{X}} \underset{\text{R}''}{\underset{}{\text{RCH}}}\underset{\text{O}}{-\overset{\|}{\text{C}}-\text{R}'}$$

Ketones,[1452] nitriles,[1453] and carboxylic esters[1454] can be alkylated in the α position in a reaction similar to **0-94,**[1411] but a stronger base must be employed, since only one activating group is present. The most common bases[1455] are Et$_2$NLi (LDA), (iso-Pr)$_2$NLi, t-BuOK, NaNH$_2$, and KH. The base lithium N-isopropyl-N-cyclohexylamide is particularly successful for carboxylic esters[1456] and nitriles.[1457] Solid KOH in Me$_2$SO has been used to methylate ketones, in high yields.[1458] Some of these bases are strong enough to convert the ketone, nitrile, or ester completely to its enolate ion conjugate base; others (especially t-BuOK) convert a significant fraction of the molecules. In the latter case, the aldol reaction (**6-39**) or Claisen condensation (**0-108**) may be side reactions, since both the free molecule and its conjugate base are present at the same time. It is therefore important to use a base strong enough to convert the starting compound completely. Protic solvents are generally not suitable because they protonate the base (though of course this is not a problem with a conjugate pair, such as t-BuOK in t-BuOH). Some common solvents are 1,2-dimethoxyethane, THF, DMF, and liquid NH$_3$. Phase transfer catalysis has been used to alkylate many nitriles, as well as some esters and ketones.[1459]

As in **0-94,** the alkyl halide may be primary or secondary. Tertiary halides give elimination. Even primary and secondary halides give predominant elimination if the enolate ion is a strong enough base (e.g., the enolate ion from Me$_3$CCOMe).[1460] Tertiary alkyl groups, as

[1448]For a review, see Consiglio; Waymouth *Chem. Rev.* **1989,** *89,* 257-276.
[1449]Trost; Lautens *Tetrahedron* **1987,** *43,* 4817, *J. Am. Chem. Soc.* **1987,** *109,* 1469.
[1450]Tamura; Kai; Kakihana; Hayashi; Tsuji; Nakamura; Oda *J. Org. Chem.* **1986,** *51,* 4375.
[1451]Hegedus; Williams; McGuire; Hayashi *J. Am. Chem. Soc.* **1980,** *102,* 4973; Hegedus, Ref. 1447, pp. 9-20.
[1452]For a review of the alkylation and acylation of ketones and aldehydes, see Caine, in Augustine *Carbon–Carbon Bond Formation,* vol. 1; Marcel Dekker: New York, 1979, pp. 85-352.
[1453]For a review, see Arseniyadis; Kyler; Watt *Org. React.* **1984,** *31,* 1-364. For a list of references, see Ref. 508, pp. 910-913.
[1454]For a review, see Petragnani; Yonashiro *Synthesis* **1982,** 521-578. For a list of references, see Ref. 508, pp. 873-890*ff.*
[1455]For a list of some bases, with references, see Ref. 508, pp. 738-740.
[1456]Rathke; Lindert *J. Am. Chem. Soc.* **1971,** *93,* 2319; Bos; Pabon *Recl. Trav. Chim. Pays-Bas* **1980,** *99,* 141. See also Cregge; Herrmann; Lee; Richman; Schlessinger *Tetrahedron Lett.* **1973,** 2425.
[1457]Watt *Tetrahedron Lett.* **1974,** 707.
[1458]Langhals; Langhals *Tetrahedron Lett.* **1990,** *31,* 859.
[1459]For reviews, see Mąkosza *Russ. Chem. Rev.* **1977,** *46,* 1151-1166, *Pure Appl. Chem.* **1975,** *43,* 439-462; Starks; Liotta, Ref. 404, pp. 170-217; Weber; Gokel *Phase Transfer Catalysis in Organic Synthesis,* Ref. 404, pp. 136-204.
[1460]Zook; Kelly; Posey *J. Org. Chem.* **1968,** *33,* 3477.

well as other groups that normally give SN1 reactions, can be introduced if the reaction is performed on a silyl enol ether[1461] of a ketone, aldehyde, or ester with a Lewis acid catalyst.[1462]

$$RCH_2-\overset{O}{\underset{||}{C}}-R' \xrightarrow{\text{2-21}} RCH=\overset{OSiMe_3}{\underset{|}{C}}-R' \xrightarrow[\text{TiCl}_4]{R_2''CX} RCH-\overset{O}{\underset{||}{C}}-R' \quad R' = R, H, \text{ or } OR$$
$$\qquad\qquad\qquad\qquad\qquad\qquad\qquad\qquad\qquad\quad\; CR_3''$$

Vinylic and aryl halides can be used to vinylate or arylate carboxylic esters (but not ketones) by the use of $NiBr_2$ as a catalyst.[1463] However, ketones have been vinylated by treating their enol acetates with vinylic bromides in the presence of a Pd compound catalyst.[1464] Also as in **0-94**, this reaction can be used to close rings.[1465] In one example of this, rings have been closed by treating a diion of a dialkyl succinate with a 1,ω-dihalide or ditosylate,[1466], e.g.:

$$\text{ROOCCH}_2\text{CH}_2\text{COOR} \xrightarrow{\text{piperidide Li}^+} \text{ROOC}-\overset{\ominus}{CH}-\overset{\ominus}{CH}-\text{COOR} \xrightarrow{\text{BrCH}_2\text{Cl}} \text{(cyclopropane)}\overset{H}{\underset{H}{<}}\overset{COOR}{\underset{COOR}{}}$$

This was applied to the synthesis of 3-, 4-, 5-, and 6-membered rings. When the R groups were chiral (e.g., menthyl) the product was formed with greater than 90% enantiomeric excess.[1466]

An efficient enantioselective alkylation has been reported:[1467]

(indanone substrate) $\xrightarrow[\text{2. MeCl}]{\text{1. aqueous OH}^--\text{PhMe}}$ (methylated indanone product)

98% yield; 94% ee

The indanone substrate was methylated in 94% enantiomeric excess, by the use of a chiral catalyst, N-(p-(trifluoromethyl)benzyl)cinchoninium bromide, under phase transfer conditions.[1468] In another method enantioselective alkylation can be achieved by using a chiral base to form the enolate.[1469]

[1461]For a list of alkylations of silyl enol ethers, see Ref. 508, pp. 750-754.
[1462]Chan; Paterson; Pinsonnault *Tetrahedron Lett.* **1977**, 4183; Reetz; Maier *Angew. Chem. Int. Ed. Engl.* **1978**, *17*, 48 [*Angew. Chem. 90*, 50]; Reetz; Schwellnus; Hübner; Massa; Schmidt *Chem. Ber.* **1983**, *116*, 3708. Lion; Dubois *Bull. Soc. Chim. Fr.* **1982**, II-375; Reetz; Sauerwald *J. Organomet. Chem.* **1990**, *382*, 121; Reetz; Chatziiosifidis; Hübner; Heimbach *Org. Synth. VII*, 424. For a review, see Reetz *Angew. Chem. Int. Ed. Engl.* **1982**, *21*, 96-108 [*Angew. Chem. 94*, 97-109].
[1463]Millard; Rathke *J. Am. Chem. Soc.* **1977**, *99*, 4833.
[1464]Kosugi; Hagiwara; Migita *Chem. Lett.* **1983**, 839. For other methods, see Negishi; Akiyoshi *Chem. Lett.* **1987**, 1007; Chang; Rosenblum; Simms *Org. Synth.* 66, 95.
[1465]For example, see Etheredge *J. Org. Chem.* **1966**, *31*, 1990; Wilcox; Whitney *J. Org. Chem.* **1967**, *32*, 2933; Bird; Stirling *J. Chem. Soc. B* **1968**, 111; Stork; Boeckman *J. Am. Chem. Soc.* **1973**, *95*, 2016; Stork; Cohen *J. Am. Chem. Soc.* **1974**, *96*, 5270. In the last case, the substrate moiety is an epoxide function.
[1466]Misumi; Iwanaga; Furuta; Yamamoto *J. Am. Chem. Soc.* **1985**, *107*, 3343; Furuta; Iwanaga; Yamamoto *Org. Synth.* 67, 76.
[1467]For reviews of stereoselective alkylation of enolates, see Nógrádi *Stereoselective Synthesis*; VCH: New York, 1986, pp. 236-245; Evans, in Morrison *Asymmetric Synthesis*, vol. 3; Academic Press: New York, 1984, pp. 1-110.
[1468]Hughes; Dolling; Ryan; Schoenewaldt; Grabowski *J. Org. Chem.* **1987**, *52*, 4745.
[1469]For example, see Murakata; Nakajima; Koga *J. Chem. Soc., Chem. Commun.* **1990**, 1657. For a review, see Cox; Simpkins *Tetrahedron: Asymmetry* **1991**, *2*, 1-26, pp. 6-13.

The reaction can be applied to aldehydes, indirectly, by alkylating an imine derivative of the aldehyde.[1470] The derivative is easily prepared (**6-14**) and the product easily hydrolyzed to the aldehyde (**6-2**). Either or both R groups may be hydrogen, so that mono-, di-, and

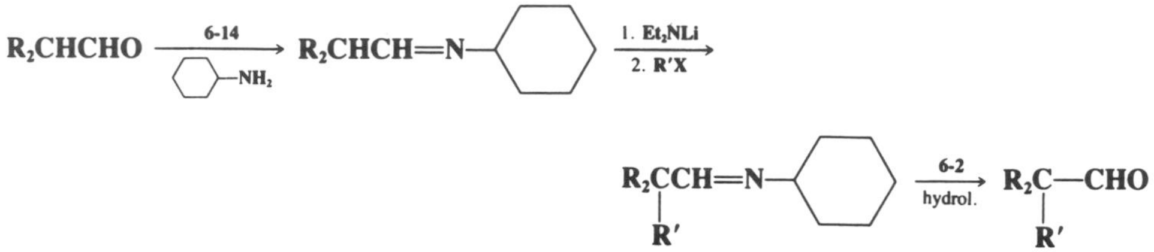

trisubstituted acetaldehydes can be prepared by this method. R' may be primary alkyl, allylic, or benzylic. Direct alkylation of aldehydes is not generally possible because base treatment of aldehydes normally gives rapid aldol reaction (**6-39**), though aldehydes bearing only one α hydrogen have been alkylated with allylic and benzylic halides in good yields by the use of the base KH to prepare the potassium enolate,[1471] or in moderate yields, by the use of a phase transfer catalyst.[1472] Hydrazones and other compounds with C=N bonds can be similarly alkylated.[1470] The use of chiral amines or hydrazines[1473] (followed by hydrolysis **6-2** of the alkylated imine) can lead to chiral alkylated ketones in high optical yields[1474] (for an example, see p. 118).

In α,β-unsaturated ketones, nitriles, and esters (e.g., **119**), the γ hydrogen assumes the acidity normally held by the position α to the carbonyl group, especially when R is not

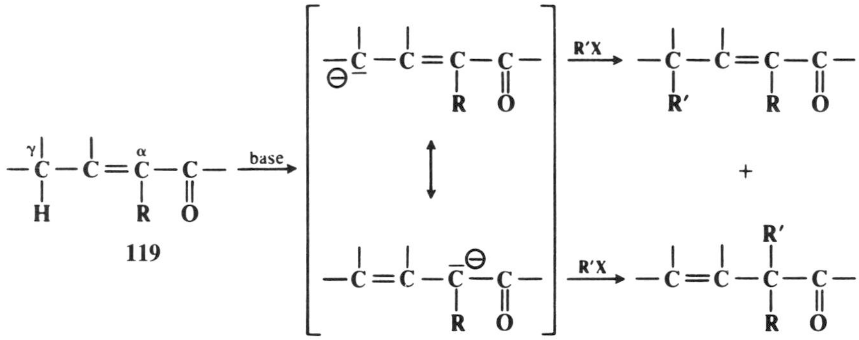

hydrogen and so cannot compete. This principle, called *vinylology*, operates because the resonance effect is transmitted through the double bond. However, because of the resonance, alkylation at the α position (with allylic rearrangement) competes with alkylation at the γ position and usually predominates.

[1470]Cuvigny; Normant *Bull. Soc. Chim. Fr.* **1970**, 3976. For reviews, see Fraser, in Buncel; Durst, Ref. 1447, pp. 65-105; Whitesell; Whitesell *Synthesis* **1983**, 517-536. For a list of references, see Ref. 508, pp. 758-761. For a method in which the metalated imine is prepared from a nitrile, see Goering; Tseng *J. Org. Chem.* **1981**, *46*, 5250.

[1471]Groenewegen; Kallenberg; van der Gen *Tetrahedron Lett.* **1978**, 491; Artaud; Torossian; Viout *Tetrahedron* **1985**, *41*, 5031.

[1472]Dietl; Brannock *Tetrahedron Lett.* **1973**, 1273; Purohit; Subramanian *Chem. Ind.* (*London*) **1978**, 731; Buschmann; Zeeh *Liebigs Ann. Chem.* **1979**, 1585.

[1473]For a review of the alkylation of chiral hydrazones, see Enders, in Morrison, Ref. 1467, pp. 275-339.

[1474]Meyers; Williams; Erickson; White; Druelinger *J. Am. Chem. Soc.* **1981**, *103*, 3081; Meyers; Williams; White; Erickson *J. Am. Chem. Soc.* **1981**, *103*, 3088; Enders; Bockstiegel *Synthesis* **1989**, 493; Enders; Kipphardt; Fey *Org. Synth.* **65**, 183.

α-Hydroxynitriles (cyanohydrins), protected by conversion to acetals with ethyl vinyl ether (**5-4**), can be easily alkylated with primary or secondary alkyl or allylic halides.[1475]

$$
\text{RCHO} \xrightarrow{\text{6-49}} \underset{\underset{\text{CN}}{|}}{\overset{\overset{\text{OH}}{|}}{\text{R--CH}}} \xrightarrow[\text{5-4}]{\text{EtOCH=CH}_2} \underset{\underset{\text{CN}}{|}}{\overset{\overset{\text{O--CHMeOEt}}{|}}{\text{R--CH}}} \xrightarrow{\text{(i-Pr)}_2\text{NLi}} \underset{\underset{\text{CN}}{|}}{\overset{\overset{\text{O--CHMeOEt}}{|}}{\text{R--C}|^{\ominus}}}
$$

120

$$
\xrightarrow{\text{R'X}} \underset{\underset{\text{CN}}{|}}{\overset{\overset{\text{O--CHMeOEt}}{|}}{\text{R--C--R'}}} \xrightarrow[\text{2. OH}^-]{\text{1. H}^+} \underset{\underset{\text{O}}{\|}}{\text{R--C--R'}}
$$

R can be aryl or saturated or unsaturated alkyl. Since the cyanohydrins[1476] are easily formed from aldehydes (**6-49**) and the product is easily hydrolyzed to a ketone, this is a method for converting an aldehyde RCHO to a ketone RCOR'[1477] (for other methods, see **0-97, 0-105**, and **8-9**).[1478] In this procedure the normal mode of reaction of a carbonyl carbon is reversed. The C atom of an aldehyde molecule is normally electrophilic and is attacked by nucleophiles (Chapter 16), but by conversion to the protected cyanohydrin this carbon atom has been induced to perform as a nucleophile.[1479] The German word *umpolung*[1480] is used to describe this kind of reversal (another example is found in **0-97**). Since the ion **120** serves as a substitute for the unavailable $\text{R--}\overline{\text{C}}\overset{\ominus}{=}\text{O}$ anion, it is often called a "masked" $\text{R--}\overline{\text{C}}\overset{\ominus}{=}\text{O}$ ion. This method fails for formaldehyde (R = H), but other masked formaldehydes have proved successful.[1481]

When the compound to be alkylated is a nonsymmetrical ketone, the question arises as to which side will be alkylated. If an α phenyl or α vinylic group is present on one side, alkylation goes predominantly on that side. When only alkyl groups are present, the reaction is generally not regioselective; mixtures are obtained in which sometimes the more alkylated and sometimes the less alkylated side is predominantly alkylated. Which product is found in higher yield depends on the nature of the substrate, the base,[1482] the cation, and the solvent. In any case, di- and trisubstitution are frequent[1483] and it is often difficult to stop with the introduction of just one alkyl group.[1484]

[1475]Stork; Maldonado *J. Am. Chem. Soc.* **1971**, *93*, 5286; Stork; Depezay; D'Angelo *Tetrahedron Lett.* **1975**, 389. See also Rasmussen; Heilmann *Synthesis* **1978**, 219; Ahlbrecht; Raab; Vonderheid *Synthesis* **1979**, 127; Hünig; Marschner; Peters; von Schnering *Chem. Ber.* **1989**, *122*, 2131, and other papers in this series.

[1476]For a review of **120**, see Albright *Tetrahedron* **1983**, *39*, 3207-3233.

[1477]For similar methods, see Stetter; Schmitz; Schreckenberg *Chem. Ber.* **1977**, *110*, 1971; Hünig; *Chimia* **1982**, *36*, 1.

[1478]For a review of methods of synthesis of aldehydes, ketones, and carboxylic acids by coupling reactions, see Martin, *Synthesis* **1979**, 633-665.

[1479]For reviews of such reversals of carbonyl group reactivity, see Block *Reactions of Organosulfur Compounds*; Academic Press: New York, 1978, pp. 56-67; Gröbel; Seebach *Synthesis* **1977**, 357-402; Lever *Tetrahedron* **1976**, *32*, 1943-1971; Seebach; Kolb *Chem. Ind. (London)* **1974**, 687-692; Seebach *Angew. Chem. Int. Ed. Engl.* **1969**, *8*, 639-649 [*Angew. Chem. 81*, 690-700]. For a compilation of references to masked acyl and formyl anions, see Hase; Koskimies *Aldrichimica Acta* **1981**, *14*, 73-77. For tables of masked reagents, see Hase, Ref. 1480, pp. xiii-xiv, 7-18, 219-317. For lists of references, see Ref. 508, pp. 709-711.

[1480]For a monograph, see Hase *Umpoled Synthons*; Wiley: New York, 1987. For a review see Seebach *Angew. Chem. Int. Ed. Engl.* **1979**, *18*, 239-258 [*Angew. Chem. 91*, 259-278].

[1481]Possel; van Leusen *Tetrahedron Lett.* **1977**, 4229; Stork; Ozorio; Leong *Tetrahedron Lett.* **1978**, 5175.

[1482]Sterically hindered bases may greatly favor one enolate over the other. See, for example, Prieto; Suarez; Larson *Synth. Commun.* **1988**, *18*, 253; Gaudemar; Bellassoued *Tetrahedron Lett.* **1989**, *30*, 2779.

[1483]For a procedure for completely methylating the α positions of a ketone, see Lissel; Neumann; Schmidt *Liebigs Ann. Chem.* **1987**, 263.

[1484]For some methods of reducing dialkylation, see Hooz; Oudenes *Synth. Commun.* **1980**, *10*, 139; Morita; Suzuki; Noyori *J. Org. Chem.* **1989**, *54*, 1785.

Several methods have been developed for ensuring that alkylation takes place regiose-lectively on the *desired* side of a ketone.[1485] Among these are:

1. Block one side of the ketone by introducing a removable group. Alkylation takes place on the other side; the blocking group is then removed. A common reaction for this purpose is formylation with ethyl formate (**0-109**); this generally blocks the less hindered side. The formyl group is easily removed by alkaline hydrolysis (**2-43**).

2. Introduce an activating group on one side; alkylation then takes place on that side (**0-94**); the activating group is then removed.

3. Prepare the desired one of the two possible enolate ions.[1486] The two ions, e.g., **121** and **122** for 2-heptanone,

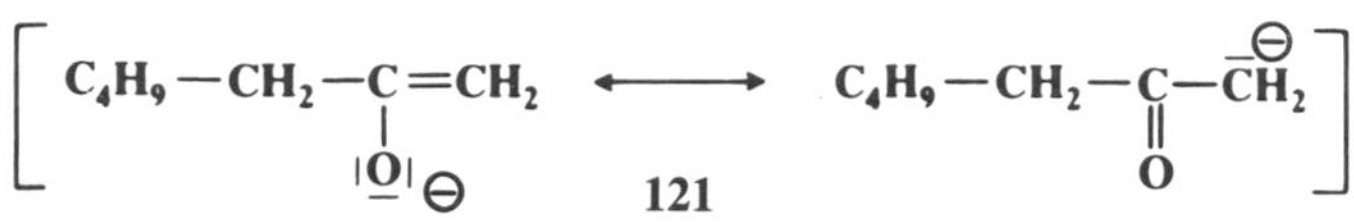

$$\left[\text{C}_4\text{H}_9-\text{CH}_2-\underset{\underset{\ominus}{|\underline{\text{O}}|}}{\text{C}}=\text{CH}_2 \quad \longleftrightarrow \quad \text{C}_4\text{H}_9-\text{CH}_2-\underset{\text{O}}{\overset{\|}{\text{C}}}-\overset{\ominus}{\text{CH}}_2 \right]$$

$$\textbf{121}$$

$$\left[\text{C}_4\text{H}_9-\text{CH}=\underset{\underset{\ominus}{|\underline{\text{O}}|}}{\text{C}}-\text{CH}_3 \quad \longleftrightarrow \quad \text{C}_4\text{H}_9-\overset{\ominus}{\text{CH}}-\underset{\text{O}}{\overset{\|}{\text{C}}}-\text{CH}_3 \right]$$

$$\textbf{122}$$

interconvert rapidly only in the presence of the parent ketone or any stronger acid.[1487] In the absence of such acids, it is possible to prepare either **121** or **122** and thus achieve selective alkylation on either side of the ketone.[1488] The desired enolate ion can be obtained by treatment of the corresponding enol acetate with two equivalents of methyllithium in 1,2-dimethoxyethane. Each enol acetate gives the corresponding enolate, e.g.,

$$\text{C}_4\text{H}_9-\text{CH}_2-\underset{\text{OAc}}{\overset{|}{\text{C}}}=\text{CH}_2 \xrightarrow{\text{MeLi}} \textbf{121} \qquad \text{C}_4\text{H}_9-\text{CH}=\underset{\text{OAc}}{\overset{|}{\text{C}}}=\text{CH}_3 \xrightarrow{\text{MeLi}} \textbf{122}$$

The enol acetates, in turn, can be prepared by treatment of the parent ketone with an appropriate reagent.[1488] Such treatment generally gives a mixture of the two enol acetates in which one or the other predominates, depending on the reagent. The mixtures are easily separable.[1488] An alternate procedure involves conversion of a silyl enol ether[1489] (see **2-23**) or a dialkylboron enol ether[1490] (an enol borinate, see p. 481) to the corresponding enolate ion. If the less hindered enolate ion is desired (e.g., **121**), it can be prepared directly from the ketone by treatment with lithium diisopropylamide in THF or 1,2-dimethoxyethane at −78°C.[1491]

[1485]For a review, see House *Rec. Chem. Prog.* **1968**, *28*, 99-120. For a review with respect to cyclohexenones, see Podraza *Org. Prep. Proced. Int.* **1991**, *23*, 217-235.
[1486]For reviews, see d'Angelo *Tetrahedron* **1976**, *32*, 2979-2990; Stork *Pure Appl. Chem.* **1975**, *43*, 553-562.
[1487]House; Trost *J. Org. Chem.* **1965**, *30*, 1341.
[1488]House; Trost *J. Org. Chem.* **1965**, *30*, 2502; Whitlock; Overman *J. Org. Chem.* **1969**, *34*, 1962; House; Gall; Olmstead *J. Org. Chem.* **1971**, *36*, 2361. For an improved procedure, see Liotta; Caruso *Tetrahedron Lett.* **1985**, *26*, 1599.
[1489]Stork; Hudrlik *J. Am. Chem. Soc.* **1968**, *90*, 4462, 4464. For reviews, see Kuwajima; Nakamura *Acc. Chem. Res.* **1985**, *18*, 181-187; Fleming *Chimia* **1980**, *34*, 265-271; Rasmussen *Synthesis* **1977**, 91-110.
[1490]Pasto; Wojtkowski *J. Org. Chem.* **1971**, *36*, 1790.
[1491]House; Gall; Olmstead, Ref. 1488. See also Corey; Gross *Tetrahedron Lett.* **1984**, *25*, 495.

4. Begin not with the ketone itself, but with an α,β-unsaturated ketone in which the double bond is present on the side where alkylation is desired. Upon treatment with lithium in liquid NH_3, such a ketone is reduced to an enolate ion. When the alkyl halide is added,

$$-\overset{|}{C}=\overset{|}{C}-\overset{|}{\underset{\underset{\displaystyle \underline{O}}{\|}}{C}}- \xrightarrow{\text{Li}} \left[-\overset{|}{\underset{\bullet}{C}}-\overset{|}{C}=\overset{|}{C}- \longleftrightarrow _{\ominus}\overset{|}{\underline{O}}\overset{|}{C}-\overset{|}{C}=\overset{|}{C}- \right] \xrightarrow{\text{NH}_3}$$

$$-\overset{|}{\underset{H}{C}}-\overset{|}{C}=\overset{|}{\underset{\underline{O}\bullet}{C}}- \xrightarrow{\text{Li}} \left[-\overset{|}{\underset{H}{C}}-\overset{|}{C}=\overset{|}{\underset{\underset{\displaystyle \underline{O}}{|}\ominus}{C}}- \longleftrightarrow -\overset{|}{\underset{H}{C}}-\overset{|}{\underset{\ominus}{C}}-\overset{|}{\underset{\underline{O}}{C}}- \right] \xrightarrow{\text{RX}} -\overset{|}{\underset{H}{C}}-\overset{|}{\underset{R}{C}}-\overset{|}{\underset{O}{C}}-$$

it must react with the enolate ion on the side where the double bond was.[1492] Of course, this method is not actually an alkylation of the ketone, but of the α,β-unsaturated ketone, though the product is the same as if the saturated ketone had been alkylated on the desired side.

Both sides of acetone have been alkylated with different alkyl groups, in one operation, by treatment of the N,N-dimethylhydrazone of acetone with *n*-BuLi, followed by a primary alkyl, benzylic, or allylic bromide or iodide; then another mole of *n*-BuLi, a second halide, and finally hydrolysis of the hydrazone.[1493]

Among other methods for the preparation of alkylated ketones are: (1) the Stork enamine reaction (**2-19**), (2) the acetoacetic ester synthesis (**0-94**), (3) alkylation of β-keto sulfones or sulfoxides (**0-94**), (4) acylation of $CH_3SOCH_2^-$ followed by reductive cleavage (**0-109**), (5) treatment of α-halo ketones with lithium dialkylcopper reagents (**0-87**), and (6) treatment of α-halo ketones with trialkylboranes (**0-99**).

Sulfones[1494] and sulfonic esters can also be alkylated in the α position if strong enough bases are used.[1495] Alkylation at the α position of selenoxides allows the formation of alkenes, since selenoxides easily undergo elimination (**7-12**).[1496]

$$\underset{\displaystyle Ph-Se-\overset{\ominus}{\underset{\displaystyle}{C}H}-CHR_2'}{\overset{\displaystyle O}{\underset{\displaystyle \|}{}}} \xrightarrow{\text{RX}} \underset{\displaystyle Ph-Se-\underset{\displaystyle}{C}H-CHR_2'}{\overset{\displaystyle OR}{\underset{\displaystyle \||}{}}} \xrightarrow{\text{7-12}} RCH{=}CR_2'$$

OS **III**, 44, 219, 221, 223, 397; **IV**, 278, 597, 641, 962; **V**, 187, 514, 559, 848; **VI**, 51, 115, 121, 401, 818, 897, 958, 991; **VII**, 153, 208, 241, 424; **65**, 32, 183; **66**, 87, 95; **67**, 76, 141; **69**, 55.

[1492]Stork; Rosen; Goldman; Coombs; Tsuji *J. Am. Chem. Soc.* **1965**, *87*, 275. For a review, see Caine *Org. React.* **1976**, *23*, 1-258. For similar approaches, see Coates; Sowerby *J. Am. Chem. Soc.* **1971**, *93*, 1027; Näf; Decorzant *Helv. Chim. Acta* **1974**, *57*, 1317; Wender; Eissenstat *J. Am. Chem. Soc.* **1978**, *100*, 292.

[1493]Yamashita; Matsuyama; Tanabe; Suemitsu *Bull. Chem. Soc. Jpn.* **1985**, *58*, 407.

[1494]For a review, see Magnus *Tetrahedron* **1977**, *33*, 2019-2045, pp. 2022-2025. For alkylation of sulfones containing the F_3CSO_2 group, see Hendrickson; Sternbach; Bair *Acc. Chem. Res.* **1977**, *10*, 306-312.

[1495]For examples, see Truce; Hollister; Lindy; Parr *J. Org. Chem.* **1968**, *33*, 43; Julia; Arnould *Bull. Soc. Chim. Fr.* **1973**, 743, 746; Bird; Stirling, Ref. 1465.

[1496]Reich; Shah *J. Am. Chem. Soc.* **1975**, *97*, 3250.

0-96 Alkylation of Carboxylic Acid Salts
α-**Carboxyalkyl-de-halogenation**

$$RCH_2COO^- \xrightarrow{\text{(i-Pr)}_2NLi} R\bar{C}HCOO^{\ominus} \xrightarrow{R'X} R-CH-COO^-$$
$$\underset{R'}{|}$$

Carboxylic acids can be alkylated in the α position by conversion of their salts to dianions [which actually have the enolate structures $RCH\!=\!C(O^-)_2$[1497]] by treatment with a strong base such as lithium diisopropylamide.[1498] The use of Li^+ as the counterion is important, because it increases the solubility of the dianionic salt. The reaction has been applied[1499] to primary alkyl, allylic, and benzylic halides, and to carboxylic acids of the form RCH_2COOH and $RR''CHCOOH$.[1454] This method, which is an example of the alkylation of a dianion at its more nucleophilic position (see p. 368), is an alternative to the malonic ester synthesis (**0-94**) as a means of preparing carboxylic acids and has the advantage that acids of the form $RR'R''CCOOH$ can also be prepared. In a related reaction, methylated aromatic acids can be alkylated at the methyl group by a similar procedure.[1500]

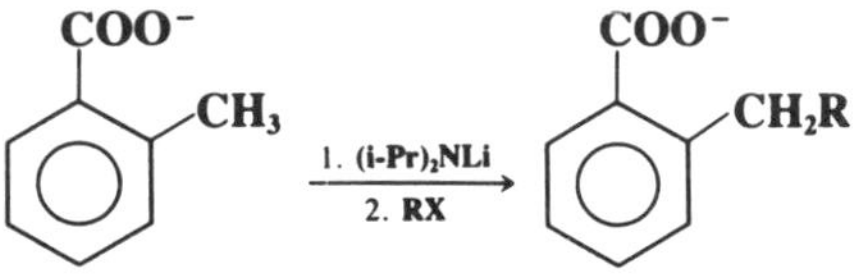

OS **V**, 526; **VI**, 517; **VII**, 249. See also OS **VII**, 164.

0-97 Alkylation at a Position α to a Hetero Atom. Alkylation of 1,3-Dithianes
2-(2-Alkyl-1,3-dithianyl)-de-halogenation

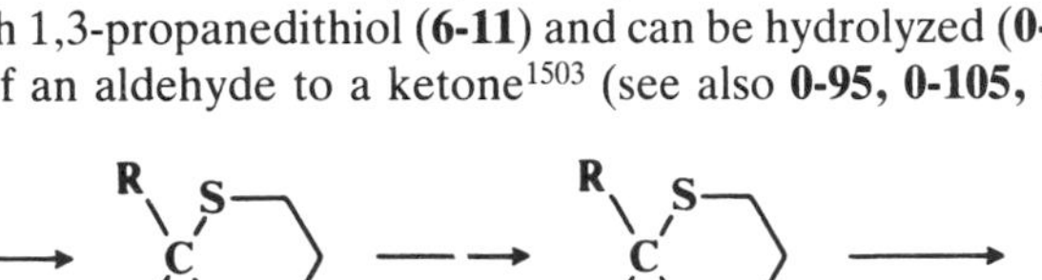

1,3-Dithianes can be alkylated[1501] if a proton is first removed by treatment with butyllithium in THF.[1502] Since 1,3-dithianes can be prepared by treatment of an aldehyde or its acetal (see OS **VI**, 556) with 1,3-propanedithiol (**6-11**) and can be hydrolyzed (**0-6**), this is a method for the conversion of an aldehyde to a ketone[1503] (see also **0-95, 0-105,** and **8-9**):

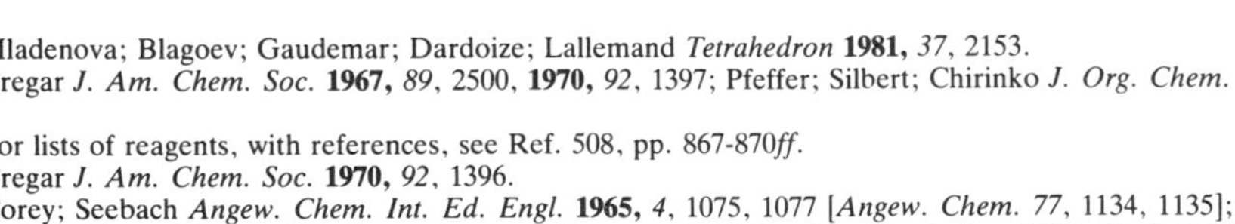

[1497]Mladenova; Blagoev; Gaudemar; Dardoize; Lallemand *Tetrahedron* **1981**, *37*, 2153.
[1498]Cregar *J. Am. Chem. Soc.* **1967**, *89*, 2500, **1970**, *92*, 1397; Pfeffer; Silbert; Chirinko *J. Org. Chem.* **1972**, *37*, 451.
[1499]For lists of reagents, with references, see Ref. 508, pp. 867-870*ff*.
[1500]Cregar *J. Am. Chem. Soc.* **1970**, *92*, 1396.
[1501]Corey; Seebach *Angew. Chem. Int. Ed. Engl.* **1965**, *4*, 1075, 1077 [*Angew. Chem.* 77, 1134, 1135]; Seebach; Corey *J. Org. Chem.* **1975**, *40*, 231. For reviews, see Page; van Niel; Prodger *Tetrahedron* **1989**, *45*, 7643-7677; Ager, in Hase, Ref. 1480, pp. 19-37; Seebach *Synthesis* **1969**, 17-36, especially pp. 24-27; Olsen; Currie, in Patai, Ref. 744, pt. 2, pp. 536-547.
[1502]For an improved method of removing the proton, see Lipshutz; Garcia *Tetrahedron Lett.* **1990**, *31*, 7261.
[1503]For examples of the use of this reaction, with references, see Ref. 508, pp. 721-725.

This is another example of umpolung (see **0-95**);[1478] the normally electrophilic carbon of the aldehyde is made to behave as a nucleophile. The reaction can be applied to the unsubstituted dithiane (R = H) and one or two alkyl groups can be introduced, so a wide variety of aldehydes and ketones can be made starting with formaldehyde.[1504] R' may be primary or secondary alkyl or benzylic. Iodides give the best results. The reaction has been used to close rings.[1505] A similar synthesis of aldehydes can be performed starting with ethyl ethyl-thiomethyl sulfoxide $EtSOCH_2SEt$.[1506]

The group **A** may be regarded as a structural equivalent for the carbonyl group **B**, since introduction of **A** into a molecule is actually an indirect means of introducing **B**. It is

| A | B | C | D |

convenient to have a word for units within molecules; such a word is *synthon*, introduced by Corey,[1507] which is defined as a structural unit within a molecule that can be formed and/or assembled by known or conceivable synthetic operations. There are many other synthons equivalent to **A** and **B**, for example, **C** (by reactions **6-25** and **9-3**) and **D** (by reactions **0-2** and **6-24**).[1508]

Carbanions generated from 1,3-dithianes also react with epoxides[1509] to give the expected products.

Another useful application of this reaction stems from the fact that dithianes can be desulfurated with Raney nickel (**4-36**). Aldehydes can therefore be converted to chain-extended hydrocarbons:[1510]

$$RCHO \longrightarrow \quad \xrightarrow{\text{Raney Ni}} \quad RCH_2R'$$

Similar reactions have been carried out with other thioacetals, as well as with compounds containing three thioether groups on a carbon.[1511]

The carbanion derived from a 1,3-dithiane is stabilized by two thioether groups. If a strong enough base is used, it is possible to alkylate at a position adjacent to only one such group. For example, benzylic and allylic thioethers ($RSCH_2Ar$ and $RSCH_2CH=CH_2$) and thioethers of the form $RSCH_3$ (R = tetrahydrofuranyl or 2-tetrahydropyranyl)[1512] have been successfully alkylated at the carbon adjacent to the sulfur atom.[1513] In the case of the $RSCH_3$

[1504]For a direct conversion of RX to RCHO, see **0-102**.
[1505]For example, see Seebach; Jones; Corey *J. Org. Chem.* **1968**, *33*, 300; Hylton; Boekelheide *J. Am. Chem. Soc.* **1968**, *90*, 6887; Ogura; Yamashita; Suzuki; Tsuchihashi *Tetrahedron Lett.* **1974**, 3653.
[1506]Richman; Herrmann; Schlessinger *Tetrahedron Lett.* **1973**, 3267. See also Ogura; Tsuchihashi *Tetrahedron Lett.* **1971**, 3151; Schill; Jones *Synthesis* **1974**, 117; Hori; Hayashi; Midorikawa *Synthesis* **1974**, 705.
[1507]Corey *Pure Appl. Chem.* **1967**, *14*, 19-37, pp. 20-23.
[1508]For a long list of synthons for RCO, with references, see Hase; Koskimies *Aldrichimica Acta* **1982**, *15*, 35-41.
[1509]For example, see Corey; Seebach, Ref. 1501; Jones; Grayshan *Chem. Commun.* **1970**, 141, 741.
[1510]For examples, see Hylton; Boekelheide, Ref. 1505; Jones; Grayshan, Ref. 1509.
[1511]For example, see Seebach *Angew. Chem. Int. Ed. Engl.* **1967**, *6*, 442 [*Angew. Chem. 79*, 468]; Olsson *Acta Chem. Scand.* **1968**, *22*, 2390; Mori; Hashimoto; Takenaka; Takigawa *Synthesis* **1975**, 720; Lissel *Liebigs Ann. Chem.* **1982**, 1589.
[1512]Block; Aslam *J. Am. Chem. Soc.* **1985**, *107*, 6729.
[1513]Biellmann; Ducep *Tetrahedron Lett.* **1968**, 5629, **1969**, 3707, *Tetrahedron* **1971**, *27*, 5861. See also Narasaka; Hayashi; Mukaiyama *Chem. Lett.* **1972**, 259.

compounds, alkylation took place at the methyl group. Stabilization by one thioether group has also been used in a method for the homologization of primary halides.[1514] Thioanisole is treated with BuLi to give the corresponding anion[1515] which reacts with the halide to give

$$\text{PhSCH}_3 \xrightarrow{\text{BuLi}} \text{PhS}\overset{\ominus}{\text{CH}}_2 \text{ Li}^+ \xrightarrow{\text{RX}} \text{RCH}_2\text{SPh} \xrightarrow[\text{DMF}]{\text{CH}_3\text{I}} \left[\begin{array}{c} \text{RCH}_2\overset{\oplus}{\text{SPh}} \\ | \\ \text{CH}_3 \end{array} \right] \xrightarrow[\text{DMF}]{\text{NaI}} \text{RCH}_2\text{I}$$

Thioanisole   **123**

the thioether **123**. **123** is then refluxed with a mixture of methyl iodide and sodium iodide in dimethylformamide. By this sequence an alkyl halide RX is converted to its homolog RCH$_2$X by a pathway involving two laboratory steps (see also **0-92**).

Vinylic sulfides containing an α hydrogen can also be alkylated[1516] by alkyl halides or epoxides. In one application, the ion **124,** which can be prepared in three steps from epichlorohydrin, reacts with alkyl halides to give the bis(methylthio) compound **125,**[1517] which

$$\text{MeS}\overset{\ominus}{-\text{CH}}-\text{CH}=\text{CHSMe Li}^+ \xrightarrow{\text{RX}} \underset{\underset{\text{SMe}}{|}}{\text{RCH}}-\text{CH}=\text{CHSMe} \xrightarrow[\text{H}_2\text{O} - \text{MeCN}]{\text{HgCl}_2} \text{RCH}=\text{CHCHO}$$

124   **125**   **126**

is easily hydrolyzed[1518] with HgCl$_2$ in aqueous MeCN. This is a method for converting an alkyl halide RX to an α,β-unsaturated aldehyde (**126**) using **124,** which is the synthetic equivalent of the unknown $\text{H}\overset{\ominus}{\text{C}}=\text{CH}-\text{CHO}$ ion.[1519] Even simple alkyl aryl sulfides RCH$_2$SAr and RR′CHSAr have been alkylated α to the sulfur.[1520]

Alkylation can also be carried out, in certain compounds, at positions α to other hetero atoms,[1521] for example, at a position α to the nitrogen of tertiary amines.[1522] Alkylation α to the nitrogen of primary or secondary amines is not generally feasible because an NH hydrogen is usually more acidic than a CH hydrogen. It has been accomplished, however, by replacing the NH hydrogens with other (removable) groups.[1523] In one example, a secondary amine is converted to its N-nitroso derivative (**2-51**).[1524] The N-nitroso product is

$$\underset{}{\text{R}_2\text{CH}}\overset{\overset{\text{R}'}{|}}{-\text{NH}} \xrightarrow{\text{2-51}} \text{R}_2\text{CH}\overset{\overset{\text{R}'}{|}}{-\text{N}}-\text{NO} \xrightarrow{(\text{i-Pr})_2\text{NLi}} \text{R}_2\overset{\ominus}{\text{C}}\overset{\overset{\text{R}'}{|}}{-\text{N}}-\text{NO} \xrightarrow{\text{R}''\text{X}}$$

$$\text{R}_2\text{C}\overset{\overset{\text{R}'' \text{R}'}{| \ |}}{-\text{N}}-\text{NO} \xrightarrow[\text{2. OH}^-]{\text{1. H}^+} \text{R}_2\text{C}\overset{\overset{\text{R}'' \text{R}'}{| \ |}}{-\text{NH}}$$

[1514]Corey; Jautelat *Tetrahedron Lett.* **1968**, 5787.

[1515]Corey; Seebach *J. Org. Chem.* **1966**, *31*, 4097.

[1516]Oshima; Shimoji; Takahashi; Yamamoto; Nozaki *J. Am. Chem. Soc.* **1973**, *95*, 2694.

[1517]Corey; Erickson; Noyori *J. Am. Chem. Soc.* **1971**, *93*, 1724.

[1518]Corey; Shulman *J. Org. Chem.* **1970**, *35*, 777. See, however, Mura; Majetich; Grieco; Cohen *Tetrahedron Lett.* **1975**, 4437.

[1519]For references to other synthetic equivalents of this ion, see Funk; Bolton *J. Am. Chem. Soc.* **1988**, *110*, 1290.

[1520]Dolak; Bryson *Tetrahedron Lett.* **1977**, 1961.

[1521]For a review of anions α to a selenium atom on small rings, see Krief *Top. Curr. Chem.* **1987**, *135*, 1-75. For alkylation α to boron, see Pelter; Smith; Brown *Borane Reagents*; Academic Press: New York, 1988, pp. 336-341.

[1522]Lepley; Khan *J. Org. Chem.* **1966**, *31*, 2061, 2064, *Chem. Commun.* **1967**, 1198; Lepley; Giumanini *J. Org. Chem.* **1966**, *31*, 2055; Ahlbrecht; Dollinger *Tetrahedron Lett.* **1984**, *25*, 1353.

[1523]For a review, see Beak; Zajdel; Reitz *Chem. Rev.* **1984**, *84*, 471-523.

[1524]Seebach; Enders; Renger *Chem. Ber.* **1977**, *110*, 1852; Renger; Kalinowski; Seebach *Chem. Ber.* **1977**, *110*, 1866. For a review, see Seebach; Enders *Angew. Chem. Int. Ed. Engl.* **1975**, *14*, 15-32 [*Angew. Chem. 87*, 1-17].

easily hydrolyzed to the product amine **(9-53)**.[1525] Alkylation of secondary and primary amines has also been accomplished with more than ten other protecting groups, involving conversion of amines to amides, carbamates,[1526] formamidines,[1527] and phosphoramides.[1523] In the case of formamidines **(127)** use of a chiral R′ leads to a chiral amine, in high enantiomeric excess, even when R is not chiral.[1528]

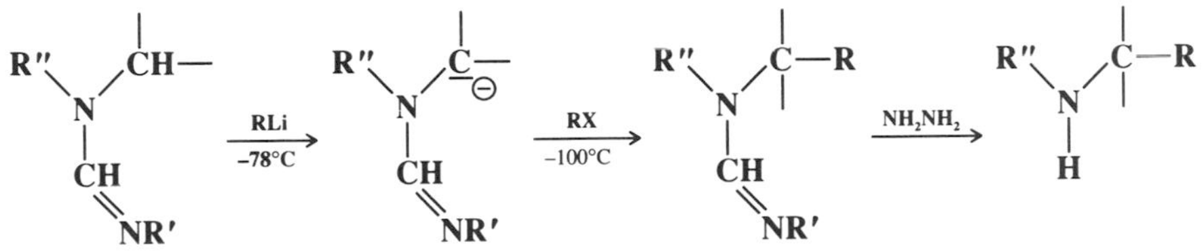

127

A proton can be removed from an allylic ether by treatment with an alkyllithium at about −70°C (at higher temperatures the Wittig rearrangement—**8-23**—takes place) to give the ion **128**, which reacts with alkyl halides to give the two products shown.[1529] Similar

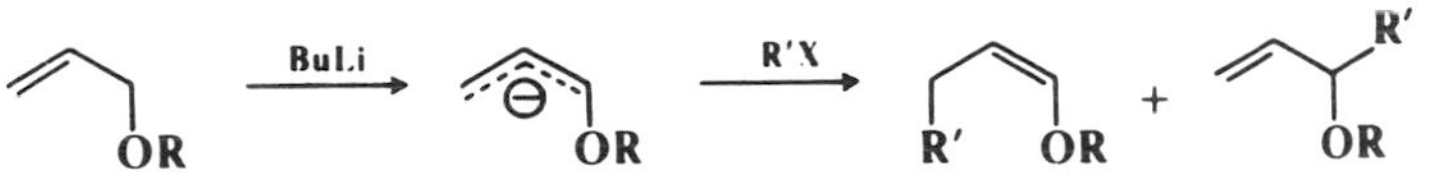

128

reactions[1530] have been reported for allylic[1531] and vinylic tertiary amines. In the latter case, enamines **129**, treated with a strong base, are converted to anions that are then alkylated, generally at C-3.[1532] (For direct alkylation of enamines at C-2, see **2-19**.)

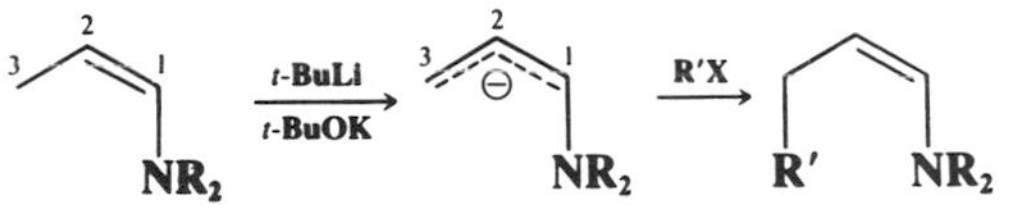

129

It is also possible to alkylate a methyl, ethyl, or other primary group of an aryl ester ArCOOR, where Ar is a 2,4,6-trialkylphenyl group.[1533] Since esters can be hydrolyzed to alcohols, this constitutes an indirect alkylation of primary alcohols. Methanol has also been alkylated by converting it to $^{\ominus}CH_2O^{\ominus}$.[1534]

OS **VI**, 316, 364, 542, 704, 869; **67**, 60.

[1525]Fridman; Mukhametshin; Novikov *Russ. Chem. Rev.* **1971**, *40*, 34-50, pp. 41-42.

[1526]For the use of *t*-butyl carbamates, see Beak; Lee *Tetrahedron Lett.* **1989**, *30*, 1197.

[1527]For a review, see Meyers *Aldrichimica Acta* **1985**, *18*, 59-68.

[1528]Meyers; Fuentes; Kubota *Tetrahedron* **1984**, *40*, 1361; Gawley; Hart; Goicoechea-Pappas; Smith *J. Org. Chem.* **1986**, *51*, 3076; Meyers; Dickman *J. Am. Chem. Soc.* **1987**, *109*, 1263; Gawley *J. Am. Chem. Soc.* **1987**, *109*, 1265; Meyers; Miller; White *J. Am. Chem. Soc.* **1988**, *110*, 4778; Gonzalez; Meyers *Tetrahedron Lett.* **1989**, *30*, 43, 47.

[1529]Evans; Andrews; Buckwalter *J. Am. Chem. Soc.* **1974**, *96*, 5560; Still; Macdonald *J. Am. Chem. Soc.* **1974**, *96*, 5561; Ref. 1519. For a similar reaction with triple-bond compounds, see Hommes; Verkruijsse; Brandsma *Recl. Trav. Chim. Pays-Bas* **1980**, *99*, 113, and references cited therein.

[1530]For a review of allylic and benzylic carbanions substituted by hetero atoms, see Biellmann; Ducep *Org. React.* **1982**, *27*, 1-344.

[1531]Martin; DuPriest *Tetrahedron Lett.* **1977**, 3925 and references cited therein.

[1532]For a review, see Ahlbrecht *Chimia* **1977**, *31*, 391-403.

[1533]Beak; McKinnie *J. Am. Chem. Soc.* **1977**, *99*, 5213; Beak; Carter *J. Org. Chem.* **1981**, *46*, 2363.

[1534]Seebach; Meyer *Angew. Chem. Int. Ed. Engl.* **1976**, *15*, 438 [*Angew. Chem.* **88**, 484].

0-98   Alkylation of Dihydro-1,3-Oxazine. The Meyers Synthesis of Aldehydes, Ketones, and Carboxylic Acids

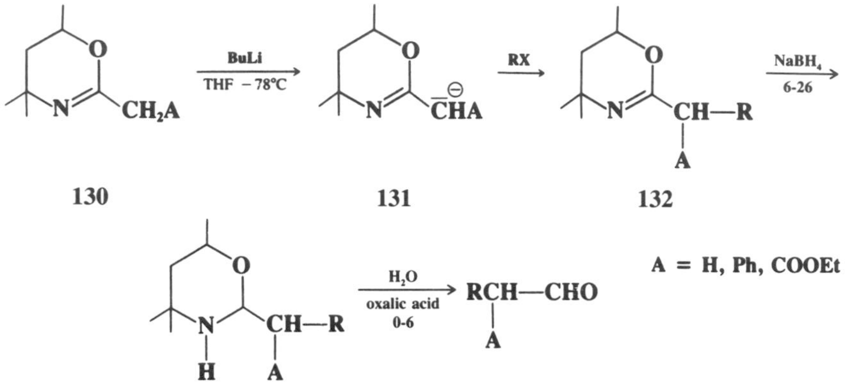

130   131   132

A = H, Ph, COOEt

$$\xrightarrow[\text{oxalic acid}]{\text{H}_2\text{O}} \quad \underset{\text{A}}{\text{RCH}}\text{—CHO}$$

0-6

A synthesis of aldehydes[1535] developed by Meyers[1536] begins with the commercially available dihydro-1,3-oxazine derivatives **130** (A = H, Ph, or COOEt).[1537] Though the ions (**131**) prepared from **130** are ambident, they are regioselectively alkylated at carbon by a wide variety of alkyl bromides and iodides. R can be primary or secondary alkyl, allylic, or benzylic and can carry another halogen or a CN group.[1538] The alkylated oxazine **132** is then reduced and hydrolyzed to give an aldehyde containing two more carbons than the starting RX. This method thus complements **0-97** which converts RX to an aldehyde containing one more carbon. Since A can be H, mono- or disubstituted acetaldehydes can be produced by this method.

The ion **131** also reacts with epoxides, to form γ-hydroxy aldehydes after reduction and hydrolysis,[1539] and with aldehydes and ketones (**6-41**). Similar aldehyde synthesis has also been carried out with thiazoles[1540] and thiazolines[1541] (five-membered rings containing N and S in the 1 and 3 positions).

The reaction has been extended to the preparation of ketones:[1542] treatment of a dihydro-1,3-oxazine (**133**) with methyl iodide forms the iminium salt **134** (**0-43**) which, when treated with a Grignard reagent or organolithium compound (**6-35**), produces **135** which can be

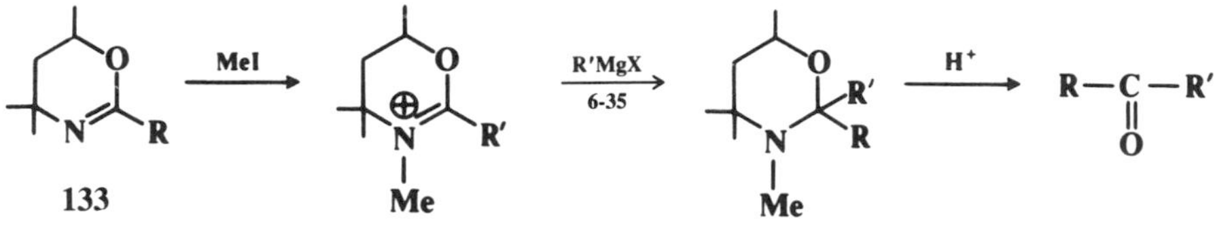

133   134   135

[1535]For examples of the preparation of aldehydes and ketones by the reactions in this section, see Ref. 508, pp. 729-732.

[1536]Meyers; Nabeya; Adickes; Politzer; Malone; Kovelesky; Nolen; Portnoy *J. Org. Chem.* **1973**, *38*, 36.

[1537]For reviews of the preparation and reactions of **130** see Schmidt *Synthesis* **1972**, 333-350; Collington *Chem. Ind. (London)* **1973**, 987-991.

[1538]Meyers; Malone; Adickes *Tetrahedron Lett.* **1970**, 3715.

[1539]Adickes; Politzer; Meyers *J. Am. Chem. Soc.* **1969**, *91*, 2155.

[1540]Altman; Richheimer *Tetrahedron Lett.* **1971**, 4709.

[1541]Meyers; Durandetta *J. Org. Chem.* **1975**, *40*, 2021.

[1542]Meyers; Smith *J. Am. Chem. Soc.* **1970**, *92*, 1084, *J. Org. Chem.* **1972**, *37*, 4289.

hydrolyzed to a ketone. R can be alkyl, cycloalkyl, aryl, benzylic, etc., and R' can be alkyl, aryl, benzylic, or allylic. **130, 132,** and **133** themselves do not react with Grignard reagents. In another procedure, 2-oxazolines[1543] (**136**) can be alkylated to give **137,**[1544] which are easily

$$
RCH_2 \underset{N}{\overset{O}{\diagdown}}\!\!\!\diagup \quad \xrightarrow[2.\ R'X]{1.\ BuLi} \quad RCH \underset{\underset{R'}{|}}{\overset{O}{\diagdown}}\!\!\!\diagup \quad \xrightarrow[EtOH]{H^+} \quad RCH-COOEt
$$

136	**137**	**138**

converted directly to the esters **138** by heating in 5 to 7% ethanolic sulfuric acid. **136** and **137** are thus synthons for carboxylic acids; this is another indirect method for the α alkylation of a carboxylic acid,[1545] representing an alternative to the malonic ester synthesis (**0-94**) and to **0-96** and **0-99**. The method can be adapted to the preparation of optically active carboxylic acids by the use of a chiral reagent.[1546] Note that, unlike **130, 136** can be alkylated even if R is alkyl. However, the C=N bond of **136** and **137** cannot be effectively reduced, so that aldehyde synthesis is not feasible here.[1547]

OS **VI,** 905.

0-99   Alkylation with Trialkylboranes
Alkyl-de-halogenation

$$
BrCH_2-\underset{O}{\overset{\displaystyle |}{\underset{\|}{C}}}-R' \; + \; R_3B \quad \xrightarrow[\text{THF, 0°C}]{} \quad RCH_2-\underset{O}{\overset{\displaystyle |}{\underset{\|}{C}}}-R'
$$

Trialkylboranes react rapidly and in high yields with α-halo ketones,[1548] α-halo esters,[1549] α-halo nitriles,[1550] and α-halo sulfonyl derivatives (sulfones, sulfonic esters, sulfonamides)[1551] in the presence of a base to give, respectively, alkylated ketones, esters, nitriles, and sulfonyl derivatives.[1552] Potassium *t*-butoxide is often a suitable base, but potassium 2,6-di-*t*-butyl-phenoxide at 0°C in THF gives better results in most cases, possibly because the large bulk of the two *t*-butyl groups prevents the base from coordinating with the R_3B.[1553] The tri-alkylboranes are prepared by treatment of 3 moles of an alkene with 1 mole of BH_3

[1543]For a review, see Meyers; Mihelich *Angew. Chem. Int. Ed. Engl.* **1976,** *15,* 270-281 [*Angew. Chem. 88,* 321-332].

[1544]Meyers; Temple; Nolen; Mihelich *J. Org. Chem.* **1974,** *39,* 2778; Meyers; Mihelich; Nolen *J. Org. Chem.* **1974,** *39,* 2783; Meyers; Mihelich; Kamata *J. Chem. Soc., Chem. Commun.* **1974,** 768.

[1545]For reviews, see Meyers, *Pure Appl. Chem.* **1979,** *51,* 1255-1268, *Acc. Chem. Res.* **1978,** *11,* 375-381. See also Hoobler; Bergbreiter; Newcomb *J. Am. Chem. Soc.* **1978,** *100,* 8182; Meyers; Snyder; Ackerman *J. Am. Chem. Soc.* **1978,** *100,* 8186.

[1546]For a review of asymmetric synthesis via chiral oxazolines, see Lutomski; Meyers, in Morrison, Ref. 1467, pp. 213-274.

[1547]Meyers; Temple *J. Am. Chem. Soc.* **1970,** *92,* 6644, 6646.

[1548]Brown; Rogić; Rathke *J. Am. Chem. Soc.* **1968,** *90,* 6218.

[1549]Brown; Rogić; Rathke; Kabalka *J. Am. Chem. Soc.* **1968,** *90,* 818.

[1550]Brown; Nambu; Rogić *J. Am. Chem. Soc.* **1969,** *91,* 6854.

[1551]Truce; Mura; Smith; Young *J. Org. Chem.* **1974,** *39,* 1449.

[1552]For reviews, see Negishi; Idacavage *Org. React.* **1985,** *33,* 1-246, pp. 42-43, 143-150; Weill-Raynal *Synthesis* **1976,** 633-651; Brown; Rogić *Organomet. Chem. Synth.* **1972,** *1,* 305-327; Rogić *Intra-Sci. Chem. Rep.* **1973,** *7*(2), 155-167; Brown *Boranes in Organic Chemistry*; Cornell University Press: Ithaca, NY, 1972, pp. 372-391, 404-409; Cragg, Ref. 1167, pp. 275-278, 283-287.

[1553]Brown; Nambu; Rogić *J. Am. Chem. Soc.* **1969,** *91,* 6852, 6854, 6855.

(5-12).[1554] With appropriate boranes, the R group transferred to α-halo ketones, nitriles, and esters can be vinylic,[1555] or (for α-halo ketones and esters) aryl.[1556]

The reaction can be extended to α,α-dihalo esters[1557] and α,α-dihalo nitriles.[1558] It is possible to replace just one halogen or both. In the latter case the two alkyl groups can be the same or different. When dialkylation is applied to dihalo nitriles, the two alkyl groups can be primary or secondary, but with dihalo esters, dialkylation is limited to primary R. Another extension is the reaction of boranes with γ-halo-α,β-unsaturated esters.[1559] Alkylation takes place in the γ position, but the double bond migrates, e.g.,

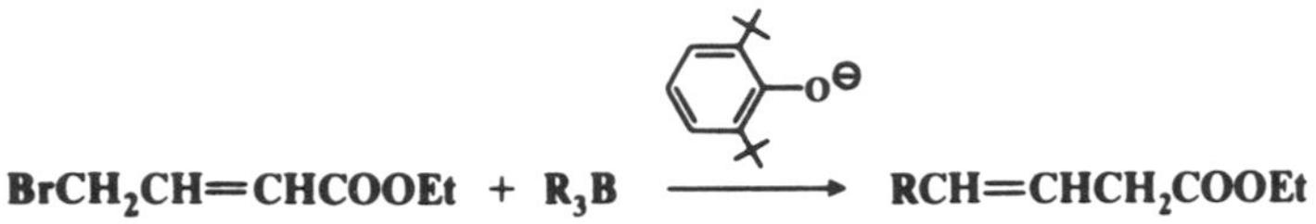

$$\text{BrCH}_2\text{CH}=\text{CHCOOEt} + \text{R}_3\text{B} \longrightarrow \text{RCH}=\text{CHCH}_2\text{COOEt}$$

In this case, however, double-bond migration is an advantage, because nonconjugated β,γ-unsaturated esters are usually much more difficult to prepare than their α,β-unsaturated isomers.

The alkylation of activated halogen compounds is one of several reactions of trialkyl-boranes developed by H. C. Brown[1560] (see also **5-12, 5-19, 8-24** to **8-28**, etc.). These compounds are extremely versatile and can be used for the preparation of many types of compounds. In this reaction, for example, an alkene (through the BR$_3$ prepared from it) can be coupled to a ketone, a nitrile, a carboxylic ester, or a sulfonyl derivative. Note that this is still another indirect way to alkylate a ketone (see **0-95**) or a carboxylic acid (see **0-96**), and provides an additional alternative to the malonic ester and acetoacetic ester syntheses (**0-94**).

Although superficially this reaction resembles **0-87** it is likely that the mechanism is quite different, involving migration of an R group from boron to carbon (see also **8-24** to **8-28**). The mechanism is not known with certainty,[1561] but it may be tentatively shown as (illustrated for an α-halo ketone):

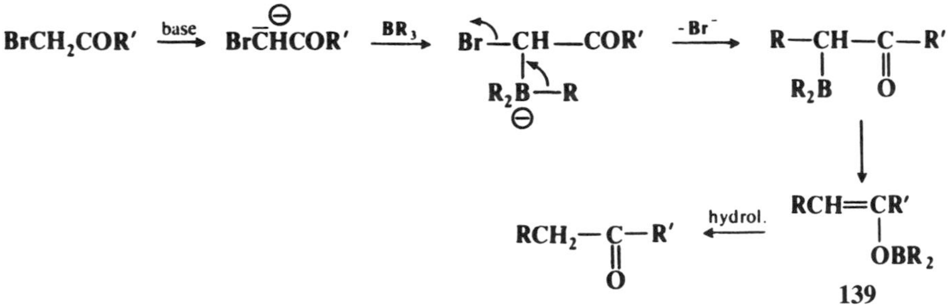

$$\text{BrCH}_2\text{COR}' \xrightarrow{\text{base}} \text{Br}\overset{\ominus}{\text{CH}}\text{COR}' \xrightarrow{\text{BR}_3} \cdots \xrightarrow{-\text{Br}^-} \cdots$$

139

$$\text{RCH}_2-\overset{\text{O}}{\underset{\|}{\text{C}}}-\text{R}'$$

[1554]For an improved procedure, with B-R-9-BBN (see p. 785), see Brown; Rogić *J. Am. Chem. Soc.* **1969**, *91*, 2146; Brown; Rogić; Nambu; Rathke *J. Am. Chem. Soc.* **1969**, *91*, 2147; Katz; Dubois; Lion *Bull. Soc. Chim. Fr.* **1977**, 683.

[1555]Brown; Bhat; Campbell *J. Org. Chem.* **1986**, *51*, 3398.

[1556]Brown; Rogić *J. Am. Chem. Soc.* **1969**, *91*, 4304.

[1557]Brown; Rogić; Rathke; Kabalka *J. Am. Chem. Soc.* **1968**, *90*, 1911.

[1558]Nambu; Brown *J. Am. Chem. Soc.* **1970**, *92*, 5790.

[1559]Brown; Nambu *J. Am. Chem. Soc.* **1970**, *92*, 1761.

[1560]Brown *Organic Syntheses via Boranes*; Wiley: New York, 1975, *Hydroboration*; W.A. Benjamin: New York, 1962, *Boranes in Organic Chemistry*, Ref. 1552; Pelter; Smith; Brown, Ref. 1521.

[1561]See Prager; Reece *Aust. J. Chem.* **1975**, *28*, 1775.

The first step is removal of the acidic proton by the base to give an enolate ion which combines with the borane (Lewis acid-base reaction). An R group then migrates, displacing the halogen leaving group.[1562] Another migration follows, this time of BR_2 from carbon to oxygen to give the enol borinate **139**[1563] which is hydrolyzed. Configuration at R is retained.[1564]

The reaction has also been applied to compounds with other leaving groups. Diazo ketones, diazo esters, diazo nitriles, and diazo aldehydes[1565] react with trialkylboranes in a similar manner, e.g.,

$$\text{H}-\overset{\overset{\text{O}}{\|}}{\text{C}}-\text{CHN}_2 \xrightarrow[\text{THF-H}_2\text{O}]{\text{R}_3\text{B}} \text{H}-\overset{\overset{\text{O}}{\|}}{\text{C}}-\text{CH}_2\text{R}$$

The mechanism is probably also similar. In this case a base is not needed, since the carbon already has an available pair of electrons. The reaction with diazo aldehydes[1566] is especially notable, since successful reactions cannot be obtained with α-halo aldehydes.[1567]

OS **VI**, 919.

0-100   Alkylation at an Alkynyl Carbon
Alkynyl-de-halogenation

$$\text{RX} + \text{R}'\text{C}\equiv\text{C}^- \longrightarrow \text{RC}\equiv\text{CR}'$$

The reaction between alkyl halides and acetylide ions is useful but of limited scope.[1568] Only primary halides unbranched in the β position give good yields, though allylic halides can be used if CuI is present.[1569] If acetylene is the reagent, two different groups can be successively attached. Sulfates, sulfonates, and epoxides[1570] are sometimes used as substrates. The acetylide ion is often prepared by treatment of an alkyne with a strong base such as $NaNH_2$. Magnesium acetylides (ethynyl Grignard reagents; prepared as in **2-21**) are also frequently used, though they react only with active substrates, such as allylic, benzylic, and propargylic halides, and not with primary alkyl halides. Alternatively, the alkyl halide can be treated with a lithium acetylide–ethylenediamine complex.[1571] If 2 moles of a very strong base are used, alkylation can be effected at a carbon α to a terminal triple bond: $RCH_2C\equiv CH +$

$2BuLi \rightarrow R\overset{\ominus}{C}HC\equiv\overset{}{C}{}^{\ominus} + R'Br \rightarrow RR'CHC\equiv C^{\ominus}$.[1572] For another method of alkylating at an alkynyl carbon, see **8-28**.

OS **IV**, 117; **VI**, 273, 564, 595; **67**, 193. Also see OS **IV**, 801; **VI**, 925.

[1562]It has been shown that this migration occurs stereospecifically with inversion in the absence of a solvent, but nonstereospecifically in the presence of a solvent such as THF or dimethyl sulfide: Midland; Zolopa; Halterman *J. Am. Chem. Soc.* **1979**, *101*, 248. See also Midland; Preston *J. Org. Chem.* **1980**, *45*, 747.

[1563]Pasto; Wojtkowski *Tetrahedron Lett.* **1970**, 215, Ref. 1490.

[1564]Brown; Rogić; Rathke; Kabalka *J. Am. Chem. Soc.* **1969**, *91*, 2150.

[1565]Hooz; Linke *J. Am. Chem. Soc.* **1968**, *90*, 5936, 6891; Hooz; Gunn; Kono *Can. J. Chem.* **1971**, *49*, 2371; Mikhailov; Gurskii *Bull. Acad. Sci. USSR, Div. Chem. Sci.* **1973**, *22*, 2588.

[1566]Hooz; Morrison *Can J. Chem.* **1970**, *48*, 868.

[1567]For an improved procedure, see Hooz; Bridson; Calzada; Brown; Midland; Levy *J. Org. Chem.* **1973**, *38*, 2574.

[1568]For reviews, see Ben-Efraim, in Patai *The Chemistry of the Carbon–Carbon Triple Bond*; Wiley: New York, 1978, pp. 790-800; Ziegenbein, in Vieh *Acetylenes*; Marcel Dekker: New York, 1969, pp. 185-206, 241-244. For a discussion of the best ways of preparing various types of alkyne, see Bernadou; Mesnard; Miginiac *J. Chem. Res. (S)* **1978**, 106, **1979**, 190.

[1569]Bourgain; Normant *Bull. Soc. Chim. Fr.* **1973**, 1777; Jeffery *Tetrahedron Lett.* **1989**, *30*, 2225.

[1570]For example, see Fried; Lin; Ford *Tetrahedron Lett.* **1969**, 1379; Krause; Seebach *Chem. Ber.* **1988**, *121*, 1315.

[1571]Smith; Beumel *Synthesis* **1974**, 441.

[1572]Bhanu; Scheinmann *J. Chem. Soc., Perkin Trans.1* **1979**, 1218; Quillinan; Scheinmann *Org. Synth. VI*, 595.

0-101   Preparation of Nitriles
Cyano-de-halogenation

$$RX + CN^- \longrightarrow RCN$$

The reaction between cyanide ion (isoelectronic with $HC\equiv\overline{C}^{\ominus}$ and of similar geometry) and alkyl halides is a convenient method for the preparation of nitriles.[1573] Primary, benzylic, and allylic halides give good yields of nitriles; secondary halides give moderate yields. The reaction fails for tertiary halides, which give elimination under these conditions. Many other groups on the molecule do not interfere. Though a number of solvents have been used, the high yields and short reaction times observed with dimethyl sulfoxide make it a very good solvent for this reaction.[1574] Other ways to obtain high yields under mild conditions are to use a phase transfer catalyst[1575] or ultrasound.[1576] This is an important way of increasing the length of a carbon chain by one carbon, since nitriles are easily hydrolyzed to carboxylic acids (**6-5**).

The cyanide ion is an ambident nucleophile and isocyanides may be side products. If the preparation of isocyanides is desired, they can be made the main products by the use of silver or copper(I) cyanide[1577] (p. 368). Vinylic bromides can be converted to vinylic cyanides with CuCN,[1578] with KCN, a crown ether, and a Pd(0) complex,[1579] with KCN and a Ni(0) catalyst,[1580] or with $K_4Ni_2(CN)_6$.[1581] Tertiary halides can be converted to the corresponding nitriles by treatment with trimethylsilyl cyanide in the presence of catalytic amounts of $SnCl_4$: $R_3CCl + Me_3SiCN \rightarrow R_3CCN$.[1582]

The cyanide nucleophile also reacts with compounds containing other leaving groups. Esters of sulfuric and sulfonic acids behave like halides. Vinylic triflates give vinylic cyanides when treated with LiCN, a crown ether, and a palladium catalyst.[1583] Epoxides give β-hydroxy nitriles. Primary, secondary, and tertiary alcohols are converted to nitriles in good yields by treatment with NaCN, Me_3SiCl, and a catalytic amount of NaI inDMF–MeCN.[1584] One alkoxy group of acetals is replaced by CN $[R_2C(OR')_2 \rightarrow R_2C(OR')CN]$ with Me_3SiCN and a catalyst[1585] or with t-BuNC and $TiCl_4$.[1586] NaCN in HMPA selectively cleaves methyl esters in the presence of ethyl esters: $RCOOMe + CN^- \rightarrow MeCN + RCOO^-$.[1587]

OS **I**, 46, 107, 156, 181, 254, 256, 536; **II**, 292, 376; **III**, 174, 372, 557; **IV**, 438, 496, 576; **V**, 578, 614.

[1573]For reviews, see, in Patai; Rappoport, Ref. 353, the articles by Fatiadi, pt. 2, pp. 1057-1303, and Friedrich, pt. 2, pp. 1343-1390; Friedrich; Wallenfels, in Rappoport *The Chemistry of the Cyano Group*; Wiley: New York, 1970, pp. 77-86.

[1574]Smiley; Arnold *J. Org. Chem.* **1960**, *25*, 257; Friedman; Shechter *J. Org. Chem.* **1960**, *25*, 877.

[1575]For reviews, see Starks; Liotta, Ref. 404, pp. 94-112; Weber; Gokel *Phase Transfer Catalysis in Organic Synthesis*, Ref. 404, pp. 96-108. See also Bram; Loupy; Pedoussaut *Tetrahedron Lett.* **1986**, *27*, 4171, *Bull. Soc. Chim. Fr.* **1986**, 124.

[1576]Ando; Kawate; Ichihara; Hanafusa *Chem. Lett.* **1984**, 725.

[1577]For an example, see Jackson; McKusick *Org. Synth.* IV, 438.

[1578]For example, see Koelsch *J. Am. Chem. Soc.* **1936**, *58*, 1328; Newman; Boden *J. Org. Chem.* **1961**, *26*, 2525; Lapouyade; Daney; Lapenue; Bouas-Laurent *Bull. Soc. Chim. Fr.* **1973**, 720.

[1579]Yamamura; Murahashi *Tetrahedron Lett.* **1977**, 4429.

[1580]Sakakibara; Yadani; Ibuki; Sakai; Uchino *Chem. Lett.* **1982**, 1565; Procházka; Široký *Collect. Czech. Chem. Commun.* **1983**, *48*, 1765.

[1581]Corey; Hegedus *J. Am. Chem. Soc.* **1969**, *91*, 1233. See also Stuhl *J. Org. Chem.* **1985**, *50*, 3934.

[1582]Reetz; Chatziiosifidis *Angew. Chem. Int. Ed. Engl.* **1981**, *20*, 1017 [*Angew. Chem. 93*, 1075].

[1583]Piers; Fleming *J. Chem. Soc., Chem. Commun.* **1989**, 756.

[1584]Davis; Untch *J. Org. Chem.* **1981**, *46*, 2985. See also Mizuno; Hamada; Shioiri *Synthesis* **1980**, 1007; Manna; Falck; Mioskowski *Synth. Commun.* **1985**, *15*, 663; Camps; Gasol; Guerrero *Synth. Commun.* **1988**, *18*, 445.

[1585]Torii; Inokuchi; Kobayashi *Chem. Lett.* **1984**, 897; Soga; Takenoshita; Yamada; Mukaiyama *Bull. Chem. Soc. Jpn.* **1990**, *63*, 3122.

[1586]Ito; Imai; Segoe; Saegusa *Chem. Lett.* **1984**, 937.

[1587]Müller; Siegfried *Helv. Chim. Acta* **1974**, *57*, 987.

0-102   Direct Conversion of Alkyl Halides to Aldehydes and Ketones
Formyl-de-halogenation

$$RX + Na_2Fe(CO)_4 \xrightarrow{Ph_3P} RCOFe(CO)_3PPh_3^- \xrightarrow{HOAc} RCHO$$
140

The direct conversion of alkyl bromides to aldehydes, with an increase in the chain length by one carbon, can be accomplished[1588] by treatment with sodium tetracarbonylferrate($-$II)[1589] (*Collman's reagent*) in the presence of triphenylphosphine and subsequent quenching of **140** with acetic acid. The reagent $Na_2Fe(CO)_4$ can be prepared by treatment of iron pentacarbonyl $Fe(CO)_5$ with sodium amalgam in THF. Good yields are obtained from primary alkyl bromides; secondary bromides give lower yields. The reaction is not satisfactory for benzylic bromides. The initial species produced from RX and $Na_2Fe(CO)_4$ is the ion $RFe(CO)_4^-$ (**141**) (which can be isolated[1590]); it then reacts with Ph_3P to give **140**.[1591]

The synthesis can be extended to the preparation of ketones in six distinct ways.[1592]

1. Instead of quenching **140** with acetic acid, the addition of a second alkyl halide at this point gives a ketone: **140** + R′X → RCOR′.

2. Treatment of $Na_2Fe(CO)_4$ with an alkyl halide in the absence of Ph_3P gives rise to a solution of **141**. Addition of a second alkyl halide produces a ketone: **141** + R′X → RCOR′.

3. Treatment of $Na_2Fe(CO)_4$ with an alkyl halide in the presence of CO results in an

$$RX + Na_2Fe(CO)_4 \xrightarrow{CO} RCOFe(CO)_4^- \xrightarrow{R'X} RCOR'$$
142

acylated iron complex (**142**) that can be isolated.[1590] Treatment of this with a second alkyl halide gives a ketone.

4. Treatment of $Na_2Fe(CO)_4$ with an acyl halide produces **142** which, when treated with an alkyl halide, gives a ketone or, when treated with an epoxide, gives an α,β-unsaturated ketone.[1593]

5. Alkyl halides and tosylates react with $Na_2Fe(CO)_4$ in the presence of ethylene to give alkyl ethyl ketones.[1594] The reaction was not successful for higher alkenes, except that where the double bond and the tosylate group are in the same molecule, 5- and 6-membered rings can be closed.[1595]

6. If 1,4-dihalides are treated with $K_2Fe(CO)_4$, 5-membered cyclic ketones are prepared.[1596]

In the first stage of methods 1, 2, and 3, primary bromides, iodides, and tosylates and secondary tosylates can be used. The second stage of the first four methods requires more active substrates, such as primary iodides or tosylates or benzylic halides. Method 5 has been applied to primary and secondary substrates.

[1588]Cooke *J. Am. Chem. Soc.* **1970**, *92*, 6080.
[1589]For a review of this reagent, see Collman *Acc. Chem. Res.* **1975**, *8*, 342-347. For a review of the related tetracarbonylhydridoferrates MHFe(CO)₄, see Brunet *Chem. Rev.* **1990**, *90*, 1041-1059.
[1590]Siegl; Collman *J. Am. Chem. Soc.* **1972**, *94*, 2516.
[1591]For the mechanism of the conversion **141** → **140**, see Collman; Finke; Cawse; Brauman *J. Am. Chem. Soc.* **1977**, *99*, 2515, **1978**, *100*, 4766.
[1592]For the first four of these methods, see Collman; Winter; Clark *J. Am. Chem. Soc.* **1972**, *94*, 1788; Collman; Hoffman *J. Am. Chem. Soc.* **1973**, *95*, 2689.
[1593]Yamashita; Yamamura; Kurimoto; Suemitsu *Chem. Lett.* **1979**, 1067.
[1594]Cooke; Parlman *J. Am. Chem. Soc.* **1975**, *97*, 6863.
[1595]McMurry; Andrus *Tetrahedron Lett.* **1980**, *21*, 4687, and references cited therein.
[1596]Yamashita; Uchida; Tashika; Suemitsu *Bull. Chem. Soc. Jpn.* **1989**, *62*, 2728.

Aryl, benzylic, vinylic, and allylic halides have been converted to aldehydes by treatment with CO and Bu_3SnH, with a Pd(0) catalyst.[1597] Various other groups do not interfere. Symmetrical ketones R_2CO can be prepared by treatment of a primary alkyl or benzylic halide with $Fe(CO)_5$ and a phase transfer catalyst,[1598] or from a halide RX (R = primary alkyl, aryl, allylic, or benzylic) and CO by an electrochemical method involving a nickel complex.[1599] Several procedures for the preparation of ketones are catalyzed by palladium complexes, among them the following: Alkyl aryl ketones are formed in good yields by treatment of a mixture of an aryl iodide, an alkyl iodide, and a Zn–Cu couple with CO (ArI + RI + CO → RCOAr);[1600] vinylic halides react with vinylic tin reagents in the presence of CO to give unsymmetrical divinyl ketones;[1601] and aryl, vinylic, and benzylic halides can be converted to methyl ketones (RX → RCOMe) by reaction with (α-ethoxy-vinyl)tributyltin $Bu_3SnC(OEt)\!\!=\!\!CH_2$.[1602]

The conversion of alkyl halides to aldehydes and ketones can also be accomplished indirectly (**0-97**). See also **2-32**.

OS **VI**, 807.

0-103 Conversion of Alkyl Halides, Alcohols, or Alkanes to Carboxylic Acids and Their Derivatives
Alkoxycarbonyl-de-halogenation

$$RX + CO + R'OH \xrightarrow[-70°C]{SbCl_5-SO_2} RCOOR'$$

Several methods, all based on carbon monoxide or metal carbonyls, have been developed for converting an alkyl halide to a carboxylic acid or an acid derivative with the chain extended by one carbon.[1603] When an alkyl halide is treated with $SbCl_5$–SO_2 at $-70°C$, it dissociates into the corresponding carbocation (p. 166). If carbon monoxide and an alcohol are present, a carboxylic ester is formed by the following route:[1604]

$$RX \xrightarrow[-70°C]{SbCl_5-SO_2} R^+ \ X^- \xrightarrow{CO} \underset{\underset{O\text{---}SbCl_5}{\|}}{R-C-X} \xrightarrow{R'OH} \underset{\underset{O}{\|}\ \underset{H}{|}}{R-C-\overset{\oplus}{O}-R'} \xrightarrow{-H^+} RCOOR'$$

This has also been accomplished with concentrated H_2SO_4 saturated with CO.[1605] Not surprisingly, only tertiary halides perform satisfactorily; secondary halides give mostly rearrangement products. An analogous reaction takes place with alkanes possessing a tertiary hydrogen, e.g.,[1606]

$$EtCHMe_2 \xrightarrow[\text{2. } H_2O]{\text{1. } HF-SbF_5-CO} 75\% \ EtCMe_2COOH$$

[1597]Baillargeon; Stille *J. Am. Chem. Soc.* **1986**, *108*, 452. See also Kasahara; Izumi; Yanai *Chem. Ind.* (*London*) **1983**, 898; Pri-Bar; Buchman *J. Org. Chem.* **1984**, *49*, 4009; Takeuchi; Tsuji; Watanabe *J. Chem. Soc., Chem. Commun.* **1986**, 351; Ben-David; Portnoy; Milstein *J. Chem. Soc., Chem. Commun* **1989**, 1816.
[1598]Kimura; Tomita; Nakanishi; Otsuji *Chem. Lett.* **1979**, 321; des Abbayes; Clément; Laurent; Tanguy; Thilmont *Organometallics* **1988**, *7*, 2293.
[1599]Garnier; Rollin; Périchon *J. Organomet. Chem.* **1989**, *367*, 347.
[1600]Tamaru; Ochiai; Yamada; Yoshida *Tetrahedron Lett.* **1983**, *24*, 3869.
[1601]Goure; Wright; Davis; Labadie; Stille *J. Am. Chem. Soc.* **1984**, *106*, 6417. For a similar preparation of diallyl ketones, see Merrifield; Godschalx; Stille *Organometallics* **1984**, *3*, 1108.
[1602]Kosugi; Sumiya; Obara; Suzuki; Sano; Migita *Bull. Chem. Soc. Jpn.* **1987**, *60*, 767.
[1603]For discussions of most of the reactions in this section, see Colquhoun; Holton; Thompson; Twigg *New Pathways for Organic Synthesis*; Plenum: New York, 1984, pp. 199-204, 212-220, 234-235. For lists of reagents, with references, see Ref. 508, pp. 850-851, 855-856, 859-860.
[1604]Yoshimura; Nojima; Tokura *Bull. Chem. Soc. Jpn.* **1973**, *46*, 2164; Puzitskii; Pirozhkov; Ryabova; Myshenkova; Éidus *Bull. Acad. Sci. USSR, Div. Chem. Sci.* **1974**, *23*, 192.
[1605]Takahashi; Yoneda *Synth. Commun.* **1989**, *19*, 1945.
[1606]Paatz; Weisgerber *Chem. Ber.* **1967**, *100*, 984.

Carboxylic acids or esters are the products, depending on whether the reaction mixture is solvolyzed with water or an alcohol. Alcohols with more than 7 carbons are cleaved into smaller fragments by this procedure.[1607] Similarly, tertiary alcohols[1608] react with H_2SO_4 and CO (which is often generated from HCOOH and the H_2SO_4 in the solution) to give trisubstituted acetic acids in a process called the *Koch–Haaf reaction* (see also **5-23**).[1609] If a primary or secondary alcohol is the substrate, the carbocation initially formed rearranges to a tertiary ion before reacting with the CO. Better results are obtained if trifluoromethanesulfonic acid F_3CSO_2OH is used instead of H_2SO_4.[1610]

Another method[1611] for the conversion of alkyl halides to carboxylic esters is treatment of a halide with nickel carbonyl $Ni(CO)_4$ in the presence of an alcohol and its conjugate

$$RX + Ni(CO)_4 \xrightarrow[R'OH]{R'O^-} RCOOR'$$

base.[1612] When R′ is primary, RX may only be a vinylic or an aryl halide; retention of configuration is observed at a vinylic R. Consequently, a carbocation intermediate is not involved here. When R′ is tertiary, R may be primary alkyl as well as vinylic or aryl. This is thus one of the few methods for preparing esters of tertiary alcohols. Alkyl iodides give the best results, then bromides. In the presence of an amine, an amide can be isolated directly, at least in some instances.

Still another method for the conversion of halides to acid derivatives makes use of $Na_2Fe(CO)_4$. As described in **0-102**, primary and secondary alkyl halides and tosylates react with this reagent to give the ion $RFe(CO)_4^-$ (**141**) or, if CO is present, the ion $RCOFe(CO)_4^-$ (**142**). Treatment of **141** or **142** with oxygen or sodium hypochlorite gives, after hydrolysis, a carboxylic acid.[1613] Alternatively, **141** or **142** reacts with a halogen (for example, I_2) in the

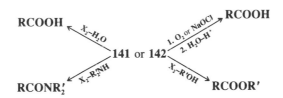

presence of an alcohol to give a carboxylic ester,[1614] or in the presence of a secondary amine or water to give, respectively, the corresponding amide or free acid. **141** and **142** prepared from primary R give high yields. With secondary R, the best results are obtained in the solvent THF by the use of **142** prepared from secondary tosylates. Ester and keto groups may be present in R without being affected. Carboxylic esters RCO_2R' have also been

[1607]Yoneda; Takahashi; Fukuhara; Suzuki *Bull. Chem. Soc. Jpn.* **1986**, *59*, 2819.

[1608]For reviews of other carbonylation reactions of alcohols and other saturated oxygenated compounds, see Bahrmann; Cornils, in Falbe *New Syntheses with Carbon Monoxide*; Springer: New York, 1980, pp. 226-241; Piacenti; Bianchi, in Wender; Pino *Organic Syntheses via Metal Carbonyls*, vol. 2; Wiley: New York, 1977, pp. 1-42.

[1609]For a review, see Bahrmann, in Falbe, Ref. 1608, pp. 372-413.

[1610]Booth; El-Fekky *J. Chem. Soc., Perkin Trans. 1* **1979**, 2441.

[1611]For reviews of methods involving transition metals, see Collman et al., Ref. 1266, pp. 749-768; Anderson; Davies, in Hartley; Patai, Ref. 1403, vol. 3, pp. 335-359, pp. 348-356; Heck *Adv. Catal.* **1977**, *26*, 323-349, pp. 323-336; Cassar; Chiusoli; Guerrieri *Synthesis* **1973**, 509-523.

[1612]Corey; Hegedus *J. Am. Chem. Soc.* **1969**, *91*, 1233. See also Crandall; Michaely *J. Organomet. Chem.* **1973**, *51*, 375.

[1613]Collman; Winter; Komoto *J. Am. Chem. Soc.* **1973**, *95*, 249.

[1614]Ref. 1613; Masada; Mizuno; Suga; Watanabe; Takegami *Bull. Chem. Soc. Jpn.* **1970**, *43*, 3824.

prepared by treating primary alkyl halides RX with alkoxides $R'O^-$ in the presence of $Fe(CO)_5$.[1615] **142** is presumably an intermediate.

Palladium complexes also catalyze the carbonylation of halides.[1616] Aryl (see **3-15**), vinylic,[1617] benzylic, and allylic halides (especially iodides) can be converted to carboxylic esters with CO, an alcohol or alkoxide, and a palladium complex.[1618] Use of an amine instead of the alcohol or alkoxide leads to an amide.[1619] Benzylic and allylic halides were converted to carboxylic acids electrocatalytically, with CO and a cobalt imine complex.[1620] Vinylic halides were similarly converted with CO and nickel cyanide, under phase-transfer conditions.[1621]

Rhodium catalysts have also been used. Benzylic halides were converted to carboxylic esters with CO in the presence of a rhodium complex. In this case, the R' could come from an ether R'_2O,[1622] a borate ester $B(OR')_3$,[1623] or an Al, Ti, or Zr alkoxide.[1624]

A number of double carbonylations have been reported. In these reactions, two molecules of CO are incorporated in the product, leading to α-keto acids or their derivatives.[1625] When the catalyst is a palladium complex, best results are obtained in the formation of α-keto amides.[1626]

$$RX + CO + 2R'_2NH \xrightarrow{\text{Pd catalyst}} R-\underset{\underset{O}{\|}}{C}-\underset{\underset{O}{\|}}{C}-NR'_2 + R'_2NH_2^+ \; X^-$$

R is usually aryl or vinylic.[1627] The formation of α-keto acids[1628] or esters[1629] requires more severe conditions. α-Hydroxy acids were obtained from aryl iodides when the reaction was carried out in the presence of an alcohol, which functioned as a reducing agent.[1630] Cobalt catalysts have also been used and require lower CO pressures.[1625]

OS **V**, 20, 739.

[1615]Yamashita; Mizushima; Watanabe; Mitsudo; Takegami *Chem. Lett.* **1977**, 1355. See also Tanguy; Weinberger; des Abbayes *Tetrahedron Lett.* **1983**, *24*, 4005.

[1616]For reviews, see Gulevich; Bumagin; Beletskaya *Russ. Chem. Rev.* **1988**, *57*, 299-315, pp. 303-309; Heck *Palladium Reagents in Organic Synthesis*, Ref. 1308, pp. 348-356, 366-370.

[1617]For conversion of vinylic triflates to carboxylic esters and amides, see Cacchi; Morera; Ortar *Tetrahedron Lett.* **1985**, *26*, 1109.

[1618]Tsuji; Kishi; Imamura; Morikawa *J. Am. Chem. Soc.* **1964**, *86*, 4350; Schoenberg; Bartoletti; Heck *J. Org. Chem.* **1974**, *39*, 3318; Hidai; Hikita; Wada; Fujikura; Uchida *Bull. Chem. Soc. Jpn.* **1975**, *48*, 2075; Bumagin; Gulevich; Beletskaya *J. Organomet. Chem.* **1985**, *285*, 415; Milstein *J. Chem. Soc., Chem. Commun.* **1986**, 817; Kiji; Okano; Nishiumi; Konishi *Chem. Lett.* **1988**, 957, **1989**, 1873; Adapa; Prasad *J. Chem. Soc., Perkin Trans. 1* **1989**, 1706.

[1619]Schoenberg; Heck *J. Org. Chem.* **1974**, *39*, 3327. See also Lindsay; Widdowson *J. Chem. Soc., Perkin Trans. 1* **1988**, 569. For a review of some methods of amide formation that involve transition metals, see Screttas; Steele *Org. Prep. Proced. Int.* **1990**, *22*, 271-314, pp. 288-314.

[1620]Folest; Duprilot; Perichon; Robin; Devynck *Tetrahedron Lett.* **1985**, *26*, 2633. For other procedures involving a cobalt catalyst, see Francalanci; Gardano; Foà *J. Organomet. Chem.* **1985**, *282*, 277; Satyanarayana; Periasamy *Tetrahedron Lett.* **1987**, *28*, 2633; Miura; Okuro; Hattori; Nomura *J. Chem. Soc., Perkin Trans. 1* **1989**, 73; Urata; Goto; Fuchikami *Tetrahedron Lett.* **1991**, *32*, 3091.

[1621]Alper; Amer; Vasapollo *Tetrahedron Lett.* **1989**, *30*, 2615. See also Amer; Alper *J. Am. Chem. Soc.* **1989**, *111*, 927.

[1622]Buchan; Hamel; Woell; Alper *Tetrahedron Lett.* **1985**, 26, 5743.

[1623]Woell; Alper *Tetrahedron Lett.* **1984**, *25*, 3791; Alper; Hamel; Smith; Woell *Tetrahedron Lett.* **1985**, *26*, 2273.

[1624]Woell; Fergusson; Alper *J. Org. Chem.* **1985**, *50*, 2134.

[1625]For a review, see Collin *Bull. Soc. Chim. Fr.* **1988**, 976-981.

[1626]Kobayashi; Tanaka *J. Organomet. Chem.* **1982**, *233*, C64; Ozawa; Sugimoto; Yuasa; Santra; Yamamoto; Yamamoto *Organometallics* **1984**, *3*, 683.

[1627]Son; Yanagihara; Ozawa; Yamamoto *Bull. Chem. Soc. Jpn.* **1988**, *61*, 1251.

[1628]Tanaka; Kobayashi; Sakakura *J. Chem. Soc., Chem. Commun.* **1985**, 837.

[1629]See Ozawa; Kawasaki; Okamoto; Yamamoto; Yamamoto *Organometallics* **1987**, *6*, 1640.

[1630]Kobayashi; Sakakura; Tanaka *Tetrahedron Lett.* **1987**, *28*, 2721.

B. Attack at an Acyl Carbon[1631]

0-104 The Conversion of Acyl Halides to Ketones with Organometallic Compounds[1632]
Alkyl-de-halogenation

$$R'-\underset{\underset{O}{\|}}{C}-X + R_2CuLi \longrightarrow R'-\underset{\underset{O}{\|}}{C}-R$$

Acyl halides react cleanly and under mild conditions with lithium dialkylcopper reagents[1633] to give high yields of ketones.[1634] R' may be primary, secondary, or tertiary alkyl or aryl and may contain iodo, keto, ester, nitro, or cyano groups. R groups that have been used successfully are methyl, primary alkyl, and vinylic. Secondary and tertiary alkyl groups can be introduced by the use of PhS(R)CuLi (p. 451) instead of R_2CuLi,[1635] or by the use of either the mixed homocuprate $(R'SO_2CH_2CuR)^- Li^+$,[1636] or a magnesium dialkylcopper reagent "RMeCuMgX."[1637] Secondary alkyl groups can also be introduced with the copper–zinc reagents RCu(CN)ZnI.[1638] R may be alkynyl if a cuprous acetylide $R''C{\equiv}CCu$ is the reagent.[1639] Organocopper reagents generated in situ from highly reactive copper, and containing such functional groups as cyano, chloro, and ester, react with acyl halides to give ketones.[1640]

Another type of organometallic reagent[1641] that gives good yields of ketones when treated with acyl halides are organocadmiums R_2Cd (prepared from Grignard reagents, **2-21**). In this case R may be aryl or primary alkyl. In general, secondary and tertiary alkylcadmium reagents are not stable enough to be useful in this reaction.[1642] An ester group may be present in either R'COX or R_2Cd. Organozinc compounds behave similarly, but are used less often.[1643] Organomercury compounds[1644] and tetraalkylsilanes[1645] also give the reaction if an AlX_3 catalyst is present.[1646] Organotin reagents R_4Sn react with acyl halides to give high yields of ketones, if a Pd complex is present.[1647] Various other groups, for example, nitrile, ester, and aldehyde can be present in the acyl halide without interference. Still

[1631]For a discussion of many of the reactions in this section, see House, Ref. 1411, pp. 691-694, 734-765.

[1632]For a review, see Cais; Mandelbaum, in Patai, Ref. 446, vol. 1, pp. 303-330.

[1633]For examples of the use of this reaction in the synthesis of natural products, see Posner, Ref. 1352, pp. 81-85. See also Ref. 1268.

[1634]Vig; Sharma; Kapur *J. Indian Chem. Soc.* **1969**, *46*, 167; Jukes; Dua; Gilman *J. Organomet. Chem.* **1970**, *21*, 241; Posner; Whitten; McFarland *J. Am. Chem. Soc.* **1972**, *94*, 5106; Luong-Thi; Rivière *J. Organomet. Chem.* **1974**, *77*, C52.

[1635]Ref. 1276; Bennett; Nadelson; Alden; Jani *Org. Prep. Proced. Int.* **1976**, *8*, 13.

[1636]Johnson; Dhanoa *J. Org. Chem.* **1987**, *52*, 1885.

[1637]Bergbreiter; Killough *J. Org. Chem.* **1976**, *41*, 2750.

[1638]Knochel; Yeh; Berk; Talbert *J. Org. Chem.* **1988**, *53*, 2390.

[1639]Castro; Havlin; Honwad; Malte; Mojé *J. Am. Chem. Soc.* **1969**, *91*, 6464. For methods of preparing acetylenic ketones, see Verkruijsse; Heus-Kloos; Brandsma *J. Organomet. Chem.* **1988**, *338*, 289.

[1640]Wehmeyer; Rieke *Tetrahedron Lett.* **1988**, *29*, 4513.

[1641]For a list of reagents, with references, see Ref. 508, pp. 686-691.

[1642]Cason; Fessenden *J. Org. Chem.* **1960**, *25*, 477.

[1643]For examples, see Grey *J. Org. Chem.* **1984**, *49*, 2288; Tamaru; Ochiai; Nakamura; Yoshida *Org. Synth. 67*, 98.

[1644]Kurts; Beletskaya; Savchenko; Reutov *J. Organomet. Chem.* **1969**, *17*, P21; Larock; Lu *Tetrahedron Lett.* **1988**, *29*, 6761. See also Bumagin; Kalinovskii; Beletskaya *J. Org. Chem. USSR* **1982**, *18*, 1152.

[1645]For a review, see Parnes; Bolestova *Synthesis* **1984**, 991-1008, pp. 991-996.

[1646]In the case of organomercury compounds a palladium catalyst can also be used: Bumagin; More; Beletskaya *J. Organomet. Chem.* **1989**, *365*, 379.

[1647]Kosugi; Shimizu; Migita *Chem. Lett.* **1977**, 1423; Labadie; Stille *J. Am. Chem. Soc.* **1983**, *105*, 669, 6129; Labadie; Tueting; Stille *J. Org. Chem.* **1983**, *48*, 4634. For the use of R_4Pb see Yamada; Yamamoto *J. Chem. Soc., Chem. Commun.* **1987**, 1302. See also Verlhac; Quintard *Tetrahedron Lett.* **1986**, *27*, 2361.

other reagents are organomanganese compounds[1648] (R can be primary, secondary, or tertiary alkyl, vinylic, alkynyl, or aryl), organothallium compounds (R can be primary alkyl or aryl),[1649] lithium aryltrialkylborates[1650] $ArBR_3^-$ Li^+ (which transfer an aryl group), and the alkylrhodium(I) complexes bis(triphenylphosphine)carbonylalkylrhodium(I) $Rh^IR(CO)(Ph_3P)_2$. The latter, generated in situ from $Rh^ICl(CO)(Ph_3P)_2$ (**143**) and a Grignard reagent or organolithium compound, react with acyl halides in THF at $-78°C$ to give good yields of ketones.[1651] R may be primary alkyl or aryl. An advantage of the rhodium reagents is that they do not react with aldehydes, esters, or nitriles, so that these groups may be present in R'. Another advantage is that the complex **143** is regenerated in reusable form at the end of the reaction.

When the organometallic compound is a Grignard reagent,[1652] ketones are generally not obtained because the initially formed ketone reacts with a second molecule of RMgX to give the salt of a tertiary alcohol (**6-32**). Ketones *have* been prepared in this manner by the use of low temperatures, inverse addition (i.e., addition of the Grignard reagent to the acyl halide rather than the other way), excess acyl halide, etc., but the yields are usually low, though high yields have been reported in THF at $-78°C$.[1653] Some ketones are unreactive toward Grignard reagents for steric or other reasons; these can be prepared in this way.[1654] Other methods involve running the reaction in the presence of Me_3SiCl[1655] (which reacts with the initial adduct **67** in the tetrahedral mechanism, p. 331), and the use of a combined Grignard–lithium diethylamide reagent.[1656] Also, certain metallic halides, notably ferric and cuprous halides, are catalysts that improve the yields of ketone at the expense of tertiary alcohol.[1657] For these catalysis, both free-radical and ionic mechanisms have been proposed.[1658] The reactions with R_2CuLi, R_2Cd, and the rhodium complexes are successful because these compounds do not generally react with ketones.

Grignard reagents react with ethyl chloroformate to give carboxylic esters $EtOCOCl + RMgX \rightarrow EtOCOR$. Acyl halides can also be converted to ketones by treatment with $Na_2Fe(CO)_4$ followed by R'X (**0-102**, method 4).

OS **II**, 198; **III**, 601; **IV**, 708; **VI**, 248, 991; **VII**, 226, 334; **65**, 47; **66**, 87, 116; **67**, 86, 98.

0-105 The Conversion of Anhydrides, Carboxylic Esters, or Amides to Ketones with Organometallic Compounds[1659]

Alkyl-de-acyloxy-substitution

$$R-\underset{\underset{O}{\|}}{C}-W + R'MgX \longrightarrow R-\underset{\underset{O}{\|}}{C}-R' \qquad W = OCOR'', OR'', NR''_2$$

[1648]Friour; Alexakis; Cahiez; Normant *Tetrahedron* **1984**, *40*, 683; Friour; Cahiez; Normant *Synthesis* **1985**, 50; Cahiez; Laboue *Tetrahedron Lett.* **1989**, *30*, 7369.

[1649]Markó; Southern *J. Org. Chem.* **1990**, *55*, 3368.

[1650]Negishi; Abramovitch; Merrill *J. Chem. Soc., Chem. Commun.* **1975**, 138, Negishi; Chiu; Yoshida *J. Org. Chem.* **1975**, *40*, 1676. See also Miyaura; Sasaki; Itoh; Suzuki *Tetrahedron Lett.* **1977**, 173.

[1651]Hegedus; Kendall; Lo; Sheats *J. Am. Chem. Soc.* **1975**, *97*, 5448. See also Pittman; Hanes *J. Org. Chem.* **1977**, *42*, 1194.

[1652]For a review, see Kharasch; Reinmuth, Ref. 1287, pp. 712-724.

[1653]Sato; Inoue; Oguro; Sato *Tetrahedron Lett.* **1979**, 4303; Eberle; Kahle *Tetrahedron Lett.* **1980**, *21*, 2303; Föhlisch; Flogaus *Synthesis* **1984**, 734.

[1654]For example, see Lion; Dubois; Bonzougou *J. Chem. Res., (S)* **1978**, 46; Dubois; Lion; Arouisse *Bull. Soc. Chim. Belg.* **1984**, *93*, 1083.

[1655]Cooke *J. Org. Chem.* **1986**, *51*, 951.

[1656]Fehr; Galindo *Helv. Chim. Acta* **1986**, *69*, 228; Fehr; Galindo; Perret *Helv. Chim. Acta* **1987**, *70*, 1745.

[1657]For examples, see Cason; Kraus *J. Org. Chem.* **1961**, *26*, 1768, 1772; MacPhee; Dubois *Tetrahedron Lett.* **1972**, 467; Cardellicchio; Fiandanese; Marchese; Ronzini *Tetrahedron Lett.* **1987**, *28*, 2053; Fujisawa; Sato *Org. Synth. 66*, 116; Babudri; D'Ettole; Fiandanese; Marchese; Naso *J. Organomet. Chem.* **1991**, *405*, 53.

[1658]For examples, see Dubois; Boussu *Tetrahedron Lett.* **1970**, 2523, *Tetrahedron* **1973**, *29*, 3943; MacPhee; Boussu; Dubois *J. Chem. Soc., Perkin Trans. 2* **1974**, 1525.

[1659]For a review, see Kharasch; Reinmuth, Ref. 1287, pp. 561-562, 846-908.

As is the case with acyl halides (**0-104**), anhydrides and carboxylic esters give tertiary alcohols (**6-32**) when treated with Grignard reagents. Low temperatures,[1660] the solvent HMPA,[1661] and inverse addition have been used to increase the yields of ketone.[1662] Amides give better yields of ketone at room temperature, but still not very high.[1663] Thiol esters RCOSR' give good yields of ketones when treated with lithium dialkylcopper reagents R''_2CuLi (R'' = primary or secondary alkyl or aryl).[1664] Ketones can also be prepared by treatment of thioamides with organolithium compounds (alkyl or aryl).[1665] Organocadmium reagents are less successful with these substrates than with acyl halides (**0-104**). Esters of formic acid, dialkylformamides, and lithium or sodium formate[1666] give good yields of aldehydes, when treated with Grignard reagents.

Alkyllithium compounds have been used to give ketones from carboxylic esters. The reaction must be carried out in a high-boiling solvent such as toluene, since reaction at lower temperatures gives tertiary alcohols.[1667] Alkyllithiums also give good yields of carbonyl compounds with N,N-disubstituted amides.[1668] Dialkylformamides give aldehydes and other disubstituted amides give ketones.

$$H—\underset{\underset{O}{\|}}{C}—NR''_2 + R'Li \longrightarrow H—\underset{\underset{O}{\|}}{C}—R'$$

$$R—\underset{\underset{O}{\|}}{C}—NR''_2 + R'Li \longrightarrow R—\underset{\underset{O}{\|}}{C}—R'$$

N,N-Disubstituted amides can be converted to alkynyl ketones by treatment with alkynyl-boranes: $RCONR''_2 + (R'C≡C)_3B \rightarrow RCOC≡CR'$.[1669] Alkynyl ketones are also obtained by treatment of anhydrides with lithium alkynyltrifluoroborates $Li(RC≡C—BF_3)$.[1670] N,N-Disubstituted carbamates ($X = OR''$) and carbamoyl chlorides ($X = Cl$) react with 2 moles of an alkyl- or aryllithium or Grignard reagent to give symmetrical ketones, in which both R groups are derived from the organometallic compound: $R'_2NCOX + 2RMgX \rightarrow R_2CO$.[1671] N,N-Disubstituted amides give ketones in high yields when treated with alkyl-lanthanum triflates $RLa(OTf)_2$.[1672]

By the use of the compound N-methoxy-N,N',N'-trimethylurea **144**, it is possible to add

$$Me_2N—\underset{\underset{O}{\|}}{C}—\underset{\underset{Me}{|}}{N}—OMe \xrightarrow{RLi} R—\underset{\underset{O}{\|}}{C}—\underset{\underset{Me}{|}}{N}—OMe \xrightarrow{R'Li} R—\underset{\underset{O}{\|}}{C}—R'$$

$$\qquad\qquad\textbf{144}\qquad\qquad\qquad\qquad\textbf{145}$$

[1660]See, for example, Newman; Smith *J. Org. Chem.* **1948**, *13*, 592; Edwards; Kammann *J. Org. Chem.* **1964**, *29*, 913; Araki; Sakata; Takei; Mukaiyama *Chem. Lett.* **1974**, 687.

[1661]Huet; Emptoz; Jubier *Tetrahedron* **1973**, *29*, 479; Huet; Pellet; Conia *Tetrahedron Lett.* **1976**, 3579.

[1662]For a list of preparations of ketones by the reaction of organometallic compounds with carboxylic esters, salts, anhydrides, or amides, with references, see Ref. 508, pp. 685-686, 693-700.

[1663]For an improved procedure with amides, see Olah; Prakash; Arvanaghi *Synthesis* **1984**, 228.

[1664]Anderson; Henrick; Rosenblum *J. Am. Chem. Soc.* **1974**, *96*, 3654. See also Kim; Lee *J. Org. Chem.* **1983**, *48*, 2608.

[1665]Tominaga; Kohra; Hosomi *Tetrahedron Lett.* **1987**, *28*, 1529.

[1666]Bogavac; Arsenijević; Pavlov; Arsenijević *Tetrahedron Lett.* **1984**, *25*, 1843.

[1667]Petrov; Kaplan; Tsir *J. Gen. Chem. USSR* **1962**, *32*, 691.

[1668]Evans *J. Chem. Soc.* **1956**, 4691. For a review, see Wakefield *Organolithium Methods*; Academic Press: New York, 1988, pp. 82-88.

[1669]Yamaguchi; Waseda; Hirao *Chem. Lett.* **1983**, 35.

[1670]Brown; Racherla; Singh *Tetrahedron Lett.* **1984**, *25*, 2411.

[1671]Michael; Hörnfeldt *Tetrahedron Lett.* **1970**, 5219; Scilly, *Synthesis* **1973**, 160.

[1672]Collins; Hong *Tetrahedron Lett.* **1987**, *28*, 4391.

two R groups, the same or different, to a CO group. Both reactions can be done in the same vessel without the isolation of **145**.[1673]

Hydrogen has been reported to be a leaving group in this reaction: Aromatic aldehydes are converted to methyl ketones (ArCHO → ArCOCH$_3$) with Al(OAr)Me$_2$ (Ar = 2,6-di-*t*-butyl-4-methylphenyl).[1674]

Carboxylic esters can be converted to their homologs (RCOOEt → RCH$_2$COOEt) by treatment with Br$_2$CHLi followed by BuLi at −90°C. The ynolate RC≡COLi is an intermediate.[1675] If the ynolate is treated with 1,3-cyclohexadiene, followed by NaBH$_4$, the product is the alcohol RCH$_2$CH$_2$OH.[1676]

Ketones can also be obtained by treatment of the lithium salt of a carboxylic acid with an alkyllithium reagent (**6-31**). For an indirect way to convert carboxylic esters to ketones, see **6-33**.

OS **II**, 282; **III**, 353; **IV**, 285; **VI**, 611; **VII**, 323, 451.

0-106   The Coupling of Acyl Halides
De-halogen-coupling

$$2RCOCl \xrightarrow{\text{pyrophoric Pb}} RCOCOR$$

Acyl halides can be coupled with pyrophoric lead to give symmetrical α-diketones in a Wurtz-type reaction.[1677] The reaction has been performed with R = Me and Ph. Other reagents that give the same reaction are samarium iodide SmI$_2$[1678] and hexaethyldistannane Et$_6$Sn$_2$ (with palladium catalysts and under CO pressure).[1679] Benzoyl chloride was coupled to give benzil by subjecting it to ultrasound in the presence of Li wire: 2PhCOCl + Li → PhCOCOPh.[1247]

Unsymmetrical α-diketones RCOCOR′ have been prepared by treatment of an acyl halide RCOCl with an acyltin reagent RCOSnBu$_3$, with a palladium-complex catalyst.[1680]

0-107   Acylation at a Carbon Bearing an Active Hydrogen
Bis(ethoxycarbonyl)methyl-de-halogenation, etc.

$$RCOCl + Z—\overset{\ominus}{CH}—Z' \longrightarrow Z—\underset{\underset{COR}{|}}{CH}—Z'$$

This reaction is similar to **0-94**, though many fewer examples have been reported.[1681] Z and Z′ may be any of the groups listed in **0-94**.[1682] Anhydrides react similarly but are used less often. The product contains three Z groups, since RCO is a Z group. One or two of these can be cleaved (**2-40, 2-43**). In this way a compound ZCH$_2$Z′ can be converted to ZCH$_2$Z″ or an acyl halide RCOCl to a methyl ketone RCOCH$_3$. O-Acylation is sometimes a side

[1673]Hlasta; Court *Tetrahedron Lett.* **1989**, *30*, 1773. See also Nahm; Weinreb *Tetrahedron Lett.* **1981**, *22*, 3815.
[1674]Power; Barron *Tetrahedron Lett.* **1990**, *31*, 323.
[1675]Kowalski; Haque; Fields *J. Am. Chem. Soc.* **1985**, *107*, 1429; Kowalski; Haque *J. Org. Chem.* **1985**, *50*, 5140.
[1676]Kowalski; Haque *J. Am. Chem. Soc.* **1986**, *108*, 1325.
[1677]Mészáros *Tetrahedron Lett.* **1967**, 4951.
[1678]Souppe; Namy; Kagan *Tetrahedron Lett.* **1984**, *25*, 2869. See also Collin; Namy; Dallemer; Kagan *J. Org. Chem.* **1991**, *56*, 3118.
[1679]Bumagin; Gulevich; Beletskaya *J. Organomet. Chem.* **1985**, *282*, 421.
[1680]Verlhac; Chanson; Jousseaume; Quintard *Tetrahedron Lett.* **1985**, *26*, 6075. For another procedure, see Olah; Wu *J. Org. Chem.* **1991**, *56*, 902.
[1681]For examples of reactions in this section, with references, see Ref. 508, pp. 742, 764-767.
[1682]For an improved procedure, see Rathke; Cowan *J. Org. Chem.* **1985**, *50*, 2622.

reaction.[1683] When thallium(I) salts of ZCH_2Z' are used, it is possible to achieve regioselective acylation at either the C or the O position. For example, treatment of the thallium(I) salt of $MeCOCH_2COMe$ with acetyl chloride at $-78°C$ gave >90% O-acylation, while acetyl fluoride at room temperature gave >95% C-acylation.[1684] The use of an alkyl chloroformate gives triesters.[1685]

The application of this reaction to simple ketones[1452] (in parallel with **0-95**) requires a strong base, such as $NaNH_2$ or Ph_3CNa, and is often complicated by O-acylation, which in many cases becomes the principal pathway because acylation at the oxygen is usually much faster. It is possible to increase the proportion of C-acylated product by employing an excess (2 to 3 equivalents) of enolate ion (and adding the substrate to this, rather than vice versa), by the use of a relatively nonpolar solvent and a metal ion (such as Mg^{2+}) which is tightly associated with the enolate oxygen atom, by the use of an acyl halide rather than an anhydride,[1686] and by working at low temperatures.[1687] In cases where the use of an excess of enolate ion results in C-acylation, it is because O-acylation takes place first, and the O-acylated product (an enol ester) is then C-acylated. Simple ketones can also be acylated by treatment of their silyl enol ethers with an acyl chloride in the presence of $ZnCl_2$ or $SbCl_3$.[1688] Ketones can be acylated by anhydrides to give β-diketones, with BF_3 as catalyst.[1689] Simple esters RCH_2COOEt can be acylated at the α carbon (at $-78°C$) if a strong base such as lithium N-isopropylcyclohexylamide is used to remove the proton.[1690]

OS **II**, 266, 268, 594, 596; **III**, 16, 390, 637; **IV**, 285, 415, 708; **V**, 384, 937; **VI**, 245; **VII**, 213, 359; **66**, 108; **69**, 44, 173. See also OS **VI**, 620; **65**, 146.

0-108   Acylation of Carboxylic Esters by Carboxylic Esters. The Claisen and Dieckmann Condensations
Alkoxycarbonylalkyl-de-alkoxy-substitution

$$2R-CH_2-\underset{\substack{\|\\O}}{C}-OR' \overset{OEt^-}{\rightleftharpoons} R-CH_2-\underset{\substack{\|\\O}}{C}-\overset{\overset{\textstyle R}{|}}{CH}-\underset{\substack{\|\\O}}{C}-OR'$$

When carboxylic esters containing an α hydrogen are treated with a strong base such as sodium ethoxide, a condensation occurs to give a β-keto ester. This reaction is called the *Claisen condensation*. When it is carried out with a mixture of two different esters, each of which possesses an α hydrogen, a mixture of all four products is generally obtained and the reaction is seldom useful synthetically.[1691] However, if only one of the esters has an α hydrogen, the mixed reaction is frequently satisfactory. Among esters lacking α hydrogens

---

[1683]When phase transfer catalysts are used, O-acylation becomes the main reaction: Jones; Nokkeo; Singh *Synth. Commun.* **1977**, 7, 195.

[1684]Taylor; Hawks; McKillop *J. Am. Chem. Soc.* **1968**, 90, 2421.

[1685]See, for example, Skarżewski *Tetrahedron* **1989**, 45, 4593. For a review of triesters, see Newkome; Baker *Org. Prep. Proced. Int.* **1986**, 19, 117-144.

[1686]See House, Ref. 1411, pp. 762-765; House; Auerbach; Gall; Peet *J. Org. Chem.* **1973**, 38, 514.

[1687]Seebach; Weller; Protschuk; Beck; Hoekstra *Helv. Chim. Acta* **1981**, 64, 716.

[1688]Tirpak; Rathke *J. Org. Chem.* **1982**, 47, 5099.

[1689]For a review, see Hauser; Swamer; Adams *Org. React.* **1954**, 8, 59-196, pp. 98-106.

[1690]For example, see Rathke; Deitch *Tetrahedron Lett.* **1971**, 2953; Logue *J. Org. Chem.* **1974**, 39, 3455; Couffignal; Moreau *J. Organomet. Chem.* **1977**, 127, C65; Ohta; Shimabayashi; Hayakawa; Sumino; Okamoto *Synthesis* **1985**, 45; Hayden; Pucher; Griengl *Monatsh. Chem.* **1987**, 118, 415.

[1691]For a method of allowing certain crossed-Claisen reactions to proceed with good yields, see Tanabe *Bull. Chem. Soc. Jpn.* **1989**, 62, 1917.

(hence acting as the substrate ester) that are commonly used in this way are esters of aromatic acids, and ethyl carbonate and ethyl oxalate. Ethyl carbonate gives malonic esters.

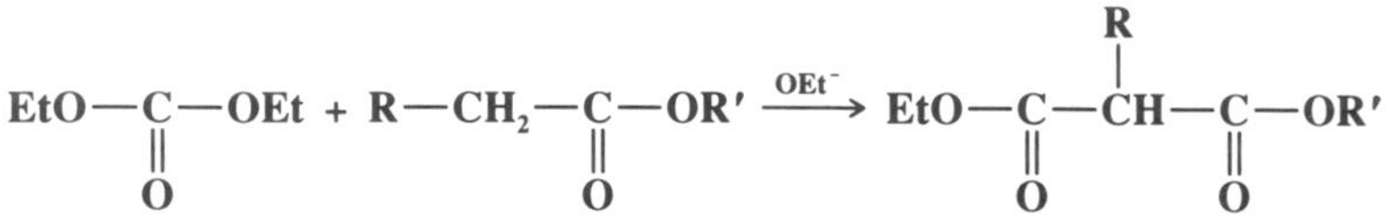

$$EtO-\underset{\underset{O}{\|}}{C}-OEt + R-CH_2-\underset{\underset{O}{\|}}{C}-OR' \xrightarrow{OEt^-} EtO-\underset{\underset{O}{\|}}{C}-\underset{\overset{R}{|}}{C}H-\underset{\underset{O}{\|}}{C}-OR'$$

Ethyl formate serves to introduce the formyl group:

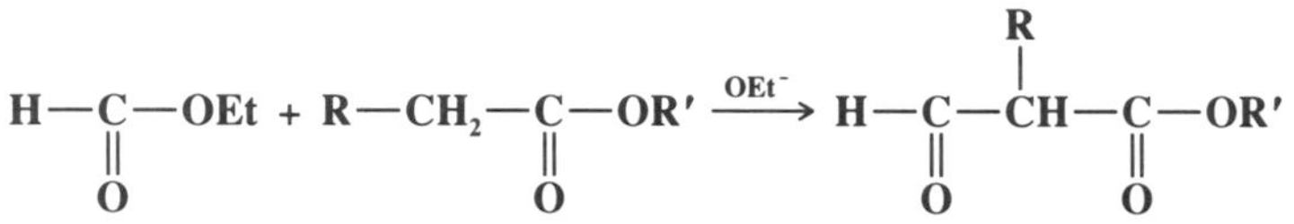

$$H-\underset{\underset{O}{\|}}{C}-OEt + R-CH_2-\underset{\underset{O}{\|}}{C}-OR' \xrightarrow{OEt^-} H-\underset{\underset{O}{\|}}{C}-\underset{\overset{R}{|}}{C}H-\underset{\underset{O}{\|}}{C}-OR'$$

When the two ester groups involved in the condensation are in the same molecule, the product is a cyclic β-keto ester and the reaction is called the *Dieckmann condensation.*[1692]

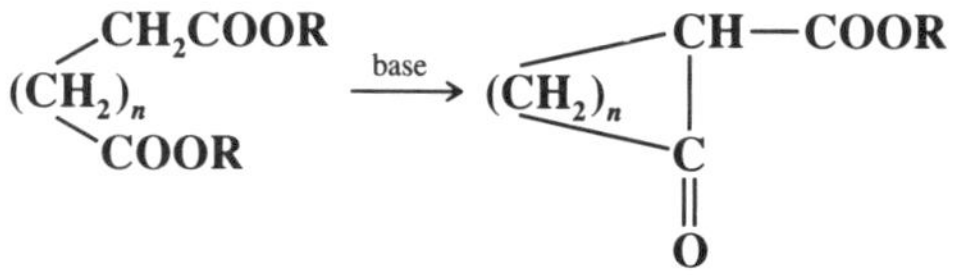

$$(CH_2)_n\begin{matrix}CH_2COOR\\ \\COOR\end{matrix} \xrightarrow{base} (CH_2)_n\begin{matrix}CH-COOR\\ |\\ \underset{\underset{O}{\|}}{C}\end{matrix}$$

The Dieckmann condensation is most successful for the formation of 5-, 6-, and 7-membered rings. Yields for rings of 9 to 12 members are very low or nonexistent; larger rings can be closed with high-dilution techniques. Reactions in which large rings are to be closed are generally assisted by high dilution, since one end of the molecule has a better chance of finding the other end than of finding another molecule. Dieckmann condensation of unsymmetrical substrates can be made regioselective (unidirectional) by the use of solid-phase supports.[1693]

The mechanism of the Claisen and Dieckmann reactions is the ordinary tetrahedral mechanism,[1694] with one molecule of ester being converted to a nucleophile by the base and the other serving as the substrate.

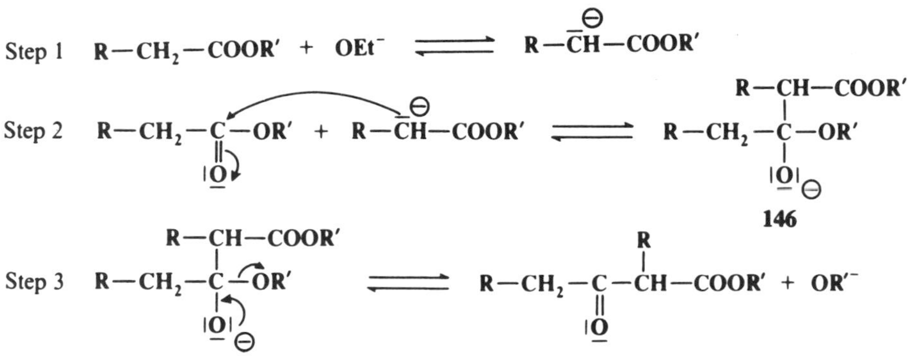

Step 1 $R-CH_2-COOR' + OEt^- \rightleftharpoons R-\overset{\ominus}{C}H-COOR'$

Step 2 $R-CH_2-\underset{\underset{|O|}{\|}}{C}-OR' + R-\overset{\ominus}{C}H-COOR' \rightleftharpoons R-CH_2-\underset{\underset{|O|_\ominus}{|}}{\overset{\overset{R-CH-COOR'}{|}}{C}}-OR'$

146

Step 3 $R-CH_2-\underset{\underset{|O|_\ominus}{|}}{\overset{\overset{R-CH-COOR'}{|}}{C}}-OR' \rightleftharpoons R-CH_2-\underset{\underset{|O}{\|}}{C}-\overset{\overset{R}{|}}{C}H-COOR' + OR'^-$

[1692]For a review, see Schaefer; Bloomfield *Org. React.* **1967**, *15*, 1-203.

[1693]Crowley; Rapoport *J. Org. Chem.* **1980**, *45*, 3215. For another method, see Yamada; Ishii; Kimura; Hosaka *Tetrahedron Lett.* **1981**, *22*, 1353.

[1694]There is evidence that, at least in some cases, an SET mechanism is involved: Ashby; Park *Tetrahedron Lett.* **1983**, 1667.

This reaction illustrates the striking difference in behavior between carboxylic esters on the one hand and aldehydes and ketones on the other. When a carbanion such as an enolate ion is added to the carbonyl group of an aldehyde or ketone (**6-41**), the H or R is not lost, since these groups are much poorer leaving groups than OR. Instead the intermediate similar to **146** adds a proton at the oxygen to give a hydroxy compound.

In contrast to **0-94** ordinary esters react quite well, that is, two Z groups are not needed. A lower degree of acidity is satisfactory because it is not necessary to convert the attacking ester entirely to its ion. Step 1 is an equilibrium that lies well to the left. Nevertheless, the small amount of enolate ion formed is sufficient to attack the readily approachable ester substrate. All the steps are equilibria. The reaction proceeds because the product is converted to its conjugate base by the base present (that is, a β-keto ester is a stronger acid than an alcohol):

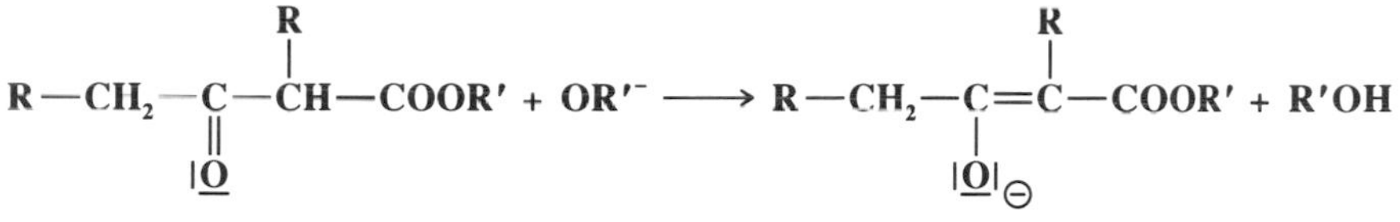

$$R-CH_2-\underset{\underline{|O|}}{\overset{R}{\underset{||}{C}}}-\overset{}{CH}-COOR' + OR'^- \longrightarrow R-CH_2-\underset{\underline{|O|}_\ominus}{\overset{R}{C}}=\overset{}{C}-COOR' + R'OH$$

The use of a stronger base, such as $NaNH_2$, NaH, or KH,[1695] often increases the yield. For some esters stronger bases *must* be used, since sodium ethoxide is ineffective. Among these are esters of the type $R_2CHCOOEt$, the products of which ($R_2CHCOCR_2COOEt$) lack an acidic hydrogen, so that they cannot be converted to enolate ions by sodium ethoxide.[1696]

OS **I**, 235; **II**, 116, 194, 272, 288; **III**, 231, 300, 379, 510; **IV**, 141; **V**, 288, 687, 989; **66**, 52.

0-109 Acylation of Ketones and Nitriles by Carboxylic Esters
α-**Acylalkyl-de-alkoxy-substitution**

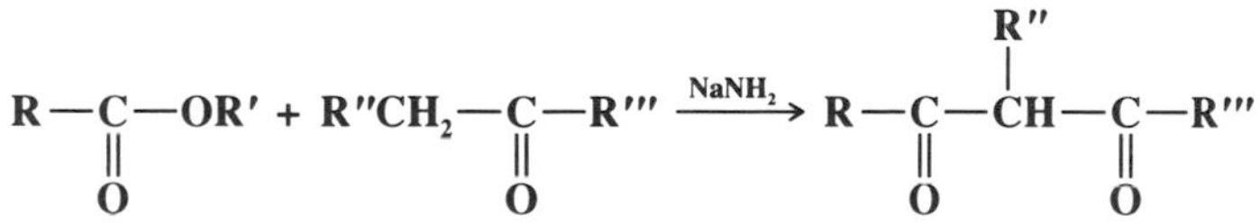

$$R-\underset{O}{\overset{||}{C}}-OR' + R''CH_2-\underset{O}{\overset{||}{C}}-R''' \xrightarrow{NaNH_2} R-\underset{O}{\overset{||}{C}}-\overset{R''}{\underset{}{CH}}-\underset{O}{\overset{||}{C}}-R'''$$

Carboxylic esters can be treated with ketones to give β-diketones in a reaction that is essentially the same as **0-108**. The reaction is so similar that it is sometimes also called the Claisen condensation, though this usage is unfortunate. A fairly strong base, such as sodium amide or sodium hydride, is required. Yields can be increased by the catalytic addition of crown ethers.[1697] Esters of formic acid (R = H) give β-keto aldehydes. Ethyl carbonate gives β-keto esters.

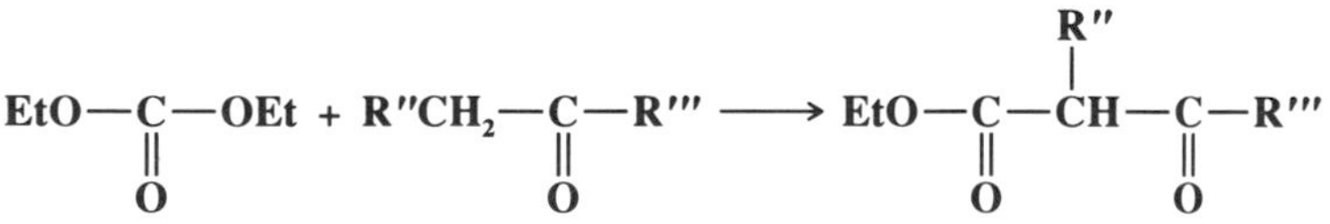

$$EtO-\underset{O}{\overset{||}{C}}-OEt + R''CH_2-\underset{O}{\overset{||}{C}}-R''' \longrightarrow EtO-\underset{O}{\overset{||}{C}}-\overset{R''}{\underset{}{CH}}-\underset{O}{\overset{||}{C}}-R'''$$

[1695]Brown *Synthesis* **1975**, 326.
[1696]For a discussion, see Garst *J. Chem. Educ.* **1979**, *56*, 721.
[1697]Popik; Nikolaev *J. Org. Chem. USSR* **1989**, *25*, 1636.

β-Keto esters can also be obtained by treating the lithium enolates of ketones with methyl cyanoformate $MeOCOCN$[1698] (in this case CN is the leaving group) and by treating ketones with KH and diethyl dicarbonate $(EtOCO)_2O$.[1699]

In the case of unsymmetrical ketones, the attack usually comes from the less highly substituted side, so that CH_3 is more reactive than RCH_2, and the R_2CH group rarely attacks. As in the case of **0-108**, this reaction has been used to effect cyclization, especially to prepare 5- and 6-membered rings. Nitriles are frequently used instead of ketones, the products being β-keto nitriles.

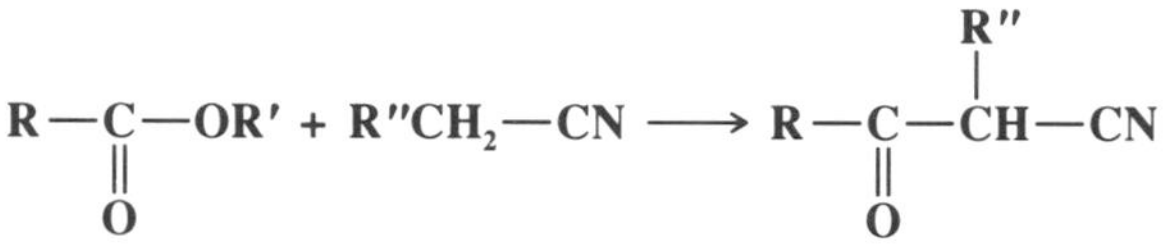

$$R-\underset{\underset{O}{\|}}{C}-OR' + R''CH_2-CN \longrightarrow R-\underset{\underset{O}{\|}}{C}-\overset{\overset{R''}{|}}{CH}-CN$$

Other carbanionic groups, such as acetylide ions, and ions derived from α-methylpyridines have also been used as nucleophiles. A particularly useful nucleophile is the methylsulfinyl carbanion $CH_3SOCH_2^-$,[1700] the conjugate base of dimethyl sulfoxide, since the β-keto sulfoxide produced can easily be reduced to a methyl ketone (p. 465). The methylsulfonyl carbanion $CH_3SO_2CH_2^-$, the conjugate base of dimethyl sulfone, behaves similarly,[1701] and the product can be similarly reduced. Certain carboxylic esters, acyl halides, and dimethylformamide acylate 1,3-dithianes[1702] (see **0-97**) to give, after oxidative hydrolysis with N-bromo- or N-chlorosuccinimide, α-keto aldehydes or α-diketones,[482] e.g.,

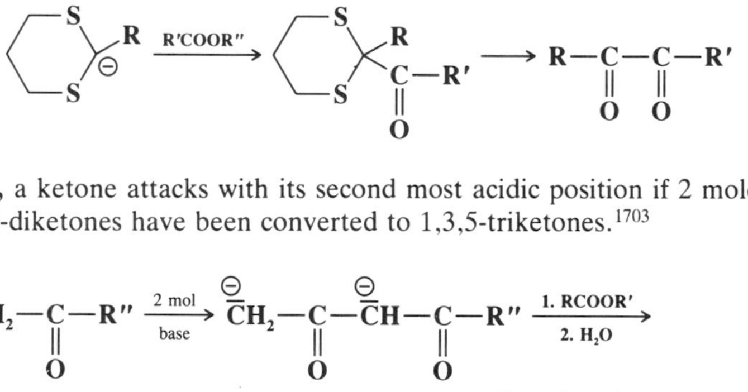

$$\underset{\ominus}{\overset{S}{\underset{S}{\diagdown}}}R \xrightarrow{R'COOR''} \overset{S}{\underset{S}{\diagdown}}\underset{\underset{O}{\|}}{\overset{R}{\underset{C-R'}{\diagdown}}} \longrightarrow R-\underset{\underset{O}{\|}}{C}-\underset{\underset{O}{\|}}{C}-R'$$

As in **0-94**, a ketone attacks with its second most acidic position if 2 moles of base are used. Thus, β-diketones have been converted to 1,3,5-triketones.[1703]

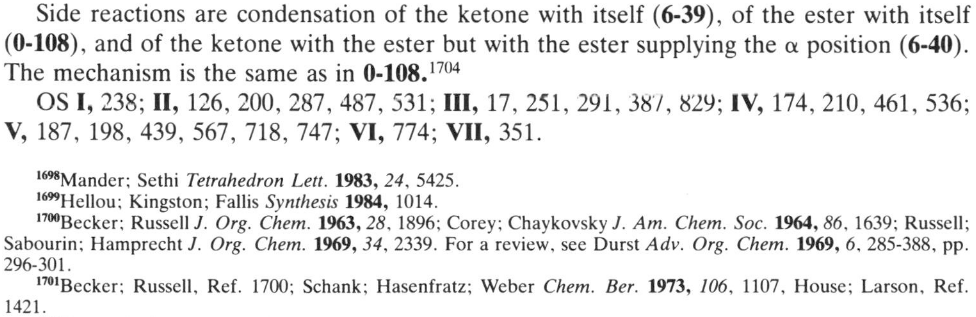

$$CH_3-\underset{\underset{O}{\|}}{C}-CH_2-\underset{\underset{O}{\|}}{C}-R'' \xrightarrow[\text{base}]{2\text{ mol}} \overset{\ominus}{C}H_2-\underset{\underset{O}{\|}}{C}-\overset{\ominus}{C}H-\underset{\underset{O}{\|}}{C}-R'' \xrightarrow[\text{2. H}_2\text{O}]{\text{1. RCOOR}'}$$

$$R-\underset{\underset{O}{\|}}{C}-CH_2-\underset{\underset{O}{\|}}{C}-CH_2-\underset{\underset{O}{\|}}{C}-R''$$

Side reactions are condensation of the ketone with itself (**6-39**), of the ester with itself (**0-108**), and of the ketone with the ester but with the ester supplying the α position (**6-40**). The mechanism is the same as in **0-108**.[1704]

OS **I**, 238; **II**, 126, 200, 287, 487, 531; **III**, 17, 251, 291, 387, 829; **IV**, 174, 210, 461, 536; **V**, 187, 198, 439, 567, 718, 747; **VI**, 774; **VII**, 351.

[1698]Mander; Sethi *Tetrahedron Lett.* **1983**, *24*, 5425.

[1699]Hellou; Kingston; Fallis *Synthesis* **1984**, 1014.

[1700]Becker; Russell *J. Org. Chem.* **1963**, *28*, 1896; Corey; Chaykovsky *J. Am. Chem. Soc.* **1964**, *86*, 1639; Russell; Sabourin; Hamprecht *J. Org. Chem.* **1969**, *34*, 2339. For a review, see Durst *Adv. Org. Chem.* **1969**, *6*, 285-388, pp. 296-301.

[1701]Becker; Russell, Ref. 1700; Schank; Hasenfratz; Weber *Chem. Ber.* **1973**, *106*, 1107, House; Larson, Ref. 1421.

[1702]Corey; Seebach, Ref. 1501.

[1703]Miles; Harris; Hauser *J. Org. Chem.* **1965**, *30*, 1007.

[1704]Hill; Burkus; Hauser *J. Am. Chem. Soc.* **1959**, *81*, 602.

0-110 Acylation of Carboxylic Acid Salts
α-Carboxyalkyl-de-alkoxy-substitution

$$\text{RCH}_2\text{COO}^- \xrightarrow{\text{(i-Pr)}_2\text{NLi}} \text{R}\overset{\ominus}{\text{C}}\text{HCOO}^{\ominus} \xrightarrow{\text{R'COOMe}} \underset{\underset{\text{O}}{\parallel}}{\text{R'}-\text{C}}-\underset{\underset{\text{R}}{|}}{\text{CH}}-\text{COO}^-$$

We have previously seen (**0-96**) that dianions of carboxylic acids can be alkylated in the α position. These ions can also be acylated on treatment with a carboxylic ester[1705] to give salts of β-keto acids. As in **0-96**, the carboxylic acid can be of the form RCH$_2$COOH or RR″CHCOOH. Since β-keto acids are so easily converted to ketones (**2-40**), this is also a method for the preparation of ketones R′COCH$_2$R and R′COCHRR″, where R′ can be primary, secondary, or tertiary alkyl, or aryl. If the ester is ethyl formate, an α-formyl carboxylate salt (R′ = H) is formed, which on acidification spontaneously decarboxylates into an aldehyde.[1706] This is a method, therefore, for achieving the conversion RCH$_2$COOH → RCH$_2$CHO, and as such is an alternative to the reduction methods discussed in **0-83**. When the carboxylic acid is of the form RR″CHCOOH, better yields are obtained by acylating with acyl halides rather than esters.[1707]

0-111 Preparation of Acyl Cyanides
Cyano-de-halogenation

$$\text{RCOX} + \text{CuCN} \longrightarrow \text{RCOCN}$$

Acyl cyanides[1708] can be prepared by treatment of acyl halides with copper cyanide. The mechanism is not known and might be free-radical or nucleophilic substitution. The reaction has also been accomplished with thallium(I) cyanide,[1709] with Me$_3$SiCN and an SnCl$_4$ catalyst,[1710] and with Bu$_3$SnCN,[1711] but these reagents are successful only when R = aryl or tertiary alkyl. KCN has also been used, along with ultrasound,[1712] as has NaCN with phase transfer catalysts.[1713]

OS **III**, 119.

0-112 Preparation of Diazo Ketones
Diazomethyl-de-halogenation

$$\text{RCOX} + \text{CH}_2\text{N}_2 \longrightarrow \text{RCOCHN}_2$$

The reaction between acyl halides and diazomethane is of wide scope and is the best way to prepare diazo ketones.[1714] Diazomethane must be present in excess or the HX produced will react with the diazo ketone (**0-71**). This reaction is the first step of the Arndt–Eistert synthesis (**8-8**). Diazo ketones can also be prepared directly from a carboxylic acid and diazomethane or diazoethane in the presence of dicyclohexylcarbodiimide.[1715]

OS **III**, 119; **VI**, 386, 613; **69**, 180.

[1705]Kuo; Yahner; Ainsworth *J. Am. Chem. Soc.* **1971**, *93*, 6321; Angelo *C.R. Seances Acad. Sci., Ser. C* **1973**, *276*, 293.
[1706]Pfeffer; Silbert *Tetrahedron Lett.* **1970**, 699; Koch; Kop *Tetrahedron Lett.* **1974**, 603.
[1707]Krapcho; Kashdan; Jahngen; Lovey *J. Org. Chem.* **1977**, *42*, 1189; Lion; Dubois *J. Chem. Res., (S)* **1980**, 44.
[1708]For a review of acyl cyanides, see Hünig; Schaller *Angew. Chem. Int. Ed. Engl.* **1982**, *21*, 36-49 [*Angew. Chem.* *94*, 1-15].
[1709]Taylor; Andrade; John; McKillop *J. Org. Chem.* **1978**, *43*, 2280.
[1710]Olah; Arvanaghi; Prakash *Synthesis* **1983**, 636.
[1711]Tanaka *Tetrahedron Lett.* **1980**, *21*, 2959. See also Tanaka; Koyanagi *Synthesis* **1981**, 973.
[1712]Ando; Kawate; Yamawaki; Hanafusa *Synthesis* **1983**, 637.
[1713]Koenig; Weber *Tetrahedron Lett.* **1974**, 2275. See also Sukata *Bull. Chem. Soc. Jpn.* **1987**, *60*, 1085.
[1714]For reviews, see Fridman; Ismagilova; Zalesov; Novikov *Russ. Chem. Rev.* **1972**, *41*, 371-389; Ried; Mengler *Fortshr. Chem. Forsch* **1965**, *5*, 1-88.
[1715]Hodson; Holt; Wall *J. Chem. Soc. C* **1970**, 971.

0-113   Ketonic Decarboxylation[1716]
Alkyl-de-hydroxylation

$$2RCOOH \xrightarrow[\text{ThO}_2]{400\text{-}500^\circ\text{C}} RCOR + CO_2$$

Carboxylic acids can be converted to symmetrical ketones by pyrolysis in the presence of thorium oxide. In a mixed reaction, formic acid and another acid heated over thorium oxide give aldehydes. Mixed alkyl aryl ketones have been prepared by heating mixtures of ferrous salts.[1717] When the R group is large, the methyl ester rather than the acid can be decarb-methoxylated over thorium oxide to give the symmetrical ketone.

The reaction has been performed on dicarboxylic acids, whereupon cyclic ketones are obtained:

$$\begin{matrix} \diagup COOH \\ (CH_2)_n \\ \diagdown COOH \end{matrix} \xrightarrow[\Delta]{ThO_2} (CH_2)_n \; CO$$

This process, called *Ruzicka cyclization*, is good for the preparation of rings of 6 and 7 members and, with lower yields, of C_8 and C_{10} to C_{30} cyclic ketones.[1718]

Not much work has been done on the mechanism of this reaction. However, a free-radical mechanism has been suggested on the basis of a thorough study of all the side products.[1719]

OS **I**, 192; **II**, 389; **IV**, 854; **V**, 589. Also see OS **IV**, 55, 560.

Nucleophilic Substitution at a Sulfonyl Sulfur Atom[1720]

Nucleophilic substitution at RSO_2X is similar to attack at RCOX. Many of the reactions are essentially the same, though sulfonyl halides are less reactive than halides of carboxylic acids.[1721] The mechanisms[1722] are not identical, because a "tetrahedral" intermediate in this case (**147**) would have five groups on the central atom. Though this is possible (since sulfur

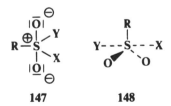

147   **148**

can accommodate up to 12 electrons in its valence shell) it seems more likely that these mechanisms more closely resemble the SN2 mechanism, with a trigonal bipyramidal transition state (**148**). There are two major experimental results leading to this conclusion.

[1716]For a review, see Kwart; King, in Patai, Ref. 197, pp. 362-370.

[1717]Granito; Schultz *J. Org. Chem.* **1963**, *28*, 879.

[1718]See, for example, Ruzicka; Stoll; Schinz *Helv. Chim. Acta* **1926**, *9*, 249, **1928**, *11*, 1174; Ruzicka; Brugger; Seidel; Schinz *Helv. Chim. Acta* **1928**, *11*, 496.

[1719]Hites; Biemann *J. Am. Chem. Soc.* **1972**, *94*, 5772. See also Bouchoule; Blanchard; Thomassin *Bull. Soc. Chim. Fr.* **1973**, 1773.

[1720]For a review of mechanisms of nucleophilic substitutions at di-, tri-, and tetracoordinated sulfur atoms, see Ciuffarin; Fava *Prog. Phys. Org. Chem.* **1968**, *6*, 81-109.

[1721]For a comparative reactivity study, see Hirata; Kiyan; Miller *Bull. Soc. Chim. Fr.* **1988**, 694.

[1722]For a review of mechanisms of nucleophilic substitution at a sulfonyl sulfur, see Gordon; Maskill; Ruasse *Chem. Soc. Rev.* **1989**, *18*, 123-151.

1. The stereospecificity of this reaction is more difficult to determine than that of nucleophilic substitution at a saturated carbon, where chiral compounds are relatively easy to prepare, but it may be recalled (p. 98) that optical activity is possible in a compound of the form RSO_2X if one oxygen is ^{16}O and the other ^{18}O. When a sulfonate ester possessing this type of chirality was converted to a sulfone with a Grignard reagent (**0-119**), inversion of configuration was found.[1723] This is not incompatible with an intermediate such as **147** but it is also in good accord with an S_N2-like mechanism with backside attack.

2. More direct evidence against **147** (though still not conclusive) was found in an experiment involving acidic and basic hydrolysis of aryl arenesulfonates, where it has been shown by the use of ^{18}O that an intermediate like **147** is not reversibly formed, since ester recovered when the reaction was stopped before completion contained no ^{18}O when the hydrolysis was carried out in the presence of labeled water.[1724]

Other evidence favoring the S_N2-like mechanism comes from kinetics and substituent effects.[1725] However, evidence for the mechanism involving **147** is that the rates did not change much with changes in the leaving group[1726] and the ρ values were large, indicating that a negative charge builds up in the transition state.[1727]

In certain cases in which the substrate carries an α hydrogen, there is strong evidence[1728] that at least some of the reaction takes place by an elimination–addition mechanism (E1cB, similar to the one shown on p. 382), going through a *sulfene* intermediate,[1729] e.g., the reaction between methanesulfonyl chloride and aniline.

$$CH_3-SO_2Cl \xrightarrow{\text{base}} CH_2\!\!=\!\!SO_2 \xrightarrow{PhNH_2} CH_3-SO_2-NHPh$$

A sulfene

In the special case of nucleophilic substitution at a sulfonic ester RSO_2OR', where R' is alkyl, R'—O cleavage is much more likely than S—O cleavage because the OSO_2R group is such a good leaving group (p. 353).[1730] Many of these reactions have been considered previously (e.g., **0-4, 0-14**, etc.), because they are nucleophilic substitutions at an alkyl carbon atom and not at a sulfur atom. However, when R' is aryl, then the S—O bond is much more likely to cleave because of the very low tendency aryl substrates have for nucleophilic substitution.[1731]

[1723]Sabol; Andersen *J. Am. Chem. Soc.* **1969**, *91*, 3603. See also Jones; Cram *J. Am. Chem. Soc.* **1974**, *96*, 2183.
[1724]Christman; Oae *Chem. Ind.* (*London*) **1959**, 1251; Oae; Fukumoto; Kiritani *Bull. Chem. Soc. Jpn.* **1963**, *36*, 346; Kaiser; Zaborsky *J. Am. Chem. Soc.* **1968**, *90*, 4626.
[1725]See, for example, Robertson; Rossall *Can. J. Chem.* **1971**, *49*, 1441; Rogne *J. Chem. Soc. B* **1971**, 1855, *J. Chem. Soc., Perkin Trans 2.* **1972**, 489; Gnedin; Ivanov; Spryskov *J. Org. Chem. USSR* **1976**, *12*, 1894; Banjoko; Okwuiwe *J. Org. Chem.* **1980**, *45*, 4966; Ballistreri; Cantone; Maccarone; Tomaselli; Tripolone *J. Chem. Soc., Perkin Trans. 2* **1981**, 438; Suttle; Williams *J. Chem. Soc., Perkin Trans. 2* **1983**, 1563; D'Rozario; Smyth; Williams *J. Am. Chem. Soc.* **1984**, *106*, 5027; Lee; Kang; Lee *J. Am. Chem. Soc.* **1987**, *109*, 7472; Arcoria; Ballistreri; Spina; Tomaselli; Maccarone *J. Chem. Soc., Perkin Trans. 2* **1988**, 1793; Gnedin; Ivanov; Shchukina *J. Org. Chem. USSR* **1988**, *24*, 731.
[1726]Ciuffarin; Senatore; Isola *J. Chem. Soc., Perkin Trans. 2* **1972**, 468.
[1727]Ciuffarin; Senatore *Tetrahedron Lett.* **1974**, 1635.
[1728]For a review, see Opitz *Angew. Chem. Int. Ed. Engl.* **1967**, *6*, 107-123 [*Angew. Chem. 79*, 161-177]. See also King; Lee *J. Am. Chem. Soc.* **1969**, *91*, 6524; Skrypnik; Bezrodnyi *Doklad. Chem.* **1982**, *266*, 341; Farng; Kice *J. Am. Chem. Soc.* **1981**, *103*, 1137; Thea; Guanti; Hopkins; Williams *J. Am. Chem. Soc.* **1982**, *104*, 1128, *J. Org. Chem.* **1985**, *50*, 5592; Bezrodnyi; Skrypnik *J. Chem. Soc. USSR* **1984**, *20*, 1660, 2349; King; Skonieczny *Tetrahedron Lett.* **1987**, *28*, 5001; Pregel; Buncel *J. Chem. Soc., Perkin Trans. 2* **1991**, 307.
[1729]For reviews of sulfenes, see King *Acc. Chem. Res.* **1975**, *8*, 10-17; Nagai; Tokura *Int. J. Sulfur Chem., Part B* **1972**, 207-216; Truce; Liu *Mech. React. Sulfur Compd.* **1969**, *4*, 145-154; Opitz *Angew. Chem. Int. Ed. Engl.* **1967**, *6*, 107-123 [*Angew. Chem. 79*, 161-177]; Wallace *Q. Rev. Chem. Soc.* **1966**, *20*, 67-74.
[1730]A number of sulfonates in which R contains α branching, e.g., $Ph_2C(CF_3)SO_2OR'$, can be used to ensure that there will be no S—O cleavage: Netscher; Prinzbach *Synthesis* **1987**, 683.
[1731]See, for example, Oae; Fukumoto; Kiritani *Bull. Chem. Soc. Jpn.* **1963**, *36*, 346; Tagaki; Kurusu; Oae *Bull. Chem. Soc. Jpn.* **1969**, *42*, 2894.

The order of nucleophilicity toward a sulfonyl sulfur has been reported as $OH^- >$ $RNH_2 > N_3^- > F^- > AcO^- > Cl^- > H_2O > I^-$.[1732] This order is similar to that at a carbonyl carbon (p. 351). Both of these substrates can be regarded as relatively hard acids, compared to a saturated carbon which is considerably softer and which has a different order of nucleophilicity (p. 350).

0-114   Attack by OH. Hydrolysis of Sulfonic Acid Derivatives
S-Hydroxy-de-chlorination, etc.

$$RSO_2Cl \xrightarrow{H_2O} RSO_2OH$$

$$RSO_2OR' \xrightarrow[H+]{H_2O} RSO_2OH$$

$$RSO_2NR_2' \xrightarrow[H+]{H_2O} RSO_2OH$$

Sulfonyl chlorides as well as esters and amides of sulfonic acids can be hydrolyzed to the corresponding acids. Sulfonyl chlorides can by hydrolyzed with water or with an alcohol in the absence of acid or base. Basic catalysis is also used, though of course the salt is the product obtained. Esters are readily hydrolyzed, many with water or dilute alkali. This is the same reaction as **0-4**, and usually involves R'—O cleavage, except when R' is aryl. However, in some cases retention of configuration has been shown at alkyl R', indicating S—O cleavage in these cases.[1733] Sulfonamides are generally not hydrolyzed by alkaline treatment, not even with hot concentrated alkali. Acids, however, do hydrolyze them, though less readily than they do sulfonyl halides or sulfonic esters. Of course, ammonia or the amine appears as the salt. However, sulfonamides can be hydrolyzed with base if the solvent is HMPA.[1734]

OS **I**, 14; **II**, 471; **III**, 262; **IV**, 34; **V**, 406; **VI**, 652, 727. Also see OS **V**, 673; **VI**, 1016.

0-115   Attack by OR. Formation of Sulfonic Esters
S-Alkoxy-de-chlorination, etc.

$$RSO_2Cl + R'OH \xrightarrow{base} RSO_2OR'$$

$$RSO_2NR_2'' + R'OH \xrightarrow{base} RSO_2OR' + NHR_2''$$

Sulfonic esters are most frequently prepared by treatment of the corresponding halides with alcohols in the presence of a base. The method is much used for the conversion of alcohols to tosylates, brosylates, and similar sulfonic esters. Both R and R' may be alkyl or aryl. The base is often pyridine, which functions as a nucleophilic catalyst,[1735] as in the similar alcoholysis of carboxylic acyl halides (**0-20**). Primary alcohols react the most rapidly, and it is often possible to sulfonate selectively a primary OH group in a molecule that also contains secondary or tertiary OH groups. The reaction with sulfonamides has been much less frequently used and is limited to N,N-disubstituted sulfonamides; that is, R'' may not be hydrogen. However, within these limits it is a useful reaction. The nucleophile in this case is actually $R'O^-$. However, R'' may be hydrogen (as well as alkyl) if the nucleophile is a phenol, so that the product is RSO_2OAr. Acidic catalysts are used in this case.[1736] Sulfonic acids have been converted directly to sulfonates by treatment with triethyl or trimethyl

[1732]Kice; Kasperek; Patterson *J. Am. Chem. Soc.* **1969**, *91*, 5516; Rogne *J. Chem. Soc. B* **1970**, 1056; Ref. 330.
[1733]Chang *Tetrahedron Lett.* **1964**, 305.
[1734]Cuvigny; Larchevêque *J. Organomet. Chem.* **1974**, *64*, 315.
[1735]Rogne *J. Chem. Soc. B* **1971**, 1334. See also Litvinenko; Shatskaya; Savelova *Doklad. Chem.* **1982**, *265*, 199.
[1736]Klamann; Fabienke *Chem. Ber.* **1960**, *93*, 252.

orthoformate $HC(OR)_3$, without catalyst or solvent;[1737] and with a trialkyl phosphite $P(OR)_3$.[1738]

OS **I**, 145; **III**, 366; **IV**, 753; **VI**, 56, 482, 587, 652; **VII**, 117; **66**, 1; **68**, 188. Also see OS **IV**, 529; **VI**, 324, 757; **VII**, 495; **66**, 185.

0-116 Attack by Nitrogen. Formation of Sulfonamides
S-Amino-de-chlorination

$$RSO_2Cl + NH_3 \longrightarrow RSO_2NH_2$$

The treatment of sulfonyl chlorides with ammonia or amines is the usual way of preparing sulfonamides. Primary amines give N-alkyl sulfonamides, and secondary amines give N,N-dialkyl sulfonamides. The reaction is the basis of the *Hinsberg test* for distinguishing between primary, secondary, and tertiary amines. N-Alkyl sulfonamides, having an acidic hydrogen, are soluble in alkali, while N,N-dialkyl sulfonamides are not. Since tertiary amines are usually recovered unchanged, primary, secondary, andtertiary amines can be told apart. However, the test is limited for at least two reasons.[1739] (1) Many N-alkyl sulfonamides in which the alkyl group has six or more carbons are insoluble in alkali, despite their acidic hydrogen,[1740] so that a primary amine may appear to be a secondary amine. (2) If the reaction conditions are not carefully controlled, tertiary amines may not be recovered unchanged.[1739]

A primary or a secondary amine can be protected by reaction with phenacyl-sulfonyl chloride ($PhCOCH_2SO_2Cl$) to give a sulfonamide $RNHSO_2CH_2COPh$ or $R_2NSO_2CH_2COPh$.[1741] The protecting group can be removed when desired with zinc and acetic acid. Sulfonyl chlorides react with azide ion to give sulfonyl azides RSO_2N_3.[1742]

OS **IV**, 34, 943; **V**, 39, 179, 1055; **VI**, 78, 652; **VII**, 501; **69**, 158. See also OS **VI**, 788.

0-117 Attack by Halogen. Formation of Sulfonyl Halides
S-Halo-de-hydroxylation

$$RSO_2OH + PCl_5 \longrightarrow RSO_2Cl$$

This reaction, parallel with **0-74**, is the standard method for the preparation of sulfonyl halides. Also used are PCl_3 and $SOCl_2$, and sulfonic acid salts can also serve as substrates. Sulfonyl bromides and iodides have been prepared from sulfonyl hydrazides ($ArSO_2NHNH_2$, themselves prepared by **0-116**) by treatment with bromine or iodine.[1743] Sulfonyl fluorides are generally prepared from the chlorides, by halogen exchange.[1744]

OS **I**, 84; **IV**, 571, 693, 846, 937; **V**, 196. See also OS **VII**, 495.

0-118 Attack by Hydrogen. Reduction of Sulfonyl Chlorides
S-Hydro-de-chlorination or **S-Dechlorination**

$$2RSO_2Cl + Zn \longrightarrow (RSO_2)_2Zn \xrightarrow{H^+} 2RSO_2H$$

Sulfinic acids can be prepared by reduction of sulfonyl chlorides. Though mostly done on aromatic sulfonyl chlorides, the reaction has also been applied to alkyl compounds. Besides

[1737]Padmapriya; Just; Lewis *Synth. Commun.* **1985**, *15*, 1057.
[1738]Karaman; Leader; Goldblum; Breuer *Chem. Ind. (London)* **1987**, 857.
[1739]For directions for performing and interpreting the Hinsberg test, see Gambill; Roberts; Shechter *J. Chem. Educ.* **1972**, *49*, 287.
[1740]Fanta; Wang *J. Chem. Educ.* **1964**, *41*, 280.
[1741]Hendrickson; Bergeron *Tetrahedron Lett.* **1970**, 345.
[1742]For an example, see Regitz; Hocker; Liedhegener *Org. Synth.* **V**, 179.
[1743]Poshkus; Herweh; Magnotta *J. Org. Chem.* **1963**, *28*, 2766; Litvinenko; Dadali; Savelova; Krichevtsova *J. Gen. Chem. USSR* **1964**, *34*, 3780.
[1744]See Bianchi; Cate *J. Org. Chem.* **1977**, *42*, 2031, and references cited therein.

zinc, sodium sulfite, hydrazine, sodium sulfide, and other reducing agents have been used. For reduction of sulfonyl chlorides to thiols, see **9-54.**

OS **I,** 7, 492; **IV,** 674.

0-119   Attack by Carbon. Preparation of Sulfones
S-Aryl-de-chlorination

$$\text{ArSO}_2\text{Cl} + \text{Ar}'\text{MgX} \longrightarrow \text{ArSO}_2\text{Ar}'$$

Grignard reagents convert aromatic sulfonyl chlorides or aromatic sulfonates to sulfones. Aromatic sulfonates have also been converted to sulfones with organolithium compounds.[1745] Vinylic and allylic sulfones have been prepared by treatment of sulfonyl chlorides with a vinylic or allylic stannane and a palladium–complex catalyst.[1746] Alkynyl sulfones can be prepared by treatment of sulfonyl chlorides with trimethylsilylalkynes, with an AlCl$_3$ catalyst.[1747]

$$\text{ArSO}_2\text{Cl} + \text{RC}{\equiv}\text{CSiMe}_3 \xrightarrow{\text{AlCl}_3} \text{ArSO}_2{-}\text{C}{\equiv}\text{CR}$$

OS **67,** 149.

[1745]Baarschers *Can. J. Chem.* **1976,** *54,* 3056.
[1746]Labadie *J. Org. Chem.* **1989,** *54,* 2496.
[1747]See Waykole; Paquette *Org. Synth. 67,* 149.

11

AROMATIC
ELECTROPHILIC
SUBSTITUTION

Most substitutions at an aliphatic carbon are nucleophilic. In aromatic systems the situation is reversed, because the high electron density at the aromatic ring attracts positive species and not negative ones. In electrophilic substitutions the attacking species is a positive ion or the positive end of a dipole or induced dipole. The leaving group (the electrofuge) must necessarily depart without its electron pair. In nucleophilic substitutions, the chief leaving groups are those best able to carry the unshared pair: Br^-, H_2O, OTs^-, etc., that is, the weakest bases. In electrophilic substitutions the most important leaving groups are those that can best exist without the pair of electrons necessary to fill the outer shell, that is, the weakest Lewis acids. The most common leaving group in electrophilic aromatic substitutions is the proton.

MECHANISMS

Electrophilic aromatic substitutions are unlike nucleophilic substitutions in that the large majority proceed by just one mechanism with respect to the substrate.[1] In this mechanism, which we call the *arenium ion mechanism*, the electrophile attacks in the first step, giving rise to a positively charged intermediate (the arenium ion), and the leaving group departs in the second step, so there is a resemblance to the tetrahedral mechanism of Chapter 10, but with the charges reversed. The IUPAC designation for this mechanism is $A_E + D_E$. Another mechanism, much less common, consists of the opposite behavior: the leaving group departs *before* the electrophile arrives. This mechanism, the SE1 mechanism, corresponds to the SN1 mechanism of nucleophilic substitution. Simultaneous attack and departure mechanisms (corresponding to SN2) are not found at all. An addition–elimination mechanism has been postulated in one case (see **1-6**).

The Arenium Ion Mechanism[2]

In the arenium ion mechanism the attacking species may be produced in various ways, but what happens to the aromatic ring is basically the same in all cases. For this reason most attention in the study of this mechanism centers around the identity of the attacking entity and how it is produced.

---

[1] For monographs, see Taylor *Electrophilic Aromatic Substitution*; Wiley: New York, 1990; Katritzky; Taylor *Electrophilic Substitution of Heterocycles: Quantitative Aspects* (Vol. 47 of *Adv. Heterocycl. Chem.*); Academic Press: New York, 1990. For a review, see Taylor, in Bamford; Tipper *Comprehensive Chemical Kinetics*, vol. 13; Elsevier: New York, 1972, pp. 1-406.

[2] This mechanism is sometimes called the SE2 mechanism because it is bimolecular, but in this book we reserve that name for aliphatic substrates (see Chapter 12).

The electrophile may be a positive ion or a dipole. If it is a positive ion, it attacks the ring, removing a pair of electrons from the sextet to give a carbocation, which is a resonance hybrid, as shown in **1**, and is frequently represented as in **2**. Ions of this type are called[3]

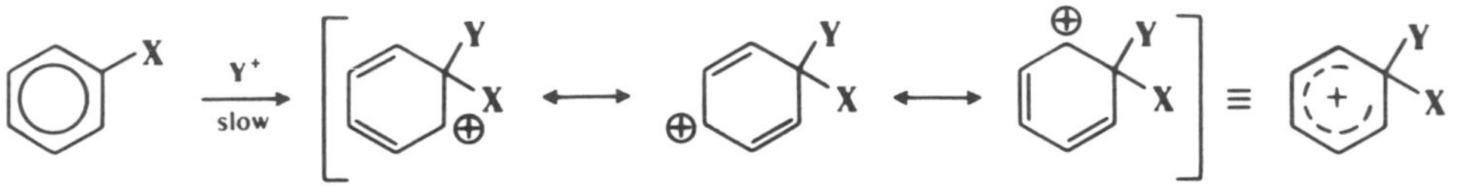

Wheland intermediates, σ complexes, or *arenium ions.*[4] In the case of benzenoid systems they are cyclohexadienyl cations. It is easily seen that the great stability associated with an aromatic sextet is no longer present in **1**, though the ion is stabilized by resonance of its own. The arenium ion is generally a highly reactive intermediate and must stabilize itself by a further reaction, although it has been isolated (see p. 504).

Carbocations can stabilize themselves in various ways (see p. 174), but for this type of ion the most likely way[5] is by loss of either X^+ or Y^+. The aromatic sextet is then restored, and in fact this is the second step of the mechanism:

The second step is nearly always faster than the first, so the first is rate-determining and the reaction is second order (unless the formation of the attacking species is slower still, in which case the aromatic compound does not take part in the rate expression at all). If Y^+ is lost, there is no net reaction, but if X^+ is lost, an aromatic substitution has taken place. If X^+ is a proton, a base is necessary to help remove it.

If the attacking species is not an ion but a dipole, the product must have a negative charge unless part of the dipole, with its pair of electrons, is broken off somewhere in the process, e.g.,

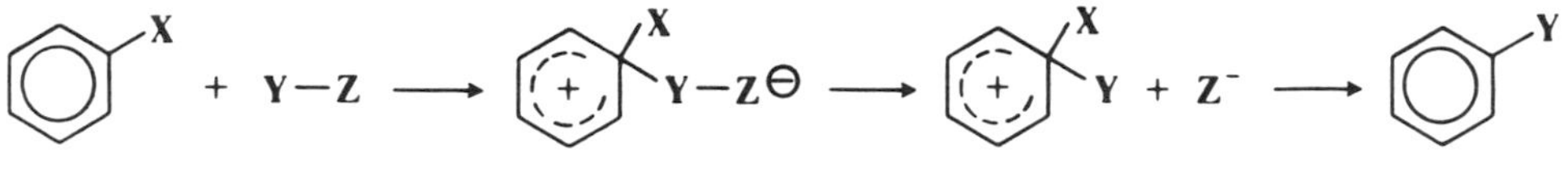

The attacking entity in each case and how it is formed are discussed for each reaction in the reactions section of this chapter.

The evidence for the arenium ion mechanism is mainly of two kinds:

1. *Isotope effects.*   If the hydrogen ion departs before the arrival of the electrophile (S_E1 mechanism) or if the arrival and departure are simultaneous, there should be a substantial isotope effect (i.e., deuterated substrates should undergo substitution more slowly than

[3]General agreement on what to call these ions has not yet been reached. The term σ complex is a holdover from the time when much less was known about the structure of carbocations and it was thought they might be complexes of the type discussed in Chapter 3. Other names have also been used. We will call them arenium ions, following the suggestion of Olah *J. Am. Chem. Soc.* **1971,** *94,* 808.

[4]For reviews of arenium ions formed by addition of a proton to an aromatic ring, see Brouwer; Mackor; MacLean, in Olah; Schleyer *Carbonium Ions,* vol. 2; Wiley: New York, 1970, pp. 837-897; Perkampus *Adv. Phys. Org. Chem.* **1966,** *4,* 195-304.

[5]For a discussion of cases in which **1** stabilizes itself in other ways, see de le Mare *Acc. Chem. Res.* **1974,** *7,* 361-368.

nondeuterated compounds) because, in each case, the C—H bond is broken in the rate-determining step. However, in the arenium ion mechanism, the C—H bond is not broken in the rate-determining step, so no isotope effect should be found. Many such studies have been carried out and, in most cases, especially in the case of nitrations, there is no isotope effect.[6] This result is incompatible with either the SE1 or the simultaneous mechanism.

However, in many instances, isotope effects have been found. Since the values are generally much lower than expected for either the SE1 or the simultaneous mechanisms (e.g., 1 to 3 for k_H/k_D instead of 6 to 7), we must look elsewhere for the explanation. For the case where hydrogen is the leaving group, the arenium ion mechanism can be summarized:

Step 1
$$ArH + Y^+ \underset{k_{-1}}{\overset{k_1}{\rightleftharpoons}} \overset{\oplus}{Ar}\begin{smallmatrix} H \\ \diagup \\ \diagdown \\ Y \end{smallmatrix}$$

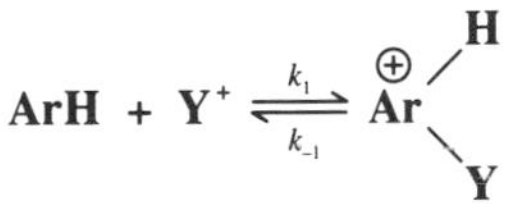

Step 2
$$\overset{\oplus}{Ar}\begin{smallmatrix} H \\ \diagup \\ \diagdown \\ Y \end{smallmatrix} \overset{k_2}{\longrightarrow} ArY + H^+$$

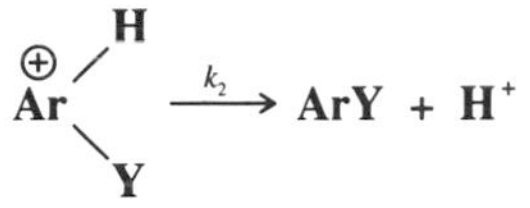

The small isotope effects found most likely arise from the reversibility of step 1 by a *partitioning effect*.[7] The rate at which ArHY$^+$ reverts to ArH should be essentially the same as that at which ArDY$^+$ (or ArTY$^+$) reverts to ArD (or ArT), since the Ar—H bond is not cleaving. However, ArHY$^+$ should go to ArY faster than either ArDY$^+$ or ArTY$^+$, since the Ar—H bond is broken in this step. If $k_2 \gg k_{-1}$, this does not matter; since a large majority of the intermediates go to product, the rate is determined only by the slow step ($k_1[ArH][Y^+]$) and no isotope effect is predicted. However, if $k_2 \gtrsim k_{-1}$, reversion to starting materials is important. If k_2 for ArDY$^+$ (or ArTY$^+$) is less than k_2 for ArHY$^+$, but k_{-1} is the same, then a larger proportion of ArDY$^+$ reverts to starting compounds. That is, k_2/k_{-1} (the *partition factor*) for ArDY$^+$ is less than that for ArHY$^+$. Consequently, the reaction is slower for ArD than for ArH and an isotope effect is observed.

One circumstance that could affect the k_2/k_{-1} ratio is steric hindrance. Thus, diazonium coupling of **3** gave no isotope effect, while coupling of **5** gave a k_H/k_D ratio of 6.55.[8] For

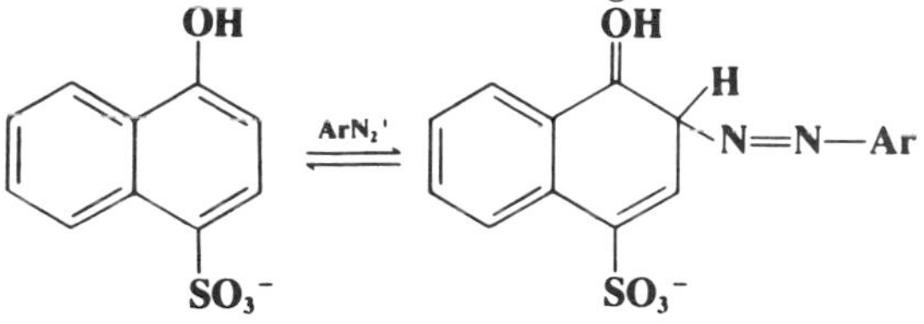

3   **4** 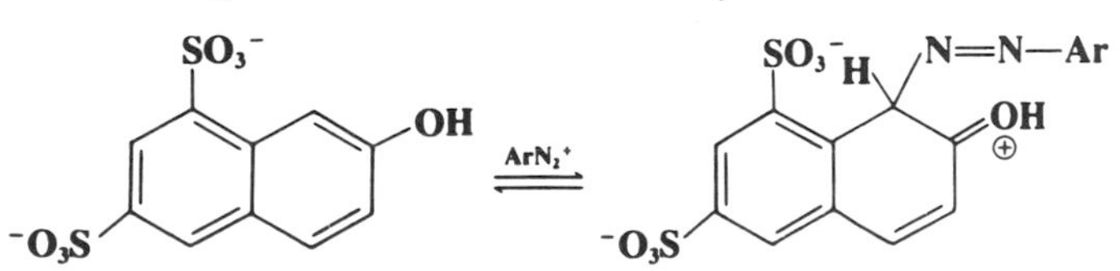   **5**   **6**

---

[6]The pioneering studies were by Melander; Melander *Ark. Kemi* **1950**, *2*, 211; Berglund-Larsson; Melander *Ark. Kemi* **1953**, *6*, 219. See also Zollinger, *Adv. Phys. Org. Chem.* **1964**, *2*, 163-200.

[7]For a discussion, see Hammett *Physical Organic Chemistry*, 2nd ed.; McGraw-Hill: New York, 1970, pp. 172-182.

[8]Zollinger *Helv. Chim. Acta* **1955**, *38*, 1597, 1617, 1623.

steric reasons it is much more difficult for **6** to lose a proton (it is harder for a base to approach) than it is for **4**, so k_2 is greater for the latter. Since no base is necessary to remove ArN_2^+, k_{-1} does not depend on steric factors[9] and is about the same for each. Thus the partition factor k_2/k_{-1} is sufficiently different for **4** and **6** that **5** exhibits a large isotope effect and **3** exhibits none.[10] Base catalysis can also affect the partition factor, since an increase in base concentration increases the rate at which the intermediate goes to product without affecting the rate at which it reverts to starting materials. In some cases, isotope effects can be diminished or eliminated by a sufficiently high concentration of base.

Evidence for the arenium ion mechanism has also been obtained from other kinds of isotope-effect experiments, involving substitutions of the type

$$ArMR_3 + H_3O^+ \longrightarrow ArH + R_3MOH_2^+$$

where M is Si, Ge, Sn, or Pb, and R is methyl or ethyl. In these reactions the proton is the electrophile. If the arenium ion mechanism is operating, then the use of D_3O^+ should give rise to an isotope effect, since the D—O bond would be broken in the rate-determining step. Isotope effects of 1.55 to 3.05 were obtained,[11] in accord with the arenium ion mechanism.

2. *Isolation of arenium ion intermediates.* Very strong evidence for the arenium ion mechanism comes from the isolation of arenium ions in a number of instances.[12] For example, **7** was isolated as a solid with melting point $-15°C$ from treatment of mesitylene with ethyl

mesitylene   **7**   **8**

fluoride and the catalyst BF_3 at $-80°C$. When **7** was heated, the normal substitution product **8** was obtained.[13] Even the simplest such ion, the benzenonium ion (**9**) has been prepared in $HF–SbF_5–SO_2ClF–SO_2F_2$ at $-134°C$, where it could be studied spectrally.[14] ^{13}C nmr

9

[9]Snyckers; Zollinger *Helv. Chim. Acta* **1970**, *53*, 1294.
[10]For some other examples of isotope effects caused by steric factors, see Helgstrand *Acta Chem. Scand.* **1965**, *19*, 1583; Nilsson *Acta Chem. Scand.* **1967**, *21*, 2423; Baciocchi; Illuminati; Sleiter, Stegel *J. Am. Chem. Soc.* **1967**, *89*, 125; Myhre; Beug; James *J. Am. Chem. Soc.* **1968**, *90*, 2105; Dubois; Uzan *Bull. Soc. Chim. Fr.* **1968**, 3534; Márton *Acta Chem. Scand.* **1969**, *23*, 3321, 3329.
[11]Bott; Eaborn; Greasley *J. Chem. Soc.* **1964**, 4803.
[12]For reviews, see Koptyug *Top. Curr. Chem.* **1984**, *122*, 1-245, *Bull. Acad. Sci. USSR, Div. Chem. Sci.* **1974**, *23*, 1031-1045. For a review of polyfluorinated arenium ions, see Shteingarts *Russ. Chem. Rev.* **1981**, *50*, 735-749. For a review of the protonation of benzene and simple alkylbenzenes, see Fǎrcaşiu *Acc. Chem. Res.* **1982**, *15*, 46-51.
[13]Olah; Kuhn *J. Am. Chem. Soc.* **1958**, *80*, 6541. For some other examples, see Ershov; Volod'kin *Bull. Acad. Sci. USSR, Div. Chem. Sci.* **1962**, 680; Farrell; Newton; White *J. Chem. Soc. B* **1967**, 637; Kamshii; Koptyug *Bull. Acad. Sci. USSR, Div. Chem. Sci.* **1974**, *23*, 232; Olah; Spear; Messina; Westerman *J. Am. Chem. Soc.* **1975**, *97*, 4051; Nambu; Hiraoka; Shigemura; Hamanaka; Ogawa *Bull. Chem. Soc. Jpn.* **1976**, *49*, 3637; Chikinev; Bushmelev; Shakirov; Shubin *J. Org. Chem. USSR* **1986**, *22*, 1311; Knoche; Schoeller; Schomäcker; Vogel *J. Am. Chem. Soc.* **1988**, *110*, 7484; Effenberger *Acc. Chem. Res.* **1989**, *22*, 27-35.
[14]Olah; Schlosberg; Porter; Mo; Kelly; Mateescu *J. Am. Chem. Soc.* **1972**, *94*, 2034.

spectra of the benzenonium ion[15] and the pentamethylbenzenonium ion[16] give graphic evidence for the charge distribution shown in **1**. According to this, the 1, 3, and 5 carbons, each of which bears a charge of about $+\frac{1}{3}$, should have a greater chemical shift in the nmr than the 2 and 4 carbons, which are uncharged. The spectra bear this out. For example, ^{13}C nmr chemical shifts for **9** are C-3: 178.1; C-1 and C-5: 186.6; C-2 and C-4: 136.9, and C-6: 52.2.[15]

In Chapter 3 it was mentioned that positive ions can form addition complexes with π systems. Since the initial step of electrophilic substitution involves attack by a positive ion on an aromatic ring, it has been suggested[17] that such a complex, called a π *complex* (represented as **10**), is formed first and then is converted to the arenium ion **11**. Stable solutions of arenium ions or π complexes (e.g., with Br_2, I_2, picric acid, Ag^+, or HCl) can

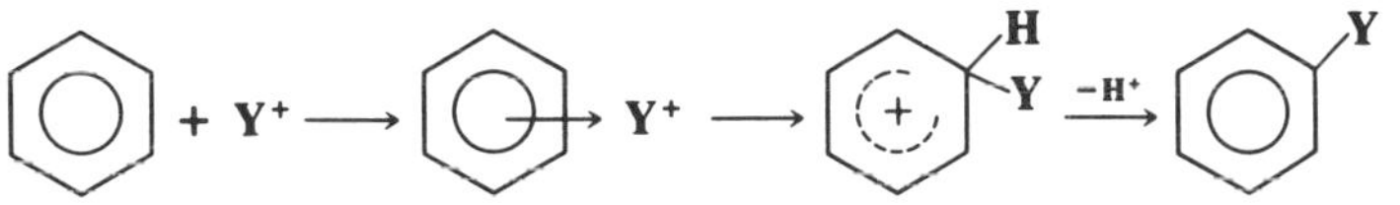

be formed at will. For example, π complexes are formed when aromatic hydrocarbons are treated with HCl alone, but the use of HCl plus a Lewis acid (e.g., $AlCl_3$) gives arenium ions. The two types of solution have very different properties. For example, a solution of an arenium ion is colored and conducts electricity (showing positive and negative ions are present), while a π complex formed from HCl and benzene is colorless and does not conduct a current. Furthermore, when DCl is used to form a π complex, no deuterium exchange takes place (because there is no covalent bond between the electrophile and the ring), while formation of an arenium ion with DCl and $AlCl_3$ gives deuterium exchange. The relative stabilities of some methylated arenium ions and π complexes are shown in Table 11.1. The arenium ion stabilities listed were determined by the relative basicity of the substrate toward HF.[18] The π complex stabilities are relative equilibrium constants for the reaction[19] between

TABLE 11.1 Relative stabilities of arenium ions and π complexes and relative rates of chlorination and nitration

In each case, p-xylene = 1.00

Substituents	Relative arenium ion stability[18]	Relative π-complex stability[18]	Rate of chlorination[19]	Rate of nitration[23]
None (benzene)	0.09	0.61	0.0005	0.51
Me	0.63	0.92	0.157	0.85
p-**Me₂**	1.00	1.00	1.00	1.00
o-**Me₂**	1.1	1.13	2.1	0.89
m-**Me₂**	26	1.26	200	0.84
1,2,4-**Me₃**	63	1.36	340	
1,2,3-**Me₃**	69	1.46	400	
1,2,3,4-**Me₄**	400	1.63	2000	
1,2,3,5-**Me₄**	16,000	1.67	240,000	
Me₅	29,000		360,000	

[15]Olah; Staral; Asencio; Liang; Forsyth; Mateescu *J. Am. Chem. Soc.* **1978**, *100*, 6299.
[16]Lyerla; Yannoni; Bruck; Fyfe *J. Am. Chem. Soc.* **1979**, *101*, 4770.
[17]Dewar *Electronic Theory of Organic Chemistry*; Clarendon Press: Oxford, 1949.
[18]Kilpatrick; Luborsky *J. Am. Chem. Soc.* **1953**, *75*, 577.
[19]Brown; Brady *J. Am. Chem. Soc.* **1952**, *74*, 3570.

the aromatic hydrocarbon and HCl. As shown in Table 11.1, the relative stabilities of the two types of species are very different: the π complex stability changes very little with methyl substitution, but the arenium ion stability changes a great deal.

How can we tell if **10** is present on the reaction path? If it is present, there are two possibilities: (1) The formation of **10** is rate-determining (the conversion of **10** to **11** is much faster), or (2) the formation of **10** is rapid, and the conversion **10** to **11** is rate-determining. One way to ascertain which species is formed in the rate-determining step in a given reaction is to use the stability information given in Table 11.1. We measure the relative rates of reaction of a given electrophile with the series of compounds listed in Table 11.1. If the relative rates resemble the arenium ion stabilities, we conclude that the arenium ion is formed in the slow step; but if they resemble the stabilities of the π complexes, the latter are formed in the slow step.[20] When such experiments are carried out, it is found in most cases that the relative rates are similar to the arenium ion and not to the π complex stabilities. For example, Table 11.1 lists chlorination rates.[19] Similar results were obtained in room-temperature bromination with Br_2 in acetic acid[21] and in acetylation with CH_3CO^+ SbF_6^-.[22] It is clear that in these cases the π complex either does not form at all, or if it does, its formation is not rate-determining (unfortunately, it is very difficult to distinguish between these two possibilities).

On the other hand, in nitration with the powerful electrophile NO_2^+ (in the form of NO_2^+ BF_4^-), the relative rates resembled π complex stabilities much more than arenium ion stabilities (Table 11.1).[23] Similar results were obtained for bromination with Br_2 and $FeCl_3$ in nitromethane. These results were taken to mean[24] that in these cases π complex formation is rate-determining. However, graphical analysis of the NO_2^+ data showed that a straight line could not be drawn when the nitration rate was plotted against π complex stability,[25] which casts doubt on the rate-determining formation of a π complex in this case.[26] There is other evidence, from positional selectivities (discussed on p. 520), that *some* intermediate is present before the arenium ion is formed, whose formation can be rate-determining with powerful electrophiles. Not much is known about this intermediate, which is given the nondescriptive name *encounter complex* and generally depicted as **12.** The arenium complex mechanism is therefore written as[27]

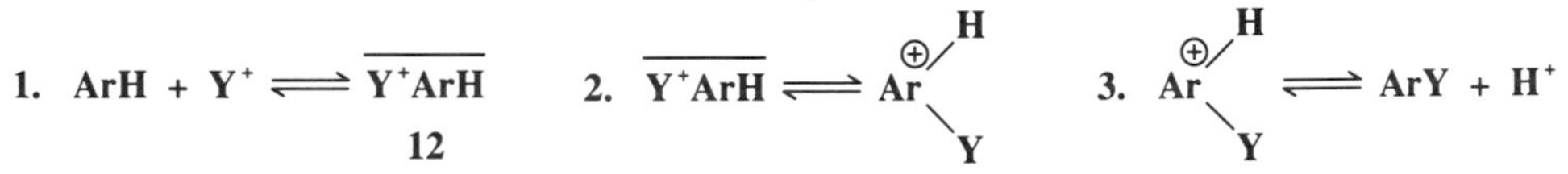

1. $ArH + Y^+ \rightleftharpoons \overline{Y^+ArH}$ 2. $\overline{Y^+ArH} \rightleftharpoons Ar\overset{\oplus}{\underset{Y}{\diagdown}}{}^{H}$ 3. $Ar\overset{\oplus}{\underset{Y}{\diagdown}}{}^{H} \rightleftharpoons ArY + H^+$

          12

[20]Condon *J. Am. Chem. Soc.* **1952**, *74*, 2528.

[21]Brown; Stock *J. Am. Chem. Soc.* **1957**, *79*, 1421.

[22]Olah; Kuhn; Flood; Hardie *J. Am. Chem. Soc.***1964**, *86*, 2203.

[23]Olah; Kuhn; Flood *J. Am. Chem. Soc.* **1961**, *83*, 4571, 4581.

[24]Olah; Kuhn; Flood; Hardie *J. Am. Chem. Soc.* **1964**, *86*, 1039, 1044; Ref. 23.

[25]Rys; Skrabal; Zollinger *Angew. Chem. Int. Ed. Engl.* **1972**, *11*, 874-883 [*Angew. Chem. 84*, 921-930]. See also DeHaan; Covey; Delker; Baker; Feigon; Miller; Stelter *J. Am. Chem. Soc.* **1979**, *101*, 1336; Santiago; Houk; Perrin *J. Am. Chem. Soc.* **1979**, *101*, 1337.

[26]For other evidence against π complexes, see Tolgyesi *Can. J. Chem.* **1965**, *43*, 343; Caille; Corriu *Chem. Commun.* **1967**, 1251, *Tetrahedron* **1969**, *25*, 2005; Coombes; Moodie; Schofield *J. Chem. Soc. B* **1968**, 800; Hoggett; Moodie; Schofield *J. Chem. Soc. B* **1969**, 1; Christy; Ridd; Stears *J. Chem. Soc. B* **1970**, 797; Ridd *Acc. Chem. Res.* **1971**, *4*, 248-253; Taylor; Tewson *J. Chem. Soc., Chem. Commun.* **1973**, 836; Naidenov; Guk; Golod *J. Org. Chem. USSR* **1982**, *18*, 1731. For further support for π complexes, see Olah; Overchuk *Can. J. Chem.* **1965**, *43*, 3279; Olah *Acc. Chem. Res.* **1971**, *4*, 240-248; Olah; Lin *J. Am. Chem. Soc.* **1974**, *96*, 2892; Koptyug; Rogozhnikova; Detsina *J. Org. Chem. USSR* **1983**, *19*, 1007; El-Dusouqui; Mahmud; Sulfab *Tetrahedron Lett.* **1987**, *28*, 2417; Sedaghat-Herati; Sharifi *J. Organomet. Chem.* **1989**, *363*, 39. For an excellent discussion of the whole question, see Banthorpe *Chem. Rev.* **1970**, *70*, 295-322, especially sections VI and IX.

[27]For discussions, see Stock *Prog. Phys. Org. Chem.* **1976**, *12*, 21-47; Ridd *Adv. Phys. Org. Chem.* **1978**, *16*, 1-49.

For the reason given above and for other reasons, it is unlikely that the encounter complex is a π complex, but just what kind of attraction exists between Y^+ and ArH is not known, other than the presumption that they are together within a solvent cage (see also p. 520). There is evidence (from isomerizations occurring in the alkyl group, as well as other observations) that π complexes are present on the pathway from substrate to arenium ion in the gas phase protonation of alkylbenzenes.[28]

The SE1 Mechanism

The SE1 mechanism (*substitution electrophilic unimolecular*) is rare, being found only in certain cases in which carbon is the leaving atom (see **1-38, 1-39**) or when a very strong base is present (see **1-1, 1-11,** and **1-42**).[29] It consists of two steps with an intermediate carbanion. The IUPAC designation is $D_E + A_E$.

Reactions **2-41, 2-45,** and **2-46** also take place by this mechanism when applied to aryl substrates.

ORIENTATION AND REACTIVITY

Orientation and Reactivity in Monosubstituted Benzene Rings[30]

When an electrophilic substitution reaction is performed on a monosubstituted benzene, the new group may be directed primarily to the ortho, meta, or para position and the substitution may be slower or faster than with benzene itself. The group already on the ring determines which position the new group will take and whether the reaction will be slower or faster than with benzene. Groups that increase the reaction rate are called *activating* and those that slow it *deactivating*. Some groups are predominantly meta-directing; all of these are deactivating. Others are mostly ortho-para directing; some of these are deactivating too, but most are activating. Groups direct *predominantly*, but usually not *exclusively*. For example, nitration of nitrobenzene gave 93% *m*-dinitrobenzene, 6% of the ortho, and 1% of the para isomer.

The orientation and reactivity effects of each group are explained on the basis of resonance and field effects on the stability of the intermediate arenium ion. To understand why we can use this approach, it is necessary to know that in these reactions the product is usually kinetically and not thermodynamically controlled (see p. 214). Some of the reactions are irreversible and the others are usually stopped well before equilibrium is reached. Therefore, which of the three possible intermediates is formed is dependent not on the thermodynamic stability of the products but on the activation energy necessary to form each of the three

[28]Holman; Gross *J. Am. Chem. Soc.* **1989,** *111,* 3560.
[29]It has also been found with a metal (SnMe₃) as electrofuge: Eaborn; Hornfeld; Walton *J. Chem. Soc. B* **1967,** 1036.
[30]For a review of orientation and reactivity in benzene and other aromatic rings, see Hoggett; Moodie; Penton; Schofield *Nitration and Aromatic Reactivity;* Cambridge University Press: Cambridge, 1971, pp. 122-145, 163-220.

intermediates. It is not easy to predict which of the three activation energies is lowest, but we make the assumption that the free-energy profile resembles either Figure 6.2(a) or (b). In either case, the transition state is closer in energy to the arenium ion intermediate than to the starting compounds. Invoking the Hammond postulate (p. 215), we can then assume that the geometry of the transition state also resembles that of the intermediate and that anything that increases the stability of theintermediate will also lower the activation energy necessary to attain it. Since the intermediate, once formed, is rapidly converted to products, we can use the relative stabilities of the three intermediates as guides to predict which products will predominantly form. Of course, if reversible reactions are allowed to proceed to equilibrium, we may get product ratios that are quite different. For example, the sulfonation of naphthalene at 80°C, where the reaction does not reach equilibrium, gives mostly α-naphthalenesulfonic acid,[31] while at 160°C, where equilibrium is attained, the β isomer predominates[32] (the α isomer is thermodynamically less stable because of steric interaction between the SO$_3$H group and the hydrogen at the 8 position).

These are the three possible ions:

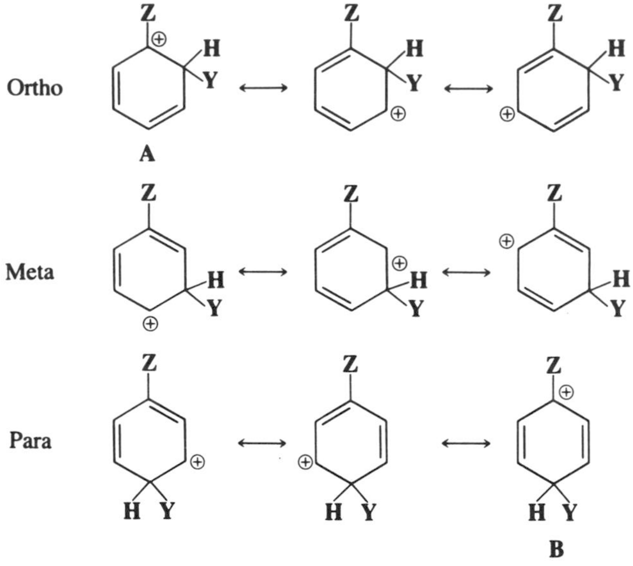

For each ion we see that the ring has a positive charge. We can therefore predict that any group Z that has an electron-donating field effect ($+I$) should stabilize all three ions (relative to **1**), but that electron-withdrawing groups, which increase the positive charge on the ring, should destabilize them. We can also make a further prediction concerning field effects. These taper off with distance and are thus strongest at the carbon connected to the group Z. Of the three arenium ions, only the ortho and para have any positive charge at this carbon. None of the canonical forms of the meta ion has a positive charge there and so the hybrid has none either. Therefore, $+I$ groups should stabilize all three ions but mostly the ortho and para, so they should be not only activating but ortho-para-directing as well. On the other hand, $-I$ groups, by removing electron density, should destabilize all three ions but mostly the ortho and para, and should be not only deactivating but also meta-directing.

These conclusions are correct as far as they go, but they do not lead to the proper results in all cases. In many cases there is *resonance interaction* between Z and the ring; this also

[31]Fierz; Weissenbach *Helv. Chim. Acta* **1920**, *3*, 312.
[32]Witt, *Ber.* **1915**, *48*, 743.

affects the relative stability, in some cases in the same direction as the field effect, in others differently.

Some substituents have a pair of electrons (usually unshared) that may be contributed *toward* the ring. The three arenium ions would then look like this:

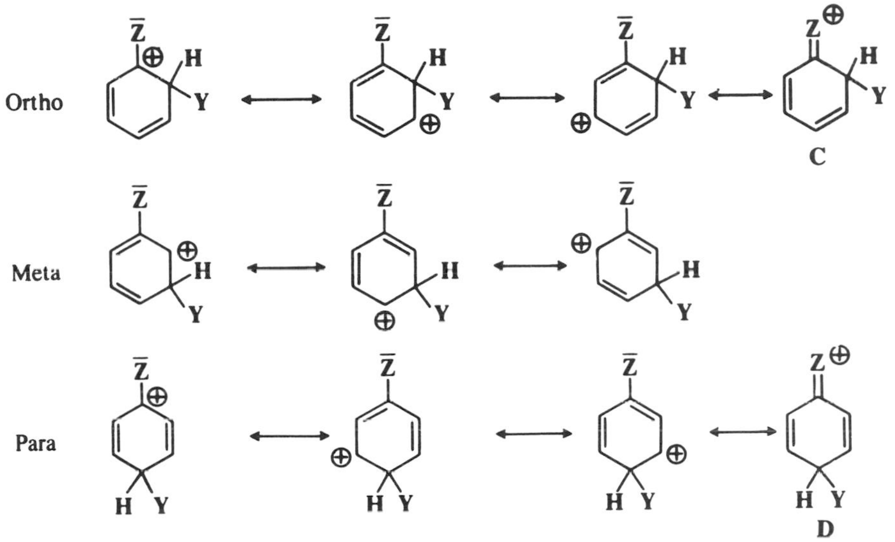

For each ion the same three canonical forms can be drawn as before, but now we can draw an extra form for the ortho and para ions. The stability of these two ions is increased by the extra form not only because it is another canonical form, but because it is more stable than the others and makes a greater contribution to the hybrid. Every atom (except of course hydrogen) in these forms (**C** and **D**) has a complete octet, while all the other forms have one carbon atom with a sextet. No corresponding form can be drawn for the meta isomer. The inclusion of this form in the hybrid lowers the energy not only because of rule 6 (p. 35), but also because it spreads the positive charge over a larger area—out onto the group Z. Groups with a pair of electrons to contibute would be expected, then, in the absence of field effects, not only to direct ortho and para, but also to activate these positions for electrophilic attack.

On the basis of these discussions, we can distinguish three types of groups.

1. Groups that contain an unshared pair of electrons on the atom connected to the ring. In this category are O⁻, NR₂, NHR, NH₂,[33] OH, OR, NHCOR, OCOR, SR, and the four halogens.[34] The SH group would probably belong here too, except that in the case of thiophenols electrophiles usually attack the sulfur rather than the ring, and ring substitution is not feasible with these substrates.[35] The resonance explanation predicts that all these

[33]It must be remembered that in acid solution amines are converted to their conjugate acids, which for the most part are meta-directing (type 2). Therefore in acid (which is the most common medium for electrophilic substitutions) amino groups may direct meta. However, unless the solution is highly acidic, there will be a small amount of free amine present, and since amino groups are activating and the conjugate acids deactivating, ortho-para direction is often found even under acidic conditions.

[34]For a review of the directing and orienting effects of amino groups, see Chuchani, in Patai *The Chemistry of the Amino Group*; Wiley: New York, 1968, pp. 250-265; for ether groups see Kohnstam; Williams, in Patai *The Chemistry of the Ether Linkage*; Wiley: New York, 1967, pp. 132-150.

[35]Tarbell; Herz *J. Am. Chem. Soc.* **1953**, *75*, 4657. Ring substitution *is* possible if the SH group is protected. For a method of doing this, see Walker *J. Org. Chem.* **1966**, *31*, 835.

groups should be ortho-para-directing, and they are, though all except O$^-$ are electron-withdrawing by the field effect (p. 18). Therefore, for these groups, resonance is more important than the field effect. This is especially true for NR$_2$, NHR, NH$_2$, and OH, which are *strongly* activating, as is O$^-$. The other groups are mildly activating, except for the halogens, which are deactivating. Fluorine is the least deactivating, and fluorobenzenes usually show a reactivity approximating that of benzene itself. The other three halogens deactivate about equally. In order to explain why chlorine, bromine, and iodine deactivate the ring, even though they direct ortho-para, we must assume that the canonical forms **C** and **D** make such great contributions to the respective hybrids that they make the ortho and para arenium ions more stable than the meta, even though the $-I$ effect of the halogen is withdrawing sufficient electron density from the ring to deactivate it. The three halogens make the ortho and para ions more stable than the meta, but less stable than the unsubstituted arenium ion (**1**). For the other groups that contain an unshared pair, the ortho and para ions are more stable than either the meta ion or the unsubstituted ion. For most of the groups in this category, the meta ion is more stable than **1**, so that groups such as NH$_2$, OH, etc. activate the meta positions too, but not as much as the ortho and para positions (see also the discussion on pp. 516-517).

2. Groups that lack an unshared pair on the atom connected to the ring and that are $-I$. In this category are, in approximate order of decreasing deactivating ability, NR$_3^+$, NO$_2$, CF$_3$, CN, SO$_3$H, CHO, COR, COOH, COOR, CONH$_2$, CCl$_3$, and NH$_3^+$. Also in this category are all other groups with a positive charge on the atom directly connected to the ring[36] (SR$_2^+$, PR$_3^+$, etc.) and many groups with positive charges on atoms farther away, since often these are still powerful $-I$ groups. The field-effect explanation predicts that these should all be meta-directing and deactivating, and (except for NH$_3^+$) this is the case. The NH$_3^+$ group is an anomaly, since this group directs para about as much as or a little more than it directs meta.[37] The NH$_2$Me$^+$, NHMe$_2^+$, and NMe$_3^+$ groups all give more meta than para substitution, the percentage of para product decreasing with the increasing number of methyl groups.[38]

3. Groups that lack an unshared pair on the atom connected to the ring and that are ortho-para-directing. In this category are alkyl groups, aryl groups, and the COO$^-$ group,[39] all of which activate the ring. We shall discuss them separately. Since aryl groups are $-I$ groups, they might seem to belong to category 2. They are nevertheless ortho-para-directing and activating. This can be explained in a similar manner as in category 1, with a pair of electrons from the aromatic sextet playing the part played by the unshared pair, so that we have forms like **E**. The effect of negatively charged groups like COO$^-$ is easily explained

E   F

[36]For discussions, see Gastaminza; Modro; Ridd; Utley *J. Chem. Soc. B* **1968**, 534; Gastaminza; Ridd; Roy *J. Chem. Soc. B* **1969**, 684; Gilow; De Shazo; Van Cleave *J. Org. Chem.* **1971**, *36*, 1745; Hoggett; Moodie; Penton; Schofield, Ref. 30, pp. 167-176.

[37]Brickman; Ridd *J. Chem. Soc.* **1965**, 6845; Hartshorn; Ridd *J. Chem. Soc. B.* **1968**, 1063. For a discussion, see Ridd, in *Aromaticity*, Chem. Soc. Spec. Publ. no. 21, 1967, pp. 149-162.

[38]Brickman; Utley; Ridd *J. Chem. Soc* **1965**, 6851.

[39]Spryskov; Golubkin *J. Gen. Chem. USSR* **1961**, *31*, 833. Since the COO$^-$ group is present only in alkaline solution, where electrophilic substitution is not often done, it is seldom met with.

by the field effect (negatively charged groups are of course electron-donating), since there is no resonance interaction between the group and the ring. The effect of alkyl groups can be explained in the same way, but, in addition, we can also draw canonical forms, even though there is no unshared pair. These of course are hyperconjugation forms like **F.** This effect, like the field effect, predicts activation and ortho-para direction, so that it is not possible to say how much each effect contributes to the result. Another way of looking at the effect of alkyl groups (which sums up both field and hyperconjugation effects) is that (for Z = R) the ortho and para arenium ions are more stable because each contains a form (**A** and **B**) that is a tertiary carbocation, while all the canonical forms for the meta ion and for **1** are secondary carbocations. In activating ability, alkyl groups usually follow the Baker–Nathan order (p. 68), but not always.[40]

The Ortho/Para Ratio[41]

When an ortho-para-directing group is on a ring, it is usually difficult to predict how much of the product will be the ortho isomer and how much the para isomer. Indeed, these proportions can depend greatly on the reaction conditions. For example, chlorination of toluene gives an ortho/para ratio anywhere from 62:38 to 34:66.[42] Nevertheless, certain points can be made. On a purely statistical basis there would be 67% ortho and 33% para, since there are two ortho positions and only one para. However, the phenonium ion **9**,

9

which arises from protonation of benzene, has the approximate charge distribution shown.[43] If we accept this as a model for the arenium ion in aromatic substitution, a para substituent would have a greater stabilizing effect on the adjacent carbon than an ortho substituent. If other effects are absent, this would mean that more than 33% para and less than 67% ortho substitution would be found. In hydrogen exchange (reaction **1-1**), where other effects are absent, it has been found for a number of substituents that the average ratio of the logarithms of the partial rate factors for these positions (see p. 516 for a definition of partial rate factor) was close to 0.865,[44] which is not far from the value predicted from the ratio of charge densities in **9**. This picture is further supported by the fact that meta-directing groups, which destabilize a positive charge, give ortho/para ratios greater than 67:33[45] (of course the total amount of ortho and para substitution with these groups is small, but the *ratios* are generally greater than 67:33). Another important factor is the steric effect. If either the group on the ring or the attacking group is large, steric hindrance inhibits formation of the ortho product and increases the amount of the para isomer. An example may be seen in the nitration, under the same conditions, of toluene and *t*-butylbenzene. The former gave 58% of the ortho compound and 37% of the para, while the more bulky *t*-butyl group gave 16% of the

[40]For examples of situations where the Baker–Nathan order is not followed, see Eaborn; Taylor, *J. Chem. Soc.* **1961,** 247; Stock *J. Org. Chem.* **1961,** *26,* 4120; Utley; Vaughan *J. Chem. Soc. B* **1968,** 196; Schubert; Gurka *J. Am. Chem. Soc.* **1969,** *91,* 1443; Himoe; Stock *J. Am. Chem. Soc.* **1969,** *91,* 1452.
[41]For a discussion, see Pearson; Buehler *Synthesis* **1971,** 455-477, pp. 455-464.
[42]Stock; Himoe *J. Am. Chem. Soc.* **1961,** *83,* 4605.
[43]Olah *Acc. Chem. Res.* **1970,** *4,* 240, p. 248.
[44]Bailey; Taylor *J. Chem. Soc. B* **1971,** 1446; Ansell; Le Guen; Taylor *Tetrahedron Lett.* **1973,** 13.
[45]Hoggett; Moodie; Penton; Schofield, Ref. 30, pp. 176-180.

ortho product and 73% of the para.[46] Some groups are so large that they direct almost entirely para.

When the ortho-para-directing group is one with an unshared pair (this of course applies to most of them), there is another effect that increases the amount of para product at the expense of the ortho. A comparison of the intermediates involved (p. 509) shows that **C** is a canonical form with an ortho-quinonoid structure, while **D** has a para-quinonoid structure. Since we know that *para*-quinones are more stable than the ortho isomers, it seems reasonable to assume that **D** is more stable than **C** and therefore contributes more to the hybrid and increases its stability compared to the ortho intermediate.

It has been shown that it is possible to compel regiospecific para substitution by enclosing the substrate molecules in a cavity from which only the para position projects. Anisole was chlorinated in solutions containing a cyclodextrin, a molecule in which the anisole is almost entirely enclosed (see Fig. 3.4). With a high enough concentration of cyclodextrin, it was possible to achieve a para/ortho ratio of 21.6[47] (in the absence of the cyclodextrin the ratio was only 1.48). This behavior is a model for the regioselectivity found in the action of enzymes.

Ipso Attack

We have discussed orientation in the case of monosubstituted benzenes entirely in terms of attack at the ortho, meta, and para positions, but attack at the position bearing the substituent (called the *ipso position*[48]) can also be important. Ipso attack has mostly been studied for nitration.[49] When NO_2^+ attacks at the ipso position there are at least five possible fates for the resulting arenium ion (**13**).

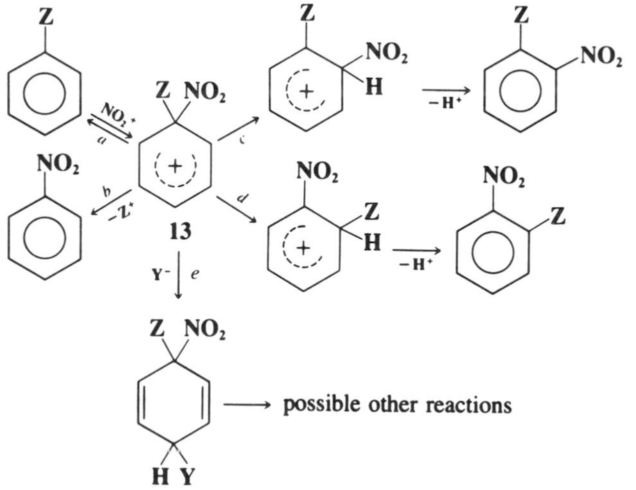

[46]Nelson; Brown *J. Am. Chem. Soc.* **1951**, *73*, 5605. For product ratios in the nitration of many monoalkylbenzenes, see Baas; Wepster *Recl. Trav. Chim. Pays-Bas* **1971**, *90*, 1081, 1089, **1972**, *91*, 285, 517, 831.

[47]Breslow; Campbell *J. Am. Chem. Soc.* **1969**, *91*, 3085, *Bioorg. Chem.* **1971**, *1*, 140. See also Chen; Kaeding; Dwyer *J. Am. Chem. Soc.* **1979**, *101*, 6783; Konishi; Yokota; Ichihashi; Okano; Kiji *Chem. Lett.* **1980**, 1423; Komiyama; Hirai *J. Am. Chem. Soc.* **1983**, *105*, 2018, **1984**, *106*, 174; Chênevert; Ampleman *Can. J. Chem.* **1987**, *65*, 307; Komiyama *Polym. J. (Tokyo)* **1988**, *20*, 439.

[48]Perrin; Skinner *J. Am. Chem. Soc.* **1971**, *93*, 3389. For a review of ipso substitution, see Traynham *J. Chem. Educ.* **1983**, *60*, 937-941.

[49]For a review, see Moodie; Schofield *Acc. Chem. Res.* **1976**, *9*, 287-292. See also Fischer; Henderson; RayMahasay *Can. J. Chem.* **1987**, *65*, 1233, and other papers in this series.

Path a. The arenium ion can lose NO_2^+ and revert to the starting compounds. This results in no net reaction and is often undetectable.

Path b. The arenium ion can lose Z^+, in which case this is simply aromatic substitution with a leaving group other than H (see **1-37** to **1-44**).

Path c. The electrophilic group (in this case NO_2^+) can undergo a 1,2-migration, followed by loss of the proton. The product in this case is the same as that obtained by direct attack of NO_2^+ at the ortho position of PhZ. It is not always easy to tell how much of the ortho product in any individual case arises from this pathway,[50] though there is evidence that it can be a considerable proportion. Because of this possibility, many of the reported conclusions about the relative reactivity of the ortho, meta, and para positions are cast into doubt, since some of the product may have arisen not from direct attack at the ortho position, but from attack at the ipso position followed by rearrangement.[51]

Path d. The ipso substituent (Z) can undergo 1,2-migration, which also produces the ortho product (though the rearrangement would become apparent if there were other substituents present). The evidence is that this pathway is very minor, at least when the electrophile is NO_2^+.[52]

Path e. Attack of a nucleophile on **13**. In some cases the products of such an attack (cyclohexadienes) have been isolated[53] (this is 1,4-addition to the aromatic ring), but further reactions are also possible.

Orientation in Benzene Rings with More than One Substituent[54]

It is often possible in these cases to predict the correct isomer. In many cases the groups already on the ring reinforce each other. Thus, 1,3-dimethylbenzene is substituted at the 4 position (ortho to one group and para to the other), but not at the 5 position (meta to both). Likewise the incoming group in *p*-chlorobenzoic acid goes to the position ortho to the chloro and meta to the carboxyl group.

When the groups oppose each other, predictions may be more difficult. In a case such as

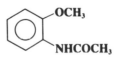

where two groups of about equal directing ability are in competing positions, all four products can be expected, and it is not easy to predict the proportions, except that steric hindrance should probably reduce the yield of substitution ortho to the acetamido group, especially for large electrophiles. Mixtures of about equal proportions are frequent in such cases. Nevertheless, even when groups on a ring oppose each other, there are some regularities.

1. If a strong activating group competes with a weaker one or with a deactivating group, the former controls. Thus *o*-cresol gives substitution mainly ortho and para to the *hydroxyl* group and not to the methyl. For this purpose we can arrange the groups in the following

[50]For methods of doing so, see Gibbs; Moodie; Schofield *J. Chem. Soc., Perkin Trans. 2* **1978**, 1145.

[51]This was first pointed out by Myhre *J. Am. Chem. Soc.* **1972**, *94*, 7921.

[52]For examples of such migration, where Z = Me, see Hartshorn; Readman; Robinson; Sies; Wright *Aust. J. Chem.* **1988**, *41*, 373.

[53]For examples, see Banwell; Morse; Myhre; Vollmar *J. Am. Chem. Soc.* **1977**, *99*, 3042; Fischer; Greig *Can. J. Chem.* **1978**, *56*, 1063.

[54]For a quantitative discussion, see pp. 516-517.

order: NH_2, OH, NR_2, O^- > OR, OCOR, NHCOR > R, Ar > halogen > meta-directing groups.

2. All other things being equal, a third group is least likely to enter between two groups in the meta relationship. This is the result of steric hindrance and increases in importance with the size of the groups on the ring and with the size of the attacking species.[55]

3. When a meta-directing group is meta to an ortho-para-directing group, the incoming group primarily goes ortho to the meta-directing group rather than para. For example, chlorination of **14** gives mostly **15**. The importance of this effect is underscored by the fact that **16,** which is in violation of the preceding rule, is formed in smaller amounts, but **17** is

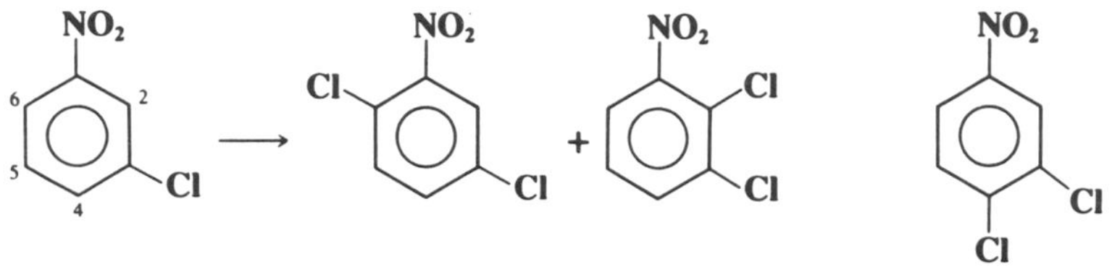

| 14 | 15 | 16 | 17 |

not formed at all. This is called the *ortho effect*,[56] and many such examples are known.[57] Another is the nitration of *p*-bromotoluene, which gives 2,3-dinitro-4-bromotoluene. In this case, once the first nitro group came in, the second was directed ortho to it rather than para, even though this means that the group has to come in between two groups in the meta position. There is no good explanation yet for the ortho effect, though possibly there is intramolecular assistance from the meta-directing group.

It is interesting that chlorination of **14** illustrates all three rules. Of the four positions open to the electrophile, the 5 position violates rule 1, the 2 position rule 2, and the 4 position rule 3. The principal attack is therefore at position 6.

Orientation in Other Ring Systems[58]

In fused ring systems the positions are not equivalent and there is usually a preferred orientation, even in the unsubstituted hydrocarbon. The preferred positions may often by predicted as for benzene rings. Thus it is possible to draw more canonical forms for the arenium ion when naphthalene is attacked at the α position than when it is attacked at the β position, and the α position is the preferred site of attack,[59] though, as previously mentioned (p. 508), the isomer formed by substitution at the β position is thermodynamically more stable and is the product if the reaction is reversible and equilibrium is reached. Because of the more extensive delocalization of charges in the corresponding arenium ions, naphthalene is more reactive than benzene and substitution is faster at both positions. Similarly,

[55]In some cases, an electrophile preferentially attacks the position between two groups in the meta relationship. For a list of some of these cases and a theory to explain them, see Kruse; Cha *J. Chem. Soc., Chem. Commun.* **1982,** 1333.

[56]This is not the same as the ortho effect mentioned on p. 286.

[57]See Hammond; Hawthorne, in Newman *Steric Effects in Organic Chemistry*; Wiley: New York, 1956, pp. 164-200, 178-182.

[58]For a review of substitution on nonbenzenoid aromatic systems, see Hafner; Moritz, in Olah *Friedel-Crafts and Related Reactions*, vol. 4; Wiley: New York, 1965, pp. 127-183. For a review of aromatic substitution on ferrocenes, see Bublitz; Rinehart, *Org. React.* **1969,** *17*, 1-154.

[59]For a discussion on the preferred site of attack for many ring systems, see de la Mare; Ridd *Aromatic Substitution—Nitration and Halogenation*; Academic Press: New York, 1959, pp. 169-209.

anthracene, phenanthrene, and other fused polycyclic aromatic hydrocarbons are also substituted faster than benzene.

Heterocyclic compounds, too, have nonequivalent positions, and the principles are similar.[60] Furan, thiophene, and pyrrole are chiefly substituted at the 2 position, and all are substituted faster than benzene.[61] Pyrrole is particularly reactive, with a reactivity approximating that of aniline or the phenoxide ion. For pyridine[62] it is not the free base that is attacked but the conjugate acid, pyridinium ion.[63] The 3 position is most reactive, but the reactivity in this case is much less than that of benzene, being similar to that of nitrobenzene. However, groups can be introduced into the 4 position of a pyridine ring indirectly, by performing the reaction on the corresponding pyridine N-oxide.[64]

When fused ring systems contain substituents, successful predictions can often be made by using a combination of the above principles. Thus, ring A of 2-methylnaphthalene (**18**)

18 **19** **20**

is activated by the methyl group; ring B is not (though the presence of a substituent in a fused ring system affects all the rings,[65] the effect is generally greatest on the ring to which it is attached). We therefore expect substitution in ring A. The methyl group activates positions 1 and 3, which are ortho to itself, but not position 4, which is meta to it. However, substitution at the 3 position gives rise to an arenium ion for which it is impossible to write a low-energy canonical form in which ring B has a complete sextet. All we can write are forms like **19**, in which the sextet is no longer intact. In contrast, substitution at the 1 position gives rise to a more stable arenium ion, for which two canonical forms (one of them is **20**) can be written in which ring B is benzenoid. We thus predict predominant substitution at C-1, and that is what is generally found.[66] However, in some cases predictions are much harder to make. For example, chlorination or nitration of **21** gives mainly the 4 derivative, but bromination yields chiefly the 6 compound.[67]

21 indole quinoline

[60]For a monograph, see Katritzky; Taylor, Ref. 1.

[61]For a review of electrophilic substitution on five-membered aromatic heterocycles, see Marino *Adv. Heterocycl. Chem.* **1971**, *13*, 235-314.

[62]For reviews of substitution on pyridines and other six-membered nitrogen-containing aromatic rings, see Comins; O'Connor *Adv. Heterocycl. Chem.* **1988**, *44*, 199-267; Aksel'rod; Berezovskii *Russ. Chem. Rev.* **1970**, *39*, 627-643; Katritzky; Johnson *Angew. Chem. Int. Ed. Engl.* **1967**, *6*, 608-615 [*Angew. Chem.* **79**, 629-636]; Abramovitch; Saha *Adv. Heterocycl. Chem.* **1966**, *6*, 229-345. For a review of methods of synthesizing 3-substituted pyrroles, see Anderson; Loader *Synthesis* **1985**, 353-364.

[63]Olah; Olah; Overchuk *J. Org. Chem.* **1965**, *30*, 3373; Katritzky; Kingsland *J. Chem. Soc. B* **1968**, 862.

[64]Jaffé *J. Am. Chem.* **1954**, *76*, 3527.

[65]See, for example, Ansell; Sheppard; Simpson; Stroud; Taylor *J. Chem. Soc., Perkin Trans 2* **1979**, 381.

[66]For example, see Alcorn; Wells *Aust. J. Chem.* **1965**, *18*, 1377, 1391; Eaborn; Golborn; Spillett; Taylor *J. Chem. Soc. B* **1968**, 1112; Kim, Chen, Krieger, Judd, Simpson, Berliner *J. Am. Chem. Soc.* **1970**, *92*, 910. For discussions, see Taylor *Chimia* **1968**, *22*, 1-8; Gore; Siddiquei; Thorburn *J. Chem. Soc., Perkin Trans 1* **1972**, 1781.

[67]Bell *J. Chem. Soc.* **1959**, 519.

For fused heterocyclic systems too, we can often make predictions based on the above principles, though many exceptions are known. Thus, indole is chiefly substituted in the pyrrole ring (at position 3) and reacts faster than benzene, while quinoline generally reacts in the benzene ring, at the 5 and 8 positions, and slower than benzene, though faster than pyridine.

In alternant hydrocarbons (p. 50) the reactivity at a given position is similar for electrophilic, nucleophilic, and free-radical substitution, because the same kind of resonance can be shown in all three types of intermediate (compare **20, 22,** and **23**). Attack at the position

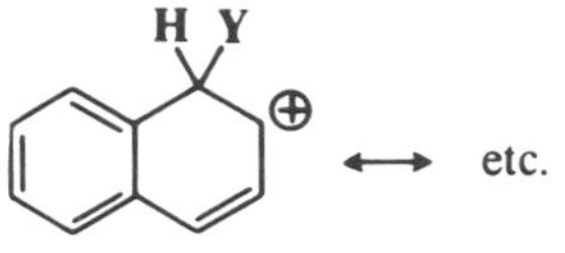

20 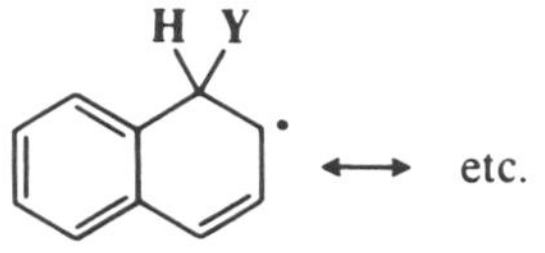   **22** 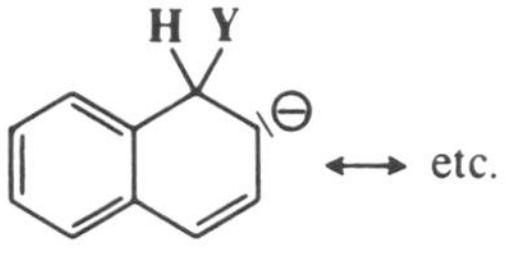   **23**

that will best delocalize a positive charge will also best delocalize a negative charge or an unpaired electron. Most results are in accord with these predictions. For example, naphthalene is attacked primarily at the 1 position by NO_2^+, NH_2^-, and Ph•, and always more readily than benzene.

Quantitative Treatments of Reactivity in the Substrate

Quantitative rate studies of aromatic substitutions are complicated by the fact that there are usually several hydrogens that can leave, so that measurements of overall rate ratios do not give a complete picture as they do in nucleophilic substitutions, where it is easy to compare substrates that have only one possible leaving group in a molecule. What is needed is not, say, the overall rate ratio for acetylation of toluene vs. that for benzene, but the *rate ratio at each position*. These can be calculated from the overall rates and a careful determination of the proportion of isomers formed, provided that the products are kinetically controlled, as is usually the case. We may thus define the *partial rate factor* for a given group and a given reaction as the rate of substitution at a single position relative to a single position in benzene. For example, for acetylation of toluene the partial rate factors are: for the ortho position o_f^{Me} = 4.5, for the meta m_f^{Me} = 4.8, and for the para p_f^{Me} = 749.[68] This means that toluene is acetylated at the ortho position 4.5 times as fast as a single position in benzene, or 0.75 times as fast as the overall rate of acetylation of benzene. A partial rate factor greater than 1 for a given position indicates that the group in question activates that position for the given reaction. Partial rate factors differ from one reaction to another and are even different, though less so, for the same reaction under different conditions.

Once we know the partial rate factors, we can predict the proportions of isomers to be obtained when two or more groups are present on a ring, *if we make the assumption that the effect of substituents is additive.* For example, if the two methyl groups in *m*-xylene have the same effect as the methyl group in toluene, we can calculate the theoretical partial rate factors at each position by multiplying those from toluene, so they should be as indicated:

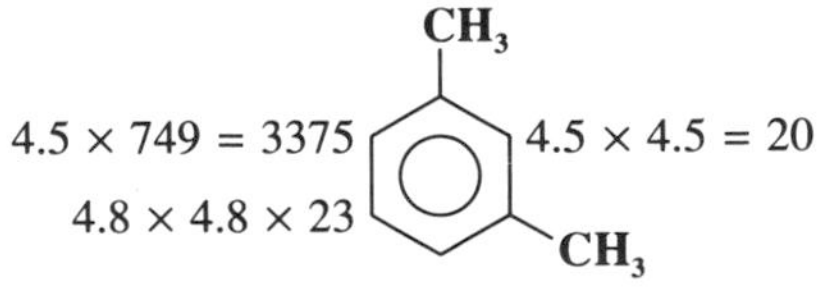

$$4.5 \times 749 = 3375 \qquad 4.5 \times 4.5 = 20$$
$$4.8 \times 4.8 \times 23$$

[68]Brown; Marino; Stock *J. Am. Chem. Soc.* **1959,** *81*, 3310.

TABLE 11.2 Calculated and experimental isomer distributions in the acetylation of m-xylene[69]

| Position | Isomer distribution, % | |
	Calculated	Observed
2	0.30	0
4	99.36	97.5
5	0.34	2.5

From this it is possible to calculate the overall theoretical rate ratio for acetylation of m-xylene relative to benzene, since this is one-sixth the sum of the partial rate factors (in this case 1130), and the isomer distribution if the reaction is kinetically controlled. The overall rate ratio actually is 347[69] and the calculated and observed isomer distributions are listed in Table 11.2.[69] In this case, and in many others, agreement is fairly good, but many cases are known where the effects are not additive.[70] For example, the treatment predicts that for 1,2,3-trimethylbenzene there should be 35% 5 substitution and 65% 4 substitution, but acetylation gave 79% 5 substitution and 21% of the 4 isomer. The treatment is thrown off by steric effects, such as those mentioned earlier (p. 511), by products arising from ipso attack (p. 512) and by resonance interaction *between* groups (for example, **24**), which must make the results deviate from simple additivity of the effects of the groups.

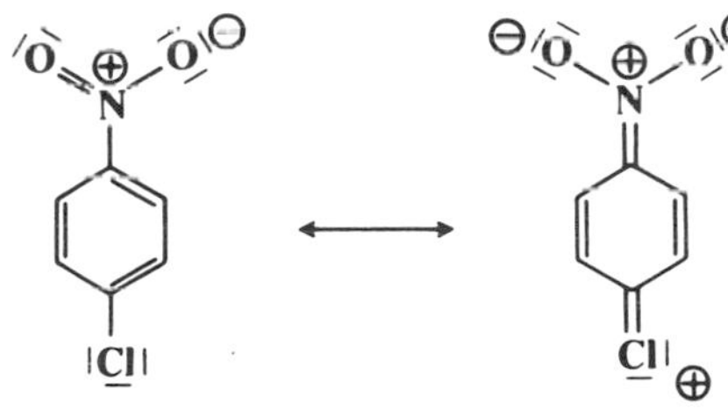

24

Another approach that avoids the problem created by having competing leaving groups present in the same substrate is the use of substrates that contain only one leaving group. This is most easily accomplished by the use of a leaving group other than hydrogen. By this means overall rate ratios can be measured for specific positions.[71] Results obtained in this way[72] give a reactivity order quite consistent with that for hydrogen as leaving group.

A quantitative scale of reactivity for aromatic substrates (fused, heterocyclic, and substituted rings) has been devised, based on the hard-soft concept (p. 261).[73] From molecular orbital theory, a quantity, called *activation hardness,* can be calculated for each position of an aromatic ring. The smaller the activation hardness, the faster the attack at that position; hence the treatment predicts the most likely orientations for incoming groups.

[69]Marino; Brown *J. Am. Chem. Soc.* **1959**, *81*, 5929.
[70]For some examples where additivity fails, see Fischer; Vaughan; Wright *J. Chem. Soc. B* **1967**, 368; Coombes; Crout; Hoggett; Moodie; Schofield *J. Chem. Soc. B* **1970**, 347; Richards; Wilkinson; Wright *Aust. J. Chem.* **1972**, *25*, 2369; Cook; Phillips; Ridd *J. Chem. Soc., Perkin Trans. 2* **1974**, 1166. For a theoretical treatment of why additivity fails, see Godfrey *J. Chem. Soc. B* **1971**, 1545.
[71]For a review of aryl–silicon and related cleavages, see Eaborn *J. Organomet. Chem.* **1975**, *100*, 43-57.
[72]See, for example, Deans and Eaborn *J. Chem. Soc.* **1959**, 2299; Eaborn; Jackson *J. Chem. Soc. B* **1969**, 21.
[73]Zhou; Parr *J. Am. Chem. Soc.* **1990**, *112*, 5720.

A Quantitative Treatment of Reactivity of the Electrophile. The Selectivity Relationship

Not all electrophiles are equally powerful. The nitronium ion attacks not only benzene but also aromatic rings that contain a strongly deactivating group. On the other hand, diazonium ions couple only with rings containing a powerful activating group. Attempts have been made to correlate the influence of substituents with the power of the attacking group. The most obvious way to do this is with the Hammett equation (p. 278):

$$\log \frac{k}{k_0} = \rho\sigma$$

For aromatic substitution, k_0 is divided by 6 and, for meta substitution, k is divided by 2, so that comparisons are made for only one position (consequently, k/k_0 for, say, the methyl group at a para position is identical to the partial rate factor p_f^{Me}). It was soon found that, while this approach worked fairly well for electron-withdrawing groups, it failed for those that are electron-donating. However, if the equation is modified by the insertion of the Brown σ^+ values instead of the Hammett σ values (because a positive charge develops during the transition state), more satisfactory correlations can be made, even for electron-donating groups (see Table 9.4 for a list of σ^+ values).[74] Groups with a negative value of σ_p^+ or σ_m^+ are activating for that position; groups with a positive value are deactivating. The ρ values correspond to the susceptibility of the reaction to stabilization or destabilization by the Z group and to the reactivity of the electrophile. The ρ values vary not only with the electrophile but also with conditions. A large negative value of ρ means an electrophile of relatively low reactivity. Of course, this approach is completely useless for ortho substitution, since the Hammett equation does not apply there.

A modification of the Hammett approach, suggested by Brown, called the *selectivity relationship*,[75] is based on the principle that reactivity of a species varies inversely with selectivity. Table 11.3 shows how electrophiles can be arranged in order of selectivity as measured by two indexes: (1) their selectivity in attacking toluene rather than benzene, and (2) their selectivity between the meta and para positions in toluene.[76] As the table shows, an electrophile more selective in one respect is also more selective in the other. In many

TABLE 11.3 Relative rates and product distributions in some electrophilic substitutions on toluene and benzene[76]

Reaction	Relative rate $k_{toluene}/k_{benzene}$	Product distribution, %	
		m	*p*
Bromination	605	0.3	66.8
Chlorination	350	0.5	39.7
Benzoylation	110	1.5	89.3
Nitration	23	2.8	33.9
Mercuration	7.9	9.5	69.5
Isopropylation	1.8	25.9	46.2

[74]For a discussion of the limitations of the Hammett equation approach, see Koptyug; Salakhutdinov; Detsina *J. Org. Chem. USSR* **1984**, *20*, 1039.
[75]Stock; Brown *Adv. Phys. Org. Chem.* **1963**, *1*, 35-154.
[76]Ref. 75, p. 45.

cases, electrophiles known to be more stable (hence less reactive) than others show a higher selectivity, as would be expected. For example, the *t*-butyl cation is more stable and more selective than the isopropyl (p. 166), and Br_2 is more selective than Br^+. However, deviations from the relationship are known.[77] Selectivity depends not only on the nature of the electrophile but also on the temperature. As expected, it normally decreases with increasing temperature.

Brown assumed that a good measurement of selectivity was the ratio of the para and meta partial rate factors in toluene. He defined the selectivity S_f of a reaction as

$$S_f = \log \frac{p_f^{Me}}{m_f^{Me}}$$

That is, the more reactive an attacking species, the less preference it has for the para position compared to the meta. If we combine the Hammett–Brown $\sigma^+\rho$ relationship with the linearity between $\log S_f$ and $\log p_f^{Me}$ and between $\log S_f$ and $\log m_f^{Me}$, it is possible to derive the following expressions:

$$\log p_f^{Me} = \frac{\sigma_p^+}{\sigma_p^+ - \sigma_m^+} S_f$$

$$\log m_f^{Me} = \frac{\sigma_m^+}{\sigma_p^+ - \sigma_m^+} S_f$$

S_f is related to ρ by

$$S_f = \rho(\sigma_p^+ - \sigma_m^+)$$

The general validity of these equations is supported by a great deal of experimental data on aromatic substitution reactions of toluene. Examples of values for some reactions obtained from these equations are given in Table 11.4.[78] For other substituents, the treatment works well with groups that, like methyl, are not very polarizable. For more polarizable groups the correlations are sometimes satisfactory and sometimes not, probably because each electrophile in the transition state makes a different demand on the electrons of the substituent group.

Not only are there substrates for which the treatment is poor, but it also fails with very powerful electrophiles; this is why it is necessary to postulate the encounter complex mentioned on p. 506. For example, relative rates of nitration of *p*-xylene, 1,2,4-trimethylbenzene, and 1,2,3,5-tetramethylbenzene were 1.0, 3.7, and 6.4,[79] though the extra methyl groups

TABLE 11.4   Values of m_f^{Me}, p_f^{Me}, s_f, and ρ for three reactions of toluene[78]

Reaction		m_f^{Me}	p_f^{Me}	S_f	ρ
PhMe + EtBr	$\xrightarrow[\text{benzene. 25°C}]{\text{GaBr}_3}$	1.56	6.02	0.587	−2.66
PhMe + HNO₃	$\xrightarrow[\text{45°C}]{\text{90\% HOAc}}$	2.5	58	1.366	−6.04
PhMe + Br₂	$\xrightarrow[\text{25°C}]{\text{85\% HOAc}}$	5.5	2420	2.644	−11.40

[77]At least some of these may arise from migration of groups already on the ring; see Olah; Olah; Ohyama *J. Am. Chem. Soc.* **1984**, *106*, 5284.

[78]Stock; Brown *J. Am. Chem. Soc.* **1959**, *81*, 3323. Ref. 75 presents many tables of these kinds of data. See also DeHaan; Chan; Chang; Ferrara; Wainschel *J. Org. Chem.* **1986**, *51*, 1591, and other papers in this series.

[79]Olah; Lin, Ref. 26.

should enhance the rates much more (*p*-xylene itself reacted 295 times faster than benzene). The explanation is that with powerful electrophiles the reaction rate is so rapid (reaction taking place at virtually every encounter[80] between an electrophile and substrate molecule)[81] that the presence of additional activating groups can no longer increase the rate.[82]

Given this behavior (little selectivity in distinguishing between different substrate molecules), the selectivity relationship would predict that positional selectivity should also be very small. However, it is not. For example, under conditions where nitration of *p*-xylene and 1,2,4-trimethylbenzene takes place at about equal rates, there was no corresponding lack of selectivity at positions *within* the latter.[83] Though steric effects are about the same at both positions, more than 10 times as much 5-nitro product was formed as 6-nitro product.

relative rate ratios

It is clear that the selectivity relationship has broken down and it becomes necessary to explain why such an extremely rapid reaction should occur with positional selectivity. The explanation offered is that the rate-determining step is formation of an encounter complex (**12**, p. 506).[84] Since the position of attack is not determined in the rate-determining step, the 5/6 ratio is not related to the reaction rate. Essentially the same idea was suggested earlier[85] and for the same reason (failure of the selectivity relationship in some cases), but the earlier explanation specifically pictured the complex as a π complex, and we have seen (p. 506) that there is evidence against this.

One interesting proposal[86] is that the encounter pair is a radical pair $\overline{NO_2\bullet\ ArH\bullet^+}$ formed by an electron transfer (SET), which would explain why the electrophile, once in the encounter complex, can acquire the selectivity that the free NO_2^+ lacked (it is not proposed that a radical pair is present in all aromatic substitutions; only in those that do not obey the selectivity relationship). The radical pair subsequently collapses to the arenium ion. There is evidence[87] both for and against this proposal.[88]

The Effect of the Leaving Group

In the vast majority of aromatic electrophilic substitutions, the leaving group is H^+ (it is certainly one of the best), and very little work has been done on the relative electrofugal

[80]See Coombes; Moodie; Schofield, Ref. 29; Moodie; Schofield; Thomas *J. Chem. Soc., Perkin Trans. 2* **1978**, 318.

[81]For a review of diffusion control in electrophilic aromatic substitution, see Ridd, Ref. 27.

[82]Coombes; Moodie; Schofield, Ref. 26; Hoggett; Moodie; Schofield, Ref. 26; Hartshorn; Moodie; Schofield; Thompson *J. Chem. Soc. B* **1971**, 2447; Manglik; Moodie; Schofield; Dedeoglu; Dutly; Rys *J. Chem. Soc., Perkin Trans 2* **1981**, 1358.

[83]Barnett; Moodie; Schofield; Weston *J. Chem. Soc., Perkin Trans. 2* **1975**, 648; Barnett; Moodie; Schofield; Taylor; Weston *J. Chem. Soc., Perkin Trans. 2* **1979**, 747.

[84]For kinetic evidence in favor of encounter complexes, see Sheats; Strachan *Can. J. Chem.* **1978**, *56*, 1280. For evidence for such complexes in the gas phase, see Attinà; Cacace; de Petris *Angew. Chem. Int. Ed. Engl.* **1987**, *26*, 1177 [*Angew. Chem. 99*, 1174].

[85]Olah, Ref. 26.

[86]Perrin *J. Am. Chem. Soc.* **1977**, *99*, 5516.

[87]For evidence in favor of the proposal, see Reents; Freiser *J. Am. Chem. Soc.* **1980**, *102*, 271; Morkovnik; Dobaeva; Panov; Okhlobystin *Doklad. Chem.* **1980**, *251*, 116; Sankararaman; Haney; Kochi *J. Am. Chem. Soc.* **1987**, *109*, 5235; Keumi; Hamanaka; Hasegawa; Minamide; Inoue; Kitajima *Chem. Lett.* **1988**, 1285; Johnston; Ridd; Sandall *J. Chem. Soc., Chem. Commun.* **1989**, 244. For evidence against it, see Barnes; Myhre *J. Am. Chem. Soc.* **1978**, *100*, 975; Eberson; Radner *Acc. Chem. Res.* **1987**, *20*, 53-59; Baciocchi; Mandolini *Tetrahedron* **1987**, *43*, 4035.

[88]For a review, see Morkovnik *Russ. Chem. Rev.* **1988**, *57*, 144-160.

ability of other leaving groups. However, the following orders of leaving-group ability have been suggested:[89] (1) for leaving groups that depart without assistance (SN1 process with respect to the leaving group), NO_2^{+}[90] $<$ iso-Pr^+ $\sim$ SO_3 $<$ t-Bu^+ $\sim$ ArN_2^+ $<$ $ArCHOH^+$ $<$ NO^+ $<$ CO_2; (2) for leaving groups that depart with assistance from an outside nucleophile (SN2 process), Me^+ $<$ Cl^+ $<$ Br^+ $<$ D^+ $\sim$ RCO^+ $<$ H^+ $\sim$ I^+ $<$ Me_3Si^+. We can use this kind of list to help predict which group, X or Y, will cleave from an arenium ion **1** once it has been formed, and so obtain an idea of which electrophilic substitutions are feasible. However, a potential leaving group can also affect a reaction in another way: by influencing the rate at which the original electrophile attacks directly at the ipso position. Partial rate factors for electrophilic attack at a position substituted by a group other than hydrogen are called ipso partial rate factors (i_f^X).[48] Such factors for the nitration of p-haloanisoles are 0.18, 0.08, and 0.06, for p-iodo, p-bromo-, and p-chloroanisole, respectively.[91] This means, for example, that the electrophile in this case attacks the 4 position of 4-iodoanisole 0.18 times as fast as a single position of benzene. Note that this is far slower than it attacks the 4 position of anisole itself so that the presence of the iodo group greatly slows the reaction at that position. A similar experiment on p-cresol showed that ipso attack at the methyl position was 6.8 times slower than attack at the para position of phenol.[92] Thus, in these cases, both an iodo and a methyl group deactivate the ipso position.[93]

REACTIONS

The reactions in this chapter are classified according to leaving group. Hydrogen replacements are treated first, then rearrangements in which the attacking entity is first cleaved from another part of the molecule (hydrogen is also the leaving group in these cases), and finally replacements of other leaving groups.

Hydrogen as the Leaving Group in Simple Substitution Reactions

A. Hydrogen as the Electrophile

1-1 Hydrogen Exchange
Deuterio-de-hydrogenation or **Deuteriation**

$$ArH + D^+ \rightleftharpoons ArD + H^+$$

Aromatic compounds can exchange hydrogens when treated with acids. The reaction is used chiefly to study mechanic questions[94] (including substituent effects), but can also be useful to deuterate or tritate aromatic rings selectively. The usual directive effects apply and, for example, phenol treated with D_2O gives slow exchange on heating, with only ortho and para hydrogens being exchanged.[95] Strong acids, of course, exchange faster with aromatic substrates, and this exchange must be taken into account when studying the mechanism of any aromatic substitution catalyzed by acids. There is a great deal of evidence that exchange

[89]Perrin *J. Org. Chem.* **1971**, *36*, 420.
[90]For examples where NO_2^+ is a leaving group (in a migration), see Bullen; Ridd; Sabek *J. Chem. Soc., Perkin Trans. 2* **1990**, 1681, and other papers in this series.
[91]Ref. 48. See also Fischer; Zollinger *Helv. Chim. Acta* **1972**, *55*, 2139.
[92]Tee; Iyengar; Bennett *J. Org. Chem.* **1986**, *51*, 2585.
[93]For other work on ipso reactivity, see Baciocchi; Illuminati *J. Am. Chem. Soc.* **1967**, *89*, 4017; Berwin *J. Chem. Soc., Chem. Commun.* **1972**, 237; Galley; Hahn *J. Am. Chem. Soc.* **1974**, *96*, 4337; Clemens; Hartshorn; Richards; Wright *Aust. J. Chem.* **1977**, *30*, 103, 113.
[94]For a review, see Taylor, in Bamford; Tipper, Ref. 1, pp. 194-277.
[95]Small; Wolfenden *J. Chem. Soc.* **1936**, 1811.

takes place by the ordinary arenium ion mechanism. Among the evidence are the orientation effects noted above and the finding that the reaction is general-acid-catalyzed, which means that a proton is transferred in the slow step[96] (p. 259). Furthermore, many examples have been reported of stable solutions of arenium ions formed by attack of a proton on an aromatic ring.[4] Simple aromatic compounds can be extensively deuterated in a convenient fashion by treatment with D_2O and BF_3.[97] It has been shown that tritium exchange takes place readily at the 2 position of **25,** despite the fact that this position is hindered by the bridge. The

$$n = 7, 8, 10$$

25

rates were not very different from the comparison compound 1,3-dimethylnaphthalene.[98]

Hydrogen exchange can also be effected with strong bases,[99] such as NH_2^-. In these cases the slow step is the proton transfer:

$$\text{ArH} + \text{B} \longrightarrow \text{Ar}^- + \text{BH}^+$$

so the SE1 mechanism and not the usual arenium ion mechanism is operating.[100] Aromatic rings can also be deuterated by treatment with D_2O and a rhodium(III) chloride[101] or platinum[102] catalyst or with C_6D_6 and an alkylaluminum dichloride catalyst,[103] though rearrangements may take place during the latter procedure. Tritium can be introduced by treatment with T_2O and an alkylaluminum dichloride catalyst.[103] Tritiation at specific sites (e.g. more than 90% para in toluene) has been achieved with T_2 gas and a microporous aluminophosphate catalyst.[104]

B. Nitrogen Electrophiles

1-2 Nitration or Nitro-de-hydrogenation

$$\text{ArH} + \text{HNO}_3 \xrightarrow{\text{H}_2\text{SO}_4} \text{ArNO}_2$$

Most aromatic compounds, whether of high or low reactivity, can be nitrated, because a wide variety of nitrating agents is available.[105] For benzene, the simple alkylbenzenes, and less reactive compounds, the most common reagent is a mixture of concentrated nitric and

[96]For example, see Challis; Long *J. Am. Chem. Soc.* **1963,** *85,* 2524; Batts; Gold *J. Chem. Soc.* **1964,** 4284; Kresge; Chiang; Sato *J. Am. Chem. Soc.* **1967,** *89,* 4418; Gruen; Long *J. Am. Chem. Soc.* **1967,** *89,* 1287; Butler; Hendry *J. Chem. Soc. B* **1970,** 852.

[97]Larsen; Chang *J. Org. Chem.* **1978,** *43,* 3602.

[98]Laws; Neary; Taylor *J. Chem. Soc., Perkin Trans. 2* **1987,** 1033.

[99]For a review of base-catalyzed hydrogen exchange on heterocycles, see Elvidge; Jones; O'Brien; Evans; Sheppard *Adv. Heterocycl. Chem.* **1974,** *16,* 1-31.

[100]Shatenshtein *Tetrahedron* **1962,** *18,* 95.

[101]Lockley *Tetrahedron Lett.* **1982,** *23,* 3819, *J. Chem. Res. (S)* **1985,** 178.

[102]See, for example, Leitch *Can. J. Chem.* **1954,** *32,* 813; Fraser; Renaud *J. Am. Chem. Soc.* **1966,** *88,* 4365; Fischer; Puza *Synthesis* **1973,** 218; Blake; Garnett; Gregor; Hannan; Hoa; Long *J. Chem. Soc., Chem. Commun.* **1975,** 930. See also Parshall *Acc. Chem. Res.* **1975,** *8,* 113-117.

[103]Garnett; Long; Vining; Mole *J. Am. Chem. Soc.* **1972,** *94,* 5913, 8632; Long; Garnett; West *Tetrahedron Lett.* **1978,** 4171.

[104]Garnett; Kennedy; Long; Than; Watson *J. Chem. Soc., Chem. Commun.* **1988,** 763.

[105]For monographs, see Olah; Malhotra; Narang *Nitration: Methods and Mechanisms;* VCH: New York, 1989; Schofield *Aromatic Nitration;* Cambridge University Press: Cambridge, 1980; Hoggett; Moodie; Penton; Schofield, Ref. 30. For reviews, see Weaver, in Feuer *Chemistry of the Nitro and Nitroso Groups,* pt. 2; Wiley: New York, 1970, pp. 1-48; de la Mare; Ridd, Ref. 59, pp. 48-93. See also Ref. 1. For a review of side reactions, see Suzuki *Synthesis* **1977,** 217-238.

sulfuric acids, but for active substrates, the reaction can be carried out with nitric acid alone, or in water, acetic acid, or acetic anhydride. In fact, these milder conditions are necessary for active compounds such as amines, phenols, and pyrroles, since reaction with mixed nitric and sulfuric acids would oxidize these substrates. If anhydrous conditions are required, nitration can be effected with N_2O_5[106] in CCl_4 in the presence of P_2O_5, which removes the water formed in the reaction.[107] Nitration in alkaline media can be accomplished with esters of nitric acid such as ethyl nitrate ($EtONO_2$). These reagents can also be used with proton or Lewis-acid catalysts. Other nitrating agents are $NaNO_2$ and trifluoroacetic acid,[108] N_2O_4 (which gives good yields with polycyclic hydrocarbons[109]), and nitronium salts[110] such as $NO_2^+ BF_4^-$, $NO_2^+ PF_6^-$, and $NO_2^+ CF_3SO_3^-$. The last-mentioned salt gives a very high yield of products at low temperatures.[111] Aromatic hydrocarbons and halobenzenes are nitrated in high yields with clay-supported cupric nitrate (claycop),[112] with predominant para regioselectivity.[113] With active substrates such as amines and phenols, nitration can be accomplished by nitrosation under oxidizing conditions with a mixture of dilute nitrous and nitric acids.[114] Active substrates can also be nitrated, conveniently and under mild conditions, with nitrocyclohexadienones such as 2,3,5,6-tetrabromo-4-methyl-4-nitro-1,4-cyclohexadienone.[115]

When amines are nitrated under strong-acid conditions, meta orientation is generally observed, because the species undergoing nitration is actually the conjugate acid of the amine. If the conditions are less acidic, the free amine is nitrated and the orientation is ortho-para. Although the free base may be present in much smaller amounts than the conjugate acid, it is far more susceptible to aromatic substitution (see also p. 510). Because of these factors and because they are vulnerable to oxidation by nitric acid, primary aromatic amines are often protected before nitration by treatment with acetyl chloride (**0-52**) or acetic anhydride (**0-53**). Nitration of the resulting acetanilide derivative avoids all these problems. There is evidence that when the reaction takes place on the free amine, it is the nitrogen that is attacked to give an N-nitro compound Ar—NH—NO_2 which rapidly undergoes rearrangement (see **1-32**) to give the product.[116]

Since the nitro group is deactivating, it is usually easy to stop the reaction after one group has entered the ring, but a second and a third group can be introduced if desired, especially when an activating group is also present. Even m-dinitrobenzene can be nitrated if vigorous conditions are applied. This has been accomplished with $NO_2^+ BF_4^-$ in FSO_3H at 150°C.[117]

[106]For a review of N_2O_5 see Fischer, in Feuer; Nielsen *Nitro Compounds, Recent Advances in Synthesis and Chemistry*; VCH: New York, 1990, pp. 267-365.

[107]For another method, see Olah; Krishnamurthy; Narang *J. Org. Chem.* **1982**, *47*, 596.

[108]Uemura; Toshimitsu; Okano *J. Chem. Soc., Perkin Trans. 1.* **1978**, 1076.

[109]Radner *Acta Chem. Scand., Ser. B* **1983**, *37*, 65.

[110]Olah; Kuhn *J. Am. Chem. Soc.* **1962**, *84*, 3684. These have also been used together with crown ethers: Masci *J. Chem. Soc., Chem. Commun.* **1982**, 1262, *J. Org. Chem.* **1985**, *50*, 4081. For a review of nitronium salts in organic chemistry, see Guk; Ilyushin; Golod; Gidaspov *Russ. Chem. Rev.* **1983**, *52*, 284-297.

[111]Coon; Blucher; Hill *J. Org. Chem.* **1973**, *38*, 4243; Effenberger; Geke *Synthesis* **1975**, 40.

[112]For reviews of clay-supported nitrates, see Cornélis; Laszlo *Synthesis* **1985**, 909-918; Laszlo *Acc. Chem. Res.* **1986**, 121-127; Laszlo; Cornélis *Aldrichimica Acta* **1988**, *21*, 97-103.

[113]Laszlo; Pennetreau *J. Org. Chem.* **1987**, *52*, 2407; Cornélis; Delaude; Gerstmans; Laszlo *Tetrahedron Lett.* **1988**, *29*, 5657; Cornélis; Gerstmans; Laszlo *Chem. Lett.* **1988**, 1839; Laszlo; Vandormael *Chem. Lett.* **1988**, 1843. See also Smith; Fry; Butters; Nay *Tetrahedron Lett.* **1989**, *30*, 5333. For similar nitrations of phenols, see Cornélis; Laszlo; Pennetreau *Bull. Soc. Chim. Belg.* **1984**, *93*, 961; Poirier; Vottero *Tetrahedron* **1989**, *45*, 1415. For a method of nitrating phenols in the ortho position, see Pervez; Onyiriuka; Rees; Rooney; Suckling *Tetrahedron* **1988**, *44*, 4555.

[114]For discussions of the mechanism in this case, see Giffney; Ridd *J. Chem. Soc., Perkin Trans. 2* **1979**, 618; Bazanova; Stotskii *J. Org. Chem. USSR* **1980**, *16*, 2070, 2075; Ross; Moran; Malhotra *J. Org. Chem.* **1983**, *48*, 2118; Dix; Moodie *J. Chem. Soc., Perkin Trans. 2* **1986**, 1097; Leis; Peña; Ridd *Can. J. Chem.* **1989**, *67*, 1677. For a review, see Ridd, Ref. 122a.

[115]Lemaire; Guy; Roussel; Guette *Tetrahedron* **1987**, *43*, 835.

[116]Ridd; Scriven *J. Chem. Soc., Chem. Commun.* **1972**, 641. See also Helsby; Ridd *J. Chem. Soc., Perkin Trans. 2* **1983**, 1191.

[117]Olah; Lin *Synthesis* **1974**, 444.

With most of the reagents mentioned, the attacking species is the nitronium ion NO_2^+. Among the ways in which this ion is formed are:

1. In concentrated sulfuric acid, by an acid–base reaction in which nitric acid is the base:

$$HNO_3 + 2H_2SO_4 \rightleftharpoons NO_2^+ + H_3O^+ + 2HSO_4^-$$

This ionization is essentially complete.

2. In concentrated nitric acid alone,[118] by a similar acid-base reaction in which one molecule of nitric acid is the acid and another the base:

$$2HNO_3 \rightleftharpoons NO_2^+ + NO_3^- + H_2O$$

This equilibrium lies to the left (about 4% ionization), but enough NO_2^+ is formed for nitration to occur.

3. The equilibrium just mentioned occurs to a small extent even in organic solvents.

4. With N_2O_5 in CCl_4, there is spontaneous dissociation:

$$N_2O_5 \rightleftharpoons NO_2^+ + NO_3^-$$

but in this case there is evidence that some nitration also takes place with undissociated N_2O_5 as the electrophile.

5. When nitronium salts are used, NO_2^+ is of course present to begin with. Esters and acyl halides of nitric acid ionize to form NO_2^+. Nitrocyclohexadienones are converted to NO_2^+ and the corresponding phenol.[115]

There is a great deal of evidence that NO_2^+ is present in most nitrations and that it is the attacking entity,[119] e.g.,

1. Nitric acid has a peak in the Raman spectrum. When nitric acid is dissolved in concentrated sulfuric acid, the peak disappears and two new peaks appear, one at 1400 cm^{-1} attributable to NO_2^+ and one at 1050 cm^{-1} due to HSO_4^-.[120]

2. On addition of nitric acid, the freezing point of sulfuric acid is lowered about four times the amount expected if no ionization has taken place.[121] This means that the addition of one molecule of nitric acid results in the production of four particles, which is strong evidence for the ionization reaction between nitric and sulfuric acids given above.

3. The fact that nitronium salts in which nitronium ion is known to be present (by x-ray studies) nitrate aromatic compounds shows that this ion does attack the ring.

4. The rate of the reaction with most reagents is proportional to the concentration of NO_2^+, not to that of other species.[122] When the reagent produces this ion in small amounts, the attack is slow and only active substrates can be nitrated. In concentrated and aqueous mineral acids the kinetics are second order: first order each in aromatic substrate and in nitric acid (unless pure nitric acid is used in which case there are pseudo-first-order kinetics). But in organic solvents such as nitromethane, acetic acid, and CCl_4, the kinetics are first order in nitric acid alone and zero order in aromatic substrate, because the rate-determining step is formation of NO_2^+ and the substrate does not take part in this.

In a few cases, depending on the substrate and solvent, there is evidence that the arenium ion is not formed directly, but via the intermediacy of a radical pair (see p. 520):[122a]

[118]See Belson; Strachan *J. Chem. Soc., Perkin Trans. 2* **1989**, 15.

[119]For an exhaustive study of this reaction, see Hughes; Ingold; and co-workers *J. Chem. Soc.* **1950**, 2400-2684.

[120]Ingold; Millen; Poole *J. Chem. Soc.* **1950**, 2576.

[121]Gillespie; Graham; Hughes; Ingold; Peeling *J. Chem. Soc.* **1950**, 2504.

[122]This is not always strictly true. See Ross; Kuhlmann; Malhotra *J. Am. Chem. Soc.* **1983**, *105*, 4299.

[122a]For a review of radical processes in aromatic nitration, see Ridd *Chem. Soc. Rev.* **1991**, *20*, 149-165. For a review of aromatic substitutions involving radical cations, see Kochi *Adv. Free Radical Chem. (Greenwich, Conn.)* **1990**, *1*, 53-119.

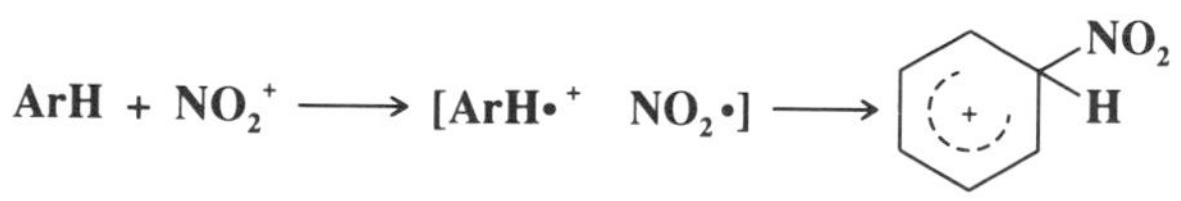

ArH + NO₂⁺ ⟶ [ArH•⁺ NO₂•] ⟶

OS **I**, 372, 396, 408 (see also OS **53**, 129); **II**, 254, 434, 438, 447, 449, 459, 466; **III**, 337, 644, 653, 658, 661, 837; **IV**, 42, 364, 654, 711, 722, 735; **V**, 346, 480, 829, 1029, 1067.

1-3 Nitrosation or **Nitroso-de-hydrogenation**

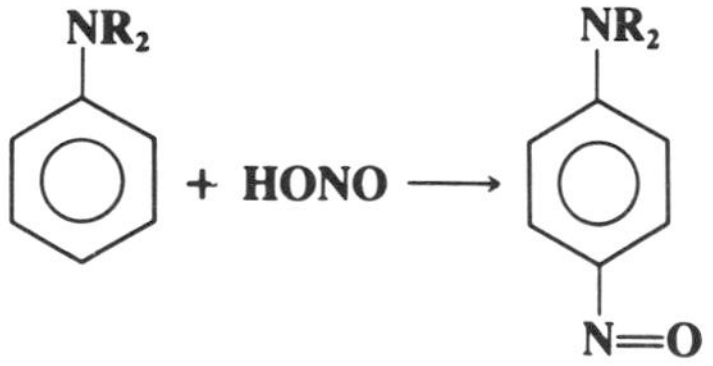

Ring nitrosation[123] with nitrous acid is normally carried out only with active substrates such as amines and phenols. However, primary aromatic amines give diazonium ions (**2-49**) when treated with nitrous acid,[124] and secondary amines tend to give N-nitroso rather than C-nitroso compounds (**2-51**); hence this reaction is normally limited to phenols and tertiary aromatic amines. Nevertheless secondary aromatic amines can be C-nitrosated in two ways. The N-nitroso compound first obtained can be isomerized to a C-nitroso compound (**1-33**), or it can be treated with another mole of nitrous acid to give an N,C-dinitroso compound. Also, a successful nitrosation of anisole has been reported, where the solvent was CF₃COOH–CH₂Cl₂.[125]

Much less work has been done on the mechanism of this reaction than on the preceding one.[126] In some cases the attacking entity is NO⁺, but in others it is apparently NOCl, NOBr, N₂O₃, etc., in each of which there is a carrier of NO⁺. NOCl and NOBr are formed during the normal process of making nitrous acid—the treatment of sodium nitrite with HCl or HBr. Nitrosation requires active substrates because NO⁺ is much less reactive than NO₂⁺. Kinetic studies have shown that NO⁺ is at least 10¹⁴ times less reactive than NO₂⁺.[127] A consequence of the relatively high stability of NO⁺ is that this species is easily cleaved from the arenium ion, so that k_{-1} competes with k_2 (p. 503) and isotope effects are found.[128] With phenols, there is evidence that nitrosation may first take place at the OH group, after which the nitrite ester thus formed rearranges to the C-nitroso product.[129] Tertiary aromatic amines substituted in the ortho position generally do not react with HONO, probably because the ortho substituent prevents planarity of the dialkylamino group, without which the ring is no longer activated. This is an example of steric inhibition of resonance (p. 36).

OS **I**, 214, 411, 511; **II**, 223; **IV**, 247.

1-4 Diazonium Coupling
Arylazo-de-hydrogenation

ArH + Ar′N₂⁺ ⟶ Ar—N=N—Ar′

[123]For a review, see Williams *Nitrosation*; Cambridge University Press: Cambridge, 1988, pp. 58-76.
[124]For examples of formation of C-nitroso compounds from primary and secondary amines, see Hoefnagel; Wepster *Recl. Trav. Chim. Pays-Bas* **1989**, *108*, 97.
[125]Radner; Wall; Loncar *Acta Chem. Scand.* **1990**, *44*, 152.
[126]For a review of nitrosation mechanisms at C and other atoms, see Williams *Adv. Phys. Org. Chem.* **1983**, *19*, 381-428. See also Ref. 123.
[127]Challis; Higgins; Lawson *J. Chem. Soc., Perkin Trans. 2*, **1972**, 1831; Challis; Higgins *J. Chem. Soc., Perkin Trans. 2* **1972**, 2365.
[128]Challis; Lawson *J. Chem. Soc. B* **1971**, 770; Challis; Higgins *J. Chem. Soc., Perkin Trans. 2* **1973**, 1597.
[129]Gosney; Page *J. Chem. Soc., Perkin Trans. 2* **1980**, 1783.

Aromatic diazonium ions normally couple only with active substrates such as amines and phenols.[130] Many of the products of this reaction are used as dyes (*azo dyes*).[131] Presumably because of the size of the attacking species, substitution is mostly para to the activating group, unless that position is already occupied, in which case ortho substitution takes place. The pH of the solution is important both for phenols and amines. For amines, the solutions may be mildly acidic or neutral. The fact that amines give ortho and para products shows that even in mildly acidic solution they react in their un-ionized form. If the acidity is too high, the reaction does not occur, because the concentration of free amine becomes too small. Phenols must be coupled in slightly alkaline solution where they are converted to the more reactive phenoxide ions, because phenols themselves are not active enough for the reaction. However, neither phenols nor amines react in moderately alkaline solution, because the diazonium ion is converted to a diazo hydroxide Ar—N=N—OH. Primary and secondary amines face competition from attack at the nitrogen.[132] However, the resulting N-azo compounds (aryl triazenes) can be isomerized to C-azo compounds (**1-34**). In at least some cases, even when the C-azo compound is isolated, it is the result of initial N-azo compound formation followed by isomerization. It is therefore possible to synthesize the C-azo compound directly in one laboratory step.[133] Acylated amines and phenolic ethers and esters are ordinarily not active enough for this reaction, though it is sometimes possible to couple them (as well as such polyalkylated benzenes as mesitylene and pentamethylbenzene) to diazonium ions containing electron-withdrawing groups in the para position, since such groups increase the concentration of the positive charge and thus the electrophilicity of the ArN_2^+. Some coupling reactions which are otherwise very slow (in cases where the coupling site is crowded) are catalyzed by pyridine for reasons discussed on p. 504. Phase transfer catalysis has also been used.[134] Coupling of a few aliphatic diazonium compounds to aromatic rings has been reported. All the examples reported so far involve cyclopropanediazonium ions and brigehead diazonium ions, in which loss of N_2 would lead to very unstable carbocations.[135]

OS **I**, 49, 374; **II**, 35, 39, 145.

1-5 Direct Introduction of the Diazonium Group
Diazoniation or Diazonio-de-hydrogenation

$$ArH \xrightarrow[HX]{2HONO} ArN_2^+ \ X^-$$

Diazonium salts can be prepared directly by replacement of an aromatic hydrogen without the necessity of going through the amino group.[136] The reaction is essentially limited to active substrates (amines and phenols), since otherwise poor yields are obtained. Since the reagents and the substrate are the same as in reaction **1-3**, the first species formed is the nitroso compound. In the presence of excess nitrous acid, this is converted to the diazonium ion.[137] The reagent (azidochloromethylene)dimethylammonium chloride $Me_2\overset{\oplus}{N}{=}C(Cl)N_3$ Cl^- can also introduce the diazonium group directly into a phenol.[138]

[130]For reviews, see Szele; Zollinger *Top. Curr. Chem.* **1983**, *112*, 1-66; Hegarty, in Patai *The Chemistry of Diazonium and Diazo Groups*, pt. 2; Wiley: New York, 1978, pp. 545-551.
[131]For reviews of azo dyes, see Zollinger *Color Chemistry*; VCH: New York, 1987, pp. 85-148; Gordon; Gregory *Organic Chemistry in Colour*; Springer: New York, 1983, pp. 95-162.
[132]See Penton; Zollinger *Helv. Chim. Acta* **1981**, *64*, 1717, 1728.
[133]Kelly; Penton; Zollinger *Helv. Chim. Acta* **1982**, *65*, 122.
[134]Hashida; Kubota; Sekiguchi *Bull. Chem. Soc. Jpn.* **1988**, *61*, 905.
[135]See Szele; Zollinger, Ref. 130, pp. 3-6.
[136]Tedder *J. Chem. Soc.* **1957**, 4003.
[137]Tedder; Theaker *Tetrahedron* **1959**, *5*, 288; Kamalova; Nazarova; Solodova; Yaskova *J. Org. Chem. USSR* **1988**, *24*, 1004.
[138]Kokel; Viehe *Angew. Chem. Int. Ed. Engl.* **1980**, *19*, 716 [*Angew. Chem. 92*, 754].

1-6   Amination or Amino-de-hydrogenation[139]

$$\text{ArH} + \text{HN}_3 \xrightarrow{\text{AlCl}_3} \text{ArNH}_2$$

Aromatic compounds can be converted to primary aromatic amines, in 10 to 65% yields, by treatment with hydrazoic acid HN_3 in the presence of $AlCl_3$ or H_2SO_4.[140] Higher yields (> 90%) have been reported with trimethylsilyl azide Me_3SiN_3 and triflic acid F_3CSO_2OH.[141] Tertiary amines have been prepared in fairly good yields (about 50 to 90%) by treatment of aromatic hydrocarbons with N-chlorodialkylamines, by heating in 96% sulfuric acid; or with $AlCl_3$ or $FeCl_3$ in nitroalkane solvents; or by irradiation.[142]

Tertiary (and to a lesser extent, secondary) aromatic amines can also be prepared in moderate to high yields by amination with an N-chlorodialkylamine (or an N-chloroalkyl-amine) and a metallic-ion catalyst (e.g., Fe^{2+}, Ti^{3+}, Cu^+, Cr^{2+}) in the presence of sulfuric acid.[143] The attacking species in this case is the aminium radical ion $R_2\overset{\oplus}{N}H\cdot$ formed by[144]

$$R_2\overset{\oplus}{N}HCl + M^+ \longrightarrow R_2\overset{\oplus}{N}H\cdot + M^{2+} + Cl^-$$

Because attack is by a positive species (even though it is a free radical), orientation is similar to that in other electrophilic substitutions (e.g., phenol and acetanilide give ortho and para substitution, mostly para). When an alkyl group is present, attack at the benzylic position competes with ring substitution. Aromatic rings containing only meta-directing groups do not give the reaction at all. Fused ring systems react well.[145]

Unusual orientation has been reported for amination with halomines and with NCl_3 in the presence of $AlCl_3$. For example, toluene gave predominately meta amination.[146] It has been suggested that initial attack in this case is by Cl^+ and that a nitrogen nucleophile (whose structure is not known but is represented here as NH_2^- for simplicity) adds to the resulting arenium ion, so that the initial reaction is addition to a carbon-carbon double bond followed by elimination of HCl:[147]

According to this suggestion, the electrophilic attack is at the para position (or the ortho, which leads to the same product) and the meta orientation of the amino group arises indirectly. This mechanism is called the σ-*substitution mechanism*.

Aromatic compounds that do not contain meta-directing groups can be converted to diarylamines by treatment with aryl azides in the presence of phenol at $-60°C$: ArH +

[139]For a review, see Kovacic, in Olah, Ref. 58, vol. 3, 1964, pp. 1493-1506.

[140]Kovacic; Russell; Bennett *J. Am. Chem. Soc.* **1964,** *86,* 1588.

[141]Olah; Ernst *J. Org. Chem.* **1989,** *54,* 1203.

[142]Bock; Kompa *Angew. Chem. Int. Ed. Engl.* **1965,** *4,* 783 [*Angew. Chem. 77,* 807], *Chem. Ber.* **1966,** *99,* 1347, 1357, 1361.

[143]For reviews, see Minisci *Top. Curr. Chem.* **1976,** *62,* 1-48, pp. 6-16, *Synthesis* **1973,** 1-24, pp. 2-12, Sosnovsky; Rawlinson *Adv. Free-Radical Chem.* **1972,** *4,* 203-284, pp. 213-238.

[144]For a review of aminium radical ions, see Chow *React. Intermed. (Plenum)* **1980,** *1,* 151-262.

[145]The reaction has been extended to the formation of primary aromatic amines, but the scope is narrow: Citterio; Gentile; Minisci; Navarrini; Serravalle; Ventura *J. Org. Chem.* **1984,** *49,* 4479.

[146]See Kovacic; Lange; Foot; Goralski; Hiller; Levisky *J. Am. Chem. Soc.* **1964,** *86,* 1650; Strand; Kovacic *J. Am. Chem. Soc.* **1973,** *95,* 2977.

[147]Kovacic; Levisky *J. Am. Chem. Soc.* **1966,** *88,* 1000.

Ar'N$_3$ → ArNHAr'.[148] Diarylamines are also obtained by the reaction of N-arylhydroxyl-amines with aromatic compounds (benzene, toluene, anisole) in the presence of F$_3$CCOOH: ArH + Ar'NHOH → ArNHAr'.[149]

Direct *amidation* can be carried out if an aromatic compound is heated with a hydroxamic acid in polyphosphoric acid, though the scope is essentially limited to phenolic ethers.[150]

$$ArH + R—\overset{\displaystyle O}{\underset{\displaystyle ||}{C}}—NH—OH \longrightarrow Ar—NH—\overset{\displaystyle O}{\underset{\displaystyle ||}{C}}—R$$

Also see **3-18** and **3-19**.

C. Sulfur Electrophiles

1-7 Sulfonation or Sulfo-de-hydrogenation

$$ArH + H_2SO_4 \longrightarrow ArSO_2OH$$

The sulfonation reaction is very broad in scope and many aromatic hydrocarbons (including fused ring systems), aryl halides, ethers, carboxylic acids, amines,[151] acylated amines, ke-tones, nitro compounds, and sulfonic acids have been sulfonated.[152] Phenols can also be successfully sulfonated, but attack at oxygen may compete.[153] Sulfonation is often accom-plished with concentrated sulfuric acid, but it can also be done with fuming sulfuric acid, SO$_3$, ClSO$_2$OH, or other reagents. As with nitration (**1-2**), reagents of a wide variety of activity are available to suit both highly active and highly inactive substrates. Since this is a reversible reaction (see **1-41**), it may be necessary to drive the reaction to completion. However, at low temperatures the reverse reaction is very slow and the forward reaction is practically irreversible.[154] SO$_3$ reacts much more rapidly than sulfuric acid—with benzene it is nearly instantaneous. Sulfones are often side products. When sulfonation is carried out on a benzene ring containing four or five alkyl and/or halogen groups, rearrangements usually occur (see **1-40**).

A great deal of work has been done on the mechanism,[155] chiefly by Cerfontain and co-workers. Mechanistic study is made difficult by the complicated nature of the solutions. Indications are that the electrophile varies with the reagent, though SO$_3$ is involved in all cases, either free or combined with a carrier. In aqueous H$_2$SO$_4$ solutions the electrophile is thought to be H$_3$SO$_4^+$ (or a combination of H$_2$SO$_4$ and H$_3$O$^+$) at concentrations below about 80 to 85% H$_2$SO$_4$, and H$_2$S$_2$O$_7$ (or a combination of H$_2$SO$_4$ and SO$_3$) at concentrations higher than this[156] (the changeover point varies with the substrate[157]). Evidence for a change

[148]Nakamura; Ohno; Oka *Synthesis* **1974**, 882. See also Takeuchi; Takano *J. Chem. Soc., Perkin Trans. 1* **1986**, 611.

[149]Shudo; Ohta; Okamoto *J. Am. Chem. Soc.* **1981**, *103*, 645.

[150]Wassmundt; Padegimas *J. Am. Chem. Soc.* **1967**, *89*, 7131; March; Engenito *J. Org. Chem.* **1981**, *46*, 4304.

[151]See Khelevin *J. Org. Chem. USSR* **1984**, *20*, 339, 1173, 1723, **1987**, *23*, 1709, **1988**, *24*, 535.

[152]For reviews, see Nelson, in Olah, Ref. 58, vol. 3, 1964, pp. 1355-1392; Gilbert, *Sulfonation and Related Reactions*; Wiley: New York, 1965, pp. 62-83, 87-124.

[153]See, for example de Wit; Woldhuis; Cerfontain *Recl. Trav. Chim. Pays-Bas* **1988**, *107*, 668.

[154]Spryskov *J. Gen. Chem. USSR* **1960**, *30*, 2433.

[155]For a monograph, see Cerfontain *Mechanistic Aspects in Aromatic Sulfonation and Desulfonation*; Wiley: New York, 1968. For reviews, see Cerfontain *Recl. Trav. Chim. Pays-Bas* **1985**, *104*, 153-165; Cerfontain; Kort *Int. J. Sulfur Chem. C* **1971**, *6*, 123-136; Taylor, in Bamford; Tipper, Ref. 1, pp. 56-77.

[156]Kort; Cerfontain *Recl. Trav. Chim. Pays-Bas* **1968**, *87*, 24, **1969**, *88*, 860; Maarsen; Cerfontain *J. Chem. Soc., Perkin Trans. 2* **1977**, 1003; Cerfontain; Lambrechts; Schaasberg-Nienhuis; Coombes; Hadjigeorgiou; Tucker *J. Chem. Soc., Perkin Trans. 2* **1985**, 659.

[157]See, for example, Kaandorp; Cerfontain *Recl. Trav. Chim. Pays-Bas* **1969**, *88*, 725.

in electrophile is that in the dilute and in the concentrated solutions the rate of the reaction was proportional to the activity of $H_3SO_4^+$ and $H_2S_2O_7$, respectively. Further evidence is that with toluene as substrate the two types of solution gave very different ortho/para ratios. The mechanism is essentially the same for both electrophiles and may be shown as:[156]

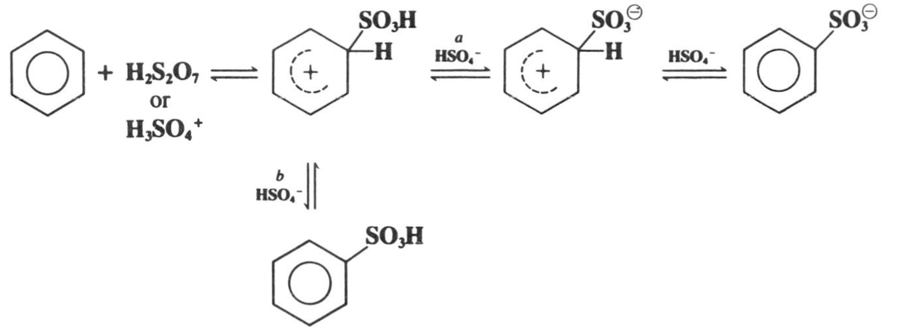

The other product of the first step is HSO_4^- or H_2O from $H_2S_2O_7$ or $H_3SO_4^+$, respectively. Path a is the principal route, except at very high H_2SO_4 concentrations, when path b becomes important. With $H_3SO_4^+$ the first step is rate-determining under all conditions, but with $H_2S_2O_7$ the first step is the slow step only up to about 96% H_2SO_4, when a subsequent proton transfer becomes partially rate-determining.[158] $H_2S_2O_7$ is more reactive than $H_3SO_4^+$. In fuming sulfuric acid (H_2SO_4 containing excess SO_3), the electrophile is thought to be $H_3S_2O_7^+$ (protonated $H_2S_2O_7$) up to about 104% H_2SO_4 and $H_2S_4O_{13}$ ($H_2SO_4 + 3SO_3$) beyond this concentration.[159] Finally, when pure SO_3 is the reagent in aprotic solvents, SO_3 itself is the actual electrophile.[160] Free SO_3 is the most reactive of all these species, so that attack here is generally fast and a subsequent step is usually rate-determining, at least in some solvents.

OS **II**, 42, 97, 482, 539; **III**, 288, 824; **IV**, 364; **VI**, 976.

1-8 Halosulfonation or Halosulfo-de-hydrogenation

$$ArH + ClSO_2OH \longrightarrow ArSO_2Cl$$

Aromatic sulfonyl chlorides can be prepared directly, by treatment of aromatic rings with chlorosulfuric acid.[161] Since sulfonic acids can also be prepared by the same reagent (**1-7**), it is likely that they are intermediates, being converted to the halides by excess chlorosulfuric acid.[162] The reaction has also been effected with bromo- and fluorosulfuric acids.

OS **I**, 8, 85.

1-9 Sulfurization

$$ArH + SCl_2 \xrightarrow{AlCl_3} ArSAr$$

[158]Kort; Cerfontain *Recl. Trav. Chim. Pays-Bas* **1967**, *86*, 865.
[159]Kort; Cerfontain *Recl. Trav. Chim. Pays-Bas* **1969**, *88*, 1298; Koeberg-Telder; Cerfontain *J. Chem. Soc., Perkin Trans. 2* **1973**, 633.
[160]Koeberg-Telder; Cerfontain *Recl. Trav. Chim. Pays-Bas* **1971**, *90*, 193, **1972**, *91*, 22; Lammertsma; Cerfontain *J. Chem. Soc., Perkin Trans. 2* **1980**, 28.
[161]For a review, see Gilbert, Ref. 152, pp. 84-87.
[162]For a discussion of the mechanism with this reagent, see van Albada; Cerfontain *J. Chem. Soc., Perkin Trans. 2* **1977**, 1548, 1557.

Diaryl sulfides can be prepared by treating aromatic compounds with SCl_2 and a Friedel–Crafts catalyst. Other reagents that can bring about the same result are S_2Cl_2, thionyl chloride, and even sulfur itself. A catalyst is not always necessary. The reaction has been used for ring closure:

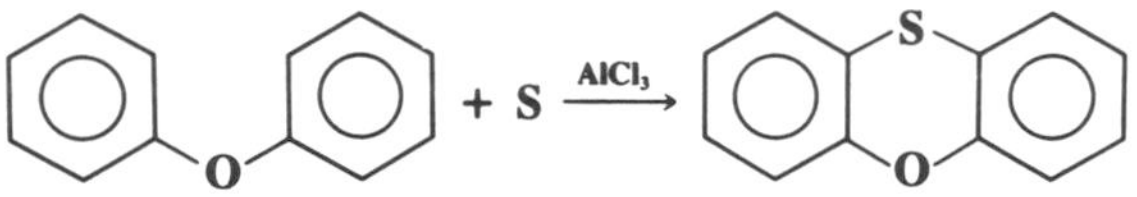

When thionyl chloride is used, diaryl sulfoxides are usually the main products.[163] Unsymmetrical diaryl sulfides can be obtained by treatment of an aromatic compound with an aryl sulfenyl chloride (ArSCl) in the presence of a trace amount of iron powder.[164] Aromatic amines and phenols can be alkylthiolated (giving mostly ortho product) by treatment with an alkyl disulfide and a Lewis acid catalyst.[165]

With certain substrates (primary amines with a chloro group, or a group not replaceable by chloro, in the para position), treatment with S_2Cl_2 and NaOH gives thiophenolate salts:

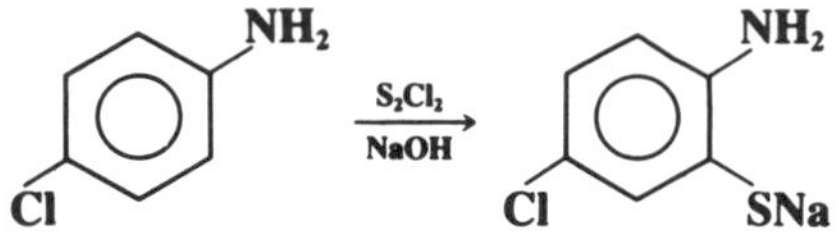

This is called the *Herz reaction*.[166]

OS **II**, 242, 485. Also see OS **I**, 574; **III**, 76.

1-10 Sulfonylation
Alkylsulfonylation or **Alkylsulfo-de-hydrogenation**

$$ArH + Ar'SO_2Cl \xrightarrow{AlCl_3} ArSO_2Ar'$$

Diaryl sulfones can be formed by treatment of aromatic compounds with aryl sulfonyl chlorides and a Friedel–Crafts catalyst.[167] This reaction is analogous to Friedel–Crafts acylation with carboxylic acid halides (**1-14**). In a better procedure, the aromatic compound is treated with an aryl sulfonic acid and P_2O_5 in polyphosphoric acid.[168] Still another method uses an arylsulfonic trifluoromethanesulfonic anhydride $ArSO_2OSO_2CF_3$ (generated in situ from $ArSO_2Br$ and CF_3SO_3Ag) without a catalyst.[169]

The reaction can be extended to the preparation of alkyl aryl sulfones by the use of a sulfonyl fluoride.[170]

[163]Nikolenko; Krizhechkovskaya *J. Gen. Chem. USSR* **1963**, *33*, 3664; Oae; Zalut *J. Am. Chem. Soc.* **1960**, *82*, 5359.
[164]Fujisawa; Kobori; Ohtsuka; Tsuchihashi *Tetrahedron Lett.* **1968**, 5071.
[165]Ranken; McKinnie *Synthesis* **1984**, 117, *J. Org. Chem.* **1989**, *54*, 2985.
[166]For a review, see Warburton *Chem. Rev.* **1957**, *57*, 1011-1020.
[167]For reviews, see Taylor, in Bamford; Tipper, Ref. 1, pp. 77-83; Jensen; Goldman, in Olah, Ref. 58, vol. 3, 1964, pp. 1319-1347.
[168]Graybill *J. Org. Chem.* **1967**, *32*, 2931; Sipe; Clary; White *Synthesis* **1984**, 283. See also Ueda; Uchiyama; Kano *Synthesis* **1984**, 323.
[169]Effenberger; Huthmacher *Chem. Ber.* **1976**, *109*, 2315. For similar methods, see Hancock; Tyobeka; Weigel *J. Chem. Res., (S)* **1980**, 270; Ono; Nakamura; Sato; Itoh *Chem. Lett.* **1988**, 395.
[170]Hyatt; White *Synthesis* **1984**, 214.

D. Halogen Electrophiles

1-11 Halogenation[171] or Halo-de-hydrogenation

$$ArH + Br_2 \xrightarrow{\text{Fe}} ArBr$$

1. *Chlorine and bromine.*   Aromatic compounds can be brominated or chlorinated by treatment with bromine or chlorine in the presence of a catalyst, most often iron. However, the real catalyst is not the iron itself, but the ferric bromide or ferric chloride formed in small amounts from the reaction between iron and the reagent. Ferric chloride and other Lewis acids are often directly used as catalysts, as is iodine. When thallium(III) acetate is the catalyst, many substrates are brominated with high regioselectivity para to an ortho-para-directing group.[172] For active substrates, including amines, phenols, naphthalene, and polyalkylbenzenes[173] such as mesitylene and isodurene, no catalyst is needed. Indeed, for amines and phenols the reaction is so rapid that it is carried out with a dilute solution of Br_2 or Cl_2 in water at room temperature. Even so, with amines it is not possible to stop the reaction before all the available ortho and para positions are substituted, because the initially formed haloamines are weaker bases than the original amines and are less likely to be protonated by the liberated HX.[174] For this reason, primary amines are often converted to the corresponding anilides if monosubstitution is desired. With phenols it is possible to stop after one group has entered.[175] The rapid room-temperature reaction with amines and phenols is often used as a test for these compounds. Chlorine is a more active reagent than bromine. Phenols can be brominated exclusively in the ortho position (disubstitution of phenol gives 2,6-dibromophenol) by treatment about $-70°C$ with Br_2 in the presence of t-butylamine or triethylenediamine, which precipitates out the liberated HBr.[176] Predominant ortho chlorination[177] of phenols has been achieved with chlorinated cyclohexadienes,[178] while para chlorination of phenols, phenolic ethers, and amines can be accomplished with N-chloroamines[179] and with N-chlorodimethylsulfonium chloride $Me_2\overset{\oplus}{S}Cl\ Cl^-$.[180] The last method is also successful for bromination. On the other hand, certain alkylated phenols can be brominated in the meta positions with Br_2 in the super-acid solution SbF_5–HF.[181] It is likely that the meta orientation is the result of conversion by the super acid of the OH group

[171]For a monograph, see de la Mare *Electrophilic Halogenation*; Cambridge University Press: Cambridge, 1976. For reviews, see Buehler; Pearson *Survey of Organic Synthesis*; Wiley: New York, 1970, pp. 392-404; Braendlin; McBee, in Olah, Ref. 58, vol. 3, 1964, pp. 1517-1593. For a review of the halogenation of heterocyclic compounds, see Eisch *Adv. Heterocycl. Chem.* **1966**, *7*, 1-37. For a list of reagents, with references, see Larock *Comprehensive Organic Transformations*; VCH: New York, 1989, pp. 315-318.

[172]McKillop; Bromley; Taylor *J. Org. Chem.* **1972**, *37*, 88.

[173]For a review of aromatic substitution on polyalkylbenzenes, see Baciocchi; Illuminati *Prog. Phys. Org. Chem.* **1967**, *5*, 1-79.

[174]Monobromination (para) of aromatic amines has been achieved with tetrabutylammonium tribromide: Berthelot; Guette; Desbène; Basselier; Chaquin; Masure *Can. J. Chem.* **1989**, *67*, 2061. For another procedure, see Onaka; Izumi *Chem. Lett.* **1984**, 2007.

[175]For a review of the halogenation of phenols, see Brittain; de la Mare, in Patai; Rappoport *The Chemistry of Functional Groups, Supplement D*, pt. 1; Wiley: New York, 1983, pp. 522-532.

[176]Pearson; Wysong; Breder *J. Org. Chem.* **1967**, *32*, 2358.

[177]For other methods of regioselective chlorination or bromination, see Schmitz; Pagenkopf *J. Prakt. Chem.* **1985**, *327*, 998; Watson *J. Org. Chem.* **1985**, *50*, 2145; Smith; Butters; Paget; Nay *Synthesis* **1985**, 1157, *Tetrahedron Lett.* **1988**, *29*, 1319; Kodomari; Takahashi; Yoshitomi *Chem. Lett.* **1987**, 1901; Kamigata; Satoh; Yoshida; Matsuyama; Kameyama *Bull. Chem. Soc. Jpn.* **1988**, *61*, 2226; de la Vega; Sasson *J. Chem. Soc., Chem. Commun.* **1989**, 653.

[178]Guy; Lemaire; Guette *Tetrahedron* **1982**, *38*, 2339, 2347; Lemaire; Guy; Guette *Bull. Soc. Chim. Fr.* **1985**, 477.

[179]Lindsay Smith; McKeer; Taylor *J. Chem. Soc., Perkin Trans. 2* **1987**, 1533, **1988**, 385, **1989**, 1529, 1537. See also Minisci; Vismara; Fontana; Platone; Faraci *J. Chem. Soc., Perkin Trans. 2* **1989**, 123.

[180]Olah; Ohannesian; Arvanaghi *Synthesis* **1986**, 868.

[181]Jacquesy; Jouannetaud; Makani *J. Chem. Soc., Chem. Commun.* **1980**, 110.

to the OH_2^+ group, which should be meta-directing because of its positive charge. Bromination and the Sandmeyer reaction (**4-25**) can be carried out in one laboratory step by treatment of an aromatic primary amine with $CuBr_2$ and t-butyl nitrite, e.g.,[182]

Other reagents have been used, among them HOCl,[183] HOBr, and N-chloro and N-bromo amides (especially N-bromosuccinimide and tetraalkylammonium polyhalides[184]). In all but the last of these cases the reaction is catalyzed by the addition of acids. Dibromoisocyanuric acid in H_2SO_4 is a very good brominating agent[185] for substrates with strongly deactivating substituents.[186] Two particularly powerful reagents consist of (1) S_2Cl_2 and $AlCl_3$ in sulfuryl chloride (SO_2Cl_2) (the *BMC reagent*)[187] and (2) dichlorine oxide Cl_2O and a strong acid such as sulfuric.[188] If the substrate contains alkyl groups, side-chain halogenation (**4-1**) is possible with most of the reagents mentioned, including chlorine and bromine. Since side-chain halogenation is catalyzed by light, the reactions should be run in the absence of light wherever possible.

For reactions in the absence of a catalyst, the attacking entity is simply Br_2 or Cl_2 that has been polarized by the ring.[189]

Evidence for molecular chlorine or bromine as the attacking species in these cases is that acids, bases, and other ions, especially chloride ion, accelerate the rate about equally, though if chlorine dissociated into Cl^+ and Cl^-, the addition of chloride should decrease the rate and the addition of acids should increase it. The conjugate base of **26** (4-bromo-2,5-cyclohexadienone) has been detected spectrally in the aqueous bromination of phenol.[190]

When a Lewis-acid catalyst is used with chlorine or bromine, the attacking entity may be Cl^+ or Br^+, formed by $FeCl_3 + Br_2 \rightarrow FeCl_3Br^- + Br^+$, or it may be Cl_2 or Br_2, polarized by the catalyst. With other reagents, the attacking entity in brominations may be Br^+ or a species such as H_2OBr^+ (the conjugate acid of HOBr), in which H_2O is a carrier of Br^+.[191]

[182]Doyle; Van Lente; Mowat; Fobare *J. Org. Chem.* **1980**, *45*, 2570.

[183]For the use of calcium hypochlorite, see Nwaukwa; Keehn *Synth. Commun.* **1989**, *19*, 799.

[184]See Kajigaeshi; Moriwaki; Tanaka; Fujisaki; Kakinami; Okamoto *J. Chem. Soc., Perkin Trans. 1* **1990**, 897, and other papers in this series.

[185]Nitrobenzene is pentabrominated in 1 min with this reagent in 15% oleum at room temperature.

[186]Gottardi *Monatsh. Chem.* **1968**, *99*, 815, **1969**, *100*, 42.

[187]Ballester; Molinet; Castañer *J. Am. Chem. Soc.* **1960**, *82*, 4254; Andrews, Glidewell; Walton *J. Chem. Res. (S)* **1978**, 294.

[188]Marsh; Farnham; Sam; Smart *J. Am. Chem. Soc.* **1982**, *104*, 4680.

[189]For reviews of the mechanism of halogenation, see de le Mare, Ref. 171; de la Mare; Swedlund, in Patai *The Chemistry of the Carbon–Halogen Bond*, pt. 1; Wiley: New York, 1973; pp. 490-536; Taylor, in Bamford; Tipper, Ref. 1, pp. 83-139; Berliner *J. Chem. Educ.* **1966**, *43*, 124-133. See also Schubert; Dial *J. Am. Chem. Soc.* **1975**, *97*, 3877; Keefer; Andrews *J. Am. Chem. Soc.* **1977**, *99*, 5693; Briggs; de la Mare; Hall *J. Chem. Soc., Perkin Trans. 2* **1977**, 106; Tee; Paventi; Bennett *J. Am. Chem. Soc.* **1989**, *111*, 2233.

[190]Tee; Iyengar; Paventi *J. Org. Chem.* **1983**, *48*, 759. See also Tee; Iyengar *J. Am. Chem. Soc.* **1985**, *107*, 455, *Can. J. Chem.* **1990**, *68*, 1769.

[191]For discussions, see Gilow; Ridd *J. Chem. Soc., Perkin Trans. 2* **1973**, 1321; Rao; Mali; Dangat *Tetrahedron* **1978**, *34*, 205.

With HOCl in water the electrophile may be Cl_2O, Cl_2, or H_2OCl^+; in acetic acid it is generally AcOCl. All these species are more reactive than HOCl itself.[192] It is extremely doubtful that Cl^+ is a significant electrophile in chlorinations by HOCl.[192] It has been demonstrated in the reaction between N-methylaniline and calcium hypochlorite that the chlorine attacking entity attacks the *nitrogen* to give N-chloro-N-methylaniline, which rearranges (as in **1-35**) to give a mixture of ring-chlorinated N-methylanilines in which the ortho isomer predominates.[193]

$FeCl_3$ itself, and also $CuCl_2$, $SbCl_5$, etc.,[194] can give moderate yields of aryl chlorides.[195] The electrophile might be a species such as $FeCl_2^+$, but the reactions can also take place by a free-radical mechanism.[196]

When chlorination or bromination is carried out at high temperatures (e.g., 300 to 400°C), ortho-para-directing groups direct meta and vice versa.[197] A different mechanism operates here, which is not completely understood. It is also possible for bromination to take place by the SE1 mechanism, e.g., in the *t*-BuOK-catalyzed bromination of 1,3,5-tribromobenzene.[198]

2. Iodine. Iodine is the least reactive of the halogens in aromatic substitution.[199] Except for active substrates, an oxidizing agent must normally be present to oxidize I_2 to a better electrophile.[200] Examples of such oxidizing agents are HNO_3, HIO_3, SO_3, peracetic acid, and H_2O_2.[201] ICl is a better iodinating agent than iodine itself.[202] Among other reagents used have been IF (prepared directly from the elements),[203] benzyltrimethylammonium dichloroiodate (which iodinates phenols, aromatic amines, and N-acylated aromatic amines),[204] and the combination of iodine cyanide ICN and a Lewis acid, which is a good reagent for active substrates.[205] Iodination can also be accomplished by treatment of the substrate with I_2 in the presence of copper salts,[206] $SbCl_5$,[207] silver trifluoromethanesulfonate CF_3SO_3Ag,[208] $HgO-BF_4$,[209] Al_2O_3,[210] $AgNO_3$,[211] Ag_2SO_4,[212] or thallium(I) acetate.[213] The TlOAc method is regioselective for ortho iodination.

The actual attacking species is less clear than with bromine or chlorine. Iodine itself is too unreactive, except for active species such as phenols, where there is good evidence that

[192]Swain; Crist *J. Am. Chem. Soc.* **1972**, *94*, 3195.

[193]Haberfield; Paul *J. Am. Chem. Soc.* **1965**, *87*, 5502; Gassman; Campbell *J. Am. Chem. Soc.* **1972**, *94*, 3891; Paul; Haberfield *J. Org. Chem.* **1976**, *41*, 3170.

[194]Kovacic; Wu; Stewart *J. Am. Chem. Soc.* **1960**, *82*, 1917; Ware; Borchert *J. Org. Chem.* **1961**, *26*, 2267; Commandeur; Mathais; Raynier; Waegell *Nouv. J. Chim.* **1979**, *3*, 385; Makhon'kov; Cheprakov; Rodkin; Beletskaya *J. Org. Chem. USSR* **1988**, *24*, 211; Kodomari; Satoh; Yoshitomi *J. Org. Chem.* **1988**, *53*, 2093.

[195]For a review of halogenations with metal halides, see Kovacic, in Olah, Ref. 58, vol. 4, 1965, pp. 111-126.

[196]Nonhebel *J. Chem. Soc.* **1963**, 1216; Nonhebel; Russell *Tetrahedron* **1969**, *25*, 3493.

[197]For a review of this type of reaction, see Kooyman *Pure. Appl. Chem.* **1963**, *7*, 193-202.

[198]Mach; Bunnett *J. Am. Chem. Soc.* **1974**, *96*, 936.

[199]For reviews of I_2 as an electrophilic reagent, see Pizey, in Pizey *Synthetic Reagents*, vol. 3; Wiley: New York, 1977, pp. 227-276. For reviews of aromatic iodination, see Merkushev *Synthesis* **1988**, 923-937, *Russ. Chem. Rev.* **1984**, *53*, 343-350.

[200]Butler *J. Chem. Educ.* **1971**, *48*, 508.

[201]For a discussion, see Makhon'kov; Cheprakov; Beletskaya *J. Org. Chem. USSR* **1989**, *24*, 2029.

[202]For a review of ICl, see McCleland, in Pizey, Ref. 199, vol. 5, 1983, pp. 85-164.

[203]Rozen; Zamir *J. Org. Chem.* **1990**, *55*, 3552.

[204]See Kajigaeshi; Kakinami; Watanabe; Okamoto *Bull. Chem. Soc. Jpn.* **1989**, *62*, 1349, and references cited therein.

[205]Radner *Acta Chem. Scand.* **1989**, *43*, 481. For another method, see Edgar; Falling *J. Org. Chem.* **1990**, *55*, 5287.

[206]Baird; Surridge *J. Org. Chem.* **1970**, *35*, 3436; Horiuchi; Satoh *Bull. Chem. Soc. Jpn.* **1984**, *57*, 2691; Makhon'kov; Cheprakov; Rodkin; Beletskaya *J. Org. Chem. USSR* **1986**, *22*, 1003.

[207]Uemura; Onoe; Okano *Bull. Chem. Soc. Jpn.* **1974**, *47*, 147.

[208]Kobayashi; Kumadaki; Yoshida *J. Chem. Res. (S)* **1977**, 215. For a similar procedure, see Merkushev; Simakhina; Koveshnikova *Synthesis* **1980**, 486.

[209]Barluenga; Campos; González; Asensio *J. Chem. Soc., Perkin Trans. 1* **1984**, 2623.

[210]Pagni; Kabalka; Boothe; Gaetano; Stewart; Conaway; Dial; Gray; Larson; Luidhart *J. Org. Chem.* **1988**, *53*, 4477.

[211]Sy; Lodge *Tetrahedron Lett.* **1989**, *30*, 3769.

[212]Sy; Lodge; By *Synth. Commun.* **1990**, *20*, 877.

[213]Cambie; Rutledge; Smith-Palmer; Woodgate *J. Chem. Soc., Perkin Trans. 1* **1976**, 1161.

I_2 is the attacking entity.[214] There is evidence that AcOI may be the attacking entity when peroxyacetic acid is the oxidizing agent,[215] and I_3^+ when SO_3 or HIO_3 is the oxidizing agent.[216] I^+ has been implicated in several procedures.[216a] For an indirect method for accomplishing aromatic iodination, see **2-30**.

3. *Fluorine.* Direct fluorination of aromatic rings with F_2 is not feasible at room temperature, because of the extreme reactivity of F_2.[217] It has been accomplished at low temperatures (e.g., -70 to $-20°C$, depending on the substrate),[218] but the reaction is not yet of preparative significance. Fluorination has also been reported with silver difluoride AgF_2,[219] with cesium fluoroxysulfate $CsSO_4F$,[220] with acetyl hypofluorite CH_3COOF (generated from F_2 and sodium acetate),[221] with XeF_2,[222] with an N-fluoroperfluoroalkyl sulfonamide, e.g., $(CF_3SO_2)_2NF$,[223] and with fluoroxytrifluoromethane CF_3OF[224] under various conditions and with various yields, in some cases by electrophilic and in other cases by free-radical mechanisms. However, none of these methods seems likely to displace the Schiemann reaction (**3-24**) as the most common method for introducing fluorine into aromatic rings.

The overall effectiveness of reagents in aromatic substitution is $Cl_2 > BrCl > Br_2 > ICl > I_2$.

OS **I**, 111, 121, 123, 128, 207, 323; **II**, 95, 97, 100, 173, 196, 343, 347, 349, 357, 592; **III**, 132, 134, 138, 262, 267, 575, 796; **IV**, 114, 166, 256, 545, 547, 872, 947; **V**, 117, 147, 206, 346; **VI**, 181, 700; **67**, 222. Also see OS **II**, 128.

E. Carbon Electrophiles In the reactions in this section, a new carbon–carbon bond is formed. With respect to the aromatic ring, they are electrophilic substitutions, because a positive species attacks the ring. We treat them in this manner because it is customary. However, with respect to the electrophile, most of these reactions are nucleophilic substitutions, and what was said in Chapter 10 is pertinent to them.

1-12 Friedel–Crafts Alkylation
Alkylation or **Alkyl-de-hydrogenation**

$$ArH + RCl \xrightarrow{\text{AlCl}_3} ArR$$

[214]Grovenstein; Aprahamian; Bryan; Gnanapragasam; Kilby; McKelvey; Sullivan *J. Am. Chem. Soc.* **1973**, *95*, 4261.

[215]Ogata; Urasaki *J. Chem. Soc. C* **1970**, 1689.

[216]Arotsky; Butler; Darby *J. Chem. Soc. C* **1970**, 1480.

[216a]Galli *J. Org. Chem.* **1991**, *56*, 3238.

[217]For a monograph on fluorinating agents, see German; Zemskov *New Fluorinating Agents in Organic Synthesis*; Springer: New York, 1989. For reviews of F_2 in organic synthesis, see Purrington; Kagen; Patrick *Chem. Rev.* **1986**, *86*, 997-1018; Grakauskas, *Intra-Sci. Chem. Rep.* **1971**, *5*, 85-104. For a review of fluoroaromatic compounds, see Hewitt; Silvester *Aldrichimica Acta* **1988**, *21*, 3-10.

[218]Grakauskas *J. Org. Chem.* **1970**, *35*, 723; Cacace; Giacomello; Wolf *J. Am. Chem. Soc.* **1980**, *102*, 3511; Stavber; Zupan *J. Org. Chem.* **1983**, *48*, 2223. See also Purrington; Woodard *J. Org. Chem.* **1991**, *56*, 142.

[219]Zweig; Fischer; Lancaster *J. Org. Chem.* **1980**, *45*, 3597.

[220]Ip; Arthur; Winans; Appelman *J. Am. Chem. Soc.* **1981**, *103*, 1964; Stavber; Zupan *J. Org. Chem.* **1985**, *50*, 3609; Appelman; Basile; Hayatsu *Tetrahedron* **1984**, *40*, 189; Patrick; Darling *J. Org. Chem.* **1986**, *51*, 3242.

[221]See Hebel; Lerman; Rozen *Bull. Soc. Chim. Fr.* **1986**, 861; Visser; Bakker; van Halteren; Herscheid; Brinkman; Hoekstra *J. Org. Chem.* **1986**, *51*, 1886.

[222]Shaw; Hyman; Filler *J. Am. Chem. Soc.* **1969**, *91*, 1563, **1970**, *92*, 6498, *J. Org. Chem.* **1971**, *36*, 2917; Mackenzie; Fajer *J. Am. Chem. Soc.* **1970**, *92*, 4994; Filler *Isr. J. Chem.* **1978**, *17*, 71.

[223]Singh; DesMarteau; Zuberi; Witz; Huang *J. Am. Chem. Soc.* **1987**, *109*, 7194.

[224]Barton; Ganguly; Hesse; Loo; Pechet *Chem. Commun.* **1968**, 806; Kollonitsch; Barash; Doldouras *J. Am. Chem. Soc.* **1970**, *92*, 7494; Patrick; Cantrell; Chang *J. Am. Chem. Soc.* **1979**, *101*, 7434; Fifolt; Olczak; Mundhenke; Bieron *J. Org. Chem.* **1985**, *50*, 4576. For a review of this reagent, see Barton *Pure. Appl. Chem.* **1977**, *49*, 1241-1249.

The alkylation of aromatic rings, called *Friedel–Crafts alkylation*, is a reaction of very broad scope.[225] The most important reagents are alkyl halides, olefins, and alcohols, but many other types of reagent have also been employed.[225] When alkyl halides are used, the reactivity order is $F > Cl > Br > I$[226]; e.g., $FCH_2CH_2CH_2Cl$ reacts with benzene to give $PhCH_2CH_2CH_2Cl$[227] when the catalyst is BCl_3. By the use of this catalyst, it is therefore possible to place a haloalkyl group on a ring (see also **1-24**).[228] Di- and trihalides, when all the halogens are the same, usually react with more than one molecule of aromatic compound; it is usually not possible to stop the reaction earlier.[229] Thus, benzene with CH_2Cl_2 gives not $PhCH_2Cl$, but Ph_2CH_2; benzene with $CHCl_3$ gives Ph_3CH. With CCl_4, however, the reaction stops when only three rings have been substituted to give Ph_3CCl.

Olefins are especially good alkylating agents. With respect to them the reaction is addition of ArH to a C=C double bond:

$$Ar{-}H \; + \; {-}\underset{|}{\overset{|}{C}}{=}\underset{|}{\overset{|}{C}}{-} \;\; \xrightarrow[\text{H}^+]{\text{AlCl}_3} \;\; {-}\underset{|}{\overset{\overset{\displaystyle Ar}{|}}{C}}{-}\underset{|}{\overset{\overset{\displaystyle H}{|}}{C}}{-}$$

Acetylene reacts with 2 moles of aromatic compound to give 1,1-diarylethanes, but other alkynes react poorly, if at all. Alcohols are more active than alkyl halides, though if a Lewis-acid catalyst is used, more catalyst is required, since the catalyst complexes with the OH group. However, proton acids, especially H_2SO_4, are often used to catalyze alkylation with alcohols. When carboxylic esters are the reagents, there is competition between alkylation and acylation (**1-14**). Though this competition can often be controlled by choice of catalyst, and alkylation is usually favored, carboxylic esters are not often employed in Friedel–Crafts reactions. Other alkylating agents are ethers, thiols, sulfates, sulfonates, alkyl nitro compounds,[230] and even alkanes and cycloalkanes, under conditions where these are converted to carbocations. Notable here are ethylene oxide, which puts the CH_2CH_2OH group onto the ring, and cyclopropane. For all types of reagent the reactivity order is allylic ~ benzylic > tertiary > secondary > primary.

Regardless of which reagent is used, a catalyst is nearly always required.[231] Aluminum chloride and boron trifluoride are the most common, but many other Lewis acids have been used, and also proton acids such as HF and H_2SO_4.[232] For active halides a trace of a less

---

[225]For a monograph, see Roberts; Khalaf *Friedel–Crafts Alkylation Chemistry*; Marcel Dekker: New York, 1984. For a treatise on Friedel–Crafts reactions in general, see Olah *Friedel–Crafts and Related Reactions*; Wiley: New York, 1963-1965. Volume 1 covers general aspects, such as catalyst activity, intermediate complexes, etc. Volume 2 covers alkylation and related reactions. In this volume the various reagents are treated by the indicated authors as follows: alkenes and alkanes, Patinkin, Friedman, pp. 1-288; dienes and substituted alkenes, Koncos; Friedman, pp. 289-412; alkynes, Franzen, pp. 413-416; alkyl halides, Drahowzal, pp. 417-475; alcohols and ethers, Schriesheim, pp. 477-595; sulfonates and inorganic esters, Drahowzal, pp. 641-658. For a monograph in which five chapters of the above treatise are reprinted and more recent material added, see Olah *Friedel–Crafts Chemistry*; Wiley: New York, 1973.

[226]For example, see Calloway *J. Am. Chem. Soc.* **1937**, *59*, 1474; Brown; Jungk *J. Am. Chem. Soc.* **1955**, *77*, 5584.

[227]Olah; Kuhn *J. Org. Chem.* **1964**, *29*, 2317.

[228]For a review of selectivity in this reaction, i.e., which group preferentially attacks when the reagent contains two or more, see Olah, in Olah, Ref. 225, vol. 1, pp. 881-905. This review also covers the case of alkylation vs. acylation.

[229]It has proven possible in some cases. Thus, arenes ArH have been converted to $ArCCl_3$ with CCl_4 and excess $AlCl_3$: Raabe; Hörhold *J. Prakt. Chem.* **1987**, *329*, 1131; Belen'kii; Brokhovetsky; Krayushkin *Chem. Scr.* **1989**, *29*, 81.

[230]Bonvino; Casini; Ferappi; Cingolani; Pietroni *Tetrahedron* **1981**, *37*, 615.

[231]There are a few exceptions. Certain alkyl and vinylic triflates alkylate aromatic rings without a catalyst; see Gramstad; Haszeldine *J. Chem. Soc.* **1957**, 4069; Olah; Nishimura *J. Am. Chem. Soc.* **1974**, *96*, 2214; Stang; Anderson *Tetrahedron Lett.* **1977**, 1485, *J. Am. Chem. Soc.* **1978**, *100*, 1520.

[232]For a review of catalysts and solvents in Friedel–Crafts reactions, see Olah, in Olah, Ref. 225, vol. 1, pp. 201-366, 853-881.

active catalyst, e.g., $ZnCl_2$, may be enough. For an unreactive halide, such as chloromethane, a more powerful catalyst is needed, for example, $AlCl_3$, and in larger amounts. In some cases, especially with olefins, a Lewis-acid catalyst causes reaction only if a small amount of proton-donating cocatalyst is present. Catalysts have been arranged in the following order of overall reactivity: $AlBr_3 > AlCl_3 > GaCl_3 > FeCl_3 > SbCl_5{}^{233} > ZrCl_4$, $SnCl_4 > BCl_3$, BF_3, $SbCl_3$;[234] but the reactivity order in each case depends on the substrate, reagent, and conditions. Nafion-H, a superacidic perfluorinated resinsulfonic acid, is a very good catalyst for gas phase alkylations with alkyl halides, alcohols, or olefins.[235]

Friedel–Crafts alkylation is unusual among the principal aromatic substitutions in that the entering group is activating so that di- and polyalkylation are frequently observed. However, the activating effect of simple alkyl groups (e.g., ethyl, isopropyl) is such that compounds with these groups as substituents are attacked in Friedel–Crafts alkylations only about 1.5 to 3 times as fast as benzene,[236] so it is often possible to obtain high yields of monoalkyl product. Actually, the fact that di- and polyalkyl derivatives are frequently obtained is not due to the small difference in reactivity but to the circumstance that alkyl-benzenes are preferentially soluble in the catalyst layer, where the reaction actually takes place.[237] This factor can be removed by the use of a suitable solvent, by high temperatures, or by high-speed stirring.

Also unusual is the fact that the OH, OR, NH_2, etc., groups do not facilitate the reaction, since the catalyst coordinates with these basic groups. Although phenols give the usual Friedel–Crafts reactions, orienting ortho and para, the reaction is very poor for amines. However, amines can undergo the reaction if olefins are used as reagents and aluminum anilides as catalysts.[238] In this method the catalyst is prepared by treating the amine to be alkylated with $\frac{1}{3}$ mole of $AlCl_3$. A similar reaction can be performed with phenols, though here the catalyst is $Al(OAr)_3$.[239] Primary aromatic amines (and phenols) can be methylated regioselectively in the ortho position by an indirect method (see **1-26**). For an indirect method for regioselective ortho methylation of phenols, see p. 872.

Naphthalene and other fused ring compounds generally give poor yields in Friedel–Crafts alkylation, because they are so reactive that they react with the catalyst. Heterocyclic rings are usually also poor substrates for the reaction. Although some furans and thiophenes have been alkylated, a true alkylation of a pyridine or a quinoline has never been described.[240] However, alkylation of pyridine and other nitrogen heterocycles can be accomplished by a free radical (**4-23**) and by a nucleophilic method (**3-17**).

In most cases, meta-directing groups make the ring too inactive for alkylation. Nitro-benzene cannot be alkylated, and there are only a few reports of successful Friedel–Crafts alkylations when electron-withdrawing groups are present.[241] This is not because the at-tacking species is not powerful enough; indeed we have seen (p. 518) that alkyl cations are among the most powerful of electrophiles. The difficulty is caused by the fact that, with inactive substrates, degradation and polymerization of the electrophile occurs before it can attack the ring. However, if an activating and a deactivating group are both present on a

[233]For a review of $SbCl_5$ as a Friedel–Crafts catalyst, see Yakobson; Furin *Synthesis* **1980**, 345-364.
[234]Russell *J. Am. Chem. Soc.* **1959**, *81*, 4834.
[235]For a review of Nafion-H in organic synthesis, see Olah; Iyer; Prakash *Synthesis* **1986**, 513-531.
[236]Condon *J. Am. Chem. Soc.* **1948**, *70*, 2265; Olah; Kuhn; Flood *J. Am. Chem. Soc.* **1962**, *84*, 1688.
[237]Francis *Chem. Rev.* **1948**, *43*, 257.
[238]For a review, see Stroh; Ebersberger; Haberland; Hahn *Newer Methods Prep. Org. Chem.* **1963**, *2*, 227-252. This article also appeared in *Angew. Chem.* **1957**, *69*, 124-131.
[239]Koshchii; Kozlikovskii; Matyusha *J. Org. Chem. USSR* **1988**, *24*, 1358; Laan; Giesen; Ward *Chem. Ind. (London)* **1989**, 354. For a review, see Stroh; Seydel; Hahn *Newer Methods Prep. Org. Chem.* **1963**, *2*, 337-359. This article also appeared in *Angew. Chem.* **1957**, *69*, 669-706.
[240]Drahowzal, in Olah, Ref. 225, vol. 2, p. 433.
[241]Campbell; Spaeth *J. Am. Chem. Soc.* **1959**, *81*, 5933; Yoneda; Fukuhara; Takahashi; Suzuki *Chem. Lett.* **1979**, 1003; Shen; Liu; Chen *J. Org. Chem.* **1990**, *55*, 3961.

ring, Friedel–Crafts alkylation can be accomplished.[242] Aromatic nitro compounds can be methylated by a nucleophilic mechanism (**3-17**).

An important synthetic limitation of Friedel–Crafts alkylation is that rearrangement frequently takes place in the reagent. For example, benzene treated with *n*-propyl bromide gives mostly isopropylbenzene (cumene) and much less *n*-propylbenzene. Rearrangement is usually in the order primary → secondary → tertiary and occurs mostly by migration of H^- but also of R^- (see discussion of rearrangement mechanisms in Chapter 18). It is therefore not usually possible to put a primary alkyl group (other than methyl and ethyl) onto an aromatic ring by Friedel–Crafts alkylation. Because of these rearrangements, *n*-alkylbenzenes are often prepared by *acylation* (**1-14**), followed by reduction (**9-37**).

An important use of the Friedel–Crafts alkylation reaction is to effect ring closure.[243] The most common method is to heat with aluminum chloride an aromatic compound having a halogen, hydroxy, or olefinic group in the proper position, as, for example, in the preparation of tetralin:

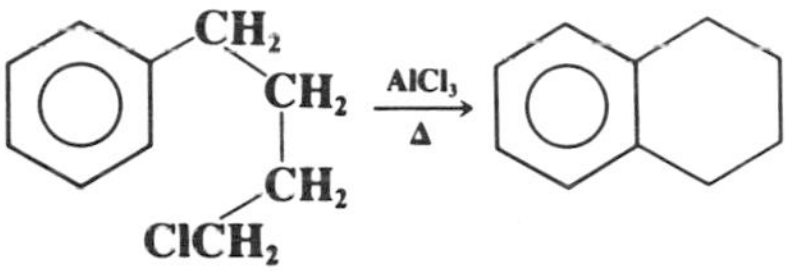

Another way of effecting ring closure through Friedel–Crafts alkylation is to use a reagent containing two groups, e.g.,

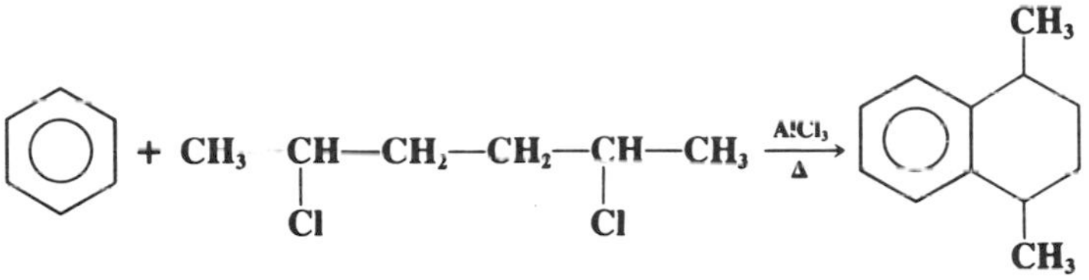

These reactions are most successful for the preparation of 6-membered rings,[244] though 5- and 7-membered rings have also been closed in this manner. For other Friedel–Crafts ring-closure reactions, see **1-13**, **1-14**, and **1-23**.

From what has been said thus far it is evident that the electrophile in Friedel–Crafts alkylation is a carbocation, at least in most cases.[245] This is in accord with the knowledge that carbocations rearrange in the direction primary → secondary → tertiary (see Chapter 18). In each case the cation is formed from the attacking reagent and the catalyst. For the three most important types of reagent these reactions are:

From alkyl halides: $\qquad$ $RCl + AlCl_3 \longrightarrow R^+ + AlCl_4^-$

From alcohols and Lewis acids:

$$ROH + AlCl_3 \longrightarrow ROAlCl_2 \longrightarrow R^+ + {}^-OAlCl_2$$

From alcohols and proton acids:

$$ROH + H^+ \longrightarrow ROH_2^+ \longrightarrow R^+ + H_2O$$

[242]Olah, in Olah, Ref. 225, vol. 1, p. 34.
[243]For a review, see Barclay, in Olah, Ref. 225, vol. 2, pp. 785-977.
[244]See Khalaf; Roberts *J. Org. Chem.* **1966**, *31*, 89.
[245]For a discussion of the mechanism see Taylor *Electrophilic Aromatic Substitution*, Ref. 1, pp. 188-213.

From olefins (a supply of protons is always required):

$$-C{=}C- + H^+ \longrightarrow H-\overset{|}{\underset{|}{C}}-\overset{\oplus}{\underset{|}{C}}-$$

There is direct evidence, from ir and nmr spectra, that the *t*-butyl cation is quantitatively formed when *t*-butyl chloride reacts with $AlCl_3$ in anhydrous liquid HCl.[246] In the case of olefins, Markovnikov's rule (p. 750) is followed. Carbocation formation is particularly easy from some reagents, because of the stability of the cations. Triphenylmethyl chloride[247] and 1-chloroadamantane[248] alkylate activated aromatic rings (e.g., phenols, amines) with no catalyst or solvent. Ions as stable as this are less reactive than other carbocations and often attack only active substrates. The tropylium ion, for example, alkylates anisole but not benzene.[249] It was noted on p. 337 that relatively stable vinylic cations can be generated from certain vinylic compounds. These have been used to introduce vinylic groups into aryl substrates.[250]

However, there is much evidence that many Friedel–Crafts alkylations, especially with primary reagents, do not go through a completely free carbocation. The ion may exist as a tight ion pair with, say, $AlCl_4^-$ as the counterion or as a complex. Among the evidence is that methylation of toluene by methyl bromide and methyl iodide gave different ortho/para/meta ratios,[251] though if the same species attacked in each case we would expect the same ratios. Other evidence is that, in some cases, the reaction kinetics are third order; first order each in aromatic substrate, attacking reagent, and catalyst.[252] In these instances a mechanism in which the carbocation is slowly formed and then rapidly attacks the ring is ruled out since, in such a mechanism, the substrate would not appear in the rate expression. Since it is known that free carbocations, once formed, rapidly attack the ring, there are no free carbocations here. Another possibility (with alkyl halides) is that some alkylations take place by an SN2 mechanism (with respect to the halide), in which case no carbocations would be involved at all. However, a completely SN2 mechanism requires inversion of configuration. Most investigations of Friedel–Crafts stereochemistry, even where an SN2 mechanism might most be expected, have resulted in total racemization, or at best a few percent inversion. A few exceptions have been found,[253] most notably where the reagent was optically active propylene oxide, in which case 100% inversion was reported.[254]

Rearrangement is possible even with a noncarbocation mechanism. The rearrangement could occur *before* the attack on the ring takes place. It has been shown that treatment of $CH_3{}^{14}CH_2Br$ with $AlBr_3$ in the absence of any aromatic compound gave a mixture of the starting material and $^{14}CH_3CH_2Br$.[255] Similar results were obtained with $PhCH_2{}^{14}CH_2Br$, in which case the rearrangement was so fast that the rate could be measured only below

[246]Kalchschmid; Mayer *Angew. Chem. Int. Ed. Engl.* **1976**, *15*, 773 [*Angew. Chem. 88*, 849].
[247]See, for example, Chuchani *J. Chem. Soc.* **1960**, 325; Hart; Cassis *J. Am. Chem. Soc.* **1954**, *76*, 1634; Hickinbottom *J. Chem. Soc.* **1934**, 1700; Chuchani and Zabicky *J. Chem. Soc. C* **1966**, 297.
[248]Takaku; Taniguchi; Inamoto *Synth. Commun.* **1971**, *1*, 141.
[249]Bryce-Smith; Perkins *J. Chem. Soc.* **1962**, 5295.
[250]Kitamura; Kobayashi; Taniguchi; Rappoport *J. Org. Chem.* **1982**, *47*, 5503.
[251]Brown; Jungk *J. Am. Chem. Soc.* **1956**, *78*, 2182.
[252]For examples, see Brown; Grayson *J. Am. Chem. Soc.* **1953**, *75*, 6285; Jungk; Smoot; Brown *J. Am. Chem. Soc.* **1956**, *78*, 2185; Choi; Brown *J. Am. Chem. Soc.* **1963**, *85*, 2596.
[253]Some instances of retention of configuration have been reported; a neighboring-group mechanism is likely in these cases: see Masuda; Nakajima; Suga *Bull. Chem. Soc. Jpn.* **1983**, *56*, 1089; Effenberger; Weber *Angew. Chem. Int. Ed. Engl.* **1987**, *26*, 142 [*Angew. Chem. 99*, 146].
[254]Nakajima; Suga; Sugita; Ichikawa *Tetrahedron* **1969**, *25*, 1807. For cases of almost complete inversion, with acyclic reagents, see Piccolo; Spreafico; Visentin; Valoti *J. Org. Chem.* **1985**, *50*, 3945; Piccolo; Azzena; Melloni; Delogu; Valoti *J. Org. Chem.* **1991**, *56*, 183.
[255]Sixma; Hendriks *Recl. Trav. Chim. Pays-Bas* **1956**, *75*, 169; Adema; Sixma *Recl. Trav. Chim. Pays-Bas* **1962**, *81*, 323, 336.

−70°C.[256] Rearrangement could also occur *after* formation of the product, since alkylation is reversible (see **1-37**).[257]

See **4-21** and **4-23** for free-radical alkylation.

OS **I**, 95, 548; **II**, 151, 229, 232, 236, 248; **III**, 343, 347, 504, 842; **IV**, 47, 520, 620, 665, 702, 898, 960; **V**, 130, 654; **VI**, 109, 744.

1-13   Friedel–Crafts Arylation. The Scholl Reaction
De-hydrogen-coupling

$$2ArH \xrightarrow[\text{H}^+]{\text{AlCl}_3} Ar\!-\!Ar + H_2$$

The coupling of two aromatic molecules by treatment with a Lewis acid and a proton acid is called the *Scholl reaction*.[258] Yields are low and the synthesis is seldom useful. High temperatures and strong-acid catalysts are required, and the reaction fails for substrates that are destroyed by these conditions. Because the reaction becomes important with large fused-ring systems, ordinary Friedel–Crafts reactions (**1-12**) on these systems are rare. For example, naphthalene gives binaphthyl under Friedel–Crafts conditions. Yields can be increased by the addition of a salt such as $CuCl_2$ or $FeCl_3$, which acts as an oxidant.[259]

Intramolecular Scholl reactions, e.g.,

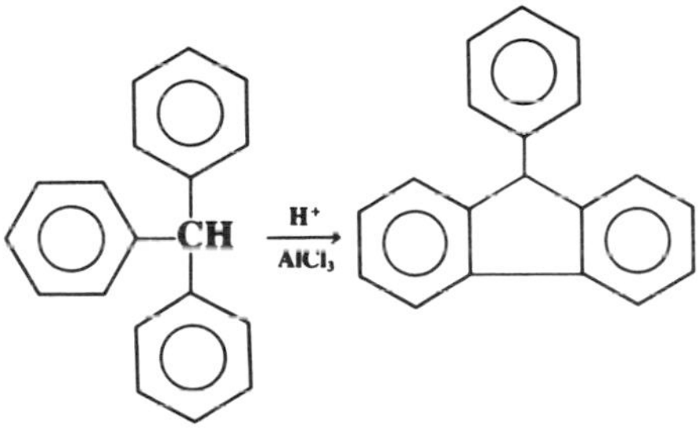

are much more successful than the intermolecular kind. The mechanism is not clear, but it may involve attack by a proton to give an arenium ion of the type **9** (p. 504), which would be the electrophile that attacks the other ring.[260] Sometimes arylations have been accomplished by treating aromatic substrates with particularly active aryl halides, especially fluorides. For free-radical arylations, see reactions **4-18** to **4-22**.

OS **IV**, 482. Also see OS **V**, 102, 952.

1-14   Friedel–Crafts Acylation
Acylation or **Acyl-de-hydrogenation**

$$ArH + RCOCl \xrightarrow{\text{AlCl}_3} ArCOR$$

The most important method for the preparation of aryl ketones is known as *Friedel–Crafts acylation*.[261] The reaction is of wide scope. Reagents used[262] are not only acyl halides but

---

[256]For a review of the use of isotopic labeling to study Friedel–Crafts reactions, see Roberts; Gibson *Isot. Org. Chem.* **1980**, *5*, 103-145.

[257]For an example, see Lee; Hamblin; Uthe *Can. J. Chem.* **1964**, *42*, 1771.

[258]For reviews, see Kovacic; Jones *Chem. Rev.* **1987**, *87*, 357-79; Balaban; Nenitzescu, in Olah, Ref. 225, vol. 2, pp. 979-1047.

[259]Kovacic; Koch *J. Org. Chem.* **1963**, *28*, 1864, **1965**, *30*, 3176; Kovacic; Wu *J. Org. Chem.* **1961**, *26*, 759, 762. For examples, with references, see Larock, Ref. 171, pp. 45-46.

[260]For a discussion, see Clowes *J. Chem. Soc. C* **1968**, 2519.

[261]For reviews of Friedel–Crafts acylation, see Olah *Friedel–Crafts and Related Reactions*; Wiley: New York, 1963-1964, as follows: vol. 1, Olah, pp. 91-115; vol. 3, Gore, pp. 1-381; Peto, pp. 535-910; Sethna, pp. 911-1002; Jensen; Goldman, pp. 1003-1032. For another review, see Gore *Chem. Ind. (London)* **1974**, 727-731.

[262]For a list of reagents, with references, see Larock, Ref. 171, pp. 703-704.

also carboxylic acids, anhydrides, and ketenes. Carboxylic esters usually give predominant alkylation (see **1-12**). R may be aryl as well as alkyl. The major disadvantages of Friedel–Crafts alkylation are not present here. Rearrangement of R is never found, and, because the RCO group is deactivating, the reaction stops cleanly after one group is introduced. All four acyl halides can be used, though chlorides are most commonly employed. The order of activity is usually, but not always, I > Br > Cl > F.[263] Catalysts are Lewis acids, similar to those in reaction **1-12**, but in acylation a little more than 1 mole of catalyst is required per mole of reagent, because the first mole coordinates with the oxygen of the reagent.[264]

$$\begin{array}{ccc} R-C-Cl \;+\; AlCl_3 & \longrightarrow & R-C-Cl \\ \;\;\;|| & & \overset{\oplus}{\;\;\;||}\;\;\overset{\ominus}{} \\ \;\;|O| & & \;\;|O-AlCl_3 \end{array}$$

Proton acids can be used as catalysts when the reagent is a carboxylic acid. The mixed carboxylic sulfonic anhydrides $RCOOSO_2CF_3$ are extremely reactive acylating agents and can smoothly acylate benzene without a catalyst.[265] With active substrates (e.g., aryl ethers, fused-ring systems, thiophenes), Friedel–Crafts acylation can be carried out with very small amounts of catalyst, often just a trace, or even sometimes with no catalyst at all. Ferric chloride, iodine, zinc chloride, and iron are the most common catalysts when the reactions is carried out in this manner.[266]

The reaction is quite successful for many types of substrate, including fused ring systems, which give poor results in **1-12**. Compounds containing ortho-para-directing groups, including alkyl, hydroxy, alkoxy, halogen, and acetamido groups, are easily acylated and give mainly or exclusively the para products, because of the relatively large size of the acyl group. However, aromatic amines give poor results. With amines and phenols there may be competition from N- or O-acylation; however, O-acylated phenols can be converted to C-acylated phenols by the Fries rearrangement (**1-30**). Friedel–Crafts acylation is usually prevented by meta-directing groups. Indeed, nitrobenzene is often used as a solvent for the reaction. Many heterocyclic systems, including furans, thiophenes, pyrans, and pyrroles but not pyridines or quinolines, can be acylated in good yield (however, pyridines and quinolines can be acylated by a free-radical mechanism, reaction **4-23**). Gore, in Ref. 261 (pp. 36–100; with tables, pp. 105–321), presents an exhaustive summary of the substrates to which this reaction has been applied.

When a mixed anhydride RCOOCOR′ is the reagent, two products are possible—ArCOR and ArCOR′. Which product predominates depends on two factors. If R contains electron-withdrawing groups, then ArCOR′ is chiefly formed, but if this factor is approximately constant in R and R′, the ketone with the larger R group predominantly forms.[267] This means that *formylations* of the ring do not occur with mixed anhydrides of formic acid HCOOCOR.

An important use of the Friedel–Crafts acylation is to effect ring closure.[268] This can be done if an acyl halide, anhydride, or acid group is in the proper position. An example is

[263]Yamase *Bull. Chem. Soc. Jpn.* **1961**, *34*, 480; Corriu *Bull. Soc. Chim. Fr.* **1965**, 821.

[264]The crystal structures of several of these complexes have been reported: Rasmussen; Broch *Acta Chem. Scand.* **1966**, *20*, 1351; Chevrier; Le Carpentier; Weiss *J. Am. Chem. Soc.* **1972**, *94*, 5718. For a review of these complexes, see Chevrier; Weiss *Angew. Chem. Int. Ed. Engl.* **1974**, *13*, 1-10 [*Angew. Chem. 86*, 12-21].

[265]Effenberger; Sohn; Epple *Chem. Ber.* **1983**, *116*, 1195. See also Keumi; Yoshimura; Shimada; Kitajima *Bull. Chem. Soc. Jpn.* **1988**, *44*, 455.

[266]For a review, see Pearson; Buehler *Synthesis* **1972**, 533-542.

[267]Edwards; Sibelle *J. Org. Chem.* **1963**, *28*, 674.

[268]For a review, see Sethna, Ref. 261. For examples, with references, see Larock, Ref. 171, pp. 704-708.

The reaction is used mostly to close 6-membered rings, but has also been done for 5- and 7-membered rings, which close less readily. Even larger rings can be closed by high-dilution techniques.[269] Tricyclic and larger systems are often made by using substrates containing one of the acyl groups on a ring. An example is the formation of acridone:

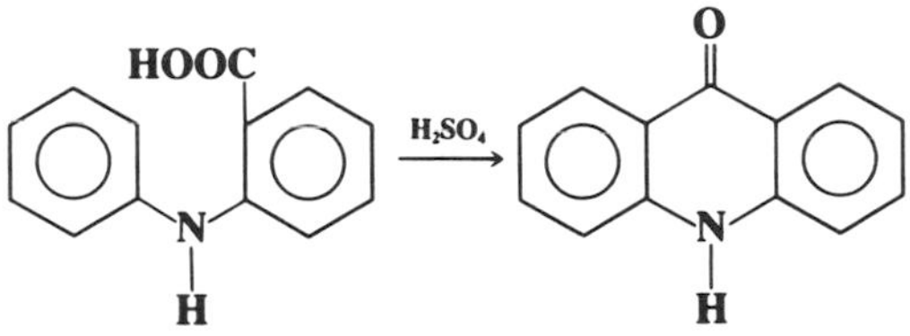

Many fused ring systems are made in this manner. If the bridging group is CO, the product is a quinone.[270] One of the most common catalysts for intramolecular Friedel–Crafts acylation is polyphosphoric acid[271] (because of its high potency), but $AlCl_3$, H_2SO_4, and other Lewis and proton acids are also used, though acylations with acyl halides are not generally catalyzed by proton acids.

Friedel–Crafts acylation can be carried out with cyclic anhydrides,[272] in which case the product contains a carboxyl group in the side chain. When succinic anhydride is used, the product is $ArCOCH_2CH_2COOH$. This can be reduced (**9-37**) to $ArCH_2CH_2CH_2COOH$, which can then be cyclized by an internal Friedel–Crafts acylation. The total process is called the *Haworth reaction*:[273]

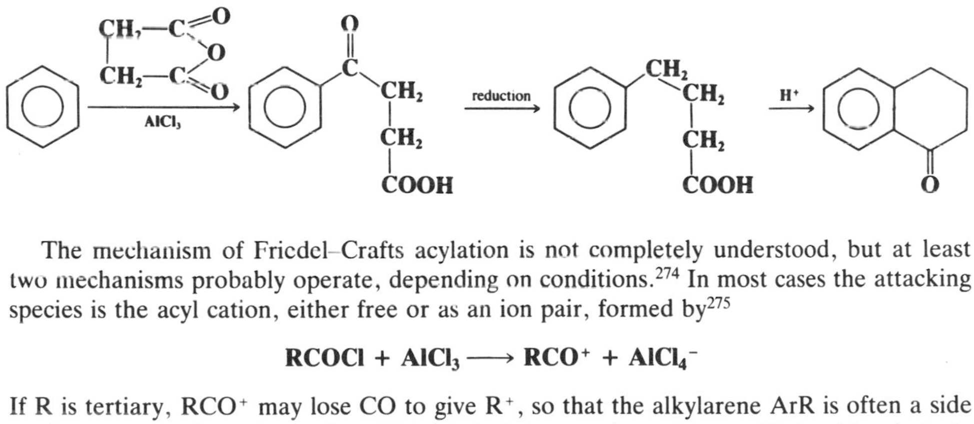

The mechanism of Friedel–Crafts acylation is not completely understood, but at least two mechanisms probably operate, depending on conditions.[274] In most cases the attacking species is the acyl cation, either free or as an ion pair, formed by[275]

$$RCOCl + AlCl_3 \longrightarrow RCO^+ + AlCl_4^-$$

If R is tertiary, RCO^+ may lose CO to give R^+, so that the alkylarene ArR is often a side product or even the main product. This kind of cleavage is much more likely with relatively unreactive substrates, where the acylium ion has time to break down. For example, pivaloyl chloride Me_3CCOCl gives the normal acyl product with anisole, but the alkyl product Me_3CPh with benzene. In the other mechanism an acyl cation is not involved, but the 1:1 complex attacks directly.[276]

[269]For example, see Schubert; Sweeney; Latourette *J. Am. Chem. Soc.* **1954**, *76*, 5462.

[270]For discussions, see Naruta; Maruyama, in Patai; Rappoport *The Chemistry of the Quinonoid Compounds*, vol. 2, pt. 1; Wiley: New York, 1988, pp. 325-332; Thomson, in Patai *The Chemistry of the Quinonoid Compounds*, vol. 1, pt. 1; Wiley: New York, 1974; pp. 136-139.

[271]For a review of polyphosphoric acid, see Rowlands, in Pizey, Ref. 199, vol. 6, 1985, pp. 156-414.

[272]See Peto, Ref. 261.

[273]See Agranat; Shih *J. Chem. Educ.* **1976**, *53*, 488.

[274]For a review of the mechanism see Taylor *Electrophilic Aromatic Substitution*, Ref. 1, pp. 222-237.

[275]After 2 min, exchange between PhCOCl and $Al(^{36}Cl)_3$ is complete: Oulevey; Susz *Helv. Chim. Acta* **1964**, *47*, 1828.

[276]For example, see Corriu; Coste *Bull. Soc. Chim. Fr.* **1967**, 2562, 2568, 2574; **1969**, 3272; Corriu; Dore; Thomassin *Tetrahedron* **1971**, *27*, 5601, 5819; Tan; Brownstein *J. Org. Chem.* **1983**, *48*, 302.

$$\text{ArH} + \overset{\overset{\oplus}{O}-\text{AlCl}_3}{\underset{\underset{Cl}{|}}{C}}-R \longrightarrow \left[\text{benzene} \cdot \overset{\overset{|\overset{\ominus}{O}-\text{AlCl}_3}{|}}{\underset{}{C}}\overset{R}{\underset{H}{\diagdown}}Cl\right] \xrightarrow{-\text{HCl}} \text{Ar}-\overset{\overset{\oplus}{O}-\overset{\ominus}{\text{AlCl}_3}}{\underset{}{C}}-R$$

Free-ion attack is more likely for sterically hindered R.[277] The ion CH_3CO^+ has been detected (by ir spectroscopy) in the liquid complex between acetyl chloride and aluminum chloride, and in polar solvents such as nitrobenzene; but in nonpolar solvents such as chloroform, only the complex and not the free ion is present.[278] In any event, 1 mole of catalyst certainly remains complexed to the product at the end of the reaction. When the reaction is performed with $RCO^+\ SbF_6^-$, no catalyst is required and the free ion[279] (or ion pair) is undoubtedly the attacking entity.[280]

OS **I**, 109, 353, 476, 517; **II**, 3, 8, 15, 81, 156, 169, 304, 520, 569; **III**, 6, 14, 23, 53, 109, 183, 248, 272, 593, 637, 761, 798; **IV**, 8, 34, 88, 898, 900; **V**, 111; **VI**, 34, 618, 625.

Reactions **1-15** through **1-18** are direct formylations of the ring.[281] Reaction **1-14** has not been used for formylation, since neither formic anhydride nor formyl chloride is stable at ordinary temperatures. Formyl chloride has been shown to be stable in chloroform solution for 1 hr at $-60°C$,[282] but it is not useful for formylating aromatic rings under these conditions. Formic anhydride has been prepared in solution, but has not been isolated.[283] Mixed anhydrides of formic and other acids are known[284] and can be used to formylate amines (see **0-53**) and alcohols, but no formylation takes place when they are applied to aromatic rings. See **3-17** for a nucleophilic method for the formylation of aromatic rings.

1-15 Formylation with Disubstituted Formamides
Formylation or **Formyl-de-hydrogenation**

$$\text{ArH} + \text{Ph}-\overset{}{\underset{\underset{Me}{|}}{N}}-\overset{\overset{O}{||}}{C}-H \xrightarrow{\text{POCl}_3} \text{ArCHO} + \text{PhNHMe}$$

The reaction with disubstituted formamides and phosphorus oxychloride, called the *Vilsmeier* or the *Vilsmeier–Haack reaction*, is the most common method for the formylation of aromatic rings.[285] However, it is applicable only to active substrates, such as amines and phenols. Aromatic hydrocarbons and heterocycles can also be formylated, but only if they are much more active than benzene (e.g., azulenes, ferrocenes). Though N-phenyl-N-methylform-

[277]Yamase *Bull. Chem. Soc. Jpn.* **1961**, *34*, 484; Gore *Bull. Chem. Soc. Jpn.* **1962**, *35*, 1627; Satchell *J. Chem. Soc.* **1961**, 5404.

[278]Cook *Can. J. Chem.* **1959**, *37*, 48; Cassimatis; Bonnin; Theophanides *Can. J. Chem.* **1970**, *48*, 3860.

[279]Crystal structures of solid $RCO^+\ SbF_6^-$ salts have been reported: Boer *J. Am. Chem. Soc.* **1968**, *90*, 6706; Chevrier; Le Carpentier; Weiss *Acta Crystallogr., Sect. B* **1972**, *28*, 2673, *J. Am. Chem. Soc.* **1972**, *94*, 5718.

[280]Olah; Kuhn; Flood; Hardie *J. Am. Chem. Soc.* **1964**, *86*, 2203; Olah; Lin; Germain *Synthesis* **1974**, 895. For a review of acylium salts in organic synthesis, see Al-Talib; Tashtoush *Org. Prep. Proced. Int.* **1990**, *22*, 1-36.

[281]For a review, see Olah; Kuhn, in Olah, Ref. 261, vol. 3, 1964, pp. 1153-1256. For a review of formylating agents, see Olah; Ohannesian; Arvanaghi *Chem. Rev.* **1987**, *87*, 671-686. For a list of reagents, with references, see Larock, Ref. 171, pp. 702-703.

[282]Staab; Datta *Angew. Chem. Int. Ed. Engl.* **1964**, *3*, 132 [*Angew. Chem.* **1963**, *75*, 1203].

[283]Olah; Vankar; Arvanaghi; Sommer *Angew. Chem. Int. Ed. Engl.* **1979**, *18*, 614 [*Angew. Chem. 91*, 649]; Schijf; Scheeren; van Es; Stevens *Recl. Trav. Chim. Pays-Bas* **1965**, *84*, 594.

[284]Stevens; van Es *Recl. Trav. Chim. Pays-Bas* **1964**, *83*, 863.

[285]For a review, see Jutz *Adv. Org. Chem.* **1976**, *9*, pt. 1, 225-342.

amide is a common reagent, other arylalkyl amides and dialkyl amides are also used.[286] Phosgene $COCl_2$ has been used in place of $POCl_3$. The reaction has also been carried out with other amides to give ketones (actually an example of **1-14**), but not often. The attacking species[287] is **27**,[288] and the mechanism is probably:

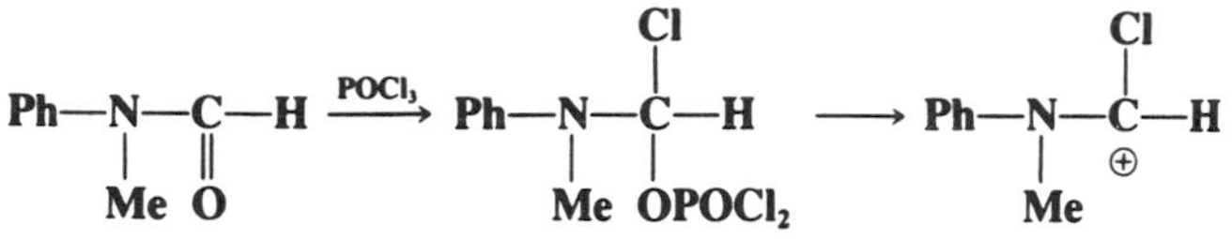

$$Ph-\underset{Me}{\underset{|}{N}}-\underset{O}{\underset{\|}{C}}-H \xrightarrow{POCl_3} Ph-\underset{Me}{\underset{|}{N}}-\underset{OPOCl_2}{\underset{|}{\overset{Cl}{\underset{|}{C}}}}-H \longrightarrow Ph-\underset{Me}{\underset{|}{N}}-\underset{\oplus}{\overset{Cl}{\underset{|}{C}}}-H$$

27

$$\underset{Me}{\underset{|}{\overset{|Z}{\bigcirc}}} + Ph-\underset{Me}{\underset{|}{N}}-\underset{\oplus}{\overset{Cl}{\underset{|}{C}}}-H \longrightarrow \overset{\oplus Z}{\bigcirc}\underset{H \ \ Me}{\overset{\overset{Cl}{\underset{|}{C}}H-N-Ph}{}} \longrightarrow \underset{Cl \ \ Me}{\overset{|Z}{\bigcirc}\overset{CH-N-Ph}{}}$$

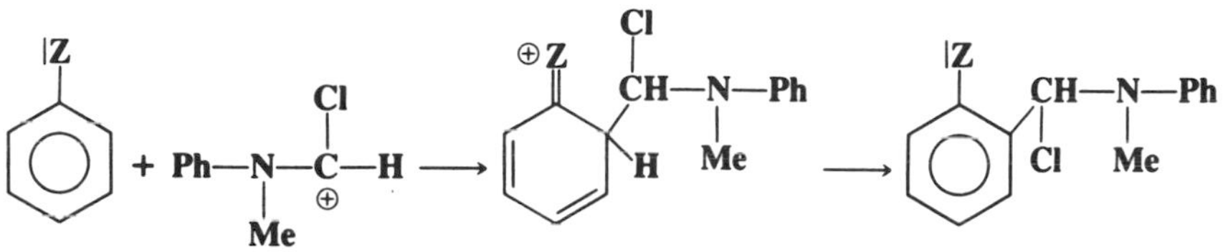

28

28 is unstable and easily hydrolyzes to the product. Either formation of **27** or the reaction of **27** with the substrate can be rate-determining, depending on the reactivity of the substrate.[289]

When $(CF_3SO_2)_2O$ was used instead of $POCl_3$, the reaction was extended to some less-active compounds, including naphthalene and phenanthrene.[290]

OS **I**, 217; **III**, 98, **IV**, 331, 539, 831, 915.

1-16 Formylation with Zinc Cyanide and HCl. The Gatterman Reaction
Formylation or **Formyl-de-hydrogenation**

$$ArH + Zn(CN)_2 \xrightarrow{HCl} ArCH{=}NH_2^+ \ Cl^- \xrightarrow{H_2O} ArCHO$$

Formylation with $Zn(CN)_2$ and HCl is called the *Gatterman reaction*.[291] It can be applied to alkylbenzenes, phenols and their ethers, and many heterocyclic compounds. However, it cannot be applied to aromatic amines. In the original version of this reaction the substrate was treated with HCN, HCl, and $ZnCl_2$, but the use of $Zn(CN)_2$ and HCl (HCN and $ZnCl_2$ are generated in situ) makes the reaction more convenient to carry out and does not reduce yields. The mechanism of the Gatterman reaction has not been investigated very much, but there is an initial nitrogen-containing product that is normally not isolated but is hydrolyzed to aldehyde. The above structure is presumed for this product. When benzene was treated with NaCN under super acidic conditions (F_3CSO_2OH—SbF_5), a good yield of product was obtained, leading to the conclusion that the electrophile in this case was $H\overset{\oplus}{C}{=}\overset{\oplus}{N}H_2$.[292] The Gatterman reaction may be regarded as a special case of **1-27**.

[286]For a review of dimethylformamide, see Pizey, Ref. 199, vol. 1, 1974, pp. 1-99.
[287]For a review of such species, see Kantlehner *Adv. Org. Chem.* **1979**, *9*, pt. 2, 5-172.
[288]See Arnold; Holý *Collect. Czech. Chem. Commun.* **1962**, *27*, 2886; Martin; Martin *Bull. Soc. Chim. Fr.* **1963**, 1637; Fritz; Oehl *Liebigs Ann. Chem.* **1971**, *749*, 159; Jugie; Smith; Martin *J. Chem. Soc., Perkin Trans. 2* **1975**, 925.
[289]Alunni; Linda; Marino; Santini; Savelli *J. Chem. Soc., Perkin Trans. 2* **1972**, 2070.
[290]Martínez; Alvarez; Barcina; Cerero; Vilar; Fraile; Hanack; Subramanian *J. Chem. Soc., Chem. Commun.* **1990**, 1571.
[291]For a review, see Truce *Org. React.* **1957**, *9*, 37-72.
[292]Yato; Ohwada; Shudo *J. Am. Chem. Soc.* **1991**, *113*, 691.

Another method, formylation with CO and HCl in the presence of AlCl₃ and CuCl[293] (the *Gatterman–Koch reaction*), is limited to benzene and alkylbenzenes.[294]

OS **II**, 583; **III**, 549.

1-17 Formylation with Chloroform. The Reimer–Tiemann Reaction
Formylation or **Formyl-de-hydrogenation**

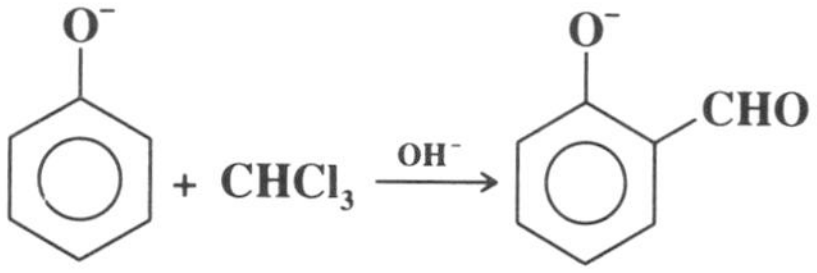

In the *Reimer–Tiemann reaction* chloroform and hydroxide ion are used to formylate aromatic rings.[295] The method is useful only for phenols and certain heterocyclic compounds such as pyrroles and indoles. Unlike the previous formylation methods (**1-15** and **1-16**), this one is conducted in basic solution. Yields are generally low, seldom rising above 50%.[296] The incoming group is directed ortho, unless both ortho positions are filled, in which case the attack is para.[297] Certain substrates have been shown to give abnormal products instead of or in addition to the normal ones. For example, **29** and **31** gave, respectively, **30** and **32** as well as the normal aldehyde products. From the nature of the reagents and from the kind

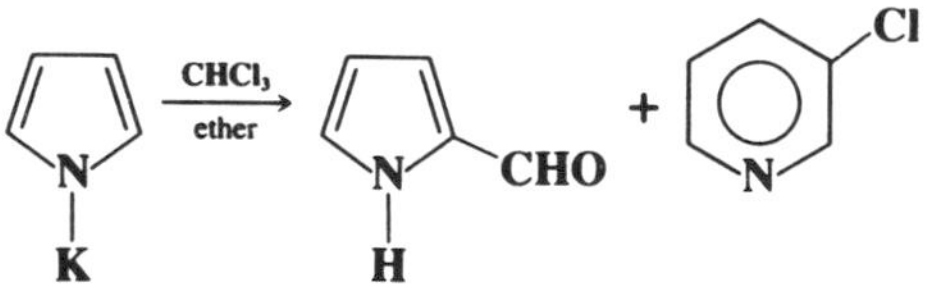

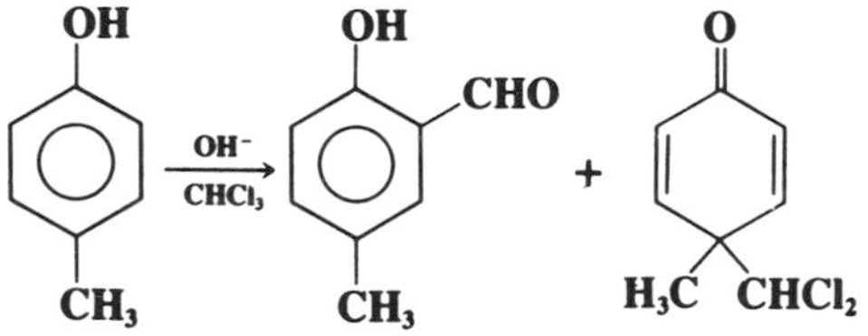

of abnormal products obtained, it is clear that the attacking entity in this reaction is dichlorocarbene CCl₂.[298] This in known to be produced by treatment of chloroform with bases (p. 371); it is an electrophilic reagent and is known to give ring expansion of aromatic rings

---

[293]The CuCl is not always necessary: see Toniolo; Graziani *J. Organomet. Chem.* **1980**, *194*, 221.
[294]For a review, see Crounse *Org. React.* **1949**, *5*, 290-300.
[295]For a review, see Wynberg; Meijer *Org. React.* **1982**, *28*, 1-36.
[296]For improved procedures, see Thoer; Denis; Delmas; Gaset *Synth. Commun.* **1988**, *18*, 2095; Cochran; Melville *Synth. Commun.* **1990**, *20*, 609.
[297]Increased para selectivity has been achieved by the use of polyethylene glycol: Neumann; Sasson *Synthesis* **1986**, 569.
[298]For a review of carbene methods for introducing formyl and acyl groups into organic molecules, see Kulinkovich *Russ. Chem. Rev.* **1989**, *58*, 711-719.

(see **5-50**), accounting for products like **30**. The mechanism of the normal reaction is thus something like this.[299]

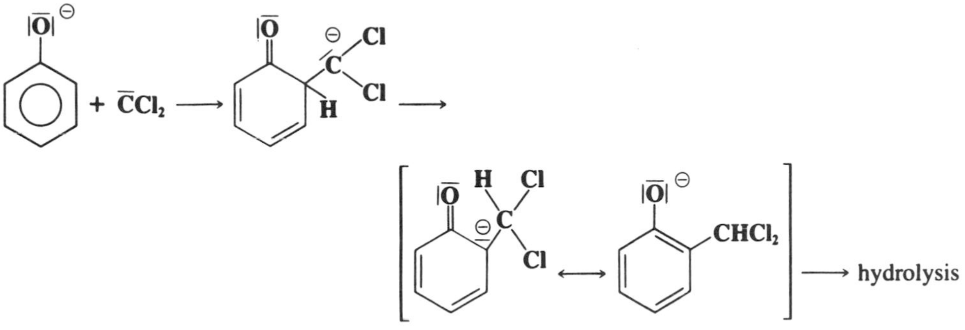

The formation of **32** in the case of **31** can be explained by attack of some of the CCl_2 ipso to the CH_3 group. Since this position does not contain a hydrogen, normal proton loss cannot take place and the reaction ends when the CCl_2^- moiety acquires a proton.

A method closely related to the Reimer–Tiemann reaction is the *Duff reaction*, in which hexamethylenetetramine $(CH_2)_6N_4$ is used instead of chloroform. This reaction can be applied only to phenols and amines; ortho substitution is generally observed and yields are low. A mechanism[300] has been proposed that involves initial aminoalkylation(**1-25**) to give $ArCH_2NH_2$, followed by dehydrogenation to $ArCH=NH$ and hydrolysis of this to the aldehyde product. When $(CH_2)_6N_4$ is used in conjunction with F_3CCOOH, the reaction can be applied to simple alkylbenzenes; yields are much higher and a high degree of regioselectively para substitution is found.[301] In this case too an imine seems to be an intermediate.

OS **III**, 463; **IV**, 866

1-18 Other Formylations
Formylation or **Formyl-de-hydrogenation**

$$ArH + Cl_2CHOMe \xrightarrow{AlCl_3} ArCHO$$

Besides **1-15** to **1-17**, several other formylation methods are known.[302] In one of these, dichloromethyl methyl ether formylates aromatic rings with Friedel–Crafts catalysts.[303] ArCHClOMe is probably an intermediate. Orthoformates have also been used.[304] In another method, aromatic rings are formylated with formyl fluoride HCOF and BF_3.[305] Unlike formyl chloride, formyl fluoride is stable enough for this purpose. This reaction was successful for benzene, alkylbenzenes, PhCl, PhBr, and naphthalene. Phenols can be regioselectively formylated in the ortho position in high yields by treatment with two equivalents of paraformaldehyde in aprotic solvents in the presence of $SnCl_4$ and a tertiary amine.[306] Phenols

[299]Robinson *J. Chem. Soc.* **1961**, 1663; Hine; van der Veen *J. Am. Chem. Soc.* **1959**, *81*, 6446. See also Langlois *Tetrahedron Lett.* **1991**, *32*, 3691.
[300]Ogata; Kawasaki; Sugiura *Tetrahedron* **1968**, *24*, 5001.
[301]Smith *J. Org. Chem.* **1972**, *37*, 3972.
[302]For methods other than those described here, see Smith; Manas *Synthesis* **1984**, 166; Olah; Laali; Farooq *J. Org. Chem.* **1985**, *50*, 1483; Nishino; Tsunoda; Kurosawa *Bull. Chem. Soc. Jpn.* **1989**, *62*, 545.
[303]Rieche; Gross; Höft *Chem. Ber.* **1960**, *93*, 88; Lewin; Parker; Fleming; Carroll *Org. Prep. Preced. Int.* **1978**, *10*, 201.
[304]Gross; Rieche; Matthey *Chem. Ber.* **1963**, *96*, 308.
[305]Olah; Kuhn *J. Am. Chem. Soc.* **1960**, *82*, 2380.
[306]Casiraghi; Casnati; Puglia; Sartori; Terenghi *J. Chem. Soc., Perkin Trans. 1* **1980**, 1862.

have also been formylated indirectly with 2-ethoxy-1,3-dithiolane.[307] See also the indirect method mentioned at **1-26.**

OS **V**, 49; **VII**, 162.

Reactions **1-19** and **1-20** are direct carboxylations[308] of aromatic rings.[309]

1-19 Carboxylation with Carbonyl Halides
Carboxylation or **Carboxy-de-hydrogenation**

$$ArH + COCl_2 \xrightarrow{AlCl_3} ArCOOH$$

Phosgene, in the presence of Friedel–Crafts catalysts, can carboxylate the ring. This process is analogous to **1-14,** but the ArCOCl initially produced hydrolyzes to the carboxylic acid. However, in most cases the reaction does not take this course, but instead the ArCOCl attacks another ring to give a ketone ArCOAr. A number of other reagents have been used to get around this difficulty, among them oxalyl chloride, urea hydrochloride, chloral Cl_3CCHO,[310] carbamoyl chloride H_2NCOCl, and N,N-diethylcarbamoyl chloride.[311] With carbamoyl chloride the reaction is called the *Gatterman amide synthesis* and the product is an amide. Among compounds carboxylated by one or another of these reagents are benzene, alkylbenzenes, and fused ring systems.[312]

OS **V**, 706; **VII**, 420.

1-20 Carboxylation with Carbon Dioxide. The Kolbe–Schmitt Reaction
Carboxylation or **Carboxy-de-hydrogenation**

Sodium phenoxides can be carboxylated, mostly in the ortho position, by carbon dioxide (*the Kolbe–Schmitt reaction*). The mechanism is not clearly understood, but apparently some kind of a complex is formed between the reactants,[313] making the carbon of the CO_2 more

[307]Jo; Tanimoto; Sugimoto; Okano *Bull. Chem. Soc. Jpn.* **1981,** *54,* 2120.
[308]For other carboxylation methods, one of which leads to the anhydride, see Sakakibara; Odaira *J. Org. Chem.* **1976,** *41*, 2049; Fujiwara; Kawata; Kawauchi; Taniguchi *J. Chem. Soc., Chem. Commun.* **1982,** 132.
[309]For a review, see Olah; Olah, in Olah, Ref. 261, vol. 3, 1964, pp. 1257-1273.
[310]Menegheli; Rezende; Zucco *Synth. Commun.* **1987,** *17,* 457.
[311]Naumov; Isakova; Kost; Zakharov; Zvolinskii; Moiseikina; Nikeryasova *J. Org. Chem. USSR* **1975,** *11*, 362.
[312]For the use of phosgene to carboxylate phenols, see Sartori; Casnati; Bigi; Bonini *Synthesis* **1988,** 763.
[313]Hales; Jones; Lindsey *J. Chem. Soc.* **1954,** 3145.

positive and putting it in a good position to attack the ring. Potassium phenoxide, which is less likely to form such a complex,[314] is chiefly attacked in the para position.[315] Carbon tetrachloride can be used instead of CO_2 under Reimer–Tiemann (**1-17**) conditions.

Sodium or potassium phenoxide can be carboxylated regioselectively in the para position in high yield by treatment with sodium or potassium carbonate and carbon monoxide.[316] [14]C labeling showed that it is the carbonate carbon that appears in the *p*-hydroxybenzoic acid product.[317] The CO is converted to sodium or potassium formate. Carbon monoxide has also been used to carboxylate aromatic rings with palladium compounds as catalysts.[318] In addition, a palladium-catalyzed reaction has been used directly to prepare acyl fluorides $ArH \rightarrow ArCOF$.[319]

OS **II**, 557.

1-21   Amidation with Isocyanates
N-Alkylcarbamoyl-de-hydrogenation

$$ArH + RNCO \xrightarrow{\text{AlCl}_3} ArCONHR$$

N-Substituted amides can be prepared by direct attack of isocyanates on aromatic rings.[320] R may be alkyl or aryl, but if the latter, dimers and trimers are also obtained. Isothiocyanates similarly give thioamides.[321] The reaction has been carried out intramolecularly both with aralkyl isothiocyanates and acyl isothiocyanates.[322] In the latter case, the product is easily hydrolyzable to a dicarboxylic acid; this is a way of putting a carboxyl group on a ring ortho

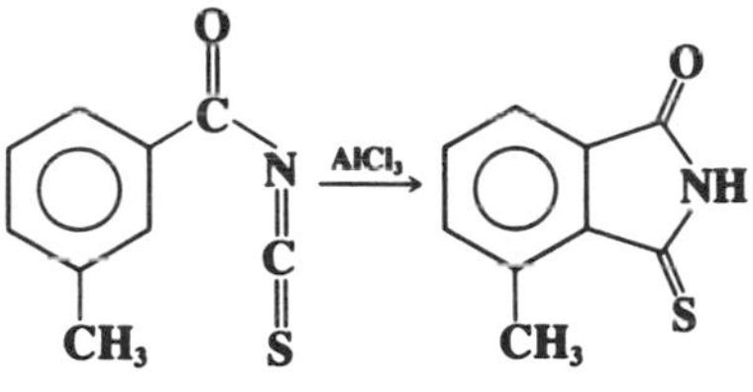

33

to one already there (**33** is prepared by treatment of the acyl halide with lead thiocyanate). The reaction gives better yields with substrates of the type $ArCH_2CONCS$, where six-membered rings are formed. Ethyl carbamate NH_2COOEt (with P_2O_5 in xylene)[323] and biscarbamoyl diselenides $R_2NCOSeSeCONR_2$[324] (with $HgBr_2$ or $SnCl_4$) have also been used to amidate aromatic rings.

OS **V**, 1051; **VI**, 465.

[314]There is evidence that, in the complex formed from potassium salts, the bonding is between the aromatic compound and the carbon atom of CO_2: Hirao; Kito *Bull. Chem. Soc. Jpn.* **1973**, *46*, 3470.

[315]Actually, the reaction seems to be more complicated than this. At least part of the potassium *p*-hydroxybenzoate that forms comes from a rearrangement of initially formed potassium salicylate. Sodium salicylate does not rearrange. See Shine, Ref. 375, pp. 344-348. See also Ota *Bull. Chem. Soc. Jpn.* **1974**, *47*, 2343.

[316]Yasuhara; Nogi *J. Org. Chem.* **1968**, *33*, 4512, *Chem. Ind.* (*London*) **1967**, 229, **1969**, 77.

[317]Yasuhara; Nogi; Saishō *Bull. Chem. Soc. Jpn.* **1969**, *42*, 2070.

[318]See Sakakibara; Odaira, Ref. 308; Jintoku; Taniguchi; Fujiwara *Chem. Lett.* **1987**, 1159; Ugo; Chiesa *J. Chem. Soc., Perkin Trans. 1* **1987**, 2625.

[319]Sakakura; Chaisupakitsin; Hayashi; Tanaka *J. Organomet. Chem.* **1987**, *334*, 205.

[320]Effenberger; Gleiter *Chem. Ber.* **1964**, *97*, 472; Effenberger; Gleiter; Heider; Niess *Chem. Ber.* **1968**, *101*, 502; Piccolo; Filippini; Tinucci; Valoti; Citterio *Tetrahedron* **1986**, *42*, 885.

[321]Jagodziński *Synthesis* **1988**, 717.

[322]Smith; Kan *J. Am. Chem. Soc.* **1960**, *82*, 4753, *J. Org. Chem.* **1964**, *29*, 2261.

[323]Chakraborty; Mandal; Roy *Synthesis* **1981**, 977.

[324]Fujiwara; Ogawa; Kambe; Ryu; Sonoda *Tetrahedron Lett.* **1988**, *29*, 6121.

Reactions **1-22** to **1-26** involve the introduction of a CH_2Z group, where Z is halogen, hydroxyl, amino, or alkylthio. They are all Friedel–Crafts reactions of aldehydes and ketones and, with respect to the carbonyl compound, additions to the C=O double bond. They follow mechanisms discussed in Chapter 16.

1-22 Hydroxyalkylation or Hydroxyalkyl-de-hydrogenation

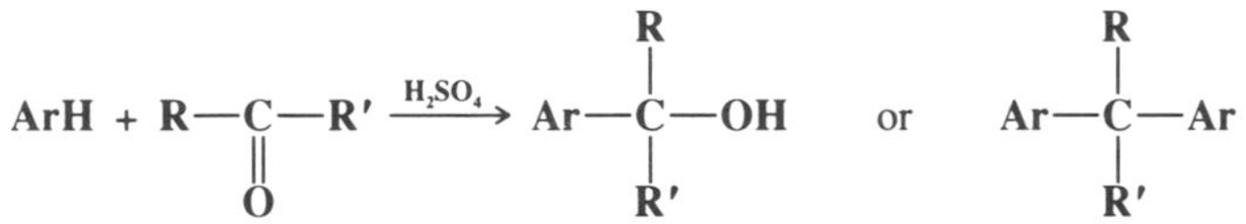

$$ArH + R—\underset{\underset{O}{\|}}{C}—R' \xrightarrow{H_2SO_4} Ar—\underset{\underset{R'}{|}}{\overset{\overset{R}{|}}{C}}—OH \quad \text{or} \quad Ar—\underset{\underset{R'}{|}}{\overset{\overset{R}{|}}{C}}—Ar$$

The condensation of aromatic rings with aldehydes or ketones is called *hydroxyalkylation*.[325] The reaction can be used to prepare alcohols,[326] though more often the alcohol initially produced reacts with another molecule of aromatic compound (**1-12**) to give diarylation. For this the reaction is quite useful, an example being the preparation of DDT:

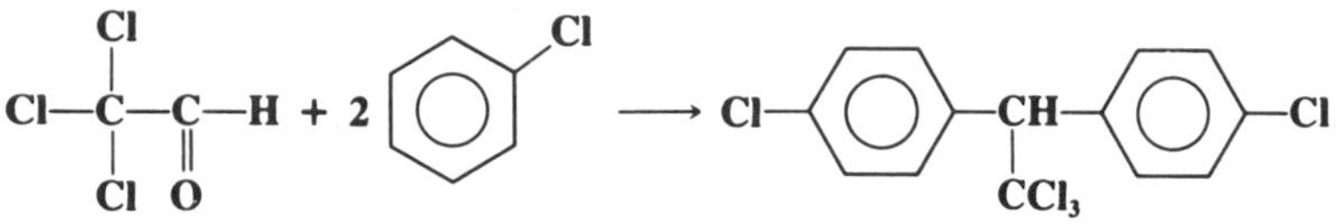

The diarylation reaction is especially common with phenols (the diaryl product here is called a *bisphenol*). The reaction is normally carried out in alkaline solution on the phenolate ion.[327] The hydroxymethylation of phenols with formaldehyde is called the *Lederer-Manasse reaction*. This reaction must be carefully controlled,[328] since it is possible for the para and both ortho positions to be substituted and for each of these to be rearylated, so that a polymeric structure is produced:

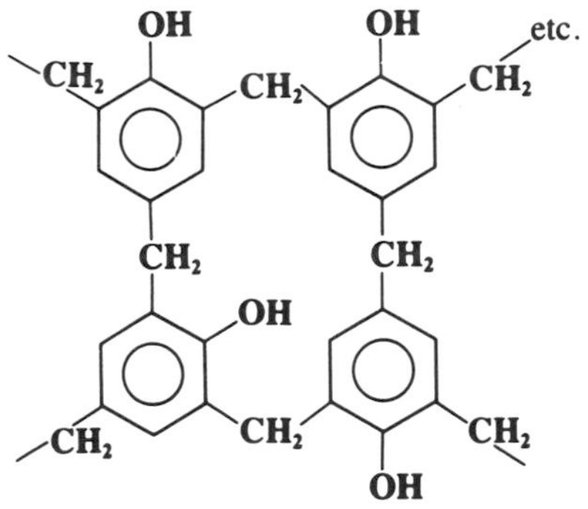

However, such polymers, which are of the Bakelite type (phenol–formaldehyde resins), are of considerable commercial importance.

The attacking species is the carbocation, $R—\overset{\oplus}{\underset{\underset{OH}{|}}{C}}—R'$, formed from the aldehyde or ketone and the acid catalyst, except when the reaction is carried out in basic solution.

---

[325]For a review, see Hofmann; Schriesheim, in Olah. Ref. 261, vol. 2, pp. 597-640.
[326]See, for example, Casiraghi; Casnati; Puglia; Sartori *Synthesis* **1980**, 124.
[327]For a review, see Schnell; Krimm *Angew. Chem. Int. Ed. Engl.* **1963**, 2, 373-379 [*Angew. Chem.* 75, 662-668].
[328]See, for example, Casiraghi; Casnati; Pochini; Puglia; Ungaro; Sartori *Synthesis* **1981**, 143.

When an aromatic ring is treated with diethyl oxomalonate $(EtOOC)_2C=O$, the product is an arylmalonic acid derivative $ArC(OH)(COOEt)_2$, which can be converted to an aryl-malonic acid $ArCH(COOEt)_2$.[329] This is therefore a way of applying the malonic ester synthesis (**0-94**) to an aryl group (see also **3-14**). Of course, the opposite mechanism applies here: the aryl species is the nucleophile.

Two methods, both involving boron-containing reagents, have been devised for the regioselective ortho hydroxymethylation of phenols or aromatic amines.[330]

OS **III**, 326; **V**, 422; **VI**, 471, 856; **68**, 234, 238, 243. Also see OS **I**, 214.

1-23   Cyclodehydration of Aldehydes and Ketones

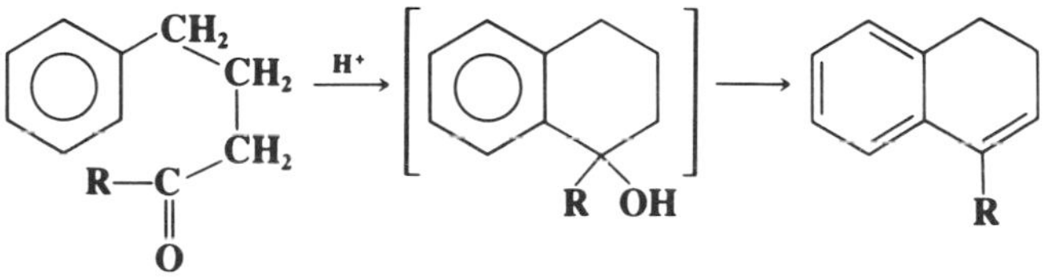

When an aromatic compound contains an aldehyde or ketone function in a position suitable for closing a six-membered ring, treatment with acid results in cyclodehydration. The reaction is a special case of **1-22**, but in this case dehydration almost always takes place to give a double bond conjugated with the aromatic ring.[331] The method is very general and is widely used to close both carbocyclic and heterocyclic rings.[332] Polyphosphoric acid is a common reagent, but other acids have also been used. In a variation known as the *Bradsher reaction*,[333]

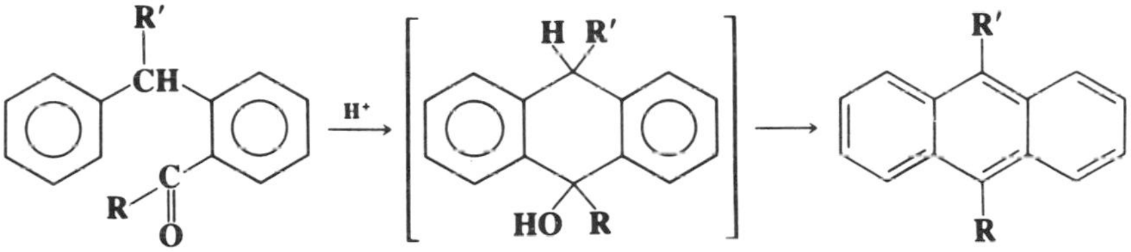

diarylmethanes containing a carbonyl group in the ortho position can be cyclized to anthracene derivatives. In this case 1,4-dehydration takes place, at least formally.

Among the many applications of cyclodehydration to the formation of heterocyclic systems is the *Bischler–Napieralski reaction*.[334] In this reaction amides of the type **34** are cyclized with phosphorous oxychloride:

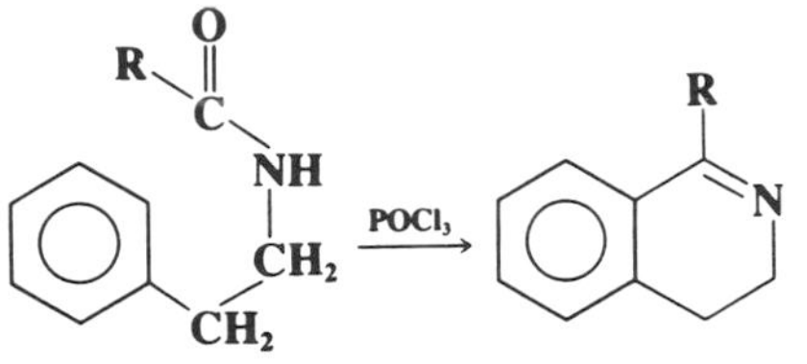

34

[329]Ghosh; Pardo; Salomon *J. Org. Chem.* **1982**, *47*, 4692.

[330]Sugasawa; Toyoda; Adachi; Sasakura *J. Am. Chem. Soc.* **1978**, *100*, 4842; Nagata; Okada; Aoki *Synthesis* **1979**, 365.

[331]For examples where the hydroxy compound was the principal product (with R = CF_3), see Fung; Abraham; Bellini; Sestanj *Can. J. Chem.* **1983**, *61*, 368; Bonnet-Delpon; Charpentier-Morize; Jacquot *J. Org. Chem.* **1988**, *53*, 759.

[332]For a review, see Bradsher *Chem. Rev.* **1987**, *87*, 1277-1297.

[333]For examples, see Bradsher *J. Am. Chem. Soc.* **1940**, *62*, 486; Saraf; Vingiello *Synthesis* **1970**, 655; Ref. 332, pp. 1287-1294.

[334]For a review of the mechanism, see Fodor; Nagubandi *Tetrahedron* **1980**, *36*, 1279-1300.

If the starting compound contains a hydroxyl group in the α position, an additional dehydration takes place and the product is an isoquinoline. Higher yields can be obtained if the amide is treated with PCl_5 to give an imino chloride $ArCH_2CH_2N{=}CR{-}Cl$, which is isolated and then cyclized by heating.[335] The nitrilium ion $ArCH_2CH_2\overset{\oplus}{N}{\equiv}CR$ is an intermediate.

OS **I**, 360, 478; **II**, 62, 194; **III**, 281, 300, 329, 568, 580, 581; **IV**, 590; **V**, 550; **VI**, 1. Also see OS **I**, 54.

1-24 Haloalkylation or Haloalkyl-de-hydrogenation

$$\text{ArH} + \text{HCHO} + \text{HCl} \xrightarrow{\text{ZnCl}_2} \text{ArCH}_2\text{Cl}$$

When certain aromatic compounds are treated with formaldehyde and HCl, the CH_2Cl group is introduced into the ring in a reaction called *chloromethylation*. The reaction has also been carried out with other aldehydes and with HBr and HI. The more general term *haloalkylation* covers these cases.[336] The reaction is successful for benzene, and alkyl-, alkoxy-, and halobenzenes. It is greatly hindered by meta-directing groups, which reduce yields or completely prevent the reactions. Amines and phenols are too reactive and usually give polymers unless deactivating groups are also present, but phenolic ethers and esters successfully undergo the reaction. Compounds of lesser reactivity can often be chloromethylated with chloromethyl methyl ether $ClCH_2OMe$, bis(chloromethyl) ether $(ClCH_2)_2O$,[337] methoxyacetyl chloride $MeOCH_2COCl$,[338] or 1-chloro-4-(chloromethoxy)butane.[339] Zinc chloride is the most common catalyst, but other Friedel–Crafts catalysts are also employed. As with reaction **1-22** and for the same reason, an important side product is the diaryl compound Ar_2CH_2 (from formaldehyde).

Apparently, the initial step involves reaction of the aromatic compound with the aldehyde to form the hydroxyalkyl compound, exactly as in **1-22,** and then the HCl converts this to the chloroalkyl compound.[340] The acceleration of the reaction by $ZnCl_2$ has been attributed[341] to the raising of the acidity of the medium, causing an increase in the concentration of $HOCH_2{}^+$ ions.

OS **III**, 195, 197, 468, 557; **IV**, 980.

1-25 Aminoalkylation and Amidoalkylation
Dialkylaminoalkylation or Dialkylamino-de-hydrogenation

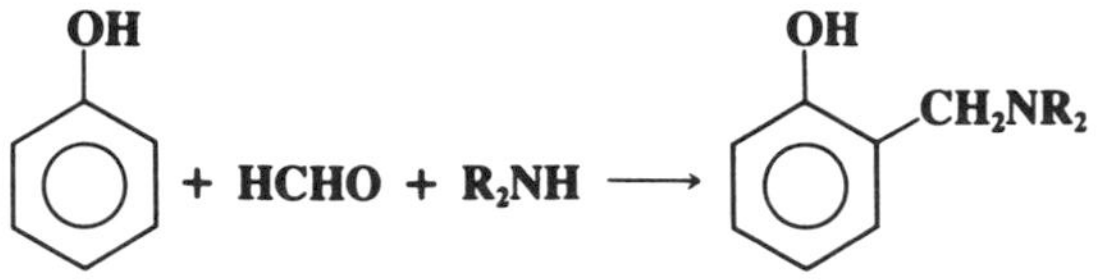

Phenols, secondary and tertiary aromatic amines,[342] pyrroles, and indoles can be aminomethylated by treatment with formaldehyde and a secondary amine. Other aldehydes have

[335]Fodor; Gal; Phillips *Angew. Chem. Int. Ed. Engl.* **1972**, *11*, 919 [*Angew. Chem. 84*, 947].
[336]For reviews, see Belen'kii; Vol'kenshtein; Karmanova *Russ. Chem. Rev.* **1977**, *46*, 891-903; Olah; Tolgyesi, in Olah, Ref. 261, vol. 2, pp. 659-784.
[337]Suzuki *Bull. Chem. Soc. Jpn.* **1970**, *43*, 3299; Kuimova; Mikhailov *J. Org. Chem. USSR* **1971**, *7*, 1485.
[338]McKillop; Madjdabadi; Long *Tetrahedron Lett.* **1983**, *24*, 1933.
[339]Olah; Beal; Olah *J. Org. Chem.* **1976**, *41*, 1627.
[340]Ziegler; Hontschik; Milowiz *Monatsh. Chem.* **1948**, *79*, 142; Ogata; Okano *J. Am. Chem. Soc.* **1956**, *78*, 5423. See also Olah; Yu *J. Am. Chem. Soc.* **1975**, *97*, 2293.
[341]Lyushin; Mekhtiev; Guseinova *J. Org. Chem. USSR* **1970**, *6*, 1445.
[342]Miocque; Vierfond *Bull. Soc. Chim. Fr.* **1970**, 1896, 1901, 1907.

sometimes been employed. Aminoalkylation is a special case of the Mannich reaction (**6-16**). When phenols and other activated aromatic compounds are treated with N-hydroxymethylchloroacetamide, *amidomethylation* takes place[343] to give **35**, which is often hydro-

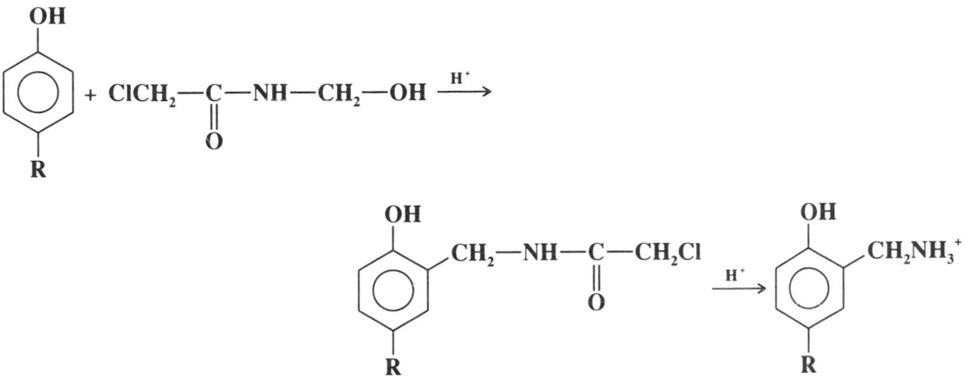

35

lyzed in situ to the aminoalkylated product. Other N-hydroxyalkyl and N-chlorinated compounds have also been used.[343]

OS **I**, 381; **IV**, 626; **V**, 434; **VI**, 965; **VII**, 162.

1-26 Thioalkylation
Alkylthioalkylation or **Alkylthioalkyl-de-hydrogenation**

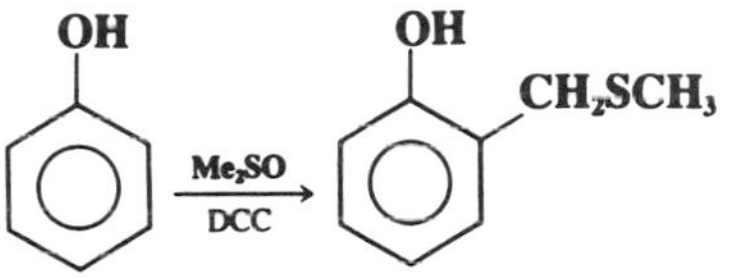

A methylthiomethyl group can be inserted into the ortho position of phenols by heating with dimethyl sulfoxide and dicyclohexylcarbodiimide (DCC).[344] Other reagents can be used instead of DCC, among them pyridine–SO_3,[345] $SOCl_2$,[346] and acetic anhydride.[347] Alternatively, the phenol can be treated with dimethyl sulfide and N-chlorosuccinimide, followed by triethylamine.[348] The reaction can be applied to amines (to give o-$NH_2C_6H_4CH_2SMe$) by treatment with *t*-BuOCl, Me_2S, and NaOMe in CH_2Cl_2.[349] It is possible to convert the CH_2SMe group to the CHO group,[350] so that this becomes an indirect method for the preparation of ortho-amino and ortho-hydroxy aromatic aldehydes; or to the CH_3 group (with Raney nickel—reaction **4-36**), which makes this an indirect method[351] for the intro-

[343]For a review, see Zaugg *Synthesis* **1984**, 85-110.
[344]Burdon; Moffatt *J. Am. Chem. Soc.* **1966**, *88*, 5855, **1967**, *89*, 4725; Olofson; Marino *Tetrahedron* **1971**, *27*, 4195.
[345]Claus *Monatsh. Chem.* **1971**, *102*, 913.
[346]Sato; Inoue; Ozawa; Tazaki *J. Chem. Soc., Perkin Trans. 1* **1984**, 2715.
[347]Hayashi; Oda *J. Org. Chem.* **1967**, *32*, 457; Pettit; Brown *Can. J. Chem.* **1967**, *45*, 1306; Claus *Monatsh. Chem.* **1968**, *99*, 1034.
[348]Gassman; Amick *J. Am. Chem. Soc.* **1978**, *100*, 7611.
[349]Gassman; Gruetzmacher *J. Am. Chem. Soc.* **1973**, *95*, 588; Gassman; van Bergen *J. Am. Chem. Soc.* **1973**, *95*, 590, 591.
[350]Gassman; Drewes *J. Am. Chem. Soc.* **1978**, *100*, 7600; Ref. 348.
[351]For another indirect method, in this case for alkylation ortho to an amino group, see Gassman; Parton *Tetrahedron Lett.* **1977**, 2055.

duction of a CH_3 group ortho to an OH or NH_2 group.[349] Aromatic hydrocarbons have been thioalkylated with ethyl α-(chloromethylthio)acetate $ClCH_2SCH_2COOEt$ (to give Ar-CH_2SCH_2COOEt)[352] and with methyl methylsulfinylmethyl sulfide $MeSCH_2SOMe$ or methylthiomethyl p-tolyl sulfone $MeSCH_2SO_2C_6H_4Me$ (to give $ArCH_2SMe$),[353] in each case with a Lewis acid catalyst.

OS **VI**, 581, 601.

1-27   Acylation with Nitriles. The Hoesch Reaction
Acylation or **Acyl-de-hydrogenation**

$$ArH + RCN \xrightarrow[ZnCl_2]{HCl} ArCOR$$

Friedel–Crafts acylation with nitriles and HCl is called the *Hoesch* or the *Houben–Hoesch reaction*.[354] In most cases, a Lewis acid is necessary; zinc chloride is the most common. The reaction is generally useful only with phenols, phenolic ethers, and some reactive heterocyclic compounds, e.g., pyrrole, but it can be extended to aromatic amines by the use of BCl_3.[355] Acylation in the case of amines is regioselectively ortho. Monohydric phenols, however, generally do not give ketones[356] but are attacked at the oxygen to produce imino esters.

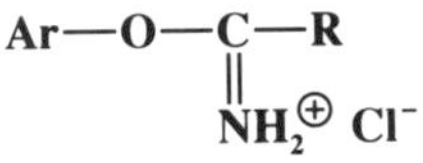

$$Ar-O-\underset{\underset{NH_2^{\oplus}\ Cl^-}{\|}}{C}-R$$

An imino ester

Many nitriles have been used. Even aryl nitriles give good yields if they are first treated with HCl and $ZnCl_2$ and then the substrate added at 0°C.[357] In fact, this procedure increases yields with any nitrile. If thiocyanates RSCN are used, thiol esters ArCOSR can be obtained. The Gatterman reaction (**1-16**) is a special case of the Hoesch synthesis.

The reaction mechanism is complex and not completely settled.[358] The first stage consists of an attack on the substrate by a species containing the nitrile and HCl (and the Lewis acid, if present) to give an imine salt (**38**). Among the possible attacking species are **36** and **37**. In the second stage, the salts are hydrolyzed to the products:

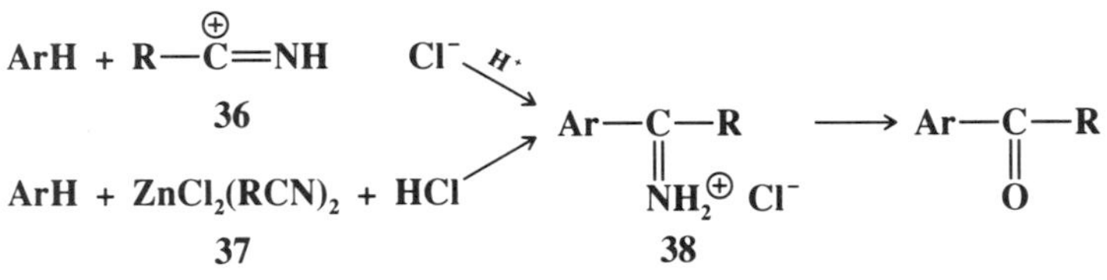

$$ArH + R-\overset{\oplus}{C}{=}NH \qquad Cl^-$$
$$\textbf{36}$$
$$ArH + ZnCl_2(RCN)_2 + HCl$$
$$\textbf{37}$$
$$\xrightarrow{H^\cdot} Ar-\underset{\underset{NH_2^{\oplus}\ Cl^-}{\|}}{C}-R \longrightarrow Ar-\underset{\underset{O}{\|}}{C}-R$$
$$\textbf{38}$$

Ketones can also be obtained by treating phenols or phenolic ethers with a nitrile in the presence of F_3CSO_2OH.[359] The mechanism in this case is different.

OS **II**, 522.

[352]Tamura; Tsugoshi; Annoura; Ishibashi *Synthesis* **1984**, 326.
[353]Torisawa; Satoh; Ikegami *Tetrahedron Lett.* **1988**, *29*, 1729.
[354]For a review, see Ruske, in Olah, Ref. 261, vol. 3, 1964, pp. 383-497.
[355]Sugasawa et al., Ref. 330; Sugasawa; Adachi; Sasakura; Kitagawa *J. Org. Chem.* **1979**, *44*, 578.
[356]For an exception, see Toyoda; Sasakura; Sugasawa *J. Org. Chem.* **1981**, *46*, 189.
[357]Zil'berman; Rybakova *J. Gen. Chem. USSR* **1960**, *30*, 1972.
[358]For discussions, see Ref. 354 and Jeffery; Satchell *J. Chem. Soc. B.* **1966**, 579.
[359]Booth; Noori *J. Chem. Soc., Perkin Trans. 1* **1980**, 2894; Amer; Booth; Noori; Proença *J. Chem. Soc., Perkin Trans. 1* **1983**, 1075.

1-28 Cyanation or Cyano-de-hydrogenation

$$\text{ArH} + \text{Cl}_3\text{CCN} \xrightarrow{\text{HCl}} \text{Ar}\!-\!\overset{\displaystyle\|}{\underset{\overset{\displaystyle |}{\overset{\oplus}{\text{NH}_2}}\ \text{Cl}^-}{\text{C}}}\!-\!\text{CCl}_3 \xrightarrow{\text{NaOH}} \text{ArCN}$$

Aromatic hydrocarbons (including benzene), phenols, and phenolic ethers can be cyanated with trichloroacetonitrile, BrCN, or mercury fulminate Hg(ONC)_2.[360] In the case of Cl_3CCN, the actual attacking entity is probably $\text{Cl}_3\text{C}\!-\!\overset{\oplus}{\text{C}}\!\!=\!\!\text{NH}$, formed by addition of a proton to the cyano nitrogen. Secondary aromatic amines ArNHR, as well as phenols, can be cyanated in the ortho position with Cl_3CCN and BCl_3.[361]

OS **III**, 293.

F. Oxygen Electrophiles Oxygen electrophiles are very uncommon, since oxygen does not bear a positive charge very well. However, there is one reaction that can be mentioned.

1-29 Hydroxylation or Hydroxy-de-hydrogenation

$$\text{ArH} + \text{F}_3\text{C}\!-\!\overset{\displaystyle\|}{\underset{\text{O}}{\text{C}}}\!-\!\text{O}\!-\!\text{OH} \xrightarrow{\text{BF}_3} \text{ArOH}$$

There have been only a few reports of direct hydroxylation[362] by an electrophilic process (see, however, **2-26** and **4-5**).[363] In general, poor results are obtained, partly because the introduction of an OH group activates the ring to further attack. Quinone formation is common. However, alkyl-substituted benzenes such as mesitylene or durene can be hydroxylated in good yield with trifluoroperacetic acid and boron trifluoride.[364] In the case of mesitylene, the product is not subject to further attack:

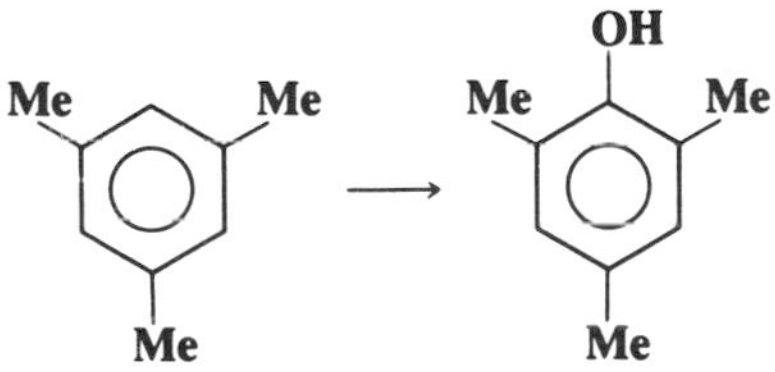

In a related procedure, even benzene and substituted benzenes (e.g., PhMe; PhCl; xylenes) can be converted to phenols in good yields with sodium perborate–$\text{F}_3\text{CSO}_2\text{OH}$.[365] Low to moderate yields of phenols can be obtained by treatment of simple alkylbenzenes with H_2O_2

[360]Olah, in Olah, Ref. 225, vol. 1, 1963, pp. 119-120.
[361]Adachi; Sugasawa *Synth. Commun.* **1990**, *20*, 71.
[362]For a list of hydroxylation reagents, with references, see Larock, Ref. 171, pp. 485-486.
[363]For reviews of electrophilic hydroxylation, see Jacquesy; Gesson; Jouannetaud *Rev. Chem. Intermed.* **1988**, *9*, 1-26, pp. 5-10; Haines *Methods for the Oxidation of Organic Compounds*; Academic Press: New York, 1985, pp. 173-176, 347-350.
[364]Hart; Buehler *J. Org. Chem.* **1964**, *29*, 2397. See also Hart *Acc. Chem. Res.* **1971**, *4*, 337-343.
[365]Prakash; Krass; Wang; Olah *Synlett* **1991**, 39.

in HF–BF$_3$[366] or H$_2$O$_2$ catalyzed by AlCl$_3$[367] or liquid HF, in some cases under CO$_2$ pressure.[368] With the last procedure even benzene could be converted to phenol in 37% yield (though 37% hydroquinone and 16% catechol were also obtained). Aromatic amines, N-acyl amines, and phenols were hydroxylated with H$_2$O$_2$ in SbF$_5$–HF.[369] Pyridine and quinoline were converted to their 2-acetoxy derivatives in high yields with acetyl hypofluorite AcOF at −75°C.[370]

Another hydroxylation reaction is the *Elbs reaction*.[371] In this method phenols can be oxidized to *p*-diphenols with K$_2$S$_2$O$_8$ in alkaline solution.[372] Primary, secondary, or tertiary aromatic amines give predominant or exclusive ortho substitution unless both ortho positions are blocked, in which case para substitution is found. The reaction with amines is called the *Boyland–Sims oxidation*. Yields are low with either phenols or amines, generally under 50%. The mechanisms are not clear,[373] but for the Boyland–Sims oxidation there is evidence that the S$_2$O$_8^{2-}$ ion attacks at the ipso position, and then a migration follows.[374]

G. Metal Electrophiles Reactions in which a metal replaces the hydrogen of an aromatic ring are considered along with their aliphatic counterparts in Chapter 12 (**2-21** and **2-22**).

Hydrogen as the Leaving Group in Rearrangement Reactions

In these reactions a group is detached from a *side chain* and then attacks the ring, but in other aspects they resemble the reactions already treated in this chapter.[375] Since a group moves from one position to another in a molecule, these are rearrangements. In all these reactions the question arises as to whether the group that cleaves from a given molecule attacks the same molecule or another one, i.e., is the reaction intramolecular or intermolecular? For intermolecular reactions the mechanism is the same as ordinary aromatic substitution, but for intramolecular cases the migrating group could never be completely free, or else it would be able to attack another molecule. Since the migrating species in intramolecular rearrangements is thus likely to remain near the atom from which it cleaved, it has been suggested that intramolecular reactions are more likely to lead to ortho products than are the intermolecular type. This characteristic has been used, among others, to help decide whether a given rearrangement is inter- or intramolecular, though there is evidence that at least in some cases, an intermolecular mechanism can still result in a high degree of ortho migration.[376]

[366]Olah; Fung; Keumi *J. Org. Chem.* **1981**, *46*, 4305. See also Gesson; Jacquesy; Jouannetaud *Nouv. J. Chem.* **1982**, *6*, 477.

[367]Kurz; Johnson *J. Org. Chem.* **1971**, *36*, 3184.

[368]Vesely; Schmerling *J. Org. Chem.* **1970**, *35*, 4028. For other hydroxylations, see Chambers; Goggin; Musgrave *J. Chem. Soc.* **1959**, 1804; Hamilton; Friedman *J. Am. Chem. Soc.* **1963**, *85*, 1008; Kovacic; Kurz *J. Am. Chem. Soc.* **1965**, *87*, 4811, *J. Org. Chem.* **1966**, *31*, 2011, 2549; Walling; Camaioni *J. Am. Chem. Soc.* **1975**, *97*, 1603; So; Miller *Synthesis* **1976**, 468; Ogata; Sawaki; Tomizawa; Ohno *Tetrahedron* **1981**, *37*, 1485; Galliani; Rindone *Tetrahedron* **1981**, *37*, 2313.

[369]Jacquesy; Joannetaud; Morellet; Vidal *Tetrahedron Lett.* **1984**, *25*, 1479; Berrier; Carreyre; Jacquesy; Joannetaud *New J. Chem.* **1990**, *14*, 283, and references cited in these papers.

[370]Rozen; Hebel; Zamir *J. Am. Chem. Soc.* **1987**, *109*, 3789.

[371]For a review of the Elbs and Boyland–Sims reactions, see Behrman *Org. React.* **1988**, *35*, 421-511.

[372]For a method for the ortho hydroxylation of phenols, see Capdevielle; Maumy *Tetrahedron Lett.* **1982**, *23*, 1573, 1577.

[373]Behrman *J. Am. Chem. Soc.* **1967**, *89*, 2424; Ogata; Akada *Tetrahedron* **1970**, *26*, 5945; Walling; Camaioni; Kim *J. Am. Chem. Soc.* **1978**, *100*, 4814.

[374]Srinivasan; Perumal; Arumugam *J. Chem. Soc., Perkin Trans. 2* **1985**, 1855.

[375]For a monograph, see Shine *Aromatic Rearrangements*; Elsevier: New York, 1967. For reviews, see Williams; Buncel *Isot. Org. Chem.* **1980**, *5*, 147-230; Williams, in Bamford; Tipper, Ref. 1, pp. 433-486.

[376]See Dawson; Hart; Littler *J. Chem. Soc., Perkin Trans. 2* **1985**, 1601.

The Claisen (**8-35**) and benzidine (**8-38**) rearrangements, which superficially resemble those in this section, have different mechanisms and are treated in Chapter 18.

A. Groups Cleaving From Oxygen

1-30   The Fries Rearrangement
1/C-Hydro,5/O-acyl-interchange[377]

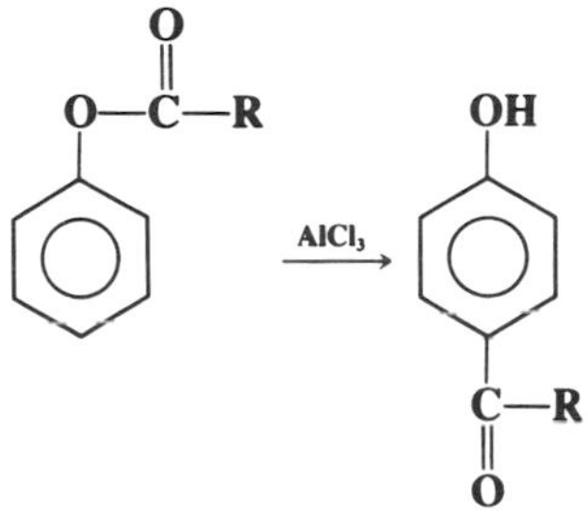

Phenolic esters can be rearranged by heating with Friedel–Crafts catalysts in a synthetically useful reaction known as the *Fries rearrangement*.[378] Both *o*- and *p*-acylphenols can be produced, and it is often possible to select conditions so that either one predominates. the ortho/para ratio is dependent on the temperature, solvent, and amount of catalyst used. Though exceptions are known, low temperatures generally favor the para product and high temperatures the ortho product. R may be aliphatic or aromatic. Any meta-directing substituent on the ring interferes with the reactions, as might be expected for a Friedel–Crafts process. In the case of aryl benzoates treated with F_3CSO_2OH, the Fries rearrangement was shown to be reversible and an equilibrium was established.[379]

The exact mechanism has still not been completely worked out. Opinions have been expressed that it is completely intermolecular,[380] completely intramolecular,[381] and partially inter- and intramolecular.[382] One way to decide between inter- and intramolecular processes is to run the reaction of the phenolic ester in the presence of another aromatic compound, say, toluene. If some of the toluene is acylated, the reaction must be, at least in part, intermolecular. If the toluene is not acylated, the presumption is that the reaction is intramolecular, though this is not certain, for it may be that the toluene is not attacked because it is less active than the other. A number of such experiments (called *crossover experiments*) have been carried out; sometimes crossover products have been found and sometimes not. As in **1-14,** an initial complex (**39**) is formed between the substrate and the catalyst, so that a catalyst/substrate molar ratio of at least 1:1 is required.

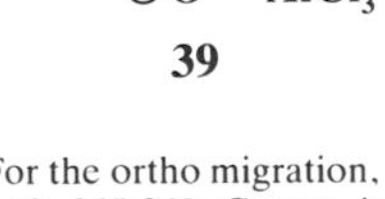

39

---

[377]This is the name for the para migration. For the ortho migration, the name is 1/C-hydro,3/O-acyl-interchange.
[378]For reviews, see Shine, Ref. 375, pp. 72-82, 365-368; Gerecs, in Olah, Ref. 261, vol. 3, 1964, pp. 499-533. For a list of references, see Larock, Ref. 171, pp. 642.
[379]Effenberger; Gutmann *Chem. Ber.* **1982,** *115,* 1089.
[380]Krausz; Martin *Bull. Soc. Chim. Fr.* **1965,** 2192; Martin *Bull. Soc. Chim. Fr.* **1974,** 983, **1979,** II-373; Martin; Gavard; Delfly; Demerseman; Tromelin *Bull. Soc. Chim. Fr.* **1986,** 659.
[381]Ogata; Tabuchi *Tetrahedron* **1964,** *20,* 1661.
[382]Munavilli *Chem. Ind.* (*London*) **1972,** 293; Warshawsky; Kalir; Patchornik *J. Am. Chem. Soc.* **1978,** *100,* 4544; Ref. 376.

The Fries rearrangement can also be carried out with uv light, in the absence of a catalyst.[383] This reaction, called the *photo-Fries rearrangement*,[384] is predominantly an intramolecular free-radical process. Both ortho and para migration are observed.[385] Unlike the Lewis-acid-catalyzed Fries rearrangement, the photo-Fries reaction can be accomplished, though often in low yields, when meta-directing groups are on the ring. The available evidence strongly suggests the following mechanism[386] for the photo-Fries rearrangement[387] (illustrated for para attack):

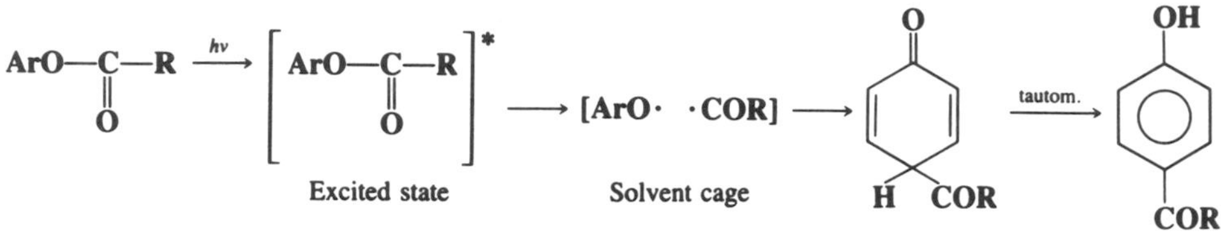

The phenol ArOH is always a side product, resulting from some ArO• that leaks from the solvent cage and abstracts a hydrogen atom from a neighboring molecule. When the reaction was performed on phenyl acetate in the gas phase, where there are no solvent molecules to form a cage (but in the presence of isobutane as a source of abstractable hydrogens), phenol was the chief product and virtually no *o*- or *p*-hydroxyacetophenone was found.[388] Other evidence[389] for the mechanism is that CIDNP has been observed during the course of the reaction[390] and that the ArO• radical has been detected by flash photolysis[391] and by nanosecond time-resolved Raman spectroscopy.[392]

OS **II**, 543; **III**, 280, 282.

1-31   Rearrangement of Phenolic Ethers
1/C-Hydro,5/O-alkyl-interchange

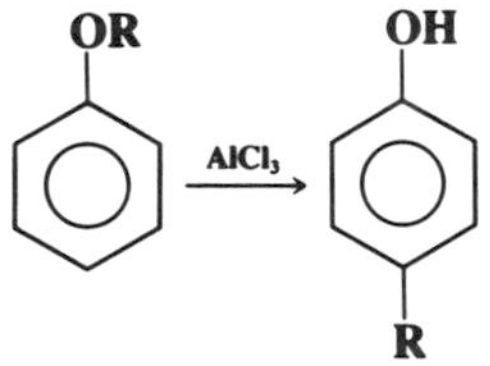

This reaction bears the same relationship to **1-30** that **1-12** bears to **1-14**.[393] However, yields are generally low and this reaction is much less useful synthetically. Isomerization of the R

[383]Kobsa *J. Org. Chem.* **1962**, *27*, 2293; Anderson; Reese *J. Chem. Soc.* **1963**, 1781; Finnegan; Matice *Tetrahedron* **1965**, *21*, 1015.
[384]For reviews, see Belluš *Adv. Photochem.* **1971**, *8* 109-159; Belluš; Hrdlovič *Chem. Rev.* **1967**, *67*, 599-609; Stenberg *Org. Photochem.* **1967**, *1*, 127-153.
[385]The migration can be made almost entirely ortho by cyclodextrin encapsulation (see p. 91): Syamala; Rao; Ramamurthy *Tetrahedron* **1988**, *44*, 7234. See also Veglia; Sanchez, de Rossi *J. Org. Chem.* **1990**, *55*, 4083.
[386]Proposed by Kobsa, Ref. 383.
[387]It has been suggested that a second mechanism, involving a four-center transition state, is also possible: Belluš; Schaffner; Hoigné *Helv. Chim. Acta* **1968**, *51*, 1980; Sander; Hedaya; Trecker *J. Am. Chem. Soc.* **1968**, *90*, 7249; Belluš Ref. 384.
[388]Meyer; Hammond *J. Am. Chem. Soc.* **1970**, *92*, 2187, **1972**, *94*, 2219.
[389]For evidence from isotope effect studies, see Shine; Subotkowski *J. Org. Chem.* **1987**, *52*, 3815.
[390]Adam; Arce de Sanabia; Fischer *J. Org. Chem.* **1973**, *38*, 2571; Adam *J. Chem. Soc., Chem. Commun.* **1974**, 289.
[391]Kalmus; Hercules *J. Am. Chem. Soc.* **1974**, *96*, 449.
[392]Beck; Brus *J. Am. Chem. Soc.* **1982**, *104*, 1805.
[393]For reviews, see Dalrymple; Kruger; White, in Patai *The Chemistry of the Ether Linkage*, Ref. 34, pp. 628-635; Shine, Ref. 375, pp. 82-89, 368-370.

group is usually found when that is possible. Evidence has been found for both inter- and intramolecular processes.[394] The fact that dialkylphenols can often be isolated shows that at least some intermolecular processes occur. Evidence for intramolecular reaction is that conversion of optically active *p*-tolyl *sec*-butyl ether to 2-*sec*-butyl-4-methylphenol proceeded with some retention of configuration.[395] The mechanism is probably similar to that of **1-14**.

B. Groups Cleaving from Nitrogen[396] It has been shown that PhN̄H$_2$D rearranges to *o*- and *p*-deuterioaniline.[397] The migration of OH, formally similar to reactions **1-32** to **1-36**, is a nucleophilic substitution and is treated in Chapter 13 (**3-27**).

1-32   Migration of the Nitro Group
  1/C-Hydro,3/N-nitro-interchange

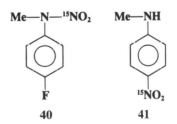

N-Nitro aromatic amines rearrange on treatment with acids to *o*- and *p*-nitroamines with the ortho compounds predominating.[398] Aside from this indication of an intramolecular process, there is also the fact that virtually no meta isomer is produced in this reaction,[399] though direct nitration of an aromatic amine generally gives a fair amount of meta product. Thus a mechanism in which NO$_2^+$ is dissociated from the ring and then attacks another molecule must be ruled out. Further results indicating an intramolecular process are that rearrangement of several substrates in the presence of K^{15}NO$_3$ gave products containing no ^{15}N[400] and that rearrangement of a mixture of PhNH^{15}NO$_2$ and unlabeled *p*-MeC$_6$H$_4$NHNO$_2$ gave 2-nitro-4-methylaniline containing no ^{15}N.[401] On the other hand, rearrangement of **40**

$$\text{Me--N--}^{15}\text{NO}_2 \qquad \text{Me--NH}$$

F	^{15}NO$_2$
40	**41**

in the presence of unlabeled PhNMeNO$_2$ gave labeled **41**, which did not arise by displacement of F.[402] R may be hydrogen or alkyl. Two principal mechanisms have been suggested, one

[394]For mechanistic discussions, see Tarbell; Petropoulos *J. Am. Chem. Soc.* **1952**, *74*, 244; Hart; Waddington *J. Chem. Soc., Perkin Trans. 2* **1985**, 1607.
  [395]Sprung; Wallis *J. Am. Chem. Soc.* **1934**, *56*, 1715. See also Hart; Elia *J. Am. Chem. Soc.* **1954**, *76*, 3031.
  [396]For a review, see Stevens; Watts *Selected Molecular Rearrangements*; Van Nostrand–Reinhold: Princeton, 1973, pp. 192-199.
  [397]Okazaki; Okumura *Bull. Chem. Soc. Jpn.* **1961**, *34*, 989.
  [398]For reviews, see Williams, in Patai *The Chemistry of Functional Groups, Supplement F*, pt. 1; Wiley: New York, 1982, pp. 127-153; White, *Mech. Mol. Migr.* **1971**, *3*, 109-143; Shine, Ref. 375, pp. 235-249.
  [399]Hughes; Jones *J. Chem. Soc.* **1950**, 2678.
  [400]Brownstein; Bunton; Hughes *J. Chem. Soc.* **1958**, 4354; Banthorpe; Thomas; Williams *J. Chem. Soc.* **1965**, 6135.
  [401]Geller; Dubrova *J. Gen. Chem. USSR* **1960**, *30*, 2627.
  [402]White; Golden *J. Org. Chem.* **1970**, *35*, 2759.

involving cyclic attack by the oxygen of the nitro group at the ortho position before the group cleaves,[403] and the other involving a cleavage into a radical and a radical ion held together in a solvent cage.[404] Among the evidence for the latter view[405] are the effects of

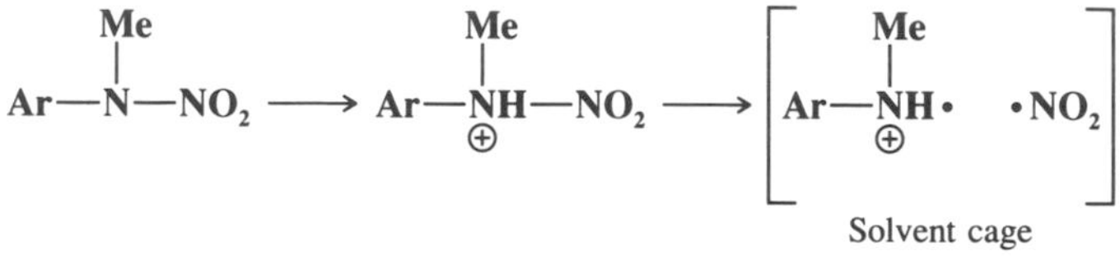

$$\text{Ar—N—NO}_2 \longrightarrow \overset{\oplus}{\text{Ar—NH—NO}_2} \longrightarrow \left[\overset{\oplus}{\text{Ar—NH}\cdot} \quad \cdot\text{NO}_2 \right]$$

Solvent cage

substituents on the rate of the reaction,[406] [15]N and ^{14}C kinetic isotope effects that show nonconcertedness,[407] and the fact that both N-methylaniline and nitrous acid are produced in sizable and comparable amounts in addition to the normal products o- and p-nitro-N-methylaniline.[408] These side products are formed when the radicals escape from the solvent cage.

1-33 Migration of the Nitroso Group. The Fischer–Hepp Rearrangement
1/C-Hydro-5/N-nitroso-interchange

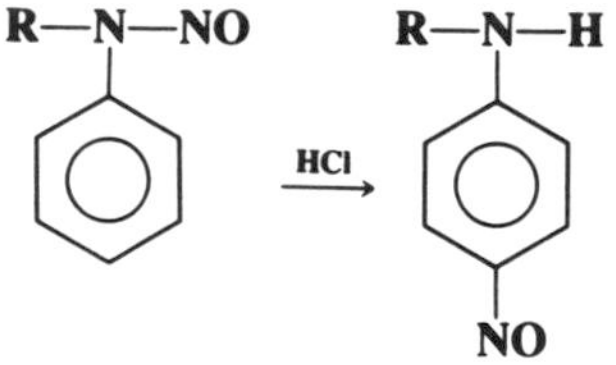

R—N—NO →(HCl) R—N—H (NO para)

The migration of a nitroso group, formally similar to **1-32**, is important because p-nitroso secondary aromatic amines cannot generally be prepared by direct C-nitrosation of secondary aromatic amines (see **2-51**). The reaction, known as the *Fischer–Hepp rearrangement*,[409] is brought about by treatment of N-nitroso secondary aromatic amines with HCl. Other acids give poor or no results. In benzene systems the para product is usually formed exclusively.[410] The mechanism of the rearrangement is not completely understood. The fact that the reaction takes place in a large excess of urea[411] shows that it is intramolecular[412] since, if NO+, NOCl,

---

[403]Banthorpe; Hughes; Williams *J. Chem. Soc.* **1964**, 5349; Banthorpe; Thomas *J. Chem. Soc.* **1965**, 7149, 7158. Also see Ref. 400.

[404]White; Lazdins; White *J. Am. Chem. Soc.* **1964**, *86*, 1517; White; White; Fentiman *J. Org. Chem.* **1976**, *41*, 3166.

[405]For additional evidence, see White; Hathaway; Huston *J. Org. Chem.* **1970**, *35*, 737; White; Golden; Lazdins *J. Org. Chem.* **1970**, *35*, 2048; White; Klink *J. Org. Chem.* **1977**, *42*, 166; Ridd; Sandall *J. Chem. Soc., Chem. Commun.* **1982**, 261.

[406]White; Klink *J. Org. Chem.* **1970**, *35*, 965.

[407]Shine; Zygmunt; Brownawell; San Filippo *J. Am. Chem. Soc.* **1984**, *106*, 3610.

[408]White; White *J. Org. Chem.* **1970**, *35*, 1803.

[409]For reviews, see Williams, Ref. 123, pp. 113-128; Williams, Ref. 398; Shine, Ref. 375, pp. 231-235.

[410]For a report of formation of about 15% ortho product in the case of N,N-diaryl-N-nitroso amides, see Titova; Arinich; Gorelik *J. Org. Chem. USSR* **1986**, *22*, 1407.

[411]Aslapovskaya; Belyaev; Kumarev; Porai–Koshits *Org. React. USSR* **1968**, *5*, 189; Morgan; Williams *J. Chem. Soc., Perkins Trans. 2* **1972**, 74.

[412]See also Belyaev; Nikulicheva *Org. React. USSR* **1971**, *7*, 165; Williams; Wilson *J. Chem. Soc., Perkin Trans. 2* **1974**, 13; Williams *Tetrahedron* **1975**, *31*, 1343, *J. Chem. Soc., Perkin Trans 2* **1975**, 655, **1982**, 801.

or some similar species were free in the solution, it would be captured by the urea, preventing the rearrangement.

1-34 Migration of an Arylazo Group
         1/C-Hydro-5/N-arylazo-interchange

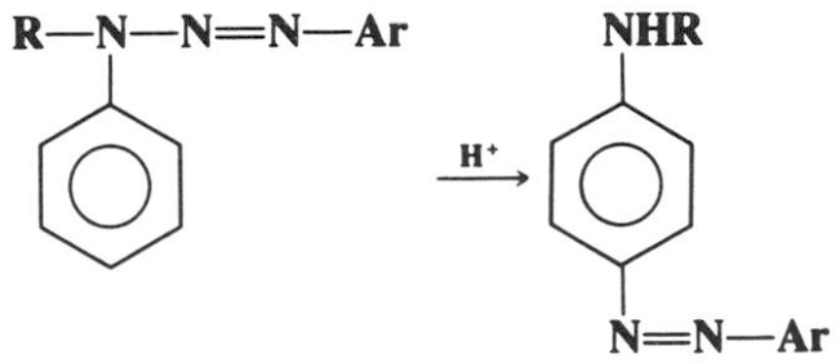

R—N—N=N—Ar   NHR

$\xrightarrow{\text{H}^+}$

N=N—Ar

Rearrangement of aryl triazenes can be used to prepare azo derivatives of primary and secondary aromatic amines.[413] These are first diazotized at the amino group (see **1-4**) to give triazenes, which are then rearranged by treatment with acid. The rearrangement always gives the para isomer, unless that position is occupied.

1-35 Migration of Halogen. The Orton Rearrangement
         1/C-Hydro-5/N-halo-interchange

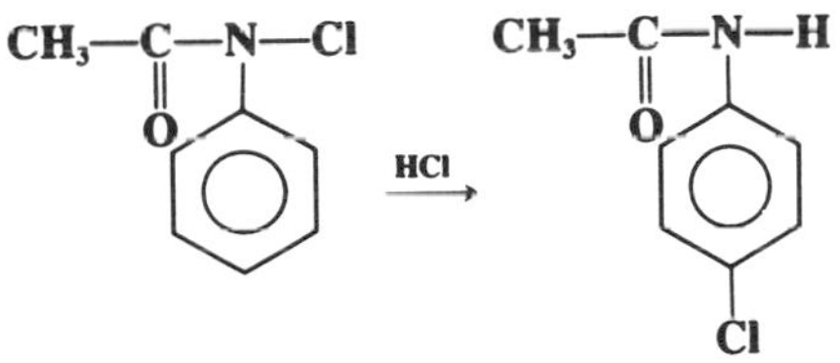

CH₃—C—N—Cl   CH₃—C—N—H
       ‖   ‖
       O   O
              $\xrightarrow{\text{HCl}}$

Cl

Migration of a halogen from a nitrogen side chain to the ring by treatment with HCl is called the *Orton rearrangement*.[414] The main product is the para isomer, though some ortho product may also be formed. The reaction has been carried out with N-chloro- and N-bromoamines and less often with N-iodo compounds. The amine must be acylated, except that PhNCl₂ gives 2,4-dichloroaniline. The reaction is usually performed in water or acetic acid. There is much evidence (cross-halogention, labeling, etc.) that this is an intermolecular process.[415] First the HCl reacts with the starting material to give ArNHCOCH₃ and Cl₂; then the chlorine halogenates the ring as in **1-11**. Among the evidence is that chlorine has been isolated from the reaction mixture. The Orton rearrangement can also be brought about photochemically[416] and by heating in the presence of benzoyl peroxide.[417] These are free-radical processes.

---

[413]For a review, see Shine, Ref. 375, pp. 212-221.
[414]For reviews, see Shine, Ref. 375, pp. 221-230, 362-364: Bieron; Dinan, in Zabicky *The Chemistry of Amides*; Wiley: New York, 1970, pp. 263-269.
[415]The reaction has been found to be intramolecular in aprotic solvents: Golding; Reddy; Scott; White; Winter *Can. J. Chem.* **1981**, *59*, 839.
[416]For example, see Hodges *J. Chem. Soc.* **1933**, 240.
[417]For example, Ayad; Beard; Garwood; Hickinbottom *J. Chem. Soc.* **1957**, 2981; Coulson; Williams; Johnston *J. Chem. Soc. B* **1967**, 174.

1-36 Migration of an Alkyl Group[418]
1/C-Hydro-5/N-alkyl-interchange

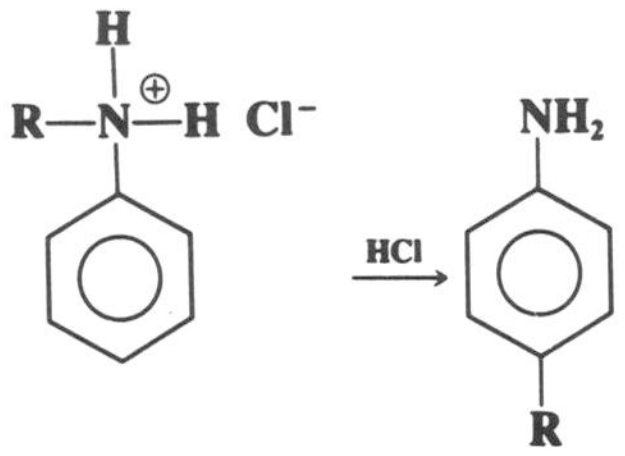

When HCl salts of arylalkylamines are heated at about 200 to 300°C, migration occurs. This is called the *Hofmann–Martius reaction*. It is an intermolecular reaction, since crossing is found. For example, methylanilinium bromide gave not only the normal products *o*- and *p*-toluidine but also aniline and di- and trimethylanilines.[419] As would be expected for an intermolecular process, there is isomerization when R is primary.

With primary R, the reaction probably goes through the alkyl halide formed initially in an S$_N$2 reaction:

$$\overset{\oplus}{R}NH_2Ar + Cl^- \longrightarrow RCl + ArNH_2$$

Evidence for this view is that alkyl halides have been isolated from the reaction mixture and that Br$^-$, Cl$^-$, and I$^-$ gave different ortho/para ratios, which indicates that the halogen is involved in the reaction.[419] Further evidence is that the alkyl halides isolated are unrearranged (as would be expected if they are formed by an S$_N$2 mechanism), even though the alkyl groups in the ring are rearranged. Once the alkyl halide is formed, it reacts with the substrate by a normal Friedel–Crafts alkylation process (**1-12**), accounting for the rearrangement. When R is secondary or tertiary, carbocations may be directly formed so that the reaction does not go through the alkyl halides.[420]

It is also possible to carry out the reaction by heating the amine (not the salt) at a temperature between 200 and 350°C with a metal halide such as CoCl$_2$, CdCl$_2$, or ZnCl$_2$. When this is done, the reaction is called the *Reilly–Hickinbottom rearrangement*. Primary R groups larger than ethyl give both rearranged and unrearranged products.[421] The reaction is not generally useful for secondary and tertiary R groups, which are usually cleaved to olefins under these conditions.

When acylated arylamines are photolyzed, migration of an acyl group takes place[422] in a process that resembles the photo-Fries reaction (**1-30**).

[418]For reviews, see Grillot *Mech. Mol. Migr.* **1971**, *3* 237-270; Shine, Ref. 375, pp. 249-257.
[419]Ogata; Tabuchi; Yoshida *Tetrahedron* **1964**, *20*, 2717.
[420]Hart; Kosak *J. Org. Chem.* **1962**, *27*, 116.
[421]For example, see Birchall; Clark; Goldwhite; Thorpe *J. Chem. Soc., Perkin Trans. 1* **1972**, 2579.
[422]For examples, see Elad; Rao; Stenberg *J. Org. Chem.* **1965**, *30*, 3252; Shizuka; Tanaka *Bull. Chem. Soc. Jpn.* **1968**, *41*, 2343, **1969**, *42*, 909; Fischer *Tetrahedron Lett.* **1968**, 4295; Hageman *Recl. Trav. Chim. Pays-Bas* **1972**, *91*, 1447; Chênevert; Plante *Can. J. Chem.* **1983**, *61*, 1092; Abdel-Malik; de Mayo *Can. J. Chem.* **1984**, *62*, 1275; Nassetta; de Rossi; Cosa *Can. J. Chem.* **1988**, *66*, 2794.

Other Leaving Groups

Three types of reactions are considered in this section.

1. Reactions in which hydrogen replaces another leaving group:

$$ArX + H^+ \longrightarrow ArH$$

2. Reactions in which an electrophile other than hydrogen replaces another leaving group:

$$ArX + Y^+ \longrightarrow ArY$$

3. Reactions in which a group (other than hydrogen) migrates from one position in a ring to another. Such migrations can be either inter- or intramolecular:

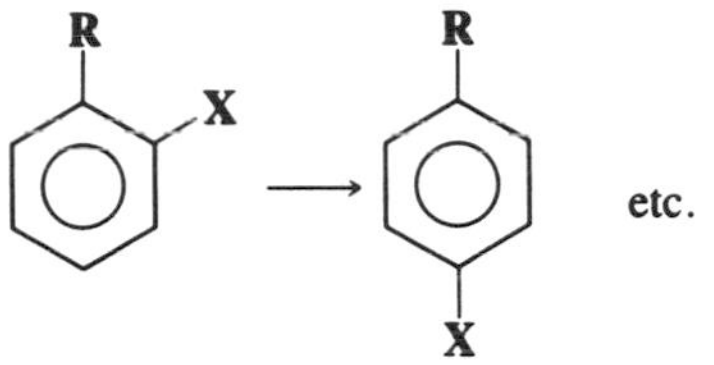

etc.

The three types are not treated separately, but reactions are classified by leaving group.

A.   Carbon Leaving Groups

1-37   Reversal of Friedel–Crafts Alkylation
Hydro-de-alkylation or **Dealkylation**

$$ArR + H^+ \xrightarrow{\;AlCl_3\;} ArH$$

Alkyl groups can be cleaved from aromatic rings by treatment with proton and/or Lewis acids. Tertiary R groups are the most easily cleaved; because this is true, the *t*-butyl group is occasionally introduced into a ring, used to direct another group, and then removed.[423] For example,[424]

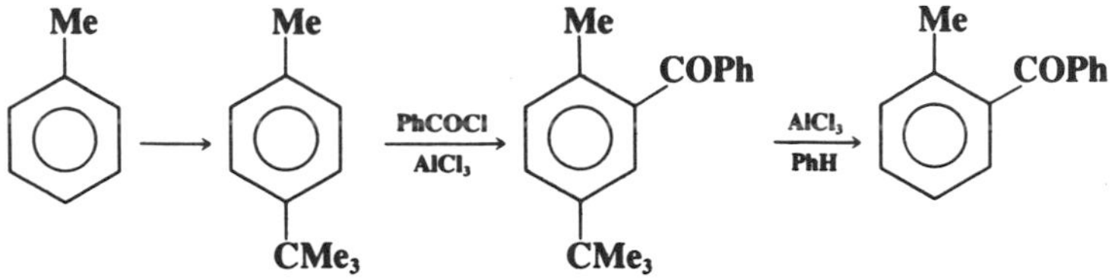

Secondary R groups are harder to cleave, and primary R harder still. Because of this reaction, care must be taken when using Friedel–Crafts catalysts (Lewis or proton acids) on aromatic compounds containing alkyl groups. True cleavage, in which the R becomes an olefin, occurs only at high temperatures—above 400°C.[425] At ordinary temperatures, the R group attacks

[423]For reviews of such reactions, where the blocking group is *t*-butyl, benzyl, or a halogen, see Tashiro, *Synthesis* **1979,** 921-936; Tashiro; Fukata *Org. Prep. Proceed. Int.* **1976,** *8*, 51-74.
[424]Hofman; Reiding; Nauta *Recl. Trav. Chim. Pays-Bas* **1960,** *79*, 790.
[425]Olah, in Olah, Ref. 261, vol. 1, 1963, pp. 36-38.

another ring, so that the bulk of the product may be dealkylated, but there is a residue of heavily alkylated material. The isomerization reaction, in which a group migrates from one position in a ring to another or to a different ring, is therefore more important than true cleavage. In these reactions, the meta isomer is generally the most favored product among the dialkylbenzenes; and the 1,3,5 product the most favored among the trialkylbenzenes, because they have the highest thermodynamic stabilities. Alkyl migrations can be inter- or intramolecular, depending on the conditions and on the R group. The following experiments can be cited: Ethylbenzene treated with HF and BF_3 gave, almost completely, benzene and diethylbenzenes[426] (entirely intermolecular); propylbenzene labeled in the β position gave benzene, propylbenzene, and di- and tripropylbenzenes, but the propylbenzene recovered was partly labeled in the α position and not at all in the γ position[427] (both intra- and intermolecular); o-xylene treated with HBr and $AlBr_3$ gave a mixture of o- and m- but no p-xylene, while p-xylene gave p- and m- but no o-xylene, and no trimethyl compounds could be isolated in these experiments[428] (exclusively intramolecular rearrangement). Apparently, methyl groups migrate only intramolecularly, while other groups may follow either path.[429]

The mechanism[430] of intermolecular rearrangement can involve free alkyl cations, but there is much evidence to show that this is not necessarily the case. For example, many of them occur without rearrangement within the alkyl group. The following mechanism has been proposed for intermolecular rearrangement without the involvement of carbocations that are separated from the ring.[431]

$$Ar-CH_2-CH_3 + AlCl_3 \longrightarrow Ar-\overset{\oplus}{CH}-CH_3 \xrightarrow{Ar'H} Ar-\underset{\underset{Ar'}{|}}{CH}-CH_3 \xrightarrow{H^\cdot}$$

$$ArH + \overset{\oplus}{\underset{\underset{Ar'}{|}}{CH}}-CH_3 \xrightarrow{ArCH_2CH_3} \underset{\underset{Ar'}{|}}{CH_2}-CH_3 + Ar\overset{\oplus}{CH}CH_3$$

Evidence for this mechanism is that optically active $PhCHDCH_3$ labeled in the ring with ^{14}C and treated with $GaBr_3$ in the presence of benzene gave ethylbenzene containing no deuterium and two deuteriums and that the rate of loss of radioactivity was about equal to the rate of loss of optical activity.[431] The mechanism of intramolecular rearrangement is not very clear. 1,2 shifts of this kind have been proposed:[432]

There is evidence from ^{14}C labeling that intramolecular migration occurs only through 1,2 shifts.[433] Any 1,3 or 1,4 migration takes place by a series of two or more 1,2 shifts.

[426]McCaulay; Lien *J. Am. Chem. Soc.* **1953**, *75*, 2407. For similar results, see Roberts; Roengsumran *J. Org. Chem.* **1981**, *46*, 3689; Bakoss; Roberts; Sadri *J. Org. Chem.* **1982**, *47*, 4053.
[427]Roberts; Brandenberger *J. Am. Chem. Soc.* **1957**, *79*, 5484; Roberts; Douglass *J. Org. Chem.* **1963**, *28*, 1225.
[428]Brown; Jungk *J. Am. Chem. Soc.* **1955**, *77*, 5579; Allen; Yats *J. Am. Chem. Soc.* **1959**, *81*, 5289.
[429]Allen; Alfrey; Yats *J. Am. Chem. Soc.* **1959**, *81*, 42; Allen *J. Am. Chem. Soc.* **1960**, *82*, 4856.
[430]For a review of the mechanism of this and closely related reactions, see Shine, Ref. 375, pp. 1-55.
[431]Streitwieser; Reif *J. Am. Chem. Soc.* **1964**, *86*, 1988.
[432]Olah; Meyer; Overchuk *J. Org. Chem.* **1964**, *29*, 2313.
[433]See, for example, Steinberg; Sixma, *Recl. Trav. Chim. Pays-Bas* **1962**, *81*, 185; Koptyug; Isaev; Vorozhtsov *Doklad. Akad. Nauk SSSR* **1963**, *149*, 100.

Phenyl groups have also been found to migrate. Thus *o*-terphenyl, heated with AlCl₃–H₂O, gave a mixture containing 7% *o*-, 70% *m*-, and 23% *p*-terphenyl.[434] Alkyl groups have also been replaced by groups other than hydrogen, e.g., nitro groups.

Unlike alkylation, Friedel–Crafts *acylation* has been generally considered to be irreversible, but a number of instances of electrofugal acyl groups have been reported,[435] especially where there are two ortho substituents, for example, the hydro-de-benzoylation of **42**.[436]

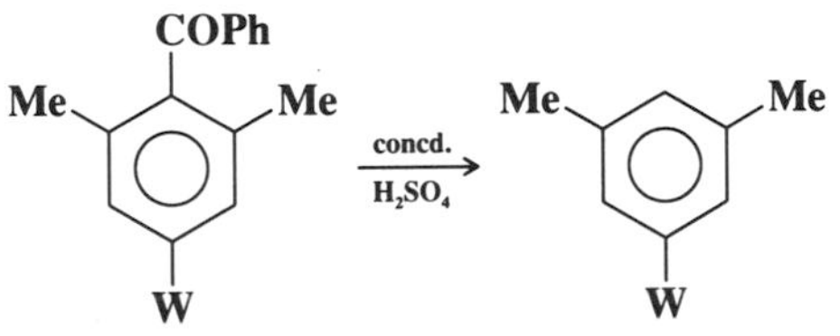

$$W = Me, Cl, OH, \text{etc.}$$

OS **V**, 332. Also see OS **III**, 282, 653; **V**, 598.

1-38 Decarbonylation of Aromatic Aldehydes
Hydro-de-formylation or Deformylation

$$\text{ArCHO} \xrightarrow{\text{H}_2\text{SO}_4} \text{ArH} + \text{CO}$$

The decarbonylation of aromatic aldehydes with sulfuric acid[437] is the reverse of the Gatterman–Koch reaction (**1-16**). It has been carried out with trialkyl- and trialkoxybenzaldehydes. The reaction takes place by the ordinary arenium ion mechanism: the attacking species is H⁺ and the leaving group is HCO⁺, which can lose a proton to give CO or combine with OH⁻ from the water solvent to give formic acid.[438] Aromatic aldehydes have also been decarbonylated with basic catalysts.[439] When basic catalysts are used, the mechanism is probably similar to the SE1 process of **1-39**. See also **4-41**.

1-39 Decarboxylation of Aromatic Acids
Hydro-de-carboxylation or Decarboxylation

$$\text{ArCOOH} \xrightarrow[\text{quinoline}]{\text{Cu}} \text{ArH} + \text{CO}_2$$

The decarboxylation of aromatic acids is most often carried out by heating with copper and quinoline. However, two other methods can be used with certain substrates. In one method the salt of the acid (ArCOO⁻) is heated, and in the other the carboxylic acid is heated with a strong acid, often sulfuric. The latter method is accelerated by the presence of electron-donating groups in ortho and para positions and by the steric effect of groups in the ortho positions; in benzene systems it is generally limited to substrates that contain such groups.

[434]Olah; Meyer *J. Org. Chem.* **1962**, *27*, 3682.

[435]For some other examples, see Agranat; Bentor; Shih *J. Am. Chem. Soc.* **1977**, *99*, 7068; Bokova; Buchina *J. Org. Chem. USSR* **1984**, *20*, 1199; Benedikt; Traynor *Tetrahedron Lett.* **1987**, *28*, 763; Gore; Moonga; Short *J. Chem. Soc.*, *Perkin Trans. 2* **1988**, 485; Keumi; Morita; Ozawa; Kitajima *Bull. Chem. Soc. Jpn.* **1989**, *62*, 599; Giordano; Villa; Annunziata *Synth. Commun.* **1990**, *20*, 383.

[436]Al-Ka'bi; Farooqi; Gore; Moonga; Waters *J. Chem. Res. (S)* **1989**, 80.

[437]For reviews of the mechanism, see Taylor, in Bamford; Tipper, Ref. 1, pp. 316-323; Schubert; Kintner, in Patai *The Chemistry of the Carbonyl Group*, vol. 1; Wiley: New York, 1966, pp. 695-760.

[438]Burkett; Schubert; Schultz; Murphy; Talbott *J. Am. Chem. Soc.* **1959**, *81*, 3923.

[439]Bunnett; Miles; Nahabedian *J. Am. Chem. Soc.* **1961**, *83*, 2512; Forbes; Gregory *J. Chem. Soc. B* **1968**, 205.

In this method decarboxylation takes place by the arenium ion mechanism,[440] with H^+ as the electrophile and CO_2 as the leaving group.[441] Evidently, the order of electrofugal ability

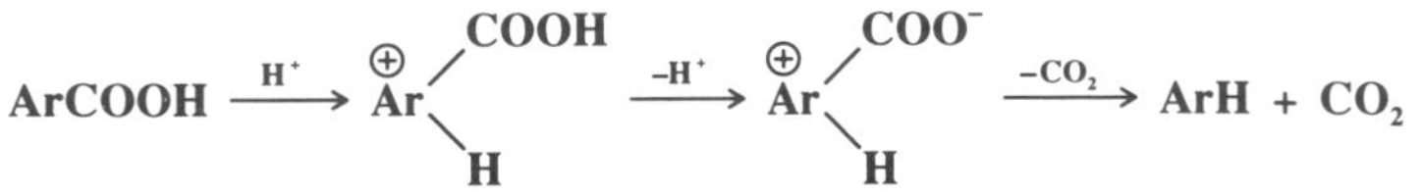

$$\text{ArCOOH} \xrightarrow{\text{H}^+} \overset{\oplus}{\text{Ar}}\!\!\diagup_{\!\!\!\!\text{H}}^{\!\!\text{COOH}} \xrightarrow{-\text{H}^+} \overset{\oplus}{\text{Ar}}\!\!\diagup_{\!\!\!\!\text{H}}^{\!\!\text{COO}^-} \xrightarrow{-\text{CO}_2} \text{ArH} + \text{CO}_2$$

is $CO_2 > H^+ > COOH^+$, so that it is necessary, at least in most cases, for the COOH to lose a proton before it can cleave.

When carboxylate *ions* are decarboxylated, the mechanism is entirely different, being of the SE1 type. Evidence for this mechanism is that the reaction is first order and that electron-withdrawing groups, which would stabilize a carbanion, facilitate the reaction.[442]

Step 1

Step 2

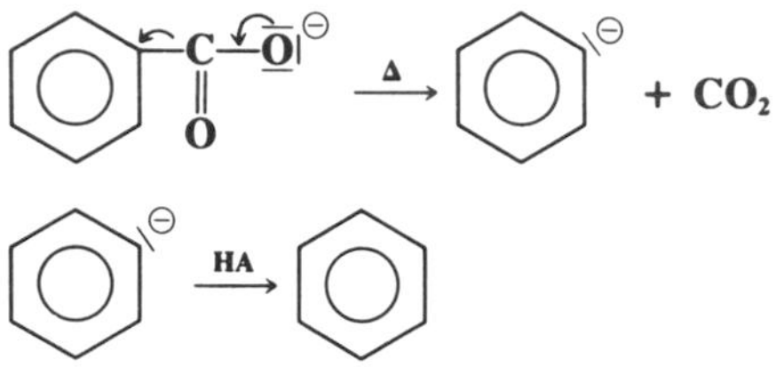

Despite its synthetic importance, the mechanism of the copper–quinoline method has been studied very little, but it has been shown that the actual catalyst is cuprous ion.[443] In fact, the reaction proceeds much faster if the acid is heated in quinoline with cuprous oxide instead of copper, provided that atmospheric oxygen is rigorously excluded. A mechanism has been suggested in which it is the cuprous salt of the acid that actually undergoes the decarboxylation.[443] It has been shown that cuprous salts of aromatic acids are easily decarboxylated by heating in quinoline[444] and that arylcopper compounds are intermediates that can be isolated in some cases.[445] Metallic silver has been used in place of copper, with higher yields.[446]

In certain cases the carboxyl group can be replaced by electrophiles other than hydrogen, e.g., NO,[446] I,[447] Br,[448] or Hg.[449]

Rearrangements are also known to take place. For example, when the phthalate ion is heated with a catalytic amount of cadmium, the terphthalate ion (**43**) is produced:[450]

[440]For a review, see Taylor, in Bamford; Tipper, Ref. 1, pp. 303-316. For a review of isotope effect studies of this reaction, see Willi *Isot. Org. Chem.* **1977**, *3*, 257-267.

[441]See, for example, Los; Rekker; Tonsbeeck *Recl. Trav. Chim. Pays-Bas* **1967**, *86*, 622; Huang; Long *J. Am. Chem. Soc.* **1969**, *91*, 2872; Willi; Cho; Won *Helv. Chim. Acta* **1970**, *53*, 663.

[442]See, for example, Segura; Bunnett; Villanova *J. Org. Chem.* **1985**, *50*, 1041.

[443]Cohen; Schambach *J. Am. Chem. Soc.* **1970**, *92*, 3189. See also Aalten; van Koten; Tromp; Stam; Goubitz; Mak *Recl. Trav. Chim. Pays-Bas* **1989**, *108*, 295.

[444]Cairncross; Roland; Henderson; Sheppard *J. Am. Chem. Soc.* **1970**, *92*, 3187; Cohen; Berninger; Wood *J. Org. Chem.* **1978**, *43*, 37.

[445]For example, see Ibne-Rasa *J. Am. Chem. Soc.* **1962**, *84*, 4962; Tedder; Theaker *J. Chem. Soc.* **1959**, 257.

[446]Chodowska-Palicka; Nilsson *Acta Chem. Scand.* **1970**, *24*, 3353.

[447]Singh; Just *Synth. Commun.* **1988**, *18*, 1327.

[448]For example, see Grovenstein; Ropp *J. Am. Chem. Soc.* **1956**, *78*, 2560.

[449]For a review, see Larock *Organomercury Compounds in Organic Synthesis*; Springer: New York, 1985, pp. 101-105.

[450]Raecke *Angew. Chem.* **1958**, *70*, 1; Riedel; Kienitz *Angew. Chem.* **1960**, *72*, 738; McNelis *J. Org. Chem.* **1965**, *30*, 1209; Ogata; Nakajima *Tetrahedron* **1965**, *21*, 2393; Ratuský; Šorm *Chem. Ind. (London)* **1966**, 1798.

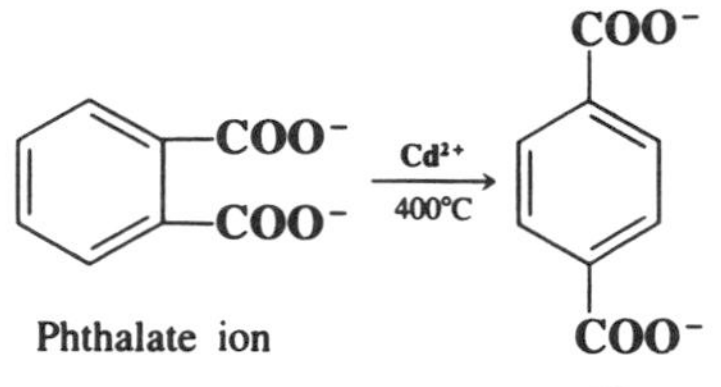

In a similar process, potassium benzoate heated with cadmium salts disproportionates to benzene and **43**. The term *Henkel reaction* (named for the company that patented the process) is used for these rearrangements.[451] An S_E1 mechanism has been suggested.[452] The terphthalate is the main product because it crystallizes from the reaction mixture, driving the equilibrium in that direction.[453]

For aliphatic decarboxylation, see **2-40**.

OS **I**, 274, 455, 541; **II**, 100, 214, 217, 341; **III**, 267, 272, 471, 637; **IV**, 590, 628; **V**, 635, 813, 982, 985. Also see OS **I**, 56.

1-40   The Jacobsen Reaction

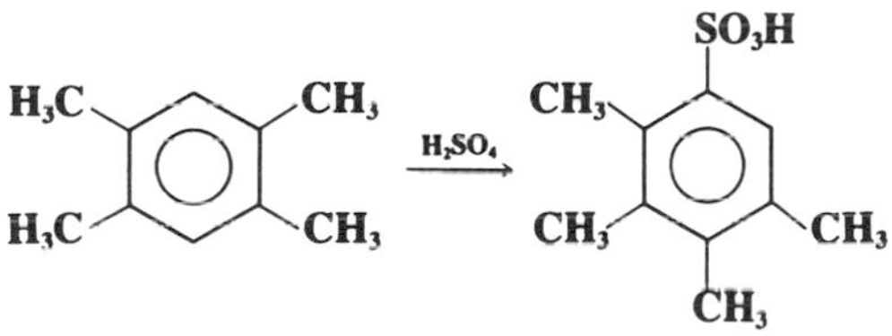

When polalkyl- or polyhalobenzenes are treated with sulfuric acid, the ring is sulfonated, but rearrangement also takes place. The reaction, known as the *Jacobsen reaction*, is limited to benzene rings that have at least four substituents, which can be any combination of alkyl and halogen groups, where the alkyl groups can be ethyl or methyl and the halogen iodo, chloro, or bromo. When isopropyl or *t*-butyl groups are on the ring, these groups are cleaved to give olefins. Since a sulfo group can later be removed (**1-41**), the Jacobsen reaction can be used as a means of rearranging polyalkylbenzenes. The rearrangement always brings the alkyl or halo groups closer together than they were originally. Side products in the case illustrated above are pentamethylbenzenesulfonic acid, 2,4,5-trimethylbenzenesulfonic acid, etc., indicating an intermolecular process, at least partially.

The mechanism of the Jacobsen reaction is not established,[454] but there is evidence, at least for polymethylbenzenes, that the rearrangement is intermolecular, and that the species to which the methyl group migrates is a polymethylbenzene, not a sulfonic acid. Sulfonation takes place after the migration.[455] It has been shown by labeling that ethyl groups migrate without internal rearrangement.[456]

---

[451]For a review, see Ratuský, in Patai *The Chemistry of Acid Derivatives*, pt. 1; Wiley: New York, 1979, pp. 915-944.

[452]See, for example, Ratuský *Collect. Czech. Chem. Commun.* **1967**, *32*, 2504, **1972**, *37*, 2436, **1973**, *38*, 74, 87.

[453]Ratuský *Chem. Ind. (London)* **1967**, 1093, *Collect. Czech. Chem. Commun.* **1968**, *33*, 2346.

[454]For discussions, see Suzuki *Bull. Chem. Soc. Jpn.* **1963**, *36*, 1642; Koeberg-Telder; Cerfontain *J. Chem. Soc., Perkin Trans. 2* **1977**, 717; Cerfontain, *Mechanistic Aspects in Aromatic Sulfonation and Desulfonation*, Ref. 155, pp. 214-226; Taylor, in Bamford; Tipper, Ref. 1, pp. 22-32, 48-55.

[455]Koeberg-Telder; Cerfontain *Recl. Trav. Chim. Pays-Bas* **1987**, *106*, 85; Cerfontain; Koeberg-Telder *Can. J. Chem.* **1988**, *66*, 162.

[456]Marvell; Webb *J. Org. Chem.* **1962**, *27*, 4408.

B. Sulfur Leaving Groups

1-41 Desulfonation or Hydro-de-sulfonation

$$ArSO_3H \xrightarrow[\text{dil. } H_2SO_4]{135-200°C} ArH + H_2SO_4$$

The cleavage of sulfo groups from aromatic rings is the reverse of **1-7**.[457] By the principle of microscopic reversibility, the mechanism is also the reverse.[458] Dilute H_2SO_4 is generally used, as the reversibility of sulfonation decreases with increasing H_2SO_4 concentration. The reaction permits the sulfo group to be used as a blocking group to direct meta and then to be removed. The sulfo group has also been replaced by nitro and halogen groups. Sulfo groups have also been removed from the ring by heating with an alkaline solution of Raney nickel.[459] In another catalytic process, aromatic sulfonyl bromides or chlorides are converted to aryl bromides or chlorides, respectively, on heating with chloro-tris(triphenylphosphine)rhodium(I).[460] This reaction is similar to the decarbonylation of aromatic acyl halides mentioned in **4-41**.

$$ArSO_2Br \xrightarrow{RhCl(PPh_3)_3} ArBr$$

OS **I**, 388; **II**, 97; **III**, 262; **IV**, 364. Also see OS **I**, 519; **II**, 128; **V**, 1070.

C. Halogen Leaving Groups

1-42 Dehalogenation or Hydro-de-halogenation

$$ArX \xrightarrow{AlCl_3} ArH$$

Aryl halides can be dehalogenated by Friedel–Crafts catalysts. Iodine is the most easily cleaved. Dechlorination is seldom performed and defluorination apparently never. The reaction is most successful when a reducing agent, say, Br^- or I^- is present to combine with the I^+ or Br^+ coming off.[461] Except for deiodination, the reaction is seldom used for preparative purposes. Migration of halogen is also found,[462] both intramolecular[463] and intermolecular.[464] The mechanism is probably the reverse of that of **1-11**.[465]

Rearrangement of polyhalobenzenes can also be catalyzed by very strong bases; e.g., 1,2,4-tribromobenzene is converted to 1,3,5-tribromobenzene by treatment with PhNHK.[466]

---

[457]For reviews, see Cerfontain, Ref. 454, pp. 185-214; Taylor, in Bamford; Tipper, Ref. 1, pp. 349-355; Gilbert, Ref. 152, pp. 427-442. See also Krylov *J. Org. Chem. USSR* **1988**, *24*, 709.

[458]For a discussion, see Kozlov; Bagrovskaya *J. Org. Chem. USSR* **1989**, *25*, 1152.

[459]Feigl *Angew. Chem.* **1961**, *73*, 113.

[460]Blum; Scharf *J. Org. Chem.* **1970**, *35*, 1895.

[461]Pettit; Piatak *J. Org. Chem.* **1960**, *25*, 721.

[462]Olah; Tolgyesi; Dear *J. Org. Chem.* **1962**, *27*, 3441, 3449, 3455; De Valois; Van Albada; Veenland *Tetrahedron* **1968**, *24*, 1835; Olah; Meidar; Olah *Nouv. J. Chim.* **1979**, *3*, 275.

[463]Koptyug; Isaev; Gershtein; Berezovskii *J. Gen. Chem. USSR* **1964**, *34*, 3830; Erykalov; Becker; Belokurova *J. Org. Chem. USSR* **1968**, *4*, 2054; Jacquesy; Jouannetaud *Tetrahedron Lett.* **1982**, *23*, 1673.

[464]Kooyman; Louw *Recl. Trav. Chim. Pays-Bas* **1962**, *81*, 365; Augustijn; Kooyman; Louw *Recl. Trav. Chim. Pays-Bas* **1963**, *82*, 965.

[465]Choguill; Ridd *J. Chem. Soc.* **1961**, 822; Ref. 430; Ref. 462.

[466]Moyer; Bunnett *J. Am. Chem. Soc.* **1963**, *85*, 1891.

This reaction, which involves aryl carbanion intermediates (S_E1 mechanism), has been called the *halogen dance*.[467]

Removal of halogen from aromatic rings can also be accomplished by various reducing agents, among them Ph_3SnH,[468] HI, Sn and HBr, Ph_3P,[469] Zn and an acid or base,[470] catalytic hydrogenolysis,[471] catalytic transfer hydrogenolysis,[472] Zn–Ag couple,[473] Na–Hg in liquid NH_3,[474] $LiAlH_4$,[475] $LiAlH_4$ irradiated with light[476] or with ultrasound,[477] $NaAlH_4$,[478] $NaBH_4$ and a catalyst,[479] NaH,[480] and Raney nickel in alkaline solution,[481] the last method being effective for fluorine as well as for the other halogens. Carbon monoxide, with potassium tetracarbonylhydridoferrate $KHFe(CO)_4$ as a catalyst, specifically reduces aryl iodides.[482] Not all these reagents operate by electrophilic substitution mechanisms. Some are nucleophilic substitutions and some are free-radical processes. Photochemical[483] and electrochemical[484] reduction are also known. Halogen can also be removed from aromatic rings indirectly by conversion to Grignard reagents (**2-38**) followed by hydrolysis (**1-44**).

OS **III**, 132, 475, 519; **V**, 149, 346, 998; **VI**, 82, 821.

1-43   Formation of Organometallic Compounds

$$ArBr + M \longrightarrow ArM$$

$$ArBr + RM \longrightarrow ArM + RBr$$

These reactions are considered along with their aliphatic counterparts at reactions **2-38** and **2-39**.

D. Metal Leaving Groups

1-44 Hydrolysis of Organometallic Compounds
Hydro-de-metallation or **Demetallation**

$$ArM + H^+ \longrightarrow ArH + M^+$$

Organometallic compounds can be hydrolyzed by acid treatment. For active metals such as Mg, Li, etc., water is sufficiently acidic. The most important example of this reaction is

[467]Bunnett; McLennan *J. Am. Chem. Soc.* **1968**, *90*, 2190; Bunnett *Acc. Chem. Res.* **1972**, *5*, 139-147; Mach; Bunnett *J. Org. Chem.* **1980**, *45*, 4660; Sauter; Fröhlich; Kalt *Synthesis*, **1989**, 771.

[468]Lorenz; Shapiro; Stern; Becker *J. Org. Chem.* **1963**, *28*, 2332; Neumann; Hillgärtner *Synthesis* **1971**, 537.

[469]Hoffmann; Michael *Chem. Ber.* **1962**, *95*, 528.

[470]Tashiro; Fukuta *J. Org. Chem.* **1977**, *42*, 835. See also Colon *J. Org. Chem.* **1982**, *47*, 2622.

[471]For example, see Subba Rao; Mukkanti; Choudary *J. Organomet. Chem.* **1989**, *367*, C29.

[472]Anwer; Spatola *Tetrahedron Lett.* **1985**, *26*, 1381.

[473]Chung; Ho; Lun; Wong; Wong; Tam *Synth. Commun.* **1988**, *18*, 507.

[474]Austin; Alonso; Rossi *J. Chem. Res. (S)* **1990**, 190.

[475]Karabatsos; Shone *J. Org. Chem.* **1968**, *33*, 619; Brown; Krishnamurthy *J. Org. Chem.* **1969**, *34*, 3918; Virtanen; Jaakkola *Tetrahedron Lett.* **1969**, 1223; Ricci; Danieli; Pirazzini *Gazz. Chim. Ital.* **1975**, *105*, 37; Chung; Chung *Tetrahedron Lett.* **1979**, 2473. Evidence for a free-radical mechanism has been found in this reaction; see Chung; Filmore *J. Chem. Soc., Chem Commun.* **1983**, 358; Beckwith; Goh *J. Chem. Soc., Chem Commun.* **1983**, 905.

[476]Beckwith; Goh *J. Chem. Soc., Chem. Commun.* **1983**, 907.

[477]Han; Baudjouk *Tetrahedron Lett.* **1982**, *23*, 1643.

[478]Zakharkin; Gavrilenko; Rukasov *Dokl. Chem.* **1972**, *205*, 551.

[479]Egli *Helv. Chim. Acta* **1968**, *51*, 2090; Bosin; Raymond; Buckpitt *Tetrahedron Lett.* **1974**, 4699; Lin; Roth *J. Org. Chem.* **1979**, *44*, 309; Narisada; Horibe; Watanabe; Takeda *J. Org. Chem.* **1989**, *54*, 5308. See also Epling; Florio *J. Chem. Soc., Perkin Trans. 1* **1988**, 703.

[480]Nelson; Gribble *J. Org. Chem.* **1974**, *39*, 1425.

[481]Buu-Hoï; Xuong; van Bac *Bull. Soc. Chim. Fr.* **1963**, 2442; de Koning *Org. Prep. Proced. Int.* **1975**, *7*, 31.

[482]Brunet; Taillefer *J. Organomet. Chem.* **1988**, *348*, C5.

[483]See, for example, Pinhey; Rigby *Tetrahedron Lett.* **1969**, 1267, 1271; Barltrop; Bradbury *J. Am. Chem. Soc.* **1973**, *95*, 5085.

[484]See Fry *Synthetic Organic Electrochemistry*, 2nd ed.; Wiley: New York, 1989, pp. 142-143.

hydrolysis of Grignard reagents, but M may be many other metals or metalloids. Examples are SiR_3, HgR, Na, and $B(OH)_2$. Since aryl Grignard and aryllithium compounds are fairly easy to prepare, they are often used to prepare salts of weak acids, e.g.,

$$\textbf{PhMgBr} + \textbf{H—C} \equiv \textbf{C—H} \longrightarrow \textbf{H—C} \equiv \textbf{C}^- \quad \textbf{MgBr}^+ + \textbf{PhH}$$

Where the bond between the metal and the ring is covalent, the usual arenium ion mechanism operates.[485] Where the bonding is essentially ionic, this is a simple acid–base reaction. For the aliphatic counterpart of this reaction, see reaction **2-24.**

Other reactions of aryl organometallic compounds are treated with their aliphatic analogs: reactions **2-25** through **2-36.**

[485]For a discussion of the mechanism, see Taylor, in Bamford; Tipper, Ref. 1, pp. 278-303, 324-349.

12
ALIPHATIC ELECTROPHILIC SUBSTITUTION

In Chapter 11 it was pointed out that the most important leaving groups in electrophilic substitution are those that can best exist with an outer shell that is deficient in a pair of electrons. For aromatic systems the most common leaving group is the proton. The proton is also a leaving group in aliphatic systems, but the reactivity depends on the acidity. Protons in saturated alkanes are very unreactive, but electrophilic substitutions are often easily carried out at more acidic positions, e.g., α to a carbonyl group, or at an alkynyl position (RC≡CH). Since metallic ions are easily able to bear positive charges, we might expect that organometallic compounds would be especially susceptible to electrophilic substitution, and this is indeed the case.[1] Another important type of electrophilic substitution, known as *anionic cleavage*, involves the breaking of C—C bonds; in these reactions there are carbon leaving groups (**2-40** to **2-46**). A number of electrophilic substitutions at a nitrogen atom are treated at the end of the chapter.

Since a carbanion is what remains when a positive species is removed from a carbon atom, the subject of carbanion structure and stability (Chapter 5) is inevitably related to the material in this chapter. So is the subject of very weak acids and very strong bases (Chapter 8), because the weakest acids are those in which the hydrogen is bonded to carbon.

MECHANISMS

For aliphatic electrophilic substitution, we can distinguish at least four possible major mechanisms,[2] which we call SE1, SE2 (front), SE2 (back), and SEi. The SE1 is unimolecular; the other three are bimolecular.

Bimolecular Mechanisms. SE2 and SEi

The bimolecular mechanisms for electrophilic aliphatic substitution are analogous to the SN2 mechanism in that the new bond forms as the old one breaks. However, in the SN2

[1]For books on the preparation and reactions of organometallic compounds, see Hartley; Patai *The Chemistry of the Metal–Carbon Bond*, 5 vols.; Wiley: New York, 1984-1990; Haiduc; Zuckerman *Basic Organometallic Chemistry*; Walter de Gruyter: New York, 1985; Negishi *Organometallics in Organic Synthesis*; Wiley: New York, 1980; Aylett *Organometallic Compounds*, 4th ed., vol. 1, pt. 2; Chapman and Hall: New York, 1979; Coates; Green; Wade *Organometallic Compounds*, 3rd ed., 2 vols.; Methuen: London, 1967-1968; Eisch *The Chemistry of Organometallic Compounds*; Macmillan: New York, 1967. For reviews, see Maslowsky *Chem. Soc. Rev.* **1980**, *9*, 25-40, and in Tsutsui *Characterization of Organometallic Compounds*; Wiley: New York, 1969-1971, the articles by Cartledge; Gilman, pt. 1, pp. 1-33, and by Reichle, pt. 2, pp. 653-826.

[2]For monographs, see Abraham *Comprehensive Chemical Kinetics*, Bamford; Tipper, Eds., vol. 12; Elsevier: New York, 1973; Jensen; Rickborn *Electrophilic Substitution of Organomercurials*; McGraw-Hill: New York, 1968; Reutov; Beletskaya *Reaction Mechanisms of Organometallic Compounds*; North-Holland Publishing Company: Amsterdam, 1968. For reviews, see Abraham; Grellier, in Hartley; Patai, Ref. 1, vol. 2, pp. 25-149; Beletskaya *Sov. Sci. Rev.*, *Sect. B* **1979**, *1*, 119-204; Reutov *Pure Appl. Chem.* **1978**, *50*, 717-724, **1968**, *17*, 79-94, *Tetrahedron*, **1978**, *34*, 2827-2855, *J. Organomet. Chem.* **1975**, *100*, 219-235, *Russ. Chem. Rev.* **1967**, *36*, 163-174, *Fortschr. Chem. Forsch.* **1967**, *8*, 61-90; Matteson *Organomet. Chem. Rev.*, *Sect. A* **1969**, *4*, 263-305; Dessy; Kitching *Adv. Organomet. Chem.* **1966**, *4*, 267-351.

mechanism the incoming group brings with it a pair of electrons, and this orbital can overlap with the central carbon only to the extent that the leaving group takes away *its* electrons; otherwise the carbon would have more than eight electrons at once in its outer shell. Since electron clouds repel, this means also that the incoming group attacks backside, at a position 180° from the leaving group, resulting in inversion of configuration. When the attacking species is an electrophile, which brings to the substrate only a vacant orbital, it is obvious that this consideration does not apply and we cannot a priori predict from which direction the attack must come. We can imagine two main possibilities: attack from the front, which we call SE2 (front), and attack from the rear, which we call SE2 (back). The possibilities can be pictured (charges not shown):

$$\text{SE2 (front)} \qquad\qquad \text{SE2 (back)}$$

Both the SE2 (front) and SE2 (back) mechanisms are designated D_EA_E in the IUPAC system. With substrates in which we can distinguish the possibility, the former mechanism should result in retention of configuration and the latter in inversion. When the electrophile attacks from the front, there is a third possibility. A portion of the electrophile may assist in the removal of the leaving group, forming a bond with it at the same time that the new C—Y bond is formed:

$$\text{SEi}$$

This mechanism, which we call the SEi mechanism[3] (IUPAC designation: cyclo-$D_EA_ED_nA_n$), also results in retention of configuration.[4] Plainly, where a second-order mechanism involves this kind of internal assistance, backside attack is impossible.

It is evident that these three mechanisms are not easy to distinguish. All three give second-order kinetics, and two result in retention of configuration.[5] In fact, although much work has been done on this question, there are few cases in which we can unequivocally say that one of these three and not another is actually taking place. Clearly, a study of the stereochemistry can distinguish between SE2 (back) on the one hand and SE2 (front) or SEi on the other. Many such investigations have been made. In the overwhelming majority of second-order electrophilic substitutions, the result has been retention of configuration or some other indication of frontside attack, indicating an SE2 (front) or SEi mechanism. For example, when *cis*-**1** was treated with labeled mercuric chloride, the **2** produced was 100% cis. The bond between the mercury and the ring must have been broken (as well as the other Hg—C bond), since each of the products contained about half of the labeled mercury.[6] Another indication of frontside attack is that second-order electrophilic substitutions proceed very easily at *bridgehead* carbons (see p. 296).[7] Still another indication is the behavior of

[3]The names for these mechanisms vary throughout the literature. For example, the SEi mechanism has also been called the SF2, the SE2 (closed), and the SE2 (cyclic) mechanism. The original designations, SE1, SE2, etc., were devised by the Hughes-Ingold school.
[4]It has been contended that the SEi mechanism violates the principle of conservation of orbital symmetry (p. 846), and that the SE2 (back) mechanism partially violates it: Slack; Baird *J. Am. Chem. Soc.* **1976**, *98*, 5539.
[5]For a review of the stereochemistry of reactions in which a carbon-transition metal σ bond is formed or broken, see Flood *Top. Stereochem.* **1981**, *12*, 37-117. See also Ref. 10.
[6]Winstein; Traylor; Garner *J. Am. Chem. Soc.* **1955**, *77*, 3741.
[7]Winstein; Traylor *J. Am. Chem. Soc.* **1956**, *78*, 2597; Schöllkopf *Angew. Chem.* **1960**, *72*, 147-159. For a discussion, see Fort; Schleyer *Adv. Alicyclic Chem.* **1966**, *1*, 283-370, pp. 353-370.

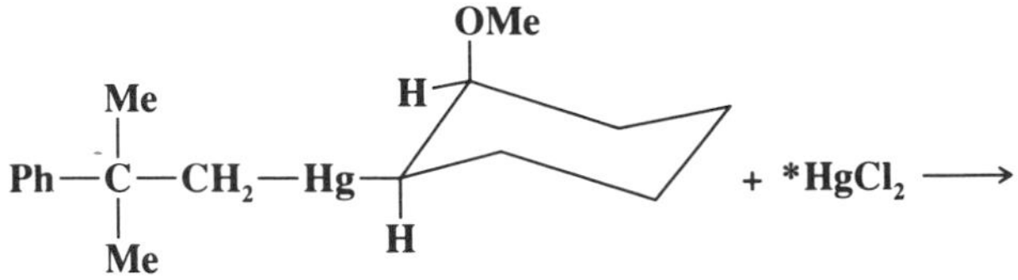

$$\text{Ph}-\overset{\overset{\displaystyle\text{Me}}{|}}{\underset{\underset{\displaystyle\text{Me}}{|}}{\text{C}}}-\text{CH}_2-\text{Hg}\!\!-\!\!\left[\begin{array}{c}\text{OMe}\\\text{H}\\\\\text{H}\end{array}\right] \;+\; {}^*\text{HgCl}_2 \;\longrightarrow$$

1

$$\text{Ph}-\overset{\overset{\displaystyle\text{Me}}{|}}{\underset{\underset{\displaystyle\text{Me}}{|}}{\text{C}}}-\text{CH}_2-{}^*\text{HgCl} \;+\; \text{Cl}{}^*\text{Hg}\!\!-\!\!\left[\begin{array}{c}\text{OMe}\\\text{H}\\\\\text{H}\end{array}\right]$$

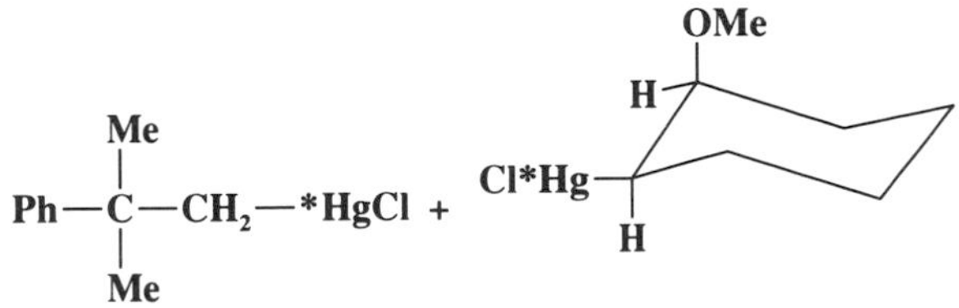

2

neopentyl as a substrate. S$_N$2 reactions at neopentyl are extremely slow (p. 339), because attack from the rear is blocked. The fact that neopentyl systems undergo electrophilic substitution only slightly more slowly than ethyl[8] is further evidence for frontside attack. One final elegant experiment may be noted. The compound di-*sec*-butylmercury was prepared with one *sec*-butyl group optically active and the other racemic.[9] This was accomplished by treatment of optically active *sec*-butylmercuric bromide with racemic *sec*-butylmagnesium bromide. The di-*sec*-butyl compound was then treated with mercuric bromide to give 2 moles of *sec*-butylmercuric bromide. The steric course of the reaction could then be predicted by the following analysis, assuming that the bonds between the mercury and each carbon have a 50% chance of breaking.

If inversion,

$$\text{attack here} \longrightarrow \overset{S}{\text{RHgX}} + \overset{RS}{\text{RHgX}}$$

$$\overset{R}{\downarrow}$$

$$\overset{R}{\text{R}}-\text{Hg}-\overset{RS}{\text{R}}$$

$$\uparrow$$

$$\text{attack here} \longrightarrow \underline{\overset{R}{\text{RHgX}} + \overset{RS}{\text{RHgX}}}$$

The sum is a racemic mixture

If retention,

$$\text{attack here} \longrightarrow \overset{R}{\text{RHgX}} + \overset{RS}{\text{RHgX}}$$

$$\overset{R}{\downarrow}$$

$$\overset{R}{\text{R}}-\text{Hg}-\overset{RS}{\text{R}}$$

$$\uparrow$$

$$\text{attack here} \longrightarrow \underline{\overset{R}{\text{RHgX}} + \overset{RS}{\text{RHgX}}}$$

The sum has one-half of the original activity

---

[8]Hughes; Volger *J. Chem. Soc.* **1961**, 2359.
[9]Jensen *J. Am. Chem. Soc.* **1960**, *82*, 2469; Ingold *Helv. Chim. Acta* **1964**, *47*, 1191.

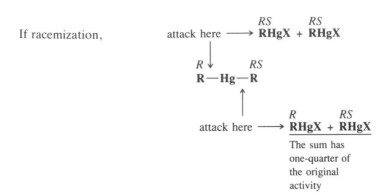

If racemization,   attack here ⟶ $\overset{RS}{R}HgX$ + $\overset{RS}{R}HgX$

The sum has
one-quarter of
the original
activity

The original activity referred to is the activity of the optically active *sec*-butylmercuric bromide used to make the dialkyl compound. The actual result was that, under several different sets of conditions, the product had one-half of the original activity, demonstrating retention of configuration.

However, inversion of configuration has been found in certain cases, demonstrating that the SE2 (back) mechanism can take place. For example, the reaction of optically active *sec*-butyltrineopentyltin with bromine (**2-30**) gives inverted *sec*-butyl bromide.[10] A number of

$$\textit{sec-}\textbf{BuSnR}_3 + \textbf{Br}_3 \longrightarrow \textit{sec-}\textbf{BuBr} \qquad \textbf{R} = \text{neopentyl}$$

other organometallic compounds have also been shown to give inversion when treated with halogens,[11] although others do not.[12] So far, no inversion has been found with an organo-mercury substrate. It may be that still other examples of backside attack exist[13] but have escaped detection because of the difficulty in preparing compounds with a configurationally stable carbon–metal bond. Compounds that are chiral because of an asymmetric carbon at which a carbon–metal bond is located[13a] are often difficult to resolve and once resolved are often easily racemized. The resolution has been accomplished most often with organo-mercury compounds,[14] and most stereochemical investigations have therefore been made with these substrates. Only a few optically active Grignard reagents have been prepared[15] (i.e., in which the only asymmetric center is the carbon bonded to the magnesium). Because of this, the steric course of electrophilic substitutions at the C—Mg bond has not often been determined. However, in one such case, the reaction of both the exo and endo isomers of the 2-norbornyl Grignard reagent with HgBr$_2$ (to give 2-norbornylmercuric bromide) has

[10]Jensen; Davis *J. Am. Chem. Soc.* **1971**, *93*, 4048. For a review of the stereochemistry of SE2 reactions with organotin substrates, see Fukuto; Jensen *Acc. Chem. Res.* **1983**, *16*, 177-184.

[11]For example, See Applequist; Chmurny *J. Am. Chem. Soc.* **1967**, *89*, 875; Glaze; Selman; Ball; Bray *J. Org. Chem.* **1969**, *34*, 641; Brown; Lane *Chem. Commun.* **1971**, 521; Jensen; Madan; Buchanan *J. Am. Chem. Soc.* **1971**, *93*, 5283; Espenson; Williams *J. Am. Chem. Soc.* **1974**, *96*, 1008; Bock; Boschetto; Rasmussen; Demers; Whitesides *J. Am. Chem. Soc.* **1974**, *96*, 2814; Magnuson; Halpern; Levitin; Vol'pin *J. Chem. Soc., Chem. Commun.* **1978**, 44.

[12]See, for example, Rahm; Pereyre *J. Am. Chem. Soc.* **1977**, *99*, 1672; McGahey; Jensen *J. Am. Chem. Soc.* **1979**, *101*, 4397. Electrophilic bromination of certain organotin compounds was found to proceed with inversion favored for equatorial and retention for axial C—Sn bonds: Olszowy; Kitching *Organometallics* **1984**, *3*, 1676. For a similar result, see Rahm; Grimeau; Pereyre *J. Organomet. Chem.* **1985**, *286*, 305.

[13]Cases of inversion involving replacement of a metal by a metal have been reported. See Tada; Ogawa *Tetrahedron Lett.* **1973**, 2639; Fritz; Espenson; Williams; Molander *J. Am. Chem. Soc.* **1974**, *96*, 2378; Gielen; Fosty *Bull. Soc. Chim. Belg.* **1974**, *83*, 333; Bergbreiter; Rainville *J. Organomet. Chem.* **1976**, *121*, 19.

[13a]For a monograph, see Sokolov *Chirality and Optical Activity in Organometallic Compounds*; Gordon and Breach: New York, 1990.

[14]Organomercury compounds were first resolved by three groups: Jensen; Whipple; Wedegaertner; Landgrebe *J. Am. Chem. Soc.* **1959**, *81*, 1262; Charman; Hughes; Ingold *J. Chem. Soc.* **1959**, 2523, 2530; Reutov; Uglova *Bull. Acad. Sci. USSR, Div. Chem. Sci.* **1959**, 735.

[15]This was done first by Walborsky; Young *J. Am. Chem. Soc.* **1964**, *86*, 3288.

been shown to proceed with retention of configuration.[16] It is likely that inversion takes place only when steric hindrance prevents frontside attack and when the electrophile does not carry a Z group (p. 570).

The S_E2 (back) mechanism can therefore be identified in certain cases (if inversion of configuration is found), but it is plain that stereochemical investigations cannot distinguish between the S_E2 (front) and the S_Ei mechanisms and that, in the many cases where configurationally stable substrates cannot be prepared, such investigations are of no help at all in distinguishing among all three of the second-order mechanisms. Unfortunately, there are not many other methods that lead to unequivocal conclusions. One method that has been used in an attempt to distinguish between the S_Ei mechanism on the one hand and the S_E2 pathways on the other involves the study of salt effects on the rate. It may be recalled (p. 358) that reactions in which neutral starting molecules acquire charges in the transition state are aided by an increasing concentration of added ions. Thus the S_Ei mechanism would be less influenced by salt effects than would either of the S_E2 mechanisms. On this basis Abraham and co-workers[17] concluded that the reactions $R_4Sn + HgX_2 \rightarrow RHgX + R_3SnX$ (X – Cl or I) take place by S_E2 and not by S_Ei mechanisms. Similar investigations involve changes in solvent polarity[18] (see also p. 580). In the case of the reaction

$$sec\text{-}\textbf{BuSnR}_2\textbf{R}' + \textbf{Br}_2 \longrightarrow sec\text{-}\textbf{BuBr}$$

(where R = R′ = iso-Pr and R = iso-Pr, R′ = neopentyl), the use of polar solvents gave predominant inversion, while nonpolar solvents gave predominant retention.[19]

On the basis of evidence from reactivity studies, it has been suggested[20] that a variation of the S_Ei mechanism is possible in which the group Z becomes attached to X before the latter becomes detached:

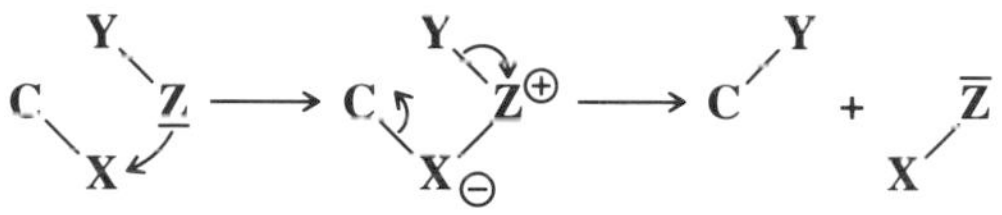

This process has been called the S_EC[20] or S_E2 (co-ord)[21] mechanism (IUPAC designation A_n + cyclo-$D_EA_ED_n$).

It has been shown that in certain cases (e.g., $Me_4Sn + I_2$) the reactants in an S_E2 reaction, when mixed, give rise to an immediate charge-transfer spectrum (p. 79), showing that an electron donor–acceptor (EDA) complex has been formed.[22] In these cases it is likely that the EDA complex is an intermediate in the reaction.

The S_E1 Mechanism

The S_E1 mechanism is analogous to the S_N1. It involves two steps—a slow ionization and a fast combination.

Step 1 $R\text{—}X \xrightarrow{\text{slow}} \overline{R}^- + X^+$

Step 2 $\overline{R}^- + Y^+ \longrightarrow R\text{—}Y$

[16]Jensen; Nakamaye *J. Am. Chem. Soc.* **1966,** *88,* 3437.
[17]Abraham; Spalding *J. Chem. Soc. A* **1969,** 784; Abraham; Johnston *J. Chem. Soc. A* **1970,** 188.
[18]See, for example, Abraham; Dorrell *J. Chem. Soc., Perkin Trans. 2* **1973,** 444.
[19]Fukuto; Newman; Jensen *Organometallics* **1987,** *6,* 415.
[20]Abraham; Hill *J. Organomet. Chem.* **1967,** *7,* 11.
[21]Abraham, Ref. 2, p. 15.
[22]Fukuzumi; Kochi *J. Am. Chem. Soc.* **1980,** *102,* 2141, 7290.

The IUPAC designation is $D_E + A_E$. First-order kinetics are predicted and many such examples have been found. Other evidence for the SE1 mechanism was obtained in a study of base-catalyzed tautomerization. In the reaction

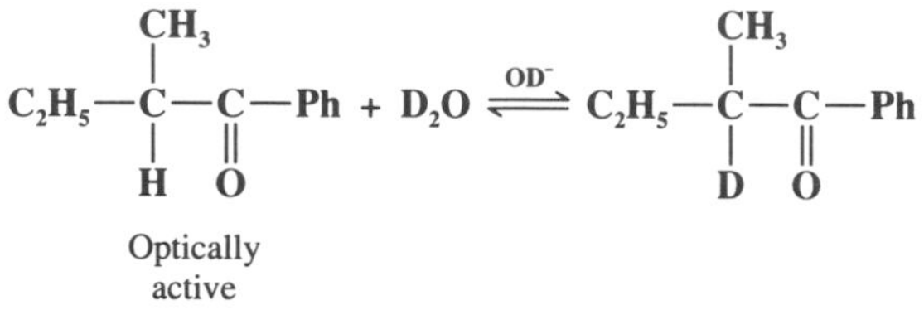

$$C_2H_5\!-\!\underset{\underset{H}{|}}{\overset{\overset{CH_3}{|}}{C}}\!-\!\underset{\underset{O}{\|}}{C}\!-\!Ph + D_2O \underset{OD^-}{\rightleftharpoons} C_2H_5\!-\!\underset{\underset{D}{|}}{\overset{\overset{CH_3}{|}}{C}}\!-\!\underset{\underset{O}{\|}}{C}\!-\!Ph$$

Optically
active

the rate of deuterium exchange was the same as the rate of racemization[23] and there was an isotope effect.[24]

SN1 reactions do not proceed at bridgehead carbons in [2.2.1] bicyclic systems (p. 300) because planar carbocations cannot form at these carbons. However, carbanions not stabilized by resonance are probably not planar; SE1 reactions should readily occur with this type of substrate. This is the case. Indeed, the question of carbanion structure is intimately tied into the problem of the stereochemistry of the SE1 reaction. If a carbanion is planar, racemization should occur. If it is pyramidal and *can hold its structure*, the result should be retention of configuration. On the other hand, even a pyramidal carbanion will give racemization if it cannot hold its structure, i.e., if there is pyramidal inversion as with amines (p. 98). Unfortunately, the only carbanions that can be studied easily are those stabilized by resonance, which makes them planar, as expected (p. 181). For simple alkyl carbanions, the main approach to determining structure has been to study the stereochemistry of SE1 reactions rather than the other way around. What is found is almost always racemization. Whether this is caused by planar carbanions or by oscillating pyramidal carbanions is not known. In either case racemization occurs whenever a carbanion is completely free or is symmetrically solvated.

However, even planar carbanions need not give racemization. Cram found that retention and even inversion can occur in the alkoxide cleavage reaction (**2-41**):

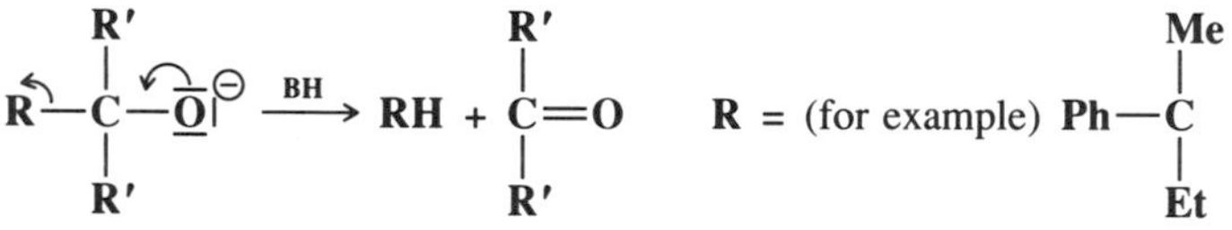

$$R\!-\!\underset{\underset{R'}{|}}{\overset{\overset{R'}{|}}{C}}\!-\!\overset{\ominus}{\underset{}{O}} \xrightarrow{BH} RH + \underset{\underset{R'}{|}}{\overset{\overset{R'}{|}}{C}}\!=\!O \qquad R = \text{(for example) } Ph\!-\!\underset{\underset{Et}{|}}{\overset{\overset{Me}{|}}{C}}$$

which is a first-order SE1 reaction involving resonance-stabilized planar carbanions (here designated R⁻).[25] By changing the solvent Cram was able to produce products ranging from 99% retention to 60% inversion and including complete racemization. These results are explained by a carbanion that is not completely free but is solvated. In nondissociating, nonpolar solvents such as benzene or dioxane, the alkoxide ion exists as an ion pair, solvated by the solvent BH:

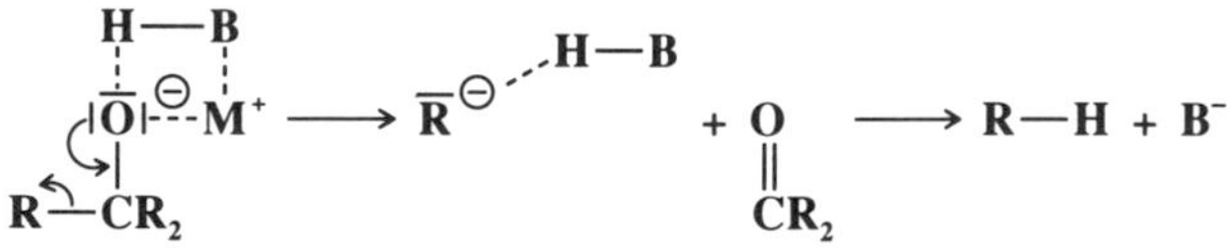

$$\underset{R\!-\!CR_2}{\overset{H\!-\!B}{\underset{|}{\overset{|}{O}}\!\ominus\!M^+}} \longrightarrow \overline{R}\overset{\ominus}{} \cdots H\!-\!B + \underset{CR_2}{\overset{O}{\|}} \longrightarrow R\!-\!H + B^-$$

[23]Hsu; Ingold; Wilson *J. Chem. Soc.* **1938**, 78.
[24]Wilson *J. Chem. Soc.* **1936**, 1550.
[25]See Cram; Langemann; Allinger; Kopecky *J. Am. Chem. Soc.* **1959**, *81*, 5740; Hoffman; Cram *J. Am. Chem. Soc.* **1969**, *91*, 1009. For a discussion, see Cram *Fundamentals of Carbanion Chemistry*; Academic Press: New York, 1965, pp. 138-158.

In the course of the cleavage, the proton of the solvent moves in to solvate the newly forming carbanion. As is easily seen, this solvation is asymmetrical since the solvent molecule is already on the front side of the carbanion. When the carbanion actually bonds with the proton, the result is retention of the original configuration. In protic solvents, such as diethylene glycol, a good deal of inversion is found. In these solvents, the *leaving group* solvates the carbanion, so the solvent can solvate it only from the opposite side:

$$
\begin{array}{c}
\overset{\displaystyle |\overline{O}|\overset{\ominus}{\cdots}H-B\cdots M^+}{\underset{R\curvearrowleft CR_2}{\big|}} \qquad\longrightarrow\qquad B-H\cdots\overline{R}\overset{\ominus}{\underset{\underset{CR_2}{\|}}{\cdots}}O \qquad\longrightarrow\qquad B^- + H-R
\end{array}
$$

Solvent-separated
ion pair

When C—H bond formation occurs, the result is inversion. Racemization results in polar aprotic solvents such as dimethyl sulfoxide. In these solvents the carbanions are relatively long-lived (because the solvent has no proton to donate) and symmetrically solvated.

Similar behavior was found for carbanions generated by base-catalyzed hydrogen exchange (reaction **2-1**):[26]

$$
R-H + B-D \underset{B^-}{\rightleftharpoons} R-D + B-H
$$

B^- = base

R = (for example) $Ph-\overset{\overset{\displaystyle CN}{|}}{C}-Et$

In this case information was obtained from measurement of the ratio of k_e (rate constant for isotopic exchange) to k_α (rate constant for racemization). A k_e/k_α ratio substantially greater than 1 means retention of configuration, since many individual isotopic exchanges are not producing a change in configuration. A k_e/k_α ratio of about 1 indicates racemization and a ratio of $\frac{1}{2}$ corresponds to inversion (see p. 296). All three types of steric behavior were found, depending on R, the base, and the solvent. As with the alkoxide cleavage reaction, retention was generally found in solvents of low dielectric constant, racemization in polar aprotic solvents, and inversion in protic solvents. However, in the proton exchange reactions, a fourth type of behavior was encountered. In aprotic solvents, with aprotic bases like tertiary amines, the k_e/k_n ratio was found to be *less* than 0.5, indicating that racemization took place *faster* than isotopic exchange (this process is known as *isoracemization*). Under these conditions, the conjugate acid of the amine remains associated with the carbanion as an ion pair. Occasionally, the ion pair dissociates long enough for the carbanion to turn over and recapture the proton:

$$
\overset{c}{\underset{a}{b\!\!\succ\!\!C}}\!-D + \overline{N}Et_3 \rightleftharpoons \overset{b\;c}{\underset{a}{C\cdots D\overset{\oplus}{N}Et_3}} \rightleftharpoons \overset{c\;b}{\underset{a}{C\cdots D\overset{\oplus}{N}Et_3}} \rightleftharpoons \overset{b}{\underset{a}{c\!\!\prec\!\!C}}\!-D + \overline{N}Et_3
$$

Thus, inversion (and hence racemization, which is produced by repeated acts of inversion) occurs without exchange. A single act of inversion without exchange is called *isoinversion*.

The isoinversion process can take place by a pathway in which a positive species migrates in a stepwise fashion around a molecule from one nucleophilic position to another. For example, in the exchange reaction of 3-carboxamido-9-methylfluorene (**3**) with Pr$_3$N in *t-*

[26]See Cram; Kingsbury; Rickborn *J. Am. Chem. Soc.* **1961**, *83*, 3688; Cram; Gosser *J. Am. Chem. Soc.* **1963**, *85*, 3890, **1964**, *86*, 5445, 5457; Roitman; Cram *J. Am. Chem. Soc.* **1971**, *93*, 2225, 2231; Cram; Cram *Intra-Sci. Chem. Rep.* **1973**, *7*(3), 1-17. For a discussion, see Cram, Ref. 25, pp. 85-105.

BuOH, it has been proposed that the amine removes a proton from the 9 position of **3** and conducts the proton out to the C=O oxygen (**5**), around the molecule, and back to C-9 on

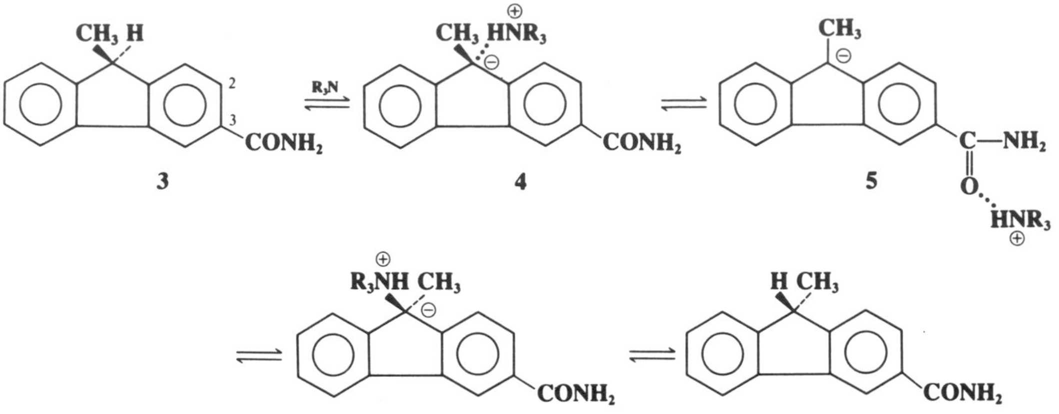

the opposite face of the anion. Collapse of **6** gives the inverted product **7**. Of course **5** could also go back to **3**, but a molecule that undergoes the total process **3 → 4 → 5 → 6 → 7** has experienced an inversion without an exchange. Evidence for this pathway, called the *conducted tour mechanism*,[27] is that the 2-carboxamido isomer of **3** does not give isoracemization. In this case the negative charge on the oxygen atom in the anion corresponding to **5** is less, because a canonical form in which oxygen acquires a full negative charge (**8**) results in

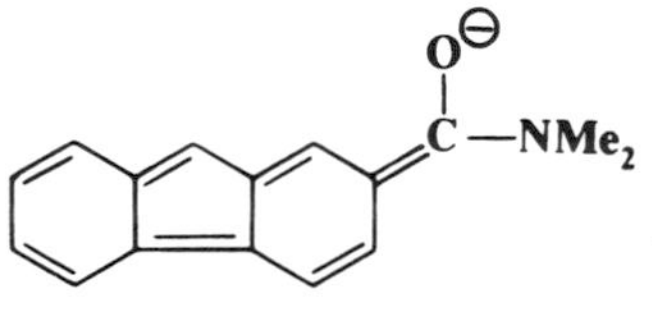

disruption of the aromatic sextet in both benzene rings (compare **9** where one benzene ring is intact). Whether the isoracemization process takes place by the conducted tour mechanism or a simple nonstructured contact ion-pair mechanism depends on the nature of the substrate (e.g., a proper functional group is necessary for the conducted tour mechanism) and of the base.[28]

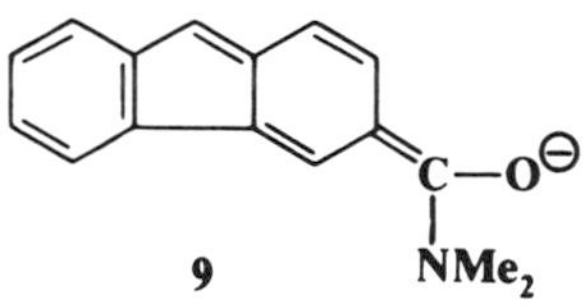

It is known that vinylic carbanions *can* maintain configuration, so that SE1 mechanisms should produce retention there. This has been found to be the case. For example, *trans*-2-bromo-2-butene was converted to 64-74% angelic acid:[29]

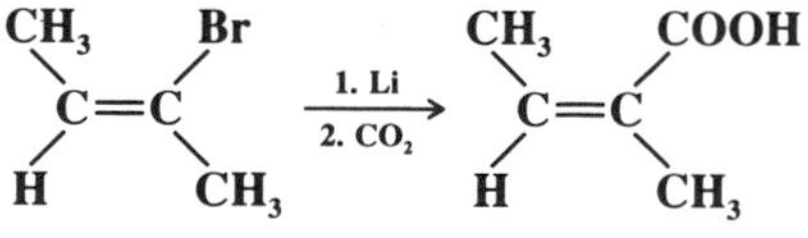

[27]Cram; Ford; Gosser *J. Am. Chem. Soc.* **1968**, *90*, 2598; Ford; Cram *J. Am. Chem. Soc.* **1968**, *90*, 2606, 2612. See also Wong; Fischer; Cram *J. Am. Chem. Soc.* **1971**, *93*, 2235; Buchholz; Harms; Massa; Boche *Angew. Chem. Int. Ed. Engl.* **1989**, *28*, 73 [*Angew. Chem. 101*, 58].
[28]Chu; Cram *J. Am. Chem. Soc.* **1972**, *94*, 3521; Almy; Hoffman; Chu; Cram *J. Am. Chem. Soc.* **1973**, *95*, 1185.
[29]Dreiding; Pratt *J. Am. Chem. Soc.* **1954**, *76*, 1902. See also Walborsky; Turner *J. Am. Chem. Soc.* **1972**, *94*, 2273.

Only about 5% of the cis isomer, tiglic acid, was produced. In addition, certain carbanions in which the negative charge is stabilized by d-orbital overlap can maintain configuration (p. 181) and S_E1 reactions involving them proceed with retention of configuration.

Electrophilic Substitution Accompanied by Double-Bond Shifts

When electrophilic substitution is carried out at an allylic substrate, the product may be rearranged:

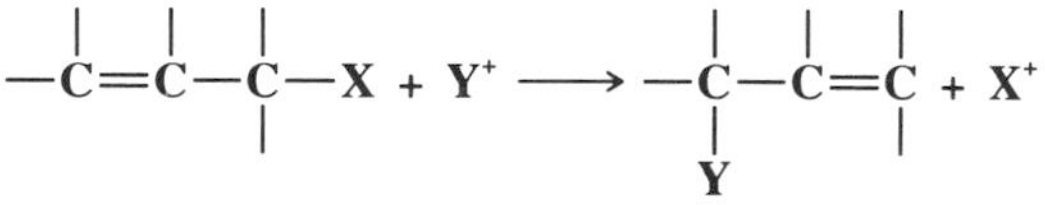

$$-\overset{|}{C}=\overset{|}{C}-\overset{|}{C}-X + Y^+ \longrightarrow -\overset{|}{C}-\overset{|}{C}=\overset{|}{C} + X^+$$
$$\underset{Y}{|}$$

This type of process is analogous to the nucleophilic allylic rearrangements discussed in Chapter 10 (p. 327). There are two principal pathways. The first of these is analogous to the S_E1 mechanism in that the leaving group is first removed, giving a resonance-stabilized allylic carbanion, and then the electrophile attacks.

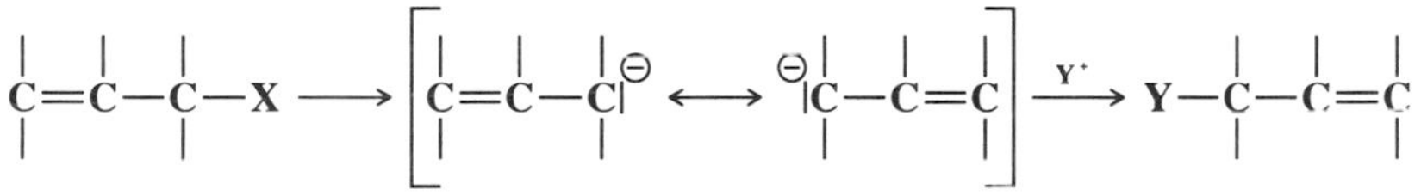

$$\overset{|}{C}=\overset{|}{C}-\overset{|}{C}-X \longrightarrow \left[\overset{|}{C}=\overset{|}{C}-\overset{|}{C}|^{\ominus} \longleftrightarrow {}^{\ominus}|\overset{|}{C}-\overset{|}{C}=\overset{|}{C}\right] \overset{Y^+}{\longrightarrow} Y-\overset{|}{C}-\overset{|}{C}=\overset{|}{C}$$

In the other pathway the Y group first attacks, giving a carbocation, which then loses X.

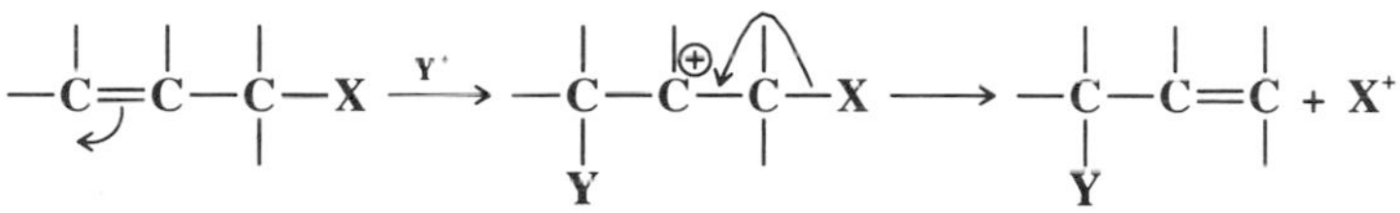

$$-\overset{|}{C}\underset{\curvearrowleft}{=}\overset{|}{C}-\overset{|}{C}-X \overset{Y^+}{\longrightarrow} -\overset{|}{C}-\overset{\overset{\oplus}{|}}{C}\overset{\curvearrowright}{-}\overset{|}{C}-X \longrightarrow -\overset{|}{C}-\overset{|}{C}=\overset{|}{C} + X^+$$
$$\underset{Y}{|} \qquad\qquad \underset{Y}{|}$$

These mechanisms are more fully discussed under reaction **2-2**.

Most electrophilic allylic rearrangements involve hydrogen as the leaving group, but they have also been observed with metallic leaving groups.[30] Sleezer, Winstein, and Young found that crotylmercuric bromide reacted with HCl about 10^7 times faster than n-butylmercuric bromide and the product was more than 99% 1-butene.[31] These facts point to an S_Ei' mechanism (IUPAC designation cyclo-$1/3/D_EA_ED_nA_n$):

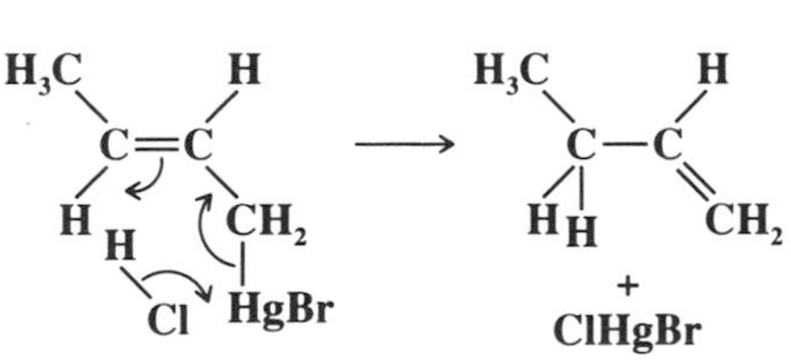

$$\underset{\overset{|}{Cl}\ \ HgBr}{\overset{H_3C}{\diagdown}\underset{\overset{H}{H}}{\overset{\diagup}{C}}=\underset{\overset{\uparrow}{CH_2}}{\overset{H}{\diagup}}\overset{\diagdown}{C}} \longrightarrow \underset{\underset{ClHgBr}{+}}{\overset{H_3C}{\diagdown}\underset{\overset{|}{H}H}{\overset{\diagup}{C}}-\underset{CH_2}{\overset{H}{\diagup}}\overset{\diagdown}{C}}$$

The reaction of the same compound with acetic acid–perchloric acid seems to proceed by an S_E2' mechanism (IUPAC designation $1/3/D_EA_E$):[31]

[30]For a review of reactions of allylic organometallic compounds, see Courtois; Miginiac *J. Organomet. Chem.* **1974**, *69*, 1-44.

[31]Sleezer; Winstein; Young *J. Am. Chem. Soc.* **1963**, *85*, 1890. See also Cunningham; Overton *J. Chem. Soc., Perkin Trans. 1* **1975**, 2140; Kashin; Bakunin; Khutoryanskii; Beletskaya; Reutov *J. Org. Chem. USSR* **1979**, *15*, 12, *J. Organomet. Chem.* **1979**, *171*, 309.

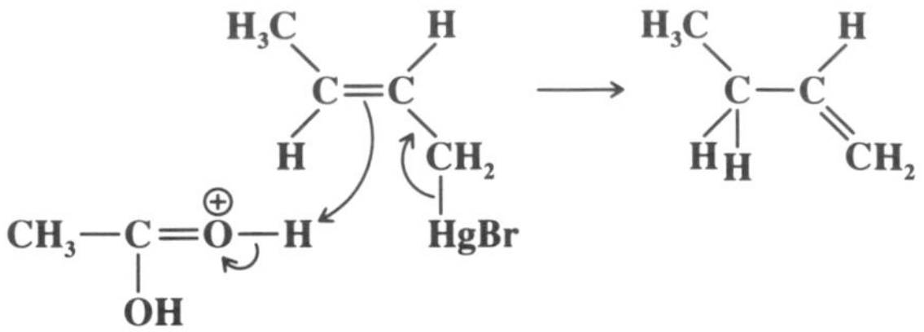

The geometry of electrophilic allylic rearrangement has not been studied very much (compare the nucleophilic case, p. 329), but in most cases the rearrangement takes place with anti stereoselectivity,[32] though syn stereoselectivity has also been demonstrated.[33] In one case, use of the electrophile H[+] and the leaving group $SnMe_3$ gave both syn and anti stereoselectivity, depending on whether the substrate was cis or trans.[34]

Other Mechanisms

Addition–elimination (**2-15**) and cyclic mechanisms (**2-40**) are also known.

Much less work has been done on electrophilic aliphatic substitution mechanisms than on nucleophilic substitutions, and the exact mechanisms of many of the reactions in this chapter are in doubt. For many of them, not enough work has been done to permit us to decide which of the mechanisms described in this chapter is operating, if indeed any is. There may be other electrophilic substitution mechanisms, and some of the reactions in this chapter may not even be electrophilic substitutions at all.

REACTIVITY

Only a small amount of work has been done in this area, compared to the vast amount done for aliphatic nucleophilic substitution and aromatic electrophilic substitution. Only a few conclusions, most of them sketchy or tentative, can be drawn.[35]

1. *Effect of substrate.*   For S_E1 reactions electron-donating groups decrease rates and electron-withdrawing groups increase them. This is as would be expected from a reaction in which the rate-determining step is analogous to the cleavage of a proton from an acid. For the S_E2 (back) mechanism, Jensen and Davis[10] showed that the reactivity of alkyl groups is similar to that for the S_N2 mechanism (i.e., Me > Et > Pr > iso-Pr > neopentyl), as would be expected, since both involve backside attack and both are equally affected by steric hindrance. In fact, this pattern of reactivity can be regarded as evidence for the occurrence of the S_E2 (back) mechanism in cases where stereochemical investigation is not feasible.[36] For S_E2 reactions that proceed with retention, several studies have been made with varying results, depending on the reaction.[37] One such study, which examined the

[32]Hayashi; Ito; Kumada *Tetrahedron Lett.* **1982**, *23*, 4605; Wetter; Scherer *Helv. Chim. Acta* **1983**, *66*, 118; Wickham; Kitching *J. Org. Chem.* **1983**, *48*, 612; Fleming; Kindon; Sarkar *Tetrahedron Lett.* **1987**, *28*, 5921; Hayashi; Matsumoto; Ito *Chem. Lett.* **1987**, 2037, *Organometallics* **1987**, *6*, 885; Matassa; Jenkins; Kümin; Damm; Schreiber; Felix; Zass; Eschenmoser *Isr. J. Chem.* **1989**, *29*, 321.

[33]Wetter; Scherer; Schweizer *Helv. Chim. Acta* **1979**, *62*, 1985; Young; Kitching *J. Org. Chem.* **1983**, *48*, 614, *Tetrahedron Lett.* **1983**, *24*, 5793.

[34]Kashin; Bakunin; Beletskaya; Reutov *J. Org. Chem. USSR* **1982**, *18*, 1973. See also Wickham; Young; Kitching *Organometallics* **1988**, *7*, 1187.

[35]For a discussion, see Abraham, Ref. 2, pp. 211-241.

[36]Another method involves measurement of the susceptibility of the rate to increased pressure: See Isaacs; Javaid *Tetrahedron Lett.* **1977**, 3073; Isaacs; Laila *Tetrahedron Lett.* **1984**, *25*, 2407.

TABLE 12.1   Relative rates of the reaction of RHgBr with Br_2 and Br^-[38]

R	Relative rate	R	Relative rate
Me	1	**Et**	10.8
Et	10.8	iso-**Bu**	1.24
iso-**Pr**	780	neopentyl	0.173
t-**Bu**	3370		

reaction $RHgBr + Br_2 \rightarrow RBr$ catalyzed by Br^-, gave the results shown in Table 12.1.[38] As can be seen, α branching increased the rates, while β branching decreased them. Sayre and Jensen attributed the decreased rates to steric hindrance, though attack here was definitely frontside, and the increased rates to the electron-donating effect of the alkyl groups, which stabilized the electron-deficient transition state.[39] Of course, steric hindrance should also be present with the α branched groups, so these workers concluded that if it were not, the rates would be even greater. The Br electrophile is rather a large one and it is likely that smaller steric effects are present with smaller electrophiles. The rates of certain second-order substitutions of organotin compounds have been found to increase with increasing electron withdrawal by substituents. This behavior has been ascribed[40] to an SE2 mechanism involving ion pairs, analogous to Sneen's ion-pair mechanism for nucleophilic substitution (p. 305).

2. *Effect of leaving group.*   For both SE1 and second-order mechanisms, the more polar the C—X bond, the easier it is for the electrofuge to cleave. For metallic leaving groups in which the metal has a valence greater than 1, the nature of the other group or groups attached to the metal thus has an effect on the reaction. For example, consider a series of organomercurials RHgW. Because a more electronegative W decreases the polarity of the C—Hg bond and furthermore results in a less stable HgW^+, the electrofugal ability of HgW decreases with increasing electronegativity of W. Thus, HgR′ (from RHgR′) is a better leaving group than HgCl (from RHgCl). Also in accord with this is the leaving-group order Hg-*t*-Bu > Hg-iso-Pr > HgEt > HgMe, reported for acetolysis of R_2Hg,[39] since the more highly branched alkyl groups better help to spread the positive charge. It might be expected that, when metals are the leaving groups, SE1 mechanisms would be favored, while with carbon leaving groups, second-order mechanisms would be found. However, the results so far reported have been just about the reverse of this. For carbon leaving groups the mechanism is usually SE1, while for metallic leaving groups the mechanism is almost always SE2 or SEi. A number of reports of SE1 reactions with metallic leaving groups have appeared,[41] but the mechanism is not easy to prove and many of these reports have been challenged.[42] Reutov and co-workers[41] have expressed the view that in such reactions a nucleophile (which

[37]For some of these, see Abraham; Grellier *J. Chem. Soc., Perkin Trans. 2* **1973**, 1132; Dessy; Reynolds; Kim *J. Am. Chem. Soc.* **1959**, *81*, 2683; Minato; Ware; Traylor *J. Am. Chem. Soc.* **1963**, *85*, 3024; Boué; Gielen; Nasielski *J. Organomet. Chem.* **1967**, *9*, 443; Abraham; Broadhurst; Clark; Koenigsberger; Dadjour *J. Organomet. Chem.* **1981**, *209*, 37.

[38]Sayre; Jensen *J. Am. Chem. Soc.* **1979**, *101*, 6001.

[39]A similar conclusion, that steric and electronic effects are both present, was reached for a different system by Nugent; Kochi *J. Am. Chem. Soc.* **1976**, *98*, 5979.

[40]Beletskaya; Kashin; Reutov *J. Organomet. Chem.* **1978**, *155*, 31; Reutov *J. Organomet. Chem.* **1983**, *250*, 145-156. See also Butin; Magdesieva *J. Organomet. Chem.* **1985**, *292*, 47; Beletskaya, Ref. 2.

[41]For discussions, see Reutov *Bull. Acad. Sci. USSR, Div. Chem. Sci.* **1980**, *29*, 1461-1477; Beletskaya; Butin; Reutov *Organomet. Chem. Rev., Sect. A* **1971**, *7*, 51-79. See also Deacon; Smith *J. Org. Chem. USSR* **1982**, *18*, 1584; Dembech; Eaborn; Seconi *J. Chem. Soc., Chem. Commun.* **1985**, 1289.

[42]For a discussion, see Kitching *Rev. Pure Appl. Chem.* **1969**, *19*, 1-16.

may be the solvent) must assist in the removal of the electrofuge and refer to such processes as SE1(N) reactions.

3. *Effect of solvent.*[43] In addition to the solvent effects on certain SE1 reactions, mentioned earlier (p. 574), solvents can influence the mechanism that is preferred. As with nucleophilic substitution (p. 356), an increase in solvent polarity increases the possibility of an ionizing mechanism, in this case SE1, in comparison with the second-order mechanisms, which do not involve ions. As previously mentioned (p. 573), the solvent can also exert an influence between the SE2 (front or back) and SEi mechanisms in that the rates of SE2 mechanisms should be increased by an increase in solvent polarity, while SEi mechanisms are much less affected.

REACTIONS

The reactions in this chapter are arranged in order of leaving group: hydrogen, metals, halogen, and carbon. Electrophilic substitutions at a nitrogen atom are treated last.

Hydrogen as Leaving Group

A. Hydrogen as the Electrophile

2-1 Hydrogen Exchange
Deuterio-de-hydrogenation or **Deuteriation**

$$R—H + D^+ \rightleftharpoons R—D + H^+$$

Hydrogen exchange can be accomplished by treatment with acids or bases. As with **1-1,** the exchange reaction is mostly used to study mechanistic questions such as relative acidities, but it can be used synthetically to prepare deuterated or tritiated molecules. When ordinary strong acids such as H_2SO_4 are used, only fairly acidic protons exchange, e.g., acetylenic, allylic, etc. However, primary, secondary, and tertiary hydrogens of alkanes can be exchanged by treatment with super-acids (p. 249).[44] The order of hydrogen reactivity is tertiary > secondary > primary. Where C—C bonds are present, they may be cleaved also (**2-47**). The mechanism of the exchange (illustrated for methane) has been formulated as involving attack of H^+ on the C—H bond to give the pentavalent methanonium ion which loses H_2 to give a tervalent carbocation.[45] The methanonium ion CH_5^+ has a three-center,

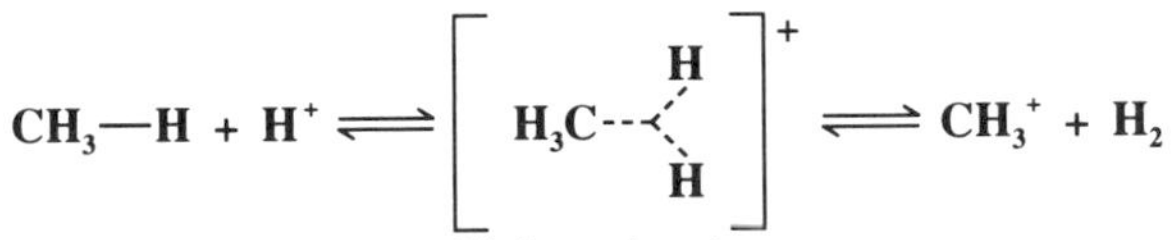

$$CH_3—H + H^+ \rightleftharpoons \left[H_3C \cdots \overset{H}{\underset{H}{\langle}} \right]^+ \rightleftharpoons CH_3^+ + H_2$$

Methanonium ion

[43]For a discussion of solvent effects on organotin alkyl exchange reactions, see Petrosyan *J. Organomet. Chem.* **1983,** *250,* 157-170.

[44]Hogeveen; Bickel *Chem. Commun.* **1967,** 635; *Recl. Trav. Chim. Pays-Bas* **1969,** *88,* 371; Hogeveen; Gaasbeek *Recl. Trav. Chim. Pays-Bas* **1968,** *87,* 319; Olah; Klopman; Schlosberg *J. Am. Chem. Soc.* **1969,** *91,* 3261; Olah; Halpern; Shen; Mo *J. Am. Chem. Soc.* **1973,** *95,* 4960. For reviews, see Olah; Prakash; Sommer *Superacids*; Wiley: New York, 1985, pp. 244-249; Olah *Angew. Chem. Int. Ed. Engl.* **1973,** *12,* 173-212 [*Angew. Chem. 85,* 183-225], *CHEMTECH* **1971,** *1,* 566-573; Brouwer; Hogeveen *Prog. Phys. Org. Chem.* **1972,** *9,* 179-240, pp. 180-203.

[45]The mechanism may not be this simple in all cases. For discussions, see McMurry; Lectka *J. Am. Chem. Soc.* **1990,** *112,* 869; Culmann; Sommer *J. Am. Chem. Soc.* **1990,** *112,* 4057.

two-electron bond.[46] It is not known whether the methanonium ion is a transition state or a true intermediate, but an ion CH_5^+ has been detected in mass spectra.[47] The ir spectrum of the ethanonium ion $C_2H_7^+$ has been measured in the gas phase.[48] Note that the two electrons in the three-center, two-electron bond can move in three directions, in accord with the threefold symmetry of such a structure. The electrons can move to unite the two hydrogens, leaving the CH_3^+ free (the forward reaction), or they can unite the CH_3 with either of the two hydrogens, leaving the other hydrogen as a free H^+ ion (the reverse reaction). Actually, the methyl cation is not stable under these conditions. It can go back to CH_4 by the route shown (leading to H^+ exchange) or it can react with additional CH_4 molecules (**2-18**) to yield, eventually, the *t*-butyl cation, which is stable in these super-acid solutions. Hydride ion can also be removed from alkanes (producing tervalent carbocations) by treatment with pure SbF_5 in the absence of any source of H^+.[49] Complete or almost complete perdeuteriation of cyclic alkenes has been achieved by treatment with dilute DCl/D_2O in sealed Pyrex tubes at 165-280°C.[50]

Exchange with bases involves an SE1 mechanism.

Step 1 $RH + B^- \longrightarrow R^- + BH$

Step 2 $R^- + BD \longrightarrow RD + B^-$

Of course, such exchange is most successful for relatively acidic protons, such as those α to a carbonyl group, but even weakly acidic protons can exchange with bases if the bases are strong enough (see p. 176).

Alkanes and cycloalkanes, of both low and high molecular weight, can be fully perdeuteriated treatment with D_2 gas and a catalyst such as Rh, Pt, or Pd.[51]

OS **VI**, 432.

2-2 Migration of Double Bonds
3/Hydro-de-hydrogenation

$$C_5H_{11}-CH_2-CH{=}CH_2 \xrightarrow[Me_2SO]{KNH_2} C_5H_{11}-CH{=}CH-CH_3$$

The double bonds of many unsaturated compounds are shifted[52] on treatment with strong bases.[53] In many cases equilibrium mixtures are obtained and the thermodynamically most stable isomer predominates.[54] Thus, if the new double bond can be in conjugation with one already present or with an aromatic ring, it goes that way.[55] If the choice is between an

[46]For a monograph on this type of species, see Olah; Prakash; Williams; Field; Wade *Hypercarbon Chemistry*; Wiley: New York, 1987.

[47]See, for example, Sefcik; Henis; Gaspar *J. Chem. Phys.* **1974**, *61*, 4321.

[48]Yeh; Price; Lee *J. Am. Chem. Soc.* **1989**, *111*, 5597.

[49]Lukas; Kramer; Kouwenhoven *Recl. Trav. Chim. Pays-Bas* **1973**, *92*, 44.

[50]Werstiuk; Timmins *Can. J. Chem.* **1985**, *63*, 530, **1986**, *64*, 1564.

[51]See, for example, Atkinson; Luke; Stuart *Can. J. Chem.* **1967**, *45*, 1511

[52]For a list of methods used to shift double and triple bonds, with references, see Larock *Comprehensive Organic Transformations*; VCH: New York, 1989, pp. 110-114, 287.

[53]For reviews of double-bond migrations, see Pines; Stalick *Base-Catalyzed Reactions of Hydrocarbons and Related Compounds*; Academic Press: New York, 1977, pp. 25-123; DeWolfe, in Bamford; Tipper *Comprehensive Chemical Kinetics*, vol. 9; Elsevier, New York, 1973, pp. 437-449; Yanovskaya; Shakhidayatov *Russ. Chem. Rev.* **1970**, *39*, 859-874; Hubert; Reimlinger *Synthesis* **1969**, 97-112, **1970**, 405-430; Mackenzie, in *The Chemistry of Alkenes*, vol. 1, Patai, Ed., pp. 416-436, vol. 2, Zabicky, Ed., pp. 132-148; Wiley: New York, 1964, 1970; Broaddus, *Acc. Chem. Res.* **1968**, *1*, 231-238; Cram, Ref. 25, pp. 175-210.

[54]For lists of which double bonds are more stable in conversions of XCH₂CH=CHY to XCH=CHCH₂Y, see Hine; Skoglund *J. Org. Chem.* **1982**, *47*, 4766. See also Hine; Linden *J. Org. Chem.* **1983**, *48*, 584.

[55]For a review of conversions of β,γ enones to α,β enones, see Pollack; Bounds; Bevins, in Patai; Rappoport *The Chemistry of Enones*, pt. 1; Wiley: New York, 1989, pp. 559-597.

exocyclic and an endocyclic double bond (in a six-membered ring), it chooses the latter. In the absence of considerations like these, Zaitsev's rule (p. 998) applies and the double bond goes to the carbon with the fewest hydrogens. All these considerations lead us to predict that terminal olefins can be isomerized to internal ones, nonconjugated olefins to conjugated, exo six-membered-ring olefins to endo, etc., and not the other way around. This is indeed usually the case.

This reaction, for which the term *prototropic rearrangement* is sometimes used, is an example of electrophilic substitution with accompanying allylic rearrangement. The mechanism involves abstraction by the base to give a resonance-stabilized carbanion, which then combines with a proton at the position that will give the more stable olefin:[56]

Step 1 $\quad$ R—CH$_2$—CH=CH$_2$ + B $\longrightarrow$

$$\left[R-\overset{\ominus}{C}H-CH=CH_2 \longleftrightarrow R-CH=CH-\overset{\ominus}{C}H_2 \right] + BH^+$$

Step 2 $\quad \left[R-\overset{\ominus}{C}H-CH=CH_2 \longleftrightarrow R-CH=CH-\overset{\ominus}{C}H_2 \right] \xrightarrow{BH^+}$

$$R-CH=CH-CH_3 + B$$

This mechanism is exactly analogous to the allylic-rearrangement mechanism for nucleophilic substitution (p. 327). Uv spectra of allylbenzene and 1-propenylbenzene in solutions containing NH$_2^-$ are identical, which shows that the same carbanion is present in both cases, as required by this mechanism.[57] The acid BH$^+$ protonates the position that will give the more stable product, though the ratio of the two possible products can vary with the identity of BH$^+$.[58] It has been shown that base-catalyzed double-bond shifts are partially intramolecular, at least in some cases.[59] The intramolecularity has been ascribed to a conducted tour mechanism (p. 576) in which the base leads the proton from one carbanionic site to the other:[60]

$$RCH_2CH=CH_2 + B \rightleftharpoons R-\underset{\underset{CH}{\diagup\diagup}}{\overset{\overset{H-B}{\vdots}}{C}H} \;\; CH_2 \rightleftharpoons R-\underset{\underset{CH}{\diagdown\diagdown}}{CH} \;\; \overset{\overset{H-B}{\vdots}}{C}H_2 \rightleftharpoons RCH=CHCH_3 + B$$

Triple bonds can also migrate in the presence of bases,[61] but through the allene intermediate:[62]

$$R-CH_2-C\equiv CH \rightleftharpoons R-CH=C=CH_2 \rightleftharpoons R-C\equiv C-CH_3$$

[56]See, for example, Hassan; Nour; Satti; Kirollos *Int. J. Chem. Kinet.* **1982**, *14*, 351; Pollack; Mack; Eldin *J. Am. Chem. Soc.* **1987**, *109*, 5048.

[57]Rabinovich; Astaf'ev; Shatenshtein *J. Gen. Chem. USSR* **1962**, *32*, 746.

[58]Hünig; Klaunzer; Schlund *Angew. Chem. Int. Ed. Engl.* **1987**, *26*, 1281 [*Angew. Chem. 99*, 1322].

[59]See, for example, Cram; Uyeda *J. Am. Chem. Soc.* **1964**, *86*, 5466; Bank; Rowe; Schriesheim *J. Am. Chem. Soc.* **1963**, *85*, 2115; Doering; Gaspar *J. Am. Chem. Soc.* **1963**, *85*, 3043; Ohlsson; Wold; Bergson *Ark. Kemi.* **1968**, *29*, 351.

[60]Almy; Cram *J. Am. Chem. Soc.* **1969**, *91*, 4459; Hussénius; Matsson; Bergson *J. Chem. Soc., Perkin Trans. 2* **1989**, 851.

[61]For reviews, see Pines; Stalick, Ref. 53, pp. 124-204; Théron; Verny; Vessière, in Patai *The Chemistry of Carbon-Carbon Triple Bond*, pt. 1; Wiley: New York, 1978, pp. 381-445; Bushby *Q. Rev. Chem. Soc.* **1970**, *24*, 585-600; Iwai *Mech. Mol. Migr.* **1969**, *2*, 73-116; Wotiz, in Viehe *Acetylenes*; Marcel Dekker: New York, 1969, pp. 365-424; Vartanyan; Babanyan *Russ. Chem. Rev.* **1967**, *36*, 670.

[62]For a review of rearrangements involving allenes, see Huntsman, in Patai *The Chemistry of Ketenes, Allenes, and Related Compounds*, pt. 2; Wiley: New York, 1980, pp. 521-667.

In general, strong bases such as $NaNH_2$ convert internal alkynes to terminal alkynes (a particularly good base for this purpose is potassium 3-aminopropylamide $NH_2CH_2CH_2CH_2NHK$[63]), because the equilibrium is shifted by formation of the acetylide ion; with weaker bases such as NaOH (which are not strong enough to remove the acetylenic proton), the internal alkynes are favored because of their greater thermodynamic stability. In some cases the reaction can be stopped at the allene stage. The reaction then becomes a method for the preparation of allenes.[64]

Double-bond rearrangements can also take place on treatment with acids. Both proton and Lewis[65] acids can be used. The mechanism in the case of proton acids is the reverse of the previous one; first a proton is gained, giving a carbocation, and then another is lost:

Step 1 CH_3—CH_2—CH=CH_2 + H^+ $\longrightarrow$ CH_3—CH_2—$\overset{\oplus}{C}H$—CH_3

Step 2 CH_3—CH_2—$\overset{\oplus}{C}H$—CH_3 $\longrightarrow$ CH_3—CH=CH—CH_3 + H^+

As in the case of the base-catalyzed reaction, the thermodynamically most stable olefin is the one predominantly formed. However, the acid-catalyzed reaction is much less synthetically useful because carbocations give rise to many side products. If the substrate has several possible locations for a double bond, mixtures of all possible isomers are usually obtained. Isomerization of 1-decene, for example, gives a mixture that contains not only 1-decene and *cis*- and *trans*-2-decene but also the cis and trans isomers of 3-, 4-, and 5-decene as well as branched alkenes resulting from rearrangement of carbocations. It is true that the most stable olefins predominate, but many of them have stabilities that are close together. Acid-catalyzed migration of triple bonds (with allene intermediates) can be accomplished if very strong acids (e.g., HF–PF_5) are used.[66] If the mechanism is the same as that for double bonds, vinyl cations are intermediates.

Double-bond isomerization can also take place in other ways. Nucleophilic allylic rearrangements were discussed in Chapter 10 (p. 327). Electrocyclic and sigmatropic rearrangements are treated at **8-29** to **8-37**. Double-bond migrations have also been accomplished photochemically,[67] and by means of metallic ion (most often complex ions containing Pt, Rh, or Ru) or metal carbonyl catalysts.[68] In the latter case there are at least two possible mechanisms. One of these, which requires external hydrogen, is called the *metal hydride addition–elimination mechanism*:

$$RCH_2CH{=}CH_2 \overset{MH}{\rightleftharpoons} RCH_2\overset{\overset{\displaystyle M}{|}}{C}HCH_3 \overset{-MH}{\rightleftharpoons} RCH{=}CHCH_3$$

[63]Brown; Yamashita *J. Am. Chem. Soc.* **1975,** *97,* 891; Macaulay *J. Org. Chem.* **1980,** *45,* 734; Abrams *Can. J. Chem.* **1984,** *62,* 1333.
[64]For example, see Enomoto; Katsuki; Yamaguchi *Tetrahedron Lett.* **1986,** *27,* 4599.
[65]For an example of a Lewis-acid catalyzed rearrangement, see Cameron; Stimson *Aust. J. Chem.* **1977,** *30,* 923.
[66]Barry; Beale; Carr; Hei; Reid *J. Chem. Soc., Chem. Commun.* **1973,** 177.
[67]Schönberg *Preparative Organic Photochemistry*; Springer: New York, 1968, pp. 22-24.
[68]For reviews, see Rodriguez; Brun; Waegell *Bull. Soc. Chim. Fr.* **1989,** 799-823; Jardine, in Harley; Patai, Ref. 1, vol. 4, pp. 733-818, pp. 736-740; Otsuka; Tani, in Morrison *Asymmetric Synthesis*, vol. 5; Academic Press: New York, 1985, pp. 171-191 (enantioselective); Colquhoun; Holton; Thompson; Twigg *New Pathways for Organic Synthesis*; Plenum: New York, 1984, pp. 173-193; Khan; Martell *Homogeneous Catalysis by Metal Complexes*; Academic Press: New York, 1974, pp. 9-37; Heck *Organotransition Metal Chemistry*; Academic Press: New York, 1974, pp. 76-82; Jira; Freiesleben, *Organomet. React.* **1972,** *3,* 1-190, pp. 133-149; Biellmann; Hemmer; Levisalles, in Zabicky, Ref. 53, vol. 2, pp. 224-230; Bird *Transition Metal Intermediates in Organic Synthesis*; Academic Press: New York, 1967, pp. 69-87; Davies *Rev. Pure Appl. Chem.* **1967,** *17,* 83-93; Orchin *Adv. Catal.* **1966,** *16,* 1-47.

The other mechanism, called the π-*allyl complex mechanism*, does not require external hydrogen:

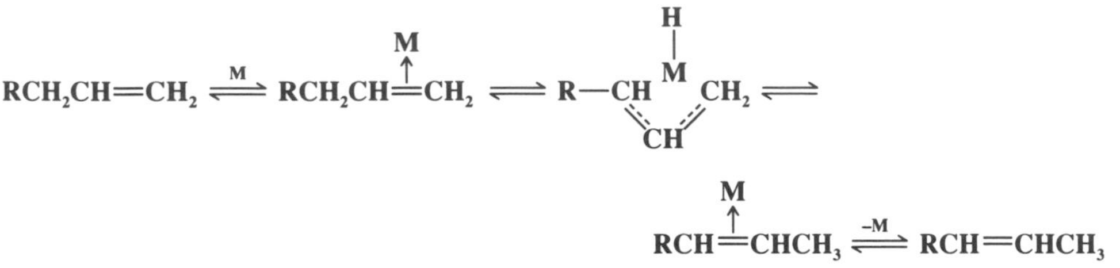

$$RCH_2CH{=}CH_2 \underset{}{\overset{M}{\rightleftharpoons}} RCH_2CH{\overset{M}{\underset{\uparrow}{=}}}CH_2 \rightleftharpoons R{-}CH{\underset{CH}{\overset{H}{\underset{\diagdown\diagup}{\overset{|}{M}}}}}CH_2 \rightleftharpoons$$

$$RCH{\overset{M}{\underset{\uparrow}{=}}}CHCH_3 \overset{-M}{\rightleftharpoons} RCH{=}CHCH_3$$

Another difference between the two mechanisms is that the former involves 1,2 and the latter 1,3 shifts. The isomerization of 1-butene by rhodium(I) is an example of a reaction that takes place by the metal hydride mechanism,[69] while an example of the π-allyl complex mechanism is found in the $Fe_3(CO)_{12}$-catalyzed isomerization of 3-ethyl-1-pentene.[70] A palladium acetate or palladium complex catalyst was used to convert alkynones $RCOC{\equiv}CCH_2CH_2R'$ to 2,4-alkadien-1-ones $RCOCH{=}CHCH{=}CHCHR'$.[71]

The metal catalysis method has been used for the preparation of simple enols, by isomerization of allylic alcohols, e.g.,[71a]

$$CH_2{=}CH{-}\underset{OH}{\underset{|}{CH}}{-}CH_3 \xrightarrow[\text{catalyst}]{\text{rhodium complex}} cis \text{ and } trans \ CH_3{-}CH{=}CH{-}\underset{OH}{\underset{|}{CH}}{-}CH_3$$

These enols are stable enough for isolation (see p. 72), but slowly tautomerize to the aldehyde or ketone, with half-lives ranging from 40-50 minutes to several days.[71a]

No matter which of the electrophilic methods of double-bond shifting is employed, the thermodynamically most stable olefin is usually formed in the largest amount in most cases, though a few anomalies are known. However, there is another, indirect, method of double-bond isomerization, by means of which migration in the other direction can often be carried out. This involves conversion of the olefin to a borane (**5-12**), rearrangement of the borane (**8-11**), oxidation and hydrolysis of the newly formed borane to the alcohol (**2-28**), and dehydration of the alcohol (**7-1**):

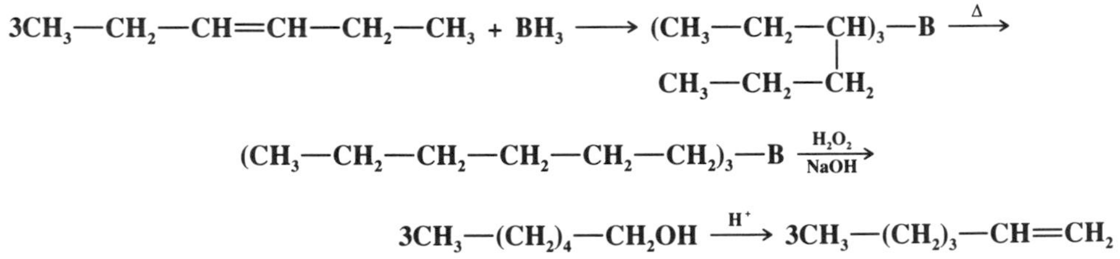

$$3CH_3{-}CH_2{-}CH{=}CH{-}CH_2{-}CH_3 + BH_3 \longrightarrow (CH_3{-}CH_2{-}\underset{CH_3{-}CH_2{-}CH_2}{\underset{|}{CH}})_3{-}B \overset{\Delta}{\longrightarrow}$$

$$(CH_3{-}CH_2{-}CH_2{-}CH_2{-}CH_2{-}CH_2)_3{-}B \xrightarrow[\text{NaOH}]{H_2O_2}$$

$$3CH_3{-}(CH_2)_4{-}CH_2OH \overset{H^+}{\longrightarrow} 3CH_3{-}(CH_2)_3{-}CH{=}CH_2$$

Since the migration reaction is always toward the end of a chain, terminal olefins can be produced from internal ones, so the migration is often opposite to that with the other methods. Alternatively, the rearranged borane can be converted directly to the olefin by heating with an alkene of molecular weight higher than that of the product (**7-15**). Photochemical isomerization can also lead to the thermodynamically less stable isomer.[72]

[69]Cramer *J. Am. Chem. Soc.* **1966**, *88*, 2272.
[70]Casey; Cyr *J. Am. Chem. Soc.* **1973**, *95*, 2248.
[71]Trost; Schmidt *J. Am. Chem. Soc.* **1988**, *110*, 2301.
[71a]Bergens; Bosnich *J. Am. Chem. Soc.* **1991**, *113*, 958.
[72]For example, see Kropp; Krauss *J. Am. Chem. Soc.* **1967**, *89*, 5199; Reardon; Krauss *J. Am. Chem. Soc.* **1971**, *93*, 5593; Duhaime; Lombardo; Skinner; Weedon *J.Org. Chem.* **1985**, *50*, 873.

If a hydroxy group is present in the chain, *it* may lose a proton, so that a ketone is the product, for example,[73]

$$R_2C{=}CHCH_2CH_2CHOHCH_3 \xrightarrow[\text{acid}]{\text{polyphosphoric}} R_2CHCH_2CH_2CH_2COCH_3$$

Similarly, α-hydroxy triple-bond compounds have given α,β-unsaturated ketones.[74]

OS **II**, 140; **III**, 207; **IV**, 189, 192, 195, 234, 398, 683; **VI**, 68, 87, 815, 925; **VII**, 249; **65**, 224; **66**, 22, 127; **68**, 162; **69**, 180.

2-3 Keto-Enol Tautomerization
3/*O*-Hydro-de-hydrogenation

$$R{-}CH_2{-}\underset{\underset{O}{\|}}{C}{-}R' \underset{OH^-}{\overset{H^+ \text{ or}}{\rightleftharpoons}} R{-}CH{=}\underset{\underset{OH}{|}}{C}{-}R'$$

The tautomeric equilibrium between enols and ketones or aldehydes is not normally a preparative reaction, though for some ketones both forms can be prepared (see p. 69 for a discussion of this and other aspects of tautomerism). For most ketones and aldehydes only the keto form is detectable under ordinary conditions, though the equilibrium must occur, since aldehydes and ketones often react through their enol forms.

Neither the forward nor the reverse reaction can take place without at least a trace of acid or base,[75] ruling out a direct shift of a hydrogen from carbon to oxygen or vice versa. The mechanisms are identical to those in **2-2**.[76]

Acid-catalyzed

$$R{-}CH_2{-}\underset{\underset{O}{\|}}{C}{-}R' \underset{\text{slow}}{\overset{H^+,\text{ fast}}{\rightleftharpoons}} R{-}CH_2{-}\overset{\oplus}{\underset{\underset{OH}{|}}{C}}{-}R' \underset{H^+,\text{ fast}}{\overset{\text{slow}}{\rightleftharpoons}} R{-}CH{=}\underset{\underset{OH}{|}}{C}{-}R'$$

Base-catalyzed[77]

$$R{-}CH_2{-}\underset{\underset{O}{\|}}{C}{-}R' \underset{\text{fast}}{\overset{B,\text{ slow}}{\rightleftharpoons}}$$

$$\left[R{-}\overset{\ominus}{CH}{-}\underset{\underset{|\underline{O}}{\|}}{C}{-}R' \longleftrightarrow R{-}CH{=}\underset{\underset{|\underline{O}|_{\ominus}}{|}}{C}{-}R' \right] \underset{B,\text{ slow}}{\overset{\text{fast}}{\rightleftharpoons}} R{-}CH{=}\underset{\underset{OH}{|}}{C}{-}R'$$

$$\textbf{10}$$

[73]Colonge; Brunie *Bull. Soc. Chim. Fr.* **1963**, 1799. For an example with basic catalysis, see Hoffmann; Köver; Pauluth *J. Chem. Soc., Chem. Commun.* **1985**, 812. For an example with a ruthenium complex catalyst, see Trost; Kulawiec *Tetrahedron Lett.* **1991**, *32*, 3039.

[74]For example, see Chabardes *Tetrahedron Lett.* **1988**, *29*, 6253.

[75]In the case of the "uncatalyzed" ketonization of CH₂=C(Ph)OH, it was shown that water functions as the basic catalyst: Chiang; Kresge; Santaballa; Wirz *J. Am. Chem. Soc.* **1988**, *110*, 5506.

[76]For reviews of the mechanism, see Keeffe; Kresge, in Rappoport *The Chemistry of Enols*; Wiley: New York, 1990, pp. 399-480; Toullec *Adv. Phys. Org. Chem.* **1982**, *18*, 1-77; Lamaty *Isot. Org. Chem.* **1976**, *2*, 33-88. For discussions, see Ingold *Structure and Mechanism in Organic Chemistry*, 2nd ed.; Cornell University Press: Ithaca, NY, 1969, pp. 794-837; Bell *The Proton in Chemistry*, 2nd ed.; Cornell University Press: Ithaca, NY, 1973, pp. 171-181; Bruice; Bruice, *J. Am. Chem. Soc.* **1976**, *98*, 844; Shelly; Venimadhavan; Nagarajan; Stewart *Can. J. Chem.* **1989**, *67*, 1274. For a review of stereoelectronic control in this mechanism, see Pollack *Tetrahedron* **1989**, *45*, 4913-4938.

[77]Another mechanism for base-catalyzed enolization has been reported when the base is a tertiary amine: See Bruice, *J. Am. Chem. Soc.* **1983**, *105*, 4982, **1989**, *111*, 962, **1990**, *112*, 7361.

For each catalyst, the mechanism for one direction is the exact reverse of the other, by the principle of microscopic reversibility.[78] As expected from mechanisms in which the C—H bond is broken in the rate-determining step, substrates of the type RCD_2COR show deuterium isotope effects (of about 5) in both the basic-[79] and the acid[80]-catalyzed processes.

Although the conversion of an aldehyde or a ketone to its enol tautomer is not generally a preparative procedure, the reactions do have their preparative aspects. If a full mole of base per mole of ketone is used, the enolate ion (**10**) is formed and can be isolated[81] (see, for example, **0-95**).[82] When enol ethers or esters are hydrolyzed, the enols initially formed immediately tatuomerize to the aldehydes or ketones. In addition, the overall processes (forward plus reverse reactions) are often used for equilibration purposes. When an optically active compound in which the chirality is due to an asymmetric carbon α to a carbonyl group (as in **11**) is treated with acid or base, racemization results.[83] If there is another asymmetric

11   *cis*-decalone   *trans*-decalone

center in the molecule, the less stable epimer can be converted to the more stable one in this manner, and this is often done. For example, *cis*-decalone can be equilibrated to the trans isomer. Isotopic exchange can also be accomplished at the α position of an aldehyde or ketone in a similar manner. For the acid-catalyzed process, exchange or equilibration is accomplished only if the carbonyl compound is completely converted to the enol and then back, but in the base-catalyzed process exchange or equilibration can take place if only the first step (conversion to the enolate ion) takes place. The difference is usually academic.

In the case of the ketone **12**, a racemic mixture was converted to an optically active mixture (optical yield 46%) by treatment with the chiral base **13**.[84] This happened because

racemic   **12**   **13**   **14**   optically active **12**

[78]It has been proposed that the acid-catalyzed ketonization of simple enols is concerted; that is, both of the processes shown in the equation take place simultaneously. This would mean that in these cases the forward reaction is also concerted. For evidence in favor of this proposal, see Capon; Siddhanta; Zucco *J. Org. Chem.* **1985**, *50*, 3580. For evidence against it, see Chiang; Kresge; Walsh *J. Am. Chem. Soc.* **1986**, *108*, 6314; Chiang; Hojatti; Keeffe; Kresge; Schepp; Wirz **1987**, *109*, 4000.

[79]Riley, Long *J. Am. Chem. Soc.* **1962**, *84*, 522; Beutelman; Xie; Saunders *J. Org. Chem.* **1989**, *54*, 1703; Xie; Saunders *J. Am. Chem. Soc.* **1991**, *113*, 3123.

[80]Swain; Stivers; Reuwer; Schaad *J. Am. Chem. Soc.* **1958**, *80*, 5885; Lienhard; Wang *J. Am. Chem. Soc.* **1969**, *91*, 1146. See also Toullec; Dubois *J. Am. Chem. Soc.* **1974**, *96*, 3524.

[81]For nmr studies of the Li enolate of acetaldehyde in solution, see Wen; Grutzner *J. Org. Chem.* **1986**, *51*, 4220.

[82]For a review of the preparation and uses of enolates, see d'Angelo *Tetrahedron* **1976**, *32*, 2979-2990.

[83]For an exception, see Guthrie; Nicolas *J. Am. Chem. Soc.* **1981**, *103*, 4637.

[84]Eleveld; Hogeveen *Tetrahedron Lett.* **1986**, *27*, 631. See also Shirai; Tanaka; Koga *J. Am. Chem. Soc.* **1986**, *108*, 543; Simpkins *J. Chem. Soc., Chem. Commun.* **1986**, 88; Cain; Cousins; Coumbarides; Simpkins *Tetrahedron* **1990**, *46*, 523.

13 reacted with one enantiomer of **12** faster than with the other (an example of kinetic resolution). The enolate **14** must remain coordinated with the chiral amine, and it is the amine that reprotonates **14**, not an added proton donor.

Enolizable hydrogens can be replaced by deuterium (and ^{16}O by ^{18}O) by passage of a sample through a deuterated (or ^{18}O-containing) gas-chromatography column.[85]

There are many enol–keto interconversions and acidifications of enolate ions to the keto forms listed in *Organic Syntheses*. No attempt is made to list them here.

B. Halogen Electrophiles

2-4   Halogenation of Aldehydes and Ketones
Halogenation or **Halo-de-hydrogenation**

$$-\overset{|}{C}H-\overset{O}{\underset{\|}{C}}-R + Br_2 \xrightarrow[OH^-]{H^+ \text{ or}} -\overset{|}{\underset{Br}{C}}-\overset{O}{\underset{\|}{C}}-R$$

Aldehydes and ketones can be halogenated in the α position with bromine, chlorine, or iodine.[86] The reaction is not successful with fluorine,[87] but active compounds, such as β-keto esters and β-diketones, have been fluorinated with XeF_2 in the presence of a resin,[88] with an N-fluoro-N-alkylsulfonamide[89] (this can result in enantioselective fluorination, if an optically active N-fluorosulfonamide is used[90]), with cesium fluoroxysulfate,[91] with N-fluoroquinuclidium fluoride,[92] and with acetyl hypofluorite.[93] The last reagent also fluorinates simple ketones in the form of their lithium enolates.[94] In another method, enolate ions of β-keto esters are fluorinated with perchloryl fluoride $FClO_3$.[95] (However, $FClO_3$ can be a dangerous reagent. Several explosions have been reported.[96]) If the carbon attacked with $FClO_3$ has two hydrogens, the reaction cannot be stopped until two fluorines have entered. Monofluorination can be accomplished indirectly by treating an enamine, enol ether, or similar ketone derivative with $FClO_3$.[97] Fluoroxytrifluoromethane CF_3OF and similar compounds behave similarly.[98] Silyl enol ethers can also be fluorinated, with XeF_2[99] or with 5%

[85]Senn; Richter; Burlingame *J. Am. Chem. Soc.* **1965**, *87*, 680; Richter; Senn; Burlingame *Tetrahedron Lett.* **1965**, 1235.

[86]For a review, see House *Modern Synthetic Reactions*, 2nd ed.; W.A. Benjamin: New York, 1972, pp. 459-478. For lists of reagents, with references, see Ref. 52, pp. 369-372. For a monograph, see De Kimpe; Verhé *The Chemistry of α Haloketones, α Haloaldehydes, and α Haloimines;* Wiley: New York, 1988.

[87]For a review of the preparation of α-fluoro carbonyl compounds, see Rozen; Filler *Tetrahedron* **1985**, *41*, 1111-1153. For a monograph, see German; Zemskov *New Fluorinating Agents in Organic Chemistry*; Springer: New York, 1989.

[88]Zajc; Zupan *J. Chem. Soc., Chem. Commun.* **1980**, 759; *J. Org. Chem.* **1982**, *47*, 573.

[89]Barnette *J. Am. Chem. Soc.* **1984**, *106*, 452.

[90]Differding; Lang *Tetrahedron* **1988**, *29*, 6087.

[91]Stavber; Šket; Zajc; Zupan *Tetrahedron* **1989**, *45*, 6003.

[92]Banks; Du Boisson; Morton; Tsiliopoulos *J. Chem. Soc., Perkin Trans. 1* **1988**, 2805.

[93]Lerman; Rozen *J. Org. Chem.* **1983**, *48*, 724. See also Purrington; Jones *J. Org. Chem.* **1983**, *48*, 761.

[94]Rozen; Brand *Synthesis* **1985**, 665. For another reagent, see Davis; Han *Tetrahedron Lett.* **1991**, *32*, 1631.

[95]Inman; Oesterling; Tyczkowski *J. Am. Chem. Soc.* **1958**, *80*, 6533; Machleidt; Hartmann *Liebigs Ann. Chem.* **1964**, *679*, 9; Kamlet; Adolph *J. Org. Chem.* **1968**, *33*, 3073; Sheppard *Tetrahedron Lett.* **1969**, 83. For reviews of perchloryl fluoride, see Sharts; Sheppard *Org. React.* **1974**, *21*, 125-406, pp. 225-236; Sheppard; Sharts *Organic Fluorine Chemistry*; W.A. Benjamin: New York, 1969, pp. 136-148; Khutoretskii; Okhlobystina; Fainzil'berg *Russ. Chem. Rev.* **1967**, *36*, 145-155.

[96]See Peet; Rockett *J. Organomet. Chem.* **1974**, *82*, C57; Adcock; Khor *J. Organomet. Chem.* **1975**, *91*, C20.

[97]For example, see Gabbard; Jensen *J. Org. Chem.* **1958**, *23*, 1406; Nakanishi; Jensen *J. Org. Chem.* **1962**, *27*, 702.

[98]Barton; Godinho; Hesse; Pechet *Chem. Commun.* **1968**, 804; Barton *Pure Appl. Chem.* **1970**, *21*, 285-293; Hesse *Isr. J. Chem.* **1978**, *17*, 60; Middleton; Bingham *J. Am. Chem. Soc.* **1980**, *102*, 4845. See also Sharts; Sheppard, Ref. 95, pp. 243-256; Rozen; Menahem *Tetrahedron Lett.* **1979**, 725.

[99]Tsushima; Kawada; Tsuji *Tetrahedron Lett.* **1982**, *23*, 1165.

F_2 in N_2 at $-78°C$ in $FCCl_3$.[100] Electrochemical fluorination has also been reported.[101] Sulfuryl chloride,[102] trichloroisocyanuric acid,[103] $Me_3SiCl-Me_2SO$,[104] $Me_3SiCl-MnO_2$,[105] $TiCl_3$,[106] and cupric chloride[107] have been used as reagents for chlorination, and N-bromosuccinimide (see **4-2**), t-BuBr-Me_2SO,[108] $Me_3SiBr-Me_2SO$,[109] and tetrabutylammonium tribromide,[110] for bromination. Iodination has been accomplished with I_2-HgCl_2[111] and with I_2-cerium(IV) ammonium nitrate.[112]

For unsymmetrical ketones the preferred position of halogenation is usually a CH group, then a CH_2 group, and then CH_3;[113] however, mixtures are frequent. With aldehydes the aldehydic hydrogen is sometimes replaced (see **4-3**). It is also possible to prepare di- and polyhalides. When basic catalysts are used, one α position of a ketone is completely halogenated before the other is attacked, and the reaction cannot be stopped until all the hydrogens of the first carbon have been replaced (see below). If one of the groups is methyl, the haloform reaction (**2-44**) takes place. With acid catalysts, it is easy to stop the reaction after only one halogen has entered, though a second halogen can be introduced by the use of excess reagent. In chlorination the second halogen generally appears on the same side as the first,[114] while in bromination the α,α'-dibromo product is found.[115] Actually, with both halogens it is the α,α-dihalo ketone that is formed first, but in the case of bromination this compound isomerizes under the reaction conditions to the α,α' isomer.[114] Aryl methyl ketones can be dibrominated ($ArCOCH_3 \rightarrow ArCOCHBr_2$) in high yields with benzyltrimethylammonium tribromide.[116]

It is not the aldehyde or ketone itself that is halogenated, but the corresponding enol or enolate ion. The purpose of the catalyst is to provide a small amount of enol or enolate. The reaction is often done without addition of acid or base, but traces of acid or base are always present, and these are enough to catalyze formation of the enol or enolate. With acid catalysis the mechanism is

Step 1

$$R_2CH-\underset{\underset{O}{\|}}{C}-R' \xrightarrow[\text{slow}]{H^+} R_2C=\underset{\underset{OH}{|}}{C}-R'$$

Step 2

$$R_2C=\underset{\underset{OH}{|}}{C}-R' + Br-Br \longrightarrow R_2C-\overset{\oplus}{\underset{\underset{Br}{|}\ \underset{OH}{|}}{C}}-R' + Br^-$$

Step 3

$$R_2C-\overset{\oplus}{\underset{\underset{Br}{|}\ \underset{OH}{|}}{C}}-R' \longrightarrow R_2C-\underset{\underset{Br}{|}\ \underset{O}{\|}}{C}-R'$$

[100]Purrington; Bumgardner; Lazaridis; Singh *J. Org. Chem.* **1987**, *52*, 4307.

[101]Laurent; Marquet; Tardivel *Tetrahedron* **1989**, *45*, 4431.

[102]For a review of sulfuryl chloride, see Tabushi; Kitaguchi, in Pizey *Synthetic Reagents*, vol. 4; Wiley: New York, 1981, pp. 336-396.

[103]Hiegel; Peyton *Synth. Commun.* **1985**, *15*, 385.

[104]Bellesia; Ghelfi; Grandi; Pagnoni *J. Chem. Res. (S)* **1986**, 426; Fraser; Kong *Synth. Commun.* **1988**, *18*, 1071.

[105]Bellesia; Ghelfi; Pagnoni; Pinetti *J. Chem. Res. (S)* **1990**, 188.

[106]Glaser; Toth *J. Chem. Soc., Chem. Commun.* **1986**, 1336.

[107]For a review, see Nigh, in Trahanovsky *Oxidation in Organic Chemistry*, pt. B; Academic Press: New York, 1973, pp. 67-81. Cupric chloride has been used to chlorinate α,β-unsaturated aldehydes and ketones in the γ position: Dietl; Normark; Payne; Thweatt; Young *Tetrahedron Lett.* **1973**, 1719.

[108]Armani; Dossena; Marchelli; Casnati *Tetrahedron* **1984**, *40*, 2035.

[109]Bellesia; Ghelfi; Grandi; Pagnoni *J. Chem. Res. (S)* **1986**, 426.

[110]Kajigaeshi; Kakinami; Okamoto; Fujisaki *Bull. Chem. Soc. Jpn.* **1987**, *60*, 1159.

[111]Barluenga; Martinez-Gallo; Najera; Yus *Synthesis* **1986**, 678.

[112]Horiuchi; Kiji *Chem. Lett.* **1988**, 31. For another reagent, see Šket; Zupet; Zupan; Dolenc *Bull. Chem. Soc. Jpn.* **1989**, *62*, 3406.

[113]For chlorination this is reversed if the solvent is methanol: Gallucci; Going *J. Org. Chem.* **1981**, *46*, 2532.

[114]Rappe *Ark. Kemi.* **1965**, *24*, 321. But see also Teo; Warnhoff *J. Am. Chem. Soc.* **1973**, *95*, 2728.

[115]Rappe; Schotte *Acta Chem. Scand.* **1962**, *16*, 2060; Rappe *Ark. Kemi* **1964**, *21*, 503; Garbisch *J. Org. Chem.* **1965**, *30*, 2109.

[116]Kajigaeshi; Kakinami; Tokiyama; Hirakawa; Okamoto *Bull. Chem. Soc. Jpn.* **1987**, *60*, 2667.

The first step, as we have already seen (**2-3**), actually consists of two steps. The second step is very similar to the first step in electrophilic addition to double bonds (p. 734). There is a great deal of evidence for this mechanism: (1) the rate is first order in substrate; (2) bromine does not appear in the rate expression at all,[117] a fact consistent with a rate-determining first step;[118] (3) the reaction rate is the same for bromination, chlorination, and iodination under the same conditions;[119] (4) the reaction shows an isotope effect; and (5) the rate of the step 2-step 3 sequence has been independently measured (by starting with the enol) and found to be very fast.[120]

With basic catalysts the mechanism may be the same as that given above (since bases also catalyze formation of the enol), or the reaction may go directly through the enolate ion without formation of the enol:

Step 1

$$R_2CH-\underset{\underset{O}{\|}}{C}-R' \xrightarrow{OH^-} R_2C=\underset{\underset{O_{\ominus}}{|}}{C}-R'$$

Step 2

$$R_2C=\underset{\underset{|\underline{O}|_\ominus}{|}}{C}-R' + Br-Br \longrightarrow R_2C-\underset{\underset{O}{\|}}{\underset{Br}{|}}C-R' + Br^-$$

It is difficult to distinguish the two possibilities. It was mentioned above that in the base-catalyzed reaction, if the substrate has two or three α halogens on the same side of the C=O group, it is not possible to stop the reaction after just one halogen atom has entered. The reason is that the electron-withdrawing field effect of the first halogen increases the acidity of the remaining hydrogens, i.e., a CHX group is more acidic than a CH_2 group, so that initially formed halo ketone is converted to enolate ion (and hence halogenated) more rapidly than the original substrate.

Regioselectivity in the halogenation of unsymmetrical ketones can be attained by treatment of the appropriate enol borinate of the ketone with N-bromo- or N-chlorosuccinimide.[121]

$$R'-\underset{\underset{OBR_2''}{|}}{C}=CH-R + NBS \longrightarrow R'-\underset{\underset{O}{\|}}{C}-\underset{\underset{Br}{|}}{CH}-R$$

The desired halo ketone is formed in high yield. Another method for achieving the same result involves bromination of the appropriate lithium enolate at a low temperature[122] (see p. 472 for the regioselective formation of enolate ions). In a similar process, α-halo aldehydes have been prepared in good yield by treatment of silyl enol ethers $R_2C=CHOSiMe_3$ with Br_2 or Cl_2,[123] with sulfuryl chloride SO_2Cl_2;[124] or with I_2 and silver acetate.[125] Enol acetates have been regioselectively iodinated with I_2 and either thallium(I) acetate[126] or copper(II)

---

[117]When the halogenating species is at low concentration or has a low reactivity, it can appear in the rate expression. The reaction becomes first order in the halogenating species. See, for example, Tapuhi; Jencks *J. Am. Chem. Soc.* **1982**, *104*, 5758. For a case in which the reaction is first order in bromine, even at relatively high Br_2 contentration, see Pinkus; Gopalan *J. Am. Chem. Soc.* **1984**, *106*, 2630. For a study of the kinetics of iodination, see Pinkus; Gopalan *Tetrahedron* **1986**, *42*, 3411.

[118]Under some conditions it is possible for step 2 to be rate-determining: Deno; Fishbein *J. Am. Chem. Soc.* **1973**, *95*, 7445.

[119]Bell; Yates *J. Chem. Soc.* **1962**, 1927.

[120]Hochstrasser; Kresge; Schepp; Wirz *J. Am. Chem. Soc.* **1988**, *110*, 7875.

[121]Hooz; Bridson *Can. J. Chem.* **1972**, *50*, 2387.

[122]Stotter; Hill *J. Org. Chem.* **1973**, *38*, 2576.

[123]Reuss; Hassner *J. Org. Chem.* **1974**, *39*, 1785; Blanco; Amice; Conia *Synthesis* **1976**, 194.

[124]Olah; Ohannesian; Arvanaghi; Prakash *J. Org. Chem.* **1984**, *49*, 2032.

[125]Rubottom; Mott *J. Org. Chem.* **1979**, *44*, 1731.

[126]Cambie; Hayward; Jurlina; Rutledge; Woodgate *J. Chem. Soc., Perkin Trans 1.* **1978**, 126.

acetate.[127] α,β-Unsaturated ketones can be converted to α-halo-α,β-unsaturated ketones by treatment with phenylselenium bromide or chloride,[128] and to α-halo-β,γ-unsaturated ketones by two-phase treatment with HOCl.[129]

OS **I**, 127; **II**, 87, 88, 244, 480; **III**, 188, 343, 538; **IV**, 110, 162, 590; **V**, 514; **VI**, 175, 193, 368, 401, 512, 520, 711, 991; **VII**, 271; **69**, 129. See also OS **VI**, 1033; **66**, 194.

2-5 Halogenation of Carboxylic Acids and Acyl Halides
Halogenation or **Halo-de-hydrogenation**

$$\text{R—CH}_2\text{—COOH} + \text{Br}_2 \xrightarrow{\text{PBr}_3} \text{R—CH—COOH} \atop \text{Br}$$

The α hydrogens of carboxylic acids can be replaced by bromine or chlorine with a phosphorus halide as catalyst.[130] The reaction, known as the *Hell–Volhard–Zelinskii reaction*, is not applicable to iodine or fluorine. When there are two α hydrogens, one or both may be replaced, though it is often hard to stop with just one. The reaction actually takes place on the acyl halide formed from the carboxylic acid and the catalyst. The acids alone are inactive, except for those with relatively high enol content, such as malonic. Less than one full mole of catalyst (per mole of substrate) is required, because of the exchange reaction between carboxylic acids and acyl halides (see **0-74**). Each molecule of acid is α halogenated while it is in the acyl halide stage. The halogen from the catalyst does not enter the α position. For example, the use of Cl_2 and PBr_3 results in α chlorination, not bromination. As expected from the foregoing, acyl halides undergo α halogenation without a catalyst. So do anhydrides and many compounds that enolize easily, e.g., malonic ester, aliphatic nitro compounds, etc. The mechanism is usually regarded as proceeding through the enol as in **2-4**.[131] If chlorosulfuric acid $ClSO_2OH$ is used as a catalyst, carboxylic acids can be α iodinated,[132] as well as chlorinated or brominated.[133]

A number of other methods exist for the α halogenation of carboxylic acids or their derivatives.[134] The acids or their chlorides or anhydrides can be α chlorinated by treatment with $CuCl_2$ in polar inert solvents (e.g., sulfolane).[135] Acyl halides can be α brominated or chlorinated by use of N-bromo- or N-chlorosuccinimide and HBr or HCl.[136] The latter is an ionic, not a free-radical halogenation (see **4-2**). Direct iodination of carboxylic acids has been achieved with I_2-Cu(II) acetate in HOAc.[137] Acyl chlorides can be α iodinated with I_2 and a trace of HI.[138] Carboxylic esters can be α halogenated by conversion to their enolate ions with lithium N-isopropylcyclohexylamide in THF and treatment of this solution at −78° with I_2[138] or with a carbon tetrahalide.[139] Carboxylic acids, esters, and amides have been α fluorinated at −78°C with F_2 diluted in N_2.[140]

OS **I**, 115, 245; **II**, 74, 93; **III**, 347, 381, 495, 523, 623, 705, 848; **IV**, 254, 348, 398, 608, 616; **V**, 255; **VI**, 90, 190, 403. Also see OS **IV**, 877; **VI**, 427.

[127]Horiuchi; Satoh *Synthesis* **1981**, 312.
[128]Ley; Whittle *Tetrahedron Lett.* **1981**, *22*, 3301.
[129]Hegde; Wolinsky *Tetrahedron Lett.* **1981**, *22*, 5019.
[130]For a review, see Harwood, *Chem. Rev.* **1962**, *62*, 99-154, pp. 102-103.
[131]See, however, Kwart; Scalzi *J. Am. Chem. Soc.* **1964**, *86*, 5496.
[132]Ogata; Watanabe *J. Org. Chem.* **1979**, *44*, 2768, **1980**, *45*, 2831.
[133]Ogata; Sugimoto *J. Org. Chem.* **1978**, *43*, 3684; Ogata; Adachi *J. Org. Chem.* **1982**, *47*, 1182.
[134]For a list of reagents, with references, see Ref. 52, pp. 378-380.
[135]Louw *Chem. Commun.* **1966**, 544.
[136]Gleason; Harpp *Tetrahedron Lett.* **1970**, 3431; Harpp; Bao; Black; Gleason; Smith *J. Org. Chem.* **1975**, *40*, 3420.
[137]Horiuchi; Satoh *Chem. Lett.* **1984**, 1509.
[138]Rathke; Lindert *Tetrahedron Lett.* **1971**, 3995.
[139]Arnold; Kulenovic *J. Org. Chem.* **1978**, *43*, 3687.
[140]Purrington; Woodard *J. Org. Chem.* **1990**, *55*, 3423.

2-6 Halogenation of Sulfoxides and Sulfones
Halogenation or **Halo-de-hydrogenation**

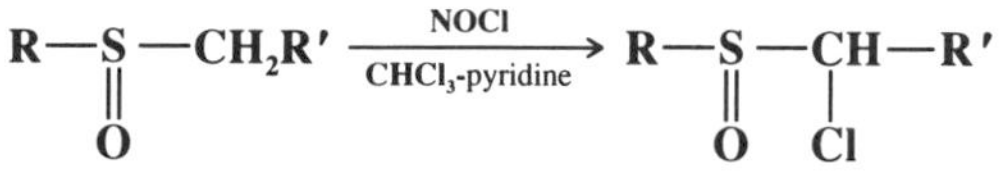

$$R—\underset{\underset{O}{\|}}{S}—CH_2R' \xrightarrow[\text{CHCl}_3\text{-pyridine}]{\text{NOCl}} R—\underset{\underset{O}{\|}}{S}—\underset{\underset{Cl}{|}}{C}H—R'$$

Sulfoxides can be chlorinated in the α position[141] by treatment with Cl_2,[142] $TsCl$,[143] N-chlorosuccinimide,[144] or $PhICl_2$,[145] all in the presence of pyridine, or with *t*-BuOCl and KOAc (or pyridine).[146] All these methods involve basic conditions. The reaction can also be accomplished in the absence of base with SO_2Cl_2 in CH_2Cl_2.[147] The bromination of sulfoxides with bromine[145] and with N-bromosuccinimide–bromine[148] have also been reported. Sulfones have been chlorinated by treatment of their conjugate bases $RSO_2\overset{\ominus}{C}HR'$ with various reagents, among them SO_2Cl_2, CCl_4,[149] N-chlorosuccinimide,[150] and hexachloroethane.[151] The α fluorination of sulfoxides has been accomplished in a two-step pro-

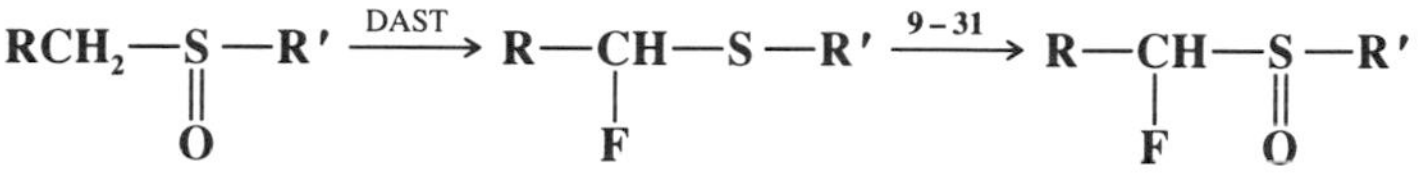

$$RCH_2—\underset{\underset{O}{\|}}{S}—R' \xrightarrow{\text{DAST}} R—\underset{\underset{F}{|}}{C}H—S—R' \xrightarrow{9-31} R—\underset{\underset{F}{|}}{C}H—\underset{\underset{O}{\|}}{S}—R'$$

cedure. Treatment with diethylaminosulfur trifluoride Et_2NSF_3 (DAST) produces an α-fluoro thioether, usually in high yield. Oxidation of this compound with *m*-chloroperbenzoic acid gives the sulfoxide.[152]

C. Nitrogen Electrophiles

2-7 Aliphatic Diazonium Coupling
Arylhydrazono-de-dihydro-bisubstitution

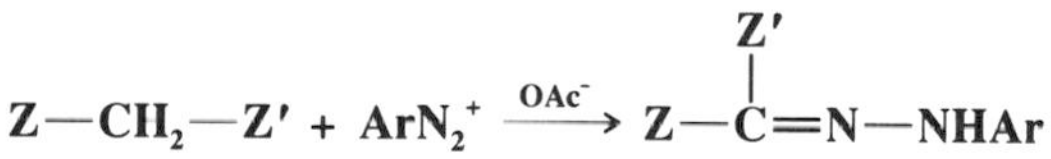

$$Z—CH_2—Z' + ArN_2^+ \xrightarrow{\text{OAc}^-} Z—\overset{\overset{\displaystyle Z'}{|}}{C}=N—NHAr$$

If a C—H bond is acidic enough, it couples with diazonium salts in the presence of a base, most often aqueous sodium acetate.[153] The reaction is commonly carried out on compounds of the form Z—CH$_2$—Z', where Z and Z' are as defined on p. 464, e.g., β-keto esters, β-keto amides, malonic ester.

[141]For a review, see Venier; Barager *Org. Prep. Proceed. Int.* **1974**, 6, 77-102, pp. 81-84.
[142]Tsuchithashi; Iriuchijima *Bull. Chem. Soc. Jpn.* **1970**, 43, 2271.
[143]Hojo; Yoshida *J. Am. Chem. Soc.* **1968**, 90, 4496.
[144]Ogura; Imaizumi; Iida; Tsuchihasi *Chem. Lett.* **1980**, 1587.
[145]Cinquini; Colonna *J. Chem. Soc., Perkin Trans. 1* **1972**, 1883. See also Cinquini; Colonna *Synthesis* **1972**, 259.
[146]Iriuchijima; Tsuchihashi *Tetrahedron Lett.* **1969**, 5259.
[147]Tin; Durst *Tetrahedron Lett.* **1970**, 4643.
[148]Iriuchijima; Tsuchihashi *Synthesis* **1970**, 588.
[149]Regis; Doweyko *Tetrahedron Lett.* **1982**, 23, 2539.
[150]Paquette; Houser *J. Am. Chem. Soc.* **1969**, 91, 3870, *J. Org. Chem.* **1971**, 36, 1015.
[151]Kattenberg; de Waard; Huisman *Tetrahedron* **1973**, 29, 4149, **1974**, 30, 463.
[152]McCarthy; Peet; LeTourneau; Inbasekaran *J. Am. Chem. Soc.* **1985**, 107, 735. See also Umemoto; Tomizawa *Bull. Chem. Soc. Jpn.* **1986**, 59, 3625.
[153]For a review, see Parmerter *Org. React.* **1959**, 10, 1-142.

The mechanism is probably of the simple SE1 type:

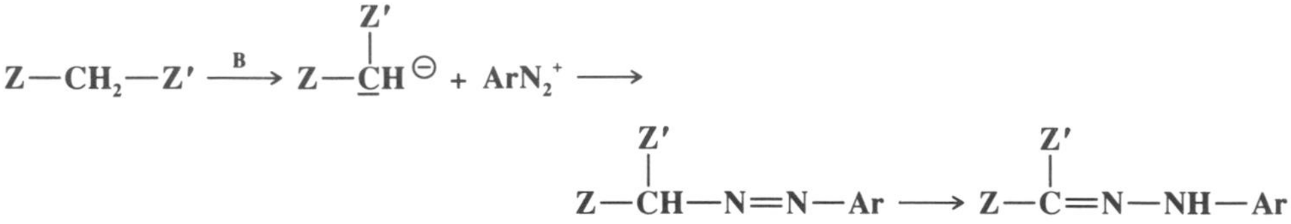

$$Z-CH_2-Z' \xrightarrow{B} Z-\underset{|}{\overset{Z'}{C}}H^{\ominus} + ArN_2^+ \longrightarrow$$

$$Z-\underset{|}{\overset{Z'}{C}}H-N=N-Ar \longrightarrow Z-\underset{|}{\overset{Z'}{C}}=N-NH-Ar$$

15

Aliphatic azo compounds in which the carbon containing the azo group is attached to a hydrogen are unstable and tautomerize to the isomeric hydrazones (**15**), which are therefore the products of the reaction.

When the reaction is carried out on a compound of the form Z—CHR—Z', so that the azo compound does not have a tautomerizable hydrogen, if at least one Z is acyl or carboxyl, this group usually cleaves:

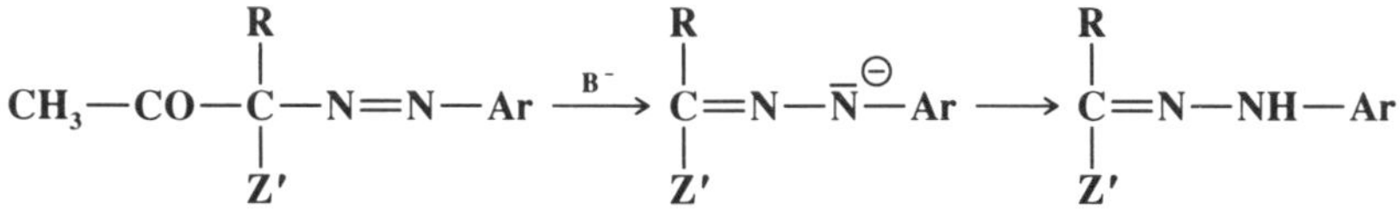

$$CH_3-CO-\underset{\underset{Z'}{|}}{\overset{\overset{R}{|}}{C}}-N=N-Ar \xrightarrow{B^-} \underset{\underset{Z'}{|}}{\overset{\overset{R}{|}}{C}}=N-\overset{\ominus}{N}-Ar \longrightarrow \underset{\underset{Z'}{|}}{\overset{\overset{R}{|}}{C}}=N-NH-Ar$$

16

so the product in this case too is the hydrazone, and not the azo compound. In fact, compounds of the type **16** are seldom isolable from the reaction, though this has been accomplished.[154] The cleavage step shown is an example of **2-43** and, when a carboxyl group cleaves, of **2-40**. The overall process in this case is called the *Japp–Klingemann reaction*[155] and involves conversion of a ketone (**17**) or a carboxylic acid (**18**) to a hydrazone (**19**). When

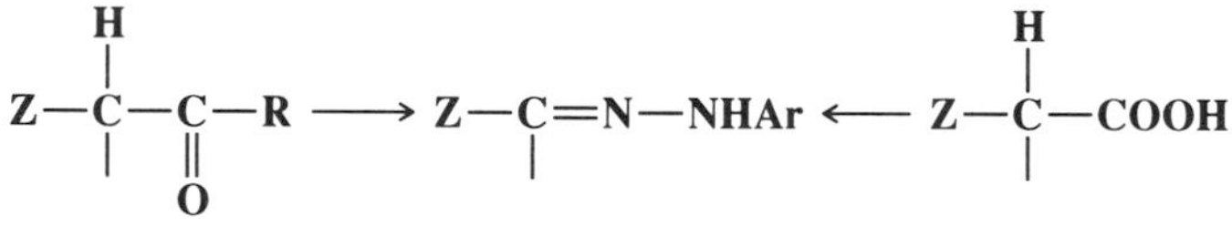

$$Z-\underset{\underset{H}{|}}{\overset{\overset{H}{|}}{C}}-\underset{\overset{||}{O}}{\overset{}{C}}-R \longrightarrow Z-\underset{|}{C}=N-NHAr \longleftarrow Z-\underset{\underset{H}{|}}{\overset{\overset{H}{|}}{C}}-COOH$$

17   **19**   **18**

an acyl and a carboxyl group are both present, the leaving group order has been reported to be MeCO > COOH > PhCO.[156] When there is no acyl or carboxyl group present, the aliphatic azo compound is stable.

OS **III**, 660; **IV**, 633.

2-8 Nitrosation at a Carbon Bearing an Active Hydrogen

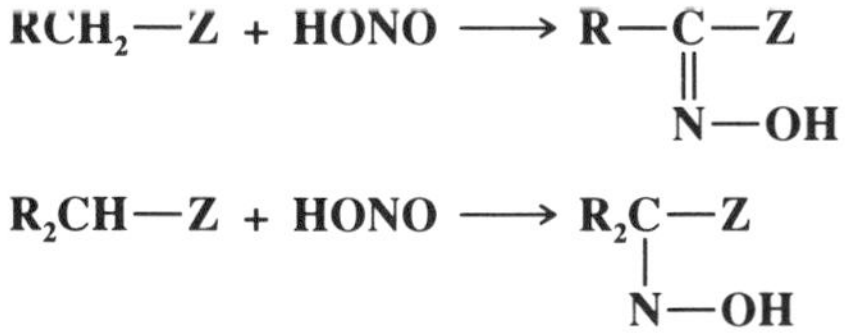

$$RCH_2-Z + HONO \longrightarrow R-\underset{\underset{N-OH}{||}}{C}-Z \qquad \textbf{Hydroxyimino-de-dihydro-bisubstitution}$$

$$R_2CH-Z + HONO \longrightarrow R_2\underset{\underset{N-OH}{|}}{C}-Z \qquad \textbf{Nitrosation or Nitroso-de-hydrogenation}$$

[154]See, for example, Yao; Resnick *J. Am. Chem. Soc.* **1962**, *84*, 3514.
[155]For a review, see Phillips, *Org. React.* **1959**, *10*, 143-178.
[156]Neplyuev; Bazavova; Lozinskii *J. Org. Chem. USSR* **1989**, *25*, 2011. This paper also includes a sequence of leaving group ability for other Z groups.

Carbons adjacent to a Z group (as defined on p. 464) can be nitrosated with nitrous acid or alkyl nitrites.[157] The initial product is the C-nitroso compound, but these are stable only when there is no tautomerizable hydrogen. When there is, the product is the more stable oxime. The situation is analogous to that with azo compounds and hydrazones (**2-7**). The mechanism is similar to that in **2-7**:[158] $R—H \rightarrow R^- + {}^+N{=}O \rightarrow R—N{=}O$. The attacking species is either NO^+ or a carrier of it. When the substrate is a simple ketone, the mechanism goes through the enol (as in halogenation **2-4**):

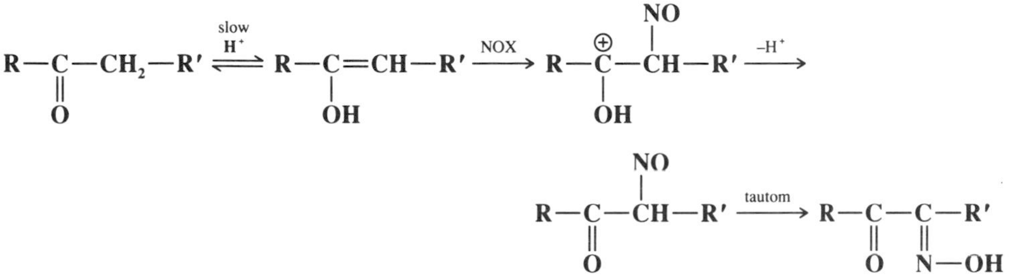

$$R—\underset{\overset{\|}{O}}{C}—CH_2—R' \underset{\overset{H^+}{slow}}{\rightleftharpoons} R—\underset{\overset{|}{OH}}{C}{=}CH—R' \xrightarrow{NOX} R—\underset{\overset{|}{OH}}{\overset{\oplus}{C}}—CH—R' \xrightarrow{-H^+}$$

$$R—\underset{\overset{\|}{O}}{C}—\underset{\overset{|}{NO}}{CH}—R' \xrightarrow{tautom} R—\underset{\overset{\|}{O}}{C}—\underset{\overset{\|}{N—OH}}{C}—R'$$
$$\textbf{20}$$

Evidence is that the reaction, in the presence of X^- (Br^-, Cl^-, or SCN^-) was first order in ketone and in H^+, but zero order in HNO_2 and X^-.[159] Furthermore, the rate of the nitrosation was about the same as that for enolization of the same ketones. The species NOX is formed by $HONO + X^- + H^+ \rightarrow HOX + H_2O$. In the cases of $F_3CCOCH_2COCF_3$ and malononitrile the nitrosation went entirely through the enolate ion rather than the enol.[160]

As in the Japp–Klingemann reaction, when Z is an acyl or carboxyl group (in the case of $R_2CH—Z$), it can be cleaved. Since oximes and nitroso compounds can be reduced to primary amines, this reaction often provides a route to amino acids. As in the case of **2-4**, the silyl enol ether of a ketone can be used instead of the ketone itself.[161] Good yields of α-oximinoketones (**20**) can be obtained by treating ketones with *t*-butyl thionitrate.[162]

Imines can be prepared in a similar manner by treatment of an active hydrogen compound with a nitroso compound:

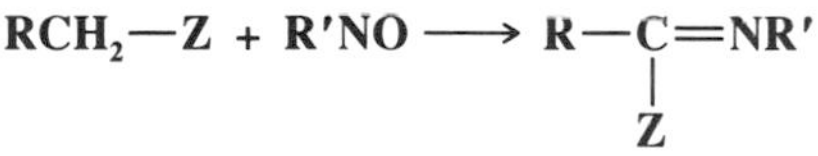

$$RCH_2—Z + R'NO \longrightarrow R—\underset{\overset{|}{Z}}{C}{=}NR'$$

Alkanes can be nitrosated photochemically, by treatment with NOCl and uv light.[163] For nitration at an activated carbon, see **4-13**.

OS **II**, 202, 204, 223, 363; **III**, 191, 513; **V**, 32, 373; **VI**, 199, 840. Also see OS **V**, 650.

2-9   Direct Formation of Diazo Compounds
Diazo-de-dihydro-bisubstitution

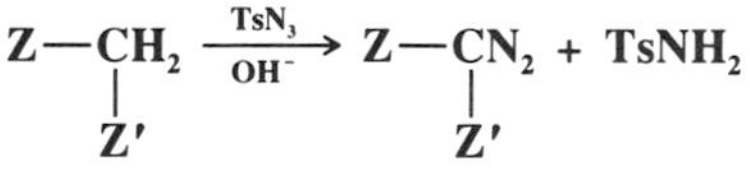

$$Z—\underset{\overset{|}{Z'}}{C}H_2 \xrightarrow[OH^-]{TsN_3} Z—\underset{\overset{|}{Z'}}{C}N_2 + TsNH_2$$

[157]For a review, see Williams *Nitrosation*; Cambridge University Press: Cambridge, 1988, pp. 1-45.
[158]For a review, see Williams *Adv. Phys. Org. Chem.* **1983**, *19*, 381-428. See also Ref. 157.
[159]Leis; Peña; Williams; Mawson *J. Chem. Soc., Perkin Trans. 2* **1988**, 157.
[160]Iglesias; Williams *J. Chem. Soc., Perkin Trans. 2* **1989**, 343; Crookes; Roy; Williams *J. Chem. Soc., Perkin Trans. 2* **1989**, 1015. See also Graham; Williams *J. Chem. Soc., Chem. Commun.* **1991**, 407.
[161]Rasmussen; Hassner *J. Org. Chem.* **1974**, *39*, 2558.
[162]Kim; Park; Kim *Tetrahedron Lett.* **1989**, *30*, 2833.
[163]For a review, see Pape *Fortschr. Chem. Forsch.* **1967**, *7*, 559-604.

Compounds containing a CH_2 bonded to two Z groups (as defined on p. 464) can be converted to diazo compounds on treatment with tosyl azide in the presence of a base.[164] The use of phase transfer catalysis increases the convenience of the method.[165] *p*-Dodecylbenzenesulfonyl azide,[166] methanesulfonyl azide,[167] and *p*-acetamidobenzenesulfonyl azide[168] also give the reaction. The reaction, which is called the *diazo transfer reaction*, can also be applied to other reactive positions, e.g., the 5 position of cyclopentadiene.[169] The mechanism is probably as follows:

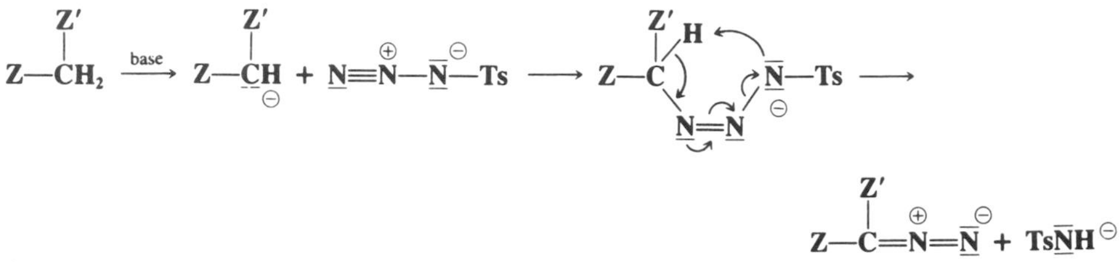

$$
\begin{array}{c}
Z'\\
|\\
Z—CH_2
\end{array}
\xrightarrow{\text{base}}
\begin{array}{c}
Z'\\
|\\
Z—\overset{\ominus}{C}H
\end{array}
+ \overset{\oplus}{N}{\equiv}N{-}\overset{\ominus}{N}{-}Ts
\longrightarrow
Z{-}\overset{Z'}{\underset{\substack{N{=}N}}{\overset{\overset{\text{H}}{|}}{C}}}

\overset{\ominus}{N}{-}Ts
\longrightarrow
$$

$$
\begin{array}{c}
Z'\\
|\\
Z—\overset{\oplus}{C}{=}N{=}\overset{\ominus}{N}
\end{array}
+ Ts\overset{\ominus}{N}H
$$

A diazo group can be introduced adjacent to a single carbonyl group indirectly by first converting the ketone to an α-formyl ketone (**0-108**) and then treating it with tosyl azide.

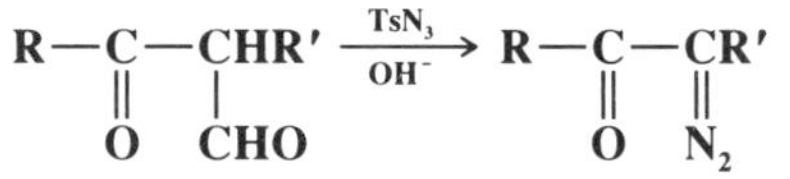

$$
\underset{\underset{O}{\|}}{R—C}—\underset{\underset{CHO}{|}}{CHR'}
\xrightarrow[OH^-]{TsN_3}
\underset{\underset{O}{\|}}{R—C}—\underset{\underset{N_2}{\|}}{CR'}
$$

As in the similar cases of **2-7** and **2-8,** the formyl group is cleaved during the reaction.[170]
OS **V,** 179; **VI,** 389, 414.

2-10 Conversion of Amides to α-Azido Amides
Azidation or **Azido-de-hydrogenation**

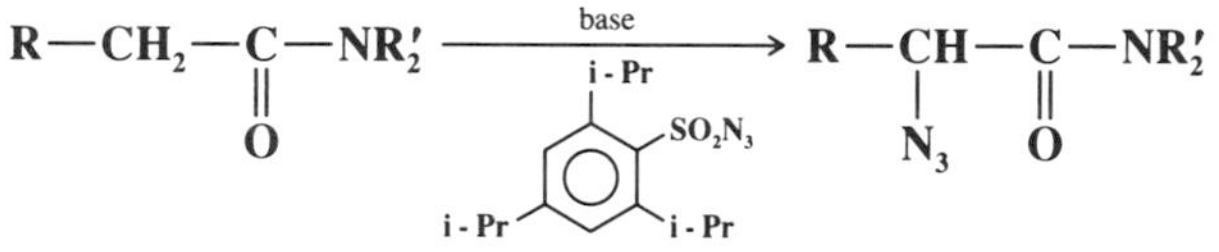

$$
\underset{\underset{O}{\|}}{R—CH_2—C}—NR'_2
\xrightarrow[\text{i-Pr}]{\text{base}}
\underset{\underset{N_3}{|}}{R—CH}—\underset{\underset{O}{\|}}{C}—NR'_2
$$

In reaction **2-9** treatment of Z—CH₂—Z' with tosyl azide gives diazo transfer. When this reaction is performed on a compound with a single Z group, formation of the azide becomes a competing process.[171] Factors favoring azide formation rather than diazo transfer include

[164]For reviews, see Regitz; Maas *Diazo Compounds*; Academic Press: New York, 1986, pp. 326-435; Regitz *Synthesis* **1972**, 351-373, *Angew. Chem. Int. Ed. Engl.* **1967**, *6*, 733-749 [*Angew. Chem. 79*, 786-801], *Newer Methods Prep. Org. Chem.* **1971**, *6*, 81-126. See also Hünig *Angew. Chem. Int. Ed. Engl.* **1968**, *7*, 335-344 [*Angew. Chem. 80*, 343-352]; Koskinen; Muñoz *J. Chem. Soc., Chem. Commun.* **1990**, 652.
[165]Ledon *Synthesis* **1974**, 347, *Org. Synth. VI*, 414. For another convenient method, see Ghosh; Datta *Synth. Commun.* **1991**, *21*, 191.
[166]Hazen; Weinstock; Connell; Bollinger *Synth. Commun.* **1981**, *11*, 947.
[167]Taber; Ruckle; Hennessy *J. Org. Chem.* **1986**, *51*, 4077.
[168]Baum; Shook; Davies; Smith *Synth. Commun.* **1987**, *17*, 1709.
[169]Doering; DePuy *J. Am. Chem. Soc.* **1953**, *75*, 5955.
[170]For a similar approach, see Danheiser; Miller; Brisbois; Park *J. Org. Chem.* **1990**, *55*, 1959.
[171]Evans; Britton *J. Am. Chem. Soc.* **1987**, *109*, 6881, and references cited therein.

K^+ as the enolate counterion rather than Na^+ or Li^+ and the use of 2,4,6-triisopropylbenzenesulfonyl azide rather than TsN_3. When the reaction was applied to amides with a chiral R', it was highly stereoselective, and the product could be converted to an optically active amino acid.[171]

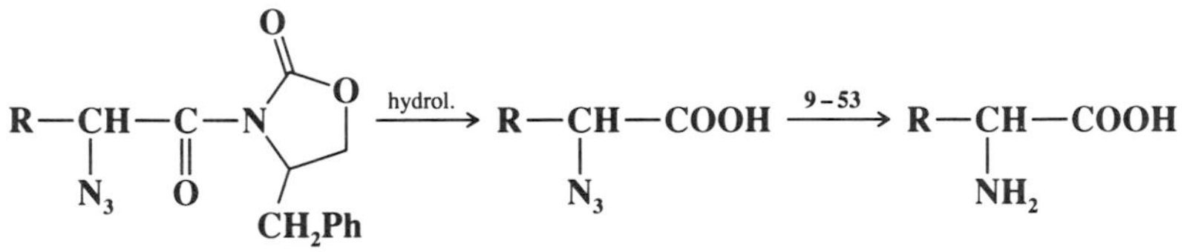

$$R-\underset{\underset{N_3}{|}}{CH}-\underset{\underset{O}{||}}{C}-N\overset{O}{\underset{CH_2Ph}{\diagdown}} \xrightarrow{\text{hydrol.}} R-\underset{\underset{N_3}{|}}{CH}-COOH \xrightarrow{\textbf{9-53}} R-\underset{\underset{NH_2}{|}}{CH}-COOH$$

2-11 Direct Amination at an Activated Position
Alkyamino-de-hydrogenation, etc.

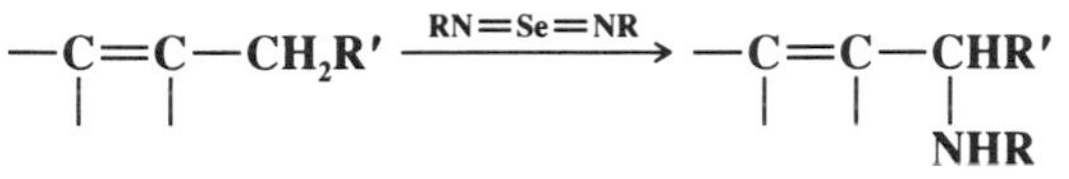

$$-\underset{|}{C}{=}\underset{|}{C}-CH_2R' \xrightarrow{\text{RN}=\text{Se}=\text{NR}} -\underset{|}{C}{=}\underset{|}{C}-\underset{\underset{NHR}{|}}{CH}R' \qquad R = t\text{-Bu, Ts}$$

Alkenes can be aminated[172] in the allylic position by treatment with solutions of imido selenium compounds R—N=Se=N—R.[173] The reaction, which is similar to the allylic oxidation of alkenes with SeO_2 (see **4-4**), has been performed with $R = t$-Bu and $R = $ Ts. The imido sulfur compound TsN=S=NTs has also been used.[174] In another reaction, compounds containing an active hydrogen can be converted to primary amines in moderate yields by treatment with O-(2,4-dinitrophenyl)hydroxylamine.[175]

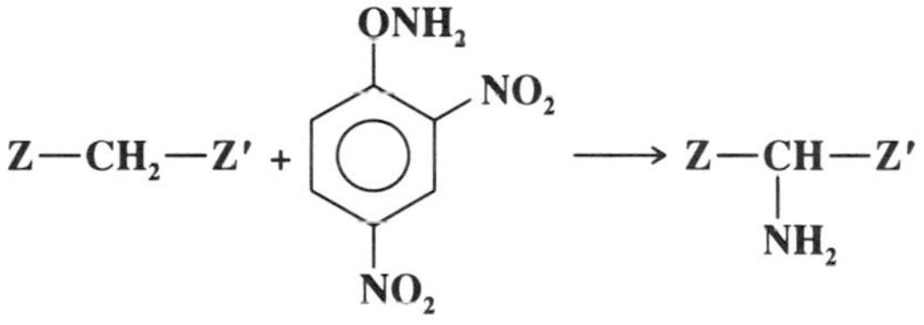

$$Z-CH_2-Z' + \underset{\underset{NO_2}{}}{\overset{\overset{ONH_2}{\underset{}{\bigcirc}}\,NO_2}{}} \longrightarrow Z-\underset{\underset{NH_2}{|}}{CH}-Z'$$

In an indirect amination process, acyl halides are enantioselectively converted to amino acids.[176] The key step involves addition to the N=N bond of a dialkyl azodicarboxylate **22.**

[172]For a review of direct aminations, see Sheradsky, in Patai *The Chemistry of Functional Groups, Supplement F*, pt. 1; Wiley: New York, 1982, pp. 395-416.
[173]Sharpless; Hori; Truesdale; Dietrich *J. Am. Chem. Soc.* **1976**, *98*, 269. For another method, see Kresze; Münsterer *J. Org. Chem.* **1983**, *48*, 3561. For a review, see Cheikh; Chaabouni; Laurent; Mison; Nafti *Synthesis* **1983**, 685-700, pp. 691-696.
[174]Sharpless; Hori, *J. Org. Chem.* **1979**, *41*, 176; Singer; Sharpless *J. Org. Chem.* **1978**, *43*, 1448. For other reagents, see Mahy; Bedi; Battioni; Mansuy *Tetrahedron Lett.* **1988**, *29*, 1927; Tsushima; Yamada; Onami; Oshima; Chaney; Jones; Swartzendruber *Bull. Chem. Soc. Jpn.* **1989**, *62*, 1167.
[175]Sheradsky; Salemnick; Nir *Tetrahedron* **1972**, *28* 3833; Radhakrishna; Loudon; Miller *J. Org. Chem.* **1979**, *44*, 4836.
[176]Trimble; Vederas *J. Am. Chem. Soc.* **1986**, *108*, 6397; Evans; Britton; Dorow; Dellaria *J. Am. Chem. Soc.* **1986**, *108*, 6395, *Tetrahedron* **1988**, *44*, 5525; Gennari; Colombo; Bertolini *J. Am. Chem. Soc.* **1986**, *108*, 6394; Oppolzer; Moretti *Helv. Chim. Acta* **1986**, *69*, 1923, *Tetrahedron* **1988**, *44*, 5541; Guanti; Banfi; Narisano *Tetrahedron* **1988**, *44*, 5523.

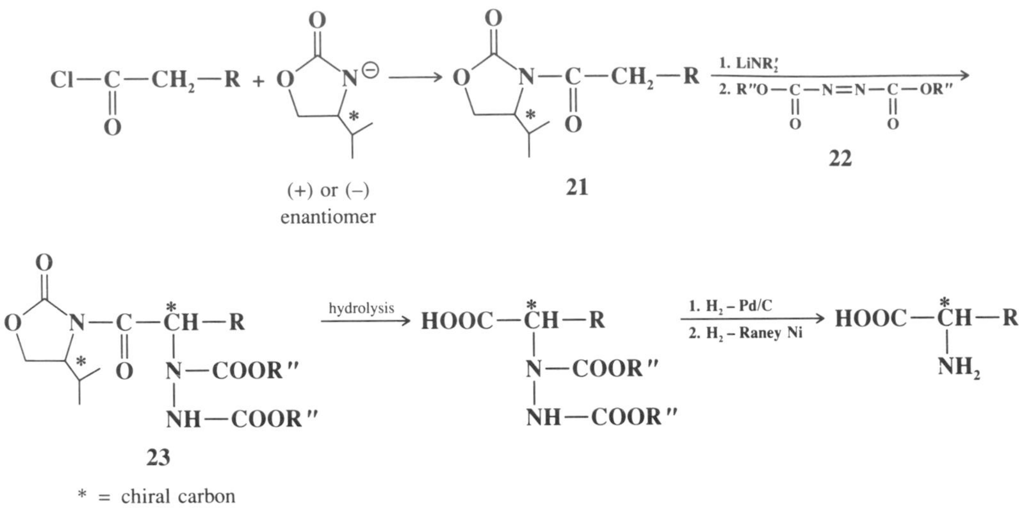

(+) or (−)
enantiomer

21

22

23

* = chiral carbon

In this process the presence of a chiral carbon in **21** induces chirality at the newly formed C—N bond in **23**.

See also **0-50.**

2-12 Insertion by Nitrenes

CH-[Acylimino]-insertion, etc.

$$RH + \overline{N}-C-W \longrightarrow R-NH-C-W$$

Carbonylnitrenes NCOW (W = R′, Ar, or OR′) are very reactive species (p. 202) and insert into the C—H bonds of alkanes to give amides (W = R′ or Ar) or carbamates (W = OR′).[177] The nitrenes are generated as discussed on p. 202. The order of reactivity among alkane C—H bonds is tertiary > secondary > primary.[178] Indications are that in general it is only singlet and not triplet nitrenes that insert.[179] Retention of configuration is found at a chiral carbon.[180] The mechanism is presumably similar to the simple one-step mechanism for insertion of carbenes (**2-20**). Other nitrenes (e.g., cyanonitrene NCN[181] and arylnitrenes NAr[182]) can also insert into C—H bonds, but alkylnitrenes usually undergo rearrangement before they can react with the alkane. The insertion reactions are not generally useful synthetically, since they usually lead to mixtures of products, but exceptions

[177]For a review, see Lwowski, in Lwowski *Nitrenes*; Wiley: New York, 1970, pp. 199-207.

[178]For example, see Maslak *J. Am. Chem. Soc.* **1989**, *111*, 8201. Nitrenes are much more selective (and less reactive) in this reaction than carbenes (**2-20**). For a discussion, see Alewood; Kazmaier; Rauk *J. Am. Chem. Soc.* **1973**, *95*, 5466.

[179]For example, see Simson; Lwowski *J. Am. Chem. Soc.* **1969**, *91*, 5107; Inagaki; Shingaki; Nagai *Chem. Lett.* **1981**, 1419.

[180]Smolinsky; Feuer *J. Am. Chem. Soc.* **1964**, *86*, 3085.

[181]For a review of cyanonitrenes, see Anastassiou; Shepelavy; Simmons; Marsh, in Lwowski, Ref. 177, pp. 305-344.

[182]For a review of arylnitrenes, see Scriven *Azides and Nitrenes*; Academic Press: New York, 1984, pp. 95-204.

are known,[183] chiefly in cyclizations.[184] For example, heating of 2-(2-methylbutyl)phenyl azide gave about 60% 2-ethyl-2-methylindoline.[180]

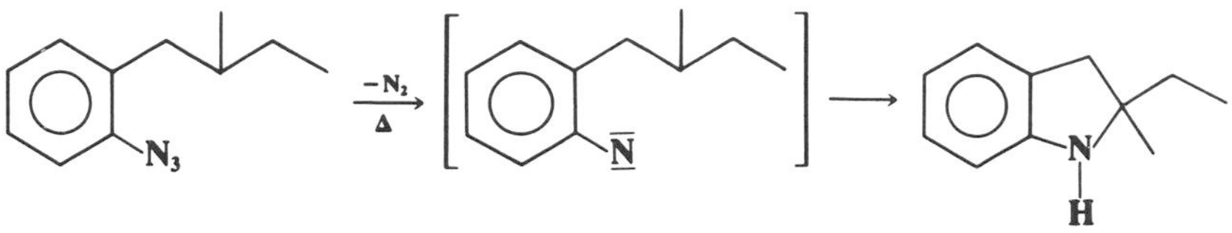

D. Sulfur Electrophiles

2-13 Sulfenylation and Selenylation of Ketones and Carboxylic Esters
Alkylthio-de-hydrogenation, etc.

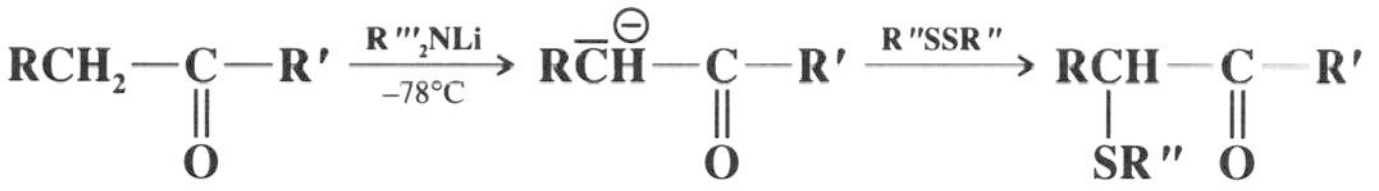

Ketones, carboxylic esters (including lactones),[185] and amides (including lactams)[186] can be sulfenylated in the α position by conversion to the enolate ion with a base such as lithium N-isopropylcyclohexylamide and treatment of this with a disulfide.[187] The reaction, shown above for ketones, involves nucleophilic substitution at sulfur. Analogously, α-phenylseleno ketones RCH(SePh)COR' and α-phenylseleno esters RCH(SePh)COOR' can be prepared[188] by treatment of the corresponding enolates with PhSeBr,[189] PhSeSePh,[190] or benzeneseleninic anhydride PhSe(O)OSe(O)Ph.[191] Another method for the introduction of a phenylseleno group into the α position of a ketone involves simple treatment of an ethyl acetate solution of the ketone with PhSeCl (but not PhSeBr) at room temperature.[192] This procedure is also successful for aldehydes but not for carboxylic esters. In another method that avoids the use of PhSeX reagents, a ketone enolate is treated with selenium to give an R'COCHRSe ion, which is treated with MeI, producing the α-methylseleno ketone R'COCHRSeMe.[193] This method has also been applied to carboxylic esters.

The α-seleno and α-sulfenyl carbonyl compounds prepared by this reaction can be converted to α,β-unsaturated carbonyl compounds (**7-12**). The sulfenylation reaction has also

[183]For a synthetically useful noncyclization example, see Meinwald; Aue *Tetrahedron Lett.* **1967**, 2317.

[184]For a list of examples, with references, see Ref. 52, p. 564.

[185]Trost; Salzmann *J. Am. Chem. Soc.* **1973**, *95*, 6840; Seebach; Teschner *Tetrahedron Lett.* **1973**, 5113. For discussions, see Trost *Pure Apple. Chem.* **1975**, *43*, 563-585, pp. 572-578; Caine, in Augustine *Carbon–Carbon Bond Formation*, vol. 1; Marcel Dekker: New York, 1979, pp. 278-282.

[186]Zoretic; Soja *J. Org. Chem.* **1976**, *41*, 3587; Gassman; Balchunis *J. Org. Chem.* **1977**, *42*, 3236.

[187]For another reagent, see Scholz *Synthesis* **1983**, 944.

[188]For reviews of selenylations, see Back, in Liotta *Organoselenium Chemistry*; Wiley: New York, 1987, pp. 1-125; Paulmier *Selenium Reagents and Intermediates in Organic Synthesis*; Pergamon: Elmsford, NY, 1986, pp. 95-98.

[189]Reich; Reich; Renga *J. Am. Chem. Soc.* **1973**, *95*, 5813; Clive *J. Chem. Soc., Chem. Commun.* **1973**, 695; Brocksom; Petragnani; Rodrigues *J. Org. Chem.* **1974**, *39*, 2114; Schwartz; Hayasi *Tetrahedron Lett.* **1980**, *21*, 1497. See also Liotta *Acc. Chem. Res.* **1984**, *17*, 28-34.

[190]Grieco; Miyashita *J. Org. Chem.* **1974**, *39*, 120. α Phenylselenation can also be accomplished with PhSeSePh, SeO$_2$, and an acid catalyst: Miyoshi; Yamamoto; Kambe; Murai; Sonoda *Tetrahedron Lett.* **1982**, *23*, 4813.

[191]Barton; Lester; Ley *J. Chem. Soc., Perkin Trans. 1* **1980**, 2209; Barton; Morzycki; Motherwell; Ley *J. Chem. Soc., Chem. Commun.* **1981**, 1044.

[192]Sharpless; Lauer; Teranishi *J. Am. Chem. Soc.* **1973**, *95*, 6137.

[193]Liotta; Zima; Barnum; Saindane *Tetrahedron Lett.* **1980**, *21*, 3643; Liotta; Saindane; Barnum; Ensley; Balakrishnan *Tetrahedron Lett.* **1981**, *22*, 3043; Liotta, Ref. 189.

been used[194] as a key step in a sequence for moving the position of a carbonyl group to an adjacent carbon.[195]

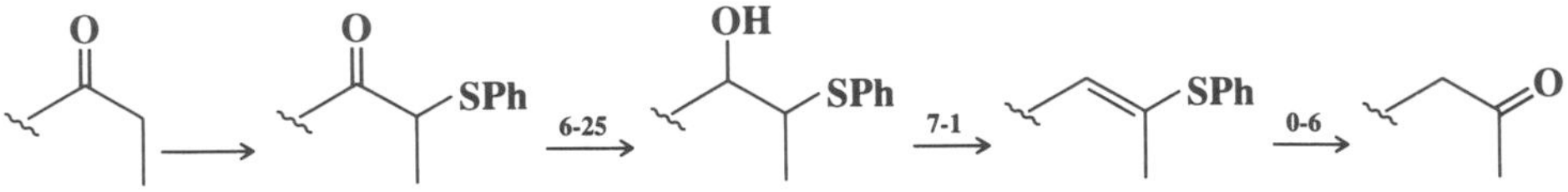

$$\text{(structures)} \xrightarrow{} \text{(structures)} \xrightarrow{6\text{-}25} \text{(structures)} \xrightarrow{7\text{-}1} \text{(structures)} \xrightarrow{0\text{-}6} \text{(structures)}$$

OS **VI,** 23, 109; **68,** 8.

2-14 Sulfonation of Aldehydes, Ketones, and Carboxylic Acids
Sulfonation or **Sulfo-de-hydrogenation**

$$R-CH_2-CO-R' \xrightarrow{SO_3} R-CH-CO-R'$$
$$\qquad\qquad\qquad\qquad\qquad\qquad\quad |$$
$$\qquad\qquad\qquad\qquad\qquad\qquad\quad SO_3H$$

Aldehydes, ketones, and carboxylic acids containing α hydrogens can be sulfonated with sulfur trioxide.[196] The mechanism is presumably similar to that of **2-4.** Sulfonation has also been accomplished at vinylic hydrogen.
OS **IV,** 846, 862.

E.   Carbon Electrophiles. With respect to the attacking molecule, these are nucleophilic substitutions.

2-15 Acylation at an Aliphatic Carbon
Acylation or **Acyl-de-hydrogenation**

$$\overset{\backslash}{\underset{/}{C}}=\overset{H}{\underset{\backslash}{C}} + RCOCl \xrightarrow{AlCl_3} \overset{\backslash}{\underset{/}{C}}=\overset{COR}{\underset{\backslash}{C}}$$

Olefins can be acylated with an acyl halide and a Lewis-acid catalyst in what is essentially a Friedel–Crafts reaction at an aliphatic carbon.[197] The product can arise by two paths. The initial attack is by the acyl cation RCO$^+$ (or by the acyl halide free or complexed; see **1-14**) at the double bond to give a carbocation:

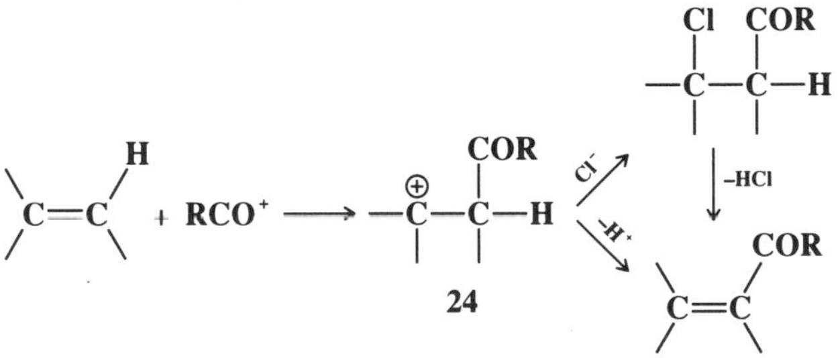

$$\overset{\backslash}{\underset{/}{C}}=\overset{H}{\underset{\backslash}{C}} + RCO^+ \longrightarrow -\overset{\oplus}{\underset{|}{C}}-\overset{COR}{\underset{|}{C}}-H$$

24

with two paths leading to products:

—C—C—H (Cl, COR)
$$\downarrow -HCl$$
$$\overset{\backslash}{\underset{/}{C}}=\overset{COR}{\underset{\backslash}{C}}$$

[194]Trost; Hiroi; Kurozumi *J. Am. Chem. Soc.* **1975,** *97,* 438.
[195]There are numerous other ways of achieving this conversion. For reviews, see Morris *Chem. Soc. Rev.* **1982,** *11,* 397-434; Kane; Singh; Martin; Doyle *Tetrahedron* **1983,** *39,* 345-394.
[196]For a review, see Gilbert *Sulfonation and Related Reactions*; Wiley: New York, 1965, pp. 33-61.
[197]For reviews, see Groves *Chem. Soc. Rev.* **1972,** *1,* 73-97; Satchell; Satchell in Patai *The Chemistry of the Carbonyl Group,* vol. 1; Wiley: New York, 1966, pp. 259-266, 270-273; Nenitzescu; Balaban, in Olah *Friedel–Crafts and Related Reactions,* vol. 3; Wiley: New York, 1964, pp. 1033-1152.

Ion **24** can either lose a proton or combine with chloride ion. If it loses a proton, the product is an unsaturated ketone; the mechanism is similar to the tetrahedral mechanism of Chapter 10, but with the charges reversed. If it combines with chloride, the product is a β-halo ketone, which can be isolated, so that the result is addition to the double bond (see **5-34**). On the other hand, the β-halo ketone may, under the conditions of the reaction, lose HCl to give the unsaturated ketone, this time by an addition–elimination mechanism. In the case of unsymmetrical olefins, the attacking ion prefers the position at which there are more hydrogens, following Markovnikov's rule (p. 750). Anhydrides and carboxylic acids (the latter with a proton acid such as anhydrous HF, H_2SO_4, or polyphosphoric acid as a catalyst) are sometimes used instead of acyl halides. With some substrates and catalysts double-bond migrations are occasionally encountered so that, for example, when 1-methylcyclohexene was acylated with acetic anhydride and zinc chloride, the major product was 6-acetyl-1-methylcyclohexene.[198]

Conjugated dienes can be acylated by treatment with acyl- or alkylcobalt tetracarbonyls, followed by base-catalyzed cleavage of the resulting π-allyl carbonyl derivatives.[199] The

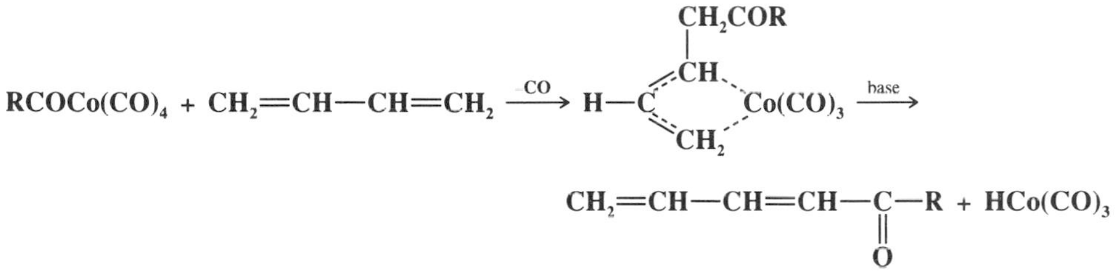

$$\text{RCOCo(CO)}_4 + \text{CH}_2{=}\text{CH}{-}\text{CH}{=}\text{CH}_2 \xrightarrow{\text{CO}} \text{H}{-}\underset{\substack{\\ \text{CH}_2}}{\overset{\substack{\text{CH}_2\text{COR} \\ | \\ \text{CH}}}{\text{C}}}{\cdots}\text{Co(CO)}_3 \xrightarrow{\text{base}}$$

$$\text{CH}_2{=}\text{CH}{-}\text{CH}{=}\text{CH}{-}\underset{\substack{|| \\ \text{O}}}{\text{C}}{-}\text{R} + \text{HCo(CO)}_3$$

reaction is very general. With unsymmetrical dienes, the acyl group generally substitutes most readily at a cis double bond, next at a terminal olefinic group, and least readily at a trans double bond. The most useful bases are strongly basic, hindered amines such as dicyclohexylethylamine. The use of an alkylcobalt tetracarbonyl $RCo(CO)_4$ gives the same product as that shown above. Acylation of vinylic ethers has been accomplished with aromatic acyl chlorides, a base, and a palladium catalyst: $ROCH{=}CH_2 \rightarrow ROCH{=}CHCOAr$.[200]

Formylation of olefins can be accomplished with N-disubstituted formamides and $POCl_3$.[201] This is an aliphatic Vilsmeier reaction (see **1-15**). Vilsmeier formylation can also be performed on the α position of acetals and ketals, so that hydrolysis of the products gives keto aldehydes or dialdehydes:[202]

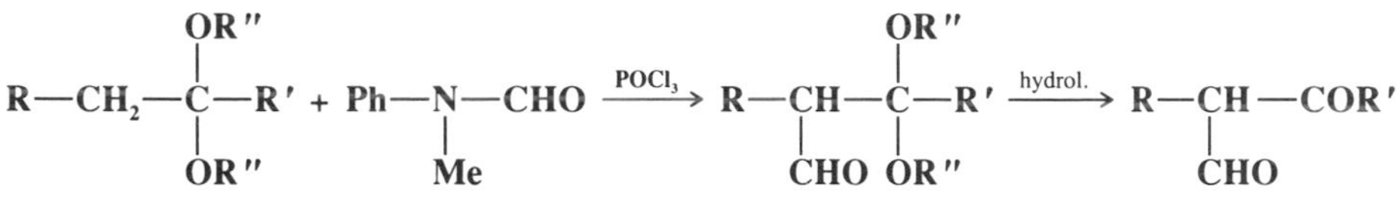

$$\text{R}{-}\text{CH}_2{-}\underset{\substack{| \\ \text{OR}''}}{\overset{\substack{\text{OR}'' \\ |}}{\text{C}}}{-}\text{R}' + \text{Ph}{-}\underset{\substack{| \\ \text{Me}}}{\text{N}}{-}\text{CHO} \xrightarrow{\text{POCl}_3} \text{R}{-}\underset{\substack{| \\ \text{CHO}}}{\overset{\substack{\text{OR}'' \\ |}}{\text{CH}}}{-}\underset{\substack{| \\ \text{OR}''}}{\text{C}}{-}\text{R}' \xrightarrow{\text{hydrol.}} \text{R}{-}\underset{\substack{| \\ \text{CHO}}}{\text{CH}}{-}\text{COR}'$$

Acetylation of acetals or ketals can be accomplished with acetic anhydride and BF_3–etherate.[203] The mechanism with acetals or ketals also involves attack at an olefinic carbon,

[198]Deno; Chafetz *J. Am. Chem. Soc.* **1952,** *74*, 3940. For other examples, see Beak; Berger *J. Am. Chem. Soc.* **1980,** *102*, 3848; Dubois; Saumtally; Lion *Bull. Soc. Chim. Fr.* **1984,** II-133; Grignon-Dubois; Cazaux *Bull. Soc. Chim. Fr.* **1986,** 332.
[199]For a review, see Heck, in Wender; Pino *Organic Syntheses via Metal Carbonyls*, vol. 1; Wiley: New York, 1968, pp. 388-397.
[200]Andersson; Hallberg *J. Org. Chem.* **1988,** *53*, 4257.
[201]For reviews, see Burn *Chem. Ind.* (*London*) **1973,** 870-873; Satchell; Satchell, Ref. 197, pp. 281-282.
[202]Youssefyeh *Tetrahedron Lett.* **1964,** 2161.
[203]Youssefyeh *J. Am. Chem. Soc.* **1963,** *85*, 3901.

since enol ethers are intermediates.[203] Ketones can be formylated in the α position by treatment with CO and a strong base.[204]

OS **IV**, 555, 560; **VI**, 744. Also see OS **VI**, 28.

2-16 Conversion of Aldehydes to β-Keto Esters or Ketones
Alkoxycarbonylalkylation or Alkoxycarbonylalkyl-de-hydrogenation

$$\text{R−C−H} + \text{EtOOC−CHN}_2 \xrightarrow[\text{CH}_2\text{Cl}_2]{\text{SnCl}_2} \text{R−C−CH}_2\text{−COOEt}$$
$$\text{O} \qquad\qquad\qquad\qquad\qquad \text{O}$$

β-Keto esters have been prepared in moderate to high yields by treatment of aldehydes with diethyl diazoacetate in the presence of a catalytic amount of a Lewis acid such as $SnCl_2$, BF_3, or $GeCl_2$.[205] The reaction was successful for both aliphatic and aromatic aldehydes, but the former react more rapidly than the latter, and the difference is great enough to allow selective reactivity. In a similar process, aldehydes react with certain carbanions stabilized by boron, in the presence of $(F_3CCO)_2O$ or N-chlorosuccinimide, to give ketones.[206]

$$\text{R−C−H} + \overset{\ominus}{\text{C}}\text{H−R}' \xrightarrow[\text{or NCS}]{(\text{F}_3\text{CCO})_2\text{O}} \text{R−C−CH}_2\text{R}' \qquad \text{Ar = mesityl}$$
$$\text{O} \qquad \text{BAr}_2 \qquad\qquad\qquad \text{O}$$

2-17 Cyanation or Cyano-de-hydrogenation

$$\text{R−C−CH} \xrightarrow[\text{2. TsCN}]{\text{1. LDA−THF}} \text{R−C−C−CN}$$
$$\text{O} \qquad\qquad\qquad\qquad \text{O}$$

Introduction of a cyano group α to the carbonyl group of a ketone can be accomplished by prior formation of the enolate with lithium diisopropylamide (LDA) in THF and addition of this solution to p-TsCN at −78°C.[207] The products are formed in moderate to high yields. The reaction is not applicable to methyl ketones. In a different kind of reaction, nitro compounds are α cyanated by treatment with CN^- and $K_3Fe(CN)_6$.[208] The mechanism probably involves ion radicals. In still another reaction, secondary amines are converted to α-cyanoamines by treatment with phenylseleninic anhydride and NaCN or Me_3SiCN.[209] Me_3SiCN has also been used in a reaction that cyanates benzylic positions.[210]

2-18 Alkylation of Alkanes
Alkylation or Alkyl-de-hydrogenation

$$\text{RH} + \text{R}'^+ \longrightarrow \text{R−R}' + \text{H}^+$$

Alkanes can be alkylated by treatment with solutions of stable carbocations[211] (p. 166), though the reaction is not generally useful for synthesis. Mixtures are usually obtained. In

[204]See, for example, van der Zeeuw; Gersmann *Recl. Trav. Chim. Pays-Bas* **1965**, *84*, 1535.
[205]Holmquist; Roskamp *J. Org. Chem.* **1989**, *54*, 3258.
[206]Pelter; Smith; Elgendy; Rowlands *Tetrahedron Lett.* **1989**, *30*, 5643.
[207]Kahne; Collum *Tetrahedron Lett.* **1981**, *22*, 5011.
[208]Matacz; Piotrowska; Urbanski *Pol. J. Chem.* **1979**, *53*, 187; Kornblum; Singh; Kelly *J. Org. Chem.* **1983**, *48*, 332.
[209]Barton; Billion; Boivin *Tetrahedron Lett.***1985**, *26*, 1229.
[210]Lemaire; Doussot; Guy *Chem. Lett.* **1988**, 1581. See also Hayashi; Mukaiyama *Chem. Lett.* **1987**, 1811.
[211]Olah; Mo; Olah *J. Am. Chem. Soc.* **1973**, *95*, 4939. For reviews, see Olah; Farooq; Prakash, in Hill *Activation and Functionalization of Alkanes*; Wiley: New York, 1989, pp. 27-78; Olah; Prakash; Sommer, Ref. 44, pp. 270-277. For a review of the thermodynamic behavior of alkanes in super-acid media, see Fabre; Devynck; Trémillon *Chem. Rev.* **1982**, *82*, 591-614. See also Ref. 46.

a typical experiment, the treatment of propane with isopropyl fluoroantimonate (Me_2C^+ SbF_6^-) gave 26% 2,3-dimethylbutane, 28% 2-methylpentane, 14% 3-methylpentane, and 32% *n*-hexane, as well as some butanes, pentanes (formed by **2-47**), and higher alkanes. Mixtures arise in part because intermolecular hydrogen exchange ($RH + R'^+ \rightleftharpoons R^+ + R'H$) is much faster than alkylation, so that alkylation products are also derived from the new alkanes and carbocations formed in the exchange reaction. Furthermore, the carbocations present are subject to rearrangement (Chapter 18), giving rise to new carbocations. Products result from all the hydrocarbons and carbocations present in the system. As expected from their relative stabilities, secondary alkyl cations alkylate alkanes more readily than tertiary alkyl cations (the *t*-butyl cation does not alkylate methane or ethane). Stable primary alkyl cations are not available, but alkylation has been achieved with complexes formed between CH_3F or C_2H_5F and SbF_5.[212] The mechanism of alkylation can be formulated (similar to that shown in hydrogen exchange with super acids, **2-1**) as

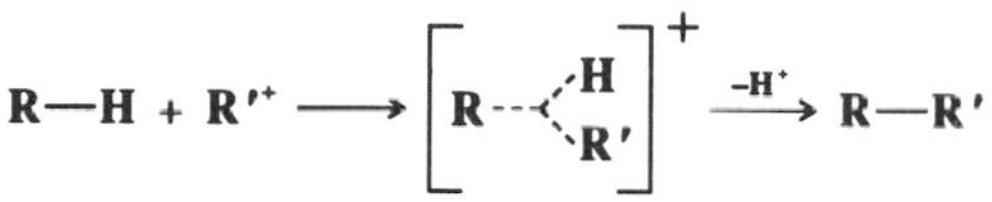

$$R{-}H + R'^+ \longrightarrow \left[R\cdots \overset{H}{\underset{R'}{<}} \right]^+ \xrightarrow{-H^+} R{-}R'$$

It is by means of successive reactions of this sort that simple alkanes like methane and ethane give *t*-butyl cations in super-acid solutions (p. 168).[213]

Intramolecular insertion has been reported. The positively charged carbon of the carbocation **26,** generated from the diazonium salt of the triptycene compound **25,** reacted with the CH_3 group in close proximity with it.[214]

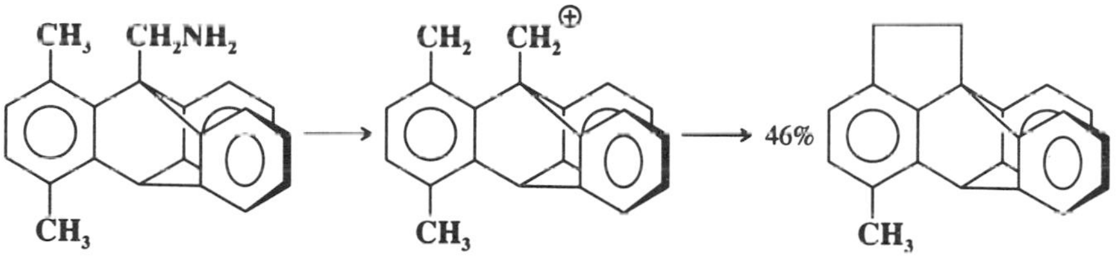

    25   **26**

2-19   The Stork Enamine Reaction
   α-**Acylalkyl-de-halogenation**[215]

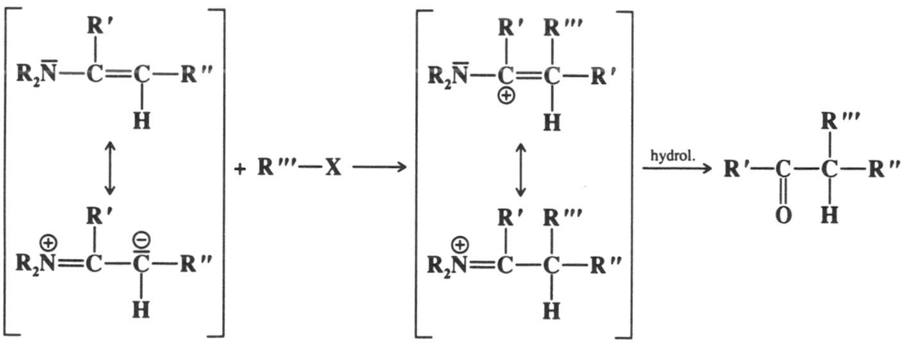

---

[212]Olah; DeMember; Shen *J. Am. Chem. Soc.* **1973,** *95,* 4952. See also Sommer; Muller; Laali *Nouv. J. Chem.* **1982,** *6,* 3.
[213]For example, see Hogeveen; Roobeek *Recl. Trav. Chim. Pays-Bas* **1972,** *91,* 137.
[214]Yamamoto; Ōki *Chem. Lett.* **1987,** 1163.
[215]This is the IUPAC name with respect to the halide as substrate.

When enamines are treated with alkyl halides, an alkylation occurs that is analogous to the first step of **2-15**. Hydrolysis of the imine salt gives a ketone. Since the enamine is normally formed from a ketone (**6-14**), the net result is alkylation of the ketone at the α position. The method, known as the *Stork enamine reaction*,[216] is an alternative to the ketone alkylation considered at **0-95**. The Stork method has the advantage that it generally leads almost exclusively to monoalkylation of the ketone, while **0-95**, when applied to ketones, is difficult to stop with the introduction of just one alkyl group. Alkylation usually takes place on the less substituted side of the original ketone. The most commonly used amines are the cyclic amines piperidine, morpholine, and pyrrolidine.

The method is quite useful for particulary active alkyl halides such as allylic, benzylic, and propargylic halides, and for α-halo ethers and esters, but is not very serviceable for ordinary primary and secondary halides. Tertiary halides do not give the reaction at all since, with respect to the halide, this is nucleophilic substitution and elimination predominates. The reaction can also be applied to activated aryl halides (such as 2,4-dinitrochlorobenzene; see Chapter 13), to epoxides,[217] and to activated olefins such as acrylonitrile, e.g.,

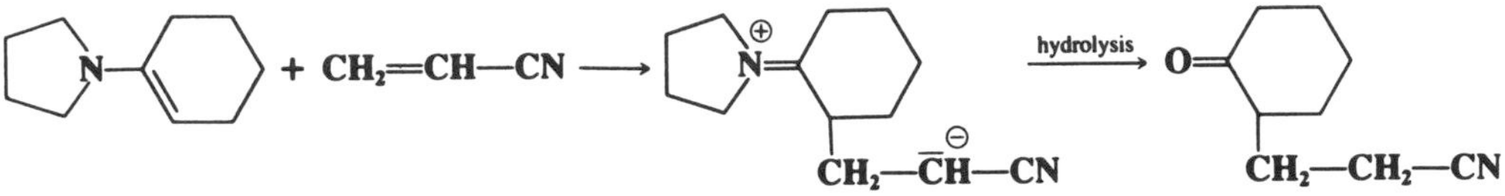

The latter is a Michael-type reaction (p. 742) with respect to the olefin.

Acylation[218] can be accomplished with acyl halides:

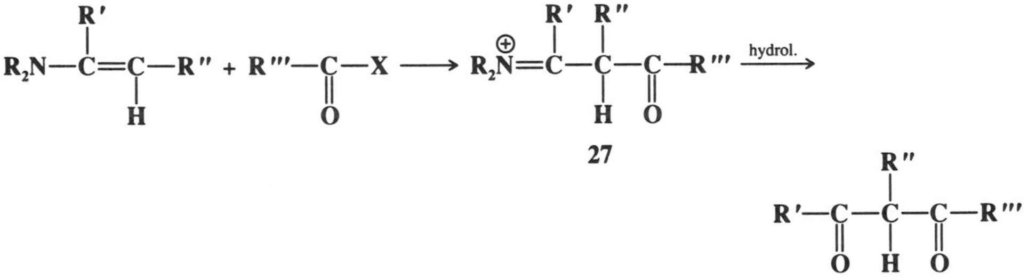

or with anhydrides. A COOEt group can be introduced by treatment of the enamine with ethyl chloroformate ClCOOEt,[219] a CN group with cyanogen chloride[220] (not cyanogen bromide or iodide, which leads to halogenation of the enamine), a CHO group with the mixed anhydride of formic and acetic acids[219] or with DMF and phosgene,[221] and a

[216]Stork; Brizzolara; Landesman; Szmuszkovicz; Terrell *J. Am. Chem. Soc.* **1963**, *85*, 207. For general reviews of enamines, see Hickmott *Tetrahedron* **1984**, *40*, 2989-3051, **1982**, *38*, 1975-2050, 3363-3446; Granik *Russ. Chem. Rev.* **1984**, *53*, 383-400. For reviews of this reaction, see in Cook *Enamines*, 2nd ed.; Marcel Dekker: New York, 1988, the articles by Alt; Cook pp. 181-246, and Gadamasetti; Kuehne, pp. 531-689; Whitesell; Whitesell *Synthesis* **1983**, 517-536; Kuehne *Synthesis* **1970**, 510-537; House, Ref. 86, pp. 570-582, 766-772; Bláha; Červinka *Adv. Heterocycl. Chem.* **1966**, *6*, 147-227, pp. 186-204.
[217]Britten; Owen; Went *Tetrahedron* **1969**, *25*, 3157.
[218]For reviews, see Hickmott *Chem. Ind. (London)* **1974**, 731; Hünig; Hoch *Fortschr. Chem. Forsch.* **1970**, *14*, 235.
[219]Stork et al., Ref. 216.
[220]Kuehne *J. Am. Chem. Soc.* **1959**, *81*, 5400.
[221]Ziegenbein *Angew. Chem. Int. Ed. Engl.* **1965**, *4*, 358 [*Angew. Chem. 77*, 380].

$C(R)\!=\!NR'$ group with a nitrilium salt $RC\!\!\equiv\!\!\overset{\oplus}{N}R'$.[222] The acylation of the enamine can take place by the same mechanism as alkylation, but another mechanism is also possible, if the acyl halide has an α hydrogen and if a tertiary amine is present, as it often is (it is added to neutralize the HX given off). In this mechanism, the acyl halide is dehydrohalogenated by the tertiary amine, producing a ketene (**7-14**) which adds to the enamine to give a cyclobutanone (**5-49**). This compound can be cleaved in the solution to form the same acylated imine salt (**27**) that would form by the more direct mechanism, or it can be isolated (in the case of enamines derived from aldehydes), or it may cleave in other ways.[223]

Primary and secondary halides do not perform well, mostly because N-alkylation becomes important, particularly with enamines derived from aldehydes. An alternative method, which gives good yields of alkylation with primary and secondary halides, is alkylation of enamine *salts*, which are prepared by treating an imine with ethylmagnesium bromide in THF:[224]

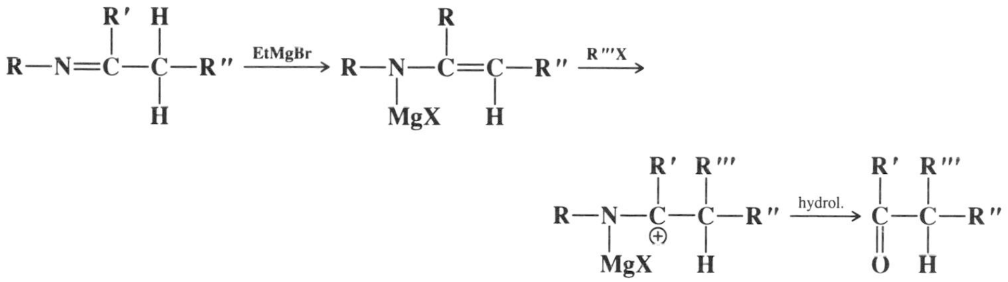

The imines are prepared by **6-14.** The enamine salt method has also been used to give good yields of mono α alkylation of α,β-unsaturated ketones.[225] Enamines prepared from aldehydes and butylisobutylamine can be alkylated by simple primary alkyl halides in good yields.[226] N-alkylation in this case is presumably prevented by steric hindrance.

When the nitrogen of the substrate contains a chiral R group, both the Stork enamine synthesis and the enamine salt method can be used to perform enantioselective syntheses, and this has often been done.[227]

OS **V,** 533, 869; **VI,** 242, 496, 526; **VII,** 473.

2-20   Insertion by Carbenes
    CH-**Methylene-insertion**

$$RH + \overline{C}H_2 \longrightarrow RCH_3$$

The highly reactive species methylene inserts into C—H bonds,[228] both aliphatic and aromatic,[229] though with aromatic compounds ring expansion is also possible (see **5-50**). The reaction is useless for synthetic purposes because of its nonselectivity (see p. 199). Alkyl-

[222]Baudoux; Fuks *Bull. Soc. Chim. Belg.* **1984,** *93,* 1009.
[223]See Alt; Cook, Ref. 216, pp. 204-215.
[224]Stork; Dowd *J. Am. Chem. Soc.* **1963,** *85,* 2178.
[225]Stork; Benaim *J. Am. Chem. Soc.* **1971,** *93,* 5938.
[226]Curphey; Hung; Chu *J. Org. Chem.* **1975,** *40,* 607. See also Ho; Wong *Synth. Commun.* **1974,** *4,* 147.
[227]For reviews, see Nóagrádi *Stereoselective Synthesis*; VCH: New York, 1986, pp. 248-255; Whitesell *Acc. Chem. Res.* **1985,** *18,* 280-284; Bergbreiter; Newcomb, in Morrison, Ref. 68, vol. 2, 1983, pp. 243-273.
[228]First reported by Meerwein; Rathjen; Werner *Ber.* **1942,** *75,* 1610. For reviews, see Bethell, in McManus *Organic Reactive Intermediates*; Academic Press: New York, 1973, pp. 92-101; Kirmse *Carbene Chemistry*, 2nd ed.; Academic Press: New York, 1971, pp. 209-266.
[229]Terao; Shida *Bull. Chem. Soc. Jpn.* **1964,** *37,* 687.

carbenes usually rearrange rather than give insertion (p. 201), but, when this is impossible, *intramolecular* insertion[230] is found rather than intermolecular.[231]

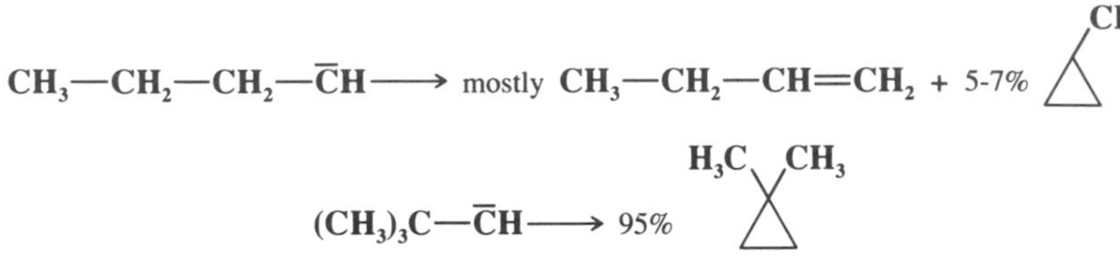

$$CH_3-CH_2-CH_2-\overline{C}H \longrightarrow \text{ mostly } CH_3-CH_2-CH=CH_2 + \text{5-7\%}$$

$$(CH_3)_3C-\overline{C}H \longrightarrow \text{95\%}$$

CH_2 generated by photolysis of CH_2N_2 in the liquid phase is indiscriminate—totally non-selective—in its reactivity (p. 199). CH_2 generated in other ways and other carbenes are less reactive and insert in the order tertiary > secondary > primary.[232] Halocarbenes insert much less readily, though a number of instances have been reported.[233] Nevertheless, even for less reactive carbenes, the insertion reaction has seldom been used for synthetic purposes.[234] The carbenes can be generated in any of the ways mentioned in Chapter 5 (p. 198). For the similar insertion of nitrenes, see **2-12**.

The mechanism[235] of the insertion reaction is not known with certainty, but there seem to be at least two possible pathways.

1. A simple one-step process involving a three-center cyclic transition state:

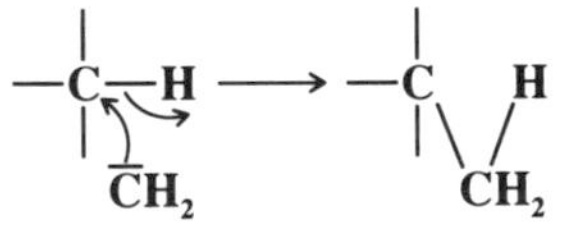

$$-\underset{\underset{CH_2}{|}}{\overset{|}{C}}\text{-} H \longrightarrow -\underset{\underset{CH_2}{|}}{\overset{|}{C}} \quad H$$

The most convincing evidence for this mechanism is that in the reaction between isobutene-1-^{14}C and carbene the product 2-methyl-1-butene was labeled only in the 1 position.[236] This rules out a free radical or other free intermediate such as a carbocation or carbanion. If **28** (or a corresponding ion) were an intermediate, resonance would ensure that some carbene attacked at the 1 position:

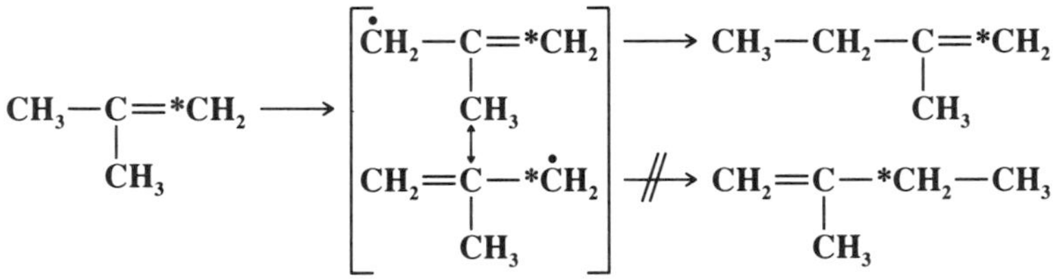

$$CH_3-C=*CH_2 \longrightarrow \left[\begin{array}{c} \overset{.}{C}H_2-C=*CH_2 \\ | \\ CH_3 \\ \uparrow\downarrow \\ CH_2=\overset{.}{C}-*\overset{.}{C}H_2 \\ | \\ CH_3 \end{array} \right] \begin{array}{c} \longrightarrow CH_3-CH_2-C=*CH_2 \\ | \\ CH_3 \\ \\ \nrightarrow CH_2=C-*CH_2-CH_3 \\ | \\ CH_3 \end{array}$$

$$\underset{\text{28}}{} \qquad \qquad \underset{\text{Not found}}{}$$

[230]Kirmse; Doering *Tetrahedron* **1960**, *11*, 266; Friedman; Berger *J. Am. Chem. Soc.* **1961**, *83*, 492, 500.

[231]For a review of the intramolecular insertions of carbenes or carbenoids generated from diazocarbonyl compounds, see Burke; Grieco *Org. React.* **1979**, *26*, 361-475.

[232]Doering; Knox *J. Am. Chem. Soc.* **1961**, *83*, 1989.

[233]For example, see Parham; Koncos *J. Am. Chem. Soc.* **1961**, *83*, 4034; Fields *J. Am. Chem. Soc.* **1962**, *82*, 1744; Anderson; Lindsay; Reese *J. Chem. Soc.* **1964**, 4874; Seyferth; Cheng *J. Am. Chem. Soc.* **1973**, *95*, 6763, *Synthesis* **1974**, 114; Steinbeck *Tetrahedron Lett.* **1978**, 1103; Boev *J. Org. Chem. USSR* **1981**, *17*, 1190.

[234]For some examples of intramolecular carbene insertions used synthetically, see Gilbert; Giamalva; Weerasooriya *J. Org. Chem.* **1983**, *48*, 5251; Taber; Ruckle *J. Am. Chem. Soc.* **1986**, *108*, 7686; Paquette; Kobayashi; Gallucci *J. Am. Chem. Soc.* **1988**, *110*, 1305; Adams; Poupart; Grenier; Schaller; Ouimet; Frenette *Tetrahedron Lett.* **1989**, *30*, 1749; Doyle; Bagheri; Pearson; Edwards *Tetrahedron Lett.* **1989**, *30*, 7001.

[235]For a discussion, see Bethell, *Adv. Phys. Org. Chem.* **1969**, *7*, 153-209, pp. 190-194.

[236]Doering; Prinzbach *Tetrahedron* **1959**, *6*, 24.

Other evidence is that retention of configuration, which is predicted by this mechanism, has been found in a number of instances.[237]

2. A free-radical process in which the carbene directly abstracts a hydrogen from the substrate to generate a pair of free radicals:

$$\textbf{RH} + \textbf{CH}_2 \longrightarrow \textbf{R} \cdot + \textbf{CH}_3 \cdot$$

$$\textbf{R} \cdot + \textbf{CH}_3 \cdot \longrightarrow \textbf{RCH}_3$$

One fact supporting this mechanism is that among the products obtained (beside butane and isobutane) on treatment of propane with CH_2 (generated by photolysis of diazomethane and ketene) were propene and ethane,[238] which could arise, respectively, by

$$\textbf{2CH}_3\textbf{CH}_2\textbf{CH}_2 \cdot \longrightarrow \textbf{CH}_3\textbf{CH}{=}\textbf{CH}_2 + \textbf{CH}_3\textbf{CH}_2\textbf{CH}_3 \quad \text{(disproportionation)}$$

and

$$\textbf{CH}_3\textbf{CH}_2\textbf{CH}_3 + \overline{\textbf{C}}\textbf{H}_2 \longrightarrow \textbf{CH}_3\textbf{CH}_2\textbf{CH}_2 \cdot + \textbf{CH}_3 \cdot$$

$$\textbf{2CH}_3 \cdot \longrightarrow \textbf{CH}_3\textbf{CH}_3$$

That this mechanism can take place under suitable conditions has been demonstrated by isotopic labeling[239] and by other means.[240] However, the obtention of disproportionation and dimerization products does not always mean that the free-radical abstraction process takes place. In some cases these products arise in a different manner.[241] We have seen that the product of the reaction between a carbene and a molecule may have excess energy (p. 197). Therefore it is possible for the substrate and the carbene to react by mechanism 1 (the direct-insertion process) and for the excess energy to cause the compound thus formed to cleave to free radicals. When this pathway is in operation, the free radicals are formed *after* the actual insertion reaction.

It has been suggested[242] that singlet carbenes insert by the one-step direct-insertion process and triplets (which, being free radicals, are more likely to abstract hydrogen) by the free-radical process. In support of this suggestion is that CIDNP signals[243] (p. 187) were observed in the ethylbenzene produced from toluene and triplet CH_2, but not from the same reaction with singlet CH_2.[244] Carbenoids (e.g., compounds of the form R_2CMCl—see **2-39**) can insert into a C—H bond by a different mechanism, similar to pathway 2, but involving abstraction of a hydride ion rather than a hydrogen atom.[245]

The reaction in which aldehydes are converted to methyl ketones, $RCHO + CH_2N_2 \rightarrow RCOCH_3$, while apparently similar, does not involve a free carbene intermediate. It is considered in Chapter 18 (**8-9**).

OS **VII**, 200.

---

[237]See, for example, Kirmse; Buschhoff *Chem. Ber.* **1969**, *102*, 1098; Seyferth; Cheng *J. Am. Chem. Soc.* **1971**, *93*, 4072.

[238]Frey *Proc. Chem. Soc.* **1959**, 318.

[239]Halberstadt; McNesby *J. Chem. Phys.* **1966**, *45*, 1666; McNesby; Kelly *Int. J. Chem. Kinet.* **1971**, *3*, 293.

[240]Ring; Rabinovitch *J. Am. Chem. Soc.* **1966**, *88*, 4285; *Can J. Chem.* **1968**, *46*, 2435.

[241]Bell *Prog. Phys. Org. Chem.* **1964**, *2*, 1-61, pp. 30-43.

[242]Richardson; Simmons; Dvoretzky *J. Am. Chem. Soc.* **1961**, *83*, 1934.

[243]For a review of the use of CIDNP to study carbene mechanisms, see Roth *Acc. Chem. Res.* **1977**, *10*, 85-91.

[244]Roth *J. Am. Chem. Soc.* **1972**, *94*, 1761. See also Closs; Closs *J. Am. Chem. Soc.* **1969**, *91*, 4549; Bethell; McDonald *J. Chem. Soc., Perkin Trans. 2* **1977**, 671.

[245]See Harada; Nozaki; Yamaura; Oku *J. Am. Chem. Soc.* **1985**, *107*, 2189; Oku; Yamaura; Harada *J. Org. Chem.* **1986**, *51*, 3730; Ritter; Cohen *J. Am. Chem. Soc.* **1986**, *108*, 3718.

F. Metal Electrophiles

2-21   Metallation with Organometallic Compounds
Metallation or Metallo-de-hydrogenation

$$RH + R'M \longrightarrow RM + R'H$$

Many organic compounds can be metallated by treatment with an organometallic compound.[246] Since the reaction involves a proton transfer, the equilibrium lies on the side of the weaker acid. For example, fluorene reacts with butyllithium to give butane and 9-fluoryllithium. Since aromatic hydrocarbons are usually stronger acids than aliphatic ones, R is most often aryl. The most common reagent is butyllithium.[247] Normally, only active aromatic rings react with butyllithium. Benzene itself is not reactive enough, though benzene can be metallated by butyllithium either in the presence of t-BuOK[248] or coordinated with various diamines.[249] Metallation of aliphatic RH is most successful when the carbanions are stabilized by resonance (allylic, benzylic, propargylic,[250] etc.) or when the negative charge is at an sp carbon (at triple bonds). Very good reagents for allylic metallation are trimethylsilylmethyl potassium Me_3SiCH_2K[251] and a combination of an organolithium compound with a bulky alkoxide (LICKOR superbase).[252] The former is also useful for benzylic positions. A combination of BuLi, t-BuOK, and tetramethylethylenediamine has been used to convert ethylene to vinylpotassium.[253] In certain cases *gem*-dialkali metal or 1,1,1-trialkali metal compounds can be prepared.[254] Examples are the conversion of phenylacetonitrile to 1,1-dilithiophenylacetonitrile $PhCLi_2CN$[255] and propyne to tetralithiopropyne $Li_3CC\equiv CLi$[256] in each case by treatment with excess butyllithium. The reaction can be used to determine relative acidities of very weak acids by allowing two R—H compounds to compete for the same R'M and to determine which proton in a molecule is the most acidic.[257]

In general, the reaction can be performed only with organometallics of active metals such as lithium, sodium, and potassium, but Grignard reagents abstract protons from a sufficiently acidic C—H bond, as in R—C≡C—H → R—C≡C—MgX. This is the best method for the preparation of alkynyl Grignard reagents.[258]

[246]For reviews, see Wardell, in Zuckerman *Inorganic Reactions and Methods*, vol. 11; VCH: New York, 1988, pp. 44-107; Wardell, in Hartley, Patai, Ref. 1, vol. 4, pp. 1-157, pp. 27-71; Narasimhan; Mali *Synthesis* **1983**, 957-986; Biellmann; Ducep *Org. React.* **1982**, *27*, 1-344; Gschwend; Rodriguez *Org. React.* **1979**, *26*, 1-360; Mallan; Bebb *Chem. Rev.* **1969**, *69*, 693-755.

[247]For a review, see Durst, in Buncel; Durst *Comprehensive Carbanion Chemistry*, vol. 5, pt. B; Elsevier: New York, 1984, pp. 239-291, pp. 265-279. For an article on the safe handling of RLi compounds, see Anderson *Chem. Ind. (London)* **1984**, 205.

[248]Schlosser *J. Organomet. Chem.* **1967**, *8*, 9. See also Schlosser; Katsoulos; Takagishi *Synlett* **1990**, 747.

[249]Eberhardt; Butte *J. Org. Chem.* **1964**, *29*, 2928; Langer *Trans. N.Y. Acad. Sci.* **1965**, *27*, 741; Eastham; Screttas *J. Am. Chem. Soc.* **1965**, *87*, 3276; Rausch; Ciappenelli *J. Organomet. Chem.* **1967**, *10*, 127.

[250]For a review of directive effects in allylic and benzylic metallation, see Klein *Tetrahedron* **1983**, *39*, 2733-2759. For a review of propargylic metallation, see Klein, in Patai *The Chemistry of the Carbon–Carbon Triple Bond*, pt. 1; Wiley: New York, 1978, pp. 343-379.

[251]Hartmann; Schlosser *Helv. Chim. Acta* **1976**, *59*, 453.

[252]Schlosser *Pure Appl. Chem.* **1988**, *60*, 1627. For sodium analogs, see Schlosser; Hartmann; Stähle; Kramař; Walde; Mordini *Chimia* **1986**, *40*, 306.

[253]Brandsma; Verkruijsse; Schade; Schleyer *J. Chem. Soc., Chem. Commun.* **1986**, 260.

[254]For a review of di and polylithium compounds, see Maercker; Theis *Top. Curr. Chem.* **1987**, *138*, 1-61.

[255]Kaiser; Solter; Schwartz; Beard; Hauser *J. Am. Chem. Soc.* **1971**, *93*, 4237. See also Kowalski; O'Dowd; Burke; Fields *J. Am. Chem. Soc.* **1980**, *102*, 5411.

[256]Priester; West *J. Am. Chem. Soc.* **1976**, *98*, 8421, 8426 and references cited therein.

[257]For examples, see Broaddus; Logan; Flautt *J. Org. Chem.* **1963**, *28*, 1174; Finnegan; McNees *J. Org. Chem.* **1964**, *29*, 3234; Shirley; Hendrix *J. Organomet. Chem.* **1968**, *11*, 217.

[258]For a review of the synthetic applications of metallation by Grignard reagents at positions other than at triple bonds, see Blagoev; Ivanov *Synthesis* **1970**, 615-628.

When a hetero atom, such as N, O, S,[259] or a halogen,[260] is present in a molecule containing an aromatic ring or a double bond, lithiation is usually quite regioselective.[261] The lithium usually bonds with the sp^2 carbon closest to the hetero atom, probably because the attacking species coordinates with the hetero atom.[262] In the case of aromatic rings this means attack at the ortho position.[263] Two examples are

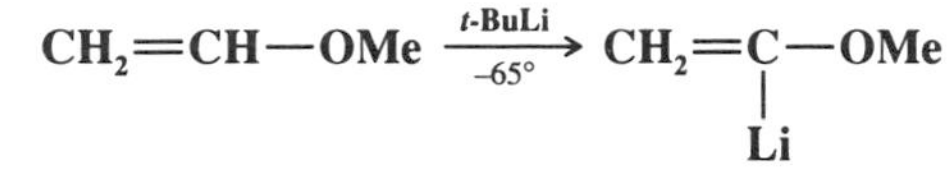

$$CH_2{=}CH{-}OMe \xrightarrow[-65°]{\textit{t}-BuLi} CH_2{=}\overset{|}{\underset{Li}{C}}{-}OMe \qquad\qquad \text{Ref. 264}$$

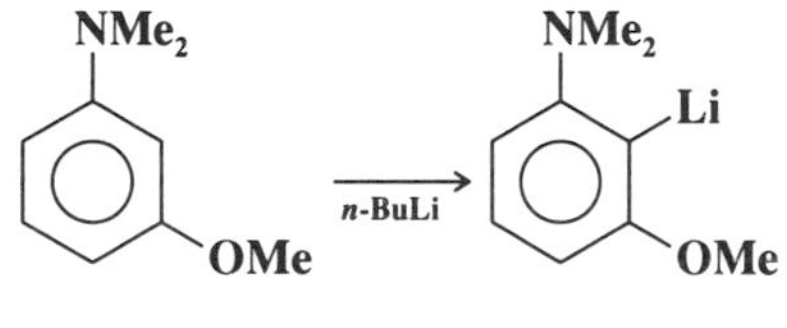

Ref. 265

In the second example, the lithium goes into the 2 position so as to be ortho to both substituents.[266] This regioselectivity can be quite valuable synthetically. In the case of γ,δ-unsaturated disubstituted amides (**29**),the lithium does not go to the closest position, but in

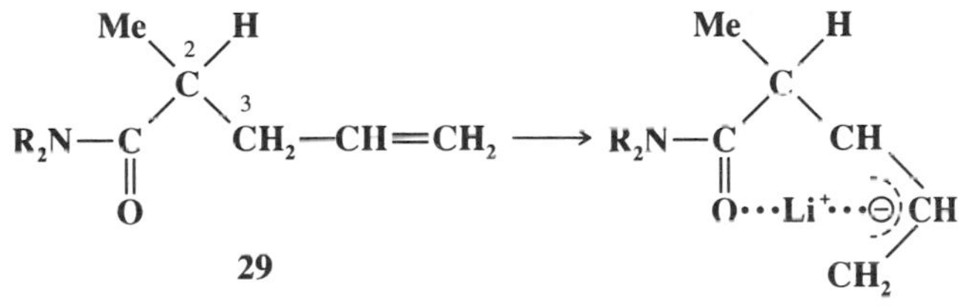

this case too the regiochemistry is controlled by coordination to the oxygen.[267] The 2 position is much more acidic than the 3 position (Table 8.1), but a negative charge at C-3 is in a more favorable position to be stabilized by the Li^+. Ortho magnesiation has been accomplished with bases of the form $(R_2N)_2Mg$.[268]

The mechanism involves a nucleophilic attack by R'^- (or a polar R') on the *hydrogen*.[269] Evidence is that resonance effects of substituents in R seem to make little difference. When

[259]For example, see Figuly; Loop; Martin *J. Am. Chem. Soc.* **1989**, *111*, 654; Block; Eswarakrishnan; Gernon; Ofori-Okai; Saha; Tang; Zubieta *J. Am. Chem. Soc.* **1989**, *111*, 658; Smith; Lindsay; Pritchard *J. Am. Chem. Soc.* **1989**, *111*, 665.

[260]Fluorine is an especially powerful ortho director in lithiation of aromatic systems: Gilday; Negri; Widdowson *Tetrahedron* **1989**, *45*, 4605.

[261]For a review of regioselective lithiation of heterocycles, see Katritzky; Lam; Sengupta *Prog. Heterocycl. Chem.* **1989**, *1*, 1-29.

[262]For many examples with references, see Ref. 246; Beak; Meyers *Acc. Chem. Res.* **1986**, *19*, 356-363; Beak; Snieckus *Acc. Chem. Res.* **1982**, *15*, 306-312; Snieckus *Bull. Soc. Chim. Fr.* **1988**, 67-78; Narasimhan; Mali *Top. Curr. Chem.* **1987**, *138*, 63-147; Reuman; Meyers *Tetrahedron* **1985**, *41*, 837-860; and the papers in *Tetrahedron* **1983**, *39*, 1955-2091.

[263]For reviews of ortho metallation, see Snieckus *Chem. Rev.* **1990**, *90*, 879-933, *Pure Appl. Chem.* **1990**, *62*, 2047-2056. For a discussion of the mechanism, see Bauer; Schleyer *J. Am. Chem. Soc.* **1989**, *111*, 7191.

[264]Baldwin; Höfle; Lever *J. Am. Chem. Soc.* **1974**, *96*, 7125.

[265]Slocum; Jennings *J. Org. Chem.* **1976**, *41*, 3653.

[266]However, the regioselectivity can depend on reaction conditions: See Meyers; Avila *Tetrahedron Lett.* **1980**, 3335.

[267]Beak; Hunter; Jun; Wallin *J. Am. Chem. Soc.* **1987**, *109*, 5403. See also Stork; Polt; Li; Houk *J. Am. Chem. Soc.* **1988**, *110*, 8360; Barluenga; Foubelo; Fañanas; Yus *J. Chem. Res. (S)* **1989**, 200.

[268]Eaton; Lee; Xiong *J. Am. Chem. Soc.* **1989**, *111*, 8016.

[269]Benkeser; Trevillyan; Hooz *J. Am. Chem. Soc.* **1962**, *84*, 4971.

R is aryl, OMe and CF₃ *both* direct ortho, while isopropyl directs meta and para (mostly meta).[270] These results are exactly what would be expected from pure field effects, with no contribution from resonance effects, which implies that attack occurs at the hydrogen and not at R. Other evidence for the involvement of H in the rate-determining step is that there are large isotope effects.[271] The nature of R' also has an effect on the rate. In the reaction between triphenylmethane and R'Li, the rate decreased in the order R' = allyl > Bu > Ph > vinyl > Me, though this order changed with changing concentration of R'Li, because of varying degrees of aggregation of the R'Li.[272]

With respect to the reagent, this reaction is a special case of **2-24**.

A closely related reaction is formation of nitrogen ylides from quaternary ammonium salts (see **7-7**):

$$CH_3-\overset{\underset{\displaystyle CH_3}{|}}{\overset{\displaystyle CH_3}{|}}{N^\oplus}-CH_3 \quad Cl^- + PhLi \longrightarrow CH_3-\overset{\underset{\displaystyle CH_3}{|}}{\overset{\displaystyle CH_3}{|}}{N^\oplus}-\overset{\ominus}{C}H_2 + PhH + LiCl$$

Phosphonium salts undergo a similar reaction (see **6-47**).

OS **II**, 198; **III**, 413, 757; **IV**, 792; **V**, 751; **VI**, 436, 478, 737, 979; **VII**, 172, 334, 456, 524; **65**, 61; **68**, 14, 25, 162.

2-22 Metallation with Metals and Strong Bases
Metallation or Metallo-de-hydrogenation

$$2RH + M \longrightarrow 2RM + H_2$$

Organic compounds can be metallated at suitably acidic positions by active metals and by strong bases.[273] The reaction has been used to study the acidities of very weak acids (see p. 176). Synthetically, the most important use of the method is to convert ketones, carboxylic esters, and similar compounds to their enolate forms,[274] e.g.,

$$CH_3-\underset{\underset{\displaystyle O}{\|}}{C}-CH_2-\underset{\underset{\displaystyle O}{\|}}{C}-OEt \xrightarrow{\text{NaOEt}} CH_3-\underset{\underset{\displaystyle O}{\|}}{C}-\overset{\ominus}{C}H-\underset{\underset{\displaystyle O}{\|}}{C}-OEt + HOEt$$

for use in nucleophilic substitutions (**0-94, 0-95,** and **3-14**) and in additions to multiple bonds (**5-17** and **6-41**). Another important use is the conversion of terminal alkynes to acetylide ions.[275] For very weak acids, the most common reagents for synthetic purposes are lithium amides, especially lithium diisopropylamide (LDA) (i-Pr)₂NLi.[276]

It has been shown that lithiation with lithium amides can also be regioselective (see **2-21**).[277] In the case of the cubane derivative **30**, a saturated unactivated position was regioselectively lithiated.[278]

---

[270]Bryce-Smith *J. Chem. Soc.* **1963**, 5983; Benkeser; Hooz; Liston; Trevillyan *J. Am. Chem Soc.* **1963**, *85*, 3984.
[271]Bryce-Smith; Gold; Satchell *J. Chem. Soc.* **1954**, 2743; Pocker; Exner *J. Am. Chem. Soc.* **1968**, *90*, 6764.
[272]West; Waack; Purmort *J. Am. Chem. Soc.* **1970**, *92*, 840.
[273]For a review, see Durst, Ref. 247, pp. 239-291. For reviews with respect to lithium, see Wardell, Ref. 246; Wakefield *Organolithium Methods*; Academic Press: New York, 1988, pp. 32-44.
[274]For a review, see Caine, Ref. 185, vol. 1, pp. 95-145, 284-291.
[275]For a review, see Ziegenbein, in Viehe *Acetylenes*; Marcel Dekker: New York, 1969, pp. 170-185. For an improved method, see Fisch; Coisne; Figeys *Synthesis* **1982**, 211.
[276]The alkali metal hydrides, LiH, NaH, and KH, when prepared in a special way, are very rapid metallation agents: Klusener; Brandsma; Verkruijsse; Schleyer; Friedl; Pi *Angew. Chem. Int. Ed. Engl.* **1986**, *25*, 465 [*Angew. Chem.* **98**, 458].
[277]For example, see Comins; Killpack *J. Org. Chem.* **1987**, *52*, 104.
[278]Eaton; Castaldi *J. Am. Chem. Soc.* **1985**, *107*, 724; Jayasuriya; Alster; Politzer *J. Org. Chem.* **1987**, *52*, 2306.

30

Mercuration of aromatic compounds[279] can be accomplished with mercuric salts, most often $Hg(OAc)_2$[280] or $Hg(ClO_4)_2$ (to give $ArHgOAc$ or $ArHgClO_4$, respectively). This is ordinary electrophilic aromatic substitution and takes place by the arenium ion mechanism (p. 501).[281] Aromatic compounds can also be converted to arylthallium bis(trifluoroacetates) $ArTl(OOCCF_3)_2$ by treatment with thallium (III) trifluoroacetate[282] in trifluoroacetic acid.[283] These arylthallium compounds can be converted to phenols (**2-26**), aryl iodides or fluorides (**2-30**), aryl cyanides (**2-33**), aryl nitro compounds,[284] or aryl esters (**2-32**). The mechanism of thallation appears to be complex, with electrophilic and electron-transfer mechanisms both taking place.[285]

OS **I**, 70, 161, 490; **IV**, 473; **VI**, 468, 542, 611, 683, 709; **VII**, 229, 339. Conversions of ketones or esters to enolates are not listed.

2-23 Conversion of Enolates to Silyl Enol Ethers
3/*O*-Trimethylsilyl-de-hydrogenation

Silyl enol ethers,[286] important reagents with a number of synthetic uses (see, for example, **0-95, 2-4, 5-17, 5-50, 6-40**), can be prepared by base treatment of a ketone (converting it to its enolate) followed by addition of a trialkylchlorosilane. Other silylating agents have also been used.[287] Both strong bases, e.g., lithium diisopropylamide (LDA), and weaker bases, e.g. Et_3N, have been used for this purpose. In some cases, the base and the silylating agent can be present at the same time.[288] Enolates prepared in other ways (e.g., as shown

[279]For reviews, see Larock *Organomercury Compounds in Organic Synthesis*; Springer: New York, 1985, pp. 60-97; Wardell, in Zuckerman, Ref. 246, pp. 308-318.

[280]For a review of mercuric acetate, see Butler, in Pizey, Ref. 102, vol. 4, 1981, pp. 1-145.

[281]For a review, see Taylor, in Bamford; Tipper, Ref. 53, vol. 13, 1972, pp. 186-194. An alternative mechanism, involving radial cations, has been reported: Courtneidge; Davies; McGuchan; Yazdi *J. Organomet. Chem.* **1988**, *341*, 63.

[282]For a review of this reagent, see Uemura, in Pizey, Ref. 102, vol. 5, 1983, pp. 165-241.

[283]McKillop; Hunt; Zelesko; Fowler; Taylor; McGillivray; Kienzle *J. Am. Chem. Soc.* **1971**, *93*, 4841; Taylor; Kienzle; McKillop *Org. Synth.* VI, 709; Al-Azzawi; Roberts *J. Chem. Soc., Perkin Trans. 2* **1982**, 677; Taylor; Katz; Alvarado; McKillop *J. Organomet. Chem.* **1985**, *285*, C9. For reviews, see Usyatinskii; Bregadze *Russ. Chem. Rev.* **1988**, *57*, 1054-1068; Uemura, in Hartley; Patai, Ref. 1, vol. 4, pp. 473-538.

[284]Uemura; Toshimitsu; Okano *Bull. Chem. Soc. Jpn.* **1976**, *49*, 2582.

[285]Lau; Kochi *J. Am. Chem. Soc.* **1984**, *106*, 7100, **1986**, *108*, 6720.

[286]For reviews of these compounds, see Poirier *Org. Prep. Proced. Int.* **1988**, *20*, 319-369; Brownbridge *Synthesis* **1983**, 1-28, 85-104; Rasmussen *Synthesis* **1977**, 91-110. See also references given in Rubottom; Mott; Krueger *Synth. Commun* **1977**, *7*, 327. For monographs on silicon reagents in organic synthesis, see Colvin *Silicon Reagents in Organic Synthesis*; Academic Press: New York, 1988; Weber *Silicon Reagents for Organic Synthesis*; Springer: New York, 1983; Colvin *Silicon in Organic Synthesis*; Butterworth: London, 1981 [reprinted, with revisions: Krieger: Melbourne, FL, 1985]. For reviews, see Colvin, in Hartley; Patai, Ref. 1, vol. 4, pp. 539-621; Ager *Chem. Soc. Rev.* **1982**, *11*, 493-522; Colvin *Chem. Soc. Rev.* **1978**, *7*, 15-64, pp. 43-50.

[287]For a review of silylating agents, see Mizhiritskii; Yuzhelevskii *Russ. Chem. Rev.* **1987**, *56*, 355-365. For a list, with references, see Ref. 52, pp. 746-748.

[288]Corey; Gross *Tetrahedron Lett.* **1984**, *25*, 495.

for **112** on p. 452) also give the reaction. The reaction can be applied to aldehydes by the use of the base KH in 1,2-dimethoxyethane.[289] A particularly mild method for conversion of ketones or aldehydes to silyl enol ethers uses Me_3SiI and the base hexamethyldisilazane $(Me_3Si)_2NH$.[290] Cyclic ketones can be converted to silyl enol ethers in the presence of acyclic ketones, by treatment with Me_3SiBr, tetraphenylstibonium bromide Ph_4SbBr, and an aziridine.[291]

OS **VI**, 327, 445; **VII**, 282, 312, 424, 512; **65**, 1; **67**, 141; **69**, 129. See also OS **VII**, 66, 266. For the conversion of ketones to vinylic triflates, see OS **68**, 116, 138.

Metals as Leaving Groups

A. Hydrogen as the Electrophile

2-24   Replacement of Metals by Hydrogen
Hydro-de-metallation or **Demetallation**

$$RM + HA \longrightarrow RH + MA$$

Organometallic compounds react with acids in reactions in which the metal is replaced by hydrogen.[292] R may be aryl (see **1-44**). The reaction is often used to introduce deuterium or tritium into susceptible positions. For Grignard reagents, water is usually a strong enough acid, but stronger acids are also used. An important method for the reduction of alkyl halides consists of the process $RX \rightarrow RMgX \rightarrow RH$.

Other organometallic compounds that are hydrolyzed by water are those of sodium, potassium, lithium, zinc, etc.—the ones high in the electromotive series. When the metal is less active, stronger acids are required. For example, R_2Zn compounds react explosively with water, R_2Cd slowly, and R_2Hg not at all, though the latter can be cleaved with concentrated HCl. However, this general statement has many exceptions, some hard to explain. For example, BR_3 compounds are completely inert to water, and GaR_3 at room temperature cleave just one R group, but AlR_3 react violently with water. However, BR_3 can be converted to RH with carboxylic acids.[293] For less active metals it is often possible to cleave just one R group from a multivalent metal. For example,

$$R_2Hg + HCl \longrightarrow RH + RHgCl$$

Organometallic compounds of less active metals and metalloids, such as silicon,[294] antimony, bismuth, etc., are quite inert to water. Organomercury compounds (RHgX or R_2Hg) can be reduced to RH by H_2, $NaBH_4$, or other reducing agents.[295] The reduction with $NaBH_4$

[289]Ladjama; Riehl *Synthesis* **1979**, 504. This base has also been used for ketones: See Orban; Turner; Twitchin *Tetrahedron Lett.* **1984**, *25*, 5099.
[290]Miller; McKean *Synthesis* **1979**, 730, *Synth. Commun.* **1982**, *12*, 319. See also Cazeau; Duboudin; Moulines; Babot; Dunogues *Tetrahedron* **1987**, *43*, 2075, 2089; Ahmad; Khan; Iqbal *Synth. Commun.* **1988**, *18*, 1679.
[291]Fujiwara; Baba; Matsuda *Chem. Lett.* **1989**, 1247.
[292]For reviews, see Abraham; Grellier, in Hartley; Patai, Ref. 1, vol. 2, pp. 25-149, pp. 105-136; Abraham, Ref. 2, pp. 107-134; Jensen; Rickborn, Ref. 2. pp. 45-74; Schlosser *Angew. Chem. Int. Ed. Engl.* **1964**, *3*, 287-306, 362-373 [*Angew. Chem. 76*, 124-143, 258-269], *Newer Methods Prep. Org. Chem.* **1968**, *5*, 238-311.
[293]Brown; Hébert *J. Organomet. Chem.* **1983**, *255*, 135; Brown; Murray *Tetrahedron* **1986**, *42*, 5497; Pelter; Smith; Brown *Borane Reagents*; Academic Press: New York, 1988, pp. 242-244.
[294]For a review of hydro-de-silylation of allylic and vinylic silanes, see Fleming; Dunoguès; Smithers *Org. React.* **1989**, *37*, 57-575, pp. 89-97, 194-243.
[295]For a review, see Makarova *Organomet. React.* **1970**, *1*, 119-348, pp. 251-270, 275-300.

takes place by a free-radical mechanism.[296] Alkyl–silicon bonds can be cleaved by H_2SO_4, e.g., $HOOCCH_2CH_2SiMe_3 \rightarrow 2CH_4 + (HOOCCH_2CH_2SiMe_2)_2O$.[297]

When the hydrogen of the HA is attached to carbon, this reaction is the same as **2-21**.

We do not list the many hydrolyses of sodium or potassium enolates, etc. found in *Organic Syntheses*. The hydrolysis of a Grignard reagent to give an alkane is found at OS **II**, 478; the reduction of a vinylic tin compound at OS **66**, 75; and the reduction of an alkynylsilane at OS **67**, 149.

B. Oxygen Electrophiles

2-25 The Reaction between Organometallic Reagents and Oxygen[298]
Hydroperoxy-de-metallation; Hydroxy-de-metallation

$$RMgX + O_2 \longrightarrow R-O-O-MgX \begin{array}{c} \nearrow \quad R-O-O-H \\ \\ \searrow \quad 2R-O-MgX \xrightarrow{H^+} 2R-OH \end{array}$$

Oxygen reacts with Grignard reagents to give either hydroperoxides or alcohols. The reaction can be used to convert alkyl halides to alcohols without side reactions. With aryl Grignard reagents yields are lower and only phenols are obtained, not hydroperoxides. It is because of the possibility of this reaction that oxygen should be excluded when Grignard reagents are desired for other purposes. A better procedure for the conversion of aryl Grignard reagents to phenols involves the use of trimethyl borate followed by oxidation with H_2O_2 in acetic acid[299] (see **2-28**).

$$ArMgX \xrightarrow{B(OMe)_3} ArB(OMe)_2 \xrightarrow[H_2O_2]{H^+} ArOH$$

Most other organometallic compounds also react with oxygen. Aryllithiums have been converted to phenols by treatment with oxygen.[300] Trialkylboranes and alkyldichloroboranes $RBCl_2$ can be conveniently converted to hydroperoxides by treatment with oxygen followed by hydrolysis.[301] Dilithiated carboxylic acids (see **0-96**) react with oxygen to give (after hydrolysis) α-hydroxy carboxylic acids.[302] There is evidence that the reaction between Grignard reagents and oxygen involves a free-radical mechanism.[303]

The 1,1-dimetallic compounds $R_2C(SnMe_3)ZnBr$ were oxidized by dry air at −10 to 0°C in the presence of Me_3SiCl to give aldehydes or ketones $R_2C{=}O$.[304]

OS **V**, 918. See also OS **69**, 96.

[296]For a review of this and other free radical reactions of organomercury compounds, see Barluenga; Yus *Chem. Rev.* **1988**, *88*, 487-509.

[297]Sommer; Marans; Goldberg; Rockett; Pioch *J. Am. Chem. Soc.* **1951**, *73*, 882. See also Abraham; Grellier, Ref. 292, p. 117.

[298]For a monograph, see Brilkina; Shushunov *Reactions of Organometallic Compounds with Oxygen and Peroxides*; CRC Press: Boca Raton, FL, 1969. For a review, see Wardell; Paterson, in Hartley; Patai, Ref. 1, vol. 2, 1985, pp. 219-338, pp. 311-316.

[299]Hawthorne *J. Org. Chem.* **1957**, *22*, 1001. For other procedures, see Lewis; Gabhe *Aust. J. Chem.* **1978**, *31*, 2091; Hoffmann; Ditrich *Synthesis* **1983**, 107.

[300]Parker; Koziski *J. Org. Chem.* **1987**, *52*, 674. For other reagents, see Taddei; Ricci *Synthesis* **1986**, 633; Einhorn; Luche; Demerseman *J. Chem. Soc., Chem. Commun.* **1988**, 1350.

[301]Brown; Midland *Tetrahedron* **1987**, *43*, 4059.

[302]Moersch; Zwiesler *Synthesis* **1971**, 647; Adam; Cueto *J. Org. Chem.* **1977**, *42*, 38.

[303]Lamb; Ayers; Toney; Garst *J. Am. Chem. Soc.* **1966**, *88*, 4261; Davies; Roberts *J. Chem. Soc. B* **1969**, 317; Walling; Cioffari *J. Am. Chem. Soc.* **1970**, *92*, 6609; Garst; Smith; Farrar *J. Am. Chem. Soc.* **1972**, *94*, 7707. For a review, see Davies *J. Organomet. Chem.* **1980**, *200*, 87-99.

[304]Knochel; Xiao; Yeh *Tetrahedron Lett.* **1988**, *29*, 6697.

2-26   Conversion of Arylthallium Compounds to Phenols
        Hydroxy-de-(bistrifluoroacetoxy)thallation

$$\text{ArTl(OOCCF}_3)_2 \xrightarrow[\substack{\text{2. PPh}_3 \\ \text{3. dil. NaOH}}]{\text{1. Pb(OAc)}_4} \text{ArOH}$$

Arythallium bis(trifluoroacetates) (prepared by **2-22**) can be converted to phenols by treatment with lead tetraacetate followed by triphenylphosphine and then dilute NaOH.[305] The entire process, including the thallation reaction, can be carried out in a single reaction vessel without isolation of any of the intermediate products, so that this is a method of accomplishing the conversion ArH → ArOH. Diarylthallium trifluoroacetates undergo the same reaction.[306]

2-27   Reaction Between Organometallic Reagents and Peroxides
        t-Butoxy-de-metallation

$$\text{RMgX} + t\text{-Bu}-\text{O}-\text{O}-\underset{\underset{\text{O}}{\|}}{\text{C}}-\text{R}' \longrightarrow \text{R}-\text{O}-t\text{-Bu} + \text{R'COOMgX}$$

A convenient method of preparation of t-butyl ethers consists of treating Grignard reagents with t-butyl acyl peroxides.[307] Both alkyl and aryl Grignard reagents can be used. The application of this reaction to Grignard reagents prepared from cyclopropyl halides permits cyclopropyl halides to be converted to t-butyl ethers of cyclopropanols,[308] which can then be easily hydrolyzed to the cyclopropanols. The direct conversion of cyclopropyl halides to cyclopropanols by **0-1** is not generally feasible, because cyclopropyl halides do not generally undergo nucleophilic substitutions without ring opening.

Vinyllic lithium reagents (**31**) react with silyl peroxides to give high yields of silyl enol ethers with retention of configuration.[309] Since the preparation of **31** from vinylic halides

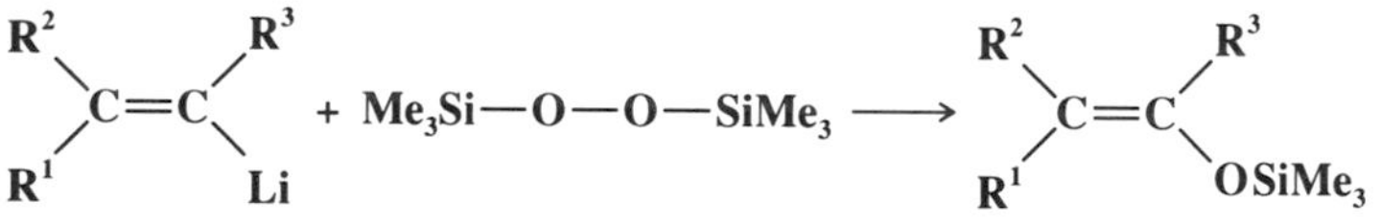

$$\underset{\underset{\text{R}^1}{\big\backslash}}{\overset{\overset{\text{R}^2}{\big\diagup}}{}}\text{C}=\underset{\underset{\text{Li}}{\big\backslash}}{\overset{\overset{\text{R}^3}{\big\diagup}}{}} + \text{Me}_3\text{Si}-\text{O}-\text{O}-\text{SiMe}_3 \longrightarrow \underset{\underset{\text{R}^1}{\big\backslash}}{\overset{\overset{\text{R}^2}{\big\diagup}}{}}\text{C}=\underset{\underset{\text{OSiMe}_3}{\big\backslash}}{\overset{\overset{\text{R}^3}{\big\diagup}}{}}$$

31

(**2-39**) also proceeds with retention, the overall procedure is a method for the stereospecific conversion of a vinylic halide to a silyl enol ether. In a related reaction, alkynyl esters can be prepared from lithium acetylides and phenyliodine(III) dicarboxylates.[310]

$$\text{RC}\equiv\text{CLi} + \text{PhI(O}_2\text{CR')}_2 \longrightarrow \text{RC}\equiv\text{C}-\text{O}-\underset{\underset{\text{O}}{\|}}{\text{C}}-\text{R}'$$

OS **V**, 642, 924.

[305]Taylor; Altland; Danforth; McGillivray; McKillop *J. Am. Chem. Soc.* **1970**, *92*, 3520.
[306]Taylor; Altland; McKillop *J. Org. Chem.* **1975**, *40*, 2351.
[307]Lawesson; Yang *J. Am. Chem. Soc.* **1959**, *81*, 4230; Lawesson; Frisell; Denney; Denney *Tetrahedron* **1963**, *19*, 1229. For a monograph on the reactions of organometallic compounds with peroxides, see Ref. 298. For a review, see Razuvaev; Shushunov; Dodonov; Brilkina, in Swern *Organic Peroxides*, vol. 3; Wiley: New York, 1972, pp. 141-270.
[308]Longone; Miller *Tetrahedron Lett.* **1967**, 4941.
[309]Davis; Lal; Wei *Tetrahedron Lett.* **1988**, *29*, 4269.
[310]Stang; Boehshar; Wingert; Kitamura *J. Am. Chem. Soc.* **1988**, *110*, 3272.

2-28 Oxidation of Trialkylboranes to Borates

$$R_3B \xrightarrow[NaOH]{H_2O_2} (RO)_3B \longrightarrow 3ROH + B(OH)_3$$

Treatment with alkaline H_2O_2 oxidizes trialkylboranes to esters of boric acid.[311] This reaction does not affect double or triple bonds, aldehydes, ketones, halides, or nitriles. The R group does not rearrange, and this reaction is a step in the hydroboration method of converting olefins to alcohols (**5-12**). The mechanism has been formulated as involving a rearrangement from boron to oxygen:[311]

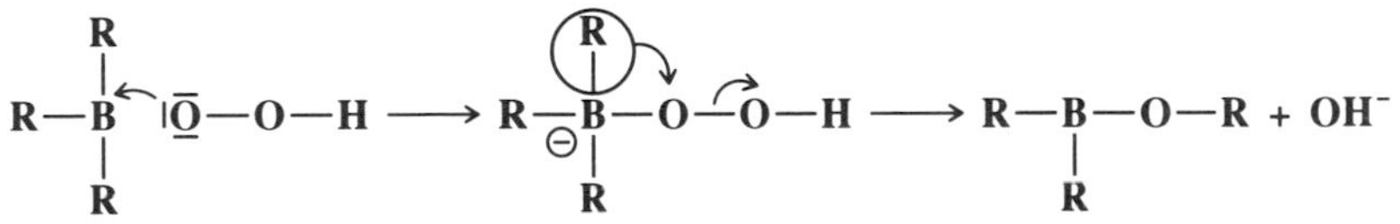

The other two R groups then similarly migrate. Retention of configuration is observed in R. Boranes can also be oxidized to borates in good yields with oxygen,[312] with sodium perborate $NaBO_3$,[313] with sodium percarbonate ($Na_2CO_3 \cdot \frac{3}{2}H_2O_2$),[314] and with trimethylamine oxide, either anhydrous[315] or in the form of the dihydrate.[316] The reaction with oxygen is free radical in nature.[317]

OS **V,** 918; **VI,** 719, 852, 919.

C. Sulfur Electrophiles

2-29 Conversion of Grignard Reagents to Sulfur Compounds

$$RMgX + S \longrightarrow R{-}S{-}MgX \xrightarrow{RMgX} R{-}S{-}R$$ **Thio-de-dimetallo-*aggre*-substitution**

$$R{-}S{-}MgX \xrightarrow{H^+} R{-}SH$$ **Mercapto-de-metallation**

Thiols and sulfides are occasionally prepared by treatment of Grignard reagents with sulfur.[318] Analogous reactions are known for selenium and tellurium compounds. Grignard reagents

$$RMgX + SO_2Cl_2 \longrightarrow RSO_2Cl$$

$$RMgX + R'SO{-}OR'' \longrightarrow RSOR'$$

$$RMgX + R'SSR' \longrightarrow RSR'$$

$$RMgX + SO_2 \longrightarrow RSO{-}OMgX \xrightarrow{H^+} RSO_2H$$

$$RSO{-}OMgX \xrightarrow{X_2} RSO_2X$$

[311]For reviews, see Pelter; Smith; Brown, Ref. 293, pp. 244-249; Brown *Boranes in Organic Chemistry*; Cornell University Press: Ithaca, NY, 1972, pp. 321-325; Matteson in Hartley; Patai, Ref. 1, vol. 4, pp. 307-409, pp. 337-340. See also Brown; Snyder; Subba Rao; Zweifel *Tetrahedron* **1986,** *42*, 5505.

[312]Brown; Midland; Kabalka *J. Am. Chem. Soc.* **1971,** *93*, 1024, *Tetrahedron* **1986,** *42*, 5523.

[313]Kabalka; Shoup; Goudgaon *J. Org. Chem.* **1989,** *54*, 5930.

[314]Kabalka; Wadgaonkar; Shoup *Organometallics* **1990,** *9*, 1316.

[315]Köster; Morita *Justus Liebigs Ann. Chem.* **1967,** *704*, 70; Köster; Arora; Binger *Angew. Chem. Int. Ed. Engl.* **1969,** *8*, 205 [*Angew. Chem. 81*, 185].

[316]Kalbalka; Hedgecock *J. Org. Chem.* **1975,** *40*, 1776, *J. Chem. Educ.* **1975,** *52*, 745; Kabalka; Slayden *J. Organomet. Chem.* **1977,** *125*, 273.

[317]Mirviss *J. Am. Chem. Soc.* **1961,** *83*, 3051, *J. Org. Chem.* **1967,** *32*, 1713; Davies; Roberts *Chem. Commun.* **1966,** 298; Midland; Brown *J. Am. Chem. Soc.* **1971,** *93*, 1506.

[318]For reviews of the reactions in this section, see Wardell; Paterson, Ref. 298, pp. 316-323; Wardell, in Patai *The Chemistry of the Thiol Group*, pt. 1; Wiley: New York, 1974, pp. 211-215; Wakefield, Ref. 273, pp. 135-142.

and other organometallic compounds[319] react with sulfuryl chloride to give sulfonyl chlorides,[320] with esters of sulfinic acids to give (stereospecifically) sulfoxides,[321] with disulfides to give sulfides,[322] and with SO_2 to give sulfinic acid salts[323] which can be hydrolyzed to sulfinic acids or treated with halogens to give sulfonyl halides.[324]

OS **III**, 771; **IV**, 667; **VI**, 533, 979.

D. Halogen Electrophiles

2-30 Halo-de-metallation

$$RMgX + I_2 \longrightarrow RI + MgIX$$

Grignard reagents react with halogens to give alkyl halides. The reaction is useful for the preparation of iodo compounds from the corresponding chloro or bromo compounds. The reaction is not useful for preparing chlorides, since the reagents RMgBr and RMgI react with Cl_2 to give mostly RBr and RI, respectively.[325] Alkyl, aryl, and vinylic Grignard reagents and lithium compounds can be converted to fluorides in moderate to high yields with perchloryl fluoride $FClO_3$[326] (but see **2-4** for the explosive nature of this reagent).

Most organometallic compounds, both alkyl and aryl, also react with halogens to give alkyl or aryl halides.[327] The reaction can be used to convert acetylide ions to 1-haloalkynes.[328] Since acetylide ions are easily prepared from alkynes (**2-22**), this provides a means of making the conversion $RC{\equiv}CH \rightarrow RC{\equiv}CX$. Trialkylboranes react rapidly with I_2[329] or Br_2[330] in the presence of NaOMe in methanol, or with $FeCl_3$ or other reagents[331] to give alkyl iodides, bromides, or chlorides, respectively. Combined with the hydroboration reaction (**5-12**), this is an indirect way of adding HBr, HI, or HCl to a double bond to give products with an anti-Markovnikov orientation (see **5-1**). Trialkylboranes can also be converted to alkyl iodides by treatment with allyl iodide and air in a free radical process.[332]

trans-1-Alkenylboronic acids **33**, prepared by hydroboration of terminal alkynes with catecholborane[333] (**5-12**) followed by hydrolysis, react with I_2 in the presence of NaOH at 0°C in ethereal solvents to give trans vinylic iodides.[334] This is an indirect way of accom-

---

[319]For a discussion of conversions of organomercury compounds to sulfur-containing compounds, see Larock, Ref. 279, pp. 210-216.

[320]Bhattacharya; Eaborn; Walton *J. Chem. Soc. C* **1968**, 1265. For similar reactions with organolithiums, see Quast; Kees *Synthesis* **1974**, 489; Hamada; Yonemitsu *Synthesis*, **1986**, 852.

[321]Harpp; Vines; Montillier; Chan *J. Org. Chem.* **1976**, *41*, 3987.

[322]For a discussion, see Negishi, Ref. 1, pp. 243-247.

[323]For a review of the reactions of organometallic compounds with SO_2, see Kitching; Fong *Organomet. Chem. Rev., Sect. A* **1970**, *5*, 281-321.

[324]Asinger; Laue; Fell; Gubelt *Chem. Ber.* **1967**, *100*, 1696.

[325]Zakharkin; Gavrilenko; Paley *J. Organomet. Chem.* **1970**, *21*, 269.

[326]Schlosser; Heinz *Chem. Ber.* **1969**, *102*, 1944. See also Satyamurthy; Bida; Phelps; Barrio *J. Org. Chem.* **1990**, *55*, 3373.

[327]For a review, see Abraham; Grellier, Ref. 292, pp. 72-105. For reviews with respect to organomercury compounds, see Larock, Ref. 279, pp. 158-178; Makarova, Ref. 295, pp. 325-348.

[328]For a review, see Delavarenne; Viehe, in Viehe, Ref. 275, pp. 665-688. For a list of reagents, with references, see Ref. 52, pp. 333-334. For an improved procedure, see Brandsma; Verkruijsse *Synthesis* **1990**, 984.

[329]Brown; Rathke; Rogić; De Lue *Tetrahedron* **1988**, *44*, 2751.

[330]Brown; Lane *Tetrahedron* **1988**, *44*, 2763; Brown; Lane; De Lue *Tetrahedron* **1988**, *44*, 2273. For another reagent, see Nelson; Soundararajan *J. Org. Chem.* **1989**, *54*, 340.

[331]Nelson; Soundararajan *J. Org. Chem.* **1988**, *53*, 5664. For other reagents, see Jigajinni; Paget; Smith *J. Chem. Res., (S)* **1981**, 376; Brown; De Lue *Tetrahedron* **1988**, *44*, 2785.

[332]Suzuki; Nozawa; Harada; Itoh; Brown; Midland *J. Am. Chem. Soc.* **1971**, *93*, 1508. For reviews, see Brown; Midland *Angew. Chem. Int. Ed. Engl.* **1972**, *11*, 692-700, pp. 699-700 [*Angew. Chem. 84*, 702-710]; Brown, Ref. 311, pp. 442-446.

[333]For a review of this reagent, see Kabalka *Org. Prep. Proced. Int.* **1977**, *9*, 131-147.

[334]Brown; Hamaoka; Ravindran; Subrahmanyam; Somayaji; Bhat *J. Org. Chem.* **1989**, *54*, 6075. See also Kabalka; Gooch; Hsu *Synth. Commun.* **1981**, *11*, 247.

plishing the anti-Markovnikov addition of HI to a terminal triple bond. The reaction cannot be applied to alkenylboronic acids prepared from internal alkynes. However, alkenylboronic

acids prepared from both internal and terminal alkynes react with Br_2 (2 moles of Br_2 must be used) followed by base to give the corresponding vinylic bromide, but in this case with *inversion* of configuration; so the product is the cis vinylic bromide.[335] Alkenylboronic acids also give vinylic bromides and iodides when treated with a mild oxidizing agent and NaBr or NaI, respectively.[336] Treatment of **33** (prepared from terminal alkynes) with Cl_2 gave vinylic chlorides with inversion.[337] Vinylic halides can also be prepared from vinylic silanes[338] and from vinylic aluminum[339] or vinylic copper reagents. The latter react with I_2 to give iodides,[340] and with N-chloro- or N-bromosuccinimide at $-45°C$ to give chlorides or bromides.[341]

Aryl iodides[342] and fluorides can be prepared from arylthallium bis(trifluoroacetates) (see **2-22**), indirectly achieving the conversions ArH → ArI and ArH → ArF. The bis(trifluoroacetates) react with KI to give ArI in high yields.[343] The reaction with KF gives arylthallium(III) difluorides ArTlF$_2$, but these react with BF_3 to give ArF in moderate overall yields.[344] Aryllead triacetates ArPb(OAc)$_3$ can be converted to aryl fluorides by treatment with BF_3–etherate.[345] Aryl fluorides have also been prepared in low-to-moderate yields by treatment of arylmetal compounds such as Ph$_4$Sn and Ph$_2$Hg with F_2[346] and with fluoroxytrifluoromethane CF$_3$OF or cesium fluoroxy sulfate CsSO$_4$F.[347]

For the reaction of lithium enolates of esters with I_2 or CX_4 see **2-5**.

It is unlikely that a single mechanism suffices to cover all conversions of organometallic compounds to alkyl halides.[348] In a number of cases the reaction has been shown to involve

[335]Brown; Hamaoka; Ravindran *J. Am. Chem. Soc.* **1973**, *95*, 6456. See also Brown; Bhat; Rajagopalan *Synthesis* **1986**, 480; Brown; Bhat *Tetrahedron Lett.* **1988**, *29*, 21.
[336]See Kabalka; Sastry; Knapp; Srivastava *Synth. Commun.* **1983**, *13*, 1027.
[337]Kunda; Smith; Hylarides; Kabalka *Tetrahedron Lett.* **1985**, *26*, 279.
[338]See, for example Chou; Kuo; Wang; McKillop *J. Org. Chem.* **1989**, *54*, 868.
[339]Zweifel; Whitney *J. Am. Chem. Soc.* **1967**, *89*, 2753.
[340]Normant; Chaiez; Chuit; Villieras *J. Organomet. Chem.* **1974**, 77, 269, *Synthesis* **1974**, 803.
[341]Westmijze; Meijer; Vermeer *Recl. Trav. Chim. Pays-Bas* **1977**, *96*, 168; Levy; Talley; Dunford *Tetrahedron Lett.* **1977**, 3545.
[342]For reviews of the synthesis of aryl iodides, see Merkushev *Synthesis* **1988**, 923-937, *Russ. Chem. Rev.* **1984**, *53*, 343-350.
[343]Ref. 283. See also Ishikawa; Sekiya *Bull. Chem. Soc. Jpn.* **1974**, *47*, 1680 and Ref. 306.
[344]Taylor; Bigham; Johnson; McKillop *J. Org. Chem.* **1977**, *42*, 362.
[345]De Meio; Pinhey *J. Chem. Soc., Chem. Commun.* **1990**, 1065.
[346]Adam; Berry; Hall; Pate; Ruth *Can. J. Chem.* **1983**, *61*, 658. See also Adam; Ruth; Jivan; Pate *J. Fluorine Chem.* **1984**, *25*, 329; Speranza; Shiue; Wolf; Wilbur; Angelini *J. Fluorine Chem.* **1985**, *30*, 97.
[347]Bryce; Chambers; Mullins; Parkin *Bull. Soc. Chim. Fr.* **1986**, 930. See also Clough; Diorazio; Widdowson *Synlett* **1990**, 761.
[348]For reviews of the mechanisms, see Abraham; Grellier, Ref. 327; Abraham, Ref. 2, pp. 135-177; Jensen; Rickborn, Ref. 2, pp. 75-97.

inversion of configuration (see p. 572), indicating an S$_E$2 (back) mechanism, while in other cases retention of configuration has been shown,[349] implicating an S$_E$2 (front) or S$_E$i mechanism. In still other cases complete loss of configuration as well as other evidence have demonstrated the presence of a free-radical mechanism.[350]

OS **I**, 125, 325, 326; **III**, 774, 813; **V**, 921; **VI**, 709; **VII**, 290; **65**, 108. Also see OS **II**, 150.

E. Nitrogen Electrophiles

2-31   The Conversion of Organometallic Compounds to Amines
Amino-de-metallation

$$RLi \xrightarrow[\text{MeLi}]{\text{CH}_3\text{ONH}_2} RNH_2$$

There are several methods for conversion of alkyl- or aryllithium compounds to primary amines.[351] The two most important are treatment with hydroxylamine derivatives and with certain azides.[352] In the first of these methods, treatment of RLi with methoxyamine and MeLi in ether at $-78°C$ gives RNH$_2$.[353] Grignard reagents give lower yields. The reaction can be extended to give secondary amines by the use of N-substituted methoxyamines CH$_3$ONHR'.[354] There is evidence[355] that the mechanism involves the direct displacement of OCH$_3$ by R on an intermediate CH$_2$O$\overline{\text{N}}$R'$^-$ (CH$_3$O$\overline{\text{N}}$R'$^-$ Li$^+$ + RLi → CH$_3$OLi + R$\overline{\text{N}}$R'$^-$ Li$^+$). The most useful azide is tosyl azide TsN$_3$.[356] The initial product is usually RN$_3$, but this is easily reducible to the amine (**9-53**). With some azides, such as azidomethyl phenyl sulfide PhSCH$_2$N$_3$, the group attached to the N$_3$ is a poor leaving group, so the inital product is a triazene (in this case ArNH=NHCH$_2$SPh from ArMgX), which can be hydrolyzed to the amine.[357]

Organoboranes react with a mixture of aqueous NH$_3$ and NaOCl to produce primary amines.[358] It is likely that the actual reagent is chloramine NH$_2$Cl. Chloramine itself,[359]

$$R_3B \xrightarrow{\text{NH}_3\text{-NaOCl}} 2RNH_2 + RB(OH)_2$$

[349]For example, see Jensen; Gale *J. Am. Chem. Soc.* **1960**, *82*, 148.

[350]See, for example, Ref. 349; Beletskaya; Reutov; Gur'yanova *Bull. Acad. Sci. USSR, Div. Chem. Sci.* **1961**, 1483; Beletskaya; Ermanson; Reutov *Bull. Acad. Sci. USSR, Div. Chem. Sci.* **1965**, 218; de Ryck; Verdonck; Van der Kelen *Bull. Soc. Chim. Belg.* **1985**, *94*, 621.

[351]For a review of methods for achieving the conversion RM → RNH$_2$, see Erdik; Ay *Chem. Rev.* **1989**, *89*, 1947-1980.

[352]For some other methods of converting organolithium or Grignard reagents to primary amines, see Alvernhe; Laurent *Tetrahedron Lett.* **1972**, 1007; Hagopian; Therien; Murdoch *J. Am. Chem. Soc.* **1984**, *106*, 5753; Genet; Mallart; Greck; Piveteau *Tetrahedron Lett.* **1991**, *32*, 2359.

[353]Beak; Kokko *J. Org. Chem.* **1982**, *47*, 2822. For other hydroxylamine derivatives, see Colvin; Kirby; Wilson *Tetrahedron Lett.* **1982**, *23*, 3835; Boche; Bernheim; Schrott *Tetrahedron Lett.* **1982**, *23*, 5399; Boche; Schrott *Tetrahedron Lett.* **1982**, *23*, 5403.

[354]Kokko; Beak *Tetrahedron Lett.* **1983**, *24*, 561.

[355]Beak; Basha; Kokko; Loo *J. Am. Chem. Soc.* **1986**, *108*, 6016.

[356]See, for example, Spagnolo; Zanirato; Gronowitz *J. Org. Chem.* **1982**, *47*, 3177; Reed; Snieckus *Tetrahedron Lett.* **1983**, *24*, 3795. For other azides, see Hassner; Munger; Belinka *Tetrahedron Lett.* **1982**, *23*, 699; Mori; Aoyama; Shioiri *Tetrahedron Lett.* **1984**, *25*, 429.

[357]Trost; Pearson *J. Am. Chem. Soc.* **1981**, *103*, 2483; **1983**, *105*, 1054.

[358]Kabalka; Sastry; McCollum; Yoshioka *J. Org. Chem.* **1981**, *46*, 4296; Kabalka; Wang; Goudgaon *Synth. Commun.* **1989**, *19*, 2409. For the extension of this reaction to the preparation of secondary amines, see Kabalka; Wang *Organometallics* **1989**, *8*, 1093, *Synth. Commun.* **1990**, *20*, 231.

[359]Brown; Heydkamp; Breuer; Murphy *J. Am. Chem. Soc.* **1964**, *86*, 3565.

hydroxylamine-O-sulfonic acid in diglyme,[360] and trimethylsilyl azide[361] also give the reaction. Since the boranes can be prepared by the hydroboration of alkenes (**5-12**), this is an indirect method for the addition of NH_3 to a double bond with anti-Markovnikov orientation. Secondary amines can be prepared[362] by the treatment of alkyl- or aryldichloroboranes or dialkylchloroboranes with alkyl or aryl azides.

$$\text{RBCl}_2 + \text{R}'\text{N}_3 \longrightarrow \text{RR}'\text{NBCl}_2 \xrightarrow[\text{OH}^-]{\text{H}_2\text{O}} \text{RNHR}'$$

$$\text{R}_2\text{BCl} + \text{R}'\text{N}_3 \xrightarrow[\text{2.H}_2\text{O}]{\text{1.Et}_2\text{O}} \text{RNHR}'$$

The use of an optically active $RBCl_2$ gave secondary amines of essentially 100% optical purity.[363] In other methods, trialkylboranes R_3B gave secondary amines RR'NH upon treatment with N-chloroamines R'NHCl,[364] and aryllead triacetates ArPb(OAc)$_3$ give secondary amines ArNHAr' when treated with primary aromatic amines Ar'NH$_2$ and Cu(OAc)$_2$.[365]

An indirect method for the conversion of aldehydes to N,N-disubstituted amides is based on the conversion of an O-(trimethylsilyl)aldehyde cyanohydrin **34** to the amine **35**.[366]

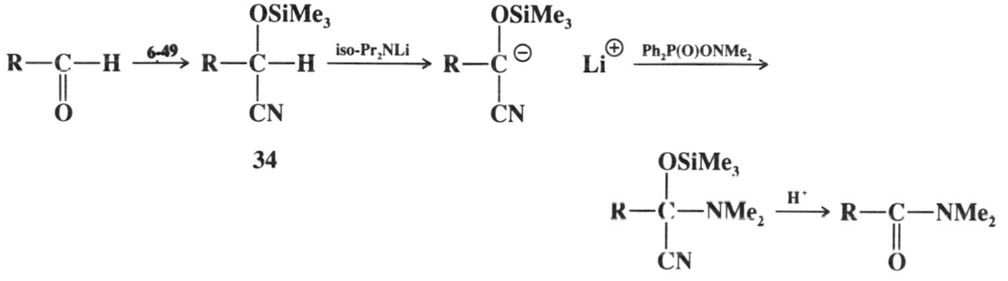

Secondary amines have been converted to tertiary amines by treatment with dialkylcopperlithium reagents: $R_2CuLi + NHR \rightarrow RNR'_2$.[367] The reaction was also used to convert primary amines to secondary, but yields were lower.[367] However, primary aromatic amines ArNH$_2$ were converted to diaryl amines ArNHPh by treatment with Ph$_3$Bi(OAc)$_2$[368] and a copper powder catalyst.[369]

Molecular nitrogen (N_2) reacts with aryllithium compounds in the presence of compounds of such transition metals as titanium, chromium, molybdenum, or vanadium (e.g., TiCl$_4$) to give (after hydrolysis) primary aromatic amines.[370]

$$\text{ArLi} + \text{N}_2 \xrightarrow[\text{2.H}_2\text{O}]{\text{1.MX}_n} \text{ArNH}_2 + \text{NH}_3$$

OS **VI**, 943.

[360]Rathke; Inouc; Varma; Brown *J. Am. Chem. Soc.* **1966**, *88*, 2870; Brown; Kim; Srebnik; Singaram *Tetrahedron* **1987**, *43*, 4071. For a method of using this reaction to prepare optically pure chiral amines, see Brown; Kim; Cole; Singaram *J. Am. Chem. Soc.* **1986**, *106*, 6761.

[361]Kabalka; Goudgaon; Liang *Synth. Commun.* **1988**, *18*, 1363.

[362]Brown; Midland; Levy; Suzuki; Sono; Itoh *Tetrahedron* **1987**, *43*, 4079; Carboni; Vaultier; Courgeon; Carrié *Bull. Soc. Chim. Fr.* **1989**, 844.

[363]Brown; Salunkhe; Singaram *J. Org. Chem.* **1991**, *56*, 1170.

[364]Kabalka; McCollum; Kunda *J. Org. Chem.* **1984**, *49*, 1656.

[365]Barton; Donnelly; Finet; Guiry *Tetrahedron Lett.* **1989**, *30*, 1377.

[366]Boche; Bosold; Niessner *Tetrahedron Lett.* **1982**, *23*, 3255.

[367]Yamamoto; Maruoka *J. Org. Chem.* **1980**, *45*, 2739.

[368]For a review of arylations with bismuth reagents, see Finet *Chem. Rev.* **1989**, *89*, 1487-1501.

[369]Dodonov; Gushchin; Brilkina *Zh. Obshch. Khim.* **1985**, *55*, 466 [*Chem. Abstr. 103*, 22218z]; Barton; Finet; Khamsi *Tetrahedron Lett.* **1986**, *27*, 3615; Barton; Yadav-Bhatnagar; Finet; Khamsi *Tetrahedron Lett.* **1987**, *28*, 3111.

[370]Vol'pin *Pure Appl. Chem.* **1972**, *30*, 607.

F. Carbon Electrophiles

2-32 The Conversion of Organometallic Compounds to Ketones, Aldehydes, Carboxylic Esters, or Amides
Acyl-de-metallation, etc.

$$\text{RHgX} + \text{Co}_2(\text{CO})_8 \xrightarrow{\text{THF}} \text{R}-\underset{\underset{O}{\|}}{\text{C}}-\text{R}$$

Symmetrical ketones[371] can be prepared in good yields by the reaction of organomercuric halides[372] with dicobalt octacarbonyl in THF,[373] or with nickel carbonyl in DMF or certain other solvents.[374] R may be aryl or alkyl. However, when R is alkyl, rearrangements may intervene in the $\text{Co}_2(\text{CO})_8$ reaction, though the Ni(CO)_4 reaction seems to be free from such rearrangements.[374] Divinylic ketones have been prepared in high yields by treatment of vinylic mercuric halides with CO and a rhodium catalyst.[375] When arylmercuric halides are treated with nickel carbonyl in the presence of Ar'I, unsymmetrical diaryl ketones can be obtained.[374] In a more general synthesis of unsymmetrical ketones, tetraalkyltin compounds R_4Sn are treated with a halide R'X (R' = aryl, vinylic, benzylic), CO, and a Pd complex catalyst.[376] Similar reactions use Grignard reagents, Fe(CO)_5, and an alkyl halide;[377] and an organoaluminum compound, an aryl halide, CO, and a palladium catalyst.[378] Aryl ketones can be prepared from aryltrimethylsilanes ArSiMe_3 and acyl chlorides in the presence of AlCl_3.[379]

Grignard reagents react with formic acid to give good yields of aldehydes. Two moles of RMgX are used; the first converts HCOOH to HCOO^-, which reacts with the second mole to give RCHO.[380] Aryllithiums and Grignard reagents react with iron pentacarbonyl to give aldehydes ArCHO,[381] while alkyllithium reagents react with CO to give symmetrical ketones.[382] α,β-Unsaturated aldehydes can be prepared by treatment of vinylic silanes with dichloromethyl methyl ether and TiCl_4 at $-90°\text{C}$.[383] Vinylic aluminum compounds react with methyl chloroformate ClCOOMe to give α,β-unsaturated esters directly.[384] The latter compounds can also be prepared by treating boronic esters **32** with CO, PdCl_2, and NaOAc in MeOH.[385] The synthesis of α,β-unsaturated esters has also been accomplished by treat-

[371]For reviews of the reactions in this section, and related reactions, see Narayana; Periasamy *Synthesis* **1985**, 253-268; Gulevich; Bumagin; Beletskaya *Russ. Chem. Rev.* **1988**, *57*, 299-315.
[372]For a monograph on the synthetic uses of organomercury compounds, see Larock, Ref. 279. For reviews, see Larock *Tetrahedron* **1982**, *38*, 1713-1754; *Angew. Chem. Int. Ed. Engl.* **1978**, *17*, 27-37 [*Angew. Chem. 90*, 28-38].
[373]Seyferth; Spohn *J. Am. Chem. Soc.* **1969**, *91*, 3037.
[374]Hirota; Ryang; Tsutsumi *Tetrahedron Lett.* **1971**, 1531; Ryu; Ryang; Rhee; Omura; Murai; Sonoda *Synth. Commun.* **1984**, *14*, 1175. For another method, see Hatanaka; Hiyama *Chem. Lett.* **1989**, 2049.
[375]Larock; Hershberger *J. Org. Chem.* **1980**, *45*, 3840.
[376]Tanaka *Tetrahedron Lett.* **1979**, 2601.
[377]Yamashita; Suemitsu *Tetrahedron Lett.* **1978**, 761. See also Vitale; Doctorovich; Nudelman *J. Organomet. Chem.* **1987**, *332*, 9.
[378]Bumagin; Ponomarev; Beletskaya *Doklad. Chem.* **1986**, *291*, 471.
[379]Dey; Eaborn; Walton *Organomet. Chem. Synth.* **1971**, *1*, 151-160.
[380]Sato; Oguro; Watanabe; Sato *Tetrahedron Lett.* **1980**, *21*, 2869. For another method of converting RMgX to RCHO, see Meyers; Comins *Tetrahedron Lett.* **1978**, 5179; Comins; Meyers *Synthesis* **1978**, 403; Amaratunga; Fréchet *Tetrahedron Lett.* **1983**, *24*, 1143.
[381]Ryang; Rhee; Tsutsumi *Bull. Chem. Soc. Jpn.* **1964**, *37*, 341; Giam; Ueno *J. Am. Chem. Soc.* **1977**, *99*, 3166; Yamashita; Miyoshi; Nakazono; Suemitsu *Bull. Chem. Soc. Jpn.* **1982**, *55*, 1663. For another method, see Gupton; Polk *Synth. Commun.* **1981**, *11*, 571.
[382]Ryang; Tsutsumi *Bull. Chem. Soc. Jpn.* **1962**, *35*, 1121; Ryang; Sawa; Hasimoto; Tsutsumi *Bull. Chem. Soc. Jpn.* **1964**, *37*, 1704; Trzupek; Newirth; Kelly; Sbarbati; Whitesides *J. Am. Chem. Soc.* **1973**, *95*, 8118.
[383]Yamamoto; Nunokawa; Tsuji *Synthesis* **1977**, 721; Yamamoto; Yohitake; Qui; Tsuji *Chem. Lett.* **1978**, 859.
[384]Zweifel; Lynd *Synthesis* **1976**, 625.
[385]Miyaura; Suzuki *Chem. Lett.* **1981**, 879. See also Yamashina; Hyuga; Hara; Suzuki *Tetrahedron Lett.* **1989**, *30*, 6555.

ment of vinylic mercuric chlorides with CO at atmospheric pressure and a Pd catalyst in an alcohol as solvent, e.g.,[386]

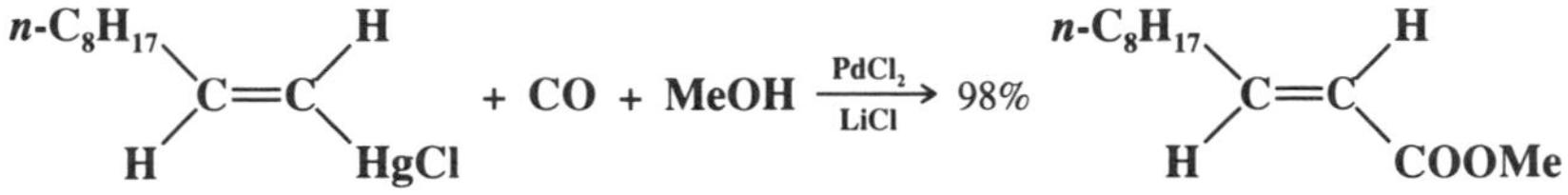

$$n\text{-}C_8H_{17}\diagdown \atop H \diagup C=C\diagdown {H \atop HgCl} + CO + MeOH \xrightarrow[\text{LiCl}]{\text{PdCl}_2} 98\% \quad n\text{-}C_8H_{17}\diagdown \atop H \diagup C=C\diagdown {H \atop COOMe}$$

Arylthallium bis(trifluoroacetates) (see **2-22**) can be carbonylated with CO, an alcohol, and a PdCl$_2$ catalyst to give esters:[387]

$$\text{ArTl(OOCCF}_3)_2 + ROH + CO \xrightarrow{\text{PdCl}_2} \text{ArCOOR}$$

Organomercury compounds undergo a similar reaction.[388] Alkyl and aryl Grignard reagents can be converted to carboxylic esters with Fe(CO)$_5$ instead of CO.[389]

Amides have been prepared by the treatment of trialkyl or triarylboranes with CO and an imine, in the presence of catalytic amounts of cobalt carbonyl:[390]

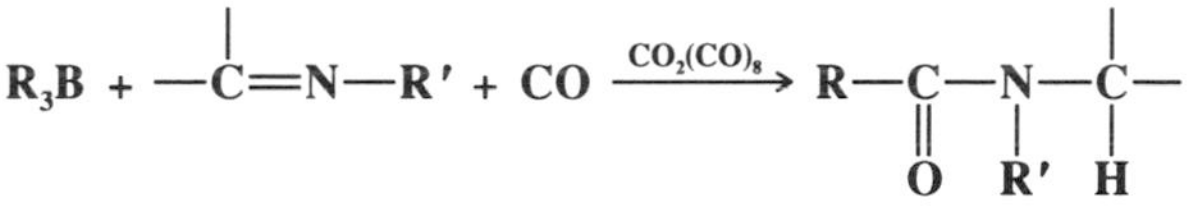

$$R_3B + \overset{|}{-}C=N-R' + CO \xrightarrow{\text{CO}_2(\text{CO})_8} R-\overset{\overset{|}{}}{\underset{\underset{O}{\parallel}}{C}}-\overset{|}{\underset{R'}{N}}-\overset{|}{\underset{H}{C}}-$$

In another method for the conversion RM → RCONR$_2'$, Grignard reagents and organolithium compounds are treated with a formamide HCONR$_2'$ to give the intermediate RCH(OM)NR$_2'$, which is not isolated, but treated with PhCHO or Ph$_2$CO to give the product RCONR$_2'$.[391]

See also reactions **0-102**, **5-21**, **6-70**, and **8-24** to **8-26**.

OS **68**, 116.

2-33   Cyano-de-metallation

$$\text{ArTl(OOCCF}_3)_2 + CuCN \longrightarrow \text{ArCN}$$

Arylthallium bis(trifluoroacetates) (see **2-22**) can be converted to aryl nitriles by treatment with copper(I) cyanide in acetonitrile.[392] Another procedure uses excess aqueous KCN followed by photolysis of the resulting complex ion ArTl(CN)$_3^-$ in the presence of excess KCN.[305] Alternatively, arylthallium acetates react with Cu(CN)$_2$ or CuCN to give aryl nitriles, e.g.[393]

$$\text{PhTl(OAc)(ClO}_4) + Cu(CN)_2 \xrightarrow[\text{pyridine}]{115°C, 5 \text{ hr}} 75\% \text{ PhCN}$$

Yields from this procedure are variable, ranging from almost nothing to 90 or 100%.

Vinylic copper reagents react with ClCN to give vinyl cyanides, though BrCN and ICN give the vinylic halide instead.[394] Vinylic cyanides have also been prepared by the reaction

[386]Larock *J. Org. Chem.* **1975**, *40*, 3237.
[387]Larock; Fellows *J. Am. Chem. Soc.* **1982**, *104*, 1900.
[388]Baird; Hartgerink; Surridge *J. Org. Chem.***1985**, *50*, 4601.
[389]Yamashita; Suemitsu *Tetrahedron Lett.* **1978**, 1477.
[390]Alper; Amaratunga *J. Org. Chem.* **1982**, *47*, 3593.
[391]Screttas; Steele *J. Org. Chem.* **1988**, *53*, 5151.
[392]Taylor; Katz; McKillop *Tetrahedron Lett.* **1984**, *25*, 5473.
[393]Uemura; Ikeda; Ichikawa *Tetrahedron* **1972**, *28*, 3025.
[394]Westmijze; Vermeer *Synthesis* **1977**, 784.

between vinylic lithium compounds and phenyl cyanate PhOCN.[395] Alkyl cyanides RCN have been prepared, in varying yields, by treatment of sodium trialkylcyanoborates with NaCN and lead tetraacetate.[396]

For other electrophilic substitutions of the type RM → RC, see **0-86** to **0-107,** which are discussed under nucleophilic substitutions in Chapter 10. See also **6-69.**

G.   Metal Electrophiles

2-34   Transmetallation with a Metal
Metallo-de-metallation

$$RM + M' \rightleftharpoons RM' + M$$

Many organometallic compounds are best prepared by this reaction, which involves replacement of a metal in an organometallic compound by another metal. RM' can be successfully prepared only when M' is above M in the electromotive series, unless some other way is found to shift the equilibrium. That is, RM is usually an unreactive compound and M' is a metal more active than M. Most often, RM is R_2Hg, since mercury alkyls[372] are easy to prepare and mercury is far down in the electromotive series.[397] Alkyls of Li, Na, K, Be, Mg, Al, Ga, Zn, Cd, Te, Sn, etc. have been prepared this way. An important advantage of this method over **2-38** is that it ensures that the organometallic compound will be prepared free of any possible halide. This method can be used for the isolation of solid sodium and potassium alkyls.[398] If the metals lie too close together in the series, it may not be possible to shift the equilibrium. For example, alkylbismuth compounds cannot be prepared in this way from alkylmercury compounds.

OS **V,** 1116.

2-35   Transmetallation with a Metal Halide
Metallo-de-metallation

$$RM + M'X \rightleftharpoons RM' + MX$$

In contrast to **2-34** the reaction between an organometallic compound and a metal *halide* is successful only when M' is *below* M in the electromotive series.[399] The two reactions considered together therefore constitute a powerful tool for preparing all kinds of organometallic compounds. In this reaction the most common substrates are Grignard reagents and organolithium compounds.[400] Among others, alkyls of Be, Zn,[401] Cd, Hg, Al, Sn, Pb, Co, Pt, and Au have been prepared by treatment of Grignard reagents with the appropriate halide.[402] The reaction has been used to prepare alkyls of almost all nontransition metals and even of some transition metals. Alkyls of metalloids and of nonmetals, including Si, B,[403] Ge, P,

[395]Murray; Zweifel *Synthesis* **1980,** 150.

[396]Masuda; Hoshi; Yamada; Arase *J. Chem. Soc., Chem. Commun.* **1984,** 398.

[397]For a review of the reaction when M is Hg, see Makarova, Ref. 295, pp. 190-226. For a review where M' is Li, see Wardell, in Zuckerman, Ref. 246, pp. 31-44.

[398]BuNa and BuK have also been prepared by exchange of BuLi with *t*-BuONa or *t*-AmOK: Pi; Bauer; Brix; Schade; Schleyer *J. Organomet. Chem.* **1986,** *306,* C1.

[399]For reviews of the mechanism, see Abraham; Grellier, Ref. 292, pp. 25-149; Abraham, Ref. 2, pp. 39-106; Jensen; Rickborn, Ref. 2, pp. 100-192. Also see Schlosser, Ref. 292.

[400]For monographs on organolithium compounds, see Wakefield, Ref. 273; Wakefield *The Chemistry of Organolithium Compounds*; Pergamon: Elmsford, NY, 1974.

[401]For a review of the use of activated zinc, see Erdik *Tetrahedron* **1987,** *43,* 2203-2212.

[402]For a review, see Noltes *Bull. Soc. Chim. Fr.* **1972,** 2151-2160.

[403]For a method of preparing organoboranes from RMgX and BF_3, where the RMgX is present only in situ, see Brown; Racherla *Tetrahedron Lett.* **1985,** *26,* 4311.

As, Sb, and Bi, can also be prepared in this manner.[404] Except for alkali-metal alkyls and Grignard reagents, the reaction between RM and M'X is the most common method for the preparation of organometallic compounds.[405]

Lithium dialkylcopper reagents can be prepared by mixing 2 moles of RLi with 1 mole of a cuprous halide in ether at low temperatures:[406]

$$\text{2RLi} + \text{CuX} \longrightarrow \text{R}_2\text{CuLi} + \text{LiX}$$

Another way is to dissolve an alkylcopper compound in an alkyllithium solution.

If M' has a valence higher than 1, it is often possible to stop the reaction before all the halogens have been replaced, e.g.,

$$\text{RMgX} + \text{SiCl}_4 \longrightarrow \text{RSiCl}_3$$

However, it is not always possible: $\text{RMgX} + \text{BF}_3$ gives only BR_3, although BRCl_2 can be prepared from R_2Zn and BCl_3.

Metallocenes (see p. 47) are usually made by this method:

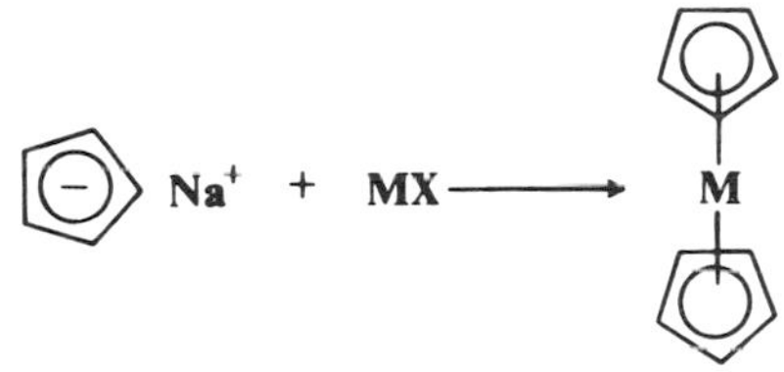

Among others, metallocenes of Sc, Ti, V, Cr, Mn, Fe, Co, and Ni have been prepared in this manner.[407]

Metal nitrates are sometimes used instead of halides.

OS **I**, 231, 550; **III**, 601; **IV**, 258, 473, 881; **V**, 211, 496, 727, 918, 1001; **VI**, 776, 875, 1033; **VII**, 236, 290, 524; **65**, 61, 108; **67**, 20, 86, 125; **68**, 104, 182. Also see OS **IV**, 476

2-36   Transmetallation with an Organometallic Compound
Metallo-de-metallation

$$\text{RM} + \text{R'M'} \rightleftharpoons \text{RM'} + \text{R'M}$$

This type of metallic exchange is used much less often than **2-34** and **2-35**. It is an equilibrium reaction and is useful only if the equilibrium lies in the desired direction. Usually the goal is to prepare a lithium compound that is not prepared easily in other ways,[408] e.g., a vinylic or an allylic lithium, most commonly from an organotin substrate. Examples are the preparation of vinyllithium from phenyllithium and tetravinyltin and the formation of α-dialkylamino organolithium compounds from the corresponding organotin compounds[409]

$$\text{RR'NCH}_2\text{SnBu}_3 + \text{BuLi} \xrightarrow{0°C} \text{RR'NCH}_2\text{Li} + \text{Bu}_4\text{Sn}$$

[404]For reviews as applied to Si, B, and P, see Wakefield, Ref. 273, pp. 149-158; Kharasch; Reinmuth *Grignard Reactions of Nonmetallic Substances*; Prentice-Hall: Englewood Cliffs, NJ, 1954, pp. 1306-1345.

[405]For a review with respect to Al, see Mole *Organomet. React.* **1970**, *1*, 1-54, pp. 31-43; to Hg, see Larock, Ref. 279, pp. 9-26; Makarova, Ref. 295, pp. 129-178, 227-240; to Cu, Ag, or Au, see van Koten, in Zuckerman, Ref. 246, pp. 219-232; to Zn, Cd, or Hg, see Wardell, in Zuckerman, Ref. 246, pp. 248-270.

[406]House; Chu; Wilkins; Umen *J. Org. Chem.* **1975**, *40*, 1460. But see also Lipshutz; Whitney; Kozlowski; Breneman *Tetrahedron Lett.* **1986**, *27*, 4273; Bertz; Dabbagh *Tetrahedron* **1989**, *45*, 425.

[407]For reviews of the preparation of metallocenes, see Bublitz; Rinehart *Org. React.* **1969**, *17*, 1-154; Birmingham *Adv. Organomet. Chem.* **1965**, *2*, 365-413, pp. 375-382.

[408]For reviews, see Wardell, in Hartley; Patai, Ref. 1, vol. 4, pp. 1-157, pp. 81-89; Kauffmann *Top. Curr. Chem.* **1980**, *92*, 109-147, pp. 130-136.

[409]Peterson *J. Am. Chem. Soc.* **1971**, *93*, 4027; Peterson; Ward *J. Organomet. Chem.* **1974**, *66*, 209; Pearson; Lindbeck *J. Org. Chem.* **1989**, *54*, 5651.

The reaction has also been used to prepare 1,3-dilithiopropanes[410] and 1,1-dilithiomethylenecyclohexane[411] from the corresponding mercury compounds. In general, the equilibrium lies in the direction in which the more electropositive metal is bonded to that alkyl or aryl group that is the more stable carbanion (p. 176). The reaction proceeds with retention of configuration;[412] an SEi mechanism is likely.[413]

"Higher order" cuprates (see Ref. 1277 in Chapter 10) have been produced by this reaction starting with a vinylic tin compound:[414]

$$\textbf{RSnR}_3' + \textbf{Me}_2\textbf{Cu(CN)Li}_2 \longrightarrow \textbf{RCuMe(CN)Li}_2 + \textbf{MeSnR}_3' \quad \textbf{R} = \text{a vinylic group}$$

These compounds are not isolated, but used directly in situ for conjugate addition reactions (**5-18**). Another method for the preparation of such reagents (but with Zn instead of Li) allows them to be made from α-acetoxy halides:[415]

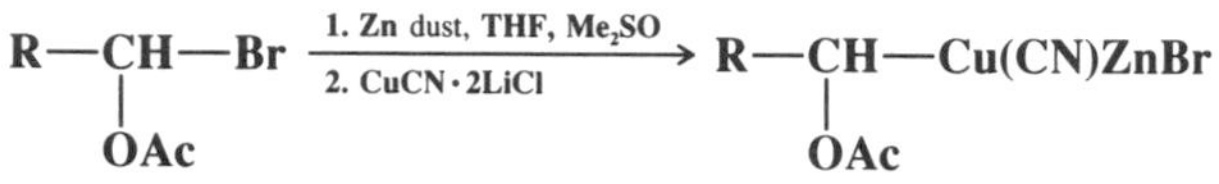

$$\underset{\overset{|}{\text{OAc}}}{\text{R—CH—Br}} \xrightarrow[\text{2. CuCN·2LiCl}]{\text{1. Zn dust, THF, Me}_2\text{SO}} \underset{\overset{|}{\text{OAc}}}{\text{R—CH—Cu(CN)ZnBr}}$$

OS **V**, 452; **VI**, 815; **68**, 116.

Halogen as Leaving Group

A. Hydrogen as the Electrophile

2-37   Reduction of Alkyl Halides

Although this reaction can proceed by an electrophilic substitution mechanism, it is considered in Chapter 10 (**0-76**).

B. Metal Electrophiles

2-38 Metallo-de-halogenation

$$\textbf{RX} + \textbf{M} \longrightarrow \textbf{RM}$$

Alkyl halides react directly with certain metals to give organometallic compounds.[416] The most common metal is magnesium, and of course this is by far the most common method for the preparation of Grignard reagents.[417] The order of halide activity is I > Br > Cl. The reaction can be applied to many alkyl halides—primary, secondary, and tertiary—and to aryl halides, though aryl *chlorides* require the use of THF or another higher-boiling solvent instead of the usual ether, or special entrainment methods.[418] Aryl iodides and bromides can be treated in the usual manner. Allylic Grignard reagents can also be prepared

[410]Seetz; Schat; Akkerman; Bickelhaupt *J. Am. Chem. Soc.* **1982**, *104*, 6848.
[411]Maercker; Dujardin *Angew. Chem. Int. Ed. Engl.* **1984**, *23*, 224 [*Angew. Chem. 96*, 222].
[412]Seyferth; Vaughan *J. Am. Chem. Soc.* **1964**, *86*, 883; Sawyer; Kucerovy; Macdonald; McGarvey *J. Am. Chem. Soc.* **1988**, *110*, 842.
[413]Dessy; Kaplan; Coe; Salinger *J. Am. Chem. Soc.* **1963**, *85*, 1191.
[414]Behling; Babiak; Ng; Campbell; Moretti; Koerner; Lipshutz *J. Am. Chem. Soc.* **1988**, *110*, 2641.
[415]Chou; Knochel *J. Org. Chem.* **1990**, *55*, 4791.
[416]For reviews, see Massey; Humphries *Aldrichimica Acta* **1989**, *22*, 31-38; Negishi, Ref. 1, pp. 30-37; Rochow *J. Chem. Educ.* **1966**, *43*, 58-62.
[417]For reviews, see Raston; Salem, in Hartley; Patai, Ref. 1, vol. 4, pp. 159-306, pp. 162-175; Kharasch; Reinmuth, Ref. 404, pp. 5-91.
[418]Pearson; Cowan; Beckler *J. Org. Chem.* **1959**, *24*, 504.

in the usual manner (or in THF),[419] though in the presence of excess halide these may give Wurtz-type coupling products (see **0-87**).[420] Like aryl chlorides, vinylic halides require higher-boiling solvents (see OS **IV**, 258). A good procedure for benzylic and allylic halides is to use magnesium anthracene (prepared from Mg and anthracene in THF)[421] instead of ordinary magnesium,[422] though activated magnesium turnings have also been used.[423] Alkynyl Grignard reagents are not generally prepared by this method at all. For these, **2-21** is used.

Dihalides[424] can be converted to Grignard reagents if the halogens are different and are at least three carbons apart. If the halogens are the same, it is possible to obtain dimagnesium compounds, e.g., $BrMg(CH_2)_4MgBr$.[425] 1,2-Dihalides give elimination[426] instead of Grignard reagent formation (**7-29**), and the reaction is seldom successful with 1,1-dihalides, though the preparation of *gem*-disubstituted compounds, such as $CH_2(MgBr)_2$, has been accomplished with these substrates.[427] α-Halo Grignard reagents and α-halolithium reagents can be prepared by the method given in **2-39**.[428] Alkylmagnesium fluorides can be prepared by refluxing alkyl fluorides with Mg in the presence of appropriate catalysts (e.g., I_2 or EtBr) in THF for several days.[429]

The presence of other functional groups in the halide usually affects the preparation of the Grignard reagent. Groups that contain active hydrogen (defined as any hydrogen that will react with a Grignard reagent), such as OH, NH_2, and COOH, can be present in the molecule, but only if they are converted to the salt form (O^-, NH^-, COO^-, respectively). Groups that react with Grignard reagents, such as C=O, C≡N, NO_2, COOR, etc., inhibit Grignard formation entirely. In general, the only functional groups that may be present in the halide molecule without any interference at all are double and triple bonds (except terminal triple bonds) and OR and NR_2 groups. However, β-halo ethers generally give β elimination when treated with magnesium (see **7-31**), and Grignard reagents from α-halo ethers[430] can only be formed in THF or dimethoxymethane at a low temperature, e.g.,[431]

$$\text{EtOCH}_2\text{Cl} + \text{Mg} \xrightarrow[-30°\text{C}]{\text{THF or CH}_2(\text{OMe})_2} \text{EtOCH}_2\text{MgCl}$$

because such reagents immediately undergo α elimination (see **2-39**) at room temperature in ether solution.

[419]For a review of allyl and crotyl Grignard reagents, see Benkeser *Synthesis* **1971,** 347-358.
[420]For a method of reducing coupling in the formation of allylic Grignard reagents, see Oppolzer; Schneider *Tetrahedron Lett.* **1984,** *25,* 3305.
[421]Freeman; Hutchinson *J. Org. Chem.* **1983,** *48,* 879; Bogdanović; Janke; Kinzelmann *Chem. Ber.* **1990,** *123,* 1507, and other papers in this series.
[422]Gallagher; Harvey; Raston; Sue *J. Chem. Soc., Chem. Commun.* **1988,** 289.
[423]Baker; Brown; Hughes; Skarnulis; Sexton *J. Org. Chem.* **1991,** *56,* 698. For a review of the use of activated magnesium, see Lai *Synthesis* **1981,** 585-604.
[424]For reviews of the preparation of Grignard reagents from dihalides, see Raston; Salem, Ref. 417, pp. 187-193; Heaney *Organomet. Chem. Rev.* **1966,** *1,* 27-42. For a review of di-Grignard reagents, see Bickelhaupt *Angew. Chem. Int. Ed. Engl.* **1987,** *26,* 990-1005 [*Angew. Chem. 99,* 1020-1036].
[425]For example, see Denise; Ducom; Fauvarque *Bull. Soc. Chim. Fr.* **1972,** 990; Seetz; Hartog; Böhm; Blomberg; Akkerman; Bickelhaupt *Tetrahedron Lett.* **1982,** *23,* 1497.
[426]For formation of 1,2-dilithio compounds and 1,2-di-Grignard reagents, but not by this method, see van Eikkema Hommes; Bickelhaupt; Klumpp *Recl. Trav. Chim. Pays-Bas* **1988,** *107,* 393, *Angew. Chem. Int. Ed. Engl.* **1988,** 27, 1083 [*Angew. Chem. 100,* 1100].
[427]For example, see Bertini; Grasselli; Zubiani; Cainelli *Tetrahedron* **1970,** *26,* 1281; Bruin; Schat; Akkerman; Bickelhaupt *J. Organomet. Chem.* **1985,** *288,* 13. For the synthesis of *gem*-dilithio and 1,1,1-trilithio compounds, see Landro; Gurak; Chinn; Newman; Lagow *J. Am. Chem. Soc.* **1982,** 104, 7345; Baran; Lagow *J. Am. Chem. Soc.* **1990,** *112,* 9415.
[428]For a review of compounds containing both carbon–halogen and carbon–metal bonds, see Chivers *Organomet. Chem. Rev., Sect.* **1970,** *6,* 1-64.
[429]Yu; Ashby *J. Org. Chem.* **1971,** *36,* 2123.
[430]For a review of organometallic compounds containing a hetero atom (N, O, P, S, or Si), see Peterson *Organomet. Chem. Rev., Sect. A* **1972,** *7,* 295-358.
[431]For example, see Normant; Castro, *C. R. Acad. Sci.* **1963,** *257,* 2115, **1964,** *259,* 830; Castro *Bull. Soc. Chim. Fr.* **1967,** 1533, 1540, 1547; Taeger; Kahlert; Walter *J. Prakt. Chem.* **1965,** [4] *28,* 13.

Because Grignard reagents react with water (**2-24**) and with oxygen (**2-25**), it is generally best to prepare them in an anhydrous nitrogen atmosphere. Grignard reagents are generally neither isolated nor stored; solutions of Grignard reagents are used directly for the required synthesis. Grignard reagents can also be prepared in benzene or toluene, if a tertiary amine is added to complex with the RMgX.[432] This method eliminates the need for an ether solvent. With certain primary alkyl halides it is even possible to prepare alkylmagnesium compounds in hydrocarbon solvents in the absence of an organic base.[433] It is also possible to obtain Grignard reagents in powdered form, by complexing them with the chelating agent tris(3,6-dioxaheptyl)amine $N(CH_2CH_2OCH_2CH_2OCH_3)_3$.[434]

Next to the formation of Grignard reagents, the most important application of this reaction is the conversion of alkyl and aryl halides to organolithium compounds,[435] but it has also been carried out with many other metals, e.g., Na, Be, Zn, Hg, As, Sb, and Sn. With sodium, the Wurtz reaction (**0-86**) is an important side reaction. In some cases where the reaction between a halide and a metal is too slow, an alloy of the metal with potassium or sodium can be used instead. The most important example is the preparation of tetraethyllead from ethyl bromide and a Pb–Na alloy.

The efficiency of the reaction can often be improved by use of the metal in its powdered[435a] or vapor[436] form. These techniques have permitted the preparation of some organometallic compounds that cannot be prepared by the standard procedures. Among the metals produced in an activated form are Mg,[437] Ca,[438] Zn,[439] Al, Sn, Cd,[440] Ni, Fe, Ti, Cu,[441] Pd, and Pt.[442]

The mechanism of Grignard reagent formation involves free radicals.[443] There is much evidence for this, from CIDNP[444] (p. 187) and from stereochemical, rate, and product studies.[445] Further evidence is that free radicals have been trapped,[446] and that experiments that studied the intrinsic reactivity of MeBr on a magnesium single-crystal surface showed that Grignard reagent formation does not take place by a single-step insertion mechanism.[447] The following SET mechanism has been proposed:[444]

$$\text{R—X} + \overline{\text{Mg}} \longrightarrow \text{R—X} \overset{\cdot}{\underset{}{\text{}}} + \text{Mg}_s \overset{\cdot}{\text{}^+}$$

$$\text{R—X} \overset{\cdot}{\underset{}{\text{}}} \longrightarrow \text{R·} + \text{X}^-$$

$$\text{X}^- + \text{Mg}_s \overset{\cdot}{\text{}^+} \longrightarrow \text{XMg}_s$$

$$\text{R·} + \text{XMg}_s \longrightarrow \text{RMgX}$$

[432]Ashby; Reed *J. Org. Chem.* **1966**, *31*, 971; Gitlitz; Considine *J. Organomet. Chem.* **1970**, *23*, 291.

[433]Smith *J. Organomet. Chem.* **1974**, *64*, 25.

[434]Boudin; Cerveau; Chuit; Corriu; Reye *Tetrahedron* **1989**, *45*, 171.

[435]For reviews, see Wakefield, Ref. 273, pp. 21-32; Wardell, in Hartley; Patai, vol. 4, pp. 1-157, pp. 5-27; Newcomb, in Zuckerman, Ref. 246, pp. 3-14.

[435a]For a review, see Rieke *Science* **1989**, *246*, 1260-1264.

[436]For reviews, see Klabunde *React. Intermed.* (Plenum) **1980**, *1*, 37-149, *Acc. Chem. Res.* **1975**, *8*, 393-399; Skell; Havel; McGlinchey *Acc. Chem. Res.* **1973**, *6*, 97-105; Timms *Adv. Inorg. Radiochem.* **1972**, *14*, 121.

[437]Burns; Rieke *J. Org. Chem.* **1987**, *52*, 3674; Ebert; Rieke *J. Org. Chem.* **1988**, *53*, 4482. See also Ref. 423.

[438]Wu; Xiong; Rieke *J. Org. Chem.* **1990**, *55*, 5045.

[439]Rieke; Li; Burns; Uhm *J. Org. Chem.* **1981**, *46*, 4323. See also Grondin; Sebban; Vottero; Blancou; Commeyras *J. Organomet. Chem.* **1989**, *362*, 237; Berk; Yeh; Jeong; Knochel *Organometallics* **1990**, *9*, 3053; Zhu; Wehmeyer; Rieke *J. Org. Chem.* **1991**, *56*, 1445.

[440]Burkhardt; Rieke *J. Org. Chem.* **1985**, *50*, 416.

[441]Stack; Dawson; Rieke *J. Am. Chem. Soc.* **1991**, *113*, 4672, and references cited therein.

[442]For reviews, see Lai, Ref. 423; Rieke *Acc. Chem. Res.* **1977**, *10*, 301-306, *Top. Curr. Chem.* **1975**, *59*, 1-31.

[443]For a review, see Blomberg *Bull. Soc. Chim. Fr.* **1972**, 2143.

[444]Bodewitz; Blomberg; Bickelhaupt *Tetrahedron Lett.* **1972**, 281, **1975**, 2003, *Tetrahedron* **1973**, *29*, 719, **1975**, *31*, 1053. See also Lawler; Livant *J. Am. Chem. Soc.* **1976**, *98*, 3710; Schaart; Blomberg; Akkerman; Bickelhaupt *Can. J. Chem.* **1980**, *58*, 932.

[445]See, for example, Walborsky; Aronoff *J. Organomet. Chem.* **1973**, *51*, 31; Czernecki; Georgoulis; Gross; Prevost *Bull. Soc. Chim. Fr.* **1968**, 3720; Rogers; Hill; Fujiwara; Rogers; Mitchell; Whitesides *J. Am. Chem. Soc.* **1980**, *102*, 217; Barber; Whitesides *J. Am. Chem. Soc.* **1980**, *102*, 239.

[446]Root; Hill; Lawrence; Whitesides *J. Am. Chem. Soc.* **1989**, *111*, 5405.

[447]Nuzzo; Dubois *J. Am. Chem. Soc.* **1986**, *108*, 2881.

The species R—X$^{\bullet}$ and Mg$^{\bullet}_{s}$ are radical ions.[448] The subscript "s" is meant to indicate that the species so marked are bound to the surface of the magnesium. It has been suggested that some of the R• radicals diffuse from the magnesium surface into the solution and then return to the surface to react with the XMg•. There is evidence both for[449] and against[450] this suggestion. Another proposal is that the fourth step is not the one shown here, but that the R• is reduced by Mg$^+$ to the carbanion R$^-$, which combines with MgX$^+$ to give RMgX.[451]

There are too many preparations of Grignard reagents in *Organic Syntheses* for us to list here. Use of the reaction to prepare other organometallic compounds can be found in OS **I**, 228; **II**, 184, 517, 607; **III**, 413, 757; **VI**, 240; **VII**, 346; **65**, 42. The preparation of unsolvated butylmagnesium bromide is described at OS **V**, 1141. The preparation of highly reactive (powdered) magnesium is given at OS **VI**, 845.

2-39   Replacement of a Halogen by a Metal from an Organometallic Compound
Metallo-de-halogenation

$$\text{RX} + \text{R}'\text{M} \longrightarrow \text{RM} + \text{R}'\text{X}$$

The exchange reaction between halides and organometallic compounds is almost entirely limited to the cases where M is lithium and X is bromide or iodide,[452] though it has been shown to occur with magnesium.[453] R′ is usually, though not always, alkyl, and often butyl; R is usually aromatic.[454] Alkyl halides are generally not reactive enough, while allylic and benzylic halides usually give Wurtz coupling. Of course, the R that becomes bonded to the halogen is the one for which RH is the weaker acid. Vinylic halides react with retention of configuration.[455] The reaction can be used to prepare α-halo organolithium and α-halo organomagnesium compounds,[456] e.g.,[457]

$$\text{CCl}_4 + \text{BuLi} \xrightarrow[-105°\text{C}]{\text{THF}} \text{Cl}_3\text{C—Li}$$

Such compounds can also be prepared by hydrogen–metal exchange, e.g.,[458]

$$\text{Br}_3\text{CH} + \text{iso-PrMgCl} \xrightarrow[-95°\text{C}]{\text{THF–HMPA}} \text{Br}_3\text{C—MgCl} + \text{C}_3\text{H}_8$$

[448]For additional evidence for this mechanism, see Vogler; Stein; Hayes *J. Am. Chem. Soc.* **1978**, *100*, 3163; Sergeev; Zagorsky; Badaev *J. Organomet. Chem.* **1983**, *243*, 123. However, there is evidence that the mechanism may be more complicated: de Souza-Barboza; Luche; Pétrier *Tetrahedron Lett.* **1987**, *28*, 2013.

[449]Garst; Deutch; Whitesides *J. Am. Chem. Soc.* **1986**, *108*, 2490; Ashby; Oswald *J. Org. Chem.* **1988**, *53*, 6068; Garst; Swift *J. Am. Chem. Soc.* **1989**, *111*, 241; Garst *Acc. Chem. Res.* **1991**, *24*, 95; Garst; Ungváry; Batlaw; Lawrence *J. Am. Chem. Soc.* **1991**, *113*, 5392. For a discussion, see Walling *Acc. Chem. Res.* **1991**, *24*, 255.

[450]Walborsky; Rachon *J. Am. Chem. Soc.* **1989**, *111*, 1896; Rachon; Walborsky *Tetrahedron Lett.* **1989**, *30*, 7345; Walborsky *Acc. Chem. Res.* **1990**, *23*, 286-293.

[451]de Boer; Akkerman; Bickelhaupt *Angew. Chem. Int. Ed. Engl.* **1988**, *27*, 687 [*Angew. Chem. 100*, 735].

[452]For reviews, see Wardell, in Zuckerman, Ref. 246, pp. 107-129; Parham; Bradsher *Acc. Chem. Res.* **1982**, *15*, 300-305.

[453]See, for example, Zakharkin; Okhlobystin; Bilevitch *J. Organomet. Chem.* **1964**, *2*, 309; Tamborski; Moore *J. Organomet. Chem.* **1971**, *26*, 153.

[454]For the preparation of primary alkyllithiums by this reaction, see Bailey; Punzalan *J. Org. Chem.* **1990**, *55*, 5404; Negishi; Swanson; Rousset *J. Org. Chem.* **1990**, *55*, 5406.

[455]For examples of exchange where R = vinylic, see Neumann; Seebach *Chem. Ber.* **1978**, *111*, 2785; Miller; McGarvey *Synth. Commun* **1979**, *9*, 831; Sugita; Sakabe; Sasahara; Tsukuda; Ichikawa *Bull. Chem. Soc. Jpn.* **1984**, *57*, 2319.

[456]For reviews of such compounds, see Siegel *Top. Curr. Chem.* **1982**, *106*, 55-78; Negishi, Ref. 1, pp. 136-151; Kaabrich *Angew. Chem. Int. Ed. Engl.* **1972**, *11*, 473-485, **1967**, *6*, 41-52 [*Angew. Chem. 84*, 557-570, *79*, 15-27]; *Bull. Soc. Chim. Fr.* **1969**, 2712-2720; Villieras *Organomet. Chem. Rev., Sect. A* **1971**, *7*, 81-94. For related reviews, see Krief *Tetrahedron* **1980**, *36*, 2531-2640; Normant *J. Organomet. Chem.* **1975**, *100*, 189-203; Zhil'tsov; Druzhkov *Russ. Chem. Rev.* **1971**, *40*, 126-141.

[457]Hoeg; Lusk; Crumbliss *J. Am. Chem. Soc.* **1965**, *87*, 4147. See also Villieras; Tarhouni; Kirschleger; Rambaud *Bull. Soc. Chim. Fr.* **1985**, 825.

[458]Villieras *Bull. Soc. Chim. Fr.* **1967**, 1520.

This is an example of **2-21.** However, these α-halo organometallic compounds are stable (and configurationally stable as well[458a]) only at low temperatures (~ −100°C) and only in THF or mixtures of THF and other solvents (e.g., HMPA). At ordinary temperatures they lose MX (α elimination) to give carbenes (which then react further) or carbenoid reactions. The α-chloro-α-magnesio sulfones $ArSO_2CH(Cl)MgBr$ are exceptions, being stable in solution at room temperature and even under reflux.[459] Compounds in which a halogen and a transition metal are on the same carbon can be more stable than the ones with lithium.[460]

There is evidence that the mechanism[461] of the reaction of alkyllithium compounds with alkyl and aryl iodides involves free radicals.[462]

$$RX + R'M \rightleftharpoons [R\cdot, X, M, R'\cdot] \rightleftharpoons RM + R'X$$

Solvent cage

Among the evidence is the obtention of coupling and disproportionation products from R· and R'· and the observation of CIDNP.[463] However, in the degenerate exchange between PhI and PhLi the ate complex $Ph_2I^- Li^+$ has been shown to be an intermediate,[464] and there is other evidence that radicals are not involved in all instances of this reaction.[465]

In a completely different kind of process, alkyl halides can be converted to certain organometallic compounds by treatment with organometallate ions, e.g.,

$$RX + R_3'SnLi \longrightarrow RSnR_3' + LiX$$

Most of the evidence is in accord with a free radical mechanism involving electron transfer, though an SN2 mechanism can compete under some conditions.[466]

OS **VI**, 82; **VII**, 271, 326, 495; **66**, 67, 210. See also OS **VII**, 512; **66**, 95.

Carbon Leaving Groups

In these reactions (**2-40** to **2-48**) a carbon–carbon bond cleaves. We regard as the substrate that side which retains the electron pair; hence the reactions are considered electrophilic substitutions. The incoming group is hydrogen in all but one (**2-42**) of the cases. The reactions in groups A and B are sometimes called *anionic cleavages*,[467] though they do not always occur by mechanisms involving free carbanions (SE1). When they do, the reactions are facilitated by increasing stability of the carbanion.

A. Carbonyl-Forming Cleavages. These reactions follow the pattern

$$-\overset{|}{\underset{|}{C}}-\overset{|}{\underset{|}{C}}-\overline{\underset{\cdot\cdot}{O}}| \longrightarrow -\overset{|}{\underset{|}{C}}|^{\ominus} + \overset{|}{C}=\overline{O}|$$

[458a]Hoffmann; Ruhland; Bewersdorf *J. Chem. Soc., Chem. Commun.* **1991**, 195; Schmidt; Köbrich; Hoffmann *Chem. Ber.* **1991**, *124*, 1253; Hoffmann; Bewersdorf *Chem. Ber.* **1991**, *124*, 1259.

[459]Stetter; Steinbeck *Liebigs Ann. Chem.* **1972**, *766*, 89.

[460]Kauffmann; Fobker; Wensing *Angew. Chem. Int. Ed. Engl.* **1988**, *27*, 943 [*Angew. Chem. 100*, 1005].

[461]For reviews of the mechanism, see Bailey; Patricia *J. Organomet. Chem.* **1988**, *352*, 1-46; Beletskaya; Artamkina; Reutov *Russ. Chem. Rev.* **1976**, *45*, 330-347.

[462]Ward; Lawler; Cooper *J. Am. Chem. Soc.* **1969**, *91*, 746; Lepley; Landau *J. Am. Chem. Soc.* **1969**, *91*, 748; Ashby; Pham *J. Org. Chem.* **1987**, *52*, 1291. See also Bailey; Patricia; Nurmi; Wang *Tetrahedron Lett.* **1986**, *27*, 1861.

[463]Ward; Lawler; Loken *J. Am. Chem. Soc.* **1968**, *90*, 7359; Ref. 462.

[464]See Farnham; Calabrese *J. Am. Chem. Soc.* **1986**, *108*, 2449; Reich; Green; Phillips *J. Am. Chem. Soc.* **1989**, *111*, 3444.

[465]Rogers; Houk *J. Am. Chem. Soc.* **1982**, *104*, 522; Beak; Allen; Lee *J. Am. Chem. Soc.* **1990**, *112*, 1629.

[466]See San Filippo; Silbermann *J. Am. Chem. Soc.* **1982**, *104*, 2831; Ashby; Su; Pham *Organometallics* **1985**, *4*, 1493; Alnajjar; Kuivila *J. Am. Chem. Soc.* **1985**, *107*, 416.

[467]For a review, see Artamkina; Beletskaya *Russ. Chem. Rev.* **1987**, *56*, 983-1001.

The leaving group is stabilized because the electron deficiency at its carbon is satisfied by a pair of electrons from the oxygen. With respect to the leaving group the reaction is elimination to form a C=O bond. Retrograde aldol reactions (**6-39**) and cleavage of cyanohydrins (**6-49**) belong to this classification but are treated in Chapter 16 under their more important reverse reactions. Other eliminations to form C=O bonds are discussed in Chapter 17 (**7-43** and **7-44**).

2-40   Decarboxylation of Aliphatic Acids
   Hydro-de-carboxylation

$$\textbf{RCOOH} \longrightarrow \textbf{RH} + \textbf{CO}_2$$

Many carboxylic acids can be successfully decarboxylated, either as the free acid or in the salt form, but not simple fatty acids.[468] An exception is acetic acid, which as the acetate, heated with base, gives good yields of methane. Aliphatic acids that do undergo successful decarboxylation have certain functional groups or double or triple bonds in the α or β position. Some of these are shown in Table 12.2. For decarboxylation of aromatic acids, see **1-39**. Decarboxylation of an α-cyano acid can give a nitrile or a carboxylic acid, since the cyano group may or may not be hydrolyzed in the course of the reaction. In addition to the compounds listed in Table 12.2, decarboxylation can also be carried out on α,β-unsaturated and α,β-acetylenic acids. α,β-Unsaturated acids can also be decarboxylated with copper and quinoline in a manner similar to that discussed in **1-39**. Glycidic acids give aldehydes on decarboxylation. The following mechanism has been suggested:[469]

TABLE 12.2   Some acids which undergo decarboxylation fairly readily
   Others are described in the text

Acid type		Decarboxylation product	
Malonic	HOOC—C—COOH	HOOC—C—H	
α-Cyano	NC—C—COOH	NC—C—H or HOOC—C—H	
α-Nitro	O$_2$N—C—COOH	O$_2$N—C—H	
α-Aryl	Ar —C—COOH	Ar —C—H	
α,α,α-Trihalo	X$_3$C—COOH	X$_3$CH	
β-Keto	—C—C—COOH (O double bond)	—C—C—H (O double bond)	
β,γ-Unsaturated	—C=C—C—COOH	—C=C—C—H	

[468]March *J. Chem. Educ.* **1963**, *40*, 212.
[469]Singh; Kagan *J. Org. Chem.* **1970**, *35*, 2203.

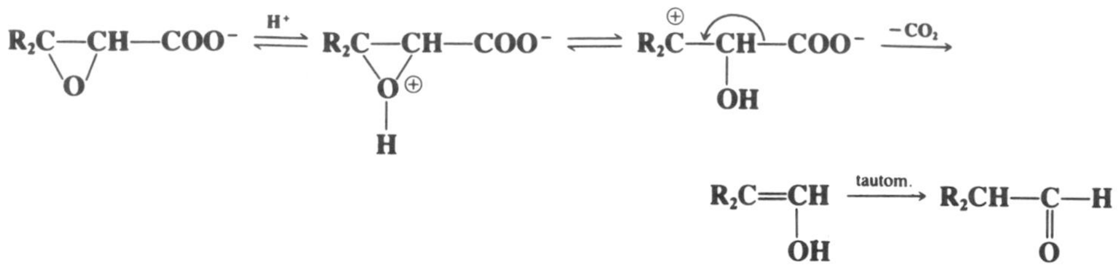

$$R_2C-CH-COO^- \underset{}{\overset{H^+}{\rightleftharpoons}} R_2C-CH-COO^- \rightleftharpoons R_2C\overset{\oplus}{-CH}-COO^- \overset{-CO_2}{\longrightarrow}$$

$$R_2C=CH \overset{tautom.}{\longrightarrow} R_2CH-C-H$$

The direct product is an enol that tautomerizes to the aldehyde.[470] This is the usual last step in the Darzens reaction (**6-45**).

Decarboxylations can be regarded as reversals of the addition of carbanions to carbon dioxide (**6-32**), but free carbanions are not always involved.[471] When the carboxylate *ion* is decarboxylated, the mechanism can be either SE1 or SE2. In the case of the SE1 mechanism, the reaction is of course aided by the presence of electron-withdrawing groups, which stabilize the carbanion.[472] Decarboxylations of carboxylate ions can be accelerated by the addition of a suitable crown ether, which in effect removes the metallic ion.[473] The reaction without the metallic ion has also been performed in the gas phase.[474] But some acids can also be decarboxylated directly and, in most of these cases, there is a cyclic, six-center mechanism:

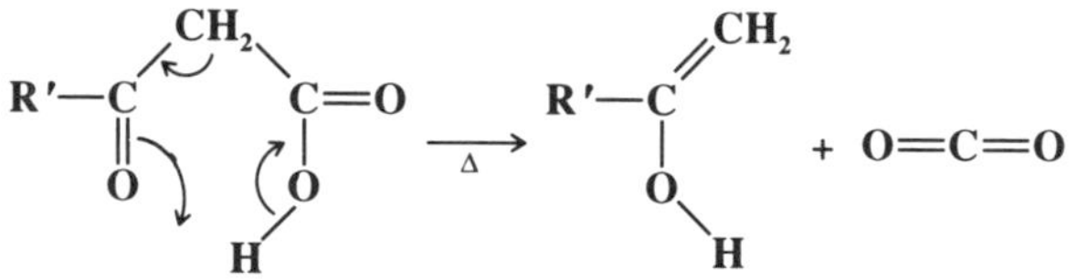

$$R'-C \overset{CH_2}{\underset{O}{\big\|}} \overset{CH_2}{\underset{O}{C=O}} \overset{\Delta}{\longrightarrow} R'-C \overset{CH_2}{\underset{O}{\diagdown}} + O=C=O$$

Here too there is an enol that tautomerizes to the product. The mechanism is illustrated for the case of β-keto acids,[475] but it is likely that malonic acids, α-cyano acids, α-nitro acids, and β,γ-unsaturated acids[476] behave similarly, since similar six-membered transition states can be written for them. Some α,β-unsaturated acids are also decarboxylated by this mechanism by isomerizing to the β,γ-isomers before they actually decarboxylate.[477] Evidence is that **36** and similar bicyclic β-keto acids resist decarboxylation.[478] In such compounds the

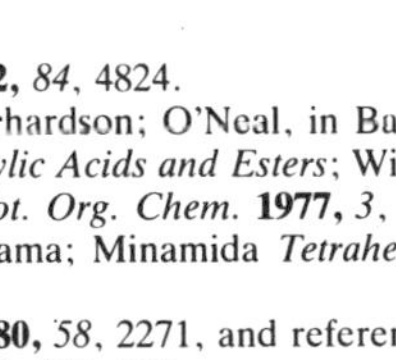

36

[470]Shiner; Martin *J. Am. Chem. Soc.* **1962**, *84*, 4824.

[471]For reviews of the mechanism, see Richardson; O'Neal, in Bamford; Tipper, Ref. 53, vol. 5, 1972, pp. 447-482; Clark, in Patai *The Chemistry of Carboxylic Acids and Esters*; Wiley: New York, 1969, pp. 589-622. For a review of carbon isotope effect studies, see Dunn *Isot. Org. Chem.* **1977**, *3*, 1-38.

[472]See, for example, Oae; Tagaki; Uneyama; Minamida *Tetrahedron* **1968**, *24*, 5283; Buncel; Venkatachalam; Menon *J. Org. Chem.* **1984**, *49*, 413.

[473]Hunter; Patel; Perry *Can. J. Chem.* **1980**, *58*, 2271, and references cited therein.

[474]Graul; Squires *J. Am. Chem. Soc.* **1988**, *110*, 607.

[475]For a review of the mechanism of the decarboxylation of β-keto acids, see Jencks *Catalysis in Chemistry and Enzmology*; McGraw-Hill: New York, 1969, pp. 116-120

[476]Bigley; Clarke *J. Chem. Soc., Perkin Trans. 2* **1982**, 1, and references cited therein. For a review, see Smith; Kelly, *Prog. Phys. Org. Chem.* **1971**, *8*, 75-234, pp. 150-153.

[477]Bigley *J. Chem. Soc.* **1964**, 3897.

[478]Wasserman, in Newman *Steric Effects in Organic Chemistry*; Wiley: New York, 1956, p. 352. See also Buchanan; Kean; Taylor *Tetrahedron* **1975**, *31*, 1583.

six-membered cyclic transition state cannot form for steric reasons, and if it could, formation of the intermediate enol would violate Bredt's rule (p. 160).[479] Some carboxylic acids that cannot form a six-membered transition state can still be decarboxylated, and these presumably react through an SE1 or SE2 mechanism.[480] Further evidence for the cyclic mechanism is that the reaction rate varies very little with a change from a nonpolar to a polar solvent (even from benzene to water[481]), and is not subject to acid catalysis.[482] The rate of decarboxylation of a β,γ-unsaturated acid was increased about 10^5-10^6 times by introduction of a β-methoxy group, indicating that the cyclic transition state has dipolar character.[483]

β-Keto acids[484] are easily decarboxylated, but such acids are usually prepared from β-keto esters, and the esters are easily decarboxylated themselves on hydrolysis without isolation of the acids.[485] This decarboxylation of β-keto esters involving cleavage on the carboxyl side of the substituted methylene group (arrow) is carried out under acidic, neutral, or

$$R-\overset{\displaystyle R}{\underset{\displaystyle O}{\overset{\displaystyle |}{\underset{\displaystyle ||}{C}}}}-\overset{\displaystyle \downarrow}{\underset{\displaystyle R}{\overset{\displaystyle |}{\underset{\displaystyle |}{C}}}}-COOR' \xrightarrow{\ H_2O\ } R-\overset{\displaystyle R}{\underset{\displaystyle O}{\overset{\displaystyle |}{\underset{\displaystyle ||}{C}}}}-\overset{\displaystyle R}{\underset{\displaystyle R}{\overset{\displaystyle |}{\underset{\displaystyle |}{CH}}}} + CO_2 + R'OH$$

slightly basic conditions to yield a ketone. When strongly basic conditions are used, cleavage occurs on the other side of the CR_2 group (**2-43**). β-Keto esters can be decarbalkoxylated without passing through the free-acid stage by treatment with boric anhydride B_2O_3 at 150°C.[486] The alkyl portion of the ester (R') is converted to an alkene or, if it lacks a β hydrogen, to an ether R'OR'. Another method for the decarbalkoxylation of β-keto esters, malonic esters, and α-cyano esters consists of heating the substrate in wet dimethyl sulfoxide containing NaCl, Na_3PO_4, or some other simple salt.[487] In this method too, the free acid is probably not an intermediate, but here the alkyl portion of the substrate is converted to the corresponding alcohol. Ordinary carboxylic acids, containing no activating groups, can be decarboxylated by conversion to esters of N-hydroxypyridine-2-thione and treatment of these with Bu_3SnH.[488] A free-radical mechanism is likely. α-Amino acids have been decarboxylated by treatment with a catalytic amount of 2-cyclohexenone.[489] Certain decarboxylations can also be accomplished photochemically.[490] See also the decarbonylation of acyl halides, mentioned in **4-41**. In some cases decarboxylations can give organometallic compounds: $RCOOM \rightarrow RM + CO_2$.[491]

[479]Sterically hindered β-keto acids decarboxylate more slowly: Meier; Wengenroth; Lauer; Krause *Tetrahedron Lett.* **1989**, *30*, 5253.
[480]For example, see Ferris; Miller *J. Am. Chem. Soc.* **1966**, *88*, 3522.
[481]Westheimer; Jones *J. Am. Chem. Soc.* **1941**, *63*, 3283; Swain; Bader; Esteve; Griffin *J. Am. Chem. Soc.* **1961**, *83*, 1951.
[482]Pedersen *Acta Chem. Scand.* **1961**, *15*, 1718; Noyce; Metesich *J. Org. Chem.* **1967**, *32*, 3243.
[483]Bigley; Al-Borno *J. Chem. Soc., Perkin Trans. 2* **1982**, 15.
[484]For a review of β-keto acids, see Oshry; Rosenfeld *Org. Prep. Proced. Int.* **1982**, *14*, 249-264.
[485]For a list of examples, with references, see Ref. 52, pp. 774-775.
[486]Lalancette; Lachance *Tetrahedron Lett.* **1970**, 3903.
[487]For a review of the synthetic applications of this method, see Krapcho *Synthesis* **1982**, 805-822, 893-914. For other methods, see Aneja; Hollis; Davies; Eaton *Tetrahedron Lett.* **1983**, *24*, 4641; Brown; Jones *J. Chem. Res. (S)* **1984**, 332; Dehmlow; Kunesch *Synthesis* **1985**, 320; Taber; Amedio; Gulino *J. Org. Chem.* **1989**, *54*, 3474.
[488]Barton; Crich; Motherwell *Tetrahedron* **1985**, *41*, 3901; Della; Tsanaktsidis *Aust. J. Chem.* **1987**, *39*, 2061. For another method of more limited scope, see Maier; Roth; Thies; Schleyer *Chem. Ber.* **1982**, *115*, 808.
[489]Hashimoto; Eda; Osanai; Iwai; Aoki *Chem. Lett.* **1986**, 893.
[490]See Davidson; Steiner *J. Chem. Soc., Perkin Trans. 2* **1972**, 1357; Kraeutler; Bard *J. Am. Chem. Soc.* **1978**, *100*, 5985; Hasebe; Tsuchiya *Tetrahedron Lett.* **1987**, *28*, 6207; Okada; Okubo; Oda *Tetrahedron Lett.* **1989**, *30*, 6733.
[491]For reviews, see Deacon *Organomet. Chem. Rev. A* **1970**, 355-372; Deacon; Faulks; Pain *Adv. Organomet. Chem.* **1986**, *25*, 237-276.

Some of the decarboxylations listed in *Organic Syntheses* are performed with concomitant ester or nitrile hydrolysis and others are simple decarboxylations.

With ester or nitrile hydrolysis: OS **I**, 290, 451, 523; **II**, 200, 391; **III**, 281, 286, 313, 326, 510, 513, 591; **IV**, 55, 93, 176, 441, 664, 708, 790, 804; **V**, 76, 288, 572, 687, 989; **VI**, 615, 781, 873, 932; **VII**, 50, 210, 319; **67**, 170.

Simple decarboxylations: OS **I**, 351, 401, 440, 473, 475; **II**, 21, 61, 93, 229, 302, 333, 368, 416, 474, 512, 523; **III**, 213, 425, 495, 705, 733, 783; **IV**, 234, 254, 278, 337, 555, 560, 597, 630, 731, 857; **V**, 251, 585; **VI**, 271, 965; **VII**, 249, 359; **65**, 98; **66**, 29; **68**, 210. Also see OS **IV**, 633.

2-41   Cleavage of Alkoxides
Hydro-de-(α-oxidoalkyl)-substitution

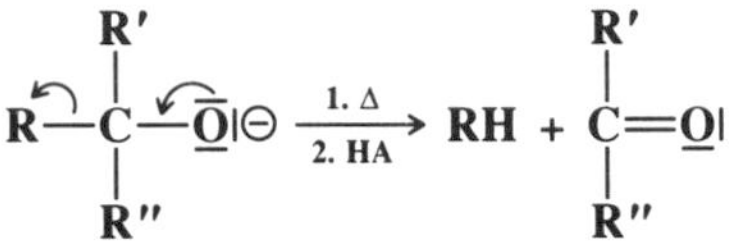

$$
\begin{array}{c}
R' \\
| \\
R\!-\!C\!-\!\underline{O}|\ominus \xrightarrow[\text{2. HA}]{\text{1. }\Delta} RH + \overset{R'}{\underset{R''}{\overset{|}{\underset{|}{C}}}}\!=\!\underline{O}|
\end{array}
$$

Alkoxides of tertiary alcohols can be cleaved in a reaction that is essentially the reverse of addition of carbanions to ketones (**6-29**).[492] The reaction is unsuccessful when the R groups are simple unbranched alkyl groups, e.g., the alkoxide of triethylcarbinol. Cleavage is accomplished with branched alkoxides such as the alkoxides of diisopropylneopentylcarbinol or tri-*t*-butylcarbinol.[493] Allylic,[494] benzylic,[495] and aryl groups also cleave; for example, the alkoxide of triphenylcarbinol gives benzene and benzophenone. Studies in the gas phase show that the cleavage is a simple one, giving the carbanion and ketone directly in one step.[496] However, with some substrates in solution, substantial amounts of dimer RR have been found, indicating a radical pathway.[497] Hindered alcohols (not the alkoxides) also lose one R group by cleavage, also by a radical pathway.[498]

The reaction has been used for extensive mechanistic studies (see p. 574).

OS **VI**, 268.

2-42   Replacement of a Carboxyl Group by an Acyl Group
Acyl-de-carboxylation

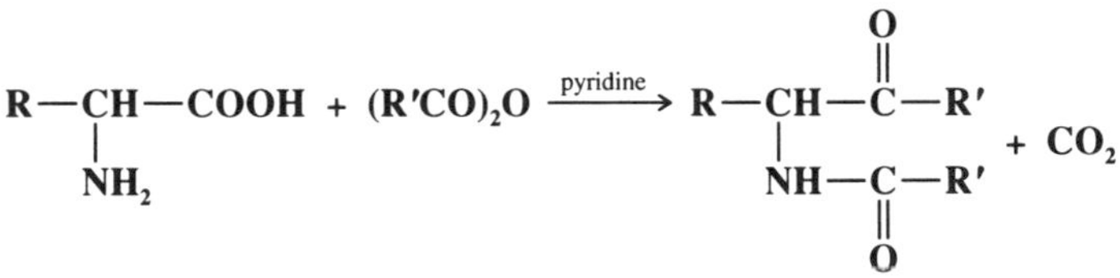

$$
R\!-\!\underset{NH_2}{\overset{|}{C}H}\!-\!COOH + (R'CO)_2O \xrightarrow{\text{pyridine}} R\!-\!\underset{\underset{O}{\overset{||}{NH-C-R'}}}{\overset{|}{C}H}\!-\!\overset{O}{\overset{||}{C}}\!-\!R' + CO_2
$$

[492]Zook; March; Smith *J. Am. Chem. Soc.* **1959**, *81*, 1617; Barbot; Miginiac *J. Organomet. Chem.* **1977**, *132*, 445; Benkeser; Siklosi; Mozdzen *J. Am. Chem. Soc.* **1978**, *100*, 2134.

[493]Arnett; Small; McIver; Miller *J. Org. Chem.* **1978**, *43*, 815. See also Lomas; Dubois *J. Org. Chem.* **1984**, *49*, 2067.

[494]See Snowden; Linder; Muller; Schulte-Elte *Helv. Chim. Acta* **1987**, *70*, 1858, 1879.

[495]Partington; Watt *J. Chem. Soc., Perkin Trans.* 2 **1988**, 983.

[496]Tumas; Foster; Brauman *J. Am. Chem. Soc.* **1988**, *110*, 2714; Ibrahim; Watt; Wilson; Moore *J. Chem. Soc., Chem. Commun.* **1989**, 161.

[497]Paquette; Gilday; Maynard *J. Org. Chem.* **1989**, *54*, 5044; Paquette; Maynard *J. Org. Chem.* **1989**, *54*, 5054.

[498]See Lomas; Fain; Briand *J. Org. Chem.* **1990**, *55*, 1052, and references cited therein.

When an α-amino acid is treated with an anhydride in the presence of pyridine, the carboxyl group is replaced by an acyl group and the NH_2 becomes acylated. This is called the *Dakin–West reaction*.[499] The mechanism involves formation of an oxazolone.[500] The reaction sometimes takes place on carboxylic acids even when an α amino group is not present. A number of N-substituted amino acids RCH(NHR')COOH give the corresponding N-alkylated products.

OS **IV**, 5; **V**, 27.

B. Acyl Cleavages. In these reactions (**2-43** to **2-46**) a carbonyl group is attacked by a hydroxide ion (or amide ion), giving an intermediate that undergoes cleavage to a carboxylic acid (or an amide). With respect to the leaving group, this is nucleophilic substitution at a carbonyl group and the mechanism is the tetrahedral one discussed in Chapter 10.

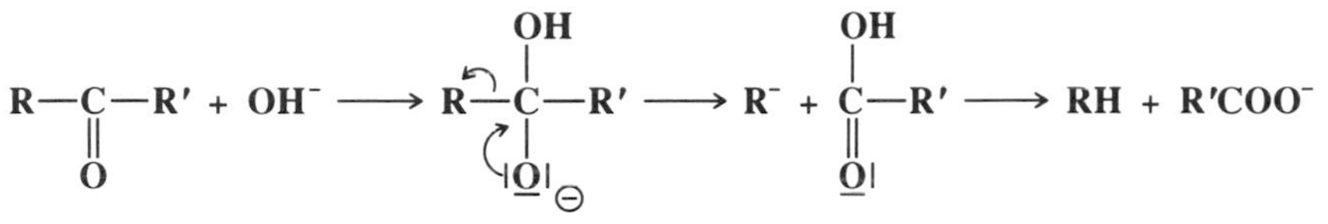

$$\underset{O}{\overset{\quad}{R-C-R'}} + OH^- \longrightarrow R\overset{OH}{\underset{\underline{|O|}_\ominus}{-C-R'}} \longrightarrow R^- + \underset{\underline{O}|}{\overset{OH}{C-R'}} \longrightarrow RH + R'COO^-$$

With respect to R this is of course electrophilic substitution. The mechanism is usually S$_E$1.

2-43 Basic Cleavage of β-Keto Esters and β-Diketones
Hydro-de-acylation

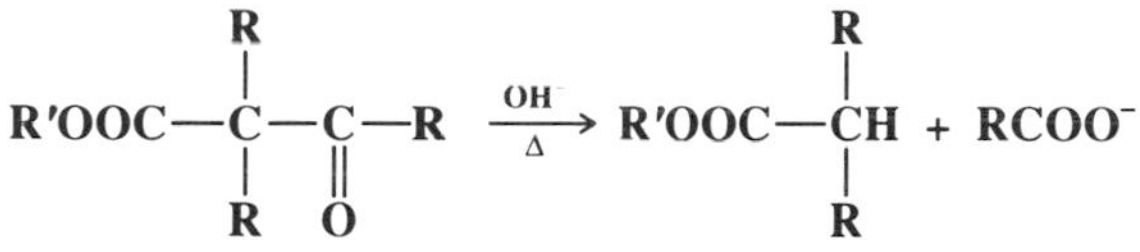

$$\underset{R\;\;O}{R'OOC-\overset{R}{\underset{|}{C}}-\overset{}{\underset{\|}{C}}-R} \xrightarrow[\Delta]{OH} \underset{R}{R'OOC-\overset{R}{\underset{|}{CH}}} + RCOO^-$$

When β-keto esters are treated with concentrated base, cleavage occurs, but is on the keto side of the CR$_2$ group (arrow) in contrast to the acid cleavage mentioned on page 629. The

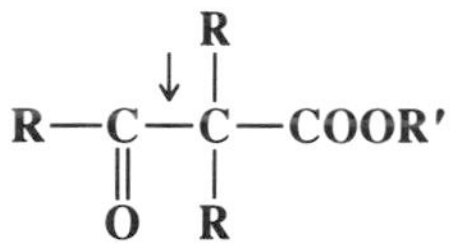

$$\underset{O\;\;R}{R-\overset{R}{\underset{\|}{C}}-\overset{\downarrow}{\underset{|}{C}}-COOR'}$$

products are a carboxylic ester and the salt of an acid. However, the utility of the reaction is somewhat limited by the fact that decarboxylation is a side reaction, even under basic conditions. β-Diketones behave similarly to give a ketone and the salt of a carboxylic acid. With both β-keto esters and β-diketones, OEt$^-$ can be used instead of OH$^-$, in which case the ethyl esters of the corresponding acids are obtained instead of the salts. In the case of β-keto esters, this is the reverse of Claisen condensation (**0-108**). The similar cleavage of

[499]For a review, see Buchanan *Chem. Soc. Rev.* **1988**, *17*, 91-109.
[500]Allinger; Wang; Dewhurst *J. Org. Chem.* **1974**, *39*, 1730.

cyclic α-cyano ketones, in an intramolecular fashion, has been used to effect a synthesis of macrocyclic lactones, e.g.,[501]

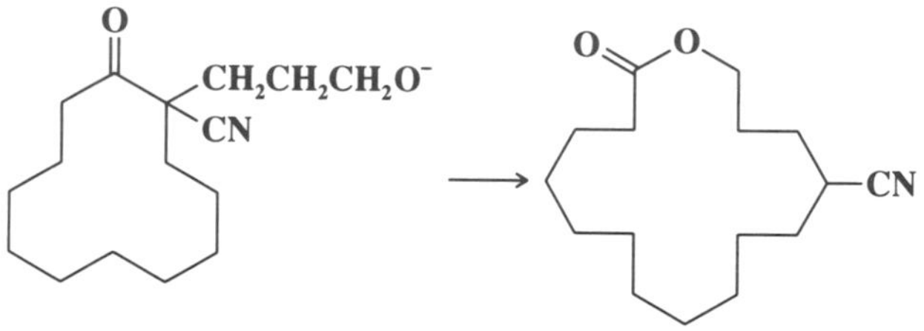

Activated F^- (from KF and a crown ether) has been used as the base to cleave an α-cyano ketone.[502]

OS **II,** 266, 531; **III,** 379; **IV,** 415, 957; **V,** 179, 187, 277, 533, 747, 767.

2-44   Haloform Reaction

$$CH_3-\underset{\underset{O}{\|}}{C}-R \xrightarrow[OH^-]{Br_2} HCBr_3 + RCOO^-$$

In the *haloform reaction*, methyl ketones (and the only methyl aldehyde, acetaldehyde) are cleaved with halogen and a base.[503] The halogen can be bromine, chlorine, or iodine. What takes place is actually a combination of two reactions. The first is an example of **2-4,** in which, under the basic conditions employed, the methyl group is trihalogenated. Then the resulting trihalo ketone is attacked by hydroxide ion:[504]

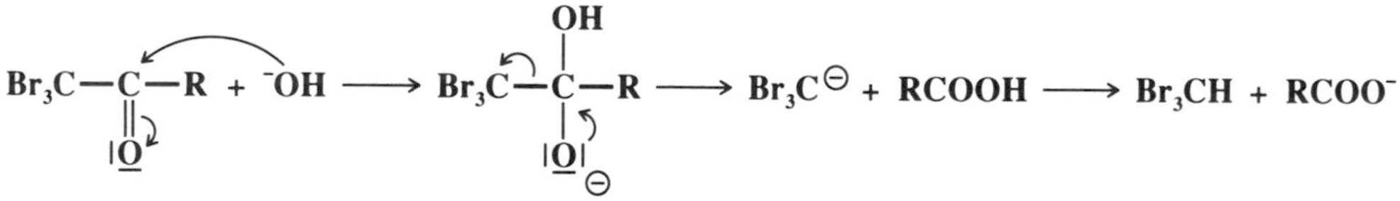

Primary or secondary methylcarbinols also give the reaction, because they are oxidized to the carbonyl compounds under the conditions employed. As with **2-4,** the rate-determining step is the preliminary enolization of the methyl ketone.[505] A side reaction is α halogenation of the nonmethyl R group. Sometimes these groups are also cleaved.[506] The reaction cannot be applied to F_2, but ketones of the form $RCOCF_3$ (R = alkyl or aryl) give fluoroform and $RCOO^-$ when treated with base.[507] Rate constants for cleavage of X_3CCOPh (X = F, Cl, Br) were found to be in the ratio $1:5.3 \times 10^{10}: 2.2 \times 10^{13}$, showing that an F_3C^- group cleaves much more slowly than the others.[508] The haloform reaction is often used as a test

[501]Milenkov; Hesse *Helv. Chim. Acta* **1987,** *70,* 308. For a similar preparation of lactams, see Wälchli; Bienz; Hesse *Helv. Chim. Acta* **1985,** *68,* 484.
[502]Beletskaya; Gulyukina; Borodkin; Solov'yanov; Reutov *Doklad. Chem.* **1984,** *276,* 202. See also Mignani; Morel; Grass *Tetrahedron Lett.* **1987,** *28,* 5505.
[503]For a review of this and related reactions, see Chakrabartty, in Trahanovsky *Oxidation in Organic Chemistry,* pt. C; Academic Press: New York, 1978, pp. 343-370.
[504]For a complete kinetic analysis of the chlorination of acetone, see Guthrie; Cossar *Can. J. Chem.* **1986,** *64,* 1250. For a discussion of the mechanism of the cleavage step, see Zucco; Lima; Rezende; Vianna; Nome *J. Org. Chem.* **1987,** *52,* 5356.
[505]Pocker *Chem. Ind. (London)* **1959,** 1383.
[506]Levine; Stephens *J. Am. Chem. Soc.* **1950,** *72,* 1642.
[507]See Hudlicky *Chemistry of Organic Fluorine Compounds,* 2nd ed.; Ellis Horwood: Chichester, 1976, pp. 276-278.
[508]Guthrie; Cossar *Can. J. Chem.* **1990,** *68,* 1640.

for methylcarbinols and methyl ketones. Iodine is most often used as the test reagent, since iodoform is an easily identifiable yellow solid. The reaction is also frequently used for synthetic purposes. Methyl ketones $RCOCH_3$ can be converted directly to methyl esters $RCOOCH_3$ by an electrochemical reaction.[509]

OS **I**, 526; **II**, 428; **III**, 302; **IV**, 345; **V**, 8. Also see OS **VI**, 618.

2-45   Cleavage of Nonenolizable Ketones
Hydro-de-acylation

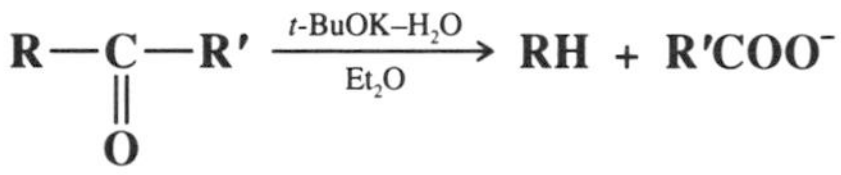

$$R-\underset{\underset{O}{\|}}{C}-R' \xrightarrow[\text{Et}_2\text{O}]{t\text{-BuOK–H}_2\text{O}} RH + R'COO^-$$

Ordinary ketones arc generally much more difficult to cleave than trihalo ketones or β-diketones, because the carbanion intermediates in these cases are more stable than simple carbanions. However, nonenolizable ketones can be cleaved by treatment with a 10:3 mixture of t-BuOK–H_2O in an aprotic solvent such as ether, dimethyl sulfoxide, 1,2-dimethoxyethane (glyme), etc.,[510] or with solid t-BuOK in the absence of a solvent.[511] When the reaction is applied to monosubstituted diaryl ketones, that aryl group preferentially cleaves that comes off as the more stable carbanion, except that aryl groups substituted in the ortho position are more readily cleaved than otherwise because of the steric effect (relief of strain).[512] In certain cases, cyclic ketones can be cleaved by base treatment, even if they are enolizable.[513]

OS **VI**, 625. See also OS **VII**, 297.

2-46   The Haller-Bauer Reaction
Hydro-de-acylation

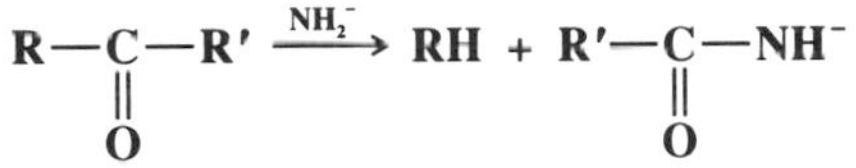

$$R-\underset{\underset{O}{\|}}{C}-R' \xrightarrow{\text{NH}_2^-} RH + R'-\underset{\underset{O}{\|}}{C}-NH^-$$

Cleavage of ketones with sodium amide is called the *Haller–Bauer reaction*.[514] As with **2-45**, which is exactly analogous, the reaction is usually applied only to nonenolizable ketones, most often to ketones of the form $ArCOCR_3$, where the products R_3CCONH_2 are not easily attainable by other methods. However, many other ketones have been used, though benzophenone is virtually unaffected. It has been shown that the configuration of optically active R is retained.[515] The NH_2 loses its proton before the R is cleaved:[516]

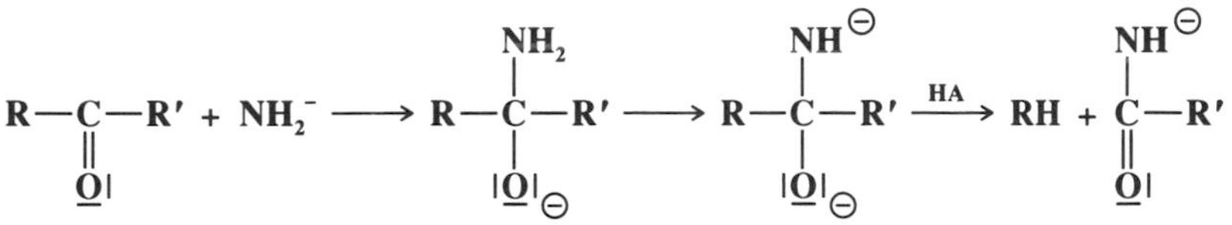

$$R-\underset{\underset{\underset{\ominus}{|}}{\overset{\|}{O}|}}{C}-R' + NH_2^- \longrightarrow R-\underset{\underset{\underset{\ominus}{|}}{\overset{|}{O}|}}{\overset{\overset{NH_2}{|}}{C}}-R' \longrightarrow R-\underset{\underset{\underset{\ominus}{|}}{\overset{|}{O}|}}{\overset{\overset{NH^\ominus}{|}}{C}}-R' \xrightarrow{HA} RH + \underset{\underset{|}{\overset{\|}{O}|}}{\overset{\overset{NH^\ominus}{|}}{C}}-R'$$

OS **V**, 384, 1074.

[509]Nikishin; Elinson; Makhova *Tetrahedron* **1991**, *47*, 895.
[510]Swan *J. Chem. Soc.* **1948**, 1408; Gassman; Lumb; Zalar *J. Am. Chem. Soc.* **1967**, *89*, 946.
[511]March; Plankl *J. Chem. Soc., Perkin Trans. 1* **1977**, 460.
[512]Davies; Derenberg; Hodge *J. Chem. Soc. C* **1971**, 455; Ref. 511.
[513]For example, see Swaminathan; Newman *Tetrahedron* **1958**, *2*, 88; Hoffman; Cram, Ref. 25.
[514]For a review, see Gilday; Paquette *Org. Prep. Proced. Int.* **1990**, *22*, 167-201. For an improved procedure, see Kaiser; Warner *Synthesis* **1975**, 395.
[515]Impastato; Walborsky *J. Am. Chem. Soc.* **1962**, *84*, 4838; Paquette; Gilday *J. Org. Chem.* **1988**, *53*, 4972; Paquette; Ra *J. Org. Chem.* **1988**, *53*, 4978.
[516]Bunnett; Hrutfiord *J. Org. Chem.* **1962**, *27*, 4152.

C. Other Cleavages

2-47 The Cleavage of Alkanes
Hydro-de-*t*-butylation, etc.

$$(CH_3)_4C \xrightarrow{\text{FSO}_3\text{H-SbF}_5} CH_4 + (CH_3)_3C^+$$

The C—C bonds of alkanes can be cleaved by treatment with super acids[44] (p. 249). For example, neopentane in FSO_3H–SbF_5 can cleave to give methane and the *t*-butyl cation. C—H cleavage (see **2-1**) is a competing reaction and, for example, neopentane can give H_2 and the *t*-pentyl cation (formed by rearrangement of the initially formed neopentyl cation) by this pathway. In general, the order of reactivity is tertiary C—H > C—C > secondary C—H ≫ primary C—H, though steric factors cause a shift in favor of C—C cleavage in such a hindered compound as tri-*t*-butylmethane. The mechanism is similar to that shown in **2-1** and **2-18** and involves attack by H^+ on the C—C bond to give a pentavalent cation.

Catalytic hydrogenation seldom breaks unactivated C—C bonds (i.e., R—R′ + H_2 → RH + R′H), but methyl and ethyl groups have been cleaved from substituted adamantanes by hydrogenation with a Ni–Al_2O_3 catalyst at about 250°C.[517] Certain C—C bonds have been cleaved by alkali metals.[518]

2-48 **Decyanation** or **Hydro-de-cyanation**

$$RCN \xrightarrow[\text{Na-Fe(acac)}_3]{\substack{\text{Na-NH}_3 \\ \text{or}}} RH$$

The cyano group of alkyl nitriles can be removed[519] by treatment with metallic sodium, either in liquid ammonia,[520] or together with tris(acetylacetonato)iron(III) Fe(acac)$_3$[521] or, with lower yields, titanocene. The two procedures are complementary. Although both can be used to decyanate many kinds of nitriles, the Na–NH_3 method gives high yields with R groups such as trityl, benzyl, phenyl, and tertiary alkyl, but lower yields (~35 to 50%) when R = primary or secondary alkyl. On the other hand, primary and secondary alkyl nitriles are decyanated in high yields by the Na–Fe(acac)$_3$ procedure. Sodium in liquid ammonia is known to be a source of solvated electrons, and the reaction may proceed through the free radical R• which would then be reduced to the carbanion R^-, which by abstraction of a proton from the solvent, would give RH. The mechanism with Fe(acac)$_3$ is presumably different. Another procedure,[522] which is successful for R = primary, secondary, or tertiary, involves the use of potassium metal and the crown ether dicyclohexano-18-crown-6 in toluene.[523]

α-Amino and α-amido nitriles $RCH(CN)NR_2'$ and $RCH(CN)NHCOR'$ can be decyanated in high yield by treatment with $NaBH_4$.[524]

[517]Grubmüller; Schleyer; McKervey *Tetrahedron Lett.* **1979**, 181.
[518]For examples and references, see Grovenstein; Bhatti; Quest; Sengupta; VanDerveer *J. Am. Chem. Soc.* **1983**, *105*, 6290.
[519]For a list of procedures, with references, see Ref. 52, pp. 42-43.
[520]Büchner; Dufaux *Helv. Chim. Acta* **1966**, *49*, 1145; Arapakos; Scott; Huber *J. Am. Chem. Soc.* **1969**, *91*, 2059; Birch; Hutchinson *J. Chem. Soc., Perkin Trans. 1* **1972**, 1546; Yamada; Tomioka; Koga *Tetrahedron Lett.* **1976**, 61.
[521]Van Tamelen; Rudler; Bjorklund *J. Am. Chem. Soc.* **1971**, *93*, 7113.
[522]For other procedures, see Cuvigny; Larcheveque; Normant *Bull. Soc. Chim. Fr.* **1973**, 1174; Berkoff; Rivard; Kirkpatrick; Ives *Synth. Commun.* **1980**, *10*, 939; Savoia; Tagliavini; Trombini; Umani-Ronchi *J. Org. Chem.* **1980**, *45*, 3227; Ozawa; Iri; Yamamoto *Chem. Lett.* **1982**, 1707.
[523]Ohsawa; Kobayashi; Mizuguchi; Saitoh; Oishi *Tetrahedron Lett.* **1985**, *26*, 6103.
[524]Yamada; Akimoto *Tetrahedron Lett.* **1969**, 3105; Fabre; Hadj Ali Salem; Welvart *Bull. Soc. Chim. Fr.* **1975**, 178. See also Ogura; Shimamura; Fujita *J. Org. Chem.* **1991**, *56*, 2920.

Electrophilic Substitution at Nitrogen

In most of the reactions in this section, an electrophile bonds with the unshared pair of a nitrogen atom. The electrophile may be a free positive ion or a positive species attached to a carrier that breaks off in the course of the attack or shortly after:

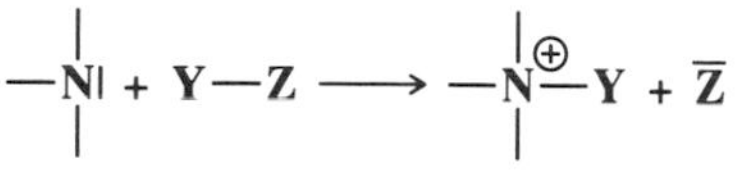

$$-\overset{|}{\underset{|}{N}}| + Y-Z \longrightarrow -\overset{|}{\underset{|}{N}}\overset{\oplus}{-}Y + \overline{Z}$$

37

Further reaction of **37** depends on the nature of Y and of the other groups attached to the nitrogen.

2-49 Diazotization

$$Ar-NH_2 + HONO \longrightarrow Ar-\overset{\oplus}{N}\equiv\overline{N}$$

When primary aromatic amines are treated with nitrous acid, diazonium salts are formed.[525] The reaction also occurs with aliphatic primary amines, but aliphatic diazonium ions are extremely unstable, even in solution (see p. 355). Aromatic diazonium ions are more stable, because of the resonance interaction between the nitrogens and the ring:

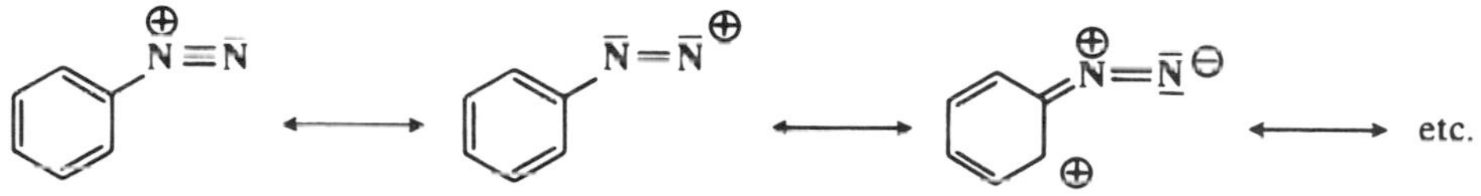

38   **39**

Incidentally, **38** contributes more to the hybrid than **39,** as shown by bond-distance measurements.[526] In benzenediazonium chloride, the C—N distance is ~1.42 Å, and the N—N distance ~1.08 Å,[527] which values fit more closely to a single and a triple bond than to two double bonds (see Table 1.5). Even aromatic diazonium salts are stable only at low temperatures, usually only below 5°C, though more stable ones, such as the diazonium salt obtained from sulfanilic acid, are stable up to 10 or 15°C. Diazonium salts are usually prepared in aqueous solution and used without isolation,[528] though it is possible to prepare solid diazonium salts if desired (see **3-24**). The stability of aryl diazonium salts can be increased by crown ether complexion.[529]

For aromatic amines, the reaction is very general. Halogen, nitro, alkyl, aldehyde, sulfonic acid, etc., groups do not interfere. Since aliphatic amines do not react with nitrous acid

---

[525]For reviews, see, in Patai, *The Chemistry of Diazonium and Diazo Groups*; Wiley: New York, 1978, the articles by Hegarty, pt. 2, pp. 511-591, and Schank, pt. 2, pp. 645-657; Godovikova; Rakitin; Khmel'nitskii *Russ. Chem. Rev.* **1983,** *52,* 440-445; Challis; Butler, in Patai *The Chemistry of the Amino Group*; Wiley: New York, 1968, pp. 305-320. For a review with respect to heterocyclic amines, see Butler *Chem. Rev.* **1975,** *75,* 241-257.

[526]For a review of diazonium salt structures, see Sorriso, in Patai *The Chemistry of Diazonium and Diazo Groups*, pt. 1, Ref. 525, pp. 95-105.

[527]Rømming *Acta Chem. Scand.* **1959,** *13,* 1260, **1963,** *17,* 1444; Sorriso, Ref. 526, p. 98; Cygler; Przybylska; Elofson *Can. J. Chem.* **1982,** *60,* 2852; Ball; Elofson *Can. J. Chem.* **1985,** *63,* 332.

[528]For a review of reactions of diazonium salts, see Wulfman, in Patai, Ref. 526, pt. 1, pp. 247-339.

[529]Korzeniowski; Leopold; Beadle; Ahern; Sheppard; Khanna; Gokel *J. Org. Chem.* **1981,** *46,* 2153, and references cited therein. For reviews, see Bartsch, in Patai; Rappoport *The Chemistry of Functional Groups, Supplement C,* pt.1; Wiley: New York, 1983, pp. 889-915; Bartsch *Prog. Macrocyclic Chem.* **1981,** *2,* 1-39.

below a pH of about 3, it is even possible, by working at a pH of about 1, to diazotize an aromatic amine without disturbing an aliphatic amino group in the same molecule.[530]

If an aliphatic amino group is α to a COOR, CN, CHO, COR, etc. and has an α hydrogen, treatment with nitrous acid gives not a diazonium salt, but a *diazo compound*.[531] Such diazo

$$\text{EtOOC—CH}_2\text{—NH}_2 + \text{HONO} \longrightarrow \text{EtOOC—CH}{=}\overset{\oplus}{\text{N}}{=}\overset{\ominus}{\underline{\text{N}}}$$

compounds can also be prepared, often more conveniently, by treatment of the substrate with isoamyl nitrite and a small amount of acid.[532] Certain heterocyclic amines also give diazo compounds rather than diazonium salts.[533]

Despite the fact that diazotization takes place in acid solution, the actual species attacked is not the salt of the amine, but the small amount of free amine present.[534] It is because aliphatic amines are stronger bases than aromatic ones that at pH values below 3 there is not enough free amine present for the former to be diazotized, while the latter still undergo the reaction. In dilute acid the actual attacking species is N_2O_3, which acts as a carrier of NO^+. Evidence is that the reaction is second order in nitrous acid and, at sufficiently low acidities, the amine does not appear in the rate expression.[535] Under these conditions the mechanism is

Step 1
$$2\text{HONO} \xrightarrow{\text{slow}} \text{N}_2\text{O}_3 + \text{H}_2\text{O}$$

Step 2
$$\text{Ar}\bar{\text{N}}\text{H}_2 + \text{N}_2\text{O}_3 \longrightarrow \text{Ar}-\overset{\text{H}}{\underset{\text{H}}{\overset{|}{\underset{|}{\text{N}}}}}\overset{\oplus}{-}\underline{\text{N}}{=}\underline{\text{O}}\text{I} + \text{NO}_2^-$$

Step 3
$$\text{Ar}-\overset{\text{H}}{\underset{\text{H}}{\overset{|}{\underset{|}{\text{N}}}}}\overset{\oplus}{-}\underline{\text{N}}{=}\underline{\text{O}}\text{I} \xrightarrow{-\text{H}^+} \text{Ar}-\overset{\bar{\text{N}}}{\underset{\text{H}}{|}}-\underline{\text{N}}{=}\underline{\text{O}}\text{I}$$
40

Step 4
$$\text{Ar}-\overset{\bar{\text{N}}}{\underset{\text{H}}{|}}{-}\text{N}{=}\underline{\text{O}}\text{I} \underset{}{\overset{\text{tautom.}}{\rightleftharpoons}} \text{Ar}-\bar{\text{N}}{=}\bar{\text{N}}-\underline{\bar{\text{O}}}-\text{H}$$

Step 5
$$\text{Ar}-\bar{\text{N}}{=}\bar{\text{N}}-\underline{\bar{\text{O}}}-\text{H} \xrightarrow{\text{H}^+} \text{Ar}-\overset{\oplus}{\text{N}}{\equiv}\bar{\text{N}} + \text{H}_2\text{O}$$

There exists other evidence for this mechanism.[536] Other attacking species can be NOCl, $H_2NO_2^+$, and at high acidities even NO^+. Nucleophiles (e.g., Cl^-, SCN^-, thiourea) catalyze the reaction by converting the HONO to a better electrophile, e.g., $HNO_2 + Cl^- + H^+ \rightarrow NOCl + H_2O$.[537]

[530]Kornblum; Iffland *J. Am. Chem. Soc.* **1949**, *71*, 2137.
[531]For a monograph on diazo compounds, see Regitz; Maas, Ref. 164. For reviews, see, in Patai, Ref. 526, the articles by Regitz, pt. 2, pp. 659-708, 751-820, and Wulfman; Linstrumelle; Cooper, pt. 2, pp. 821-976.
[532]Takamura; Mizoguchi; Koga; Yamada *Tetrahedron* **1975**, *31*, 227.
[533]Butler, Ref. 525.
[534]Challis; Ridd *J. Chem. Soc.* **1962**, 5197, 5208; Challis; Larkworthy; Ridd *J. Chem. Soc.* **1962**, 5203.
[535]Hughes; Ingold; Ridd *J. Chem. Soc.* **1958**, 58, 65, 77, 88; Hughes; Ridd *J. Chem. Soc.* **1958**, 70, 82.
[536]For discussions, see Ref. 157, pp. 95-109; Ridd, Ref. 540, pp. 422-424.
[537]Ref. 157, pp. 84-93.

There are many preparations of diazonium salts listed in *Organic Syntheses*, but they are always prepared for use in other reactions. We do not list them here, but under reactions in which they are used. The preparation of aliphatic diazo compounds can be found in OS **III**, 392; **IV**, 424. See also OS **VI**, 840.

2-50   The Conversion of Hydrazines to Azides
Hydrazine-azide transformation

$$\textbf{RNHNH}_2 + \textbf{HONO} \longrightarrow \textbf{R} \overline{\textbf{N}} {=} \overset{\oplus}{\textbf{N}} {=} \overline{\textbf{N}}{}^{\ominus}$$

Monosubstituted hydrazines treated with nitrous acid give azides in a reaction exactly analogous to the formation of aliphatic diazo compounds mentioned in **2-49**. Among other reagents used for this conversion have been N_2O_4[538] and nitrosyl tetrafluoroborate $NOBF_4$.[539]
OS **III**, 710; **IV**, 819; **V**, 157.

2-51   *N*-Nitrosation or *N*-Nitroso-de-hydrogenation

$$\textbf{R}_2\textbf{NH} + \textbf{HONO} \longrightarrow \textbf{R}_2\textbf{N} {-} \textbf{NO}$$

When secondary amines are treated with nitrous acid, N-nitroso compounds (also called nitrosamines) are formed.[540] The reaction can be accomplished with dialkyl-, diaryl-, or alkylarylamines, and even with mono-N-substituted amides: RCONHR' + HONO → RCON(NO)R'.[541] Tertiary amines have also been N-nitrosated, but in these cases one group cleaves, so that the product is the nitroso derivative of a secondary amine.[542] The group that cleaves appears as an aldehyde or ketone. Other reagents have also been used, for example NOCl, which is useful for amines or amides that are not soluble in an acidic aqueous solution or where the N-nitroso compounds are highly reactive. N-Nitroso compounds can be prepared in basic solution by treatment of secondary amines with gaseous N_2O_3, N_2O_4,[543] or alkyl nitrites,[544] and, in aqueous or organic solvents, by treatment with $BrCH_2NO_2$.[545]
The mechanism of nitrosation is essentially the same as in **2-49** up to the point where **41** (analogous to **40**) is formed. Since this species cannot lose a proton, it is stable and the

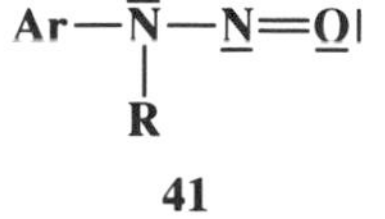

$$\text{Ar} {-} \overline{\text{N}} {-} \underline{\text{N}} {=} \underline{\text{O}}|$$
$$|$$
$$\text{R}$$

41

[538]Kim; Kim; Shim *Tetrahedron Lett.* **1986**, *27*, 4749.
[539]Pozsgay; Jennings *Tetrahedron Lett.* **1987**, *28*, 5091.
[540]For reviews, see Williams, Ref. 157, pp. 95-109; Kostyukovskii; Melamed *Russ. Chem. Rev.* **1988**, *57*, 350-366; Saavedra *Org. Prep. Proced. Int.* **1987**, *19*, 83-159; Ref. 158; Challis; Challis, in Patai; Rappoport, Ref. 172, pt. 2, pp. 1151-1223; Ridd, *Q. Rev., Chem. Soc.* **1961**, *15*, 418-441. For a review of the chemistry of aliphatic N-nitroso compounds, including methods of synthesis, see Fridman; Mukhametshin; Novikov *Russ. Chem. Rev.* **1971**, *40*, 34-50.
[541]For a discussion of the mechanism with amides, see Castro; Iglesias; Leis; Peña; Tato *J. Chem. Soc., Perkin Trans. 2* **1986**, 1725.
[542]Hein *J. Chem. Educ.* **1963**, *40*, 181. See also Verardo; Giumanini; Strazzolini *Tetrahedron* **1990**, *46*, 4303.
[543]Challis; Kyrtopoulos *J. Chem. Soc., Perkin Trans. 1* **1979**, 299.
[544]Casado; Castro; Lorenzo; Meijide *Monatsh. Chem.* **1986**, *117*, 335.
[545]Challis; Yousaf *J. Chem. Soc., Chem. Commun.* **1990**, 1598.

reaction ends there. The attacking entity can be any of those mentioned in **2-49.** The following has been suggested as the mechanism for the reaction with tertiary amines:[546]

$$R_2N-CHR'_2 \xrightarrow{\text{HONO}} \overset{\oplus}{R_2N}-CHR'_2 \longrightarrow \overset{\oplus}{R_2N}=CR'_2 + HNO$$
$$\underset{N=O}{|} \qquad\qquad\qquad \downarrow H_2O$$

$$R_2N-NO \xleftarrow{\text{HONO}} \overset{\oplus}{R_2NH_2} + O=CR'_2$$

The evidence for this mechanism includes the facts that nitrous oxide is a product (formed by $2HNO \rightarrow H_2O + N_2O$) and that quinuclidine, where the nitrogen is at a bridgehead and therefore cannot give elimination, does not react. Tertiary amines have also been converted to nitrosamines with nitric acid in Ac_2O[547] and with N_2O_4.[548]

Amines and amides can be N-*nitrated*[549] with nitric acid,[550] N_2O_5,[551] or NO_2^+,[552] and aromatic amines can be converted to triazenes with diazonium salts. Aliphatic primary amines can also be converted to triazenes if the diazonium salts contain electron-withdrawing groups.[553] C-Nitrosation is discussed at **1-3** and **2-8.**

OS **I,** 177, 399, 417; **II,** 163, 211, 290, 460, 461, 462, 464 (also see **V,** 842); **III,** 106, 244; **IV,** 718, 780, 943; **V,** 336, 650, 797, 839, 962; **VI,** 542, 981. Also see OS **III,** 711.

2-52   Conversion of Amines to Azo Compounds
N-Arylimino-de-dihydro-bisubstitution

$$ArNH_2 + Ar'NO \xrightarrow{\text{HOAc}} Ar-N=N-Ar'$$

Aromatic nitroso compounds combine with primary arylamines in glacial acetic acid to give symmetrical or unsymmetrical azo compounds (the *Mills reaction*).[554] A wide variety of substituents may be present in both aryl groups. Unsymmetrical azo compounds have also been prepared by the reaction between aromatic nitro compounds $ArNO_2$ and N-acyl aromatic amines $Ar'NHAc$.[555] The use of phase transfer catalysis increased the yields.

2-53   Conversion of Nitroso Compounds to Azoxy Compounds

$$RNO + R'NHOH \longrightarrow R-\overset{\oplus}{N}=N-R'$$
$$\underset{O_\ominus}{|}$$

[546]Smith; Loeppky *J. Am. Chem. Soc.* **1967,** *89,* 1147; Smith; Pars *J. Org. Chem.* **1959,** *24,* 1324; Gowenlock; Hutchison; Little; Pfab *J. Chem. Soc., Perkin Trans. 2* **1979,** 1110. See also Loeppky; Outram; Tomasik; Faulconer *Tetrahedron Lett.* **1983,** *24,* 4271.

[547]Boyer; Pillai; Ramakrishnan *Synthesis* **1985,** 677.

[548]Boyer; Kumar; Pillai *J. Chem. Soc., Perkin Trans. 1* **1986,** 1751.

[549]For other reagents, see Mayants; Pyreseva; Gordeichuk *J. Org. Chem. USSR* **1986,** *22,* 1900; Bottaro; Schmitt; Bedford *J. Org. Chem.* **1987,** *52,* 2292; Suri; Chapman *Synthesis* **1988,** 743; Carvalho; Iley; Norberto; Rosa *J. Chem. Res. (S)* **1989,** 260.

[550]Cherednichenko; Dmitrieva; Kuznetsov; Gidaspov *J. Org. Chem. USSR* **1976,** *12,* 2101, 2105.

[551]Emmons; Pagano; Stevens *J. Org. Chem.* **1958,** *23,* 311; Runge; Treibs *J. Prakt. Chem.* **1962,** [4]' *15,* 223; Halevi; Ron; Speiser *J. Chem. Soc.* **1965,** 2560.

[552]Ilyushin; Golod; Gidaspov *J. Org. Chem. USSR* **1977,** *13,* 8; Andreev; Lededev; Tselinskii *J. Org. Chem. USSR* **1980,** *16,* 1166, 1170, 1175, 1179.

[553]For a reveiw of alkyl traizenes, see Vaughan; Stevens *Chem. Soc. Rev.* **1978,** *7,* 377-397.

[554]For a review, see Boyer, in Feuer *The Chemistry of the Nitro and Nitroso Groups,* pt. 1; Wiley: New York, 1969, pp. 278-283.

[555]Ayyangar; Naik; Srinivasan *Tetrahedron Lett.* **1989,** *30,* 7253.

In a reaction similar to **2-52,** azoxy compounds can be prepared by the condensation of a nitroso compound with a hydroxylamine.[556] The position of the oxygen in the final product is determined by the nature of the R groups, not by which R groups came from which starting compound. Both R and R' can be alkyl or aryl, but when two different aryl groups are involved, mixtures of azoxy compounds (ArNONAr', ArNONAr', and Ar'NONAr') are obtained[557] and the unsymmetrical product (ArNONAr') is likely to be formed in the smallest amount. This behavior is probably caused by an equilibration between the starting compounds prior to the actual reaction (ArNO + Ar'NHOH → Ar'NO + ArNHOH).[558] The mechanism[559] has been investigated in the presence of base. Under these conditions both reactants are converted to radical anions, which couple:

$$\text{ArNO} + \text{ArNHOH} \longrightarrow 2\text{Ar}-\overset{\bullet}{\underset{|\underline{O}|_{\ominus}}{N}}| \longrightarrow \text{Ar}-\overset{|\overline{O}|^{\ominus}}{\underset{|\underline{O}|_{\ominus}}{N}}-\text{N}-\text{Ar} \xrightarrow[\text{H}_2\text{O}]{-2\text{OH}^-} \text{Ar}-\overset{|\overline{O}|^{\ominus}}{N}=\underset{\oplus}{N}-\text{Ar}$$

These radical anions have been detected by esr.[560] This mechanism is consistent with the following result: when nitrosobenzene and phenylhydroxylamine are coupled, ^{18}O and ^{15}N labeling show that the two nitrogens and the two oxygens become equivalent.[561] Unsymmetrical azoxy compounds can be prepared[562] by combination of a nitroso compound with an N,N-dibromoamine. Symmetrical and unsymmetrical azo and azoxy compounds are produced when aromatic nitro compounds react with aryliminodimagnesium reagents ArN(MgBr)$_2$.[563]

2-54 N-Halogenation or N-Halo-de-hydrogenation

$$\text{RNH}_2 + \text{NaOCl} \longrightarrow \text{RNHCl}$$

Treatment with sodium hypochlorite or hypobromite converts primary amines into N-halo- or N,N-dihaloamines. Secondary amines can be converted to N-halo secondary amines. Similar reactions can be carried out on unsubstituted and N-substituted amides and on sulfonamides. With unsubstituted amides the N-halogen product is seldom isolated but usually rearranges (see **8-14**); however, N-halo-N-alkyl amides and N-halo imides are quite stable. The important reagent N-bromosuccinimide is made in this manner. N-Halogenation has also been accomplished with other reagents, e.g., t-BuOCl,[564] sodium bromite NaBrO$_2$,[565] benzyltrimethylammonium tribromide PhCH$_2$NMe$_3$$^+$ Br$_3$$^-$,[566] and N-chlorosuccinimide.[567] The mechanisms of these reactions[568] involve attack by a positive halogen and are probably

[556]Boyer, Ref. 554.

[557]See, for example, Ogata; Tsuchida; Takagi *J. Am. Chem. Soc.* **1957,** *79,* 3397.

[558]Knight; Saville *J. Chem. Soc., Perkin Trans. 2* **1973,** 1550.

[559]For discussions of the mechanism in the absence of base, see Darchen; Moinet *Bull. Soc. Chim. Fr.* **1976,** 812; Becker; Sternson *J. Org. Chem.* **1980,** *45,* 1708. See also Pizzolatti; Yunes *J. Chem. Soc., Perkin Trans. 1* **1990,** 759.

[560]Russell; Geels; Smentowski; Chang; Reynolds; Kaupp *J. Am. Chem. Soc.* **1967,** *89,* 3821.

[561]Shemyakin; Maimind; Vaichunaite *Izv. Akad. Nauk SSSR, Ser. Khim.* **1957,** 1260; Oae; Fukumoto; Yamagami *Bull. Chem. Soc. Jpn.* **1963,** *36,* 728.

[562]Zawalski; Kovacic *J. Org. Chem.* **1979,** *44,* 2130. For another method, see Moriarty; Hopkins; Prakash; Vaid; Vaid *Synth. Commun.* **1990,** *20,* 2353.

[563]Okubo; Matsuo; Yamauchi *Bull. Chem. Soc. Jpn.* **1989,** *62,* 915, and other papers in this series.

[564]Altenkirk; Isrealstam *J. Org. Chem.* **1962,** *27,* 4532.

[565]Kajigaeshi; Nakagawa; Fujisaki *Chem. Lett.* **1984,** 2045.

[566]Kajigaeshi; Murakawa; Asano; Fujisaki; Kakinami *J. Chem. Soc., Perkin Trans. 1* **1989,** 1702.

[567]See Deno; Fishbein; Wyckoff *J. Am. Chem. Soc.* **1971,** *93,* 2065; Guillemin; Denis *Synthesis* **1985,** 1131.

[568]For a study of the mechanism, see Matte; Solastiouk; Merlin; Deglise *Can. J. Chem.* **1989,** *67,* 786.

similar to those of **2-49** and **2-51**.[569] N-Fluorination can be accomplished by direct treatment of amines[570] or amides[571] with F_2. Fluorination of N-alkyl-N-fluoro amides results in cleavage to N,N-difluoroamines.[572]

$$\textbf{RNFCOR'} \xrightarrow{\text{F}_2} \textbf{RNF}_2$$

OS **III,** 159; **IV,** 104, 157; **V,** 208, 663, 909; **VI,** 968; **VII,** 223; **65,** 159; **67,** 222.

2-55   The Reaction of Amines with Carbon Monoxide
***N*-Formylation** or ***N*-Formyl-de-hydrogenation,** etc.

$$\textbf{RNH}_2 + \textbf{CO} \xrightarrow{\text{catalyst}} \textbf{RNH}{-}\underset{\underset{\textbf{O}}{\|}}{\textbf{C}}{-}\textbf{H} \text{ or } \textbf{RNH}{-}\underset{\underset{\textbf{O}}{\|}}{\textbf{C}}{-}\textbf{NHR} \text{ or } \textbf{RNCO}$$

Three types of product can be obtained from the reaction of amines with carbon monoxide, depending on the catalyst. (1) Both primary and secondary amines react with CO in the presence of various catalysts [e.g., $Cu(CN)_2$, $Me_3N{-}H_2Se$, rhodium or ruthenium complexes] to give N-substituted and N,N-disubstituted formamides, respectively.[573] (2) Symmetrically substituted ureas can be prepared by treatment of a primary amine (or ammonia) with CO in the presence of selenium[574] or sulfur.[575] R can be alkyl or aryl. The same thing can be done with secondary amines, by using $Pd(OAc)_2{-}I_2{-}K_2CO_3$.[576] (3) When $PdCl_2$ is the catalyst, primary amines yield isocyanates.[577] Isocyanates can also be obtained by treatment of CO with azides: $RN_3 + CO \rightarrow RNCO$,[578] or with an aromatic nitroso or nitro compound and a rhodium complex catalyst.[579] A fourth type of product, a carbamate RNHCOOR', can be obtained from primary or secondary amines, if these are treated with CO, O_2, and an alcohol R'OH in the presence of a catalyst.[580] Carbamates can also be obtained from nitroso compounds, by treatment with CO, R'OH, $Pd(OAc)_2$, and $Cu(OAc)_2$,[581] and from nitro compounds.[582] When allylic amines $R_2C{=}CHRCHRNR'_2$ are treated with CO and a palladium–phosphine catalyst, the CO inserts to produce the β,γ-unsaturated amides $R_2C{=}CHRCHRCONR'_2$ in good yields.[583] See also **6-19**.

[569]For studies of reactivitiy in this reaction, see Thomm; Wayman *Can. J. Chem.* **1969,** *47,* 3289; Higuchi; Hussain; Pitman *J. Chem. Soc. B* **1969,** 626.
[570]Sharts *J. Org. Chem.* **1968,** *33,* 1008.
[571]Grakauskas; Baum *J. Org. Chem.* **1969,** *34,* 2840, **1970,** *35,* 1545.
[572]Ref. 571. See also Wiesboeck; Ruff *Tetrahedron* **1970,** *26,* 837; Barton; Hesse; Klose; Pechet *J. Chem. Soc., Chem. Commun.* **1975,** 97.
[573]See Tsuji; Iwamoto *Chem. Commun.* **1966,** 380; Durand; Lassau *Tetrahedron Lett.* **1969,** 2329; Saegusa; Kobayashi; Hirota; Ito *Bull. Chem. Soc. Jpn.* **1969,** *42,* 2610; Nefedov; Sergeeva; Éidus *Bull. Acad. Sci. USSR, Div. Chem. Sci.* **1973,** *22,* 784; Kondo; Sonoda; Sakurai *J. Chem. Soc., Chem. Commun.* **1973,** 853; Yoshida; Asano; Inoue *Chem. Lett.* **1984,** 1073; Bitsi; Jenner *J. Organomet. Chem.* **1987,** *330,* 429.
[574]Sonoda; Yasuhara; Kondo; Ikeda; Tsutsumi *J. Am. Chem. Soc.* **1971,** *93,* 6344.
[575]Franz; Applegath; Morriss; Baiocchi; Bolze *J. Org. Chem.* **1961,** *26,* 3309.
[576]Pri-Bar; Alper *Can. J. Chem.* **1990,** *68,* 1544.
[577]Stern; Spector *J. Org. Chem.* **1966,** *31,* 596.
[578]Bennett; Hardy *J. Am. Chem. Soc.* **1968,** *90,* 3295.
[579]Unverferth; Rüger; Schwetlick *J. Prakt. Chem.* **1977,** *319,* 841; Unverferth; Tietz; Schwetlick *J. Prakt. Chem.* **1985,** *327,* 932. See also Braunstein; Bender; Kervennal *Organometallics* **1982,** *1,* 1236; Kunin; Noirot; Gladfelter *J. Am. Chem. Soc.* **1989,** *111,* 2739.
[580]Fukuoka; Chono; Kohno *J. Org. Chem.* **1984,** *49,* 1458, *J. Chem. Soc., Chem. Commun.* **1984,** 399. See also Alper; Vasapollo; Hartstock; Mlekuz; Smith; Morris *Organometallics* **1987,** *6,* 2391.
[581]Alper; Vasapollo *Tetrahedron Lett.* **1987,** *28,* 6411.
[582]Cenini; Crotti; Pizzotti; Porta *J. Org. Chem.* **1988,** *53,* 1243.
[583]Murahashi; Imada; Nishimura *J. Chem. Soc., Chem. Commun.* **1988,** 1578.

13

AROMATIC NUCLEOPHILIC SUBSTITUTION

On p. 341 it was pointed out that nucleophilic substitutions proceed so slowly at an aromatic carbon that the reactions of Chapter 10 are not feasible for aromatic substrates. There are, however, exceptions to this statement, and it is these exceptions that form the subject of this chapter.[1] Reactions that *are* successful at an aromatic substrate are largely of four kinds: (1) reactions activated by electron-withdrawing groups ortho and para to the leaving group; (2) reactions catalyzed by very strong bases and proceeding through aryne intermediates; (3) reactions initiated by electron donors; and (4) reactions in which the nitrogen of a diazonium salt is replaced by a nucleophile. However, not all the reactions discussed in this chapter fit into these categories.

MECHANISMS

There are four principal mechanisms for aromatic nucleophilic substitution.[2] Each of the four is similar to one of the aliphatic nucleophilic substitution mechanisms discussed in Chapter 10.

The SNAr Mechanism

By far the most important mechanism for aromatic nucleophilic substitution consists of two steps:

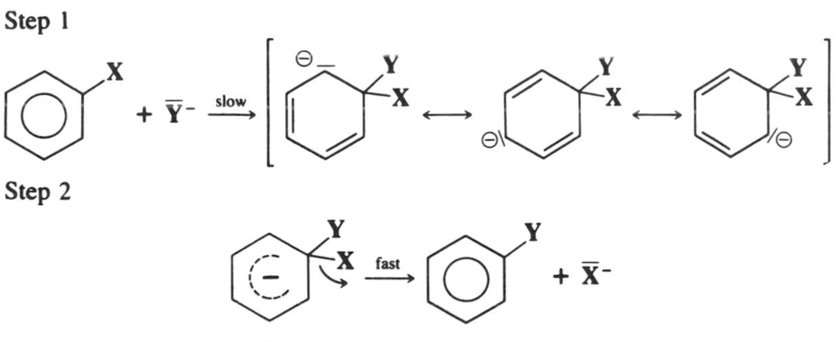

Step 1

Step 2

1

---

[1]For a review of aromatic nucleophilic substitution, see Zoltewicz *Top. Curr. Chem.* **1975**, *59*, 33-64.

[2]For a monograph on aromatic nucleophilic substitution mechanisms, see Miller *Aromatic Nucleophilic Substitution*; Elsevier: New York, 1968. For reviews, see Bernasconi *Chimia* **1980**, *34*, 1-11, *Acc. Chem. Res.* **1978**, *11*, 147-152; Bunnett *J. Chem. Educ.* **1974**, *51*, 312-315; Ross, in Bamford; Tipper *Comprehensive Chemical Kinetics*, vol. 13; Elsevier: New York, 1972, pp. 407-431; Buck *Angew. Chem. Int. Ed. Engl.* **1969**, *8*, 120-131 [*Angew. Chem. 81*, 136-148]; Buncel; Norris; Russell *Q. Rev., Chem. Soc.* **1968**, *22*, 123-146; Ref. 1.

The first step is usually, but not always, rate-determining. It can be seen that this mechanism greatly resembles the tetrahedral mechanism discussed in Chapter 10 and, in another way, the arenium ion mechanism of electrophilic aromatic substitution. In all three cases, the attacking species forms a bond with the substrate, giving an intermediate, and then the leaving group departs. We refer to this mechanism as the SNAr mechanism.[3] The IUPAC designation is $A_N + D_N$ (the same as for the tetrahedral mechanism; compare the designation $A_E + D_E$ for the arenium ion mechanism). This mechanism is generally found where activating groups are present on the ring (see p. 649).

There is a great deal of evidence for the mechanism; we shall discuss only some of it.[2] Probably the most convincing evidence was the isolation, as long ago as 1902, of the intermediate **2** in the reaction between ethyl picrate and methoxide ion.[4] Intermediates of this

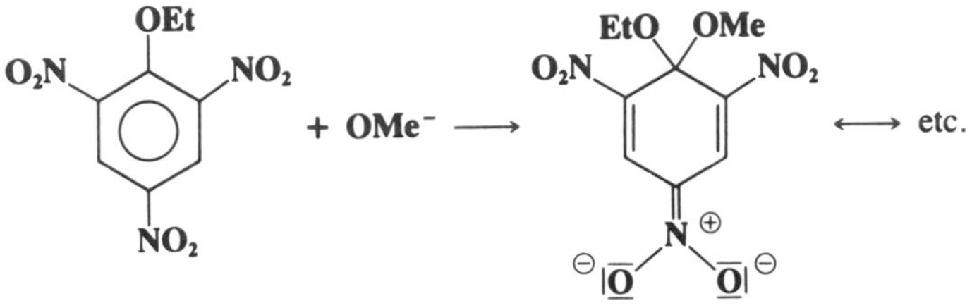

type are stable salts, called *Meisenheimer* or *Meisenheimer–Jackson salts*, and many more have been isolated since 1902.[5] The structures of several of these intermediates have been proved by nmr[6] and by x-ray crystallography.[7] Further evidence comes from studies of the effect of the leaving group on the reaction. If the mechanism were similar to either the SN1 or SN2 mechanisms described in Chapter 10, the Ar—X bond would be broken in the rate-determining step. In the SNAr mechanism this bond is not broken until after the rate-determining step (that is, if step 1 is rate-determining). We would predict from this that if the SNAr mechanism is operating, a change in leaving group should not have much effect on the reaction rate. In the reaction

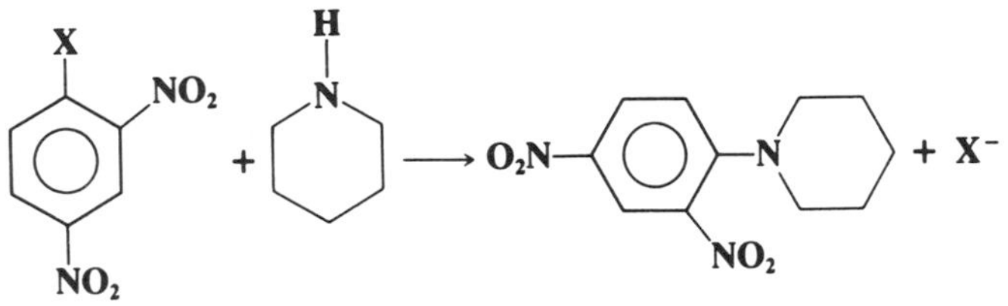

[3]The mechanism has also been called by other names, including the SN2Ar, the addition-elimination, and the intermediate complex mechanism.

[4]Meisenheimer *Liebigs Ann. Chem.* **1902**, *323*, 205. Similar salts were isolated even earlier by Jackson; see Jackson; Gazzolo *Am. Chem. J.* **1900**, *23*, 376; Jackson; Earle *Am. Chem. J.* **1903**, *29*, 89.

[5]For a monograph on Meisenheimer salts and on this mechanism, see Buncel; Crampton; Strauss; Terrier *Electron Deficient Aromatic- and Heteroaromatic-Base Interactions*; Elsevier: New York, 1984. For reviews of structural and other studies, see Illuminati; Stegel *Adv. Heterocycl. Chem.* **1983**, *34*, 305-444; Artamkina; Egorov; Beletskaya *Chem. Rev.* **1982**, *82*, 427-459; Terrier *Chem. Rev.* **1982**, *82*, 77-152; Strauss *Chem. Rev.* **1970**, *70*, 667-712. *Acc. Chem. Res.* **1974**, *7*, 181-188; Hall; Poranski, in Feuer *The Chemistry of the Nitro and Nitroso Groups*, pt. 2; Wiley: New York, 1970, pp. 329-384; Crampton, *Adv. Phys. Org. Chem.* **1969**, *7*, 211-257; Foster; Fyfe *Rev. Pure Appl. Chem.* **1966**, *16*, 61-82.

[6]First done by Crampton; Gold *J. Chem. Soc.* **1964**, 4293. *J. Chem. Soc. B* **1966**, 893. A good review of spectral studies is found in Buncel et al., Ref. 5, pp. 15-133.

[7]Destro; Gramaccioli; Simonetta *Acta Crystallogr.* **1968**, *24*, 1369; Ueda; Sakabe; Tanaka; Furusaki *Bull. Chem. Soc. Jpn.* **1968**, *41*, 2866; Messmer; Palenik *Chem. Commun.* **1969**, 470.

when X was Cl, Br, I, SOPh, SO$_2$Ph, or *p*-nitrophenoxy, the rates differed only by a factor of about 5.[8] This behavior would not be expected in a reaction in which the Ar—X bond is broken in the rate-determining step. We do not expect the rates to be *identical*, because the nature of X affects the rate at which Y attacks. An increase in the electronegativity of X causes a decrease in the electron density at the site of attack, resulting in a faster attack by a nucleophile. Thus, in the reaction just mentioned, when X = F, the relative rate was 3300 (compared with I = 1). The very fact that fluoro is the best leaving group among the halogens in most aromatic nucleophilic substitutions is good evidence that the mechanism is different from the S$_N$1 and S$_N$2 mechanisms, where fluoro is by far the poorest leaving group of the halogens. This is an example of the element effect (p. 336).

The pattern of base catalysis of reactions with amine nucleophiles provides additional evidence. These reactions are catalyzed by bases only when a relatively poor leaving group (such as OR) is present (not Cl or Br) and only when relatively bulky amines are nucleophiles.[9] Bases could not catalyze step 1, but if amines are nucleophiles, bases can catalyze step 2. Base catalysis is found precisely in those cases where the amine moiety cleaves easily

3

but X does not, so that k_{-1} is large and step 2 is rate-determining. This is evidence for the S$_N$Ar mechanism because it implies two steps. Furthermore, in cases where bases *are* catalysts, they catalyze only at low base concentrations: a plot of the rate against the base concentration shows that small increments of base rapidly increase the rate until a certain concentration of base is reached, after which further base addition no longer greatly affects the rate. This behavior, based on a partitioning effect (see p. 503), is also evidence for the S$_N$Ar mechanism. At low base concentration, each increment of base, by increasing the rate of step 2, increases the fraction of intermediate that goes to product rather than reverting to reactants. At high base concentration the process is virtually complete: there is very little reversion to reactants and the rate becomes dependent on step 1. Just how bases catalyze step 2 has been investigated. For protic solvents two proposals have been presented. One is that step 2 consists of two steps: rate-determining deprotonation of **3** followed by rapid loss of X, and that bases catalyze the reaction by increasing the rate of the deprotonation

3

step.[10] According to the other proposal, loss of X assisted by BH$^+$ is rate-determining.[11] Two mechanisms, both based on kinetic evidence, have been proposed for aprotic solvents

[8]Bunnett; Garbisch; Pruitt *J. Am. Chem. Soc.* **1957**, *79*, 385.
[9]Kirby; Jencks *J. Am. Chem. Soc.* **1965**, *87*, 3217; Bunnett; Garst *J. Am. Chem. Soc.* **1965**, *87*, 3875, 3879, *J. Org. Chem.* **1968**, *33*, 2320; Bunnett; Bernasconi *J. Org. Chem.* **1970**, *35*, 70; Bernasconi; Schmid *J. Org. Chem.* **1967**, *32*, 2953; Bernasconi; Zollinger *Helv. Chim. Acta* **1966**, *49*, 103, **1967**, *50*, 1; Pietra; Vitali *J. Chem. Soc. B* **1968**, 1200; Chiacchiera; Singh; Anunziata; Silber *J. Chem. Soc., Perkin Trans. 2* **1987**, 987.
[10]Bernasconi; de Rossi; Schmid *J. Am. Chem. Soc.* **1977**, *99*, 4090, and references cited therein.
[11]Bunnett; Sekiguchi; Smith *J. Am. Chem. Soc.* **1981**, *103*, 4865, and references cited therein.

such as benzene. In both proposals the ordinary SNAr mechanism operates, but in one the attacking species involves two molecules of the amine (the *dimer mechanism*),[12] while in the other there is a cyclic transition state.[13] Further evidence for the SNAr mechanism has been obtained from $^{18}O/^{16}O$ and $^{15}N/^{14}N$ isotope effects.[14]

Step 1 of the SNAr mechanism has been studied for the reaction between picryl chloride (as well as other substrates) and OH$^-$ ions (**3-1**), and spectral evidence has been reported[15] for two intermediates, one a π complex (p. 505), and the other a radical ion–radical pair:

| Picryl chloride | π complex | radical ion– radical pair |

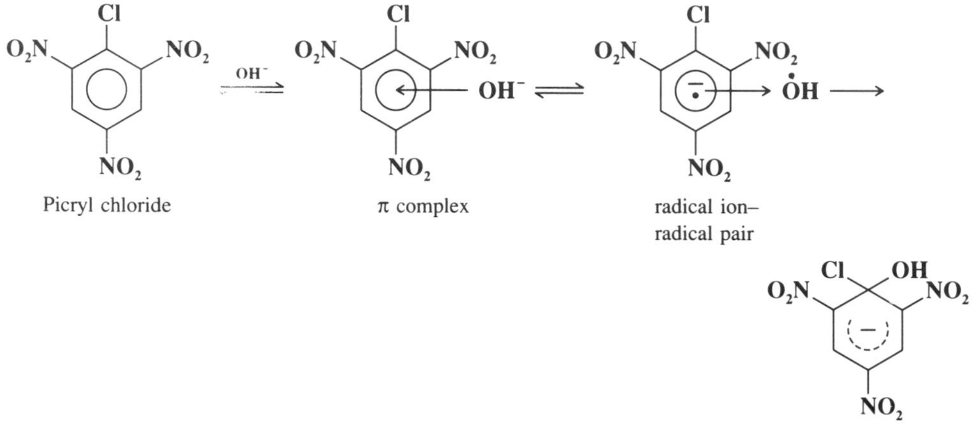

As with the tetrahedral mechanism at an acyl carbon, nucleophilic catalysis (p. 334) has been demonstrated with an aryl substrate, in certain cases.[16]

The SN1 Mechanism

For aryl halides and sulfonates, even active ones, a unimolecular SN1 mechanism (IUPAC: $D_N + A_N$) is very rare; it has only been observed for aryl triflates in which both ortho positions contain bulky groups (*t*-butyl or SiR$_3$).[17] It is in reactions with diazonium salts that this mechanism is important:[18]

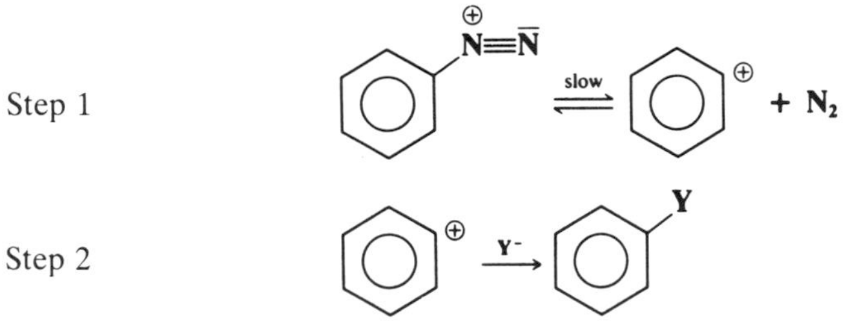

Step 1

Step 2

[12]For a review of this mechanism, see Nudelman *J. Phys. Org. Chem.* **1989**, *2*, 1-14. See also Nudelman; Montserrat *J. Chem. Soc., Perkin Trans. 2* **1990**, 1073.

[13]Banjoko; Ezcani *J. Chem. Soc., Perkin Trans. 2* **1986**, 531; Banjoko; Bayeroju *J. Chem. Soc., Perkin Trans. 2* **1988**, 1853; Jain; Gupta; Kumar *J. Chem. Soc., Perkin Trans. 2* **1990**, 11.

[14]Hart; Bourns *Tetrahedron Lett.* **1966**, 2995; Ayrey; Wylie *J. Chem. Soc. B* **1970**, 738.

[15]Bacaloglu; Blaskó; Bunton; Dorwin; Ortega; Zucco *J. Am. Chem. Soc.* **1991**, *113*, 238, and references cited therein. For earlier reports, based on kinetic data, of complexes with amine nucleophiles, see Forlani *J. Chem. Res. (S)* **1984**, 260; Hayami; Otani; Yamaguchi; Nishikawa *Chem. Lett.* **1987**, 739; Crampton; Davis; Greenhalgh; Stevens *J. Chem. Soc., Perkin Trans. 2* **1989**, 675.

[16]See Muscio; Rutherford *J. Org. Chem.* **1987**, *52*, 5194.

[17]Himeshima; Kobayashi; Sonoda *J. Am. Chem. Soc.* **1985**, *107*, 5286.

[18]Aryl iodonium salts Ar$_2$I$^+$ also undergo substitutions by this mechanism (and by a free-radical mechanism).

Among the evidence for the S_N1 mechanism[19] with aryl cations as intermediates,[20] is the following:[21]

1. The reaction rate is first order in diazonium salt and independent of the concentration of Y.

2. When high concentrations of halide salts are added, the product is an aryl halide but the rate is independent of the concentration of the added salts.

3. The effects of ring substituents on the rate are consistent with a unimolecular rate-determining cleavage.[22]

4. When reactions were run with substrate deuterated in the ortho position, isotope effects of about 1.22 were obtained.[23] It is difficult to account for such high secondary isotope effects in any other way except that an incipient phenyl cation is stabilized by hyperconjugation,[24] which is reduced when hydrogen is replaced by deuterium.

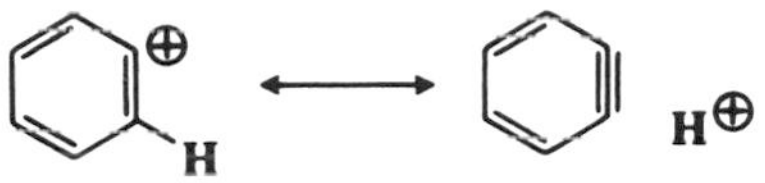

5. That the first step is reversible cleavage[25] was demonstrated by the observation that when $Ar^{15}\overset{\oplus}{N}{\equiv}N$ was the reaction species, recovered starting material contained not only $Ar^{15}\overset{\oplus}{N}{\equiv}N$ but also $Ar\overset{\oplus}{N}{\equiv}^{15}N$.[26] This could arise only if the nitrogen breaks away from the ring and then returns. Additional evidence was obtained by treating $Ph\overset{\oplus}{N}{\equiv}^{15}N$ with unlabeled N_2 at various pressures. At 300 atm the recovered product had lost about 3% of the labeled nitrogen, indicating that PhN_2^+ was exchanging with atmospheric N_2.[27]

There is kinetic and other evidence[28] that step 1 is more complicated and involves two steps, both reversible:

$$ArN_2^+ \rightleftharpoons [Ar^+ \ N_2] \rightleftharpoons Ar^+ + N_2$$
$$\mathbf{4}$$

4, which is probably some kind of tight ion–molecule pair, has been trapped with carbon monoxide.[29]

[19]For additional evidence, see Lorand *Tetrahedron Lett.* **1989**, *30*, 7337.

[20]For a review of aryl cations, see Ambroz; Kemp *Chem. Soc. Rev.* **1979**, *8*, 353-365. Also see Ref. 51 in Chapter 5.

[21]For a review, see Zollinger *Angew. Chem. Int. Ed. Engl.* **1978**, *17*, 141-150 [*Angew. Chem. 90*, 151-160]. For discussions, see Swain; Sheats; Harbison *J. Am. Chem. Soc.* **1975**, *97*, 783, 796; Burri; Wahl; Zollinger *Helv. Chim. Acta* **1974**, *57*, 2099; Richey; Richey, in Olah; Schleyer *Carbonium Ions*, vol. 2; Wiley: New York, 1970, pp. 922-931; Zollinger *Azo and Diazo Chemistry*; Wiley: New York, 1961, pp. 138-142; Miller, Ref. 2, pp. 29-40.

[22]Lewis; Miller *J. Am. Chem. Soc.* **1953**, *75*, 429.

[23]Swain; Sheats; Gorenstein; Harbison *J. Am. Chem. Soc.* **1975**, *97*, 791.

[24]See Apeloig; Arad *J. Am. Chem. Soc.* **1985**, *107*, 5285.

[25]For discussions, see Williams; Buncel *Isot. Org. Chem.* **1980**, 147-230, pp. 212-221; Zollinger *Pure Appl. Chem.* **1983**, *55*, 401-408.

[26]Lewis; Insole *J. Am. Chem. Soc.* **1964**, *86*, 32; Lewis; Kotcher *Tetrahedron* **1969**, *25*, 4873; Lewis; Holliday *J. Am. Chem. Soc.* **1969**, *91*, 426; Ref. 27; Tröndlin; Medina; Rüchardt *Chem. Ber.* **1979**, *112*, 1835.

[27]Bergstrom; Landells; Wahl; Zollinger *J. Am. Chem. Soc.* **1976**, *98*, 3301.

[28]Maurer; Szele; Zollinger *Helv. Chim. Acta* **1979**, *62*, 1079; Szele; Zollinger *Helv. Chim. Acta* **1981**, *64*, 2728.

[29]Ravenscroft; Skrabal; Weiss; Zollinger *Helv. Chim. Acta* **1988**, *71*, 515.

The Benzyne Mechanism[30]

Some aromatic nucleophilic substitutions are clearly different in character from those that occur by the SNAr mechanism (or the SN1 mechanism). These substitutions occur on aryl halides that have no activating groups; bases are required that are stronger than those normally used; and most interesting of all, the incoming group does not always take the position vacated by the leaving group. That the latter statement is true was elegantly demonstrated by the reaction of $1\text{-}^{14}C$-chlorobenzene with potassium amide:

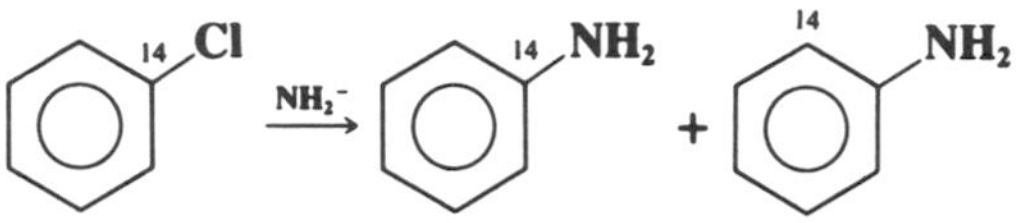

The product consisted of almost equal amounts of aniline labeled in the 1 position and in the 2 position.[31]

A mechanism that can explain all these facts involves elimination followed by addition:

Step 1

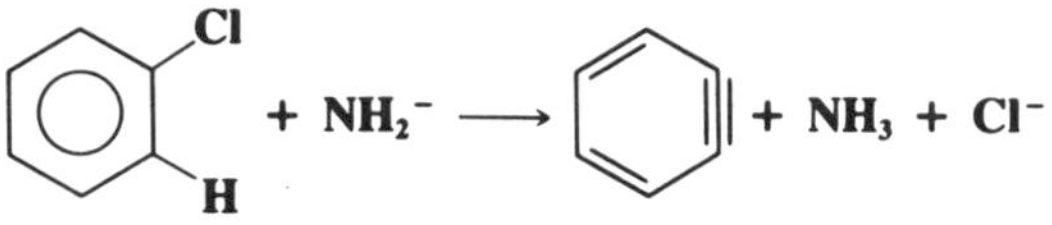

Step 2

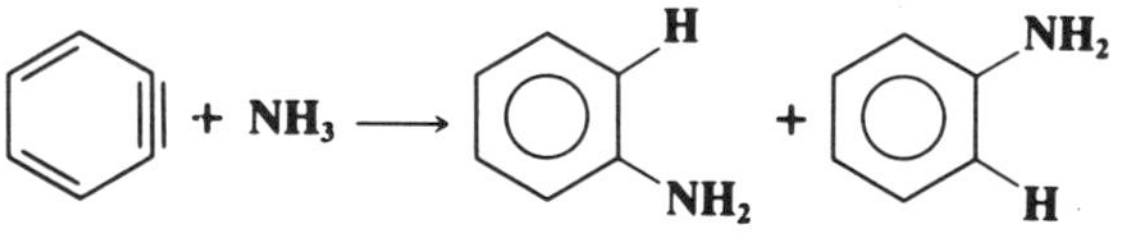

The symmetrical intermediate **5** can be attacked by the NH_3 at either of two positions, which explains why about half of the aniline produced from the radioactive chlorobenzene was labeled at the 2 position. The fact that the 1 and 2 positions were not labeled equally is the result of a small isotope effect. Other evidence for this mechanism is the following:

1. If the aryl halide contains two ortho substituents, the reaction should not be able to occur. This is indeed the case.[31]

2. It had been known many years earlier that aromatic nucleophilic substitution occasionally results in substitution at a different position. This is called *cine substitution* and can

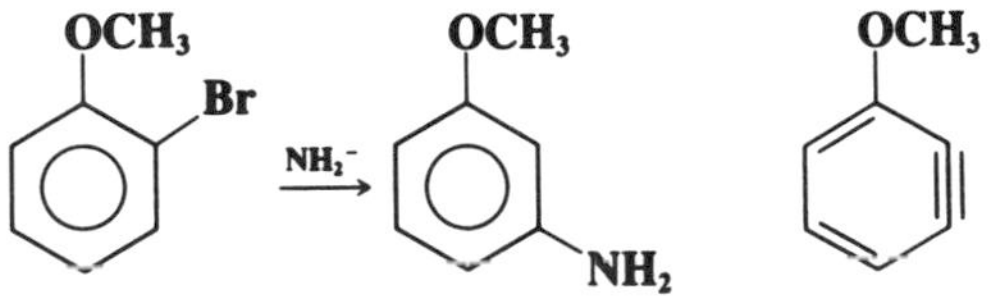

6

[30]For a monograph, see Hoffmann *Dehydrobenzene and Cycloalkynes*; Academic Press: New York, 1967. For reviews, see Gilchrist, in Patai; Rappoport *The Chemistry of Functional Groups, Supplement C*, pt. 1; Wiley: New York, 1983, pp. 383-419; Bryce; Vernon *Adv. Heterocycl. Chem.* **1981**, *28*, 183-229; Levin *React. Intermed.* (*Wiley*) **1985**, *3*, 1-18, **1981**, *2*, 1-14, **1978**, *1*, 1-26; Nefedov; D'yachenko; Prokof'ev *Russ. Chem. Rev.* **1977**, *46*, 941-966; Fields, in McManus *Organic Reactive Intermediates*; Academic Press: New York, 1973, pp. 449-508; Heaney *Fortschr. Chem. Forsch.* **1970**, *16*, 35-74, *Essays Chem.* **1970**, *1*, 95-115; Hoffmann, in Viehe *Acetylenes*; Marcel Dekker: New York, 1969, pp. 1063-1148; Fields; Meyerson *Adv. Phys. Org. Chem.* **1968**, *6*, 1-61; Wittig *Angew. Chem. Int. Ed. Engl.* **1965**, *4*, 731-737 [*Angew. Chem. 77*, 752-759].
[31]Roberts; Semenow; Simmons; Carlsmith *J. Am. Chem. Soc.* **1965**, *78*, 601.

be illustrated by the conversion of *o*-bromoanisole to *m*-aminoanisole.[32] In this particular case, only the meta isomer is formed. The reason a 1:1 mixture is not formed is that the intermediate **6** is not symmetrical and the methoxy group directs the incoming group meta but not ortho (see p. 651). However, not all cine substitutions proceed by this kind of mechanism (see **3-25**).

3. The fact that the order of halide reactivity is Br > I > Cl > F (when the reaction is performed with KNH_2 in liquid NH_3) shows that the SNAr mechanism is not operating here.[31]

In the conversion of the substrate to **6**, either proton removal or subsequent loss of halide ion can be rate-determining. In fact, the unusual leaving-group order just mentioned (Br > I > Cl) stems from a change in the rate-determining step. When the leaving group is Br or I, proton removal is rate-determining and the rate order for this step is F > Cl > Br > I. When Cl or F is the leaving group, cleavage of the C—X bond is rate-determining and the order for this step is I > Br > Cl > F. Confirmation of the latter order was found in a direct competitive study. *meta*-Dihalobenzenes in which the two halogens are different were treated with NH_2^-.[33] In such compounds, the most acidic hydrogen is the one between the two halogens; when it leaves, the remaining anion can lose either halogen. Therefore a study of which halogen is preferentially lost provides a direct measure of leaving-group ability. The order was found to be I > Br > Cl.[33]

Species such as **5** and **6** are called *benzynes* (sometimes *dehydrobenzenes*), or more generally, *arynes*, and the mechanism is known as the *benzyne mechanism*. Benzynes are very reactive. Neither benzyne nor any other aryne has yet been isolated under ordinary conditions,[31] but benzyne has been isolated in an argon matrix at 8 K,[35] where its ir spectrum could be observed. In addition, benzynes can be trapped, e.g., they undergo the Diels Alder reaction (see **5-47**). It should be noted that the extra pair of electrons does not affect the aromaticity. The original sextet still functions as a closed ring, and the two additional electrons are merely located in a π orbital that covers only two carbons. Benzynes do not have a formal triple bond, since two canonical forms (**A** and **B**) contribute to the hybrid.

A   **B**

The ir spectrum, mentioned above, indicates that **A** contributes more than **B**. Not only benzene rings but other aromatic rings[36] and even nonaromatic rings (p. 338) can react through this kind of intermediate. Of course, the nonaromatic rings do have a formal triple bond.

[32]This example is from Gilman; Avakian *J. Am. Chem. Soc.* **1945**, *67*, 349. For a table of many such examples, see Bunnett; Zahler *Chem. Rev.* **1951**, *49*, 273-412, pp. 385-386.

[33]Bunnett; Kearley *J. Org. Chem.* **1971**, *36*, 184.

[34]For the measurement of aryne lifetimes in solution, see Gaviña; Luis; Costero; Gil *Tetrahedron* **1986**, *42*, 155.

[35]Chapman; Mattes; McIntosh; Pacansky; Calder; Orr *J. Am. Chem. Soc.* **1973**, *95*, 6134. For the ir spectrum of pyridyne trapped in a matrix, see Nam; Leroi *J. Am. Chem. Soc.* **1988**, *110*, 4096. For spectra of transient arynes, see Berry; Spokes; Stiles *J. Am. Chem. Soc.* **1962**, *84*, 3570; Brown; Godfrey; Rodler *J. Am. Chem. Soc.* **1986**, *108*, 1296.

[36]For reviews of *hetarynes* (benzyne intermediates in heterocyclic rings), see van der Plas; Roeterdink, in Patai; Rappoport, Ref. 30, pt. 1, pp. 421-511; Reinecke, *React. Intermed.* (Plenum) **1982**, *2*, 367-526, *Tetrahedron* **1982**, *38*, 427-498; den Hertog; van der Plas, in Viehe, Ref. 30, pp. 1149-1197, *Adv. Heterocycl. Chem.* **1971**, *40*, 121-144; Kauffmann; Wirthwein *Angew. Chem. Int. Ed. Engl.* **1971**, *10*, 20-33 [*Angew. Chem. 83*, 21-34]; Kauffmann *Angew. Chem. Int. Ed. Engl.* **1965**, *4*, 543-557 [*Angew. Chem. 77*, 557-571]; Hoffmann, *Dehydrobenzene and Cycloalkynes* Ref. 30, pp. 275-309.

The SRN1 Mechanism

When 5-iodo-1,2,4-trimethylbenzene **7** was treated with KNH_2 in NH_3, **8** and **9** were formed in the ratio 0.63:1. From what we have already seen, the presence of an unactivated substrate,

a strong base, and the occurrence of cine along with normal substitution would be strong indications of a benzyne mechanism. Yet if that were so, the 6-iodo isomer of **7** should have given **8** and **9** in the same ratio (because the same aryne intermediate would be formed in both cases), but in this case the ratio of **8** to **9** was 5.9:1 (the chloro and bromo analogs did give the same ratio, 1.46:1, showing that the benzyne mechanism may be taking place there).

To explain the iodo result, it has been proposed[37] that besides the benzyne mechanism, this free-radical mechanism is also operating here:

$$\text{ArI} \xrightarrow[\text{donor}]{\text{electron}} \text{ArI}\cdot^- \longrightarrow \text{Ar}\cdot + \text{I}^-$$

$$\text{Ar}\cdot + \text{NH}_2^- \longrightarrow \text{ArNH}_2\cdot^- + \text{ArI} \longrightarrow \text{ArNH}_2 + \text{ArI}\cdot^-$$

Termination steps

This is called the SRN1 mechanism,[38] and many other examples are known (see **3-4, 3-5, 3-7, 3-14**). The IUPAC designation is $T + D_N + A_N$.[39] Note that the last step of the mechanism produces $\text{ArI}\cdot^-$ radical ions, so the process is a chain mechanism[40] (see p. 678). An electron donor is required to initiate the reaction. In the case above it was solvated electrons from KNH_2 in NH_3. Evidence was that the addition of potassium metal (a good producer of solvated electrons in ammonia) completely suppresssed the cine substitution. Further evidence for the SRN1 mechanism was that addition of radical scavengers (which would suppress a free-radical mechanism) led to **8:9** ratios much closer to 1.46:1. Numerous other observations of SRN1 mechanisms that were stimulated by solvated electrons and inhibited by radical scavengers have also been recorded.[41] Further evidence for the SRN1 mechanism in the case above was that some 1,2,4-trimethylbenzene was found among the products. This could easily be formed by abstraction by Ar• of H from the solvent NH_3. Besides initiation

---

[37]Kim; Bunnett *J. Am. Chem. Soc.* **1970**, *92*, 7463, 7464. For an alternative proposal, in which the first step is the same, but the radical ion reacts directly with the nucleophile, see Denney; Denney *Tetrahedron* **1991**, *47*, 6577.

[38]For a monograph, see Rossi; de Rossi *Aromatic Substitution by the SRN1 Mechanism*; American Chemical Society: Washington, **1983**. For reviews, see Savéant *Adv. Phys. Org. Chem.* **1990**, *26*, 1-130; Russell *Adv. Phys. Org. Chem.* **1987**, *23*, 271-322; Norris, in Patai; Rappoport *The Chemistry of Functional Groups, Supplement D*, pt. 1; Wiley: New York, **1983**, pp. 681-701; Chanon; Tobe *Angew. Chem. Int. Ed. Engl.* **1982**, *21*, 1-23 [*Angew. Chem. 94*, 27-49]; Rossi *Acc. Chem. Res.* **1982**, *15*, 164-170; Beletskaya; Drozd *Russ. Chem. Rev.* **1979**, *48*, 431-448; Bunnett; *Acc. Chem. Res.* **1978**, *11*, 413-420; Wolfe; Carver *Org. Prep. Proceed. Int.* **1978**, *10*, 225-253. For a review of this mechanism with aliphatic substrates, see Rossi; Pierini; Palacios *Adv. Free Radical Chem. (Greenwich, Conn.)* **1990**, *1*, 193-252.

[39]The symbol T is used for electron transfer.

[40]For a discussion, see Amatore, Pinson; Savéant; Thiébault *J. Am. Chem. Soc.* **1981**, *103*, 6930.

[41]Bunnett, Ref. 38.

by solvated electrons, S$_{RN}$1 reactions have been initiated photochemically,[42] electrochemically,[43] and even thermally.[44]

S$_{RN}$1 reactions have a fairly wide scope. There is no requirement for activating groups or strong bases. Alkyl, alkoxy, aryl, and COO$^-$ groups do not interfere, although Me$_2$N, O$^-$, and NO$_2$ groups do interfere. Cine substitution is not found.

Other Mechanisms

There is no clear-cut proof that a one-step S$_N$2 mechanism, so important at a saturated carbon, ever actually occurs with an aromatic substrate. The hypothetical aromatic S$_N$2 process is sometimes called the *one-stage* mechanism to distinguish it from the *two-stage* S$_N$Ar mechanism. Some of the reactions in this chapter operate by still other mechanisms, among them an addition–elimination mechanism (see **3-17**).

REACTIVITY

The Effect of Substrate Structure

In the discussion of electrophilic aromatic substitution (Chapter 11) equal attention was paid to the effect of substrate structure on reactivity (activation or deactivation) and on orientation. The question of orientation was important because in a typical substitution there are four or five hydrogens that could serve as leaving groups. This type of question is much less important for aromatic nucleophilic substitution, since in most cases there is only one potential leaving group in a molecule. Therefore attention is largely focused on the reactivity of one molecule compared with another and not on the comparison of the reactivity of different positions within the same molecule.

S$_N$Ar mechanism   These substitutions are accelerated by electron-withdrawing groups, especially in positions ortho and para to the leaving group[45] and hindered by electron-donating groups. This is, of course, opposite to the effects of these groups on electrophilic substitutions, and the reasons are similar to those discussed in Chapter 11 (p. 507). Table 13.1 contains a list of groups arranged approximately in order of activating or deactivating ability.[46] Hetero nitrogen atoms are also strongly activating (especially to the α and γ positions) and are even more so when quaternized.[47] Thus 2- and 4-chloropyridine, for example, are often used as substrates. Heterocyclic N-oxides are readily attacked by nucleophiles in the 2 and 4 positions, but the oxygen is generally lost in these reactions.[48] The most highly activating group, N$_2$$^+$, is seldom deliberately used to activate a reaction, but it

[42]For reviews of photochemical aromatic nucleophilic substitutions, see Cornelisse, de Gunst, Havinga *Adv. Phys. Org. Chem.* **1975**, *11*, 225-266; Cornelisse *Pure Appl. Chem.* **1975**, *41*, 433-453; Pietra *Q. Rev. Chem. Soc.* **1969**, *23*, 504-521, pp. 519-521.

[43]For a review, see Savéant *Acc. Chem. Res.* **1980**, *13* 323-329. See also Alam; Amatore; Combellas; Thiébault; Verpeaux *J. Org. Chem.* **1990**, *55*, 6347.

[44]Swartz; Bunnett *J. Org. Chem.* **1979**, *44*, 340, and references cited therein.

[45]The effect of meta substituents has been studied much less, but it has been reported that here too, electron-withdrawing groups increase the rate: See Nurgatin; Sharnin; Ginzburg *J. Org. Chem. USSR* **1983**, *19*, 343.

[46]For additional tables of this kind, see Miller, Ref. 2, pp. 61-136.

[47]For reviews of reactivity of nitrogen-containing heterocycles, see Illuminati *Adv. Heterocycl. Chem.* **1964**, *3*, 285-371; Shepherd; Fedrick *Adv. Heterocycl. Chem.* **1965**, *4*, 145-423.

[48]For reviews, see Albini; Pietra *Heterocyclic N-Oxides*; CRC Press: Boca Raton, FL, 1991, pp. 142-180; Katritzky; Lagowski *Chemistry of the Heterocyclic N-Oxides*; Academic Press: New York, 1971, pp. 258-319, 550-553.

TABLE 13.1 Groups listed in approximate descending order of activating ability in the SNAr mechanism[46]

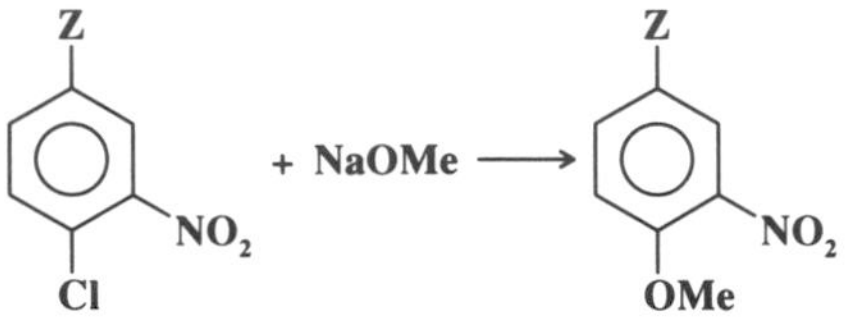

For reaction (a) the rates are relative to **H**; for (b) they are relative to **NH₂**

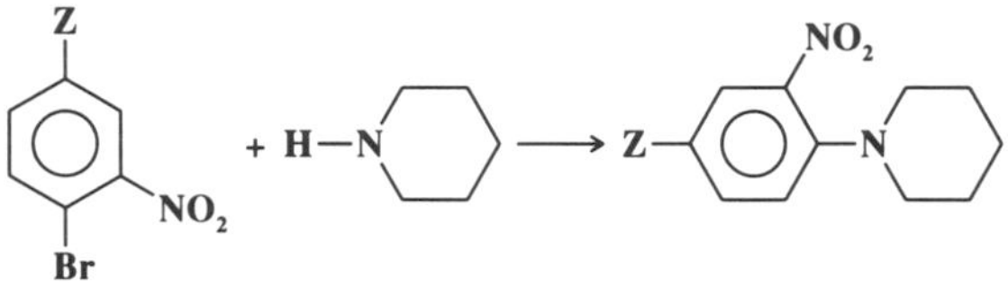

	Group Z	Relative rate of reaction	
		(*a*) H = 1[49]	(*b*) NH₂ = 1[50]
Activates halide exchange at room temperature	N₂⁺		
Activates reaction with strong nucleophiles at room temperature	N⁺—R (heterocyclic)		
Activate reactions with strong nucleophiles at 80–100°C	NO	5.22×10^6	
	NO₂	6.73×10^5	Very fast
	N (heterocyclic)		
With nitro also present, activate reactions with strong nucleophiles at room temperature	SO₂Me		
	NMe₃⁺		
	CF₃		
	CN	3.81×10^4	
	CHO	2.02×10^4	
With nitro also present, activate reactions with strong nucleophiles at 40–60°C	COR		
	COOH		
	SO₃⁻		
	Br		6.31×10^4
	Cl		4.50×10^4
	I		4.36×10^4
	COO⁻		2.02×10^4
	H		8.06×10^3
	F		2.10×10^3
	CMe₃		1.37×10^3
	Me		1.17×10^3
	OMe		145
	NMe₂		9.77
	OH		4.70
	NH₂		1

The comments on the left are from Bunnett and Zahler, Ref. 31, p. 308.

[49]Miller; Parker *Aust. J. Chem.* **1958**, *11*, 302.
[50]Berliner; Monack *J. Am. Chem. Soc.* **1952**, *74*, 1574.

sometimes happens that in the diazotization of a compound such as *p*-nitroaniline or *p*-chloroaniline the group para to the diazonium group is replaced by OH from the solvent or by X from $ArN_2{}^+ X^-$, to the surprise and chagrin of the investigator, who was trying only to replace the diazonium group and to leave the para group untouched. By far the most common activating group is the nitro group and the most common substrates are 2,4-dinitrophenyl halides and 2,4,6-trinitrophenyl halides (also called picryl halides).[51] Poly-fluorobenzenes,[52] e.g., C_6F_6, also undergo aromatic nucleophilic substitution quite well.[53] Benzene rings that lack activating substituents are generally not useful substrates for the SNAr mechanism, because the two extra electrons in **1** are in an antibonding orbital (p. 27). Activating groups, by withrawing electron density, are able to stabilize the intermediates and the transition states leading to them. Reactions taking place by the SNAr mechanism are also accelerated when the aromatic ring is coordinated with a transition metal (e.g., **7** in Chapter 3).[54]

Just as electrophilic aromatic substitutions were found more or less to follow the Hammett relationship (with σ^+ instead of σ; see p. 518), so do nucleophilic substitutions, with σ^- instead of σ for electron-withdrawing groups.[55]

Benzyne mechanism   Two factors affect the positions of the incoming group, the first being the direction in which the aryne forms.[56] When there are groups ortho or para to the leaving group, there is no choice:

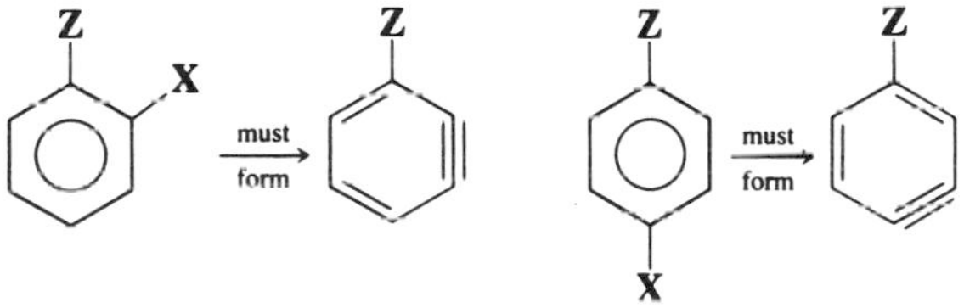

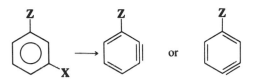

but when a meta group is present, the aryne can form in two different ways:

In such cases, the more acidic hydrogen is removed. Since acidity is related to the field effect of Z, it can be stated that an electron-attracting Z favors removal of the ortho hydrogen while an electron-donating Z favors removal of the para hydrogen. The second factor is that the aryne, once formed, can be attacked at two positions. The favored position for nucleophilic attack is the one that leads to the more stable carbanion intermediate, and this

---

[51]For a review of the activating effect of nitro groups, see de Boer; Dirkx, in Feuer, Ref. 5, pt. 1, pp. 487-612.

[52]Fluorine significantly activates ortho and meta positions, and slightly deactivates (see Table 13.1) para positions: Chambers; Seabury; Williams; Hughes *J. Chem. Soc., Perkin Trans. 1* **1988**, 255.

[53]For reviews, see Yakobson; Vlasov *Synthesis* **1976**, 652-672; Kobrina *Fluorine Chem. Rev.* **1974**, *7*, 1-114.

[54]For a review, see Balas; Jhurry; Latxague; Grelier; Morel; Hamdani; Ardoin; Astruc *Bull. Soc. Chim. Fr.* **1990**, 401-426.

[55]For a discussion of linear free-energy relationships in this reaction, see Bartoli; Todesco *Acc. Chem. Res.* **1977**, *10*, 125-132. For a list of σ^- values, see Table 9.4.

[56]This analysis is from Roberts; Vaughan; Carlsmith; Semenow *J. Am. Chem. Soc.* **1956**, *78*, 611. For a discussion, see Hoffmann *Dehydrobenzene and Cycloalkynes*, Ref. 30, pp. 134-150.

in turn also depends on the field effect of Z. For $-I$ groups, the more stable carbanion is the one in which the negative charge is closer to the substituent. These principles are illustrated by the reaction of the three dichlorobenzenes with alkali-metal amides. The predicted products are

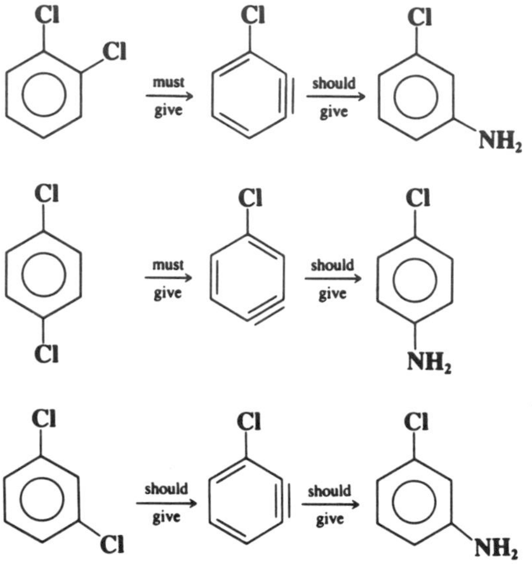

In each case the predicted product was the one chiefly formed.[57] The obtention of m-aminoanisole, mentioned on p. 647, is also in accord with these predictions.

The Effect of the Leaving Group[58]

The common leaving groups in aliphatic nucleophilic substitution (halide, sulfate, sulfonate, NR_3^+, etc.) are also common leaving groups in aromatic nucleophilic substitutions, but the groups NO_2, OR, OAr, SO_2R,[59] and SR, which are not generally lost in aliphatic systems, *are* leaving groups when attached to aromatic rings. Surprisingly, NO_2 is a particularly good leaving group.[60] An approximate order of leaving-group ability is[61] $F > NO_2 > OTs > SOPh > Cl$, Br, $I > N_3 > NR_3^+ > OAr$, OR, SR, NH_2. However, this depends greatly on the nature of the nucleophile, as illustrated by the fact that $C_6Cl_5OCH_3$ treated with NH_2^- gives mostly $C_6Cl_5NH_2$; i.e., one methoxy group is replaced in preference to five chlorines.[62] As usual, OH can be a leaving group if it is converted to an inorganic ester. Among the halogens, fluoro is generally a much better leaving group than the other halogens, which have reactivities fairly close together. The order is usually $Cl > Br > I$, but not always.[63]

[57]Wotiz; Huba *J. Org. Chem.* **1959**, *24*, 595. Eighteen other reactions also gave products predicted by these principles. See also Caubere; Lalloz *Bull. Soc. Chim. Fr.* **1974**, 1983, 1989, 1996; Biehl; Razzuk; Jovanovic; Khanapure *J. Org. Chem.* **1986**, *51*, 5157.
[58]For a review, see Miller, Ref. 2, pp. 137-179.
[59]See, for example Furukawa; Ogawa; Kawai; Oae *J. Chem. Soc., Perkin Trans. 1* **1984**, 1839.
[60]For a review, see Beck *Tetrahedron* **1978**, *34*, 2057-2068. See also Effenberger; Koch; Streicher *Chem. Ber.* **1991**, *24*, 163.
[61]Loudon; Shulman *J. Chem. Soc.* **1941**, 772; Suhr *Chem. Ber.* **1963**, *97*, 3268.
[62]Kobrina; Yakobson *J. Gen. Chem. USSR* **1963**, *33*, 3238.
[63]Reinheimer; Taylor; Rohrbaugh *J. Am. Chem. Soc.* **1961**, *83*, 835; Ross *J. Am. Chem. Soc.* **1959**, *81*, 2113; Bunnett; Garbisch; Pruitt *J. Am. Chem. Soc.* **1957**, *79*, 385; Parker; Read *J. Chem. Soc.* **1962**, 9, 3149; Litvinenko; Shpan'ko; Korostylev *Doklad. Chem.* **1982**, *266*, 309.

The leaving-group order is quite different from that for the S_N1 or S_N2 mechanisms. The most likely explanation is that the first step of the SNAr mechanism is usually rate determining, and this step is promoted by groups with strong $-I$ effects. This would explain why fluoro and nitro are such good leaving groups when this mechanism is operating. Fluoro is the poorest leaving group of the halogens when the second step of the SNAr mechanism is rate-determining or when the benzyne mechanism is operating. The four halogens, as well as SPh, NMe_3^+, and $OPO(OEt)_2$, have been shown to be leaving groups in the SRN1 mechanism.[41] The only important leaving group in the S_N1 mechanism is N_2^+.

The Effect of the Attacking Nucleophile[64]

It is not possible to construct an invariant nucleophilicity order because different substrates and different conditions lead to different orders of nucleophilicity, but an overall approximate order is $NH_2^- > PH_3C^- > PhNH^-$ (aryne mechanism) $> ArS^- > RO^- > R_2NH > ArO^- > OH^- > ArNH_2 > NH_3 > I^- > Br^- > Cl^- > H_2O > ROH$.[65] As with aliphatic nucleophilic substitution, nucleophilicity is generally dependent on base strength and nucleophilicity increases as the attacking atom moves down a column of the periodic table, but there are some surprising exceptions, e.g., OH^-, a stronger base than ArO^-, is a poorer nucleophile.[66] In a series of similar nucleophiles, such as substituted anilines, nucleophilicity *is* correlated with base strength. Oddly, the cyanide ion is not a nucleophile for aromatic systems, except for sulfonic acid salts (**3-12**) and in the von Richter (**3-25**) and Rosenmund–von Braun (**3-11**) reactions, which are special cases.

REACTIONS

In the first part of this section, reactions are classified according to attacking species, with all leaving groups considered together, except for hydrogen and N_2^+, which are treated subsequently. Finally, a few rearrangement reactions are discussed.

All Leaving Groups except Hydrogen and N_2^+

A. Oxygen Nucleophiles

3-1 Hydroxy-de-halogenation

$$ArBr + OH^- \longrightarrow ArOH$$

Aryl halides can be converted to phenols only if activating groups are present or if exceedingly strenuous conditions are employed.[67] Other leaving groups, including nitro,[68] azide, NR_3^+, etc., can also be replaced by OH groups. When the reaction is carried out at high

[64]For a review, see Miller, Ref. 2, pp. 180-233.

[65]This list is compiled from data in Bunnett; Zahler, Ref. 32, p. 340; Bunnett *Q. Rev. Chem. Soc.* **1958**, *12*, 1-16, p. 13; Sauer; Huisgen *Angew. Chem.* **1960**, *72*, 294-315, p. 311; Bunnett *Annu. Rev. Phys. Chem.* **1963**, *14*, 271-290.

[66]For studies of nucleophilicity in the SRN1 mechanism, see Amatore; Combellas; Robveille; Savéant; Thiébault *J. Am. Chem. Soc.* **1986**, *108*, 4754, and references cited therein.

[67]For a review of OH$^-$ and OR$^-$ as nucleophiles in aromatic substitution, see Fyfe, in Patai *The Chemistry of the Hydroxyl Group*, pt. 1; Wiley: New York, 1971, pp. 83-124.

[68]For a convenient way of achieving this conversion, see Knudsen; Snyder *J. Org. Chem.* **1974**, *39*, 3343.

temperatures, cine substitution is observed, indicating a benzyne mechanism.[69] Phenols have been obtained from unactivated aryl halides by treatment with borane and a metal such as lithium, followed by oxidation with alkaline H_2O_2.[70]

OS **I**, 455; **II**, 451; **V**, 632. Also see OS **V**, 918.

3-2 Replacement of an Amino Group by a Hydroxyl Group
Hydroxy-de-amination

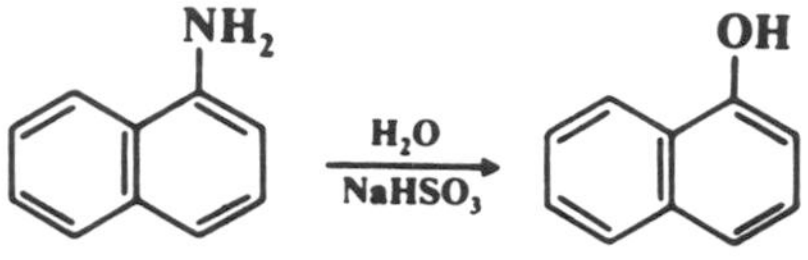

The amino group of naphthylamines can be replaced by a hydroxyl group by treatment with aqueous bisulfite.[71] The scope is greatly limited; the amino group (which may be NH_2 or NHR) must be on a naphthalene ring, with very few exceptions. The reaction is reversible (see **3-7**), and both the forward and reverse reactions are called the *Bucherer reaction*. The mechanism is completely different from any outlined in the first section of this chapter and is discussed at **3-7**.

3-3 Alkali Fusion of Sulfonate Salts
Oxido-de-sulfonato-substitution

$$ArSO_3^- \xrightarrow[300-320°C]{NaOH \text{ fus.}} ArO^-$$

Aryl sulfonic acids can be converted, through their salts, to phenols, by alkali fusion. In spite of the extreme conditions, the reaction gives fairly good yields, except when the substrate contains other groups that are attacked by alkali at the fusion temperatures. Milder conditions can be used when the substrate contains activating groups, but the presence of deactivating groups hinders the reaction. The mechanism is obscure, but a benzyne intermediate has been ruled out by the finding that cine substitution does not occur.[72]

OS **I**, 175; **III**, 288.

3-4 Replacement by OR or OAr
Alkoxy-de-halogenation

$$ArBr + OR^- \longrightarrow ArOR$$

This reaction is similar to **3-1** and, like that one, generally requires activated substrates.[67] With unactivated substrates, side reactions predominate, though aryl methyl ethers have been prepared from unactivated chlorides by treatment with MeO^- in HMPA.[73] This reaction gives better yields than **3-1** and is used more often. A good solvent is liquid ammonia. NaOMe reacted with *o*- and *p*-fluoronitrobenzenes about 10^9 times faster in NH_3 at $-70°C$

---

[69]The benzyne mechanism for this reaction is also supported by ^{14}C labeling experiments: Bottini; Roberts *J. Am. Chem. Soc.* **1957**, *79*, 1458; Dalman; Neumann *J. Am. Chem. Soc.* **1968**, *90*, 1601.

[70]Pickles; Thorpe *J. Organomet. Chem.* **1974**, *76*, C23.

[71]For reviews, see Seeboth *Angew. Chem. Int. Ed. Engl.* **1967**, *6*, 307-317 [*Angew. Chem. 79*, 329-340]; Gilbert *Sulfonation and Related Reactions*; Wiley: New York, 1965, pp. 166-169.

[72]Buzbee *J. Org. Chem.* **1966**, *31*, 3289; Oae; Furukawa; Kise; Kawanishi *Bull. Chem. Soc. Jpn.* **1966**, *39*, 1212.

[73]Shaw; Kunerth; Swanson *J. Org. Chem.* **1976**, *41*, 732; Testaferri; Tiecco; Tingoli; Chianelli; Montanucci *Tetrahedron* **1983**, *39*, 193.

than in MeOH.[74] Phase transfer catalysis has also been used.[75] In addition to halides, leaving groups can be nitro, NR_3^+, other OR, etc., even OH.[76] Acid salts, RCOO⁻, are sometimes used as nucleophiles. Good yields of aryl benzoates can be obtained by the treatment of aryl halides with cuprous benzoate in diglyme or xylene at 140 to 160°C.[77] Unactivated substrates have been converted to carboxylic esters in low-to-moderate yields under oxidizing conditions.[78] The following chain mechanism, called the SON2 mechanism,[79] has been suggested:[78]

Initiation

$$\text{ArX} \xrightarrow[\text{acceptor}]{\text{electron}} [\text{ArX}]^{+\bullet} \xrightarrow{\text{RCOO}^-} [\text{ArX(OCOR)}]^{\bullet} \xrightarrow{-X^-} [\text{Ar-OCOR}]^{+\bullet}$$

Propagation

$$[\text{Ar-OCOR}]^{+\bullet} + \text{ArX} \longrightarrow [\text{ArX}]^{+\bullet} + \text{Ar-OCOR}$$

For aroxide nucleophiles, the reaction is promoted by copper salts,[80] and when these are used, activating groups need not be present. This method of preparation of diaryl ethers is called the *Ullmann ether synthesis*[81] and should not be confused with the Ullmann biaryl synthesis (**3-16**). The reactivity order is typical of nucleophilic substitutions, despite the presence of the copper salts.[82] Because aryloxycopper(I) reagents ArOCu react with aryl halides to give ethers, it has been suggested that they are intermediates in the Ullmann ether synthesis.[83] Indeed, high yields of ethers can be obtained by reaction of ROCu or ArOCu with aryl halides.[84] Unactivated substrates also react with phenoxide ion with electrochemical catalysis in liquid NH_3–Me_2SO, to give diaryl ethers, presumably by the SRN1 mechanism.[85] Diaryl ethers can be prepared from activated aryl halides by treatment with triaryl phosphate $(ArO)_3PO$.[86]

OS **I**, 219; **II**, 445; **III**, 293, 566; **V**, 926; **VI**, 150.

B. Sulfur Nucleophiles

3-5 Replacement by SH or SR

ArBr + SH⁻ ⟶ ArSH   **Mercapto-de-halogenation**

ArBr + SR⁻ ⟶ ArSR   **Alkylthio-de-halogenation**

Aryl thiols and thioethers can be prepared in reactions similar to **3-1** and **3-4**.[87] Activated aryl halides generally give good results, but side reactions are occasionally important. Diaryl

---

[74]Kizner; Shteingarts *J. Org. Chem. USSR* **1984**, *20*, 991.
[75]Artamanova; Seregina; Shner; Salov; Kokhlova; Zhdamarova *J. Org. Chem. USSR* **1989**, *25*, 554.
[76]Oae; Kiritani *Bull. Chem. Soc. Jpn.* **1964**, *37*, 770, **1966**, *39*, 611.
[77]Cohen; Lewin *J. Am. Chem. Soc.* **1966**, *88*, 4521; Cohen; Wood; Dietz *Tetrahedron Lett.* **1974**, 3555.
[78]Eberson; Jönsson; Wistrand *Tetrahedron* **1982**, *38*, 1087; Jönsson; Wistrand *J. Org. Chem.* **1984**, *49*, 3340.
[79]First proposed by Alder *J. Chem. Soc., Chem. Commun.* **1980**, 1184.
[80]For a review of copper-assisted aromatic nucleophilic substitution, see Lindley *Tetrahedron* **1984**, *40*, 1433-1456.
[81]For a review of the Ullmann ether synthesis, see Moroz; Shvartsberg *Russ. Chem. Rev.* **1974**, *43*, 679-689.
[82]Weingarten *J. Org. Chem.* **1964**, *29*, 977, 3624.
[83]Kawaki; Hashimoto *Bull. Chem. Soc. Jpn.* **1972**, *45*, 1499.
[84]Whitesides; Sadowski; Lilburn *J. Am. Chem. Soc.* **1974**, *96*, 2829.
[85]Alam; Amatore; Combellas; Pinson; Savéant; Thiébault; Verpeaux *J. Org. Chem.* **1988**, *53*, 1496.
[86]Ohta; Iwasaki; Akita *Synthesis* **1982**, 828. For other procedures, see Bates; Janda *J. Org. Chem.* **1982**, *47*, 4374; Sammes; Thetford; Voyle *J. Chem. Soc., Perkin Trans. 1* **1988**, 3229.
[87]For a review of sulfur nucleophiles in aromatic substitution, see Peach, in Patai *The Chemistry of the Thiol Group*, pt. 2; Wiley: New York, 1974, pp. 735-744.

sulfides can be prepared by the use of SAr⁻. Even unactivated aryl halides react with SAr⁻ if polar aprotic solvents, e.g., DMF,[88] Me$_2$SO,[89] tetraglyme,[90] 1-methyl-2-pyrrolidinone,[91] or HMPA,[92] are used, though the mechanisms are still mostly or entirely nucleophilic substitution. Unactivated aryl halides also give good yields of sulfides on treatment with SAr⁻ or SR⁻ in the presence of a catalytic amount of (Ph$_3$P)$_4$Pd.[93] Copper catalysts have also been used.[94] Diaryl sulfides can also be prepared (in high yields) by treatment of unactivated aryl iodides with ArS⁻ in liquid ammonia under irradiation.[95] The mechanism in this case is probably SRN1. The reaction (with unactivated halides) has also been carried out electrolytically, with a nickel complex catalyst.[96]

Other sulfur nucleophiles also react with activated aryl halides:

$$2ArX + S_2^{2-} \longrightarrow Ar-S-S-Ar \qquad ArX + SCN^- \longrightarrow ArSCN$$

$$ArX + SO_3^{2-} \longrightarrow Ar-SO_3^- \qquad ArX + RSO_2^- \longrightarrow Ar-SO_2-R$$

Unactivated thiocyanation has been accomplished with charcoal supported copper (I) thiocyanate.[97]

OS **I**, 220; **III**, 86, 239, 667; **V**, 107, 474; **VI**, 558, 824. Also see OS **V**, 977.

C. Nitrogen Nucleophiles

3-6 Replacement by NH$_2$, NHR, or NR$_2$
Amino-de-halogenation

$$ArBr + NH_3 \longrightarrow ArNH_2$$

$$ArBr + RNH_2 \longrightarrow ArNHR$$

$$ArBr + R_2NH \longrightarrow ArNR_2$$

Activated aryl halides react quite well with ammonia and with primary and secondary amines to give the corresponding arylamines. Primary and secondary amines usually give better results than ammonia, with piperidine especially reactive. Picryl chloride (2,4,6-trinitrochlorobenzene) is often used to form amine derivatives. 2,4-Dinitrofluorobenzene is used to tag the amino end of a peptide or protein chain. Other leaving groups in this reaction may be NO$_2$, N$_3$, OSO$_2$R, OR, SR, N=NAr (where Ar contains electron-withdrawing groups)[98] and even NR$_2$.[99] Activated halides can be converted to diethylamino compounds ArX → ArNMe$_2$ by treatment with HMPA.[100]

[88]Campbell *J. Org. Chem.* **1964**, *29*, 1830; Testaferri; Tiecco; Tingoli; Chianelli; Montanucci *Synthesis* **1983**, 751. For the extension of this to selenides, see Tiecco; Testaferri; Tingoli; Chianelli; Montanucci *J. Org. Chem.* **1983**, *48*, 4289.

[89]Bradshaw; South; Hales *J. Org. Chem.* **1972**, *37*, 2381.

[90]Pastor, Hessell *J. Org. Chem.* **1985**, *50*, 4812; Pastor *Helv. Chim. Acta* **1988**, *71*, 859.

[91]Caruso; Colley; Bryant *J. Org. Chem.* **1991**, *56*, 862; Shaw *J. Org. Chem.* **1991**, *56*. 3728.

[92]Cogolli; Maiolo; Testaferri; Tingoli; Tiecco *J. Org. Chem.* **1979**, *44*, 2642. See also Testaferri; Tingoli; Tiecco *Tetrahedron Lett.* **1980**, *21*, 3099; Suzuki; Abe; Osuka *Chem. Lett.* **1980**, 1363.

[93]Migita; Shimizu; Asami; Shiobara; Kato; Kosugi *Bull. Chem. Soc. Jpn.* **1980**, *53*, 1385.

[94]Bowman; Heaney; Smith *Tetrahedron Lett.* **1984**, *25*, 5821; Yamamoto; Sekine *Can. J. Chem.* **1984**, *62*, 1544. For other catalysts, see Cristau; Chabaud; Chêne; Christol *Synthesis* **1981**, 892; Takagi *Chem. Lett.* **1985**, 1307. **1986**, 1379. **1987**, 2221.

[95]Bunnett; Creary *J. Org. Chem.* **1974**, *39*, 3173, 3611.

[96]Meyer; Troupel *J. Organomet. Chem.* **1988**, *354*, 249.

[97]Clark; Jones; Duke; Miller *J. Chem. Soc., Chem. Commun.* **1989**, 81.

[98]Kazankov; Ginodman *J. Org. Chem. USSR* **1975**, *11*, 451.

[99]Sekiguchi; Horie; Suzuki *J. Chem. Soc., Chem. Commun.* **1988**, 698.

[100]See, for example, Gupton; Idoux; Baker; Colon; Crews; Jurss; Rampi *J. Org. Chem.* **1983**, *48*, 2933.

Unactivated aryl halides can be converted to amines by the use of $NaNH_2$, NaNHR, or $NaNR_2$.[101] With these reagents, the benzyne mechanism generally operates, so cine substitution is often found. Ring closure has been effected by this type of reaction,[102] e.g.,

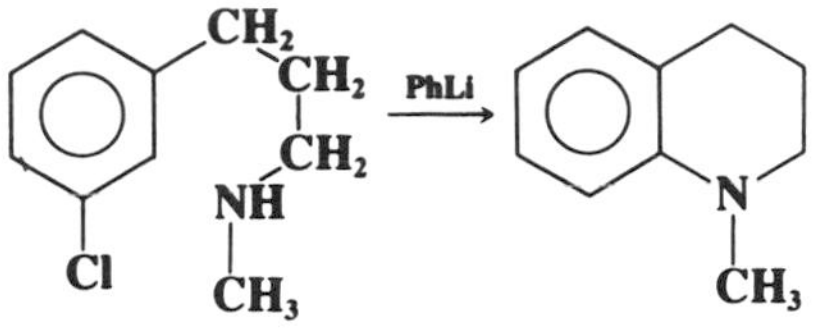

It has also proved possible to close larger rings in this manner: eight- and even twelve-membered. Triarylamines have been prepared in a similar manner from ArI and $Ar_2'NLi$, even with unactivated ArI.[103] In the *Goldberg reaction*, an aryl bromide reacts with an acetanilide in the presence of K_2CO_3 and CuI to give an N-acetyldiarylamine, which can be hydrolyzed to a diarylamine: ArBr + Ar'NHAc → ArAr'NAc.[104]

The reaction with ammonia or amines, which undoubtedly proceeds by the SNAr mechanism, is catalyzed by copper[80] and nickel[105] salts, though these are normally used only with rather unreactive halides.[106] This reaction, with phase transfer catalysis, has been used to synthesize triarylamines.[107] Copper ion catalysts (especially cuprous oxide or iodide) also permit the Gabriel synthesis (**0-58**) to be applied to aromatic substrates. Aryl bromides or iodides are refluxed with potassium phthalimide and Cu_2O or CuI in dimethylacetamide to give N-aryl phthalimides, which can be hydrolyzed to primary aryl amines.[108]

In certain cases the SRN1 mechanism has been found (p. 648). When the substrate is a heterocyclic aromatic nitrogen compound, still a different mechanism [the SN(ANRORC) mechanism], involving opening and reclosing of the aromatic ring, has been shown to take place.[109]

OS **I**, 544; **II**, 15, 221, 228; **III**, 53, 307, 573; **IV**, 336, 364; **V**, 816, 1067; **VII**, 15.

3-7   Replacement of a Hydroxy Group by an Amino Group
Amino-de-hydroxylation

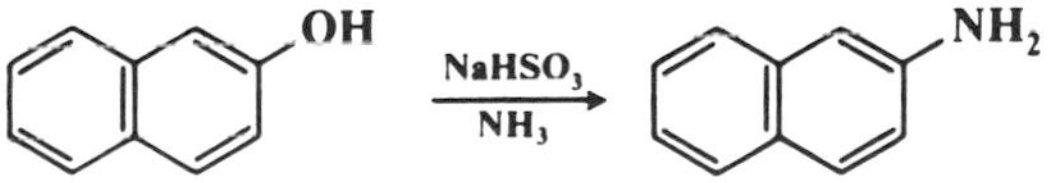

The reaction of naphthols with ammonia and sodium bisulfite is the reverse of **3-2** and has a similar scope.[71] It is also called the *Bucherer reaction*. Primary amines can be used instead

[101]For a review, see Heaney *Chem Rev.* **1962**, *62*, 81-97, pp. 83-89.
[102]Huisgen; König; Lepley *Chem. Ber.* **1960**, *93*, 1496; Bunnett; Hrutfiord *J. Am. Chem. Soc.* **1961**, *83*, 1691. For a review of ring closures by the benzyne mechanism, see Hoffmann *Dehydrobenzene and Cycloalkynes*, Ref. 30, pp. 150-164.
[103]Neunhoeffer; Heitmann *Chem. Ber.* **1961**, *94*, 2511.
[104]See Freeman; Butler; Freedman, *J. Org. Chem.* **1978**, *43*, 4975; Renger *Synthesis* **1985**, 856.
[105]See Cramer; Coulson *J. Org. Chem.* **1975**, *40*, 2267.
[106]For discussions of the mechanism, see Bethell; Jenkins; Quan *J. Chem. Soc., Perkin Trans. 1* **1985**, 1789; Tuong; Hida *J. Chem. Soc., Perkin Trans. 2* **1974**, 676; Kondratov; Shein *J. Org. Chem. USSR* **1979**, *15*, 2160; Paine *J. Am. Chem. Soc.* **1987**, *109*, 1496.
[107]Gauthier; Fréchet *Synthesis* **1987**, 383.
[108]Bacon; Karim *Chem. Commun.* **1969**, 578, *J. Chem. Soc., Perkin Trans. 1* **1973**, 272, 278; Sato; Ebine; Akabori *Synthesis* **1981**, 472. See also Yamamoto; Kurata *Can. J. Chem.* **1983**, *61*, 86.
[109]For reviews, see van der Plas *Tetrahedron* **1985**, *41*, 237-281, *Acc. Chem. Res.* **1978**, *11*, 462-468.

of ammonia, in which case N-substituted naphthylamines are obtained. In addition, primary naphthylamines can be converted to secondary, by a transamination reaction:

$$\text{ArNH}_2 + \text{RNH}_2 \xrightarrow{\text{NaHSO}_3} \text{ArNHR}$$

The mechanism of the Bucherer reaction amounts to a kind of overall addition–elimination:[110]

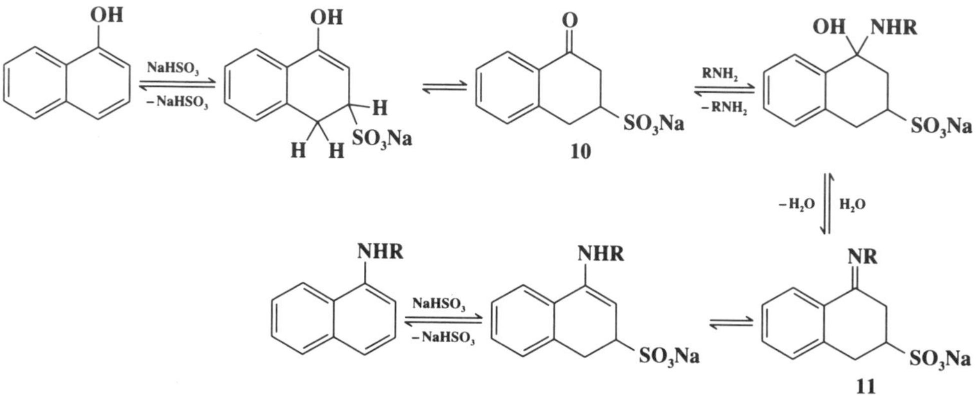

The first step in either direction consists of addition of NaHSO_3 to one of the double bonds of the ring, which gives an enol (or enamine) that tautomerizes to the keto (or imine) form. The conversion of **10** to **11** (or vice versa) is an example of **6-14** (or **6-2**). Evidence for this mechanism was the isolation of **10**[111] and the demonstration that for β-naphthol treated with ammonia and HSO_3^-, the rate of the reaction depends only on the substrate and on HSO_3^-, indicating that ammonia is not involved in the rate-determining step.[112] If the starting compound is a β-naphthol, the intermediate is a 2-keto-4-sulfonic acid compound, so the sulfur of the bisulfite in either case attacks meta to the OH or NH_2.[113]

Hydroxy groups on benzene rings can be replaced by NH_2 groups if they are first converted to aryl diethyl phosphates. Treatment of these with KNH_2 and potassium metal in liquid

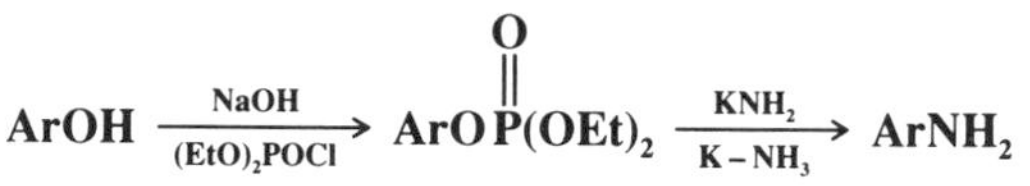

$$\text{ArOH} \xrightarrow[\text{(EtO)}_2\text{POCl}]{\text{NaOH}} \text{ArO}\overset{\displaystyle\text{O}}{\overset{\displaystyle\|}{\text{P}}}(\text{OEt})_2 \xrightarrow[\text{K}-\text{NH}_3]{\text{KNH}_2} \text{ArNH}_2$$

ammonia gives the corresponding primary aromatic amines.[114] The mechanism of the second step is SRN1.[115]

OS **III**, 78.

[110]Riechc: Secboth *Liebigs Ann. Chem.* **1960,** *638,* 66.
[111]Riechc: Secboth *Liebigs Ann. Chem.* **1960,** *638,* 43, 57.
[112]Kozlov: Vcsclovskaia *J. Gen. Chem. USSR* **1958,** *28,* 3359.
[113]Riechc: Secboth *Liebigs Ann. Chem.* **1960,** *638,* 76.
[114]Rossi: Bunnett *J. Org. Chem.* **1972,** *37,* 3570.
[115]For another method of converting phenols to amines, see Scherrer: Beatty *J. Org. Chem.* **1972,** *37,* 1681.

D. Halogen Nucleophiles

3-8 The Introduction of Halogens
Halo-de-halogenation, etc.

$$Ar—X + X'^- \longrightarrow Ar—X' + X^-$$

It is possible to replace a halogen on a ring by another halogen[116] if the ring is activated. There is an equilibrium, but it is usually possible to shift this in the desired direction by the use of an excess of added halide ion.[117] Another common leaving group is nitro, which can be replaced with chloro by use of NH_4Cl, PCl_5, $SOCl_2$, HCl, Cl_2, or CCl_4. Some of these reagents operate only at high temperatures and the mechanism is not always nucleophilic substitution. Activated aromatic nitro compounds can be converted to fluorides with F^-.[118]

A phenolic hydroxy group can be replaced by chloro with PCl_5 or $POCl_3$, but only if activated. Unactivated phenols give phosphates when treated with $POCl_3$: 3ArOH + $POCl_3 \rightarrow (ArO)_3PO$. Phenols, even unactivated ones, can be converted to aryl bromides by treatment with Ph_3PBr_2[119] (see **0-66**) and to aryl chlorides by treatment with $PhPCl_4$.[120]

Halide exchange is particularly useful for putting fluorine into a ring, since there are fewer alternate ways of doing this than for the other halogens. Activated aryl chlorides give fluorides when treated with KF in DMF, Me_2SO, or dimethyl sulfone.[121] Halide exchange can also be accomplished with copper halides. Since the leaving-group order in this case is I > Br > Cl $\gg$ F (which means that iodides cannot normally be made by this method), the S_NAr mechanism is probably not operating.[122] However, aryl iodides have been prepared from bromides, by the use of Cu supported on charcoal or Al_2O_3,[123] and by treatment with excess KI and a nickel catalyst.[124]

OS **III**, 194, 272, 475; **V**, 142, 478; **67**, 20.

E. Hydrogen as Nucleophile

3-9 Reduction of Phenols and Phenolic Esters and Ethers[125]
Hydro-de-hydroxylation or **Dehydroxylation**, etc.

$$ArOH \xrightarrow{\text{Zn}} ArH$$

[116]For a list of reagents, with references, see Larock *Comprehensive Organic Transformations*; VCH: New York, 1989. p. 340.

[117]Sauer; Huisgen *Angew. Chem.* **1960**, *72*, 294-315, p. 297.

[118]Attiná; Cacace; Wolf *J. Chem. Soc. Chem. Commun.* **1983**, 108; Clark; Smith *Tetrahedron Lett.* **1985**, *26*, 2233; Suzuki; Yazawa; Yoshida; Furusawa; Kimura *Bull. Chem. Soc. Jpn.* **1990**, *63*, 2010; Effenberger; Streicher *Chem. Ber.* **1991**, *124*, 157.

[119]Wiley; Hershkowitz; Rein; Chung *J. Am. Chem. Soc.* **1964**, *86*, 964; Wiley; Rein; Hershkowitz *Tetrahedron Lett.* **1964**, 2509; Schaefer; Higgins *J. Org. Chem.* **1967**, *32*, 1607.

[120]Bay; Bak; Timony; Leone-Bay *J. Org. Chem.* **1990**, *55*, 3415.

[121]Starr; Finger *Chem. Ind. (London)* **1962**, 1328; Shiley; Dickerson; Finger *J. Fluorine Chem.* **1972**, *2*, 19; Kimura; Suzuki *Tetrahedron Lett.* **1989**, *30*, 1271. For the use of phase transfer catalysis in this reaction, see Yoshida; Kimura *Chem. Lett.* **1988**, 1355. For a review of the preparation of aryl fluorides by halogen exchange, see Dolby-Glover *Chem. Ind. (London)* **1986**, 518-523.

[122]Bacon; Hill *J. Chem. Soc.* **1964**, 1097, 1108. See also Nefedov; Tarygina; Kryuchkova; Ryabokobylko *J. Org. Chem. USSR* **1981**, *17*, 487; Suzuki; Kondo; Ogawa *Chem. Lett.* **1985**, 411; Liedholm; Nilsson *Acta Chem. Scand., Ser. B* **1988**, *42*, 289; Clark; Jones; Duke; Miller *J. Chem. Res. (S)* **1989**, 238.

[123]Clark; Jones *J. Chem. Soc., Chem. Commun.* **1987**, 1409.

[124]Yang; Li; Cheng *J. Org. Chem.* **1987**, *52*, 691.

[125]For a list of reagents, with references, see Ref. 116, pp. 27-31ff.

Phenols can be reduced by distillation over zinc dust or with HI and red phosphorus, but these methods are quite poor and are seldom feasible. Catalytic hydrogenation has also been used, but the corresponding cyclohexanol (see **5-10**) is a side product.[126]

Much better results have been obtained by conversion of phenols to certain esters or ethers and reduction of the latter:

$$ArOSO_2CF_3 \xrightarrow[\substack{Pd(OAc)_2, \ Ph_3P \\ DMF}]{HCOOH, \ Et_3N} ArH \qquad \text{Ref. 127}$$

$$ArOTs + NH_2NH_2 \xrightarrow{Pd} ArH \qquad \text{Ref. 128}$$

$$\underset{\substack{\| \\ O}}{ArO-P(OEt)_2} \xrightarrow[THF]{Ti} ArH \qquad \text{Ref. 129}$$

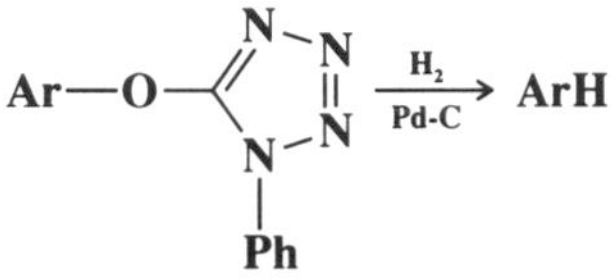

$$Ar-O-\underset{\substack{| \\ Ph}}{\overset{N-N}{\underset{N-N}{\| \ \|}}} \xrightarrow[Pd-C]{H_2} ArH \qquad \text{Ref. 130}$$

12

12 are prepared by treatment of phenols with 1-phenyl-5-chlorotetrazole in acetone containing K_2CO_3.

OS **VI**, 150. See also OS **VII**, 476.

3-10 Reduction of Halides and Nitro Compounds

The reaction ArX → ArH is treated in Chapter 11 (reaction **1-42**), although, depending on reagent and conditions, it can be nucleophilic or free-radical substitution, as well as electrophilic.

The nitro group of aromatic nitro compounds has been removed with sodium borohydride.[131] This reaction involves an addition–elimination mechanism.

F. Carbon Nucleophiles[131a]

Some formations of new aryl–carbon bonds formed from aryl substrates have been considered in Chapter 10 (see **0-87, 0-95, 0-102, 0-103**).

3-11 The Rosenmund–von Braun Reaction
Cyano-de-halogenation

$$ArBr + \text{anhydrous CuCN} \xrightarrow{200^\circ C} ArCN$$

[126]Shuikin; Erivanskaya *Russ. Chem. Rev.* **1960**, *29*, 309-320, pp. 313-315. See also Bagnell; Jeffery *Aust. J. Chem.* **1981**, *34*, 697.

[127]Cacchi; Ciattini; Morera; Ortar *Tetrahedron Lett.* **1986**, *27*, 5541. See also Peterson; Kunng; McCallum; Wulff *Tetrahedron Lett.* **1987**, *28*, 1381; Chen; He *Synthesis* **1988**, 896; Cabri; De Bernardinis; Francalanci; Penco *J. Org. Chem.* **1990**, *55*, 350.

[128]Kenner; Murray *J. Chem. Soc.* **1949**, S178; Rottendorf; Sternhell *Aust. J. Chem.* **1963**, *16*, 647.

[129]Welch; Walters *J. Org. Chem.* **1978**, *43*, 4797. See also Rossi; Bunnett *J. Org. Chem.* **1973**, *38*, 2314.

[130]Musliner; Gates *J. Am. Chem. Soc.* **1966**, *88*, 4271; Hussey; Johnstone; Entwistle *Tetrahedron* **1982**, *38*, 3775; Johnstone; Price *J. Chem. Soc., Chem. Commun.* **1984**, 845. For related methods, see Pailer; Gössinger *Monatsh. Chem.* **1969**, *100*, 1613; van Muijlwijk; Kieboom; van Bekkum *Recl. Trav. Chim. Pays-Bas* **1974**, *93*, 204.

[131]Severin; Schmitz; Temme *Chem. Ber.* **1963**, *96*, 2499; Kniel *Helv. Chim. Acta* **1968**, *51*, 371. For another method, see Ono; Tamura; Kaji *J. Am. Chem. Soc.* **1983**, *105*, 4017.

[131a]For a review of many of these reactions, see Artamkina; Kovalenko; Beletskaya; Reutov *Russ. Chem. Rev.* **1990**, *59*, 750-777.

The reaction between aryl halides and cuprous cyanide is called the *Rosenmund–von Braun reaction*.[132] Reactivity is in the order I > Br > Cl > F, indicating that the SNAr mechanism does not apply.[133] Other cyanides, e.g., KCN and NaCN, do not react with aryl halides, even activated ones. However, alkali cyanides do convert aryl halides to nitriles[134] in dipolar aprotic solvents in the presence of Pd(II) salts[135] or copper[136] or nickel[137] complexes. A nickel complex also catalyzes the reaction between aryl triflates and KCN to give aryl nitriles.[138] Aromatic ethers ArOR[139] and some nitro compounds ArNO$_2$[140] have been photochemically converted to ArCN.

OS **III**, 212, 631.

3-12 Cyanide Fusion of Sulfonate Salts
Cyano-de-sulfonato-substitution

$$ArSO_3^- \xrightarrow{\text{NaCN fus.}} ArCN$$

This reaction is very similar to **3-3.** Yields are usually low.

3-13 Coupling of Organometallic Compounds with Aryl Halides, Ethers, and Carboxylic Esters
Alkyl-de-halogenation, etc.

$$ArI + R_2CuLi \longrightarrow ArR$$

Aryl iodides, which need not be activated, couple with lithium dialkylcopper reagents. The reaction is discussed at **0-87.** Aryl halides, even when activated, generally do not couple with Grignard reagents, though certain transition-metal catalysts do effect this reaction in variable yields.[141] The reaction with Grignard reagents proceeds better when OR can be the leaving group, providing that activating groups are present in the ring. The oxazoline group activates *o*-methoxy and *o*-fluoro groups to reaction with Grignard reagents and organolithiums; the product can be hydrolyzed after coupling[142] (see **0-98**):

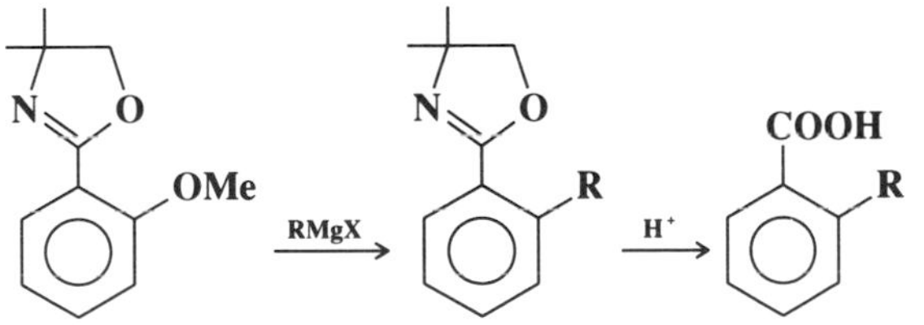

[132]For a review of cyano-de-halogenation, see Ellis; Romney-Alexander *Chem. Rev.* **1987**, *87*, 779-794.
[133]For discussions of the mechanism, see Couture; Paine *Can. J. Chem.* **1985**, *63*, 111; Connor; Leeming; Price *J. Chem. Soc., Perkin Trans. 1* **1990**, 1127.
[134]For a list of reagents that convert aryl halides to cyanides, with references, see Ref. 116, pp. 861-862.
[135]Takagi; Okamoto; Sakakibara; Ohno; Oka; Hayama *Bull Chem. Soc. Jpn.* **1975**, *48*, 3298. **1976**, *49*, 3177. See also Sekiya; Ishikawa *Chem. Lett.* **1975**, 277; Takagi; Sasaki; Sakakibara *Bull. Chem. Soc. Jpn.* **1991**, *64*, 1118.
[136]Connor; Gibson; Price *J. Chem. Soc., Perkin Trans. 1* **1987**, 619.
[137]Cassar; Foà; Montanari; Marinelli *J. Organomet. Chem.* **1979**, *173*, 335; Sakakibara; Okuda; Shimobayashi; Kirino; Sakai; Uchino; Takagi *Bull. Chem. Soc. Jpn.* **1988**, *61*, 1985.
[138]Chambers; Widdowson *J. Chem. Soc., Perkin Trans. 1* **1989**, 1365; Takagi; Sakakibara *Chem. Lett.* **1989**, 1957.
[139]Letsinger; Colb *J. Am. Chem. Soc.* **1972**, *94*, 3665.
[140]See, for example, Vink; Verheijdt; Cornelisse; Havinga *Tetrahedron* **1972**, *28*, 5081.
[141]See, for example, Sekiya; Ishikawa *J. Organomet. Chem.* **1976**, *118*, 349, **1977**, *125*, 281; Negishi; Matsushita; Kobayashi; Rand *Tetrahedron Lett.* **1983**, *24*, 3823; Tiecco; Testaferri; Tingoli; Chianelli; Wenkert *Tertrahedron Lett.* **1982**, *23*, 4629; Eapen; Dua; Tamborski *J. Org. Chem.* **1984**, *49*, 478; Bell; Hu; Patel *J. Org. Chem.* **1987**, *52*, 3847; Bumagin; Andryukhova; Beletskaya *Doklad. Chem.* **1987**, *297*, 524; Ozawa; Kurihara; Fujimori; Hidaka; Toyoshima; Yamamoto *Organometallics* **1989**, *8*, 180.
[142]For a review of oxazolines in aromatic substitutions, see Reuman; Meyers *Tetrahedron* **1985**, *41*, 837-860. For the similar use of oxazoles, see Cram; Bryant; Doxsee *Chem. Lett.* **1987**, 19.

Unactivated aryl halides couple with alkyllithium reagents in THF[143] and with organotin compounds and a Pd complex catalyst[144] to give moderate-to-good yields of alkyl arenes. Unactivated aryl triflates[145] $ArOSO_2CF_3$ react to give ArR in good yields when treated with $R_2Cu(CN) Li_2$,[146] with RZnX,[147] with R_3Al,[148] or with $R_3'SnR$ and a Pd complex catalyst.[149] The coupling reaction between aryl halides and alkenes, with a Pd catalyst, is treated at **4-20**.

Unactivated aryl halides react with copper acetylides to give good yields of arylacetylenes (*Stephens–Castro coupling*).[150]

$$ArI + RC{\equiv}CCu \longrightarrow ArC{\equiv}CR$$

R many be alkyl or aryl. A wide variety of aryl iodides has been used and the reaction is of considerable synthetic importance.

Unactivated aryl iodides undergo the conversion ArI $\rightarrow$ $ArCH_3$ when treated with tris(diethylamino)sulfonium difluorotrimethylsilicate and a palladium catalyst.[151] A number of methods, all catalyzed by palladium complexes, have been used to prepare unsymmetrical biaryls (see also **3-16**). In these methods, aryl bromides or iodides are coupled with aryl Grignard reagents,[152] with arylboronic acids $ArB(OH)_2$,[153] with aryltin compounds Ar-SnR_3,[154] and with arylmercury compounds.[155] Unsymmetrical binaphthyls were synthesized by photochemically stimulated reaction of naphthyl iodides with naphthoxide ions in an S$_{RN}$1 reaction.[156] Grignard reagents also couple with aryl halides without a palladium catalyst, by the benzyne mechanism.[157]

OS **VI**, 916; **65**, 108; **66**, 67.

3-14   Arylation at a Carbon Containing Active Hydrogen
Bis(ethoxycarbonyl)methyl-de-halogenation, etc.

$$ArBr + Z{-}\overset{\ominus}{CH}{-}Z' \longrightarrow Z{-}\underset{\underset{Ar}{|}}{CH}{-}Z'$$

[143]Merrill; Negishi *J. Org. Chem.* **1974**, *39*, 3452. For another method, see Hallberg; Westerlund *Chem. Lett.* **1982**, 1993.

[144]Bumagin; Bumagina; Beletskaya *Doklad. Chem.* **1984**, *274*, 39; Bumagin; Ponomarev; Beletskaya *J. Org. Chem. USSR* **1987**, *23*, 1215, 1222; Kosugi; Sumiya; Ohhashi; Sano; Migita *Chem. Lett.* **1985**, 997; McKean; Parrinello; Renaldo; Stille *J. Org. Chem.* **1987**, *52*, 422.

[145]For another coupling reaction of aryl triflates, see Aoki; Fujimura; Nakamura; Kuwajima *J. Am. Chem. Soc.* **1988**, *110*, 3296.

[146]McMurry; Mohanraj *Tetrahedron Lett.* **1983**, *24*, 2723.

[147]Chen; He *Tetrahedron Lett.* **1987**, *28*, 2387.

[148]Hirota; Isobe; Maki *J. Chem. Soc., Perkin Trans. 1* **1989**, 2513.

[149]Echevarren; Stille *J. Am. Chem. Soc.* **1987**, *109*, 5478. For a similar reaction with aryl fluorosulfonates, see Roth; Fuller *J. Org. Chem.* **1991**, *56*, 3493.

[150]Castro; Stephens *J. Org. Chem.* **1963**, *28*, 2163; Stephens; Castro *J. Org. Chem.* **1963**, *28*, 3313; Sladkov; Ukhin; Korshak *Bull. Acad. Sci. USSR., Div. Chem. Sci.* **1963**, 2043. For a review, see Sladkov; Gol'ding *Russ. Chem. Rev.* **1979**, *48*, 868-896. For an improved procedure, see Bumagin; Kalinovskii; Ponomarev; Beletskaya *Doklad. Chem.* **1982**, *265*, 262.

[151]Hatanaka; Hiyama *Tetrahedron Lett.* **1988**, *29*, 97.

[152]Widdowson; Zhang *Tetrahedron* **1986**, *42*, 2111. See also Ikoma; Taya; Ozaki; Higuchi; Naoi; Fuji-i *Synthesis* **1990**, 147.

[153]Miyaura; Yanagi; Suzuki *Synth. Commun.* **1981**, *11*, 513; Miller; Dugar *Organometallics* **1984**, *3*, 1261; Sharp; Cheng; Snieckus *Tetrahedron Lett.* **1987**, *28*, 5093; Cheng; Snieckus *Tetrahedron Lett.* **1987**, *28*, 5097.

[154]Bailey *Tetrahedron Lett.* **1986**, *27*, 4407.

[155]Bumagin; More; Beletskaya *J. Organomet. Chem.* **1989**, *364*, 231.

[156]Beugelmans; Bois-Choussy; Tang *Tetrahedron Lett.* **1988**, *29*, 1705. For other preparations of biaryls via S$_{RN}$1 processes, see Alam; Amatore; Combellas; Thiébault; Verpeaux *Tetrahedron Lett.* **1987**, *28*, 6171; Picrini; Baumgartner; Rossi *Tetrahedron Lett.* **1988**, *29*, 3429.

[157]Du; Hart; Ng *J. Org. Chem.* **1986**, *51*, 3162.

The arylation of compounds of the form ZCH_2Z' is analogous to **0-94**, and Z is as defined there. Activated aryl halides generally give good results.[158] Even unactivated aryl halides can be employed if the reaction is carried out in the presence of a strong base such as $NaNH_2$[159] or lithium diisopropylamide (LDA). Compounds of the form ZCH_2Z' and even simple ketones[160] and carboxylic esters have been arylated in this manner. The reaction with unactivated halides proceeds by the benzyne mechanism and represents a method for extending the malonic ester (and similar) syntheses to aromatic compounds. The base performs two functions: it removes a proton from ZCH_2Z' and catalyzes the benzyne mechanism. The reaction has been used for ring closure:[161]

The reaction on unactivated halides can also be done with copper halide catalysts[80] (the *Hurtley reaction*),[162] and with palladium complex catalysts.[163]

Compounds of the form CH_3Z can be arylated by treatment with an aryl halide in liquid ammonia containing Na or K, e.g.,[164]

The same products are obtained (though in different proportions) when Na or K is omitted but the solution is irradiated with near-uv-light.[165] In either case other leaving groups can be used instead of halogens (e.g., NR_3^+, SAr) and the mechanism is the $S_{RN}1$ mechanism. Iron(II) salts have also been used to initiate this reaction.[166] The reaction can also take place without an added initiator: Enolate ions of ketones react with PhI in the dark.[167] In this case, it has been suggested[167] that intiation takes place by

[158] There is evidence for both SNAr (see Leffek; Matinopoulos-Scordou *Can. J. Chem.* **1977**, *55*, 2656, 2664) and SRN1 (see Zhang; Yang; Liu; Chen; Cheng *Res. Chem. Intermed.* **1989**, *11*, 281) mechanisms.

[159] Leake; Levine *J. Am. Chem. Soc.* **1959**, *81*, 1169, 1627.

[160] For example, see Caubere; Guillaumet *Bull. Soc. Chim. Fr.* **1972**, 4643, 4649.

[161] Bunnett; Hrutfiord *J. Am. Chem. Soc.* **1961**, *83*, 1691; Bunnett; Kato; Flynn; Skorcz *J. Org. Chem.* **1963**, *28*, 1. For reviews, see Biehl; Khanapure *Acc. Chem. Res.* **1989**, *22*, 275-281; Hoffmann, Ref. 102, pp. 150-164. See also Kessar, *Acc. Chem. Res.* **1978**, *11*, 283-288.

[162] For discussions and procedures, see Bruggink; McKillop, *Tetrahedron* **1975**, *31*, 2607; McKillop; Rao *Synthesis* **1977**, 759; Setsune; Matsukawa; Wakemoto; Kitao *Chem. Lett.* **1981**, 367; Osuka; Kobayashi; Suzuki *Synthesis* **1983**, 67; Suzuki; Kobayashi; Yoshida; Osuka *Chem. Lett.* **1983**, 193; Aalten; van Koten; Vrieze; van der Kerk-van Hoof *Recl. Trav. Chim. Pays-Bas* **1990**, *109*, 46.

[163] Uno; Seto; Ueda; Masuda; Takahashi *Synthesis* **1985**, 506.

[164] Rossi; Bunnett *J. Am. Chem. Soc.* **1972**, *94*, 683, *J. Org. Chem.* **1973**, *38*, 3020; Bunnett; Gloor *J. Org. Chem.* **1973**, *38*, 4156, **1974**, *39*, 382.

[165] Rossi; Bunnett *J. Org. Chem.* **1973**, *38*, 1407; Hay; Hudlicky; Wolfe *J. Am. Chem. Soc.* **1975**, *97*, 374; Bunnett; Sundberg *J. Org. Chem.* **1976**, *41*, 1702; Rajan; Muralimohan *Tetrahedron Lett.* **1978**, 483; Rossi; de Rossi; Pierini *J. Org. Chem.* **1979**, *44*, 2662; Rossi; Alonso *J. Org. Chem.* **1980**, *45*, 1239; Beugelmans *Bull. Soc. Chim. Belg.* **1984**, *93*, 547.

[166] Galli; Bunnett *J. Org. Chem.* **1984**, *49*, 3041.

[167] Scamehorn; Bunnett *J. Org. Chem.* **1977**, *42*, 1449; Scamehorn; Hardacre; Lukanich; Sharpe *J. Org. Chem.* **1984**, *49*, 4881.

This is an SET mechanism (see p. 307). The photostimulated reaction has also been used for ring closure.[168] In certain instances of the intermolecular reaction there is evidence that the leaving group exerts an influence on the product ratios, even when it has already departed at the time that product selection takes place.[169] Malonic and β-keto esters can be arylated in high yields by treatment with aryllead tricarboxylates: RCOCHR'COOEt + ArPb(OAc)$_3$ → RCOCArR'COOEt,[170] and with triphenylbismuth carbonate[171] Ph$_3$BiCO$_3$ and other bismuth reagents.[172] In a related process, manganese(III) acetate was used to convert a mixture of ArH and ZCH$_2$Z' to ArCHZZ'.[173]

OS **V**, 12, 263; **VI**, 36, 873, 928; **VII**, 229.

3-15 Conversion of Aryl Substrates to Carboxylic Acids, Their Derivatives, Aldehydes, and Ketones[174]

Alkoxycarbonyl-de-halogenation, etc.

$$ArX + CO + ROH \xrightarrow[\text{Pd complex}]{\text{base}} ArCOOR$$

Aryl bromides and iodides, when treated with carbon monoxide, an alcohol ROH, a base, and a palladium complex catalyst, give carboxylic esters. The use of H$_2$O, RNH$_2$, or an alkali metal or calcium carboxylate[175] instead of ROH, gives the carboxylic acid,[175a] amide,[176] or mixed anhydride, respectively.[177] With certain palladium catalysts, aryl chlorides[178] and aryl triflates[179] can also be substrates. Other reagents used (instead of CO) have been nickel carbonyl Ni(CO)$_4$[180] (see **0-103**) and dicobalt octacarbonyl Co$_2$(CO)$_8$.[181] Aryl chlorides have been converted to carboxylic acids by an electrochemical synthesis,[182] and aryl iodides to aldehydes by treatment with CO, Bu$_3$SnH, and NCCMe$_2$N=NCMe$_2$CN (AIBN).[183]

Lead tetraacetate has been used to convert phenols, with a hydrazone group in the ortho position, to carboxylic esters,[184] e.g.,

[168]See Semmelhack; Bargar *J. Am. Chem. Soc.* **1980**, *102*, 7765; Bard; Bunnett *J. Org. Chem.* **1980**, *45*, 1546.
[169]Bard; Bunnett; Creary; Tremelling *J. Am. Chem. Soc.* **1980**, *102*, 2852; Tremelling; Bunnett *J. Am. Chem. Soc.* **1980**, *102*, 7375.
[170]Pinhey; Rowe *Aust. J. Chem.* **1980**, *33*, 113; Kopinski; Pinhey; Rowe *Aust. J. Chem.* **1984**, *37*, 1245; Kozyrod; Morgan, Pinhey *Aust. J. Chem.* **1991**, *44*, 369.
[171]For a review of these and related reactions, see Abramovitch; Barton; Finet *Tetrahedron* **1988**, *44*, 3039-3071.
[172]Barton; Blazejewski; Charpiot; Finet; Motherwell; Papoula; Stanforth *J. Chem. Soc., Perkin Trans. 1* **1985**, 2667; O'Donnell; Bennett; Jacobsen; Ma *Tetrahedron Lett.* **1989**, *30*, 3913.
[173]Citterio; Santi; Fiorani; Strologo *J. Org. Chem.* **1989**, *54*, 2703; Citterio; Fancelli; Finzi; Pesce; Santi *J. Org. Chem.* **1989**, *54*, 2713.
[174]For a review, see Weil; Cassar; Foà, in Wender; Pino *Organic Synthesis Via Metal Carbonyls*, vol. 2; Wiley: New York, 1977, pp. 517-543.
[175]Pri-Bar; Alper *J. Org. Chem.* **1989**, *54*, 36.
[175a]For example, see Bumagin; Nikitin; Beletskaya *Doklad. Chem.* **1990**, *312*, 149.
[176]For another reagent that also gives amides, see Bumagin; Gulevich; Beletskaya *J. Organomet. Chem.* **1985**, *285*, 415.
[177]For a review, see Heck *Palladium Reagents in Organic Synthesis*; Academic Press: New York, 1985, pp. 348-358.
[178]Ben-David; Portnoy; Milstein *J. Am. Chem. Soc.* **1989**, *111*, 8742.
[179]Cacchi; Ciattini; Morera; Ortar *Tetrahedron Lett.* **1986**, *27*, 3931.
[180]Bauld *Tetrahedron Lett.* **1963**, 1841. See also Corey; Hegedus *J. Am. Chem. Soc.* **1969**, *91*, 1233; Nakayama; Mizoroki *Bull. Chem. Soc. Jpn.* **1971**, *44*, 508.
[181]Brunet; Sidot; Caubere *Tetrahedron Lett.* **1981**, *22*, 1013. *J. Org. Chem.* **1983**, *48*, 1166. See also Foà; Francalanci; Bencini; Gardano *J. Organomet. Chem.* **1985**, *285*, 293; Kudo; Shibata; Kashimura; Mori; Sugita *Chem. Lett.* **1987**, 577.
[182]Heintz; Sock; Saboureau; Périchon *Tetrahedron* **1988**, *44*, 1631.
[183]Ryu; Kusano; Masumi; Yamazaki; Ogawa; Sonoda *Tetrahedron Lett.* **1990**, *31*, 6887.
[184]Katritzky; Kotali *Tetrahedron Lett.* **1990**, *31*, 6781.

The hydrazone group is hydrolyzed (**6-2**) during the course of the reaction. Yields are high. Aryl iodides are converted to unsymmetrical diaryl ketones on treatment with arylmercury halides and nickel carbonyl: $ArI + Ar'HgX + Ni(CO)_4 \rightarrow ArCOAr'$.[185]

3-16   The Ullmann Reaction
De-halogen-coupling

$$2ArI \xrightarrow[\Delta]{Cu} Ar\!-\!Ar$$

The coupling of aryl halides with copper is called the *Ullmann reaction*.[186] The reaction is of broad scope and has been used to prepare many symmetrical and unsymmetrical biaryls.[187] When a mixture of two different aryl halides is used, there are three possible products, but often only one is obtained. For example, picryl chloride and iodobenzene gave only 2,4,6-trinitrobiphenyl.[188] The best leaving group is iodo, and the reaction is most often done on aryl iodides, but bromides, chlorides, and even thiocyanates have been used.

The effects of other groups on the ring are ususual. The nitro group is strongly activating, but only in the ortho (not meta or para) position.[189] R and OR are active in all positions. Not only do OH, NH_2, NHR, and NHCOR inhibit the reaction, as would be expected for aromatic nucleophilic substitution, but so do COOH (but not COOR), SO_2NH_2, and similar groups for which the reaction fails completely. These groups inhibit the coupling reaction by causing side reactions.

The mechanism is not known with certainty. It seems likely that it is basically a two-step process, similar to that of the Wurtz reaction (**0-86**), which can be represented schematically by:

Step 1 $ArI + Cu \longrightarrow ArCu$

Step 2 $ArCu + ArI \longrightarrow ArAr$

Organocopper compounds have been trapped by coordination with organic bases.[190] In addition, arylcopper compounds (ArCu) have been independently prepared and shown to give biaryls (ArAr′) when treated with aryl iodides Ar′I.[191]

A similar reaction has been used for ring closure:[192]

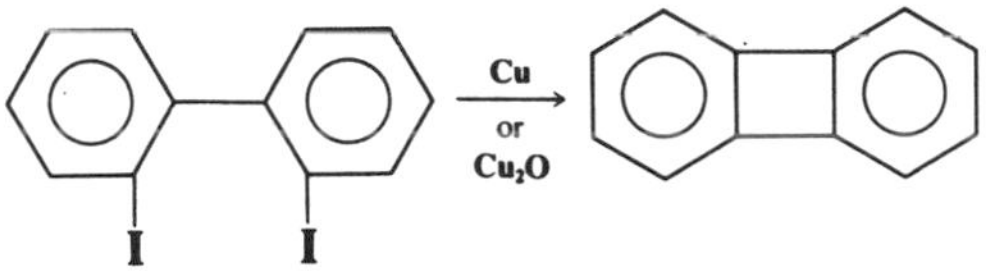

[185]Rhee; Ryang; Watanabe; Omura; Murai; Sonoda *Synthesis* **1977**, 776. For other acylation reactions, see Tanaka; *Synthesis* **1981**, 47. *Bull Chem. Soc. Jpn.* **1981**, *54*, 637; Bumagin; Ponomaryov; Beletskaya *Tetrahedron Lett.* **1985**, *26*, 4819; Koga; Makinouchi; Okukado *Chem. Lett.* **1988**, 1141; Echavarren; Stille *J. Am. Chem. Soc.* **1988**, *110*, 1557.

[186]For reviews, see Fanta *Synthesis* **1974**, 9-21; Goshaev; Otroshchenko; Sadykov *Russ. Chem. Rev.* **1972**, *41*, 1046-1059.

[187]For reviews of methods of aryl–aryl bond formation, see Bringmann; Walter; Weirich *Angew. Chem. Int. Ed. Engl.* **1990**, *29*, 977-991 [*Angew. Chem. 102*, 1006-1019]; Sainsbury *Tetrahedron* **1980**, *36*, 3327-3359.

[188]Rule; Smith *J. Chem. Soc.* **1937**, 1096.

[189]Forrest *J. Chem. Soc.* **1960**, 592.

[190]Lewin; Cohen *Tetrahedron Lett.* **1965**, 4531.

[191]For examples, see Nilsson *Tetrahedron Lett.* **1966**, 675; Cairncross; Sheppard *J. Am. Chem. Soc.* **1968**, *90*, 2186; Ullenius *Acta Chem. Scand.* **1972**, *26*, 3383; Mack; Suschitzky; Wakefield *J. Chem. Soc., Perkin Trans. 1* **1980**, 1682.

[192]Salfeld; Baume *Tetrahedron Lett.* **1966**, 3365; Lothrop *J. Am. Chem. Soc.* **1941**, *63*, 1187.

An important alternative to the Ullmann method is the use of certain nickel complexes.[193] This method has also been used intramolecularly.[194] Aryl halides ArX can also be converted to Ar—Ar[195] by treatment with activated Ni metal,[196] with Zn and nickel complexes,[197] with aqueous alkaline sodium formate, Pd–C, and a phase transfer catalyst,[198] and in an electrochemical process catalyzed by a nickel complex.[199]

For other methods of coupling aromatic rings, see **3-13**, **3-17**, **4-18**, **4-21**, and **4-22**. **OS III**, 339; **V**, 1120.

Hydrogen as Leaving Group[200]

3-17   Alkylation and Arylation
Alkylation or **Alkyl-de-hydrogenation**, etc.

The alkylation of heterocyclic nitrogen compounds[201] with alkyllithiums is called *Ziegler alkylation*. Aryllithiums give arylation. The reaction occurs by an addition–elimination mechanism and the adduct can be isolated.[202] Upon heating of the adduct, elimination of LiH occurs (see **7-16**) and an alkylated product is obtained. With respect to the 2-carbon the first step is the same as that of the SNAr mechanism. The difference is that the unshared pair of electrons on the nitrogen combines with the lithium, so the extra pair of ring electrons has a place to go: it becomes the new unshared pair on the nitrogen.

The reaction has been applied to nonheterocyclic aromatic compounds: Benzene, naphthalene, and phenanthrene have been alkylated with alkyllithiums, though the usual reaction with these reagents is **2-21**,[203] and Grignard reagents have been used to alkylate naphthalene.[204] The addition–elimination mechanism apparently applies in these cases too.

Aromatic nitro compounds can be methylated with dimethyloxosulfonium methylide[205] or the methylsulfinyl carbanion (obtained by treatment of dimethyl sulfoxide with a strong base):[206]

[193]See, for example Semmelhack; Helquist; Jones *J. Am. Chem. Soc.* **1971**, *93*, 5908; Clark; Norman; Thomas *J. Chem. Soc., Perkin Trans. 1* **1975**, 121; Tsou; Kochi *J. Am. Chem. Soc.* **1979**, *101*, 7547; Colon; Kelsey *J. Org. Chem.* **1986**, *51*, 2627; Lourak; Vanderesse; Fort; Caubere *J. Org. Chem.* **1989**, *54*, 4840, 4844; Iyoda; Otsuka; Sato; Nisato; Oda *Bull. Chem. Soc. Jpn.* **1990**, *63*, 80. For a review of the mechanism, see Amatore; Jutand *Acta Chem. Scand.* **1990**, *44*, 755-764.
[194]See for example, Karimipour; Semones; Asleson; Heldrich *Synlett* **1990**, 525.
[195]For a list of reagents, with references, see Ref. 116, pp. 46-47.
[196]Inaba; Matsumoto; Rieke *Tetrahedron Lett.* **1982**, *23*, 4215; Matsumoto; Inaba; Rieke *J. Org. Chem.* **1983**, *48*, 840; Chao; Cheng; Chang *J. Org. Chem.* **1983**, *48*, 4904.
[197]Takagi; Hayama; Sasaki *Bull. Chem. Soc. Jpn.* **1984**, *57*, 1887.
[198]Bamfield; Quan *Synthesis* **1978**, 537.
[199]Meyer; Rollin; Perichon *J. Organomet. Chem.* **1987**, *333*, 263.
[200]For a review, see Chupakhin; Postovskii *Russ. Chem. Rev.* **1976**, *45*, 454-468. For a review of reactivity and mechanism in these cases, see Chupakhin; Charushin; van der Plas *Tetrahedron* **1988**, *44*, 1-34.
[201]For a review of substitution by carbon groups on a nitrogen heterocycle, see Vorbrüggen; Maas *Heterocycles* **1988**, *27*, 2659-2776. For a related review, see Comins; O'Connor *Adv. Heterocycl. Chem.* **1988**, *44*, 199-267.
[202]See, for example, Armstrong; Mulvey; Barr; Snaith; Reed *J. Organomet. Chem.* **1988**, *350*, 191.
[203]Dixon; Fishman *J. Am. Chem. Soc.* **1963**, *85*, 1356; Eppley; Dixon *J. Am. Chem. Soc.* **1968**, *90*, 1606.
[204]Bryce-Smith; Wakefield *Tetrahedron Lett.* **1964**, 3295.
[205]Traynelis; McSweeney *J. Org. Chem.* **1966**, *31*, 243.
[206]Russell; Weiner *J. Org. Chem.* **1966**, *31*, 248.

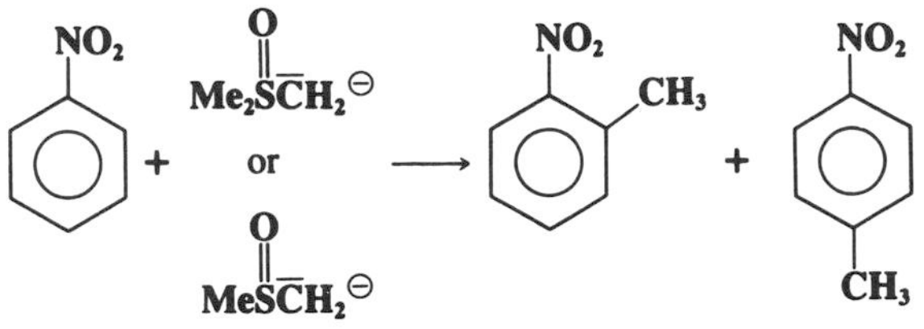

The latter reagent also methylates certain heterocyclic compounds, e.g., quinoline, and certain fused aromatic compounds, e.g., anthracene, phenanthrene.[207] The reactions with the sulfur carbanions are especially useful, since none of these substrates can be methylated by the Friedel–Crafts procedure (**1-12**). It has been reported[208] that aromatic nitro compounds can also be alkylated, not only with methyl but with other alkyl and substituted alkyl groups as well, in ortho and para positions, by treatment with an alkyllithium compound (or, with lower yields, a Grignard reagent), followed by an oxidizing agent such as Br_2 or DDQ (p. 1163). Trinitrobenzene was alkylated (ArH → ArR) by treatment with a silane $RSiMe_3$ in the presence of KF and a crown ether.[209] In this reaction, R was not a simple alkyl group, but a group such as CH_2COOMe, COMe, CH_2Ph, $CH_2CH=CH_2$, etc.

A different kind of alkylation of nitro compounds uses carbanion nucleophiles that have a chlorine at the carbanionic carbon. The following process takes place:[210]

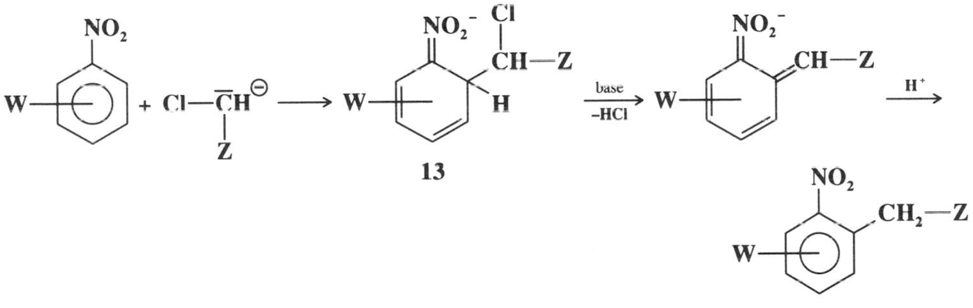

This type of process is called *vicarious nucleophilic substitution of hydrogen.*[211] Z is an electron-withdrawing group such as SO_2R, SO_2OR, SO_2NR_2, COOR, or CN; it stabilizes the negative charge. The carbanion attacks the activated ring ortho or para to the nitro group. Hydride ion H^- is not normally a leaving group, but in this case the presence of the adjacent Cl allows the hydrogen to be replaced. Hence, Cl is a "vicarious" leaving group. Other leaving groups have been used, e.g., OMe, SPh, but Cl is generally the best. Many groups W in ortho, meta, or para positions do not interfere. The reaction is also successful for di- and trinitro compounds, for nitronaphthalenes,[212] and for many nitro heterocycles.

$Z-\overset{\ominus}{C}R-Cl$ may also be used.[213] When Br_3C^- or Cl_3C^- is the nucleophile the product is $ArCHX_2$, which can easily be hydrolyzed to ArCHO.[214] This is therefore an indirect way

[207]Ref. 206; Argabright; Hofmann; Schriesheim *J. Org. Chem.* **1965**, *30*, 3233; Trost *Tetrahedron Lett.* **1966**, 5761; Yamamoto; Nisimura; Nozaki *Bull. Chem. Soc. Jpn.* **1971**, *44*, 541.

[208]Kienzle *Helv. Chim. Acta* **1978**, *61*, 449.

[209]Artamkina; Kovalenko; Beletskaya; Reutov *J. Organomet. Chem.* **1987**, *329*, 139, *J. Org. Chem. USSR* **1990**, 26, 801. See also RajanBabu; Reddy; Fukunaga *J. Am. Chem. Soc.* **1985**, *107*, 5473.

[210]In some cases intermediate **13** has been isolated: Stahly; Stahly; Maloney *J. Org. Chem.* **1988**, *53*, 690.

[211]Goliński; Mąkosza *Tetrahedron Lett.* **1978**, 3495. For reviews, see Mąkosza *Synthesis* **1991**, 103-111, *Russ. Chem. Rev.* **1989**, *58*, 747-757; Mąkosza; Winiarski *Acc. Chem. Res.* **1987**, *20*, 282-289.

[212]Mąkosza; Danikiewicz; Wojciechowski *Liebigs Ann. Chem.* **1987**, 711.

[213]See Mudryk; Mąkosza *Tetrahedron* **1988**, *44*, 209.

[214]Mąkosza; Owczarczyk *J. Org. Chem.* **1989**, *54*, 5094. See also Mąkosza; Winiarski *Chem. Lett.* **1984**, 1623.

of formylating an aromatic ring containing one or more NO_2 groups, which cannot be done by any of the formylations mentioned in Chapter 11 (**1-16** to **1-18**).

For the introduction of CH_2SR groups into phenols, see **1-26**. See also **4-23**.

OS **II**, 517.

3-18   Amination of Nitrogen Heterocycles
Amination or **Amino-de-hydrogenation**

$$\text{pyridine} + NH_2^- \xrightarrow{100–200°C} \text{2-aminopyridine anion} + H_2$$

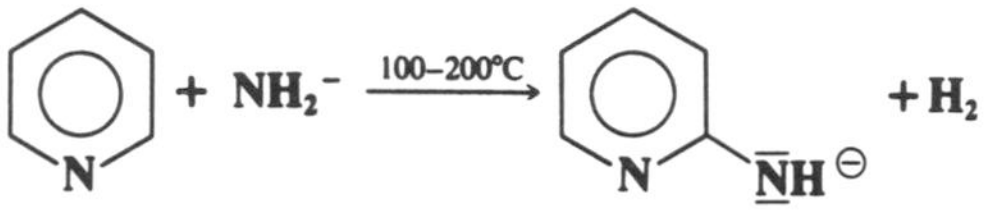

Pyridine and other heterocyclic nitrogen compounds can be aminated with alkali-metal amides in a process called the *Chichibabin reaction*.[215] The attack is always in the 2 position unless both such positions are filled, in which case the 4 position is attacked. Substituted alkali-metal amides, e.g., RNH^- and R_2N^-, have also been used. The mechanism is probably similar to that of **3-17**. The existence of intermediate ions such as **14**

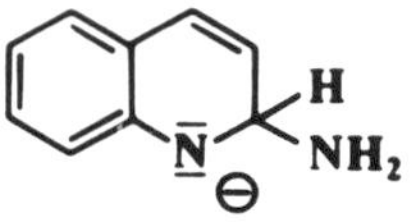

14

(from quinoline) has been demonstrated by nmr spectra.[216] A pyridyne type of intermediate was ruled out by several observations including the facts that 3-ethylpyridine gave 2-amino-3-ethylpyridine[217] and that certain heterocycles that cannot form an aryne could nevertheless be successfully aminated. Nitro compounds do not give this reaction,[218] but they have been aminated ($ArH \rightarrow ArNH_2$ or $ArNHR$) via the vicarious substitution principle (see **3-17**), using 4-amino- or 4-alkylamino-1,2,4-triazoles as nucleophiles.[219] The vicarious leaving group in this case is the triazole ring.

Analogous reactions have been carried out with hydrazide ions, R_2NNH^-.[220] For other methods of aminating aromatic rings, see **1-6** and **3-19**.

There are no *Organic Syntheses* references, but see OS **V**, 977, for a related reaction.

3-19   Amination by Hydroxylamine
Amination or **Amino-de-hydrogenation**

$$\text{1-nitronaphthalene} + NH_2OH \xrightarrow{OEt^-} \text{1-nitro-4-aminonaphthalene}$$

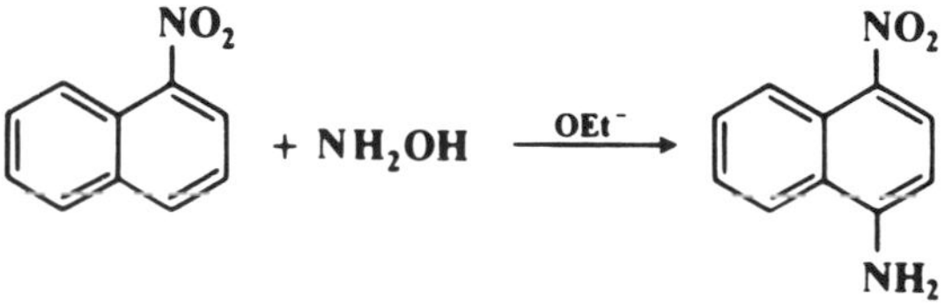

[215]For reviews, see Vorbrüggen *Adv. Heterocycl. Chem.* **1990**, *49*, 117-192; McGill; Rappa *Adv. Heterocycl. Chem.* **1988**, *44*, 1-79; Pozharskii; Simonov; Doron'kin *Russ. Chem. Rev.* **1978**, *47*, 1042-1060.

[216]Zoltewicz; Helmick; Oestreich; King; Kandetzki *J. Org. Chem.* **1973**, *38*, 1947; Woźniak; Baránski; Nowak; van der Plas *J. Org. Chem.* **1987**, *52*, 5643.

[217]Ban; Wakamatsu *Chem. Ind.* (*London*) **1964**, 710.

[218]See, for example, Levitt; Levitt *Chem. Ind.* (*London*) **1975**, 520.

[219]Katritzky; Laurenzo *J. Org. Chem.* **1986**, *51*, 5039, **1988**, *53*, 3978.

[220]Kauffmann; Hansen; Kosel; Schoeneck *Liebigs Ann. Chem.* **1962**, *656*, 103.

Activated aromatic compounds can be directly aminated with hydroxylamine in the presence of strong bases.[221] Conditions are mild and yields are high. Ions of the type **15** are intermediates:

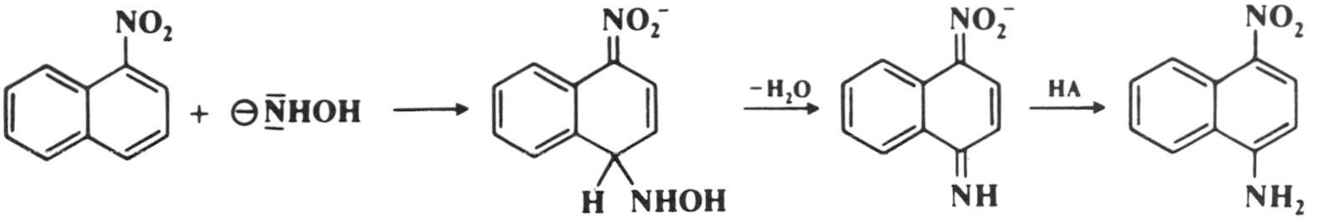

OS **III**, 664.

N_2^+ as Leaving Group

The diazonium group can be replaced by a number of groups.[222] Some of these are nucleophilic substitutions, with SN1 mechanisms (p. 644), but others are free-radical reactions and are treated in Chapter 14. The solvent in all these reactions is usually water. With other solvents it has beeen shown that the SN1 mechanism is favored by solvents of low nucleophilicity, while those of high nucleophilicity favor free-radical mechanisms.[223] (For *formation* of diazonium ions, see 2-49.) The N_2^+ group can be replaced by Cl^-, Br^-, and CN^-, by a nucleophilic mechanism (see OS **IV**, 182), but the Sandmeyer reaction is much more useful (**4-25** and **4-28**). As mentioned on p. 651 it must be kept in mind that the N_2^+ group can activate the removal of another group on the ring.

3-20 Hydroxy-de-diazoniation

$$ArN_2^+ + H_2O \longrightarrow ArOH$$

Water is usually present whenever diazonium salts are made, but at these temperatures (0 to 5°C) the reaction proceeds very slowly. When it is *desired* to have OH replace the diazonium group, the excess nitrous acid is destroyed and the solution is usually boiled. Some diazonium salts require even more vigorous treatment, e.g., boiling with aqueous sulfuric acid or with trifluoroacetic acid containing potassium trifluoroacetate.[224] The reaction can be performed on solutions of any diazonium salts, but hydrogen sulfates are preferred to chlorides or nitrates, since in these cases there is competition from the nucleophiles Cl^- or NO_3^-. A better method, which is faster, avoids side reactions, takes place at room temperature, and gives higher yields consists of adding Cu_2O to a dilute solution of the diazonium salt dissolved in a solution containing a large excess of $Cu(NO_3)_2$.[225] Aryl radicals are intermediates when this method is used. It has been shown that aryl radicals are at least partly involved when ordinary hydroxy-de-diazoniation is carried out in weakly alkaline

[221]See Chupakhin; Postovskii, Ref. 200, p. 456.
[222]For a review of such reactions, see Wulfman, in Patai *The Chemistry of Diazonium and Diazo Groups*, pt. 1; Wiley: New York, 1978, pp. 286-297.
[223]Szele; Zollinger *Helv. Chim. Acta* **1978**, *61*, 1721.
[224]Horning; Ross; Muchowski *Can. J. Chem.* **1973**, *51*, 2347.
[225]Cohen; Dietz; Miser *J. Org. Chem.* **1977**, *42*, 2053.

aqueous solution.[226] Decomposition of arenediazonium tetrafluoroborates in F_3CSO_2OH gives aryl triflates directly, in high yields.[226a]

OS **I**, 404; **III**, 130, 453, 564; **V**, 1130.

3-21 Replacement by Sulfur-Containing Groups
Mercapto-de-diazoniation, etc.

$$ArN_2^+ + HS^- \longrightarrow ArSH$$

$$ArN_2^+ + S^{2-} \longrightarrow ArSAr$$

$$ArN_2^+ + RS^- \longrightarrow ArSR$$

$$ArN_2^+ + SCN^- \longrightarrow ArSCN + ArNCS$$

These reactions are convenient methods for putting sulfur-containing groups onto an aromatic ring. With Ar'S$^-$, diazosulfides Ar—N=N—S—Ar' are intermediates,[227] which can in some cases be isolated.[228] Thiophenols can be made as shown above, but more often the diazonium ion is treated with EtO—CSS$^-$ or S_2^{2-}, which give the expected products, and these are easily convertible to thiophenols. See also **4-27**.

OS **II**, 580; **III**, 809 (but see OS **V**, 1050). Also see OS **II**, 238.

3-22 Azido-de-diazoniation

$$ArN_2^+ + N_3^- \longrightarrow ArN_3$$

Diazonium salts can be converted to aryl azides by the addition of sodium azide to the acidic diazonium salt solution.[229]

OS **IV**, 75; **V**, 829.

3-23 Iodo-de-diazoniation

$$ArN_2^+ + I^- \longrightarrow ArI$$

One of the best methods for the introduction of iodine into aromatic rings is the reaction of diazonium salts with iodide ions. Analogous reactions with chloride, bromide, and fluoride ions give poorer results, and **4-25** and **3-24** are preferred for the preparation of aryl chlorides, bromides, and fluorides. However, when other diazonium reactions are carried out in the presence of these ions, halides are usually side products.

The actual attacking species is probably not only I$^-$, if it is I$^-$ at all. The iodide ion is oxidized (by the diazonium ion, nitrous acid, or some other oxidizing agent) to iodine, which in a solution containing iodide ions is converted to I_3^-; this is the actual attacking species, at least partly. This was shown by isolation of $ArN_2^+ I_3^-$ salts, which, on standing, gave ArI.[230] From this, it can be inferred that the reason the other halide ions give poor results

[226]Dreher; Niederer; Rieker; Schwarz; Zollinger *Helv. Chim. Acta* **1981**, *64*, 488.
[226a]Yoneda; Fukuhara; Mizokami; Suzuki *Chem. Lett.* **1991**, 459.
[227]Abeywickrema; Beckwith *J. Am. Chem. Soc.* **1986**, *108*, 8227, and references cited therein.
[228]See, for example Price; Tsunawaki *J. Org. Chem.* **1963**, *28*, 1867.
[229]Smith; Brown *J. Am. Chem. Soc.* **1951**, *73*, 2438. For a review, see Biffin; Miller; Paul, in Patai *The Chemistry of the Azido Group*; Wiley: New York, 1971, pp. 147-176.
[230]Carey; Millar *Chem. Ind. (London)* **1960**, 97.

is not that they are poor nucleophiles but that they are poor reducing agents (compared with iodide). There is also evidence for a free radical mechanism.[231]

OS **II**, 351, 355, 604; **V**, 1120.

3-24 The Schiemann Reaction
Fluoro-de-diazoniation (overall tranformation)

$$ArN_2^+ BF_4^- \xrightarrow{\Delta} ArF + N_2 + BF_3$$

Heating of diazonium fluoroborates (the *Schiemann* or *Balz–Schiemann reaction*) is by far the best way of introducing fluorine into an aromatic ring.[232] In the most common procedure, the fluoroborate salts are prepared by diazotizing as usual with nitrous acid and HCl and then adding a cold aqueous solution of $NaBF_4$, HBF_4, or NH_4BF_4. A precipitate forms, which is dried, and the salt is heated in the dry state. These salts are unusually stable for diazonium salts, and the reaction is usually successsful. In general, any aromatic amine that can be diazotized will form a BF_4^- salt, usually with high yields. The diazonium fluoroborates can be formed directly from primary aromatic amines with *t*-butyl nitrate and BF_3–etherate.[233] The reaction has also been carried out on $ArN_2^+ PF_6^-$, $ArN_2^+ SbF_6^-$, and $ArN_2^+ AsF_6^-$ salts, in many cases with better yields.[234] The reaction has been extended to $ArN_2^+ BCl_4^-$ and $ArN_2^+ BBr_4^-$,[235] but aryl chlorides and bromides are more commonly prepared by the Sandmeyer reaction (**4-25**). In an alternative procedure, aryl fluorides have been prepared by treatment of aryltriazenes Ar—N=N—NR$_2$ with 70% HF in pyridine.[236]

The mechanism is of the SN1 type. That aryl cations are intermediates was shown by the following experiments:[237] aryl diazonium chlorides are known to arylate other aromatic rings by a free-radical mechanism (see **4-18**). In radical arylation it does not matter whether the other ring contains electron-withdrawing or electron-donating groups; in either case a mixture of isomers is obtained, since the attack is not by a charged species. If an aryl radical were an intermediate in the Schiemann reaction and the reaction were run in the presence of other rings, it should not matter what kinds of groups were on these other rings: mixtures of biaryls should be obtained in all cases. But if an aryl cation is an intermediate in the Schiemann reaction, compounds containing meta-directing groups, i.e., meta-directing for *electrophilic* substitutions, should be meta-arylated and those containing ortho-para-directing groups should be ortho- and para-arylated, since an aryl cation should behave in this respect like any electrophile (see Chapter 11). Experiments are shown[238] that such orientation is observed, demonstrating that the Schiemann reaction has a positively charged intermediate. The attacking species, in at least some instances, is not F^- but BF_4^-.[239]

OS **II**, 188, 295, 299; **V**, 133.

[231]Singh; Kumar *Aust. J. Chem.* **1972**, *25*, 2133; Kumar; Singh *Tetrahedron Lett.* **1972**, 613; Meyer; Rössler; Stöcklin *J. Am. Chem. Soc.* **1979**, *101*, 3121; Packer; Taylor *Aust. J. Chem.* **1985**, *38*, 991; Abeywickrema; Beckwith *J. Org. Chem.* **1987**, *52*, 2568.
[232]For a review, see Suschitzky *Adv. Fluorine Chem.* **1965**, *4*, 1-30.
[233]Doyle; Bryker *J. Org. Chem.* **1979**, *44*, 1572.
[234]Rutherford; Redmond; Rigamonti *J. Org. Chem.* **1961**, *26*, 5149; Sellers; Suschitzky *J. Chem. Soc. C* **1968**, 2317.
[235]Olah; Tolgyesi *J. Org. Chem.* **1961**, *26*, 2053.
[236]Rosenfeld; Widdowson *J. Chem. Soc., Chem. Commun.* **1979**, 914. For another alternative procedure, see Yoneda; Fukuhara; Kikuchi; Suzuki *Synth. Commun.* **1989**, *19*, 865.
[237]See also Swain; Sheats; Harbison, Ref. 21; Becker; Israel *J. Prakt. Chem.* **1979**, *321*, 579.
[238]Makarova; Matveeva *Bull. Acad. Sci. USSR, Div. Chem. Sci.* **1958**, 548; Makarova; Matveeva; Gribchenko *Bull. Acad. Sci. USSR, Div. Chem. Sci.* **1958**, 1399.
[239]Swain; Rogers *J. Am. Chem. Soc.* **1975**, *97*, 799.

Rearrangements

3-25   The von Richter Rearrangement
Hydro-de-nitro-*cine*-substitution

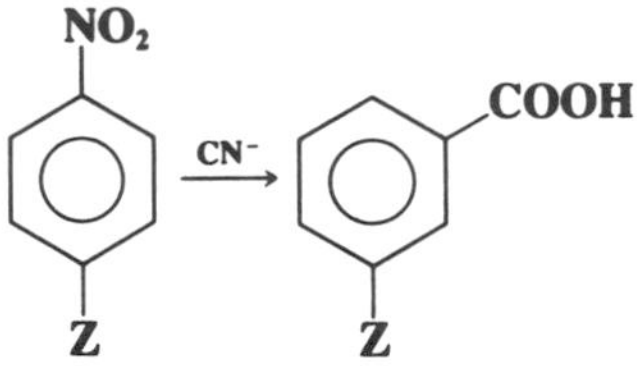

When aromatic nitro compounds are treated with cyanide ion, the nitro group is displaced and a carboxyl group enters with cine substitution (p. 646), always ortho to the displaced group, never meta or para. The scope of this reaction, called the *von Richter rearrangement*, is variable.[240] As with other nucleophilic aromatic substitutions, the reaction gives best results when electron-withdrawing groups are in ortho and para positions, but yields are low, usually less than 20% and never more than 50%.

At one time it was believed that a nitrile, ArCN, was an intermediate, since cyanide is the reagent and nitriles are hydrolyzable to carboxylic acids under the reaction conditions (**6-5**). However, a remarkable series of results proved this belief to be in error. Bunnett and Rauhut demonstrated[241] that α-naphthyl cyanide is *not* hydrolyzable to α-naphthoic acid under conditions at which β-nitronaphthalene undergoes the von Richter rearrangement to give α-naphthoic acid. This proved that the nitrile cannot be intermediate. It was subsequently demonstrated that N_2 is a major product of the reaction.[242] It had previously been assumed that all the nitrogen in the reaction was converted to ammonia, which would be compatible with a nitrile intermediate, since ammonia is a hydrolysis product of nitriles. At the same time it was shown that NO_2^- is not a major product. The discovery of nitrogen indicated that a nitrogen–nitrogen bond must be formed during the course of the reaction. A mechanism in accord with all the facts was proposed by Rosenblum:[242]

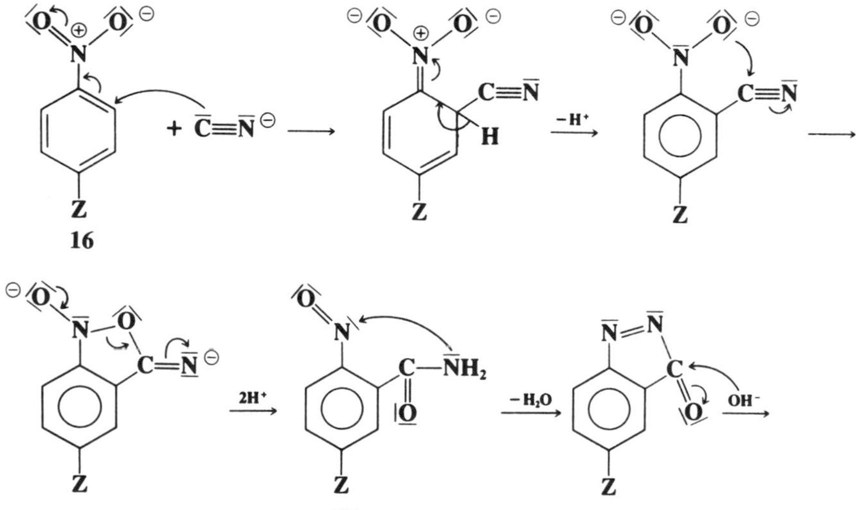

[240]For a review, see Shine *Aromatic Rearrangements*; Elsevier: New York, 1967, pp. 326-335.
[241]Bunnett; Rauhut *J. Org. Chem.* **1956**, *21*, 934, 944.
[242]Rosenblum *J. Am. Chem. Soc.* **1960**, *82*, 3796.

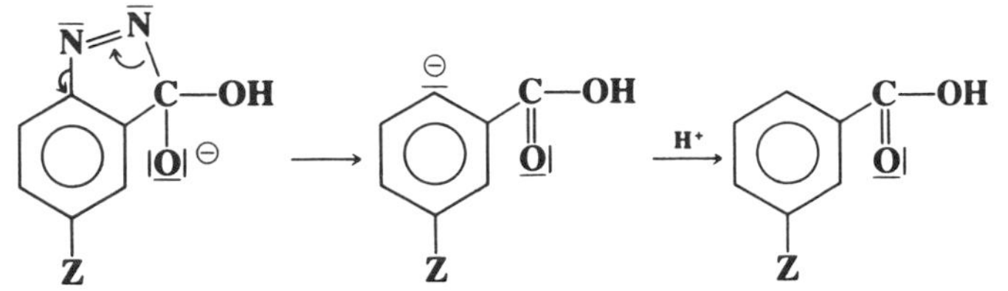

It may be noted that **17** are stable compounds; hence it should be possible to prepare them independently and to subject them to the conditions of the von Richter rearrangement. This was done and the correct products are obtained.[243] Further evidence is that when **16** (Z = Cl or Br) was treated with cyanide in $H_2{}^{18}O$, half the oxygen in the product was labeled, showing that one of the oxygens of the carboxyl group came from the nitro group and one from the solvent, as required by this mechanism.[244]

3-26 The Sommelet–Hauser Rearrangement

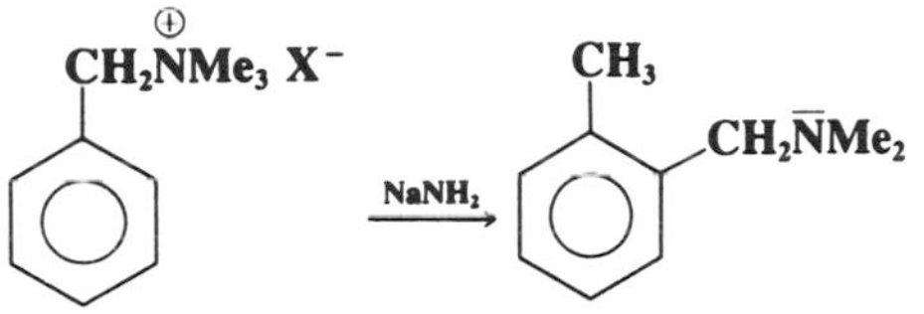

Benzylic quaternary ammonium salts, when treated with alkali-metal amides, undergo a rearrangement called the *Sommelet–Hauser rearrangement*.[245] Since the product is a benzylic tertiary amine, it can be further alkylated and the product again subjected to the rearrangement. This process can be continued around the ring until an ortho position is blocked.[246]

The rearrangement occurs with high yields and can be performed with various groups present in the ring.[247] The reaction is most often carried out with three methyl groups on the nitrogen, but other groups can also be used, though if a β hydrogen is present, Hofmann elimination (**7-6**) often competes. The Stevens rearrangement (**8-22**) is also a competing process.[248] When both rearrangements are possible, the Stevens is favored at high temperatures and the Sommelet–Hauser at low temperatures.[249] The mechanism is

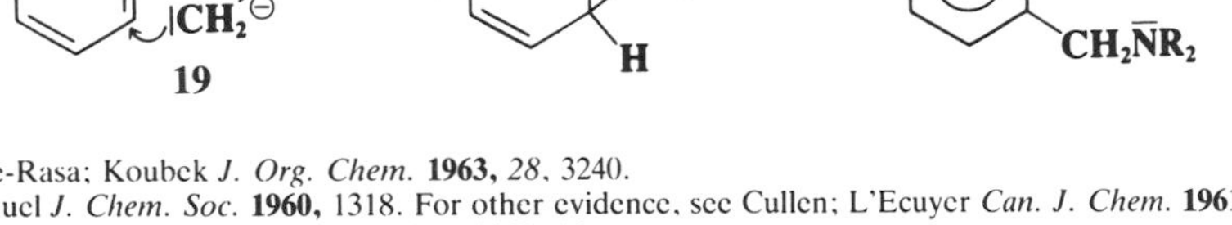

[243]Ibne-Rasa; Koubek *J. Org. Chem.* **1963**, *28*, 3240.

[244]Samuel *J. Chem. Soc.* **1960**, 1318. For other evidence, see Cullen; L'Ecuyer *Can. J. Chem.* **1961**, *39*, 144, 155, 382; Ullman; Bartkus *Chem. Ind. (London)* **1962**, 93.

[245]For reviews, see Pine, *Org. React.* **1970**, *18*, 403-464; Lepley; Giumanini *Mech. Mol. Migr.* **1971**, *3*, 297-440; Wittig *Bull. Soc. Chim. Fr.* **1971**, 1921-1924; Stevens; Watts *Selected Molecular Rearrangements*; Van Nostrand-Reinhold: Princeton, 1973, pp. 81-88; Shine, Ref. 240, pp. 316-326.

[246]Beard; Hauser *J. Org. Chem.* **1960**, *25*, 334.

[247]Beard; Hauser *J. Org. Chem.* **1961**, *26*, 371; Jones; Beard; Hauser *J. Org. Chem.* **1963**, *28*, 199.

[248]For a method that uses nonbasic conditions, and gives high yields of the Sommelet–Hauser product, with little or no Stevens rearrangement, see Nakano; Sato *J. Org. Chem.* **1987**, *52*, 1844; Shirai; Sato *J. Org. Chem.* **1988**, *53*, 194.

[249]Wittig; Streib *Liebigs Ann. Chem.* **1953**, *584*, 1.

The benzylic hydrogen is most acidic and is the one that first loses a proton to give the ylide **18**. However, **19**, which is present in smaller amount, is the species that undergoes the rearrangement, shifting the equilibrium in its favor. This mechanism is an example of a [2,3] sigmatropic rearrangment (see **8-37**). Another mechanism that might be proposed is one in which a methyl group actually breaks away (in some form) from the nitrogen and then attaches itself to the ring. That this is not so was shown by a product study.[250] If the second mechanism were true, **20** should give **21**, but the first mechanism predicts the formation of **22**, which is what was actually obtained.[251]

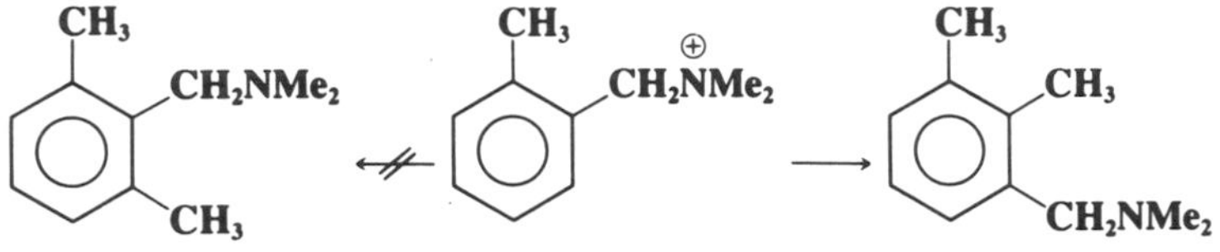

The mechanism as we have pictured it can lead only to an ortho product. However, a small amount of para product has been obtained in some cases.[252] A mechanism[253] in which there is a dissociation of the ArC—N bond (similar to the ion-pair mechanism of the Stevens rearrangement, p. 1101) has been invoked to explain the obtention of the para products.

Sulfur ylides containing a benzylic group (analogous to **19**) undergo an analogous rearrangement.[254]

OS **IV**, 585.

3-27 Rearrangement of Aryl Hydroxylamines
1/C-Hydro-5/N-hydroxy-interchange

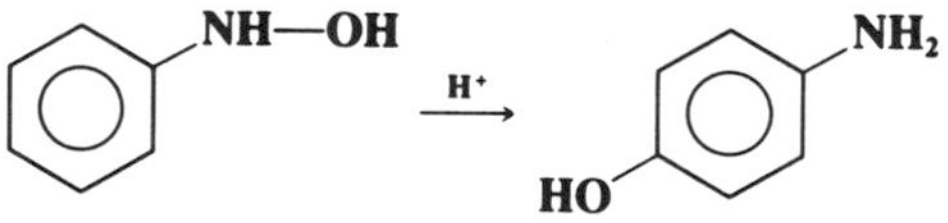

Aryl hydroxylamines treated with acids rearrange to aminophenols.[255] Although this reaction (known as the *Bamberger rearrangement*) is similar in appearance to **1-32** to **1-36**, the attack on the ring is not electrophilic but nucleophilic. The rearrangement is intermolecular, with the following mechanism:

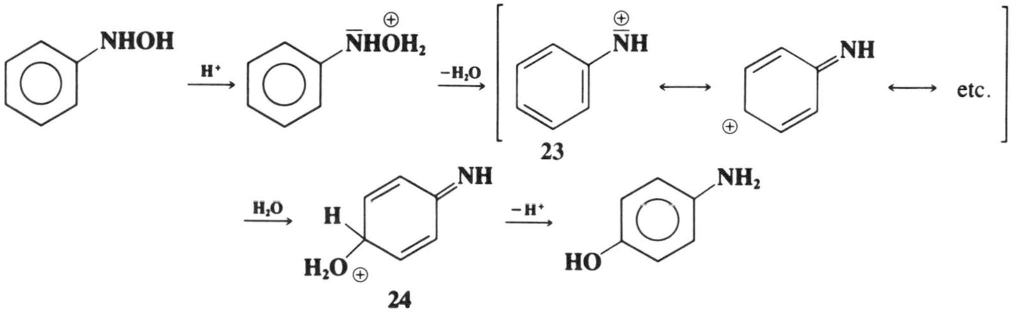

[250]For other evidence for the mechanism given, see Hauser; Van Eenam *J. Am. Chem. Soc.* **1957**, *79*, 5512; Jones; Hauser *J. Org. Chem.* **1961**, *26*, 2979; Puterbaugh; Hauser *J. Am. Chem. Soc.* **1964**, *86*, 1105; Pine; Sanchez *Tetrahedron Lett.* **1969**, 1319; Shirai; Watanabe; Sato *J. Org. Chem.* **1990**, *55*, 2767.
[251]Kantor; Hauser *J. Am. Chem. Soc.* **1951**, *73*, 4122.
[252]Pine *Tetrahedron Lett.* **1967**, 3393; Pine, Ref. 245, p. 418.
[253]Bumgardner *J. Am. Chem. Soc.* **1963**, *85*, 73.
[254]See Block *Reactions of Organosulfur Compounds*; Academic Press: New York, 1978, pp. 118-124.
[255]For a review, see Ref. 240, pp. 182-190.

Among the evidence[256] for this mechanism are the facts that other products are obtained when the reaction is run in the presence of competing nucleophiles, e.g., *p*-ethoxyaniline when ethanol is present, and that when the para position is blocked, compounds similar to **24** are isolated. In the case of 2,6-dimethylphenylhydroxylamine, the intermediate nitrenium ion **23** was trapped, and its lifetime in solution was measured.[257] The reaction of **23** with water was found to be diffusion controlled.[257]

OS **IV,** 148.

3-28   The Smiles Rearrangement

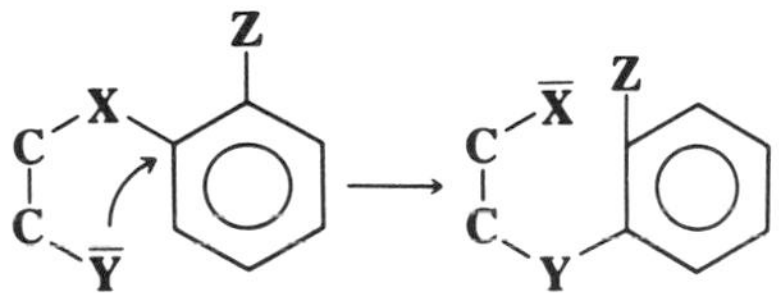

The *Smiles rearrangement* actually comprises a group of rearrangements that follow the pattern given above.[258] A specific example is

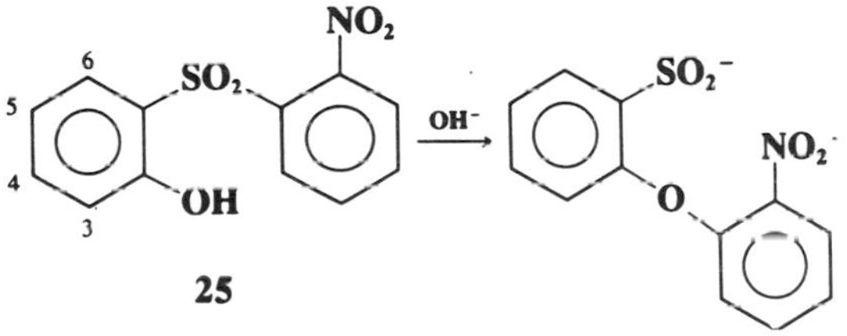

25

Smiles rearrangements are simply intramolecular nucleophilic substitutions. In the example given, SO_2Ar is the leaving group and ArO^- the nucleophile, and the nitro group serves to activate its ortho position. The ring at which the substitution takes place is nearly always activated, usually by ortho or para nitro groups. X is usually S, SO, SO_2,[259] O, or COO. Y is usually the conjugate base of OH, NH_2, NHR, or SH. The reaction has even been carried out with $Y = CH_2^-$ (phenyllithium was the base here).[260]

The reaction rate is greatly enhanced by substitution in the 6 position of the attacking ring, for steric reasons. For example, a methyl, chloro, or bromo group in the 6 position of **25** caused the rate to be about 10^5 times faster than when the same groups were in the 4 position,[261] though electrical effects should be similar at these positions. The enhanced rate comes about because the most favorable conformation the molecule can adopt to suit the bulk of the 6-substituent is also the conformation required for the rearrangement. Thus, less entropy of activation is required.

[256]For additional evidence, see Sone; Hamamoto; Seiji; Shinkai; Manabe *J. Chem. Soc., Perkin Trans. 2* **1981,** 1596; Kohnstam; Petch; Williams *J. Chem. Soc., Perkin Trans. 2* **1984,** 423; Sternson; Chandrasakar *J. Org. Chem.* **1984,** *49.* 4295. and references cited in these papers.

[257]Fishbein; McClelland *J. Am. Chem. Soc.* **1987,** *109.* 2824.

[258]For reviews, see Truce; Kreider; Brand *Org. React.* **1971,** *18,* 99-215; Shine. Ref. 240, pp. 307-316; Stevens; Watts. Ref. 245. pp. 120-126.

[259]For a review for the case of X = SO_2, see Cerfontain *Mechanistic Aspects in Aromatic Sulfonation and Desulfonation:* Wiley: New York, 1968, pp. 262-274.

[260]Truce *J. Am. Chem. Soc.* **1959,** *81.* 481; Truce; Robbins; Kreider **1966,** *88,* 4027; Drozd; Nikonova *J. Org. Chem. USSR* **1969,** *5.* 313.

[261]Bunnett; Okamoto *J. Am. Chem. Soc.* **1956,** *78,* 5363.

Although the Smiles rearrangement is usually carried out on compounds containing two rings, this need not be the case; e.g.,[262]

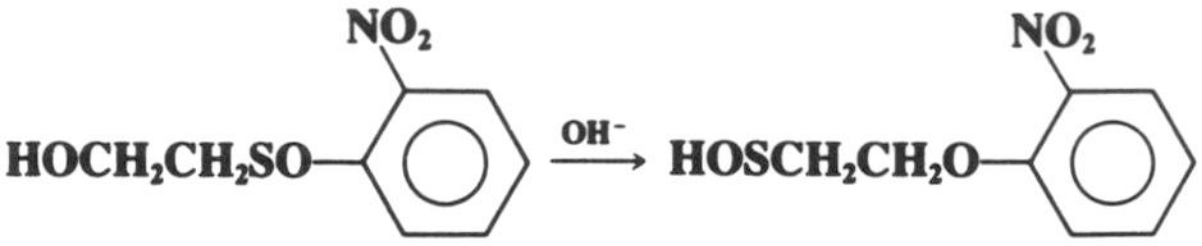

26

In this case the sulfenic acid (**26**) is unstable[263] and the actual products isolated were the corresponding sulfinic acid (RSO_2H) and disulfide (R_2S_2).

[262]Kent; Smiles *J. Chem. Soc.* **1934**, 422.
[263]For a stable sulfenic acid, see Nakamura *J. Am. Chem. Soc.* **1983**, *105*, 7172.

14

FREE-RADICAL SUBSTITUTION

MECHANISMS

Free-Radical Mechanisms in General[1]

A free-radical process consists of at least two steps. The first step involves the *formation* of free radicals, usually by homolytic cleavage of bond, i.e., a cleavage in which each fragment retains one electron:

$$A—B \longrightarrow A• + B•$$

This is called an *initiation* step. It may happen spontaneously or may be induced by heat or light (see the discussion on p. 193), depending on the type of bond. Peroxides, including hydrogen peroxide, dialkyl, diacyl, and alkyl acyl peroxides, and peracids are the most common source of free radicals induced spontaneously or by heat, but other organic compounds with low-energy bonds, such as azo compounds, are also used. Molecules that are cleaved by light are most often chlorine, bromine, and various ketones (see Chapter 7). Radicals can also be formed in another way, by a one-electron transfer (loss or gain), e.g., $A^+ + e^- \rightarrow A•$. One-electron transfers usually involve inorganic ions or electrochemical processes.

The second step involves the *destruction* of free radicals. This usually happens by a process opposite to the first, namely, a combination of two like or unlike radicals to form a new bond:[2]

$$A• + B• \longrightarrow A—B$$

This type of step is called *termination*, and it ends the reaction as far as these particular radicals are concerned.[3] However, it is not often that termination follows *directly* upon initiation. The reason is that most radicals are very reactive and will react with the first available species with which they come in contact. In the usual situation, in which the concentration of radicals is low, this is much more likely to be a molecule than another radical. When a radical (which has an odd number of electrons) reacts with a molecule

---

[1]For books on free-radical mechanisms, see Nonhebel; Tedder; Walton *Radicals*; Cambridge University Press: Cambridge, 1979; Nonhebel; Walton *Free-Radical Chemistry*; Cambridge University Press: London, 1974; Huyser *Free-Radical Chain Reactions*; Wiley: New York, 1970; Pryor *Free Radicals*; McGraw-Hill: New York, 1966; For reviews, see Huyser, in McManus *Organic Reactive Intermediates*; Academic Press: New York, 1973, pp. 1-59; Lloyd, *CHEMTECH* **1971**, 176-180, 371-381, 687-696, **1972**, 182-188. For monographs on the use of free-radical reactions in synthesis, see Giese *Radicals in Organic Synthesis: Formation of Carbon–Carbon Bonds*; Pergamon: Elmsford, NY, 1986; Davies; Parrott *Free Radicals in Organic Synthesis*; Springer: New York, 1978. For reviews, see Curran *Synthesis* **1988**, 417-439, 489-513; Ramaiah *Tetrahedron* **1987**, *43*, 3541-3676.
[2]For a review of the stereochemistry of this type of combination reaction, see Porter; Krebs *Top. Stereochem.* **1988**, *18*, 97-127.
[3]Another type of termination is disproportionation (see p. 194).

(which has an even number), the total number of electrons in the products must be odd. The product in a particular step of this kind may be one particle, e.g.,

$$R\cdot + -\underset{|}{C}=\underset{|}{C}- \longrightarrow -\underset{|}{\overset{R}{\underset{|}{C}}}-\overset{\bullet}{\underset{|}{C}}-$$

in which case it may be another free radical; or it may consist of two particles, e.g.,

$$R\cdot + R'H \longrightarrow RH + R'\cdot$$

in which case one must be a molecule and one a free radical, but in any case a *new radical is generated*. This type of step is called *propagation*, since the newly formed radical can now react with another molecule and produce another radical, and so on, until two radicals do meet each other and terminate the sequence. The process just described is called a *chain reaction*,[4] and there may be hundreds or thousands of propagation steps between an initiation and a termination. Two other types of propagation reactions do not involve a molecule at all. These are (1) cleavage of a radical into, necessarily, a radical and a molecule and (2) rearrangement of one radical to another (see Chapter 18). When radicals are highly reactive, e.g., alkyl radicals, chains are long, since reactions occur with many molecules; but with radicals of low reactivity, e.g., aryl radicals, the radical may be unable to react with anything until it meets another radical, so that chains are short, or the reaction may be a nonchain process. In any particular chain process there is usually a wide variety of propagation and termination steps. Because of this, these reactions lead to many products and are often difficult to treat kinetically.[5]

The following are some general characteristics of free-radical reactions:

1. Reactions are fairly similar whether they are occurring in the vapor or liquid phase, though solvation of free radicals in solution does cause some differences.[6]

2. They are largely unaffected by the presence of acids or bases or by changes in the polarity of solvents, except that nonpolar solvents may suppress competing ionic reactions.

3. They are initiated or accelerated by typical free-radical sources, such as the peroxides referred to, or by light. In the latter case the concept of quantum yield applies (p. 247). Quantum yields can be quite high, e.g., 1000, if each quantum generates a long chain, or low, in the case of nonchain processes.

4. Their rates are decreased or the reactions are suppressed entirely by substances that scavenge free radicals, e.g., nitric oxide, molecular oxygen, or benzoquinone. These substances are called *inhibitors*.[7]

In this chapter are discussed free-radical substitution reactions. Free-radical additions to unsaturated compounds and rearrangements are discussed in Chapters 15 and 18, respectively. In addition, many of the oxidation-reduction reactions considered in Chapter 19 involve free-radical mechanisms. Several important types of free-radical reactions do not usually lead to reasonable yields of pure products and are not generally treated in this book. Among these are polymerizations and high-temperature pyrolyses.

[4]For a discussion of radical chain reactions from a synthetic point of view, see Walling *Tetrahedron* **1985,** *41*, 3887.

[5]For a discussion of the kinetic aspects of radical chain reactions, see Huyser *Free-Radical Chain Reactions*, Ref. 1, pp. 39-65.

[6]For a discussion, see Mayo *J. Am. Chem. Soc.* **1967,** *89*, 2654.

[7]For a review of the action of inhibitors, see Denisov; Khudyakov *Chem. Rev.* **1987,** *87*, 1313-1357.

Free-Radical Substitution Mechanisms[8]

In a free-radical substitution reaction

$$R\text{—}X \longrightarrow R\text{—}Y \tag{1}$$

there must first be a cleavage of the substrate RX so that R• radicals are produced. This can happen by a spontaneous cleavage

$$R\text{—}X \longrightarrow R\bullet + X\bullet \tag{2}$$

or it can be caused by light or heat, or, more often, there is no actual cleavage, but R• is produced by an *abstraction*

$$R\text{—}X + W\bullet \longrightarrow R\bullet + W\text{—}X \tag{3}$$

W• is produced by adding a compound, such as a peroxide, that spontaneously forms free radicals. Such a compound is called an *initiator*. Once R• is formed, it can go to product in two ways, by abstraction

$$R\bullet + Y\text{—}W \longrightarrow R\text{—}Y + W\bullet \tag{4}$$

or by coupling with another radical

$$R\bullet + Y\bullet \longrightarrow R\text{—}Y \tag{5}$$

In a reaction with a moderately long chain, much more of the product will be produced by abstraction (4) than by coupling (5). Cleavage steps like (2) have been called S_H1 (H for homolytic), and abstraction steps like (3) and (4) have been called S_H2; reactions can be classified as S_H1 or S_H2 on the basis of whether RX is converted to R by (2) or (3).[9] Most chain substitution mechanisms follow the pattern (3), (4), (3), (4) . . . . Chains are long and reactions go well where both (3) and (4) are energetically favored (no worse that slightly endothermic, see pp. 683, 693). The IUPAC designation of a chain reaction that follows the pattern (3), (4) . . . is $A_rD_R + A_RD_r$ (R stands for radical).

With certain radicals the transition state in an abstraction reaction has some polar character. For example, consider the abstraction of hydrogen from the methyl group of toluene by a bromine atom. Since bromine is more electronegative than carbon, it is reasonable to assume that in the transition state there is a separation of charge, with a partial negative charge on the halogen and a partial positive charge on the carbon:

$$\overset{\delta+}{Ph\text{—}CH_2}\text{-----}H\text{-----}\overset{\delta-}{Br}$$

Evidence for the polar character of the transition state is that electron-withdrawing groups in the para position of toluene (which would destabilize a positive charge) decrease the rate of hydrogen abstraction by bromine while electron-donating groups increase it.[10] However, as we might expect, substituents have a smaller effect here ($\rho \approx -1.4$) than they do in reactions where a completely ionic intermediate is involved, e.g., the S_N1 mechanism (see p. 344). Other evidence for polar transition states in radical abstraction reactions is mentioned on p. 685. For abstraction by radicals such as methyl or phenyl, polar effects are

---

[8]For a review, see Poutsma, in Kochi *Free Radicals*, vol. 2; Wiley: New York, 1973, pp. 113-158.
[9]Eliel, in Newman *Steric Effects in Organic Chemistry*; Wiley: New York, 1956, pp. 142-143.
[10]For example, see Pearson; Martin *J. Am. Chem. Soc.* **1963,** *85,* 354, 3142; Kim; Choi; Kang *J. Am. Chem. Soc.* **1985,** *107,* 4234.

very small or completely absent. For example, rates of hydrogen abstraction from ring-substituted toluenes by the methyl radical were relatively unaffected by the presence of electron-donating or electron-withdrawing substituents.[11] Those radicals (e.g., Br•) that have a tendency to abstract electron-rich hydrogen atoms are called *electrophilic radicals*.

When the reaction step R—X → R• takes place at a chiral carbon, racemization is almost always observed because free radicals do not retain configuration. Exceptions to this rule are found at cyclopropyl substrates, where both inversion[12] and retention[13] of configuration have been reported, and in the reactions mentioned on p. 682.

Mechanisms at an Aromatic Substrate[14]

When R in reaction (1) is aromatic, the simple abstraction mechanism just discussed may be operating, especially in gas-phase reactions. However, mechanisms of this type cannot account for all reactions of aromatic substrates. In processes such as the following (see **4-18**, **4-21**, and **4-22**):

$$\text{Ar•} + \text{ArH} \longrightarrow \text{Ar—Ar} \tag{6}$$

which occur in solution, the coupling of two rings cannot be explained on the basis of a simple abstraction

$$\text{Ar•} + \text{ArH} \longrightarrow \text{Ar—Ar} + \text{H•} \tag{7}$$

since, as discussed on p. 683, abstraction of an entire group such as phenyl by a free radical is very unlikely. The products can be explained by a mechanism similar to that of electrophilic and nucleophilic aromatic substitution. In the first step, the radical attacks the ring in much the same way as would an electrophile or a nucleophile:

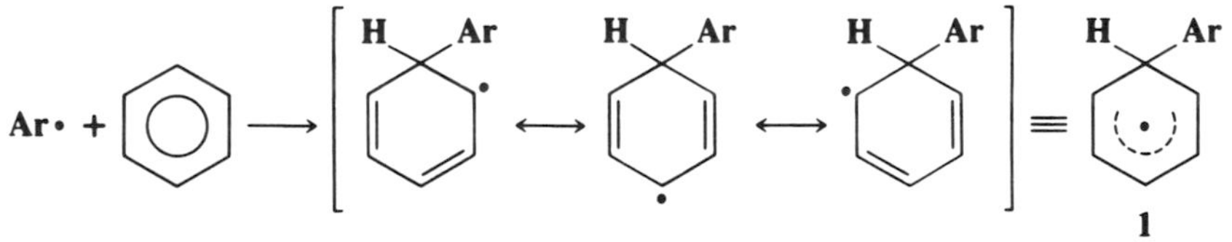

$$\tag{8}$$

The intermediate is relatively stable because of the resonance. The reaction can terminate in three ways: by simple coupling, or by disproportionation

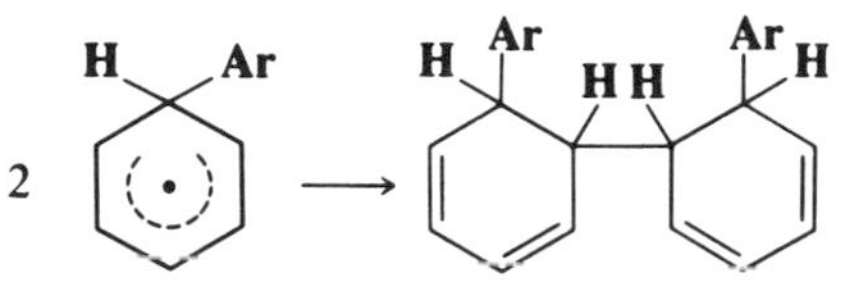

$$\tag{9}$$

---

[11]For example, see Kalatzis; Williams *J. Chem. Soc. B* **1966**, 1112; Pryor; Tonellato; Fuller; Jumonville *J. Org. Chem.* **1969**, *34*, 2018.

[12]Altman; Nelson *J. Am. Chem. Soc.* **1969**, *91*, 5163.

[13]Jacobus; Pensak *Chem. Commun.* **1969**, 400.

[14]For reviews, see Kobrina *Russ. Chem. Rev.* **1977**, *46*, 348-360; Perkins, in Kochi, Ref. 8, vol. 2, 231-271; Bolton; Williams, *Adv. Free-Radical Chem.* **1975**, *5*, 1-25; Nonhebel; Walton, Ref. 1, pp. 417-469; Minisci; Porta *Adv. Heterocycl. Chem.* **1974**, *16*, 123-180; Bass; Nababsing *Adv. Free-Radical Chem.* **1972**, *4*, 1-47; Hey *Bull. Soc. Chim. Fr.* **1968**, 1591.

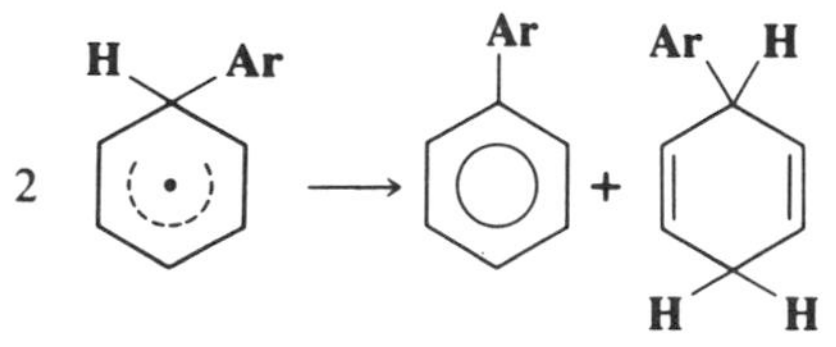

$$2 \quad \boxed{\cdot} \quad \longrightarrow \quad \boxed{} \quad + \quad \boxed{}$$

$$\tag{10}$$

$$\mathbf{3}$$

or, if a species (R'•) is present which abstracts hydrogen, by abstraction[15]

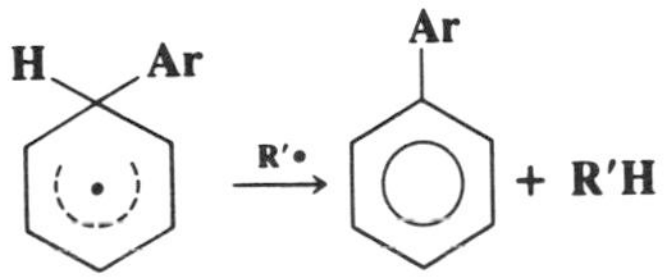

$$\boxed{\cdot} \quad \xrightarrow{\text{R'•}} \quad \boxed{} \quad + \quad \text{R'H}$$

$$\tag{11}$$

2 is a partially hydrogenated quaterphenyl. Of course, the coupling need not be ortho–ortho, and other isomers can also be formed. Among the evidence for steps (9) and (10) was isolation of compounds of types **2** and **3**,[16] though normally under the reaction conditions dihydrobiphenyls like **3** are oxidized to the corresponding biphenyls. Other evidence for this mechanism is the detection of the intermediate **1** by CIDNP[17] and the absence of isotope effects, which would be expected if the rate-determining step were (7), which involves cleavage of the Ar—H bond. In the mechanism just given, the rate-determining step (8) does not involve loss of hydrogen. The reaction between aromatic rings and the OH• radical takes place by the same mechanism. A similar mechanism has been shown for substitution at some vinylic and acetylenic substrates, e.g.:[18]

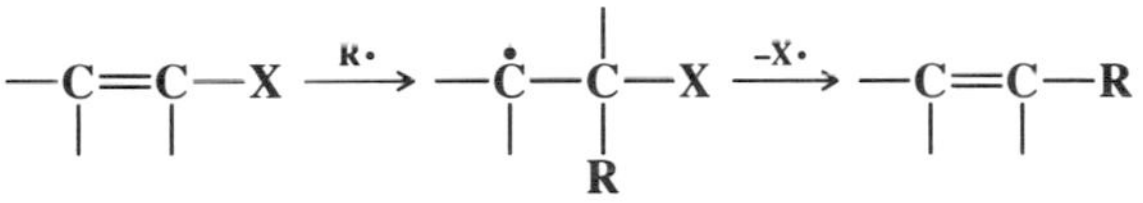

$$-\text{C}{=}\text{C}-\text{X} \xrightarrow{\text{R•}} -\overset{\cdot}{\underset{\underset{\text{R}}{|}}{\text{C}}}-\text{C}-\text{X} \xrightarrow{-\text{X•}} -\text{C}{=}\text{C}-\text{R}$$

This is reminiscent of the nucleophilic tetrahedral mechanism at a vinylic carbon (p. 336)

Neighboring-Group Assistance in Free-Radical Reactions

In a few cases it has been shown that cleavage steps (2) and abstraction steps (3) have been accelerated by the presence of neighboring groups. Photolytic halogenation (**4-1**) is a process that normally leads to mixtures of many products. However, bromination of carbon chains containing a bromine atom occurs with high regioselectivity. Bromination of alkyl bromides gave 84 to 94% substitution at the carbon adjacent to the bromine already in the molecule.[19] This result is especially surprising because, as we shall see (p. 685), positions close to a polar group such as bromine should actually be *deactivated* by the electron-withdrawing field effect

[15]**1** can also be oxidized to the arene ArPh by atmospheric O_2. For a discussion of the mechanism of this oxidation, see Narita; Tezuka *J. Am. Chem. Soc.* **1982**, *104*, 7316.

[16]De Tar; Long *J. Am. Chem. Soc.* **1958**, *80*, 4742. See also Ref. 334.

[17]Fahrenholtz; Trozzolo *J. Am. Chem. Soc.* **1972**, *94*, 282.

[18]Russell; Ngoviwatchai *Tetrahedron Lett.* **1986**, *27*, 3479, and references cited therein.

[19]Thaler *J. Am. Chem. Soc.* **1963**, *85*, 2607. See also Traynham; Hines *J. Am. Chem. Soc.* **1968**, *90*, 5208; Ucciani; Pierri; Naudet *Bull. Soc. Chim. Fr.* **1970**, 791; Hargis *J. Org. Chem.* **1973**, *38*, 346.

of the bromine. The unusual regioselectivity is explained by a mechanism in which abstraction (3) is assisted by a neighboring bromine atom:[20]

In the normal mechanism, Br• abstracts a hydrogen from RH, leaving R•. When a bromine is present in the proper position, it assists this process, giving a cyclic intermediate (a *bridged free radical*, **4**).[21] In the final step (very similar to R• + $Br_2 \rightarrow$ RBr + Br•) the ring is broken. If this mechanism is correct, the configuration at the substituted carbon (marked *) should be retained. This has been shown to be the case: optically active 1-bromo-2-methylbutane gave 1,2-dibromo-2-methylbutane with retention of configuration.[20] Furthermore, when this reaction was carried out in the presence of DBr, the "recovered" 1-bromo-2-methylbutane was found to be deuterated in the 2 position, and its configuration was retained.[22] This is just what would be predicted if some of the **4** present abstracted D from DBr. There is evidence that Cl can form bridged radicals,[23] though esr spectra show that the bridging is not necessarily symmetrical.[24] Still more evidence for bridging by Br has been found in isotope effect and other studies.[25] However, evidence from CIDNP shows that the methylene protons of the β-bromoethyl radical are not equivalent, at least while the radical is present in the radical pair [PhCOO• •CH_2CH_2Br] within a solvent cage.[26] This evidence indicates that under these conditions $BrCH_2CH_2$• is not a symmetrically bridged radical, but it could be unsymmetrically bridged. A bridged intermediate has also been invoked, when a bromo group is in the proper position, in the Hunsdiecker reaction[27] (**4-39**), and in abstraction of iodine atoms by the phenyl radical.[28] Participation by other neighboring groups, e.g. SR, SiR_3, SnR_3, has also been reported.[29]

[20]Skell; Tuleen; Readio *J. Am. Chem. Soc.* **1963**, *85*, 2849. For other stereochemical evidence, see Huyser; Feng *J. Org. Chem.* **1971**, *36*, 731. For another explanation, see Lloyd; Wood *J. Am. Chem. Soc.* **1975**, *97*, 5986.

[21]For a monograph, see Kaplan *Bridged Free Radicals*; Marcel Dekker: New York, 1972. For reviews, see Skell; Traynham *Acc. Chem. Res.* **1984**, *17*, 160-166; Skell; Shea, in Kochi, Ref. 8, vol. 2, pp. 809-852.

[22]Shea; Skell *J. Am. Chem. Soc.* **1973**, *95*, 283.

[23]Everly; Schweinsberg; Traynham *J. Am. Chem. Soc.* **1978**, *100*, 1200; Wells; Franke *Tetrahedron Lett.* **1979**, 4681.

[24]Bowles; Hudson; Jackson *Chem. Phys. Lett.* **1970**, *5*, 552; Cooper; Hudson; Jackson *Tetrahedron Lett.* **1973**, 831; Chen; Elson; Kochi *J. Am. Chem. Soc.* **1973**, *95*, 5341.

[25]Skell; Readio *J. Am. Chem. Soc.* **1964**, *86*, 3334; Skell; Pavlis; Lewis; Shea *J. Am. Chem. Soc.* **1973**, *95*, 6735; Juneja; Hodnett *J. Am. Chem. Soc.* **1967**, *89*, 5685, Lewis; Kozuka *J. Am. Chem. Soc.* **1973**, *95*, 282; Cain; Solly *J. Chem. Soc., Chem. Commun.* **1974**, 148; Chenier; Tremblay; Howard *J. Am. Chem. Soc.* **1975**, *97*, 1618; Howard; Chenier; Holden *Can. J. Chem.* **1977**, *55*, 1463. See however Tanner; Blackburn; Kosugi; Ruo *J. Am. Chem. Soc.* **1977**, *99*, 2714.

[26]Hargis; Shevlin *J. Chem. Soc., Chem. Commun.* **1973**, 179.

[27]Applequist; Werner *J. Org. Chem.* **1963**, *28*, 48.

[28]Danen; Winter *J. Am. Chem. Soc.* **1971**, *93*, 716.

[29]Tuleen; Bentrude; Martin *J. Am. Chem. Soc.* **1963**, *85*, 1938; Fisher; Martin *J. Am. Chem. Soc.* **1966**, *88*, 3382; Jackson; Ingold; Griller; Nazran *J. Am. Chem. Soc.* **1985**, *107*, 208. For a review of neighboring-group participation in cleavage reactions, especially those involving SiR_3 as a neighboring group, see Reetz *Angew. Chem. Int. Ed. Engl.* **1979**, *18*, 173-180 [*Angew. Chem. 91*, 185-192].

REACTIVITY

Reactivity for Aliphatic Substrates[30]

In a chain reaction, the step that determines what the product will be is most often an abstraction step. What is abstracted by a free radical is almost never a tetra-[31] or tervalent atom[32] (except in strained systems, see p. 757)[33] and seldom a divalent one.[34] Nearly always it is univalent, and so, for organic compounds, it is hydrogen or halogen. For example, a reaction between a chlorine atom and ethane gives an ethyl radical, not a hydrogen atom:

$$CH_3CH_3 + Cl\cdot \begin{cases} \nearrow H\!-\!Cl + CH_3CH_2\cdot \quad \Delta H = -3 \text{ kcal/mol}, -13 \text{ kJ/mol} \\ \searrow\!\!\!\!\! / \quad CH_3CH_2\!-\!Cl + H\cdot \quad \Delta H = +18 \text{ kcal/mol}, +76 \text{ kJ/mol} \end{cases}$$

The principal reason for this is steric. A univalent atom is much more exposed to attack by the incoming radical than an atom with a higher valence. Another reason is that in many cases abstraction of a univalent atom is energetically more favored. For example, in the reaction given above, a C_2H_5—H bond is broken ($D = 100$ kcal/mol, 419 kJ/mol, from Table 5.3) whichever pathway is taken, but in the former case an H—Cl bond is formed ($D = 103$ kcal/mol, 432 kJ/mol) while in the latter case it is a C_2H_5—Cl bond ($D = 82$ kcal/mol, 343 kJ/mol). Thus the first reaction is favored because it is exothermic by 3 kcal/mol (100 − 103) [13 kJ/mol (419 − 432)], while the latter is endothermic by 18 kcal/mol (100 − 82) [76 kJ/mol (419 − 343)].[35] However, the steric reason is clearly more important, because even in cases where ΔH is not very different for the two possibilities, the univalent atom is chosen.

Most studies of aliphatic reactivity have been made with hydrogen as the leaving atom and chlorine atoms as the abstracting species.[36] In these reactions, every hydrogen in the substrate is potentially replaceable and mixtures are usually obtained. However, the abstracting radical is not totally unselective, and some positions on a molecule lose hydrogen more easily than others. We discuss the position of attack under several headings:[37]

1. *Alkanes.* The tertiary hydrogens of an alkane are the ones preferentially abstracted by almost any radical, with secondary hydrogens being next preferred. This is in the same order as D values for these types of C—H bonds (Table 5.3). The extent of the preference

---

[30]For a review of the factors involved in reactivity and regioselectivity in free-radical substitutions and additions, see Tedder *Angew. Chem. Int. Ed. Engl.* **1982**, *21*, 401-410 [*Angew. Chem. 94*, 433-442].

[31]Abstraction of a tetravalent carbon has been seen in the abstraction by F• of R from RCl: Firouzbakht; Ferrieri; Wolf; Rack *J. Am. Chem. Soc.* **1987**, *109*, 2213.

[32]See, for example, Back *Can. J. Chem.* **1983**, *61*, 916.

[33]For an example of an abstraction occurring to a small extent at an unstrained carbon atom, see Jackson; Townson *J. Chem. Soc., Perkin Trans. 2* **1980**, 1452. See also Johnson *Acc. Chem. Res.* **1983**, *16*, 343-349.

[34]For a monograph on abstractions of divalent and higher-valent atoms, see Ingold; Roberts *Free-Radical Substitution Reactions*; Wiley: New York, 1971.

[35]ΔH for a free-radical abstraction reaction can be regarded simply as the difference in D values for the bond being broken and the one formed.

[36]For a review that lists many rate constants for abstraction of hydrogen at various positions of many molecules, see Hendry; Mill; Piszkiewicz; Howard; Eigenmann *J. Phys. Chem. Ref. Data* **1974**, *3*, 937-978.

[37]For reviews, see Tedder *Tetrahedron* **1982**, *38*, 313-329; Kerr, in Bamford; Tipper *Comprehensive Chemical Kinetics*, vol. 18; Elsevier: New York, 1976, pp. 39-109; Russell, in Kochi, Ref. 8, vol. 2, pp. 275-331; Rüchardt *Angew. Chem. Int. Ed. Engl.* **1970**, *9*, 830-843 [*Angew. Chem. 82*, 845-858]; Poutsma *Methods Free-Radical Chem.* **1969**, *1*, 79-193; Davidson *Q. Rev., Chem. Soc.* **1967**, *21*, 249-258; Pryor; Fuller; Stanley *J. Am. Chem. Soc.* **1972**, *94*, 1632.

TABLE 14.1   Relative susceptibility to attack by Cl• of primary, secondary, and tertiary positions at 100 and 600°C in the gas phase[38]

Temp., °C	Primary	Secondary	Tertiary
100	1	4.3	7.0
600	1	2.1	2.6

depends on the selectivity of the abstracting radical and on the temperature. Table 14.1 shows[38] that at high temperatures selectivity decreases, as might be expected.[39] An example of the effect of radical selectivity may be noted in a comparison of fluorine atoms with bromine atoms. For the former, the ratio of primary to tertiary abstraction (of hydrogen) is 1:1.4, while for the less reactive bromine atom this ratio is 1:1600. With certain large radicals there is a steric factor that may change the selectivity pattern. For example, in the photochemical chlorination of isopentane in H_2SO_4 with N-chloro-di-t-butylamine and N-chloro-t-butyl-t-pentylamine, the primary hydrogens are abstracted 1.7 times *faster* than the tertiary hydrogen.[40] In this case the attacking radicals (the radical ions $R_2NH•^+$, see p. 692) are bulky enough for steric hindrance to become a major factor.

2. *Olefins.*   When the substrate molecule contains a double bond, treatment with chlorine or bromine usually leads to addition rather than substitution. However, for other radicals (and even for chlorine or bromine atoms when they do abstract a hydrogen) the position of attack is perfectly clear. Vinylic hydrogens are practically never abstracted, and allylic hydrogens are greatly preferred to other positions of the molecule. This is generally attributed[41] to resonance stabilization of the allylic radical:

$$\overset{|}{C}=\overset{|}{C}-\overset{|}{C}-H \longrightarrow \left[-\overset{|}{C}=\overset{|}{C}-\overset{|}{C}• \longleftrightarrow •\overset{|}{C}=\overset{|}{C}-\overset{|}{C} \right]$$

As might be expected, allylic rearrangements (see p. 327) are common in these cases.[42]

3. *Alkyl side chains of aromatic rings.*   The preferential position of attack on a side chain is usually the one α to the ring. Both for active radicals such as chlorine and phenyl and for more selective ones such as bromine such attack is faster than that at a primary carbon, but for the active radicals benzylic attack is slower than for tertiary positions, while for the selective ones it is faster. Two or three aryl groups on a carbon activate its hydrogens even more, as would be expected from the resonance involved. These statements can be illustrated by the following abstraction ratios:[43]

	Me–H	MeCH$_2$–H	Me$_2$CH–H	Me$_3$C–H	PhCH$_2$–H	Ph$_2$CH–H	Ph$_3$C–H
Br	0.0007	1	220	19,400	64,000	1.1×10^6	6.4×10^6
Cl	0.004	1	4.3	6.0	1.3	2.6	9.5

[38]Hass; McBee; Weber *Ind. Eng. Chem.* **1936**, *28*, 333.
[39]For a similar result with phenyl radicals, see Kopinke; Zimmermann; Anders *J. Org. Chem.* **1989**, *54*, 3571.
[40]Deno; Fishbein; Wyckoff *J. Am. Chem. Soc.* **1971**, *93*, 2065. Similar steric effects, though not a reversal of primary–tertiary reactivity, were found by Dneprovskii; Mil'tsov *J. Org. Chem. USSR* **1988**, *24*, 1836.
[41]See however Kwart; Brechbiel; Miles; Kwart *J. Org. Chem.* **1982**, *47*, 4524.
[42]For reviews, see Wilt, in Kochi, Ref. 8, vol. 1, pp. 458-466.
[43]Russell, Ref. 37, p. 289.

However, many anomalous results have been reported for these substrates. The benzylic position is not always the most favored. One thing certain is that *aromatic* hydrogens are seldom abstracted if there are aliphatic ones to compete (note from Table 5.3, that D for Ph—H is higher than that for any alkyl H bond). Several $\sigma^{\bullet}$ scales (similar to the σ, σ^{+} and σ^{-} scales discussed in Chapter 9) have been developed for benzylic radicals.[44]

4. *Compounds containing electron-withdrawing substituents.* In halogenations electron-withdrawing groups greatly deactivate adjacent positions. Compounds of the type Z—CH$_2$—CH$_3$ are attacked predominantly or exclusively at the β position when Z is COOH, COCl, COOR, SO$_2$Cl, or CX$_3$. Such compounds as acetic acid and acetyl chloride are not attacked at all. This is in sharp contrast to electrophilic halogenations (**2-4** to **2-6**), where *only* the α position is substituted. This deactivation of α positions is also at variance with the expected stability of the resulting radicals, since they would be expected to be stabilized by resonance similar to that for allylic and benzylic radicals. This behavior is a result of the polar transition states discussed on p. 679. Halogen atoms are electrophilic radicals and look for positions of high electron density. Hydrogens on carbon atoms next to electron-withdrawing groups have low electron densities (because of the field effect of Z) and are therefore shunned. Radicals that are not electrophilic do not display this behavior. For example, the methyl radical is essentially nonpolar and does not avoid positions next to electron-withdrawing groups; relative rates of abstraction at the α and β carbons of propionic acid are:[45]

	CH$_3$—	CH$_2$—COOH
Me•	1	7.8
Cl•	1	0.03

Some radicals, e.g., *t*-butyl,[46] benzyl,[47] and cyclopropyl,[48] are *nucleophilic* (they tend to abstract electron-poor hydrogen atoms). The phenyl radical appears to have a very small degree of nucleophilic character.[49] For longer chains, the field effect continues, and the β position is also deactivated to attack by halogen, though much less so than the α position. We have already mentioned (p. 679) that abstraction of an α hydrogen atom from ring-substituted toluenes can be correlated by the Hammett equation.

5. *Stereoelectronic effects.* On p. 334 we saw an example of a stereoelectronic effect. It has been shown that such effects are important where a hydrogen is abstracted from a carbon adjacent to a C—O or C—N bond. In such cases hydrogen is abstracted from C—H bonds that have a relatively small dihedral angle ($\sim 30°$) with the unshared orbitals of the O or N much more easily than from those with a large angle ($\sim 90°$). For example, the starred hydrogen of **5** was abstracted about 8 times faster than the starred hydrogen of **6**.[50]

[44]See, for example, Dinçtürk; Jackson *J. Chem. Soc., Perkin Trans. 2* **1981**, 1127; Dust; Arnold *J. Am. Chem. Soc.* **1983**, *105*, 1221, 6531; Creary; Mehrsheikh-Mohammadi; McDonald *J. Org. Chem.* **1987**, *52*, 3254, **1989**, *54*, 2904; Fisher; Dershem; Prewitt *J. Org. Chem.* **1990**, *55*, 1040.

[45]Russell, Ref. 37, p. 311.

[46]Pryor; Davis; Stanley *J. Am. Chem. Soc.* **1973**, *95*, 4754; Pryor; Tang; Tang; Church *J. Am. Chem. Soc.* **1982**, *104*, 2885; Dütsch; Fischer *Int. J. Chem. Kinet.* **1982**, *14*, 195.

[47]Clerici; Minisci; Porta *Tetrahedron* **1973**, *29*, 2775.

[48]Stefani; Chuang; Todd *J. Am. Chem. Soc.* **1970**, *92*, 4168.

[49]Suehiro; Suzuki; Tsuchida; Yamazaki *Bull. Chem. Soc. Jpn.* **1977**, *50*, 3324.

[50]Hayday; McKelvey *J. Org. Chem.* **1976**, *41*, 2222. For additional examples, see Malatesta; Ingold *J. Am. Chem. Soc.* **1981**, *103*, 609; Beckwith; Easton *J. Am. Chem. Soc.* **1981**, *103*, 615; Beckwith; Westwood *Aust. J. Chem.* **1983**, *36*, 2123; Griller; Howard; Marriott; Scaiano *J. Am. Chem. Soc.* **1981**, *103*, 619. For a stereoselective abstraction step, see Dneprovskii; Pertsikov; Temnikova *J. Org. Chem. USSR* **1982**, *18*, 1951. See also Bunce; Cheung; Langshaw *J. Org. Chem.* **1986**, *51*, 5421.

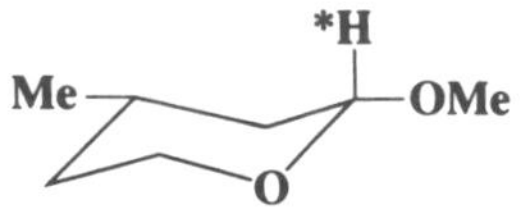

Abstraction of a halogen has been studied much less,[51] but the order of reactivity is RI > RBr > RCl ≫ RF.

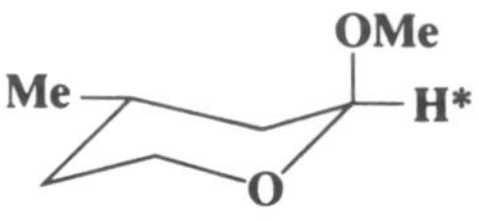

Reactivity at a Bridgehead[52]

Many free-radical reactions have been observed at bridgehead carbons, e.g. (see **4-39**),[53]

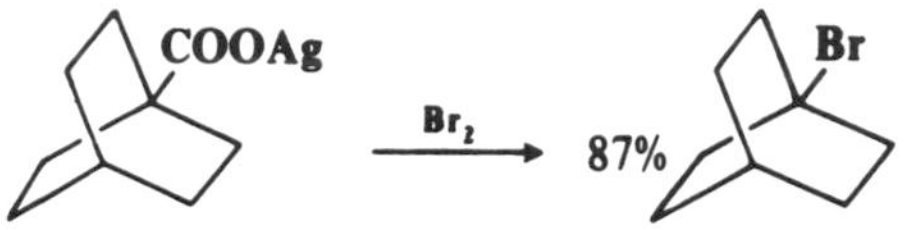

demonstrating that the free radical need not be planar. However, treatment of norbornane with sulfuryl chloride and benzoyl peroxide gave mostly 2-chloronorbornane, though the bridgehead position is tertiary.[54] So, while bridgehead free-radical substitution is possible, it is not preferred, presumably because of the strain involved.[55]

Reactivity in Aromatic Substrates

Free-radical substitution at an aromatic carbon seldom takes place by a mechanism in which a hydrogen is abstracted to give an aryl radical. Reactivity considerations here are similar to those in Chapters 11 and 13; i.e., we need to know which position on the ring will be attacked to give the intermediate

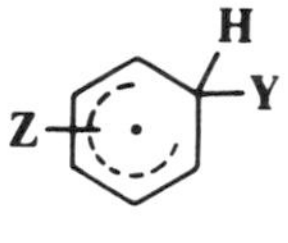

The obvious way to obtain this information is to carry out reactions with various Z groups and to analyze the products for percent ortho, meta, and para isomers, as has so often been done for electrophilic substitution. However, this procedure is much less accurate in the case of free-radical substitutions because of the many side reactions. It may be, for example, that in a given case the ortho position is more reactive than the para, but the intermediate from the para attack may go on to product while that from ortho attack gives a side reaction. In such a case, analysis of the three products does not give a true picture of which position

---

[51]For a review, see Danen *Methods Free-Radical Chem.* **1974**, *5*, 1-99.
[52]For reviews, see Bingham; Schleyer *Fortschr. Chem. Forsch.* **1971**, *18*, 1-102, pp. 79-81; Fort; Schleyer *Adv. Alicyclic Chem.* **1966**, *1*, 283-370, pp. 337-352.
[53]Grob; Ohta; Renk; Weiss *Helv. Chim. Acta* **1958**, *41*, 1191.
[54]Roberts; Urbanek; Armstrong *J. Am. Chem. Soc.* **1949**, *71*, 3049. See also Kooyman; Vegter *Tetrahedron* **1958**, *4*, 382; Walling; Mayahi *J. Am. Chem. Soc.* **1959**, *81*, 1485.
[55]See, for example, Koch; Gleicher *J. Am. Chem. Soc.* **1971**, *93*, 1657.

is most susceptible to attack. The following generalizations can nevertheless be drawn, though there has been much controversy over just how meaningful such conclusions are:[56]

1. All substituents increase reactivity at ortho and para positions over that of benzene. There is no great difference between electron-donating and electron-withdrawing groups.

2. Reactivity at meta positions is usually similar to that of benzene, perhaps slightly higher or lower. This fact, coupled with the preceding one, means that all substituents are activating and ortho-para-directing; none are deactivating or (chiefly) meta-directing.

3. Reactivity at ortho positions is usually somewhat greater than at para positions, except where a large group decreases ortho reactivity for steric reasons.

4. In direct competition, electron-withdrawing groups exert a somewhat greater influence than electron-donating groups. Arylation of para-disubstituted compounds XC_6H_4Y showed that substitution ortho to the group X became increasingly preferred as the electron-withdrawing character of X increases (with Y held constant).[57] The increase could be correlated with the Hammett σ_p values for X.

5. Substituents have a much smaller effect than in electrophilic or nucleophilic substitution; hence the partial rate factors (see p. 516) are not great.[58] Partial rate factors for a few groups are given in Table 14.2.[59]

6. Although hydrogen is the leaving group in most free-radical aromatic substitutions, ipso attack (p. 512) and ipso substitution (e.g., with Br, NO_2, or CH_3CO as the leaving group) have been found in certain cases.[60]

Reactivity in the Attacking Radical[61]

We have already seen that some radicals are much more selective than others (p. 684). The bromine atom is so selective that when only primary hydrogens are available, as in neo-

TABLE 14.2 Partial rate factors for attack of substituted benzenes by phenyl radicals generated from Bz_2O_2 (reaction **4-21**)[59]

	Partial rate factor		
Z	o	m	p
H	1	1	1
NO₂	5.50	0.86	4.90
CH₃	4.70	1.24	3.55
CMe₃	0.70	1.64	1.81
Cl	3.90	1.65	2.12
Br	3.05	1.70	1.92
MeO	5.6	1.23	2.31

[56]De Tar *J. Am. Chem. Soc.* **1961**, *83*, 1014 (book review); Dickerman; Vermont *J. Am. Chem. Soc.* **1962**, *84*, 4150; Morrison; Cazes; Samkoff; Howe *J. Am. Chem. Soc.* **1962**, *84*, 4152; Ohta; Tokumaru *Bull. Chem. Soc. Jpn.* **1971**, *44*, 3218; Vidal; Court; Bonnier *J. Chem. Soc. Perkin Trans. 2* **1973**, 2071; Tezuka; Ichikawa; Marusawa; Narita *Chem. Lett.* **1983**, 1013.
[57]Davies; Hey; Summers *J. Chem. Soc. C* **1970**, 2653.
[58]For a quantitative treatment, see Charton; Charton *Bull. Soc. Chim. Fr.* **1988**, 199.
[59]Davies; Hey; Summers *J. Chem. Soc. C* **1971**, 2681.
[60]For reviews, see Traynham *J. Chem. Educ.* **1983**, *60*, 937-941, *Chem. Rev.* **1979**, *79*, 323-330; Tiecco *Acc. Chem. Res.* **1980**, *13*, 51-57; *Pure Appl. Chem.* **1981**, *53*, 239-258.
[61]For reviews with respect to CH_3· and CF_3·, see Trotman-Dickenson *Adv. Free-Radical Chem.* **1965**, *1*, 1-38; Spirin *Russ. Chem. Rev.* **1969**, *38*, 529-539; Gray; Herod; Jones *Chem. Rev.* **1971**, *71*, 247-294.

pentane or *t*-butylbenzene, the reaction is slow or nonexistent; and isobutane can be selectively brominated to give *t*-butyl bromide in high yields. However, toluene reacts with bromine atoms instantly. Bromination of other alkylbenzenes, e.g., ethylbenzene and cumene, takes place exclusively at the α position,[62] emphasizing the selectivity of Br•. The dissociation energy *D* of the C—H bond is more important for radicals of low reactivity than for highly reactive radicals, since bond breaking in the transition state is greater. Thus, bromine shows a greater tendency than chlorine to attack α to an electron-withdrawing group because the energy of the C—H bond there is lower than in other places in the molecule.

Some radicals, e.g., triphenylmethyl, are so unreactive that they abstract hydrogens very poorly if at all. Table 14.3 lists some common free radicals in approximate order of reactivity.[63]

It has been mentioned that some free radicals, e.g., chloro, are electrophilic and some, e.g., *t*-butyl, are nucleophilic. It must be borne in mind that these tendencies are relatively slight compared with the electrophilicity of a positive ion or the nucleophilicity of a negative ion. The predominant character of a free radical is neutral, whether it has slight electrophilic or nucleophilic tendencies.

The Effect of Solvent on Reactivity[65]

As has been noted earlier, the solvent usually has little effect on free-radical substitutions in contrast to ionic ones: indeed, reactions in solution are often quite similar in character to those in the gas phase, where there is no solvent at all. However, in certain cases the solvent *can* make an appreciable difference. Chlorination of 2,3-dimethylbutane in aliphatic solvents gave about 60% $(CH_3)_2CHCH(CH_3)CH_2Cl$ and 40% $(CH_3)_2CHCCl(CH_3)_2$, while in aromatic solvents the ratio became about 10:90.[66] This result is attributed to complex

TABLE 14.3   Some common free radicals in decreasing order of activity
The E values represent activation energies for the reaction

$$X\bullet + C_2H_6 \longrightarrow X—H + C_2H_5\bullet^{63}$$

iso-Pr• *is less active than* Me• *and* t-Bu• *still less so*[64]

Radical	E kcal/mol	kJ/mol	Radical	E kcal/mol	kJ/mol
F•	0.3	1.3	H•	9.0	38
Cl•	1.0	4.2	Me•	11.8	49.4
MeO•	7.1	30	Br•	13.2	55.2
CF₃•	7.5	31			

[62]Huyser *Free-Radical Chain Reactions*, Ref. 1, p. 97.
[63]Trotman-Dickenson, Ref. 61.
[64]Kharasch; Hambling; Rudy *J. Org. Chem.* **1959,** *24,* 303.
[65]For reviews, see Reichardt *Solvent Effects in Organic Chemistry*; Verlag Chemie: Deerfield Beach, FL, 1979, pp. 110-123; Martin, in Kochi, Ref. 8, vol. 2, pp. 493-524; Huyser *Adv. Free-Radical Chem.* **1965,** *1,* 77-135.
[66]Russell *J. Am. Chem. Soc.* **1958,** *80,* 4987, 4997, 5002, *J. Org. Chem.* **1959,** *24,* 300.

formation between the aromatic solvent and the chlorine atom which makes the chlorine more selective.[67] This type of effect is not found in cases where the differences in abstract-

7

ability are caused by field effects of electron-withdrawing groups (p. 685). In such cases aromatic solvents make little difference.[68] The complex **7** has been detected[69] as a very short-lived species by observation of its visible spectrum in the pulse radiolysis of a solution of benzene in CCl_4.[70] Differences caused by solvents have also been reported in reactions of other radicals.[71] Some of the anomalous results obtained in the chlorination of aromatic side chains (p. 685) can also be explained by this type of complexing, in this case not with the solvent but with the reacting species.[72] Much smaller, though real, differences in selectivity have been found when the solvent in the chlorination of 2,3-dimethylbutane is changed from an alkane to CCl_4.[73] However, these differences are not caused by formation of a complex between Cl• and the solvent.

REACTIONS

The reactions in this chapter are classified according to leaving group. The most common leaving groups are hydrogen and nitrogen (from the diazonium ion); these are considered first.

Hydrogen as Leaving Group

A.   Substitution by Halogen

4-1   Halogenation at an Alkyl Carbon[74]
Halogenation or **Halo-de-hydrogenation**

$$R{-}H + Cl_2 \xrightarrow{h\nu} R{-}Cl$$

[67]See also Soumillion; Bruylants *Bull. Soc. Chim. Belg.* **1969**, *78*, 425; Potter; Tedder *J. Chem. Soc., Perkin Trans. 2* **1982**, 1689; Aver'yanov; Ruban; Shvets *J. Org. Chem. USSR* **1987**, *23*, 782; Aver'yanov; Ruban *J. Org. Chem. USSR* **1987**, *23*, 1119; Raner; Lusztyk; Ingold *J. Am. Chem. Soc.* **1989**, *111*, 3652; Ingold; Lusztyk; Raner *Acc. Chem. Res.* **1990**, *23*, 219-225.

[68]Russell *Tetrahedron* **1960**, *8*, 101; Nagai; Horikawa; Ryang; Tokura *Bull. Chem. Soc. Jpn.* **1971**, *44*, 2771.

[69]It has been contended that another species, a chlorocyclohexadienyl radical (the structure of which is the same as **1**, except that Cl replaces Ar), can also be attacking when the solvent is benzene: Skell; Baxter; Taylor *J. Am. Chem. Soc.* **1983**, *105*, 120; Skell; Baxter; Tanko; Chebolu *J. Am. Chem. Soc.* **1986**, *108*, 6300. For arguments against this proposal, see Bunce; Ingold; Landers; Lusztyk; Scaiano *J. Am. Chem. Soc.* **1985**, *107*, 5464; Walling *J. Org. Chem.* **1988**, *53*, 305; Aver'yanov; Shvets; Semenov *J. Org. Chem. USSR* **1990**, *26*, 1261.

[70]Bühler *Helv. Chim. Acta* **1968**, *51*, 1558. For other spectral observations, see Raner; Lusztyk; Ingold *J. Phys. Chem.* **1989**, *93*, 564.

[71]Walling; Azar *J. Org. Chem.* **1968**, *33*, 3885; Walling; Wagner *J. Am. Chem. Soc.* **1963**, *85*, 2333; Ito; Matsuda *J. Am. Chem. Soc.* **1982**, *104*, 568; Minisci; Vismara; Fontana; Morini; Serravalle; Giordano *J. Org. Chem.* **1987**, *52*, 730.

[72]Russell; Ito; Hendry *J. Am. Chem. Soc.* **1963**, *85*, 2976; Corbiau; Bruylants *Bull. Soc. Chim. Belg.* **1970**, *79*, 203, 211; Newkirk; Gleicher *J. Am. Chem. Soc.* **1974**, *96*, 3543.

[73]See Raner; Lusztyk; Ingold *J. Org. Chem.* **1988**, *53*, 5220.

[74]For lists of reagents, with references, see Larock *Comprehensive Organic Transformations*; VCH: New York, 1989, pp. 311-313.

Alkanes can be chlorinated or brominated by treatment with chlorine or bromine in the presence of visible or uv light.[75] The reaction can also be applied to alkyl chains containing many functional groups. The chlorination reaction is usually not useful for preparative purposes precisely because it is so general: not only does substitution take place at virtually every alkyl carbon in the molecule, but di- and polychloro substitution almost invariably occur even if there is a large molar ratio of substrate to halogen. When functional groups are present, the principles are those outlined on p. 684; favored positions are those α to aromatic rings, while positions α to electron-withdrawing groups are least likely to be substituted. Tertiary carbons are most likely to be attacked and primary least. Positions α to an OR group are very readily attacked. Nevertheless, mixtures are nearly always obtained. This can be contrasted to the regioselectivity of electrophilic halogenation (**2-4** to **2-6**), which always takes place α to a carbonyl group (except when the reaction is catalyzed by $AgSbF_6$; see following). Of course, if a *mixture* of chlorides is wanted, the reaction is usually quite satisfactory. For obtaining pure compounds, the chlorination reaction is essentially limited to substrates with only one type of replaceable hydrogen, e.g., ethane, cyclohexane, neopentane. The most common are methylbenzenes and other substrates with methyl groups on aromatic rings, since few cases are known where halogen atoms substitute at an aromatic position.[76] Of course, ring substitution *does* take place in the presence of a positive-ion-forming catalyst (**1-11**). In addition to mixtures of various alkyl halides, traces of other products are obtained. These include H_2, olefins, higher alkanes, lower alkanes, and halogen derivatives of these compounds.

The bromine atom is much more selective than the chlorine atom. As indicated on p. 688, it is often possible to brominate tertiary and benzylic positions selectively. High regioselectivity can also be obtained where the neighboring-group mechanism (p. 681) can operate.

As already mentioned, halogenation can be performed with chlorine or bromine. Fluorine has also been used,[77] but seldom, because it is too reactive and hard to control.[78] It often breaks carbon chains down into smaller units, a side reaction that sometimes becomes troublesome in chlorinations too. Fluorination[78a] has been achieved by the use of chlorine trifluoride ClF_3 at $-75°C$.[79] For example, cyclohexane gave 41% fluorocyclohexane and methylcyclohexane gave 47% 1-fluoro-1-methylcyclohexane. Fluoroxytrifluoromethane CF_3OF fluorinates tertiary positions of certain molecules in good yields with high regioselectivity.[80] For example, adamantane gave 75% 1-fluoroadamantane. F_2 at $-70°C$, diluted with N_2,[81] and bromine trifluoride at 25-35°C[82] are also highly regioselective for tertiary

---

[75]For reviews, see Poutsma, in Kochi, Ref. 8, vol. 2, pp. 159-229; Huyser, in Patai *The Chemistry of the Carbon–Halogen Bond*, pt. 1; Wiley: New York, 1973, pp. 549-607; Poutsma, Ref. 37 (chlorination); Thaler *Methods Free-Radical Chem.* **1969**, *2*, 121-227 (bromination).

[76]Dermer; Edmison *Chem. Rev.* **1957**, *57*, 77-122, pp. 110-112. An example of free-radical ring halogenation can be found in Engelsma; Kooyman *Revl. Trav. Chim. Pays-Bas* **1961**, *80*, 526, 537. For a review of aromatic halogenation in the gas phase, see Kooyman *Adv. Free-Radical Chem.* **1965**, *1*, 137-153.

[77]Rozen *Acc. Chem. Res.* **1988**, *21*, 307-312; Purrington; Kagen; Patrick *Chem. Rev.* **1986**, *86*, 997-1018, pp. 1003-1005; Gerstenberger; Haas *Angew. Chem. Int. Ed. Engl.* **1981**, *20*, 647-667 [*Angew. Chem.* *93*, 659-680]; Hudlický *The Chemistry of Organic Fluorine Compounds*, 2nd ed.; Ellis Horwood: Chichester, 1976; pp. 67-91. For descriptions of the apparatus necessary for handling F_2, see Vypel *Chimia* **1985**, *39*, 305-311.

[78]However, there are several methods by which all the C—H bonds in a molecule can be converted to C—F bonds. For reviews, see Rozhkov, in Baizer; Lund *Organic Electrochemistry*; Marcel Dekker: New York, 1983, pp. 805-825; Lagow; Margrave *Prog. Inorg. Chem.* **1979**, *26*, 161-210. See also Adcock; Horita; Renk *J. Am. Chem. Soc.* **1981**, *103*, 6937; Adcock; Evans *J. Org. Chem.* **1984**, *49*, 2719; Huang; Lagow *Bull. Soc. Chim. Fr.* **1986**, 993.

[78a]For a monograph on fluorinating agents, see German; Zemskov *New Fluorinating Agents in Organic Synthesis*; Springer: New York, 1989.

[79]Brower *J. Org. Chem.* **1987**, *52*, 798.

[80]Alker; Barton; Hesse; Lister-James; Markwell; Pechet; Rozen; Takeshita; Toh *Nouv. J. Chem.* **1980**, *4*, 239.

[81]Rozen; Gal; Faust *J. Am. Chem. Soc.* **1980**, *102*, 6860; Gal; Rozen *Tetrahedron Lett.* **1984**, *25*, 449; Rozen; Ben-Shushan *J. Org. Chem.* **1986**, *51*, 3522; Rozen; Gal *J. Org. Chem.* **1987**, *52*, 4928, **1988**, *53*, 2803; Ref. 80.

[82]Boguslavskaya; Kartashov; Chuvatkin *J. Org. Chem. USSR* **1989**, *25*, 1835.

positions. These reactions probably have electrophilic,[83] not free-radical mechanisms. In fact, the success of the F_2 reactions depends on the suppression of free radical pathways, by dilution with an inert gas, by working at low temperatures, and/or by the use of radical scavengers.

Iodine can be used if the activating light has a wavelength of 184.9 nm,[84] but iodinations are seldom attempted, largely because the HI formed reduces the alkyl iodide.

Many other halogenation agents have been employed, the most common of which is sulfuryl chloride SO_2Cl_2.[85] A mixture of Br_2 and HgO is a more active brominating agent than bromine alone.[86] The actual brominating agent in this case is believed to be bromine monoxide Br_2O. Among other agents used have been N-bromosuccinimide (see **4-2**), CCl_4,[87] dichlorine monoxide Cl_2O,[88] $BrCCl_3$,[89] PCl_5,[90] phosgene, t-butyl hypobromite[91] and hypochlorite,[92] and N-haloamines and sulfuric acid.[93] In all these cases a chain-initiating catalyst is required, usually peroxides or uv light.

When chlorination is carried out with N-haloamines and sulfuric acid (catalyzed by either uv light or metal ions), selectivity is much greater than with other reagents.[93] In particular, alkyl chains are chlorinated with high regioselectivity at the position next to the end of the chain (the $\omega - 1$ position).[94] Some typical selectivity values are[95]

$$CH_3\!-\!CH_2\!-\!CH_2\!-\!CH_2\!-\!CH_2\!-\!CH_2\!-\!CH_3 \qquad\qquad \text{Ref. 96}$$
1   56   29   14

$$CH_3\!-\!CH_2\!-\!CH_2\!-\!CH_2\!-\!CH_2\!-\!CH_2\!-\!CH_2\!-\!OH \qquad\qquad \text{Ref. 97}$$
1   92   3   1   1   2   0   0

$$CH_3\!-\!CH_2\!-\!CH_2\!-\!CH_2\!-\!CH_2\!-\!CH_2\!-\!COOMe \qquad\qquad \text{Ref. 98}$$
3   72   20   4   1   0

Furthermore, di- and polychlorination are much less prevalent. Dicarboxylic acids are predominantly chlorinated in the middle of the chain,[99] and adamantane and bicyclo[2.2.2]octane at the bridgeheads[100] by this procedure. The reasons for the high $\omega - 1$ specificity are not clearly understood.[101] Alkyl bromides can be regioselectively chlorinated

[83]See, for example, Rozen; Gal *J. Org. Chem.* **1987**, *52*, 2769.
[84]Gover; Willard *J. Am. Chem. Soc.* **1960**, *82*, 3816.
[85]For a review of this reagent, see Tabushi; Kitaguchi, in Pizey *Synthetic Reagents*, vol. 4; Wiley: New York, 1981, pp. 336-396.
[86]Bunce *Can. J. Chem.* **1972**, *50*, 3109.
[87]For a discussion of the mechanism with this reagent, see Hawari; Davis; Engel; Gilbert; Griller *J. Am. Chem. Soc.* **1985**, *107*, 4721.
[88]Marsh; Farnham; Sam; Smart *J. Am. Chem. Soc.* **1982**, *104*, 4680.
[89]Huyser *J. Am. Chem. Soc.* **1960**, *82*, 391; Baldwin; O'Neill *Synth. Commun.* **1976**, *6*, 109.
[90]Wyman; Wang; Freeman *J. Org. Chem.* **1963**, *28*, 3173.
[91]Walling; Padwa *J. Org. Chem.* **1962**, *27*, 2976.
[92]Walling; Mintz *J. Am. Chem. Soc.* **1967**, *89*, 1515.
[93]For reviews, see Minisci *Synthesis* **1973**, 1-24; Deno *Methods Free-Radical Chem.* **1972**, *3*, 135-154; Sosnovsky; Rawlinson *Adv. Free-Radical Chem.* **1972**, *4*, 203-284.
[94]The $\omega - 1$ regioselectivity diminishes when the chains are longer than 10 carbons; see Deno; Jedziniak *Tetrahedron Lett.* **1976**, 1259; Konen; Maxwell; Silbert *J. Org. Chem.* **1979**, *44*, 3594.
[95]The $\omega - 1$ selectivity values shown here may actually be lower than the true values because of selective solvolysis of the $\omega - 1$ chlorides in concentrated H_2SO_4: see Deno; Pohl *J. Org. Chem.* **1975**, *40*, 380.
[96]Bernardi; Galli; Minisci *J. Chem. Soc. B* **1968**, 324. See also Deno; Gladfelter; Pohl *J. Org. Chem.* **1979**, *44*, 3728; Fuller; Lindsay Smith; Norman; Higgins *J. Chem. Soc., Perkin Trans. 2* **1981**, 545.
[97]Deno; Billups; Fishbein; Pierson; Whalen; Wyckoff *J. Am. Chem. Soc.* **1971**, *93*, 438.
[98]Minisci; Galli; Galli; Bernardi *Tetrahedron Lett.* **1967**, 2207; Minisci; Gardini; Bertini *Can. J. Chem.* **1970**, *48*, 544.
[99]Kämper; Schäfer; Luftmann *Angew. Chem. Int. Ed. Engl.* **1976**, *15*, 306 [*Angew. Chem. 88*, 334].
[100]Smith; Billups *J. Am. Chem. Soc.* **1974**, *96*, 4307.
[101]It has been reported that the selectivity in one case is in accord with a pure electrostatic (field effect) explanation: Dneprovskii; Mil'tsov; Arbuzov *J. Org. Chem. USSR* **1988**, *24*, 1826. See also Tanner; Arhart; Meintzer *Tetrahedron* **1985**, *41*, 4261; Ref. 95.

one carbon away from the bromine (to give *vic*-bromochlorides) by treatment with PCl_5.[102] Alkyl chlorides can be converted to *vic*-dichlorides by treatment with $MoCl_5$.[103] Enhanced selectivity at a terminal position of *n*-alkanes has been achieved by absorbing the substrate onto a pentasil zeolite.[104] In another regioselective chlorination, alkanesulfonamides $RCH_2CH_2CH_2SO_2NHR'$ are converted primarily to $RCHClCH_2CH_2SO_2NHR'$ by sodium peroxydisulfate $Na_2S_2O_8$ and $CuCl_2$.[105] For regioselective chlorination at certain positions of the steroid nucleus, see **9-2**.

In almost all cases, the mechanism involves a free-radical chain:

Initiation	$X_2 \xrightarrow{h\nu} 2X\cdot$
Propagation	$RH + X\cdot \longrightarrow R\cdot + XH$
	$R\cdot + X_2 \longrightarrow RX + X\cdot$
Termination	$R\cdot + X\cdot \longrightarrow RX$

When the reagent is halogen, initiation occurs as shown above.[106] When it is another reagent, a similar cleavage occurs (catalyzed by light or, more commonly, peroxides), followed by propagation steps that do not necessarily involve abstraction by halogen. For example, the propagation steps for chlorination by *t*-BuOCl have been formulated as[107]

$$RH + t\text{-}BuO\cdot \longrightarrow R\cdot + t\text{-}BuOH$$
$$R\cdot + t\text{-}BuOCl \longrightarrow RCl + t\text{-}BuO\cdot$$

and the abstracting radicals in the case of N-haloamines are the aminium radical cations $R_2NH\cdot^+$ (p. 527), with the following mechanism (in the case of initiation by Fe^{2+}):[93]

Initiation	$R_2NCl \xrightarrow{H^+} R_2\overset{\oplus}{N}HCl \xrightarrow{Fe^{2+}} R_2NH\cdot^+ + \overset{\oplus}{F}eCl$
Propagation	$R_2NH\cdot^+ + RH \longrightarrow R_2\overset{\oplus}{N}H_2 + R\cdot$
	$R\cdot + R_2\overset{\oplus}{N}HCl \longrightarrow RCl + R_2NH\cdot^+$

This mechanism is similar to that of the Hofmann–Löffler reaction (**8-42**).

The two propagation steps shown above for X_2 are those that lead directly to the principal products (RX and HX), but many other propagation steps are possible and many occur. Similarly, the only termination step shown is the one that leads to RX, but any two radicals may combine. Thus, products like H_2, higher alkanes, and higher alkyl halides can be

[102]Luche; Bertin; Kagan *Tetrahedron Lett.* **1974**, 759.

[103]San Filippo; Sowinski; Romano *J. Org. Chem.* **1975**, *40*, 3463.

[104]Turro; Fehlner; Hessler; Welsh; Ruderman; Firnberg; Braun *J. Org. Chem.* **1988**, *53*, 3731.

[105]Nikishin; Troyansky; Lazareva *Tetrahedron Lett.* **1985**, *26*, 3743.

[106]There is evidence (unusually high amounts of multiply chlorinated products) that under certain conditions in the reaction of RH with Cl_2, the products of the second propagation step (RX + X·) are enclosed within a solvent cage. See Skell; Baxter *J. Am. Chem. Soc.* **1985**, *107*, 2823; Raner; Lusztyk; Ingold *J. Am. Chem. Soc.* **1988**, *110*, 3519; Tanko; Anderson *J. Am. Chem. Soc.* **1988**, *110*, 3525.

[107]Carlsson; Ingold *J. Am. Chem. Soc.* **1967**, *89*, 4885, 4891; Walling; Kurkov *J. Am. Chem. Soc.* **1967**, *89*, 4895; Walling; McGuiness *J. Am. Chem. Soc.* **1969**, *91*, 2053. See also Zhulin; Rubinshtein *Bull. Acad. Sci. USSR, Div. Chem. Sci.* **1977**, *26*, 2082.

accounted for by steps like these (these are for chlorination of methane, but analogous steps can be written for other substrates):

$$Cl\cdot + HCl \longrightarrow Cl\!\!-\!\!Cl + H\cdot$$

$$H\cdot + H\cdot \longrightarrow H\!\!-\!\!H$$

$$CH_3\cdot + CH_3\cdot \longrightarrow CH_3CH_3$$

$$CH_3CH_3 + Cl\cdot \longrightarrow CH_3CH_2\cdot + HCl$$

$$CH_3CH_2\cdot + Cl_2 \longrightarrow CH_3CH_2Cl + Cl\cdot$$

$$CH_3CH_2\cdot + CH_3CH_2\cdot \longrightarrow CH_3CH_3 + CH_2\!\!=\!\!CH_2$$

$$CH_3CH_2\cdot + CH_3CH_2\cdot \longrightarrow CH_3CH_2CH_2CH_3$$

$$CH_3CH_2\cdot + HCl \longrightarrow CH_3CH_3 + Cl\cdot$$

etc.

At least when methane is the substrate, the rate-determining step is

$$CH_4 + Cl\cdot \longrightarrow CH_3\cdot + HCl$$

since an isotope effect of 12.1 was observed at 0°C.[108] For chlorinations, chains are very long, typically 10^4 to 10^6 propagations before a termination step takes place.

The order of reactivity of the halogens can be explained by energy considerations. For the substrate methane, ΔH values for the two principal propagation steps are

	kcal/mol				kJ/mol			
	F_2	Cl_2	Br_2	I_2	F_2	Cl_2	Br_2	I_2
$CH_4 + X\cdot \rightarrow CH_3\cdot + HX$	-31	$+2$	$+17$	$+34$	-132	$+6$	$+72$	$+140$
$CH_3\cdot + X_2 \rightarrow CH_3X + X\cdot$	-70	-26	-24	-21	-293	-113	-100	-87

In each case D for CH_3—H is 105 kcal/mol (438 kJ/mol), while D values for the other bonds involved are given in Table14.4.[109] F_2 is so reactive[110] that neither uv light nor any other initiation is needed (total $\Delta H = -101$ kcal/mol; -425 kJ/mol);[111] while Br_2 and I_2 essentially do not react with methane. The second step is exothermic in all four cases, but it cannot take place before the first, and it is this step that is very unfavorable for Br_2 and I_2. It is apparent that the most important single factor causing the order of halogen reactivity

---

[108]Wiberg; Motell *Tetrahedron* **1963**, *19*, 2009.

[109]Kerr, in Weast *Handbook of Chemistry and Physics*, 69th ed.; CRC Press: Boca Raton, FL, 1988, pp. F174-F189.

[110]It has been reported that the reaction of F atoms with CH_4 at 25 K takes place with practically zero activation energy: Johnson; Andrews *J. Am. Chem. Soc.* **1980**, *102*, 5736.

[111]For F_2 the following initiation step is possible: $F_2 + RH \rightarrow R\cdot + F\cdot + HF$ [first demonstrated by Miller; Koch; McLafferty *J. Am. Chem. Soc.* **1956**, *78*, 4992]. ΔH for this reaction is equal to the small positive value of 5 kcal/mol (21 kJ/mol). The possibility of this reaction (which does not require an initiator) explains why fluorination can take place without uv light (which would otherwise be needed to furnish the 38 kcal/mol (159 kJ/mol) necessary to break the F—F bond). Once the reaction has been initiated, the large amount of energy given off by the propagation steps is ample to cleave additional F_2 molecules. Indeed, it is the magnitude of this energy that is responsible for the cleavage of carbon chains by F_2.

TABLE 14.4 Some D values[109]

Bond	D	
	kcal/mol	kJ/mol
H–F	136	570
H–Cl	103	432
H–Br	88	366
H–I	71	298
F–F	38	159
Cl–Cl	59	243
Br–Br	46	193
I–I	36	151
CH₃–F	108	452
CH₃–Cl	85	356
CH₃–Br	70	293
CH₃–I	57	238

to be $F_2 > Cl_2 > Br_2 > I_2$ is the decreasing strength of the HX bond in the order HF > HCl > HBr > HI. The increased reactivity of secondary and tertiary positions is in accord with the decrease in D values for R—H in the order primary > secondary > tertiary (Table 5.3). (Note that for chlorination step 1 is exothermic for practically all substrates other than CH_4, since most other aliphatic C—H bonds are weaker than those in CH_4.)

Bromination and chlorination of alkanes and cycloalkanes can also take place by an electrophilic mechanism if the reaction is catalyzed by $AgSbF_6$.[112] Direct chlorination at a vinylic position by an electrophilic mechanism has been achieved with benzeneseleninyl chloride PhSe(O)Cl and $AlCl_3$ or $AlBr_3$.[113] However, while some substituted alkenes give high yields of chloro substitution products, others (such as styrene) undergo addition of Cl_2 to the double bond (**5-26**).[113] Electrophilic fluorination has already been mentioned (p. 690).

OS **II**, 89, 133, 443, 549; **III**, 737, 788; **IV**, 807, 921, 984; **V**, 145, 221, 328, 504, 635, 825; **VI**, 271, 404, 715; **VII**, 491; **65**, 68.

4-2   Allylic Halogenation
Halogenation or Halo-de-hydrogenation

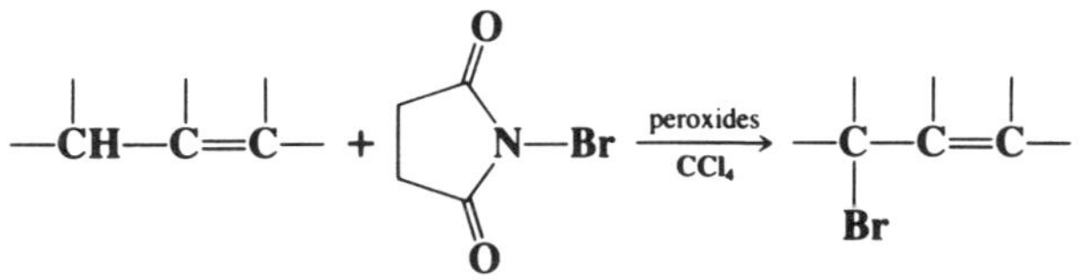

This reaction is a special case of **4-1**, but is important enough to be treated separately.[114] Olefins can be halogenated in the allylic position by a number of reagents, of which N-bromosuccinimide (NBS)[115] is by far the most common. When this reagent is used, the

---

[112]Olah; Renner; Schilling; Mo *J. Am. Chem. Soc.* **1973**, *95*, 7686. See also Olah; Wu; Farooq *J. Org. Chem.* **1989**, *54*, 1463.
[113]Kamigata; Satoh; Yoshida *Bull. Chem. Soc. Jpn.* **1988**, *44*, 449.
[114]For a review, see Nechvatal *Adv. Free-Radical Chem.* **1972**, *4*, 175-201.
[115]For a review of this reagent, see Pizey, Ref. 85, vol. 2, pp. 1-63, 1974.

reaction is known as *Wohl–Ziegler bromination*. A nonpolar solvent is used, most often CCl_4. Other N-bromo amides have also been used. To a much lesser extent, allylic chlorination has been carried out, with N-chlorosuccinimide, N-chloro-N-cyclohexylbenzenesulfonamide,[116] or t-butyl hypochlorite.[117] With any reagent an initiator is needed; this is usually a peroxide or, less often, uv light.

The reaction is usually quite specific at the allylic position and good yields are obtained. However, when the allylic radical intermediate is unsymmetrical, allylic rearrangements can take place, so that mixtures of both possible products are obtained, e.g.,

$$CH_3—CH_2—CH{=}CH_2 + NBS \longrightarrow CH_3—\underset{\underset{Br}{|}}{CH}—CH{=}CH_2 + CH_3—CH{=}CH—\underset{\underset{Br}{|}}{CH_2}$$

When a double bond has two different allylic positions, e.g., $CH_3CH{=}CHCH_2CH_3$, a secondary position is substituted more readily than a primary. The relative reactivity of tertiary hydrogen is not clear, though many substitutions at allylic tertiary positions have been performed.[118] It is possible to brominate both sides of the double bond.[119] Because of the electron-withdrawing nature of bromine, the second bromine substitutes on the other side of the double bond rather than α to the first bromine.

NBS is also a highly regioselective brominating agent at other positions, including positions α to a carbonyl group, to a $C{\equiv}C$ triple bond, and to an aromatic ring (benzylic position). When both a double and a triple bond are in the same molecule, the preferred position is α to the triple bond.[120]

That the mechanism of allylic bromination is of the free-radical type was demonstrated by Dauben and McCoy,[121] who showed that the reaction is very sensitive to free-radical initiators and inhibitors and indeed does not proceed at all unless at least a trace of initiator is present. Subsequent work indicated that the species that actually abstracts hydrogen from the substrate is the bromine atom. The reaction is initiated by small amounts of Br•. Once it is formed, the main propagation steps are

Step 1 $Br• + RH \longrightarrow R• + HBr$

Step 2 $R• + Br_2 \longrightarrow RBr + Br•$

The source of the Br_2 is a fast ionic reaction between NBS and the HBr liberated in step 1:

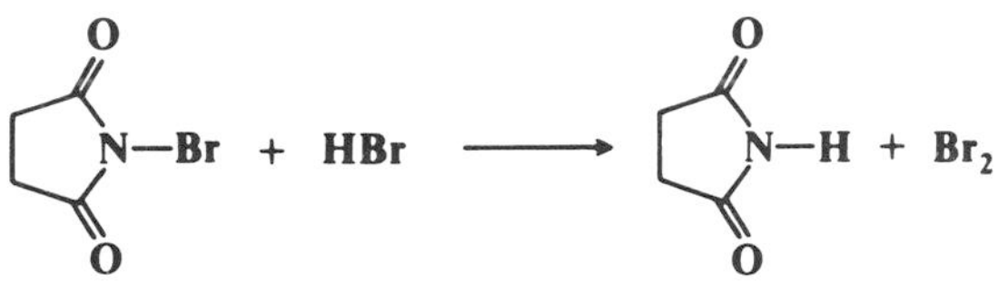

The function of the NBS is therefore to provide a source of Br_2 in a low, steady-state concentration and to use up the HBr liberated in step 1.[122] The main evidence for this

[116]Theilacker; Wessel *Liebigs Ann. Chem.* **1967**, *703*, 34.
[117]Walling; Thaler *J. Am. Chem. Soc.* **1961**, *83*, 3877.
[118]Dauben; McCoy *J. Org. Chem.* **1959**, *24*, 1577.
[119]Ucciani; Naudet *Bull. Soc. Chim. Fr.* **1962**,871.
[120]Peiffer *Bull. Soc. Chim. Fr.* **1963**, 537.
[121]Dauben; McCoy *J. Am. Chem. Soc.* **1959**, *81*, 4863.
[122]This mechanism was originally suggested by Adam; Gosselain; Goldfinger *Nature* **1953**, *171*, 704, *Bull. Soc. Chim. Belg.* **1956**, *65*, 533.

mechanism is that NBS and Br_2 show similar selectivity[123] and that the various N-bromo amides also show similar selectivity,[124] which is consistent with the hypothesis that the same species is abstracting in each case.[125]

It may be asked why, if Br_2 is the reacting species, it does not add to the double bond, either by an ionic or by a free-radical mechanism (see **5-26**). Apparently the concentration is too low. In bromination of a double bond, only one atom of an attacking bromine molecule becomes attached to the substrate, whether the addition is electrophilic or free-radical:

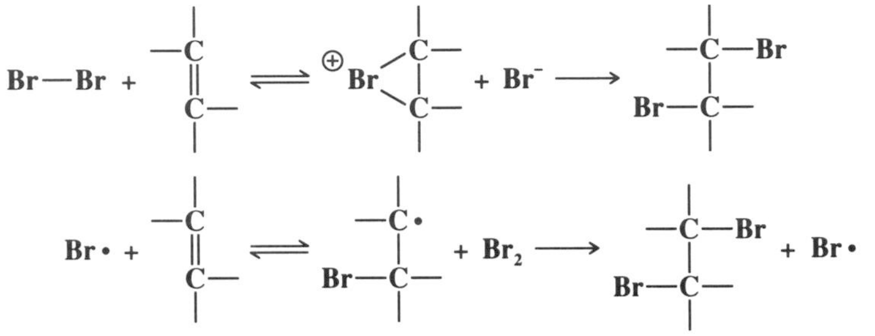

The other bromine atom comes from another bromine-containing molecule or ion. If the concentration is sufficiently low, there is a low probability that the proper species will be in the vicinity once the intermediate forms. The intermediate in either case reverts to the initial species and the allylic substitution competes successfully. If this is true, it should be possible to brominate an olefin in the allylic position without competition from addition, even in the absence of NBS or a similar compound, if a very low concentration of bromine is used and if the HBr is removed as it is formed so that it is not available to complete the addition step. This has indeed been demonstrated.[126]

When NBS is used to brominate non-olefinic substrates such as alkanes, another mechanism, involving abstraction of the hydrogen of the substrate by the succinimidyl radical[127] **8** can operate.[128] This mechanism is facilitated by solvents (such as CH_2Cl_2, $CHCl_3$, or

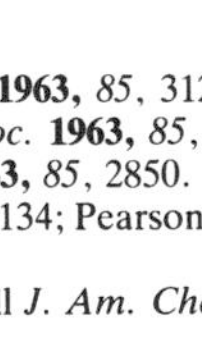

8

[123]Walling; Rieger; Tanner *J. Am. Chem. Soc.* **1963**, *85*, 3129; Russell; Desmond *J. Am. Chem. Soc.* **1963**, *85*, 3139; Russell; DeBoer; Desmond *J. Am. Chem. Soc.* **1963**, *85*, 365; Pearson; Martin *J. Am. Chem. Soc.* **1963**, *85*, 3142; Skell; Tuleen; Readio *J. Am. Chem. Soc.* **1963**, *85*, 2850.

[124]Walling; Rieger *J. Am. Chem. Soc.* **1963**, *85*, 3134; Pearson; Martin, Ref. 123; Incremona; Martin *J. Am. Chem. Soc.* **1970**, *92*, 627.

[125]For other evidence, see Day; Lindstrom; Skell *J. Am. Chem. Soc.* **1974**, *96*, 5616.

[126]McGrath; Tedder *Proc. Chem. Soc.* **1961**, 80.

[127]For a review of this radical, see Chow; Naguib *Rev. Chem. Intermed.* **1984**, *5*, 325-345.

[128]Skell; Day *Acc. Chem. Res.* **1978**, *11*, 381; Walling; El-Taliawi; Zhao *J. Am. Chem. Soc.* **1983**, *105*, 5119; Tanner; Reed; Tan; Meintzer; Walling; Sopchik *J. Am. Chem. Soc.* **1985**, *107*, 6576; Lüning; Skell *Tetrahedron* **1985**, *41*, 4289; Skell; Lüning; McBain; Tanko *J. Am. Chem. Soc.* **1986**, *108*, 121; Lüning; Seshadri; Skell *J. Org. Chem.* **1986**, *51*, 2071; Chow; Zhao *J. Org. Chem.* **1987**, *52*, 1931, **1989**, *54*, 530; Zhang; Dong; Jiang; Chow *Can. J. Chem.* **1990**, *68*, 1668.

MeCN) in which NBS is more soluble, and by the presence of small amounts of an alkene that lacks an allylic hydrogen (e.g., ethene). The alkene serves to scavenge any Br• that forms from the reagent. Among the evidence for the mechanism involving **8** are abstraction selectivities similar to those of Cl• atoms and the isolation of β-bromopropionyl isocyanate BrCH₂CH₂CONCO, which is formed by ring-opening of **8.**

Allylic chlorination has also been carried out[129] with N-chlorosuccinimide and either arylselenyl chlorides ArSeCl, aryl diselenides ArSeSeAr, or TsNSO as catalysts. Use of the selenium catalysts produces almost entirely the allylically rearranged chlorides in high yields. With TsNSO the products are the unrearranged chlorides in lower yields. Dichlorine monoxide Cl_2O, with no catalyst, also gives allylically rearranged chlorides in high yields.[130] A free-radical mechanism is unlikely in these reactions.

OS **IV**, 108; **V**, 825; **VI**, 462.

4-3 Halogenation of Aldehydes
Halogenation or **Halo-de-hydrogenation**

$$RCHO + Cl_2 \longrightarrow RCOCl$$

Aldehydes can be directly converted to acyl chlorides by treatment with chlorine; however, the reaction operates only when the aldehyde does not contain an α hydrogen and even then it is not very useful. When there is an α hydrogen, α halogenation (**2-4**) occurs instead. Other sources of chlorine have also been used, among them SO_2Cl_2[131] and t-BOCl.[132] The mechanisms are probably of the free-radical type. NBS, with AIBN (p. 664) as a catalyst, has been used to convert aldehydes to acyl bromides.[133]

OS **I**, 155.

B. Substitution by Oxygen

4-4 Hydroxylation at an Aliphatic Carbon
Hydroxylation or **Hydroxy-de-hydrogenation**

$$R_3CH \xrightarrow[\text{silica gel}]{O_3} R_3COH$$

Compounds containing susceptible C—H bonds can be oxidized to alcohols.[134] Nearly always, the C—H bond involved is tertiary, so the product is a tertiary alcohol. This is partly because tertiary C—H bonds are more susceptible to free-radical attack than primary and secondary bonds and partly because the reagents involved would oxidize primary and secondary alcohols further. In the best method the reagent is ozone and the substrate is absorbed on silica gel.[135] Yields as high as 99% have been obtained by this method. Other reagents, which often give much lower yields, are chromic acid,[136] alkaline permanganate,[137] potassium

[129]Hori; Sharpless *J. Org. Chem.* **1979**, *44*, 4204.
[130]Torii; Tanaka; Tada; Nagao; Sasaoka *Chem. Lett.* **1984**, 877.
[131]Arai *Bull. Chem. Soc. Jpn.* **1964**, *37*, 1280, **1965**, *38*, 252.
[132]Walling; Mintz, Ref. 92.
[133]Markó; Mekhalfia *Tetrahedron Lett.* **1990**, *31*, 7237. For a related procedure, see Cheung *Tetrahedron Lett.* **1979**, 3809.
[134]For reviews, see Chinn *Selection of Oxidants in Synthesis*; Marcel Dekker: New York, 1971, pp. 7-11; Lee, in Augustine *Oxidation*, vol. 1; Marcel Dekker: New York, 1969, pp. 2-6. For a monograph on all types of alkane activation, see Hill *Activation and Functionalization of Alkanes*; Wiley: New York, 1989.
[135]Cohen; Keinan; Mazur; Varkony *J. Org. Chem.* **1975**, *40*, 2141, *Org. Synth. VI*, 43; Keinan; Mazur *Synthesis* **1976**, 523; McKillop; Young *Synthesis* **1979**, 401-422, pp. 418-419.
[136]For a review, see Cainelli; Cardillo *Chromium Oxidations in Organic Chemistry*; Springer: New York, 1984, pp. 8-23.
[137]Eastman; Quinn *J. Am. Chem. Soc.* **1969**, *82*, 4249.

hydrogen persulfate $KHSO_5$,[138] methyl(trifluoromethyl)dioxirane,[139] ruthenium tetroxide RuO_4,[140] F_2 in $MeCN–H_2O$,[141] sodium chlorite $NaClO_2$ with a metalloporphyrin catalyst,[142] and certain perbenzoic acids.[143] Alkanes and cycloalkanes have been oxidized at secondary positions, to a mixture of alcohols and trifluoroacetates, by 30% aqueous H_2O_2 in trifluoroacetic acid.[144] This reagent does not oxidize the alcohols further and ketones are not found. As in the case of chlorination with N-haloamines and sulfuric acid (see **4-1**), the $\omega - 1$ position is the most favored. Another reagent[145] that oxidizes secondary positions is iodosylbenzene, catalyzed by Fe(III)–porphyrin catalysts.[146] Use of an optically active Fe(III)–porphyrin gave enantioselective hydroxylation, with moderate enantiomeric excesses.[147]

When chromic acid is the reagent, the mechanism is probably as follows: a Cr^{6+} species abstracts a hydrogen to give $R_3C•$, which is held in a solvent cage near the resulting Cr^{5+} species. The two species then combine to give R_3COCr^{4+}, which is hydrolyzed to the alcohol. This mechanism predicts retention of configuration; this is largely observed.[148] The oxidation by permanganate also involves predominant retention of configuration, and a similar mechanism has been proposed.[149]

Treatment of double-bond compounds with selenium dioxide introduces an OH group into the allylic position (see also **9-16**).[150] Allylic rearrangements are common. There is evidence that the mechanism does not involve free radicals but includes two pericyclic steps (**A** and **B**):[151]

The step marked **A** is similar to the ene synthesis (**5-16**). The step marked **B** is a [2,3] sigmatropic rearrangement (see **8-37**). The reaction can also be accomplished with

[138]De Poorter; Ricci; Meunier *Tetrahedron Lett.* **1985**, *26*, 4459.
[139]Mello; Fiorentino; Fusco; Curci *J. Am. Chem. Soc.* **1989**, *111*, 6749. For a review of dioxiranes as oxidizing agents, see Adam; Curci; Edwards *Acc. Chem. Res.* **1989**, *22*, 205-211. See also Murray; Jeyaraman; Mohan *J. Am. Chem. Soc.* **1986**, *108*, 2470.
[140]Bakke; Lundquist *Acta Chem. Scand., Ser. B* **1986**, *40*, 430; Tenaglia; Terranova; Waegell *Tetrahedron Lett.* **1989**, *30*, 5271; Bakke; Braenden *Acta Chem. Scand.* **1991**, *45*, 418.
[141]Rozen; Brand; Kol *J. Am. Chem. Soc.* **1989**, *111*, 8325.
[142]Collman; Tanaka; Hembre; Brauman *J. Am. Chem. Soc.* **1990**, *112*, 3689.
[143]Schneider; Müller *Angew. Chem. Int. Ed. Engl.* **1982**, *21*, 146 [*Angew. Chem. 94*, 153], *J. Org. Chem.* **1985**, *50*, 4609; Takaishi; Fujikura; Inamoto *Synthesis* **1983**, 293; Tori; Sono; Asakawa *Bull. Chem. Soc. Jpn.* **1985**, *58*, 2669. See also Querci; Ricci *Tetrahedron Lett.* **1990**, *31*, 1779.
[144]Deno; Jedziniak; Messer; Meyer; Stroud; Tomezsko *Tetrahedron* **1977**, *33*, 2503.
[145]For other procedures, see Sharma; Sonawane; Dev *Tetrahedron* **1985**, *41*, 2483; Nam; Valentine *New J.Chem.* **1989**, *13*, 677.
[146]See Groves; Nemo *J. Am. Chem. Soc.* **1983**, *105*, 6243.
[147]Groves; Viski *J. Org. Chem.* **1990**, *55*, 3628.
[148]Wiberg; Foster *J. Am. Chem. Soc.* **1961**, *83*, 423, *Chem. Ind. (London)* **1961**, 108; Wiberg; Eisenthal *Tetrahedron* **1964**, *20*, 1151.
[149]Wiberg; Fox *J. Am. Chem. Soc.* **1963**, *85*, 3487; Brauman; Pandell *J. Am. Chem. Soc.* **1970**, *92*, 329; Stewart; Spitzer *Can. J. Chem.* **1978**, *56*, 1273.
[150]For reviews, see Rabjohn, *Org. React.* **1976**, *24*, 261-415; Jerussi *Sel. Org. Transform.* **1970**, *1*, 301-326; Trachtenberg, in Augustine, Ref. 134, pp. 123-153.
[151]Sharpless; Lauer *J. Am. Chem. Soc.* **1972**, *94*, 7154; Arigoni; Vasella; Sharpless; Jensen *J. Am. Chem. Soc.* **1973**, *95*, 7917; Woggon; Ruther; Egli *J. Chem. Soc., Chem. Commun.* **1980**, 706. For other mechanistic proposals, see Schaefer; Horvath; Klein *J. Org. Chem.* **1968**, *33*, 2647; Trachtenberg; Nelson; Carver *J. Org. Chem.* **1970**, *35*, 1653; Bhalerao; Rapoport *J. Am. Chem. Soc.* **1971**, *93*, 4835; Stephenson; Speth *J. Org. Chem.* **1979**, *44*, 4683.

t-butyl hydroperoxide, if SeO$_2$ is present in catalytic amounts (the *Sharpless method*).[152] The SeO$_2$ is the actual reagent; the peroxide reoxidizes the Se(OH)$_2$.[153] This method makes work-up easier, but gives significant amounts of side products when the double bond is in a ring.[154] Alkynes generally give α,α' dihydroxylation.[155]

Ketones and carboxylic esters can be α hydroxylated by treatment of their enolate forms (prepared by adding the ketone or ester to lithium diisopropylamide) with a molybdenum peroxide reagent (MoO$_5$–pyridine–HMPA) in THF–hexane at −70°C.[156] The enolate forms of amides and esters[157] and the enamine derivatives of ketones[158] can similarly be converted to their α hydroxy derivatives by reaction with molecular oxygen. The MoO$_5$ method can also be applied to certain nitriles.[156] Ketones have also been α hydroxylated by treating the corresponding silyl enol ethers with *m*-chloroperbenzoic acid,[159] or with certain other oxidizing agents.[160] When the silyl enol ethers are treated with iodosobenzene in the presence of trimethylsilyl trifluoromethyl sulfonate, the product is the α-keto triflate.[161]

Ketones can be α hydroxylated in good yields, without conversion to the enolates, by treatment with the hypervalent iodine reagents[162] *o*-iodosobenzoic acid[163] or phenyliodoso acetate PhI(OAc)$_2$ in methanolic NaOH.[164] The latter reagent has also been used on carboxylic esters.[165] O$_2$ and a chiral phase transfer catalyst gave enantioselective α hydroxylation of ketones, if the α position was tertiary.[166]

A different method for the conversion of ketones to α-hydroxy ketones consists of treating the enolate with a 2-sulfonyloxaziridine (such as **9**).[167] This is not a free-radical process; the following mechanism is likely:

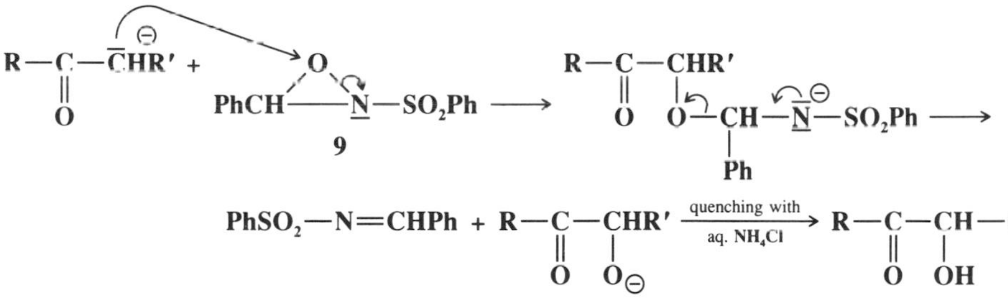

[152]Umbreit; Sharpless *J. Am. Chem. Soc.* **1977**, *99*, 5526. See also Uemura; Fukuzawa; Toshimitsu; Okano *Tetrahedron Lett.* **1982**, *23*, 87; Singh; Sabharwal; Sayal; Chhabra *Chem. Ind. (London)* **1989**, 533.

[153]For the use of the peroxide with O$_2$ instead of SeO$_2$, see Sabol; Wiglesworth; Watt *Synth. Commun.* **1988**, *18*, 1.

[154]Warpehoski; Chabaud; Sharpless *J. Org. Chem.* **1982**, *47*, 2897.

[155]Chabaud; Sharpless *J. Org. Chem.* **1979**, *44*, 4202.

[156]Vedejs *J. Am. Chem. Soc.* **1974**, *96*, 5944; Vedejs; Telschow *J. Org. Chem.* **1976**, *41*, 740; Vedejs; Larsen *Org. Synth. VII*, 277; Gamboni; Tamm *Tetrahedron Lett.* **1986**, *27*, 3999; *Helv. Chim. Acta* **1986**, *69*, 615. See also Anderson; Smith *Synlett* **1990**, 107.

[157]Wasserman; Lipshutz *Tetrahedron Lett.* **1975**, 1731. For another method, see Pohmakotr; Winotai *Synth. Commun.* **1988**, *18*, 2141.

[158]Cuvigny; Valette; Larcheveque; Normant *J. Organomet. Chem.* **1978**, *155*, 147.

[159]Rubottom; Vazquez; Pelegrina *Tetrahedron Lett.* **1974**, 4319; Rubottom; Gruber *J. Org. Chem.* **1978**, *43*, 1599; Hassner; Reuss; Pinnick *J. Org. Chem.* **1975**, *40*, 3427; Andriamialisoa; Langlois; Langlois *Tetrahedron Lett.* **1985**, *26*, 3563; Rubottom; Gruber; Juve; Charleson *Org. Synth. VII*, 282. See also Horiguchi; Nakamura; Kuwajima *Tetrahedron Lett.* **1989**, *30*, 3323.

[160]McCormick; Tomasik; Johnson *Tetrahedron Lett.* **1981**, *22*, 607; Moriarty; Prakash; Duncan *Synthesis* **1985**, 943; Iwata; Takemoto; Nakamura; Imanishi *Tetrahedron Lett.* **1985**, *26*, 3227; Davis; Sheppard *J. Org. Chem.* **1987**, *52*, 954; Takai; Yamada; Rhode; Mukaiyama *Chem. Lett.* **1991**, 281.

[161]Moriarty; Epa; Penmasta; Awasthi *Tetrahedron Lett.* **1989**, *30*, 667.

[162]For a review, see Moriarty; Prakash *Acc. Chem. Res.* **1986**, *19*, 244-250.

[163]Moriarty; Hou *Tetrahedron Lett.* **1984**, *25*, 691; Moriarty; Hou; Prakash; Arora *Org. Synth. VII*, 263.

[164]Moriarty; Hu; Gupta *Tetrahedron Lett.* **1981**, *22*, 1283.

[165]Moriarty; Hu *Tetrahedron Lett.* **1981**, *22*, 2747.

[166]Masui; Ando; Shioiri *Tetrahedron Lett.* **1988**, *29*, 2835.

[167]Davis; Vishwakarma; Billmers; Finn *J. Org. Chem.* **1984**, *49*, 3241.

The method is also successful for carboxylic esters[167] and N,N-disubstituted amides,[168] and can be made enantioselective by the use of a chiral oxaziridine.[169] Dimethyldioxirane also oxidizes ketones (through their enolate forms) to α-hydroxy ketones.[169a]

$$\underset{\substack{\big| \\ Me}}{\overset{\displaystyle O}{O\!\!-\!\!-\!\!-\!\!C\!\!-\!\!Me}}$$

Dimethyldioxirane

Tetrahydrofuran was converted to the hemiacetal 2-hydroxytetrahydrofuran (which was relatively stable under the conditions used) by electrolysis in water[170] (see also **4-7**).

OS **IV**, 23; **VI**, 43, 946; **VII**, 263, 277, 282.

4-5 Hydroxylation at an Aromatic Carbon[171]

Hydroxylation or **Hydroxy-de-hydrogenation**

$$ArH + H_2O_2 + FeSO_4 \longrightarrow ArOH$$

A mixture of hydrogen peroxide and ferrous sulfate,[172] called *Fenton's reagent*,[173] can be used to hydroxylate aromatic rings, though yields are usually not high.[174] Biaryls are usually side products.[175] Among other reagents used have been H_2O_2 and titanous ion; O_2 and Cu(I)[176] or Fe(III),[177] a mixture of ferrous ion, oxygen, ascorbic acid, and ethylenetetra-aminetetraacetic acid (*Udenfriend's reagent*);[178] α-azo hydroperoxides ArN=NCHPhOOH;[179] O_2 and KOH in liquid NH_3;[180] and peracids such as pernitrous and trifluoroperacetic acids.

Much work has been done on the mechanism of the reaction with Fenton's reagent, and it is known that free aryl radicals (formed by a process such as HO• + ArH → Ar• + H_2O) are not intermediates. The mechanism is essentially that outlined on p. 680, with HO• as the attacking species,[181] formed by

$$Fe^{2+} + H_2O_2 \longrightarrow Fe^{3+} + OH^- + HO•$$

[168]Davis; Vishwakarma *Tetrahedron Lett.* **1985**, *26*, 3539.

[169]Evans; Morrissey; Dorow *J. Am. Chem. Soc.* **1985**, *107*, 4346; Davis; Ulatowski; Haque *J. Org. Chem.* **1987**, *52*, 5288; Enders; Bhushan *Tetrahedron Lett.* **1988**, *29*, 2437; Davis; Sheppard; Chen; Haque *J. Am. Chem. Soc.* **1990**, *112*, 6679; Davis; Weismiller *J. Org. Chem.* **1990**, *55*, 3715.

[169a]Guertin; Chan *Tetrahedron Lett.* **1991**, *32*, 715.

[170]Wermeckes; Beck; Schulz *Tetrahedron* **1987**, *43*, 577.

[171]For reviews, see Vysotskaya *Russ. Chem. Rev.* **1973**, *42*, 851-856; Sangster, in Patai *The Chemistry of the Hydroxyl Group*, pt. 1; Wiley: New York, 1971, pp. 133-191; Metelitsa *Russ. Chem. Rev.* **1971**, *40*, 563-580; Enisov; Metelitsa *Russ. Chem. Rev.* **1968**, *37*, 656-665; Loudon *Prog. Org. Chem.* **1961**, *5*, 47-72.

[172]For a review of reactions of H_2O_2 and metal ions with all kinds of organic compounds, including aromatic rings, see Sosnovsky; Rawlinson, in Swern *Organic Peroxides*, vol. 2; Wiley: New York, 1970, pp. 269-336. See also Sheldon; Kochi *Metal-Catalyzed Oxidations of Organic Compounds*; Academic Press: New York, 1981.

[173]For a discussion of Fenton's reagent, see Walling *Acc. Chem. Res.* **1975**, *8*, 125-131.

[174]Yields can be improved with phase transfer catalysis: Karakhanov; Narin; Filippova; Dedov *Doklad. Chem.* **1987**, *292*, 81.

[175]See the discussion of the aromatic free-radical substitution mechanism on pp. 680–681.

[176]See Karlin; Hayes; Gultneh; Cruse; McKown; Hutchinson; Zubieta *J. Am. Chem. Soc.* **1984**, *106*, 2121; Cruse; Kaderli; Meyer; Zuberbühler; Karlin *J. Am. Chem. Soc.* **1988**, *110*, 5020; Ito; Kunai; Okada; Sasaki *J. Org. Chem.* **1988**, *53*, 296.

[177]Funabiki; Tsujimoto; Ozawa; Yoshida *Chem. Lett.* **1989**, 1267.

[178]Udenfriend; Clark; Axelrod; Brodie *J. Biol. Chem.* **1954**, *208*, 731; Brodie; Shore; Udenfriend *J. Biol. Chem.* **1954**, *208*, 741. See also Tamagaki; Suzuki; Tagaki *Bull. Chem. Soc. Jpn.* **1989**, *62*, 148, 153, 159.

[179]Tezuka; Narita; Ando; Oae *J. Am. Chem. Soc.* **1981**, *103*, 3045.

[180]Malykhin; Kolesnichenko; Shteingarts *J. Org. Chem. USSR* **1986**, *22*, 720.

[181]Jefcoate; Lindsay Smith; Norman; *J. Chem. Soc. B* **1969**, 1013; Brook; Castle; Lindsay Smith; Higgins; Morris *J. Chem. Soc., Perkin Trans. 2* **1982**, 687; Lai; Piette *Tetrahedron Lett.* **1979**, 775; Kunai; Hata; Ito; Sasaki *J. Am. Chem. Soc.* **1986**, *108*, 6012.

The rate-determining step is formation of HO• and not its reaction with the aromatic substrate.

See also **1-29.**

4-6 Oxidation of Aldehydes to Carboxylic Acids
Hydroxylation or **Hydroxy-de-hydrogenation**

$$R-\underset{\underset{O}{\|}}{C}-H \xrightarrow{MnO_4^-} R-\underset{\underset{O}{\|}}{C}-OH$$

Oxidation of aldehydes to carboxylic acids is one of the most common oxidation reactions in organic chemistry[182] and has been carried out with many oxidizing agents, the most popular of which is permanganate in acid, basic, or neutral solution.[183] Chromic acid[184] and bromine are other reagents frequently employed. Silver oxide is a fairly specific oxidizing agent for aldehydes and does not readily attack other groups. Benedict's and Fehling's solutions oxidize aldehydes,[185] and a test for aldehydes depends on this reaction, but the method is seldom used for preparative purposes and in any case gives very poor results with aromatic aldehydes. α,β-Unsaturated aldehydes can be oxidized by sodium chlorite without disturbing the double bond.[186] Aldehydes are also oxidized to carboxylic acids by atmospheric oxygen, but the actual direct oxidation product in this case is the peroxy acid RCO_3H,[187] which with another molecule of aldehyde disproportionates to give two molecules of acid (see **4-9**).[188]

Mechanisms of aldehyde oxidation[189] are not firmly established, but there seem to be at least two main types—a free-radical mechanism and an ionic one. In the free-radical process, the aldehydic hydrogen is abstracted to leave an acyl radical, which obtains OH from the oxidizing agent. In the ionic process, the first step is addition of a species OZ^- to the carbonyl bond to give **10** in alkaline solution and **11** in acid or neutral solution. The aldehydic hydrogen of **10** or **11** is then lost as a proton to a base, while Z leaves with its electron pair.

$$R-\underset{\underset{\underline{|}\underline{O}}{\|}}{C}-H + OZ^- \longrightarrow R-\underset{\underset{|\underline{O}|_-}{|}}{\overset{O-Z}{\underset{|}{C}}}H \xrightarrow{B^-} R-\underset{\underset{|\underline{O}|_-}{|}}{\overset{O}{\underset{\|}{C}}} + BH + Z^-$$

10

[182]For reviews, see Haines *Methods for the Oxidation of Organic Compounds*; Academic Press: New York, 1988, pp. 241-263, 423-428; Chinn, Ref. 134, pp. 63-70; Lee, Ref. 134, pp. 81-86.

[183]For lists of some of the oxidizing agents used, with references, see Hudlicky *Oxidations in Organic Chemistry*; American Chemical Society: Washington, 1990, pp. 174-180; Ref. 74, pp. 838-840; Srivastava; Venkataramani *Synth. Commun.* **1988**, *18*, 2193. See also Haines, Ref. 182.

[184]For a review, see Cainelli; Cardillo, Ref. 136, pp. 217-225.

[185]For a review, see Nigh, in Trahanovsky *Oxidation in Organic Chemistry*, pt. B; Academic Press: New York, 1973, pp. 31-34.

[186]Bal; Childers; Pinnick *Tetrahedron* **1981**, *37*, 2091; Dalcanale; Montanari *J. Org. Chem.* **1986**, *51*, 567. See also Bayle; Perez; Courtieu *Bull. Soc. Chim. Fr.* **1990**, 565.

[187]For a review of the preparation of peroxy acids by this and other methods, see Swern, in Swern, Ref. 172, vol. 1, pp. 313-516.

[188]For reviews of the autoxidation of aldehydes, see Vardanyan; Nalbandyan *Russ. Chem. Rev.* **1985**, *54*, 532-543 (gas phase); Sajus; Sérée de Roch, in Bamford; Tipper, Ref. 37, vol. 16, 1980, pp. 89-124 (liquid phase); Maslov; Blyumberg *Russ. Chem. Rev.* **1976**, *45*, 155-167 (liquid phase). For a review of photochemical oxidation of aldehydes by O_2, see Niclause; Lemaire; Letort *Adv. Photochem.* **1966**, *4*, 25-48. For a discussion of the mechanism of catalyzed atmospheric oxidation of aldehydes, see Larkin *J. Org. Chem.* **1990**, *55*, 1563.

[189]For a review, see Roček, in Patai *The Chemistry of the Carbonyl Group*, vol. 1; Wiley: New York, 1966, pp. 461-505.

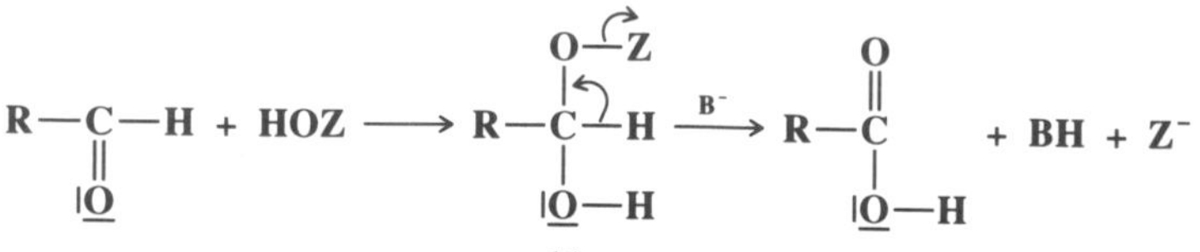

$$R-\underset{\underset{\underline{|}O}{\|}}{C}-H + HOZ \longrightarrow R-\underset{\underset{\underline{|}O-H}{|}}{\overset{\overset{O-Z}{|}}{C}}-H \xrightarrow{B^-} R-\underset{\underset{\underline{|}O-H}{|}}{\overset{\overset{O}{\|}}{C}} + BH + Z^-$$

11

For oxidation with acid dichromate the picture seems to be quite complex, with several processes of both types going on:[190]

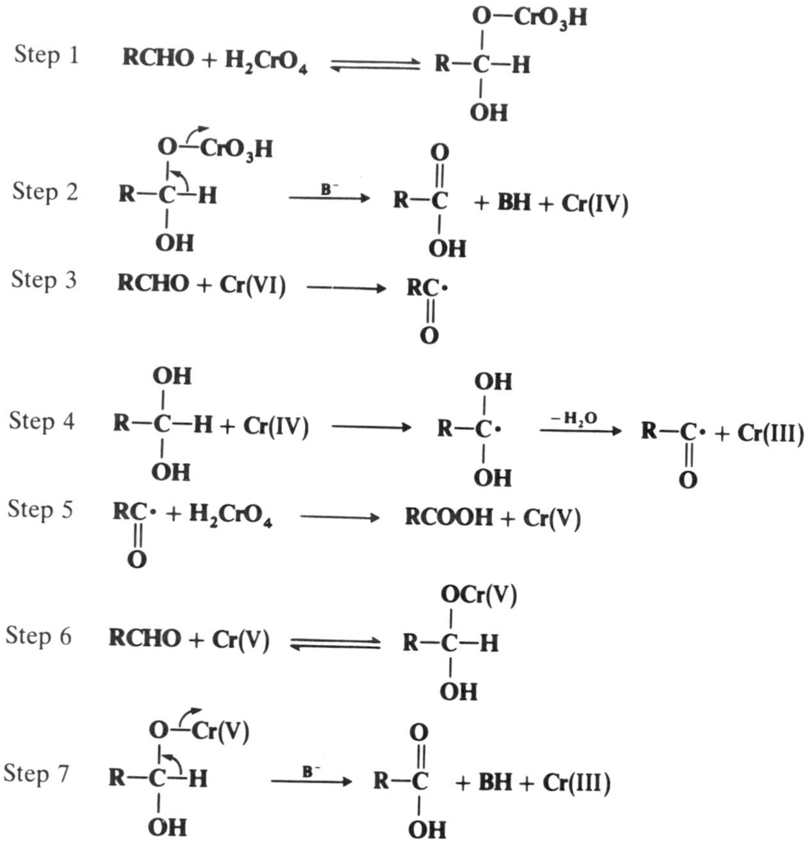

Step 1 $$RCHO + H_2CrO_4 \rightleftharpoons R-\underset{\underset{OH}{|}}{\overset{\overset{O-CrO_3H}{|}}{C}}-H$$

Step 2 $$R-\underset{\underset{OH}{|}}{\overset{\overset{O-CrO_3H}{|}}{C}}-H \xrightarrow{B^-} R-\underset{\underset{OH}{|}}{\overset{\overset{O}{\|}}{C}} + BH + Cr(IV)$$

Step 3 $$RCHO + Cr(VI) \longrightarrow R\underset{\overset{\|}{O}}{C}\cdot$$

Step 4 $$R-\underset{\underset{OH}{|}}{\overset{\overset{OH}{|}}{C}}-H + Cr(IV) \longrightarrow R-\underset{\underset{OH}{|}}{\overset{\overset{OH}{|}}{C}}\cdot \xrightarrow{-H_2O} R-\underset{\overset{\|}{O}}{C}\cdot + Cr(III)$$

Step 5 $$R\underset{\overset{\|}{O}}{C}\cdot + H_2CrO_4 \longrightarrow RCOOH + Cr(V)$$

Step 6 $$RCHO + Cr(V) \rightleftharpoons R-\underset{\underset{OH}{|}}{\overset{\overset{OCr(V)}{|}}{C}}-H$$

Step 7 $$R-\underset{\underset{OH}{|}}{\overset{\overset{O-Cr(V)}{|}}{C}}-H \xrightarrow{B^-} R-\underset{\underset{OH}{|}}{\overset{\overset{O}{\|}}{C}} + BH + Cr(III)$$

Steps 1 and 2 constitute an oxidation by the ionic pathway by Cr(VI), and steps 6 and 7 a similar oxidation by Cr(V), which is produced by an electron-transfer process. Either Cr(VI) (step 3) or Cr(IV) (step 4) [Cr(IV) is produced in step 2] may abstract a hydrogen and the resulting acyl radical is converted to carboxylic acid in step 5. Thus, chromium in three oxidation states is instrumental in oxidizing aldehydes. Still another possible process has

[190]Wiberg; Richardson *J. Am. Chem. Soc.* **1962**, *84*, 2800; Wiberg; Szeimies *J. Am. Chem. Soc.* **1974**, *96*, 1889. See also Roček; Ng *J. Am. Chem. Soc.* **1974**, *96*, 1522, 2840; Sen Gupta; Dey; Sen Gupta *Tetrahedron* **1990**, *46*, 2431.

been proposed in which the chromic acid ester decomposes as follows:[191]

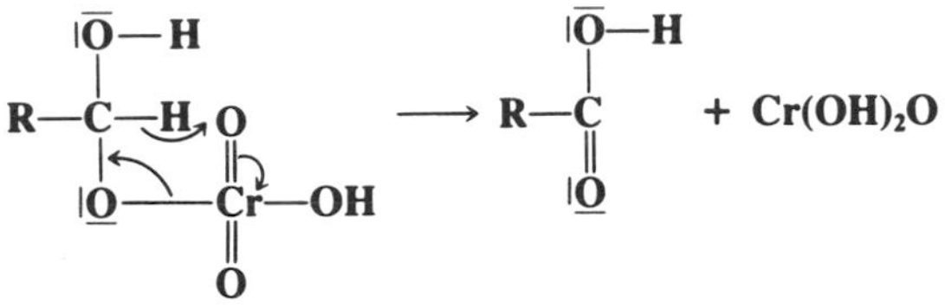

The mechanism with permanganate is less well-known, but an ionic mechanism has been proposed[192] for neutral and acid permanganate, similar to steps 1 and 2 for dichromate:

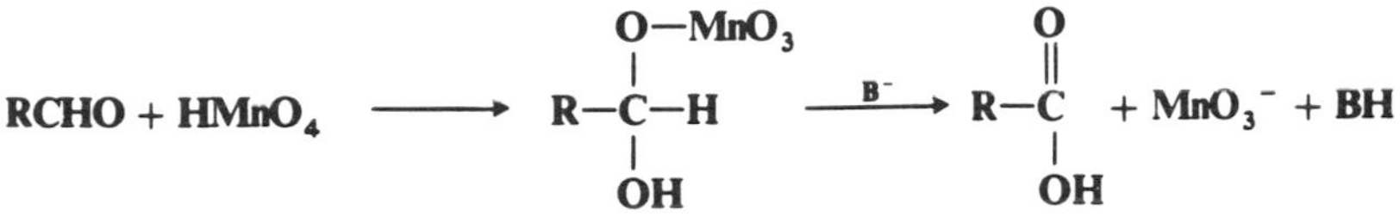

For alkaline permanganate, the following mechanism has been proposed:[193]

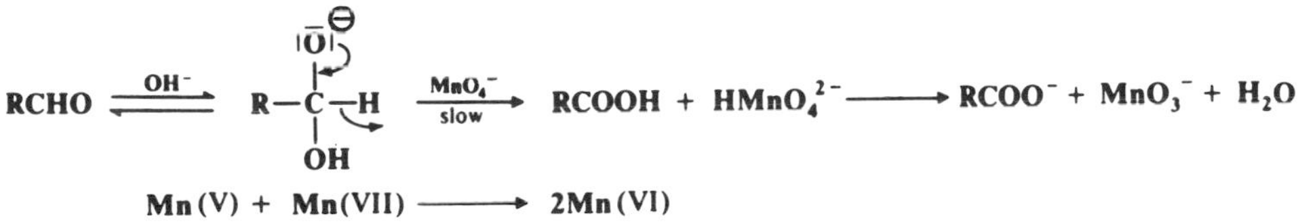

$$Mn(V) + Mn(VII) \longrightarrow 2Mn(VI)$$

OS **I**, 166; **II**, 302, 315, 538; **III**, 745; **IV**, 302, 493, 499, 919, 972, 974.

4-7   Electrochemical Alkoxylation

Alkoxylation or **Alkoxy-de-hydrogenation**

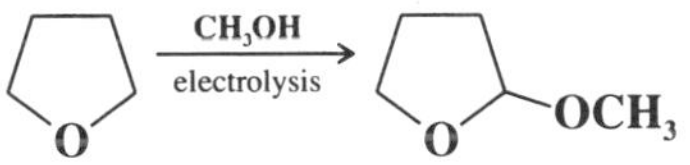

Ethers can be converted to acetals, and acetals to ortho esters, by anodic oxidation in an alcohol as solvent.[194] Yields are moderate. In a similar reaction, certain amides, carbamates, and sulfonamides can be alkoxylated α to the nitrogen, e.g., $MeSO_2NMe_2 \rightarrow MeSO_2N(Me)CH_2OCH_3$.[195]

OS **VII**, 307.

[191]See Roček; Ng *J. Org. Chem.* **1973**, *38*, 3348.
[192]See, for example, Freeman; Lin; Moore *J. Org. Chem.* **1982**, *47*, 56; Jain; Banerji *J. Chem. Res. (S)* **1983**, 60.
[193]Freeman; Brant; Hester; Kamego; Kasner; McLaughlin; Paull *J. Org. Chem.* **1970**, *35*, 982.
[194]Shono; Matsumura; Onomura; Yamada *Synthesis* **1987**, 1099; Ginzel; Steckhan *Tetrahedron* **1987**, *43*, 5797.
[195]Ross; Finkelstein; Rudd *J. Org. Chem.* **1972**, *37*, 2387. See also Moeller; Tarazi; Marzabadi *Tetrahedron Lett.* **1989**, *30*, 1213; Shono; Matsumura; Tsubata *Org. Synth. VII*, 307. For a table of compounds subjected to this reaction, see Shono *Electroorganic Chemistry as a New Tool in Organic Synthesis*; Springer: New York, 1984, pp. 63-66.

4-8   Formation of Cyclic Ethers
(5)*OC-cyclo*-Alkoxy-de-hydro-substitution

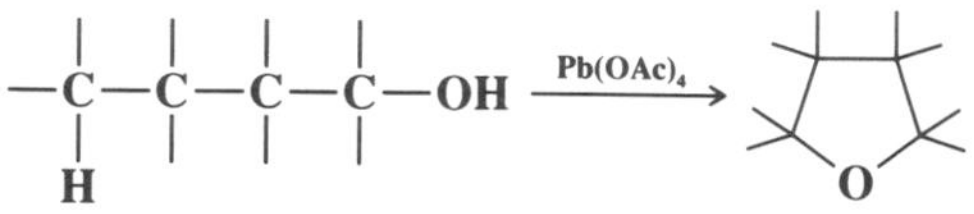

$$-\overset{|}{\underset{\underset{H}{|}}{C}}-\overset{|}{\underset{|}{C}}-\overset{|}{\underset{|}{C}}-\overset{|}{\underset{|}{C}}-OH \quad \xrightarrow{Pb(OAc)_4} \quad$$

Alcohols with a hydrogen in the δ position can be cyclized with lead tetraacetate.[196] The reaction is usually carried out at about 80°C (most often in refluxing benzene) but can also be done at room temperature if the reaction mixture is irradiated with uv light. Tetrahydrofurans are formed in high yields. Little or no four- and six-membered cyclic ethers (oxetanes and tetrahydropyrans, respectively) are obtained even when γ and ε hydrogens are present. The reaction has also been carried out with a mixture of halogen (Br$_2$ or I$_2$) and a salt or oxide of silver or mercury (especially HgO or AgOAc),[197] with iodosobenzene diacetate and I$_2$,[198] and with ceric ammonium nitrate (CAN).[199] The following mechanism is likely for the lead tetraacetate reaction:[200]

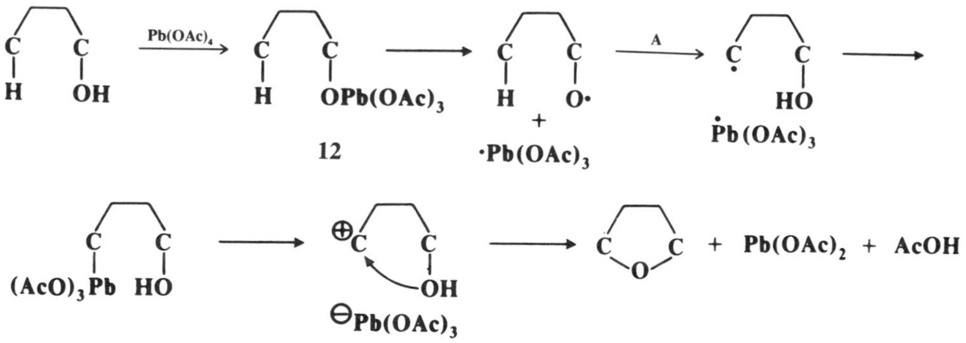

though **12** has never been isolated. The step marked **A** is a 1,5 internal hydrogen abstraction. Such abstractions are well-known (see p. 1153) and are greatly favored over 1,4 or 1,6 abstractions (the small amounts of tetrahydropyran formed result from 1,6 abstractions).[201]

Reactions that sometimes compete are oxidation to the aldehyde or acid (**9-3** and **9-22**) and fragmentation of the substrate. When the OH group is on a ring of at least seven

[196]For reviews, see Mihailović; Partch *Sel. Org. Transform.* **1972,** 2, 97-182; Milhailović; Ćeković *Synthesis* **1970,** 209-224. For a review of the chemistry of lead tetraacetate, see Butler, in Pizey, Ref. 85, vol. 3, 1977, pp. 277-419.

[197]Akhtar; Barton *J. Am. Chem. Soc.* **1964,** 86, 1528; Sneen; Matheny *J. Am. Chem. Soc.* **1964,** 86, 3905, 5503; Roscher; Shaffer *Tetrahedron* **1984,** 40, 2643. For a review, see Kalvoda; Heusler *Synthesis* **1971,** 501-526. For a list of references, see Ref. 74, p. 445.

[198]Concepción; Francisco; Hernández; Salazar; Suárez *Tetrahedron Lett.* **1984,** 25, 1953; Furuta; Nagata; Yamamoto *Tetrahedron Lett.* **1988,** 29, 2215.

[199]See, for example, Trahanovsky; Young; Nave *Tetrahedron Lett.* **1969,** 2501; Doyle; Zuidema; Bade *J. Org. Chem.* **1975,** 40, 1454.

[200]Milhailović; Ćeković; Maksimović; Jeremić; Lorenc; Mamuzić *Tetrahedron* **1965,** 21, 2799.

[201]Milhailović; Ćeković; Jeremić *Tetrahedron* **1965,** 21, 2813.

members, a transannular product can be formed, e.g.,[202]

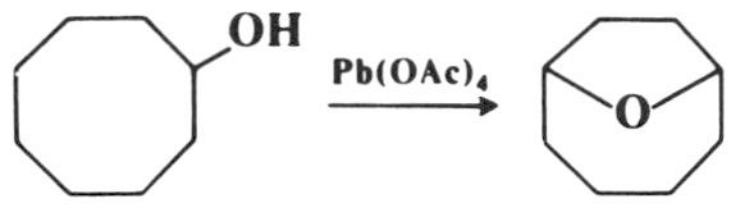

β-Hydroxy ethers can give cyclic acetals, e.g.,[203]

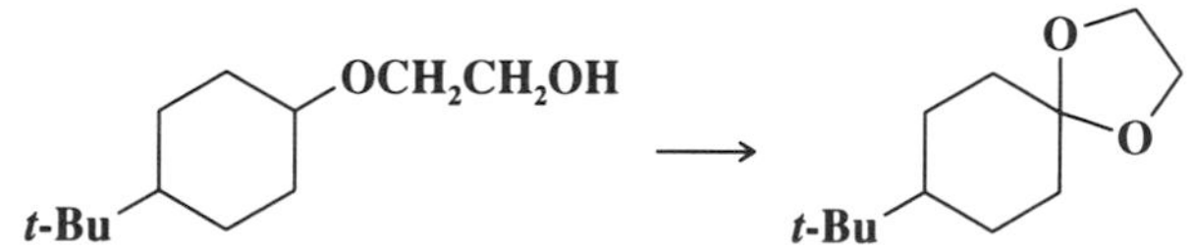

A different kind of formation of a cyclic ether was reported by Paquette and Kobayashi,[204] who found that when the epoxide **13** of secododecahedrane was treated with sodium chro-

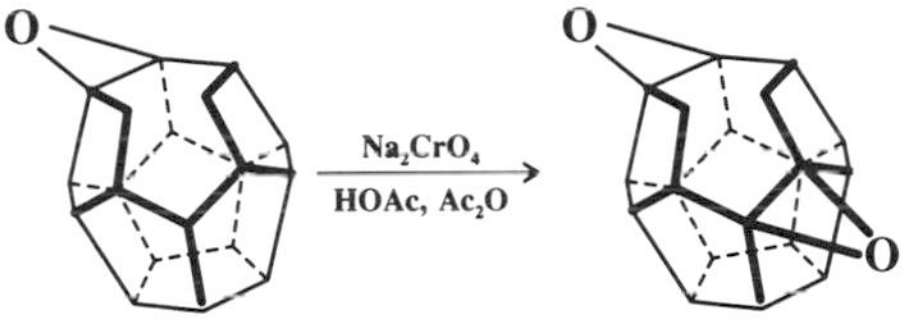

13   **14**

mate and HOAc–Ac$_2$O, the diepoxide **14** was obtained. Thus, the unusual transformation

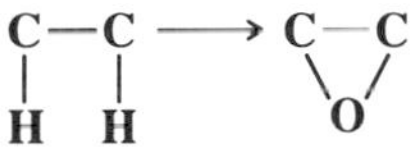

was achieved in this case. It is likely that the large degree of strain in this system was at least partially responsible for the formation of this product.

There are no references in *Organic Syntheses*, but see OS **V**, 692; **VI**, 958, for related reactions.

4-9   Formation of Hydroperoxides
Hydroperoxy-de-hydrogenation

$$RH + O_2 \longrightarrow R\!-\!O\!-\!O\!-\!H$$

The slow atmospheric oxidation (*slow* meaning without combustion) of C—H to C—O—O—H is called *autoxidation*.[205] The reaction occurs when compounds are allowed to stand in air and is catalyzed by light, so unwanted autoxidations can be greatly slowed by keeping the compounds in dark places. The hydroperoxides produced often react further

[202]Cope; Gordon; Moon; Park *J. Am. Chem. Soc.* **1965,** *87,* 3119; Moriarty; Walsh *Tetrahedron Lett.* **1965,** 465; Milhailović; Čeković; Andrejević; Matić; Jeremić *Tetrahedron* **1968,** *24,* 4947.
[203]Furuta et al., Ref. 198.
[204]Paquette; Kobayashi *Tetrahedron Lett.* **1987,** *28,* 3531.
[205]The term autoxidation actually applies to any slow oxidation with atmospheric oxygen. For reviews, see Sheldon; Kochi *Adv. Catal.* **1976,** *25,* 272-413; Howard, in Kochi, Ref. 8, vol. 2, pp. 3-62; Lloyd *Methods Free-Radical Chem.* **1973,** *4,* 1-131; Betts *Q. Rev., Chem. Soc.* **1971,** *25,* 265-288; Huyser *Free-Radical Chain Reactions,* Ref. 1, pp. 306-312; Chinn, Ref. 134, pp. 29-39; Ingold *Acc. Chem. Res.* **1969,** *2,* 1-9; Mayo *Acc. Chem. Res.* **1968,** *1,* 193-201. For monographs on these and similar reactions, see Bamford; Tipper, Ref. 37, Vol. 16, 1980; Sheldon; Kochi, Ref. 172.

to give alcohols, ketones, and more complicated products, so the reaction is not often used for preparative purposes, although in some cases hydroperoxides have been prepared in good yield.[206] It is because of autoxidation that foods, rubber, paint, lubricating oils, etc. deteriorate on exposure to the atmosphere over periods of time. On the other hand, a useful application of autoxidation is the atmospheric drying of paints and varnishes. As with other free-radical reactions of C—H bonds, some bonds are attacked more readily than others,[207] and these are the ones we have seen before (pp. 683-685), though the selectivity is very low at high temperatures and in the gas phase. The reaction can be carried out successfully at tertiary (to a lesser extent, secondary), benzylic,[208] and allylic (though allylic rearrangements are common) R.[209] The following are actual examples:

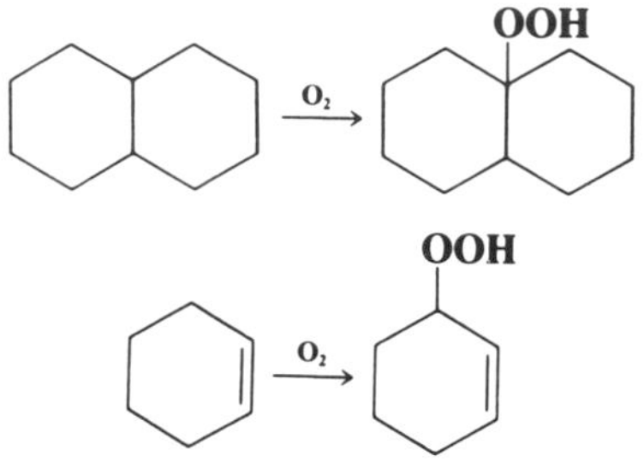

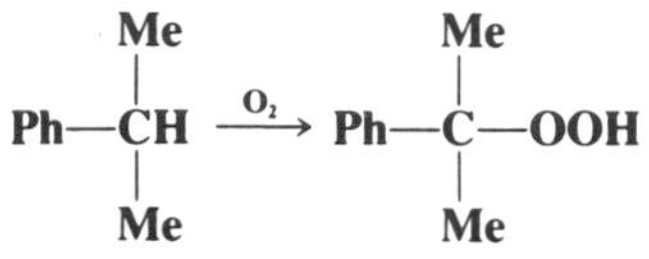

Another susceptible position is aldehydic C—H, but the peracids so produced are not easily isolated[187] since they are converted to the corresponding carboxylic acids (**4-6**). The α positions of ethers are also easily attacked by oxygen:

$$\text{RO}\!-\!\overset{|}{\underset{|}{C}}\!-\!\text{H} \xrightarrow{O_2} \text{RO}\!-\!\overset{|}{\underset{|}{C}}\!-\!\text{OOH}$$

but the resulting hydroperoxides are seldom isolated. However, this reaction constitutes a hazard in the storage of ethers since solutions of these hydroperoxides and their rearrangement products in ethers are potential spontaneous explosives.[210]

Oxygen itself (a diradical) is not reactive enough to be the species that actually abstracts the hydrogen. But if a trace of free radical (say R'•) is produced by some initiating process, *it* reacts with oxygen[211] to give R'—O—O•; since this type of radical *does* abstract hydrogen, the chain is

$$\text{R'OO•} + \text{RH} \longrightarrow \text{R•} + \text{R'OOH}$$
$$\text{R•} + O_2 \longrightarrow \text{R—O—O•}$$

etc.

[206]For a review of the synthesis of alkyl peroxides and hydroperoxides, see Sheldon, in Patai *The Chemistry of Peroxides*; Wiley: New York, 1983, pp. 161-200.
[207]For a discussion, see Korcek; Chenier; Howard; Ingold *Can. J. Chem.* **1972**, *50*, 2285, and other papers in this series.
[208]For a method that gives good yields at benzylic positions, see Santamaria; Jroundi; Rigaudy *Tetrahedron Lett.* **1989**, *30*, 4677.
[209]For a review of autoxidation at allylic and benzylic positions, see Voronenkov; Vinogradov; Belyaev *Russ. Chem. Rev.* **1970**, *39*, 944-952.
[210]For methods of detection and removal of peroxides from ether solvents, see Gordon; Ford *The Chemist's Companion*; Wiley: New York, 1972, p. 437; Burfield, *J. Org. Chem.* **1982**, *47*, 3821.
[211]See, for example Schwetlick *J. Chem. Soc., Perkin Trans. 2* **1988**, 2007.

In at least some cases (in alkaline media)[212] the radical R• can be produced by formation of a carbanion and its oxidation (by O_2) to a radical, e.g.,[213]

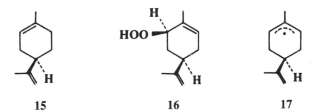

Autoxidations in alkaline media can also proceed by a different mechanism: R—H + base → R⁻ + O_2 → ROO⁻.[214]

When alkenes are treated with oxygen that has been photosensitized (p. 241), they are substituted by OOH in the allylic position in a synthetically useful reaction.[215] Although superficially similar to autoxidation, this reaction is clearly different because 100% allylic rearrangement always takes place. The reagent here is not the ground-state oxygen (a triplet) but an excited singlet state[216] (in which all electrons are paired), and the function of the photosensitization is to promote the oxygen to this singlet state. Singlet oxygen can also be produced by nonphotochemical means,[217] e.g., by the reaction between H_2O_2 and NaOCl[218] or sodium molybdate,[219] or between ozone and triphenyl phosphite.[220] The oxygen generated by either photochemical or nonphotochemical methods reacts with olefins in the same way;[221] this is evidence that singlet oxygen is the reacting species in the photochemical reaction and not some hypothetical complex between triplet oxygen and the photosensitizer, as had previously been suggested. The fact that 100% allylic rearrangement always takes place is incompatible with a free-radical mechanism, and further evidence that free radicals are not involved comes from the treatment of optically active limonene (**15**) with singlet oxygen. Among other products is the optically active hydroperoxide **16**, though if **17** were an inter-

15   **16**   **17**

[212]For a review of base-catalyzed autoxidations in general, see Sosnovsky; Zaret, in Swern, Ref. 172, vol. 1, pp. 517-560.

[213]Barton; Jones *J. Chem. Soc.* **1965**, 3563; Russell; Bemis *J. Am. Chem. Soc.* **1966**, *88*, 5491.

[214]Gersmann; Bickel *J. Chem. Soc. B* **1971**, 2230.

[215]For reviews, see Frimer; Stephenson, in Frimer, Ref. 216, vol. 2, pp. 67-91; Wasserman; Ives *Tetrahedron* **1981**, *37*, 1825-1852; Gollnick; Kuhn, in Wasserman; Murray, Ref. 216, pp. 287-427; Denny; Nickon *Org. React.* **1973**, *20*, 133-336; Adams, in Augustine, Ref. 134, vol. 2, pp. 65-112.

[216]For books on singlet oxygen, see Frimer *Singlet O₂*, 4 vols.; CRC Press: Boca Raton, FL, 1985; Wasserman; Murray *Singlet Oxygen*; Academic Press: New York, 1979. For reviews, see Frimer, in Patai, Ref. 206, pp. 201-234; Gorman; Rodgers, *Chem. Soc. Rev.* **1981**, *10*, 205-231; Shinkarenko; Aleskovskii *Russ. Chem. Rev.* **1981**, *50*, 220-231; Shlyapintokh; Ivanov *Russ. Chem. Rev.* **1976**, *45*, 99-110; Ohloff *Pure Appl. Chem.* **1975**, *43*, 481-502; Kearns *Chem. Rev.* **1971**, *71*, 395-427; Wayne *Adv. Photochem.* **1969**, *7*, 311-371.

[217]For reviews, see Turro; Ramamurthy, in de Mayo *Rearrangements in Ground and Excited States*, vol. 3; Academic Press: New York, 1980, pp. 1-23; Murray, in Wasserman; Murray, Ref. 216, pp. 59-114. For a general monograph, see Adam; Cilento, *Chemical and Biological Generation of Excited States*; Academic Press: New York, 1982.

[218]Foote; Wexler *J. Am. Chem. Soc.* **1964**, *86*, 3879.

[219]Aubry; Cazin; Duprat *J. Org. Chem.* **1989**, *54*, 726.

[220]Murray; Kaplan *J. Am. Chem. Soc.* **1969**, *91*, 5358; Bartlett; Mendenhall; Durham *J. Org. Chem.* **1980**, *45*, 4269.

[221]Foote; Wexler; Ando; Higgins *J. Am. Chem. Soc.* **1968**, *90*, 975. See also McKeown; Waters *J. Chem. Soc. B* **1966**, 1040.

mediate, it could not give an optically active product since it possesses a plane of symmetry.[222] In contrast, autoxidation of **15** gave optically inactive **16** (a mixture of four diastereomers in which the two pairs of enantiomers are present as racemic mixtures). As this example shows, singlet oxygen reacts faster with more-highly substituted than with less-highly substituted alkenes. The order of alkene reactivity is tetrasubstituted > trisubstituted > disubstituted. Electron-withdrawing substituents deactivate the olefin.[223] In simple trisubstituted olefins, there is a general preference for the hydrogen to be removed from the more highly congested side of the double bond.[224] With *cis*-alkenes of the form RCH=CHR', the hydrogen is removed from the larger R group.[225] Many functional groups in an allylic position cause the hydrogen to be removed from that side rather than the other (geminal selectivity).[226] Also, in alkyl-substituted alkenes, the hydrogen that is preferentially removed is the one geminal to the larger substituent on the double bond.[227]

Several mechanisms have been proposed for the reaction with singlet oxygen.[228] One of these is a pericyclic mechanism, similar to that of the ene synthesis (**5-16**) and to the first

step of the reaction between alkenes and SeO$_2$(**4-4**). However, there is strong evidence against this mechanism,[229] and a more likely mechanism involves addition of singlet oxygen to the double bond to give a perepoxide (**18**),[230] followed by internal proton transfer.[231]

18

Still other proposed mechanisms involve diradicals or dipolar intermediates.[232]

OS **IV**, 895.

[222]Schenck; Gollnick; Buchwald; Schroeter; Ohloff *Liebigs Ann. Chem.* **1964,** *674,* 93; Schenck; Neumüller; Ohloff; Schroeter *Liebigs Ann. Chem.* **1965,** *687,* 26.

[223]For example, see Foote; Denny *J. Am. Chem. Soc.* **1971,** *93,* 5162.

[224]Schulte-Elte; Muller; Rautenstrauch *Helv. Chim. Acta* **1978,** *61,* 2777; Orfanopoulos; Grdina; Stephenson *J. Am. Chem. Soc.* **1979,** *101,* 275; Rautenstrauch; Thommen; Schulte-Elte *Helv. Chim. Acta* **1986,** *69,* 1638.

[225]Orfanopoulos; Stratakis; Elemes *Tetrahedron Lett.* **1989,** *30,* 4875.

[226]Clennan; Chen; Koola *J. Am. Chem. Soc.* **1990,** *112,* 5193, and references cited therein.

[227]Orfanopoulos; Stratakis; Elemes *J. Am. Chem. Soc.* **1990,** *112,* 6417.

[228]For reviews of the mechanism, see Frimer; Stephenson, Ref. 215, pp. 80-87; Stephenson; Grdina; Orfanopoulos *Acc. Chem. Res.* **1980,** *13,* 419-425; Gollnick; Kuhn, Ref. 215, pp. 288-341; Frimer *Chem. Rev.* **1979,** *79,* 359-387; Foote *Acc. Chem. Res.* **1968,** *1,* 104-110, *Pure Appl. Chem.* **1971,** *27,* 635-645; Gollnick *Adv. Photochem.* **1968,** *6,* 1-122; Kearns, Ref. 216.

[229]Asveld; Kellogg *J. Org. Chem.* **1982,** *47,* 1250.

[230]For a review of perepoxides as intermediates in organic reactions, see Mitchell *Chem. Soc. Rev.* **1985,** *14,* 399-419, pp. 401-406.

[231]For evidence in favor of this mechanism, at least with some kinds of substrates, see Jefford; Rimbault *J. Am. Chem. Soc.* **1978,** *100,* 6437; Okada; Mukai *J. Am. Chem. Soc.* **1979,** *100,* 6509; Paquette; Hertel; Gleiter; Böhm *J. Am. Chem. Soc.* **1978,** *100,* 6510; Hurst; Wilson; Schuster *Tetrahedron* **1985,** *41,* 2191; Wilson; Schuster *J. Org. Chem.* **1986,** *51,* 2056; Davies; Schiesser *Tetrahedron Lett.* **1989,** *30,* 7099; Orfanopoulos; Smonou; Foote *J. Am. Chem. Soc.* **1990,** *112,* 3607.

[232]See, for example, Jefford *Helv. Chim. Acta* **1981,** *64,* 2534.

4-10   Formation of Peroxides
Alkyldioxy-de-hydrogenation

$$RH + R'OOH \xrightarrow{CuCl} ROOR'$$

Peroxy groups (ROO) can be introduced into susceptible organic molecules by treatment with a hydroperoxide in the presence of cuprous chloride or other catalysts, e.g., cobalt and manganese salts.[233] Very high yields can be obtained. The type of hydrogen replaced is similar to that with N-bromosuccinimide (**4-2**), i.e., mainly benzylic, allylic, and tertiary. The mechanism is therefore of the free-radical type, involving ROO• formed from ROOH and the metal ion. The reaction can be used to demethylate tertiary amines of the form R_2NCH_3, since the product R_2NCH_2OOR' can easily be hydrolyzed by acid (**0-6**) to give R_2NH.[234]

4-11   Acyloxylation or Acyloxy-de-hydrogenation

$$RH + Me_3C-O-O-\underset{\underset{O}{\|}}{C}-R' \xrightarrow{Cu^+/Cu^{2+}} R-O-\underset{\underset{O}{\|}}{C}-R'$$

Susceptible positions of organic compounds can be directly acyloxylated[235] by t-butyl peresters, the most frequently used being acetic and benzoic (R' = Me or Ph).[236] The reaction requires a catalyst (cuprous ion is the actual catalyst, but a trace is all that is necessary, and such traces are usually present in cupric compounds, so that these are often used) and without it is not selective. Susceptible positions are similar to those in **4-9**: benzylic, allylic, and the α position of ethers and sulfides. Terminal olefins are substituted almost entirely in the 3 position, i.e., with only a small amount of allylic rearrangement, but internal olefins generally give mixtures containing a large amount of allylic-shift product. If the reaction with olefins is carried out in an excess of another acid R''COOH, the ester produced is of *that* acid ROCOR''. Aldehydes give anhydrides:

$$R-\underset{\underset{O}{\|}}{C}-H + Me_3COOCR' \xrightarrow{Cu^+} R-\underset{\underset{O}{\|}}{C}-O-\underset{\underset{O}{\|}}{C}-R'$$

Acyloxylation has also been achieved with metallic acetates such as lead tetraacetate,[237] mercuric acetate,[238] and palladium(II) acetate.[239] In the case of the lead and mercuric acetates, not only does the reaction take place at allylic and benzylic positions and at those α to an OR or SR group but also at positions α to the carbonyl groups of aldehydes, ketones, or esters and at those α to two carbonyl groups (ZCH_2Z'). It is likely that in the latter cases

---

[233]For a review, see Sosnovsky; Rawlinson, Ref. 172, pp. 153-268. See also Murahashi; Naota; Kuwabara; Saito; Kumobayashi; Akutagawa *J. Am. Chem. Soc.* **1990**, *112*, 7820; Ref. 206.
[234]See Murahashi; Naota; Yonemura *J. Am. Chem. Soc.* **1988**, *110*, 8256.
[235]For a list of reagents, with references, see Ref. 74, pp. 823-827ff, 841-842.
[236]For reviews, see Rawlinson; Sosnovsky *Synthesis* **1972**, 1-28; Sosnovsky; Rawlinson, in Swern, Ref. 172, vol. 1, pp. 585-608; Doumaux, in Augustine, Ref. 134, vol. 2, 1971, pp. 141-185.
[237]For a review of lead tetraacetate, see Butler, Ref. 196.
[238]For reviews, see Larock *Organomercury Compounds in Organic Synthesis*; Springer: New York, 1985, pp. 190-208; Rawlinson; Sosnovsky *Synthesis* **1973**, 567-602.
[239]Hansson; Heumann; Rein; Åkermark *J. Org. Chem.* **1990**, *55*, 975; Byström; Larsson; Åkermark *J. Org. Chem.* **1990**, *55*, 5674.

it is the enol forms that react. Ketones can be α-acyloxylated indirectly by treatment of various enol derivatives with metallic acetates, for example, silyl enol ethers with silver carboxylates–iodine,[240] enol thioethers with lead tetraacetate,[241] and enamines[242] with lead tetraacetate[243] or thallium triacetate.[244] α,β-Unsaturated ketones can be acyloxylated in good yields in the α' position with manganese triacetate.[245] Palladium acetate converts alkenes to vinylic and/or allylic acetates.[246] Lead tetraacetate even acyloxylates alkanes, in a slow reaction (10 days to 2 weeks), with tertiary and secondary positions greatly favored over primary ones.[247] Yields are as high as 50%. Acyloxylation of certain alkanes has also been reported with palladium(II) acetate.[248]

Studies of the mechanism of the cuprous-catalyzed reaction show that the most common mechanism is the following:[249]

Step 1
$$\underset{\displaystyle \overset{\|}{O}}{R'COO}—t\text{-Bu} + Cu^+ \longrightarrow \underset{\displaystyle \overset{\|}{O}}{R'CO}—Cu^+(II) + t\text{-BuO}•$$

Step 2
$$RH + t\text{-BuO}• \longrightarrow R• + t\text{-BuOH}$$

Step 3
$$R• + \underset{\displaystyle \overset{\|}{O}}{R'CO}—Cu^+(II) \longrightarrow \underset{\displaystyle \overset{\|}{O}}{ROCR'} + Cu^+$$

19

This mechanism, involving a free radical R•, is compatible with the allylic rearrangements found.[250] The finding that t-butyl peresters labeled with [18]O in the carbonyl oxygen gave ester with 50% of the label in each oxygen[251] is in accord with a combination of R• with the intermediate **19,** in which the copper is ionically bound, so that the oxygens are essentially equivalent. Other evidence is that t-butoxy radicals have been trapped with dienes.[252] Much less is known about the mechanisms of the reactions with metal acetates.[253]

Free-radical acyloxylation of aromatic substrates[254] has been accomplished with a number of reagents including copper(II) acetate,[255] benzoyl peroxide–iodine,[256] silver(II) complexes,[257] and cobalt(III) trifluoroacetate.[258]

OS **III**, 3; **V**, 70, 151; **68**, 109.

[240]Rubottom; Mott; Juve *J. Org. Chem.* **1981**, *46*, 2717.
[241]Trost; Tanigawa *J. Am. Chem. Soc.* **1979**, *101*, 4413.
[242]For a review, see Cook, in Cook *Enamines*, 2nd ed.; Marcel Dekker: New York, 1988, pp. 251-258.
[243]See Butler, *Chem. Ind. (London)* **1976**, 499-500.
[244]Kuehne; Giacobbe *J. Org. Chem.* **1968**, *33*, 3359.
[245]Dunlap; Sabol; Watt *Tetrahedron Lett.* **1984**, *25*, 5839; Demir; Sayrac; Watt *Synthesis* **1990**, 1119.
[246]For reviews, see Rylander *Organic Synthesis with Noble Metal Catalysts*; Academic Press: New York, 1973, pp. 80-87; Jira; Freiesleben *Organomet. React.* **1972**, *3*, 1-190, pp. 44-84; Heck *Fortschr. Chem. Forsch.* **1971**, *16*, 221-242, pp. 231-237; Tsuji *Adv. Org. Chem.* **1969**, *6*, 109-255, pp. 132-143.
[247]Bestre; Cole; Crank *Tetrahedron Lett.* **1983**, *24*, 3891; Mosher; Cox *Tetrahedron Lett.* **1985**, *26*, 3753.
[248]This was done in trifluoroacetic acid, and the products were trifluoroacetates: Sen; Gretz; Oliver; Jiang *New J. Chem.* **1989**, *13*, 755.
[249]Kharasch; Sosnovsky; Yang *J. Am. Chem. Soc.* **1959**, *81*, 5819; Kochi; Mains *J. Org. Chem.* **1965**, *30*, 1862. See also Beckwith; Zavitsas *J. Am. Chem. Soc.* **1986**, *108*, 8230.
[250]Goering; Mayer *J. Am. Chem. Soc.* **1964**, *86*, 3753; Denney; Appelbaum; Denney *J. Am. Chem. Soc.* **1962**, *84*, 4969.
[251]Denney; Denney; Feig *Tetrahedron Lett.* **1959**, no. 15, p. 19.
[252]Kochi *J. Am. Chem. Soc.* **1962**, *84*, 2785, 3271; Story *Tetrahedron Lett.* **1962**, 401.
[253]See, for example, Jones; Mellor *J. Chem. Soc., Perkin Trans. 2* **1977**, 511.
[254]For a review, see Haines *Methods for the Oxidation of Organic Compounds*; Academic Press: New York, 1985, pp. 177-180, 351-355.
[255]Takizawa; Tateishi; Sugiyama; Yoshida; Yoshihara *J. Chem. Soc., Chem. Commun.* **1991**, 104. See also Kaeding; Kerlinger; Collins *J. Org. Chem.* **1965**, *30*, 3754.
[256]For example, see Kovacic; Reid; Brittain *J. Org. Chem.* **1970**, *35*, 2152.
[257]Nyberg; Wistrand *J. Org. Chem.* **1978**, *43*, 2613.
[258]Kochi; Tank; Bernath *J. Am. Chem. Soc.* **1973**, *95*, 7114; DiCosimo; Szabo *J. Org. Chem.* **1986**, *51*, 1365.

C. Substitution by Sulfur

4-12   Chlorosulfonation or Chlorosulfo-de-hydrogenation

$$RH + SO_2 + Cl_2 \xrightarrow{hv} RSO_2Cl$$

The chlorosulfonation of organic molecules with chlorine and sulfur dioxide is called the *Reed reaction*.[259] In scope and range of products obtained, the reaction is similar to **4-1**. The mechanism is also similar, except that there are two additional main propagation steps:

$$R\cdot + SO_2 \longrightarrow R-SO_2\cdot$$

$$R-SO_2\cdot + Cl_2 \longrightarrow R-SO_2Cl + Cl\cdot$$

Chlorosulfenation[260] can be accomplished by treatment with SCl_2 and uv light: $RH + SCl_2 \xrightarrow{hv} RSCl$.

D. Substitution by Nitrogen

4-13   Nitration of Alkanes
Nitration or Nitro-de-hydrogenation

$$RH + HNO_3 \xrightarrow{400^\circ C} RNO_2$$

Nitration of alkanes[261] can be carried out in the gas phase at about 400°C or in the liquid phase. The reaction is not practical for the production of pure products for any alkane except methane. For other alkanes, not only does the reaction produce mixtures of the mono-, di-, and polynitrated alkanes at every combination of positions, but extensive chain cleavage occurs.[262] A free-radical mechanism is involved.[263]

Activated positions (e.g., ZCH_2Z' compounds) can be nitrated by fuming nitric acid in acetic acid, by acetyl nitrate and an acid catalyst,[264] or by alkyl nitrates under alkaline conditions.[265] In the latter case it is the carbanionic form of the substrate that is actually nitrated. What is isolated under these alkaline conditions is the conjugate base of the nitro

$$-\overset{|}{\underset{|}{C}}{}^{\ominus} + MeONO_2 \longrightarrow -\overset{|}{\underset{|}{C}}-NO_2 + OMe^-$$

compound. Yields are not high. Of course, the mechanism in this case is not of the free-radical type, but is electrophilic substitution with respect to the carbon (similar to the mechanisms of **2-7** and **2-8**). Positions activated by only one electron-withdrawing group, e.g., α positions of simple ketones, nitriles, sulfones, or N,N-dialkyl amides, can be nitrated with alkyl nitrates if a very strong base, e.g., *t*-BuOK or $NaNH_2$, is present to convert the substrate to the carbanionic form.[266] Electrophilic nitration of alkanes has been performed

[259]For a review, see Gilbert *Sulfonation and Related Reactions*; Wiley: New York, 1965, pp. 126-131.
[260]Müller; Schmidt *Chem. Ber.* **1963**, *96*, 3050, **1964**, *97*, 2614. For a review of the formation and reactions of sulfenyl halides, see Kühle *Synthesis* **1970**, 561-580, **1971**, 563-586, 617-638.
[261]For reviews, see Olah; Malhotra; Narang *Nitration*; VCH: New York, 1989, pp. 219-295; Ogata, in Trahanovsky, Ref. 185, part C, 1978, pp. 295-342; Ballod; Shtern *Russ. Chem. Rev.* **1976**, *45*, 721-737.
[262]For a discussion of the mechanism of this cleavage, see Matasa; Hass *Can. J. Chem.* **1971**, *49*, 1284.
[263]Titov *Tetrahedron* **1963**, *19*, 557-580.
[264]Sifniades *J. Org. Chem.* **1975**, *40*, 3562.
[265]For a review, see Larson, in Feuer *The Chemistry of the Nitro and Nitroso Groups*, vol. 1; Wiley: New York, 1969, pp. 310-316.
[266]For examples, see Feuer; Shepherd; Savides *J. Am. Chem. Soc.* **1956**, *78*, 4364; Feuer; Lawrence *J. Org. Chem.* **1972**, *37*, 2662; Truce; Christensen *Tetrahedron* **1969**, *25*, 181; Pfeffer; Silbert *Tetrahedron Lett.* **1970**, 699; Feuer; Spinicelli *J. Org. Chem.* **1976**, *41*, 2981; Feuer; Van Buren; Grutzner *J. Org.Chem.* **1978**, *43*, 4676.

with nitronium salts, e.g., NO_2^+ PF_6^- and with HNO_3–H_2SO_4 mixtures, but mixtures of nitration and cleavage products are obtained and yields are generally low.[267]

Aliphatic nitro compounds can be α nitrated [$R_2CHNO_2 \rightarrow R_2C(NO_2)_2$] by treatment of their conjugate bases $R\overset{\ominus}{\overline{C}}NO_2$ with NO_2^- and $K_3Fe(CN)_6$.[268]

OS **I**, 390; **II**, 440, 512.

4-14   The Direct Conversion of Aldehydes to Amides
Amination or Amino-de-hydrogenation

$$ArCHO \xrightarrow[\text{NBS–AIBN}]{\text{NH}_3} ArCONH_2$$

Aliphatic and aromatic aldehydes have been converted to the corresponding amides with ammonia or a primary or secondary amine, N-bromosuccinimide, and a catalytic amount of AIBN (p. 664).[269] In a reaction of more limited scope, amides are obtained from aromatic and α,β-unsaturated aldehydes by treatment with dry ammonia gas and nickel peroxide.[270] Best yields (80 to 90%) are obtained at -25 to $-20°C$. The reaction has also been performed with MnO_2 and NaCN along with ammonia or an amine at 0°C in isopropyl alcohol,[271] and with a secondary amine and a palladium acetate catalyst.[272] In the nickel peroxide reaction the corresponding alcohols ($ArCH_2OH$) have also been used as substrates. For an indirect way of converting aldehydes to amides, see **2-31**. Thioamides $RCSNR_2'$ have been prepared in good yield from thioaldehydes (produced in situ from phosphoranes and sulfur) and secondary amines.[273]

4-15   Amidation and Amination at an Alkyl Carbon
Acylamino-de-hydrogenation

$$R_3CH + CH_3CN \xrightarrow[\text{H}_3(\text{PW}_{12}\text{O}_{40})\cdot 6\text{H}_2\text{O}]{hv} R_3C-NH-\underset{\underset{O}{\|}}{C}-CH_3 + H_2$$

When alkanes bearing a tertiary hydrogen are exposed to uv light in acetonitrile containing a heteropolytungstic acid, they are amidated.[274] The oxygen in the product comes from the tungstic acid. When the substrate bears two adjacent tertiary hydrogens, alkenes are formed (by loss of two hydrogens), rather than amides (**9-2**).

An electrochemical method for amination has been reported by Shono and co-workers.[275] Derivatives of malonic esters containing an N-tosyl group were cyclized in high yields by anodic oxidation:

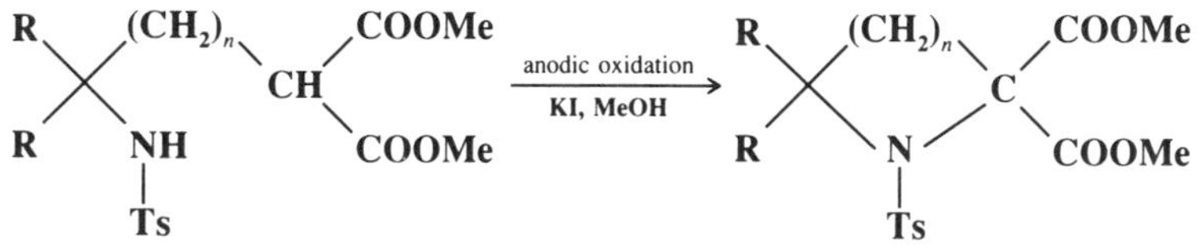

$$n = 0, 1, \text{ or } 2$$

Three-, four-, and five-membered rings were synthesized by this procedure.

[267]Olah; Lin *J. Am. Chem. Soc.* **1973**, *93*, 1259. See also Bach; Holubka; Badger; Rajan *J. Am. Chem. Soc.* **1979**, *101*, 4416.

[268]Matacz; Piotrowska; Urbanski *Pol. J. Chem.* **1979**, *53*, 187; Kornblum; Singh; Kelly *J. Org. Chem.* **1983**, *48*, 332; Garver; Grakauskas; Baum *J. Org. Chem.* **1985**, *50*, 1699.

[269]Markó; Mekhalfia, Ref. 133.

[270]Nakagawa; Onouc; Minami *Chem. Commun.* **1966**, 17.

[271]Gilman *Chem. Commun.* **1971**, 733.

[272]Tamaru; Yamada; Yoshida *Synthesis* **1983**, 474.

[273]Okuma; Komiya; Ohta *Chem. Lett.* **1988**, 1145.

[274]Rennckc; Hill *J. Am. Chem. Soc.* **1986**, *108*, 3528.

[275]Shono; Matsumura; Katoh; Ohshita *Chem. Lett.* **1988**, 1065.

E. Substitution by Carbon In these reactions a new carbon–carbon bond is formed, and they may be given the collective title *coupling reactions*. In each case an alkyl or aryl radical is generated and then combines with another radical (a termination process) or attacks an aromatic ring or olefin to give the coupling product.[276]

4-16 Simple Coupling at a Susceptible Position
De-hydrogen-coupling

$$2RH \xrightarrow[\text{Hg}]{h\nu} R\text{—}R + H_2$$

Alkanes can be dimerized by vapor-phase mercury photosensitization[277] in a synthetically useful process. Best results are obtained for coupling at tertiary positions, but compounds lacking tertiary hydrogens (e.g., cyclohexane) also give good yields. Dimerization of *n*-alkanes gives secondary–secondary coupling in a nearly statistical distribution, with primary positions essentially unaffected. Alcohols and ethers dimerize at the position α to the oxygen [e.g., 2EtOH → MeCH(OH)CH(OH)Me].

When a mixture of compounds is treated, cross-dimerization and homodimerization take place statistically, e.g.:

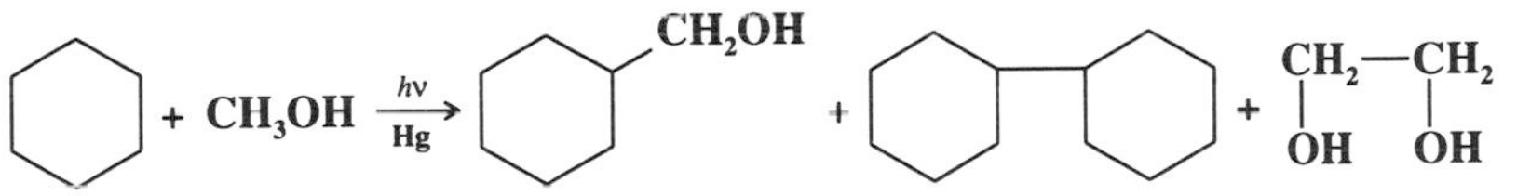

Even with the limitation on yield implied by the statistical process, cross-dimerization is still useful when one of the reactants is an alkane, because the products are easy to separate, and because of the few other ways to functionalize an alkane. The cross-coupling of an alkane with trioxane is especially valuable, because hydrolysis of the product (**0-6**) gives an

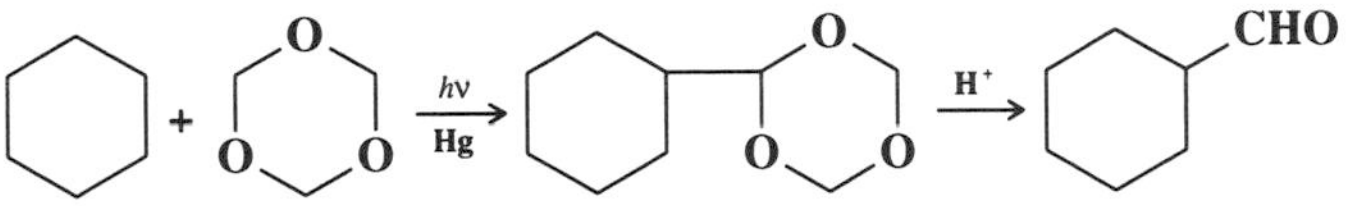

aldehyde, thus achieving the conversion RH → RCHO. The mechanism probably involves abstraction of H by the excited Hg atom, and coupling of the resulting radicals.

The reaction has been extended to ketones, carboxylic acids and esters (all of which couple α to the C=O group), and amides (which couple α to the nitrogen) by running it in the presence of H_2.[278] Under these conditions it is likely that the excited Hg abstracts H• from H_2, and that the remaining H• abstracts H from the substrate.

In an older reaction, substrates RH are treated with peroxides, which decompose to give a radical that abstracts a hydrogen from RH to give R•, which dimerizes. Dialkyl and diacyl peroxides have been used, as well as Fenton's reagent (p. 700). This reaction is far from general, though in certain cases respectable yields have been obtained. Among susceptible positions are those at a tertiary carbon,[279] as well as those α to a phenyl group (especially if there is also an α-alkyl or α-chloro group),[280] an ether group,[281] a carbonyl group,[282] a

[276]For a monograph on the formation of C—C bonds by radical reactions, see Giese, Ref. 1. For a review of arylation at carbon, see Abramovitch; Barton; Finet *Tetrahedron* **1988**, *44*, 3039-3071. For a review of aryl–aryl coupling, see Sainsbury *Tetrahedron* **1980**, *36*, 3327-3359.
[277]Brown; Crabtree *J. Am. Chem. Soc.* **1989**, *111*, 2935, 2946, *J. Chem. Educ.* **1988**, *65*, 290.
[278]Boojamra; Crabtree; Ferguson; Muedas *Tetrahedron Lett.* **1989**, *30*, 5583.
[279]Meshcheryakov; Érzyutova *Bull. Acad. Sci. USSR, Div. Chem. Sci.* **1966**, 94.
[280]McBay; Tucker; Groves *J. Org. Chem.* **1959**, *24*, 536; Johnston; Williams *J. Chem. Soc.* **1960**, 1168.
[281]Pfordte; Leuschner *Liebigs. Ann. Chem.* **1961**, *643*, 1.
[282]Kharasch; McBay; Urry *J. Am. Chem. Soc.* **1948**, *70*, 1269; Leffingwell *Chem. Commun.* **1970**, 357; Hawkins; Large *J. Chem. Soc., Perkin Trans. 1* **1974**, 280.

cyano group,[283] a dialkylamino group,[284] or a carboxylic ester group, either the acid or alcohol side.[285]

OS **IV**, 367; **V**, 1026; **VII**, 482.

4-17 Coupling of Alkynes
De-hydrogen-coupling

$$2R-C{\equiv}C-H \xrightarrow[\text{pyridine}]{CuX_2} R-C{\equiv}C-C{\equiv}C-R$$

Terminal alkynes can be coupled by heating with stoichiometric amounts of cupric salts in pyridine or a similar base. This reaction, which produces symmetrical diynes in high yields, is called the *Eglinton reaction*.[286] The large-ring annulenes of Sondheimer et al. (see p. 62) were prepared by rearrangement and hydrogenation of cyclic polyynes,[287] prepared by Eglinton coupling of terminal diynes, e.g.,[288]

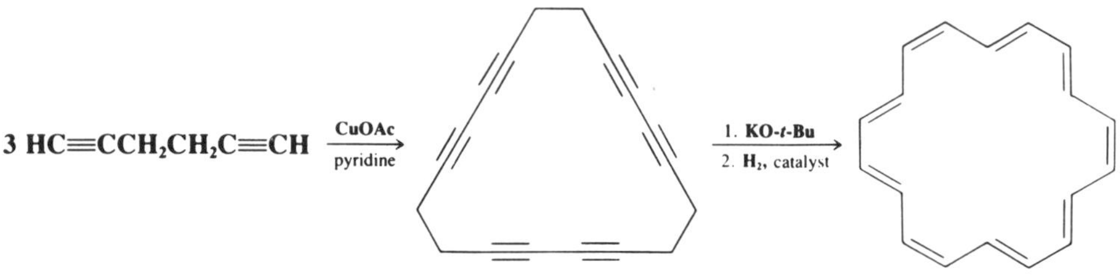

20

20 is a cyclic trimer of 1,5-hexadiyne. The corresponding tetramers (C_{24}), pentamers (C_{30}), and hexamers (C_{36}) were also formed.

The Eglinton reaction is of wide scope. Many functional groups can be present on the alkyne. The oxidation is usually quite specific for triple-bond hydrogen. Another common procedure is the use of catalytic amounts of cuprous salts in the presence of ammonia or ammonium chloride (this method is called the *Glaser reaction*). Atmospheric oxygen or some other oxidizing agent such as permanganate or hydrogen peroxide is required in the latter procedure. This method is not satisfactory for cyclic coupling. Unsymmetrical diynes can be prepared by *Cadiot–Chodkiewicz coupling*:[289]

$$R-C{\equiv}C-H + R'-C{\equiv}C-Br \xrightarrow{Cu^+} R-C{\equiv}C-C{\equiv}C-R' + HBr$$

This may be regarded as a variation of **0-100** but it must have a different mechanism since acetylenic halides give the reaction but ordinary alkyl halides do not, which is hardly compatible with a nucleophilic mechanism. However, the mechanism is not fully understood. Propargyl halides also give the reaction.[290] A variation of the Cadiot–Chodkiewicz method

[283]Kharasch; Sosnovsky *Tetrahedron* **1958**, *3*, 97.

[284]Schwetlick; Jentzsch; Karl; Wolter *J. Prakt. Chem.* **1964**, [4] *25*, 95.

[285]Boguslavskaya; Razuvaev *J. Gen. Chem. USSR* **1963**, *33*, 1967.

[286]For reviews, see Simándi, in Patai; Rappoport *The Chemistry of Functional Groups, Supplement C*, pt. 1; Wiley: New York, 1983, pp. 529-534; Nigh, Ref. 185, pp. 11-31; Cadiot; Chodkiewicz, in Viehe *Acetylenes*; Marcel Dekker: New York; 1969, pp. 597-647.

[287]For a review of cyclic alkynes, see Nakagawa, in Patai *The Chemistry of the Carbon–Carbon Triple Bond*, pt. 2; Wiley: New York, 1978, pp. 635-712.

[288]Sondheimer; Wolovsky *J. Am. Chem. Soc.* **1962**, *84*, 260; Sondheimer; Wolovsky; Amiel *J. Am. Chem. Soc.* **1962**, *84*, 274.

[289]Chodkiewicz *Ann. Chim. (Paris)* **1957**, [13] *2*, 819.

[290]Sevin; Chodkiewicz; Cadiot *Bull. Soc. Chim. Fr.* **1974**, 913.

consists of treating a haloalkyne (R'C≡CX) with a copper acetylide (RC≡CCu).[291] The Cadiot–Chodkiewicz procedure can be adapted to the preparation of diynes in which R' = H by the use of BrC≡CSiEt$_3$ and subsequent cleavage of the SiEt$_3$ group.[292] This protecting group can also be used in the Eglinton or Glaser methods.[293]

The mechanism of the Eglinton and Glaser reactions probably begins with loss of a proton

$$\text{R—C≡C—H} \xrightarrow{\text{base}} \text{R—C≡C}^-$$

since there is a base present and acetylenic protons are acidic. The last step is probably the coupling of two radicals:

$$\text{2R—C≡C·} \longrightarrow \text{R—C≡C—C≡C—R}$$

but just how the carbanion becomes oxidized to the radical and what part the cuprous ion plays (other than forming the acetylide salt) are matters of considerable speculation,[294] and depend on the oxidizing agent. It is known, of course, that cuprous ion can form complexes with triple bonds.

OS **V**, 517; **VI**, 68, 925; **65**, 52.

4-18 Arylation of Aromatic Compounds by Diazonium Salts
Arylation or **Aryl-de-hydrogenation**

$$\text{ArH} + \text{Ar'N}_2{}^+ \text{ X}^- \xrightarrow{\text{OH}^-} \text{Ar—Ar'}$$

When the normally acidic solution of a diazonium salt is made alkaline, the aryl portion of the diazonium salt can couple with another aromatic ring. Known as the *Gomberg* or *Gomberg–Bachmann reaction*,[295] it has been performed on several types of aromatic rings and on quinones. Yields are not high (usually under 40%) because of the many side reactions undergone by diazonium salts, though higher yields have been obtained under phase transfer conditions.[296] The conditions of the Meerwein reaction (**4-19**), treatment of the solution with a copper–ion catalyst, have also been used, as has the addition of sodium nitrite in Me$_2$SO (to benzene diazonium fluoroborate in Me$_2$SO).[297] When the Gomberg–Bachmann reaction is performed intramolecularly, either by the alkaline solution or by the copper–ion procedure,

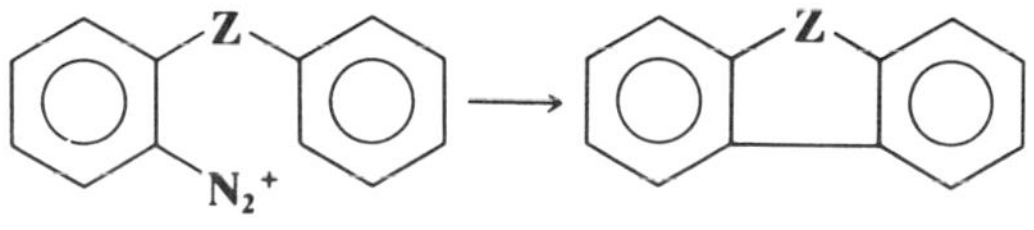

it is called the *Pschorr ring closure*[298] and yields are usually somewhat higher. Still higher yields have been obtained by carrying out the Pschorr reaction electrochemically.[299] The Pschorr reaction has been carried out for Z = CH=CH, CH$_2$CH$_2$, NH, C=O, CH$_2$, and quite a few others. A rapid and convenient way to carry out the Pschorr synthesis is to

[291]Curtis; Taylor *J. Chem. Soc. C* **1971**, 186.
[292]Eastmond; Walton *Tetrahedron* **1972**, *28*, 4591; Ghose; Walton *Synthesis* **1974**, 890.
[293]Johnson; Walton *Tetrahedron* **1972**, *28*, 5221.
[294]See the discussions in Nigh, Ref. 185, pp. 27-31; Fedenok; Berdnikov; Shvartsberg *J. Org. Chem. USSR* **1973**, *9*, 1806; Clifford; Waters *J. Chem. Soc.* **1963**, 3056.
[295]For reviews, see Bolton; Williams *Chem. Soc. Rev.* **1986**, *15*, 261-289; Hey *Adv. Free-Radical Chem.* **1966**, *2*, 47-86. For a review applied to heterocyclic substrates, see Vernin; Dou; Metzger *Bull. Soc. Chim. Fr.* **1972**, 1173-1203.
[296]Beadle; Korzeniowski; Rosenberg; Garcia-Slanga; Gokel *J. Org. Chem.* **1984**, *49*, 1594.
[297]Kamigata; Kurihara; Minato; Kobayashi *Bull. Chem. Soc. Jpn.* **1971**, *44*, 3152.
[298]For a review, see Abramovitch *Adv. Free-Radical Chem.* **1966**, *2*, 87-138.
[299]Elofson; Gadallah *J. Org. Chem.* **1971**, *36*, 1769.

diazotize the amine substrate with isopropyl nitrite in the presence of sodium iodide, in which case the ring-closed product is formed in one step.[300]

Other compounds with nitrogen–nitrogen bonds have been used instead of diazonium salts. Among these are N-nitroso amides [ArN(NO)COR], triazenes,[301] and azo compounds. Still another method involves treatment of an aromatic primary amine directly with an alkyl nitrite in an aromatic substrate as solvent.[302]

In each case the mechanism involves generation of an aryl radical from a covalent azo compound. In acid solution diazonium salts are ionic and their reactions are polar. When they cleave, the product is an aryl cation (see p. 644). However, in neutral or basic solution, diazonium ions are converted to covalent compounds, and these cleave to give free radicals:

$$\text{Ar}-\text{N}{=}\text{N}-\text{Z} \longrightarrow \text{Ar}\cdot + \text{N}{\equiv}\text{N} + \text{Z}\cdot$$

Under Gomberg–Bachmann conditions, the species that cleaves is the anhydride:[303]

$$\text{Ar}-\text{N}{=}\text{N}-\text{O}-\text{N}{=}\text{N}-\text{Ar} \longrightarrow \text{Ar}\cdot + \text{N}_2 + \cdot\text{O}-\text{N}{=}\text{N}-\text{Ar}$$

$$\mathbf{21}$$

The aryl radical thus formed attacks the substrate to give the intermediate **1** (p. 680), from which the radical **21** abstracts hydrogen to give the product. N-Nitroso amides probably rearrange to N-acyloxy compounds, which cleave to give aryl radicals:[304]

$$2\text{Ar}-\underset{\underset{\text{O}}{\|}}{\overset{\overset{\text{N}{=}\text{O}}{|}}{\text{N}}}-\text{C}-\text{R} \longrightarrow 2\text{Ar}-\text{N}{=}\text{N}-\text{O}-\text{COR} \longrightarrow$$

$$\text{Ar}\cdot + \text{Ar}-\text{N}{=}\text{N}-\text{O}\cdot + \text{N}_2 + (\text{RCO})_2\text{O}$$

There is evidence that the reaction with alkyl nitrites also involves attack by aryl radicals.[305]

The Pschorr reaction can take place by two different mechanisms, depending on conditions: (1) attack by an aryl radical (as in the Gomberg–Bachmann reaction) or (2) attack by an aryl cation (similar to the S$_N$1 mechanism discussed on p. 644).[306] Under certain conditions the ordinary Gomberg–Bachmann reaction can also involve attack by aryl cations.[307]

OS **I**, 113; **IV**, 718.

4-19 Arylation of Activated Olefins by Diazonium Salts. Meerwein Arylation
Arylation or **Aryl-de-hydrogenation**

$$\underset{\underset{\text{H}}{|}}{\overset{|}{\text{Z}-\text{C}}}{=}\overset{|}{\text{C}} \xrightarrow[\text{ArN}_2{}^+\text{Cl}^-]{\text{CuCl}_2} \overset{|}{\text{Z}-\text{C}}{=}\overset{|}{\text{C}}-\text{Ar}$$

[300]Chauncy; Gellert *Aust. J. Chem.* **1969**, *22*, 993. See also Duclos; Tung; Rapoport *J. Org. Chem.* **1984**, *49*, 5243.
[301]See, for example, Patrick; Willaredt; DeGonia *J. Org. Chem.* **1985**, *50*, 2232; Butler; O'Shea; Shelly *J. Chem. Soc., Perkin Trans. 1* **1987**, 1039.
[302]Cadogan *J. Chem. Soc.* **1962**, 4257; Fillipi; Vernin; Dou; Metzger; Perkins *Bull. Soc. Chim. Fr.* **1974**, 1075.
[303]Rüchardt; Merz *Tetrahedron Lett.* **1964**, 2431; Eliel; Saha; Meyerson *J. Org. Chem.* **1965**, *30*, 2451.
[304]Cadogan; Murray; Sharp *J. Chem. Soc., Perkin Trans. 2* **1976**, 583, and references cited therein.
[305]Gragerov; Levit *J. Org. Chem. USSR* **1968**, *4*, 7.
[306]For an alternative to the second mechanism, see Gadallah; Cantu; Elofson *J. Org. Chem.* **1973**, *38*, 2386.
[307]For examples; see Kobori; Kobayashi; Minato *Bull. Chem. Soc., Jpn.* **1970**, *43*, 223; Cooper; Perkins *Tetrahedron Lett.* **1969**, 2477; Burri; Zollinger *Helv. Chim. Acta* **1973**, *56*, 2204; Eustathopoulos; Rinaudo; Bonnier *Bull. Soc. Chim. Fr.* **1974**, 2911. For a discussion, see Zollinger *Acc. Chem. Res.* **1973**, *6*, 335-341, pp. 338-339.

Olefins activated by an electron-withdrawing group (Z may be C=C, halogen, C=O, Ar, CN, etc.) can be arylated by treatment with a diazonium salt and a cupric chloride[308] catalyst. This is called the *Meerwein arylation reaction.*[309] Addition of ArCl to the double bond (to give Z—C—C—Ar) is a side reaction (**5-33**). In an improved procedure, an arylamine is

$$Z\!-\!\!\overset{|}{\underset{\underset{Cl}{|}}{C}}\!-\!\!\overset{|}{\underset{\underset{H}{|}}{C}}\!-\!Ar$$

treated with an alkyl nitrite (generating ArN_2^+ in situ) and a copper(II) halide in the presence of the olefin.[310]

The mechanism is probably of the free-radical type, with Ar• forming as in **4-25** and then[311]

The radical **22** can react with cupric chloride by two pathways, one of which leads to addition and the other to substitution. Even when the addition pathway is taken, however, the substitution product may still be formed by subsequent elimination of HCl.

OS **IV**, 15.

4-20   Arylation and Alkylation of Olefins by Organopalladium Compounds
The Heck Reaction

Alkylation or **Alkyl-de-hydrogenation**, etc.

$$R_2C{=}CH_2 + \text{``ArPdX''} \longrightarrow R_2C{=}CH{-}Ar$$

Arylation of olefins can also be achieved[312] by treatment with an "arylpalladium" reagent that can be generated in situ by several[313] methods: (1) by treatment of an aryl bromide with a palladium-triarylphosphine complex (ArBr → "ArPdBr");[314] (2) by treatment of an aryl iodide[315] with palladium acetate[316] in the presence of a base such as tributylamine or

---

[308]$FeCl_2$ is also effective: Ganushchak; Obushak; Luka *J. Org. Chem. USSR* **1981**, *17*, 765.

[309]For reviews, see Dombrovskii *Russ. Chem. Rev.* **1984**, *53*, 943-955; Rondestvedt *Org. React.* **1976**, *24*, 225-259.

[310]Doyle; Siegfried; Elliott; Dellaria *J. Org. Chem.* **1977**, *42*, 2431.

[311]Dickerman; Vermont *J. Am. Chem. Soc.* **1962**, *84*, 4150; Morrison; Cazes; Samkoff; Howe *J. Am. Chem. Soc.* **1962**, *84*, 4152.

[312]For reviews of this and related reactions, see Heck *Palladium Reagents in Organic Syntheses*; Academic Press: New York, 1985, pp. 179-321; Ryabov *Synthesis* **1985**, 233-252; Heck *Org. React.* **1982**, *27*, 345-390, *Adv. Catal.* **1977**, *26*, 323-349; Volkova; Levitin; Vol'pin *Russ. Chem. Rev.* **1975**, *44*, 552-560; Moritani; Fujiwara *Synthesis* **1973**, 524-533; Jira; Freiesleben *Organomet. React.* **1972**, *3*, 1-190, pp. 84-105.

[313]For other methods, see Murahashi; Yamamura; Mita *J. Org. Chem.* **1977**, *42*, 2870; Luong-Thi; Riviere *J. Chem. Soc., Chem. Commun.* **1978**, 918; Akiyama; Miyazaki; Kaneda; Teranishi; Fujiwara; Abe; Taniguchi *J. Org. Chem.* **1980**, *45*, 2359; Tsuji; Nagashima *Tetrahedron* **1984**, *40*, 2699; Kikukawa; Naritomi; He; Wada; Matsuda *J. Org. Chem.* **1985**, *50*, 299; Chen; Yang *Tetrahedron Lett.* **1986**, *27*, 1171; Kasahara; Izumi; Miyamoto; Sakai *Chem. Ind. (London)* **1989**, 192; Miura; Hashimoto; Itoh; Nomura *Tetrahedron Lett.* **1989**, *30*, 975.

[314]For reviews, see Heck *Acc. Chem. Res.* **1979**, *12*, 146-151, *Pure Appl. Chem.* **1978**, *50*, 691-701. See also Bender; Stakem; Heck *J. Org. Chem.* **1982**, *47*, 1278; Spencer *J. Organomet. Chem.* **1983**, *258*, 101.

[315]For a method that uses an aryl chloride, but converts it to an aryl iodide in situ, see Bozell; Vogt *J. Am. Chem. Soc.* **1988**, *110*, 2655.

[316]For a more efficient palladium reagent, see Andersson; Karabelas; Hallberg; Andersson *J. Org. Chem.* **1985**, *50*, 3891. See also Merlic; Semmelhack *J. Organomet. Chem.* **1990**, *391*, C23.

potassium acetate (ArI → "ArPdI");[317] (3) by treatment of an arylmercury compound (either Ar₂Hg or ArHgX) with LiPdCl₃ (ArHgX → "ArPdX")[318] (in some cases other noble metal salts have been used); or (4) by the reaction of an aromatic compound with palladium acetate or palladium metal and silver acetate in acetic acid [in this case an aryl *hydrogen* is replaced (ArH → "ArPdOAc")].[319] Whichever of these methods is used, the reaction is known as *the Heck reaction*.

Unlike **4-19**, the Heck reaction is not limited to activated substrates. The substrate can be a simple olefin, or it can contain a variety of functional groups, such as ester, ether,[319a] carboxyl, phenolic, or cyano groups.[320] Primary and secondary allylic alcohols (and even nonallylic unsaturated alcohols[321]) give aldehydes or ketones that are products of double-bond migration,[322] e.g.,

$$CH_2=CH-CH_2OH \xrightarrow{\text{"PhPdCl"}} PhCH_2CH_2CHO$$

Ethylene is the most reactive olefin. Increasing substitution lowers the reactivity. Substitution therefore takes place at the less highly substituted side of the double bond.[323] Alkylation can also be accomplished, but only if the alkyl group lacks a β hydrogen, e.g., the reaction is successful for the introduction of methyl, benzyl, and neopentyl groups.[324] However, vinylic groups, even those possessing β hydrogens, have been successfully introduced (to give 1,3-dienes) by the reaction of the olefin with a vinylic halide in the presence of a trialkylamine and a catalyst composed of palladium acetate and a triarylphosphine at 100 to 150°C.[325] The reaction has also been done with terminal alkynes as substrates.[326]

The evidence is in accord with an addition-elimination mechanism (addition of ArPdX followed by elimination of HPdX) in most cases.[327] The reactions are stereospecific, yielding products expected from syn addition followed by syn elimination.[328] Because the product is formed by an elimination step, with suitable substrates the double bond can go the other way, resulting in allylic rearrangement, e.g.,[329]

$$PhI + \bigcirc \xrightarrow[\text{NaOAc, } n\text{-Bu}_4\text{N}^+\text{Cl}^-]{Pd(OAc)_2} 89\% \ Ph-\bigcirc$$

The Heck reaction has also been performed intramolecularly.[330]
OS **VI**, 815; **VII**, 361.

[317]Mizoroki; Mori; Ozaki *Bull. Chem. Soc. Jpn.* **1971**, *44*, 581; Mori; Mizoroki; Ozaki *Bull. Chem. Soc. Jpn.* **1973**, *46*, 1505; Heck; Nolley *J. Org. Chem.* **1972**, *37*, 2320; Ziegler; Heck *J. Org. Chem.* **1978**, *43*, 2941; Hirao; Enda; Ohshiro; Agawa *Chem. Lett.* **1981**, 403; Jeffery *J. Chem. Soc., Chem. Commun.* **1984**, 1287; Bumagin; More; Beletskaya *J. Organomet. Chem.* **1989**, *371*, 397; Larock; Johnson *J. Chem. Soc., Chem. Commun.* **1989**, 1368.
[318]Heck *J. Am. Chem. Soc.* **1968**, *90*, 5518, 5526, 5535. For a review, see Larock, Ref. 238, pp. 273-292.
[319]See, for example, Fujiwara; Moritani; Matsuda *Tetrahedron* **1968**, *24*, 4819; Fujiwara; Maruyama; Yoshidomi; Taniguchi *J. Org. Chem.* **1981**, *46*, 851. For a review, see Kozhevnikov *Russ. Chem. Rev.* **1983**, *52*, 138-151.
[319a]For a review pertaining to enol ethers, see Daves *Adv. Met.-Org. Chem.* **1991**, *2*, 59-99.
[320]For a review of cases where the olefin contains an α hetero atom, see Daves; Hallberg *Chem. Rev.* **1989**, *89*, 1433-1445.
[321]Larock; Leung; Stolz-Dunn *Tetrahedron Lett.* **1989**, *30*, 6629.
[322]See, for example, Melpolder; Heck *J. Org. Chem.* **1976**, *41*, 265; Chalk; Magennis *J. Org. Chem.* **1976**, *41*, 273, 1206.
[323]Heck *J. Am. Chem. Soc.* **1969**, *91*, 6707, **1971**, *93*, 6896.
[324]Heck *J. Organomet. Chem.* **1972**, *37*, 389; Heck; Nolley, Ref. 317.
[325]Dieck; Heck *J. Org. Chem.* **1975**, *40*, 1083; Kim; Patel; Heck *J. Org. Chem.* **1981**, *46*, 1067; Heck *Pure Appl. Chem.* **1981**, *53*, 2323-2332. See also Luong-Thi; Riviere *Tetrahedron Lett.* **1979**, 4657; Jeffery *Tetrahedron Lett.* **1985**, *26*, 2667, *J. Chem. Soc., Chem. Commun.* **1991**, 324; Scott; Peña; Swärd; Stoessel; Stille *J. Org. Chem.* **1985**, *50*, 2302; Larock; Gong *J. Org. Chem.* **1989**, *54*, 2047.
[326]Cassar *J. Organomet. Chem.* **1975**, *93*, 253; Dieck; Heck *J. Organomet. Chem.* **1975**, *93*, 259; Sonogashira; Tohda; Hagihara *Tetrahedron Lett.* **1975**, 4467; Singh; Just *J. Org. Chem.* **1989**, *54*, 4453. See also Heck *Palladium Reagents in Organic Syntheses*, Ref. 312, pp. 299-306.
[327]Heck *J. Am. Chem. Soc.* **1969**, *91*, 6707; Shue *J. Am. Chem. Soc.* **1971**, *93*, 7116; Heck; Nolley, Ref. 317.
[328]Heck, Ref. 327; Moritani; Danno; Fujiwara; Teranishi *Bull. Chem. Soc. Jpn.* **1971**, *44*, 578.
[329]Larock; Baker *Tetrahedron Lett.* **1988**, *29*, 905. Also see Larock; Gong; Baker *Tetrahedron Lett.* **1989**, *30*, 2603.
[330]See, for example, Abelman; Oh; Overman *J. Org. Chem.* **1987**, *52*, 4130; Negishi; Zhang; O'Connor *Tetrahedron Lett.* **1988**, *29*, 2915; Larock; Song; Baker; Gong *Tetrahedron Lett.* **1988**, *29*, 2919.

4-21 Alkylation and Arylation of Aromatic Compounds by Peroxides
Alkylation or **Alkyl-de-hydrogenation**

$$\text{ArH} + \text{R}-\underset{\underset{\text{O}}{\|}}{\text{C}}-\text{O}-\text{O}-\underset{\underset{\text{O}}{\|}}{\text{C}}-\text{R} \longrightarrow \text{Ar}-\text{R}$$

This reaction is most often carried out with R = aryl, so the net result is the same as in **4-18**, though the reagent is different.[331] It is used less often than **4-18**, but the scope is similar. When R = alkyl, the scope is more limited.[332] Only certain aromatic compounds, particularly benzene rings with two or more nitro groups, and fused ring systems, can be alkylated by this procedure. 1,4-Quinones can be alkylated with diacyl peroxides or with lead tetraacetate (methylation occurs with this reagent).

The mechanism is as shown on p. 680 (CIDNP has been observed[333]); the radicals are produced by

$$\text{R}-\underset{\underset{\text{O}}{\|}}{\text{C}}-\text{O}-\text{O}-\underset{\underset{\text{O}}{\|}}{\text{C}}-\text{R} \longrightarrow 2\text{R}-\underset{\underset{\text{O}}{\|}}{\text{C}}-\text{O}\cdot \longrightarrow 2\text{R}\cdot + 2\text{CO}_2$$

Since no relatively stable free radical is present (such as **21** in **4-18**), most of the product arises from dimerization and disproportionation.[334] The addition of a small amount of nitrobenzene increases the yield of arylation product because the nitrobenzene is converted to diphenyl nitroxide, which abstracts the hydrogen from **1** and reduces the extent of side reactions.[335]

Aromatic compounds can also be arylated by aryllead tricarboxylates.[336] Best yields (~70 to 85%) are obtained when the substrate contains alkyl groups; an electrophilic mechanism

$$\text{ArH} + \text{Ar}'\text{Pb}(\text{OAc})_3 \longrightarrow \text{ArAr}'$$

is likely. Phenols are phenylated ortho to the OH group (and enols are α phenylated) by triphenylbismuth dichloride or by certain other Bi(V) reagents.[337] O-Phenylation is a possible side reaction. As with the aryllead tricarboxylate reactions, a free-radical mechanism is unlikely.[338]

OS **V**, 51. See also OS **V**, 952; **VI**, 890.

4-22 Photochemical Arylation of Aromatic Compounds
Arylation or **Aryl-de-hydrogenation**

$$\text{ArH} + \text{Ar}'\text{I} \xrightarrow{h\nu} \text{ArAr}'$$

Another free-radical arylation method consists of the photolysis of aryl iodides in an aromatic solvent.[339] Yields are generally higher than in **4-18** or **4-21**. The aryl iodide may contain OH

[331]For reviews, see Ref. 295.

[332]For reviews of the free-radical alkylation of aromatic compounds, see Tiecco; Testaferri *React. Intermed.* (*Plenum*) **1983**, *3*, 61-11; Dou; Vernin; Metzger *Bull. Soc. Chim. Fr.* **1971**, 4593.

[333]Kaptein; Freeman; Hill; Bargon *J. Chem. Soc., Chem. Commun.* **1973**, 953.

[334]We have given the main steps that lead to biphenyls. The mechanism is actually more complicated than this and includes more than 100 elementary steps resulting in many side products, including those mentioned on p. 681: DeTar; Long; Rendleman; Bradley; Duncan *J. Am. Chem. Soc.* **1967**, *89*, 4051; DeTar *J. Am. Chem. Soc.* **1967**, *89*, 4058. See also Jandu; Nicolopoulou; Perkins *J. Chem. Res. (S)* **1985**, 88.

[335]Chalfont; Hey; Liang; Perkins *J. Chem. Soc. B* **1971**, 233.

[336]Bell; Kalman; May; Pinhey; Sternhell *Aust. J. Chem.* **1979**, *32*, 1531.

[337]For a review, see Abramovitch; Barton; Finet, Ref. 276, pp. 3040-3047.

[338]Barton; Finet; Giannotti; Halley *J. Chem. Soc., Perkin Trans. 1* **1987**, 241.

[339]Wolf; Kharasch *J. Org. Chem.* **1965**, *30*, 2493. For a review, see Sharma; Kharasch *Angew. Chem. Int. Ed. Engl.* **1968**, *7*, 36-44 [*Angew. Chem. 80*, 69-77].

or COOH groups. The mechanism is similar to that of **4-18**. The aryl radicals are generated by the photolytic cleavage $ArI \rightarrow Ar\cdot + I\cdot$. The reaction has been applied to intramolecular arylation (analogous to the Pschorr reaction).[340] A similar reaction is photolysis of an arylthallium bis(trifluoroacetate)(**2-22**) in an aromatic solvent. Here too, an unsymmetrical biaryl is produced in good yields.[341]

$$Ar'Tl(OCOCF_3)_2 \xrightarrow[ArH]{h\nu} ArAr'$$

In this case it is the C—Tl bond that is cleaved to give aryl radicals.

4-23 Alkylation, Acylation, and Carbalkoxylation of Nitrogen Heterocycles[342]
Alkylation or Alkyl-de-hydrogenation, etc.

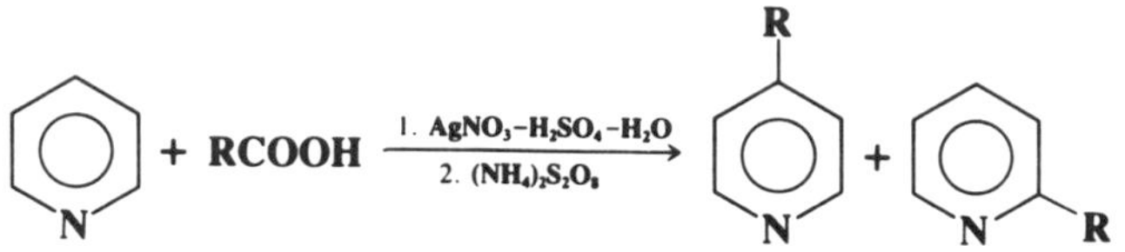

Alkylation of protonated nitrogen heterocycles (e.g., pyridines, quinolines) can be accomplished by treatment with a carboxylic acid, silver nitrate, sulfuric acid, and ammonium peroxydisulfate.[343] R can be primary, secondary, or tertiary. The attacking species is R·, formed by[344]

$$2Ag^+ + S_2O_8^{2-} \longrightarrow 2Ag^{2+} + 2SO_4^{2-}$$

$$RCOOH + Ag^{2+} \longrightarrow RCOO\cdot + H^+ + Ag^+$$

$$RCOO\cdot \longrightarrow R\cdot + CO_2$$

A hydroxymethyl group can be introduced ($ArH \rightarrow ArCH_2OH$) by several variations of this method.[345] Alkylation of these substrates can also be accomplished by generating the alkyl radicals in other ways: from hydroperoxides and $FeSO_4$,[346] from alkyl iodides and H_2O_2–Fe(II),[347] from carboxylic acids and lead tetraacetate, or from the photochemically induced decarboxylation of carboxylic acids by iodosobenzene diacetate.[348] The reaction has also been applied to acetophenone and ferrocene.[349]

[340]See, for example, Kupchan; Wormser *J. Org. Chem.* **1965**, *30*, 3792; Jeffs; Hansen *J. Am. Chem. Soc.* **1967**, *89*, 2798; Thyagarajan; Kharasch; Lewis; Wolf *Chem. Commun.* **1967**, 614.
[341]Taylor; Kienzle; McKillop *J. Am. Chem. Soc.* **1970**, *92*, 6088.
[342]For reviews; see Heinisch *Heterocycles* **1987**, *26*, 481-496; Minisci; Vismara; Fontana *Heterocycles* **1989**, *28*, 489-519; Minisci *Top. Curr. Chem.* **1976**, *62*, 1-48, pp. 17-46, *Synthesis* **1973**, 1-24, pp. 12-19. For a review of substitution of carbon groups on nitrogen heterocycles, see Vorbrüggen; Maas *Heterocycles* **1988**, *27*, 2659-2776.
[343]Minisci; Mondelli; Gardini; Porta *Tetrahedron* **1972**, *28*, 2403; Citterio; Minisci; Franchi *J. Org. Chem.* **1980**, *45*, 4752; Fontana; Minisci; Barbosa; Vismara *Tetrahedron* **1990**, *46*, 2525.
[344]Anderson; Kochi *J. Am. Chem. Soc.* **1970**, *92*, 1651.
[345]See Citterio; Gentile; Minisci; Serravalle; Ventura *Tetrahedron* **1985**, *41*, 617; Katz; Mistry; Mitchell *Synth. Commun.* **1989**, *19*, 317.
[346]Minisci; Selva; Porta; Barilli; Gardini *Tetrahedron* **1972**, *28*, 2415.
[347]Fontana; Minisci; Barbosa; Vismara *Acta Chem. Scand.* **1989**, *43*, 995.
[348]Minisci; Vismara; Fontana; Barbosa *Tetrahedron Lett.* **1989**, *30*, 4569.
[349]Din; Meth-Cohn; Walshe *Tetrahedron Lett.* **1979**, 4783.

Protonated nitrogen heterocycles can be acylated by treatment with an aldehyde, *t*-butyl hydroperoxide, sulfuric acid, and ferrous sulfate, e.g.,[350]

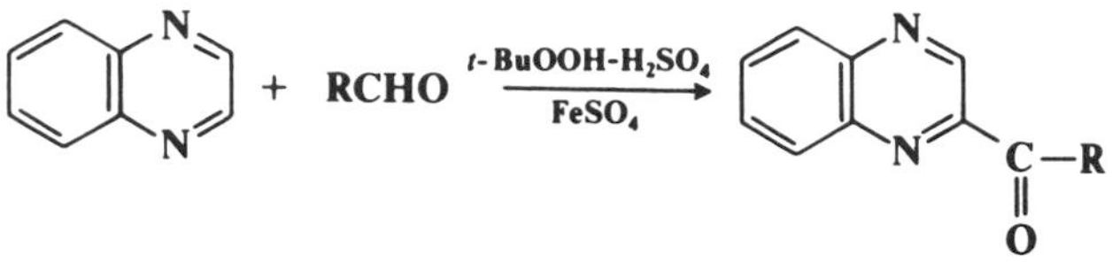

These alkylation and acylation reactions are important because Friedel-Crafts alkylation and acylation (**1-12, 1-14**) cannot be applied to most nitrogen heterocycles. See also **3-17**.

Protonated nitrogen heterocycles can be carbalkoxylated[351] by treatment with esters of α-keto acids and Fenton's reagent:

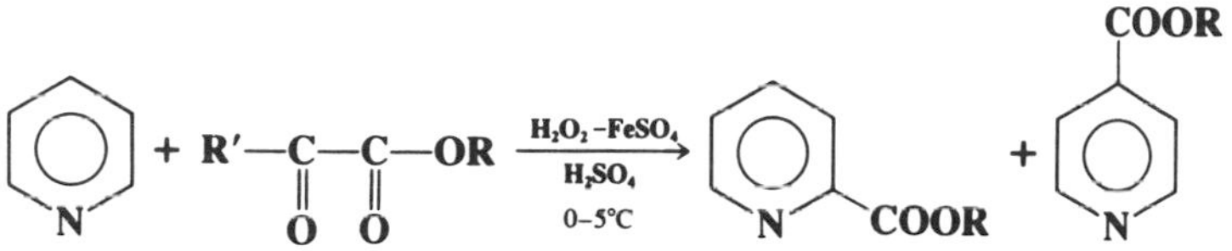

The attack is by •COOR radicals generated from the esters:

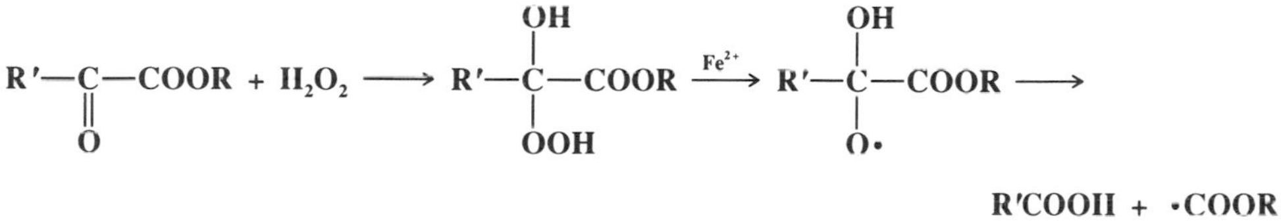

$$R'COOH + \cdot COOR$$

Similarly, a carbamoyl group can be introduced[352] by the use of the radicals $H_2N\overset{\underset{\|}{O}}{C}\cdot$ or $Me_2N\overset{\underset{\|}{O}}{C}\cdot$ generated from formamide or dimethylformamide and H_2SO_4, H_2O_2, and $FeSO_4$ or other oxidants.

N_2 as Leaving Group[353]

In these reactions diazonium salts are cleaved to aryl radicals,[354] in most cases with the assistance of copper salts. Reactions **4-18** and **4-19** may also be regarded as belonging to this category with respect to the attacking compound. For nucleophilic substitutions of diazonium salts, see **3-20** to **3-24**.

4-24   Replacement of the Diazonium Group by Hydrogen
Dediazoniation or **Hydro-de-diazoniation**

$$ArN_2^+ + H_3PO_2 \longrightarrow ArH$$

[350]Caronna; Gardini; Minisci *Chem. Commun.* **1969,** 201; Arnoldi; Bellatti; Caronna; Citterio; Minisci; Porta; Sesana *Gazz. Chim. Ital.* **1977,** *107,* 491.
[351]Bernardi; Caronna; Galli; Minisci; Perchinunno *Tetrahedron Lett.* **1973,** 645; Heinisch; Lötsch *Angew. Chem. Int. Ed. Engl.* **1985,** *24,* 692 [*Angew. Chem. 97,* 694].
[352]Minisci; Gardini; Galli; Bertini *Tetrahedron Lett.* **1970,** 15; Minisci; Citterio; Vismara; Giordano *Tetrahedron* **1985,** *41,* 4157.
[353]For a review, see Wulfman, in Patai *The Chemistry of Diazonium and Diazo Groups,* pt. 1; Wiley: New York, 1978, pp. 286-297.
[354]For reviews, see Galli *Chem. Rev.* **1988,** *88,* 765-792; Zollinger *Acc. Chem. Res.* **1973,** *6,* 355-341, pp. 339-341.

Reduction of the diazonium group (*dediazoniation*) provides an indirect method for the removal of an amino group from an aromatic ring.[355] The best and most common way of accomplishing this is by use of hypophosphorous acid H_3PO_2, though many other reducing agents[356] have been used, among them ethanol, HMPA,[356] thiophenol,[357] and sodium stannite. Ethanol was the earliest reagent used, and it frequently gives good yields, but often ethers (ArOEt) are side products. When H_3PO_2 is used, 5 to 15 moles of this reagent are required per mole of substrate. Diazonium salts can be reduced in nonaqueous media by several methods,[358] including treatment with Bu_3SnH or Et_3SiH in ethers or $MeCN$[359] and by isolation as the BF_4^- salt and reduction of this with $NaBH_4$ in DMF.[360] Aromatic amines can be deaminated ($ArNH_2 \rightarrow ArH$) in one laboratory step by treatment with an alkyl nitrite in DMF[361] or boiling THF.[362] The corresponding diazonium salt is an intermediate.

Not many investigations of the mechanism have been carried out. It is generally assumed that the reaction of diazonium salts with ethanol to produce ethers takes place by an ionic (SN1) mechanism while the reduction to ArH proceeds by a free-radical process.[363] The reduction with H_3PO_2 is also believed to have a free-radical mechanism.[364] In the reduction with $NaBH_4$, an aryldiazene intermediate ($ArN=NH$) has been demonstrated,[365] arising from nucleophilic attack by BH_4^- on the β nitrogen. Such diazenes can be obtained as moderately stable (half-life of several hours) species in solution.[366] It is not entirely clear how the aryldiazene decomposes, but there are indications that either the aryl radical Ar• or the corresponding anion Ar^- may be involved.[367]

An important use of the dediazoniation reaction is to remove an amino group after it has been used to direct one or more other groups to ortho and para positions. For example, the compound 1,3,5-tribromobenzene cannot be prepared by direct bromination of benzene because the bromo group is ortho-para-directing; however, this compound is easily prepared by the following sequence:

$$C_6H_6 \xrightarrow[\substack{H_2SO_4 \\ 1\text{-}2}]{HNO_3} PhNO_2 \xrightarrow[9\text{-}47]{Sn-HCl} PhNH_2 \xrightarrow[1\text{-}11]{3Br_2}$$

[355]For a review, see Zollinger in Patai; Rappoport, Ref. 286, pp. 603-669.
[356]For lists of some of these, with references, see Ref.74, p. 25; Tröndlin; Rüchardt *Chem. Ber.* **1977**, *110*, 2494.
[357]Shono; Matsumura; Tsubata *Chem. Lett.* **1979**, 1051.
[358]For a list of some of these, with references, see Korzeniowski; Blum; Gokel *J. Org. Chem.* **1977**, *42*, 1469.
[359]Nakayama; Yoshida; Simamura *Tetrahedron* **1970**, *26*, 4609.
[360]Hendrickson *J. Am. Chem. Soc.* **1961**, *83*, 1251. See also Threadgill; Gledhill *J. Chem. Soc., Perkin Trans. 1* **1986**, 873.
[361]Doyle; Dellaria; Siegfried; Bishop *J. Org. Chem.* **1977**, *42*, 3494.
[362]Cadogan; Molina *J. Chem. Soc., Perkin Trans. 1* **1973**, 541.
[363]For examples, see DeTar; Turetzky *J. Am. Chem. Soc.* **1955**, 77, 1745, **1956**, *78*, 3925, 3928; DeTar; Kosuge *J. Am. Chem. Soc.* **1958**, *80*, 6072; Lewis; Chambers *J. Am. Chem. Soc.* **1971**, *93*, 3267; Broxton; Bunnett; Paik *J. Org. Chem.* **1977**, *42*, 643.
[364]See, for example, Kornblum; Cooper; Taylor *J. Am. Chem. Soc.* **1950**, 72, 3013; Beckwith *Aust. J. Chem.* **1972**, 25, 1887; Levit; Kiprianova; Gragerov *J. Org. Chem. USSR* **1975**, *11*, 2395.
[365]Bloch; Musso; Záhorszky *Angew. Chem. Int. Ed. Engl.* **1969**, *8*, 370 [*Angew. Chem. 81*, 392]; König; Musso; Záhorszky *Angew. Chem. Int. Ed. Engl.* **1972**, *11*, 45 [*Angew. Chem. 84*, 33]; McKenna; Traylor *J. Am. Chem. Soc.* **1971**, *93*, 2313.
[366]Huang; Kosower *J. Am. Chem. Soc.* **1968**, *90*, 2354, 2362, 2367; Smith; Hillhouse *J. Am. Chem. Soc.* **1988**, *110*, 4066.
[367]Rieker; Niederer; Leibfritz *Tetrahedron Lett.* **1969**, 4287; Kosower; Huang; Tsuji *J. Am. Chem. Soc.* **1969**, *91*, 2325; König; Musso; Záhorszky, Ref. 365; Broxton; McLeish *Aust. J. Chem.* **1983**, *36*, 1031.

Many other compounds that would otherwise be difficult to prepare are easily synthesized with the aid of the dediazoniation reaction.

Unwanted dediazoniation can be suppressed by using hexasulfonated calix[6]arenes (see p. 84).[368]

OS **I**, 133, 415; **II**, 353, 592; **III**, 295; **IV**, 947; **VI**, 334.

4-25 Replacement of the Diazonium Group by Chlorine or Bromine
Chloro-de-diazoniation, etc.

$$ArN_2^+ + CuCl \longrightarrow ArCl$$

Treatment of diazonium salts with cuprous chloride or bromide leads to aryl chlorides or bromides, respectively. In either case the reaction is called the *Sandmeyer reaction*. The reaction can also be carried out with copper and HBr or HCl, in which case it is called the *Gatterman reaction* (not to be confused with **1-16**). The Sandmeyer reaction is not useful for the preparation of fluorides or iodides, but for bromides and chlorides it is of wide scope and is probably the best way of introducing bromine or chlorine into an aromatic ring. The yields are usually high.

The mechanism is not known with certainty but is believed to take the following course:[369]

$$ArN_2^+ X^- + CuX \longrightarrow Ar\cdot + N_2 + CuX_2$$

$$Ar\cdot + CuX_2 \longrightarrow ArX + CuX$$

The first step involves a reduction of the diazonium ion by the cuprous ion, which results in the formation of an aryl radical. In the second step, the aryl radical abstracts halogen from cupric chloride, reducing it. CuX is regenerated and is thus a true catalyst.

Aryl bromides and chlorides can be prepared from primary aromatic amines in one step by several procedures,[370] including treatment of the amine (1) with *t*-butyl nitrite and anhydrous $CuCl_2$ or $CuBr_2$ at 65°C,[371] and (2) with *t*-butyl thionitrite or *t*-butyl thionitrate and $CuCl_2$ or $CuBr_2$ at room temperature.[372] These procedures are, in effect, a combination of **2-49** and the Sandmeyer reaction. A further advantage is that cooling to 0°C is not needed.

For the preparation of fluorides and iodides from diazonium salts, see **3-24** and **3-23**.

OS **I**, 135, 136, 162, 170; **II**, 130; **III**, 185; **IV**, 160. Also see OS **III**, 136; **IV**, 182.

4-26 **Nitro-de-diazoniation**

$$ArN_2^+ + NaNO_2 \xrightarrow{\text{Cu}^+} ArNO_2$$

Nitro compounds can be formed in good yields by treatment of diazonium salts with sodium nitrite in the presence of cuprous ion. The reaction occurs only in neutral or alkaline solution. This is not usually called the Sandmeyer reaction, although, like **4-25** and **4-28**, it was discovered by Sandmeyer. BF_4^- is often used as the negative ion to avoid competition from the chloride ion. The mechanism is probably like that of **4-25**.[373] If electron-withdrawing

[368]Shinkai; Mori; Araki; Manabe *Bull. Chem. Soc. Jpn.* **1987**, *60*, 3679.
[369]Dickerman; Weiss; Ingberman *J. Org. Chem.* **1956**, *21*, 380, *J. Am. Chem. Soc.* **1958**, *80*, 1904; Kochi *J. Am. Chem. Soc.* **1957**, *79*, 2942; Dickerman; DeSouza; Jacobson *J. Org. Chem.* **1969**, *34*, 710; Galli *J. Chem. Soc., Perkin Trans. 2* **1981**, 1459, **1982**, 1139, **1984**, 897. See also Hanson; Jones; Gilbert; Timms *J. Chem. Soc., Perkin Trans. 2* **1991**, 1009.
[370]For other procedures, see Brackman; Smit *Recl. Trav. Chim. Pays-Bas* **1966**, *85*, 857; Cadogan; Roy; Smith *J. Chem. Soc. C* **1966**, 1249.
[371]Doyle; Siegfried; Dellaria *J. Org. Chem.* **1977**, *42*, 2426.
[372]Oae; Shinhama; Kim *Chem. Lett.* **1979**, 939, *Bull. Chem. Soc. Jpn.* **1980**, *53*, 1065.
[373]For discussions, see Opgenorth; Rüchardt *Liebigs Ann. Chem.* **1974**, 1333; Singh; Kumar; Khanna *Tetrahedron Lett.* **1982**, *23*, 5191.

groups are present, the catalyst is not needed; $NaNO_2$ alone gives nitro compounds in high yields.[374]

OS **II**, 225; **III**, 341.

4-27 Replacement of the Diazonium Group by Sulfur-containing Groups
Chlorosulfo-de-diazoniation

$$ArN_2^+ + SO_2 \xrightarrow[HCl]{CuCl_2} ArSO_2Cl$$

Diazonium salts can be converted to sulfonyl chlorides by treatment with sulfur dioxide in the presence of cupric chloride.[375] The use of $FeSO_4$ and copper metal instead of $CuCl_2$ gives sulfinic acids $ArSO_2H$.[376] See also **3-21**.

OS **V**, 60; **VII**, 508.

4-28 Cyano-de-diazoniation

$$ArN_2^+ + CuCN \longrightarrow ArCN$$

This reaction, also called the *Sandmeyer reaction*, is similar to **4-25** in scope and mechanism. It is usually conducted in neutral solution to avoid liberation of HCN.

OS **I**, 514.

4-29 Aryl Dimerization with Diazonium Salts
De-diazonio-coupling; Arylazo-de-diazonio-substitution

$$2ArN_2^+ \xrightarrow[\text{or } Cu+H^+]{Cu^+} Ar-Ar + 2N_2 \quad \text{or} \quad Ar-N=N-Ar + N_2$$

When diazonium salts are treated with cuprous ion (or with copper and acid, in which case it is called the *Gatterman method*), two products are possible. If the ring contains electron-withdrawing groups, the main product is the biaryl, but the presence of electron-donating groups leads mainly to the azo compound. This reaction is different from **4-18** (and from **1-4**) in that *both* aryl groups in the product originate from ArN_2^+, i.e., hydrogen is not a leaving group in this reaction. The mechanism probably involves free radicals.[377]

OS **I**, 222; **IV**, 872. Also see OS **IV**, 273.

4-30 Methylation and Vinylation of Diazonium Salts
Methyl-de-diazoniation, etc.

$$ArN_2^+ + Me_4Sn \xrightarrow[MeCN]{Pd(OAc)_2} ArMe$$

A methyl group can be introduced into an aromatic ring by treatment of diazonium salts with tetramethyltin and a palladium acetate catalyst.[378] The reaction has been performed with Me, Cl, Br, and NO_2 groups on the ring. A vinylic group can be introduced with $CH_2=CHSnBu_3$.

[374]Bagal; Pevzner; Frolov *J. Org. Chem. USSR* **1969**, *5*, 1767.
[375]Gilbert *Synthesis* **1969**, 1-10, p. 6.
[376]Wittig; Hoffmann *Org. Synth. V*, 60.
[377]See Cohen; Lewarchik; Tarino *J. Am. Chem. Soc.* **1974**, *96*, 7753.
[378]Kikukawa; Kono; Wada; Matsuda *J. Org. Chem.* **1983**, *48*, 1333.

4-31 Conversion of Diazonium Salts to Aldehydes, Ketones, or Carboxylic Acids
Acyl-de-diazoniation, etc.

$$ArN_2^+ + RCH{=}NOH \xrightarrow[Na_2SO_3]{CuSO_4} Ar{-}\underset{R}{\overset{}{C}}{=}NOH \xrightarrow[6-2]{hydrol.} Ar{-}\underset{O}{\overset{}{C}}{-}R$$

Diazonium salts react with oximes to give aryl oximes, which are easily hydrolyzed to aldehydes (R = H) or ketones.[379] A copper sulfate–sodium sulfite catalyst is essential. In most cases higher yields (40 to 60%) are obtained when the reaction is used for aldehydes than for ketones. In another method[380] for achieving the conversion $ArN_2^+ \rightarrow ArCOR$, diazonium salts are treated with R_4Sn and CO with palladium acetate as catalyst.[381] In a different kind of reaction, silyl enol ethers of aryl ketones $Ar'C(OSiMe_3){=}CHR$ react with solid diazonium fluoroborates $ArN_2^+ BF_4^-$ to give ketones $ArCHRCOAr'$.[382] This is, in effect, an α arylation of the aryl ketone.

Carboxylic acids can be prepared in moderate-to-high yields by treatment of diazonium fluoroborates with carbon monoxide and palladium acetate[383] or copper(II) chloride.[384] The mixed anhydride ArCOOCOMe is an intermediate that can be isolated. Other mixed anhydrides can be prepared by the use of other salts instead of sodium acetate.[385] An aryl-palladium compound is probably an intermediate.[385]

OS **V**, 139.

4-32 Replacement of the Diazonium Group by a Metal
Metallo-de-diazoniation

$$ArN_2^+ BF_4^- + M \longrightarrow ArM$$

Aromatic organometallic compounds can be prepared by the treatment of diazonium salts (most often fluoroborates) with metals.[386] Among the metals used have been Hg, Tl, Sn, Pb, Sb, and Bi. Another method consists of treating the double salt of the diazonium salt and a metal chloride with a metallic powder, e.g.,

$$ArN_2Cl{\cdot}HgCl_2 \xrightarrow{Cu} ArHgCl + CuCl_2$$

Organometallic compounds of Hg,[387] Ge, Sn, and As have been among those prepared by this method. The mechanisms are not clear and may be either homolytic or heterolytic.
OS **II**, 381, 432, 494; **III**, 665.

Metals as Leaving Groups

4-33 Coupling of Grignard Reagents
De-metallo-coupling

$$2RMgX \xrightarrow{TlBr} RR$$

[379]Beech *J. Chem. Soc.* **1954**, 1297.
[380]For still another method, see Citterio; Serravalle; Vismara *Tetrahedron Lett.* **1982**, *23*, 1831.
[381]Kikukawa; Idemoto; Katayama; Kono; Wada; Matsuda *J. Chem. Soc., Perkin Trans. 1* **1987**, 1511.
[382]Sakakura; Hara; Tanaka *J. Chem. Soc., Chem. Commun.* **1985**, 1545.
[383]Nagira; Kikukawa; Wada; Matsuda *J. Org. Chem.* **1980**, *45*, 2365.
[384]Olah; Wu; Bagno; Prakash *Synlett* **1990**, 596.
[385]Kikukawa; Kono; Nagira; Wada; Matsuda *Tetrahedron Lett.* **1980**, *21*, 2877, *J. Org. Chem.* **1981**, *46*, 4413.
[386]For a review, see Reutov; Ptitsyna *Organomet. React.* **1972**, *4*, 73-162.
[387]For reviews with respect to Hg, see Wardell, in Zuckerman *Inorganic Reactions and Methods*, vol. 11; VCH: New York, 1988, pp. 320-323; Larock, Ref. 238, pp. 97-101.

Grignard reagents can be coupled to give symmetrical dimers[388] by treatment with either thallium(I) bromide[389] or with a transition-metal halide such as $CrCl_2$, $CrCl_3$, $CoCl_2$, $CoBr_2$, or $CuCl_2$.[390] The metallic halide is an oxidizing agent and becomes reduced. Both aryl and alkyl Grignard reagents can be dimerized by either procedure, though the TlBr method cannot be applied to R = primary alkyl or to aryl groups with ortho substituents. Aryl Grignard reagents can also be dimerized by treatment with 1,4-dichloro-2-butene, 1,4-dichloro-2-butyne, or 2,3-dichloropropene.[391] Vinylic and alkynyl Grignard reagents can be coupled (to give 1,3-dienes and 1,3-diynes, respectively) by treatment with thionyl chloride.[392] Primary alkyl, vinylic, aryl, and benzylic Grignard reagents give symmetrical dimers in high yield (~90%) when treated with a silver(I) salt, e.g., $AgNO_3$, $AgBr$, $AgClO_4$, in the presence of a nitrogen-containing oxidizing agent such as lithium nitrate, methyl nitrate, or NO_2.[393] This method has been used to close rings of 4, 5, and 6 members.[394]

The mechanisms of the reactions with metal halides, at least in some cases, probably begin with conversion of RMgX to the corresponding RM (**2-35**), followed by its decomposition to free radicals.[395]

OS **VI**, 488.

4-34 Coupling of Boranes
Alkyl-de-dialkylboration

$$\text{R—B—} + \text{R'—B—} \xrightarrow[\text{NaOH}]{\text{AgNO}_3} \text{R—R'}$$

Alkylboranes can be coupled by treatment with silver nitrate and base.[396] Since alkylboranes are easily prepared from olefins (**5-12**), this is essentially a way of coupling and reducing olefins; in fact, olefins can be hydroborated and coupled in the same flask. For symmetrical coupling (R = R') yields range from 60 to 80% for terminal olefins and from 35 to 50% for internal ones. Unsymmetrical coupling has also been carried out,[397] but with lower yields. Arylboranes react similarly, yielding biaryls.[398] The mechanism is probably of the free-radical type.

Vinylic dimerization can be achieved by treatment of divinylchloroboranes (prepared by addition of BH_2Cl to alkynes; see **5-12**) with methylcopper. (E,E)-1,3-Dienes are prepared in high yields.[399]

$$\text{RC} \equiv \text{CR'} \xrightarrow{\text{BH}_2\text{Cl}} \underset{H}{\overset{R}{\diagdown}} C = C \underset{\overset{|}{2}\;\text{BCl}}{\overset{R'}{\diagup}} \xrightarrow{\text{3MeCu}} \underset{H}{\overset{R}{\diagdown}} C = C \underset{\underset{R'}{\diagup} C = C \underset{R}{\overset{H}{\diagup}}}{\overset{R'}{\diagup}}$$

[388]For a list of reagents, with references, see Ref. 74, pp. 48-49.
[389]McKillop; Elsom; Taylor *J. Am. Chem. Soc.* **1968**, *90*, 2423, *Tetrahedron* **1970**, *26*, 4041.
[390]For reviews, see Kauffmann *Angew. Chem. Int. Ed. Engl.* **1974**, *13*, 291-305 [*Angew. Chem. 86*, 321-335]; Elsom; Hunt; McKillop *Organomet. Chem. Rev., Sect. A* **1972**, *8*, 135-152; Nigh, Ref. 185, pp. 85-91.
[391]Taylor; Bennett; Heinz; Lashley *J. Org. Chem.* **1981**, *46*, 2194; Cheng; Luo *Tetrahedron Lett.* **1988**, *29*, 1293.
[392]Uchida; Nakazawa; Kondo; Iwata; Matsuda *J. Org. Chem.* **1972**, *37*, 3749.
[393]Tamura; Kochi *Bull. Chem. Soc. Jpn.* **1972**, *45*, 1120.
[394]Whitesides; Gutowski *J. Org. Chem.* **1976**, *41*, 2882.
[395]For a review of the mechanism, see Kashin; Beletskaya *Russ. Chem. Rev.* **1982**, *51*, 503-526.
[396]Pelter; Smith; Brown *Borane Reagents*; Academic Press: New York, 1988, pp. 306-308.
[397]Brown; Verbrugge; Snyder *J. Am. Chem. Soc.* **1961**, *83*, 1001.
[398]Breuer; Broster *Tetrahedron Lett.* **1972**, 2193.
[399]Yamamoto; Yatagai; Maruyama; Sonoda; Murahashi *J. Am. Chem. Soc.* **1977**, *99*, 5652, *Bull. Chem. Soc. Jpn.* **1977**, *50*, 3427. For other methods of dimerizing vinylic boron compounds, see Rao; Kumar; Devaprabhakara *J. Organomet. Chem.* **1979**, *179*, C7; Campbell; Brown *J. Org. Chem.* **1980**, *45*, 549.

In a similar reaction, symmetrical conjugated diynes RC≡C—C≡CR can be prepared by reaction of lithium dialkyldialkynylborates Li$^+$ [R'$_2$B(C≡CR)$_2$]$^-$ with iodine.[400]

4-35   Coupling of Other Organometallic Reagents[388]
De-metallo-coupling

$$R_2CuLi \xrightarrow[-78°C, THF]{O_2} RR$$

Lithium dialkylcopper reagents can be oxidized to symmetrical dimers by O$_2$ at $-78°C$ in THF.[401] The reaction is successful for R = primary and secondary alkyl, vinylic, or aryl. Other oxidizing agents, e.g., nitrobenzene, can be used instead of O$_2$. Vinylic copper reagents dimerize on treatment with oxygen, or simply on standing at 0°C for several days or at 25°C for several hours, to yield 1,3-dienes.[402] The finding of retention of configuration for this reaction demonstrates that free-radical intermediates are not involved. Lithium organoaluminates LiAlR$_4$ are dimerized to RR by treatment with Cu(OAc)$_2$.[403] Terminal vinylic alanes (prepared by **5-13**) can be dimerized to 1,3-dienes with CuCl in THF.[404] Symmetrical 1,3-dienes can also be prepared in high yields by treatment of vinylic mercury chlorides[405] with LiCl and a rhodium catalyst[406] and by treatment of vinylic tin compounds with a palladium catalyst.[407] Arylmercuric salts are converted to biaryls by treatment with copper and a catalytic amount of PdCl$_2$.[408] Vinylic, alkynyl, and aryl tin compounds were dimerized with Cu(NO$_3$)$_2$.[409] Alkyl- and aryllithium compounds can be dimerized by transition-metal halides in a reaction similar to **4-33**.[410] Triarylbismuth compounds Ar$_3$Bi react with palladium(0) complexes to give biaryls ArAr.[411] Unsymmetrical coupling of vinylic, alkynyl, and arylmercury compounds was achieved in moderate-to-good yields by treatment with alkyl and vinylic dialkylcopper reagents, e.g., PhCH=CHHgCl + Me$_2$CuLi → PhCH=CHMe.[412] Unsymmetrical biaryls were prepared by treating a cyanocuprate ArCu(CN)Li (prepared from ArLi and CuCN) with an aryllithium Ar'Li.[412a]

Halogen as Leaving Group

The conversion of RX to RH can occur by a free-radical mechanism but is treated at **0-76.**

[400]Pelter; Smith; Tabata *J. Chem. Soc., Chem. Commun.* **1975**, 857. For extensions to unsymmetrical conjugated diynes, see Pelter; Hughes; Smith; Tabata *Tetrahedron Lett.* **1976**, 4385; Sinclair; Brown *J. Org. Chem.* **1976**, *41*, 1078.

[401]Whitesides; SanFilippo; Casey; Panek *J. Am. Chem. Soc.* **1967**, *89*, 5302. See also Kauffmann; Kuhlmann; Sahm; Schrecken *Angew. Chem. Int. Ed. Engl.* **1968**, 7, 541 [*Angew. Chem. 80*, 566]; Bertz; Gibson *J. Am. Chem. Soc.* **1986**, *108*, 8286.

[402]Whitesides; Casey; Krieger *J. Am. Chem. Soc.* **1971**, *93*, 1379; Walborsky; Banks; Banks; Duraisamy *Organometallics* **1982**, *1*, 667; Rao; Periasamy *J. Chem. Soc., Chem. Commun.* **1987**, 495. See also Lambert; Duffley; Dalzell; Razdan *J. Org. Chem.* **1982**, *47*, 3350.

[403]Sato; Mori; Sato *Chem. Lett.* **1978**, 1337.

[404]Zweifel; Miller, *J. Am. Chem. Soc.* **1970**, *92*, 6678.

[405]For reviews of coupling with organomercury compounds, see Russell *Acc. Chem. Res.* **1989**, *22*, 1-8; Larock, Ref. 238, pp. 240-248.

[406]Larock; Bernhardt *J. Org. Chem.* **1977**, *42*, 1680. For extension to unsymmetrical 1,3-dienes, see Larock; Riefling *J. Org. Chem.* **1978**, *43*, 1468.

[407]Tolstikov; Miftakhov; Danilova; Vel'der; Spirikhin *Synthesis* **1989**, 633.

[408]Kretchmer; Glowinski *J. Org. Chem.* **1976**, *41*, 2661. See also Bumagin; Kalinovskii; Beletskaya *J. Org. Chem. USSR* **1982**, *18*, 1151; Larock; Bernhardt, Ref. 406.

[409]Ghosal; Luke; Kyler *J. Org. Chem.* **1987**, *52*, 4296.

[410]Morizur *Bull. Soc. Chim. Fr.* **1964**, 1331.

[411]Barton; Ozbalik; Ramesh *Tetrahedron* **1988**, *44*, 5661.

[412]Larock; Leach *Tetrahedron Lett.* **1981**, *22*, 3435. *Organometallics* **1982**, *1*, 74. For another method, see Larock; Hershberger *Tetrahedron Lett.* **1981**, *22*, 2443.

[412a]Lipshutz; Siegmann; Garcia *J. Am. Chem. Soc.* **1991**, *113*, 8161.

Sulfur as Leaving Group

4-36 Desulfurization with Raney Nickel
Hydro-de-mercapto-substitution, etc.

$$RSH \xrightarrow[\text{Ni}]{H_2} RH$$

$$RSR' \xrightarrow[\text{Ni}]{H_2} RH + R'H$$

Thiols and thioethers,[413] both alkyl and aryl, can be desulfurized by hydrogenolysis with Raney nickel.[414] The hydrogen is usually not applied externally, since Raney nickel already contains enough hydrogen for the reaction. Other sulfur compounds can be similarly desulfurized, among them:

Disulfides	$RSSR' \longrightarrow RH + R'H$
Thiono esters[415]	$RCSOR' \longrightarrow RCH_2OR'$
Thioamides	$RCSNHR' \longrightarrow RCH_2NHR$
Sulfoxides	$RSOR' \longrightarrow RH + R'H$
Thioacetals	$R\overset{\textstyle |}{\underset{\textstyle |}{C}}SR' \longrightarrow \overset{\textstyle |}{\underset{\textstyle |}{C}}H_2$

The last reaction, which is an indirect way of accomplishing reduction of a carbonyl to a methylene group (see **9-37**), can also give the olefin if an α hydrogen is present.[416] In most of the examples given, R can also be aryl. Other reagents[417] have also been used.[418]

An important special case of RSR reduction is desulfurization of thiophene derivatives. This proceeds with concomitant reduction of the double bonds. Many compounds have been made by alkylation of thiophene, followed by reduction:

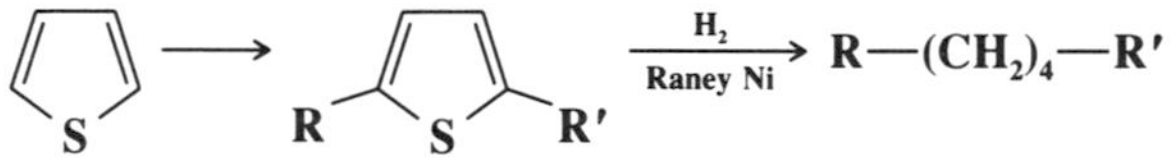

23

Thiophenes can also be desulfurized to alkenes ($RCH_2CH{=}CHCH_2R'$ from **23**) with a nickel boride catalyst prepared from nickel(II) chloride and $NaBH_4$ in methanol.[419] It is possible to reduce just one SR group of a dithioacetal by treatment with borane–pyridine

---

[413]For a review of the reduction of thioethers, see Block, in Patai *The Chemistry of Functional Groups, Supplement E*, pt. 1; Wiley: New York, 1980, pp. 585-600.

[414]For reviews, see Belen'kii, in Belen'kii *Chemistry of Organosulfur Compounds*; Ellis Horwood: Chichester, 1990, pp. 193-228; Pettit; van Tamelen *Org. React.* **1962**, *12*, 356-529; Hauptmann; Walter *Chem. Rev.* **1962**, *62*, 347-404.

[415]See Baxter; Bradshaw *J. Org. Chem.* **1981**, *46*, 831.

[416]Fishman; Torigoe; Guzik *J. Org. Chem.* **1963**, *28*, 1443.

[417]For lists of reagents, with references, see Ref. 74, pp. 31-35. For a review with respect to transition-metal reagents, see Luh; Ni *Synthesis* **1990**, 89-103. For some very efficient nickel-containing reagents, see Becker; Fort; Vanderesse; Caubère *J. Org. Chem.* **1989**, *54*, 4848.

[418]For example, diphosphorus tetraiodide by Suzuki; Tani; Takeuchi *Bull. Chem. Soc. Jpn.* **1985**, *58*, 2421; Shigemasa; Ogawa; Sashiwa; Saimoto *Tetrahedron Lett.* **1989**, *30*, 1277; $NiBr_2$–Ph_3P–$LiAlH_4$ by Ho; Lam; Luh *J. Org. Chem.* **1989**, *54*, 4474.

[419]Schut; Engberts; Wynberg *Synth. Commun.* **1972**, *2*, 415.

in trifluoroacetic acid or in CH_2Cl_2 in the presence of $AlCl_3$.[420] Phenyl selenides RSePh can be reduced to RH with Ph_3SnH[421] and with nickel boride.[422]

The exact mechanisms of the Raney nickel reactions are still in doubt, though they are probably of the free-radical type.[423] It has been shown that reduction of thiophene proceeds through butadiene and butene, not through 1-butanethiol or other sulfur compounds, i.e., the sulfur is removed before the double bonds are reduced. This was demonstrated by isolation of the olefins and the failure to isolate any potential sulfur-containing intermediates.[424]

OS **IV,** 638; **V,** 419; **VI,** 109, 581, 601. See also OS **VII,** 124, 476.

4-37   Conversion of Sulfides to Organolithium Compounds
Lithio-de-phenylthio-substitution

$$\text{RSPh} \xrightarrow[\text{THF}]{\text{Li naphthalenide}} \text{RLi}$$

Sulfides can be cleaved, with a phenylthio group replaced by a lithium,[425] by treatment with lithium or lithium naphthalenide in THF.[426] Good yields have been obtained with R = primary, secondary, or tertiary alkyl, or allylic,[427] and containing groups such as double bonds or halogens. Dilithio compounds can be made from compounds containing two separated SPh groups, but it is also possible to replace just one SPh from a compound with two such groups on a single carbon, to give an α-lithio sulfide.[428] The reaction has also been used to prepare α-lithio ethers and α-lithio organosilanes.[425] For some of these compounds lithium 1-(dimethylamino)naphthalenide is a better reagent than either Li or lithium naphthalenide.[429] The mechanism is presumably of the free-radical type.

Carbon as Leaving Group

4-38   Decarboxylative Dimerization. The Kolbe Reaction
De-carboxylide-coupling

$$2\text{RCOO}^- \xrightarrow{\text{electrol.}} \text{R—R}$$

Electrolysis of carboxylate ions, which results in decarboxylation and combination of the resulting radicals, is called the *Kolbe reaction*.[430] It is used to prepare symmetrical RR, where R is straight- or branched-chained, except that little or no yield is obtained when there is α branching. The reaction is not successful for R = aryl. Many functional groups

[420]Kikugawa J. Chem. Soc., Perkin Trans. 1 **1984**, 609.
[421]Clive; Chittattu; Wong J. Chem. Soc., Chem. Commun. **1978**, 41.
[422]Back J. Chem. Soc., Chem. Commun. **1984**, 1417.
[423]For a review, see Bonner; Grimm, in Kharasch; Meyers The Chemistry of Organic Sulfur Compounds, vol. 2; Pergamon: New York, 1966, pp. 35-71, 410-413. For a review of the mechanism of desulfurization on molybdenum surfaces, see Friend; Roberts Acc. Chem. Res. **1988**, 21, 394-400.
[424]Owens; Ahmberg Can. J. Chem. **1962**, 40, 941.
[425]For a review, see Cohen; Bhupathy Acc. Chem. Res. **1989**, 22, 152-161.
[426]Screttas; Micha-Screttas J. Org. Chem. **1978**, 43, 1064, **1979**, 44, 713.
[427]See Cohen; Guo Tetrahedron **1986**, 42, 2803.
[428]See, for example; Cohen; Sherbine; Matz; Hutchins; McHenry; Willey J. Am. Chem. Soc. **1984**, 106, 3245; Ager J. Chem. Soc., Perkin Trans. 1 **1986**, 183; Ref. 426.
[429]See Cohen; Matz Synth. Commun. **1980**, 10, 311.
[430]For reviews, see Schäfer Top. Curr. Chem. **1990**, 152, 91-151, Angew. Chem. Int. Ed. Engl. **1981**, 20, 911-934 [Angew. Chem. 93, 978-1000]; Fry Synthetic Organic Electrochemistry, 2nd ed.; Wiley: New York, 1989, pp. 238-253; Eberson; Utley, in Baizer; Lund Organic Electrochemistry; Marcel Dekker: New York, 1983, pp. 435-462; Gilde Methods Free-Radical Chem. **1972**, 3, 1-82; Eberson, in Patai The Chemistry of Carboxylic Acids and Esters; Wiley: New York, 1969, pp. 53-101; Vijh; Conway Chem. Rev. **1967**, 67, 623-664.

may be present, though many others inhibit the reaction.[430] Unsymmetrical RR' have been made by coupling mixtures of acid salts.

A free-radical mechanism is involved:

$$RCOO^- \xrightarrow[\text{oxidation}]{\text{electrolytic}} RCOO\cdot \xrightarrow{-CO_2} R\cdot \longrightarrow R\text{—}R$$

There is much evidence[431] for this mechanism, including side products (RH, alkenes) characteristic of free-radical intermediates and the fact that electrolysis of acetate ion in the presence of styrene caused some of the styrene to polymerize to polystyrene (such polymerizations can be initiated by free radicals, see p. 744). Other side products (ROH, RCOOR) are sometimes found; these stem from further oxidation of the radical R· to the carbocation R+.[432]

When the reaction is conducted in the presence of 1,3-dienes, additive dimerization can occur:[433]

$$2RCOO^- + CH_2\!\!=\!\!CH\text{—}CH\!\!=\!\!CH_2 \longrightarrow RCH_2CH\!\!=\!\!CHCH_2CH_2CH\!\!=\!\!CHCH_2R$$

The radical R· adds to the conjugated system to give $RCH_2CH\!\!=\!\!CHCH_2\cdot$, which dimerizes. Another possible product is $RCH_2CH\!\!=\!\!CHCH_2R$, from coupling of the two kinds of radicals.[434]

In a non-electrolytic reaction, which is limited to R = primary alkyl, the thiohydroxamic esters 24 give dimers when irradiated at $-64°C$ in an argon atmosphere:[435]

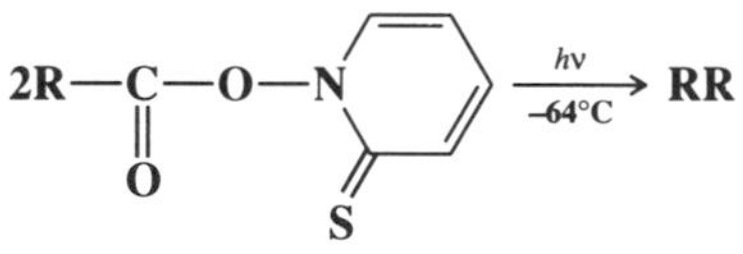

24

In another non-electrolytic process, arylacetic acids are converted to vic-diaryl compounds $2ArCR_2COOH \to ArCR_2CR_2Ar$ by treatment with sodium persulfate $Na_2S_2O_8$ and a catalytic amount of $AgNO_3$.[436] Both of these reactions involve dimerization of free radicals. In still another process, electron-deficient aromatic acyl chlorides are dimerized to biaryls ($2ArCOCl \to ArAr$) by treatment with a disilane R_3SiSiR_3 and a palladium catalyst.[437]

OS **III**, 401; **V**, 445, 463; **VII**, 181.

4-39   The Hunsdiecker Reaction

Bromo-de-carboxylation

$$RCOOAg + Br_2 \longrightarrow RBr + CO_2 + AgBr$$

Reaction of a silver salt of a carboxylic acid with bromine is called the *Hunsdiecker reaction*[438] and is a way of decreasing the length of a carbon chain by one unit.[439] The reaction is of

[431]For other evidence, see Kraeutler; Jaeger; Bard *J. Am. Chem. Soc.* **1978**, *100*, 4903.
[432]See Corey; Bauld; La Londe; Casanova; Kaiser *J. Am. Chem. Soc.* **1960**, *82*, 2645.
[433]Lindsey; Peterson *J. Am. Chem. Soc.* **1959**, *81*, 2073; Khrizolitova; Mirkind; Fioshin *J. Org. Chem. USSR* **1968**, *4*, 1640; Bruno; Dubois *Bull. Soc. Chim. Fr.* **1973**, 2270.
[434]Smith; Gilde *J. Am. Chem. Soc.* **1959**, *81*, 5325, **1961**, *83*, 1355; Schäfer; Pistorius *Angew. Chem. Int. Ed. Engl.* **1972**, *11*, 841 [*Angew. Chem. 84*, 893].
[435]Barton; Bridon; Fernandez-Picot; Zard *Tetrahedron* **1987**, *43*, 2733.
[436]Fristad; Klang *Tetrahedron Lett.* **1983**, *24*, 2219.
[437]Krafft; Rich; McDermott *J. Org. Chem.* **1990**, *55*, 5430.
[438]This reaction was first reported by the Russian composer–chemist Alexander Borodin: *Liebigs Ann. Chem.* **1861**, *119*, 121.
[439]For reviews, see Wilson *Org. React.* **1957**, *9*, 332-388; Johnson; Ingham *Chem. Rev.* **1956**, *56*, 219-269.

wide scope, giving good results for *n*-alkyl R from 2 to 18 carbons and for many branched R too, producing primary, secondary, and tertiary bromides. Many functional groups may be present as long as they are not α substituted. R may also be aryl. However, if R contains unsaturation, the reaction seldom gives good results. Although bromine is the most often used halogen, chlorine and iodine have also been used.

When iodine is the reagent, the ratio between the reactants is very important and determines the products. A 1:1 ratio of salt to iodine gives the alkyl halide, as above. A 2:1 ratio, however, gives the ester RCOOR. This is called the *Simonini reaction* and is sometimes used to prepare carboxylic esters. The Simonini reaction can also be carried out with lead salts of acids.[440] A more convenient way to perform the Hunsdiecker reaction is by use of a mixture of the acid and mercuric oxide instead of the salt, since the silver salt must be very pure and dry and such pure silver salts are often not easy to prepare.[441]

Other methods for accomplishing the conversion RCOOH → RX are:[442] (1) treatment of thallium(I) carboxylates[443] with bromine;[444] (2) treatment of carboxylic acids with lead tetraacetate and halide *ions* (Cl^-, Br^-, or I^-);[445] (3) reaction of the acids with lead tetraacetate and N-chlorosuccinimide, which gives tertiary and secondary chlorides in good yields but is not good for R = primary alkyl or phenyl;[446] (4) the reaction between a diacyl peroxide and $CuCl_2$, $CuBr_2$, or CuI_2[447] [this reaction also takes place with $Cu(SCN)_2$, and $Cu(CN)_2$]; (5) treatment of thiohydroxamic esters (**24**) with CCl_4, $BrCCl_3$ (which gives bromination), CHI_3, or CH_2I_2 in the presence of a radical initiator;[448] (6) photolysis of benzophenone oxime esters of carboxylic acids in CCl_4 ($RCON{=}CPh_2$ → RCl).[449] Alkyl fluorides can be prepared in moderate to good yields by treating carboxylic acids RCOOH with XeF_2.[450] This method works best for R = primary and tertiary alkyl, and benzylic. Aromatic and vinylic acids do not react.

The mechanism of the Hunsdiecker reaction is believed to be as follows:

Step 1
$$RCOOAg + X_2 \longrightarrow R-\underset{\underset{O}{\|}}{C}-O-X + AgX$$

25

Step 2
$$R\underset{\underset{O}{\|}}{C}-O-X \longrightarrow RCOO\cdot + X\cdot \text{ (initiation)}$$

Step 3
$$RCOO\cdot \longrightarrow R\cdot + CO_2$$

Step 4
$$R\cdot + RCOOX \longrightarrow RX + RCOO\cdot \text{ (propagation)}$$

etc.

[440]Bachman; Kite; Tuccarbasu; Tullman *J. Org. Chem.* **1970**, *35*, 3167.
[441]Cristol; Firth *J. Org. Chem.* **1961**, *26*, 280. See also Meyers; Fleming *J. Org. Chem.* **1979**, *44*, 3405, and references cited therein.
[442]For a list of reagents, with references, see Ref. 74, pp. 381-382.
[443]These salts are easy to prepare and purify; see Ref. 444.
[444]McKillop; Bromley; Taylor *J. Org. Chem.* **1969**, *34*, 1172; Cambie; Hayward; Jurlina; Rutledge; Woodgate *J. Chem. Soc., Perkin Trans. 1* **1981**, 2608.
[445]Kochi *J. Am. Chem. Soc.* **1965**, *87*, 2500, *J. Org. Chem.* **1965**, *30*, 3265. For a review, see Sheldon; Kochi *Org. React.* **1972**, *19*, 279-421, pp. 326-334, 390-399.
[446]Becker; Geisel; Grob; Kuhnen *Synthesis* **1973**, 493.
[447]Jenkins; Kochi *J. Org. Chem.* **1971**, *36*, 3095, 3103.
[448]Barton; Crich; Motherwell *Tetrahedron Lett.* **1983**, *24*, 4979; Barton; Lacher; Zard *Tetrahedron* **1987**, *43*, 4321; Stofer; Lion *Bull. Soc. Chim. Belg.* **1987**, *96*, 623; Della; Tsanaktsidis *Aust. J. Chem.* **1989**, *42*, 61.
[449]Hasebe; Tsuchiya *Tetrahedron Lett.* **1988**, *29*, 6287.
[450]Patrick; Johri; White; Bertrand; Mokhtar; Kilbourn; Welch **1986**, *Can. J. Chem. 64*, 138. For another method, see Grakauskas *J. Org. Chem.* **1969**, *34*, 2446.

The first step is not a free-radical process, and its actual mechanism is not known.[451] **25** is an acyl hypohalite and is presumed to be an intermediate, though it has never been isolated from the reaction mixture. Among the evidence for the mechanism is that optical activity at R is lost (except when a neighboring bromine atom is present, see p. 682); if R is neopentyl, there is no rearrangement, which would certainly happen with a carbocation; and the side products, notably RR, are consistent with a free-radical mechanism. There is evidence that the Simonini reaction involves the same mechanism as the Hunsdiecker reaction but that the alkyl halide formed then reacts with excess RCOOAg (**0-24**) to give the ester.[452] See also **9-13**.

OS **III**, 578; **V**, 126; **VI**, 179. See also OS **VI**, 403.

4-40 Decarboxylative Allylation
Allyl-de-carboxylation

$$R-\underset{\underset{O}{\|}}{C}-\underset{|}{C}-COOH + H_2C=CHCH_2-O-\underset{\underset{O}{\|}}{C}-CH_3 \xrightarrow{Pd(PPh_3)_4}$$

$$R-\underset{\underset{O}{\|}}{C}-\underset{|}{C}-CH_2-CH=CH_2 + CO_2 + CH_3COOH$$

The COOH group of a β-keto acid is replaced by an allylic group when the acid is treated with an allylic acetate and a palladium catalyst at room temperature.[453] The reaction is successful for various substituted allylic groups. The less-highly-substituted end of the allylic group forms the new bond. Thus, both $CH_2=CHCHMeOAc$ and $MeCH=CHCH_2OAc$

gave $RCO-\underset{|}{C}-CH_2CH=CHMe$ as the product.

4-41 Decarbonylation of Aldehydes and Acyl Halides
Carbonyl-extrusion

$$\textbf{RCHO} \xrightarrow{\text{RhCl(Ph}_3\text{P)}_3} \textbf{RH}$$

Aldehydes, both aliphatic and aromatic, can be decarbonylated[454] by heating with chloro-tris(triphenylphosphine)rhodium[455] or other catalysts such as palladium.[456] RhCl(Ph$_3$P)$_3$ is often called *Wilkinson's catalyst.*[457] In an older reaction aliphatic (but not aromatic) aldehydes are decarbonylated by heating with di-*t*-peroxide or other peroxides,[458] usually in a solution

[451]When Br$_2$ reacts with aryl R, at low temperature in inert solvents, it is possible to isolate a complex containing both Br$_2$ and the silver carboxylate: see Bryce-Smith; Isaacs; Tumi *Chem. Lett.* **1984**, 1471.

[452]Oae; Kashiwagi; Kozuka *Bull. Chem. Soc. Jpn.* **1966**, *39*, 2441; Bunce; Murray *Tetrahedron* **1971**, 27, 5323.

[453]Tsuda; Okada; Nishi; Saegusa *J. Org. Chem.* **1986**, *51*, 421.

[454]For reviews, see Collman; Hegedus; Norton; Finke *Principles and Applications of Organotransition Metal Chemistry*; University Science Books: Mill Valley, CA, 1987, pp. 768-775; Baird, in Patai *The Chemistry of Functional Groups, Supplement B*, pt. 2; Wiley: New York, 1979, pp. 825-857; Tsuji, in Wender; Pino *Organic Syntheses Via Metal Carbonyls*, vol. 2; Wiley: New York, 1977, pp. 595-654; Tsuji; Ohno *Synthesis* **1969**, 157-169; Bird *Transition Metal Intermediates in Organic Synthesis*; Academic Press: New York, 1967, pp. 239-247.

[455]Tsuji; Ohno *Tetrahedron Lett.* **1965**, 3969; Ohno; Tsuji *J. Am. Chem. Soc.* **1968**, *90*, 99; Baird; Nyman; Wilkinson *J. Chem. Soc. A* **1968**, 348.

[456]For a review, see Rylander, Ref. 246, pp. 260-267.

[457]For a review of this catalyst, see Jardine *Prog. Inorg. Chem.* **1981**, *28*, 63-202.

[458]For reviews of free-radical aldehyde decarbonylations, see Vinogradov; Nikishin *Russ. Chem. Rev.* **1971**, *40*, 916-932; Schubert; Kintner, in Patai, Ref. 189, pp. 711-735.

containing a hydrogen donor, such as a thiol. The reaction has also been initiated with light, and thermally (without an initiator) by heating at about 500°C.

Wilkinson's catalyst has also been reported to decarbonylate aromatic acyl halides at 180°C (ArCOX → ArX).[459] This reaction has been carried out with acyl iodides,[460] bromides, and chlorides. Aliphatic acyl halides that lack an α hydrogen also give this reaction,[461] but if an α hydrogen is present, elimination takes place instead (**7-19**). Aromatic acyl cyanides give aryl cyanides (ArCOCN → ArCN).[462] Aromatic acyl chlorides and cyanides can also be decarbonylated with palladium catalysts.[463]

It is possible to decarbonylate acyl halides in another way, to give alkanes (RCOCl → RH). This is done by heating the substrate with tripropylsilane Pr_3SiH in the presence of t-butyl peroxide.[464] Yields are good for R = primary or secondary alkyl and poor for R = tertiary alkyl or benzylic. There is no reaction when R = aryl. (See also the decarbonylation ArCOCl → ArAr mentioned in **4-38**.)

The mechanism of the peroxide- or light-induced reaction seems to be as follows (in the presence of thiols):[465]

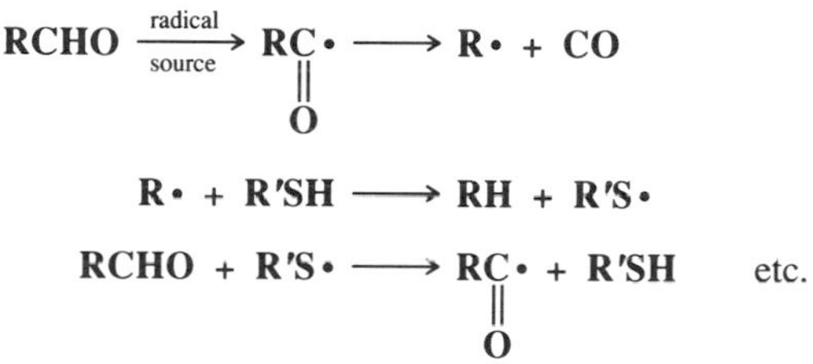

$$RCHO \xrightarrow[\text{source}]{\text{radical}} RC\cdot \longrightarrow R\cdot + CO$$
$$\overset{\|}{O}$$

$$R\cdot + R'SH \longrightarrow RH + R'S\cdot$$

$$RCHO + R'S\cdot \longrightarrow RC\cdot + R'SH \qquad \text{etc.}$$
$$\overset{\|}{O}$$

The reaction of aldehydes with Wilkinson's catalyst goes through complexes of the form **26** and **27**, which have been trapped.[466] The reaction has been shown to give retention of

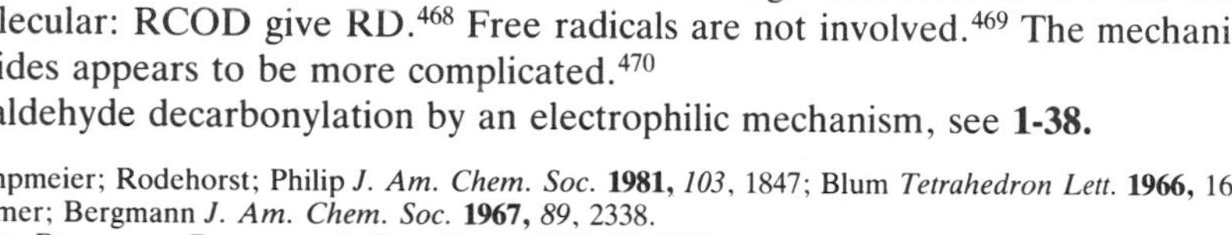

26   **27**

configuration at a chiral R;[467] and deuterium labeling demonstrates that the reaction is intramolecular: RCOD give RD.[468] Free radicals are not involved.[469] The mechanism with acyl halides appears to be more complicated.[470]

For aldehyde decarbonylation by an electrophilic mechanism, see **1-38**.

[459]Kampmeier; Rodehorst; Philip *J. Am. Chem. Soc.* **1981**, *103*, 1847; Blum *Tetrahedron Lett.* **1966**, 1605; Blum; Oppenheimer; Bergmann *J. Am. Chem. Soc.* **1967**, *89*, 2338.

[460]Blum; Rosenman; Bergmann *J. Org. Chem.* **1968**, *33*, 1928.

[461]Tsuji; Ohno *Tetrahedron Lett.* **1966**, 4713, *J. Am. Chem. Soc.* **1966**, *88*, 3452.

[462]Blum; Oppenheimer; Bergmann, Ref. 459.

[463]Verbicky; Dellacoletta; Williams *Tetrahedron Lett.* **1982**, *23*, 371; Murahashi; Naota; Nakajima *J. Org. Chem.* **1986**, *51*, 898.

[464]Billingham; Jackson; Malek *J. Chem. Soc., Perkin Trans. 1* **1979**, 1137.

[465]Slaugh *J. Am. Chem. Soc.* **1959**, *81*, 2262; Berman; Stanley; Sherman; Cohen *J. Am. Chem. Soc.* **1963**, *85*, 4010.

[466]Suggs *J. Am. Chem. Soc.* **1978**, *100*, 640; Kampmeier; Harris; Mergelsberg *J. Org. Chem.* **1984**, *49*, 621.

[467]Walborsky; Allen *J. Am. Chem. Soc.* **1971**, *93*, 5465. See also Tsuji; Ohno *Tetrahedron Lett.* **1967**, 2173.

[468]Prince; Raspin *J. Chem. Soc. A* **1969**, 612; Walborsky; Allen, Ref. 467. See, however, Baldwin; Barden; Pugh; Widdison *J. Org. Chem.* **1987**, *52*, 3303.

[469]Kampmeier; Harris; Wedegaertner *J. Org. Chem.* **1980**, *45*, 315.

[470]Kampmeier; Rodehorst; Philip, Ref. 459; Kampmeier; Mahalingam; Liu *Organometallics* **1986**, *5*, 823; Kampmeier; Liu *Organometallics* **1989**, *8*, 2742.

15

ADDITION TO CARBON–CARBON MULTIPLE BONDS

There are basically four ways in which addition to a double or triple bond can take place. Three of these are two-step processes, with initial attack by a nucleophile, an electrophile, or a free radical. The second step consists of combination of the resulting intermediate with, respectively, a positive species, a negative species, or a neutral entity. In the fourth type of mechanism, attack at the two carbon atoms of the double or triple bond is simultaneous. Which of the four mechanisms is operating in any given case is determined by the nature of the substrate, the reagent, and the reaction conditions. Some of the reactions in this chapter can take place by all four mechanistic types.

MECHANISMS

Electrophilic Addition[1]

In this mechanism a positive species approaches the double or triple bond and in the first step forms a bond by converting the π pair of electrons into a σ pair:

Step 1

$$-\overset{|}{\underset{|}{C}}\!\!=\!\!\overset{|}{\underset{|}{C}}- \; + \; Y^+ \; \xrightarrow{\;slow\;} \; -\overset{\oplus}{\underset{|}{C}}-\overset{\overset{\displaystyle Y}{|}}{\underset{|}{C}}-$$

Step 2

$$-\overset{\oplus}{\underset{|}{C}}-\overset{\overset{\displaystyle Y}{|}}{\underset{|}{C}}- \; + \; \overline{W} \; \longrightarrow \; -\overset{\overset{\displaystyle W}{|}}{\underset{|}{C}}-\overset{\overset{\displaystyle Y}{|}}{\underset{|}{C}}-$$

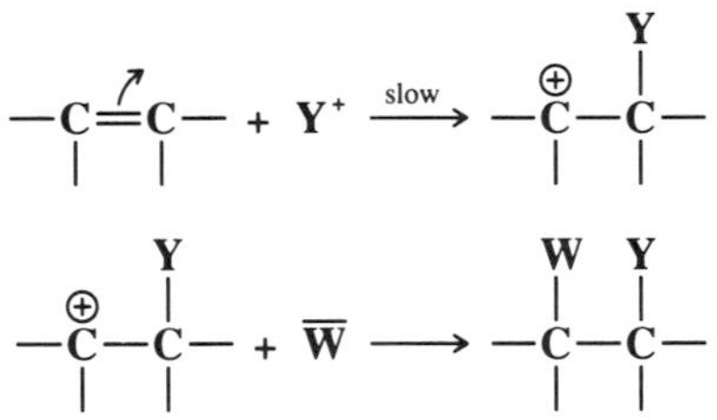

1

The IUPAC designation for this mechanism is $A_E + A_N$ (or $A_H + A_N$ if $Y^+ = H^+$). As in electrophilic substitution (p. 502), Y need not actually be a positive ion but can be the

---

[1]For a monograph, see de la Mare; Bolton *Electrophilic Additions to Unsaturated Systems*, 2nd ed.; Elsevier: New York, 1982. For reviews, see Schmid, in Patai *Supplement A: The Chemistry of Double-bonded Functional Groups*, vol. 2, pt. 1; Wiley: New York, 1989, pp. 679-731; Smit *Sov. Sci. Rev. Sect. B* **1985**, *7*, 155-236; V'yunov; Ginak *Russ. Chem. Rev.* **1981**, *50*, 151-163; Schmid; Garratt, in Patai *Supplement A: The Chemistry of Double-bonded Functional Groups*, vol. 1, pt. 2; Wiley: New York, 1977, pp. 725-912; Freeman *Chem. Rev.* **1975**, *75*, 439-490; Bolton, in Bamford; Tipper *Comprehensive Chemical Kinetics*, vol. 9; Elsevier: New York, 1973, pp. 1-86; Dolbier *J. Chem. Educ.* **1969**, *46*, 342-344.

positive end of a dipole or an induced dipole, with the negative part breaking off either during the first step or shortly after. The second step is a combination of **1** with a species carrying an electron pair and often bearing a negative charge. This step is the same as the second step of the S_N1 mechanism. Not all electrophilic additions follow the simple mechanism given above. In many brominations it is fairly certain that **1**, if formed at all, very rapidly cyclizes to a bromonium ion (**2**):

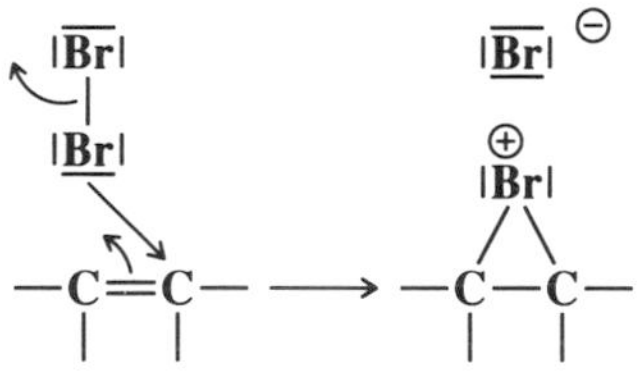

2

This intermediate is similar to those encountered in the neighboring-group mechanism of nucleophilic substitution (see p. 308). The attack of $\overline{W}$ on an intermediate like 2 is an S_N2 step. Whether the intermediate is **1** or **2**, the mechanism is called Ad_E2 (electrophilic addition, bimolecular).

In investigating the mechanism of addition to a double bond, perhaps the most useful type of information is the stereochemistry of the reaction.[2] The two carbons of the double bond and the four atoms immediately attached to them are all in a plane (p. 8); there are thus three possibilities. Y and W may enter from the same side of the plane, in which case the addition is stereospecific and *syn*; they may enter from opposite sides for stereospecific *anti* addition; or the reaction may be nonstereospecific. In order to determine which of these possibilities is occurring in a given reaction, the following type of experiment is often done: YW is added to the cis and trans isomers of an olefin of the form ABC=CBA. We may use the cis olefin as an example. If the addition is syn, the product will be the erythro *dl* pair, because each carbon has a 50% chance of being attacked by Y:

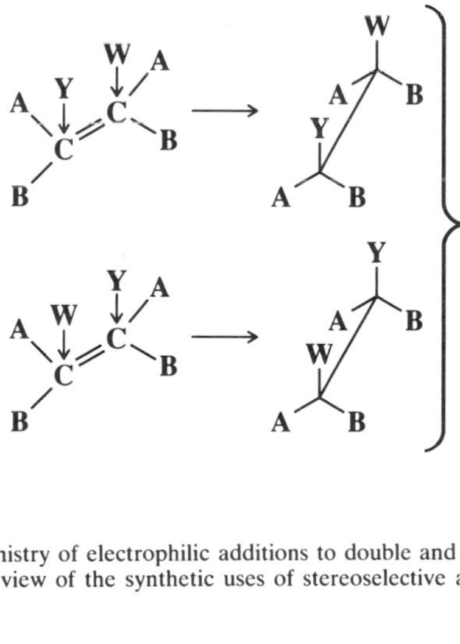

---

[2]For a review of the stereochemistry of electrophilic additions to double and triple bonds, see Fahey *Top. Stereochem.* **1968**, *3*, 237-342. For a review of the synthetic uses of stereoselective additions, see Bartlett *Tetrahedron* **1980**, *36*, 2-72, pp. 3-15.

On the other hand, if the addition is anti, the threo *dl* pair will be formed:

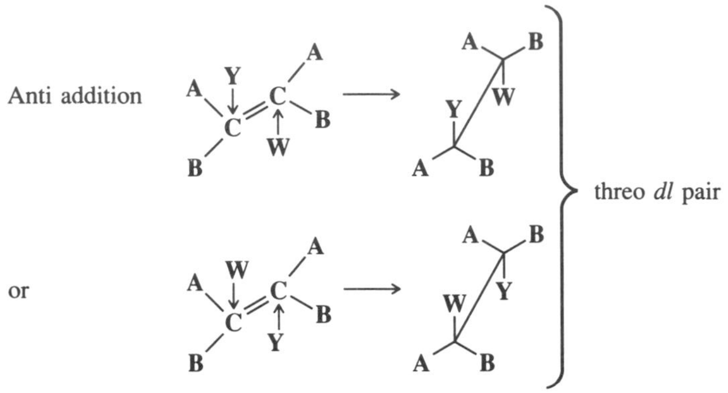

Anti addition

or

threo *dl* pair

Of course, the trans isomer will give the opposite results: the threo pair if the addition is syn and the erythro pair if it is anti. The threo and erythro isomers have different physical properties. In the special case where Y = W (as in the addition of Br_2), the "erythro pair" is a meso compound. In addition to triple-bond compounds of the type AC≡CA, syn addition results in a cis olefin and anti addition in a trans olefin. By the definition given on p. 137, addition to triple bonds cannot be stereospecific, though it can be, and often is, stereoselective.

It is easily seen that in reactions involving cyclic intermediates like **2** addition must be anti, since the second step is an SN2 step and must occur from the back side. It is not so easy to predict the stereochemistry for reactions involving **1**. If **1** has a relatively long life, the addition should be nonstereospecific, since there will be free rotation about the single bond. On the other hand, there may be some factor that maintains the configuration, in which case W may come in from the same side or the opposite side, depending on the circumstances. For example, the positive charge might be stabilized by an attraction for Y that does not involve a full bond:

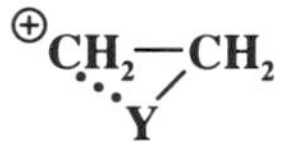

$$\overset{\oplus}{CH_2}-CH_2$$
$$\vdots$$
$$Y$$

3

The second group would then come in anti. A circumstance that would favor syn addition would be the formation of an ion pair after the addition of Y:[3]

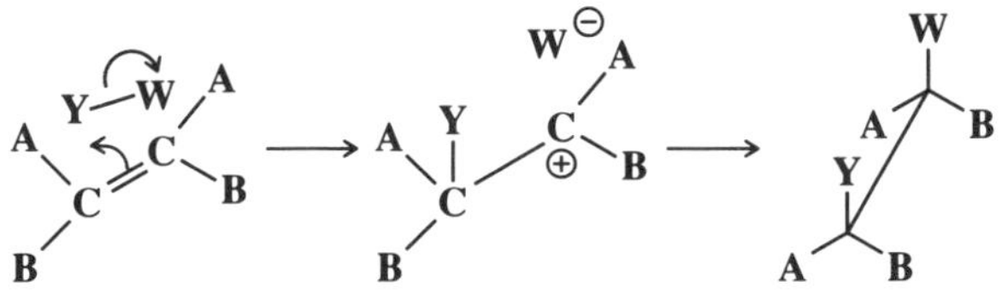

Since W is already on the same side of the plane as Y, collapse of the ion pair leads to syn addition.

[3]Dewar *Angew. Chem. Int. Ed. Engl.* **1964**, *3*, 245-249 [*Angew. Chem.* *76*, 320-325]; Heasley; Bower; Dougharty; Easdon; Heasley; Arnold; Carter; Yaeger; Gipe; Shellhamer *J. Org. Chem.* **1980**, *45*, 5150.

Another possibility is that anti addition might, at least in some cases, be caused by the operation of a mechanism in which attack by W and Y are essentially simultaneous but from opposite sides:

$$\begin{array}{c} \text{Y} \\ \diagdown \text{C}=\text{C} \diagdown \\ \diagup \quad \uparrow \diagdown \\ \overline{\text{W}} \end{array} \longrightarrow \begin{array}{c} \text{Y} \\ | \quad | \\ -\text{C}-\text{C}- \\ | \quad | \\ \text{W} \end{array}$$

This mechanism, called the AdE3 mechanism (*termolecular addition*, IUPAC $A_N A_E$),[4] has the disadvantage that three molecules must come together in the transition state. However, it is the reverse of the E2 mechanism for elimination, for which the transition state is known to possess this geometry (p. 983).

There is much evidence that when the attack is by Br⁺ (or a carrier of it), the bromonium ion **2** is often an intermediate and the addition is anti. As long ago as 1911, McKenzie and Fischer independently showed that treatment of maleic acid with bromine gave the *dl* pair of 2,3-dibromosuccinic acid, while fumaric acid (the trans isomer) gave the meso compound.[5] Many similar experiments have been performed since with similar results. For triple bonds, stereoselective anti addition was shown even earlier. Bromination of dicarboxyacetylene gave 70% of the trans isomer.[6]

$$\text{HOOC}-\text{C}\equiv\text{C}-\text{COOH} + \text{Br}_2 \longrightarrow 70\% \text{ trans} \quad \begin{array}{c} \text{HOOC} \diagdown \quad \diagup \text{Br} \\ \text{C}=\text{C} \\ \text{Br} \diagup \quad \diagdown \text{COOH} \end{array}$$

There is other evidence for mechanisms involving **2.** We have already mentioned (p. 312) that bromonium ions have been isolated in stable solutions in nucleophilic substitution reactions involving bromine as a neighboring group. Such ions have also been isolated in reactions involving addition of a Br⁺ species to a double bond.[7] The following is further evidence. If the two bromines approach the double bond from opposite sides, it is very unlikely that they could come from the same bromine molecule. This means that if the reaction is performed in the presence of nucleophiles, some of these will compete in the second step with the bromide liberated from the bromine. It has been found, indeed, that treatment of ethylene with bromine in the presence of chloride ions gives some 1-chloro-2-bromoethane along with the dibromoethane.[8] Similar results are found when the reaction is carried out in the presence of water (**5-27**) or of other nucleophiles.[9] Ab initio molecular

[4]For evidence for this mechanism, see, for example, Hammond; Nevitt *J. Am. Chem. Soc.* **1954**, *76*, 4121; Bell; Pring *J. Chem. Soc. B* **1966**, 1119; Pincock; Yates *J. Am. Chem. Soc.* **1968**, *90*, 5643; Fahey; Lee *J. Am. Chem. Soc.* **1967**, *89*, 2780, **1968**, *90*, 2124; Fahey; Monahan *J. Am. Chem. Soc.* **1970**, *92*, 2816; Fahey; Payne; Lee *J. Org. Chem.* **1974**, *39*, 1124; Roberts *J. Chem. Soc., Perkin Trans. 2* **1976**, 1374; Pasto; Gadberry *J. Am. Chem. Soc.* **1978**, *100*, 1469; Naab; Staab *Chem. Ber.* **1978**, *111*, 2982.
[5]This was done by Fischer *Liebigs Ann. Chem.* **1911**, *386*, 374; McKenzie *Proc. Chem. Soc.* **1911**, 150, *J. Chem. Soc.* **1912**, *101*, 1196.
[6]Michael *J. Prakt. Chem.* **1892**, *46*, 209.
[7]Strating; Wieringa; Wynberg *Chem. Commun.* **1969**, 907; Olah *Angew. Chem. Int. Ed. Engl.* **1973**, *12*, 173-212, p. 207 [*Angew. Chem. 85*,183-225]; Slebocka-Tilk; Ball; Brown *J. Am. Chem. Soc.* **1985**, *107*, 4504.
[8]Francis *J. Am. Chem. Soc.* **1925**, *47*, 2340.
[9]See, for example, Zefirov; Koz'min; Dan'kov; Zhdankin; Kirin *J. Org. Chem. USSR* **1984**, *20*, 205.

orbital studies show that **2** is more stable than its open isomer **1** (Y = Br).[10] There is evidence that formation of **2** is reversible.[11]

However, a number of examples have been found where addition of bromine is not stereospecifically anti. For example, the addition of Br_2 to *cis*- and *trans*-1-phenylpropenes in CCl_4 was nonstereospecific.[12] Furthermore, the stereospecificity of bromine addition to stilbene depends on the dielectric constant of the solvent. In solvents of low dielectric constant, the addition was 90 to 100% anti, but with an increase in dielectric constant, the reaction became less stereospecific, until, at a dielectric constant of about 35, the addition was completely nonstereospecific.[13] Likewise in the case of triple bonds, stereoselective anti addition was found in bromination of 3-hexyne, but both cis and trans products were obtained in bromination of phenylacetylene.[14] These results indicate that a bromonium ion is not formed where the open cation can be stabilized in other ways (e.g., addition of Br^+ to 1-phenylpropene gives the ion $Ph\overset{\oplus}{C}HCHBrCH_3$, which is a relatively stable benzylic cation) and that there is probably a spectrum of mechanisms between complete bromonium ion (**2**, no rotation) formation and completely open-cation (**1**, free rotation) formation, with partially bridged bromonium ions (**3**, restricted rotation) in between.[15] We have previously seen cases (e.g., p. 315) where cations require more stabilization from outside sources as they become intrinsically less stable themselves.[16] Further evidence for the open cation mechanism where aryl stabilization is present was reported in an isotope effect study of addition of Br_2 to ArCH=CHCHAr′ (Ar = *p*-nitrophenyl, Ar′ = *p*-tolyl). The ^{14}C isotope effect for one of the double bond carbons (the one closer to the NO_2 group) was considerably larger than for the other one.[17]

Attack by Cl^+,[18] I^+,[19] and RS^{+20} is similar to that by Br^+; there is a spectrum of mechanisms between cyclic intermediates and open cations. As might be expected from our discussion in Chapter 10 (p. 312), iodonium ions compete with open carbocations more effectively than bromonium ions, while chloronium ions compete less effectively. There is

[10]Hamilton; Schaefer *J. Am. Chem. Soc.* **1990,** *112,* 8260.

[11]Brown; Gedye; Slebocka-Tilk; Buschek; Kopecky *J. Am. Chem. Soc.* **1984,** *106,* 4515; Bellucci; Bianchini; Chiappe; Marioni; Spagna *J. Am. Chem. Soc.* **1988,** *110,* 546; Ruasse; Motallebi; Galland *J. Am. Chem. Soc.* **1991,** *113,* 3440; Bellucci; Bianchini; Chiappe; Brown; Slebocka-Tilk; *J. Am. Chem. Soc.* **1991,** *113,* 8012; Bennet; Brown; McClung; Klobukowski; Aarts; Santarsiero; Bellucci; Bianchini; *J. Am. Chem. Soc.* **1991,** *113,* 8532.

[12]Fahey; Schneider *J. Am. Chem. Soc.* **1968,** *90,* 4429. See also Rolston; Yates *J. Am. Chem. Soc.* **1969,** *91,* 1469, 1477, 1483.

[13]Buckles; Bader; Thurmaier *J. Org. Chem.* **1962,** *27,* 4523; Heublein *J. Prakt. Chem.* **1966,** [4] *31,* 84. See also Buckles; Miller; Thurmaier *J. Org. Chem.* **1967,** *32,* 888; Heublein; Lauterbach *J. Prakt. Chem.* **1969,** *311,* 91; Ruasse; Dubois *J. Am. Chem. Soc.* **1975,** *97,* 1977. For the dependence of stereospecificity in this reaction on the solvent concentration, see Bellucci; Bianchini; Chiappe; Marioni *J. Org. Chem.* **1990,** *55,* 4094.

[14]Pincock; Yates *Can. J. Chem.* **1970,** *48,* 3332.

[15]For other evidence for this concept, see Pincock; Yates *Can. J. Chem.* **1970,** *48,* 2944; Heasley; Chamberlain *J. Org. Chem.* **1970,** *35,* 539; Dubois; Toullec; Barbier *Tetrahedron Lett.* **1970,** 4485; Dalton; Davis *Tetrahedron Lett.* **1972,** 1057; Wilkins; Regulski *J. Am. Chem. Soc.* **1972,** *94,* 6016; Sisti; Meyers *J. Org. Chem.* **1973,** *38,* 4431; McManus; Peterson *Tetrahedron Lett.* **1975,** 2753; Abraham; Monasterios *J. Chem. Soc., Perkin Trans. 1* **1973,** 1446; Ruasse; Argile; Dubois *J. Am. Chem. Soc.* **1978,** *100,* 7645, *J. Org. Chem.* **1979,** *44,* 1173; Schmid; Modro; Yates *J. Org. Chem.* **1980,** *45,* 665; Ruasse; Argile *J. Org. Chem.* **1983,** *48,* 202; Cadogan; Cameron; Gosney; Highcock; Newlands *J. Chem. Soc., Chem. Commun.* **1985,** 1751. For a review, see Ruasse *Acc. Chem.Res.* **1990,** *23,* 87-93.

[16]In a few special cases, stereospecific syn addition of Br_2 has been found, probably caused by an ion pair mechanism as shown on p. 736: Naae *J. Org. Chem.* **1980,** *45,* 1394.

[17]Kokil; Fry *Tetrahedron Lett.* **1986,** *27,* 5051.

[18]Fahey, Ref. 2, pp. 273-277.

[19]Hassner; Boerwinkle; Levy *J. Am. Chem. Soc.* **1970,** *92,* 4879.

[20]For reviews of thiiranium and/or thiirenium ions, see Capozzi; Modena, in Bernardi; Csizmadia; Mangini *Organic Sulfur Chemistry;* Elsevier: New York, 1985, pp. 246-298; Smit, Ref. 1, pp. 180-202; Dittmer; Patwardhan, in Stirling *The Chemistry of the Sulphonium Group,* pt. 1; Wiley: New York, 1981, pp. 387-412; Capozzi; Lucchini; Modena; *Rev. Chem. Intermed.* **1979,** *2,* 347-375; Schmid *Top. Sulfur Chem.* **1977,** *3,* 102-117; Mueller *Angew. Chem. Int. Ed. Engl.* **1969,** *8,* 482-492 [*Angew. Chem. 81,* 475-484]. The specific nature of the 3-membered sulfur-containing ring is in dispute; see Smit; Zefirov; Bodrikov; Krimer *Acc. Chem. Res.* **1979,** *12,* 282-288; Bodrikov; Borisov; Chumakov; Zefirov; Smit *Tetrahedron Lett.* **1980,** *21,* 115; Schmid; Garratt; Dean *Can. J. Chem.* **1987,** *65,* 1172; Schmid; Strukelj; Dalipi *Can. J. Chem.* **1987,** *65,* 1945.

kinetic and spectral evidence that at least in some cases, for example in the addition of Br_2 or ICl, the electrophile forms a π complex with the alkene before a covalent bond is formed.[21]

When the electrophile is a proton,[22] a cyclic intermediate is not possible, and the mechanism is the simple $A_H + A_N$ process shown before

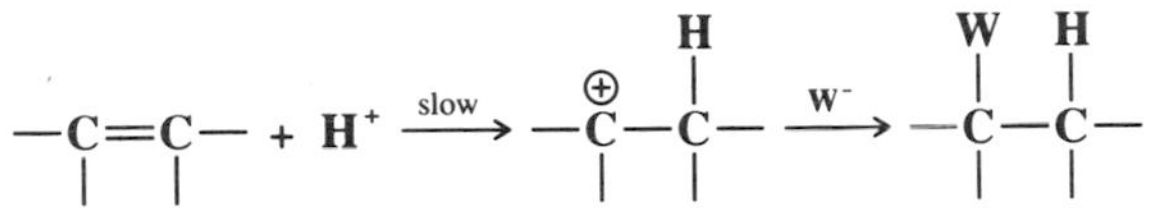

$$-\overset{|}{C}=\overset{|}{C}- + H^+ \xrightarrow{slow} -\overset{|}{\underset{|}{C}}-\overset{\overset{H}{|}}{\underset{|}{\overset{\oplus}{C}}}- \xrightarrow{w^-} -\overset{\overset{W}{|}}{\underset{|}{C}}-\overset{\overset{H}{|}}{\underset{|}{C}}-$$

This is an A-SE2 mechanism (p. 374). There is a great deal of evidence[23] for it, including:

1. The reaction is general-acid, not specific-acid-catalyzed, implying rate-determining proton transfer from the acid to the double bond.[24]

2. The existence of open carbocation intermediates is supported by the contrast in the pattern of alkyl substituent effects[25] with that found in brominations, where cyclic intermediates are involved. In the latter case substitution of alkyl groups on $H_2C{=}CH_2$ causes a cumulative rate acceleration until all four hydrogens have been replaced by alkyl groups,

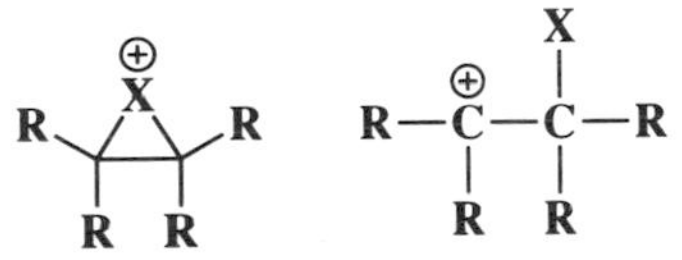

because each group helps to stabilize the positive charge.[26] In addition of HX the effect is not cumulative. Replacement of the two hydrogens on one carbon causes great rate increases (primary → secondary → tertiary carbocation), but additional substitution on the other carbon produces little or no acceleration.[27] This is evidence for open cations when a proton is the electrophile.[28]

3. Open carbocations are prone to rearrange (Chapter 18). Many rearrangements have been found to accompany additions of HX and H_2O).[29]

---

[21]See Norlander; Haky; Landino *J. Am. Chem. Soc.* **1980,** *102,* 7487; Fukuzumi; Kochi *Int. J. Chem. Kinet.* **1983,** *15,* 249; Schmid; Gordon *Can. J. Chem.* **1984,** *62,* 2526, **1986,** *64,* 2171; Bellucci; Bianchini; Ambrosetti *J. Am. Chem. Soc.* **1985,** *107,* 2464; Bellucci; Bianchini; Chiappe; Marioni; Ambrosetti; Brown; Slebocka-Tilk *J. Am. Chem. Soc.* **1989,** *111,* 2640.

[22]For a review of the addition of HCl, see Sergeev; Smirnov; Rostovshchikova *Russ. Chem. Rev.* **1983,** *52,* 259-274.

[23]For other evidence, see Baliga; Whalley *Can. J. Chem.* **1964,** *42,* 1019, **1965,** *43,* 2453; Gold; Kessick *J. Chem. Soc.* **1965,** 6718; Corriu; Guenzet *Tetrahedron* **1970,** *26,* 671; Simandoux; Torck; Hellin; Coussemant *Bull. Soc. Chim. Fr.* **1972,** 4402, 4410; Bernasconi; Boyle *J. Am. Chem. Soc.* **1974,** *96,* 6070; Hampel; Just; Pisanenko; Pritzkow *J. Prakt. Chem.* **1976,** *318,* 930; Allen; Tidwell, *J. Am. Chem. Soc.* **1983,** *104,* 3145.

[24]Kresge; Chiang; Fitzgerald; McDonald; Schmid *J. Am. Chem. Soc.* **1971,** *93,* 4907; Loudon; Noyce *J. Am. Chem. Soc.* **1969,** *91,* 1433; Schubert; Keeffe *J. Am. Chem. Soc.* **1972,** *94,* 559; Chiang; Kresge *J. Am. Chem. Soc.* **1985,** *107,* 6363.

[25]Bartlett; Sargent *J. Am. Chem. Soc.* **1965,** *87,* 1297; Schmid; Garratt *Can. J. Chem.* **1973,** *51,* 2463.

[26]See, for example, Anantakrishnan; Ingold *J. Chem. Soc.* **1935,** 1396; Swern, in Swern *Organic Peroxides,* vol. 2; Wiley: New York, 1971, pp. 451-454; Nowlan; Tidwell *Acc. Chem. Res.* **1977,** *10,* 252-258.

[27]Bartlett; Sargent, Ref. 25; Riesz; Taft; Boyd *J. Am. Chem. Soc.* **1957,** *79,* 3724.

[28]A similar result (open cations) was obtained with carbocations Ar_2CH^+ as electrophiles: Mayr; Pock *Chem. Ber.* **1986,** *119,* 2473.

[29]For example, see Whitmore; Johnston *J. Am. Chem. Soc.* **1933,** *55,* 5020; Fahey; McPherson *J. Am. Chem. Soc.* **1969,** *91,* 3865; Bundel'; Ryabstev; Sorokin; Reutov *Bull. Acad. Sci. USSR, Div. Chem. Sci.* **1969,** 1311; Pocker; Stevens *J. Am. Chem. Soc.* **1969,** *91,* 4205; Staab; Wittig; Naab *Chem. Ber.* **1978,** *111,* 2965; Stammann; Griesbaum *Chem. Ber.* **1980,** *113,* 598.

It may also be recalled that vinylic ethers react with proton donors in a similar manner (see **0-6**).

The stereochemistry of HX addition is varied. Examples are known of predominant syn, anti, and nonstereoselective addition. It was found that treatment of 1,2-dimethylcyclohexene (**4**) with HBr gave predominant anti addition,[30] while addition of water to **4** gave equal amounts of the cis and trans alcohols:[31]

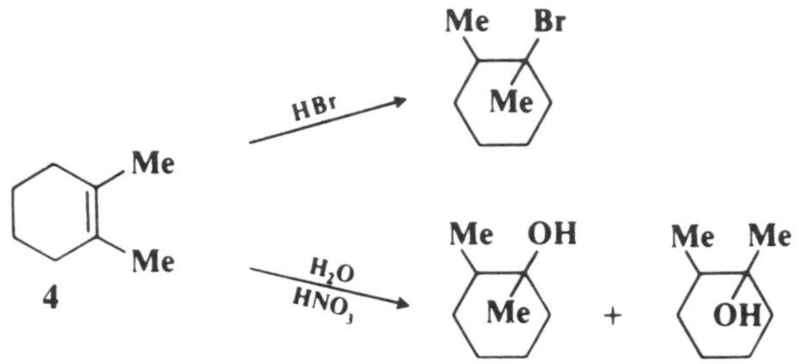

On the other hand, addition of DBr to acenaphthylene (**5**) and to indene and 1-phenylpropene gave predominant syn addition.[32]

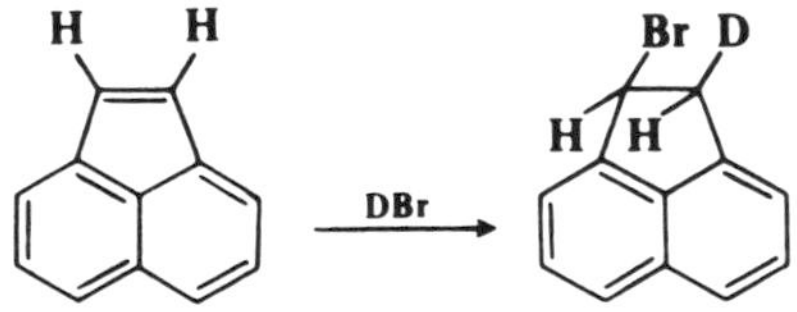

In fact it has been shown that the stereoselectivity of HCl addition can be controlled by changing the reaction conditions. Addition of HCl to **4** in CH_2Cl_2 at $-98°C$ gave predominantly syn addition, while in ethyl ether at $0°C$, the addition was mostly anti.[33]

Addition of HX to triple bonds has the same mechanism, though the intermediate in this case is a vinylic cation:[34]

In all these cases (except for the AdE3 mechanism) we have assumed that formation of the intermediate (**1, 2,** or **3**) is the slow step and attack by the nucleophile on the intermediate

---

[30]Hammond; Nevitt, Ref. 4; See also Fahey; Monahan, Ref. 4; Pasto; Meyer; Lepeska *J. Am. Chem. Soc.* **1974,** *96,* 1858.

[31]Collins; Hammond *J. Org. Chem.* **1960,** *25,* 911.

[32]Dewar; Fahey *J. Am. Chem. Soc.* **1963,** *85,* 2245, 2248. For a review of syn addition of HX, see Ref. 3.

[33]Becker; Grob *Synthesis* **1973,** 789. See also Marcuzzi; Melloni; Modena *Tetrahedron Lett.* **1974,** 413; Naab; Staab, Ref. 4.

[34]For reviews of electrophilic addition to alkynes, including much evidence, see Rappoport *React. Intermed.* (*Plenum*) **1983,** *3,* 427-615, pp. 428-440; Stang; Rappoport; Hanack; Subramanian *Vinyl Cations*; Academic Press: New York, 1979, pp. 24-151; Stang *Prog. Phys. Org. Chem.* **1973,** *10,* 205-325; Modena; Tonellato *Adv. Phys. Org. Chem.* **1971,** *9,* 185-280, pp. 187-231; Richey; Richey, in Olah; Schleyer *Carbonium Ions,* vol. 2; Wiley: New York, 1970, pp. 906-922.

is rapid, and this is probably true in most cases. However, some additions have been found in which the second step is rate-determining.[35]

Nucleophilic Addition[36]

In the first step of nucleophilic addition a nucleophile brings its pair of electrons to one carbon atom of the double or triple bond, creating a carbanion. The second step is combination of this carbanion with a positive species:

Step 1

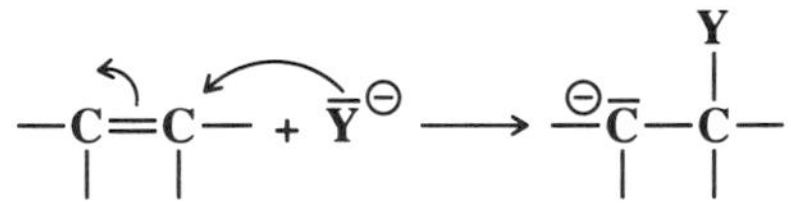

$$-\overset{|}{C}=\overset{|}{C}- + \overline{Y}^{\ominus} \longrightarrow {}^{\ominus}\overline{C}-\overset{Y}{\underset{|}{C}}-$$

Step 2

$$^{\ominus}\overline{C}-\overset{Y}{\underset{|}{C}}- + W^{\oplus} \longrightarrow -\overset{W}{\underset{|}{C}}-\overset{Y}{\underset{|}{C}}-$$

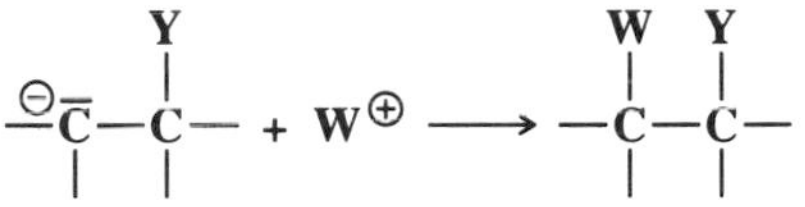

This mechanism is the same as the simple electrophilic one shown on p. 734 except that the charges are reversed (IUPAC $A_N + A_E$ or $A_N + A_H$). When the olefin contains a good leaving group (as defined for nucleophilic substitution), substitution is a side reaction (this is nucleophilic substitution at a vinylic substrate, see p. 335).

In the special case of addition of HY to a substrate of the form —C=C—Z, where

Z = CHO, COR[37] (including quinones[38]), COOR, CONH$_2$, CN, NO$_2$, SOR, SO$_2$R,[39] etc., addition nearly always follows a nucleophilic mechanism,[40] with Y$^-$ bonding with the carbon *away* from the Z group, e.g.,

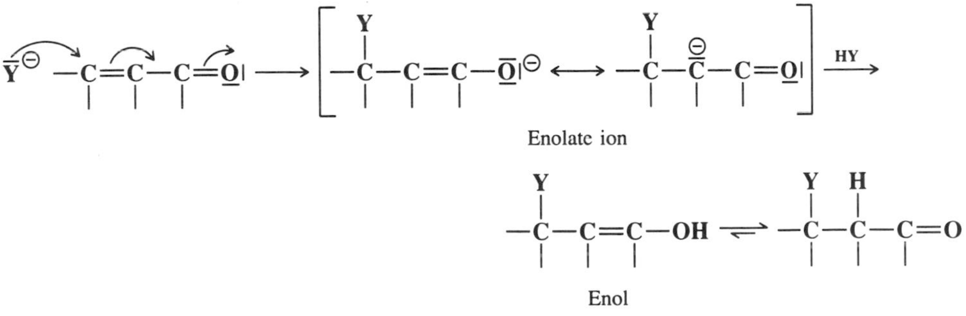

$$\overline{Y}^{\ominus} \ -\overset{|}{C}=\overset{|}{C}-\overset{|}{C}=\overline{O}| \longrightarrow \left[-\overset{Y}{\underset{|}{C}}-\overset{|}{C}=\overset{|}{C}-\overline{\underline{O}}|^{\ominus} \longleftrightarrow -\overset{Y}{\underset{|}{C}}-\overset{\ominus}{\overline{C}}-\overset{|}{C}=\overline{O}| \right] \xrightarrow{\text{HY}}$$

Enolate ion

$$-\overset{Y}{\underset{|}{C}}-\overset{|}{C}=\overset{|}{C}-OH \rightleftharpoons -\overset{Y}{\underset{|}{C}}-\overset{H}{\underset{|}{C}}-\overset{|}{C}=O$$

Enol

[35]See, for example, Rau; Alcais; Dubois *Bull. Soc. Chim. Fr.* **1972**, 3336; Bellucci; Berti; Ingrosso; Mastrorilli *Tetrahedron Lett.* **1973**, 3911.

[36]For a review, see Patai; Rappoport, in Patai *The Chemistry of Alkenes*, vol. 1; Wiley: New York, 1964, pp. 469-584.

[37]For reviews of reactions of C=C—C=O compounds, see, in Patai; Rappoport *The Chemistry of Enones*, pt. 1; Wiley: New York, 1989, the articles by Boyd, pp. 281-315; Duval; Géribaldi, pp. 355-469.

[38]For reviews of addition reactions of quinones, see Kutyrev; Moskva *Russ. Chem. Rev.* **1991**, *60*, 72-88; Finley, in Patai; Rappoport *The Chemistry of the Quinonoid Compounds*, vol. 2, pt. 1; Wiley: New York, 1988, pp. 537-717, pp. 539-589; Finley, in Patai *The Chemistry of the Quinonoid Compounds*, pt. 2; Wiley: New York, 1974, pp. 877-1144.

[39]For a review of vinylic sulfones, see Simpkins *Tetrahedron* **1990**, *46*, 6951-6984. For a review of conjugate addition to cycloalkenyl sulfones, see Fuchs; Braish *Chem. Rev.* **1986**, *86*, 903-917.

[40]For a review of the mechanism with these substrates, see Bernasconi *Tetrahedron* **1989**, *45*, 4017-4090.

Protonation of the enolate ion is chiefly at the oxygen, which is more negative than the carbon, but this produces the enol, which tautomerizes. So although the net result of the reaction is addition to a carbon–carbon double bond, the *mechanism* is 1,4 nucleophilic addition to the C=C—C=O (or similar) system and is thus very similar to the mechanism of addition to carbon–oxygen double and similar bonds (see Chapter 16). When Z is CN or a C=O group, it is also possible for Y⁻ to attack at *this* carbon, and this reaction sometimes competes. When it happens, it is called 1,2 addition. 1,4 addition to these substrates is also known as *conjugate addition*. Y⁻ almost never attacks at the 3 position, since the resulting carbanion would have no resonance stabilization:[41]

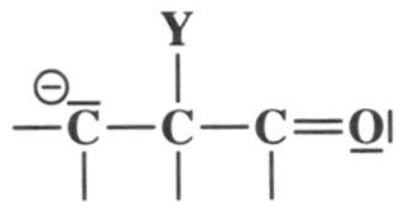

$$-\overset{\ominus}{\underset{|}{C}}-\underset{|}{\overset{|}{C}}-\underset{|}{\overset{Y}{C}}=\underline{O}|$$

An important substrate of this type is acrylonitrile, and 1,4 addition to it is called *cyanoethylation* because the Y is cyanoethylated:

$$\text{H}_2\text{C}=\text{CH}—\text{CN} + \text{HY} \longrightarrow \text{Y}—\text{CH}_2—\text{CH}_2—\text{CN}$$

With any substrate, when Y is an ion of the type $\text{Z}—\overset{\ominus}{\text{C}}\text{R}_2$ (Z is as defined above; R may be alkyl, aryl, hydrogen, or another Z), the reaction is called the *Michael reaction* (see **5-17**). In this book we will call all other reactions that follow this mechanism *Michael-type additions*. Systems of the type C=C—C=C—Z can give 1,2, 1,4, or 1,6 addition.[42] Michael-type reactions are reversible, and compounds of the type YCH₂CH₂Z can often be decomposed to YH and CH₂=CHZ by heating, either with or without alkali.

If the mechanism for nucleophilic addition is the simple carbanion mechanism outlined on p. 741, the addition should be nonstereospecific, though it might well be stereoselective (see p. 137 for the distinction). For example, the *E* and *Z* forms of an olefin ABC=CDE would give **6** and **7**:

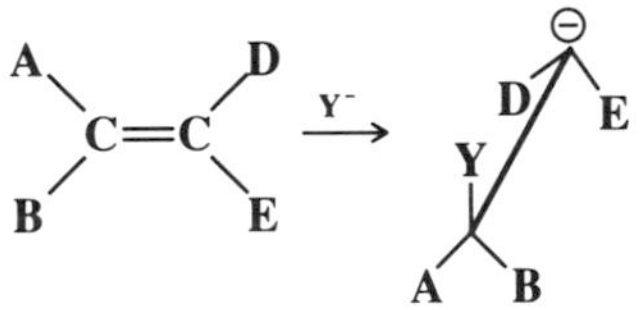

6

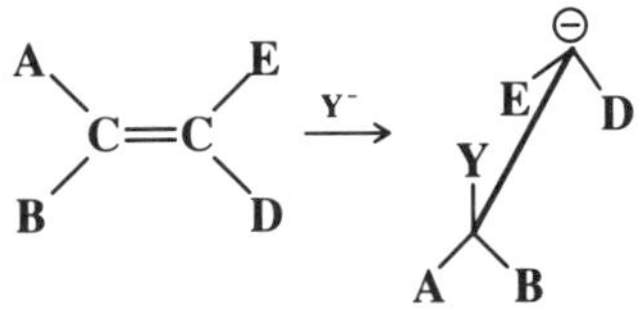

7

If the carbanion has even a short lifetime, **6** and **7** will assume the most favorable conformation before the attack of W. This is of course the same for both, and when W attacks, the same product will result from each. This will be one of two possible diastereomers, so the reaction will be stereoselective; but since the cis and trans isomers do not give rise to

[41]For 1,8 addition to a trienone, see Barbot; Kadib-Elban; Miginiac *J. Organomet. Chem.* **1988,** *345,* 239.
[42]However, attack at the 3 position has been reported when the 4 position contains one or two carbanion-stabilizing groups such as SiMe₃: Klumpp; Mierop; Vrielink; Brugman; Schakel *J. Am. Chem. Soc.* **1985,** *107,* 6740.

different isomers, it will not be stereospecific. Unfortunately, this prediction has not been tested on open-chain olefins. Except for Michael-type substrates, the stereochemistry of nucleophilic addition to double bonds has been studied only in cyclic systems, where only the cis isomer exists. In these cases the reaction has been shown to be stereoselective, with syn addition reported in some cases[43] and anti addition in others.[44] When the reaction is performed on a Michael-type substrate, C=C—Z, the hydrogen does not arrive at the carbon directly but only through a tautomeric equilibrium. The product naturally assumes the most thermodynamically stable configuration, without relation to the direction of original attack of Y. In one such case (the addition of EtOD and of Me_3CSD to *trans*-MeCH=CHCOOEt) predominant anti addition was found; there is evidence that the stereoselectivity here results from the final protonation of the enolate, and not from the initial attack.[45] For obvious reasons, additions to triple bonds cannot be stereospecific. As with electrophilic additions, nucleophilic additions to triple bonds are usually stereoselective and anti,[46] though syn addition[47] and nonstereoselective addition[48] have also been reported.

Free-Radical Addition

The mechanism of free-radical addition[49] follows the pattern discussed in Chapter 14 (pp. 677-678). A radical is generated by

$$YW \xrightarrow[\text{dissociation}]{h\nu \text{ or spontaneous}} Y\cdot + W\cdot$$

or $R\cdot$ (from some other source) $+ YW \longrightarrow RW + Y\cdot$

Propagation then occurs by

Step 1

Step 2

[43]For example, Truce; Levy *J. Org. Chem.* **1963**, *28*, 679.

[44]For example, Truce; Levy *J. Am. Chem. Soc.* **1961**, *83*, 4641; Zefirov; Yur'ev; Prikazchikova; Bykhovskaya *J. Gen. Chem. USSR* **1963**, *33*, 2100.

[45]Mohrig; Fu; King; Warnet; Gustafson *J. Am. Chem. Soc.* **1990**, *112*, 3665.

[46]Truce; Simms *J. Am. Chem. Soc.* **1956**, *78*, 2756; Shostakovskii; Chekulaeva; Kondrat'eva; Lopatin *Bull. Acad. Sci. USSR, Div. Chem. Sci* **1962**, 2118; Théron; Vessière *Bull. Soc. Chim. Fr.* **1968**, 2994; Bowden; Price *J. Chem. Soc. B* **1970**, 1466. 1472; Raunio; Frey *J. Org. Chem.* **1971**, *36*, 345; Truce; Tichenor *J. Org. Chem.* **1972**, *37*, 2391.

[47]Truce; Goldhamer; Kruse *J. Am. Chem. Soc.* **1959**, *81*, 4931; Dolfini *J. Org. Chem.* **1965**, *30*, 1298; Winterfeldt; Preuss *Chem. Ber.* **1966**, *99*, 450; Hayakawa; Kamikawaji; Wakita; Kanematsu *J. Org. Chem.* **1984**, *49*, 1985.

[48]Gracheva; Laba; Kul'bovskaya; Shostakovskii *J. Gen. Chem. USSR* **1963**, *33*, 2431; Truce; Brady *J. Org. Chem.* **1966**, *31*, 3543; Prilezhaeva; Vasil'ev; Mikhaleshvili; Bogdanov *Bull. Acad. Sci., USSR, Div. Chem. Sci.* **1970**, 1820.

[49]For a monograph on this subject, see Huyser *Free-Radical Chain Reactions*; Wiley: New York, 1970. Other books with much of interest in this field are Nonhebel; Walton *Free-Radical Chemistry*; Cambridge University Press: London, 1974; Pyor *Free Radicals*; McGraw-Hill: New York, 1965. For reviews, see Giese *Rev. Chem. Intermed.* **1986**, *7*, 3-11; *Angew. Chem. Int. Ed. Engl.* **1983**, *22*, 753-764 [*Angew. Chem. 95*, 771-782]; Amiel, in Patai; Rappoport *The Chemistry of Functional Groups, Supplement C*, pt. 1; Wiley: New York, 1983, pp. 341-382; Abell, in Bamford; Tipper *Comprehensive Chemical Kinetics*, vol. 18; Elsevier: New York, 1976, pp. 111-165; Abell, in Kochi *Free Radicals*, vol. 2; Wiley: New York, 1973, pp. 63-112; Minisci *Acc. Chem. Res.* **1975**, *8*, 165-171; Julia, in Viehe *Acetylenes*; Marcel Dekker: New York, 1969, pp. 335-354; Elad *Org. Photochem.* **1969**, *2*, 168-212; Schönberg *Preparative Organic Photochemistry*; Springer: New York, 1968, pp. 155-181; Cadogan; Perkins, in Patai, Ref. 36, pp. 585-632.

Step 2 is an abstraction, so W is nearly always univalent, either hydrogen or halogen (p. 683). Termination of the chain can occur in any of the ways discussed in Chapter 14. If **8** adds to another olefin molecule,

$$-\overset{\underset{\mid}{Y}}{\underset{\mid}{C}}-\overset{\mid}{\underset{\mid}{C}}-\;+\;-C=C-\;\longrightarrow\;-\overset{\underset{\mid}{}}{\underset{\mid}{C}}-\overset{\underset{\mid}{Y}}{\underset{\mid}{C}}-\overset{\mid}{\underset{\mid}{C}}-\overset{\underset{\mid}{Y}}{\underset{\mid}{C}}-$$

8

a dimer is formed. This can add to still another, and chains, long or short, may be built up. This is the mechanism of free-radical polymerization. Short polymeric molecules (called *telomers*), formed in this manner, are often troublesome side products in free-radical addition reactions.

When free radicals are added to 1,5- or 1,6-dienes, the initially formed radical can add intramolecularly to the other bond, leading to a cyclic product, e.g.,[50]

9

Radicals of the type **9**, generated in other ways, also undergo these cyclizations. Both five- and six-membered rings can be formed in these reactions (see p. 752).

The free-radical addition mechanism just outlined predicts that the addition should be nonstereospecific, at least if **8** has any but an extremely short lifetime. However, the reactions may be stereoselective, for reasons similar to those discussed for nucleophilic addition on p. 742. Not all free-radical additions have been found to be stereoselective, but many are. For example, addition of HBr to 1-bromocyclohexene gave only *cis*-1,2-dibromocyclohexane and none of the trans isomer (anti addition),[51] and propyne (at -78 to $-60°C$) gave only *cis*-1-bromopropene (anti addition).[52] However, stereospecificity has been found only in a few cases. The most important of these is addition of HBr to 2-bromo-2-butene under free-radical conditions at $-80°C$. Under these conditions, the cis isomer gave 92% of the meso product, while the trans isomer gave mostly the *dl* pair.[53] This stereospecificity disappeared at room temperature, where both olefins gave the same mixture of products (about 78% of the *dl* pair and 22% of the meso compound), so the addition was still stereoselective but no longer stereospecific. The stereospecificity at low temperatures is probably caused by a stabilization of the intermediate radical through the formation of a bridged bromine radical, of the type mentioned on p. 682:

$$-\overset{\underset{\mid}{Br}}{\underset{\mid}{C}}-\overset{\underset{\mid}{}}{\underset{\mid}{C}}-\;\longrightarrow\;-\overset{Br}{\underset{\mid}{C}}\diagdown\overset{}{\underset{\mid}{C}}-$$

---

[50]For reviews of these and other free-radical cyclization reactions, see RajanBabu *Acc. Chem. Res.* **1991**, *24*, 139-145; Beckwith *Rev. Chem. Intermed.* **1986**, *7*, 143-154; Giese *Radicals in Organic Synthesis: Formation of Carbon-Carbon Bonds*; Pergamon: Elmsford, NY, 1986, pp. 141-209; Surzur *React. Intermed. (Plenum)* **1982**, *2*, 121-295; Julia *Acc. Chem. Res.* **1972**, *4*, 386-392, *Pure Appl. Chem.* **1974**, *40*, 553-567, **1967**, *15*, 167-183; Nonhebel; Walton, Ref. 49, pp. 533-544; Wilt, in Kochi, Ref. 49, vol. 1, pp. 418-446. For a review of cyclizations in general, see Thebtaranonth; Thebtaranonth *Tetrahedron* **1990**, *46*, 1385-1489.
[51]Goering; Abell; Aycock *J. Am. Chem. Soc.* **1952**, *74*, 3588. See also LeBel; Czaja; DeBoer *J. Org. Chem.* **1969**, *34*, 3112.
[52]Skell; Allen *J. Am. Chem. Soc.* **1958**, *80*, 5997.
[53]Goering; Larsen *J. Am. Chem. Soc.* **1957**, *79*, 2653, **1959**, *81*, 5937. Also see Skell; Allen *J. Am. Chem. Soc.* **1959**, *81*, 5383; Skell; Freeman *J. Org. Chem.* **1964**, *29*, 2524.

This species is similar to the bromonium ion that is responsible for stereospecific anti addition in the electrophilic mechanism. Further evidence for the existence of such bridged radicals was obtained by addition of Br• to olefins at 77 K. Esr spectra of the resulting species were consistent with bridged structures.[54]

For many radicals step 1 (C=C + Y• → •C—C—Y) is reversible. In such cases free radicals can cause cis → trans isomerization of a double bond by the pathway[55]

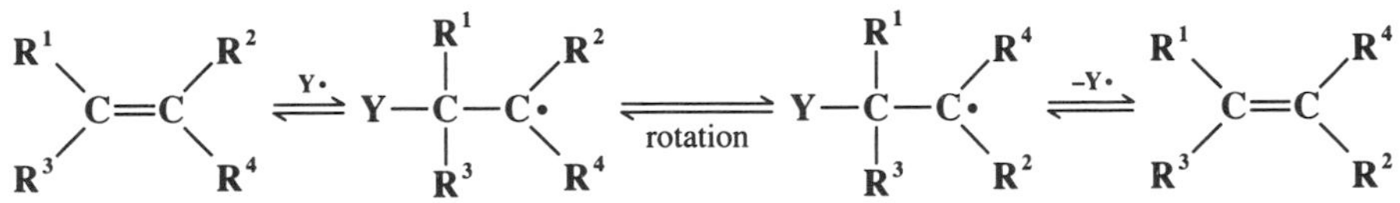

$$
\begin{array}{ccccccc}
R^1\diagdown\diagup R^2 & & R^1R^2 & & R^1R^4 & & R^1\diagdown\diagup R^4 \\
C{=}C & \overset{Y\bullet}{\rightleftharpoons} & Y{-}\overset{|}{C}{-}\overset{|}{C}\bullet & \underset{\text{rotation}}{\rightleftharpoons} & Y{-}\overset{|}{C}{-}\overset{|}{C}\bullet & \overset{-Y\bullet}{\rightleftharpoons} & C{=}C \\
R^3\diagup\diagdown R^4 & & R^3R^4 & & R^3R^2 & & R^3\diagup\diagdown R^2
\end{array}
$$

Cyclic Mechanisms

There are some addition reactions where the initial attack is not at one carbon of the double bond, but both carbons are attacked simultaneously. Some of these are four-center mechanisms, which follow this pattern:

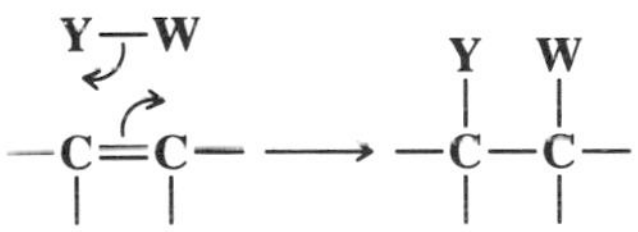

$$
\begin{array}{ccc}
Y{-}W & & YW \\
\downarrow & & || \\
{-}C{=}C{-} & \longrightarrow & {-}C{-}C{-} \\
|| & & ||
\end{array}
$$

In others there is a five- or a six-membered transition state. In these cases the addition to the double or triple bond must be syn. The most important reaction of this type is the Diels-Alder reaction (5-47).

Addition to Conjugated Systems

When electrophilic addition is carried out on a compound with two double bonds in conjugation, a 1,2-addition product (10) is often obtained, but in most cases there is also a 1,4-addition product (11), often in larger yield:[55a]

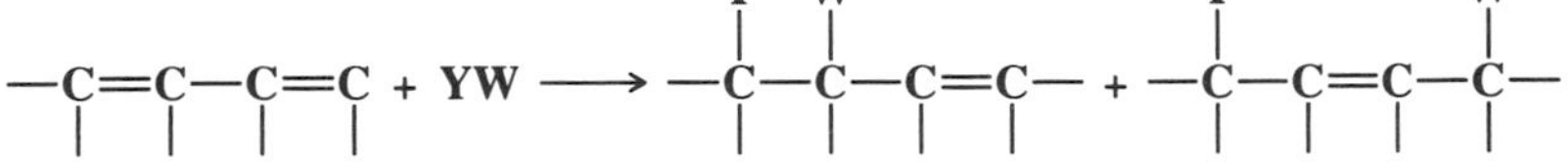

$$
\begin{array}{ccccc}
& & YW & & YW \\
& & || & & || \\
{-}C{=}C{-}C{=}C{-} + YW & \longrightarrow & {-}C{-}C{-}C{=}C{-} & + & {-}C{-}C{=}C{-}C{-} \\
& & & & \\
& & \mathbf{10} & & \mathbf{11}
\end{array}
$$

If the diene is unsymmetrical, there may be two 1,2-addition products. The competition between two types of addition product comes about because the carbocation resulting from

[54]Abell; Piette J. Am. Chem. Soc. 1962, 84, 916. See also Leggett; Kennerly; Kohl J. Chem. Phys. 1974, 60, 3264.

[55]Benson; Egger; Golden J. Am. Chem. Soc. 1965, 87, 468; Golden; Furuyama; Benson Int. J. Chem. Kinet. 1969, 1, 57.

[55a]For a review of electrophilic addition to conjugated dienes, see Khristov; Angelov; Petrov Russ. Chem. Rev. 1991, 60, 39-56.

attack by Y^+ is a resonance hybrid, with partial positive charges at the 2 and 4 positions:

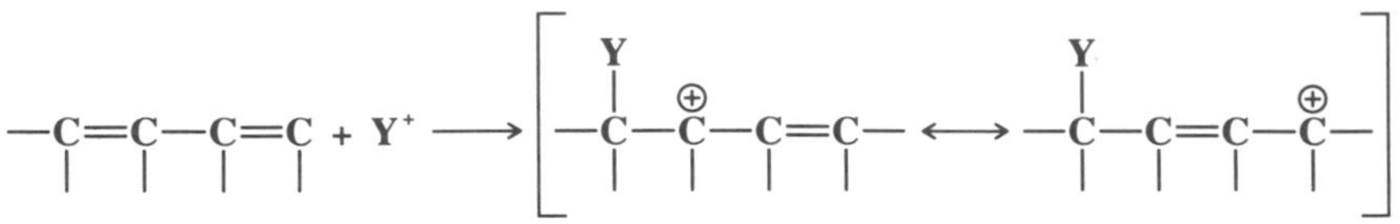

$$-C=C-C=C + Y^+ \longrightarrow \left[-\overset{Y}{\underset{}{C}}-\overset{\oplus}{\underset{}{C}}-C=C- \longleftrightarrow -\overset{Y}{\underset{}{C}}-C=C-\overset{\oplus}{\underset{}{C}}- \right]$$

W^- may then attack either position. The original attack of Y^+ is always at the end of the conjugated system because an attack at a middle carbon would give a cation unstabilized by resonance:

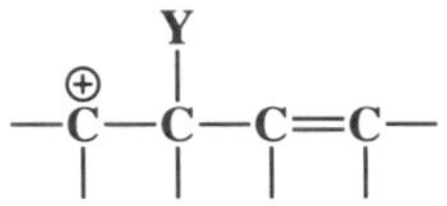

$$\overset{\oplus}{\underset{}{C}}-\overset{Y}{\underset{}{C}}-C=C-$$

In the case of electrophiles like Br^+, which can form cyclic intermediates, both 1,2- and 1,4-addition products can be rationalized as stemming from an intermediate like **12**. Direct nucleophilic attack by W^- would give the 1,2-product, while the 1,4-product could be formed by attack at the 4 position, by an S_N2'-type mechanism (see p. 329). Intermediates like **13**

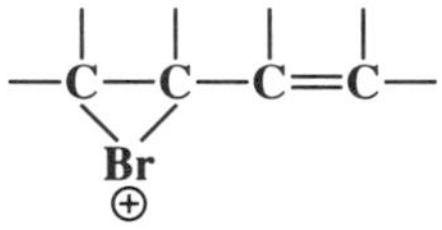

$$-\overset{|}{C}-\overset{|}{C}-\overset{|}{C}=\overset{|}{C}- \qquad \text{[cyclopentene ring with Br]}$$

$$\underset{\underset{\oplus}{Br}}{\diagdown\diagup} \qquad\qquad\qquad \underset{\underset{\oplus}{Br}}{}$$

12    **13**

have been postulated but ruled out for Br and Cl by the observation that chlorination or bromination of butadiene gives trans 1,4-products.[56] If an ion like **13** were the intermediate, the 1,4-products would have to have the cis configuration.

In most cases more 1,4- than 1,2-addition product is obtained. This may be a consequence of thermodynamic control of products, as against kinetic. In most cases, under the reaction conditions, **10** is converted to a mixture of **10** and **11** which is richer in **11**. That is, either isomer gives the same mixture of both, which contains more **11**. It was found that at low temperatures, butadiene and HCl gave only 20 to 25% 1,4-adduct, while at high temperatures, where attainment of equilibrium is more likely, the mixture contained 75% 1,4-product.[57] 1,2-Addition predominated over 1,4- in the reaction between DCl and 1,3-pentadiene, where the intermediate was the symmetrical (except for the D label)

$CH_3CH\overset{\overset{\oplus}{\cdots}}{=}CH\overset{\cdots}{=}CHCH_2D$.[58] Ion pairs were invoked to explain this result, since a free ion would be expected to be attacked by Cl^- equally well at both positions, except for the very small isotope effect.

Addition to conjugated systems can also be accomplished by any of the other three mechanisms. In each case there is competition between 1,2 and 1,4 addition. In the case of

[56]Mislow; Hellman *J. Am. Chem. Soc.* **1951**, *73*, 244; Mislow *J. Am. Chem. Soc.* **1953**, *75*, 2512.
[57]Kharasch; Kritchevsky; Mayo *J. Org. Chem.* **1938**, *2*, 489.
[58]Nordlander; Owuor; Haky *J. Am. Chem. Soc.* **1979**, *101*, 1288.

nucleophilic or free-radical attack,[59] the intermediates are resonance hybrids and behave

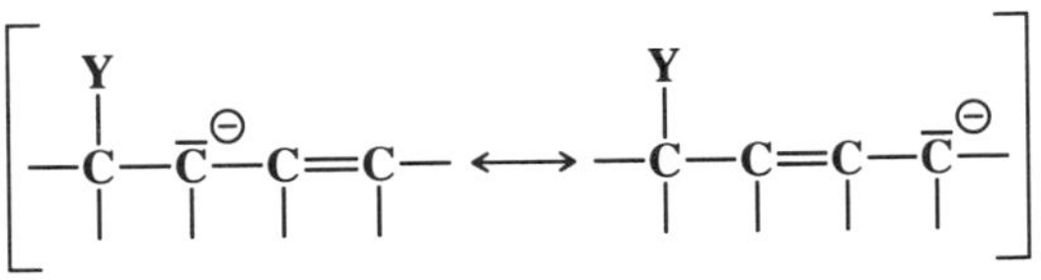

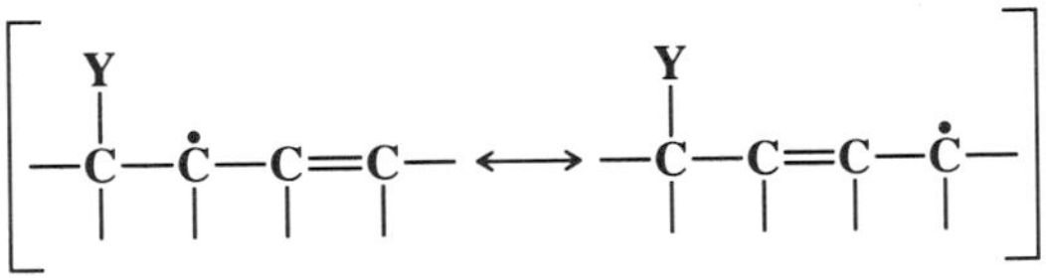

like the intermediate from electrophilic attack. Dienes can give 1,4 addition by a cyclic mechanism in this way:

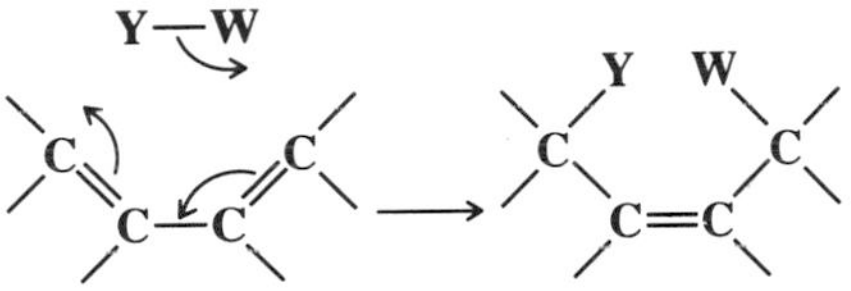

Other conjugated systems, including trienes, enynes, diynes, etc., have been studied much less but behave similarly. 1,4 addition to enynes is an important way of making allenes:

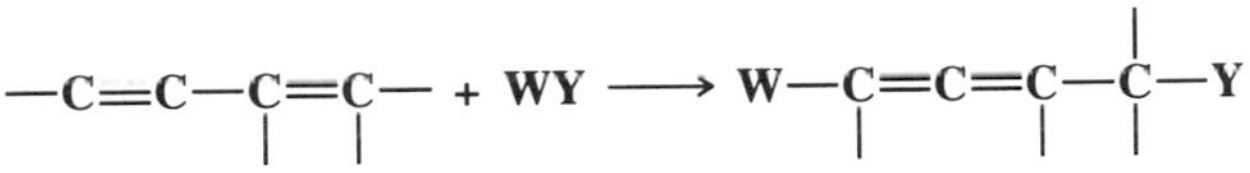

ORIENTATION AND REACTIVITY

Reactivity

As with electrophilic aromatic substitution (Chapter 11), electron-donating groups increase the reactivity of a double bond toward electrophilic addition and electron-withdrawing groups decrease it. This is illustrated in Tables 15.1 and 15.2.[60] As a further illustration it may be mentioned that the reactivity toward electrophilic addition of a group of olefins increased in the order $CCl_3CH=CH_2 < Cl_2CHCH=CH_2 < ClCH_2CH=CH_2 < CH_3CH_2=CH_2$.[61] For nucleophilic addition the situation is reversed. These reactions are best carried out on substrates containing three or four electron-withdrawing groups, two of the most common being $F_2C=CF_2$[62] and $(NC)_2C=C(CN)_2$.[63] The effect of substituents is so great that it is

---

[59]For a review of free-radical addition to conjugated dienes, see Afanas'ev; Samokhvalov *Russ. Chem. Rev.* **1969,** *38,* 318-329.

[60]Table 15.1 is from de la Mare *Q. Rev., Chem. Soc.* **1949,** *3,* 126-145, p. 145. Table 15.2 is from Dubois; Mouvier *Tetrahedron Lett.* **1963,** 1325. See also Dubois; Mouvier *Bull. Soc. Chim. Fr.* **1968,** 1426; Grosjean; Mouvier; Dubois *J. Org. Chem.* **1976,** *41,* 3869, 3872.

[61]Shelton; Lee *J. Org. Chem.* **1960,** *25,* 428.

[62]For a review of additions to $F_2C=CF_2$ and other fluoroolefins, see Chambers; Mobbs *Adv. Fluorine Chem.* **1965,** *4,* 51-112.

[63]For reviews of additions to tetracyanoethylene, see Fatiadi *Synthesis* **1987,** 249-284, 749-789; Dhar *Chem. Rev.* **1967,** *67,* 611-622.

TABLE 15.1 Relative reactivity of some olefins toward bromine in acetic acid at 24°C[60]

Olefin	Relative rate
PhCH=CH$_2$	Very fast
PhCH=CHPh	18
CH$_2$=CHCH$_2$Cl	1.6
CH$_2$=CHCH$_2$Br	1.0
PhCH=CHBr	0.11
CH$_2$=CHBr	0.0011

TABLE 15.2 Relative reactivity of some olefins toward bromine in methanol[60]

Olefin	Relative rate
CH$_2$=CH$_2$	3.0×10^1
CH$_3$CH$_2$CH=CH$_2$	2.9×10^3
cis-CH$_3$CH$_2$CH=CHCH$_3$	1.3×10^5
(CH$_3$)$_2$C=C(CH$_3$)$_2$	2.8×10^7

possible to make the statement that *simple olefins do not react by the nucleophilic mechanism, and polyhalo or polycyano olefins do not generally react by the electrophilic mechanism.*[64] There are some reagents that attack only as nucleophiles, e.g., ammonia, and these add only to substrates susceptible to nucleophilic attack. Other reagents attack only as electrophiles, and, for example, F$_2$C=CF$_2$ does not react with these. In still other cases, the same reagent reacts with a simple olefin by the electrophilic mechanism and with a polyhalo olefin by a nucleophilic mechanism. For example, Cl$_2$ and HF are normally electrophilic reagents, but it has been shown that Cl$_2$ adds to (NC)$_2$C=CHCN with initial attack by Cl$^-$[65] and that HF adds to F$_2$C=CClF with initial attack by F$^-$.[66] Compounds that have a double bond conjugated with a Z group (as defined on p. 741) nearly always react by a nucleophilic mechanism.[67] These are actually 1,4 additions, as discussed on p. 742. A number of studies have been made of the relative activating abilities of various Z groups.[68] On the basis of these studies, the following order of decreasing activating ability has been suggested: Z = NO$_2$, COAr, CHO, COR, SO$_2$Ar, CN, COOR, SOAr, CONH$_2$, CONHR.[69]

It seems obvious that electron-withdrawing groups enhance nucleophilic addition and inhibit electrophilic addition because they lower the electron density of the double bond. This is probably true, and yet similar reasoning does not always apply to a comparison between double and triple bonds.[70] There is a higher concentration of electrons between the carbons of a triple bond than in a double bond, and yet triple bonds are *less* subject to electrophilic attack and *more* subject to nucleophilic attack than double bonds.[71] This statement is not universally true, but it does hold in most cases. In compounds containing both double and triple bonds (nonconjugated), bromine, an electrophilic reagent, always adds

---

[64]Such reactions can take place under severe conditions. For example, electrophilic addition could be accomplished withF$_2$C=CHF in super-acid solutions [Olah; Mo *J. Org. Chem.* **1972**, *37*, 1028] although F$_2$C=CF$_2$ did not react under these conditions. For reviews of electrophilic additions to fluoroolefins, see Belen'kii; German *Sov. Sci. Rev. Sect. B* **1984**, *5*, 183-218; Dyatkin; Mochalina; Knunyants *Russ. Chem. Rev.* **1966**, *35*, 417-427, *Fluorine Chem. Rev.* **1969**, *3*, 45-71; Ref. 62, pp. 77-81.

[65]Dickinson; Wiley; McKusick *J. Am. Chem. Soc.* **1960**, *82*, 6132. For another example, see Atkinson; de la Mare; Larsen *J. Chem. Soc., Perkin Trans. 2* **1983**, 271.

[66]Miller; Fried; Goldwhite *J. Am. Chem. Soc.* **1960**, *82*, 3091.

[67]For a review of electrophilic reactions of such compounds, see Müllen; Wolf, in Patai; Rappoport, Ref. 37, pp. 513-558.

[68]See, for example, Friedman; Wall *J. Org. Chem.* **1966**, *31*, 2888; Ring; Tesoro; Moore *J. Org. Chem.* **1967**, *32*, 1091.

[69]Shenhav; Rappoport; Patai *J. Chem. Soc. B* **1970**, 469.

[70]For reviews of ionic additions to triple bonds, see, in Patai *The Chemistry of the Carbon–Carbon Triple Bond*; Wiley: New York, 1978, the articles by Schmid, pt. 1, pp. 275-341, and by Dickstein; Miller, pt. 2, pp. 813-955; Miller; Tanaka *Sel. Org. Transform.* **1970**, *1*, 143-238; Winterfeldt, in Viehe, Ref. 49, pp. 267-334. For comparisons of double and triple bond reactivity, see Melloni; Modena; Tonellato *Acc. Chem. Res.* **1981**, *14*, 227-233; Allen; Chiang; Kresge; Tidwell *J. Org. Chem.* **1982**, *47*, 775.

[71]For discussions, see Daniels; Bauer *J. Chem. Educ.* **1958**, *35*, 444; DeYoung; Ehrlich; Berliner *J. Am. Chem. Soc.* **1977**, *99*, 290; Strozier; Caramella; Houk *J. Am. Chem. Soc.* **1979**, *101*, 1340.

to the double bond.[72] In fact, all reagents that form bridged intermediates like **2** react faster with double than with triple bonds. On the other hand, addition of electrophilic H^+ (acid-catalyzed hydration, **5-2**; addition of hydrogen halides, **5-1**) takes place at about the same rates for alkenes as for corresponding alkynes.[73] Furthermore, the presence of electron-withdrawing groups lowers the alkene/alkyne rate ratio. For example, while styrene $PhCH=CH_2$ was brominated 3000 times faster than $PhC\equiv CH$, the addition of a second phenyl group ($PhCH=CHPh$ vs. $PhC\equiv CPh$) lowered the rate ratio to about 250.[74] In the case of *trans*-$MeOOCCH=CHCOOMe$ vs. $MeOOCC\equiv CCOOMe$, the triple bond compound was actually brominated faster.[75]

Still, it is true that in general triple bonds are more susceptible to nucleophilic and less to electrophilic attack than double bonds, in spite of their higher electron density. One explanation is that the electrons in the triple bond are held more tightly because of the smaller carbon–carbon distance; it is thus harder for an attacking electrophile to pull out a pair. There is evidence from far-uv spectra to support this conclusion.[76] Another possible explanation has to do with the availability of the unfilled orbital in the alkyne. It has been shown that a π^* orbital of bent alkynes (such as cyclooctyne) has a lower energy than the π^* orbital of alkenes, and it has been suggested[77] that linear alkynes can achieve a bent structure in their transition states when reacting with an electrophile. Where electrophilic addition involves bridged-ion intermediates, those arising from triple bonds (**14**) are more strained than the corresponding **15** and furthermore are antiaromatic systems (see p. 56),

14   **15**

which **15** are not. This may be a reason why electrophilic addition by such electrophiles as Br, I, SR, etc., is slower for triple than for double bonds.[78] As might be expected, triple bonds connected to a Z group ($C\equiv C-Z$) undergo nucleophilic addition especially well.[79]

Although alkyl groups in general increase the rates of electrophilic addition, we have already mentioned (p. 739) that there is a different pattern depending on whether the intermediate is a bridged ion or an open carbocation. For brominations and other electrophilic additions in which the first step of the mechanism is rate-determining, the rates for substituted alkenes correlate well with the ionization potentials of the alkenes, which means that steric effects are not important.[80] Where the second step is rate-determining [e.g., oxymercuration (**5-2**), hydroboration (**5-13**)], steric effects are important.[80]

Free-radical additions can occur with any type of substrate. The determining factor is the presence of a free-radical attacking species. Some reagents, e.g., HBr, RSH, attack by ionic mechanisms if no initiator is present, but in the presence of a free-radical initiator, the mechanism changes and the addition is of the free-radical type. Nucleophilic radicals

[72]Petrov *Russ. Chem. Rev.* **1960**, *29*, 489-509.

[73]Melloni; Modena; Tonellato, Ref. 70, p. 228.

[74]Robertson; Dasent; Milburn; Oliver *J. Chem. Soc.* **1950**, 1628.

[75]Wolf; Ganguly; Berliner *J. Am. Chem. Soc.* **1985**, *50*, 1053.

[76]Walsh *Q. Rev., Chem. Soc.* **1948**, *2*, 73-91.

[77]Ng; Jordan; Krebs; Rüger *J. Am. Chem. Soc.* **1982**, *104*, 7414.

[78]Nevertheless, bridged ions **14** have been implicated in some additions to triple bonds. See, for example, Pincock; Yates, Ref. 14; Mauger; Berliner *J. Am. Chem. Soc.* **1972**, *94*, 194; Bassi; Tonellato *J. Chem. Soc., Perkin Trans. 1* **1973**, 669; Schmid; Modro; Lenz; Garratt; Yates *J. Org. Chem.* **1976**, *41*, 2331.

[79]For a review of additions to these substrates, see Winterfeldt *Angew. Chem. Int. Ed. Engl.* **1967**, *6*, 423-434 [*Angew. Chem. 79*, 389-400], *Newer Methods Prep. Org. Chem.* **1971**, *6*, 243-279.

[80]Nelson; Cooper; Soundararajan *J. Am. Chem. Soc.* **1989**, *111*, 1414; Nelson; Soundararajan *Tetrahedron Lett.* **1988**, *29*, 6207.

(see p. 679) behave like nucleophiles in that the rate is increased by the presence of electron-withdrawing groups in the substrate. The reverse is true for electrophilic radicals.[81] However, nucleophilic radicals react with alkynes more slowly than with the corresponding alkenes,[82] which is contrary to what might have been expected.[83]

Steric influences are important in some cases. In catalytic hydrogenation, where the substrate must be adsorbed onto the catalyst surface, the reaction becomes more difficult with increasing substitution. The hydrocarbon **16,** in which the double bond is entombed

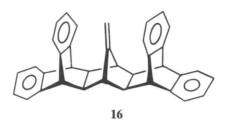

16

between the benzene rings, does not react with Br_2, H_2SO_4, O_3, BH_3, CBr_2, or other reagents that react with most double bonds.[84] A similarly inactive compound is tetra-t-butylallene $(t\text{-Bu})_2C{=}C{=}C(t\text{-Bu})_2$, which is inert to Br_2, Cl_2, O_3, and catalytic hydrogenation.[85]

Orientation

When an unsymmetrical reagent is added to an unsymmetrical substrate, the question arises: Which side of the reagent goes to which side of the double or triple bond? For electrophilic attack, the answer is given by *Markovnikov's rule: the positive portion of the reagent goes to the side of the double or triple bond that has more hydrogens.*[86] A number of explanations have been suggested for this regioselectivity, but the most probable is that Y^+ adds to that side that will give the more stable carbocation. Thus, when an alkyl group is present, secondary carbocations are more stable than primary:

$$R{-}C{=}C{-}H + Y^+ \longrightarrow R{-}\overset{\oplus}{\underset{H}{C}}{-}\underset{H}{\overset{Y}{C}}{-}H \text{ or } R{-}\underset{H}{\overset{Y}{C}}{-}\overset{\oplus}{\underset{H}{C}}{-}H$$

More stable

We may ask: How does Y^+ "know" which side will give the more stable carbocation? As in the similar case of electrophilic aromatic substitution (p. 508), we invoke the Hammond postulate and say that the lower energy carbocation is preceded by the lower energy transition state. Markovnikov's rule also applies for halogen substituents because the halogen stabilizes

[81]For reviews of reactivity in free-radical additions, see Tedder *Angew. Chem. Int. Ed. Engl.* **1982**, *21*, 401-410 [*Angew. Chem. 94*, 433-442]; Tedder; Walton *Tetrahedron* **1980**, *36*, 701-707.

[82]Giese; Lachhein *Angew. Chem. Int. Ed. Engl.* **1982**, *21*, 768 [*Angew. Chem. 94*, 780].

[83]For a discussion of reactivity and orientation of polar radicals, see Volovik; Dyadyusha; Staninets *J. Org. Chem. USSR* **1986**, *22*, 1224.

[84]Butler; Gupta; Ng; Nyburg *J. Chem. Soc., Chem. Commun.* **1980**, 596.

[85]Bolze; Eierdanz; Schlüter; Massa; Grahn; Berndt *Angew. Chem. Int. Ed. Engl.* **1982**, *21*, 924 [*Angew. Chem. 94*, 927].

[86]For discussions of Markovnikov's rule, see Isenberg; Grdinic *J. Chem. Educ.* **1969**, *46*, 601; Grdinic; Isenberg, *Intra-Sci. Chem. Rep.* **1970**, *4*, 145-162.

the carbocation by resonance:

$$Cl-\overset{\underset{\displaystyle |}{H}}{C}=\overset{\underset{\displaystyle |}{H}}{C}-H + Y^+ \longrightarrow \left[|\underline{Cl}-\overset{\oplus}{\underset{\underset{\displaystyle H}{|}}{C}}-\overset{\overset{\displaystyle Y}{|}}{\underset{\underset{\displaystyle H}{|}}{C}}-H \longleftrightarrow |\overset{\oplus}{\underline{Cl}}=\overset{\overset{\displaystyle Y}{|}}{\underset{\underset{\displaystyle H}{|}}{C}}-H \right] \text{ or } Cl-\overset{\overset{\displaystyle Y}{|}}{\underset{\underset{\displaystyle H}{|}}{C}}-\overset{\oplus}{\underset{\underset{\displaystyle H}{|}}{C}}-H$$

More stable

Markovnikov's rule is also usually followed where bromonium ions or other three-membered rings are intermediates.[87] This means that in these cases attack by W must resemble the S_N1 rather than the S_N2 mechanism (see p. 369), though the overall stereospecific anti

$$\overset{\overset{\displaystyle \oplus}{\underset{\displaystyle Y}{\diagdown}}}{\underset{\displaystyle CH_2-CHR}{}} \overset{w}{\longrightarrow} \overset{\overset{\displaystyle Y}{|} \quad \overset{\displaystyle W}{|}}{CH_2-CHR}$$

addition in these reactions means that the nucleophilic substitution step is taking place with inversion of configuration.

Olefins containing strong electron-withdrawing groups may violate Markovnikov's rule. For example, attack at the Markovnikov position of $Me_3\overset{\oplus}{N}-CH=CH_2$ would give an ion with positive charges on adjacent atoms. The compound CF_3CH-CH_2 has been reported to give electrophilic addition with acids in an anti-Markovnikov direction, but it has been shown[88] that, when treated with acids, this compound does not give simple electrophilic addition at all; the apparently anti-Markovnikov products are formed by other pathways.

For nucleophilic addition the direction of attack has been studied very little, except for Michael-type addition, with compounds of the type C=C—Z. Here the negative part of the reagent almost always attacks regioselectively at the carbon that does not carry the Z (see p. 742).

In free-radical addition[89] the main effect seems to be steric.[90] All substrates $CH_2=CHX$ are preferentially attacked at the CH_2, regardless of the identity of X or of the attacking radical. With a regent such as HBr, this means that the addition is anti-Markovnikov:

$$R-\overset{\underset{\displaystyle |}{H}}{C}=\overset{\underset{\displaystyle |}{H}}{C}-H + Br\cdot \longrightarrow R-\overset{\overset{\displaystyle Br}{|}}{\underset{\underset{\displaystyle H}{|}}{\overset{\displaystyle \centerdot}{C}}}-\overset{\underset{\displaystyle H}{|}}{C}-H \longrightarrow R-\overset{\overset{\displaystyle H}{|}}{\underset{\underset{\displaystyle H}{|}}{C}}-\overset{\overset{\displaystyle Br}{|}}{\underset{\underset{\displaystyle H}{|}}{C}}-H$$

preferentially   product
formed
intermediate

Thus the observed orientation in both kinds of HBr addition (Markovnikov electrophilic and anti-Markovnikov free radical) is caused by formation of the secondary intermediate.

[87]This has been graphically demonstrated by direct treatment of stabilized bromonium ions by nucleophiles: Dubois; Chrétien *J. Am. Chem. Soc.* **1978**, *100*, 3506.

[88]Myhre; Andrews *J. Am. Chem. Soc.* **1970**, *92*, 7595, 7596. See also Newton *J. Chem. Educ.* **1987**, *64*, 531.

[89]For reviews of orientation in free-radical additions, see Tedder; Walton *Tetrahedron* **1980**, *36*, 701-707, *Adv. Phys. Org. Chem.* **1978**, *16*, 51-86, *Acc. Chem. Res.* **1976**, *9*, 183-191. See also Giese, Ref. 49; Tedder *J. Chem. Educ.* **1984**, *61*, 237.

[90]See, however, Riemenschneider; Bartels; Dornow; Drechsel-Grau; Eichel; Luthe; Matter; Michaelis; Boldt *J. Org. Chem.* **1987**, *52*, 205; Gleicher; Mahiou; Aretakis *J. Org. Chem.* **1989**, *54*, 308.

In the electrophilic case it forms because it is more stable than the primary; in the free-radical case because it is sterically preferred. The stability order of the free-radical intermediates is also usually in the same direction: $3° > 2° > 1°$ (p. 188), but this factor is apparently less important than the steric factor. Internal olefins with no groups present to stabilize the radical usually give an approximately 1:1 mixture.

In *intramolecular* additions of radicals containing a 5,6 double bond,[50] both five- and six-membered rings can be formed, but in most cases[91] the five-membered rings are greatly preferred kinetically, even (as in the case shown) where five-membered ring closure means

Favored

generating a primary radical and six-membered ring closure a secondary radical. This phenomenon may be caused by more favorable entropy factors leading to a five-membered ring, as well as by stereoelectronic factors, but other explanations have also been offered.[92] Similar behavior is found when the double bond is in other positions (from the 3,4 to the 7,8 position). In each case the smaller ring (*Exo–Trig* addition) is preferred to the larger (*Endo–Trig* addition)[93] (see the Baldwin rules, p. 212). However, when a radical that is unsaturated in the 5,6 position contains an alkyl group in the 5 position, formation of the 6-membered ring is generally favored.[94]

For conjugated dienes, attack by a positive ion, a negative ion, or a free radical is almost always at the *end* of the conjugated system, since in each case this gives an intermediate stabilized by resonance. In the case of an unsymmetrical diene, the more stable ion is formed. For example, isoprene CH_2=$CMeCH$=CH_2, treated with HCl gives only Me_2CClCH=CH_2 and Me_2C=$CHCH_2Cl$, with none of the product arising from attack at the other end. $PhCH$=$CHCH$=CH_2 gives only $PhCH$=$CHCHClCH_3$ since it is the only one of the eight possible products that has a double bond in conjugation with the ring and that results from attack by H^+ at an end of the conjugated system.

When allenes are attacked by electrophilic reagents,[95] Markovnikov's rule would predict that the attack should be at the end of the system, since there are no hydrogens in the middle. Attack at the center gives a carbocation stabilized by resonance, but not immediately.

[91]For an exception, see Wilt *Tetrahedron* **1985**, *41*, 3979.

[92]For discussions, see Beckwith *Tetrahedron* **1981**, *37*, 3073-3100; Verhoeven *Revl. Trav. Chim. Pays-Bas* **1980**, *99*, 143. For molecular mechanics force-field approaches to this problem, see Beckwith; Schiesser *Tetrahedron* **1985**, *41*, 3925; Spellmeyer; Houk *J. Org. Chem.* **1987**, *52*, 959.

[93]See Beckwith; Easton; Serelis *J. Chem. Soc., Chem. Commun.* **1980**, 482.

[94]See Chuang; Gallucci; Hart; Hoffman *J. Org. Chem.* **1988**, *53*, 3218, and references cited therein.

[95]For a monograph on addition to allenes, see Schuster; Coppola *Allenes in Organic Synthesis*; Wiley: New York, 1984. For reviews, see Pasto *Tetrahedron* **1984**, *40*, 2805-2827; Smadja *Chem. Rev.* **1983**, *83*, 263-320; in Landor *The Chemistry of Allenes*, vol. 2; Academic Press: New York, 1982, articles by Landor, Jacobs, and Hopf, pp. 351-577; Stang; Rappoport; Hanack; Subramanian, Ref. 34, pp. 152-167; Blake, in Patai *The Chemistry of Ketenes, Allenes and Related Compounds*, pt. 1; Wiley: New York, 1980; pp. 342-357; Modena; Tonellato, Ref. 34, pp. 215-231; Richey; Richey Ref. 34, pp. 917-922; Caserio *Sel. Org. Transform.* **1970**, *1*, 239-299; Taylor *Chem. Rev.* **1967**, *67*, 317-359, pp. 338-346; Mavrov; Kucherov *Russ. Chem. Rev.* **1967**, *36*, 233-249; Griesbaum *Angew. Chem. Int. Ed. Engl.* **1966**, *5*, 933-946 [*Angew. Chem. 78*, 953-966].

In order for such stabilization to be in effect the three p orbitals must be parallel, and it requires a rotation about the C—C bond for this to happen.[96] Therefore, the stability of the allylic cation has no effect on the transition state, which still has a geometry similar to that of the original allene (p. 102). Probably because of this, attack on the unsubstituted $CH_2=C=CH_2$ is most often at the end carbon, to give a vinylic cation, though center attack has also been reported. However, as alkyl or aryl groups are substituted on the allene carbons, attack at the middle carbon becomes more favorable because the resulting cation is stabilized by the alkyl or aryl groups (it is now a secondary, tertiary, or benzylic cation). For example, allenes of the form $RCH=C=CH_2$ are still attacked most often at the end, but with $RCH=C=CHR'$ center attack is more prevalent. Tetramethylallene is also attacked predominantly at the center carbon.[97] Free radicals[98] attack allenes most often at the end,[99] though attack at the middle has also been reported.[100] As with electrophilic attack and for the same reason, the stability of the allylic radical has no effect on the transition state of the reaction between a free radical and an allene. Again, as with electrophilic attack, the presence of alkyl groups increases the extent of attack by a radical at the middle carbon.[101]

Stereochemical Orientation

It has already been pointed out that some additions are syn, with both groups, approaching from the same side, and that others are anti, with the groups approaching from opposite sides of the double or triple bond. For cyclic compounds there are further aspects of steric orientation. In syn addition to an unsymmetrical cyclic olefin, the two groups can come in from the more-hindered face or from the less-hindered face of the double bond. The rule is that syn addition is usually, though not always, from the less-hindered face. For example, epoxidation of 4-methylcyclopentene gave 76% addition from the less-hindered and 24% from the more-hindered face.[102]

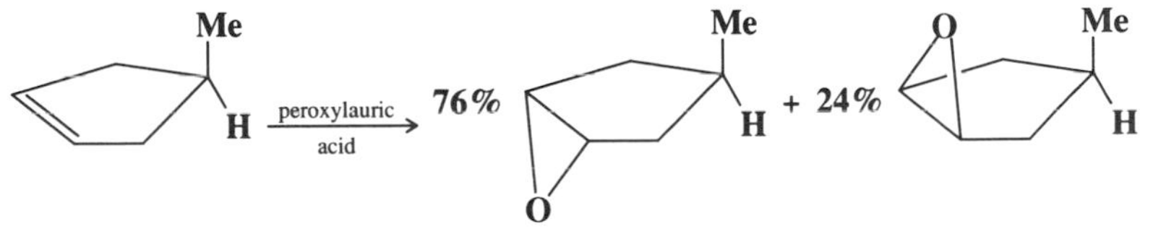

In anti addition to a cyclic substrate, the initial attack by the electrophile is also from the less-hindered face. However, many (though not all) electrophilic additions to norbornene and similar strained bicycloalkenes are syn additions.[103] In these cases attack is always from

---

[96]For evidence that this is so, see Okuyama; Izawa; Fueno *J. Am. Chem. Soc.* **1973**, *95*, 6749.

[97]For example, see Bianchini; Guillemonat *Bull. Soc. Chim. Fr.* **1968**, 2120; Pittman *Chem. Commun.* **1969**, 122; Poutsma; Ibarbia *J. Am. Chem. Soc.* **1971**, *93*, 440.

[98]For a review, see Jacobs, in Landor, Ref. 95, vol. 2.pp. 399-415.

[99]Griesbaum; Oswald; Quiram; Naegele *J. Org. Chem.* **1963**, *28*, 1952.

[100]See, for example, Pasto; L'Hermine *J. Org. Chem.* **1990**, *55*, 685.

[101]For example, see Byrd; Caserio *J. Org. Chem.* **1972**, *37*, 3881; Pasto; Warren; Morrison *J. Org. Chem.* **1981**, *46*, 2837. See however Bartels; Boldt *Liebigs Ann. Chem.* **1981**, 40.

[102]Henbest; McCullough *Proc. Chem. Soc.* **1962**, 74.

[103]For a discussion, see Traylor *Acc. Chem. Res.* **1969**, *2*, 152-160.

the exo side, e.g.,[104]

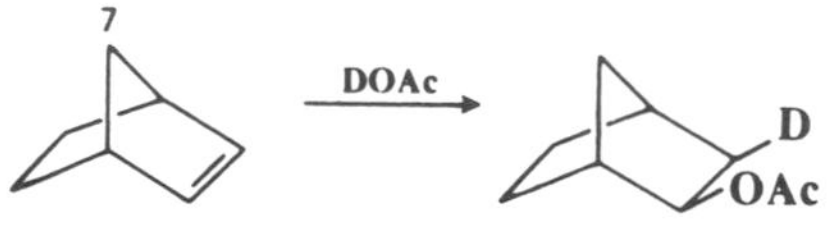

unless the exo side is blocked by substituents in the 7 position, in which case endo attack may predominate; e.g., 7,7-dimethylnorbornene undergoes syn-endo epoxidation (**5-36**) and hydroboration[105] (**5-12**). However, addition of DCl and F$_3$CCOOD to, and oxymercuration (**5-2**) of, 7,7-dimethylnorbornene proceed syn-exo in spite of the methyl groups in the 7 position.[106] Similarly, free-radical additions to norbornene and similar molecules are often syn-exo, though anti additions and endo attacks are also known.[107]

Electronic effects can also play a part in determining which face is attacked. In the adamantane derivative **17** steric effects are about the same for each face of the double bond. Yet epoxidation, dibromocarbene attack (**5-50**), and hydroboration (**5-12**) all predominantly

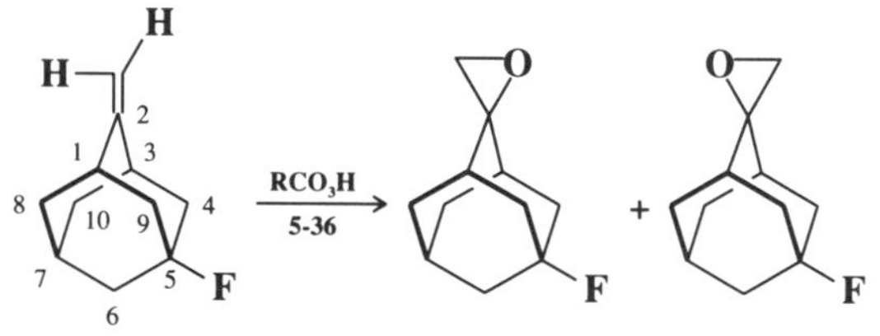

take place from the face that is syn to the electron-withdrawing fluorine.[108] In the case shown, about twice as much **18** was formed, compared to **19**. Similar results have been obtained on other substrates:[109] groups that are electron-withdrawing by the field effect ($-I$) direct attack from the syn face; $+I$ groups from the anti face, for both electrophilic and nucleophilic attack. These results are attributed[110] to hyperconjugation: For the adamantane case, there is overlap between the σ^* orbital of the newly-forming bond (between the attacking species and C-2 in **17**) and the filled σ orbitals of the C$_\alpha$—C$_\beta$ bonds on the opposite side. The four possible bonds are C-3—C-4 and C-1—C-9 on the syn side and C-3—C-10 and C-1—C-8 on the anti side. The preferred pathway is the one where the incoming group has the more electron-rich bonds on the side *opposite* to it (these are the ones it overlaps with). Since the electron-withdrawing F has its greatest effect on the bonds closest to it, the C-1—C-8 and C-3—C-10 bonds are more electron rich, and the group comes in on the face syn to the F.

[104]Cristol; Morrill; Sanchez *J. Org. Chem.* **1966**, *31*, 2719; Brown; Kawakami; Liu *J. Am. Chem. Soc.* **1970**, *92*, 5536; Alvernhe; Anker; Laurent; Haufe; Beguin *Tetrahedron* **1988**, *44*, 3551; Koga; Ozawa; Morokuma *J. Phys. Org. Chem.* **1990**, *3*, 519.

[105]Brown; Kawakami *J. Am. Chem. Soc.* **1970**, *92*, 201, 1990; Brown; Kawakami; Liu *J. Am. Chem. Soc.* **1973**, *95*, 2209.

[106]Brown; Liu *J. Am. Chem. Soc.* **1975**, *97*, 600, 2469; Brown; Kawakami *J. Am. Chem. Soc.* **1973**, *95*, 8665; Tidwell; Traylor *J. Org. Chem.* **1968**, *33*, 2614.

[107]For a review of free-radical addition to these systems, see Azovskaya; Prilezhaeva *Russ. Chem. Rev.* **1972**, *41*, 516-528.

[108]Srivastava; le Noble *J. Am. Chem. Soc.* **1987**, *109*, 5874. See also Bodepudi; le Noble *J. Am. Chem. Soc.* **1991**, *113*, 2001.

[109]Johnson; Tait; Cieplak *J. Am. Chem. Soc.* **1987**, *109*, 5875; Cieplak; Tait; Johnson *J. Am. Chem. Soc.* **1989**, *111*, 8447.

[110]Cieplak *J. Am. Chem. Soc.* **1981**, *103*, 4540. See also Jorgensen *Chemtracts: Org. Chem.* **1988**, *1*, 71.

It has been mentioned that additions of Br_2 and HOBr are often anti because of formation of bromonium ions and that free-radical addition of HBr is also anti. When the substrate in any of these additions is a cyclohexene, the addition is not only anti but the initially formed product is conformationally specific too, being mostly diaxial.[111] This is so because diaxial opening of the three-membered ring preserves a maximum coplanarity of the participating centers in the transition state; indeed, on opening, epoxides also give diaxial products.[112] However, the initial diaxial product may then pass over to the diequatorial conformer unless other groups on the ring render the latter less stable than the former. In free-radical additions to cyclohexenes in which cyclic intermediates are not involved, the initial attack by the radical is also usually from the axial direction,[113] resulting in a diaxial initial product if the overall addition is anti. The direction from which unsymmetrical radicals attack has also been studied.[114] For example, when the radical **20** adds to a double bond it preferentially does so anti to the OH group, leading to a diaxial trans product.[114]

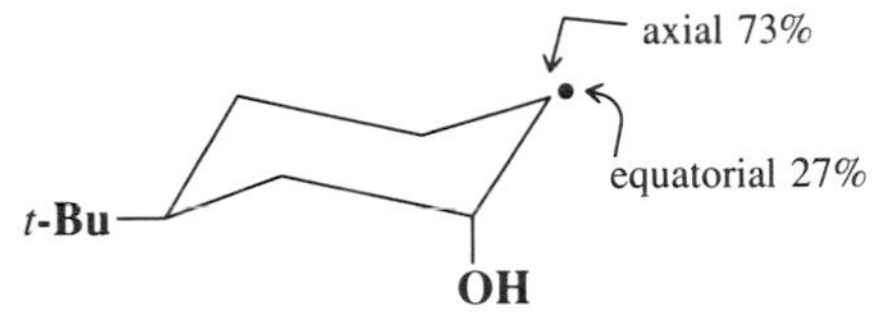

20

Addition to Cyclopropane Rings[115]

We have previously seen (p. 152) that in some respects, cyclopropane rings resemble double bonds.[116] It is not surprising, therefore, that cyclopropanes undergo addition reactions analogous to those undergone by double-bond compounds, resulting in the opening of the three-membered rings, e.g. (the reaction numbers of the analogous addition reactions are given in parentheses),

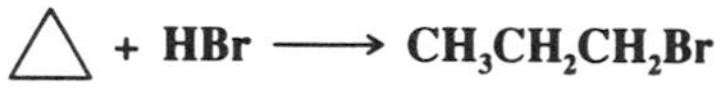

$$\triangle + HBr \longrightarrow CH_3CH_2CH_2Br \quad \text{(5-1)}$$

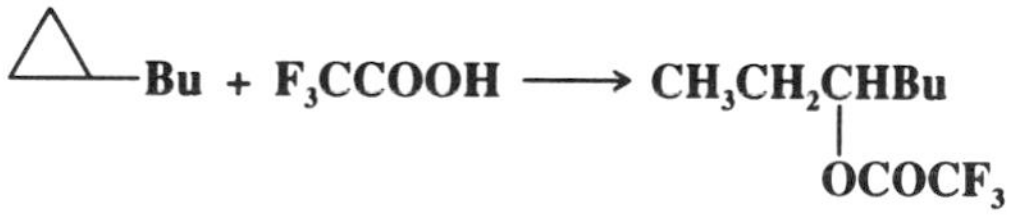

$$\triangle\!\!-Bu + F_3CCOOH \longrightarrow CH_3CH_2CHBu \quad \text{(5-5)} \qquad \text{Ref. 117}$$
$$\qquad\qquad\qquad\qquad\qquad\qquad\quad OCOCF_3$$

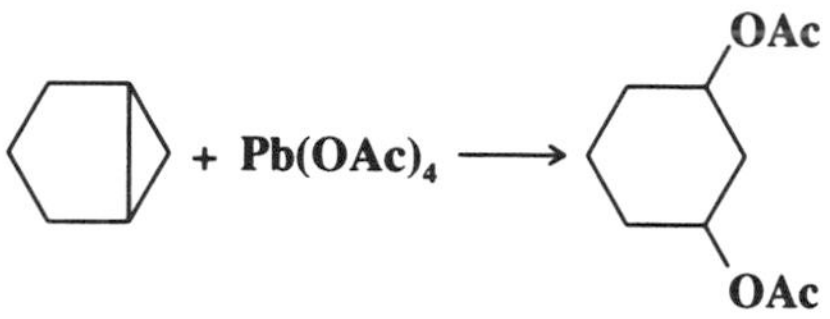

(5-34)   Ref. 118

[111]Barton, in *Theoretical Organic Chemistry, The Kekulé Symposium*; Butterworth: London, 1959, pp. 127-143; Goering; Abell; Aycock *J. Am. Chem. Soc.* **1952**, *74*,3588; Goering; Sims *J. Am. Chem. Soc.* **1955**, *77*, 3465; Shoppee; Akhtar; Lack *J. Chem. Soc.* **1964**, 877; Readio; Skell *J. Org. Chem.* **1966**, *31*, 753, 759.
[112]For example, see Anselmi; Berti; Catelani; Lecce; Monti *Tetrahedron* **1977**, *33*, 2771.
[113]Huyser; Benson; Sinnige *J. Org. Chem.* **1967**, *32*, 622; LeBel; Czaja; DeBoer, Ref. 51.
[114]For a review, see Giese *Angew. Chem. Int. Ed. Engl.* **1989**, *28*, 969-980 [*Angew. Chem. 101*, 993-1004].
[115]For a review, see Charton, in Zabicky *The Chemistry of Alkenes*, vol 2.; Wiley: New York, 1970, pp. 569-592. For reviews of the use of cyclopropanes in organic synthesis, see Reissig *Top. Curr. Chem.* **1988**, *144*, 73-135; Wong; Hon; Tse; Yip; Tanko; Hudlicky *Chem. Rev.* **1989**, *89*, 165-198.
[116]The analogies are by no means complete: see Gordon *J. Chem. Educ.* **1967**, *44*, 461.
[117]Peterson; Thompson *J. Org. Chem.* **1968**, *33*, 968.
[118]Moon *J. Org. Chem.* **1964**, *39*, 3456.

Other examples are discussed at **5-2, 5-11,** and **5-49.**

Additions to cyclopropanes can take place by any of the four mechanisms already discussed in this chapter, but the most important type involves electrophilic attack.[119] For substituted cyclopropanes, these reactions usually follow Markovnikov's rule, though exceptions are known and the degree of regioselectivity is often small. The application of Markovnikov's rule to these substrates can be illustrated by the reaction of 1,1,2-trimethylcyclopropane with HX.[120] The rule predicts that the electrophile (in this case H+)

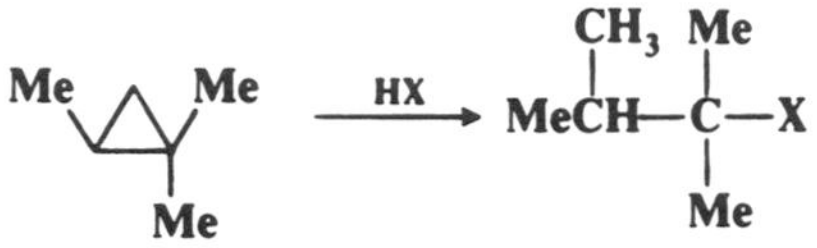

goes to the carbon with the most hydrogens and the nucleophile goes to the carbon that can best stabilize a positive charge (in this case the tertiary rather than the secondary carbon). The stereochemistry of the reaction can be investigated at two positions—the one that becomes connected to the electrophile and the one that becomes connected to the nucleophile. The results at the former position are mixed. Additions have been found to take place with 100% retention,[121] 100% inversion,[122] and with mixtures of retention and inversion.[123] At the carbon that becomes connected to the nucleophile the result is usually inversion, though retention has also been found,[124] and elimination, rearrangement, and racemization processes often compete, indicating that in many cases a positively charged carbon is generated at this position.

At least three mechanisms have been proposed for electrophilic addition (these mechanisms are shown for attack by HX, but analogous mechanisms can be written for other electrophiles).

Mechanism *a*

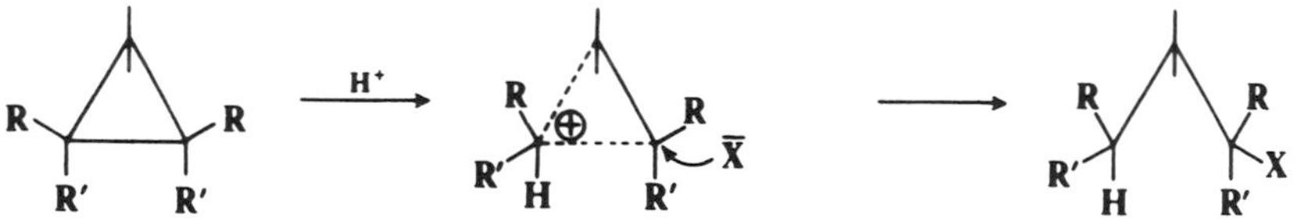

Mechanism *b*

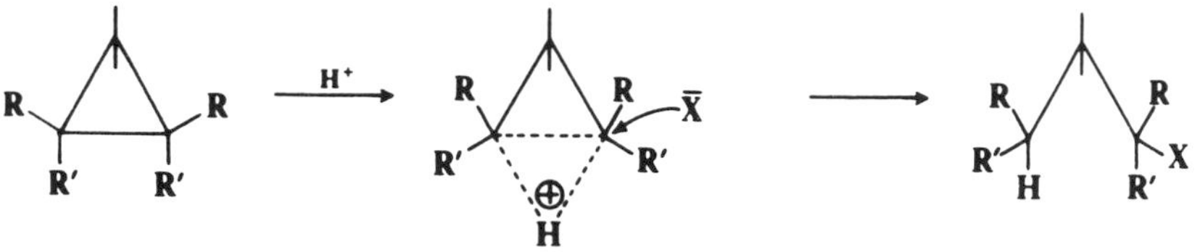

[119]For a review, see DePuy *Top. Curr. Chem.* **1973,** *40,* 73-101. For a list of references to pertinent mechanistic studies, see Wiberg; Kass *J. Am. Chem. Soc.* **1985,** *107,* 988
[120]Kramer *J. Am. Chem. Soc.* **1970,** *92,* 4344.
[121]For example, see DePuy; Breitbeil; DeBruin *J. Am. Chem. Soc.* **1966,** *88,* 3347; Hendrickson; Boeckman *J. Am. Chem. Soc.* **1969,** *91,* 3269.
[122]For example, see LaLonde; Ding; Tobias *J. Am. Chem. Soc.* **1967,** *89,* 6651; Warnet; Wheeler *Chem. Commun.* **1971,** 547; Hogeveen; Roobeek; Volger *Tetrahedron Lett.* **1972,** 221; Battiste; Mackiernan *Tetrahedron Lett.* **1972,** 4095. See also Jensen; Patterson; Dinizo *Tetrahedron Lett.* **1974,** 1315; Coxon; Steel; Whittington *J. Org. Chem.* **1990,** *55,* 4136.
[123]Nickon; Hammons *J. Am. Chem. Soc.* **1964,** *86,* 3322; Hammons; Probasco; Sanders; Whalen *J. Org. Chem.* **1968,** *33,* 4493; DePuy; Fünfschilling; Andrist; Olson *J. Am. Chem. Soc.* **1977,** *99,* 6297.
[124]Cristol; Lim; Dahl *J. Am. Chem. Soc.* **1970,** *92,* 4013; Hendrickson; Boeckman *J. Am. Chem. Soc.* **1971,** *93,* 4491.

Mechanism *c*

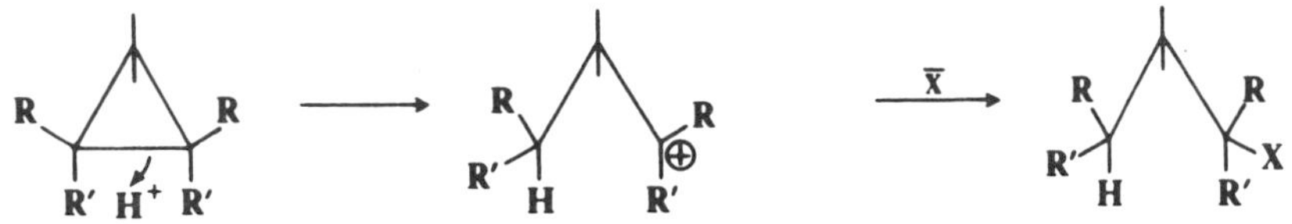

23

Mechanism *a* involves a corner-protonated cyclopropane[125] (**21**); we have already seen examples of such ions in the 2-norbornyl and 7-norbornenyl cations (pp. 321, 314). Mechanism *b* involves an edge-protonated cyclopropane (**22**). Mechanism *c* consists of a one-step SE2-type attack by H$^+$ to give the classical cation **23**, which then reacts with the nucleophile. Although the three mechanisms as we have drawn them show retention of configuration at the carbon that becomes attached to the proton, mechanisms *a* and *c* at least can also result in inversion at this carbon. Unfortunately, the evidence on hand at present does not allow us unequivocally to select any of these as the exclusive mechanism in all cases. Matters are complicated by the possibility that more than one edge-protonated cyclopropane is involved, at least in some cases. There is strong evidence for mechanism *b* with the electrophiles Br$^+$ and Cl$^+$;[126] and for mechanism *a* with D$^+$ and Hg^{2+}.[127] Ab initio studies show that the corner-protonated **21** is slightly more stable (about 1.4 kcal/mole, 6 kJ/mol) than the edge-protonated **22**.[128] There is some evidence against mechanism *c*.[129]

Free-radical additions to cyclopropanes have been studied much less, but it is known that Br$_2$ and Cl$_2$ add to cyclopropanes by a free-radical mechanism in the presence of uv light. The addition follows Markovnikov's rule, with the initial radical attacking the least-substituted carbon and the second group going to the most-substituted position. Several investigations have shown that the reaction is stereospecific at one carbon, taking place with inversion there, but nonstereospecific at the other carbon.[130] A mechanism that accounts for this behavior is[131]

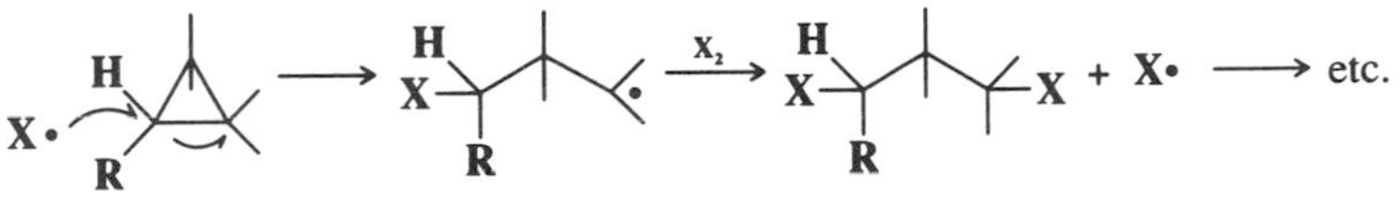

In some cases conjugate addition has been performed on systems where a double bond is "conjugated" with a cyclopropyl ring. An example is[132]

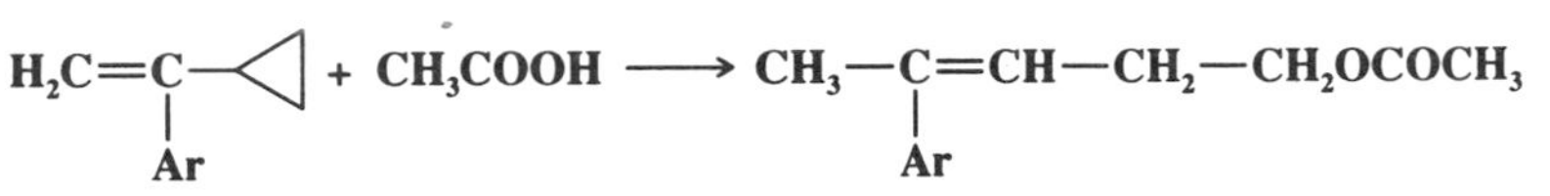

$$H_2C=C-\triangleleft + CH_3COOH \longrightarrow CH_3-C=CH-CH_2-CH_2OCOCH_3 \qquad (5\text{-}5)$$

with Ar substituents.

[125]For reviews of protonated cyclopropanes, see Collins *Chem. Rev.* **1969**, *69*, 543-550; Lee *Prog. Phys. Org. Chem.* **1970**, *7*, 129-187.
[126]Coxon; Steel; Whittington; Battiste *J. Org. Chem.* **1989**, *54*, 1383; Coxon; Steel; Whittington *J. Org. Chem.* **1989**, *54*, 3702.
[127]Lambert; Chelius; Schulz; Carpenter *J. Am. Chem. Soc.* **1990**, *112*, 3156; Lambert; Chelius; Bible; Hadju *J. Am. Chem. Soc.* **1991**, *113*, 1331.
[128]Koch; Liu; Schleyer *J. Am. Chem. Soc.* **1989**, *111*, 3479, and references cited therein.
[129]Wiberg; Kass, Ref. 119
[130]Maynes; Applequist *J. Am. Chem. Soc.* **1973**, *95*, 856; Incremona; Upton *J. Am. Chem. Soc.* **1972**, *94*, 301; Shea; Skell *J. Am. Chem. Soc.* **1973**, *95*, 6728; Poutsma *J. Am. Chem. Soc.* **1965**, *87*, 4293; Jarvis *J. Org. Chem.* **1970**, *35*, 924; Upton; Incremona *J. Org. Chem.* **1976**, *41*, 523.
[131]For free-radical addition to [1.1.1]propellane and bicyclo[1.1.0]butane, see Wiberg; Waddell; Laidig *Tetrahedron Lett.* **1986**, *27*, 1553.
[132]Sarel; Ben-Shoshan *Tetrahedron Lett.* **1965**, 1053. See also Danishefsky *Acc. Chem. Res.* **1979**, *12*, 66-72.

REACTIONS

Reactions are classified by type of reagent. All reactions where hydrogen adds to one side of the double bond are treated first.

Reactions in Which Hydrogen Adds to One Side

A.   Halogen on the Other Side

5-1   Addition of Hydrogen Halides
Hydro-halo-addition

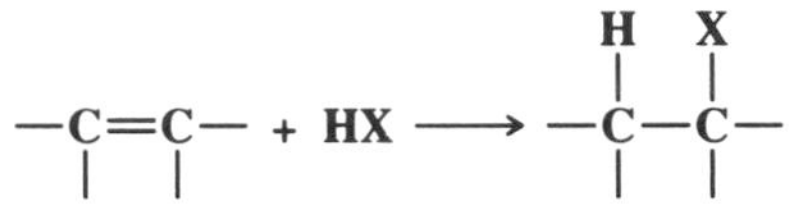

$$-\overset{|}{\underset{|}{C}}=\overset{|}{\underset{|}{C}}-\ +\ HX\ \longrightarrow\ -\overset{\overset{\displaystyle H}{|}}{\underset{|}{C}}-\overset{\overset{\displaystyle X}{|}}{\underset{|}{C}}-$$

Any of the four hydrogen halides can be added to double bonds.[133] HI, HBr, and HF[134] add at room temperature. The addition of HCl is more difficult and usually requires heat.[22] The reaction has been carried out with a large variety of double-bond compounds, including conjugated systems, where both 1,2 and 1,4 addition are possible. A convenient method for the addition of HF involves the use of a polyhydrogen fluoride–pyridine solution.[135] When the substrate is mixed with this solution in a solvent such as THF at 0°C, alkyl fluorides are obtained in moderate-to-high yields.

The addition of hydrogen halides to simple olefins, in the absence of peroxides, takes place by an electrophilic mechanism, and the orientation is in accord with Markovnikov's rule.[136] When peroxides are added, the addition of HBr occurs by a free-radical mechanism and the orientation is anti-Markovnikov (p. 751).[137] It must be emphasized that this is true only for HBr. Free-radical addition of HF and HI has never been observed, even in the presence of peroxides, and of HCl only rarely. In the rare cases where free-radical addition of HCl was noted, the orientation was still Markovnikov, presumably because the more stable *product* was formed.[138] Free-radical addition of HF, HI, and HCl is energetically unfavorable (see the discussions on pp. 683, 693). It has often been found that anti-Markovnikov addition of HBr takes place even when peroxides have not been added. This happens because the substrate alkenes absorb oxygen from the air, forming small amounts of peroxides (**4-9**). Markovnikov addition can be ensured by rigorous purification of the substrate, but in practice this is not easy to achieve, and it is more common to add inhibitors, e.g., phenols or quinones, which suppress the free-radical pathway. The presence of free-radical precursors such as peroxides does not inhibit the ionic mechanism, but the radical reaction, being a chain process, is much more rapid than the electrophilic reaction. In most cases it is possible to control the mechanism (and hence the orientation) by adding peroxides

---

[133]For a list of references, see Larock *Comprehensive Organic Transformations*; VCH: New York, 1989, pp. 322-323.
[134]For reviews of addition of HF, see Sharts; Sheppard *Org. React.* **1974**, *21*, 125-406, pp. 192-198, 212-214; Hudlický *The Chemistry of Organic Fluorine Compounds*, 2nd ed.; Ellis Horwood: Chichester, 1976, pp. 36-41.
[135]Olah; Welch; Vankar; Nojima; Kerekes; Olah *J. Org. Chem.* **1979**, *44*, 3872. For related methods, see Yoneda; Abe; Fukuhara; Suzuki *Chem. Lett.* **1983**, 1135; Olah; Li *Synlett* **1990**, 267.
[136]For reviews of electrophilic addition of HX, see Ref. 22, and Dewar, Ref. 3.
[137]For reviews of free-radical addition of HX, see Thaler *Methods Free-Radical Chem.* **1969**, *2*, 121-227, pp. 182-195.
[138]Mayo *J. Am. Chem. Soc.* **1962**, *84*, 3964.

to achieve complete free-radical addition, or inhibitors to achieve complete electrophilic addition, though there are some cases where the ionic mechanism is fast enough to compete with the free-radical mechanism and complete control cannot be attained. Markovnikov addition of HBr, HCl, and HI has also been accomplished, in high yields, by the use of phase transfer catalysis.[139] For alternative methods of adding HBr (or HI) with anti-Markovnikov orientation, see **2-30.**

It is also possible to add one[140] or two moles of any of the four hydrogen halides to triple bonds.

$$-C{\equiv}C- \xrightarrow{\text{HX}} -CH{=}CX- \xrightarrow{\text{HX}} -CH_2-CX_2-$$

Markovnikov's rule ensures that *gem*-dihalides and not *vic*-dihalides are the products of the addition of two moles.

HX are electrophilic reagents, and many polyhalo and polycyano alkenes, e.g., $Cl_2C{=}CHCl$, do not react with them at all in the absence of free-radical conditions. When such reactions do occur, however, they take place by a nucleophilic addition mechanism, i.e., initial attack is by X^-. This type of mechanism also occurs with Michael-type substrates $C{=}C-Z$,[141] where the orientation is always such that the halogen goes to the carbon that does not bear the Z, so the product is of the form $X-C-CH-Z$, even in the presence of free-radical initiators. HI adds 1,4 to conjugated dienes in the gas phase by a pericyclic mechanism:[142]

HX can be added to ketenes[143] to give acyl halides:

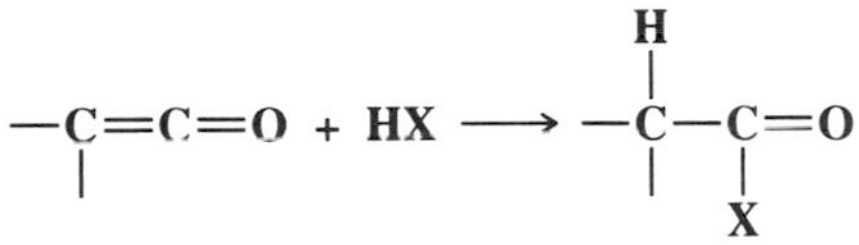

$$-C{=}C{=}O + HX \longrightarrow \overset{\overset{\displaystyle H}{|}}{-C}-C{=}O \atop {\underset{\displaystyle X}{|}}$$

OS **I,** 166; **II,** 137, 336; **III,** 576; **IV,** 238, 543; **VI,** 273; **VII,** 59.

B.   Oxygen on the Other Side

5-2   Hydration of Double Bonds
Hydro-hydroxy-addition

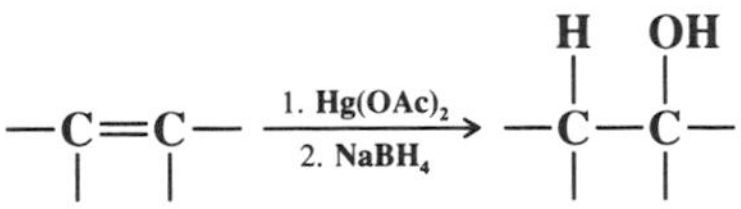

$$-C{=}C- \xrightarrow[\text{2. NaBH}_4]{\text{1. Hg(OAc)}_2} \overset{\overset{\displaystyle H \quad OH}{| \quad \;\; |}}{-C}-C-$$

---

[139]Landini; Rolla *J. Org. Chem.* **1980,** *45,* 3527.

[140]For a convenient method of adding one mole of HCl or HBr to a triple bond, see Cousseau; Gouin *J. Chem. Soc., Perkin Trans. 1* **1977,** 1797; Cousseau *Synthesis* **1980,** 805. For the addition of one mole of HI, see Kamiya; Chikami; Ishii *Synlett* **1990,** 675.

[141]For an example, see Marx *Tetrahedron* **1983,** *39,* 1529.

[142]Gorton; Walsh *J. Chem. Soc., Chem. Commun.* **1972,** 782. For evidence that a pericyclic mechanism may be possible, even for an isolated double bond, see Sergeev; Stepanov; Leenson; Smirnov; Pupyshev; Tyurina; Mashyanov *Tetrahedron* **1982,** *38,* 2585.

[143]For reviews of additions to ketenes, and their mechanisms, see Tidwell *Acc. Chem. Res.* **1990,** *23,* 273-279; Seikaly; Tidwell *Tetrahedron* **1986,** *42,* 2587-2613; Satchell; Satchell *Chem. Soc. Rev.* **1975,** *4,* 231-250.

Olefins can be hydrated quickly under mild conditions in high yields without rearrangement products by the use of *oxymercuration*[144] (addition of oxygen and mercury) followed by in situ treatment with sodium borohydride[145] (**2-24**). For example, 2-methyl-1-butene treated with mercuric acetate,[146] followed by NaBH$_4$, gave 2-methyl-2-butanol:

$$\underset{\text{CH}_3\text{CH}_2\text{C}=\text{CH}_2}{\overset{\overset{\displaystyle\text{CH}_3}{|}}{}} \xrightarrow[\text{2. NaBH}_4]{\text{1. Hg(OAc)}_2} 90\% \quad \underset{\underset{|}{\text{OH}}}{\overset{\overset{\displaystyle\text{CH}_3}{|}}{\text{CH}_3\text{CH}_2\text{CCH}_3}}$$

This method, which is applicable to mono-, di-, tri-, and tetraalkyl as well as phenyl-substituted olefins, gives almost complete Markovnikov addition. Hydroxy, methoxy, acetoxy, halo, and other groups may be present in the substrate without, in general, causing difficulties.[147] When two double bonds are present in the same molecule, the use of ultrasound allows oxymercuration of the less-substituted one without affecting the other.[148]

Double bonds can also be hydrated by treatment with water and an acid catalyst. The most common catalyst is sulfuric acid, but other acids, such as nitric or perchloric can also be used. The mechanism is electrophilic and begins with attack by a proton (see p. 739). The negative attacking species may be HSO$_4^-$ (or similar ion in the case of other acids) to give the initial product **24** which can be isolated, but under the conditions of the reaction,

$$\underset{\overset{|}{}}{\overset{\overset{\displaystyle\text{H}}{|}}{-\text{C}-}}\underset{\overset{|}{}}{\overset{\overset{\displaystyle\text{OSO}_2\text{OH}}{|}}{\text{C}-}}$$

24

is usually hydrolyzed to the alcohol (**0-4**). However, the conjugate base of the acid is not the only possible species that attacks the initial carbocation. The attack can also be by water:

$$-\underset{|}{\overset{|}{\text{C}}}=\underset{|}{\overset{|}{\text{C}}}- + \text{H}^+ \longrightarrow -\underset{\overset{|}{}}{\overset{\overset{\displaystyle\text{H}}{|}}{\text{C}}}-\overset{\oplus}{\underset{|}{\text{C}}}- + \text{H}_2\text{O} \longrightarrow -\underset{\overset{|}{}}{\overset{\overset{\displaystyle\text{H}}{|}}{\text{C}}}-\underset{\overset{|}{}}{\overset{\overset{\displaystyle\overset{\oplus}{\text{OH}_2}}{|}}{\text{C}}}- \xrightarrow{-\text{H}^+} -\underset{\overset{|}{}}{\overset{\overset{\displaystyle\text{H}}{|}}{\text{C}}}-\underset{\overset{|}{}}{\overset{\overset{\displaystyle\text{OH}}{|}}{\text{C}}}-$$

When the reaction proceeds by this pathway, **24** and similar intermediates are not involved and the mechanism is exactly (by the principle of microscopic reversibility) the reverse of E1 elimination of alcohols (**7-1**).[149] It is likely that the mechanism involves both pathways.

[144]For a monograph, see Larock *Solvation/Demercuration Reactions in Organic Synthesis*; Springer: New York, 1986. For reviews of this and other oxymetallation reactions, see Kitching *Organomet. React.* **1972**, *3*, 319-398, *Organomet. Chem. Rev.* **1968**, *3*, 61-134; Oullette, in Trahanovsky *Oxidation in Organic Chemistry*, pt. B; Academic Press: New York, 1973, pp. 140-166; House *Modern Synthetic Reactions*, 2nd ed.; W.A. Benjamin: New York, 1972, pp. 387-396; Zefirov *Russ. Chem. Rev.* **1965**, *34*, 527-536.
[145]Brown; Geoghegan *J. Am. Chem. Soc.* **1967**, *89*, 1522, *J. Org. Chem.* **1970**, *35*, 1844, **1972**, *37*, 1937; Brown; Geoghegan; Lynch; Kurek; *J. Org. Chem.* **1972**, *37*, 1941; Moon; Waxman *Chem. Commun.* **1967**, 1283; Moon; Takakis; Waxman *J. Org. Chem.* **1969**, *34*, 2951; Moon; Ganz; Waxman *Chem. Commun.* **1969**, 866; Johnson; Rickborn *Chem. Commun.* **1968**, 1073; Klein; Levene *Tetrahedron Lett.* **1969**, 4833; Chamberlain; Whitham *J. Chem. Soc. B* **1970**, 1382; Barrelle; Apparu *Bull. Soc. Chim. Fr.* **1972**, 2016.
[146]For a review of this reagent, see Butler, in Pizey *Synthetic Reagents*, vol. 4; Wiley: New York, 1981, pp. 1-145.
[147]See the extensive tables in Larock, Ref. 144, pp. 4-71.
[148]Einhorn; Einhorn; Luche *J. Org. Chem.* **1989**, *54*, 4479.
[149]For discussions of the mechanism, see Vinnik; Obraztsov *Russ. Chem. Rev.* **1990**, *59*, 63-77; Liler *Reaction Mechanisms in Sulphuric Acid*; Academic Press: New York, 1971, pp. 210-225.

The initial carbocation occasionally rearranges to a more stable one. For example, hydration of CH_2=CHCH(CH$_3$)$_2$ gives $CH_3CH_2COH(CH_3)_2$. With ordinary olefins the addition predominantly follows Markovnikov's rule. Another method for Markovnikov addition of water consists of simultaneously adding an oxidizing agent (O$_2$) and a reducing agent (either Et$_3$SiH[150] or a secondary alcohol such as 2-propanol[151]) to the olefin in the presence of a cobalt-complex catalyst. No rearrangement is observed with this method. The corresponding alkane and ketone are usually side products.

Water can be added indirectly, with anti-Markovnikov orientation, by treatment of the alkene with a 1:1 mixture of PhCH$_2$NEt$_3$$^+$ BH$_4$$^-$ and Me$_3$SiCl, followed by addition of an aqueous solution of K$_2$CO$_3$.[152] For another method of anti-Markovnikov hydration, see **5-12.** With substrates of the type C=C—Z (Z is as defined on p. 741) the product is almost always HO—C—CH—Z and the mechanism is usually nucleophilic,[153] though electrophilic addition gives the same product[154] since a cation CH—$\overset{\oplus}{C}$—Z would be destabilized by the positive charges (full or partial) on two adjacent atoms. However, the other product, HC—CH(OH)Z, was obtained by treatment of the substrate with O$_2$, PhSiH$_3$, and a manganese-complex catalyst.[155] When the substrate is of the type RCH=CZZ', addition of water may result in cleavage of the adduct, to give an aldehyde and CH$_2$ZZ', e.g.,[156]

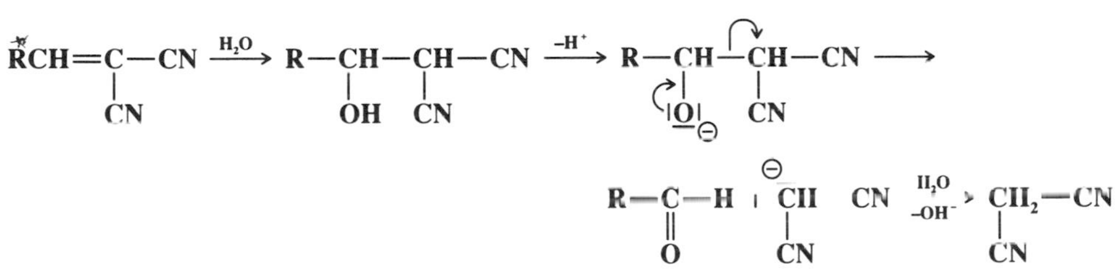

The cleavage step is an example of **2-41.**

Conjugated dienes are seldom hydrated.

The addition of water to enol ethers causes hydrolysis to aldehydes or ketones (**0-6**). Ketenes add water to give carboxylic acids in a reaction catalyzed by acids:[157]

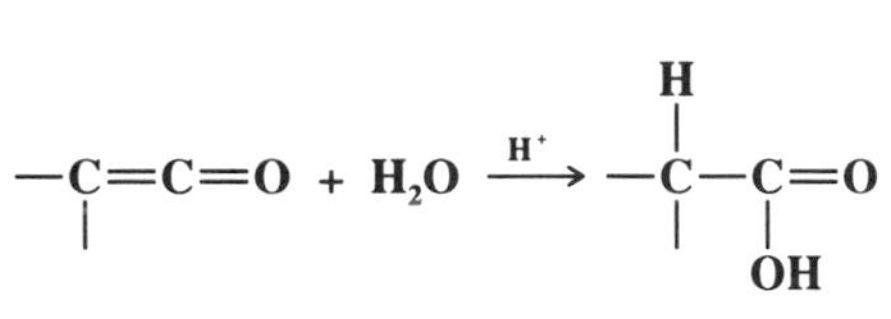

[150]Isayama; Mukaiyama *Chem. Lett.* **1989,** 569.

[151]Inoki; Kato; Takai; Isayama; Yamada; Mukaiyama *Chem. Lett.* **1989,** 515.

[152]Baskaran; Gupta; Chidambaram; Chandrasekaran *J. Chem. Soc., Chem. Commun.* **1989,** 903.

[153]For example, see Fedor; De; Gurwara *J. Am. Chem. Soc.* **1973,** *95,* 2905; Jensen; Hashtroudi *J. Org. Chem.* **1976,** *41,* 3299; Bernasconi; Leonarduzzi *J. Am. Chem. Soc.* **1982,** *104,* 5133, 5143.

[154]For example, see Noyce; DeBruin *J. Am. Chem. Soc.* **1968,** *90,* 372.

[155]Inoki; Kato; Isayama; Mukaiyama *Chem. Lett.* **1990,** 1869.

[156]Bernasconi; Fox; Kanavarioti; Panda *J. Am. Chem. Soc.* **1986,** *108,* 2372; Bernasconi; Paschalis *J. Am. Chem. Soc.* **1989,** *111,* 5893, and other papers in this series.

[157]For discussions of the mechanism, see Poon; Satchell *J. Chem. Soc., Perkin Trans. 2* **1983,** 1381; **1986,** 1485; Allen; Tidwell *J. Am. Chem. Soc.* **1987,** *109,* 2774; Allen; Stevenson; Tidwell *J. Org. Chem.* **1989,** *54,* 2843; Ref. 143.

The oxymercuration procedure (with mercuric trifluoroacetate) has been used to open cyclopropane rings, e.g.,[158]

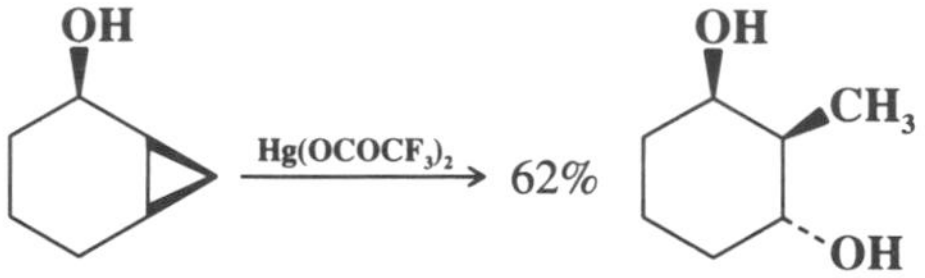

OH   OH
⌂▷ —Hg(OCOCF₃)₂→ 62% ⌂ CH₃
                                              OH

OS **IV**, 555, 560; **VI**, 766. Also see OS **V**, 818.

5-3 Hydration of Triple Bonds
Dihydro-oxo-biaddition

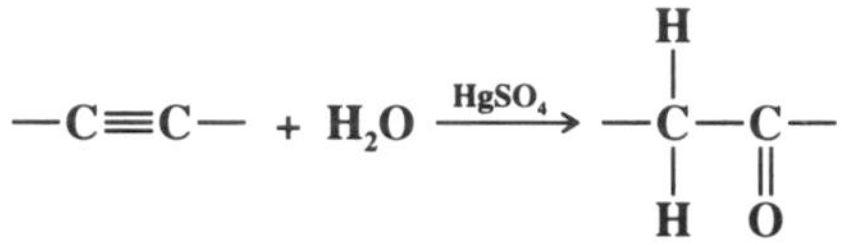

$$-C\equiv C- + H_2O \xrightarrow{HgSO_4} -\underset{\underset{H}{|}}{\overset{\overset{H}{|}}{C}}-\underset{\underset{O}{||}}{C}-$$

The hydration of triple bonds is generally carried out with mercuric ion salts (often the sulfate or acetate) as catalysts.[159] Mercuric oxide in the presence of an acid is also a common reagent. Since the addition follows Markovnikov's rule, only acetylene gives an aldehyde. All other triple-bond compounds give ketones (for a method of reversing the orientation for terminal alkynes, see **5-12**). With alkynes of the form $RC\equiv CH$ methyl ketones are formed almost exclusively, but with $RC\equiv CR'$ both possible products are usually obtained. The reaction can be conveniently carried out with a catalyst prepared by impregnating mercuric oxide onto Nafion-H (a superacidic perfluorinated resinsulfonic acid).[160]

The first step of the mechanism is formation of a complex (**25**) (ions like Hg^{2+} form complexes with alkynes—p. 80). Water then attacks in an S_N2-type process to give the intermediate **26**, which loses a proton to give **27**. Hydrolysis of **27** (an example of **2-24**)

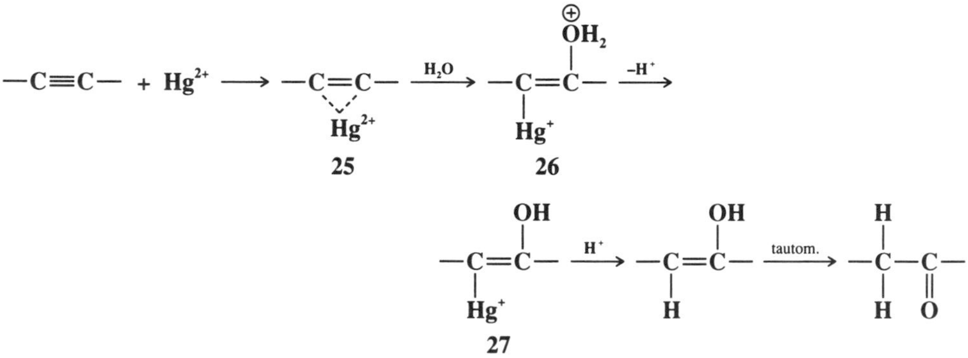

$$-C\equiv C- + Hg^{2+} \longrightarrow -\underset{\underset{Hg^{2+}}{|}}{C}=C- \xrightarrow{H_2O} -C=\underset{\underset{Hg^+}{|}}{\overset{\overset{\oplus}{OH_2}}{C}}- \xrightarrow{-H^+}$$

                        25   **26**

$$-\underset{\underset{Hg^+}{|}}{\overset{\overset{OH}{|}}{C}}=C- \xrightarrow{H^+} -\underset{\underset{H}{|}}{\overset{\overset{OH}{|}}{C}}=C- \xrightarrow{tautom.} -\underset{\underset{H}{|}}{\overset{\overset{H}{|}}{C}}-\underset{\underset{O}{||}}{C}-$$

                27

gives the enol, which tautomerizes to the product. A spectrum of the enol was detected by flash photolysis when phenylacetylene was hydrated photolytically.[161]

[158]Collum; Mohamadi; Hallock *J. Am. Chem. Soc.* **1983**, *105*, 6882; Collum; Still; Mohamadi *J. Am. Chem. Soc.* **1986**, *108*, 2094.
[159]For reviews, see Larock, Ref. 144, pp. 123-148; Khan; Martell *Homogeneous Catalysis by Metal Complexes*, vol. 2; Academic Press: New York, 1974, pp. 91-95. For a list of reagents, with references, see Ref. 133, pp. 596-597.
[160]Olah; Meidar *Synthesis* **1978**, 671.
[161]Chiang; Kresge; Capponi; Wirz *Helv. Chim. Acta* **1986**, *69*, 1331.

Carboxylic esters, thiol esters, and amides can be made, respectively, by acid-catalyzed hydration of acetylenic ethers, thioethers,[162] and ynamines, without a mercuric catalyst:[163]

$$-C\equiv C-A + H_2O \xrightarrow{H^+} -CH_2-\underset{\underset{O}{\parallel}}{C}-A \qquad A = OR, SR, NR_2$$

This is ordinary electrophilic addition, with rate-determining protonation as the first step.[164] Certain other alkynes have also been hydrated to ketones with strong acids in the absence of mercuric salts.[165] Simple alkynes can also be converted to ketones by heating with formic acid, without a catalyst.[166]

Allenes can also be hydrolyzed to ketones, with an acid catalyst.[167]

$$-\underset{|}{\overset{|}{C}}=C=\underset{|}{\overset{|}{C}}- \xrightarrow[H_2O]{H^+} -\underset{|}{\overset{H}{\underset{|}{C}}}-\underset{}{\overset{OH}{\underset{}{C}}}=\underset{|}{\overset{|}{C}}- \xrightarrow{tautom.} -\underset{|}{\overset{H}{\underset{|}{C}}}-\underset{}{\overset{O}{\underset{}{\overset{\parallel}{C}}}}-\underset{|}{\overset{H}{\underset{|}{C}}}-$$

OS **III**, 22; **IV**, 13; **V**, 1024.

5-4 Addition of Alcohols and Phenols
Hydro-alkoxy-addition

$$-\underset{|}{\overset{|}{C}}=\underset{|}{\overset{|}{C}}- + ROH \longrightarrow -\underset{|}{\overset{H}{\underset{|}{C}}}-\underset{|}{\overset{OR}{\underset{|}{C}}}-$$

The addition of alcohols and phenols to double bonds is catalyzed by acids or bases. When the reactions are acid-catalyzed, the mechanism is electrophilic, with H^+ as the attacking species. The resulting carbocation combines with a molecule of alcohol:

$$-\underset{|}{\overset{|}{C}}=\underset{|}{\overset{|}{C}}- + H^+ \longrightarrow -\underset{|}{\overset{H}{\underset{|}{C}}}-\underset{|}{\overset{\oplus}{\underset{|}{C}}}- + ROH \longrightarrow -\underset{|}{\overset{H}{\underset{|}{C}}}-\underset{|}{\overset{\overset{\oplus}{ORH}}{\underset{|}{C}}}- \xrightarrow{-H^+} -\underset{|}{\overset{H}{\underset{|}{C}}}-\underset{|}{\overset{OR}{\underset{|}{C}}}-$$

The addition, therefore, follows Markovnikov's rule. Primary alcohols give better results than secondary, and tertiary alcohols are very inactive. This is a convenient method for the preparation of tertiary ethers by the use of a suitable olefin such as $Me_2C=CH_2$.

For those substrates more susceptible to nucleophilic attack, e.g., polyhalo olefins and olefins of the type C=C—Z, it is better to carry out the reaction in basic solution, where the attacking species is RO^-.[168] The reactions with C=C—Z are of the Michael type, and OR goes to the side away from the Z. Since triple bonds are more susceptible to nucleophilic

---

[162]For a review of acetylenic ethers and thioethers, see Brandsma; Bos; Arens, in Viehe, Ref. 49, pp. 751-860.
[163]Arens *Adv. Org. Chem.* **1960**, *2*, 163; Ref. 162, pp. 774-775.
[164]Hogeveen; Drenth *Recl. Trav. Chim. Pays-Bas* **1963**, *82*, 375, 410; Verhelst; Drenth *J. Am. Chem. Soc.* **1974**, *96*, 6692; Banait; Hojatti; Findlay; Kresge *Can. J. Chem.* **1987**, *65*, 441.
[165]See, for example, Noyce; Schiavelli *J. Org. Chem.* **1968**, *33*, 845, *J. Am. Chem. Soc.* **1968**, *90*, 1020, 1023.
[166]Menashe; Reshef; Shvo *J. Org. Chem.* **1991**, *56*, 2912.
[167]For example, see Fedorova; Petrov *J. Gen. Chem. USSR* **1962**, *32*, 1740; Mühlstadt; Graefe *Chem. Ber.* **1967**, *100*, 223; Cramer; Tidwell *J. Org. Chem.* **1981**, *46*, 2683.
[168]For a review with respect to fluoroolefins, see Ref. 62, pp. 53-61.

attack than double bonds, it might be expected that bases would catalyze addition to triple bonds particularly well. This is the case, and enol ethers and acetals can be produced by this reaction:[169]

$$-C\equiv C- + ROH \xrightarrow{OH^-} \underset{\underset{H}{|}}{-C} = \underset{\underset{OR}{|}}{C} - + ROH \xrightarrow{OH^-} \underset{\underset{H}{|}}{-C} - \underset{\underset{OR}{|}}{C} -$$

with H, OR substituents on the carbons.

Because enol ethers are more susceptible than triple bonds to electrophilic attack, the addition of alcohols to enol ethers can also be catalyzed by acids.[170] One utilization of this reaction involves the compound dihydropyran (**28**), which is often used to protect the OH

$$\mathbf{28} \qquad\qquad \mathbf{29}$$

groups of primary and secondary[171] alcohols and phenols.[172] The tetrahydropyranyl acetal formed by this reaction (**29**) is stable to bases, Grignard reagents, LiAlH$_4$, and oxidizing agents, any of which can be used to react with functional groups located within the R group. When the reactions are completed, **29** is easily cleaved by treatment with dilute acids (**0-6**). The addition of alcohols to enol ethers is also catalyzed by CoCl$_2$.[173]

In base-catalyzed addition to triple bonds the rate falls in going from a primary to a tertiary alcohol, and phenols require more severe conditions. Other catalysts, namely, BF$_3$ and mercuric salts, have also been used in addition of ROH to triple bonds.

Alcohols can be added to certain double-bond compounds (cyclohexenes, cycloheptenes) photochemically[174] in the presence of a photosensitizer such as benzene. The mechanism is electrophilic and Markovnikov orientation is found. The olefins react in their first excited triplet states.[175]

The oxymercuration–demercuration procedure mentioned in **5-2** can be adapted to the preparation of ethers (Markovnikov orientation) if the oxymercuration is carried out in an alcohol ROH as solvent,[176] e.g., 2-methyl-1-butene in ethanol gives EtMe$_2$COEt.[177] Primary alcohols give good yields when mercuric acetate is used, but for secondary and tertiary alcohols it is necessary to use mercuric trifluoroacetate.[178] However, even with this reagent the method fails where the product would be a ditertiary ether. Alkynes generally give acetals. If the oxymercuration is carried out in the presence of a hydroperoxide instead of an alcohol, the product (after demercuration with NaBH$_4$) is an alkyl peroxide (peroxy-mercuration).[179]

[169]For a review, see Shostakovskii; Trofimov; Atavin; Lavrov *Russ. Chem. Rev.* **1968**, *37*, 907-919.
[170]For discussions of the mechanism, see Toullec; El-Alaoui; Bertrand *J. Chem. Soc., Perkin Trans. 2* **1987**, 1517; Kresge; Yin *J. Phys. Org. Chem.* **1989**, *2*, 43.
[171]Tertiary alcohols can also be protected in this way if triphenylphosphine hydrobromide is used as a catalyst: Bolitt; Mioskowski; Shin; Falck *Tetrahedron Lett.* **1988**, *29*, 4583.
[172]For useful catalysts for this reaction, some of which are also applicable to tertiary alcohols, see Miyashita; Yoshikoshi; Grieco *J. Org. Chem.* **1977**, *42*, 3772; Olah; Husain; Singh *Synthesis* **1985**, 703; Johnston; Marston; Krieger; Goe *Synthesis* **1988**, 393.
[173]Iqbal; Srivastava; Gupta; Khan *Synth. Commun.* **1989**, *19*, 901.
[174]For a review of the photochemical protonation of double and triple bonds, see Wan; Yates *Rev. Chem. Intermed.* **1984**, *5*, 157-181.
[175]Marshall *Acc. Chem. Res.* **1969**, *2*, 33-40.
[176]For a review, with tables of many examples, see Larock, Ref. 144, pp. 162-345.
[177]Brown; Rei *J. Am. Chem. Soc.* **1969**, *91*, 5646.
[178]Brown; Kurek; Rei; Thompson *J. Org. Chem.* **1984**, *49*, 2551, **1985**, *50*, 1171.
[179]Ballard; Bloodworth *J. Chem. Soc. C* **1971**, 945; Sokolov; Reutov *J. Org. Chem. USSR* **1969**, *5*, 168. For a review, see Larock, Ref. 144, pp. 346-366.

Both alcohols and phenols add to ketenes to give carboxylic esters:[180]

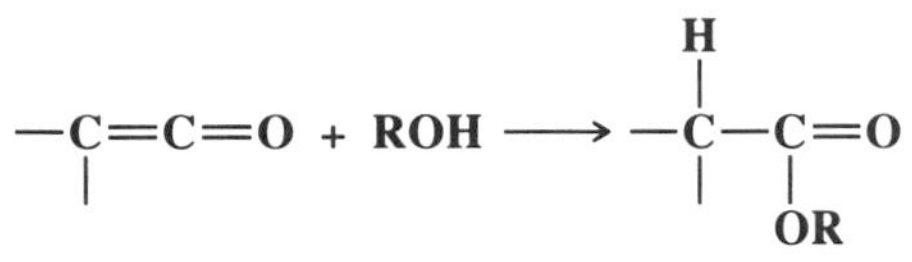

$$-C=C=O + ROH \longrightarrow \overset{\overset{\displaystyle H}{|}}{-C}-C=O \atop \overset{|}{OR}$$

This has been done intramolecularly (with the ketene end of the molecule generated and used in situ) to form medium- and large-ring lactones.[181] In the presence of a strong acid, ketene reacts with aldehydes or ketones (in their enol forms) to give enol acetates:

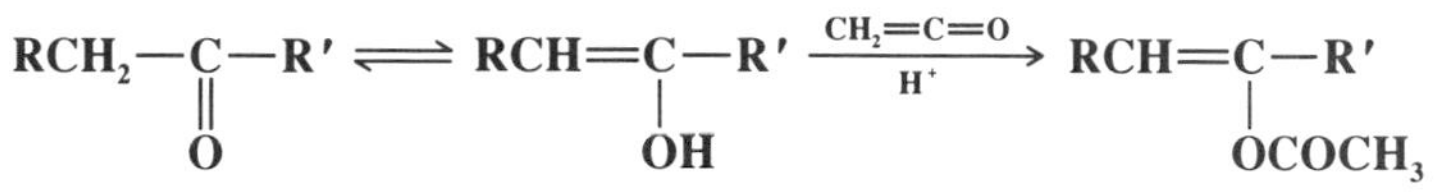

$$RCH_2-\overset{\overset{\displaystyle O}{||}}{C}-R' \rightleftharpoons RCH=\overset{\overset{\displaystyle OH}{|}}{C}-R' \xrightarrow[H^+]{CH_2=C=O} RCH=\overset{\overset{\displaystyle OCOCH_3}{|}}{C}-R'$$

Alcohols can also add to olefins in a different way (see **5-22**).
OS **III**, 371, 774, 813; **IV**, 184, 558; **VI**, 916; **VII**, 66, 160, 304, 334, 381; **67**, 52; **69**, 238.

5-5 Addition of Carboxylic Acids
Hydro-acyloxy-addition

$$-C=C- + RCOOH \xrightarrow{H^+} \overset{\overset{\displaystyle H \quad OCOR}{|\quad\quad|}}{-C}-\overset{|\quad\quad|}{C}-$$

Carboxylic esters are produced by the addition of carboxylic acids to olefins, a reaction that is usually acid-catalyzed (by proton or Lewis acids[182]) and similar in mechanism to **5-4**. Since Markovnikov's rule is followed, hard-to-get esters of tertiary alcohols can be prepared from olefins of the form $R_2C=CHR$.[183] When a carboxylic acid that contains a double bond in the chain is treated with a strong acid, the addition occurs internally and the product is a γ- and/or a δ-lactone, regardless of the original position of the double bond in the chain, since strong acids catalyze double bond shifts (**2-2**).[184] The double bond always migrates to a position favorable for the reaction, whether this has to be toward or away from the carboxyl group. Carboxylic esters have also been prepared by the acyloxymercuration–demercuration of olefins (similar to the procedures mentioned in **5-2** and **5-4**).[185]

Triple bonds can give enol esters or acylals when treated with carboxylic acids. Mercuric

[180]Quadbeck *Newer Methods Prep. Org. Chem.* **1963**, *2*, 133-161. See also Chihara; Teratini; Ogawa *J. Chem. Soc., Chem. Commun.* **1981**, 1120. For discussions of the mechanism see Tille; Pracejus *Chem. Ber.* **1967**, *100*, 196-210; Brady; Vaughn; Hoff *J. Org. Chem.* **1969**, *34*, 843; Ref. 143; Jähme; Rüchardt *Tetrahedron Lett.* **1982**, *23*, 4011; Poon; Satchell *J. Chem. Soc., Perkin Trans. 2* **1984**, 1083; **1985**, 1551.
[181]Boeckman; Pruitt *J. Am. Chem. Soc.* **1989**, *111*, 8286.
[182]See, for example, Guenzet; Camps *Bull. Soc. Chim. Fr.* **1973**, 3167, *Tetrahedron* **1974**, *30*, 849; Ballantine; Davies; Purnell; Rayanakorn; Thomas; Williams *J. Chem. Soc., Chem. Commun.* **1981**, 8.
[183]See, for example, Peterson; Tao *J. Org. Chem.* **1964**, *29*, 2322.
[184]For a review of such lactonizations, see Ansell; Palmer *Q. Rev., Chem. Soc.* **1964**, *18*, 211-225.
[185]For a review, see Larock, Ref. 144, pp. 367-442.

salts are usually catalysts,[186] and vinylic mercury compounds $-\underset{\underset{HgX}{|}}{C}=\underset{|}{C}-OCOR$ are inter-

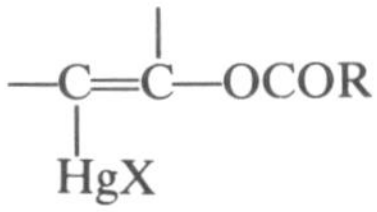

mediates.[187] Terminal alkynes $RC\equiv CH$ react with CO_2, a secondary amine $R_2'NH$, and a

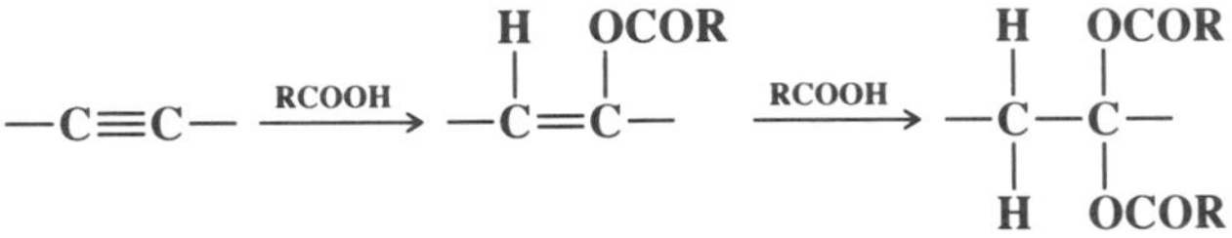

$$-C\equiv C- \xrightarrow{RCOOH} \underset{\underset{H}{|}}{-C}=\underset{\underset{OCOR}{|}}{C}- \xrightarrow{RCOOH} \underset{\underset{H}{|}}{-C}-\underset{\underset{OCOR}{|}}{C}-$$

ruthenium complex catalyst, to give enol carbamates $RCH=CHOC(=O)NR_2'$.[188] This reaction has also been performed intramolecularly, to produce unsaturated lactones.[189] With ketenes, carboxylic acids give anhydrides[190] and acetic anhydride is prepared industrially in this manner:

$$CH_2=C=O + CH_3COOH \longrightarrow CH_3-\underset{\underset{OCOCH_3}{|}}{C}=O$$

Carboxylic esters can also be obtained by the addition to olefins of diacyl peroxides.[191] These reactions are catalyzed by copper and are free-radical processes.

OS **III**, 853; **IV**, 261, 417, 444; **V**, 852, 863; **VII**, 30, 411. Also see OS **I**, 317.

C. Sulfur on the Other Side

5-6 Addition of H_2S and Thiols
Hydro-alkylthio-addition

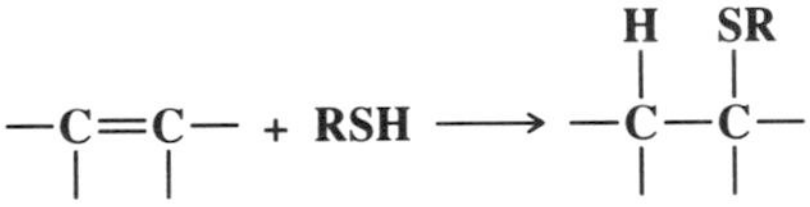

$$-\underset{|}{C}=\underset{|}{C}- + RSH \longrightarrow \underset{\underset{H}{|}}{-C}-\underset{\underset{SR}{|}}{C}-$$

H_2S and thiols add to olefins by electrophilic, nucleophilic, or free-radical mechanisms.[192] In the absence of initiators the addition to simple olefins is by an electrophilic mechanism, similar to that in **5-4**, and Markovnikov's rule is followed. However, this reaction is usually very slow and often cannot be done or requires very severe conditions unless a proton or Lewis acid catalyst is used. For example, the reaction can be performed in concentrated

[186]For the use of rhodium complex catalysts, see Bianchini; Meli; Peruzzini; Zanobini; Bruneau; Dixneuf *Organometallics* **1990**, *9*, 1155.
[187]See for example, Bach; Woodard; Anderson; Glick *J. Org. Chem.* **1982**, *47*, 3707; Alekseeva; Chalov; Temkin *J. Org. Chem. USSR* **1983**, *19*, 431; Bassetti; Floris *J. Org. Chem.* **1986**, *51*, 4140, *J. Chem. Soc., Perkin Trans. 2* **1988**, 227; Grishin; Bazhenov; Ustynyuk; Zefirov; Kartashov; Sokolova; Skorobogatova; Chernov *Tetrahedron Lett.* **1988**, *29*, 4631; Camps; Monthéard; Benzaïd *Bull. Soc. Chim. Fr.* **1989**, 123. Ruthenium complexes have also been used as catalysts: Rotem; Shvo *Organometallics*. **1983**, *2*, 1689; Ruppin; Dixneuf *Tetrahedron Lett.* **1986**, *27*, 6323; Mitsudo; Hori; Yamakawa; Watanabe *J. Org. Chem.* **1987**, *52*, 2230.
[188]Mitsudo; Hori; Yamakawa; Watanabe *Tetrahedron Lett.* **1987**, *28*, 4417; Mahé; Sasaki; Bruneau; Dixneuf *J. Org. Chem.* **1989**, *54*, 1518.
[189]See, for example, see Sofia; Katzenellenbogen *J. Org. Chem.* **1985**, *50*, 2331. For a list of other examples, see Ref. 133, p. 950.
[190]For discussions of the mechanism, see Briody; Lillford; Satchell *J. Chem. Soc. B* **1968**, 885; Corriu; Guenzet; Camps; Reye *Bull. Soc. Chim. Fr.* **1970**, 3679; Blake; Vayjooee *J. Chem. Soc., Perkin Trans. 2* **1976**, 1533.
[191]Kharasch; Fono *J. Org. Chem.* **1959**, *24*, 606; Kochi *J. Am. Chem. Soc.* **1962**, *84*, 1572.
[192]For a review, see Wardell, In Patai *The Chemistry of the Thiol Group*, pt. 1; Wiley: New York, 1974, pp. 169-178.

H_2SO_4[193] or together with $AlCl_3$.[194] In the presence of free-radical initiators, H_2S and thiols add to double and triple bonds by a free-radical mechanism and the orientation is anti-Markovnikov.[195] In fact, the orientation can be used as a diagnostic tool to indicate which mechanism is operating. Free-radical addition can be done with H_2S, RSH (R may be primary, secondary, or tertiary), ArSH, or RCOSH.[196] R may contain various functional groups. The olefins may be terminal, internal, contain branching, be cyclic, and have various functional groups including OH, COOH, COOR, NO_2, RSO_2, etc. With alkynes it is possible to add 1 or 2 moles of RSH.

When thiols are added to substrates susceptible to nucleophilic attack, bases catalyze the reaction and the mechanism is nucleophilic. These substrates may be of the Michael type[197] or may be polyhalo olefins or alkynes.[169] As with the free-radical mechanism, alkynes can give either vinylic thioethers or dithioacetals:

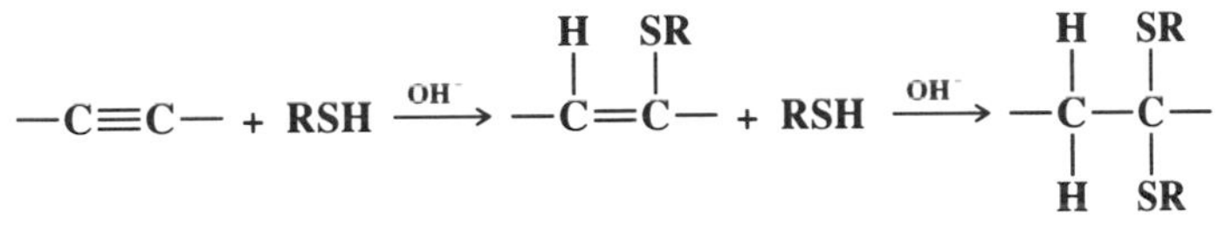

$$-C{\equiv}C- + RSH \xrightarrow{OH^-} \underset{\overset{|}{H}}{-}\overset{\overset{H}{|}}{C}{=}\overset{\overset{SR}{|}}{C}- + RSH \xrightarrow{OH^-} \overset{\overset{H}{|}}{\underset{\overset{|}{H}}{C}}-\overset{\overset{SR}{|}}{\underset{\overset{|}{SR}}{C}}-$$

By any mechanism, the initial product of addition of H_2S to a double bond is a thiol, which is capable of adding to a second molecule of olefin, so that sulfides are often produced:

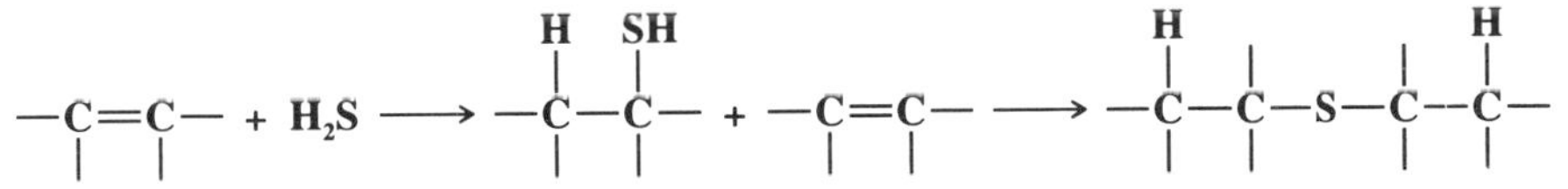

$$-C{=}C- + H_2S \longrightarrow \overset{\overset{H}{|}}{\underset{\overset{|}{}}{C}}-\overset{\overset{SH}{|}}{\underset{\overset{|}{}}{C}}- + -C{=}C- \longrightarrow \overset{\overset{H}{|}}{\underset{\overset{|}{}}{C}}-\overset{\overset{|}{}}{\underset{\overset{|}{}}{C}}-S-\overset{\overset{H}{|}}{\underset{\overset{|}{}}{C}}-\overset{\overset{H}{|}}{\underset{\overset{|}{}}{C}}-$$

Ketenes add thiols to give thiol esters:

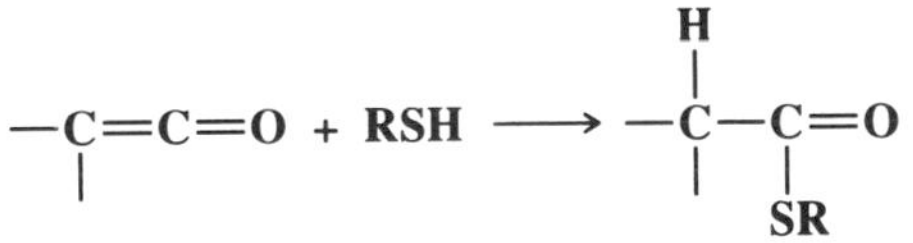

$$-C{=}C{=}O + RSH \longrightarrow \overset{\overset{H}{|}}{\underset{\overset{|}{}}{C}}-\overset{\overset{|}{}}{\underset{\overset{|}{SR}}{C}}{=}O$$

OS **III**, 458; **IV**, 669; **65**, 215. See also OS **69**, 169.

---

[193]Shostakovskii; Kul'bovskaya; Gracheva; Laba; Yakushina *J. Gen. Chem. USSR* **1962**, *32*, 707.

[194]Belley; Zamboni *J. Org. Chem.* **1989**, *54*, 1230.

[195]For reviews of free-radical addition of H_2S and RSH, see Voronkov; Martynov; Mirskova *Sulfur Rep.* **1986**, *6*, 77-95; Griesbaum *Angew. Chem. Int. Ed. Engl.* **1970**, *9*, 273-287 [*Angew. Chem. 82*, 276-290]; Oswald; Griesbaum, in Kharasch; Meyers *Organic Sulfur Compounds*, vol. 2; Pergamon: Elmsford, NY, 1966, pp. 233-256; Stacey; Harris *Org. React.* **1963**, *13*, 150-376, pp. 165-196, 247-324.

[196]For a review of the addition of thio acids, see Janssen, in Patai *The Chemistry of Carboxylic Acids and Esters*; Wiley: New York, 1969, pp. 720-723.

[197]Michael substrates usually give the expected orientation. For a method of reversing the orientation for RS groups (the RS group goes α to the C=O bond of a C=C—C=O system), see Gassman; Gilbert; Cole *J. Org. Chem.* **1977**, *42*, 3233.

D. Nitrogen on the Other Side

5-7   Addition of Ammonia and Amines
Hydro-amino-addition

$$-\text{C}=\text{C}- + \text{NH}_3 \longrightarrow -\text{CH}-\overset{|}{\underset{|}{\text{C}}}\text{NH}_2 + (-\text{CH}-\overset{|}{\underset{|}{\text{C}}})_2\text{NH} + (-\text{CH}-\overset{|}{\underset{|}{\text{C}}})_3\text{N}$$

$$-\text{C}=\text{C}- + \text{RNH}_2 \longrightarrow -\text{CH}-\overset{|}{\underset{|}{\text{C}}}\text{NHR} + (-\text{CH}-\overset{|}{\underset{|}{\text{C}}})_2\text{NR}$$

$$-\text{C}=\text{C}- + \text{R}_2\text{NH} \longrightarrow -\text{CH}-\overset{|}{\underset{|}{\text{C}}}\text{NR}_2$$

Ammonia and primary and secondary amines add to olefins that are susceptible to nucleophilic attack.[198] Ammonia gives three possible products, since the initial product is a primary amine, which may add to a second molecule of olefin, etc. Similarly, primary amines give both secondary and tertiary products. In practice it is usually possible to control which product predominates. Since ammonia and amines are much weaker acids than water, alcohols, and thiols (see **5-2, 5-4, 5-6**) and since acids could hardly catalyze the reaction (because they would turn NH_3 into NH_4^+), this reaction does not occur by an electrophilic mechanism and so gives very low yields, if any, with ordinary olefins, unless extreme conditions are used (e.g., 178-200°C, 800-1000 atm, and the presence of metallic Na, for the reaction between NH_3 and ethylene[199]). The mechanism is nearly always nucleophilic, and the reaction is generally performed on polyhalo olefins,[200] Michael-type substrates, and alkynes. As expected, on Michael-type substrates the nitrogen goes to the carbon that does not carry the Z. With substrates of the form RCH=CZZ', the same type of cleavage of the adduct can take place as in **5-2**.[201]

Other nitrogen compounds, among them hydroxylamine, hydrazines, amides (RCONH_2 and RCONHR' including imides and lactams), and sulfonamides, also add to olefins. In the case of amides, basic catalysts are required, since amides are not good enough nucleophiles for the reaction and must be converted to RCONH^-. Even with amines, basic catalysts are sometimes used, so that RNH^- or R_2N^- is the actual nucleophile. Tertiary amines (except those that are too bulky) add to Michael-type substrates in a reaction that is catalyzed by acids like HCl or HNO_3 to give the corresponding quaternary ammonium salts.[202]

$$\text{Z}-\text{C}=\text{C}- + \text{R}_3\text{NH}^+ \ \text{Cl}^- \xrightarrow{\text{HCl}} \text{Z}-\overset{|}{\underset{|}{\text{C}}}-\overset{|}{\underset{|}{\text{C}}}-$$
$$\text{H}\quad \text{NR}_3^+ \ \text{Cl}^-$$

The tertiary amine can be aliphatic, cycloalkyl, or heterocyclic (including pyridine).

[198]For reviews, see Gasc; Lattes; Périé *Tetrahedron* **1983**, *39*, 703-731; Pines; Stalick *Base-Catalyzed Reactions of Hydrocarbons and Related Compounds*; Academic Press: New York, 1977, pp. 423-454; Suminov; Kost *Russ. Chem. Rev.* **1969**, *38*, 884-899; Gibson, in Patai *The Chemistry of the Amino Group*; Wiley: New York, 1968, pp. 61-65.
[199]Howk; Little; Scott; Whitman *J. Am. Chem. Soc.* **1954**, *76*, 1899.
[200]For a review with respect to fluoroolefins, see Chambers; Mobbs *Adv. Fluorine Chem.* **1965**, *4*, 51-112, pp. 62-68.
[201]See, for example, Bernasconi; Murray *J. Am. Chem. Soc.* **1986**, *108*, 5251, 5257; Bernasconi; Bunnell *J. Org. Chem.* **1988**, *53*, 2001.
[202]Le Berre; Delacroix *Bull. Soc. Chim. Fr.* **1973**, 640, 647. See also Vogel; Büchi *Org. Synth. 66*, 29.

Primary amines add to triple bonds[203] to give enamines that have a hydrogen on the nitrogen and (analogously to enols) tautomerize to the more stable imines:

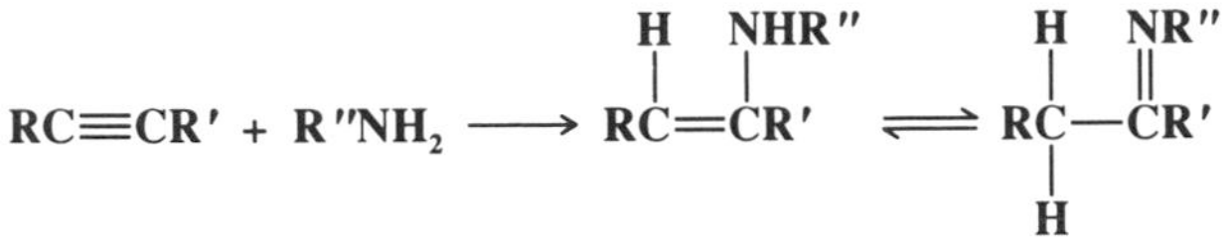

$$RC{\equiv}CR' + R''NH_2 \longrightarrow \underset{\overset{|}{H}}{RC}{=}\underset{\overset{|}{NHR''}}{CR'} \; \rightleftharpoons \; \underset{\overset{|}{H}}{RC}{-}\underset{\overset{\|}{NR''}}{CR'}$$

These are often stable enough for isolation.[204] When ammonia is used instead of a primary amine, the corresponding $RCH_2{-}\overset{\overset{\displaystyle NH}{\|}}{C}R'$ is not stable enough for isolation, but polymerizes. Ammonia and primary amines (aliphatic and aromatic) add to conjugated diynes to give pyrroles:[205]

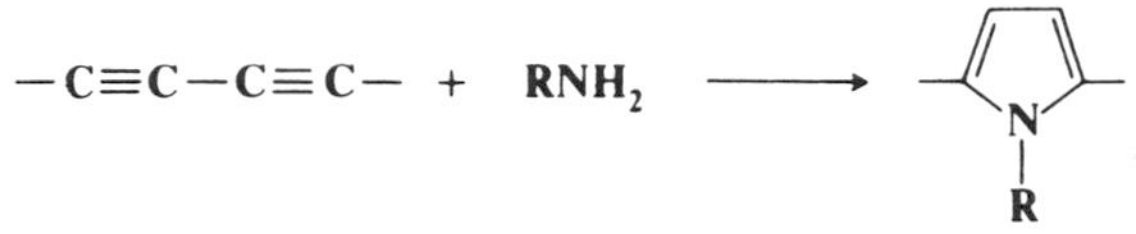

$$-C{\equiv}C{-}C{\equiv}C- \; + \; RNH_2 \; \longrightarrow \; \underset{\overset{|}{R}}{\boxed{}}$$

This is not 1,4 addition but 1,2 addition twice.

Primary and secondary amines add to ketenes to give, respectively, N-substituted and N,N-disubstituted amides:[206]

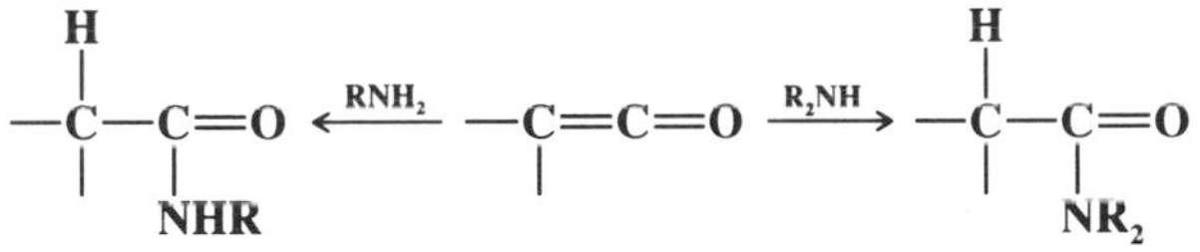

$$\underset{\overset{|}{NHR}}{-\overset{\overset{\displaystyle H}{|}}{C}}-C{=}O \; \xleftarrow{RNH_2} \; -C{=}C{=}O \; \xrightarrow{R_2NH} \; \underset{\overset{|}{NR_2}}{-\overset{\overset{\displaystyle H}{|}}{C}}-C{=}O$$

and to ketenimines to give amidines:[207]

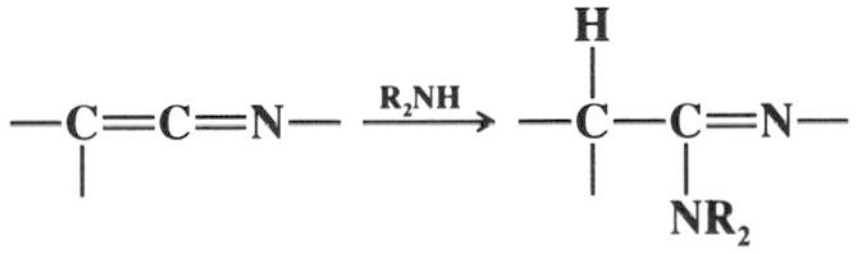

$$-C{=}C{=}N- \; \xrightarrow{R_2NH} \; \underset{\overset{|}{NR_2}}{-\overset{\overset{\displaystyle H}{|}}{C}}-C{=}N- \qquad \mathbf{R = H \text{ or alkyl}}$$

Secondary amines can be added to certain nonactivated olefins if palladium(II) complexes are used as catalysts.[208] The complexation lowers the electron density of the double bond, facilitating nucleophilic attack.[209] Markovnikov orientation is observed and the addition is anti.[210]

[203]For a review of addition of ammonia and amines to triple bonds, see Chekulaeva; Kondrat'eva *Russ. Chem. Rev.* **1965**, *34*, 669-680.

[204]For example, see Kruse; Kleinschmidt *J. Am. Chem. Soc.* **1961**, *83*, 213, 216.

[205]Schulte; Reisch; Walker *Chem. Ber.* **1965**, *98*, 98.

[206]For discussions of the mechanism of this reaction, see Briody; Satchell *Tetrahedron* **1966**, *22*, 2649; Lillford; Satchell *J. Chem. Soc. B* **1967**, 360, **1968**, 54; Ref. 143.

[207]Stevens; Freeman; Noll *J. Org. Chem.* **1965**, *30*, 3718.

[208]See, for example, Walker; Manyik; Atkins; Farmer *Tetrahedron Lett.* **1970**, 3817; Takahashi; Miyake; Hata *Bull. Chem. Soc. Jpn.* **1972**, *45*, 1183; Baker; Cook; Halliday; Smith *J. Chem. Soc., Perkin Trans. 2* **1974**, 1511; Hegedus; Allen; Waterman *J. Am. Chem. Soc.* **1976**, *98*, 2674. For a review, see Gasc et al., Ref. 198. For a review of metal-catalyzed nucleophilic addition, see Bäckvall *Adv. Met.-Org. Chem.* **1989**, *1*, 135-175.

[209]For a discussion of the mechanism, see Hegedus; Åkermark; Zetterberg; Olsson *J. Am. Chem. Soc.* **1984**, *106*, 7122.

[210]Åkermark; Zetterberg *J. Am. Chem. Soc.* **1984**, *106*, 5560.

NH$_3$ can be added to double bonds (even ordinary double bonds) in an indirect manner by the use of hydroboration (**5-12**) followed by treatment with NH$_2$Cl or NH$_2$OSO$_2$OH (**2-31**). This produces a primary amine with anti-Markovnikov orientation. An indirect way of adding a primary or secondary amine to a double bond consists of aminomercuration followed by reduction (see **5-2** for the analogous oxymercuration–demercuration procedure), e.g.,[211]

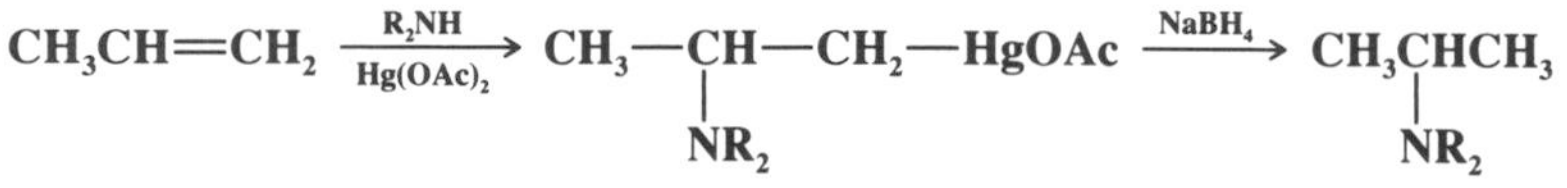

$$CH_3CH{=}CH_2 \xrightarrow[Hg(OAc)_2]{R_2NH} CH_3{-}\underset{\underset{\textbf{30}}{\underset{|}{NR_2}}}{\overset{|}{CH}}{-}CH_2{-}HgOAc \xrightarrow{NaBH_4} CH_3\underset{\underset{|}{NR_2}}{\overset{|}{CH}}CH_3$$

The addition of a secondary amine (shown above) produces a tertiary amine, while addition of a primary amine gives a secondary amine. The overall orientation follows Markovnikov's rule. Amido- and sulfamidomercuration–demercuration[212] and nitromercuration[213] have also been accomplished (see also **6-55**). For conversion of **30** to other products, see **5-40** and **5-41**.

OS **I**, 196; **III**, 91, 93, 244, 258; **IV**, 146, 205; **V**, 39, 575, 929; **VI**, 75, 943; **66**, 29; **67**, 44, 48. See also OS **VI**, 932.

5-8 Addition of Hydrazoic Acid
Hydro-azido-addition

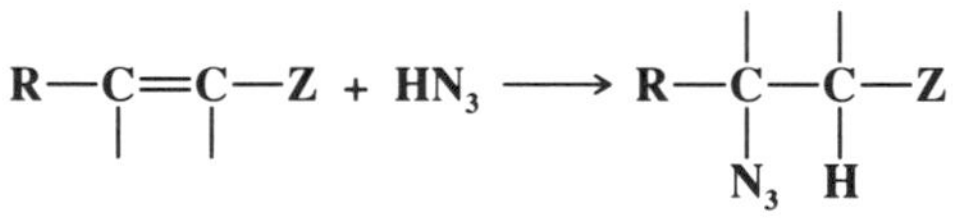

$$R{-}\overset{|}{C}{=}\overset{|}{C}{-}Z + HN_3 \longrightarrow R{-}\underset{\underset{|}{N_3}}{\overset{|}{C}}{-}\underset{\underset{|}{H}}{\overset{|}{C}}{-}Z$$

Hydrazoic acid can be added to certain Michael-type substrates (Z is as defined on p. 741) to give β-azido compounds.[214] The reaction apparently fails if R is phenyl. HN$_3$ also adds to enol ethers CH$_2$=CHOR to give CH$_3$—CH(OR)N$_3$, and to silyl enol ethers,[215] but it does not add to ordinary alkenes unless a Lewis acid catalyst, such as TiCl$_4$, is used, in which case good yields of azide can be obtained.[215] HN$_3$ can also be added indirectly to ordinary olefins by azidomercuration, followed by demercuration,[216] analogous to the similar

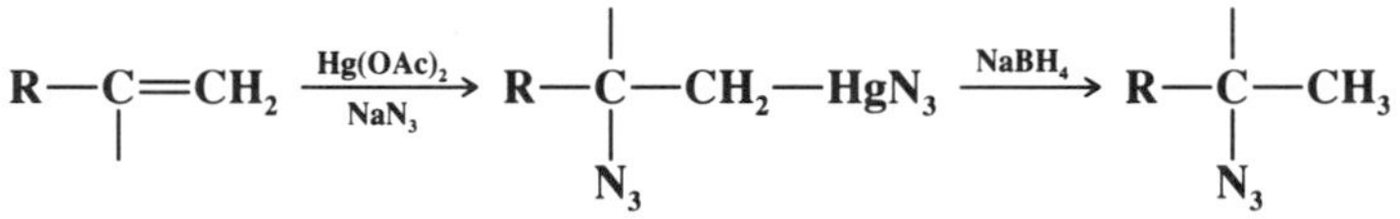

$$R{-}\overset{|}{C}{=}CH_2 \xrightarrow[NaN_3]{Hg(OAc)_2} R{-}\underset{\underset{|}{N_3}}{\overset{|}{C}}{-}CH_2{-}HgN_3 \xrightarrow{NaBH_4} R{-}\underset{\underset{|}{N_3}}{\overset{|}{C}}{-}CH_3$$

procedures mentioned in **5-2**, **5-4**, **5-5**, and **5-7**. The method can be applied to terminal alkenes or strained cycloalkenes (e.g., norbornene) but fails for unstrained internal alkenes.

[211]For a review, see Larock, Ref. 144, pp. 443-504. See also Barluenga; Perez-Prieto; Asensio *Tetrahedron* **1990**, *46*, 2453.

[212]For a review, see Larock, Ref. 144, pp. 505-521.

[213]Bachman; Whitehouse *J. Org. Chem.* **1967**, *32*, 2303. For a review, see Larock, Ref. 144, pp. 528-531.

[214]Boyer *J. Am. Chem. Soc.* **1951**, *73*, 5248; Harvey; Ratts *J. Org. Chem.* **1966**, *31*, 3907. For a review, see Biffin; Miller; Paul, in Patai *The Chemistry of the Azido Group*; Wiley: New York, 1971, pp. 120-136.

[215]Hassner; Fibiger; Andisik *J. Org. Chem.* **1984**, *49*, 4237.

[216]Heathcock *Angew. Chem. Int. Ed. Engl.* **1969**, *8*, 134 [*Angew. Chem. 81*, 148]. For a review, see Larock, Ref. 144, pp. 522-527.

E. Hydrogen on Both Sides

5-9 Hydrogenation of Double and Triple Bonds[217]
Dihydro-addition

$$-\overset{|}{\underset{|}{C}}=\overset{|}{\underset{|}{C}}- \;+\; H_2 \;\xrightarrow{\text{cat.}}\; -\overset{H}{\underset{|}{C}}-\overset{H}{\underset{|}{C}}-$$

Most carbon–carbon double bonds, whether substituted by electron-donating or electron-withdrawing substituents, can be catalytically hydrogenated, usually in quantitative or near-quantitative yields.[218] Almost all known alkenes added hydrogen at temperatures between 0 and 275°C. Many functional groups may be present in the molecule, e.g., OH, COOH, NH$_2$, CHO, COR, COOR, or CN. Some of these groups are also susceptible to catalytic reduction, but it is usually possible to find conditions under which double bonds can be reduced selectively[219] (see Table 19.2). The catalysts used can be divided into two broad classes, both of which mainly consist of transition metals and their compounds: (1) catalysts insoluble in the reaction medium (*heterogeneous catalysts*). Among the most effective are Raney nickel,[220] palladium-on-charcoal (perhaps the most common), NaBH$_4$-reduced nickel[221] (also called nickel boride), platinum metal or its oxide, rhodium, ruthenium, and zinc oxide,[222] (2) Catalysts soluble in the reaction medium (*homogeneous catalysts*).[223] The most important is chlorotris(triphenylphosphine)rhodium RhCl(PH$_3$P)$_3$,[224] (*Wilkinson's catalyst*),[225] which catalyzes the hydrogenation of many olefinic compounds without disturbing such groups as COOR, NO$_2$, CN, or COR present in the same molecule.[226] Even unsaturated

[217]For a review, see Mitsui; Kasahara, in Zabicky, Ref. 115, vol. 2, pp. 175-214.

[218]For books on catalytic hydrogenation, see Rylander *Hydrogenation Methods*; Academic Press: New York, 1985, *Catalytic Hydrogenation in Organic Synthesis*; Academic Press: New York, 1979, *Catalytic Hydrogenation over Platinum Metals*, Academic Press: New York, 1967; Červeny *Catalytic Hydrogenation*; Elsevier: New York, 1986 (this book deals mostly with industrial aspects); Freifelder *Catalytic Hydrogenation in Organic Synthesis*; Wiley: New York, 1978, *Practical Catalytic Hydrogenation*; Wiley: New York, 1971; Augustine *Catalytic Hydrogenation*; Marcel Dekker: New York, 1965. For reviews, see Parker, in Hartley *The Chemistry of the Metal–carbon Bond*, vol. 4; Wiley: New York, 1987, pp. 979-1047; Carruthers *Some Modern Methods of Organic Synthesis*, 3rd ed.; Cambridge University Press: Cambridge, 1986, pp. 411-431; Colquhoun; Holton; Thompson; Twigg *New Pathways for Organic Synthesis*; Plenum: New York, 1984, pp. 266-300, 325-334; Kalinkin; Kolomnikova; Parnes; Kursanov *Russ. Chem. Rev.* **1979**, *48*, 332-342; Candlin; Rennie, in Bentley; Kirby *Elucidation of Organic Structures by Physical and Chemical Methods*, 2nd ed. (vol. 4 of Weissberger *Techniques of Chemistry*), pt. 2; Wiley: New York, 1973, pp. 97-117; House, Ref. 144, pp, 1-34.

[219]For a discussion, see Rylander *Catalytic Hydrogenation over Platinum Metals*, Ref. 218, pp. 59-120.

[220]For a review of Raney nickel, see Pizey, Ref. 146, vol. 2, 1974, pp. 175-311. Double bonds have been reduced with Raney nickel alone; with no added H$_2$. The hydrogen normally present in this reagent was sufficient: Pojer *Chem. Ind. (London)* **1986**, 177.

[221]Paul; Buisson; Joseph *Ind. Eng. Chem.* **1952**, *44*, 1006; Brown *Chem. Commun.* **1969**, 952, *J. Org. Chem.* **1970**, *35*, 1900. For a review of reductions with nickel boride and related catalysts, see Ganem; Osby *Chem. Rev.* **1986**, *86*, 763-780.

[222]For reviews of hydrogenation with metal oxides, see Minachev; Khodakov; Nakhshunov *Russ. Chem. Rev.* **1976**, *45*, 142-154; Kokes; Dent *Adv. Catal.* **1972**, *22*, 1-50 (ZnO).

[223]For a monograph, see James *Homogeneous Hydrogenation*; Wiley: New York, 1973. For reviews, see Collman; Hegedus; Norton; Finke *Principles and Applications of Organotransition Metal Chemistry*; University Science Books: Mill Valley, CA, 1987, pp. 523-564; Birch; Williamson *Org. React.* **1976**, *24*, 1-186; James *Adv. Organomet. Chem.* **1979**, *17*, 319-405; Harmon; Gupta; Brown *Chem. Rev.* **1973**, *73*, 21-52; Strohmeier *Fortschr. Chem. Forsch.* **1972**, *25*, 71-104; Heck *Organotransition Metal Chemistry*; Academic Press: New York, 1974, pp. 55-65; Rylander *Organic Syntheses with Noble Metal Catalysts*; Academic Press: New York, 1973, pp. 60-76; Lyons; Rennick; Burmeister *Ind. Eng. Chem., Prod. Res. Dev.* **1970**, *9*, 2-20; Vol'pin; Kolomnikov *Russ. Chem. Rev.* **1969**, *38*, 273-289.

[224]Young; Osborn; Jardine; Wilkinson *Chem. Commun.* **1965**, 131; Osborn; Jardine; Young; Wilkinson *J. Chem. Soc. A* **1966**, 1711; Osborn; Wilkinson *Inorg. Synth.* **1967**, *10*, 67; Biellmann *Bull. Soc. Chim. Fr.* **1968**, 3055; van Bekkum; van Rantwijk; van de Putte *Tetrahedron Lett.* **1969**, 1.

[225]For a review of Wilkinson's catalyst, see Jardine, *Prog. Inorg. Chem.* **1981**, *28*, 63-202.

[226]Harmon; Parsons; Cooke; Gupta; Schoolenberg *J. Org. Chem.* **1969**, *34*, 3684. See also Mohrig; Dabora; Foster; Schultz *J. Org. Chem.* **1984**, *49*, 5179.

aldehydes can be reduced to saturated aldehydes,[227] though in this case decarbonylation (**4-41**) may be a side reaction. Among other homogeneous catalysts are chlorotris(triphenylphosphine)hydridoruthenium(II) $(Ph_3P)_3RuClH$,[228] which is specific for terminal double bonds (other double bonds are hydrogenated slowly or not at all), and pentacyanocobaltate(II) $Co(CN)_5{}^{3-}$, which is effective for double and triple bonds only when they are part of conjugated systems[229] (the conjugation may be with $C=C$, $C=O$, or an aromatic ring). Homogeneous catalysts often have the advantages of better catalyst reproducibility and better selectivity. They are also less susceptible to catalyst poisoning[230] (heterogeneous catalysts are usually poisoned by small amounts of sulfur, often found in rubber stoppers, or by sulfur-containing compounds such as thiols and sulfides).[231] On the other hand, heterogeneous catalysts are usually easier to separate from the reaction mixture.

Optically active homogeneous (as well as heterogeneous) catalysts have been used to achieve partially asymmetric (enantioselective) hydrogenations of certain prochiral substrates.[232] For example,[233] hydrogenation of **31** with a suitable catalyst gives $(+)$ or $(-)$ **32**

(depending on which enantiomer of the catalyst is used) with an enantiomeric excess as high as 96%.[234] Prochiral substrates that give such high optical yields generally contain functional groups similar to those in **31**.[235] The catalyst in such cases[236] is usually a ruthenium- or rhodium-phosphine in which the phosphine is optically active either because of an asymmetric phosphorus atom, e.g., **33**,[237] or because of a chiral group connected to the phosphorus,

[227]Jardine; Wilkinson *J. Chem. Soc. C* **1967**, 270.

[228]Hallman; Evans; Osborn; Wilkinson *Chem. Commun.* **1967**, 305; Hallman; McGarvey; Wilkinson *J. Chem. Soc. A* **1968**, 3143; Jardine; McQuillin *Tetrahedron Lett.* **1968**, 5189.

[229]Kwiatek; Mador; Seyler *J. Am. Chem. Soc.* **1962**, *84*, 304; Jackman; Hamilton; Lawlor *J. Am. Chem. Soc.* **1968**, *90*, 1914; Funabiki; Matsumoto; Tarama *Bull. Chem. Soc. Jpn.* **1972**, *45*, 2723; Reger; Habib; Fauth *Tetrahedron Lett.* **1979**, 115.

[230]Birch; Walker *Tetrahedron Lett.* **1967**, 1935.

[231]For a review of catalyst poisoning by sulfur, see Barbier; Lamy-Pitara; Marecot; Boitiaux; Cosyns; Verna *Adv. Catal.* **1990**, *37*, 279-318.

[232]For reviews, see, in Morrison *Asymmetric Synthesis*, vol. 5; Academic Press: New York, 1985, the reviews by Halpern, pp. 41-69, Koenig, pp. 71-101, Harada, pp. 345-383; Ojima; Clos; Bastos *Tetrahedron* **1989**, *45*, 6901-6939, pp. 6902-6916; Jardine, in Hartley, Ref. 218, pp. 751-775; Nógrádi *Stereoselective Synthesis*; VCH: New York, 1986, pp. 53-87; Knowles *Acc. Chem. Res.* **1983**, *16*, 106-112; Brunner *Angew. Chem. Int. Ed. Engl.* **1983**, *22*, 897-907 [*Angew. Chem. 95*, 921-931]; Klabunovskii *Russ. Chem. Rev.* **1982**, *51*, 630-643; Čaplar; Comisso; Šunjić *Synthesis* **1981**, 85-116; Morrison; Masler; Neuberg *Adv. Catal.* **1976**, *25*, 81-124; Kagan *Pure Appl. Chem.* **1975**, *43*, 401-421; Bogdanović *Angew. Chem. Int. Ed. Engl.* **1973**, *12*, 954-964 [*Angew. Chem. 85*, 1013-1023]. See also Ref. 94 in Chapter 4.

[233]For some other recent examples, see Hayashi; Kawamura; Ito *Tetrahedron Lett.* **1988**, *29*, 5969; Muramatsu; Kawano; Ishii; Saburi; Uchida *J. Chem. Soc., Chem. Commun.* **1989**, 769; Amrani; Lecomte; Sinou; Bakos; Toth; Heil *Organometallics* **1989**, *8*, 542; Yamamoto; Ikeda; Lin *J. Organomet. Chem.* **1989**, *370*, 319; Waymouth; Pino *J. Am. Chem. Soc.* **1990**, *112*, 4911; Ohta; Takaya; Noyori *Tetrahedron Lett.* **1990**, *31*, 7189; Ashby; Halpern *J. Am. Chem. Soc.* **1991**, *113*, 589; Heiser; Broger; Crameri *Tetrahedron: Asymmetry* **1991**, *2*, 51; Burk *J. Am. Chem. Soc.* **1991**, *113*, 8518.

[234]Koenig, in Morrison, Ref. 232, p. 74.

[235]For tables of substrates that have been enantioselectively hydrogenated, see Koenig, in Morrison, Ref. 232, pp. 83-101.

[236]For a list of these, with references, see Ref. 133, p. 7. For reviews of optically active nickel catalysts, see Izumi *Adv. Catal.* **1983**, *32*, 215-271, *Angew. Chem. Int. Ed. Engl.* **1971**, *10*, 871-881 [*Angew. Chem. 83*, 956-966]. For a review of the synthesis of some of these phosphines, see Mortreux; Petit; Buono; Peiffer *Bull. Soc. Chim. Fr.* **1987**, 631-639.

[237]Knowles; Sabacky; Vineyard *J. Chem. Soc., Chem. Commun.* **1972**, 10. See also Vineyard; Knowles; Sabacky; Bachman; Weinkauff *J. Am. Chem. Soc.* **1977**, *99*, 5946.

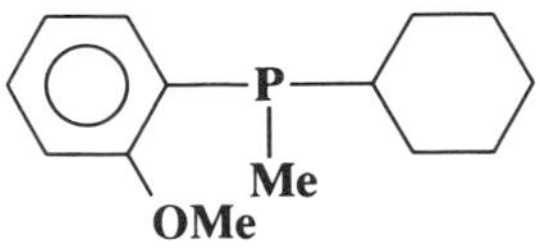

33 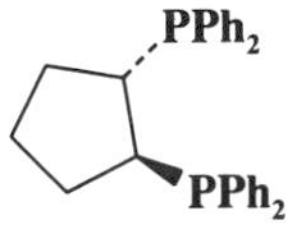   34

e.g., **34**.[238] Other types of catalysts, for example, titanocenes with chiral cyclopentadienyl ligands, have given enantioselective hydrogenation of olefins that lack functional groups such as COOH or NHCOCH$_3$, for example, 2-phenyl-1-butene.[239] Enantioselective reduction of certain olefins has also been achieved by reducing with baker's yeast.[240]

Hydrogenations in most cases are carried out at room temperature and just above atmospheric pressure, but some double bonds are more resistant and require higher temperatures and pressures. The resistance is usually a function of increasing substitution and is presumably caused by steric factors. Trisubstituted double bonds require, say, 25°C and 100 atm, while tetrasubstituted double bonds may require 275°C and 1000 atm. Among the double bonds most difficult to hydrogenate or which cannot be hydrogenated at all are those common to two rings, as in the steroid shown. Hydrogenations, even at about atmospheric

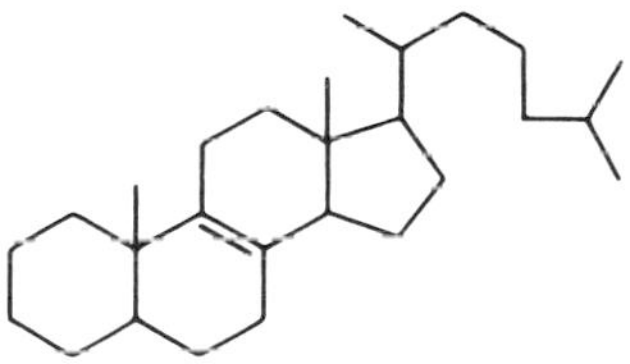

pressure, are ordinarily performed in a special hydrogenator, but this is not always necessary. Both the catalyst and the hydrogen can be generated in situ, by treatment of H$_2$PtCl$_6$ or RhCl$_3$ with NaBH$_4$;[241] ordinary glassware can then be used. The great variety of catalysts available often allows an investigator to find one that is highly selective. For example, the catalyst Pd(salen) encapsulated in zeolites permitted the catalytic hydrogenation of 1-hexene in the presence of cyclohexene.[242]

Although catalytic hydrogenation is the method most often used, double bonds can be reduced by other reagents, as well. Among these are sodium in ethanol, sodium and t-butyl alcohol in HMPA,[243] lithium and aliphatic amines[244] (see also **5-10**), chromous ion,[245] zinc and acids, sodium hypophosphate and Pd-C,[246] (EtO)$_3$SiH–Pd(OAc)$_2$,[247] trifluoroacetic acid

[238]Allen; Gibson; Green; Skinner; Bashkin; Grebenik *J. Chem. Soc., Chem. Commun.* **1983**, 895.
[239]Halterman; Vollhardt; Welker; Bläser; Boese *J. Am. Chem. Soc.* **1987**, *109*, 8105.
[240]See, for example, Gramatica; Manitto; Monti; Speranza *Tetrahedron* **1988**, *44*, 1299; Ohta; Kobayashi; Ozaki *J. Org. Chem.* **1989**, *54*, 1802. For reviews of baker's yeast, see Csuk; Glänzer *Chem. Rev.* **1991**, *91*, 49-97; Servi *Synthesis* **1990**, 1-25.
[241]Brown; Sivasankaran *J. Am. Chem. Soc.* **1962**, *84*, 2828; Brown; Brown *J. Am. Chem. Soc.* **1962**, *84*, 1494, 1945, 2829, *J. Org. Chem.* **1966**, *31*, 3989.
[242]Kowalak; Weiss; Balkus *J. Chem. Soc., Chem. Commun.* **1991**, 57.
[243]Angibeaud; Larchevêque; Normant; Tchoubar *Bull. Soc. Chim. Fr.* **1968**, 595; Whitesides; Ehmann *J. Org. Chem.* **1970**, *35*, 3565.
[244]Benkeser; Schroll; Sauve *J. Am. Chem. Soc.* **1955**, *77*, 3378.
[245]For example, see Castro; Stephens *J. Am. Chem. Soc.* **1964**, *86*, 4358; Castro; Stephens; Mojé *J. Am. Chem. Soc.* **1966**, *88*, 4964.
[246]Sala; Doria; Passarotti *Tetrahedron Lett.* **1984**, *25*, 4565.
[247]Tour; Pendalwar *Tetrahedron Lett.* **1990**, *31*, 4719.

and triethylsilane Et_3SiH,[248] hydrazine (if a small amount of oxidizing agent, such as air, H_2O_2, or cupric ion is present),[249] hydroxylamine and ethyl acetate,[250] and NH_2OSO_3H.[251] However, metallic hydrides, such as lithium aluminum hydride and sodium borohydride, do not in general reduce carbon–carbon double bonds, although this can be done in special cases where the double bond is polar, as in 1,1-diarylethenes[252] and in enamines.[253] In certain cases[254] these metallic hydride reagents may also reduce double bonds in conjugation with $C=O$ bonds, as well as reducing the $C=O$ bonds, e.g.,[255]

$NaBH_4$ has a greater tendency than $LiAlH_4$ to effect this double reduction, though even with $NaBH_4$ the product of single reduction (of the $C=O$ bond) is usually formed in larger amount than the doubly reduced product. $LiAlH_4$ gives significant double reduction only in cinnamyl systems, e.g., with $PhCH=CHCOOH$.[256]

Reduction of only the $C=C$ bond of conjugated $C=C-C=O$ and $C=C-C\equiv N$ systems[257] has been achieved by many reducing agents,[258] a few of which are H_2 and a Rh catalyst,[259] $Bu_3SnH-Pd(PPh_3)_4$,[260] $Bu_3SnH-CuI-LiCl$,[261] Zn–Cu in boiling MeOH,[262] Mg–MeOH,[263] diisobutylaluminum hydride–MeCu–HMPA,[264] $PhSiH_3-Mo(CO)_6$,[265] $[(Ph_3P)CuH]_6$,[266] Zn–$NiCl_2$ in the presence of ultrasound,[267] Al–$NiCl_2$,[268] potassium triphenylborohydride,[269] $CO-Se-H_2O$,[270] and catecholborane.[271] See **6-25** for methods of re-

[248]Kursanov; Parnes; Bassova; Loim; Zdanovich *Tetrahedron* **1967**, *23*, 2235; Doyle; McOsker *J. Org. Chem.* **1978**, *43*, 693. For a monograph, see Kursanov; Parnes; Kalinkin; Loim *Ionic Hydrogenation and Related Reactions*; Harwood Academic Publishers: Chur, Switzerland, 1985. For a review, see Kursanov; Parnes; Loim *Synthesis* **1974**, 633-651.

[249]Corey; Mock; Pasto *Tetrahedron Lett.* **1961**, 347; Hünig; Müller; Thier *Tetrahedron Lett.* **1961**, 353; Furst; Berlo; Hooton *Chem. Rev.* **1965**, *65*, 51-68, pp. 64-65; Kondo; Murai; Sonoda *Tetrahedron Lett.* **1977**, 3727.

[250]Wade; Amin *Synth. Commun.* **1982**, *12*, 287.

[251]Appel; Büchner *Liebigs Ann. Chem.* **1962**, *654*, 1; Dürckheimer *Liebigs Ann. Chem.* **1969**, *721*, 240. For a review of the reagent hydroxylamine-O-sulfonic acid, see Wallace *Org. Prep. Proced. Int.* **1982**, *14*, 265-307.

[252]See Granoth; Segall; Leader; Alkabets *J. Org. Chem.* **1976**, *41*, 3682.

[253]For a review of the reduction of enamines and indoles with $NaBH_4$ and a carboxylic acid, see Gribble; Nutaitis *Org. Prep. Proced. Int.* **1985**, *17*, 317-384. Enamines can also be reduced by formic acid; see Nilsson; Carlson *Acta Chem. Scand. Sect. B* **1985**, *39*, 187.

[254]For discussion, see Meyer *J. Chem. Educ.* **1981**, *58*, 628.

[255]Brown; Hess *J. Org. Chem.* **1969**, *34*, 2206. For other methods of reducing both double bonds, see Ref. 133, p. 540.

[256]Nystrom; Brown *J. Am. Chem. Soc.* **1947**, *69*, 2548, **1948**, *70*, 3738; Gammill; Gold; Mizsak *J. Am. Chem. Soc.* **1980**, *102*, 3095.

[257]For a review of the reduction of α,β-unsaturated carbonyl compounds, see Keinan; Greenspoon, in Patai; Rappoport, Ref. 37, pt. 2, pp. 923-1022. For a review of the stereochemistry of catalytic hydrogenation of α,β-unsaturated ketones, see Augustine *Adv. Catal.* **1976**, *25*, 56-80.

[258]For a long list of these, with references, see Ref. 133, pp. 8-17.

[259]Djerassi; Gutzwiller *J. Am. Chem. Soc.* **1966**, *88*, 4537; Cabello; Campelo; Garcia; Luna; Marinas *J. Org. Chem.* **1986**, *51*, 1786; Ref. 226.

[260]Keinan; Gleize *Tetrahedron Lett.* **1982**, *23*, 477; Four; Guibe *Tetrahedron Lett.* **1982**, *23*, 1825.

[261]Lipshutz; Ung; Sengupta *Synlett* **1989**, 64.

[262]Sondengam; Fomum; Charles; Akam *J. Chem. Soc., Perkin. Trans. 1* **1983**, 1219.

[263]Youn; Yon; Pak *Tetrahedron Lett.* **1986**, *27*, 2409; Hudlicky; Sinai-Zingde; Natchus *Tetrahedron Lett.* **1987**, *28*, 5287.

[264]Tsuda; Hayashi; Satomi; Kawamoto; Saegusa *J. Org. Chem.* **1986**, *51*, 537.

[265]Keinan; Perez *J. Org. Chem.* **1987**, *52*, 2576.

[266]Mahoney; Brestensky; Stryker *J. Am. Chem. Soc.* **1988**, *110*, 291.

[267]Petrier; Luche *Tetrahedron Lett.* **1987**, *28*, 2347, 2351.

[268]Hazarika; Barua *Tetrahedron Lett.* **1989**, *30*, 6567.

[269]Kim; Park; Yoon *Synth. Commun.* **1988**, *18*, 89.

[270]Nishiyama; Makino; Hamanaka; Ogawa; Sonoda *Bull. Chem. Soc. Jpn.* **1989**, *62*, 1682.

[271]Evans; Fu *J. Org. Chem.* **1990**, *55*, 5678.

ducing C=O bonds in the presence of conjugated C=C bonds. LiAlH$_4$ also reduces the double bonds of allylic alcohols[272] and NaBH$_4$ in MeOH–THF reduces α,β-unsaturated nitro compounds to nitroalkanes.[273] Furthermore, both LiAlH$_4$ and NaBH$_4$, as well as NaH, reduce ordinary alkenes and alkynes when complexed with transition metal salts, such as FeCl$_2$ or CoBr$_2$.[274]

The inertness of ordinary double bonds toward metallic hydrides is quite useful, since it permits reduction of, say, a carbonyl or nitro group, without disturbing a double bond in the same molecule (see Chapter 19 for a discussion of selectivity in reduction reactions). Sodium in liquid ammonia also does not reduce ordinary double bonds,[275] although it does reduce alkynes, allenes, conjugated dienes,[276] and aromatic rings (**5-10**).

Another hydrogenation method is called *transfer hydrogenation*.[277] In this method the hydrogen comes from another organic molecule, which is itself oxidized. A transition-metal catalyst, heterogeneous or homogeneous, is frequently employed. A common reducing agent is cyclohexene, which, when a palladium catalyst is used, is oxidized to benzene, losing 2 moles of hydrogen.

Triple bonds can be reduced, either by catalytic hydrogenation or by the other methods mentioned. The comparative reactivity of triple and double bonds depends on the catalyst. With most catalysts, e.g., Pd, triple bonds are hydrogenated more easily, and therefore it is possible to add just 1 mole of hydrogen and reduce a triple bond to a double bond (usually a stereoselective syn addition) or to reduce a triple bond without affecting a double bond present in the same molecule.[278] A particularly good catalyst for this purpose is the Lindlar catalyst (Pd–CaCO$_3$–PbO).[279] Triple bonds can also be selectively reduced to double bonds with diisobutylaluminum hydride (DIBALH),[280] with tetramethyldihydrodisiloxane–HOAc and a Pd(0) catalyst,[281] with activated zinc (see **2-38**),[282] with a zinc–copper couple,[283] or (internal triple bonds only) with alkali metals (Na, Li) in liquid ammonia or a low-molecular-weight amine.[284] Terminal alkynes are not reduced by the Na–NH$_3$ procedure because they are converted to acetylide ions under these conditions. However, terminal triple bonds can be reduced to double bonds by the addition to the Na–NH$_3$ solution of (NH$_4$)$_2$SO$_4$, which liberates the free ethynyl group.[285]

An indirect method[286] of double-bond reduction involves hydrolysis of boranes (prepared

[272]For discussions of the mechanism of this reaction, see Snyder *J. Org. Chem.* **1967**, *32*, 3531; Borden *J. Am. Chem. Soc.* **1968**, *90*, 2197; Blunt; Hartshorn; Soong; Munro *Aust. J. Chem.* **1982**, *35*, 2519; Vincens; Fadel; Vidal *Bull. Soc. Chim. Fr.* **1987**, 462.

[273]Varma; Kabalka *Synth. Commun.* **1985**, *15*, 151.

[274]See for example Sato; Sato; Sato *J. Organomet. Chem.* **1976**, *122*, C25, **1977**, *131*, C26; Fujisawa; Sugimoto; Ohta *Chem. Lett.* **1976**, 581; Ashby; Lin *J. Org. Chem.* **1978**, *43*, 2567; Chung *J. Org. Chem.* **1979**, *44*, 1014. See also Osby; Heinzman; Ganem *J. Am. Chem. Soc.* **1986**, *108*, 67.

[275]There are some exceptions. See, for example, Butler *Synth. Commun.* **1977**, *7*, 441, and references cited therein.

[276]For a review of reductions of α,β-unsaturated carbonyl compounds with metals in liquid NH$_3$, see Caine, *Org. React.* **1976**, *23*, 1-258.

[277]For reviews, see Johnstone; Wilby; Entwistle *Chem. Rev.* **1985**, *85*, 129; Brieger; Nestrick *Chem. Rev.* **1974**, *74*, 567-580.

[278]For reviews of the hydrogenation of alkynes, see Hutchins; Hutchins, in Patai; Rappoport, Ref. 49, pt. 1, pp. 571-601; Marvell; Li *Synthesis* **1973**, 457-468; Gutmann; Lindlar, in Viehe, Ref. 70, pp. 355-363.

[279]Lindlar; Dubuis *Org. Synth. V*, 880. See also Rajaram; Narula; Chawla; Dev *Tetrahedron* **1983**, *39*, 2315; McEwen; Guttieri; Maier; Laine; Shvo *J. Org. Chem.* **1983**, *48*, 4436.

[280]Wilke; Müller *Chem. Ber.* **1956**, *89*, 444, *Liebigs Ann. Chem.* **1960**, *629*, 224; Gensler; Bruno *J. Org. Chem.* **1963**, *28*, 1254; Eisch; Kaska *J. Am. Chem. Soc.* **1966**, *88*, 2213. For a catalyst with even better selectivity for triple bonds, see Ulan; Maier; Smith *J. Org. Chem.* **1987**, *52*, 3132.

[281]Trost; Braslau *Tetrahedron Lett.* **1989**, *30*, 4657.

[282]Aerssens; van der Heiden; Heus; Brandsma *Synth. Commun.* **1990**, *20*, 3421; Chou; Clark; White *Tetrahedron Lett.* **1991**, *32*, 299.

[283]Sondengam; Charles; Akam *Tetrahedron Lett.* **1980**, *21*, 1069.

[284]For a list of methods of reducing triple to double bonds, with syn or anti addition, see Ref. 133, pp. 212-214.

[285]Henne; Greenlee *J. Am. Chem. Soc.* **1943**, *65*, 2020.

[286]For a review, see Zweifel *Intra-Sci. Chem. Rep.* **1973**, *7*(2), 181-189.

by **5-12**). Trialkylboranes can be hydrolyzed by refluxing with carboxylic acids,[287] while monoalkylboranes RBH_2 can be hydrolyzed with base.[288] Triple bonds can be similarly reduced, to cis olefins.[289]

Conjugated dienes can add hydrogen by 1,2 or 1,4 addition. Selective 1,4 addition can be achieved by hydrogenation in the presence of carbon monoxide, with bis(cyclopentadienyl)chromium as catalyst.[290] With allenes[291] catalytic hydrogenation usually reduces both double bonds, but reduction of just one double bond, to give an olefin, has been accomplished by treatment with $Na–NH_3$[292] or with DIBALH,[293] and by hydrogenation with $RhCl(PPh_3)_3$ as catalyst.[294]

Most catalytic reductions of double or triple bonds, whether heterogeneous or homogeneous, have been shown to be syn, with the hydrogens entering from the less-hindered side of the molecule.[295] Stereospecificity can be investigated only for tetrasubstituted olefins (except when the reagent is D_2), which are the hardest to hydrogenate, but the results of these investigations show that the addition is usually 80 to 100% syn, though some of the anti addition product is normally also found and in some cases predominates. Catalytic hydrogenation of alkynes is nearly always stereoselective, giving the cis olefin (usually at least 80%), even when it is thermodynamically less stable. For example, **35** gave **36**, even though the steric hindrance is such that a planar molecule is impossible.[296] This is thus a

useful method for preparing cis olefins.[297] However, when steric hindrance is too great, the trans olefin may be formed. One factor that complicates the study of the stereochemistry of heterogeneous catalytic hydrogenation is that exchange of hydrogens takes place, as can be shown by hydrogenation with deuterium.[298] Thus deuterogenation of ethylene produced all the possible deuterated ethylenes and ethanes (even C_2H_6), as well as HD.[299] With 2-butene, it was found that double-bond migration, cis–trans isomerization, and even exchange of hydrogen with groups not on the double bond could occur; e.g., $C_4H_2D_8$ and C_4HD_9 were detected on treatment of cis-2-butene with deuterium and a catalyst.[300] Indeed, *alkanes*

[287]Brown; Murray *Tetrahedron* **1986**, *42*, 5497; Kabalka; Newton; Jacobus *J. Org. Chem.* **1979**, *44*, 4185.
[288]Weinheimer; Marisco *J. Org. Chem.* **1962**, *27*, 1926.
[289]Brown; Zweifel *J. Am. Chem. Soc.* **1959**, *81*, 1512.
[290]Miyake; Kondo *Angew. Chem. Int. Ed. Engl.* **1968**, *7*, 631 [*Angew. Chem. 80*, 663]. For other methods, with references, see Ref. 133, p. 211.
[291]For a review, see Schuster; Coppola, Ref. 95, pp. 57-61.
[292]Gardner; Narayana *J. Org. Chem.* **1961**, *26*, 3518; Vaidyanathaswamy; Joshi; Devaprabhakara *Tetrahedron Lett.* **1971**, 2075.
[293]Montury; Goré *Tetrahedron Lett.* **1980**, *21*, 51.
[294]Bhagwat; Devaprabhakara *Tetrahedron Lett.* **1972**, 1391.
[295]For a review of homogeneous hydrogenation directed to only one face of a substrate molecule, see Brown *Angew. Chem. Int. Ed. Engl.* **1987**, *26*, 190-203 [*Angew. Chem. 99*, 169-182].
[296]Holme; Jones; Whiting *Chem. Ind. (London)* **1956**, 928.
[297]For a catalyst that leads to trans olefins, see Burch; Muetterties; Teller; Williams *J. Am. Chem. Soc.* **1982**, *104*, 4257.
[298]For a review of the use of deuterium to study the mechanism of heterogeneous organic catalysis, see Gudkov *Russ. Chem. Rev.* **1986**, *55*, 259-270.
[299]Turkevich; Schissler; Irsa *J. Phys. Chem.* **1951**, *55*, 1078.
[300]Wilson; Otvos; Stevenson; Wagner *Ind. Eng. Chem.* **1953**, *45*, 1480.

have been found to exchange with deuterium over a catalyst,[301] and even without deuterium, e.g., $CH_4 + CD_4 \rightarrow CHD_3 + CH_3D$ in the gas phase, with a catalyst. All this makes it difficult to investigate the stereochemistry of heterogeneous catalytic hydrogenation.

Catalytic hydrogenation of triple bonds and the reaction with DIBALH usually give the cis olefin. Most of the other methods of triple-bond reduction lead to the more thermodynamically stable trans olefin. However, this is not the case with the method involving hydrolysis of boranes or with the reductions with activated zinc, hydrazine, or NH_2OSO_3H, which also give the cis products.

The mechanism of the heterogenous catalytic hydrogenation of double bonds is not thoroughly understood because it is a very difficult reaction to study.[302] Because the reaction is heterogeneous, kinetic data, though easy to obtain (measurement of decreasing hydrogen pressure), are difficult to interpret. Furthermore, there are the difficulties caused by the aforementioned hydrogen exchange. The currently accepted mechanism for the common two-phase reaction was originally proposed in 1934.[303] According to this, the olefin is adsorbed onto the surface of the metal, though the nature of the actual bonding is unknown,[304]

$$CH_2{=}CH_2 \; \rightleftharpoons \; \underset{*\quad *}{CH_2{-}CH_2} \qquad H{-}H \; \rightleftharpoons \; \underset{*\;\;\; *}{H \;\; H}$$

$$\underset{*\quad *}{CH_2{-}CH_2} \; + \; \underset{*}{H} \; \rightleftharpoons \; \underset{*\quad *\quad *}{CH_3{-}CH_2}$$

$$\underset{*\quad *}{CH_3{-}CH_2} \; + \; \underset{*}{H} \; \rightleftharpoons \; \underset{*\quad *}{CH_3{-}CH_3}$$

despite many attempts to elucidate it.[305] The metallic site is usually indicated by an asterisk. For steric reasons it is apparent that adsorption of the olefin takes place with its less-hindered side attached to the catalyst surface. The fact that addition of hydrogen is generally also from the less-hindered side indicates that the hydrogen too is probably adsorbed on the catalyst surface before it reacts with the olefin. It is likely that the H_2 molecule is cleaved to hydrogen atoms in the act of being adsorbed. It has been shown that platinum catalyzes homolytic cleavage of hydrogen molecules.[306] In the second step one of the adsorbed hydrogen atoms becomes attached to a carbon atom, creating in effect, an alkyl radical (which is still bound to the catalyst though only by one bond) and two vacant catalyst sites. Finally, another hydrogen atom (not necessarily the one originally connected to the first hydrogen) combines with the radical to give the reaction product, freed from the catalyst surface, and two more vacant sites. All the various side reactions, including hydrogen exchange and isomerism, can be explained by this type of process. For example, Figure 15.1 shows the

[301]For a review, see Gudkov; Balandin *Russ. Chem. Rev.* **1966**, *35*, 756-761. For an example of intramolecular exchange, see Lebrilla; Maier *Tetrahedron Lett.* **1983**, *24*, 1119. See also Poretti; Gäumann *Helv. Chim. Acta* **1985**, *68*, 1160.

[302]For reviews, see Webb, in Bamford; Tipper *Comprehensive Chemical Kinetics*, vol. 20; Elsevier: New York, 1978, pp. 1-121; Clarke; Rooney *Adv. Catal.* **1976**, *25*, 125-183; Siegel *Adv. Catal.* **1966**, *16*, 123-177; Burwell *Chem. Eng. News* **1966**, *44*(34), 56-67.

[303]Horiuti; Polanyi *Trans. Faraday Soc.* **1934**, *30*, 1164.

[304]See, for example, Burwell; Schrage *J. Am. Chem. Soc.* **1965**, *87*, 5234.

[305]See, for example, McKee *J. Am. Chem. Soc.* **1962**, *84*, 1109; Ledoux *Nouv. J. Chim.* **1978**, *2*, 9; Bautista; Campelo; Garcia; Guardeño; Luna; Marinas *J. Chem. Soc., Perkin Trans. 2* **1989**, 493.

[306]Krasna *J. Am. Chem. Soc.* **1961**, *83*, 289.

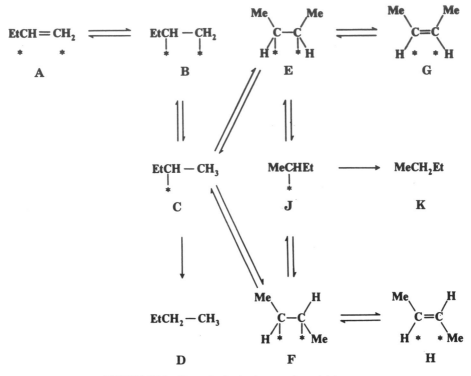

FIGURE 15.1 Steps in the hydrogenation of 1-butene.

steps that may be occurring in hydrogenation of 1-butene.[307] In this scheme the normal reaction is represented by **A → B → C → D,** double-bond migration by **A → B → C → E → G,** cis–trans isomerization by **H → F → C → E → G,** and hydrogen exchange by **A → B → C → E → J → K.** Although this mechanism is satisfactory as far as it goes,[308] there are still questions it does not answer, among them questions[309] involving the nature of the asterisk, the nature of the bonding, and the differences caused by the differing nature of each catalyst.[310]

The mechanism of homogeneous hydrogenation[311] catalyzed by $RhCl(PH_3P)_3$[312] involves

$$P-\underset{\underset{Cl}{|}}{\overset{\overset{P}{|}}{Rh}}-P \rightleftharpoons \underset{P}{\overset{P}{\diagdown}}Rh-Cl \xrightarrow{H_2} \underset{\underset{Cl}{|}}{\overset{\overset{P\diagdown\;H}{\diagup|}}{Rh}}-H \xrightarrow{C=C}$$

37

[307]Smith; Burwell *J. Am. Chem. Soc.* **1962,** *84,* 925.

[308]A different mechanism has been proposed by Zaera; Somorjai *J. Am. Chem. Soc.* **1984,** *106,* 2288, but there is evidence against it: Beebe; Yates *J. Am. Chem. Soc.* **1986,** *108,* 663. See alsoThomson; Webb *J. Chem. Soc., Chem. Commun.* **1976,** 526.

[309]For discussions, see Augustine; Yaghmaie; Van Peppen *J. Org. Chem.* **1984,** *49,* 1865; Maier *Angew. Chem. Int. Ed. Engl.* **1989,** *28,* 135-145 [*Angew. Chem. 101,* 135-146].

[310]For a study of the detailed structure of Lindlar catalysts (which were shown to consist of seven distinct chemical phases), see Schlögl; Noack; Zbinden; Reller *Helv. Chim. Acta* **1987,** *70,* 627.

[311]For reviews, see Crabtree *Organometallic Chemistry of the Transition Metals*; Wiley: New York, 1988, pp. 190-200; Jardine, in Hartley, Ref. 218, vol. 4, pp. 1049-1071.

[312]Osborn; Jardine; Young; Wilkinson, Ref. 224; Jardine; Osborn; Wilkinson *J. Chem. Soc. A* **1967,** 1574; Montelatici; van der Ent; Osborn; Wilkinson *J. Chem. Soc. A* **1968,** 1054; Wink; Ford *J. Am. Chem. Soc.* **1985,** *107,* 1794; Koga; Daniel; Han; Fu; Morokuma *J. Am. Chem. Soc.* **1987,** *109,* 3455.

38   P = PPh$_3$

reaction of the catalyst with hydrogen to form a metal hydride $(PPh_3)_2RhH_2Cl$ (**37**), which rapidly transfers two hydrogen atoms to the alkene. The intermediate **37** can be isolated. If a mixture of H_2 and D_2 is used, the product contains only dideuterated and nondeuterated compounds; no monodeuterated products are found, indicating that (unlike the case of heterogeneous catalysis) H_2 or D_2 has been added to one olefin molecule and that no exchange takes place.[312] Although conversion of **38** to the products takes place in two steps,[313] the addition of H_2 to the double bond is syn.

In the above-mentioned reactions with hydrazine and hydroxylamine, the actual reducing species is diimide NH=NH, which is formed from N_2H_4 by the oxidizing agent and from NH_2OH by the ethyl acetate.[314] Although both the syn and anti forms of diimide are produced, only the syn form reduces the double bond,[315] at least in part by a cyclic mechanism:[316]

The addition is therefore stereospecifically syn[317] and, like catalytic hydrogenation, generally takes place from the less-hindered side of a double bond, though not much discrimination in this respect is observed where the difference in bulk effects is small.[318] Diimide reductions are most successful with symmetrical multiple bonds (C=C, C≡C, N=N) and are not useful for those inherently polar (C≡N, C=N, C=O. etc.). Diimide is not stable enough for isolation at ordinary temperatures, though it has been prepared[319] as a yellow solid at −196°C.

When double bonds are reduced by lithium in ammonia or amines, the mechanism is similar to that of the Birch reduction (**5-10**).[320] The reduction with trifluoroacetic acid and Et$_3$SiH has an ionic mechanism, with H$^+$ coming in from the acid and H$^-$ from the silane.[248] In accord with this mechanism, the reaction can be applied only to those olefins which when

[313]Biellmann; Jung *J. Am. Chem. Soc.* **1968**, *90*, 1673; Hussey; Takeuchi *J. Am. Chem. Soc.* **1969**, *91*, 672; Heathcock; Poulter *Tetrahedron Lett.* **1969**, 2755; Smith; Shuford *Tetrahedron Lett.* **1970**, 525; Atkinson; Luke *Can. J. Chem.* **1970**, *48*, 3580.

[314]For reviews of hydrogenations with diimide, see Pasto; Taylor *Org. React.* **1991**, *40*, 91-155; Miller *J. Chem. Educ.* **1965**, *42*, 254-259; House, Ref. 144, pp. 248-256. For reviews of diimides, see Back *Rev. Chem. Intermed.* **1984**, *5*, 293-323; Hünig; Müller; Thier *Angew. Chem. Int. Ed. Engl.* **1965**, *4*, 271-280 [*Angew. Chem. 77*, 368-377].

[315]Aylward; Sawistowska *J. Chem. Soc.* **1964**, 1435.

[316]Ref. 249; van Tamelen; Dewey; Lease; Pirkle *J. Am. Chem. Soc.* **1961**, *83*, 4302; Willis; Back; Parsons; Purdon *J. Am. Chem. Soc.* **1977**, *99*, 4451.

[317]Corey; Pasto; Mock *J. Am. Chem. Soc.* **1961**, *83*, 2957.

[318]van Tamelen; Timmons *J. Am. Chem. Soc.* **1962**, *84*, 1067.

[319]Wiberg; Fischer; Bachhuber *Chem. Ber.* **1974**, *107*, 1456, *Angew. Chem. Int. Ed. Engl.* **1977**, *16*, 780 [*Angew. Chem. 89*, 828]. See also Trombetti *Can. J. Phys.* **1968**, *46*, 1005; Bondybey; Nibler *J. Chem. Phys.* **1973**, *58*, 2125; Craig; Kliewer; Shih *J. Am. Chem. Soc.* **1979**, *101*, 2480.

[320] For a review of the steric course of this reaction, see Toromanoff *Bull. Soc. Chim. Fr.* **1987**, 893-901. For a review of this reaction as applied to α,β-unsaturated ketones, see Russell, in Patai; Rappoport, Ref. 37, pt. 2, pp. 471-512.

protonated can form a tertiary carbocation or one stabilized in some other way, e.g., by α OR substitution.[321] It has been shown, by the detection of CIDNP, that reduction of α-methylstyrene by hydridopentacarbonylmanganese(I) $HMn(CO)_5$ involves free-radical addition.[322]

The occurrence of hydrogen exchange and double-bond migration in heterogeneous catalytic hydrogenation means that the hydrogenation does not necessarily take place by straightforward addition of two hydrogen atoms at the site of the original double bond. Consequently, this method is not synthetically useful for adding D_2 to a double or triple bond in a regioselective or stereospecific manner. However, this objective can be achieved (with syn addition) by a homogeneous catalytic hydrogenation, which usually adds D_2 without scrambling[323] or by the use of one of the diimide methods.[317] Deuterium can also be regioselectively added by the hydroboration-reduction procedure previously mentioned.

Reductions of double and triple bonds are found at OS **I**, 101, 311; **II**, 191, 491; **III**, 385, 586, 742, 794; **IV**, 136, 298, 302, 304, 408, 887; **V**, 16, 96, 277, 281, 993; **VI**, 68, 459; **VII**, 226, 287, 524; **68**, 64, 182.

Catalysts and apparatus for hydrogenation are found at OS **I**, 61, 463; **II**, 142; **III**, 176, 181, 685; **V**, 880.

5-10 Hydrogenation of Aromatic Rings

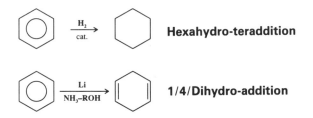

Aromatic rings can be reduced by catalytic hydrogenation,[324] but higher temperatures (100 to 200°C) are required than for ordinary double bonds.[325] Though the reaction is usually carried out with heterogeneous catalysts, homogeneous catalysts have also been used; conditions are much milder with these.[326] Mild conditions are also successful in hydrogenations with phase transfer catalysts.[327] Many functional groups, such as OH, O⁻, COOH, COOR, NH_2, etc., do not interfere with the reaction, but some groups may be preferentially reduced. Among these are CH_2OH groups, which undergo hydrogenolysis to CH_3 (**0-78**). Phenols may be reduced to cyclohexanones, presumably through the enol. Heterocyclic compounds are often reduced. Thus furan gives tetrahydrofuran. With benzene rings it is usually impossible to stop the reaction after only one or two bonds have been reduced, since olefins are more easily reduced than aromatic rings.[328] Thus, 1 mole of benzene, treated with 1

[321]Parnes; Bolestova; Kursanov *Bull. Acad. Sci. USSR, Div. Chem. Sci.* **1972**, *21*, 1927.

[322]Sweany; Halpern *J. Am. Chem. Soc.* **1977**, *99*, 8335. See also Thomas; Shackleton; Wright; Gillis; Colpa; Baird *J. Chem. Soc., Chem. Commun.* **1986**, 312; Garst; Bockman; Batlaw *J. Am. Chem. Soc.* **1986**, *108*, 1689; Bullock; Samsel *J. Am. Chem. Soc.* **1987**, *109*, 6542.

[323]Biellmann; Liesenfelt *Bull. Soc. Chim. Fr.* **1966**, 4029; Birch; Walker *Tetrahedron Lett.* **1966**, 4939, *J. Chem. Soc. C* **1966**, 1894; Morandi; Jensen *J. Org. Chem.* **1969**, *34*, 1889. See, however, Atkinson; Luke, Ref. 313.

[324]For reviews, see Karakhanov; Dedov; Loktev *Russ. Chem. Rev.* **1985**, *54*, 171-184; Weitkamp *Adv. Catal.* **1968**, *18*, 1-110 (for naphthalenes).

[325]For a highly active heterogeneous Rh catalyst, see Timmer; Thewissen; Meinema; Bulten *Recl. Trav. Chim. Pays-Bas* **1990**, *109*, 87.

[326]For reviews, see Bennett *CHEMTECH* **1980**, *10*, 444-446; Muetterties; Bleeke *Acc. Chem. Res.* **1979**, *12*, 324-331.

[327]Januszkiewicz; Alper *Organometallics* **1983**, *2*, 1055.

[328]For an indirect method of hydrogenating benzene to cyclohexene, see Harman; Taube *J. Am. Chem. Soc.* **1988**, *110*, 7906.

mole of hydrogen, gives no cyclohexadiene or cyclohexene but $\frac{1}{3}$ mole of cyclohexane and $\frac{2}{3}$ mole of recovered benzene. This is not true for all aromatic systems. With phenanthrene, for example, it is easy to stop after only the 9,10-bond has been reduced (see p. 43).

When aromatic rings are reduced by lithium (or potassium or sodium) in liquid ammonia (such reductions are known as *dissolving metal reductions*), usually in the presence of an alcohol (often ethyl, isopropyl, or *t*-butyl alcohol), 1,4 addition of hydrogen takes place and nonconjugated cyclohexadienes are produced.[329] This reaction is called the *Birch reduction*.[330] Ammonia obtained commercially often has iron salts as impurities that lower the yield in the Birch reduction. Therefore it is often necessary to distill the ammonia. When substituted aromatic compounds are subjected to the Birch reduction, electron-donating groups such as alkyl or alkoxyl decrease the rate of the reaction and are generally found on the nonreduced positions of the product. For example, anisole gives 1-methoxy-1,4-cyclohexadiene, not 3-methoxy-1,4-cyclohexadiene. On the other hand, electron-withdrawing groups such as COOH or CONH$_2$ increase the reaction rate and are found on the reduced positions of the product.[331] The mechanism involves solvated electrons,[332] which are transferred from the metal to the solvent and thence to the ring:[333]

The sodium becomes oxidized to Na$^+$ and creates a radical ion (**39**).[334] There is a great deal of evidence from esr spectra for these species.[335] The radical ion accepts a proton from the alcohol to give a radical, which is reduced to a carbanion by another sodium atom. Finally, **40** accepts another proton. Thus the function of the alcohol is to supply protons, since with most substrates ammonia is not acidic enough for this purpose. In the absence of the alcohol, products arising from dimerization of **39** are frequently obtained. There is evidence [336] at least with some substrates, e.g., biphenyl, that the radical ion corresponding to **39** is converted to the carbanion corresponding to **40** by a different pathway, in which the order of the steps is reversed: first a second electron is gained to give a dianion,[334] which then acquires a proton, producing the intermediate corresponding to **40**.

[329]For a procedure that converts benzene to pure 1,4-cyclohexadiene, see Brandsma; van Soolingen; Andringa *Synth. Commun.* **1990**, *20*, 2165.

[330]For a monograph, see Akhrem; Reshotova; Titov *Birch Reduction of Aromatic Compounds*; Plenum: New York, 1972. For reviews, see Rabideau *Tetrahedron* **1989**, *45*, 1579-1603; Birch; Subba Rao *Adv. Org. Chem.* **1972**, *8*, 1-65; Kaiser *Synthesis* **1972**, 391-415; Harvey *Synthesis* **1970**, 161-172; House, Ref. 144, pp. 145-150, 173-209; Hückel *Fortschr. Chem. Forsch* **1966**, *6*, 197-250; Smith, in Augustine *Reduction Techniques and Applications in Organic Synthesis*; Marcel Dekker: New York, 1968, pp. 95-170.

[331]These regioselectivities have generally been explained by molecular-orbital considerations regarding the intermediates involved. For example, see Birch; Hinde; Radom *J. Am. Chem. Soc.* **1980**, *102*, 3370, 4074, 6430; **1981**, *103*, 284; Zimmerman; Wang *J. Am. Chem. Soc.* **1990**, *112*, 1280. For methods of reversing the regioselectivities, see Epling; Florio *Tetrahedron Lett.* **1986**, *27*, 1469; Rabideau; Karrick *Tetrahedron Lett.* **1987**, *28*, 2481.

[332]For reviews of solvated electrons and related topics, see Dye *Prog. Inorg. Chem.* **1984**, *32*, 327-441; Alpatova; Krishtalik; Pleskov *Top. Curr. Chem.* **1987**, *138*, 149-219.

[333]Birch; Nasipuri *Tetrahedron* **1959**, *6*, 148.

[334]For a review of radical ions and diions generated from aromatic compounds, see Holy *Chem. Rev.* **1974**, *74*, 243-277.

[335]For example, see Jones, in Kaiser; Kevan *Radical Ions*; Wiley: New York, 1968, pp. 245-274; Bowers *Adv. Magn. Reson.* **1965**, *1*, 317-396; Carrington *Q. Rev., Chem. Soc.* **1963**, *17*, 67-99.

[336]Lindow; Cortez; Harvey *J. Am. Chem. Soc.* **1972**, *94*, 5406; Rabideau; Peters; Huser *J. Org. Chem.* **1981**, *46*, 1593.

Ordinary olefins are usually unaffected by Birch-reduction conditions, and double bonds may be present in the molecule if they are not conjugated with the ring. However, phenylated olefins, internal alkynes (p. 775), and conjugated olefins (with C=C or C=O) are reduced under these conditions.

It may be noted that **40** is a resonance hybrid; i.e., we can write two additional canonical forms:

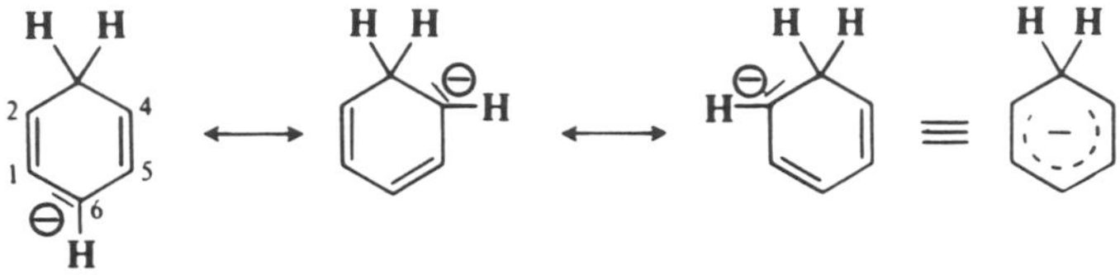

The question therefore arises: Why does the carbanion pick up a proton at the 6 position to give the 1,4-diene? Why not at the 2 position to give the 1,3-diene?[337] An answer to this question has been proposed by Hine, who has suggested that this case is an illustration of the operation of the *principle of least motion*.[338] According to this principle, "those elementary reactions will be favored that involve the least change in atomic position and electronic configuration."[338] The principle can be applied to the case at hand in the following manner (simplified): The valence-bond bond orders (p. 26) for the six carbon–carbon bonds (on the assumption that each of the three forms contributes equally) are (going around the ring) $1\frac{2}{3}$, 1, 1, $1\frac{2}{3}$, $1\frac{1}{3}$, and $1\frac{1}{3}$. When the carbanion is converted to the diene, these bond orders change as follows:

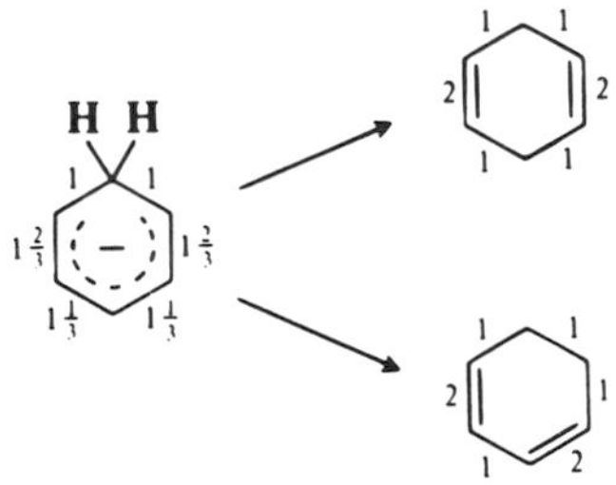

It can be seen that the two bonds whose bond order is 1 are unchanged in the two products, but for the other four bonds there is a change. If the 1,4-diene is formed, the change is $\frac{1}{3} + \frac{1}{3} + \frac{1}{3} + \frac{1}{3}$, while formation of the 1,3-diene requires a change of $\frac{1}{3} + \frac{2}{3} + \frac{2}{3} + \frac{1}{3}$. Since a greater change is required to form the 1,3-diene, the principle of least motion predicts formation of the 1,4-diene. This may not be the only factor, because the ^{13}C nmr spectrum of **40** shows that the 6 position has a somewhat greater electron density than the 2 position, which presumably would make the former more attractive to a proton.[339]

Reduction of aromatic rings with lithium[340] or calcium[341] in amines (instead of ammonia)

[337]For a discussion of this question, see Rabideau; Huser *J. Org. Chem.* **1983**, *48*, 4266.

[338]Hine *J. Org. Chem.* **1966**, *31*, 1236. For a review of this principle, see Hine *Adv. Phys. Org. Chem.* **1977**, *15*, 1-61. See also Tee *J. Am. Chem. Soc.* **1969**, *91*, 7144; Jochum; Gasteiger; Ugi *Angew. Chem. Int. Ed. Engl.* **1980**, *19*, 495-505 [*Angew. Chem. 92*, 503-513].

[339]Bates; Brenner; Cole; Davidson; Forsythe; McCombs; Roth *J. Am. Chem. Soc.* **1973**, *95*, 926.

[340]Benkeser; Robinson; Sauve; Thomas *J. Am. Chem. Soc.* **1955**, *77*, 3230; Reggel; Friedel; Wender *J. Org. Chem.* **1957**, *22*, 891; Benkeser; Agnihotri; Burrous; Kaiser; Mallan; Ryan *J. Org. Chem.* **1964**, *29*, 1313; Kwart; Conley *J. Org. Chem.* **1973**, *38*, 2011.

[341]Benkeser; Kang *J. Org. Chem.* **1979**, *44*, 3737; Benkeser; Belmonte; Kang *J. Org. Chem.* **1983**, *48*, 2796. See also Benkeser; Laugal; Rappa *Tetrahedron Lett.* **1984**, *25*, 2089.

proceeds further and cyclohexenes are obtained. It is thus possible to reduce a benzene ring, by proper choice of reagent, so that one, two, or all three double bonds are reduced.[342]

OS **I**, 99, 499; **II**, 566; **III**, 278, 742; **IV**, 313, 887, 903; **V**, 398, 400, 467, 591, 670, 743, 989; **VI**, 371, 395, 461, 731, 852, 856, 996; **VII**, 249.

5-11   Reductive Cleavage of Cyclopropanes

$$\triangle \xrightarrow[\text{cat.}]{\text{H}_2} CH_3CH_2CH_3$$

Cyclopropanes can be cleaved by catalytic hydrogenolysis.[343] Among the catalysts used have been Ni, Pd, and Pt. The reaction can often be run under mild conditions.[344] Certain cyclopropane rings, especially cyclopropyl ketones and aryl-substituted cyclopropanes,[345] can be reductively cleaved by an alkali metal (generally Na or Li) in liquid ammonia.[346]

F.   A Metal on the Other Side

5-12   Hydroboration

$$3\!-\!\overset{|}{C}\!=\!\overset{|}{C}\!- \; + \; BH_3 \; \longrightarrow \; \left(\!-\!\overset{\overset{\displaystyle H}{|}}{\underset{|}{C}}\!-\!\overset{|}{\underset{|}{C}}\!-\!\right)_{\!3}\!B$$

When olefins are treated with borane[347] in ether solvents, BH_3 adds across the double bond.[348] Borane cannot be prepared as a stable pure compound[349] (it dimerizes to diborane B_2H_6), but it is commercially available in the form of complexes with THF, Me_2S,[350] phosphines, or tertiary amines. The olefins can be treated with a solution of one of these complexes (THF–BH_3 reacts at 0°C and is the most convenient to use; R_3N–BH_3 generally require temperatures of about 100°C; however, the latter can be prepared as air-stable liquids or solids, while the former can only be used as relatively dilute solutions in THF and are decomposed by moisture in air) or with a mixture of $NaBH_4$ and BF_3 etherate, which generates borane in situ.[351] Ordinarily, the process cannot be stopped with the addition of one molecule of BH_3 because the resulting RBH_2 adds to another molecule of olefin to give R_2BH, which in turn adds to a third olefin molecule, so that the isolated product is a trialkylborane R_3B. The reaction can be performed on alkenes with one to four substituents,

---

[342]One, two, or all three double bonds of certain aromatic nitrogen heterocycles can be reduced with metallic hydrides such as $NaBH_4$ or $LiAlH_4$. For a review, see Keay *Adv. Heterocycl. Chem.* **1986**, *39*, 1-77.

[343]For reviews, see Charton, Ref. 115, pp. 588-592; Newham *Chem. Rev.* **1963**, *63*, 123-137; Rylander *Catalytic Hydrogenation over Platinum Metals*, Ref. 218, pp. 469-474.

[344]See, for example, Woodworth; Buss; Schleyer *Chem. Commun.* **1968**, 569.

[345]See, for example, Walborsky; Pierce *J. Org. Chem.* **1968**, *33*, 4102; Walborsky; Aronoff; Schulman *J. Org. Chem.* **1970**, *36*, 1036.

[346]For a review, see Staley *Sel. Org. Transform.* **1972**, *2*, 309-348.

[347]For a review of this reagent, see Lane, in Pizey, Ref. 146, vol. 3, 1977, pp. 1-191.

[348]For books on this reaction and its manifold applications, see Pelter; Smith; Brown *Borane Reagents*; Academic Press: New York, 1988; Brown *Boranes in Organic Chemistry*; Cornell University Press: Ithaca, NY, 1972, *Organic Syntheses Via Boranes*; Wiley: New York, 1975; Cragg *Organoboranes in Organic Synthesis*; Marcel Dekker: New York, 1973. For reviews, see Matteson, in Hartley, Ref. 218, vol. 4, pp. 307-409, pp. 315-337; Smith *Chem. Ind. (London)* **1987**, 603-611; Brown; Vara Prasad *Heterocycles* **1987**, *25*, 641-567; Suzuki; Dhillon *Top. Curr. Chem.* **1986**, *130*, 23-88.

[349]Mappes; Fehlner *J. Am. Chem. Soc.* **1970**, *92*, 1562; Fehlner *J. Am. Chem. Soc.* **1971**, *93*, 6366.

[350]For a review of $BH_3{\cdot}SMe_2$, see Hutchins; Cistone *Org. Prep. Proced. Int.* **1981**, *13*, 225-240.

[351]For a list of hydroborating reagents, with references, see Ref. 133, pp. 497-499.

including cyclic olefins, but when the olefin is moderately hindered, the product *is* the dialkylborane R_2BH or even the monalkylborane RBH_2.[352] For example, **41** (*disiamylborane*), **42** (*thexylborane*),[353] and **44** have been prepared in this manner. Monoalkylboranes

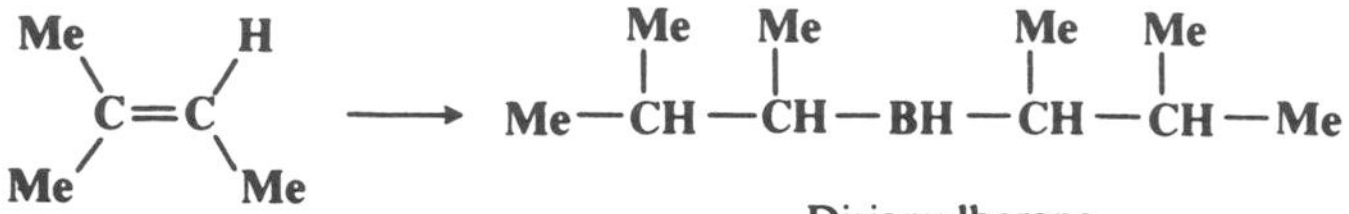

$$\underset{Me}{\overset{Me}{>}}C=C\underset{Me}{\overset{H}{<}} \longrightarrow Me-\underset{}{\overset{Me}{\underset{|}{CH}}}-\underset{}{\overset{Me}{\underset{|}{CH}}}-BH-\underset{}{\overset{Me}{\underset{|}{CH}}}-\underset{}{\overset{Me}{\underset{|}{CH}}}-Me$$

Disiamylborane

41

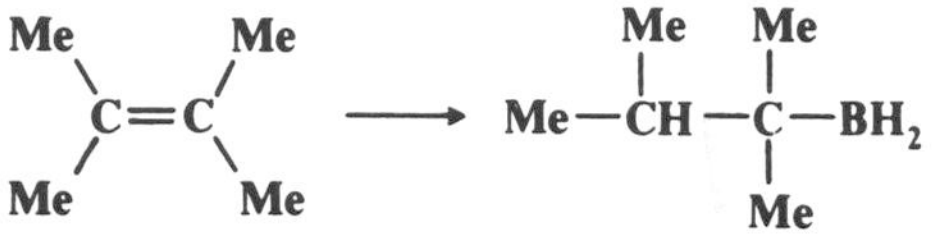

$$\underset{Me}{\overset{Me}{>}}C=C\underset{Me}{\overset{Me}{<}} \longrightarrow Me-\underset{}{\overset{Me}{\underset{|}{CH}}}-\underset{Me}{\overset{Me}{\underset{|}{C}}}-BH_2$$

Thexylborane

42

$$\underset{H}{\overset{Me_3C}{>}}C=C\underset{Me}{\overset{Me}{<}} \longrightarrow Me_3C-\underset{BH_2}{\overset{|}{CH}}-CHMe_2$$

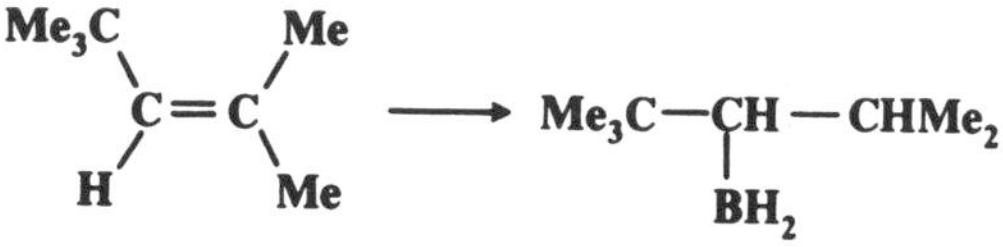

43

RBH_2 (which can be prepared from hindered olefins, as above) and dialkylboranes R_2BH also add to olefins, to give the mixed trialkylboranes $RR_2'B$ and $R_2R'B$, respectively. Surprisingly, when methylborane $MeBH_2$,[354] which is not a bulky molecule, adds to olefins in the solvent THF, the reaction can be stopped with one addition to give the dialkylboranes $RMeBH$.[355] Reaction of this with a second olefin produces the trialkylborane $RR'MeB$.[356] Other monoalkylboranes, i-$PrBH_2$, n-$BuBH_2$, s-$BuBH_2$, and t-$BuBH_2$, behave similarly with internal olefins, but not with olefins of the type $RCH=CH_2$.[357]

In all cases the boron goes to the side of the double bond that has more hydrogens, whether the substituents are aryl or alkyl.[358] Thus the reaction of **43** with BH_3 gives 98% **44** and only 2% of the other product. This actually follows Markovnikov's rule, since boron is more positive than hydrogen. However, the regioselectivity is caused mostly by steric factors, though electronic factors also play a part. Studies of the effect of ring substituents on rates and on the direction of attack in hydroboration of substituted styrenes showed that the attack by boron has electrophilic character.[359] When both sides of the double bond are

---

[352]Unless coordinated with a strong Lewis base such as a tertiary amine, mono and dialkylboranes actually exist

as dimers e.g., $R_2B\overset{\overset{\displaystyle H}{\diamond}}{\underset{\underset{\displaystyle H}{}}{}}BR_2$: Brown; Klender *Inorg. Chem.* **1962**, *1*, 204.

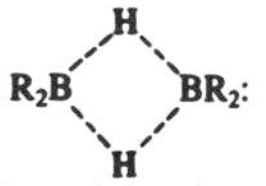

[353]For a review of the chemistry of thexylborane, see Negishi; Brown *Synthesis* **1974**, 77-89.
[354]Prepared from lithium methylborohydride and HCl: Brown; Cole; Srebnik; Kim *J. Org. Chem.* **1986**, *51*, 4925.
[355]Srebnik; Cole; Brown *Tetrahedron Lett.* **1987**, *28*, 3771, *J. Org. Chem.* **1990**, *55*, 5051.
[356]For a method of synthesis of RR'R"B, see Kulkarni; Basavaiah; Zaidlewicz; Brown *Organometallics* **1982**, *1*, 212.
[357]Srebnik; Cole; Ramachandran; Brown *J. Org. Chem.* **1989**, *54*, 6085.
[358]For a thorough discussion of the regioselectivity with various types of substrate and hydroborating agents, see Cragg, Ref. 348, pp.63-84, 137-197. See also Brown; Vara Prasad; Zee *J. Org. Chem.* **1986**, *51*, 439.
[359]Brown; Sharp *J. Am. Chem. Soc.* **1966**, *88*, 5851; Klein; Dunkelblum; Wolff *J. Organomet. Chem.* **1967**, *7*, 377. See also Marshall; Prager *Aust. J. Chem.* **1979**, *32*, 1251.

monosubstituted or both disubstituted, about equal amounts of each isomer are obtained. However, it is possible in such cases to make the addition regioselective by the use of a large attacking molecule. For example, treatment of iso-PrCH=CHMe with borane gave 57% of product with boron on the methyl-bearing carbon and 43% of the other, while treatment with **41** gave 95% **45** and only 5% of the other isomer.[360]

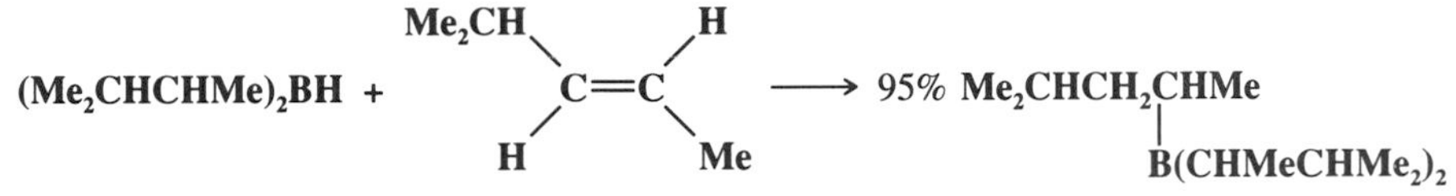

$$(Me_2CHCHMe)_2BH \; + \quad \underset{H}{\overset{Me_2CH}{\diagdown}}C=C\underset{Me}{\overset{H}{\diagup}} \quad \longrightarrow \quad 95\% \; \underset{B(CHMeCHMe_2)_2}{\overset{}{Me_2CHCH_2CHMe}}$$

$$\textbf{41} \qquad\qquad\qquad\qquad\qquad\qquad\qquad\qquad\qquad \textbf{45}$$

Another reagent with high regioselectivity is 9-borabicyclo[3.3.1]nonane (9-BBN), which is prepared by hydroboration of 1,5-cyclooctadiene:[361]

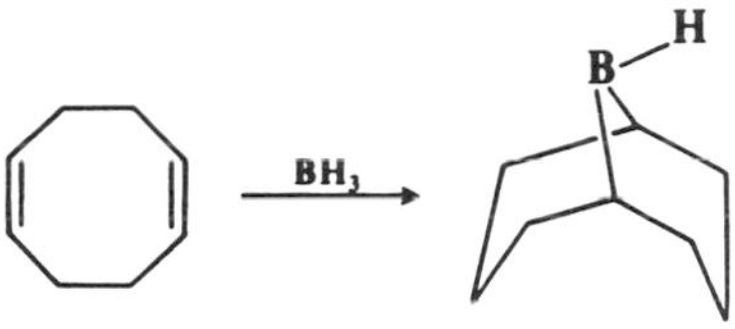

9-BBN

9-BBN has the advantage that it is stable in air. Borane is quite unselective and attacks all sorts of double bonds. Disiamylborane, 9-BBN, and similar molecules are far more selective and preferentially attack less-hindered bonds, so it is often possible to hydroborate one double bond in a molecule and leave others unaffected or to hydroborate one olefin in the presence of a less reactive olefin.[362] For example, 1-pentene can be removed from a mixture of 1- and 2-pentenes, and a cis olefin can be selectively hydroborated in a mixture of the cis and trans isomers.

A hydroboration reagent with greater regioselectivity than BH_3 (for terminal alkenes or those of the form $R_2C=CHR$) is monochloroborane[363] BH_2Cl coordinated with dimethyl sulfide (the hydroboration product is a dialkylchloroborane R_2BCl).[364] For example, 1-hexene gave 94% of the anti-Markovnikov product with BH_3-THF, but 99.2% with BH_2Cl-SMe_2. Treatment of alkenes with dichloroborane–dimethyl sulfide $BHCl_2$-SMe_2 in the presence of BF_3[365] or with BCl_3 and Me_3SiH[366] gives alkyldichloroboranes $RBCl_2$.

An important use of the hydroboration reaction is that alkylboranes, when oxidized with hydrogen peroxide and NaOH, are converted to alcohols (**2-28**). This is therefore an indirect way of adding H_2O across a double bond in an anti-Markovnikov manner. However, boranes undergo many other reactions as well. Among other things, they react with α-halo carbonyl compounds to give alkylated products (**0-99**), with α,β-unsaturated carbonyl compounds to give Michael-type addition of R and H (**5-19**), with CO to give alcohols and ketones (**8-24**

[360]Brown; Zweifel *J. Am. Chem. Soc.* **1961**, *83*, 1241.
[361]See Knights; Brown *J. Am. Chem. Soc.* **1968**, *90*, 5280, 5281; Brown; Chen *J. Org. Chem.* **1981**, *46*, 3978; Soderquist; Brown *J. Org. Chem.* **1981**, *46*, 4599.
[362]Brown; Moerikofer *J. Am. Chem. Soc.* **1963**, *85*, 2063; Zweifel; Brown *J. Am. Chem. Soc.* **1963**, *85*, 2066; Zweifel; Ayyangar; Brown *J. Am. Chem. Soc.* **1963**, *85*, 2072; Ref. 359.
[363]For a review of haloboranes, see Brown; Kulkarni *J. Organomet. Chem.* **1982**, *239*, 23-41.
[364]Brown; Ravindran; Kulkarni *J. Org. Chem.* **1979**, *44*, 2417.
[365]Brown; Ravindran; Kulkarni *J. Org. Chem.* **1980**, *45*, 384; Brown; Racherla *J. Org. Chem.* **1986**, *51*, 895.
[366]Soundararajan; Matteson *J. Org. Chem.* **1990**, *55*, 2274.

to **8-26**); they can be reduced with carboxylic acids, providing an indirect method for reduction of double bonds (**5-9**), or they can be oxidized with chromic acid or pyridinium chlorochromate to give ketones[367] or aldehydes (from terminal olefins),[368] dimerized with silver nitrate and NaOH (**4-34**), isomerized (**8-11**), or converted to amines (**2-31**) or halides (**2-30**). They are thus useful intermediates for the preparation of a wide variety of compounds.

Such functional groups as OR, OH, NH_2, SMe, halogen, and COOR may be present in the molecule,[369] but not groups that are reducible by borane. Hydroboration of enamines with 9-BBN provides an indirect method for reducing an aldehyde or ketone to an alkene, e.g.,[370]

$$R^1CH_2 - \underset{\underset{O}{\parallel}}{C} - R^2 \xrightarrow{\text{6-14}} R^1CH = \underset{\underset{NR^3}{|}}{C} - R^2 \xrightarrow[\text{2. MeOH}]{\text{1. 9-BBN}} R^1CH = \underset{\underset{H}{|}}{C} - R^2 + R^3N - \underset{\underset{|}{|}}{B} -$$

Use of the reagent diisopinocampheylborane **46** (prepared by treating optically active α-pinene with BH_3) results in enantioselective hydroboration–oxidation.[371] Alcohols with op-

α-Pinene
Optically active

46

Optically active

tical purities as high as 98% have been obtained in this way.[372] However, **46** does not give good results with even moderately hindered alkenes; a better reagent for these compounds is isopinocampheylborane[373] though optical yields are lower. Limonylborane,[374] 2- and 4-dicaranylboranes,[375] a myrtanylborane,[376] and dilongifolylborane[377] have also been used. The method has been improved[378] by synthesizing the chiral isopinocampheylborane in the presence of tetramethylenediamine (TMED), whereupon a TMED–isopinocampheylborane adduct is formed. This adduct,[379] in Et_2O, reacts with a prochiral alkene to give a dialkylborane RBHR′ (R′ = isocampheyl). The RBHR′ crystallizes from THF in 99–100% optical

[367]Brown; Garg *J. Am. Chem. Soc.* **1961**, *83*, 2951; *Tetrahedron* **1986**, *42*, 5511; Rao; Devaprabhakara; Chandrasekaran *J. Organomet. Chem.* **1978**, 162, C9; Parish; Parish; Honda *Synth. Commun.* **1990**, *20*, 3265.

[368]Brown; Kulkarni; Rao; Patil *Tetrahedron* **1986**, *42*, 5515.

[369]See, for example, Brown; Unni *J. Am. Chem. Soc.* **1968**, *90*, 2902; Brown; Gallivan *J. Am. Chem. Soc.* **1968**, *90*, 2906; Brown; Sharp *J. Am. Chem. Soc.* **1968**, *90*, 2915.

[370]Singaram; Rangaishenvi; Brown; Goralski; Hasha *J. Org. Chem.* **1991**, *56*, 1543.

[371]Brown; Ayyangar; Zweifel *J. Am. Chem. Soc.* **1964**, *86*, 397; Brown; Singaram *J. Org. Chem.* **1984**, *49*, 945; Brown; Vara Prasad *J. Am. Chem. Soc.* **1986**, *108*, 2049.

[372]For reviews of enantioselective syntheses with organoboranes, see Brown *Chemtracts: Org. Chem.* **1988**, *1*, 77-88; Brown; Singaram *Acc. Chem. Res.* **1988**, *21*, 287-293, *Pure Appl. Chem.* **1987**, *59*, 879-894; Srebnik; Ramachandran *Aldrichimica Acta* **1987**, *20*, 9-24; Matteson, Ref. 348, pp. 381-395; Brown; Jadhav; Singaram *Mod. Synth. Methods* **1986**, *4*, 307-356; Matteson *Synthesis* **1986**, 973-985; Brown; Jadhav, in Morrison, Ref. 232, vol. 2, 1983, pp. 1-43; Brown; Jadhav; Mandal *Tetrahedron* **1981**, *37*, 3547-3587.

[373]Brown; Jadhav; Mandal *J. Org. Chem.* **1982**, *47*, 5074. See also Brown; Weissman; Perumal; Dhokte *J. Org. Chem.* **1990**, *55*, 1217.

[374]Jadhav; Kulkarni *Heterocycles* **1982**, *18*, 169.

[375]Brown; Vara Prasad; Zaidlewicz *J. Org. Chem.* **1988**, *53*, 2911.

[376]Kiesgen de Richter; Bonato; Follet; Kamenka *J. Org. Chem.* **1990**, *55*, 2855.

[377]Jadhav; Brown *J. Org. Chem.* **1981**, *46*, 2988.

[378]Brown; Singaram *J. Am. Chem. Soc.* **1984**, *106*, 1797; Brown; Gupta; Vara Prasad *Bull. Chem. Soc. Jpn.* **1988**, *61*, 93.

[379]For the crystal structure of this adduct, see Soderquist; Hwang-Lee; Barnes *Tetrahedron Lett.* **1988**, *29*, 3385.

purity (the other diastereomer remains in solution). The optically pure RBHR′ is treated with acetaldehyde to produce α-pinene and optically pure R_2BH, which can be converted to optically pure alcohols or to other products.[380] Since both (+) and (−) α-pinene are readily available, both enantiomers can be prepared. The chiral cyclic boranes *trans*-2,5-dimethylborolanes (**47** and **48**) also add enantioselectively to olefins (except olefins of the

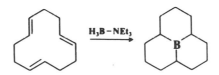

47   **48**

(*R*, *R*)   (*S*, *S*)

form RR′C=CH₂) to give boranes of high optical purity.[381] When chiral boranes are added to trisubstituted olefins of the form RR′C=CHR″, two new chiral centers are created, and, with **47** or **48**, only one of the four possible diastereomers is predominantly produced, in yields greater than 90%.[381] This has been called *double-asymmetric synthesis*.[382]

The double bonds in a conjugated diene are hydroborated separately, i.e., there is no 1,4 addition. However, it is not easy to hydroborate just one of a conjugated system, since conjugated double bonds are less reactive than isolated ones. Thexylborane[353] (**42**) is particularly useful for achieving the cyclic hydroboration of dienes, conjugated or nonconjugated,[383] e.g.,

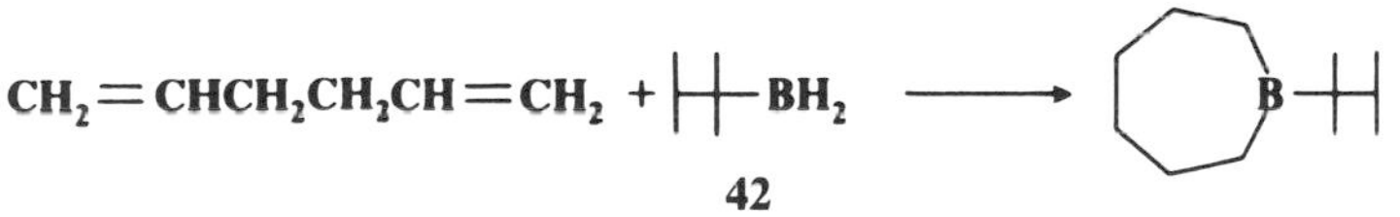

$$CH_2=CHCH_2CH_2CH=CH_2 + \text{⊢BH}_2 \longrightarrow$$

42

Rings of five, six, or seven members can be formed in this way. Similar cyclization can also be accomplished with other monoalkylboranes and, in some instances, with BH_3 itself.[384] One example is the formation of 9-BBN, shown above. Another is conversion of 1,5,9-cyclododecatriene to perhydro-9*b*-boraphenalene:[385]

$$\xrightarrow{\text{H}_3\text{B}-\text{NEt}_3}$$

Triple bonds[386] can be monohydroborated to give vinylic boranes, which can be reduced with carboxylic acids to cis alkenes or oxidized and hydrolyzed to aldehydes or ketones. Terminal alkynes give aldehydes by this method, in contrast to the mercuric or acid-catalyzed addition of water discussed at **5-3**. However, terminal alkynes give vinylic boranes[387] (and

[380]For another method of preparing optically pure mono- and dialkylboranes, see Brown; Singaram; Cole *J. Am. Chem. Soc.* **1985**, *107*, 460.
[381]Masamune; Kim; Petersen; Sato; Veenstra; Imai *J. Am. Chem. Soc.* **1985**, *107*, 4549.
[382]For another enantioselective hydroboration method, see p. 788.
[383]Brown; Pfaffenberger *J. Am. Chem. Soc.* **1967**, *89*, 5475; Brown; Negishi *J. Am. Chem. Soc.* **1972**, *94*, 3567.
[384]For a review of cyclic hydroboration, see Brown; Negishi *Tetrahedron* **1977**, *33*, 2331-2357. See also Brown; Pai; Naik *J. Org. Chem.* **1984**, *49*, 1072.
[385]Rotermund; Köster *Liebigs Ann. Chem.* **1965**, *686*, 153; Brown; Negishi; Dickason *J. Org.Chem.* **1985**, *50*, 520.
[386]For a review of hydroboration of triple bonds, see Hudrlik; Hudrlik, in Patai, Ref. 70, pt. 1, pp. 203-219.
[387]For a review of the preparation and reactions of vinylic boranes, see Brown; Campbell *Aldrichimica Acta* **1981**, *14*, 1-11.

hence aldehydes) only when treated with a hindered borane such as **41, 42,** or catechol-borane (p. 615),[388] or with $BHBr_2$–SMe_2.[389] The reaction between terminal alkynes and BH_3 produces 1,1-dibora compounds, which can be oxidized either to primary alcohols (with

$$RC{\equiv}CH \xrightarrow{BH_3} RCH_2-\underset{\underset{\displaystyle -B}{|}}{CH}-B \underset{\xrightarrow{m\text{-}ClC_6H_4COOH}}{\overset{\xrightarrow{H_2O_2/NaOH}}{}} \begin{array}{l} RCH_2CH_2OH \\[2em] RCH_2COOH \end{array}$$

$NaOH$–H_2O_2) or to carboxylic acids (with m-chloroperbenzoic acid).[390] Double bonds can be hydroborated in the presence of triple bonds if the reagent is 9-BBN.[391] On the other hand, dimesitylborane selectively hydroborates triple bonds in the presence of double bonds.[392] Furthermore, it is often possible to hydroborate selectively one particular double bond of a nonconjugated diene.[393] When the reagent is catecholborane, hydroboration is catalyzed by rhodium complexes, such as Wilkinson's catalyst.[394] Enantioselective hydro-boration-oxidation has been achieved by the use of optically active rhodium complexes.[395]

For most substrates, the addition in hydroboration is stereospecific and syn, with attack taking place from the less-hindered side.[396] The mechanism[397] may be a cyclic four-center one:[398]

$$-\underset{|}{\overset{|}{C}}=\underset{|}{\overset{|}{C}}- \longrightarrow -\underset{|}{\overset{|}{C}}-\underset{|}{\overset{|}{C}}-$$
$$-\underset{|}{B}-H \qquad\qquad -\underset{|}{B}\quad H$$

When the substrate is an allylic alcohol or amine, the addition is generally anti,[399] though the stereoselectivity can be changed to syn by the use of catecholborane and the rhodium complexes mentioned above.[400] Because the mechanism is different, use of this pro-

[388]Brown; Gupta *J. Am. Chem. Soc.* **1972,** *94,* 4370, **1975,** *97,* 5249. For a review of catecholborane, see Lane; Kabalka *Tetrahedron* **1976,** *32,* 981-990.

[389]Brown; Campbell *J. Org. Chem.* **1980,** *45,* 389.

[390]Zweifel; Arzoumanian *J. Am. Chem. Soc.* **1967,** *89,* 291.

[391]Brown; Coleman *J. Org. Chem.* **1979,** *44,* 2328.

[392]Pelter; Singaram; Brown *Tetrahedron Lett.* **1983,** *24,* 1433.

[393]For a list of references, see Gautam; Singh; Dhillon *J. Org. Chem.* **1988,** *53,* 187. See also Suzuki; Dhillon, Ref. 348.

[394]Männig; Nöth *Angew. Chem. Int. Ed. Engl.* **1985,** *24,* 878 [*Angew. Chem. 97,* 854]. For a review, see Burgess; Ohlmeyer *Chem. Rev.* **1991,** *91,* 1179-1191.

[395]Burgess; Ohlmeyer *J. Org. Chem.* **1988,** *53,* 5178; Hayashi; Matsumoto; Ito *J. Am. Chem. Soc.* **1989,** *111,* 3426; Sato; Miyaura; Suzuki *Tetrahedron Lett.* **1990,** *31,* 231; Brown; Lloyd-Jones *Tetrahedron: Assymmetry* **1990,** *1,* 869.

[396]Kabalka; Bowman *J. Org. Chem.* **1973,** *38,* 1607; Brown; Zweifel *J. Am. Chem. Soc.* **1961,** *83,* 2544; Bergbreiter; Rainville *J. Org. Chem.* **1976,** *41,* 3031; Kabalka; Newton; Jacobus *J. Org. Chem.* **1978,** *43,* 1567.

[397]For kinetic studies, see Wang; Brown *J. Org. Chem.* **1980,** *45,* 5303, *J. Am. Chem. Soc.* **1982,** *104,* 7148; Vishwakarma; Fry *J. Org. Chem.* **1980,** *45,* 5306; Brown; Chandrasekharan; Wang *J. Org. Chem.* **1983,** *48,* 2901, *Pure Appl. Chem.* **1983,** *55,* 1387-1414; Chandrasekharan; Brown *J. Org. Chem.* **1985,** *50,* 518; Nelson; Cooper *Tetrahedron Lett.* **1986,** *27,* 4693; Brown; Chandrasekharan *J. Org. Chem.* **1988,** *53,* 4811.

[398]Brown; Zweifel *J. Am. Chem. Soc.* **1959,** *81,* 247; Pasto; Lepeska; Balasubramaniyan *J. Am. Chem. Soc.* **1972,** *94,* 6090; Pasto; Lepeska; Cheng *J. Am. Chem. Soc.* **1972,** *94,* 6083; Narayana; Periasamy *J. Chem. Soc., Chem. Commun.* **1987,** 1857. See, however, Jones *J. Org. Chem.* **1972,** *37,* 1886.

[399]See Still; Barrish *J. Am. Chem. Soc.* **1983,** *105,* 2487.

[400]See Evans; Fu; Hoveyda *J. Am. Chem. Soc.* **1988,** *110,* 6917; Burgess; Cassidy; Ohlmeyer *J. Org. Chem.* **1991,** *56,* 1020; Burgess; Ohlmeyer *J. Org. Chem.* **1991,** *56,* 1027.

cedure can result in a change in regioselectivity as well, e.g., styrene $PhCH{=}CH_2$ gave $PhCH(OH)CH_3$.[401]

OS **VI**, 719, 852, 919, 943; **VII**, 164, 339, 402, 427; **68**, 130.

5-13 Other Hydrometalation
Hydro-metallo-addition

$$
\underset{\displaystyle |}{\overset{\displaystyle |}{-}}C{=}C\underset{\displaystyle |}{\overset{\displaystyle |}{-}} + HM \longrightarrow \underset{\displaystyle |}{\overset{\displaystyle H}{\overset{|}{-C}}}\underset{\displaystyle |}{\overset{\displaystyle M}{\overset{|}{C-}}}
$$

Metal hydrides of groups 13 and 14 of the periodic table (e.g., AlH_3, GaH_3) as well as many of their alkyl and aryl derivatives (e.g., R_2AlH, Ar_3SnH) add to double bonds to give organometallic compounds.[402] The hydroboration reaction (**5-12**) is the most important example, but other important metals in this reaction are aluminum,[403] silicon, tin,[404] and zirconium[405] (a group 4 metal). Some of these reactions are uncatalyzed, but in other cases various types of catalyst have been used.[406] Hydrozirconation is most commonly carried out with Cp_2ZrHCl (Cp = cyclopentadienyl),[407] known as *Schwartz's reagent*. The mechanism with group 13 hydrides seems to be electrophilic (or four-centered pericyclic with some electrophilic characteristics) while with group 14 hydrides a mechanism involving free radicals seems more likely. Dialkylmagnesiums have been obtained by adding MgH_2 to double bonds.[408] RMgX can be added to an alkene $R'CH{=}CH_2$ to give $R'CH_2CH_2MgX$, with $TiCl_4$ as a catalyst (see also **8-12**).[409] With some reagents triple bonds[410] can add 1 or 2 moles, e.g.,[411]

$$
RC{\equiv}CH \xrightarrow{\ R'_2AlH\ } RCH{=}CH{-}AlR'_2 \xrightarrow{\ R'_2AlH\ } RCH_2\underset{\displaystyle \underset{|}{AlR'_2}}{CH}{-}AlR'_2
$$

[401]Hayashi; Matsumoto; Ito, Ref. 395; Zhang; Lou; Guo; Dai *J. Org. Chem.* **1991**, *56*, 1670.

[402]Negishi *Adv. Met.-Org. Chem.* **1989**, *1*, 177-207; Eisch *The Chemistry of Organometallic Compounds*; Macmillan: New York, 1967, pp. 107-111. See also Eisch; Fichter *J. Organomet. Chem.* **1983**, *250*, 63.

[403]For reviews of organoaluminums in organic synthesis, see Dzhemilev; Vostrikova; Tolstikov *Russ. Chem. Rev.* **1990**, *59*, 1157-1173; Maruoka; Yamamoto *Tetrahedron* **1988**, *44*, 5001-5032.

[404]For a review with respect to Al, Si, and Sn, see Negishi *Organometallics in Organic Synthesis*, vol. 1; Wiley: New York, 1980, pp. 45-48, 357-363, 406-412. For reviews of hydrosilylation, see Ojima, in Patai; Rappoport *The Chemistry of Organic Silicon Compounds*, pt. 2; Wiley: New York, 1989, pp. 1479-1526; Alberti; Pedulli *Rev. Chem. Intermed.* **1987**, *8*, 207-246; Speier *Adv. Organomet. Chem.* **1979**, *17*, 407-447; Andrianov; Souček; Khananashvili *Russ. Chem. Rev.* **1979**, *48*, 657-668.

[405]For reviews of hydrozirconation, and the uses of organozirconium compounds, see Negishi; Takahashi *Synthesis* **1988**, 1-19; Dzhemilev; Vostrikova; Tolstikov *J. Organomet. Chem.* **1986**, *304*, 17-39; Schwartz; Labinger *Angew. Chem. Int. Ed. Engl.* **1976**, *15*, 333-340 [*Angew. Chem.* **88**, 402-409].

[406]See, for example, Oertle; Wetter *Tetrahedron Lett.***1985**, *26*, 5511; Randolph; Wrighton *J. Am. Chem. Soc.* **1986**, *108*, 3366; Maruoka; Sano; Shinoda; Nakai; Yamamoto *J. Am. Chem. Soc.* **1986**, *108*, 6036; Miyake; Yamamura *Chem. Lett.* **1989**, 981; Doyle; High; Nesloney; Clayton; Lin *Organometallics* **1991**, *10*, 1225.

[407]For a method of preparing this reagent (which is also available commercially), see Buchwald; LaMaire; Nielsen; Watson; King *Tetrahedron Lett.* **1987**, *28*, 3895. It can also be generated in situ: Lipshutz; Keil; Ellsworth *Tetrahedron Lett.* **1990**, *31*, 7257.

[408]For a review, see Bogdanović *Angew. Chem. Int. Ed. Engl.* **1985**, *24*, 262-273 [*Angew. Chem.* **97**, 253-264].

[409]For a review, see Sato *J. Organomet. Chem.* **1985**, *285*, 53-64. For another catalyst, see Hoveyda; Xu *J. Am. Chem. Soc.* **1991**, *113*, 5079.

[410]For a review of the hydrometalation of triple bonds, see Ref. 386, pp. 219-232.

[411]Wilke; Müller *Liebigs Ann. Chem.* **1960**, *629*, 222; Eisch; Kaska *J. Am. Chem. Soc.* **1966**, *88*, 2213; Eisch; Rhee *Liebigs Ann. Chem.* **1975**, 565.

When 2 moles are added, electrophilic addition generally gives 1,1-dimetallic products (as with hydroboration), while free-radical addition usually gives the 1,2-dimetallic products.

OS **VII**, 456; **66**, 60; **67**, 86; **69**, 106. See also OS **66**, 43, 75.

G. Carbon on the Other Side

5-14 Addition of Alkanes
Hydro-alkyl-addition

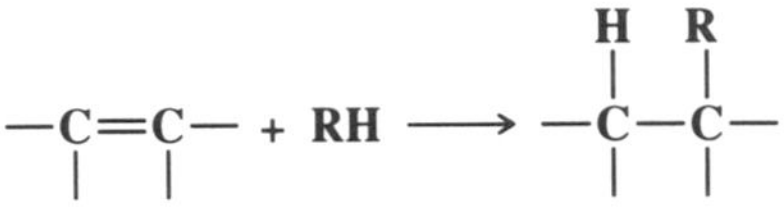

$$-C=C- + RH \longrightarrow -\overset{\overset{\displaystyle H}{|}}{C}-\overset{\overset{\displaystyle R}{|}}{C}-$$

There are two important ways of adding alkanes to olefins—the thermal method and the acid-catalysis method.[412] Both give chiefly mixtures, and neither is useful for the preparation of relatively pure compounds in reasonable yields. However, both are useful industrially. In the thermal method the reactants are heated to high temperatures (about 500°C) at high pressures (150 to 300 atm) without a catalyst. As an example, propane and ethylene gave 55.5% isopentane, 7.3% hexanes, 10.1% heptanes, and 7.4% alkenes.[413] The mechanism is undoubtedly of a free-radical type and can be illustrated by one possible sequence in the reaction between propane and ethylene:

Step 1 $\qquad CH_3CH_2CH_3 + CH_2{=}CH_2 \overset{\Delta}{\longrightarrow} CH_3{-}\overset{\bullet}{C}H{-}CH_3 + CH_3CH_2\bullet$

Step 2 $\qquad CH_3\overset{\bullet}{C}HCH_3 + CH_2{=}CH_2 \longrightarrow (CH_3)_2CHCH_2CH_2\bullet$

Step 3 $\quad (CH_3)_2CHCH_2CH_2\bullet + CH_3CH_2CH_3 \longrightarrow (CH_3)_2CHCH_2CH_3 + CH_3\overset{\bullet}{C}HCH_3$

There is kinetic evidence that the initiation takes place primarily by steps like 1, which are called *symproportionation* steps[414] (the opposite of disproportionation, p. 194).

In the acid-catalysis method, a proton or Lewis acid is used as the catalyst and the reaction is carried out at temperatures between -30 and 100°C. This is a Friedel–Crafts process with a carbocation mechanism[415] (illustrated for a proton acid catalyst):

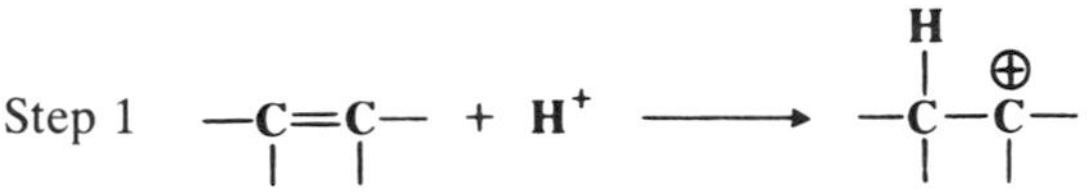

Step 1 $\quad -C=C- + H^+ \longrightarrow -\overset{\overset{\displaystyle H}{|}}{C}-\overset{\overset{\displaystyle \oplus}{|}}{C}-$

49

Step 2 $\quad -\overset{\overset{\displaystyle H}{|}}{C}-\overset{\overset{\displaystyle \oplus}{|}}{C}- + RH \longrightarrow -\overset{\overset{\displaystyle H}{|}}{C}-\overset{\overset{\displaystyle H}{|}}{C}- + R^+ \quad (H^- \text{ abstraction})$

Step 3 $\quad R^+ + -C=C- \longrightarrow -\overset{\overset{\displaystyle R}{|}}{C}-\overset{\overset{\displaystyle \oplus}{|}}{C}-$

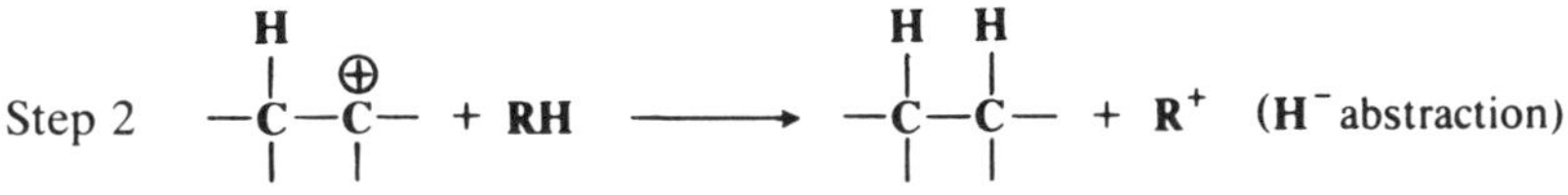

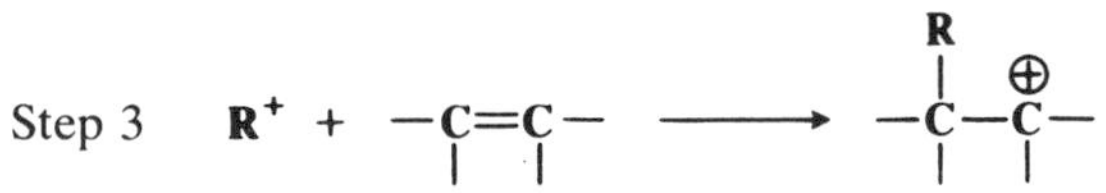

[412]For reviews, see Shuikin; Lebedev *Russ. Chem. Rev.* **1966**, *35*, 448-455; Schmerling, in Olah *Friedel-Crafts and Related Reactions*, vol. 2; Wiley: New York, 1964, pp. 1075-1111, 1121-1122.
[413]Frey; Hepp *Ind. Eng. Chem.* **1936**, *28*, 1439.
[414]Metzger *Angew. Chem. Int. Ed. Engl.* **1983**, *22*, 889 [*Angew. Chem. 95*, 914]; Hartmanns; Klenke; Metzger *Chem. Ber.* **1986**, *119*, 488.
[415]For a review, see Mayr *Angew. Chem. Int. Ed. Engl.* **1990**, *29*, 1371-1384 [*Angew. Chem. 102*, 1415-1428].

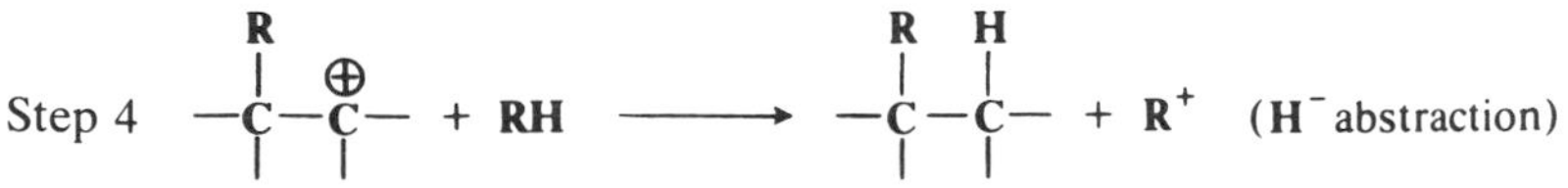

Step 4
$$\underset{\mathbf{50}}{-\overset{\mathbf{R}}{\underset{|}{C}}-\overset{\oplus}{\underset{|}{C}}-} + \mathbf{RH} \longrightarrow -\overset{\mathbf{R}}{\underset{|}{C}}-\overset{\mathbf{H}}{\underset{|}{C}}- + \mathbf{R^+} \quad (\mathrm{H^-\ abstraction})$$

50 often rearranges before it abstracts a hydride ion, explaining, for example, why the principal product from the reaction between isobutane and ethylene is 2,3-dimethylbutane. It is also possible for **49** (or **50**) instead of abstracting a hydride ion, to add to another mole of olefin, so that not only rearrangement products but also dimeric and polymeric products are frequent. If the tri- or tetrasubstituted olefins are treated with Me_4Si, HCl, and $AlCl_3$, they become protonated to give a tertiary carbocation, which reacts with the Me_4Si to give a product that is the result of addition of H and Me to the original alkene.[416] (For a free-radical hydro-methyl-addition, see **5-20**.)

The addition of secondary or tertiary cations (generated from the corresponding alcohols, esters, or alkenes) to 1,1-dichloroethene gives carboxylic acids by hydrolysis of the intermediate ions (see **0-3**):[417]

$$\mathbf{R_3C^+ + CH_2{=}CCl_2 \longrightarrow R_3C{-}CH_2{-}\overset{\oplus}{C}Cl_2 \xrightarrow{\ H_2O\ } R_3CCH_2COOH}$$

The reaction can also be base-catalyzed, in which case there is nucleophilic addition and a carbanion mechanism.[418] Carbanions most often used are those stabilized by one or more α-aryl groups. For example, toluene adds to styrene in the presence of sodium to give 1,3-diphenylpropane:[419]

$$\mathbf{PhCH_3 \xrightarrow{\ Na\ } Ph\bar{C}H_2{}^{\ominus} + PhCH{=}CH_2 \longrightarrow}$$

$$\mathbf{Ph\bar{C}H{-}CH_2CH_2Ph \xrightarrow{\ solvent\ } PhCH_2CH_2CH_2Ph}$$

Conjugated dienes give 1,4 addition.[420] This reaction has also been performed with salts of carboxylic acids in what amounts to a method of alkylation of carboxylic acids[421] (see also **0-96**).

$$\mathbf{CH_3COOK \xrightarrow{\ NaNH_2\ } {}^{\ominus}CH_2COOK + CH_2{=}CH_2 \longrightarrow {}^{\ominus}CH_2{-}CH_2CH_2COOK}$$

OS **I**, 229; **IV**, 665; **VII**, 479.

5-15 Addition of Alkenes and/or Alkynes to Alkenes and/or Alkynes
Hydro-alkenyl-addition

$$\mathbf{CH_2{=}CH_2 + CH_2{=}CH_2 \xrightarrow{\ H^+\ } CH_2{=}CHCH_2CH_3}$$

[416]Bolestova; Parnes; Kursanov *J. Org. Chem. USSR* **1983**, *19*, 2175.
[417]For reviews, see Bott *Angew. Chem. Int. Ed. Engl.* **1980**, *19*, 171-178 [*Angew. Chem. 92*, 169-176]; Bott; Hellmann *Angew. Chem. Int. Ed. Engl.* **1966**, *5*, 870-874 [*Angew. Chem. 78*, 932-936], *Newer Methods Prep. Org. Chem.* **1971**, *6*, 67-80.
[418]For reviews, see Pines; Stalick, Ref. 198, pp. 240-422; Pines *Acc. Chem. Res.* **1974**, *7*, 155-162; Pines; Schaap *Adv. Catal.* **1960**, *12*, 117-148, pp. 126-146.
[419]Pines; Wunderlich *J. Am. Chem. Soc.* **1958**, *80*, 6001.
[420]Eberhardt; Peterson *J. Org. Chem.* **1965**, *30*, 82; Pines; Stalick *Tetrahedron Lett.* **1968**, 3723.
[421]Schmerling; Toekelt *J. Am. Chem. Soc.* **1962**, *84*, 3694.

With certain substrates, alkenes can be dimerized by acid catalysts, so that the product is a dimer that contains one double bond.[422] This reaction is more often carried out internally, e.g.,

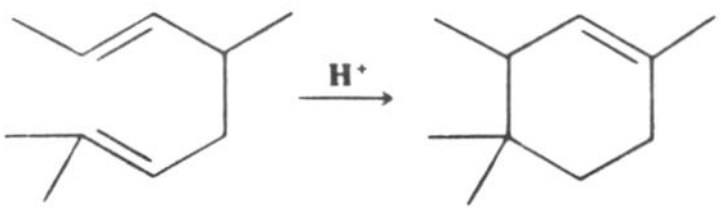

Processes of this kind are important in the biosynthesis of steroids and tetra- and pentacyclic terpenes. For example, squalene 2,3-oxide is converted by enzymic catalysis to dammaradienol.

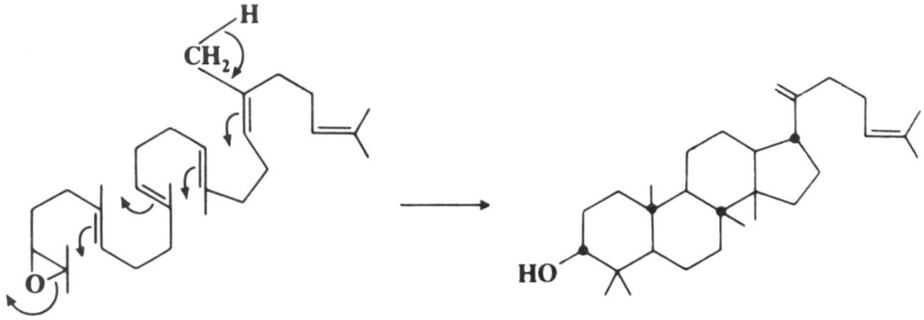

Squalene 2,3-oxide

Dammaradienol

The squalene → lanosterol biosynthesis (which is a key step in the biosynthesis of cholesterol) is similar. The idea that the biosynthesis of such compounds involves this type of multiple ring closing was proposed in 1955 and is known as the *Stork–Eschenmoser hypothesis*.[423] Such reactions can also be carried out in the laboratory, without enzymes.[424] By putting cation-stabilizing groups at positions at which positive charges develop, Johnson and co-workers have been able to close as many as four rings stereoselectively and in high yield, in one operation.[425] An example is

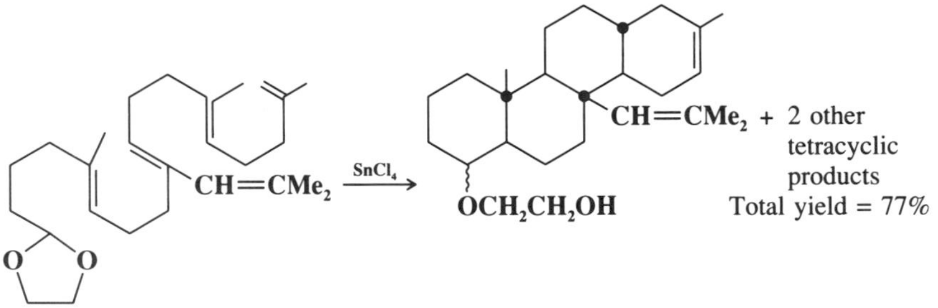

$CH=CMe_2$ + 2 other tetracyclic products Total yield = 77%

which uses the $CH=CMe_2$ group as the cation-stabilizing auxilliary.

[422]For a review, see Onsager; Johansen, in Hartley; Patai *The Chemistry of the Metal–Carbon Bond*, vol. 3; Wiley: New York, 1985, pp. 205-257.

[423]Stork; Burgstahler *J. Am. Chem. Soc.* **1955**, *77*, 5068; Eschenmoser; Ruzicka; Jeger; Arigoni *Helv. Chim. Acta* **1955**, *38*, 1890.

[424]For reviews, see Gnonlonfoun *Bull. Soc. Chim. Fr.* **1988**, 862-869; Sutherland *Chem. Soc. Rev.* **1980**, *9*, 265-280; Johnson *Angew. Chem. Int. Ed. Engl.* **1976**, *15*, 9-17 [*Angew. Chem. 88*, 33-40], *Bioorg. Chem.* **1976**, *5*, 51-98, *Acc. Chem. Res.* **1968**, *1*, 1-8; van Tamelen *Acc. Chem. Res.* **1975**, *8*, 152-158. For a review of the stereochemical aspects, see Bartlett, in Morrison, Ref. 232, vol. 3, pp. 341-409.

[425]Johnson; Telfer; Cheng; Schubert *J. Am. Chem. Soc.* **1987**, *109*, 2517; Johnson; Lindell; Steele *J. Am. Chem. Soc.* **1987**, *109*, 5852; Guay; Johnson; Schubert *J. Org. Chem.* **1989**, *54*, 4731.

The addition of olefins to olefins[426] can also be accomplished by bases[427] as well as by the use of catalyst systems[428] consisting of nickel complexes and alkylaluminum compounds (known as *Ziegler catalysts*),[429] catalysts derived from rhodium chloride,[430] and other transition metal catalysts. These and similar catalysts also catalyze the 1,4-addition of olefins to conjugated dienes,[431] e.g.,

$$CH_2=CH_2 + CH_2=CH-CH=CH_2 \xrightarrow{RhCl_3} \overset{\overset{\displaystyle H}{|}}{CH_2}-CH=CH-\overset{\overset{\displaystyle CH=CH_2}{|}}{CH_2}$$

and the dimerization of 1,3-butadienes to octatrienes.[432]

In the presence of cuprous chloride and ammonium chloride, acetylene adds to another molecule of itself to give vinylacetylene.

$$HC\equiv CH + HC\equiv CH \xrightarrow[NH_4Cl]{CuCl} HC\equiv C-CH=CH_2$$

This type of alkyne dimerization is also catalyzed by certain nickel complexes, as well as other catalysts[433] and has been carried out internally to convert diynes to large-ring cycloalkynes with an exocyclic double bond.[434]

In another type of alkyne dimerization, two molecules of alkyne, the same or different, can be coupled to give a 1,3-diene[435]

$$R^1C\equiv CR^2 + R^3C\equiv CR^4 \longrightarrow \underset{\overset{\displaystyle |}{H}}{\overset{\overset{\displaystyle R^1}{|}}{C}}\underset{}{=}\overset{\overset{\displaystyle R^2}{|}}{C}-\overset{\overset{\displaystyle R^3}{|}}{C}\underset{\overset{\displaystyle |}{H}}{\overset{\overset{\displaystyle R^4}{}}{C}}$$

In this method, one alkyne is treated with Schwartz's reagent (see **5-13**) to produce a vinylic zirconium intermediate. Addition of MeLi or MeMgBr, followed by the second alkyne, gives another intermediate, which, when treated with aqueous acid, gives the diene in moderate-to-good yields. The stereoisomer shown is the one formed in usually close to 100% purity. If the second intermediate is treated with I_2 instead of aqueous acid, the 1,4-diiodo-1,3-diene is obtained instead, in comparable yield and isomeric purity. This reaction can

[426]For a review of olefin dimerization and oligomerization with all catalysts, see Fel'dblyum; Obeshchalova *Russ. Chem. Rev.* **1968**, *37*, 789-797.

[427]For a review, see Pines *Synthesis* **1974**, 309-327.

[428]For reviews, see Pillai; Ravindranathan; Sivaram *Chem. Rev.* **1986**, *86*, 353-399; Jira; Freiesleben *Organomet. React.* **1972**, *3*, 1-190, pp. 117-130; Heck, Ref. 223, pp. 84-94, 150-157; Khan; Martell, Ref. 159, vol. 2, pp. 135-15; Rylander Ref. 223, pp. 175-196; Tsuji *Adv. Org. Chem.* **1969**, *6*, 109-255, pp. 213-220.

[429]See for example, Onsager; Wang; Blindheim *Helv. Chim. Acta* **1969**, *52*, 187, 230; Fischer; Jonas; Misbach; Stabba; Wilke *Angew. Chem. Int. Ed. Engl.* **1973**, *12*, 943 [*Angew. Chem. 85*, 1002].

[430]Cramer *J. Am. Chem. Soc.* **1965**, *87*, 4717, *Acc. Chem. Res.* **1968**, *1*, 186-191; Kobayashi; Taira *Tetrahedron* **1968**, *24*, 5763; Takahashi; Okura; Keii *J. Am. Chem. Soc.* **1975**, *97*, 7489.

[431]Alderson; Jenner; Lindsey *J. Am. Chem. Soc.* **1965**, *87*, 5638. For a review see Su *Adv. Organomet. Chem.* **1979**, *17*, 269-318.

[432]See, for example, Denis; Jean; Croizy; Mortreux; Petit *J. Am. Chem. Soc.* **1990**, *112*, 1292.

[433]See for example, Carlton; Read *J. Chem. Soc., Perkin Trans. 1* **1978**, 1631; Schmitt; Singer *J. Organomet. Chem.* **1978**, *153*, 165; Selimov; Rutman; Dzhemilev *J. Org. Chem. USSR* **1983**, *19*, 1621.

[434]Trost; Matsubara; Carinji *J. Am. Chem. Soc.* **1989**, *111*, 8745.

[435]Buchwald; Nielsen *J. Am. Chem. Soc.* **1989**, *111*, 2870.

also be done intramolecularly: Diynes **51** can be cyclized to E,E exocyclic dienes **52** by treatment with a zirconium complex.[436]

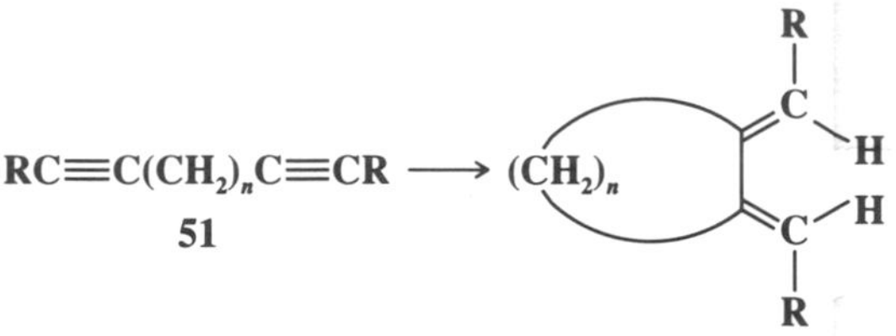

$$RC\equiv C(CH_2)_nC\equiv CR \longrightarrow$$
51

52

Rings of 4, 5, and 6 members were obtained in high yield; 7-membered rings in lower yield. When the reaction is applied to enynes, compounds similar to **52** but with only one double bond are obtained.[437]

In a conversion that is formally similar, substituted alkenes (CH_2=CH—Y; Y = R, COOMe, OAc, CN, etc.) can be dimerized to substituted alkanes $CH_3CHYCHYCH_3$ by photolysis in an H_2 atmosphere, using Hg as a photosensitizer.[438] Still another procedure involves palladium-catalyzed addition of vinylic halides to triple bonds to give 1,3-dienes.[439]

Olefins and alkynes can also add to each other to give cyclic products in other ways (see **5-49** and **5-51**).

OS **65**, 42; **66**, 52, 75; **67**, 48.

5-16   The Ene Synthesis
Hydro-allyl-addition

Olefins can add to double bonds in a reaction different from those discussed in **5-15**, which, however, is still formally the addition of RH to a double bond. This reaction is called the *ene synthesis*[440] and bears a certain similarity to the Diels–Alder reaction (**5-47**). For the reaction to proceed without a catalyst, one of the components must be a reactive dienophile (see **5-47** for a definition of this word) such as maleic anhydride, but the other (which supplies the hydrogen) may be a simple alkene such as propene. There has been much discussion of the mechanism of this reaction, and both concerted pericyclic (as shown above) and stepwise mechanisms have been suggested. The reaction between maleic anhydride and optically active PhCHMeCH=CH_2 gave an optically active product,[441] which is strong evi-

[436]Nugent; Thorn; Harlow *J. Am. Chem. Soc.* **1987**, *109*, 2788. See also Trost; Lee *J. Am. Chem. Soc.* **1988**, *110*, 7255; Tamao; Kobayashi; Ito *J. Am. Chem. Soc.* **1989**, *111*, 6478.

[437]RajanBabu; Nugent; Taber; Fagan *J. Am. Chem. Soc.* **1988**, *110*, 7128.

[438]Muedas; Ferguson; Crabtree *Tetrahedron Lett.* **1989**, *30*, 3389.

[439]Arcadi; Bernocchi; Burini; Cacchi; Marinelli; Pietroni *Tetrahedron Lett.* **1989**, *30*, 3465.

[440]Alder; Brachel *Liebigs Ann. Chem.* **1962**, *651*, 141. For a monograph, see Carruthers *Cycloaddition Reactions in Organic Synthesis*; Pergamon: Elmsford, NY, 1990. For reviews, see Boyd, in Patai *Supplement A: The Chemistry of Double-bonded Functional Groups*, vol. 2, pt. 1; Wiley: New York, 1989, pp. 477-525; Keung; Alper *J. Chem. Educ.* **1972**, *49*, 97-100; Hoffmann *Angew. Chem. Int. Ed. Engl.* **1969**, *8*, 556-577 [*Angew. Chem. 81*, 597-618]. For reviews of intramolecular ene reactions see Taber *Intramolecular Diels–Alder and Alder Ene Reactions*; Springer: New York, 1984; pp. 61-94; Oppolzer; Snieckus *Angew. Chem. Int. Ed. Engl.* **1978**, *17*, 476-486 [*Angew. Chem. 90*, 506-516]; Conia; Le Perchec *Synthesis* **1975**, 1-19. For a review of ene reactions in which one of the reactants bears a Si or Ge atom, see Dubac; Laporterie *Chem. Rev.* **1987**, *87*, 319-334.

[441]Hill; Rabinovitz *J. Am. Chem. Soc.* **1964**, *86*, 965. See also Garsky; Koster; Arnold *J. Am. Chem. Soc.* **1974**, *96*, 4207; Stephenson; Mattern *J. Org. Chem.* **1976**, *41*, 3614; Nahm; Cheng *J. Org. Chem.* **1986**, *51*, 5093.

dence for a concerted rather than a stepwise mechanism.[442] The reaction can be extended

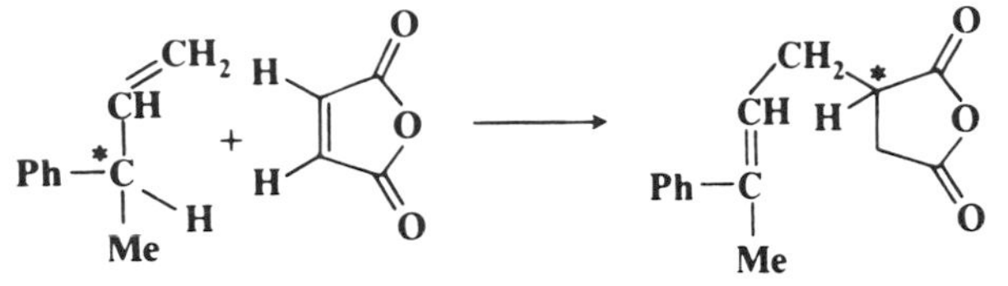

to less-reactive enophiles by the use of Lewis-acid catalysts, especially alkylaluminum halides.[443] The Lewis-acid catalyzed reaction probably has a stepwise mechanism.[444]

OS **IV**, 766; **V**, 459. See also OS **65**, 159.

5-17 The Michael Reaction
Hydro-bis(ethoxycarbonyl)methyl-addition, etc.

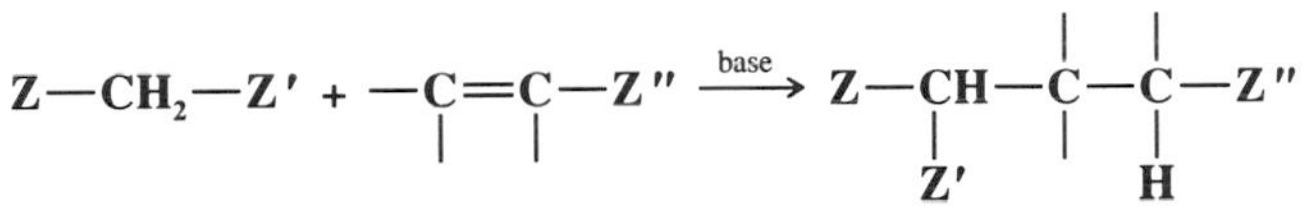

$$Z-CH_2-Z' + \begin{array}{c}|\\-C\\|\end{array}=\begin{array}{c}|\\C-Z''\\|\end{array} \xrightarrow{\text{base}} Z-CH-\begin{array}{c}|\\C\\|\end{array}-\begin{array}{c}|\\C-Z''\\|\end{array}$$
$$\qquad\qquad\qquad\qquad\qquad\qquad Z' \qquad H$$

Compounds containing electron-withdrawing groups (Z is defined on p. 741) add, in the presence of bases, to olefins of the form C=C—Z (including quinones). This is called the *Michael reaction* and involves conjugate addition.[445] The base removes the acidic proton and then the mechanism is as outlined on p. 741. The reaction has been carried out with malonates, cyanoacetates, acetoacetates, other β-keto esters, and compounds of the form ZCH₃, ZCH₂R, ZCHR₂, and ZCHRZ′, including carboxylic esters, ketones, aldehydes, nitriles, nitro compounds,[446] and sulfones, as well as other compounds with relatively acidic hydrogens, such as indenes and fluorenes. These reagents do not add to ordinary double bonds, except in the presence of free-radical initiators (**5-22**). 1,2 addition (to the C=O or C≡N group) often competes and sometimes predominates (**6-41**).[447] In particular, α,β-unsaturated *aldehydes* seldom give 1,4 addition.[448] The Michael reaction has traditionally been performed in protic solvents, with catalytic amounts of base, but more recently better yields with fewer side reactions have been obtained in some cases by using an equimolar amount of base to convert the nucleophile to its enolate form (*preformed enolate*). In particular, preformed enolates are often used where stereoselective reactions are desired.[449]

---

[442]For other evidence for a concerted mechanism see Benn; Dwyer; Chappell *J. Chem. Soc., Perkin Trans.* 2 **1977**, 533; Jenner; Salem; El'yanov; Gonikberg *J. Chem. Soc., Perkin Trans.* 2 **1989**, 1671.
[443]For reviews, see Chaloner, in Hartley, Ref. 218, vol. 4, pp. 456-460; Snider *Acc. Chem. Res.* **1980**, *13*, 426-432.
[444]See Snider; Ron *J. Am. Chem. Soc.* **1985**, *107*, 8160.
[445]For reviews, see Yanovskaya; Kryshtal; Kulganek *Russ. Chem. Rev.* **1984**, *53*, 744-756; Bergmann; Ginsburg; Pappo *Org. React.* **1959**, *10*, 179-560; House, Ref. 144, pp. 595-623. The subject is also discussed at many places in Stowell *Carbanions in Organic Synthesis*; Wiley: New York, 1979.
[446]For reviews of Michael reactions where Z or Z′ is nitro see Yoshikoshi; Miyashita *Acc. Chem. Res.* **1985**, *18*, 284-290; Baer; Urbas, in Feuer *The Chemistry of the Nitro and Nitroso Groups*, pt. 2; Wiley: New York, 1970, pp. 130-148.
[447]For a discussion of 1,2 vs. 1,4 addition, see Oare; Heathcock, *Top. Stereochem.* **1989**, Ref. 449, pp. 232-236.
[448]For reports of successful 1,4 additions to α,β-unsaturated aldehydes, see Kryshtal; Kulganek; Kucherov; Yanovskaya *Synthesis* **1979**, 107; Yamaguchi; Yokota; Minami *J. Chem. Soc., Chem. Commun.* **1991**, 1088.
[449]For reviews of stereoselective Michael additions, see Oare; Heathcock *Top. Stereochem.* **1991**, *20*, 87-170, **1989**, *19*, 227-407.

In a Michael reaction with suitably different R groups, two new chiral centers are created:

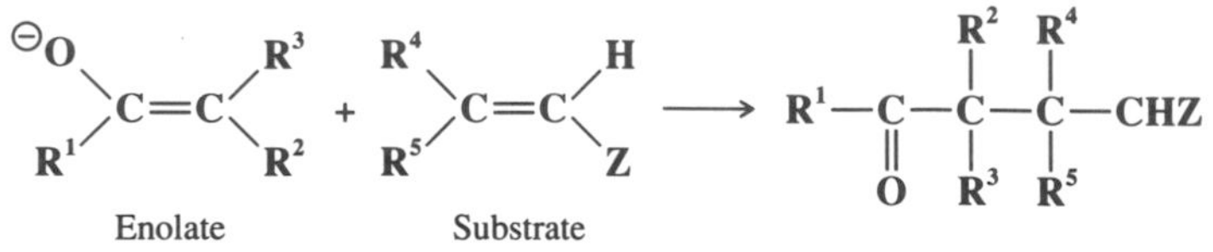

Thus the product in such cases can exist as two pairs of enantiomers.[450] In a diastereoselective process one of the two pairs is formed exclusively or prodominantly, as a racemic mixture. Many such examples have been reported.[449] In many of these cases, both the enolate and substrate can exist as *Z* or *E* isomers. With enolates derived from ketones or carboxylic esters, *E* enolates gave the syn pair of enantiomers (p. 115), while *Z* enolates gave the anti pair.[451]

When either or both of the reaction components has a chiral substituent, the reaction can be enantioselective (only one of the four diastereomers formed predominantly), and this has been accomplished a number of times.[452] Enantioselective addition has also been achieved by the use of a chiral catalyst[453] and by using optically active enamines instead of enolates.[454]

Mannich bases (see **6-16**) and β-halo carbonyl compounds can also be used as substrates; these are converted to the C=C—Z compounds in situ by the base (**6-16, 7-13**).[455] Substrates of this kind are especially useful in cases where the C=C—Z compound is unstable. The reaction of C=C—Z compounds with enamines (**2-19**) can also be considered a Michael reaction. Michael reactions are reversible (**7-20**).

When the substrate contains *gem*-Z groups, e.g., **54**, bulky groups can be added, if the

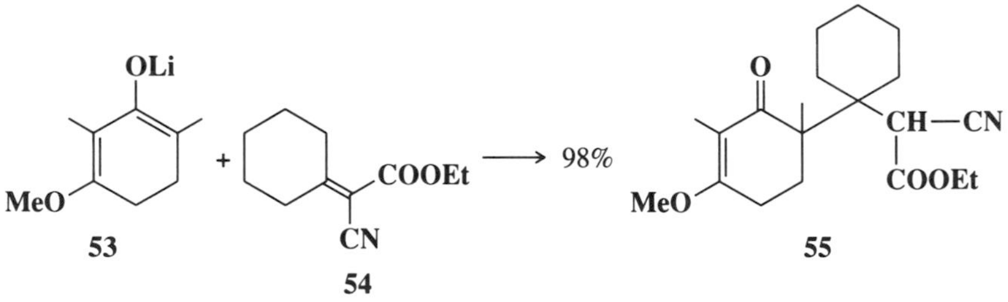

reaction is carried out under aprotic conditions. For example, addition of enolate **53** to **54** gave **55** in which two adjacent quaternary centers have been formed.[456]

In certain cases, Michael reactions can take place under acidic conditions.[457]

[450]For a more extended analysis, see Oare; Heathcock *Top. Stereochem.* **1989**, Ref. 449, pp. 237-242.

[451]For example, see Oare; Heathcock *J. Org. Chem.* **1990**, *55*, 157.

[452]See, for example, Corey; Peterson *Tetrahedron Lett.* **1985**, *26*, 5025; Calderari; Seebach *Helv. Chim. Acta* **1985**, *68*, 1592; Tomioka; Ando; Yasuda; Koga *Tetrahedron Lett.* **1986**, *27*, 715; Posner; Switzer *J. Am. Chem. Soc.* **1986**, *108*, 1239; Enders; Demir; Rendenbach *Chem. Ber.* **1987**, *120*, 1731.

[453]Yura; Iwasaka; Mukaiyama *Chem. Lett.* **1988**, 1021; Yura; Iwasaka; Narasaka; Mukaiyama *Chem. Lett.* **1988**, 1025; Desimoni; Quadrelli; Righetti *Tetrahedron* **1990**, *46*, 2927.

[454]See d'Angelo; Revial; Volpe; Pfau *Tetrahedron Lett.* **1988**, *29*, 4427.

[455]Mannich bases react with ketones *without* basic catalysts to give 1,5-diketones, but this process, known as the *thermal-Michael reaction*, has a different mechanism: Brown; Buchanan; Curran; McLay *Tetrahedron* **1968**, *24*, 4565; Gill; James; Lions; Potts *J. Am. Chem. Soc.* **1952**, *74*, 4923.

[456]Holton; Williams; Kennedy *J. Org. Chem.* **1986**, *51*, 5480.

[457]See Hajos; Parrish *J. Org. Chem.* **1974**, *39*, 1612, *Org. Synth.* *VII*, 363.

Michael reactions are sometimes applied to substrates of the type C≡C—Z, e.g.,

$$HC\equiv C-COOEt + CH_3COC\overset{\ominus}{H}COOEt \longrightarrow CH_3COCHCOOEt$$
$$\underset{CH=CHCOOEt}{|}$$

Indeed, because of the greater susceptibility of triple bonds to nucleophilic attack, it is even possible for nonactivated alkynes, e.g., acetylene, to be substrates in this reaction.[458]

In a closely related reaction, silyl enol ethers add to α,β-unsaturated ketones and esters when catalyzed[459] by TiCl$_4$, e.g.,[460]

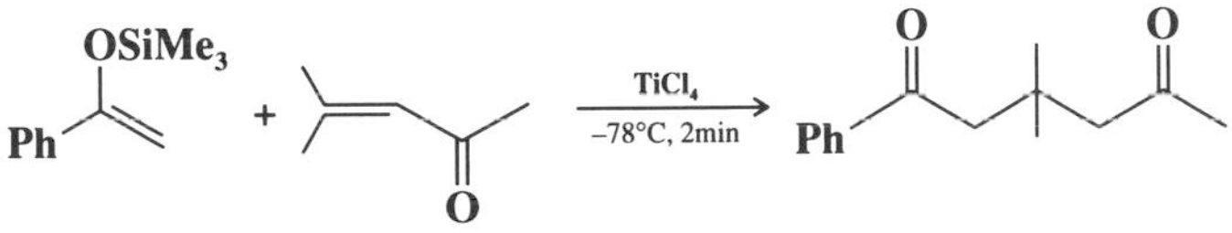

This reaction, also, has been performed diastereoselectively.[461] Allylic silanes R$_2$C=CHCH$_2$SiMe$_3$ can be used instead of silyl enol ethers (the *Sakurai reaction*).[462] Similarly, silyl ketene acetals, e.g., **56**, give δ-keto esters, in MeNO$_2$ as solvent, for example,[463]

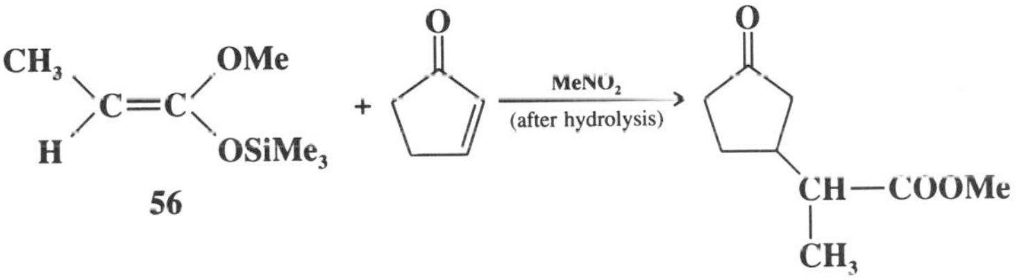

OS **I**, 272; **II**, 200; **III**, 286; **IV**, 630, 652, 662, 776; **V**, 486, 1135; **VI**, 31, 648, 666, 940; **VII**, 50, 363, 368, 414, 443; **65**, 12, 98; **66**, 37; **69**, 173, 226. See also OS **65**, 236.

5-18 1,4 Addition of Organometallic Compounds to Activated Double Bonds
Hydro-alkyl-addition

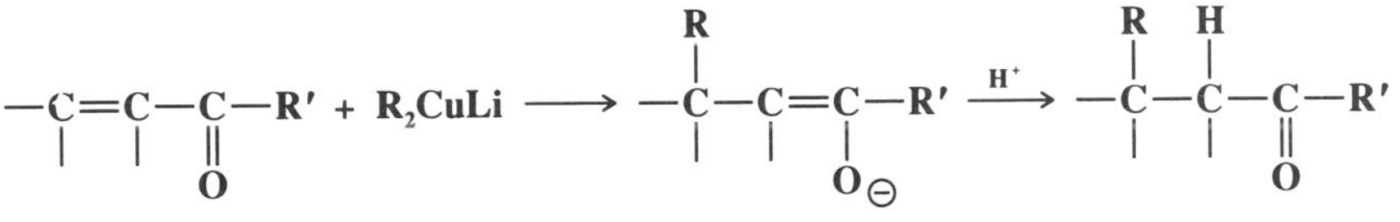

57

[458]See, for example, Makosza *Tetrahedron Lett.* **1966**, 5489.

[459]Other catalysts have also been used. For a list of catalysts, with references, see Ref. 133, pp. 793-795. See also Mukaiyama; Kobayashi; Tamura; Sagawa *Chem. Lett.* **1987**, 491; Mukaiyama; Kobayashi *J. Organomet. Chem.* **1990**, *382*, 39.

[460]Narasaka; Soai; Aikawa; Mukaiyama *Bull. Chem. Soc. Jpn.* **1976**, *49*, 779; Saigo; Osaki; Mukaiyama *Chem. Lett.* **1976**, 163; Matsuda *J. Organomet. Chem.* **1987**, *321*, 307; Narasaka *Org. Synth.* 65, 12. See also Yoshikoshi; Miyashita, Ref. 446.

[461]See Heathcock; Uehling *J. Org. Chem.* **1986**, *51*, 279; Mukaiyama; Tamura; Kobayashi *Chem. Lett.* **1986**, 1017, 1817, 1821, **1987**, 743.

[462]Hosomi; Sakurai *J. Am. Chem. Soc.* **1977**, *99*, 1673; Jellal; Santelli *Tetrahedron Lett.* **1980**, *21*, 4487; Sakurai; Hosomi; Hayashi *Org. Synth.* VII, 443. For a review, see Fleming; Dunoguès; Smithers *Org. React.* **1989**, *37*, 57-575, pp. 127-132, 335-370. For a review of intramolecular additions, see Schinzer *Synthesis* **1988**, 263-273.

[463]RajanBabu *J. Org. Chem.* **1984**, *49*, 2083.

Lithium dialkylcopper reagents (see **0-87**) add to α,β-unsaturated aldehydes[464] and ketones (R' = H, R, Ar) to give conjugate addition products[465] in a reaction closely related to the Michael reaction. α,β-Unsaturated esters are less reactive,[466] and the corresponding acids do not react at all. R can be primary alkyl, vinylic, or aryl. If Me_3SiCl is present, the reaction takes place much faster and with higher yields; in this case the product is the silyl enol ether of **57** (see **2-23**).[467] The use of Me_3SiCl also permits good yields with allylic R groups.[468]

Various functional groups such as OH and unconjugated C=O groups may be present in the substrate.[469] A characteristic of the reaction is that only one of the R groups of R_2CuLi adds to the substrate; the other is wasted. This can be a limitation where the precursor (RLi or RCu, see **2-35**) is expensive or available in limited amounts. The difficulty can be overcome by using one of the mixed reagents $R(R'C≡C)CuLi$,[470] $R(O—t-Bu)CuLi$,[471] or $R(PhS)CuLi$,[472] each of which transfers only the R group. These reagents are easily prepared by the reaction of RLi with R'C≡CCu (R' = n-Pr or t-Bu), t-BuOCu, or PhSCu, respectively. A further advantage of the mixed reagents is that good yields of addition product are achieved when R is tertiary, so that use of one of them permits the introduction of a tertiary alkyl group. The mixed reagents $R(CN)CuLi$[473] (prepared from RLi and CuCN) and $R_2Cu(CN)Li_2$[474] also selectively transfer the R group.[475] The reaction has also been carried out with α,β-acetylenic ketones, esters, and nitriles.[476] Conjugate addition to α,β-unsaturated and acetylenic acids and esters, as well as ketones, can be achieved by the use of the coordinated reagents $RCu·BF_3$ (R = primary).[477] Alkylcopper compounds RCu (R = primary or secondary alkyl) have also been used with tetramethylethylenediamine and Me_3SiCl to give silyl enol ethers from α,β-unsaturated ketones in high yield.[478]

There is generally little or no competition from 1,2 addition (to the C=O). However, when R is allylic, 1,4 addition is observed with some substrates and 1,2 addition with others.[479] R_2CuLi also add to α,β-unsaturated sulfones[480] but not to simple α,β-unsaturated nitriles.[481]

[464]Chuit; Foulon; Normant *Tetrahedron* **1980**, *36*, 2305, **1981**, *37*, 1385. For a review, see Alexakis; Chuit; Commerçon-Bourgain; Foulon; Jabri; Mangeney; Normant *Pure Appl. Chem.* **1984**, *56*, 91-98. A better reagent for the addition of a methyl group to an α,β-unsaturated aldehyde is $Me_5Cu_3Li_2$: Clive; Farina; Beaulieu, *J. Org. Chem.* **1982**, *47*, 2572.

[465]House; Respess; Whitesides *J. Org. Chem.* **1966**, *31*, 3128. For reviews, see Posner *Org. React.* **1972**, *19*, 1-113; House *Acc. Chem. Res.* **1976**, *9*, 59-67. For examples of the use of this reaction in the synthesis of natural products, see Posner *An Introduction to Synthesis Using Organocopper Reagents*; Wiley: New York 1980, pp. 10-67. For a list of organocopper reagents that give this reaction, with references, see Ref. 133, pp. 805-809, 916-920.

[466]R_2CuLi also add to N-tosylated α,β-unsaturated amides: Nagashima; Ozaki; Washiyama; Itoh *Tetrahedron Lett.* **1985**, *26*, 657.

[467]Corey; Boaz *Tetrahedron Lett.* **1985**, *26*, 6019; Alexakis; Berlan; Besace *Tetrahedron Lett.* **1986**, *27*, 1047; Matsuza; Horiguchi; Nakamura; Kuwajima *Tetrahedron* **1989**, *45*, 349; Horiguchi; Komatsu; Kuwajima *Tetrahedron Lett.* **1989**, *30*, 7087; Linderman; McKenzie *J. Organomet. Chem.* **1989**, *361*, 31; Bertz; Smith *Tetrahedron* **1990**, *46*, 4091. For a list of references, see Ref. 133, p. 748.

[468]Lipshutz; Ellsworth; Dimock; Smith *J. Am. Chem.Soc.* **1990**, *112*, 4404.

[469]For the use of enol tosylates of 1,2-diketones as substrates, see Charonnat; Mitchell; Keogh *Tetrahedron Lett.* **1990**, *31*, 315.

[470]Corey; Beames *J. Am. Chem. Soc.* **1972**, *94*, 7210; House; Umen *J. Org. Chem.* **1973**, *38*, 3893; Corey; Floyd; Lipshutz *J. Org. Chem.* **1978**, *43*, 3419.

[471]Posner; Whitten *Tetrahedron Lett.* **1973**, 1815.

[472]Posner; Whitten; Sterling *J. Am. Chem. Soc.* **1973**, *95*, 7788.

[473]Gorlier; Hamon; Levisalles; Wagnon *J. Chem. Soc., Chem. Commun.* **1973**, 88. For another useful mixed reagent see Ledlie; Miller *J. Org. Chem.* **1979**, *44*, 1006.

[474]Lipshutz; Wilhelm; Kozlowski *Tetrahedron Lett.* **1982**, *23*, 3755; Lipshutz *Tetrahedron Lett.* **1983**, *24*, 127.

[475]When the two R groups of $R_2Cu(CN)Li_2$ are different, one can be selectively transferred: Lipshutz; Wilhelm; Kozlowski *J. Org. Chem.* **1984**, *49*, 3938.

[476]For a list of references, see Ref. 133, pp. 237-238.

[477]For a review, see Yamamoto *Angew. Chem. Int. Ed. Engl.* **1986**, *25*, 947-959 [*Angew. Chem. 98*, 945-957]. For a discussion of the role of the BF_3, see Lipshutz; Ellsworth; Siahaan *J. Am. Chem. Soc.* **1988**, *110*, 4834, **1989**, *111*, 1351.

[478]Johnson; Marren *Tetrahedron Lett.* **1987**, *28*, 27.

[479]House; Fischer *J. Org. Chem.* **1969**, *34*, 3615. See also Daviaud; Miginiac *Tetrahedron Lett.* **1973**, 3345.

[480]Posner; Brunelle *Tetrahedron Lett.* **1973**, 935.

[481]House; Umen Ref. 470.

Organocopper reagents RCu (as well as certain R₂CuLi) add to α,β-unsaturated and acetylenic sulfoxides.[482]

Usually, after an enolate ion is generated from an α,β-unsaturated ketone, it is converted to the β-alkylated product as shown above. But it is often possible to have the enolate react with some other electrophile (*tandem vicinal difunctionalization*), in some cases at the O and in other cases at the C.[483] For example, if an alkyl halide R″X is present (R″ = primary alkyl or allylic), and the solvent is 1,2-dimethoxyethane, the enolate **57** can be alkylated

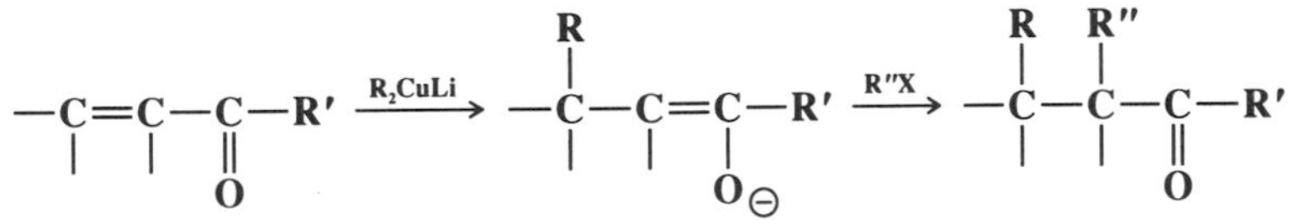

directly.[484] Thus, by this method, both the α and β positions of a ketone are alkylated in one synthetic operation (see also **5-53**).

Grignard reagents also add to these substrates, but with these reagents, 1,2 addition may seriously compete:[485] The product is often controlled by steric factors. Thus **58** with phenylmagnesium bromide gives 100% 1,4 addition, while **59** gives 100% 1,2 addition. In general,

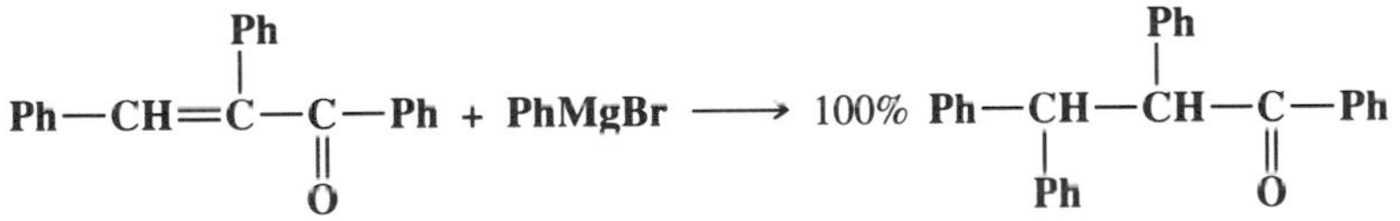

substitution at the carbonyl group increases 1,4 addition, while substitution at the double bond increases 1,2 addition. In most cases both products are obtained, but α,β-unsaturated *aldehydes* nearly always give exclusive 1,2 addition when treated with Grignard reagents. However, the extent of 1,4 addition of Grignard reagents can be increased by the use of a copper ion catalyst, e.g., CuCl, Cu(OAc)₂.[486] It is likely that alkylcopper reagents, formed from RMgX and Cu⁺ (cupric acetate is reduced to cuprous ion by excess RMgX), are the actual attacking species in these cases.[465] Alkyllithiums,[487] treated with compounds of the form C=C—COCH₃ and C=C—COOC₂H₅, gave only 1,2 addition,[488] but 1,4 addition was achieved with esters of the form C=C—COOAr, where Ar was a bulky group such as 2,6-

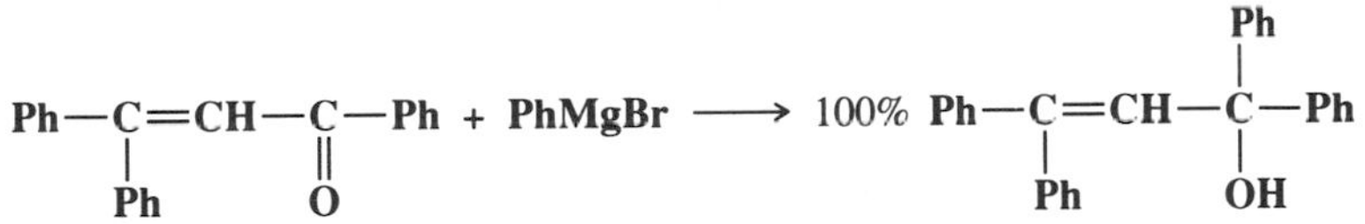

[482]Truce; Lusch *J. Org. Chem.* **1974**, *39*,3174, **1978**, *43*, 2252.

[483]For reviews of such reactions, see Chapdelaine; Hulce *Org. React.* **1990**, *38*, 225-653; Taylor *Synthesis* **1985**, 364-392. For a list of references, see Ref. 133, pp. 810-811, 922.

[484]Coates; Sandefur *J. Org. Chem.* **1974**, *39*, 275; Posner; Lentz *Tetrahedron Lett.* **1977**, 3215.

[485]For a discussion of the factors affecting 1,2 vs. 1,4 addition, see Negishi, Ref. 404, pp. 127-133.

[486]Posner, Ref. 465.

[487]For a review of addition of organolithium compounds to double bonds, see Hunt *Org. Prep. Proceed. Int.* **1989**, *21*, 705-749.

[488]Rozhkov; Makin *J. Gen. Chem. USSR* **1964**, *34*, 57. For a discussion of 1,2 vs. 1,4 addition with organolithiums, see Cohen; Abraham; Myers *J. Am. Chem. Soc.* **1987**, *109*, 7923.

di-*t*-butyl-4-methoxyphenyl.[489] Also, alkyllithiums can be made to give 1,4 addition with α,β-unsaturated ketones[490] and aldehydes[491] if the reactions are conducted in the presence of HMPA. Among alkyllithiums that have been found to add 1,4 in this manner are 2-lithio-1,3-dithianes (see **0-97**).[492] 1,4 Addition of alkyllithiums to α,β-unsaturated aldehydes can also be achieved by converting the aldehyde to a benzothiazole derivative (masking the aldehyde function),[493] from which the aldehyde group can be regenerated.

However, neither Grignard reagents nor lithium dialkylcopper reagents generally add to ordinary C=C double bonds.[494] Grignard reagents in general add only to double bonds susceptible to nucleophilic attack, e.g., fluoroolefins and tetracyanoethylene.[495] However, active Grignard reagents (benzylic, allylic) also add to the double bonds of allylic amines,[496] and of allylic and homoallylic alcohols,[497] as well as to the triple bonds of propargyl alcohols and certain other alkynols.[498] It is likely that cyclic intermediates are involved in these cases, in which the magnesium coordinates with the hetero atom. Organolithium reagents (primary, secondary, and tertiary alkyl and in some cases aryl) also add to the double and triple bonds of allylic and propargylic alcohols[499] (in this case tetramethylethylenediamine is a catalyst) and to certain other olefins containing hetero groups such as OR, NR_2, or SR. Allylic, benzylic, and tertiary alkyl Grignard reagents also add to 1-alkenes and strained internal alkenes, e.g., norbornene, if the reaction is carried out not in ether but in a hydrocarbon solvent such as pentane or in the alkene itself as solvent, heated, under pressure if necessary, to 60 to 130°C.[500] Yields are variable. *Intramolecular* addition of RMgX to completely unactivated double and triple bonds has been demonstrated,[501] e.g., refluxing of 6-chloro-1-heptene with Mg for 5 hr gave, after hydrolysis, an 88% yield of 1,2-dimethylcyclopentane.[502]

An alkynyl group can be added to the double bond of an α,β-unsaturated ketone by use of the diethylalkynylalane reagents $Et_2AlC≡CR$.[503] In a similar manner, the alkenyl reagents $R_2AlCH=CR_2'$ transfer an alkenyl group.[504] Trialkylalanes R_3Al also add 1,4 to such ketones

[489]Cooke *J. Org. Chem.* **1986**, *51*, 1637.
[490]Sauvetre; Seyden-Penne *Tetrahedron Lett.* **1976**, 3949; Roux; Wartski; Seyden-Penne *Tetrahedron* **1981**, *37*, 1927, *Synth. Commun.* **1981**, *11*, 85.
[491]El-Bouz; Wartski *Tetrahedron Lett.* **1980**, *21*, 2897.
[492]Lucchetti; Dumont; Krief *Tetrahedron Lett.* **1979**, 2695; Brown; Yamaichi *J. Chem. Soc., Chem. Commun.* **1979**, 100; Ref. 491. See also Bürstinghaus; Seebach *Chem. Ber.* **1977**, *110*, 841.
[493]Corey; Boger *Tetrahedron Lett.* **1978**, 9. For another indirect method, see Sato; Okazaki; Otera; Nozaki *Tetrahedron Lett.* **1988**, *29*, 2979.
[494]For reviews of the addition of RM to isolated double bonds see Wardell; Paterson, in Hartley; Patai, Ref. 422, vol. 2, 1985, pp. 219-338, pp. 268-296; Vara Prasad; Pillai *J. Organomet. Chem.* **1983**, *259*, 1-30.
[495]Gardner; Kochi *J. Am. Chem. Soc.* **1976**, *98*, 558.
[496]Richey; Moses; Domalski; Erickson; Heyn *J. Org. Chem.* **1981**, *46*, 3773.
[497]Eisch; Husk *J. Am. Chem. Soc.* **1965**, *87*, 4194; Felkin; Kaeseberg *Tetrahedron Lett.* **1970**, 4587; Richey; Szucs *Tetrahedron Lett.* **1971**, 3785; Eisch; Merkley *J. Am. Chem. Soc.* **1979**, *101*, 1148; Kang *Organometallics* **1984**, *3*, 525.
[498]Eisch; Merkley Ref. 497; Von Rein; Richey *Tetrahedron Lett.* **1971**, 3777; Miller; Reichenbach *Synth. Commun.* **1976**, *6*, 319. See also Duboudin; Jousseaume *J. Organomet. Chem.* **1979**, *168*, 1, *Synth. Commun.* **1979**, *9*, 53.
[499]For a review of the addition of organolithium compounds to double or triple bonds, see Wardell, in Zuckerman *Inorganic Reactions and Methods*, vol. 11; VCH: New York, 1988, pp. 129-142.
[500]Lehmkuhl; Reinehr *J. Organomet. Chem.* **1970**, *25*, C47; **1973**, *57*, 29; Lehmkulhl; Janssen *Liebigs Ann. Chem.* **1978**, 1854. This is actually a type of ene reaction. For a review of the intramolecular version of this reaction, see Oppolzer *Angew. Chem. Int. Ed. Engl.* **1989**, *28*, 38-52 [*Angew. Chem. 101*, 39-53].
[501]See, for example, Richey; Rees *Tetrahedron Lett.* **1966**, 4297; Drozd; Ustynyuk; Tsel'eva; Dmitriev *J. Gen. Chem. USSR* **1969**, *39*, 1951; Felkin; Umpleby; Hagaman; Wenkert *Tetrahedron Lett.* **1972**, 2285; Hill; Myers *J. Organomet. Chem.* **1979**, *173*, 1.
[502]For intramolecular addition of RLi and R_2CuLi, see Wender; White *J. Am. Chem. Soc.* **1988**, *110*, 2218; Bailey; Nurmi; Patricia; Wang *J. Am. Chem. Soc.* **1987**, *109*, 2442.
[503]Hooz; Layton *J. Am. Chem. Soc.* **1971**, *93*, 7320; Schwartz; Carr; Hansen; Dayrit *J. Org. Chem.* **1980**, *45*, 3053.
[504]Hooz; Layton *Can. J. Chem.* **1973**, *51*, 2098. For a similar reaction with an alkenylzirconium reagent, see Schwartz; Loots; Kosugi *J. Am. Chem. Soc.* **1980**, *102*, 1333; Dayrit; Schwartz *J. Am. Chem. Soc.* **1981**, *103*, 4466.

in the presence of nickel acetylacetonate.[505] Also used for 1,4 addition to these ketones are trialkylzinclithium reagents R_3ZnLi (reagents of the type RMe_2ZnLi transfer only R),[506] alkyl- and arylmanganese chlorides, catalyzed by CuCl (this reagent is succesful for α,β-unsaturated aldehydes and esters also),[507] arylpalladium compounds,[508] and arylmercury compounds with phase transfer catalysts.[509] Diarylzinc compounds (prepared with the aid of ultrasound) in the presence of nickel acetylacetonate, undergo 1,4 addition not only to α,β-unsaturated ketones, but also to α,β-unsaturated aldehydes.[510] An allyl group can be added, to α,β-unsaturated carboxylic esters, amides and nitriles, with CH_2=$CHCH_2SiMe_3$ and F^- ion.[511] This reagent gave better results than lithium diallylcuprate. Functionalized allylic groups can be added to terminal alkynes with allylic halides, zinc, and ultrasound, to give 1,4-dienes.[512] An alkyl group can be added to nitroolefins with $RCu(CN)ZnI$[513] or with a trialkylalane; when one of the R groups of the latter is alkenyl, it is the one transferred.[514] Trialkylalanes and dialkylzinc compounds add to triple bonds in the presence of a zirconium complex.[515] An aryl group can be added to a triple bond with an aryl iodide and a Pd–HCOOH–R_3N catalyst.[516]

As with the Michael reaction (**5-17**) the 1,4 addition of organometallic compounds has been performed diastereoselectively[517] and enantioselectively.[518] In one example of the latter,[519] α,β-unsaturated sulfoxides that are optically active because of chirality at sulfur (p. 100) have given high enantiomeric excesses, e.g.,[520]

[505]Jeffery; Meisters; Mole *J. Organomet. Chem.* **1974**, *74*, 365; Bagnell; Meisters; Mole *Aust. J. Chem.* **1975**, *28*, 817, Ashby; Heinsohn *J. Org. Chem.***1974**, *39*, 3297. See also Sato; Oikawa; Sato *Chem. Lett.* **1979**, 167; Kunz; Pees *J. Chem. Soc., Perkin Trans. 1* **1989**, 1168.
[506]Isobe; Kondo; Nagasawa; Goto *Chem. Lett.* **1977**, 679; Watson; Kjonaas *Tetrahedron Lett.* **1986**, *27*, 1437; Tückmantel; Oshima; Nozaki *Chem. Ber.* **1986**, *119*, 1581; Kjonaas; Vawter *J. Org. Chem.* **1986**, *51*, 3993.
[507]Cahiez; Alami *Tetrahedron Lett.* **1989**, *30*, 3541, 7365, **1990**, *31*, 7423.
[508]Cacchi; Arcadi *J. Org. Chem.* **1983**, *48*, 4236.
[509]Cacchi; Misiti; Palmieri *Tetrahedron* **1981**, *37*, 2941.
[510]de Souza Barboza; Pétrier; Luche *Tetrahedron Lett.* **1985**, *26*, 829; Pétrier; de Souza Barboza; Dupuy; Luche *J. Org. Chem.* **1985**, *50*, 5761.
[511]Majetich; Casares; Chapman; Behnke *J. Org. Chem.* **1986**, *51*, 1745.
[512]Knochel; Normant *J. Organomet. Chem.* **1986**, *309*, 1.
[513]Retherford; Yeh; Schipor; Chen; Knochel *J. Org. Chem.* **1989**, *54*, 5200.
[514]Pecunioso; Menicagli *Tetrahedron Lett.* **1987**, *43*, 5411, *J. Org. Chem.* **1988**, *53*, 45.
[515]Negishi; Van Horn; Yoshida; Rand *Organometallics* **1983**, *2*, 563.
[516]Cacchi; Felici; Pietroni *Tetrahedron Lett.* **1984**, *25*, 3137.
[517]For some examples, see Isobe; Funabashi; Ichikawa; Mio; Goto *Tetrahedron Lett.* **1984**, *25*, 2021; Kawasaki; Tomioka; Koga *Tetrahedron Lett.* **1985**, *26*, 3031; Yamamoto; Nishii; Ibuka *J. Chem. Soc., Chem. Commun.* **1987**, 464, 1572; Smith; Dunlap; Sulikowski *Tetrahedron Lett.* **1988**, *29*, 439; Smith; Trumper *Tetrahedron Lett.* **1988**, *29*, 443; Alexakis; Sedrani; Mangeney; Normant *Tetrahedron Lett.* **1988**, *29*, 4411; Larchevêque; Tamagnan; Petit *J. Chem. Soc., Chem. Commun.* **1989**, 31; Page; Prodger; Hursthouse; Mazid *J. Chem. Soc., Perkin Trans. 1* **1990**, 167; Corey; Hannon *Tetrahedron Lett.* **1990**, *31*, 1393.
[518]For reviews, see Posner *Acc. Chem. Res.* **1987**, *20*, 72-78; in Morrison, Ref. 232, vol. 2, 1983, the articles by Tomioka; Koga pp. 201-224; Posner, pp. 225-241.
[519]For other examples, see Oppolzer; Moretti; Godel; Meunier; Löhrer *Tetrahedron Lett.* **1983**,*24*, 4971; Helmchen; Wegner *Tetrahedron Lett.* **1985**, *26*, 6051; Corey; Naef; Hannon *J. Am. Chem. Soc.* **1986**, *108*, 7114; Dieter; Tokles *J. Am. Chem. Soc.* **1987**,*109*, 2040; Ahn; Klassen; Lippard *Organometallics* **1990**, *9*, 3178; Alexakis; Sedrani; Mangeney *Tetrahedron Lett.* **1990**, *31*, 345; Rossiter; Eguchi *Tetrahedron Lett.* **1990**, *31*, 965; Bolm; Ewald *Tetrahedron Lett.* **1990**, *31*, 5011; Jansen; Feringa *J. Org. Chem.* **1990**, *55*, 4168; Soai; Okudo; Okamoto *Tetrahedron Lett.* **1991**, *32*, 95.
[520]Posner; Kogan; Hulce *Tetrahedron Lett.* **1984**, *25*, 383.

In certain cases, Grignard reagents add 1,4 to *aromatic* systems, e.g.,[521]

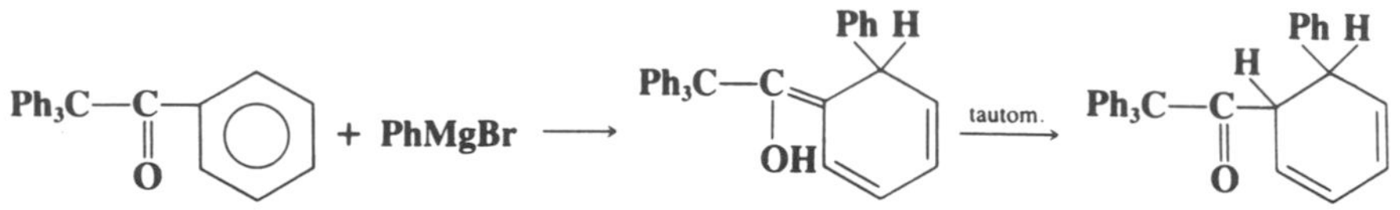

Such cyclohexadienes are easily oxidizable to benzenes (often by atmospheric oxygen), so this reaction becomes a method of alkylating and arylating suitably substituted (usually hindered) aryl ketones. A similar reaction has been reported for aromatic nitro compounds:[522]

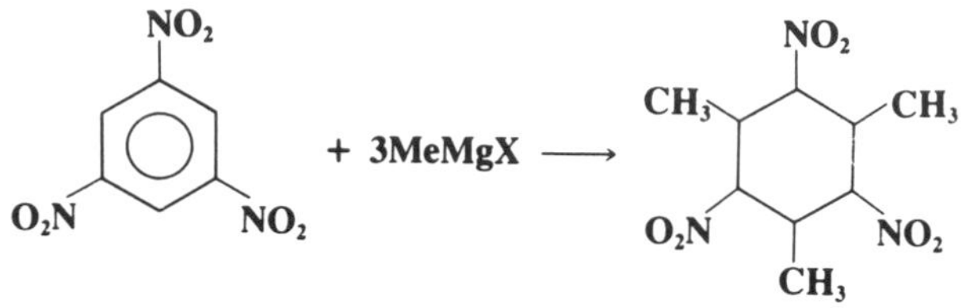

Both Grignard and R_2CuLi reagents[523] have also been added to triple-bond systems of the form $C \equiv C - C = O$.[524]

The mechanisms of most of these reactions are not well known. The 1,4 uncatalyzed Grignard reaction has been postulated to proceed by a cyclic mechanism

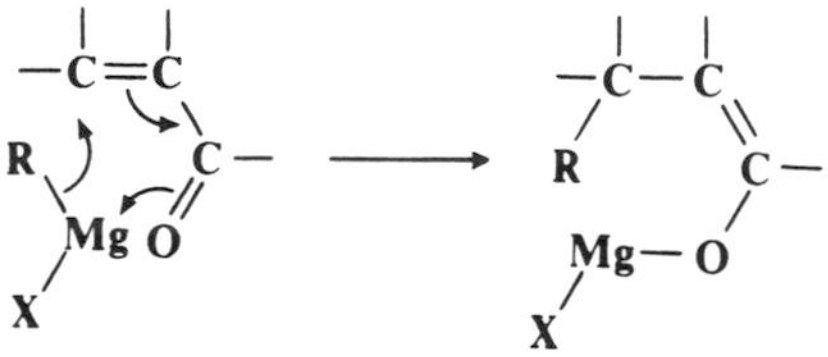

but there is evidence against it.[525] The R_2CuLi and copper-catalyzed Grignard additions may involve a number of mechanisms, since the actual attacking species and substrates are so diverse.[526] A free-radical mechanism of some type (perhaps SET) has been suggested[527]

---

[521]This example is from Schmidlin; Wohl *Ber.* **1910**, *43*, 1145; Mosher; Huber *J. Am. Chem. Soc.* **1953**, *75*, 4604. For a review of such reactions see Fuson *Adv. Organomet. Chem.* **1964**, *1*, 221-238.

[522]Severin; Schmitz *Chem. Ber.* **1963**, *96*, 3081. See also Bartoli; Bosco; Baccolini *J. Org. Chem.* **1980**, *45*, 522; Bartoli *Acc. Chem. Res.* **1984**, *17*, 109-115; Bartoli; Dalpozzo; Grossi *J. Chem. Soc., Perkin Trans. 2* **1989**, 573. For a study of the mechanism, see Bartoli; Bosco; Cantagalli; Dalpozzo; Ciminale *J. Chem. Soc., Perkin Trans. 2* **1985**, 773.

[523]For example see Corey; Kim; Chen; Takeda *J. Am. Chem. Soc.* **1972**, *94*, 4395; Anderson; Corbin; Cotterrell; Cox; Henrick; Schaub; Siddall *J. Am. Chem. Soc.* **1975**, *97*, 1197.

[524]For a review of the addition of organometallic reagents to conjugated enynes see Miginiac *J. Organomet. Chem.* **1982**, *238*, 235-266.

[525]House; Thompson *J. Org. Chem.* **1963**, *28*, 360; Klein *Tetrahedron* **1964**, *20*, 465. See however Marets; Rivière *Bull. Soc. Chim. Fr.* **1970**, 4320.

[526]For some mechanistic investigations see Berlan; Battioni; Koosha *J. Organomet. Chem.* **1978**, *152*, 359, *Bull Soc. Chim. Fr.* **1979**, II-183; Four; Riviere; Tang *Tetrahedron Lett.* **1977**, 3879; Casey; Cesa *J. Am. Chem. Soc.* **1979**, *101*, 4236; Smith; Hannah *Tetrahedron* **1979**, *35*, 1183; Krauss; Smith *J. Am. Chem. Soc.* **1981**, *103*, 141; Bartoli; Bosco; Dal Pozzo; Ciminale *J. Org. Chem.* **1982**, *47*, 5227; Corey; Boaz *Tetrahedron Lett.* **1985**, *26*, 6015; Yamamoto; Yamada; Uyehara *J. Am. Chem. Soc.* **1987**, *109*, 5820; Ullenius; Christenson *Pure Appl. Chem.* **1988**, *60*, 57; Christenson; Olsson; Ullenius *Tetrahedron* **1989**, *45*, 523; Krause *Tetrahedron Lett.* **1989**, *30*, 5219.

[527]See, for example, House; Umen *J. Am. Chem. Soc.* **1972**, *94*, 5495; Ruden; Litterer *Tetrahedron Lett.* **1975**, 2043; House; Snoble *J. Org. Chem.* **1976**, *41*, 3076; Wigal; Grunwell; Hershberger; *J. Org. Chem.* **1991**, *56*, 3759.

though the fact that retention of configuration at R has been demonstrated in several cases rules out a completely free R• radical.[528] For simple α,β-unsaturated ketones, such as 2-cyclohexenone, and Me_2CuLi, there is evidence[529] for this mechanism:

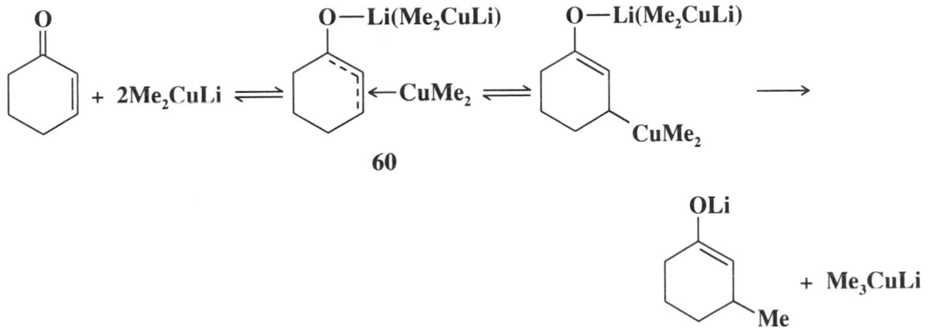

60 is a d,π^* complex, with bonding between copper, as a base supplying a pair of d electrons, and the enone as a Lewis acid using the π^* orbital of the allylic system.[529] The ^{13}C nmr spectrum of an intermediate similar to **60** has been reported.[530] The addition of R_3Al takes place by a free-radical mechanism.[505]

For the addition of organocopper reagents to alkynes and conjugated dienes, see **5-53**.
OS **IV**, 93; **V**, 762; **VI**, 442, 666, 762, 786; **65**, 203; **66**, 43, 52, 95.

5-19 The Addition of Boranes to Activated Double Bonds
     Hydro-alkyl-addition (overall transformation)

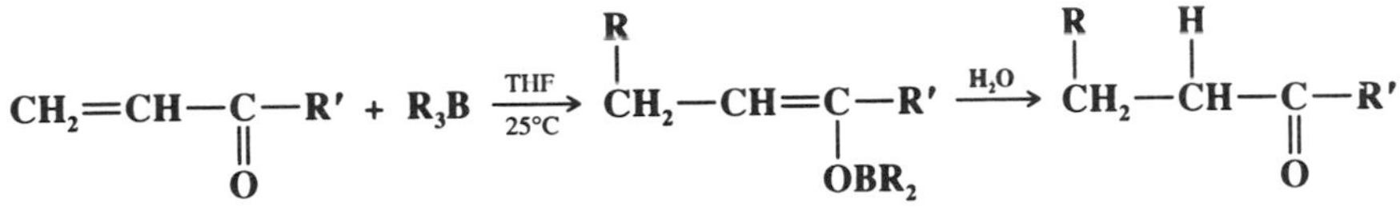

Trialkylboranes rapidly add to the double bonds of acrolein, methyl vinyl ketone, and certain of their derivatives in THF at 25°C to give enol borinates, which can be hydrolyzed to aldehydes or ketones.[531] The water may be present from the beginning, so the reaction can be run in one laboratory step. Since the boranes can be prepared from olefins (**5-12**), this reaction provides a means of lengthening a carbon chain by three or four carbons, respectively. Compounds containing a terminal alkyl group, such as crotonaldehyde $CH_3CH{=}CHCHO$ and 3-penten-2-one, fail to react under these conditions, as does acrylonitrile, but these compounds can be induced to react by the slow and controlled addition of O_2 or by initiation with peroxides or uv light.[532] A disadvantage is that only one of the

[528]Näf; Degen *Helv. Chim. Acta* **1971**, *54*, 1939; Whitesides; Kendall *J. Org. Chem.* **1972**, *37*, 3718. See also Ref. 465.
[529]Corey; Hannon; Boaz *Tetrahedron* **1989**, *45*, 545.
[530]Bertz; Smith *J. Am. Chem. Soc.* **1989**, *111*, 8276.
[531]Suzuki; Arase; Matsumoto; Itoh; Brown; Rogić; Rathke *J. Am. Chem. Soc.* **1967**, *89*, 5708; Köster; Zimmermann; Fenzl *Liebigs Ann. Chem.* **1976**, 1116. For reviews see Pelter; Smith; Brown; Ref. 348, pp. 301-305, 318-323; Brown; Midland *Angew. Chem. Int. Ed. Engl.* **1972**, *11*, 692-700, pp. 694-698 [*Angew. Chem. 84*, 702-710]; Kabalka *Intra-Sci. Chem. Rep.* **1973**, 7(1), 57-64; Brown *Boranes in Organic Chemistry*, Ref. 348, pp. 413-433.
[532]Brown; Kabalka *J. Am. Chem. Soc.* **1970**, *92*, 712, 714. See also Utimoto; Tanaka; Furubayashi; Nozaki *Tetrahedron Lett.* **1973**, 787; Miyaura; Kashiwagi; Itoh; Suzuki *Chem. Lett.* **1974**, 395.

three R groups of R_3B adds to the substrate, so that the other two are wasted. This difficulty is overcome by the use of a B-alkyl borinate such as **61**,[533] which can be prepared as shown.

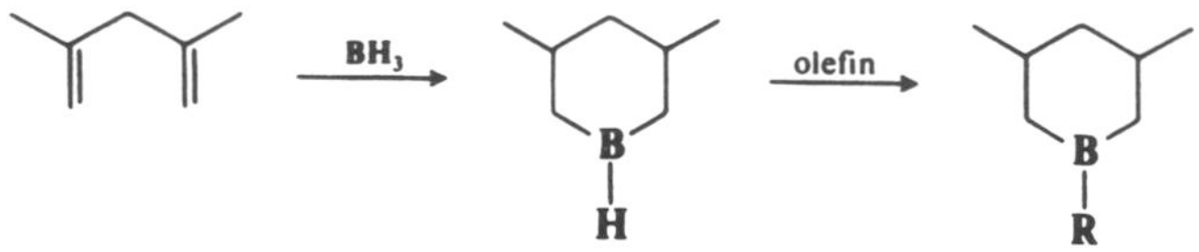

61

61 (R = *t*-butyl) can be made by treatment of **61** (R = OMe) with *t*-BuLi. The use of this reagent permits *t*-butyl groups to be added. B-1-Alkenyl-9-BBN compounds B-RCH=CR′-9-BBN (prepared by treatment of alkynes with 9-BBN or of RCH=CR′Li with B-methoxy-9-BBN[534]) add to methyl vinyl ketones to give, after hydrolysis, γ,δ-unsaturated ketones,[535] though B-R-9-BBN, where R = a saturated group, are not useful here, because the R group of these reagents does not preferentially add to the substrate.[533] The corresponding B-1-alkynyl-9-BBN compounds also give the reaction.[536] Like the three substrates mentioned above, 3-butyn-2-one fails to react in the absence of air but undergoes the reaction when exposed to a slow stream of air:[537]

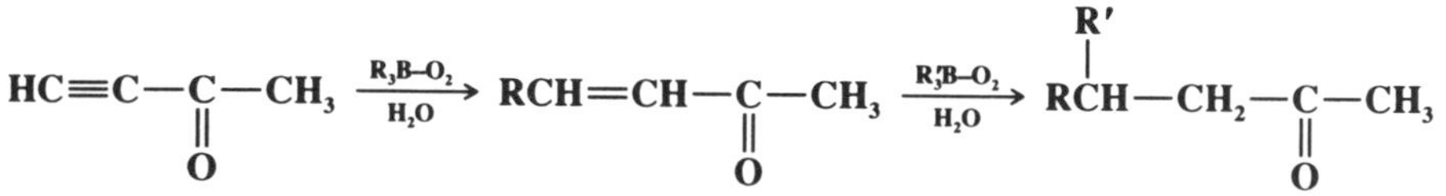

Since the product, **62**, is an α,β-unsaturated ketone, it can be made to react with another BR_3, the same or different, to produce a wide variety of ketones **63**.

The fact that these reactions are catalyzed by free-radical initiators and inhibited by galvinoxyl[538] (a free-radical inhibitor) indicates that free-radical mechanisms are involved.

5-20 The Addition of Tin and Mercury Hydrides to Activated Double Bonds
Hydro-alkyl-addition

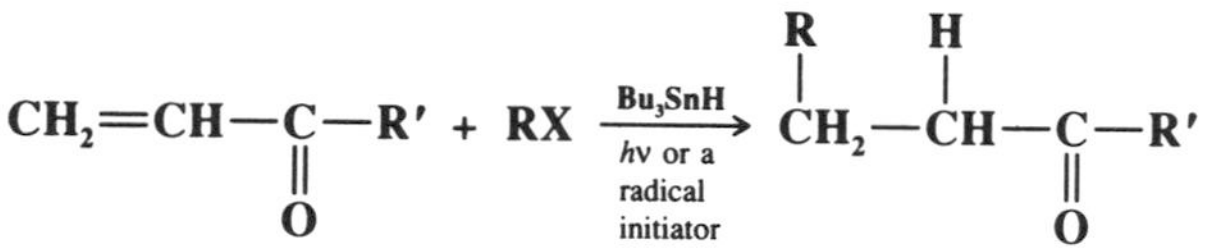

In a reaction similar to **5-19**, alkyl groups can be added to olefins activated by such groups as COR′, COOR′, CN, and even Ph.[539] In the method illustrated above, the R group comes

[533]Brown; Negishi *J. Am. Chem. Soc.* **1971**, *93*, 3777.
[534]Brown; Bhat; Rajagopalan *Organometallics* **1986**, *5*, 816.
[535]Jacob; Brown *J. Am. Chem. Soc.* **1976**, *98*, 7832; Satoh; Serizawa; Hara; Suzuki *J. Am. Chem. Soc.* **1985**, *107*, 5225. See also Molander; Singaram; Brown *J. Org. Chem.* **1984**, *49*, 5024. Alkenyldialkoxyboranes, together with BF_3-etherate, also transfer vinylic groups: Hara; Hyuga; Aoyama; Sato; Suzuki *Tetrahedron Lett.* **1990**, *31*, 247.
[536]Sinclair; Molander; Brown *J. Am. Chem. Soc.* **1977**, *99*, 954. See also Molander; Brown *J. Org. Chem.* **1977**, *42*, 3106.
[537]Suzuki; Nozawa; Itoh; Brown; Kabalka; Holland *J. Am. Chem. Soc.* **1970**, *92*, 3503.
[538]Kabalka; Brown; Suzuki; Honma; Arase; Itoh *J. Am. Chem. Soc.* **1970**, *92*, 710. See also Arase; Masuda; Suzuki *Bull. Chem. Soc. Jpn.* **1976**, *49*, 2275.
[539]For reviews, see Giese, Ref. 50, pp. 36-68; Giese *Angew. Chem. Int. Ed. Engl.* **1985**, *24*, 553-565 [*Angew. Chem. 97*, 555-567]; Larock *Organomercury Compounds in Organic Synthesis*; Springer: New York, 1985, pp. 263-273. The last review includes a table with many examples of the mercury method. For a list of reagents, with references, see Ref. 133, pp. 915-916.

from an alkyl halide (R = primary, secondary, or tertiary alkyl; X = Br or I) and the hydrogen from the tin hydride. Organomercury hydrides RHgH, generated in situ from RHgX and $NaBH_4$, can also be used.[540] When the tin method is used, Bu_3SnH can also be generated in a similar way, from R_3SnX and $NaBH_4$. The tin method has a broader scope (e.g., it can be used on $CH_2=CCl_2$), but the mercury method uses milder reaction conditions. Like **5-19**, these additions have free-radical mechanisms. The reaction has been used for free-radical cyclizations of the type discussed on p. 752.[541] Such cyclizations normally give predominant formation of 5-membered rings, but large rings (11 to 20 members) have also been synthesized by this reaction.[542]

Free-radical addition of an aryl group and a hydrogen has been achieved by treatment of activated olefins with a diazonium salt and $TiCl_3$.[543]

In a related reaction, a methyl group and a hydrogen can be added indirectly to the double bond of an otherwise unactivated allylic alcohol in this manner:[544]

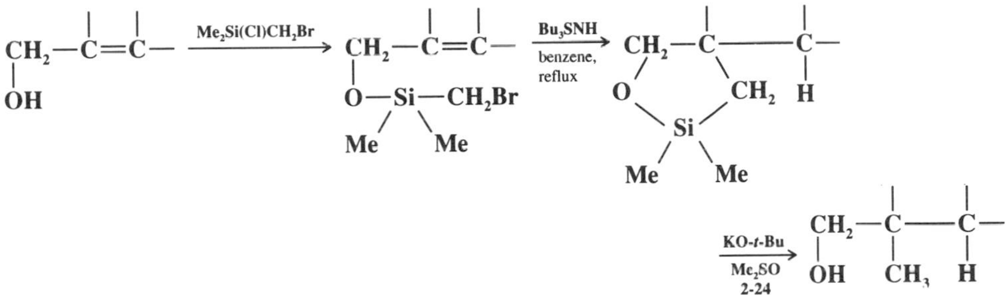

This procedure has been used to introduce angular methyl groups (methyl groups at the bridgeheads of fused rings).[545]

OS **VII**, 105.

5-21 Acylation of Activated Double Bonds and of Triple Bonds
Hydro-acyl-addition

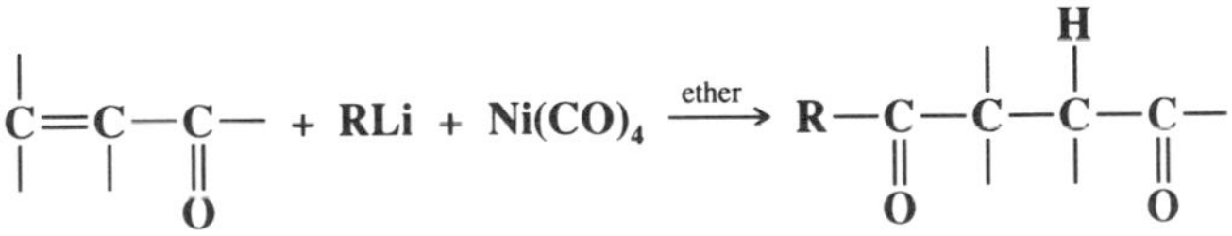

An acyl group can be introduced into the 4 position of an α,β-unsaturated ketone by treatment with an organolithium compound and nickel carbonyl.[546] The product is a 1,4-

[540]For the use of tris(trimethylsilyl)silane instead, see Giese; Koppong; Chatgilialoglu *Tetrahedron Lett.* **1989**, *30*, 681.

[541]For reviews, see Jasperse; Curran; Fevig *Chem. Rev.* **1991**, *91*, 1237-1286; Curran *Adv. Free Radical Chem.* (*Greenwich, Conn.*) **1990**, *1*, 121-157; Giese, Ref. 50, pp. 151-169. For a list of references, see Ref. 133, pp. 215-216.

[542]See Porter; Chang *J. Am. Chem. Soc.* **1987**, *109*, 4976.

[543]Citterio; Vismara *Synthesis* **1980**, 291. For other methods of adding an alkyl or aryl group and a hydrogen to activated double bonds by free-radical processes, see Cacchi; Palmieri *Synthesis* **1984**, 575; Lebedev; Lopatina; Berestova; Petrov; Beletskaya *J. Org. Chem. USSR* **1986**, *22*, 1238; Barton; Crich *J. Chem. Soc., Perkin Trans. 1* **1986**, 1603; Luche; Allavena *Tetrahedron Lett.* **1988**, *29*, 5369; Varea; González-Núñez; Rodrigo-Chiner; Asensio *Tetrahedron Lett.* **1989**, *30*, 4709; Barton; Sarma *Tetrahedron Lett.* **1990**, *31*, 1965.

[544]Stork; Sofia *J. Am. Chem. Soc.* **1986**, *108*, 6826. See also Stork *Bull. Chem. Soc. Jpn.* **1988**, *61*, 149.

[545]Stork; Mah *Tetrahedron Lett.* **1989**, *30*, 3609.

[546]Corey; Hegedus *J. Am. Chem. Soc.* **1969**, *91*, 4926.

diketone. R may be aryl or primary alkyl. The reaction can also be applied to alkynes (which need not be activated), in which case 2 moles add and the product is also a 1,4-diketone, e.g., $R'C{\equiv}CH \rightarrow RCOCHR'CH_2COR$.[547] In a different procedure, α,β-unsaturated ketones and aldehydes are acylated by treatment at $-110°C$ with $R_2(CN)CuLi_2$ and CO. This method is successful for R = primary, secondary, and tertiary alkyl.[548] For secondary and tertiary groups, $R(CN)CuLi$ (which does not waste an R group) can be used instead.[549]

Another method involves treatment with an aldehyde and cyanide ion (see **6-49**) in a polar aprotic solvent such as DMF or Me_2SO.[550]

$$RCHO + CN^- \rightleftharpoons R\overset{\overset{\displaystyle O^\ominus}{|}}{\underset{\underset{\displaystyle CN}{|}}{C}}H \rightleftharpoons R\overset{\ominus}{\underset{\underset{\displaystyle CN}{|}}{C}} + \overset{\displaystyle |}{C}{=}\overset{\displaystyle |}{C}-\overset{\overset{\displaystyle |}{|}}{\underset{\underset{\displaystyle O}{\|}}{C}}- \longrightarrow$$

$$\textbf{65}$$

$$R-\overset{\overset{\displaystyle OH}{|}}{\underset{\underset{\displaystyle CN}{|}}{C}}-\overset{\overset{\displaystyle H}{|}}{\underset{|}{C}}-\overset{|}{\underset{|}{C}}-\overset{\underset{\underset{\displaystyle O}{\|}}{C}}{-} \xrightarrow{-HCN} \textbf{64}$$

This method has been applied to α,β-unsaturated ketones, esters, and nitriles to give the corresponding 1,4-diketones, γ-keto esters, and γ-keto nitriles, respectively (see also **6-54**). The ion **65** is a synthon for the unavailable $R\overset{\ominus}{\overline{C}}{=}O$ anion (see also p. 471); it is a masked $R\overset{\ominus}{\overline{C}}{=}O$ anion. Other masked carbanions that have been used in this reaction are the $R\overset{\ominus}{\overline{C}}(CN)NR_2'$ ion,[551] the $EtS\overset{\ominus}{\overline{C}}RSOEt$ ion[552] (see p. 475), the $CH_2{=}\overset{\ominus}{\overline{C}}OEt$ ion,[553] $CH_2{=}C(OEt)Cu_2Li$,[554] $CH_2{=}CMe(SiMe_3)$,[554] and the $R\overset{\ominus}{\overline{C}}(OCHMeOEt)CN$ ion[555] (see p. 471). In the last case, best results are obtained when R is a vinylic group. Anions of 1,3-dithianes (**0-97**) do not give 1,4 addition to these substrates (except in the presence of HMPA, see **5-18**) but add 1,2 to the $C{=}O$ group instead (**6-41**).

In another procedure, acyl radicals derived from phenyl selenoesters ArCOSePh (by treatment of them with Bu_3SnH) add to α,β-unsaturated esters and nitriles to give γ-keto esters and γ-keto nitriles, respectively.[556] Hydroacylation has also been done by electrochemical reaction of the substrate with an anhydride.[557]

OS **VI**, 866; **65**, 26.

[547]Sawa; Hashimoto; Ryang; Tsutsumi *J. Org. Chem.* **1968**, *33*, 2159.
[548]Seyferth; Hui *J. Am. Chem. Soc.* **1985**, *107*, 4551. See also Lipshutz; Elworthy *Tetrahedron Lett.* **1990**, *31*, 477.
[549]Seyferth; Hui *Tetrahedron Lett.* **1986**, *27*, 1473.
[550]For reviews, see Stetter; Kuhlmann *Org. React.* **1991**, *40*, 407-496; Stetter *Angew. Chem. Int. Ed. Engl.* **1976**, *15*, 639-647 [*Angew. Chem.* **88**, 695-704]. For a similar method involving thiazolium salts, see Stetter; Kuhlmann *Chem. Ber.* **1976**, *109*, 2890; Stetter; Skobel *Chem. Ber.* **1987**, *120*, 643; Stetter; Kuhlmann; Haese *Org. Synth.* 65, 26.
[551]Enders; Gerdes; Kipphardt *Angew. Chem. Int. Ed. Engl.* **1990**, *29*, 179 [*Angew. Chem. 102*, 226].
[552]Herrmann; Richman; Schlessinger *Tetrahedron Lett.* **1973**, 3271, 3275.
[553]Beockman; Bruza; Baldwin; Lever *J. Chem. Soc., Chem. Commun.* **1975**, 519.
[554]Boeckman; Bruza *J. Org. Chem.* **1979**, *44*, 4781.
[555]Stork; Maldonado *J. Am. Chem. Soc.* **1974**, *96*, 5272.
[556]Boger; Mathvink *J. Org. Chem.* **1989**, *54*, 1777.
[557]Shono; Nishiguchi; Ohmizu *J. Am. Chem. Soc.* **1977**, *99*, 7396; Lund; Degrand *Tetrahedron Lett.* **1977**, 3593.

5-22   Addition of Alcohols, Amines, Carboxylic Esters, Aldehydes, etc.
      Hydro-acyl-addition, etc.

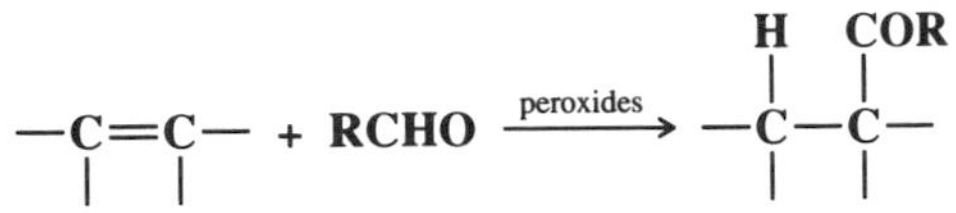

$$-C{=}C- + RCHO \xrightarrow{\text{peroxides}} \begin{array}{cc} H & COR \\ | & | \\ -C{-}C{-} \\ | & | \end{array}$$

Aldehydes, formates, primary, and secondary alcohols, amines, ethers, alkyl halides, compounds of the type Z—CH$_2$—Z', and a few other compounds add to double bonds in the presence of free-radical initiators.[558] This is formally the addition of RH to a double bond, but the "R" is not just any carbon but one connected to an oxygen or a nitrogen, a halogen, or to two Z groups (defined as on p. 464). The addition of aldehydes is illustrated above. Formates and formamides[559] add similarly:

$$-C{=}C- + HCOW \longrightarrow \begin{array}{cc} H & COW \\ | & | \\ -C{-}C{-} \\ | & | \end{array} \qquad W = OR, NH_2$$

Alcohols, ethers, amines, and alkyl halides add as follows (shown for alcohols):

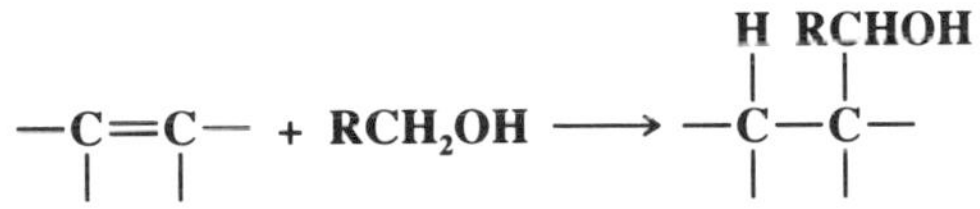

$$-C{=}C- + RCH_2OH \longrightarrow \begin{array}{cc} H & RCHOH \\ | & | \\ -C{-}C{-} \\ | & | \end{array}$$

ZCH$_2$Z' compounds react at the carbon bearing the active hydrogen:[560]

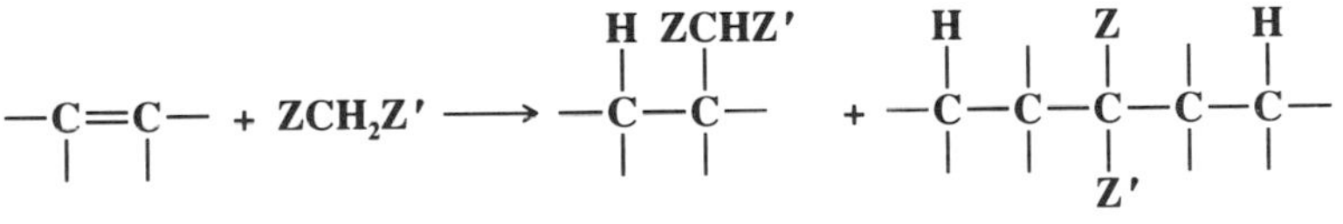

$$-C{=}C- + ZCH_2Z' \longrightarrow \begin{array}{cc} H & ZCHZ' \\ | & | \\ -C{-}C{-} \\ | & | \end{array} + \begin{array}{ccccc} H & & Z & & H \\ | & & | & & | \\ -C{-}C{-}C{-}C{-}C{-} \\ | & & | & & | \\ & & Z' & & \end{array}$$

Similar additions have been successfully carried out with carboxylic acids, anhydrides,[561] acyl halides, carboxylic esters, nitriles, and other types of compounds.[562]
      These reactions are not successful when the olefin contains electron-withdrawing groups such as halo or carbonyl groups. A free-radical initiator is required, usually peroxides or

[558]For reviews see Giese, Ref. 50, pp. 69-77; Vogel *Synthesis* **1970**, 99-140; Huyser, Ref. 49, pp.152-159; Elad *Fortschr. Chem. Forsch.* **1967**, *7*, 528-558.
      [559]Elad, Ref. 558, pp. 530-543.
      [560]For example, see Cadogan; Hey; Sharp *J. Chem. Soc. C* **1966**, 1743, *J. Chem. Soc. B* **1967**, 803; Hájek; Málek *Coll. Czech. Chem. Commun.* **1979**, *44*, 3695.
      [561]de Klein *Recl. Trav. Chim. Pays-Bas* **1975**, *94*, 48.
      [562]Allen; Cadogan; Hey *J. Chem. Soc.* **1965**, 1918; Cadogan *Pure Appl. Chem.* **1967**, *15*, 153-165, pp. 153-158. See also Giese; Zwick *Chem. Ber.* **1982**, *115*, 2526; Giese; Erfort *Chem. Ber.* **1983**, *116*, 1240.

uv light. The mechanism is illustrated for aldehydes but is similar for the other compounds:

$$R-\underset{\underset{O}{\|}}{C}-H \xrightarrow{\text{initiator}} R-\underset{\underset{O}{\|}}{C}\cdot \;+\; -\underset{\underset{|}{|}}{C}=\underset{\underset{|}{|}}{C}- \longrightarrow R-\underset{\underset{O}{\|}}{C}-\underset{\underset{|}{|}}{C}-\underset{\underset{|}{|}}{C}\cdot$$

$$R-\underset{\underset{O}{\|}}{C}-\underset{\underset{|}{|}}{C}-\underset{\underset{|}{|}}{C}\cdot \;+\; R-\underset{\underset{O}{\|}}{C}-H \longrightarrow R-\underset{\underset{O}{\|}}{C}-\underset{\underset{|}{|}}{C}-\underset{\underset{|}{|}}{C}-H \;+\; R-\underset{\underset{O}{\|}}{C}\cdot$$

etc.

Polymers are often side products.

Similar reactions have been carried out on acetylene.[563] In a cyclic version of the addition of aldehydes, 4-pentenal was converted to cyclopentanone with a rhodium-complex catalyst.[564]

OS **IV**, 430; **V**, 93; **VI**, 587, 615.

5-23 Hydrocarboxylation
Hydro-carboxy-addition

$$-\underset{\underset{|}{|}}{C}=\underset{\underset{|}{|}}{C}- \;+\; CO \;+\; H_2O \xrightarrow[\text{pressure}]{H^+} -\underset{\underset{|}{|}}{\overset{\overset{H}{|}}{C}}-\underset{\underset{|}{|}}{\overset{\overset{COOH}{|}}{C}}-$$

The acid-catalyzed hydrocarboxylation of olefins (the *Koch reaction*) can be performed in a number of ways.[565] In one method, the olefin is treated with carbon monoxide and water at 100 to 350°C and 500 to 1000 atm pressure with a mineral-acid catalyst. However, the reaction can also be performed under milder conditions. If the olefin is first treated with CO and catalyst and then water added, the reaction can be accomplished at 0 to 50°C and 1 to 100 atm. If formic acid is used as the source of both the CO and the water, the reaction can be carried out at room temperature and atmospheric pressure.[566] The formic acid procedure is called the *Koch–Haaf reaction* (the Koch–Haaf reaction can also be applied to alcohols, see **0-103**). Nearly all olefins can be hydrocarboxylated by one or more of these procedures. However, conjugated dienes are polymerized instead.

Hydrocarboxylation can also be accomplished under mild conditions (160°C and 50 atm) by the use of nickel carbonyl as catalyst. This is more often applied to triple bonds to give α,β-unsaturated acids, in which cases the conditions are milder still. Acid catalysts are used along with the nickel carbonyl, but basic catalysts can also be employed.[567] Other metallic

---

[563]For example, see Cywinski; Hepp *J. Org. Chem.* **1965**, *31*, 3814; DiPietro; Roberts *Angew. Chem. Int. Ed. Engl.* **1966**, *5*, 415 [*Angew. Chem. 78*, 388].

[564]Fairlie; Bosnich *Organometallics* **1988**, *7*, 936, 946.

[565]For reviews of hydrocarboxylation of double and triple bonds catalyzed by acids or metallic compounds, see Lapidus; Pirozhkov *Russ. Chem. Rev.* **1989**, *58*, 117-137; Anderson; Davies, in Hartley; Patai, Ref. 422, vol. 3, pp. 335-359, pp. 335-348; in Falbe *New Syntheses with Carbon Monoxide*; Springer: New York, 1980, the articles by Mullen, pp. 243-308; and Bahrmann, pp. 372-413; in Wender; Pino *Organic Syntheses via Metal Carbonyls*, vol. 2; Wiley: New York, 1977, the articles by Pino; Piacenti; Bianchi, pp. 233-296; and Pino; Braca pp. 419-516; Eidus; Lapidus; Puzitskii; Nefedov *Russ. Chem. Rev.* **1973**, *42*, 199-213, *Russ. Chem. Rev.* **1971**, *40*, 429-440; Falbe *Carbon Monoxide in Organic Synthesis*; Springer: Berlin, 1970, pp. 78-174.

[566]Koch; Haaf *Liebigs Ann. Chem.* **1958**, *618*, 251; Haaf *Chem. Ber.* **1966**, *99*, 1149; Christol; Solladié *Bull. Soc. Chim. Fr.* **1966**, 1307.

[567]Sternberg; Markby; Wender *J. Am. Chem. Soc.* **1960**, *82*, 3638.

salts and complexes, e.g., bis(triphenylphosphine)palladium dichloride $(Ph_3P)_2PdCl_2$,[568] have also been used. This has been done enantioselectively, with moderate-to-high optical yields, by the use of an optically active palladium-complex catalyst.[569] Triple bonds give unsaturated acids and saturated dicarboxylic acids when treated with carbon dioxide and an electrically reduced nickel complex catalyst.[570]

With any method, if the olefin contains a functional group such as OH, NH_2, or $CONH_2$, the corresponding lactone (**0-22**), lactam (**0-54**), or cyclic imide may be the product.[571]

When acid catalysts are employed, in the absence of nickel carbonyl, the mechanism[572] involves initial attack by a proton, followed by attack of the resulting carbocation on carbon monoxide to give an acyl cation, which, with water, gives the product:

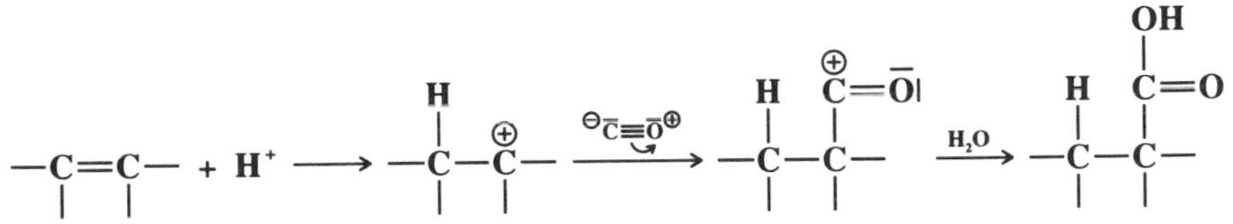

$$-C=C- \;+\; H^+ \;\longrightarrow\; -\overset{|}{\underset{|}{C}}-\overset{\oplus}{\underset{|}{C}}- \;\xrightarrow{\;^\ominus\overline{C}\equiv\overline{O}^\oplus\;}\; -\overset{H\;\;\overset{\oplus}{C}=\overline{O}|}{\underset{|}{C}}-\overset{|}{\underset{|}{C}}- \;\xrightarrow{\;H_2O\;}\; -\overset{H\;\;\overset{OH}{C}=O}{\underset{|}{C}}-\overset{|}{\underset{|}{C}}-$$

Therefore, Markovnikov's rule is followed, and carbon skeleton rearrangements and double-bond isomerizations (prior to attack by CO) are frequent.

For the nickel carbonyl reaction, the addition is syn for both alkenes and alkynes.[573] The following is the accepted mechanism:[573]

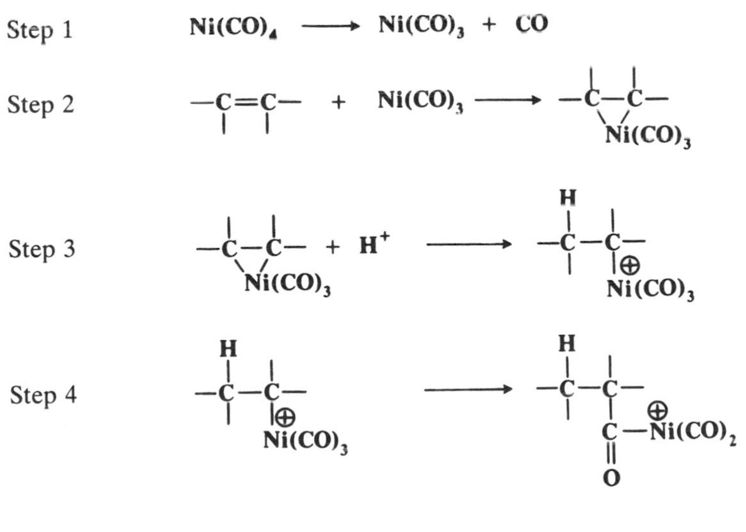

Step 1　　$Ni(CO)_4 \longrightarrow Ni(CO)_3 + CO$

Step 2　　$-C=C- \;+\; Ni(CO)_3 \longrightarrow -\overset{|}{C}-\overset{|}{C}- \atop \underset{Ni(CO)_3}{\diagdown\diagup}$

Step 3　　$-\overset{|}{C}-\overset{|}{C}- \atop \underset{Ni(CO)_3}{\diagdown\diagup} \;+\; H^+ \longrightarrow -\overset{H}{\underset{|}{C}}-\overset{|}{\underset{\overset{|}{\oplus Ni(CO)_3}}{C}}-$

Step 4　　$-\overset{H}{\underset{|}{C}}-\overset{|}{\underset{\overset{|}{\oplus Ni(CO)_3}}{C}}- \longrightarrow -\overset{H}{\underset{|}{C}}-\overset{|}{\underset{\overset{|}{\underset{\overset{\|}{O}}{C}-\overset{\oplus}{Ni}(CO)_2}}{C}}-$

[568]For reviews, see Heck *Palladium Reagents in Organic Synthesis*; Academic Press: New York, 1985, pp. 381-395; Bittler; Kutepow; Neubauer; Reis *Angew. Chem. Int. Ed. Engl.* **1968**, *7*, 329-335 [*Angew. Chem. 80*, 329-335]. For a review with respect to fluoroolefins, see Ojima *Chem. Rev.* **1988**, *88*, 1011-1030, pp. 1016-1019. See also Fenton *J. Org. Chem.* **1973**, *38*, 3192; Knifton *J. Org. Chem.* **1976**, *41*, 2885; Alper; Woell; Despeyroux; Smith *J. Chem. Soc., Chem. Commun.* **1983**, 1270; Lin; Alper *J. Chem. Soc., Chem. Commun.* **1989**, 248; Amer; Alper *J. Organomet. Chem.* **1990**, *383*, 573; Inomata; Toda; Kinoshita *Chem. Lett.* **1990**, 1567.

[569]Alper; Hamel *J. Am. Chem. Soc.* **1990**, *112*, 2803.

[570]Duñach; Dérien; Périchon *J. Organomet. Chem.* **1989**, *364*, C33.

[571]For reviews of these ring closures see Ohshiro; Hirao *Heterocycles* **1984**, *22*, 859-873; Falbe, Ref. 565, pp. 147-174, *Angew. Chem. Int. Ed. Engl.* **1966**, *5*, 435-446 [*Angew. Chem. 78*, 532-544], *Newer Methods Prep. Org. Chem.* **1971**, *6*, 193-222. See also Krafft; Wilson; Onan *Tetrahedron Lett.* **1989**, *30*, 539.

[572]For a review, see Hogeveen *Adv. Phys. Org. Chem.* **1973**, *10*, 29-52.

[573]Bird; Cookson; Hudec; Williams *J. Chem. Soc.* **1963**, 410.

Step 5

$$
\underset{\underset{O}{\overset{\parallel}{C}}-\overset{\oplus}{\text{Ni(CO)}_2}}{-\overset{\overset{H}{|}}{\underset{|}{C}}-\overset{|}{\underset{|}{C}}-} \quad \xrightarrow{\text{H}_2\text{O}} \quad -\overset{\overset{H}{|}}{\underset{|}{C}}-\overset{|}{\underset{\underset{O}{\overset{\parallel}{C}-\text{OH}}}{C}}-
$$

Step 3 is an electrophilic substitution. The principal step of the mechanism, step 4, is a rearrangement.

In either the acid catalysis or the nickel carbonyl (or other metallic catalyst) method, if alcohols, thiols, amines, etc. are used instead of water, the product is the corresponding ester, thiol ester, or amide, instead of the carboxylic acid.

5-24   Hydroformylation
Hydro-formyl-addition

$$
-\overset{|}{\underset{|}{C}}=\overset{|}{\underset{|}{C}}- + \text{CO} + \text{H}_2 \xrightarrow[\text{[Co(CO)}_4]_2]{\text{pressure}} -\overset{\overset{H}{|}}{\underset{|}{C}}-\overset{\overset{CHO}{|}}{\underset{|}{C}}-
$$

Olefins can be hydroformylated[574] by treatment with carbon monoxide and hydrogen over a catalyst. The most common catalysts are cobalt carbonyls and rhodium complexes[575] [e.g., hydridocarbonyltris(triphenylphosphine)rhodium], but other transition metal compounds have also been used. Cobalt catalysts are less active than the rhodium type, and catalysts of other metals are less active still.[576] Commercially, this is called the *oxo process*, but it can be carried out in the laboratory in an ordinary hydrogenation apparatus. The order of reactivity is straight-chain terminal olefins > straight-chain internal olefins > branched-chain olefins. Conjugated dienes give dialdehydes when rhodium catalysts are used[577] but saturated monoaldehydes (the second double bond is reduced) with cobalt carbonyls. 1,4- and 1,5-dienes may give cyclic ketones.[578] Many functional groups, e.g., OH, CHO, COOR, CN, can be present in the molecule, though halogens usually interfere. Hydroformylation of triple bonds proceeds very slowly, and few examples have been reported.[579] Among the side

[574]For reviews, see Kalck; Peres; Jenck *Adv. Organomet. Chem.* **1991**, *32*, 121-146; Davies, in Hartley; Patai, Ref. 422, vol. 3, pp. 361-389; Pino; Piacenti; Bianchi, in Wender; Pino, Ref. 565, pp. 43-231; Cornils, in Falbe *New Syntheses with Carbon Monoxide*, Ref. 565, pp. 1-225; Collman et al., Ref. 223, pp. 621-632; Pino *J. Organomet. Chem.* **1980**, *200*, 223-242; Pruett *Adv. Organomet. Chem.* **1979**, *17*, 1-60; Stille; James, in Patai, Ref. 1, pt. 2, pp. 1099-1166; Heck, Ref. 223, pp. 215-224; Khan; Martell, Ref. 159, vol. 2, pp. 39-60; Falbe *Carbon Monoxide in Organic Synthesis*, Ref. 565, pp. 3-77; Chalk; Harrod *Adv. Organomet. Chem.* **1968**, *6*, 119-170. For a review with respect to fluoroolefins, see Ohshiro; Hirao, Ref. 571.
[575]For example, see Osborn; Wilkinson; Young *Chem. Commun.* **1965**, 17; Brown; Wilkinson *Tetrahedron Lett.* **1969**, 1725, *J. Chem. Soc. A* **1970**, 2753; Stefani; Consiglio; Botteghi; Pino *J. Am. Chem. Soc.* **1973**, *95*, 6504; Bott *Chem. Ber.* **1975**, *108*, 997; van Leeuwen; Roobeek *J. Organomet. Chem.* **1983**, *258*, 343; Salvadori; Vitulli; Raffaelli; Lazzaroni *J. Organomet. Chem.* **1983**, *258*, 351; Collman; Belmont; Brauman *J. Am. Chem. Soc.* **1983**, *105*, 7288; Brown; Kent *J. Chem. Soc., Perkin Trans. 1* **1987**, 1597; Hanson; Davis *J. Chem. Educ.* **1987**, *64*, 928; Jackson; Perlmutter; Suh *J. Chem. Soc., Chem. Commun.* **1987**, 724; Hendriksen; Oswald; Ansell; Leta; Kastrup *Organometallics* **1989**, *8*, 1153; Lazzaroni; Uccello-Barretta; Benetti *Organometallics* **1989**, *8*, 2323; Amer; Alper *J. Am. Chem. Soc.* **1990**, *112*, 3674. For a review of the rhodium-catalyzed process, see Jardine, in Hartley, Ref. 218, vol. 4, pp. 733-818, pp. 778-784.
[576]Collman et al., Ref. 223, p. 630.
[577]Fell; Rupilius *Tetrahedron Lett.* **1969**, 2721.
[578]For a review of ring closure reactions with CO, see Mullen, in Falbe *New Syntheses with Carbon Monoxide*, Ref. 565, pp. 414-439. See also Eilbracht; Hüttmann; Deussen *Chem. Ber.* **1990**, *123*, 1063, and other papers in this series.
[579]For examples with rhodium catalysts, see Fell; Beutler *Tetrahedron Lett.* **1972**, 3455; Botteghi; Salomon *Tetrahedron Lett.* **1974**, 4285. For an indirect method, see Campi; Fitzmaurice; Jackson; Perlmutter; Smallridge *Synthesis* **1987**, 1032.

reactions are the aldol reaction (**6-39**), acetal formation, the Tishchenko reaction (**9-70**), and polymerization. Stereoselective syn addition has been reported.[580] Asymmetric hydroformylation has been accomplished with a chiral catalyst.[581]

When dicobalt octacarbonyl $[Co(CO)_4]_2$ is the catalyst, the species that actually adds to the double bond is tricarbonylhydrocobalt $HCo(CO)_3$.[582] Carbonylation $RCo(CO)_3$ + $CO \rightarrow RCo(CO)_4$ takes place, followed by a rearrangement and a reduction of the C—Co bond, similar to steps 4 and 5 of the nickel carbonyl mechanism shown in **5-23**. The reducing agent in the reduction step is tetracarbonylhydrocobalt $HCo(CO)_4$,[583] or, under some conditions, H_2.[584] When $HCo(CO)_4$ was the agent used to hydroformylate styrene, the observation of CIDNP indicated that the mechanism is different, and involves free radicals.[585] Alcohols can be obtained by allowing the reduction to continue after all the carbon monoxide is used up. It has been shown[586] that the formation of alcohols is a second step, occurring after the formation of aldehydes, and that $HCo(CO)_3$ is the reducing agent.

An indirect method for the hydroformylation of olefins involves formation of the trialkylborane (**5-12**) and treatment of this with carbon monoxide and a reducing agent (see **8-26**). *Hydroacylation* of alkenes has been accomplished, in variable yields, by treatment with an acyl halide and a rhodium complex catalyst, e.g.,[587]

$$CH_2{=}CH_2 \ + \ PhCOCl \ \xrightarrow{\text{HRh(CO)(PPh}_3)_2} \ CH_3CH_2{-}\underset{\underset{O}{\|}}{C}{-}Ph$$

OS **VI**, 338.

5-25   Addition of HCN
Hydro-cyano-addition

$$-\underset{|}{\overset{|}{C}}{=}\underset{|}{\overset{|}{C}}{-} \ + \ HCN \ \longrightarrow \ -\underset{|}{\overset{\overset{\displaystyle H}{|}}{C}}{-}\underset{|}{\overset{\overset{\displaystyle CN}{|}}{C}}{-}$$

Ordinary olefins do not react with HCN, but polyhalo olefins and olefins of the form C=C—Z add HCN to give nitriles.[588] The reaction is therefore a nucleophilic addition and is base-

[580]See, for example, Haelg; Consiglio; Pino *Helv. Chim. Acta* **1981**, *64*, 1865.

[581]For reviews, see Ojima; Hirai, in Morrison, Ref. 232, vol. 5, 1985, pp. 103-145, pp. 125-139; Consiglio; Pino *Top. Curr. Chem.* **1982**, *105*, 77-123. See also Kollár; Bakos; Tóth; Heil *J. Organomet. Chem.* **1988**, *350*, 277, **1989**, *370*, 257; Pottier; Mortreux; Petit *J. Organomet. Chem.* **1989**, *370*, 333; Stille; Su; Brechot; Parrinello; Hegedus *Organometallics* **1991**, *10*, 1183; Consiglio; Nefkens; Borer *Organometallics* **1991**, *10*, 2046.

[582]Heck; Breslow, *Chem. Ind. (London)* **1960**, 467, *J. Am. Chem. Soc.* **1961**, *83*, 4023; Karapinka; Orchin *J. Org. Chem.* **1961**, *26*, 4187; Whyman *J. Organomet. Chem.* **1974**, *81*, 97; Mirbach *J. Organomet. Chem.* **1984**, *265*, 205. For discussions of the mechanism see Orchin *Acc. Chem. Res.* **1981**, *14*, 259-266; Versluis; Ziegler; Baerends; Ravenek *J. Am. Chem. Soc.* **1989**, *111*, 2018.

[583]Alemdaroğlu; Penninger; Oltay *Monatsh. Chem.* **1976**, *107*, 1153; Ungváry; Markó *Organometallics* **1982**, *1*, 1120.

[584]See Kovács; Ungváry; Markó *Organometallics* **1986**, *5*, 209.

[585]Bockman; Garst; King; Markó; Ungváry *J.Organomet. Chem.* **1985**, *279*, 165.

[586]Aldridge; Jonassen *J. Am. Chem. Soc.* **1963**, *85*, 886.

[587]Schwartz; Cannon *J. Am. Chem. Soc.* **1974**, *96*, 4721. For some other hydroacylation methods see Cooke; Parlman *J. Am. Chem. Soc.* **1977**, *99*, 5222; Larock; Bernhardt *J. Org. Chem.* **1978**, *43*, 710; Suggs *J. Am. Chem. Soc.* **1979**, *101*, 489; Isnard; Denise; Sneeden; Cognion; Durual *J. Organomet. Chem.* **1982**, *240*, 285; Zudin; Il'inich; Likholobov; Yermakov *J. Chem. Soc., Chem. Commun.* **1984**, 545; Kondo; Akazome; Tsuji; Watanabe *J. Org. Chem.* **1990**, *55*, 1286.

[588]For reviews see Friedrich, in Patai; Rappoport, Ref. 49, pt. 2, pp. 1345-1390; Nagata; Yoshioka *Org. React.* **1977**, *25*, 255-476; Brown, in Wender; Pino, Ref. 565, pp. 655-672; Friedrich; Wallenfels, in Rappoport *The Chemistry of the Cyano Group*; Wiley: New York, 1970, pp. 68-72.

catalyzed. When Z is COR or, more especially, CHO, 1,2 addition (**6-51**) is an important competing reaction and may be the only reaction. Triple bonds react very well when catalyzed by an aqueous solution of CuCl, NH$_4$Cl, and HCl or by Ni or Pd compounds.[589] The HCN can be generated in situ from acetone cyanohydrin (see **6-49**), avoiding the use of the poisonous HCN.[590] One or two moles of HCN can be added to a triple bond, since the initial product is a Michael-type substrate. Acrylonitrile is commercially prepared this way, by the addition of HCN to acetylene. Alkylaluminum cyanides, e.g., Et$_2$AlCN, or mixtures of HCN and trialkylalanes R$_3$Al are especially good reagents for conjugate addition of HCN[591] to α,β-unsaturated ketones and α,β-unsaturated acyl halides. HCN can be added to ordinary olefins in the presence of dicobalt octacarbonyl[592] or certain other transition-metal compounds.[593] An indirect method for the addition of HCN to ordinary olefins uses an isocyanide RNC and Schwartz's reagent (see **5-13**); this method gives anti-Markovnikov addition.[594] *t*-Butyl isocyanide and TiCl$_4$ have been used to add HCN to C=C—Z olefins.[595]

OS **I**, 451; **II**, 498; **III**, 615; **IV**, 392, 393, 804; **V**, 239, 572; **VI**, 14.

For addition of ArH, see **1-12** (Friedel–Crafts alkylation).

Reactions in Which Hydrogen Adds to Neither Side

Some of these reactions are *cycloadditions* (reactions **5-36**, **5-37**, **5-42**, and **5-45** to **5-52**). In such cases addition to the multiple bond closes a ring:

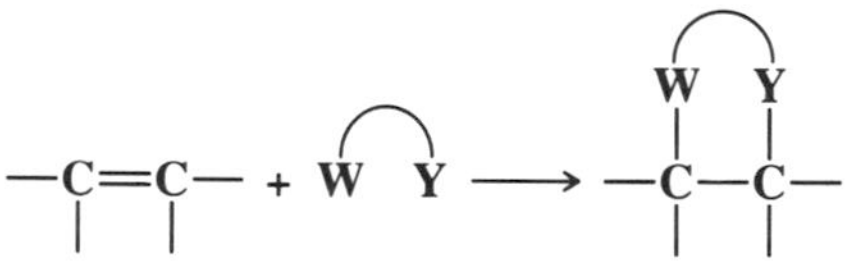

$$-\underset{|}{C}=\underset{|}{C}- \; + \; \overset{\frown}{W \quad Y} \longrightarrow -\overset{\overset{\displaystyle W}{|}}{\underset{|}{C}}-\overset{\overset{\displaystyle Y}{|}}{\underset{|}{C}}-$$

A.   Halogen on One or Both Sides

5-26   Halogenation of Double and Triple Bonds (Addition of Halogen, Halogen)
Dihalo-addition

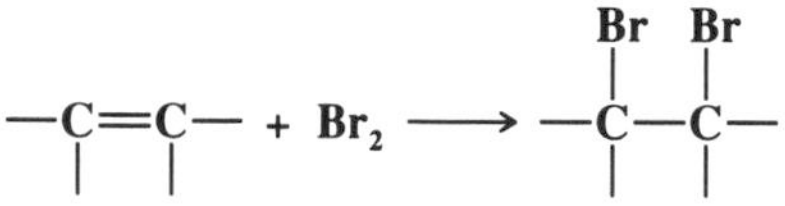

$$-\underset{|}{C}=\underset{|}{C}- \; + \; Br_2 \longrightarrow -\overset{\overset{\displaystyle Br}{|}}{\underset{|}{C}}-\overset{\overset{\displaystyle Br}{|}}{\underset{|}{C}}-$$

[589]Jackson; Lovel *Aust. J. Chem.* **1983**, *36*, 1975.
[590]Jackson; Permutter *Chem. Br.* **1986**, 338.
[591]For a review, see Nagata; Yoshioka Ref. 588.
[592]Arthur; England; Pratt; Whitman *J. Am. Chem. Soc.* **1954**, *76*, 5364.
[593]For a review, see Brown, Ref. 588, pp. 658-667. For a review of the nickel-catalyzed process, see Tolman; McKinney; Seidel; Druliner; Stevens *Adv. Catal.* **1985**, *33*, 1-46. For studies of the mechanism see Tolman; Seidel; Druliner; Domaille *Organometallics* **1984**, *3*, 33; Druliner *Organometallics* **1984**, *3*, 205; Bäckvall; Andell *Organometallics* **1986**, *5*, 2350; McKinney; Roe *J. Am. Chem. Soc.* **1986**, *108*, 5167; Funabiki; Tatsami; Yoshida *J. Organomet. Chem.* **1990**, *384*, 199. See also Jackson; Lovel; Perlmutter; Smallridge *Aust. J. Chem.* **1988**, *41*, 1099.
[594]Buchwald; LeMaire *Tetrahedron Lett.* **1987**, *28*, 295.
[595]Ito; Kato; Imai; Saegusa *J. Am. Chem. Soc.* **1982**, *104*, 6449.

Most double bonds are easily halogenated[596] with bromine, chlorine, or interhalogen compounds.[597] Iodination has also been accomplished, but the reaction is slower.[598] Under free-radical conditions, iodination proceeds more easily.[599] However, *vic*-diiodides are generally unstable and tend to revert to iodine and the olefin. The order of activity for some of the reagents is $BrCl > ICl$[600] $> Br_2 > IBr > I_2$.[601] Mixed halogenations have also been achieved by other methods. Mixtures of Br_2 and Cl_2 have been used to give bromochlorination,[602] as has tetrabutylammonium dichlorobromate Bu_4NBrCl_2;[603] iodochlorination has been achieved with $CuCl_2$ and either I_2, HI, CdI_2, or other iodine donors;[604] iodofluorination[605] with mixtures of AgF and I_2;[606] and mixtures of N-bromo amides in anhydrous HF give bromofluorination.[607] Bromo-, iodo-, and chlorofluorination have also been achieved by treatment of the substrate with a solution of Br_2, I_2, or an N-halo amide in polyhydrogen fluoride–pyridine;[608] while addition of I along with Br, Cl, or F has been accomplished with the reagent bis(pyridine)iodo(I) tetrafluoroborate $I(Py)_2BF_4$ and Br^-, Cl^-, or F^-, respectively.[609] This reaction (which is also successful for triple bonds[610]) can be extended to addition of I and other nucleophiles, e.g., NCO, OH, OAc, and NO_2.[609]

Under ordinary conditions fluorine itself is too reactive to give simple addition; it attacks other bonds and mixtures are obtained.[611] However, F_2 has been successfully added to certain double bonds in an inert solvent at low temperatures ($-78°C$), usually by diluting the F_2 gas with Ar or N_2.[612] Addition of fluorine has also been accomplished with other reagents, e.g., CoF_3,[613] XeF_2,[614] and a mixture of PbO_2 and SF_4.[615]

[596]For a list of reagents that have been used for di-halo-addition, with references, see Ref. 133, pp. 319-321.
[597]For a monograph, see de la Mare *Electrophilic Halogenation*; Cambridge University Press: Cambridge, 1976. For a review, see House, Ref. 144, pp. 422-431.
[598]Sumrell; Wyman; Howell; Harvey *Can. J. Chem.* **1964**, *42*, 2710; Zanger; Rabinowitz *J. Org. Chem.* **1975**, *40*, 248.
[599]Skell; Pavlis *J. Am. Chem. Soc.* **1964**, *86*, 2956; Ayres; Michejda; Rack *J. Am. Chem. Soc.* **1971**, *93*, 1389.
[600]For a review of ICl, see McCleland, in Pizey, Ref. 146, vol. 5, 1983, pp. 85-164.
[601]White; Robertson *J. Chem. Soc.* **1939**, 1509.
[602]Buckles; Forrester; Burham; McGee *J. Org. Chem.* **1960**, *25*, 24.
[603]Negoro; Ikeda *Bull. Chem. Soc. Jpn.* **1986**, *59*, 3519.
[604]Baird; Surridge; Buza *J. Org. Chem.* **1971**, *36*, 2088, 3324.
[605]For a review of mixed halogenations where one side is fluorine, see Sharts; Sheppard *Org. React.* **1974**, *21*, 125-406, pp. 137-157. See also German; Zemskov, Ref. 612. For a review of halogen fluorides in organic synthesis, see Boguslavskaya *Russ. Chem. Rev.* **1984**, *53*, 1178-1194.
[606]Hall; Jones *Can. J. Chem.* **1973**, *51*, 2902. See also Zupan; Pollak *J. Org. Chem.* **1976**, *41*, 2179; *J. Chem. Soc., Perkin Trans. 1* **1976**, 1745; Rozen; Brand *Tetrahedron Lett.* **1980**, *21*, 4543; Evans; Schauble *Synthesis* **1987**, 551; Kuroboshi; Hiyama *Synlett* **1991**, 185.
[607]Robinson; Finckenor; Oliveto; Gould *J. Am. Chem. Soc.* **1959**, *81*, 2191; Bowers *J. Am. Chem. Soc.* **1959**, *81*, 4107; Pattison; Peters; Dean *Can. J. Chem.* **1965**, *43*, 1689. For other methods, see Boguslavskaya; Chuvatkin; Kartashov; Ternovskoi *J. Org. Chem. USSR* **1987**, *23*, 230; Shimizu; Nakahara; Yoshioka *J. Chem. Soc., Chem. Commun.* **1989**, 1881.
[608]Olah; Nojima; Kerekes *Synthesis* **1973**, 780; Ref. 135. For other halofluorination methods, see Rozen; Brand *J. Org. Chem.* **1985**, *50*, 3342, **1986**, *51*, 222; Alvernhe; Laurent; Haufe *Synthesis* **1987**, 562; Camps; Chamorro; Gasol; Guerrero *J. Org. Chem.* **1989**, *54*, 4294; Ichihara; Funabiki; Hanafusa *Tetrahedron Lett.* **1990**, *31*, 3167.
[609]Barluenga; González; Campos; Asensio *Angew. Chem. Int. Ed. Engl.* **1985**, *24*, 319 [*Angew. Chem. 97*, 341].
[610]Barluenga; Rodríguez; González; Campos; Asensio *Tetrahedron Lett.* **1986**, *27*, 3303.
[611]See, for example, Fuller; Stacey; Tatlow; Thomas *Tetrahedron* **1962**, *18*, 123.
[612]Merritt; Stevens *J. Am. Chem. Soc.* **1966**, *88*, 1822; Merritt *J. Am. Chem. Soc.* **1967**, *89*, 609; Barton; Lister-James; Hesse; Pechet; Rozen *J. Chem. Soc., Perkin Trans. 1* **1982**, 1105; Rozen; Brand *J. Org. Chem.* **1986**, *51*, 3607. For reviews of the use of F_2 in organic synthesis, see Haas; Lieb *Chimia* **1985**, *39*, 134-140; Purrington; Kagen; Patrick *Chem. Rev.* **1986**, *86*, 997-1018. See also German; Zemskov *New Fluorinating Agents in Organic Synthesis*; Springer: New York, 1989.
[613]Rausch; Davis; Osborne *J. Org. Chem.* **1963**, *28*, 494.
[614]Zupan; Pollak *J. Org. Chem.* **1974**, *39*, 2646, **1976**, *41*, 4002, **1977**, *42*, 1559, *Tetrahedron Lett.* **1974**, 1015; Gregorčič; Zupan *J. Org. Chem.* **1979**, *44*, 1255; Shackelford *J. Org. Chem.* **1979**, *44*, 3485; Filler *Isr. J. Chem.* **1978**, *17*, 71-79. For a review of fluorination with xenon fluorides see Zupan, in Patai; Rappoport *The Chemistry of Functional Groups, Supplement D*, pt. 1; Wiley: New York, 1983, pp. 657-679.
[615]Bissell; Fields *J. Org. Chem.* **1964**, 29, 1591.

The reaction with bromine is very rapid and is easily carried out at room temperature. Bromine is often used as a test, qualitative or quantitative, for unsaturation.[616] The vast majority of double bonds can be successfully brominated. Even when aldehyde, ketone, amine, etc. functions are present in the molecule, they do not interfere, since the reaction with double bonds is faster.

Several other reagents add Cl_2 to double bonds, among them SO_2Cl_2,[617] PCl_5,[618] Me_3SiCl–MnO_2,[619] $MoCl_5$,[620] $KMnO_4$–oxalyl chloride,[620a] and iodobenzene dichloride $PhICl_2$.[621] A convenient reagent for the addition of Br_2 to a double bond on a small scale is the commercially available pyridinium bromide perbromide $C_5H_5NH^+$ Br_3^-.[622] Br_2 or Cl_2 can also be added with $CuBr_2$ or $CuCl_2$ in the presence of a compound such as acetonitrile, methanol, or triphenylphosphine.[623]

The mechanism is usually electrophilic (see p. 737), but when free-radical initiators (or uv light) are present, addition can occur by a free-radical mechanism.[624] Once Br· or Cl· radicals are formed, however, substitution may compete (**4-1** and **4-2**). This is especially important when the olefin has allylic hydrogens. Under free-radical conditions (uv light) bromine or chlorine adds to the benzene ring to give, respectively, hexabromo- and hexachlorocyclohexane. These are mixtures of stereoisomers (see p. 131).[625]

Conjugated systems give both 1,2 and 1,4 addition.[625] Triple bonds add bromine, though generally more slowly than double bonds (see p. 748). Molecules that contain both double and triple bonds are preferentially attacked at the double bond. Two moles of bromine can be added to triple bonds to give tetrabromo products. There is evidence that the addition of the first mole of bromine to a triple bond may take place by a nucleophilic mechanism.[626] I_2 on Al_2O_3 adds to triple bonds to give good yields of 1,2-diiodoalkenes.[627] With allenes it is easy to stop the reaction after only 1 mole has added, to give X—C—CX=C.[628] Addition of halogen to ketenes gives α-halo acyl halides, but the yields are not good.

OS **I**, 205, 521; **II**, 171, 177, 270, 408; **III**, 105, 123, 127, 209, 350, 526, 531, 731, 785; **IV**, 130, 195, 748, 851, 969; **V**, 136, 370, 403, 467; **VI**, 210, 422, 675, 862, 954.

5-27 Addition of Hypohalous Acids and Hypohalites (Addition of Halogen, Oxygen)
Hydroxy-chloro-addition, etc.[629]

$$—C{=}C— + HOCl \longrightarrow \overset{\displaystyle Cl}{\underset{\displaystyle |}{—C}}\!\!-\!\!\overset{\displaystyle OH}{\underset{\displaystyle |}{C—}}$$

[616]For a review of this, see Kuchar, in Patai, Ref. 36, pp. 273-280.
[617]Kharasch; Brown *J. Am. Chem. Soc.* **1939**, *61*, 3432.
[618]Spiegler; Tinker *J. Am. Chem. Soc.* **1939**, *61*, 940.
[619]Bellesia; Ghelfi; Pagnoni; Pinetti *J. Chem. Res. (S)* **1989**, *108*, 360.
[620]Uemura; Onoe; Okano *Bull. Chem. Soc. Jpn.* **1974**, *47*, 3121; San Filippo; Sowinski; Romano *J. Am. Chem. Soc.* **1975**, *97*, 1599. See also Nugent *Tetrahedron Lett.* **1978**, 3427.
[620a]Markó; Richardson *Tetrahedron Lett.* **1991**, *32*, 1831.
[621]See, for example, Tanner; Gidley *J. Org. Chem.* **1968**, *33*, 38; Masson; Thuillier *Bull. Soc. Chim. Fr.* **1969**, 4368; Lasne; Thuillier *Bull Soc. Chim. Fr.* **1974**, 249.
[622]Fieser; Fieser *Reagents for Organic Synthesis*, vol. 1; Wiley: New York, 1967, pp. 967-970. For a discussion of the mechanism with Br_3^-, see Bellucci; Bianchini; Vecchiani *J. Org. Chem.* **1986**, *51*, 4224.
[623]Koyano *Bull. Chem. Soc. Jpn.* **1970**, *43*, 1439, 3501; Uemura; Tabata; Kimura; Ichikawa *Bull. Chem. Soc. Jpn.* **1971**, *44*, 1973; Or; Levy; Asscher; Vofsi *J. Chem. Soc., Perkin Trans. 2* **1974**, 857; Uemura; Okazaki; Onoe; Okano *J. Chem. Soc., Perkin Trans. 1* **1977**, 676; Ref. 604.
[624]For example, see Poutsma *J. Am. Chem. Soc.* **1965**, *87*, 2161, 2172, *J. Org. Chem.* **1966**, *31*, 4167; Dessau, *J. Am. Chem. Soc.* **1979**, *101*, 1344.
[625]For a review, see Cais, in Patai, Ref. 36, pp. 993-999.
[626]Sinn; Hopperdietzel; Sauermann *Monatsh. Chem.* **1965**, *96*, 1036.
[627]Hondrogiannis; Lee; Kabalka; Pagni *Tetrahedron Lett.* **1989**, *30*, 2069.
[628]For a review of additions of halogens to allenes, see Jacobs, in Landor, Ref. 95, vol. 2, pp. 466-483.
[629]Addends are listed in order of priority in the Cahn–Ingold–Prelog system (p. 109).

HOCl, HOBr, and HOI can be added to olefins[630] to produce halohydrins.[631] HOBr and HOCl are often generated in situ by the reaction between water and Br_2 or Cl_2 respectively. HOI, generated from I_2 and H_2O, also adds to double bonds, if the reaction is carried out in tetramethylene sulfone–$CHCl_3$[632] or if an oxidizing agent such as HIO_3 is present.[633] HOF has also been added, but this reagent is difficult to prepare in a pure state and detonations occur.[634] HOBr can also be conveniently added by the use of a reagent consisting of an N-bromo amide [e.g., N-bromosuccinimide (NBS) or N-bromoacetamide] and a small amount of water in a solvent such as Me_2SO or dioxane.[635] An especially powerful reagent for HOCl addition is t-butyl hydroperoxide (or di-t-butyl peroxide) along with $TiCl_4$. This reaction is generally complete within 15 min at $-78°C$.[636] Chlorohydrins can be conveniently prepared by treatment of the alkene with Chloramine T ($TsNCl^- Na^+$)[637] in acetone–water.[638] HOI can be added by treatment of alkenes with periodic acid and $NaHSO_3$.[639]

The mechanism of HOX addition is electrophilic, with initial attack by the positive halogen end of the HOX dipole. Following Markovnikov's rule, the positive halogen goes to the side of the double bond that has more hydrogens. The resulting carbocation (or bromonium or iodonium ion) reacts with OH^- or H_2O to give the product. If the substrate is treated with Br_2 or Cl_2 (or another source of positive halogen such as NBS) in an alcohol or a carboxylic acid solvent, it is possible to obtain, directly X—C—C—OR or X—C—C—OCOR, respectively (see also **5-35**).[640] Even the weak nucleophile $CF_3SO_2O^-$ can participate in the second step: The addition of Cl_2 or Br_2 to olefins in the presence of this ion resulted in the formation of some β-haloalkyl triflates.[641] There is evidence that the mechanism with Cl_2 and H_2O is different from that with HOCl.[642] HOCl and HOBr can be added to triple bonds to give dihalo carbonyl compounds —CX_2—CO—.

t-Butyl hypochlorite, hypobromite, and hypoiodite[643] add to double bonds to give halogenated t-butyl ethers, e.g.,

$$—C{=}C— + Me_3COCl \longrightarrow —\overset{\overset{\textstyle Cl}{|}}{C}—\overset{\overset{\textstyle OCMe_3}{|}}{C}—$$

This is a convenient method for the preparation of tertiary ethers. When Me_3COCl or Me_3COBr is added to olefins in the presence of excess ROH, the ether produced is

[630]For a list of reagents used to accomplish these additions, with references, see Ref. 133, pp. 325-327.

[631]For a review, see Boguslavskaya *Russ. Chem. Rev.* **1972**, *41*, 740-749.

[632]Cambie; Noall; Potter; Rutledge; Woodgate *J. Chem. Soc., Perkin Trans. 1* **1977**, 266.

[633]See, for example, Cornforth; Green *J. Chem. Soc. C* **1970**, 846; Furrow *Int. J. Chem. Kinet.* **1982**, *14*, 927; Antonioletti; D'Auria; De Mico; Piancatelli; Scettri *Tetrahedron* **1983**, *39*, 1765.

[634]Migliorese; Appelman; Tsangaris *J. Org. Chem.* **1979**, *44*, 1711.

[635]For examples, see Dalton; Hendrickson; Jones *Chem. Commun.* **1966**, 591; Dalton; Dutta *J. Chem. Soc. B* **1971**, 85; Sisti *J. Org. Chem.* **1970**, *35*, 2670.

[636]Klunder; Caron; Uchiyama; Sharpless *J. Org. Chem.* **1985**, *50*, 912.

[637]For reviews of this reagent, see Bremner, in Pizey, Ref. 146, vol. 6, 1985, pp. 9-59; Campbell; Johnson *Chem. Rev.* **1978**, *78*, 65-79.

[638]Damin; Garapon; Sillion *Synthesis* **1981**, 362.

[639]Ohta; Sakata; Takeuchi; Ishii *Chem. Lett.* **1990**, 733.

[640]For a list of reagents that accomplish alkoxy-halo-addition, with references, see Ref. 133, pp. 327-328.

[641]Zefirov; Koz'min; Sorokin; Zhdankin *J. Org. Chem. USSR* **1982**, *18*, 1546. For reviews of this and related reactions, see Zefirov; Koz'min *Acc. Chem. Res.* **1985**, *18*, 154, *Sov. Sci. Rev., Sect. B* **1985**, *7*, 297-339.

[642]Buss; Rockstuhl; Schnurpfeil *J. Prakt. Chem.* **1982**, *324*, 197.

[643]Glover; Goosen *Tetrahedron Lett.* **1980**, *21*, 2005.

X—C—C—OR.[644] Vinylic ethers give β-halo acetals.[645] A mixture of Cl_2 and SO_3 at $-78°C$ converts alkenes to 2-chloro chlorosulfates $ClCHRCHROSO_2Cl$, which are stable compounds.[646] Chlorine acetate [solutions of which are prepared by treating Cl_2 with $Hg(OAc)_2$ in an appropriate solvent] adds to olefins to give acetoxy chlorides.[647] The latter are also produced[648] by treatment of olefins with a mixture of $PdCl_2$ and $CuCl_2$ in acetic acid[649] or with chromyl chloride CrO_2Cl_2 in acetyl chloride.[650] Acetoxy fluorides have been obtained by treatment of olefins with CH_3COOF.[651]

An internal example of the addition of X and OCOR is called *halolactonization*;[652] e.g.,

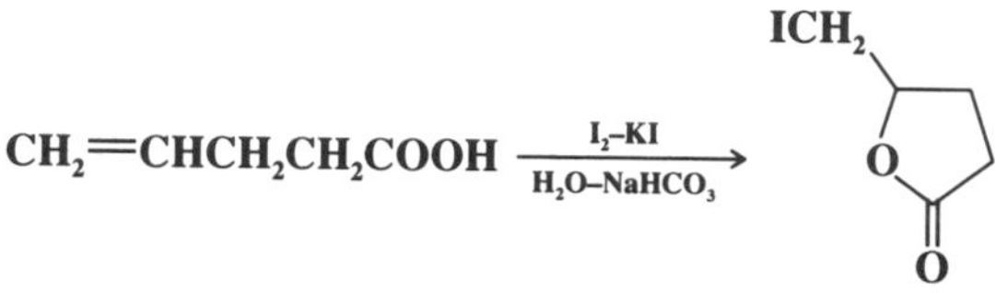

$$CH_2=CHCH_2CH_2COOH \xrightarrow[H_2O-NaHCO_3]{I_2-KI}$$

This reaction has been used mostly to prepare iodo lactones, but bromo lactones and, to a lesser extent, chloro lactones, have also been prepared. In the case of γ,δ-unsaturated acids, 5-membered rings (γ-lactones) are predominantly formed (as shown above; note that Markovnikov's rule is followed), but 6-membered and even 4-membered lactones have also been made by this procedure. Thallium reagents, along with the halogen, have also been used.[653]

For a method of iodoacetyl addition, see **5-35**.

OS **I**, 158; **IV**, 130, 157; **VI**, 184, 361, 560; **VII**, 164; **67**, 105; **69**, 38.

5-28 Addition of Sulfur Compounds (Addition of Halogen, Sulfur)
Alkylsulfonyl-chloro-addition, etc.[654]

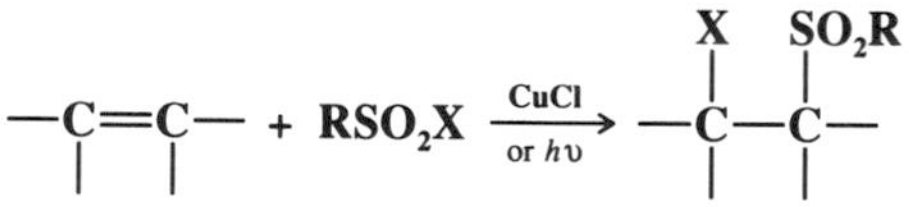

$$—C=C— + RSO_2X \xrightarrow[or\ h\upsilon]{CuCl} —\overset{X}{\underset{|}{C}}—\overset{SO_2R}{\underset{|}{C}}—$$

Sulfonyl halides add to double bonds, to give β-halo sulfones, in the presence of free-radical initiators or uv light. A particularly good catalyst is cuprous chloride.[655] Triple bonds behave

[644]Bresson; Dauphin; Geneste; Kergomard; Lacourt *Bull. Soc. Chim. Fr.* **1970**, 2432, **1971**, 1080.
[645]Weissermel; Lederer *Chem. Ber.* **1963**, *96*, 77.
[646]Zefirov; Koz'min; Sorokin *J. Org. Chem.* **1984**, *49*, 4086.
[647]de la Mare; Wilson; Rosser *J. Chem. Soc., Perkin Trans. 2* **1973**, 1480; de la Mare; O'Connor; Wilson *J. Chem. Soc., Perkin Trans. 2* **1975**, 1150. For the addition of bromine acetate see Wilson; Woodgate *J. Chem. Soc., Perkin Trans. 2* **1976**, 141.
[648]For a list of reagents that accomplish acyloxy-halo-addition, with references, see Ref. 133, pp. 328-329.
[649]Henry *J. Org. Chem.* **1967**, *32*, 2575, **1973**, *38*, 1681. For a 1,4 example of this addition, see Bäckvall; Nystrom; Nordberg *J. Am. Chem. Soc.* **1985**, *107*, 3676; Nyström; Rein; Bäckvall *Org. Synth.* 67, 105.
[650]Bäckvall Young; Sharpless *Tetrahedron Lett.* **1977**, 3523.
[651]Rozen; Lerman; Kol; Hebel *J. Org. Chem.* **1985**, *50*, 4753.
[652]For reviews, see Cardillo; Orena *Tetrahedron* **1990**, *46*, 3321-3408; Dowle; Davies *Chem. Soc. Rev.* **1979**, *8*, 171-197. For a list of reagents that accomplish this, with references, see Ref. 133, pp. 945-946. For a review with respect to the stereochemistry of the reaction, see Bartlett, in Morrison, Ref. 232, vol. 3, 1984, pp. 411-454, pp. 416-425.
[653]See Cambie; Rutledge; Somerville; Woodgate *Synthesis* **1988**, 1009, and references cited therein.
[654]When a general group (such as halo) is used, its priority is that of the lowest member of its group (see footnote 629). Thus the general name for this transformation is halo-alkylsulfonyl-addition because "halo" has the same priority as "fluoro," its lowest member.
[655]Asscher; Vofsi *J. Chem. Soc.* **1964**, 4962; Truce; Goralski; Christensen; Bavry *J. Org. Chem.* **1970**, *35*, 4217; Sinnreich; Asscher *J. Chem. Soc., Perkin Trans. 1* **1972**, 1543.

similarly, to give β-halo-α,β-unsaturated sulfones.[656] In a similar reaction, sulfenyl chlorides, RSCl, give β-halo thioethers.[657] The latter may be free-radical or electrophilic additions, depending on conditions. The addition of MeS and Cl has also been accomplished by treating the olefin with Me_3SiCl and Me_2SO.[658] The use of Me_3SiBr and Me_2SO does not give this result; dibromides (**5-26**) are formed instead. MeS and F have been added by treatment of the olefin with dimethyl(methylthio)sulfonium fluoroborate $Me_2\overset{\oplus}{S}SMe$ BF_4^- and triethylamine tris-hydrofluoride Et_3N-3HF.[659] β-Iodo thiocyanates can be prepared from alkenes by treatment with I_2 and isothiocyanatotributylstannane Bu_3SnNCS.[660] Bromothiocyanation can be accomplished with Br_2 and thallium(I) thiocyanate.[661]

β-Halo disulfides, formed by addition of arenethiosulfenyl chlorides to double-bond compounds, are easily converted to thiiranes by treatment with sodium amide or sodium sulfide.[662]

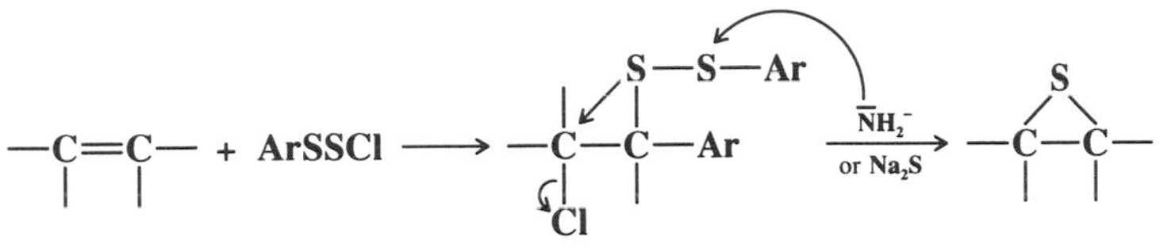

The overall episulfidation is a stereospecific syn addition.

OS **65**, 90. See also OS **VII**, 251.

5-29 Addition of Halogen and an Amino Group (Addition of Halogen, Nitrogen)
Dialkylamino-chloro-addition

The groups R_2N and Cl can be added directly to olefins, allenes, conjugated dienes, and alkynes, by treatment with dialkyl-N-chloroamines and acids.[663] These are free-radical additions, with initial attack by the $R_2NH^{\bullet+}$ radical ion.[664] N-Halo amides RCONHX add RCONH and X to double bonds under the influence of uv light or chromous chloride.[665] For an indirect way of adding NH_2 and I to a double bond, see **5-32.**

[656]Truce; Wolf *J. Org. Chem.* **1971**, *36*, 1727; Amiel *J. Org. Chem.* **1971**, *36*, 3691, 3697, **1974**, *39*, 3867; Zakharin; Zhigareva *J. Org. Chem. USSR* **1973**, *9*, 918; Okuyama; Izawa; Fueno *J. Org. Chem.* **1974**, *39*, 351.

[657]For reviews, see Rasteikiene; Greiciute; Lin'kova; Knunyants *Russ. Chem. Rev.* **1977**, *46*, 548-564; Kühle *Synthesis* **1971**, 563-586.

[658]Bellesia; Ghelfi; Pagnoni; Pinetti *J. Chem. Res. (S)* **1987**, 238. See also Liu; Nyangulu *Tetrahedron Lett.* **1988**, *29*, 5467.

[659]Haufe; Alvernhe; Anker; Laurent; Saluzzo *Tetrahedron Lett.* **1988**, *29*, 2311.

[660]Woodgate; Janssen; Rutledge; Woodgate; Cambie *Synthesis* **1984**, 1017, and references cited therein. See also Watanabe; Uemura; Okano *Bull. Chem. Soc. Jpn.* **1983**, *56*, 2458.

[661]Cambie; Larsen; Rutledge; Woodgate *J. Chem. Soc., Perkin Trans. 1* **1981**, 58.

[662]Fujisawa; Kobori *Chem. Lett.* **1972**, 935. For another method of olefin-thiirane conversion, see Capozzi; Capozzi; Menichetti *Tetrahedron Lett.* **1988**, *29*, 4177.

[663]Neale; Hinman *J. Am. Chem. Soc.* **1963**, *85*, 2666; Neale; Marcus *J. Org. Chem.* **1967**, *32*, 3273; Minisci; Galli; Cecere *Tetrahedron Lett.* **1966**, 3163. For reviews see Mirskova; Drozdova; Levkovskaya; Voronkov *Russ. Chem. Rev.* **1989**, *58*, 250-271; Neale *Synthesis* **1971**, 1-15; Sosnovsky; Rawlinson *Adv. Free-Radical Chem.* **1972**, *4*, 203-284, pp. 238-249.

[664]For a review of these species, see Chow; Danen; Nelson; Rosenblatt *Chem. Rev.* **1978**, *78*, 243-274.

[665]Tuaillon; Couture; Lessard *Can. J. Chem.* **1987**, *65*, 2194, and other papers in this series. For a review, see Labeish; Petrov *Russ. Chem. Rev.* **1989**, *58*, 1048-1061.

5-30   Addition of NOX and NO$_2$X (Addition of Halogen, Nitrogen)
Nitroso-chloro-addition

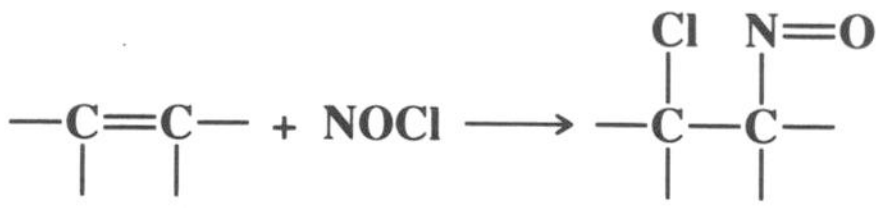

$$
-C=C- + NOCl \longrightarrow \underset{\underset{Cl}{|}}{-C}-\underset{\underset{N=O}{|}}{C}-
$$

There are three possible products when NOCl is added to olefins.[666] The initial product is always the β-halo nitroso compound, but these are stable only if the carbon bearing the nitrogen has no hydrogen. If it has, the nitroso compound tautomerizes to the oxime:

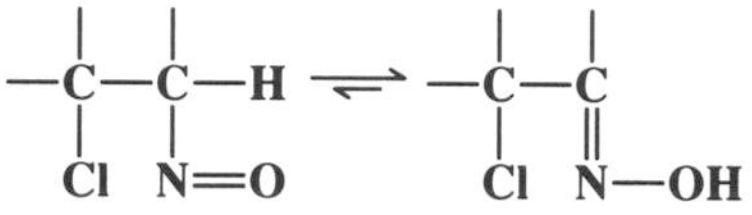

$$
-\underset{\underset{Cl}{|}}{C}-\underset{\underset{N=O}{|}}{C}-H \;\rightleftharpoons\; -\underset{\underset{Cl}{|}}{C}-\underset{\underset{N-OH}{\|}}{C}
$$

With some olefins, the initial β-halo nitroso compound is oxidized by the NOCl to a β-halo nitro compound.[667] Many functional groups can be present without interference, e.g., COOH, COOR, CN, OR. The mechanism in most cases is probably simple electrophilic addition, and the addition is usually anti, though syn addition has been reported in some cases.[668] Markovnikov's rule is followed, the positive NO going to the carbon that has more hydrogens.

Nitryl chloride NO$_2$Cl also adds to olefins, to give β-halo nitro compounds, but this is a free-radical process. The NO$_2$ goes to the less-substituted carbon.[669] Nitryl chloride also adds to triple bonds to give the expected 1-nitro-2-chloro olefins.[670] FNO$_2$ can be added to olefins[671] by treatment with HF in HNO$_3$[672] or by addition of the olefin to a solution of nitronium tetrafluoroborate NO$_2^+$ BF$_4^-$ (see **1-2**) in 70% polyhydrogen fluoride–pyridine solution[673] (see also **5-26**).

OS **IV**, 711; **V**, 266, 863.

5-31   Addition of XN$_3$ (Addition of Halogen, Nitrogen)
Azido-iodo-addition

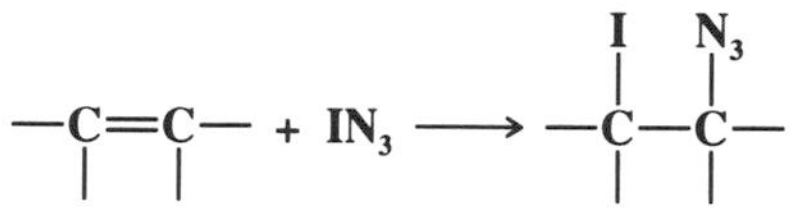

$$
-C=C- + IN_3 \longrightarrow \underset{\underset{I}{|}}{-C}-\underset{\underset{N_3}{|}}{C}-
$$

The addition of iodine azide to double bonds gives β-iodo azides.[674] The addition is stereospecific and anti, suggesting that the mechanism involves a cyclic iodonium ion interme-

---

[666]For a review, see Kadzyauskas; Zefirov *Russ. Chem. Rev.* **1968**, *37*, 543-550.

[667]For a review of the preparation of halo nitro compounds see Shvekhgeimer; Smirnyagin; Sadykov; Novikov *Russ. Chem. Rev.* **1968**, *37*, 351-363.

[668]For example, see Meinwald; Meinwald; Baker *J. Am. Chem. Soc.* **1964**, *86*, 4074.

[669]Shechter *Rec. Chem. Prog.* **1964**, *25*, 55-76.

[670]Schlubach; Braun *Liebigs Ann. Chem.* **1959**, *627*, 28.

[671]For a review, see Sharts; Sheppard *Org. React.* **1974**, *21*, 125-406, pp. 236-243.

[672]Knunyants; German; Rozhkov *Bull Acad. Sci. USSR, Div. Chem. Sci.* **1963**, 1794.

[673]Olah; Nojima *Synthesis* **1973**, 785.

[674]For reviews, see Dehnicke *Angew. Chem. Int. Ed. Engl.* **1979**, *18*, 507-514 [*Angew. Chem. 91*, 527-534]; Hassner *Acc. Chem. Res.* **1971**, *4*, 9-16; Biffin; Miller; Paul, Ref. 214, pp. 136-147.

diate.[675] The reaction has been performed on many double-bond compounds, including allenes[676] and α,β-unsaturated ketones. Similar reactions can be performed with BrN_3[677] and ClN_3. 1,4 addition has been found with acylic conjugated dienes.[678] In the case of BrN_3 both electrophilic and free-radical mechanisms are important,[679] while with ClN_3 the additions are chiefly free-radical.[680] IN_3 also adds to triple bonds to give β-iodo-α,β-unsaturated azides.[681]

β-Iodo azides can be reduced to aziridines with $LiAlH_4$[682] or converted to N-alkyl- or N-arylaziridines by treatment with an alkyl- or aryldichloroborane followed by a base.[683] In

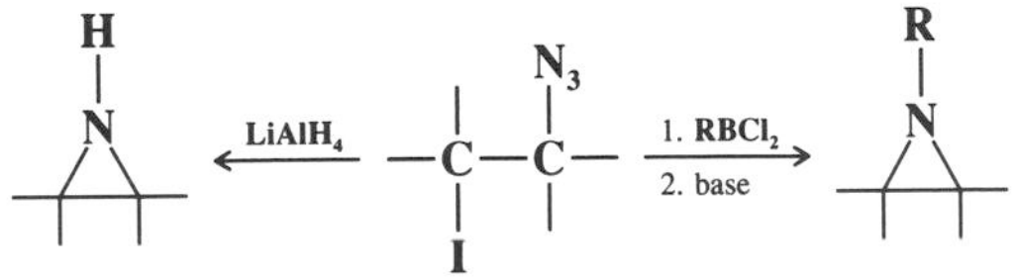

both cases the azide is first reduced to the corresponding amine (primary or secondary, respectively) and ring closure (**0-43**) follows.

OS **VI**, 893.

5-32 Addition of INCO (Addition of Halogen, Nitrogen)
Isocyanato-iodo-addition

In a reaction similar to **5-31**, iodine isocyanate adds to double bonds to give β-iodo isocyanates.[684] The addition is stereospecific and anti; the mechanism similar to that shown in **5-31**. The reaction has been applied to mono-, di-, and some trisubstituted olefins. The orientation generally follows Markovnikov's rule, the positive iodine adding to the less highly substituted side. α,β-Unsaturated carbonyl compounds do not react. Triple bonds give β-iodo-α,β-unsaturated isocyanates.[685] Allenes add 1 mole of INCO to give β-iodo-β,γ-unsaturated isocyanates.[686] Since an isocyanate group can be hydrolyzed to an amino group (RNCO $\rightarrow$ RNH_2, **6-3**), the method is an indirect way of adding H_2N and I to double bonds.

OS **VI**, 795.

[675]See, however, Cambie; Hayward; Rutledge; Smith-Palmer; Swedlund; Woodgate *J. Chem. Soc., Perkin Trans. 1* **1979**, 180.
[676]Hassner; Keogh *J. Org. Chem.* **1986**, *51*, 2767.
[677]Azido-bromo-addition has also been done with another reagent: Olah; Wang; Li; Prakash *Synlett* **1990**, 487.
[678]Hassner; Keogh *Tetrahedron Lett.* **1975**, 1575.
[679]Hassner; Boerwinkle *J. Am. Chem. Soc.* **1968**, *90*, 217; Hassner; Teeter *J. Org. Chem.* **1971**, *36*, 2176.
[680]Even IN_3 can be induced to add by a free-radical mechanism [see, for example, Cambie; Jurlina; Rutledge; Swedlund; Woodgate *J. Chem. Soc., Perkin Trans. 1* **1982**, 327]. For a review of free-radical additions of XN_3, see Hassner *Intra-Sci. Chem. Rep.* **1970**, *4*, 109-114.
[681]Hassner; Isbister; Friederang *Tetrahedron Lett.* **1969**, 2939.
[682]Hassner; Matthews; Fowler *J. Am. Chem. Soc.* **1969**, *91*, 5046.
[683]Levy; Brown *J. Am. Chem. Soc.* **1973**, *95*, 4067.
[684]Heathcock; Hassner *Angew. Chem. Int. Ed. Engl.* **1963**, *2*, 213 [*Angew. Chem. 75*, 344]; Birckenbach; Linhard *Ber.* **1931**, *64B*, 961, 1076; Drehfahl; Ponsold *Chem. Ber.* **1960**, *93*, 519; Hassner; Hoblitt; Heathcock; Kropp; Lorber *J. Am. Chem. Soc.* **1970**, *92*, 1326; Gebelein; Rosen; Swern *J. Org. Chem.* **1969**, *34*, 1677; Cambie; Hume; Rutledge; Woodgate *Aust. J. Chem.* **1983**, *36*, 2569.
[685]Grimwood; Swern *J. Org. Chem.* **1967**, *32*, 3665.
[686]Greibrokk *Acta Chem. Scand.* **1973**, *27*, 3368.

5-33   Addition of Alkyl Halides (Addition of Halogen, Carbon)
Alkyl-halo-addition[654]

$$-\overset{|}{C}=\overset{|}{C}- + RX \xrightarrow{AlCl_3} -\overset{\overset{\displaystyle R}{|}}{\underset{|}{C}}-\overset{\overset{\displaystyle X}{|}}{\underset{|}{C}}-$$

Alkyl halides can be added to olefins in the presence of a Friedel–Crafts catalyst, most often AlCl₃.[687] The yields are best for tertiary R. Secondary R can also be used, but primary R give rearrangement products (as with **1-12**). Methyl and ethyl halides, which cannot rearrange, give no reaction at all. The attacking species is the carbocation formed from the alkyl halide and the catalyst (see **1-12**).[688] The addition therefore follows Markovnikov's rule, with the cation going to the carbon with more hydrogens. Substitution is a side reaction, arising from loss of hydrogen from the carbocation formed when the original carbocation attacks the double bond:

$$H-\overset{|}{C}=\overset{|}{C}- + R^+ \longrightarrow H-\overset{\overset{\displaystyle R}{|}}{\underset{|}{C}}-\overset{\oplus}{\underset{|}{C}}- \begin{array}{c} \xrightarrow[X^-]{addition} \\ \\ \xrightarrow[-H^+]{substitution} \end{array} \begin{array}{c} H-\overset{\overset{\displaystyle R}{|}}{\underset{|}{C}}-\overset{\overset{\displaystyle X}{|}}{\underset{|}{C}}- \\ \\ \overset{\overset{\displaystyle R}{|}}{C}=\overset{|}{C}- \end{array}$$

Conjugated dienes can add 1,4.[689] Triple bonds also undergo the reaction, to give vinylic halides.[690]

CCl₄, BrCCl₃, ICF₃, and similar simple polyhalo alkanes add to olefins in good yield.[691] These are free-radical additions and require initiation, e.g.,[692] by peroxides, metal halides (e.g., FeCl₂, CuCl),[693] or uv light. The initial attack is by the carbon, and it goes to the carbon with more hydrogens, as in most free-radical attack:

$$RCH{=}CH_2 + \cdot CX_3 \longrightarrow R\overset{\cdot}{C}H-CH_2CX_3 \xrightarrow{CX_4} R\overset{\overset{\displaystyle X}{|}}{C}H-CH_2CX_3 + \cdot CX_3$$

This type of polyhalo alkane adds to halogenated olefins in the presence of AlCl₃ by an electrophilic mechanism. This is called the *Prins reaction* (not to be confused with the other Prins reaction, **6-53**).[694]

ArX can be added across double bonds, in a free-radical process, by treatment of olefins

[687]For a review, see Schmerling, in Olah, Ref. 412, vol. 2, pp. 1133-1174. See also Mayr; Striepe *J. Org. Chem.* **1983**, *48*, 1159; Mayr; Schade; Rubow; Schneider *Angew. Chem. Int. Ed. Engl.* **1987**, *26*, 1029 [*Angew.Chem. 99*, 1059]. For a list of references, see Ref. 133, p. 342.

[688]For a discussion of the mechanism, see Pock; Mayr; Rubow; Wilhelm *J. Am. Chem. Soc.* **1986**, *108*, 7767.

[689]Kolyaskina; Petrov *J. Gen. Chem. USSR* **1962**, *32*, 1067.

[690]See, for example, Maroni; Melloni; Modena *J. Chem. Soc., Perkin Trans. 1* **1973**, 2491, **1974**, 353.

[691]For reviews, see Freidlina; Velichko *Synthesis* **1977**, 145-154; Freidlina; Chukovskaya *Synthesis* **1974**, 477-488.

[692]For other initiators, see Matsumoto; Nakano; Takasu; Nagai *J. Org. Chem.* **1978**, *43*, 1734; Tsuji; Sato; Nagashima *Tetrahedron* **1985**, *41*, 393; Bland; Davis; Durrant *J. Organomet. Chem.* **1985**, *280*, 397; Phelps; Bergbreiter; Lee; Villani; Weinreb *Tetrahedron Lett.* **1989**, *30*, 3915.

[693]For example, see Asscher; Vofsi *J. Chem. Soc.* **1963**, 1887, 3921, *J. Chem. Soc. B* **1968**, 947; Murai; Tsutsumi *J. Org. Chem.* **1966**, *31*, 3000; Martin; Steiner; Streith; Winkler; Belluš *Tetrahedron* **1985**, *41*, 4057. For the addition of CH₂Cl₂ and PhBr, see Mitani; Nakayama; Koyama *Tetrahedron Lett.* **1980**, *21*, 4457.

[694]For a review with respect to fluoroolefins, see Paleta *Fluorine Chem. Rev.* **1977**, *8*, 39-71.

with diazonium salts, although Meerwein arylation (substitution) (**4-19**) competes.[695] This addition can be either 1,2 or 1,4 with conjugated dienes.[696] Addition of ArX can also be accomplished by treatment with an arylmercury halide ArHgX in the presence of CuX_2, LiX, and a palladium compound catalyst, usually Li_2PdCl_4.[697] In this case also, substitution (**4-20**) is a side reaction. Yields of addition product are increased by increasing the concentration of CuX_2. Palladium compounds also catalyze the addition of allylic halides to alkynes.[698]

A variant of the free-radical addition method has been used for ring closure. For example, treatment of **66** with the free-radical initiator hexamethylditin gave a mixture of cis- and

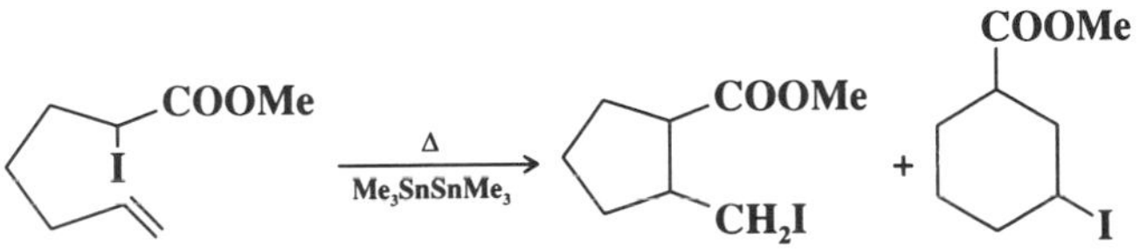

66   cis and trans   cis and trans
               67   68

trans-**67**, with a small amount of cis- and trans-**68** (total yield 83%).[699] The reaction has been performed with α-iodo esters, ketones, and malonates.

For another method of adding R and I to a triple bond, see **5-53**.

OS **II**, 312; **IV**, 727; **V**, 1076; **VI**, 21; **VII**, 290.

5-34 Addition of Acyl Halides (Addition of Halogen, Carbon)
Acyl-halo-addition

Acyl halides have been added to many olefins, in the presence of Friedel–Crafts catalysts. The reaction has been applied to straight-chain, branched, and cyclic olefins, but to very few containing functional groups, other than halogen.[700] The mechanism is similar to that of **5-33**, and, as in that case, substitution competes (**2-15**). Increasing temperature favors substitution,[701] and good yields of addition products can be achieved if the temperature is kept under 0°C. The reaction usually fails with conjugated dienes, since polymerization predominates.[702] The reaction can be performed on triple-bond compounds, producing com-

[695]For example, see Iurkevich; Dombrovskii; Terent'ev *J. Gen. Chem. USSR* **1958**, *28*, 226; Fedorov; Pribytkova; Kanishchev; Dombrovskii *J. Org. Chem. USSR* **1973**, *9*, 1517; Cleland *J. Org. Chem.* **1961**, *26*, 3362, **1969**, *34*, 744; Doyle; Siegfried; Elliott; Dellaria *J. Org. Chem.* **1977**, *42*, 2431; Ganushchak; Obushak; Polishchuk *J. Org. Chem. USSR* **1986**, *22*, 2291.
[696]For example, see Dombrovskii; Ganushchak *J. Gen. Chem. USSR* **1961**, *31*, 1191, **1962**, *32*, 1867; Ganushchak; Golik; Migaichuk *J. Org. Chem. USSR* **1972**, *8*, 2403.
[697]Heck *J. Am. Chem. Soc.* **1968**, *90*, 5538. See also Bäckvall; Nordberg *J. Am. Chem. Soc.* **1980**, *102*, 393.
[698]Kaneda; Uchiyama; Fujiwara; Imanaka; Teranishi *J. Org. Chem.* **1979**, *44*, 55.
[699]Curran; Chang *J. Org. Chem.* **1989**, *54*, 3140; Curran; Chen; Spletzer; Seong; Chang *J. Am. Chem. Soc.* **1989**, *111*, 8872. See also Ichinose; Matsunaga; Fugami; Oshima; Utimoto *Tetrahedron Lett.* **1989**, *30*, 3155.
[700]For reviews, see Groves *Chem. Soc. Rev.* **1972**, *1*, 73-97; House, Ref. 144, pp. 786-797; Nenitzescu; Balaban, in Olah, Ref. 412, vol. 3, 1964, pp. 1033-1152.
[701]Jones; Taylor; Rudd *J. Chem. Soc.* **1961**, 1342.
[702]For examples of 1,4 addition at low temperatures, see Melikyan; Babayan; Atanesyan; Badanyan *J. Org. Chem. USSR* **1984**, *20*, 1884.

pounds of the form RCO—C=C—Cl.[703] A *formyl* group and a halogen can be added to triple bonds by treatment with N,N-disubstituted formamides and $POCl_3$ (Vilsmeier conditions, see **1-15**).[704]

OS **IV**, 186; **VI**, 883; **69**, 238.

B. Oxygen, Nitrogen, or Sulfur on One or Both Sides

5-35 Hydroxylation (Addition of Oxygen, Oxygen)
Dihydroxy-addition

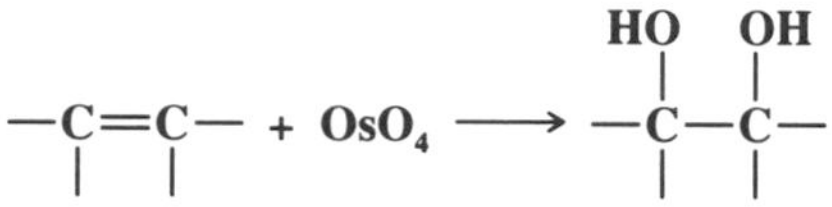

$$-C{=}C- \;+\; OsO_4 \;\longrightarrow\; \overset{\displaystyle HO \quad OH}{-\underset{|}{C}-\underset{|}{C}-}$$

There are many reagents that add two OH groups to a double bond.[705] OsO_4[706] and alkaline $KMnO_4$[707] give syn addition, from the less-hindered side of the double bond. Osmium tetroxide adds rather slowly but almost quantitatively. The cyclic ester **69** is an intermediate and can be isolated,[708] but is usually decomposed in solution, with sodium sulfite in ethanol

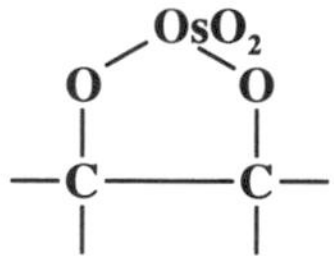

69 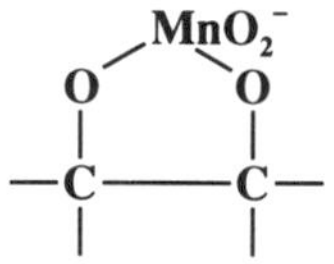   **70**

or other reagents. Bases catalyze the reaction by coordinating with the ester. The chief drawback to this reaction is that OsO_4 is expensive and highly toxic, so that its use has been limited to small-scale preparations of scarce materials. However, the same result (syn addition) can be accomplished more economically by the use of H_2O_2, with OsO_4 present in catalytic amounts.[709] t-Butyl hydroperoxide in alkaline solution,[710] N-methylmorpholine-N-oxide,[711] and $K_3Fe(CN)_6$[712] have been substituted for H_2O_2 in this procedure. Another method uses polymer-bound OsO_4.[713]

[703]For example see Nifant'ev; Grachev; Bakinovskii; Kara-Murza; Kochetkov *J. Appl. Chem. USSR* **1963**, *36*, 646; Savenkov; Khokhlov; Nazarova; Mochalkin *J. Org. Chem. USSR* **1973**, *9*, 914; Martens; Janssens; Hoornaert *Tetrahedron* **1975**, *31*, 177; Brownstein; Morrison; Tan *J. Org. Chem.* **1985**, *50*, 2796.

[704]Yen *Ann. Chim. (Paris)* **1962**, [13] 7, 785.

[705]For reviews, see Hudlický *Oxidations in Organic Chemistry*; American Chemical Society: Washington, 1990, pp. 67-73; Haines *Methods for the Oxidation of Organic Compounds*; Academic Press: New York, 1985, pp. 73-98, 278-294; Sheldon; Kochi *Metal-Catalyzed Oxidations of Organic Compounds*; Academic Press: New York, 1981, pp. 162-171, 294-296. For a list of reagents, with references, see Ref. 133, pp. 494-496.

[706]For a review, see Schröder *Chem. Rev.* **1980**, *80*, 187-213. OsO_4 was first used for this purpose by Criegee *Liebigs Ann. Chem.* **1936**, *522*, 75.

[707]For a review, see Fatiadi *Synthesis* **1987**, 85-127, pp. 86-96.

[708]For a molecular orbital study of the formation of **69**, see Jørgensen; Hoffmann *J. Am. Chem. Soc.* **1986**, *108*, 1867.

[709]Milas; Sussman *J. Am. Chem. Soc.* **1936**, *58*, 1302, **1937**, *59*, 2345. For a review, see Rylander, Ref. 223, pp. 121-133. For another procedure that uses H_2O_2, see Venturello; Gambaro *Synthesis* **1989**, 295.

[710]Akashi; Palermo; Sharpless *J. Org. Chem.* **1978**, *43*, 2063.

[711]VanRheenen; Kelly; Cha *Tetrahedron Lett.* **1976**, 1973; Iwasawa; Kato; Narasaka *Chem. Lett.* **1988**, 1721. See also Ray; Matteson *Tetrahedron Lett.* **1980**, 449.

[712]Minato; Yamamoto; Tsuji *J. Org. Chem.* **1990**, *55*, 766.

[713]Cainelli; Contento; Manescalchi; Plessi *Synthesis* **1989**, 45.

Potassium permanganate is a strong oxidizing agent and can oxidize the glycols[714] that are the products of this reaction (see **9-7** and **9-10**). In acid and neutral solution it always does so; hence it is not feasible to prepare glycols in this manner. Glycols can be prepared with alkaline[715] permanganate, but the conditions must be mild. Even so, yields are seldom above 50%, though they can be improved with phase transfer catalysis[716] or increased stirring.[717] There has been much speculation that, as with OsO_4, cyclic esters (**70**) are intermediates, and there is evidence for them.[718] This reaction is the basis of the *Baeyer test* for the presence of double bonds.

Anti hydroxylation can be achieved by treatment with H_2O_2 and formic acid. In this case, epoxidation (**5-36**) occurs first, followed by an SN2 reaction, which results in overall anti addition:

$$-\overset{|}{C}=\overset{|}{C}- + H_2O_2 \longrightarrow -\overset{|}{\underset{|}{C}}\overset{O}{\diagdown}\overset{|}{\underset{|}{C}}- \overset{H^+}{\longrightarrow} -\overset{|}{\underset{|}{C}}\overset{\overset{H}{\underset{|}{O}}{}^{\oplus}}{\diagdown}\overset{|}{\underset{|}{C}}- \overset{H_2O}{\longrightarrow}$$

$$-\overset{OH}{\underset{\underset{\oplus}{OH_2}}{\overset{|}{C}}}\overset{|}{\underset{|}{C}}- \overset{}{\underset{-H^+}{\longrightarrow}} -\overset{OH}{\underset{OH}{\overset{|}{C}}}\overset{|}{\underset{|}{C}}-$$

The same result can be achieved in one step with *m*-chloropcroxybenzoic acid and water.[719] Overall anti addition can also be achieved by the method of Prevost. In this method the olefin is treated with iodine and silver benzoate in a 1:2 molar ratio. The initial addition is anti and results in a β-halo benzoate (**71**). These can be isolated, and this represents a method of addition of IOCOPh. However, under the normal reaction conditions, the iodine is replaced by a second PhCOO group. This is a nucleophilic substitution reaction, and it operates by the neighboring-group mechanism (p. 308), so the groups are still anti:

$$-\overset{|}{C}=\overset{|}{C}- \overset{I_2}{\underset{PhCOOAg}{\longrightarrow}} -\overset{I}{\underset{OCOPh}{\overset{|}{C}}}\overset{|}{\underset{|}{C}}- \longrightarrow -\overset{PhCOO}{\underset{OCOPh}{\overset{|}{C}}}\overset{|}{\underset{|}{C}}- \longrightarrow -\overset{OH}{\underset{OH}{\overset{|}{C}}}\overset{|}{\underset{|}{C}}-$$

Hydrolysis of the ester does not change the configuration. Woodward's method is similar, but results in overall syn hydroxylation. The olefin is treated with iodine and silver acetate in a 1:1 molar ratio in acetic acid containing water. Here again, the initial product is a β-

[714]Or give more-highly-oxidized products such as α-hydroxy ketones without going through the glycols. See, for example, Wolfe; Ingold; Lemieux *J. Am. Chem. Soc.* **1981**, *103*, 938; Wolfe; Ingold *J. Am. Chem. Soc.* **1981**, *103*, 940.

[715]The role of the base seems merely to be to inhibit acid-promoted oxidations. The base does not appear to play any part in the mechanism: Taylor; Green *Can. J. Chem.* **1985**, *63*, 2777.

[716]See, for example, Weber; Shepherd *Tetrahedron Lett.* **1972**, 4907; Ogino; Mochizuki *Chem. Lett.* **1979**, 443.

[717]Taylor; Williams; Edwards; Otonnaa; Samanich *Can. J. Chem.* **1984**, *62*, 11; Taylor *Can. J. Chem.* **1984**, *62*, 2641.

[718]For some recent evidence, see Ogino; Kikuiri *J. Am. Chem. Soc.* **1989**, *111*, 6174; Lee; Chen *J. Am. Chem. Soc.* **1989**, *111*, 7534; Ogino; Hasegawa; Hoshino *J. Org. Chem.* **1990**, *55*, 2653. See however Freeman; Chang; Kappos; Sumarta *J. Org. Chem.* **1987**, *52*, 1461; Freeman; Kappos *J. Org. Chem.* **1989**, *54*, 2730, and other papers in this series; Perez-Benito; Lee *Can. J. Chem.* **1985**, *63*, 3545.

[719]Fringuelli; Germani; Pizzo; Savelli *Synth. Commun.* **1989**, *19*, 1939.

halo ester; the addition is anti and a nucleophilic replacement of the iodine occurs. However, in the presence of water, neighboring-group participation is prevented or greatly decreased by solvation of the ester function, and the mechanism is the normal S$_N$2 process,[720] so the monoacetate is syn and hydrolysis gives the glycol that is the product of overall syn addition. Although the Woodward method results in overall syn addition, the product may be different from that with OsO$_4$ or KMnO$_4$, since the overall syn process is from the more-hindered side of the olefin.[721] Both the Prevost and the Woodward methods[722] have also been carried out in high yields with thallium(I) acetate and thallium(I) benzoate instead of the silver carboxylates.[723]

With suitable substrates, addition of two OH groups creates either one or two new chiral centers:

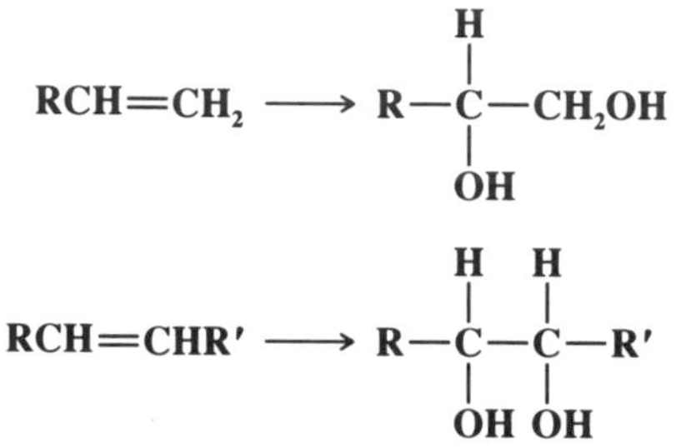

$$RCH{=}CH_2 \longrightarrow R{-}\overset{\overset{\displaystyle H}{|}}{\underset{\underset{\displaystyle OH}{|}}{C}}{-}CH_2OH$$

$$RCH{=}CHR' \longrightarrow R{-}\overset{\overset{\displaystyle H}{|}}{\underset{\underset{\displaystyle OH}{|}}{C}}{-}\overset{\overset{\displaystyle H}{|}}{\underset{\underset{\displaystyle OH}{|}}{C}}{-}R'$$

Addition to olefins of the form RCH=CH$_2$ has been made enantioselective, and addition to RCH=CHR' both diastereoselective[724] and enantioselective, by using optically active amines, such as **72**, **73** (derivatives of the naturally occurring quinine and quinidine),

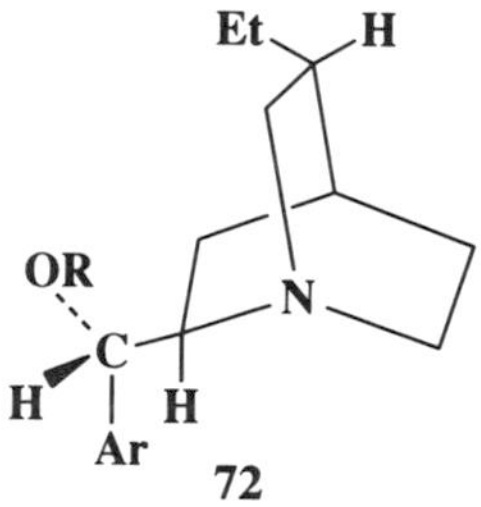

72
9′-Phenanthryl ether of
dihydroquinidine

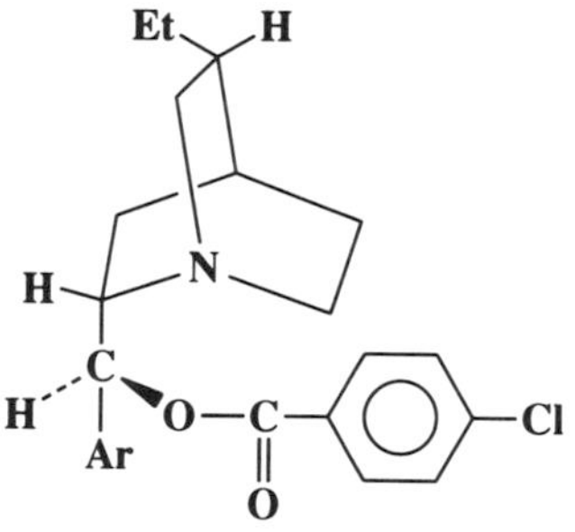

73
dihydroquinine
p-chlorobenzoate

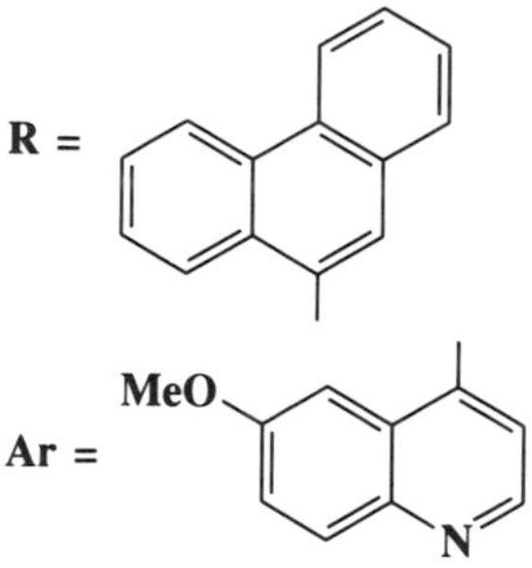

---

[720]For another possible mechanism that accounts for the stereochemical result of the Woodward method, see Woodward; Brutcher *J. Am. Chem. Soc.* **1958**, *80*, 209.

[721]For another method of syn hydroxylation, which can be applied to either face, see Corey; Das *Tetrahedron Lett.* **1982**, *23*, 4217.

[722]For some related methods, see Jasserand; Girard; Rossi; Granger *Tetrahedron Lett.* **1976**, 1581; Ogata; Aoki *J. Org. Chem.* **1966**, *31*, 1625; Mangoni; Adinolfi; Barone; Parrilli *Tetrahedron Lett.* **1973**, 4485, *Gazz. Chim. Ital.* **1975**, *105*, 377; Horiuchi; Satoh *Chem. Lett.* **1988**, 1209; Campi; Deacon; Edwards; Fitzroy; Giunta; Jackson; Trainor *J. Chem. Soc., Chem. Commun.* **1989**, 407.

[723]Cambie; Hayward; Roberts; Rutledge *J. Chem. Soc., Chem. Commun.* **1973**, 359, *J. Chem. Soc., Perkin Trans. 1* **1974**, 1858, 1864; Cambie; Rutledge *Org. Synth. VI*, 348.

[724]For diastereoselective, but not enantioselective, addition of OsO$_4$, see Cha; Christ; Kishi *Tetrahedron Lett.* **1983**, *24*, 3943, 3947, *Tetrahedron* **1984**, *40*, 2247; Stork; Kahn *Tetrahedron Lett.* **1983**, *24*, 3951; Vedejs; McClure *J. Am. Chem. Soc.* **1986**, *108*, 1094; Evans; Kaldor *J. Org. Chem.* **1990**, *55*, 1698.

$$\begin{array}{cc} & \text{Ph} \quad \text{H} \\ & | \quad | \\ \text{ArCH}_2-\text{NH}-\text{C}-\text{C}-\text{NH}-\text{CH}_2\text{Ar} \\ & | \quad | \\ & \text{H} \quad \text{Ph} \quad\quad \text{Ar = mesityl} \\ & \mathbf{74} \end{array}$$

$$\text{Ph} \cdots \overset{\text{Ph}}{\underset{\text{Ph}}{\bigtriangleup}} \text{N}-\text{CH}_2\text{CH}_2-\text{N} \overset{\text{Ph}}{\underset{\text{Ph}}{\bigtriangleup}} \cdots \text{Ph}$$

75

74,[725] or **75**,[726] along with OsO$_4$.[727] These amines bind to the OsO$_4$ in situ as chiral ligands, causing it to add asymmetrically.[728] This has been done both with the stoichiometric and with the catalytic method.[729] The catalytic method has been extended to conjugated dienes, which give tetrahydroxy products diastereoselectively.[730]

Ligands **72** and **73** not only cause enantioselective addition, but also accelerate the reaction, so that they may be useful even where enantioselective addition is not required.[731] Although **72** and **73** are not enantiomers, they give enantioselective addition to a given olefin in the opposite sense; e.g., styrene predominantly gave the (R) diol with **72**, and the (S) diol with **73**.[732] Enantioselective and diastereoselective addition have also been achieved by using preformed derivatives of OsO$_4$, already containing chiral ligands,[733] and by the use of OsO$_4$ on olefins that have a chiral group elsewhere in the molecule.[734]

Olefins can also be oxidized with metallic acetates such as lead tetraacetate[735] or thallium(III) acetate[736] to give bisacetates of glycols.[737] Oxidizing agents such as benzoquinone, MnO$_2$, or O$_2$, along with palladium acetate, have been used to convert conjugated dienes to 1,4-diacetoxy-2-alkenes (1,4 addition).[738]

OS **II**, 307; **III**, 217; **IV**, 317; **V**, 647; **VI**, 196, 342, 348.

[725]Corey; Jardine; Virgil; Yuen; Connell *J. Am. Chem. Soc.* **1989**, *111*, 9243; Corey; Lotto *Tetrahedron Lett.* **1990**, *31*, 2665.

[726]Tomioka; Nakajima; Koga *J. Am. Chem. Soc.* **1987**, *109*, 6213, *Tetrahedron Lett.* **1990**, *31*, 1741; Tomioka; Nakajima; Iitaka; Koga *Tetrahedron Lett.* **1988**, *29*, 573.

[727]Wai; Marko; Svendsen; Finn; Jacobsen; Sharpless *J. Am. Chem. Soc.* **1989**, *111*, 1123; Lohray; Kalantar; Kim; Park; Shibata; Wai; Sharpless *Tetrahedron Lett.* **1989**, *30*, 2041; Kwong; Sorato; Ogino; Chen; Sharpless *Tetrahedron Lett.* **1990**, *31*, 2999; Shibata; Gilheany; Blackburn; Sharpless *Tetrahedron Lett.* **1990**, *31*, 3817; Sharpless et al., *J. Org. Chem.* **1991**, *56*, 4585.

[728]For discussions of the mechanism of the enantioselectivity, see Jørgensen *Tetrahedron Lett.* **1990**, *31*, 6417; Ogino; Chen; Kwong; Sharpless *Tetrahedron Lett.* **1991**, *32*, 3965.

[729]For other examples of asymmetric dihydroxylation, see Yamada; Narasaka *Chem. Lett.* **1986**, 131; Tokles; Snyder *Tetrahedron Lett.* **1986**, *27*, 3951; Annunziata; Cinquini; Cozzi; Raimondi; Stefanelli *Tetrahedron Lett.* **1987**, *28*, 3139; Hirama; Oishi; Itô *J. Chem. Soc., Chem. Commun.* **1989**, 665.

[730]Park; Kim; Sharpless *Tetrahedron Lett.* **1991**, *32*, 1003.

[731]Sharpless et al., Ref. 727. See also Jacobsen; Marko; France; Svendsen; Sharpless *J. Am. Chem. Soc.* **1989**, *111*, 737.

[732]Jacobsen; Marko; Mungall; Schröder; Sharpless *J. Am. Chem. Soc.* **1988**, *110*, 1968.

[733]Kokubo; Sugimoto; Uchida; Tanimoto; Okano *J. Chem. Soc., Chem. Commun.* **1983**, 769.

[734]Hauser; Ellenberger; Clardy; Bass *J. Am. Chem. Soc.* **1984**, *106*, 2458; Johnson; Barbachyn *J. Am. Chem. Soc.* **1984**, *106*, 2459.

[735]For a review, see Moriarty *Sel Org. Transform.* **1972**, *2*, 183-237.

[736]See, for example, Uemura; Miyoshi; Tabata; Okano *Tetrahedron* **1981**, *37*, 291. For a review of the reactions of thallium(III) compounds with olefins, see Uemura, in Hartley, Ref. 218, vol. 4, pp. 473-538, pp. 497-513. For a review of thallium(III) acetate and trifluoroacetate, see Uemura, in Pizey, Ref. 146, vol. 5, 1983, pp. 165-187.

[737]For another method see Fristad; Peterson *Tetrahedron* **1984**, *40*, 1469.

[738]See Bäckvall; Nordberg *J. Am. Chem. Soc.* **1981**, *103*, 4959; Bäckvall; Byström; Nordberg *J. Org. Chem.* **1984**, *49*, 4619; Bäckvall; Awasthi; Renko *J. Am. Chem. Soc.* **1987**, *109*, 4750. For articles on this and related reactions, see Bäckvall *Bull. Soc. Chim. Fr.* **1987**, 665-670, *New. J. Chem.* **1990**, *14*, 447-452. For another method, see Uemura; Fukuzawa; Patil; Okano *J. Chem. Soc., Perkin Trans. 1* **1985**, 499.

5-36   Epoxidation (Addition of Oxygen, Oxygen)
epi-Oxy-addition

$$-\overset{|}{C}=\overset{|}{C}- \;+\; \text{PhCOOH} \longrightarrow -\overset{|}{\underset{\underset{O}{\diagup\;\diagdown}}{C}}-\overset{|}{C}-$$

Olefins can be epoxidized with any of a number of peracids,[739] of which *m*-chloroperbenzoic has been the most often used. The reaction, called the *Prilezhaev reaction*, has wide utility.[740] Alkyl, aryl, hydroxyl, ester, and other groups may be present, though not amino groups, since these are affected by the reagent. Electron-donating groups increase the rate, and the reaction is particularly rapid with tetraalkyl olefins. Conditions are mild and yields are high. Other peracids, especially peracetic and perbenzoic, are also used; trifluoroperacetic acid[741] and 3,5-dinitroperoxybenzoic acid[742] are particularly reactive ones.

The following one-step mechanism[743] was proposed by Bartlett:[744]

Evidence for this mechanism is as follows:[745] (1) The reaction is second order. If ionization were the rate-determining step, it would be first order in peracid. (2) The reaction readily takes place in nonpolar solvents, where formation of ions is inhibited. (3) Measurements of the effect on the reaction rate of changes in the substrate structure show that there is no carbocation character in the transition state.[746] (4) The addition is stereospecific (i.e., a trans olefin gives a trans epoxide and a cis olefin a cis epoxide) even in cases where electron-donating substituents would stabilize a hypothetical carbocation intermediate. However, where there is an OH group in the allylic or homoallylic position, the stereospecificity diminishes or disappears, with both cis and trans isomers giving predominantly or exclusively the product where the incoming oxygen is syn to the OH group. This probably indicates a transition state in which there is hydrogen bonding between the OH group and the peroxy acid.[747]

[739]For a list of reagents, including peracids and others, used for epoxidation, with references, see Ref. 133, pp. 456-461.

[740]For reviews, see Hudlický, Ref. 705, pp. 60-64; Haines, Ref. 705, pp. 98-117, 295-303; Dryuk *Russ. Chem. Rev.* **1985**, *54*, 986-1005; Plesničar, in Trahanovsky *Oxidation in Organic Chemistry*, pt. C; Academic Press: New York, 1978, pp. 211-252; Swern, in Swern *Organic Peroxides*, vol. 2; Wiley: New York, 1971, pp. 355-533; Metelitsa *Russ. Chem. Rev.* **1972**, *41*, 807-821; Hiatt, in Augustine; Trecker *Oxidation*, vol. 2; Marcel Dekker: New York, 1971; pp. 113-140; House, Ref. 144, pp. 292-321. For a review pertaining to the stereochemistry of the reaction, see Berti *Top. Stereochem.* **1973**, *7*, 93-251, pp. 95-187.

[741]Emmons; Pagano *J. Am. Chem. Soc.* **1955**, *77*, 89.

[742]Rastetter; Richard; Lewis *J. Org. Chem.* **1978**, *43*, 3163.

[743]For discussions of the mechanism, see Dryuk *Tetrahedron* **1976**, *32*, 2855-2866; Finn; Sharpless, in Morrison, Ref. 232, vol. 5, pp. 247-308. For a review of polar mechanisms involving peroxides, see Plesnicar in Patai *The Chemistry of Peroxides*; Wiley: New York, 1983, pp. 521-584.

[744]Bartlett *Rec. Chem. Prog.* **1957**, *18*, 111. For other proposed mechanisms see Kwart; Hoffman *J. Org. Chem.* **1966**, *31*, 419; Hanzlik; Shearer *J. Am. Chem. Soc.* **1975**, *97*, 5231.

[745]Ogata; Tabushi *J. Am. Chem. Soc.* **1961**, *83*, 3440. See also Woods; Beak *J. Am. Chem. Soc.* **1991**, *113*, 6281.

[746]Khalil; Pritzkow *J. Prakt. Chem.* **1973**, *315*, 58; Schneider; Becker; Philippi *Chem. Ber.* **1981**, *114*, 1562; Batog; Savenko; Batrak; Kucher *J. Org. Chem. USSR* **1981**, *17*, 1860.

[747]See Berti, Ref. 740, pp. 130-162.

Conjugated dienes can be epoxidized (1,2 addition), though the reaction is slower than for corresponding olefins, but α,β-unsaturated ketones do not generally give epoxides when treated with peracids.[748] However, α,β-unsaturated esters react normally, to give glycidic esters.[749] When a carbonyl group is in the molecule but not conjugated with the double bond, the Baeyer–Villiger reaction (**8-20**) may compete. Allenes[750] are converted by peracids to allene oxides[751] or spiro dioxides, both of which species can in certain cases be isolated[752]

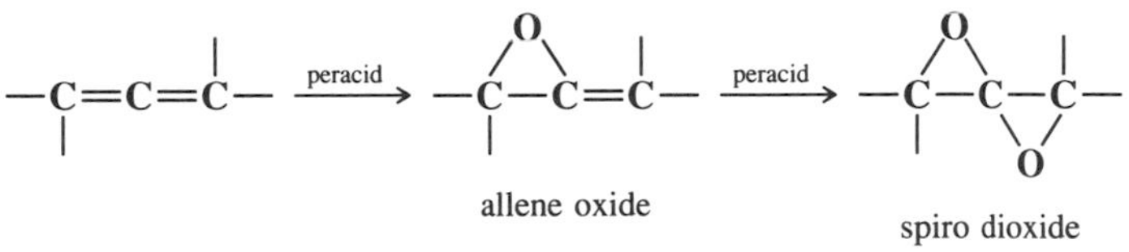

but more often are unstable under the reaction conditions and react further to give other products.[753]

α,β-Unsaturated ketones (including quinones), aldehydes, and sulfones can be epoxidized with alkaline H_2O_2.[754] This is a nucleophilic addition by a Michael-type mechanism, involving attack by HO_2^-:[755]

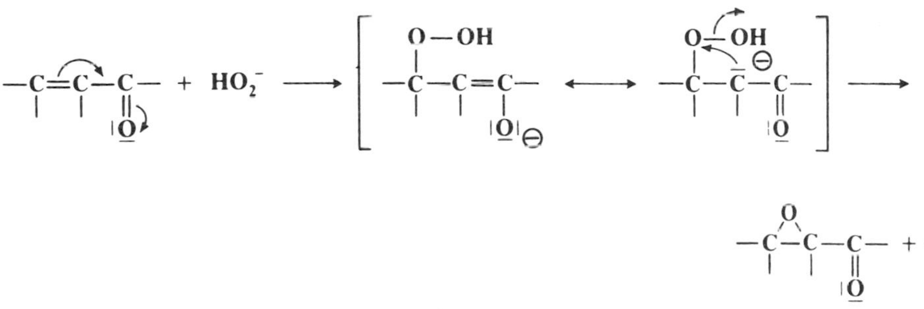

α,β-Unsaturated carboxylic acids can be epoxidized with H_2O_2 and heteropoly acids,[756] and α,β-unsaturated esters, amides, and sulfones with t-BuOOH and an alkyllithium in THF.[757] Epoxides can also be prepared by treating olefins with oxygen or with an alkyl peroxide,[758]

[748]A few exceptions are known. For example see Hart; Verma; Wang *J. Org. Chem.* **1973**, *38*, 3418.

[749]MacPeek; Starcher; Phillips *J. Am. Chem. Soc.* **1959**, *81*, 680.

[750]For a review of epoxidation of allenes, see Jacobs, in Landor, Ref. 95, vol. 2, pp. 417-510, pp. 483-491.

[751]For a review of allene oxides see Chan; Ong *Tetrahedron* **1980**, *36*, 2269-2289.

[752]Crandall; Machleder; Thomas *J. Am. Chem. Soc.* **1968**, *90*, 7346; Camp; Greene *J. Am. Chem. Soc.* **1968**, *90*, 7349; Crandall; Conover; Komin; Machleder *J. Org. Chem.* **1974**, *39*, 1723; Crandall; Batal *J. Org. Chem.* **1988**, *53*, 1338.

[753]For example see Crandall; Machleder; Sojka *J. Org. Chem.* **1973**, *38*, 1149; Crandall; Rambo *J. Org. Chem.* **1990**, *55*, 5929.

[754]For example, see Payne *J. Am. Chem. Soc.* **1959**, *81*, 4901; Payne; Williams *J. Org. Chem.* **1961**, *26*, 651; Zwanenburg; ter Wiel *Tetrahedron Lett.* **1970**, 935.

[755]Bunton; Minkoff *J. Chem. Soc.* **1949**, 665; Temple *J. Org. Chem.* **1970**, *35*, 1275; Apeloig; Karni; Rappoport *J. Am. Chem. Soc.* **1983**, *105*, 2784. For a review of allene oxides, see Patai; Rappoport in Patai, Ref. 36, pt. 1, pp. 512-517.

[756]Oguchi; Sakata; Takeuchi; Kaneda; Ishii; Ogawa *Chem. Lett.* **1989**, 2053.

[757]Meth-Cohn; Moore; Taljaard *J. Chem. Soc., Perkin Trans. 1* **1988**, 2663; Bailey; Clegg; Jackson; Meth-Cohn *J. Chem. Soc., Perkin Trans. 1* **1990**, 200.

[758]For example, see Gould; Hiatt; Irwin *J. Am. Chem. Soc.* **1968**, *90*, 4573; Sharpless; Michaelson *J. Am. Chem. Soc.* **1973**, *95*, 6136; Hart; Lavrik *J. Org. Chem.* **1974**, *39*, 1793; Beg; Ahmad *J. Org. Chem.* **1977**, *42*, 1590; Kochi *Organometallic Mechanisms and Catalysis*; Academic Press: New York, 1978, pp. 69-73; Mihelich *Tetrahedron Lett.* **1979**, 4729; Ledon; Durbut; Varescon *J. Am. Chem. Soc.* **1981**, *103*, 3601; Mimoun; Mignard; Brechot; Saussine *J. Am. Chem. Soc.* **1986**, *108*, 3711; Kato; Ota; Matsukawa; Endo *Tetrahedron Lett.* **1988**, *29*, 2843; Laszlo; Levart; Singh *Tetrahedron Lett.* **1991**, *32*, 3167.

catalyzed by a complex of a transition metal such as V, Mo, Ti, or Co.[759] The reaction with oxygen, which can also be carried out without a catalyst, is probably a free-radical process.[760] In the *Sharpless asymmetric epoxidation*,[761] allylic alcohols are converted to optically active epoxides in better than 90% enantiomeric excess, by treatment with *t*-BuOOH, titanium tetraisopropoxide and optically active diethyl tartrate.[762] The $Ti(OCHMe_2)_4$ and diethyl tartrate can be present in catalytic amounts (5-10 mole %) if molecular sieves are present.[763] Since both (+) and (−) diethyl tartrate are readily available, and the reaction is stereospecific, either enantiomer of the product can be prepared. The method has been successful for a wide range of primary allylic alcohols, where the double bond is mono-, di-, tri-, and tetrasubstituted.[764] This procedure, in which an optically active catalyst is used to induce asymmetry, has proved to be one of the most important methods of asymmetric synthesis, and has been used to prepare a large number of optically active natural products and other compounds. Among these are the 8 L-aldohexoses (the unnatural isomers), which were totally synthesized with the aid of this method, starting from a common precursor, 4-benzhydryloxy-(*E*)-but-2-en-ol.[765] The mechanism of the Sharpless epoxidation is believed to involve attack on the substrate by a compound[766] formed from the titanium alkoxide and the diethyl tartrate to produce a complex that also contains the substrate and the *t*-BuOOH.[767] Ordinary alkenes (without an allylic OH group) have been enantioselectively epoxidized with sodium hypochlorite (commercial bleach) and an optically active manganese-complex catalyst.[768]

Among other reagents for converting olefins to epoxides[769] are H_2O_2, catalyzed by tungstic acid or its derivatives,[770] F_2–H_2O–MeCN,[771] and magnesium monoperoxyphthalate.[772] The last reagent, which is commercially available, has been shown to be a good substitute for *m*-chloroperbenzoic acid in a number of reactions.[773]

---

[759]For a review, see Jørgensen *Chem. Rev.* **1989**, *89*, 431-458.

[760]For reviews, see Van Santen; Kuipers *Adv. Catal.* **1987**, *35*, 265-321; Filippova; Blyumberg *Russ. Chem. Rev.* **1982**, *51*, 582-591.

[761]For reviews, see Pfenninger *Synthesis* **1986**, 89-116; Rossiter, in Morrison, Ref. 232, vol. 5, pp. 193-246. For histories of its discovery, see Sharpless *Chem. Br.* **1986**, 38-44, *CHEMTECH* **1985**, 692-700.

[762]Katsuki; Sharpless *J. Am. Chem. Soc.* **1980**, *102*, 5974; Rossiter; Katsuki; Sharpless *J. Am. Chem. Soc.* **1981**, *103*, 464; Sharpless; Woodard; Finn *Pure Appl. Chem.* **1983**, *55*, 1823-1836.

[763]Gao; Hanson; Klunder; Ko; Masamune; Sharpless *J. Am. Chem. Soc.* **1987**, *109*, 5765. For another improvement, see Wang; Zhou *Tetrahedron* **1987**, *43*, 2935.

[764]See the table in Finn; Sharpless, Ref. 743, pp. 249-250. See also Schweiter; Sharpless *Tetrahedron Lett.* **1985**, *26*, 2543.

[765]Ko; Lee; Masamune; Reed; Sharpless; Walker *Tetrahedron* **1990**, *46*, 245. For other stereospecific syntheses of monosaccharides, see Mukaiyama; Suzuki; Yamada; Tabusa *Tetrahedron* **1990**, *46*, 265, and references cited therein.

[766]Very similar compounds have been prepared and isolated as solids whose structures have been determined by x-ray crystallography: Williams; Pedersen; Sharpless; Lippard *J. Am. Chem. Soc.* **1984**, *106*, 6430.

[767]For a review of the mechanism, see Finn; Sharpless, Ref. 743. For other mechanistic studies, see Jørgensen; Wheeler; Hoffmann *J. Am. Chem. Soc.* **1987**, *109*, 3240; Hawkins; Sharpless *Tetrahedron Lett.* **1987**, *28*, 2825; Carlier; Sharpless *J. Org. Chem.* **1989**, *54*, 4016; Corey *J. Org. Chem.* **1990**, *55* 1693; Woodard; Finn; Sharpless *J. Am. Chem. Soc.* **1991**, *113*, 106; Finn; Sharpless *J. Am. Chem. Soc.* **1991**, *113*, 113; Takano; Iwebuchi; Ogasawara *J. Am. Chem. Soc.* **1991**, *113*, 2786.

[768]Jacobsen; Zhang; Muci; Ecker; Deng *J. Am. Chem. Soc.* **1991**, *113*, 7063. See also Irie; Noda; Ito; Katsuki *Tetrahedron Lett.* **1991**, *32*, 1055; Irie; Ito; Katsuki *Synlett* **1991**, 265; Halterman; Jan *J. Org. Chem.* **1991**, *56*, 5253.

[769]For other methods of converting olefins to epoxides, see Balavoine; Eskenazi; Meunier; Rivière *Tetrahedron Lett.* **1984**, *25*, 3187; Tezuka; Iwaki *J. Chem. Soc.*, *Perkin Trans. 1* **1984**, 2507; Samsel; Srinivasan; Kochi *J. Am. Chem. Soc.* **1985**, *107*, 7606; Xie; Xu; Hu; Ma; Hou; Tao *Tetrahedron Lett.* **1988**, *29*, 2967; Bruice *Aldrichimica Acta* **1988**, *21*, 87-94; Adam; Curci; Edwards *Acc. Chem. Res.* **1989**, *22*, 205-211; Troisi; Cassidei; Lopez; Mello; Curci *Tetrahedron Lett.* **1989**, *30*, 257; Rodriguez; Dulcère *J. Org. Chem.* **1991**, *56*, 469.

[770]See, for example, Bortolini; Di Furia; Modena; Seraglia *J. Org. Chem.* **1985**, *50*, 2688; Prat; Lett *Tetrahedron Lett.* **1986**, *27*, 707; Prandi; Kagan; Mimoun *Tetrahedron Lett.* **1986**, *27*, 2617; Venturello; D'Aloisio *J. Org. Chem.* **1988**, *53*, 1553.

[771]Rozen; Kol *J. Org. Chem.* **1990**, *55*, 5155.

[772]Brougham; Cooper; Cummerson; Heaney; Thompson *Synthesis* **1987**, 1015; Querci; Ricci *J. Chem. Soc.*, *Chem. Commun.* **1989**, 889.

[773]Brougham et al., Ref. 772.

It would be useful if triple bonds could be similarly epoxidized to give oxirenes. However, oxirenes are not stable compounds.[774] Two of them have been trapped in solid argon matrices

oxirene

at very low temperatures, but they decayed on warming to 35 K.[775] Oxirenes probably form in the reaction,[776] but react further before they can be isolated. Note that oxirenes bear the same relationship to cyclobutadiene that furan does to benzene and may therefore be expected to be antiaromatic (see p. 55).

In a different type of reaction, olefins are photooxygenated (with singlet O_2, see **4-9**) in the presence of a Ti, V, or Mo complex to give epoxy alcohols formally derived from allylic hydroxylation followed by epoxidation, e.g.,[777]

VO(acac)$_2$ / O$_2$, hv → 67%

OS **I**, 494; **IV**, 552, 860; **V**, 191, 414, 467, 100/; **VI**, 39, 320, 679, 862; **VII**, 121, 126, 461; **66**, 203.

5-37 Photooxidation of Dienes (Addition of Oxygen, Oxygen)
(2 + 4)*OC,OC-cyclo*-Peroxy-1/4/addition

$$-\overset{|}{\underset{|}{C}}=\overset{|}{\underset{|}{C}}-\overset{|}{\underset{|}{C}}=\overset{|}{\underset{|}{C}}- \ + \ O_2 \ \xrightarrow{hv} \ $$

76

Conjugated dienes react with oxygen under the influence of light to give cyclic peroxides **76**.[778] The reaction has mostly[779] been applied to cyclic dienes.[780] The scope extends to certain aromatic compounds,[781] e.g.,

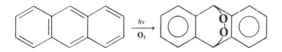

[774]For a review of oxirenes, see Lewars *Chem. Rev.* **1983**, *83*, 519-534.

[775]Torres; Bourdelande; Clement; Strausz *J. Am. Chem. Soc.* **1983**, *105*, 1698. See also Laganis; Janik; Curphey; Lemal *J. Am. Chem. Soc.* **1983**, *105*, 7457.

[776]McDonald; Schwab *J. Am. Chem. Soc.* **1964**, *86*, 4866; Ibne-Rasa; Pater; Ciabattoni; Edwards *J. Am. Chem. Soc.* **1973**, *95*, 7894; Ogata; Sawaki; Inoue *J. Org. Chem.* **1973**, *38*, 1044.

[777]Adam; Braun; Griesbeck; Lucchini; Staab; Will *J. Am. Chem. Soc.* **1989**, *111*, 203.

[778]For reviews, see Clennan *Tetrahedron* **1991**, *47*, 1343-1382, *Adv. Oxygenated Processes* **1988**, *1*, 85-122; Wasserman; Ives *Tetrahedron* **1981**, *37*, 1825-1852; Denny; Nickon *Org. React.* **1973**, *20*, 133-336; Adams in Augustine; Trecker, Ref. 740, vol. 2, pp. 65-112; Gollnick *Adv. Photochem.* **1968**, *6*, 1-122; Schönberg, Ref. 49, pp. 382-397; Gollnick; Schenck in Hamer *1,4-Cycloaddition Reactions*; Academic Press: New York, 1967, pp. 255-344; Arbuzov *Russ. Chem. Rev.* **1965**, *34*, 558-574.

[779]For many examples with acyclic dienes, see Matsumoto; Dobashi; Kuroda; Kondo *Tetrahedron* **1985**, *41*, 2147.

[780]For reviews of cyclic peroxides, see Saito; Nittala, in Patai, Ref. 743, pp. 311-374; Balci *Chem. Rev.* **1981**, *81*, 91-108; Adam; Bloodworth *Top. Curr. Chem.* **1981**, *97*, 121-158.

[781]For reviews, see in Wasserman; Murray *Singlet Oxygen*; Academic Press: New York, 1979, the articles by Wasserman; Lipshutz, pp. 429-509; Saito; Matsuura, pp. 511-574; Rigaudy *Pure Appl. Chem.* **1968**, *16*, 169-186.

Besides those dienes and aromatic rings that can be photooxidized directly, there is a larger group that give the reaction in the presence of a photosensitizer such as eosin (see p. 241). Among these is α-terpinene, which is converted to ascaridole:

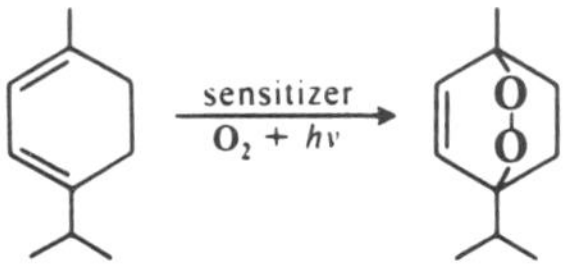

As in **4-9**, it is not the ground-state oxygen (the triplet), that reacts, but the excited singlet state,[782] so the reaction is actually a Diels-Alder reaction (see **5-47**) with singlet oxygen as dienophile:[783]

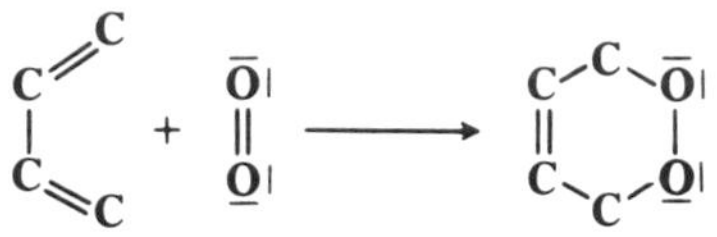

Like **5-47**, this reaction is reversible.

We have previously discussed the reaction of singlet oxygen with double-bond compounds to give hydroperoxides (**4-9**), but singlet oxygen can also react with double bonds in another way to give a dioxetane intermediate[784] (**77**), which usually cleaves to aldehydes or ketones[785]

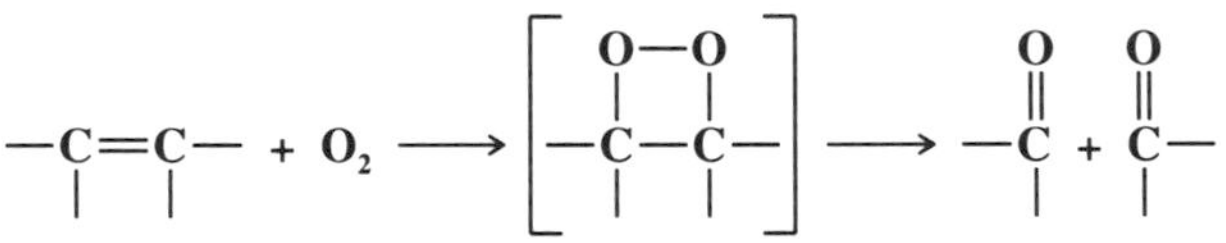

77

but has been isolated.[786] Both the 6-membered cyclic peroxides **76**[787] and the 4-membered **77**[788] have been formed from oxygenation reactions that do not involve singlet oxygen. If **77** are desired, better reagents[789] are triphenyl phosphite ozonide $(PhO)_3PO_3$ and triethylsilyl hydrotrioxide $Et_3SiOOOH$, though yields are not high.[790]

[782]For books and reviews on singlet oxygen, see Ref. 216 in Chapter 14.

[783]Foote; Wexler *J. Am. Chem. Soc.* **1964**, *86*, 3880; Corey; Taylor *J. Am. Chem. Soc.* **1964**, *86*, 3881; Foote; Wexler; Ando *Tetrahedron Lett.* **1965**, 4111; Monroe *J. Am. Chem. Soc.* **1981**, *103*, 7253. See also Hathaway; Paquette *Tetrahedron Lett.* **1985**, *41*, 2037; O'Shea; Foote *J. Am. Chem. Soc.* **1988**, *110*, 7167.

[784]For reviews, see Adam; Cilento *Angew. Chem. Int. Ed. Engl.* **1983**, *22*, 529-542 [*Angew. Chem. 95*, 525-538]; Schaap; Zaklika in Wasserman; Murray, Ref. 781, pp. 173-242; Bartlett *Chem. Soc. Rev.* **1976**, *5*, 149-163. For discussions of the mechanisms see Frimer *Chem. Rev.* **1979**, *79*, 359-387; Clennan; Nagraba *J. Am. Chem. Soc.* **1988**, *110*, 4312.

[785]For discussions see Kearns *Chem. Rev.* **1971**, *71*, 395-427, pp. 422-424; Foote *Pure Appl. Chem.* **1971**, *27*, 635-645.

[786]For reviews of 1,2-dioxetanes see Adam, in Patai, Ref. 743, pp. 829-920; Bartlett; Landis, in Wasserman; Murray Ref. 781, pp. 243-286; Adam *Adv. Heterocycl. Chem.* **1977**, *21*, 437-481. See also Inoue; Hakushi; Turro *Kokagaku Toronkai Koen Yoshishu* **1979**, 150 [*C.A. 92*, 214798q]; Adam; Encarnación *Chem. Ber.* **1982**, *115*, 2592; Adam; Baader *Angew. Chem. Int. Ed. Engl.* **1984**, *23*, 166 [*Angew. Chem 96*, 156].

[787]See Nelson; Teasley; Kapp *J. Am. Chem. Soc.* **1986**, *108*, 5503.

[788]For a review, see Nelson *Acc. Chem. Res.* **1987**, *20*, 269-276.

[789]For another reagent, see Curci; Lopez; Troisi; Rashid; Schaap *Tetrahedron Lett.* **1987**, *28*, 5319.

[790]Posner; Weitzberg; Nelson; Murr; Seliger *J. Am. Chem. Soc.* **1987**, *109*, 278.

5-38 Hydroxysulfenylation (Addition of Oxygen, Sulfur)
Hydroxy-arylthio-addition (overall transformation)

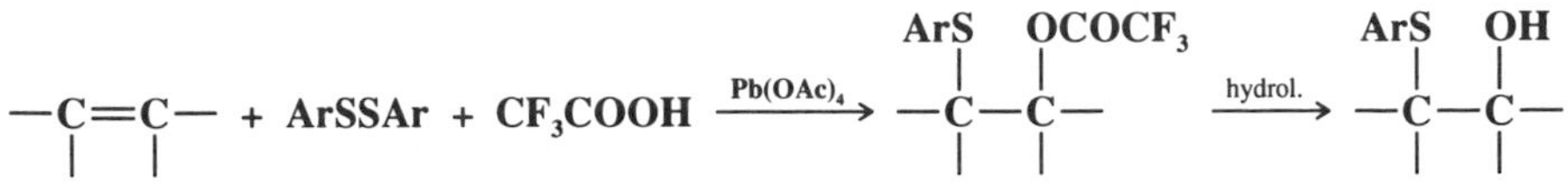

$$-\overset{|}{C}=\overset{|}{C}- \ + \ ArSSAr \ + \ CF_3COOH \ \xrightarrow{Pb(OAc)_4} \ \overset{ArS}{\underset{|}{-\overset{|}{C}}}\overset{OCOCF_3}{\underset{|}{-\overset{|}{C}}}- \ \xrightarrow{hydrol.} \ \overset{ArS}{\underset{|}{-\overset{|}{C}}}\overset{OH}{\underset{|}{-\overset{|}{C}}}-$$

A hydroxy and an arylthio group can be added to a double bond by treatment with an aryl disulfide and lead tetraacetate in the presence of trifluoroacetic acid.[791] Manganese and copper acetates have been used instead of $Pb(OAc)_4$.[792] Addition of the groups OH and RSO has been achieved by treatment of olefins with O_2 and a thiol RSH.[793] Two RS groups were added, to give *vic*-dithiols, by treatment of the alkene with a disulfide RSSR and BF_3–etherate.[794]

In a number of cases, addition of an ether group and a thioether group has been carried

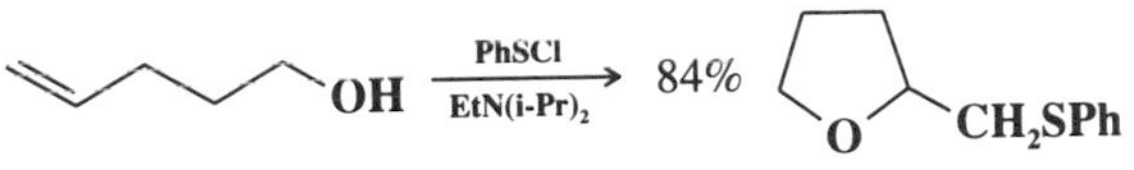

$$\text{(alkenyl alcohol)}\underset{OH}{} \ \xrightarrow[EtN(i\text{-}Pr)_2]{PhSCl} \ 84\% \ \text{(tetrahydrofuran)}CH_2SPh$$

78

out internally. For example, 4-penten-1-ol, treated with benzenesulfenyl chloride and ethyl-diisopropylamine, gave the tetrahydrofuran **78**.[795]

5-39 Oxyamination (Addition of Oxygen, Nitrogen)
Tosylamino-hydroxy-addition

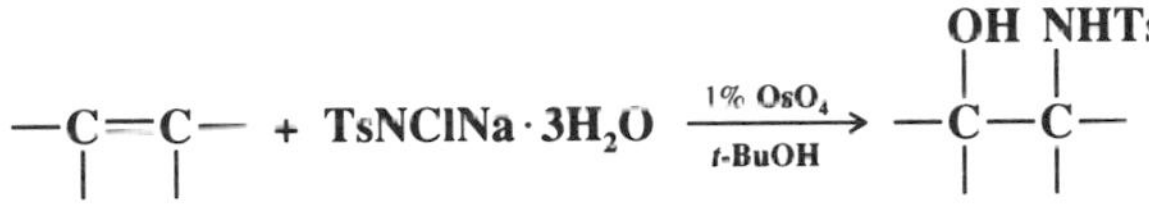

$$-\overset{|}{C}=\overset{|}{C}- \ + \ TsNClNa \cdot 3H_2O \ \xrightarrow[t\text{-BuOH}]{1\% \ OsO_4} \ \overset{OH}{\underset{|}{-\overset{|}{C}}}\overset{NHTs}{\underset{|}{-\overset{|}{C}}}-$$

N-Tosylated β-hydroxy alkylamines (which can be easily hydrolyzed to β-hydroxyamines[796]) can be prepared[797] by treatment of alkenes with the trihydrate of Chloramine-T[637] and a catalytic amount of OsO_4. In some cases yields can be improved by the use of phase-transfer catalysis.[798] The reaction has been carried out enantioselectively.[799] In another procedure, certain β-hydroxy secondary alkylamines can be prepared by treatment of alkenes with the osmium compounds t-Bu—N=OsO_3, followed by reductive cleavage with $LiAlH_4$ of the

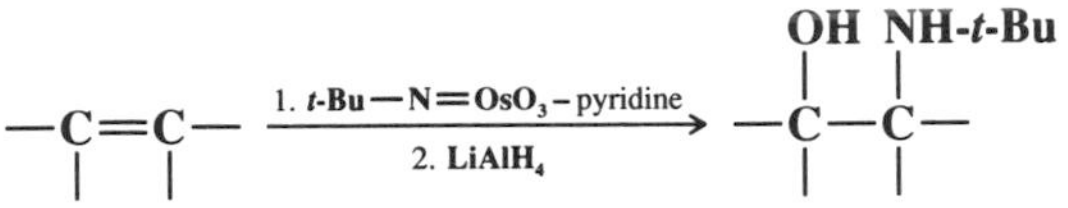

$$-\overset{|}{C}=\overset{|}{C}- \ \xrightarrow[\text{2. LiAlH}_4]{1. \ t\text{-Bu}-N=OsO_3-\text{pyridine}} \ \overset{OH}{\underset{|}{-\overset{|}{C}}}\overset{NH\text{-}t\text{-Bu}}{\underset{|}{-\overset{|}{C}}}-$$

[791]Trost; Ochiai; McDougal *J. Am. Chem. Soc.* **1978,** *100,* 7103. For a related reaction, see Zefirov; Zyk; Kuta-teladze; Kolbasenko; Lapin *J. Org. Chem. USSR* **1986,** *22,* 190.
[792]Bewick; Mellor; Owton *J. Chem. Soc., Perkin Trans. 1* **1985,** 1039; Bewick; Mellor; Milano; Owton *J. Chem. Soc., Perkin Trans. 1* **1985,** 1045; Samii; Ashmawy; Mellor *Tetrahedron Lett.* **1986,** *27,* 5289.
[793]Chung; D'Souza; Szmant *J. Org. Chem.* **1987,** *52,* 1741, and other papers in this series.
[794]Caserio; Fisher; Kim *J. Org. Chem.* **1985,** *50,* 4390.
[795]Tuladhar; Fallis *Tetrahedron Lett.* **1987,** *28,* 523. For a list of other examples, with references, see Ref. 133, pp. 451-452.
[796]For some reactions of the oxyamination products, see Bäckvall; Oshima; Palermo; Sharpless *J. Org. Chem.* **1979,** *44,* 1953.
[797]Sharpless; Chong; Oshima *J. Org. Chem.* **1976,** *41,* 177.
[798]Herranz; Sharpless *J. Org. Chem.* **1978,** *43,* 2544.
[799]Hassine; Gorsane; Pecher; Martin *Bull. Soc. Chim. Belg.* **1985,** *94,* 759.

initially formed osmic esters.[800] It is presumed that Ts—N=OsO$_3$ is an intermediate in the Chloramine-T reaction. Another oxyamination reaction involves treatment of a palladium complex of the olefin with a secondary or primary amine, followed by lead tetraacetate or another oxidant.[801]

β-Amino alcohols can be prepared by treatment of an olefin with a reagent prepared from HgO and HBF$_4$ along with aniline to give an aminomercurial compound

$$\text{PhNH}-\overset{\displaystyle |}{\underset{\displaystyle |}{C}}-\overset{\displaystyle |}{\underset{\displaystyle |}{C}}-\text{HgBF}_4 \quad \text{(aminomercuration; see } \textbf{5-7}\text{) which is hydrolyzed to}$$

$$\text{PhNH}-\overset{\displaystyle |}{\underset{\displaystyle |}{C}}-\overset{\displaystyle |}{\underset{\displaystyle |}{C}}-\text{OH.}$$

[802] The use of an alcohol instead of water gives the corresponding amino ether.

OS **VII**, 223, 375.

5-40 Addition of N$_2$O$_4$ and Related Reactions (Addition of Nitrogen, Nitrogen or Nitrogen, Oxygen)
Dinitro-addition; Nitro-nitrosooxy-addition

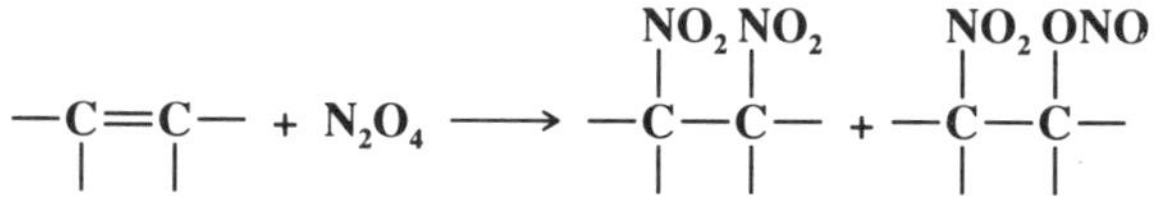

$$-\text{C}=\text{C}- + \text{N}_2\text{O}_4 \longrightarrow -\overset{\text{NO}_2}{\underset{|}{\text{C}}}-\overset{\text{NO}_2}{\underset{|}{\text{C}}}- + -\overset{\text{NO}_2}{\underset{|}{\text{C}}}-\overset{\text{ONO}}{\underset{|}{\text{C}}}-$$

When olefins are treated with N$_2$O$_4$ in an ether, ester, or alkane as solvent, *vic*-dinitro compounds and β-nitro alkyl nitrites are produced.[803] The reaction can be successfully performed with all kinds of olefins and acetylenes. Generally, both products are produced. The dinitro compound is usually stable, but the ester is quite reactive. Upon addition of water or alcohol it is hydrolyzed to a β-nitro alcohol. If oxygen is added, it is oxidized to a β-nitro alkyl nitrate or an α-nitro aldehyde or ketone.

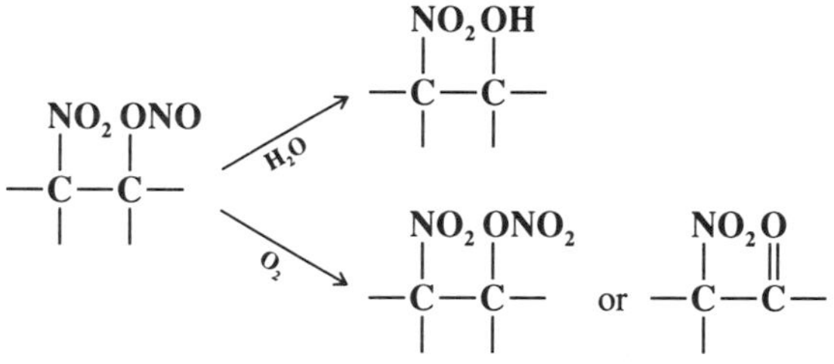

$$-\overset{\text{NO}_2}{\underset{|}{\text{C}}}-\overset{\text{ONO}}{\underset{|}{\text{C}}}- \quad \xrightarrow{\text{H}_2\text{O}} \quad -\overset{\text{NO}_2}{\underset{|}{\text{C}}}-\overset{\text{OH}}{\underset{|}{\text{C}}}-$$

$$\xrightarrow{\text{O}_2} \quad -\overset{\text{NO}_2}{\underset{|}{\text{C}}}-\overset{\text{ONO}_2}{\underset{|}{\text{C}}}- \quad \text{or} \quad -\overset{\text{NO}_2}{\underset{|}{\text{C}}}-\overset{\text{O}}{\underset{\|}{\text{C}}}-$$

The nitrate is stable. Even without deliberate addition of oxygen, it is not uncommon to find some nitrate or ketone. It is therefore possible to prepare four types of compound in this reaction, not counting the nitrite.

[800]Sharpless; Patrick; Truesdale; Biller *J. Am. Chem. Soc.* **1975**, *97*, 2305; Hentges; Sharpless *J. Org. Chem.* **1980**, *45*, 2257. For another method, in which the NH in the product is connected to an easily removable protecting group, see Herranz; Biller; Sharpless *J. Am. Chem. Soc.* **1978**, *100*, 3596; Herranz; Sharpless *J. Org. Chem.* **1980**, *45*, 2710.

[801]Bäckvall; Björkman *J. Org. Chem.* **1980**, *45*, 2893, *Acta Chem. Scand., Ser. B* **1984**, *38*, 91; Bäckvall; Bystrom *J. Org. Chem.* **1982**, *47*, 1126.

[802]Barluenga; Alonso-Cires; Asensio *Synthesis* **1981**, 376.

[803]For reviews, see Ogata, in Trahanovsky, Ref. 740, pt. C pp. 309-313; Larson, in Feuer, Ref. 446, pt. 1, 1969, pp. 316-323.

The mechanism is probably of the free-radical type,[804] with initial attack by NO_2 to give

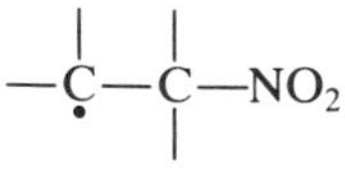

$$-\overset{|}{\underset{\bullet}{C}}-\overset{|}{\underset{|}{C}}-NO_2$$ as the intermediate for both products. In accord with this, the nitro group (in the nitrite derivatives) is found on the side with more hydrogens. An NO_2 and an acetamido (AcNH) group can be added to arylalkenes (with the NHAc going to the side closer to the aryl group) with nitronium tetrafluoroborate in MeCN.[805] Unsubstituted alkenes give poor yields.

OS **VI**, 837.

5-41   Diamination (Addition of Nitrogen, Nitrogen)
   Di(alkylarylamino)-addition

$$-\overset{|}{C}=\overset{|}{C}- \ + \ PhNHR \ \xrightarrow{\text{Ti(OAc)}_3} \ -\overset{\overset{\displaystyle PhNR}{|}}{\underset{\underset{\displaystyle RNPh}{|}}{C}}-\overset{|}{\underset{|}{C}}-$$

Primary (R = H) and secondary aromatic amines react with alkenes in the presence of thallium(III) acetate to give *vic*-diamines in good yields.[806] The reaction is not successful for primary aliphatic amines. In another procedure, olefins can be diaminated by treatment with the osmium compounds R_2NOsO_2 and R_3NOsO (R = *t*-Bu),[807] analogous to the osmium compound mentioned at **5-39**. The palladium-promoted method of **5-39** has also been extended to diamination.[808] Alkenes can also be diaminated[809] indirectly by treatment of the aminomercurial compound mentioned in **5-39** with a primary or secondary aromatic amine.[810]

Two azido groups can be added to double bonds by treatment with sodium azide and iodosobenzene in acetic acid.[811]

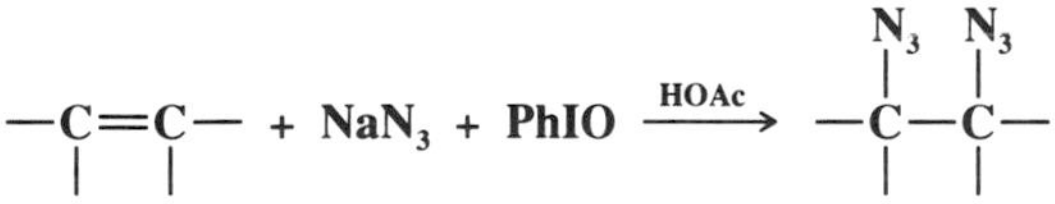

$$-\overset{|}{C}=\overset{|}{C}- \ + \ NaN_3 \ + \ PhIO \ \xrightarrow{\text{HOAc}} \ -\overset{\overset{\displaystyle N_3 \ \ N_3}{| \ \ \ |}}{C}-\overset{|}{\underset{|}{C}}-$$

5-42   Formation of Aziridines (Addition of Nitrogen, Nitrogen)
   epi-**Arylimino-addition,** etc.

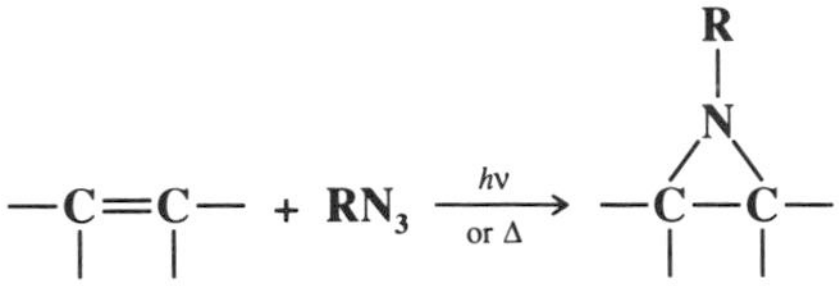

$$-\overset{|}{C}=\overset{|}{C}- \ + \ RN_3 \ \xrightarrow[\text{or } \Delta]{h\nu} \ \overset{\displaystyle R}{\underset{\displaystyle -C-C-}{\overset{|}{N}}}$$

[804]Shechter; Gardikes; Pagano *J. Am. Chem. Soc.* **1959**, *81*, 5420; Shechter; Gardikes; Cantrell; Tiers *J. Am. Chem. Soc.* **1967**, *89*, 3005.

[805]Bloom; Fleischmann; Mellor *J. Chem. Soc., Perkin Trans. 1* **1984**, 2357.

[806]Gómez Aranda; Barluenga; Aznar *Synthesis* **1974**, 504.

[807]Chong; Oshima; Sharpless *J. Am. Chem. Soc.* **1977**, *99*, 3420. See also Sharpless; Singer *J. Org Chem.* **1976**, *41*, 2504.

[808]Bäckvall *Tetrahedron Lett.* **1978**, 163.

[809]For other diamination methods, see Michejda; Campbell *J. Am. Chem. Soc.* **1979**, *101*, 7687; Becker; White; Bergman *J. Am. Chem. Soc.* **1980**, *102*, 5676; Becker; Bergman *Organometallics* **1983**, *2*, 787; Jung; Kohn *Tetrahedron Lett.* **1984**, *25*, 399, *J. Am. Chem. Soc.* **1985**, *107*, 2931; Osowska-Pacewicka; Zwierzak *Synthesis* **1990**, 505.

[810]Barluenga; Alonso-Cires; Asensio *Synthesis* **1979**, 962.

[811]Moriarty; Khosrowshahi *Tetrahedron Lett.* **1986**, *27*, 2809. For other methods, see Minisci; Galli *Tetrahedron Lett.* **1962**, 533; Fristad; Brandvold; Peterson; Thompson *J. Org. Chem.* **1985**, *50*, 3647.

Aziridines can be prepared directly from double-bond compounds by photolysis or thermolysis of a mixture of the substrate and an azide.[812] The reaction has been carried out with R = aryl, cyano, EtOOC, and RSO_2, as well as other groups. The reaction can take place by at least two pathways. In one, the azide is converted to a nitrene, which adds to the double bond in a manner analogous to that of carbene addition (**5-50**). In the other pathway a 1,3 dipolar addition (**5-46**) takes place to give a triazoline (which can be isolated), followed by extrusion of nitrogen (**7-46**). Evidence for the nitrene pathway is most compelling for

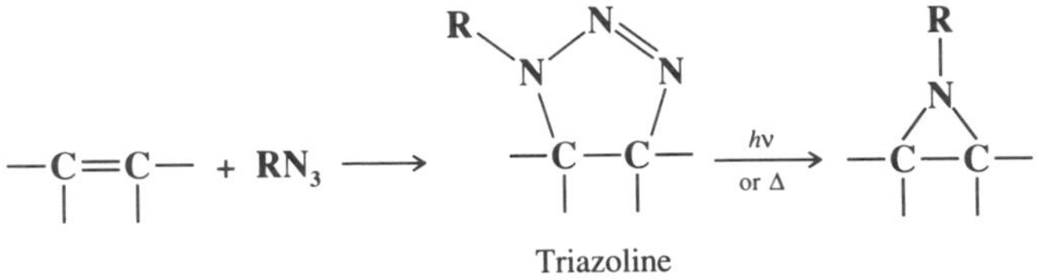

$$-\overset{|}{C}=\overset{|}{C}- \ + \ RN_3 \ \longrightarrow \ \underset{\text{Triazoline}}{\overset{\displaystyle R\diagdown N\diagdown N\diagup N}{-\overset{|}{C}-\overset{|}{C}-}} \ \xrightarrow[\text{or } \Delta]{h\nu} \ \overset{\displaystyle R}{\underset{\displaystyle}{-\overset{|}{C}-\overset{|}{C}-}}$$

R = acyl groups. As discussed on p. 202, singlet nitrenes add stereospecifically while triplet nitrenes do not. Diphenyl sulfimide Ph_2SNH converts Michael-type substrates to the corresponding aziridines.[813] Aminonitrenes R_2NN have been shown to add to triple bonds to

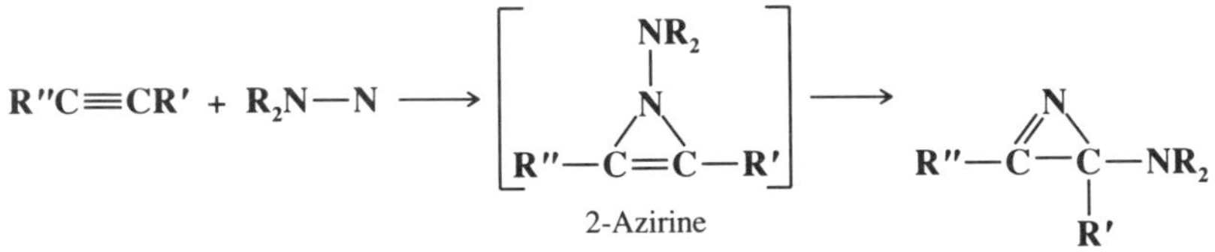

$$R''C\equiv CR' \ + \ R_2N{-}N \ \longrightarrow \ \left[\underset{\text{2-Azirine}}{\overset{\displaystyle NR_2}{R''{-}\overset{\displaystyle \diagup N\diagdown}{C}{=}C{-}R'}}\right] \ \longrightarrow \ \underset{\text{1-Azirine}}{\overset{\displaystyle N}{R''{-}\overset{\diagup\diagdown}{C}{-}\underset{\displaystyle R'}{\overset{|}{C}}{-}NR_2}}$$

give 1-azirines, which arise from rearrangement of the initially formed 2-azirines.[814] Like oxirenes (see **5-36**), 2-azirines are unstable, probably because of antiaromaticity.

Nitrenes can also add to aromatic rings to give ring-expansion products analogous to those mentioned in **5-50**.[815]

OS **VI**, 56.

5-43 Aminosulfenylation (Addition of Nitrogen, Sulfur)
Arylamino-arylthio-addition

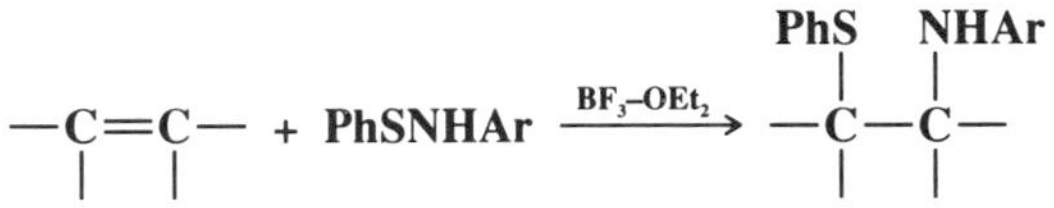

$$-\overset{|}{C}=\overset{|}{C}- \ + \ PhSNHAr \ \xrightarrow{BF_3{-}OEt_2} \ \overset{\displaystyle PhS \quad NHAr}{-\overset{|}{C}-\overset{|}{C}-}$$

An amino group and an arylthio group can be added to a double bond by treatment with a sulfenanilide PhSNHAr in the presence of BF_3–etherate.[816] The addition is anti, and the

[812]For reviews, see Dermer; Ham *Ethylenimine and Other Aziridines*; Academic Press: New York, 1969, pp. 68-79; Muller; Hamer *1,2-Cycloaddition Reactions*; Wiley: New York, 1967.

[813]Furukawa; Yoshimura; Ohtsu; Akasaka; Oae *Tetrahedron* **1980**, *36*, 73. For other methods see Groves; Takahashi *J. Am. Chem. Soc.* **1983**, *105*, 2073; Mahy; Bedi; Battioni; Mansuy *J. Chem. Soc., Perkin Trans. 2* **1988**, 1517; Atkinson; Kelly *J. Chem. Soc., Perkin Trans. 1* **1989**, 1515.

[814]Anderson; Gilchrist; Rees *Chem. Commun.* **1969**, 147.

[815]For example, see Hafner; König *Angew. Chem. Int. Ed. Engl.* **1963**, *2*, 96 [*Angew. Chem. 75*, 89]; Lwowski; Johnson *Tetrahedron Lett.* **1967**, 891.

[816]Benati; Montavecchi; Spagnolo *Tetrahedron Lett.* **1984**, *25*, 2039. See also Brownbridge *Tetrahedron Lett.* **1984**, *25*, 3759.

mechanism probably involves a thiiranium ion.[817] In another aminosulfenylation procedure, the substrate is treated with dimethyl(methylthio)sulfonium fluoroborate MeS$\overset{\oplus}{S}$Me$_2$ BF$_4^-$ and ammonia or an amine,[818] the latter acting as a nucleophile. This reaction was extended

to other nucleophiles:[819] N$_3^-$, NO$_2^-$, CN$^-$, OH$^-$, and OAc$^-$ to give MeS—$\overset{|}{\underset{|}{C}}$—$\overset{|}{\underset{|}{C}}$—A, where

A = N$_3$, NO$_2$, CN, OH, and OAc, respectively. An RS (R = alkyl or aryl) and an NHCOMe group have been added in an electrochemical procedure.[820]

5-44 Acylacyloxylation and Acylamidation (Addition of Oxygen, Carbon, or Nitrogen, Carbon)
  Acyl-acyloxy-addition

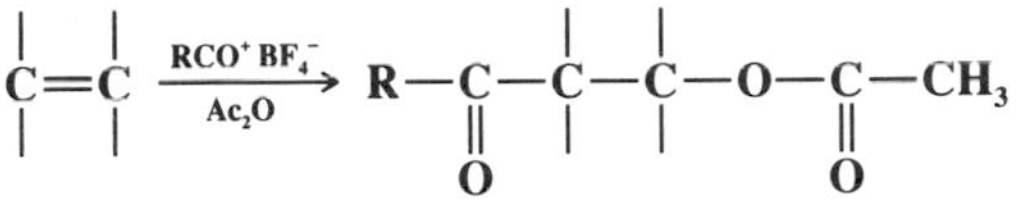

$$\overset{|}{\underset{|}{C}}=\overset{|}{\underset{|}{C}} \xrightarrow[\text{Ac}_2\text{O}]{\text{RCO}^+\text{BF}_4^-} R-\overset{O}{\overset{\|}{C}}-\overset{|}{\underset{|}{C}}-\overset{|}{\underset{|}{C}}-O-\overset{O}{\overset{\|}{C}}-CH_3$$

An acyl and an acyloxy group can be added to a double bond by treatment with an acyl fluoroborate and acetic anhydride.[821] As expected, the addition follows Markovnikov's rule, with the electrophile Ac$^+$ going to the carbon with more hydrogens. In an analogous reaction, an acyl and an amido group can be added, if a nitrile is used in place of the anhydride:

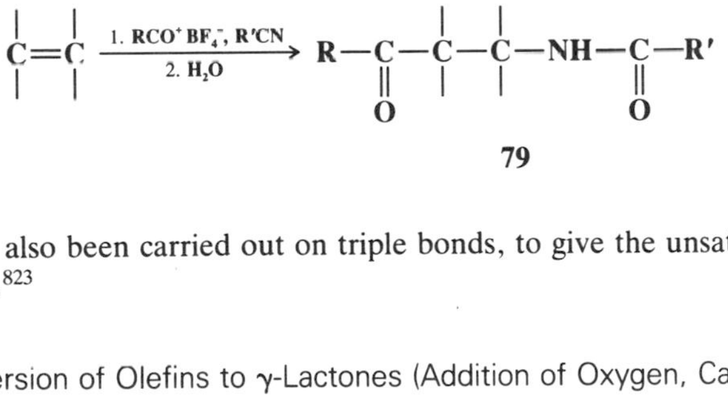

$$\overset{|}{\underset{|}{C}}=\overset{|}{\underset{|}{C}} \xrightarrow[\text{2. H}_2\text{O}]{\text{1. RCO}^+\text{BF}_4^-, \text{R'CN}} R-\overset{O}{\overset{\|}{C}}-\overset{|}{\underset{|}{C}}-\overset{|}{\underset{|}{C}}-NH-\overset{O}{\overset{\|}{C}}-R'$$

79

This reaction has also been carried out on triple bonds, to give the unsaturated analogs of **79** (syn addition).[823]

5-45 The Conversion of Olefins to γ-Lactones (Addition of Oxygen, Carbon)

$$-\overset{|}{\underset{|}{C}}=\overset{|}{\underset{|}{C}}- + \text{Mn(OAc)}_3 \xrightarrow{\text{HOAc}} \begin{array}{c} -\overset{|}{C}-\overset{|}{C}- \\ O \quad\;\; CH_2 \\ \diagdown C \diagup \\ \| \\ O \end{array}$$

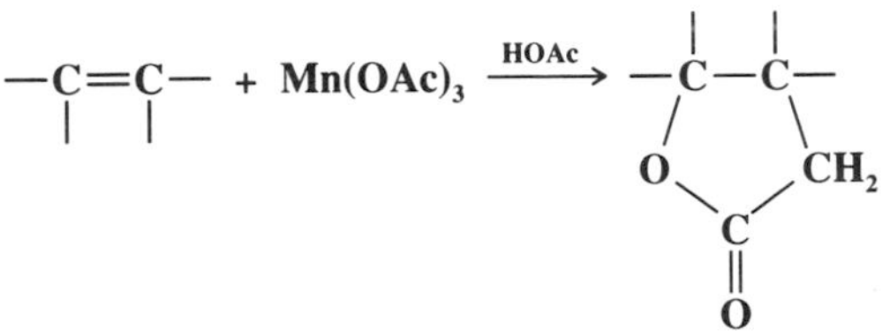

[817]See Ref. 20.
[818]Trost; Shibata *J. Am. Chem. Soc.* **1982**, *104*, 3225; Caserio; Kim. *J. Am. Chem. Soc.* **1982**, *104*, 3231.
[819]Trost; Shibata; Martin *J. Am. Chem. Soc.* **1982**, *104*, 3228; Trost; Shibata, Ref. 818. For an extension that allows A to be C≡CR, see Trost; Martin *J. Am. Chem. Soc.* **1984**, *106*, 4263.
[820]Bewick; Coe; Mellor; Owton *J. Chem. Soc., Perkin Trans. 1* **1985**, 1033.
[821]Shastin; Balenkova *J. Org. Chem. USSR* **1984**, *20*, 870.
[822]Shastin; Balenkova *J. Org. Chem. USSR* **1984**, *20*, 1235; Gridnev; Shastin; Balenkova *J. Org. Chem. USSR* **1987**, *23*, 1389; Gridnev; Buevich; Sergeyev; Balenkova *Tetrahedron Lett.* **1989**, *30*, 1987.
[823]Gridnev; Balenkova *J. Org. Chem. USSR* **1988**, *24*, 1447.

Olefins react with manganese(III) acetate to give γ-lactones.[824] The mechanism is probably free-radical, involving addition of $\cdot CH_2COOH$ to the double bond. Lactone formation has also been accomplished by treatment of olefins with lead tetraacetate,[825] with α-bromo carboxylic acids in the presence of benzoyl peroxide as catalyst,[826] and with dialkyl malonates and iron(III) perchlorate $Fe(ClO_4)_3 \cdot 9H_2O$.[827] Olefins can also be converted to γ-lactones by indirect routes.[828]

OS **VII**, 400.

For addition of aldehydes and ketones, see the Prins reaction (**6-53**), and reactions **6-63** and **6-64**.

5-46 1,3-Dipolar Addition (Addition of Oxygen, Nitrogen, Carbon)

Azides add to double bonds to give triazolines. This is one example of a large group of reactions (2 + 3 cycloadditions) in which five-membered heterocyclic compounds are prepared by addition of 1,3-dipolar compounds to double bonds (see Table 15.3).[829] These are compounds that have a sequence of three atoms a—b—c, of which a has a sextet of electrons in the outer shell and c an octet with at least one unshared pair. The reaction can then be formulated as

---

[824]Bush; Finkbeiner *J. Am. Chem. Soc.* **1968**, *90*, 5903; Heiba; Dessau; Koehl *J. Am. Chem. Soc.* **1968**, *90*, 5905; Heiba; Dessau; Rodewald *J. Am. Chem. Soc.* **1974**, *96*, 7977; Midgley; Thomas *J. Chem. Soc., Perkin Trans. 2* **1984**, 1537; Ernst; Fristad *Tetrahedron Lett.* **1985**, *26*, 3761; Shundo; Nishiguchi; Matsubara; Hirashima *Tetrahedron* **1991**, *47*, 831. See also Corey; Gross *Tetrahedron Lett.* **1985**, *26*, 4291.

[825]Heiba; Dessau; Koehl *J. Am. Chem. Soc.* **1968**, *90*, 2706.

[826]Nakano; Kayama; Nagai *Bull. Chem. Soc. Jpn.* **1987**, *60*, 1049. See also Kraus; Landgrebe *Tetrahedron Lett.* **1984**, *25*, 3939.

[827]Citterio; Sebastiano; Nicolini; Santi *Synlett* **1990**, 42.

[828]See, for example, Boldt; Thielecke; Etzemüller *Chem. Ber.* **1969**, *102*, 4157; Das Gupta; Felix, Kempe; Eschenmoser *Helv. Chim. Acta* **1972**, *55*, 2198; Bäuml; Tscheschlok; Pock; Mayr *Tetrahedron Lett.* **1988**, *29*, 6925.

[829]For a treatise, see Padwa *1,3-Dipolar Cycloaddition Chemistry*, 2 vols.; Wiley: New York, 1984. For general reviews, see Carruthers, Ref. 440; Drygina; Garnovskii *Russ. Chem. Rev.* **1986**, *55*, 851-866; Samuilov; Konovalov *Russ. Chem. Rev.* **1984**, *53*, 332-342; Beltrame, in Bamford; Tipper, Ref. 1, vol. 9, pp. 117-131; Huisgen; Grashey; Sauer, in Patai, Ref. 36, vol. 1, pp. 806-878; Huisgen *Helv. Chim. Acta* **1967**, *50*, 2421-2439, *Bull. Soc. Chim. Fr.* **1965**, 3431-3440, *Angew. Chem. Int. Ed. Engl.* **1963**, *2*, 565-598, 633-645 [*Angew. Chem. 75*, 604-637, 742-754]. For specific monographs and reviews, see Torssell *Nitrile Oxides, Nitrones, and Nitronates in Organic Synthesis*; VCH: New York, 1988; Scriven *Azides and Nitrenes*; Academic Press: New York, 1984; Stanovnik *Tetrahedron* **1991**, *47*, 2925-2945 (diazoalkanes); Kanemasa; Tsuge *Heterocycles* **1990**, *30*, 719-736 (nitrile oxides); Paton *Chem. Soc. Rev.* **1989**, *18*, 33-52 (nitrile sulfides); Terao; Aono; Achiwa *Heterocycles* **1988**, *27*, 981-1008 (azomethine ylides); Vedejs *Adv. Cycloaddit.* **1988**, *1*, 33-51 (azomethine ylides); DeShong; Lander; Leginus; Dicken *Adv. Cycloaddit.* **1988**, *1*, 87-128 (nitrones); Balasubramanian *Org. Prep. Proced. Int.* **1985**, *17*, 23-47 (nitrones); Confalone; Huie *Org. React.* **1988**, *36*, 1-173 (nitrones); Padwa, in Horspool *Synthetic Organic Photochemistry*; Plenum: New York, 1984, pp. 313-374 (nitrile ylides); Bianchi; Gandolfi; Grünanger in Patai; Rappoport, Ref. 49, pp. 752-784 (nitrile oxides); Black; Crozier; Davis *Synthesis* **1975**, 205-221 (nitrones); Stuckwisch *Synthesis* **1973**, 469-483 (azomethine ylides, azomethine imines). For reviews of intramolecular 1,3-dipolar additions see Padwa, in Padwa, treatise cited above, vol. 2, pp. 277-406; Padwa; Schoffstall *Adv. Cycloaddit.* **1990**, *2*, 1-89; Tsuge; Hatta; Hisano, in Patai *Supplement A: The Chemistry of Double-bonded Functional Groups*, vol. 2, pt. 1; Wiley: New York, 1989, pp. 345-475; Padwa *Angew. Chem. Int. Ed. Engl.* **1976**, *15*, 123-136 [*Angew. Chem. 88*, 131-144]. For a review of azomethine ylides, see Tsuge; Kanemasa *Adv. Heterocycl. Chem.* **1989**, *45*, 231-349. For reviews of 1,3-dipolar cycloreversions, see Bianchi; Gandolfi, in Padwa, treatise cited above, vol. 2, pp. 451-542; Bianchi; De Micheli; Gandolfi *Angew. Chem. Int. Ed. Engl.* **1979**, *18*, 721-738 [*Angew. Chem. 91*, 781-798]. For a related review, see Petrov; Petrov *Russ. Chem. Rev.* **1987**, *56*, 152-162. For the use of this reaction to synthesize natural products, see papers in *Tetrahedron* **1985**, *41*, 3447-3568.

TABLE 15.3 Some common 1,3-dipolar compounds

Type 1

Azide	$R-\overset{\ominus}{\underline{N}}-N=\overset{\oplus}{\underline{N}} \longleftrightarrow R-\overset{\ominus}{\underline{N}}-\overset{\oplus}{N}\equiv N$
Diazoalkane	$R_2\overset{\ominus}{C}-N=\overset{\oplus}{\underline{N}} \longleftrightarrow R_2\overset{\ominus}{C}-\overset{\oplus}{N}\equiv N$
Nitrous oxide	$I\overset{\ominus}{\underline{O}}-N=\overset{\oplus}{\underline{N}} \longleftrightarrow I\overset{\ominus}{\underline{O}}-\overset{\oplus}{N}\equiv N$
Nitrile imine	$R-\overset{\ominus}{\underline{N}}-N=\overset{\oplus}{C}R' \longleftrightarrow R-\overset{\ominus}{\underline{N}}-\overset{\oplus}{N}\equiv CR'$
Nitrile ylide	$R_2\overset{\ominus}{C}-N=\overset{\oplus}{C}R' \longleftrightarrow R_2\overset{\ominus}{C}-\overset{\oplus}{N}\equiv CR'$
Nitrile oxide	$I\overset{\ominus}{\underline{O}}-\underline{N}=\overset{\oplus}{C}R \longleftrightarrow I\overset{\ominus}{\underline{O}}-\overset{\oplus}{N}\equiv CR$

Type 2

Azomethine imine	$R_2\overset{\ominus}{\underline{C}}-\overset{\oplus}{\underset{\underset{R''}{\vert}}{N}}-\overset{\oplus}{N}R' \longleftrightarrow R_2\overset{\ominus}{C}-\overset{\oplus}{\underset{\underset{R''}{\vert}}{N}}=\underline{N}R'$
Azoxy compound	$I\overset{\ominus}{\underline{O}}-\overset{}{\underset{\underset{R}{\vert}}{\underline{N}}}-\overset{\oplus}{N}R' \longleftrightarrow I\overset{\ominus}{\underline{O}}-\overset{\oplus}{\underset{\underset{R}{\vert}}{N}}=\underline{N}R'$
Azomethine ylide	$R_2\overset{\ominus}{\underline{C}}-\overset{}{\underset{\underset{R''}{\vert}}{\underline{N}}}-\overset{\oplus}{C}R'_2 \longleftrightarrow R_2\overset{\ominus}{\underline{C}}-\overset{\oplus}{\underset{\underset{R''}{\vert}}{N}}=CR'_2$
Nitrone	$I\overset{\ominus}{\underline{O}}-\overset{}{\underset{\underset{R'}{\vert}}{\underline{N}}}-\overset{\oplus}{C}R_2 \longleftrightarrow I\overset{\ominus}{\underline{O}}-\overset{\oplus}{\underset{\underset{R'}{\vert}}{N}}=CR_2$
Carbonyl oxide	$I\overset{\ominus}{\underline{O}}-\underline{O}-\overset{\oplus}{C}R_2 \longleftrightarrow I\overset{\ominus}{\underline{O}}-\overset{\oplus}{O}=CR_2$
Ozone	$I\overset{\ominus}{\underline{O}}-\underline{O}-\overset{\oplus}{\underline{O}}I \longleftrightarrow I\overset{\ominus}{\underline{O}}-\overset{\oplus}{O}=\underline{O}I$

Since compounds with six electrons in the outer shell of an atom are usually not stable, the a—b—c system is actually one canonical form of a resonance hybrid, for which at least one other form can be drawn (see Table 15.3). 1,3-Dipolar compounds can be divided into two main types:

1. Those in which the dipolar canonical form has a double bond on the sextet atom and the other canonical form has a triple bond on that atom:

$$\underset{}{\overset{\ominus}{\underline{a}}}-\underline{b}\overset{\oplus}{=}c-\longleftrightarrow \overset{\ominus}{\underline{a}}-\underline{b}\overset{\oplus}{\equiv}c-$$

If we limit ourselves to the first row of the periodic table, b can only be nitrogen, c can be carbon or nitrogen, and a can be carbon, oxygen, or nitrogen; hence there are six types. Among these are azides (a = b = c = N), illustrated above, and diazoalkanes.

2. Those in which the dipolar canonical form has a single bond on the sextet atom and the other form has a double bond:

$$\overset{\ominus}{\underline{a}}-\underline{b}\overset{\oplus}{-}c-\longleftrightarrow \overset{\ominus}{\underline{a}}-\underline{b}\overset{\oplus}{=}c-$$

Here b can be nitrogen or oxygen, and a and c can be nitrogen, oxygen, or carbon, but there are only 12 types, since, for example, N—N—C is only another form of C—N—N. Examples are shown in Table 15.3.

Of the 18 systems, some of which are unstable and must be generated in situ,[830] the reaction has been accomplished for at least 15, though not in all cases with a carbon–carbon double bond (the reaction also can be carried out with other double bonds[831]). Not all olefins undergo 1,3-dipolar addition equally well. The reaction is most successful for those that are good dienophiles in the Diels–Alder reaction (**5-47**). The addition is stereospecific and syn, and the mechanism is probably a one-step concerted process, as illustrated above.[832] As expected for this type of mechanism, the rates do not vary much with changes in solvent.[833] There are no simple rules covering orientation in 1,3-dipolar additions. Regioselectivities are complicated but have been explained by molecular-orbital treatments.[834] When the 1,3-dipolar compound is a thiocarbonyl ylide ($R_2C\overset{\oplus}{=}S—\overset{\ominus}{C}H_2$) the addition has been shown to be nonstereospecific with certain substrates (though stereospecific with others), indicating a nonsynchronous mechanism in these cases, and in fact, a diionic intermediate (see mechanism c on p. 857) has been trapped in one such case.[835]

Conjugated dienes generally give exclusive 1,2 addition, though 1,4 addition (a 3 + 4 cycloaddition) has been reported.[836]

Carbon–carbon triple bonds can also undergo 1,3-dipolar addition.[837] For example, azides give triazoles:

$$-C\equiv C- + RN_3 \longrightarrow \quad \text{(triazole ring)}$$

The 1,3-dipolar reagent can in some cases be generated by the in situ opening of a suitable three-membered ring system. For example, aziridines can add to activated double bonds to give pyrrolidines, e.g.,[838]

[830]For a review of some aspects of this, see Grigg *Chem. Soc. Rev.* **1987**, *16*, 89-121.

[831]For a review of 1,3-dipolar addition to other double bonds, see Bianchi; De Micheli; Gandolfi, in Patai, Ref. 1, pt. 1, pp. 369-532. For a review of such addition to the C=S bond, see Dunn; Rudorf *Carbon Disulfide in Organic Chemistry*; Wiley: New York, 1989, pp. 97-119.

[832]For a review, see Huisgen *Adv. Cycloaddit.* **1988**, *1*, 1-31. For discussions, see Huisgen *J. Org. Chem.* **1976**, *41*, 403; Firestone *Tetrahedron* **1977**, *33*, 3009-3039; Harcourt *Tetrahedron* **1978**, *34*, 3125; Haque *J. Chem. Educ.* **1984**, *61*, 490; Al-Sader; Kadri *Tetrahedron Lett.* **1985**, *26*, 4661; Houk; Firestone; Munchausen; Mueller; Arison; Garcia *J. Am. Chem. Soc.* **1985**, *107*, 7227; Majchrzak; Warkentin *J. Phys. Org. Chem.* **1990**, *3*, 339.

[833]For a review of the role of solvents in this reaction, see Kadaba *Synthesis* **1973**, 71-84.

[834]For a review, see Houk; Yamaguchi, in Padwa *1,3-Dipolar Cycloaddition Chemistry*, Ref. 829, vol. 2, pp. 407-450. See also Burdisso; Gandolfi; Quartieri; Rastelli *Tetrahedron* **1987**, 159.

[835]Huisgen; Mloston; Langhals *J. Am. Chem. Soc.* **1986**, *108*, 6401, *J. Org. Chem.* **1986**, *51*, 4085; Mloston; Langhals; Huisgen *Tetrahedron Lett.* **1989**, *30*, 5373; Huisgen; Mloston *Tetrahedron Lett.* **1989**, *30*, 7041.

[836]Baran; Mayr *J. Am. Chem. Soc.* **1987**, *109*, 6519.

[837]For reviews, see Bastide; Hamelin; Texier; Quang *Bull. Soc. Chim. Fr.* **1973**, 2555-2579; 2871-2887; Fuks; Viehe in Viehe Ref. 49, p. 460-477.

[838]For a review, see Lown, in Padwa, Ref. 834, vol 1. pp. 653-732.

Aziridines also add to C≡C triple bonds as well as to other unsaturated linkages, including C=O, C=N, and C≡N.[839] In some of these reactions it is a C—N bond of the aziridine that opens rather than the C—C bond.

For other 2 + 3 cycloadditions, see **5-48.**

OS **V**, 957, 1124; **VI**, 592, 670; **67**, 133. Also see OS **IV**, 380.

C.   Carbon on Both Sides.   Reactions **5-47** to **5-52** are cycloaddition reactions.[840]

5-47   The Diels–Alder Reaction

(2 + 4)*cyclo***-Ethylene-1/4/addition** or **(4 + 2)***cyclo***-[But-2-ene-1,4-diyl]-1/2/addition,** etc.

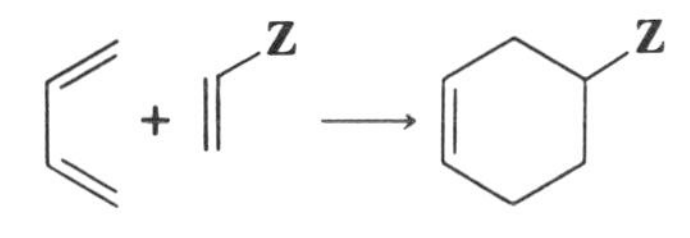

In the *Diels–Alder reaction* a double bond adds 1,4 to a conjugated diene (a 2 + 4 cycloaddition),[841] so the product is always a six-membered ring. The double-bond compound is called a *dienophile*. The reaction is easy and rapid and of very broad scope.[842] Ethylene and simple olefins make poor dienophiles, although the reaction has been carried out with these compounds. Most dienophiles are of the form —C=C—Z or Z—C=C—Z', where Z and Z' are CHO, COR,[843] COOH, COOR, COCl, COAr, CN,[844] NO$_2$,[845] Ar, CH$_2$OH, CH$_2$Cl, CH$_2$NH$_2$, CH$_2$CN, CH$_2$COOH, halogen, or C=C. In the last case, the dienophile is itself a diene:

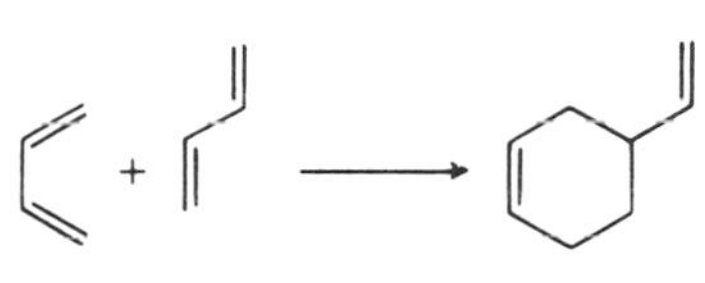

When two dienes react, mixtures are quite possible. Thus, butadiene and isoprene (CH$_2$=CH—CMe=CH$_2$) gave all nine possible Diels–Alder adducts, as well as eight-mem-

---

[839]For reviews, see Lown *Rec. Chem. Prog.* **1971,** *32,* 51-83; Gladysheva; Sineokov; Etlis *Russ. Chem. Rev.* **1970,** *39,* 118-129.

[840]For a system of classification of cycloaddition reactions, see Huisgen *Angew. Chem. Int. Ed. Engl.* **1968,** *7,* 321-328 [*Angew Chem. 80,* 329-337]. For a review of certain types of cycloadditions leading to 3- to 6-membered rings involving 2, 3, or 4 components, see Posner *Chem. Rev.* **1986,** *86,* 831-844. See also the series *Advances in Cycloaddition.*

[841]For a monograph, see Wasserman *Diels–Alder Reactions;* Elsevier: New York, 1965. For reviews, see Roush *Adv. Cycloaddit.* **1990,** *2,* 91-146; Carruthers, Ref. 440; Brieger; Bennett *Chem. Rev.* **1980,** *80,* 63-97; Oppolzer *Angew. Chem. Int. Ed. Engl.* **1977,** *16,* 10-23 [*Angew. Chem. 89,* 10-24]; Beltrame, in Bamford; Tipper, Ref. 1, vol. 9, pp. 94-117; Huisgen; Grashey; Sauer, in Patai, Ref. 36, vol. 1, pp. 878-929; Carruthers, Ref. 218, pp. 183-244; Sauer *Angew. Chem. Int. Ed. Engl.* **1966,** *5,* 211-230, **1967,** *6,* 16-33 [*Angew Chem. 78,* 233-252, *79,* 76-94]. For a monograph on intramolecular Diels–Alder reactions see Taber, Ref. 440. For reviews, see Deslongchamps *Aldrichimica Acta* **1991,** *24,* 43-56; Craig *Chem. Soc. Rev.* **1987,** *16,* 187-238; Salakhov; Ismailov *Russ. Chem. Rev.* **1986,** *55,* 1145-1163; Fallis *Can. J. Chem.* **1984,** *62,* 183-234. For a long list of references to various aspects of the Diels–Alder reaction, see Ref. 133, pp. 263-272.

[842]For a review of reactivity in the Diels–Alder reaction, see Konovalov *Russ. Chem. Rev.* **1983,** *52,* 1064-1080.

[843]For a review of Diels–Alder reactions with cyclic enones, see Fringuelli; Taticchi; Wenkert *Org. Prep. Proced. Int.* **1990,** *22,* 131-165.

[844]For a review of the Diels–Alder reaction with acrylonitrile, see Butskus *Russ. Chem. Rev.* **1962,** *31,* 283-294. For a review of tetracyanoethylene as a dienophile, see Ciganek; Linn; Webster, in Rappoport, Ref. 588, pp. 449-453.

[845]For a review of the Diels–Alder reaction with nitro compounds, see Novikov; Shuekhgeimer; Dudinskaya *Russ. Chem. Rev.* **1960,** *29,* 79-94.

bered rings and trimers.[846] Particularly common dienophiles are maleic anhydride[847] and quinones.[848] Triple bond compounds (—C≡C—Z or Z—C≡C—Z') may be dienophiles[849]

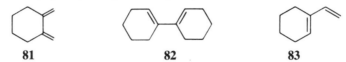

as may allenes, though allenes without activating groups are very poor dienophiles.[850] Ketenes, however, do not undergo Diels–Alder reactions.[851] Benzynes, although not isolable, act as dienophiles and can be trapped with dienes,[852] e.g.,

The low reactivity of ethylene can be overcome by using phenyl vinyl sulfone Ph-SO$_2$CH=CH$_2$ instead.[853] The PhSO$_2$ group can be easily removed with Na–Hg after the ring-closure reaction. Similarly, phenyl vinyl sulfoxide PhSOCH=CH$_2$ can be used as a synthon for acetylene.[854] In this case PhSOH is lost from the sulfoxide product (**7-12**).

Besides carbon–carbon multiple bonds, other double- and triple-bond compounds can be dienophiles, giving rise to heterocyclic compounds. Among these are N≡C—, —N=C—, —N=N—, O=N—, and —C=O compounds[855] and, as we have seen (**5-37**), even molecular oxygen.

Dienes can be open-chain, inner-ring (e.g., **80**), outer-ring[856] (e.g., **81**), across rings (e.g., **82**), or inner–outer (e.g., **83**), except that they may not be frozen into a transoid conformation

81   **82**   **83**

[846]Johnstone; Quan *J. Chem. Soc.* **1963**, 935.

[847]For a review of Diels–Alder reactions with maleic anhydride see Kloetzel *Org. React.* **1948**, *4*, 1-59.

[848]For reviews of Diels–Alder reactions with quinones, see Finley, in Patai Ref. 38, vol. 1, pt. 2, pp. 986-1018, vol. 2, pt. 1 (edited by Patai; Rappoport), 1988, pp. 537-717, pp. 614-645. For a review of the synthesis of quinones using Diels–Alder reactions, see Naruta; Maruyama, in the same treatise, vol. 2, pt. 1, pp. 241-402, pp. 277-303.

[849]For reviews of triple bonds in cycloaddition reactions, see Bastide; Henri-Rousseau, in Patai, Ref. 70, pt. 1, pp. 447-522, Fuks; Viehe, in Viehe, Ref. 49, pp. 477-508.

[850]For a review of allenes as dienes or dienophiles, see Hopf, in Landor, Ref. 95, vol. 2, pp. 563-577.

[851]Ketenes react with conjugated dienes to give 1,2 addition (see **5-49**).

[852]For a review of benzynes as dienophiles, see Hoffmann *Dehydrobenzene and Cycloalkynes*; Academic Press: New York, 1967, pp. 200-239. For a review of the reactions of benzynes with heterocyclic compounds see Bryce; Vernon *Adv. Heterocycl. Chem.* **1981**, *28*, 183-229.

[853]Carr; Williams; Paquette *J. Org. Chem.* **1983**, *48*, 4976; Kinney; Crouse; Paquette *J. Org. Chem.* **1983**, *48*, 4986.

[854]Paquette; Moerck; Harirchian; Magnus *J. Am. Chem. Soc.* **1978**, *100*, 1597. For other acetylene synthons see De Lucchi; Lucchini; Pasquato; Modena *J. Org. Chem.* **1984**, *49*, 596; Hermeling; Schäfer *Angew. Chem. Int. Ed. Engl.* **1984**, *23*, 233 [*Angew. Chem. 96*, 238]. For a review, see De Lucchi; Modena *Tetrahedron* **1984**, *40*, 2585-2632. For a review of 2 + 2 and 2 + 4 cycloadditions of vinylic sulfides, sulfoxides, and sulfones, see De Lucchi; Pasquato *Tetrahedron* **1988**, *44*, 6755-6794.

[855]For monographs on dienes and dienophiles with hetero atoms, see Boger; Weinreb *Hetero Diels–Alder Methodology in Organic Synthesis*; Academic Press: New York, 1987; Hamer, Ref. 778. For reviews, see Weinreb; Scola *Chem. Rev.* **1989**, *89*, 1525-1534; Boger, in Lindberg *Strategies and Tactics in Organic Synthesis*, vol. 2; Academic Press: New York, 1989, pp. 1-56; Kametani; Hibino *Adv. Heterocycl. Chem.* **1987**, *42*, 245-333; Boger *Tetrahedron* **1983**, *39*, 2869-2939; Weinreb; Staib *Tetrahedron* **1982**, *38*, 3087-3128; Weinreb; Levin *Heterocycles* **1979**, *12*, 949-975; Desimoni; Tacconi *Chem. Rev.* **1975**, *75*, 651-692; Kresze; Firl *Fortschr. Chem. Forsch.* **1969**, *11*, 245-284. See also Ref. 862.

[856]For reviews of Diels–Alder reactions of some of these compounds, see Charlton; Alauddin *Tetrahedron* **1987**, *43*, 2873-2889; Oppolzer *Synthesis* **1978**, 793-802.

(see p. 842). They need no special activating groups, and nearly all conjugated dienes undergo the reaction with suitable dienophiles.[857]

Aromatic compounds can also behave as dienes.[858] Benzene is very unreactive toward dienophiles; very few dienophiles (one of them is benzyne) have been reported to give Diels–Alder adducts with it.[859] Naphthalene and phenanthrene are also quite resistant, though naphthalene has given Diels–Alder addition at high pressures.[860] However, anthracene and other compounds with at least three linear benzene rings give Diels–Alder reactions readily. The interesting compound triptycene can be prepared by a Diels–Alder reaction between benzyne and anthracene:[861]

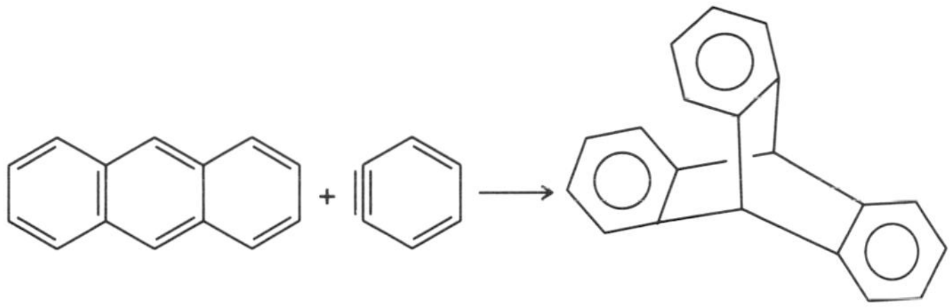

Certain heterocyclic aromatic rings (among them furans)[862] can also behave as dienes in the Diels–Alder reaction. Some hetero dienes that give the reaction are —C—C—C—O, O=C—C=O, and N=C—C=N.[854] For both all-carbon and hetero systems, the "diene" can be a conjugated enyne. If the geometry of the molecule is suitable, the diene can even be nonconjugated, e.g.,[863]

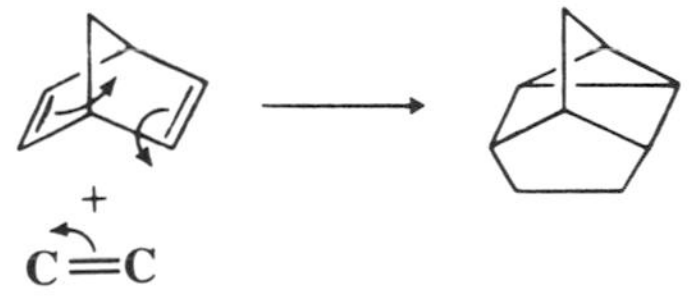

This last reaction is known as the *homo-Diels–Alder reaction*.

---

[857]For a monograph on dienes, with tables showing more than 800 types, see Fringuelli; Taticchi *Dienes in the Diels–Alder Reaction*; Wiley: New York, 1990. For a review of Diels–Alder reactions with 2-pyrones, see Shusherina *Russ. Chem. Rev.* **1974**, *43*, 851-861. For reviews of dienes with hetero substituents, see Danishefsky *Chemtracts: Org. Chem.* **1989**, *2*, 273-297; Petrzilka; Grayson *Synthesis* **1981**, 753-786. For a review of dienes containing a 1-CONR₂ group, see Smith *Org. Prep. Proced. Int.* **1990**, *22*, 315-397.

[858]For a review, see Wagner–Jauregg *Synthesis* **1980**, 165-214, 769-798. See also Balaban; Biermann; Schmidt *Nouv. J. Chim.* **1985**, *9*, 443.

[859]Miller; Stiles *J. Am. Chem. Soc.* **1963**, *85*, 1798; Meyerson; Fields *Chem. Ind. (London)* **1966**, 1230; Ciganek *Tetrahedron Lett.* **1967**, 3321; Friedman *J. Am. Chem. Soc.* **1967**, *89*, 3071; Liu; Krespan *J. Org. Chem.* **1969**, *34*, 1271.

[860]Jones; Mangold; Plieninger *Tetrahedron* **1962**, *18*, 267; Plieninger; Wild; Westphal *Tetrahedron* **1969**, *25*, 5561.

[861]Wittig; Niethammer *Chem. Ber.* **1960**, *93*, 944; Wittig; Härle; Knauss; Niethammer *Chem. Ber.* **1960**, *93*, 951. For a review of triptycene, see Skvarchenko; Shalaev; Klabunovskii *Russ. Chem. Rev.* **1974**, *43*, 951-966.

[862]For reviews, see Katritzky; Dennis *Chem. Rev.* **1989**, *89*, 827-861; Schmidt *Acc. Chem. Res.* **1986**, *19*, 250-259; Boger *Chem. Rev.* **1986**, *86*, 781-793.

[863]See, for example, Fickes; Metz *J. Org. Chem.* **1978**, *43*, 4057; Paquette; Kesselmayer; Künzer *J. Org. Chem.* **1988**, *53*, 5183.

When an unsymmetrical diene adds to an unsymmetrical dienophile, there are two possible products (not counting stereoisomers):

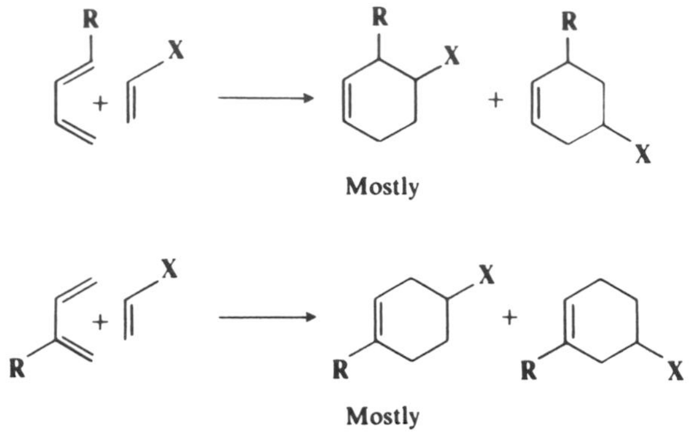

Mostly

Mostly

Although mixtures are often obtained, usually one predominates, the one indicated above. This regioselectivity, in which the "ortho" or "para" product is favored over the "meta," has been explained by molecular-orbital considerations.[864] When X = NO$_2$, regioselectivity to give the "ortho" or "para" product was very high at room temperature, and this method, combined with subsequent removal of the NO$_2$ (see **0-82**) has been used to perform regioselective Diels–Alder reactions.[865]

The stereochemistry of the Diels–Alder reaction can be considered from several aspects:

1. With respect to the dienophile, the addition is stereospecifically syn, with very few exceptions.[866] This means that groups that are cis in the olefin will be cis in the cyclohexene ring:

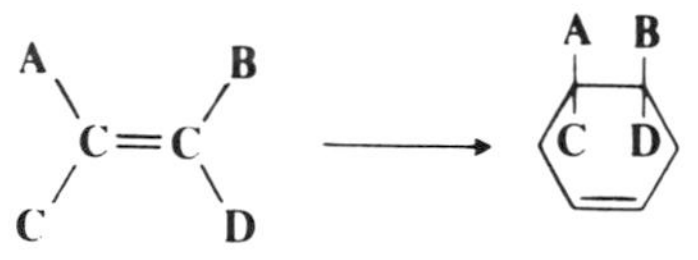

2. With respect to 1,4-disubstituted dienes, fewer cases have been investigated, but here too the reaction is stereospecific and syn. Thus, *trans,trans*-1,4-diphenylbutadiene gives *cis*-1,4-diphenylcyclohexene derivatives.

3. The diene must be in the cisoid conformation. If it is frozen into the transoid conformation, as in **84,** the reaction does not take place. The diene either must be frozen into the cisoid conformation or must be able to achieve it during the reaction.

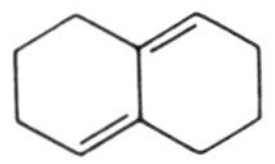

84

[864]Feuer; Herndon; Hall *Tetrahedron* **1968,** *24,* 2575; Inukai; Sato; Kojima *Bull. Chem. Soc. Jpn.* **1972,** *45,* 891; Epiotis *J. Am. Chem. Soc.* **1973,** *95,* 5624; Sustmann *Pure Appl. Chem.* **1974,** *40,* 569-593; Trost; Vladuchick; Bridges *J. Am. Chem. Soc.* **1980,** *102,* 3554; Alston; Gordon; Ottenbrite; Cohen *J. Org. Chem.* **1983,** *48,* 5051; Kahn; Pau; Overman; Hehre *J. Am. Chem. Soc.* **1986,** *108,* 7381.

[865]Danishefsky; Hershenson *J. Org. Chem.* **1979,** *44,* 1180; Ono; Miyake; Kamimura; Kaji *J. Chem. Soc., Perkin Trans. 1* **1987,** 1929. For another method of controlling regioselectivity, see Kraus; Liras *Tetrahedron Lett.* **1989,** *30,* 1907.

[866]For an exception, see Meier; Eckes; Niedermann; Kolshorn *Angew. Chem. Int. Ed. Engl.* **1987,** *26,* 1046 [*Angew. Chem. 99,* 1040.

4. When the diene is cyclic, there are two possible ways in which addition can occur if the dienophile is not symmetrical. The larger side of the dienophile may be under the ring (*endo addition*), or it may be the smaller side (*exo addition*):

Endo addition   Exo addition

Most of the time, the addition is predominantly endo; i.e., the more bulky side of the olefin is under the ring, and this is probably true for open-chain dienes also.[867] However, exceptions are known, and in many cases mixtures of exo and endo addition products are found.[868]

5. In some cases, the Diels–Alder reaction can be made enantioselective.[869] Most such work has used a chiral dienophile (e.g., **85**) and an achiral diene,[870] along with a Lewis acid catalyst (see below). In such cases addition of the diene to the two faces of **85** takes place at different rates, and **86** and **87** are formed in different amounts.[871] In the case illustrated,

86   **85**   **87**

R = (+) or (–)-phenylmenthyl

hydrolysis of the product removes the chiral R group, making it a chiral auxiliary in this reaction. Asymmetric Diels–Alder reactions have also been carried out with achiral dienes and dienophiles, but with an optically active catalyst.[872]

Electron-donating substituents in the diene accelerate the reaction; electron-withdrawing groups retard it. For the dienophile it is just the reverse: donating groups decrease the rate,

[867]See, for example, Baldwin; Reddy *J. Org. Chem.* **1989**, *54*, 5264.

[868]See, for example, Alder; Günzl *Chem. Ber.* **1960**, *93*, 809; Stockmann *J. Org. Chem.* **1961**, *26*, 2025; Jones; Wife *J. Chem. Soc., Chem. Commun.* **1973**, 421; Lindsay Smith; Norman; Stillings *Tetrahedron* **1978**, *34*, 1381; Müller; Bernardinelli; Rodriguez; Pfyffer; Schaller *Chimia* **1987**, *41*, 244.

[869]For reviews, see Taschner *Org. Synth: Theory Appl.* **1989**, *1*, 1-101; Helmchen; Karge; Weetman *Mod. Synth. Methods* **1986**, *4*, 261-306; Paquette, in Morrison, Ref. 232, vol. 3, pp. 455-501; Oppolzer *Angew. Chem. Int. Ed. Engl.* **1984**, *23*, 876-889 [*Angew. Chem. 96*, 840-854]. See also the list of references in Macaulay; Fallis *J. Am. Chem. Soc.* **1990**, *112*, 1136.

[870]For the use of chiral dienes, see Fisher; Hehre; Kahn; Overman *J. Am. Chem. Soc.* **1988**, *110*, 4625; Menezes; Zezza; Sheu; Smith *Tetrahedron Lett.* **1989**, *30*, 3295; Charlton; Plourde; Penner *Can. J. Chem.* **1989**, *67*, 1010; Tripathy; Carroll; Thornton *J. Am. Chem. Soc.* **1990**, *112*, 6743, **1991**, *113*, 7630; Rieger; Breitmaier *Synthesis* **1990**, 697.

[871]Oppolzer; Kurth; Reichlin; Moffatt *Tetrahedron Lett.* **1981**, *22*, 2545. See also Walborsky; Barash; Davis *Tetrahedron* **1963**, *19*, 2333; Furuta; Iwanaga; Yamamoto *Tetrahedron Lett.* **1986**, *27*, 4507; Evans; Chapman; Bisaha *J. Am. Chem. Soc.* **1988**, *110*, 1238; Mattay; Mertes; Maas *Chem. Ber.* **1989**, *122*, 327; Alonso; Carretero; Garcia Ruano *Tetrahedron Lett.* **1989**, *30*, 3853; Tomioka; Hamada; Suenaga; Koga *J. Chem. Soc., Perkin Trans. 1* **1990**, 426; Cativiela; López; Mayoral *Tetrahedron: Asymmetry* **1990**, *1*, 61.

[872]For a review, see Narasaka *Synthesis* **1991**, 1-11. For some recent examples, see Bir; Kaufmann *J. Organomet. Chem.* **1990**, *390*, 1; Rebiere; Riant; Kagan *Tetrahedron: Asymmetry* **1990**, *1*, 199; Terada; Mikami; Nakai *Tetrahedron Lett.* **1991**, *32*, 935; Corey; Imai; Zhang *J. Am. Chem. Soc.* **1991**, *113*, 728; Narasaka; Tanaka; Kanai *Bull. Chem. Soc. Jpn.* **1991**, *64*, 387; Hawkins; Loren *J. Am. Chem. Soc.* **1991**, *113*, 7794.

and withdrawing groups increase it. Cyclic dienes, in which the cisoid conformation is built in, usually react faster than the corresponding open-chain compounds, which have to achieve the cisoid conformation by rotation.[873]

As should be apparent from the foregoing, many interesting compounds can be prepared by the Diels–Alder reaction, some of which would be hard to make in any other way. It has thus been exceedingly useful. Competing reactions are polymerization of the diene or dienophile, or both, and 1,2 cycloaddition (**5-49**). However, yields are usually quite high. No catalyst is needed, though it has been found that Lewis acids catalyze some Diels–Alder reactions,[874] usually those in which Z in the dienophile is a C=O or C=N group. A Lewis acid catalyst usually increases both the regioselectivity of the reaction (in the sense given above) and the extent of endo addition,[875] and, in the case of enantioselective reactions, the extent of enantioselectivity. Some Diels–Alder reactions can also be catalyzed by the addition of a stable cation radical, e.g., tris(4-bromophenyl)aminium hexachloroantimonate $Ar_3N \cdot^+ SbCl_6^-$.[876]

A number of other methods have been reported for the acceleration of Diels–Alder reactions, including the use of a microwave oven,[877] water as a solvent (a hydrophobic effect),[878] 5 M $LiClO_4$ in Et_2O as solvent,[879] absorption of the reactants on chromatographic absorbents,[880] and the use of an ultracentrifuge[881] (one of several ways to achieve reaction at high pressures).[882]

The Diels–Alder reaction is usually reversible and has been used to protect double bonds.[883] A convenient substitute for butadiene in the Diels–Alder reaction is the compound

3-sulfolene

3-sulfolene since the latter is a solid which is easy to handle while the former is gas.[884] Butadiene is generated in situ by a reverse Diels–Alder reaction (see **7-25**).

There are, broadly speaking, three possible mechanisms that have been considered for

[873]Sauer; Lang; Mielert *Angew. Chem. Int. Ed. Engl.* **1962**, *1*, 268 [*Angew. Chem. 74*, 352]; Sauer; Wiest *Angew. Chem. Int. Ed. Engl.* **1962**, *1*, 269 [*Angew. Chem. 74*, 353]. See, however, Scharf; Plum; Fleischhauer; Schleker *Chem. Ber.* **1979**, *112*, 862.

[874]Yates; Eaton *J. Am. Chem. Soc.* **1960**, *82*, 4436; Fray; Robinson *J. Am. Chem. Soc.* **1961**, *83*, 249; Inukai; Kojima *J. Org. Chem.* **1967**, *32*, 869, 872; Laszlo; Lucchetti *Tetrahedron Lett.* **1984**, *25*, 4387; Bonnesen; Puckett; Honeychuck; Hersh *J. Am. Chem. Soc.* **1989**, *111*, 6070. For review of the role of the catalyst in increasing reactivity, see Kiselev; Konovalov *Russ. Chem. Rev.* **1989**, *58*, 230-249.

[875]For discussions see Houk; Strozier *J. Am. Chem. Soc.* **1973**, *95*, 4094; Alston; Ottenbrite *J. Org. Chem.* **1975**, *40*, 1111.

[876]For a review, see Bauld; *Tetrahedron* **1989**, *45*, 5307-5363.

[877]Giguere; Bray; Duncan; Majetich *Tetrahedron Lett.* **1986**, *27*, 4945; Berlan; Giboreau; Lefeuvre; Marchand *Tetrahedron Lett.* **1991**, *32*, 2363.

[878]Rideout; Breslow *J. Am. Chem. Soc.* **1980**, *102*, 7816. For a review, see Breslow *Acc. Chem. Res.* **1991**, *24*, 159-164. See also Grieco; Garner; He *Tetrahedron Lett.* **1983**, 1897; Blokzijl; Blandamer; Engberts *J. Am. Chem. Soc.* **1991**, *113*, 4241; Breslow; Rizzo *J. Am. Chem. Soc.* **1991**, *113*, 4340.

[879]Grieco; Nunes; Gaul *J. Am. Chem. Soc.* **1990**, *112*, 4595. See also Braun; Sauer *Chem. Ber.* **1986**, *119*, 1269; Forman; Dailey *J. Am. Chem. Soc.* **1991**, *113*, 2761.

[880]Veselovsky; Gybin; Lozanova; Moiseenkov; Smit; Caple *Tetrahedron Lett.* **1988**, *29*, 175.

[881]Dolata; Bergman *Tetrahedron Lett.* **1987**, *28*, 707.

[882]For reviews, see Isaacs; George *Chem. Br.* **1987**, 47-54; Asano; le Noble *Chem. Rev.* **1978**, *78*, 407-489. See also Firestone; Smith *Chem. Ber.* **1989**, *122*, 1089.

[883]For reviews of the reverse Diels–Alder reaction, see Ichihara *Synthesis* **1987**, 207-222; Lasne; Ripoll *Synthesis* **1985**, 121-143; Ripoll; Rouessac; Rouessac *Tetrahedron* **1978**, *34*, 19-40; Brown *Pyrolytic Methods in Organic Chemistry*; Academic Press: New York, 1980, pp. 259-281; Kwart; King *Chem. Rev.* **1968**, *68*, 415-447.

[884]Sample; Hatch *Org. Synth. VI*, 454. For a review, see Chou; Tso *Org. Prep. Proced. Int.* **1989**, *21*, 257-296.

the uncatalyzed Diels–Alder reaction.[885] In mechanism a there is a cyclic six-centered transition state and no intermediate. The reaction is concerted and occurs in one step. In

Mechanism a

Mechanism b

mechanism b one end of the diene fastens to one end of the dienophile first to give a diradical, and then, in a second step, the other ends become fastened. A diradical formed in this manner must be a singlet; i.e., the two unpaired electrons must have opposite spins, by an argument similar to that outlined on p. 196. The third mechanism (c, not shown) is similar to mechanism b, but the initial bond and the subsequent bond are formed by movements of electron pairs and the intermediate is a diion. There have been many mechanistic investigations of the Diels–Alder reaction. The bulk of the evidence suggests that most Diels–Alder reactions take place by the one-step cyclic mechanism a,[886] although it is possible that a diradical[887] or even a diion[888] mechanism may be taking place in some cases. The main evidence in support of mechanism a is as follows: (1) The reaction is stereospecific in both the diene and dienophile. A completely free diradical or diion probably would not be able to retain its configuration. (2) In general, the rates of Diels–Alder reactions depend very little on the nature of the solvent. This would rule out a diion intermediate because polar solvents increase the rates of reactions that develop charges in the transition state. (3) It was shown that, in the decomposition of **88**, the isotope effect k_I/k_{II} was equal to 1.00 within experimental error.[889] If bond x broke before bond y, there should surely be a

I: R = H, R′ = D
II: R = D, R′ = H

88

secondary isotope effect. This result strongly indicates that the bond breaking of x and y is simultaneous. This is the reverse of a Diels–Alder reaction, and by the principle of microscopic reversibility, the mechanism of the forward reaction should involve simultaneous formation of bonds x and y. Subsequently, a similar experiment was carried out on the forward reaction[890] and the result was the same. There is also other evidence for mechanism

---

[885]For reviews, see Sauer; Sustmann *Angew. Chem. Int. Ed. Engl.* **1980**, *19*, 779-807 [*Angew. Chem. 92*, 773-801]; Houk *Top. Curr. Chem.* **1979**, *79*, 1-40; Seltzer *Adv. Alicyclic Chem.* **1968**, *2*, 1-57; Ref. 841. For a review of the application of quantum-chemical methods to the study of this reaction, see Babichev; Kovtunenko; Voitenko; Tyltin *Russ. Chem. Rev.* **1988**, *57*, 397-405.

[886]For a contrary view, see Dewar; Pierini *J. Am. Chem. Soc.* **1984**, *106*, 203; Dewar; Olivella; Stewart *J. Am. Chem. Soc.* **1986**, *108*, 5771. For arguments against this view, see Houk; Lin; Brown *J. Am. Chem. Soc.* **1986**, *108*, 554; Hancock; Wood *J. Chem. Soc., Chem. Commun.* **1988**, 351; Gajewski; Peterson; Kagel; Huang *J. Am. Chem. Soc.* **1989**, *111*, 9078.

[887]See, for example, Bartlett; Mallet *J. Am. Chem. Soc.* **1976**, *98*, 143; Jenner; Rimmelin *Tetrahedron Lett.* **1980**, *21*, 3039; Van Mele; Huybrechts *Int. J. Chem. Kinet.* **1987**, *19*, 363, **1989**, *21*, 967.

[888]For a reported example, see Gassman; Gorman *J. Am. Chem. Soc.* **1990**, *112*, 8624.

[889]Seltzer *J. Am. Chem. Soc.* **1963**, *85*, 1360, **1965**, *87*, 1534. For a review of isotope effect studies of Diels–Alder and other pericyclic reactions, see Gajewski *Isot. Org. Chem.* **1987**, *7*, 115-176.

[890]Van Sickle; Rodin *J. Am. Chem. Soc.* **1964**, *86*, 3091.

a.[891] However, the fact that the mechanism is concerted does not necessarily mean that it is synchronous. In the transition state of a synchronous reaction both new σ bonds would be formed to the same extent, but a Diels–Alder reaction with non-symmetrical components might very well be non-synchronous; i.e., it could have a transition state in which one bond has been formed to a greater degree than the other.[671]

In another aspect of the mechanism, the effects of electron-donating and electron-withdrawing substituents (p. 843) indicate that the diene is behaving as a nucleophile and the dienophile as an electrophile. However, this can be reversed. Perchlorocyclopentadiene reacts better with cyclopentene than with maleic anhydride and not at all with tetracyanoethylene, though the latter is normally the most reactive dienophile known. It is apparent, then, that this diene is the electrophile in its Diels–Alder reactions.[893] Reactions of this type are said to proceed with *inverse electron demand*.[894]

We have emphasized that the Diels–Alder reaction generally takes place rapidly and conveniently. In sharp contrast, the apparently similar dimerization of olefins to cyclobutanes (**5-49**) gives very poor results in most cases, except when photochemically induced. Fukui, Woodward, and Hoffmann have shown that these contrasting results can be explained by the *principle of conservation of orbital symmetry*,[895] which predicts that certain reactions are allowed and others forbidden. The orbital-symmetry rules (also called the Woodward–Hoffmann rules) apply *only to concerted reactions*, e.g., mechanism *a*, and are based on the principle that reactions take place in such a way as to maintain maximum bonding throughout the course of the reaction. There are several ways of applying the orbital-symmetry principle to cycloaddition reactions, three of which are used more frequently than others.[896] Of these three we will discuss two: the frontier-orbital method and the Möbius–Hückel method. The third, called the correlation diagram method,[897] is less convenient to apply than the other two.

[891]See, for example, Dewar; Pyron *J. Am. Chem. Soc.* **1970**, *92*, 3098; Brun; Jenner *Tetrahedron* **1972**, *28*, 3113; Doering; Franck-Neumann; Hasselmann; Kaye *J. Am. Chem. Soc.* **1972**, *94*, 3833; McCabe; Eckert *Acc. Chem. Res.* **1974**, *7*, 251-257; Berson; Dervan; Malherbe; Jenkins *J. Am. Chem. Soc.* **1976**, *98*, 5937; Rücker; Lang; Sauer; Friege; Sustmann *Chem. Ber.* **1980**, *113*, 1663; Tolbert; Ali *J. Am. Chem. Soc.* **1981**, *103*, 2104.

[892]Woodward; Katz *Tetrahedron* **1959**, *5*, 70; Liu; Schmidt *Tetrahedron* **1971**, *27*, 5289; Dewar; Pyron, Ref. 891; Papadopoulos; Jenner *Tetrahedron Lett.* **1982**, *23*, 1889; Houk; Loncharich; Blake; Jorgensen *J. Am. Chem. Soc.* **1989**, *111*, 9172; Lehd; Jensen *J. Org. Chem.* **1990**, *55*, 1034.

[893]Sauer; Wiest *Angew. Chem. Int. Ed. Engl.* **1962**, *1*, 269 [*Angew. Chem. 74*, 353].

[894]For a review, see Boger; Patel *Prog. Heterocycl. Chem.* **1989**, *1*, 30-64.

[895]For monographs, see Gilchrist; Storr *Organic Reactions and Orbital Symmetry*, 2nd ed.; Cambridge University Press: Cambridge, 1979; Fleming *Frontier Orbitals and Organic Chemical Reactions*; Wiley: New York, 1976; Woodward; Hoffmann *The Conservation of Orbital Symmetry*; Academic Press: New York, 1970 [the text of this book also appears in *Angew. Chem. Int. Ed. Engl.* **1969**, *8*, 781-853; *Angew. Chem. 81*, 797-869]; Lehr; Marchand *Orbital Symmetry*; Academic Press, New York, 1972. For reviews, see Pearson *J. Chem. Educ.* **1981**, *58*, 753-757; in Klopman *Chemical Reactivity and Reaction Paths*; Wiley: New York, 1974, the articles by Fujimoto; Fukui, pp. 23-54, Klopman, pp. 55-165, Herndon; Feuer; Giles; Otteson; Silber, pp. 275-299; Michl, pp. 301-338; Simonetta *Top. Curr. Chem.* **1973**, *42*, 1-47; Houk *Surv. Prog. Chem.* **1973**, *6*, 113-208; Vollmer; Servis *J. Chem. Educ.* **1970**, *47*, 491-500; Gill *Essays Chem.* **1970**, *1*, 43-76, *Q. Rev., Chem. Soc.* **1968**, *22*, 338-389; Seebach *Fortschr. Chem. Forsch.* **1969**, *11*, 177-215; Miller *Adv. Phys. Org. Chem.* **1968**, *6*, 185-332; Millie *Bull. Soc. Chim. Fr.* **1966**, 4031-4038. For a review of applications to inorganic chemistry, see Pearson *Top Curr. Chem.* **1973**, *41*, 75-112.

[896]For other approaches see Epiotis *Theory of Organic Reactions*; Springer: New York, 1978; Epiotis; Shaik *J. Am. Chem. Soc.* **1978**, *100*, 1, 9; Halevi *Angew. Chem. Int. Ed. Engl.* **1976**, *15*, 593-607 [*Angew. Chem. 88*, 664-679]; Shen *J. Chem. Educ.* **1973**, *50*, 238-242; Salem *J. Am. Chem. Soc.* **1968**, *90*, 543, 553; Trindle *J. Am. Chem. Soc.* **1970**, *92*, 3251, 3255; Mulder; Oosterhoff *Chem. Commun.* **1970**, 305, 307; Goddard *J. Am. Chem. Soc.* **1970**, *92*, 7520, **1972**, *94*, 793; Herndon *Chem. Rev.* **1972**, *72*, 157-179; Perrin *Chem. Br.* **1972**, *8*, 163-173; Langlet; Malrieu *J. Am. Chem. Soc.* **1972**, *94*, 7254; Pearson *J. Am. Chem. Soc.* **1972**, *94*, 8287; Mathieu, *Bull. Soc. Chim. Fr.* **1973**, 807; Silver; Karplus *J. Am. Chem. Soc.* **1975**, *97*, 2645; Day *J. Am. Chem. Soc.* **1975**, *97*, 2431; Mok; Nye *J. Chem. Soc., Perkin Trans. 2* **1975**, 1810; Ponec *Collect. Czech. Chem. Commun.* **1984**, *49*, 455, **1985**, *50*, 1121; Hua-ming; De-xiang *Tetrahedron* **1986**, *42*, 515; Bernardi; Olivucci; Robb *Res. Chem. Intermed.* **1989**, *12*, 217, *Acc. Chem. Res.* **1990**, *23*, 405.

[897]For excellent discussions of this method see Woodward; Hoffmann, Ref. 895; Jones *Physical and Mechanistic Organic Chemistry*, 2nd ed.; Cambridge University Press: Cambridge, 1984, pp. 352-366; Klumpp *Reactivity in Organic Chemistry*; Wiley: New York, 1982, pp. 378-389; Yates *Hückel Molecular Orbital Theory*; Academic Press: New York, 1978, pp. 263-276.

The Frontier–Orbital Method[898]

As applied to cycloaddition reactions the rule is that *reactions are allowed only when all overlaps between the highest-occupied molecular orbital (HOMO) of one reactant and the lowest-unoccupied molecular orbital (LUMO) of the other are such that a positive lobe overlaps only with another positive lobe and a negative lobe only with another negative lobe.* We may recall that monoolefins have two π molecular orbitals (p. 9) and that conjugated dienes have four (p. 31), as shown in Figure 15.2. A concerted cyclization of two monoolefins (a 2 + 2 reaction) is not allowed because it would require that a positive lobe overlap with a negative lobe (Figure 15.3). On the other hand, the Diels–Alder reaction (a 2 + 4 reaction) is allowed, whether considered from either direction (Figure 15.4).

These considerations are reversed when the ring closures are photochemically induced since in such cases an electron is promoted to a vacant orbital before the reaction occurs. Obviously, the 2 + 2 reaction is now allowed (Figure 15.5) and the 2 + 4 reaction disallowed. The reverse reactions follow the same rules, by the principle of microscopic reversibility. In fact, Diels–Alder adducts are usually cleaved quite readily, while cyclobutanes, despite the additional strain, require more strenuous conditions.

The Möbius–Hückel Method[899]

In this method, the orbital symmetry rules are related to the Hückel aromaticity rule discussed in Chapter 2. Hückel's rule, which states that a cyclic system of electrons is aromatic (hence, stable) when it consists of $4n + 2$ electrons, applies of course to molecules in their ground states. In applying the orbital symmetry principle we are not concerned with ground states, but with transition states. In the present method we do not examine the molecular

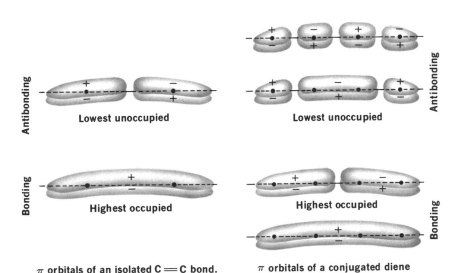

π orbitals of an isolated C=C bond.   π orbitals of a conjugated diene

FIGURE 15.2 Schematic drawings of the π orbitals of an isolated C=C bond and a conjugated diene.

[898]Fukui; Fujimoto *Bull Chem. Soc. Jpn.* **1967**, *40*, 2018, **1969**, *42*, 3399; Fukui *Fortschr. Chem. Forsch.* **1970**, *15*, 1-85, *Acc. Chem. Res.* **1971**, *4*, 57-64; Houk *Acc. Chem. Res.* **1975**, *8*, 361-369. See also Chu *Tetrahedron* **1978**, *34*, 645. For a monograph on frontier orbitals see Fleming, Ref. 895. For reviews, see Fukui *Angew. Chem. Int. Ed. Engl.* **1982**, *21*, 801-809 [*Angew Chem.* **94**, 852-861]; Houk, in Marchand; Lehr, *Pericyclic Reactions*, vol. 2; Academic Press: New York, 1977, pp. 181-271.

[899]Zimmerman, in Marchand; Lehr, Ref. 898, pp. 53-107, *Acc. Chem. Res.* **1971**, *4*, 272-280, *J. Am. Chem. Soc.* **1966**, *88*, 1564, 1566; Dewar *Angew. Chem. Int. Ed. Engl.* **1971**, *10*, 761-775 [*Angew Chem.* **83**, 859-875]; Jefford; Burger *Chimia* **1971**, *25*, 297-307; Herndon *J. Chem. Educ.* **1981**, *58*, 371-376.

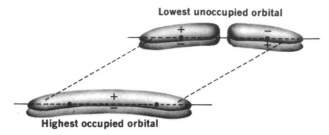

Lowest unoccupied orbital

Highest occupied orbital

FIGURE 15.3   Overlap of orbitals in a thermal 2 + 2 cycloaddition.

orbitals themselves, but rather the *p* orbitals before they overlap to form the molecular orbitals. Such a set of *p* orbitals is called a *basis set* (Figure 15.6). In investigating the possibility of a concerted reaction, we put the basis sets into the position they would occupy in the transition state. Figure 15.7 shows this for both the 2 + 2 and the 2 + 4 ring closures. What we look for are *sign inversions*. In Figure 15.7 we can see that there are no sign inversions in either case. That is, the dashed line connects only lobes with a minus sign. Systems with *zero or an even number* of sign inversions are called *Hückel systems*. Because they have no sign inversions, both of these systems are Hückel systems. Systems with *an odd number* of sign inversions are called *Möbius systems* (because of the similarity to the Möbius strip, which is a mathematical surface, shown in Figure 15.8). Möbius systems do not enter into either of these reactions, but an example of such a system is shown on p. 1114.

The rule may then be stated: *A thermal pericyclic reaction involving a Hückel system is allowed only if the total number of electrons is 4n + 2. A thermal pericyclic reaction involving a Möbius system is allowed only if the total number of electrons is 4n.* For photochemical reactions these rules are reversed. Since both the 2 + 4 and 2 + 2 cycloadditions are Hückel systems, the Möbius–Hückel method predicts that the 2 + 4 reaction, with 6 electrons, is thermally allowed, but the 2 + 2 reaction is not. One the other hand, the 2 + 2 reaction is allowed photochemically, while the 2 + 4 reaction is forbidden.

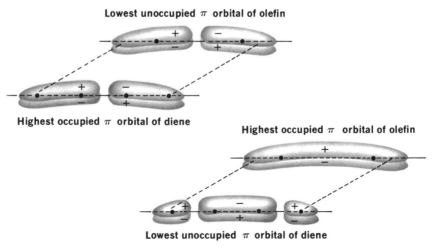

Lowest unoccupied π orbital of olefin

Highest occupied π orbital of diene

Highest occupied π orbital of olefin

Lowest unoccupied π orbital of diene

FIGURE 15.4   Two ways for orbitals to overlap in a thermal 2 + 4 cycloaddition.

Lowest unoccupied π orbital
of an unexcited olefin

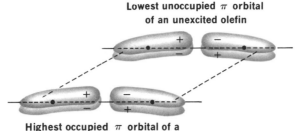

Highest occupied π orbital of a
photochemically excited olefin

FIGURE 15.5 Overlap of orbitals in a photochemical 2 + 2 cycloaddition.

Note that both the 2 + 2 and 2 + 4 transition states are Hückel systems no matter what basis sets we chose. For example, Figure 15.9 shows other basis sets we might have chosen. In every case there will be zero or an even number of sign inversions.

Thus, the frontier-orbital and Hückel–Möbius methods (and the correlation-diagram method as well) lead to the same conclusions: thermal 2 + 4 cycloadditions and photochemical 2 + 2 cycloadditions (and the reverse ring openings) are allowed, while photochemical 2 + 4 and thermal 2 + 2 ring closings (and openings) are forbidden. Application of the same procedures to other ring closures shows that 4 + 4 and 2 + 6 ring closures and openings require photochemical induction while the 4 + 6 and 2 + 8 reactions can take place only thermally (see **5-52**). In general, cycloaddition reactions allowed thermally are those with $4n + 2$ electrons, while those allowed photochemically have $4n$ electrons.

It must be emphasized once again that the rules apply only to cycloaddition reactions that take place by cyclic mechanisms, i.e., where two σ bonds are formed (or broken) at about the same time.[900] The rule does not apply to cases where one bond is clearly formed (or broken) before the other. It must further be emphasized that the fact that the thermal Diels–Alder reaction (mechanism *a*) is allowed by the principle of conservation of orbital symmetry does not constitute proof that any given Diels–Alder reaction proceeds by this mechanism. The principle merely says the mechanism is allowed, not that it must go by this pathway. However, the principle does say that thermal 2 + 2 cycloadditions in which the molecules assume a face-to-face geometry cannot[901] take place by a cyclic mechanism because

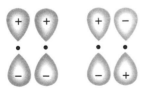

Two basis sets for an
isolated double bond

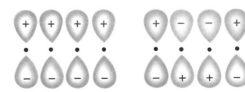

Two basis sets for a conjugated diene

FIGURE 15.6 Some basis sets.

[900]For a discussion of concertedness in these reactions see Lehr; Marchand, in Marchand; Lehr, Ref. 898, vol. 1, pp. 1-51.

[901]The possibility has been raised that some disallowed reactions may nevertheless proceed by concerted mechanisms: see Schmidt *Helv. Chim. Acta* **1971**, *54*, 862, *Tetrahedron Lett.* **1972**, 581; Muszkat; Schmidt *Helv. Chim. Acta* **1971**, *54*, 1195; Baldwin; Andrist; Pinschmidt *Acc. Chem. Res.* **1972**, *5*, 402-406; Berson *Acc. Chem. Res.* **1972**, *5*, 406-414; Baldwin, in Marchand; Lehr, Ref. 898, vol. 2, pp. 273-302.

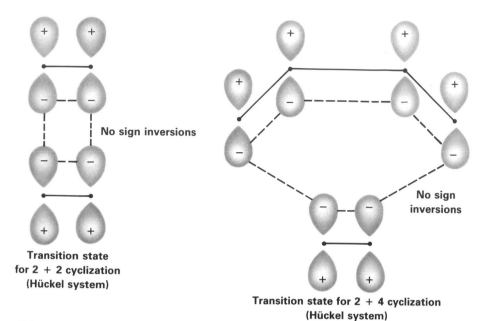

Transition state
for 2 + 2 cyclization
(Hückel system)

No sign inversions

Transition state for 2 + 4 cyclization
(Hückel system)

No sign inversions

FIGURE 15.7 Transition states illustrating Hückel–Möbius rules for cycloaddition reactions.

their activation energies would be too high (however, see below). As we shall see (**5-49**), such reactions largely occur by two-step mechanisms. Similarly, 2 + 4 photochemical cycloadditions are also known, but the fact that they are not stereospecific indicates that they also take place by the two-step diradical mechanism (mechanism *b*).[902]

In all of the above discussion we have assumed that a given molecule forms both the new σ bonds from the same face of the π system. This manner of bond formation, called *suprafacial*, is certainly most reasonable and almost always takes place. The subscript s is used to designate this geometry, and a normal Diels–Alder reaction would be called a $[_\pi 2_s + _\pi 4_s]$ cycloaddition (the subscript π indicates that π electrons are involved in the

FIGURE 15.8 A Möbius strip. Such a strip is easily constructed by twisting a thin strip of paper 180° and fastening the ends together.

[902]For example, see Sieber; Heimgartner; Hansen; Schmid *Helv. Chim. Acta* **1972**, *55*, 3005. For discussions see Bartlett; Helgeson; Wersel *Pure Appl. Chem.* **1968**, *16*,187-200; Seeley *J. Am. Chem. Soc.* **1972**, *94*, 4378; Kaupp *Angew. Chem. Int. Ed. Engl.* **1972**, *11*, 313, 718 [*Angew Chem. 84*, 259, 718].

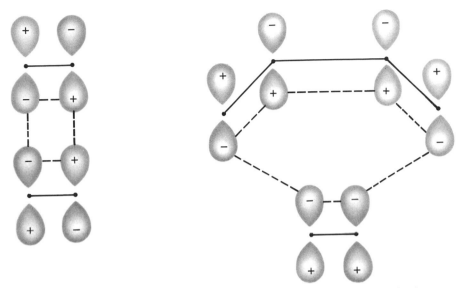

· **FIGURE 15.9** Transition states for 2 + 2 and 2 + 4 cyclizations involving other basis sets.

cycloaddition). However, we can conceive of another approach in which the newly forming bonds of the diene lie on *opposite* faces of the π system, i.e., they point in opposite directions.

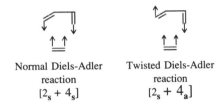

Normal Diels-Adler   Twisted Diels-Adler
reaction   reaction
$[2_s + 4_s]$ $[2_s + 4_a]$

This type of orientation of the newly formed bonds is called *antarafacial*, and the reaction would be a $[_\pi 2_s + _\pi 4_a]$ cycloaddition (a stands for antarafacial). We can easily show by the frontier-orbital method that this reaction (and consequently the reverse ring-opening reactions) are thermally forbidden and photochemically allowed. Thus in order for a $[_\pi 2_s + _\pi 4_a]$ reaction to proceed, overlap between the highest occupied π orbital of the olefin and the lowest unoccupied π orbital of the diene would have to occur as shown in Figure 15.10, with a + lobe overlapping a − lobe. Since like signs are no longer overlapping, the thermal reaction is now forbidden. Similarly, thermal $[_\pi 2_a + _\pi 4_s]$ and $[_\pi 2_a + _\pi 2_a]$ cyclizations are forbidden, while thermal $[_\pi 2_a + _\pi 4_a]$ and $[_\pi 2_s + _\pi 2_a]$ cyclizations are allowed, and these considerations are reversed for the corresponding photochemical processes. Of course, an antarafacial approach is highly unlikely in a 2 + 4 cyclization,[903] but larger ring closures could take place by such a pathway, and 2 + 2 thermal cyclizations, where the $[_\pi 2_s + _\pi 2_s]$ pathway is forbidden, can also do so in certain cases (see **5-49**). We therefore see that whether a given cycloaddition is allowed or forbidden depends on the geometry of approach of the two molecules involved.

Symmetry considerations have also been advanced to explain predominant endo addi-

[903]A possible photochemical $[_\pi 2_a + _\pi 4_s]$ cycloaddition has been reported: Hart; Miyashi; Buchanan; Sasson *J. Am. Chem. Soc.* **1974**, *96*, 4857.

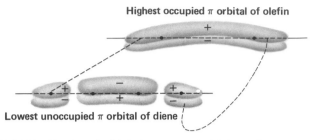

FIGURE 15.10 Overlap of orbitals in an antarafacial thermal 2 + 4 cycloaddition.

tion.[904] In the case of 2 + 4 addition of butadiene to itself, the approach can be exo or endo. It can be seen (Figure 15.11) that whether the highest-occupied molecular orbital of the diene overlaps with the lowest-unoccupied molecular orbital of the olefin or vice versa, the endo orientation is stabilized by additional secondary overlap of orbitals[905] of like sign (dashed lines between heavy dots). Addition from the exo direction has no such stabilization. Evidence for secondary orbital overlap as the cause of predominant endo orientation, at least in some cases, is that 4 + 6 cycloaddition is predicted by similar considerations to proceed with predominant exo orientation, and that is what is found.[906] However, this explanation does not account for endo orientation in cases where the dienophile does not possess additional π orbitals, and a number of alternative explanations have been offered.[907]

OS **II**, 102; **III**, 310, 807; **IV**, 238, 311, 738, 890, 964; **V**, 60, 96, 414, 424, 604, 985, 1037; **VI**, 82, 196, 422, 427, 445, 454; **VII**, 4, 312, 485; **65**, 98; **66**, 142; **67**, 163; **68**, 198, 206; **69**, 31. For a reverse Diels–Alder reaction, see OS **VII**, 339. See also OS **VII**, 326.

5-48 All-Carbon 2 + 3 Cycloadditions

$$\underset{\textbf{89}}{\underset{\text{AcO} \qquad \text{SiMe}_3}{\bigwedge}} + \| \quad \xrightarrow[]{\text{Z}} \quad \xrightarrow{\text{Pd complex}} \quad \underset{\text{Z}}{\bigotimes}$$

Several methods have been reported for the formation of cyclopentanes by 2 + 3 cycloadditions.[908] One type involves reagents that produce intermediates **90** or **91**.[909] A synthetically useful example[910] uses 2-[(trimethylsilyl)methyl]-2-propen-1-yl acetate (**89**) (which is com-

[904]Hoffmann; Woodward *J. Am. Chem. Soc.* **1965**, *87*, 4388.

[905]For reviews of secondary orbital interactions, see Ginsburg *Tetrahedron* **1983**, *39*, 2095-2135; Gleiter; Paquette *Acc. Chem. Res.* **1983**, *16*, 328-334.

[906]See, for example, Cookson; Drake; Hudec; Morrison *Chem. Commun.* **1966**, 15; Itô; Fujise; Okuda; Inoue *Bull. Chem. Soc. Jpn.* **1966**, *39*, 1351; Paquette; Barrett *J. Am. Chem. Soc.* **1966**, *88*, 2590; Paquette; Barrett; Kuhla *J. Am. Chem. Soc.* **1969**, *91*, 3616; Houk; Woodward *J. Am. Chem. Soc.* **1970**, *92*, 4143, 4145; Jones; Kneen *J. Chem. Soc., Chem. Commun.* **1973**, 420.

[907]See, for example, Houk; Luskus *J. Am. Chem. Soc.* **1971**, *93*, 4606; Kobuke; Sugimoto; Furukawa; Fueno *J. Am. Chem. Soc.* **1972**, *94*, 3633; Jacobson *J. Am. Chem. Soc.* **1973**, *95*, 2579; Mellor; Webb *J. Chem. Soc., Perkin Trans. 2* **1974**, 17, 26; Fox; Cardona; Kiwiet *J. Org. Chem.* **1987**, *52*, 1469.

[908]For a list of methods, with references, see Trost; Seoane; Mignani; Acemoglu *J. Am. Chem. Soc.* **1989**, *111*, 7487.

[909]For reviews, see Trost *Pure Appl. Chem.* **1988**, *60*, 1615-1626, *Angew. Chem. Int. Ed. Engl.* **1986**, *25*, 1-20 [*Angew. Chem. 98*, 1-20].

[910]See, for example, Trost; Lynch; Renaut; Steinman *J. Am. Chem. Soc.* **1986**, *108*, 284.

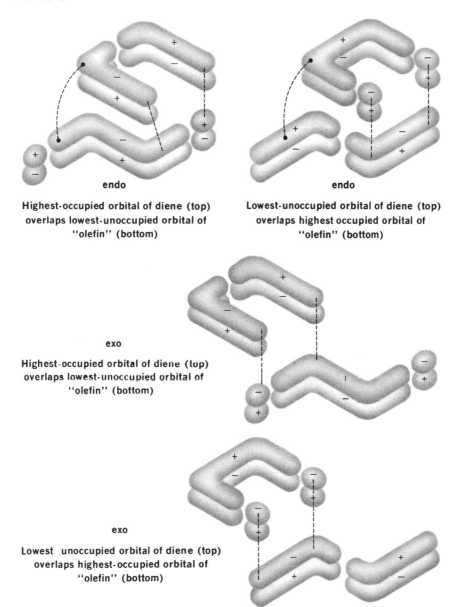

endo

Highest-occupied orbital of diene (top)
overlaps lowest-unoccupied orbital of
"olefin" (bottom)

endo

Lowest-unoccupied orbital of diene (top)
overlaps highest occupied orbital of
"olefin" (bottom)

exo

Highest-occupied orbital of diene (top)
overlaps lowest-unoccupied orbital of
"olefin" (bottom)

exo

Lowest unoccupied orbital of diene (top)
overlaps highest-occupied orbital of
"olefin" (bottom)

FIGURE 15.11 Overlap of orbitals in 2 + 4 cycloaddition of dienes.

90   **91**

mercially available) and a palladium or other transition metal catalyst to generate **90** or **91**, which adds to double bonds, to give, in good yields, cyclopentanes with an exocyclic double bond. Similar or identical intermediates generated from bicyclo azo compounds **92** (see

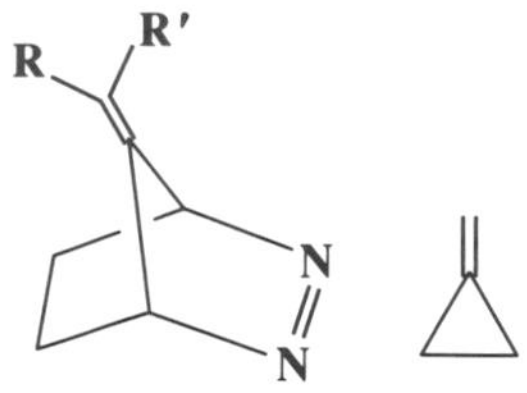

92   **93**

7-46)[911] or methylenecyclopropane **93**[912] also add to activated double bonds. With suitable substrates the addition can be enantioselective.[913] The reagent **94**, similar to **89**, forms cis 5-membered cyclic unsaturated diols when treated with α-diketones in the presence of SnF$_2$.

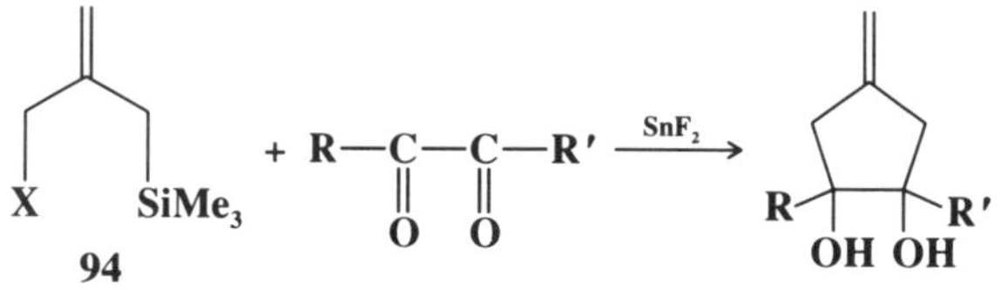

94

X = Br or I

In a different type of procedure, 2 + 3 cycloadditions are performed with allylic anions. Such reactions are called 1,3-anionic cycloadditions.[915] For example, α-methylstyrene adds to stilbene on treatment with the strong base lithium diisopropylamide.[916]

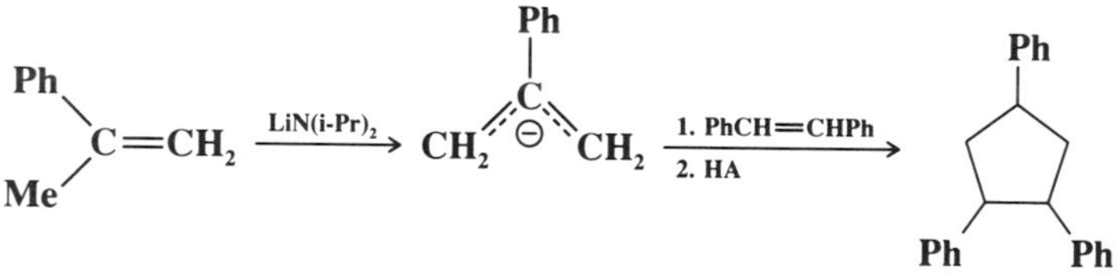

The mechanism can be outlined as

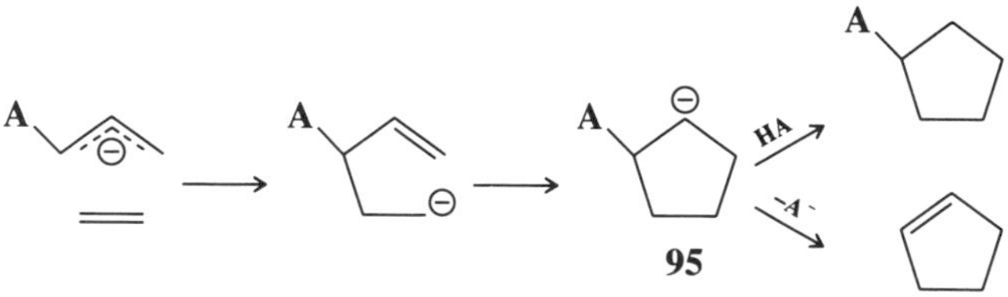

95

[911]For a review, see Little *Chem. Rev.* **1986**, *86*, 875-884.
[912]See Yamago; Nakamura *J. Am. Chem. Soc.* **1989**, *111*, 7285.
[913]See Binger; Schäfer *Tetrahedron Lett.* **1988**, *29*, 529; Chaigne; Gotteland; Malacria *Tetrahedron Lett.* **1989**, *30*, 1803.
[914]Molander; Shubert *J. Am. Chem. Soc.* **1986**, *108*, 4683.
[915]For reviews, see Kauffmann *Top. Curr. Chem.* **1980**, *92*, 109-147, pp. 111-116; *Angew. Chem. Int. Ed. Engl.* **1974**, *13*, 627-639 [*Angew. Chem. 86*, 715-727].
[916]Eidenschink; Kauffmann *Angew. Chem. Int. Ed. Engl.* **1972**, *11*, 292 [*Angew. Chem. 84*, 292].

In the case above, **95** is protonated in the last step by the acid HA, but if the acid is omitted and a suitable nucleofuge is present, it may leave, resulting in a cyclopentene.[917] In these cases the reagent is an allylic anion, but similar 2 + 3 cycloadditions involving allylic cations have also been reported.[918]

In a third type of procedure,[919] cyclopropene ketal **96** reacts with olefins bearing two

electron-withdrawing groups Z to give cyclopentenes.[920]

OS **65**, 32; **66**, 8.

5-49   Dimerization of Olefins
    (2 + 2)cyclo-Ethylene-1/2/addition

The thermal reaction between two molecules of olefin to give cyclobutane derivatives (a 2 + 2 cycloaddition) can be carried out where the olefins are the same or different, but the reaction is not a general one for olefins.[921] Dimerization of like olefins occurs with the following compounds: $F_2C=CX_2$ (X – F or Cl) and certain other fluorinated alkenes (though not $F_2C=CH_2$), allenes (to give derivatives of **97**),[922] benzynes (to give biphenylene deriv-

97   Biphenylene

atives), activated olefins (e.g., styrene, acrylonitrile, butadiene), and certain methylene-cyclopropanes.[923] Substituted ketenes dimerize to give cyclobutene derivatives (**98**) as the

[917]See, for example, Padwa; Yeske *J. Am. Chem. Soc.* **1988**, *110*, 1617; Beak; Burg *J. Org. Chem.* **1989**, *54*, 1647.

[918]For example, see Hoffmann; Vathke-Ernst *Chem. Ber.* **1981**, *114*, 2208, 2898; Klein; Mayr *Angew. Chem. Int. Ed. Engl.* **1981**, *20*, 1027 [*Angew. Chem. 93*, 1069]; Noyori; Hayakawa *Tetrahedron* **1985**, *41*, 5879.

[919]For some other methods of making cyclopentanes or cyclopentenes by 2 + 3 cycloadditions, see Danheiser; Carini; Fink; Basak *Tetrahedron* **1983**, *39*, 935; Shimizu; Ohashi; Tsuji *Tetrahedron Lett.* **1985**, *26*, 3825; Marino; Laborde *J. Org. Chem.* **1987**, *52*, 1; Curran; Chen *J. Am. Chem. Soc.* **1987**, *109*, 6558; Feldman; Romanelli; Ruckle; Miller *J. Am. Chem. Soc.* **1988**, *110*, 3300; Herndon; Tumer; Schnatter *J. Am. Chem. Soc.* **1988**, *110*, 3334; Ghera; Yechezkel; Hassner *Tetrahedron Lett.* **1990**, *31*, 3653; Crimmins; Nantermet *J. Org. Chem.* **1990**, *55*, 4235. For a review of a 2 + 2 + 1 method (the *Pauson-Khand reaction*), see Schore *Org. React.* **1991**, *40*, 1-90.

[920]Boger; Brotherton *J. Am. Chem. Soc.* **1986**, *108*, 6695, 6713; Boger; Wysocki *J. Org. Chem.* **1988**, *53*, 3408.

[921]For reviews, see Carruthers, Ref. 440; Reinhoudt *Adv. Heterocycl. Chem.* **1977**, *21*, 253-321; Roberts; Sharts *Org. React.* **1962**, *12*, 1-56; Gilchrist; Storr, Ref. 895, pp. 173-212; Beltrame, in Bamford; Tipper, Ref. 1, vol. 9, pp. 131-152; Huisgen; Grashey; Sauer, in Patai, Ref. 36, pp. 779-802. For a review of the use of 2 + 2 cycloadditions in polymerization reactions see Dilling. *Chem. Rev.* **1983**, *83*, 1-47. For a list of references, see Ref. 133, pp. 82-83, 659-660.

[922]For a review, see Fischer, in Patai, Ref. 36, pp. 1064-1067.

[923]Dolbier; Lomas; Garza; Harmon; Tarrant *Tetrahedron* **1972**, *28*, 3185.

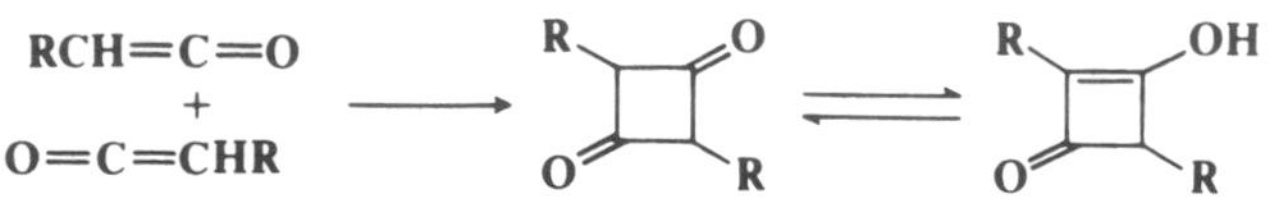

98

major primary products, though ketene itself dimerizes in a different manner, to give an unsaturated β-lactone (**6-63**).[924]

Different olefins combine as follows:

1. $F_2C=CX_2$ (X = F or Cl), especially $F_2C=CF_2$, form cyclobutanes with many olefins. Compounds of this type even react with conjugated dienes to give four-membered rings rather than undergoing normal Diels–Alder reactions.[925]

2. Allenes[926] and ketenes[927] react with activated olefins and alkynes. Ketenes give 1,2 addition, even with conjugated dienes.[928] Ketenes also add to unactivated olefins if sufficiently long reaction times are used.[929] Allenes and ketenes also add to each other.[930]

3. Enamines[931] form four-membered rings with Michael-type olefins[932] and ketenes.[933] In both cases, only enamines from aldehydes give stable four-membered rings:

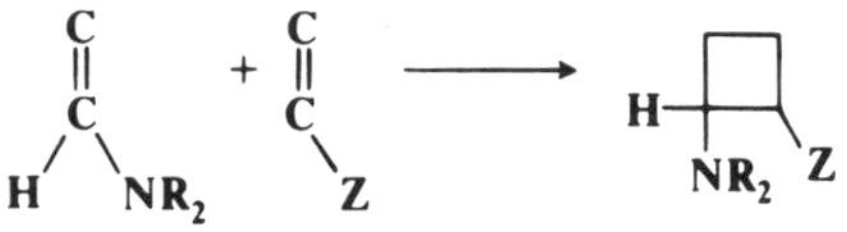

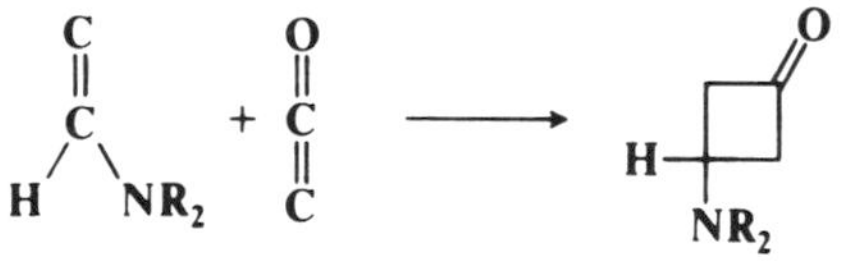

The reaction of enamines with ketenes can be conveniently carried out by generating the ketene in situ from an acyl halide and a tertiary amine.

4. Olefins with electron-withdrawing groups may form cyclobutanes with olefins containing electron-donating groups. The enamine reactions, mentioned above, are examples

[924]Farnum; Johnson; Hess; Marshall; Webster *J. Am. Chem. Soc.* **1965**, *87*, 5191; Dehmlow; Pickardt; Slopianka; Fastabend; Drechsler; Soufi *Liebigs Ann. Chem.* **1987**, 377.

[925]Bartlett; Montgomery; Seidel *J. Am. Chem. Soc.* **1964**, *86*, 616; De Cock; Piettre; Lahousse; Janousek; Merényi; Viehe *Tetrahedron* **1985**, *41*, 4183.

[926]For reviews of 2 + 2 cycloadditions of allenes, see Schuster; Coppola, Ref. 95, pp. 286-317; Hopf, in Landor, Ref. 95, vol. 2, pp. 525-562; Ghosez; O'Donnell, in Marchand; Lehr, Ref. 898, vol. 2, pp. 79-140; Baldwin; Fleming *Fortschr. Chem. Forsch.* **1970**, *15*, 281-310.

[927]For reviews of cycloadditions of ketenes, see Ghosez; O'Donnell, Ref. 926; Brady *Synthesis* **1971**, 415-422; Luknitskii; Vovsi *Russ. Chem. Rev.* **1969**, *38*, 487-494; Ulrich *Cycloaddition Reactions of Heterocumulenes*; Academic Press: New York, 1967, pp. 38-121; Holder *J. Chem. Educ.* **1976**, *53*, 81-85. For a review of intramolecular cycloadditions of ketenes to alkenes, see Snider *Chem. Rev.* **1988**, *88*, 793-811.

[928]See, for example, Martin; Gott; Goodlett; Hasek *J. Org. Chem.* **1965**, *30*, 4175; Brady; O'Neal *J. Org. Chem.* **1967**, *32*, 2704; Huisgen; Feiler; Otto *Tetrahedron Lett.* **1968**, 4491, *Chem. Ber.* **1969**, *102*, 3475. For indirect methods of the 1,4 addition of the elements of ketene to a diene see Freeman; Balls; Brown *J. Org. Chem.* **1968**, *33*, 2211; Corey; Ravindranathan; Terashima *J. Am. Chem. Soc.* **1971**, *93*, 4326. For a review of ketene equivalents see Ranganathan; Ranganathan; Mehrotra *Synthesis* **1977**, 289-296.

[929]Huisgen; Feiler *Chem. Ber.* **1969**, *102*, 3391; Brady; Patel *J. Org. Chem.* **1973**, *38*, 4106; Bak; Brady *J. Org. Chem.* **1979**, *44*, 107.

[930]Bampfield; Brook; McDonald *J. Chem. Soc., Chem. Commun.* **1975**, 132; Gras; Bertrand *Nouv. J. Chim.* **1981**, *5*, 521.

[931]For a review of cycloaddition reactions of enamines, see Cook, in Cook *Enamines*, 2nd ed.; Marcel Dekker: New York, 1988, pp. 347-440.

[932]Brannock; Bell; Goodlett; Thweatt *J. Org. Chem.* **1964**, *29*, 813.

[933]Berchtold; Harvey; Wilson *J. Org. Chem.* **1961**, *26*, 4776; Opitz; Kleeman *Liebigs Ann. Chem.* **1963**, *665*, 114; Hasek; Gott; Martin *J. Org. Chem.* **1966**, *31*, 1931.

of this, but it has also been accomplished with tetracyanoethylene and similar molecules, which give substituted cyclobutanes when treated with olefins of the form C=C—A, where A may be OR,[934] SR (enol and thioenol ethers),[935] cyclopropyl,[936] or certain aryl groups.[937]

Solvents are not necessary for 2 + 2 cycloadditions. They are usually carried out at 100 to 225°C under pressure, although the reactions in group 4 occur under milder conditions.

The reaction is similar to the Diels–Alder (in action, not in scope), and if dienes are involved, the latter reaction may compete, though most olefins react with a diene either entirely by 1,2 or entirely by 1,4 addition. Three mechanisms can be proposed[938] analogous to those proposed for the Diels–Alder reaction. Mechanism *a* is a concerted pericyclic process, and mechanisms *b* and *c* are two-step reactions involving, respectively, a diradical (**99**) and a diion (**100**) intermediate. As in **5-47**, a diradical intermediate must be a singlet.

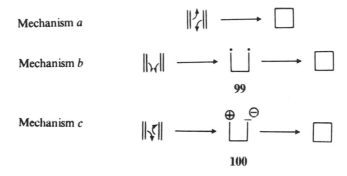

In searching for ways to tell which mechanism is operating in a given case, we would expect mechanism *c* to be sensitive to changes in solvent polarity, while mechanisms *a* and *b* should be insensitive. We would also expect mechanism *a* to be stereospecific, while mechanisms *b* and *c* probably would not be stereospecific, though if the second step of these processes takes place very rapidly, before **99** or **100** has a chance to rotate about the newly formed single bond, stereospecificity might be observed. Because of entropy considerations such rapid ring closure might be more likely here than in a 2 + 4 cycloaddition.

There is evidence that the reactions can take place by all three mechanisms, depending on the structure of the reactants. A thermal $[_\pi 2_s + _\pi 2_s]$ mechanism is ruled out for most of these substrates by the orbital symmetry rules, but a $[_\pi 2_s + _\pi 2_a]$ mechanism is allowed (p. 851), and there is much evidence that ketenes and certain other linear molecules[939] in which the steric hindrance to such an approach is minimal can and often do react by this mechanism. In a $[_\pi 2_s + _\pi 2_a]$ cycloaddition the molecules must approach each other in such a way (Figure 15.12a) that the + lobe of the HOMO of one molecule (I) overlaps with both + lobes of the LUMO of the other (II), even though these lobes are on opposite sides of the nodal plane of II. The geometry of this approach requires that the groups S and U of molecule II project *into* the plane of molecule I. This has not been found to happen for ordinary

[934]For a review with ketene acetals $R_2C=C(OR')_2$, see Scheeren *Recl. Trav. Chim. Pays-Bas* **1986**, *105*, 71-84.
[935]Williams; Wiley; McKusick *J. Am. Chem. Soc.* **1962**, *84*, 2210.
[936]Nishida; Moritani; Teraji *J. Org. Chem.* **1973**, *38*, 1878.
[937]Nagata; Shirota; Nogami; Mikawa *Chem. Lett.* **1973**, 1087; Shirota; Yoshida; Nogami; Mikawa *Chem. Lett.* **1973**, 1271.
[938]For a review, see Bartlett *Q. Rev., Chem. Soc.* **1970**, *24*, 473-497.
[939]There is evidence that a cyclopentyne (generated in situ) also adds to a double bond by an antarafacial process: Gilbert; Baze *J. Am. Chem. Soc.* **1984**, *106*, 1885.

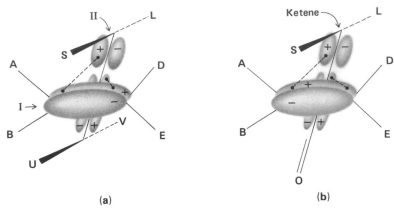

FIGURE 15.12 Orbital overlap in $_\pi2_s + _\pi2_a$ cycloaddition between (a) two olefin molecules and (b) a ketene and an olefin. S and L stand for small and large.

alkenes,[940] but if molecule II is a ketene (Figure 15.12b), the group marked U is not present and the $[_\pi2_s + _\pi2_a]$ reaction can take place. Among the evidence[941] for this mechanism[942] is the following: (1) The reactions are stereospecific.[943] (2) The isomer that forms is the *more-hindered one*. Thus methylketene plus cyclopentadiene gave only the endo product (**101**, A = H, R = CH₃).[944] Even more remarkably, when haloalkyl ketenes RXC=C=O were treated with cyclopentadiene, the endo-exo ratio of the product (**101, 102**, A = halogen)

$$\underset{R'}{\overset{A}{\diagdown}}C=C=O \ + \ \text{⬡} \ \longrightarrow \ \text{endo } \mathbf{101} \qquad \text{exo } \mathbf{102} \qquad \begin{array}{l} A = H \text{ or} \\ \text{halogen} \end{array}$$

actually *increased* substantially when R was changed from Me to iso-Pr to *t*-Bu![945] One would expect preferential formation of the exo products (**102**) from $[_\pi2_s + _\pi2_s]$ cycloadditions where the molecules approach each other face-to-face, but a $[_\pi2_s + _\pi2_a]$ process leads to endo products because the ketene molecule (which for steric reasons would approach with its smaller group directed toward the olefin) must twist as shown in Figure 15.13 (L = larger;

[940]See, for example, Padwa; Koehn; Masaracchia; Osborn; Trecker *J. Am. Chem. Soc.* **1971**, *93*, 3633; Bartlett; Cohen; Elliott; Hummel; Minns; Sharts; Fukunaga *J. Am. Chem. Soc.* **1972**, *94*, 2899.

[941]For other evidence, see Baldwin; Kapecki *J. Am. Chem. Soc.* **1970**, *92*, 4874; Brook; Griffiths *Chem. Commun.* **1970**, 1344; Frey; Isaacs *J. Chem. Soc. B* **1970**, 830; Egger *Int. J. Chem. Kinet.* **1973**, *5*, 285; Moon; Kolesar *J. Org. Chem.* **1974**, *39*, 995; Isaacs; Hatcher *J. Chem. Soc., Chem. Commun.* **1974**, 593; Hassner; Cory; Sartoris *J. Am. Chem. Soc.* **1976**, *98*, 7698; Gheorghiu; Pârvulescu; Drâghici; Elian *Tetrahedron* **1981**, *37* Suppl., 143. See, however, Holder; Graf; Duesler; Moss *J. Am. Chem. Soc.* **1983**, *105*, 2929.

[942]On the other hand, molecular orbital calculations predict that the cycloaddition of ketenes to olefins does not take place by a $[_\pi2_s + _\pi2_a]$ mechanism: Wang; Houk *J. Am. Chem. Soc.* **1990**, *112*, 1754; Bernardi; Bottoni; Robb; Venturini *J. Am. Chem. Soc.* **1990**, *112*, 2106; Valentí; Pericàs; Moyano *J. Org. Chem.* **1990**, *55*, 3582.

[943]Huisgen; Feiler; Binsch *Angew. Chem. Int. Ed. Engl.* **1964**, *3*, 753 [*Angew. Chem. 76*, 892], *Chem. Ber.* **1969**, *102*, 3460; Martin; Goodlett; Burpitt *J. Org. Chem.* **1965**, *30*, 4309; Montaigne; Ghosez *Angew. Chem. Int. Ed. Engl.* **1968**, *7*, 221 [*Angew. Chem. 80*, 194]; Bertrand; Gras; Gore *Tetrahedron* **1975**, *31*, 857; Marchand-Brynaert; Ghosez *J. Am. Chem. Soc.* **1972**, *94*, 2870; Huisgen; Mayr *Tetrahedron Lett.* **1975**, 2965, 2969.

[944]Brady; Hoff; Roe; Parry *J. Am. Chem. Soc.* **1969**, *91*, 5679; Rey; Roberts; Dieffenbacher; Dreiding *Helv. Chim. Acta* **1970**, *53*, 417. See also Brady; Parry; Stockton *J. Org. Chem.* **1971**, *36*, 1486; DoMinh; Strausz *J. Am. Chem. Soc.* **1970**, *92*, 1766; Isaacs; Stanbury *Chem. Commun.* **1970**, 1061; Brook; Harrison; Duke *Chem. Commun.* **1970**, 589; Dehmlow, *Tetrahedron Lett.* **1973**, 2573; Rey; Roberts; Dreiding; Roussel; Vanlierde; Toppet; Ghosez *Helv. Chim. Acta* **1982**, *65*, 703.

[945]Brady; Roe *J. Am. Chem. Soc.* **1970**, *92*, 4618.

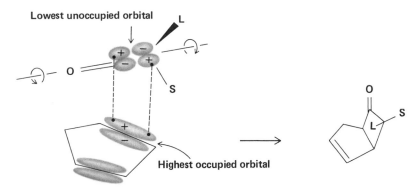

FIGURE 15.13 Orbital overlap in the reaction of a ketene with cyclopentadiene. S and L stand for small and large.

S = smaller group) in order for the + lobes to interact and this swings the larger group into the endo position.[946] The experimental results in which the amount of endo isomer increases with the increasing size of the R group would seem to be contrary to what would be expected from considerations of steric hindrance (we may call them *masochistic steric effects*), but they are just what is predicted for a $[_\pi 2_s + _\pi 2_a]$ reaction. (3) There is only moderate polar solvent acceleration.[947] (4) The rate of the reaction is not very sensitive to the presence of electron-withdrawing or electron-donating substituents.[948] Because cycloadditions involving allenes are often stereospecific, it has been suggested that these also take place by the $[_\pi 2_s + _\pi 2_a]$ mechanism,[949] but the evidence in these cases is more consistent with the diradical mechanism *b*.[950]

The diradical mechanism *b* is most prominent in the reactions involving fluorinated alkenes.[951] These reactions are generally not stereospecific[952] and are insensitive to solvent effects. Further evidence that a diion is not involved is that head-to-head coupling is found when an unsymmetrical molecule is dimerized. Thus dimerization of $F_2C=CFCl$ gives **103**,

104   **103**

[946]Brook; Harrison; Duke, Ref. 944.

[947]Brady; O'Neal *J. Org. Chem.* **1967**, *32*, 612; Huisgen; Feiler; Otto *Tetrahedron Lett.* **1968**, 4485, *Chem. Ber.* **1969**, *102*, 3444; Sterk *Z. Naturforsch., Teil B* **1972**, *27*, 143.

[948]Baldwin; Kapecki *J. Am. Chem. Soc.* **1970**, *92*, 4868; Isaacs; Stanbury *J. Chem. Soc., Perkin Trans. 2* **1973**, 166.

[949]For example, see Kiefer; Okamura *J. Am. Chem. Soc.* **1968**, *90*, 4187; Baldwin; Roy *Chem. Commun.* **1969**, 1225; Moore; Bach; Ozretich *J. Am. Chem. Soc.* **1969**, *91*, 5918.

[950]Muscio; Jacobs *Tetrahedron Lett.* **1969**, 2867; Taylor; Warburton; Wright *J. Chem. Soc. C* **1971**, 385; Dai; Dolbier *J. Am. Chem. Soc.* **1972**, *94*, 3946; Duncan; Weyler; Moore *Tetrahedron Lett.* **1973**, 4391; Grimme; Rother *Angew. Chem. Int. Ed. Engl.* **1973**, *12*, 505 [*Angew. Chem. 85*, 512]; Levek; Kiefer *J. Am. Chem. Soc.* **1976**, *98*, 1875; Pasto; Heid; Warren *J. Am. Chem. Soc.* **1982**, *104*, 3676; Pasto; Yang *J. Org. Chem.* **1986**, *51*, 1676; Dolbier; Seabury *Tetrahedron Lett.* **1987**, *28*, 1491, *J. Am. Chem. Soc.* **1987**, *109*, 4393; Dolbier; Weaver *J. Org. Chem.* **1990**, *55*, 711.

[951]It has been argued that the mechanism here is not the diradical mechanism, but the $[_\pi 2_s + _\pi 2_a]$ mechanism: Roberts *Tetrahedron* **1985**, *41*, 5529.

[952]Montgomery; Schueller; Bartlett *J. Am. Chem. Soc.* **1964**, *86*, 621; Bartlett; Hummel; Elliott; Minns *J. Am. Chem. Soc.* **1972**, *94*, 2898.

not **104**. If one pair of electrons moved before the other, the positive end of one molecule would be expected to attack the negative end of the other.[953]

The diion mechanism[954] c has been reported for at least some of the reactions[955] in categories 3 and 4,[956] as well as some ketene dimerizations.[957] For example, the rate of the reaction between 1,2-bis(trifluoromethyl)-1,2-dicyanoethene and ethyl vinyl ether was strongly influenced by changes in solvent polarity.[958] Some of these reactions are nonstereospecific, but others are stereospecific.[959] As previously indicated, it is likely that in the latter cases the diionic intermediate closes before rotation can take place. Such rapid ring closure is more likely for a diion than for a diradical because of the attraction between the opposite charges. Other evidence for the diion mechanism in these cases is that reaction rates are greatly dependent on the presence of electron-donating and electron-withdrawing groups and that it is possible to trap the diionic intermediates.

Whether a given olefin reacts by the diradical or diion mechanism depends, among other things, on the groups attached to it. For example, phenyl and vinyl groups at the α positions of **99** or **100** help to stabilize a diradical, while donors such as oxygen and nitrogen favor a diion (they stabilize the positively charged end).[960] A table on p. 451 of reference 960 shows which mechanism is more likely for 2 + 2 cycloadditions of various pairs of olefins.

Thermal cleavage of cyclobutanes[961] to give two olefin molecules (*cycloreversion*,[962] the reverse of 2 + 2 cycloaddition) operates by the diradical mechanism, and the $[_\sigma 2_s + _\sigma 2_a]$ pathway has not been found[963] (the subscripts σ indicate that σ bonds are involved in this reaction).

In some cases, double bonds add to triple bonds to give cyclobutenes, apparently at about the same rate that they add to double bonds, e.g.,

About equal amounts

The addition of triple bonds to triple bonds would give cyclobutadienes, and this has not been observed, except where these rearrange before they can be isolated (see **5-51**)[964] or in the presence of a suitable coordination compound, so that the cyclobutadiene is produced in the form of a complex (p. 55).[965]

---

[953]For additional evidence based on radical stabilities, see Silversmith; Kitahara; Caserio; Roberts *J. Am. Chem. Soc.* **1958**, *80*, 5840; Ref. 925; Doering; Guyton *J. Am. Chem. Soc.* **1978**, *100*, 3229.

[954]For reviews of this mechanism, see Huisgen *Acc. Chem. Res.* **1977**, *10*, 117-124, 199-206; Huisgen; Schug; Steiner *Bull. Soc. Chim. Fr.* **1976**, 1813-1820.

[955]For a review of cycloadditions with polar intermediates, see Gompper *Angew. Chem. Int. Ed. Engl.* **1969**, *8*, 312-327 [*Angew. Chem. 81*, 348-363].

[956]The reactions of ketenes with enamines are apparently not concerted but take place by the diionic mechanism: Otto; Feiler; Huisgen *Angew. Chem. Int. Ed. Engl.* **1968**, *7*, 737 [*Angew. Chem. 80*, 759].

[957]See Moore; Wilbur *J. Am. Chem. Soc.* **1978**, *100*, 6523.

[958]Proskow; Simmons; Cairns *J. Am. Chem. Soc.* **1966**, *88*, 5254. See also Huisgen *Pure Appl. Chem.* **1980**, *52*, 2283-2302.

[959]Proskow; Simmons; Cairns, Ref. 958; Huisgen; Steiner *J. Am. Chem. Soc.* **1973**, *95*, 5054, 5055.

[960]Hall *Angew. Chem. Int. Ed. Engl.* **1983**, *22*, 440-455 [*Angew. Chem. 95*, 448-464].

[961]See Frey *Adv. Phys. Org. Chem.* **1966**, *4*, 147-193, pp. 170-175, 180-183.

[962]For reviews of 2 + 2 cycloreversions, see Schaumann; Ketcham *Angew. Chem. Int. Ed. Engl.* **1982**, *21*, 225-247 [*Angew. Chem. 94*, 231-253]; Brown, Ref. 883, pp. 247-259.

[963]See, for example, Cocks; Frey; Stevens *Chem. Commun.* **1969**, 458; Srinivasan; Hsu *J. Chem. Soc., Chem. Commun.* **1972**, 1213; Paquette; Kukla *Tetrahedron Lett.* **1973**, 1241; Paquette; Carmody *J. Am. Chem. Soc.* **1976**, *98*, 8175. See however Cant; Coxon; Hartshorn *Aust. J. Chem.* **1975**, *28*, 391; Doering; Roth; Breuckmann; Figge; Lennartz; Fessner; Prinzbach *Chem. Ber.* **1988**, *121*, 1.

[964]For a review of these cases, and of cycloadditions of triple bonds to double bonds, see Fuks; Viehe, in Viehe, Ref. 49, pp. 435-442.

[965]D'Angelo; Ficini; Martinon; Riche; Sevin *J. Organomet. Chem.* **1979**, *177*, 265. For a review, see Hogeveen; Kok in Patai; Rappoport, Ref. 49, pt. 2, pp. 981-1013.

Although thermal 2 + 2 cycloaddition reactions are essentially limited to the cases described above, many (though by no means all) double-bond compounds undergo such reactions *when photochemically excited* (either directly or by a photosensitizer—see p. 241), even if they are not in the above categories.[966] Simple alkenes absorb in the far uv (p. 235), which is difficult to reach experimentally, though this problem can sometimes be overcome by the use of suitable photosensitizers. The reaction has been applied to simple alkenes[967] (especially to strained compounds such as cyclopropenes and cyclobutenes), but more often the double-bond compounds involved are conjugated dienes,[968] α,β-unsaturated ketones,[969] acids, or acid derivatives, or quinones, since these compounds, because they are conjugated, absorb at longer wavelengths (p. 234). Both dimerizations and mixed additions are common, some examples being (see also the example on p. 246):

Diels-Alder product

Ref. 970

Ref. 971

Ref. 972

ratio ~ 9 : 1

[966]For reviews, see Demuth; Mikhail *Synthesis* **1989**, 145-162; Ninomiya; Naito *Photochemical Synthesis*; Academic Press: New York, 1989, pp. 58-109; Ramamurthy; Venkatesan *Chem. Rev.* **1987**, *87*, 433-481; Lewis *Adv. Photochem.* **1986**, *13*, 165-235; Wender, in Coyle *Photochemistry in Organic Synthesis*; Royal Society of Chemistry: London, 1986, pp. 163-188; Schreiber *Science* **1985**, *227*, 857-863; Neckers; Tinnemans, in Horspool *Synthetic Organic Photochemistry*; Plenum: New York, 1984, pp. 285-311; Baldwin *Org. Photochem.* **1981**, *5*, 123-225; Turro *Modern Molecular Photochemistry*; W.A. Benjamin: New York, 1978, pp. 417-425, 458-465; Kricka; Ledwith *Synthesis* **1974**, 539-549; Herndon *Top. Curr. Chem.* **1974**, *46*, 141-179; Sammes *Q. Rev., Chem. Soc.* **1970**, *24*, 37-68, pp. 46-55; Crowley; Mazzocchi, in Zabicky, Ref. 115, pp. 297-316; Turro; Dalton; Weiss *Org. Photochem.* **1969**, *2*, 1-62; Trecker *Org. Photochem.* **1969**, *2*, 63-116; Scharf *Fortschr. Chem. Forsch.* **1969**, *11*, 216-244; Steinmetz *Fortschr. Chem. Forsch.* **1967**, *7*, 445-527; Fonken *Org. Photochem.* **1967**, *1*, 197-246; Chapman; Lenz *Org. Photochem.* **1967**, *1*, 283-321; Schönberg, Ref. 49, pp. 70-96, 109-117; Warrener; Bremner *Rev. Pure Appl. Chem.* **1966**, *16*, 117-173, pp. 122-128.
[967]For examples of nonphotosensitized dimerization of simple alkenes, see Arnold; Abraitys *Chem. Commun.* **1967**, 1053; Yamazaki; Cvetanović *J. Am. Chem. Soc.* **1969**, *91*, 520.
[968]For a review, see Dilling *Chem. Rev.* **1969**, *69*, 845-877.
[969]For reviews of various aspects of this subject, see Cossy; Carrupt; Vogel, in Patai *Supplement A: The Chemistry of Double-bonded Functional Groups*, vol. 2, pt. 2; Wiley: New York, 1989, pp. 1369-1565; Kemernitskii; Ignatov; Levina *Russ. Chem. Rev.* **1988**, *57*, 270-282; Weedon, in Horspool, Ref. 966, pp. 61-143; Lenz *Rev. Chem. Intermed.* **1981**, *4*, 369-404; Margaretha *Chimia* **1975**, *29*, 203-209; Bauslaugh *Synthesis* **1970**, 287-300; Eaton *Acc. Chem. Res.* **1968**, *1*, 50-57.
[970]Hammond; Turro; Fischer *J. Am. Chem. Soc.* **1961**, *83*, 4674; Liu; Turro; Hammond *J. Am. Chem. Soc.* **1965**, *87*, 3406; Cundall; Griffiths *Trans. Faraday Soc.* **1965**, *61*, 1968; DeBoer; Turro; Hammond *Org. Synth. V*, 528.
[971]Cookson; Cox; Hudec *J. Chem. Soc.* **1961**, 4499.
[972]Stensen; Svendsen; Hofer; Sydnes *Acta Chem. Scand., Ser. B* **1988**, *42*, 259.

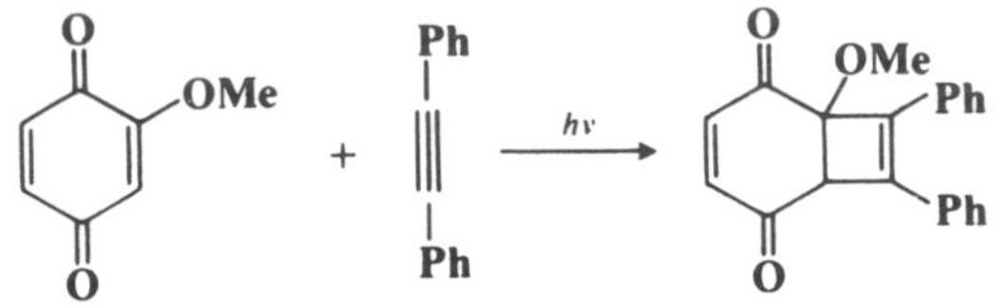

Ref. 973

Photochemical 2 + 2 cycloadditions can also take place intramolecularly if a molecule has two double bonds that are properly oriented.[974] The cyclization of the quinone dimer shown above is one example. Other examples are

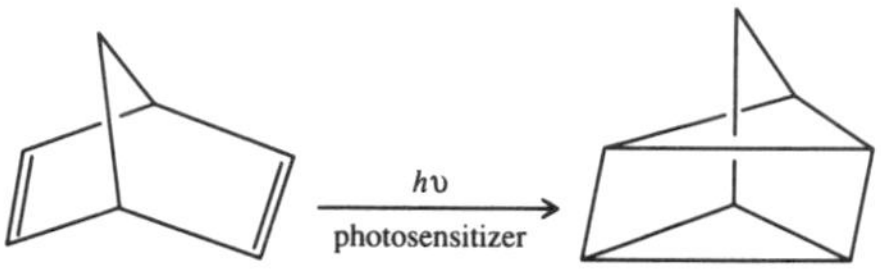

Quadricyclane

Ref. 975

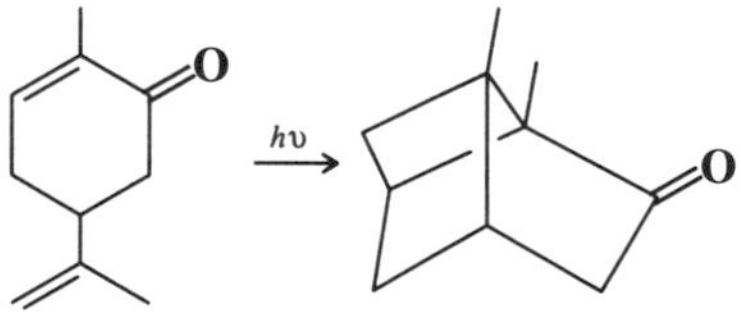

Carvone   Carvonecamphor

Ref. 976

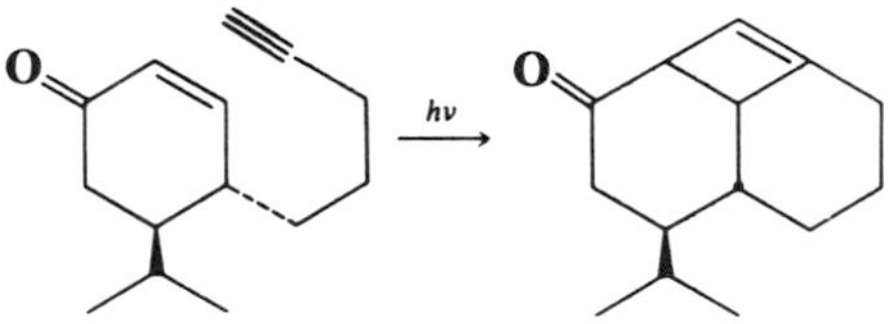

Ref. 977

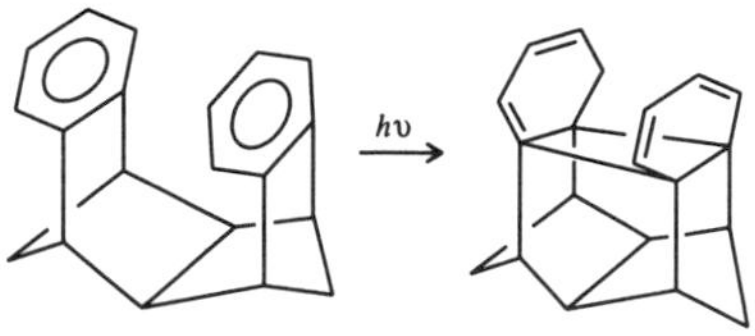

Ref. 978

[973]Pappas; Pappas *Tetrahedron Lett.* **1967,** 1597.

[974]For reviews, see Becker; Haddad *Org. Photochem.* **1989,** *10,* 1-162; Crimmins *Chem. Rev.* **1988,** *88,* 1453-1473; Oppolzer *Acc. Chem. Res.* **1982,** *15,* 135-141; Prinzbach *Pure Appl. Chem.* **1968,** *16,* 17-46; Dilling *Chem. Rev.* **1966,** *66,* 373-393.

[975]Hammond; Turro; Fischer, Ref. 970; Dauben; Cargill *Tetrahedron* **1961,** *15,* 197. See also Cristol; Snell *J. Am. Chem. Soc.* **1958,** *80,* 1950.

[976]Ciamician; Silber *Ber.* **1908,** *41,* 1928; Büchi; Goldman *J. Am. Chem. Soc.* **1957,** *79,* 4741.

[977]Koft; Smith *J. Am. Chem. Soc.* **1984,** *106,* 2115.

[978]Fessner; Prinzbach; Rihs *Tetrahedron Lett.* **1983,** *24,* 5857.

Ref. 979

It is obvious that many molecules can be constructed in this way that would be difficult to make by other procedures. However, attempted cyclizations of this kind are not always successful. In many cases polymeric or other side products are obtained instead of the desired product.

It is possible that some of these photochemical cycloadditions take place by a $[_\pi2_s + _\pi2_s]$ mechanism (which is of course allowed by orbital symmetry); when and if they do, one of the molecules must be in the excited singlet state (S_1) and the other in the ground state.[980] The nonphotosensitized dimerizations of *cis*- and *trans*-2-butene are stereospecific,[981] making it likely that the $[_\pi2_s + _\pi2_s]$ mechanism is operating in these reactions. However, in most cases it is a triplet excited state that reacts with the ground-state molecule; in these cases the diradical (or in certain cases, the diionic) mechanism is taking place. In one intramolecular case, the intermediate diradical has been trapped.[982] Photosensitized $2\pi + 2\pi$ cycloadditions almost always involve the triplet state and hence a diradical (or diionic) mechanism.

The photochemical diradical mechanism is not quite the same as the thermal diradical mechanism. In the thermal mechanism the initially formed diradical must be a singlet, but in the photochemical process a triplet excited state is adding to a ground state (which is of course a singlet). Thus, in order to conserve spin,[983] the initially formed diradical must be a triplet; i.e., the two electrons must have the same spin. Consequently the second, or ring-closing, step of the mechanism cannot take place at once, because a new bond cannot form from a combination of two electrons with the same spin, and the diradical has a reasonably long lifetime before collisions with molecules in the environment allow a spin inversion to take place and the diradical to cyclize. We would therefore predict nonstereospecificity, and that is what is found.[984] It has been believed that at least some 2 + 2 photocycloadditions take place by way of exciplex intermediates[985] [an *exciplex*[986] is an excited EDA complex (p. 79) which is dissociated in the ground state; in this case one double bond is the donor and the other the acceptor], but there is evidence against this.[987]

It has been found that certain 2 + 2 cycloadditions which do not occur thermally can be made to take place without photochemical initiation by the use of certain catalysts, usually

[979]Mehta; Padma; Ōsawa; Barbiric; Mochizuki *Tetrahedron Lett.* **1987**, *28*, 1295.

[980]We have previously seen (p. 242) that reactions between two excited molecules are extremely rare.

[981]Yamazaki; Cvetanović, Ref. 967; Yamazaki; Cvetanović; Irwin *J. Am. Chem. Soc.* **1976**, *98*, 2198. For other likely examples, see Lewis; Hoyle; Johnson *J. Am. Chem. Soc.* **1975**, *97*, 3267; Lewis; Kojima *J. Am. Chem. Soc.* **1988**, *110*, 8660.

[982]Becker; Haddad; Sahali *Tetrahedron Lett.* **1989**, *30*, 2661.

[983]This is an example of the Wigner spin conservation rule (p. 241). Note that spin conservation is something entirely different from symmetry conservation.

[984]See, for example, Liu; Hammond *J. Am. Chem. Soc.* **1967**, *89*, 4936; Kramer; Bartlett *J. Am. Chem. Soc.* **1972**, *94*, 3934.

[985]See, for example, Farid; Doty; Williams *J. Chem. Soc., Chem. Commun.* **1972**, 711; Mizuno; Pac; Sakurai *J. Am. Chem. Soc.* **1974**, *96*, 2993; Caldwell; Creed *Acc. Chem. Res.* **1980**, *13*, 45-50; Mattes; Farid *Acc. Chem. Res.* **1982**, *15*, 80-86; Swapna; Lakshmi; Rao; Kunwar *Tetrahedron* **1989**, *45*, 1777.

[986]For a review of exciplexes, see Davidson *Adv. Phys. Org. Chem.* **1983**, *19*, 1-130.

[987]Schuster; Heibel; Brown; Turro; Kumar *J. Am. Chem. Soc.* **1988**, *110*, 8261.

transition-metal compounds.[988] Examples are:

$$\text{(structure)} + \text{(structure)} \xrightarrow{\text{Ti(CH}_2\text{Ph)}_4} \text{(structure)} \qquad \text{Ref. 989}$$

$$2 \ \triangle \xrightarrow{\text{zeolites}} \text{(structure)} \qquad \text{Ref. 990}$$

Among the catalysts used are Lewis acids[991] and phosphine-nickel complexes.[992] Certain of the reverse cyclobutane ring openings can also be catalytically induced (**8-40**). The role of the catalyst is not certain and may be different in each case. One possibility is that the presence of the catalyst causes a forbidden reaction to become allowed, through coordination of the catalyst to the π or σ bonds of the substrate.[993] In such a case the reaction would of course be a concerted $2_s + 2_s$ process. However, the available evidence is more consistent with nonconcerted mechanisms involving metal–carbon σ-bonded intermediates, at least in most cases.[994] For example, such an intermediate was isolated in the dimerization of norbornadiene, catalyzed by iridium complexes.[995]

Thermal cycloadditions leading to four-membered rings can also take place between a cyclopropane ring and an alkene or alkyne[996] bearing electron-withdrawing groups.[997] These reactions are $_\pi 2 + _\sigma 2$ cycloadditions. Ordinary cyclopropanes do not undergo the reaction, but it has been accomplished with strained systems such as bicyclo[1.1.0]butanes[998] and bicyclo[2.1.0]pentanes. For example, bicyclo[2.1.0]pentane reacts with maleonitrile (or fu-

$$\text{(structure)} + \underset{H}{\overset{NC}{C}}=\underset{H}{\overset{CN}{C}} \longrightarrow \text{(structure, CN, CN)} + \text{(structure, CN, CN)} + \text{(structure, CN, CN)}$$

maronitrile) to give all three isomers of 2,3-dicyanonorbornane, as well as four other products.[999] The lack of stereospecificity and the negligible effect of solvent on the rate indicate

[988]For reviews, see Dzhemilev; Khusnutdinov; Tolstikov *Russ. Chem. Rev.* **1987**, *56*, 36-51; Kricka; Ledwith, Ref. 966.

[989]Cannell *J. Am. Chem. Soc.* **1972**, *94*, 6867.

[990]Schipperijn; Lukas *Tetrahedron Lett.* **1972**, 231.

[991]West; Kwitowski *J. Am. Chem. Soc.* **1968**, *90*, 4697; Lukas; Baardman; Kouwenhoven *Angew. Chem. Int. Ed. Engl.* **1976**, *15*, 369 [*Angew Chem.* **88**, 412].

[992]See, for example, Hoover; Lindsey *J. Org. Chem.* **1969**, *34*, 3051; Noyori; Ishigami; Hayashi; Takaya *J. Am. Chem. Soc.* **1973**, *95*, 1674; Yoshikawa; Aoki; Kiji; Furukawa *Tetrahedron* **1974**, *30*, 405.

[993]For discussions, see Labunskaya; Shebaldova; Khidekel' *Russ. Chem. Rev.* **1974**, *43*, 1-16; Mango *Top. Curr. Chem.* **1974**, *45*, 39-91, *Tetrahedron Lett.* **1973**, 1509, *Intra-Sci. Chem. Rep.* **1972**, *6* (3), 171-187, *CHEMTECH* **1971**, *1*, 758-765, *Adv. Catal.* **1969**, *20*, 291-325; Mango; Schachtschneider *J. Am. Chem. Soc.* **1971**, *93*, 1123, **1969**, *91*, 2484; van der Lugt *Tetrahedron Lett.* **1970**, 2281; Wristers; Brener; Pettit *J. Am. Chem. Soc.* **1970**, *92*, 7499.

[994]See, for example, Cassar; Halpern *Chem. Commun.* **1970**, 1082; Doyle; McMeeking; Binger *J. Chem. Soc., Chem. Commun.* **1976**, 376; Grubbs; Miyashita; Liu; Burk *J. Am. Chem. Soc.* **1977**, *99*, 3863.

[995]Fraser; Bird; Bezman; Shapley; White; Osborn *J. Am. Chem. Soc.* **1973**, *95*, 597.

[996]Gassman; Mansfield *J. Am. Chem. Soc.* **1968**, *90*, 1517, 1524.

[997]For a review, see Gassman *Acc. Chem. Res.* **1971**, *4*, 128-136.

[998]Cairncross; Blanchard *J. Am. Chem. Soc.* **1966**, *88*, 496.

[999]Gassman; Mansfield; Murphy *J. Am. Chem. Soc.* **1969**, *91*, 1684.

a diradical mechanism. Photochemical[1000] and metal-catalyzed[1001] $_{\pi}2 + _{\sigma}2$ cycloadditions have also been reported.

In **5-47** we used the principle of conservation of orbital symmetry to explain why certain reactions take place readily and others do not. The orbital-symmetry principle can also explain why certain molecules are stable though highly strained. For example, quadricyclane and hexamethylprismane[1002] are thermodynamically much less stable (because much more strained) than their corresponding isomeric dienes, norbornadiene and hexamethylbicyclo[2.2.0]hexadiene (**105**).[1003] Yet the former two compounds can be kept indefinitely at

			105
Quadricyclane	Norbornadiene	Hexamethyl-	(A Dewar benzene)
		prismane	

room temperature, although in the absence of orbital-symmetry considerations it is not easy to understand why the electrons simply do not move over to give the more stable diene isomers. The reason is that both these reactions involve the conversion of a cyclobutane ring to a pair of double bonds (a $_{\sigma}2 + _{\sigma}2$ process) and, as we have seen, a thermal process of this sort is forbidden by the Woodward–Hoffmann rules. The process is allowed photochemically, and we are not surprised to find that both quadricyclane and hexamethylprismane are photochemically converted to the respective dienes at room temperature or below.[1004] It is also possible to conceive of simple bond rearrangements whereby hexamethylprismane is converted to hexamethylbenzene, which of course is far more stable than either hexa-

methylprismane or **105**. It has been calculated that hexamethylbenzene is at least 90 kcal/mol (380 kJ/mol) more stable than hexamethylprismane. The fact that hexamethylprismane does not spontaneously undergo this reaction has prompted the observation[1005] that the prismane has "the aspect of an angry tiger unable to break out of a paper cage." However, a correlation diagram for this reaction[1005] discloses that it too is a symmetry-forbidden process. All three of these "forbidden" reactions do take place when the compounds are heated, but the diradical mechanism is likely under these conditions.[1006]

[1000]Freeman; Balls *J. Org. Chem.* **1967**, *32*, 2354; Wiskott; Schleyer *Angew. Chem. Int. Ed. Engl.* **1967**, *6*, 694 [*Angew. Chem. 79*, 680]; Prinzbach; Eberbach *Chem. Ber.* **1968**, *101*, 4083; Prinzbach; Sedelmeier; Martin *Angew. Chem. Int. Ed. Engl.* **1977**, *16*, 103 [*Angew. Chem. 89*, 111].
[1001]See, for example, Volger; Hogeveen; Gaasbeek *J. Am. Chem. Soc.* **1969**, *91*, 218; Katz; Cerefice *J. Am. Chem. Soc.* **1969**, *91*, 2405, 6519.
[1002]This compound can be prepared by photolysis of **105**, another example of an intramolecular photochemical 2 + 2 cycloaddition: Lemal; Lokensgard *J. Am. Chem. Soc.* **1966**, *88*, 5934; Schäfer; Criegee; Askani; Grüner *Angew. Chem. Int. Ed. Engl.* **1967**, *6*, 78 [*Angew. Chem. 79*, 54].
[1003]For a review of this compound, see Schäfer; Hellmann *Angew. Chem. Int. Ed. Engl.* **1967**, *6*, 518-525 [*Angew. Chem. 79*, 566-573].
[1004]These conversions can also be carried out by the use of transition metal catalysts: Hogeveen; Volger *Chem. Commun.* **1967**, 1133, *J. Am. Chem. Soc.* **1967**, *89*, 2486; Kaiser; Childs; Maitlis *J. Am. Chem. Soc.* **1971**, *93*, 1270; Landis; Gremaud; Patrick *Tetrahedron Lett.* **1982**, *23*, 375; Maruyama; Tamiaki *Chem. Lett.* **1987**, 683.
[1005]Woodward; Hoffmann Ref. 895, pp. 107-112.
[1006]See, for example, Oth *Recl. Trav. Chim. Pays-Bas* **1968**, *87*, 1185.

Bicyclo[2.2.0]hexadienes and prismanes are *valence isomers* of benzenes.[1007] These compounds actually have the structures that were proposed for benzenes in the nineteenth century. Prismanes have the Ladenburg formula, and bicyclo[2.2.0]hexadienes have the Dewar formula. Because of this bicyclo[2.2.0]hexadiene is often called Dewar benzene. On p. 26 it was mentioned that Dewar formulas are canonical forms (though not very important) of benzenes. Yet they also exist as separate compounds in which the positions of the nuclei are different from those of benzenes.

OS **V**, 54, 235, 277, 297, 370, 393, 424, 459, 528; **VI**, 378, 571, 962, 1002, 1024, 1037; **VII**, 177, 256, 315; **68**, 32, 41; **69**, 199, 205. For the reverse reaction, see OS **V**, 734.

5-50 The Addition of Carbenes and Carbenoids to Double and Triple Bonds
epi-Methylene-addition

$$-\overset{|}{C}=\overset{|}{C}- \; + \; \overline{C}H_2 \; \longrightarrow \; \overset{H \quad H}{\bigtriangleup}$$

Carbenes and substituted carbenes add to double bonds to give cyclopropane derivatives (1 + 2 cycloaddition).[1008] Many derivatives of carbene, e.g., PhCH, ROCH,[1009] Me$_2$C=C, C(CN)$_2$, have been added to double bonds, but the reaction is most often performed with CH$_2$ itself, with halo and dihalocarbenes,[1010] and with carbalkoxycarbenes[1011] (generated from diazoacetic esters). Alkylcarbenes HCR have been added to olefins,[1012] but more often these rearrange to give olefins (p. 201). The carbene can be generated in any of the ways normally used (p. 198). However, most reactions in which a cyclopropane is formed by treatment of an olefin with a carbene "precursor" do not actually involve free carbene intermediates. In some cases it is certain that free carbenes are not involved, and in other cases there is doubt. Because of this, the term *carbene transfer* is often used to cover all reactions in which a double bond is converted to a cyclopropane, whether a carbene or a carbenoid (p. 199) is actually involved.

Carbene itself is extremely reactive and gives many side reactions, especially insertion reactions (**2-20**), which greatly reduce yields. When it is desired to add CH$_2$ for preparative purposes, free carbene is not used, but the Simmons–Smith procedure (p. 870) or some other method that does not involve free carbenes is employed instead. Halocarbenes are less active than carbenes, and this reaction proceeds quite well, since insertion reactions do not interfere.[1013] A few of the many ways[1014] in which halocarbenes or carbenoids are

[1007]For reviews of valence isomers of benzene, see Kobayashi; Kumadaki *Adv. Heterocycl. Chem.* **1982**, *31*, 169-206, *Acc. Chem. Res.* **1981**, *14*, 76-82; van Tamelen *Acc. Chem. Res.* **1972**, *5*, 186-192, *Angew. Chem. Int. Ed. Engl.* **1965**, *4*, 738-745 [*Angew. Chem.* **77**, 759-767]; Bolesov *Russ. Chem. Rev.* **1968**, *37*, 666-670; Viehe *Angew. Chem. Int. Ed. Engl.* **1965**, *4*, 746-751 [*Angew. Chem.* **77**, 768-773]; Ref. 1003.

[1008]For reviews, see, in Rappoport *The Chemistry of the Cyclopropyl Group*; Wiley: New York, 1987, the reviews by Tsuji; Nishida, pt. 1, pp. 307-373; Verhé; De Kimpe, pt. 1, pp. 445-564; Marchand, in Patai, Ref. 1, pt. 1, pp. 534-607, 625-635; Bethell, in McManus *Organic Reactive Intermediates*; Academic Press: New York, 1973, pp. 101-113; in Patai, Ref. 36, the articles by Cadogan; Perkins, pp. 633-671; Huisgen; Grashey; Sauer, pp. 755-776; Kirmse *Carbene Chemistry*, 2nd ed.; Academic Press: New York, 1971, pp. 85-122, 267-406. For a review of certain intramolecular additions, see Burke; Grieco *Org. React.* **1979**, *26*, 361-475. For a list of reagents, with references, see Ref. 133, pp. 71-79.

[1009]For a review, see Schöllkopf *Angew. Chem. Int. Ed. Engl.* **1968**, *7*, 588-598 [*Angew. Chem.* **80**, 603-613].

[1010]For a review of the addition of halocarbenes, see Parham; Schweizer *Org. React.* **1963**, *13*, 55-90.

[1011]For a review, see Dave; Warnhoff *Org. React.* **1970**, *18*, 217-401.

[1012]For example see Frey *J. Chem. Soc.* **1962**, 2293.

[1013]For reviews of carbene selectivity in this reaction, see Moss *Acc. Chem. Res.* **1989**, *22*, 15-21, **1980**, *13*, 58-64. For a review with respect to halocarbenes, see Kostikov; Molchanov; Khlebnikov *Russ. Chem. Rev.* **1989**, *58*, 654-666.

[1014]Much of the work in this field has been carried out by Seyferth and co-workers; see, for example, Seyferth; Burlitch; Minasz; Mui; Simmons; Treiber; Dowd *J. Am. Chem. Soc.* **1965**, *87*, 4259; Seyferth; Haas *J. Organomet. Chem.* **1972**, *46*, C33, *J. Org. Chem.* **1975**, *40*, 1620; Seyferth; Hopper *J. Org. Chem.* **1972**, *37*, 4070, *J. Organomet. Chem.* **1973**, *51*, 77; Seyferth; Haas; Dagani *J. Organomet. Chem.* **1976**, *104*, 9.

generated for this reaction are the following,[1015] most of which involve formal elimination (the first two steps of the SN1cB mechanism, p. 356):

$$CH_2Cl_2 + RLi \longrightarrow CHCl$$

$$N_2CHBr \xrightarrow{h\nu} CHBr$$

$$CHCl_3 + OH^- \longrightarrow CCl_2$$

$$PhHgCCl_2Br \xrightarrow{\Delta} CCl_2 \qquad\qquad\qquad\qquad \text{Ref. 1016}$$

$$Me_3SnCF_3 + NaI \longrightarrow CF_2 \qquad\qquad\qquad\qquad \text{Ref. 1017}$$

$$CHBr_2F + RLi \longrightarrow CFBr \qquad\qquad\qquad\qquad \text{Ref. 1017}$$

$$CFCl_3 + TiCl_4 + LiAlH_4 \longrightarrow CFCl \qquad\qquad\qquad \text{Ref. 1018}$$

The reaction between $CHCl_3$ and OH^- is often carried out under phase transfer conditions.[1019] It has been shown that the reaction between $PhCHCl_2$ and *t*-BuOK produces a carbenoid, but when the reaction is run in the presence of a crown ether, the free $PhCCl$ is formed instead.[1020] Dihalocyclopropanes are very useful compounds[1021] that can be reduced to cyclopropanes, treated with magnesium or sodium to give allenes (**8-3**), or converted to a number of other products.

Olefins of all types can be converted to cyclopropane derivatives by this reaction (though difficulty may be encountered with sterically hindered ones).[1022] Even tetracyanoethylene, which responds very poorly to electrophilic attack, gives cyclopropane derivatives with carbenes.[1023] Conjugated dienes give 1,2 addition:[1024]

Addition of a second mole gives bicyclopropyl derivatives.[1025] 1,4 addition is rare but has been reported in certain cases.[1026] Carbene adds to ketene to give cyclopropanone.[1027]

$$CH_2{=}C{=}O + CH_2N_2 \xrightarrow[-78°C]{CH_2Cl_2} \triangleright{=}O$$

[1015]A much longer list, with references, is given in Kirmse, *Carbene Chemistry*, Ref. 1008, pp. 313-319. See also Ref. 133, pp. 73-75.

[1016]For a review of the use of phenyl(trihalomethyl)mercury compounds as dihalocarbene or dihalocarbenoid precursors, see Seyferth *Acc. Chem. Res.* **1972**, *5*, 65-74. For a review of the synthesis of cyclopropanes with the use of organomercury reagents, see Larock, Ref. 539, pp. 341-380.

[1017]For reviews of flourinated carbenes, see Seyferth in Moss; Jones *Carbenes*, vol. 2; Wiley: New York, 1975, pp. 101-158; Sheppard; Sharts *Organic Fluorine Chemistry*; W. A. Benjamin: New York, 1969, pp. 237-270.

[1018]Dolbier; Burkholder *Tetrahedron Lett.* **1988**, *29*, 6749.

[1019]For reviews of the use of phase-transfer catalysis in the addition of dihalocarbenes to C=C bonds, see Starks; Liotta *Phase Transfer Catalysis*; Academic Press: New York, 1978, pp. 224-268; Weber; Gokel *Phase Transfer Catalysis in Organic Synthesis*; Springer: New York, 1977, pp. 18-43, 58-62. For a discussion of the mechanism, see Gol'dberg; Shimanskaya *J. Org. Chem. USSR* **1984**, *20*, 1212.

[1020]Moss; Pilkiewicz *J. Am. Chem. Soc.* **1974**, *96*, 5632; Moss; Lawrynowicz *J. Org. Chem.* **1984**, *49*, 3828.

[1021]For reviews of dihalocyclopropanes, see Banwell; Reum *Adv. Strain Org. Chem.* **1991**, *1*, 19-64; Kostikov; Molchanov; Hopf *Top. Curr. Chem.* **1990**, *155*, 41-80; Weyerstahl, in Patai; Rappoport, Ref. 614, pt. 2, pp. 1451-1497; Barlet; Vo-Quang *Bull. Soc. Chim. Fr.* **1969**, 3729-3760.

[1022]Dehmlow; Eulenberger *Liebigs Ann. Chem.* **1979**, 1112.

[1023]Cairns; McKusick *Angew. Chem.* **1961**, *73*, 520.

[1024]Woodworth; Skell *J. Am. Chem. Soc.* **1957**, *79*, 2542.

[1025]Orchin; Herrick *J. Org. Chem.* **1959**, *24*, 139; Nakhapetyan; Safonova; Kazanskii *Bull. Acad. Sci. USSR Div. Chem. Sci.* **1962**, 840; Skattebøl *J. Org. Chem.* **1964**, *29*, 2951.

[1026]Anastassiou; Cellura; Ciganek *Tetrahedron Lett.* **1970**, 5267; Jefford; Mareda; Gehret; Kabengele; Graham; Burger *J. Am. Chem. Soc.* **1976**, *98*, 2585; Mayr; Heigl *Angew. Chem. Int. Ed. Engl.* **1985**, *24*, 579 [*Angew. Chem. 97*, 567]; Jenneskens; de Wolf; Bickelhaupt *Angew. Chem. Int. Ed. Engl.* **1985**, *24*, 585 [*Angew. Chem. 97*, 568]; Le; Jones; Bickelhaupt; de Wolf *J. Am. Chem. Soc.* **1989**, *111*, 8491; Kraakman; de Wolf; Bickelhaupt *J. Am. Chem. Soc.* **1989**, *111*, 8534; Hudlicky; Seoane; Price; Gadamasetti *Synlett* **1990**, 433; Lambert; Ziemnicka-Merchant *J. Org. Chem.* **1990**, *55*, 3460.

[1027]Turro; Hammond *Tetrahedron* **1968**, *24*, 6017; Rothgery; Holt; McGee *J. Am. Chem. Soc.* **1975**, *97*, 4971. For a review of cyclopropanones, see Wasserman; Berdahl; Lu, in Rappoport, Ref. 1008, pt. 2, pp. 1455-1532.

Allenes react with carbenes to give cyclopropanes with exocyclic unsaturation:[1028]

$$=C= \quad \xrightarrow{\bar{C}H_2} \quad \rhd\mathbin{=} \quad \xrightarrow{\bar{C}H_2} \quad \bowtie$$

A second mole gives spiropentanes. In fact, any size ring with an exocyclic double bond can be converted by a carbene to a spiro compound.[1029]

Triple-bond compounds[1030] react with carbenes to give cyclopropenes, except that in the case of acetylene itself, the cyclopropenes first formed cannot be isolated because they rearrange to allenes.[1031] Cyclopropenones (p. 53) are obtained by hydrolysis of dihalocyclopropenes.[1032] It has proved possible to add 2 moles of a carbene to an alkyne to give a bicyclobutane:[1033]

$$Me-C\equiv C-Me \;+\; 2\bar{C}H_2 \quad\longrightarrow\quad Me\!\!-\!\!\langle\rangle\!\!-\!\!Me$$

Most carbenes are electrophilic, and, in accord with this, electron-donating substituents on the olefin increase the rate of the reaction, and electron-withdrawing groups decrease it,[1034] though the range of relative rates is not very great.[1035] As discussed on p. 196, carbenes in the singlet state (which is the most common state) react stereospecifically and syn,[1036] probably by a one-step mechanism,[1037] similar to mechanism *a* of **5-47** and **5-49**:

$$-\underset{\overset{\displaystyle\curvearrowright}{\underset{\displaystyle\bar{C}}{\curvearrowleft}}}{C}=C- \quad\longrightarrow\quad \bigtimes$$

Infrared spectra of a carbene and the cyclopropane product have been observed in an argon matrix at 12 to 45 K.[1038] Carbenes in the triplet state react nonstereospecifically,[1039] probably by a diradical mechanism, similar to mechanism *b* of **5-47** and **5-49**:

$$\begin{array}{c} -\overset{|}{C}=\overset{|}{C}- \\[4pt] -\overset{\cdot}{\underset{\cdot}{C}}- \end{array} \quad\longrightarrow\quad -\overset{|}{C}-\overset{|}{\underset{\diagdown}{C}}{\cdot} \quad\longrightarrow\quad \bigtimes$$

[1028]For reviews of the addition of carbenes and carbenoids to allenes, see Landor, in Landor, Ref. 95, vol. 2, pp. 351-360; Bertrand *Bull. Soc. Chim. Fr.* **1968,** 3044-3054. For a review of the synthetic uses of methylenecyclopropanes and cyclopropenes, see Binger; Büch *Top. Curr. Chem.***1987,** *135,* 77-151.

[1029]For a review of the preparation of spiro compounds by this reaction, see Krapcho *Synthesis* **1978,** 77-126.

[1030]For reviews, see Fuks; Viehe, in Viehe, Ref. 49, pp. 427-434; Closs *Adv. Alicyclic Chem.* **1966,** *1,* 53-127, pp. 58-65.

[1031]Frey *Chem. Ind. (London)* **1960,** 1266.

[1032]Vol'pin; Koreshkov; Kursanov *Bull. Acad. Sci. USSR Div. Chem. Sci.* **1959,** 535.

[1033]Doering; Coburn *Tetrahedron Lett.* **1965,** 991. Also see Mahler *J. Am. Chem. Soc.* **1962,** *84,* 4600.

[1034]Skell; Garner *J. Am. Chem. Soc.* **1956,** *78,* 5430; Doering; Henderson *J. Am. Chem. Soc.* **1958,** *80,* 5274; Mitsch; Rodgers *Int. J. Chem. Kinet.* **1969,** *1,* 439.

[1035]For a review of reactivity in this reaction, with many comprehensive tables of data, see Moss in Jones; Moss *Carbenes,* vol. 1; Wiley: New York, 1973, pp. 153-304. See also Cox; Gould; Hacker; Moss; Turro *Tetrahedron Lett.* **1983,** *24,* 5313.

[1036]Woodworth; Skell *J. Am. Chem. Soc.* **1959,** *81,* 3383; Jones; Ando; Hendrick; Kulczycki; Howley; Hummel; Malament *J. Am. Chem. Soc.* **1972,** *94,* 7469.

[1037]For evidence that at least some singlet carbenes add by a two-step mechanism, see Giese; Lee; Neumann *Angew. Chem. Int. Ed. Engl.* **1982,** *21,* 310 [*Angew. Chem. 94,* 320].

[1038]Nefedov; Zuev; Maltsev; Tomilov *Tetrahedron Lett.* **1989,** *30,* 763.

[1039]Skell; Klebe *J. Am. Chem. Soc.* **1960,** *82,* 247. See also Jones; Tortorelli; Gaspar; Lambert *Tetrahedron Lett.* **1978,** 4257.

For carbenes or carbenoids of the type R—C—R' there is another aspect of stereochemistry.[1040] When these species are added to all but symmetrical olefins, two isomers are possible, even if the four groups originally on the double-bond carbons maintain their configurations:

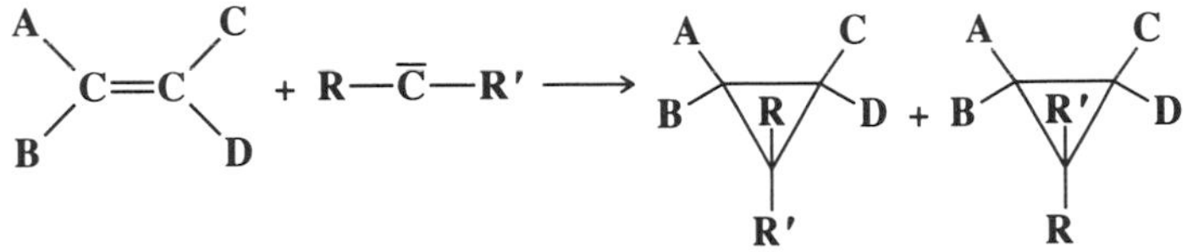

Which isomer is predominantly formed depends on R, R', and on the method by which the carbene or carbenoid is generated. Most studies have been carried out on monosubstituted species (R' $-$ H), and in these studies it is found that aryl groups generally prefer the more substituted side (syn addition) while carbethoxy groups usually show anti stereoselectivity. When R = halogen, free halocarbenes show little or no stereochemical preference, while halocarbenoids exhibit a preference for syn addition. Beyond this, it is difficult to make simple generalizations.

Carbenes are so reactive that they add to the "double bonds" of aromatic rings. The products are usually not stable and rearrange to give ring expansion. Carbene reacts with benzene to give cycloheptatriene:[1041]

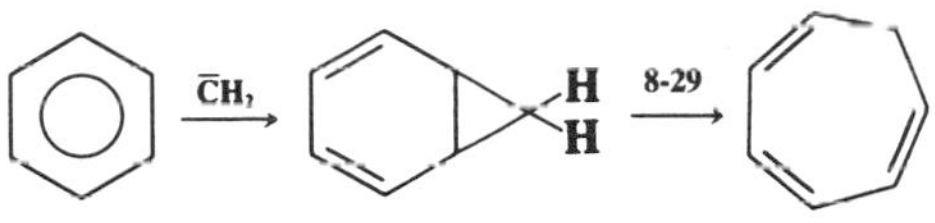

Norcaradiene

but not all carbenes are reactive enough to add to benzene. The norcaradiene intermediate cannot be isolated in this case[1042] (it undergoes an electrocyclic rearrangement, **8-29**), though certain substituted norcaradienes, e.g., the product of addition of $C(CN)_2$ to benzene,[1043] have been isolated.[1044] With CH_2, insertion is a major side reaction, and, for example, benzene gives toluene as well as cycloheptatriene. A method of adding CH_2 to benzene rings without the use of free carbene is the catalytic decomposition of CH_2N_2 in the aromatic compound as solvent with CuCl or CuBr.[1045] By this method better yields of cycloheptatrienes are obtained without insertion side products. CHCl is active enough to add to benzene, but dihalocarbenes do not add to benzene or toluene, only to rings with greater electron density.

[1040]For reviews of the stereochemistry of carbene and carbenoid addition to double bonds, see Moss *Sel. Org. Transform.* **1970**, *1*, 35-88; Closs *Top Stereochem.* **1968**, *3*, 193-235. For a discussion of enantioselectivity in this reaction, see Nakamura *Pure App. Chem.* **1978**, *50*, 37.

[1041]Doering; Knox *J. Am. Chem. Soc.* **1951**, *75*, 297.

[1042]It has been detected by uv spectroscopy: Rubin *J. Am. Chem. Soc.* **1981**, *103*, 7791.

[1043]Ciganek *J. Am. Chem. Soc.* **1967**, *89*, 1454.

[1044]See, for example, Mukai; Kubota; Toda *Tetrahedron Lett.* **1967**, 3581; Maier; Heep *Chem. Ber.* **1968**, *101*, 1371; Ciganek *J. Am. Chem. Soc.* **1971**, *93*, 2207; Dürr; Kober *Tetrahedron Lett.* **1972**, 1255, 1259; Vogel; Wiedemann; Roth; Eimer; Günther *Liebigs Ann. Chem.* **1972**, *759*, 1; Bannerman; Cadogan; Gosney; Wilson *J. Chem. Soc., Chem. Commun.* **1975**, 618; Takeuchi; Kitagawa; Senzaki; Okamoto *Chem. Lett.* **1983**, 73; Kawase; Iyoda; Oda *Angew. Chem. Int. Ed. Engl.* **1987**, *26*, 559 [*Angew. Chem.* **99**, 572].

[1045]Wittig; Schwarzenbach *Liebigs Ann. Chem.* **1961**, *650*, 1; Müller; Fricke *Liebigs Ann. Chem.* **1963**, *661*, 38; Müller; Kessler; Fricke; Kiedaisch *Liebigs Ann. Chem.* **1961**, *675*, 63.

Pyrroles and indoles can be expanded, respectively, to pyridines and quinolines by treatment with halocarbenes,[1046] e.g.,

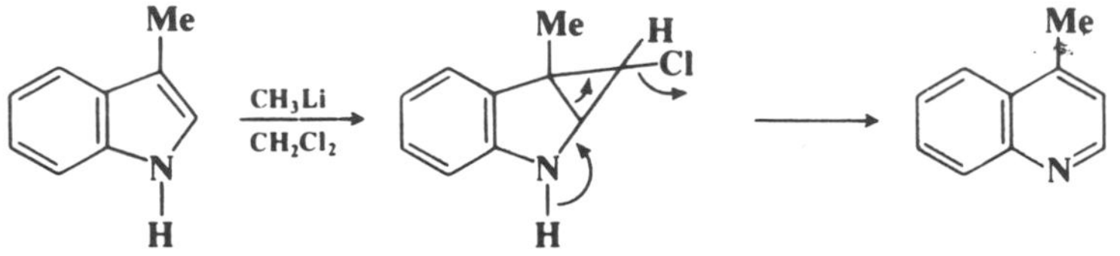

In such cases a side reaction that sometimes occurs is expansion of the *six-membered* ring. Ring expansion can occur even with nonaromatic compounds, when the driving force is supplied by relief of strain,[1047] e.g.,

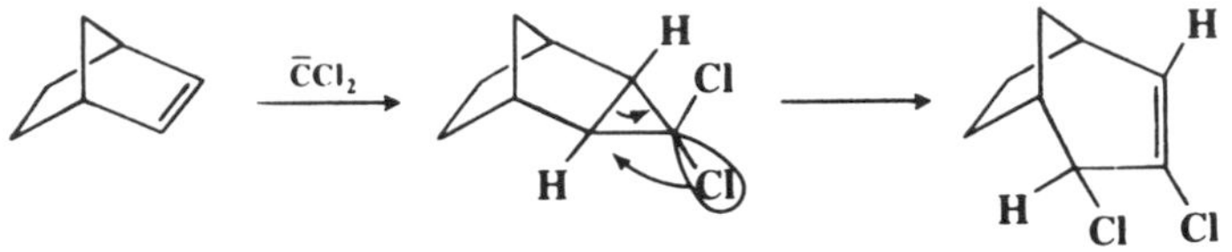

As previously mentioned, free carbene is not very useful for additions to double bonds since it gives too many side products. The *Simmons–Smith procedure* accomplishes the same result without a free carbene intermediate and without insertion side products.[1048] This procedure involves treatment of the double-bond compound with CH_2I_2 and a Zn–Cu couple and leads to cyclopropane derivatives in good yields.[1049] The Zn–Cu couple can be prepared in several ways,[1050] of which heating Zn dust with CuCl in ether under nitrogen[1051] is particularly convenient. The reaction has also been done with unactivated zinc and ultrasound.[1052] When $TiCl_4$ is used along with Zn and CuCl, CH_2I_2 can be replaced by the cheaper CH_2Br_2.[1053] The actual attacking species is an organozinc intermediate, probably $(ICH_2)_2Zn \cdot ZnI_2$. This intermediate is stable enough for solutions of it to be isolable.[1054] An x-ray crystallographic investigation of the intermediate, complexed with a diether, has been reported.[1055] The addition is stereospecifically syn, and a concerted mechanism is likely, perhaps[1056]

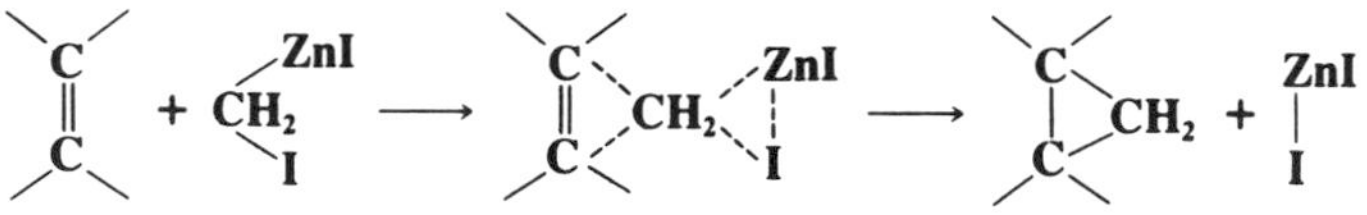

[1046]For a review of the reactions of heterocyclic compounds with carbenes, see Rees; Smithen *Adv. Heterocycl. Chem.* **1964**, *3*, 57-78.

[1047]Jefford; Gunsher; Hill; Brun; Le Gras; Waegell *Org. Synth. VI*, 142. For a review of the addition of halocarbenes to bridged bicyclic olefins see Jefford *Chimia* **1970**, *24*, 357-363.

[1048]For reviews, see Simmons; Cairns; Vladuchick; Hoiness *Org. React.* **1973**, *20*, 1-131; Furukawa; Kawabata *Adv. Organomet. Chem.* **1974**, *12*, 83-134, pp. 84-103.

[1049]Simmons; Smith *J. Am. Chem. Soc.* **1959**, *81*, 4256.

[1050]Shank; Shechter *J. Org. Chem.* **1959**, *24*, 1525; LeGoff *J. Org. Chem.* **1964**, *29*, 2048. For the use of a Zn–Ag couple, see Denis; Girard; Conia *Synthesis* **1972**, 549.

[1051]Rawson; Harrison *J. Org. Chem.* **1970**, *35*, 2057.

[1052]Repič; Vogt *Tetrahedron Lett.* **1982**, *23*, 2729; Repič; Lee; Giger *Org. Prep. Proced. Int.* **1984**, *16*, 25.

[1053]Friedrich; Lunetta; Lewis *J. Org. Chem.* **1989**, *54*, 2388.

[1054]Blanchard; Simmons *J. Am. Chem. Soc.* **1964**, *86*, 1337.

[1055]Denmark; Edwards; Wilson *J. Am. Chem. Soc.* **1991**, *113*, 723.

[1056]Simmons; Blanchard; Smith *J. Am. Chem. Soc.* **1964**, *86*, 1347.

With the Simmons–Smith procedure, as with free carbenes, conjugated dienes give 1,2 addition,[1057] and allenes give methylenecyclopropanes or spiropentanes. An alternative way of carrying out the Simmons–Smith reaction is by treatment of the substrate with CH_2I_2 or another dihalomethane and Et_2Zn in ether. This method can be adapted to the introduction of RCH and ArCH by the use of $RCHI_2$ or $ArCHI_2$ instead of the dihalomethane.[1058] In another method, CH_2I_2 or $MeCHI_2$ is used along with an alane R_3Al to transfer CH_2 or MeCH.[1059] For the conversion of enolates to cyclopropanols, CH_2I_2 has been used along with SmI_2.[1060]

Free carbenes can also be avoided by using transition metal–carbene complexes $L_nM{=}CRR'$ (L = a ligand, M = a metal), which add the group CRR' to double bonds.[1061] An example is[1062]

These complexes can be isolated in some cases; in others they are generated in situ from appropriate precursers, of which diazo compounds are among the most important. These compounds, including CH_2N_2 and others, react with metals or metal salts (copper, palladium, and rhodium are most commonly used) to give the carbene complexes that add CRR' to double bonds.[1063] Optically active complexes have been used for enantioselective cyclopropane synthesis.[1064]

The Simmons–Smith reaction has been used as the basis of a method for the indirect α methylation of a ketone.[1065] The ketone (illustrated for cyclohexanone) is first converted to an enol ether, e.g., by **6-6**, or to an enamine (**6-14**) or silyl enol ether (**2-23**). Application of the Simmons–Smith reaction gives the norcarane derivative **106**, which is then cleaved

A = OR, NR$_2$, SiR$_3$ **106**   **107**

[1057]Overberger; Halek *J. Org. Chem.* **1963**, *28*, 867.

[1058]Furukawa; Kawabata; Nishimura *Tetrahedron* **1968**, *24*, 53, *Tetrahedron Lett.* **1968**, 3495; Nishimura; Kawabata; Furukawa *Tetrahedron* **1969**, *25*, 2647; Miyano; Hashimoto *Bull. Chem. Soc. Jpn.* **1973**, *46*, 892; Friedrich; Biresaw *J. Org. Chem.* **1982**, *47*, 1615.

[1059]Maruoka; Fukutani; Yamamoto *J. Org. Chem.* **1985**, *50*, 4412, *Org. Synth.* **67**, 176.

[1060]Imamoto; Takiyama *Tetrahedron Lett.* **1987**, *28*, 1307. See also Molander; Harring *J. Org. Chem.* **1989**, *54*, 3525.

[1061]For reviews, see Helquist *Adv. Met.-Org. Chem.* **1991**, *2*, 143-194; Brookhart; Studabaker *Chem. Rev.* **1987**, *87*, 411-432; Syatkovskii; Babitskii *Russ. Chem. Rev.* **1984**, *53*, 672-682.

[1062]Brookhart; Tucker; Husk *J. Am. Chem. Soc.* **1983**, *105*, 258.

[1063]For reviews, see Adams; Spero *Tetrahedron* **1991**, *47*, 1765-1808; Collman et al., Ref. 223, pp. 800-806; Maas *Top. Curr. Chem.* **1987**, *137*, 75-253; Doyle *Chem. Rev.* **1986**, *86*, 919-939, *Acc. Chem. Res.* **1986**, *19*, 348-356; Heck, Ref. 568, pp. 401-407; Wulfman; Poling *React. Intermed. (Plenum)* **1980**, *1*, 321-512; Müller; Kessler; Zeeh *Fortschr. Chem. Forsch.* **1966**, *7*, 128-171.

[1064]Brookhart; Liu *Organometallics* **1989**, *8*, 1572, *J. Am. Chem. Soc.* **1991**, *113*, 939; Brookhart; Liu; Goldman; Timmers; Williams *J. Am. Chem. Soc.* **1991**, *113*, 927; Lowenthal; Abiko; Masamune *Tetrahedron Lett.* **1990**, *31*, 6005; Evans; Woerpel; Hinman; Faul *J. Am. Chem. Soc.* **1991**, *113*, 726. For asymmetric Simmons–Smith reactions, see Mori; Arai; Yamamoto *Tetrahedron* **1986**, *42*, 6447; Mash; Nelson; Heidt *Tetrahedron Lett.* **1987**, *28*, 1865; Sugimura; Futagawa; Yoshikawa; Tai *Tetrahedron Lett.* **1989**, *30*, 3807. See also Ojima; Clos; Bastos, Ref. 232, pp. 6919-6921.

[1065]See Wenkert; Mueller; Reardon; Sathe; Scharf; Tosi *J. Am. Chem. Soc.* **1970**, *92*, 7428 for the enol ether procedure; Kuehne; King *J. Org. Chem.* **1973**, *38*, 304 for the enamine procedure; Conia *Pure Appl. Chem.* **1975**, *43*, 317-326 for the silyl ether procedure.

(addition of water to a cyclopropane ring) to an intermediate **107,** which loses ROH, RNH$_2$, or R$_3$SiH, producing the methylated ketone. Cleavage of **106** is carried out by acid hydrolysis if A is OR, by basic hydrolysis if A is SiR$_3$,[1066] and by neutral hydrolysis in aqueous methanol if A is NR$_2$.

In another variation, phenols can be ortho-methylated in one laboratory step, by treatment with Et$_2$Zn and CH$_2$I$_2$.[1067] The following mechanism was proposed:

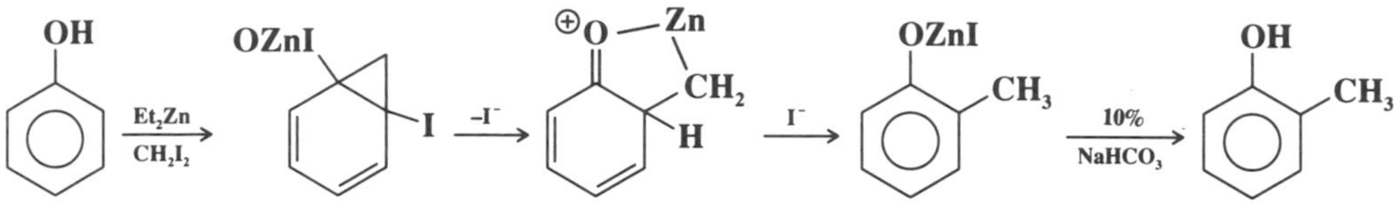

Double-bond compounds that undergo the Michael reaction (**5-17**) can be converted to cyclopropane derivatives with sulfur ylides.[1068] Among the most common of these is dimethyloxosulfonium methylide (**108**),[1069] which is widely used to transfer CH$_2$ to activated

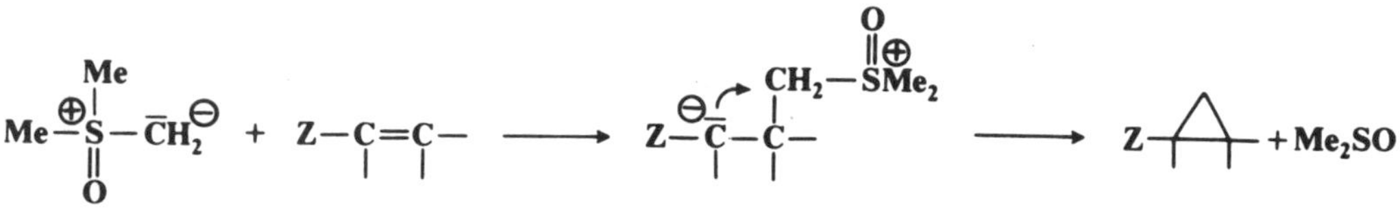

108

double bonds, but other sulfur ylides, e.g., **109** (A = acyl,[1070] carbethoxy[1071]), **110,**[1072] and **111,**[1073] which transfer CHA, CH—vinyl, and CMe$_2$, respectively, have also been used. CHR

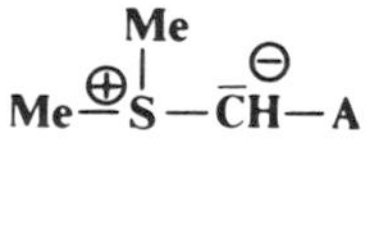

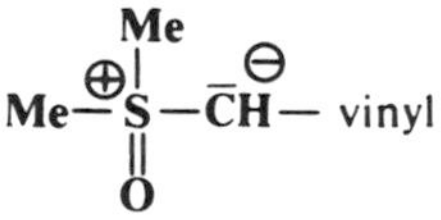

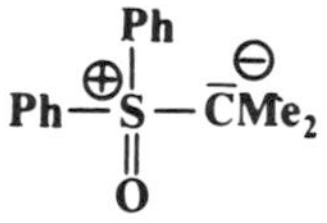

and CR$_2$ can be added in a similar manner with certain nitrogen-containing compounds. For example, the ylides[1074] **112** and **113** and the carbanion **114** can be used, respectively, to add

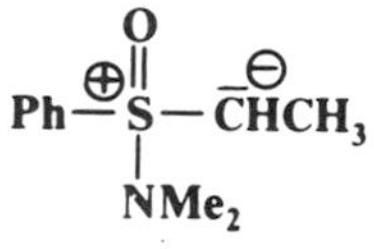

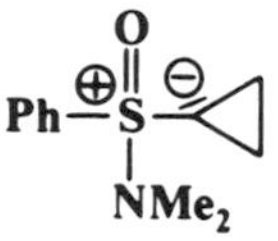

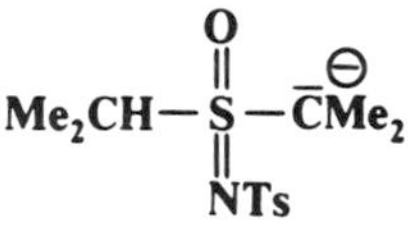

[1066]In the case of silyl enol ethers the inner bond can be cleaved with FeCl$_3$, giving a ring-enlarged β-chloro ketone: Ito; Fujii; Saegusa *J. Org. Chem.* **1976,** *41,* 2073; *Org. Synth. VI.* 327.

[1067]Lehnert; Sawyer; Macdonald *Tetrahedron Lett.* **1989,** *30,* 5215.

[1068]For a monograph and reviews on sulfur ylides, see Chapter 2, Ref. 53.

[1069]Truce; Badiger *J. Org. Chem.* **1964,** *29,* 3277; Corey; Chaykovsky *J. Am. Chem. Soc.* **1965,** *87,* 1353; Agami; Prevost *Bull. Soc. Chim. Fr.* **1967,** 2299. For a review of this reagent, see Gololobov; Nesmeyanov; Lysenko; Boldeskul *Tetrahedron* **1987,** *43,* 2609-2651.

[1070]Trost *J. Am. Chem. Soc.* **1967,** *89,* 138. See also Nozaki; Takaku; Kondô *Tetrahedron* **1966,** *22,* 2145.

[1071]Payne *J. Org. Chem.* **1967,** *32,* 3351.

[1072]LaRochelle; Trost; Krepski *J. Org. Chem.* **1971,** *36,* 1126; Marino; Kaneko *Tetrahedron Lett.* **1973,** 3971, 3975.

[1073]Corey; Jautelat *J. Am. Chem. Soc.* **1967,** *89,* 3912.

[1074]For a review of sulfoximides R$_2$S(O)NR$_2$ and ylides derived from them, see Kennewell; Taylor *Chem. Soc. Rev.* **1980,** *9,* 477-498.

CHMe, cyclopropylidene, and CMe_2 to activated double bonds.[1075] Similar reactions have been performed with phosphorus ylides,[1076] with pyridinium ylides,[1077] and with the compounds $(PhS)_3CLi$ and $Me_3Si(PhS)_2CLi$.[1078] The reactions with ylides are of course nucleophilic addition.

OS **V**, 306, 855, 859, 874; **VI**, 87, 142, 187, 327, 731, 913, 974; **VII**, 12, 200, 203; **67**, 176; **68**, 220; **69**, 144, 180.

5-51 Trimerization and Tetramerization of Alkynes

When acetylene is heated with nickel cyanide, other Ni(II) or Ni(0) compounds, or similar catalysts, it gives benzene and cyclooctatetraene.[1079] It is possible to get more of either product by a proper choice of catalyst. Substituted acetylenes give substituted benzenes. This reaction has been used to prepare very crowded molecules. Diisopropylacetylene was trimerized over $Co_2(CO)_8$ and over $Hg[Co(CO)_4]_2$ to hexaisopropylbenzene.[1080] The six isopropyl groups are not free to rotate but are lined up perpendicular to the plane of the benzene ring. Even more interesting was the *spontaneous* (no catalyst) trimerization of *t*-BuC≡CF to give 1,2,3-tri-*t*-butyl-4,5,6-trifluorobenzene (**116**), the first time three adjacent *t*-butyl groups had been put onto a benzene ring.[1081] The fact that this is a head-to-head joining makes the following sequence likely:

R = *t*-butyl   **115**   **116**

The fact that **115** (a dewar benzene) was also isolated lends support to this scheme.[1082]

[1075]For reviews, see Johnson *Aldrichimica Acta* **1985**, *18*, 1-10, *Acc. Chem. Res.* **1973**, *6*, 341-347; Kennewell; Taylor *Chem. Soc. Rev.* **1975**, *4*, 189-209; Trost *Acc. Chem. Res.* **1974**, *7*, 85-92.

[1076]Bestmann; Seng *Angew. Chem. Int. Ed. Engl.* **1962**, *1*, 116 [*Angew. Chem. 74*, 154]; Grieco; Finkelhor *Tetrahedron Lett.* **1972**, 3781.

[1077]Shestopalov; Sharanin; Litvinov; Nefedov *J. Org. Chem. USSR* **1989**, *25*, 1000.

[1078]Cohen; Myers *J. Org. Chem.* **1988**, *53*, 457.

[1079]For reviews, see Winter, in Hartley; Patai, Ref. 422, vol. 3, pp. 259-294; Vollhardt *Angew. Chem. Int. Ed. Engl.* **1984**, *23*, 539-556 [*Angew. Chem. 96*, 525-541], *Acc. Chem. Res.* **1977**, *10*, 1-8; Maitlis *J. Organomet. Chem.* **1980**, *200*, 161-176, *Acc. Chem. Res.* **1976**, *9*, 93-99, *Pure Appl. Chem.* **1972**, *30*, 427-448; Yur'eva *Russ. Chem. Rev.* **1974**, *43*, 48-68; Khan; Martell, Ref. 159, pp. 163-168; Reppe; Kutepow; Magin *Angew. Chem. Int. Ed. Engl.* **1969**, *8*, 727-733 [*Angew. Chem. 81*, 717-723]; Fuks; Viehe, in Viehe, Ref. 49, pp. 450-460; Hoogzand; Hübel, in Wender; Pino *Organic Syntheses Via Metal Carbonyls*, vol. 1; Wiley: New York, 1968, pp. 343-371; Reikhsfel'd; Makovetskii *Russ. Chem. Rev.* **1966**, *35*, 510-523. For a list of reagents, with references, see Ref. 133, pp. 100-101. For a review of metal-catalyzed cycloadditions of alkynes to give rings of all sizes, see Schore *Chem. Rev.* **1988**, *88*, 1081-1119.

[1080]Arnett; Bollinger *J. Am. Chem. Soc.* **1964**, *86*, 4729; Hopff *Chimia* **1964**, *18*, 140; Hopff; Gati *Helv. Chim. Acta* **1965**, *48*, 509.

[1081]Viehe; Merényi; Oth; Valange *Angew. Chem. Int. Ed. Engl.* **1964**, *3*, 746 [*Angew. Chem. 76*, 888]; Viehe; Merényi; Oth; Senders; Valange *Angew. Chem. Int. Ed. Engl.* **1964**, *3*, 755 [*Angew. Chem. 76*, 923].

[1082]For other reactions between cyclobutadienes and triple bonds to give dewar benzenes, see Wingert; Regitz *Chem. Ber.* **1986**, *119*, 244.

In contrast to the spontaneous reaction, the catalyzed process seldom gives the 1,2,3-trisubstituted benzene isomer from an acetylene $RC\equiv CH$. The chief product is usually the 1,2,4-isomer, with lesser amounts of the 1,3,5-isomer also generally obtained, but little if any of the 1,2,3-isomer. The mechanism of the catalyzed reaction to form benzenes[1083] is believed to go through a species **117** in which two molecules of alkyne coordinate with the

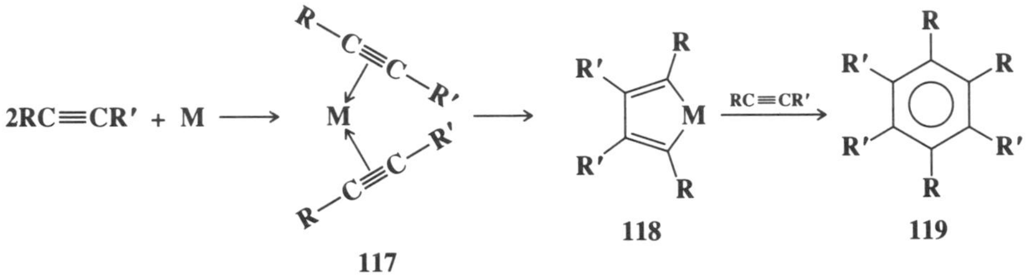

$$2RC\equiv CR' + M \longrightarrow \quad \mathbf{117} \quad \longrightarrow \quad \mathbf{118} \quad \xrightarrow{RC\equiv CR'} \quad \mathbf{119}$$

metal, and another species **118**, a five-membered heterocyclic intermediate.[1084] Such intermediates (where M = Rh, Ir, or Ni) have been isolated and shown to give benzenes when treated with alkynes.[1085] Note that this pathway accounts for the predominant formation of the 1,2,4-isomer. Two possiblities for the last step are a Diels–Alder reaction, and a ring expansion, each followed by extrusion of the metal:[1085a]

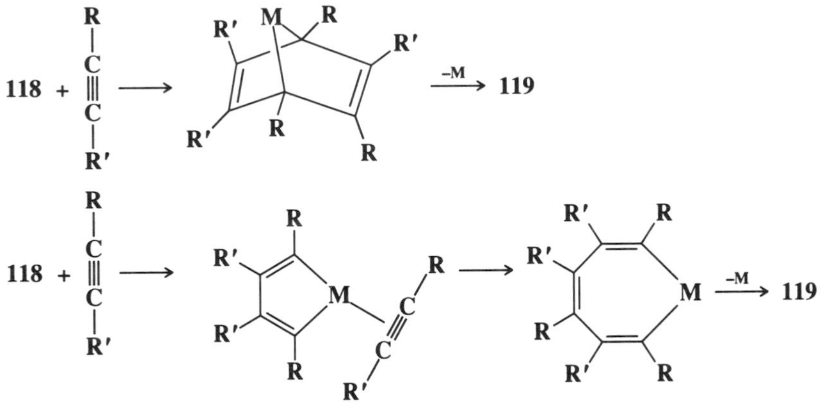

$$\mathbf{118} + \quad \longrightarrow \quad \xrightarrow{-M} \quad \mathbf{119}$$

$$\mathbf{118} + \quad \longrightarrow \quad \longrightarrow \quad \xrightarrow{-M} \quad \mathbf{119}$$

In at least one case the mechanism is different, going through a cyclobutadiene–nickel complex (p. 55), which has been isolated.[1086]

When benzene, in the gas phase, was adsorbed onto a surface of 10% rhodium-on-alumina, the reverse reaction took place, and acetylene was formed.[1087]

[1083]For studies of the mechanism of the reaction that produces cyclooctatetraenes, see Diercks; Stamp; Kopf; tom Dieck *Angew. Chem. Int. Ed. Engl.* **1984**, *23*, 893 [*Angew. Chem. 96*, 891]; Colborn; Vollhardt *J. Am. Chem. Soc.* **1986**, *108*, 5470; Lawrie; Gable; Carpenter *Organometallics* **1989**, *8*, 2274.

[1084]See, for example, Colborn; Vollhardt *J. Am. Chem. Soc.* **1981**, *103*, 6259; Kochi *Organometallic Mechanisms and Catalysis*; Academic Press: New York, 1978, pp. 428-432; Collman et al., Ref. 223, pp. 870-877; Eisch; Sexsmith *Res. Chem. Intermed.* **1990**, *13*, 149-192.

[1085]See, for example, Collman; Kang *J. Am. Chem. Soc.* **1967**, *89*, 844; Collman *Acc. Chem. Res.* **1968**, *1*, 136-143; Yamazaki; Hagihara *J. Organomet. Chem.* **1967**, *7*, P22; Wakatsuki; Kuramitsu; Yamazaki *Tetrahedron Lett.* **1974**, 4549; Moseley; Maitlis *J. Chem. Soc., Dalton Trans.* **1974**, 169; Müller *Synthesis* **1974**, 761-774; Eisch; Galle *J. Organomet. Chem.* **1975**, *96*, C23; McAlister; Bercaw; Bergman *J. Am. Chem. Soc.* **1977**, *99*, 1666.

[1085a]There is evidence that the mechanism of the last step more likely resembles the Diels-Alder pathway than the ring expansion pathway: Bianchini et al. *J. Am. Chem. Soc.* **1991**, *113*, 5127.

[1086]Mauret; Alphonse *J. Organomet. Chem.* **1984**, *276*, 249. See also Pepermans; Willem; Gielen; Hoogzand *Bull. Soc. Chim. Belg.* **1988**, *97*, 115.

[1087]Parker; Hexter; Siedle *J. Am. Chem. Soc.* **1985**, *107*, 4584.

For addition of triple bonds to triple bonds, but not with ring formation, see **5-15**.
OS **VII**, 256.

5-52 Other Cycloaddition Reactions
cyclo-[But-2-en-1,4-diyl]-1/4/addition, etc.

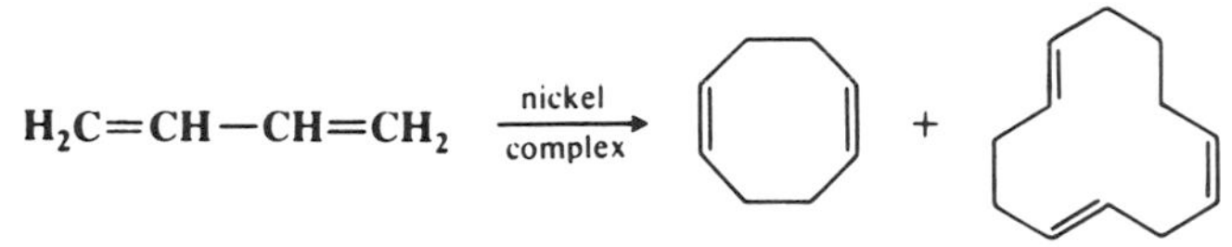

$$H_2C{=}CH{-}CH{=}CH_2 \xrightarrow[\text{complex}]{\text{nickel}} \quad + \quad$$

Conjugated dienes can be dimerized or trimerized at their 1,4 positions (formally, 4 + 4
and 4 + 4 + 4 cycloadditions) by treatment with certain complexes or other transition-metal compounds.[1088] Thus butadiene gives 1,5-cyclooctadiene and 1,5,9-cyclodo-
decatriene.[1089] The relative amount of each product can be controlled by use of the
proper catalyst. For example, Ni:P(OC$_6$H$_4$-*o*-Ph)$_3$ gives predominant dimerization, while
Ni(cyclooctadiene)$_2$ gives mostly trimerization. The products arise, not by direct 1,4 to 1,4
attack, but by stepwise mechanisms involving metal–olefin complexes.[1090]

As we saw in **5-47**, the Woodward–Hoffmann rules allow suprafacial concerted cycloadditions to take place thermally if the total number of electrons is $4n + 2$ and photochemically
if the number is $4n$. Furthermore, forbidden reactions become allowed if one molecule reacts
antarafacially. It would thus seem that syntheses of many large rings could easily be achieved.
However, when the newly formed ring is eight-membered or greater, concerted mechanisms,
though allowed by orbital symmetry for the cases stated, become difficult to achieve because
of the entropy factor (the two ends of one system must simultaneously encounter the two
ends of the other), unless one or both components are cyclic, in which case the molecule
has many fewer possible conformations. There have been a number of reports of cycloaddition reactions leading to eight-membered and larger rings, some thermally and some
photochemically induced, but (apart from the dimerization and trimerization of butadienes
mentioned above, which are known not to involve direct 4 + 4 or 4 + 4 + 4 cycloaddition)
in most cases evidence is lacking to indicate whether they are concerted or stepwise processes.

Some examples are

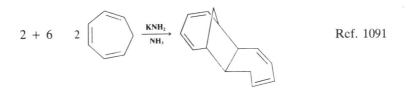

$$2 + 6 \qquad 2 \qquad \xrightarrow[\text{NH}_3]{\text{KNH}_2} \qquad\qquad\qquad\qquad \text{Ref. 1091}$$

[1088]For reviews, see Wilke *Angew. Chem. Int. Ed. Engl.* **1988,** 27, 186-206 [*Angew. Chem. 100*, 189-211], *J.
Organomet. Chem.* **1980,** 200, 349-364; Tolstikov; Dzhemilev *Sov. Sci. Rev., Sect. B* **1985,** 7, 237-295, pp. 278-290;
Heimbach; Schenkluhn *Top Curr. Chem.* **1980,** 92, 45-108; Heimbach *Angew. Chem. Int. Ed. Engl.* **1973,** 12, 975-
989 [*Angew. Chem. 85*, 1035-1049]; Baker *Chem. Rev.* **1973,** 73, 487-530, pp. 489-512; Semmelhack *Org. React.* **1972,**
19, 115-198, pp. 128-143; Heimbach; Jolly; Wilke *Adv. Organomet. Chem.* **1970,** 8, 29-86, pp. 48-83; Khan; Martell,
Ref. 159, pp. 159-163; Heck, Ref. 223, pp. 157-164.
[1089]For a review of the 1,5,9-cyclododecatrienes (there are four stereoisomers, of which the *ttt* is shown above),
see Rona *Intra-Sci. Chem. Rep.* **1971,** 5, 105-148.
[1090]For example, see Heimbach; Wilke *Liebigs Ann. Chem.* **1969,** 727, 183; Barnett; Büssemeier; Heimbach; Jolly;
Krüger; Tkatchenko; Wilke *Tetrahedron Lett.* **1972,** 1457; Barker; Green; Howard; Spencer; Stone *J. Am. Chem.
Soc.* **1976,** 98, 3373; Graham; Stephenson *J. Am. Chem. Soc.* **1977,** 99, 7098.
[1091]Staley; Orvedal *J. Am. Chem. Soc.* **1974,** 96, 1618. In this case the reagent converted one molecule of cycloheptatriene to the cycloheptatrienyl anion (p. 46), which then added stepwise to the other molecule.

$4 + 4$ Ref. 1092

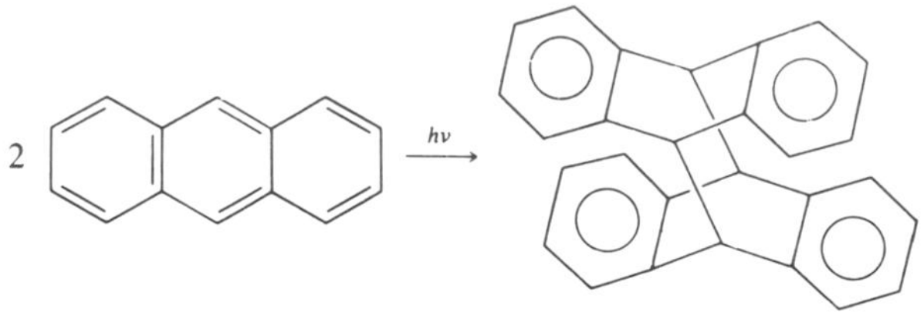

$2 + 8$ Ref. 1093

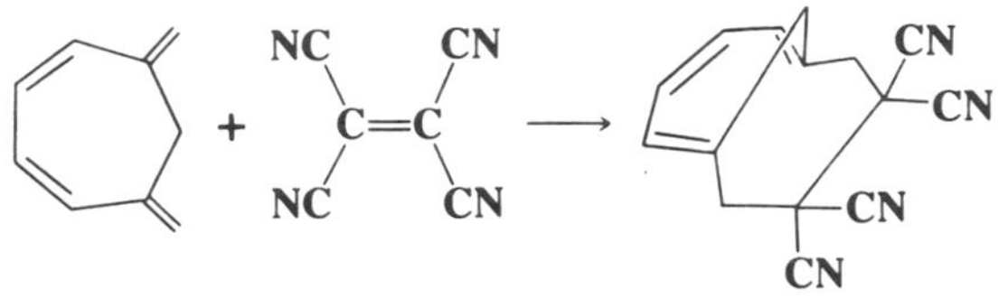

$4 + 6$ Ref. 1094

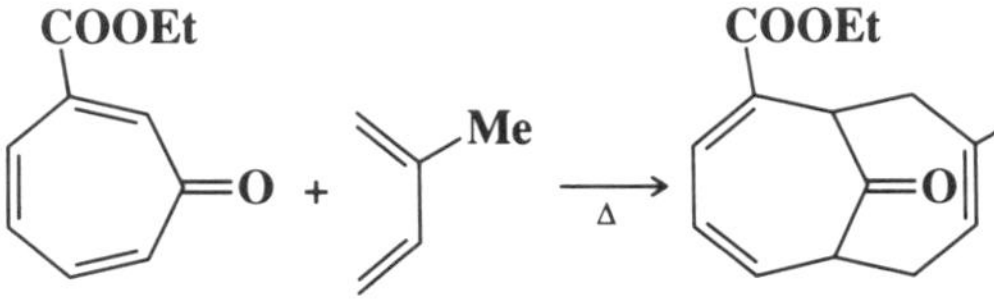

The suprafacial thermal addition of an allylic cation to a diene (a 3 + 4 cycloaddition) is allowed by the Woodward–Hoffmann rules (this reaction would be expected to follow the same rules as the Diels–Alder reaction[1095]). Such cycloadditions can be carried out[1096] by treatment of a diene with an allylic halide in the presence of a suitable silver salt, e.g.,[1097]

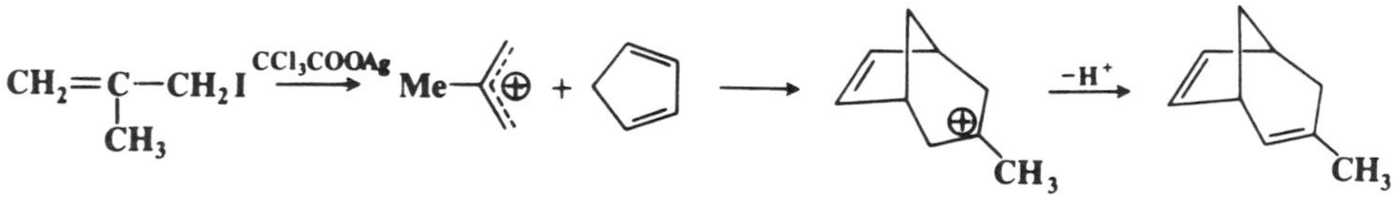

This reaction has even been carried out with benzene assuming the role of the diene.[1098]

Benzene rings can undergo photochemical cycloaddition with olefins.[1099] The major product is usually the 1,3 addition product **120** (in which a 3-membered ring has also been formed), though some of the 1,2 product **121** (**5-49**) is sometimes formed as well. (**121** is

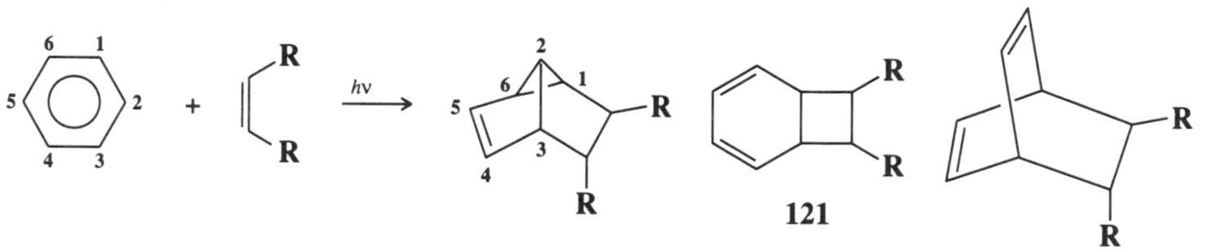

120   **121**   **122**

[1092]Shönberg, Ref. 49, pp. 97-99.
[1093]Farrant; Feldmann *Tetrahedron Lett.* **1970**, 4979.
[1094]Garst; Roberts; Houk; Rondan *J. Am. Chem. Soc.* **1984**, *106*, 3882.
[1095]Because the HOMO of an allylic cation, p. 33, has the same symmetry as the HOMO of an alkene.
[1096]For reviews of 3 + 4 cycloadditions see Mann *Tetrahedron* **1986**, *42*, 4611-4659; Hoffmann *Angew. Chem. Int. Ed. Engl.* **1984**, *23*, 1-19, **1973**, *12*, 819-835 [*Angew. Chem. 96*, 29-48, *85*, 877-894]; Noyori *Acc. Chem. Res.* **1979**, *12*, 61-66.
[1097]Hoffmann; Joy; Suter *J. Chem. Soc. B* **1968**, 57.
[1098]Hoffmann; Hill *Angew. Chem. Int. Ed. Engl.* **1974**, *13*, 136 [*Angew. Chem. 86*, 127].
[1099]For reviews, see Wender; Ternansky; deLong; Singh; Olivero; Rice *Pure Appl. Chem.* **1990**, *62*, 1597-1602; Wender; Siggel; Nuss *Org. Photochem.* **1989**, *10*, 357-473; Gilbert, in Horspool, Ref. 966, pp. 1-60. For a review of this and related reactions, see McCullough *Chem. Rev.* **1987**, *87*, 811-860.

usually the main product where the olefin bears electron-withdrawing groups and the aromatic compound electron-donating groups, or vice versa.) The 1,4 product **122** is rarely formed. The reaction has also been run with benzenes substituted with alkyl, halo, OR, CN, and other groups, and with acyclic and cyclic olefins bearing various groups.[1100]

OS **VI**, 512; **VII**, 485.

5-53   The Addition of Two Alkyl Groups to an Alkyne
Dialkyl-addition

$$RC{\equiv}CH + R'Cu{\cdot}MgBr_2 + R''I \xrightarrow[\text{ether-HMPA}]{(EtO)_3P} \begin{array}{c} R \\ \diagdown \\ C{=}C \\ \diagup \qquad \diagdown \\ R' \qquad\quad R'' \end{array} \begin{array}{c} H \\ \diagup \end{array}$$

Two different alkyl groups can be added to a terminal alkyne[1101] in one laboratory step by treatment with an alkylcopper–magnesium bromide reagent (called *Normant reagents*)[1102] and an alkyl iodide in ether–HMPA containing triethyl phosphite.[1103] The groups add stereoselectively syn. The reaction, which has been applied to primary[1104] R' and to primary, allylic, benzylic, vinylic, and α-alkoxyalkyl R'', involves initial addition of the alkylcopper reagent,[1105] followed by a coupling reaction (**0-87**):

$$RC{\equiv}CH \xrightarrow{R'Cu{\cdot}MgBr_2} \begin{array}{c} R \qquad\quad H \\ \diagdown \qquad \diagup \\ C{=}C \\ \diagup \qquad \diagdown \\ R' \qquad Cu, MgBr_2 \end{array} \xrightarrow{R''I} \begin{array}{c} R \qquad H \\ \diagdown \quad \diagup \\ C{=}C \\ \diagup \quad \diagdown \\ R' \qquad R'' \end{array}$$

$$\text{\textbf{123}}$$

Acetylene itself (R = H) undergoes the reaction with R_2CuLi instead of the Normant reagent.[1106] The use of R' containing functional groups has been reported.[1107] If the alkyl iodide is omitted, the vinylic copper intermediate **123** can be converted to a carboxylic acid by the addition of CO_2 (see **6-34**) or to an amide by the addition of an isocyanate

$$\begin{array}{c} R \qquad H \\ \diagdown \quad \diagup \\ C{=}C \\ \diagup \quad \diagdown \\ R' \qquad CONHR'' \end{array} \xleftarrow[\text{P(OEt)}_3, \text{HMPA}]{RNCO} \begin{array}{c} R \qquad H \\ \diagdown \quad \diagup \\ C{=}C \\ \diagup \quad \diagdown \\ R' \qquad Cu, MgBr_2 \end{array} \xrightarrow[\text{P(OEt)}_3, \text{HMPA}]{CO_2} \begin{array}{c} R \qquad H \\ \diagdown \quad \diagup \\ C{=}C \\ \diagup \quad \diagdown \\ R' \qquad COOH \end{array}$$

(see **6-36**), in either case in the presence of HMPA and a catalytic amount of triethyl phosphite.[1108] The use of I_2 results in a vinylic iodide.[1109]

[1100]See the table in Wender; Siggel; Nuss, Ref. 1099, pp. 384-415.

[1101]For reviews of this and related reactions, see Raston; Salem, in Hartley, Ref. 218, vol. 4, pp. 159-306, pp. 233-248; Normant; Alexakis *Synthesis* **1981**, 841-870; Hudrlik; Hudrlik, in Patai, Ref. 70, pt. 1 pp. 233-238. For a list of reagents and references for this and related reactions, see Ref. 133, pp. 233-238.

[1102]For the composition of these reagents see Ashby; Smith; Goel *J. Org. Chem.* **1981**, *46*, 5133; Ashby; Goel *J. Org. Chem.* **1983**, *48*, 2125.

[1103]Normant; Cahiez; Chuit; Alexakis; Villieras *J. Organomet. Chem.* **1972**, *40*, C49; Alexakis; Cahiez; Normant; Villieras *Bull. Soc. Chim. Fr.* **1977**, 693; Gardette; Alexakis; Normant *Tetrahedron* **1985**, *41*, 5887. For an extensive list of references see Marfat; McGuirk; Helquist *J. Org. Chem.* **1979**, *44*, 3888.

[1104]For a method of using secondary and tertiary R, see Rao; Periasamy *Tetrahedron Lett.* **1988**, *29*, 4313.

[1105]The initial product, **123**, can be hydrolyzed with acid to give RR'C=CH_2. See Westmijze; Kleijn; Meijer; Vermeer *Recl. Trav. Chim. Pays-Bas* **1981**, *100*, 98, and references cited therein.

[1106]Alexakis; Normant; Villieras *Tetrahedron Lett.* **1976**, 3461; Alexakis; Cahiez; Normant *Synthesis* **1979**, 826, *Tetrahedron* **1980**, *36*, 1961; Furber; Taylor; Burford *J. Chem. Soc., Perkin Trans. 1* **1986**, 1809.

[1107]Rao; Knochel *J. Am. Chem. Soc.* **1991**, *113*, 5735.

[1108]Normant; Cahiez; Chuit; Villieras *J. Organomet. Chem.* **1973**, *54*, C53.

[1109]Alexakis; Cahiez; Normant *Org. Synth. VII*, 290.

Similar reactions, in which two alkyl groups are added to a triple bond, have been carried out with trialkylalanes R_3Al, with zirconium complexes as catalysts.[1110]

Allylic zinc bromides add to vinylic Grignard and lithium reagents to give the *gem-*

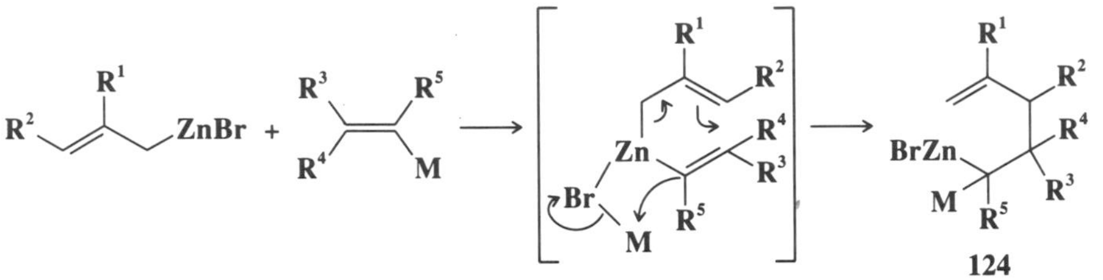

dimetallo compounds **124**. The two metallo groups can be separately reacted with various nucleophiles.[1111]

OS **VII**, 236, 245, 290.

5-54 Dicarbalkoxylation of Olefins and Acetylenes
Dicarboxy-addition

$$\underset{\displaystyle |\quad|}{C\!\!=\!\!C} + CO + ROH \xrightarrow[\text{HgCl}_2]{\text{PdCl}_2} ROOC\!\!-\!\!\underset{\displaystyle |}{C}\!\!-\!\!\underset{\displaystyle |}{C}\!\!-\!\!COOR$$

Alkenes can be converted to succinic esters by reaction with carbon monoxide, an alcohol, and palladium chloride in the presence of mercuric chloride.[1112] The addition is mostly syn. In similar reaction, both terminal and internal alkynes can be converted to esters of maleic acid.

5-55 The Conversion of Diphenylacetylene to a Butadiene
Dimethylene-biaddition

$$PhC\!\!\equiv\!\!CPh + CH_3SOCH_2^{\ominus}Na^+ \longrightarrow Ph\!\!-\!\!\underset{\displaystyle \underset{CH_2}{\|}}{C}\!\!-\!\!\underset{\displaystyle \underset{CH_2}{\|}}{C}\!\!-\!\!Ph$$

Diphenylacetylene reacts with methylsulfinyl carbanion to give 2,3-diphenylbutadiene.[1113] Neither the scope nor the mechanism of the reaction seems to have been investigated.

OS **VI**, 531.

[1110]Yoshida; Negishi *J. Am. Chem. Soc.* **1981**, *103*, 4985; Rand; Van Horn; Moore; Negishi *J. Org. Chem.* **1981**, *46*, 4093; Negishi; Van Horn; Yoshida *J. Am. Chem. Soc.* **1985**, *107*, 6639. For reviews, see Negishi *Acc. Chem. Res.* **1987**, *20*, 65-72, *Pure Appl. Chem.* **1981**, *53*, 2333-2356; Negishi; Takahashi *Aldrichimica Acta* **1985**, *18*, 31-47.
[1111]Knochel; Normant *Tetrahedron Lett.* **1986**, *27*, 1039, 1043, 4427, 4431, 5727.
[1112]Heck *J. Am. Chem. Soc.* **1972**, *94*, 2712. See also Fenton; Steinwand *J. Org. Chem.* **1972**, *37*, 2034; Stille; Divakaruni *J. Org. Chem.* **1979**, *44*, 3474; Catellani; Chiusoli; Peloso *Tetrahedron Lett.* **1983**, *24*, 813; Deprés; Coelho; Greene *J. Org. Chem.* **1985**, *50*, 1972.
[1113]Iwai; Ide *Org. Synth. VI*, 531.

16

ADDITION TO CARBON–HETERO MULTIPLE BONDS

MECHANISM AND REACTIVITY

The reactions considered in this chapter involve addition to the carbon–oxygen, carbon–nitrogen, and carbon–sulfur double bonds and the carbon–nitrogen triple bond. The mechanistic study of these reactions is much simpler than that of the additions to carbon–carbon multiple bonds considered in Chapter 15.[1] Most of the questions that concerned us there either do not arise here or can be answered very simply. Since C=O, C=N, and C≡N bonds are strongly polar, with the carbon always the positive end (except for isocyanides, see p. 979), there is never any doubt about the *orientation* of unsymmetrical addition to these bonds. Nucleophilic attacking species always go to the carbon and electrophilic ones to the oxygen or nitrogen. Additions to C=S bonds are much less common,[2] but in these cases the addition can be in the other direction.[3] For example, thiobenzophenone Ph$_2$C=S, when treated with phenyllithium gives, after hydrolysis, benzhydryl phenyl sulfide Ph$_2$CHSPh.[4] The *stereochemistry* of addition is not generally a factor because it is not normally possible to determine whether the addition is syn or anti. In addition of YH to a ketone, e.g.,

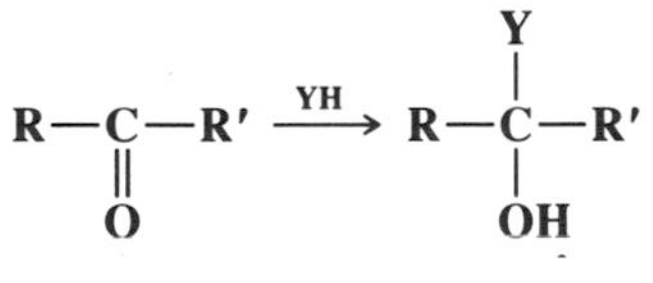

$$R - \overset{\displaystyle ||}{\underset{\displaystyle O}{C}} - R' \xrightarrow{\text{YH}} R - \overset{\displaystyle Y}{\underset{\displaystyle OH}{\overset{|}{\underset{|}{C}}}} - R'$$

the product has a chiral carbon, but unless there is chirality in R or R' or YH is optically active, the product must be a racemic mixture and there is no way to tell from its steric nature whether the addition of Y and H was syn or anti. The same holds true for C=N and C=S bonds, since in none of these cases can chirality be present at the hetero atom. The stereochemistry of addition of a single YH to the carbon–nitrogen triple bond could be investigated, since the product can exist in *E* and *Z* forms (p. 127), but these reactions are not very important. Of course, if R or R' *is* chiral, a racemic mixture will not always arise

---

[1]For a discussion, see Jencks *Prog. Phys. Org. Chem.* **1964**, *2*, 63-118.

[2]For reviews of thioketones and other compounds with C=S bonds, see Schaumann, in Patai *Supplement A: The Chemistry of Double-bonded Functional Groups*, vol. 2, pt. 2; Wiley: New York, 1989, pp. 1269-1367; Ohno, in Oae *Organic Chemistry of Sulfur*; Plenum: New York, 1977, pp. 189-229; Mayer, in Janssen *Organosulfur Chemistry*; Wiley: New York, 1967, pp. 219-240; Campaigne, in Patai *The Chemistry of the Carbonyl Group*, pt. 1; Wiley: New York, 1966, pp. 917-959.

[3]For a review of additions of organometallic compounds to C=S bonds, both to the sulfur (*thiophilic addition*) and to the carbon (*carbophilic addition*), see Wardell; Paterson, in Hartley; Patai *The Chemistry of the Metal–Carbon Bond*, vol. 2; Wiley: New York, 1985, pp. 219-338, pp. 261-267.

[4]Beak; Worley *J. Am. Chem. Soc.* **1972**, *94*, 597. For some other examples, see Schaumann; Walter *Chem. Ber.* **1974**, *107*, 3562; Metzner; Vialle; Vibet *Tetrahedron* **1978**, *34*, 2289.

and the stereochemistry of addition can be studied in such cases. Cram's rule (p. 117) allows us to predict the direction of attack of Y in many cases.[5] However, even in this type of study, the relative directions of attack of Y and H are not determined, but only the direction of attack of Y with respect to the rest of the substrate molecule.

On p. 754 it was mentioned that electronic effects can play a part in determining which face of a carbon–carbon double bond is attacked. The same applies to additions to carbonyl groups. For example, in 5-substituted adamantanones:

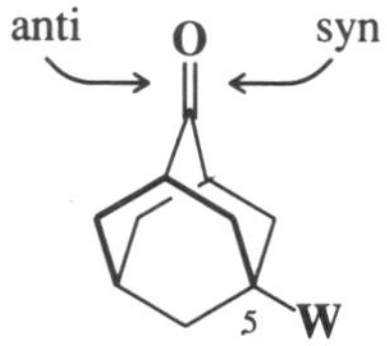

electron-withdrawing $(-I)$ groups W cause the attack to come from the syn face, while electron-donating groups cause it to come from the anti face.[6]

The mechanistic picture is further simplified by the fact that free-radical additions to carbon-hetero double bonds are rare.[7] The principal question remaining is which attacks first, the nucleophile or electrophile. In most cases it is the nucleophile, and these reactions are regarded as *nucleophilic additions*, which can be represented thus (for the C=O bond, analogously for the others):

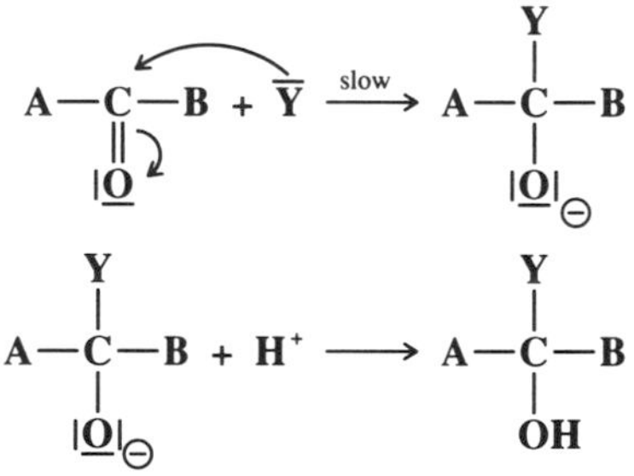

The electrophile shown in step 2 is the proton. In almost all the reactions considered in this chapter the electrophilic attacking atom is either hydrogen or carbon. It may be noted that step 1 is exactly the same as step 1 of the tetrahedral mechanism of nucleophilic substitution at a carbonyl carbon (p. 331), and it might be expected that substitution would compete with addition. However, this is seldom the case. When A and B are H, R, or Ar, the substrate is an aldehyde or ketone and these almost never undergo substitution, owing to the extremely poor nature of H, R, and Ar as leaving groups. For carboxylic acids and their

---

[5]For a discussion of such rules, see Eliel *The Stereochemistry of Carbon Compounds*; McGraw-Hill: New York, 1962, pp. 68-74. For reviews of the stereochemistry of addition to carbonyl compounds, see Bartlett *Tetrahedron* **1980,** *36,* 2-72, pp. 22-28; Ashby; Laemmle *Chem. Rev.* **1975,** *75,* 521-546; Goller *J. Chem. Educ.* **1974,** *51,* 182-185; Toromanoff *Top. Stereochem.* **1967,** *2,* 157-198.

[6]Cheung; Tseng; Lin; Srivastava; le Noble *J. Am. Chem. Soc.* **1986,** *108,* 1598; Laube; Stilz *J. Am. Chem. Soc.* **1987,** *109,* 5876.

[7]An example is found in **6-35.** For other examples, see Kaplan *J. Am. Chem. Soc.* **1966,** *88,* 1833; Drew; Kerr *Int. J. Chem. Kinet.* **1983,** *15,* 281; Fraser-Reid; Vite; Yeung; Tsang *Tetrahedron Lett.* **1988,** *29,* 1645; Beckwith; Hay *J. Am. Chem. Soc.* **1989,** *111,* 2674; Clerici; Porta *J. Org. Chem.* **1989,** *54,* 3872; Cossy; Pete; Portella *Tetrahedron Lett.* **1989,** *30,* 7361.

derivatives (B = OH, OR, NH$_2$, etc.) addition is seldom found, because these are much better leaving groups. It is thus the nature of A and B that determines whether a nucleophilic attack at a carbon-hetero multiple bond will lead to substitution or addition.

As is the case in the tetrahedral mechanism, it is also possible for the electrophilic species to attack first, in which case it goes to the hetero atom. This species is most often a proton and the mechanism is

Step 1

$$A-\underset{\underset{|\underline{O}|}{\overset{\|}{C}}}{C}-B \; + \; H^+ \; \xrightarrow{\;fast\;} \; A-\overset{\oplus}{\underset{\underset{OH}{|}}{C}}-B$$

1

Step 2

$$A-\overset{\oplus}{\underset{\underset{OH}{|}}{C}}-B \; + \; \overline{Y} \; \xrightarrow{\;slow\;} \; A-\underset{\underset{OH}{|}}{\overset{\overset{Y}{|}}{C}}-B$$

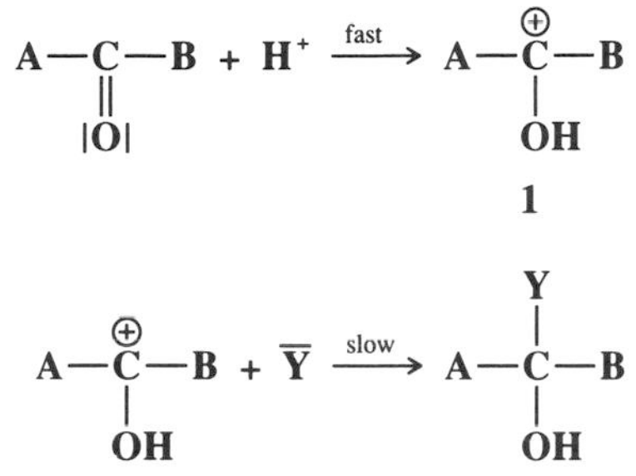

No matter which species attacks first, the rate-determining step is usually the one involving nucleophilic attack. It may be observed that many of these reactions can be catalyzed by both acids and bases.[8] Bases catalyze the reaction by converting a reagent of the form YH to the more powerful nucleophile Y$^-$ (see p. 348). Acids catalyze it by converting the substrate to an ion (e.g., **1**) in which the positive charge on the carbon is greatly increased, thus making it more attractive to nucleophilic attack. Similar catalysis can also be found with metallic ions, such as Ag$^+$, which act here as Lewis acids.[9] We have mentioned before (p. 170) that ions of type **1** are comparatively stable carbocations because the positive charge is spread by resonance.

Reactivity factors in additions to carbon-hetero multiple bonds are similar to those for the tetrahedral mechanism of nucleophilic substitution.[10] If A and/or B are electron-donating groups, rates are decreased. Electron-attracting substituents increase rates. This means that aldehydes are more reactive than ketones. Aryl groups are somewhat deactivating compared to alkyl, because of resonance that stabilizes the substrate molecule but is lost on going to the intermediate:

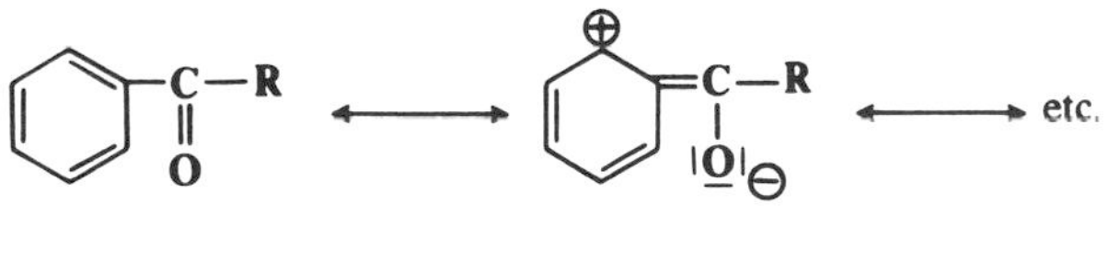

Double bonds in conjugation with the carbon-hetero multiple bond also lower addition rates, for similar reasons but, more important, may provide competition from 1,4 addition (p. 742). Steric factors are also quite important and contribute to the decreased reactivity of ketones compared with aldehydes. Highly hindered ketones like hexamethylacetone and dineopentyl ketone either do not undergo many of these reactions or require extreme conditions.

---

[8]For a discussion of acid and base catalysis in these reactions, see Jencks; Gilbert *Pure Appl. Chem.* **1977**, *49*, 1021-1027.

[9]Toromanoff *Bull. Soc. Chim. Fr.* **1962**, 1190.

[10]For a review of the reactivity of nitriles, see Schaefer, in Rappoport *The Chemistry of the Cyano Group*; Wiley: New York, 1970, pp. 239-305.

REACTIONS

Many of the reactions in this chapter are simple additions to carbon-hetero multiple bonds, with the reaction ending when the two groups have been added. But in many other cases subsequent reactions take place. We shall meet a number of such reactions, but most are of two types:

$$
\text{Type } A \qquad \underset{\underset{\text{O}}{\|}}{\text{A}-\text{C}-\text{B}} + \text{YH}_2 \longrightarrow \underset{\underset{\text{OH}}{|}}{\overset{\overset{\text{YH}}{|}}{\text{A}-\text{C}-\text{B}}} \xrightarrow{-\text{H}_2\text{O}} \underset{}{\overset{\overset{\text{Y}}{\|}}{\text{A}-\text{C}-\text{B}}}
$$

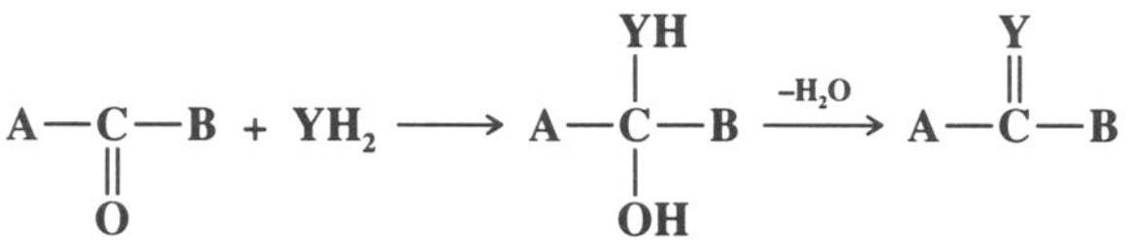

$$
\text{Type } B \qquad \underset{\underset{\text{O}}{\|}}{\text{A}-\text{C}-\text{B}} + \text{YH}_2 \longrightarrow \underset{\underset{\text{OH}}{|}}{\overset{\overset{\text{YH}}{|}}{\text{A}-\text{C}-\text{B}}} \xrightarrow{z} \underset{\underset{\text{Z}}{|}}{\overset{\overset{\text{YH}}{|}}{\text{A}-\text{C}-\text{B}}}
$$

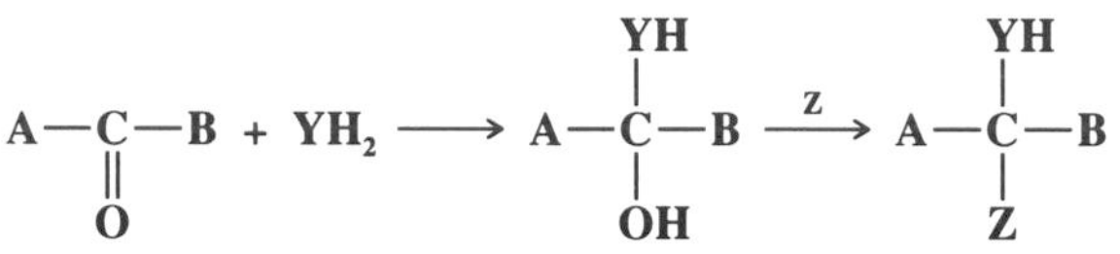

In type A, the adduct loses water (or, in the case of addition to C=NH, ammonia, etc.), and the net result of the reaction is the substitution of C=Y for C=O (or C=NH, etc.). In type B there is a rapid substitution, and the OH (or NH$_2$, etc.) is replaced by another group Z, which is often another YH moiety. This substitution is in most cases nucleophilic, since Y usually has an unshared pair and SN1 reactions occur very well on this type of compound (see p. 342), even when the leaving group is as poor as OH or NH$_2$. In this chapter we shall classify reactions according to what is initially adding to the carbon-hetero multiple bond, even if subsequent reactions take place so rapidly that it is not possible to isolate the initial adduct.

Most of the reactions considered in this chapter can be reversed. In many cases we shall consider the reverse reactions with the forward ones, in the same section. The reverse of some of the other reactions are considered in other chapters. In still other cases, one of the reactions in this chapter is the reverse of another, e.g., **6-2** and **6-14**. For reactions that are reversible, the principle of microscopic reversibility (p. 215) applies.

We shall discuss first reactions in which hydrogen or a metallic ion (or in one case phosphorus or sulfur) adds to the hetero atom and then reactions in which carbon adds to the hetero atom. Within each group, the reactions are classified by the nature of the nucleophile. Additions to isocyanides, which are different in character, are treated at the end.

Reactions in Which Hydrogen or a Metallic Ion Adds to the Hetero Atom

A. Attack by OH (Addition of H₂O)

6-1 The Addition of Water to Aldehydes and Ketones. Formation of Hydrates
_O_-Hydro-_C_-hydroxy-addition

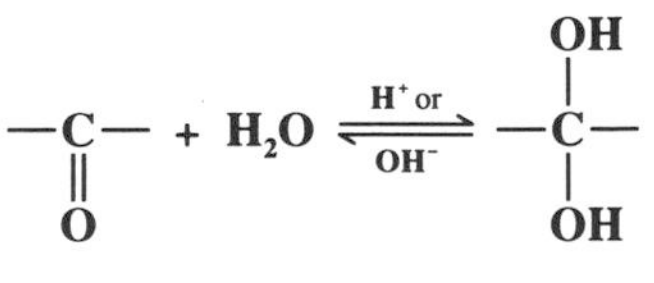

$$
\underset{\underset{\text{O}}{\|}}{-\text{C}-} + \text{H}_2\text{O} \underset{\text{OH}^-}{\overset{\text{H}^+ \text{ or}}{\rightleftharpoons}} \underset{\underset{\text{OH}}{|}}{\overset{\overset{\text{OH}}{|}}{-\text{C}-}}
$$

The adduct formed upon addition of water to an aldehyde or ketone is called a hydrate or *gem*-diol.[11] These compounds are usually stable only in water solution and decompose on distillation; i.e., the equilibrium shifts back toward the carbonyl compound. The position of the equilibrium is greatly dependent on the structure of the hydrate. Thus, formaldehyde in water at 20°C exists 99.99% in the hydrated form, while for acetaldehyde this figure is 58%, and for acetone the hydrate concentration is negligible.[12] It has been found, by exchange with ^{18}O, that the reaction with acetone is quite rapid when catalyzed by acid or base, but the equilibrium lies on the side of acetone and water.[13] Since methyl, a $+ I$ group, inhibits hydrate formation, it may be expected that electron-attracting groups would have the opposite effect, and this is indeed the case. The hydrate of chloral[14] is a stable crystalline substance. In order for it to revert to chloral, OH^- or H_2O must leave; this is made difficult by the electron-withdrawing character of the Cl_3C group. Some other[15] polychlorinated and

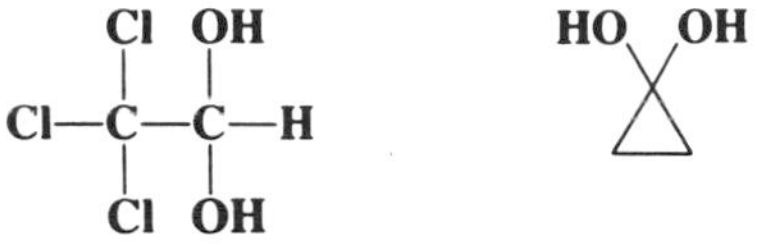

chloral hydrate   hydrate of cyclopropanone

polyfluorinated aldehydes and ketones[16] and α-keto aldehydes also form stable hydrates, as do cyclopropanones.[17] In the last case[18] formation of the hydrate relieves some of the I strain (p. 276) of the parent ketone.

The reaction is subject to both general-acid and general-base catalysis; the following mechanisms can be written for basic (B) and acidic (BH) catalysis, respectively:[19]

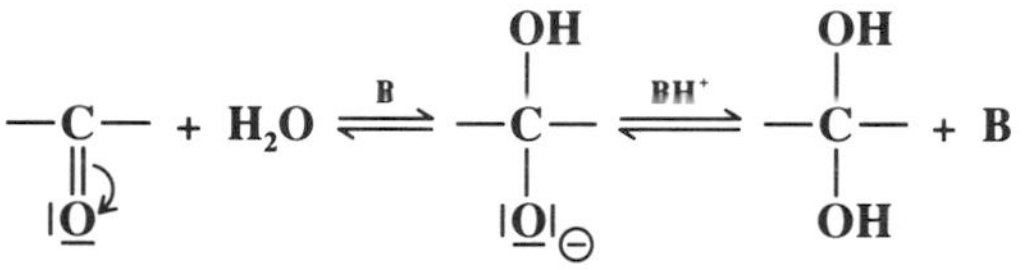

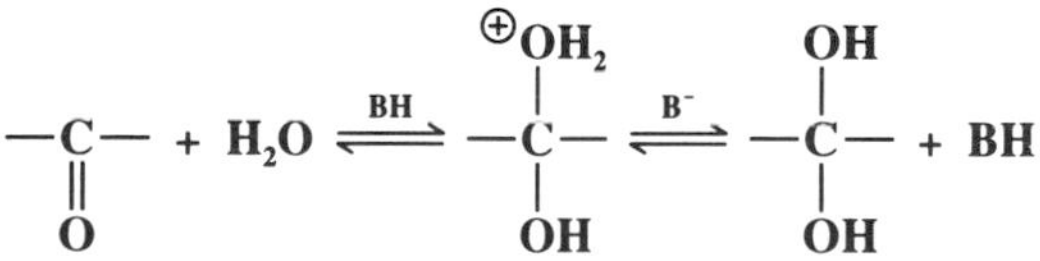

[11]For reviews, see Bell *The Proton in Chemistry*, 2nd ed.; Cornell University Press: Ithaca, NY, 1973, pp. 183-187, *Adv. Phys. Org. Chem.* **1966**, *4*, 1-29; Le Hénaff *Bull. Soc. Chim. Fr.* **1968**, 4687-4700.
[12]Bell; Clunie *Trans. Faraday Soc.* **1952**, *48*, 439. See also Bell; McDougall *Trans. Faraday Soc.* **1960**, *56*, 1281.
[13]Cohn; Urey *J. Am. Chem. Soc.* **1938**, *60*, 679.
[14]For a review of chloral, see Luknitskii *Chem. Rev.* **1975**, *75*, 259-289.
[15]For a discussion, see Schulman; Bonner; Schulman; Laskovics *J. Am. Chem. Soc.* **1976**, *98*, 3793.
[16]For a review of addition to fluorinated ketones, see Gambaryan; Rokhlin; Zeifman; Ching-Yun; Knunyants *Angew. Chem. Int. Ed. Engl.* **1966**, *5*, 947-956 [*Angew. Chem. 78*, 1008-1017].
[17]For other examples, see Krois; Langer; Lehner *Tetrahedron* **1980**, *36*, 1345; Krois; Lehner *Monatsh. Chem.* **1982**, *113*, 1019.
[18]Turro; Hammond *J. Am. Chem. Soc.* **1967**, *89*, 1028; Schaafsma; Steinberg; de Boer *Recl. Trav. Chim. Pays-Bas* **1967**, *86*, 651. For a review of cyclopropanone chemistry, see Wasserman; Clark; Turley *Top. Curr. Chem.* **1974**, *47*, 73-156.
[19]Bell; Rand; Wynne-Jones *Trans. Faraday Soc.* **1956**, *52*, 1093; Pocker *Proc. Chem. Soc.* **1960**, 17; Funderburk; Aldwin; Jencks *J. Am. Chem. Soc.* **1978**, *100*, 5444; Sørensen; Jencks *J. Am. Chem. Soc.* **1987**, *109*, 4675. For a comprehensive treatment, see Lowry; Richardson *Mechanism and Theory in Organic Chemistry*, 3rd ed.; Harper and Row: New York, 1987, pp. 662-680.

In mechanism *a*, as the H_2O attacks, the base pulls off a proton, and the net result is addition of OH^-. This can happen because the base is already hydrogen-bonded to the H_2O molecule before the attack. In mechanism *b*, because HB is already hydrogen-bonded to the oxygen of the carbonyl group, it gives up a proton to the oxygen as the water attacks. In this way B and HB accelerate the reaction even beyond the extent that they form OH^- or H_3O^+ by reaction with water. Reactions in which the catalyst donates a proton to the electrophilic reagent (in this case the aldehyde or ketone) in one direction and removes it in the other are called class e reactions. Reactions in which the catalyst does the same to the nucleophilic reagent are called class n reactions.[20] Thus the acid-catalyzed process here is a class e reaction, while the base catalyzed process is a class n reaction.

For the reaction between ketones and H_2O_2, see **7-49**.

There are no OS references, but see OS **66**, 142, for the reverse reaction.

6-2 Hydrolysis of the Carbon–Nitrogen Double Bond
Oxo-de-alkylimino-bisubstitution, etc.

$$\begin{array}{ccc} -\overset{\displaystyle \|}{\underset{\displaystyle N-W}{C}}- & \xrightarrow{H_2O} & -\overset{\displaystyle \|}{\underset{\displaystyle O}{C}}- \ + \ W-NH_2 \end{array}$$

Compounds containing carbon–nitrogen double bonds can be hydrolyzed to the corresponding aldehydes or ketones. For imines (W = R or H) the hydrolysis is easy and can be carried out with water. When W = H, the imine is seldom stable enough for isolation, and hydrolysis usually occurs in situ, without isolation. The hydrolysis of Schiff bases (W = Ar) is more difficult and requires acid or basic catalysis. Oximes (W = OH), arylhydrazones (W = NHAr), and, most easily, semicarbazones (W = $NHCONH_2$) can also be hydrolyzed. Often a reactive aldehyde, e.g., formaldehyde, is added to combine with the liberated amine.

A number of other reagents[21] have been used to cleave C=N bonds, especially those not easily hydrolyzable with acidic or basic catalysts or which contain other functional groups that are attacked under these conditions. In particular, oximes have been converted to the corresponding aldehyde or ketone by treatment with, among other reagents, thallium(III) nitrate,[22] aqueous $TiCl_3$ and acetic acid,[23] aqueous $NaHSO_3$,[24] benzeneseleninic anhydride $(PhSeO)_2O$,[25] N_2O_4,[26] $Me_3SiCl–NaNO_2$,[27] $LiAlH_4$–HMPA,[28] Amberlyst 15 and acetone,[29] pyridinium dichromate–*t*-BuOOH,[30] alkaline H_2O_2,[31] and by treatment of the O-acetate of the oxime with chromium(II) acetate.[32] Tosylhydrazones can be hydrolyzed to the corresponding ketones with NaOCl,[33] aqueous acetone and BF_3–etherate,[34] $CuSO_4 \cdot 5H_2O$,[35] so-

[20]Jencks *Acc. Chem. Res.* **1976**, *9*, 425-432.
[21]For a list of many of these reagents, with references,see Ranu; Sarkar *J. Org. Chem.* **1988**, *53*, 878.
[22]McKillop; Hunt; Naylor; Taylor *J. Am. Chem. Soc.* **1971**, *93*, 4918.
[23]Timms; Wildsmith *Tetrahedron Lett.* **1971**, 195. See also McMurry; Silvestri *J. Org. Chem.* **1975**, *40*, 1502; Balicki; Kaczmarek; Malinowski *Liebigs Ann. Chem.* **1989**, 1139.
[24]Pine; Chemerda; Kozlowski *J. Org. Chem.* **1966**, *31*, 3446.
[25]Barton; Lester; Ley *J. Chem. Soc., Perkin Trans. 1* **1980**, 1212.
[26]Shim; Kim; Kim *Tetrahedron Lett.* **1987**, *28*, 645.
[27]Lee; Kwak; Hwang *Tetrahedron Lett.* **1990**, *31*, 6677.
[28]Wang; Sukenik *J. Org. Chem.* **1985**, *50*, 5448.
[29]Ballini; Petrini *J. Chem. Soc., Perkin Trans. 1* **1988**, 2563.
[30]Chidambaram; Satyanarayana; Chandrasekaran *Synth. Commun.* **1989**, *19*, 1727.
[31]Ho *Synth. Commun.* **1980**, *10*, 465.
[32]Corey; Richman *J. Am. Chem. Soc.* **1970**, *92*, 5276.
[33]Ho; Wong *J. Org. Chem.* **1974**, *39*, 3453.
[34]Sacks; Fuchs *Synthesis* **1976**, 456.
[35]Attanasi; Gasperoni *Gazz. Chim. Ital.* **1978**, *108*, 137.

dium peroxide,[36] as well as with other reagents.[37] Among other reagents that have been used to cleave C=N bonds are nitrous acid (as well as nitrosonium salts such as $NO^+ BF_4^-$)[38] and ozone[39] (see **9-9**).

The hydrolysis of carbon–nitrogen double bonds involves initial addition of water and elimination of a nitrogen moiety:

$$-\overset{\overset{\displaystyle \oplus OH_2}{|}}{\underset{\underset{\displaystyle |\overset{\ominus}{N}-W}{||}}{C}}- \xrightarrow{H_2O} -\overset{\overset{\displaystyle OH_2\oplus}{|}}{\underset{\underset{\displaystyle \ominus \underset{}{N}-W}{|}}{C}}- \longrightarrow -\overset{\overset{\displaystyle OH}{|}}{\underset{\underset{\displaystyle NHW}{|}}{C}}- \longrightarrow -\overset{\overset{\displaystyle OH}{|}}{\underset{\underset{\displaystyle \oplus}{|}}{C}}- \xrightarrow{-H^+} -\overset{\overset{\displaystyle O}{||}}{C}-$$

It is thus an example of reaction type A (p. 882). The sequence shown is generalized.[40] In specific cases there are variations in the sequence of the steps, depending on acid or basic catalysis or other conditions.[41] Which step is rate-determining also depends on acidity and on the nature of W and of the groups connected to the carbonyl.[42] Iminium ions (**2**)[43] would

$$\left[\begin{array}{ccc} -\overset{\overset{\displaystyle}{|}}{\underset{\underset{\underset{\displaystyle R \quad\quad R}{}}{N\oplus}}{C}}- & \longleftrightarrow & -\overset{\overset{\displaystyle}{|}}{\underset{\underset{\underset{\displaystyle R \quad\quad R}{}}{N}}{\overset{\oplus}{C}}}- \\ & \mathbf{2} & \end{array} \right]$$

be expected to undergo hydrolysis quite readily, since there is a contributing form with a positive charge on the carbon. Indeed, they react with water at room temperature.[44] Acid-catalyzed hydrolysis of enamines (the last step of the Stork reaction, **2-19**) involves conversion to iminium ions:[45]

$$-\overset{}{\underset{\underset{\displaystyle |NR_2}{}}{C}}=\overset{}{\underset{}{C}}- \underset{H^+}{\rightleftarrows} -\overset{\overset{\displaystyle H}{|}}{\underset{\underset{\displaystyle NR_2}{||\oplus}}{C}}-\overset{}{\underset{}{C}}- \xrightarrow{H_2O} -\overset{\overset{\displaystyle H}{|}}{\underset{\underset{\displaystyle \underset{\oplus}{HNR_2}}{|}}{C}}-\overset{\overset{\displaystyle OH}{|}}{\underset{}{C}}- \underset{-R_2NH}{\rightleftarrows} -\overset{\overset{\displaystyle H}{|}}{\underset{}{C}}-\overset{\overset{\displaystyle}{||}}{\underset{\underset{\displaystyle \underset{\oplus}{OH}}{}}{C}}- \underset{-H^+}{\rightleftarrows} -\overset{\overset{\displaystyle H}{|}}{\underset{}{C}}-\overset{\overset{\displaystyle}{||}}{\underset{\underset{\displaystyle O}{}}{C}}-$$

The mechanism of enamine hydrolysis is thus similar to that of vinyl ether hydrolysis (**0-6**).

OS **I**, 217, 298, 318, 381; **II**, 49, 223, 234, 284, 310, 333, 395, 519, 522; **III**, 20, 172, 626, 818; **IV**, 120; **V**, 139, 277, 736, 758; **VI**, 1, 358, 640, 751, 901, 932; **VII**, 8; **65**, 108, 183; **67**, 33.

[36]Ho; Olah *Synthesis* **1976**, 611.

[37]For references, see Jiricny; Orere; Reese *Synthesis* **1970, 919.**

[38]Doyle; Wierenga; Zaleta *J. Org. Chem.* **1972,** *37,* 1597; Doyle; Zaleta; DeBoer; Wierenga *J. Org. Chem.* **1973,** *38,* 1663; Olah; Ho *Synthesis* **1976,** 610.

[39]For example, see Erickson; Andrulis; Collins; Lungle; Mercer *J. Org. Chem.* **1969,** *34,* 2961.

[40]For reviews of the mechanism, see Bruylants; Feytmants-de Medicis, in Patai *The Chemistry of the Carbon–Nitrogen Double Bond*; Wiley: New York, 1970, pp. 465-504; Salomaa, in Patai, Ref. 2, pt. 1, pp. 199-205.

[41]For example, see Reeves *J. Am. Chem. Soc.* **1962,** *82,* 3332; Sayer; Conlon *J. Am. Chem Soc.* **1980,** *102,* 3592.

[42]Cordes; Jencks *J. Am. Chem. Soc.* **1963,** *85,* 2843.

[43]For a review of iminium ions, see Böhme; Haake *Adv. Org. Chem.* **1976,** *9,* pt. 1, 107-223.

[44]Hauser; Lednicer *J. Org. Chem.* **1959,** *24,* 46. For a study of the mechanism, see Gopalakrishnan; Hogg *J. Org. Chem.* **1989,** *54,* 768.

[45]Stamhuis; Maas *J. Org. Chem.* **1965,** *30,* 2156; Maas; Janssen; Stamhuis; Wynberg *J. Org. Chem.* **1967,** *32,* 1111; Sollenberger; Martin *J. Am. Chem. Soc.* **1970,** *92,* 4261. For a review of enamine hydrolysis, see Stamhuis; Cook, in Cook *Enamines,* 2nd ed.; Marcel Dekker: New York, 1988, pp. 165-180.

6-3   Hydrolysis of Isocyanates and Isothiocyanates
Oxo-de-alkylimino-bisubstitution

$$R—N=C=O + H_2O \xrightarrow{\text{OH}^- \text{ or } H^+} R—NH_2 + CO_2$$

A common method for the preparation of primary amines involves the hydrolysis of iso-cyanates or isothiocyanates.[46] The latter react more slowly and more vigorous conditions are required. The reaction is catalyzed by acids or bases. In this case simple addition of water to the carbon–nitrogen double bond would give an N-substituted carbamic acid (**3**). Such compounds are unstable and break down to carbon dioxide (or COS in the case of isothiocyanates) and the amine:

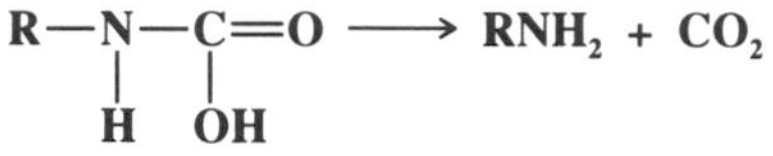

$$\underset{\substack{| \quad | \\ H \quad OH \\ \textbf{3}}}{R—N—C=O} \longrightarrow RNH_2 + CO_2$$

OS **II**, 24; **IV**, 819; **V**, 273; **VI**, 910.

6-4   Hydrolysis of Aliphatic Nitro Compounds
Oxo-de-hydro,nitro-bisubstitution

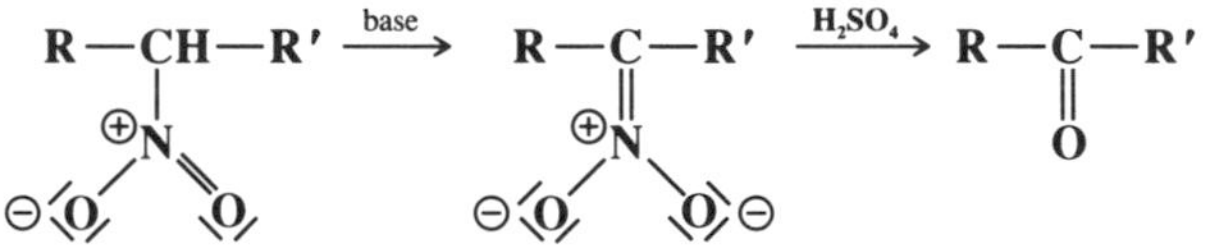

Primary or secondary aliphatic nitro compounds can be hydrolyzed, respectively, to alde-hydes or ketones, by treatment of their conjugate bases with sulfuric acid. This is called the *Nef reaction.*[47] Tertiary aliphatic nitro compounds do not give the reaction because they cannot be converted to their conjugate bases. Like **6-2**, this reaction involves hydrolysis of a C=N double bond. A possible mechanism is[48]

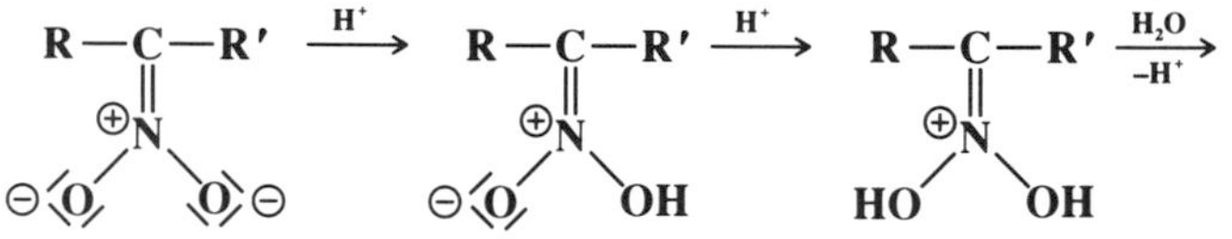

Aci form of the nitro compound

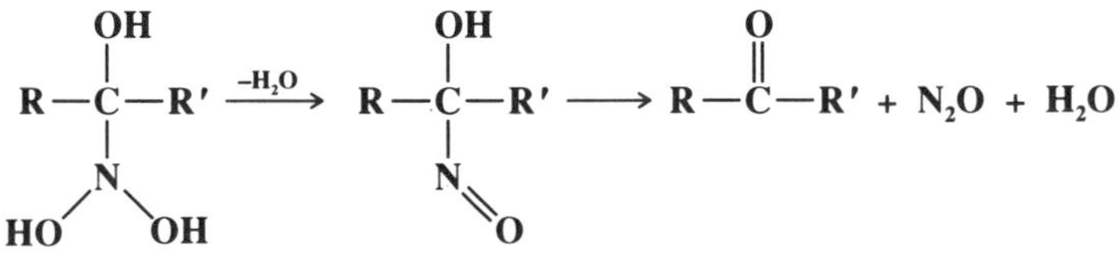

Intermediates of type **4** have been isolated in some cases.[49]

---

[46]For a study of the mechanism, see Castro; Moodie; Sansom *J. Chem. Soc., Perkin Trans. 2* **1985**, 737. For a review of the mechanisms of reactions of isocyanates with various nucleophiles, see Satchell; Satchell *Chem. Soc. Rev.* **1975**, *4*, 231-250.

[47]For reviews, see Pinnick *Org. React.* **1990**, *38*, 655-792; Haines *Methods for the Oxidation of Organic Compounds*; Academic Press: New York, 1988, pp. 220-231, 416-419.

[48]Hawthorne *J. Am. Chem. Soc.* **1957**, *79*, 2510. A similar mechanism, but with some slight differences, was suggested earlier by van Tamelen; Thiede *J. Am. Chem. Soc.* **1952**, *74*, 2615. See also Sun; Folliard *Tetrahedron* **1971**, *27*, 323.

[49]Feuer; Spinicelli *J. Org. Chem.* **1977**, *42*, 2091.

The conversion of nitro compounds to aldehydes or ketones has been carried out with better yields and fewer side reactions by several alternative methods. Among these are treatment of the nitro compound with aqueous $TiCl_3$,[50] cetyltrimethylammonium permanganate,[51] tin complexes and $NaHSO_3$,[52] activated dry silica gel,[53] or 30% H_2O_2–K_2CO_3,[54] and treatment of the conjugate base[55] of the nitro compound with $KMnO_4$,[56] t-BuOOH and a catalyst,[57] ceric ammonium nitrate (CAN),[58] MoO_5–pyridine–HMPA,[59] or ozone.[60]

When *primary* nitro compounds are treated with sulfuric acid without previous conversion to the conjugate bases, they give carboxylic acids. Hydroxamic acids are intermediates and can be isolated, so that this is also a method for preparing them.[61] Both the Nef reaction and the hydroxamic acid process involve the aci form; the difference in products arises from higher acidity, e.g., a difference in sulfuric acid concentration from 2 M to 15.5 M changes the product from the aldehyde to the hydroxamic acid.[62] The mechanism of the hydroxamic acid reaction is not known with certainty, but if higher acidity is required, it may be that the protonated aci form of the nitro compound is further protonated.

OS **VI**, 648; **VII**, 414. See also OS **IV**, 573.

6-5 Hydrolysis of Nitriles

$$R\!-\!C\!\equiv\!N + H_2O \xrightarrow{\text{H}^+ \text{ or OH}^-} R\!-\!\underset{\underset{O}{\|}}{C}\!-\!NH_2$$

NN-Dihydro-*C*-oxo-biaddition

$$R\!-\!C\!\equiv\!N + H_2O \xrightarrow{\text{H}^+ \text{ or OH}^-} R\!-\!\underset{\underset{O}{\|}}{C}\!-\!OH \text{ or } R\!-\!\underset{\underset{O}{\|}}{C}\!-\!O^-$$

Hydroxy,oxo-de-nitrilo-tersubstitution

Nitriles can be hydrolyzed to give either amides or carboxylic acids.[63] The amide is formed initially, but since amides are also hydrolyzed with acid or basic treatment, the carboxylic acid is the more common product. When the acid is desired,[64] the reagent of choice is

[50]McMurry; Melton *J. Org. Chem.* **1973**, *38*, 4367; McMurry *Acc. Chem. Res.* **1974**, *7*, 281-286, pp. 282-284. See also Kirchhoff *Tetrahedron Lett.* **1976**, 2533.

[51]Vankar; Rathore; Chandrasekaran *Synth. Commun.* **1987**, *17*, 195.

[52]Urpí; Vilarrasa *Tetrahedron Lett.* **1990**, *31*, 7499.

[53]Keinan; Mazur *J. Am. Chem. Soc.* **1977**, *99*, 3861.

[54]Olah; Arvanaghi; Vankar; Prakash *Synthesis* **1980**, 662.

[55]For other methods, see Barton; Motherwell; Zard *Tetrahedron Lett.* **1983**, *24*, 5227; Yano; Ohshima; Sutoh *J. Chem. Soc., Chem. Commun.* **1984**, 695.

[56]Shechter; Williams *J. Org. Chem.* **1962**, *27*, 3699; Freeman; Yeramyan *J. Org. Chem.* **1970**, *35*, 2061; Freeman; Lin *J. Org. Chem.* **1971**, *36*, 1335; Kornblum; Erickson; Kelly; Henggeler *J. Org. Chem.* **1982**, *47*, 4534; Steliou; Poupart *J. Org. Chem.* **1985**, *50*, 4971.

[57]Bartlett; Green; Webb *Tetrahedron Lett.* **1977**, 331.

[58]Olah; Gupta *Synthesis* **1980**, 44.

[59]Galobardes; Pinnick *Tetrahedron Lett.* **1981**, *22*, 5235.

[60]McMurry; Melton; Padgett *J. Org. Chem.* **1974**, *39*, 259. See Williams; Unger; Moore *J. Org. Chem.* **1978**, *43*, 1271, for the use of singlet oxygen instead of ozone.

[61]Hydroxamic acids can also be prepared from primary nitro compounds with SeO_2 and Et_3N: Sosnovsky; Krogh *Synthesis* **1980**, 654.

[62]Kornblum; Brown *J. Am. Chem. Soc.* **1965**, *87*, 1742. See also Cundall; Locke *J. Chem. Soc. B* **1968**, 98; Edward; Tremaine *Can J. Chem.* **1971**, *49*, 3483, 3489, 3493.

[63]For reviews, see Zil'berman *Russ. Chem. Rev.* **1984**, *53*, 900-912; Compagnon; Miocque *Ann. Chim. (Paris)* **1970**, [14] *5*, 11-22, 23-37.

[64]For a list of reagents, with references, see Larock *Comprehensive Organic Transformations*; VCH: New York, 1989, p. 993.

aqueous NaOH containing about 6 to 12% H_2O_2, though acid-catalyzed hydrolysis is also frequently carried out. However, there are a number of procedures for stopping at the amide stage,[65] among them the use of concentrated H_2SO_4; formic acid and HCl or HBr;[66] acetic acid and BF_3; H_2O_2 and OH^-;[67] 30% H_2O_2 in Me_2SO,[68] sodium percarbonate,[69] and dry HCl followed by H_2O. The same result can also be obtained by use of water and certain metal ions or complexes;[70] MnO_2 in methylene chloride[71] or on silica gel;[72] sodium perborate in aqueous MeOH;[73] $Hg(OAc)_2$ in HOAc;[74] 2-mercaptoethanol in a phosphate buffer;[75] $KF–Al_2O_3$;[76] or $TiCl_4$ and water.[77] Nitriles can be hydrolyzed to the carboxylic acids without disturbing carboxylic ester functions also present, by the use of tetrachloro- or tetrafluorophthalic acid.[78]

The hydrolysis of nitriles to carboxylic acids is one of the best methods for the preparation of these compounds. Nearly all nitriles give the reaction, with either acidic or basic catalysts. The sequences

$$RX + NaCN \longrightarrow RCN \longrightarrow RCOOH \qquad (0\text{-}101)$$

$$RCHO + HCN \longrightarrow RCH(OH)CN \longrightarrow RCH(OH)COOH \qquad (6\text{-}49)$$

$$RCHO + NaCN + NH_4Cl \longrightarrow RCH(NH_2)CN \longrightarrow RCH(NH_2)COOH \qquad (6\text{-}50)$$

are very common. The last two sequences are often carried out without isolation of the cyanide intermediates. Hydrolysis of cyanohydrins $RCH(OH)CN$ is usually carried out under acidic conditions, because basic solutions cause competing reversion of the cyanohydrin to the aldehyde and CN^-. However, cyanohydrins have been hydrolyzed under basic conditions with borax or alkaline borates.[79]

The first addition product is **5**, which tautomerizes to the amide.

$$R—C{\equiv}N \longrightarrow \underset{OH}{R—C{=}NH} \rightleftharpoons \underset{O}{R—C—NH_2}$$

5

Thiocyanates can be converted to thiocarbamates, in a similar reaction:[80] $R—S—C{\equiv}N + H_2O \xrightarrow{H^+} R—S—CO—NH_2$. Hydrolysis of cyanamides gives amines, produced by the breakdown of the unstable carbamic acid intermediates: $R_2NCN \rightarrow [R_2NCOOH] \rightarrow R_2NH$.

OS **I**, 21, 131, 201, 289, 298, 321, 336, 406, 436, 451; **II**, 29, 44, 292, 376, 512, 586 (see, however, **V**, 1054), 588; **III**; 34, 66, 84, 88, 114, 221, 557, 560, 615, 851; **IV**, 58, 93, 496, 506, 664, 760, 790; **V**, 239; **VI**, 932. Also see OS **III**, 609; **IV**, 359, 502; **66**, 142.

[65]For a discussion, see Beckwith, in Zabicky *The Chemistry of Amides*; Wiley: New York, 1970, pp. 119-125. For a list of reagents, with references, see Ref. 64, p. 994.
[66]Becke; Fleig; Pässler *Liebigs Ann. Chem.* **1971**, *749*, 198.
[67]For an example with phase transfer catalysis, see Cacchi; Misiti; La Torre *Synthesis* **1980**, 243.
[68]Katritzky; Pilarski; Urogdi *Synthesis* **1989**, 949.
[69]Kabalka; Deshpande; Wadgaonkar; Chatla *Synth. Commun.* **1990**, *20*, 1445.
[70]For example, see Watanabe *Bull. Chem. Soc. Jpn.* **1959**, *32*, 1280, **1964**, *37*, 1325; Bennett; Yoshida *J. Am. Chem. Soc.* **1973**, *95*, 3030; Paraskewas *Synthesis* **1974**, 574; McKenzie; Robson *J. Chem. Soc., Chem. Commun.* **1988**, 112.
[71]Cook; Forbes; Kahn *Chem. Commun.* **1966**, 121.
[72]Liu; Shih; Huang; Hu *Synthesis* **1988**, 715.
[73]McKillop; Kemp *Tetrahedron* **1989**, *45*, 3299; Reed; Gupton; Solarz *Synth. Commun.* **1990**, *20*, 563.
[74]Plummer; Menendez; Songster *J. Org. Chem.* **1989**, *54*, 718.
[75]Lee; Goo; Lee; Lee *Tetrahedron Lett.* **1989**, *30*, 7439.
[76]Rao *Synth. Commun.* **1982**, *12*, 177.
[77]Mukaiyama; Kamio; Kobayashi; Takei *Chem. Lett.* **1973**, 357.
[78]Rounds; Eaton; Urbanowicz; Gribble *Tetrahedron Lett.* **1988**, *29*, 6557.
[79]Jammot; Pascal; Commeyras *Tetrahedron Lett.* **1989**, *30*, 563.
[80]Zil'berman; Lazaris *J. Gen. Chem. USSR* **1963**, *33*, 1012.

B. Attack by OR (Addition of ROH)

6-6 The Addition of Alcohols to Aldehydes and Ketones
Dialkoxy-de-oxo-bisubstitution

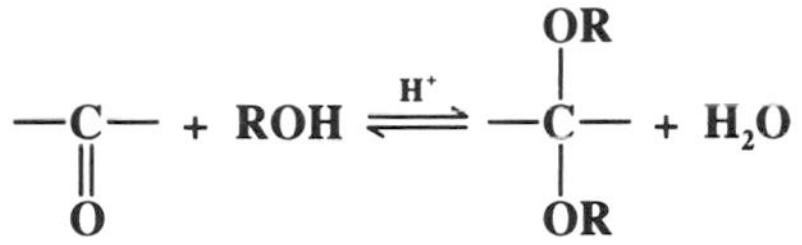

$$-\overset{\displaystyle ||}{\underset{\displaystyle O}{C}}- + \text{ROH} \underset{H^+}{\rightleftharpoons} -\overset{\displaystyle OR}{\underset{\displaystyle OR}{C}}- + H_2O$$

Acetals and ketals are formed by treatment of aldehydes and ketones, respectively, with alcohols in the presence of acid catalysts.[81] This reaction is reversible, and acetals and ketals can be hydrolyzed by treatment with acid (**0-6**). With small unbranched aldehydes the equilibrium lies to the right. If it is desired to prepare ketals, or acetals of larger molecules, the equilibrium must be shifted, usually by removal of water. This can be done by azeotropic distillation, ordinary distillation, or the use of a drying agent such as Al_2O_3 or a molecular sieve.[82] The reaction in neither direction is catalyzed by bases, so most acetals and ketals are quite stable to bases, though they are easily hydrolyzed by acids. This makes this reaction a useful method of protection of aldehyde or ketone functions from attack by bases. The reaction is of wide scope. Most aldehydes are easily converted to acetals.[83] With ketones the process is more difficult, presumably for steric reasons, and the reaction often fails, though many ketals, especially from cyclic ketones, have been made in this manner.[84] Many functional groups may be present without being affected. 1,2-Glycols and 1,3-glycols form cyclic acetals and ketals, e.g.,

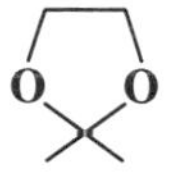

and these are often used to protect aldehydes and ketones.

The mechanism, which involves initial formation of a *hemiacetal*,[85] is the reverse of that given for acetal hydrolysis (**0-6**):

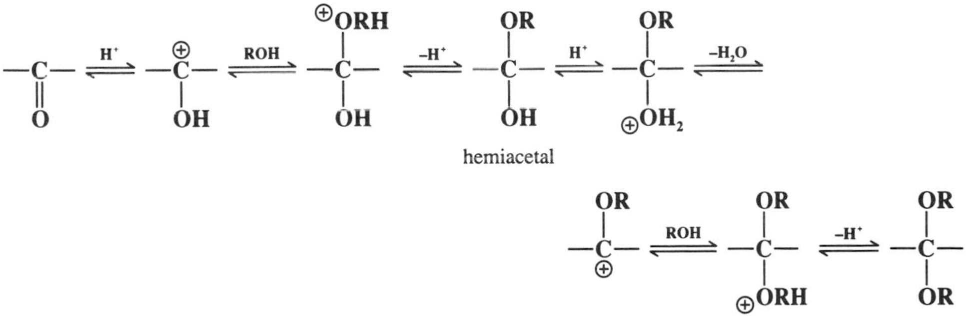

hemiacetal

[81]For reviews, see Meskens *Synthesis* **1981,** 501-522; Schmitz; Eichhorn, in Patai *The Chemistry of the Ether Linkage*; Wiley: New York, 1967, pp. 309-351.
[82]For many examples of each of these methods, see Meskens, Ref. 81, pp. 502-505.
[83]For other methods, see Caputo; Ferreri; Palumbo *Synthesis* **1987,** 386; Ott; Tombo; Schmid; Venanzi; Wang; Ward *Tetrahedron Lett.* **1989,** *30,* 6151, *New J. Chem.* **1990,** *14,* 495; Liao; Huang; Zhu *J. Chem. Soc., Chem. Commun.* **1990,** 493; Chan; Brook; Chaly *Synthesis* **1983,** 203.
[84]High pressure has been used to improve the results with ketones: Dauben; Gerdes; Look *J. Org. Chem.* **1986,** *51,* 4964. For other methods, see Otera; Mizutani; Nozaki *Organometallics* **1989,** *8,* 2063; Thurkauf; Jacobson; Rice *Synthesis* **1988,** 233.
[85]For a review of hemiacetals, see Hurd *J. Chem. Educ.* **1966,** *43,* 527-531.

In a study of the acid-catalyzed formation of the hemiacetal, Grunwald has shown[86] that the data best fit a mechanism in which the three steps shown here are actually all concerted; that is, the reaction is simultaneously catalyzed by acid and base, with water acting as the base:[87]

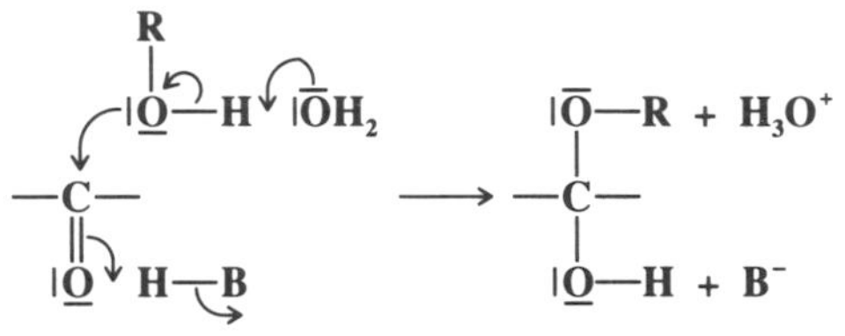

$$
\begin{array}{c}
\text{R} \\
| \\
\underset{|}{\overset{\curvearrowleft}{|\underline{O}}\text{—}\overset{\curvearrowleft}{H}}\; |\overline{O}H_2 \\
-\!C\!- \\
\| \\
|\underline{O}\overset{}{} \; H\text{—}B
\end{array}
\qquad
\begin{array}{c}
|\overline{O}\text{—}R + H_3O^+ \\
| \\
-\!C\!- \\
| \\
|\underline{O}\text{—}H + B^-
\end{array}
$$

If the original aldehyde or ketone has an α hydrogen, it is possible for water to split out in that way and enol ethers can be prepared in this manner:

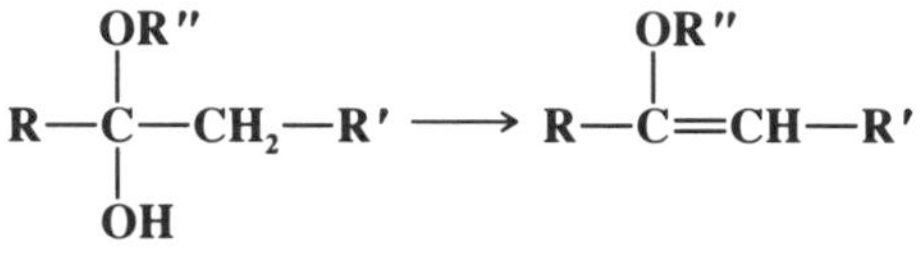

$$
\begin{array}{c}
OR'' \\
| \\
R\text{—}C\text{—}CH_2\text{—}R' \\
| \\
OH
\end{array}
\longrightarrow
\begin{array}{c}
OR'' \\
| \\
R\text{—}C\!=\!CH\text{—}R'
\end{array}
$$

Similarly, treatment with an anhydride and a catalyst can give an enol ester.[88]

Hemiacetals themselves are no more stable than the corresponding hydrates (6-1). As with hydrates, hemiacetals of cyclopropanones[89] and of polychloro and polyfluoro aldehydes and ketones may be quite stable.

When acetals or ketals are treated with an alcohol of higher molecular weight than the one already there, it is possible to get a transacetalation (see 0-17). In another type of transacetalation, aldehydes or ketones can be converted to acetals or ketals by treatment with another acetal or ketal or with an ortho ester,[90] in the presence of an acid catalyst (shown for an ortho ester):

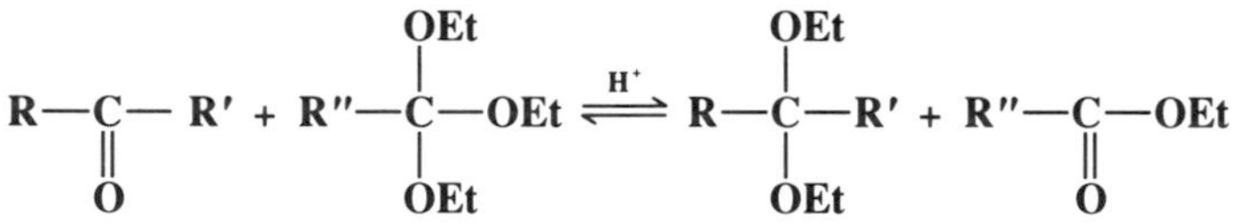

$$
\begin{array}{c}
\text{R}\text{—}\underset{\overset{\|}{O}}{C}\text{—}R'
\end{array}
+
\begin{array}{c}
OEt \\
| \\
R''\text{—}C\text{—}OEt \\
| \\
OEt
\end{array}
\overset{H^+}{\rightleftharpoons}
\begin{array}{c}
OEt \\
| \\
R\text{—}C\text{—}R' \\
| \\
OEt
\end{array}
+
\begin{array}{c}
R''\text{—}\underset{\overset{\|}{O}}{C}\text{—}OEt
\end{array}
$$

This method is especially useful for the conversion of ketones to ketals, since the direct reaction of a ketone with an alcohol often gives poor results. In another method, the substrate is treated with an alkoxysilane $ROSiMe_3$ in the presence of trimethylsilyl trifluoromethanesulfonate.[91]

---

[86]Grunwald *J. Am. Chem. Soc.* **1985**, *107*, 4715.

[87]Grunwald also studied the mechanism of the base-catalyzed formation of the hemiacetal, and found it to be the same as that of base-catalyzed hydration (6-1, mechanism *a*): Grunwald *J. Am. Chem. Soc.* **1985**, *107*, 4710. See also Sørensen; Pedersen; Pedersen; Kanagasabapathy; McClelland *J. Am. Chem. Soc.* **1988**, *110*, 5118; Leussing *J. Org. Chem.* **1990**, *55*, 666.

[88]For a list of catalysts, with references, see Ref. 64, p. 743.

[89]For a review, see Salaun *Chem. Rev.* **1983**, *83*, 619-632.

[90]For a review with respect to ortho esters, see DeWolfe *Carboxylic Ortho Ester Derivatives*; Academic Press: New York, 1970, pp. 154-164.

[91]Tsunoda; Suzuki; Noyori *Tetrahedron Lett.* **1980**, *21*, 1357; Kato; Iwasawa; Mukaiyama *Chem. Lett.* **1985**, 743. See also Torii; Takagishi; Inokuchi; Okumoto *Bull. Chem. Soc. Jpn.* **1987**, *60*, 775.

1,4-Diketones give furans when treated with acids. This is actually an example of an intramolecular addition of an alcohol to a ketone, since it is the enol form that adds:

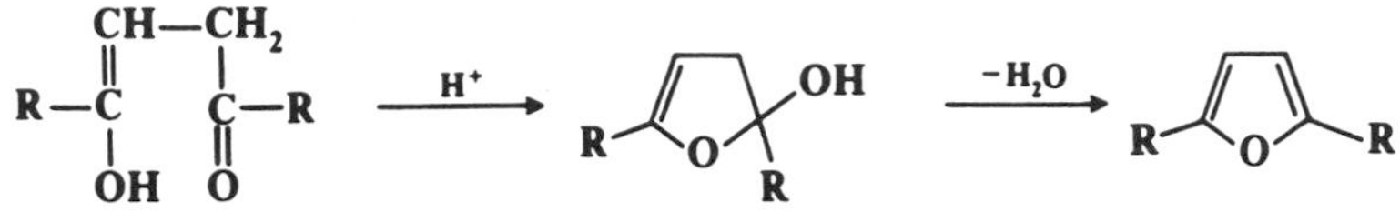

Similarly, 1,5-diketones give pyrans. Formic acid reacts with alcohols to give orthoformates.

OS **I**, 1, 298, 364, 381; **II**, 137; **III**, 123, 387, 502, 536, 644, 731, 800; **IV**, 21, 479, 679; **V**, 5, 292, 303, 450, 539; **VI**, 567, 666, 954; **VII**, 59, 149, 168, 177, 241, 271, 297; **67**, 202. Also see OS **IV**, 558, 588; **V**, 25; **67**, 193.

6-7 Reductive Alkylation of Alcohols
   C-Hydro-*O*-alkyl-addition

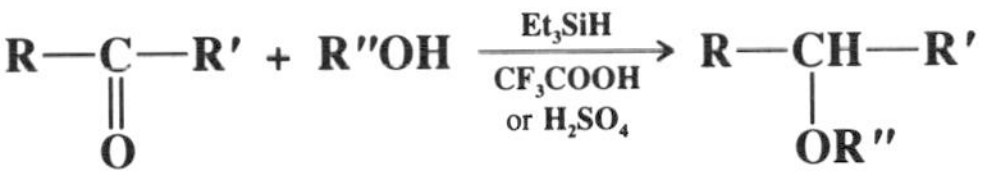

Aldehydes and ketones can be converted to ethers by treatment with an alcohol and triethylsilane in the presence of a strong acid[92] or by hydrogenation in alcoholic acid in the presence of platinum oxide.[93] The process can formally be regarded as addition of ROH to give a hemiacetal RR'C(OH)OR", followed by reduction of the OH. In this respect it is similar to **6-15**. In a similar reaction, ketones can be converted to carboxylic esters (reductive acylation of ketones) by treatment with an acyl chloride and triphenyltin hydride.[94]

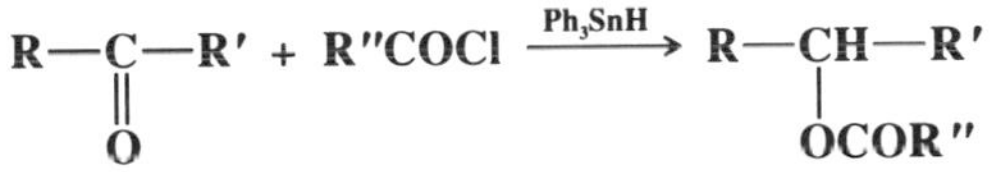

Ethers have also been prepared by the reductive dimerization of two molecules of an aldehyde or ketone (e.g., cyclohexanone → dicyclohexyl ether). This was accomplished by treatment of the substrate with a trialkylsilane and a catalyst.[95]

6-8 The Addition of Alcohols to Isocyanates
   N-Hydro-*C*-alkoxy-addition

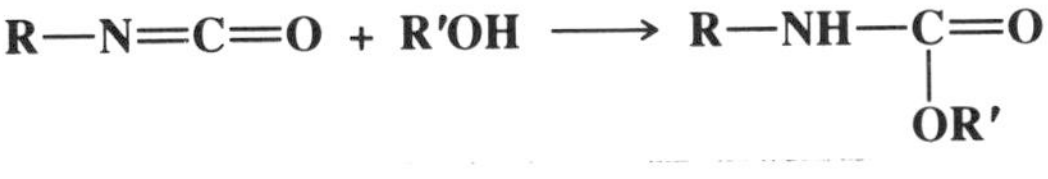

$$ R-N=C=O + R'OH \longrightarrow R-NH-\underset{\underset{OR'}{|}}{C}=O $$

Carbamates (substituted urethans) are prepared when isocyanates are treated with alcohols. This is an excellent reaction, of wide scope, and gives good yields. Isocyanic acid HNCO gives unsubstituted carbamates. Addition of a second mole of HNCO gives *allophanates*.

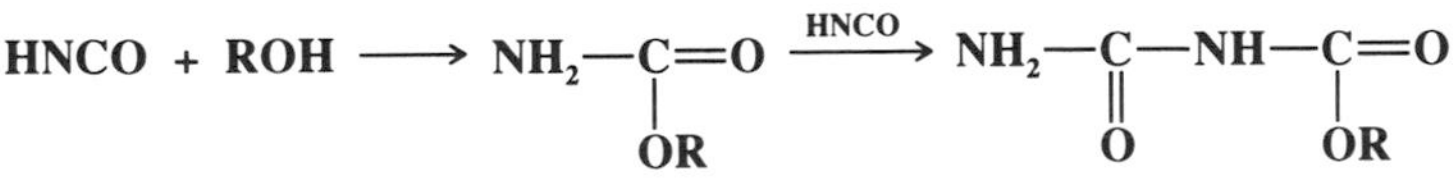

Allophanate

[92]Doyle; DeBruyn; Kooistra *J. Am. Chem. Soc.* **1972**, *94*, 3659.
[93]Verzele; Acke; Anteunis *J. Chem. Soc.* **1963**, 5598. For still another method, see Loim; Parnes; Vasil'eva; Kursanov *J. Org. Chem. USSR* **1972**, *8*, 902.
[94]Kaplan *J. Am. Chem. Soc.* **1966**, *88*, 4970.
[95]Sassaman; Kotian; Prakash; Olah *J. Org. Chem.* **1987**, *52*, 4314. See also Kikugawa *Chem. Lett.* **1979**, 415.

Polyurethans are made by combining compounds with two NCO groups with compounds containing two OH groups. Isothiocyanates similarly give thiocarbamates[96] RNHCSOR′, though they react slower than the corresponding isocyanates.

The details of the mechanism are poorly understood,[97] though the oxygen of the alcohol is certainly attacking the carbon of the isocyanate. Hydrogen bonding complicates the kinetic picture.[98] The addition of ROH to isocyanates can also be catalyzed by metallic compounds,[99] by light,[100] or, for tertiary ROH, by lithium alkoxides[101] or n-butyllithium.[102]

OS **I**, 140; **V**, 162; **VI**, 95, 226, 788, 795.

6-9 Alcoholysis of Nitriles
Alkoxy,oxo-de-nitrilo-tersubstitution

$$R\!-\!C\!\equiv\!N + R'OH \xrightarrow{\text{HCl}} R\!-\!\underset{\underset{OR'}{|}}{C}\!=\!NH_2^+ \ Cl^- \xrightarrow[\text{H}^+]{\text{H}_2\text{O}} R\!-\!\underset{\underset{OR'}{|}}{C}\!=\!O$$

The addition of dry HCl to a mixture of a nitrile and an alcohol in the absence of water leads to the hydrochloride salt of an imino ester (imino esters are also called imidates and imino ethers). This reaction is called the *Pinner synthesis*.[103] The salt can be converted to the free imino ester by treatment with a weak base such as sodium bicarbonate, or it can be hydrolyzed with water and an acid catalyst to the corresponding carboxylic ester. If the latter is desired, water may be present from the beginning, in which case aqueous HCl can be used and the need for gaseous HCl is eliminated. Imino esters can also be prepared from nitriles with basic catalysts.[104]

This reaction is of broad scope and is good for aliphatic, aromatic, and heterocyclic R and for nitriles with oxygen-containing functional groups. The application of the reaction to nitriles containing a carboxyl group constitutes a good method for the synthesis of mono esters of dicarboxylic acids with the desired group esterified and with no diester or diacid present.

Cyanogen chloride reacts with alcohols in the presence of an acid catalyst such as dry HCl or $AlCl_3$ to give carbamates:[105]

$$\textbf{ClCN} + \textbf{2ROH} \xrightarrow[\text{or AlCl}_3]{\text{HCl}} \textbf{ROCONH}_2 + \textbf{RCl}$$

ROH can also be added to nitriles in another manner (**6-55**).
OS **I**, 5, 270; **II**, 284, 310; **IV**, 645; **VI**, 507; **67**, 193.

[96]For a review of thiocarbamates, see Walter; Bode *Angew. Chem. Int. Ed. Engl.* **1967**, 6, 281-293 [*Angew. Chem.* 79, 285-297].

[97]For reviews, see Satchell; Satchell, Ref. 46; Entelis; Nesterov *Russ. Chem. Rev.* **1966**, 35, 917-930.

[98]See for example, Robertson; Stutchbury *J. Chem. Soc.* **1964**, 4000; Lammiman; Satchell *J. Chem. Soc., Perkin Trans. 2* **1972**, 2300, **1974**, 877; Donohoe; Satchell; Satchell *J. Chem. Soc., Perkin Trans. 2* **1990**, 1671. See also Sivakamasundari; Ganesan *J. Org. Chem.* **1984**, 49, 720.

[99]For example, see Davies; Puddephatt *J. Chem. Soc. C* **1967**, 2663, **1968**, 1479; Hazzard; Lammiman; Poon; Satchell; Satchell *J. Chem. Soc., Perkin Trans. 2* **1985**, 1029; Duggan; Imagire *Synthesis* **1989**, 131.

[100]McManus; Bruner; Coble; Ortiz *J. Org. Chem.* **1977**, 42, 1428.

[101]Bailey; Griffith *J. Org. Chem.* **1978**, 43, 2690.

[102]Nikoforov; Jirovetz; Buchbauer *Liebigs Ann. Chem.* **1989**, 489.

[103]For a review, see Compagnon; Miocque *Ann. Chim. (Paris)* [14] 5, 23-27, pp. 24-26. For a review of imino esters, see Neilson, in Patai *The Chemistry of Amidines and Imidates*; Wiley: New York, 1975, pp. 385-489.

[104]Schaefer; Peters *J. Org. Chem.* **1961**, 26, 412.

[105]Bodrikov; Danova *J. Org. Chem. USSR* **1968**, 4, 1611, **1969**, 5, 1558; Fuks; Hartemink *Bull. Soc. Chim. Belg.* **1973**, 82, 23.

6-10   The Formation of Xanthates
  S-Metallo-C-alkoxy-addition

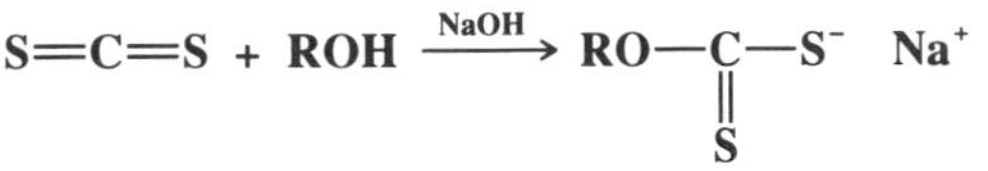

$$S{=}C{=}S + ROH \xrightarrow{\text{NaOH}} RO{-}\underset{\underset{S}{\|}}{C}{-}S^- \ Na^+$$

The addition of alcohols to carbon disulfide in the presence of a base produces xanthates.[106] The base is often OH⁻, but in some cases better results can be obtained by using methyl-sulfinyl carbanion MeSOCH₂⁻.[107] If an alkyl halide RX is present, the xanthate ester ROCSSR' can be produced directly. In a similar manner, alkoxide ions add to CO_2 to give carbonate ester salts ROCOO⁻.

OS **V**, 439; **VI**, 207, 418; **VII**, 139.

C. Sulfur Nucleophiles

6-11   The Addition of H₂S and Thiols to Carbonyl Compounds
  O-Hydro-C-mercapto-addition[108]

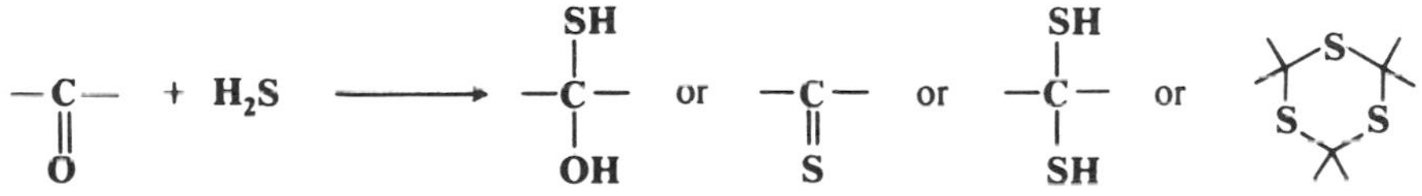

The addition of H₂S to an aldehyde or ketone can result in a variety of products. The most usual product is the trithiane **9**.[109] α-Hydroxy thiols (**6**) can be prepared from polychloro and polyfluoro aldehydes and ketones.[110] Apparently **6** are stable only when prepared from these compounds, and not even for all of them. Thioketones² (**7**) can be prepared from certain ketones, such as diaryl ketones, by treatment with H₂S and an acid catalyst, usually HCl. They are often unstable and tend to trimerize (to **9**) or to react with air. Thioaldehydes[111] are even less stable and simple ones[112] apparently have never been isolated, though t-BuCHS has been prepared in solution, where it exists for several hours at 20°C.[113] A high-yield synthesis of thioketones involves treatment of acyclic[114] ketones with 2,4-bis(4-methoxyphenyl)-1,3,2,4-dithiadiphosphetane-2,4-disulfide **10** (known as *Lawesson's*

---

[106]For a review of the formation and reactions of xanthates, see Dunn; Rudorf *Carbon Disulphide in Organic Chemistry*; Ellis Horwood: Chichester, 1989, pp. 316-367.

[107]Meurling; Sjöberg; Sjöberg *Acta Chem. Scand.* **1972**, *26*, 279.

[108]This name applies to formation of **6**. Names for formation of **7**, **8**, and **9**, are, respectively, **thioxo-de-oxo-bisubstitution, dimercapto-de-oxo-bisubstitution**, and **carbonyl-trithiane transformation**.

[109]Campaigne; Edwards *J. Org. Chem.* **1962**, *27*, 3760.

[110]Harris *J. Org. Chem.* **1960**, *25*, 2259.

[111]For a review of thioaldehydes, see Usov; Timokhina; Voronkov *Russ. Chem. Rev.* **1990**, *59*, 378-395.

[112]For the preparation and reactions of certain substituted thioaldehydes, see Hofstra; Kamphuis; Bos *Tetrahedron Lett.* **1984**, *25*, 873; Okazaki; Ishii; Inamoto *J. Am. Chem. Soc.* **1987**, *109*, 279; Adelaere; Guemas; Quiniou *Bull. Soc. Chim. Fr.* **1987**, 517; Muraoka; Yamamoto; Enomoto; Takeshima *J. Chem. Soc., Perkin Trans. 1* **1989**, 1241, and references cited in these papers.

[113]Vedejs; Perry *J. Am. Chem. Soc.* **1983**, *105*, 1683. See also Baldwin; Lopez *J. Chem. Soc., Chem. Commun.* **1982**, 1029.

[114]Cyclopentanone and cyclohexanone gave different products: Scheibye; Shabana; Lawesson; Rømming *Tetrahedron* **1982**, *38*, 993.

reagent[115]).[116] **10** also converts the C=O groups of amides and carboxylic esters[117] to C=S groups.[118] In similar reactions, bis(tricyclohexyltin)sulfide $(R_3Sn)_2S$ [R = cyclohexyl] and

$$MeO-\langle \bigcirc \rangle-P\underset{S}{\overset{S}{\underset{}{\parallel}}}\underset{S}{\overset{S}{\underset{\parallel}{}}}P-\langle \bigcirc \rangle-OMe$$

10

BCl_3 convert C=O groups of ketones, lactones, and lactams to C=S groups[119] and H_2S–Me_3SiCl–i-Pr_2NLi converts carboxylic esters to thiono esters.[120] Carboxylic acids RCOOH can be converted directly to dithiocarboxylic esters RCSSR',[120a] in moderate yield, with P_4S_{10} and a primary alcohol R'OH.[121] Thioketones can also be prepared by treatment of ketones with P_4S_{10},[122] and from oximes or various types of hydrazone (overall conversion C=N— → C=S).[123]

gem-Dithiols (**8**) are much more stable than the corresponding hydrates or α-hydroxy thiols.[124] They have been prepared by the treatment of ketones with H_2S under pressure[125] and under mild conditions with HCl as a catalyst.[126]

Thiols add to aldehydes and ketones to give hemimercaptals and dithioacetals. Hemimercaptals are ordinarily unstable,[127] though they are more stable than the corresponding

$$-\overset{\displaystyle |}{\underset{\displaystyle \parallel}{C}}- + RSH \longrightarrow -\overset{\displaystyle SR}{\underset{\displaystyle OH}{\underset{\displaystyle |}{C}}}- \quad \text{or} \quad -\overset{\displaystyle SR}{\underset{\displaystyle SR}{\underset{\displaystyle |}{C}}}-$$

hemimercaptal   dithioacetal

hemiacetals and can be isolated in certain cases.[128] Dithioacetals, like acetals, are stable in the presence of bases, except that a strong base can remove the aldehyde proton, if there is one[129] (see **0-97**). A common method for the protection of ketones involves treatment

$$R-\overset{\displaystyle R'}{\underset{\displaystyle O}{\underset{\displaystyle \parallel}{C}}}-R' + HSCH_2CH_2SH \xrightarrow{\text{BF}_3\text{-etherate}} \overset{R\ R'}{\underset{\displaystyle S\ \ S}{\times}}$$

[115]For reviews of this and related reagents, see Cava; Levinson *Tetrahedron* **1985**, *41*, 5061-5087; Cherkasov; Kutyrev; Pudovik *Tetrahedron* **1985**, *41*, 2567-2624. For the preparation of **10**, see Thomsen; Clausen; Scheibye; Lawesson *Org. Synth. VII*, 372.

[116]Pedersen; Scheibye; Nilsson; Lawesson *Bull. Soc. Chim. Belg.* **1978**, *87*, 223. For a study of the mechanism, see Rauchfuss; Zank *Tetrahedron Lett.* **1986**, *27*, 3445.

[117]For a review of thiono esters RC(=S)OR', see Jones; Bradshaw *Chem. Rev.* **1984**, *84*, 17-30.

[118]Scheibye; Pedersen; Lawesson *Bull. Soc. Chim. Belg.* **1978**, *87*, 229; Ghattas; El-Khrisy; Lawesson *Sulfur Lett.* **1982**, *1*, 69; Yde; Yousif; Pedersen; Thomsen; Lawesson *Tetrahedron* **1984**, *40*, 2047; Thomsen et al., Ref. 115.

[119]Steliou; Mrani *J. Am. Chem. Soc.* **1982**, *104*, 3104.

[120]Corey; Wright *Tetrahedron Lett.* **1984**, *25*, 2639.

[120a]For a review of dithiocarboxylic esters, see Kato; Ishida *Sulfur Rep.* **1988**, *8*, 155-323.

[121]Davy; Metzner *Chem. Ind. (London)* **1985**, 824.

[122]See, for example, Scheeren; Ooms; Nivard *Synthesis* **1973**, 149.

[123]See for example, Kimura; Niwa; Motoki *Bull. Chem. Soc. Jpn.* **1977**, *50*, 2751; de Mayo; Petrašiūnas; Weedon *Tetrahedron Lett.* **1978**, 4621; Okazaki; Inoue; Inamoto *Tetrahedron Lett.* **1979**, 3673.

[124]For a review of the preparation of *gem*-dithiols, see Mayer; Hiller; Nitzschke; Jentzsch *Angew. Chem. Int. Ed. Engl.* **1963**, *2*, 370-373 [*Angew. Chem. 75*, 1011-1014].

[125]Cairns; Evans; Larchar; McKusick *J. Am. Chem. Soc.* **1952**, *74*, 3982.

[126]Ref. 109; Demuynck; Vialle *Bull. Soc. Chim. Fr.* **1967**, 1213.

[127]See, for example, Fournier; Lamaty; Nata; Roque *Tetrahedron* **1975**, *31*, 809.

[128]For example, see Field; Sweetman *J. Org. Chem.* **1969**, *34*, 1799.

[129]Truce; Roberts *J. Org. Chem.* **1963**, *28*, 961.

with ethanedithiol to give a cyclic dithioketal.[130] After subsequent reactions involving the R or R' group, the protecting group can then be removed by **0-6.** Alternatively, the dithioketal can be desulfurized with Raney nickel (**4-36**), giving the overall conversion $C=O \rightarrow CH_2$. Dithioacetals can also be prepared from aldehydes or ketones by treatment with thiols in the presence of $TiCl_4$,[131] $SiCl_4$,[132] or polyphosphoric acid trimethylsilyl ester;[133] with a disulfide RSSR (R = alkyl or aryl),[134] or with methylthiotrimethylsilane $MeSSiMe_3$.[135]

If an aldehyde or ketone possesses an α hydrogen, it can be converted to the corresponding enol thioether by treatment with a thiol in the presence of $TiCl_4$:[136]

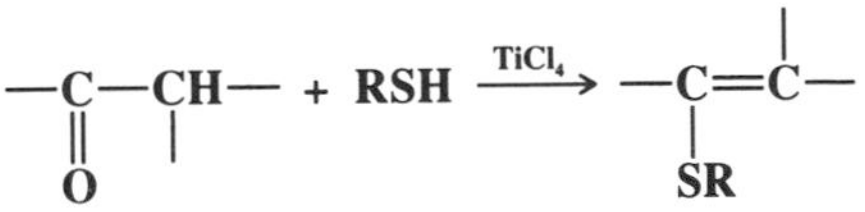

$$-\overset{\|}{\underset{O}{C}}-\overset{|}{C}H- + RSH \xrightarrow{TiCl_4} -C=\overset{|}{\underset{SR}{C}}-$$

Aldehydes and ketones have been converted to sulfides by treatment with thiols and pyridine–borane, $RCOR' + R''SH \xrightarrow{BH_3} RR'CHSR''$,[137] in a reductive alkylation reaction, analogous to **6-7.**

OS **II**, 610; **IV**, 927; **VI**, 109; **VII**, 124, 372. Also see OS **III**, 332; **IV**, 967; **V**, 780; **VI**, 556; **65**, 215.

6-12   Formation of Bisulfite Addition Products
O-Hydro-*C*-sulfonato-addition

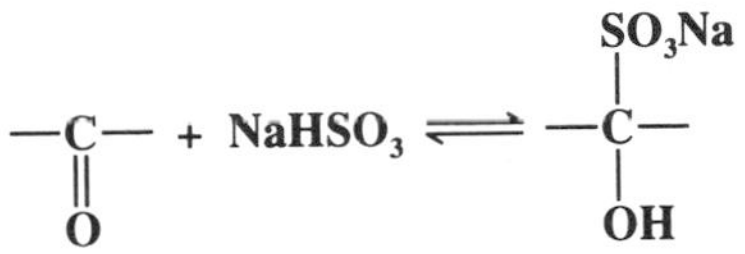

$$-\overset{\|}{\underset{O}{C}}- + NaHSO_3 \rightleftharpoons -\overset{\overset{SO_3Na}{|}}{\underset{\underset{OH}{|}}{C}}-$$

Bisulfite addition products are formed from aldehydes, methyl ketones, cyclic ketones (generally seven-membered and smaller rings), α-keto esters, and isocyanates, upon treatment with sodium bisulfite. Most other ketones do not undergo the reaction, probably for steric reasons. The reaction is reversible (by treatment of the addition product with either acid or base[138])[139] and is useful for the purification of the starting compounds, since the addition products are soluble in water and many of the impurities are not.[140]

OS **I**, 241, 336; **III**, 438; **IV**, 903; **V**, 437.

[130]For a review, see Olsen; Currie, in Patai *The Chemistry of the Thiol Group*, pt. 2; Wiley: New York, 1974, pp. 521-532.
[131]Kumar; Dev *Tetrahderon Lett.* **1983**, *24*, 1289.
[132]Ku; Oh *Synth. Commun.* **1989**, 433.
[133]Kakimoto; Seri; Imai *Synthesis* **1987**, 164.
[134]Tazaki; Takagi *Chem. Lett.* **1979**, 767.
[135]Evans; Grimm; Truesdale *J. Am. Chem. Soc.* **1975**, *97*, 3229.
[136]Mukaiyama; Saigo *Chem. Lett.* **1973**, 479.
[137]Kikugawa *Chem. Lett.* **1981**, 1157.
[138]For cleavage with ion-exchange resins, see Khusid; Chizhova *J. Org. Chem. USSR* **1985**, *21*, 37.
[139]For a discussion of the mechanism, see Young; Jencks *J. Am. Chem. Soc.* **1978**, *100*, 1228.
[140]The reaction has also been used to protect an aldehyde group in the presence of a keto group: Chihara; Wakabayashi; Taya *Chem. Lett.* **1981**, 1657.

D. Attack by NH_2, NHR, or NR_2 (Addition of NH_3, RNH_2, R_2NH)

6-13   The Addition of Ammonia to Aldehydes and Ketones
Formaldehyde–hexamethylenetetramine transformation

$$HCHO + NH_3 \longrightarrow \text{[structure 11]}$$

11

The addition of ammonia[141] to aldehydes or ketones does not generally give useful products. According to the pattern followed by analogous nucleophiles, the initial products would be expected to be *hemiaminals*[142] (also called "aldehyde ammonias") (**12**) and/or imines (**13**):

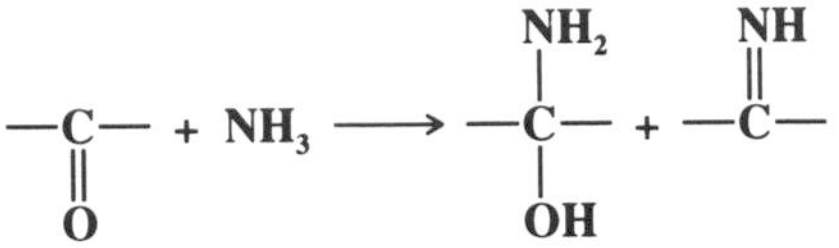

$$-\underset{\underset{O}{\|}}{C}- + NH_3 \longrightarrow -\underset{\underset{OH}{|}}{\overset{\overset{NH_2}{|}}{C}}- + -\overset{\overset{NH}{\|}}{C}-$$

12   **13**

However, these compounds are generally unstable. Most imines with a hydrogen on the nitrogen spontaneously polymerize.[143] Stable hemiaminals can be prepared from polychlorinated and polyfluorinated aldehydes and ketones, and diaryl ketones do give stable imines Ar_2C=NH.[144] Aside from these, when stable compounds *are* prepared in this reaction, they are the result of combinations and condensations of one or more molecules of **12** and/or **13** with each other or with additional molecules of ammonia or carbonyl compound. The most important example of such a product is hexamethylenetetramine[145] (**11**), prepared from ammonia and formaldehyde.[146] Aromatic aldehydes give hydrobenzamides $ArCH(N$=$CHAr)_2$ derived from three molecules of aldehyde and two of ammonia.[147]

OS **II**, 214, 219; **IV**, 451; **VI**, 664, 976. Also see OS **III**, 471; **V**, 897.

6-14   The Addition of Amines to Aldehydes and Ketones
Alkylimino-de-oxo-bisubstitution

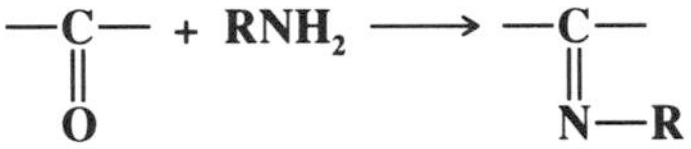

$$-\underset{\underset{O}{\|}}{C}- + RNH_2 \longrightarrow -\underset{\underset{N-R}{\|}}{C}-$$

Primary, secondary, and tertiary amines can add to aldehydes[148] and ketones to give different kinds of products. Primary amines give imines.[149] In contrast to imines in which the nitrogen

---

[141]For a review of this reagent in organic synthesis, see Jeyaraman, in Pizey *Synthetic Reagents*, vol. 5; Wiley: New York, 1983, pp. 9-83.

[142]These compounds have been detected by ^{13}C nmr: Chudek; Foster; Young *J. Chem. Soc., Perkin Trans. 2* **1985,** 1285.

[143]Methanimine CH_2=NH is stable in solution for several hours at $-95°C$, but rapidly decomposes at $-80°C$: Braillon; Lasne; Ripoll; Denis *Nouv. J. Chim.* **1982,** 6, 121. See also Bock; Dammel *Chem. Ber.* **1987,** 120, 1961.

[144]Verardo; Giumanini; Strazzolini; Poiana *Synth. Commun.* **1988,** 18, 1501.

[145]For a review of this compound, see Blažević; Kolbah; Belin; Šunjić; Kajfež *Synthesis* **1979,** 161-176.

[146]For a discussion of the mechanism, see Nielsen; Moore; Ogan; Atkins *J. Org. Chem.* **1979,** 44, 1678.

[147]Ogata; Kawasaki; Okumura *J. Org. Chem.* **1964,** 29, 1985; Crowell; McLeod *J. Org. Chem.* **1967,** 32, 4030.

[148]For a review of the reactions between amines and formaldehyde, see Farrar *Rec. Chem. Prog.* **1968,** 29, 85-101.

[149]For reviews of reactions of carbonyl compounds leading to the formation of C=N bonds, see Dayagi; Degani, in Patai *The Chemistry of the Carbon–Nitrogen Double Bond*; Ref. 40, pp. 64-83; Reeves, in Patai, Ref. 2, pp. 600-614.

is attached to a hydrogen (**6-13**), these imines are stable enough for isolation. However, in some cases, especially with simple R groups, they rapidly decompose or polymerize unless there is at least one aryl group on the nitrogen or the carbon. When there is an aryl group, the compounds are quite stable. They are usually called *Schiff bases*, and this reaction is the best way to prepare them. The reaction is straightforward and proceeds in high yields. The initial N-substituted hemiaminals[150] lose water to give the stable Schiff bases:

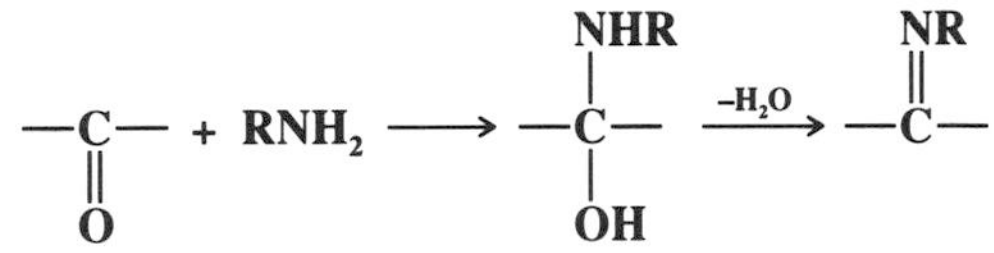

$$\underset{O}{\overset{\displaystyle |}{-C-}} \;+\; RNH_2 \;\longrightarrow\; \underset{OH}{\overset{NHR}{\underset{|}{\overset{|}{-C-}}}} \;\xrightarrow{-H_2O}\; \overset{NR}{\overset{\|}{-C-}}$$

In general, ketones react more slowly than aldehydes, and higher temperatures and longer reaction times are often required.[151] In addition, the equilibrium must often be shifted, usually by removal of the water, either azeotropically by distillation, or with a drying agent such as $TiCl_4$,[152] or with a molecular sieve.[153]

The reaction is often used to effect ring closure.[154] The *Friedländer quinoline synthesis*[155] is an example:

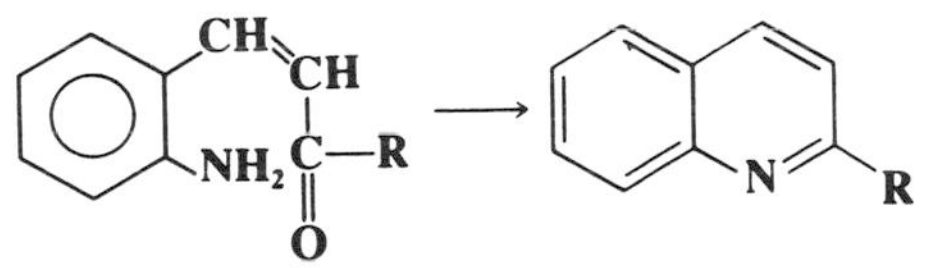

Pyrylium ions react with ammonia or primary amines to give pyridinium ions[156] (see p. 354).

When secondary amines are added to aldehydes or ketones, the initially formed N,N-disubstituted hemiaminals (**14**) cannot lose water in the same way, and it is possible to isolate them.[157] However, they are generally unstable, and under the reaction conditions

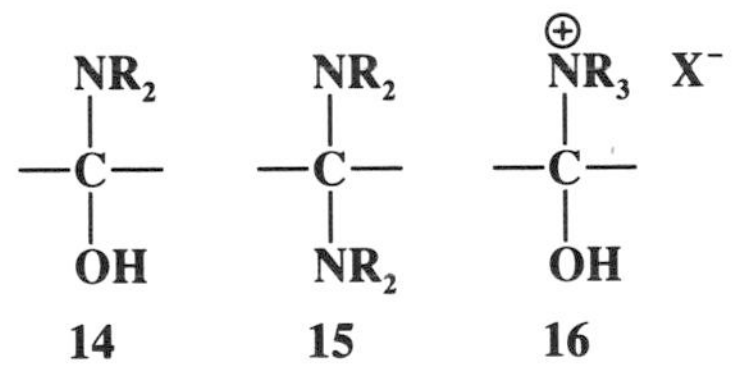

$$\underset{OH}{\overset{NR_2}{\underset{|}{\overset{|}{-C-}}}} \qquad \underset{NR_2}{\overset{NR_2}{\underset{|}{\overset{|}{-C-}}}} \qquad \underset{OH}{\overset{\overset{\oplus}{N}R_3 \;\; X^-}{\underset{|}{\overset{|}{-C-}}}}$$

$$\quad\;\; \textbf{14} \qquad\qquad\quad \textbf{15} \qquad\qquad\;\; \textbf{16}$$

$$\text{aminal}$$

usually react further. If no α hydrogen is present, **14** is converted to the more stable *aminal* (**15**).[158] However, if an α hydrogen is present, water (from **14**) or RNH_2 (from **15**) can be lost in that direction to give an enamine:[159]

[150]Some of these have been observed spectrally; see Forlani; Marianucci; Todesco *J. Chem. Res. (S)* **1984**, 126.

[151]For improved methods, see Morimoto; Sekiya *Chem. Lett.* **1985**, 1371; Eisch; Sanchez *J. Org. Chem.* **1986**, *51*, 1848.

[152]Weingarten; Chupp; White *J. Org. Chem.* **1967**, *32*, 3246.

[153]Bonnett; Emerson *J. Chem. Soc.* **1965**, 4508; Roelofsen; van Bekkum *Recl. Trav. Chim. Pays-Bays* **1972**, *91*, 605.

[154]For a review of such ring closures, see Katritzky; Ostercamp; Yousaf *Tetrahedron* **1987**, *43*, 5171-5186.

[155]For a review, see Cheng; Yan *Org. React.* **1982**, *28*, 37-201.

[156]For a review, see Zvezdina; Zhadonva; Dorofeenko *Russ. Chem. Rev.* **1982**, *51*, 469-484.

[157]For example, see Duhamel; Cantacuzène *Bull. Soc. Chim. Fr.* **1962**, 1843.

[158]For a review of aminals, see Duhamel, in Patai *The Chemistry of Functional Groups, Supplement F*, pt. 2; Wiley: New York, 1982, pp. 849-907.

[159]For reviews of the preparation of enamines, see Haynes; Cook, in Cook, Ref. 45, pp. 103-163; Pitacco; Valentin, in Patai, Ref. 158, pt. 1, pp. 623-714.

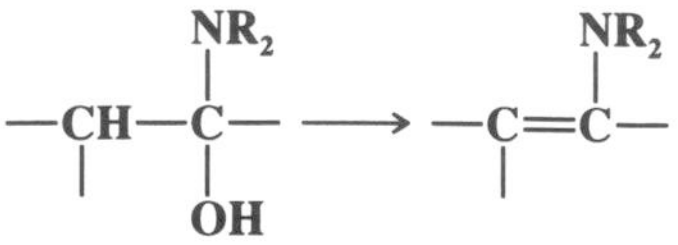

$$-\underset{|}{\overset{|}{C}}H-\underset{|}{\overset{NR_2}{\underset{|}{\overset{|}{C}}}}- \longrightarrow -\underset{|}{C}=\underset{|}{\overset{NR_2}{\overset{|}{C}}}-$$

This is the most common method[160] for the preparation of enamines and usually takes place when an aldehyde or ketone containing an α hydrogen is treated with a secondary amine. The water is usually removed azeotropically or with a drying agent,[161] but molecular sieves can also be used.[162] Secondary amine perchlorates react with aldehydes and ketones to give iminium salts (**2**, p. 885).[163] Tertiary amines can only give salts (**16**).

Amides can add to aldehydes in the presence of bases (so the nucleophile is actually $RCONH^-$) or acids to give acylated amino alcohols, which often react further to give alkylidene or arylidene bisamides:[164]

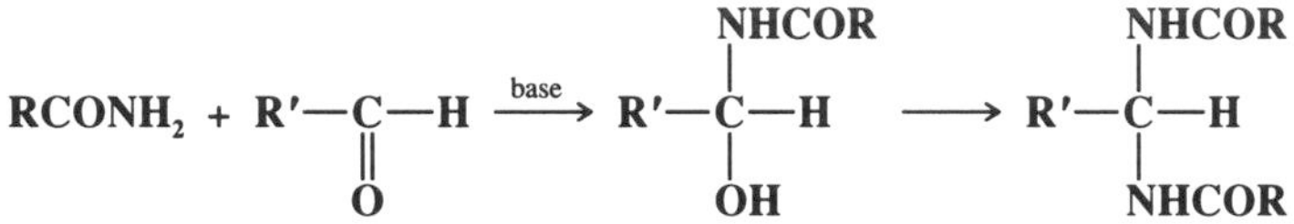

$$RCONH_2 + R'-\underset{\overset{\|}{O}}{C}-H \xrightarrow{\text{base}} R'-\underset{\overset{|}{OH}}{\overset{NHCOR}{\overset{|}{C}}}-H \longrightarrow R'-\underset{\overset{|}{NHCOR}}{\overset{NHCOR}{\overset{|}{C}}}-H$$

If the R' group contains an α hydrogen, water may split out.

OS **I**, 80, 355, 381; **II**, 31, 49, 65, 202, 231, 422; **III**, 95, 328, 329, 332, 358, 374, 513, 753, 827; **IV**, 210, 605, 638, 824; **V**, 191, 277, 533, 567, 627, 703, 716, 736, 758, 808, 941, 1070; **VI**, 5, 448, 474, 496, 520, 526, 592, 601, 818, 901, 1014; **VII**, 8, 135, 144, 473; **65**, 108, 119, 146, 183; **66**, 133, 142, 203; **68**, 206. Also see OS **IV**, 283, 464; **VII**, 197; **66**, 52; **69**, 55, 158.

6-15 Reductive Alkylation of Ammonia or Amines
Hydro,dialkylamino-de-oxo-bisubstitution

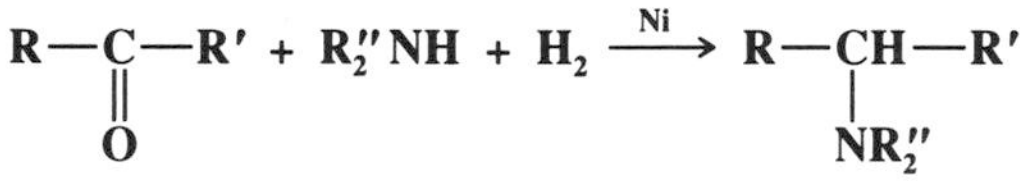

$$R-\underset{\overset{\|}{O}}{C}-R' + R_2''NH + H_2 \xrightarrow{\text{Ni}} R-\underset{\overset{|}{NR_2''}}{\overset{|}{C}}H-R'$$

When an aldehyde or a ketone is treated with ammonia or a primary or secondary amine in the presence of hydrogen and a hydrogenation catalyst (heterogeneous or homogeneous), *reductive alkylation* of ammonia or the amine (or *reductive amination* of the carbonyl compound) takes place.[165] The reaction can formally be regarded as occurring in the following manner (shown for a primary amine), which probably does correspond to the actual sequence of steps:[166]

---

[160]For another method, see Katritzky; Long; Lue; Jozwiak *Tetrahedron* **1990**, *46*, 8153.

[161]For example, TiCl₄: White; Weingarten *J. Org. Chem.* **1967**, *32*, 213; Kuo; Daly *J. Org. Chem.* **1970**, *35*, 1861; Nilsson; Carlson *Acta Chem. Scand. Sect. B* **1984**, *38*, 523.

[162]Brannock; Bell; Burpitt; Kelly *J. Org. Chem.* **1964**, *29*, 801; Taguchi; Westheimer *J. Org. Chem.* **1971**, *36*, 1570; Roelofsen; van Bekkum, Ref. 153; Carlson; Nilsson; Strömqvist *Acta Chem. Scand., Ser. B* **1983**, *37*, 7.

[163]Leonard; Paukstelis *J. Org. Chem.* **1964**, *28*, 3021.

[164]For reviews, see Challis; Challis, in Zabicky, Ref. 65, pp. 754-759; Zaugg; Martin *Org. React.* **1965**, *14*, 52-269, pp. 91-95, 104-112. For a discussion, see Gilbert *Synthesis* **1972**, 30.

[165]For reviews, see Rylander *Hydrogenation Methods*; Academic Press: New York, 1985, pp. 82-93; Klyuev; Khidekel *Russ. Chem. Rev.* **1980**, *49*, 14-27; Rylander, *Catalytic Hydrogenation over Platinum Metals;* Academic Press: New York, 1967, pp. 291-303.

[166]See, for example, Le Bris; Lefebvre; Coussemant *Bull. Soc. Chim. Fr.* **1964**, 1366, 1374, 1584, 1594.

$$
\begin{array}{ccc}
 & \overset{\displaystyle \text{NR}}{\underset{\displaystyle \|}{-\text{C}-}} & \\
 & \overset{\nearrow}{} \quad \underset{\text{6-26}}{} \quad \searrow \text{hydrogenation} & \\
\overset{\displaystyle\text{NHR}}{} & & \overset{\displaystyle\text{NHR}}{}
\end{array}
$$

$$
\underset{\underset{\displaystyle \text{O}}{\|}}{-\text{C}-} \; + \; \text{RNH}_2 \longrightarrow \underset{\underset{\displaystyle \text{OH}}{|}}{-\text{C}-} \xrightarrow[\text{0-78}]{\text{hydrogenol}} \underset{\underset{\displaystyle \text{H}}{|}}{-\text{C}-}
$$

For ammonia and primary amines there are two possible pathways, but when secondary amines are involved, only the hydrogenolysis pathway is possible. Other reducing agents[167] can be used instead of hydrogen and a catalyst, among them zinc and HCl, sodium cyanoborohydride $NaBH_3CN$,[168] sodium triacetoxyborohydride,[169] sodium borohydride,[170] iron pentacarbonyl and alcoholic KOH,[171] BH_3–pyridine,[172] and formic acid. When the last is used, the process is called the *Wallach reaction.* In the particular case where primary or secondary amines are reductively methylated with formaldehyde and formic acid, the method is called the *Eschweiler–Clarke procedure.* It is possible to use ammonium (or amine) salts of formic acid,[173] or formamides, as a substitute for the Wallach conditions. This method is called the *Leuckart reaction,*[174] and in this case the products obtained are often the N-formyl derivatives of the amines instead of the free amines. Primary and secondary amines can be N-ethylated (e.g., ArNHR → ArNREt) by treatment with $NaBH_4$ in acetic acid.[175]

When the reagent is ammonia, it is possible for the initial product to react again and for this product to react again, so that secondary and tertiary amines are usually obtained as side products:

$$
\underset{\underset{\displaystyle \text{O}}{\|}}{-\text{C}-} \longrightarrow \underset{\underset{\displaystyle \text{NH}_2}{|}}{-\text{CH}-} \longrightarrow \underset{\underset{\underset{\displaystyle -\text{CH}-}{|}}{\text{NH}}}{\overset{|}{-\text{CH}-}} \longrightarrow \underset{\underset{\underset{\displaystyle -\text{CH}}{|}}{\text{N}-\text{CH}-}}{\overset{|}{-\text{CH}-}}
$$

Similarly, primary amines give tertiary as well as secondary amines. In order to minimize this, the aldehyde or ketone is treated with an excess of ammonia or primary amine (unless of course the higher amine is desired).

Primary amines have been prepared from many aldehydes with at least five carbons and from many ketones by treatment with ammonia and a reducing agent. Smaller aldehydes are usually too reactive to permit isolation of the primary amine. Secondary amines have

---

[167]For a list of many of these, with references, see Ref. 64, pp. 421-423.

[168]Borch; Bernstein; Durst *J. Am. Chem. Soc.* **1971**, *93*, 2897; Mattson; Pham; Leuck; Cowen *J. Org. Chem.* **1990**, *55*, 2552. See also Barney; Huber; McCarthy *Tetrahedron Lett.* **1990**, *31*, 5547. For reviews of $NaBH_3CN$, see Hutchins; Natale *Org. Prep. Proced. Int.* **1979**, *11*, 201-246; Lane *Synthesis* **1975**, 135-146.

[169]Abdel-Magid; Maryanoff; Carson *Tetrahedron Lett.* **1990**, *31*, 5595.

[170]Schellenberg *J. Org. Chem.* **1963**, *28*, 3259; Gribble; Nutaitis *Synthesis* **1987**, 709.

[171]Watanabe; Yamashita; Mitsudo; Tanaka; Takegami *Tetrahedron Lett.* **1974**, 1879; Watanabe; Mitsudo; Yamashita; Shim; Takegami *Chem. Lett.* **1974**, 1265.

[172]Pelter; Rosser; Mills *J. Chem. Soc., Perkin Trans. 1* **1984**, 717.

[173]For a review of ammonium formate in organic synthesis, see Ram; Ehrenkaufer *Synthesis* **1988**, 91-95.

[174]For a review, see Moore, *Org. React.* **1949**, *5*, 301-330; for discussions of the mechanism, see Lukasiewicz *Tetrahedron* **1963**, *19*, 1789; Ito; Oba; Sekiya *Bull. Chem. Soc. Jpn.* **1976**, *49*, 2485; Awachie; Agwada *Tetrahedron* **1990**, *46*, 1899.

[175]Gribble; Lord; Skotnicki; Dietz; Eaton; Johnson *J. Am. Chem. Soc.* **1974**, *96*, 7812; Gribble; Jasinski; Pellicone; Panetta *Synthesis* **1978**, 766. See also Marchini; Liso; Reho; Liberatore; Moracci *J. Org. Chem.* **1975**, *40*, 3453. For a review, see Gribble; Nutaitis *Org. Prep. Proced. Int.* **1985**, *17*, 317-384, pp. 336-350.

been prepared by both possible procedures: 2 moles of ammonia and 1 mole of aldehyde or ketone, and 1 mole of primary amine and 1 mole of carbonyl compound, the latter method being better for all but aromatic aldehydes. Tertiary amines can be prepared in three ways, but the method is seldom carried out with 3 moles of ammonia and 1 mole of carbonyl compound. Much more often they are prepared from primary or secondary amines.[176] The most common method for this purpose is the Eschweiler-Clarke procedure, i.e., treatment of the primary or secondary amine with formaldehyde and formic acid. Amines of the form $RNMe_2$ and R_2NMe are prepared in this manner.[177] Another method for accomplishing the conversions $RNH_2 \rightarrow RNMe_2$ and $R_2NH \rightarrow R_2NMe$ is to treat the amine with aqueous formaldehyde and $NaBH_4$[178] or $NaBH_3CN$.[179]

Reductive alkylation has also been carried out on nitro, nitroso, azo, and other compounds that are reduced in situ to primary or secondary amines.

OS **I**, 347, 528, 531; **II**, 503; **III**, 328, 501, 717, 723; **IV**, 603; **V**, 552; **VI**, 499; **VII**, 27.

6-16 The Mannich Reaction
Acyl,amino-de-oxo-bisubstitution, etc.

$$H-\underset{\underset{O}{\|}}{C}-H \; + \; NH_4Cl \; + \; CH_3-\underset{\underset{O}{\|}}{C}-R \; \xrightarrow[OH^-]{H^+ \, or} \; H_2N-CH_2-CH_2-\underset{\underset{O}{\|}}{C}-R$$

In the *Mannich reaction*, formaldehyde (or sometimes another aldehyde) is condensed with ammonia, in the form of its salt, and a compound containing an active hydrogen.[180] This can formally be considered as an addition of ammonia to give H_2NCH_2OH, followed by a nucleophilic substitution. Instead of ammonia, the reaction can be carried out with salts of primary or secondary amines,[181] or with amides,[182] in which cases the product is substituted on the nitrogen with R, R_2, and RCO, respectively. Arylamines do not normally give the reaction. The product is referred to as a *Mannich base*. Many active hydrogen compounds give the reaction. Among these are the following types, with the active hydrogen underlined:

$$-\underset{|}{C}\underline{H}-COR \text{ (or H)} \quad -\underset{|}{C}\underline{H}-COOR \text{ (or H)} \quad -\underset{|}{C}\underline{H}-NO_2 \quad -\underset{|}{C}\underline{H}-CN$$

RC≡C$\underline{H}$ $\underline{H}$CN   RO$\underline{H}$ RS$\underline{H}$

See **6-50**

See **1-25**

---

[176]For a review of the preparation of tertiary amines by reductive alkylation, see Spialter; Pappalardo *The Acyclic Aliphatic Tertiary Amines*; Macmillan: New York, 1965, pp. 44-52.

[177]For a discussion, see Pine; Sanchez *J. Org. Chem.* **1971**, *36*, 829.

[178]Sondengam; Hentchoya Hémo; Charles *Tetrahedron Lett.* **1973**, 261.

[179]Borch; Hassid *J. Org. Chem.* **1972**, *37*, 1673; Kapnang; Charles; Sondengam; Hentchoya Hémo *Tetrahedron Lett.* **1977**, 3469. See also Ref. 168.

[180]For reviews, see Tramontini; Angiolini *Tetrahedron* **1990**, *46*, 1791-1837; Gevorgyan; Agababyan; Mndzhoyan *Russ. Chem. Rev.* **1984**, *53*, 561-581; Tramontini *Synthesis* **1973**, 703-775; House *Modern Synthetic Reactions*, 2nd ed.; W.A. Benjamin: New York, 1972, pp. 654-660. For reviews of Mannich reactions in which the active-hydrogen component is a thiol, see Massy *Synthesis* **1987**, 589-603; Dronov; Nikitin *Russ. Chem. Rev.* **1985**, *54*, 554-561; in which it is a nitro compound, see Baer; Urbas, in Feuer *The Chemistry of the Nitro and Nitroso Groups*; Wiley: New York, 1970, pp. 117-130. For reviews on the reactions of Mannich Bases, see Tramontini; Angeloni, cited above; Gevorgyan; Agababyan; Mndzhoyan *Russ. Chem. Rev.* **1985**, *54*, 495-514.

[181]For a review where the amine component is an amino acid, see Agababyan; Gevorgyan; Mndzhoyan *Russ. Chem. Rev.* **1982**, *51*, 387-396.

[182]Hellmann, *Angew. Chem.* **1957**, *69*, 463, *Newer Methods Prep. Org. Chem.* **1963**, *2*, 277-302.

The Mannich base can react further in three ways. If it is a primary or secondary amine, it may condense with one or two additional molecules of aldehyde and active compound, e.g.,

$$\text{H}_2\text{NCH}_2\text{CH}_2\text{COR} \xrightarrow[\text{CH}_3\text{COR}]{\text{HCHO}} \text{HN}(\text{CH}_2\text{CH}_2\text{COR})_2 \xrightarrow[\text{CH}_3\text{COR}]{\text{HCHO}} \text{N}(\text{CH}_2\text{CH}_2\text{COR})_3$$

If the active hydrogen compound has two or three active hydrogens, the Mannich base may condense with one or two additional molecules of aldehyde and ammonia or amine, e.g,

$$\text{H}_2\text{NCH}_2\text{CH}_2\text{COR} \xrightarrow[\text{NH}_3]{\text{HCHO}} (\text{H}_2\text{NCH}_2)\text{CHCOR} \xrightarrow[\text{NH}_3]{\text{HCHO}} (\text{H}_2\text{NCH}_2)_3\text{CCOR}$$

Another further reaction consists of condensation of the Mannich base with excess formaldehyde:

$$\text{H}_2\text{NCH}_2\text{CH}_2\text{COR} + \text{HCHO} \longrightarrow \text{H}_2\text{C}=\text{NCH}_2\text{CH}_2\text{COR}$$

Sometimes it is possible to obtain these products of further condensation as the main products of the reaction. At other times they are side products.

When the Mannich base contains an amino group β to a carbonyl (and it usually does), ammonia is easily eliminated. This is a route to α,β-unsaturated aldehydes, ketones, esters, etc.:

$$\text{H}_2\text{NCH}_2\text{CH}_2\text{COR} \xrightarrow{\Delta} \text{H}_2\text{C}=\text{CHCOR}$$

The Mannich reaction is an important biosynthetic route to natural products, mainly alkaloids, and some of these routes have been duplicated in the laboratory. A classic example is the synthesis of tropinone (**17**) by Robinson in 1917. Robinson synthesized tropinone by a Mannich reaction involving succindialdehyde, methylamine, and acetone:[183]

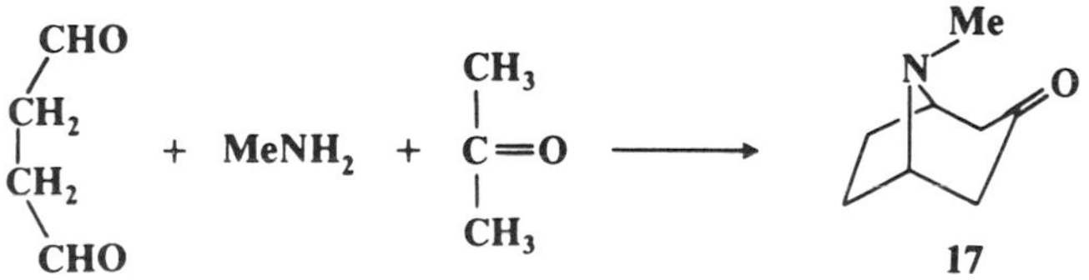

Studies of the reaction kinetics have led to the following proposals for the mechanism of the Mannich reaction.[184]

The base-catalyzed reaction

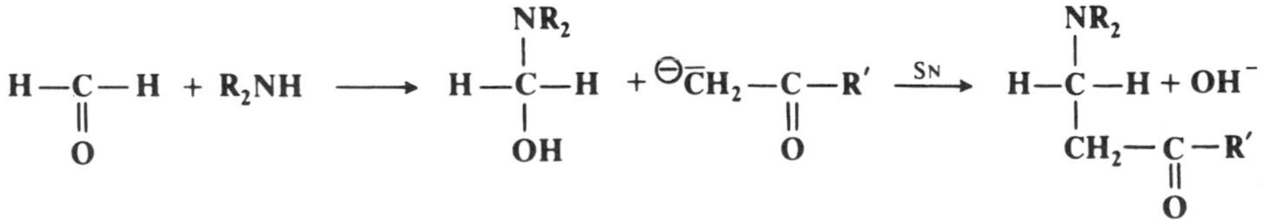

[183]Robinson *J. Chem. Soc.* **1917**, *111*, 762.
[184]Cummings; Shelton *J. Org. Chem.* **1960**, *25*, 419.

The acid-catalyzed reaction

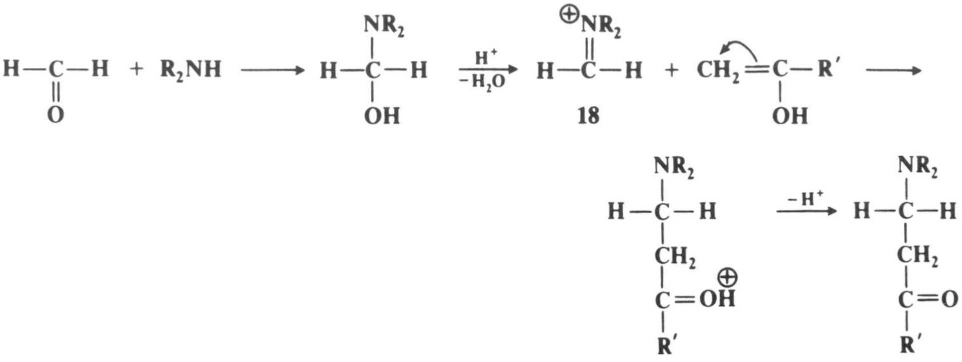

$$H-\underset{\underset{O}{\|}}{C}-H + R_2NH \longrightarrow H-\underset{\underset{OH}{|}}{C}-H \xrightarrow[-H_2O]{H^+} H-\underset{\underset{\mathbf{18}}{|}}{\overset{\overset{\oplus NR_2}{\|}}{C}}-H + \overset{\curvearrowleft}{CH_2}\!\!=\!\!\underset{\underset{OH}{|}}{C}-R' \longrightarrow$$

$$H-\underset{\underset{\underset{\underset{R'}{|}}{C=\overset{\oplus}{O}H}}{|}}{\overset{\overset{NR_2}{|}}{C}}-H \xrightarrow{-H^+} H-\underset{\underset{\underset{\underset{R'}{|}}{C=O}}{|}}{\overset{\overset{NR_2}{|}}{C}}-H$$

According to this mechanism, it is the free amine, not the salt that reacts, even in acid solution; and the active-hydrogen compound (in the acid-catalyzed process) reacts as the enol when that is possible. This latter step is similar to what happens in **2-4**. There is kinetic evidence for the intermediacy of the iminium ion (**18**).[185]

When it is desired to use an unsymmetrical ketone as the active-hydrogen component, it is possible to get two products. Regioselectivity has been obtained by treatment of the ketone with preformed iminium ions:[186] the use of $Me_2\overset{\oplus}{N}\!\!=\!\!CH_2$ CF_3COO^- in CF_3COOH gives substitution at the more highly substituted position, while with iso-$Pr_2\overset{\oplus}{N}\!\!=\!\!CH_2$ ClO_4^- the reaction takes place at the less highly substituted position.[187] The preformed iminium compound dimethyl(methylene)ammonium iodide $CH_2\!\!=\!\!NMe_2^+$ I^-, called *Eschenmoser's salt*,[188] has also been used in Mannich reactions.[189]

Another type of preformed reagent (**20**) has been used to carry out diastereoselective Mannich reactions. The lithium salts **19** are treated with $TiCl_4$ to give **20**, which is then treated with the enolate of a ketone.[190]

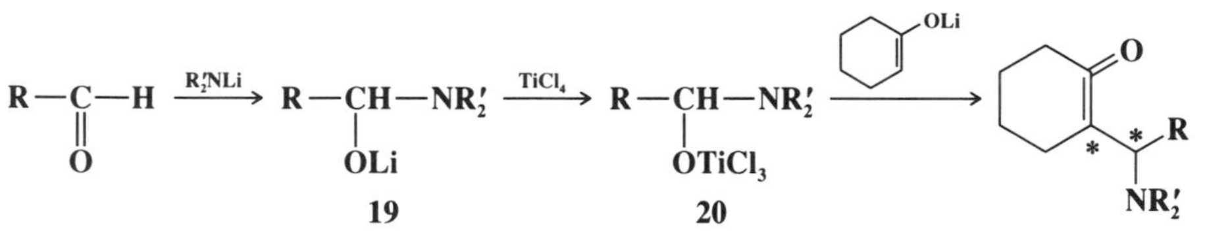

$$R-\underset{\underset{O}{\|}}{C}-H \xrightarrow{R_2'NLi} R-\underset{\underset{OLi}{|}}{C}H-NR_2' \xrightarrow{TiCl_4} R-\underset{\underset{OTiCl_3}{|}}{C}H-NR_2'$$
$$\qquad\qquad\qquad\qquad \mathbf{19} \qquad\qquad\qquad\qquad \mathbf{20}$$

Also see **6-50** and **1-25**.

OS **III**, 305; **IV**, 281, 515, 816; **VI**, 474, 981, 987; **VII**, 34. See also OS **68**, 188.

[185]Benkovic; Benkovic; Comfort *J. Am. Chem. Soc.* **1969**, *91*, 1860.

[186]For earlier use of preformed iminium ions in the Mannich reaction, see Ahond; Cavé; Kan-Fan; Husson; de Rostolan; Potier *J. Am. Chem. Soc.* **1968**, *90*, 5622; Ahond; Cavé; Kan-Fan; Potier *Bull. Soc. Chim. Fr.* **1970**, 2707; Ref. 188.

[187]Jasor; Luche; Gaudry; Marquet *J. Chem. Soc., Chem. Commun.* **1974**, 253; Gaudry; Jasor; Khac *Org. Synth.* *VI*, 474.

[188]Schreiber; Maag; Hashimoto; Eschenmoser *Angew. Chem. Int. Ed. Engl.* **1971**, *10*, 330 [*Angew. Chem. 83*, 355].

[189]See Holy; Fowler; Burnett; Lorenz *Tetrahedron* **1979**, *35* 613; Bryson; Bonitz; Reichel; Dardis *J. Org. Chem.* **1980**, *45*, 524, and references cited in these papers.

[190]Seebach; Betschart; Schweizer *Helv. Chim. Acta* **1984**, *67*, 1593; Seebach; Schiess; Schweizer *Chimia* **1985**, *39*, 272. See also Heaney; Papageorgiou; Wilkins *J. Chem. Soc., Chem. Commun.* **1988**, 1161; Katritzky; Harris *Tetrahedron* **1990**, *46*, 987.

6-17   The Addition of Amines to Isocyanates
N-Hydro-C-alkylamino-addition

$$R-N=C=O \ + \ R'NH_2 \ \longrightarrow \ R-NH-\underset{\underset{NHR'}{|}}{C}=O$$

Ammonia and primary and secondary amines can be added to isocyanates[191] to give substituted ureas.[192] Isothiocyanates give thioureas. This is an excellent method for the preparation of ureas and thioureas, and these compounds are often used as derivatives for primary
and secondary amines. Isocyanic acid HNCO also gives the reaction; usually its salts, e.g.,
NaNCO, are used. Wöhler's famous synthesis of urea involved the addition of ammonia to
a salt of this acid.[193]

OS **II**, 79; **III**, 76, 617, 735; **IV**, 49, 180, 213, 515, 700; **V**, 555, 801, 802, 967; **VI**, 936,
951; **65**, 173.

6-18   The Addition of Ammonia or Amines to Nitriles
N-Hydro-C-amino-addition

$$R-C\equiv N \ + \ NH_3 \ \xrightarrow[\text{pressure}]{NH_4Cl} \ R-\underset{\underset{NH_2}{|}}{C}=NH_2^+ \ \ Cl^-$$

Unsubstituted amidines (in the form of their salts) can be prepared by addition of ammonia
to nitriles.[194] Many amidines have been made in this way. Dinitriles of suitable chain length
can give imidines:[195]

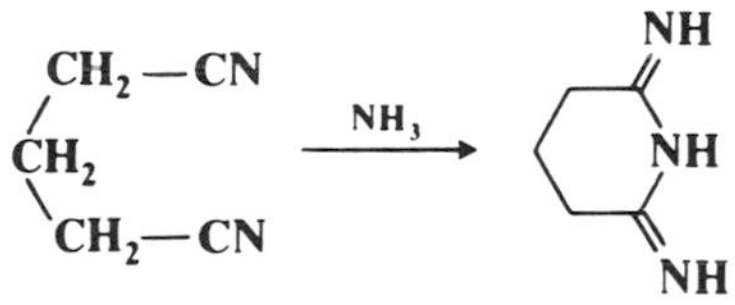

$$\begin{matrix} CH_2-CN \\ | \\ CH_2 \\ | \\ CH_2-CN \end{matrix} \xrightarrow{NH_3} \text{(cyclic imidine structure with NH groups)}$$

Primary and secondary amines can be used instead of ammonia, to give substituted
amidines, but only if the nitrile contains electron-withdrawing groups; e.g., Cl_3CCN gives
the reaction. Ordinary nitriles do not react, and, in fact, acetonitrile is often used as a
solvent in this reaction.[196] However, ordinary nitriles can be converted to amidines by
treatment with an alkylchloroaluminum amide $MeAl(Cl)NR_2$ (R = H or Me).[197] The addition of ammonia to cyanamide NH_2CN gives guanidine $(NH_2)_2C=NH$.

If water is present, and a ruthenium complex catalyst is used, the addition of a

---

[191]For a review of the mechanism, see Satchell; Satchell, Ref. 46.
[192]For a review of substituted ureas, see Vishnyakova; Golubeva; Glebova *Russ. Chem. Rev.* **1985**, *54*, 249-261.
[193]For a history of the investigation of the mechanism of the Wöhler synthesis, see Shorter, *Chem. Soc. Rev.* **1978**,
7, 1-14. See also Williams; Jencks *J. Chem. Soc., Perkin Trans. 2* **1974**, 1753, 1760; Hall; Watts *Aust. J. Chem.* **1977**,
30, 781, 903.
[194]For reviews of amidines, see Granik *Russ. Chem. Rev.* **1983**, *52*, 377-393; Gautier; Miocque; Farnoux, in Patai,
Ref. 103, pp. 283-348.
[195]Elvidge; Linstead; Salaman *J. Chem. Soc.* **1959**, 208.
[196]Grivas; Taurins *Can. J. Chem.* **1961**, *39*, 761.
[197]Garigipati *Tetrahedron Lett.* **1990**, *31*, 1969.

primary or secondary amine to a nitrile gives an amide: $RCN + R'NHR'' + H_2O \rightarrow$ $RCONR'R'' + NH_3$ (R'' may be H).[198]

OS **I**, 302 [but also see OS **V**, 589]; **IV**, 245, 247, 515, 566, 769. See also OS **V**, 39.

6-19 The Addition of Amines to Carbon Disulfide and Carbon Dioxide
S-Metallo-C-alkylamino-addition

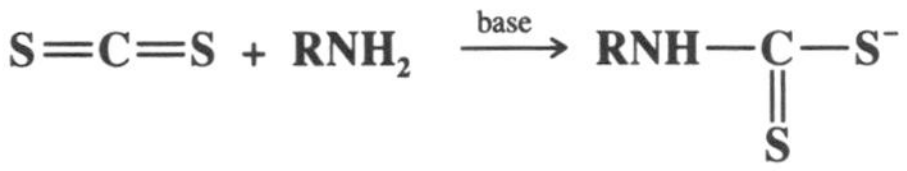

$$S{=}C{=}S + RNH_2 \xrightarrow{\text{base}} RNH{-}\underset{\underset{S}{\|}}{C}{-}S^-$$

Salts of dithiocarbamic acid can be prepared by the addition of primary or secondary amines to carbon disulfide.[199] This reaction is similar to **6-10**. H_2S can be eliminated from the product, directly or indirectly, to give isothiocyanates RNCS. Isothiocyanates can be obtained directly by the reaction of primary amines and CS_2 in pyridine in the presence of dicyclohexylcarbodiimide.[200] In the presence of diphenyl phosphite and pyridine, primary amines add to CO_2 and to CS_2 to give, respectively, symmetrically substituted ureas and thioureas:[201]

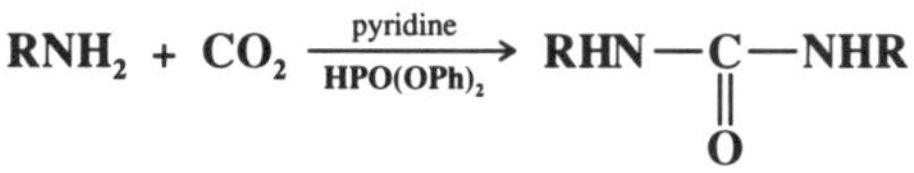

$$RNH_2 + CO_2 \xrightarrow[\text{HPO(OPh)}_2]{\text{pyridine}} RHN{-}\underset{\underset{O}{\|}}{C}{-}NHR$$

OS **I**, 447; **III**, 360, 394, 599, 763; **V**, 223.

E. Other Nitrogen Nucleophiles

6-20 The Addition of Hydrazine Derivatives to Carbonyl Compounds
Hydrazono-de-oxo-bisubstitution

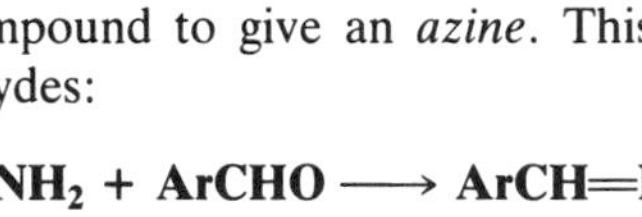

$$\underset{\underset{O}{\|}}{-}C{-} \xrightarrow{RNHNH_2} \underset{\underset{N{-}NHR}{\|}}{-}C{-}$$

The product of condensation of a hydrazine and an aldehyde or ketone is called a *hydrazone*. Hydrazine itself gives hydrazones only with aryl ketones. With other aldehydes and ketones, either no useful product can be isolated, or the remaining NH_2 group condenses with a second mole of carbonyl compound to give an *azine*. This type of product is especially important for aromatic aldehydes:

$$ArCH{=}N{-}NH_2 + ArCHO \longrightarrow ArCH{=}N{-}N{=}CHAr$$

An azine

[198]Murahashi; Naota; Saito *J. Am. Chem. Soc.* **1986**, *108*, 7846.

[199]For reviews, see Ref. 106, pp. 226-315; Katritzky; Faid-Allah; Marson *Heterocycles* **1987**, *26*, 1657-1670; Yokoyama; Imamoto *Synthesis* **1984**, 797-824, pp. 804-812. For a review of the addition of heterocyclic amines to CO_2 to give, e.g., salts of pyrrole-1-carboxylic acids, see Katritzky; Marson; Faid-Allah *Heterocycles* **1987**, *26*, 1333-1344.

[200]Jochims *Chem. Ber.* **1968**, *101*, 1746. For other methods, see Sakai; Fujinami; Aizawa *Bull. Chem. Soc. Jpn.* **1975**, *48*, 2981; Gittos; Davies; Iddon; Suschitzky *J. Chem. Soc., Perkin Trans. 1* **1976**, 141; Shibanuma; Shiono; Mukaiyama *Chem. Lett.* **1977**, 573; Molina; Alajarin; Arques *Synthesis* **1982**, 596.

[201]Yamazaki; Higashi; Iguchi *Tetrahedron Lett.* **1974**, 1191. For other methods for the conversion of amines and CO_2 to ureas, see Ogura; Takeda; Tokue; Kobayashi *Synthesis* **1978**, 394; Fournier; Bruneau; Dixneuf; Lécolier *J. Org. Chem.* **1991**, *56*, 4456.

However, in some cases azines can be converted to hydrazones by treatment with excess hydrazine and NaOH.[202] Arylhydrazines, especially phenyl, *p*-nitrophenyl, and 2,4-dinitrophenyl,[203] are used much more often and give the corresponding hydrazones with most aldehydes and ketones.[204] Since these are usually solids, they make excellent derivatives and are commonly employed for this purpose. α-Hydroxy aldehydes and ketones and α-dicarbonyl compounds give *osazones*, in which two adjacent carbons have carbon–nitrogen double bonds:

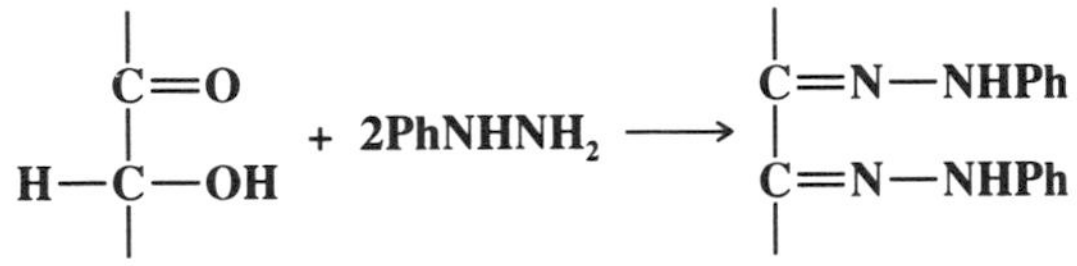

An osazone

Osazones are particularly important in carbohydrate chemistry. In contrast to this behavior, β-diketones and β-keto esters give *pyrazoles* and *pyrazolones*, respectively (illustrated for β-keto esters):

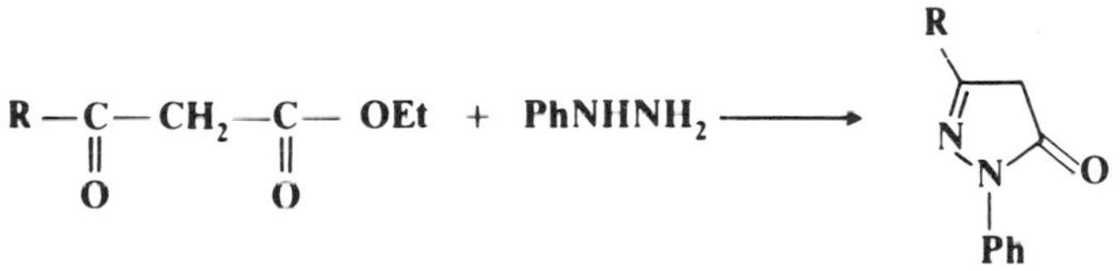

Other hydrazine derivatives frequently used to prepare the corresponding hydrazone are semicarbazide $NH_2NHCONH_2$, in which case the hydrazone is called a semicarbazone, and *Girard's reagents T and P*, in which case the hydrazone is water-soluble because of the ionic

Girard's reagent T   Girard's reagent P

group. Girard's reagents are often used for purification of carbonyl compounds.[205]

Simple N-unsubstituted hydrazones can be obtained by an exchange reaction. The N,N-dimethylhydrazone is prepared first and then treated with hydrazine:[206]

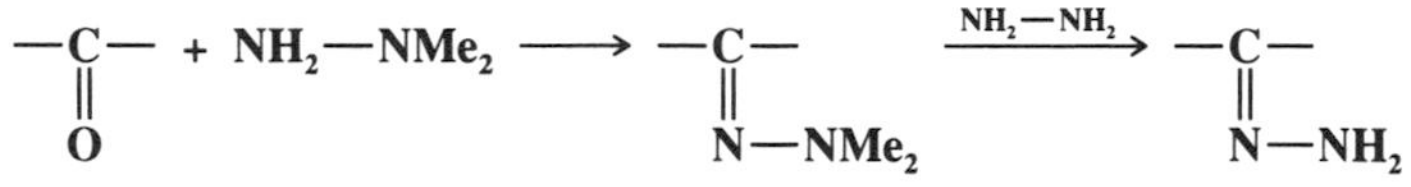

No azines are formed under these conditions.

[202]For example, see Day; Whiting *Org. Synth. VI*, 10.
[203]For an improved procedure for the preparation of 2,4-dinitrophenylhydrazones, see Behforouz; Bolan; Flynt *J. Org. Chem.* **1985**, *50*, 1186.
[204]For a review of arylhydrazones, see Buckingham *Q. Rev., Chem. Soc.* **1969**, *23*, 37-56.
[205]For a study of the mechanism with Girard's reagent T, see Stachissini; do Amaral *J. Org. Chem.* **1991**, *56*, 1419.
[206]Newkome; Fishel *J. Org. Chem.* **1966**, *31*, 677.

OS **II**, 395; **III**, 96, 351; **IV**, 351, 377, 536, 884; **V**, 27, 258, 747, 929; **VI**, 10, 12, 62, 242, 293, 679, 791; **VII**, 77, 438. Also see OS **III**, 708; **VI**, 161; **66**, 142.

6-21   The Formation of Oximes
Hydroxyimino-de-oxo-bisubstitution

$$\underset{\displaystyle O}{\overset{\displaystyle |}{-C-}} \, \| \quad + \ NH_2OH \ \longrightarrow \ \underset{\displaystyle N-OH}{\overset{\displaystyle |}{-C-}} \, \|$$

In a reaction very much like **6-20,** oximes can be prepared by the addition of hydroxylamine to aldehydes or ketones. Derivatives of hydroxylamine, e.g., H_2NOSO_3H and $HON(SO_3Na)_2$, have also been used. For hindered ketones, such as hexamethylacetone, high pressures, e.g., 10,000 atm, may be necessary.[207]

It has been shown[208] that the rate of formation of oximes is at a maximum at a pH which depends on the substrate but is usually about 4, and that the rate decreases as the pH is either raised or lowered from this point. We have previously seen (p. 332) that bell-shaped curves like this are often caused by changes in the rate-determining step. In this case, at low pH values step 2 is rapid (because it is acid-catalyzed), and step 1 is slow (and rate-

$$\underset{\displaystyle O}{\overset{\displaystyle |}{-C-}} \, \| \ + \ NH_2OH \ \xrightarrow{\ 1\ } \ \underset{\displaystyle OH}{\overset{\displaystyle NHOH}{-C-}} \ \xrightarrow{\ 2\ } \ \underset{}{\overset{\displaystyle N-OH}{-C-}} \, \|$$

21

determining), because under these acidic conditions most of the NH_2OH molecules have been converted to the conjugate NH_3OH^+ ions, which cannot attack the substrate. As the pH is slowly increased, the fraction of free NH_2OH molecules increases and consequently so does the reaction rate, until the maximum rate is reached at about pH = 4. As the rising pH has been causing an increase in the rate of step 1, it has also been causing a *decrease* in the rate of the acid-catalyzed step 2, although this latter process has not affected the overall rate since step 2 was still faster than step 1. However, when the pH goes above about 4, step 2 becomes rate-determining, and although the rate of step 1 is still increasing (as it will until essentially all the NH_2OH is unprotonated), it is now step 2 that determines the rate, and this step is slowed by the decrease in acid concentration. Thus the overall rate decreases as the pH rises beyond about 4. It is likely that similar considerations apply to the reaction of aldehydes and ketones with amines, hydrazines, and other nitrogen nucleophiles.[209] There is evidence that when the nucleophile is 2-methylthiosemicarbazide, there is a second change in the rate-determining step: above pH about 10 *basic* catalysis of step 2 has increased the rate of this step to the point where step 1 is again rate-determining.[210] Still a third change in the rate-determining step has been found at about pH = 1, showing

[207]Jones; Tristram; Benning *J. Am. Chem. Soc.* **1959,** *81,* 2151.
[208]Jencks *J. Am. Chem. Soc.* **1959,** *81,* 475, *Prog. Phys. Org. Chem.* **1964,** *2,* 63-128.
[209]For reviews of the mechanism of such reactions, see Cockerill; Harrison, in Patai *The Chemistry of Functional Groups: Supplement A,* pt. 1; Wiley: New York, 1977, pp. 288-299; Sollenberger; Martin, in Patai *The Chemistry of the Amino Group*; Wiley: New York, 1968, pp. 367-392. For isotope effect studies, see Rossi; Stachissini; do Amaral *J. Org. Chem.* **1990,** *55,* 1300.
[210]Sayer; Jencks *J. Am. Chem. Soc.* **1972,** *94,* 3262.

that at least in some cases step 1 actually consists of two steps: formation of a zwitterion,

e.g., $\overset{\oplus}{HO\overset{|}{N}H_2}\overset{|}{-}\overset{|}{C}\overset{|}{-}O^{\ominus}$ in the case shown above, and conversion of this to **21**.[211] The intermediate **21** has been detected by nmr in the reaction between NH_2OH and acetaldehyde.[212]

In another type of process, oximes can be obtained by passing a mixture of ketone vapor, NH_3, and O_2 over a silica-gel catalyst.[213] Ketones can also be converted to oximes by treatment with other oximes, in a transoximation reaction.[214]

OS **I**, 318, 327; **II**, 70, 204, 313, 622; **III**, 690, **IV**, 229; **V**, 139, 1031; **VII**, 149. See also OS **VI**, 670.

6-22 The Conversion of Aldehydes to Nitriles
Nitrilo-de-hydro,oxo-tersubstitution

$$R-\underset{\underset{O}{\|}}{C}-H \ + \ NH_2OH\cdot HCl \ \xrightarrow{\text{HCOOH}} \ R-C\equiv N$$

Aldehydes can be converted to nitriles in one step by treatment with hydroxylamine hydrochloride and either formic acid,[215] concentrated HCl,[216] SeO_2,[217] $MeNO_2$–polyphosphoric acid,[218] or pyridine–toluene.[219] The reaction is a combination of **6-21** and **7-37**. Direct nitrile formation has also been accomplished with certain derivatives of NH_2OH, notably, N,O-bistrifluoroacetylhydroxylamine $F_3CCONHOCOCF_3$[220] and NH_2OSO_2OH.[221] Another method involves treatment with hydrazoic acid, though the Schmidt reaction (**8-17**) may compete.[222] Aromatic aldehydes have been converted to nitriles in good yield with $NH_4H_2PO_4$ and nitropropane in acetic acid,[223] with trimethylsilyl azide,[224] with S,S-dimethylsulfurdiimide,[225] with NH_4Cl–O_2–Cu in pyridine,[226] with hydroxylamine hydrochloride, $MgSO_4$, and TsOH,[227] and with ammonia and iodine or lead tetraacetate.[228]

[211]Rosenberg; Silver; Sayer; Jencks *J. Am. Chem. Soc.* **1974**, *96*, 7986; Sayer; Pinsky; Schonbrunn; Washtien *J. Am. Chem. Soc.* **1974**, *96*, 7998; Sayer; Edman *J. Am. Chem. Soc.* **1979**, *101*, 3010.
[212]Cocivera; Fyfe; Effio; Vaish; Chen *J. Am. Chem. Soc.* **1976**, *98*, 1573; Cocivera; Effio *J. Am. Chem. Soc.* **1976**, *98*, 7371.
[213]Armor *J. Am. Chem. Soc.* **1980**, *102*, 1453.
[214]For example, see Block; Newman *Org. Synth. V*, 1031.
[215]Olah; Keumi *Synthesis* **1979**, 112.
[216]Findlay; Tang *Can. J. Chem.* **1967**, *45*, 1014.
[217]Sosnovsky; Krogh; Umhoefer *Synthesis* **1979**, 722.
[218]Ganboa; Palomo *Synth. Commun.* **1983**, *13*, 999.
[219]Saednya *Synthesis* **1982**, 190.
[220]Pomeroy; Craig *J. Am. Chem. Soc.* **1959**, *81*, 6340.
[221]Streith; Fizet; Fritz *Helv. Chim. Acta* **1976**, *59*, 2786.
[222]For additional methods, see Glass; Hoy *Tetrahedron Lett.* **1976**, 1781; Ikeda; Machii; Okahara *Synthesis* **1978**, 301; Nakagawa; Mineo; Kawamura; Horikawa; Tokumoto; Mori *Synth. Commun.* **1979**, *9*, 529; Furukawa; Fukumura; Akasaka; Yoshimura; Oae *Tetrahedron Lett.* **1980**, *21*, 761; Gelas-Mialhe; Vessière *Synthesis* **1980**, 1005; Arques; Molina; Soler *Synthesis* **1980**, 702; Sato; Itoh; Itoh; Nishina; Goto; Saito *Chem. Lett.* **1984**, 1913; Reddy; Reddy *Synth. Commun.* **1988**, *18*, 2179; Neunhoeffer; Diehl; Karafiat *Liebigs Ann. Chem.* **1989**, 105; Said; Skarżewski; Młochowski *Synthesis* **1989**, 223.
[223]Blatter; Lukaszewski; de Stevens, *J. Am. Chem. Soc.* **1961**, *83*, 2203. See also Dauzonne; Demerseman; Royer *Synthesis* **1981**, 739; Karmarkar; Kelkar; Wadia *Synthesis* **1985**, 510.
[224]Nishiyama; Oba; Watanabe *Tetrahedron* **1987**, *43*, 693.
[225]Georg; Pfeifer; Haake *Tetrahedron Lett.* **1985**, *26*, 2739.
[226]Capdevielle; Lavigne; Maumy *Synthesis* **1989**, 451. See also Yamazaki; Yamazaki *Chem. Lett.* **1990**, 571.
[227]Ganboa; Palomo *Synth. Commun.* **1983**, *13*, 219.
[228]Misono; Osa; Koda *Bull. Chem. Soc. Jpn.* **1966**, *39*, 854, **1967**, *40*, 2875; Parameswaran; Friedman *Chem. Ind.* (*London*) **1965**, 988.

On treatment with two equivalents of dimethylaluminum amide Me_2AlNH_2, carboxylic esters can be converted to nitriles: $RCOOR' \rightarrow RCN$.[229] This is very likely a combination of **0-55** and **7-39**.

See also **9-5**.

OS **V**, 656.

F. Halogen Nucleophiles

6-23   The Formation of α-Halo Ethers
Alkoxy, halo-de-oxo-bisubstitution

$$-\overset{\displaystyle ||}{\underset{\displaystyle O}{C}}- + ROH + HCl \longrightarrow -\overset{\displaystyle Cl}{\underset{\displaystyle OR}{C}}-$$

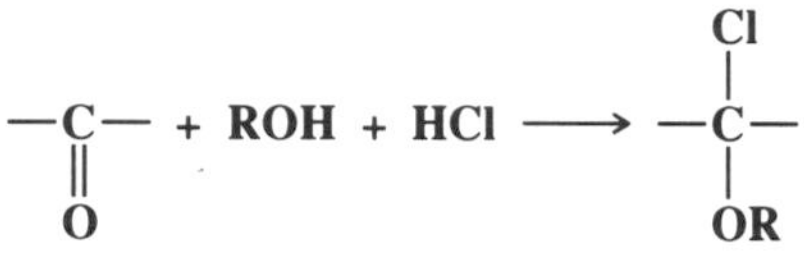

α-Halo ethers can be prepared by treatment of aldehydes and ketones with an alcohol and HX. The reaction is applicable to aliphatic aldehydes and ketones and to primary and secondary alcohols. Aromatic aldehydes and ketones react poorly.[230]

The addition of HX to an aldehyde or ketone gives α-halo alcohols, which are usually unstable, though exceptions are known, especially with perfluoro and perchloro species.[231] Unstable α-halo alcohols may be quite stable in the dimeric form $2XCR_2OH \rightarrow XCR_2OCR_2X$.

OS **I**, 377; **IV**, 101 (see, however, OS **V**, 218), 748; **VI**, 101.

6-24   The Formation of *gem*-Dihalides from Aldehydes and Ketones
Dihalo-de-oxo-bisubstitution

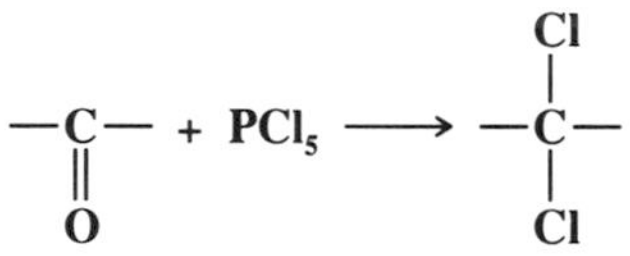

$$-\overset{\displaystyle ||}{\underset{\displaystyle O}{C}}- + PCl_5 \longrightarrow -\overset{\displaystyle Cl}{\underset{\displaystyle Cl}{C}}-$$

Aliphatic aldehydes and ketones can be converted to *gem*-dichlorides[232] by treatment with PCl_5. The reaction fails for perhalo ketones.[233] If the aldehyde or ketone has an α hydrogen, elimination of HCl may follow and a vinylic chloride is a frequent side product:[234]

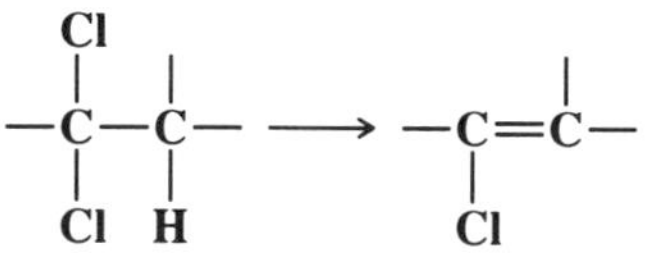

$$-\overset{\displaystyle Cl}{\underset{\displaystyle Cl}{C}}-\overset{\displaystyle |}{\underset{\displaystyle H}{C}}- \longrightarrow -\overset{\displaystyle |}{C}=\overset{\displaystyle |}{\underset{\displaystyle Cl}{C}}-$$

[229]Wood; Khatri; Weinreb *Tetrahedron Lett.* **1979,** 4907.
[230]Klages; Mühlbauer *Chem. Ber.* **1959,** *92*, 1818.
[231]For example, see Andreades; England *J. Am Chem. Soc.* **1961,** *83*, 4670; Clark; Emsley; Hibbert *J. Chem. Soc., Perkin Trans. 2* **1988,** 1107.
[232]For a list of reagents that convert aldehydes and ketones to *gem*-dihalides or vinylic halides, with references, see Ref. 64, pp. 372-375.
[233]Farah; Gilbert *J. Org. Chem.* **1965,** *30*, 1241.
[234]See, for example, Nikolenko; Popov *J. Gen. Chem. USSR* **1962,** *32*, 29.

or even the main product.[235] PBr_5 does not give good yields of *gem*-dibromides,[236] but these can be obtained from aldehydes, by the use of Br_2 and triphenyl phosphite.[237]

The mechanism of *gem*-dichloride formation involves initial attack of PCl_4^+ (which is present in solid PCl_5) at the oxygen, followed by addition of Cl^- to the carbon:[238]

$$-\overset{\|}{\underset{O}{C}}- + PCl_4^+ \longrightarrow -\overset{\oplus}{\underset{OPCl_4}{C}}- \xrightarrow{Cl^-} -\underset{OPCl_4}{\overset{Cl}{C}}- \longrightarrow -\overset{Cl}{\underset{\oplus}{C}}- \xrightarrow{Cl^-} -\underset{Cl}{\overset{Cl}{C}}-$$
$$\textbf{22}$$

This chloride ion may come from PCl_6^- (which is also present in solid PCl_5). There follows a two-step SN1 process. Alternatively, **22** can be converted to the product without going through the chlorocarbocation, by an SNi process.

This reaction has sometimes been performed on carboxylic esters, though these compounds very seldom undergo any addition to the C=O bond. An example is the conversion of $F_3CCOOPh$ to F_3CCCl_2OPh.[239] However, formates commonly give the reaction.

Many aldehydes and ketones have been converted to *gem*-difluoro compounds with sulfur tetrafluoride SF_4,[240] including quinones, which give 1,1,4,4-tetrafluorocyclohexadiene derivatives. With ketones, yields can be raised and the reaction temperature lowered, by the addition of anhydrous HF.[241] Carboxylic acids, acyl chlorides, and amides react with SF_4 to give 1,1,1-trifluorides. In these cases the first product is the acyl fluoride, which then undergoes the *gem*-difluorination reaction:

$$R-\overset{\|}{\underset{O}{C}}-W + SF_4 \longrightarrow R-\overset{\|}{\underset{O}{C}}-F + SF_4 \longrightarrow R-\underset{F}{\overset{F}{C}}-F \qquad W = OH, Cl, NH_2, NHR$$

The acyl fluoride can be isolated. Carboxylic esters also give trifluorides, though more vigorous conditions are required, but in this case the carbonyl group of the ester is attacked first, and RCF_2OR' can be isolated from $RCOOR'$[242] and then converted to the trifluoride. Anhydrides can react in either manner, and both types of intermediate are isolable under the right conditions. SF_4 even converts carbon dioxide to CF_4. A disadvantage of reactions with SF_4 is that they require a pressure vessel lined with stainless steel. Selenium tetrafluoride SeF_4 gives similar reactions, but atmospheric pressure and ordinary glassware can be used.[243] Another reagent that is often used to convert aldehydes and ketones to *gem*-difluorides is the commercially available diethylaminosulfur trifluoride (DAST) Et_2NSF_3.[244] Among other

[235]See, for example, Newman; Fraenkel; Kirn *J. Org. Chem.* **1963**, *28*, 1851.
[236]For an indirect method of converting ketones to *gem*-dibromides, see Napolitano; Fiaschi; Mastrorilli *Synthesis* **1986**, 122.
[237]Hoffmann; Bovicelli *Synthesis* **1990**, 657. See also Lansinger; Ronald *Synth. Commun.* **1979**, *9*, 341.
[238]Newman; Wood *J. Am. Chem. Soc.* **1959**, *81*, 4300; Newman *J. Org. Chem.* **1969**, *34*, 741.
[239]Kirsanov; Molosnova *J. Gen. Chem. USSR* **1958**, *28*, 31; Clark; Simons *J. Org. Chem.* **1961**, *26*, 5197.
[240]For reviews, see Wang *Org. React.* **1985**, *34*, 319-400; Boswell; Ripka; Scribner; Tullock *Org. React.* **1974**, *21*, 1-124.
[241]Muratov; Mohamed; Kunshenko; Burmakov; Alekseeva; Yagupol'skii *J. Org. Chem. USSR* **1985**, *21*, 1292.
[242]For methods of converting $RCOOR'$ to RCF_2OR', see Boguslavaskaya; Panteleeva; Chuvatkin *J. Org. Chem. USSR* **1982**, *18*, 198; Bunnelle; McKinnis; Narayanan *J. Org. Chem.* **1990**, *55*, 768.
[243]Olah; Nojima; Kerekes *J. Am. Chem. Soc.* **1974**, *96*, 925.
[244]Markovskij; Pashinnik; Kirsanov *Synthesis* **1973**, 787; Middleton *J. Org. Chem.* **1975**, *40*, 574. For a review of DAST and related reagents, see Hudlický *Org. React.* **1988**, *35*, 513-637.

reagents[245] used have been phenylsulfur trifluoride PhSF$_3$,[246] and molybdenum hexafluoride MoF$_6$.[247]

The mechanism with SF$_4$ is probably similar in general nature, if not in specific detail, to that with PCl$_5$.

Aromatic aldehydes, ketones, and carboxylic acids and esters can be halogenated and reduced in one operation (e.g., ArCHO → ArCH$_2$Br), by treatment with LiAlH$_4$ followed by HBr.[248]

OS **II**, 549; **V**, 365, 396, 1082; **VI**, 505, 845; **66**, 173. Also see OS **I**, 506.

G. Attack by Hydrogen

6-25 Reduction of Aldehydes and Ketones to Alcohols
C,O-Dihydro-addition

$$-\overset{\|}{\underset{O}{C}}- + \text{LiAlH}_4 \longrightarrow -\overset{H}{\underset{OH}{\overset{|}{C}}}-$$

Aldehydes can be reduced to primary alcohols, and ketones to secondary alcohols, by a number of reducing agents,[249] of which lithium aluminum hydride and other metallic hydrides are the most commonly used.[250] These reagents have two main advantages over many other reducing agents: they do not reduce carbon–carbon double (or triple) bonds, and they generally contain a lot of hydrogen in a small amount of reagent—with LiAlH$_4$, all four hydrogens are usable for reduction. The reaction is broad and general. LiAlH$_4$ easily reduces aliphatic, aromatic, alicyclic, and heterocyclic aldehydes, containing double or triple bonds and/or nonreducible groups such as NR$_3$, OH, OR, F, etc. If the molecule contains a group reducible by LiAlH$_4$ (e.g., NO$_2$, CN, COOR), then it is also reduced. LiAlH$_4$ reacts readily with water and alcohols, so these compounds must be excluded. Common solvents are ether and THF. NaBH$_4$ has a similar scope but is more selective and so may be used with NO$_2$, Cl, COOR, CN, etc. in the molecule. Another advantage of NaBH$_4$ is that it can be used in water or alcoholic solvents and so reduces compounds such as sugars that are not soluble in ethers.[251] The scope of these reagents with ketones is similar to that with aldehydes. LiAlH$_4$ reduces even sterically hindered ketones.

The double bonds that are generally not affected by metallic hydrides may be isolated or conjugated, but double bonds that are conjugated with the C=O group may or may not be reduced, depending on the substrate, reagent, and reaction conditions.[252] Some reagents that reduce only the C=O bonds of α,β-unsaturated aldehydes and ketones are

---

[245]For some indirect methods, see Sondej; Katzenellenbogen *J. Org. Chem.* **1986**, *51*, 3508; Prakesh; Reddy; Li; Olah *Synlett* **1990**, 594; Rozen; Zamir *J. Org. Chem.* **1991**, *56*, 4695.

[246]Sheppard *J. Am. Chem. Soc.* **1962**, *84*, 3058.

[247]Mathey; Bensoam *Tetrahedron* **1971**, *27*, 3965, **1975**, *31*, 391.

[248]Bilger; Royer; Demerseman *Synthesis* **1988**, 902.

[249]For a review, see Hudlický *Reductions in Organic Chemistry*; Ellis Horwood: Chichester, 1984, pp. 96-129. For a list of reagents, with references, see Ref. 64, pp. 527-547.

[250]For books on metal hydrides, see Seyden-Penne *Reductions by the Alumino- and Borohydrides*; VCH: New York, 1991; Hajos *Complex Hydrides*; Elsevier: New York, 1979. For reviews, see House, Ref. 180, pp. 49-71; Wheeler, in Patai, Ref. 2, pp. 507-566.

[251]NaBH$_4$ reduces solid ketones in the absence of any solvent (by mixing the powders): Toda; Kiyoshige; Yagi *Angew. Chem. Int. Ed. Engl.* **1989**, *28*, 320 [*Angew. Chem.* **101**, 329].

[252]For a review of the reduction of α,β-unsaturated carbonyl compounds, see Keinan; Greenspoon, in Patai; Rappoport *The Chemistry of Enones*, pt. 2; Wiley: New York, 1989, pp. 923-1022.

AlH$_3$,[253] NaBH$_4$, or LiAlH$_4$ in the presence of lanthanide salts (e.g., LaCl$_3$, CeBr$_3$),[254] NaBH$_3$(OAc),[255] Et$_3$SiH,[256] lithium *n*-butylborohydride,[257] and diisobutylaluminium hydride (DIBALH).[258] Also, both LiAlH$_4$[259] and NaBH$_4$[260] predominantly reduce only the C=O bonds of C=C—C=O systems in most cases, though substantial amounts of fully saturated alcohols have been found in some cases[259] (p. 774). For some reagents that reduce only the C=C bonds of conjugated aldehydes and ketones, see **5-9**.

When a functional group is selectively attacked in the presence of a different functional group, the reaction is said to be *chemoselective*. A number of reagents have been found to reduce aldehydes much faster than ketones. Among these[261] are NaBH$_4$ in isopropyl alcohol,[262] sodium triacetoxyborohydride,[263] lithium tris[(3-ethyl-3-pentyl)oxy]aluminum hydride Li(Et$_3$CO)$_3$AlH,[264] zinc borohydride in THF,[264a] and tributyltin hydride.[265] On the other hand, ketones can be chemoselectively reduced in the presence of aldehydes with NaBH$_4$ in aqueous EtOH at −15°C in the presence of cerium trichloride CeCl$_3$.[266] The reagent lithium N-dihydropyridylaluminum hydride reduces diaryl ketones much better than dialkyl or alkyl aryl ketones.[267] Most other hydrides reduce diaryl ketones more slowly than other types of ketones. Saturated ketones can be reduced in the presence of α,β-unsaturated ketones with NaBH$_4$–50% MeOH–CH$_2$Cl$_2$ at −78°C[268] and with zinc borohydride.[269] In general, NaBH$_4$ reduces carbonyl compounds in this order: aldehydes > α,β-unsaturated aldehydes > ketones > α,β-unsaturated ketones, and a carbonyl group of one type can be selectively reduced in the presence of a carbonyl group of a less reactive type.[270] Potassium triphenylborohydride KPh$_3$BH shows 99.4:0.6 selectivity between cyclohexanone and 4-heptanone, and 97:3 selectivity between cyclohexanone and cyclopentanone.[271] A number of reagents will preferentially reduce the less sterically hindered of two carbonyl compounds, but by the use of DIBALH in the presence of the Lewis acid methylaluminum bis(2,6-di-*t*-butyl-4-methyl-phenoxide), it was possible selectively to reduce the *more hindered* of a mixture of two ketones.[272] It is obvious that reagents can often be found to reduce one kind of carbonyl

[253]Jorgenson *Tetrahedron Lett.* **1962**, 559; Dilling; Plepys *J. Org. Chem.* **1970**, *35*, 2971.

[254]Gemal; Luche *J. Am. Chem. Soc.* **1981**, *103*, 5454; Fukuzawa; Fujinami; Yamauchi; Sakai *J. Chem. Soc., Perkin Trans. 1* **1986**, 1929. See also Chênevert; Ampleman *Chem. Lett.* **1985**, 1489; Varma; Kabalka *Synth. Commun.* **1985**, *15*, 985.

[255]Nutaitis; Bernardo *J. Org. Chem.* **1989**, *54*, 5629.

[256]Ojima; Kogure *Organometallics* **1982**, *1*, 1390.

[257]Kim; Moon; Ahn *J. Org. Chem.* **1982**, *47*, 3311.

[258]Wilson; Seidner; Masamune *Chem. Commun.* **1970**, 213.

[259]Johnson; Rickborn *J. Org. Chem.* **1970**, *35*, 1041.

[260]Chaikin; Brown *J. Am. Chem. Soc.* **1949**, *71*, 122.

[261]For some others (not all of them metal hydrides) see Hutchins; Kandasamy *J. Am. Chem. Soc.* **1973**, *95*, 6131; Risbood; Ruthven *J. Org. Chem.* **1979**, *44*, 3969; Babler; Invergo *Tetrahedron Lett.* **1981**, *22*, 621; Fleet; Harding *Tetrahedron Lett.* **1981**, *22*, 675; Yamaguchi; Kabuto; Yasuhara *Chem. Lett.* **1981**, 461; Kim; Kang; Yang *Tetrahedron Lett.* **1984**, *25*, 2985; Kamitori; Hojo; Masuda; Yamamoto *Chem. Lett.* **1985**, 253; Borbaruah; Barua; Sharma *Tetrahedron Lett.* **1987**, *28*, 5741.

[262]Brown; Wheeler; Ichikawa *Tetrahedron* **1957**, *1*, 214; Adams *Synth. Commun.* **1984**, *14*, 1349.

[263]Gribble; Ferguson *J. Chem. Soc., Chem. Commun.* **1975**, 535. See also Nutaitis; Gribble *Tetrahedron Lett.* **1983**, *24*, 4287.

[264]Krishnamurthy *J. Org. Chem.* **1981**, *46*, 4628.

[264a]Ranu; Chakraborty *Tetrahedron Lett.* **1990**, *31*, 7663.

[265]Fung; Mayo; Schauble; Weedon *J. Org. Chem.* **1978**, *43*, 3977; Shibata; Yoshida; Baba; Matsuda *Chem. Lett.* **1989**, 619; Adams; Schemenaur *Synth. Commun.* **1990**, *20*, 2359. For a review, see Kuivila *Synthesis* **1970**, 499-509.

[266]Luche; Gemal *J. Am. Chem. Soc.* **1979**, *101*, 5848. See also Gemal; Luche *Tetrahedron Lett.* **1981**, *22*, 4077. For other methods, see Paradisi; Zecchini; Ortar *Tetrahedron Lett.* **1980**, *21*, 5085; Bordoloi; Sarmah *Chem. Ind. (London)* **1987**, 459.

[267]Lansbury; Peterson *J. Am. Chem. Soc.* **1962**, *84*, 1756.

[268]Ward; Rhee; Zoghaib *Tetrahedron Lett.* **1988**, *29*, 517.

[269]Sarkar; Das; Ranu *J. Org. Chem.* **1990**, *55*, 5799.

[270]Ward; Rhee *Can. J. Chem.* **1989**, *67*, 1206.

[271]Yoon; Kim; Kang *J. Org. Chem.* **1986**, *51*, 226.

[272]Maruoka; Araki; Yamamoto *J. Am. Chem. Soc.* **1988**, *110*, 2650.

function in the presence of another.[273] For a discussion of selectivity in reduction reactions, see p. 1206.

Quinones are reduced to hydroquinones by $LiAlH_4$, $SnCl_2$–HCl, or sodium hydrosulfite $Na_2S_2O_4$, as well as by other reducing agents.

The reagent lithium tri-*sec*-butylborohydride $LiBH(sec-Bu)_3$ reduces cyclic and bicyclic ketones in a highly stereoselective manner, giving the less stable isomer.[274] For example, 2-methylcyclohexanone gave *cis*-2-methylcyclohexanol with an isomeric purity greater than 99%. The more usual reagents, e.g., $LiAlH_4$, $NaBH_4$, reduce relatively unhindered cyclic ketones either with little or no stereoselectivity[275] or give predominant formation of the more stable isomer (axial attack).[276] The less stable alcohol is also predominantly formed when cyclohexanones are reduced with (among other reagents) AlH_3 in ether at $-70°C$[277] and with triethyl phosphite and iridium tetrachloride in aqueous isopropyl alcohol.[278] Cyclohexanones that have a large degree of steric hindrance near the carbonyl group usually give predominant formation of the less stable alcohol, even with $LiAlH_4$ and $NaBH_4$.

Among other reagents that reduce aldehydes and ketones to alcohols[279] are the following:

1. *Hydrogen and a catalyst.*[280] The most common catalysts are platinum and ruthenium, but homogeneous catalysts have also been used.[281] Before the discovery of the metal hydrides this was one of the most common ways of effecting this reduction, but it suffers from the fact that C=C, C≡C, C=N and C≡N bonds are more susceptible to attack than C=O bonds.[282] For aromatic aldehydes and ketones, reduction to the hydrocarbon (**9-37**) is a side reaction, stemming from hydrogenolysis of the alcohol initially produced (**0-78**).

---

[273]For lists of some of these chemoselective reagents, with references, see Ref. 64, pp. 535-537, and references given in Ref. 270.

[274]Brown; Krishnamurthy *J. Am. Chem. Soc.* **1972**, *94*, 7159; Krishnamurthy; Brown *J. Am. Chem. Soc.* **1976**, *98*, 3383.

[275]For reviews of the stereochemistry and mechanism, see Caro; Boyer; Lamaty; Jaouen *Bull. Soc. Chim. Fr.* **1983**, II-281-II-303; Boone; Ashby *Top. Stereochem.* **1979**, *11*, 53-95; Wigfield *Tetrahedron* **1979**, *35*, 449-462. For a review of stereoselective synthesis of amino alcohols by this method, see Tramontini *Synthesis* **1982**, 605-644.

[276]For a discussion of why this isomer is predominantly formed, see Mukherjee; Wu; Fronczek; Houk *J. Am. Chem. Soc.* **1988**, *110*, 3328.

[277]Ayres; Sawdaye *J. Chem. Soc. B* **1967**, 581; Ayres; Kirk; Sawdaye *J. Chem. Soc. B* **1970**, 505.

[278]Henbest; Mitchell *J. Chem. Soc. C* **1970**, 785; Eliel; Doyle; Hutchins; Gilbert *Org. Synth. VI*, 215. See also Henbest; Zurqiyah *J. Chem. Soc., Perkin Trans. 1* **1974**, 604.

[279]This can also be done electrochemically. For a review, see Feoktistov; Lund, in Baizer; Lund *Organic Electochemistry*; Marcel Dekker: New York, 1983, pp. 315-358, pp. 315-326. See also Coche; Moutet *J. Am. Chem. Soc.* **1987**, *109*, 6887.

[280]For reviews, see Parker, in Hartley *The Chemistry of the Metal-Carbon Bond*, vol. 4; Wiley: New York, 1987, pp. 979-1047; Tanaka, in Červený *Catalytic Hydrogenation*; Elsevier: New York, 1986, pp. 79-104; Rylander *Hydrogenation Methods*, Ref. 165, pp. 66-77; Rylander *Catalytic Hydrogenation over Platinum Metals*, Ref. 165, pp. 238-290.

[281]For a review, see Heck *Organotransition Metal Chemistry*; Academic Press: New York, 1974, pp. 65-70.

[282]For catalysts that allow hydrogenation of only the C=O bond of α,β-unsaturated aldehydes, see Galvagno; Poltarzewski; Donato; Neri; Pietropaolo *J. Chem. Soc., Chem. Commun.* **1986**, 1729; Farnetti; Pesce; Kašpar; Spogliarich; Graziani *J. Chem. Soc., Chem. Commun.* **1986**, 746; Narasimhan; Deshpande; Ramnarayan *J. Chem. Soc., Chem. Commun.* **1988**, 99.

2. *Sodium in ethanol.*[283]   This is called the *Bouveault–Blanc procedure* and was more popular for the reduction of carboxylic esters **(9-42)** than of aldehydes or ketones before the discovery of LiAlH$_4$.

3. *Isopropyl alcohol and aluminum isopropoxide.*   This is called the *Meerwein–Ponndorf–Verley reduction*. It is reversible, and the reverse reaction is known as the *Oppenauer oxidation* (see **9-3**):

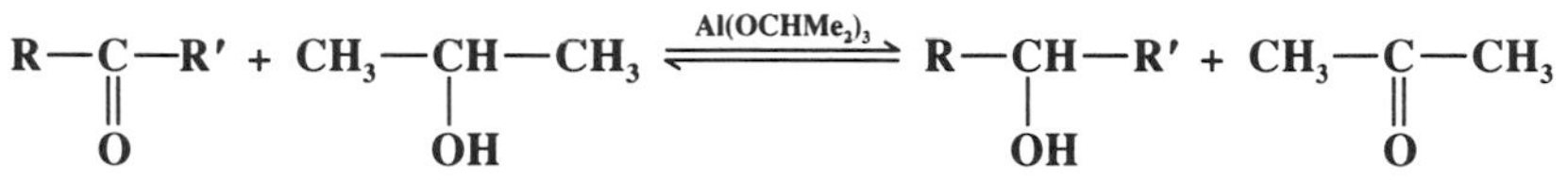

$$\underset{O}{\overset{\parallel}{R-C-R'}} + CH_3-\underset{OH}{\overset{}{CH}}-CH_3 \xrightarrow[]{Al(OCHMe_2)_3} \underset{OH}{\overset{}{R-CH-R'}} + CH_3-\underset{O}{\overset{\parallel}{C}}-CH_3$$

The equilibrium is shifted by removal of the acetone by distillation. The reaction takes place under very mild conditions and is highly specific for aldehydes and ketones, so that C=C bonds (including those conjugated with the C=O bonds) and many other functional groups can be present without themselves being reduced.[284] This includes acetals, so that one of two carbonyl groups in a molecule can be specifically reduced if the other is first converted to an acetal. β-Keto esters, β-diketones, and other ketones and aldehydes with a relatively high enol content do not give this reaction.

4. Borane BH$_3$ and substituted boranes reduce aldehydes and ketones in a manner similar to their addition to C=C bonds **(5-12)**.[285] That is, the boron adds to the oxygen and the hydrogen to the carbon:[286]

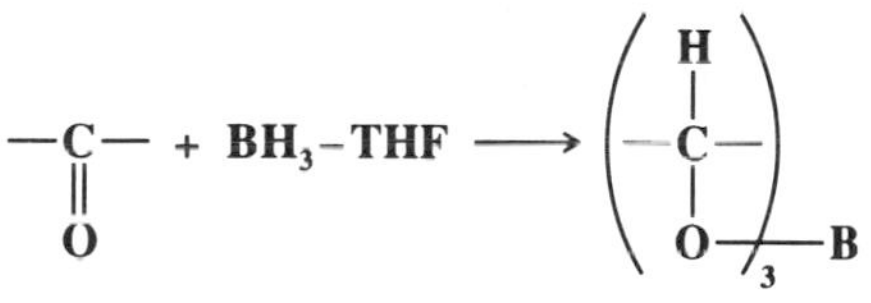

$$-\underset{O}{\overset{\parallel}{C}}- + BH_3-THF \longrightarrow \left(-\underset{O}{\overset{\overset{\displaystyle H}{|}}{\underset{|}{C}}}- \right)_3 B$$

The borate is then hydrolyzed to the alcohol. 9-BBN[287] (p. 785) and BH$_3$–Me$_2$S[288] reduce only the C=O group of conjugated aldehydes and ketones.

5. *Diimide*   (N$_2$H$_2$, see p. 779) reduces aromatic aldehydes[289] and ketones, but aliphatic carbonyl compounds react very poorly.[290]

6. A single carbonyl group of an α-diketone can be reduced (to give an α-hydroxy ketone) by heating with zinc powder in aqueous DMF.[291] This has also been accomplished with aqueous VCl$_2$[292] and with Zn–ZnCl$_2$–EtOH.[293]

7. In the Cannizzaro reaction **(9-69)** aldehydes without an α hydrogen are reduced to alcohols.

[283]For a discussion, see House, Ref. 180, pp. 152-160.
[284]Diisobornyloxyaluminum isopropoxide gives higher yields under milder conditions than aluminum isopropoxide: Hutton, *Synth. Commun.* **1979**, *9*, 483. For other substitutes for aluminum isopropoxide, see Namy; Souppe; Collin; Kagan *J. Org. Chem.* **1984**, *49*, 2045; Okano; Matsuoka; Konishi; Kiji *Chem. Lett.* **1987**, 181.
[285]For a review, see Cragg *Organoboranes in Organic Synthesis*; Marcel Dekker: New York, 1973, pp. 324-335.
[286]Brown; Subba Rao *J. Am. Chem. Soc.* **1960**, *82*, 681; Brown; Korytnyk *J. Am. Chem. Soc.* **1960**, *J. Am. Chem. Soc.* **1960**, *82*, 3866.
[287]Krishnamurthy; Brown *J. Org. Chem.* **1975**, *40*, 1864; Lane *Aldrichimica Acta* **1976**, *9*, 31.
[288]Mincione *J. Org. Chem.* **1978**, *43*, 1829.
[289]Curry; Uff; Ward *J. Chem. Soc. C.* **1967**, 1120.
[290]van Tamelen; Davis; Deem *Chem. Commun.* **1965**, 71.
[291]Kreiser *Liebigs Ann. Chem.* **1971**, *745*, 164.
[292]Ho; Olah *Synthesis* **1976**, 815.
[293]Toda; Tanaka; Tange *J. Chem. Soc., Perkin Trans. 1* **1989**, 1555.

Unsymmetrical ketones are prochiral (p. 135); that is, reduction creates a new chiral center:

$$
\underset{\underset{O}{\parallel}}{R-C-R'} \longrightarrow \underset{OH}{R-\overset{H}{\underset{|}{\overset{*}{C}}}-R'}
$$

Much effort has been put into finding optically active reducing agents that will produce one enantiomer of the alcohol enantioselectively, and considerable success has been achieved,[294] both with biologically-derived reducing agents[295] such as baker's yeast,[296] and with synthetic reagents. Each reagent is more effective for certain types of ketones than for others.[297] H.C. Brown and co-workers reduced various types of ketone with a number of reducing agents,[298] and reported in 1987 that of the reagents available at that time, the highest enantiomeric excesses (ee) for acyclic ketones were obtained with (*R,R*)- or (*S,S*)-2,5-dimethylborolane (**47** and **48** on p. 787).[299] For cyclic ketones the best reagents were diisopinocampheyl-chloroborane (**23**),[300] (*S*)-2-amino-1,1-diphenylbutan-1-ol–BH$_3$,[301] and K-Glucoride, a boron derivative of a carbohydrate.[302] These workers also determined the relative effectiveness of

23

various reagents for reduction of 8 other types of ketone, including heterocyclic, aralkyl, β-keto esters, etc.[298] In most cases, ee values of greater than 90% can be obtained with the proper reagent.[303]

Asymmetric reduction with very high ee values has also been achieved with achiral reducing agents and optically active catalysts. The two most important examples are (1) homogeneous catalytic hydrogenation with the catalyst 2,2'-bis(diphenylphosphino)-1,1'-

[294]For reviews, see Midland *Chem. Rev.* **1989**, *89*, 1553-1561; Nógrádi *Stereoselective Synthesis*; VCH: New York, 1986, pp. 105-130; in Morrison *Asymmetric Synthesis*; Academic Press: New York, 1983, the articles by Midland, vol. 2, pp. 45-69, and Grandbois; Howard; Morrison, vol. 2, pp. 71-90; Haubenstock *Top. Stereochem.* **1983**, *14*, 231-300.

[295]For a review, see Sih; Chen *Angew. Chem. Int. Ed. Engl.* **1984**, *23*, 570-578 [*Angew. Chem. 96*, 556-565].

[296]See, for example, Fujisawa; Hayashi; Kishioka *Chem. Lett.* **1987**, 129; Nakamura; Kawai; Ohno *Tetrahedron Lett.* **1990**, *31*, 267; Spiliotis; Papahatjis; Ragoussis *Tetrahedron Lett.* **1990**, *31*, 1615.

[297]For a list of many of these reducing agents, with references, see Ref. 64, pp. 540-547.

[298]Brown; Park; Cho; Ramachandran *J. Org. Chem.* **1987**, *52*, 5406.

[299]First used in this way by Imai; Tamura; Yamamuro; Sato; Wollmann; Kennedy; Masamune *J. Am. Chem. Soc.* **1986**, *108*, 7402; Masamune; Kennedy; Petersen; Houk; Wu *J. Am. Chem. Soc.* **1986**, *108*, 7404.

[300]Chandrasekharan; Ramachandran; Brown *J. Org. Chem.* **1985**, *50*, 5446; Brown; Chandrasekharan; Ramachandran *J. Org. Chem.* **1986**, *51*, 3394, *J. Am. Chem. Soc.* **1988**, *110*, 1539; Srebnik; Ramachandran; Brown *J. Org. Chem.* **1988**, *53*, 2916. See also Brown; Srebnik; Ramachandran *J. Org. Chem.* **1989**, *54*, 1577.

[301]For the preparation and use of this and related reagents, see Itsuno; Nakano; Miyazaki; Masuda; Ito; Hirao; Nakahama *J. Chem. Soc., Perkin Trans. 1* **1985**, 2039, and other papers in this series.

[302]Brown; Park; Cho *J. Org. Chem.* **1986**, *51*, 1934, 3278; Brown; Cho; Park *J. Org. Chem.* **1986**, *51*, 3396, **1988**, *53*, 1231.

[303]For some recent examples, see Youn; Lee; Pak *Tetrahedron Lett.* **1988**, *29*, 4453; Meyers; Brown *Tetrahedron Lett.* **1988**, *29*, 5617; Brown; Ramachandran; Weissman; Swaminathan *J. Org. Chem.* **1990**, *55*, 6328; Rama Rao; Gurjar; Sharma; Kaiwar *Tetrahedron Lett.* **1990**, *31*, 2341; Midland; Kazubski; Woodling *J. Org. Chem.* **1991**, *56*, 1068.

binaphthyl–ruthenium acetate [BINAP–Ru(OAc)$_2$],[304] which reduces β-keto esters in >98% ee,[305] and (2) reduction with BH$_3$–THF or catecholborane, using an oxazaborolidine

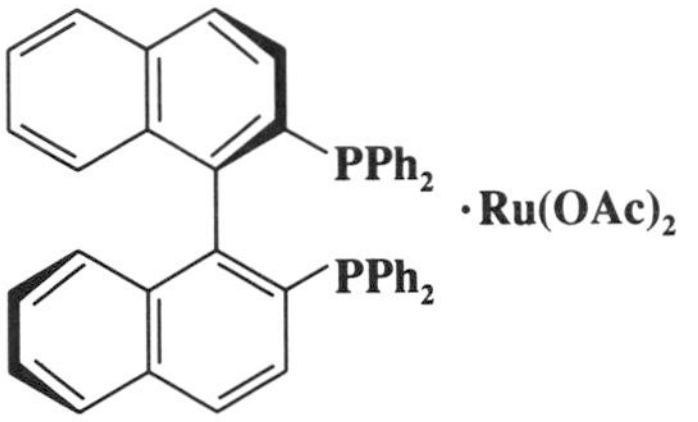

R(+)-BINAP

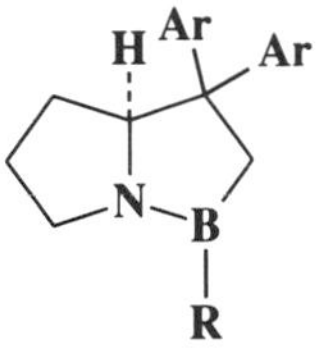

24

24 (R – H, Me, or *n*-Bu; Ar = Ph or β-naphthyl) as a catalyst.[306] This method gives high ee values with various types of ketone, especially α,β-unsaturated ketones.

Enantioselective reduction is not possible for aldehydes, since the products are primary alcohols in which the reduced carbon is not chiral, but deuterated aldehydes RCDO give a chiral product, and these have been reduced enantioselectively with B-(3-pinanyl)-9-bora-bicyclo[3.3.1]nonane (Alpine–Borane) with almost complete optical purity.[307]

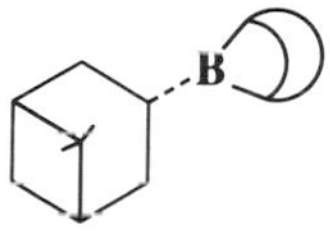

Alpine – Borane

In the above cases an optically active reducing agent or catalyst interacts with a prochiral substrate. Asymmetric reduction of ketones has also been achieved with an achiral reducing agent, if the ketone is complexed to an optically active transition metal Lewis acid.[308]

There are other stereochemical aspects to the reduction of aldehydes and ketones. If there is a chiral center α to the carbonyl group,[309] even an achiral reducing agent can give

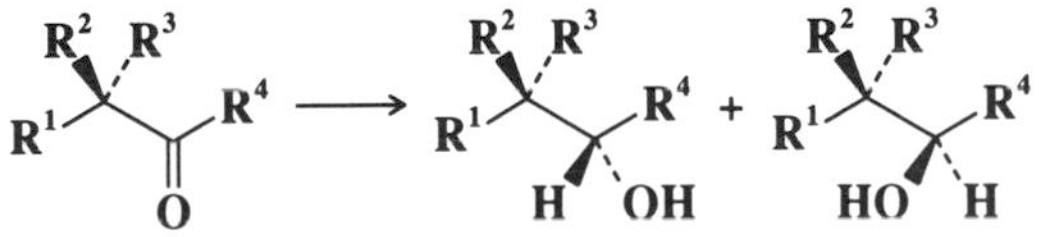

diastereomers

[304]For reviews of BINAP, see Noyori *Science* **1990,** *248,* 1194-1199; Noyori; Takaya *Acc. Chem. Res.* **1990,** *23,* 345-350. For the synthesis of BINAP, see Takaya; Akutagawa; Noyori *Org. Synth. 67,* 20.

[305]Noyori; Ohkuma; Kitamura; Takaya; Sayo; Kumobayashi; Akutagawa *J. Am. Chem. Soc.* **1987,** *109,* 5856; Taber; Silverberg *Tetrahedron Lett.* **1991,** *32,* 4227. See also Kitamura; Ohkuma; Inoue; Sayo; Kumobayashi; Akutagawa; Ohta; Takaya; Noyori; *J. Am. Chem. Soc.* **1988,** *110,* 629.

[306]Corey; Bakshi; Shibata *J. Am. Chem. Soc.* **1987,** *109,* 5551; Corey; Bakshi; Shibata; Chen; Singh *J. Am. Chem. Soc.* **1987,** *109,* 7924; Corey; Link; *Tetrahedron Lett.* **1989,** *30,* 6275; Corey; Bakshi *Tetrahedron Lett.* **1990,** *31,* 611.

[307]Midland; Greer; Tramontano; Zderic *J. Am. Chem. Soc.* **1979,** *101,* 2352. See also Noyori; Tomino; Tanimoto *J. Am. Chem. Soc.* **1979,** *101,* 3129; Brown; Jadhav; Mandal *Tetrahedron* **1981,** *37,* 3547-3587; Midland; Zderic *J. Am. Chem. Soc.* **1982,** *104,* 525.

[308]Dalton; Gladysz *J. Organomet. Chem.* **1989,** *370,* C17.

[309]In theory, the chiral center can be anywhere in the molecule, but in practice, reasonable diastereoselectivity is most often achieved when it is in the α position. For examples of high diastereoselectivity when the chiral center is further away, especially in reduction of β-hydroxy ketones, see Narasaka; Pai *Tetrahedron* **1984,** *40,* 2233; Hassine; Gorsane; Pecher; Martin *Bull. Soc. Chim. Belg.* **1985,** *94,* 597; Bloch; Gilbert; Girard *Tetrahedron Lett.* **1988,** *53,* 1021; Evans; Chapman; Carreira *J. Am. Chem. Soc.* **1988,** *110,* 3560.

more of one diastereomer than of the other. Such diastereoselective reductions have been carried out with considerable success.[310] In most such cases Cram's rule (p. 117) is followed, but exceptions are known.[311]

With most reagents there is an initial attack on the carbon of the carbonyl group by H⁻ or some carrier of it, though with BH_3[312] the initial attack is on the oxygen. Detailed mechanisms are not known in most cases.[275] With AlH_4^- (or BH_4^-) compounds, the attacking species is the AlH_4^- (or BH_4^-) ion, which, in effect, transfers H⁻ to the carbon. The following mechanism has been proposed for $LiAlH_4$:[313]

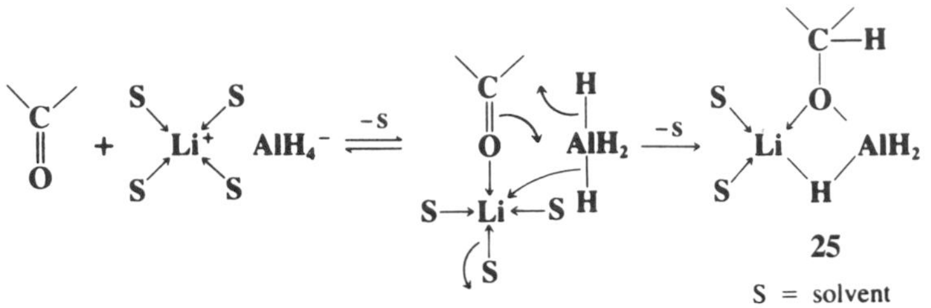

S = solvent molecules

Evidence that the cation plays an essential role, at least in some cases, is that when the Li⁺ was effectively removed from $LiAlH_4$ (by the addition of a crown ether), the reaction did not take place.[314] The complex **25** must now be hydrolyzed to the alcohol. For $NaBH_4$ the Na⁺ does not seem to participate in the transition state, but kinetic evidence shows that an OR group from the solvent does participate and remains attached to the boron:[315]

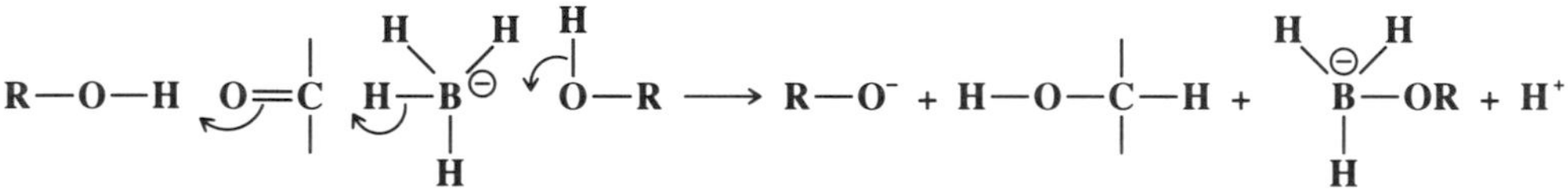

Free H⁻ cannot be the attacking entity in most reductions with boron or aluminum hydrides because the reactions are frequently sensitive to the size of the MH_4^- [or $MR_mH_n^-$ or $M(OR)_mH_n^-$, etc.].

There has been much controversy about whether the initial complex in the $LiAlH_4$ reduction (**25**, which can be written as $H\!-\!\overset{|}{\underset{|}{C}}\!-\!OAlH_3^-$, **26**) can reduce another carbonyl to give $(H\!-\!\overset{|}{\underset{|}{C}}\!-\!O)_2AlH_2^-$, and so on. It has been shown[316] that this is probably not the

---

[310]For reviews, see Nógrádi, Ref. 294, pp. 131-148; Oishi; Nakata *Acc. Chem. Res.* **1984**, *17*, 338-344.

[311]One study showed that the Cram's rule product predominates with metal hydride reducing agents, but the other product with Bouveault-Blanc and dissolving metal reductions: Yamamoto; Matsuoka; Nemoto *J. Am. Chem. Soc.* **1988**, *110*, 4475.

[312]For a discussion of the mechanism with boranes, see Brown, Wang, Chandrasekharan *J. Am. Chem. Soc.* **1983**, *105*, 2340.

[313]Ashby; Boone *J. Am. Chem. Soc.* **1976**, *98*, 5524.

[314]Pierre; Handel *Tetrahedron Lett.* **1974**, 2317. See also Loupy, Seyden-Penne; Tchoubar *Tetrahedron Lett.* **1976**, 1677; Ref. 313.

[315]Wigfield; Gowland *J. Org. Chem.* **1977**, *42*, 1108, *Tetrahedron Lett.* **1976**, 3373. See however Adams; Gold; Reuben *J. Chem. Soc., Chem. Commun.* **1977**, 182, *J. Chem. Soc., Perkin Trans 2* **1977**, 1466, 1472; Kayser; Eliev; Eisenstein *Tetrahedron Lett.* **1983**, *24*, 1015.

[316]Haubenstock; Eliel *J. Am. Chem. Soc.* **1962**, *84*, 2363; Malmvik; Obenius; Henriksson *J. Chem. Soc., Perkin Trans. 2* **1986**, 1899, 1905.

case but that, more likely, **26** disproportionates to $(H-\overset{|}{\underset{|}{C}}-O)_4Al^-$ and AlH_4^-, which is the

only attacking species. Disproportionation has also been reported in the $NaBH_4$ reaction.[317]

26 is essentially $LiAlH_4$ with one of the hydrogens replaced by an alkoxy group, i.e., $LiAlH_3OR$. The fact that **26** and other alkoxy derivatives of $LiAlH_4$ are less reactive than $LiAlH_4$ itself has led to the use of such compounds as reducing agents that are less reactive and more selective than $LiAlH_4$.[318] We have already met some of these, e.g., $LiAlH(O-t-Bu)_3$ (reactions **0-83** to **0-85**; see also Table 19.5). As an example of chemoselectivity in this reaction it may be mentioned that $LiAlH(O-t-Bu)_3$ has been used to reduce only the keto group in a molecule containing both keto and carboxylic ester groups.[319] However, the use of such reagents is sometimes complicated by the disproportionation mentioned above, which may cause $LiAlH_4$ to be the active species, even if the reagent is an alkoxy derivative. Another highly selective reagent (reducing aldehydes and ketones, but not other functional groups), which does not disproportionate, is potassium triisopropoxyborohydride.[320]

The Meerwein–Ponndorf–Verley reaction usually[321] involves a cyclic transition state:[322]

but in some cases 2 moles of aluminum alkoxide are involved—one attacking the carbon and the other the oxygen, a conclusion that stems from the finding that in these cases the reaction was 1.5 order in alkoxide.[323] Although, for simplicity, we have shown the alkoxide as a monomer, it actually exists as trimers and tetramers, and it is these that react.[324]

For the reaction with sodium in ethanol the following mechanism[325] has been suggested:[326]

A ketyl

The ketyl intermediate can be isolated.[327]

[317]Malmvik; Obenius; Henriksson *J. Org. Chem.* **1988**, *53*, 221.

[318]For reviews of reductions with alkoxyaluminum hydrides, see Málek *Org. React.* **1988**, *36*, 249-590, **1985**, *34*, 1-317; Málek; Černý *Synthesis* **1972**, 217-234.

[319]Levine; Eudy *J. Org. Chem.* **1970**, *35*, 549; Heusler; Wieland; Meystre *Org. Synth.* V, 692.

[320]Brown; Krishnamurthy; Kim *J. Chem. Soc., Chem. Commun.* **1973**, 391.

[321]It has been that shown in some cases reduction with metal alkoxides, including aluminum isopropoxide, involves free-radical intermediates (SET mechanism): Screttas; Cazianis *Tetrahedron* **1978**, *34*, 933; Ashby; Goel; Argyropoulos *Tetrahedron Lett.* **1982**, *23*, 2273; Nasipuri; Gupta; Banerjee *Tetrahedron Lett.* **1984**, *25*, 5551; Ashby; Argyropoulos *Tetrahedron Lett.* **1986**, 27, 465, *J. Org. Chem.* **1986**, *51*, 3593; Yamataka; Hanafusa *Chem. Lett.* **1987**, 643.

[322]See, for example, Shiner; Whittaker *J. Am. Chem. Soc.* **1963**, *85*, 2337; Warnhoff; Reynolds-Warnhoff; Wong *J. Am. Chem. Soc.* **1980**, *102*, 5956.

[323]Moulton; Van Atta; Ruch *J. Org. Chem.* **1961**, *26*, 290.

[324]Williams; Krieger; Day *J. Am. Chem. Soc.* **1953**, *75*, 2404; Shiner; Whittaker *J. Am. Chem. Soc.*, **1969**, *91*, 394.

[325]For reviews of the mechanisms of these reactions, see Pradhan *Tetrahedron* **1986**, *42*, 6351-6388; Huffman *Acc. Chem. Res.* **1983**, *16*, 399-405. For discussions of the mechanism in the absence of protic solvents, see Huffman; Liao; Wallace *Tetrahedron Lett.* **1987**, *28*, 3315; Rautenstrauch *Tetrahedron* **1988**, *44*, 1613; Song; Dewald *J. Chem. Soc., Perkin Trans. 2* **1989**, 269. For a review of the stereochemistry of these reactions in liquid NH_3, see Rassat *Pure Appl. Chem.* **1977**, *49*, 1049-1058.

[326]House, Ref. 180, p. 151. See, however Giordano; Perdoncin; Castaldi *Angew. Chem. Int. Ed. Engl.* **1985**, *24*, 499 [*Angew. Chem. 97*, 510].

[327]For example, see Rautenstrauch; Geoffroy *J. Am. Chem. Soc.* **1976**, *98*, 5035, **1977**, *99*, 6280.

The mechanism of catalytic hydrogenation of aldehydes and ketones is probably similar to that of reaction **5-9**, though not much is known about it.[328]

For other reduction reactions of aldehydes and ketones, see **9-37, 9-62**, and **9-69**.

OS **I**, 90, 304, 554; **II**, 317, 545, 598; **III**, 286; **IV**, 15, 25, 216, 660; **V**, 175, 294, 595, 692; **VI**, 215, 769, 887; **VII**, 129, 215, 241, 402, 417; **65**, 203, 215; **68**, 56; **69**, 44.

6-26   Reduction of the Carbon–Nitrogen Double Bond
   C,N-Dihydro-addition

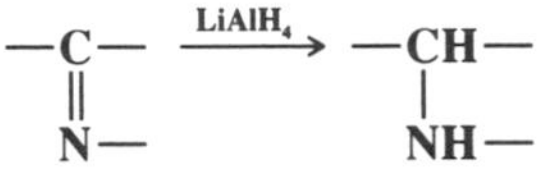

$$-\underset{\underset{N-}{\|}}{C}- \xrightarrow{\text{LiAlH}_4} -\underset{\underset{NH-}{|}}{CH}-$$

Imines, Schiff bases, hydrazones, and other C=N compounds can be reduced with LiAlH$_4$, NaBH$_4$, Na–EtOH, hydrogen and a catalyst, as well as with other reducing agents.[329] Iminium salts are also reduced by LiAlH$_4$, though here there is no "addition" to the nitrogen:[330]

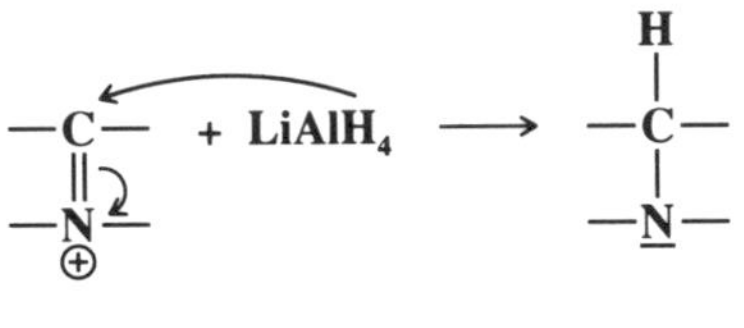

$$-\underset{\underset{\oplus}{\overset{\|}{N}-}}{C}- + \text{LiAlH}_4 \longrightarrow -\underset{\underset{N-}{|}}{\overset{\overset{H}{|}}{C}}-$$

Reduction of imines has been carried out enantioselectively.[331]

Isocyanates have been catalytically hydrogenated to N-substituted formamides: RNCO → R—NH—CHO.[332]

Oximes are generally reduced to amines (**9-51**), but simple addition of H$_2$ to give hydroxylamines can be accomplished with borane[333] or sodium cyanoborohydride.[168]

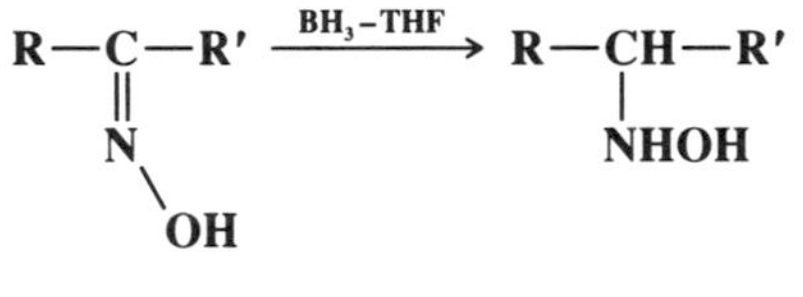

$$\underset{\underset{\underset{OH}{\diagdown}}{\overset{|}{N}}}{R-\overset{\|}{C}-R'} \xrightarrow{\text{BH}_3-\text{THF}} \underset{NHOH}{R-CH-R'}$$

OS **III**, 328, 827; **VI**, 905; **66**, 185; **69**, 154. Also see OS **IV**, 283.

6-27   The Reduction of Nitriles to Amines
   CC,NN-Tetrahydro-biaddition

$$R-C\equiv N + \text{LiAlH}_4 \longrightarrow R-CH_2-NH_2$$

[328]For a review of the mechanism of gas-phase hydrogenation, see Pavlenko *Russ. Chem. Rev.* **1989**, *58*, 453-469.

[329]For a review, see Harada, in Patai *The Chemistry of the Carbon–Nitrogen Double Bond*, Ref. 40, pp. 276-293. For a review with respect to catalytic hydrogenation, see Rylander, *Catalytic Hydrogenation over Platinum Metals*, Ref. 165, pp. 123-138.

[330]For a review of nucleophilic addition to iminium salts, see Paukstelis; Cook, in Cook, Ref. 45, pp. 275-356.

[331]See Cho; Chun *J. Chem. Soc., Perkin Trans. 1* **1990**, 3200; Chan; Osborn *J. Am. Chem. Soc.* **1990**, *112*, 9400, and references cited in these papers.

[332]Howell *Synth. Commun.* **1983**, *13*, 635.

[333]Feuer; Vincent *J. Am. Chem. Soc.* **1962**, *84*, 3771; Feuer; Vincent; Bartlett *J. Org. Chem.* **1965**, *30*, 2877; Ioffe; Tartakovskii; Medvedeva; Novikov *Bull. Acad. Sci. USSR, Div. Chem. Sci.* **1964**, 1446; Kawase; Kikugawa, *J. Chem. Soc., Perkin Trans. 1* **1979**, 643.

Nitriles can be reduced to primary amines with many reducing agents,[334] including LiAlH$_4$, BH$_3$–Me$_2$S,[335] NaOEt, and hydrogen and a catalyst.[336] NaBH$_4$ does not generally reduce nitriles but does so in alcoholic solvents when a CoCl$_2$ catalyst is added[337] or in the presence of Raney nickel.[338] The reaction is of wide scope and has been applied to many nitriles. When catalytic hydrogenation is used, secondary amines (RCH$_2$)$_2$NH are often side products.[339] These can be avoided by adding a compound such as acetic anhydride, which removes the primary amine as soon as it is formed,[340] or by the use of excess ammonia to drive the equilibria backward.[341]

It is not possible to stop with the addition of only 1 mole of hydrogen, i.e., to convert the nitrile to an imine, except where the imine is subsequently hydrolyzed (**6-28**).

N-Alkylnitrilium ions are reduced to secondary amines by NaBH$_4$.[342]

$$RCN \xrightarrow{R_3'O^+ \ BF_4^-} R-C\overset{\oplus}{=}N-R' \xrightarrow[\text{diglyme}]{NaBH_4} RCH_2-NH-R'$$

Since nitrilium salts can be prepared by treatment of nitriles with trialkyloxonium salts (see **6-9**), this is a method for the conversion of nitriles to secondary amines.

OS **III,** 229, 358, 720; **VI,** 223.

6-28 The Reduction of Nitriles to Aldehydes
Hydro,oxy-de-nitrilo-tersubstitution

$$R-C\equiv N \xrightarrow[\text{2. hydrolysis}]{\text{1. HCl, SnCl}_2} R-CH=O$$

There are two principal methods for the reduction of nitriles to aldehydes.[343] In one of these, known as the *Stephen reduction*, the nitrile is treated with HCl to form

$$RCCl=NH_2^+ \ Cl^-$$

27

This is reduced with anhydrous SnCl$_2$ to RCH=NH, which precipitates as a complex with SnCl$_4$ and is then hydrolyzed (**6-2**) to the aldehyde. The Stephen reduction is most successful when R is aromatic, but it can be done for aliphatic R up to about six carbons.[344] It is also possible to prepare **27** in a different way, by treating ArCONHPh with PCl$_5$. The **27** obtained in this way can then be converted to the aldehyde. This is known as the *Sonn–Müller method*.

The other way of reducing nitriles to aldehydes involves using a metal hydride reducing agent to add 1 mole of hydrogen and hydrolysis, in situ, of the resulting imine (which is undoubtedly coordinated to the metal). This has been carried out with LiAlH$_4$,

[334]For a review, see Rabinovitz, in Rappoport *The Chemistry of the Cyano Group*; Wiley: New York, 1970, pp. 307-340. For a list of reagents, with references, see Ref. 64, pp. 437-438.
[335]See Brown; Choi; Narasimhan *Synthesis* **1981**, 605.
[336]For reviews of catalytic hydrogenation of nitriles, see Volf; Pašek, in Červený, Ref. 280, pp. 105-144; Rylander, Ref. 329, pp. 203-226; Freidlin; Sladkova *Russ. Chem. Rev.* **1964**, *33*, 319-330.
[337]Satoh; Suzuki *Tetrahedron Lett.* **1969**, 4555. For a discussion of the mechanism, see Heinzman; Ganem *J. Am. Chem. Soc.* **1982**, *104*, 6801.
[338]Egli *Helv. Chim. Acta* **1970**, *53*, 47.
[339]For a method of making secondary amines the main products, see Galán; de Mendoza; Prados; Rojo; Echavarren *J. Org. Chem.* **1991**, *56*, 452.
[340]For example, see Carothers; Jones *J. Am. Chem. Soc.* **1925**, *47*, 3051; Gould; Johnson; Ferris *J. Org. Chem.* **1960**, *25*, 1658.
[341]For example, see Freifelder *J. Am. Chem. Soc.* **1960**, *82*, 2386.
[342]Borch *Chem. Commun.* **1968**, 442.
[343]For a review, see Rabinovitz, Ref. 334. For a list of reagents, with references, see Ref. 64, pp. 624-625.
[344]Zil'berman; Pyryalova *J. Gen. Chem. USSR* **1963**, *33*, 3348.

LiAlH(OEt)$_3$,[345] DIBALH,[346] and NaAlH$_4$.[347] The metal hydride method is useful for aliphatic and aromatic nitriles. Reduction to the aldehyde has also been accomplished by treatment of the nitrile with sodium hypophosphate and Raney nickel in aqueous acetic acid–pyridine or formic acid,[348] and with zinc and a Cob(I)alamin catalyst in aqueous acetic acid.[349]

OS **III**, 626, 818; **VI**, 631.

H. Carbon Attack by Organometallic Compounds[350]

6-29 The Addition of Organometallic Compounds to Aldehydes and Ketones
O-Hydro-C-alkyl-addition

$$
\begin{array}{ccc}
-\overset{\displaystyle\|}{\underset{\displaystyle O}{C}}- \;+\; RMgX & \longrightarrow & -\overset{\displaystyle R}{\underset{\displaystyle OMgX}{C}}- \quad\xrightarrow{\text{hydrol.}}\quad -\overset{\displaystyle R}{\underset{\displaystyle OH}{C}}-
\end{array}
$$

The addition of Grignard reagents to aldehydes and ketones is known as the *Grignard reaction*.[351] Formaldehyde gives primary alcohols; other aldehydes give secondary alcohols; and ketones give tertiary alcohols. The reaction is of very broad scope, and hundreds of alcohols have been prepared in this manner. R may be alkyl or aryl. In many cases the hydrolysis step is carried out with dilute HCl or H$_2$SO$_4$, but this cannot be done for tertiary alcohols in which at least one R group is alkyl because such alcohols are easily dehydrated under acidic conditions (**7-1**). In such cases (and often for other alcohols as well) an aqueous solution of ammonium chloride is used instead of a strong acid. Other organometallic compounds can also be used,[352] but in general only of active metals; e.g., alkylmercurys do not react. In practice, the only organometallic compounds used to any extent, besides Grignard reagents, are alkyl- and aryllithiums,[353] and alkylzinc reagents[354] where enantioselective addition is desired (see below). For the addition of acetylenic groups, sodium may be the metal used: RC≡CNa (**6-41**); while vinylic alanes (prepared as in **5-13**) are the reagents of choice for the addition of vinylic groups.[355] Many methods have been reported

[345]Brown; Shoaf *J. Am. Chem. Soc.* **1964**, *86*, 1079. For a review of reductions with this and related reagents, see Málek *Org. React.* **1988**, *36*, 249-590, pp. 287-289, 438-448.

[346]Miller; Biss; Schwartzman *J. Org. Chem.* **1959**, *24*, 627; Marshall; Andersen; Schlicher *J. Org. Chem.* **1970**, *35*, 858.

[347]Zakharkin; Maslin; Gavrilenko *Bull. Acad. Sci. USSR, Div. Chem. Sci.* **1964**, 1415.

[348]Backeberg; Staskun *J. Chem. Soc.* **1962**, 3951; van Es; Staskun *J. Chem. Soc.* **1965**, 5775, *Org. Synth. VI*, 631. For a related method, see Khai; Arcelli *J. Org. Chem.* **1989**, *54*, 949.

[349]Fischli *Helv. Chim. Acta* **1978**, *61*, 2560.

[350]Discussions of most of the reactions in this section are found in Hartley; Patai *The Chemistry of the Metal–Carbon Bond*, vols. 2, 3 and 4; Wiley: New York, 1985-1987.

[351]For reviews of the addition of organometallic compounds to carbonyl groups, see Eicher, in Patai, Ref. 2, pp. 621-693; Kharasch; Reinmuth *Grignard Reactions of Nonmetallic Substances*; Prentice-Hall: Englewood Cliffs, NJ, 1954, pp. 138-528. For a review of reagents that extend carbon chains by 3 carbons, with some functionality at the new terminus, see Stowell *Chem. Rev.* **1984**, *84*, 409-435.

[352]For a list of reagents, with references, see Ref. 64, pp. 559-567.

[353]For a discussion, see Wakefield *Organolithium Methods*; Academic Press: New York, 1988, pp. 67-75.

[354]For a review with respect to organozinc compounds, see Furukawa; Kawabata *Adv. Organomet. Chem.* **1974**, *12*, 103-112. For a review with respect to organocadmium compounds, see Jones; Desio *Chem. Rev.* **1978**, *78*, 491-516.

[355]Newman *Tetrahedron Lett.* **1971**, 4571. Vinylic groups can also be added with 9-vinylic-9-BBN compounds: Jacob; Brown *J. Org. Chem.* **1977**, *42*, 579.

for the addition of allylic groups.[356] Among these are the use of allyltrialkyltin compounds (in the presence of BF_3–etherate),[357] allyltrialkylsilanes (in the presence of a Lewis acid),[358] as well as other allylic metal compounds.[359] Although organoboranes do not generally add to aldehydes and ketones,[360] allylic boranes are exceptions.[361] When they add, an allylic rearrangement always takes place, e.g.,

$$\text{RCH}{=}\text{CH}{-}\text{CH}_2\text{Br}_2' \; + \; \text{R}''{-}\underset{\underset{\displaystyle O}{\|}}{C}{-}\text{H} \longrightarrow \text{R}''{-}\text{CH}{-}\underset{\underset{\displaystyle OH}{|}}{\overset{\overset{\displaystyle R}{|}}{\text{CH}}}{-}\text{CH}{=}\text{CH}_2$$

indicating a cyclic mechanism:

Allylic rearrangements sometimes take place with the other reagents as well.

Certain functional groups (COOEt, CONMe$_2$, CN) can be present in the R group when organotin reagents RSnEt$_3$ are added to aldehydes.[362] A trifluoromethyl group can be added with Me$_3$SiCF$_3$, with Bu$_4$NF as a catalyst, in THF.[363]

The reaction with alkyl- and aryllithium reagents has also been carried out without preliminary formation of RLi: a mixture of RX and the carbonyl compound was added to a suspension of lithium pieces in THF.[364] Yields were generally satisfactory. The magnesium analog of this process is called the *Barbier reaction*.[365] Lithium dimethylcopper Me$_2$CuLi

[356]For a list of reagents and references, see Ref. 64, pp. 567-572.

[357]Naruta; Ushida; Maruyama *Chem. Lett.* **1979**, 919. For a review, see Yamamoto *Aldrichimica Acta* **1987**, *20*, 45-49.

[358]For reviews, see Fleming; Dunoguès; Smithers *Org. React.* **1989**, *37*, 57-575, pp. 113-125, 290-328; Parnes; Bolestova *Synthesis* **1984**, 991-1008, pp. 997-1000. For studies of the mechanism, see Denmark; Wilson; Willson *J. Am. Chem. Soc.* **1988**, *110*, 984; Denmark; Weber; Wilson; Willson *Tetrahedron* **1989**, *45*, 1053; Keck; Andrus; Castellino *J. Am. Chem. Soc.* **1989**, *111*, 8136.

[359]See, for example, Furuta; Ikeda; Meguriya; Ikeda; Yamamoto *Bull. Chem. Soc. Jpn.* **1984**, *57*, 2781; Pétrier; Luche *J. Org. Chem.* **1985**, *50*, 910; Tanaka; Yamashita; Hamatani; Ikemoto; Torii *Chem. Lett.* **1986**, 1611; *Synth. Commun.* **1987**, *17*, 789; Guo; Doubleday; Cohen *J. Am. Chem. Soc.* **1987**, *109*, 4710; Hosomi *Acc. Chem. Res.* **1988**, *21*, 200-206; Araki; Butsugan *Chem. Lett.* **1988**, 457; Minato; Tsuji *Chem. Lett.* **1988**, 2049; Coxon; van Eyk; Steel *Tetrahedron* **1989**, *45*, 1029; Knochel; Rao *J. Am. Chem. Soc.* **1990**, *112*, 6146; Wada; Ohki; Akiba *Bull. Chem. Soc. Jpn.* **1990**, *63*, 1738; Marton; Tagliavini; Zordan; Wardell *J. Organomet. Chem.* **1990**, *390*, 127; Wang; Shi; Xu; Huang *J. Chem. Soc., Perkin Trans. 1* **1990**, 424; Shono; Ishifune; Kashimura *Chem. Lett.* **1990**, 449.

[360]For another exception, involving a vinylic borane, see Satoh; Tayano; Hara; Suzuki *Tetrahedron Lett.* **1989**, *30*, 5153.

[361]For reviews, see Hoffmann; Niel; Schlapbach *Pure Appl. Chem.* **1990**, *62*, 1993-1998; Pelter; Smith; Brown *Borane Reagents*; Academic Press: New York, 1988, pp. 310-318. For a review of allylic boranes, see Bubnov *Pure Appl. Chem.* **1987**, *21*, 895-906.

[362]Kashin; Tulchinsky; Beletskaya *J. Organomet. Chem.* **1985**, *292*, 205.

[363]Prakash; Krishnamurti; Olah *J. Am. Chem. Soc.* **1989**, *111*, 393.

[364]Pearce; Richards; Scilly *J. Chem. Soc., Perkin Trans 1* **1972**, 1655; de Souza-Barboza; Pétrier; Luche; *J. Org. Chem.* **1988**, *53*, 1212.

[365]For a review, with Mg, Li, and other metals, see Blomberg; Hartog *Synthesis* **1977**, 18-30. For a discussion of the mechanism, see Molle; Bauer *J. Am. Chem. Soc.* **1982**, *104*, 3481. For a list of Barbier-type reactions, with references, see Ref. 64, pp. 553-555.

reacts with aldehydes[366] and with certain ketones[367] to give the expected alcohols. The similar reagents RCu(CN)ZnI also react with aldehydes, in the presence of BF_3–etherate, to give secondary alcohols. Carboxylic ester, nitrile, and imide groups in the R are not affected by the reaction conditions.[368]

Trimethylaluminum[369] and dimethyltitanium dichloride[370] exhaustively methylate ketones to give *gem*-dimethyl compounds[371] (see also **0-90**):

$$
\begin{array}{c}
\text{R}-\overset{\displaystyle \|}{\underset{\displaystyle \text{O}}{\text{C}}}-\text{R}' \quad \xrightarrow[\substack{\text{or} \\ \text{Me}_2\text{TiCl}_2}]{\text{Me}_3\text{Al}} \quad \text{R}-\overset{\displaystyle \text{Me}}{\underset{\displaystyle \text{Me}}{\overset{\displaystyle |}{\underset{\displaystyle |}{\text{C}}}}}-\text{R}'
\end{array}
$$

The titanium reagent also dimethylates aromatic aldehydes.[372]

α,β-Unsaturated aldehydes or ketones can give 1,4-addition as well as normal 1,2 addition (see **5-18**). In general, alkyllithiums give less 1,4 addition than the corresponding Grignard reagents.[373] Quinones add Grignard reagents on one or both sides or give 1,4 addition. In a compound containing both an aldehyde and a ketone function it is possible to add RMgX chemoselectively to the aldehyde function without significantly disturbing the ketonic group[374] (see also p. 927). On the other hand, chemoselective addition to a ketonic group can be carried out if the aldehyde is protected with a titanium tetrakis(dialkylamide).[375]

As with the reduction of aldehydes and ketones (**6-25**), the addition of organometallic compounds to these substrates can be carried out enantioselectively and diastereoselectively.[376] Chiral secondary alcohols have been obtained with high ee values by addition to aromatic aldehydes of Grignard and organolithium compounds in the presence of optically active amino alcohols as ligands.[377] High ee values have also been obtained with other organometallics,[378] including organotitanium compounds (methyl, aryl, allylic) in which an optically active ligand is coordinated to the titanium,[379] allylic boron compounds, and organozinc compounds.

[366]Barreiro; Luche; Zweig; Crabbé *Tetrahedron Lett.* **1975**, 2353; Zweig; Luche; Barreiro; Crabbé *Tetrahedron Lett.* **1975**, 2355.

[367]House; Prabhu; Wilkins; Lee *J. Org. Chem.* **1976**, *41*, 3067; Matsuzawa; Isaka; Nakamura; Kuwajima *Tetrahedron Lett.* **1989**, *30*, 1975.

[368]Yeh; Knochel; Santa *Tetrahedron Lett.* **1988**, *29*, 3887.

[369]Meisters; Mole *Aust. J. Chem.* **1974**, *27*, 1655. See also Jeffery; Meisters; Mole *Aust. J. Chem.* **1974**, *27*, 2569. For discussions of the mechanism of this reaction, see Ashby; Goel *J. Organomet. Chem.* **1981**, *221*, C15; Ashby; Smith *J. Organomet. Chem.* **1982**, *225*, 71. For a review of organoaluminum compounds in organic synthesis, see Maruoka; Yamamoto *Tetrahedron* **1988**, *44*, 5001-5032.

[370]Reetz; Westermann; Kyung *Chem. Ber.* **1985**, *118*, 1050.

[371]For the *gem*-diallylation of anhydrides, with an indium reagent, see Araki; Katsumura; Ito; Butsugan *Tetrahedron Lett.* **1989**, *30*, 1581.

[372]Reetz; Kyung *Chem. Ber.* **1987**, *120*, 123.

[373]An example was given on p. 799.

[374]Vaskan; Kovalev *J. Org. Chem. USSR* **1973**, *9*, 501.

[375]Reetz; Wenderoth; Peter *J. Chem. Soc., Chem. Commun.* **1983**, 406. For another method, see Maruoka; Araki; Yamamoto *Tetrahedron Lett.* **1988**, *29*, 3101.

[376]For reviews, see Solladié, in Morrison, Ref. 294, vol. 2, pp. 157-199, pp. 158-183; Nógrádi, Ref. 294, pp. 160-193; Noyori; Kitamura *Angew. Chem. Int. Ed. Engl.* **1991**, *30*, 49-69 [*Angew. Chem. 103*, 34-55].

[377]Mukaiyama; Soai; Shimizu; Suzuki *J. Am. Chem. Soc.* **1979**, *101*, 1455; Mazaleyrat; Cram *J. Am. Chem. Soc.* **1981**, *103*, 4585; Eleveld; Hogeveen *Tetrahedron Lett.* **1984**, *25*, 5187.

[378]For examples involving other organometallic compounds, see Abenhaïm; Boireau; Deberly *J. Org. Chem.* **1985**, *50*, 4045; Minowa; Mukaiyama *Bull. Chem. Soc. Jpn.* **1987**, *60*, 3697; Takai; Kataoka; Utimoto *J. Org. Chem.* **1990**, *55*, 1707.

[379]Reetz; Kükenhöhner; Weinig *Tetrahedron Lett.* **1986**, *27*, 5711; Wang; Fan; Feng; Quian *Synthesis* **1989**, 291; Riediker; Duthaler *Angew. Chem. Int. Ed. Engl.* **1989**, *28*, 494 [*Angew. Chem. 101*, 488]; Riediker; Hafner; Piantini; Rihs; Togni *Angew. Chem. Int. Ed. Engl.* **1989**, *30*, 499 [*Angew. Chem. 101*, 493].

A number of optically active allylic boron compounds have been used, including[380] B-allylbis(2-isocaranyl)borane (**28**),[381] *E*- and *Z*-crotyl-(*R*,*R*)-2,5-dimethylborolanes (**29**),[382]

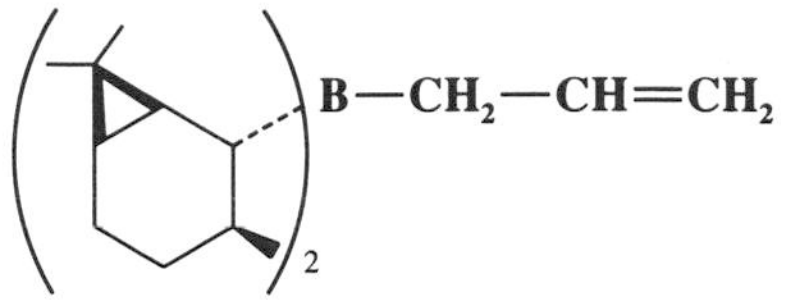

and the borneol derivative **30**,[383] all of which allylate aldehydes with ee values of 90% or more. Where the substrate possesses an aryl group or a triple bond, enantioselectivity is enhanced by using a a metal carbonyl complex of the substrate.[384]

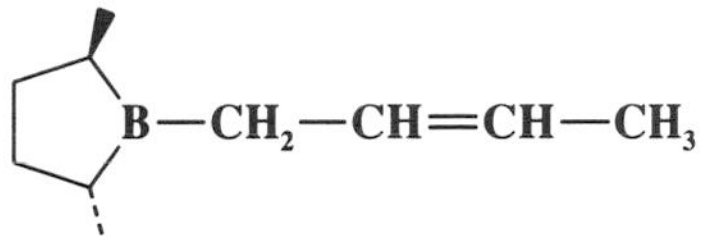

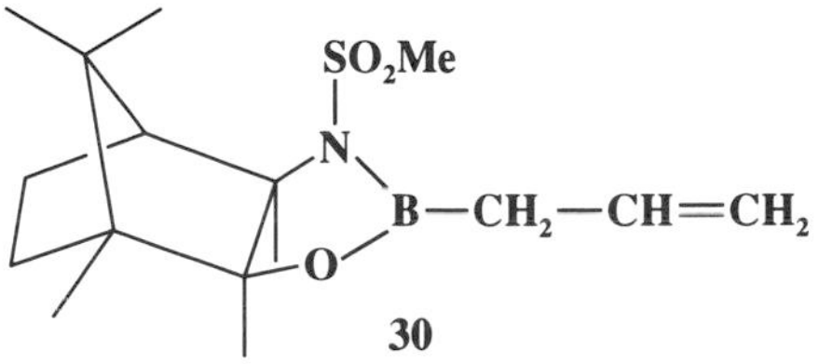

As for the organozinc reagents, very high ee values (90-98%) were obtained from R_2Zn reagents (R = alkyl) and aromatic[385] aldehydes by the use of a small amount (2 mole percent) of the catalyst[386] (−)-3-*exo*-(dimethylamino)isoborneol (DAIB).[387] High ee values

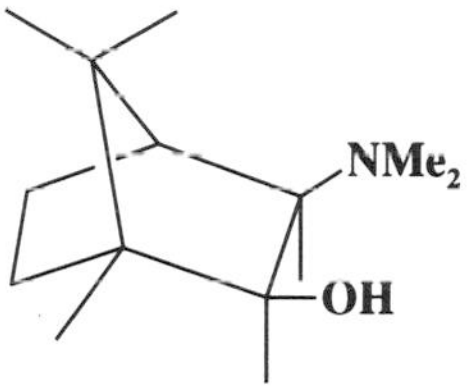

were also achieved with divinylzinc and both aromatic and aliphatic aldehydes, with other optically active amino alcohols as catalysts.[388] When benzaldehyde was treated with Et_2Zn

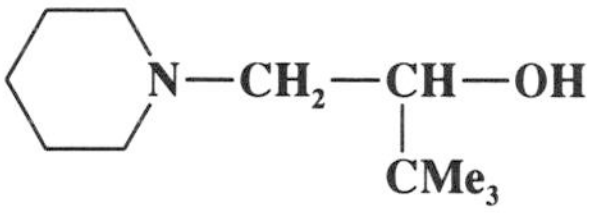

[380]For some others, see Hoffmann *Pure Appl. Chem.* **1988**, *60*, 123; Corey; Yu; Kim *J. Am. Chem. Soc.* **1989**, *111*, 5495; Roush; Ando; Powers, Palkowitz; Halterman *J. Am. Chem. Soc.* **1990**, *112*, 6339; Brown; Randad *Tetrahedron Lett.* **1990**, *31*, 455; Stürmer; Hoffmann *Synlett* **1990**, 759.

[381]Brown; Randad *Tetrahedron* **1990**, *46*, 4457; Racherla; Brown *J. Org. Chem.* **1991**, *56*, 401, and references cited in these papers.

[382]Garcia; Kim; Masamune *J. Org. Chem.* **1987**, *52*, 4831.

[383]Reetz; Zierke *Chem. Ind. (London)* **1988**, 663.

[384]Roush; Park *J. Org. Chem.* **1990**, *55*, 1143.

[385]For catalysts that are also successful for aliphatic aldehydes, see Takahashi; Kawakita; Yoshioka; Kobayashi; Ohno *Tetrahedron Lett.* **1989**, *30*, 7095; Tanaka; Ushio; Suzuki *J. Chem. Soc., Chem. Commun.* **1989**, 1700; Soai; Yokoyama; Hayasaka *J. Org. Chem.* **1991**, *56*, 4264.

[386]For some other optically active catalysts used with R_2Zn and ArCHO, see Smaardijk; Wynberg *J. Org. Chem.* **1987**, *52*, 135; Joshi; Srebnik; Brown *Tetrahedron Lett.* **1989**, *30*, 5551; Soai; Watanabe; Yamamoto *J. Org. Chem.* **1990**, *55*, 4832; Soai; Hori; Kawahara *Tetrahedron: Asymmetry* **1990**, *1*, 769; Chelucci; Falorni; Giacomelli *Tetrahedron: Asymmetry* **1990**, *1*, 843; Chaloner; Langadianou *Tetrahedron Lett.* **1990**, *31*, 5185; Corey; Yuen; Hannon; Wierda *J. Org. Chem.* **1990**, *55*, 784.

[387]Kitamura; Okada; Suga; Noyori *J. Am. Chem. Soc.* **1989**, *111*, 4028; Noyori; Suga; Kawai; Okada; Kitamura; Oguni; Hayashi; Kaneko; Matsuda *J. Organomet. Chem.* **1990**, *382*, 19.

[388]Oppolzer; Radinov *Tetrahedron Lett.* **1988**, *29*, 5645; Watanabe; Araki; Butsugan; Uemura *J. Org. Chem.* **1991**, *56*, 2218; Soai; Watanabe *Tetrahedron: Asymmetry* **1991**, *2*, 97; Asami; Inoue *Chem. Lett.* **1991**, 685.

in the presence of the optically active catalyst 1-piperidino-3,3-dimethyl-2-butanol (**31**), a surprising result was obtained. Although the catalyst had only 10.7% excess of one enantiomer, the product PhCH(OH)Me had an ee of 82%.[389] When the catalyst ee was increased to 20.5%, the product ee rose to 88%. The question is, how could a catalyst produce a product with an ee much higher than itself? One possible explanation[390] is that *R* and *S* molecules of the catalyst form a complex with each other, and that only the uncomplexed molecules are actually involved in the reaction. Since initially the number of *R* and *S* molecules was not the same, the *R:S* ratio of the uncomplexed molecules must be considerably higher (or lower) than that of the initial mixture.

Diastereoselective addition[391] has been carried out with achiral reagents and chiral substrates,[392] similar to the reduction shown on p. 915,[393] but because the attacking atom in this case is carbon, not hydrogen, it is also possible to get diastereoselective addition with an achiral substrate and an optically active reagent.[394] Use of suitable reactants creates, in the most general case, two new chiral centers, so the product can exist as two pairs of enantiomers:

$$R^1{-}CHM + R^3{-}\underset{\underset{O}{\|}}{C}{-}R^4 \longrightarrow$$

diastereomers

Even if the organometallic compound is racemic, it still may be possible to get a diastereoselective reaction; that is, one pair of enantiomers is formed in greater amount than the other.[395]

In some cases the Grignard reaction can be performed intramolecularly.[396] For example, treatment of 5-bromo-2-pentanone with magnesium and a small amount of mercuric chloride in THF produced 1-methyl-1-cyclobutanol in 60% yield.[397] Other four- and five-membered

$$BrCH_2CH_2CH_2{-}\underset{\underset{O}{\|}}{C}{-}CH_3 \xrightarrow[THF]{Mg} 60\%$$

[389]Oguni; Matsuda; Kaneko *J. Am. Chem. Soc.* **1988**, *110*, 7877.

[390]See Wynberg *Chimia* **1989**, *43*, 150.

[391]For a review, see Yamamoto; Maruyama *Heterocycles* **1982**, *18*, 357-386.

[392]For a review of cases in which the substrate bears a group that can influence the diastereoselectivity by chelating with the metal, see Reetz *Angew. Chem. Int. Ed. Engl.* **1984**, *23*, 556-569 [*Angew. Chem. 96*, 542-555]. See also Keck; Castellino *J. Am. Chem. Soc.* **1986**, *108*, 3847.

[393]See, for example, Eliel; Morris-Natschke *J. Am. Chem. Soc.* **1984**, *106*, 2937; Reetz; Steinbach; Westermann; Peter; Wenderoth *Chem. Ber.* **1985**, *118*, 1441; Yamamoto; Matsuoka *J. Chem. Soc., Chem. Commun.* **1987**, 923; Boireau; Deberly; Abenhaïm *Tetrahedron Lett.* **1988**, *29*, 2175; Page; Westwood; Slawin; Williams *J. Chem. Soc., Perkin Trans. 1* **1989**, 1158; Soai; Niwa; Hatanaka *Bull. Chem. Soc. Jpn.* **1990**, *63*, 2129. For examples in which both reactants were chiral, see Roush; Halterman *J. Am. Chem. Soc.* **1986**, *108*, 294; Hoffmann; Dresely; Hildebrandt *Chem. Ber.* **1988**, *121*, 2225; Paquette; Learn; Romine; Lin *J. Am. Chem. Soc.* **1988**, *110*, 879; Brown; Bhat; Randad *J. Org. Chem.* **1989**, *54*, 1570.

[394]For a review of such reactions with crotylmetallic reagents, see Hoffmann *Angew. Chem. Int. Ed. Engl.* **1982**, *21*, 555-566 [*Angew. Chem. 94*, 569-580]. For a discussion of the mechanism, see Denmark; Weber *J. Am. Chem. Soc.* **1984**, *106*, 7970. For some examples, see Hoffmann; Landmann *Chem. Ber.* **1986**, *119*, 2013; Zweifel; Shoup *J. Am. Chem. Soc.* **1988**, *110*, 5578; Gung; Smith; Wolf *Tetrahedron Lett.* **1991**, *32*, 13.

[395]For examples, see Coxon; van Eyk; Steel *Tetrahedron Lett.* **1985**, *26*, 6121; Mukaiyama; Ohshima; Miyoshi *Chem. Lett.* **1987**, 1121; Masuyama; Takahara; Kurusu *Tetrahedron Lett.* **1989**, *30*, 3437.

[396]For a list of reagents, with references, see Ref. 64, p. 557.

[397]Leroux *Bull. Soc. Chim. Fr.* **1968**, 359.

ring compounds were also prepared by this procedure. Similar closing of five- and six-membered rings was achieved by treatment of a δ- or ε-halocarbonyl compound, not with a metal, but with a dianion derived from nickel tetraphenyporphine.[398] An interesting organometallic ring closure is

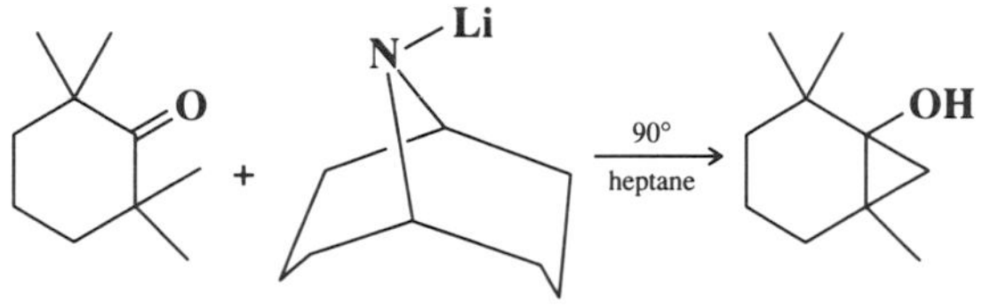

base

In this case, because the ketone has no α hydrogen, the base removed a β hydrogen (from a CH_3 group), and the intramolecular addition to the C=O followed.[399]

The *gem*-disubstituted magnesium compounds formed from CH_2Br_2 or CH_2I_2 (**2-38**) react with aldehydes or ketones to give olefins in moderate-to-good yields.[400] The reaction could

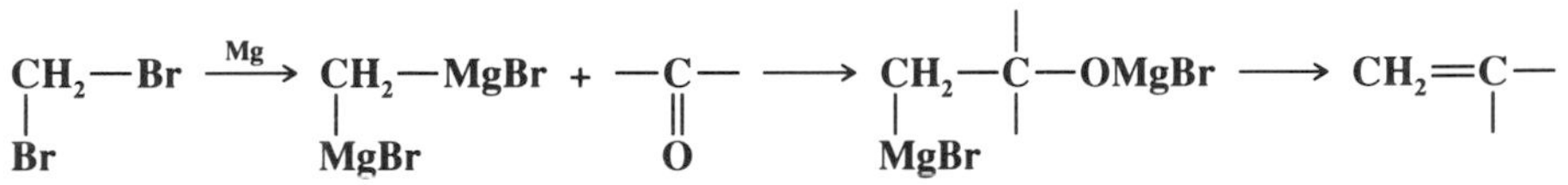

not be extended to other *gem*-dihalides. Similar reactions with *gem*-dimetallic compounds prepared with metals other than magnesium have also produced olefins.[401] The α,α-dimetallic derivatives of phenyl sulfones $PhSO_2CM_2R$ (M = Li or Mg) react with aldehydes or ketones R'COR" to give good yields of the α,β-unsaturated sulfones $PhSO_2CR=CR'R"$,[402] which can be reduced with aluminum amalgam (see **0-94**) or with $LiAlH_4$–$CuCl_2$ to give the olefins CHR=CR'R".[403] Olefins can also be obtained from organolithium compounds R^1R^2CHLi, by treating them with ketones R^3COR^4, followed by $SOCl_2$, a procedure which gives $R^1R^2C=CR^3R^4$.[404] These reactions are closely related to the Wittig reaction (**6-47**) and, like it, provide a means of achieving the conversion $R_2C=O \rightarrow R_2C=CR'R"$. On the other hand, *gem*-dihalides treated with a carbonyl compound and Li or BuLi give epoxides[405] (see also **6-61**).

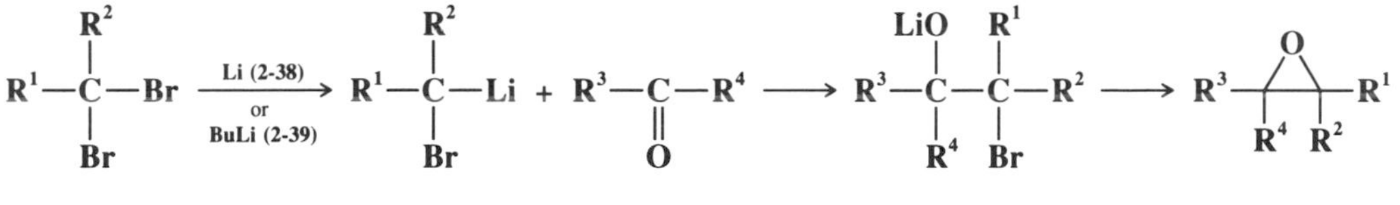

[398]Corey; Kuwajima *J. Am. Chem. Soc.* **1970**, *92*, 395. For another method, see Molander; Etter; Zinke *J. Am. Chem. Soc.* **1987**, *109*, 453; Molander; McKie *J. Org. Chem.* **1991**, *56*, 4112.

[399]Shiner; Berks; Fisher *J. Am. Chem. Soc.* **1988**, *110*, 957.

[400]Bertini; Grasselli; Zubiani; Cainelli *Tetrahedron* **1970**, *26*, 1281.

[401]For example, see Zweifel; Steele *Tetrahedron Lett.* **1966**, 6021; Cainelli; Bertini; Grasselli; Zubiani *Tetrahedron Lett.* **1967**, 1581; Takai; Hotta; Oshima; Nozaki *Bull. Chem. Soc. Jpn.* **1980**, *53*, 1698; Knochel; Normant *Tetrahedron Lett.* **1986**, *27*, 1039; Barluenga; Fernández-Simón; Concellón; Yus *J. Chem. Soc., Chem. Commun.* **1986**, 1665; Okazoe; Takai; Utimoto *J. Am. Chem. Soc.* **1987**, *109*, 951; Piotrowski; Malpass; Boleslawski; Eisch *J. Org. Chem.* **1988**, *53*, 2829; Tour; Bedworth; Wu *Tetrahedron Lett.* **1989**, *30*, 3927; Lombardo *Org. Synth.* 65, 81.

[402]Pascali; Tangari; Umani-Ronchi *J. Chem. Soc., Perkin Trans. 1* **1973**, 1166.

[403]Pascali; Umani-Ronchi *J. Chem. Soc., Chem. Commun.* **1973**, 351.

[404]Olah; Wu; Farooq *J. Org. Chem.* **1989**, *54*, 1375.

[405]Cainelli; Umani-Ronchi; Bertini; Grasselli; Zubiani *Tetrahedron* **1971**, *27*, 6109; Cainelli; Tangari; Umani-Ronchi *Tetrahedron* **1972**, *28*, 3009.

In other uses of *gem*-dihalo compounds, aldehydes and ketones add the CH_2I group ($R_2CO \rightarrow R_2C(OH)CH_2I$) when treated with CH_2I_2 in the presence of SmI_2[406] and the CHX_2 group when treated with methylene halides and lithium dicyclohexylamide at low temperatures.[407]

$$CH_2X_2 + \underset{\underset{O}{\|}}{-C-} \xrightarrow[\text{2. H}_2\text{O}]{\substack{\text{1. LiN(C}_6\text{H}_{11}\text{)}_2 \\ -78°C}} \underset{\underset{OH}{|}}{-C-}CHX_2 \quad X = Cl, Br, I$$

A hydroxymethyl group can be added to an aldehyde or ketone with the masked reagent Me_2(i-PrO)$SiCH_2MgCl$, which with R_2CO gives $R_2C(OH)CH_2Si(O-i-Pr)Me_2$, which, with H_2O_2, give 1,2-diols $R_2C(OH)CH_2OH$.[408]

It is possible to add an acyl group to a ketone to give (after hydrolysis) an α-hydroxy ketone.[409] This can be done by adding RLi and CO to the ketone at $-110°C$:[410]

$$\underset{\underset{O}{\|}}{-C-} + RLi + CO \xrightarrow{-110°} R-\underset{\underset{O}{\|}}{C}-\underset{\underset{OLi}{|}}{C}- \xrightarrow{H_2O^+} R-\underset{\underset{O}{\|}}{C}-\underset{\underset{OH}{|}}{C}-$$

When the same reaction is carried out with carboxylic esters R'COOR'', α-diketones RCOCOR' are obtained.[410] Another way to add RCO to aldehydes and ketones is to treat the substrate with ArCOLi, generated by treating ArCOTeBu with BuLi.[411]

Although most aldehydes and ketones react very nicely with most Grignard reagents, there are several types of side reaction that occur mostly with hindered ketones and with bulky Grignard reagents. The two most important of these are *enolization* and *reduction*. The former requires that the aldehyde or ketone have an α hydrogen, and the latter requires that the Grignard reagent have a β hydrogen:

Enolization

$$RMgX + \underset{\underset{H}{|}}{-C}-\underset{\underset{O}{\|}}{C}-R' \longrightarrow RH + \underset{\underset{\underset{|\underline{O}|^{\ominus}}{|}}{}}{-C=C}-R \xrightarrow{\text{hydrol.}} \underset{\underset{OH}{|}}{-C=C}-R' \rightleftharpoons \underset{\underset{H}{|}}{-C}-\underset{\underset{O}{\|}}{C}-R'$$

Reduction

$$\underset{\underset{H}{|}}{-C}-\underset{|}{C}-MgX + \underset{\underset{O}{\|}}{-C-} \longrightarrow -C=C- + \underset{\underset{OMgX}{|}}{-\overset{\overset{H}{|}}{C}-} \xrightarrow{\text{hydrol.}} \underset{\underset{OH}{|}}{-\overset{\overset{H}{|}}{C}-}$$

Enolization is an acid–base reaction (**2-24**) in which a proton is transferred from the α carbon to the Grignard reagent. The carbonyl compound is converted to its enolate ion form, which, on hydrolysis, gives the original ketone or aldehyde. Enolization is important not only for hindered ketones but also for those that have a relatively high percentage of enol form, e.g., β-keto esters, etc. In reduction, the carbonyl compound is reduced to an alcohol (**6-25**)

[406]Imamoto; Takeyama; Koto *Tetrahedron Lett.* **1986,** *27,* 3243.
[407]Taguchi; Yamamoto; Nozaki *J. Am. Chem. Soc.* **1974,** *96,* 3010 *Bull. Chem. Soc. Jpn.* **1977,** *50,* 1588.
[408]Tamao; Ishida *Tetrahedron Lett.* **1984,** *25,* 4245. For another method, see Imamoto; Takeyama; Yokoyama *Tetrahedron Lett.* **1984,** *25,* 3225.
[409]For a review, see Seyferth; Weinstein; Wang; Hui; Archer *Isr. J. Chem.* **1984,** *24,* 167-175.
[410]Seyferth; Weinstein; Wang *J. Org. Chem.* **1983,** *48,* 1144; Seyferth; Weinstein; Wang; Hui; *Tetrahedron Lett.* **1983,** *24,* 4907.
[411]Hiiro; Morita; Inoue; Kambe; Ogawa; Ryu; Sonoda *J. Am. Chem. Soc.* **1990,** *112,* 455.

by the Grignard reagent, which itself undergoes elimination to give an olefin. Two other side reactions are condensation (between enolate ion and excess ketone) and Wurtz-type coupling (**0-92**). Such highly hindered tertiary alcohols as triisopropylcarbinol, tri-*t*-butylcarbinol, and diisopropylneopentylcarbinol cannot be prepared (or can be prepared only in extremely low yields) by the addition of Grignard reagents to ketones, because reduction and/or enolization become prominent.[412] However, these carbinols can be prepared by the use of alkyllithiums at $-80°C$,[413] under which conditions enolization and reduction are much less important.[414] Other methods of increasing the degree of addition at the expense of reduction consist of complexing the Grignard reagent with $LiClO_4$ or $Bu_4N^+ Br^-$,[415] or using benzene or toluene instead of ether as solvent.[416] Both reduction and enolization can be avoided by adding $CeCl_3$ to the Grignard reagent.[417]

Another way to avoid complications is to add $(RO)_3TiCl$, $TiCl_4$,[418] $(RO)_3ZrCl$, or $(R_2N)_3TiX$ to the Grignard or lithium reagent. This produces organotitanium or organozirconium compounds that are much more selective than Grignard or organolithium reagents.[419] An important advantage of these reagents is that they do not react with NO_2 or CN functions that may be present in the substrate, as Grignard and organolithium reagents do. Furthermore, organotitanium reagents can be made to add chemoselectively to aldehydes in presence of ketones.[420] Organomanganese compounds are also chemoselective in this way.[421]

There has been much controversy regarding the mechanism of addition of Grignard reagents to aldehydes and ketones.[422] The reaction is difficult to study because of the variable nature of the species present in the Grignard solution (p. 183) and because the presence of small amounts of impurities in the magnesium seems to have a great effect on the kinetics of the reaction, making reproducible experiments difficult.[423] There seem to be two basic mechanisms, depending on the reactants and the reaction conditions. In one of these, the R group is transferred to the carbonyl carbon with its electron pair. A detailed mechanism of this type has been proposed by Ashby and co-workers,[424] based on the discovery that this reaction proceeds by two paths—one first order in MeMgBr and the other first order in Me₂Mg.[425] According to this proposal, both MeMgBr and Me₂Mg add to the carbonyl

[412]Whitmore; George *J. Am. Chem. Soc.* **1942**, *64*, 1239.

[413]Bartlett; Lefferts *J. Am. Chem. Soc.* **1955**, 77, 2804; Zook; March; Smith *J. Am. Chem. Soc.* **1959**, *81*, 1617; Bartlett; Tidwell *J. Am. Chem. Soc.* **1968**, *90*, 4421. See also Lomas *Nouv. J. Chim.* **1984**, 8, 365; Molle; Briand; Bauer; Dubois *Tetrahedron* **1984**, *40*, 5113.

[414]Buhler *J. Org. Chem.* **1973**, *38*, 904.

[415]Chastrette; Amouroux *Chem. Commun.* **1970**, 470, *Bull. Soc. Chim. Fr.* **1970**, 4348. See also Richey; DeStephano *J. Org. Chem.* **1990**, 55, 3281.

[416]Canonne; Foscolos; Caron; Lemay *Tetrahedron* **1982**, *38*, 3563.

[417]Imamoto; Takiyama; Nakamura; Hatajima; Kamiya *J. Am. Chem. Soc.* **1989**, *111*, 4392.

[418]See Reetz; Kyung; Hüllmann *Tetrahedron* **1986**, *42*, 2931.

[419]For a monograph, see Reetz *Organotitanium Reagents in Organic Synthesis*; Springer: New York, 1986. For reviews, see Weidmann; Seebach *Angew. Chem. Int. Ed. Engl.* **1983**, *22*, 31-45 [*Angew. Chem. 95*, 12-26]; Reetz *Top. Curr. Chem.* **1982**, *106*, 1-54.

[420]Reetz, Ref. 419 (monograph), pp. 75-86. See also Reetz; Maus *Tetrahedron* **1987**, *43*, 101.

[421]Cahiez; Figadere *Tetrahedron Lett.* **1986**, *27*, 4445. For other organometallic reagents with high selectivity towards aldehyde functions, see Kauffmann; Hamsen; Beirich *Angew. Chem. Int. Ed. Engl.* **1982**, *21*, 144 [*Angew. Chem. 94*, 145]; Takai; Kimura; Kuroda; Hiyama; Nozaki *Tetrahedron Lett.* **1983**, *24*, 5281; Soai; Watanabe; Koyano *Bull. Chem. Soc. Jpn.* **1989**, *62*, 2124.

[422]For reviews, see Holm *Acta Chem. Scand., Ser. B* **1983**, *37*, 567-584; Ashby *Pure Appl. Chem.* **1980**, *52*, 545-569, *Bull Soc. Chim. Fr.* **1972**, 2133-2142, *Q. Rev. Chem. Soc.* **1967**, *21*, 259-285; Ashby; Laemmle; Neumann *Acc. Chem. Res.* **1974**, 7, 272-280; Blomberg *Bull. Soc. Chim. Fr.* **1972**, 2143-2149. For a review of the stereochemistry of the reaction, see Ashby; Laemmle, Ref. 5. For a review of the effects of the medium and the cation, see Solv'yanov; Beletskaya *Russ. Chem. Rev.* **1987**, *56*, 465-476.

[423]See, for example, Ashby; Walker; Neumann *Chem. Commun.* **1970**, 330; Ashby; Neumann; Walker; Laemmle; Chao *J. Am. Chem. Soc.* **1973**, *95*, 3330.

[424]Ashby; Laemmle; Neumann *J. Am. Chem. Soc.* **1972**, *94*, 5421.

[425]Ashby; Laemmle; Neumann *J. Am. Chem. Soc.* **1971**, *93*, 4601; Laemmle; Ashby; Neumann *J. Am. Chem. Soc.* **1971**, *93*, 5120.

carbon, though the exact nature of the step by which MeMgBr or Me$_2$Mg reacts with the substrate is not certain. One possibility is a four-centered cyclic transition state:[426]

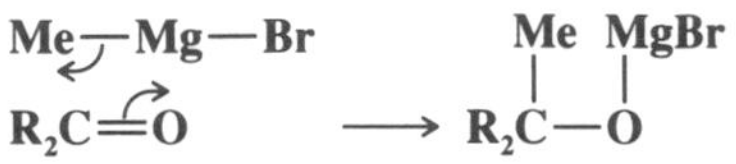

$$\begin{array}{cc} \text{Me} \overset{\longleftarrow}{\rightharpoondown} \text{Mg} - \text{Br} & \text{Me MgBr} \\ \underset{\overset{\curvearrowright}{\longrightarrow}}{R_2C \overset{\textstyle\backslash}{=\!\!=} O} & \begin{array}{c} | \quad | \\ R_2C - O \end{array} \end{array}$$

The other type of mechanism is a single electron transfer (SET)·process[427] with a ketyl intermediate:[428]

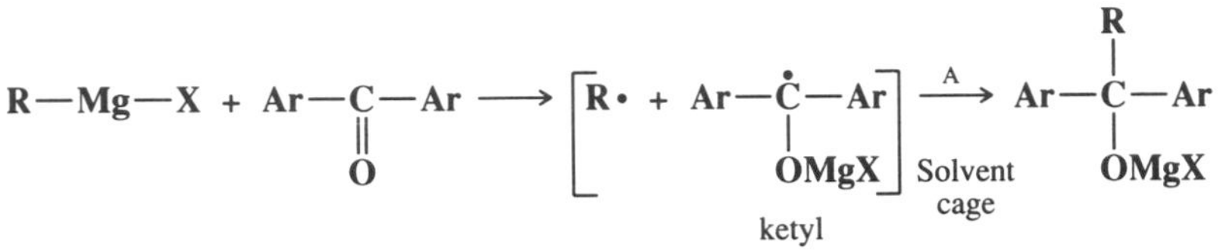

$$R - Mg - X \; + \; \underset{\overset{\|}{O}}{Ar - C - Ar} \longrightarrow \left[R\cdot \; + \; \underset{\overset{|}{OMgX}}{Ar - \overset{\bullet}{C} - Ar} \right] \xrightarrow[\substack{\text{Solvent} \\ \text{cage}}]{A} \underset{\overset{|}{OMgX}}{\overset{\overset{\textstyle R}{|}}{Ar - C - Ar}}$$

ketyl

This mechanism, which has been mostly studied with diaryl ketones, is more likely for aromatic and other conjugated aldehydes and ketones than it is for strictly aliphatic ones. Among the evidence[429] for the SET mechanism are esr spectra[430] and the obtention of Ar$_2$C—CAr$_2$ side products (from dimerization of the ketyl).[431] In the case of addition of
| |
OH OH
RMgX to benzil PhCOCOPh, esr spectra of two different ketyl radicals were observed, both reported to be quite stable at room temperature.[432] Carbon isotope effect studies with Ph^{14}COPh showed that the rate-determining step with most Grignard reagents is the carbon–carbon bond-forming step (marked A), though with allylmagnesium bromide it is the initial electron transfer step.[433]

Mechanisms for the addition of organolithium reagents have been investigated much less.[434] Addition of a cryptand that binds Li$^+$ inhibited the normal addition reaction, showing that the lithium is necessary for the reaction to take place.[435]

There is general agreement that the mechanism leading to reduction[436] is usually as follows:

[426]Tuulmets *Org. React. (USSR)* **1967**, *4*, 5; House; Oliver *J. Org. Chem.* **1968**, *33*, 929; Ashby; Yu; Roling; *J. Org. Chem.* **1972**, *37*, 1918. See also Billet; Smith *J. Am. Chem. Soc.* **1968**, *90*, 4108; Lasperas; Perez-Rubalcaba; Quiroga-Feijoo *Tetrahedron* **1980**, *36*, 3403.

[427]For a review, see Dagonneau *Bull. Soc. Chim. Fr.* **1982**, II-269-II-280.

[428]There is kinetic evidence that the solvent cage shown may not be necessary: Walling *J. Am. Chem. Soc.* **1988**, *110*, 6846.

[429]For other evidence, see Savin; Kitaev *J. Org. Chem. USSR* **1975**, *11*, 2622; Ōkubo *Bull. Chem. Soc. Jpn.* **1977**, *50*, 2379; Ashby; Bowers *J. Am. Chem. Soc.* **1981**, *103*, 2242; Holm *Acta. Chem. Scand., Ser. B* **1982**, *36*, 266, **1988**, *42*, 685; Liotta; Saindane; Waykole *J. Am. Chem. Soc.* **1983**, *105*, 2922; Zhang; Wenderoth; Su; Ashby *J. Organomet. Chem.* **1985**, *292*, 29; Yamataka; Miyano; Hanafusa *J. Org. Chem.* **1991**, *56*, 2573.

[430]Fauvarque; Rouget, *C. R. Acad. Sci., Ser C* **1968**, *267*, 1355; Maruyama; Katagiri *Chem. Lett.* **1987**, 731, 735, *J. Phys. Org. Chem.* **1988**, *1*, 21.

[431]Blomberg; Mosher *J. Organomet. Chem.* **1968**, *13*, 519; Holm; Crossland *Acta Chem. Scand.* **1971**, *25*, 59.

[432]Maruyama; Katagiri *J. Am. Chem. Soc.* **1986**, *108*, 6263, *J. Phys. Org. Chem.* **1989**, *2*, 205. See also Holm *Acta Chem. Scand., Ser. B* **1987**, *41*, 278; Maruyama; Katagiri *J. Phys. Org. Chem.* **1991**, *4*, 158.

[433]Yamataka; Matsuyama; Hanafusa *J. Am. Chem. Soc.* **1989**, *111*, 4912.

[434]See, for example, Al-Aseer; Smith *J. Org. Chem.* **1984**, *49*, 2608; Yamataka; Kawafuji; Nagareda; Miyano; Hanafusa *J. Org. Chem.* **1989**, *54*, 4706.

[435]Perraud; Handel; Pierre *Bull. Soc. Chim. Fr.* **1980**, II-283.

[436]For discussions of the mechanism of reduction, see Singer; Salinger; Mosher *J. Org. Chem.* **1967**, *32*, 3821; Denise; Fauvarque; Ducom *Tetrahedron Lett.* **1970**, 335; Cabaret; Welvart *J. Organomet. Chem.* **1974**, *80*, 199; Holm *J. Organomet. Chem.* **1971**, *29*, C45, *Acta Chem. Scand.* **1973**, *27*, 1552; Morrison; Tomaszewski; Mosher; Dale; Miller; Elsenbaumer *J. Am. Chem. Soc.* **1977**, *99*, 3167; Okuhara *J. Am. Chem. Soc.* **1980**, *102*, 244.

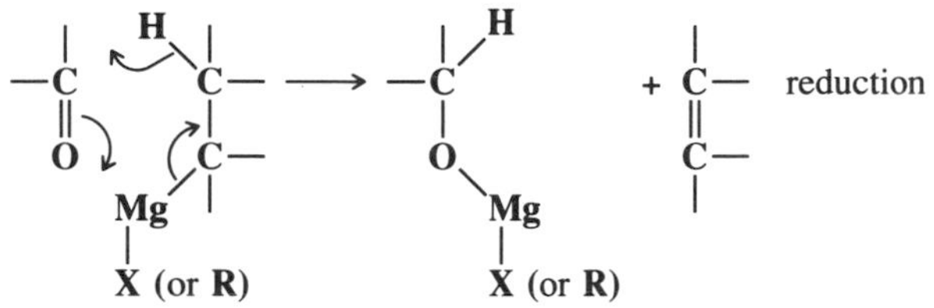

There is evidence that the mechanism leading to enolization is also cyclic, but involves prior coordination with magnesium:[437]

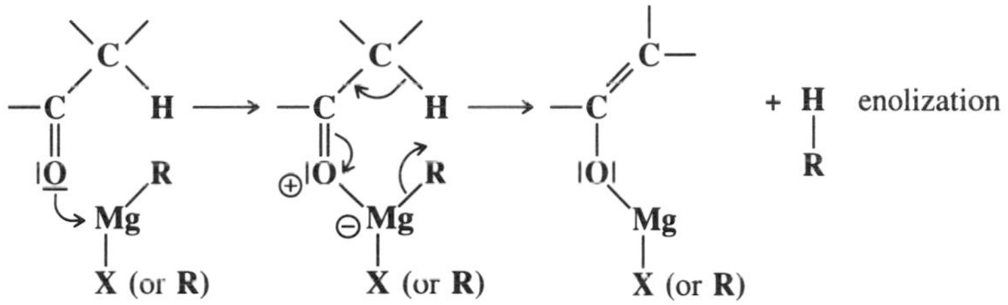

Aromatic aldehydes and ketones can be alkylated and reduced in one reaction vessel by treatment with an alkyl- or aryllithium, followed by lithium and ammonia and then by ammonium chloride.[438]

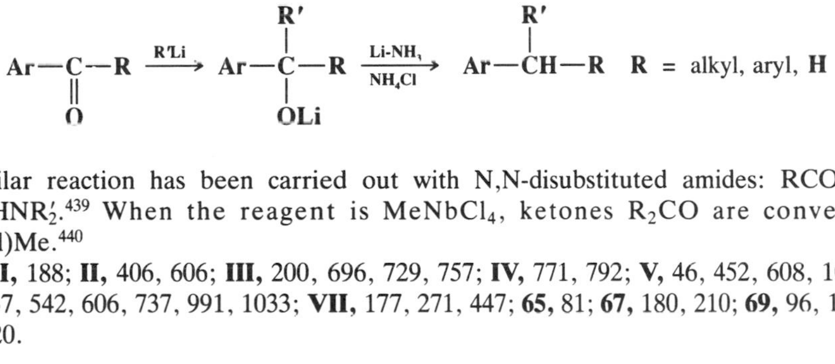

A similar reaction has been carried out with N,N-disubstituted amides: $RCONR'_2 \rightarrow$ $RR''CHNR'_2$.[439] When the reagent is $MeNbCl_4$, ketones R_2CO are converted to $R_2C(Cl)Me$.[440]

OS **I**, 188; **II**, 406, 606; **III**, 200, 696, 729, 757; **IV**, 771, 792; **V**, 46, 452, 608, 1058; **VI**, 478, 537, 542, 606, 737, 991, 1033; **VII**, 177, 271, 447; **65**, 81; **67**, 180, 210; **69**, 96, 106, 114, 120, 220.

6-30 The Reformatsky Reaction
O-Hydro-*C*-α-ethoxycarbonylalkyl-addition

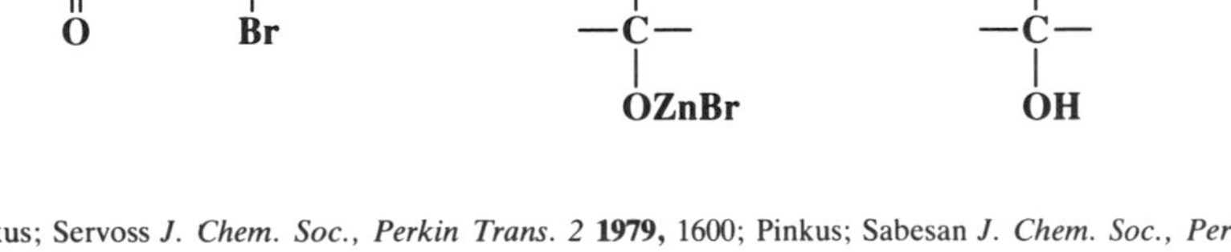

[437]Pinkus; Servoss *J. Chem. Soc., Perkin Trans. 2* **1979,** 1600; Pinkus; Sabesan *J. Chem. Soc., Perkin Trans. 2.* **1981,** 273.
[438]Hall; Lipsky *J. Org. Chem.* **1973,** *38,* 1735; Lipsky; Hall *Org. Synth. VI,* 537; McEnroe; Sha; Hall *J. Org. Chem.* **1976,** *41,* 3465.
[439]Hwang; Chu; Fowler *J. Org. Chem.* **1985,** *50,* 3885.
[440]Kauffmann; Abel; Neiteler; Schreer *Tetrahedron Lett.* **1990,** 503.

The *Reformatsky reaction* is very similar to **6-29**.[441] An aldehyde or ketone is treated with zinc and a halide; the halide is usually an α-halo ester or a vinylog of an α-halo ester (e.g., RCHBrCH=CHCOOEt), though α-halo nitriles,[442] α-halo ketones,[443] α-halo N,N-disubstituted amides, and the zinc salts of α-halo carboxylic acids[444] have also been used. With the last reagent the product is a β-hydroxy acid. Especially high reactivity can be achieved with activated zinc,[445] with zinc/silver–graphite,[446] and with zinc and ultrasound.[447] The reaction has also been carried out with other metals instead of zinc (e.g., In,[448] Mn[449]) and with certain other compounds, including SmI_2,[450] Bu_2Te,[451] and Bu_3Sb.[452] The aldehyde or ketone can be aliphatic, aromatic, or heterocyclic or contain various functional groups. Solvents used are generally ethers, including Et_2O, THF, and 1,4-dioxane.

Formally, the reaction can be regarded as if it were analogous to the Grignard reaction

(6-29), with EtOOC—$\overset{|}{\underset{|}{C}}$—ZnBr (**32**) as an intermediate analogous to RMgX. There *is* an

intermediate derived from zinc and the ester, the structure of which has been shown to be **33**, by x-ray crystallography of the solid intermediate prepared from *t*-$BuOCOCH_2Br$ and

33

Zn. As can be seen, it has some of the characteristics of **32**.

Usually, after hydrolysis, the alcohol is the product, but sometimes (especially with aryl aldehydes) elimination follows directly and the product is an olefin. By the use of Bu_3P along with Zn, the olefin can be made the main product,[454] making this an alternative to the Wittig reaction (**6-47**). Since Grignard reagents cannot be formed from α-halo esters, the method is quite useful, though there are competing reactions and yields are sometimes low. A similar reaction (called the *Blaise reaction*) has been carried out on nitriles:[455]

[441]For reviews, see Fürstner *Synthesis* **1989**, 571-590; Rathke *Org. React.* **1975**, *22*, 423-460; Gaudemar *Organomet. Chem. Rev., Sect A* **1972**, *8*, 183-233.

[442]Vinograd; Vul'fson *J. Gen. Chem. USSR* **1959**, *29*, 248, 1118, 2656, 2659; Palomo; Aizpurua; López; Aurrekoetxea *Tetrahedron Lett.* **1990**, *31*, 2205; Zheng; Yu; Shen *Synth. Commun.* **1990**, *20*, 3277.

[443]For examples (with R_3Sb and $CrCl_2$, respectively, instead of Zn), see Huang; Chen; Shen *J. Chem. Soc., Perkin Trans. 1* **1988**, 2855; Dubois; Axiotis; Bertounesque *Tetrahedron Lett.* **1985**, *26*, 4371.

[444]Bellassoued; Gaudemar *J. Organomet. Chem.* **1975**, *102*, 1.

[445]Rieke; Uhm *Synthesis* **1975**, 452; Bouhlel; Rathke *Synth. Commun.* **1991**, *21*, 133.

[446]Csuk; Fürstner; Weidmann *J. Chem. Soc., Chem. Commun.* **1986**, 775. See also Bortolussi; Seyden-Penne *Synth. Commun.* **1989**, *19*, 2355.

[447]Han; Boudjouk *J. Org. Chem.* **1982**, *47*, 5030.

[448]Chao; Rieke *J. Org. Chem.* **1975**, *40*, 2253; Araki; Ito; Butsugan *Synth. Commun.* **1988**, *18*, 453.

[449]Cahiez; Chavant *Tetrahedron Lett.* **1989**, *30*, 7373.

[450]Kagan; Namy; Girard *Tetrahedron Suppl.* **1981**, *37*, 175; Tabuchi; Kawamura; Inanaga; Yamaguchi *Tetrahedron Lett.* **1986**, *27*, 3889; Molander; Etter *J. Am. Chem. Soc.* **1987**, *109*, 6556.

[451]Huang; Xie; Wu *Tetrahedron Lett.* **1987**, *28*, 801.

[452]Chen; Huang; Shen; Liao *Heteroat. Chem.* **1990**, *1*, 49.

[453]Dekker; Budzelaar; Boersma; van der Kerk; Spek *Organometallics* **1984**, *3*, 1403.

[454]Shen; Xin; Zhao *Tetrahedron Lett.* **1988**, *29*, 6119. For another method, see Huang; Shi; Li; Wen *J. Chem. Soc., Perkin Trans. 1* **1989**, 2397.

[455]See Cason; Rinehart; Thornton *J. Org. Chem.* **1953**, *18*, 1594; Bellassoued; Gaudemar *J. Organomet. Chem.* **1974**, *81*, 139; Hannick; Kishi *J. Org. Chem.* **1983**, *48*, 3833.

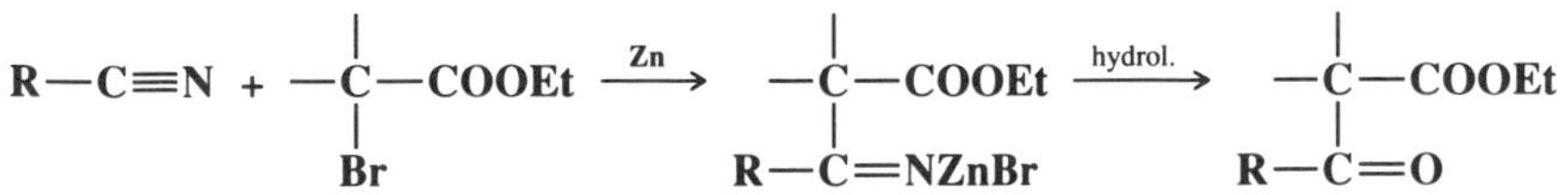

$$R-C\equiv N + \underset{\underset{Br}{|}}{\overset{|}{-}}C-COOEt \xrightarrow{Zn} \underset{\underset{R-C=NZnBr}{}}{\overset{|}{-}}C-COOEt \xrightarrow{hydrol.} \underset{\underset{R-C=O}{}}{\overset{|}{-}}C-COOEt$$

Carboxylic esters have also been used as substrates, but then, as might be expected (p. 881), the result is substitution and not addition:

$$R-\underset{\underset{O}{\|}}{C}-OR' + \underset{\underset{Br}{|}}{\overset{|}{-}}C-COOEt \xrightarrow{Zn} \underset{\underset{R-C=O}{}}{\overset{|}{-}}C-COOEt$$

The product in this case is the same as with the corresponding nitrile, though the pathways are different.

Addition of *t*-butyl acetate to lithium diisopropylamide (LDA) in hexane at $-78°C$ gives the lithium salt of *t*-butyl acetate[456] (2-22) as a stable solid. The nmr and ir spectra of this

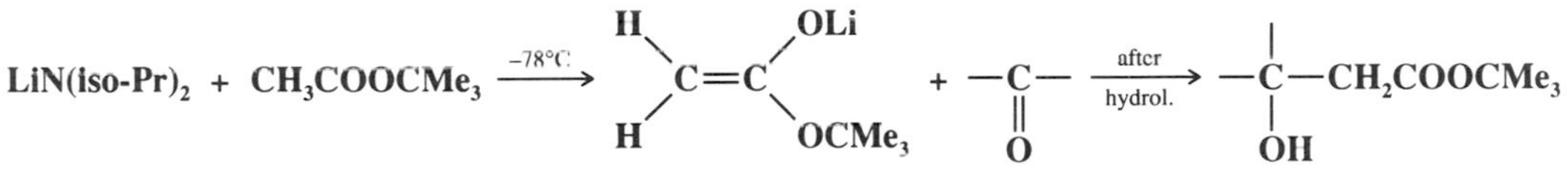

$$LiN(iso-Pr)_2 + CH_3COOCMe_3 \xrightarrow{-78°C} \underset{H}{\overset{H}{\diagdown}}C=C\underset{OCMe_3}{\overset{OLi}{\diagup}} + \underset{\underset{O}{\|}}{-}C- \xrightarrow[hydrol.]{after} \underset{\underset{OH}{|}}{-}C-CH_2COOCMe_3$$

34

salt in benzene show it to have the enolate structure **34.** Reaction of **34** with a ketone provides a simple rapid alternative to the Reformatsky reaction as a means of preparing β-hydroxy *t*-butyl esters. A similar reaction involves treatment of a ketone with a silyl ketene acetal $R_2C=C(OSiMe_3)OR'$ in the presence of $TiCl_4$[457] (see also the reaction between silyl enol ethers and aldehydes and ketones, in **6-39**).

OS **III**, 408; **IV**, 120, 444.

6-31 The Conversion of Carboxylic Acid Salts to Ketones with Organometallic Compounds

Alkyl-de-oxido-substitution

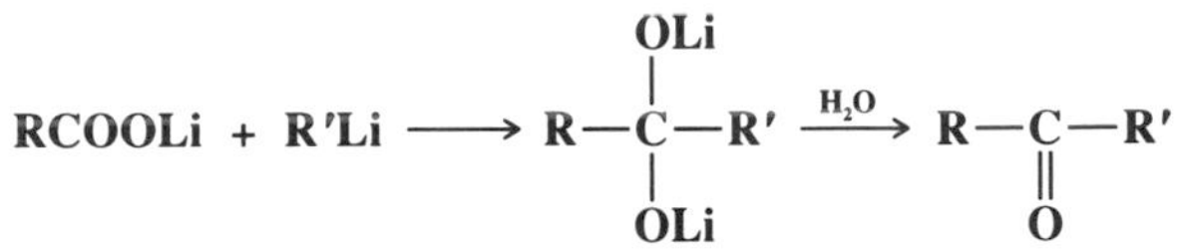

$$RCOOLi + R'Li \longrightarrow R-\underset{\underset{OLi}{|}}{\overset{\overset{OLi}{|}}{C}}-R' \xrightarrow{H_2O} R-\underset{\underset{O}{\|}}{C}-R'$$

Good yields of ketones can often be obtained by treatment of the lithium salt of a carboxylic acid with an alkyllithium reagent, followed by hydrolysis.[458] R' may be aryl or primary, secondary, or tertiary alkyl. MeLi and PhLi have been employed most often. R may be

---

[456]Rathke; Sullivan *J. Am. Chem. Soc.* **1973**, *95*, 3050.

[457]See for example, Saigo; Osaki; Mukaiyama *Chem. Lett.* **1975**, 989; Palazzi; Colombo; Gennari *Tetrahedron Lett.* **1986**, *27*, 1735; Oppolzer; Marco-Contelles *Helv. Chim. Acta* **1986**, *69*, 1699; Hara; Mukaiyama *Chem. Lett.* **1989**, 1909. For a list of references, see Ref. 64, pp. 885-887. For methods of preparing silyl ketene acetals, see Revis; Hilty *Tetrahedron Lett.* **1987**, *28*, 4809, and references cited therein.

[458]For a review, see Jorgenson *Org. React.* **1970**, *18*, 1-97. For an improved procedure, see Rubottom; Kim *J. Org. Chem.* **1983**, *48*, 1550.

alkyl or aryl, though lithium acetate generally gives low yields. Tertiary alcohols are side products.

OS **V**, 775.

6-32   The Addition of Grignard Reagents to Acid Derivatives
Dialkyl,hydroxy-de-alkoxy,oxo-tersubstitution

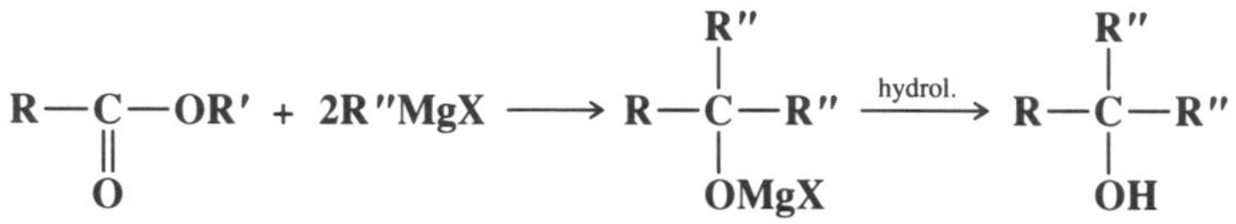

$$\underset{\underset{O}{\|}}{R-C-OR'} + 2R''MgX \longrightarrow \underset{\underset{OMgX}{|}}{R-\overset{\overset{R''}{|}}{C}-R''} \overset{hydrol.}{\longrightarrow} \underset{\underset{OH}{|}}{R-\overset{\overset{R''}{|}}{C}-R''}$$

When carboxylic esters are treated with Grignard reagents, there is usually concomitant addition to the carbonyl (**6-29**) and substitution of R″ for OR′ (**0-104**), so that tertiary alcohols are formed in which two R groups are the same. Formates give secondary alcohols and carbonates give tertiary alcohols in which all three R groups are the same: $(EtO)_2C=O + RMgX \rightarrow R_3COMgX$. Acyl halides and anhydrides behave similarly, though these substrates are employed less often.[459] There are many side reactions possible, especially when the acid derivative or the Grignard reagent is branched: enolizations, reductions (not for esters, but for halides), condensations, and cleavages, but the most important is simple substitution (**0-104**), which in some cases can be made to predominate. When 1,4-diagnesium compounds are used, carboxylic esters are converted to cyclopentanols.[460]

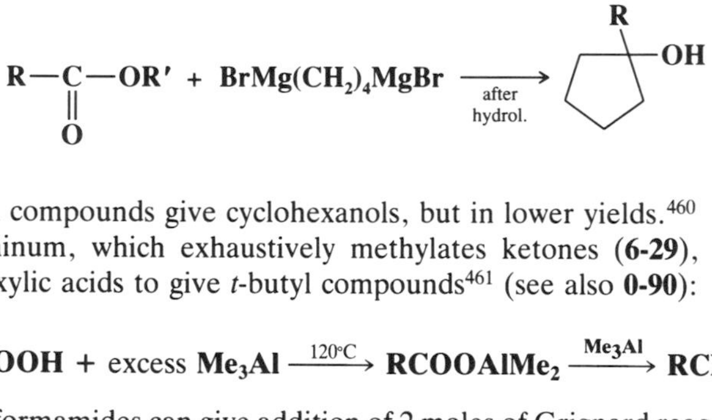

$$\underset{\underset{O}{\|}}{R-C-OR'} + BrMg(CH_2)_4MgBr \xrightarrow[\text{hydrol.}]{\text{after}} \text{(cyclopentane with R and OH)}$$

1,5-Dimagnesium compounds give cyclohexanols, but in lower yields.[460]

Trimethylaluminum, which exhaustively methylates ketones (**6-29**), also exhaustively methylates carboxylic acids to give *t*-butyl compounds[461] (see also **0-90**):

$$RCOOH + \text{excess } Me_3Al \xrightarrow{120^\circ C} RCOOAlMe_2 \xrightarrow{Me_3Al} RCMe_3$$

Disubstituted formamides can give addition of 2 moles of Grignard reagent. The products of this reaction (called *Bouveault reaction*) are an aldehyde and a tertiary amine.[462] The use

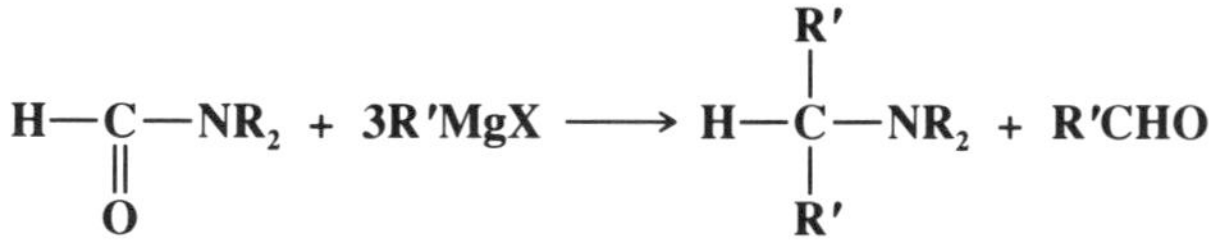

$$\underset{\underset{O}{\|}}{H-C-NR_2} + 3R'MgX \longrightarrow \underset{\underset{R'}{|}}{H-\overset{\overset{R'}{|}}{C}-NR_2} + R'CHO$$

of an amide other than a formamide can give a ketone instead of an aldehyde, but yields are generally low. It has proven possible to add two different R groups by sequential addition

[459]For a review of these reactions, see Kharasch; Reinmuth, Ref. 351, pp. 549-766, 846-869.
[460]Canonne; Bernatchez *J. Org. Chem.* **1986**, *51*, 2147, **1987**, *52*, 4025.
[461]Meisters; Mole *Aust. J. Chem.* **1974**, *27*, 1665.
[462]For a review, see Ref. 176, pp. 59-63.

of two Grignard reagents.[463] Alternatively, if R' contains an α hydrogen, the product may be an enamine, and enamines have been synthesized in goods yields by this method.[464]

$$H-\underset{\underset{O}{\|}}{C}-NR_2 + R'-\underset{\underset{R''}{|}}{CH}-MgX \longrightarrow \left[R'-\underset{\underset{R''}{|}}{CH}-\underset{\underset{OMgX}{|}}{CH}-NR_2\right] \xrightarrow{-MgXOH} R'-\underset{\underset{R''}{|}}{C}=CHNR_2$$

OS **I,** 226; **II,** 179, 602; **III,** 237, 831, 839; **IV,** 601; **VI,** 240, 278; **65,** 42; **67,** 125.

6-33 Conversion of Carboxylic Esters to Enol Ethers
Methylene-de-oxo-bisubstitution

$$R-\underset{\underset{OR'}{|}}{C}=O + Cp_2Ti\underset{Cl}{\overset{CH_2}{\diagdown}}AlMe_2 \longrightarrow R-\underset{\underset{OR'}{|}}{C}=CH_2 \qquad Cp = cyclopentadienide$$

35

Carboxylic esters and lactones can be converted in good yields to the corresponding enol ethers by treatment with the titanium cyclopentadienide complex **35** (Tebbe's reagent) in toluene-THF containing a small amount of pyridine.[465] **35** is prepared from dicyclopentadienyltitaniumdichloride and trimethylaluminum.[466] Dimethyltitanocene has been used instead of **35.**[467] There are several methods for the conversion C=O to C=CH$_2$ when the substrate is an aldehyde or ketone (see **6-29, 6-30, 6-39 to 6-44, 6-47**), but very few ways to make the same conversion for a carboxylic ester. (Tebbe's reagent also gives good results with ketones.[468]) The enol ether can be hydrolyzed to a ketone (**0-6**), so this is also an indirect method for making the conversion RCOOR' → RCOCH$_3$ (see also **0-105**).

Carboxylic esters undergo the conversion C=O → C=CHR (R = primary or secondary alkyl) when treated with RCHBr$_2$, Zn, and TiCl$_4$ in the presence of N,N,N',N'-tetramethylethylenediamine.[469] Metal carbene complexes[470] R$_2$C=ML$_n$ (L = ligand), where M is a transition metal such as Zr, W, or Ta, have also been used to convert the C=O of carboxylic esters and lactones to CR$_2$.[471] It is likely that the complex Cp$_2$Ti=CH$_2$ is an intermediate in the reaction with Tebbe's reagent.

OS **69,** 72.

6-34 The Addition of Organometallic Compounds to CO$_2$
C-Alkyl-O-halomagnesio-addition

$$O=C=O + RMgX \longrightarrow R-\underset{\underset{OMgX}{|}}{C}=O$$

[463]Comins; Dernell *Tetrahedron Lett.* **1981,** *22,* 1085.
[464]Hansson; Wickberg *J. Org. Chem.* **1973,** *38,* 3074.
[465]Tebbe; Parshall; Reddy *J. Am. Chem. Soc.* **1978,** *100,* 3611; Pine; Pettit; Geib; Cruz; Gallego; Tijerina; Pine *J. Org. Chem.* **1985,** *50,* 1212. See also Clawson; Buchwald; Grubbs *Tetrahedron Lett.* **1984,** *25,* 5733; Clift; Schwartz *J. Am. Chem. Soc.* **1984,** *106,* 8300.
[466]For a method of generating this reagent in situ, see Cannizzo; Grubbs *J. Org. Chem.* **1985,** *50,* 2386.
[467]Petasis; Bzowej *J. Am. Chem. Soc.* **1990,** *112,* 6392.
[468]Pine; Shen; Hoang *Synthesis* **1991,** 165.
[469]Okazoe; Takai; Oshima; Utimoto *J. Org. Chem.* **1987,** *52,* 4410. This procedure is also successful for silyl esters, to give silyl enol ethers: Takai; Kataoka; Okazoe; Utimoto *Tetrahedron Lett.* **1988,** *29,* 1065.
[470]For a review of the synthesis of such complexes, see Aguero; Osborn *New J. Chem.* **1988,** *12,* 111-118.
[471]See, for example, Schrock *J. Am. Chem. Soc.* **1976,** *98,* 5399; Aguero; Kress; Osborn *J. Chem. Soc., Chem. Commun.* **1986,** 531; Hartner; Schwartz; Clift *J. Am. Chem. Soc.* **1990,** *105,* 640.

Grignard reagents add to one C=O bond of CO_2 exactly as they do to an aldehyde or a ketone.[472] Here, of course, the product is the salt of a carboxylic acid. The reaction is usually performed by adding the Grignard reagent to dry ice. Many carboxylic acids have been prepared in this manner, and, along with the sequence **0-101—6-5** and reaction **8-8**, this constitutes an important way of increasing a carbon chain by one unit. Since labeled CO_2 is commercially available, this is a good method for the preparation of carboxylic acids labeled in the carboxyl group. Other organometallic compounds have also been used (RLi, RNa, RCaX, etc.), but much less often. The formation of the salt of a carboxylic acid after the addition of CO_2 to a reaction mixture is regarded as a positive test for the presence of a carbanion or of a reactive organometallic intermediate in that reaction mixture (see also **6-43**).

OS **I**, 361, 524; **II**, 425; **III**, 413, 553, 555; **V**, 890, 1043; **VI**, 845.

6-35 The Addition of Organometallic Compounds to C=N Compounds
N-Hydro-C-alkyl-addition

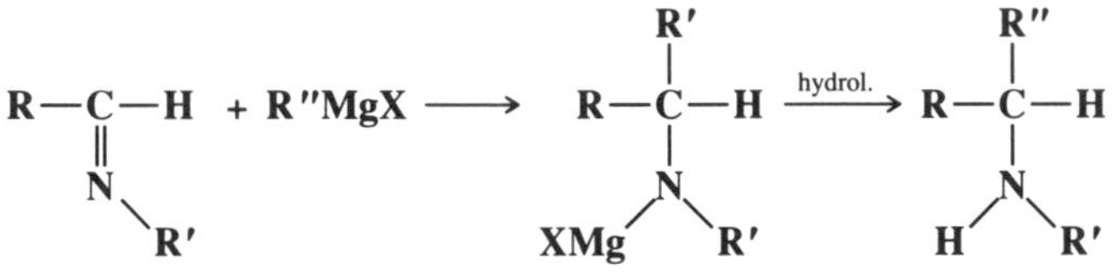

Aldimines can be converted to secondary amines by treatment with Grignard reagents.[473] Ketimines generally give reduction instead of addition. However, organolithium compounds give the normal addition product with both aldimines and ketimines.[474] Other organometallic compounds,[475] including RCu–BF_3,[476] allylic boranes[477] (see **6-29**), and allylic stannanes[478] also add to aldimines in the same manner. The addition of organolithiums has been done enantioselectively, with an optically active amino ether as catalyst.[478a] Many other C=N systems (phenylhydrazones, oxime ethers, etc.) give normal addition when treated with Grignard reagents; others give reductions; others give miscellaneous reactions. Oximes can be converted to hydroxylamines by treatment with 2 moles of an alkyllithium reagent, followed by methanol.[479]

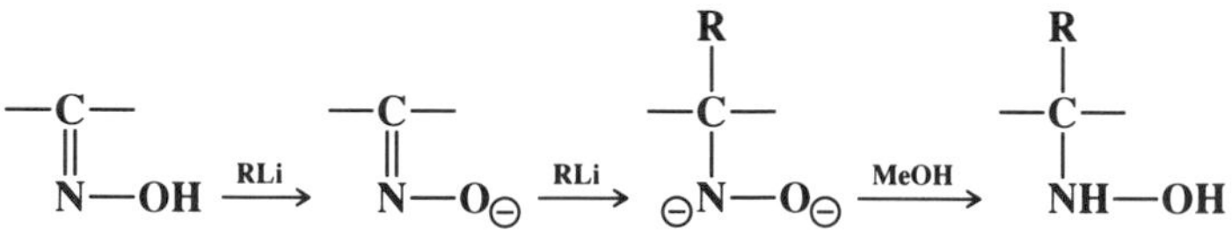

[472]For reviews of the reaction between organometallic compounds and CO_2, see Volpin; Kolomnikov; *Organomet. React.* **1975**, *5*, 313-386; Sneeden, in Patai *The Chemistry of Carboxylic Acids and Esters*; Wiley: New York, 1969, pp. 137-173; Kharasch; Reinmuth, Ref. 351, pp. 913-948. For a more general review, see Lapidus; Ping *Russ. Chem. Rev.* **1981**, *50*, 63-75.

[473]For reviews of the addition of organometallic reagents to C=N bonds, see Harada, in Patai *The Chemistry of the Carbon–Nitrogen Double Bond*, Ref. 40, pp. 266-272; Kharasch; Reinmuth, Ref. 451, pp. 1204-1227.

[474]Huet *Bull. Soc. Chim. Fr.* **1964**, 952, 960, 967, 973.

[475]For a list of reagents, with references, see Ref. 64, pp. 425-427.

[476]Wada; Sakurai; Akiba *Tetrahedron Lett.* **1984**, *25*, 1079.

[477]Yamamoto; Nishii; Maruyama; Komatsu; Ito *J. Am. Chem. Soc.* **1986**, *108*, 7778. See also Yamamoto *Acc. Chem. Res.* **1987**, *20*, 243-249.

[478]Keck; Enholm *J. Org. Chem.* **1985**, *50*, 146.

[478a]Tomioka; Inoue; Shindo; Koga *Tetrahedron Lett.* **1991**, *32*, 3095.

[479]Richey; McLane; Phillips *Tetrahedron Lett.* **1976**, 233.

The conjugate bases of nitro compounds (formed by treatment of the nitro compound with BuLi) react with Grignard reagents in the presence of $ClCH=NMe_2^+ Cl^-$ to give oximes: $RCH=N(O)OLi + R'MgX \rightarrow RR'C=NOH$.[480]

For the addition of an organometallic compound to an imine to give a primary amine, R' in $RCH=NR'$ would have to be H, and such compounds are seldom stable (**6-13**). However, the conversion has been done, for R = aryl, by the use of the masked reagents $(ArCH=N)_2SO_2$ [prepared from an aldehyde RCHO and sulfamide $(NH_2)_2SO_2$]. Addition of $R''MgX$ or $R''Li$ to these compounds gives $ArCHR''NH_2$ after hydrolysis.[481]

Iminium salts[330] give tertiary amines directly, with just R adding:

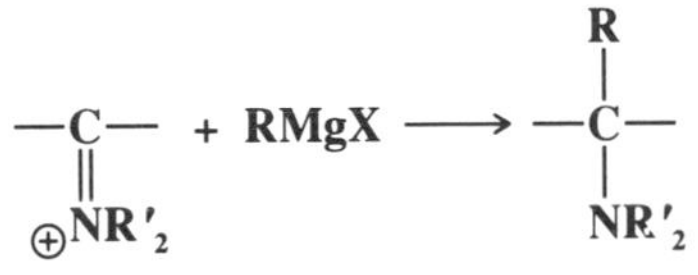

$$-\overset{\underset{\displaystyle \oplus NR'_2}{\|}}{C}- \ + \ RMgX \longrightarrow -\overset{\underset{\displaystyle NR'_2}{|}}{\overset{\overset{\displaystyle R}{|}}{C}}-$$

Chloroiminium salts $ClCH=\overset{\oplus}{N}R'_2Cl^-$ (generated in situ from an amide $HCONR'_2$ and phosgene $COCl_2$) react with 2 moles of a Grignard reagent RMgX, one adding to the C=N and the other replacing the Cl, to give tertiary amines $R_2CHNR'_2$.[482]

An alkyl group (primary, secondary, or tertiary) can be added to the oxime ether $CH_2=NOCH_2Ph$ by treatment with the appropriate alkyl halide and an equimolar amount of bis(trimethylstannyl)benzopicolinate.[483] This reaction, which is a free radical addition, is another way to extend a chain by one carbon.

OS **IV**, 605; **VI**, 64. Also see OS **III**, 329.

6-36   The Addition of Grignard Reagents to Isocyanates
N-Hydro-C-alkyl-addition

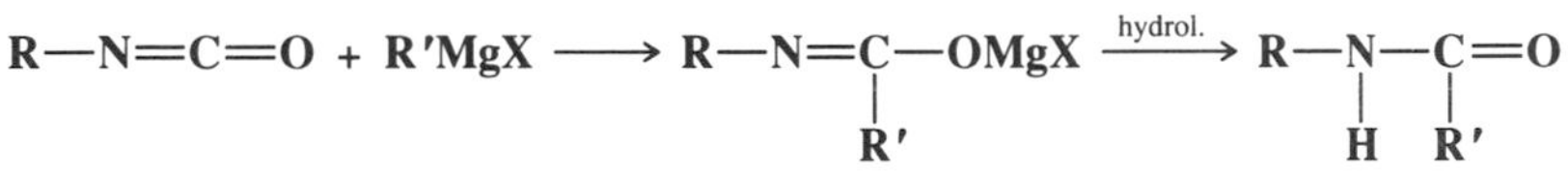

$$R-N=C=O + R'MgX \longrightarrow R-N=\overset{\underset{\displaystyle R'}{|}}{C}-OMgX \xrightarrow{\text{hydrol.}} R-\overset{\underset{\displaystyle H}{|}}{N}-\overset{\underset{\displaystyle R'}{|}}{C}=O$$

The addition of Grignard reagents to isocyanates gives, after hydrolysis, N-substituted amides.[484] The reaction is written above as involving addition to C=O, but the ion is a resonance hybrid and the addition might just as well have been shown as occurring on the C=N. Hydrolysis gives the amide. This is a very good reaction and can be used to prepare derivatives of alkyl and aryl halides. The reaction has also been performed with alkyllithium compounds.[485] Isothiocyanates give N-substituted thioamides.

6-37   The Addition of Grignard Reagents to Nitriles
Alkyl,oxo-de-nitrilo-tersubstitution (Overall transformation)

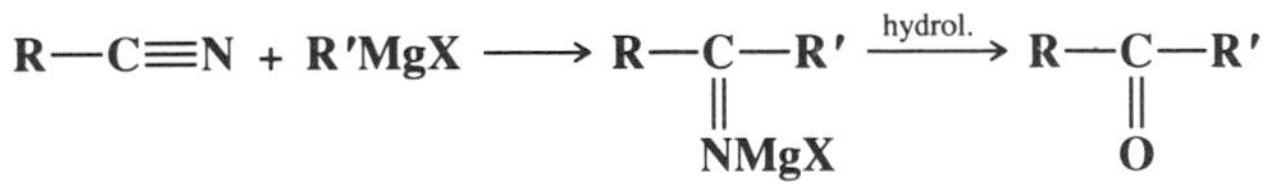

$$R-C\equiv N + R'MgX \longrightarrow R-\overset{\underset{\displaystyle NMgX}{\|}}{C}-R' \xrightarrow{\text{hydrol.}} R-\overset{\underset{\displaystyle O}{\|}}{C}-R'$$

[480]Fujisawa; Kurita; Sato *Chem. Lett.* **1983**, 1537.
[481]Davis; Giangiordano; Starner *Tetrahedron Lett.* **1986**, 27, 3957.
[482]Wieland; Simchen *Liebigs Ann. Chem.* **1985**, 2178.
[483]Hart; Seely *J. Am. Chem. Soc.* **1988**, *110*, 1631.
[484]For a review of this and related reactions, see Screttas; Steele *Org. Prep. Proceed. Int.* **1990**, *22*, 271-314.
[485]LeBel; Cherluck; Curtis *Synthesis* **1973**, 678. For another method, see Einhorn; Luche *Tetrahedron Lett.* **1986**, 27, 501.

Ketones can be prepared by addition of Grignard reagents to nitriles and subsequent hydrolysis. Many ketones have been made in this manner, though when both R groups are alkyl, yields are not high.[486] Yields can be improved by the use of Cu(I) salts[487] or by using benzene containing one equivalent of ether as the solvent, rather than ether alone.[488] The ketimine salt does not in general react with Grignard reagents: hence tertiary alcohols or tertiary alkyl amines are not often side products.[489] By careful hydrolysis of the salt it is sometimes possible to isolate ketimines R—C—R′,[490] especially when R and R′ = aryl.
$$\underset{NH}{\overset{\|}{}}$$

The addition of Grignard reagents to the C≡N group is normally slower than to the C=O group, and CN-containing aldehydes add the Grignard reagent without disturbing the CN group.[491] In a similar reaction,[492] triethylaluminum[493] reacts with nitriles (in a 2:1 ratio) to give, after hydrolysis, ethyl ketones.[494]

The following mechanism has been proposed for the reaction of the methyl Grignard reagent with benzonitrile:[495]

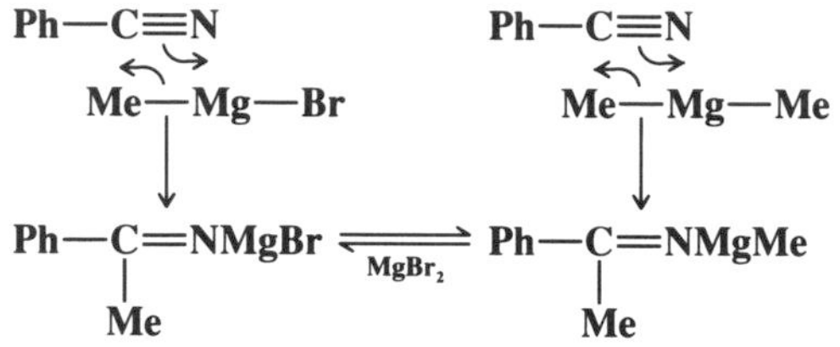

$$Ph—C≡N \qquad\qquad Ph—C≡N$$
$$Me—Mg—Br \qquad\qquad Me—Mg—Me$$
$$\downarrow \qquad\qquad\qquad \downarrow$$
$$Ph—\underset{Me}{\overset{\|}{C}}=NMgBr \underset{MgBr_2}{\rightleftharpoons} Ph—\underset{Me}{\overset{\|}{C}}=NMgMe$$

OS **III**, 26, 562; **V**, 520.

6-38 The Addition of Organometallic Reagents to the C=S Bond
C-Alkyl-S-halomagnesio-addition

$$S=C=S + RMgX \longrightarrow S—\underset{SMgX}{\overset{\|}{C}}=S$$

Grignard reagents add to CS_2 to give salts of dithiocarboxylic acids (analogous to **6-34**).[496] Two other reactions are worthy of note. (1) Lithium dialkylcopper reagents react with dithiocarboxylic esters to give tertiary thiols[497] (analogous to **6-32**):

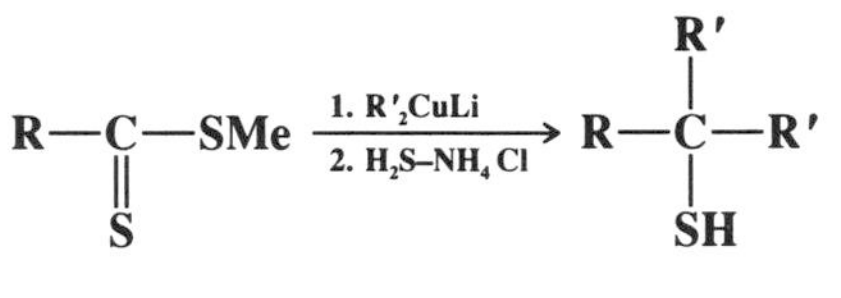

$$R—\underset{S}{\overset{\|}{C}}—SMe \xrightarrow[\text{2. } H_2S–NH_4\,Cl]{\text{1. } R'_2CuLi} R—\underset{SH}{\overset{R'}{\underset{|}{\overset{|}{C}}}}—R'$$

[486]For a review, see Kharasch; Reinmuth, Ref. 351, pp. 767-845.
[487]Weiberth; Hall *J. Org. Chem.* **1987**, *52*, 3901.
[488]Canonne; Foscolos; Lemay *Tetrahedron Lett.* **1980**, 155.
[489]For examples where tertiary amines have been made the main products, see Alvernhe; Laurent *Tetrahedron Lett.* **1973**, 1057; Gauthier; Axiotis; Chastrette *J. Organomet. Chem.* **1977**, *140*, 245.
[490]Pickard; Toblert *J. Org. Chem.* **1961**, *26*, 4886.
[491]Cason; Kraus; McLeod *J. Org. Chem.* **1959**, *24*, 392.
[492]For some other reagents, with references, see Ref. 64, p. 701.
[493]For a review of the reactions of organoaluminum compounds, see Reinheckel; Haage; Jahnke *Organomet. Chem. Rev., Sect. A* **1969**, *4*, 47-136.
[494]Reinheckel; Jahnke *Chem. Ber.* **1964**, *97*, 2661. See also Bagnell; Jeffery; Meisters; Mole *Aust. J. Chem.* **1974**, *27*, 2577.
[495]Ashby; Chao; Neumann *J. Am. Chem. Soc.* **1973**, *95*, 4896, 5186.
[496]For a review of the addition of Grignard reagents to C=S bonds, see Paquer *Bull Soc. Chim. Fr.* **1975**, 1439-1449. For a review of the synthesis of dithiocarboxylic acids and esters, see Ramadas; Srinivasan; Ramachandran; Sastry *Synthesis* **1983**, 605-622.
[497]Bertz; Dabbagh; Williams *J. Org. Chem.* **1985**, *50*, 4414.

(2) Thiono lactones can be converted to cyclic ethers,[498] e.g.:

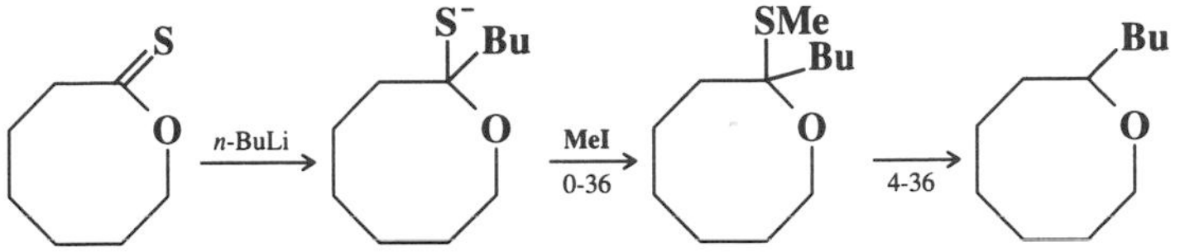

This is a valuable procedure because medium and large ring ethers are not easily made, while the corresponding thiono lactones can be prepared from the readily available lactones (see, for example, **0-22**) by reaction **6-11.**

I. Carbon Attack by Active Hydrogen Compounds. Reactions **6-39** through **6-48** are base-catalyzed condensations (though some of them are also catalyzed to acids).[499] In **6-39** through **6-47,** a base removes a C—H proton to give a carbanion, which then adds to a C=O. The oxygen acquires a proton, and the resulting alcohol may or may not be dehydrated, depending on whether an α hydrogen is present and on whether the new double bond would be in conjugation with double bonds already present:

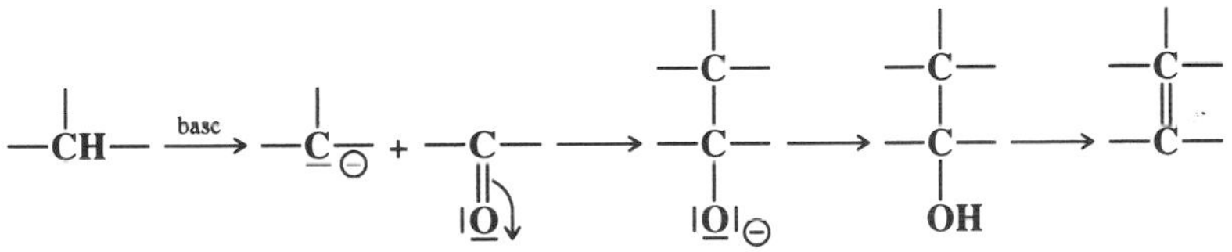

The reactions differ in the nature of the active hydrogen component and the carbonyl component. Table 16.1 illustrates the differences. Reaction **6-48** is an analogous reaction involving addition to C≡N.

6-39 The Aldol Reaction
O-Hydro-*C*-(α-acylalkyl)-addition; α-Acylalkylidene-de-oxo-bisubstitution

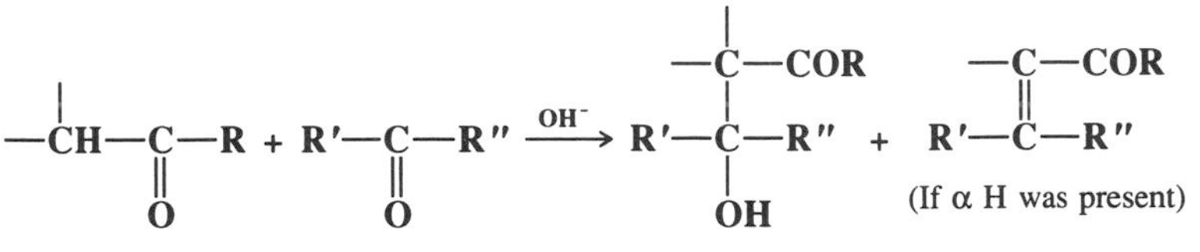

In the *aldol reaction*[500] the α carbon of one aldehyde or ketone molecule adds to the carbonyl carbon of another.[501] The base most often used is OH⁻, though stronger bases, e.g., alu-

[498]Nicolaou; McGarry; Somers; Veale; Furst *J. Am. Chem. Soc.* **1987,** *109,* 2504.
[499]For reviews, see House, Ref. 180, pp. 629-682; Reeves, in Patai, Ref. 2, pp. 567-619. See also Stowell *Carbanions in Organic Synthesis*; Wiley: New York, 1979.
[500]This reaction is also called the *aldol condensation*, though, strictly speaking, this term applies to the formation only of the α,β-unsaturated product, and not the aldol.
[501]For reviews, see Thebtaranonth; Thebtaranonth, in Patai, Ref. 252, pt. 1, pp. 199-280, pp. 199-212; Hajos, in Augustine *Carbon–Carbon Bond Formation*, vol. 1; Marcel Dekker: New York, 1979; pp. 1-84; Nielsen; Houlihan, *Org. React.* **1968,** *16,* 1-438.

TABLE 16.1 Base-catalyzed condensations showing the active-hydrogen components and the carbonyl components

Reaction	Active-hydrogen component	Carbonyl component	Subsequent reactions
6-39 Aldol reaction	Aldehyde $-\overset{\|}{C}H-CHO$ Ketone $-\overset{\|}{C}H-COR$	Aldehyde, ketone	Dehydration may follow
6-40	Ester $-\overset{\|}{C}H-COOR$	Aldehyde, ketone (usually without α-hydrogens)	Dehydration may follow
6-41 Knoevenagel reaction	$Z-CH_2-Z'$, $Z-CHR-Z'$, and similar molecules	Aldehyde, ketone (usually without α-hydrogens)	Dehydration usually follows
6-42 Peterson reaction	$Me_3Si-\overset{\|}{\underset{\ominus}{C}}H$	Aldehyde, ketone	Dehydration may follow
6-43	$-\overset{\|}{C}H-Z$ $\quad\begin{array}{l}\mathbf{Z = COR,}\\ \mathbf{COOR, NO_2}\end{array}$	CO_2, CS_2	
6-44 Perkin reaction	Anhydride $-\overset{\|}{C}H-COOCOR$	Aromatic aldehyde	Dehydration usually follows
6-45 Darzen's reaction	α-Halo ester $XC\overset{\|}{H}-COOR$	Aldehyde, ketone	Epoxidation (S$_N$ reaction) follows
6-46 Tollens' reaction	Aldehyde $-\overset{\|}{C}H-CHO$ Ketone $-\overset{\|}{C}H-COR$	Formaldehyde	Crossed Cannizzaro reaction follows
6-47 Wittig reaction	Phosphorous ylide $Ph_3\overset{\oplus}{P}-\overset{\|}{\underset{\ominus}{C}}$	Aldehyde, ketone	"Dehydration" always follows
6-48 Thorpe reaction	Nitrile $-\overset{\|}{C}H-CN$	Nitrile	

minum *t*-butoxide, are sometimes employed. Hydroxide ion is not a strong enough base to convert substantially all of an aldehyde or ketone molecule to the corresponding enolate

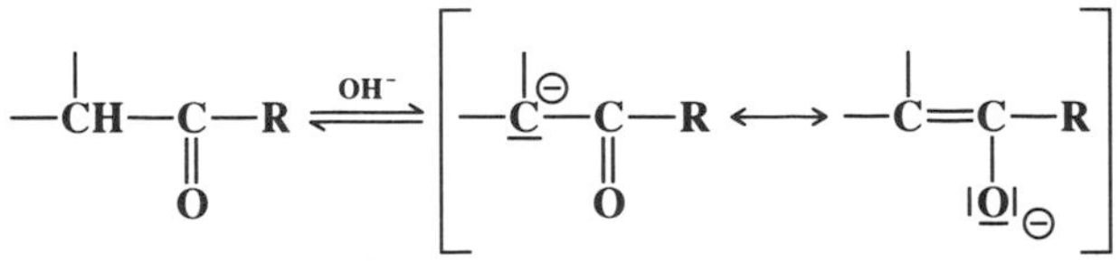

$$-\overset{\|}{C}H-\underset{\overset{\|}{O}}{C}-R \;\overset{OH^-}{\rightleftharpoons}\; \left[-\underset{\overset{\|}{O}}{\overset{\ominus}{\underline{C}}}-C-R \longleftrightarrow -\overset{\|}{C}=\underset{\underset{\ominus}{|\underline{O}|}}{C}-R \right]$$

ion, i.e., the equilibrium lies well to the left, for both aldehydes and ketones. Nevertheless, enough enolate ion is present for the reaction to proceed:

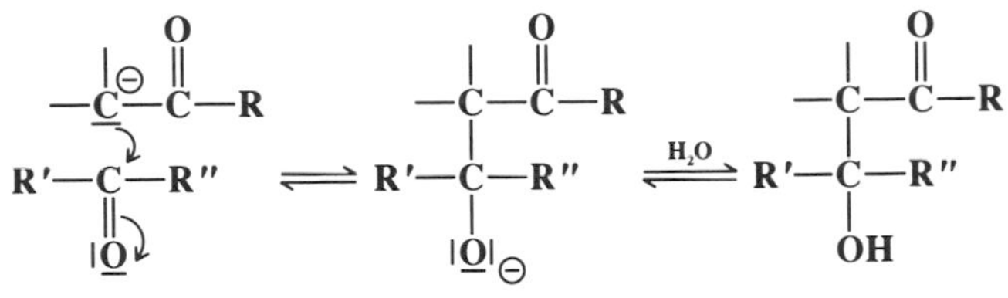

The product is a β-hydroxy aldehyde (called an *aldol*) or ketone, which in some cases is dehydrated during the course of the reaction. Even if the dehydration is not spontaneous, it can usually be done easily, since the new double bond is in conjugation with the C=O bond; so that this is a method of preparing α,β-unsaturated aldehydes and ketones as well as β-hydroxy aldehydes and ketones. The entire reaction is an equilibrium (including the dehydration step), and α,β-unsaturated and β-hydroxy aldehydes and ketones can be cleaved by treatment with OH⁻ (the *retrograde aldol reaction*). There is evidence that an SET mechanism can intervene when the substrate is an aromatic ketone.[502]

Under the principle of vinylology, the active hydrogen can be one in the γ position of an α,β-unsaturated carbonyl compound:

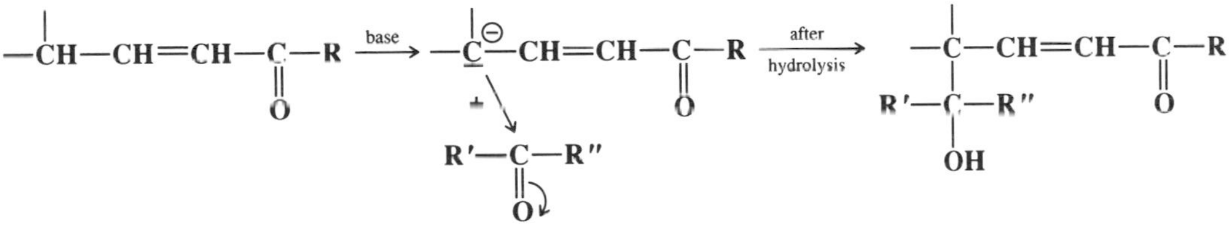

The scope of the aldol reaction may be discussed under five headings:

1. *Reaction between two molecules of the same aldehyde.* The equilibrium lies far to the right,[503] and the reaction is quite feasible. Many aldehydes have been converted to aldols and/or their dehydration products in this manner. The most effective catalysts are basic ion-exchange resins. Of course, the aldehyde must possess an α hydrogen.

2. *Reaction between two molecules of the same ketone.* In this case the equilibrium lies well to the left,[504] and the reaction is feasible only if the equilibrium can be shifted. This can often be done by allowing the reaction to proceed in a Soxhlet extractor (for example, see OS **I**, 199). In this method the ketone is refluxed in such a way that the condensate drips into a separate chamber, in which the base is present. In this chamber the reaction proceeds to the small extent permitted by the unfavorable equilibrium. When the chamber is full, the mixture of the ketone and its dimer is siphoned back into the original flask, out of contact with the base. Since the boiling point of the dimer is higher than that of the ketone, only the ketone is volatilized back to the chamber containing the base, where a little more of it is converted to dimer, and the process is repeated until a reasonable yield of dimer is obtained. Two molecules of the same ketone can also be condensed without a

[502]Ashby; Argyropoulos; Meyer; Goel *J. Am. Chem. Soc.* **1982**, *104*, 6788; Ashby; Argyropoulos *J. Org. Chem.* **1986**, *51*, 472.
[503]For discussions of equilibrium constants in aldol reactions, see Guthrie; Wang *Can. J. Chem.* **1991**, *69*, 339; Guthrie *J. Am. Chem. Soc.* **1991**, *113*, 7249, and references cited in these papers.
[504]The equilibrium concentration of the product from acetone in pure acetone was determined to be 0.01%: Maple; Allerhand *J. Am. Chem. Soc.* **1987**, *109*, 6609.

Soxhlet extractor,[505] by treatment with basic Al_2O_3.[506] Unsymmetrical ketones condense on the side that has more hydrogens. (An exception is butanone, which reacts at the CH_2 group with acid catalysts, though with basic catalysts, it too reacts at the CH_3 group.)

3. *Reaction between two different aldehydes.* In the most general case, this will produce a mixture of four products (eight, if the olefins are counted). However, if one aldehyde does not have an α hydrogen, only two aldols are possible, and in many cases the crossed product is the main one. The crossed aldol reaction is often called the *Claisen–Schmidt reaction*.

4. *Reaction between two different ketones.* This is seldom attempted (except with the use of preformed enolates, see below), but similar considerations apply.

5. *Reaction between an aldehyde and a ketone.* This is usually feasible, especially when the aldehyde has no α hydrogen, since there is no competition from ketone condensing with itself.[507] This is also called the *Claisen–Schmidt reaction*. Even when the aldehyde has an α hydrogen, it is the α carbon of the ketone that adds to the carbonyl of the aldehyde, not the other way around. The reaction can be made regioselective by preparing an enol derivative of the ketone separately[508] and then adding this to the aldehyde (or ketone), which assures that the coupling takes place on the desired side of an unsymmetrical ketone. A number of these preformed enolates have been used, the most common of which is the silyl enol ether of the ketone. This can be combined with an aldehyde or ketone, with $TiCl_4$[509]

$$R^1-\underset{\underset{OSiMe_3}{|}}{C}=CHR^2 \ + \ R^3-\underset{\underset{O}{\|}}{C}-R^4 \ \xrightarrow{\underset{2. \ H_2O}{1. \ TiCl_4}} \ R^1-\underset{\underset{O}{\|}}{C}-\underset{\underset{R^2}{|}}{CH}-\overset{\overset{R^3}{|}}{\underset{\underset{OH}{|}}{C}}-R^4$$

36

(the *Mukaiyama reagent*), with various other catalysts, and even in aqueous solution, with no catalyst at all.[510] The large number of catalysts reported[511] testify to the importance of this method. This reaction can also be run with the aldehyde or ketone in the form of its acetal $R^3R^4C(OR')_2$, in which case the product is the ether $R^1COCHR^2CR^3R^4OR'$ instead of **36**.[512] Enol acetates and enol ethers also give this product when treated with acetals and $TiCl_4$ or a similar catalyst.[513] When the catalyst is dibutyltin bis(triflate) $Bu_2Sn(OTf)_2$, al-

[505]For another method, see Barot; Sullins; Eisenbraun *Synth. Commun.* **1984**, *14*, 397.

[506]Muzart *Synthesis* **1982**, 60, *Synth. Commun.* **1985**, 285.

[507]For a study of the rate and equilibrium constants in the reaction between acetone and benzaldehyde, see Guthrie; Cossar; Taylor *Can. J. Chem.* **1984**, *62*, 1958.

[508]For some other aldol reactions with preformed enol derivatives, see Schulz; Steglich *Angew. Chem. Int. Ed. Engl.* **1977**, *16*, 251 [*Angew. Chem. 89*, 255]; Paterson; Fleming *Tetrahedron Lett.* **1979**, 2179; Itoh; Ozawa; Oshima; Nozaki *Bull. Chem. Soc. Jpn.* **1981**, *54*, 274; Yamamoto; Yatagai; Maruyama *J. Chem. Soc., Chem. Commun.* **1981**, 162; Kowalski; Fields *J. Am. Chem. Soc.* **1982**, *104*, 1777; Fujita; Schlosser *Helv. Chim. Acta* **1982**, *65*, 1258; Kato; Mukaiyama *Chem. Lett.* **1983**, 1727; Dubois; Axiotis *Tetrahedron Lett.* **1984**, *25*, 2143. For reviews of this subject, see Mukaiyama *Isr. J. Chem.* **1984**, *24*, 162-166; Caine, in Augustine, Ref. 501, pp. 264-276.

[509]Mukaiyama; Banno; Narasaka *J. Am. Chem. Soc.* **1974**, *96*, 7503; Mukaiyama *Pure Appl. Chem.* **1983**, *55*, 1749-1758; Kohler *Synth. Commun.* **1985**, *15*, 39; Mukaiyama; Narasaka *Org. Synth.* 65, 6. For a discussion of the mechanism, see Gennari; Colombo; Bertolini; Schimperna *J. Org. Chem.* **1987**, *52*, 2754. For a review of this and other applications of $TiCl_4$ in organic synthesis, see Mukaiyama *Angew. Chem. Int. Ed. Engl.* **1977**, *16*, 817-826 [*Angew. Chem. 89*, 856-866]. See also Reetz, Ref. 419.

[510]Lubineau; Meyer *Tetrahedron* **1988**, *44*, 6065.

[511]See, for example, Noyori; Nishida; Sakata *J. Am. Chem. Soc.* **1981**, *103*, 2106; Nakamura; Shimizu; Kuwajima; Sakata; Yokoyama; Noyori *J. Org. Chem.* **1983**, *48*, 932; Naruse; Ukai; Ikeda; Yamamoto *Chem. Lett.* **1985**, 1451; Sato; Matsuda; Izumi *Tetrahedron Lett.* **1986**, *27*, 5517; Reetz; Vougioukas *Tetrahedron Lett.* **1987**, *28*, 793; Vougioukas; Kagan *Tetrahedron Lett.* **1987**, *28*, 5513; Mukaiyama; Kobayashi; Tamura; Sagawa *Chem. Lett.* **1987**, 491; Iwasawa; Mukaiyama *Chem. Lett.* **1987**, 463; Kawai; Onaka; Izumi *Bull. Chem. Soc. Jpn.* **1988**, *61*, 1237; Ohki; Wada; Akiba *Tetrahedron Lett.* **1988**, *29*, 4719; Mukaiyama; Matsui; Kashiwagi *Chem. Lett.* **1989**, 993.

[512]Mukaiyama; Hayashi *Chem. Lett.* **1974**, 15; Mukaiyama; Kobayashi; Murakami *Chem. Lett.* **1984**, 1759; Murata; Suzuki; Noyori *Tetrahedron* **1988**, *44*, 4259. For a review of cross-coupling reactions of acetals, see Mukaiyama; Murakami *Synthesis* **1987**, 1043-1054.

[513]Mukaiyama; Izawa; Saigo *Chem. Lett.* **1974**, 323; Kitazawa; Imamura; Saigo; Mukaiyama *Chem. Lett.* **1975**, 569.

dehydes react, but not their acetals, while acetals of ketones react, but not the ketones themselves.[514] Other types of preformed derivatives that react with aldehydes and ketones are enamines (with a Lewis acid catalyst),[515] and enol borinates $R'CH=CR''—OBR_2$[516] (which can be synthesized by **5-19,** or directly from an aldehyde or ketone[517]). Preformed metallic enolates are also used. For example lithium enolates[518] (prepared by **2-22**) react with the substrate in the presence of $ZnCl_2$;[519] in this case the aldol product is stabilized by chelation of its two oxygen atoms with the zinc ion.[520] Among other metallic enolates used for aldol reactions are those of Ti,[521] Zr,[522] and Pd,[523] all of which give products regioselectively. α-Alkoxy ketones react with lithium enolates particularly rapidly.[524]

The reactions with preformed enol derivatives provide a way to control the stereoselectivity of the aldol reaction.[525] As with the Michael reaction (**5-17**), the aldol reaction creates two new chiral centers, and, in the most general case, there are four stereoisomers of the aldol product, which can be represented as

syn (or erythro) ($\pm$) pair   anti (or threo) ($\pm$) pair

Among the preformed enol derivatives used in this way have been enolates of magnesium, lithium,[526] titanium,[527] rhodium,[528] zirconium,[522] and tin,[529] silyl enol ethers,[530] enol borinates,[531] and enol borates $R'CH=CR''—OB(OR)_2$.[532] In general, metallic Z enolates give

[514]Sato; Otera; Nozaki *J. Am. Chem. Soc.* **1990**, *112*, 901.

[515]Takazawa; Kogami; Hayashi *Bull. Chem. Soc. Jpn.* **1985**, *58*, 2427.

[516]Inoue; Mukaiyama *Bull. Chem. Soc. Jpn.* **1980**, *53*, 174; Kuwajima; Kato; Mori *Tetrahedron Lett.* **1980**, *21*, 4291; Wada *Chem. Lett.* **1981**, 153; Hooz; Oudenes; Roberts; Benderly *J. Org. Chem.* **1987**, *52*, 1347; Nozaki; Oshima; Utimoto *Tetrahedron Lett.* **1988**, *29*, 1041. For a review, see Pelter; Smith; Brown, Ref. 361, pp. 324-333.

[517]For conversion of ketones to either *Z* or *E* enol borinates, see, for example, Evans; Nelson; Vogel; Taber *J. Am. Chem. Soc.* **1981**, *103*, 3099; Brown; Dhar; Bakshi; Pandiarajan; Singaram *J. Am. Chem. Soc.* **1989**, *111*, 3441.

[518]For a complete structure–energy analysis of one such reaction, see Arnett; Fisher; Nichols; Ribeiro *J. Am. Chem. Soc.* **1990**, *112*, 801.

[519]House; Crumrine; Teranishi; Olmstead *J. Am. Chem. Soc.* **1973**, *95*, 3310.

[520]It has been contended that such stabilization is not required: Mulzer; Brüntrup; Finke; Zippel *J. Am. Chem. Soc.* **1979**, *101*, 7723.

[521]Stille; Grubbs *J. Am. Chem. Soc.* **1983**, *105*, 1664.

[522]Evans; McGee *Tetrahedron Lett.* **1980**, *21*, 3975, *J. Am. Chem. Soc.* **1981**, *103*, 2876.

[523]Nokami; Mandai; Watanabe; Ohyama; Tsuji *J. Am. Chem. Soc.* **1989**, *111*, 4126.

[524]Das; Thornton *J. Am. Chem. Soc.* **1990**, *112*, 5360.

[525]For reviews, see Heathcock *Aldrichimica Acta* **1990**, *23*, 99-111; *Science* **1981**, *214*, 395-400; Nógrádi, Ref. 294, pp. 193-220; Heathcock, in Morrison, Ref. 294, vol. 3, 1984, pp. 111-212; Heathcock, in Buncel; Durst *Comprehensive Carbanion Chemistry*, pt. B, Elsevier: New York, 1984, pp. 177-237; Evans; Nelson; Taber *Top. Stereochem.* **1982**, *13*, 1-115; Evans *Aldrichimica Acta* **1982**, *15*, 23-32.

[526]Fellmann; Dubois *Tetrahedron* **1978**, *34*, 1349; Heathcock; Pirrung; Montgomery; Lampe *Tetrahedron* **1981**, *37*, 4087; Masamune; Ellingboe; Choy *J. Am. Chem. Soc.* **1982**, *104*, 5526; Ertas; Seebach *Helv. Chim. Acta* **1985**, *68*, 961.

[527]Siegel; Thornton *Tetrahedron Lett.* **1986**, *27*, 457; Nerz-Stormes; Thornton *Tetrahedron* **1986**, 897; Evans; Rieger; Bilodeau; Urpí *J. Am. Chem. Soc.* **1991**, *113*, 1047.

[528]Slough; Bergman; Heathcock *J. Am. Chem. Soc.* **1989**, *111*, 938.

[529]Mukaiyama; Iwasawa; Stevens; Haga *Tetrahedron* **1984**, *40*, 1381; Labadie; Stille *Tetrahedron* **1984**, *40*, 2329; Yura; Iwasawa; Mukaiyama *Chem. Lett.* **1986**, 187. See also Nakamura; Kuwajima *Tetrahedron Lett.* **1983**, *24*, 3347.

[530]Matsuda; Izumi *Tetrahedron Lett.* **1981**, *22*, 1805; Yamamoto; Maruyama; Matsumoto *J. Am. Chem. Soc.* **1983**, *105*, 6963; Sakurai; Sasaki; Hosomi; *Bull. Chem. Soc. Jpn.* **1983**, *56*, 3195; Hagiwara; Kimura; Uda *J. Chem. Soc., Chem. Commun.* **1986**, 860.

[531]Masamune; Mori; Van Horn; Brooks *Tetrahedron Lett.* **1979**, 1665; Evans et al., Ref. 517; Evans; Bartroli; Shih *J. Am. Chem. Soc.* **1981**, *103*, 2127; Masamune; Choy; Kerdesky; Imperiali *J. Am. Chem. Soc.* **1981**, *103*, 1566; Heathcock; Arseniyadis *Tetrahedron Lett.* **1985**, *26*, 6009; Paterson; Goodman; Lister; Schumann; McClure; Norcross *Tetrahedron* **1990**, *46*, 4663; Walker; Heathcock *J. Org. Chem.* **1991**, *56*, 5747. For reviews, see Paterson *Chem. Ind. (London)* **1988**, 390-394; Pelter; Smith; Brown, Ref. 516.

[532]Hoffmann; Ditrich; Fröch *Liebigs Ann. Chem.* **1987**, 977.

the syn (or erythro) pair, and this reaction is highly useful for the diastereoselective synthesis of these products.[533] The *E* isomers generally react nonstereoselectively. However, anti (or threo) stereoselectivity has been achieved in a number of cases, with titanium enolates,[534] with germanium enolates,[535] with magnesium enolates,[535a] with certain enol borinates,[536] and with lithium enolates at −78°C.[537] High diastereoselectivity was also achieved, without a preformed enolate, in the reaction between ethyl ketones and aldehydes, by performing the reaction in the presence of PhBCl$_2$ and Et$_3$N.[538]

These reactions can also be made enantioselective (in which case only one of the four isomers predominates) by using[539] chiral enol derivatives,[540] chiral aldehydes or ketones,[541] or both.[542] Since both new chiral centers are formed enantioselectively, this kind of process is called *double asymmetric synthesis*.[543] A single one of the four stereoisomers has also been produced where both the enolate derivative and substrate were achiral, by carrying out the reaction in the presence of an optically active boron compound[544] or a diamine coordinated with a tin compound.[545]

It is possible to make the α carbon of the aldehyde add to the carbonyl carbon of the ketone, by using an imine instead of an aldehyde, and LiN(iso-Pr)$_2$ as the base:[546]

$$\text{LiCH}_2\text{—C—H} + \text{CH}_3\text{—C—Ph} \xrightarrow{-70°C} \quad \xrightarrow{\text{hydrol.}} \text{CH}_3\text{—C—Ph}$$

[533]For discussion of transition state geometries in this reaction, see Hoffmann; Ditrich; Froech; Cremer *Tetrahedron* **1985,** *41,* 5517; Anh; Thanh *Nouv. J. Chim.* **1986,** *10,* 681; Li; Paddon-Row; Houk *J. Org. Chem.* **1990,** *55,* 481; Denmark; Henke *J. Am. Chem. Soc.* **1991,** *113,* 2177.

[534]See Murphy; Procter; Russell *Tetrahedron Lett.* **1987,** *28,* 2037; Shirodkar; Nerz-Stormes; Thornton *Tetrahedron Lett.* **1990,** *31,* 4699; Nerz-Stormes; Thornton *J. Org. Chem.* **1991,** *56,* 2489.

[535]Yamamoto; Yamada *J. Chem. Soc., Chem. Commun.* **1988,** 802.

[535a]Swiss; Choi; Liotta; Abdel-Magid; Maryanoff *J. Org. Chem.* **1991,** *56,* 5978.

[536]Masamune; Sato; Kim; Wollmann *J. Am. Chem. Soc.* **1986,** *108,* 8279; Danda; Hansen; Heathcock *J. Org. Chem.* **1990,** *55,* 173. See also Corey; Kim *Tetrahedron Lett.* **1990,** *31,* 3715.

[537]Hirama; Noda; Takeishi; Itô *Bull. Chem. Soc. Jpn.* **1988,** *61,* 2645; Majewski; Gleave *Tetrahedron Lett.* **1989,** *30,* 5681.

[538]Hamana; Sasakura; Sugasawa *Chem. Lett.* **1984,** 1729.

[539]For reviews, see Klein, in Patai *Supplement A: The Chemistry of Double-bonded Functional Groups,* vol. 2, pt. 1; Wiley: New York, 1989, pp. 567-677; Braun *Angew. Chem. Int. Ed. Engl.* **1987,** *26,* 24-37 [*Angew. Chem. 99,* 24-37].

[540]For examples, see Eichenauer; Friedrich; Lutz; Enders *Angew. Chem. Int. Ed. Engl.* **1978,** *17,* 206 [*Angew. Chem. 90,* 219]; Meyers; Yamamoto *Tetrahedron* **1984,** *40,* 2309; Ando; Shioiri *J. Chem. Soc., Chem. Commun.* **1987,** 1620; Muraoka; Kawasaki; Koga *Tetrahedron Lett.* **1988,** *29,* 337; Paterson; Goodman *Tetrahedron Lett.* **1989,** *30,* 997; Siegel; Thornton *J. Am. Chem. Soc.* **1989,** *111,* 5722; Gennari; Molinari; Cozzi; Oliva *Tetrahedron Lett.* **1989,** *30,* 5163; Faunce; Grisso; Mackenzie *J. Am. Chem. Soc.* **1991,** *113,* 3418.

[541]For example, see Ojima; Yoshida; Inaba *Chem. Lett.* **1977,** 429; Heathcock; Flippin *J. Am. Chem. Soc.* **1983,** *105,* 1667; Reetz; Kesseler; Jung *Tetrahedron Lett.* **1984,** *40,* 4327.

[542]For example, see Heathcock; White; Morrison; VanDerveer *J. Org. Chem.* **1981,** *46,* 1296; Short; Masamune *Tetrahedron Lett.* **1987,** *28,* 2841.

[543]For a review, see Masamune; Choy; Petersen; Sita *Angew. Chem. Int. Ed. Engl.* **1985,** *24,* 1-30 [*Angew. Chem. 97,* 1-31].

[544]Corey; Imwinkelried; Pikul; Xiang *J. Am. Chem. Soc.* **1989,** *111,* 5493; Corey; Kim *J. Am. Chem. Soc.* **1990,** *112,* 4976; Furuta; Maruyama; Yamamoto *J. Am. Chem. Soc.* **1991,** *113,* 1041; Kiyooka; Kaneko; Komura; Matsuo; Nakano *J. Org. Chem.* **1991,** *56,* 2276.

[545]Mukaiyama; Uchiro; Kobayashi *Chem. Lett.* **1990,** 1147.

[546]Wittig; Frommeld; Suchanek *Angew. Chem. Int. Ed. Engl.* **1963,** *2,* 683 [*Angew. Chem. 75,* 303]. For reviews, see Mukaiyama *Org. React.* **1982,** *28,* 203-331; Wittig *Top. Curr. Chem.* **1976,** *67,* 1-14; *Rec. Chem. Prog.* **1967,** *28,* 45-60; Wittig; Reiff *Angew. Chem. Int. Ed. Engl.* **1968,** *7,* 7-14; [*Angew. Chem. 80,* 8-15]; Reiff *Newer Methods Prep. Org. Chem.* **1971,** *6,* 48-66.

This is known as the *directed aldol reaction*. Similar reactions have been performed with α-lithiated dimethylhydrazones of aldehydes or ketones[547] and with α-lithiated aldoximes.[548]

The aldol reaction can also be performed with acid catalysts, in which case dehydration usually follows. Here there is initial protonation of the carbonyl group, which attacks the α carbon of the *enol* form of the other molecule:[549]

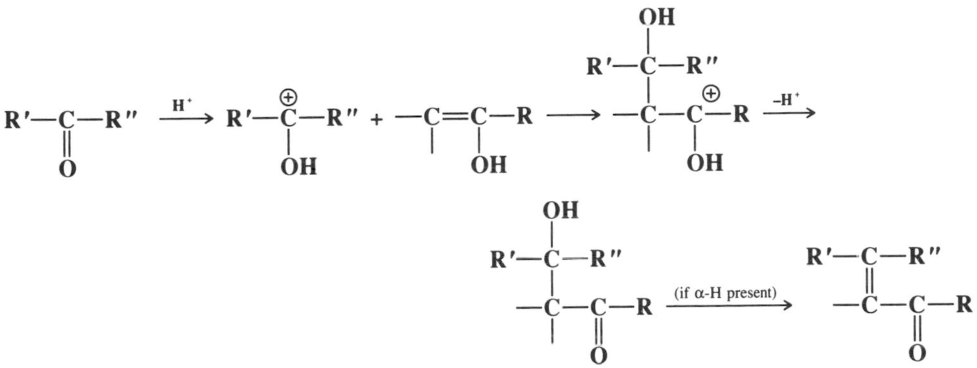

With respect to the enol, this mechanism is similar to that of halogenation (**2-4**).

A side reaction that is sometimes troublesome is further condensation, since the product of an aldol reaction is still an aldehyde or ketone.

Aldol reactions are often used to close five- and six-membered rings. Because of the favorable entropy (p. 211), such ring closures generally take place with ease, even where a ketone condenses with a ketone. An important example is the *Robinson annulation reaction*,[550] which has often been used in the synthesis of steroids and terpenes. In this reaction a cyclic ketone is converted to another cyclic ketone, with one additional six-membered ring containing a double bond. The substrate is treated with methyl vinyl ketone (or a simple derivative of methyl vinyl ketone) and a base.[551] The enolate ion of the substrate adds to the methyl vinyl ketone in a Michael reaction (**5-17**) to give a diketone that undergoes or

is made to undergo an internal aldol reaction and subsequent dehydration to give the product.[552] Because methyl vinyl ketone has a tendency to polymerize, precursors are often used instead, i.e., compounds that will give methyl vinyl ketone when treated with a base. One common example, $MeCOCH_2CH_2NEt_2Me^+$ I^- (see **7-8**), is easily prepared by quaternization of $MeCOCH_2CH_2NEt_2$, which itself is prepared by a Mannich reaction (**6-16**)

[547]Corey; Enders *Tetrahedron Lett.* **1976**, 11. See also Beam; Thomas; Sandifer; Foote; Hauser *Chem. Ind. (London)* **1976**, 487; Sugasawa; Toyoda; Sasakura *Synth. Commun.* **1979**, *9*, 515; Depezay; Le Merrer *Bull. Soc. Chim. Fr.* **1981**, II-306.

[548]Hassner; Näumann *Chem. Ber.* **1988**, *121*, 1823.

[549]There is evidence (in the self-condensation of acetaldehyde) that a water molecule acts as a base (even in concentrated H_2SO_4) in assisting the addition of the enol to the protonated aldehyde: Baigrie; Cox; Slebocka-Tilk; Tencer; Tidwell *J. Am. Chem. Soc.* **1985**, *107*, 3640.

[550]For reviews of this and related reactions, see Gawley *Synthesis* **1976**, 777-794; Jung *Tetrahedron* **1976**, *32*, 1-31; Mundy *J. Chem. Educ.* **1973**, *50*, 110-113. For a list of references, see Ref. 64, pp. 668-670.

[551]Acid catalysis has also been used: see Heathcock; Ellis; McMurry; Coppolino *Tetrahedron Lett.* **1971**, 4995.

[552]For improved procedures, see Sato; Wakahara; Otera; Nozaki *Tetrahedron Lett.* **1990**, *31*, 1581, and references cited therein.

involving acetone, formaldehyde, and diethylamine. The Robinson annulation reaction has also been carried out with 3-butyn-2-one, in which case the new ring of the product contains two double bonds.[553] α-Silylated vinyl ketones $RCOC(SiMe_3)\!\!=\!\!CH_2$ have also been used successfully in annulation reactions.[554] The $SiMe_3$ group is easily removed. 1,5-Diketones prepared in other ways are also frequently cyclized by internal aldol reactions. When the ring closure of a 1,5-diketone is catalyzed by the amino acid (S)-proline, the product is optically active with high enantiomeric excess.[555]

OS **I**, 77, 78, 81, 199, 283, 341; **II**, 167, 214; **III**, 317, 353, 367, 747, 806, 829; **V**, 486, 869; **VI**, 496, 666, 692, 781, 901; **VII**, 185, 190, 332, 363, 368, 473; **65**, 6, 26; **67**, 121; **68**, 83; **69**, 55, 226. Also see OS **65**, 146.

6-40   Aldol-type Reactions between Carboxylic Esters and Aldehydes or Ketones
O-Hydro-C-(α-alkoxycarbonylalkyl)-addition; α-Alkoxycarbonylalkylidene-de-oxo-bisubstitution

$$
-\overset{|}{C}H-COOR + R'-\overset{\overset{\displaystyle ||}{}}{\underset{\displaystyle O}{C}}-R'' \xrightarrow{\text{base}} R'-\overset{\overset{\displaystyle -\overset{|}{C}-COOR}{|}}{\underset{\displaystyle OH}{C}}-R'' + R'-\overset{\overset{\displaystyle -\overset{||}{C}-COOR}{}}{C}-R''
$$

(If α-H was present)

In the presence of a strong base, the α carbon of a carboxylic ester can condense with the carbonyl carbon of an aldehyde or ketone to give a β-hydroxy ester,[556] which may or may not be dehydrated to the α,β-unsaturated ester. This reaction is sometimes called the Claisen condensation,[557] an unfortunate usage since that name is more firmly connected to **0-108**. It is also possible for the α carbon of an aldehyde or ketone to add to the carbonyl carbon of a carboxylic ester, but this is a different reaction (**0-109**) involving nucleophilic substitution and not addition to a C=O bond. It can, however, be a side reaction if the aldehyde or ketone has an α hydrogen.

Besides ordinary esters (containing an α hydrogen), the reaction can also be carried out with lactones and, as in **6-39**, with the γ position of α,β-unsaturated esters (vinylology).

For most esters, a much stronger base is needed than for aldol reactions; (i-Pr)$_2$NLi, Ph$_3$CNa and LiNH$_2$ are among those employed. However, one type of ester reacts more easily, and such strong bases are not needed: diethyl succinate and its derivatives condense with aldehydes and ketones in the presence of bases such as NaOEt, NaH, or KOCMe$_3$. This reaction is called the *Stobbe condensation*.[558] One of the ester groups (sometimes both) is hydrolyzed in the course of the reaction. The following mechanism accounts for (1) the fact the succinic esters react so much better than others; (2) one ester group is always cleaved; and (3) the alcohol is not the product but the olefin. In addition, intermediate lactones **37** have been isolated from the mixture:[559]

[553]For example, see Woodward; Singh *J. Am. Chem. Soc.* **1950**, *72*, 494.
[554]Stork; Ganem *J. Am. Chem. Soc.* **1973**, *95*, 6152; Stork; Singh *J. Am. Chem. Soc.* **1974**, *96*, 6181; Boeckman *J. Am. Chem. Soc.* **1974**, *96*, 6179.
[555]Eder; Sauer; Wiechert *Angew. Chem. Int. Ed. Engl.* **1971**, *10*, 496 [*Angew. Chem. 83*, 492]; Hajos; Parrish *J. Org. Chem.* **1974**, *39*, 1615. For a review of the mechanism, see Agami *Bull. Soc. Chim. Fr.* **1988**, 499-507.
[556]If the reagent is optically active because of the presence of a chiral sulfoxide group, the reaction can be enantioselective. For a review of such cases, see Solladié *Chimia* **1984**, *38*, 233-243.
[557]Because it was discovered by Claisen: *Ber.* **1890**, *23*, 977.
[558]For a review, see Johnson; Daub *Org. React.* **1951**, *6*, 1-73.
[559]Robinson; Seijo *J. Chem. Soc.* **1941**, 582.

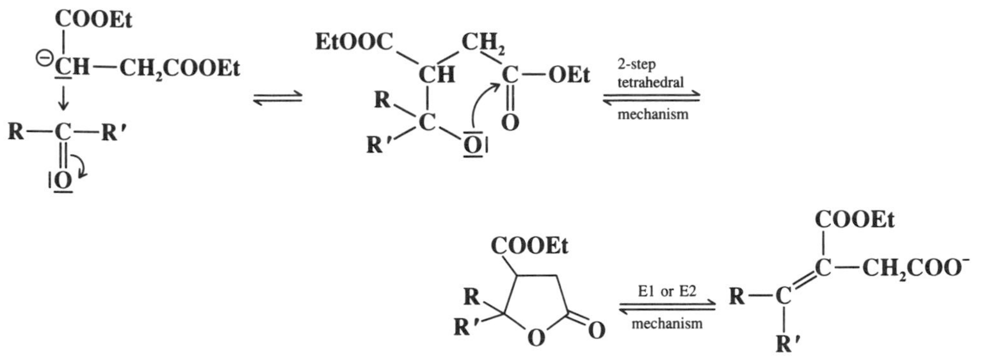

37

The Stobbe condensation has been extended to di-*t*-butyl esters of glutaric acid.[560]

This reaction is one step in an annulation sequence that also features two Michael (5-17) steps. An α,β-unsaturated ester is treated with a lithium enolate:

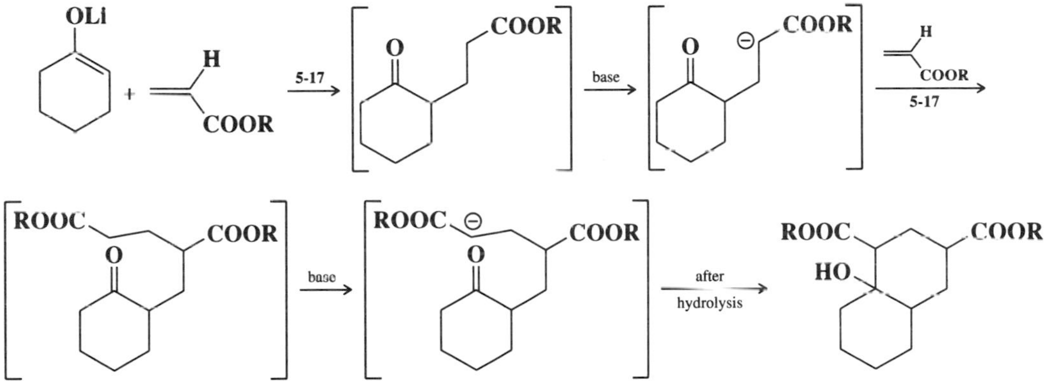

The entire sequence takes place in one laboratory step.[561]

OS **I**, 252; **III**, 132; **V**, 80, 564. Also see OS **IV**, 278, 478; **V**, 251.

6-41 The Knoevenagel Reaction
Bis(ethoxycarbonyl)methylene-de-oxo-bisubstitution, etc.

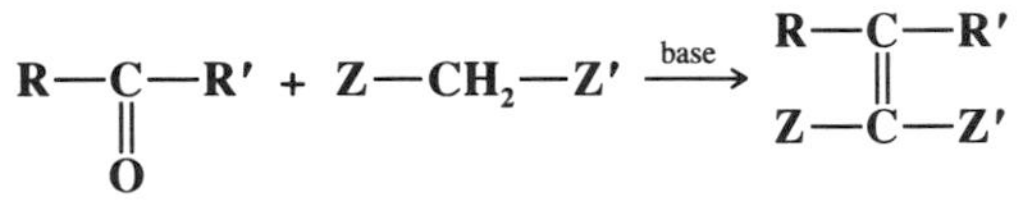

$$ R-\underset{\underset{O}{\|}}{C}-R' + Z-CH_2-Z' \xrightarrow{\text{base}} \underset{\underset{Z-C-Z'}{\underset{\|}{}}}{R-C-R'} $$

The condensation of aldehydes or ketones, usually not containing an α hydrogen, with compounds of the form Z—CH₂—Z' or Z—CHR—Z' is called the *Knoevenagel reaction*.[562]

[560]Puterbaugh *J. Org. Chem.* **1962**, *27*, 4010. See also El-Newaihy; Salem; Enayat; El-Bassiouny *J. Prakt. Chem.* **1982**, *324*, 379.

[561]Posner; Lu; Asirvatham; Silversmith; Shulman *J. Am. Chem. Soc.* **1986**, *108*, 511. For an extension of this work to the coupling of four components, see Posner; Webb; Asirvatham; Jew; Degl'Innocenti *J. Am. Chem. Soc.* **1988**, *110*, 4754.

[562]For a review, see Jones *Org. React.* **1967**, *15*, 204-599.

Z and Z' may be CHO, COR, COOH, COOR, CN, NO_2,[563] SOR, SO_2R, SO_2OR, or similar groups. When Z = COOH, decarboxylation of the product often takes place in situ.[564] If a strong enough base is used, the reaction can be performed on compounds possessing only a single Z, e.g., CH_3Z or RCH_2Z. Other active hydrogen compounds can also be employed, among them $CHCl_3$, 2-methylpyridines, terminal acetylenes, cyclopentadienes, etc.; in fact any compound that contains a C—H bond the hydrogen of which can be removed by a base. The following examples illustrate the wide scope of the reaction:

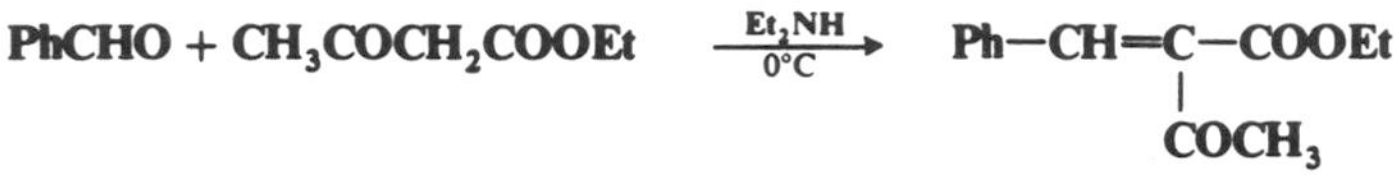

$$PhCHO + CH_3COCH_2COOEt \xrightarrow[0°C]{Et_2NH} Ph-CH=C-COOEt$$
$$\underset{COCH_3}{|}$$
38

$$PhCHO + CH_3NO_2 \xrightarrow{n\text{-}C_5H_{11}NH_2} Ph-CH=CH-NO_2 \qquad \text{Ref. 563}$$

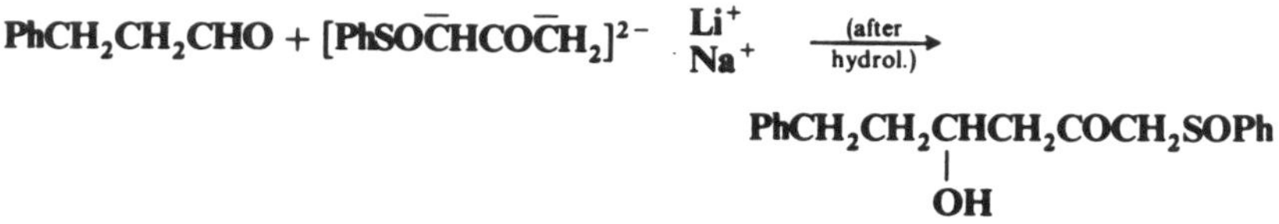

$$PhCH_2CH_2CHO + [PhSO\overline{C}HCO\overline{C}H_2]^{2-} \quad \underset{Na^+}{Li^+} \quad \xrightarrow[\text{hydrol.})]{\text{(after}}$$
$$PhCH_2CH_2CHCH_2COCH_2SOPh \qquad \text{Ref. 565}$$
$$\underset{OH}{|}$$
39

$$PrCHO + HOOCCH_2COOH \xrightarrow[\text{piperidine}]{\text{pyridine-}} PrCH=CHCOOH$$

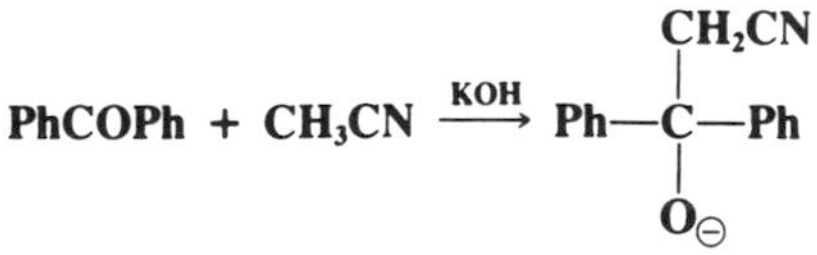

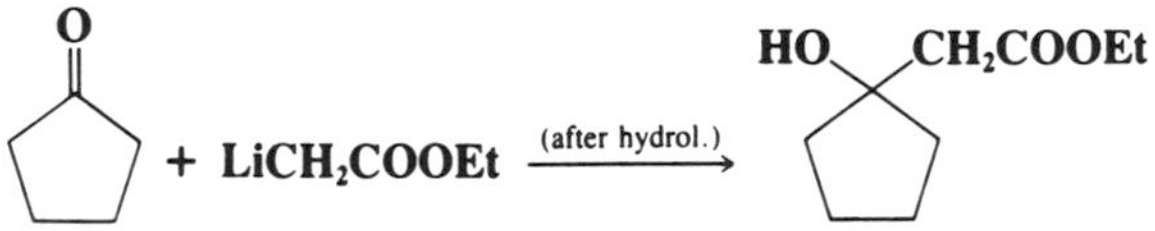

$$PhCOPh + CH_3CN \xrightarrow{KOH} \underset{\underset{O_\ominus}{|}}{\overset{\overset{CH_2CN}{|}}{Ph-C-Ph}} \qquad \text{Ref. 566}$$

$$\text{(cyclopentanone)} + LiCH_2COOEt \xrightarrow{\text{(after hydrol.)}} \text{(1-hydroxy-1-(CH}_2\text{COOEt)-cyclopentane)} \qquad \text{Ref. 567}$$

$$EtCOEt + Me_2\overline{C}-COO^\ominus \xrightarrow{\text{(after hydrol.)}} \underset{\underset{OH\ Me}{|\ |}}{\overset{\overset{Et\ Me}{|\ |}}{Et-C-C-COOH}} \qquad \text{Ref. 568}$$

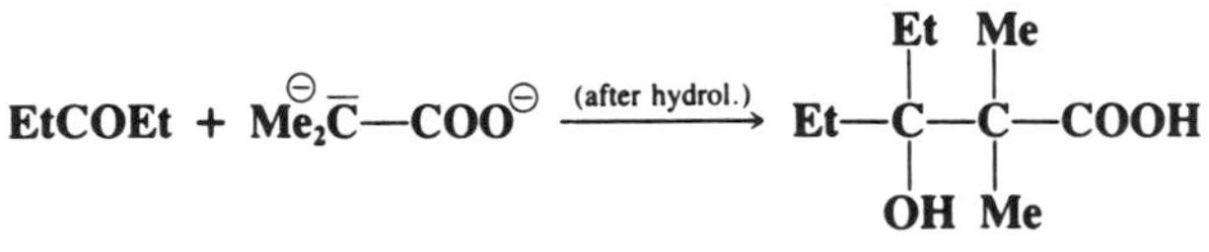

[563]For a review of this reaction with respect to nitroalkanes (often called the *Henry reaction*), see Baer; Urbas, in Feuer, Ref. 180, pp. 76-117. See also Rosini; Ballini; Sorrenti *Synthesis* **1983**, 1014; Matsumoto *Angew. Chem. Int. Ed. Engl.* **1984**, *23*, 617 [*Angew. Chem. 96*, 599]; Eyer; Seebach *J. Am. Chem. Soc.* **1985**, *107*, 3601. For reviews of the nitroalkenes that are the products of this reaction, see Barrett; Graboski *Chem. Rev.* **1986**, *86*, 751-762; Kabalka; Varma *Org. Prep. Proced. Int.* **1987**, *19*, 283-328.

[564]For a discussion of the mechanism when the reaction is accompanied by decarboxylation, see Tanaka; Oota; Hiramatsu; Fujiwara *Bull. Chem. Soc. Jpn.* **1988**, *61*, 2473.

[565]Kuwajima; Iwasawa *Tetrahedron Lett.* **1974**, 107. See also Huckin; Weiler *Can. J. Chem.* **1974**, *52*, 2157.

[566]DiBiase; Lipisko; Haag; Wolak; Gokel *J. Org. Chem.* **1979**, *44*, 4640. For a review of addition of the conjugate bases of nitriles, see Arseniyadis; Kyler; Watt *Org. React.* **1984**, *31*, 1-364.

[567]Rathke *J. Am. Chem. Soc.* **1970**, *92*, 3222; van der Veen; Geenevasen; Cerfontain *Can. J. Chem.* **1984**, *62*, 2202.

[568]Moersch; Burkett *J. Org. Chem.* **1971**, *36*, 1149. See also Cainelli; Cardillo; Contento; Umani-Ronchi *Gazz. Chim. Ital.* **1974**, *104*, 625. When the nucleophile is $Ph\overline{C}HCOO^\ominus$, the reaction is known as the *Ivanov reaction*. For a discussion of the mechanism, see Toullec; Mladenova; Gaudemar-Bardone; Blagoev *J. Org. Chem.* **1985**, *50*, 2563.

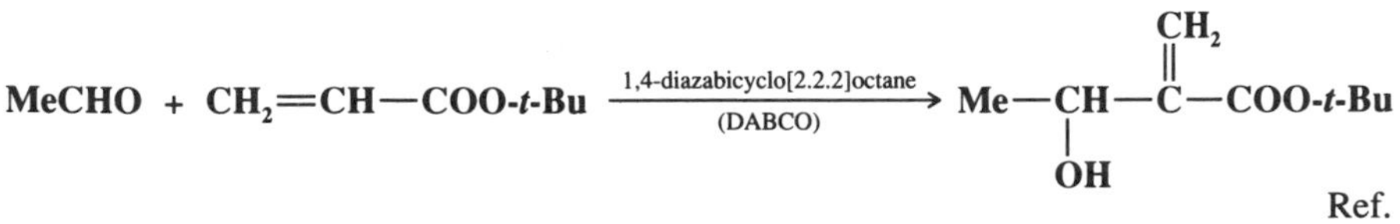

$$\text{MeCHO} + \text{CH}_2{=}\text{CH}{-}\text{COO-}t\text{-Bu} \xrightarrow[\text{(DABCO)}]{\text{1,4-diazabicyclo[2.2.2]octane}} \overset{\overset{\displaystyle\text{CH}_2}{\underset{\displaystyle|}{\|}}}{\text{Me}{-}\underset{\underset{\displaystyle\text{OH}}{|}}{\text{CH}}{-}\text{C}{-}\text{COO-}t\text{-Bu}}$$

Ref. 569

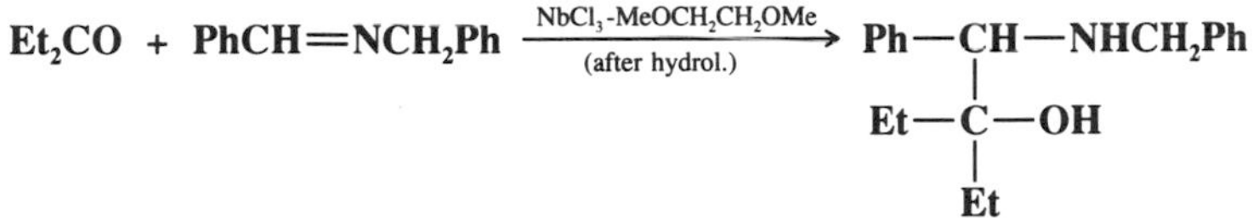

$$\text{Et}_2\text{CO} + \text{PhCH}{=}\text{NCH}_2\text{Ph} \xrightarrow[\text{(after hydrol.)}]{\text{NbCl}_3\text{-MeOCH}_2\text{CH}_2\text{OMe}} \text{Ph}{-}\underset{\underset{\displaystyle\text{Et}}{|}}{\overset{\overset{\displaystyle\text{Et}}{|}}{\text{C}}}{-}\text{OH} \quad\text{with}\quad \text{Ph}{-}\text{CH}{-}\text{NHCH}_2\text{Ph}$$

Ref. 570

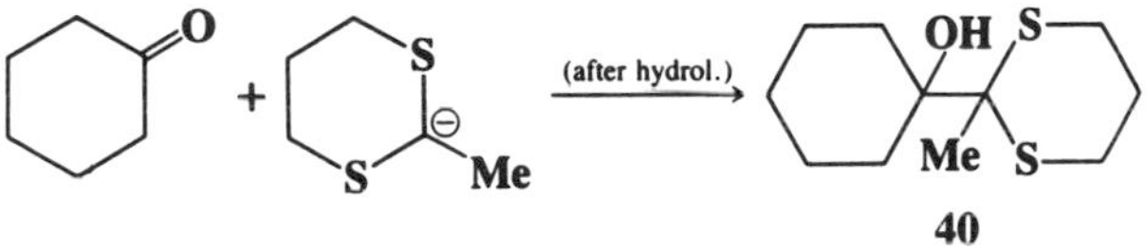

40

Ref. 571

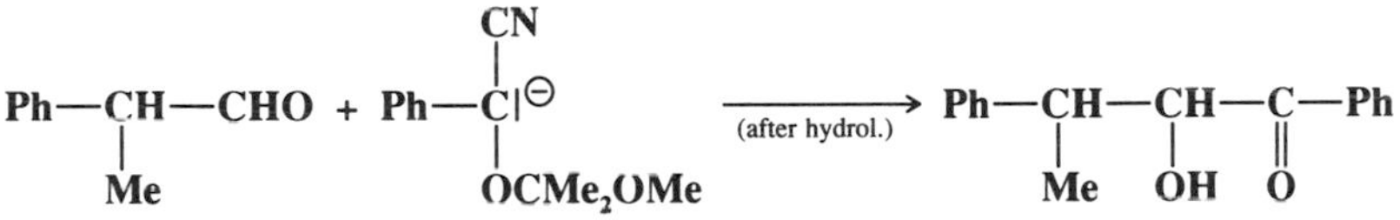

$$\text{Ph}{-}\underset{\underset{\displaystyle\text{Me}}{|}}{\text{CH}}{-}\text{CHO} + \text{Ph}{-}\underset{\underset{\displaystyle\text{OCMe}_2\text{OMe}}{|}}{\overset{\overset{\displaystyle\text{CN}}{|}}{\text{C}}}\text{I}^{\ominus} \xrightarrow[\text{(after hydrol.)}]{} \text{Ph}{-}\underset{\underset{\displaystyle\text{Me}}{|}}{\text{CH}}{-}\underset{\underset{\displaystyle\text{OH}}{|}}{\text{CH}}{-}\overset{\overset{\displaystyle\text{O}}{\|}}{\text{C}}{-}\text{Ph}$$

Ref. 572

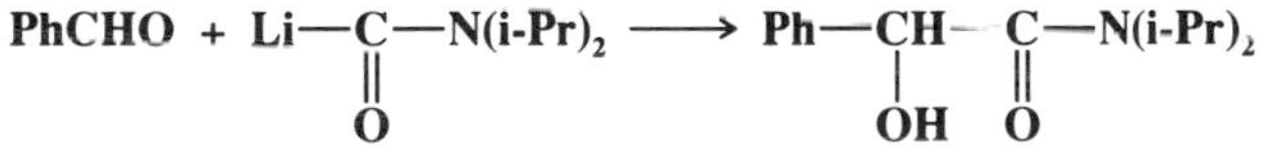

$$\text{PhCHO} + \text{Li}{-}\overset{\overset{\displaystyle\text{O}}{\|}}{\text{C}}{-}\text{N}(i\text{-Pr})_2 \longrightarrow \text{Ph}{-}\underset{\underset{\displaystyle\text{OH}}{|}}{\text{CH}}{-}\overset{\overset{\displaystyle\text{O}}{\|}}{\text{C}}{-}\text{N}(i\text{-Pr})_2$$

Ref. 573

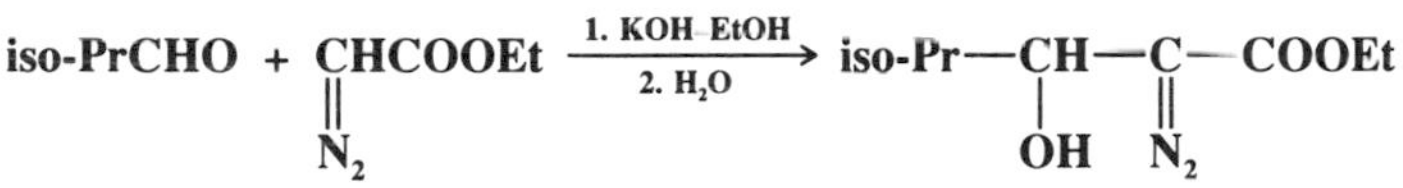

$$\text{iso-PrCHO} + \overset{\overset{\displaystyle}{}}{\underset{\underset{\displaystyle\text{N}_2}{\|}}{\text{CHCOOEt}}} \xrightarrow[\text{2. H}_2\text{O}]{\text{1. KOH-EtOH}} \text{iso-Pr}{-}\underset{\underset{\displaystyle\text{OH}}{|}}{\text{CH}}{-}\underset{\underset{\displaystyle\text{N}_2}{\|}}{\text{C}}{-}\text{COOEt}$$

Ref. 574

$$\text{PrCHO} + \quad \xrightarrow{\text{BuLi}}$$

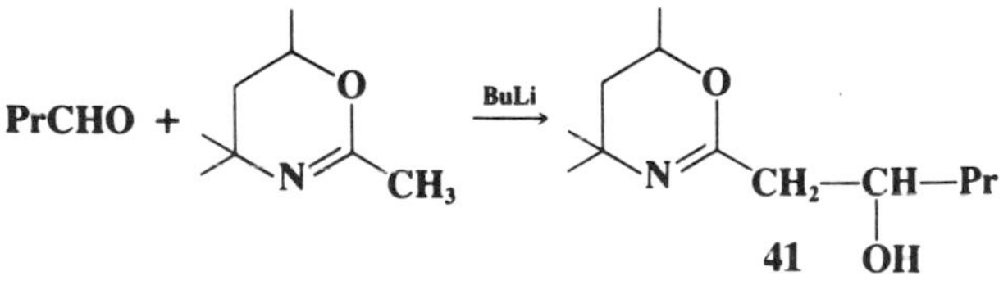

41

Ref. 575

[569]Hoffmann; Rabe *Angew. Chem. Int. Ed. Engl.* **1983,** *22,* 795 [*Angew. Chem.* **95,** 795]; Basavaiah; Gowriswari *Tetrahedron Lett.* **1986,** *27,* 2031. For a review of reactions of vinylic carbanions with aldehydes, see Drewes; Roos *Tetrahedron* **1988,** *44,* 4653-4670.

[570]Roskamp; Pedersen *J. Am. Chem. Soc.* **1987,** *109,* 6551.

[571]Corey; Seebach *Angew. Chem. Int. Ed. Engl.* **1965,** *4,* 1075 [*Angew. Chem.* **77,** 1134]. For other examples of the addition of 1,3-dithianes and similar reagents to aldehydes, ketones, and compounds containing C=N bonds, see Seebach *Synthesis* **1969,** 17-36, pp. 27-29; Corey; Crouse *J. Org. Chem.* **1968,** *33,* 298; Duhamel; Duhamel; Mancelle *Bull. Soc. Chim. Fr.* **1974,** 331; Gröbel; Bürstinghaus; Seebach, *Synthesis* **1976,** 121; Meyers; Tait; Comins *Tetrahedron Lett.* **1978,** 4657; Blatcher; Warren *J. Chem. Soc., Perkin Trans. 1* **1979,** 1074; Ogura *Pure Appl. Chem.* **1987,** *59,* 1033.

[572]Hünig; Marschner *Chem. Ber.* **1989,** *122,* 1329.

[573]Smith; Swaminathan *J. Chem. Soc., Chem. Commun.* **1976,** 387.

[574]Wenkert; McPherson *J. Am. Chem. Soc.* **1972,** *94,* 8084. See also Schöllkopf; Bánhidai; Frasnelli; Meyer; Beckhaus *Liebigs Ann. Chem.* **1974,** 1767.

[575]Meyers; Nabeya; Adickes; Fitzpatrick; Malone; Politzer *J. Am. Chem. Soc.* **1969,** *91,* 764. For other examples, see Meyers; Temple *J. Am. Chem. Soc.* **1970,** *92,* 6644; Meyers; Nabeya; Adickes; Politzer; Malone; Kovelesky; Nolen; Portnoy *J. Org. Chem.* **1973,** *38,* 36.

Ref. 576

We see from these examples that many of the carbon nucleophiles we encountered in Chapter 10 are also nucleophiles toward aldehydes and ketones (compare reactions **0-94** through **0-98** and **0-100**). As we saw in Chapter 10, the initial products in many of these cases, e.g., **38** through **41,** can be converted by relatively simple procedures (hydrolysis, reduction, decarboxylation, etc.) to various other products. In the reaction with terminal acetylenes,[577] sodium acetylides are the most common reagents (when they are used, the reaction is often called the *Nef reaction*), but lithium,[578] magnesium, and other metallic acetylides have also been used. A particularly convenient reagent is lithium acetylide–ethylenediamine complex,[579] a stable, free-flowing powder that is commercially available. Alternatively, the substrate may be treated with the alkyne itself in the presence of a base, so that the acetylide is generated in situ. This procedure is called the *Favorskii reaction*, not to be confused with the Favorskii rearrangement (**8-7**).[580] 1,4-Diols can be prepared by the treatment of aldehydes with dimetalloacetylenes $MC\equiv CM$.[581]

With most of these reagents the alcohol is not isolated (only the olefin) if the alcohol has a hydrogen in the proper position.[582] However, in some cases the alcohol is the major product. With suitable reactants, the Knoevenagel reaction, like the aldol (**6-39**), has been carried out diastereoselectively[583] and enantioselectively.[584] When the reactant is of the form ZCH_2Z', aldehydes react much better than ketones and few successful reactions with ketones have been reported. However, it is possible to get good yields of olefin from the condensation of diethyl malonate $CH_2(COOEt)_2$ with ketones, as well as with aldehydes, if the reaction is run with $TiCl_4$ and pyridine in THF.[585] In reactions with ZCH_2Z', the catalyst is most often a secondary amine (piperidine is the most common), though many other catalysts have been used. When the catalyst is pyridine (to which piperidine may or may not be added) the reaction is known as the *Doebner modification* of the Knoevenagel reaction. Alkoxides are also common catalysts.

As with **6-39**, these reactions have sometimes been performed with acid catalysts.[586]

[576]Dimroth; Berndt; Reichardt *Org. Synth.* V 1128. See also Dimroth *Angew. Chem.* **1960,** 72, 331-342; Dimroth; Wolf *Newer Methods Prep. Org. Chem.* **1964,** 3, 357-423.

[577]For reviews, see Ziegenbein, in Viehe *Acetylenes*; Marcel Dekker: New York, 1969, pp. 207-241; Ried *Newer Methods Prep. Org. Chem.* **1968,** 4, 95-138.

[578]See Midland *J. Org. Chem.* **1975,** 40, 2250, for the use of amine-free monolithium acetylide.

[579]Beumel; Harris *J. Org. Chem.* **1963,** 28, 2775.

[580]For a discussion of the mechanism of the Favorskii addition reaction, see Kondrat'eva; Potapova; Grigina; Glazunova; Nikitin *J. Org. Chem. USSR* **1976,** 12, 948.

[581]Sudwecks; Broadbent *J. Org. Chem.* **1975,** 40, 1131.

[582]For lists of reagents (with references) that condense with aldehydes and ketones to give olefin products, see Ref. 64, pp. 167-171, 180-184. For those that give the alcohol product, see Ref. 64, pp. 575, 773, 868-871, 875, 878-880, 901, 910-911.

[583]See, for example, Trost; Florez; Jebaratnam *J. Am. Chem. Soc.* **1987,** 109, 613; Mahler; Devant; Braun *Chem. Ber.* **1988,** 121, 2035; Ronan; Marchalin; Samuel; Kagan *Tetrahedron Lett.* **1988,** 29, 6101; Barrett; Robyr; Spilling *J. Org. Chem.* **1989,** 54, 1233; Pyne; Boche *J. Org. Chem.* **1989,** 54, 2663.

[584]See, for example, Enders; Lotter; Maigrot; Mazaleyrat; Welvart *Nouv. J. Chim.* **1984,** 8, 747; Ito; Sawamura; Hayashi *J. Am. Chem. Soc.* **1986,** 108, 6405; Togni; Pastor *J. Org. Chem.* **1990,** 55, 1649; Pastor; Togni *Tetrahedron Lett.* **1990,** 31, 839; Sakuraba; Ushiki *Tetrahedron Lett.* **1990,** 31, 5349; Niwa; Soai *J. Chem. Soc., Perkin Trans. 1* **1990,** 937.

[585]Lehnert *Tetrahedron Lett.* **1970,** 4723, *Tetrahedron* **1972,** 28, 663, **1973,** 29, 635, *Synthesis* **1974,** 667.

[586]For example, see Rappoport; Patai *J. Chem. Soc.* **1962,** 731.

Imines can be employed instead of aldehydes or ketones; the products are the same—an amine is lost instead of water.[587]

A number of special applications of the Knoevenagel reaction follow:

1. The dilithio derivative of N-methanesulfinyl-p-toluidine[588] (**42**) adds to aldehydes and ketones to give, after hydrolysis, the hydroxysulfinamides **43**, which, upon heating, undergo stereospecifically syn eliminations to give olefins.[589] The reaction is thus a method for achieving the conversion RR'CO → RR'C=CH$_2$ and represents an alternative to the Wittig reaction.[590]

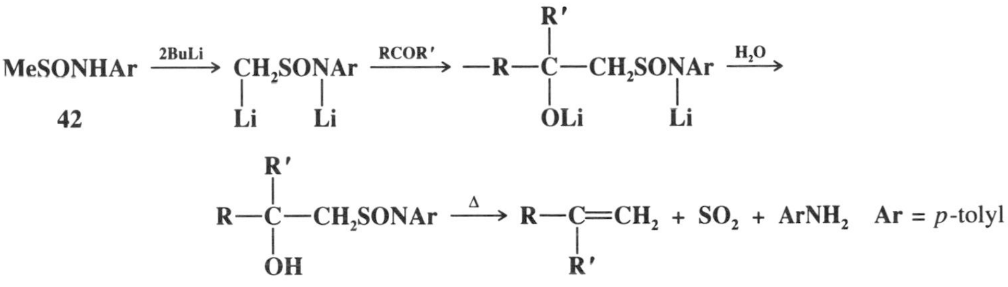

2. The reaction of ketones with tosylmethylisocyanide (**44**) gives different products,[591] depending on the reaction conditions.

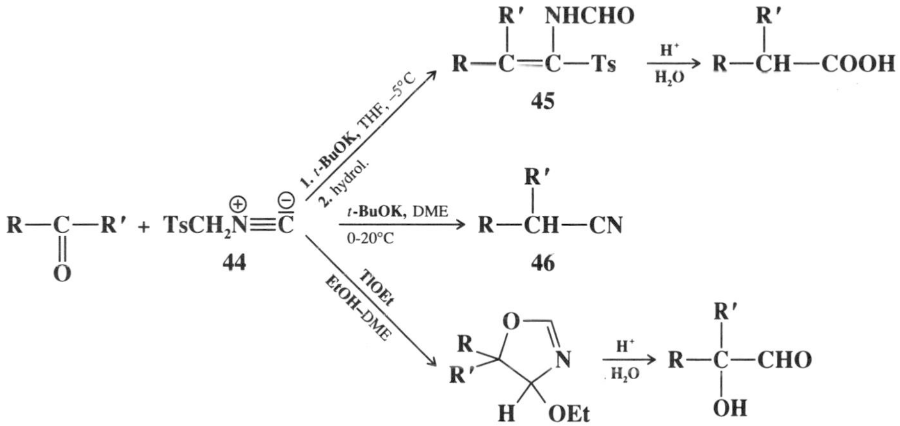

[587]Charles *Bull. Soc. Chim. Fr.* **1963**, 1559, 1566, 1573, 1576; Siegrist; Liechti; Meyer; Weber *Helv. Chim. Acta* **1969**, *52*, 2521. For the use of iminium salts, see Nair; Jahnke *Synthesis* **1984**, 424. For a review as applied to heterocyclic compounds, see Fletcher; Siegrist *Adv. Heterocycl. Chem.* **1978**, *23*, 171-261.

[588]For a method of preparing **42**, see Bowlus; Katzenellenbogen *Synth. Commun.* **1974**, *4*, 137.

[589]Corey; Durst *J. Am. Chem. Soc.* **1968**, *90*, 5548, 5553.

[590]For similar reactions, see Jung; Sharma; Durst *J. Am. Chem. Soc.* **1973**, *95*, 3420; Kuwajima; Uchida *Tetrahedron Lett.* **1972**, 649; Johnson; Shanklin; Kirchhoff *J. Am. Chem. Soc.* **1973**, *95*, 6462; Lau; Chan *Tetrahedron Lett.* **1978**, 2383; Yamamoto; Tomo; Suzuki *Tetrahedron Lett.* **1980**, *21*, 2861; Martin; Phillips; Puckette; Colapret *J. Am. Chem. Soc.* **1980**, *102*, 5866; Arenz; Vostell; Frauenrath *Synlett* **1991**, 23.

[591]For reviews of α-metalated isocyanides, see Schöllkopf *Pure Appl. Chem.* **1979**, *51*, 1347-1355, *Angew. Chem. Int. Ed. Engl.* **1977**, *16*, 339-348 [*Angew. Chem. 89*, 351-360]; Hoppe *Angew. Chem. Int. Ed. Engl.* **1974**, *13*, 789-804 [*Angew. Chem. 86*, 878-893].

When the reaction is run with potassium t-butoxide in THF at $-5°C$, one obtains (after hydrolysis) the normal Knoevenagel product **45**, except that the isocyano group has been hydrated (**6-65**).[592] With the same base but with 1,2-dimethoxyethane (DME) as solvent the product is the nitrile **46**.[593] When the ketone is treated with **44** and thallium(I) ethoxide in a 4:1 mixture of absolute ethanol and DME at room temperature, the product is a 4-ethoxy-2-oxazoline **47**.[594] Since **46** can be hydrolyzed[595] to a carboxylic acid[592] and **47** to an α-hydroxy aldehyde,[594] this versatile reaction provides a means for achieving the conversion of RCOR' to RCHR'COOH, RCHR'CN, or RCR'(OH)CHO. The conversions to RCHR'COOH and to RCHR'CN[596] have also been carried out with certain aldehydes (R' = H).

3. Aldehydes and ketones RCOR' react with α-methoxyvinyllithium CH_2=C(Li)OMe to give hydroxy enol ethers RR'C(OH)C(OMe)=CH_2, which are easily hydrolyzed to acyloins RR'C(OH)COMe.[597] In this reaction, the CH_2=C(Li)OMe is a synthon for the unavailable CH_3—$\overset{\ominus}{C}$=O ion.[598] The reagent also reacts with esters RCOOR' to give RC(OH)(COMe=CH_2)$_2$. A synthon for the Ph—$\overset{\ominus}{C}$=O ion is PhC(CN)OSiMe$_3$, which adds to aldehydes and ketones RCOR' to give, after hydrolysis, the α-hydroxy ketones RR'C(OH)COPh.[599]

4. Lithiated allylic carbamates (**48**) (prepared as shown) react with aldehydes or ketones (R⁶COR⁷), in a reaction accompanied by an allylic rearrangement, to give (after hydrolysis) γ-hydroxy aldehydes or ketones.[600] The reaction is called *the homoaldol reaction*, since the

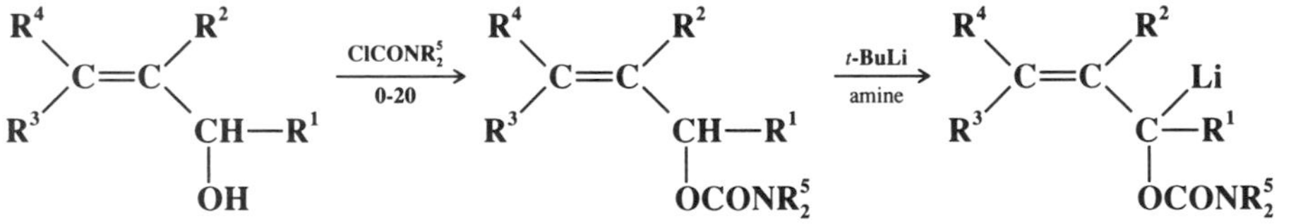

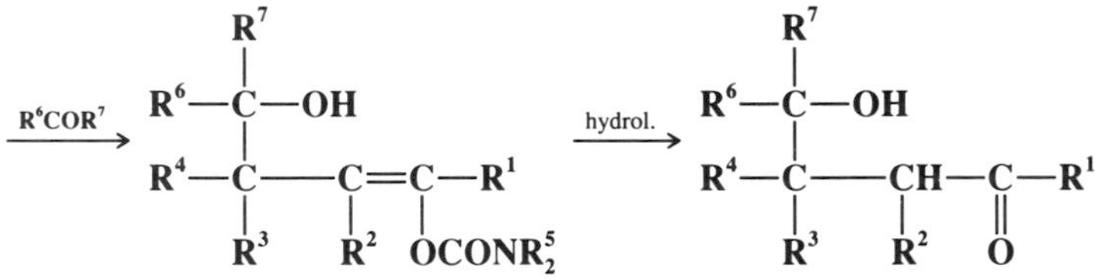

product is a homolog of the product of **6-39**. The reaction has been performed enantioselectively.[601]

[592]Schöllkopf; Schröder; Blume *Liebigs Ann. Chem.* **1972**, *766*, 130; Schöllkopf; Schröder *Angew. Chem. Int. Ed. Engl.* **1972**, *11*, 311 [*Angew. Chem. 84*, 289].
[593]Oldenziel; van Leusen; van Leusen *J. Org. Chem.* **1977**, *42*, 3114.
[594]Oldenziel; van Leusen *Tetrahedron Lett.* **1974**, 163, 167. For conversions to α,β-unsaturated ketones and diketones, see, respectively, Moskal; van Leusen *Tetrahedron Lett.* **1984**, *25*, 2585; van Leusen; Oosterwijk; van Echten; van Leusen *Recl. Trav. Chim. Pays-Bas* **1985**, *104*, 50.
[595]**45** can also be converted to a nitrile; see **7-38**.
[596]van Leusen; Oomkes *Synth. Commun.* **1980**, *10*, 399.
[597]Baldwin; Höfle; Lever *J. Am. Chem. Soc.* **1974**, *96*, 7125. For a similar reaction, see Tanaka; Nakai; Ishikawa *Tetrahedron Lett.* **1978**, 4809.
[598]For a synthon for the $\ominus$COCOOEt ion, see Reetz; Heimbach; Schwellnus *Tetrahedron Lett.* **1984**, *25*, 511.
[599]Hünig; Wehner *Synthesis* **1975**, 391.
[600]For a review, see Hoppe *Angew. Chem. Int. Ed. Engl.* **1984**, *23*, 932-948 [*Angew. Chem. 96*, 930-946].
[601]Krämer; Hoppe *Tetrahedron Lett.* **1987**, *28*, 5149.

5. A procedure for converting an aldehyde or ketone RR'CO to the homologous aldehyde RR'CHCHO consists of treating the substrate with lithium bis(ethylenedioxyboryl)methide, followed by oxidation with aqueous H_2O_2:[602]

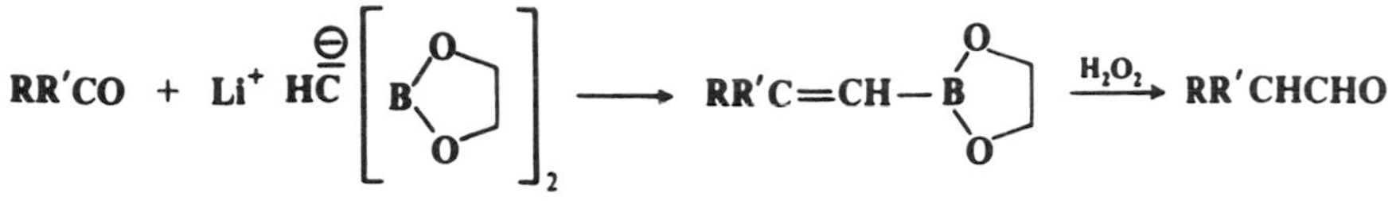

$$RR'CO + Li^+ \; H\overset{\ominus}{C}\left[B\overset{O}{\underset{O}{\bigcirc}}\right]_2 \longrightarrow RR'C=CH-B\overset{O}{\underset{O}{\bigcirc}} \xrightarrow{H_2O_2} RR'CHCHO$$

6. A method for the stereoselective synthesis of 1,2-diols consists of treating aromatic aldehydes with carbanions stabilized by an adjacent dimesitylboron group at $-120°C$, followed by oxidation with H_2O_2.[603]

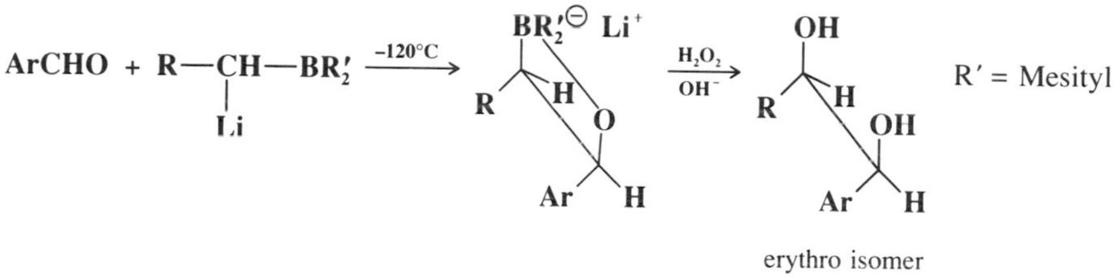

$$ArCHO + R-\underset{Li}{\underset{|}{CH}}-BR_2' \xrightarrow{-120°C} \quad \xrightarrow{H_2O_2} \quad R' = \text{Mesityl}$$

erythro isomer

The erythro–threo ratio of the product was greater than 9:1.

7. The lithium salt of an active hydrogen compound adds to the lithium salt of the tosylhydrazone of an aldehyde to give product **49**. If X = CN, SPh, or SO_2R, **49** spontaneously loses N_2 and LiX to give the alkene **50**. The entire process is done in one reaction

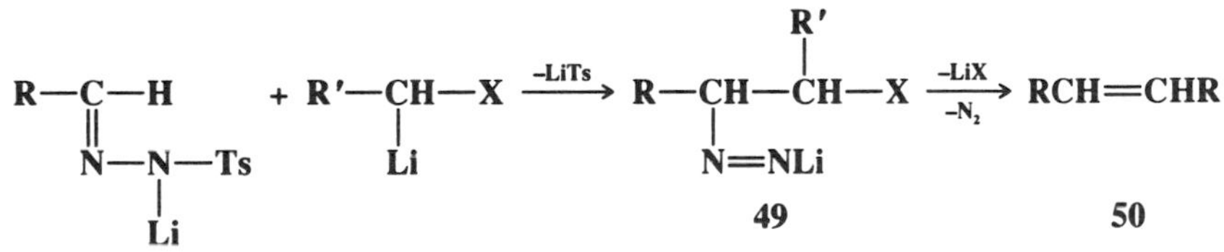

$$R-\underset{\underset{Li}{\underset{|}{N-N-Ts}}}{\overset{\overset{}{\parallel}}{C}}-H \quad + \quad R'-\underset{Li}{\underset{|}{CH}}-X \xrightarrow{-LiTs} R-\underset{\underset{N=NLi}{\underset{|}{CH}}}{CH}-\overset{\overset{R'}{|}}{CH}-X \xrightarrow[-N_2]{-LiX} RCH=CHR'$$

                                                                    49   **50**

vessel: The active hydrogen compound is mixed with the tosylhydrazone and the mixture is treated with $(i\text{-Pr})_2$NLi to form both salts at once.[604] This process is another alternative to the Wittig reaction for forming double bonds.

OS **I**, 181, 290, 413; **II**, 202; **III**, 39, 165, 317, 320, 377, 385, 399, 416, 425, 456, 479, 513, 586, 591, 597, 715, 783; **IV**, 93, 210, 221, 234, 293, 327, 387, 392, 408, 441, 463, 471, 549, 573, 730, 731, 777; **V**, 130, 381, 572, 585, 627, 833, 1088, 1128; **VI**, 41, 95, 442, 598, 683; **VII**, 50, 108, 142, 276, 381, 386, 456; **66**, 220; **67**, 205; **68**, 14, 64; **69**, 19, 31. Also see OS **III**, 395; **V**, 450.

[602]Matteson; Moody *J. Org. Chem.* **1980**, *45*, 1091. For other methods of achieving this conversion, see Corey; Tius *Tetrahedron Lett.* **1980**, *21*, 3535, 1980; Huang; Zhang *Synthesis* **1989**, 42.
[603]Pelter; Buss; Pitchford *Tetrahedron Lett.* **1985**, *26*, 5093.
[604]Vedejs; Dolphin; Stolle *J. Am. Chem. Soc.* **1979**, *101*, 249.

6-42 The Peterson Olefination Reaction
Alkylidene-de-oxo-bisubstitution

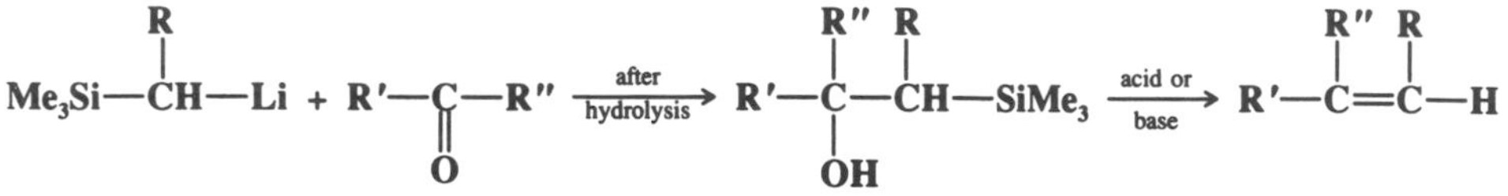

$$\text{Me}_3\text{Si}\overset{\overset{R}{|}}{\underset{}{\text{CH}}}\text{—Li} + \text{R'}\overset{}{\underset{\overset{\|}{O}}{\text{—C—R''}}} \xrightarrow[\text{hydrolysis}]{\text{after}} \text{R'}\overset{\overset{R''}{|}}{\underset{\overset{|}{\text{OH}}}{\text{—C}}}\overset{\overset{R}{|}}{\underset{}{\text{—CH}}}\text{—SiMe}_3 \xrightarrow[\text{base}]{\text{acid or}} \text{R'}\overset{\overset{R''}{|}}{\underset{}{\text{—C}}}{=}\overset{\overset{R}{|}}{\underset{}{\text{C}}}\text{—H}$$

In the *Peterson olefination reaction*[605] the lithio (or sometimes magnesio) derivative of a trialkylsilane adds to an aldehyde or ketone to give a β-hydroxysilane, which spontaneously eliminates water, or can be made to do so by treatment with acid or base, to produce an olefin. This reaction is still another alternative to the Wittig reaction, and is sometimes called the *silyl-Wittig reaction*.[606] R can also be a COOR group, in which case the product is an α,β-unsaturated ester,[607] or an SO_2Ph group, in which case the product is a vinylic sulfone.[608] The stereochemistry of the product can often be controlled by whether an acid or a base is used to achieve elimination. Use of a base generally gives syn elimination (Ei mechanism, see p. 1006), while an acid usually results in anti elimination (E2 mechanism, see p. 983).[609]

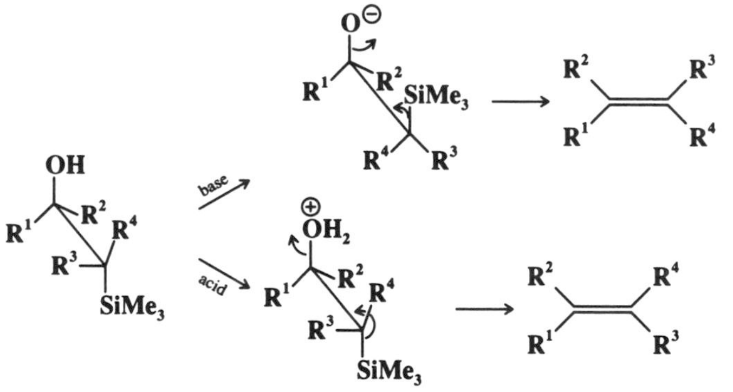

When aldehydes or ketones are treated with reagents of the form **51**, the product is an epoxy silane (**6-61**), which can be hydrolyzed to a methyl ketone.[610] For aldehydes, this is a method for converting RCHO to a methyl ketone RCH_2COMe.

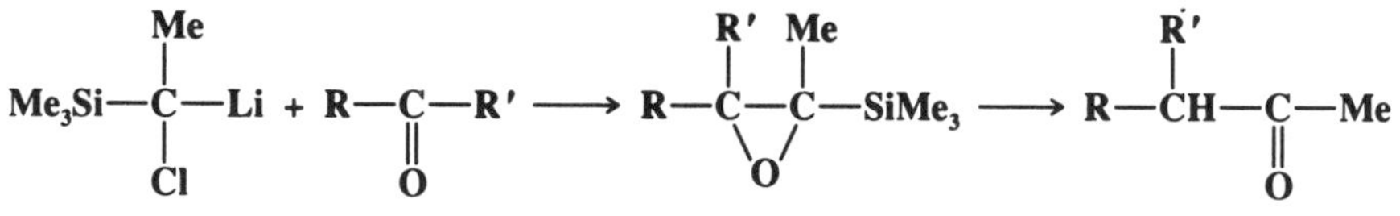

$$\text{Me}_3\text{Si}\overset{\overset{\text{Me}}{|}}{\underset{\overset{|}{\text{Cl}}}{\text{—C}}}\text{—Li} + \text{R}\overset{}{\underset{\overset{\|}{O}}{\text{—C—R'}}} \longrightarrow \text{R}\overset{\overset{\text{R'}}{|}}{\underset{\diagdown\!\underset{O}{}\!\diagup}{\text{—C}}}\overset{\overset{\text{Me}}{|}}{\underset{}{\text{—C}}}\text{—SiMe}_3 \longrightarrow \text{R}\overset{\overset{\text{R'}}{|}}{\underset{}{\text{—CH}}}\overset{}{\underset{\overset{\|}{O}}{\text{—C}}}\text{—Me}$$

51

[605]Peterson *J. Org. Chem.* **1968**, *33*, 780. For reviews, see Ager *Org. React.* **1990**, *38*, 1-223, *Synthesis* **1984**, 384-398; Colvin *Silicon Reagents in Organic Synthesis*; Academic Press: New York, 1988, pp. 63-75; Weber *Silicon Reagents for Organic Synthesis*; Springer: New York, 1983, pp. 58-78; Magnus *Aldrichimica Acta* **1980**, *13*, 43-51; Chan *Acc. Chem. Res.* **1977**, *10*, 442-448. For a list of references, see Ref. 64, pp. 178-180. For books and reviews on silicon reagents in organic synthesis, see Chapter 12, Ref. 286.

[606]For discussions of the mechanism, see Bassindale; Ellis; Lau; Taylor *J. Chem. Soc., Perkin Trans. 2* **1986**, 593; Hudrlik; Agwaramgbo; Hudrlik *J. Org. Chem.* **1989**, *54*, 5613.

[607]Hartzell; Sullivan; Rathke *Tetrahedron Lett* **1974**, 1403; Shimoji; Taguchi; Oshima; Yamamoto; Nozaki *J. Am. Chem. Soc.* **1974**, *96*, 1620; Chan; Moreland *Tetrahedron Lett.* **1978**, 515; Strekowski; Visnick; Battiste *Tetrahedron Lett.* **1984**, *25*, 5603.

[608]Craig; Ley; Simpkins; Whitham; Prior *J. Chem. Soc., Perkin Trans. 1* **1985**, 1949.

[609]See Colvin, Ref. 605, pp. 65-69.

[610]Cooke; Roy; Magnus *Organometallics* **1982**, *1*, 893.

The reagents Me$_3$SiCHRM (M = Li or Mg) are often prepared from Me$_3$SiCHRCl[611] (by **2-38** or **2-39**), but they have also been made by **2-21** and by other procedures.[612]

There are no references in *Organic Syntheses*, but see OS **69**, 89, for a related reaction.

6-43 The Addition of Active Hydrogen Compounds to CO_2 and CS_2

  α-**Acylalkyl-de-methoxy-substitution** (overall reaction)

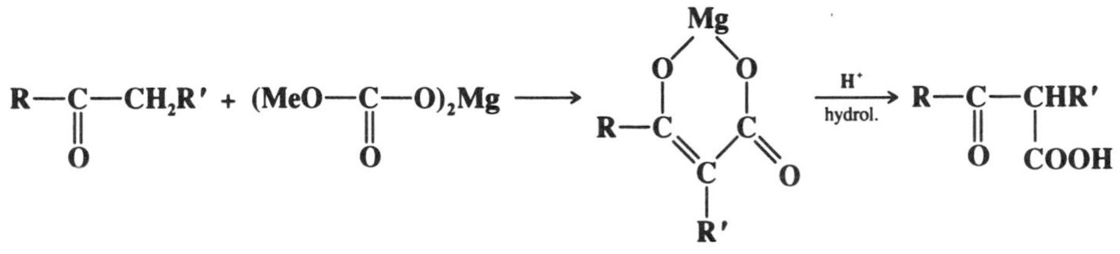

Ketones of the form $RCOCH_3$ and $RCOCH_2R'$ can be carboxylated indirectly by treatment with magnesium methyl carbonate **52**.[613] Because formation of the chelate **53** provides the driving force of the reaction, carboxylation cannot be achieved at a disubstituted α position. The reaction has also been performed on CH_3NO_2 and compounds of the form RCH_2NO_2[614] and on certain lactones.[615] Direct carboxylation has been reported in a number of instances. Ketones have been carboxylated in the α position to give β-keto acids.[616] The base here was lithium 4-methyl-2,6-di-*t*-butylphenoxide.

Ketones $RCOCH_2R'$ (as well as other active hydrogen compounds) undergo base-catalyzed addition to CS_2[617] to give a dianion intermediate $RCOCHR'CSS^-$, which can be dialkylated with a halide $R''X$ to produce α-dithiomethylene ketones $RCOCR'=C(SR'')_2$.[618] Compounds of the form ZCH_2Z' also react with bases and CS_2 to give analogous dianions.[619]

OS **VII**, 476. See also OS **65**, 17.

6-44 The Perkin Reaction

  α-**Carboxyalkylidene-de-oxo-bisubstitution**

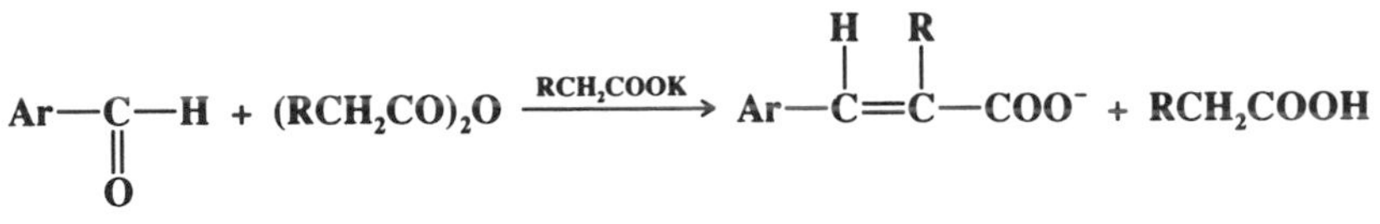

[611]For a review of these reagents, see Anderson *Synthesis* **1985**, 717-734.

[612]See, for example, Ager *J. Chem. Soc., Perkin Trans. 1* **1986**, 183; Barrett; Flygare *J. Org. Chem.* **1991**, *56*, 638.

[613]Stiles *J. Am. Chem. Soc.* **1959**, *81*, 2598, *Ann. N.Y. Acad. Sci* **1960**, *88*, 332; Crombie; Hemesley; Pattenden *Tetrahedron Lett.* **1968**, 3021.

[614]Finkbeiner; Stiles *J. Am. Chem. Soc.* **1963**, *85*, 616; Finkbeiner; Wagner *J. Org. Chem.* **1963**, *28*, 215.

[615]Martin; Watts; Johnson *Chem. Commun.* **1970**, 27.

[616]Corey; Chen *J. Org. Chem.* **1973**, *38*, 4086; Tirpak; Olsen; Rathke *J. Org. Chem.* **1985**, *50*, 4877. For an enantioselective version, see Hogeveen; Menge *Tetrahedron Lett.* **1986**, *27*, 2767.

[617]For reviews of the reactions of CS_2 with carbon nucleophiles, see Ref. 106, pp. 120-225; Yokoyama; Imamoto *Synthesis* **1984**, 797-824, pp. 797-804.

[618]See, for example Corey; Chen *Tetrahedron Lett.* **1973**, 3817.

[619]Jensen; Dalgaard; Lawesson *Tetrahedron* **1974**, *30*, 2413; Konen; Pfeffer; Silbert *Tetrahedron* **1976**, *32*, 2507, and references cited in these papers.

The condensation of aromatic aldehydes with anhydrides is called the *Perkin reaction*.[620] When the anhydride has two α hydrogens (as shown), dehydration always occurs; the β-hydroxy acid salt is never isolated. In some cases, anhydrides of the form $(R_2CHCO)_2O$ have been used, and then the hydroxy compound is the product since dehydration cannot take place. The base in the Perkin reaction is nearly always the salt of the acid corresponding to the anhydride. Although the Na and K salts have been most frequently used, higher yields and shorter reaction times have been reported for the Cs salt.[621] Besides aromatic aldehydes, their vinylogs $ArCH=CHCHO$ also give the reaction. Otherwise, the reaction is not suitable for aliphatic aldehydes.[622]

OS **I**, 398; **II**, 61, 229; **III**, 426.

6-45   Darzens Glycidic Ester Condensation
(2 + 1)*OC,CC-cyclo*-α-Alkoxycarbonylmethylene-addition

$$-\overset{\displaystyle R}{\underset{\displaystyle \|}{\underset{O}{C}}}- \;+\; Cl-\overset{\displaystyle R}{\underset{\displaystyle |}{CH}}-COOEt \xrightarrow{\text{NaOEt}} -\overset{}{C}-\overset{\displaystyle R}{\underset{\displaystyle |}{C}}-COOEt$$

Aldehydes and ketones condense with α-halo esters in the presence of bases to give α,β-epoxy esters, called *glycidic esters*. This is called *the Darzens condensation*.[623] The reaction consists of an initial Knoevenagel-type reaction (**6-41**), followed by an internal S$_N$2 reaction (**0-13**):[624]

$$-\overset{}{\underset{\|O}{C}}- \;+\; Cl-\overset{}{\underset{R}{CH}}-COOEt \xrightarrow{\text{OEt}^-} \; -\overset{}{C}-\overset{Cl}{\underset{R}{C}}-COOEt \longrightarrow -\overset{R}{\underset{O}{C}}-\overset{R}{\underset{}{C}}-COOEt$$

Although the intermediate halo alkoxide is generally not isolated, it has been done, not only with α-fluoro esters (since fluorine is such a poor leaving group in nucleophilic substitutions) but also with α-chloro esters.[625] This is only one of several types of evidence that rule out a carbene intermediate.[626] Sodium ethoxide is often used as the base, though other bases, including sodium amide, are sometimes used. Aromatic aldehydes and ketones give good yields, but aliphatic aldehydes react poorly. However, the reaction can be made to give good yields (~80%) with simple aliphatic aldehydes as well as with aromatic aldehydes and ketones by treatment of the α-halo ester with the base lithium bis(trimethylsilyl)amide LiN(SiMe$_3$)$_2$ in THF at −78°C (to form the conjugate base of the ester) and addition of the aldehyde or ketone to this solution.[627] If a preformed dianion of an α-halo carboxylic

[620]For a review, see Johnson, *Org. React.* **1942**, *1*, 210-266.
[621]Koepp; Vögtle *Synthesis* **1987**, 177.
[622]Crawford; Little *J. Chem. Soc.* **1959**, 722.
[623]For a review, see Berti *Top. Stereochem.* **1973**, 7, 93-251, pp. 210-218.
[624]For discussions of the mechanism of the reaction, and especially of the stereochemistry, see Roux-Schmitt; Seyden-Penne; Wolfe *Tetrahedron* **1972**, *28*, 4965; Bansal; Sethi *Bull. Chem. Soc. Jpn.* **1980**, *53*, 1197.
[625]Ballester; Pérez-Blanco *J. Org. Chem.* **1958**, *23*, 652; Martynov; Titov *J. Gen. Chem. USSR* **1960**, *30*, 4072, **1962**, *32*, 716, **1963**, *33*, 1350, **1964**, *34*, 2139; Elkik; Francesch *Bull. Soc. Chim. Fr.* **1973**, 1277, 1281.
[626]Another, based on the stereochemistry of the products, is described by Zimmerman; Ahramjian *J. Am. Chem. Soc.* **1960**, *82*, 5459.
[627]Borch *Tetrahedron Lett.* **1972**, 3761.

acid Cl—$\overset{\ominus}{\underset{}{\overline{C}}}$R—COO$^\ominus$ is used instead, α,β-epoxy acids are produced directly.[628] The Darzens reaction has also been carried out on α-halo ketones, α-halo nitriles,[629] α-halo sulfoxides[630] and sulfones,[631] α-halo N,N-disubstituted amides,[632] α-halo ketimines,[633] and even on allylic[634] and benzylic halides. Phase transfer catalysis has been used.[635] The Darzens reaction has been performed enantioselectively, by coupling optically active α-bromo-β-hydroxy esters with aldehydes.[635a]

Glycidic esters can easily be converted to aldehydes (**2-40**). The reaction has been extended to the formation of analogous aziridines by treatment of an imine with an α-halo ester or an α-halo N,N-disubstituted amide and t-BuOK in the solvent 1,2-dimethoxyethane.[636] However, yields were not high. Acid-catalyzed Darzens reactions have also been reported.[637] See also **6-61**.

OS **III**, 727; **IV**, 459, 649.

6-46 Tollens' Reaction
O-Hydro-C-(β-hydroxyalkyl)-addition

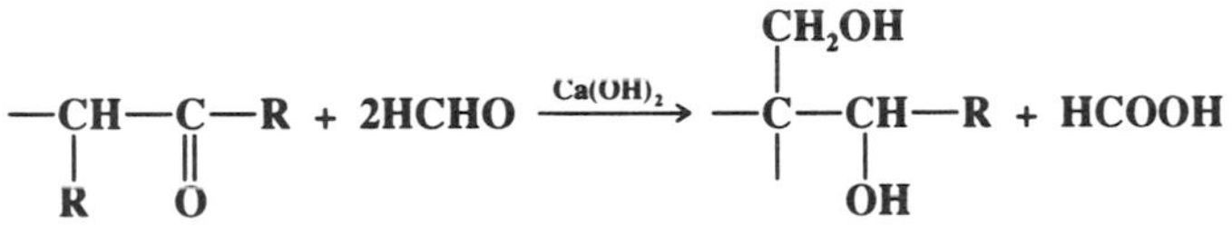

$$-\underset{\underset{R}{|}}{CH}-\underset{\underset{O}{\|}}{C}-R + 2HCHO \xrightarrow{Ca(OH)_2} -\underset{\underset{OH}{|}}{\overset{\overset{CH_2OH}{|}}{C}}-\underset{}{CH}-R + HCOOH$$

In *Tollens' reaction* an aldehyde or ketone containing an α hydrogen is treated with formaldehyde in the presence of $Ca(OH)_2$ or a similar base. The first step is a mixed aldol reaction (**6-39**).

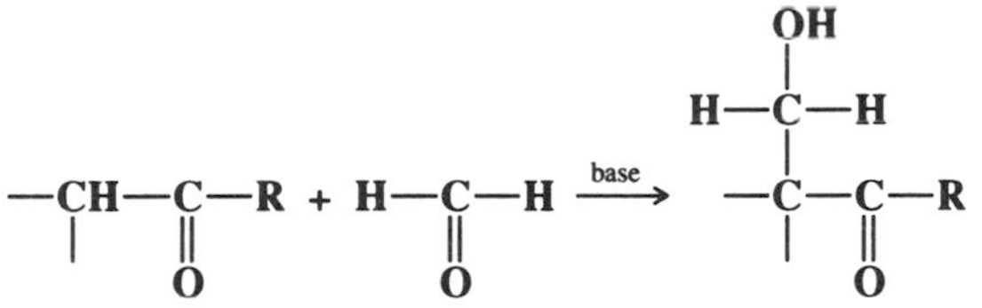

$$-\underset{\underset{O}{\|}}{CH}-\underset{}{C}-R + H-\underset{\underset{O}{\|}}{C}-H \xrightarrow{base} -\underset{}{C}-\underset{\underset{O}{\|}}{C}-R$$
(with $\overset{\overset{OH}{|}}{H-C-H}$ group)

The reaction can be stopped at this point, but more often a second mole of formaldehyde is permitted to reduce the newly formed aldol to a 1,3-diol, in a crossed Cannizzaro reaction (**9-69**). If the aldehyde or ketone has several α hydrogens, they can all be replaced. An important use of the reaction is to prepare pentaerythritol from acetaldehyde:

$$CH_3CHO + 4HCHO \longrightarrow C(CH_2OH)_4 + HCOOH$$

[628]Johnson; Bade *J. Org. Chem.* **1982**, *47*, 1205.
[629]See White; Wu *J. Chem. Soc., Chem. Commun.* **1974**, 988.
[630]Satoh; Sugimoto; Itoh; Yamakawa *Tetrahedron Lett.* **1989**, *30*, 1083.
[631]Vogt; Tavares *Can. J. Chem.* **1969**, *47*, 2875.
[632]Tung; Speziale; Frazier *J. Org. Chem.* **1963**, *28*, 1514.
[633]Mauzé *J. Organomet. Chem.* **1979**, *170*, 265.
[634]Sulmon; De Kimpe; Schamp; Declercq; Tinant *J. Org. Chem.* **1988**, *53*, 4457.
[635]See Jończyk; Kwast; Makosza *J. Chem. Soc., Chem. Commun.* **1977**, 902; Gladiali; Soccolini *Synth. Commun.* **1982**, *12*, 355; Starks; Liotta *Phase Transfer Catalysis*; Academic Press: New York, 1978, pp. 197-198.
[635a]Corey; Choi *Tetrahedron Lett.* **1991**, *32*, 2857.
[636]Deyrup *J. Org. Chem.* **1969**, *34*, 2724.
[637]Sipos; Schöbel; Baláspiri *J. Chem. Soc. C* **1970**, 1154; Sipos; Schöbel; Sirokmán *J. Chem. Soc., Perkin Trans. 2* **1975**, 805.

When aliphatic nitro compounds are used instead of aldehydes or ketones, no reduction occurs, and the reaction is essentially a Knoevenagel reaction, though it is usually also called a Tollens' reaction:

$$CH_3NO_2 + HCHO \xrightarrow{OH^-} HOCH_2CH_2NO_2$$

OS **I**, 425; **IV**, 907; **V**, 833.

6-47 The Wittig Reaction
Alkylidene-de-oxo-bisubstitution

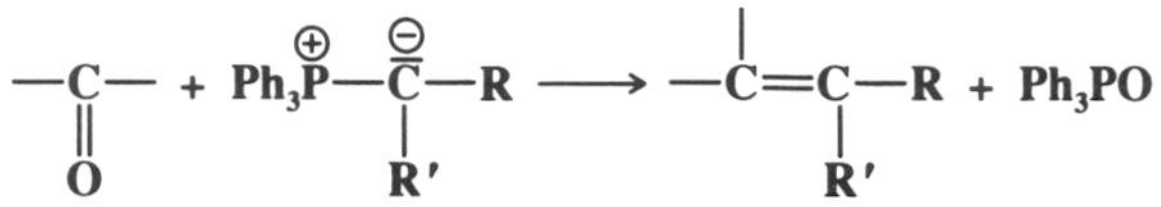

In the *Wittig reaction* an aldehyde or ketone is treated with a *phosphorus ylide* (also called a *phosphorane*) to give an olefin.[638] Phosphorus ylides are usually prepared by treatment of a phosphonium salt with a base,[639] and phosphonium salts are usually prepared from the phosphine and an alkyl halide (**0-43**):

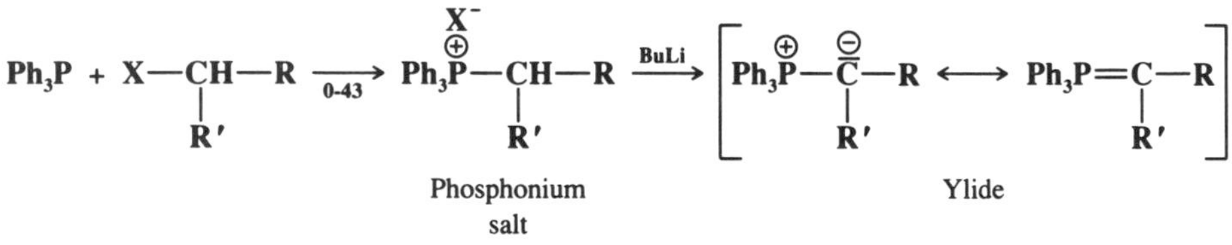

The overall sequence of three steps may be called the Wittig reaction, or only the final step. Phosphonium salts are also prepared by addition of phosphines to Michael olefins (like **5-7**) and in other ways. The phosphonium salts are most often converted to the ylides by treatment with a strong base such as butyllithium, sodium amide,[640] sodium hydride, or a sodium alkoxide, though weaker bases can be used if the salt is acidic enough. For $(Ph_3P^+)_2CH_2$, sodium carbonate is a strong enough base.[641] When the base used does not contain lithium, the ylide is said to be prepared under "salt-free" conditions.[642]

---

[638]For a general treatise, see Cadogan *Organophosphorus Reagents in Organic Synthesis*; Academic Press: New York, 1979. For a monograph on the Wittig reaction, see Johnson *Ylid Chemistry*; Academic Press: New York, 1966. For reviews, see Maryanoff; Reitz *Chem. Rev.* **1989**, *89*, 863-927; Bestmann; Vostrowsky *Top. Curr. Chem.* **1983**, *109*, 85-164; Pommer; Thieme *Top. Curr. Chem.* **1983**, *109*, 165-188; Pommer *Angew. Chem. Int. Ed. Engl.* **1977**, *16*, 423-429 [*Angew. Chem. 89*, 437-443]; Maercker *Org. React.* **1965**, *14*, 270-490; House, Ref. 180, pp. 682-709; Lowe *Chem. Ind. (London)* **1970**, 1070-1079; Bergelson; Shemyakin, in Patai, Ref. 472, pp. 295-340, *Newer Methods Prep. Org. Chem.* **1968**, *5*, 154-175. For related reviews, see Tyuleneva; Rokhlin; Knunyants *Russ. Chem. Rev.* **1981**, *50*, 280-290; Starks; Liotta, Ref. 635, pp. 288-297; Weber; Gokel *Phase Transfer Catalysis in Organic Synthesis*; Springer: New York, 1977; pp. 234-241; Zbiral *Synthesis* **1974**, 775-797; Bestmann *Bull. Soc. Chim. Fr.* **1971**, 1619-1634, *Angew. Chem. Int. Ed. Engl.* **1965**, *4*, 583-587, 645-660, 830-838 [*Angew. Chem. 77*, 609-613, 651-666, 850-858], *Newer Methods Prep. Org. Chem.* **1968**, *5*, 1-60; Horner *Fortschr. Chem. Forsch.* **1966**, *7*, 1-61. For a historical background, see Wittig, *Pure Appl. Chem.* **1964**, *9*, 245-254. For a list of reagents and references for the Wittig and related reactions, see Ref. 64, pp. 173-178.

[639]When phosphonium *fluorides* are used, no base is necessary, as these react directly with the substrate to give the olefin: Schiemenz; Becker; Stöckigt *Chem. Ber.* **1970**, *103*, 2077.

[640]For a convenient method of doing this that results in high yields, see Schlosser; Schaub *Chimia* **1982**, *36*, 396.

[641]Ramirez; Pilot; Desai; Smith; Hansen; McKelvie *J. Am. Chem. Soc.* **1967**, *89*, 6273.

[642]Bestmann *Angew. Chem. Int. Ed. Engl.* **1965**, *4*, 586 [*Angew. Chem. 77*, 612].

In the overall Wittig reaction, an olefin is formed from the aldehyde or ketone and an alkyl halide in which the halogen-bearing carbon contains at least one hydrogen:

$$
-\overset{|}{\underset{|}{C}}=O \ + \ Br-\overset{\overset{\displaystyle H}{|}}{\underset{\underset{\displaystyle R'}{|}}{C}}-R \longrightarrow -\overset{|}{\underset{|}{C}}=\overset{}{\underset{\underset{\displaystyle R'}{|}}{C}}-R
$$

This result is similar to that obtained in the Reformatsky reaction (**6-30**), but this is more general since no ester or other group is required to be α to the halogen. Another important advantage of the Wittig reaction is that the *position* of the new double bond is always certain, in contrast to the result in the Reformatsky reaction and in most of the base-catalyzed condensations (**6-39** to **6-46**). Examples of this are given below.

The reaction is very general. The aldehyde or ketone may be aliphatic, alicyclic, or aromatic (including diaryl ketones); it may contain double or triple bonds; it may contain various functional groups, such as OH, OR, NR$_2$, aromatic nitro or halo, acetal, or even ester groups.[643] Double or triple bonds *conjugated* with the carbonyl also do not interfere, the attack being at the C=O carbon.

The phosphorus ylide may also contain double or triple bonds and certain functional groups. Simple ylides (R, R′ = hydrogen or alkyl) are highly reactive, reacting with oxygen, water, hydrohalic acids, and alcohols, as well as carbonyl compounds and carboxylic esters, so the reaction must be run under conditions where these materials are absent. When an electron-withdrawing group, e.g., COR, CN, COOR, CHO, is present in the α position, the ylides are much more stable, because the charge on the carbon is spread by resonance:

$$
\overset{\oplus}{Ph_3P}-\overset{\ominus}{\underset{\underset{\displaystyle \underset{|}{\underline{\underline{O}}|}}{\|}}{CH}}-C-R \longleftrightarrow \overset{\oplus}{Ph_3P}-CH=\underset{\underset{\displaystyle |\underline{\underline{O}}|_{\ominus}}{|}}{C}-R
$$

These ylides react readily with aldehydes, but slowly or not at all with ketones.[644] In extreme cases, e.g., **54,** the ylide does not react with ketones *or* aldehydes. Besides these groups,

54

the ylide may contain one or two α halogens[645] or an α OR or OAr group. In the latter case the product is an enol ether, which can be hydrolyzed (**0-6**) to an aldehyde,[646] so that

$$
R''OCH_2Cl \xrightarrow{Ph_3P} R''OCH_2\overset{\oplus}{PPh_3} \xrightarrow[\text{2. RCOR'}]{\text{1. base}} R''OCH=\underset{\underset{\displaystyle R}{|}}{C}-R' \xrightarrow{\text{hydrol.}} R'-\underset{\underset{\displaystyle R}{|}}{CH}-CHO
$$

[643]Although phosphorus ylides also react with esters, that reaction is too slow to interfere: Greenwald; Chaykovsky; Corey *J. Org. Chem.* **1963**, *28*, 1128.

[644]For successful reactions of stabilized ylides with ketones, under high pressure, see Isaacs; El-Din *Tetrahedron Lett.* **1987**, *28*, 2191. See also Dauben; Takasugi *Tetrahedron Lett.* **1987**, 4377.

[645]Seyferth; Grim; Read *J. Am. Chem. Soc.* **1960**, *82*, 1510, **1961**, *83*, 1617; Seyferth; Heeren; Singh; Grim; Hughes *J. Organomet. Chem.* **1966**, *5*, 267; Schlosser; Zimmermann *Synthesis* **1969**, 75; Burton; Greenlimb *J. Fluorine Chem.* **1974**, *3*, 447; Smithers *J. Org. Chem.* **1978**, *43*, 2833; Miyano; Izumi; Fujii; Ohno; Hashimoto *Bull. Chem. Soc. Jpn.* **1979**, *52*, 1197; Stork; Zhao *Tetrahedron Lett.* **1989**, *30*, 2173.

[646]For references to the use of the Wittig reaction to give enol ethers or enol thioethers, which are then hydrolyzed, see Ref. 64, pp. 715-716, 726.

this reaction is a means of achieving the conversion RCOR′ → RR′CHCHO.[647] However, the ylide may not contain an α nitro group. If the phosphonium salt contains a potential leaving group, such as Br or OMe, in the β position, treatment with a base gives elimination, instead of the ylide:

$$Ph_3\overset{\oplus}{P}CH_2CH_2Br \xrightarrow{\text{base}} Ph_3\overset{\oplus}{P}CH=CH_2$$

However, a β COO⁻ group may be present, and the product is a β,γ-unsaturated acid:[648]

$$Ph_3\overset{\oplus}{P}-\overset{\ominus}{C}HCH_2COO^- + -\overset{|}{C}=O \longrightarrow -\overset{|}{C}=CHCH_2COO^-$$

This is the only convenient way to make these compounds, since elimination by any other route gives the thermodynamically more stable α,β-unsaturated isomers. This is an illustration of the utility of the Wittig method for the specific location of a double bond. Another illustration is the conversion of cyclohexanones to olefins containing double bonds, e.g.,[649]

$$\bigcirc\!\!=\!O + Ph_3P=CH_2 \longrightarrow \bigcirc\!\!=\!CH_2$$

Still another example is the easy formation of anti-Bredt bicycloalkenones[650] (see p. 160). As indicated above, α,α′-dihalophosphoranes can be used to prepare 1,1-dihaloalkenes. Another way to prepare such compounds[651] is to treat the carbonyl compound with a mixture of CX₄ (X = Cl, Br, or I) and triphenylphosphine, either with or without the addition of zinc dust (which allows less Ph₃P to be used).[652]

The Wittig reaction has been carried out with polymer-supported ylides[653] (see p. 421).

Ylides are usually prepared from triphenylphosphine, but other triarylphosphines,[654] trialkylphosphines,[655] and triphenylarsine[656] have also been used. The Wittig reaction has also been carried out with other types of ylides, the most important being prepared from phosphonates:[657]

$$(RO)_2\underset{\underset{O}{\|}}{P}-\underset{\underset{R'}{|}}{C}H-R'' \xrightarrow{\text{base}} (RO)_2\underset{\underset{O}{\|}}{P}-\underset{\underset{R'}{|}}{\overset{\ominus}{C}}-R'' \xrightarrow{-\overset{|}{C}=O} -\underset{\underset{R'}{|}}{C}=\overset{|}{C}-R'' + (RO)_2PO_2^-$$

[647]For other methods of achieving this conversion via Wittig-type reactions, see Ceruti; Degani; Fochi *Synthesis* **1987**, 79; Moskal; van Leusen *Recl. Trav. Chim. Pays-Bas* **1987**, *106*, 137; Doad *J. Chem. Res. (S)* **1987**, 370.

[648]Corey; McCormick; Swensen *J. Am. Chem. Soc.* **1964**, *86*, 1884.

[649]Wittig; Schöllkopf *Chem. Ber.* **1954**, *87*, 1318.

[650]Bestmann; Schade *Tetrahedron Lett.* **1982**, *23*, 3543.

[651]For a list of references to the preparation of haloalkenes by Wittig reactions, with references, see Ref. 64, pp. 376-377.

[652]See, for example, Rabinowitz; Marcus *J. Am. Chem. Soc.* **1962**, *84*, 1312; Ramirez; Desai; McKelvie *J. Am. Chem. Soc.* **1962**, *84*, 1745; Corey; Fuchs *Tetrahedron Lett.* **1972**, 3769; Posner; Loomis; Sawaya *Tetrahedron Lett.* **1975**, 1373; Suda; Fukushima *Tetrahedron Lett.* **1981**, *22*, 759; Gaviña; Luis; Ferrer; Costero; Marco *J. Chem. Soc., Chem. Commun.* **1985**, 296; Li; Alper *J. Org. Chem.* **1986**, *51*, 4354.

[653]Bernard; Ford; Nelson *J. Org. Chem.* **1983**, *48*, 3164.

[654]Schiemenz; Thobe *Chem. Ber.* **1966**, *99*, 2663.

[655]For example, see Johnson; LaCount *Tetrahedron* **1960**, *9*, 130; Bestmann; Kratzer *Chem. Ber.* **1962**, *95*, 1894.

[656]An arsenic ylide has been used in a catalytic version of the Wittig reaction; that is, the R₃AsO product is constantly regenerated to produce more arsenic ylide: Shi; Wang; Wang; Huang *J. Org. Chem.* **1989**, *54*, 2027.

[657]Horner; Hoffmann; Wippel *Chem. Ber.* **1958**, *91*, 61; Horner; Hoffmann; Wippel; Klahre *Chem. Ber.* **1959**, *92*, 2499; Wadsworth; Emmons *J. Am. Chem. Soc.* **1961**, *83*, 1733.

This method, sometimes called the *Horner–Emmons, Wadsworth–Emmons,* or *Wittig–Horner reaction*,[658] has several advantages over the use of phosphoranes.[659] These ylides are more reactive than the corresponding phosphoranes, and when R' is an electron-withdrawing group, these compounds often react with ketones that are inert to phosphoranes. In addition, the phosphorus product is a phosphate ester and hence soluble in water, unlike Ph_3PO, which makes it easy to separate it from the olefin product. Phosphonates are also cheaper than phosphonium salts and can easily be prepared by the *Arbuzov reaction*:[660]

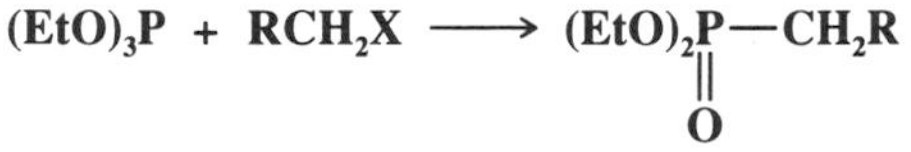

$$(EtO)_3P + RCH_2X \longrightarrow (EtO)_2\underset{\underset{O}{\|}}{P}-CH_2R$$

Ylides formed from phosphinoxides $Ar_2\underset{\underset{O}{\|}}{P}CHRR'$, phosphonic acid bisamides $(R_2''N)_2POCHRR'$,[661] and alkyl phosphonothionates $(McO)_2PSCHRR'$[662] share some of these advantages. Phosphonates $Ph_2POCH_2NR_2'$ react with aldehydes or ketones R^2COR^3 to give good yields of enamines $R^2R^3C=CHNR_2'$.[663]

The mechanism[664] of the key step of the Wittig reaction is as follows:[665]

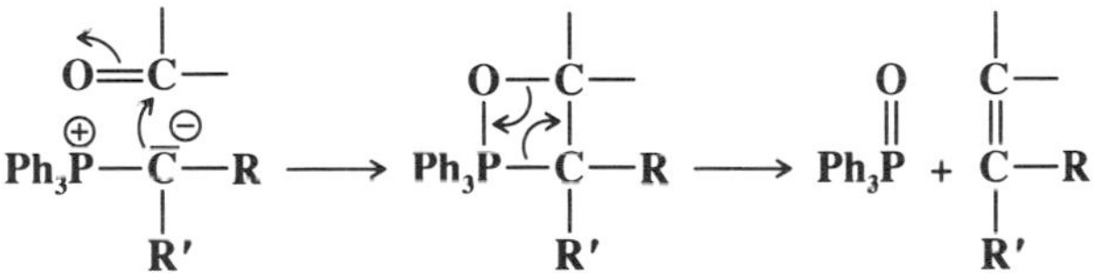

$$\underset{\underset{R'}{|}}{\overset{\overset{|}{O=C-}}{Ph_3\overset{\oplus}{P}-\overset{\ominus}{C}-R}} \longrightarrow \underset{\underset{R'}{|}}{\overset{\overset{|}{O-C-}}{Ph_3P-C-R}} \longrightarrow \underset{\underset{R'}{|}}{\overset{\overset{|}{O \quad C-}}{Ph_3P + C-R}}$$

Oxaphosphetane

For many years it was assumed that a diionic compound, called a *betaine*, is an intermediate on the pathway from the starting compounds to the oxaphosphetane, and in fact it may be

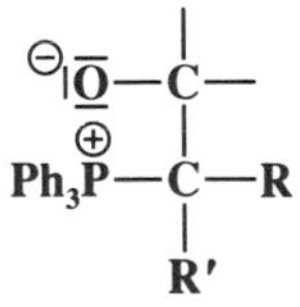

$$\underset{\underset{R'}{|}}{\overset{\overset{|}{^\ominus|\overline{O}-C-}}{Ph_3\overset{\oplus}{P}-C-R}}$$

Betaine

[658]For reviews, see Wadsworth *Org. React.* **1977**, *25*, 73-253; Stec *Acc. Chem. Res.* **1983**, *16*, 411-417; Walker, in Cadogan, Ref. 638, pp. 156-205; Dombrovskii; Dombrovskii *Russ. Chem. Rev.* **1966**, *35*, 733-741; Boutagy; Thomas *Chem. Rev.* **1974**, *74*, 87-99.

[659]For a convenient method of carrying out this reaction, see Seguineau; Villieras *Tetrahedron Lett.* **1988**, *29*, 477, and other papers in this series.

[660]Also known as the *Michaelis–Arbuzov rearrangement*. For reviews, see Petrov; Dogadina; Ionin; Garibina; Leonov *Russ. Chem. Rev.* **1983**, *52*, 1030-1035; Bhattacharya; Thyagarajan *Chem. Rev.* **1981**, *81*, 415-430. For related reviews, see Shokol; Kozhushko *Russ. Chem. Rev.* **1985**, *53*, 98-104; Brill; Landon *Chem. Rev.* **1984**, *84*, 577-585.

[661]Corey; Kwiatkowski *J. Am. Chem. Soc.* **1968**, *90*, 6816; Corey; Cane *J. Org. Chem.* **1969**, *34*, 3053.

[662]Corey; Kwiatkowski *J. Am. Chem. Soc.* **1966**, *88*, 5654.

[663]Broekhof; van der Gen *Recl. Trav. Chim. Pays-Bas* **1984**, *103*, 305; Broekhof; van Elburg; Hoff; van der Gen *Recl. Trav. Chim. Pays-Bas* **1984**, *103*, 317.

[664]For a review of the mechanism, see Cockerill; Harrison, Ref. 209, pp. 232-240. For a thorough discussion, see Vedejs; Marth *J. Am. Chem. Soc.* **1988**, *110*, 3948.

[665]It has been contended that another mechanism, involving single electron transfer, may be taking place in some cases: Olah; Krishnamurthy *J. Am. Chem. Soc.* **1982**, *104*, 3987; Yamataka; Nagareda; Hanafusa; Nagase *Tetrahedron Lett.* **1989**, *30*, 7187. A diradical mechanism has also been proposed for certain cases: Ward; McEwen *J. Org. Chem.* **1990**, *55*, 493.

so, but there is little or no evidence for it,[666] though many attempts have been made to find it. "Betaine" precipitates have been isolated in certain Wittig reactions,[667] but these are betaine–lithium halide adducts, and might just as well have been formed from the oxaphosphetane as from a true betaine.[668] In contrast, there is much evidence for the presence of the oxaphosphetane intermediates, at least with unstable ylides. For example,[31]P nmr spectra taken of the reaction mixtures at low temperatures[669] are compatible with an oxaphosphetane structure that persists for some time but not with a tetracoordinated phosphorus species. Since a betaine, an ylide, and a phosphine oxide all have tetracoordinated phosphorus, these species could not be causing the spectra, leading to the conclusion that an oxaphosphetane intermediate is present in the solution. In certain cases oxaphosphetanes have been isolated.[670] It has even been possible to detect cis and trans isomers of the intermediate oxaphosphetanes by nmr spectroscopy.[671] According to this mechanism, an optically active phosphonium salt $RR'R''\overset{\oplus}{P}CHR_2$ should retain its configuration all the way through the reaction, and it should be preserved in the phosphine oxide $RR'R''PO$. This has been shown to be the case.[672]

The proposed betaine intermediates can be formed, in a completely different manner, by nucleophilic substitution by a phosphine on an epoxide (**0-49**):

$$Ph_3\overset{\frown}{P} + \underset{O}{\overset{|}{\underset{}{C}}}\!-\!\overset{|}{C}\!- \longrightarrow \overset{Ph_3P^{\oplus}}{\underset{\underset{|\underline{O}|_{\ominus}}{|}}{\overset{|}{C}}}\!-\!\overset{|}{\underset{|}{C}}\!-$$

Betaines formed in this way can then be converted to the olefin, and this is one reason why betaine intermediates were long accepted in the Wittig reaction.

Some Wittig reactions give the Z olefin; some the E, and others give mixtures, and the question of which factors determine the stereoselectivity has been much studied.[673] It is generally found that ylides containing stabilizing groups or formed from trialkylphosphines give E olefins. However, ylides formed from triarylphosphines and not containing stabilizing groups often give Z or a mixture of Z and E olefins.[674] One explanation for this[669] is that the reaction of the ylide with the carbonyl compound is a 2 + 2 cycloaddition, which in order to be concerted must adopt the $[_\pi 2_s + _\pi 2_a]$ pathway. As we have seen earlier (p. 858), this pathway leads to the formation of the more sterically crowded product, in this case the Z olefin. If this explanation is correct, it is not easy to explain the predominant formation of E products from stable ylides, but E compounds are of course generally thermodynamically more stable than the Z isomers, and the stereochemistry seems to depend on many factors.

[666]See Vedejs; Marth *J. Am. Chem. Soc.* **1990**, *112*, 3905.

[667]Wittig; Weigmann; Schlosser *Chem. Ber.* **1961**, *94*, 676; Schlosser; Christmann *Liebigs Ann. Chem.* **1967**, *708*, 1.

[668]Maryanoff; Reitz, Ref. 638, p. 865.

[669]Vedejs; Snoble *J. Am. Chem. Soc.* **1973**, *95*, 5778; Vedejs; Meier; Snoble *J. Am. Chem. Soc.* **1981**, *103*, 2823. See also Nesmayanov; Binshtok; Reutov *Doklad. Chem.* **1973**, *210*, 499.

[670]Birum; Matthews *Chem. Commun.* **1967**, 137; Mazhar-Ul-Haque; Caughlan; Ramirez; Pilot; Smith *J. Am. Chem. Soc.* **1971**, *93*, 5229.

[671]Maryanoff; Reitz; Mutter; Inners; Almond; Whittle; Olofson *J. Am. Chem. Soc.* **1986**, *108*, 7664. See also Pískala; Rchan; Schlosser *Coll. Czech. Chem. Commun.* **1983**, *48*, 3539.

[672]McEwen; Kumli; Bladé-Font; Zanger; VanderWerf *J. Am. Chem. Soc.* **1964**, *86*, 2378.

[673]For reviews of the stereochemistry of the Wittig reactions, see Maryanoff; Reitz, Ref. 638; Gosney; Rowley, in Cadogan, Ref. 638, pp. 17-153; Reucroft; Sammes *Q. Rev., Chem. Soc.* **1971**, *25*, 135-169, pp. 137-148, 169; Schlosser *Top. Stereochem.* **1970**, *5*, 1-30.

[674]For cases where such an ylide gave E olefins, see Maryanoff; Reitz; Duhl-Emswiler *J. Am. Chem. Soc.* **1985**, *107*, 217; Le Bigot; El Gharbi; Delmas; Gaset *Tetrahedron* **1986**, *42*, 3813. For guidance in how to obtain the maximum yields of the Z product, see Schlosser; Schaub; de Oliveira-Neto; Jeganathan *Chimia* **1986**, *40*, 244.

The E:Z ratio of the product can often be changed by a change in solvent or by the addition of salts.[675] Another way of controlling the stereochemistry of the product is by use of the aforementioned phosphonic acid bisamides. In this case the betaine (**55**) does form

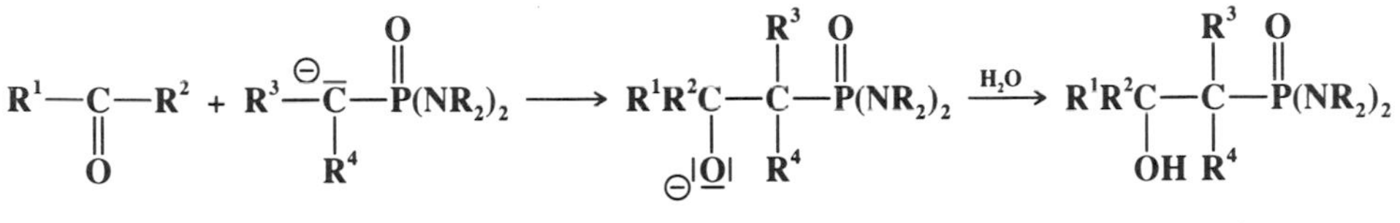

$$R^1-\overset{\overset{\displaystyle O}{\|}}{C}-R^2 \;+\; R^3-\overset{\ominus}{\underset{\underset{\displaystyle R^4}{|}}{C}}-\overset{\overset{\displaystyle O}{\|}}{P}(NR_2)_2 \longrightarrow R^1R^2C-\overset{\overset{\displaystyle R^3}{|}}{\underset{\underset{\displaystyle \ominus|\underset{\displaystyle R^4}{\overset{O}{|}}|}{}}{C}}-\overset{\overset{\displaystyle O}{\|}}{P}(NR_2)_2 \;\xrightarrow{H_2O}\; R^1R^2C-\overset{\overset{\displaystyle R^3}{|}}{\underset{\underset{\displaystyle \overset{|}{OH}\;\underset{\displaystyle R^4}{|}}{}}{C}}-\overset{\overset{\displaystyle O}{\|}}{P}(NR_2)_2$$

55   **56**

and when treated with water gives the β-hydroxyphosphonic acid bisamides **56,** which can be crystallized and then cleaved to $R^1R^2C{=}CR^3R^4$ by refluxing in benzene or toluene in the presence of silica gel.[661] **56** are generally formed as mixtures of diastereomers, and these mixtures can be separated by recrystallization. Cleavage of the two diastereomers gives the two isomeric olefins. Optically active phosphonic acid bisamides have been used to give optically active olefins.[676] Another method of controlling the stereochemistry of the olefin (to obtain either the Z or E isomer) starting with a phosphine oxide Ph_2POCH_2R, has been reported.[677]

In reactions where the betaine–lithium halide intermediate is present, it is possible to extend the chain further if a hydrogen is present α to the phosphorus. For example, reaction of ethylidenetriphenylphosphorane with heptanal at $-78°C$ gave **57,** which with butyllithium gave the ylide **58.** Treatment of this with an aldehyde R'CHO gave the intermediate **59,**

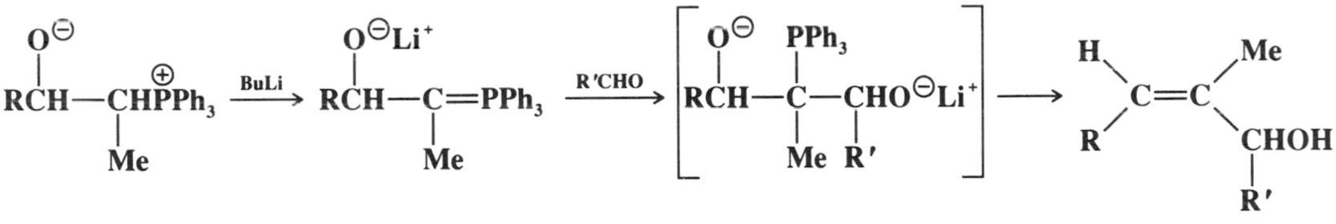

$$R = n\text{-}C_6H_{13}$$

which after workup gave **60.**[678] This reaction gives the unsaturated alcohols **60** stereoselectively. **58** also reacts with other electrophiles. For example, treatment of **58** with N-chlorosuccinimide or $PhICl_2$ gives the vinylic chloride RCH=CMeCl stereoselectively: NCS giving the cis and $PhICl_2$ the trans isomer.[679] The use of Br_2 and $FClO_3$ (see **2-4** for the explosive nature of this reagent) gives the corresponding bromides and fluorides, respectively.[680] Reactions of **58** with electrophiles have been called *scoopy* reactions (α substitution plus carbonyl *o*lefination via β-*o*xido *p*hosphorus *y*lides).[681]

[675]See, for example, Reitz; Nortey; Jordan; Mutter; Maryanoff *J. Org. Chem.* **1986,** *51,* 3302.

[676]Hanessian; Delorme; Beaudoin; Leblanc *J. Am. Chem. Soc.* **1984,** *106,* 5754.

[677]Buss; Warren *J. Chem. Soc., Perkin Trans. 1* **1985,** 2307; Ayrey; Warren *Tetrahedron Lett.* **1989,** *30,* 4581.

[678]Corey; Yamamoto *J. Am. Chem. Soc.* **1970,** *92,* 226; Schlosser; Christmann; Piskala; Coffinet *Synthesis* **1971,** 29; Schlosser; Coffinet *Synthesis* 380, **1971, 1972,** 575; Corey; Ulrich; Venkateswarlu *Tetrahedron Lett.* **1977,** 3231; Schlosser; Tuong; Respondek; Schaub *Chimia* **1983,** *37,* 10.

[679]Schlosser; Christmann *Synthesis* **1969,** 38; Corey; Shulman; Yamamoto *Tetrahedron Lett.* **1970,** 447.

[680]Schlosser; Christmann, Ref. 679.

[681]Schlosser, Ref. 673, p. 22.

The Wittig reaction has been carried out intramolecularly, to prepare rings containing from 5 to 16 carbons,[682] both by single ring closure

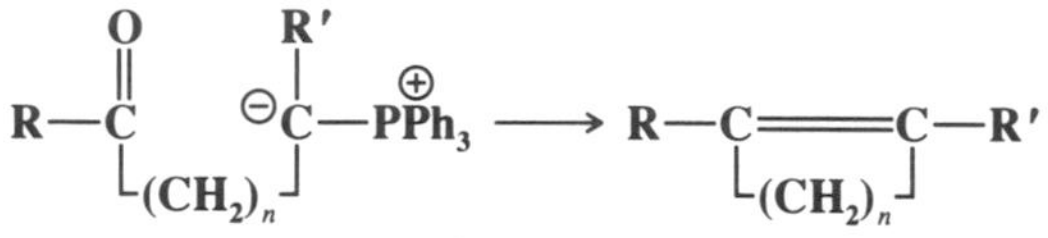

$$R-\underset{\underset{(CH_2)_n}{\big|}}{\overset{\overset{O}{\|}}{C}} \quad \overset{\oplus}{\underset{\underset{(CH_2)_n}{\big|}}{\overset{R'}{C}}}-PPh_3 \longrightarrow R-\underset{\underset{(CH_2)_n}{\big|}}{C}\!=\!=\!\overset{R'}{C}-R'$$

and double ring closure.[683]

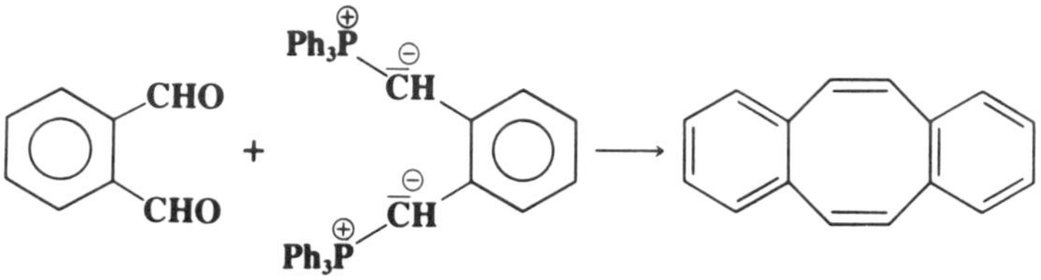

The Wittig reaction has proved very useful in the synthesis of natural products, some of which are quite difficult to prepare in other ways.[684] One example out of many is the synthesis of β-carotene:[685]

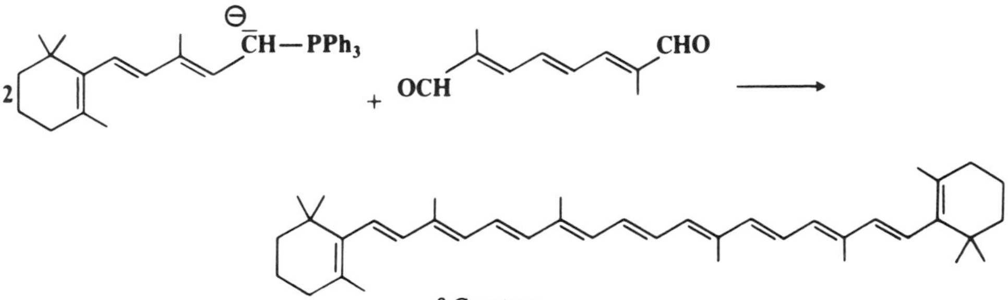

β-Carotene

Phosphorus ylides also react in a similar manner with the C=O bonds of ketenes,[686] isocyanates,[687] and certain anhydrides[688] and imides,[689] the N=O of nitroso groups, and the C=N of imines.[690]

[682]For a review, see Becker *Tetrahedron* **1980**, *36*, 1717-1745.
[683]For a review of these double ring closures, see Vollhardt *Synthesis* **1975**, 765-780.
[684]For a review of applications of the Wittig reaction to the synthesis of natural products, see Bestmann; Vostrowsky, Ref. 638.
[685]Wittig; Pommer; German patent **1956**, 954,247, *CA* **1959**, *53*, 2279.
[686]For example, see Aksnes; Frøyen *Acta Chem. Scand.* **1968**, *22*, 2347.
[687]For example, see Frøyen *Acta Chem. Scand., Ser. B* **1974**, *28*, 586.
[688]See, for example, Abell; Massy-Westropp *Aust. J. Chem.* **1982**, *35*, 2077; Kayser; Breau *Can. J. Chem.* **1989**, *67*, 1401. For a study of the mechanism, see Abell; Clark; Robinson *Aust. J. Chem.* **1988**, *41*, 1243.
[689]For a review of the reactions with anhydrides and imides (and carboxylic esters, thiol esters, and amides), see Murphy; Brennan *Chem. Soc. Rev.* **1988**, *17*, 1-30. For a review with respect to imides, see Flitsch; Schindler *Synthesis* **1975**, 685-700.
[690]Bestmann; Seng *Tetrahedron* **1965**, *21*, 1373.

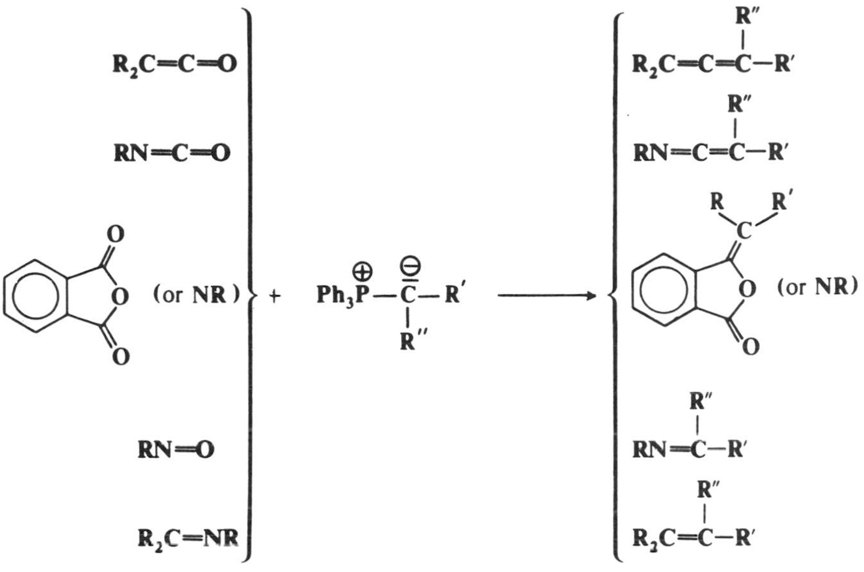

Phosphorus ylides react with carbon dioxide to give the isolable salts **61**,[691] which can be hydrolyzed to the carboxylic acids **62** (thus achieving the conversion RR'CHX →

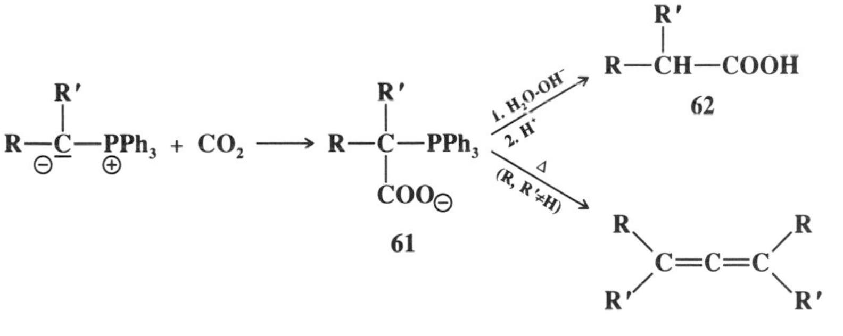

RR'CHCOOH) or (if neither R nor R' is hydrogen) dimerized to allenes.

OS **V**, 361, 390, 499, 509, 547, 751, 949, 985; **VI**, 358; **VII**, 164, 232; **65**, 119; **66**, 220.

6-48   The Thorpe Reaction
N-Hydro-C-(α-cyanoalkyl)-addition

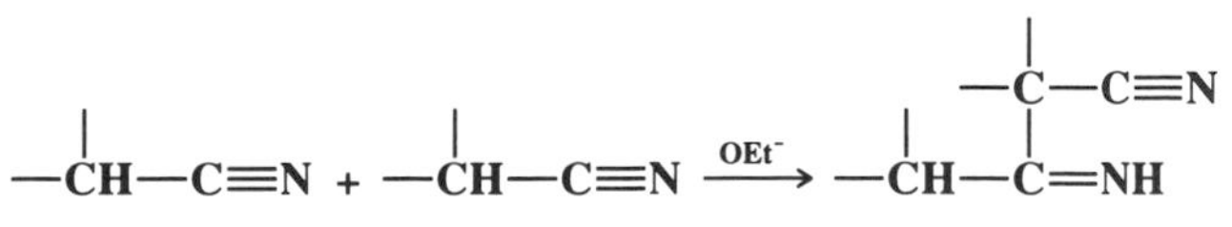

In the *Thorpe reaction*, the α carbon of one nitrile molecule is added to the CN carbon of another, so this reaction is analogous to the aldol reaction (**6-39**). The C=NH bond is, of

---

[691]Bestmann; Denzel; Salbaum *Tetrahedron Lett* **1974**, 1275.

course, hydrolyzable (**6-2**), so β-keto nitriles can be prepared in this manner. The Thorpe reaction can be done internally, in which case it is called the *Thorpe–Ziegler reaction*.[692] This is a useful method for closing large rings. Yields are high for five- to eight-membered rings, fall off to about zero for rings of nine to thirteen members, but are high again for fourteen-membered and larger rings, if high-dilution techniques are employed. The product

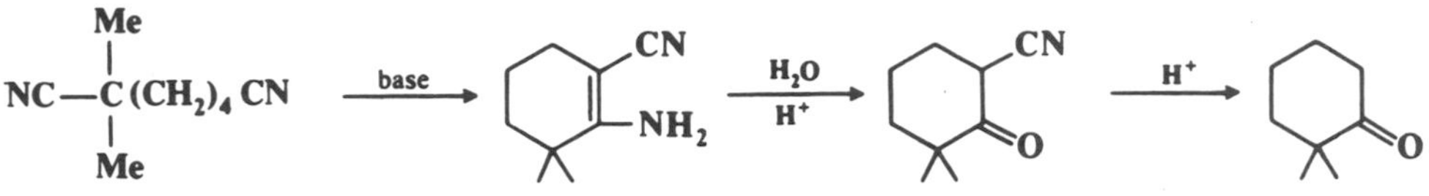

63

in the Thorpe–Ziegler reaction is not the imine, but the tautomeric enamine, e.g., **63**; if desired this can be hydrolyzed to an α-cyano ketone (**6-2**), which can in turn be hydrolyzed and decarboxylated (**6-5, 2-40**). Other active-hydrogen compounds can also be added to nitriles.[693]

OS **VI**, 932.

J.   Other Carbon Nucleophiles

6-49   The Formation of Cyanohydrins
       O-**Hydro-*C*-cyano-addition**

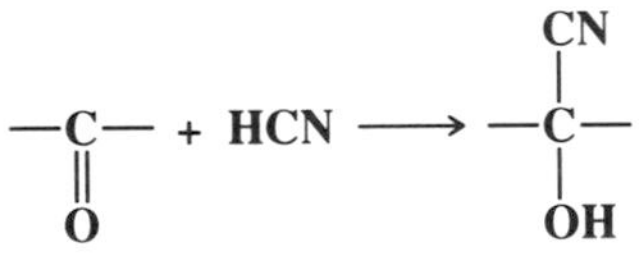

The addition of HCN to aldehydes or ketones produces cyanohydrins.[694] This is an equilibrium reaction. For aldehydes and aliphatic ketones the equilibrium lies to the right; therefore the reaction is quite feasible, except with sterically hindered ketones such as diisopropyl ketone. However, ketones ArCOR give poor yields, and the reaction cannot be carried out with ArCOAr since the equilibrium lies too far to the left. With aromatic aldehydes the benzoin condensation (**6-54**) competes. With α,β-unsaturated aldehydes and ketones, 1,4 addition competes (**5-25**). Ketones of low reactivity, such as ArCOR, can be converted to cyanohydrins by treatment with diethylaluminum cyanide Et_2AlCN (see OS **VI**, 307) or, indirectly, with cyanotrimethylsilane Me_3SiCN[695] in the presence of a Lewis acid or base,[695a] followed by hydrolysis of the resulting O-trimethylsilyl cyanohydrin **64**. When $TiCl_4$ is used,

[692]For a monograph, see Taylor; McKillop *The Chemistry of Cyclic Enaminonitriles and ortho-Amino Nitriles*; Wiley: New York, 1970. For a review, see Schaefer; Bloomfield, *Org. React.* **1967**, *15*, 1-203.

[693]See for example, Josey *J. Org. Chem.* **1964**, *29*, 707; Barluenga; Fustero; Rubio; Gotor *Synthesis* **1977**, 780; Hiyama; Kobayashi *Tetrahedron Lett.* **1982**, *23*, 1597; Gewald; Bellmann; Jänsch *Liebigs Ann. Chem.* **1984**, 1702; Page; van Niel; Westwood *J. Chem. Soc., Perkin Trans. 1* **1988**, 269.

[694]For reviews, see Friedrich, in Patai; Rappoport *The Chemistry of Functional Groups, Supplement C*, pt. 2; Wiley: New York, 1983, pp. 1345-1390; Friedrich; Wallenfels, in Rappoport, Ref. 334, pp. 72-77.

[695]For reviews of Me₃SiCN and related compounds, see Rasmussen; Heilmann; Krepski *Adv. Silicon Chem.* **1991**, *1*, 65-187; Groutas; Felker *Synthesis* **1980**, 861-868. For procedures using Me₃SiCl and CN⁻ instead of Me₃SiCN, see Yoneda; Santo; Harusawa; Kurihara *Synthesis* **1986**, 1054; Sukata *Bull. Chem. Soc. Jpn.* **1987**, *60*, 3820.

[695a]Kobayashi; Tsuchiya; Mukaiyama *Chem. Lett.* **1991**, 537.

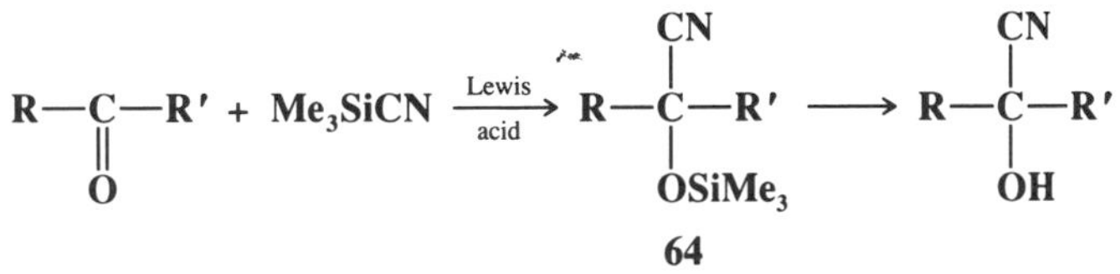

$$\underset{\underset{O}{\|}}{R-C-R'} + Me_3SiCN \xrightarrow[\text{acid}]{\text{Lewis}} \underset{\underset{OSiMe_3}{|}}{\overset{\overset{CN}{|}}{R-C-R'}} \longrightarrow \underset{\underset{OH}{|}}{\overset{\overset{CN}{|}}{R-C-R'}}$$

64

the reaction between Me₃SiCN and aromatic aldehydes or ketones gives α-chloro nitriles Cl—CRR'—CN.[696]

Frequently it is the bisulfite addition product that is treated with CN⁻. This method is especially useful for aromatic aldehydes, since it avoids competition from the benzoin condensation. If desired, it is possible to hydrolyze the cyanohydrin in situ to the corresponding α-hydroxy acid. This reaction is important in the *Kiliani–Fischer* method of extending the carbon chain of a sugar.

The addition is nucleophilic and the actual nucleophile is CN⁻, so the reaction rate is increased by the addition of base.[697] This was demonstrated by Lapworth in 1903, and consequently this was one of the first organic mechanisms to be known.[698]

The reaction has been carried out enantioselectively: optically active cyanohydrins were prepared with the aid of optically active catalysts.[699]

OS **I**, 336; **II**, 7, 29, 387; **III**, 436; **IV**, 58, 506; **VI**, 307; **VII**, 20, 381, 517, 521. For the reverse reaction, see OS **III**, 101.

6-50 The Strecker Synthesis
Cyano,amino-de-oxo-bisubstitution

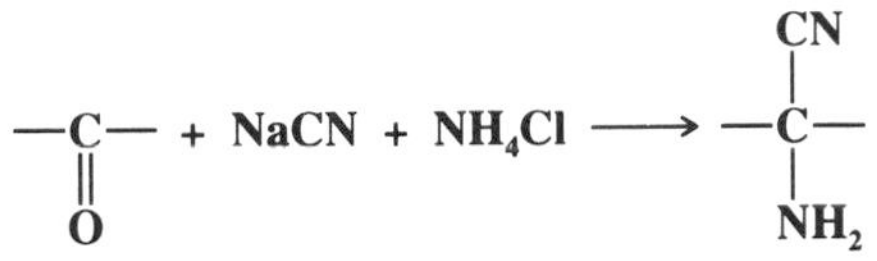

$$\underset{\underset{O}{\|}}{-C-} + NaCN + NH_4Cl \longrightarrow \underset{\underset{NH_2}{|}}{\overset{\overset{CN}{|}}{-C-}}$$

α-Amino nitriles[700] can be prepared in one step by the treatment of an aldehyde or ketone with NaCN and NH₄Cl. This is called the *Strecker synthesis*;[700a] it is a special case of the Mannich reaction (**6-16**). Since the CN is easily hydrolyzed to the acid, this is a convenient method for the preparation of α-amino acids. The reaction has also been carried out with NH₃ + HCN and with NH₄CN. Salts of primary and secondary amines can be used instead of NH₄⁺ to obtain N-substituted and N,N-disubstituted α-amino nitriles. Unlike **6-49**, the Strecker synthesis is useful for aromatic as well as aliphatic ketones. As in **6-49**, the Me₃SiCN method has been used; **64** is converted to the product with ammonia or an amine.[701]

OS **I**, 21, 355; **III**, 66, 84, 88, 275; **IV**, 274; **V**, 437; **VI**, 334.

[696]Kiyooka; Fujiyama; Kawaguchi *Chem. Lett.* **1984**, 1979.

[697]For a review, see Ogata; Kawasaki, in Zabicky *The Chemistry of the Carbonyl Group*, vol. 2, Wiley: New York, 1970, pp. 21-32. See also Okano; do Amaral; Cordes *J. Am. Chem. Soc.* **1976**, *98*, 4201; Ching; Kallen *J. Am. Chem. Soc.* **1978**, *100*, 6119.

[698]Lapworth *J. Chem. Soc.* **1903**, *83*, 998.

[699]See Minamikawa; Hayakawa; Yamada; Iwasawa; Narasaka *Bull. Chem. Soc. Jpn.* **1988**, *61*, 4379; Jackson; Jayatilake; Matthews; Wilshire *Aust. J. Chem.* **1988**, *41*, 203; Garner; Fernández; Gladysz *Tetrahedron Lett.* **1989**, *30*, 3931; Mori; Ikeda; Kinoshita; Inoue *Chem. Lett.* **1989**, 2119; Kobayashi; Tsuchiya; Mukaiyama *Chem. Lett.* **1991**, 541, and references cited in these papers.

[700]For a review of α-amino nitriles, see Shafran; Bakulev; Mokrushin *Russ. Chem. Rev.* **1989**, *58*, 148-162.

[700a]For a review of asymmetric Strecker syntheses, see Williams *Synthesis of Optically Active α-Amino Acids*; Pergamon: Elmsford, NY, 1989, pp. 208-229.

[701]See Mai; Patil *Tetrahedron Lett.* **1984**, *25*, 4583, *Synth. Commun.* **1985**, *15*, 157.

6-51 The Addition of HCN to C=N and C≡N Bonds
N-Hydro-C-cyano-addition

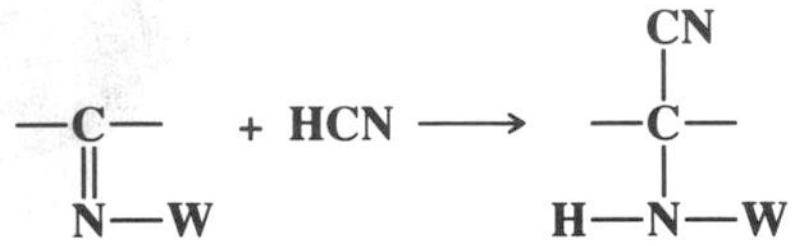

$$
\overset{\displaystyle |}{\underset{\displaystyle \underset{\displaystyle |}{N-W}}{-C-}} + \text{HCN} \longrightarrow \overset{\displaystyle \overset{\displaystyle CN}{|}}{\underset{\displaystyle \underset{\displaystyle |}{H-N-W}}{-C-}} \qquad \text{W = H, R, Ar, OH, NHAr, etc.}
$$

HCN adds to imines, Schiff bases, hydrazones, oximes, and similar compounds. CN⁻ can be added to iminium ions:[330]

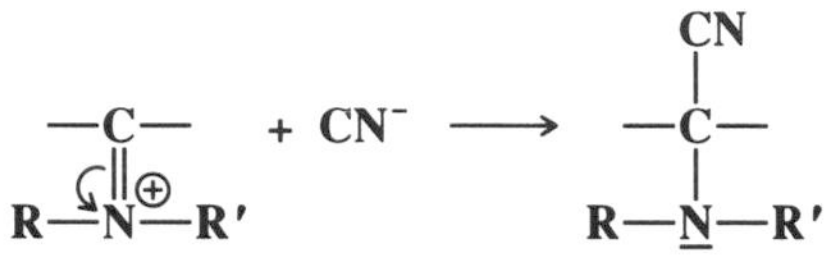

$$
\overset{\displaystyle |}{\underset{\displaystyle \underset{\oplus}{R-N-R'}}{-C-}} + \text{CN}^- \longrightarrow \overset{\displaystyle \overset{\displaystyle CN}{|}}{\underset{\displaystyle \underset{\displaystyle |}{R-N-R'}}{-C-}}
$$

As in **6-48**, the addition to imines has been carried out enantioselectively.[702]

The addition of KCN to triisopropylbenzenesulfonyl hydrazones **65** provides an indirect method for achieving the conversion RR′CO → RR′CHCN.[703] The reaction is successful for hydrazones of aliphatic aldehydes and ketones.

$$
\underset{\textbf{65}}{\text{RR'C=NNHSO}_2\text{Ar}} + \text{KCN} \xrightarrow{\text{MeOH}} \text{RR'CHCN} \qquad \text{Ar = 2,4,6-(i-Pr)}_3\text{C}_6\text{H}_2
$$

HCN can also be added to the C≡N bond to give iminonitriles or α-aminomalononitriles.[704]

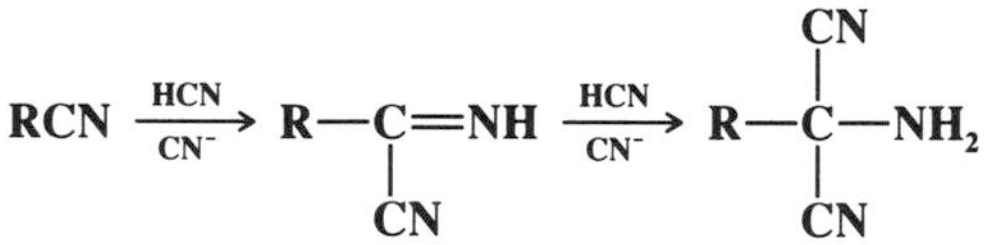

$$
\text{RCN} \xrightarrow[\text{CN}^-]{\text{HCN}} \underset{\underset{\displaystyle CN}{|}}{\text{R-C=NH}} \xrightarrow[\text{CN}^-]{\text{HCN}} \underset{\underset{\displaystyle CN}{|}}{\overset{\overset{\displaystyle CN}{|}}{\text{R-C-NH}_2}}
$$

OS **V**, 344. See also OS **V**, 269.

6-52 The Addition of CO₂ to Aldehydes and Ketones
O-Hydro-C-carboxyl-addition

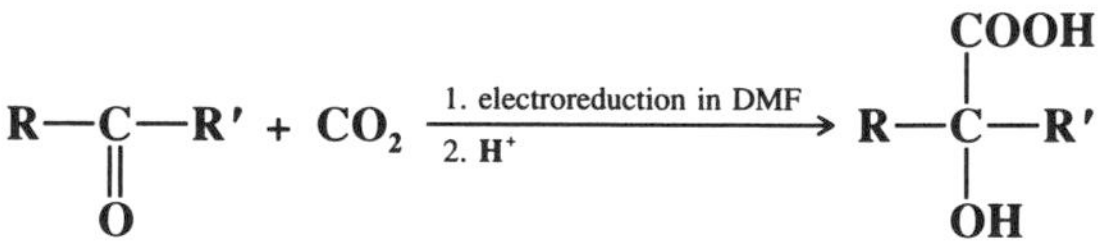

$$
\underset{\underset{\displaystyle O}{\parallel}}{\text{R-C-R'}} + \text{CO}_2 \xrightarrow[\text{2. H}^+]{\text{1. electroreduction in DMF}} \underset{\underset{\displaystyle OH}{|}}{\overset{\overset{\displaystyle COOH}{|}}{\text{R-C-R'}}}
$$

[702]Saito; Harada *Tetrahedron Lett.* **1989**, *30*, 4535.

[703]Jiricny; Orere; Reese *J. Chem. Soc., Perkin Trans. 1* **1980**, 1487. For other methods of achieving this conversion, see Ziegler; Wender *J. Org. Chem.* **1977**, *42*, 2001; Cacchi; Caglioti; Paolucci *Synthesis* **1975**, 120; Yoneda; Harusawa; Kurihara *Tetrahedron Lett.* **1989**, *30*, 3681; Okimoto; Chiba *J. Org. Chem.* **1990**, *55*, 1070.

[704]For an example, see Ferris; Sanchez *Org. Synth.* V. 344.

Aromatic aldehydes and ketones have been converted to α-hydroxy acids by electrolysis carried out in the presence of CO_2 in DMF, followed by hydrolysis.[705] Yields were moderate to high.

Addition of ArH to C=O, C=N, and C≡N bonds is discussed under aromatic substitution: **1-16**, **1-20** to **1-25**, **1-27**, and **1-28**.

6-53 The Prins Reaction

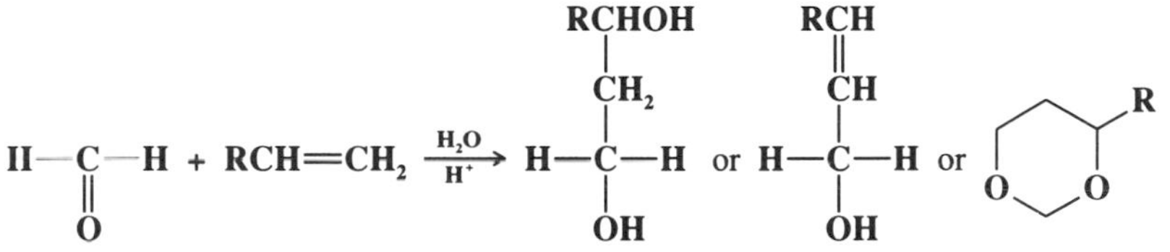

$$
\text{II—C—H} + \text{RCH=CH}_2 \xrightarrow[\text{H}^+]{\text{H}_2\text{O}}
\begin{array}{c}\text{RCHOH}\\ \| \\ \text{CH}_2 \\ \text{H—C—H} \\ \text{OH}\end{array}
\text{ or }
\begin{array}{c}\text{RCH}\\ \| \\ \text{CH} \\ \text{H—C—H} \\ \text{OH}\end{array}
\text{ or }
\text{(dioxane with R)}
$$

The addition of an olefin to formaldehyde in the presence of an acid[706] catalyst is called the *Prins reaction*.[707] Three main products are possible; which one predominates depends on the olefin and the conditions. When the product is the 1,3-diol or the dioxane,[708] the reaction involves addition to the C=C as well as to the C=O. The mechanism is one of electrophilic attack on both double bonds. The acid first protonates the C=O, and the resulting carbocation attacks the C=C:

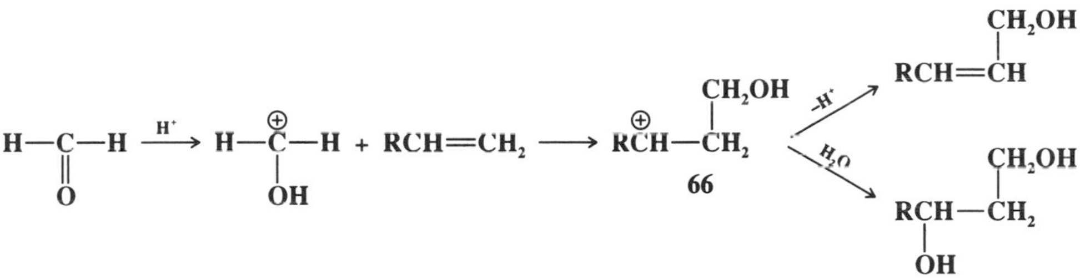

$$
\text{H—C—H} \xrightarrow{\text{H}^+} \text{H—C—H} + \text{RCH=CH}_2 \longrightarrow \overset{\oplus}{\text{RCH—CH}_2} \quad \mathbf{66}
$$

66 can undergo loss of H^+ to give the olefin or add water to give the diol.[709] It has been proposed that **66** is stabilized by neighboring-group attraction, with either the oxygen[710] or

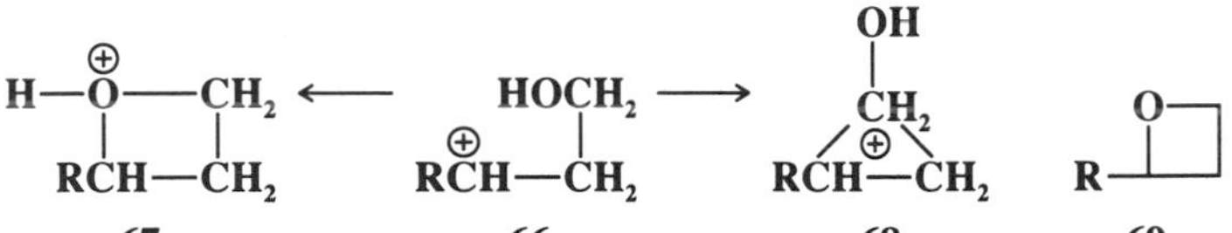

$$
\begin{array}{ccccc}
\overset{\oplus}{\text{H—O—CH}_2} & \leftarrow & \text{HOCH}_2 & \rightarrow & \text{CH}_2 \\
| \quad\quad | & & \overset{\oplus}{|} \quad\quad | & & \\
\text{RCH—CH}_2 & & \text{RCH—CH}_2 & & \text{RCH—CH}_2 \\
\mathbf{67} & & \mathbf{66} & & \mathbf{68} \quad\quad \mathbf{69}
\end{array}
$$

[705]Mcharek; Heintz; Troupel; Perichon *Bull. Soc. Chim. Fr.* **1989**, 95.

[706]The Prins reaction has also been carried out with basic catalysts: Griengl; Sieber *Monatsh. Chem.* **1973**, *104*, 1008, 1027.

[707]For reviews, see Adams; Bhatnagar *Synthesis* **1977**, 661-672; Isagulyants; Khaimova; Melikyan; Pokrovskaya *Russ. Chem. Rev.* **1968**, *37*, 17-25. For a list of references, see Ref. 64, p. 125.

[708]The reaction to produce dioxanes has also been carried out with equimolar mixtures of formaldehyde and another aldehyde RCHO. The R appears in the dioxane on the carbon between the two oxygens: Safarov; Nigmatullin; Ibatullin; Rafikov *Doklad. Chem.* **1977**, *236*, 507.

[709]Hellin; Davidson; Coussemant *Bull. Soc. Chim. Fr.* **1966**, 1890, 3217.

[710]Blomquist; Wolinsky *J. Am. Chem. Soc.* **1957**, *79*, 6025; Schowen; Smissman; Schowen *J. Org. Chem.* **1968**, *33*, 1873.

a carbon[711] stabilizing the charge (**67** and **68,** respectively). This stabilization is postulated to explain the fact that with 2-butenes[712] and with cyclohexenes the addition is anti. A backside attack of H_2O on the three- or four-membered ring would account for it. Other products are obtained too, which can be explained on the basis of **67** or **68.**[710,711] Additional evidence for the intermediacy of **67** is the finding that oxetanes (**69**) subjected to the reaction conditions (which would protonate **69** to give **67**) give essentially the same product ratios as the corresponding alkenes.[713] An argument against the intermediacy of **67** and **68** is that not all alkenes show the anti stereoselectivity mentioned above. Indeed, the stereochemical results are often quite complex, with syn, anti, and nonstereoselective addition reported, depending on the nature of the reactants and the reaction conditions.[714] Since addition to the C=C bond is electrophilic, the reactivity of the olefin increases with alkyl substitution and Markovnikov's rule is followed. The dioxane product may arise from a reaction between the 1,3-diol and formaldehyde[715] (**6-6**) or between **66** and formaldehyde.

Lewis acids such as $SnCl_4$ also catalyze the reaction, in which case the species that adds to the olefins is $H_2\overset{\oplus}{C}-O-\overset{\ominus}{SnCl_4}$.[716] The reaction can also be catalyzed by peroxides, in which case the mechanism is probably a free-radical one.

A closely related reaction has been performed with other aldehydes and even with ketones; without a catalyst, but with heat.[717] The aldehydes and ketones here are active ones, such as chloral and acetoacetic ester. The product in these cases is a β-hydroxy olefin, and the mechanism is pericyclic:[718]

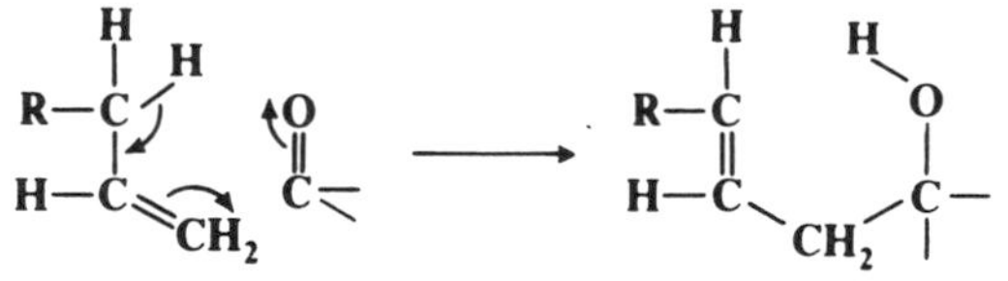

This reaction is reversible and suitable β-hydroxy olefins can be cleaved by heat (**7-43**). There is evidence that the cleavage reaction occurs by a cyclic mechanism (p. 1043), and, by the principle of microscopic reversibility, the addition mechanism should be cyclic too.[719] Note that this reaction is an oxygen analog of the ene synthesis (**5-16**). This reaction can also be done with unactivated aldehydes[720] and ketones[721] if Lewis-acid catalysts such as

[711]Dolby; Lieske; Rosencrantz; Schwarz *J. Am. Chem. Soc.* **1963,** *85,* 47; Dolby; Schwarz *J. Org. Chem.* **1963,** *28,* 1456; Safarov; Isagulyants; Nigmatullin *J. Org. Chem. USSR* **1974,** *10,* 1378.

[712]Fremaux; Davidson; Hellin; Coussemant *Bull. Soc. Chim. Fr.* **1967,** 4250.

[713]Meresz; Leung; Denes *Tetrahedron Lett.* **1972,** 2797.

[714]For example, see LeBel, Liesemer; Mehmedbasich *J. Org. Chem.* **1963,** *28,* 615; Portoghese; Smissman *J. Org. Chem.* **1962,** *27,* 719; Wilkins; Marianelli *Tetrahedron* **1970,** *26,* 4131; Karpaty; Hellin; Davidson; Coussemant *Bull. Soc. Chim. Fr.* **1971,** 1736; Coryn; Anteunis *Bull. Soc. Chim. Belg.* **1974,** *83,* 83.

[715]Ref. 709; Isagulyants; Isagulyants; Khairudinov; Rakhmankulov *Bull. Acad. Sci. USSR. Div. Chem. Sci* **1973,** *22,* 1810; Sharf; Kheifets; Freidlin *Bull. Acad. Sci. USSR, Div. Chem. Sci* **1974,** *23,* 1681.

[716]Yang; Yang; Ross *J. Am. Chem. Soc.* **1959,** *81,* 133.

[717]Arnold; Veeravagu *J. Am. Chem. Soc.* **1960,** *82,* 5411; Klimova; Abramov; Antonova; Arbuzov *J. Org. Chem. USSR* **1969,** *5,* 1308; Klimova; Antonova; Arbuzov *J. Org. Chem. USSR* **1969,** *5,* 1312, 1315.

[718]See for example, Achmatowicz; Szymoniak *J. Org. Chem.* **1980,** *45,* 1228; Ben Salem; Jenner *Tetrahedron Lett.* **1986,** *27,* 1575. There is evidence that the mechanism is somewhat more complicated than shown here: Kwart; Brechbiel *J. Org. Chem.* **1982,** *47,* 3353.

[719]For other evidence, see Ref. 718; Papadopoulos; Jenner *Tetrahedron Lett.* **1981,** *22,* 2773.

[720]Snider *Acc. Chem. Res.* **1980,** *13,* 426-432; Snider; Phillips *J. Org. Chem.* **1983,** *48,* 464; Cartaya-Marin; Jackson; Snider *J. Org. Chem.* **1984,** *49,* 2443.

[721]Jackson; Goldman; Snider *J. Org. Chem.* **1984,** *49,* 3988.

dimethylaluminum chloride Me_2AlCl or ethylaluminum dichloride $EtAlCl_2$ are used.[722] Lewis acid catalysts also increase rates with activated aldehydes.[723] The use of optically active catalysts has given optically active products with high enantiomeric excesses.[724]

In a related reaction, alkenes can be added to aldehydes and ketones to give reduced alcohols **70**. This has been accomplished by several methods,[725] including treatment with

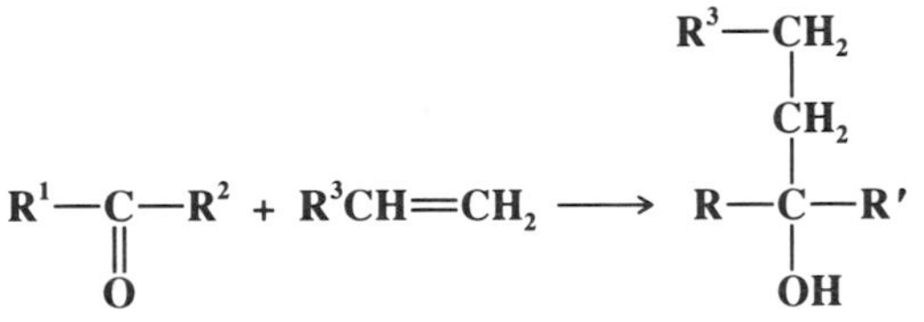

$$R^1\!-\!\underset{\underset{O}{\|}}{C}\!-\!R^2 + R^3CH\!=\!CH_2 \longrightarrow R\!-\!\underset{\underset{OH}{|}}{\overset{\overset{R^3\!-\!CH_2}{|}\atop\overset{CH_2}{|}}{C}}\!-\!R'$$

70

SmI_2[726] or Zn and Me_3SiCl,[727] and by electrochemical[728] and photochemical[729] methods. Most of these methods have been used for intramolecular addition and most or all involve free radical intermediates.

OS **IV**, 786. See also OS **VII**, 102.

6-54   The Benzoin Condensation
Benzoin aldehyde condensation

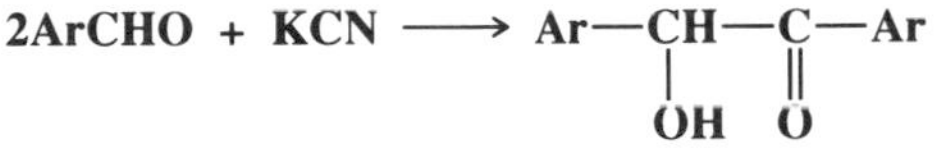

$$2ArCHO + KCN \longrightarrow Ar\!-\!\underset{\underset{OH}{|}}{C}H\!-\!\underset{\underset{O}{\|}}{C}\!-\!Ar$$

When certain aldehydes are treated with cyanide ion, *benzoins* are produced in a reaction called the *benzoin condensation*. The condensation can be regarded as involving the addition of one molecule of aldehyde to the C=O group of another. The reaction can be accomplished only for aromatic aldehydes, though not for all of them,[730] and for glyoxals RCOCHO. The two molecules of aldehyde obviously perform different functions. The one that no longer has a C—H bond in the product is called the *donor*, because it has "donated" its hydrogen to the oxygen of the other molecule, the *acceptor*. Some aldehydes can perform only one of these functions and hence cannot be self-condensed, though they can often be condensed with a different aldehyde. For example, *p*-dimethylaminobenzaldehyde is not an acceptor but only a donor. Thus it cannot condense with itself, but it can condense with benzaldehyde, which can perform both functions, but is a better acceptor than it is a donor.

---

[722]For discussions of the mechanism with Lewis-acid catalysts, see Stephenson; Orfanopoulos *J. Org. Chem.* **1981**, *46*, 2200; Kwart; Brechbiel *J. Org. Chem.* **1982**, *47*, 5409; Song; Beak *J. Org. Chem.* **1990**, *112*, 8126.
[723]Benner; Gill; Parrott; Wallace *J. Chem. Soc., Perkin Trans. 1* **1984**, 291, 315, 331.
[724]Maruoka; Hoshino; Shirasaka; Yamamoto *Tetrahedron Lett.* **1988**, *29*, 3967; Mikami; Terada; Nakai *J. Am. Chem. Soc.* **1990**, *112*, 3949.
[725]For references, see Ujikawa; Inanaga; Yamaguchi *Tetrahedron Lett.* **1989**, *30*, 2837; Ref. 64, pp. 575-576.
[726]Ujikawa et al., Ref. 725.
[727]Corey; Pyne *Tetrahedron Lett.* **1983**, *24*, 2821.
[728]See Shono; Kashimura; Mori; Hayashi; Soejima; Yamaguchi *J. Org. Chem.* **1989**, *54*, 6001.
[729]See Belotti; Cossy; Pete; Portella *J. Org. Chem.* **1986**, *51*, 4196.
[730]For a review, see Ide; Buck *Org. React.* **1948**, *4*, 269-304.

The following is the accepted mechanism,[731] which was originally proposed by Lapworth in 1903:[732]

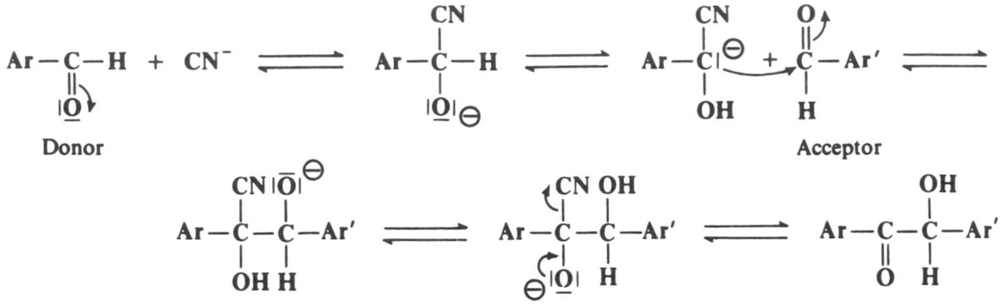

$$
\begin{array}{c}
\text{Ar-C-H + CN}^- \rightleftharpoons \text{Ar-C-H} \rightleftharpoons \text{Ar-C}|^{\ominus} + \text{C-Ar}' \rightleftharpoons \\
\end{array}
$$

Donor   Acceptor

$$
\text{Ar-C-C-Ar}' \rightleftharpoons \text{Ar-C-C-Ar}' \rightleftharpoons \text{Ar-C-C-Ar}'
$$

The reaction is reversible. The key step, the loss of the aldehydic proton, can take place because the acidity of this C—H bond is increased by the electron-withdrawing power of the CN group. Thus, CN⁻ is a highly specific catalyst for this reaction, because, almost uniquely, it can perform three functions: (1) It acts as a nucleophile; (2) its electron-withdrawing ability permits loss of the aldehydic proton; and (3) having done this, it then acts as a leaving group. Certain thiazolium salts can also catalyze the reaction.[733] In this case aliphatic aldehydes can also be used[734] (the products are called *acyloins*), and mixtures of aliphatic and aromatic aldehydes give mixed α-hydroxy ketones.[735] The reaction has also been carried out without CN⁻, by using the benzoylated cyanohydrin as one of the components in a phase-transfer catalyzed process. By this means products can be obtained from aldehydes that normally fail to self-condense.[736]

OS **I**, 94; **VII**, 95.

Reactions in Which Carbon Adds to the Hetero Atom

A.   Oxygen Adding to the Carbon

6-55   The Ritter Reaction
N-Hydro,N-alkyl-C-oxo-biaddition

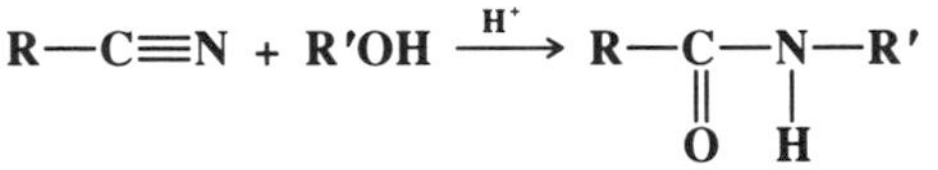

$$
\text{R-C}\equiv\text{N} + \text{R'OH} \xrightarrow{\text{H}^+} \text{R-C-N-R'}
$$

Alcohols can be added to nitriles in an entirely different manner from that of reaction **6-9**. In this reaction, the alcohol is converted by a strong acid to a carbocation, which adds to the negative nitrogen, water adding to the carbon:

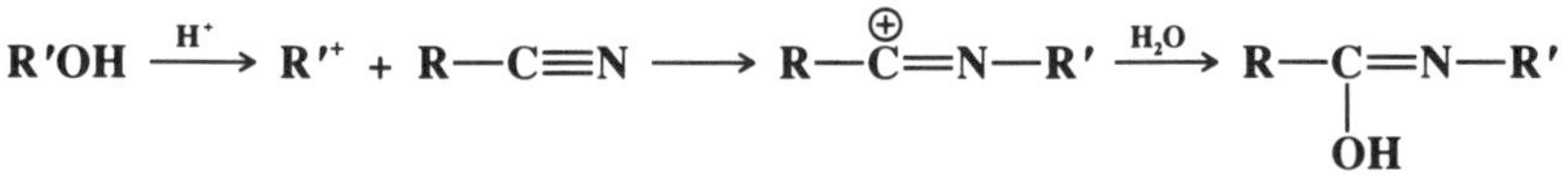

$$
\text{R'OH} \xrightarrow{\text{H}^+} \text{R'}^+ + \text{R-C}\equiv\text{N} \longrightarrow \text{R-}\overset{\oplus}{\text{C}}=\text{N-R'} \xrightarrow{\text{H}_2\text{O}} \text{R-C}=\text{N-R'}
$$

[731]For a discussion, See Kuebrich; Schowen; Wang; Lupes *J. Am. Chem. Soc.* **1971**, *93*, 1214.

[732]Lapworth *J. Chem. Soc.* **1903**, *83*, 995, **1904**, *85*, 1206.

[733]See Ugai; Tanaka; Dokawa *J. Pharm. Soc. Jpn.* **1943**, *63*, 296 [*CA 45*, 5148]; Breslow *J. Am. Chem. Soc.* **1958**, *80*, 3719; Breslow; Kool *Tetrahedron Lett.* **1988**, *29*, 1635; Castells; López-Calahorra; Domingo *J. Org. Chem.* **1988**, *53*, 4433; Diederich; Lutter *J. Am. Chem. Soc.* **1989**, *111*, 8438. For another catalyst, see Lappert; Maskell *J. Chem. Soc., Chem. Commun.* **1982**, 580.

[734]Stetter; Rämsch; Kuhlmann *Synthesis* **1976**, 733; Stetter; Kuhlmann *Org. Synth. VII*, 95; Matsumoto; Ohishi; Inouc *J. Org. Chem.* **1985**, *50*, 603.

[735]Stetter; Dämbkes *Synthesis* **1977**, 403.

[736]Rozwadowska *Tetrahedron* **1985**, *41*, 3135.

The immediate product tautomerizes to the N-alkyl amide. Only alcohols that give rise to fairly stable carbocations react (secondary, tertiary, benzylic, etc.); primary alcohols do not give the reaction. The carbocation need not be generated from an alcohol but may come from protonation of an olefin or from other sources. In any case, the reaction is called the *Ritter reaction*.[737] HCN also gives the reaction, the product being a formamide. Since the amides (especially the formamides) are easily hydrolyzable to amines, the Ritter reaction provides a method for achieving the conversions R'OH → R'NH$_2$ (see **0-46**) and alkene → R'NH$_2$ (see **5-7**) in those cases where R' can form a relatively stable carbocation. The reaction is especially useful for the preparation of tertiary alkyl amines because there are few alternate ways of preparing these compounds. The reaction can be extended to primary alcohols by treatment with triflic anhydride[738] or Ph$_2$CCl$^+$ SbCl$_6^-$ or a similar salt[739] in the presence of the nitrile.

Olefins of the form RCH=CHR' and RR'C=CH$_2$ add to nitriles in the presence of mercuric nitrate to give, after treatment with NaBH$_4$, the same amides that would be obtained by the Ritter reaction.[740] This method has the advantage of avoiding strong acids.

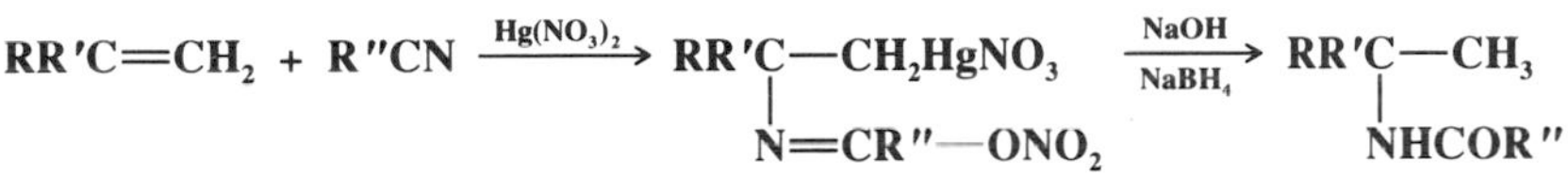

$$RR'C{=}CH_2 + R''CN \xrightarrow{Hg(NO_3)_2} \underset{\underset{N=CR''-ONO_2}{|}}{RR'C-CH_2HgNO_3} \xrightarrow[NaBH_4]{NaOH} \underset{\underset{NHCOR''}{|}}{RR'C-CH_3}$$

The Ritter reaction can be applied to cyanamides RNHCN to give ureas RNHCONHR'.[741]

OS **V**, 73, 471.

6-56 Acylation of Aldehydes and Ketones
   O-Acyl-C-acyloxy-addition

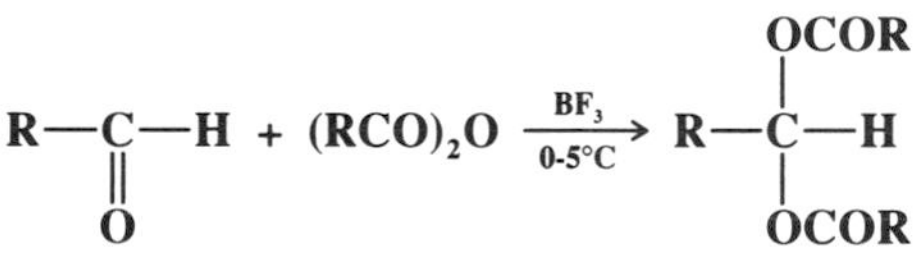

$$R-\underset{\underset{O}{\|}}{C}-H + (RCO)_2O \xrightarrow[0\text{-}5°C]{BF_3} R-\underset{\underset{OCOR}{|}}{\overset{\overset{OCOR}{|}}{C}}-H$$

Aldehydes can be converted to *acylals* by treatment with an anhydride in the presence of BF$_3$, other Lewis acids,[742] proton acids,[743] or PCl$_3$.[744] The reaction cannot normally be applied to ketones, though an exception has been reported when the reagent is trichloroacetic anhydride, which gives acylals with ketones without a catalyst.[745]

OS **IV**, 489.

---

[737]Ritter; Minieri *J. Am. Chem. Soc.* **1948**, *70*, 4045. For reviews, see Krimen; Cota *Org. React.* **1969**, *17*, 213-325; Beckwith, in Zabicky, Ref. 65, pp. 125-130; Johnson; Madroñero *Adv. Heterocycl. Chem.* **1966**, *6*, 95-146.
[738]Martinez; Alvarez; Vilar; Fraile; Hanack; Subramanian *Tetrahedron Lett.* **1989**, *30*, 581.
[739]Barton; Magnus; Garbarino; Young *J. Chem. Soc., Perkin Trans. 1* **1974**, 2101. See also Top; Jaouen *J. Org. Chem.* **1981**, *46*, 78.
[740]Sokolov; Reutov *Bull. Acad. Sci. USSR, Div. Chem. Sci.* **1968**, 225; Brown; Kurek *J. Am. Chem. Soc.* **1969**, *91*, 5647; Chow; Robson; Wright *Can. J. Chem.* **1965**, *43*, 312; Fry; Simon *J. Org. Chem.* **1982**, *47*, 5032.
[741]Anatol; Berecoechea *Bull. Soc. Chim. Fr.* **1975**, 395, *Synthesis* **1975**, 111.
[742]For example, FeCl$_3$: Kochhar; Bal; Deshpande; Rajadhyaksha; Pinnick *J. Org. Chem.* **1983**, *48*, 1765.
[743]For example, see Olah; Mehrotra *Synthesis* **1982**, 962.
[744]See Michie; Miller *Synthesis* **1981**, 824.
[745]Libman; Sprecher; Mazur *Tetrahedron* **1969**, *25*, 1679.

6-57   The Addition of Aldehydes to Aldehydes

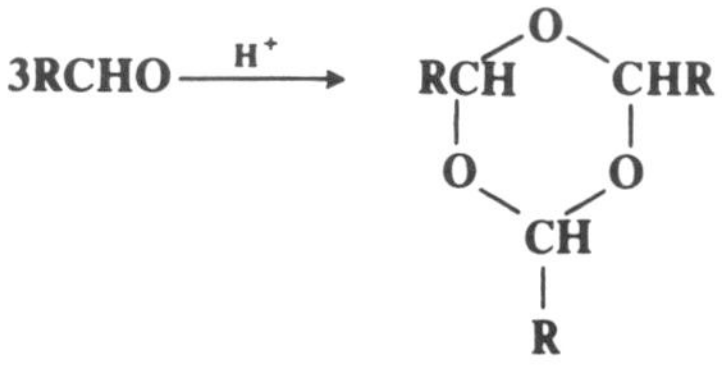

$$3RCHO \xrightarrow{H^+} \begin{array}{c} RCH \quad CHR \\ \end{array}$$

When catalyzed by acids, low-molecular-weight aldehydes add to each other to give cyclic acetals, the most common product being the trimer.[746] The cyclic trimer of formaldehyde is called *trioxane*, and that of acetaldehyde is known as *paraldehyde*. Under certain conditions, it is possible to get tetramers[747] or dimers. Aldehydes can also polymerize to linear polymers, but here a small amount of water is required to form hemiacetal groups at the ends of the chains. The linear polymer formed from formaldehyde is called *paraformalde-hyde*. Since trimers and polymers of aldehydes are acetals, they are stable to bases but can be hydrolyzed by acids. Because formadehyde and acetaldehyde have low boiling points, it is often convenient to use them in the form of their trimers or polymers.

B. Nitrogen Adding to the Carbon

6-58   The Addition of Isocyanates to Isocyanates
Alkylimino-de-oxo-bisubstitution

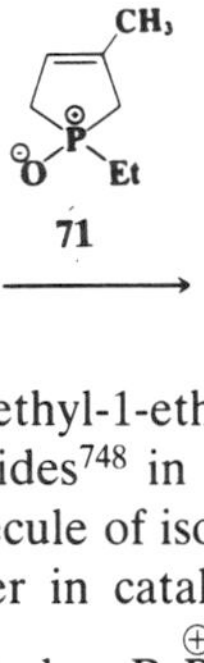

71

$$2R—N=C=O \longrightarrow R—N=C=N—R$$

The treatment of isocyanates with 3-methyl-1-ethyl-3-phospholene-1-oxide (**71**) is a useful method for the synthesis of carbodiimides[748] in good yields.[749] The mechanism does not simply involve the addition of one molecule of isocyanate to another, since the kinetics are first order in isocyanate and first order in catalyst. The following mechanism has been proposed (the catalyst is here represented as $R_3\overset{\oplus}{P}—\overset{\ominus}{O}$):[750]

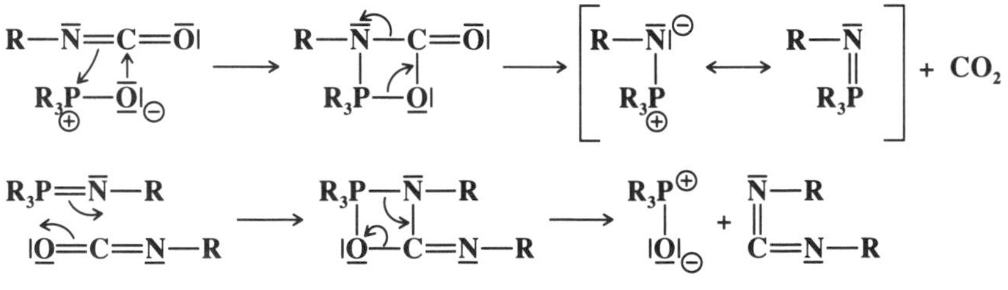

[746]For a review, see Bevington *Q. Rev., Chem. Soc.* **1952**, *6*, 141-156.
[747]Barón; Manderola; Westerkamp *Can. J. Chem.* **1963**, *41*, 1893.
[748]For reviews of the chemistry of carbodiimides, see Williams; Ibrahim *Chem. Rev.* **1981**, *81*, 589-636; Mikołajczyk; Kiełbasiński *Tetrahedron* **1981**, *37*, 233-284; Kurzer; Douraghi-Zadeh *Chem. Rev.* **1967**, *67*, 107-152.
[749]Campbell; Monagle; Foldi *J. Am. Chem. Soc.* **1962**, *84*, 3673.
[750]Monagle; Campbell; McShane *J. Am. Chem. Soc.* **1962**, *84*, 4288.

According to this mechanism, one molecule of isocyanate undergoes addition to C=O, and the other addition to C=N. Evidence is that ^{18}O labeling experiments have shown that each molecule of CO_2 produced contains one oxygen atom derived from the isocyanate and one from **71**,[751] precisely what is predicted by this mechanism. Certain other catalysts are also effective.[752]

OS **V**, 501.

6-59   The Conversion of Carboxylic Acid Salts to Nitriles
        Nitrilo-de-oxido,oxo-tersubstitution

$$RCOO^- + BrCN \xrightarrow{\text{250-300°C}} RCN + CO_2$$

Salts of aliphatic or aromatic carboxylic acids can be converted to the corresponding nitriles by heating with BrCN or ClCN. Despite appearances, this is not a substitution reaction. When $R^{14}COO^-$ was used, the label appeared in the nitrile, not in the CO_2,[753] and optical activity in R was retained.[754] The acyl isocyanate RCON=C=O could be isolated from the reaction mixture; hence the following mechanism was proposed:[753]

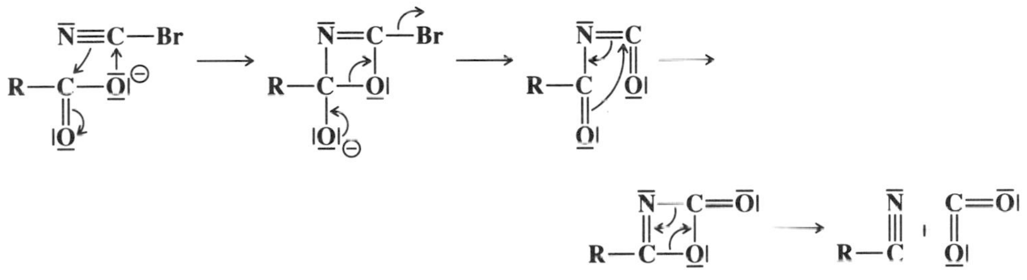

6-60   The Trimerization of Nitriles

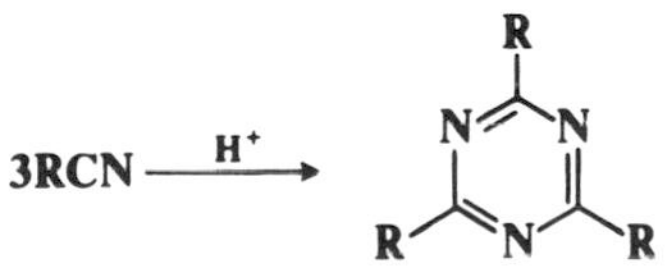

Nitriles can be trimerized with various acids, bases, or other catalysts to give triazines.[755] HCl is most often used, and then the reaction is similar to reaction **6-57.** However, most nitriles with an α hydrogen do not give the reaction.

OS **III**, 71.

C.   Carbon Adding to the Carbon. The reactions in this group (**6-61** to **6-64**) are cycloadditions.

[751]Monagle; Mengenhauser *J. Org. Chem.* **1966**, *31*, 2321.
[752]Monagle *J. Org. Chem.* **1962**, *27*, 3851; Appleman; DeCarlo *J. Org. Chem.* **1967**, *32*, 1505; Ulrich; Tucker; Sayigh *J. Org. Chem.* **1967**, *32*, 1360, *Tetrahedron Lett.* **1967**, 1731; Ostrogovich; Kerek; Buzás; Doca *Tetrahedron* **1969**, *25*, 1875.
[753]Douglas; Eccles; Almond *Can. J. Chem.* **1953**, *31*, 1127; Douglas; Burditt *Can. J. Chem.* **1958**, *36*, 1256.
[754]Barltrop; Day; Bigley *J. Chem. Soc.* **1961**, 3185.
[755]For a review, see Martin; Bauer; Pankratov *Russ. Chem. Rev.* **1978**, *47*, 975-990. For a review with respect to cyanamides RNH—CN, see Pankratov; Chesnokova *Russ. Chem. Rev.* **1989**, *58*, 879-890.

6-61   The Formation of Epoxides from Aldehydes and Ketones
(1 + 2)*OC,CC-cyclo*-Methylene-addition

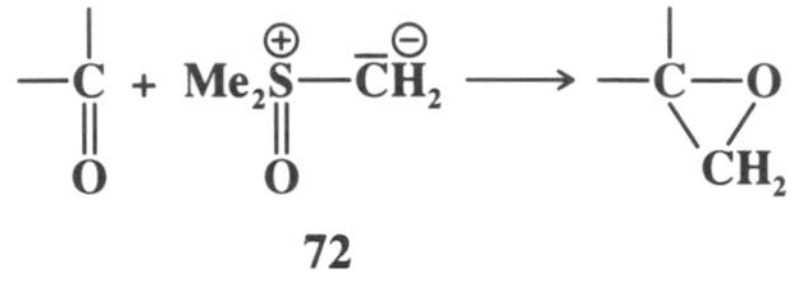

Aldehydes and ketones can be converted to epoxides[756] in good yields with the sulfur ylides dimethyloxosulfonium methylide (**72**) and dimethylsulfonium methylide (**73**).[757] For most purposes, **72** is the reagent of choice, because **73** is much less stable and ordinarily must be

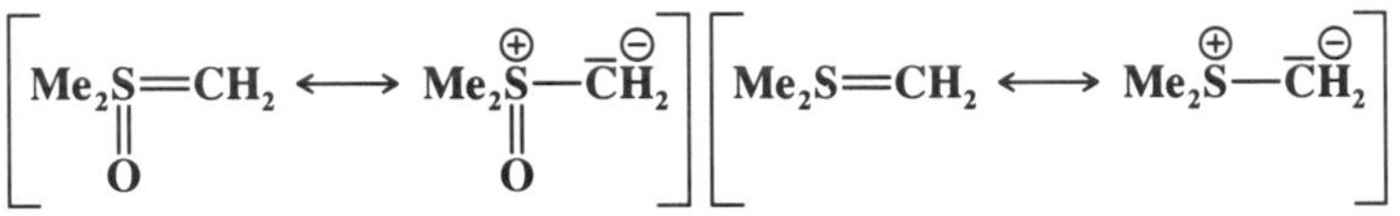

used as soon as it is formed, while **72** can be stored several days at room temperature. However, when diastereomeric epoxides can be formed, **73** usually attacks from the more hindered and **72** from the less-hindered side. Thus, 4-*t*-butylcyclohexanone, treated with **72** gave exclusively **75** while **73** gave mostly **74**.[758] Another difference in behavior between the

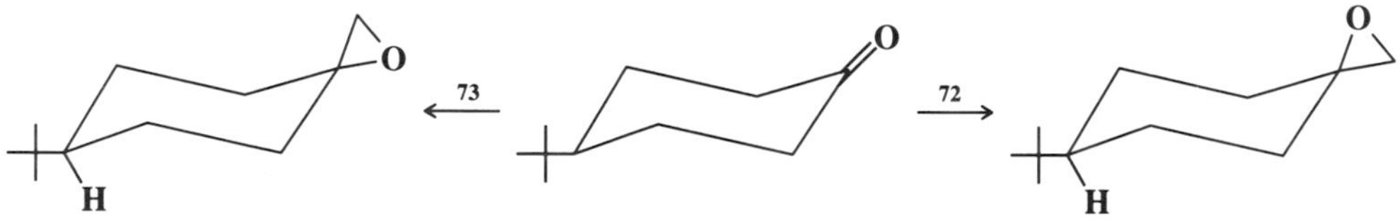

74		75
New bond is axial		New bond is equatorial

two reagents is that with α,β-unsaturated ketones, **72** gives only cyclopropanes (reaction **5-50**), while **73** gives oxirane formation. Other sulfur ylides have been used in an analogous manner, to transfer CHR or CR$_2$. Among these are Me$_2$S=CHCOO$^-$,[759] Me$_2$S=CHPh,[760] Me$_2$S=CH—vinyl,[761] and **111** on p. 872,[762] which transfer CHCOO$^-$, CHPh, CH—vinyl, and CPh$_2$, respectively. Nitrogen-containing sulfur ylides, such as **112** on p. 872 and Ph(Me$_2$N)SO=CH$_2$, as well as carbanions like **114** on p. 872 and sulfonium salts such as trimethylsulfonium bromide Me$_3$S$^+$ Br$^-$ (with a phase-transfer catalyst)[763] have also been

[756]For reviews, see Block *Reactions of Organosulfur Compounds*; Academic Press: New York, 1978, pp. 101-105; Berti *Top. Stereochem.* **1973**, *7*, 93-251, pp. 218-232. For a list of reagents, with references, see Ref. 64, pp. 468-470.
[757]For reviews, see House, Ref. 180, pp. 709-733; Durst *Adv. Org. Chem.* **1969**, *6*, 285-388, pp. 321-330; Johnson, Ref. 638, pp. 328-351. For a monograph on sulfur ylides, see Trost; Melvin *Sulfur Ylides*; Academic Press: New York, 1975.
[758]Corey; Chaykovsky *J. Am. Chem. Soc.* **1965**, *87*, 1353.
[759]Adams; Hoffman; Trost *J. Org. Chem.* **1970**, *35*, 1600.
[760]Yoshimine; Hatch *J. Am. Chem. Soc.* **1967**, *89*, 5831.
[761]Braun; Huber; Kresze *Tetrahedron Lett.* **1973**, 4033.
[762]Corey; Jautelat; Oppolzer *Tetrahedron Lett.* **1967**, 2325.
[763]Borredon; Delmas; Gaset *Tetrahedron Lett.* **1982**, *23*, 5283, *Tetrahedron* **1987**, *43*, 3945, **1988**, *44*, 1073; Mosset; Grée *Synth. Commun.* **1985**, *15*, 749; Bouda; Borredon; Delmas; Gaset *Synth. Commun.* **1987**, *17*, 503.

used.[764] High yields have been achieved by the use of sulfonium ylides anchored to insoluble polymers under phase transfer conditions.[765]

The generally accepted mechanism for the reaction between sulfur ylides and aldehydes or ketone is

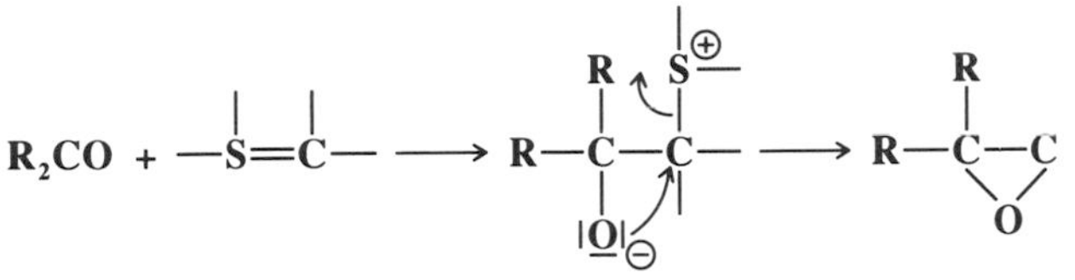

76

which is similar to that of the reaction of sulfur ylides with C=C double bonds (**5-50**).[766] The stereochemical difference in the behavior of **72** and **73** has been attributed to formation of the betaine **76** being reversible for **72** but not for the less stable **73,** so that the more-hindered product is the result of kinetic control and the less-hindered of thermodynamic control.[767]

Phosphorus ylides do not give this reaction, but give **6-47** instead.

Aldehydes and ketones can also be converted to epoxides by treatment with a diazoalkane,[768] most commonly diazomethane, but an important side reaction is the formation of an aldehyde or ketone with one more carbon than the starting compound (reaction **8-9**). The reaction can be carried out with many aldehydes, ketones, and quinones. A mechanism that accounts for both products is

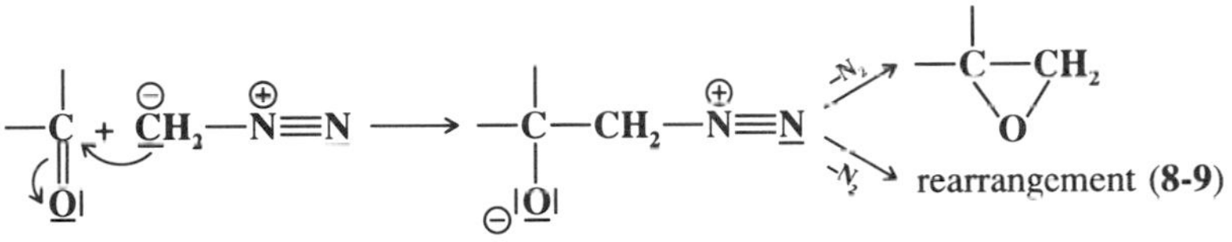

77

Compound **77** or nitrogen-containing derivatives of it have sometimes been isolated.

Dihalocarbenes and carbenoids, which readily add to C=C bonds (**5-50**), do not generally add to the C=O bonds of ordinary aldehydes and ketones.[769]

Symmetrical epoxides can be prepared by treatment of aromatic aldehydes with hexamethylphosphorus triamide.[770]

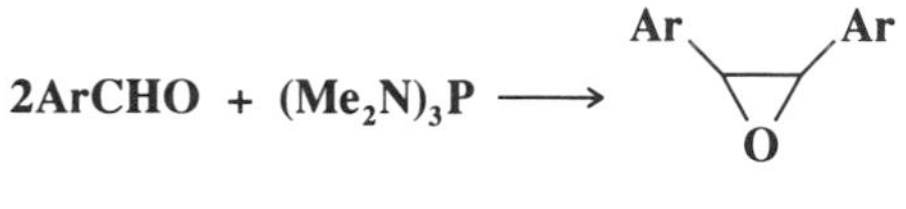

See also **6-45.**
OS **V**, 358, 755.

[764]Johnson; Haake; Schroeck *J. Am. Chem. Soc.* **1970,** *92,* 6594; Johnson; Janiga *J. Am. Chem. Soc.* **1973,** *95,* 7692; Johnson *Acc. Chem. Res.* **1973,** *6,* 341-347; Tamura; Matsushima; Ikeda; Sumoto *Synthesis* **1976,** 35.

[765]Farrall; Furst; Fréchet *Tetrahedron Lett.* **1979,** 203.

[766]See, for example, Townsend; Sharpless *Tetrahedron Lett.* **1972,** 3313; Johnson; Schroeck; Shanklin *J. Am. Chem. Soc.* **1973,** *95,* 7424.

[767]Johnson et al., Ref. 766.

[768]For a review, see Gutsche, *Org. React.* **1954,** *8,* 364-429.

[769]For exceptions, see Greuter; Winkler; Belluš *Helv. Chim. Acta* **1979,** *62,* 1275; Sadhu; Matteson *Tetrahedron Lett.* **1986,** *27,* 795; Araki; Butsugan *J. Chem. Soc., Chem. Commun.* **1989,** 1286.

[770]Mark *J. Am. Chem. Soc.* **1963,** *85,* 1884; *Org. Synth.* V, 358; Newman; Blum *J. Am. Chem. Soc.* **1964,** *86,* 5598.

6-62   The Formation of Episulfides and Episulfones[771]

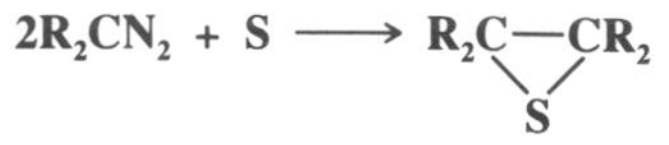

$$2R_2CN_2 + S \longrightarrow R_2C\overset{\diagdown}{\underset{S}{\diagup}}CR_2$$

Diazoalkanes, treated with sulfur, give episulfides.[772] It is likely that $R_2C{=}S$ is an intermediate, which is attacked by another molecule of diazoalkane, in a process similar to that shown in **6-61**. Thioketones *do* react with diazoalkanes to give episulfides.[773] Thioketones have also been converted to episulfides with sulfur ylides.[758]

Alkanesulfonyl chlorides, when treated with diazomethane in the presence of a base (usually a tertiary amine), give episulfones (**79**).[774] The base removes HCl from the sulfonyl

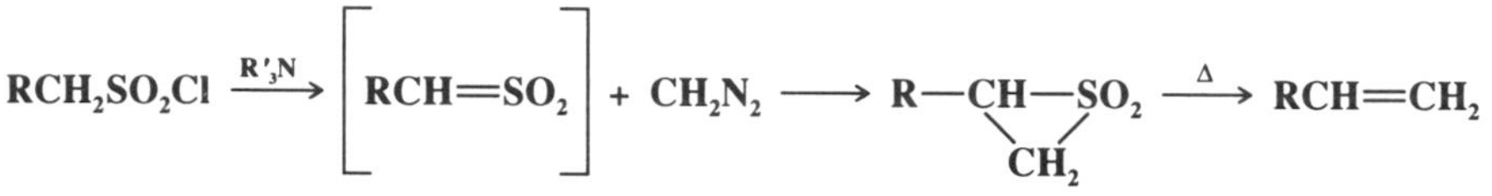

$$RCH_2SO_2Cl \xrightarrow{R'_3N} \left[RCH{=}SO_2 \right] + CH_2N_2 \longrightarrow R{-}CH{-}SO_2 \xrightarrow{\Delta} RCH{=}CH_2$$

78   **79**

halide to produce the highly reactive sulfene (**78**) (**7-14**), which then adds CH_2. The episulfone can then be heated to give off SO_2 (**7-25**), making the entire process a method for achieving the conversion $RCH_2SO_2Cl \rightarrow RCH{=}CH_2$.[775]

OS **V**, 231, 877.

6-63   The Formation of β-Lactones and Oxetanes
(2 + 2)OC,CC-cyclo-[oxoethylene]-1/2/addition

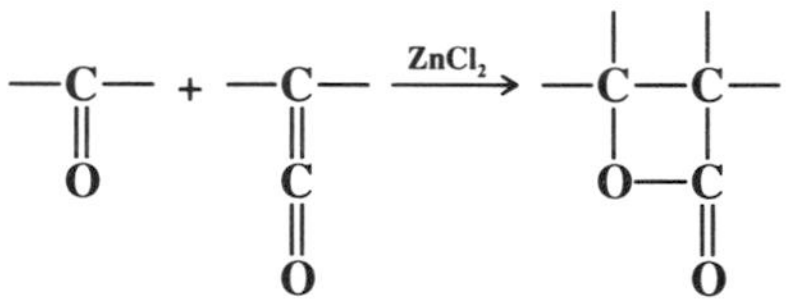

$$-\underset{O}{\overset{\|}{C}}- + -\underset{\overset{\|}{O}}{\overset{|}{\underset{C}{\overset{\|}{C}}}}- \xrightarrow{ZnCl_2} -\underset{O{-}C}{\overset{|}{\underset{\|}{\overset{|}{C}}}}\overset{|}{\underset{O}{\overset{|}{C}}}-$$

Aldehydes, ketones, and quinones react with ketenes to give β-lactones, diphenylketene being used most often.[776] The reaction is catalyzed by Lewis acids, and without them most ketenes do not give adducts because the adducts decompose at the high temperatures necessary when no catalyst is used. When ketene was added to chloral Cl_3CCHO in the presence of the chiral catalyst (+)-quinidine, one enantiomer of the β-lactone was produced in 98% enantiomeric excess.[777] Other di- and trihalo aldehydes and ketones also give the reaction enantioselectively, with somewhat lower ee values.[778] Ketene adds to another molecule of itself:

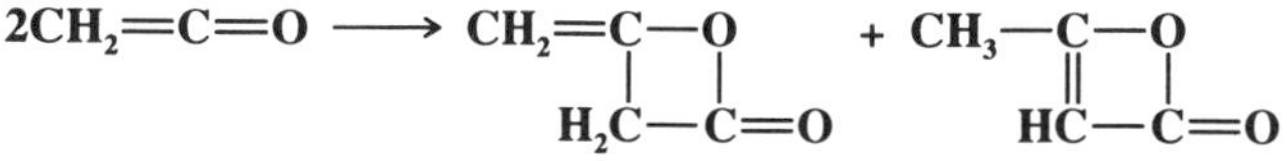

$$2CH_2{=}C{=}O \longrightarrow \underset{H_2C{-}C{=}O}{\overset{|}{\underset{|}{CH_2{=}C{-}O}}} + \underset{HC{-}C{=}O}{\overset{\|}{\underset{|}{CH_3{-}C{-}O}}}$$

[771]For a review, see Muller; Hamer *1,2-Cycloaddition Reactions*; Wiley: New York, 1967, pp. 57-86.
[772]Schönberg; Frese *Chem. Ber.* **1962**, *95*, 2810.
[773]For example, see Beiner; Lecadet; Paquer; Thuillier *Bull. Soc. Chim. Fr.* **1973**, 1983.
[774]Opitz; Fischer *Angew. Chem. Int. Ed. Engl.* **1965**, *4*, 70 [*Angew. Chem. 77*, 41].
[775]For a review of this process, see Fischer *Synthesis* **1970**, 393-404.
[776]For reviews, see Ref. 771, pp. 139-168; Ulrich *Cycloaddition Reactions of Heterocumulenes*; Academic Press: New York, 1967, pp. 39-45, 64-74.
[777]Wynberg; Staring *J. Am. Chem. Soc.* **1982**, *104*, 166, *J. Chem. Soc., Chem. Commun.* **1984**, 1181.
[778]Wynberg; Staring *J. Org. Chem.* **1985**, *50*, 1977.

This dimerization is so rapid that ketene does not form β-lactones with aldehydes or ketones, except at low temperatures. Other ketenes dimerize more slowly. In these cases the major dimerization product is not the β-lactone, but a cyclobutenone (see **5-49**). However, the proportion of ketene that dimerizes to β-lactone can be increased by the addition of catalysts such as triethylamine or triethyl phosphite.[779] Ketene acetals $R_2C=C(OR')_2$ add to aldehydes and ketones in the presence of $ZnCl_2$ to give the corresponding oxetanes.[780]

Ordinary aldehydes and ketones can add to olefins, under the influence of uv light, to give oxetanes. This reaction, called the *Paterno–Büchi reaction*,[781] is similar to the photo-chemical dimerization of olefins discussed at **5-49**. In general, the mechanism consists of the

$$-\overset{|}{\underset{\overset{\|}{O}}{C}}- \quad + \quad -\overset{|}{\underset{\overset{\|}{\underset{|}{C}}}{C}}- \quad \xrightarrow{h\upsilon} \quad -\overset{|}{\underset{\overset{|}{O}}{C}}-\overset{|}{\underset{|}{C}}- \\ \qquad\qquad\qquad\qquad\qquad\qquad\quad \overset{|}{O}-\overset{|}{\underset{|}{C}}-$$

addition of an excited state of the carbonyl compound to the ground state of the olefin. Both singlet (S_1)[782] and n,π^* triplet[783] states have been shown to add to olefins to give oxetanes. A diradical intermediate[784] $\overset{\cdot}{O}-\overset{|}{\underset{|}{C}}-\overset{|}{\underset{|}{C}}-\overset{\cdot}{\underset{}{}}$ has been detected spectrally.[785] Yields in the Paterno-Büchi reaction are variable, ranging from very low to fairly high (90%). There are several side reactions. When the reaction proceeds through a triplet state, it can in general be successful only when the alkene possesses a triplet energy comparable to, or higher than, the carbonyl compound; otherwise energy transfer from the excited carbonyl group to the ground-state alkene can take place (triplet–triplet photosensitization, see p. 241). In most cases quinones react normally with alkenes, giving oxetane products, but other α,β-unsaturated ketones usually give preferential cyclobutane formation (**5-49**). Aldehydes and ketones also add photochemically to allenes to give the corresponding alkylideneoxe-tanes and dioxaspiro compounds:[786]

$$C=O \; + \; C=C=C \quad \xrightarrow{h\nu} \quad \text{[structures]}$$

OS **III**, 508; **V**, 456. For the reverse reaction, see OS **V**, 679.

[779]Farnum; Johnson; Hess; Marshall; Webster *J. Am. Chem. Soc.* **1965**, *87*, 5191; Elam; *J. Org. Chem.* **1967**, *32*, 215.

[780]Aben; Hofstraat; Scheeren *Recl. Trav. Chim. Pays-Bas* **1981**, *100*, 355.

[781]For reviews, see Ninomiya; Naito *Photochemical Synthesis*; Academic Press: New York, 1989, pp. 138-152; Carless, in Coyle *Photochemistry in Organic Synthesis*; Royal Society of Chemistry: London, 1986, pp. 95-117; Carless, in Horspool *Synthetic Organic Photochemistry*; Plenum: New York, 1984, pp. 425-487; Jones *Org. Photochem.* **1981**, *5*, 1-122; Arnold *Adv. Photochem.* **1968**, *6*, 301-423; Chapman; Lenz *Org. Photochem.* **1967**, *1*, 283-321, pp. 283-294; Ref. 771, pp. 111-139.

[782]See, for example, Turro *Pure Appl. Chem.* **1971**, *27*, 679-705; Yang; Kimura; Eisenhardt *J. Am. Chem. Soc.* **1973**, *95*, 5058; Singer; Davis; Muralidharan *J. Am. Chem. Soc.* **1969**, *91*, 897; Barltrop; Carless *J. Am. Chem. Soc.* **1972**, *94*, 1951, 8761.

[783]Arnold; Hinman; Glick *Tetrahedron Lett.* **1964**, 1425; Yang; Nussim; Jorgenson; Murov *Tetrahedron Lett.* **1964**, 3657.

[784]For other evidence for these diradical intermediates, see references cited in Griesbeck; Stadtmüller *J. Am. Chem. Soc.* **1990**, *112*, 1281.

[785]Freilich; Peters *J. Am. Chem. Soc.* **1981**, *103*, 6255, **1985**, *107*, 3819.

[786]Arnold; Glick *Chem. Commun.* **1966**, 813; Gotthardt; Steinmetz; Hammond *Chem. Commun.* **1967**, 480, *J. Org. Chem.* **1968**, *33*, 2774. For a review of the formation of heterocycles by cycloadditions of allenes, see Schuster; Coppola *Allenes in Organic Synthesis*; Wiley: New York, 1984, pp. 317-326.

6-64   The Formation of β-Lactams

(2 + 2)NC,CC-cyclo-**[oxoethylene]-1/2/addition**

$$-\overset{|}{C}- \;+\; -\overset{|}{C}- \longrightarrow -\overset{|}{\underset{R-N-C}{C}}-\overset{|}{\underset{\parallel}{C}}-$$

Ketenes add to imines to give β-lactams.[787] The reaction is generally carried out with ketenes of the form $R_2C=C=O$. It has not been successfully applied to $RCH=C=O$, except when these are generated is situ by decomposition of a diazo ketone (the Wolff rearrangement, **8-8**) in the presence of the imine. It has been done with ketene, but the more usual course with this reagent is an addition to the enamine tautomer of the substrate. Thioketenes[788] $R_2C=C=S$ give β-thiolactams.[789] Imines also form β-lactams when treated with (1) zinc (or another metal) and an α-bromo ester (Reformatsky conditions—**6-30**),[790] or (2) the chromium carbene complexes $(CO)_5Cr=C(Me)OMe$.[791] The latter method has been used to prepare optically active β-lactams.[792] Ketenes have also been added to certain hydrazones (e.g., $PhCH=NNMe_2$) to give N-amino β-lactams.[793]

Like the similar cycloaddition of ketenes to olefins (**5-49**), most of these reactions probably take place by the diionic mechanism c (p. 857).[794] β-Lactams have also been prepared in the opposite manner: by the addition of enamines to isocyanates:[795]

$$\underset{NR_2''}{\overset{R'}{\underset{|}{\overset{|}{R-C}}}}\;\;+\;\;\underset{Ar}{\overset{O}{\underset{\parallel}{\underset{N}{\overset{C}{\parallel}}}}} \longrightarrow \underset{NR_2''}{\overset{R'}{\underset{|}{\overset{|}{R-C-C=O}}}}$$

The reactive compound chlorosulfonyl isocyanate[796] $ClSO_2NCO$ forms β-lactams even with unactivated alkenes,[797] as well as with allenes,[798] conjugated dienes,[799] and cyclopropenes.[800] OS **V**, 673; **65**, 135, 140.

[787]For a list of references, see Ref. 64, pp. 961-962. For reviews of the formation of β-lactams, see Brown *Heterocycles* **1989**, *29*, 2225-2294; Isaacs *Chem. Soc. Rev.* **1976**, *5*, 181-202; Mukerjee; Srivastava *Synthesis* **1973**, 327-346; Ref. 771, pp. 173-206; Ulrich, Ref. 776, pp. 75-83, 135-152; Anselme, in Patai *The Chemistry of the Carbon–Nitrogen Double Bond*, Ref. 40, pp. 305-309. For a review of cycloaddition reactions of imines, see Sandhu; Sain *Heterocycles* **1987**, *26*, 777-818.

[788]For a review of thioketenes, see Schaumann *Tetrahedron* **1988**, *44*, 1827-1871.

[789]Schaumann *Chem. Ber.* **1976**, *109*, 906.

[790]For a review, see Hart; Ha *Chem. Rev.* **1989**, *89*, 1447-1465.

[791]Hegedus; McGuire; Schultze; Yijun; Anderson *J. Am. Chem. Soc.* **1984**, *106*, 2680; Hegedus; McGuire; Schultze *Org. Synth.* **65**, 140.

[792]Hegedus; Imwinkelried; Alarid-Sargent; Dvorak; Satoh *J. Am. Chem. Soc.* **1990**, *112*, 1109.

[793]Sharma; Pandhi *J. Org. Chem.* **1990**, *55*, 2196.

[794]See Moore; Hernandez; Chambers *J. Am. Chem. Soc.* **1978**, *100*, 2245; Pacansky; Chang; Brown; Schwarz *J. Org. Chem.* **1982**, *47*, 2233; Brady; Shieh *J. Org. Chem.* **1983**, *48*, 2499.

[795]For example, see Perelman; Mizsak *J. Am. Chem. Soc.* **1962**, *84*, 4988; Opitz; Koch *Angew. Chem. Int. Ed. Engl.* **1963**, *2*, 152 [*Angew. Chem.* **75**, 167].

[796]For reviews of this compound, see Kamal; Sattur *Heterocycles* **1987**, *26*, 1051-1076; Szabo *Aldrichimica Acta* **1977**, *10*, 23-29; Rasmussen; Hassner *Chem. Rev.* **1976**, *76*, 389-408; Graf *Angew. Chem. Int. Ed. Engl.* **1968**, *7*, 172-182 [*Angew. Chem. 80*, 179-189].

[797]Graf *Liebigs Ann. Chem.* **1963**, *661*, 111; Bestian *Pure Appl. Chem.* **1971**, *27*, 611-634. See also Barrett; Betts; Fenwick *J. Org. Chem.* **1985**, *50*, 169.

[798]Moriconi; Kelly *J. Am. Chem. Soc.* **1966**, *88*, 3657, *J. Org. Chem.* **1968**, *33*, 3036. See also Martin; Carter; Chitwood *J. Org. Chem.* **1971**, *36*, 2225.

[799]Moriconi; Meyer *J. Org. Chem.* **1971**, *36*, 2841; Malpass; Tweddle *J. Chem. Soc. Perkin Trans. 1* **1977**, 874.

[800]Moriconi; Kelly; Salomone *J. Org. Chem.* **1968**, *33*, 3448.

Addition to Isocyanides[801]

Addition to R—N≡C (with ⊕ on N and ⊖ on C) is not a matter of a species with an electron pair adding to one atom and a species without a pair adding to the other, as is addition to the other types of double and triple bonds in this chapter and Chapter 15. In these additions the electrophile and the nucleophile *both add to the carbon*. No species add to the nitrogen, which, however, loses its positive charge by obtaining as an unshared pair one of the triple-bond pairs of electrons:

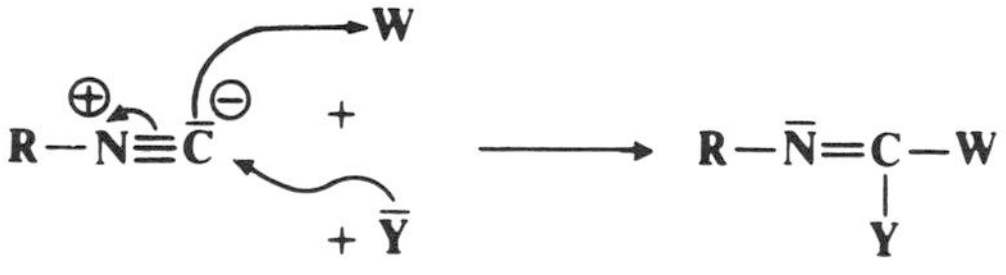

$$R-\overset{\oplus}{N}\equiv\overset{\ominus}{C} \underset{+\ \bar{Y}}{\overset{\longrightarrow W}{}} + \longrightarrow R-\bar{N}=\underset{\underset{Y}{|}}{C}-W$$

80

In most of the reactions considered below, **80** undergoes a further reaction, so the product is of the form $R-\bar{N}H-\overset{|}{\underset{|}{C}}-$. See also **9-30**.

6-65 The Addition of Water to Isocyanides
1/N,2/C-Dihydro-2/C-oxo-biaddition

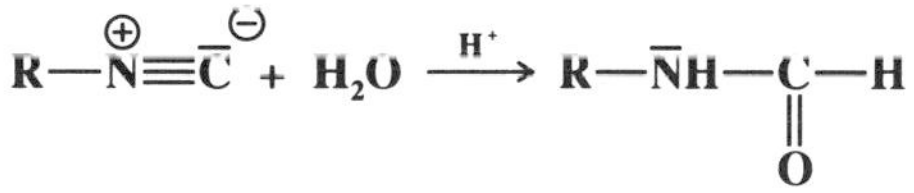

$$R-\overset{\oplus}{N}\equiv\overset{\ominus}{C} + H_2O \xrightarrow{H^+} R-\bar{N}H-\underset{\underset{O}{\|}}{C}-H$$

Formamides can be prepared by the acid-catalyzed addition of water to isocyanides. The mechanism is probably[802]

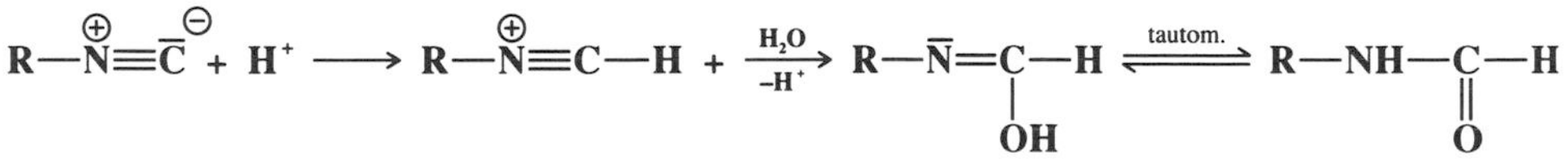

$$R-\overset{\oplus}{N}\equiv\overset{\ominus}{C} + H^+ \longrightarrow R-\overset{\oplus}{N}\equiv C-H + \xrightarrow[-H^+]{H_2O} R-\bar{N}=\underset{\underset{OH}{|}}{C}-H \underset{}{\overset{\text{tautom.}}{\rightleftharpoons}} R-NH-\underset{\underset{O}{\|}}{C}-H$$

The reaction has also been carried out under alkaline conditions, with OH^- in aqueous dioxane.[803] The mechanism here involves nucleophilic attack by OH^- at the carbon atom.

6-66 The Reduction of Isocyanides
1/N,2,2,2/C-Tetrahydro-biaddition

$$R-\overset{\oplus}{N}\equiv\overset{\ominus}{C} + LiAlH_4 \longrightarrow R-NH-CH_3$$

Isocyanides have been reduced to N-methylamines with lithium aluminum hydride as well as with other reducing agents.

[801]For a monograph, see Ugi *Isonitrile Chemistry*; Academic Press: New York, 1971. For reviews, see Walborsky; Periasamy, in Patai; Rappoport, Ref. 694, pt. 2, pp. 835-887; Hoffmann; Marquarding; Kliimann; Ugi, in Rappoport, Ref. 334, pp. 853-883.
[802]Drenth; *Recl. Trav. Chim. Pays-Bas* **1962**, *81*, 319; Lim; Stein *Can. J. Chem.* **1971**, *49*, 2455.
[803]Cunningham; Buist; Arkle *J. Chem. Soc., Perkin Trans. 2* **1991**, 589.

6-67 The Passerini and Ugi Reactions[804]
1/N-Hydro-2/C-(α-acyloxyalkyl),2/C-oxo-biaddition

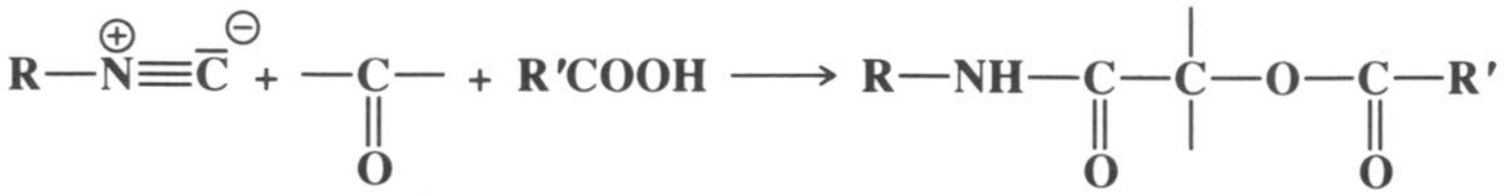

$$R-\overset{\oplus}{N}\equiv\overset{\ominus}{C} + \underset{\underset{O}{\|}}{-C-} + R'COOH \longrightarrow R-NH-\underset{\underset{O}{\|}}{C}-\overset{|}{\underset{|}{C}}-O-\underset{\underset{O}{\|}}{C}-R'$$

When an isocyanide is treated with a carboxylic acid and an aldehyde or ketone, an α-acyloxy amide is prepared. This is called the *Passerini reaction*. The following mechanism has been postulated:

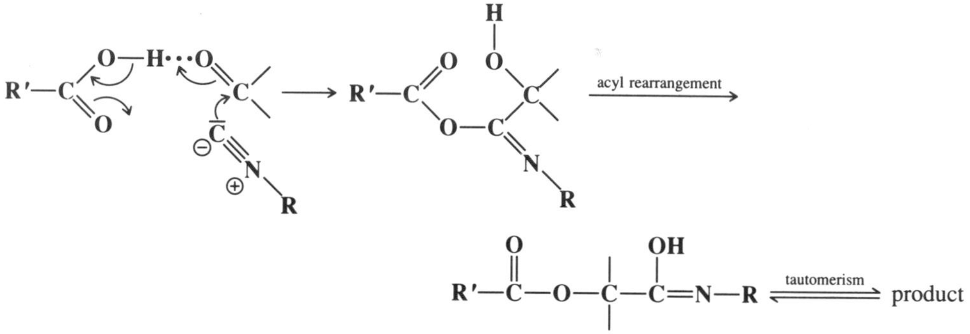

$$R'-\overset{\overset{O}{\|}}{C}-O-\overset{|}{\underset{|}{C}}-\overset{OH}{\underset{|}{C}}=N-R \xrightarrow{\text{tautomerism}} \text{product}$$

If ammonia or an amine is also added to the mixture (in which case the reaction is known as the *Ugi reaction*, or the *Ugi four-component condensation*, abbreviated 4 CC), the product is the corresponding bisamide $R'-\overset{\overset{O}{\|}}{C}-NH-\overset{|}{\underset{|}{C}}-\overset{\overset{O}{\|}}{C}-NH-R$ (from NH_3) or

$R'-\overset{\overset{O}{\|}}{C}-NR''-\overset{|}{\underset{|}{C}}-\overset{\overset{O}{\|}}{C}-NH-R$ (from a primary amine $R''NH_2$). This product probably arises from a reaction between the carboxylic acid, the isocyanide, and the *imine* formed from the aldehyde or ketone and ammonia or the primary amine. The use of an N-protected amino acid or peptide as the carboxylic acid component and/or the use of an isocyanide containing a C-protected carboxyl group allows the reaction to be used for peptide synthesis.[805]

6-68 The Addition of O- and N-Halides to Isocyanides

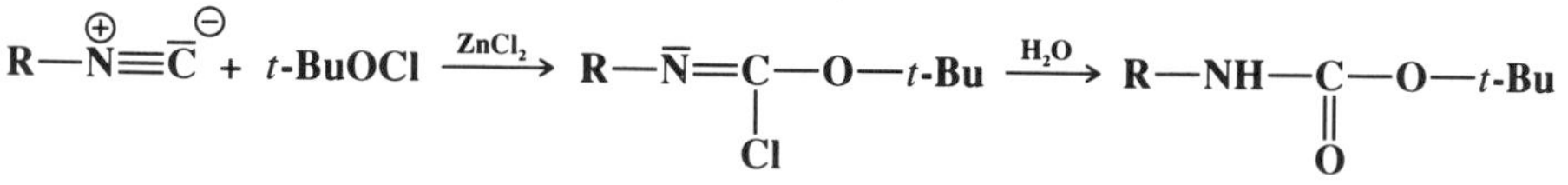

$$R-\overset{\oplus}{N}\equiv\overset{\ominus}{C} + t\text{-BuOCl} \xrightarrow{\text{ZnCl}_2} R-\overline{N}=\underset{Cl}{\overset{|}{C}}-O-t\text{-Bu} \xrightarrow{H_2O} R-NH-\underset{\underset{O}{\|}}{C}-O-t\text{-Bu}$$

1/N-Hydro-2/C-butoxy,2/C-oxo-biaddition

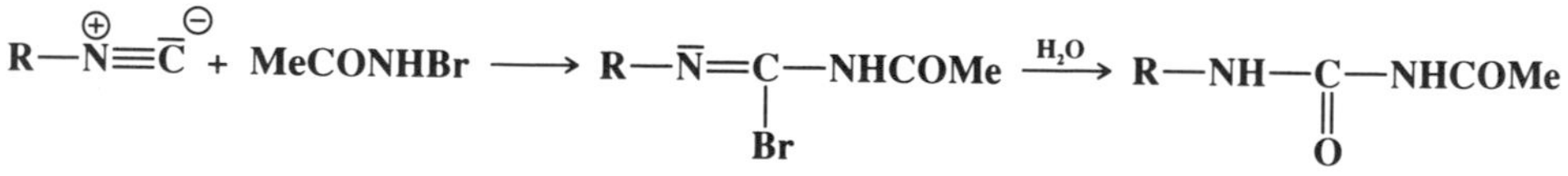

$$R-\overset{\oplus}{N}\equiv\overset{\ominus}{C} + \text{MeCONHBr} \longrightarrow R-\overline{N}=\underset{Br}{\overset{|}{C}}-NHCOMe \xrightarrow{H_2O} R-NH-\underset{\underset{O}{\|}}{C}-NHCOMe$$

1/N-Hydro-2/C-acylamino,2/C-oxo-biaddition

---

[804]For reviews, see Ugi *Angew. Chem. Int. Ed. Engl.* **1982**, *21*, 810-819 [*Angew. Chem. 94*, 826-836]; Marquarding; Gokel; Hoffmann; Ugi, in Ugi, Ref. 801, pp. 133-143, Gokel; Lüdke; Ugi, in Ugi, Ref. 801, pp. 145-199, 252-254.

[805]For reviews, see Ugi, in Gross; Meienhofer *The Peptides*, vol. 2; Academic Press: New York, 1980, pp. 365-381, *Intra-Sci. Chem. Rep.* **1971**, *5*, 229-261, *Rec. Chem. Prog.* **1969**, *30*, 289-311; Gokel; Hoffmann; Kleimann; Klusacek; Lüdke; Marquarding; Ugi, in Ugi, Ref. 801, pp. 201-215. See also Kunz; Pfrengle *J. Am. Chem. Soc.* **1988**, *110*, 651.

Alkyl hypochlorites and N-halo amides add to isocyanides to give, after hydrolysis, carbamates and N-acylureas (ureides), respectively.[806]

6-69 The Formation of Metalated Aldimines
1/1/Lithio-alkyl-addition

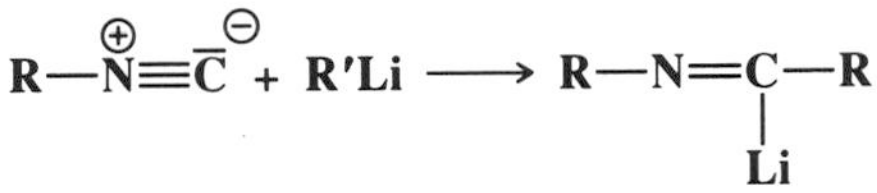

$$R-\overset{\oplus}{N}\equiv\overset{\ominus}{C} + R'Li \longrightarrow R-N=\underset{\underset{Li}{|}}{C}-R'$$

Isocyanides that do not contain an α hydrogen react with alkyllithium compounds,[807] as well as with Grignard reagents, to give lithium (or magnesium) aldimines.[808] These metalated aldimines are versatile nucleophiles and react with various substrates as follows (see also **8-25**):

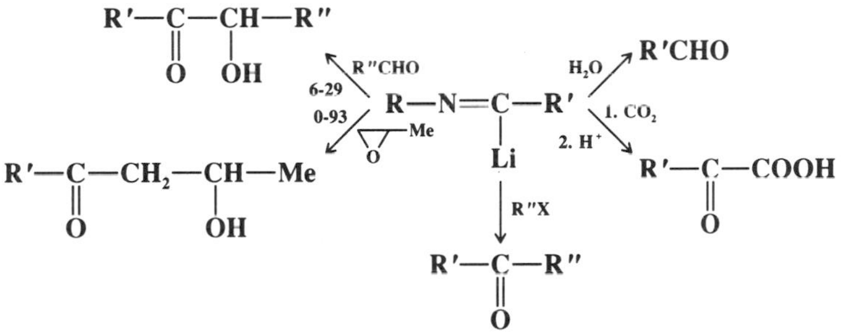

The reaction therefore constitutes a method for converting an organometallic compound R'M to an aldehyde R'CHO (see also **2-32**), an α-keto acid,[809] a ketone R'COR (see also **2-32**), an α-hydroxy ketone, or a β-hydroxy ketone. In each case the C=N bond is hydrolyzed to a C=O bond (**6-2**).

In a related reaction, isocyanides can be converted to aromatic aldimines by treatment with an iron complex followed by irradiation in benzene solution: RNC + C_6H_6 → PhCH=NR.[810]

OS **VI**, 751.

[806]Okano; Ito; Shono; Oda *Bull Chem. Soc. Jpn.* **1963**, *36*, 1314. See also Yamada; Wada; Tanimoto; Okano *Bull. Chem. Soc. Jpn.* **1982**, *55*, 2480.

[807]For a review of other metallation reactions of isocyanides, see Ito; Murakami *Synlett* **1990**, 245-250.

[808]Niznik; Morrison; Walborsky *J. Org. Chem.* **1974**, *39*, 600; Marks; Walborsky *J. Org. Chem.* **1981**, *46*, 5405, **1982**, *47*, 52. See also Walborsky; Ronman *J. Org. Chem.* **1978**, *43*, 731. For the formation of zinc aldimines, see Murakami; Ito; Ito *J. Org. Chem.* **1988**, *53*, 4158.

[809]For a review of the synthesis and properties of α-keto acids, see Cooper; Ginos; Meister *Chem. Rev.* **1983**, *83*, 321-358.

[810]Jones; Foster; Putinas *J. Am. Chem. Soc.* **1987**, *109*, 5047.

17

ELIMINATIONS

When two groups are lost from adjacent atoms so that a new double (or triple) bond is

$$-\overset{|}{\underset{W}{A}}-\overset{|}{\underset{X}{B}}- \longrightarrow -A{=}B-$$

formed the reaction is called β *elimination*; one atom is the α, the other the β atom. In an α elimination both groups are lost from the same atom to give a carbene (or a nitrene):

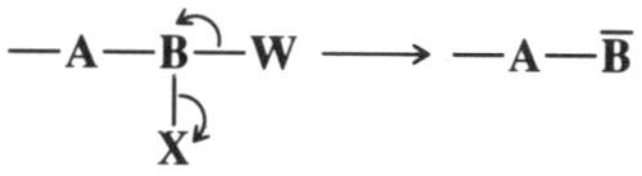

$$-A-\overset{\frown}{\underset{\underset{X^{\downarrow}}{\overset{|}{\downarrow}}}{B}}-W \longrightarrow -A-\bar{B}$$

In a γ elimination, a three-membered ring is formed:

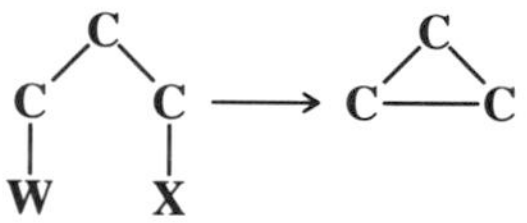

$$\underset{\underset{W}{\overset{|}{C}}}{C}\overset{C}{\underset{\underset{X}{\overset{|}{C}}}{\diagdown}} \longrightarrow C{-\!\!-\!\!-}C$$

Some of these processes were discussed in Chapter 10. Another type of elimination involves the expulsion of a fragment from within a chain or ring (X—Y—Z → X—Z + Y). Such reactions are called *extrusion reactions*. This chapter discusses β elimination and (beginning on p. 1045) extrusion reactions; however, β elimination in which both X and W are hydrogens are oxidation reactions and are treated in Chapter 19.

MECHANISMS AND ORIENTATION

β elimination reactions may be divided into two types; one type taking place largely in solution, the other (pyrolytic eliminations) mostly in the gas phase. In the reactions in solution one group leaves with its electrons and the other without, the latter most often being hydrogen. In these cases we refer to the former as the leaving group or nucleofuge. For pyrolytic eliminations there are two principal mechanisms, one pericyclic and the other a free-radical pathway. A few photochemical eliminations are also known (the most important is Norrish type II cleavage of ketones, p. 243), but these are not generally of synthetic importance[1] and will not be discussed further. In most β eliminations the new bonds are

---

[1] For synthetically useful examples of Norrish type II cleavage, see Neckers; Kellogg; Prins; Schoustra *J. Org. Chem.* **1971**, *36*, 1838.

C=C or C≡C; our discussion of mechanisms is largely confined to these cases.[2] Mechanisms in solution (E2, E1, E1cB) are discussed first.

The E2 Mechanism

In the E2 mechanism (elimination, bimolecular), the two groups depart simultaneously, with the proton being pulled off by a base:

$$\beta \overset{|}{\underset{\nearrow H}{C}}-\overset{\overset{X}{|}}{\underset{|}{C}}\alpha \longrightarrow -C=C- \; + \; X^- \; + \; BH$$

$$\underset{B}{}$$

The mechanism thus takes place in one step and kinetically is second order: first order in substrate and first order in base. The IUPAC designation is $A_{xH}D_HD_N$, or more generally (to include cases where the electrofuge is not hydrogen), $A_nD_ED_N$. It is analogous to the SN2 mechanism (p. 294) and often competes with it. With respect to the substrate, the difference between the two pathways is whether the species with the unshared pair attacks the carbon (and thus acts as a nucleophile) or the hydrogen (and thus acts as a base). As in the case of the SN2 mechanism, the leaving group may be positive or neutral and the base may be negatively charged or neutral.

Among the evidence for the existence of the E2 mechanism are: (1) the reaction displays the proper second-order kinetics; (2) when the hydrogen is replaced by deuterium in second-order eliminations, there is an isotope effect of from 3 to 8, consistent with breaking of this bond in the rate-determining step.[3] However, neither of these results alone could prove an E2 mechanism, since both are compatible with other mechanisms also (e.g., see E1cB p. 991). The most compelling evidence for the E2 mechanism is found in stereochemical studies.[4] As will be illustrated in the examples below, the E2 mechanism is stereospecific: the five atoms involved (including the base) in the transition state must be in one plane. There are two ways for this to happen. The H and X may be trans to one another (**A**) with a dihedral angle of 180°, or they may be cis (**B**) with a dihedral angle of 0°.[5] Conformation

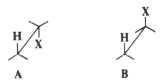

A   B

[2]For a monograph on elimination mechanisms, see Saunders; Cockerill *Mechanisms of Elimination Reactions*; Wiley: New York, 1973. For reviews, see Gandler, in Patai *Supplement A: The Chemistry of Double-bonded Functional Groups*, vol. 2, pt. 1; Wiley: New York, 1989, pp. 733-797; Aleskerov; Yufit; Kucherov *Russ. Chem. Rev.* **1978**, 47, 134-147; Cockerill; Harrison, in Patai *The Chemistry of Functional Groups, Supplement A*, pt. 1; Wiley: New York, 1977, pp. 153-221; Willi *Chimia* **1977**, *31*, 93-101; More O'Ferrall, in Patai *The Chemistry of the Carbon-Halogen Bond*, pt. 2; Wiley: New York, 1973, pp. 609-675; Cockerill, in Bamford; Tipper *Comprehensive Chemical Kinetics*, vol. 9; Elsevier: New York, 1973, pp. 163-372; Saunders *Acc. Chem. Res.* **1976**, *9*, 19-25; Stirling *Essays Chem.* **1973**, *5*, 123-149; Bordwell *Acc. Chem. Res.* **1972**, *5*, 374-381; Fry *Chem. Soc. Rev.* **1972**, *1*, 163-210; LeBel *Adv. Alicyclic Chem.* **1971**, *3*, 195-290; Bunnett *Surv. Prog. Chem.* **1969**, *5*, 53-93; in Patai *The Chemistry of Alkenes*, vol. 1; Wiley: New York, 1964, the articles by Saunders, pp. 149-201 (eliminations in solution); and by Maccoll, pp. 203-240 (pyrolytic eliminations); Köbrich *Angew. Chem. Int. Ed. Engl.* **1965**, *4*, 49-68, pp. 59-63 [*Angew. Chem. 77*, 75-94] (for the formation of triple bonds).
[3]See, for example, Saunders; Edison *J. Am. Chem. Soc.* **1960**, *82*, 138; Shiner; Smith *J. Am. Chem. Soc.* **1958**, *80*, 4095, **1961**, *83*, 593. For a review of isotope effects in elimination reactions, see Fry, Ref. 2.
[4]For reviews, see Bartsch; Závada *Chem. Rev.* **1980**, *80*, 453-494; Coke *Sel. Org. Transform.* **1972**, *2*, 269-307; Sicher *Angew. Chem. Int. Ed. Engl.* **1972**, *11*, 200-214 [*Angew. Chem. 84*, 177-191], *Pure Appl. Chem.* **1971**, *25*, 655-666; Saunders; Cockerill, Ref. 2, pp. 105-163; Cockerill, Ref. 2, pp. 217-235; More O'Ferrall, Ref. 2, pp. 630-640.
[5]DePuy; Morris; Smith; Smat *J. Am. Chem. Soc.* **1965**, *87*, 2421.

A is called *anti-periplanar*, and this type of elimination, in which H and X depart in opposite directions, is called *anti elimination*. Conformation **B** is *syn-periplanar*, and this type of elimination, with H and X leaving in the same direction, is called *syn elimination*. Many examples of both kinds have been discovered. In the absence of special effects (discussed below) anti elimination is usually greatly favored over syn elimination, probably because **A** is a staggered conformation (p. 139) and the molecule requires less energy to reach this transition state than it does to reach the eclipsed transition state **B**. A few of the many known examples of predominant or exclusive anti elimination follow.

1. Elimination of HBr from *meso*-1,2-dibromo-1,2-diphenylethane gave *cis*-2-bromostilbene, while the (+) or (−) isomer gave the trans olefin. This stereospecific result, which

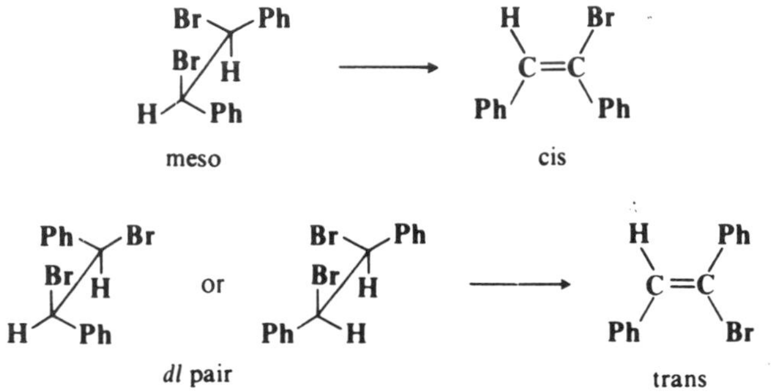

was obtained in 1904,[6] demonstrates that in this case elimination is anti. Many similar examples have been discovered since. Obviously, this type of experiment need not be restricted to compounds that have a meso form. Anti elimination requires that an erythro *dl* pair (or either isomer) give the cis olefin, and the threo *dl* pair (or either isomer) give the trans isomer, and this has been found many times. Anti elimination has also been demonstrated in cases where the electrofuge is not hydrogen. In the reaction of 2,3-dibromobutane with iodide ion, the two bromines are removed (**7-29**). In this case the meso compound gave the trans olefin and the *dl* pair the cis:[7]

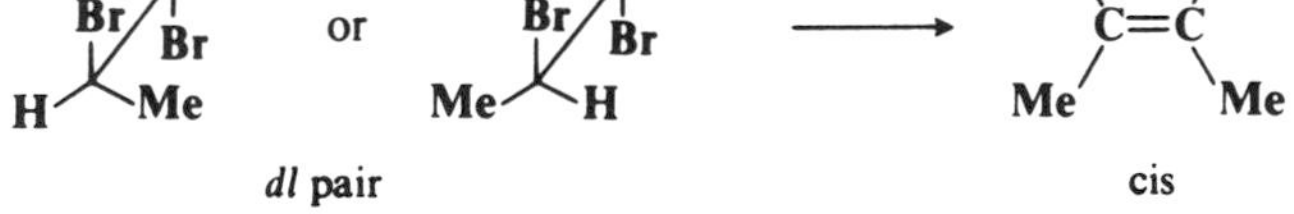

2. In open-chain compounds the molecule can usually adopt that conformation in which H and X are anti-periplanar. However, in cyclic systems this is not always the case. There

[6]Pfeiffer *Z. Phys. Chem.* **1904**, *48*, 40.
[7]Winstein; Pressman; Young *J. Am. Chem. Soc.* **1939**, *61*, 1645.

are nine stereoisomers of 1,2,3,4,5,6-hexachlorocyclohexane: seven meso forms and a *dl* pair (see p. 131). Four of the meso compounds and the *dl* pair (all that were then known) were subjected to elimination of HCl. Only one of these (**1**) has no Cl trans to an H. Of

the other isomers, the fastest elimination rate was about three times as fast as the slowest, but the rate for **1** was 7000 times slower than that of the slowest of the other isomers.[8] This result demonstrates that with these compounds anti elimination is greatly favored over syn elimination, though the latter must be taking place on **1**, very slowly, to be sure.

3. The preceding result shows that elimination of HCl in a six-membered ring proceeds best when the H and X are trans to each other. However, there is an additional restriction. Adjacent trans groups on a six-membered ring can be diaxial or diequatorial (p. 144) and the molecule is generally free to adopt either conformation, though one may have a higher energy than the other. Anti-periplanarity of the leaving groups requires that they be diaxial, even if this is the conformation of higher energy. The results with menthyl and neomenthyl chlorides are easily interpretable on this basis. Menthyl chloride has two chair conformations,

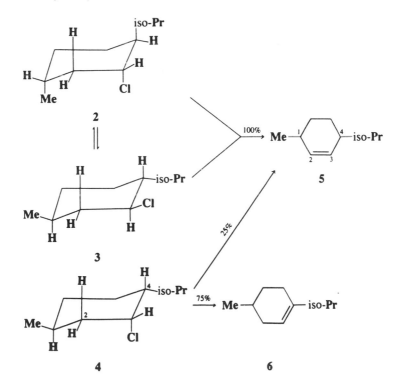

[8]Cristol *J. Am. Chem. Soc.* **1947**, *69*, 338; Cristol; Hause; Meek *J. Am. Chem. Soc.* **1951**, *73*, 674.

2 and **3. 3,** in which the three substituents are all equatorial, is the more stable. The more stable chair conformation of neomenthyl chloride is **4,** in which the chlorine is axial; there are axial hydrogens on both C-2 and C-4. The results are: neomenthyl chloride gives rapid E2 elimination and the olefin produced is predominantly **6** (**6/5** ratio is about 3:1) in accord with Zaitsev's rule (p. 998). Since an axial hydrogen is available on both sides, this factor does not control the direction of elimination and Zaitsev's rule is free to operate. However, for menthyl chloride, elimination is much slower and the product is entirely the anti-Zaitsev **5.** It is slow because the unfavorable conformation **2** has to be achieved before elimination can take place, and the product is **5** because only on this side is there an axial hydrogen.[9]

4. That anti elimination also occurs in the formation of triple bonds is shown by elimination from *cis-* and *trans-*HOOC—CH=CCl—COOH. In this case the product in both cases is HOOCC≡CCOOH, but the trans isomer reacts about 50 times faster than the cis compound.[10]

Some examples of syn elimination have been found in molecules where H and X could not achieve an anti-periplanar conformation.

1. The deuterated norbornyl bromide (**7,** X = Br) gave 94% of the product containing no deuterium.[11] Similar results were obtained with other leaving groups and with bicy-

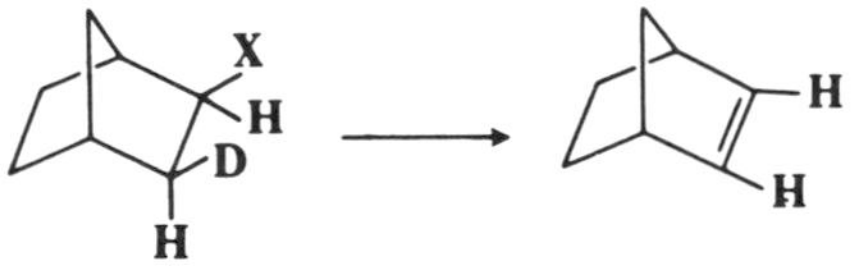

clo[2.2.2] compounds.[12] In these cases the exo X group cannot achieve a dihedral angle of 180° with the endo β hydrogen because of the rigid structure of the molecule. The dihedral angle here is about 120°. These leaving groups prefer syn elimination with a dihedral angle of about 0° to anti elimination with an angle of about 120°.

2. The molecule **8** is a particularly graphic example of the need for a planar transition state. In **8** each Cl has an adjacent hydrogen trans to it, and if planarity of leaving groups

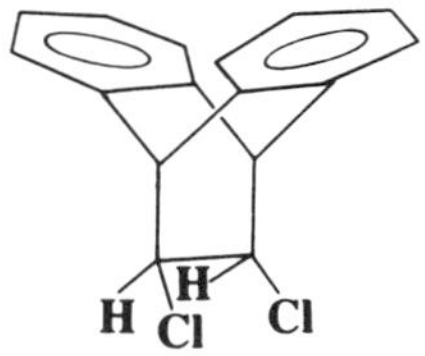

were not required, anti elimination could easily take place. However, the crowding of the rest of the molecule forces the dihedral angle to be about 120°, and elimination of HCl from

[9]Hughes; Ingold; Rose *J. Chem. Soc.* **1953,** 3839.
[10]Michael *J. Prakt. Chem.* **1895,** *52,* 308. See also Marchese; Naso; Modena *J. Chem. Soc. B* **1968,** 958.
[11]Kwart; Takeshita; Nyce *J. Am. Chem. Soc.* **1964,** *86,* 2606.
[12]For example, see Bird; Cookson; Hudec; Williams *J. Chem. Soc.* **1963,** 410; Stille; Sonnenberg; Kinstle *J. Am. Chem. Soc.* **1966,** *88,* 4922; Coke; Cooke *J. Am. Chem. Soc.* **1967,** *89,* 6701; DePuy; Naylor; Beckman *J. Org. Chem.* **1970,** *35,* 2750; Brown; Liu *J. Am. Chem. Soc.* **1970,** *92,* 200; Sicher; Pánkova; Závada; Knieżo; Orahovats *Collect. Czech. Chem. Commun.* **1971,** *36,* 3128; Bartsch; Lee *J. Org. Chem.* **1991,** *56,* 212, 2579.

8 is much slower than from corresponding nonbridged compounds.[13] (Note that syn elimination from **8** is even less likely than anti elimination.) Syn elimination can take place from the trans isomer of **8** (dihedral angle about 0°); this isomer reacted about eight times faster than **8**.[13]

The examples so far given illustrate two points. (1) Anti elimination *requires* a dihedral angle of 180°. When this angle cannot be achieved, anti elimination is greatly slowed or prevented entirely. (2) For the simple systems so far discussed syn elimination is not found to any significant extent unless anti elimination is greatly diminished by failure to achieve the 180° angle.

As noted in Chapter 4 (p. 156), six-membered rings are the only ones among rings of four to thirteen members in which strain-free anti-periplanar conformations can be achieved. It is not surprising, therefore, that syn elimination is least common in six-membered rings. Cooke and Coke subjected cycloalkyltrimethylammonium hydroxides to elimination (**7-6**) and found the following percentages of syn elimination with ring size: four-membered, 90%; five-membered, 46%; six-membered, 4% seven-membered, 31 to 37%.[14] It should be noted that the NMe_3^+ group has a greater tendency to syn elimination than do other common leaving groups such as OTs, Cl, and Br.

Other examples of syn elimination have been found in medium-ring compounds, where both cis and trans olefins are possible (p. 128). As an illustration, we can look at experiments performed by Závada, Svoboda, and Sicher.[15] These workers subjected 1,1,4,4-tetramethyl-7-cyclodecyltrimethylammonium chloride (**9**) to elimination and obtained mostly *trans*- but

also some *cis*-tetramethylcyclodecenes as products. (Note that *trans*-cyclodecenes, though stable, are less stable than the cis isomers). In order to determine the stereochemistry of the reaction, they repeated the elimination, this time using deuterated substrates. They found that when **9** was deuterated in the trans position (H_t = D), there was a substantial isotope effect in the formation of *both* cis and trans olefins, but when **9** was deuterated in the cis position (H_c = D), there was *no* isotope effect in the formation of either olefin. Since an isotope effect is expected for an E2 mechanism,[16] these results indicated that *only the trans hydrogen* (H_t) was lost, whether the product was the cis or the trans isomer.[17] This in turn means that the cis isomer must have been formed by anti elimination and the trans isomer by syn elimination. (Anti elimination could take place from approximately the conformation shown, but for syn elimination the molecule must twist into a conformation in which the $C-H_t$ and $C-NMe_3^+$ bonds are syn-periplanar.) This remarkable result, called the *syn-anti dichotomy*, has also been demonstrated by other types of evidence.[18] The fact

[13]Cristol; Hause *J. Am. Chem. Soc.* **1952**, *74*, 2193.

[14]Cooke; Coke *J. Am. Chem. Soc.* **1968**, *90*, 5556. See also Coke; Smith; Britton *J. Am. Chem. Soc.* **1975**, *97*, 4323.

[15]Závada; Svoboda; Sicher *Tetrahedron Lett.* **1966**, 1627, *Collect. Czech. Chem. Commun.* **1968**, *33*, 4027.

[16]Other possible mechanisms, such as E1cB (p. 991) or α′,β elimination (p. 1018), were ruled out in all these cases by other evidence.

[17]This conclusion has been challenged by Coke, Ref. 4.

[18]Sicher; Závada; Krupička *Tetrahedron Lett.* **1966**, 1619; Sicher; Závada *Collect. Czech. Chem. Commun.* **1967**, *32*, 2122; Závada; Sicher *Collect. Czech. Chem. Commun.* **1967**, *32*, 3701. For a review, see Bartsch; Závada, Ref. 4.

that syn elimination in this case predominates over anti (as indicated by the formation of trans isomer in greater amounts than cis) has been explained by conformational factors.[19] The syn-anti dichotomy has also been found in other medium-ring systems (8- to 12-membered),[20] though the effect is greatest for 10-membered rings. With leaving groups,[21] the extent of this behavior decreases in the order NMe_3^+ > OTs > Br > Cl, which parallels steric requirements. When the leaving group is uncharged, syn elimination is favored by strong bases and by weakly ionizing solvents.[22]

Syn elimination and the syn-anti dichotomy have also been found in open-chain systems, though to a lesser extent than in medium-ring compounds. For example, in the conversion of 3-hexyl-4-d-trimethylammonium ion to 3-hexene with potassium sec-butoxide, about 67% of the reaction followed the syn-anti dichotomy.[23] In general syn elimination in open-chain systems is only important in cases where certain types of steric effect are present. One such type is compounds in which substituents are found on both the β' and the γ carbons (the unprimed letter refers to the branch in which the elimination takes place). The factors that cause these results are not completely understood, but the following conformational effects have been proposed as a partial explanation.[24] The two anti- and two syn-periplanar conformations are, for a quaternary ammonium salt:

C	D	E	F
anti ⟶ trans	anti ⟶ cis	syn ⟶ trans	syn ⟶ cis

In order for an E2 mechanism to take place a base must approach the proton marked *. In **C** this proton is shielded on both sides by R and R'. In **D** the shielding is on only one side. Therefore, when anti elimination does take place in such systems, it should give more cis product than trans. Also, when the normal anti elimination pathway is hindered sufficiently to allow the syn pathway to compete, the anti → trans route should be diminished more than the anti → cis route. When syn elimination begins to appear, it seems clear that **E,** which is less eclipsed than **F,** should be the favored pathway and syn elimination should generally give the trans isomer. In general, deviations from the syn-anti dichotomy are greater on the trans side than on the cis. Thus, trans olefins are formed partly or mainly by syn elimination, but cis olefins are formed entirely by anti elimination. Predominant syn

[19]For discussions, see Ref. 4.
[20]For example, see Coke; Mourning *J. Am. Chem. Soc.* **1968**, *90*, 5561, where the experiment was performed on cyclooctyltrimethylammonium hydroxide, and *trans*-cyclooctene was formed by a 100% syn mechanism, and *cis*-cyclooctene by a 51% syn and 49% anti mechanism.
[21]For examples with other leaving groups, see Závada; Krupička; Sicher *Chem. Commun.* **1967**, *66*, *Collect. Czech. Chem. Commun.* **1968**, *33*, 1393; Sicher; Jan; Schlosser *Angew. Chem. Int. Ed. Engl.* **1971**, *10*, 926 [*Angew. Chem. 83*, 1012]; Závada; Pánková *Collect. Czech. Chem. Commun.* **1980**, *45*, 2171.
[22]See, for example, Sicher; Závada *Collect. Czech. Chem. Commun.* **1968**, *33*, 1278.
[23]Bailey; Saunders *Chem. Commun.* **1968**, 1598, *J. Am. Chem. Soc.* **1970**, *92*, 6904. For other examples of syn elimination and the syn-anti dichotomy in open-chain systems, see Pánková; Sicher; Závada *Chem. Commun.* **1967**, 394; Pánková; Vítek; Vašíčková; Řeřicha; Závada *Collect. Czech. Chem. Commun.* **1972**, *37*, 3456; Schlosser; An *Helv. Chim. Acta* **1979**, *62*, 1194; Sugita; Nakagawa; Nishimoto; Kasai; Ichikawa *Bull. Chem. Soc. Jpn.* **1979**, *52*, 871; Pánková; Kocián; Krupička; Závada *Collect. Czech. Chem. Commun.* **1983**, *48*, 2944.
[24]Bailey; Saunders, Ref. 23; Chiao; Saunders *J. Am. Chem. Soc.* **1977**, *99*, 6699.

elimination has also been found in compounds of the form $R^1R^2CHCHDNMe_3{}^+$, where R^1 and R^2 are both bulky.[25] In this case also the conformation leading to syn elimination (**H**) is less strained than **G,** which gives anti elimination. **G** has three bulky groups (including $NMe_3{}^+$) in the gauche position to each other.

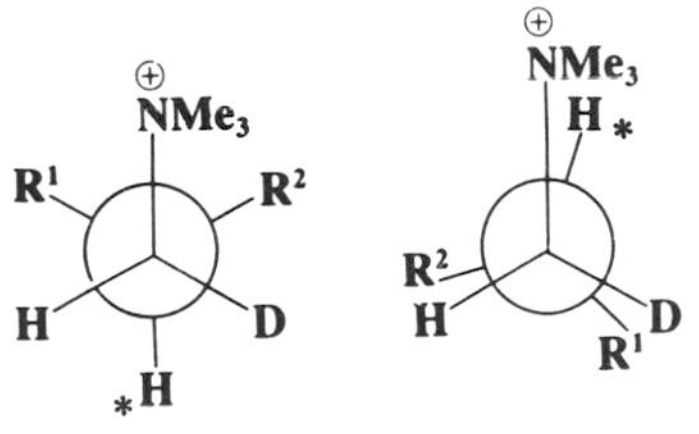

It was mentioned above that weakly ionizing solvents promote syn elimination when the leaving group is uncharged. This is probably caused by ion pairing, which is greatest in nonpolar solvents.[26] Ion pairing can cause syn elimination with an uncharged leaving group by means of the transition state shown in **10.** This effect was graphically illustrated by

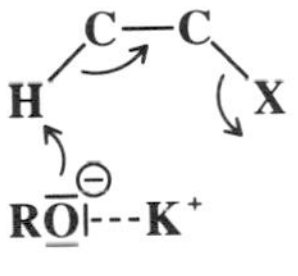

10

elimination from 1,1,4,4-tetramethyl-7-cyclodecyl bromide.[27] The ratio of syn to anti elimination when this compound was treated with t-BuOK in the nonpolar benzene was 55.0. But when the crown ether dicyclohexano-18-crown-6 was added (this compound selectively removes K^+ from the t-BuO$^-$ K^+ ion pair and thus leaves t-BuO$^-$ as a free ion), the syn/anti ratio decreased to 0.12. Large decreases in the syn/anti ratio on addition of the crown ether were also found with the corresponding tosylate and with other nonpolar solvents.[28] However, with positively charged leaving groups the effect is reversed. Here, ion pairing *increases* the amount of anti elimination.[29] In this case a relatively free base (e.g., PhO$^-$) can be attracted to the leaving group, putting it in a favorable position for attack on the syn β hydrogen, while ion pairing would reduce this attraction.

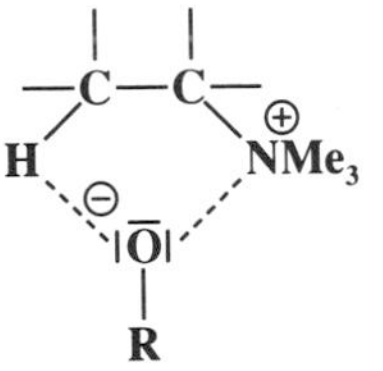

[25]Tao; Saunders *J. Am. Chem. Soc.* **1983,** *105,* 3183; Dohner; Saunders *J. Am. Chem. Soc.* **1986,** *108,* 245.

[26]For reviews of ion pairing in this reaction, see Bartsch; Závada, Ref. 4; Bartsch *Acc. Chem. Res.* **1975,** *8,* 239-245.

[27]Svoboda; Hapala; Závada *Tetrahedron Lett.* **1972,** 265.

[28]For other examples of the effect of ion pairing, see Bayne; Snyder *Tetrahedron Lett.* **1971,** 571; Bartsch; Wiegers *Tetrahedron Lett.* **1972,** 3819; Fiandanese; Marchese; Naso; Sciacovelli *J. Chem. Soc., Perkin Trans. 2* **1973,** 1336; Borchardt; Swanson; Saunders *J. Am. Chem. Soc.* **1974,** *96,* 3918; Mano; Sera; Maruyama *Bull. Chem. Soc. Jpn.* **1974,** *47,* 1758; Závada; Pánková; Svoboda *Collect. Czech. Chem. Commun.* **1976,** *41,* 3778; Baciocchi; Ruzziconi; Sebastiani *J. Org. Chem.* **1979,** *44,* 3718; Croft; Bartsch *Tetrahedron Lett.* **1983,** *24,* 2737; Kwart; Gaffney; Wilk *J. Chem. Soc., Perkin Trans. 2* **1984,** 565.

[29]Borchardt; Saunders *J. Am. Chem. Soc.* **1974,** *96,* 3912.

We can conclude that anti elimination is generally favored in the E2 mechanism, but that steric (inability to form the anti-periplanar transition state), conformational, ion-pairing, and other factors cause syn elimination to intervene (and even predominate) in some cases.

The E1 Mechanism

The E1 mechanism is a two-step process in which the rate-determining step is ionization of the substrate to give a carbocation that rapidly loses a β proton to a base, usually the solvent:

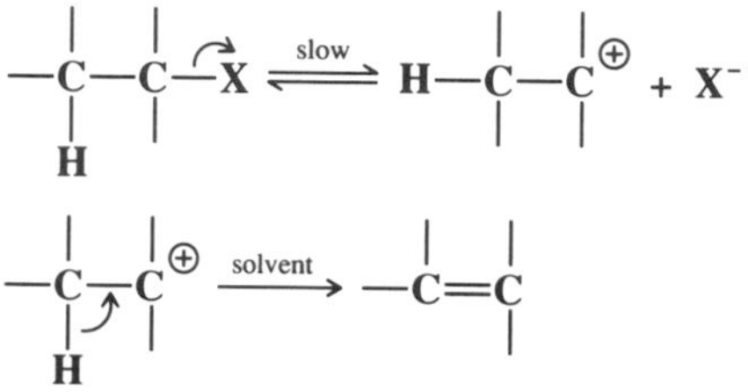

Step 1
$$-\overset{\mid}{\underset{\mid}{C}}-\overset{\mid}{\underset{\overset{\mid}{H}}{C}}\overset{\frown}{-}X \overset{slow}{\rightleftharpoons} H-\overset{\mid}{\underset{\mid}{C}}-\overset{\mid}{\underset{\mid}{\overset{\oplus}{C}}} + X^-$$

Step 2
$$-\overset{\mid}{\underset{\overset{\mid}{H}}{C}}\overset{\frown}{\underset{}{}}\overset{\mid}{\underset{\mid}{\overset{\oplus}{C}}} \overset{solvent}{\longrightarrow} -\overset{\mid}{C}=\overset{\mid}{C}-$$

The IUPAC designation is $D_N + D_E$ (or $D_N + D_H$). This mechanism normally operates without an *added* base. Just as the E2 mechanism is analogous to and competes with the SN2, so is the E1 mechanism related to the SN1. In fact, the first step of the E1 is exactly the same as that of the SN1 mechanism. The second step differs in that the solvent pulls a proton from the β carbon of the carbocation rather than attacking it at the positively charged carbon, as in the SN1 process. In a pure E1 reaction (i.e., without ion pairs, etc.) the product should be completely nonstereospecific, since the carbocation is free to adopt its most stable conformation before giving up the proton.

Some of the evidence for the E1 mechanism is as follows:

1. The reaction exhibits first-order kinetics (in substrate) as expected. Of course the solvent is not expected to appear in the rate equation, even if it were involved in the rate-determining step (p. 222), but this point can be easily checked by adding a small amount of the conjugate base of the solvent. It is generally found that such an addition does not increase the rate of the reaction. If this more powerful base does not enter into the rate-determining step, it is unlikely that the solvent does. An example of an E1 mechanism with a rate-determining second step (proton transfer) has been reported.[30]

2. If the reaction is performed on two molecules that differ only in the leaving group (for example, *t*-BuCl and *t*-BuSMe$_2^+$), the rates should obviously be different, since they depend on the ionizing ability of the molecule. However, once the carbocation is formed, if the solvent and the temperature are the same, it should suffer the same fate in both cases, since the nature of the leaving group does not affect the second step. This means that *the ratio of elimination to substitution should be the same*. The compounds mentioned in the example were solvolyzed at 65.3°C in 80% aqueous ethanol with the following results:[31]

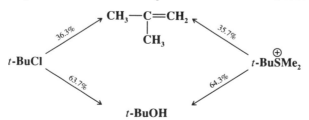

[30]Baciocchi; Clementi; Sebastiani; Ruzziconi *J. Org. Chem.* **1979**, *44*, 32.
[31]Cooper; Hughes; Ingold; MacNulty *J. Chem. Soc.* **1948**, 2038.

Although the rates were greatly different (as expected with such different leaving groups), the product ratios were the same, within 1%. If this had taken place by a second-order mechanism, the nucleophile would not be expected to have the same ratio of preference for attack at the β hydrogen compared to attack at a *neutral* chloride as for attack at the β hydrogen compared to attack at a *positive* SMe$_2$ group.

3. Many reactions carried out under first-order conditions on systems where E2 elimination is anti proceed quite readily to give olefins where a cis hydrogen must be removed, often in preference to the removal of a trans hydrogen. For example, menthyl chloride (**2**, p. 985), which by the E2 mechanism gave only **5**, under E1 conditions gave 68% **6** and 32% **5**, since the steric nature of the hydrogen is no longer a factor here, and the more stable olefin (Zaitsev's rule, p. 998) is predominantly formed.

4. If carbocations are intermediates, we should expect rearrangements with suitable substrates. These have often been found in elimination reactions performed under E1 conditions.

E1 reactions can involve ion pairs, just as is true for S$_N$1 reactions (p. 302).[32] This effect is naturally greatest for nondissociating solvents: it is least in water, greater in ethanol, and greater still in acetic acid. It has been proposed that the ion-pair mechanism (p. 305) extends to elimination reactions too, and that the S$_N$1, S$_N$2, E1, and E2 mechanisms possess in common an ion-pair intermediate, at least occasionally.[33]

The E1cB Mechanism[34]

In the E1 mechanism, X leaves first and then H. In the E2 mechanism the two groups leave at the same time. There is a third possibility: the H leaves first and then the X. This is a two-step process, called the *E1cB mechanism*, or the *carbanion mechanism*, since the intermediate is a carbanion:

Step 1

$$-\overset{\overset{\displaystyle H}{|}}{\underset{|}{C}}-\overset{|}{\underset{|}{C}}-X \;\overset{base}{\rightleftharpoons}\; {}^{\ominus}\overset{|}{\underset{|}{C}}-\overset{|}{\underset{|}{C}}-X$$

11

Step 2

$$ {}^{\ominus}\overset{|}{\underset{|}{C}}\overset{\frown}{}\overset{|}{\underset{|}{C}}\overset{\frown}{}X \longrightarrow -\overset{|}{C}=\overset{|}{C}- $$

The name E1cB comes from the fact that it is the conjugate base of the substrate that is giving up the leaving group (see the S$_N$1cB mechanism, p. 356). The IUPAC designation is A$_n$D$_E$ + D$_N$ or A$_{xh}$D$_H$ + D$_N$ (see p. 290). We can distinguish three limiting cases: (1) The carbanion returns to starting material faster than it forms product: step 1 is reversible;

[32]Cocivera; Winstein *J. Am. Chem. Soc.* **1963**, *85*, 1702; Smith; Goon *J. Org. Chem.* **1969**, *34*, 3127; Bunnett; Eck *J. Org. Chem.* **1971**, *36*, 897; Sridharan; Vitullo *J. Am. Chem. Soc.* **1977**, *99*, 8093; Seib; Shiner; Sendijarevič; Humski *J. Am. Chem. Soc.* **1978**, *100*, 8133; Jansen; Koshy; Mangru; Tidwell *J. Am. Chem. Soc.* **1981**, *103*, 3863; Coxon; Simpson; Steel; Whiteling *Tetrahedron* **1984**, *40*, 3503; Thibblin *J. Am. Chem. Soc.* **1987**, *109*, 2071, *J. Phys. Org. Chem.* **1989**, *2*, 15.
[33]Sneen; Robbins *J. Am. Chem. Soc.* **1969**, *91*, 3100, Sneen *Acc. Chem. Res.* **1973**, *6*, 46-53. See, however, McLennan *J. Chem. Soc., Perkin Trans. 2* **1972**, 1577.
[34]For reviews, see Cockerill; Harrison, Ref. 2, pp. 158-178; Hunter *Intra-Sci. Chem. Rep.* **1973**, *7*(3), 19-26; McLennan *Q. Rev., Chem. Soc.* **1967**, *21*, 490-506. For a general discussion, see Koch *Acc. Chem. Res.* **1984**, *17*, 137-144.

step 2 is slow. (2) Step 1 is the slow step, and formation of product is faster than return of the carbanion to starting material. In this case step 1 is essentially irreversible. (3) Step 1 is rapid, and the carbanion goes slowly to product. This case occurs only with the most stable carbanions. Here, too, step 1 is essentially irreversible. These cases have been given the designations: (1) $(E1cB)_R$, (2) $(E1cB)_I$ (or $E1cB_{irr}$), and (3) $(E1)_{anion}$. Their characteristics are listed in Table 17.1.[35] Investigations of the reaction order are generally not very useful (except for case 3 which is first order), because cases 1 and 2 are second order and thus difficult or impossible to distinguish from the E2 mechanism by this procedure.[36] We would expect the greatest likelihood of finding the E1cB mechanism in substrates that have (a) a poor nucleofuge and (b) an acidic hydrogen, and most investigations have concerned such substrates. The following is some of the evidence in support of the E1cB mechanism.

1. The first step of the $(E1cB)_R$ mechanism involves a reversible exchange of protons between the substrate and the base. In that case, if deuterium is present in the base, recovered starting material should contain deuterium. This was found to be the case in the treatment of $Cl_2C=CHCl$ with NaOD to give $ClC\equiv CCl$. When the reaction was stopped before completion, there *was* deuterium in the recovered olefin.[37] A similar result was found for pentahaloethanes.[38] These substrates are relatively acidic. In both cases the electron-withdrawing halogens increase the acidity of the hydrogen, and in the case of trichloroethylene there is the additional factor that a hydrogen on an sp^2 carbon is more acidic than one on an sp^3 carbon (p. 269). Thus, the E1cB mechanism is more likely to be found in eliminations yielding triple bonds than in those giving double bonds. Another likely place for the E1cB mechanism should be in reaction of a substrate like $PhCH_2CH_2Br$, since the carbanion is stabilized by resonance with the phenyl group. Nevertheless, no deuterium exchange was found here.[39] If this type of evidence is a guide, then it may be inferred that the $(E1cB)_R$ mechanism is quite rare, at least for eliminations with common leaving groups such as Br, Cl, or OTs, which yield C=C double bonds.

2. When the reaction

$$p\text{-NO}_2\text{C}_6\text{H}_4\text{—CH}_2\text{—CH}_2\text{—NR}_3{}^+ + \text{B}^- \longrightarrow p\text{-NO}_2\text{C}_6\text{H}_4\text{—CH}=\text{CH}_2 + \text{BH} + \text{NR}_3$$

was carried out in water containing acetohydroxamate buffers, a plot of the rate against the buffer concentration was curved and the rate leveled off at high buffer concentrations, indicating a change in rate-determining step.[40] This rules out an E2 mechanism, which has only one step. When D_2O was used instead of H_2O as solvent, there was an initial inverse solvent isotope effect of 7.7 (the highest inverse solvent isotope effect yet reported). That is, the reaction took place faster in D_2O than in H_2O. This is compatible only with an E1cB mechanism in which the proton-transfer step is not entirely rate-determining. The isotope effect arises from a partitioning of the carbanion intermediate **11**. This intermediate either can go to product or it can revert to starting compound, which requires taking a proton from the solvent. In D_2O the latter process is slower (because the O—D bond of D_2O cleaves less easily than the O—H bond of H_2O), reducing the rate at which **11** returns to

[35]This table, which appears in Cockerill; Harrison, Ref. 2, p. 161, was adapted from a longer one in Bordwell, Ref. 2, p. 375.

[36](E1cB)$_I$ cannot be distinguished from E2 by this means, because it has the identical rate law: Rate = k[substrate][B$^-$]. The rate law for (E1cB)$_R$ is different: Rate = k[substrate][B$^-$]/[BH], but this is often not useful because the only difference is that the rate is also dependent (inversely) on the concentration of the conjugate acid of the base, and this is usually the solvent, so that changes in its concentration cannot be measured.

[37]Houser; Bernstein; Miekka; Angus *J. Am. Chem. Soc.* **1955**, *77*, 6201.

[38]Hine; Wiesboeck; Ghirardelli *J. Am. Chem. Soc.* **1961**, *83*, 1219; Hine; Wiesboeck; Ramsay *J. Am. Chem. Soc.* **1961**, *83*, 1222.

[39]Skell; Hauser *J. Am. Chem. Soc.* **1945**, *67*, 1661.

[40]Keeffe; Jencks *J. Am. Chem. Soc.* **1983**, *105*, 265.

TABLE 17.1 Kinetic predictions for base-induced β-eliminations[35]

$$\overline{B} + (D)H-\overset{\displaystyle\diagup}{\underset{\displaystyle\diagdown}{C}}_\beta-\overset{\displaystyle\diagup}{\underset{\displaystyle\diagdown}{C}}_\alpha-X \longrightarrow BH + \overset{\displaystyle\diagup}{C}=\overset{\displaystyle\diagdown}{C} + \overline{X}$$

Mechanism	Kinetic[a] order	β-hydrogen exchange faster than elimination	General or specific base catalysis	k_H/k_D	Electron withdrawal at $C_\beta{}^d$	Electron release at $C_\alpha{}^d$	Leaving-group isotope effect or element effect
$(E1)_{anion}$	1	Yes	General[c]	1.0	Rate decrease	Rate increase	Substantial
$(E1cB)_R$	2	Yes	Specific	1.0	Small rate increase	Small rate increase	Substantial
$(E1cB)_{ip}$	2	No	General[e]	$1.0 \rightarrow 1.2$	Small rate increase	Small rate increase	Substantial
$(E1cB)_I$	2	No	General	$2 \rightarrow 8$	Rate increase	Little effect	Small to negligible
$E2^b$	2	No	General	$2 \rightarrow 8$	Rate increase	Small rate increase	Small

[a]All mechanisms exhibit first-order kinetics in substrate.
[b]Only transition states with considerable carbanion character considered in this table.
[c]Specific base catalysis predicted if extent of substrate ionization reduced from almost complete.
[d]Effect on rate assuming no change in mechanism is caused; steric factors upon substitution at C_α and C_β have not been considered. The rate predictions are geared to substituent effects such as those giving rise to Hammett reaction constants on β- and α-aryl substitution.
[e]Depends on whether ion pair assists in removal of leaving group.

starting compound. With the return reaction competing less effectively, the rate of conversion of **11** to product is increased.

3. We have predicted that the E1cB mechanism would be most likely to be found with substrates containing acidic hydrogens and poor leaving groups. Compounds of the type ZCH_2CH_2OPh, where Z is an electron-withdrawing group (e.g., NO_2, $SMe_2{}^+$, $ArSO_2$, CN, COOR, etc.), belong to this category, because OPh is a very poor leaving group (p. 352). There is much evidence to show that the mechanism here is indeed E1cB.[41] Isotope effects, measured for $MeSOCD_2CH_2OPh$ and $Me_2\overset{\oplus}{S}CD_2CH_2OPh$ with NaOD in D_2O, are about 0.7. This is compatible with an $(E1cB)_R$ mechanism, but not with an E2 mechanism for which an isotope effect of perhaps 5 might be expected (of course, an E1 mechanism is precluded by the extremely poor nucleofugal ability of OPh). The fact that k_H/k_D is less than the expected value of 1 is attributable to solvent and secondary isotope effects. Among other evidence for an E1cB mechanism in these systems is that changes in the identity of Z had a dramatic effect on the relative rates: a span of 10^{11} between NO_2 and COO^-. Note that elimination from substrates of the type $RCOCH_2CH_2Y$ is the reverse of Michael-type addition to C=C bonds. We have seen (p. 741) that such addition involves initial attack by a nucleophile Y and subsequent attack by a proton. Thus the initial loss of a proton from substrates of this type (i.e., an E1cB mechanism) is in accord with the principle of microscopic reversibility.[42] It may also be recalled that benzyne formation (p. 647) can occur by such a

[41]Crosby; Stirling *J. Chem. Soc. B* **1970**, 671, 679; Redman; Stirling *Chem. Commun.* **1970**, 633; Cann; Stirling *J. Chem. Soc., Perkin Trans. 2* **1974**, 820. For other examples; see Fedor *J. Am. Chem. Soc.* **1969**, *91*, 908; More O'Ferrall; Slae *J. Chem. Soc. B* **1970**, 260; Kurzawa; Leffek *Can. J. Chem.* **1977**, 55, 1696.
[42]Patai; Weinstein; Rappoport *J. Chem. Soc.* **1962**, 1741. See also Hilbert; Fedor *J. Org. Chem.* **1978**, *43*, 452.

process. It has been suggested that all base-initiated eliminations wherein the proton is activated by a strong electron-withdrawing group are E1cB reactions,[43] but there is evidence that this is not the case—that when there is a good nucleofuge, the mechanism is E2 even when strong electron-withdrawing groups are present.[44] On the other hand, Cl⁻ has been found to be a leaving group in an E1cB reaction.[45]

Of the three cases of the E1cB mechanism, the one most difficult to distinguish from E2 is (E1cB)₁. One way to make this distinction is to study the effect of a change in leaving group. This was done in the case of the three acenaphthylenes **12,** where it was found that (1) the three rates were fairly similar, the largest being only about four times that of the

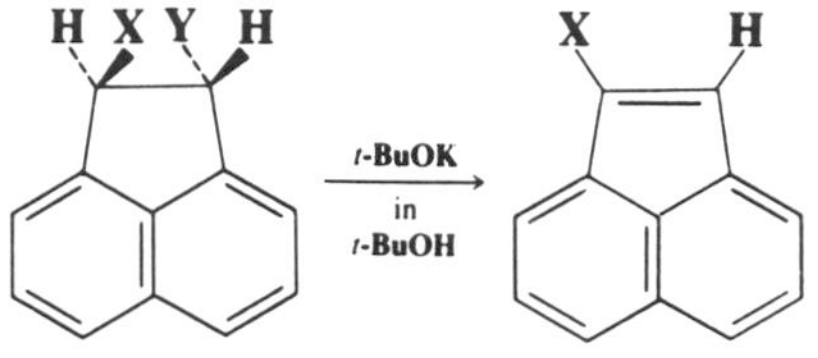

	X	Y
a	Br	Cl
b	Cl	Cl
c	Cl	F

smallest, and (2) in compound *c* (X = Cl, Y = F), the only product contained Cl and no F, i.e., only the poorer nucleofuge F departed while Cl remained.[46] Result (1) rules out all the E1cB mechanisms except (E1cB)₁, because the others should all have considerable leaving group effects (Table 17.1). An ordinary E2 mechanism should also have a large leaving group effect, but an E2 mechanism with substantial carbanionic character (see the next section) might not. However, no E2 mechanism can explain result (2), which can be explained by the fact that an α Cl is more effective than an α F in stabilizing the planar carbanion that remains when the proton is lost. Thus (as in the somewhat similar case of aromatic nucleophilic substitution, see p. 653), when X⁻ leaves in the second step, the one that leaves is not determined by which is the better nucleofuge, but by which has had its β hydrogen removed.[47] Additional evidence for the existence of the (E1cB)₁ mechanism was the observation of a change in the rate-determining step in the elimination reaction of N-(2-cyanoethyl)pyridinium ions **13,** treated with base, when X was changed.[48] Once again,

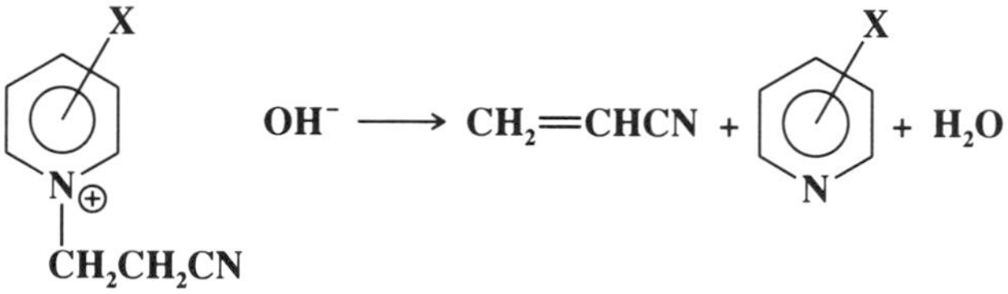

the demonstration that two steps are involved precludes the one-step E2 mechanism.

[43]Bordwell; Vestling; Yee *J. Am. Chem. Soc.* **1970,** *92,* 5950; Bordwell, Ref. 2.
[44]Marshall; Thomas; Stirling *J. Chem. Soc., Perkin Trans. 2,* **1977,** 1898, 1914; Fishbein; Jencks *J. Am. Chem. Soc.* **1988,** *110,* 5075, 5087; Banait; Jencks *J. Am. Chem. Soc.* **1990,** *112,* 6950.
[45]Ölwegård; McEwen; Thibblin; Ahlberg *J. Am. Chem. Soc.* **1985,** *107,* 7494.
[46]Baciocchi; Ruzziconi; Sebastiani *J. Org. Chem.* **1982,** *47,* 3237.
[47]For other evidence for the existence of the (E1cB)₁ mechanism, see Bordwell; Vestling; Yee, Ref. 43; Fedor; Glave *J. Am. Chem. Soc.* **1971,** *93,* 985; Redman; Thomas; Stirling *J. Chem. Soc., Perkin Trans. 2* **1978,** 1135; Thibblin *Chem. Scr.* **1980,** *15,* 121; Carey; More O'Ferrall; Vernon *J. Chem. Soc., Perkin Trans. 2* **1982,** 1581; Baciocchi; Ruzziconi *J. Org. Chem.* **1984,** *49,* 3395; Jarczewski; Waligorska; Leffek *Can. J. Chem.* **1985,** *63,* 1194; Gula; Vitale; Dostal; Trometer; Spencer *J. Am. Chem. Soc.* **1988,** *110,* 4400; Garay; Cabaleiro *J. Chem. Res. (S)* **1988,** 388; Gandler; Storer; Ohlberg *J. Am. Chem. Soc.* **1990,** *112,* 7756.
[48]Bunting; Toth; Heo; Moors *J. Am. Chem. Soc.* **1990,** *112,* 8878. See also Bunting; Kanter *J. Am. Chem. Soc.* **1991,** *113,* 6950.

4. An example of an $(E1)_{anion}$ mechanism has been found with the substrate **14**, which when treated with methoxide ion undergoes elimination to **16**, which is unstable under the

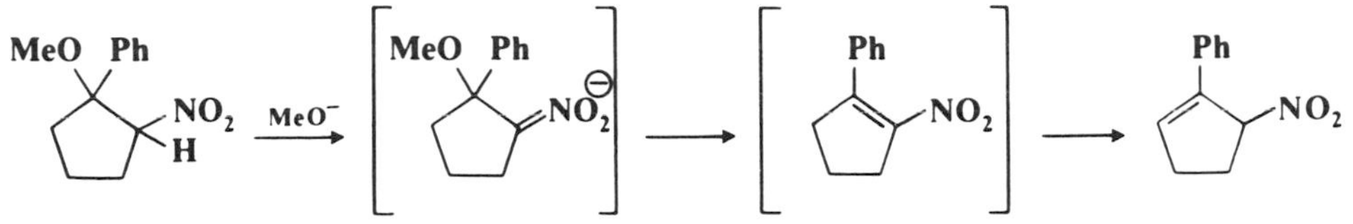

reaction conditions and rearranges as shown.[49] Among the evidence for the proposed mechanism in this case were kinetic and isotope-effect results, as well as the spectral detection of **15**.[50]

5. In many eliminations to form C=O and C≡N bonds the initial step is loss of a positive group (normally a proton) from the oxygen or nitrogen. These may also be regarded as E1cB processes.

There is evidence that some E1cB mechanisms can involve carbanion ion pairs, e.g.,[51]

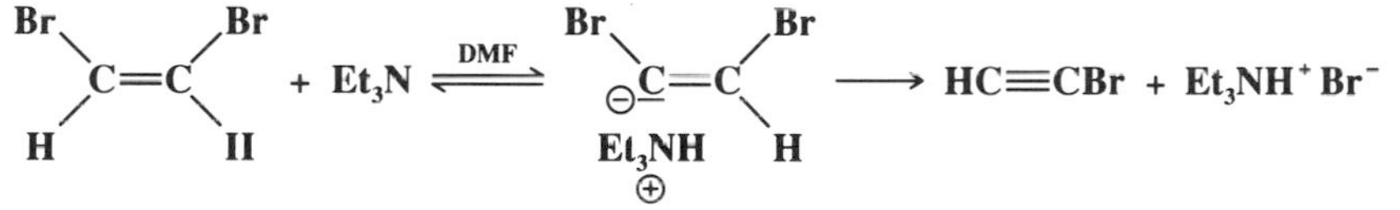

This case is designated $(E1cB)_{ip}$; its characteristics are shown in Table 17.1.

The E1–E2–E1cB Spectrum

In the three mechanisms so far considered the similarities are greater than the differences. In each case there is a leaving group that comes off with its pair of electrons and another group (usually hydrogen) that comes off without them. The only difference is in the order of the steps. It is now generally accepted that there is a spectrum of mechanisms ranging from one extreme, in which the leaving group departs well before the proton (pure E1), to the other extreme, in which the proton comes off first and then, after some time, the leaving group follows (pure E1cB). The *pure* E2 case would be somewhere in the middle, with both groups leaving simultaneously. However, most E2 reactions are not exactly in the middle, but somewhere to one side or the other. For example, the nucleofuge might depart just before the proton. This case may be described as an E2 reaction with a small amount of E1 character. The concept can be expressed by the question: In the transition state, which bond (C—H or C—X) has undergone more cleavage?[52]

[49]Bordwell; Yee; Knipe *J. Am. Chem. Soc.* **1970**, *92*, 5945.

[50]For other examples of this mechanism, see Rappoport *Tetrahedron Lett.* **1968**, 3601; Berndt *Angew. Chem. Int. Ed. Engl.* **1969**, *8*, 613 [*Angew. Chem. 81*, 567]; Albeck; Hoz; Rappoport *J. Chem. Soc., Perkin Trans. 2* **1972**, 1248, **1975**, 628.

[51]Kwok; Lee; Miller *J. Am. Chem. Soc.* **1969**, *91*, 468. See also Lord; Naan; Hall *J. Chem. Soc. B* **1971**, 220; Rappoport; Shohamy *J. Chem. Soc. B* **1971**, 2060; Fiandanese; Marchese; Naso *J. Chem. Soc., Chem. Commun.* **1972**, 250; Koch; Dahlberg; Toczko; Solsky *J. Am. Chem. Soc.* **1973**, *95*, 2029; Hunter; Shearing *J. Am. Chem. Soc.* **1973**, *95*, 8333; Thibblin; Ahlberg *J. Am. Chem. Soc.* **1977**, *99*, 7926, **1979**, *101*, 7311; Thibblin; Bengtsson; Ahlberg *J. Chem. Soc., Perkin Trans. 2* **1977**, 1569; Petrillo; Novi; Garbarino; Dell'Erba; Mugnoli *J. Chem. Soc., Perkin Trans. 2* **1985**, 1291.

[52]For discussions, see Cockerill; Harrison, Ref. 2, pp. 178-189; Saunders *Acc. Chem. Res.*, Ref. 2; Bunnett, Ref. 2; Saunders; Cockerill, Ref. 2, pp. 47-104; Bordwell, Ref. 2.

One way to determine just where a given reaction stands on the E1–E2–E1cB spectrum is to study isotope effects, which ought to tell something about the behavior of bonds in the transition state.[53] For example, $CH_3CH_2NMe_3^+$ showed a nitrogen isotope effect (k^{14}/k^{15}) of 1.017, while $PhCH_2CH_2NMe_3^+$ gave a corresponding value of 1.009.[54] It would be expected that the phenyl group would move the reaction toward the E1cB side of the line, which means that for this compound the C—N bond is not as greatly broken in the transition state as it is for the unsubstituted one. The isotope effect bears this out, for it shows that in the phenyl compound, the mass of the nitrogen has less effect on the reaction rate than it does in the unsubstituted compound. Similar results have been obtained with SR_2^+ leaving groups by the use of $^{32}S/^{34}S$ isotope effects[55] and with Cl ($^{35}Cl/^{37}Cl$).[56] The position of reactions along the spectrum has also been studied from the other side of the newly forming double bond by the use of H/D and H/T isotope effects,[57] though interpretation of these results is clouded by the fact that β hydrogen isotope effects are expected to change smoothly from small to large to small again as the degree of transfer of the β hydrogen from the β carbon to the base increases[58] (recall—p. 227—that isotope effects are greatest when the proton is half-transferred in the transition state), by the possibility of secondary isotope effects (e.g., the presence of a β deuterium or tritium may cause the leaving group to depart more slowly), and by the possibility of tunneling[59] (see footnote 55 in Chapter 6). Other isotope-effect studies have involved labeled α or β carbon, labeled α hydrogen, or labeled base.[53]

Another way to study the position of a given reaction on the spectrum involves the use of β aryl substitution. Since a positive Hammet ρ value is an indication of a negatively charged transition state, the ρ value for substituted β aryl groups should increase as a reaction moves from E1-like to E1cB-like along the spectrum. This has been shown to be the case in a number of studies;[60] e.g., ρ values of $ArCH_2CH_2X$ increase as the leaving-group ability of X decreases. A typical set of ρ values was: X = I, 2.07; Br, 2.14; Cl, 2.61; SMe_2^+, 2.75; F, 3.12.[61] As we have seen, decreasing leaving-group ability correlates with increasing E1cB character.

Still another method measures volumes of activation.[62] These are negative for E2 and positive for E1cB mechanisms. Measurement of the activation volume therefore provides a continuous scale for deciding just where a reaction lies on the spectrum.

[53]For a review, see Fry, Ref. 2. See also Hasan; Sims; Fry. *J. Am. Chem. Soc.* **1983**, *105*, 3967; Pulay; Fry *Tetrahedron Lett.* **1986**, *27*, 5055.

[54]Ayrey; Bourns; Vyas *Can. J. Chem.* **1963**, *41*, 1759. Also see Simon; Müllhofer *Chem. Ber.* **1963**, *96*, 3167, **1964**, *97*, 2202, *Pure Appl. Chem.* **1964**, *8*, 379, 536; Smith; Bourns *Can. J. Chem.* **1970**, *48*, 125.

[55]Saunders; Zimmerman *J. Am. Chem. Soc.* **1964**, *86*, 3789; Wu; Hargreaves; Saunders *J. Org. Chem.* **1985**, *50*, 2392.

[56]Grout; McLennan; Spackman *J. Chem. Soc., Perkin Trans. 2* **1977**, 1758.

[57]For example, see Saunders; Edison *J. Am. Chem. Soc.* **1960**, *82*, 138; Hodnett; Sparapany *Pure Appl. Chem.* **1964**, *8*, 385, 537; Finley; Saunders *J. Am. Chem. Soc.* **1967**, *89*, 898; Ghanbarpour; Willi *Liebigs Ann. Chem.* **1975**, 1295; Simon; Müllhofer, Ref. 54; Thibblin *J. Am. Chem. Soc.* **1988**, *110*, 4582; Smith; Amin *Can. J. Chem.* **1989**, *67*, 1457.

[58]There is controversy as to whether such an effect has been established in this reaction: See Cockerill *J. Chem. Soc. B* **1967**, 964; Blackwell *J. Chem. Soc., Perkin Trans. 2* **1976**, 488.

[59]For examples of tunneling in elimination reactions, see Miller; Saunders *J. Org. Chem.* **1981**, *46*, 4247 and previous papers in this series. See also Shiner; Smith, Ref. 3; McLennan *J. Chem. Soc., Perkin Trans. 2* **1977**, 1753; Fouad; Farrell *Tetrahedron Lett.* **1978**, 4735; Koth; Tumas; Dobson; Koch *J. Am. Chem. Soc.* **1983**, *105*, 1930; Kwart; Wilk *J. Org. Chem.* **1985**, *50*, 817; Amin; Price; Saunders *J. Am. Chem. Soc.* **1990**, *112*, 4467.

[60]Saunders; Bushman; Cockerill *J. Am. Chem. Soc.* **1968**, *90*, 1775; Oae; Yano *Tetrahedron* **1968**, *24*, 5721; Yano; Oae *Tetrahedron* **1970**, *26*, 27, 67; Blackwell; Buckley; Jolley; MacGibbon *J. Chem. Soc., Perkin Trans. 2* **1973**, 169; Smith; Tsui *J. Am. Chem. Soc.* **1973**, *95*, 4760, *Can. J. Chem.* **1974**, *52*, 749.

[61]DePuy; Froemsdorf *J. Am. Chem. Soc.* **1957**, *79*, 3710; DePuy; Bishop *J. Am. Chem. Soc.* **1960**, *82*, 2532, 2535.

[62]Brower; Muhsin; Brower *J. Am. Chem. Soc.* **1976**, *98*, 779. For a review, see van Eldik; Asano; le Noble *Chem. Rev.* **1989**, *89*, 549-688.

The E2C Mechanism[63]

Certain alkyl halides and tosylates undergo E2 eliminations faster when treated with such weak bases as Cl^- in polar aprotic solvents or PhS^- than with the usual E2 strong bases such as RO^- in ROH.[64] In order to explain these results Parker and co-workers proposed[65] that there is a spectrum[66] of E2 transition states in which the base can interact in the transition state with the α carbon as well as with the β hydrogen. At one end of this spectrum is a mechanism (called E2C) in which, in the transition state, the base interacts mainly with the

| E2C | **17** | E2H | **18** |

carbon. The E2C mechanism is characterized by strong nucleophiles that are weak bases. At the other extreme is the normal E2 mechanism, here called E2H to distinguish it from E2C, characterized by strong bases. **17** represents a transition state between these extremes. Additional evidence[67] for the E2C mechanism is derived from Brønsted equation considerations (p. 258), from substrate effects, from isotope effects, and from the effects of solvents on rates.

However, the E2C mechanism has been criticized, and it has been contended that all the experimental results can be explained by the normal E2 mechanism.[68] McLennan has suggested that the transition state is that shown as **18**.[69] An ion-pair mechanism has also been proposed.[70] Although the actual mechanisms involved may be a matter of controversy, there is no doubt that a class of elimination reactions exists that is characterized by second-order attack by weak bases.[71] These reactions also have the following general characteris-

---

[63]For reviews, see McLennan *Tetrahedron* **1975**, *31*, 2999-3010; Ford *Acc. Chem. Res.* **1973**, *6*, 410-415; Parker *CHEMTECH* **1971**, 297-303.

[64]For example, see Winstein; Darwish; Holness *J. Am. Chem. Soc.* **1956**, *78*, 2915; de la Mare; Vernon *J. Chem. Soc.* **1956**, 41; Eliel; Ro *Tetrahedron* **1958**, *2*, 353; Bunnett; Davis; Tanida *J. Am. Chem. Soc.* **1962**, *84*, 1606; McLennan *J. Chem. Soc. B* **1966**, 705, 709; Hayami; Ono; Kaji *Bull. Chem. Soc. Jpn.* **1971**, *44*, 1628.

[65]Parker; Ruane; Biale; Winstein *Tetrahedron Lett.* **1968**, 2113.

[66]This is apart from the E1–E2–E1cB spectrum.

[67]Lloyd; Parker *Tetrahedron Lett.* **1968**, 5183, **1970**, 5029; Cook; Parker; Ruane *Tetrahedron Lett.* **1968**, 5715; Alexander; Ko; Parker; Broxton *J. Am. Chem. Soc.* **1968**, *90*, 5049; Ko; Parker *J. Am. Chem. Soc.* **1968**, *90*, 6447; Parker; Ruane; Palmer; Winstein *J. Am. Chem. Soc.* **1972**, *94*, 2228; Biale; Parker; Stevens; Takahashi; Winstein *J. Am. Chem. Soc.* **1972**, *94*, 2235; Cook; Hutchinson; Parker *J. Org. Chem.* **1974**, *39*, 3029; Cook; Hutchinson; MacLeod; Parker *J. Org. Chem.* **1974**, *39*, 534; Cook *J. Org. Chem.* **1976**, *41*, 2173; Muir; Parker *Aust. J. Chem.* **1983**, *36*, 1667; Kwart; Wilk *J. Org. Chem.* **1985**, *50*, 3038.

[68]Anderson; Ang; England; McCann; McLennan *Aust. J. Chem.* **1969**, *22*, 1427; Bunnett; Baciocchi *J. Org. Chem.* **1967**, *32*, 11, **1970**, *35*, 76; Jackson; McLennan; Short; Wong *J. Chem. Soc., Perkin Trans. 2* **1972**, 2308; McLennan; Wong *Tetrahedron Lett.* **1970**, 881, *J. Chem. Soc., Perkin Trans. 2* **1972**, 279, **1974**, 1818; Bunnett; Eck *J. Am. Chem. Soc.* **1973**, *95*, 1897, 1900; Ford; Pietsek *J. Am. Chem. Soc.* **1975**, *97*, 2194; Loupy *Bull. Soc. Chim. Fr.* **1975**, 2662; Miller; Saunders *J. Am. Chem. Soc.* **1979**, *101*, 6749; Bunnett; Sridharan; Cavin *J. Org. Chem.* **1979**, *44*, 1463; Bordwell; Mrozack *J. Org. Chem.* **1982**, *47*, 4813; Bunnett; Migdal *J. Org. Chem.* **1989**, *54*, 3037, 3041.

[69]McLennan, Ref. 63, *J. Chem. Soc., Perkin Trans. 2* **1977**, 293, 298; McLennan; Lim *Aust. J. Chem.* **1983**, *36*, 1821. For an opposing view, see Kwart; Gaffney *J. Org. Chem.* **1983**, *48*, 4502.

[70]Ford, Ref. 63.

[71]For convenience, we will refer to this class of reactions as E2C reactions, though the actual mechanism is in dispute.

tics:[72] (1) they are favored by good leaving groups; (2) they are favored by polar aprotic solvents; (3) the reactivity order is tertiary > secondary > primary, the opposite of the normal E2 order (p. 1003); (4) the elimination is always anti (syn elimination is not found), but in cyclohexyl systems, a diequatorial anti elimination is about as favorable as a diaxial anti elimination (unlike the normal E2 reaction, p. 985); (5) they follow Zaitsev's rule (see below), where this does not conflict with the requirement for anti elimination.

Orientation of the Double Bond

With some substrates, a β hydrogen is present on only one carbon and (barring rearrangements) there is no doubt as to the identity of the product. For example, $PhCH_2CH_2Br$ can give only $PhCH{=}CH_2$. However, in many other cases two or three olefinic products are possible. In the simplest such case, a *sec*-butyl compound can give either 1-butene or 2-butene. There are a number of rules that enable us to predict, in many instances, which product will predominantly form.[73]

1. No matter what the mechanism, a double bond does not go to a bridgehead carbon unless the ring sizes are large enough (Bredt's rule, see p. 160). This means, for example, not only that **19** gives only **20** and not **21** (indeed **21** is not a known compound), but also that **22** does not undergo elimination.

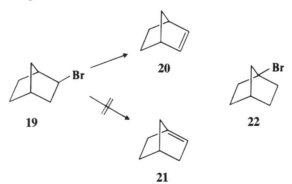

2. No matter what the mechanism, if there is a double bond (C=C or C=O) or an aromatic ring already in the molecule that can be in conjugation with the new double bond, the conjugated product usually predominates, sometimes even when the stereochemistry is unfavorable (for an exception, see p. 1001).

3. In the E1 mechanism the leaving group is gone before the choice is made as to which direction the new double bond takes. Therefore the direction is determined almost entirely by the relative stabilities of the two (or three) possible olefins. In such cases *Zaitsev's rule*[74] *operates*. This rule states that *the double bond goes mainly toward the most highly substituted carbon*. That is, a *sec*-butyl compound gives more 2-butene than 1-butene, and 3-bromo-

[72]Biale; Parker; Smith; Stevens; Winstein *J. Am. Chem. Soc.* **1970**, *92*, 115; Lloyd; Muir; Parker *Tetrahedron Lett.* **1971**, 3015; Beltrame; Biale; Lloyd; Parker; Ruane; Winstein *J. Am. Chem. Soc.* **1972**, *94*, 2240; Beltrame; Ceccon; Winstein *J. Am. Chem. Soc.* **1972**, *94*, 2315.

[73]For a review of orientation in cycloalkyl systems, see Hückel; Hanack *Angew. Chem. Int. Ed. Engl.* **1967**, *6*, 534-544 [*Angew. Chem. 79*, 555-565].

[74]Often given the German spelling: Saytzeff.

2,3-dimethylpentane gives more 2,3-dimethyl-2-pentene than either 3,4-dimethyl-2-pentene or 2-ethyl-3-methyl-1-butene. Thus Zaitsev's rule predicts that the olefin predominantly formed will be the one with the largest possible number of alkyl groups on the C=C carbons, and in most cases this is what is found. From heat of combustion data (see p. 23) it is known that olefin stability increases with alkyl substitution, though just why this should be is a matter of conjecture. The most common explanation is hyperconjugation. For E1 eliminations Zaitsev's rule governs the orientation whether the leaving group is neutral or positive, since, as already mentioned, the leaving group is not present when the choice of direction is made. This statement does not hold for E2 eliminations, and it may be mentioned here, for contrast with later results, that E1 elimination of $Me_2CHCHMeSMe_2^+$ gave 91% of the Zaitsev product and 9% of the other.[75] However, there *are* cases in which the leaving group affects the direction of the double bond in E1 eliminations.[76] This may be attributed to ion pairs; that is, the leaving group is not completely gone when the hydrogen departs. Zaitsev's rule breaks down in cases where the non-Zaitsev product is more stable for steric reasons. For example, E1 or E1-like eliminations of 1,2-diphenyl-2-X-propanes $PhMeCXCH_2Ph$ were reported to give about 50% $CH_2=CPhCH_2Ph$, despite the fact that the double bond of the Zaitsev product ($PhMeC=CHPh$) is conjugated with two benzene rings.[77]

4. For the anti E2 mechanism a trans β proton is necessary; if this is available in only one direction, that is the way the double bond will form. Because of the free rotation in acyclic systems (except where steric hindrance is great), this is a factor only in cyclic systems. Where trans β hydrogens are available on two or three carbons, two types of behavior are found, depending on substrate structure and the nature of the leaving group. Some compounds follow Zaitsev's rule and give predominant formation of the most highly substituted olefin, but others follow *Hofmann's rule: the double bond goes mainly toward the least highly substituted carbon.* Though many exceptions are known, the following general statements can be made: In most cases, compounds containing uncharged nucleofuges (those that come off as negative ions) follow Zaitsev's rule, just as they do in E1 elimination, no matter what the structure of the substrate. However, elimination from compounds with charged nucleofuges, e.g., NR_3^+, SR_2^+ (those that come off as neutral molecules), follow Hofmann's rule if the substrate is acyclic,[78] but Zaitsev's rule if the leaving group is attached to a six-membered ring.[79]

Much work has been devoted to searching for the reasons for the differences in orientation. Since Zaitsev orientation almost always gives the thermodynamically more stable isomer, what needs to be explained is why in some cases the less stable Hofmann product predominates. Three explanations have been offered for the change in orientation in acyclic systems with a change from uncharged to charged nucleofuges. The first of these, by Hughes and Ingold,[80] is that Hofmann orientation is caused by the fact that the acidity of the β hydrogen is decreased by the presence of the electron-donating alkyl groups. For example, under E2 conditions $Me_2CHCHMeSMe_2^+$ gives more of the Hofmann product; it is the more acidic hydrogen that is removed by the base.

---

[75]de la Mare *Prog. Stereochem.* **1954,** *1*, 112.

[76]Cram; Sahyun *J. Am. Chem. Soc.* **1963,** *85*, 1257; Silver *J. Am. Chem. Soc.* **1961,** *83*, 3482.

[77]Ho; Smith *Tetrahedron* **1970,** *26*, 4277.

[78]An example of an acyclic quaternary ammonium salt that follows Zaitsev's rule is found in Feit; Saunders *J. Am. Chem. Soc.* **1970,** *92*, 5615.

[79]For examples where Zaitsev's rule is followed with charged leaving groups in cyclohexyl systems, see Gent; McKenna *J. Chem. Soc* **1959,** 137; Hughes; Wilby *J. Chem. Soc.* **1960,** 4094; Brownlee; Saunders *Proc. Chem. Soc.* **1961,** *314*; Booth; Franklin; Gidley *J. Chem. Soc. C* **1968,** 1891. For a discussion of the possible reasons for this, see Saunders; Cockerill, Ref. 2, pp. 192-193.

[80]For summaries of this position, see Ingold *Proc. Chem. Soc.* **1962,** 265-274; Banthorpe; Hughes; Ingold *J. Chem. Soc.* **1960,** 4054.

Of course, the CH$_3$ hydrogens would still be more acidic than the Me$_2$CH hydrogen even if a neutral leaving group were present, but the explanation of Hughes and Ingold is that acidity matters with charged and not with neutral leaving groups, because the charged groups exert a strong electron-withdrawing effect, making differences in acidity greater than they are with the less electron-withdrawing neutral groups.[80] The explanation of Bunnett[81] is similar. According to this, the change to a positive leaving group causes the mechanism to shift toward the E1cB end of the spectrum, where there is more C—H bond breaking in the rate-determining step and where, consequently, acidity is more important. In this view, when there is a neutral leaving group, the mechanism is more E1-like, C—X bond breaking is more important, and olefin stability determines the direction of the new double bond. The third explanation, by H. C. Brown, is completely different. In this picture, field effects are unimportant, and the difference in orientation is largely a steric effect caused by the fact that charged groups are usually larger than neutral ones. A CH$_3$ group is more open to attack than a CH$_2$R group and a CHR$_2$ group is still less easily attacked. Of course, these considerations also apply when the leaving group is neutral, but, according to Brown, they are much less important here because the neutral groups are smaller and do not block access to the hydrogens as much. Brown showed that Hofmann elimination increases with the size of the leaving group. Thus the percentage of 1-ene obtained from CH$_3$CH$_2$CH$_2$CHXCH$_3$ was as follows (X listed in order of increasing size): Br, 31%; I, 30%; OTs, 48%; SMe$_2$$^+$, 87%; SO$_2$Me, 89%; NMe$_3$$^+$, 98%.[82] Hofmann elimination was also shown to increase with increase in bulk of the substrate.[83] With large enough compounds, Hofmann orientation can be obtained even with halides, e.g., *t*-amyl bromide gave 89% of the Hofmann product. Even those who believe in the acidity explanations concede that these steric factors operate in extreme cases.[84]

There is one series of results incompatible with the steric explanation—E2 elimination from the four 2-halopentanes gave the following percentages of 1-pentene: F, 83%; Cl, 37%; Br, 25%; I, 20%.[85] The same order was found for the four-2-halohexanes.[86] Although there is some doubt about the relative steric requirements of Br, Cl, and I, there is no doubt that F is the smallest of the halogens, and if the steric explanation were the only valid one, the fluoroalkanes could not give predominant Hofmann orientation. Another result that argues against the steric explanation is the effect of changing the nature of the base. An experiment in which the effective size of the base was kept constant while its basicity was increased (by

[81]Bunnett, Ref. 2.
[82]Brown; Wheeler *J. Am. Chem. Soc.* **1956**, *78*, 2199.
[83]Brown; Moritani; Nakagawa *J. Am. Chem. Soc.* **1956**, *78*, 2190; Brown; Moritani *J. Am. Chem. Soc.* **1956**, *78*, 2203; Bartsch *J. Org. Chem.* **1970**, *35*, 1334. See also Charton *J. Am. Chem. Soc.* **1975**, *97*, 6159.
[84]For example, see Banthorpe; Hughes; Ingold *J. Chem. Soc.* **1960**, 4054.
[85]Saunders; Fahrenholtz; Caress; Lowe; Schreiber *J. Am. Chem. Soc.* **1965**, *87*, 3401. Similar results were obtained by Brown; Klimisch *J. Am. Chem. Soc.* **1966**, *88*, 1425.
[86]Bartsch; Bunnett *J. Am. Chem. Soc.* **1968**, *90*, 408.

using as bases a series of $XC_6H_4O^-$ ions) showed that the percentage of Hofmann elimination increased with increasing base strength, though the size of the base did not change.[87] These results are in accord with the explanation of Bunnett, since an increase in base strength moves an E2 reaction closer to the E1cB end of the spectrum. In further experiments, a large series of bases of different kinds was shown to obey linear free-energy relationships between basicity and percentage of Hofmann elimination,[88] though certain very large bases (e.g., 2,6-di-t-butyl-phenoxide) did not obey the relationships, steric effects becoming important in these cases. How large the base must be before steric effects are observed depends on the pattern of alkyl substitution in the substrate, but not on the nucleofuge.[89] One further result may be noted. In the gas phase, elimination of H and BrH^+ or H and ClH^+ using Me_3N as the base predominantly followed Hofmann's rule,[90] although BrH^+ and ClH^+ are not very large.

5. Only a few investigations on the orientation of syn E2 eliminations have been carried out, but these show that Hofmann orientation is greatly favored over Zaitsev.[91]

6. In the E1cB mechanism the question of orientation seldom arises because the mechanism is generally found only where there is an electron-withdrawing group in the β position, and that is where the double bond goes.

7. As already mentioned, E2C reactions show a strong preference for Zaitsev orientation.[92] In some cases this can be put to preparative use. For example, the compound $PhCH_2CHOTsCHMe_2$ gave about 98% $PhCH=CHCHMe_2$ under the usual E2 reaction conditions (t-BuOK in t-BuOH). In this case the double bond goes to the side with more hydrogens because on that side it will be able to conjugate with the benzene ring. However, with the weak base Bu_4N^+ Br^- in acetone the Zaitsev product $PhCH_2CH=CMe_2$ was formed in 90% yield.[93]

Steric Orientation of the Double Bond

When elimination takes place on a compound of the form CH_3—CABX or CHAB—CGGX, the new olefin does not have cis–trans isomerism, but for compounds of the form CHEG—CABX (E and G *not* H) (**23**) and CH_2E—CABX (**24**), cis and trans isomers are possible. When the anti E2 mechanism is in operation, **23** gives the isomer arising from

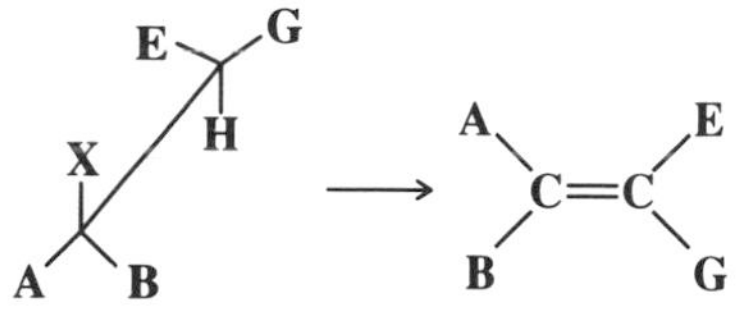

23

[87]Froemsdorf; Robbins *J. Am. Chem. Soc.* **1967**, *89*, 1737. See also Froemsdorf; Dowd; Leimer *J. Am. Chem. Soc.* **1966**, *88*, 2345; Bartsch; Kelly; Pruss *Tetrahedron Lett.* **1970**, 3795; Feit; Breger; Capobianco; Cooke; Gitlin *J. Am. Chem. Soc.* **1975**, *97*, 2477; Ref. 78.

[88]Bartsch; Pruss; Bushaw; Wiegers *J. Am. Chem. Soc.* **1973**, *95*, 3405; Bartsch; Roberts; Cho *J. Org. Chem.* **1979**, *44*, 4105.

[89]Bartsch; Read; Larsen; Roberts; Scott; Cho *J. Am. Chem. Soc.* **1979**, *101*, 1176.

[90]Angelini; Lilla; Speranza *J. Am. Chem. Soc.* **1989**, *111*, 7393.

[91]Sicher; Svoboda; Pánková; Závada *Collect. Czech. Chem. Commun.* **1971**, *36*, 3633; Bailey; Saunders *J. Am. Chem. Soc.* **1970**, *92* 6904.

[92]For example; see Ono *Bull. Chem. Soc. Jpn.* **1971**, *44*, 1369; Bailey; Saunders *J. Org. Chem.* **1973**, *38*, 3363; Muir; Parker *J. Org. Chem.* **1976**, *41*, 3201.

[93]Lloyd; Muir; Parker, Ref. 72.

trans orientation of X and H and, as we have seen before (p. 984), an erythro compound gives the cis olefin and a threo compound the trans. For **24** two conformations are possible for the transition state; these lead to different isomers and often both are obtained. However, the one that predominates is often determined by an eclipsing effect.[94] For example, Zaitsev elimination from 2-bromopentane can occur as follows:

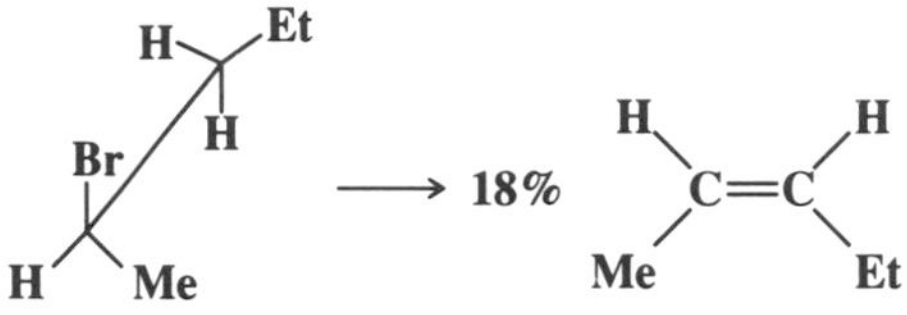

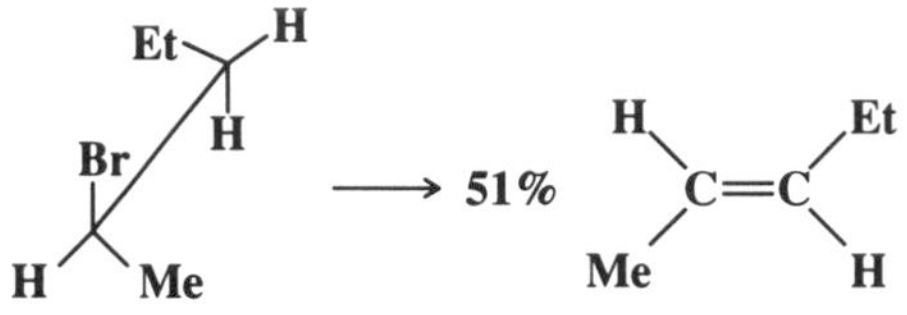

In conformation **J** the ethyl group is between Br and Me, while in **K** it is between Br and H. This means that **K** is more stable, and most of the elimination should occur from this conformation. This is indeed what happens, and 51% of the trans isomer is formed (with KOEt) compared to 18% of the cis (the rest is the Hofmann product).[95] These effects become larger with increasing size of A, B, and E.

However, eclipsing effects are not the only factors that affect the cis/trans ratio in anti E2 eliminations. Other factors are the nature of the leaving group, the base, the solvent, and the substrate. Not all these effects are completely understood.[96]

For E1 eliminations, if there is a free carbocation (**25**), it is free to rotate, and no matter

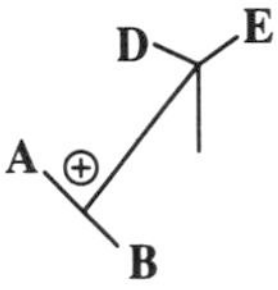

what the geometry of the original compound, the more stable situation is the one where the larger of the D–E pair is opposite the smaller of the A–B pair and the corresponding olefin should form. If the carbocation is not completely free, then to that extent, E2-type products are formed. Similar considerations apply in E1cB eliminations.[97]

---

[94]See Cram; Greene; DePuy *J. Am. Chem. Soc.* **1956**, *78*, 790; Cram, in Newman *Steric Effects in Organic Chemistry*; Wiley: New York, 1956, pp. 338-345.

[95]Brown; Wheeler *J. Am. Chem. Soc.* **1956**, *78* 2199.

[96]For discussions, see Bartsch; Bunnett *J. Am. Chem. Soc.* **1969**, *91*, 1376, 1382; Feit; Saunders *J. Am. Chem. Soc.* **1970**, *92*, 1630, 5615; Alunni; Baciocchi *J. Chem. Soc., Perkin Trans. 2* **1976**, 877; Saunders; Cockerill, Ref. 2, pp. 165-193.

[97]See, for example, Redman; Thomas; Stirling *J. Chem. Soc., Chem. Commun.* **1978**, 43.

REACTIVITY

In this section we examine the effects of changes in the substrate, base, leaving group, and medium on (1) overall reactivity, (2) E1 vs. E2 vs. E1cB,[98] and (3) elimination vs. substitution.

Effect of Substrate Structure

1. *Effect on reactivity.* We refer to the carbon containing the nucleofuge (X) as the α carbon and to the carbon that loses the positive species as the β carbon. Groups attached to the α or β carbons can exert at least four kinds of influence:

a. They can stabilize or destabilize the incipient double bond (both α and β groups).

b. They can stabilize or destabilize an incipient negative charge, affecting the acidity of the proton (β groups only).

c. They can stabilize or destabilize an incipient positive charge (α groups only).

d. They can exert steric effects (e.g., eclipsing effects) (both α and β groups).

Effects a and d can apply in all three mechanisms, though steric effects are greatest for the E2 mechanism. Effect b does not apply in the E1 mechanism, and effect c does not apply in the E1cB mechanism. Groups such as Ar and C=C increase the rate by any mechanism, whether they are α or β (effect a). Electron-withdrawing groups increase the acidity when in the β position, but have little effect in the α position unless they also conjugate with the double bond. Thus Br, Cl, CN, Ts, NO_2, CN, and SR in the β position all increase the rate of E2 eliminations.

2. *Effect on E1 vs. E2 vs. E1cB.* α alkyl and α aryl groups stabilize the carbocation character of the transition state, shifting the spectrum toward the E1 end. β alkyl groups also shift the mechanism toward E1, since they *decrease* the acidity of the hydrogen. However, β aryl groups shift the mechanism the other way (toward E1cB) by stabilizing the carbanion. Indeed, as we have seen (p. 993), all electron-withdrawing groups in the β position shift the mechanism toward E1cB.[99] α alkyl groups also increase the extent of elimination with weak bases (E2C reactions).

3. *Effect on elimination vs. substitution.* Under second-order conditions α branching increases elimination, to the point where tertiary substrates undergo few SN2 reactions, as we saw in Chapter 10. For example, Table 17.2 shows results on some simple alkyl bromides. Similar results were obtained with SMe_2^+ as the leaving group.[100] Two reasons can be presented for this trend. One is statistical: as α branching increases, there are usually more hydrogens for the base to attack. The other is that α branching presents steric hindrance to attack of the base at the carbon. Under first-order conditions, increased α branching also increases the amount of elimination (E1 vs. SN1), though not so much, and usually the substitution product predominates. For example, solvolysis of *t*-butyl bromide gave only 19% elimination[101] (compare with Table 17.2). β branching also increases the amount of E2 elimination with respect to SN2 substitution (Table 17.2), not because elimination is faster but because the SN2 mechanism is so greatly slowed (p. 339). Under first-order conditions too, β branching favors elimination over substitution, probably for steric reasons.[102] However, E2 eliminations from compounds with charged leaving groups are slowed

[98]For discussions, see Cockerill; Harrison, Ref. 2, pp. 178-189.

[99]For a review of eliminations with COOH, COOR, $CONH_2$, and CN groups in the β position, see Butskus; Denis *Russ. Chem. Rev.* **1966**, *35*, 839-850.

[100]Hughes; Ingold; Maw *J. Chem. Soc.* **1948**, 2072; Hughes; Ingold; Woolf *J. Chem. Soc.* **1948**, 2084.

[101]Hughes; Ingold; Maw *J. Chem. Soc.* **1948**, 2065.

[102]Brown; Berneis *J. Am. Chem. Soc.* **1953**, *75*, 10.

TABLE 17.2   The effect of α and β branching on the rate of E2 elimination and the amount of olefin formed

The reactions were between the alkyl bromide and OEt⁻. The rate for isopropyl bromide was actually greater than that for ethyl bromide, if the temperature difference is considered. Neopentyl bromide, the next compound in the β-branching series, cannot be compared because it has no β-hydrogen and cannot give an elimination product without rearrangement.

Substrate	Temperature, °C	Olefin, %	Rate × 10^5 of E2 reaction	Reference
CH$_3$CH$_2$Br	55	0.9	1.6	103
(CH$_3$)$_2$CHBr	25	80.3	0.237	104
(CH$_3$)$_3$CBr	25	97	4.17	101
CH$_3$CH$_2$CH$_2$Br	55	8.9	5.3	103
CH$_3$CHCH$_2$Br | CH$_3$	55	59.5	8.5	103

by β branching. This is related to Hofmann's rule (p. 999). Electron-withdrawing groups in the β position not only increase the rate of E2 eliminations and shift the mechanisms toward the E1cB end of the spectrum but also increase the extent of elimination as opposed to substitution.

Effect of the Attacking Base

1. *Effect on E1 vs. E2 vs. E1cB.*   In the E1 mechanism, an external base is generally not required: The solvent acts as the base. Hence, when external bases are added, the mechanism is shifted toward E2. Stronger bases and higher base concentrations cause the mechanism to move toward the E1cB end of the E1–E2–E1cB spectrum.[105] However, weak bases in polar aprotic solvents can also be effective in elimination reactions with certain substrates (the E2C reaction). Normal E2 elimination has been accomplished with the following bases:[106] H$_2$O, NR$_3$, OH⁻, OAc⁻, OR⁻, OAr⁻, NH$_2$⁻, CO$_3$$^{2-}$, LiAlH$_4$, I⁻, CN⁻, and organic bases. However, the only bases of preparative importance in the normal E2 reaction are OH⁻, OR⁻, and NH$_2$⁻, usually in the conjugate acid as solvent, and certain amines. Weak bases effective in the E2C reaction are Cl⁻, Br⁻, F⁻, OAc⁻, and RS⁻. These bases are often used in the form of their R$_4$N⁺ salts.

2. *Effect on elimination vs. substitution.*   Strong bases not only benefit E2 as against E1, but also benefit elimination as against substitution. With a high concentration of strong base in a nonionizing solvent, bimolecular mechanisms are favored and E2 predominates over S$_N$2. At low base concentrations, or in the absence of base altogether, in ionizing solvents, unimolecular mechanisms are favored, and the S$_N$1 mechanism predominates over the E1. In Chapter 10, it was pointed out that some species are strong nucleophiles though weak bases (p. 349). The use of these obviously favors substitution, except that, as we have seen, elimination can predominate if polar aprotic solvents are used. It has been shown for the

[103]Dhar; Hughes; Ingold; Masterman *J. Chem. Soc.* **1948**, 2055.

[104]Dhar; Hughes; Ingold *J. Chem. Soc.* **1948**, 2058.

[105]For a review, see Baciocchi *Acc. Chem. Res.* **1979**, *12*, 430-436. See also Baciocchi; Ruzziconi; Sebastiani *J. Org. Chem.* **1980**, *45*, 827.

[106]This list is from Banthorpe *Elimination Reactions*; Elsevier: New York, 1963, p. 4.

base CN⁻ that in polar aprotic solvents, the less the base is encumbered by its counterion in an ion pair (i.e., the freer the base), the more substitution is favored at the expense of elimination.[107]

Effect of the Leaving Group

1. *Effect on reactivity.*   The leaving groups in elimination reactions are similar to those in nucleophilic substitution. E2 eliminations have been performed with the following groups: NR_3^+, PR_3^+, SR_2^+, OHR^+, SO_2R, OSO_2R, $OCOR$, OOH OOR, NO_2,[108] F, Cl, Br, I, and CN (*not* OH_2^+). E1 eliminations have been carried out with: NR_3^+, SR_2^+, OH_2^+, OHR^+, OSO_2R, $OCOR$, Cl, Br, I, and N_2^+.[109] However, the major leaving groups for preparative purposes are OH_2^+ (always by E1) and Cl, Br, I, and NR_3^+ (usually by E2).

2. *Effect on E1 vs. E2 vs. E1cB.*   Better leaving groups shift the mechanism toward the E1 end of the spectrum, since they make ionization easier. This effect has been studied in various ways. One way already mentioned was a study of ρ values (p. 996). Poor leaving groups and positively charged leaving groups shift the mechanism toward the E1cB end of the spectrum because the strong electron-withdrawing field effects increase the acidity of the β hydrogen.[110] The E2C reaction is favored by good leaving groups.

3. *Effect on elimination vs. substitution.*   As we have already seen (p. 990), for first-order reactions the leaving group has nothing to do with the competition between elimination and substitution, since it is gone before the decision is made as to which path to take. However, where ion pairs are involved, this is not true, and results have been found where the nature of the leaving group does affect the product.[111] In second-order reactions, the elimination/substitution ratio is not greatly dependent on a halide leaving group, though there is a slight increase in elimination in the order I > Br > Cl. When OTs is the leaving group, there is usually much more substitution. For example, n-$C_{18}H_{37}Br$ treated with t-BuOK gave 85% elimination, while n-$C_{18}H_{37}OTs$ gave, under the same conditions, 99% substitution.[112] On the other hand, positively charged leaving groups increase the amount of elimination.

Effect of the Medium

1. *Effect of solvent on E1 vs. E2 vs. E1cB.*   With any reaction a more polar environment enhances the rate of mechanisms that involve ionic intermediates. For neutral leaving groups, it is expected that E1 and E1cB mechanisms will be aided by increasing polarity of solvent and by increasing ionic strength. With certain substrates, polar aprotic solvents promote elimination with weak bases (the E2C reaction).

2. *Effect of solvent on elimination vs. substitution.*   Increasing polarity of solvent favors SN2 reactions at the expense of E2. In the classical example, alcoholic KOH is used to effect elimination, while the more polar aqueous KOH is used for substitution. Charge-dispersal discussions, similar to those on p. 358,[113] only partially explain this. In most solvents SN1

[107]Loupy; Seyden-Penne *Bull. Soc. Chim. Fr.* **1971**, 2306.

[108]For a review of eliminations in which NO_2 is a leaving group, see Ono, in Feuer; Nielsen *Nitro Compounds; Recent Advances in Synthesis and Chemistry*; VCH: New York, 1990,pp. 1-135, pp. 86-126.

[109]These lists are from Banthorpe, Ref. 106, pp. 4, 7.

[110]For a discussion of leaving-group ability, see Stirling *Acc. Chem. Res.* **1979**, *12*, 198-203. See also Varma; Stirling *J. Chem. Soc., Chem. Commun.* **1981**, 553.

[111]For example, see Skell; Hall *J. Am. Chem. Soc.* **1963**, *85* 2851; Cocivera; Winstein, Ref. 32; Feit; Wright *J. Chem. Soc., Chem. Commun.* **1975**, 776. See, however, Cavazza *Tetrahedron Lett.* **1975**, 1031.

[112]Veeravagu; Arnold; Eigenmann *J. Am. Chem. Soc.* **1964**, *86*, 3072.

[113]Cooper; Dhar; Hughes; Ingold; MacNulty; Woolf *J. Chem. Soc.* **1948**, 2043.

reactions are favored over E1. E1 reactions compete best in polar solvents that are poor nucleophiles, especially dipolar aprotic solvents.[114] A study made in the gas phase, where there is no solvent, has shown that when 1-bromopropane reacts with MeO^- only elimination takes place—no substitution—even with this primary substrate.[115]

3. *Effect of temperature.* Elimination is favored over substitution by increasing temperature, whether the mechanism is first or second order.[116] The reason is that the activation energies of eliminations are higher than those of substitutions (because eliminations have greater changes in bonding).

MECHANISMS AND ORIENTATION IN PYROLYTIC ELIMINATIONS

Mechanisms[117]

Several types of compound undergo elimination on heating, with no other reagent present. Reactions of this type are often run in the gas phase. The mechanisms are obviously different from those already discussed, since all those require a base (which may be the solvent) in one of the steps, and there is no base or solvent present in pyrolytic elimination. Two mechanisms have been found to operate. One involves a cyclic transition state, which may be four-, five-, or six-membered. Examples of each size are:

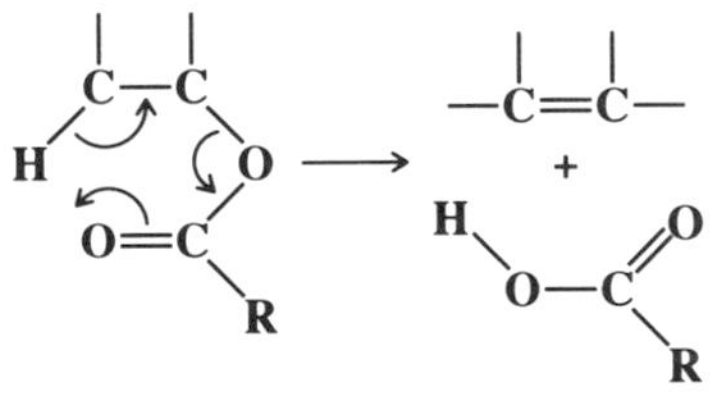

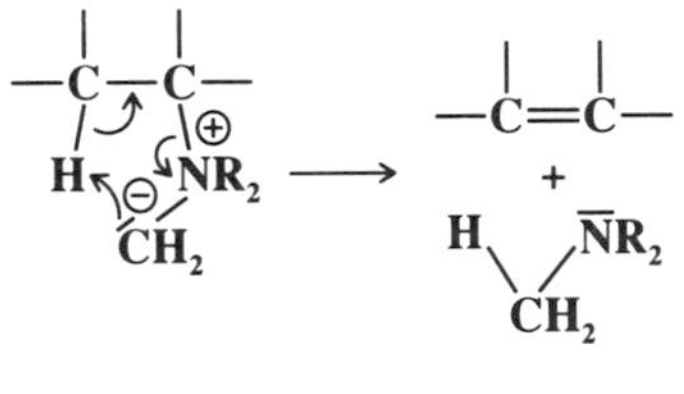

In this mechanism the two groups leave at about the same time and bond to each other as they are doing so. The designation is Ei in the Ingold terminology and cyclo-$D_ED_NA_n$ in the IUPAC system. The elimination must be syn and, for the four- and five-membered transition states, the four or five atoms making up the ring must be coplanar. Coplanarity

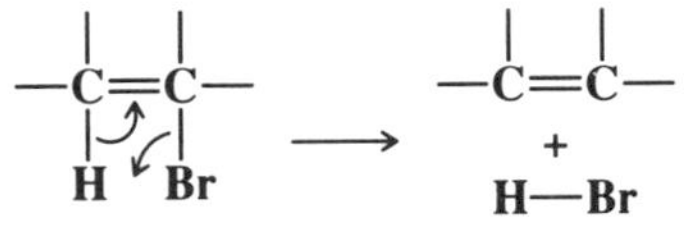

[114]Aksnes; Stensland *Acta Chem. Scand.* **1989**, *43*, 893, and references cited therein.
[115]Jones; Ellison *J. Am. Chem. Soc.* **1989**, *111*, 1645. For a different result with other reactants, see Lum; Grabowski *J. Am. Chem. Soc.* **1988**, *110*, 8568.
[116]Cooper; Hughes; Ingold; Maw; MacNulty *J. Chem. Soc.* **1948**, 2049.
[117]For reviews, see Taylor, in Patai *The Chemistry of Functional Groups, Supplement B*, pt. 2; Wiley: New York, 1979, pp. 860-914; Smith; Kelly *Prog. Phys. Org. Chem.* **1971**, *8*, 75-234, pp. 76-143, 207-234; in Bamford; Tipper, Ref. 2, vol. 5, 1972, the articles by Swinbourne, pp. 149-233 (pp. 158-188), and by Richardson; O'Neal, pp. 381-565 (pp. 381-446); Maccoll, Ref. 2, *Adv. Phys. Org. Chem.* **1965**, *3*, 91-122. For reviews of mechanisms in pyrolytic eliminations of halides, see Egger; Cocks; in Patai *The Chemistry of the Carbon–Halogen Bond*, pt. 2; Wiley: New York, 1973, pp. 677-745; Maccoll *Chem. Rev.* **1969**, *69*, 33-60.

is not required for the six-membered transition state, since there is room for the outside atoms when the leaving atoms are staggered.

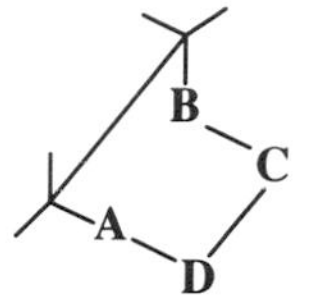

As in the E2 mechanism, it is not necessary that the C—H and C—X bond be equally broken in the transition state. In fact, there is also a spectrum of mechanisms here, ranging from a mechanism in which C—X bond breaking is a good deal more advanced than C—H bond breaking to one in which the extent of bond breaking is virtually identical for the two bonds. Evidence for the existence of the Ei mechanism is:

1. The kinetics are first order, so only one molecule of the substrate is involved in the reaction (that is, if one molecule attacked another, the kinetics would be second order in substrate).[118]

2. Free-radical inhibitors do not slow the reactions, so no free-radical mechanism is involved.[119]

3. The mechanism predicts exclusive syn elimination, and this behavior has been found in many cases.[120] The evidence is inverse to that for the anti E2 mechanism and generally involves the following facts: (1) an erythro isomer gives a trans olefin and a threo isomer gives a cis olefin; (2) the reaction takes place only when a cis β hydrogen is available; (3) if, in a cyclic compound, a cis hydrogen is available on only one side, the elimination goes in that direction. Another piece of evidence involves a pair of steroid molecules. In 3β-acetoxy-(R)-5α-methylsulfinylcholestane (**26** shows rings A and B of this compound) and in 3β-acetoxy-(S)-5α-methylsulfinylcholestane (**27**: rings A and B), the *only* difference is the

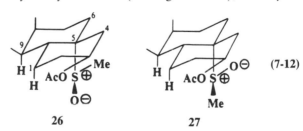

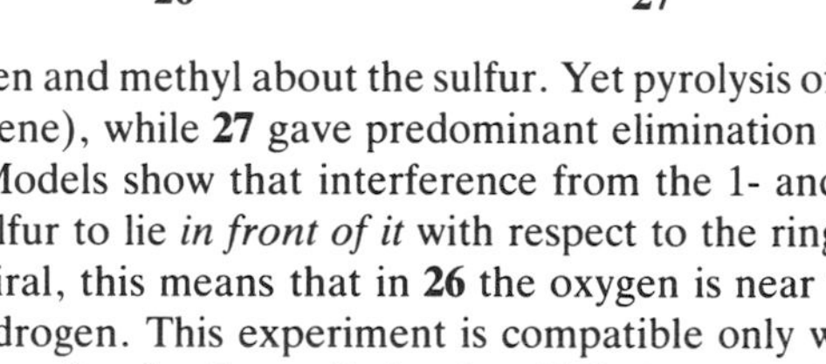

(7-12)

26 **27**

configuration of oxygen and methyl about the sulfur. Yet pyrolysis of **26** gave only elimination to the 4-side (86% 4-ene), while **27** gave predominant elimination to the 6-side (65% 5-ene and 20% 4-ene).[121] Models show that interference from the 1- and 9-hydrogens causes the two groups on the sulfur to lie *in front of* it with respect to the rings, rather than behind it. Since the sulfur is chiral, this means that in **26** the oxygen is near the 4-hydrogen, while in **27** it is near the 6 hydrogen. This experiment is compatible only with syn elimination.[122]

4. [14]C isotope effects for the Cope elimination (**7-8**) show that both the C—H and C—N bonds have been extensively broken in the transition state.[123]

[118]O'Connor; Nace *J. Am. Chem. Soc.* **1953**, *75*, 2118.
[119]Barton; Head; Williams *J. Chem. Soc.* **1953**, 1715.
[120]In a few instances anti or nonstereoselective elimination has been found; this behavior is generally ascribed to the intervention of other mechanisms. For example, see Bordwell; Landis *J. Am. Chem. Soc.* **1958**, *80*, 2450, 6383; Briggs; Djerassi *J. Org. Chem.* **1968**, *33*, 1625; Smissman; Li; Creese *J. Org. Chem.* **1970**, *35*, 1352.
[121]Jones; Saeed *Proc. Chem. Soc.* **1964**, 81. See also Goldberg; Sahli *J. Org. Chem.* **1967**, *32*, 2059.
[122]For other evidence for syn elimination, see Curtin; Kellom *J. Am. Chem. Soc.* **1953**, *75*, 6011; Skell; Hall *J. Am. Chem. Soc.* **1964**, *86*, 1557; Bailey; Bird *J. Org. Chem.* **1977**, *42*, 3895.
[123]Wright; Sims; Fry *J. Am. Chem. Soc.* **1983**, *105*, 3714.

5. Some of these reactions have been shown to exhibit negative entropies of activation, indicating that the molecules are more restricted in geometry in the transition state than they are in the starting compound.

Where a pyrolytic elimination lies on the mechanistic spectrum seems to depend mostly on the leaving group. When this is halogen, all available evidence suggests that in the transition state the C—X bond is cleaved to a much greater extent than the C—H bond, i.e., there is a considerable amount of carbocation character in the transition state. This is in accord with the fact that a completely nonpolar four-membered cyclic transition state violates the Woodward–Hoffmann rules (see the similar case of **5-49**). Evidence for the carbocation-like character of the transition state when halide is the leaving group is that relative rates are in the order $I > Br > Cl^{124}$ (see p. 352), and that the effects of substituents on reaction rates are in accord with such a transition state.[125] Rate ratios for pyrolysis of some alkyl bromides at 320°C were: ethyl bromide, 1; isopropyl bromide, 280; t-butyl bromide, 78,000. Also, α-phenylethyl bromide had about the same rate as t-butyl bromide. On the other hand, β-phenylethyl bromide was only slightly faster than ethyl bromide.[126] This indicates that C—Br cleavage was much more important in the transition state than C—H cleavage, since the incipient carbocation was stabilized by α alkyl and α aryl substitution, while there was no incipient carbanion to be stabilized by β aryl substitution. These substituent effects, as well as those for other groups, are very similar to the effects found for the SN1 mechanism and thus in very good accord with a carbocation-like transition state.

For carboxylic esters, the rate ratios were much smaller,[127] though still in the same order, so that this reaction is closer to a pure Ei mechanism, though the transition state still has some carbocationic character. Other evidence for a greater initial C—O cleavage with carboxylic esters is that a series of 1-arylethyl acetates followed σ^{+} rather than σ, showing carbocationic character at the 1 position.[128] The extent of E1 character in the transition state increases in the following order of ester types: acetate < phenylacetate < benzoate < carbamate < carbonate.[129] Cleavage of xanthates (**7-4**), cleavage of sulfoxides (**7-12**), the Cope reaction (**7-8**), and reaction **7-7** are probably very close to straight Ei mechanisms.[130]

The second type of pyrolysis mechanism is completely different and involves free radicals. Initiation occurs by pyrolytic homolytic cleavage. The remaining steps may vary, and a few are shown:

Initiation $\quad$ $R_2CHCH_2X \quad \longrightarrow R_2CHCH_2\cdot + X\cdot$

Propagation $\quad$ $R_2CHCH_2X + X\cdot \longrightarrow R_2\overset{\cdot}{C}CH_2X + HX$

$\qquad\qquad\quad R_2\overset{\cdot}{C}CH_2X \qquad \longrightarrow R_2C{=}CH_2 + X\cdot$

Termination (disproportionation)

$$2R_2\overset{\cdot}{C}CH_2X \longrightarrow R_2C{=}CH_2 + R_2CXCH_2X$$

[124]Maccoll, Ref. 2, pp. 215-216.
[125]For reviews of such studies, see Maccoll, Ref. 117.
[126]For rate studies of pyrolysis of some β-alkyl substituted ethyl bromides, see Chuchani; Rotinov; Dominguez; Martin *Int. J. Chem. Kinet.* **1987**, *19*, 781.
[127]For example, see Scheer; Kooyman; Sixma *Recl. Trav. Chim. Pays-Bas* **1963**, *82*, 1123. See also Louw; Vermeeren; Vogelzang *J. Chem. Soc., Perkin Trans. 2* **1983**, 1875.
[128]Taylor; Smith; Wetzel *J. Am. Chem. Soc.* **1962**, *84*, 4817; Smith; Jones; Brown *J. Org. Chem.* **1963**, *28*, 403; Taylor *J. Chem. Soc., Perkin Trans. 2* **1978**, 1255. See also Ottenbrite; Brockington *J. Org. Chem.* **1974**, *39*, 2463; Jordan; Thorne *J. Chem. Soc., Perkin Trans. 2* **1984**, 647; August; McEwen; Taylor *J. Chem. Soc., Perkin Trans. 2* **1987**, 1683, and other papers in this series; Al-Awadi *J. Chem. Soc., Perkin Trans. 2* **1990**, 2187.
[129]Taylor *J. Chem. Soc., Perkin Trans. 2* **1975**, 1025.
[130]For a review of the mechanisms of **7-12**, **7-8**, and the pyrolysis of sulfilimines, see Oae; Furukawa *Tetrahedron* **1977**, *33*, 2359-2367.

Free-radical mechanisms are mostly found in pyrolyses of polyhalides and of primary mo-nohalides,[131] though they also have been postulated in pyrolysis of certain carboxylic esters.[132] Much less is known about these mechanisms and we shall not consider them further. Free-radical eliminations in solution are also known but are rare.[133]

Orientation in Pyrolytic Eliminations

As in the E1–E2–E1cB mechanistic spectrum, Bredt's rule applies; and if a double bond is present, a conjugated system will be preferred, if sterically possible. Apart from these considerations, the following statements can be made for Ei eliminations:

1. In the absence of considerations mentioned below, orientation is statistical and is determined by the number of β hydrogens available (therefore *Hofmann's rule* is followed). For example, *sec*-butyl acetate gives 55 to 62% 1-butene and 38 to 45% 2-butene,[134] which is close to the 3:2 distribution predicted by the number of hydrogens available.[135]

2. A cis β hydrogen is required. Therefore in cyclic systems, if there is a cis hydrogen on only one side, the double bond will go that way. However, when there is a six-membered transition state, this does not necessarily mean that the leaving groups must be cis to each other, since such transition states need not be completely coplanar. If the leaving group is axial, then the hydrogen obviously must be equatorial (and consequently cis to the leaving group), since the transition state cannot be realized when the groups are both axial. But if the leaving group is equatorial, it can form a transition state with a β hydrogen that is either axial (hence, cis) or equatorial (hence, trans). Thus **28**, in which the leaving group is most likely axial, does not form a double bond in the direction of the carbethoxyl group, even though that would be conjugated, because there is no equatorial hydrogen on that side. Instead it gives 100% **29**.[136] On the other hand, **30**, with an equatorial leaving group, gives

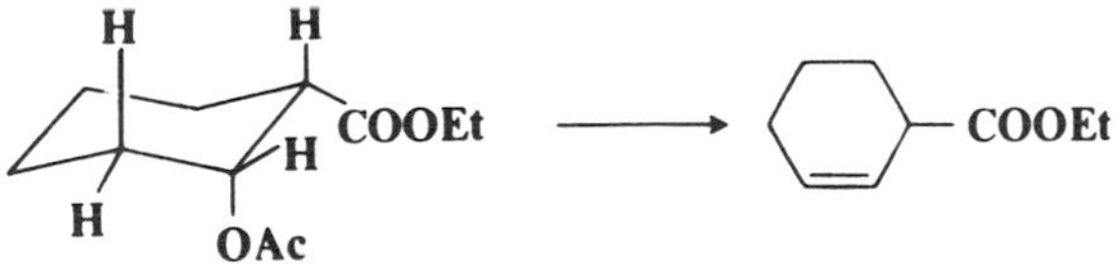

(7-3)

28   **29**

(7-4)

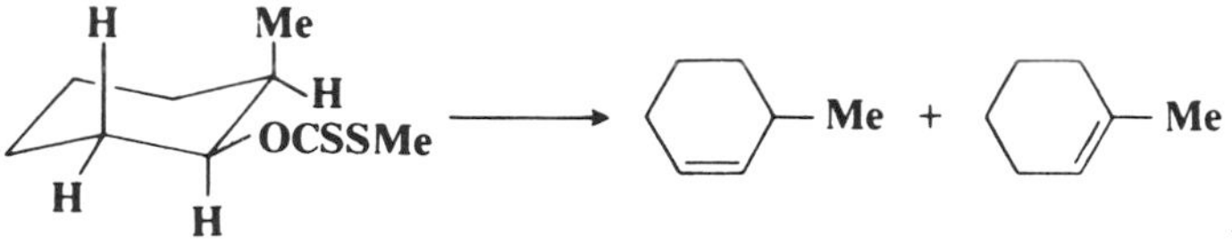

30

[131]For example, see Barton; Howlett *J. Chem. Soc.* **1949**, 155, 165.

[132]For example, see Rummens *Recl. Trav. Chim. Pays-Bas* **1964**, *83*, 901; Louw; Kooyman *Recl. Trav. Chim. Pays-Bas* **1965**, *84*, 1511.

[133]For examples; see Kampmeier; Geer; Meskin; D'Silva *J. Am. Chem. Soc.* **1966**, *88*, 1257; Kochi; Singleton; Andrews *Tetrahedron* **1968**, *24*, 3503; Boothe; Greene; Shevlin *J. Org. Chem.* **1980**, *45*, 794; Stark; Nelson; Jensen *J. Org. Chem.* **1980**, *45*, 420; Kochi *Organic Mechanisms and Catalysis*; Academic Press: New York, 1978, pp. 346-349; Kamimura; Ono *J. Chem. Soc., Chem. Commun.* **1988**, 1278.

[134]Froemsdorf; Collins; Hammond; DePuy *J. Am. Chem. Soc.* **1959**, *81*, 643; Haag; Pines *J. Org. Chem.* **1959**, *24*, 877.

[135]DePuy; King *Chem. Rev.* **1960**, *60*, 431-445, have tables showing the product distribution for many cases.

[136]Bailey; Baylouny *J. Am. Chem. Soc.* **1959**, *81*, 2126.

about 50% of each olefin, even though, for elimination to the 1-ene, the leaving group must go off with a trans hydrogen.[137]

3. In some cases, especially with cyclic compounds, the more stable olefin forms and Zaitsev's rule applies. For example, menthyl acetate gives 35% of the Hofmann product and 65% of the Zaitsev, even though a cis β hydrogen is present on both sides and the statistical distribution is the other way. A similar result was found for the pyrolysis of menthyl chloride.[138]

4. There are also steric effects. In some cases the direction of elimination is determined by the need to minimize steric interactions in the transition state or to relieve steric interactions in the ground state.

1,4 Conjugate Eliminations

1,4 eliminations of the type

$$\text{H—C—C}{=}\text{C—C—X} \longrightarrow \text{C}{=}\text{C—C}{=}\text{C}$$

are much rarer than conjugate additions (Chapter 15), but some examples are known.[140] One such is[141]

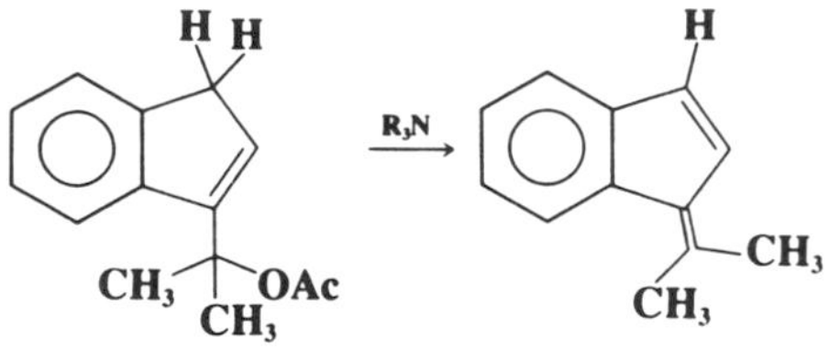

REACTIONS

First we consider reactions in which a C=C or a C≡C bond is formed. From a synthetic point of view, the most important reactions for the formation of double bonds are **7-1** (usually by an E1 mechanism), **7-6**, **7-13**, and **7-29** (usually by an E2 mechanism), and **7-3**, **7-4**, and **7-8** (usually by an Ei mechanism). The only synthetically important method for the formation of triple bonds is **7-13**.[142] In the second section we treat reactions in which C≡N bonds and C=N bonds are formed, and then eliminations that give C=O bonds and diazoalkanes. Finally, we discuss extrusion reactions.

Reactions in Which C=C and C≡C Bonds are Formed

A. Reactions in Which Hydrogen is Removed from One Side. In **7-1** to **7-5** the other leaving atom is oxygen. In **7-6** to **7-10** it is nitrogen. For reactions in which hydrogen is removed from both sides, see **9-1** to **9-6**.

[137]Botteron; Shulman *J. Org. Chem.* **1962**, *27*, 2007.

[138]Barton; Head; Williams *J. Chem. Soc.* **1952**, 453; Bamkole; Maccoll *J. Chem. Soc. B* **1970**, 1159.

[139]Taylor, Ref. 117, pp. 885-890; Smith; Mutter; Todd *J. Org. Chem.* **1977**, *42*, 44; Chuchani; Dominguez *Int. J. Chem. Kinet.* **1981**, *13*, 577; Hernández A.; Chuchani *Int. J. Chem. Kinet* **1983**, *15*, 205.

[140]For a review of certain types of 1,4 and 1,6 eliminations, see Wakselman *Nouv. J. Chem.* **1983**, *7*, 439-447.

[141]Thibblin; Onyido; Ahlberg *Chem. Scr.* **1982**, *19*, 145; Thibblin *J. Chem. Soc., Perkin Trans. 2* **1986**, 321; Ölwegård; Ahlberg *Acta Chem. Scand.* **1990**, *44*, 642. For studies of the stereochemistry of 1,4 eliminations, see Hill; Bock *J. Am. Chem. Soc.* **1978**, *100*, 637; Moss; Rickborn *J. Org. Chem.* **1986**, *51*, 1992; Ölwegård; Ahlberg *J. Chem. Soc., Chem. Commun.* **1989**, 1279.

[142]For reviews of methods for preparing alkynes, see Friedrich, in Patai; Rappoport *The Chemistry of Functional Groups, Supplement C*, pt. 2; Wiley: New York, 1983; pp. 1376-1384; Ben-Efraim, in Patai *The Chemistry of the Carbon-Carbon Triple Bond*, pt. 2; Wiley: New York, 1978, pp. 755-790. For a comparative study of various methods, see Mesnard; Bernadou; Miginiac *J. Chem. Res. (S)* **1981**, 270, and other papers in this series.

7-1 Dehydration of Alcohols
Hydro-hydroxy-elimination

$$-\overset{|}{\underset{H}{C}}-\overset{|}{\underset{OH}{C}}- \xrightarrow[\text{Al}_2\text{O}_3,\ 300°\text{C}]{\text{H}_2\text{SO}_4\ \text{or}} -\overset{|}{C}=\overset{|}{C}-$$

Dehydration of alcohols can be accomplished in several ways. H_2SO_4 and H_3PO_4 are common reagents, but in many cases these lead to rearrangement products and to ether formation (**0-16**). If the alcohol can be evaporated, vapor-phase elimination over Al_2O_3 is an excellent method since side reactions are greatly reduced. This method has even been applied to such high-molecular-weight alcohols as 1-dodecanol.[143] Other metallic oxides (e.g., Cr_2O_3, TiO_2, WO_3) have also been used, as have been sulfides, other metallic salts, and zeolites. Another method of avoiding side reactions is the conversion of alcohols to esters, and the pyrolysis of these (**7-3** to **7-5**). The ease of dehydration increases with α branching, and tertiary alcohols are dehydrated so easily with only a trace of acid that it sometimes happens even when the investigator desires otherwise. It may also be recalled that the initial alcohol products of many base-catalyzed condensations dehydrate spontaneously (Chapter 16) because the new double bond can be in conjugation with one already there. Many other dehydrating agents[144] have been used on occasion: P_2O_5, I_2, $ZnCl_2$, BF_3–etherate, dimethyl sulfoxide, $KHSO_4$, anhydrous $CuSO_4$, and phthalic anhydride, among others. Secondary and tertiary alcohols can also be dehydrated, without rearrangements, simply on refluxing in HMPA.[145] With nearly all reagents, dehydration follows Zaitsev's rule. An exception involves the passage of hot alcohol vapors over thorium oxide at 350 to 450°C, under which conditions Hofmann's rule is followed,[146] and the mechanism is probably different.

Carboxylic acids can be dehydrated by pyrolysis, the product being a ketene:

$$\text{R}-\overset{|}{\underset{H}{C}}\text{H}-\overset{|}{\underset{OH}{C}}=\text{O} \xrightarrow{\Delta} \text{R}-\text{CH}=\text{C}=\text{O}$$

Ketene itself is commercially prepared in this manner. In a similar reaction, carbon suboxide is produced by heating malonic acid with P_2O_5:

$$\text{HOOC}-\text{CH}_2-\text{COOH} \longrightarrow \text{O}=\text{C}=\text{C}=\text{C}=\text{O}$$

Carboxylic acids have also been converted to ketenes by treatment with certain reagents, among them TsCl,[147] dicyclohexylcarbodiimide,[148] and 1-methyl-2-chloropyridinium iodide (*Mukaiyama's reagent*).[149] Analogously, amides can be dehydrated with P_2O_5, pyridine, and Al_2O_3 to give ketenimines:[150]

$$\text{R}_2\text{CH}-\overset{\overset{\displaystyle O}{\|}}{\text{C}}-\text{NHR}' \xrightarrow[\text{pyridine}]{\text{P}_2\text{O}_5,\ \text{Al}_2\text{O}_3} \text{R}_2\text{C}=\text{C}=\text{NR}'$$

[143]For example, see Spitzin; Michailenko; Pirogowa *J. Prakt. Chem.* **1964**, [4] 25, 160; Bertsch; Greiner; Kretzschmar; Falk *J. Prakt. Chem.* **1964**, [4] 25, 184.
[144]For a list of reagents, with references, see Larock *Comprehensive Organic Transformations*; VCH: New York, 1989, pp. 151-152.
[145]Monson *Tetrahedron Lett.* **1971**, 567; Monson; Priest *J. Org. Chem.* **1971**, 36, 3826; Lomas; Sagatys; Dubois *Tetrahedron Lett.* **1972**, 165.
[146]Lundeen; Van Hoozer *J. Am. Chem. Soc.* **1963**, 85, 2180, *J. Org. Chem.* **1967**, 32, 3386. See also Davis *J. Org. Chem.* **1982**, 47, 900; Iimori; Ohtsuka; Oishi *Tetrahedron Lett.* **1991**, 32, 1209.
[147]Brady; Marchand; Giang; Wu *Synthesis* **1987**, 395, *J. Org. Chem.* **1987**, 52, 3457.
[148]Olah; Wu; Farooq *Synthesis* **1989**, 568.
[149]Ref. 147; Funk; Abelman; Jellison *Synlett* **1989**, 36.
[150]Stevens; Singhal *J. Org. Chem.* **1964**, 29, 34.

There is no way in which dehydration of alcohols can be used to prepare triple bonds: *gem*-diols and vinylic alcohols are not normally stable compounds and *vic*-diols[151] give either conjugated dienes or lose only 1 mole of water to give an aldehyde or ketone.

When proton acids catalyze alcohol dehydration, the mechanism is E1.[152] The principal process involves conversion of ROH to ROH_2^+ and cleavage of the latter to R^+ and H_2O, though with some acids a secondary process probably involves conversion of the alcohol to an inorganic ester and ionization of *this* (illustrated for H_2SO_4):

$$ROH \xrightarrow{H_2SO_4} ROSO_2OH \longrightarrow R^+ + HSO_4^-$$

Note that these mechanisms are the reverse of those involved in the acid-catalyzed hydration of double bonds (**5-2**), in accord with the principle of microscopic reversibility. With anhydrides (e.g., P_2O_5, phthalic anhydride) as well as with some other reagents such as HMPA,[153] it is likely that an ester is formed, and the leaving group is the conjugate base of the corresponding acid. In these cases the mechanism can be E1 or E2. The mechanism with Al_2O_3 and other solid catalysts has been studied extensively but is poorly understood.[154]

Dehydration of alcohols has also been accomplished by treating the *alkoxide* form of the alcohol with bromoform.[155] This reaction is called *deoxidation*. It is known that bromoform in basic solution gives rise to dibromocarbene, and the following mechanism is likely:

$$RO^- + \overline{C}Br_2 \xrightarrow{-Br^-} R-O-\overline{C}-Br \xrightarrow[-CO]{-Br^-} R^{\oplus} \longrightarrow \text{alkenes}$$

Note that the cleavage of the intermediate ROCBr is analogous to cleavage of RN_2^+ (p. 355) and the product distribution is similar.[156] Magnesium alkoxides (formed by ROH + $Me_2Mg \rightarrow ROMgMe$) have been decomposed thermally, by heating at 195-340°C to give the alkene, CH_4, and MgO.[157] Syn elimination is found and an Ei mechanism is likely. Similar decomposition of aluminum and zinc alkoxides has also been accomplished.[158]

OS **I**, 15, 183, 226, 280, 345, 430, 473, 475; **II**, 12, 368, 408, 606; **III**, 22, 204, 237, 312, 313, 353, 560, 729, 786; **IV**, 130, 444, 771; **V**, 294; **VI**, 307, 901; **VII**, 210, 241, 363, 368, 396; **65**, 12, 98. See also OS **VII**, 63; **67**, 125; **69**, 199. No attempt has been made to list olefin-forming dehydrations accompanying condensations or rearrangements.

7-2 Cleavage of Ethers to Olefins
Hydro-alkoxy-elimination

$$-\overset{|}{\underset{H}{C}}-\overset{|}{\underset{OR}{C}}- \xrightarrow{R'\,Na} -\overset{|}{C}=\overset{|}{C}- + RONa + R'H$$

[151]For a review on the dehydration of 1,2 and 1,3 diols, see Bartók; Molnár, in Patai *The Chemistry of Functional Groups, Supplement E*, pt. 2; Wiley: New York, 1980, pp. 721-760.

[152]For reviews of dehydration mechanisms, see Vinnik; Obraztsov *Russ. Chem. Rev.* **1990**, *59*, 63-77; Saunders; Cockerill, Ref. 2, pp. 221-274, 317-331; Knözinger, in Patai *The Chemistry of the Hydroxyl Group*, pt. 2; Wiley: New York, 1971, pp. 641-718.

[153]See, for example, Kawanisi; Arimatsu; Yamaguchi; Kimoto *Chem. Lett.* **1972**, 881.

[154]For reviews, see Beránek; Kraus; in Bamford; Tipper, Ref. 2, vol. 20, 1978, pp. 274-295; Pines *Intra-Sci. Chem. Rep.* **1972**, *6*(2), 1-42, pp. 17-21; Noller; Andréu; Hunger *Angew. Chem. Int. Ed. Engl.* **1971**, *10*, 172-181 [*Angew. Chem. 83*, 185-194]; Knözinger *Angew. Chem. Int. Ed. Engl.* **1968**, *7*, 791-805 [*Angew. Chem. 80*, 778-792]; Pines; Manassen *Adv. Catal.* **1966**, *16*, 49-93; Ref. 152. See also Berteau; Ruwet; Delmon *Bull. Soc. Chim. Belg.* **1985**, *94*, 859.

[155]Skell; Starer *J. Am. Chem. Soc.* **1959**, *81*, 4117.

[156]See, for example, Lee; Hahn *Can J. Chem.* **1967**, *45*, 2129.

[157]Ashby; Willard; Goel *J. Org. Chem.* **1979**, *44*, 1221.

[158]Ref. 157; Brieger; Watson; Barar; Shene *J. Org. Chem.* **1979**, *44*, 1340.

Olefins can be formed by the treatment of ethers with very strong bases, such as alkylsodium or alkyllithium compounds or sodium amide,[159] though there are usually side reactions too. The reaction is aided by electron-withdrawing groups in the β position, and, for example, $EtOCH_2CH(COOEt)_2$ can be converted to $CH_2\!=\!C(COOEt)_2$ without any base at all, but simply on heating.[160] t-Butyl ethers are cleaved more easily than others. Several mechanisms are possible. In many cases the mechanism is probably E1cB or on the E1cB side of the mechanistic spectrum,[161] since the base required is so strong, but it has been shown (by the use of $PhCD_2OEt$) that $PhCH_2OEt$ reacts by the five-membered Ei mechanism:[162]

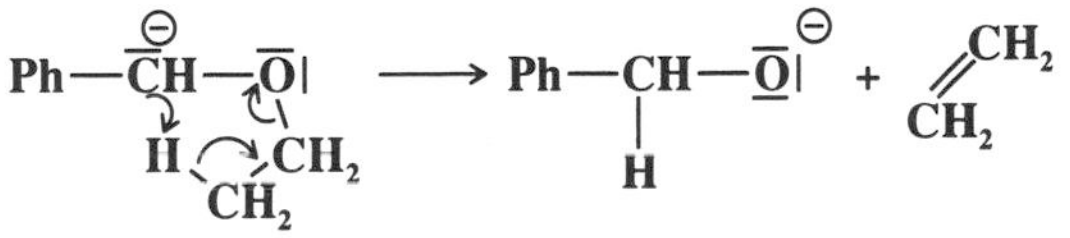

Epoxides can be converted to allylic alcohols[163] by treatment with several reagents, including lithium diethylamide,[164] t-butyldimethylsilyl iodide,[165] methylmagnesium N-cy-

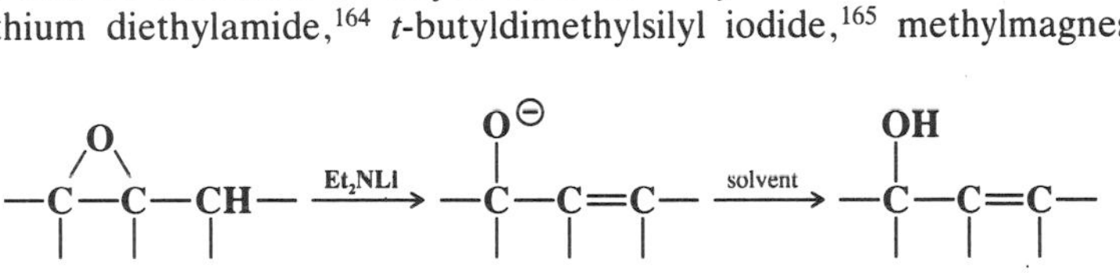

clohexylisopropylamide,[166] i-Pr$_2$NLi–t-BuOK (the *LIDAKOR reagent*),[167] and a diethylaluminum dialkylamide R_2NAlEt[168] (an alternative procedure is given in **7-12**). When an optically active reagent is used, optically active allylic alcohols can be produced from achiral epoxides.[169]

Ethers have also been converted to olefins and alcohols by passing vapors over hot P_2O_5 or Al_2O_3 (this method is similar to **7-1**), but this is not a general reaction. However, acetals can be converted to enol ethers in this manner:

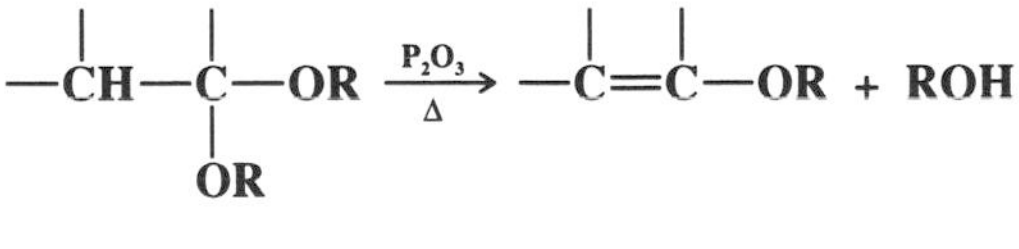

This can also be done at room temperature by treatment with trimethylsilyl triflate and a tertiary amine[170] or with Me_3SiI in the presence of hexamethyldisilazane.[171]

[159]For a review, see Maercker *Angew. Chem. Int. Ed. Engl.* **1987**, *26*, 972-989 [*Angew. Chem. 99*, 1002-1019].

[160]Feely; Boekelheide *Org. Synth. IV*, 298.

[161]For an investigation in the gas phase, see DePuy; Bierbaum *J. Am. Chem. Soc.* **1981**, *103*, 5034.

[162]Letsinger; Pollart *J. Am. Chem. Soc.* **1956**, *78*, 6079.

[163]For reviews, see Smith *Synthesis* **1984**, 629-656, pp. 637-642; Crandall; Apparu *Org. React.* **1983**, *29*, 345-443. For a list of reagents, with references, see Ref. 144, pp. 117-118.

[164]See, for example, Cope; Brown; Lee *J. Am. Chem. Soc.* **1958**, *80*, 2855; Kissel; Rickborn *J. Org. Chem.* **1972**, *37*, 2060; Crandall; Crawley *Org. Synth. VI*, 948.

[165]Detty *J. Org. Chem.* **1980**, *45*, 924. For another silyl reagent, see Murata; Suzuki; Noyori *J. Am. Chem. Soc.* **1979**, *101*, 2738.

[166]Mosset; Manna; Viala; Falck *Tetrahedron Lett.* **1986**, *27*, 299.

[167]Mordini; Ben Rayana; Margot; Schlosser *Tetrahedron* **1990**, *46*, 2401.

[168]For a review, see Yamamoto; Nozaki *Angew. Chem. Int. Ed. Engl.* **1978**, *17*, 169-175 [*Angew. Chem. 90*, 180-186]. See also Yasuda; Tanaka; Yamamoto; Nozaki *Bull. Chem. Soc. Jpn.* **1979**, *52*, 1752.

[169]Su; Walder; Zhang; Scheffold *Helv. Chim. Acta* **1988**, *71*, 1073, and references cited therein.

[170]Gassman; Burns *J. Org. Chem.* **1988**, *53*, 5574.

[171]Miller; McKean *Tetrahedron Lett.* **1982**, *23*, 323. For another method, see Marsi; Gladysz *Organometallics* **1982**, *1*, 1467.

Enol ethers can be pyrolyzed to olefins and aldehydes in a manner similar to that of **7-3**:

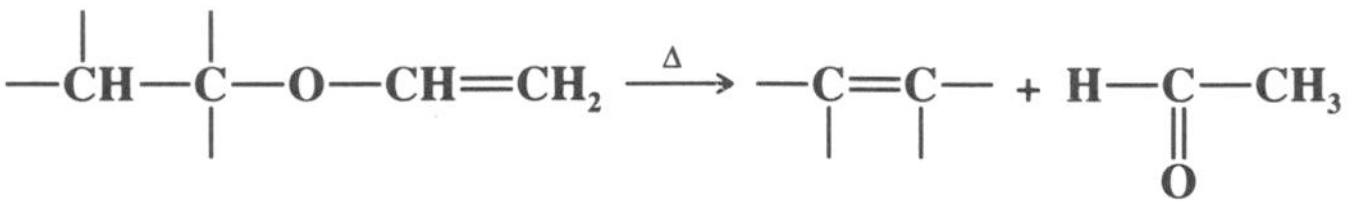

$$-\overset{|}{\underset{|}{CH}}-\overset{|}{\underset{|}{C}}-O-CH{=}CH_2 \xrightarrow{\Delta} -\overset{|}{C}{=}\overset{|}{C}- + H-\overset{O}{\underset{\|}{C}}-CH_3$$

The rate of this reaction for R—O—CH=CH$_2$ increased in the order Et < i-Pr < *t*-Bu.[172] The mechanism is similar to that of **7-3**.

OS **IV**, 298, 404; **V**, 25, 642, 859, 1145; **VI**, 491, 564, 584, 606, 683, 948; **65**, 98.

7-3 Pyrolysis of Esters of Carboxylic Acids
Hydro-acyloxy-elimination

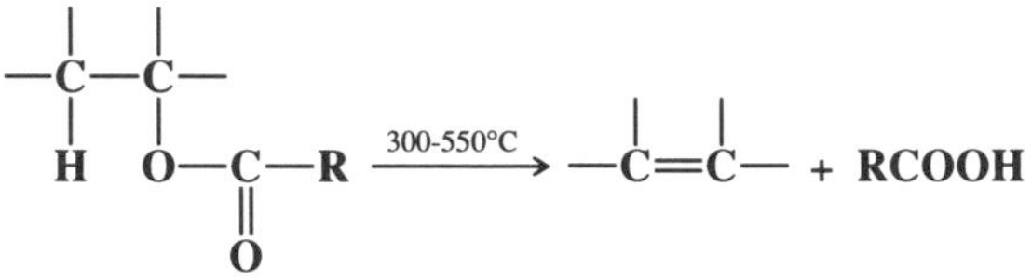

$$\begin{array}{c} -\overset{|}{\underset{|}{C}}-\overset{|}{\underset{|}{C}}- \\ \overset{}{H} \quad \overset{}{\underset{\underset{O}{\|}}{O}}-\overset{}{\underset{}{C}}-R \end{array} \xrightarrow{300\text{-}550°C} -\overset{|}{C}{=}\overset{|}{C}- + RCOOH$$

Carboxylic esters in which the alkyl group has a β hydrogen can be pyrolyzed, most often in the gas phase, to give the corresponding acid and an olefin.[173] No solvent is required. Since rearrangement and other side reactions are few, the reaction is synthetically very useful and is often carried out as an indirect method of accomplishing **7-1**. The yields are excellent and the workup is easy. Many olefins have been prepared in this manner. For higher olefins (above about C$_{10}$) a better method is to pyrolyze the alcohol in the presence of acetic anhydride.[174]

The mechanism is Ei (see p. 1006). Lactones can be pyrolyzed to give unsaturated acids, provided that the six-membered transition state required for Ei reactions is available (it is not available for five- and six-membered lactones, but it is for larger rings[175]). Amides give a similar reaction but require higher temperatures.

Allylic acetates give dienes when heated with certain palladium[176] or molybdenum[177] compounds.

OS **III**, 30; **IV**, 746; **V**, 235.

7-4 The Chugaev Reaction

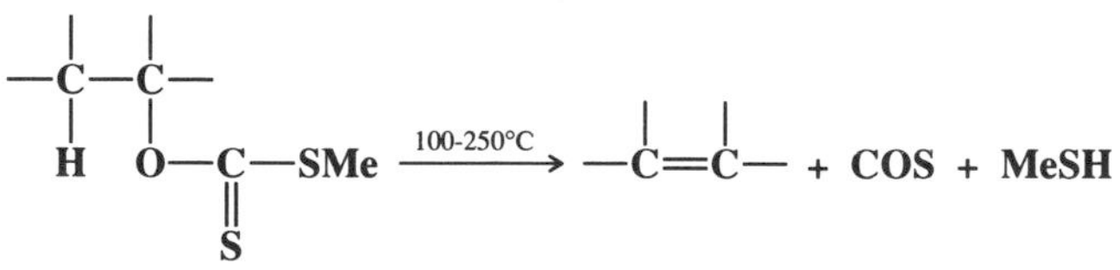

$$\begin{array}{c} -\overset{|}{\underset{|}{C}}-\overset{|}{\underset{|}{C}}- \\ \overset{}{H} \quad \overset{}{\underset{\underset{S}{\|}}{O}}-\overset{}{\underset{}{C}}-SMe \end{array} \xrightarrow{100\text{-}250°C} -\overset{|}{C}{=}\overset{|}{C}- + COS + MeSH$$

[172]McEwen; Taylor *J. Chem. Soc., Perkin Trans. 2* **1982**, 1179. See also Taylor *J. Chem. Soc., Perkin Trans. 2* **1988**, 737.

[173]For a review, see DePuy; King, Ref. 135, pp. 432-444. For some procedures, see Jenneskens; Hoefs; Wiersum *J. Org. Chem.* **1989**, *54*, 5811, and references cited therein.

[174]Aubrey; Barnatt; Gerrard *Chem. Ind. (London)* **1965**, 681.

[175]See, for example, Bailey; Bird, Ref. 122.

[176]For a review, see Heck *Palladium Reagents in Organic Synthesis*; Academic Press: New York, 1985, pp. 172-178.

[177]Trost; Lautens; Peterson *Tetrahedron Lett.* **1983**, *24*, 4525.

Methyl xanthates are prepared by treatment of alcohols with NaOH and CS_2 to give RO—CS—SNa, followed by treatment of this with methyl iodide.[178] Pyrolysis of the xanthate to give the olefin, COS, and the thiol is called the *Chugaev reaction*.[179] The reaction is thus, like **7-3**, an indirect method of accomplishing **7-1**. The temperatures required with xanthates are lower than with ordinary esters, which is advantageous because possible isomerization of the resulting olefin is minimized. The mechanism is Ei, similar to that of **7-3**. For a time there was doubt as to which sulfur atom closed the ring, but now there is much evidence, including the study of ^{34}S and ^{13}C isotope effects, to show that it is the C=S sulfur:[180]

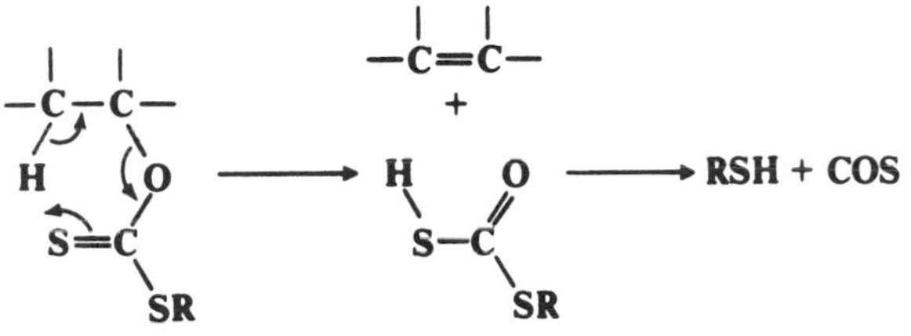

The mechanism is thus exactly analogous to that of **7-3**.
    OS **VII**, 139.

7-5 Decomposition of Other Esters
Hydro-tosyloxy-elimination

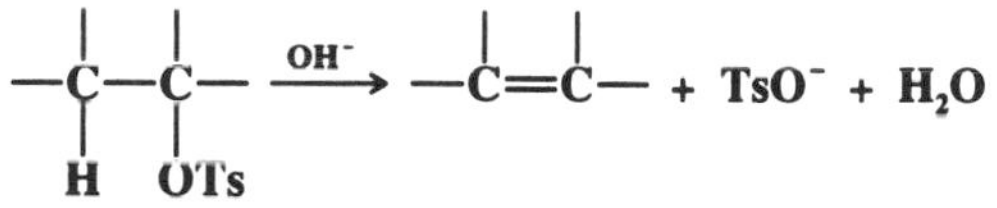

Several types of inorganic ester can be cleaved to olefins by treatment with bases. Esters of sulfuric, sulfurous, and other acids undergo elimination in solution by E1 or E2 mechanisms, as do tosylates and other esters of sulfonic acids.[181] It has been shown that bis(tetra-*n*-butylammonium) oxalate $(Bu_4N^+)_2 (COO^-)_2$ is an excellent reagent for inducing tosylates to undergo elimination rather than substitution.[182] Aryl sulfonates have also been cleaved without a base. Esters of 2-pyridinesulfonic acid and 8-quinolinesulfonic acid gave olefins in high yields simply on heating, without a solvent.[183] Esters of $PhSO_2OH$ and TsOH behaved similarly when heated in a dipolar aprotic solvent such as Me_2SO or HMPA.[184]
    OS, **VI**, 837; **VII**, 117.

7-6 Cleavage of Quaternary Ammonium Hydroxides
Hydro-trialkylammonio-elimination

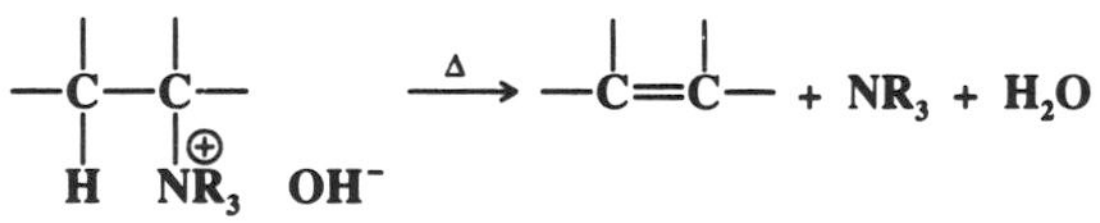

[178]For a method of preparing xanthates from alcohols in one laboratory step, see Lee; Chan; Wong; Wong *Synth. Commun.* **1989**, *19*, 547.
[179]For reviews, see DePuy; King, Ref. 135, pp. 444-448; Nace *Org. React.* **1962**, *12*, 57-100.
[180]Bader; Bourns *Can. J. Chem.* **1961**, *39*, 348.
[181]For a list of reagents used for sulfonate cleavages, with references, see Ref. 144, pp. 153-154.
[182]Corey; Terashima *Tetrahedron Lett.* **1972**, 111.
[183]Corey; Posner; Atkinson; Wingard; Halloran; Radzik; Nash *J. Org. Chem.* **1989**, *54*, 389.
[184]Nace *J. Am. Chem. Soc.* **1959**, *81*, 5428.

Cleavage of quaternary ammonium hydroxides is the final step of the process known as *Hofmann exhaustive methylation* or *Hofmann degradation*.[185] In the first step, a primary, secondary, or tertiary amine is treated with enough methyl iodide to convert it to the quaternary ammonium iodide (**0-43**). In the second step, the iodide is converted to the hydroxide by treatment with silver oxide. In the cleavage step an aqueous or alcoholic solution of the hydroxide is distilled, often under reduced pressure. The decomposition generally takes place at a temperature between 100 and 200°C. Alternatively, the solution can be concentrated to a syrup by distillation or freeze-drying.[186] When the syrup is heated at low pressures, the cleavage reaction takes place at lower temperatures than are required for the reaction in the ordinary solution, probably because the base (OH⁻ or RO⁻) is less solvated.[187] The reaction has never been an important synthetic tool, but in the 19th century and the first part of the 20th century it saw much use in the determination of the structure of unknown amines, especially alkaloids. In many of these compounds the nitrogen is in a ring, or even at a ring junction, and in such cases the olefin still contains nitrogen. Repetitions of the process are required to remove the nitrogen completely, e.g.,

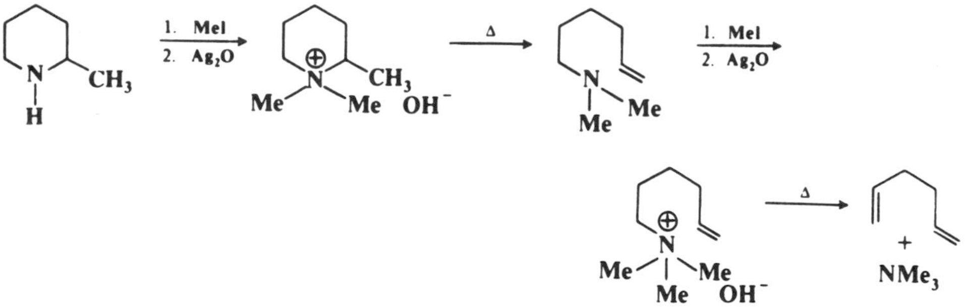

A side reaction involving nucleophilic substitution to give an alcohol (R₄N⁺ OH⁻ → ROH + R₃N) generally accompanies the normal elimination reaction[188] but seldom causes trouble. However, when none of the four groups on the nitrogen has a β hydrogen, substitution is the only reaction possible. On heating Me₄N⁺ OH⁻ in water, methanol is obtained, though without a solvent the product is not methanol but dimethyl ether.[189]

The mechanism is usually E2. Hofmann's rule is generally obeyed by acyclic and Zaitsev's rule by cyclohexyl substrates (p. 999). In certain cases, where the molecule is highly hindered, a five-membered Ei mechanism, similar to that in **7-7**, has been shown to operate. That is, the OH⁻ in these cases does not attract the β hydrogen, but instead removes one of the methyl hydrogens:

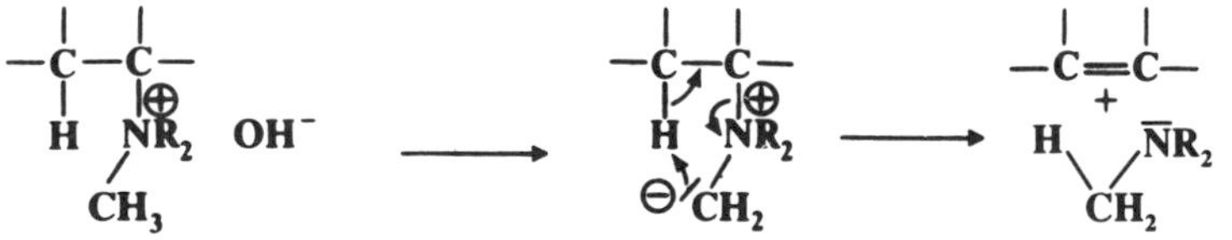

[185]For reviews, see Bentley, in Bentley; Kirby *Elucidation of Organic Structures by Physical and Chemical Methods*, 2nd ed. (vol. 4 of Weissberger *Techniques of Chemistry*), pt. 2; Wiley: New York, 1973, pp. 255-289; White; Woodcock, in Patai *The Chemistry of the Amino Group*; Wiley: New York, 1968, pp. 409-416; Cope; Trumbull *Org. React.* **1960**, *11*, 317-493.

[186]Archer *J. Chem. Soc. C* **1971**, 1327.

[187]Saunders; Cockerill, Ref. 2, pp. 4-5.

[188]Baumgarten *J. Chem. Educ.* **1968**, *45*, 122.

[189]Musker *J. Am. Chem. Soc.* **1964**, *86*, 960, *J. Chem. Educ.* **1968**, *45*, 200; Musker; Stevens *J. Am. Chem. Soc.* **1968**, *90*, 3515; Tanaka; Dunning; Carter *J. Org. Chem.* **1966**, *31*, 3431.

The obvious way to distinguish between this mechanism and the ordinary E2 mechanism is by the use of deuterium labeling. For example, if the reaction is carried out on a quaternary hydroxide deuterated on the β carbon ($R_2CDCH_2NMe_3^+ OH^-$), the fate of the deuterium indicates the mechanism. If the E2 mechanism is in operation, the trimethylamine produced would contain no deuterium (which would be found only in the water). But if the mechanism is Ei, the amine would contain deuterium. In the case of the highly hindered compound $(Me_3C)_2CDCH_2NMe_3^+ OH^-$, the deuterium did appear in the amine, demonstrating an Ei mechanism for this case.[190] With simpler compounds, the mechanism is E2, since here the amine was deuterium-free.[191]

When the nitrogen bears more than one group possessing a β hydrogen, which group cleaves? The Hofmann rule says that *within* a group the hydrogen on the least alkylated carbon cleaves. This tendency is also carried over to the choice of which group cleaves: thus ethyl with three β hydrogens cleaves more readily than any longer *n*-alkyl group, all of which have two β hydrogens. "The β hydrogen is removed most readily if it is located on a methyl group, next from RCH_2, and least readily from R_2CH."[192] In fact, the Hofmann rule as first stated[193] in 1851 applied only to which group cleaved, not to the orientation within a group; the latter could not have been specified in 1851, since the structural theory of organic compounds was not formulated until 1857-1860. Of course, the Hofmann rule (applied to which group cleaves *or* to orientation within a group) is superseded by conjugation possibilities. Thus $PhCH_2CH_2NMe_2Et^+ OH^-$ gives mostly styrene instead of ethylene.

Triple bonds have been prepared by pyrolysis of 1,2-bis salts.[194]

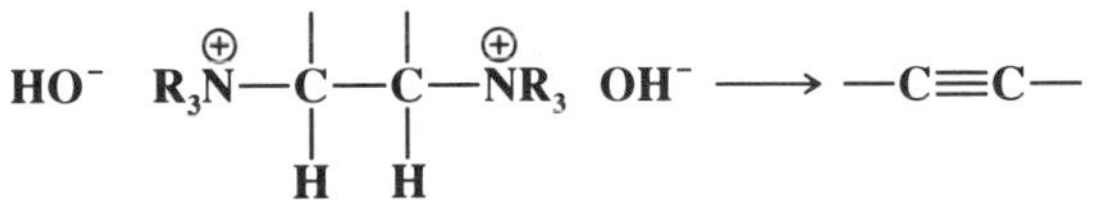

$$HO^-\quad R_3\overset{\oplus}{N}-\overset{|}{\underset{H}{C}}-\overset{|}{\underset{H}{C}}-\overset{\oplus}{N}R_3, \ OH^- \longrightarrow \ -C\equiv C-$$

OS **IV**, 980; **V**, 315, 608; **VI**, 552. Also see OS **V**, 621, 883; **VI**, 75.

7-7   Cleavage of Quaternary Ammonium Salts with Strong Bases
Hydro-trialkylammonio-elimination

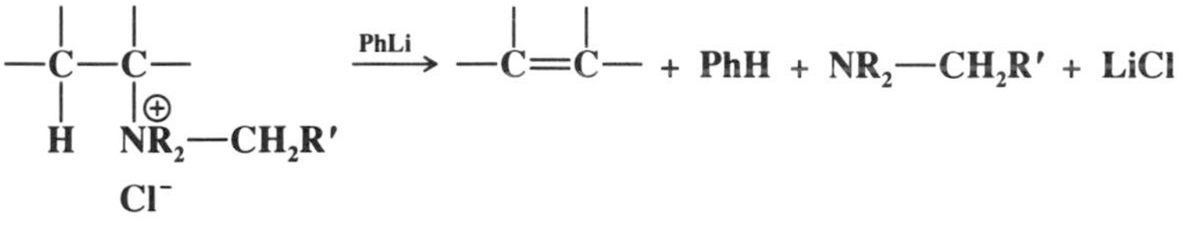

$$-\overset{|}{\underset{H}{C}}-\overset{|}{\underset{\substack{\oplus\\NR_2-CH_2R'\\Cl^-}}{C}}-\ \xrightarrow{\text{PhLi}}\ -\overset{|}{C}=\overset{|}{C}-\ +\ \text{PhH}\ +\ NR_2-CH_2R'\ +\ \text{LiCl}$$

When quaternary ammonium halides are treated with strong bases (e.g., PhLi, KNH_2 in liquid NH_3[195]), an elimination can occur that is similar in products, though not in mechanism,

[190]Cope; Mehta *J. Am. Chem Soc.* **1963**, *85*, 1949. See also Baldwin; Banthorpe; Loudon; Waller *J. Chem. Soc. B* **1967**, 509.
[191]Cope; LeBel; Moore; Moore *J. Am. Chem. Soc.* **1961**, *83*, 3861.
[192]Cope; Trumbull. Ref. 185, p. 348.
[193]Hofmann *Liebigs Ann. Chem.* **1851**, *78*, 253.
[194]For a review, see Franke; Ziegenbein; Meister *Angew. Chem.* **1960**, *72*, 391-400, pp. 397-398.
[195]Bach; Andrzejewski *J. Am. Chem. Soc.* **1971**, *93*, 7118; Bach; Bair; Andrzejewski *J. Am. Chem. Soc.* **1972**, *94*, 8608, *J. Chem. Soc., Chem. Commun.* **1974**, 819.

to **7-6.** This is an alternative to **7-6** and is done on the quaternary ammonium halide, so that it is not necessary to convert this to the hydroxide. The mechanism is Ei:

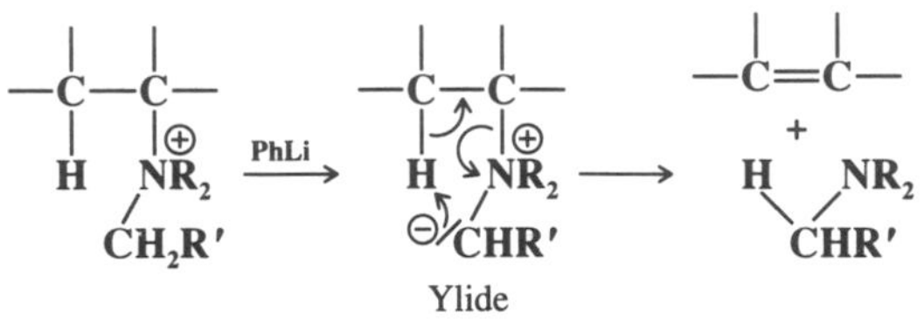

$$-\overset{|}{\underset{H}{C}}-\overset{|}{\underset{\overset{\oplus}{NR_2}}{C}}- \quad\xrightarrow{PhLi}\quad -\overset{|}{\underset{H}{C}}-\overset{|}{\underset{\overset{\oplus}{NR_2}}{C}}- \quad\longrightarrow\quad -C{=}C- \;+\; $$

Ylide

An α' hydrogen is obviously necessary in order for the ylide to be formed. This type of mechanism is called α',β elimination, since a β hydrogen is removed by the α' carbon. The mechanism has been confirmed by labeling experiments similar to those described at **7-6,**[196] and by isolation of the intermediate ylides.[197] An important synthetic difference between this and most instances of **7-6** is that syn elimination is observed here and anti elimination in **7-6,** so products of opposite configuration are formed when the olefin exhibits cis–trans isomerism.

An alternative procedure that avoids the use of a very strong base is heating the salt with KOH in polyethylene glycol monomethyl ether.[198]

7-8 Cleavage of Amine Oxides
Hydro-(Dialkyloxidoammonio)-elimination

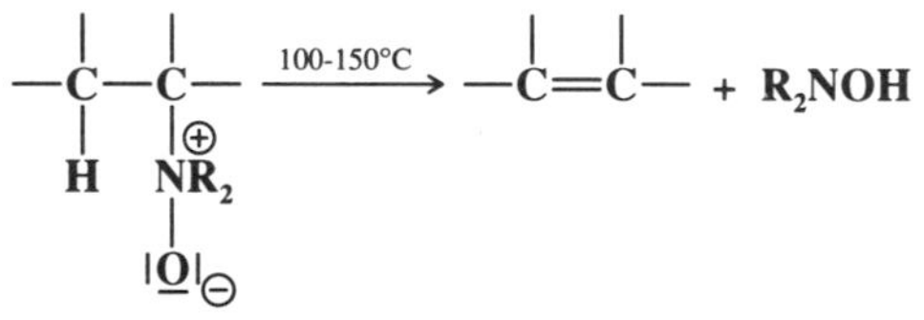

$$-\overset{|}{\underset{H}{C}}-\overset{|}{\underset{\overset{|}{\underset{\underset{|\underline{O}|\ominus}{\overset{\oplus}{NR_2}}}{}}}{C}}- \quad\xrightarrow{100\text{-}150°C}\quad -C{=}C- \;+\; R_2NOH$$

Cleavage of amine oxides to produce an alkene and a hydroxylamine is called the *Cope reaction* (not to be confused with the Cope *rearrangement,* **8-34**). It is an alternative to **7-6** and **7-7.**[199] The reaction is usually performed with a mixture of amine and oxidizing agent (see **9-28**) without isolation of the amine oxide. Because of the mild conditions side reactions are few, and the olefins do not usually rearrange. The reaction is thus very useful for the preparation of many olefins. A limitation is that it does not open 6-membered rings containing hetero nitrogen, though it does open rings of 5 and 7 to 10 members.[200] Rates of the reaction increase with increasing size of α and β substituents.[201] The reaction can be carried out at room temperature in dry Me$_2$SO or THF.[202] The elimination is a stereoselective syn process,[203] and the five-membered Ei mechanism operates:

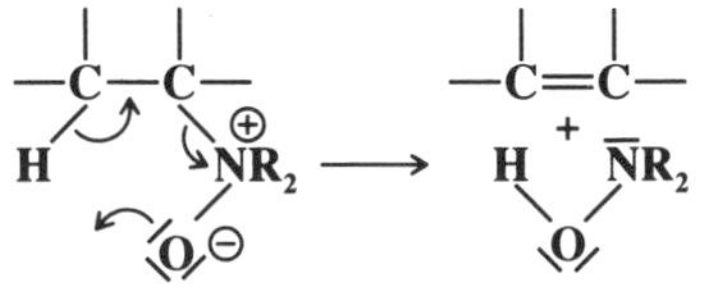

[196]Weygand; Daniel; Simon *Chem. Ber.* **1958,** *91,* 1691; Bach; Andrzejewski; Bair *J. Chem. Soc., Chem. Commun.* **1974,** 820; Bach; Knight *Tetrahedron Lett.* **1979,** 3815.
[197]Wittig; Polster *Liebigs Ann. Chem.* **1958,** *612,* 102; Wittig; Burger *Liebigs Ann. Chem.* **1960,** *632,* 85.
[198]Hünig; Öller; Wehner *Liebigs Ann. Chem.* **1979,** 1925.
[199]For reviews, see Cope; Trumbull, Ref. 185, pp. 361-370; DePuy; King, Ref. 135, pp. 448-451.
[200]Cope; LeBel *J. Am. Chem. Soc.* **1960,** *82,* 4656; Cope; Ciganek; Howell; Schweizer *J. Am. Chem. Soc.* **1960,** *82,* 4663.
[201]Závada; Pánková; Svoboda *Collect. Czech. Chem. Commun.* **1973,** *38,* 2102.
[202]Cram; Sahyun; Knox *J. Am. Chem. Soc.* **1962,** *84,* 1734.
[203]See, for example, Bach; Andrzejewski; Dusold *J. Org. Chem.* **1973,** *38,* 1742.

Almost all evidence indicates that the transition state must be planar. Deviations from planarity as in **7-3** (see p. 1006) are not found here, and indeed this is why six-membered heterocyclic nitrogen compounds do not react. Because of the stereoselectivity of this reaction and the lack of rearrangement of the products, it is useful for the formation of trans cycloolefins (eight-membered and higher).

OS **IV**, 612.

7-9   Olefins from Aliphatic Diazonium Salts
Hydro-diazonio-elimination

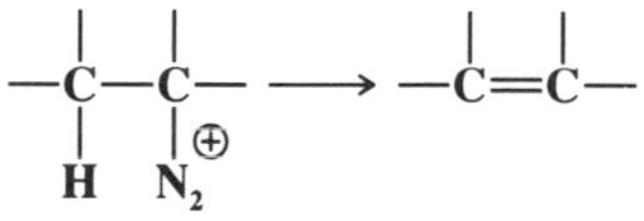

$$-\overset{|}{\underset{H}{C}}-\overset{|}{\underset{\overset{|}{N_2}}{C}}- \longrightarrow -\overset{|}{C}=\overset{|}{C}-$$

The treatment of aliphatic amines with nitrous acid is not a useful method for the preparation of olefins any more than it is for the preparation of alcohols (p. 355), though some olefin is usually formed in such reactions.

7-10   Decomposition of Toluene-*p*-sulfonylhydrazones

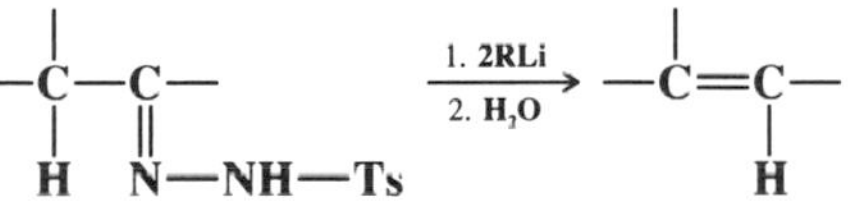

$$-\overset{|}{\underset{H}{C}}-\overset{|}{\underset{N-NH-Ts}{C}}- \xrightarrow[\text{2. H}_2\text{O}]{\text{1. 2RLi}} -\overset{|}{C}=\overset{|}{\underset{H}{C}}-$$

Treatment of the tosylhydrazone of an aldehyde or a ketone with a strong base leads to the formation of an olefin, the reaction being formally an elimination accompanied by a hydrogen shift.[204] The reaction (called the *Shapiro reaction*) has been applied to tosylhydrazones of many aldehydes and ketones. The most useful method synthetically involves treatment of the substrate with at least two equivalents of an organolithium compound[205] (usually MeLi) in ether, hexane, or tetramethylenediamine.[206] This procedure gives good yields of alkenes without side reactions and, where a choice is possible, predominantly gives the less highly substituted olefin. Tosylhydrazones of α,β-unsaturated ketones give conjugated dienes.[207] The mechanism[208] has been formulated as:

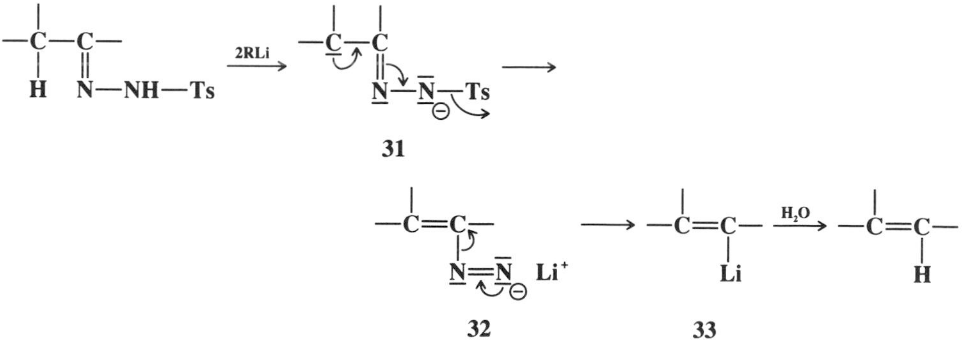

$$-\overset{|}{\underset{H}{C}}-\overset{|}{\underset{N-NH-Ts}{C}}- \xrightarrow{\text{2RLi}} -\overset{|}{C}-\overset{|}{\underset{\underset{\ominus}{N-N-Ts}}{C}}- \longrightarrow$$

31

$$-\overset{|}{C}=\overset{|}{\underset{\underset{\ominus}{N=N}}{C}}-\ \text{Li}^+ \longrightarrow -\overset{|}{C}=\overset{|}{\underset{\text{Li}}{C}}- \xrightarrow{\text{H}_2\text{O}} -\overset{|}{C}=\overset{|}{\underset{\text{H}}{C}}-$$

32   **33**

---

[204]For reviews, see Adlington; Barrett *Acc. Chem. Res.* **1983**, *16*, 55-59; Shapiro *Org. React.* **1976**, *23*, 405-507.
[205]Shapiro; Heath *J. Am. Chem. Soc.* **1967**, *89*, 5734; Kaufman; Cook; Shechter; Bayless; Friedman *J. Am. Chem. Soc.* **1967**, *89*, 5736; Shapiro *Tetrahedron Lett.* **1968**, 345; Meinwald; Uno *J. Am. Chem. Soc.* **1968**, *90*, 800.
[206]Stemke; Bond *Tetrahedron Lett.* **1975**, 1815.
[207]See Dauben; Rivers; Zimmerman *J. Am. Chem. Soc.* **1977**, *99*, 3414.
[208]For a review of the mechanism, see Casanova; Waegell *Bull. Soc. Chim. Fr.* **1975**, 922-932.

Evidence for this mechanism is: (1) two equivalents of RLi are required; (2) the hydrogen in the product comes from the water and not from the adjacent carbon, as shown by deuterium labeling;[209] and (3) the intermediates **31-33** have been trapped.[210] This reaction, when performed in tetramethylenediamine, can be a synthetically useful method[211] of generating vinylic lithium compounds (**33**), which can be trapped by various electrophiles such as D_2O (to give deuterated alkenes), CO_2 (to give α,β-unsaturated carboxylic acids—**6-34**), or DMF (to give α,β-unsaturated aldehydes—**0-105**).

The reaction also takes place with other bases (e.g., LiH,[213] Na in ethylene glycol, NaH, $NaNH_2$) or with smaller amounts of RLi, but in these cases side reactions are common and the orientation of the double bond is in the other direction (to give the more highly substituted olefin). The reaction with Na in ethylene glycol is called the *Bamford–Stevens reaction*.[214] For these reactions two mechanisms are possible—a carbenoid and a carbocation mechanism.[215] The side reactions found are those expected of carbenes and carbocations. In general, the carbocation mechanism is chiefly found in protic solvents and the carbenoid mechanism in aprotic solvents. Both routes involve formation of a diazo compound (**34**) which in some cases can be isolated.

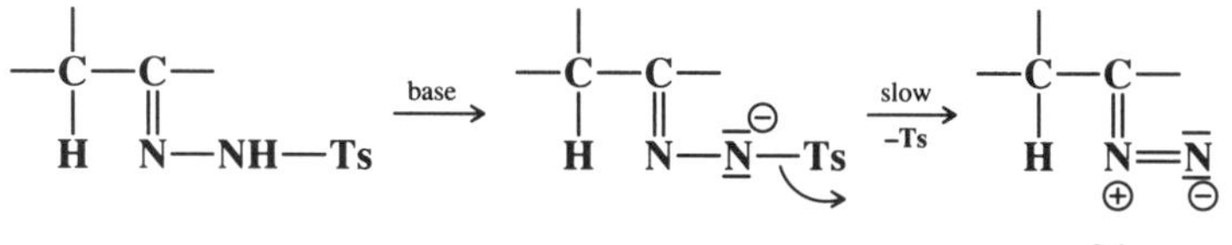

In fact, this reaction has been used as a synthetic method for the preparation of diazo compounds.[216] In the absence of protic solvents **34** loses N_2, and hydrogen migrates, to give the olefin product. The migration of hydrogen may immediately follow, or be simultaneous with, the loss of N_2. In a protic solvent, **34** becomes protonated to give the diazonium ion **35** which loses N_2 to give the corresponding carbocation which may then undergo elimination

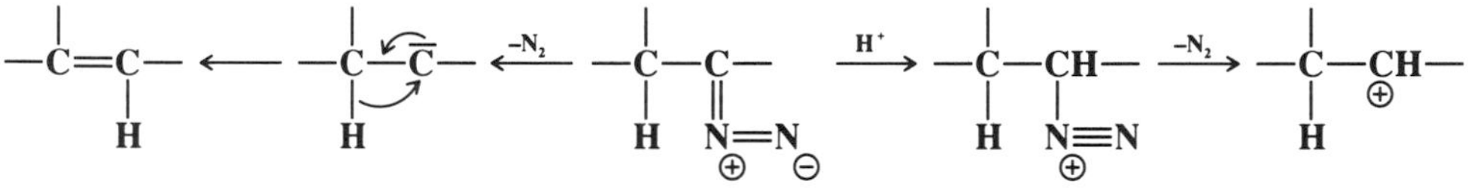

(**7-9**) or give other reactions characteristic of carbocations. A diazo compound is an intermediate in the formation of olefins by treatment of N-nitrosoamides with a rhodium(II) catalyst.[217]

[209]Ref. 205; Shapiro; Hornaman *J. Org. Chem.* **1974**, *39*, 2302.

[210]Shapiro; Lipton; Kolonko; Buswell; Capuano *Tetrahedron Lett.* **1975**, 1811, Ref. 206; Lipton; Shapiro *J. Org. Chem.* **1978**, *43*, 1409.

[211]See Traas; Boelens; Takken *Tetrahedron Lett.* **1976**, 2287; Stemke; Chamberlin; Bond *Tetrahedron Lett.* **1976**, 2947.

[212]For a review, see Chamberlin; Bloom *Org. React.* **1990**, *39*, 1-83.

[213]Biellmann; Pète *Bull. Soc. Chim. Fr.* **1967**, 675.

[214]Bamford; Stevens *J. Chem. Soc.* **1952**, 4735.

[215]Powell; Whiting *Tetrahedron* **1959**, 7, 305, **1961**, *12* 168; DePuy; Froemsdorf *J. Am. Chem. Soc.* **1960**, *82*, 634; Bayless; Friedman; Cook; Shechter *J. Am. Chem. Soc.* **1968**, *90*, 531; Nickon; Werstiuk *J. Am. Chem. Soc.* **1972**, *94*, 7081.

[216]For a review, see Regitz; Maas *Diazo Compounds*; Academic Press: New York, 1986, pp. 257-295. For an improved procedure, see Wulfman; Yousefian; White *Synth. Commun.* **1988**, *18*, 2349.

[217]Godfrey; Ganem *J. Am. Chem. Soc.* **1990**, *112*, 3717.

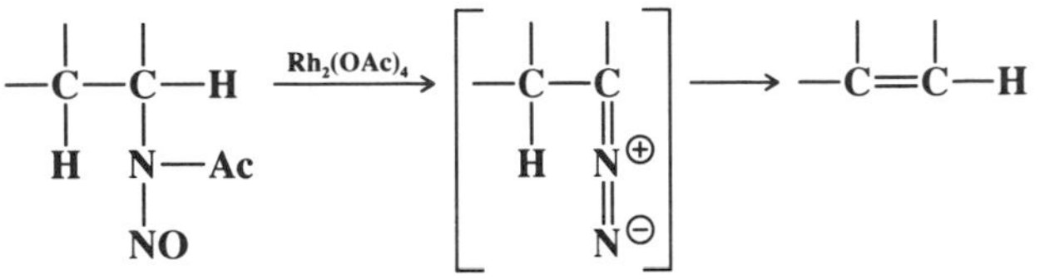

$$-\overset{|}{\underset{H}{C}}-\overset{|}{\underset{\underset{NO}{N}-Ac}{C}}-H \xrightarrow{Rh_2(OAc)_4} \left[-\overset{|}{\underset{H}{C}}-\overset{|}{\underset{\underset{N_\ominus}{N^\oplus}}{C}} \right] \longrightarrow -\overset{|}{C}=\overset{|}{C}-H$$

See also **7-28**.

OS **VI**, 172; **VII**, 77. For the preparation of a diazo compound, see OS **VII**, 438.

7-11   Cleavage of Sulfonium Compounds
Hydro-dialkylsulfonio-elimination

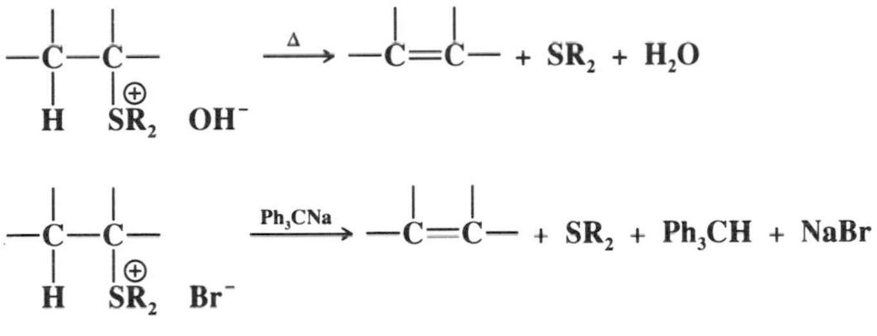

$$-\overset{|}{\underset{H}{C}}-\overset{|}{\underset{\underset{OH^-}{SR_2}}{C}}- \xrightarrow{\Delta} -\overset{|}{C}=\overset{|}{C}- + SR_2 + H_2O$$

$$-\overset{|}{\underset{H}{C}}-\overset{|}{\underset{\underset{Br^-}{SR_2}}{C}}- \xrightarrow{Ph_3CNa} -\overset{|}{C}=\overset{|}{C}- + SR_2 + Ph_3CH + NaBr$$

Sulfonium compounds undergo elimination similar to that of their ammonium counterparts (**7-6** and **7-7**) in scope and mechanism. The decomposition by heat of sulfonium hydroxides has been known for many years.[218] The ylide reaction was discovered more recently.[219] Neither is important synthetically.

7-12   Cleavage of Sulfoxides, Selenoxides, and Sulfones

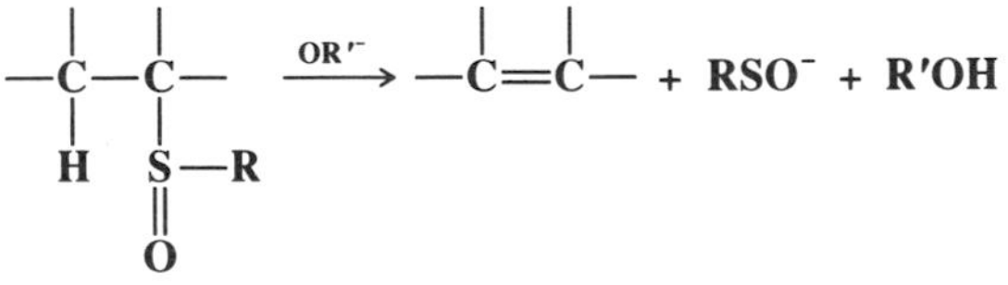

$$-\overset{|}{\underset{H}{C}}-\overset{|}{\underset{\underset{O}{\overset{\|}{S}-R}}{C}}- \xrightarrow{OR'^-} -\overset{|}{C}=\overset{|}{C}- + RSO^- + R'OH \qquad \textbf{Hydro-alkylsulfinyl-elimination}$$

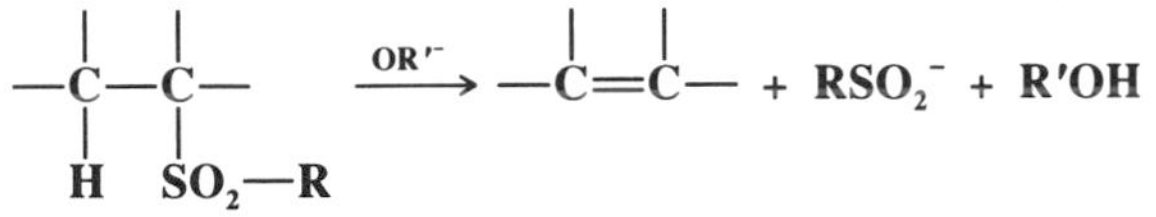

$$-\overset{|}{\underset{H}{C}}-\overset{|}{\underset{SO_2-R}{C}}- \xrightarrow{OR'^-} -\overset{|}{C}=\overset{|}{C}- + RSO_2^- + R'OH \qquad \textbf{Hydro-alkylsulfonyl-elimination}$$

Sulfones and sulfoxides with a β hydrogen undergo elimination on treatment with an alkoxide or, for sulfones,[220] even with OH$^-$.[221] In mechanism, these reactions belong on the E1–E2–E1cB spectrum.[222] Although the leaving groups are uncharged, the orientation follows Hofmann's rule, not Zaitsev's. Sulfoxides (but not sulfones) also undergo elimination on pyrolysis

[218]For a discussion, see Knipe, in Stirling *The Chemistry of the Sulphonium Group*, pt. 1; Wiley: New York, 1981, pp. 334-347.

[219]Franzen; Mertz *Chem. Ber.* **1960**, *93*, 2819. For a review, see Block *Reactions of Organosulfur Compounds*; Academic Press: New York, 1978, pp. 112-117.

[220]Certain sulfones undergo elimination with 5% HCl in THF: Yoshida; Saito *Chem. Lett.* **1982**, 165.

[221]Hofmann; Wallace; Argabright; Schriesheim *Chem. Ind. (London)* **1963**, 1234.

[222]Hofmann; Wallace; Schriesheim *J. Am. Chem. Soc.* **1964**, *86*, 1561.

at about 80°C in a manner analogous to **7-8.** The mechanism is also analogous, being the five-membered Ei mechanism with syn elimination.[223] Selenoxides[224] and sulfinate esters R_2CH—CHR—SO—OMe[225] also undergo elimination by the Ei mechanism, the selenoxide reaction taking place at room temperature. The reaction with selenoxides has been extended to the formation of triple bonds.[226]

Both the selenoxide[227] and sulfoxide[228] reactions have been used in a method for the conversion of ketones, aldehydes, and carboxylic esters to their α,β-unsaturated derivatives (illustrated for the selenoxide).

$$PhCOCH_2CH_3 \xrightarrow{2-13} \underset{\underset{SePh}{|}}{PhCOCHCH_3} \xrightarrow[9-31]{NaIO_4} \left[\underset{\underset{\underset{O}{\|}}{\overset{|}{SePh}}}{PhCOCHCH_3} \right] \longrightarrow PhCOCH=CH_2$$

Because of the mildness of the procedure, this is probably the best means of accomplishing this conversion. The selenoxide reaction has been used in a procedure for the conversion of epoxides to allylic alcohols.[229]

$$\underset{O}{\overset{R_2CH \diagdown \diagup R'}{\triangle}} + PhSe^- \xrightarrow[0-35]{EtOH} R_2CH—\underset{\underset{SePh}{|}}{CH}—\overset{\overset{OH}{|}}{CHR'} \xrightarrow[9-31]{H_2O_2} R_2CH—\underset{\underset{\underset{O}{\|}}{\overset{|}{SePh}}}{CH}—\overset{\overset{OH}{|}}{CHR'}$$

$$\xrightarrow[\text{room temp.}]{\text{10 hr}}$$

$$R_2C=CH—\overset{\overset{OH}{|}}{CHR'}$$

In another process, an olefin is converted to a rearranged allylic alcohol.[230]

$$R_2CH—CH=CR'_2 \xrightarrow[\underset{H_2O}{PhSeSePh}]{PhSeO_2H} R_2CH—\underset{\underset{SePh}{|}}{CH}—CR'_2OH \xrightarrow[\text{above}]{as} R_2C=CH—CR'_2OH$$

[223]Kingsbury; Cram *J. Am. Chem. Soc.* **1960**, *82*, 1810; Walling; Bollyky *J. Org. Chem.* **1964**, *29*, 2699; Entwistle; Johnstone *Chem. Commun.* **1965**, 29; Yoshimura; Tsukurimichi; Iizuka; Mizuno; Isaji; Shimasaki *Bull. Chem. Soc. Jpn.* **1989**, *62*, 1891.

[224]For reviews, see Back, in Patai *The Chemistry of Organic Selenium and Tellurium Compounds*, vol. 2; Wiley: New York, 1987, pp. 91-213, pp. 95-109; Paulmier *Selenium Reagents and Intermediates in Organic Synthesis*; Pergamon: Elmsford, NY, 1986, pp. 132-143; Reich *Acc. Chem. Res.* **1979**, *12*, 22-30, in Trahanovsky *Oxidation in Organic Chemistry*, pt. C; Academic Press: New York, 1978, pp. 15-101; Sharpless; Gordon; Lauer; Patrick; Singer; Young *Chem. Scr.* **1975**, *8A*; 9-13. See also Liotta *Organoselenium Chemistry*; Wiley: New York, 1987.

[225]Jones; Higgins *J. Chem. Soc. C* **1970**, 81.

[226]Reich; Willis *J. Am. Chem. Soc.* **1980**, *102*, 5967.

[227]Clive *J. Chem. Soc., Chem. Commun.* **1973**, 695; Reich; Reich; Renga *J. Am. Chem. Soc.* **1973**, *95*, 5813; Reich; Renga; Reich *J. Org. Chem.* **1974**, *39*, 2133, *J. Am. Chem. Soc.* **1975**, *97*, 5434; Sharpless; Lauer; Teranishi *J. Am. Chem. Soc.* **1973**, *95*, 6137; Grieco; Miyashita *J. Org. Chem.* **1974**, *39*, 120. For lists of reagents, with references, see Ref. 144, pp. 149-150.

[228]Trost; Salzmann; Hiroi *J. Am. Chem. Soc.* **1976**, *98*, 4887. For a review of this and related methods, see Trost *Acc. Chem. Res.* **1978**, *11*, 453-461.

[229]Sharpless; Lauer *J. Am. Chem. Soc.* **1973**, *95*, 2697.

[230]Hori; Sharpless *J. Org. Chem.* **1978**, *43*, 1689; Reich; Wollowitz; Trend; Chow; Wendelborn *J. Org. Chem.* **1978**, *43*, 1697. See also Reich *J. Org. Chem.* **1974**, *39*, 428; Clive *J. Chem. Soc., Chem. Commun.* **1974**, 100; Sharpless; Lauer *J. Org. Chem.* **1974**, *39*, 429.

See p. 473 for another application of the selenoxide reaction. Allylic sulfoxides undergo 1,4 elimination to give dienes.[231]

OS **VI**, 23, 737; **67**, 157.

7-13 Dehydrohalogenation of Alkyl Halides
Hydro-halo-elimination

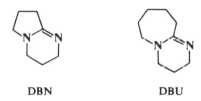

The elimination of HX from an alkyl halide is a very general reaction and can be accomplished with chlorides, fluorides, bromides, and iodides.[232] Hot alcoholic KOH is the most frequently used base, though stronger bases[233] (OR^-, NH_2^-, etc.) or weaker ones (e.g., amines) are used where warranted.[234] The bicyclic amidines 1,5-diazabicyclo[3.4.0]nonene-5 (DBN)[235] and 1,8-diazabicyclo[5.4.0]undecene-7 (DBU)[236] are good reagents for difficult cases.[237]

DBN   DBU

Dehydrohalogenation with the non-ionic base $(Me_2N)_3P{=}N{-}P(NMe_2)_2{=}NMe$ is even faster.[238] Phase transfer catalysis has been used with OH^- as base.[239] As previously mentioned (p. 997), certain weak bases in dipolar aprotic solvents are effective reagents for dehydrohalogenation. Among those most often used for synthetic purposes are LiCl or LiBr–LiCO$_3$ in DMF.[240] Dehydrohalogenation has also been effected by heating of the alkyl halide in HMPA with no other reagent present.[241] As in nucleophilic substitution (p. 352), the order of leaving group reactivity is I > Br > Cl > F.[242]

[231]de Groot; Jansen; Reuvers; Tedjo *Tetrahedron Lett.* **1981**, *22*, 4137.

[232]For a review of eliminations involving the carbon–halogen bond, see Baciocchi, in Patai; Rappoport *The Chemistry of Functional Groups, Supplement D*, pt. 2; Wiley: New York, 1983, pp. 1173-1227.

[233]Triphenylmethylpotassium rapidly dehydrohalogenates secondary alkyl bromides and iodides, in over 90% yields, at 0°C: Anton; Crabtree *Tetrahedron Lett.* **1983**, *24*, 2449.

[234]For a list of reagents, with references, see Ref. 144, pp. 131-133.

[235]Truscheit; Eiter *Liebigs Ann. Chem.* **1962**, *658*, 65; Oediger; Kabbe; Möller; Eiter *Chem. Ber.* **1966**, *99*, 2012; Vogel; Klärner *Angew. Chem. Int. Ed. Engl.* **1968**, *7*, 374 [*Angew. Chem. 80*, 402].

[236]Oediger; Möller *Angew. Chem. Int. Ed. Engl.* **1967**, *6*, 76 [*Angew. Chem. 79*, 53]; Wolkoff *J. Org. Chem.* **1982**, *47*, 1944.

[237]For a review of these reagents, see Oediger; Möller; Eiter *Synthesis* **1972**, 591.

[238]Schwesinger; Schlemper *Angew. Chem. Int. Ed. Engl.* **1987**, *26*, 1167 [*Angew. Chem. 99*, 1212].

[239]Kimura; Regen *J. Org. Chem.* **1983**, *48*, 195; Halpern; Zahalka; Sasson; Rabinovitz *J. Org. Chem.* **1985**, *50*, 5088. See also Barry; Bram; Decodts; Loupy; Pigeon; Sansoulet *J. Org. Chem.* **1984**, *49*, 1138.

[240]For a discussion, see Fieser; Fieser *Reagents for Organic Syntheses*, vol. 1; Wiley: New York, 1967, pp. 606-609. For a review of alkali-metal fluorides in this reaction, see Yakobson; Akhmetova *Synthesis* **1983**, 169-184, pp. 170-173.

[241]Hanna *Tetrahedron Lett.* **1968**, 2105; Monson *Chem. Commun.* **1971**, 113; Hutchins; Hutchins; Milewski *J. Org. Chem.* **1972**, *37*, 4190.

[242]Matsubara; Matsuda; Hamatani; Schlosser *Tetrahedron* **1988**, *44*, 2855.

Tertiary halides undergo elimination most easily. Eliminations of chlorides, bromides, and iodides follow Zaitsev's rule, except for a few cases where steric effects are important (for an example, see p. 1000). Eliminations of fluorides follow Hofmann's rule (p. 1000).

This reaction is by far the most important way of introducing a triple bond into a molecule.[243] This can be accomplished with substrates of the types:[244]

$$\text{H—}\underset{\underset{\text{H}}{|}}{\overset{|}{\text{C}}}\text{—}\underset{\underset{\text{X}}{|}}{\overset{|}{\text{C}}}\text{—X} \qquad \text{H—}\underset{\underset{\text{X}}{|}}{\overset{|}{\text{C}}}\text{—}\underset{\underset{\text{X}}{|}}{\overset{|}{\text{C}}}\text{—H} \qquad \text{—}\underset{\underset{\text{H}}{|}}{\text{C}}\text{=}\underset{\underset{\text{X}}{|}}{\text{C}}\text{—}$$

When the base is $NaNH_2$ 1-alkynes predominate (where possible), because this base is strong enough to form the salt of the alkyne, shifting any equilibrium between 1- and 2-alkynes. When the base is OH^- or OR^-, the equilibrium tends to be shifted to the internal alkyne, which is thermodynamically more stable. If another hydrogen is suitably located (e.g., —CRH—CX$_2$—CH$_2$—), allene formation can compete, though alkynes are usually more stable.

Dehydrohalogenation is generally carried out in solution, with a base, and the mechanism is usually E2, though the E1 mechanism has been demonstrated in some cases. However, elimination of HX can be accomplished by pyrolysis of the halide, in which case the mechanism is Ei (p. 1006) or, in some instances, the free-radical mechanism (p. 1008). Pyrolysis is normally performed without a catalyst at about 400°C. The pyrolysis reaction is not generally useful synthetically, because of its reversibility. Less work has been done on pyrolysis with a catalyst[245] (usually a metallic oxide or salt), but the mechanisms here are probably E1 or E2.

A combination elimination and substitution reaction has been used to synthesize alkynes. In this reaction a compound RCH=CFCl is treated with YM, where M is a metal and Y may be alkyl, aryl, NR_2, or OR:

$$\text{RCH=}\underset{\underset{\text{Cl}}{|}}{\text{C}}\text{—F} + \text{YM} \longrightarrow \text{R—C}\equiv\text{C—Y} \quad \text{Y = R, Ar, NR}_2\text{, OR}$$

Alkynes, ynamines,[246] and acetylenic ethers[247] can be prepared in this manner.[248]

In the special case of the prochiral carboxylic acids **36**, dehydrohalogenation with an

$$\underset{\textbf{36}}{\left.\begin{array}{c}\text{R} \qquad \text{X} \\ \\ \text{H} \qquad \text{CH}_2\text{COOH}\end{array}\right.} \longrightarrow \left.\begin{array}{c}\text{R} \\ \\ \text{H}\end{array}\right\}\text{=CHCOOH}$$

optically active lithium amide gave an optically active product with enantiomeric excesses as high as 82%.[249]

[243]For reviews, see Ben-Efraim, Ref. 142; Köbrich; Buck, in Viehe *Acetylenes*; Marcel Dekker: New York, 1969, pp. 100-134; Ref. 194, pp. 391-397; Köbrich, Ref. 2, pp. 50-53.

[244]For a list of reagents, with references, see Ref. 144, pp. 289-291.

[245]For a review, see Noller; Andréu; Hunger, Ref. 154.

[246]For a review of methods for the synthesis of ynamines, see Collard-Motte; Janousek *Top. Curr. Chem.* **1986**, *130*, 89-131.

[247]For a review of acetylenic ethers, see Radchenko; Petrov *Russ. Chem. Rev.* **1989**, *58*, 948-966.

[248]Viehe *Angew. Chem. Int. Ed. Engl.* **1963**, *2*, 477 [*Angew. Chem. 75*, 638]. For reviews of ynamines, see Ficini *Tetrahedron* **1976**, *32*, 1448-1486; Viehe, in Viehe, Ref. 243, pp. 861-912.

[249]Duhamel; Ravard; Plaquevent; Plé; Davoust *Bull. Soc. Chim. Fr.* **1990**, 787.

OS **I**, 191, 205, 209, 438; **II**, 10, 17, 515; **III**, 125, 209, 270, 350, 506, 623, 731, 785; **IV**, 128, 162, 398, 404, 555, 608, 616, 683, 711, 727, 748, 755, 763, 851, 969; **V**, 285, 467, 514; **VI**, 87, 210, 327, 361, 368, 427, 462, 505, 564, 862, 883, 893, 954, 991, 1037; **VII**, 126, 319, 453, 491; **65,** 32, 68, 90; **69,** 238. Also see OS **VI**, 968.

7-14 Dehydrohalogenation of Acyl Halides and Sulfonyl Halides
Hydro-halo-elimination

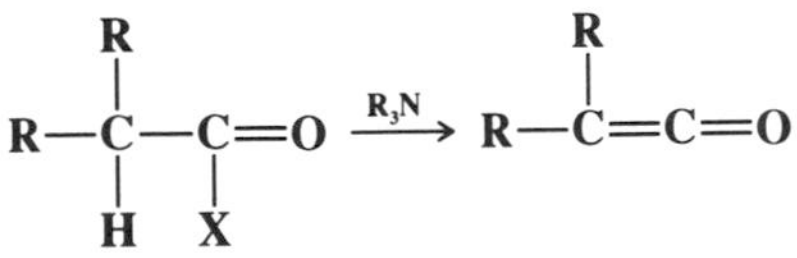

$$R-\overset{\overset{\displaystyle R}{|}}{\underset{\underset{\displaystyle H}{|}}{C}}-\overset{\underset{\underset{\displaystyle X}{|}}{}}{C}=O \xrightarrow{R_3N} R-\overset{\overset{\displaystyle R}{|}}{C}=C=O$$

Ketenes can be prepared by treatment of acyl halides with tertiary amines. The scope is broad, and most acyl halides possessing an α hydrogen give the reaction, but if at least one R is hydrogen, only the ketene dimer, not the ketene, is isolated. However, if it is desired to use a reactive ketene in a reaction with a given compound, the ketene can be generated in situ in the presence of the given compound.[250]

Closely related is the reaction of tertiary amines with sulfonyl halides that contain an α hydrogen. In this case the initial product is the highly reactive sulfene, which cannot be

$$RCH_2SO_2Cl \xrightarrow{R_3N} [RCH{=}SO_2] \longrightarrow RCH{=}CHR + \text{other products}$$

Sulfene

isolated but reacts further to give products, one of which may be the alkene that is the dimer of RCH.[251] Reactions of sulfenes in situ are also common (for example, see **6-62**).

OS **IV**, 560; **V**, 294, 877; **VI**, 549, 1037; **VII**, 232; **68**, 32.

7-15 Elimination of Boranes
Hydro-boranetriyl-elimination

$$(R_2CH{-}CH_2)_3B + 3\ 1\text{-decene} \rightleftharpoons 3R_2C{=}CH_2 + [CH_3(CH_2)_8CH_2]_3B$$

Trialkylboranes are formed from an olefin and BH_3 (**5-12**). When the resulting borane is treated with another olefin, an exchange reaction occurs.[252] This is an equilibrium process that can be shifted by using a large excess of olefin, by using an unusually reactive olefin, or by using an olefin with a higher boiling point than the displaced olefin and removing the latter by distillation. The reaction is useful for shifting a double bond in the direction opposite to that resulting from normal isomerization methods (**2-2**). This cannot be accomplished simply by treatment of a borane such as **38** with an olefin, because elimination in this reaction follows Zaitsev's rule: It is in the direction of the most stable olefin, and the product would be **37**, not **40**. However, if it is desired to convert **37** to **40**, this can be accomplished by converting **37** to **38**, isomerizing **38** to **39** (**8-11**) and then subjecting **39** to the exchange

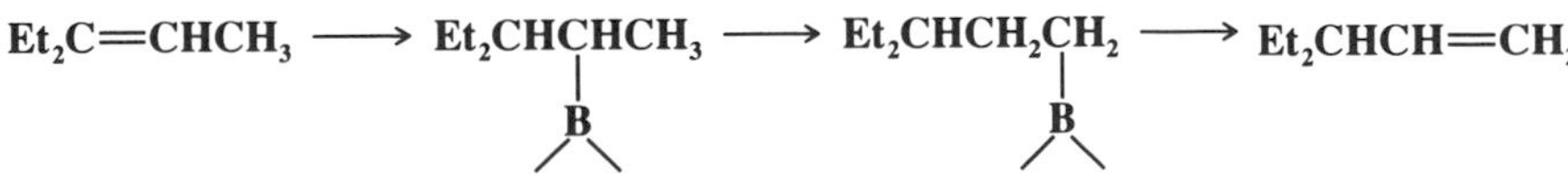

$$Et_2C{=}CHCH_3 \longrightarrow Et_2CHCHCH_3 \longrightarrow Et_2CHCH_2CH_2 \longrightarrow Et_2CHCH{=}CH_2$$

$$\qquad\qquad\qquad\qquad \overset{|}{\underset{\diagdown}{B}} \qquad\qquad\qquad\quad \overset{|}{\underset{\diagdown}{B}}$$

37	**38**	**39**	**40**

---

[250]For a review of this procedure, see Luknitskii; Vovsi *Russ. Chem. Rev.* **1969**, *38*, 487-494.
[251]For reviews of sulfenes, see Ref. 1729 in Chapter 10.
[252]Brown; Bhatt; Munekata; Zweifel *J. Am. Chem. Soc.* **1967**, *89*, 567; Taniguchi *Bull. Chem. Soc. Jpn.* **1979**, *52*, 2942.

reaction with a higher-boiling olefin, e.g., 1-decene, whereupon **40** is produced. In the usual isomerizations (**2-2**), **40** could be isomerized to **37**, but not the other way around. The reactions **38** → **39** and **39** → **40** proceed essentially without rearrangement. The mechanism is probably the reverse of borane addition (**5-12**).

A similar reaction, but irreversible, has been demonstrated for alkynes.[253]

$$(R_2CH—CH_2)_3B + R'C≡CR' \longrightarrow 3R_2C=CH_2 + (R'CH=CR')_3B$$

7-16 Pyrolysis of Alkali-Metal Organometallic Compounds
Hydro-metallo-elimination

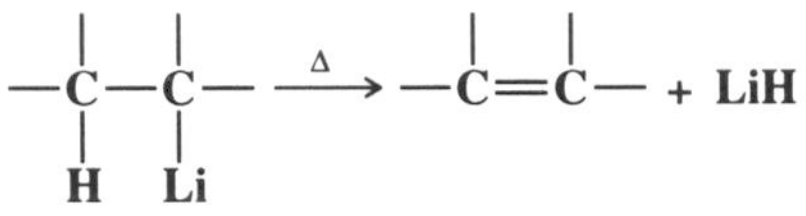

$$-\overset{|}{\underset{H}{C}}-\overset{|}{\underset{Li}{C}}- \overset{\Delta}{\longrightarrow} -\overset{|}{C}=\overset{|}{C}- + LiH$$

Solid lithium hydride and an olefin can be obtained by heating alkyllithium compounds containing a β hydrogen.[254] The reaction has also been applied to alkylsodium and alkyl-potassium compounds.[255] Grignard reagents gave olefins when thermally decomposed in nonsolvating solvents, e.g., cumene.[256] Alkenes have also been obtained from RLi and RMgX in solution, by treatment with ethylene and $NiCl_2$ or with certain other reagents.[257] Nitroalkenes have been obtained by cleavage of H and HgCl from β-nitro mercuric halides[258] (prepared by nitromercuration—see **5-7**). The mechanism is generally believed to be a four-centered pericyclic one (Ei).[259]

OS **68**, 148.

7-17 Conversion of Alkenes to Alkynes
Hydro-methyl-elimination

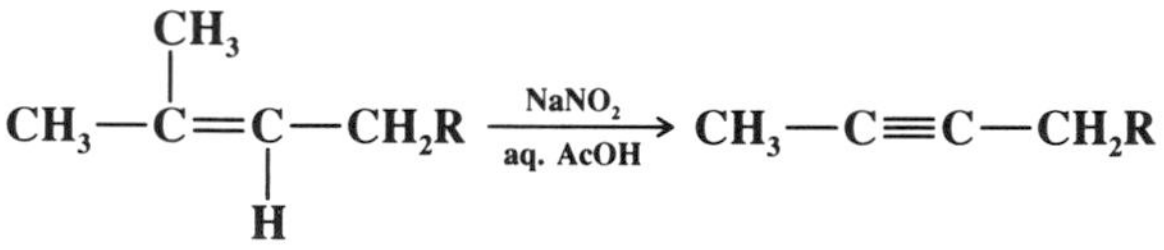

$$CH_3—\overset{\overset{\displaystyle CH_3}{|}}{C}=\overset{\underset{\displaystyle H}{|}}{C}—CH_2R \xrightarrow[\text{aq. AcOH}]{\text{NaNO}_2} CH_3—C≡C—CH_2R$$

Alkenes of the form shown lose the elements of methane when treated with sodium nitrite in acetic acid and water, to form alkynes in moderate-to-high yields.[260] The R may contain additional unsaturation as well as OH, OR, OAc, C=O, and other groups, but the $Me_2C=CHCH_2—$ portion of the substrate is necessary for the reaction to take place. The mechanism is complex, beginning with a nitration that takes place with allylic rearrangement [$Me_2C=CHCH_2R → H_2C=CMeCH(NO_2)CH_2R$], and involving several additional intermediates.[261] The CH_3 lost from the substrate appears as CO_2, as demonstrated by the trapping of this gas.[261]

[253]Hubert *J. Chem. Soc.* **1965**, 6669.
[254]Ziegler; Gellert *Liebigs Ann. Chem.* **1950**, *567*, 179.
[255]For example, see Finnegan *Chem. Ind. (London)* **1962**, 895, *Tetrahedron Lett.* **1963**, 851.
[256]Zakharkin; Okhlobystin; Strunin *J. Organomet. Chem.* **1965**, *4*, 349; Lefrançois; Gault *J. Organomet. Chem.* **1969**, *16*, 7; Dymova; Grazhulene; Kuchinskii; Kuznetsov *Bull. Acad. Sci. USSR; Div. Chem. Sci.* **1971**, *20*, 1532.
[257]Reetz; Stephan *Liebigs Ann. Chem.* **1980**, 171, and previous papers in this series. See also Laycock; Baird *Tetrahedron Lett.* **1978**, 3307; Baudin; Julia; Rolando; Verpeaux *Tetrahedron Lett.* **1984**, *25*, 3203.
[258]Corey; Estreicher *J. Am. Chem. Soc.* **1978**, *100*, 6294.
[259]See, for example, Li; San Filippo *Organometallics* **1983**, *2*, 554.
[260]Abidi *Tetrahedron Lett.* **1986**, *27*, 267, *J. Org. Chem.* **1986**, *51*, 2687.
[261]Corey; Seibel; Kappos *Tetrahedron Lett.* **1987**, *28*, 4921.

7-18   Dehydrocyanation
Hydro-cyano-elimination

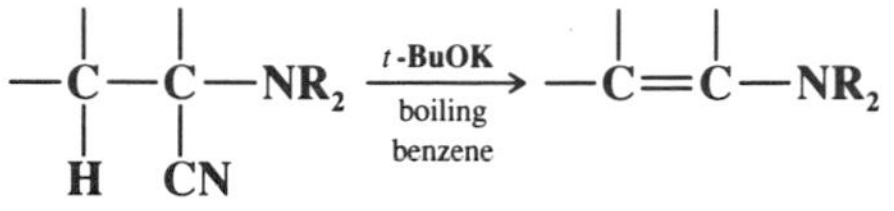

$$-\underset{\underset{\textstyle H}{|}}{\overset{|}{C}}-\underset{\underset{\textstyle CN}{|}}{\overset{|}{C}}-NR_2 \xrightarrow[\substack{boiling \\ benzene}]{t\text{-BuOK}} -\overset{|}{C}=\overset{|}{C}-NR_2$$

Enamines can be prepared from α-cyano tertiary amines by treatment with KOH or *t*-BuOK in boiling benzene or toluene, or in *t*-butyl methyl ether at room temperature.[262]

7-19   Decarbonylation of Acyl Halides
Hydro-chloroformyl-elimination

$$RCH_2CH_2COCl + RhCl(Ph_3P)_3 \xrightarrow{\Delta} RCH{=}CH_2 + HCl + RhClCO(Ph_3P)_2$$

Acyl chlorides containing an α hydrogen are smoothly converted to olefins, with loss of HCl and CO, on heating with chlorotris(triphenylphosphine)rhodium, with metallic platinum, or with certain other catalysts.[263] The mechanism probably involves conversion of RCH_2CH_2COCl to $RCH_2CH_2—RhCO(Ph_3P)_2Cl_2$ followed by a concerted syn elimination of Rh and H.[264] See also **4-41** and **9-13**.

7-20   Reversal of the Michael Reaction
Hydro-bis(ethoxycarbonyl)methyl-elimination, etc.

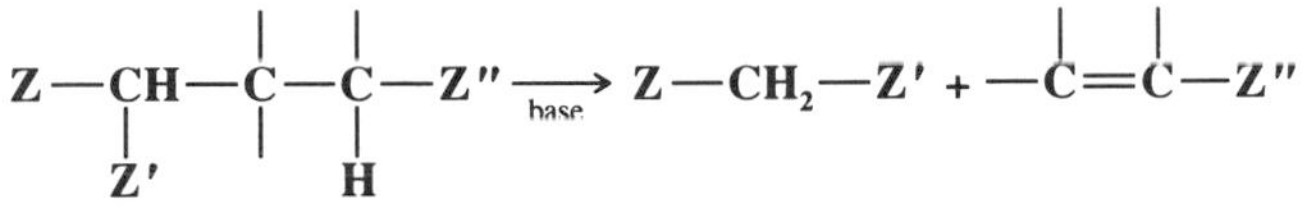

$$Z-\underset{\underset{\textstyle Z'}{|}}{\overset{|}{C}H}-\overset{|}{\underset{|}{C}}-\underset{\underset{\textstyle H}{|}}{\overset{|}{C}}-Z'' \xrightarrow{base} Z-CH_2-Z' + -\overset{|}{C}=\overset{|}{C}-Z''$$

Olefins can be formed on base cleavage of Michael adducts. (See **5-17.** Z is defined on p. 741) In some cases cleavage occurs simply on heating, without basic catalysis.

B.   Reactions in Which Neither Leaving Atom is Hydrogen

7-21   Deoxygenation of Vicinal Diols
Dihydroxy-elimination

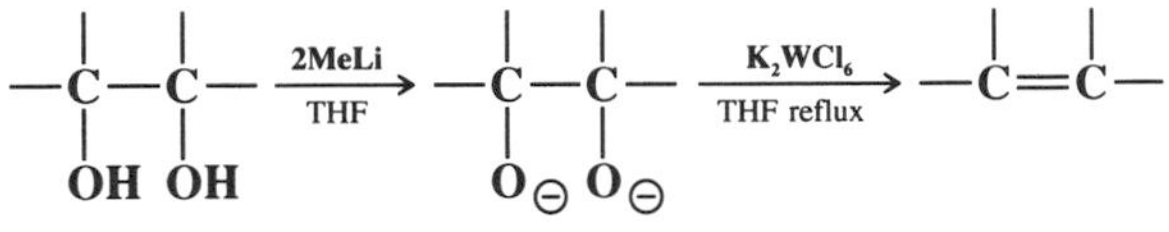

$$-\underset{\underset{\textstyle OH}{|}}{\overset{|}{C}}-\underset{\underset{\textstyle OH}{|}}{\overset{|}{C}}- \xrightarrow[THF]{2MeLi} -\underset{\underset{\textstyle O_\ominus}{|}}{\overset{|}{C}}-\underset{\underset{\textstyle O_\ominus}{|}}{\overset{|}{C}}- \xrightarrow[THF\ reflux]{K_2WCl_6} -\overset{|}{C}=\overset{|}{C}-$$

vic-Diols can be deoxygenated by treatment of the dilithium dialkoxide with the tungsten halide K_2WCl_6, or with certain other tungsten reagents, in refluxing THF.[265] Tetrasubstituted diols react most rapidly. The elimination is largely, but not entirely, syn. Several other

[262]Ahlbrecht; Raab; Vonderheid *Synthesis* **1979,** 127; Ahlbrecht; Raab *Synthesis* **1980,** 320.
[263]Tsuji; Ohno *J. Am. Chem. Soc.* **1966,** *88,* 3452, **1968,** *90,* 94; Ohno; Tsuji *J. Am. Chem. Soc.* **1968,** *90,* 99. For a review, see Tsuji; Ohno *Synthesis* **1969,** 157-169. For extensions to certain other acid derivatives, see Minami; Nisar; Yuhara; Shimizu; Tsuji *Synthesis* **1987,** 992.
[264]Lau; Becker; Huang; Baenziger; Stille *J. Am. Chem. Soc.* **1977,** *99,* 5664.
[265]Sharpless; Flood *J. Chem. Soc., Chem. Commun.* **1972,** 370; Sharpless; Umbreit; Nieh; Flood *J. Am. Chem. Soc.* **1972,** *94,* 6538.

methods have been reported,[266] in which the diol is deoxygenated directly, without conversion to the dialkoxide. These include treatment with titanium metal,[267] with TsOH–NaI,[268] with Ph_2PCl–imidazole–I_2 in toluene,[269] and with PBr_3–CuBr–ether at low temperatures, followed by zinc powder.[270]

vic-Diols can also be deoxygenated indirectly, through sulfonate ester derivatives. For example, *vic*-dimesylates and *vic*-ditosylates have been converted to alkenes by treatment, respectively, with naphthalene–sodium[271] and with NaI in dimethylformamide.[272] In another procedure, the diols are converted to bisdithiocarbonates (bis xanthates), which undergo

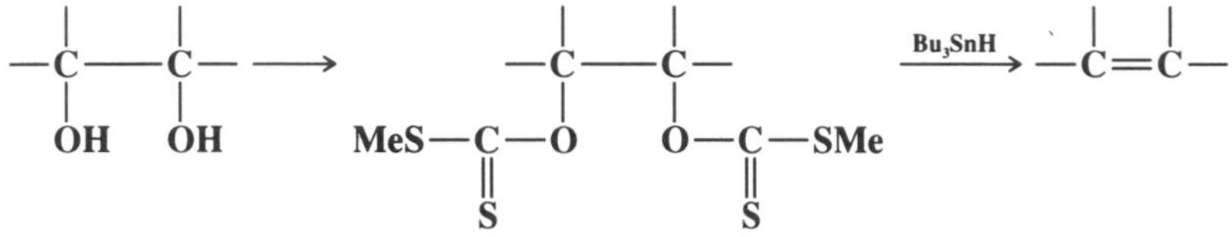

$$-\overset{|}{\underset{OH}{C}}-\overset{|}{\underset{OH}{C}}- \longrightarrow -\overset{|}{\underset{\underset{\overset{\|}{S}}{MeS-C-O}}{C}}-\overset{|}{\underset{\underset{\overset{\|}{S}}{O-C-SMe}}{C}}- \xrightarrow{Bu_3SnH} -\overset{|}{C}=\overset{|}{C}-$$

elimination (probably by a free-radical mechanism) when treated with tri-*n*-butylstannane in toluene or benzene.[273] *vic*-Diols can also be deoxygenated through cyclic derivatives (**7-22**).

7-22   Cleavage of Cyclic Thionocarbonates

$$-\overset{|}{\underset{\underset{\overset{\|}{S}}{\underset{C}{O}}}{C}}-\overset{|}{\underset{O}{C}}- + P(OMe)_3 \xrightarrow{\Delta} -\overset{|}{C}=\overset{|}{C}- + S{=}P(OMe)_3 + CO_2$$

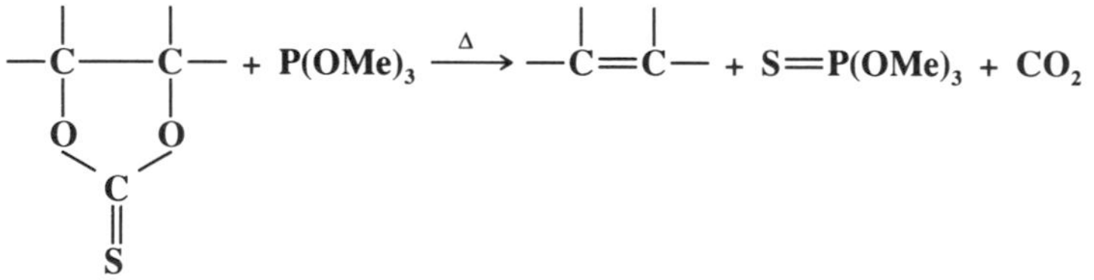

41

Cyclic thionocarbonates (**41**) can be cleaved to olefins (the *Corey–Winter reaction*)[274] by heating with trimethyl phosphite[275] or other trivalent phosphorus compounds[276] or by treatment with bis(1,5-cyclooctadiene)nickel.[277] The thionocarbonates can be prepared by treatment of 1,2-diols with thiophosgene and 4-dimethylaminopyridine (DMAP):[278]

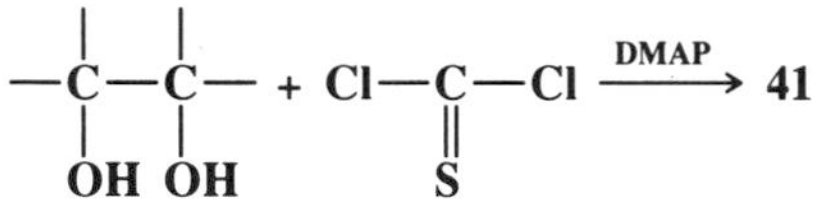

$$-\overset{|}{\underset{OH}{C}}-\overset{|}{\underset{OH}{C}}- + Cl-\overset{\overset{\|}{S}}{C}-Cl \xrightarrow{DMAP} \textbf{41}$$

[266]For a list of reagents, with references, see Ref. 144, pp. 155-156.

[267]McMurry; Fleming *J. Org. Chem.* **1976**, *41*, 896; McMurry *Acc. Chem. Res.* **1983**, *16*, 405-411.

[268]Sarma; Sharma *Chem. Ind. (London)* **1987**, 96.

[269]Liu; Classon; Samuelsson *J. Org. Chem.* **1990**, *55*, 4273.

[270]Tanaka; Yasuda; Yamamoto; Nozaki *J. Am. Chem. Soc.* **1975**, *97*, 3252.

[271]Carnahan; Closson *Tetrahedron Lett.* **1972**, 3447.

[272]Dafaye *Bull Soc. Chim. Fr.* **1968**, 2099.

[273]Barrett; Barton; Bielski *J. Chem. Soc., Perkin Trans. 1* **1979**, 2378.

[274]For reviews, see Block *Org. React.* **1984**, *30*, 457-566; Sonnet *Tetrahedron* **1980**, *36*, 557-604, pp. 593-598; Mackie, in Cadogan *Organophosphorus Reagents in Organic Synthesis*; Academic Press: New York, 1979, pp. 354-359; Block, Ref. 219, pp. 229-235.

[275]Corey; Winter *J. Am. Chem. Soc.* **1963**, *85*, 2677.

[276]Corey *Pure Appl. Chem.* **1967**, *14*, 19-37, pp. 32-33.

[277]Semmelhack; Stauffer *Tetrahedron Lett.* **1973**, 2667. For another method, see Vedejs; Wu *J. Org. Chem.* **1974**, *39*, 3641.

[278]Corey; Hopkins *Tetrahedron Lett.* **1982**, *23*, 1979.

The elimination is of course syn, so the product is sterically controlled. Olefins that are not sterically favored can be made this way in high yield, e.g., cis-PhCH$_2$CH=CHCH$_2$Ph.[279] Certain other 5-membered cyclic derivatives of 1,2-diols can also be converted to alkenes.[280]

7-23   The Conversion of Epoxides to Olefins
     epi-**Oxy-elimination**

$$-\overset{|}{\underset{\diagdown}{C}}-\overset{|}{\underset{O}{C}}-\ +\ Ph_3P\ \longrightarrow\ -\overset{|}{C}=\overset{|}{C}-\ +\ Ph_3PO$$

Epoxides can be converted to olefins[281] by treatment with triphenylphosphine[282] or triethyl phosphite P(OEt)$_3$.[283] The first step of the mechanism is nucleophilic substitution (**0-49**), followed by a four-center elimination. Since inversion accompanies the substitution, the overall elimination is anti, i.e., if two groups A and C are cis in the epoxide, they will be trans in the olefin:

betaine

Alternatively, the epoxide can be treated with lithium diphenylphosphide Ph$_2$PLi, and the product quaternized with methyl iodide.[284] Olefins have also been obtained from epoxides by reaction with a large number of reagents,[285] among them Li in THF,[286] TsOH and NaI,[287] trimethylsilyl iodide,[288] dimethyl diazomalonate,[289] PI$_3$,[290] P$_2$I$_4$,[291] AlI$_3$,[292] Mg-I$_2$-Et$_2$O,[293] F$_3$COOH-NaI,[294] 9-diazofluorene and uv light,[295] SmI$_2$,[296] titanocene dichloride-Mg,[297]

[279]Corey; Carey; Winter *J. Am. Chem. Soc.* **1965**, *87*, 934.
[280]See Hines; Peagram; Whitham; Wright *Chem. Commun.* **1968**, 1593; Josan; Eastwood *Aust. J. Chem.* **1968**, *21*, 2013; Hiyama; Nozaki *Bull. Chem. Soc. Jpn.* **1973**, *46*, 2248; Marshall; Lewellyn *J. Org. Chem.* **1977**, *42*, 1311; Breuer; Bannet *Tetrahedron* **1978**, *34*, 997; Hanessian; Bargiotti; LaRue *Tetrahedron Lett.* **1978**, 737; Hatanaka; Tanimoto; Oida; Okano *Tetrahedron Lett.* **1981**, *22*, 5195; Ando; Ohhara; Takase *Chem. Lett.* **1986**, 879; King; Posner; Mak; Yang *Tetrahedron Lett.* **1987**, *28*, 3919; Beels; Coleman; Taylor *Synlett* **1990**, 479.
[281]For reviews, see Wong; Fok; Wong *Heterocycles* **1987**, *26*, 1345-1382; Sonnet, Ref. 274, pp. 576-586.
[282]Wittig; Haag *Chem. Ber.* **1955**, *88*, 1654.
[283]Scott *J. Org. Chem.* **1957**, *22*, 1118.
[284]Vedejs; Fuchs *J. Am. Chem. Soc.* **1971**, *93*, 4070, **1973**, *95*, 822.
[285]For a list of reagents, with references, see Ref. 144, pp. 140-142.
[286]Gurudutt; Ravindranath *Tetrahedron Lett.* **1980**, *21*, 1173.
[287]Baruah; Sharma; Baruah *Chem. Ind. (London)* **1983**, 524.
[288]Denis; Magnane; Van Eenoo; Krief *Nouv. J. Chim.* **1979**, *3*, 705. For other silyl reagents, see Reetz; Plachky *Synthesis* **1976**, 199; Dervan; Shippey *J. Am. Chem. Soc.* **1976**, *98*, 1265; Caputo; Mangoni; Neri; Palumbo *Tetrahedron Lett.* **1981**, *22*, 3551.
[289]Martin; Ganem *Tetrahedron Lett.* **1984**, *25*, 251.
[290]Denis, et al., Ref. 288.
[291]Suzuki; Fuchita; Iwasa; Mishina *Synthesis* **1978**, 905; Ref. 290.
[292]Sarmah; Barua *Tetrahedron Lett.* **1988**, *29*, 5815.
[293]Chowdhury *J. Chem. Res. (S)* **1990**, 192.
[294]Sarma; Sharma *Chem. Ind. (London)* **1984**, 712.
[295]Shields; Schuster *Tetrahedron Lett.* **1987**, *28*, 853.
[296]Girard; Namy; Kagan *J. Am. Chem. Soc.* **1980**, *102*, 2693; Matsukawa; Tabuchi; Inanaga; Yamaguchi *Chem. Lett.* **1987**, 2101.
[297]Schobert *Angew. Chem. Int. Ed. Engl.* **1988**, *27*, 855 [*Angew. Chem. 100*, 869]. See also Yadav; Shekharam; Gadgil *J. Chem. Soc., Chem. Commun.* **1990**, 843.

$Fe(CO)_5$,[298] $TiCl_3$–$LiAlH_4$,[299] $FeCl_3$–BuLi,[300] the tungsten reagents mentioned in **7-21**,[265] and NaI–NaOAc–Zn–AcOH.[301] The last-mentioned method is actually a variation of **7-31**, since iodohydrins are intermediates. Some of these methods give syn elimination.

7-24 The Conversion of Episulfides to Olefins
***epi*-Thio-elimination**

$$-\overset{|}{\underset{\diagdown_S\diagup}{C}}-\overset{|}{C}- \;+\; P(OEt)_3 \longrightarrow -\overset{|}{C}=\overset{|}{C}- \;+\; S=P(OEt)_3$$

Episulfides[302] can be converted to olefins in a reaction similar in appearance to **7-23**.[303] However, in this case the elimination is syn, so the mechanism cannot be the same as that of **7-23**. The phosphite attacks not the carbon, but the sulfur. Among other reagents that convert episulfides to olefins are Bu_3SnH,[304] P_2I_4,[304] certain rhodium complexes,[305] $LiAlH_4$[306] (this compound behaves quite differently with epoxides, see **0-80**), and methyl iodide.[307] Episulfoxides can be converted to olefins and sulfur monoxide simply by heating.[308]

7-25 The Ramberg-Bäcklund Reaction
Ramberg-Bäcklund halosulfone transformation

$$R-CH_2-SO_2-\underset{\underset{Cl}{|}}{CH}-R \;+\; OH^- \longrightarrow R-\overset{\overset{H}{|}}{C}=\overset{\overset{H}{|}}{C}-R$$

The reaction of an α-halo sulfone with a base to give an olefin is called the *Ramberg–Bäcklund reaction*.[309] The reaction is quite general for α-halo sulfones with an α' hydrogen, despite the unreactivity of α-halo sulfones in normal S_N2 reactions (p. 344). Halogen reactivity is in the order I > Br ≫ Cl. Phase transfer catalysis has been used.[310] In general, mixtures of cis and trans isomers are obtained, but usually the less stable cis isomer predominates. The mechanism involves formation of an episulfone and then elimination of

$$R-\underset{\underset{SO_2}{|}}{CH_2}\;\underset{\underset{Cl}{|}}{CH}-R \xrightarrow{OH^-} R-\underset{\underset{SO_2}{|}}{\overset{\ominus}{CH}}\;\underset{\underset{Cl}{|}}{CH}-R \longrightarrow R-\underset{\underset{SO_2}{\diagdown\diagup}}{CH}-CH-R \longrightarrow RCH=CHR$$
$$+$$
$$SO_2$$

[298]Alper; Des Roches *Tetrahedron Lett.* **1977**, 4155.

[299]McMurry; Silvestri; Fleming; Hoz; Grayston *J. Org. Chem.* **1978**, *43*, 3249.

[300]Fujisawa; Sugimoto; Ohta *Chem. Lett.* **1975**, 883.

[301]Cornforth; Cornforth; Mathew *J. Chem. Soc.* **1959**, 112. See also Yamada; Goto; Nagase; Kyotani; Hirata *J. Org. Chem.* **1978**, *43*, 2076; Sonnet *Synthesis* **1980**, 828.

[302]For a review of this reaction, see Sonnet, Ref. 274, pp. 587-590. For a review of episulfides, see Goodman; Reist, in Kharasch; Meyers *The Chemistry of Organic Sulfur Compounds*, vol. 2; Pergamon: Elmsford, NY, 1966, pp. 93-113.

[303]Neureiter; Bordwell *J. Am. Chem. Soc.* **1959**, *81*, 578; Davis *J. Org. Chem.* **1957**, *23*, 1767.

[304]Schauder; Denis; Krief *Tetrahedron Lett.* **1983**, *24*, 1657.

[305]Calet; Alper *Tetrahedron Lett.* **1986**, *27*, 3573.

[306]Lightner; Djerassi *Chem. Ind. (London)* **1962**, 1236; Latif; Mishriky; Zeid *J. Prakt. Chem.* **1970**, *312*, 421.

[307]Culvenor; Davies; Heath *J. Chem. Soc.* **1949**, 282; Helmkamp; Pettitt *J. Org. Chem.* **1964**, *29*, 3258.

[308]Hartzell; Paige *J. Am. Chem. Soc.* **1966**, *88*, 2616, *J. Org. Chem.* **1967**, *32*, 459; Aalbersberg; Vollhardt *J. Am. Chem. Soc.* **1977**, *99*, 2792.

[309]For reviews, see Paquette *Org. React.* **1977**, *25*, 1-71, *Mech. Mol. Migr.* **1968**, *1*, 121-156, *Acc. Chem. Res.* **1968**, *1*, 209-216; Meyers; Matthews; Ho; Kolb; Parady, in Smith *Catalysis in Organic Synthesis*; Academic Press: New York, 1977, pp. 197-278; Rappe, in Patai *The Chemistry of the Carbon–Halogen Bond*, Ref. 2, pt. 2, pp. 1105-1110; Bordwell *Acc. Chem. Res.* **1970**, *3*, 281-290, pp. 285-286; in Janssen *Organosulfur Chemistry*; Wiley: New York, 1967, pp. 271-284.

[310]Hartman; Hartman *Synthesis* **1982**, 504.

SO_2. There is much evidence for this mechanism,[311] including the isolation of the episulfone intermediate,[312] and the preparation of episulfones in other ways and the demonstration that they give olefins under the reaction conditions faster than the corresponding α-halo sulfones.[313] Episulfones synthesized in other ways (e.g., **6-62**) are reasonably stable compounds but eliminate SO_2 to give olefins when heated or treated with base.

If the reaction is run on the unsaturated bromo sulfones $RCH_2CH=CHSO_2CH_2Br$ (prepared by reaction of $BrCH_2SO_2Br$ with $RCH_2CH=CH_2$ followed by treatment with Et_3N), the dienes $RCH=CHCH=CH_2$ are produced in moderate-to-good yields.[314] The compound mesyltriflone $CF_3SO_2CH_2SO_2CH_3$ can be used as a synthon for the tetraion $^{2-}C=C^{2-}$. Successive alkylation (**0-94**) converts it to $CF_3SO_2CR^1R^2SO_2CHR^3R^4$ (anywhere from one to four alkyl groups can be put in), which, when treated with base, gives $R^1R^2C=CR^3R^4$.[315] The nucleofuge here is the $CF_3SO_2^-$ ion.

2,5-Dihydrothiophene-1,1-dioxides (**42**) and 2,7-dihydrothiepin-1,1-dioxides (**43**)

undergo analogous 1,4 and 1,6 eliminations, respectively (see also **7-48**). These are concerted reactions and, as predicted by the orbital-symmetry rules (p. 846), the former[316] is a suprafacial process and the latter[317] an antarafacial process. The rules also predict that elimination of SO_2 from episulfones cannot take place by a concerted mechanism (except antarafacially, which is unlikely for such a small ring), and the evidence shows that this reaction occurs by a nonconcerted pathway.[318] The eliminations of SO_2 from **42** and **43** are examples of *cheletropic reactions*,[319] which are defined as reactions in which two σ bonds that terminate at a single atom (in this case the sulfur atom) are made or broken in concert.[320]

α,α-Dichlorobenzyl sulfones (**44**) react with an excess of the base triethylenediamine in

$$ArCH_2SO_2CCl_2Ar \xrightarrow[\text{2. H}_2\text{O}]{\text{1. TED, Me}_2\text{SO}} \underset{\underset{\textbf{45}}{Ar \quad Ar}}{\overset{SO_2}{\triangle}} \xrightarrow{\Delta} ArC\equiv CAr$$

44   **46**

dimethyl sulfoxide at room temperature to give 2,3-diarylthiiren-1,1-dioxides (**45**), which can be isolated.[321] Thermal decomposition of **45** gives the alkynes **46**.[322]

[311]See, for example, Bordwell; Cooper *J. Am. Chem. Soc.* **1951**, *73*, 5187; Paquette *J. Am. Chem. Soc.* **1964**, *86*, 4089; Neureiter *J. Am. Chem. Soc.* **1966**, *88*, 558; Bordwell; Wolfinger *J. Org. Chem.* **1974**, *39*, 2521; Bordwell; Doomes *J. Org. Chem.* **1974**, *39*, 2526, 2531.

[312]Sutherland; Taylor *Tetrahedron Lett.* **1989**, *30*, 3267.

[313]Bordwell; Williams; Hoyt; Jarvis *J. Am. Chem. Soc.* **1968**, *90*, 429; Bordwell; Williams *J. Am. Chem. Soc.* **1968**, *90*, 435.

[314]Block; Aslam; Eswarakrishnan; Gebreyes; Hutchinson; Iyer; Laffitte; Wall *J. Am. Chem. Soc.* **1986**, *108*, 4568.

[315]Hendrickson; Boudreaux; Palumbo *J. Am. Chem. Soc.* **1986**, *108*, 2358.

[316]Mock *J. Am. Chem. Soc.* **1966**, *88*, 2857; McGregor; Lemal *J. Am. Chem. Soc.* **1966**, *88*, 2858.

[317]Mock *J. Am. Chem. Soc.* **1969**, *91*, 5682.

[318]Ref. 313. See also Vilsmaier; Tropitzsch; Vostrowsky *Tetrahedron Lett.* **1974**, 3987.

[319]For a review, see Mock, in Marchand; Lehr *Pericyclic Reactions*, vol. 2; Academic Press: New York, 1977, pp. 141-179.

[320]Woodward; Hoffmann *The Conservation of Orbital Symmetry*; Academic Press: New York, 1970, pp. 152-163.

[321]Philips; Swisher; Haidukewych; Morales *Chem. Commun.* **1971**, 22.

[322]Carpino; McAdams; Rynbrandt; Spiewak *J. Am. Chem. Soc.* **1971**, *93*, 476; Philips; Morales *J. Chem. Soc., Chem. Commun.* **1977**, 713.

A Ramberg–Bäcklund-type reaction has been carried out on the α-halo *sulfides* ArCHClSCH₂Ar, which react with *t*-BuOK and PPh₃ in refluxing THF to give the alkenes ArCH=CHAr.[323]

The Ramberg–Bäcklund reaction can be regarded as a type of extrusion reaction (see p. 1045).

OS **V**, 877; **VI**, 454, 555; **65**, 90.

7-26 The Conversion of Aziridines to Olefins
epi-Imino-elimination

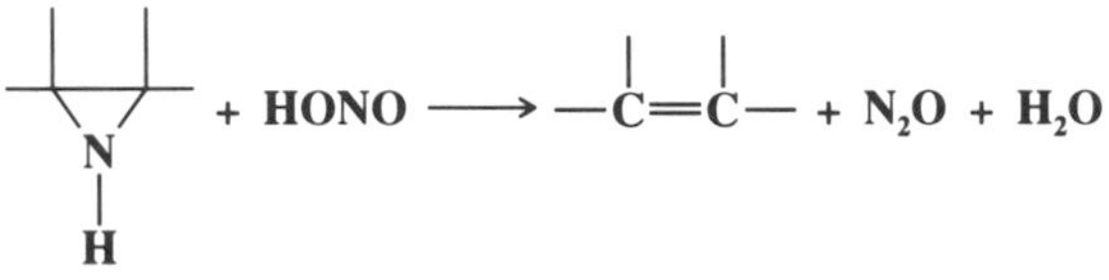

$$\text{(aziridine)} + \text{HONO} \longrightarrow -\overset{|}{\text{C}}=\overset{|}{\text{C}}- + \text{N}_2\text{O} + \text{H}_2\text{O}$$

Aziridines not substituted on the nitrogen atom react with nitrous acid to produce olefins.[324] An N-nitroso compound is an intermediate (**2-51**); other reagents that produce such intermediates also give olefins. The reaction is stereospecific: cis aziridines give cis olefins and trans aziridines give trans olefins.[325] Aziridines carrying N-alkyl substituents can be converted to olefins by treatment with ferrous iodide[326] or with *m*-chloroperbenzoic acid.[327] An N-oxide intermediate (**9-28**) is presumably involved in the latter case.

7-27 Conversion of Vicinal Dinitro Compounds to Olefins
Dinitro-elimination

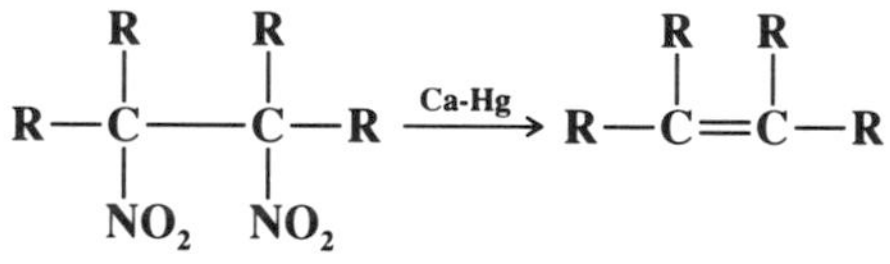

$$\text{R}-\overset{\overset{\text{R}}{|}}{\underset{\underset{\text{NO}_2}{|}}{\text{C}}}-\overset{\overset{\text{R}}{|}}{\underset{\underset{\text{NO}_2}{|}}{\text{C}}}-\text{R} \xrightarrow{\text{Ca-Hg}} \text{R}-\overset{\overset{\text{R}}{|}}{\text{C}}=\overset{\overset{\text{R}}{|}}{\text{C}}-\text{R}$$

Tetrasubstituted *vic*-dinitro compounds have been converted to olefins by treatment with amalgamated calcium.[328] Various functional groups, such as CN and COOR, did not affect the reaction. Other reagents that have been used include sodium sulfide in DMF,[329] nickel boride and ultrasound,[330] Bu₃SnH,[331] and SnCl₂.[332] Radical-ion mechanisms are likely in all these cases.

7-28 The Conversion of Dihydrazones to Alkynes
Dihydrazono-bielimination

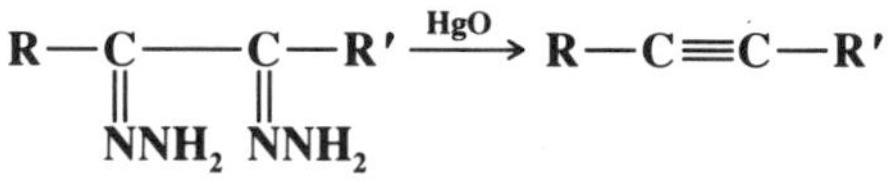

$$\text{R}-\overset{\overset{\|}{\text{NNH}_2}}{\text{C}}-\overset{\overset{\|}{\text{NNH}_2}}{\text{C}}-\text{R}' \xrightarrow{\text{HgO}} \text{R}-\text{C}\equiv\text{C}-\text{R}'$$

[323]Mitchell *Tetrahedron Lett.* **1973**, 4395. For a similar reaction without base treatment, see Pommelet; Nyns; Lahousse; Merényi; Viehe *Angew. Chem. Int. Ed. Engl.* **1981**, *20*, 585 [*Angew. Chem. 93*, 594].

[324]For reviews, see Sonnet, Ref. 274, pp. 591-592; Dermer; Ham *Ethylenimine and other Aziridines*; Academic Press: New York, 1969, pp. 293-295.

[325]Clark; Helmkamp *J. Org. Chem.* **1964**, *29*, 1316; Carlson; Lee *Tetrahedron Lett.* **1969**, 4001.

[326]Imamoto; Yukawa *Chem. Lett.* **1974**, 165.

[327]Heine; Myers; Peltzer *Angew. Chem. Int. Ed. Engl.* **1970**, *9*, 374 [*Angew. Chem. 82*, 395].

[328]Kornblum; Cheng *J. Org. Chem.* **1977**, *42*, 2944.

[329]Kornblum; Boyd; Pinnick; Smith *J. Am. Chem. Soc.* **1971**, *93*, 4316.

[330]Madjdabadi; Beugelmans; Lechavallier *Synth. Commun.* **1989**, *19*, 1631.

[331]Ono; Miyake; Tamura; Hamamoto; Kaji *Chem. Lett.* **1981**, 1139.

[332]Fukunaga; Kimura *Bull. Chem. Soc. Jpn.* **1979**, *52*, 1107.

1,2-Dihydrazones can be made to lose two moles of nitrogen to give alkynes by treatment with HgO, Ag$_2$O, CuCl$_2$–O$_2$–pyridine, or certain other reagents.[333] R and R' may be alkyl or aryl. Highly strained seven- and eight-membered cycloalkynes (see p. 159), as well as large cycloalkynes, have been obtained by this reaction.[334]

OS **IV,** 377. See also OS **VI,** 791.

7-29   Dehalogenation of Vicinal Dihalides
Dihalo-elimination

$$-\overset{|}{\underset{X}{C}}-\overset{|}{\underset{X}{C}}- \quad \xrightarrow{Zn} \quad -\overset{|}{C}=\overset{|}{C}-$$

Dehalogenation has been accomplished with many reagents, the most common being zinc, magnesium, and iodide ion.[335] Among reagents used less frequently have been phenyllithium, phenylhydrazine, CrCl$_2$, naphthalene–sodium,[336] Na–NH$_3$,[337] Na$_2$S in DMF,[338] Na$_2$Te,[339] and LiAlH$_4$.[340] Electrochemical reduction has also been used.[341] Though the reaction usually gives good yields, it is not very useful because the best way to prepare *vic*-dihalides is by the addition of halogen to a double bond (**5-26**). One useful feature of this reaction is that there is no doubt about the *position* of the new double bond, so that it can be used to give double bonds exactly where they are wanted. For example, allenes, which are not easily prepared by other methods, can be prepared from X—C—CX$_2$—C—X or X—C—CX—C— systems.[342] Cumulenes have been obtained from 1,4 elimination:

$$\text{BrCH}_2\text{—C}\equiv\text{C—CH}_2\text{Br} + \text{Zn} \longrightarrow \text{CH}_2\text{=C=C=CH}_2$$

Triple bonds can be prepared from X—C=C—X or X$_2$C—CX$_2$ systems,[343] but availability considerations are even more extreme here. 1,4 Elimination of BrC—C=C—CBr has been used to prepare conjugated dienes C=C—C=C.[344]

The reaction can be carried out for any combination of halogens, except where one is fluorine. Mechanisms are often complex and depend on the reagent and reaction conditions.[345] For different reagents, mechanisms involving carbocations, carbanions, and free-radical intermediates, as well as concerted mechanisms, have been proposed. When the reagent is zinc, anti stereospecificity has been observed in some cases,[346] but not in others.[347]

[333]For a list of reagents, with references, see Ref. 144, p. 293.
[334]For example, see Blomquist; Liu *J. Am. Chem. Soc.* **1953,** *75,* 2153; Krebs; Kimling *Tetrahedron Lett.* **1970,** 761; Tsuji; Kezuka; Toshida; Takayanagi; Yamamoto *Tetrahedron* **1983,** *39,* 3279.
[335]For a review of this reaction, see Baciocchi, in Patai; Rappoport, Ref. 232; pt. 1, pp. 161-201.
[336]Scouten; Barton; Burgess; Story; Garst *Chem. Commun.* **1969,** 78; Garst; Pacifici; Singleton; Ezzel; Morris *J. Am. Chem. Soc.* **1975,** *97,* 5242.
[337]Allred; Beck; Voorhees *J. Org. Chem.* **1974,** *39,* 1426.
[338]Fukunaga; Yamaguchi *Synthesis* **1981,** 879. See also Nakayama; Machida; Hoshino *Tetrahedron Lett.* **1983,** *24* 3001; Landini; Milesi; Quadri; Rolla *J. Org. Chem.* **1984,** *49,* 152.
[339]Suzuki; Inouye *Chem. Lett.* **1985,** 225. See also Huang; Hou *Synth. Commun.* **1988,** *18,* 2201.
[340]For a lists of reagents, with references, see Ref. 144, pp. 133-135.
[341]See Shono *Electroorganic Chemistry as a New Tool in Organic Synthesis*; Springer: New York, 1984, pp. 145-147; Fry *Synthetic Organic Electrochemistry,* 2nd ed.; Wiley: New York, 1989, pp. 151-154.
[342]For reviews of allene formation, see Schuster; Coppola *Allenes in Organic Synthesis*; Wiley: New York, 1984, pp. 9-56; Landor, in Landor *The Chemistry of the Allenes,* vol. 1; Academic Press: New York, 1982; pp. 19-233; Taylor *Chem. Rev.* **1967,** *67,* 317-359.
[343]For a review, see Köbrich; Buck; in Viehe, Ref. 243, pp. 134-138.
[344]Engman; Byström *J. Org. Chem.* **1985,** *50,* 3170.
[345]For discussion, see Saunders; Cockerill, Ref. 2, pp. 332-368; Ref. 335.
[346]For example, see House; Ro *J. Am. Chem. Soc.* **1958,** *80,* 182; Gordon; Hay *J. Org. Chem.* **1968,** *33,* 427.
[347]For example, see Stevens; Valicenti *J. Am. Chem. Soc.* **1965,** *87,* 838; Sicher; Havel; Svoboda *Tetrahedron Lett.* **1968,** 4269.

OS **III**, 526, 531; **IV**, 195, 268; **V**, 22, 255, 393, 901; **VI**, 310, **VII**, 241. Also see OS **IV**, 877, 914, 964.

7-30   Dehalogenation of α-Halo Acyl Halides
Dihalo-elimination

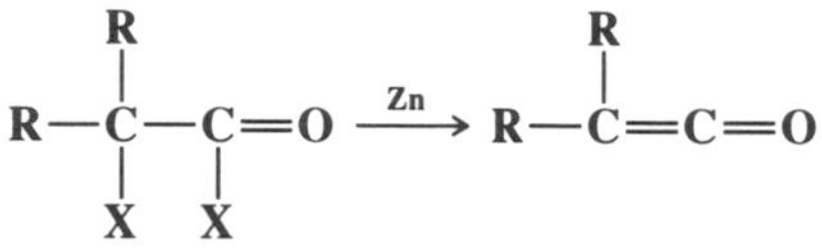

$$
\underset{\underset{X}{\overset{\overset{R}{|}}{|}}{R-\overset{}{C}}-C=O \xrightarrow{\ Zn\ } R-\overset{\overset{R}{|}}{C}=C=O
$$

Ketenes can be prepared by dehalogenation of α-halo acyl halides with zinc or with triphenylphosphine.[348] The reaction generally gives good results when the two R groups are aryl or alkyl, but not when either one is hydrogen.[349]

OS **IV**, 348; **68**, 41.

7-31   Elimination of a Halogen and a Hetero Group
Alkoxy-halo-elimination

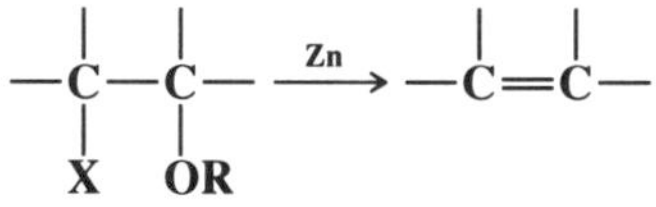

$$
-\overset{|}{\underset{\underset{X}{|}}{C}}-\overset{|}{\underset{\underset{OR}{|}}{C}}- \xrightarrow{\ Zn\ } -\overset{|}{C}=\overset{|}{C}-
$$

The elimination of OR and halogen from β-halo ethers is called the *Boord reaction*. It can be carried out with zinc, magnesium, sodium, or certain other reagents.[350] The yields are high and the reaction is of broad scope. β-Halo acetals readily yield vinylic ethers $X-\overset{|}{\underset{|}{C}}-C(OR)_2 \rightarrow -\overset{|}{C}=\overset{|}{C}-OR$. Besides β-halo ethers, the reaction can also be carried out on compounds of the formula $X-\overset{|}{\underset{|}{C}}-\overset{|}{\underset{|}{C}}-Z$, where X is halogen and Z is OCOR, OTs,[351] NR$_2$,[352] or SR.[353] Z may also be OH, but then X is limited Br and I. Like **7-29**, this method ensures that the new double bond will be in a specific position. The fact that magnesium causes elimination in these cases limits the preparation of Grignard reagents from these compounds. It has been shown that treatment of β-halo ethers and esters with zinc gives nonstereospecific elimination,[354] so the mechanism was not E2. An E1cB mechanism was postulated because of the poor leaving-group ability of OR and OCOR. Bromohydrins can be converted to olefins (elimination of Br, OH) in high yields by treatment with LiAlH$_4$–TiCl$_3$.[355]

OS **III**, 698, **IV**, 748; **VI**, 675.

[348]Darling; Kidwell *J. Org. Chem.* **1968**, *33*, 3974.
[349]For a procedure that gives 60 to 65% yields when one R = H, see McCarney; Ward *J. Chem. Soc., Perkin Trans. 1* **1975**, 1600. See also Masters; Sorensen; Ziegler *J. Org. Chem.* **1986**, *51*, 3558.
[350]See Ref. 144, pp. 136-139, for reagents that produce olefins from β-halo ethers and esters, and from halohydrins.
[351]Cristol; Rademacher *J. Am. Chem. Soc.* **1959**, *81*, 1600; Reeve; Brown; Steckel *J. Am. Chem. Soc.* **1971**, *93*, 4607.
[352]Gurien *J. Org. Chem.* **1963**, *28*, 878.
[353]Amstutz *J. Org. Chem.* **1944**, *9*, 310.
[354]House; Ro, Ref. 346.
[355]McMurry; Hoz *J. Org. Chem.* **1975**, *40*, 3797.

Fragmentations

When carbon is the positive leaving group (the electrofuge) in an elimination, the reaction is called *fragmentation*.[356] These processes occur on substrates of the form W—C—C—X, where X is a normal nucleofuge (e.g., halogen, OH_2^+, OTs, NR_3^+, etc.) and W is a positive-carbon electrofuge. In most of the cases W is HO—C— or R_2N—C—, so that the positive charge on the carbon atom is stabilized by the unshared pair of the oxygen or nitrogen, e.g.,

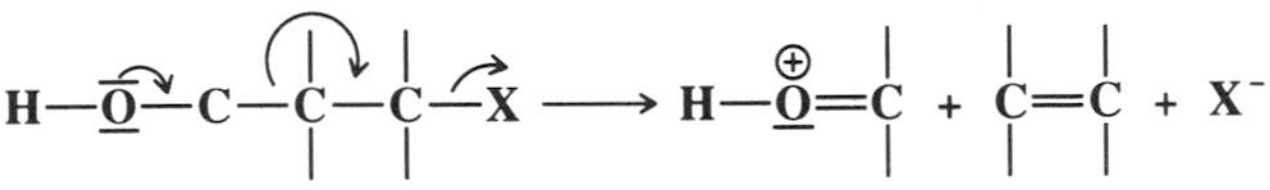

$$H-\overset{\curvearrowright}{\underline{\overline{O}}}\overset{\curvearrowright}{-}C-\overset{\curvearrowright}{\underset{|}{\overset{|}{C}}}-\overset{\curvearrowright}{\underset{|}{\overset{|}{C}}}-\overset{\nearrow}{X} \longrightarrow H-\overset{\oplus}{\underline{O}}=\overset{|}{\underset{|}{C}} + \overset{|}{\underset{|}{C}}=\overset{|}{\underset{|}{C}} + X^-$$

The mechanisms are mostly E1 or E2. We shall discuss only a few fragmentations, since many are possible and not much work has been done on most of them. Reactions **7-32** to **7-36** and **7-38** may be considered fragmentations. See also **9-13** and **9-14**.

7-32   Fragmentation of γ-Amino and γ-Hydroxy Halides
   Dialkylaminoalkyl-halo-elimination, etc.

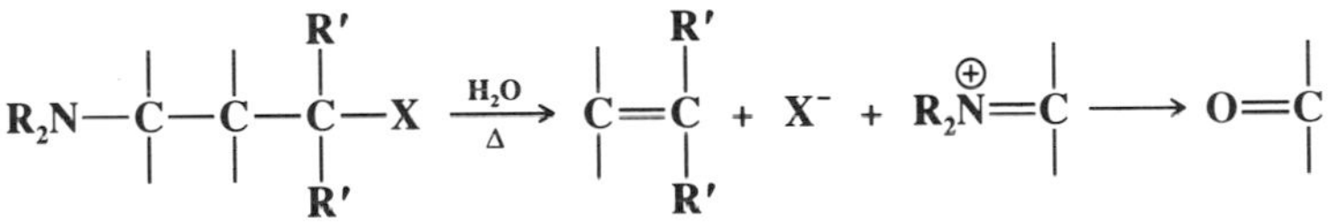

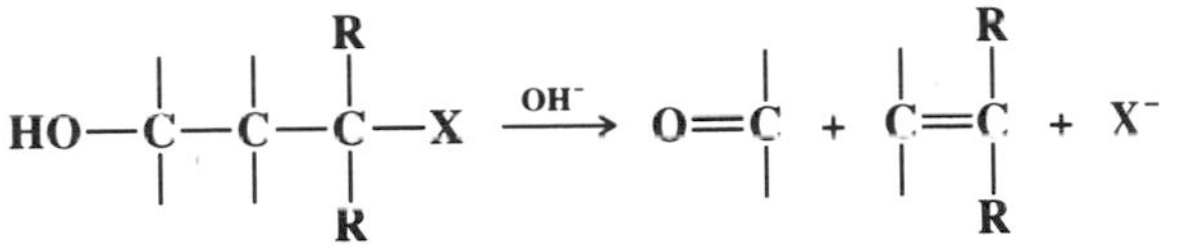

γ-Dialkylamino halides undergo fragmentation when heated with water to give an olefin and an iminium salt, which under the reaction conditions is hydrolyzed to an aldehyde or ketone (**6-2**).[357] γ-Hydroxy halides and tosylates are fragmented with base. In this instance the base does not play its usual role in elimination reactions but instead serves to remove a proton from the OH group, which enables the carbon leaving group to come off more easily:

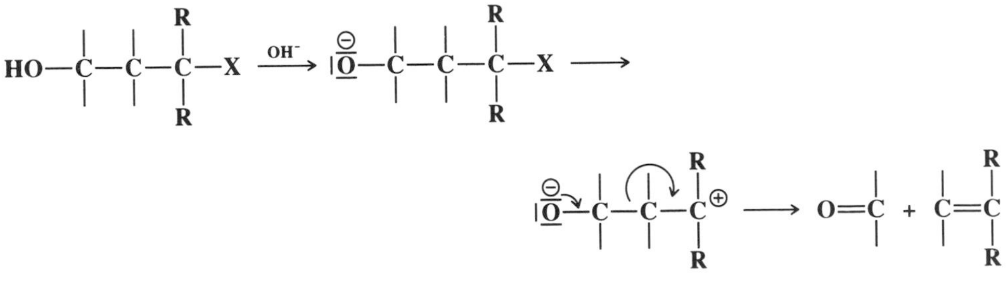

[356]For reviews, see Becker; Grob, in Patai, *The Chemistry of Functional Groups, Supplement A*, Ref. 2, pt. 2, pp. 653-723; Grob *Angew. Chem. Int. Ed. Engl.* **1969,** *8*, 535-546 [*Angew. Chem. 81*, 543-554]; Grob; Schiess *Angew. Chem. Int. Ed. Engl.* **1967,** *6*, 1-15 [*Angew. Chem. 79*, 1-14].
[357]Grob; Ostermayer; Raudenbusch *Helv. Chim. Acta* **1962,** *45*, 1672.

The mechanism of these reactions is often E1. However, in at least some cases, an E2 mechanism operates.[358] It has been shown that stereoisomers of cyclic γ-amino halides and tosylates in which the two leaving groups can assume an anti-periplanar conformation react by the E2 mechanism, while those isomers in which the groups cannot assume such a conformation either fragment by the E1 mechanism or do not undergo fragmentation at all, but in either case give rise to side products characteristic of carbocations.[359]

γ-Dialkylamino alcohols do not give fragmentation, since for ionization the OH group must be converted to OH_2^+ and this would convert NR_2 to NR_2H^+, which does not have the unshared pair necessary to form the double bond with the carbon.[360]

7-33 Fragmentation of 1,3-Diols
Hydroxyalkyl-hydroxy-elimination

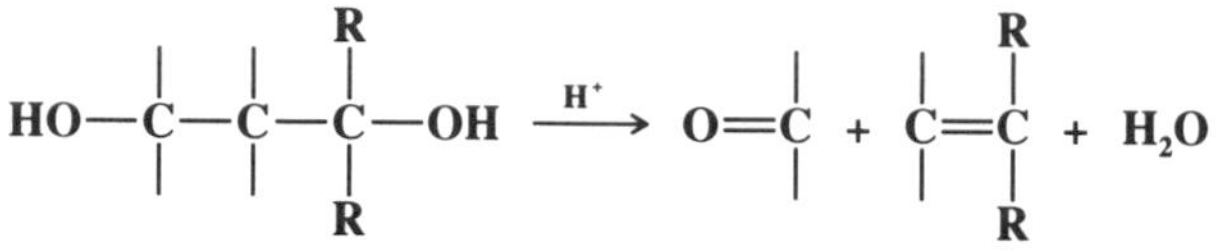

$$HO-\overset{|}{\underset{|}{C}}-\overset{|}{\underset{|}{C}}-\overset{\overset{R}{|}}{\underset{\underset{R}{|}}{C}}-OH \overset{H^+}{\longrightarrow} O=\overset{|}{\underset{|}{C}} + \overset{|}{\underset{|}{C}}=\overset{\overset{R}{|}}{\underset{\underset{R}{|}}{C}} + H_2O$$

1,3-Diols in which at least one OH group is tertiary or is located on a carbon with aryl substituents can be cleaved by acid treatment.[361] The reaction is most useful synthetically when at least one of the OH groups is on a ring.[362]

7-34 Decarboxylation of β-Hydroxy Carboxylic Acids and of β-Lactones
Carboxy-hydroxy-elimination

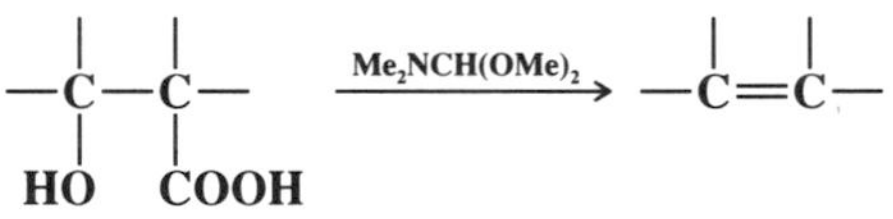

$$-\overset{|}{\underset{\underset{HO}{|}}{C}}-\overset{|}{\underset{\underset{COOH}{|}}{C}}- \overset{Me_2NCH(OMe)_2}{\longrightarrow} -\overset{|}{\underset{|}{C}}=\overset{|}{\underset{|}{C}}-$$

An OH and a COOH group can be eliminated from β-hydroxy carboxylic acids by refluxing with excess dimethylformamide dimethyl acetal.[363] Mono-, di-, tri-, and tetrasubstituted olefins have been prepared by this method in good yields.[364] There is evidence that the mechanism involves E1 or E2 elimination from the zwitterionic intermediate $^-O_2C-\overset{|}{\underset{|}{C}}-\overset{|}{\underset{|}{C}}-OCH=NMe_2^+$.[365] The reaction has also been accomplished[366] under extremely mild conditions (a few seconds at 0°C) with PPh_3 and diethyl azodicarboxylate EtOOC—N=N—COOEt.[367] In a related procedure, β-lactones undergo thermal decar-

[358]Grob; Schwarz *Helv. Chim. Acta* **1964,** *47,* 1870; Fischer; Grob *Helv. Chim. Acta* **1978,** *61,* 2336.
[359]Bottini; Grob; Schumacher; Zergenyi *Helv. Chim. Acta* **1966,** *49,* 2516; Burckhardt; Grob; Kiefer *Helv. Chim. Acta* **1967,** *50,* 231; Grob; Kiefer; Lutz; Wilkens *Helv. Chim. Acta* **1967,** *50,* 416; Geisel; Grob; Wohl *Helv. Chim. Acta* **1969,** *52,* 2206.
[360]Grob; Hoegerle; Ohta *Helv. Chim. Acta* **1962,** *45,* 1823.
[361]Zimmerman; English *J. Am. Chem. Soc.* **1954,** *76,* 2285, 2291, 2294.
[362]For a review of such cases, see Caine *Org. Prep. Proced. Int.* **1988,** *20,* 1-51.
[363]Hara; Taguchi; Yamamoto; Nozaki *Tetrahedron Lett.* **1975,** 1545.
[364]For a 1,4 example of this reaction, see Rüttimann; Wick; Eschenmoser *Helv. Chim. Acta* **1975,** *58,* 1450.
[365]Mulzer; Brüntrup *Tetrahedron Lett.* **1979,** 1909.
[366]For another method, see Tanzawa; Schwartz *Organometallics* **1990,** *9,* 3026.
[367]Mulzer; Brüntrup *Angew. Chem. Int. Ed. Engl.* **1977,** *16,* 255 [*Angew. Chem. 89,* 265]; Mulzer; Lammer *Angew. Chem. Int. Ed. Engl.* **1983,** *22,* 628 [*Angew. Chem. 95,* 629].

boxylation to give olefins in high yields. The reaction has been shown to be a stereospecific

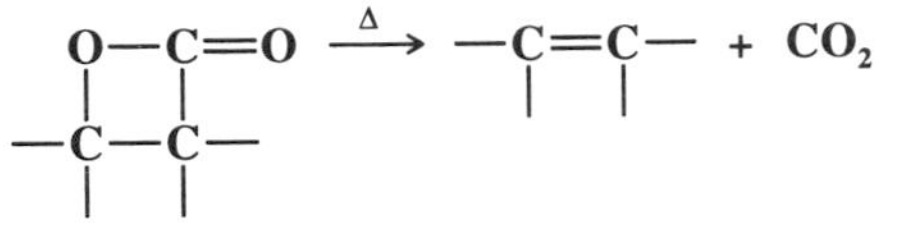

$$\text{O—C=O} \xrightarrow{\Delta} \text{—C=C—} + CO_2$$

syn elimination.[368] There is evidence that this reaction also involves a zwitterionic inter-mediate.[369]

There are no OS references, but see OS **VII**, 172, for a related reaction.

7-35 Fragmentation of α,β-Epoxy Hydrazones
Eschenmoser–Tanabe ring cleavage

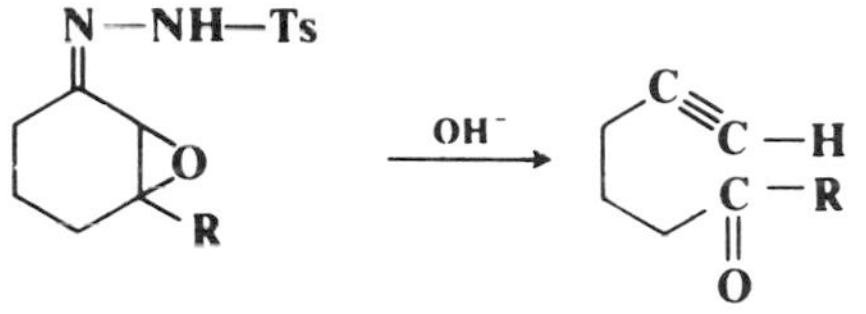

Cyclic α,β-unsaturated ketones[370] can be cleaved by treatment with base of their epoxy tosylhydrazone derivatives to give acetylenic ketones.[371] The reaction can be applied to the formation of acetylenic aldehydes (R = H) by using the corresponding, 2,4-dinitrotosylhy-drazone derivatives.[372] Hydrazones (e.g., **47**) prepared from epoxy ketones and ring-sub-

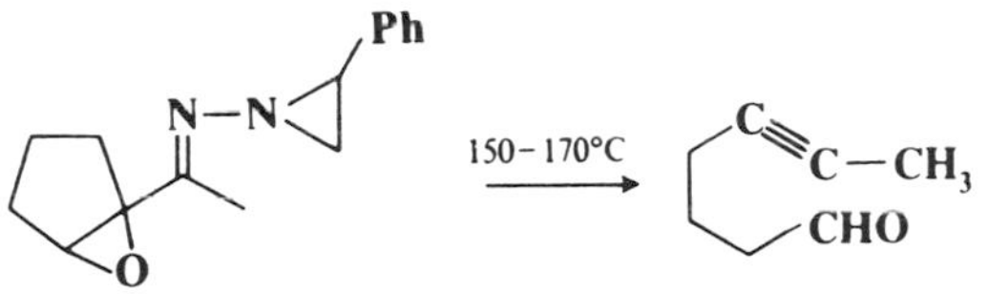

47

stituted N-aminoaziridines undergo similar fragmentation when heated.[373]

OS **VI**, 679.

7-36 Elimination of CO and CO₂ from Bridged Bicyclic Compounds
***seco*-Carbonyl-1/4/elimination**

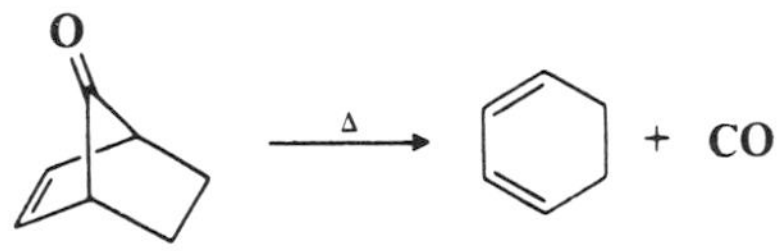

48

[368]Noyce; Banitt *J. Org. Chem.* **1966**, *31*, 4043; Adam; Baeza; Liu *J. Am. Chem. Soc.* **1972**, *94*, 2000; Krapcho; Jahngen *J. Org. Chem.* **1974**, *39*, 1322, 1650; Mageswaran; Sultanbawa *J. Chem. Soc., Perkin Trans. 1* **1976**, 884; Adam; Martinez; Thompson; Yany *J. Org. Chem.* **1981**, *46*, 3359.
[369]Mulzer; Zippel; Brüntrup *Angew. Chem. Int. Ed. Engl.* **1980**, *19*, 465 [*Angew. Chem. 92*, 469]; Mulzer; Zippel *Tetrahedron Lett.* **1980**, *21*, 751. See also Moyano; Pericaas; Valentí *J. Org. Chem.* **1989**, 573.
[370]For other methods of fragmentation of α,β-epoxy ketone derivatives, see MacAlpine; Warkentin *Can. J. Chem.* **1978**, *56*, 308, and references cited therein.
[371]Eschenmoser; Felix; Ohloff *Helv. Chim. Acta* **1967**, *50*, 708; Tanabe; Crowe; Dehn; Detre *Tetrahedron Lett.* **1967**, 3739; Tanabe; Crowe; Dehn *Tetrahedron Lett.* **1967**, 3943.
[372]Corey; Sachdev *J. Org. Chem.* **1975**, *40*, 579.
[373]Felix, Müller; Horn; Joos; Schreiber; Eschenmoser *Helv. Chim. Acta* **1972**, *55*, 1276.

On heating, bicyclo[2.2.1]hept-2,3-en-7-ones (**48**) usually lose CO to give cyclohexadienes,[374] in a type of reverse Diels–Alder reaction. Bicyclo[2.2.1]heptadienones (**49**) undergo the

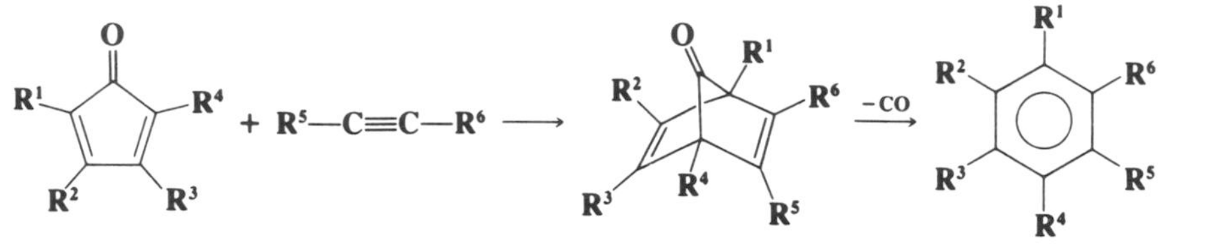

49

reaction so readily (because of the stability of the benzene ring produced) that they cannot generally be isolated. The parent **49** has been obtained at 10-15 K in an Ar matrix, where its spectrum could be studied.[375] **48** and **49** can be prepared by Diels–Alder reactions between a cyclopentadienone and an alkyne or olefin, so that this reaction is a useful method for the preparation of specifically substituted benzene rings and cyclohexadienes.[376] Unsaturated bicyclic lactones of the type **50** can also undergo the reaction, losing CO_2. See also **7-47**.

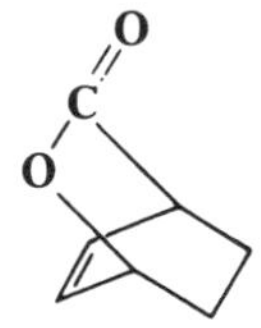

50

OS **III**, 807; **V**, 604, 1037.

Reversal of the Diels–Alder reaction may be considered a fragmentation. See **5-47**.

Reactions in Which C≡N or C=N Bonds Are Formed

7-37   Dehydration of Aldoximes and Similar Compounds
C-Hydro-N-hydroxy-elimination

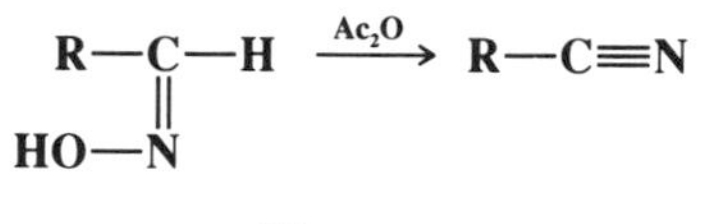

$$R—\underset{\underset{HO—N}{\|}}{C}—H \xrightarrow{Ac_2O} R—C≡N$$

Aldoximes can be dehydrated to nitriles[377] by many dehydrating agents, of which acetic anhydride is the most common. Among reagents that are effective under mild conditions[378]

[374]For a review, see Stark; Duke, Ref. 444, pp. 16-46.

[375]Birney; Berson *J. Am. Chem. Soc.* **1985**, *107*, 4553; *Tetrahedron* **1986**, *42*, 1561; LeBlanc; Sheridan *J. Am. Chem. Soc.* **1985**, *107*, 4554; Birney; Wiberg; Berson *J. Am. Chem. Soc.* **1988**, *110*, 6631.

[376]For a review with many examples; see Ogliaruso; Romanelli; Becker *Chem. Rev.* **1965**, *65*, 261-367, pp. 300-348. For references to this and related reactions, see Ref. 144, pp. 101-103.

[377]For reviews, see Friedrich, in Patai; Rappoport, Ref. 142, pt. 2, pp. 1345-1390; Friedrich; Wallenfels in Rappoport *The Chemistry of the Cyano Group*; Wiley: New York, 1970, pp. 92-96. For a review of methods of synthesizing nitriles, see Fatiadi, in Patai; Rappoport, Ref. 142, pt. 2, pp. 1057-1303.

[378]For lists of some other reagents, with references, see Molina; Alajarin; Vilaplana *Synthesis* **1982**, 1016; Aizpurua; Palomo *Nouv. J. Chim.* **1983**, *7*, 465; Attanasi; Palma; Serra-Zanetti *Synthesis* **1983**, 741; Juršić *Synth. Commun.* **1989**, *19*, 689.

(room temperature) are ethyl orthoformate and H$^+$,[379] Ph$_3$P–CCl$_4$,[380] trichloromethyl chloroformate ClCOOCCl$_3$,[381] methyl (or ethyl) cyanoformate ROCOCN,[382] trifluoromethane sulfonic anhydride,[383] P$_2$I$_4$,[291] SeO$_2$,[384] CS$_2$ under phase transfer conditions,[385] Cl$_3$COCl–Et$_3$N,[386] and chloromethylene dimethylammonium chloride Me$_2$N=CHCl$^+$ Cl$^-$.[387] Electrochemical synthesis has also been used.[388] The reaction is most successful when the H and OH are anti. Various alkyl and acyl derivatives of aldoximes, for example, RCH=NOR, RCH=NOCOR, RCH=NOSO$_2$Ar, etc., also give nitriles, as do chlorimines RCH=NCl (the latter with base treatment).[389] N,N-dichloro derivatives of primary amines give nitriles on pyrolysis: RCH$_2$NCl$_2$ → RCN.[390]

Quaternary hydrazonium salts (derived from aldehydes) give nitriles when treated with OEt$^-$[391] or DBU (p. 1023):[392]

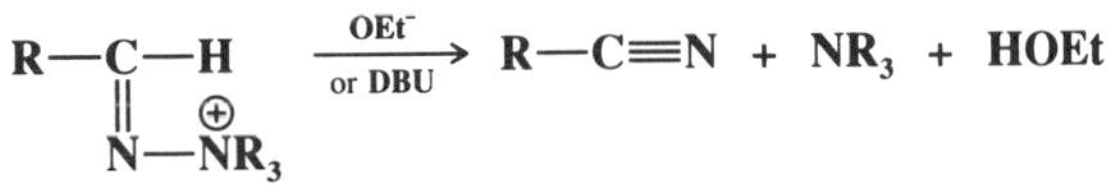

$$R-\overset{\underset{\displaystyle N-NR_3}{\|}}{\underset{\oplus}{C}}-H \xrightarrow[\text{or DBU}]{\text{OEt}^-} R-C\equiv N \;+\; NR_3 \;+\; HOEt$$

as do dimethylhydrazones RCH=NNMe$_2$ when treated with Et$_2$NLi and HMPA.[393] All these are methods of converting aldehyde derivatives to nitriles. For the conversion of aldehydes directly to nitriles, without isolation of intermediates, see **6-22**.

OS **II**, 622; **III**, 690.

7-38   The Conversion of Ketoximes to Nitriles
        C-Acyl-N-hydroxy-elimination

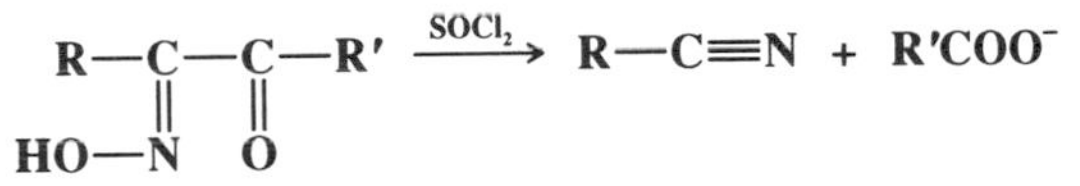

$$R-\overset{\underset{\displaystyle HO-N}{\|}}{C}-\overset{\underset{\displaystyle O}{\|}}{C}-R' \xrightarrow{\text{SOCl}_2} R-C\equiv N \;+\; R'COO^-$$

Certain ketoximes can be converted to nitriles by the action of proton or Lewis acids.[394] Among these are oximes of α-diketones (illustrated above), α-keto acids, α-dialkylamino ketones, α-hydroxy ketones, β-keto ethers, and similar compounds.[395] These are fragmen-

[379]Rogić; Van Peppen; Klein; Demmin *J. Org. Chem.* **1974**, *39*, 3424.
[380]Kim; Chung; Ryu *Synth. Commun.* **1990**, *20*, 2785.
[381]Mai; Patil *Synthesis* **1986**, 1037.
[382]Thomas; Greyn *Synthesis* **1990**, 129.
[383]Hendrickson; Blair; Keehn; *Tetrahedron Lett.* **1976**, 603.
[384]Sosnovsky; Krogh *Synthesis* **1978**, 703.
[385]Shinozaki; Imaizumi; Tajima *Chem. Lett.* **1983**, 929.
[386]Saednya *Synthesis* **1983**, 748.
[387]Dulcere *Tetrahedron Lett.* **1981**, *22*, 1599.
[388]See Shono; Matsumura; Tsubata; Kamada; Kishi *J. Org. Chem.* **1989**, *54*, 2249.
[389]Hauser; Le Maistre; Rainsford *J. Am. Chem. Soc.* **1935**, *57*, 1056.
[390]Roberts; Rittberg; Kovacic *J. Org. Chem.* **1981**, *46*, 4111.
[391]Smith; Walker *J. Org. Chem.* **1962**, *27*, 4372; Grandberg *J. Gen. Chem. USSR* **1964**, *34*, 570; Grundon; Scott *J. Chem. Soc.* **1964**, 5674; Ioffe; Zelenina *J. Org. Chem. USSR* **1968**, *4*, 1496.
[392]Moore; Stupp *J. Org. Chem.* **1990**, *55*, 3374.
[393]Cuvigny; Le Borgne; Larchevêque; Normant *Synthesis* **1976**, 237.
[394]For reviews, see Gawley *Org. React.* **1988**, *35*, 1-420; Conley; Ghosh *Mech. Mol. Migr.* **1971**, *4*, 197-308, pp. 197-251; McCarty; in Patai *The Chemistry of the Carbon–Nitrogen Double Bond*; Wiley: New York, 1970, pp. 416-439; Casanova; in Rappoport, Ref. 377, pp. 915-932.
[395]For more complete lists with references, see Olah; Vankar; Berrier *Synthesis* **1980**, 45; Conley; Ghosh, Ref. 394.

tation reactions, analogous to **7-32** and **7-33.** For example, α-dialkylamino ketoximes also give amines and aldehydes or ketones besides nitriles:[396]

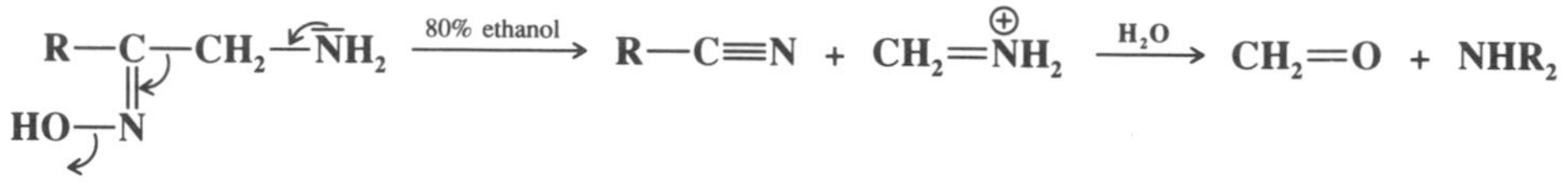

$$R-\underset{\underset{HO-N}{\parallel}}{C}-CH_2-NH_2 \xrightarrow{\text{80\% ethanol}} R-C\equiv N + CH_2=\overset{\oplus}{N}H_2 \xrightarrow{H_2O} CH_2=O + NHR_2$$

The reaction that normally occurs on treatment of a ketoxime with a Lewis or proton acid is the Beckmann rearrangement (**8-18**); fragmentations are considered side reactions, often called "abnormal" or "second-order" Beckmann rearrangements.[397] Obviously, the substrates mentioned are much more susceptible to fragmentation than are ordinary ketoximes, since in each case an unshared pair is available to assist in removal of the group cleaving from the carbon. However, fragmentation is a side reaction even with ordinary ketoximes[398] and, in cases where a particularly stable carbocation can be cleaved, may be the main reaction:[399]

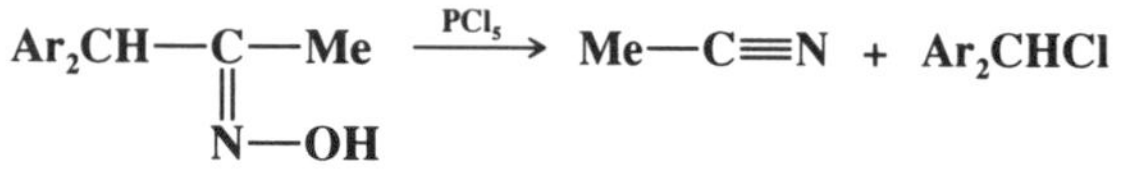

$$Ar_2CH-\underset{\underset{N-OH}{\parallel}}{C}-Me \xrightarrow{PCl_5} Me-C\equiv N + Ar_2CHCl$$

There are indications that the mechanism at least in some cases first involves a rearrangement and then cleavage. The ratio of fragmentation to Beckmann rearrangement of a series of oxime tosylates RC(=NOTs)Me was not related to the solvolysis rate but *was* related to the stability of R⁺ (as determined by the solvolysis rate of the corresponding RCl), which showed that fragmentation did not take place in the rate-determining step.[400] It may be postulated then that the first step in the fragmentation and in the rearrangement is the same and that this is the rate-determining step. The product is determined in the second step:

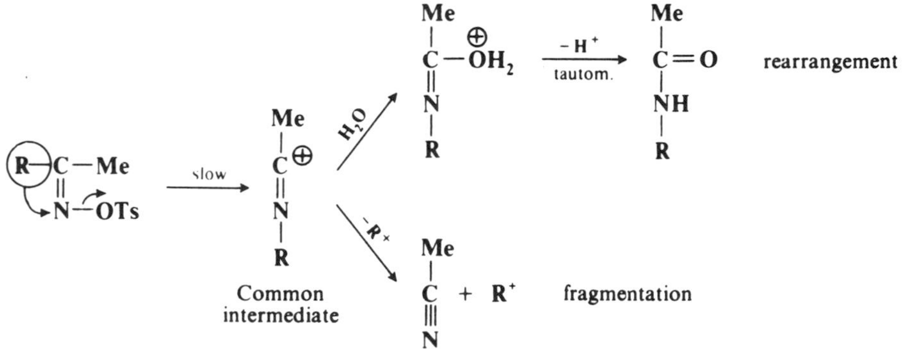

However, in other cases the simple E1 or E2 mechanisms operate.[401]

[396]Fischer; Grob; Renk *Helv. Chim. Acta* **1962**, *45*, 2539; Fischer; Grob *Helv. Chim. Acta* **1963**, *46*, 936.
[397]See the discussion in Ferris *J. Org. Chem.* **1960**, *25*, 12.
[398]See, for example, Hill; Conley *J. Am. Chem. Soc.* **1960**, *82*, 645.
[399]Hassner; Nash *Tetrahedron Lett.* **1965**, 525.
[400]Grob; Fischer; Raudenbusch; Zergenyi *Helv. Chim.Acta* **1964**, *47*, 1003.
[401]Ahmad; Spenser *Can. J. Chem.* **1961**, *39*, 1340; Ferris; Johnson; Gould *J. Org. Chem.* **1960**, *25*, 1813; Grob; Sieber *Helv. Chim. Acta* **1967**, *50*, 2520; Green; Pearson *J. Chem. Soc. B* **1969**, 593.

I-(1-tosyl-1-alkenyl)formamides (**51**) by refluxing with NaOMe in version of a ketone to a nitrile,[402] since **51** can be prepared by

$$\text{--Ts} \xrightarrow{\text{base}} \underset{\substack{| \\ \text{R--CH--C--Ts}}}{\overset{\substack{\text{R'} \quad \text{N--CHO} \\ | \qquad ||}}{}} \xrightarrow{\text{base}} \underset{\substack{| \\ \text{R--CH--CN}}}{\overset{\text{R'}}{}}$$

51

catment of ketones with TsCH$_2$NC (p. 949). The overall conversion is RR′C=O to RR′CHCN.

OS **V**, 266.

7-39 Dehydration of Unsubstituted Amides
***NN*-Dihydro-*C*-oxo-bielimination**

$$\underset{\substack{|| \\ \text{O}}}{\text{R--C--NH}_2} \xrightarrow{\text{P}_2\text{O}_5} \text{R--C} \equiv \text{N}$$

Unsubstituted amides can be dehydrated to nitriles.[403] Phosphorous pentoxide is the most common dehydrating agent for this reaction, but many others, including POCl$_3$, PCl$_5$, CCl$_4$–Ph$_3$P,[404] TiCl$_4$–base,[405] HMPA,[406] Cl$_3$COCl–Et$_3$N,[407] MeOOC$\overset{\ominus}{N}$SO$_2\overset{\oplus}{N}$Et$_3$ (the Burgess reagent),[408] nitrilium salts,[409] cyanuric chloride,[410] Me$_2$N=CHCl$^+$ Cl$^-$,[411] trimethylsilyl polyphosphate,[412] and SOCl$_2$ have also been used.[413] It is possible to convert an acid to the nitrile, without isolation of the amide, by heating its ammonium salt with the dehydrating agent,[414] or by other methods.[415] Acyl halides can also be directly converted to nitriles by heating with sulfamide (NH$_2$)$_2$SO$_2$.[416] The reaction may be formally looked on as a β elimination from the enol form of the amide RC(OH)=NH, in which case it is like **7-37**, except that H and OH have changed places. In some cases, for example, with SOCl$_2$, the mechanism probably *is* through the enol form, with the dehydrating agent forming an ester with the OH group, for example, RC(OSOCl)=NH, which undergoes elimination by the E1 or E2 mechanism.[417] N,N-Disubstituted ureas give cyanamides (R$_2$N—CO—NH$_2$ → R$_2$N—CN) when dehydrated with CHCl$_3$–NaOH under phase transfer conditions.[418]

[402]Schöllkopf; Schröder *Angew. Chem. Int. Ed. Engl.* **1973**, *12*, 407 [*Angew. Chem. 85*, 402].
[403]For reviews, see Bieron; Dinan; in Zabicky *The Chemistry of Amides*; Wiley: New York, 1970, pp. 274-283; Friedrich; Wallenfels, Ref. 377, pp. 96-103; Friedrich, Ref. 377.
[404]Yamato; Sugasawa *Tetrahedron Lett.* **1970**, 4383; Appel; Kleinstück; Ziehn *Chem. Ber.* **1971**, *104*, 1030; Harrison; Hodge; Rogers *Synthesis* **1977**, 41.
[405]Lehnert *Tetrahedron Lett.* **1971**, 1501.
[406]Monson; Priest *Can. J. Chem.* **1971**, *49*, 2897.
[407]Saednya *Synthesis* **1985**, 184.
[408]Claremon; Phillips *Tetrahedron Lett.* **1988**, *29*, 2155.
[409]Jochims; Glocker *Chem. Ber.* **1990**, *123*, 1537.
[410]Olah; Narang; Fung; Gupta *Synthesis* **1980**, 657.
[411]Barger; Riley *Synth. Commun.* **1980**, *10*, 479.
[412]Yokoyama; Yoshida; Imamoto *Synthesis* **1982**, 591. See also Rao; Rambabu; Srinivasan *Synth. Commun.* **1989**, *19*, 1431.
[413]For a list of reagents, with references, see Ref. 144, pp. 991-992.
[414]See, for example, Imamoto; Takaoka; Yokoyama *Synthesis* **1983**, 142.
[415]For a list of methods, with references, see Ref. 144, pp. 976-977.
[416]Hulkenberg; Troost *Tetrahedron Lett.* **1982**, *23*, 1505.
[417]Rickborn; Jensen *J. Org. Chem.* **1962**, *27*, 4608.
[418]Schroth; Kluge; Frach; Hodek; Schädler *J. Prakt. Chem.* **1983**, *325*, 787.

N-Alkyl-substituted amides can be converted to nitriles and alkyl chlorides by tr‌
with PCl_5. This is called the *von Braun reaction* (not to be confused with the oth‌

$$\text{R'CONHR} + \text{PCl}_5 \longrightarrow \text{R'CN} + \text{RCl}$$

Braun reaction, **0-73**). In a similar reaction, treatment of N-alkyl-substituted amides with
chlorotris(triphenylphosphine)rhodium $RhCl(PPh_3)_3$ or certain other catalysts give nitriles
and the corresponding alcohols.[419]

OS **I**, 428; **II**, 379; **III**, 493, 535, 584, 646, 768; **IV**, 62, 144, 166, 172, 436, 486, 706; **VI**,
304, 465.

7-40 Conversion of Primary Nitro Compounds to Nitriles

$$\text{RCH}_2\text{—NO}_2 \xrightarrow[\text{pyridine}]{\text{PCl}_3} \text{RC}\!\equiv\!\text{N}$$

Nitriles can be obtained in one step by treatment of primary nitro compounds with PCl_3
and pyridine.[420] R may be alkyl or aryl and may contain C=C double bonds or various
functional groups. Yields are moderate to good. The reaction has also been carried out with
Me_3N–SO_2 and with HMPA.[421] Primary azides RCH_2N_3 have been converted to nitriles
RCN with Pd metal.[422] Primary nitro compounds RCH_2NO_2 were converted to nitrile oxides
$\overset{\oplus}{R}C\overset{\ominus}{N}$—$O$ by treatment with ClCOOEt or $PhSO_2Cl$ in the presence of Et_3N.[423]

7-41 Conversion of N-Alkylformamides to Isocyanides
CN-Dihydro-C-oxo-bielimination

$$\text{H—C—NH—R} \xrightarrow[\text{R}_3\text{N}]{\text{COCl}_2} \overset{\ominus}{C}\!\equiv\!\overset{\oplus}{N}\text{—R}$$
$$\underset{\text{O}}{\overset{\|}{}}$$

Isocyanides can be prepared by elimination of water from N-alkylformamides with phosgene
and a tertiary amine.[424] Other reagents, among them TsCl in quinoline, $POCl_3$ and a tertiary
amine,[425] Me_2N=$CHCl^+$ Cl^-,[426] di-2-pyridyl sulfite,[427] triflic anhydride–(i-Pr)$_2$NEt,[428] Ph_3P–
CCl_4–Et_3N,[429] and Ph_3PBr_2–Et_3N[430] have also been employed.

OS **V**, 300, 772; **VI**, 620, 751, 987. See also OS **VII**, 27.

[419]Blum; Fisher; Greener *Tetrahedron* **1973**, *29*, 1073.
[420]Wehrli; Schaer *J. Org. Chem.* **1977**, *42*, 3956.
[421]Olah; Vankar; Gupta *Synthesis*; **1979**, 36. For another method, see Urpí; Vilarrasa *Tetrahedron Lett.* **1990**, *31*,
7497.
[422]Hayashi; Ohno; Oka *Bull. Chem. Soc. Jpn.* **1976**, *49*, 506. See also Jarvis; Nicholas *J. Org. Chem.* **1979**, *44*,
2951.
[423]Shimizu; Hayashi; Shibafuchi; Teramura *Bull. Chem. Soc. Jpn.* **1986**, *59*, 2827.
[424]For reviews, see Hoffmann; Gokel; Marquarding; Ugi, in Ugi *Isonitrile Chemistry*; Academic Press: New York,
1971, pp. 10-17; Ugi; Fetzer; Eholzer; Knupfer; Offermann *Angew. Chem. Int. Ed. Engl.* **1965**, *4*, 472-484 [*Angew.
Chem. 77*, 492-504], *Newer Methods Prep. Org. Chem.* **1968**, *4*, 37-66.
[425]See Obrecht; Herrmann; Ugi *Synthesis* **1985**, 400.
[426]Walborsky; Niznik *J. Org. Chem.* **1972**, *37*, 187.
[427]Kim; Yi *Tetrahedron Lett.* **1986**, *27*, 1925.
[428]Baldwin; O'Neil *Synlett* **1991**, 603.
[429]Appel; Kleinstück; Ziehn *Angew. Chem. Int. Ed. Engl.* **1971**, *10*, 132 [*Angew. Chem. 83*, 143].
[430]Bestmann; Lienert; Mott *Liebigs Ann. Chem.* **1968**, *718*, 24.

7-42   Dehydration of N,N′-Disubstituted Ureas and Thioureas
   1/N,3/N-Dihydro-2/C-oxo-bielimination

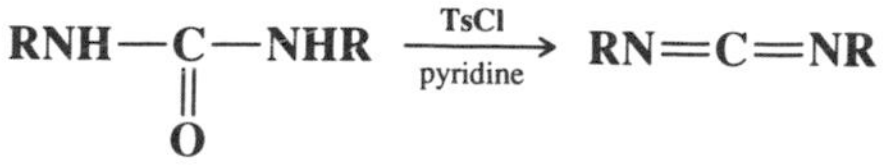

$$\text{RNH—C—NHR} \xrightarrow[\text{pyridine}]{\text{TsCl}} \text{RN}{=}\text{C}{=}\text{NR}$$
$$\underset{\text{O}}{\overset{\|}{}}$$

Carbodiimides[431] can be prepared by the dehydration of N,N′-disubstituted ureas with various dehydrating agents,[432] among which are TsCl in pyridine, $POCl_3$, PCl_5, P_2O_5–pyridine, TsCl (with phase-transfer catalysis),[433] and Ph_3PBr_2–Et_3N.[430] H_2S can be removed from the corresponding thioureas by treatment with HgO, NaOCl, phosgene,[434] or diethyl azodicarboxylate–triphenylphospine.[435]
   OS **V**, 555; **VI**, 951.

Reactions in Which C=O Bonds Are Formed

Many elimination reactions in which C=O bonds are formed were considered in Chapter 16, along with their more important reverse reactions. Also see **2-40** and **2-41**.

7-43   Pyrolysis of β-Hydroxy Olefins
   O-Hydro-C-allyl-elimination

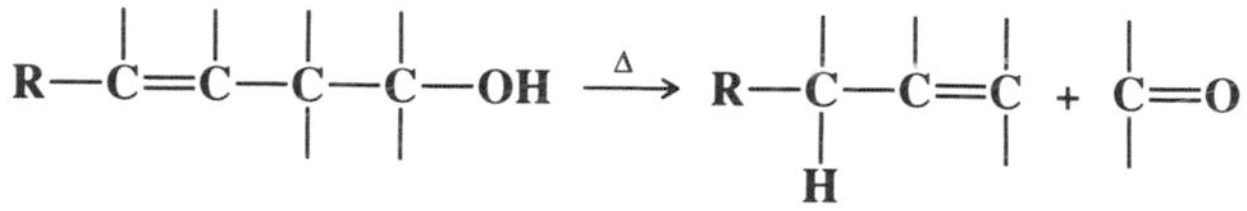

$$\text{R—C}{=}\text{C—C—C—OH} \xrightarrow{\Delta} \text{R—C—C}{=}\text{C} + \text{C}{=}\text{O}$$

When pyrolyzed, β-hydroxy olefins cleave to give olefins and aldehydes or ketones.[436] Olefins produced this way are quite pure, since there are no side reactions. The mechanism has

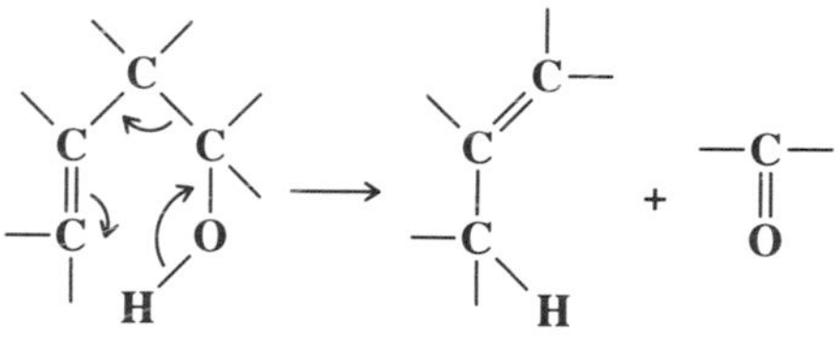

been shown to be pericyclic, primarily by observations that the kinetics are first order[437] and that, for ROD, the deuterium appeared in the allylic position of the new olefin.[438] This

[431]For a review of the reactions in this section, see Bocharov *Russ. Chem. Rev.* **1965**, *34*, 212-219. For a review of carbodiimide chemistry; see Williams; Ibrahim *Chem. Rev.* **1981**, *81*, 589-636.
   [432]For some others not mentioned here, see Sakai; Fujinami; Otani; Aizawa *Chem. Lett.* **1976**, 811; Shibanuma; Shiono; Mukaiyama *Chem. Lett.* **1977**, 575; Kim; Yi *J. Org. Chem.* **1986**, *51*, 2613, Ref. 427.
   [433]Jászay; Petneházy; Tóke; Szajáni *Synthesis* **1987**, 520.
   [434]Ulrich; Sayigh *Angew. Chem. Int. Ed. Engl.* **1966**, *5*, 704-712 [*Angew. Chem. 78*, 761-769], *Newer Methods Prep. Org. Chem.* **1971**, *6*, 223-242.
   [435]Mitsunobu; Kato; Tomari *Tetrahedron* **1970**, *26*, 5731.
   [436]Arnold; Smolinsky *J. Am. Chem. Soc.* **1959**, *81*, 6643. For a review, see Marvell; Whalley, in Patai, Ref. 152, pt. 2, pp. 729-734.
   [437]Smith; Yates *J. Chem. Soc.* **1965**, 7242; Voorhees; Smith *J. Org. Chem.* **1971**, *36*, 1755.
   [438]Arnold; Smolinsky *J. Org. Chem.* **1960**, *25*, 128; Smith; Taylor *Chem. Ind. (London)* **1961**, 949.

mechanism is the reverse of that for the oxygen analog of the ene synthesis (**6-53**). β-Hydroxyacetylenes react similarly to give the corresponding allenes and carbonyl compounds.[439] The mechanism is the same despite the linear geometry of the triple bonds.

7-44 Pyrolysis of Allylic Ethers
C-Hydro-O-allyl-elimination

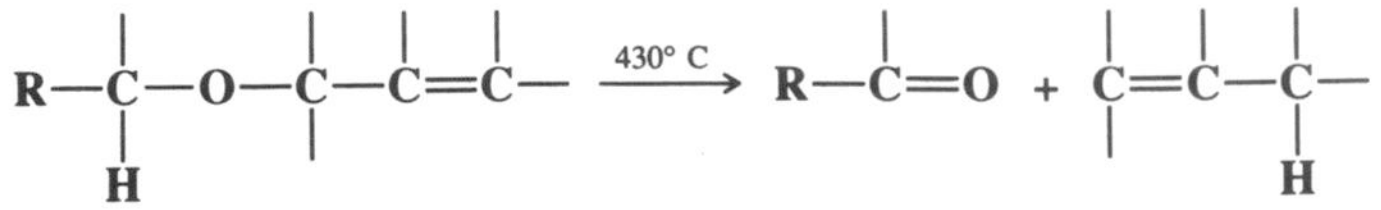

$$R-\underset{\underset{H}{|}}{\overset{|}{C}}-O-\underset{|}{\overset{|}{C}}-\overset{|}{C}=\overset{|}{C}- \xrightarrow{430° C} R-\overset{|}{C}=O \ + \ \overset{|}{C}=\overset{|}{C}-\underset{\underset{H}{|}}{\overset{|}{C}}-$$

Pyrolysis of allylic ethers that contain at least one α hydrogen gives olefins and aldehydes or ketones. The reaction is closely related to **7-43,** and the mechanism is also pericyclic[440]

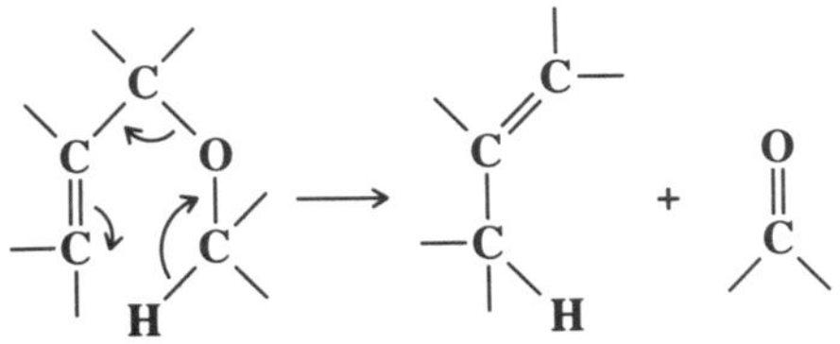

Reactions in Which N≡N Bonds Are Formed

7-45 Eliminations to Give Diazoalkanes
N-Nitrosoamine-diazoalkane transformation

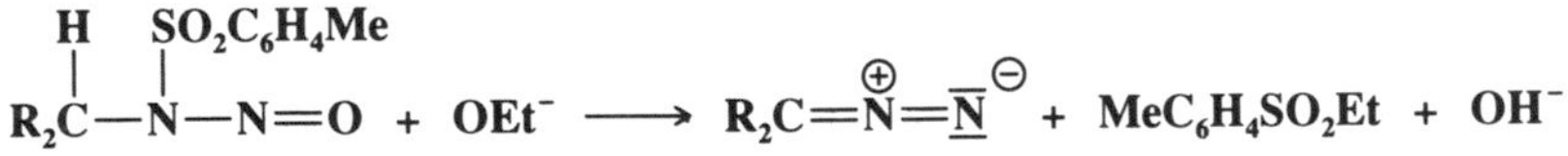

$$\underset{R_2C-N-N=O}{\overset{H \quad SO_2C_6H_4Me}{\overset{|}{\underset{}{\overset{|}{}}}}} + OEt^- \longrightarrow R_2C=\overset{\oplus}{N}=\overset{\ominus}{\overline{N}} + MeC_6H_4SO_2Et + OH^-$$

Various N-nitroso-N-alkyl compounds undergo elimination to give diazoalkanes.[441] One of the most convenient methods for the preparation of diazomethane involves base treatment of N-nitroso-N-methyl-*p*-toluenesulfonamide (illustrated above, with R = H).[442] However, other compounds commonly used are (base treatment is required in all cases):

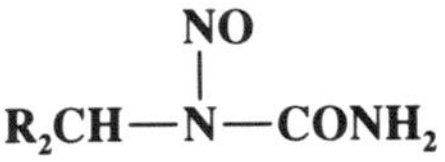

$$\underset{R_2CH-N-CONH_2}{\overset{NO}{\overset{|}{}}}$$

N-nitroso-N-alkylureas

$$\underset{R_2CH-N-COOEt}{\overset{NO}{\overset{|}{}}}$$

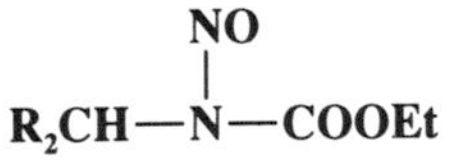

N-nitroso-N-alkylcarbamates

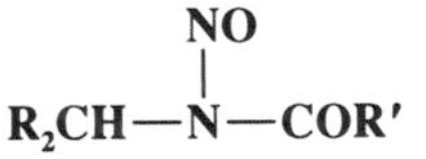

$$\underset{R_2CH-N-COR'}{\overset{NO}{\overset{|}{}}}$$

N-nitroso-N-alkyl amides

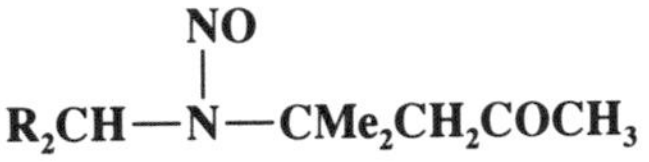

$$\underset{R_2CH-N-CMe_2CH_2COCH_3}{\overset{NO}{\overset{|}{}}}$$

N-nitroso-N-alkyl-4-amino-4-methyl-2-pentanones

[439]Viola; MacMillan; Proverb; Yates *J. Am. Chem. Soc.* **1971,** *93,* 6967; Viola; Proverb; Yates; Larrahondo *J. Am. Chem. Soc.* **1973,** *95,* 3609.

[440]Cookson; Wallis *J. Chem. Soc. B.* **1966,** 1245; Kwart; Slutsky; Sarner *J. Am. Chem. Soc.* **1973,** *95,* 5242; Egger; Vitins *Int. J. Chem. Kinet.* **1974,** *6,* 429.

[441]For a review, see Regitz; Maas *Diazo Compounds;* Academic Press: New York, 1986, pp. 296-325. For a review of the preparation and reactions of diazomethane, see Black *Aldrichimica Acta* **1983,** *16,* 3-10. For discussions, see Cowell; Ledwith *Q. Rev., Chem. Soc.* **1970,** *24,* 119-167, pp. 126-131; Smith *Open-chain Nitrogen Compounds;* W. A. Benjamin: New York, 1966, especially pp. 257-258, 474-475, in vol. 2.

[442]de Boer; Backer *Org. Synth. IV* 225, 250; Hudlicky *J. Org. Chem.* **1980,** *45,* 5377.

All these compounds can be used to prepare diazomethane, though the sulfonamide, which
is commercially available, is most satisfactory. (N-Nitroso-N-methylcarbamate and N-ni-
troso-N-methylurea give good yields, but are highly irritating and carcinogenic.[443]) For higher
diazoalkanes the preferred substrates are nitrosoalkylcarbamates.

Most of these reactions probably begin with a 1,3 nitrogen-to-oxygen rearrangement,
followed by the actual elimination (illustrated for the carbamate):

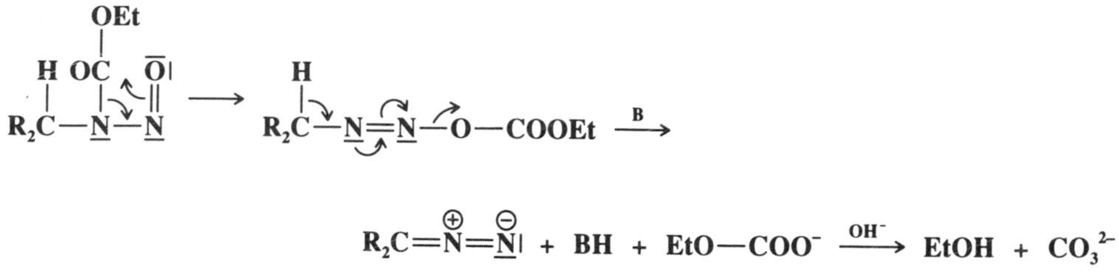

$$R_2C=\overset{\oplus}{N}=\overset{\ominus}{\underline{N}}| + BH + EtO—COO^- \xrightarrow{\text{OH}^-} EtOH + CO_3^{2-}$$

OS **II**, 165; **III**, 119, 244; **IV**, 225, 250; **V**, 351; **VI**, 981.

Extrusion Reactions

We consider an *extrusion reaction*[444] to be one in which an atom or group Y connected to
two other atoms X and Z is lost from a molecule, leading to a product in which X is bonded
directly to Z.

$$X—Y—Z \longrightarrow X—Z + Y$$

Reactions **4-41** and **7-25** also fit this definition. Reaction **7-36** does not fit the definition, but
is often also classified as an extrusion reaction. An extrusibility scale has been developed,
showing that the ease of extrusion of the common Y groups is in the order: —N=N— >
—COO— > —SO₂— > —CO—.[445]

7-46 Extrusion of N₂ from Pyrazolines, Pyrazoles, and Triazolines
      Azo-extrusion

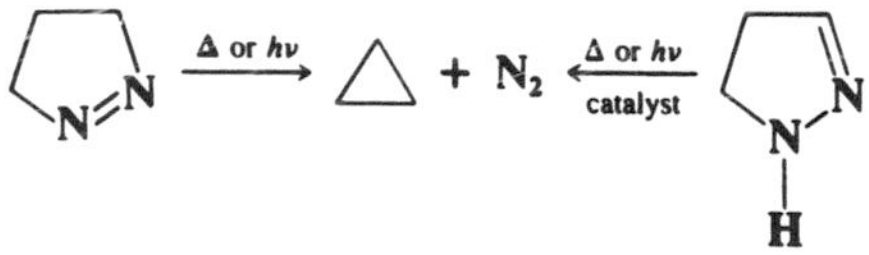

52   **53**

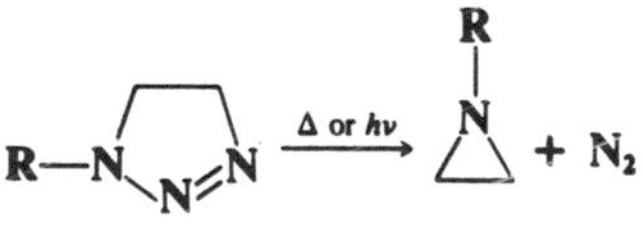

54

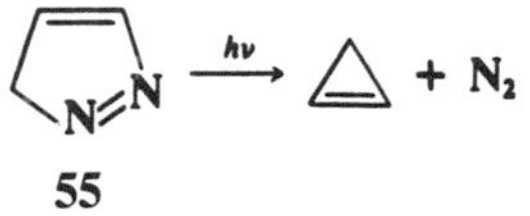

55

---

[443]Searle *Chem. Br.* **1970,** *6*, 5-10.
[444]For a monograph, see Stark; Duke *Extrusion Reactions*; Pergamon: Elmsford, NY, 1967. For a review of
extrusions that are photochemically induced, see Givens *Org.Photochem.* **1981,** *5*, 227-346.
[445]Paine; Warkentin *Can. J. Chem.* **1981,** *59*, 491.

1-Pyrazolines (**52**) can be converted to cyclopropane and N_2 on photolysis[446] or pyrolysis.[447] The tautomeric 2-pyrazolines (**53**), which are more stable than **52**, also give the reaction, but in this case an acidic or basic catalyst is required, the function of which is to convert **53** to **52**.[448] In the absence of such catalysts, **53** do not react.[449] In a similar manner, triazolines (**54**) are converted to aziridines.[450] Side reactions are frequent with both **52** and **54**, and some substrates do not give the reaction at all. However, the reaction has proved synthetically useful in many cases. In general, photolysis gives better yields and fewer side reactions than pyrolysis with both **52** and **54**. 3*H*-Pyrazoles[451] (**55**) are stable to heat, but in some cases can be converted to cyclopropenes on photolysis,[452] though in other cases other types of products are obtained.

There is much evidence that the mechanism[453] of the 1-pyrazoline reactions generally involves diradicals, though the mode of formation and detailed structure (e.g., singlet vs.

$$\underset{N^{\diagdown}N}{\diagup} \xrightarrow{-N_2} \overset{\bullet}{\underset{\bullet}{[}} \longrightarrow \triangle$$

triplet) of these radicals may vary with the substrate and reaction conditions. The reactions of the 3*H*-pyrazoles have been postulated to proceed through a diazo compound that loses N_2 to give a vinylic carbene.[454]

$$\mathbf{55} \xrightarrow{hv} -\overset{|}{C}=\overset{|}{C}-\overset{|}{C}N_2 \longrightarrow -\overset{|}{C}=\overset{|}{C}-\overset{|}{\underline{C}} \longrightarrow \triangle$$

OS **V**, 96, 929. See also OS **66**, 142.

7-47   Extrusion of CO or CO_2
Carbonyl-extrusion

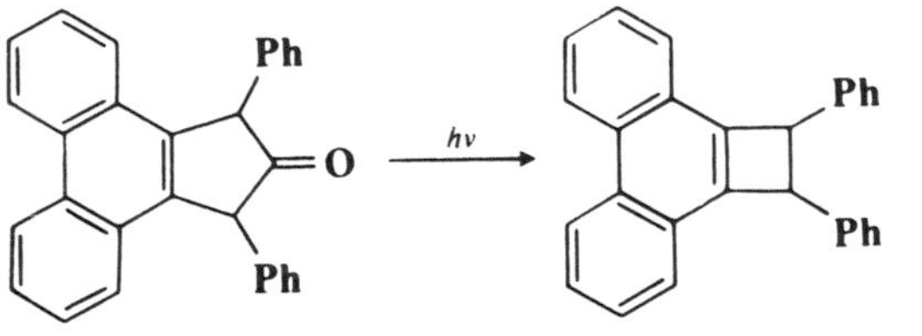

56   **57**

[446]Van Auken; Rinehart *J. Am. Chem. Soc.* **1962**, *84*, 3736.
[447]For reviews of the reactions in this section, see Adam; De Lucchi *Angew. Chem. Int. Ed. Engl.* **1980**, *19*, 762-779 [*Angew. Chem. 92*, 815-832]; Meier; Zeller *Angew. Chem. Int. Ed. Engl.* **1977**, *16*, 835-851 [*Angew. Chem. 89*, 876-890]; Stark; Duke, Ref. 444, pp. 116-151. For a review of the formation and fragmentation of cyclic azo compounds, see Mackenzie; in Patai *The Chemistry of the Hydrazo, Azo, and Azoxy Groups*, pt. 1; Wiley: New York, 1975, pp. 329-442.
[448]For example, see Jones; Sanderfer; Baarda *J. Org. Chem.* **1967**, *32*, 1367.
[449]McGreer; Wai; Carmichael *Can. J. Chem.* **1960**, *38*, 2410; Kocsis; Ferrini; Arigoni; Jeger *Helv. Chim. Acta* **1960**, *43*, 2178.
[450]For a review, see Scheiner *Sel. Org. Transform.* **1970**, *1*, 327-362.
[451]For a review of 3*H*-pyrazoles, see Sammes; Katritsky *Adv. Heterocycl. Chem.* **1983**, *34*, 2-52.
[452]Closs; Böll *J. Am. Chem. Soc.* **1963**, *85*, 3904, *Angew. Chem. Int. Ed. Engl.* **1963**, *2*, 399 [*Angew. Chem. 75*, 640]; Ege *Tetrahedron Lett.* **1963**, 1667; Closs; Böll; Heyn; Dev *J. Am. Chem. Soc.* **1968**, *90*, 173; Franck-Neumann; Buchecker *Tetrahedron Lett.* **1969**, 15; Pincock; Morchat; Arnold *J. Am. Chem. Soc.* **1973**, *95*, 7536.
[453]For a review of the mechanism; see Engel *Chem. Rev.* **1980**, *80*, 99-150. See also Engel; Nalepa *Pure Appl. Chem.* **1980**, *52*, 2621; Engel; Gerth *J. Am. Chem. Soc.* **1983**, *105*, 6849; Reedich; Sheridan *J. Am. Chem. Soc.* **1988**, *110*, 3697.
[454]Closs; Böll; Heyn; Dev, Ref. 452; Pincock; Morchat; Arnold, Ref. 452.

Though the reaction is not general, certain cyclic ketones can be photolyzed to give ring contracted products.[455] In the example above, the tetracyclic ketone **56** was photolyzed to give **57**.[456] This reaction was used to synthesize tetra-*t*-butyltetrahedrane:[457]

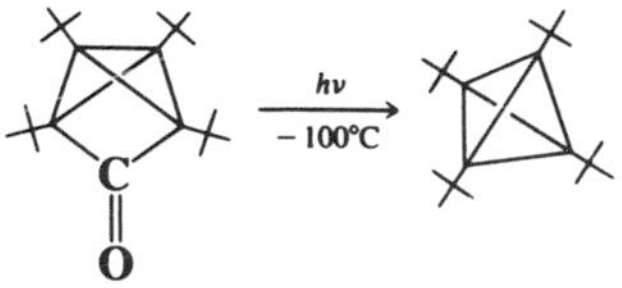

The mechanism probably involves a Norrish type I cleavage (p. 243), loss of CO from the resulting radical, and recombination of the radical fragments.

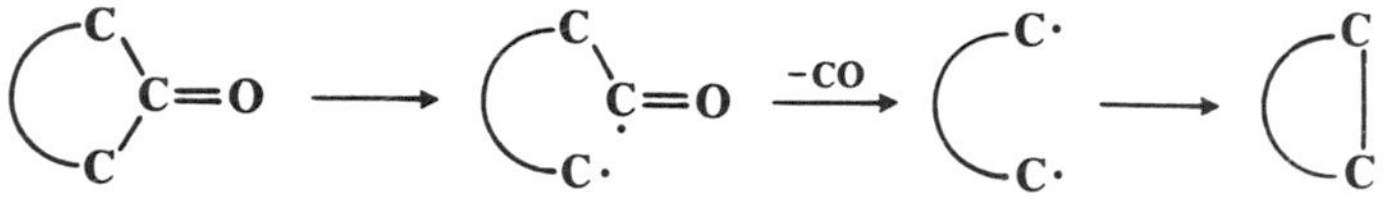

Certain lactones extrude CO_2 on heating or on irradiation, examples being pyrolysis of **58**,[458]

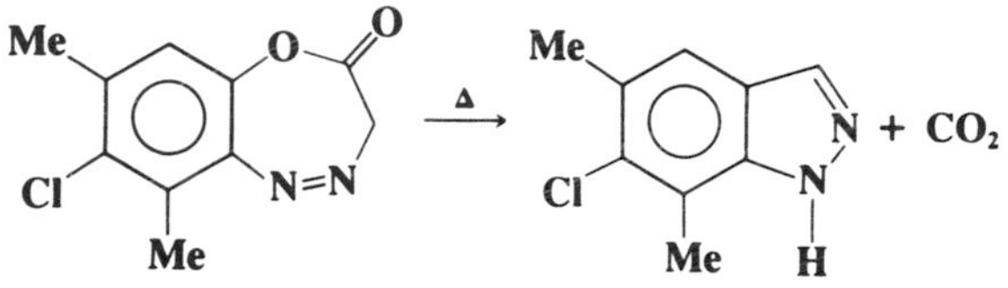

and the formation of α-lactones by photolysis of 1,2-dioxolane-3,5-diones.[459]

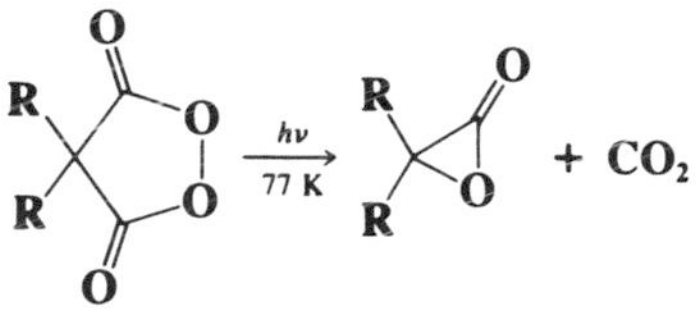

Decarboxylation of β-lactones (see **7-34**) may be regarded as a degenerate example of this reaction. Unsymmetrical diacyl peroxides RCO—OO—COR′ lose two molecules of CO_2 when photolyzed in the solid state to give the product RR′.[460] Electrolysis was also used, but yields were lower. This is an alternative to the Kolbe reaction (**4-38**). See also **7-36** and **7-51**.

There are no OS references, but see OS **VI**, 418, for a related reaction.

[455]For reviews of the reactions in this section, see Redmore; Gutsche *Adv. Alicyclic Chem.* **1971**, *3*, 1-138, pp. 91-107; Stark; Duke, Ref. 444, pp. 47-71.

[456]Cava; Mangold *Tetrahedron Lett.* **1964**, 1751.

[457]Maier; Pfriem; Schäfer; Matusch *Angew. Chem. Int. Ed. Engl.* **1978**, *17*, 520 [*Angew. Chem. 90*, 552].

[458]Ried; Dietrich *Angew. Chem. Int. Ed. Engl.* **1963**, *2*, 323 [*Angew. Chem. 75*, 476]; Ried; Wagner *Liebigs Ann. Chem.* **1965**, *681*, 45.

[459]Chapman; Wojtkowski; Adam; Rodriquez; Rucktäschel *J. Am. Chem. Soc.* **1972**, *94*, 1365.

[460]Feldhues; Schäfer *Tetrahedron* **1985**, *41*, 4195, 4213, **1986**, *42*, 1285; Lomölder; Schäfer *Angew. Chem. Int. Ed. Engl.* **1987**, *26*, 1253 [*Angew. Chem. 99*, 1282].

-48 Extrusion of SO_2
Sulfonyl-extrusion

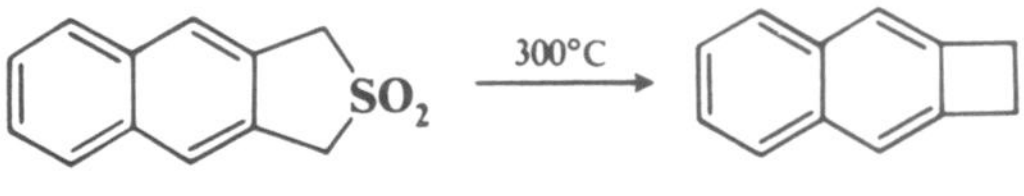

In a reaction similar to **7-47,** certain sulfones, both cyclic and acylic,[461] extrude SO_2 on heating or photolysis to give ring-contracted products.[462] An example is the preparation of naphtho(*b*)cyclobutene shown above.[463] In a different kind of reaction, five-membered cyclic sulfones can be converted to cyclobutenes by treatment with butyllithium followed by $LiAlH_4$,[464] e.g.,

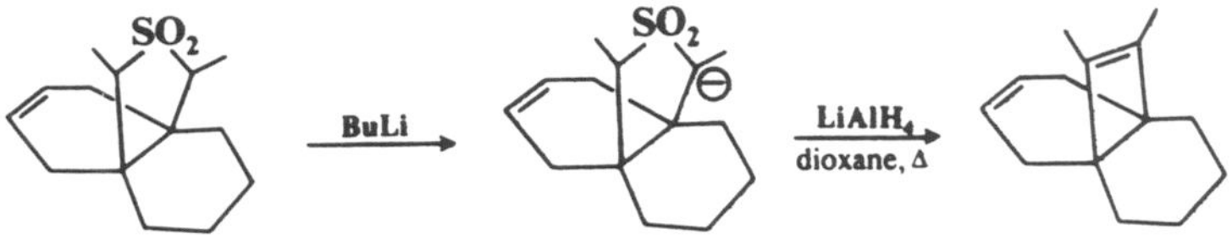

This method is most successful when both the α and α' position of the sulfone bear alkyl substituents. See also **7-25.**

OS **VI,** 482.

7-49 The Story Synthesis

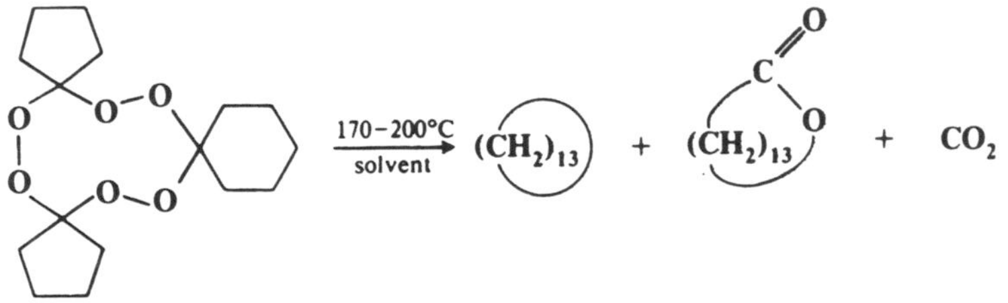

59

When cycloalkylidene peroxides (e.g., **59**) are heated in an inert solvent (e.g., decane), extrusion of CO_2 takes place; the products are the cycloalkane containing three carbon atoms less than the starting peroxide and the lactone containing two carbon atoms less[465] (the *Story synthesis*).[466] The two products are formed in comparable yields, usually about 15 to 25% each. Although the yields are low, the reaction is useful because there are not many other ways to prepare large rings. The reaction is versatile, having been used to prepare rings of every size from 8 to 33 members. The method is also applicable to dimeric

[461]See, for example, Gould; Tung; Turro; Givens; Matuszewski *J. Am. Chem. Soc.* **1984,** *106,* 1789.
[462]For reviews of extrusions of SO_2, see Vögtle; Rossa *Angew. Chem. Int. Ed. Engl.* **1979,** *18,* 515-529 [*Angew. Chem. 91,* 534-549]; Stark; Duke, Ref. 444, pp. 72-90; Kice, in Kharasch; Meyers, Ref. 302, pp. 115-136. For a review of extrusion reactions of S, Se, and Te compounds, see Guziec; SanFilippo *Tetrahedron* **1988,** *44,* 6241-6285.
[463]Cava; Shirley *J. Am. Chem. Soc.* **1960,** *82,* 654.
[464]Photis; Paquette *J. Am. Chem. Soc.* **1974,** *96,* 4715.
[465]Story; Denson; Bishop; Clark; Farine *J. Am. Chem. Soc.* **1968,** *90,* 817; Sanderson; Story; Paul *J. Org. Chem.* **1975,** *40,* 691; Sanderson; Paul; Story *Synthesis* **1975,** 275.
[466]For a review, see Story; Busch *Adv. Org. Chem.* **1972,** *8,* 67-95, pp. 79-94.

cycloalkylidene peroxides, in which case the cycloalkane and lactone products result from loss of two molecules and one molecule of CO_2, respectively, e.g.,

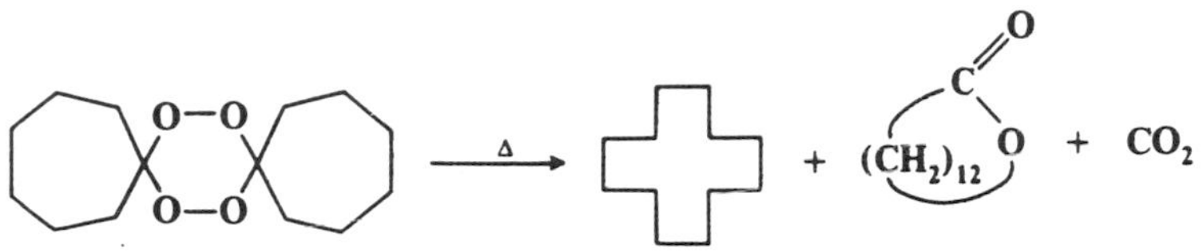

Both dimeric and trimeric cycloalkylidene peroxides can be synthesized[467] by treatment of the corresponding cyclic ketones with H_2O_2 in acid solution.[468] The trimeric peroxide is formed first and is subsequently converted to the dimeric compound.[469]

7-50   Formation of β-Dicarbonyl Compounds by Extrusion of Sulfur
Thio-extrusion

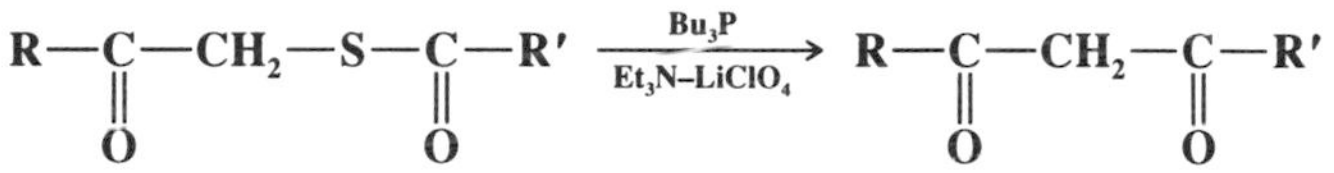

Thioesters containing a β keto group in the alkyl portion can be converted to β-diketones by treatment with a tertiary phosphine under basic conditions.[470] The starting thioesters can be prepared by the reaction between a thiol acid and an α-halo ketone (similar to **0-24**).
OS **VI**, 776.

7-51   Olefin Synthesis by Twofold Extrusion
Carbon dioxide,thio-extrusion

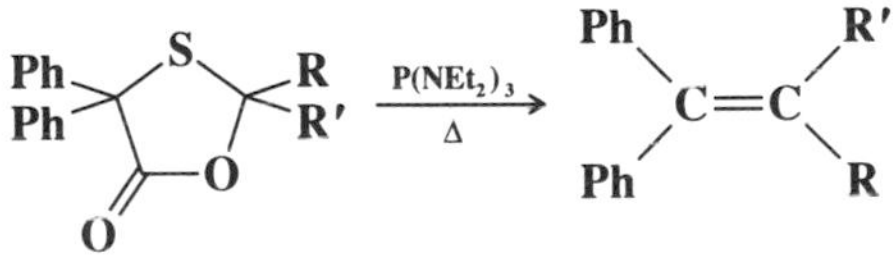

60

4,4-Diphenyloxathiolan-5-ones (**60**) give good yields of the corresponding olefins when heated with tris(diethylamino)phosphine.[471] This reaction is an example of a general type:

[467]For synthesis of mixed trimeric peroxides (e.g., **59**), see Sanderson; Zeiler *Synthesis* **1975**, 388; Paul; Story; Busch; Sanderson *J. Org. Chem.* **1976**, *41*, 1283.
[468]Kharasch; Sosnovsky *J. Org. Chem.* **1958**, *23*, 1322; Ledaal *Acta Chem. Scand.* **1967**, *21*, 1656. For another method, see Sanderson; Zeiler *Synthesis* **1975**, 125.
[469]Story; Lee; Bishop; Denson; Busch *J. Org. Chem.* **1970**, *35*, 3059. See also Sanderson; Wilterdink; Zeiler *Synthesis* **1976**, 479.
[470]Roth; Dubs; Götschi; Eschenmoser *Helv. Chim. Acta* **1971**, *54*, 710. For a review of thio-extrusion, see Williams; Harpp *Sulfur Rep.* **1990**, *10*, 103-191.
[471]Barton; Willis *J. Chem. Soc., Perkin Trans. 1* **1972**, 305.

olefin synthesis by twofold extrusion of X and Y from a molecule of the type **61**.[472] Other examples are photolysis of 1,4-diones[473] (e.g., **62**) and treatment with Ph_3P of the azo sulfide

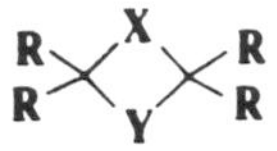

63.[474] **60** can be prepared by the condensation of thiobenzilic acid $Ph_2C(SH)COOH$ with aldehydes or ketones.

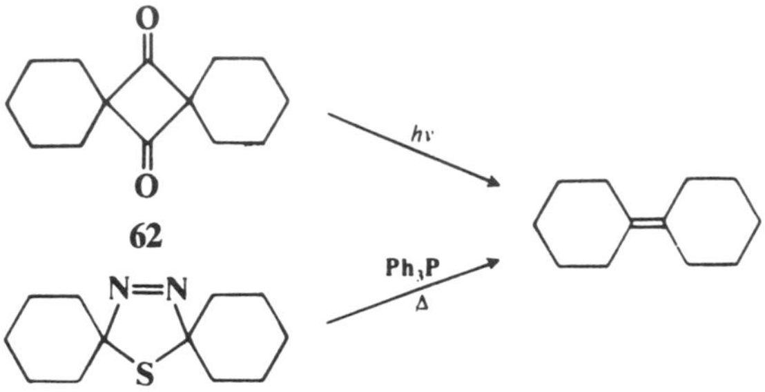

OS **V**, 297.

[472]For a review of those in which X or Y contains S, Se, or Te, see Guziec; SanFilippo, Ref. 462.

[473]Turro; Leermakers; Wilson; Neckers; Byers; Vesley *J. Am. Chem. Soc.* **1965**, *87*, 2613.

[474]Barton; Smith; Willis *Chem. Commun.* **1970**, 1226; Barton; Guziec; Shahak *J. Chem. Soc., Perkin Trans. 1* **1974**, 1794. See also Bee; Beeby; Everett; Garratt *J. Org. Chem.* **1975**, *40*, 2212; Back; Barton; Britten-Kelly; Guziec *J. Chem. Soc., Perkin Trans. 1* **1976**, 2079; Guziec; Moustakis *J. Chem. Soc., Chem. Commun.* **1984**, 63.

18

REARRANGEMENTS

In a rearrangement reaction a group moves from one atom to another in the same molecule.[1] Most are migrations from an atom to an adjacent one (called 1,2 shifts), but some are over

$$
\begin{array}{ccc}
W & & W \\
| & & | \\
A\!-\!B & \longrightarrow & A\!-\!B
\end{array}
$$

longer distances. The migrating group (W) may move with its electron pair (these can be called *nucleophilic* or *anionotropic* rearrangements; the migrating group can be regarded as a nucleophile), without its electron pair (*electrophilic* or *cationotropic* rearrangements; in the case of migrating hydrogen, *prototropic* rearrangements), or with just one electron (free-radical rearrangements). The atom A is called the *migration origin* and B is the *migration terminus*. However, there are some rearrangements that do not lend themselves to neat categorization in this manner. Among these are those with cyclic transition states (**8-29** to **8-38**).

As we shall see, nucleophilic 1,2 shifts are much more common than electrophilic or free-radical 1,2 shifts. The reason for this can be seen by a consideration of the transition states (or in some cases intermediates) involved. We represent the transition state or intermediate for all three cases by **1**, in which the two-electron A—W bond overlaps with the

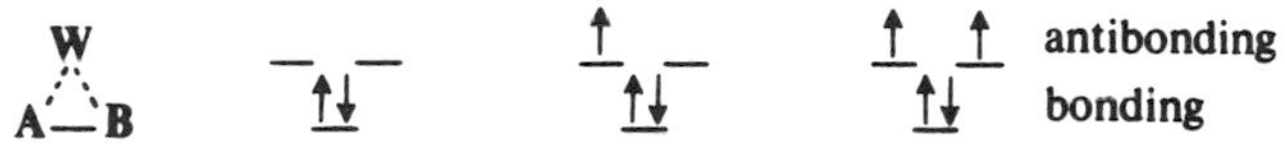

orbital on atom B, which contains zero, one, and two electrons, in the case of nucleophilic, free-radical, and electrophilic migration, respectively. The overlap of these orbitals gives rise to three new orbitals, which have an energy relationship similar to those on p. 52 (one bonding and two degenerate antibonding orbitals). In a nucleophilic migration, where only two electrons are involved, both can go into the bonding orbital and **1** is a low-energy transition state; but in a free-radical or electrophilic migration, there are, respectively, three or four electrons that must be accommodated, and antibonding orbitals must be occupied. It is not surprising therefore that, when 1,2-electrophilic or free-radical shifts are found, the migrating group W is usually aryl or some other group that can accommodate the extra one or two electrons and thus effectively remove them from the three-membered transition state or intermediate (see **37** on p. 1065).

In any rearrangement we can in principle distinguish between two possible modes of reaction: In one of these the group W becomes completely detached from A and may end

[1]For books, see Mayo *Rearrangements in Ground and Excited States*, 3 vols.; Academic Press: New York, 1980; Stevens; Watts *Selected Molecular Rearrangements*; Van Nostrand-Reinhold: Princeton, 1973. For a review of many of these rearrangements, see Collins; Eastham, in Patai *The Chemistry of the Carbonyl Group*, vol. 1; Wiley: New York, 1966, pp. 761-821. See also the series *Mechanisms of Molecular Migrations*.

up on the B atom of a different molecule (*intermolecular* rearrangement); in the other W goes from A to B in the *same* molecule (*intramolecular* rearrangement), in which case there must be some continuing tie holding W to the A—B system, preventing it from coming completely free. Strictly speaking, only the intramolecular type fits our definition of a rearrangement, but the general practice, which is followed here, is to include under the title "rearrangement" all net rearrangements whether they are inter- or intramolecular. It is usually not difficult to tell whether a given rearrangement is inter- or intramolecular. The most common method involves the use of *crossover* experiments. In this type of experiment, rearrangement is carried out on a mixture of W—A—B and V—A—C, where V is closely related to W (say, methyl vs. ethyl) and B to C. In an intramolecular process only A—B—W and A—C—V are recovered, but if the reaction is intermolecular, then not only will these two be found, but also A—B—V and A—C—W.

MECHANISMS

Nucleophilic Rearrangements[2]

Broadly speaking, such rearrangements consist of three steps, of which the actual migration is the second:

This process is sometimes called the Whitmore 1,2 shift.[3] Since the migrating group carries the electron pair with it, the migration terminus B must be an atom with only six electrons in its outer shell (an open sextet). The first step therefore is creation of a system with an open sextet. Such a system can arise in various ways, but two of these are the most important:

   1. *Formation of a carbocation.*   These can be formed in a number of ways (see p. 173), but one of the most common methods when a rearrangement is desired is the acid treatment of an alcohol:

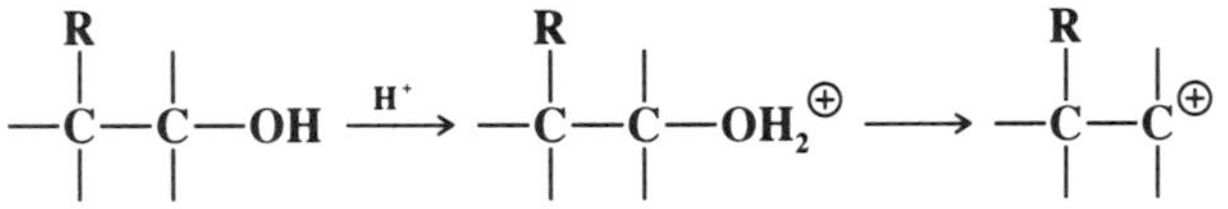

These two steps are of course the same as the first two steps of the S_N1cA or the E1 reactions of alcohols.

   2. *Formation of a nitrene.*   The decomposition of acyl azides is one of several ways in which nitrenes are formed (see p. 202):

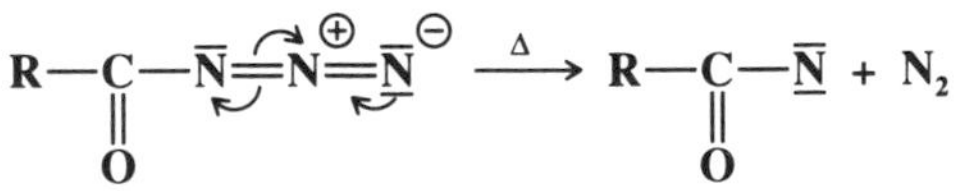

[2]For reviews, see Vogel *Carbocation Chemistry*; Elsevier: New York, 1985, pp. 323-372; Shubin *Top. Curr. Chem.* **1984,** *116/117,* 267-341; Saunders; Chandrasekhar; Schleyer, in Mayo, Ref. 1, vol. 1, pp. 1-53; Kirmse *Top. Curr. Chem.* **1979,** *80,* 89-124. For reviews of rearrangements in vinylic cations, see Shchegolev; Kanishchev *Russ. Chem. Rev.* **1981,** *50,* 553-564; Lee *Isot. Org. Chem.* **1980,** *5,* 1-44.
[3]It was first postulated by Whitmore *J. Am. Chem. Soc.* **1932,** *54,* 3274.

After the migration has taken place, the atom at the migration origin (A) must necessarily have an open sextet. In the third step this atom acquires an octet. In the case of carbocations, the most common third steps are combinations with a nucleophile (rearrangement with substitution) and loss of H⁺ (rearrangement with elimination).

Though we have presented this mechanism as taking place in three steps, and some reactions do take place in this way, in many cases two or all three steps are simultaneous. For instance, in the nitrene example above, as the R migrates, an electron pair from the nitrogen moves into the C—N bond to give a stable isocyanate:

In this example, the second and third steps are simultaneous. It is also possible for the second and third steps to be simultaneous even when the "third" step involves more than just a simple motion of a pair of electrons. Similarly, there are many reactions in which the first two steps are simultaneous; that is, there is no actual formation of a species such as **2** or **3.** In these instances it may be said that R assists in the removal of the leaving group, with migration of R and the removal of the leaving group taking place simultaneously. Many investigations have been carried out in attempts to determine, in various reactions, whether such intermediates as **2** or **3** actually form, or whether the steps are simultaneous (see, for example, the discussions on pp. 1055, 1090), but the difference between the two possibilities is often subtle, and the question is not always easily answered.[4]

Evidence for this mechanism is that rearrangements of this sort occur under conditions where we have previously encountered carbocations: SN1 conditions, Friedel–Crafts alkylation, etc. Solvolysis of neopentyl bromide leads to rearrangement products, and the rate increases with increasing ionizing power of the solvent but is unaffected by concentration of base,[5] so that the first step is carbocation formation. The same compound under SN2 conditions gave no rearrangement, but only ordinary substitution, though slowly. Thus with neopentyl bromide, formation of a carbocation leads only to rearrangement. Carbocations usually rearrange to more stable carbocations. Thus the direction of rearrangement is usually primary → secondary → tertiary. Neopentyl (Me_3CCH_2), neophyl ($PhCMe_2CH_2$), and norbornyl (e.g., **4**) type systems are especially prone to carbocation rearrangement reactions.

4

It has been shown that the rate of migration increases with the degree of electron deficiency at the migration terminus.[6]

We have previously mentioned (p. 166) that stable tertiary carbocations can be obtained, in solution, at very low temperatures. Nmr studies have shown that when these solutions are warmed, rapid migrations of hydride and of alkyl groups take place, resulting in an

[4]The IUPAC designations depend on the nature of the steps. For the rules, see Guthrie *Pure Appl. Chem.* **1989,** *61*, 23-56, pp. 44-45.
[5]Dostrovsky; Hughes *J. Chem. Soc.* **1946, 166.**
[6]Borodkin; Shakirov; Shubin; Koptyug *J. Org. Chem. USSR* **1976,** *12*, 1293, 1298, **1978,** *14*, 290, 924.

equilibrium mixture of structures.[7] For example, the *t*-pentyl cation (**5**)[8] equilibrates as follows:

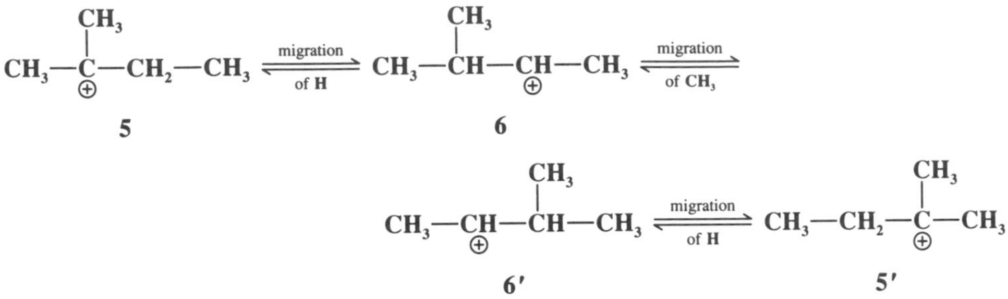

Carbocations that rearrange to give products of identical structure(e.g., **5** ⇌ **5'**, **6** ⇌ **6'**) are called *degenerate carbocations* and such rearrangements are *degenerate rearrangements*. Many examples are known.[9]

The Actual Nature of the Migration

Most nucleophilic 1,2 shifts are intramolecular. W does not become free but always remains connected in some way to the substrate. Apart from the evidence from crossover experiments, the strongest evidence is that when the group W is chiral, the configuration is *retained* in the product. For example, (+)-PhCHMeCOOH was converted to (−)-PhCHMeNH$_2$ by the Curtius (**8-15**), Hofmann (**8-14**), Lossen (**8-16**), and Schmidt (**8-17**) reactions.[10] In these reactions the extent of retention varied from 95.8 to 99.6%. Retention of configuration in the migrating group has been shown many times since.[11] Another experiment demonstrating

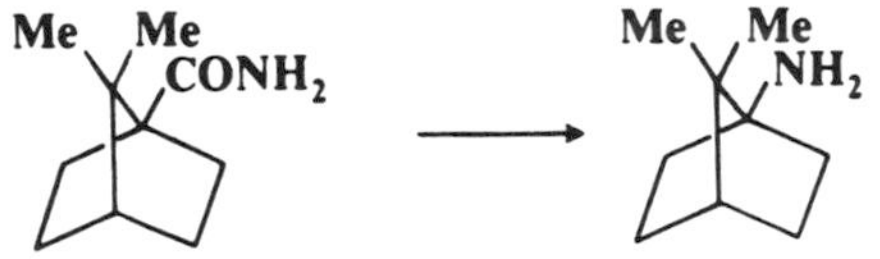

retention was the easy conversion of **7** to **8**.[11] Neither inversion nor racemization could take place at a bridgehead. There is much other evidence that retention of configuration usually occurs in W, and inversion never.[13] However, this is not the state of affairs at A and B. In

[7]For reviews, see Brouwer; Hogeveen *Prog. Phys. Org. Chem.* **1972**, *9*, 179-240, pp. 203-237; Olah; Olah, in Olah; Schleyer *Carbonium Ions*, vol. 2; Wiley: New York, 1970, pp. 751-760, 766-778. For a discussion of the rates of these reactions, see Sorensen *Acc. Chem. Res.* **1976**, *9*, 257-265.

[8]Brouwer *Recl. Trav. Chim. Pays-Bas* **1968**, *87*, 210; Saunders; Hagen *J. Am. Chem. Soc.* **1968**, *90*, 2436.

[9]For reviews, see Ahlberg; Jonsäll; Engdahl *Adv. Phys. Org. Chem.* **1983**, *19*, 223-379; Leone; Barborak; Schleyer, in Olah; Schleyer, Ref. 7, vol. 4, pp. 1837-1939; Leone; Schleyer *Angew. Chem. Int. Ed. Engl.* **1970**, *9*, 860-890 [*Angew. Chem. 82*, 889-919].

[10]Arcus; Kenyon *J. Chem. Soc.* **1939**, 916; Kenyon; Young *J. Chem. Soc.* **1941**, 263; Campbell; Kenyon *J. Chem. Soc.* **1946**, 25.

[11]For retention of migrating group configuration in the Wagner–Meerwein and pinacol rearrangements, see Beggs; Meyers *J. Chem. Soc. B* **1970**, 930; Kirmse; Gruber; Knist *Chem. Ber.* **1973**, *106*, 1376; Shono; Fujita; Kumai *Tetrahedron Lett.* **1973**, 3123; Borodkin; Panova; Shakirov; Shubin *J. Chem. Soc., Chem. Commun.* **1979**, 354, *J. Org. Chem. USSR* **1983**, *19*, 103.

[12]Barlett; Knox *J. Am. Chem. Soc.* **1939**, *61*, 3184.

[13]See Cram, in Newman *Steric Effects in Organic Chemistry*; Wiley: New York, 1956; pp. 251-254; Wheland *Advanced Organic Chemistry*, 3rd ed.; Wiley: New York, 1960, pp. 597-604.

many reactions, of course, the structure of W—A—B is such that the product has only one steric possibility at A or B or both, and in most of these cases nothing can be learned. But in cases where the steric nature of A or B can be investigated, the results are mixed. It has been shown that either inversion or racemization can occur at A or B. Thus the following conversion proceeded with inversion at B:[14]

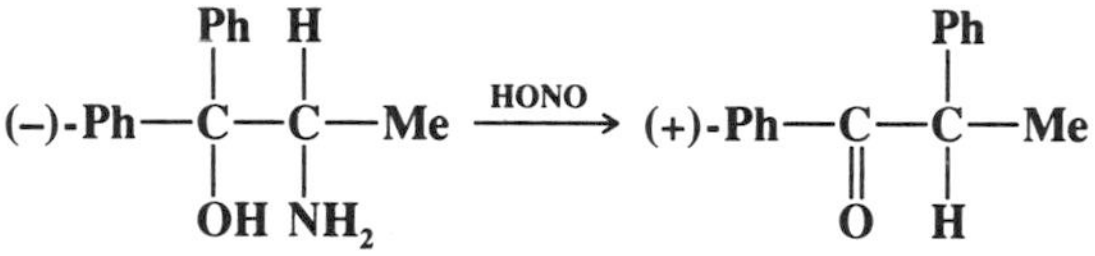

$$(-)\text{-Ph}-\overset{\overset{\displaystyle Ph}{|}}{\underset{\underset{\displaystyle OH}{|}}{C}}-\overset{\overset{\displaystyle H}{|}}{\underset{\underset{\displaystyle NH_2}{|}}{C}}-Me \xrightarrow{\text{HONO}} (+)\text{-Ph}-\overset{\overset{\displaystyle Ph}{|}}{\underset{\underset{\displaystyle O}{\|}}{C}}-\overset{\overset{\displaystyle }{|}}{\underset{\underset{\displaystyle H}{|}}{C}}-Me \qquad (8\text{-}2)$$

and inversion at A has been shown in other cases.[15] However, in many other cases, racemization occurs at A or B or both.[16] It is not always necessary for the product to have two steric possibilities in order to investigate the stereochemistry at A or B. Thus, in most Beckmann rearrangements (**8-18**), only the group trans (usually called *anti*) to the hydroxyl group migrates:

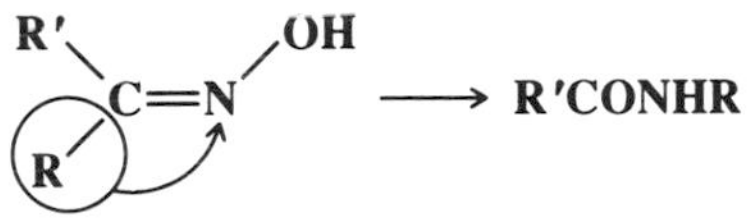

$$\underset{R'}{\overset{R'}{\diagdown}}C=N\overset{\diagup OH}{} \longrightarrow R'CONHR$$

showing inversion at B.

This information tells us about the degree of concertedness of the three steps of the rearrangement. First consider the migration terminus B. If racemization is found at B, it is probable that the first step takes place before the second and that a positively charged carbon (or other sextet atom) is present at B:

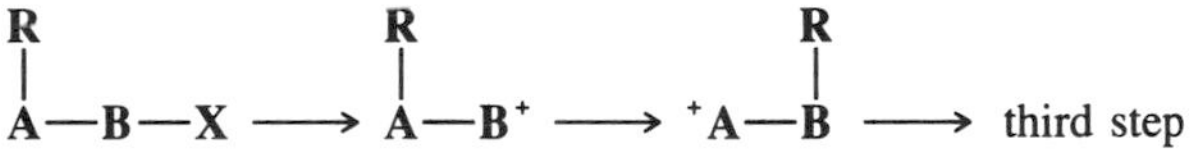

$$\overset{\overset{\displaystyle R}{|}}{A}-B-X \longrightarrow \overset{\overset{\displaystyle R}{|}}{A}-B^+ \longrightarrow {}^+\overset{\overset{\displaystyle R}{|}}{A}-B \longrightarrow \text{third step}$$

With respect to B this is an S_N1-type process. If inversion occurs at B, it is likely that the first two steps are concerted, that a carbocation is *not* an intermediate, and that the process is S_N2-like:

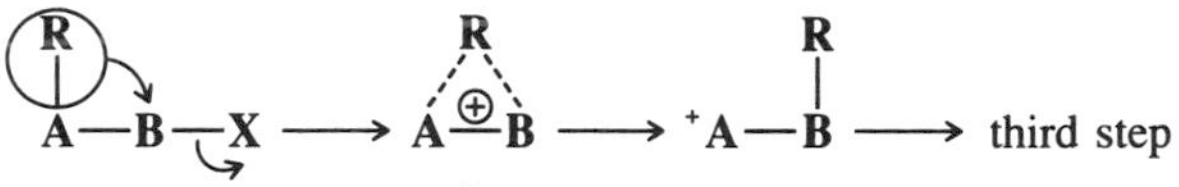

$$\overset{\overset{\displaystyle R}{\downarrow}}{A}-B\overset{\curvearrowleft}{}X \longrightarrow \overset{\overset{\displaystyle R}{\diagup\oplus\diagdown}}{A\text{---}B} \longrightarrow {}^+\overset{\overset{\displaystyle R}{|}}{A}-B \longrightarrow \text{third step}$$

$$\mathbf{9}$$

In this case participation by R assists in removal of X in the same way that neighboring groups do (p. 309). Indeed, R *is* a neighboring group here. The only difference is that, in the case of the neighboring-group mechanism of nucleophilic substitution, R never becomes detached from A, while in a rearrangement the bond between R and A is broken. In either

[14]Bernstein; Whitmore *J. Am. Chem. Soc.* **1939,** *61,* 1324. For other examples, see Tsuchihashi; Tomooka; Suzuki *Tetrahedron Lett.* **1984,** *25,* 4253.

[15]See Meerwein; van Emster *Ber.* **1920,** *53,* 1815, **1922,** *55,* 2500; Meerwein; Gérard *Liebigs Ann. Chem.* **1923,** *435,* 174.

[16]For example, see Winstein; Morse *J. Am. Chem. Soc.* **1952,** *74,* 1133.

case, the anchimeric assistance results in an increased rate of reaction. Of course, for such a process to take place, R must be in a favorable geometrical position (R and X antiperiplanar). **9** may be a true intermediate or only a transition state, depending on what migrates. In certain cases of the SN1-type process, it is possible for migration to take place with net retention of configuration at the migrating terminus because of conformational effects in the carbocation.[17]

We may summarize a few conclusions:

1. The SN1-type process occurs mostly when B is a tertiary atom or has one aryl group and at least one other alkyl or aryl group. In other cases, the SN2-type process is more likely. Inversion of configuration (indicating an SN2-type process) has been shown for a neopentyl substrate by the use of the chiral neopentyl-1-*d* alcohol.[18] On the other hand, there is other evidence that neopentyl systems undergo rearrangement by a carbocation (SN1-type) mechanism.[19]

2. The question as to whether **9** is an intermediate or a transition state has been much debated. When R is aryl or vinyl, then **9** is probably an intermediate and the migrating group lends anchimeric assistance[20] (see p. 319 for resonance stabilization of this intermediate when R is aryl). When R is alkyl, **9** is a protonated cyclopropane (edge- or corner-protonated; see p. 757). There is much evidence that in simple migrations of a methyl group, the bulk of the products formed do not arise from protonated cyclopropane *intermediates*. Evidence for this statement has already been given (p. 325). Further evidence was obtained from experiments involving labeling. Rearrangement of the neopentyl cation labeled with deuterium in the 1 position (**10**) gave only *t*-pentyl products with the label in the 3 position

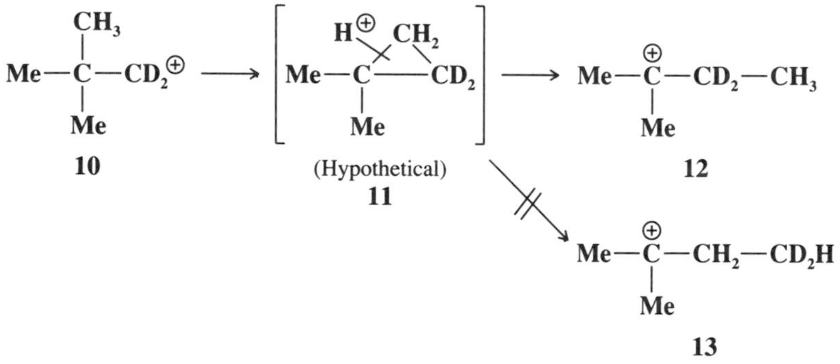

$$
\underset{\substack{|\\ \textbf{Me}\\ \textbf{10}}}{\overset{\substack{\textbf{CH}_3\\|}}{\textbf{Me}-\textbf{C}-\textbf{CD}_2^{\oplus}}} \longrightarrow \left[\underset{\substack{|\\ \textbf{Me}\\ \text{(Hypothetical)}\\ \textbf{11}}}{\overset{\textbf{H}^{\oplus}\diagdown\textbf{CH}_2}{\textbf{Me}-\textbf{C}--\textbf{CD}_2}} \right] \longrightarrow \underset{\substack{|\\ \textbf{Me}\\ \textbf{12}}}{\overset{\oplus}{\textbf{Me}-\textbf{C}-\textbf{CD}_2-\textbf{CH}_3}}
$$

$$
\underset{\substack{|\\ \textbf{Me}\\ \textbf{13}}}{\overset{\oplus}{\textbf{Me}-\textbf{C}-\textbf{CH}_2-\textbf{CD}_2\textbf{H}}}
$$

(derived from **12**), though if **11** were an intermediate, the cyclopropane ring could just as well cleave the other way to give *t*-pentyl derivatives labeled in the 4 position (derived from **13**).[21] Another experiment that led to the same conclusion was the generation, in several ways, of $Me_3C^{13}CH_2^+$. In this case the only *t*-pentyl products isolated were labeled in C-3, that is, $Me_2\overset{\oplus}{C}-{}^{13}CH_2CH_3$ derivatives; no derivatives of $Me_2\overset{\oplus}{C}-CH_2{}^{13}CH_3$ were found.[22]

[17]Benjamin; Collins *J. Am. Chem. Soc.* **1961**, *83*, 3662; Collins; Staum; Benjamin *J. Org. Chem.* **1962**, *27*, 3525; Collins; Benjamin *J. Org. Chem.* **1972**, *37*, 4358.

[18]Sanderson; Mosher *J. Am. Chem. Soc.* **1966**, *88*, 4185; Mosher *Tetrahedron* **1974**, *30*, 1733. See also Guthrie, *J. Am. Chem. Soc.* **1967**, *89*, 6718.

[19]Nordlander; Jindal; Schleyer; Fort; Harper; Nicholas *J. Am. Chem. Soc.* **1966**, *88*, 4475; Shiner; Imhoff *J. Am. Chem. Soc.* **1985**, *107*, 2121.

[20]For example, see Rachon; Goedkin; Walborsky *J. Org. Chem.* **1989**, *54*, 1006. For an opposing view, see Kirmse; Feyen *Chem. Ber.* **1975**, *108*, 71; Kirmse; Plath; Schaffrodt *Chem. Ber.* **1975**, *108*, 79.

[21]Skell; Starer; Krapcho *J. Am. Chem. Soc.* **1960**, *82*, 5257.

[22]Karabastos; Graham *J. Am. Chem. Soc.* **1960**, *82*, 5250; Karabatsos; Orzech; Meyerson *J. Am. Chem. Soc.* **1964**, *86*, 1994.

Though the bulk of the products are not formed from protonated cyclopropane intermediates, there is considerable evidence that at least in 1-propyl systems, a small part of the product can in fact arise from such intermediates.[23] Among this evidence is the isolation of 10 to 15% cyclopropanes (mentioned on p. 325). Additional evidence comes from propyl cations generated by diazotization of labeled amines ($CH_3CH_2CD_2^+$, $CH_3CD_2CH_2^+$, $CH_3CH_2^{14}CH_2^+$), where isotopic distribution in the products indicated that a small amount (about 5%) of the product had to be formed from protonated cyclopropane intermediates, e.g.,[24]

$$CH_3CH_2CD_2NH_2 \xrightarrow{HONO} \sim 1\% \ C_2H_4D\text{—}CHD\text{—}OH$$

$$CH_3CD_2CH_2NH_2 \xrightarrow{HONO} \sim 1\% \ C_2H_4D\text{—}CHD\text{—}OH$$

$$CH_3CH_2^{14}CH_2NH_2 \xrightarrow{HONO} \sim 2\% \ {}^{14}CH_3CH_2CH_2OH + \sim 2\% \ CH_3{}^{14}CH_2CH_2OH$$

Even more scrambling was found in trifluoroacetolysis of 1-propyl-1-^{14}C-mercuric perchlorate.[25] However, protonated cyclopropane intermediates accounted for less than 1% of the products from diazotization of labeled isobutylamine[26] and from formolysis of labeled 1-propyl tosylate.[27]

It is likely that protonated cyclopropane transition states or intermediates are also responsible for certain non-1,2 rearrangements. For example, in super-acid solution, the ions **14** and **16** are in equilibrium. It is not possible for these to interconvert solely by 1,2 alkyl

or hydride shifts unless primary carbocations (which are highly unlikely) are intermediates. However, the reaction can be explained[28] by postulating that (in the forward reaction) it is the 1,2 bond of the intermediate or transition state **15** that opens up rather than the 2,3 bond, which is the one that would open if the reaction were a normal 1,2 shift of a methyl group. In this case opening of the 1,2 bond produces a tertiary cation, while opening of the 2,3 bond would give a secondary cation. (In the reaction **16 → 14**, it is of course the 1,3 bond that opens).

3. There has been much discussion of H as migrating group. There is no conclusive evidence for the viewpoint that **9** in this case is or is not a true intermediate, though both positions have been argued (see p. 325).

[23]For reviews, see Saunders; Vogel; Hagen; Rosenfeld *Acc. Chem. Res.* **1973**, *6*, 53-59; Lee *Prog. Phys. Org. Chem.* **1970**, *7*, 129-187; Collins *Chem. Rev.* **1969**, *69*, 543-550. See also Cooper; Jenner; Perry; Russell-King; Storesund; Whiting *J. Chem. Soc., Perkin Trans. 2* **1982**, 605.

[24]Lee; Kruger; Wong *J. Am. Chem. Soc.* **1965**, *87*, 3985; Lee; Kruger *J. Am. Chem. Soc.* **1965**, *87*, 3986, *Tetrahedron* **1967**, *23*, 2539; Karabatsos; Orzech; Meyerson *J. Am. Chem. Soc.* **1965**, *87*, 4394; Lee; Wan *J. Am. Chem. Soc.* **1969**, *91*, 6416; Karabatsos; Orzech; Fry; Meyerson *J. Am. Chem. Soc.* **1970**, *92*, 606.

[25]Lee; Cessna; Ko; Vassie *J. Am. Chem. Soc.* **1973**, *95*, 5688. See also Lee; Chwang *Can. J. Chem.* **1970**, *48*, 1025; Lee; Law *Can. J. Chem.* **1971**, *49*, 2746; Lee; Reichle *J. Org. Chem.* **1977**, *42*, 2058.

[26]Karabatsos; Hsi; Meyerson *J. Am. Chem. Soc.* **1970**, *92*, 621. See also Karabatsos; Anand; Rickter; Meyerson *J. Am. Chem. Soc.* **1970**, *92*, 1254.

[27]Lee; Kruger *Can. J. Chem.* **1966**, *44*, 2343; Shatkina; Lovtsova; Reutov *Bull. Acad. Sci. USSR, Div. Chem. Sci.* **1967**, 2616; Karabatsos; Fry; Meyerson *J. Am. Chem. Soc.* **1970**, *92*, 614. See also Lee; Zohdi *Can. J. Chem.* **1983**, *61*, 2092.

[28]Brouwer; Oelderik *Recl. Trav. Chim. Pays-Bas* **1968**, *87*, 721; Saunders; Jaffe; Vogel *J. Am. Chem. Soc.* **1971**, *93*, 2558; Saunders; Vogel *J. Am. Chem. Soc.* **1971**, *93*, 2559, 2561; Kirmse; Loosen; Prolingheuer *Chem. Ber.* **1980**, *113*, 129.

The stereochemistry at the migration origin A is less often involved, since in most cases it does not end up as a tetrahedral atom; but when there is inversion here, there is an SN2-type process at the beginning of the migration. This may or may not be accompanied by an SN2 process at the migration terminus B:

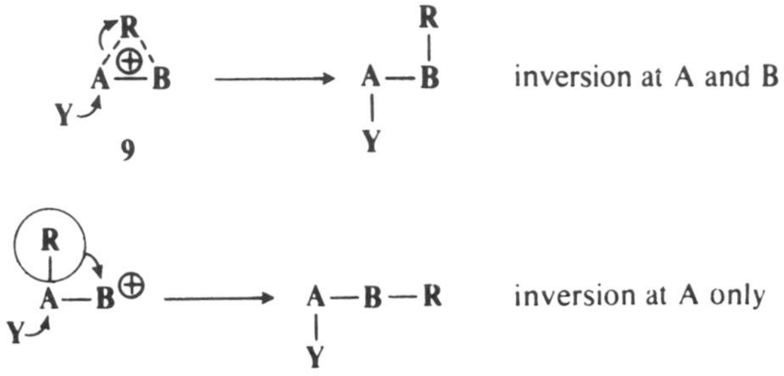

inversion at A and B

9

inversion at A only

In some cases it has been found that, when H is the migrating species, the configuration at A may be *retained*.[29]

There is evidence that the configuration of the molecule may be important even where the leaving group is gone long before migration takes place. For example, the 1-adamantyl cation (**17**) does not equilibrate intramolecularly, even at temperatures up to 130°C,[30] though open-chain (e.g., **5** ⇌ **5'**) and cyclic tertiary carbocations undergo such equilibration at 0°C

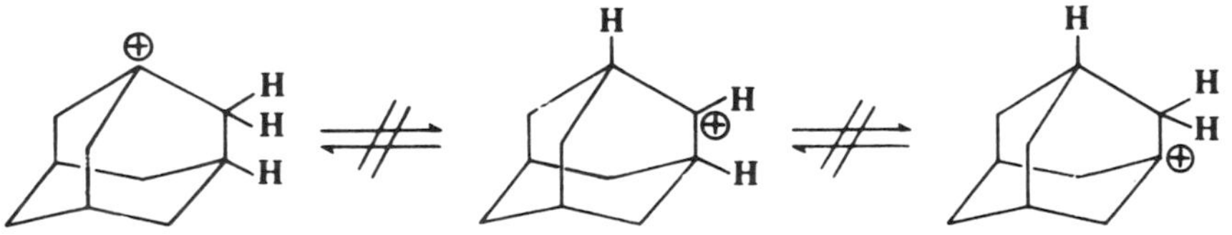

17   **17'**

or below. On the basis of this and other evidence it has been concluded that for a 1,2 shift of hydrogen or methyl to proceed as smoothly as possible, the vacant *p* orbital of the carbon bearing the positive charge and the sp^3 orbital carrying the migrating group must be coplanar,[30] which is not possible for **17**.

Migratory Aptitudes[31]

In many reactions there is no question about which group migrates. For example, in the Hofmann, Curtius, and similar reactions there is only one possible migrating group in each molecule, and one can measure migratory aptitudes only by comparing the relative rearrangement rates of different compounds. In other instances there are two or more potential migrating groups, but which migrates is settled by the geometry of the molecule. The Beckmann rearrangement (**8-18**) provides an example. As we have seen, only the group

[29]Winstein; Holness *J. Am. Chem. Soc.* **1955**, *77*, 5562; Cram; Tadanier *J. Am. Chem. Soc.* **1959**, *81*, 2737; Bundel'; Pankratova; Gordin; Reutov *Doklad. Chem.* **1971**, *199*, 700; Kirmse; Arold *Chem. Ber.* **1971**, *104*, 1800; Kirmse; Ratajczak; Rauleder *Chem. Ber.* **1977**, *110*, 2290.

[30]Brouwer; Hogeveen *Recl. Trav. Chim. Pays-Bas* **1970**, *89*, 211; Majerski; Schleyer; Wolf *J. Am. Chem. Soc.* **1970**, *92*, 5731.

[31]For discussions, see Koptyug; Shubin *J. Org. Chem. USSR* **1980**, *16*, 1685-1714; Wheland, Ref. 13, pp. 573-597.

trans to the OH migrates. In compounds whose geometry is not restricted in this manner, there still may be eclipsing effects (see p. 1002), so that the choice of migrating group is largely determined by which group is in the right place in the most stable conformation of the molecule.[32] However, in some reactions, especially the Wagner–Meerwein (**8-1**) and the pinacol (**8-2**) rearrangements, the molecule may contain several groups that, geometrically at least, have approximately equal chances of migrating, and these reactions have often been used for the direct study of relative migratory aptitudes. In the pinacol rearrangement there is the additional question of which OH group leaves and which does not, since a group can migrate only if the OH group on *the other* carbon is lost.

We deal with the second question first. To study this question, the best type of substrate to use is one of the form R_2C—CR'_2, since the only thing that determines migratory aptitude

$$\underset{\text{OH OH}}{R_2C\text{---}CR'_2}$$

is which OH group comes off. Once the OH group is gone, the migrating group is determined. As might be expected, the OH that leaves is the one whose loss gives rise to the more stable carbocation. Thus 1,1-diphenylethanediol (**18**) gives diphenylacetaldehyde (**19**), not phen-

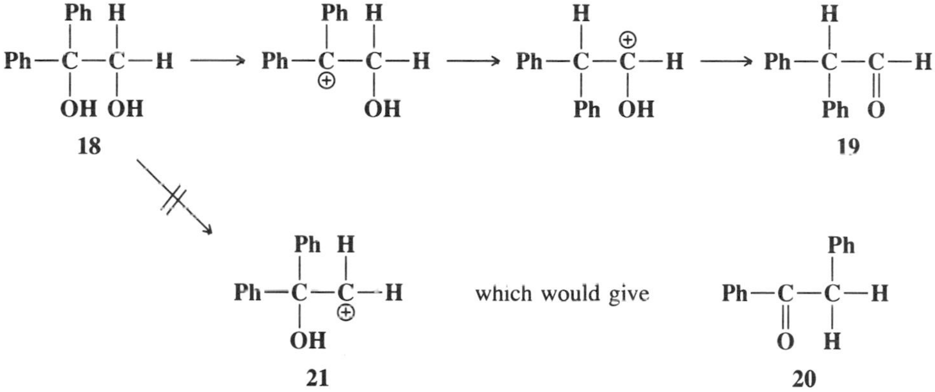

ylacetophenone (**20**). Obviously, it does not matter in this case whether phenyl has a greater inherent migratory aptitude than hydrogen or not. Only the hydrogen can migrate because **21** is not formed. As we know, carbocation stability is enhanced by groups in the order aryl > alkyl > hydrogen, and this normally determines which side loses the OH group. However, exceptions are known, and which group is lost may depend on the reaction conditions (for an example, see the reaction of **41**, p. 1073).

In order to answer the question about inherent migratory aptitudes, the obvious type of substrate to use (in the pinacol rearrangement) is $RR'C$—CRR', since the same carbocation

$$\underset{\text{OH OH}}{RR'C\text{---}CRR'}$$

is formed no matter which OH leaves, and it would seem that a direct comparison of the migratory tendencies of R and R' is possible. On closer inspection, however, we can see that several factors are operating. Apart from the question of possible conformational effects, already mentioned, there is also the fact that whether the group R or R' migrates is determined not only by the relative inherent migrating abilities of R and R' but also by whether the group that does *not* migrate is better at stabilizing the positive charge that will now be found at the migration origin.[33] Thus, migration of R gives rise to the cation

---

[32]For a discussion, see Cram, Ref. 13, pp. 270-276. For an interesting example, see Nickon; Weglein *J. Am. Chem. Soc.* **1975**, *97*, 1271.

[33]For example, see Howells; Warren *J. Chem. Soc., Perkin Trans. 2* **1973**, 1645; McCall; Townsend; Bonner *J. Am. Chem. Soc.* **1975**, *97*, 2743; Brownbridge; Hodgson; Shepherd; Warren *J. Chem. Soc., Perkin Trans. 1* **1976**, 2024.

$R'\overset{\oplus}{C}(OH)CR_2R'$, while migration of R' gives the cation $R\overset{\oplus}{C}(OH)CRR_2'$ and these cations have different stabilities. It is possible that in a given case R might be found to migrate less than R', not because it actually has a lower inherent migrating tendency, but because it is much better at stabilizing the positive charge. In addition to this factor, migrating ability of a group is also related to its capacity to render anchimeric assistance to the departure of the nucleofuge. An example of this effect is the finding that in the decomposition of the tosylate **22** only the phenyl group migrates, while in acid treatment of the corresponding

$$\underset{\textbf{22}}{Ph-\underset{\underset{Me}{|}}{\overset{\overset{Me}{|}}{C}}-\overset{14}{CH}-\underset{\underset{OTs}{|}}{Me}} \quad\xrightarrow[\text{benzene}]{\text{refluxing}}\quad Me-\underset{\underset{Me}{|}}{\overset{}{C}}=\overset{14}{CH}-\underset{\underset{Me}{|}}{Ph}$$

$$\underset{\textbf{23}}{Ph-\underset{\underset{Me}{|}}{\overset{\overset{Me}{|}}{C}}-\overset{14}{CH}=CH_2} \quad\xrightarrow{H^+}\quad Ph-\underset{\underset{Me}{|}}{\overset{}{C}}=\overset{14}{C}-Me \;+\; Me-\underset{\underset{Me}{|}}{\overset{}{C}}=\overset{14}{C}-Ph$$

alkene **23,** there is competitive migration of both methyl and phenyl (in these reactions ^{14}C labeling is necessary to determine which group has migrated).[34] **22** and **23** give the same carbocation; the differing results must be caused by the fact that in **22** the phenyl group can assist the leaving group, while no such process is possible for **23**. This example clearly illustrates the difference between migration to a relatively free terminus and one that proceeds with the migrating group lending anchimeric assistance.[35]

It is not surprising therefore that clear-cut answers as to relative migrating tendencies are not available. More often than not migratory aptitudes are in the order aryl > alkyl, but exceptions are known, and the position of hydrogen in this series is often unpredictable. In some cases migration of hydrogen is preferred to aryl migration; in other cases migration of alkyl is preferred to that of hydrogen. Mixtures are often found, and the isomer that predominates often depends on conditions. For example, the comparison between methyl and ethyl has been made many times in various systems, and in some cases methyl migration and in others ethyl migration has been found to predominate.[36] However, it can be said that among aryl migrating groups, electron-donating substituents in the para and meta positions increase the migratory aptitudes, while the same substituents in the ortho positions decrease them. Electron-withdrawing groups decrease migrating ability in all positions. The following are a few of the relative migratory aptitudes determined for aryl groups by Bachmann and Ferguson:[37] p-anisyl, 500; p-tolyl, 15.7; m-tolyl, 1.95; phenyl, 1.00; p-chlorophenyl, 0.7; o-anisyl, 0.3. For the o-anisyl group, the poor migrating ability probably has a

[34]Grimaud; Laurent *Bull. Soc. Chim. Fr.* **1967**, 3599.

[35]A number of studies of migratory aptitudes in the dienone-phenol rearrangement (**8-5**) are in accord with the above. For a discussion, see Fischer; Henderson *J. Chem. Soc., Chem. Commun.* **1979**, 279, and references cited therein. See also Palmer; Waring *J. Chem. Soc,. Perkin Trans. 2* **1979**, 1089; Marx; Hahn *J. Org. Chem.* **1988**, *53*, 2866.

[36]For examples, see Cram; Knight *J. Am. Chem. Soc.* **1952**, *74*, 5839; Stiles; Mayer *J. Am. Chem. Soc.* **1959**, *81*, 1497; Heidke; Saunders *J. Am. Chem. Soc.* **1966**, *88*, 5816; Dubois; Bauer *J. Am. Chem. Soc.* **1968**, *90*, 4510, 4511; Bundel'; Levina; Reutov *J. Org. Chem. USSR* **1970**, *6*, 1; Pilkington; Waring *J. Chem. Soc., Perkin Trans. 2* **1976**, 1349; Korchagina; Derendyaev; Shubin; Koptyug *J. Org. Chem. USSR* **1976**, *12*, 378; Wistuba; Rüchardt *Tetrahedron Lett.* **1981**, *22*, 4069; Jost; Laali; Sommer *Nouv. J. Chim.* **1983**, *7*, 79.

[37]Bachmann; Ferguson *J. Am. Chem. Soc.* **1934**, *56*, 2081.

steric cause, while for the others there is a fair correlation with activation or deactivation of electrophilic aromatic substitution, which is what the process is with respect to the benzene ring. It has been reported that at least in certain systems acyl groups have a greater migratory aptitude than alkyl groups.[38]

Memory Effects[39]

Solvolysis of the endo bicyclic compound **24** (X = ONs, p. 353, or Br) gave mostly the bicyclic allylic alcohol **27,** along with a smaller amount of the tricyclic alcohol **31,** while

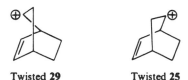

solvolysis of the exo isomers **28** gave mostly **31,** with smaller amounts of **27.**[40] Thus the two isomers gave entirely different ratios of products, though the carbocation initially formed (**25** or **29**) seems to be the same for each. In the case of **25,** a second rearrangement (a shift of the 1,7 bond) follows, while with **29** what follows is an intramolecular addition of the positive carbon to the double bond. It seems as if **25** and **29** "remember" how they were formed before they go on to give the second step. Such effects are called *memory effects* and other such cases are known.[41] The causes of these effects are not well understood, though there has been much discussion. One possible cause is differential solvation of the apparently identical ions **25** and **29.** Other possibilities are: (1) that the ions have geometrical structures that are twisted in opposite senses (e.g., a twisted **29** might have its positive carbon

Twisted 29   Twisted 25

[38]Le Drian; Vogel *Helv. Chim. Acta* **1987,** *70,* 1703, *Tetrahedron Lett.* **1987,** *28,* 1523.
[39]For a review, see Berson *Angew. Chem. Int. Ed. Engl.* **1968,** *7,* 779-791 [*Angew. Chem.* **80,** 765-777].
[40]Berson; Poonian; Libbey *J. Am. Chem. Soc.* **1969,** *91,* 5567; Berson; Donald; Libbey *J. Am. Chem. Soc.* **1969,** *91,* 5580; Berson; Wege; Clarke; Bergman *J. Am. Chem. Soc.* **1969,** *91,* 5594, 5601.
[41]For examples of memory effects in other systems, see Berson; Gajewski; Donald *J. Am. Chem. Soc.,* **1969,** *91,* 5550; Berson; Luibrand; Kundu; Morris *J. Am. Chem. Soc.* **1971,** *93,* 3075; Collins *Acc. Chem. Res.* **1971,** *4,* 315-322; Collins; Glover; Eckart; Raaen; Benjamin; Benjaminov *J. Am. Chem. Soc.* **1972,** *94,* 899; Svensson *Chem. Scr.* **1974,** *6,* 22.

closer to the double bond than a twisted **25**); (2) that ion pairing is responsible;[42] and (3) that nonclassical carbocations are involved.[43] One possibility that has been ruled out is that the steps **24 → 25 → 26** and **28 → 29 → 30** are concerted, so that **25** and **29** never exist at all. This possibility has been excluded by several kinds of evidence, including the fact that **24** gives not only **27**, but also some **31**; and **28** gives some **27** along with **31**. This means that some of the **25** and **29** ions interconvert, a phenomenon known as *leakage*.

Longer Nucleophilic Rearrangements

The question as to whether a group can migrate with its electron pair from A to C in W—A—B—C or over longer distances has been much debated. Although claims have been made that alkyl groups can migrate in this way, the evidence is that such migration is extremely rare, if it occurs at all. One experiment that demonstrated this was the generation of the 3,3-dimethyl-1-butyl cation $Me_3CCH_2CH_2{}^+$. If 1,3 methyl migrations are possible, this cation would appear to be a favorable substrate, since such a migration would convert a primary cation into the tertiary 2-methyl-2-pentyl cation $Me_2\overset{\oplus}{C}CH_2CH_2CH_3$, while the only possible 1,2 migration (of hydride) would give only a secondary cation. However, no products arising from the 2-methyl-2-pentyl cation were found, the only rearranged products being those formed by the 1,2 hydride migration.[44] 1,3 Migration of bromine has been reported.[45]

However, most of the debate over the possibility of 1,3 migrations has concerned not methyl or bromine but 1,3 hydride shifts.[46] There is no doubt that *apparent* 1,3 hydride shifts take place (many instances have been found), but the question is whether they are truly direct hydride shifts or whether they occur by another mechanism. There are at least two ways in which indirect 1,3 hydride shifts can take place: (1) by successive 1,2 shifts or (2) through the intervention of protonated cyclopropanes (see p. 1057). A direct 1,3 shift would have the transition state **A,** while the transition state for a 1,3 shift involving a

A   **B**

protonated cyclopropane intermediate would resemble **B.** The evidence is that most reported 1,3 hydride shifts are actually the result of successive 1,2 migrations,[47] but that in some cases small amounts of products cannot be accounted for in this way. For example, the reaction of 2-methyl-1-butanol with KOH and bromoform gave a mixture of olefins, nearly all of which could have arisen from simple elimination or 1,2 shifts of hydride or alkyl. However, 1.2% of the product was **32:**[48]

$$CH_3-CH_2-\underset{\underset{CH_3}{|}}{CH}-CH_2-OH \xrightarrow{\text{KOH}} 1.2\% \ CH_2{=}CH-\underset{\underset{CH_3}{|}}{CH}-CH_3$$

32

[42]See Collins *Chem. Soc. Rev.* **1975,** *4,* 251-262.

[43]See, for example, Seybold; Vogel; Saunders; Wiberg *J. Am. Chem. Soc.* **1973,** *95,* 2045; Kirmse; Günther *J. Am. Chem. Soc.* **1978,** *100,* 3619.

[44]Skell; Reichenbacher *J. Am. Chem. Soc.* **1968,** *90,* 2309.

[45]Reineke; McCarthy *J. Am. Chem. Soc.* **1970,** *92,* 6376; Smolina; Gopius; Gruzdneva; Reutov *Doklad. Chem.* **1973,** *209,* 280.

[46]For a review, see Fry; Karabatsos, in Olah; Schleyer, Ref. 7, vol. 2, pp. 527-566.

[47]For example, see Bundel'; Levina; Krzhizhevskii; Reutov *Doklad. Chem.* **1968,** *181,* 583; Fărcaşiu; Kascheres; Schwartz *J. Am. Chem. Soc.* **1972,** *94,* 180; Kirmse; Knist; Ratajczak *Chem. Ber.* **1976,** *109,* 2296.

[48]Skell; Maxwell *J. Am. Chem. Soc.* **1962,** *84,* 3963. See also Skell; Starer *J. Am. Chem. Soc.* **1962,** *84,* 3962.

Hypothetically, **32** could have arisen from a 1,3 shift (direct or through a protonated cyclopropane) or from two successive 1,2 shifts:

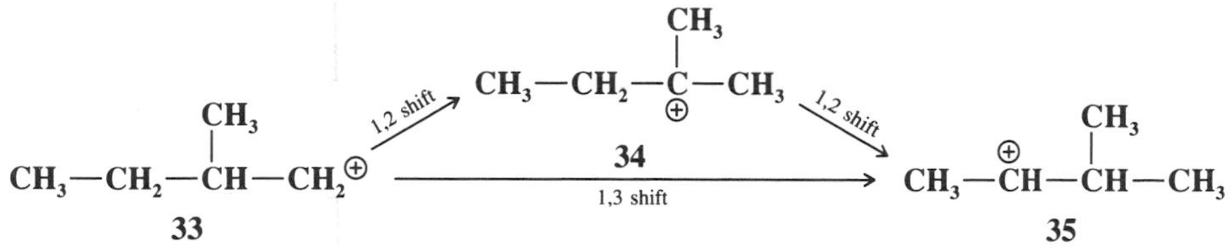

However, the same reaction applied to 2-methyl-2-butanol gave no **32**, which demonstrated that **35** was not formed from **34**. The conclusion was thus made that **35** was formed directly from **33**. This experiment does not answer the question as to whether **35** was formed by a direct shift or through a protonated cyclopropane, but from other evidence[49] it appears that 1,3 hydride shifts that do not result from successive 1,2 migrations usually take place through protonated cyclopropane intermediates (which, as we saw on p. 1056, account for only a small percentage of the product in any case). However, there is evidence that direct 1,3 hydride shifts by way of **A** may take place in super-acid solutions.[50]

Although direct nucleophilic rearrangements over distances greater than 1,2 are rare (or perhaps nonexistent) when the migrating atom or group must move along a chain, this is not so for a shift across a ring of 8 to 11 members. Many such transannular rearrangements are known.[51] Several examples are given on p. 157. This is the mechanism of one of these:[52]

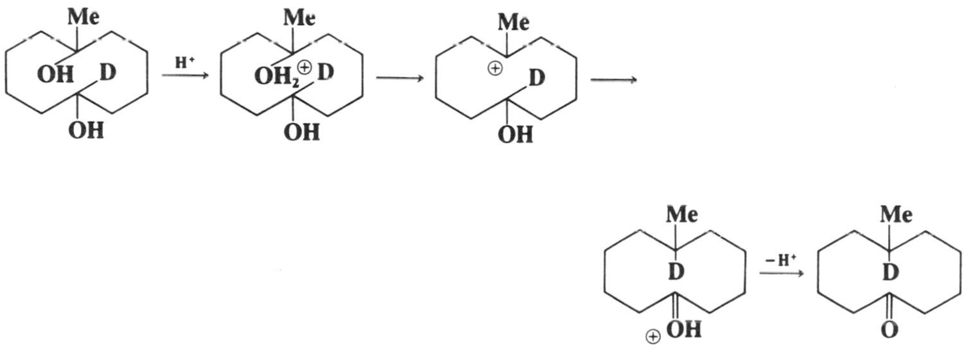

It is noteworthy that the *methyl* group does not migrate in this system. It is generally true that alkyl groups do not undergo transannular migration.[53] In most cases it is hydride that undergoes this type of migration, though a small amount of phenyl migration has also been shown.[54]

[49]For example, see Brouwer; van Doorn *Recl. Trav. Chim. Pays-Bas* **1969**, *8*, 573; Dupuy; Goldsmith; Hudson *J. Chem. Soc., Perkin Trans. 2* **1973**, 74; Hudson; Koplick; Poulton *Tetrahedron Lett.* **1975**, 1449; Fry; Karabatsos, Ref. 46.
[50]Saunders; Stofko *J. Am. Chem. Soc.* **1973**, *95*, 252.
[51]For reviews, see Cope; Martin; McKervey *Q. Rev. Chem. Soc.* **1966**, *20*, 119-152. For many references, see Blomquist; Buck *J. Am. Chem. Soc.* **1951**, *81*, 672.
[52]Prelog; Küng *Helv. Chim. Acta* **1956**, *39*, 1394.
[53]For an apparent exception, see Fărcaşiu; Seppo; Kizirian; Ledlie; Sevin *J. Am. Chem. Soc.* **1989**, *111*, 8466.
[54]Cope; Burton; Caspar *J. Am. Chem. Soc.* **1962**, *84*, 4855.

Free-Radical Rearrangements[55]

1,2-Free-radical rearrangements are much less common than the nucleophilic type previously considered, for the reasons mentioned on p. 1051. Where they do occur, the general pattern is similar. There must first be generation of a free radical, and then the actual migration in which the migrating group moves with one electron:

$$
\begin{array}{ccc}
\text{R} & & \text{R} \\
| & & | \\
\text{A} - \text{B} \cdot & \longrightarrow & \cdot \text{A} - \text{B}
\end{array}
$$

Finally, the new free radical must stabilize itself by a further reaction. The order of radical stability leads us to predict that here too, as with carbocation rearrangements, any migrations should be in the order primary → secondary → tertiary, and that the logical place to look for them should be in neopentyl and neophyl systems. The most common way of generating free radicals for the purpose of detection of rearrangements is by decarbonylation of aldehydes (**4-41**). In this manner it was found that neophyl radicals *do* undergo rearrangement. Thus, $PhCMe_2CH_2CHO$ treated with di-*t*-butyl peroxide gave about equal amounts of the normal product $PHCMe_2CH_3$ and the product arising from migration of phenyl:[56]

$$
\begin{array}{ccccccc}
\text{Ph} & & \text{Ph} & & & \text{Ph} \\
| & & | & & & | \\
\text{Me} - \text{C} - \text{CH}_2 \cdot & \longrightarrow & \text{Me} - \overset{\cdot}{\text{C}} - \text{CH}_2 & \xrightarrow[\text{of H}]{\text{abstraction}} & & \text{Me} - \text{CH} - \text{CH}_2 \\
| & & | & & & | \\
\text{Me} & & \text{Me} & & & \text{Me}
\end{array}
$$

Many other cases of free-radical migration of aryl groups have been found.[57]

It is noteworthy that the extent of migration is much less than with corresponding carbocations: thus in the example given, there was only about 50% migration, whereas the carbocation would have given much more. Also noteworthy is that there was no migration of the methyl group. In general it may be said that free-radical migration of alkyl groups does not occur at ordinary temperatures. Many attempts have been made to detect such migration on the traditional neopentyl and bornyl types of substrates. However, alkyl migration is not observed, even in substrates where the corresponding carbocations undergo facile rearrangement.[58] Another type of migration that is very common for carbocations, but not observed for free radicals, is 1,2 migration of hydrogen. We confine ourselves to a few examples of the lack of migration of alkyl groups and hydrogen:

1. 3,3-Dimethylpentanal ($EtCMe_2CH_2CHO$) gave no rearranged products on decarbonylation.[59]

[55]For reviews, see Beckwith; Ingold, in Mayo, Ref. 1, vol.1, pp. 161-310; Wilt, in Kochi *Free Radicals*, vol. 1; Wiley: New York, 1973, pp. 333-501; Stepukhovich; Babayan *Russ. Chem. Rev.* **1972**, *41*, 750; Nonhebel; Walton *Free-Radical Chemistry*; Cambridge University Press: London, 1974, pp. 498-552; Huyser *Free-Radical Chain Reactions*; Wiley: New York, 1970, pp. 235-255; Freidlina *Adv. Free-Radical Chem.* **1965**, *1*, 211-278; Pryor *Free Radicals*; McGraw-Hill: New York, 1966, pp. 266-284.

[56]Winstein; Seubold *J. Am. Chem. Soc.* **1947**, *69*, 2916; Seubold *J. Am. Chem. Soc.* **1953**, *75*, 2532. For the observation of this rearrangement by esr, see Hamilton; Fischer *Helv. Chim. Acta* **1973**, *56*, 795.

[57]For example, see Curtin; Hurwitz *J. Am. Chem. Soc.* **1952**, *74*, 5381; Wilt; Philip *J. Org. Chem.* **1959**, *24*, 441, **1960**, *25*, 891; Pines; Goetschel *J. Am. Chem. Soc.* **1964**, *87*, 4207; Goerner; Cote; Vittimberga *J. Org. Chem.* **1977**, *42*, 19; Collins; Roark; Raaen; Benjamin *J. Am. Chem. Soc.* **1979**, *101*, 1877; Walter; McBride *J. Am. Chem. Soc.* **1981**, *103*, 7069, 7074.

[58]For a summary of unsuccessful attempts, see Slaugh; Magoon; Guinn *J. Org. Chem.* **1963**, *28*, 2643.

[59]Seubold *J. Am. Chem. Soc.* **1954**, *76*, 3732.

2. Addition of RSH to norbornene gave only *exo*-norbornyl sulfides, though **36** is an

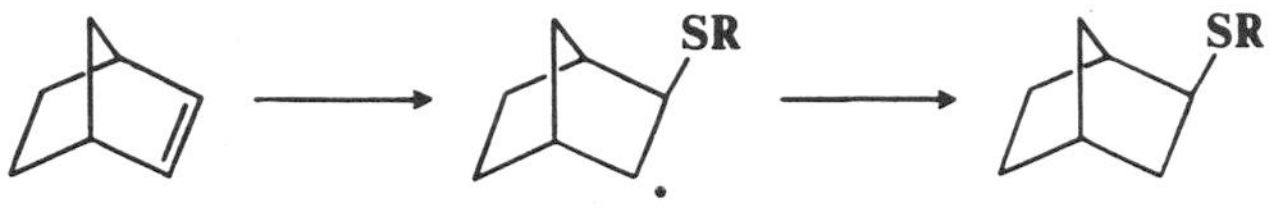

SR → SR

36

intermediate, and the corresponding carbocation cannot be formed without rearrangement.[60]

3. The cubylcarbinyl radical did not rearrange to the 1-homocubyl radical, though doing

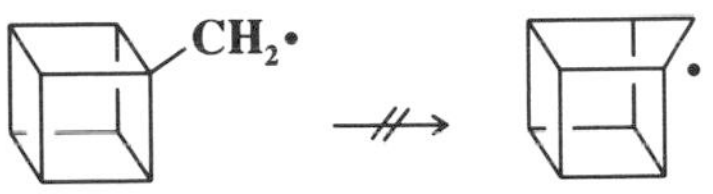

CH$_2$· $\xrightarrow{\,/\!/\,}$

cubylcarbinyl
radical

1-homocubyl
radical

so would result in a considerable decrease in strain.[60a]

4. It was shown[61] that no rearrangement of isobutyl radical to *t*-butyl radical (which would involve the formation of a more stable radical by a hydrogen shift) took place during the chlorination of isobutane.

However, 1,2 migration of alkyl groups has been shown to occur in certain *diradicals*.[62] For example, the following rearrangement has been established by tritium labeling.[63]

$$\overset{\cdot}{C}HT-\overset{\overset{\displaystyle CH_3}{|}}{C}H-\overset{\cdot}{C}H-CH_3 \longrightarrow CH_3-CHT-CH=CH-CH_3$$

In this case the fact that migration of the methyl group leads directly to a compound in which all electrons are paired undoubtedly contributes to the driving force of the reaction.

The fact that aryl groups migrate, but alkyl groups and hydrogen generally do not, leads to the proposition that **37**, in which the odd electron is not found in the three-membered

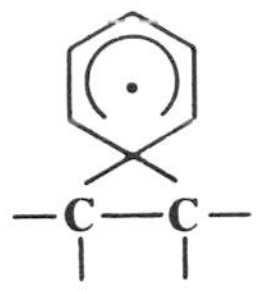

$$-C-C-$$

37

ring, may be an intermediate. There has been much controversy on this point, but the bulk of the evidence indicates that **37** is a transition state, not an intermediate.[64] Among the

[60]Cristol; Brindell *J. Am. Chem. Soc.* **1954**, *76*, 5699.
[60a]Eaton; Yip *J. Am. Chem. Soc.* **1991**, *113*, 7692.
[61]Brown; Russell *J. Am. Chem. Soc.* **1952**, *74*, 3995. See also Desai; Nechvatal; Tedder *J. Chem. Soc. B.* **1970**, 386.
[62]For a review, see Freidlina; Terent'ev *Russ. Chem. Rev* **1974**, *43*, 129-139.
[63]McKnight; Rowland *J. Am. Chem. Soc.* **1966**, *88*, 3179. For other examples, see Greene; Adam; Knudsen *J. Org. Chem.* **1966**, *31*, 2087; Gajewski; Burka *J. Am. Chem. Soc.* **1972**, *94*, 8857, 8860, 8865; Adam; Aponte *J. Am. Chem. Soc.* **1971**, *93*, 4300.
[64]For molecular orbital calculations indicating that **37** is an intermediate, see Yamabe *Chem. Lett.* **1989**, 1523.

evidence is the failure to observe **37** either by esr[65] or CIDNP.[66] Both of these techniques can detect free radicals with extremely short lifetimes (pp. 186-187).[67]

Besides aryl, vinylic[68] and acetoxy groups[69] also migrate. Vinylic groups migrate by way of a cyclopropylcarbinyl radical intermediate,[70] while the migration of acetoxy groups may involve the charge-separated structure shown.[71] In addition, migration has been observed

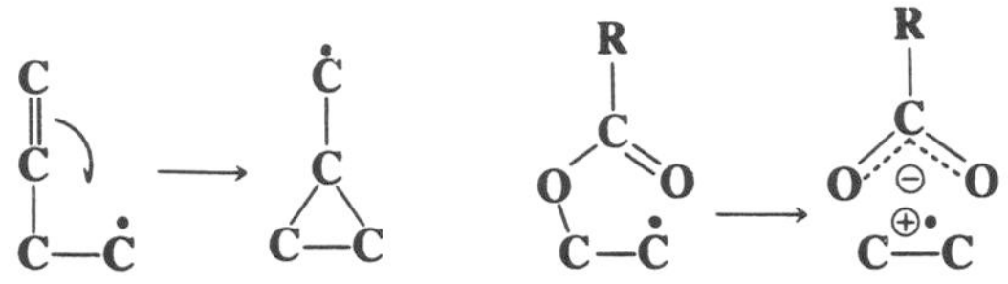

38

for chloro (and to a much lesser extent bromo) groups. For example, in the reaction of $Cl_3CCH=CH_2$ with bromine under the influence of peroxides, the products were 47% $Cl_3CCHBrCH_2Br$ (the normal addition product) and 53% $BrCCl_2CHClCH_2Br$, which arose by rearrangement:

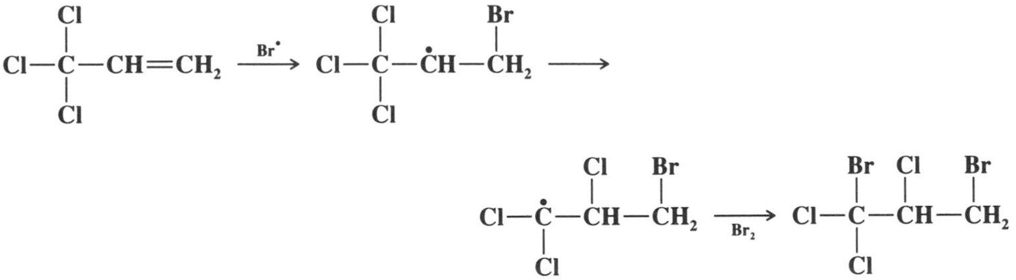

In this particular case the driving force for the rearrangement is the particular stability of dichloroalkyl free radicals. Nesmeyanov, Freidlina, and co-workers have extensively studied reactions of this sort.[72] It has been shown that the 1,2 migration of Cl readily occurs if the migration origin is tertiary and the migration terminus primary.[73] Migration of Cl and Br could take place by a transition state in which the odd electron is accommodated in a vacant *d* orbital of the halogen.

[65]Kochi; Krusic *J. Am. Chem. Soc.* **1969**, *91*, 3940; Edge; Kochi *J. Am. Chem. Soc.* **1972**, *94*, 7695.

[66]Shevlin; Hansen *J. Org. Chem.* **1977**, *42*, 3011; Olah; Krishnamurthy; Singh; Iyer *J. Org. Chem.* **1983**, *48*, 955. **37** has been detected as an intemediate in a different reaction: Effio; Griller; Ingold; Scaiano; Sheng *J. Am. Chem. Soc.* **1980**, *102*, 6063; Leardini; Nanni; Pedulli; Tundo; Zanardi; Foresti; Palmieri *J. Am. Chem. Soc.* **1989**, *111*, 7723.

[67]For other evidence, see Martin *J. Am. Chem. Soc.* **1962**, *84*, 1986; Rüchardt; Hecht *Tetrahedron Lett.* **1962**, 957, *Chem. Ber.* **1965**, *98*, 2460, 2471; Rüchardt; Trautwein *Chem. Ber.* **1965**, *98*, 2478.

[68]For cxample, see Slaugh; Mullineaux; Raley *J. Am. Chem. Soc.* **1963**, *85*, 3180; Slaugh *J. Am. Chem. Soc.* **1965**, *87*, 1522; Newcomb; Glenn; Williams *J. Org. Chem.* **1989**, *54*, 2675.

[69]Surzur; Teissier *C. R. Acad. Sci., Ser. C* **1967**, *264*, 1981, *Bull. Soc. Chim. Fr.* **1970**, 3060; Tanner; Law *J. Am. Chem. Soc.* **1969**, *91*, 7535; Julia; Lorne *C. R. Acad. Sci., Ser. C* **1971**, *273*, 174; Lewis; Miller; Winstein *J. Org. Chem.* **1972**, *37*, 1478.

[70]For evidence for this species, see Montgomery; Matt; Webster *J. Am. Chem. Soc.* **1967**, *89*, 923; Montgomery; Matt *J. Am. Chem. Soc.* **1967**, *89*, 934, 6556; Giese; Heinrich; Horler; Koch; Schwarz *Chem. Ber.* **1986**, *119*, 3528.

[71]Beckwith; Tindal *Aust. J. Chem.* **1971**, *24*, 2099; Beckwith; Thomas *J. Chem. Soc., Perkin Trans. 2* **1973**, 861; Barclay; Lusztyk; Ingold *J. Am. Chem. Soc.* **1984**, *106*, 1793.

[72]For reviews, see Freidlina; Terent'ev *Russ. Chem. Rev.* **1979**, *48*, 828-839; Freidlina, Ref. 55, pp. 231-249.

[73]See, for example, Skell; Pavlis; Lewis; Shea *J. Am. Chem. Soc.* **1973**, *95*, 6735; Chen; Tang; Montgomery; Kochi *J. Am. Chem. Soc.* **1974**, *96*, 2201.

Migratory aptitudes have been measured for the phenyl and vinyl groups, and for three
other groups, using the system $RCMe_2CH_2\cdot \rightarrow Me_2\overset{\cdot}{C}CH_2R$. These were found to be in the
order R $= H_2C=CH_2 > Me_3CC=O > Ph > Me_3C\equiv C > CN$.[74]

In summary then, 1,2 free-radical migrations are much less prevalent than the analogous
carbocation processes, and are important only for aryl, vinylic, acetoxy, and halogen migrating groups. The direction of migration is normally toward the more stable radical, but
"wrong-way" rearrangements are also known.[75]

Despite the fact that hydrogen atoms do not migrate 1,2, longer free-radical migrations
of hydrogen are known.[76] The most common are 1,5 shifts, but 1,6 and longer shifts have
also been found. The possibility of 1,3 hydrogen shifts has been much investigated, but it
is not certain if any actually occur. If they do they are rare, presumably because the most
favorable geometry for C—II—C in the transition state is linear and this geometry cannot
be achieved in a 1,3 shift. 1,4 shifts are definitely known, but are still not very common.
These long shifts are best regarded as internal abstractions of hydrogen (for reactions involving them, see **4-8** and **8-42**):

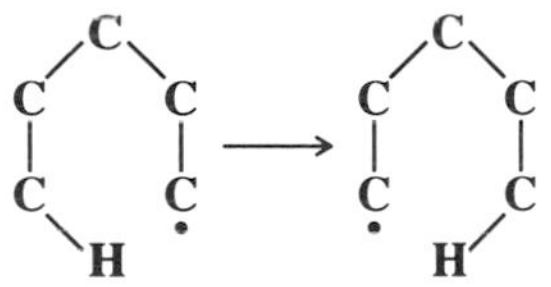

Transannular shifts of hydrogen atoms have also been observed.[77]

Electrophilic Rearrangements[78]

Rearrangements in which a group migrates without its electrons are much rarer than the
two kinds previously considered, but the general principles are the same. A carbanion (or
other negative ion) is created first, and the actual rearrangement step involves migration of
a group without its electrons:

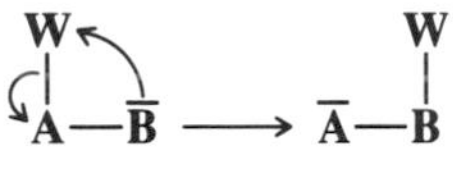

The product of the rearrangement may be stable or may react further, depending on its
nature (see also p. 1072).

[74]Lindsay; Lusztyk; Ingold *J. Am. Chem. Soc.* **1984**, *106*, 7087.
[75]Slaugh; Raley *J. Am. Chem. Soc.* **1960**, *82*, 1259; Bonner; Mango *J. Org. Chem.* **1964**, *29*, 29; Dannenberg;
Dill *Tetrahedron Lett.* **1972**, 1571.
[76]For a discussion, see Freidlina; Terent'ev. *Acc. Chem. Res.* **1977**, *10*, 9-15.
[77]Heusler; Kalvoda *Tetrahedron Lett.* **1963**, 1001; Cope; Bly; Martin; Petterson *J. Am. Chem. Soc.* **1965**, *87*, 3111;
Fisch; Ourisson *Chem. Commun.* **1965**, 407; Traynham; Couvillon *J. Am. Chem. Soc.* **1967**, *89*, 3205.
[78]For reviews, see Hunter; Stothers; Warnhoff, in Mayo, Ref. 1, vol. 1, pp. 391-470; Grovenstein *Angew. Chem.
Int. Ed. Engl.* **1978**, *17*, 313-332 [*Angew. Chem. 90*, 317-336], *Adv. Organomet. Chem.* **1977**, *16*, 167-193; Jensen;
Rickborn *Electrophilic Substitution of Organomercurials*; McGraw-Hill: New York, 1968, pp. 21-30; Cram *Funda-
mentals of Carbanion Chemistry*; Academic Press: New York, 1965, pp. 223-243.

REACTIONS

The reactions in this chapter are classified into three main groups. 1,2 shifts are considered first. Within this group, reactions are classified according to (1) the identity of the substrate atoms A and B and (2) the nature of the migrating group W. In the second group are the cyclic rearrangements. The third group consists of rearrangements that cannot be fitted into either of the first two categories.

Reactions in which the migration terminus is on an aromatic ring have been treated under aromatic substitution. These are **1-30** to **1-36**, **1-40**, **3-25** to **3-28**, and, partially, **1-37**, **1-41**, and **1-42**. Double-bond shifts have also been treated in other chapters, though they may be considered rearrangements (p. 327, p. 577, and **2-2**). Other reactions that may be regarded as rearrangements are the Pummerer (**9-71**) and Willgerodt (**9-72**) reactions.

1,2 Rearrangements

A. Carbon-to-Carbon Migrations of R, H, and Ar

8-1 Wagner–Meerwein and Related Reactions
1/Hydro,1/hydroxy-(2/→1/alkyl)-*migro*-elimination, etc.

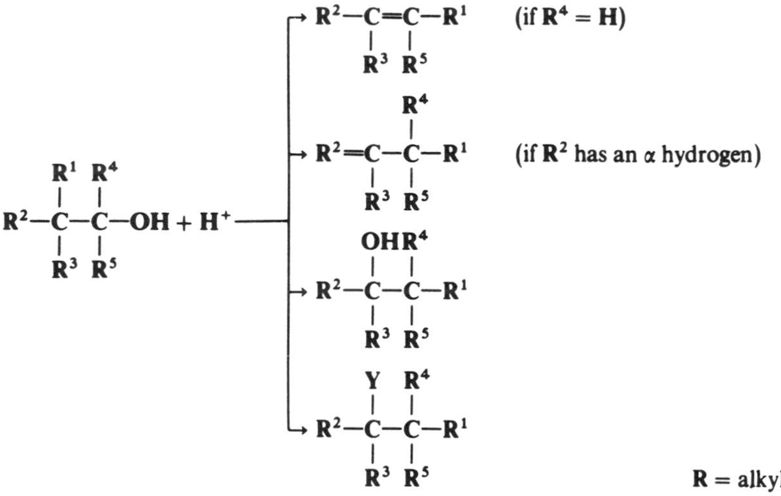

$$R^2-C\!=\!C-R^1 \quad \text{(if } R^4 = H\text{)}$$
$$\underset{R^3 \quad R^5}{|\quad\quad|}$$

$$\overset{R^4}{\underset{R^3 \quad R^5}{R^2\!=\!C-C-R^1}} \quad \text{(if } R^2 \text{ has an } \alpha \text{ hydrogen)}$$

$$\underset{R^3 \quad R^5}{\overset{R^1 \quad R^4}{R^2-C-C-OH}} + H^+ \longrightarrow$$

$$\overset{OHR^4}{\underset{R^3 \quad R^5}{R^2-C-C-R^1}}$$

$$\overset{Y \quad R^4}{\underset{R^3 \quad R^5}{R^2-C-C-R^1}}$$

R = alkyl, aryl, or hydrogen

When alcohols are treated with acids, simple substitution (e.g., **0-67**) or elimination (**7-1**) usually accounts for most or all of the products. But in many cases, especially where two or three alkyl or aryl groups are on the β carbon, some or all of the product is rearranged. These rearrangements are called *Wagner–Meerwein rearrangements*. As pointed out previously, the carbocation that is a direct product of the rearrangement must stabilize itself, and most often it does this by the loss of a hydrogen β to it, so the rearrangement product is usually an olefin.[79] The proton lost may be R^4 (if this is a hydrogen) or an α proton from R^2 (if it has one). If there is a choice of protons, Zaitsev's rule (p. 998) governs the direction, as we might expect. Sometimes a different positive group is lost instead of a proton. Less

[79]For a review of such rearrangements, see Kaupp *Top. Curr. Chem.* **1988**, *146*, 57-98.

often, the new carbocation stabilizes itself by combining with a nucleophile instead of losing a proton. The nucleophile may be the water which is the original leaving group, so that the product is a rearranged alcohol, or it may be some other species present, which we have called Y. Rearrangement is usually predominant in neopentyl and neophyl types of substrates, and with these types normal nucleophilic substitution is difficult (normal elimination is of course impossible). Under S_N2 conditions, substitution is extremely slow;[80] under S_N1 conditions, carbocations are formed that rapidly rearrange. However, free-radical substitution, unaccompanied by rearrangement, can be carried out on neopentyl systems, though, as we have seen (p. 1064), neophyl systems undergo rearrangement as well as substitution.

Wagner–Meerwein rearrangements were first discovered in the bicyclic terpenes, and most of the early development of this reaction was with these compounds.[81] An example is

Example *a*

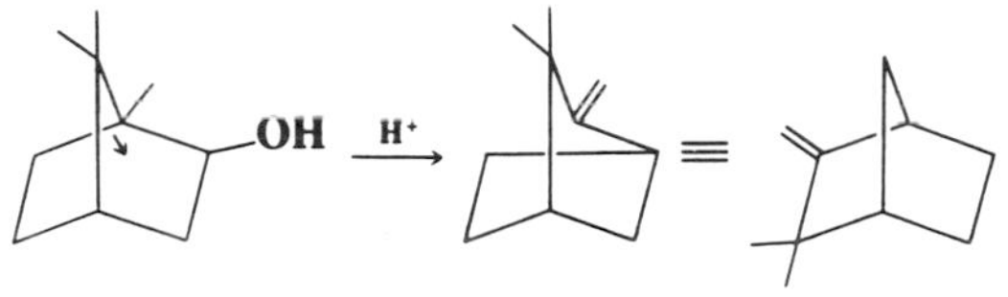

isoborneol   camphene

Examples in simpler systems are:

Example *b*

$$CH_3-\underset{\underset{CH_3}{|}}{\overset{\overset{CH_3}{|}}{C}}-CH_2Cl \xrightarrow{OH^-} CH_3-\underset{\underset{CH_3}{|}}{C}=CH-CH_3$$

Example *c*

$$CH_3CH_2CH_2Br \xrightarrow{AlBr_3} CH_3\underset{\underset{Br}{|}}{C}HCH_3$$

These examples illustrate the following points:

1. Hydride ion can migrate. In example *c*, it was hydride that shifted, not bromine:

$$CH_3CH_2CH_2Br \xrightarrow{AlBr_3} \underset{+}{CH_3CH_2CH_2^{\oplus}} \longrightarrow CH_3\overset{\oplus}{C}HCH_3 \xrightarrow{AlBr_4^-} CH_3\underset{\underset{Br}{|}}{C}HCH_3$$
$$AlBr_4^-$$

2. The leaving group does not have to be H_2O, but can be any departing species whose loss creates a carbocation, including N_2 from aliphatic diazonium ions[82] (see the section on leaving groups in nucleophilic substitution, p. 352). Also, rearrangement may follow when the carbocation is created by addition of a proton or other positive species to a double bond. Even alkanes give rearrangements when heated with Lewis acids, provided some species is initially present to form a carbocation from the alkane.

[80]See, however, Ref. 248 in Chapter 10.
[81]For a review of rearrangements in bicyclic systems, see Hogeveen; van Kruchten *Top. Curr. Chem.* **1979**, *80*, 89-124. For reviews concerning caranes and pinanes see, respectively, Arbuzov; Isaeva *Russ. Chem. Rev.* **1976**, *45*, 673-683; Banthorpe; Whittaker *Q. Rev. Chem. Soc.* **1966**, *20*, 373-387.
[82]For reviews of rearrangements arising from diazotization of aliphatic amines, see, in Patai *The Chemistry of the Amino Group*; Wiley: New York, 1968, the articles by White; Woodcock, pp. 407-497 (pp. 473-483) and by Banthorpe, pp. 585-667 (pp. 586-612).

3. Example *c* illustrates that the last step can be substitution instead of elimination.

4. Example *b* illustrates that the new double bond is formed in accord with Zaitsev's rule.

2-Norbornyl cations, besides displaying the 1,2 shifts of a CH_2 group previously illustrated for the isoborneol → camphene conversion, are also prone to rapid hydride shifts from the 3 to the 2 position (known as 3,2 shifts). These 3,2 shifts usually take place from the exo

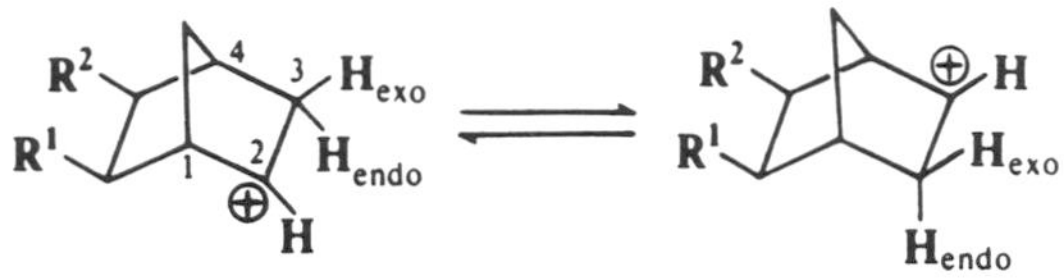

side;[83] that is, the 3-exo hydrogen migrates to the 2-exo position.[84] This stereoselectivity is analogous to the behavior we have previously seen for norbornyl systems, namely, that nucleophiles attack norbornyl cations from the exo side (p. 321) and that addition to nor-bornenes is also usually from the exo direction (p. 753).

The direction of rearrangement is usually towards the most stable carbocation (or radical), which is tertiary > secondary > primary, but rearrangements in the other direction have also been found,[85] and often the product is a mixture corresponding to an equilibrium mixture of the possible carbocations.

The term "Wagner–Meerwein rearrangement" is not precise. Some use it to refer to all the rearrangements in this section and in **8-2**. Others use it only when an alcohol is converted to a rearranged olefin. Terpene chemists call the migration of a methyl group the *Nametkin rearrangement*. The term *retropinacol rearrangement* is often applied to some or all of these. Fortunately, this disparity in nomenclature does not seem to cause much confusion.

Sometimes several of these rearrangements occur in one molecule, either simultaneously or in rapid succession. A spectacular example is found in the triterpene series. Friedelin is a triterpenoid ketone found in cork. Reduction gives 3β-friedelanol (**39**). When this compound is treated with acid, 13(18)-oleanene (**40**) is formed.[86] In this case *seven* 1,2 shifts take place. On removal of H_2O from position 3 to leave a positive charge, the following

39   **40**

shifts occur: hydride from 4 to 3; methyl from 5 to 4; hydride from 10 to 5; methyl from 9 to 10; hydride from 8 to 9; methyl from 14 to 8; and methyl from 13 to 14. This leaves a

---

[83]For example, see Kleinfelter; Schleyer *J. Am. Chem. Soc.* **1961**, *83*, 2329; Collins; Cheema; Werth; Benjamin *J. Am. Chem. Soc.* **1964**, *86*, 4913; Berson; Hammons; McRowe; Bergman; Remanick; Houston *J. Am. Chem. Soc.* **1967**, *89*, 2590.

[84]For examples of 3,2 endo shifts, see Bushell; Wilder *J. Am. Chem. Soc.* **1967**, *89*, 5721; Wilder; Hsieh *J. Org. Chem.* **1971**, *36*, 2552.

[85]See, for example, Cooper et al., Ref. 23.

[86]Corey; Ursprung *J. Am. Chem. Soc.* **1956**, *78*, 5041.

positive charge at position 13, which is stabilized by loss of the proton at the 18 position to give **40**. All these shifts are stereospecific, the group always migrating on the side of the ring system on which it is located; that is, a group above the "plane" of the ring system (indicated by a solid line in **39**) moves above the plane, and a group below the plane (dashed line) moves below it. It is probable that the seven shifts are not all concerted, though some of them may be, for intermediate products can be isolated.[87] As an illustration of point 2 (p. 1069), it may be mentioned that friedelene, derived from dehydration of **39,** also gives **40** on treatment with acid.[88]

It was mentioned above that even alkanes undergo Wagner–Meerwein rearrangements if treated with Lewis acids and a small amount of initiator. An interesting application of this reaction is the conversion of tricyclic molecules to adamantane and its derivatives.[89] It has been found that *all* tricyclic alkanes containing 10 carbons are converted to adamantane by treatment with a Lewis acid such as $AlCl_3$. If the substrate contains more than 10 carbons, alkyl-substituted adamantanes are produced. The IUPAC name for these reactions is **Schleyer adamantization.** Some examples are

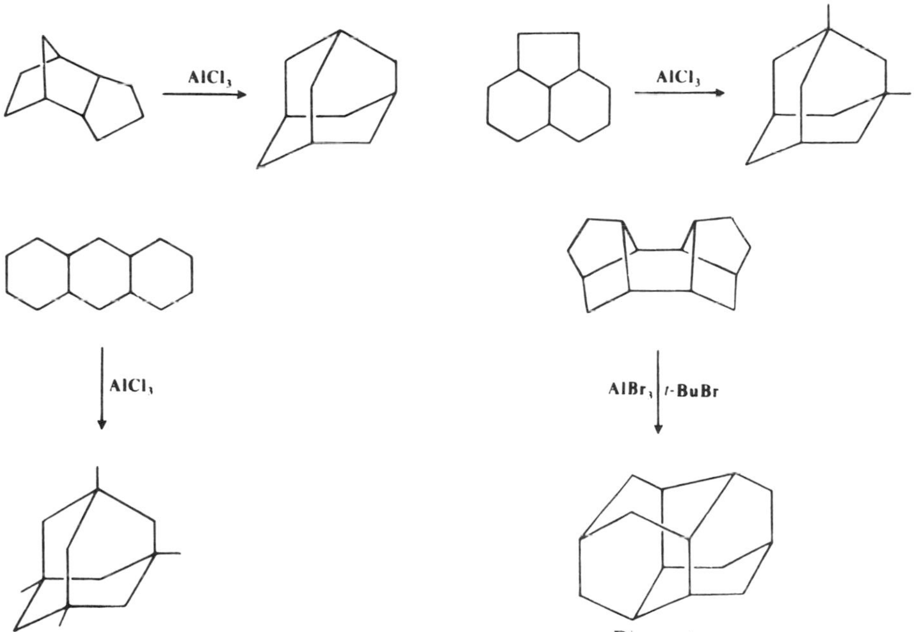

Diamantane

If 14 or more carbons are present, the product may be diamantane or a substituted diamantane.[90] These reactions are successful because of the high thermodynamic stability of adamantane, diamantane, and similar diamond-like molecules. The most stable of a set of C_nH_m isomers (called the *stabilomer*) will be the end product if the reaction reaches equi-

[87]For a discussion, see Whitlock; Olson *J. Am. Chem. Soc.* **1970,** *92,* 5383.
[88]Dutler; Jeger; Ruzicka *Helv. Chim. Acta* **1955,** *38,* 1268; Brownlie; Spring; Stevenson; Strachan *J. Chem. Soc.* **1956,** 2419; Coates *Tetrahedron Lett.* **1967,** 4143.
[89]For reviews, see McKervey; Rooney, in Olah *Cage Hydrocarbons*; Wiley: New York, 1990, pp. 39-64; McKervey *Tetrahedron* **1980,** *36,* 971-992, *Chem. Soc. Rev.* **1974,** *3,* 479-512; Greenberg; Liebman *Strained Organic Molecules*; Academic Press: New York, 1978, pp. 178-202; Bingham; Schleyer, *Fortschr. Chem. Forsch.* **1971,** *18,* 1-102, pp. 3-23.
[90]See Gund; Osawa; Williams; Schleyer *J. Org. Chem.* **1974,** *39,* 2979.

librium.[91] Best yields are obtained by the use of "sludge" catalysts[92] (i.e., a mixture of AlX_3 and *t*-butyl bromide or *sec*-butyl bromide).[93] Though it is certain that these adamantane-forming reactions take place by nucleophilic 1,2 shifts, the exact pathways are not easy to unravel because of their complexity.[94] Treatment of adamantane-2-^{14}C with $AlCl_3$ results in total carbon scrambling on a statistical basis.[95]

As already indicated, the mechanism of the Wagner–Meerwein rearrangement is usually nucleophilic. Free-radical rearrangements are also known (see the mechanism section of this chapter), though virtually only with aryl migration. However, carbanion mechanisms (electrophilic) have also been found.[78] Thus Ph_3CCH_2Cl treated with sodium gave Ph_2CHCH_2Ph along with unrearranged products.[96] This is called the *Grovenstein–Zimmerman rearrangement*. The intermediate is $Ph_3C\bar{C}H_2{}^-$, and the phenyl moves without its electron pair. Only aryl and vinylic,[97] and not alkyl, groups migrate by the electrophilic mechanism (p. 1051) and transition states or intermediates analogous to **37** and **38** are likely.[98]

OS **V**, 16, 194; **VI**, 378, 845.

8-2 The Pinacol Rearrangement
1/*O*-Hydro,3/hydroxy-(2/→3/alkyl)-*migro*-elimination

$$R^2{-}\underset{\underset{OH}{|}}{\overset{\overset{R^1}{|}}{C}}{-}\underset{\underset{OH}{|}}{\overset{\overset{R^3}{|}}{C}}{-}R^4 \xrightarrow{\ H^+\ } R^2{-}\underset{\underset{O}{\|}}{\overset{\overset{R^1}{|}}{C}}{-}\underset{\underset{R^3}{|}}{\overset{}{C}}{-}R^4 \quad R = \text{alkyl, aryl, or hydrogen}$$

When *vic*-diols (glycols) are treated with acids, they can be rearranged to give aldehydes or ketones, though elimination without rearrangement can also be accomplished. This reaction is called the *pinacol rearrangement*; the reaction gets its name from the typical compound pinacol $Me_2COHCOHMe_2$, which is rearranged to pinacolone Me_3CCOCH_3.[99] The reaction has been accomplished many times, with alkyl, aryl, hydrogen, and even ethoxycarbonyl (COOEt)[100] as migrating groups. In most cases each carbon has at least one alkyl or aryl group, and the reaction is most often carried out with tri- and tetrasubstituted glycols. As mentioned earlier, glycols in which the four R groups are not identical can give rise to more than one product, depending on which group migrates (see p. 1058 for a discussion of migratory aptitudes). Mixtures are often produced, and which group preferentially migrates

[91]For a method for the prediction of stabilomers, see Godleski; Schleyer; Ōsawa; Wipke *Prog. Phys. Org. Chem.* **1981**, *13*, 63-117.

[92]Schneider; Warren; Janoski *J. Org. Chem.* **1966**, *31*, 1617; Williams; Schleyer; Gleicher; Rodewald *J. Am. Chem. Soc.* **1966**, *88*, 3862; Robinson; Tarratt *Tetrahedron Lett.* **1968**, 5.

[93]For other methods, see Johnston; McKervey; Rooney *J. Am. Chem. Soc.* **1971**, *93*, 2798; Olah; Wu; Farooq; Prakash *J. Org. Chem.* **1989**, *54*, 1450.

[94]See, for example, Engler; Farcasiu; Sevin; Cense; Schleyer *J. Am. Chem. Soc.* **1973**, *95*, 5769; Klester; Ganter *Helv. Chim. Acta* **1983**, *66*, 1200, **1985**, *68*, 734.

[95]Majerski; Liggero; Schleyer; Wolf *Chem. Commun.* **1970**, 1596.

[96]Grovenstein *J. Am. Chem. Soc.* **1957**, *79*, 4985; Zimmerman; Smentowski *J. Am. Chem. Soc.* **1957**, *79*, 5455; Grovenstein; Williams *J. Am. Chem. Soc.* **1961**, *83*, 412; Zimmerman; Zweig *J. Am. Chem. Soc.* **1961**, *83*, 1196. See also Crimmins; Murphy; Hauser *J. Org. Chem.* **1966**, *31*, 4273; Grovenstein; Cheng *J. Am. Chem. Soc.* **1972**, *94*, 4971.

[97]See Grovenstein; Black; Goel; Hughes; Northrop; Streeter; VanDerveer *J. Org. Chem.* **1989**, *54*, 1671, and references cited therein.

[98]Grovenstein; Wentworth *J. Am. Chem. Soc.* **1967**, *89*, 2348; Bertrand; Grovenstein; Lu; VanDerveer *J. Am. Chem. Soc.* **1976**, *98*, 7835.

[99]For reviews, see Bartók; Molnár, in Patai *The Chemistry of Functional Groups, Supplement E*; Wiley: New York, 1980, pp. 722-732; Collins; Eastham, Ref. 1, pp. 762-771.

[100]Kagan; Agdeppa; Mayers; Singh; Walters; Wintermute *J. Org. Chem.* **1976**, *41*, 2355. COOH has been found to migrate in a Wagner-Meerwein reaction: Berner; Cox; Dahn *J. Am. Chem. Soc.* **1982**, *104*, 2631.

may depend on the reaction conditions as well as on the nature of the substrate. Thus the action of cold, concentrated sulfuric acid on **41** produces mainly the ketone **42** (methyl

$$\text{Me}-\underset{\underset{\textbf{O}}{\|}}{\text{C}}-\underset{\underset{\textbf{Me}}{|}}{\text{CPh}_2} \xleftarrow[\text{H}_2\text{SO}_4]{\text{cold}} \text{Ph}_2\text{C}-\underset{\underset{\textbf{HO}}{|}}{}\underset{\underset{\textbf{OH}}{|}}{\text{CMe}_2} \xrightarrow[\substack{+ \\ \text{a trace of} \\ \text{H}_2\text{SO}_4}]{\text{HOAc}} \text{Ph}-\underset{\underset{\textbf{O}}{\|}}{\text{C}}-\underset{\underset{\textbf{Ph}}{|}}{\text{CMe}_2}$$

$$\textbf{42} \qquad\qquad\qquad\qquad \textbf{41} \qquad\qquad\qquad\qquad \textbf{43}$$

migration), while treatment of **41** with acetic acid containing a trace of sulfuric acid gives mostly **43** (phenyl migration).[101] If at least one R is hydrogen, aldehydes can be produced as well as ketones. Generally, aldehyde formation is favored by the use of mild conditions (lower temperatures, weaker acids), because under more drastic conditions the aldehydes may be converted to ketones (**8-4**). The reaction has been carried out in the solid state, by treating solid substrates with HCl gas or with an organic solid acid.[102]

The mechanism involves a simple 1,2 shift. The ion **44** (where all four R groups are Me)

$$\underset{\underset{\textbf{OH OH}}{|\quad|}}{\overset{\overset{\textbf{R}^1 \;\; \textbf{R}^3}{|\quad|}}{\text{R}^2-\text{C}-\text{C}-\text{R}^4}} \xrightarrow{\text{H}^+} \underset{\underset{\textbf{OH OH}_2^\oplus}{|\quad|}}{\overset{\overset{\textbf{R}^1 \;\; \textbf{R}^3}{|\quad|}}{\text{R}^2-\text{C}-\text{C}-\text{R}^4}} \longrightarrow \underset{\underset{\textbf{OH}}{|}}{\overset{\overset{\textbf{R}^1 \;\; \textbf{R}^3}{|\quad|}}{\text{R}^2-\text{C}-\overset{\oplus}{\text{C}}-\text{R}^4}} \longrightarrow$$

$$\textbf{44}$$

$$\underset{\underset{\textbf{OH R}^3}{|\quad|}}{\overset{\overset{\textbf{R}^1}{|}}{\text{R}^2-\overset{\oplus}{\text{C}}-\text{C}-\text{R}^4}} \xrightarrow{-\text{H}^+} \underset{\underset{\textbf{O R}^3}{\|\quad|}}{\overset{\overset{\textbf{R}^1}{|}}{\text{R}^2-\text{C}-\text{C}-\text{R}^4}}$$

has been trapped by the addition of tetrahydrothiophene.[103] It may seem odd that a migration takes place when the positive charge is already at a tertiary position, but carbocations stabilized by an oxygen atom are even more stable than tertiary alkyl cations (p. 170). There is also the driving force supplied by the fact that the new carbocation can immediately stabilize itself by losing a proton.

It is obvious that other compounds in which a positive charge can be placed on a carbon α to one bearing an OH group can also give this rearrangement. This is true for β-amino alcohols, which rearrange on treatment with nitrous acid (this is called the *semipinacol rearrangement*), iodohydrins, for which the reagent is mercuric oxide or silver nitrate, β-hydroxyalkyl selenides $\text{R}^1\text{R}^2\text{C(OH)C(SeR}^5)\text{R}^3\text{R}^4$,[104] and allylic alcohols, which can rearrange on treatment with a strong acid that protonates the double bond. A similar rearrangement is given by epoxides, when treated with acidic[105] reagents such as BF_3–etherate or MgBr_2–etherate, or sometimes by heat alone.[106] It has been shown that epoxides are

$$\underset{\underset{\textbf{O}}{\diagdown\;\diagup}}{\overset{\overset{\textbf{R}^1 \;\; \textbf{R}^3}{|\quad|}}{\text{R}^2-\text{C}-\text{C}-\text{R}^4}} \xrightarrow[\substack{\text{or} \\ \text{MgBr}_2-\text{Et}_2\text{O}}]{\text{BF}_3-\text{Et}_2\text{O}} \underset{\underset{\textbf{R}^3 \;\; \textbf{O}}{|\quad\|}}{\overset{\overset{\textbf{R}^1}{|}}{\text{R}^2-\text{C}-\text{C}-\text{R}^4}} \qquad \text{R = alkyl, aryl, or hydrogen}$$

[101]Ramart-Lucas; Salmon-Legagneur *C. R. Acad. Sci.* **1928**, *188*, 1301.
[102]Toda; Shigemasa *J. Chem. Soc., Perkin Trans. 1* **1989**, 209.
[103]Bosshard; Baumann; Schetty *Helv. Chim. Acta* **1970**, *53*, 1271.
[104]For a review, see Krief; Laboureur; Dumont; Labar *Bull. Soc. Chim. Fr.* **1990**, 681-696.
[105]Epoxides can also be rearranged with basic catalysts, though the products are usually different. For a review, see Yandovskii; Ershov *Russ. Chem. Rev.* **1972**, *41*, 403, 410.
[106]For a list of reagents that accomplish this transformation, with references, see Larock *Comprehensive Organic Transformations*; VCH: New York, 1989, p. 628.

intermediates in the pinacol rearrangements of certain glycols.[107] Among the evidence for the mechanism given is that $Me_2COHCOHMe_2$, $Me_2COHCNH_2Me_2$, and $Me_2COHCClMe_2$ gave the reaction at different rates (as expected) but yielded the *same mixture* of two products—pinacol and pinacolone—indicating a common intermediate.[108]

Epoxides can also be rearranged to aldehydes or ketones on treatment with certain metallic catalysts.[109] A good way to prepare β-diketones consists of heating α,β-epoxy ketones at 80-140°C in toluene with small amounts of $(Ph_3P)_4Pd$ and 1,2-bis(diphenylphosphino)ethane.[110]

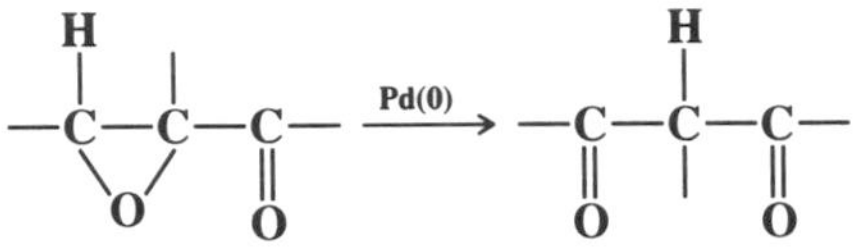

β-Hydroxy ketones can be prepared by treating the silyl ethers (**45**) of α,β-epoxy alcohols with $TiCl_4$.[111]

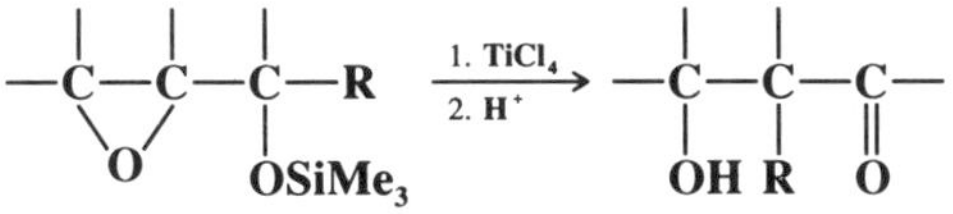

OS **I**, 462; **II**, 73, 408; **III**, 312; **IV**, 375, 957; **V**, 326, 647; **VI**, 39, 320; **VII**, 129. See also OS **VII**, 456.

8-3 Expansion and Contraction of Rings
Demyanov ring contraction; Demyanov ring expansion

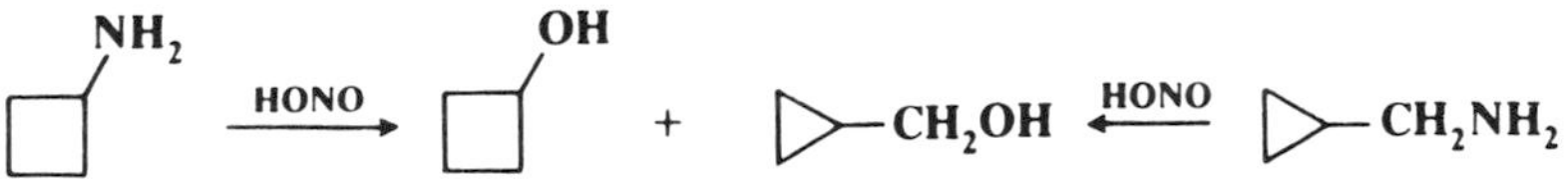

When a positive charge is formed on an alicyclic carbon, migration of an alkyl group can take place to give ring contraction, producing a ring that is one carbon smaller than the original

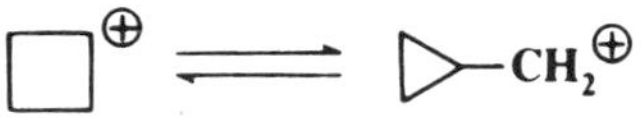

Note that this change involves conversion of a secondary to a primary carbocation. In a similar manner, when a positive charge is placed on a carbon α to an alicyclic ring, ring

---

[107]See, for example, Matsumoto *Tetrahedron* **1968**, *24*, 6851; Pocker; Ronald *J. Am. Chem. Soc.* **1970**, *92*, 3385, *J. Org. Chem.* **1970**, *35*, 3362; Tamura; Moriyoshi *Bull. Chem. Soc. Jpn.* **1974**, *47*, 2942.

[108]Pocker *Chem. Ind.* (*London*) **1959**, 332. See also Herlihy *Aust. J. Chem.* **1981**, *34*, 107.

[109]For example, see Alper; Des Roches; Durst; Legault *J. Org. Chem.* **1976**, *41*, 3611; Milstein; Buchman; Blum *J. Org. Chem.* **1977**, *42*, 2299; Prandi; Namy; Menoret; Kagan *J. Organomet. Chem.* **1985**, *285*, 449; Miyashita; Shimada; Sugawara; Nohira *Chem. Lett.* **1986**, 1323; Maruoka; Nagahara; Ooi; Yamamoto *Tetrahedron Lett.* **1989**, *30*, 5607.

[110]Suzuki; Watanabe; Noyori *J. Am. Chem. Soc.* **1980**, *102*, 2095.

[111]Maruoka; Hasegawa; Yamamoto; Suzuki; Shimazaki; Tsuchihashi *J. Am. Chem. Soc.* **1986**, *108*, 3827. For a different rearrangement of **45**, see Maruoka; Ooi; Yamamoto *J. Am. Chem. Soc.* **1989**, *111*, 6431.

expansion can take place.[112] The new carbocation, and the old one, can then give products by combination with a nucleophile (e.g., the alcohols shown above), or by elimination, so that this reaction is a special case of **8-1**. Often, both rearranged and unrearranged products are formed, so that, for example, cyclobutylamine and cyclopropylmethylamine give similar mixtures of the two alcohols shown above on treatment with nitrous acid (a small amount of 3-buten-1-ol is also produced). When the carbocation is formed by diazotization of an amine, the reaction is called the *Demyanov rearrangement*,[113] but of course similar products are formed when the carbocation is generated in other ways. The expansion reaction has been performed on rings of C_3 to C_8,[114] but yields are best with the smaller rings, where relief of small-angle strain provides a driving force for the reaction. The contraction reaction has been applied to four-membered rings and to rings of C_6 to C_8, but contraction of a cyclopentyl cation to a cyclobutylmethyl system is generally not feasible because of the additional strain involved. Strain is apparently much less of a factor in the cyclobutyl–cyclopropylmethyl interconversion (for a discussion of this interconversion, see p. 323).

An interesting example of a cascade of ring expansions, similar to the friedelin example described in **8-1**, is the conversion of 16-methylpentaspiro[2.0.2.0.2.0.2.0.2.1]hexadecan-16-ol (**46**) to 2-methylhexacyclo[12.2.0.0^{2,5}.0^{5,8}.0^{8,11}.0^{11,14}]hexadecan-1-ol (**47**) on treatment

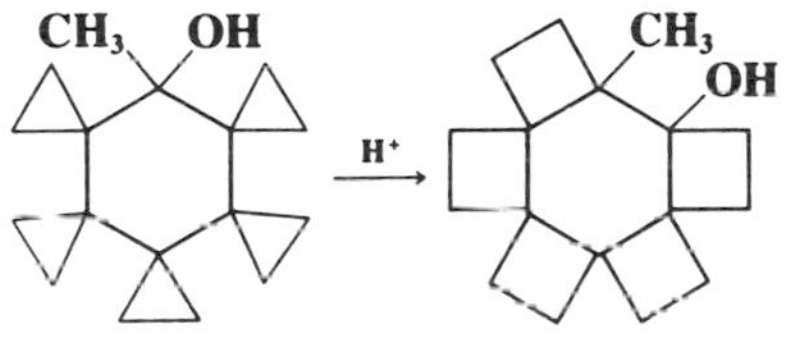

with *p*-toluenesulfonic acid in acetone–water.[115] The student may wish to write out the mechanism as an exercise. Ring expansions of certain hydroxyamines, e.g.,

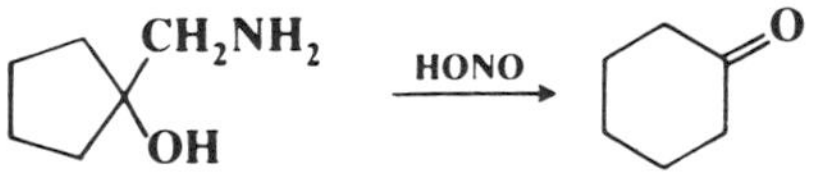

are analogous to the semipinacol rearrangement (**8-2**). This reaction is called the *Tiffeneau–Demyanov ring expansion*. These have been performed on rings of C_4 to C_8 and the yields are better than for the simple Demyanov ring expansion. A similar reaction has been used

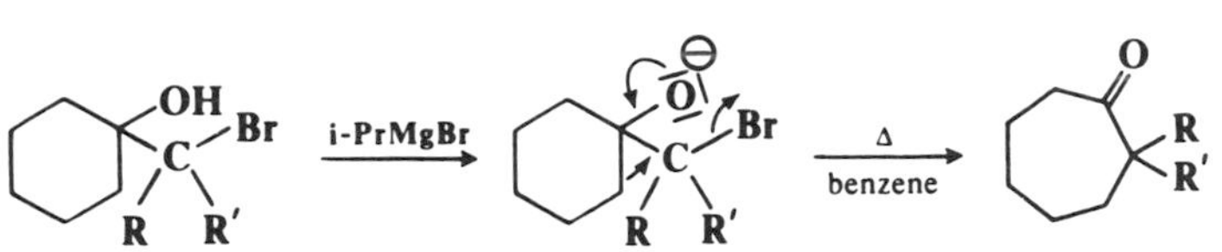

---

[112]For monographs on ring expansions, see Hesse *Ring Enlargement in Organic Chemistry*; VCH: New York, 1991; Gutsche; Redmore *Carbocyclic Ring Expansion Reactions*; Academic Press: New York, 1968. For a review of ring contractions, see Redmore; Gutsche *Adv. Alicyclic Chem.* **1971**, *3*, 1-138. For reviews of ring expansions in certain systems, see Baldwin; Adlington; Robertson *Tetrahedron* **1989**, *45*, 909-922; Stach; Hesse *Tetrahedron* **1988**, *44*, 1573-1590; Dolbier *Mech. Mol. Migr.* **1971**, *3*, 1-66. For reviews of expansions and contractions of three- and four-membered rings, see Salaün, in Rappoport *The Chemistry of the Cyclopropyl Group*, pt. 2; Wiley: New York, 1987, pp. 809-878; Conia; Robson *Angew. Chem. Int. Ed. Engl.* **1975**, *14*, 473-485 [*Angew. Chem.* **87**, 505-516]. For a list of ring expansions and contractions, with references, see Ref. 106, pp. 630-637.

[113]For a review, see Smith; Baer *Org. React.* **1960**, *11*, 157-188.

[114]For a review concerning three-membered rings, see Wong; Hon; Tse; Yip; Tanko; Hudlicky *Chem. Rev.* **1989**, *89*, 165-198, pp. 182-186. For a review concerning three- and four-membered rings, see Breslow, in Mayo *Molecular Rearrangements*, vol. 1; Wiley: New York, 1963, pp. 233-294.

[115]Fitjer; Wehle; Noltemeyer; Egert; Sheldrick *Chem. Ber.* **1984**, *117*, 203. For similar cascade rearrangements, see Giersig; Wehle; Fitjer; Schormann; Clegg *Chem. Ber.* **1988**, *121*, 525, and other papers in this series.

to expand rings of from five to eight members.[116] In this case, a cyclic bromohydrin of the form **48** is treated with a Grignard reagent which, acting as a base, removes the OH proton to give the alkoxide **49**. Refluxing of **49** brings about the ring enlargement. The reaction has been accomplished for **48** in which at least one R group is phenyl or methyl,[117] but fails when both R groups are hydrogen.[118]

A positive charge generated on a three-membered ring gives "contraction" to an allylic cation.[119]

We have previously seen (p. 345) that this is the reason nucleophilic substitutions are not feasible at a cyclopropyl substrate. The reaction is often used to convert cyclopropyl halides and tosylates to allylic products, especially for the purpose of ring expansion, an example being[120]

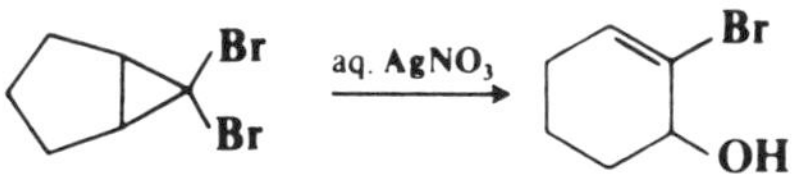

The stereochemistry of these cyclopropyl cleavages is governed by the principle of orbital symmetry conservation (for a discussion, see p. 1119).

Three-membered rings can also be cleaved to unsaturated products in at least two other ways. (1) On pyrolysis, cyclopropanes can undergo "contraction" to propenes.[121] In the simplest case, cyclopropane gives propene when heated to 400 to 500°C. The mechanism is generally regarded[122] as involving a diradical intermediate[123] (recall that free-radical 1,2

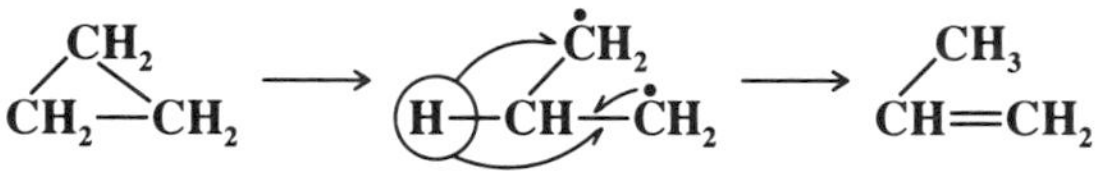

migration is possible for diradicals, p. 1065). (2) The generation of a carbene or carbenoid carbon in a three-membered ring can lead to allenes, and allenes are often prepared in this

[116]Sisti *Tetrahedron Lett.* **1967**, 5327, *J. Org. Chem.* **1968**, *33*, 453. See also Sisti; Vitale *J. Org. Chem.* **1972**, *37*, 4090.

[117]Sisti *J. Org. Chem.* **1970**, *35*, 2670, *Tetrahedron Lett.* **1970**, 3305; Sisti; Meyers *J. Org. Chem.* **1973**, *38*, 4431; Sisti; Rusch *J. Org. Chem.* **1974**, *39*, 1182.

[118]Sisti *J. Org. Chem.* **1968**, *33*, 3953.

[119]For reviews, see Marvell, Ref. 365, pp. 23-53; Sorensen; Rauk, in Marchand; Lehr *Pericyclic Reactions*, vol. 2; Academic Press: New York, 1977, pp. 1-78.

[120]Skell; Sandler *J. Am. Chem. Soc.* **1958**, *80*, 2024.

[121]For reviews, see Berson, in Mayo, Ref. 1, vol. 1, pp. 324-352, *Ann. Rev. Phys. Chem.* **1977**, *28*, 111-132; Bergman, in Kochi, Ref. 55, vol. 1, pp. 191-237; Frey *Adv. Phys. Org. Chem.* **1966**, *4*, 147-193, pp. 148-170.

[122]For evidence that diradical intermediates may not be involved, at least in some cases, see Fields; Haszeldine; Peter *Chem. Commun.* **1967**, 1081; Parry; Robinson *Chem. Commun.* **1967**, 1083; Clifford; Holbrook *J. Chem. Soc., Perkin Trans. 2* **1972**, 1972; Baldwin; Grayston *J. Am. Chem. Soc.* **1974**, *96*, 1629, 1630.

[123]We have seen before that such diradicals can close up to give cyclopropanes (7-46). Therefore, pyrolysis of cyclopropanes can produce not only propenes but also isomerized (cis → trans or optically active → inactive) cyclopropanes. See, for example, Berson; Balquist *J. Am. Chem. Soc.* **1968**, *90*, 7343; Bergman; Carter *J. Am. Chem. Soc.* **1969**, *91*, 7411.

way.[124] One way to generate such a species is treatment of a 1,1-dihalocyclopropane with an alkyllithium compound (**2-39**).[125] In contrast, the generation of a carbene or carbenoid

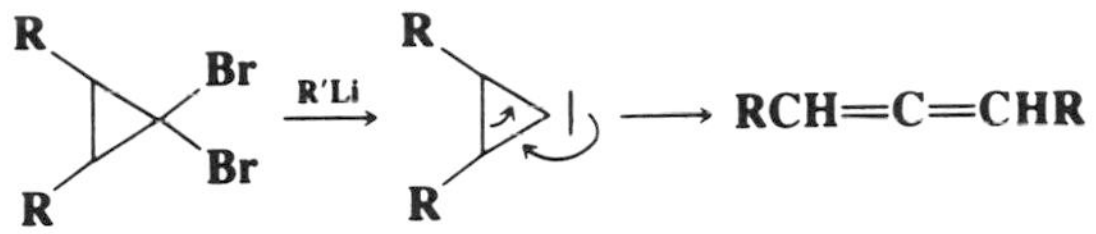

at a cyclopropylmethyl carbon gives ring expansion.[126]

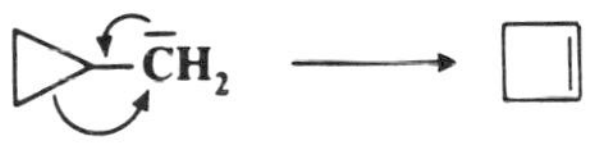

Some free-radical ring enlargements are also known, an example being:[127]

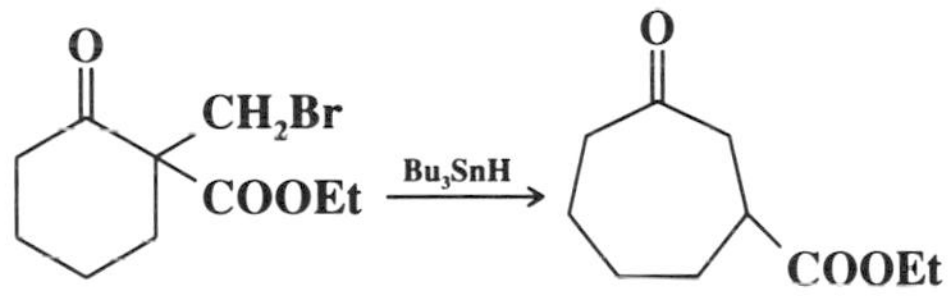

This reaction has been used to make rings of 6, 7, 8, and 13 members. A possible mechanism is:

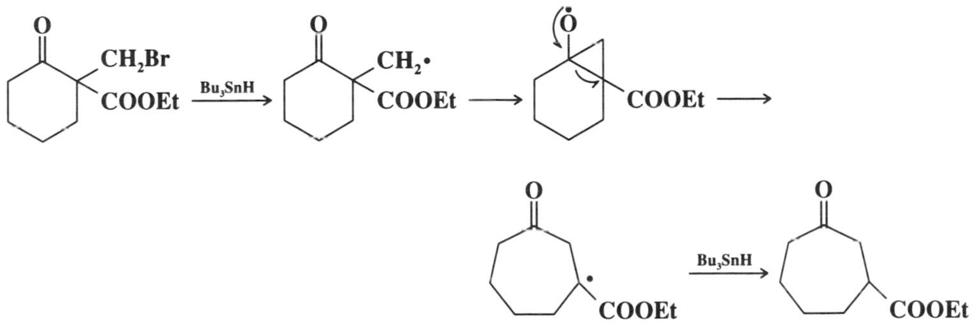

This reaction has been extended to the expansion of rings by 3 or 4 carbons, by the use of a substrate containing $(CH_2)_nX$ ($n = 3$ or 4) instead of CH_2Br.[128] By this means, 5-, 6-, and 7-membered rings were enlarged to 8- to 11-membered rings.

OS **III**, 276; **IV**, 221, 957; **V**, 306, 320; **VI**, 142, 187; **VII**, 12, 114, 117, 129, 135; **65**, 17; **67**, 210; **68**, 220; **69**, 220.

[124]For reviews, see Schuster; Coppola *Allenes in Organic Synthesis*; Wiley: New York, 1984, pp. 20-23; Kirmse *Carbene Chemistry*, 2nd ed.; Academic Press: New York, 1971, pp. 462-467.
[125]See Baird; Baxter *J. Chem. Soc., Perkin Trans. 1* **1979**, 2317, and references cited therein.
[126]For a review, see Gutsche; Redmore, Ref. 112, pp. 111-117.
[127]Dowd; Choi *J. Am. Chem. Soc.* **1987**, *109*, 3493, *Tetrahedron Lett.* **1991**, *32*, 565, *Tetrahedron* **1991**, *47*, 4847. For a related ring expansion, see Baldwin; Adlington; Robertson *J. Chem. Soc., Chem. Commun.* **1988**, 1404.
[128]Dowd; Choi *J. Am. Chem. Soc.* **1987**, *109*, 6548, *Tetrahedron Lett.* **1991**, *32*, 565.

8-4 Acid-Catalyzed Rearrangements of Aldehydes and Ketones

1/Alkyl,2/alkyl-interchange, etc.

$$
\begin{array}{c}
R^1 \\
| \\
R^2-C-C-R^4 \\
| \quad \| \\
R^3 \quad O
\end{array}
\xrightarrow{H^+}
\begin{array}{c}
R^4 \\
| \\
R^2-C-C-R^1 \\
| \quad \| \\
R^3 \quad O
\end{array}
$$

Rearrangements of this type, where a group α to a carbonyl "changes places" with a group attached to the carbonyl carbon, occur when migratory aptitudes are favorable.[129] R^2, R^3, and R^4 may be alkyl or hydrogen. Certain aldehydes have been converted to ketones, and ketones to other ketones (though more drastic conditions are required for the latter), but no rearrangement of a ketone to an aldehyde (R^1 = H) has so far been reported. There are two mechanisms,[130] each beginning with protonation of the oxygen and each involving two migrations. In one pathway, the migrations are in opposite directions:[131]

$$
\begin{array}{c}
\widehat{R^1}_{\oplus} \\
| \\
R^2-C-C-R^4 \\
| \quad | \\
R^3 \quad OH
\end{array}
\longrightarrow
\begin{array}{c}
\oplus\widehat{R^4} \\
| \\
R^2-C-C-R^1 \\
| \quad | \\
R^3 \quad OH
\end{array}
\longrightarrow
\begin{array}{c}
R^4 \; R^1 \\
| \quad | \\
R^2-C-C\oplus \\
| \quad | \\
R^3 \quad OH
\end{array}
\xrightarrow{-H^+}
\begin{array}{c}
R^4 \\
| \\
R^2-C-C-R^1 \\
| \quad \| \\
R^3 \quad O
\end{array}
$$

In the other pathway the migrations are in the same direction. The actual mechanism of this pathway is not certain, but an epoxide (protonated) intermediate[132] is one possibility:[133]

$$
\begin{array}{c}
\widehat{R^2}_{\oplus} \\
| \\
R^1-C-C-R^4 \\
| \quad | \\
R^3 \quad OH
\end{array}
\longrightarrow
\begin{array}{c}
\widehat{R^3}\; R^4 \\
| \quad | \\
R^1-C-C-R^2 \\
\quad \diagdown \diagup \\
\quad O_\oplus \\
\quad | \\
\quad H
\end{array}
\longrightarrow
\begin{array}{c}
R^3 \\
\oplus \; | \\
R^1-C-C-R^2 \\
| \quad | \\
OH \; R^4
\end{array}
\xrightarrow{-H^+}
\begin{array}{c}
R^3 \\
| \\
R^1-C-C-R^2 \\
\| \quad | \\
O \; R^4
\end{array}
$$

If the reaction is carried out with ketone labeled in the C=O group with ^{14}C, the first pathway predicts that the product will contain all the ^{14}C in the C=O carbon, while in the second pathway the label will be in the α carbon (demonstrating migration of oxygen). The results of such experiments[134] have shown that in some cases only the C=O carbon was labeled, in other cases only the α carbon, while in still others both carbons bore the label, indicating that in these cases both pathways were in operation. With α-hydroxy aldehydes and ketones, the process may stop after only one migration (this is called the α-*ketol* rearrangement).

$$
\begin{array}{c}
R^1 \\
| \\
R^2-C-C-R^3 \\
| \quad \| \\
OH \quad O
\end{array}
\xrightarrow{H^+}
\begin{array}{c}
R^1 \\
| \\
R^2-C-C-R^3 \\
\| \quad | \\
O \quad OH
\end{array}
$$

[129]For reviews, see Fry *Mech. Mol. Migr.* **1971**, *4*, 113-196; Collins; Eastham, in Patai, Ref. 1, pp. 771-790.
[130]Favorskii; Chilingaren *C. R. Acad. Sci.* **1926**, *182*, 221.
[131]Raaen; Collins *J. Am. Chem. Soc.* **1958**, *80*, 1409; Kendrick; Benjamin; Collins *J. Am. Chem. Soc.* **1958**, *80*, 4057; Rothrock; Fry *J. Am. Chem. Soc.* **1958**, *80*, 4349; Collins; Bowman *J. Am. Chem. Soc.* **1959**, *81*, 3614.
[132]Zook; Smith; Greene *J. Am. Chem. Soc.* **1957**, *79*, 4436.
[133]Some such pathway is necessary to account for the migration of oxygen that is found. It may involve a protonated epoxide, a 1,2-diol, or simply a 1,2 shift of an OH group.
[134]See, for example, Barton; Porter *J. Chem. Soc.* **1956**, 2483; Fry; Carrick; Adams *J. Am. Chem. Soc.* **1958**, *80*, 4743; Zalesskaya; Remizova *J. Gen. Chem. USSR.* **1965**, *35*, 29; Fry; Oka *J. Am. Chem. Soc.* **1979**, *101*, 6353.

The α-ketol rearrangement can also be brought about by base catalysis, but only if the alcohol is tertiary, since if R^1 or R^2 = hydrogen, enolization of the substrate is more favored than rearrangement.

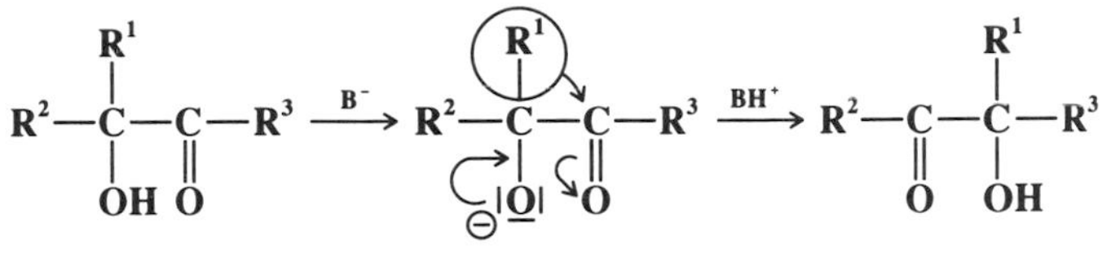

8-5 The Dienone–Phenol Rearrangement
2/C→5/O-Hydro,1/C→2/C-alkyl-*bis*-migration

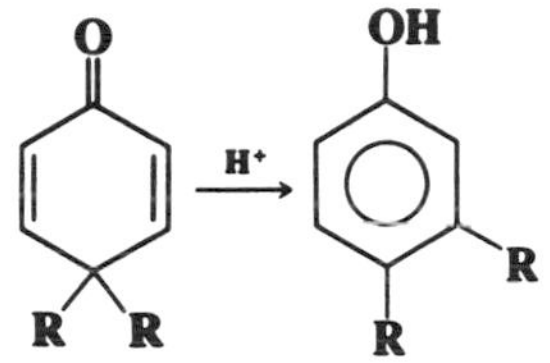

Compounds in which a cyclohexadienone has two alkyl groups in the 4 position undergo, on acid treatment,[135] 1,2 migration of one of these groups:

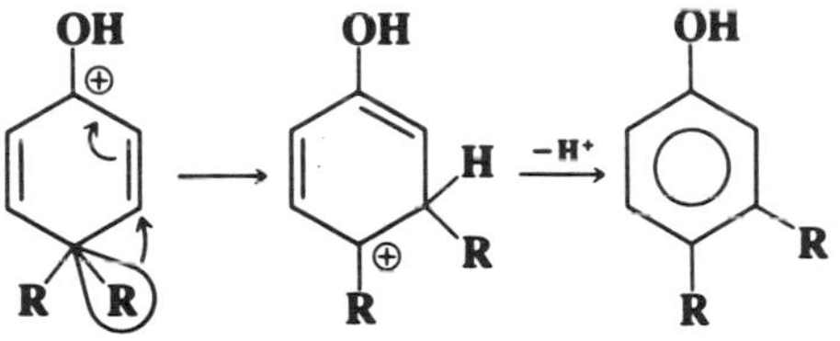

The driving force in the overall reaction (the *dienone–phenol rearrangement*) is of course creation of an aromatic system.[136] It may be noted that **50** and **51** are arenium ions (p. 502), the same as those generated by attack of an electrophile on a phenol.[137] Sometimes, in the reaction of a phenol with an electrophile, a kind of reverse rearrangement (called the *phenol–dienone rearrangement*) takes place, though without an actual migration.[138] An example is

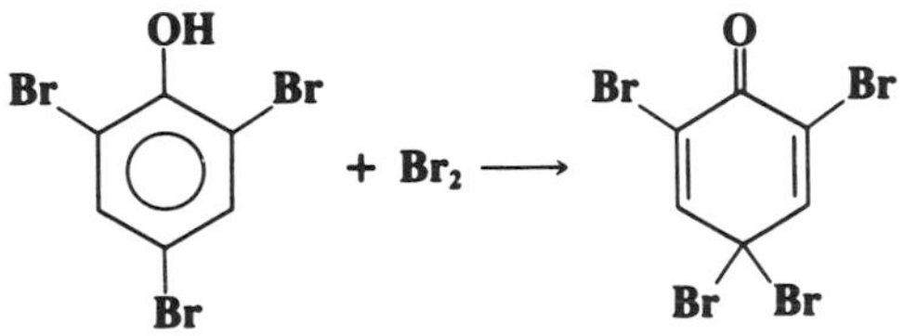

---

[135]For a reagent that greatly accelerates this reaction, see Chalais; Laszlo; Mathy *Tetrahedron Lett.* **1986**, *27*, 2627.
[136]For reviews, see Perkins; Ward *Mech. Mol. Migr.* **1971**, *4*, 55-112, pp. 90-103; Miller *Mech. Mol. Migr.* **1968**, *1*, 247-313; Shine *Aromatic Rearrangements*; Elsevier: New York, 1967, pp. 55-68; Waring *Adv. Alicyclic Chem.* **1966**, *1*, 129-256, pp. 207-223. For a review of other rearrangements of cyclohexadienones, see Miller *Acc. Chem. Res.* **1975**, *8*, 245-256.
[137]For evidence that these ions are indeed intermediates in this rearrangement, see Vitullo *J. Org. Chem.* **1969**, *34*, 224, *J. Org. Chem.* **1970**, *35*, 3976; Vitullo; Grossman *J. Am. Chem. Soc.* **1972**, *94*, 3844; Planas; Tomás; Bonet *Tetrahedron Lett.* **1987**, *28*, 471.
[138]For a review, see Ershov; Volod'kin; Bogdanov *Russ. Chem. Rev.* **1963**, *32*, 75-93.

8-6   The Benzil–Benzilic Acid Rearrangement
   1/O-Hydro,3/oxido-(1/→2/aryl)-*migro*-addition

$$\text{Ar}-\underset{\underset{O}{\|}}{C}-\underset{\underset{O}{\|}}{C}-\text{Ar}' \xrightarrow{\text{OH}^-} \text{Ar}-\underset{\underset{OH}{|}}{\overset{\overset{Ar'}{|}}{C}}-\text{COO}^-$$

When treated with base, α-diketones rearrange to give the salts of α-hydroxy acids, a reaction known as the *benzil–benzilic acid rearrangement* (benzil is PhCOCOPh; benzilic acid is Ph$_2$COHCOOH).[139] Though the reaction is usually illustrated with aryl groups, it can also be applied to aliphatic diketones[140] and to α-keto aldehydes. The use of alkoxide ion instead of OH$^-$ gives the corresponding ester directly,[141] though alkoxide ions that are readily oxidized (such as OEt$^-$ or OCHMe$_2$$^-$) are not useful here, since they reduce the benzil to a benzoin. The mechanism is similar to the rearrangements in **8-1** to **8-4**, but there is a difference: The migrating group does not move to a carbon with an open sextet. The carbon makes room for the migrating group by releasing a pair of π electrons from the C=O bond to the oxygen. The first step is attack of the base at the carbonyl group, the same as the first step of the tetrahedral mechanism of nucleophilic substitution (p. 331) and of many additions to the C=O bond (Chapter 16):

$$\underset{\underset{\underset{|O|}{\overset{\|}{}}}{\overset{\overset{Ar}{|}}{C}}-\underset{\underset{|O|}{\overset{\|}{}}}{C}-\text{Ar}' \xrightarrow{\text{OH}^-} \text{HO}-\underset{\underset{|O|}{}}{\overset{\overset{\overset{Ar}{\frown}}{|}}{C}}-\underset{\underset{|O|}{\overset{\|}{}}}{C}-\text{Ar}' \longrightarrow \text{HO}-\underset{\underset{|O|}{}}{\overset{\overset{Ar}{|}}{C}}-\underset{\underset{|O|}{}}{C}-\text{Ar}' \xrightarrow[\text{shift}]{\text{proton}} {}^-\text{O}-\underset{\underset{O}{\|}}{C}-\underset{\underset{OH}{|}}{\overset{\overset{Ar}{|}}{C}}-\text{Ar}'$$

The mechanism has been intensely studied,[139] and there is much evidence for it.[142] The reaction is irreversible.

   OS **I**, 89.

8-7   The Favorskii Rearrangement
   2/Alkoxy-de-chloro(2/→1/alkyl)-*migro*-substitution

$$\text{R}^1-\underset{\underset{O}{\|}}{C}-\underset{\underset{Cl}{|}}{\overset{\overset{R^2}{|}}{C}}-\text{R}^3 + \text{OR}^{4-} \longrightarrow \text{R}^4\text{O}-\underset{\underset{O}{\|}}{C}-\underset{\underset{R^1}{|}}{\overset{\overset{R^2}{|}}{C}}-\text{R}^3$$

The reaction of α-halo ketones (chloro, bromo, or iodo) with alkoxide ions[143] to give rearranged esters is called the *Favorskii rearrangement*.[144] The use of hydroxide ions or amines

---

[139]For a review, see Selman; Eastham *Q. Rev. Chem. Soc.* **1960**, *14*, 221-235.

[140]For an example, see Schaltegger; Bigler *Helv. Chim. Acta* **1986**, *69*, 1666.

[141]Doering; Urban *J. Am. Chem. Soc.* **1956**, *78*, 5938.

[142]However, some evidence for an SET pathway has been reported: Screttas; Micha-Screttas; Cazianis *Tetrahedron Lett.* **1983**, *24*, 3287.

[143]The reaction has also been reported to take place with BF$_3$–MeOH and Ag$^+$: Giordano; Castaldi; Casagrande; Abis *Tetrahedron Lett.* **1982**, *23*, 1385.

[144]For reviews, see Hunter; Stothers; Warnhoff, in Mayo, Ref. 1, vol. 1, pp. 437-461; Chenier *J. Chem. Educ.* **1978**, *55*, 286-291; Rappe, in Patai *The Chemistry of the Carbon-Halogen Bond*, pt. 2; Wiley: New York, 1973, pp. 1084-1101; Redmore; Gutsche, Ref. 112, pp. 46-69; Akhrem; Ustynyuk; Titov *Russ. Chem. Rev.* **1970**, *39*, 732-746.

as bases leads to the free carboxylic acid (salt) or amide, respectively, instead of the ester. Cyclic α-halo ketones give ring contraction:

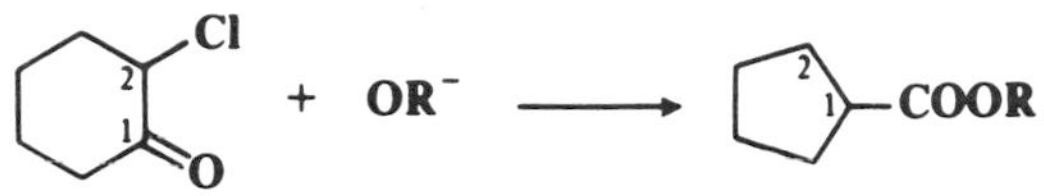

The reaction has also been carried out on α-hydroxy ketones[145] and on α,β-epoxy ketones:[146]

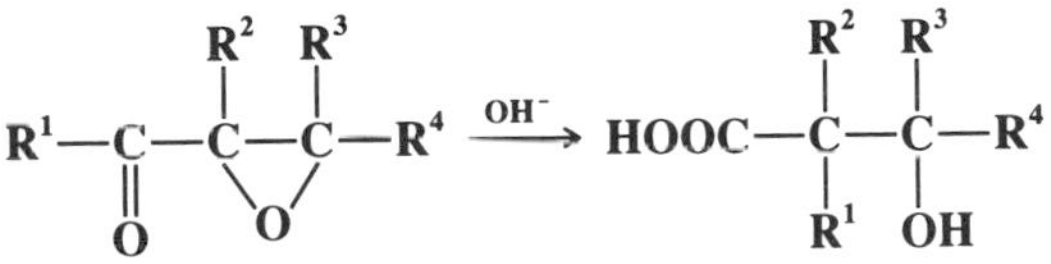

The fact that an epoxide gives a reaction analogous to a halide indicates that the oxygen and halogen are leaving groups in a nucleophilic substitution step.

Through the years, the mechanism[147] of the Favorskii rearrangement has been the subject of much investigation; at least five different mechanisms have been proposed. However, the finding[148] that **54** and **55** *both* give **56** (this behavior is typical) shows that any mechanism

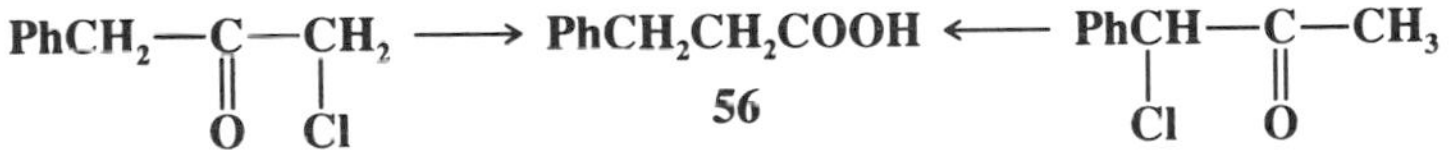

where the halogen leaves and R^1 takes its place is invalid, since in such a case **54** would be expected to give **56** (with $PhCH_2$ migrating), but **55** should give PhCHMeCOOH (with CH_3 migrating). That is, in the case of **55**, it was PhCH that migrated and not methyl. Another important result was determined by radioactive labeling. **52**, in which C-1 and C-2 were equally labeled with ^{14}C, was converted to **53**. The product was found to contain 50% of the label on the carbonyl carbon, 25% on C-1, and 25% on C-2.[149] Now the carbonyl carbon, which originally carried half of the radioactivity, still had this much, so the rearrangement did not directly affect *it*. However, if the C-6 carbon had migrated to C-2, the other half of the radioactivity would be only on C-1 of the product:

On the other hand, if the migration had gone the other way—if the C-2 carbon had migrated to C-6—then this half of the radioactivity would be found solely on C-2 of the product:

[145]Craig; Dinner; Mulligan *J. Org. Chem.* **1972**, *37*, 3539.

[146]See, for example, House; Gilmore *J. Am. Chem. Soc.* **1961**, *83*, 3972; Mouk; Patel; Reusch *Tetrahedron* **1975**, *31*, 13.

[147]For a review of the mechanism, see Baretta; Waegell *React. Intermed.* (Plenum) **1982**, *2*, 527-585.

[148]McPhee; Klingsberg *J. Am. Chem. Soc.* **1944**, *66*, 1132; Bordwell; Scamehorn; Springer *J. Am. Chem. Soc.* **1969**, *91*, 2087.

[149]Loftfield *J. Am. Chem. Soc.* **1951**, *73*, 4707.

The fact that C-1 and C-2 were found to be equally labeled showed that *both migrations occurred*, with equal probability. Since C-2 and C-6 of **52** are not equivalent, this means that there must be a symmetrical intermediate.[150] The type of intermediate that best fits the circumstances is a cyclopropanone,[151] and the mechanism (for the general case) is formulated (replacing R^1 of our former symbolism with CHR^5R^6, since it is obvious that for this mechanism an α hydrogen is required on the nonhalogenated side of the carbonyl):

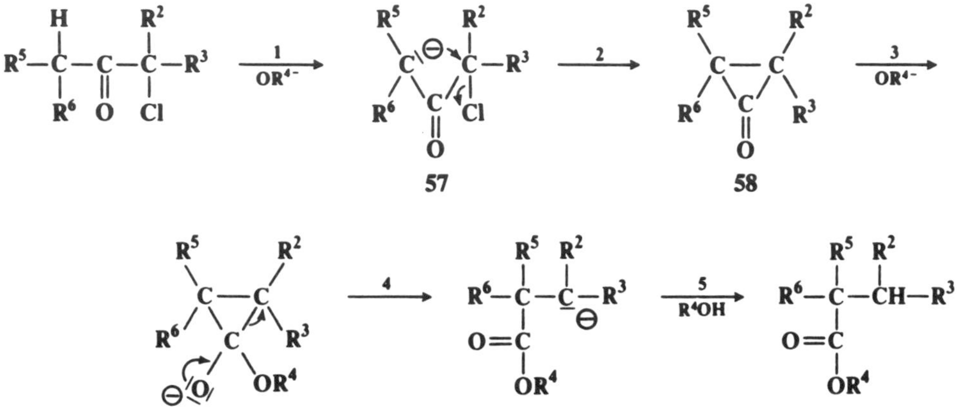

The intermediate corresponding to **58** in the case of **52** is a symmetrical compound, and the three-membered ring can be opened with equal probability on either side of the carbonyl, accounting for the results with ^{14}C. In the general case, **58** is not symmetrical and should open on the side that gives the more stable carbanion.[152] This accounts for the fact that **54** and **55** give the same product. The intermediate in both cases is **59**, which always opens to

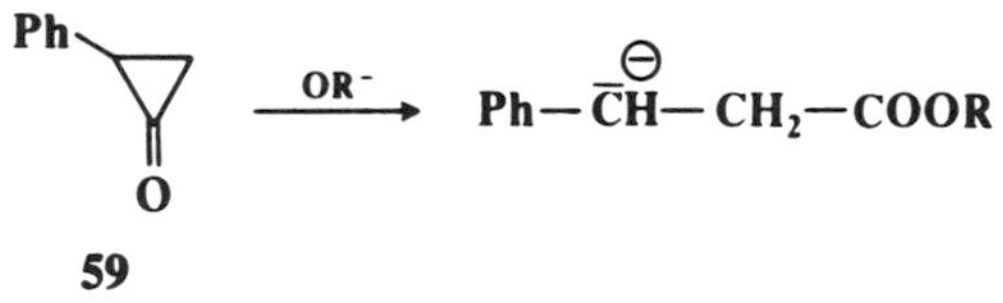

give the carbanion stabilized by resonance. The cyclopropanone intermediate (**58**) has been isolated in the case where $R^2 = R^5 = t$-Bu and $R^3 = R^6 = H$,[153] and it has also been trapped.[154] Also, cyclopropanones synthesized by other methods have been shown to give Favorskii products on treatment with NaOMe or other bases.[155]

The mechanism discussed is in accord with all the facts when the halo ketone contains an α hydrogen on the other side of the carbonyl group. However, ketones that do not have

---

[150]A preliminary migration of the chlorine from C-2 to C-6 was ruled out by the fact that recovered **52** had the same isotopic distribution as the starting **52.**

[151]Although cyclopropanones are very reactive compounds, several of them have been isolated. For reviews of cyclopropanone chemistry, see Wasserman; Clark; Turley *Top. Curr. Chem.* **1974**, *47*, 73-156; Turro *Acc. Chem. Res.* **1969**, *2*, 25-32.

[152]Factors other than carbanion stability (including steric factors) may also be important in determining which side of an unsymmetrical **58** is preferentially opened. See, for example, Rappe; Knutsson *Acta Chem. Scand.* **1967**, *21*, 2205; Rappe; Knutsson; Turro; Gagosian *J. Am. Chem. Soc.* **1970**, *92*, 2032.

[153]Pazos; Pacifici; Pierson; Sclove; Greene *J. Org. Chem.* **1974**, *39*, 1990.

[154]Fort *J. Am. Chem. Soc.* **1962**, *84*, 4979; Cookson; Nye *Proc. Chem. Soc.* **1963**, 129; Breslow; Posner; Krebs *J. Am. Chem. Soc.* **1963**, *85*, 234; Baldwin; Cardellina *Chem. Commun.* **1968**, 558.

[155]Turro; Hammond *J. Am. Chem. Soc.* **1965**, *87*, 3258; Crandall; Machleder *J. Org. Chem.* **1968**, *90*, 7347; Turro; Gagosian; Rappe; Knutsson *Chem. Commun.* **1969**, 270; Wharton; Fritzberg *J. Org. Chem.* **1972**, *37*, 1899.

a hydrogen there also rearrange to give the same type of product. This is usually called the *quasi-Favorskii rearrangement*. An example is found in the preparation of Demerol:[156]

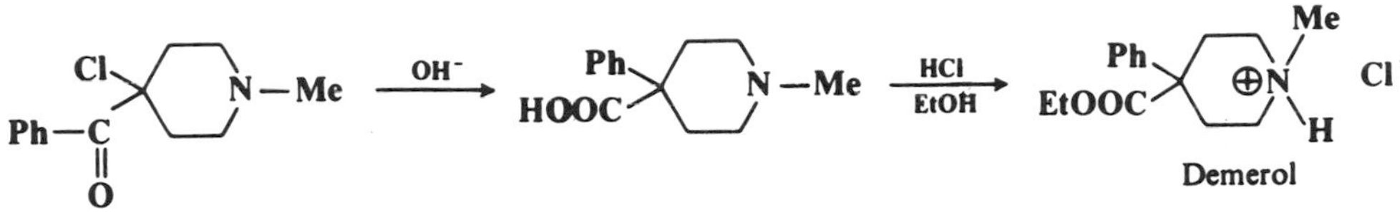

The quasi-Favorskii rearrangement obviously cannot take place by the cyclopropanone mechanism. The mechanism that is generally accepted (called the *semibenzilic mechanism*[157])

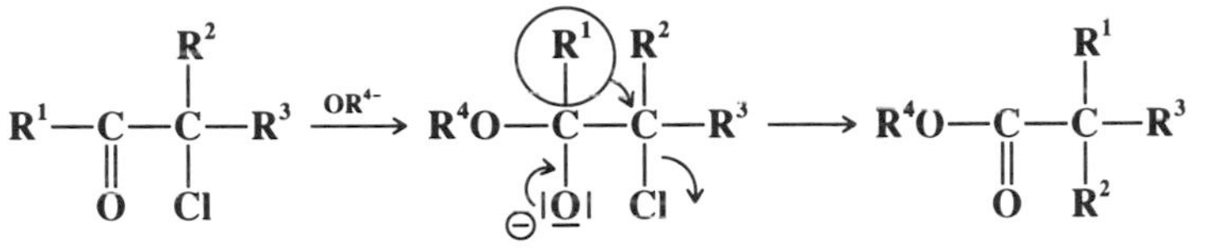

is a base-catalyzed pinacol rearrangement-type mechanism similar to that of **8-6**. This mechanism requires inversion at the migration terminus and this has been found.[158] It has been shown that even where there *is* an appropriately situated α hydrogen, the semibenzilic mechanism may still operate.[159]

OS **IV**, 594; **VI**, 368, 711.

8-8   The Arndt–Eistert Synthesis

$$\underset{O}{R-\overset{|}{\underset{\|}{C}}-Cl} + CH_2N_2 \longrightarrow \underset{O}{R-\overset{|}{\underset{\|}{C}}-CHN_2} \xrightarrow[Ag_2O]{H_2O} R-CH_2-COOH$$

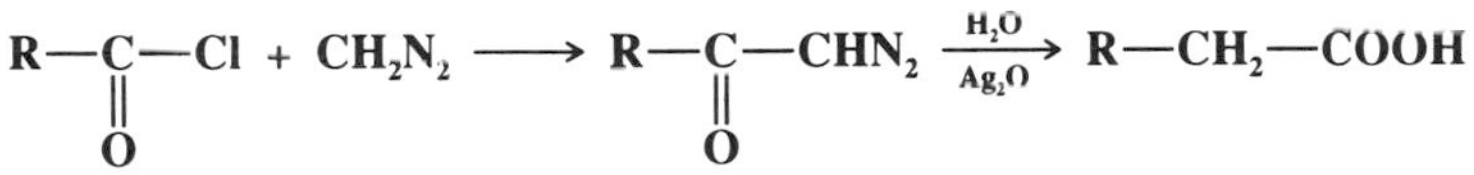

In the *Arndt–Eistert synthesis* an acyl halide is converted to a carboxylic acid with one additional carbon.[160] The first step of this process is reaction **0-112**. The actual rearrangement occurs in the second step on treatment of the diazo ketone with water and silver oxide or with silver benzoate and triethylamine. This rearrangement is called the *Wolff rearrangement*. It is the best method of increasing a carbon chain by one if a *carboxylic acid* is available (**0-101** and **6-34** begin with alkyl halides). If an alcohol R'OH is used instead of water, the ester RCH$_2$COOR' is isolated directly. Similarly, ammonia gives the amide. Other catalysts are sometimes used, e.g., colloidal platinum, copper, etc., but occasionally the diazo ketone is simply heated or photolyzed in the presence of water, an alcohol, or ammonia, with no catalyst at all.[161] The photolysis method[162] often gives better results than the silver catalysis

---

[156]Smissman; Hite *J. Am. Chem. Soc.* **1959**, *81*, 1201.

[157]Tchoubar; Sackur *C. R. Acad. Sci.* **1939**, *208*, 1020.

[158]Baudry; Bégué; Charpentier-Morize *Bull. Soc.Chim. Fr.* **1971**, 1416, *Tetrahedron Lett.* **1970**. 2147.

[159]For example, see Conia; Salaun *Tetrahedron Lett.* **1963**, 1175, *Bull. Soc. Chem. Fr.* **1964**, 1957; Salaun; Garnier; Conia *Tetrahedron* **1973**, *29*, 2895; Rappe; Knutsson *Acta Chem. Scand.* **1967**, *21*, 163; Warnhoff; Wong; Tai *J. Am. Chem. Soc.* **1968**, *90*, 514.

[160]For reviews, see Meier; Zeller *Angew. Chem. Int. Ed. Engl.* **1975**, *14*, 32-43 [*Angew. Chem.* *87*, 52-63]; Kirmse, Ref. 124, pp. 475-493; Rodina; Korobitsyna *Russ. Chem. Rev.* **1967**, *36*, 260-272; For a review of rearrangements of diazo and diazonium compounds, see Whittaker, in Patai *The Chemistry of Diazonium and Diazo Compounds*, pt. 2; Wiley: New York, 1978, pp. 593-644.

[161]For a list of methods, with references, see Ref. 106, p. 933.

[162]For reviews of the photolysis method, see Regitz; Maas *Diazo Compounds*; Academic Press: New York, 1986, pp. 185-195; Ando, in Patai, Ref. 160, pp. 458-475.

method. Of course, diazo ketones prepared in any other way also give the rearrangement.[163] The reaction is of wide scope. R may be alkyl or aryl and may contain many functional groups including unsaturation, but not including groups acidic enough to react with CH_2N_2 or diazo ketones (e.g., **0-5** and **0-26**). Sometimes the reaction is performed with other diazoalkanes (that is, R'CHN_2) to give RCHR'COOH. The reaction has often been used for ring contraction of cyclic diazo ketones,[164] e.g.,[165]

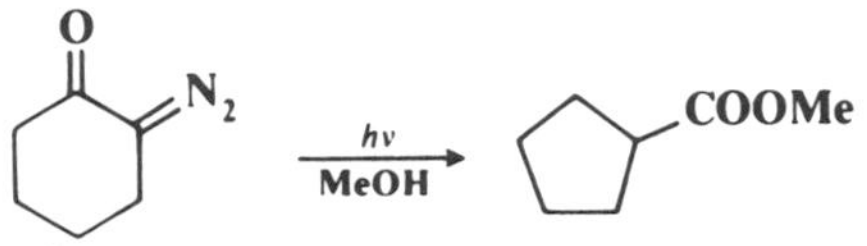

The mechanism is generally regarded as involving formation of a carbene. It is the divalent carbon that has the open sextet and to which the migrating group brings its electron pair:

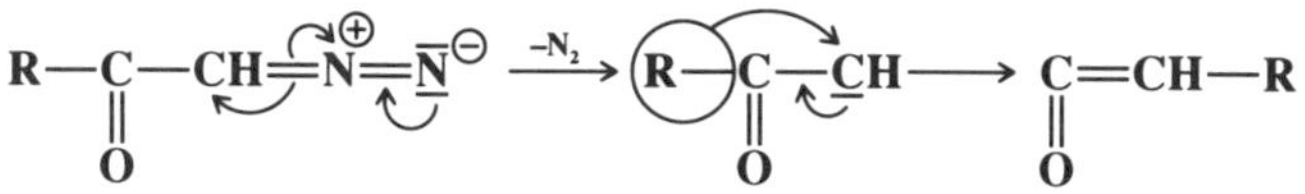

The actual product of the reaction is thus the ketene, which then reacts with water (**5-2**), an alcohol (**5-4**), or ammonia or an amine (**5-7**). Particularly stable ketenes (e.g., $Ph_2C=C=O$) have been isolated and others have been trapped in other ways (e.g., as β-lactams,[166] **6-64**). The purpose of the catalyst is not well understood, though many suggestions have been made. This mechanism is strictly analogous to that of the Curtius rearrangement (**8-15**). Although the mechanism as shown above involves a free carbene and there is much evidence to support this,[167] it is also possible that at least in some cases the two steps are concerted and a free carbene is absent.

When the Wolff rearrangement is carried out photochemically, the mechanism is basically the same,[162] but another pathway can intervene. Some of the ketocarbene originally formed can undergo a carbene–carbene rearrangement, through an oxirene intermediate.[168] This was shown by ^{14}C labeling experiments, where diazo ketones labeled in the carbonyl group

[163]For a method of conducting the reaction with trimethylsilyldiazomethane instead of CH_2N_2, see Aoyama; Shioiri *Tetrahedron Lett.* **1980**, *21*, 4461.

[164]For a review, see Redmore; Gutsche, Ref. 112, pp. 125-136.

[165]Korobitsyna; Rodina; Sushko *J. Org. Chem. USSR* **1968**, *4*, 165; Jones; Ando *J. Am. Chem. Soc.* **1968**, *90*, 2200.

[166]Kirmse; Horner *Chem. Ber.* **1956**, *89*, 2759; also see Horner; Spietschka *Chem. Ber.* **1956**, *89*, 2765.

[167]For a summary of evidence on both sides of the question, see Kirmse, Ref. 124, pp. 476-480. See also Torres; Ribo; Clement; Strausz *Can J. Chem.* **1983**, *61*, 996; Tomoika; Hayashi; Asano; Izawa *Bull. Chem. Soc. Jpn.* **1983**, *56*, 758.

[168]For a review of oxirenes, see Lewars *Chem. Rev.* **1983**, *83*, 519-534.

gave rise to ketenes that bore the label at both C=C carbons.[169] In general, the smallest degree of scrambling (and thus of the oxirene pathway) was found when R' = H. An intermediate believed to be an oxirene has been detected by laser spectroscopy.[170] The oxirene pathway is not found in the thermal Wolff rearrangement. It is likely that an excited singlet state of the carbene is necessary for the oxirene pathway to intervene.[171] In the photochemical process, ketocarbene intermediates, in the triplet state, have been isolated in an Ar matrix at 10–15 K, where they have been identified by uv-visible, ir, and esr spectra.[172] These intermediates went on to give the rearrangement via the normal pathway, with no evidence for oxirene intermediates.

The diazo ketone can exist in two conformations, called *s-E* and *s-Z*. Studies have shown

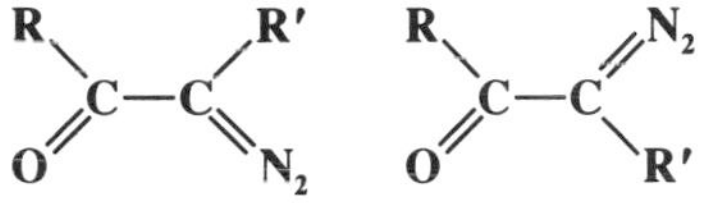

s-Z   *s-E*

that Wolff rearrangement takes place preferentially from the *s-Z* conformation.[173]

Other 1,2 alkyl migrations to a carbene or carbenoid terminus are also known,[174] e.g.,[175]

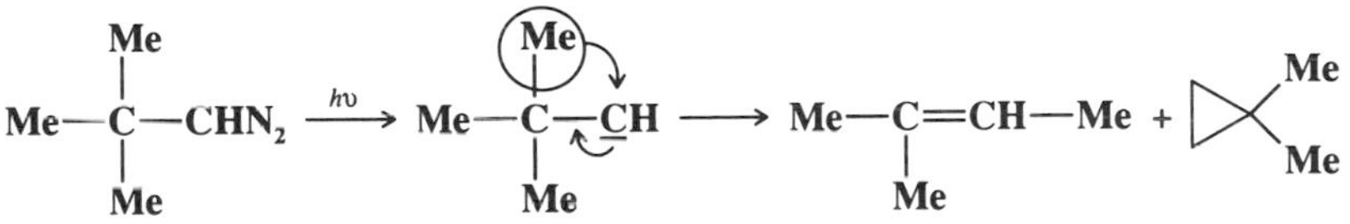

52%   47%

OS **III**, 356; **VI**, 613, 840.

8-9 Homologation of Aldehydes and Ketones
Methylene-insertion

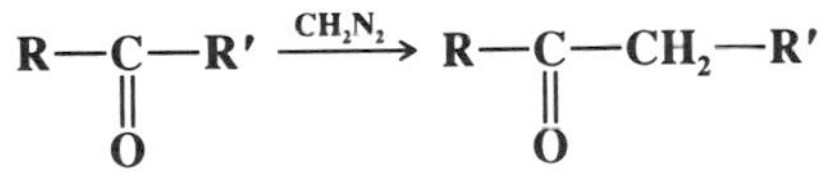

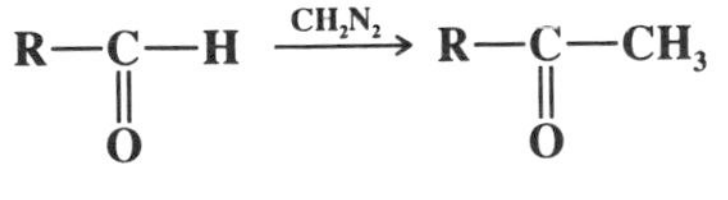

Aldehydes and ketones[176] can be converted to their homologs[177] with diazomethane.[178] Formation of the epoxide (**6-61**) is a side reaction. Although this reaction appears super-

[169]Csizmadia; Font; Strausz *J. Am. Chem. Soc.* **1968**, *90*, 7360; Fenwick; Frater; Ogi; Strausz *J. Am. Chem. Soc.* **1973**, *95*, 124; Zeller *Chem. Ber.* **1978**, *112*, 678. See also Thornton; Gosavi; Strausz *J. Am. Chem. Soc.* **1970**, *92*, 1768; Russell; Rowland *J.Am. Chem. Soc.* **1970**, *92*, 7508; Majerski; Redvanly *J. Chem. Soc. Chem. Commun.* **1972**, 694.

[170]Tanigaki; Ebbesen *J. Am. Chem. Soc.* **1987**, *109*, 5883. See also Bachmann; N'Guessan; Debû; Monnier; Pourcin; Aycard; Bodot *J. Am. Chem. Soc.* **1990**, *112*, 7488.

[171]Csizmadia; Gunning; Gosavi; Strausz *J. Am. Chem. Soc.* **1973**, *95*, 133.

[172]McMahon; Chapman; Hayes; Hess; Krimmer *J. Am. Chem. Soc.* **1985**, *107*, 7597.

[173]Kaplan; Mitchell *Tetrahedron Lett.* **1979**, 759; Tomioka; Okuno; Izawa *J. Org. Chem.* **1980**, *45*, 5278.

[174]For a review, see Kirmse, Ref. 124, pp. 457-462.

[175]Kirmse; Horn *Chem. Ber.* **1967**, *100*, 2698.

[176]For a homologation of carboxylic esters RCOOEt → RCH₂COOEt, which goes by an entirely different pathway, see Kowalski; Haque; Fields *J. Am. Chem. Soc.* **1985**, *107*, 1429.

[177]Other homologation reagents have also reported: See Taylor; Chiang; McKillop *Tetrahedron Lett.* **1977**, 1827; Villieras; Perriot; Normant *Synthesis* **1979**, 968; Hashimoto; Aoyama; Shioiri *Tetrahedron Lett.* **1980**, *21*, 4619; Aoyama; Shioiri *Synthesis* **1988**, 228.

[178]For a review, see Gutsche, *Org. React.* **1954**, *8*, 364-429.

ficially to be similar to the insertion of carbenes into C—H bonds, **2-20** (and IUPAC names it as an insertion), the mechanism is quite different. This is a true rearrangement and no free carbene is involved. The first step is an addition to the C=O bond:

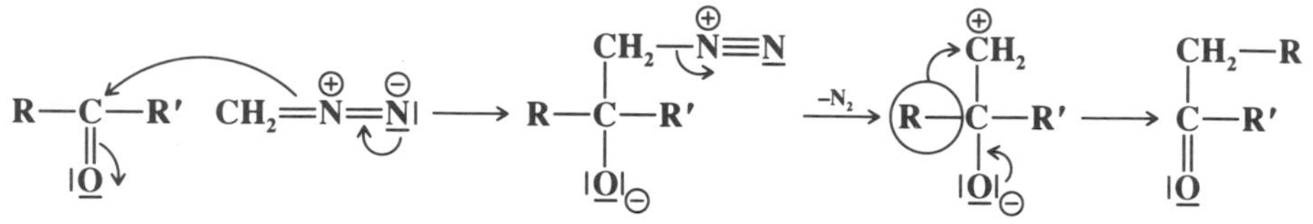

60

The betaine **60** can sometimes be isolated. As shown in **6-61, 60** can also go to the epoxide. The evidence for this mechanism is summarized in the review by Gutsche.[178] It may be noted that this mechanism is essentially the same as in the apparent "insertions" of oxygen (**8-20**) and nitrogen (**8-17**) into ketones.

Aldehydes give fairly good yields of methyl ketones; that is, hydrogen migrates in preference to alkyl. The most abundant side product is not the homologous aldehyde, but the epoxide. However, the yield of aldehyde at the expense of methyl ketone can be increased by the addition of methanol. If the aldehyde contains electron-withdrawing groups, the yield of epoxides is increased and the ketone is formed in smaller amounts, if at all. Ketones give poorer yields of homologous ketones. Epoxides are usually the predominant product here, especially when one or both R groups contain an electron-withdrawing group. The yield of ketones also decreases with increasing length of the chain. The use of BF_3[179] or $AlCl_3$[180] increases the yield of ketone.[181] Cyclic ketones,[182] three-membered[183] and larger, behave particularly well and give good yields of ketones with the ring expanded by one.[184] Aliphatic diazo compounds ($RCHN_2$ and R_2CN_2) are sometimes used instead of diazomethane, with the expected results.[185] An interesting example is the preparation of bicyclic compounds from alicyclic compounds with a diazo group in the side chain, e.g.,[186]

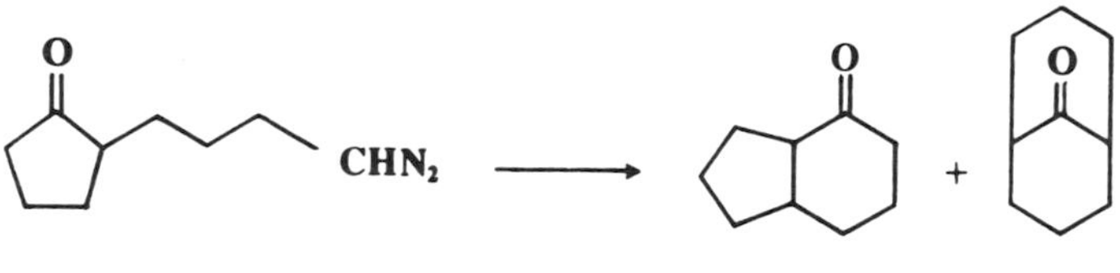

Ethyl diazoacetate can be used analogously, in the presence of a Lewis acid or of triethyloxonium fluoroborate,[187] e.g.,

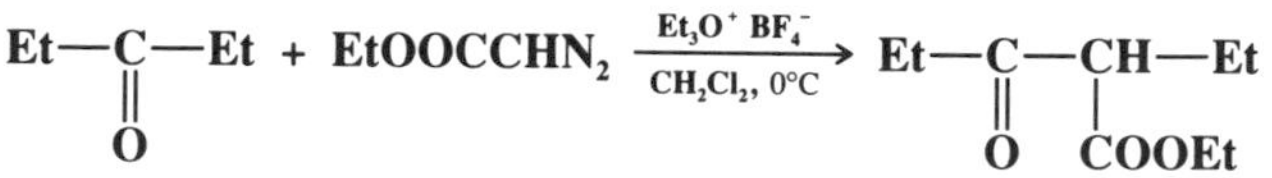

[179]House; Grubbs; Gannon *J. Am. Chem. Soc.* **1960**, *82*, 4099.

[180]Müller; Heischkeil *Tetrahedron Lett.* **1964**, 2809.

[181]For a review of homologations catalyzed by Lewis acids, see Müller; Kessler; Zeeh *Fortschr. Chem. Forsch* **1966**, *7*, 128-171, pp. 137-150.

[182]For other methods for the ring enlargement of cyclic ketones, see Krief; Laboureur *Tetrahedron Lett.* **1987**, *28*, 1545; Krief; Laboureur; Dumont *Tetrahedron Lett.* **1987**, *28*, 1549; Abraham; Bhupathy; Cohen *Tetrahedron Lett.* **1987**, *28*, 2203; Trost; Mikhail *J. Am. Chem. Soc.* **1987**, *109*, 4124.

[183]For example, see Turro; Gagosian *J. Am. Chem. Soc.* **1970**, *92*, 2036.

[184]For a review, see Gutsche; Redmore, Ref. 112, pp. 81-98. For a review pertaining to bridged bicyclic ketones, see Krow *Tetrahedron* **1987**, *43*, 3-38.

[185]For example, see Smith *J. Org. Chem.* **1960**, *25*, 453; Warner; Walsh; Smith *J. Chem. Soc.* **1962**, 1232; Loeschorn; Nakajima; Anselme *Bull. Soc. Chim. Belg.* **1981**, *90*, 985.

[186]Gutsche; Bailey *J. Org. Chem.* **1963**, *28*, 607; Gutsche; Zandstra *J. Org. Chem.* **1974**, *39*, 324.

[187]Mock; Hartman *J. Org. Chem.* **1977**, *42*, 459, 466; Baldwin; Landmesser *Synth. Commun.* **1978**, *8*, 413.

When unsymmetrical ketones were used in this reaction (with BF_3 as catalyst), the less highly substituted carbon preferentially migrated.[188] The reaction can be made regioselective by applying this method to the α-halo ketone, in which case only the other carbon migrates.[189] The ethyl diazoacetate procedure has also been applied to the acetals or ketals of α,β-unsaturated aldehydes and ketones.[190]

OS **IV**, 225, 780.

B. Carbon-to-Carbon Migrations of Other Groups

8-10 Migrations of Halogen, Hydroxyl, Amino, etc.
Hydroxy-de-bromo-*cine*-substitution, etc.

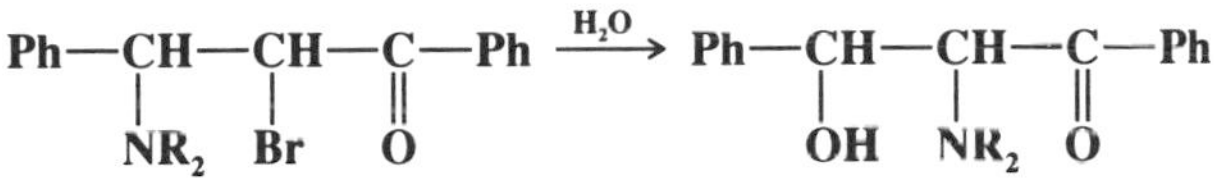

$$Ph-CH-CH-C-Ph \xrightarrow{H_2O} Ph-CH-CH-C-Ph$$
$$\quad\;\; | \quad\;\; | \quad\; || \qquad\qquad\qquad | \quad\;\; | \quad\; ||$$
$$\quad NR_2 \;\; Br \;\; O \qquad\qquad\qquad OH \;\; NR_2 \; O$$

When a nucleophilic substitution is carried out on a substrate that has a neighboring group (p. 309) on the adjacent carbon, if the cyclic intermediate is opened on the opposite side, the result is migration of the neighboring group. In the example shown above (NR_2 = morpholino),[191] the reaction took place as follows:

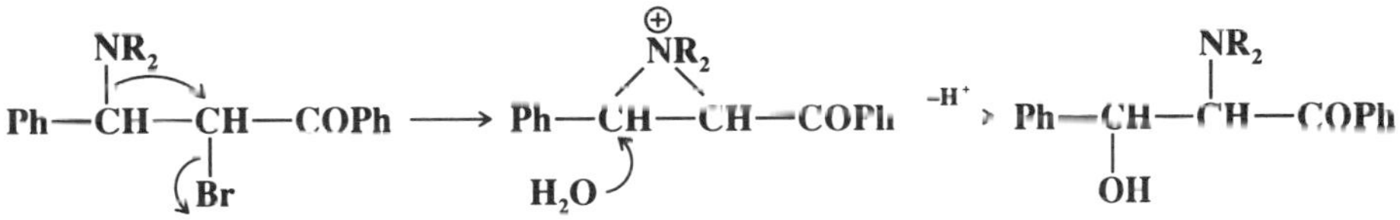

$$Ph-\overset{\overset{\displaystyle NR_2}{|}}{CH}-\overset{\overset{\displaystyle |}{CH}}{\underset{\underset{\displaystyle Br}{|}}{}}-COPh \longrightarrow Ph-\overset{\overset{\displaystyle \oplus NR_2}{|}}{CH}-CH-COPh \xrightarrow{-H^+} Ph-\overset{\overset{\displaystyle NR_2}{|}}{CH}-\overset{\overset{\displaystyle |}{CH}}{\underset{\underset{\displaystyle OH}{|}}{}}-COPh$$

Another example is[192] (ONs = nosylate, see p. 353):

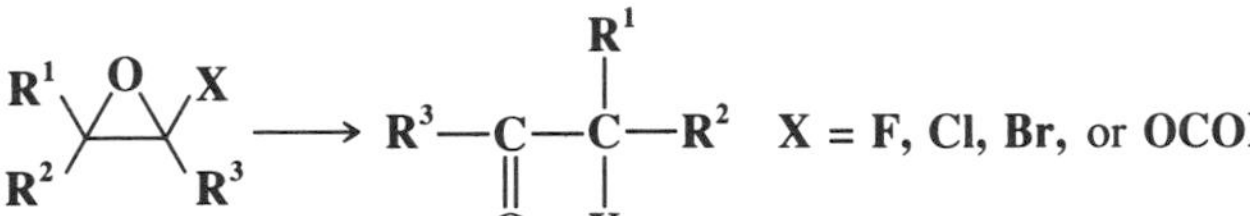

$$CH_3-\overset{\overset{\displaystyle |}{CH}}{\underset{\underset{\displaystyle Cl}{|}}{}}-CH_2-ONs \xrightarrow{F_3CCOOH} CH_3-\overset{\overset{\displaystyle |}{CH}}{\underset{\underset{\displaystyle OOCCF_3}{|}}{}}-CH_2-Cl$$

α-Halo and α-acyloxy epoxides undergo ready rearrangement to α-halo and α-acyloxy ketones, respectively.[193] These substrates are very prone to rearrange, and often do so on

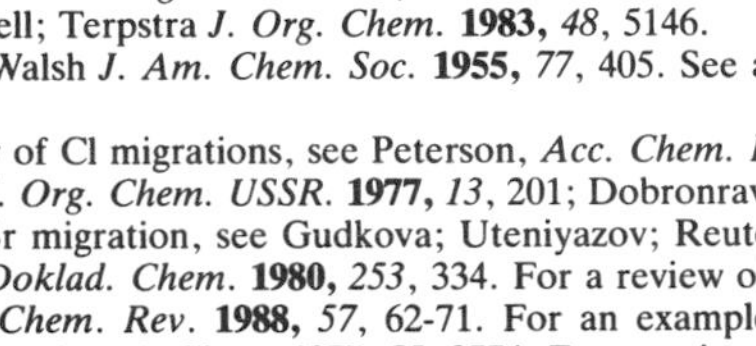

$$\overset{R^1}{\underset{R^2}{\Large\diagdown}}\overset{O}{\underset{}{\bigtriangleup}}\overset{X}{\underset{R^3}{\Large\diagup}} \longrightarrow R^3-\overset{\overset{\displaystyle R^1}{|}}{\underset{\underset{\displaystyle O}{||}}{C}}-\overset{\overset{}{|}}{\underset{\underset{\displaystyle X}{|}}{C}}-R^2 \quad X = F, Cl, Br, or OCOR$$

[188]Liu; Majumdar *Synth. Commun.* **1975**, *5*, 125.
[189]Dave; Warnhoff *J. Org. Chem.* **1983**, *48*, 2590.
[190]Doyle; Trudell; Terpstra *J. Org. Chem.* **1983**, *48*, 5146.
[191]Southwick; Walsh *J. Am. Chem. Soc.* **1955**, *77*, 405. See also Suzuki; Okano; Nakai; Terao; Sekiya *Synthesis* **1983**, 723.
[192]For a review of Cl migrations, see Peterson, *Acc. Chem. Res.* **1971**, *4*, 407-413. See also Loktev; Korchagina; Shubin; Koptyug *J. Org. Chem. USSR.* **1977**, *13*, 201; Dobronravov; Shteingarts *J. Org. Chem. USSR.* **1977**, *13*, 420. For examples of Br migration, see Gudkova; Uteniyazov; Reutov *Doklad. Chem.* **1974**, *214*, 70; Brusova; Gopius; Smolina; Reutov *Doklad. Chem.* **1980**, *253*, 334. For a review of F migration (by several mechanisms) see Kobrina; Kovtonyuk *Russ. Chem. Rev.* **1988**, *57*, 62-71. For an example OH migration, see Cathcart; Bovenkamp; Moir; Bannard; Casselman *Can. J. Chem.* **1977**, *55*, 3774. For a review of migrations of ArS and $Ar_2P(O)$, see Warren *Acc. Chem. Res.* **1978**, *11*, 403-406. See also Aggarwal; Warren *J. Chem. Soc.*, *Perkin Trans. 1* **1987**, 2579.
[193]For a review, see McDonald *Mech. Mol. Migr.* **1971**, *3*, 67-107.

standing without a catalyst, though in some cases an acid catalyst is necessary. The reaction is essentially the same as the rearrangement of epoxides shown in **8-2**, except that in this case halogen or acyloxy is the migrating group (as shown above; however, it is also possible for one of the R groups—alkyl, aryl, or hydrogen—to migrate instead, and mixtures are sometimes obtained).

8-11   Migration of Boron
Hydro,dialkylboro-interchange, etc.

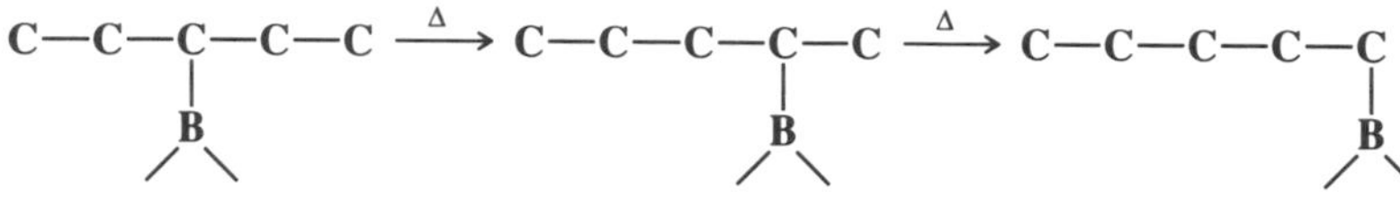

$$\text{C—C—C—C—C} \xrightarrow{\Delta} \text{C—C—C—C—C} \xrightarrow{\Delta} \text{C—C—C—C—C}$$

When a nonterminal borane is heated at temperatures ranging from 100 to 200°C, the boron moves toward the end of the chain.[194] The reaction is catalyzed by small amounts of borane or other species containing B—H bonds. The boron can move past a branch, e.g.,

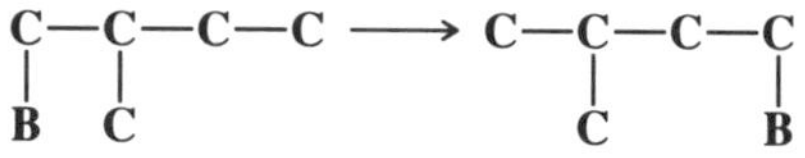

$$\text{C—C—C—C} \longrightarrow \text{C—C—C—C}$$

but not past a double branch, e.g.,

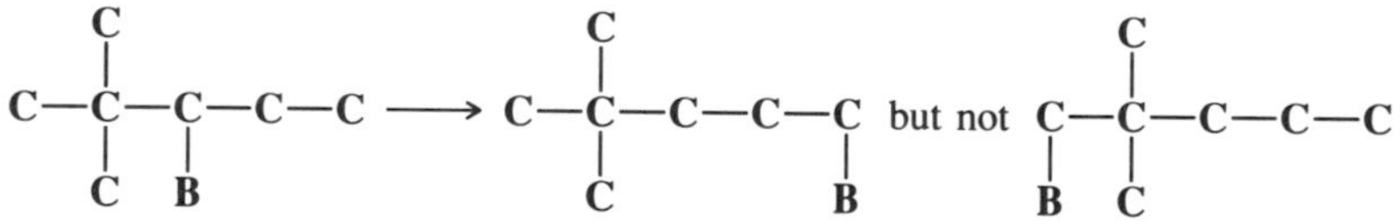

$$\text{C—C—C—C—C} \longrightarrow \text{C—C—C—C—C} \text{ but not } \text{C—C—C—C—C}$$

The reaction is an equilibrium: **61, 62,** and **63** each gave a mixture containing about 40% **61,** 1% **62,** and 59% **63.** The migration can go quite a long distance. Thus

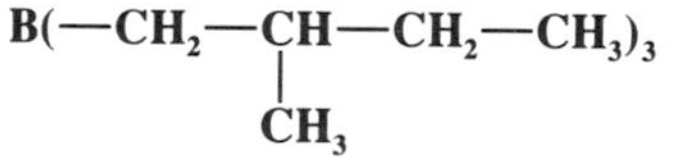

$$\text{B(—CH}_2\text{—CH—CH}_2\text{—CH}_3)_3 \quad (\text{CH}_3\text{—CH—CH—CH}_3)_3\text{B} \quad (\text{CH}_3\text{—CH—CH}_2\text{—CH}_2\text{—})_3\text{B}$$
$$\text{CH}_3 \qquad\qquad\qquad \text{CH}_3 \qquad\qquad\qquad \text{CH}_3$$
$$\textbf{61} \qquad\qquad\qquad\qquad \textbf{62} \qquad\qquad\qquad\qquad \textbf{63}$$

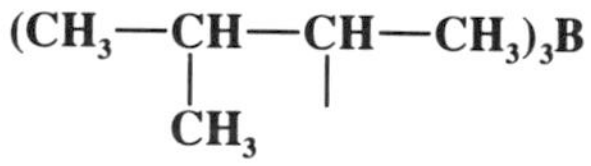

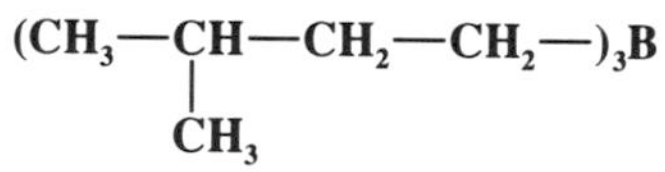

$(C_{11}H_{23}CHC_{11}H_{23})_3B$ was completely converted to $(C_{23}H_{47})_3B$, involving a migration of 11 positions.[195] If the boron is on a cycloalkyl ring, it can move around the ring; if any alkyl chain is also on the ring, the boron may move from the ring to the chain, ending up at the end of the chain.[196] The reaction is useful for the migration of double bonds in a controlled way (see **2-2**). The mechanism may involve a π complex, at least partially.[197]

8-12   Rearrangement of Grignard Reagents
Hydro,magnesio-interchange

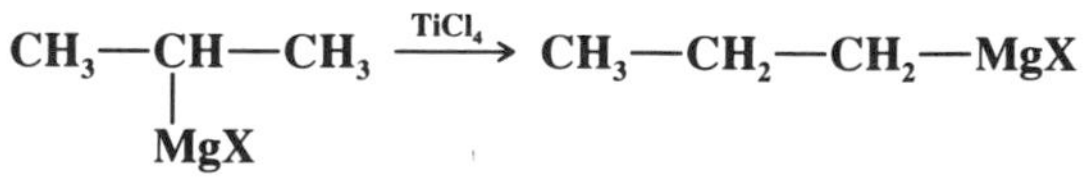

$$\text{CH}_3\text{—CH—CH}_3 \xrightarrow{\text{TiCl}_4} \text{CH}_3\text{—CH}_2\text{—CH}_2\text{—MgX}$$
$$\text{MgX}$$

[194]Brown *Hydroboration*; W. A. Benjamin: New York, 1962, pp. 136-149, Brown; Zweifel *J. Am. Chem. Soc.* **1966**, *88*, 1433. See also Brown; Racherla *J. Organomet. Chem.* **1982**, *241*, C37.
[195]Logan *J. Org. Chem.* **1961**, *26*, 3657.
[196]Brown; Zweifel *J. Am. Chem. Soc.* **1967**, *89*, 561.
[197]See Wood; Rickborn *J. Org. Chem.* **1983**, *48*, 555; Field; Gallagher *Tetrahedron Lett.* **1985**, *26*, 6125.

The MgX of Grignard reagents[198] can migrate to terminal positions in the presence of small amounts of $TiCl_4$.[199] The proposed mechanism consists of metal exchange (**2-35**), elimination–addition, and metal exchange:

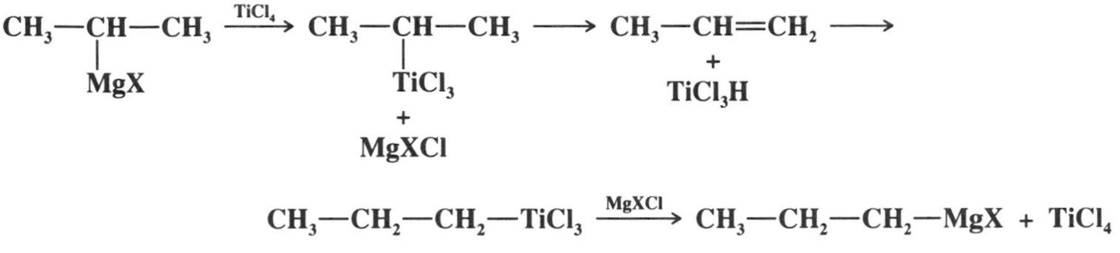

$$CH_3-\underset{\underset{MgX}{|}}{CH}-CH_3 \xrightarrow{TiCl_4} CH_3-\underset{\underset{\underset{MgXCl}{+}}{\underset{TiCl_3}{|}}}{CH}-CH_3 \longrightarrow CH_3-\underset{\underset{TiCl_3H}{+}}{CH}=CH_2 \longrightarrow$$

$$CH_3-CH_2-CH_2-TiCl_3 \xrightarrow{MgXCl} CH_3-CH_2-CH_2-MgX + TiCl_4$$

The addition step is similar to **5-12** or **5-13** and follows Markovnikov's rule, so the positive titanium goes to the terminal carbon.

8-13 The Neber Rearrangement
Neber oxime tosylate-amino ketone rearrangement

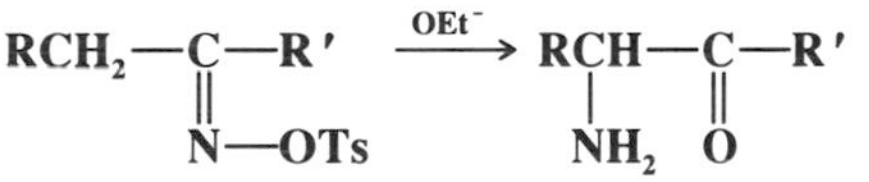

$$RCH_2-\underset{\underset{N-OTs}{||}}{C}-R' \xrightarrow{OEt^-} RCH-\underset{\underset{O}{||}}{\underset{NH_2}{|}}C-R'$$

α-Amino ketones can be prepared by treatment of ketoxime tosylates with a base such as ethoxide ion or pyridine.[200] This is called the *Neber rearrangement*. R is usually aryl, though the reaction has been carried out with R = alkyl or hydrogen. R' may be alkyl or aryl but not hydrogen. The Beckmann rearrangement (**8-18**) and the abnormal Beckmann reaction (elimination to the nitrile, **7-38**) may be side reactions, though these generally occur in acid media. A similar rearrangement is given by N,N-dichloroamines of the type $RCH_2CH(NCl_2)R'$, where the product is also $RCH(NH_2)COR'$.[201] The mechanism of the Neber rearrangement is as follows:[202]

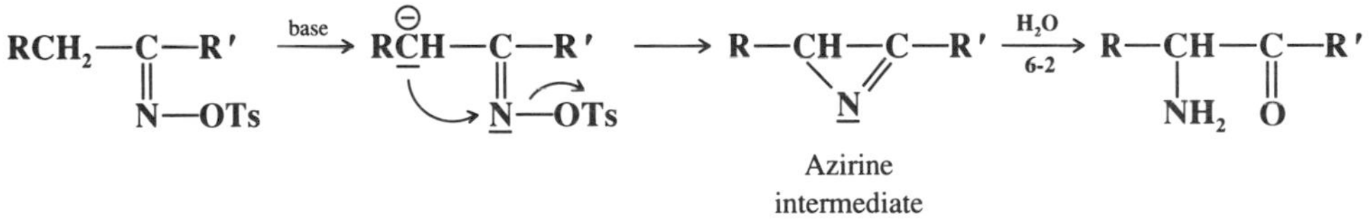

$$RCH_2-\underset{\underset{N-OTs}{||}}{C}-R' \xrightarrow{base} R\overset{\ominus}{CH}-\underset{\underset{N-OTs}{||}}{C}-R' \longrightarrow R-\underset{\underset{N}{\diagdown\diagup}}{CH-C}-R' \xrightarrow[6-2]{H_2O} R-\underset{\underset{NH_2}{|}}{CH}-\underset{\underset{O}{||}}{C}-R'$$

Azirine
intermediate

The best evidence for this mechanism is that the azirine intermediate has been isolated.[203] In contrast to the Beckmann rearrangement, this one is sterically indiscriminate:[204] Both a syn and an anti ketoxime give the same product. The mechanism as shown above consists of three steps. However, it is possible that the first two steps are concerted, and it is also possible that what is shown as the second step is actually two steps: loss of OTs to give a nitrene, and formation of the azirine. In the case of the dichloroamines, HCl is first lost to

[198]For reviews of rearrangements in organomagnesium chemistry, see Hill *Adv. Organomet. Chem.* **1977**, *16*, 131-165, *J. Organomet. Chem.* **1975**, *91*, 123-271.
[199]Cooper; Finkbeiner *J. Org. Chem.* **1962,** *27*, 1493; Fell; Asinger; Sulzbach *Chem. Ber.* **1970**, *103*, 3830. See also Ashby; Ainslie *J. Organomet. Chem.* **1983**, *250*, 1.
[200]For a review, see Conley; Ghosh *Mech. Mol. Migr.* **1971**, *4*, 197-308, pp. 289-304.
[201]Baumgarten; Petersen *J. Am. Chem. Soc.* **1960**, *82*, 459, and references cited therein.
[202]Cram; Hatch *J. Am. Chem. Soc.* **1953**, *75*, 33; Hatch; Cram *J. Am. Chem. Soc.* **1953**, *75*, 38.
[203]Neber; Burgard *Liebigs Ann. Chem.* **1932**, *493*, 281; Parcell *Chem. Ind.* (*London*) **1963**, 1396; Ref. 202.
[204]House; Berkowitz *J. Org. Chem.* **1963**, *28*, 2271.

give $RCH_2C(=NCl)R'$, which then behaves analogously.[205] N-Chloroimines prepared in other ways also give the reaction.[206]

OS **V**, 909; **VII**, 149.

C. Carbon-to-Nitrogen Migrations of R and Ar. The reactions in this group are nucleophilic migrations from a carbon to a nitrogen atom. In each case the nitrogen atom either has six electrons in its outer shell (and thus invites the migration of a group carrying an electron pair) or else loses a nucleofuge concurrently with the migration (p. 1053). Reactions **8-14** to **8-17** are used to prepare amines from acid derivatives. Reactions **8-17** and **8-18** are used to prepare amines from ketones. The mechanisms of **8-14**, **8-15**, **8-16**, and **8-17** (with carboxylic acids) are very similar and follow one of two patterns:

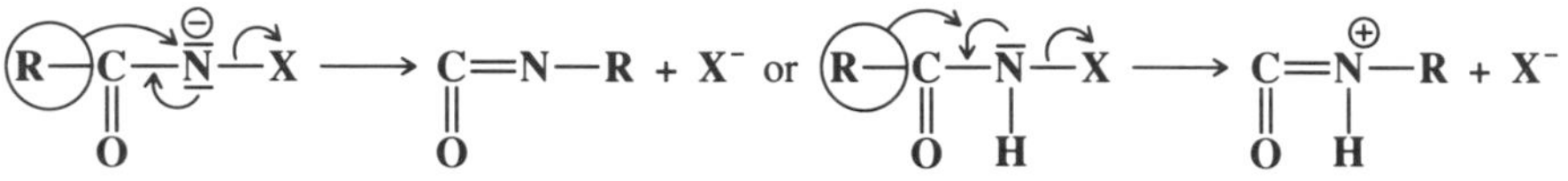

$$\text{R}-\overset{\overset{\displaystyle \|}{O}}{C}-\overset{\ominus}{N}-X \longrightarrow \overset{\overset{\displaystyle \|}{O}}{C}=N-R + X^- \quad \text{or} \quad \text{R}-\overset{\overset{\displaystyle \|}{O}}{C}-\overset{\overset{\displaystyle |}{H}}{N}-X \longrightarrow \overset{\overset{\displaystyle \|}{O}}{C}=\overset{\overset{\displaystyle |}{H}}{\overset{\oplus}{N}}-R + X^-$$

Some of the evidence[207] is: (1) configuration is retained in R (p. 1054); (2) the kinetics are first order; (3) intramolecular rearrangement is shown by labeling; and (4) no rearrangement occurs *within* the migrating group, e.g., a neopentyl group on the carbon of the starting material is still a neopentyl group on the nitrogen of the product.

In many cases it is not certain whether the nucleofuge X is lost first, creating an intermediate nitrene[208] or nitrenium ion, or whether migration and loss of the nucleofuge are simultaneous, as shown above.[209] It is likely that both possibilities can exist, depending on the substrate and reaction conditions.

8-14 The Hofmann Rearrangement

Bishydrogen-(2/→1/N-alkyl)-*migro*-detachment (formation of isocyanate)

$$RCONH_2 + NaOBr \longrightarrow R-N=C=O \xrightarrow[\text{6-3}]{\text{hydrolysis}} RNH_2$$

In the *Hofmann rearrangement*, an unsubstituted amide is treated with sodium hypobromite (or sodium hydroxide and bromine, which is essentially the same thing) to give a primary amine that has one carbon fewer than the starting amide.[210] The actual product is the isocyanate, but this compound is seldom isolated[211] since it is usually hydrolyzed under the reaction conditions (6-3). R may be alkyl or aryl, but if it is an alkyl group of more than about six or seven carbons, low yields are obtained unless Br_2 and NaOMe are used instead of Br_2 and NaOH.[212] Under these conditions the product of addition to the isocyanate is the carbamate RNHCOOMe (6-8), which is easily isolated or can be hydrolyzed to the amine. Side reactions when NaOH is the base are formation of ureas RNHCONHR and acylureas RCONHCONHR by addition, respectively, of RNH_2 and $RCONH_2$ to RNCO (6-17). If acylureas are desired, they can be made the main products by using only half the

[205]For example, see Oae; Furukawa *Bull. Chem. Soc. Jpn.* **1965**, *38*, 62; Nakai; Furukawa; Oae *Bull. Chem. Soc. Jpn.* **1969**, *42*, 2917.

[206]Baumgarten; Petersen; Wolf *J. Org. Chem.* **1963**, *28*, 2369.

[207]For a discussion of this mechanism and the evidence for it, see Smith, in Mayo, Ref. 114, vol. 1, pp. 258-550.

[208]For a review of rearrangements involving nitrene intermediates, see Boyer *Mech. Mol. Migr.* **1969**, *2*, 267-318. See also Ref. 221.

[209]The question is discussed by Lwowski, in Lwowski *Nitrenes*; Wiley: New York, 1970, pp. 217-221.

[210]For a review, see Wallis; Lane *Org. React.* **1946**, *3*, 267-306.

[211]If desired, the isocyanate can be isolated by the use of phase transfer conditions: see Sy and Raksis *Tetrahedron Lett.* **1980**, *21*, 2223.

[212]For an example of the use of this method at low temperatures, see Radlick; Brown *Synthesis* **1974**, 290.

usual quantities of Br_2 and NaOH. Another side product, though only from primary R, is the nitrile derived from oxidation of RNH_2 (**9-5**). Imides react to give amino acids, e.g., phthalimide gives *o*-aminobenzoic acid. α-Hydroxy and α-halo amides give aldehydes and ketones by way of the unstable α-hydroxy- or α-haloamines. However, a side product with an α-halo amide is a *gem*-dihalide. Ureas analogously give hydrazines.

The mechanism follows the pattern outlined on p. 1090.

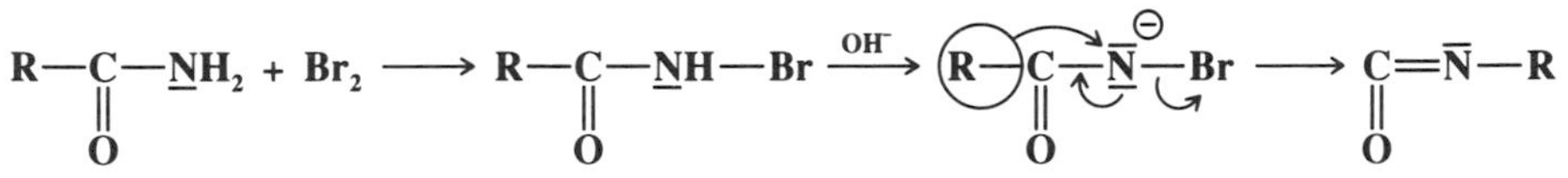

$$R-\underset{O}{\overset{\parallel}{C}}-\underline{N}H_2 + Br_2 \longrightarrow R-\underset{O}{\overset{\parallel}{C}}-\underline{N}H-Br \xrightarrow{OH^-} \left(R-\underset{O}{\overset{\parallel}{C}}-\overset{\ominus}{\underline{N}}-Br\right) \longrightarrow \underset{O}{\overset{\parallel}{C}}=\overline{N}-R$$

$$\textbf{64}$$

The first step is an example of **2-54** and intermediate N-halo amides (**64**) have been isolated. In the second step, **64** lose a proton to the base. **64** are acidic because of the presence of two electron-withdrawing groups (acyl and halo) on the nitrogen. It is possible that the third step is actually two steps: loss of bromide to form a nitrene, followed by the actual migration, but most of the available evidence favors the concerted reaction.[213]

A similar reaction can be effected by the treatment of amides with lead tetraacetate.[214] In this case the initial isocyanate and the amine formed from it react with the acetic acid liberated from the lead tetraacetate to give, respectively, ureas and amides. If the reaction is carried out in the presence of an alcohol, carbamates are formed (**6-8**).

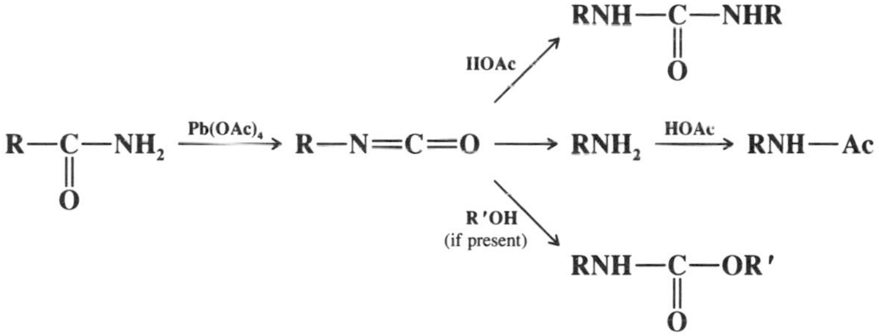

$$R-\underset{O}{\overset{\parallel}{C}}-NH_2 \xrightarrow{Pb(OAc)_4} R-N=C=O \longrightarrow RNH_2 \xrightarrow{HOAc} RNH-Ac$$

with branches:

$$\xrightarrow{HOAc} RNH-\underset{O}{\overset{\parallel}{C}}-NHR$$

$$\xrightarrow[\text{(if present)}]{R'OH} RNH-\underset{O}{\overset{\parallel}{C}}-OR'$$

Among other reagents that convert $RCONH_2$ to RNH_2 (R = alkyl, but not aryl) are phenyliodosyl bis(trifluoroacetate) $PhI(OCOCF_3)_2$[215] and hydroxy(tosyloxy)iodobenzene $PhI(OH)OTs$.[216] A mixture of N-bromosuccinimide, $Hg(OAc)_2$, and R'OH is one of several reagent mixtures that convert an amide $RCONH_2$ to the carbamate $RNHCOOR'$ (R = primary, secondary, or tertiary alkyl or aryl) in high yield.[217]

OS **II**, 19, 44, 462; **IV**, 45; **65**, 173; **66**, 132.

8-15 The Curtius Rearrangement
Dinitrogen-(2/→1/N-alkyl)-*migro*-detachment

$$RCON_3 \xrightarrow{\Delta} R-N=C=O$$

[213]See, for example, Imamoto; Tsuno; Yukawa *Bull. Chem. Soc. Jpn.* **1971**, *44*, 1632, 1639, 1644; Imamoto; Kim; Tsuno; Yukawa *Bull. Chem. Soc. Jpn.* **1971**, *44*, 2776.

[214]Acott; Beckwith *Chem. Commun.* **1965**, 161; Baumgarten; Staklis *J. Am. Chem. Soc.* **1965**, *87*, 1141; Acott; Beckwith; Hassanali *Aust. J. Chem.* **1968**, *21*, 185, 197; Baumgarten; Smith; Staklis *J. Org. Chem.* **1975**, *40*, 3554.

[215]Loudon; Radhakrishna; Almond; Blodgett; Boutin *J. Org. Chem.* **1984**, *49*, 4272; Boutin; Loudon *J. Org. Chem.* **1984**, *49*, 4277; Pavlides; Chan; Pennington; McParland; Whitehead; Coutts *Synth. Commun.* **1988**, *18*, 1615.

[216]Lazbin; Koser *J. Org. Chem.* **1986**, *51*, 2669; Vasudevan; Koser *J. Org. Chem.* **1988**, *53*, 5158.

[217]Jew; Park; Park; Park; Cho *Tetrahedron Lett.* **1990**, *31*, 1559.

The *Curtius rearrangement* involves the pyrolysis of acyl azides to yield isocyanates.[218] The reaction gives good yields of isocyanates, since no water is present to hydrolyze them to the amine. Of course, they can be subsequently hydrolyzed, and indeed the reaction *can* be carried out in water or alcohol, in which case the products are amines, carbamates, or acylureas, as in **8-14**.[219] This is a very general reaction and can be applied to almost any carboxylic acid: aliphatic, aromatic, alicyclic, heterocyclic, unsaturated, and containing many functional groups. Acyl azides can be prepared as in **0-61** or by treatment of acylhydrazines (hydrazides) with nitrous acid (analogous to **2-50**). The Curtius rearrangement is catalyzed by Lewis or protic acids, but these are usually not necessary for good results.

The mechanism is similar to that in **8-14**:

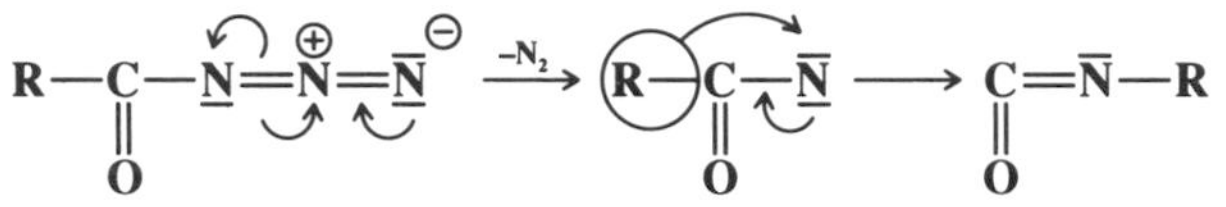

Also note the exact analogy between this reaction and **8-8**. However, in this case, there is no evidence for a free nitrene and it is probable that the steps are concerted.[220]

Alkyl azides can be similarly pyrolyzed to give imines, in an analogous reaction:[221]

$$R_3CN_3 \xrightarrow{\Delta} R_2C{=}NR$$

The R groups may be alkyl, aryl, or hydrogen, though if hydrogen migrates, the product is the unstable $R_2C{=}NH$. The mechanism is essentially the same as that of the Curtius rearrangement. However, in pyrolysis of tertiary alkyl azides, there is evidence that free alkyl nitrenes are intermediates.[222] The reaction can also be carried out with acid catalysis, in which case lower temperatures can be used, though the acid may hydrolyze the imine (**6-2**). Cycloalkyl azides give ring expansion.[223]

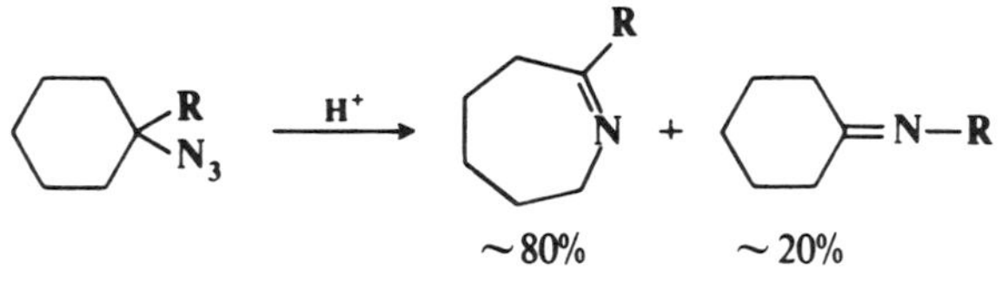

Aryl azides also give ring expansion on heating, e.g.,[224]

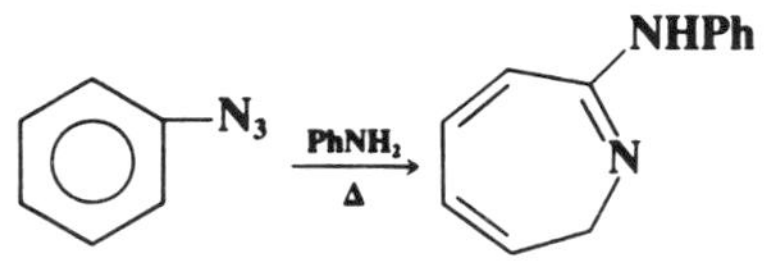

OS **III**, 846; **IV**, 819; **V**, 273; **VI**, 95, 910. Also see OS **VI**, 210.

[218]For a review, see Banthorpe, in Patai *The Chemistry of the Azido Group*; Wiley: New York, 1971, pp. 397-405.

[219]For a variation that conveniently produces the amine directly, see Pfister; Wyman *Synthesis* **1983**, 38. See also Capson; Poulter *Tetrahedron Lett.* **1984**, *25*, 3515.

[220]See, for example, Lwowski *Angew. Chem. Int. Ed. Engl.* **1967**, *6*, 897-906 [*Angew. Chem. 79*, 922-932]; Linke; Tissue; Lwowski *J. Am. Chem. Soc.* **1967**, *89*, 6308; Smalley; Bingham *J. Chem. Soc. C* **1969**, 2481.

[221]For a treatise on azides, which includes discussion of rearrangement reactions, see Scriven *Azides and Nitrenes*; Academic Press: New York, 1984. For a review of rearrangements of alkyl and aryl azides, see Stevens; Watts, Ref. 1, pp. 45-52. For reviews of the formation of nitrenes from alkyl and aryl azides, see, in Lwowski, Ref. 209, the chapters by Lewis; Saunders, pp. 47-97, pp. 47-78 and by Smith, pp. 99-162.

[222]Abramovitch; Kyba *J. Am. Chem. Soc.* **1974**, *96*, 480; Montgomery; Saunders *J. Org. Chem.* **1976**, *41*, 2368.

[223]Smith; Lakritz, cited in Smith, in Mayo, Ref. 114, vol. 1, p. 474.

[224]Huisgen; Vossius; Appl *Chem. Ber.* **1958**, *91*, 1, 12.

8-16   The Lossen Rearrangement
Hydro,acetoxy-(2/→1/N-alkyl)-*migro*-detachment

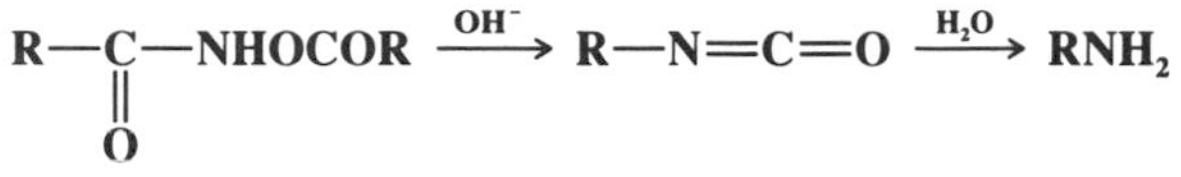

$$R\text{—}\underset{\substack{\|\\O}}{C}\text{—NHOCOR} \xrightarrow{\text{OH}^-} R\text{—N}=C=O \xrightarrow{\text{H}_2\text{O}} RNH_2$$

The O-acyl derivatives of hydroxamic acids[225] give isocyanates when treated with bases or sometimes even just on heating, in a reaction known as the *Lossen rearrangement*. The mechanism is similar to that of **8-14** and **8-15**:

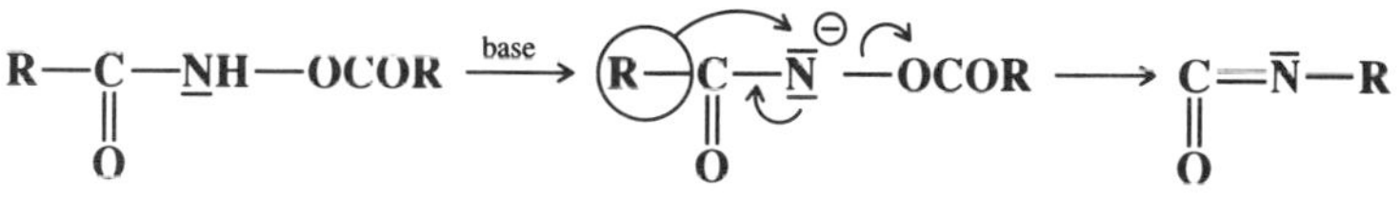

$$R\text{—}\underset{\substack{\|\\O}}{C}\text{—}\underline{\text{NH}}\text{—OCOR} \xrightarrow{\text{base}} (R)\underset{\substack{\|\\O}}{C}\text{—}\overset{\ominus}{\underline{N}}\text{—OCOR} \longrightarrow \underset{\substack{\|\\O}}{C}=\bar{N}\text{—R}$$

In a similar reaction, aromatic acyl halides are converted to amines in one laboratory step by treatment with hydroxylamine-O-sulfonic acid.[226]

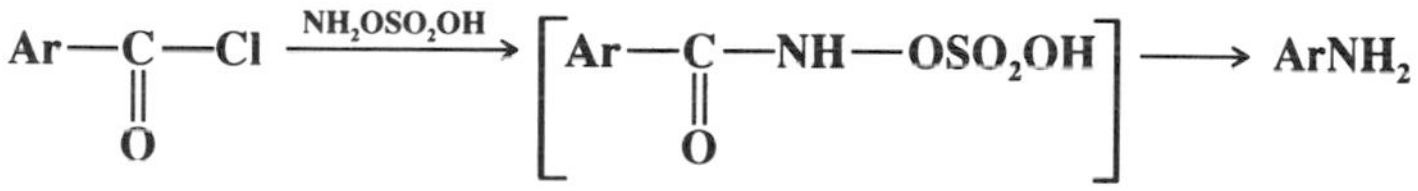

$$Ar\text{—}\underset{\substack{\|\\O}}{C}\text{—Cl} \xrightarrow{\text{NH}_2\text{OSO}_2\text{OH}} \left[Ar\text{—}\underset{\substack{\|\\O}}{C}\text{—NH—OSO}_2\text{OH}\right] \longrightarrow ArNH_2$$

8-17   The Schmidt Reaction

$$RCOOH + HN_3 \xrightarrow{\text{H}^+} R\text{—N}=C=O \xrightarrow{\text{H}_2\text{O}} RNH_2$$

There are actually three reactions called by the name *Schmidt reaction*, involving the addition of hydrazoic acid to carboxylic acids, aldehydes and ketones, and alcohols and olefins.[227] The most common is the reaction with carboxylic acids, illustrated above.[228] Sulfuric acid is the most common catalyst, but Lewis acids have also been used. Good results are obtained for aliphatic R, especially for long chains. When R is aryl, the yields are variable, being best for sterically hindered compounds like mesitoic acid. This method has the advantage over **8-14** and **8-15** that it is just one laboratory step from the acid to the amine, but conditions are more drastic.[229] Under the acid conditions employed, the isocyanate is virtually never isolated.

The reaction between a ketone and hydrazoic acid is a method for "insertion" of NH between the carbonyl group and one R group, converting a ketone into an amide.[230]

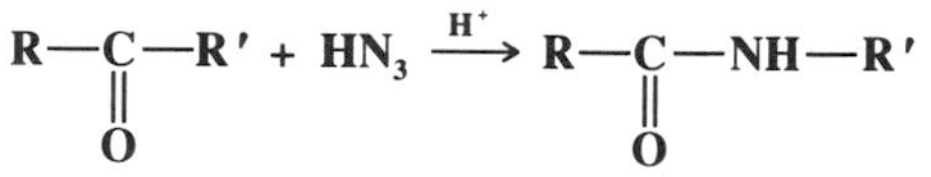

$$R\text{—}\underset{\substack{\|\\O}}{C}\text{—R}' + HN_3 \xrightarrow{\text{H}^+} R\text{—}\underset{\substack{\|\\O}}{C}\text{—NH—R}'$$

Either or both of the R groups may be aryl. In general, dialkyl ketones and cyclic ketones react more rapidly than alkyl aryl ketones, and these more rapidly than diaryl ketones. The

[225]For a review of hydroxamic acids, see Bauer; Exner *Angew. Chem. Int. Ed. Engl.* **1974**, *13*, 376-384 [*Angew. Chem.* *86*, 419-428].
[226]Wallace; Barker; Wood *Synthesis* **1990**, 1143.
[227]For a review, see Banthorpe, Ref. 218, pp. 405-434.
[228]For a review, see Koldobskii; Ostrovskii; Gidaspov *Russ. Chem. Rev.* **1978**, *47*, 1084-1094.
[229]For a comparision of reactions **8-14** to **8-17** as methods for converting an acid to an amine, see Smith, *Org. React.* **1946**, *3*, 337-449, pp. 363-366.
[230]For reviews, see Koldobskii; Tereschenko; Gerasimova; Bagal *Russ. Chem. Rev.* **1971**, *40*, 835-846; Beckwith, in Zabicky *The Chemistry of Amides*; Wiley: New York, 1970, pp. 137-145.

latter require sulfuric acid and do not react in concentrated HCl, which is strong enough for dialkyl ketones. Dialkyl and cyclic ketones react sufficiently faster than diaryl or aryl alkyl ketones—or carboxylic acids or alcohols—that these functions may be present in the same molecule without interference. Cyclic ketones give lactams:[231]

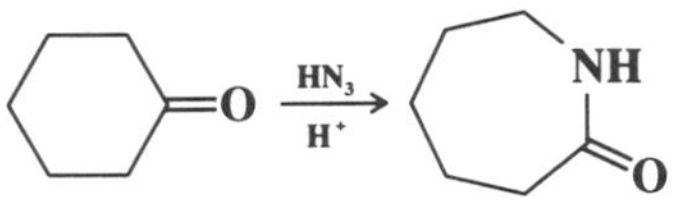

With alkyl aryl ketones, it is the aryl group that generally migrates to the nitrogen, except when the alkyl group is bulky.[232] The reaction has been applied to a few aldehydes, but rarely. With aldehydes the product is usually the nitrile (**6-22**). Even with ketones, conversion to the nitrile is often a side reaction, especially with the type of ketone that gives **7-38**.

Alcohols and olefins react with HN_3 to give alkyl azides, which in the course of reaction rearrange in the same way as discussed in reaction **8-15**.[221]

There is evidence that the mechanism with carboxylic acids[228] is similar to that of **8-15**, except that it is the protonated azide that undergoes the rearrangement:[233]

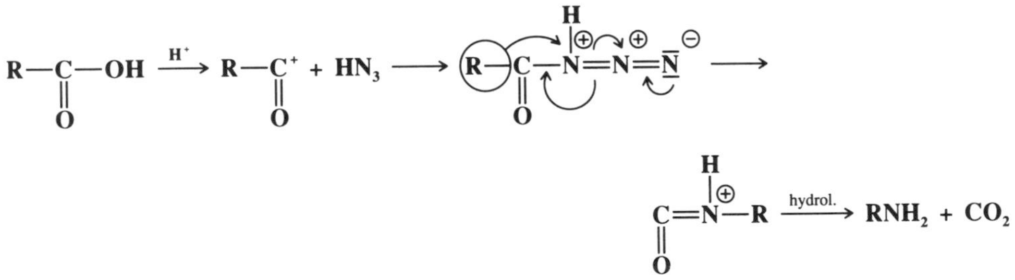

The first step is the same as that of the AAc1 mechanism (**0-10**), which explains why good results are obtained with hindered substrates. The mechanism with ketones is[234]

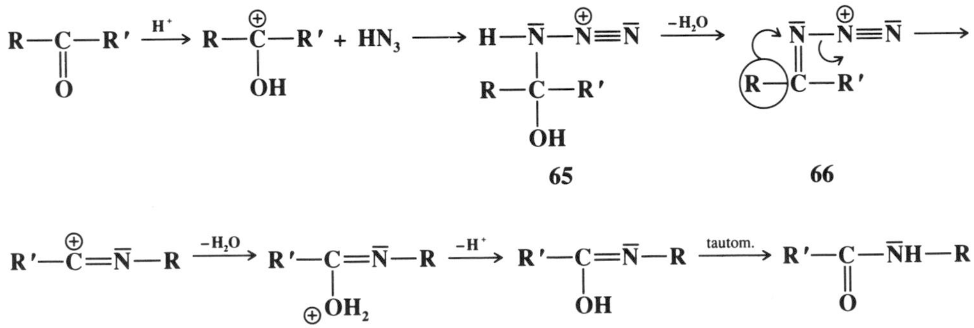

[231]For a review with respect to bicyclic ketones, see Krow, *Tetrahedron* **1981**, *37*, 1283-1307.

[232]Exceptions to this statement have been noted in the case of cyclic aromatic ketones bearing electron-donating groups in ortho and para positions: Bhalerao; Thyagarajan *Can. J. Chem.* **1968**, *46*, 3367; Tomita; Minami; Uyeo *J. Chem. Soc. C.* **1969**, 183.

[233]There has been some controversy about this mechanism. For a discussion, see Vogler; Hayes *J. Org. Chem.* **1979**, *44*, 3682.

[234]Smith *J. Am. Chem. Soc.* **1948**, *70*, 320; Smith; Antoniades *Tetrahedron* **1960**, *9*, 210. A slightly different mechanism, involving direct rearrangement of **65**, has been shown in certain cases: Fikes; Shechter *J. Org. Chem.* **1979**, *44*, 741. See also Bach; Wolber *J. Org. Chem.* **1972**, *47*, 239.

The intermediates **66** have been independently generated in aqueous solution.[235] Note the similarity of this mechanism to those of "insertion" of CH_2 (**8-9**) and of O (**8-20**). The three reactions are essentially analogous, both in products and in mechanism.[236] Also note the similarity of the latter part of this mechanism to that of the Beckmann rearrangement (**8-18**).

OS **V**, 408; **VI**, 368; **VII**, 254. See also OS **V**, 623.

8-18 The Beckmann Rearrangement
Beckmann oxime–amide rearrangement

$$R-\underset{\underset{\text{OH}}{\overset{\|}{N}}}{\overset{\|}{C}}-R' \xrightarrow{PCl_5} R'-\underset{\overset{\|}{O}}{\overset{\|}{C}}-NH-R$$

When oximes are treated with PCl_5 or a number of other reagents, they rearrange to substituted amides in a reaction called the *Beckmann rearrangement*.[237] Among other reagents used have been concentrated H_2SO_4, formic acid, liquid SO_2, HMPA,[238] $SOCl_2$,[239] silica gel,[240] P_2O_5–methanesulfonic acid,[241] HCl–HOAc–Ac_2O, and polyphosphoric acid.[242] The group that migrates is generally the one anti to the hydroxyl, and this is often used as a method of determining the configuration of the oxime. However, it is not unequivocal. It is known that with some oximes the syn group migrates and that with others, especially where R and R' are both alkyl, mixtures of the two possible amides are obtained. However, this behavior does not necessarily mean that the syn group actually undergoes migration. In most cases the oxime undergoes isomerization under the reaction conditions *before* migration takes place.[243] The scope of the reaction is quite broad. R and R' may be alkyl, aryl, or hydrogen. However, hydrogen very seldom *migrates*, so the reaction is not generally a means of converting aldoximes to unsubstituted amides $RCONH_2$. This conversion can be accomplished, though, by treatment of the aldoxime with nickel acetate under neutral conditions[244] or by heating the aldoxime for 60 hr at 100°C after it has been adsorbed onto silica gel.[245] As in the case of the Schmidt rearrangement, when the oxime is derived from an alkyl aryl ketone, it is generally the aryl group that preferentially migrates. The oximes of cyclic ketones give ring enlargement,[246] e.g.,

[235]Amyes; Richard *J. Am. Chem. Soc.* **1991**, *113*, 1867.
[236]For evidence for this mechanism, see Koldobskii; Enin; Naumov; Ostrovskii; Tereshchenko; Bagal *J. Org. Chem. USSR* **1972**, *8*, 242; Ostrovskii; Koshtaleva; Shirokova; Koldobskii; Gidaspov *J. Org. Chem. USSR* **1974**, *10*, 2365; Ref. 230.
[237]For reviews, see Gawley *Org. React.* **1988**, *35*, 1-420; McCarty, in Patai *The Chemistry of the Carbon–Nitrogen Double Bond*; Wiley: New York, 1970, pp. 408-439.
[238]Monson; Broline *Can. J. Chem.* **1973**, *51*, 942; Gupton; Idoux; Leonard; DeCrescenzo *Synth. Commun.* **1983**, *13*, 1083.
[239]Butler; O'Donoghue *J. Chem. Res. (S)* **1983**, 18.
[240]Costa; Mestres; Riego *Synth. Commun.* **1982**, *12*, 1003.
[241]Eaton; Carlson; Lee *J. Org. Chem.* **1973**, *38*, 4071.
[242]For a review of Beckmann rearrangements with polyphosphoric acid, see Beckwith, in Zabicky, Ref. 230, pp. 131-137.
[243]Lansbury; Mancuso *Tetrahedron Lett.* **1965**, 2445 have shown that some Beckmann rearrangements are *authentically* nonstereospecific.
[244]Field; Hughmark; Shumaker; Marshall *J. Am. Chem. Soc.* **1961**, *83*, 1983. See also Leusink; Meerbeek; Noltes *Recl. Trav. Chim. Pays-Bas* **1976**, *95*, 123, **1977**, *96*, 142.
[245]Chattopadhyaya; Rama Rao *Tetrahedron* **1974**, *30*, 2899.
[246]For a review of such ring enlargements, see Vinnik; Zarakhani *Russ. Chem. Rev.* **1967**, *36*, 51-64. For a review with respect to bicyclic oximes, see Ref. 231.

Not only do oximes undergo the Beckmann rearrangement, but so also do esters of oximes with many acids, organic and inorganic. A side reaction with many substrates is the formation of nitriles (the "abnormal" Beckmann rearrangement, **7-38**). Cyclic ketones can be converted directly to lactams in one laboratory step by treatment with NH_2OSO_2OH and formic acid (**6-20** takes place first, then the Beckmann rearrangement).[247]

In the first step of the mechanism, the OH group is converted by the reagent to a better leaving group, e.g., proton acids convert it to OH_2^+. After that, the mechanism follows a course analogous to that for the Schmidt reaction of ketones (**8-17**) from the formation of **66** on:[248]

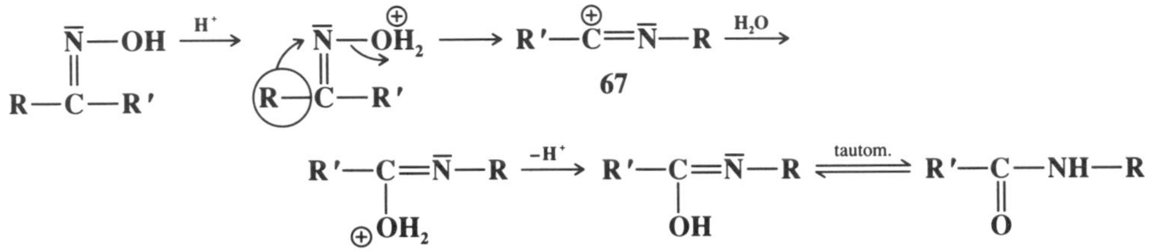

The other reagents convert OH to an ester leaving group (e.g., $OPCl_4$ from PCl_5 and OSO_2OH from concentrated H_2SO_4[249]). Alternatively, the attack on **67** can be by the leaving group, if different from H_2O. Intermediates of the form **67** have been detected by nmr and uv spectroscopy.[250] The rearrangement has also been found to take place by a different mechanism, involving formation of a nitrile by fragmentation, and then addition by a Ritter reaction (**6-55**).[251] Beckmann rearrangements have also been carried out photochemically.[252]

If the rearrangement of oxime sulfonates is induced by organoaluminum reagents,[253] the intermediate **67** is captured by the nucleophile originally attached to the Al. By this means an oxime can be converted to an imine, an imino thioether, or an imino nitrile[254] (in the

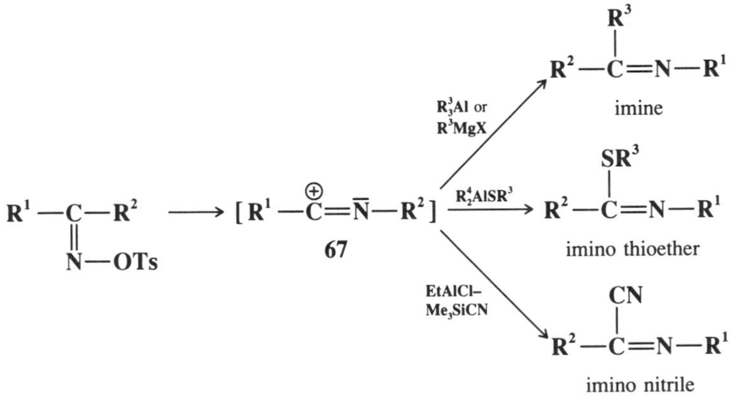

[247]Olah; Fung *Synthesis* **1979**, 537. See also Novoselov; Isaev; Yurchenko; Vodichka; Trshiska *J. Org. Chem. USSR* **1981**, *17*, 2284.

[248]For summaries of the considerable evidence for this mechanism, see Donaruma; Heldt *Org. React.* **1960**, *11*, 1-156, pp. 5-14; Smith, in Mayo, Ref. 114, vol. 1, 483-507, pp. 488-493.

[249]Gregory; Moodie; Schofield *J. Chem. Soc. B* **1970**, 338; Kim; Kawakami; Ando; Yukawa *Bull. Chem. Soc. Jpn.* **1979**, *52*, 1115.

[250]Gregory; Moodie; Schofield, Ref. 249.

[251]Hill; Conley; Chortyk *J. Am. Chem. Soc.* **1965**, *87*, 5646; Palmere; Conley; Rabinowitz *J. Org. Chem.* **1972**, *37*, 4095.

[252]See, for example, Izawa; Mayo; Tabata *Can. J. Chem.* **1969**, *47*, 51; Cunningham; Ng Lim; Just *Can. J. Chem.* **1971**, *49*, 2891; Suginome; Yagihashi *J. Chem. Soc., Perkin Trans. 1* **1977**, 2488.

[253]For a review, see Maruoka; Yamamoto *Angew. Chem. Int. Ed. Engl.* **1985**, *24*, 668-682 [*Angew. Chem. 97*, 670-683].

[254]Maruoka; Miyazaki; Ando; Matsumura; Sakane; Hattori; Yamamoto *J. Am. Chem. Soc.* **1983**, *105*, 2831; Maruoka; Nakai; Yamamoto *Org. Synth. 66*, 185.

last case, the nucleophile comes from added trimethylsilyl cyanide). The imine-producing reaction can also be accomplished with a Grignard reagent in benzene or toluene.[255]

OS **II**, 76, 371; **66,** 185.

8-19 Stieglitz and Related Rearrangements
Methoxy-de-N-chloro-(2/→1/N-alkyl)-*migro*-substitution, etc.

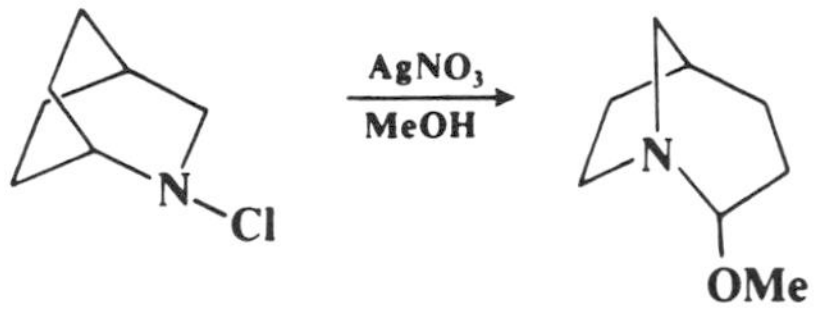

Besides the reactions discussed at **8-14** to **8-18,** a number of other rearrangements are known in which an alkyl group migrates from C to N. Certain bicyclic N-haloamines, for example N-chloro-2-azabicyclo[2.2.2]octane (above), undergo rearrangement when solvolyzed in the presence of silver nitrate.[256] This reaction is similar to the Wagner–Meerwein rearrangement (**8-1**) and is initiated by the silver-catalyzed departure of the chloride ion.[257] Similar reactions have been used for ring expansions and contractions, analogous to those discussed for reaction **8-3.**[258] An example is the conversion of 1-(N-chloroamino)cyclopropanols to β-lactams.[259]

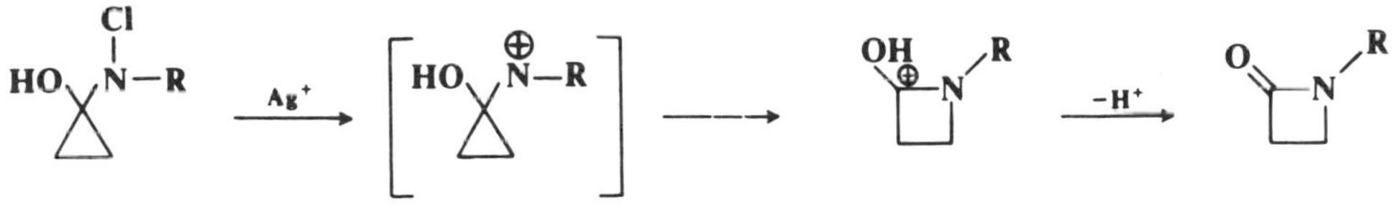

The name *Stieglitz rearrangement* is generally applied to the rearrangements of trityl N-haloamines and hydroxylamines. These reactions are similar to the rearrangements of alkyl

$$Ar_3CNHX \xrightarrow{\text{base}} Ar_2C{=}NAr$$

$$Ar_3CNHOH \xrightarrow{\text{PCl}_5} Ar_2C{=}NAr$$

azides (**8-15**), and the name Stieglitz rearrangement is also given to the rearrangement of trityl azides. Another similar reaction is the rearrangement undergone by trityl amines when treated with lead tetraacetate:[260]

$$Ar_3CNH_2 \xrightarrow{\text{Pb(OAc)}_4} Ar_2C{=}NAr$$

D. Carbon-to-Oxygen Migrations of R and Ar

[255]Hattori; Maruoka; Yamamoto *Tetrahedron Lett.* **1982,** *23,* 3395.

[256]Gassman; Fox *J. Am. Chem. Soc.* **1967,** *89,* 338. See also Schell; Ganguly *J. Org. Chem.* **1980,** *45,* 4069; Davies; Malpass; Walker *J. Chem. Soc., Chem. Commun.* **1985,** 686; Hoffman; Kumar; Buntain *J. Am. Chem. Soc.* **1985,** *107,* 4731.

[257]For C → N rearrangements induced by AlCl₃, see Kovacic; Lowery; Roskos *Tetrahedron* **1970,** *26,* 529.

[258]Gassman; Carrasquillo *Tetrahedron Lett.* **1971,** 109; Hoffman; Buntain *J. Org. Chem.* **1988,** *53,* 3316.

[259]Wasserman; Adickes; Espejo de Ochoa *J. Am. Chem. Soc.* **1971,** *93,* 5586; Wasserman; Glazer; Hearn *Tetrahedron Lett.* **1973,** 4855.

[260]Sisti *Chem. Commun.* **1968,** 1272; Sisti; Milstein *J. Org. Chem.* **1974,** *39,* 3932.

8-20   The Baeyer–Villiger Rearrangement
Oxy-insertion

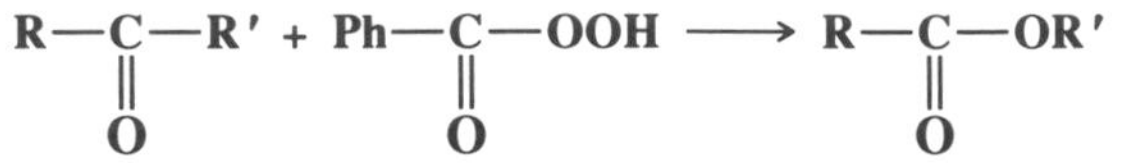

$$R—C—R' + Ph—C—OOH \longrightarrow R—C—OR'$$
$$\overset{\|}{O} \qquad\quad \overset{\|}{O} \qquad\qquad\quad \overset{\|}{O}$$

The treatment of ketones with peracids such as perbenzoic or peracetic acid, or with other peroxy compounds in the presence of acid catalysts, gives carboxylic esters by "insertion" of oxygen.[261] The reaction is called the *Baeyer–Villiger rearrangement*.[262] A particularly good reagent is peroxytrifluoroacetic acid. Reactions with this reagent are rapid and clean, giving high yields of product, though it is often necessary to add a buffer such as Na_2HPO_4 to prevent transesterification of the product with trifluoroacetic acid. The reaction is often applied to cyclic ketones to give lactones.[263] Enantioselective synthesis of chiral lactones from achiral ketones has been achieved by the use of enzymes as catalysts.[264] For acyclic compounds, R' must usually be secondary, tertiary, or vinylic, although primary R' has been rearranged with peroxytrifluoroacetic acid,[265] with BF_3–H_2O_2,[266] and with $K_2S_2O_8$–H_2SO_4.[267] For unsymmetrical ketones the approximate order of migration is tertiary alkyl > secondary alkyl, aryl > primary alkyl > methyl. Since the methyl group has a low migrating ability, the reaction provides a means of cleaving a methyl ketone R'COMe to produce an alcohol or phenol R'OH (by hydrolysis of the ester R'OCOMe). The migrating ability of aryl groups is increased by electron-donating and decreased by electron-withdrawing substituents.[268] Enolizable β-diketones do not react. α-Diketones can be converted to anhydrides.[269] With aldehydes, migration of hydrogen gives the carboxylic acid, and this is a way of accomplishing **4-6**. Migration of the other group would give formates, but this seldom happens, though aryl aldehydes have been converted to formates with H_2O_2 and a selenium compound[270] (see also the Dakin reaction in **9-12**).

The mechanism[271] is similar to those of the analogous reactions with hydrazoic acid (**8-17** with ketones) and diazomethane (**8-8**):

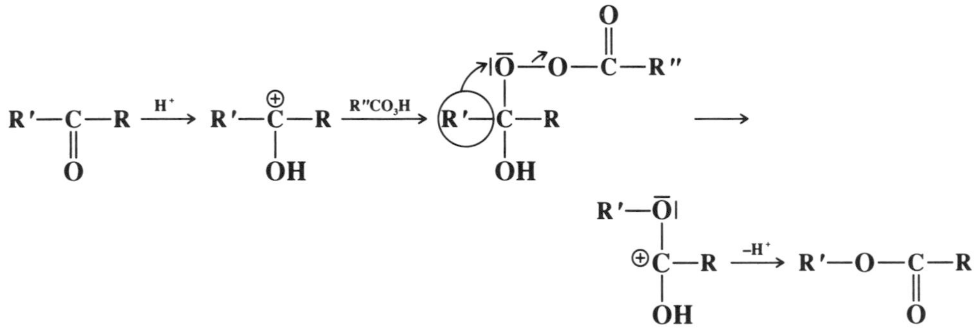

[261]For a list of reagents, with references, see Ref. 106, p. 843.
[262]For reviews, see Hudlický *Oxidations in Organic Chemistry*; American Chemical Society: Washington, 1990, pp. 186-195; Plesničar, in Trahanovsky *Oxidation in Organic Chemistry*, pt. C; Academic Press: New York, 1978, pp. 254-267; House *Modern Synthetic Reactions*, 2nd ed.; W.A. Benjamin: New York, 1972, pp. 321-329; Lewis, in Augustine *Oxidation*, vol. 1; Marcel Dekker: New York, 1969, pp. 237-244; Lee; Uff *Q. Rev. Chem. Soc.* **1967**, *21*, 429-457, pp. 449-453. For a review of enzyme-catalyzed Baeyer–Villiger rearrangements, see Walsh; Chen *Angew. Chem. Int. Ed. Engl.* **1988**, *27*, 333-343 [*Angew. Chem. 100*, 342-352].
[263]For a review of the reaction as applied to bicyclic ketones, see Krow *Tetrahedron* **1981**, *37*, 2697-2724.
[264]See Taschner; Black *J. Am. Chem. Soc.* **1988**, *110*, 6892.
[265]Emmons; Lucas *J. Am. Chem. Soc.* **1955**, *77*, 2287.
[266]McClure; Williams *J. Org. Chem.* **1962**, *27*, 24.
[267]Deno; Billups; Kramer; Lastomirsky *J. Org. Chem.* **1970**, *35*, 3080.
[268]For a report of substituent effects in the α, β, and γ positions of alkyl groups, see Noyori; Sato; Kobayashi *Bull. Chem. Soc. Jpn.* **1983**, *56*, 2661.
[269]For a study of the mechanism of this conversion, see Cullis; Arnold; Clarke; Howell; DeMira; Naylor; Nicholls *J. Chem. Soc., Chem. Commun.* **1987**, 1088.
[270]Syper *Synthesis* **1989**, 167. See also Godfrey; Sargent; Elix *J. Chem. Soc., Perkin Trans. 1* **1974**, 1353.
[271]Proposed by Criegee *Liebigs Ann. Chem.* **1948**, *560*, 127.

One important piece of evidence for this mechanism was that benzophenone–^{18}O gave ester entirely labeled in the carbonyl oxygen, with none in the alkoxyl oxygen.[272] Carbon-14 isotope-effect studies on acetophenones have shown that migration of aryl groups takes place in the rate-determining step,[273] demonstrating that migration of Ar is concerted with departure of OCOR″.[274] (It is hardly likely that migration would be the slow step if the leaving group departed first to give an ion with a positive charge on an oxygen atom, which would be a highly unstable species.)

8-21   Rearrangement of Hydroperoxides
        C-Alkyl-*O*-hydroxy-elimination

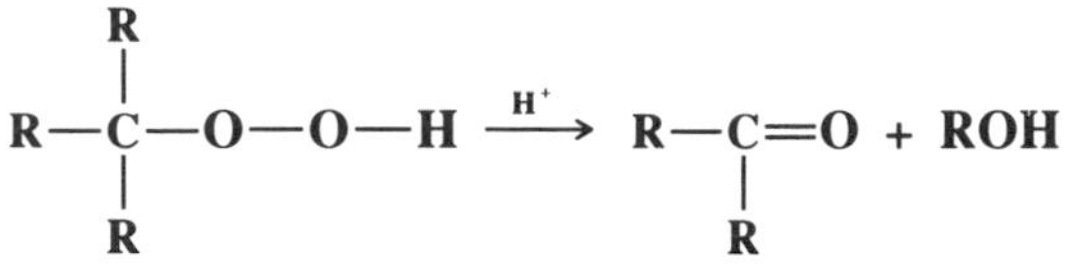

$$R-\underset{\underset{R}{|}}{\overset{\overset{R}{|}}{C}}-O-O-H \xrightarrow{\ H^+\ } R-\underset{\underset{R}{|}}{C}{=}O + ROH$$

Hydroperoxides (R = alkyl, aryl, or hydrogen) can be cleaved by proton or Lewis acids in a reaction whose principal step is a rearrangement.[275] The reaction has also been applied to peroxy esters R$_3$COOCOR′, but less often. When aryl and alkyl groups are both present, migration of aryl dominates. It is not necessary actually to prepare and isolate hydroperoxides. The reaction takes place when the alcohols are treated with H$_2$O$_2$ and acids. Migration of an alkyl group of a primary hydroperoxide provides a means for converting an alcohol to its next lower homolog (RCH$_2$OOH → CH$_2$=O + ROH).

The mechanism is as follows:[276]

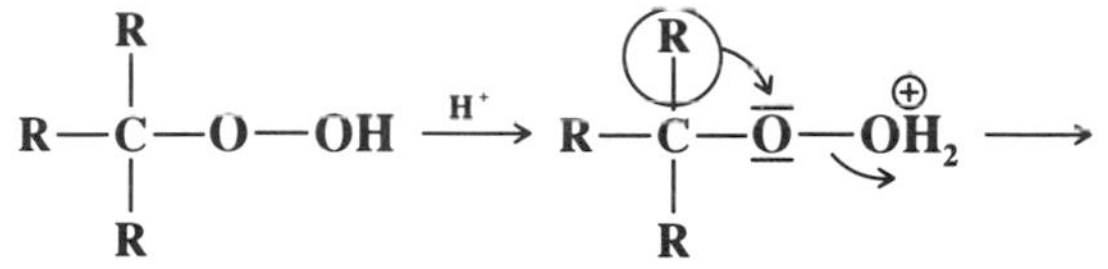

$$R-\underset{\underset{R}{|}}{\overset{\overset{R}{|}}{C}}-O-OH \xrightarrow{\ H^+\ } R-\underset{\underset{R}{|}}{\overset{\overset{\textcircled{R}}{|}}{C}}-\bar{O}-\overset{\oplus}{O}H_2 \longrightarrow$$

68

$$R-\underset{\underset{R}{|}}{\overset{\oplus}{C}}-\bar{O}-R \xrightarrow{\ H_2O\ } R-\underset{\underset{R}{|}}{\overset{\overset{\oplus}{O}H_2}{|}{C}}-O-R \longrightarrow R-\underset{\underset{R}{|}}{C}{=}O + ROH$$

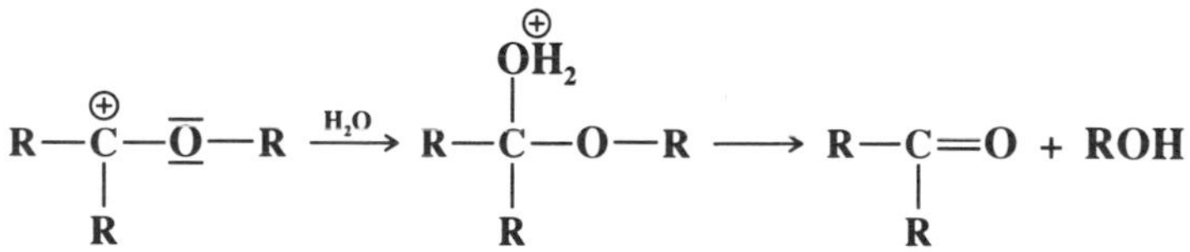

69

The last step is hydrolysis of the unstable hemiacetal. Alkoxycarbocation intermediates (**69**, R = alkyl) have been isolated in super-acid solution[277] at low temperatures, and their structures proved by nmr.[278] The protonated hydroperoxides (**68**) could not be observed in these solutions, evidently reacting immediately on formation.

OS **V**, 818.

[272]Doering; Dorfman *J. Am. Chem. Soc.* **1953**, *75*, 5595. For summaries of the other evidence, see Smith, Ref. 248, pp. 578-584.
[273]Palmer; Fry *J. Am. Chem. Soc.* **1970**, *92*, 2580. See also Mitsuhashi; Miyadera; Simamura; *Chem. Commun.* **1970**, 1301. For secondary isotope-effect studies, see Winnik; Stoute; Fitzgerald *J. Am. Chem. Soc.* **1974**, *96*, 1977.
[274]In some cases the rate-determining step has been shown to be the addition of peracid to the substrate [see, for example, Ogata; Sawaki *J. Org. Chem.* **1972**, *37*, 2953]. Even in these cases it is still highly probable that migration is concerted with departure of the nucleofuge.
[275] For reviews, see Yablokov *Russ. Chem. Rev.* **1980**, *49*, 833-842; Lee; Uff, Ref. 262, 445-449.
[276]For a discussion of the transition state involved in the migration step, see Wistuba; Rüchardt *Tetrahedron Lett.* **1981**, *22*, 3389.
[277]For a review of peroxy compounds in super acids, see Olah; Parker; Yoneda *Angew. Chem. Int. Ed. Engl.* **1978**, *17*, 909-931 [*Angew. Chem. 90*, 962-984].
[278]Sheldon; van Doorn *Tetrahedron Lett.* **1973**, 1021.

E.   Nitrogen-to-Carbon, Oxygen-to-Carbon, and Sulfur-to-Carbon Migration

8-22   The Stevens Rearrangement
Hydron-(2/N→1/alkyl)-*migro*-detachment

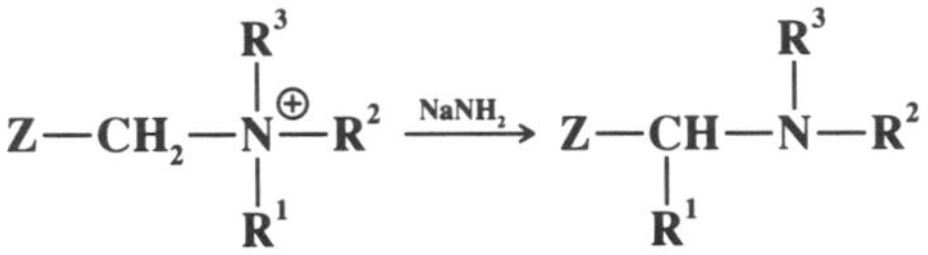

$$Z-CH_2-\overset{\overset{\displaystyle R^3}{|}}{\underset{\underset{\displaystyle R^1}{|}}{N}}{}^{\oplus}-R^2 \xrightarrow{\text{NaNH}_2} Z-\overset{\overset{\displaystyle R^3}{|}}{\underset{\underset{\displaystyle R^1}{|}}{CH}}-N-R^2$$

70

In the *Stevens rearrangement* a quaternary ammonium salt containing an electron-withdrawing group Z on one of the carbons attached to the nitrogen is treated with a strong base (such as NaOR or NaNH$_2$) to give a rearranged tertiary amine. Z is a group such as RCO, ROOC, phenyl, etc.[279] The most common migrating groups are allylic, benzylic, benzhydryl, 3-phenylpropargyl, and phenacyl, though even methyl migrates to a sufficiently negative center.[280] When an allylic group migrates, it may or may not involve an allylic rearrangement within the migrating group (see **8-37**), depending on the substrate and reaction conditions. The reaction has been used for ring enlargement,[281] e.g.:

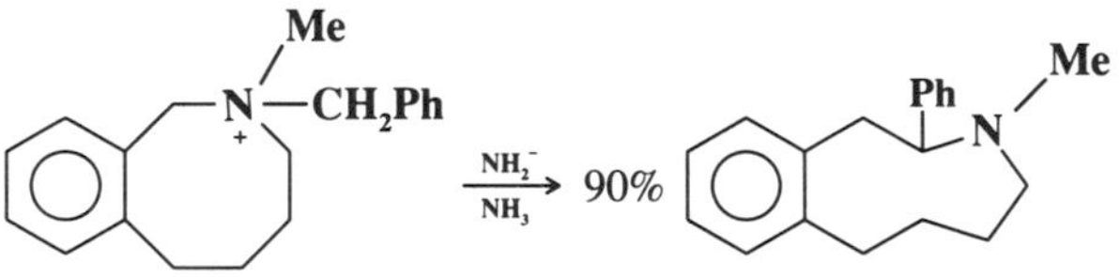

The mechanism has been the subject of much study.[282] That the rearrangement is intramolecular was shown by crossover experiments, by ^{14}C labeling,[283] and by the fact that retention of configuration is found at R^1.[284] The first step is loss of the acidic proton to give the ylide **71**, which has been isolated.[285] The finding[286] that CIDNP spectra[287] could be obtained in many instances shows that in these cases the product is formed directly from a free-radical precursor. The following radical pair mechanism was proposed:[288]

[279]For reviews of the Stevens rearrangement, see Lepley; Giumanini *Mech. Mol. Migr.* **1971**, *3*, 297-440; Pine *Org. React.* **1970**, *18*, 403-464. For reviews of the Stevens and the closely related Wittig rearrangement (**8-23**), see Stevens; Watts, Ref. 1, pp. 81-116; Wilt, in Kochi, Ref. 55, pp. 448-458; Iwai *Mech. Mol. Migr.* **1969**, *2*, 73-116, pp. 105-113; Stevens *Prog. Org. Chem.* **1968**, *7*, 48-74.

[280]Migration of aryl is rare, but has been reported: Heaney; Ward *Chem. Commun.* **1969**, 810; Truce; Heuring *Chem. Commun.* **1969**, 1499.

[281]Elmasmodi; Cotelle; Barbry; Hasiak; Couturier *Synthesis* **1989**, 327.

[282]For example, see Pine *J. Chem. Educ.* **1971**, *48*, 99-102.

[283]Stevens *J. Chem. Soc.* **1930**, 2107; Johnstone; Stevens *J. Chem. Soc.* **1955**, 4487.

[284]Brewster; Kline *J. Am. Chem. Soc.* **1952**, *74*, 5179; Schöllkopf; Ludwig; Ostermann; Patsch *Tetrahedron Lett.* **1969**, 3415.

[285]Jemison; Mageswaran; Ollis; Potter; Pretty; Sutherland; Thebtaranonth *Chem. Commun.* **1970**, 1201.

[286]Lepley *J. Am. Chem. Soc.* **1969**, *91*, 1237, *Chem. Commun.* **1969**, 1460; Lepley; Becker; Giumanini *J. Org. Chem.* **1971**, *36*, 1222; Baldwin; Brown; *J. Am. Chem. Soc.* **1969**, *91*, 3646; Jemison; Morris *Chem. Commun.* **1969**, 1226; Ref. 285; Schöllkopf et al., Ref. 284.

[287]For a review of the application of CIDNP to rearrangement reactions, see Lepley, in Lepley; Closs *Chemically Induced Magnetic Polarization*; Wiley: New York, 1973, pp. 323-384.

[288]Schöllkopf; Ludwig *Chem. Ber.* **1968**, *101*, 2224; Ollis; Rey; Sutherland *J. Chem. Soc., Perkin Trans 1.* **1983**, 1009, 1049.

Mechanism *a*

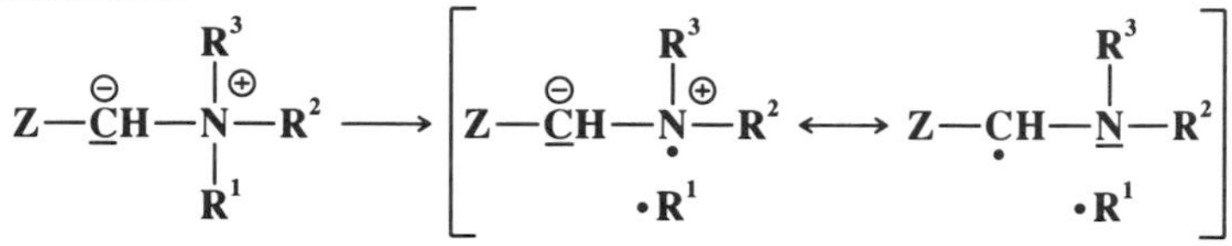

$$Z-\underset{\underset{R^1}{|}}{\overset{\overset{R^3}{|}}{\underset{}{C}H}}\overset{\oplus}{N}-R^2 \longrightarrow \left[Z-\underset{\overset{}{\bullet}}{\overset{\overset{R^3}{|}}{C}H}\overset{\oplus}{\underset{\bullet R^1}{N}}-R^2 \longleftrightarrow Z-\underset{\bullet}{C}H-\underset{\bullet R^1}{\overset{\overset{R^3}{|}}{N}}-R^2 \right]$$

71   Solvent cage

The radicals do not drift apart because they are held together by the solvent cage. According to this mechanism, the radicals must recombine rapidly in order to account for the fact that R^1 does not racemize. Other evidence in favor of mechanism *a* is that in some cases small amounts of coupling products (R^1R^1) have been isolated,[289] which would be expected if some •R^1 leaked from the solvent cage. However, not all the evidence is easily compatible with mechanism *a*.[290] It is possible that another mechanism (*b*) similar to mechanism *a*, but involving ion pairs in a solvent cage instead of radical pairs, operates in some cases. A third

Mechanism *b*

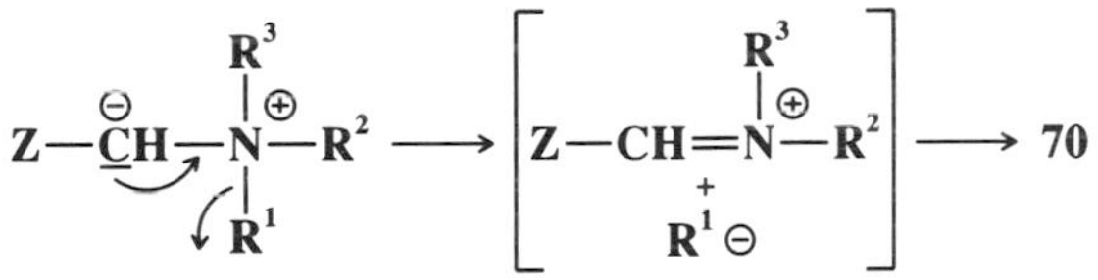

$$Z-\underset{\underset{R^1}{|}}{\overset{\overset{R^3}{|}}{\underset{}{C}H}}\overset{\oplus}{N}-R^2 \longrightarrow \left[Z-CH=\overset{\overset{R^3}{|}}{\underset{R^1\ominus}{\overset{+}{N}}}-R^2 \right] \longrightarrow \mathbf{70}$$

71   Solvent cage

possible mechanism would be a concerted 1,2-shift,[291] but the orbital symmetry principle requires that this take place with inversion at R^1.[292] (See p. 1126.) Since the actual migration takes place with retention, it cannot, according to this argument, proceed by a concerted mechanism. However, in the case where the migrating group is allylic, a concerted mechanism can also operate (**8-37**). An interesting finding compatible with all three mechanisms is that optically active allylbenzylmethylphenylammonium iodide (asymmetric nitrogen, see p. 98) gave an optically active product:[293]

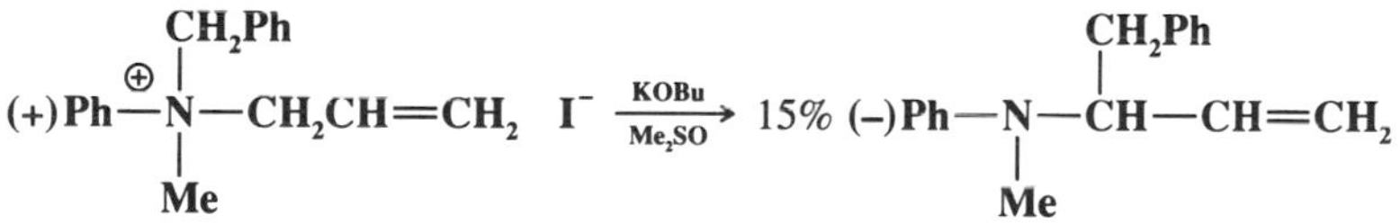

$$(+)Ph-\underset{\underset{Me}{|}}{\overset{\overset{CH_2Ph}{|}}{\overset{\oplus}{N}}}-CH_2CH=CH_2 \quad I^- \xrightarrow[\text{Me}_2\text{SO}]{\text{KOBu}} 15\% \ (-)Ph-\underset{\underset{Me}{|}}{\overset{\overset{CH_2Ph}{|}}{N}}-CH-CH=CH_2$$

The Sommelet-Hauser rearrangement competes when Z is an aryl group (see **3-26**). Hofmann elimination competes when one of the R groups contains a β hydrogen atom (**7-6** and **7-7**).

Sulfur ylides containing a Z group give an analogous rearrangement, often also referred to as a Stevens rearrangement.[294] In this case too, there is much evidence (including CIDNP)

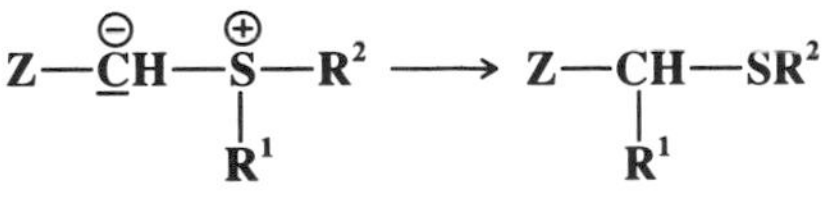

$$Z-\underset{\underset{R^1}{|}}{\overset{}{\underset{}{C}H}}\overset{\oplus}{S}-R^2 \longrightarrow Z-\underset{\underset{R^1}{|}}{\overset{}{C}H}-SR^2$$

[289]Schöllkopf et al., Ref. 284; Hennion; Shoemaker *J. Am. Chem. Soc.* **1970**, *92*, 1769.

[290]See, for example, Pine; Catto; Yamagishi *J. Org. Chem.* **1970**, *35*, 3663.

[291]For evidence against this mechanism, see Jenny; Druey *Angew. Chem. Int. Ed. Engl.* **1962**, *1*, 155 [*Angew. Chem. 74*, 152].

[292]Woodward; Hoffmann *The Conservation of Orbital Symmetry*; Academic Press: New York, 1970, p. 131.

[293]Hill; Chan *J. Am. Chem. Soc.* **1966**, *88*, 866.

[294]For a review, see Olsen; Currie, in Patai *The Chemistry of The Thiol Group*, pt. 2; Wiley: New York, 1974, pp. 561-566.

that a radical-pair cage mechanism is operating,[295] except that when the migrating group is allylic, the mechanism may be different (see **8-37**). Another reaction with a similar mechanism[296] is the *Meisenheimer rearrangement*,[297] in which certain tertiary amine oxides rearrange on heating to give substituted hydroxylamines. The migrating group R^1 is almost

$$R^2—\overset{\overset{\displaystyle R^1}{|}}{\underset{\underset{\displaystyle R^3}{|}}{N}}{}^{\oplus}\!\!=\!\!O^{\ominus} \xrightarrow{\Delta} R^2—\underset{\underset{\displaystyle R^3}{|}}{N}—O—R^1$$

always allylic or benzilic.[298] R^2 and R^3 may be alkyl or aryl, but if one of the R groups contains a β hydrogen, Cope elimination (**7-8**) often competes.

Certain tertiary benzylic amines, when treated with BuLi, undergo a rearrangement analogous to the Wittig rearrangement (**8-23**), e.g., $PhCH_2NPh_2 \rightarrow Ph_2CHNHPh$.[299] Only aryl groups migrate in this reaction.

Isocyanides, when heated in the gas phase or in nonpolar solvents, undergo a 1,2-intramolecular rearrangement to nitriles: $RNC \rightarrow RCN$.[300] In polar solvents the mechanism is different.[301]

8-23 The Wittig Rearrangement
Hydron-(2/*O*→1/alkyl)-*migro*-detachment

$$R—CH_2—O—R' \xrightarrow{R''Li} R—\underset{\underset{\displaystyle R'}{|}}{CH}—O^- \ Li^+ + R''H$$

The rearrangement of ethers with alkyllithiums is called the *Wittig rearrangement* (not to be confused with the Wittig reaction, **6-47**) and is similar to **8-22**.[279] However, a stronger base is required (e.g., phenyllithium or sodium amide). R and R' may be alkyl, aryl, or vinylic.[302] Also, one of the hydrogens may be replaced by an alkyl or aryl group, in which case the product is the salt of a tertiary alcohol. Migratory aptitudes here are allylic, benzylic > ethyl > methyl > phenyl.[303] The following radical-pair mechanism[304] (similar to

$$R—\overset{\ominus}{\underset{}{C}}H—\overset{\ominus}{O}—R' \longrightarrow \left[R—\overset{\ominus}{\underset{•R'}{C}}H—\overset{•}{\underset{}{O}}| \longleftrightarrow R—\overset{•}{\underset{•R'}{C}}H—\overset{\ominus}{\underset{}{O}}| \right] \longrightarrow R—\overset{}{\underset{\underset{\displaystyle R'}{|}}{C}}H—\overset{•}{\underset{}{O}}|{}^{\ominus}$$

Solvent cage

[295]See, for example, Baldwin; Erickson; Hackler; Scott *Chem. Commun.* **1970**, 576; Schöllkopf; Schossig; Ostermann *Liebigs Ann. Chem.* **1970**, *737*, 158; Iwamura; Iwamura; Nishida; Yoshida; Nakayama *Tetrahedron Lett.* **1971**, 63.

[296]For some of the evidence, see Schöllkopf; Ludwig *Chem. Ber.* **1968**, *101*, 2224; Ostermann; Schöllkopf *Liebigs Ann. Chem.* **1970**, *737*, 170; Lorand; Grant; Samuel; O'Connell; Zaro *Tetrahedron Lett.* **1969**, 4087.

[297]For a review, see Johnstone *Mech. Mol. Migr.* **1969**, *2*, 249-266.

[298]Migration of aryl and of certain alkyl groups has also been reported. See Khuthier; Al-Mallah; Hanna; Abdulla *J. Org. Chem.* **1987**, *52*, 1710, and references cited therein.

[299]Eisch; Dua; Kovacs *J. Org. Chem.* **1987**, *52*, 4437; Eisch; Kovacs; Chobe *J. Org. Chem.* **1989**, *54*, 1275.

[300]See Meier; Rüchardt *Chem. Ber.* **1987**, *120*, 1; Meier; Müller; Rüchardt *J. Org. Chem.* **1987**, *52*, 648; Pakusch; Rüchardt *Chem. Ber.* **1991**, *124*, 971.

[301]Meier; Rüchardt *Chimia* **1986**, *40*, 238.

[302]For migration of vinyl, see Rautenstrauch; Büchi; Wüest *J. Am. Chem. Soc.* **1974**, *96*, 2576.

[303]Wittig *Angew. Chem.* **1954**, *66*, 10; Solov'yanov; Ahmed; Beletskaya; Reutov *J. Chem. Soc., Chem. Commun.* **1987**, *23*, 1232.

[304]For a review of the mechanism, see Schöllkopf *Angew. Chem. Int. Ed. Engl.* **1970**, *9*, 763-773 [*Angew. Chem.* *82*, 795-805].

mechanism *a* of **8-22**) is likely, after removal of the proton by the base. One of the radicals in the radical pair is a ketyl. Among the evidence for this mechanism is (1) the rearrangement is largely intramolecular; (2) migratory aptitudes are in the order of free-radical stabilities, not of carbanion stabilities[305] (which rules out an ion-pair mechanism similar to mechanism *b* of **8-22**); (3) aldehydes are obtained as side products;[306] (4) partial racemization of R′ has been observed[307] (the remainder of the product retained its configuration); (5) crossover products have been detected;[308] and (6) when ketyl radicals and R• radicals from different precursors were brought together, similar products resulted.[309] However, there is evidence that at least in some cases the radical-pair mechanism accounts for only a portion of the product, and some kind of concerted mechanism can also take place.[310] Most of the above investigations were carried out with systems where R′ is alkyl, but a radical-pair mechanism has also been suggested for the case where R′ is aryl.[311] When R′ is allylic a concerted mechanism can operate (**8-37**).

When R is vinylic it is possible, by using a combination of an alkyllithium and *t*-BuOK, to get migration to the γ carbon (as well as to the α carbon), producing an enolate that, on hydrolysis, gives an aldehyde:[312]

$$CH_2=CH—CH_2—OR' \longrightarrow R'CH_2—CH=CH—OLi \longrightarrow R'CH_2CH_2CHO$$

There are no OS references, but see OS **66**, 14, for a related reaction.

F.   Boron-to-Carbon Migrations.[313] For another reaction involving boron-to-carbon migration, see **0-99**.

8-24   Conversion of Boranes to Tertiary Alcohols

$$R_3B + CO \xrightarrow[100-125°\,C]{HOCH_2CH_2OH} R_3C—B\begin{matrix} O \\ \\ O \end{matrix}\begin{matrix} CH_2 \\ | \\ CH_2 \end{matrix} \xrightarrow[NaOH]{H_2O,} R_3COH$$

72

Trialkylboranes (which can be prepared from olefins by **5-12**) react with carbon monoxide[314] at 100 to 125°C in the presence of ethylene glycol to give the 2-bora-1,3-dioxolanes **72**, which

[305]Lansbury; Pattison; Sidler; Bieber *J. Am. Chem. Soc.* **1966**, *88*, 78; Schäfer; Schöllkopf; Walter *Tetrahedron Lett.* **1968**, 2809.

[306]For example, see Hauser; Kantor *J. Am. Chem. Soc.* **1951**, *73*, 1437; Cast; Stevens; Holmes *J. Chem. Soc.* **1960**, 3521.

[307]Schöllkopf; Fabian *Liebigs Ann. Chem.* **1961**, *642*, 1; Schöllkopf; Schäfer *Liebigs Ann. Chem.* **1963**, *663*, 22; Felkin; Frajerman *Tetrahedron Lett.* **1977**, 3485; Hebert; Welvart *J. Chem. Soc., Chem. Commun.* **1980**, 1035, *Nouv. J. Chim.* **1981**, *5*, 327.

[308]Lansbury; Pattison *J. Org. Chem.* **1962**, *27*, 1933, *J. Am. Chem. Soc.* **1962**, *84*, 4295.

[309]Garst; Smith *J. Am. Chem. Soc.* **1973**, *95*, 6870.

[310]Garst; Smith *J. Am. Chem. Soc.* **1976**, *98*, 1526. For evidence against this, see Hebert; Welvart; Ghelfenstein; Szwarc *Tetrahedron Lett.* **1983**, *24*, 1381.

[311]Eisch; Kovacs; Rhee *J. Organomet. Chem.* **1974**, *65*, 289.

[312]Schlosser; Strunk *Tetrahedron* **1989**, *45*, 2649.

[313]For reviews, see Matteson, in Hartley *The Chemistry of the Metal–Carbon Bond*, vol. 4; Wiley: New York, 1984, pp. 307-409, pp. 346-387; Pelter; Smith; Brown *Borane Reagents*; Academic Press: New York, 1988, pp. 256-301; Negishi; Idacavage *Org. React.* **1985**, *33*, 1-246; Suzuki *Top. Curr. Chem.* **1983**, *112*, 67-115; Pelter, in Mayo, Ref. 1, vol. 2, pp. 95-147, *Chem. Soc. Rev.* **1982**, *11*,191-225; Cragg; Koch *Chem. Soc. Rev.* **1977**, *6*, 393-412; Weill-Raynal *Synthesis* **1976**, 633-651; Cragg *Organoboranes in Organic Synthesis*; Marcel Dekker: New York, 1973, pp. 249-300; Paetzold; Grundke *Synthesis* **1973**, 635-660.

[314]For discussions of the reactions of boranes with CO, see Negishi *Intra-Sci. Chem. Rep.* **1973**, *7*(1), 81-94; Brown *Boranes in Organic Chemistry*; Cornell University Press: Ithaca, NY, 1972, pp. 343-371, *Acc. Chem. Res.* **1969**, *2*, 65-72.

are easily oxidized (**2-28**) to tertiary alcohols.[315] The R groups may be primary, secondary, or tertiary, and may be the same or different.[316] Yields are high and the reaction is quite useful, especially for the preparation of sterically hindered alcohols such as tricyclohexylcarbinol and tri-2-norbornylcarbinol, which are difficult to prepare by **6-29.** Heterocycles in which boron is a ring atom react similarly (except that high CO pressures are required), and cyclic alcohols can be obtained from these substrates.[317] The preparation of such het-

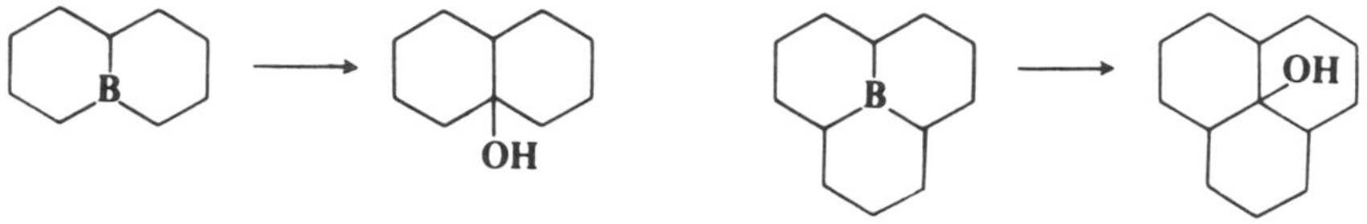

erocyclic boranes was discussed at **5-12.** The overall conversion of a diene or triene to a cyclic alcohol has been described by H. C. Brown as "stitching" with boron and "riveting" with carbon.

Though the mechanism has not been investigated thoroughly, it has been shown to be intramolecular by the failure to find crossover products when mixtures of boranes are used.[318] The following scheme, involving three boron-to-carbon migrations, has been suggested.

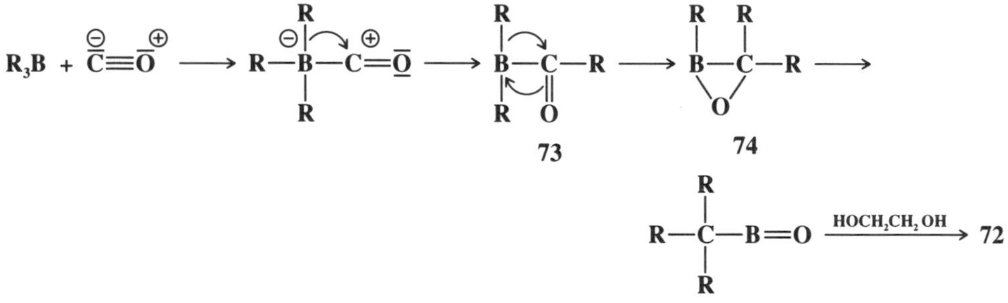

73 **74** **75**

The purpose of the ethylene glycol is to intercept the boronic anhydride **75,** which otherwise forms polymers that are difficult to oxidize. As we shall see in **8-25** and **8-26,** it is possible to stop the reaction after only one or two migrations have taken place.

There are two other methods for achieving the conversion $R_3B \rightarrow R_3COH$, which often give better results: (1) treatment with α,α-dichloromethyl methyl ether and the base lithium

$$\text{Method 1} \qquad R_3B + CHCl_2OMe \xrightarrow[2.\ H_2O_2\text{-}NaOH]{1.\ LiOCEt_3\text{-}THF} R_3COH$$

$$\text{Method 2} \qquad R_3B + CN^- \xrightarrow{THF} R_3\overset{\ominus}{B}\text{—}CN \xrightarrow[2.\ NaOH\text{-}H_2O_2]{1.\ \text{excess}\ (CF_3CO)_2O} R_3COH$$

76

[315]Hillman *J. Am. Chem. Soc.* **1962,** *84*, 4715, **1963,** *85*, 982; Brown; Rathke *J. Am. Chem. Soc.* **1967,** *89*, 2737; Puzitskii; Pirozhkov; Ryabova; Pastukhova; Eidus *Bull. Acad. Sci. USSR, Div. Chem. Sci.* **1972,** *21*, 1939, **1973,** *22*, 1760; Brown; Cole; Srebnik; Kim *J. Org. Chem.* **1986,** *51*, 4925.

[316]Brown; Negishi; Gupta *J. Am. Chem. Soc.* **1970,** *92*, 6648; Brown; Gupta *J. Am. Chem. Soc.* **1971,** *93*, 1818; Negishi; Brown *Synthesis* **1972,** 197.

[317]Brown; Negishi *J. Am. Chem. Soc.* **1967,** *89*, 5478; Knights; Brown *J. Am. Chem. Soc.* **1968,** *90*, 5283; Brown; Negishi; Dickason *J. Org. Chem.* **1985,** *50*, 520.

[318]Brown; Rathke *J. Am. Chem. Soc.* **1967,** *89*, 4528.

triethylcarboxide;[319] (2) treatment with a suspension of sodium cyanide in THF followed by reaction of the resulting trialkylcyanoborate **76** with an excess (more than 2 moles) of trifluoroacetic anhydride.[320] All the above migrations take place with retention of configuration at the migrating carbon.[321]

Several other methods for the conversion of boranes to tertiary alcohols are also known.[322] OS **VII**, 427.

8-25 Conversion of Boranes to Secondary Alcohols or Ketones

If the reaction between trialkylboranes and carbon monoxide (**8-24**) is carried out in the presence of water followed by addition of NaOH, the product is a secondary alcohol. If H_2O_2 is added along with the NaOH, the corresponding ketone is obtained instead.[323] Various functional groups (e.g., OAc, COOR, CN) may be present in R without being affected,[324] though if they are in the α or β position relative to the boron atom, difficulties may be encountered. The reaction has been extended to the formation of unsymmetrical ketones by use of a borane of the form $R_2R'B$, where one of the groups migrates much less readily than the other (migratory aptitudes are in the order primary > secondary > tertiary).[318]

The reaction follows the mechanism shown in **8-24** until formation of the borepoxide **74**. In the presence of water the third boron → carbon migration does not take place, because the water hydrolyzes **74** to the diol **77**.

Trialkylboranes can also be converted to ketones by the cyanoborate procedure, mentioned in **8-24.** In this case the procedure is similar, but use of an equimolar amount of

$$R_3\overset{\ominus}{B}-CN \xrightarrow[\text{2.}\,H_2O_2\text{-}OH^-]{\text{1.}\,(CF_3CO)_2O} RCOR$$

76

trifluoroacetic anhydride leads to the ketone rather than the tertiary alcohol.[325] By this procedure the thexylboranes $RR'R''B$ (R'' = thexyl) can be converted to unsymmetrical ketones $RCOR'$.[326] Like the carbon monoxide procedure, this method tolerates the presence of

[319]Brown; Carlson *J. Org. Chem.* **1973**, *38*, 2422; Brown; Katz; Carlson *J. Org. Chem.* **1973**, *38*, 3968.

[320]Pelter; Hutchings; Smith *J. Chem. Soc., Chem. Commun.* **1973**, 186; Pelter; Hutchings; Smith; Williams *J. Chem. Soc., Perkin Trans. 1* **1975**, 145; Pelter *Chem. Ind.* (*London*) **1973**, 206-209, *Intra-Sci. Chem. Rep.* **1973**, 7(1), 73-79.

[321]See however Pelter; Maddocks; Smith *J. Chem. Soc., Chem. Commun.* **1978**, 805.

[322]See, for example, Lane; Brown *J. Am. Chem. Soc.* **1971**, *93*, 1025; Brown; Yamamoto *Synthesis* **1972**, 699; Brown; Lane *Synthesis* **1972**, 303; Yamamoto; Brown *J. Chem. Soc., Chem. Commun.* **1973**, 801, *J. Org. Chem* **1974**, *39*, 861; Zweifel; Fisher *Synthesis* **1974**, 339; Midland; Brown; *J. Org. Chem.* **1975**, *40*, 2845; Levy; Schwartz *Tetrahedron Lett.* **1976**, 2201; Hughes; Ncube; Pelter; Smith; Negishi; Yoshida *J. Chem. Soc., Perkin Trans. 1* **1977**, 1172; Avasthi; Baba; Suzuki *Tetrahedron Lett.* **1980**, *21*, 945; Baba; Avasthi; Suzuki *Bull. Chem. Soc. Jpn.* **1983**, *56*, 1571; Pelter; Rao *J. Organomet. Chem.* **1985**, *285*, 65; Junchai; Weike; Hongxun *J. Organomet. Chem.* **1989**, *367*, C9; Junchai; Hongxun *J. Chem. Soc., Chem. Commun.* **1990**, 323.

[323]Brown; Rathke *J. Am. Chem. Soc.* **1967**, *89*, 2738.

[324]Brown; Kabalka; Rathke *J. Am. Chem. Soc.* **1967**, *89*, 4530.

[325]Pelter; Smith; Hutchings; Rowe *J. Chem. Soc., Perkin Trans. 1* **1975**, 129; Ref. 320. See also Pelter; Hutchings; Smith *J. Chem. Soc., Perkin Trans. 1* **1975**, 142; Mallison; White; Pelter; Rowe; Smith *J. Chem. Res. (S)* **1978**, 234.

[326]This has been done enantioselectively: Brown; Bakshi; Singaram *J. Am. Chem. Soc.* **1988**, *110*, 1529.

various functional groups in R. Another method involves the treatment of borinic acid esters (which can be prepared by treatment of dialkylchloroboranes with alcohols) with α,α-di-

$$R_2BCl + R'OH \longrightarrow R_2BOR' \xrightarrow[\text{2. } H_2O_2-OH^-]{\substack{\text{1. CHCl}_2\text{OMe, LiOCEt}_3 \\ \text{THF, 25°C}}} R_2CO$$

chloromethyl methyl ether and lithium triethylcarboxide.[327] This method does not waste an R group, is carried out under mild conditions, and has been made enantioselective.[328] A closely related method uses boronic esters $RB(OR')_2$ and $LiCHCl_2$. By the use of chiral R', this method has been used to prepare optically active alcohols.[329] In still another procedure ketones are prepared by the reaction between dialkylchloroboranes and lithium aldimines[330] (which can be prepared by **6-69**).

$$R_2BCl + \underset{\underset{Li}{|}}{R''{-}N{=}C{-}R'} \xrightarrow[\text{2. } H_2O_2-OH^-]{\text{1. HCN or PhCHO}} \underset{\underset{O}{\|}}{R{-}C{-}R'}$$

For another conversion of trialkylboranes to ketones, see **8-28**.[331] Other conversions of boranes to secondary alcohols are also known.[332]

OS **VI**, 137.

8-26   Conversion of Boranes to Primary Alcohols, Aldehydes, or Carboxylic Acids

$$R_3B + CO \xrightarrow[\text{2. } H_2O_2-OH^-]{\text{1. LiBH}_4} RCH_2OH$$

$$R_3B + CO \xrightarrow[\text{2. } H_2O_2-NaH_2PO_4-Na_2HPO_4]{\text{1. LiAl(OMe)}_3} RCHO$$

When the reaction between a trialkylborane and carbon monoxide (**8-24**) is carried out in the presence of a reducing agent such as lithium borohydride or potassium triisopropoxy-borohydride, the reduction agent intercepts the intermediate **73**, so that only one boron-to-carbon migration takes place, and the product is hydrolyzed to a primary alcohol or oxidized to an aldehyde.[333] This procedure wastes two of the three R groups, but this problem can be avoided by the use of B-alkyl-9-BBN derivatives (p. 785). Since only the 9-alkyl group

[327]Carlson; Brown *J. Am. Chem. Soc.* **1973**, *95*, 6876, *Synthesis* **1973**, 776.
[328]Brown; Srebnik; Bakshi; Cole *J. Am. Chem. Soc.* **1987**, *109*, 5420; Brown; Gupta; Vara Prasad; Srebnik *J. Org. Chem.* **1988**, *53*, 1391.
[329]For reviews, see Matteson *Mol. Struct. Energ.* **1988**, *5*, 343-356, *Acc. Chem. Res.* **1988**, *21*, 294-300, *Synthesis* **1986**, 973-985, pp. 980-983.
[330]Yamamoto; Kondo; Moritani *Tetrahedron Lett.* **1974**, 793; *Bull. Chem. Soc. Jpn.* **1975**, *48*, 3682. See also Yamamoto; Kondo; Moritani *J. Org. Chem.* **1975**, *40*, 3644.
[331]For still other methods, see Brown; Levy; Midland *J. Am. Chem. Soc.* **1975**, *97*, 5017; Ncube; Pelter; Smith *Tetrahedron Lett.* **1979**, 1893; Pelter; Rao, Ref. 322; Yogo; Koshino; Suzuki *Chem. Lett.* **1981**, 1059; Kulkarni; Lee; Brown *J. Org. Chem.* **1980**, *45*, 4542, *Synthesis* **1982**, 193; Brown; Bhat; Basavaiah *Synthesis* **1983**, 885; Narayana; Periasamy *Tetrahedron Lett.* **1985**, *26*, 6361.
[332]See for example, Zweifel; Fisher, Ref. 322; Brown; Yamamoto *J. Am. Chem. Soc.* **1971**, *93*, 2796, *Chem. Commun.* **1971**, 1535, *J. Chem. Soc., Chem. Commun.* **1972**, 71; Brown; DeLue *J. Am. Chem. Soc.* **1974**, *96*, 311; Hubbard; Brown *Synthesis* **1978**, 676; Uguen *Bull. Soc. Chim. Fr.* **1981**, II-99.
[333]Brown; Rathke *J. Am. Chem. Soc.* **1967**, *89*, 2740; Brown; Coleman; Rathke *J. Am. Chem. Soc.* **1968**, *90*, 499; Brown; Hubbard; Smith *Synthesis* **1979**, 701. For discussions of the mechanism, see Brown; Hubbard *J. Org. Chem.* **1979**, *44*, 467; Hubbard; Smith *J. Organomet. Chem.* **1984**, *276*, C41.

migrates, this method permits the conversion in high yield of an alkene to a primary alcohol or aldehyde containing one more carbon.[334] When B-alkyl-9-BBN derivatives are treated with CO and lithium tri-t-butoxyaluminum hydride,[335] other functional groups (e.g., CN and ester) can be present in the alkyl group without being reduced.[336] Boranes can be directly converted to carboxylic acids by reaction with the dianion of phenoxyacetic acid.[337]

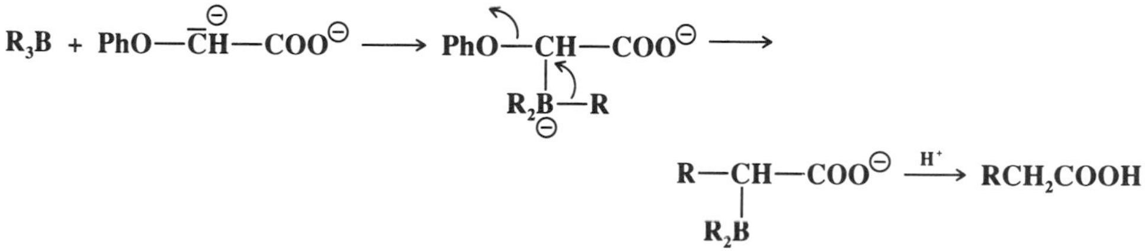

$$R_3B + PhO\overset{\ominus}{-}\overset{}{\underset{}{C}}H-COO^{\ominus} \longrightarrow PhO-\underset{\underset{\ominus}{\overset{|}{R_2B-R}}}{\overset{}{C}}H-COO^{\ominus} \longrightarrow$$

$$R-\underset{R_2B}{\overset{|}{\underset{}{C}}}H-COO^{\ominus} \xrightarrow{H^+} RCH_2COOH$$

Boronic esters $RB(OR')_2$ react with methoxy(phenylthio)methyllithium LiCH(OMe)SPh to give salts, which, after treatment with $HgCl_2$ and then H_2O_2, yield aldehydes.[338] This synthesis has been made enantioselective, with high ee values ($> 99\%$), by the use of an optically pure boronic ester,[339] e.g.:

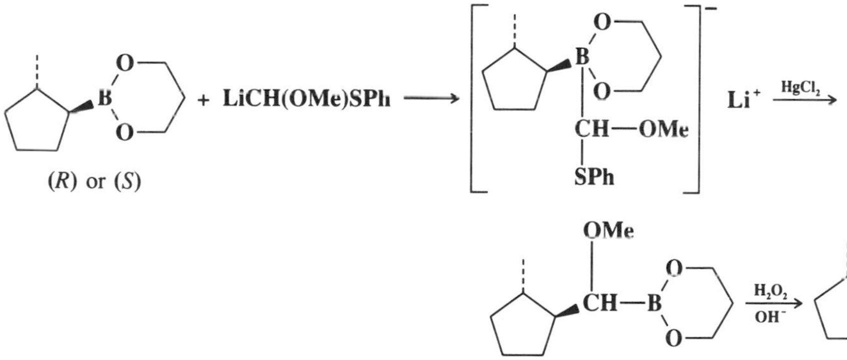

(R) or (S)

8-27 Conversion of Vinylic Boranes to Alkenes

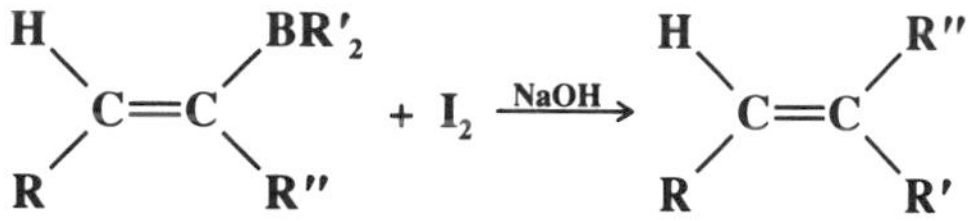

78   79

The reaction between trialkylboranes and iodine to give alkyl iodides was mentioned at **2-30**. When the substrate contains a vinylic group, the reaction takes a different course,[340]

[334]Brown; Knights; Coleman *J. Am. Chem. Soc.* **1969**, *91*, 2144.

[335]Brown; Coleman *J. Am. Chem. Soc.* **1969**, *91*, 4606.

[336]For other methods of converting boranes to aldehydes, see Yamamoto; Shiono; Mukaiyama *Chem. Lett.* **1973**, 961; Negishi; Yoshida; Silveira; Chiou *J. Org. Chem.* **1975**, *40*, 814.

[337]Hara; Kishimura; Suzuki; Dhillon *J. Org. Chem.* **1990**, *55*, 6356. See also Brown; Imai *J. Org. Chem.* **1984**, *49*, 892.

[338]Brown; Imai *J. Am. Chem. Soc.* **1983**, *105*, 6285. For a related method that produces primary alcohols, see Brown; Imai; Perumal; Singaram *J. Org. Chem.* **1985**, *50*, 4032.

[339]Brown; Imai; Desai; Singaram *J. Am. Chem. Soc.* **1985**, *107*, 4980.

[340]Zweifel; Arzoumanian; Whitney *J. Am. Chem. Soc.* **1967**, *89*, 3652; Zweifel; Fisher *Synthesis* **1975**, 376; Brown; Basavaiah; Kulkarni; Bhat; Vara Prasad *J. Org. Chem.* **1988**, *53*, 239.

with one of the R' groups migrating to the carbon, to give alkenes **79**.[341] The reaction is stereospecific in two senses: (1) if the groups R and R″ are cis in the starting compound, they will be trans in the product; (2) there is retention of configuration within the migrating group R'.[342] Since vinylic boranes can be prepared from alkynes (**5-12**), this is a method for the addition of R' and H to a triple bond. If R″ = H, the product is a Z alkene. The mechanism is believed to be

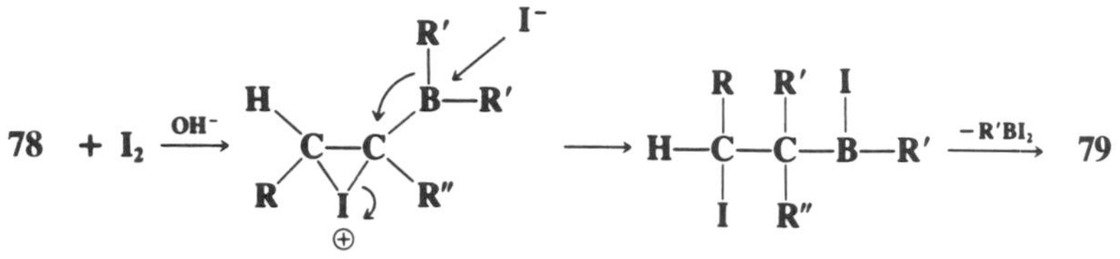

$$78 \ + \ I_2 \ \xrightarrow{\text{OH}^-} \ \ldots \ \longrightarrow \ \underset{\substack{| \\ I}}{\overset{R}{\underset{|}{H-C}}}-\underset{\substack{| \\ R''}}{\overset{R'}{\underset{|}{C}}}-\overset{I}{\underset{|}{B}}-R' \ \xrightarrow{-\text{R'BI}_2} \ 79$$

When R' is vinylic, the product is a conjugated diene.[343]

In another procedure, the addition of a dialkylborane to a 1-haloalkyne produces an α-halo vinylic borane (**80**).[344] Treatment of this with NaOMe gives the rearrangement shown,

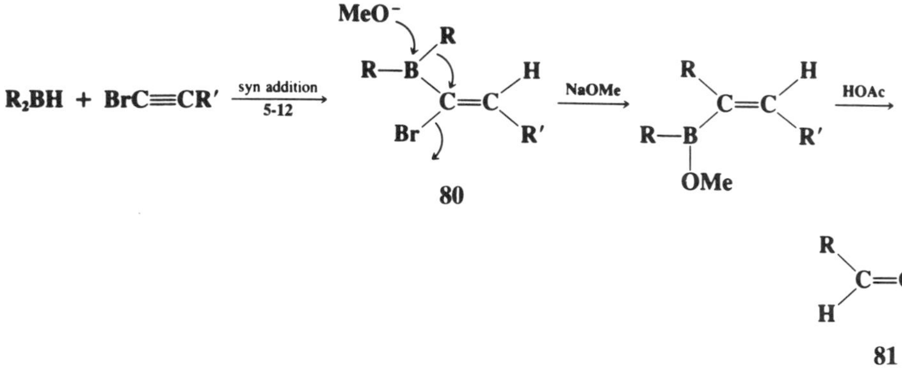

$$R_2BH \ + \ BrC{\equiv}CR' \ \xrightarrow[\text{5-12}]{\text{syn addition}} \ \mathbf{80} \ \xrightarrow{\text{NaOMe}} \ \ldots \ \xrightarrow{\text{HOAc}} \ \mathbf{81}$$

and protonolysis of the product produces the E-alkene **81**.[342] If R is a vinylic group the product is a 1,3-diene.[345] If one of the groups is thexyl, the other migrates.[346] This extends the scope of the synthesis, since dialkylboranes where one R group is thexyl are easily prepared.

A combination of both of the procedures described above results in the preparation of trisubstituted olefins.[347] The entire conversion of haloalkyne to **82** can be carried out in one

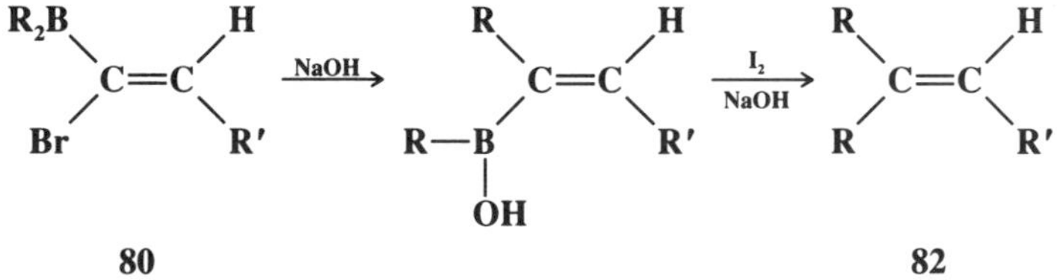

$$\mathbf{80} \ \xrightarrow{\text{NaOH}} \ \ldots \ \xrightarrow[\text{NaOH}]{\text{I}_2} \ \mathbf{82}$$

[341]For a list of methods of preparing alkenes using boron reagents, with references, see Ref. 106, pp. 218-222.

[342]Zweifel; Fisher; Snow; Whitney *J. Am. Chem. Soc.* **1971**, *93*, 6309.

[343]Zweifel; Polston; Whitney *J. Am. Chem. Soc.* **1968**, *90*, 6243; Brown; Ravindran *J. Org. Chem.* **1973**, *38*, 1617; Hyuga; Takinami; Hara; Suzuki *Tetrahedron Lett.* **1986**, *27*, 977.

[344]For improvements in this method, see Brown; Basavaiah; Kulkarni; Lee; Negishi; Katz *J. Org. Chem.* **1986**, *51*, 5270.

[345]Negishi; Yoshida *J. Chem. Soc., Chem. Commun.* **1973**, 606. See also Negishi; Yoshida; Abramovitch; Lew; Williams *Tetrahedron* **1991**, *47*, 343.

[346]Corey; Ravindranathan; *J. Am. Chem. Soc.* **1972**, *94*, 4013; Negishi; Katz; Brown *Synthesis* **1972**, 555.

[347]Zweifel; Fisher *Synthesis* **1972**, 557.

reaction vessel, without isolation of intermediates. An aluminum counterpart of the α-halo vinylic borane procedure has been reported.[348]

E-alkenes **81** can also be obtained[349] by treatment of **78** (R″ = H) with cyanogen bromide or cyanogen iodide in CH_2Cl_2[350] or with $Pd(OAc)_2$–Et_3N.[351]

8-28 Formation of Alkynes, Alkenes, and Ketones from Boranes and Acetylides
Alkyl-de-lithio-substitution

$$R'_3B + RC{\equiv}CLi \longrightarrow RC{\equiv}C{-}\overset{\ominus}{B}R'_3 \; Li^+ \xrightarrow{\;I_2\;} RC{\equiv}CR'$$
$$\textbf{83}$$

A hydrogen directly attached to a triple-bond carbon can be replaced in high yield by an alkyl or an aryl group, by treatment of the lithium acetylide with a trialkyl- or triarylborane, followed by reaction of the lithium alkynyltrialkylborate **83** with iodine.[352] R′ may be primary or secondary alkyl as well as aryl, so the reaction has a broader scope than the older reaction **0-100**.[353] R may be alkyl, aryl, or hydrogen, though in the last-mentioned case satisfactory yields are obtained only if lithium acetylide–ethylenediamine is used as the starting compound.[354] Optically active alkynes can be prepared by using optically active thexylborinates RR″BOR′ (R″ − thexyl), where R is chiral, and $LiC{\equiv}CSiMe_3$.[355] The reaction can be adapted to the preparation of alkenes[341] by treatment of **83** with an electrophile such as

$$\textbf{83} \xrightarrow{R''COOH} R{-}\underset{H}{C}{=}\overset{R'}{\underset{\ominus}{\overset{\oplus}{C}}}{-}BR'_2 \longrightarrow R{-}\underset{H}{C}{=}\overset{R'}{C}{-}BR'_2 \xrightarrow{R''COOH} R{-}\underset{H}{C}{=}\underset{H}{C}{-}R'$$

propanoic acid[356] or tributyltin chloride.[357] The reaction with Bu_3SnCl produces the Z alkene stereoselectively.

Treatment of **83** with an electrophile such as methyl sulfate, allyl bromide, or triethyloxonium borofluoride, followed by oxidation of the resulting vinylic borane gives a ketone (illustrated for methyl sulfate):[358]

$$Me{\overset{\frown}{-}}OSO_2OMe$$

$$R{-}C{\equiv}C{-}\underset{R'}{\overset{\ominus}{B}R'_2} \; Li^+ \longrightarrow R{-}\overset{Me}{\underset{R'}{C}}{=}\overset{R'}{C}{-}BR'_2 \xrightarrow[NaOH]{H_2O_2} R{-}CH{-}\underset{\parallel}{\overset{Me}{C}}{-}R'$$
$$+ \qquad\qquad\qquad\qquad O$$
$$MeOSO_2OLi$$

[348]Miller *J. Org. Chem.* **1989**, *54*, 998.
[349]For other methods of converting boranes to alkenes, see Pelter; Subrahmanyam; Laub; Gould; Harrison *Tetrahedron Lett.* **1975**, 1633; Utimoto; Uchida; Yamaya; Nozaki *Tetrahedron* **1977**, *33*, 1945; Ncube; Pelter; Smith *Tetrahedron Lett.* **1979**, 1895; Levy; Angelastro; Marinelli *Synthesis* **1980**, 945; Brown; Lee; Kulkarni *Synthesis* **1982**, 195; Pelter; Hughes; Rao *J. Chem. Soc., Perkin Trans. 1* **1982**, 719; Hoshi; Masuda; Arase *Bull. Chem. Soc. Jpn.* **1986**, *59*, 3985; Brown; Bhat *J. Org. Chem.* **1988**, *53*, 6009.
[350]Zweifel; Fisher; Snow; Whitney *J. Am. Chem. Soc.* **1972**, *94*, 6560.
[351]Yatagai *Bull. Chem. Soc. Jpn.* **1980**, *53*, 1670.
[352]Suzuki; Miyaura; Abiko; Itoh; Brown; Sinclair; Midland *J. Am. Chem. Soc.* **1973**, *95*, 3080, *J. Org. Chem.* **1986**, *51*, 4507; Sikorski; Bhat; Cole; Wang; Brown *J. Org. Chem.* **1986**, *51*, 4521. For a review of reactions of organoborates, see Suzuki *Acc. Chem. Res.* **1982**, *15*, 178-184.
[353]For a study of the relative migratory aptitudes of R′, see Slayden *J. Org. Chem.* **1981**, *46*, 2311.
[354]Midland; Sinclair; Brown *J. Org. Chem.* **1974**, *39*, 731.
[355]Brown; Mahindroo; Bhat; Singaram *J. Org. Chem.* **1991**, *56*, 1500.
[356]Pelter; Harrision; Kirkpatrick *J. Chem. Soc., Chem. Commun.* **1973**, 544; Miyaura; Yoshinari; Itoh; Suzuki *Tetrahedron Lett.* **1974**, 2961; Pelter; Gould; Harrison *Tetrahedron Lett.* **1975**, 3327.
[357]Hooz; Mortimer *Tetrahedron Lett.* **1976**, 805; Wang; Chu *J. Org. Chem.* **1984**, *49*, 5175.
[358]Pelter; Bentley; Harrison; Subrahmanyam; Laub *J. Chem. Soc., Perkin Trans. 1* **1976**, 2419; Pelter; Gould; Harrison *J. Chem. Soc., Perkin Trans. 1* **1976**, 2428; Pelter; Drake *Tetrahedron Lett.* **1988**, *29*, 4181.

Non-1,2 Rearrangements

A. Electrocyclic Rearrangements

8-29 Electrocyclic Rearrangements of Cyclobutenes and 1,3-Cyclohexadienes
(4)*seco*-1/4/Detachment; (4)*cyclo*-1/4/Attachment
(6)*seco*-1/6/Detachment; (6)*cyclo*-1/6/Attachment

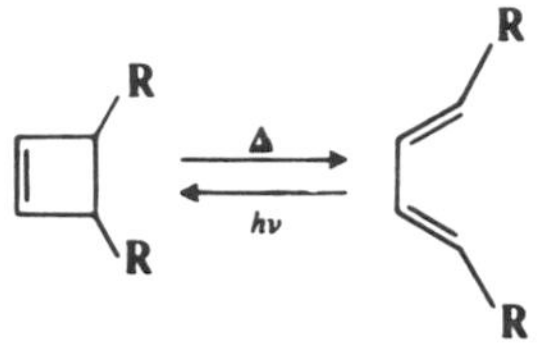

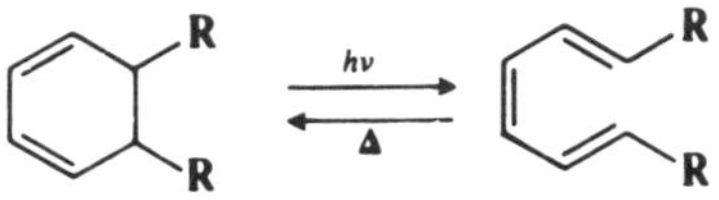

Cyclobutenes and 1,3-dienes can be interconverted by treatment with uv light or with heat. The thermal reaction is generally not reversible (though exceptions[359] are known), and many cyclobutenes have been converted to 1,3-dienes by heating at temperatures between 100 and 200°C. The photochemical conversion can in principle be carried out in either direction, but most often 1,3-dienes are converted to cyclobutenes rather than the reverse, because the dienes are stronger absorbers of light at the wave lengths used.[360] In a similar reaction, 1,3-cyclohexadienes interconvert with 1,3,5-trienes, but in this case the ring-closing process is generally favored thermally and the ring-opening process photochemically, though exceptions are known in both directions.[361]

Some examples are

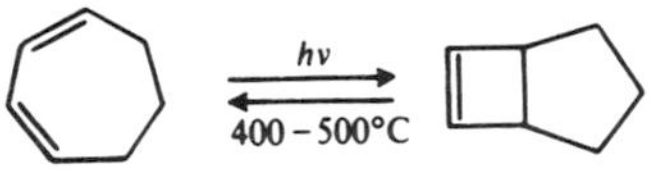

Ref. 362

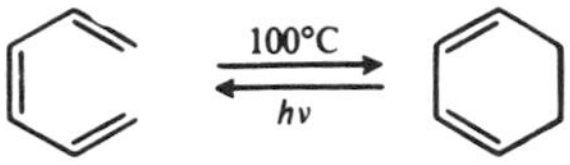

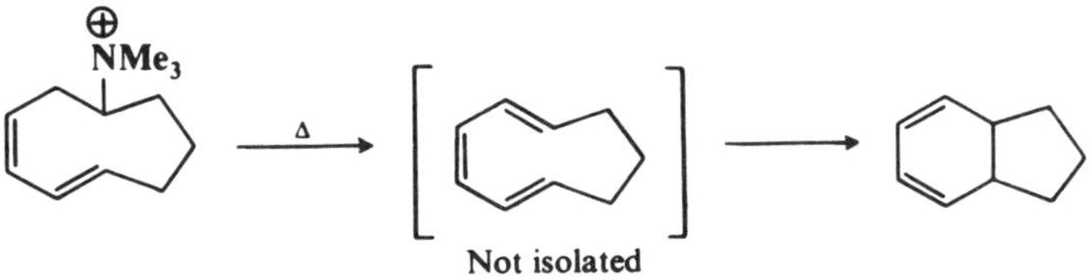

Not isolated

---

[359]For example; see Shumate; Neuman; Fonken *J. Am. Chem. Soc.* **1965**, *87*, 3996; Gil-Av; Herling *Tetrahedron Lett.* **1967**, 1; Doorakian; Freedman *J. Am. Chem. Soc.* **1968**, *90*, 3582; Brune; Schwab *Tetrahedron* **1969**, *25*, 4375; Steiner; Michl *J. Am. Chem. Soc.* **1978**, *100*, 6413.

[360]For examples of photochemical conversion of a cyclobutene to a 1,3-diene, see Scherer *J. Am. Chem. Soc.* **1968**, *90*, 7352; Saltiel; Lim *J. Am. Chem. Soc.* **1969**, *91*, 5404; Adam; Oppenländer; Zang *J. Am. Chem. Soc.* **1985**, *107*, 3921; Dauben; Haubrich *J. Org. Chem.* **1988**, *53*, 600.

[361]For a review of photochemical rearrangements in trienes, see Dauben; McInnis; Michno, in Mayo, Ref. 1, vol. 3, pp. 91-129.

[362]Dauben; Cargill *Tetrahedron* **1961**, *12*, 186; Chapman; Pasto; Borden; Griswold *J. Am. Chem. Soc.* **1962**, *84*, 1220.

An interesting example of 1,3-cyclohexadiene—1,3,5-triene interconversion is the reaction of norcaradienes to give cycloheptatrienes.[363] Norcaradienes give this reaction so readily

Norcaradiene

(because they are *cis*-1,2-divinylcyclopropanes, see p. 1131) that they cannot generally be isolated, though some exceptions are known[364] (see also p. 869).

These reactions, called *electrocyclic rearrangements*,[365] take place by pericyclic mechanisms. The evidence comes from stereochemical studies, which show a remarkable stereospecificity whose direction depends on whether the reaction is induced by heat or light. For example, it was found for the thermal reaction that *cis*-3,4-dimethylcyclobutene gave only *cis,trans*-2,4-hexadiene, while the trans isomer gave only the trans–trans diene:[366]

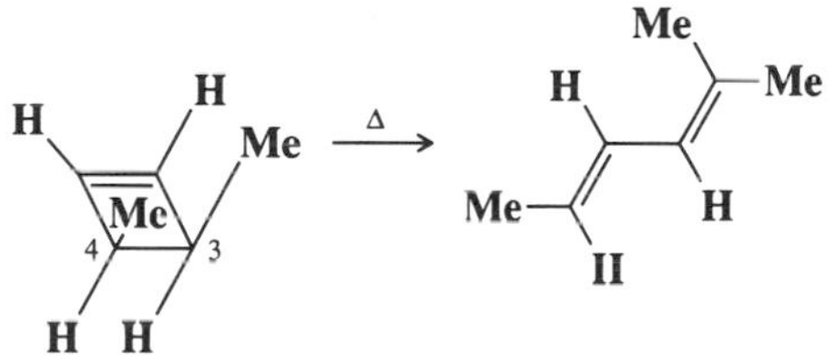

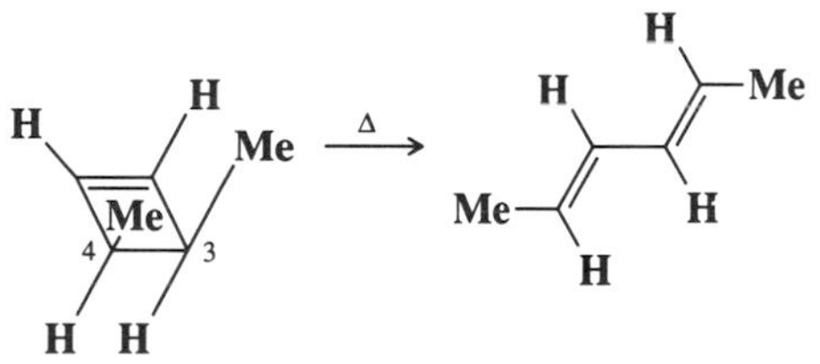

This is evidence for a four-membered cyclic transition state and arises from conrotatory motion about the C-3—C-4 bond. It is called conrotatory because both movements are

[363]For reviews of the norcaradiene–cycloheptatriene interconversion and the analogous benzene oxide–oxepin interconversion, see Maier *Angew. Chem. Int. Ed. Engl.* **1967**, *6*, 402-413 [*Angew. Chem. 79*, 446-458]; Vogel; Günther *Angew. Chem. Int. Ed. Engl.* **1967**, *6*, 385-401 [*Angew. Chem. 79*, 429-446]; Vogel *Pure Appl. Chem.* **1969**, *20*, 237-262.

[364]See Refs. 1043 and 1044 in Chapter 15.

[365]For a monograph on thermal isomerizations, which includes electrocyclic and sigmatropic rearrangements, as well as other types, see Gajewski *Hydrocarbon Thermal Isomerizations*; Academic Press: New York, 1981. For a monograph on electrocyclic reactions, see Marvell *Thermal Electrocyclic Reactions*; Academic Press: New York, 1980. For reviews, see Dolbier; Koroniak *Mol. Struct. Energ.* **1988**, *8*, 65-81; Laarhoven *Org. Photochem.* **1987**, *9*, 129-224; George; Mitra; Sukumaran *Angew. Chem. Int. Ed. Engl.* **1980**, *19*, 973-983 [*Angew. Chem. 92*, 1005-1014]; Jutz *Top. Curr. Chem.* **1978**, *73*, 125-230; Gilchrist; Storr *Organic Reactions and Orbital Symmetry*; Cambridge University Press: Cambridge, 1972, pp. 48-72; DeWolfe, in Bamford; Tipper *Comprehensive Chemical Kinetics*, vol. 9; Elsevier: New York, 1973; pp. 461-470; Crowley; Mazzocchi, in Zabicky *The Chemistry of Alkenes*, vol. 2; Wiley: New York, 1970, pp. 284-297; Criegee *Angew. Chem. Int. Ed. Engl.* **1968**, *7*, 559-565 [*Angew. Chem. 80*, 585-591]; Vollmer; Servis *J. Chem. Educ.* **1968**, *45*, 214-220. For a review of isotope effects in these reactions, see Gajewski *Isot. Org. Chem.* **1987**, *7*, 115-176. For a related review, see Schultz; Motyka *Org. Photochem.* **1983**, *6*, 1-119.

[366]Winter *Tetrahedron Lett.* **1965**, 1207. Also see Vogel *Liebigs Ann. Chem.* **1958**, *615*, 14; Criegee; Noll *Liebigs Ann. Chem.* **1959**, *627*, 1.

clockwise (or both counterclockwise). Because both rotate in the same direction, the cis isomer gives the cis–trans diene:[367]

The other possibility (*disrotatory* motion) would have one moving clockwise while the other moves counterclockwise; the cis isomer would have given the cis–cis diene (shown) or the trans–trans diene:

If the motion had been disrotatory, this would still have been evidence for a cyclic mechanism. If the mechanism were a diradical or some other kind of noncyclic process, it is likely that no stereospecificity of either kind would have been observed. The reverse reaction is also conrotatory. In contrast, the photochemical cyclobutene—1,3-diene interconversion is *disrotatory* in either direction.[368] On the other hand, the cyclohexadiene—1,3,5-triene interconversion shows precisely the opposite behavior. The thermal process is *disrotatory*, while the photochemical process is *conrotatory* (in either direction). These startling results are a consequence of the symmetry rules mentioned in Chapter 15 (p. 846).[369] As in the case of cycloaddition reactions, we will use the frontier-orbital and Möbius–Hückel approaches.[370]

The Frontier-Orbital Method[371]
As applied to these reactions, the frontier-orbital method may be expressed: *A σ bond will open in such a way that the resulting p orbitals will have the symmetry of the highest occupied π orbital of the product.* In the case of cyclobutenes, the HOMO of the product in the thermal reaction is the χ_2 orbital (Figure 18.1). Therefore, in a thermal process, the cyclo-

χ_2 χ_3^*

FIGURE 18.1   Symmetries of the χ_2 and χ_3^* orbitals of a conjugated diene.

[367]This picture is from Woodward; Hoffmann *J. Am. Chem. Soc.* **1965**, *87*, 395, who coined the terms, *conrotatory* and *disrotatory*.

[368]Photochemical ring-opening of cyclobutenes can also be nonstereospecific. See Leigh; Zheng *J. Am. Chem. Soc.* **1991**, *113*, 4019; Leigh; Zheng; Nguyen; Werstiuk; Ma *J. Am. Chem. Soc.* **1991**, *113*, 4993, and references cited in these papers.

[369]Woodward; Hoffmann, Ref. 367. Also see Longuet-Higgins; Abrahamson *J. Am. Chem. Soc.* **1965**, *87*, 2045; Fukui *Tetrahedron Lett.* **1965**, 2009.

[370]For the correlation diagram method, see Jones *Physical and Mechanistic Organic Chemistry*, 2nd ed.; Cambridge University Press: Cambridge, 1984, pp. 352-359; Yates *Hückel Molecular Orbital Theory*; Academic Press: New York, 1978, pp. 250-263; Ref. 897 in Chapter 15.

[371]See Ref. 898 in Chapter 15.

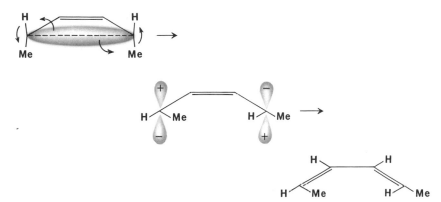

FIGURE 18.2   Thermal ring opening of 1,2-dimethylcyclobutene. The two hydrogens and two methyls are forced into conrotatory motion so that the resulting p orbitals have the symmetry of the HOMO of the diene.

butene must open so that on one side the positive lobe lies above the plane, and on the other side below it. Thus the substituents are forced into conrotatory motion (Figure 18.2). On the other hand, in the photochemical process, the HOMO of the product is now the χ_3 orbital (Figure 18.1), and in order for the p orbitals to achieve this symmetry (the two plus lobes on the same side of the plane), the substituents are forced into disrotatory motion.

We may also look at this reaction from the opposite direction (ring closing). For this direction the rule is that *those lobes of orbitals that overlap (in the HOMO) must be of the same sign.* For thermal cyclization of butadienes, this requires conrotatory motion (Figure 18.3). In the photochemical process the HOMO is the χ_3 orbital, so that disrotatory motion is required for lobes of the same sign to overlap.

The Möbius–Hückel Method[372]
As we saw on p. 848, in this method we choose a basis set of p orbitals and look for sign inversions in the transition state. Figure 18.4 shows a basis set for a 1,3-diene. It is seen that disrotatory ring closing (Figure 18.4a) results in overlap of plus lobes only, while in conrotatory closing (Figure 18.4b) there is one overlap of a plus with a minus lobe. In the first case we have zero sign inversions, while in the second there is one sign inversion. With zero (or an even number of) sign inversions, the disrotatory transition state is a Hückel

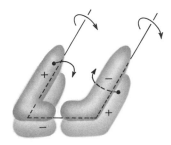

FIGURE 18.3   Thermal ring closing of a 1,3-diene. Conrotatory motion is required for two + lobes to overlap.

[372]See Ref. 899 in Chapter 15.

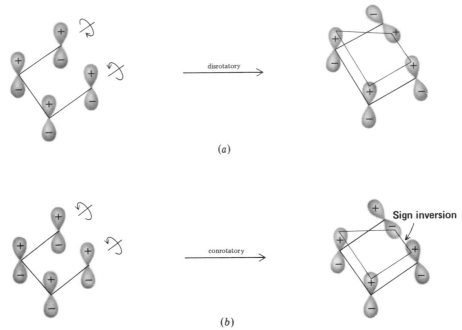

(a)

(b)

FIGURE 18.4 The 1,3-diene-cyclobutene interconversion. The orbitals shown are *not* molecular orbitals, but a basis set of *p* atomic orbitals. (*a*) Disrotatory ring closure gives zero sign inversions. (*b*) Conrotatory ring closure gives one sign inversion. We could have chosen to show any other basis set (for example, another basis set would have two plus lobes above the plane and two below, etc.). This would change the number of sign inversions, but the disrotatory mode would still have an even number of sign inversions, and the conrotatory mode an odd number, whichever basis set was chosen.

system, and so is allowed thermally only if the total number of electrons is $4n + 2$ (p. 848). Since the total here is 4, the disrotatory process is not allowed. On the other hand, the conrotatory process, with one sign inversion, is a Möbius system, which is thermally allowed if the total number is $4n$. The conrotatory process is therefore allowed thermally. For the photochemical reactions the rules are reversed: A reaction with $4n$ electrons requires a Hückel system, so only the disrotatory process is allowed.

Both the frontier-orbital and the Möbius–Hückel methods can also be applied to the cyclohexadiene—1,3,5-triene reaction; in either case the predicted result is that for the thermal process, only the disrotatory pathway is allowed, and for the photochemical process, only the conrotatory. For example, for a 1,3,5-triene, the symmetry of the HOMO is

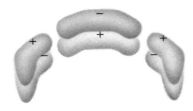

In the thermal cleavage of cyclohexadienes, then, the positive lobes must lie on the same side of the plane, requiring disrotatory motion:

Disrotatory motion is also necessary for the reverse reaction, in order that the orbitals which overlap may be of the same sign:

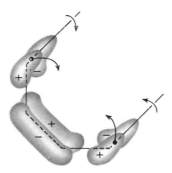

All these directions are reversed for photochemical processes, because in each case a higher orbital, with inverted symmetry, is occupied.

In the Möbius–Hückel approach, diagrams similar to Figure 18.4 can be drawn for this case. Here too, the disrotatory pathway is a Hückel system and the conrotatory pathway a Möbius system, but since six electrons are now involved, the thermal reaction follows the Hückel pathway and the photochemical reaction the Möbius pathway.

In the most general case, there are four possible products that can arise from a given cyclobutene or cyclohexadiene—two from the conrotatory and two from the disrotatory

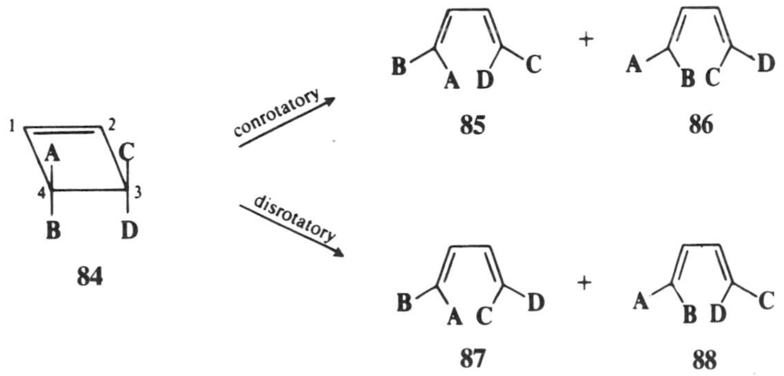

pathway. For example, conrotatory ring opening of **84** gives either **85** or **86**, while disrotatory opening gives either **87** or **88**. The orbital-symmetry rules tell us when a given reaction will operate by the conrotatory and when by the disrotatory mode, but they do not say which of the two possible conrotatory or disrotatory pathways will be followed. It is often possible, however, to make such predictions on steric grounds. For example, in the opening of **84** by the disrotatory pathway, **87** arises when groups A and C swing in toward each other (clockwise motion around C-4, counterclockwise around C-3), while **88** is formed when groups B and D swing in and A and C swing out (clockwise motion around C-3, counterclockwise around C-4). We therefore predict that when A and C are larger than B and D, the predominant or exclusive product will be **88**, rather than **87**. Predictions of this kind have largely been borne out.[373] There is evidence, however, that steric effects are not the only factor, and that electronic effects also play a role, which may be even greater.[374] An electron-donating group stabilizes the transition state when it rotates *outward*, because it mixes with the LUMO; if it rotates *inward*, it mixes with the HOMO, destabilizing the transition state.[375] The compound 3-formylcyclobutene provided a test. Steric factors would cause the CHO (an electron-withdrawing group) to rotate outward; electronic effects would cause it to rotate inward. The experiment showed inward rotation.[376]

Cyclohexadienes are of course 1,3-dienes, and in certain cases it is possible to convert them to cyclobutenes instead of to 1,3,5-trienes.[377] An interesting example is found in the pyrocalciferols. Photolysis of the syn isomer **89** (or of the other syn isomer, not shown)

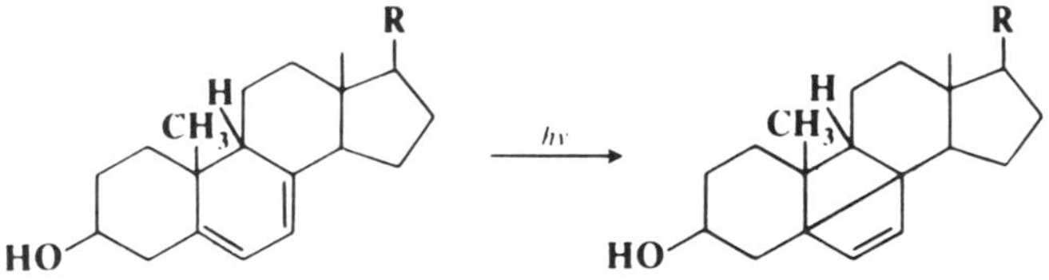

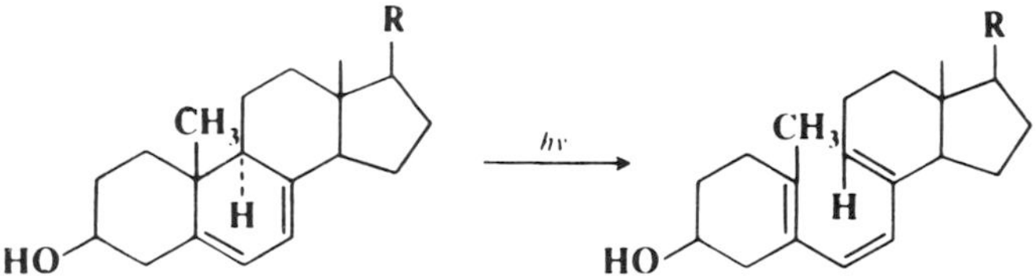

leads to the corresponding cyclobutene,[378] while photolysis of the anti isomers (one of them is **90**) gives the ring-opened 1,3,5-triene **91**. This difference in behavior is at first sight remarkable, but is easily explained by the orbital-symmetry rules. Photochemical ring opening to a 1,3,5-triene must be conrotatory. If **89** were to react by this pathway, the product would be the triene **91**, but this compound would have to contain a *trans*-cyclohexene ring (either the methyl group or the hydrogen would have to be directed inside the ring). On

[373]For example, see Baldwin; Krueger *J. Am. Chem. Soc.* **1969**, *91*, 6444; Spangler; Hennis *J. Chem. Soc., Chem. Commun.* **1972**, 24; Gesche; Klinger; Riesen; Tschamber; Zehnder; Streith *Helv. Chim. Acta* **1987**, *70*, 2087.

[374]Kirmse; Rondan; Houk *J. Am. Chem. Soc.* **1984**, *106*, 7989; Dolbier; Koroniak; Burton; Heinze; Bailey; Shaw; Hansen *J. Am. Chem. Soc.* **1987**, *109*, 219; Dolbier; Gray; Keaffaber; Celewicz; Koroniak *J. Am. Chem. Soc.* **1990**, *112*, 363; Hayes; Ingham; Saengchantara; Wallace *Tetrahedron Lett.* **1991**, *32*, 2953.

[375]For theoretical studies, see Rondan; Houk *J. Am. Chem. Soc.* **1985**, *107*, 2099; Buda; Wang; Houk *J. Org. Chem.* **1989**, *54*, 2264; Kallel; Wang; Spellmeyer; Houk *J. Am. Chem. Soc.* **1990**, *112*, 6759.

[376]Rudolf; Spellmeyer; Houk *J. Org. Chem.***1987**, *52*, 3708; Piers; Lu *J. Org. Chem.* **1989**, *54*, 2267.

[377]For a discussion of the factors favoring either direction, see Dauben; Kellogg; Seeman; Vietmeyer; Wendschuh *Pure Appl. Chem.* **1973**, *33*, 197-215.

[378]Dauben; Fonken *J. Am. Chem. Soc.* **1959**, *81*, 4060. This was the first reported example of the conversion of a 1,3-diene to a cyclobutene.

the other hand, photochemical conversion to a cyclobutene must be disrotatory, but if **90** were to give this reaction, the product would have to have a trans-fused ring junction. Compounds with such ring junctions are known (p. 132) but are very strained. Stable *trans*-cyclohexenes are unknown (p. 158). Thus, **89** and **90** give the products they do owing to a combination of orbital-symmetry rules and steric influences.

The 1,3-diene—cyclobutene interconversion can even be applied to benzene rings. For example,[379] photolysis of 1,2,4-tri-*t*-butylbenzene (**92**) gives 1,2,5-tri-*t*-butyl[2.2.0]hexadiene (**93**, a Dewar benzene).[380] The reaction owes its success to the fact that once **93** is formed,

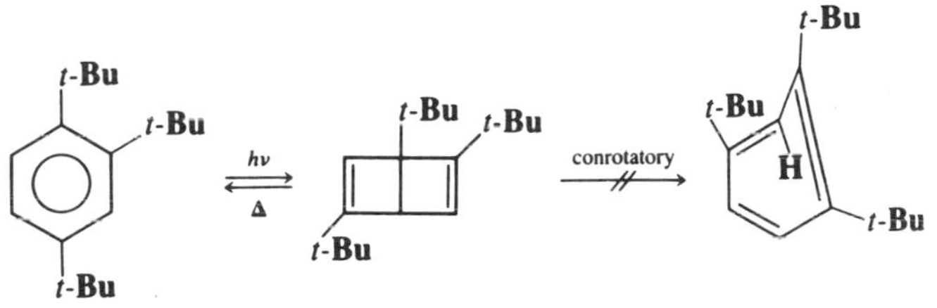

it cannot, under the conditions used, revert to **92** by either a thermal or a photochemical route. The orbital-symmetry rules prohibit thermal conversion of **93** to **92** by a pericyclic mechanism, because thermal conversion of a cyclobutene to a 1,3-diene must be conrotatory, and conrotatory reaction of **93** would result in a 1,3,5-cyclohexatriene containing one trans double bond (**94**), which is of course too strained to exist. **93** cannot revert to **92** by a photochemical pathway either, because light of the frequency used to excite **92** would not be absorbed by **93**. This is thus another example of a molecule that owes its stability to the orbital-symmetry rules (see p. 865). Pyrolysis of **93** does give **92**, probably by a diradical mechanism.[381] In the case of **95** and **96**, the Dewar benzene is actually more stable than the

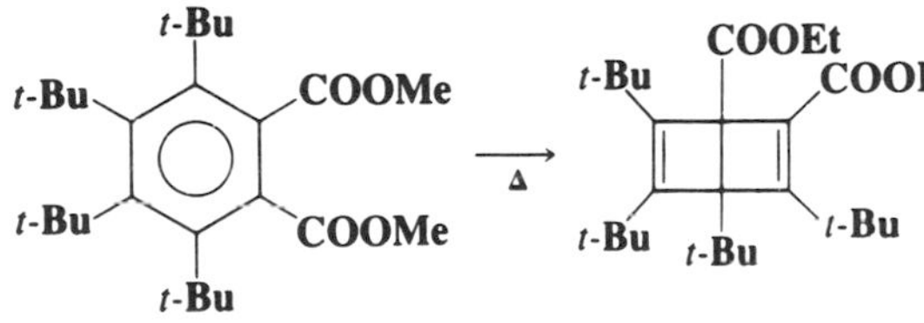

benzene. **95** rearranges to **96** in 90% yield at 120°.[382] In this case thermolysis of the benzene gives the Dewar benzene (rather than the reverse), because of the strain of four adjacent *t*-butyl groups on the ring.

---

[379]Unsubstituted Dewar benzene has been obtained, along with other photoproducts, by photolysis of benzene: Ward; Wishnok *J. Am. Chem. Soc.* **1968**, *90*, 1085; Bryce-Smith; Gilbert; Robinson *Angew. Chem. Int. Ed. Engl.* **1971**, *10*, 745 [*Angew. Chem. 83*, 803]. For other examples, see Arnett; Bollinger *Tetrahedron Lett.* **1964**, 3803; Camaggi; Gozzo; Cevidalli *Chem. Commun.* **1966**, 313; Haller *J. Am. Chem. Soc.* **1966**, *88*, 2070, *J. Chem. Phys.* **1967**, *47*, 1117; Barlow; Haszeldine; Hubbard *Chem. Commun.* **1969**, 202; Lemal; Staros; Austel *J. Am. Chem. Soc.* **1969**, *91*, 3373.

[380]van Tamelen; Pappas *J. Am. Chem. Soc.* **1962**, *84* 3789; Wilzbach; Kaplan *J. Am. Chem. Soc.* **1965**, *87*, 4004; van Tamelen; Pappas; Kirk *J. Am. Chem. Soc.* **1971**, *93*, 6092; van Tamelen *Acc. Chem. Res.* **1972**, *5*, 186-192. As mentioned on p. 865 (Ref. 1002), Dewar benzenes can be photolyzed further to give prismanes.

[381]See, for example, Oth *Recl. Trav. Chim. Pays-Bas* **1968**, *87*, 1185; Adam; Chang *Int. J. Chem. Kinet.* **1969**, *1*, 487; Lechtken; Breslow; Schmidt; Turro *J. Am. Chem. Soc.* **1973**, *95*, 3025; Wingert; Irngartinger; Kallfass; Regitz *Chem. Ber.* **1987**, *120*, 825.

[382]Maier; Schneider *Angew. Chem. Int. Ed. Engl.* **1980**, *19*, 1022 [*Angew. Chem. 95*, 1056]. See also Wingert; Maas; Regitz *Tetrahedron* **1986**, *42*, 5341.

A number of electrocyclic reactions have been carried out with systems of other sizes, e.g., conversion of the 1,3,5,7-octatetraene **97** to the cyclooctatriene **98**.[383] The stereochem-

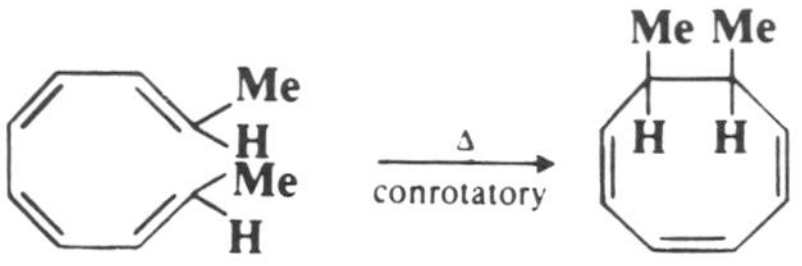

istry of these reactions can be predicted in a similar manner. The results of such predictions can be summarized according to whether the number of electrons involved in the cyclic process is of the form $4n$ or $4n + 2$ (where n is any integer including zero).

	Thermal reaction	**Photochemical reaction**
$4n$	conrotatory	disrotatory
$4n + 2$	disrotatory	conrotatory

Although the orbital-symmetry rules predict the stereochemical results in almost all cases, it is necessary to recall (p. 849) that they only say what is allowed and what is forbidden, but the fact that a reaction is allowed does not necessarily mean that that reaction takes place, and if an allowed reaction does take place, it does not *necessarily* follow that a concerted pathway is involved, since other pathways of lower energy may be available.[384] Furthermore, a "forbidden" reaction might still be made to go, if a method of achieving its high activation energy can be found. This was, in fact, done for the cyclobutene—butadiene interconversion (*cis*-3,4-dichlorocyclobutene gave the forbidden *cis,cis*- and *trans,trans*-1,4-dichloro-1,3-cyclobutadienes, as well as the allowed cis, trans isomer) by the use of ir laser light.[385] This is a thermal reaction. The laser light excites the molecule to a higher vibrational level (p. 232), but not to a higher electronic state.

As is the case for 2 + 2 cycloaddition reactions (**5-49**), certain forbidden electrocyclic reactions can be made to take place by the use of metallic catalysts.[386] An example is the silver ion-catalyzed conversion of tricyclo[4.2.0.0$^{2.5}$]octa-3,7-diene to cyclooctatetraene:[387]

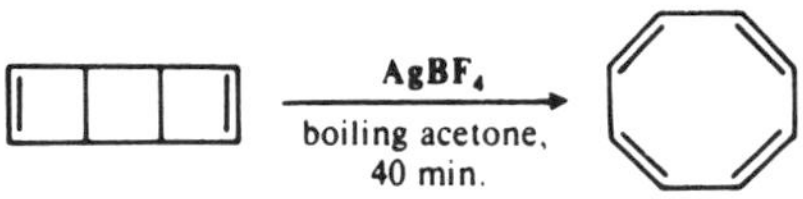

This conversion is very slow thermally (i.e., without the catalyst) because the reaction must take place by a disrotatory pathway, which is disallowed thermally.[388]

[383]Marvell; Seubert *J. Am. Chem. Soc.* **1967**, *89*, 3377; Huisgen; Dahmen; Huber *J. Am. Chem. Soc.* **1967**, *89*, 7130, *Tetrahedron Lett.* **1969**, 1461; Dahmen; Huber *Tetrahedron Lett.* **1969**, 1465.

[384]For a discussion, see Baldwin; Andrist; Pinschmidt *Acc. Chem. Res.* **1972**, *5*, 402-406.

[385]Mao; Presser; John; Moriarty; Gordon *J. Am. Chem. Soc.* **1981**, *103*, 2105.

[386]For a review, see Pettit; Sugahara; Wristers; Merk *Discuss. Faraday Soc.* **1969**, *47*, 71-78. See also Ref. 993 in Chapter 15.

[387]Merk; Pettit *J. Am. Chem. Soc.* **1967**, *89*, 4788.

[388]For discussions of how these reactions take place, see Slegeir; Case; McKennis; Pettit *J. Am. Chem. Soc.* **1974**, *96*, 287; Pinhas; Carpenter *J. Chem. Soc., Chem. Commun.* **1980**, 15.

The ring opening of cyclopropyl cations (pp. 345, 1076) is an electrocyclic reaction and is governed by the orbital symmetry rules.[389] For this case we invoke the rule that the σ bond opens in such a way that the resulting p orbitals have the symmetry of the highest occupied orbital of the product, in this case, an allylic cation. We may recall that an allylic system has three molecular orbitals (p. 32). For the cation, with only two electrons, the highest occupied orbital is the one of the lowest energy (**A**). Thus, the cyclopropyl cation must

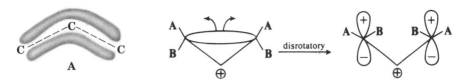

undergo a disrotatory ring opening in order to maintain the symmetry. (Note that, in contrast, ring opening of the cyclopropyl *anion* must be conrotatory,[390] since in this case it is the next orbital of the allylic system which is the highest occupied, and this has the opposite symmetry.[391]) However, it is very difficult to generate a free cyclopropyl cation (p. 345), and it is likely that in most cases, cleavage of the σ bond is concerted with departure of the leaving group in the original cyclopropyl substrate. This of course means that the σ bond provides anchimeric assistance to the removal of the leaving group (an S_N2-type process), and we would expect that such assistance should come from the back side. This has an important effect on the direction of ring opening. The orbital-symmetry rules require that the ring opening be disrotatory, but as we have seen, there are two disrotatory pathways and the rules do not tell us which is preferred. But the fact that the σ orbital provides assistance from the back side means that the two substituents which are trans to the leaving group must move *outward*, not inward.[392] Thus, the disrotatory pathway that is followed is the one shown in **B,** not the one shown in **C,** because the former puts the electrons of the σ

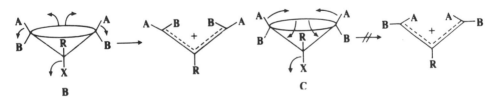

bond on the side opposite that of the leaving group.[393] Strong confirmation of this picture[394] comes from acetolysis of *endo-* (**99**) and *exo*-bicyclo[3,1,0]hexyl-6-tosylate (**100**). The groups

[389]For discussions, see DePuy *Acc. Chem. Res.* **1968,** *1*, 33-41; Schöllkopf *Angew. Chem. Int. Ed. Engl.* **1968,** 7, 588-598 [*Angew. Chem.* 80, 603-613].

[390]For a review of ring opening of cyclopropyl anions and related reactions, see Boche *Top. Curr. Chem.* **1988,** *146,* 1-56.

[391]For evidence that this is so, see Newcomb; Ford *J. Am. Chem. Soc.* **1974,** *96*, 2968; Boche; Buckl; Martens; Schneider; Wagner *Chem. Ber.* **1979,** *112*, 2961; Coates; Last *J. Am. Chem. Soc.* **1983,** *105*, 7322. For a review of the analogous ring opening of epoxides, see Huisgen *Angew. Chem. Int. Ed. Engl.* **1977,** 16, 572-585 [*Angew. Chem.* 89, 589-602].

[392]This was first proposed by DePuy; Schnack; Hausser; Wiedemann *J. Am. Chem. Soc.* **1965,** 87, 4006.

[393]It has been suggested that the pathway shown in C is possible in certain cases: Hausser; Grubber *J. Org. Chem.* **1972,** *37*, 2648; Hausser; Uchic *J. Org. Chem.* **1972,** *37*, 4087.

[394]There is much other evidence. For example, see Jefford; Medary *Tetrahedron Lett.* **1966,** 2069; Jefford; Wojnarowski *Tetrahedron Lett.* **1968,** 199; Schleyer; Van Dine; Schöllkopf; Paust *J. Am. Chem. Soc.* **1966,** *88*, 2868; Sliwinski; Su; Schleyer *J. Am. Chem. Soc.* **1972,** *94*, 133; Sandler *J. Org. Chem.* **1967,** *32*, 3876; Ghosez; Slinckx; Glineur; Hoet; Laroche *Tetrahedron Lett.* **1967,** 2773; Parham; Yong *J. Org. Chem.* **1968,** *33*, 3947; Reese; Shaw *J. Am. Chem. Soc.* **1970,** 92, 2566; Dolbier; Phanstiel *Tetrahedron Lett.* **1988,** 29, 53.

trans to the tosylate must move outward. For **99** this means that the two hydrogens can go outside the framework of the six-membered ring, but for **100** they are forced to go inside.

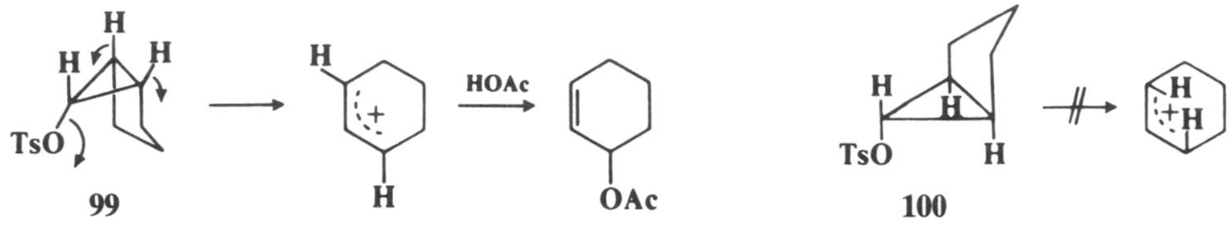

Consequently, it is not surprising that the rate ratio for solvolysis of **99:100** was found to be greater than 2.5×10^6 and that at 150°C **100** did not solvolyze at all.[395] This evidence is kinetic. Unlike the cases of the cyclobutene—1,3-diene and cyclohexadiene—1,3,5-triene interconversions, the direct product here is a cation, which is not stable but reacts with a nucleophile and loses some of its steric integrity in the process, so that much of the evidence has been of the kinetic type rather than from studies of product stereochemistry. However, it has been shown by investigations in super acids, where it is possible to keep the cations intact and to study their structures by nmr, that in all cases studied the cation that is predicted by these rules is in fact formed.[396]

OS **V**, 235, 277, 467; **VI**, 39, 145, 196, 422, 427, 862.

8-30 Conversion of Stilbenes to Phenanthrenes
(6)*cyclo*-De-hydrogen-coupling (overall transformation)

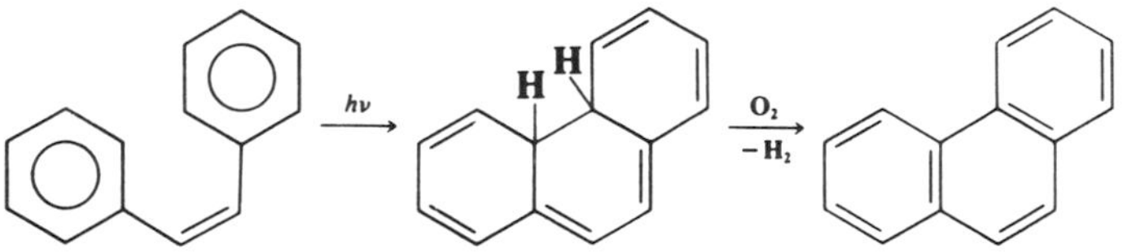

Stilbenes can be converted to phenanthrenes by irradiation with uv light[397] in the presence of an oxidizing agent such as dissolved molecular oxygen, FeCl$_3$, Pd–C,[398] or iodine.[398a] The reaction is a photochemically allowed conrotatory[399] conversion of a 1,3,5-hexatriene to a cyclohexadiene, followed by removal of two hydrogen atoms by the oxidizing agent. The intermediate dihydrophenanthrene has been isolated.[400] The use of substrates containing hetero atoms (e.g., PhN=NPh) allows the formation of heterocyclic ring systems. The actual reacting species must be the *cis*-stilbene, but *trans*-stilbenes can often be used, because they are isomerized to the cis isomers under the reaction conditions. The reaction can be extended to the preparation of many fused aromatic systems, e.g.,[401]

[395]Schöllkopf; Fellenberger; Patsch; Schleyer; Su; Van Dine *Tetrahedron Lett.* **1967**, 3639.
[396]Schleyer; Su; Saunders; Rosenfeld *J. Am. Chem. Soc.* **1969**, *91*, 5174.
[397]For reviews, see Mallory; Mallory *Org. React.* **1984**, *30*, 1-456; Laarhoven *Recl. Trav. Chim. Pays-Bas* **1983**, *102*, 185-204, 241-254; Blackburn; Timmons *Q. Rev., Chem. Soc.* **1969**, *23*, 482-503; Stermitz; *Org. Photochem.* **1967**, *1*, 247-282. For a review of electrocyclizations of conjugated aryl olefins in general, see Laarhoven *Org. Photochem.* **1989**, *10*, 163-308.
[398]Rawal; Jones; Cava *Tetrahedron Lett.* **1985**, *26*, 2423.
[398a]For the use of iodine plus propylene oxide in the absence of air, see Liu; Yang; Katz; Poindexter *J. Org. Chem.* **1991**, *56*, 3769.
[399]Cuppen; Laarhoven *J. Am. Chem. Soc.* **1972**, *94*, 5914.
[400]Doyle; Benson; Filipescu *J. Am. Chem. Soc.* **1976**, *98*, 3262.
[401]Sato; Shimada; Hata *Bull. Chem. Soc. Jpn.* **1971**, *44*, 2484.

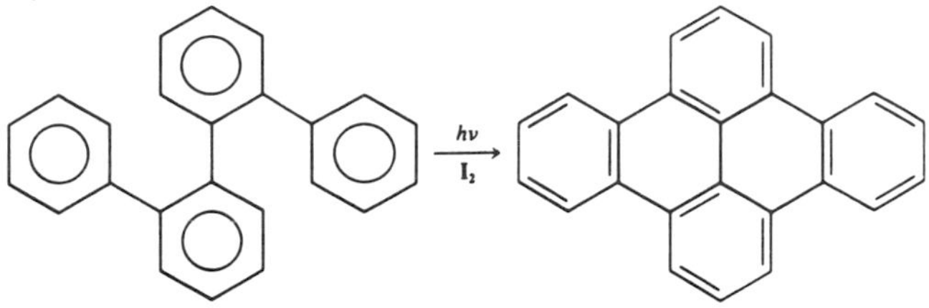

though not all such systems give reaction.[402]

B.   Sigmatropic Rearrangements. A sigmatropic rearrangement is defined[403] as migration, in an uncatalyzed intramolecular process, of a σ bond, adjacent to one or more π systems, to a new position in a molecule, with the π systems becoming reorganized in the process. Examples are

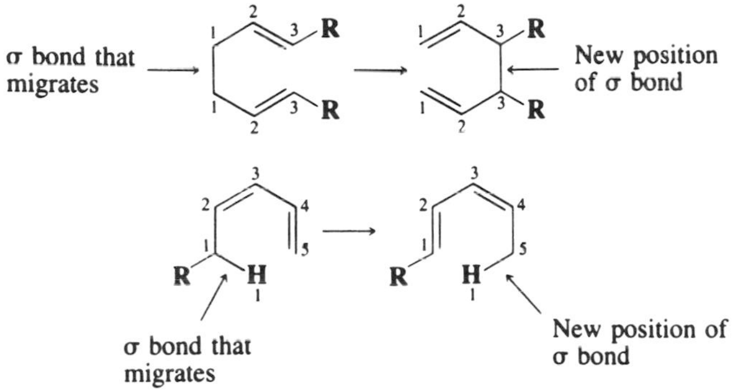

σ bond that ⟶
migrates

New position
of σ bond

Reaction **8-34**
A [3,3] sigmatropic
rearrangement

Reaction **8-31**
A [1,5] sigmatropic
rearrangement

σ bond that
migrates

New position of
σ bond

The *order* of a sigmatropic rearrangement is expressed by two numbers set in brackets: $[i,j]$. These numbers can be determined by counting the atoms over which each end of the σ bond has moved. Each of the original termini is given the number 1. Thus in the first example above, each terminus of the σ bond has migrated from C-1 to C-3, so the order is [3,3]. In the second example the carbon terminus has moved from C-1 to C-5, but the hydrogen terminus has not moved at all, so the order is [1,5].

8-31   [1,*j*] Sigmatropic Migrations of Hydrogen
       1/→3/Hydrogen-migration; 1/→5/Hydrogen-migration

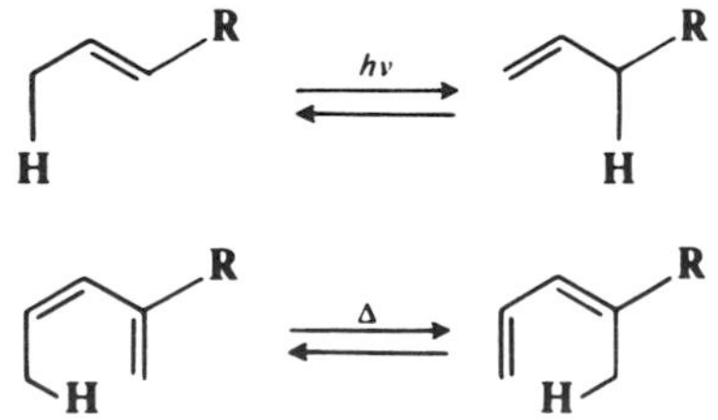

[1,3]

[1,5]

[402]For a discussion and lists of photocyclizing and nonphotocyclizing compounds, see Laarhoven *Recl. Trav. Chim. Pays-Bas*, Ref. 397, pp. 185-204.
   [403]Woodward; Hoffmann *The Conservation of Orbital Symmetry*; Academic Press: New York, 1970, p. 114.

Many examples of thermal or photochemical rearrangements in which a hydrogen atom migrates from one end of a system of π bonds to the other have been reported,[404] though the reaction is subject to geometrical conditions. Pericyclic mechanisms are involved, and the hydrogen must, in the transition state, be in contact with both ends of the chain at the same time. This means that for [1,5] and longer rearrangements, the molecule must be able to adopt the cisoid conformation. Furthermore, there are two geometrical pathways by which any sigmatropic rearrangement can take place, which we illustrate for the case of a [1,5] sigmatropic rearrangement,[405] starting with a substrate of the form **101**, where the migration origin is an asymmetric carbon atom and U $\neq$ V. In one of the two pathways,

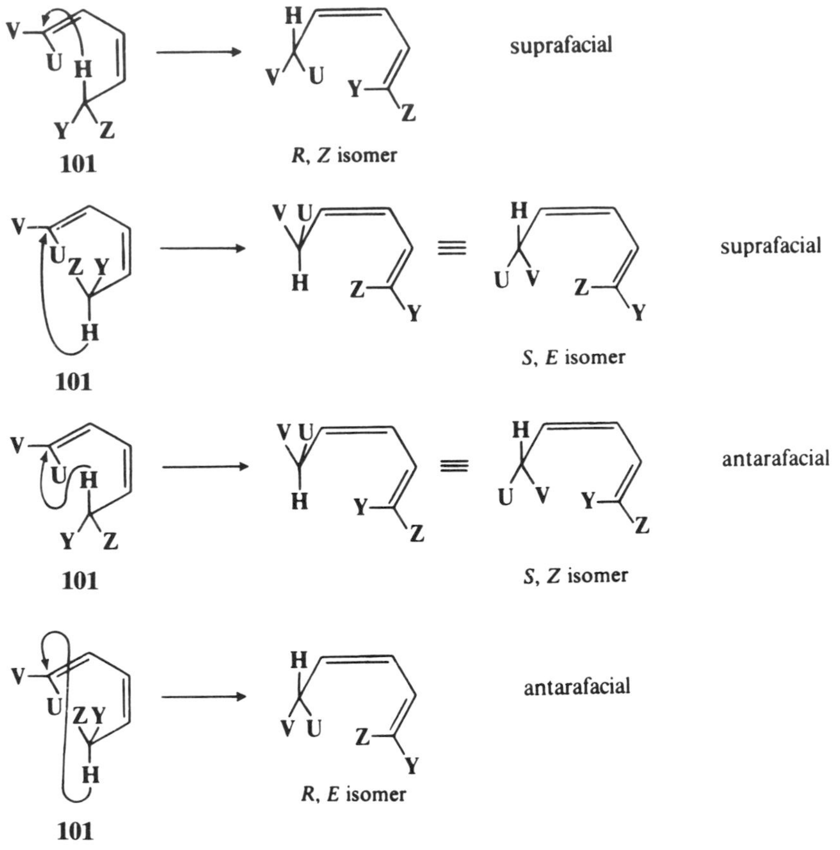

the hydrogen moves along the top or bottom face of the π system. This is called *suprafacial migration*. In the other pathway, the hydrogen moves *across* the π system, from top to bottom, or vice versa. This is *antarafacial* migration. Altogether, a single isomer like **101** can give four products. In a suprafacial migration, H can move across the top of the π system (as drawn above) to give the R, Z isomer, or it can rotate 180° and move across the

[404]For a monograph, see Gajewski, Ref. 365. For reviews, see Mironov; Fedorovich; Akhrem *Russ. Chem. Rev.* **1981**, *50*, 666-681; Spangler *Chem. Rev.* **1976**, *76*, 187-217; DeWolfe, in Bamford; Tipper, Ref. 365, pp. 474-480; Woodward; Hoffmann, Ref. 403, pp. 114-140; Hansen; Schmid *Chimia* **1970**, *24*, 89-99; Roth *Chimia* **1966**, *20*, 229-236.
[405]Note that a [1,5] sigmatropic rearrangement of hydrogen is also an internal ene synthesis (**5-16**).

bottom of the π system to give the S,E isomer.[406] The antarafacial migration can similarly lead to two diastereomers, in this case the S,Z and R,E isomers.

In any given sigmatropic rearrangement, only one of the two pathways is allowed by the orbital-symmetry rules; the other is forbidden. To analyze this situation we first use a modified frontier orbital approach.[407] We will imagine that in the transition state the migrating H atom breaks away from the rest of the system, which we may treat as if it were a free radical.

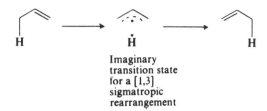

Imaginary
transition state
for a [1,3]
sigmatropic
rearrangement

Note that this is not what actually takes place; we merely imagine it in order to be able to analyze the process. In a [1,3] sigmatropic rearrangement the imaginary transition state consists of a hydrogen atom and an allyl radical. The latter species (p. 32) has three π orbitals, but the only one that concerns us here is the HOMO which, in a thermal rearrangement is **D**. The electron of the hydrogen atom is of course in a $1s$ orbital, which has only one lobe. The rule governing sigmatropic migration of hydrogen is *the H must move from a plus to a plus or from a minus to a minus lobe, of the highest occupied molecular*

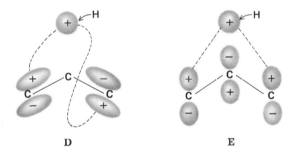

D   E

orbital; it cannot move to a lobe of opposite sign.[408] Obviously, the only way this can happen in a thermal [1,3] sigmatropic rearrangement is if the migration is antarafacial. Consequently, the rule predicts that antarafacial thermal [1,3] sigmatropic rearrangements are allowed, but the suprafacial pathway is forbidden. However, in a photochemical reaction, promotion of an electron means that **E** is now the HOMO; the suprafacial pathway is now allowed and the antarafacial pathway forbidden.

A similar analysis of [1,5] sigmatropic rearrangements shows that in this case the thermal reaction must be suprafacial and the photochemical process antarafacial. For the general case, with odd-numbered j, we can say that $[1,j]$ suprafacial migrations are allowed thermally when j is of the form $4n + 1$, and photochemically when j has the form $4n - 1$; the opposite is true for antarafacial migrations.

[406]Since we are using the arbitrary designations U, V, Y, and Z, we have been arbitrary in which isomer to call R,Z and which to call S,E.

[407]See Woodward; Hoffmann, Ref. 403, pp. 114-140.

[408]This follows from the principle that bonds are formed only by overlap of orbitals of the same sign. Since this is a concerted reaction, the hydrogen orbital in the transition state must overlap simultaneously with one lobe from the migration origin and one from the terminus. It is obvious that both of these lobes must have the same sign.

As expected, the Möbius–Hückel method leads to the same predictions. Here we look at the basis set of orbitals shown in **F** and **G** for [1,3] and [1,5] rearrangements, respectively.

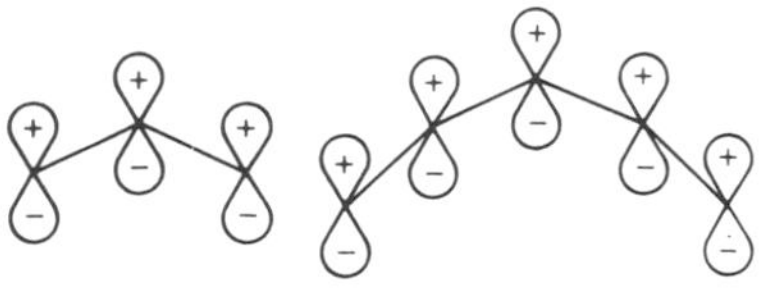

F   **G**

A [1,3] shift involves four electrons, so an allowed thermal pericyclic reaction must be a Möbius system (p. 1115) with one or an odd number of sign inversions. As can be seen in **F**, only an antarafacial migration can achieve this. A [1,5] shift, with six electrons, is allowed thermally only when it is a Hückel system with zero or an even number of sign inversions; hence it requires a suprafacial migration.

The actual reported results bear out this analysis. Thus a thermal [1,3] migration is allowed to take place only antarafacially, but such a transition state would be extremely strained, and thermal [1,3] sigmatropic migrations of hydrogen are unknown.[409] On the other hand, the photochemical pathway allows suprafacial [1,3] shifts, and a few such reactions are known, an example being[410]

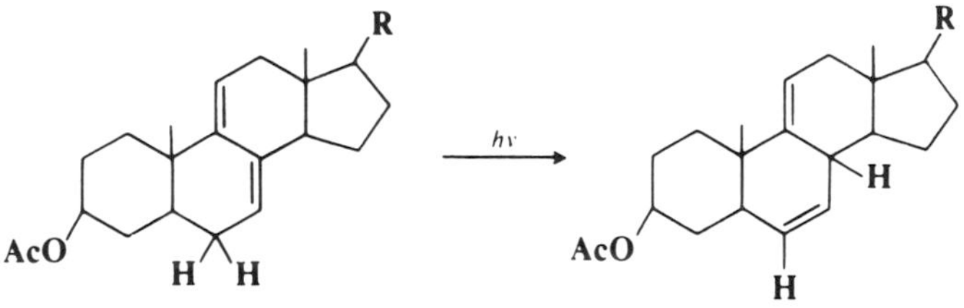

The situation is reversed for [1,5] hydrogen shifts. In this case the thermal rearrangements, being suprafacial, are quite common, while photochemical rearrangements are rare.[411] Examples of the thermal reaction are

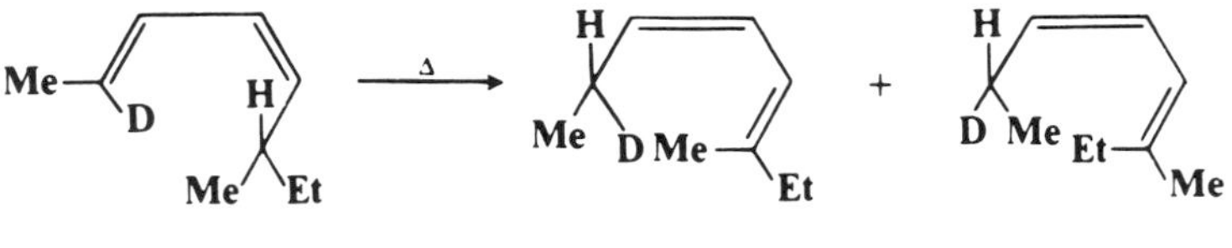

Ref. 412

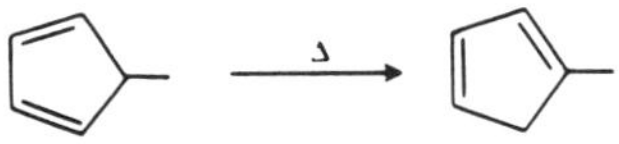

Ref. 413

[409]A possible [1,3] migration of hydrogen has been reported. See Yeh; Linder; Hoffman; Barton *J. Am. Chem. Soc.* **1986,** *108*, 7849. See also Parto; Brophy *J. Org. Chem.* **1991,** *56*, 4554.

[410]Dauben; Wipke *Pure Appl. Chem.* **1964,** *9*, 539-553, p. 546. For another example, see Kropp; Fravel; Fields *J. Am. Chem. Soc.* **1976,** *98*, 840.

[411]For examples of photochemical [1,5] antarafacial reactions, see Kiefer; Tanna *J. Am. Chem. Soc.* **1969,** *91*, 4478; Kiefer; Fukunaga *Tetrahedron Lett.* **1969,** 993; Dauben; Poulter; Suter *J. Am. Chem. Soc.* **1970,** *92*, 7408.

[412]Roth; König; Stein *Chem. Ber.* **1970,** *103*, 426.

[413]McLean; Haynes *Tetrahedron* **1965,** *21*, 2329. For a review of such rearrangements, see Klärner *Top. Stereochem.* **1984,** *15* 1-42.

Note that the first example bears out the stereochemical prediction made earlier. Only the two isomers shown were formed. In the second example, migration can continue around the ring. Migrations of this kind are called *circumambulatory rearrangements*.[414]

With respect to [1,7] hydrogen shifts, the rules predict the thermal reaction to be antarafacial. Unlike the case of [1,3] shifts, the transition state is not too greatly strained, and such rearrangements have been reported, e.g.,[415]

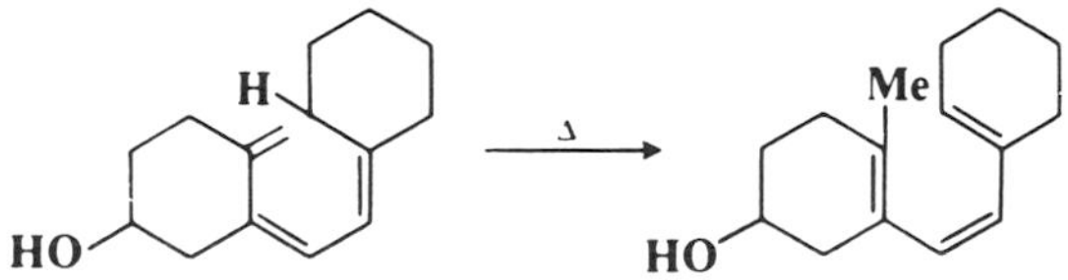

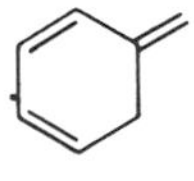

102

Photochemical [1,7] shifts are suprafacial and, not surprisingly, many of these have been observed.[416]

The orbital symmetry rules also help us to explain, as on pp. 865 and 1117, the unexpected stability of certain compounds. Thus, **102** could, by a thermal [1,3] sigmatropic rearrangement, easily convert to toluene, which of course is far more stable because it has an aromatic sextet. Yet **102** has been prepared and is stable at dry ice temperature and in dilute solutions.[417]

Analogs of sigmatropic rearrangements in which a cyclopropane ring replaces one of the double bonds are also known, e.g.,[418]

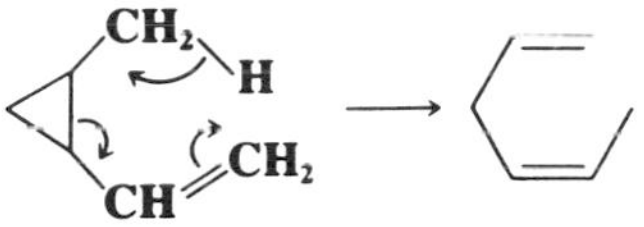

a homodienyl [1,5] shift

The reverse reaction has also been reported.[419] 2-Vinylcycloalkanols[420] undergo an analogous reaction, as do cyclopropyl ketones (see p. 1138 for this reaction).

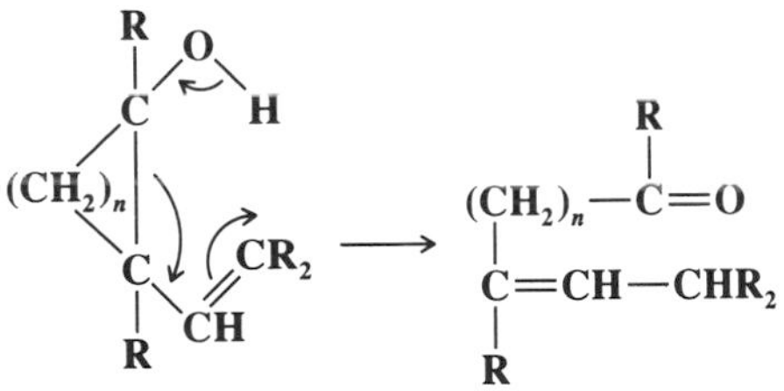

[414]For a review, see Childs *Tetrahedron* **1982**, *38*, 567-608. See also Minkin; Mikhailov; Dushenko; Yudilevich; Minyaev; Zschunke; Mügge *J. Phys. Org. Chem.* **1991**, *4*, 31.

[415]Schlatmann; Pot; Havinga *Recl. Trav. Chim. Pays-Bas* **1964**, *83*, 1173; Hoeger; Johnston; Okamura *J. Am. Chem. Soc.* **1987**, *109*, 4690; Baldwin; Reddy *J. Am. Chem. Soc.* **1987**, *109*, 8051, **1988**, *110*, 8223.

[416]See Murray; Kaplan *J. Am. Chem. Soc.* **1966**, *88*, 3527; ter Borg; Kloosterziel *Recl. Trav. Chim. Pays-Bas* **1969**, *88*, 266; Tezuka; Kimura; Sato; Mukai *Bull. Chem. Soc. Jpn.* **1970**, *43*, 1120.

[417]Bailey; Baylouny *J. Org. Chem.* **1962**, *27*, 3476.

[418]Ellis; Frey *Proc. Chem. Soc.* **1964**, 221; Frey; Solly *Int. J. Chem. Kinet.* **1969**, *1*, 473; Roth; König *Liebigs Ann. Chem.* **1965**, *688*, 28; Ohloff *Tetrahedron Lett.* **1965**, 3795; Jorgenson; Thacher *Tetrahedron Lett.* **1969**, 4651; Corey; Yamamoto; Herron; Achiwa *J. Am. Chem. Soc.* **1970**, *92*, 6635; Loncharich; Houk *J. Am. Chem. Soc.* **1988**, *110*, 2089; Parziale; Berson *J. Am. Chem. Soc.* **1990**, *112*, 1650; Pegg; Meehan *Aust. J. Chem.* **1990**, *43*, 1009, 1071.

[419]Roth; König, Ref. 418. Also see Grimme *Chem. Ber.* **1965**, *98*, 756.

[420]Arnold; Smolinsky *J. Am. Chem. Soc.* **1960**, *82*, 4918; Leriverend; Conia *Tetrahedron Lett.* **1969**, 2681; Conia; Barnier *Tetrahedron Lett.* **1969**, 2679.

8-32   [1,*j*] Sigmatropic Migrations of Carbon

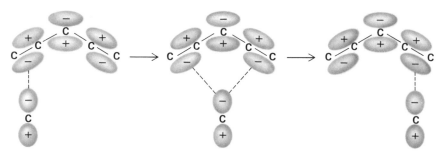

Sigmatropic migrations of alkyl or aryl groups[423] are less common than the corresponding hydrogen migrations.[424] When they do take place, there is an important difference. Unlike a hydrogen atom, whose electron is in a 1*s* orbital with only one lobe, a carbon free radical has its odd electron in a *p* orbital that has *two lobes of opposite sign*. Therefore, if we draw the imaginary transition states for this case (see p. 1123), we see that in a thermal suprafacial [1,5] process (Figure 18.5), symmetry can be conserved only if the migrating carbon moves in such a way that the lobe which was originally attached to the π system remains attached to the π system. This can happen only if configuration is *retained within the migrating group*. On the other hand, thermal suprafacial [1,3] migration (Figure 18.6) *can* take place if the migrating carbon switches lobes. If the migrating carbon was originally bonded by its minus lobe, it must now use its plus lobe to form the new C—C bond. Thus, configuration in the migrating group will be *inverted*. From these considerations we predict that suprafacial [1,*j*] sigmatropic rearrangements in which carbon is the migrating group are always allowed, both thermally and photochemically, but that thermal [1,3] migrations will proceed with inversion and thermal [1,5] migrations with retention of configuration within the migrating group.

FIGURE 18.5 Hypothetical orbital movement for a thermal [1,5] sigmatropic migration of carbon. To move from one − lobe to the other − lobe, the migrating carbon uses only its own − lobe, retaining its configuration.

[421]Roth; Friedrich *Tetrahedron Lett.* **1969**, 2607.
[422]Youssef; Ogliaruso *J. Org. Chem.* **1972**, *37*, 2601.
[423]For reviews, see Mironov; Fedorovich; Akhrem, Ref. 404; Spangler, Ref. 404.
[424]It has been shown that methyl and phenyl have lower migratory aptitudes than hydrogen in thermal sigmatropic rearrangements: Shen; McEwen; Wolf *Tetrahedron Lett.* **1969**, 827; Miller; Greisinger; Boyer *J. Am. Chem. Soc.* **1969**, *91*, 1578.

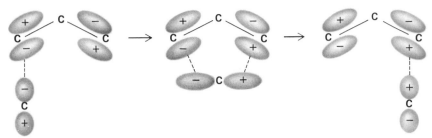

FIGURE 18.6 Hypothetical orbital movement for a thermal [1,3] sigmatropic migration of carbon. The migrating carbon moves from a − to a + lobe, requiring it to switch its own bonding lobe from − to +, inverting its configuration.

More generally, we can say that suprafacial [1,j] migrations of carbon in systems where $j = 4n - 1$ proceed with inversion thermally and retention photochemically, while systems where $j = 4n + 1$ show the opposite behavior. Where antarafacial migrations take place, all these predictions are of course reversed.

The first laboratory test of these predictions was the pyrolysis of deuterated *endo*-bicyclo[3.2.0]hept-2-en-6-yl acetate (**103**), which gave the *exo*-deuterio-*exo*-norbonyl acetate

103   **104**

104.[425] Thus, as predicted by the orbital symmetry rules, this thermal suprafacial [1,3] sigmatropic reaction took place with complete inversion at C-7. Similar results have been obtained in a number of other cases.[426] However, similar studies of the pyrolysis of the parent hydrocarbon of **103**, labeled with D at C-6 and C-7, showed that while most of the product was formed with inversion at C-7, a significant fraction (11 to 29%) was formed with retention.[427] Other cases of lack of complete inversion are also known.[428] A diradical mechanism has been invoked to explain such cases.[429] There is strong evidence for a radical mechanism for some [1,3] sigmatropic rearrangements.[430] Photochemical suprafacial [1,3] migrations of carbon have been shown to proceed with retention, as predicted.[431]

[425]Berson; Nelson *J. Am. Chem. Soc.* **1967**, *89*, 5503; Berson *Acc. Chem. Res.* **1968**, *1*, 152-160.

[426]See Ref. 421; Berson *Acc. Chem. Res.* **1972**, *5*, 406-414; Bampfield; Brook; Hunt *J. Chem. Soc., Chem. Commun.* **1976**, 146; Franzus; Scheinbaum; Waters; Bowlin *J. Am. Chem. Soc.* **1976**, *98*, 1241; Klärner; Adamsky *Angew. Chem. Int. Ed. Engl.* **1979**, *18*, 674 [*Angew. Chem. 91*, 738].

[427]Baldwin; Belfield *J. Am. Chem. Soc.* **1988**, *110*, 296; Klärner; Drewes; Hasselmann *J. Am. Chem. Soc.* **1988**, *110*, 297.

[428]See, for example, Berson; Nelson *J. Am. Chem. Soc.* **1970**, *92*, 1096; Berson; Holder *J. Am. Chem. Soc.* **1973**, *95*, 2037; Pikulin; Berson *J. Am. Chem. Soc.* **1988**, *110*, 8500.

[429]See Newman-Evans; Carpenter *J. Am. Chem. Soc.* **1984**, *106*, 7994; Pikulin; Berson, Ref. 428. See also Berson *Chemtracts: Org. Chem.* **1989**, *2*, 213-227.

[430]See, for example, Bates; Ramaswamy *Can. J. Chem.* **1985**, *63*, 745; Dolbier; Phanstiel *J. Am. Chem. Soc.* **1989**, *111*, 4907.

[431]Cookson; Hudec; Sharma *Chem. Commun.* **1971**, 107, 108.

Although allylic vinylic ethers generally undergo [3,3] sigmatropic rearrangements (**8-35**), they can be made to give the [1,3] kind, to give aldehydes, e.g.,

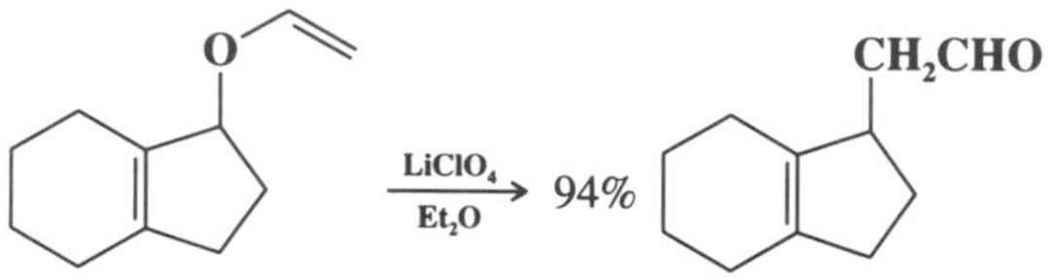

by treatment with $LiClO_4$ in diethyl ether.[431a] In this case the C—O bond undergoes a 1,3 migration from the O to the end vinylic carbon. When the vinylic ether is of the type $ROCR'{=}CH_2$, ketones RCH_2COR' are formed. There is evidence that this [1,3] sigmatropic rearrangement is not concerted, but involves dissociation of the substrate into ions.[431a]

Thermal suprafacial [1,5] migrations of carbon have been found to take place with retention,[432] but also with inversion.[433,434] A diradical mechanism has been suggested for the latter case.[433]

Simple nucleophilic, electrophilic, and free-radical 1,2 shifts can also be regarded as sigmatropic rearrangements (in this case, [1,2] rearrangements). We have already (p. 1051) applied similar principles to such rearrangements to show that nucleophilic 1,2 shifts are allowed, but the other two types are forbidden unless the migrating group has some means of delocalizing the extra electron or electron pair.

8-33 Conversion of Vinylcyclopropanes to Cyclopentenes

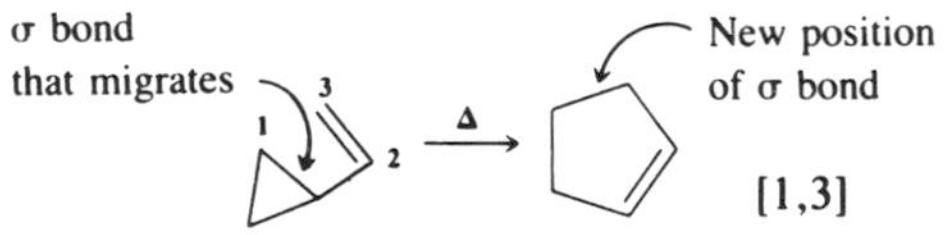

The thermal expansion of a vinylcyclopropane to a cyclopentene ring[435] is a special case of a [1,3] sigmatropic migration of carbon, though it can also be considered an internal [$_\pi2$ + $_\sigma2$] cycloaddition reaction (see **5-49**). The reaction has been carried out on many vinylcyclopropanes bearing various substituents in the ring or on the vinyl group and has been extended to 1,1-dicyclopropylethene[436]

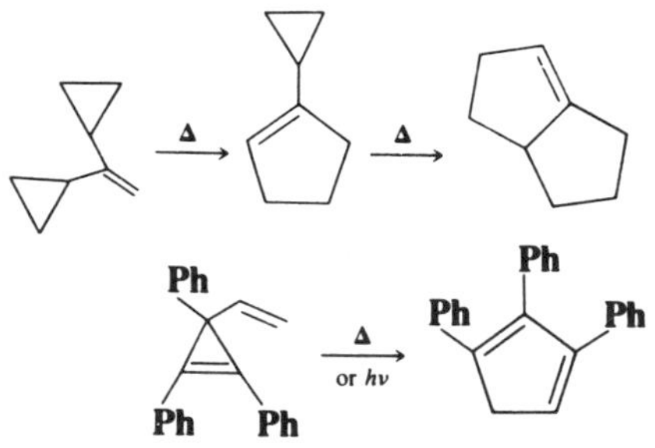

[431a]Grieco; Clarke; Jagoe *J. Am. Chem. Soc.* **1991,** *113,* 5488.
[432]Boersma; de Haan; Kloosterziel; van de Ven *Chem. Commun.* **1970,** 1168.
[433]Klärner; Yaslak; Wette *Chem. Ber.* **1979,** *112,* 1168; Klärner; Brassel *J. Am. Chem. Soc.* **1980,** *102,* 2469; Borden; Lee; Young *J. Am. Chem. Soc.* **1980,** *102,* 4841; Gajewski; Gortva; Borden *J. Am. Chem. Soc.* **1986,** *108,* 1083.
[434]Baldwin; Broline *J. Am. Chem. Soc.* **1982,** *104,* 2857.
[435]For reviews, see Wong et al., Ref. 114, pp. 169-172; Goldschmidt; Crammer *Chem. Soc. Rev.* **1988,** *17,* 229-267; Hudlický; Kutchan; Naqvi *Org. React.* **1985,** *33,* 247-335; Mil'vitskaya; Tarakanova; Plate *Russ. Chem. Rev.* **1976,** *45,* 469-478; DeWolfe, in Bamford; Tipper, Ref. 365, pp. 470-474; Gutsche; Redmore, Ref. 112, pp. 163-170; Frey *Adv. Phys. Org. Chem.* **1966,** *4,* 147-193, pp. 155-163, 175-176.
[436]Ketley *Tetrahedron Lett.* **1964,** 1687; Branton; Frey *J. Chem. Soc. A* **1966,** 1342.

and (both thermally[437] and photochemically[438]) to vinylcyclopropenes. Various heterocyclic analogs[438a] are also known, e.g.,[439]

$$\text{Ph}-\overset{\text{O}}{\underset{\|}{\text{C}}}-\text{N}\triangleleft \quad \xrightarrow{\Delta} \quad \text{Ph}-\langle\text{oxazoline}\rangle$$

Two competing reactions are the homodienyl [1,5] shift (if a suitable H is available, see **8-31**), and simple cleavage of the cyclopropane ring, leading in this case to a diene (see **8-3**).

Vinylcyclobutanes can be similarly converted to cyclohexenes,[440] but larger ring compounds do not generally give the reaction.[441] Though high temperatures (as high as 500°C)

$$\text{(vinylcyclobutane)} \quad \xrightarrow{\Delta} \quad \text{(cyclohexene)}$$

are normally required for the thermal reaction, the lithium salts of 2-vinylcyclopropanols rearrange at 25°C.[442]

$$\text{(2-vinylcyclopropanol, OLi)} \quad \longrightarrow \quad \text{(cyclopentene, OLi)}$$

Salts of 2-vinylcyclobutanols behave analogously.[443]

The reaction rate has also been greatly increased by the addition of a one-electron oxidant tris-(4-bromophenyl)aminium hexafluoroantimonate $Ar_3N^{\bullet+}$ SbF_6^- ($Ar = p$-bromophenyl).[444] This reagent converts the substrate to a cation radical, which undergoes ring expansion much faster.[445]

The mechanisms of these ring expansions are not certain. Both concerted[446] and diradical[447] pathways have been proposed, and it is possible that both pathways operate, in different systems.

[437]Small; Breslow, cited in Breslow, in Mayo, Ref. 114, vol. 1, p. 236.

[438]Padwa; Blacklock; Getman; Hatanaka; Loza *J. Org. Chem.* **1978**, *43*, 1481; Zimmerman; Aasen *J. Org. Chem.* **1978**, *43*, 1493; Zimmerman; Kreil *J. Org. Chem.* **1982**, *47*, 2060.

[438a]For a review of a nitrogen analog, see Boeckman; Walters *Adv. Heterocycl. Nat. Prod. Synth.* **1990**, *1*, 1-41.

[439]For reviews of ring expansions of aziridines, see Heine *Mech. Mol. Migr.* **1971**, *3*, 145-176; Dermer; Ham *Ethylenimine and Other Aziridines*; Academic Press: New York, 1969, pp. 282-290. See also Wong et al., Ref. 114, pp. 190-192.

[440]See, for example, Overberger; Borchert *J. Am. Chem. Soc.* **1960**, *82*, 1007; Gruseck; Heuschmann *Chem. Ber.* **1990**, *123*, 1911.

[441]For an exception, see Thies *J. Am. Chem. Soc.* **1972**, *94*, 7074.

[442]Danheiser; Martinez-Davila; Morin *J. Org. Chem.* **1980**, *45*, 1340; Danheiser; Bronson; Okano *J. Am. Chem. Soc.* **1985**, *107*, 4579.

[443]Danheiser; Martinez-Davila; Sard *Tetrahedron* **1981**, *37*, 3943.

[444]Dinnocenzo; Conlan *J. Am. Chem. Soc.* **1988**, *110*, 2324.

[445]For a review of ring expansion of vinylcyclobutane cation radicals, see Bauld *Tetrahedron* **1989**, *45*, 5307-5363.

[446]For evidence favoring the concerted mechanism, see Shields; Lepely *J. Am. Chem. Soc.* **1968**, *90*, 4749; Billups; Leavell; Lewis; Vanderpool *J. Am. Chem. Soc.* **1973**, *95*, 8096; Berson; Dervan; Malherbe; Jenkins *J. Am. Chem. Soc.* **1976**, *98*, 5937; Andrews; Baldwin *J. Am. Chem. Soc.* **1976**, *98*, 6705, 6706; Dolbier; Al-Sader; Sellers; Koroniak *J. Am. Chem. Soc.* **1981**, *103*, 2138; Gajewski; Olson *J. Am. Chem. Soc.* **1991**, *113*, 7432.

[447]For evidence favoring the diradical mechanism, see Willcott; Cargle *J. Am. Chem. Soc.* **1967**, *89*, 723; Doering; Schmidt *Tetrahedron* **1971**, *27*, 2005; Roth; Schmidt *Tetrahedron Lett.* **1971**, 3639; Simpson; Richey *Tetrahedron Lett.* **1973**, 2545; Gilbert; Higley *Tetrahedron Lett.* **1973**, 2075; Caramella; Huisgen; Schmolke *J. Am. Chem. Soc.* **1974**, *96*, 2997, 2999; Mazzocchi; Tamburin *J. Am. Chem. Soc.* **1975**, *97*, 555; Zimmerman; Fleming *J. Am. Chem. Soc.* **1983**, *105*, 622; Klumpp; Schakel *Tetrahedron Lett.* **1983**, *24*, 4595; McGaffin; de Meijere; Walsh *Chem. Ber.* **1991**, *124*, 939. A "continuous diradical transition state" has also been proposed: Doering; Sachdev *J. Am. Chem. Soc.* **1974**, *96*, 1168, **1975**, *97*, 5512; Roth; Lennartz; Doering; Birladeanu; Guyton; Kitagawa *J. Am. Chem. Soc.* **1990**, *112*, 1722.

For the conversion of a vinylcyclopropane to a cyclopentene in a different way, see OS **68,** 220.

8-34   The Cope Rearrangement
         (3/4/)↦(1/6/)-*sigma*-Migration

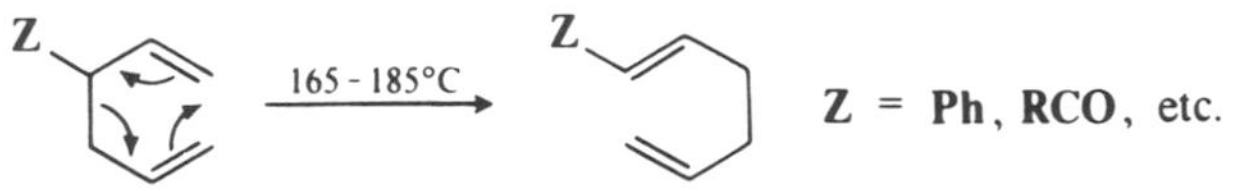

Z = **Ph**, **RCO**, etc.

When 1,5-dienes are heated, they isomerize, in a [3,3] sigmatropic rearrangement known as the *Cope rearrangement* (not to be confused with the Cope elimination reaction, **7-8**).[448] When the diene is symmetrical about the 3,4 bond, we have the unusual situation where a reaction gives a product identical with the starting material:[449]

Therefore, a Cope rearrangement can be detected only when the diene is not symmetrical about this bond. Any 1,5-diene gives the rearrangement; for example, 3-methyl-1,5-hexadiene heated to 300°C gives 1,5-heptadiene.[450] However, the reaction takes place more easily (lower temperature required) when there is a group on the 3- or 4-carbon with which the new double bond can conjugate. The reaction is obviously reversible and produces an equilibrium mixture of the two 1,5-dienes, which is richer in the thermodynamically more stable isomer. However, the reaction is not generally reversible[451] for 3-hydroxy-1,5-dienes, because the product tautomerizes to the ketone or aldehyde:

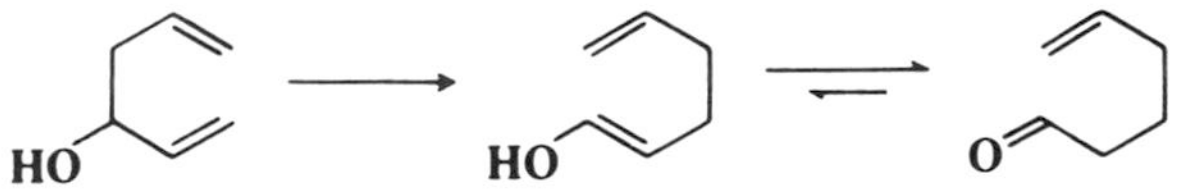

This reaction, called the *oxy-Cope rearrangement*,[452] has proved highly useful in synthesis.[453] The oxy-Cope rearrangement is greatly accelerated (by factors of 10^{10} to 10^{17}) if the alkoxide is used rather than the alcohol.[454] In this case the direct product is the enolate ion, which is hydrolyzed to the ketone.

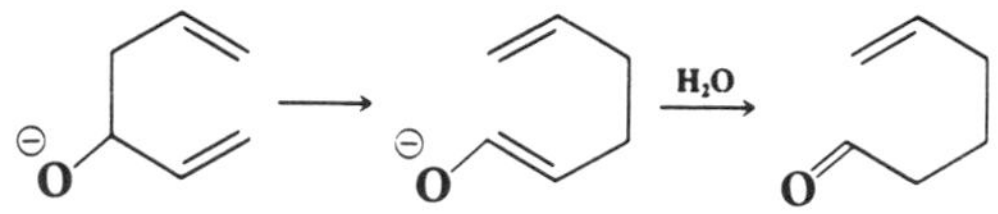

[448]For reviews, see Bartlett *Tetrahedron* **1980,** *36,* 2-72, pp. 28-39; Rhoads; Raulins *Org. React.* **1975,** *22,* 1-252; Smith; Kelly *Prog. Phys. Org. Chem.* **1971,** *8,* 75-234, pp. 153-201; DeWolfe, in Bamford; Tipper, Ref. 365, pp. 455-461.

[449]Note that the same holds true for [1,*j*] sigmatropic reactions of symmetrical substrates (**8-31, 8-32**).

[450]Levy; Cope *J. Am. Chem. Soc.* **1944,** *66,* 1684.

[451]For an exception, see Elmore; Paquette *Tetrahedron Lett.* **1991,** *32,* 319.

[452]Berson; Jones *J. Am. Chem. Soc.* **1964,** *86,* 5017, 5019; Viola; Levasseur *J. Am. Chem. Soc.* **1965,** *87,* 1150; Berson; Walsh *J. Am. Chem. Soc.* **1968,** *90,* 4729; Viola; Padilla; Lennox; Hecht; Proverb *J. Chem. Soc., Chem. Commun.* **1974,** 491; For reviews, see Paquette *Angew. Chem. Int. Ed. Engl.* **1990,** *29,* 609-626 [*Angew. Chem. 102,* 642-660], *Synlett* **1990,** 67-73; Marvell; Whalley, in Patai *The Chemistry of the Hydroxyl Group,* pt. 2; Wiley: New York, 1971, pp. 738-743.

[453]For a list of references, see Ref. 106, pp. 639-640.

[454]Evans; Golub *J. Am. Chem. Soc.* **1975,** *97,* 4765; Evans; Nelson *J. Am. Chem. Soc.* **1980,** *102,* 774; Miyashi; Hazato; Mukai *J. Am. Chem. Soc.* **1978,** *100,* 1008; Paquette; Pegg; Toops; Maynard; Rogers *J. Am. Chem. Soc.* **1990,** *112,* 277; Gajewski; Gee *J. Am. Chem. Soc.* **1991,** *113,* 967. See also Wender; Ternansky; Sieburth *Tetrahedron Lett.* **1985,** *26,* 4319.

The 1,5-diene system may be inside a ring or part of an allenic system (this example illustrates both of these situations):[455]

but the reaction does not take place when one of the double bonds is part of an aromatic system, e.g., 1-phenyl-1-butene.[456] When the two double bonds are in vinylic groups attached to adjacent ring positions, the product is a ring four carbons larger. This has been applied to divinylcyclopropanes and cyclobutanes:[457]

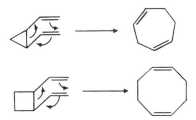

Indeed, *cis*-1,2-divinylcyclopropanes give this rearrangement so rapidly that they generally cannot be isolated at room temperature,[458] though exceptions are known.[459] When heated, 1,5-diynes are converted to 3,4-dimethylenecyclobutenes.[460] A rate-determining Cope rearrangement is followed by a very rapid electrocyclic (**8-29**) reaction. The interconversion of

1,3,5-trienes and cyclohexadienes (in **8-29**) is very similar to the Cope rearrangement, though in **8-29**, the 3,4 bond goes from a double bond to a single bond rather than from a single bond to no bond.

Like 2 + 2 cycloadditions (p. 863), Cope rearrangements of simple 1,5-dienes can be catalyzed by certain transition-metal compounds. For example, the addition of $PdCl_2(PhCN)_2$ causes the reaction to take place at room temperature.[461] This can be quite useful synthetically, because of the high temperatures required in the uncatalyzed process.

[455]Harris *Tetrahedron Lett.* **1965**, 1359.
[456]See, for example, Lambert; Fabricius; Hoard *J. Org. Chem.* **1979**, *44*, 1480; Marvell; Almond *Tetrahedron Lett.* **1979**, 2777, 2779; Newcomb; Vieta *J. Org. Chem.* **1980**, *45*, 4793. For exceptions in certain systems, see Doering; Bragole *Tetrahedron* **1966**, *22*, 385; Jung; Hudspeth *J. Am. Chem. Soc.* **1978**, *100*, 4309; Yasuda; Harano; Kanematsu *J. Org. Chem.* **1980**, *45*, 2368.
[457]Vogel; Ott; Gajek *Liebigs Ann. Chem.* **1961**, *644*, 172. For reviews, see Wong et al., Ref. 114, pp. 172-174; Mil'vitskaya et al., Ref. 435, pp. 475-476.
[458]Unsubstituted *cis*-1,2-divinylcyclopropane is fairly stable at −20°: Brown; Golding; Stofko *J. Chem. Soc., Chem. Commun.* **1973**, 319; Schneider; Rebell *J. Chem. Soc., Chem. Commun.* **1975**, 283.
[459]See, for example, Brown *Chem. Commun.* **1965**, 226; Schönleber *Chem. Ber.* **1969**, *102*, 1789; Bolesov; Ii-hsein; Levina *J. Org. Chem. USSR* **1970**, *6*, 1791; Schneider; Rau *J. Am. Chem. Soc.* **1979**, *101*, 4426.
[460]For reviews of Cope rearrangements involving triple bonds, see Viola, Collins, and Filipp *Tetrahedron* **1981**, *37*, 3765-3811; Théron; Verny; Vessière, in Patai *The Chemistry of the Carbon–Carbon Triple Bond*, pt. 1; Wiley: New York, 1978, pp. 381-445, pp. 428-430; Huntsman *Intra-Sci. Chem. Rep.* **1972**, *6*, 151-159.
[461]Overman; Knoll *J. Am. Chem. Soc.* **1980**, *102*, 865; Hamilton; Mitchell; Rooney *J. Chem. Soc., Chem. Commun.* **1981**, 456. For reviews of catalysis of Cope and Claisen rearrangements, see Overman *Angew. Chem. Int. Ed. Engl.* **1984**, *23*, 579-586 [*Angew. Chem. 96*, 565-573]; Lutz *Chem. Rev.* **1984**, *84*, 205-247. For a study of the mechanism, see Overman; Renaldo *J. Am. Chem. Soc.* **1990**, *112*, 3945.

As we have indicated with our arrows, the mechanism of the uncatalyzed Cope rearrangement is a simple six-centered pericyclic process. Since the mechanism is so simple, it has been possible to study some rather subtle points, among them the question of whether the six-membered transition state is in the boat or the chair form. For the case of 3,4-dimethyl-1,5-hexadiene it was demonstrated conclusively that the transition state is in the chair form. This was shown by the stereospecific nature of the reaction: The meso isomer gave the cis–trans product, while the (±) compound gave the trans–trans diene.[462] If the transition state is in the chair form (taking the meso isomer, for example), one methyl must be "axial" and the other "equatorial" and the product must be the cis–trans olefin:

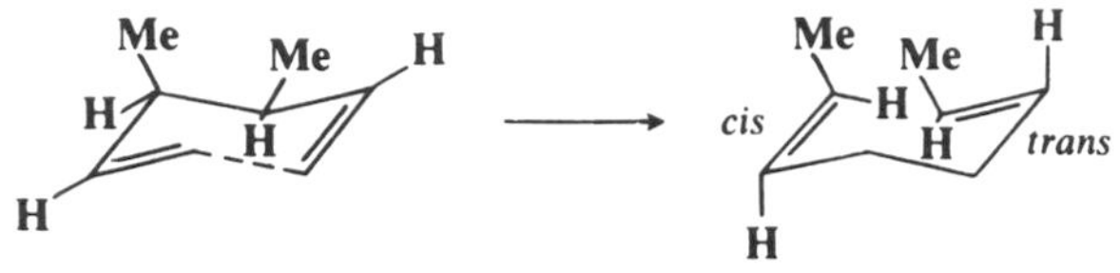

There are two possible boat forms for the transition state of the meso isomer. One leads to a trans–trans product;

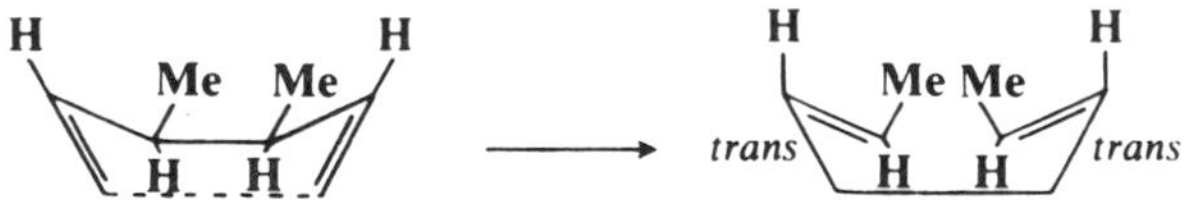

the other to a cis–cis olefin. For the (±) pair the predictions are just the opposite: There is just one boat form, and it leads to the cis–trans olefin, while one chair form ("diaxial" methyls) leads to the cis–cis product and the other ("diequatorial" methyls) predicts the trans–trans product. Thus the nature of the products obtained demonstrates that the transition state is a chair and not a boat.[463] However, 3,4-dimethyl-1,5-hexadiene is free to assume either the chair or boat (it prefers the chair), but other compounds are not so free. Thus 1,2-divinylcyclopropane (p. 1131) can react *only* in the boat form, demonstrating that such reactions are not impossible.[464]

Because of the nature of the transition state in the pericyclic mechanism, optically active substrates with a chiral carbon at C-3 or C-4 transfer the chirality to the product, making this an enantioselective synthesis (see p. 1139 for an example in the mechanistically similar Claisen rearrangement).[465]

Not all Cope rearrangements proceed by the cyclic six-centered mechanism. Thus *cis*-1,2-divinylcyclobutane (p. 1131) rearranges smoothly to 1,5-cyclooctadiene, since the geometry is favorable. The trans isomer also gives this product, but the main product is 4-vinylcyclohexene (resulting from **8-33**). This reaction can be rationalized as proceeding by

[462]Doering; Roth *Tetrahedron* **1962**, *18*, 67. See also Hill; Gilman *Chem. Commun.* **1967**, 619; Goldstein; DeCamp *J. Am. Chem. Soc.* **1974**, *96*, 7356; Hansen; Schmid *Tetrahedron* **1974**, *30*, 1959; Gajewski; Benner; Hawkins *J. Org. Chem.* **1987**, *52*, 5198; Paquette; DeRussy; Cottrell *J. Am. Chem. Soc.* **1988**, *110*, 890.

[463]Preference for the chair transition state is a consequence of orbital-symmetry relationships: Hoffmann; Woodward *J. Am. Chem. Soc.* **1965**, *87*, 4389; Fukui; Fujimoto *Tetrahedron Lett.* **1966**, 251.

[464]For other examples of Cope rearrangements in the boat form, see Goldstein; Benzon *J. Am. Chem. Soc.* **1972**, *94*, 7147; Shea; Phillips *J. Am. Chem. Soc.* **1980**, *102*, 3156; Wiberg; Matturro; Adams *J. Am. Chem. Soc.* **1981**, *103*, 1600; Gajewski; Jiminez *J. Am. Chem. Soc.* **1986**, *108*, 468.

[465]For a review of Cope and Claisen reactions as enantioselective syntheses, see Hill, in Morrison *Asymmetric Synthesis*, vol. 3; Academic Press: New York, 1984, pp. 503-572, pp. 503-545.

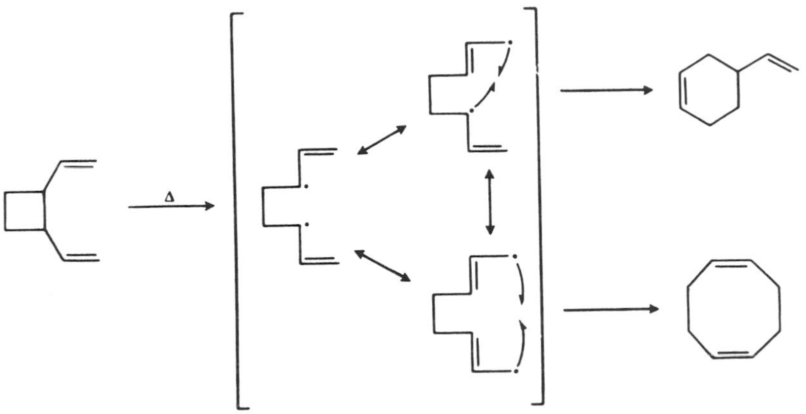

a diradical mechanism,[466] though it is possible that at least part of the cyclooctadiene produced comes from a prior epimerization of the *trans*- to the *cis*-divinylcyclobutane followed
by Cope rearrangement of the latter.[467]

It has been suggested that another type of diradical two-step mechanism may be preferred
by some substrates.[468] In this pathway,[469] the 1,6 bond is formed before the 3,4 bond breaks:

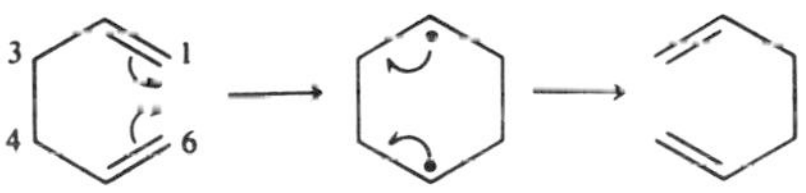

It was pointed out earlier that a Cope rearrangement of 1,5-hexadiene gives 1,5-hex-
adiene. This is a *degenerate Cope rearrangement* (p. 1054). Another molecule that undergoes
it is bicyclo[5.1.0]octadiene (**105**).[470] At room temperature the nmr spectrum of this com-

105   **105**

pound is in accord with the structure shown on the left. At 180°C it is converted by a Cope
reaction to a compound equivalent to itself. The interesting thing is that at 180°C the nmr
spectrum shows that what exists is an equilibrium mixture of the two structures. That is, at

[466]Hammond; De Boer *J. Am. Chem. Soc.* **1964**, *86*, 899; Trecker; Henry *J. Am. Chem. Soc.* **1964**, *86*, 902. Also
see Dolbier; Mancini *Tetrahedron Lett.* **1975**, 2141; Kessler; Ott *J. Am. Chem. Soc.* **1976**, *98*, 5014. For a discussion
of diradical mechanisms in Cope rearrangements, see Berson, in Mayo, Ref. 1, pp. 358-372.

[467]See, for example, Berson; Dervan *J. Am. Chem. Soc.* **1972**, *94*, 8949; Baldwin; Gilbert *J. Am. Chem. Soc.*
1976, *98*, 8283. For a similar result in the 1,2-divinylcyclopropane series, see Baldwin; Ullenius *J. Am. Chem. Soc.*
1974, *96*, 1542.

[468]Doering; Toscano; Beasley *Tetrahedron* **1971**, *27*, 5299; Dewar; Wade *J. Am. Chem. Soc.* **1977**, *99*, 4417; Padwa;
Blacklock *J. Am. Chem. Soc.* **1980**, *102*, 2797; Dollinger; Henning; Kirmse *Chem. Ber.* **1982**, *115*, 2309; Kaufmann;
de Meijere *Chem. Ber.* **1984**, *117*, 1128; Dewar; Jie *J. Am. Chem. Soc.* **1987**, *109*, 5893, *J. Chem. Soc., Chem.
Commun.* **1989**, 98. For evidence against this view, see Gajewski; Conrad *J. Am. Chem. Soc.* **1978**, *100*, 6268, 6269,
1979, *101*, 6693; Gajewski *Acc. Chem. Res.* **1980**, *13*, 142-148; Morokuma; Borden; Hrovat *J. Am. Chem. Soc.* **1988**,
110, 4474; Berson *Chemtracts: Org. Chem.* **1989**, *2*, 213-227; Halevi; Rom *Isr. J. Chem.* **1989**, *29*, 311; Owens; Berson
J. Am. Chem. Soc. **1990**, *112*, 5973.

[469]For a report of still another mechanism, featuring a diionic variant of the diradical, see Gompper; Ulrich *Angew.
Chem. Int. Ed. Engl.* **1976**, *15*, 299 [*Angew. Chem.* **88**, 298].

[470]Doering; Roth *Tetrahedron* **1963**, *19*, 715.

this temperature the molecule rapidly (faster than 10^3 times per second) changes back and forth between the two structures. This is called *valence tautomerism* and is quite distinct from resonance, even though only electrons shift.[471] The positions of the nuclei are not the same in the two structures. Molecules like **105** that exhibit valence tautomerism (in this case, at 180°C) are said to have *fluxional* structures. It may be recalled that *cis*-1,2-divinyl-cyclopropane does not exist at room temperature because it rapidly rearranges to 1,4-cycloheptadiene (p. 1131), but in **105** the *cis*-divinylcyclopropane structure is frozen into the molecule in both structures. Several other compounds with this structural feature are also known. Of these, *bullvalene* (**106**) is especially interesting. The Cope rearrangement shown

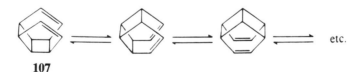

106

changes the position of the cyclopropane ring from 4,5,10 to 1,7,8. But the molecule could also have undergone rearrangements to put this ring at 1,2,8 or 1,2,7. Any of these could then undergo several Cope rearrangements. In all, there are 10!/3, or more than 1.2 million tautomeric forms, and the cyclopropane ring can be at any three carbons that are adjacent. Since each of these tautomers is equivalent to all the others, this has been called an infinitely degenerate Cope rearrangement. Bullvalene has been synthesized and its proton nmr spectrum determined.[472] At −25°C there are two peaks with an area ratio of 6:4. This is in accord with a single nontautomeric structure. The six are the vinylic protons and the four are the allylic ones. But at 100°C the compound shows only one nmr peak, indicating that we have here a truly unusual situation where the compound rapidly interchanges its structure among 1.2 million equivalent forms.[473] The ^{13}C nmr spectrum of bullvalene also shows only one peak at 100°C.[474]

Another compound for which degenerate Cope rearrangements result in equivalence for all the carbons is *hypostrophene* (**107**).[475] In the case of the compound *barbaralane* (**108**)

107

[471]For reviews of valence tautomerizations, see Decock-Le Révérend; Goudmand *Bull. Soc. Chim. Fr.* **1973**, 389-407; Gajewski *Mech. Mol. Migr.* **1971**, *4*, 1-53, pp. 32-49; Paquette *Angew. Chem. Int. Ed. Engl.* **1971**, *10*, 11-20 [*Angew. Chem. 83*, 11-20]; Domareva-Mandel'shtam; D'yakonov *Russ. Chem. Rev.* **1966**, *35*, 559, 568; Schröder; Oth; Merényi *Angew. Chem. Int. Ed. Engl.* **1965**, *4*, 752-761 [*Angew. Chem. 77*, 774-784].

[472]Schröder *Angew. Chem. Int. Ed. Engl.* **1963**, *2*, 481 [*Angew. Chem. 75*, 772], *Chem. Ber.* **1964**, *97*, 3140; Merényi; Oth; Schröder *Chem. Ber.* **1964**, *97*, 3150. For a review of bullvalenes, see Schröder; Oth *Angew. Chem. Int. Ed. Engl.* **1967**, *6*, 414-423 [*Angew. Chem. 79*, 458-467].

[473]A number of azabullvalenes (**106** containing heterocyclic nitrogen) have been synthesized. They also have fluxional structures when heated, though with fewer tautomeric forms than bullvalene itself: Paquette; Barton *J. Am. Chem. Soc.* **1967**, *89*, 5480; Wegener *Tetrahedron Lett.* **1967**, 4985; Paquette; Malpass; Krow; Barton *J. Am. Chem. Soc.* **1969**, *91*, 5296.

[474]Oth; Müllen; Gilles; Schröder *Helv. Chim. Acta* **1974**, *57*, 1415; Nakanishi; Yamamoto *Tetrahedron Lett.* **1974**, 1803; Günther; Ulmen *Tetrahedron* **1974**, *30*, 3781. For deuterium nmr spectra see Poupko; Zimmermann; Luz *J. Am. Chem. Soc.* **1984**, *106*, 5391. For a crystal structure study, see Luger; Buschmann; McMullan; Ruble; Matias; Jeffrey *J. Am. Chem. Soc.* **1986**, *108*, 7825.

[475]McKennis; Brener; Ward; Pettit *J. Am. Chem. Soc.* **1971**, *93*, 4957; Paquette; Davis; James *Tetrahedron Lett.* **1974**, 1615.

(bullvalene in which one CH=CH has been replaced by a CH_2):

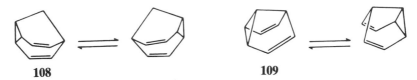

108   **109**

there are only two equivalent tautomers.[476] However, nmr spectra indicate that even at room temperature a rapid interchange of both tautomers is present, though by about $-100°C$ this has slowed to the point where the spectrum is in accord with a single structure. In the case of *semibullvalene* (**109**) (barbaralane in which the CH_2 has been removed), not only is there a rapid interchange at room temperature, but even at $-110°C$.[477] **109** has the lowest energy barrier of any known compound capable of undergoing the Cope rearrangement.[478]

The molecules taking part in a valence tautomerization need not be equivalent. Thus, nmr spectra indicate that a true valence tautomerization exists at room temperature between the cycloheptatriene **110** and the norcaradiene **111**.[479] In this case one isomer (**111**) has the

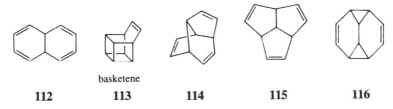

110   **111**   Oxepin   Benzene oxide

cis-1,2-divinylcyclopropane structure, while the other does not. In an analogous interconversion, benzene oxide[480] and oxepin exist in a tautomeric equilibrium at room temperature.[481]

Bullvalene and hypostrophene are members of a group of compounds all of whose formulas can be expressed by the symbol $(CH)_{10}$.[482] Many other members of this group are known, including **112** to **116** and the [10]annulenes (p. 58). All these compounds represent

		basketene		
112	**113**	**114**	**115**	**116**

[476]Barbaralane was synthesized by Biethan; Klusacek; Musso *Angew. Chem. Int. Ed. Engl.* **1967**, 6, 176 [*Angew. Chem.* 79, 152]; by Tsuruta; Kurabayashi; Mukai *Tetrahedron Lett.* **1965**, 3775; by Doering; Ferrier; Fossel; Hartenstein; Jones; Klumpp; Rubin; Saunders *Tetrahedron* **1967**, 23, 3943; and by Henkel; Hane *J. Org. Chem.* **1983**, 48, 3858.
[477]Zimmerman; Grunewald *J. Am. Chem. Soc.* **1966**, 88, 183; Meinwald; Schmidt *J. Am. Chem. Soc.* **1969**, 91, 5877; Zimmerman; Binkley; Givens; Grunewald; Sherwin *J. Am. Chem. Soc.* **1969**, 91, 3316.
[478]Cheng; Mioduski; Meinwald *J. Am. Chem. Soc.* **1974**, 96, 2887; Moskau; Aydin; Leber; Günther; Quast; Martin; Hassenrück; Miller; Grohmann *Chem. Ber.* **1989**, 122, 925.
[479]Ciganek *J. Am. Chem. Soc.* **1965**, 87, 1149. For other examples of norcaradiene–cycloheptatriene valence tautomerizations, see Görlitz; Günther *Tetrahedron* **1969**, 25, 4467; Ciganek *J. Am. Chem. Soc.* **1965**, 93, 2207; Dürr; Kober *Chem. Ber.* **1973**, 106, 1565; Betz; Daub *Chem. Ber.* **1974**, 107, 2095; Maas; Regitz *Chem. Ber.* **1976**, 109, 2039; Warner; Lu *J. Am. Chem. Soc.* **1980**, 102, 331; Neidlein; Radke *Helv. Chim. Acta* **1983**, 66, 2626; Takeuchi; Kitagawa; Ueda; Senzaki; Okamoto *Tetrahedron* **1985**, 41, 5455.
[480]For a review of arene oxides, see Shirwaiker; Bhatt *Adv. Heterocycl. Chem.* **1984**, 37, 67-165.
[481]For reviews, see Ref. 363. See also Boyd; Stubbs *J. Am. Chem. Soc.* **1983**, 105, 2554.
[482]For reviews of rearrangements and interconversions of $(CH)_n$ compounds, see Balaban; Banciu *J. Chem. Educ.* **1984**, 61, 766-770; Greenberg; Liebman, Ref. 89, pp. 203-215; Scott; Jones *Chem. Rev.* **1972**, 72, 181-202; Balaban *Rev. Roum. Chim.* **1966**, 11, 1097-1116. See also Maier; Wiegand; Baum; Wüllner *Chem. Ber.* **1989**, 122, 781.

positions of minimum energy on the $(CH)_{10}$ energy surface, and many have been interconverted by electrocyclic or Cope rearrangements. Similar groups of $(CH)_n$ compounds exist for other even-numbered values of n.[482] For example, there are 20 possible $(CH)_8$[483] compounds,[484] including semibullvalene (**109**), cubane (p. 154), cuneane (p. 1149), octabisvalene (p. 154), cyclooctatetraene (p. 57), **117** to **119**, and five possible $(CH)_6$ compounds,[485] all

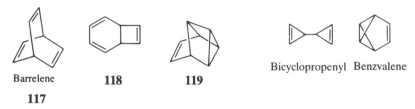

| Barrelene | **118** | **119** | Bicyclopropenyl | Benzvalene |

117

Some $(CH)_8$ compounds   Two $(CH)_6$ compounds

of which are known: benzene, prismane (p. 154), Dewar benzene (p. 1117), bicyclopropenyl,[486] and benzvalene.[487]

An interesting example of a valence tautomerism is the case of 1,2,3-tri-*t*-butylcyclobutadiene (p. 54). There are two isomers, both rectangular, and ^{13}C nmr spectra show that

they exist in a dynamic equilibrium, even at $-185°C$.[488]

8-35   The Claisen Rearrangement

Allylic aryl ethers, when heated, rearrange to *o*-allylphenols in a reaction called the *Claisen rearrangement*.[489] If both ortho positions are filled, the allylic group migrates to the para

[483]For a review of strain in $(CH)_8$ compounds, see Hassenrück; Martin; Walsh *Chem. Rev.* **1989**, *89*, 1125-1146.

[484]The structures of all possible $(CH)_n$ compounds, for $n = 4, 6, 8$, and 10, are shown in Balaban, Ref. 482. For a review of $(CH)_{12}$ compounds, see Banciu; Popa; Balaban *Chem. Scr.* **1984**, *24*, 28.

[485]For reviews of valence isomers of benzene and some related compounds, see Kobayashi; Kumadaki *Top. Curr. Chem.* **1984**, *123*, 103-150; Bickelhaupt; de Wolf *Recl. Trav. Chim. Pays-Bas* **1988**, *107*, 459-478.

[486]For a study of how this compound isomerizes to benzene, see Davis, Shea; Bergman *J. Am. Chem. Soc.* **1977**, *99*, 1499.

[487]For reviews of benzvalenes, see Christl *Angew. Chem. Int. Ed. Engl.* **1981**, *20*, 529-546 [*Angew. Chem. 93*, 515-531]; Burger *Chimia* **1979**, 147-152.

[488]Maier; Kalinowski; Euler *Angew. Chem. Int. Ed. Engl.* **1982**, *21*, 693 [*Angew. Chem. 94*, 706].

[489]For reviews, see Moody *Adv. Heterocycl. Chem.* **1987**, *42*, 203-244; Bartlett, Ref. 448, pp. 28-39; Ziegler *Acc. Chem. Res.* **1977**, *10*, 227-232; Bennett *Synthesis* **1977**, 589-606; Rhoads; Raulins, Ref. 448; Shine *Aromatic Rearrangements*; Elsevier: New York, 1969, pp. 89-120; Smith; Kelly *Prog. Phys. Org. Chem.* **1971**, *8*, 75-234, pp. 153-201; Hansen; Schmid *Chimia* **1970**, *24*, 89-99, *Chem. Br.* **1969**, *5*, 111-116; Jefferson; Scheinmann *Q. Rev., Chem. Soc.* **1968**, *22*, 391-421; Thyagarajan *Adv. Heterocycl. Chem.* **1967**, *8*, 143-163; Dalrymple; Kruger; White, in Patai *The Chemistry of the Ether Linkage*; Wiley: New York, 1967, pp. 635-660.

position (this is often called the *para-Claisen rearrangement*). There is no reaction when the para and both ortho positions are filled. Migration to the meta position has not been observed. In the ortho migration the allylic group always undergoes an allylic shift. That is, as shown above, a substituent α to the oxygen is now γ to the ring (and vice versa). On the other hand, in the para migration there is never an allylic shift: the allylic group is found exactly as it was in the original ether. Compounds with propargylic groups (i.e., groups with a triple bond in the appropriate position) do not generally give the corresponding products.

The mechanism is a concerted pericyclic [3,3] sigmatropic rearrangement[490] and accounts for all these facts. For the ortho rearrangement:

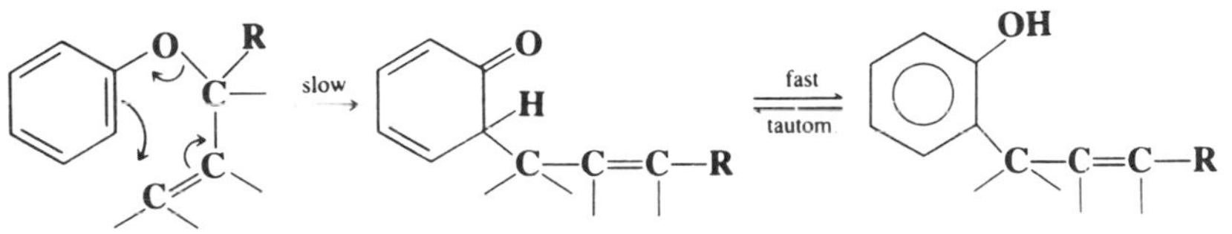

Evidence is the lack of a catalyst, the fact that the reaction is first order in the ether, the absence of crossover products when mixtures are heated, and the presence of the allylic shift, which is required by this mechanism. The allylic shift for the ortho rearrangement (and the absence of one for the para) has been demonstrated by ^{14}C labeling, even when no substituents are present. Studies of the transition-state geometry have shown that, like the Cope rearrangement, the Claisen rearrangement usually prefers a chairlike transition state.[491] When the ortho positions have no hydrogen, a second [3,3] sigmatropic migration (a Cope reaction) follows:

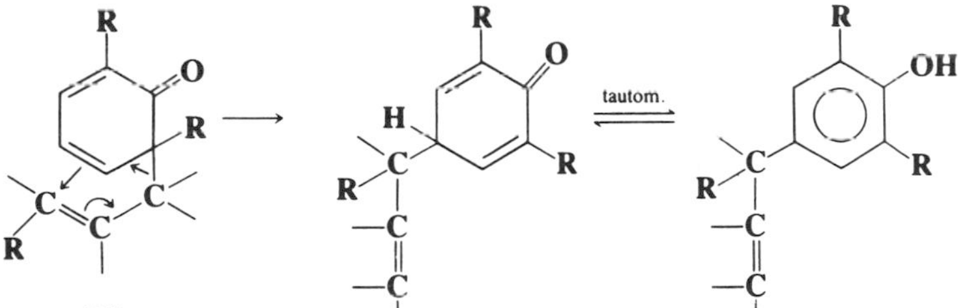

and the migrating group is restored to its original structure. Intermediates of structure **120** have been trapped by means of a Diels-Alder reaction.[492]

Ethers with an alkyl group in the γ position (ArO—C—C=C—R systems) sometimes give abnormal products, with the β carbon becoming attached to the ring:[493]

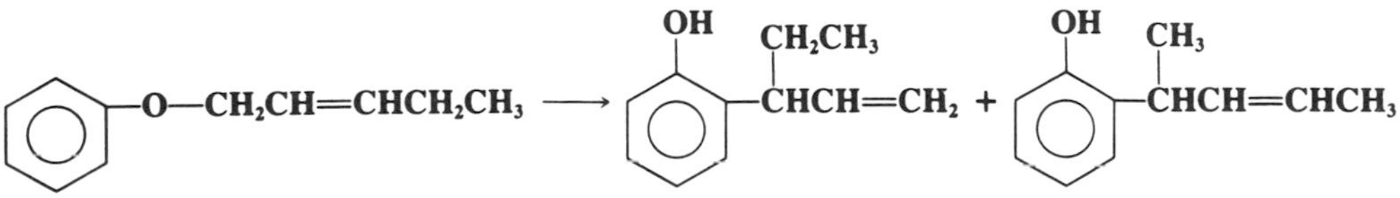

Normal product   Abnormal product

[490]For isotope effect evidence regarding the nature of the concerted transition state, see McMichael; Korver *J. Am. Chem. Soc.* **1979**, *101*, 2746; Gajewski; Conrad *J. Am. Chem. Soc.* **1979**, *101*, 2747; Kupczyk-Subotkowska; Saunders; Shine *J. Am. Chem. Soc.* **1988**, *110*, 7153.

[491]Vittorelli; Winkler; Hansen; Schmid *Helv. Chim. Acta* **1968**, *51*, 1457; Wunderli; Winkler; Hansen *Helv. Chim. Acta* **1977**, *60*, 2436; Copley; Knowles *J. Am. Chem. Soc.* **1985**, *107*, 5306.

[492]Conroy; Firestone *J. Am. Chem. Soc.* **1956**, *78*, 2290.

[493]For reviews of these abnormal Claisen rearrangements, see Hansen *Mech. Mol. Migr.* **1971**, *3*, 177-236; Marvell; Whalley, in Patai, Ref. 452, pt. 2, pp. 743-750.

It has been established that these abnormal products do not arise directly from the starting ether but are formed by a further rearrangement of the normal product:[494]

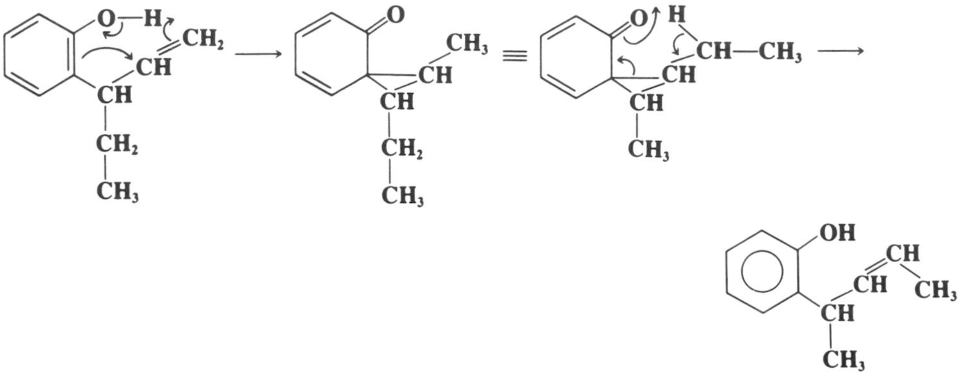

This rearrangement, which has been called an *enolene rearrangement*, a *homodienyl* [1,5] *sigmatropic hydrogen shift* (see **8-31**), and a [1,5] *homosigmatropic rearrangement*, involves a shift of three electron pairs over *seven* atoms. It has been found that this "abnormal" Claisen rearrangement is general and can interconvert the enol forms of systems of the types **121** and **123** through the cyclopropane intermediate **122**.[495]

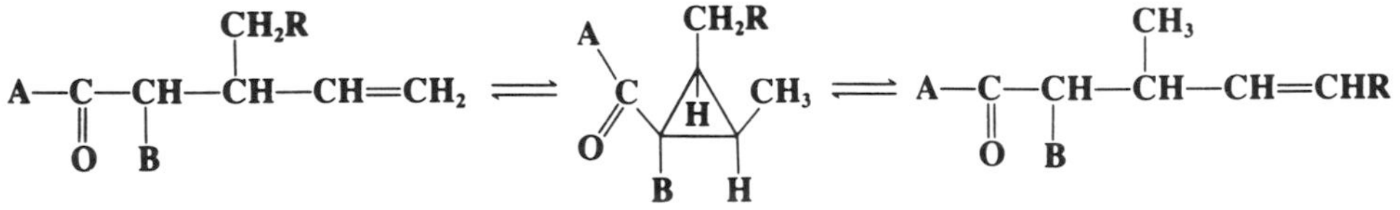

A = H, R, Ar, Or, etc.
B = H, R, Ar, COR, COAr, COOR, etc.

Since the Claisen rearrangement mechanism does not involve ions, it should not be greatly dependent on the presence or absence of substituent groups on the ring. This is the case. Electron-donating groups increase the rate and electron-withdrawing groups decrease it, but the effect is small, with the *p*-amino compound reacting only about 10 to 20 times faster than the *p*-nitro compound.[496] However, solvent effects are greater: rates varied over a 300-fold range when the reaction was run in 17 different solvents.[497] An especially good solvent is trifluoroacetic acid, in which the reaction can be carried out at room temperature.[498] Most Claisen rearrangements are performed without a catalyst, but $AlCl_3$ or BF_3 are sometimes

[494]Marvell; Anderson; Ong *J. Org. Chem.* **1962**, *27*, 1109; Habich; Barner; Roberts; Schmid *Helv. Chim. Acta* **1962**, *45*, 1943; Lauer; Johnson *J. Org. Chem.* **1963**, *28*, 2913; Fráter; Schmid *Helv. Chim. Acta* **1966**, *49*, 1957; Marvell; Schatz *Tetrahedron Lett.* **1967**, 67.

[495]Roberts; Landolt; Greene; Heyer *J. Am. Chem. Soc.* **1967**, *89*, 1404; Watson; Irvine; Roberts *J. Am. Chem. Soc.* **1973**, *95*, 3348.

[496]Goering; Jacobson *J. Am. Chem. Soc.* **1958**, *80*, 3277; White; Gwynn; Schlitt; Girard; Fife *J. Am. Chem. Soc.* **1958**, *80*, 3271; White; Slater *J. Org. Chem.* **1962**, *27*, 2908; Zahl; Kosbahn; Kresze *Liebigs Ann. Chem.* **1975**, 1733. See also Desimoni; Faita; Gamba; Righetti; Tacconi; Toma *Tetrahedron* **1990**, *46*, 2165; Gajewski; Gee; Jurayj *J. Org. Chem.* **1990**, *55*, 1813.

[497]White; Wolfarth *J. Org. Chem.* **1970**, *35*, 2196. See also Brandes; Greico; Gajewski *J. Org. Chem.* **1989**, *54*, 515.

[498]Svanholm; Parker *J. Chem. Soc., Perkin Trans. 2* **1974**, 169.

used.[499] In this case it may become a Friedel–Crafts reaction, with the mechanism no longer cyclic,[500] and ortho, meta, and para products may be obtained.

Allylic ethers of enols (allylic vinylic ethers) also undergo the Claisen rearrangement;[501] in fact, it was discovered with these compounds first:[502]

$$\begin{array}{ccc}
\text{OCH}_2\text{CH}=\text{CH}_2 & \xrightarrow{\;\Delta\;} & \text{O}\quad\text{CH}_2\text{CH}=\text{CH}_2 \\
\mid & & \parallel\;\;\mid \\
\text{R}-\text{C}=\text{CR}'_2 & & \text{R}-\text{C}-\text{CR}'_2
\end{array}$$

In these cases of course the final tautomerization does not take place even when R′ = H, since there is no aromaticity to restore, and ketones are more stable than enols.[503] The use of water as solvent accelerates the reaction.[504] The mechanism is similar to that with allylic aryl ethers.[505] One experiment that demonstrated this was the conversion of optically active

$$(+)\;\text{cyclopentene ring with }O\text{—allyl group} \quad \longrightarrow \quad (-)\;\text{cyclopentene ring with CH}_2\text{CHO group}$$

124   **125**

124 to **125**, which was still optically active.[506] This is another example of asymmetric induction (p. 117).[465]

It is possible to treat ketones with allyl alcohol and an acid catalyst to give γ,δ-unsaturated ketones directly, presumably by initial formation of the vinylic ethers, and then Claisen rearrangement.[507] In an analogous procedure, the enolates (**126**) of allylic esters [formed by treatment of the esters with lithium isopropylcyclohexylamide (ICA)] rearrange to γ,δ-unsaturated acids.[508]

$$\begin{array}{ccccccc}
\text{OCH}_2\text{CH}=\text{CHR}^1 & & \xrightarrow[\text{THF, }-78°\text{C}]{\text{Li–ICA}} & \text{OCH}_2\text{CH}=\text{CHR}^1 & \xrightarrow[\text{temp.}]{\text{room}} & \text{O}\quad\text{CHR}^1\text{CH}=\text{CH}_2 \\
\mid & & & \mid & & \parallel\;\;\mid \\
\text{O}=\text{C}-\text{CHR}^2\text{R}^3 & & & {}^{\ominus}\text{O}-\text{C}=\text{CR}^2\text{R}^3 & & {}^{\ominus}\text{O}-\text{C}-\text{CR}^2\text{R}^3
\end{array}$$

126

[499]For a review, see Lutz, Ref. 461.

[500]For example, crossover experiments have demonstrated that the ZnCl₂-catalyzed reaction is intermolecular: Yagodin; Bunina-Krivorukova; Bal'yan *J. Org. Chem. USSR* **1971**, *7*, 1491.

[501]For a review, see Ziegler *Chem. Rev.* **1988**, *88*, 1423-1452

[502]Claisen *Ber.* **1912**, *45*, 3157.

[503]However, it has proved possible to reverse the reaction, with a Lewis acid catalyst. See Boeckman; Flann; Poss *J. Am. Chem. Soc.* **1985**, *107*, 4359.

[504]Grieco; Brandes; McCann; Clark *J. Org. Chem.* **1989**, *54*, 5849.

[505]For discussions of the transition state, see Burrows; Carpenter *J. Am. Chem. Soc.* **1981**, *103*, 6983, 6984; Gajewski; Jurayj; Kimbrough; Gande; Ganem; Carpenter *J. Am. Chem. Soc.* **1987**, *109*, 1170. For mo calculations, see Vance; Rondan; Houk; Jensen; Borden; Komornicki; Wimmer *J. Am. Chem. Soc.* **1988**, *110*, 2314; Dewar; Jie *J. Am. Chem. Soc.* **1989**, *111*, 511.

[506]Hill; Edwards *Tetrahedron Lett* **1964**, 3239.

[507]Lorette *J. Org. Chem.* **1961**, *26*, 4855. See also Saucy; Marbet *Helv. Chim. Acta* **1967**, *50*, 2091; Marbet; Saucy *Helv. Chim. Acta* **1967**, *50*, 2095; Thomas *J. Am. Chem. Soc.* **1969**, *91*, 3281; Johnson; Werthemann; Bartlett; Brocksom; Li; Faulkner; Petersen *J. Am. Chem. Soc.* **1970**, *92*, 741; Pitteloud; Petrzilka; *Helv. Chim. Acta* **1979**, *62*, 1319; Daub; Sanchez; Cromer; Gibson *J. Org. Chem.* **1982**, *47*, 743; Bartlett; Tanzella; Barstow *J. Org. Chem.* **1982**, *47*, 3941.

[508]Ireland; Mueller; Willard *J. Am. Chem. Soc.* **1976**, *98*, 2868; Gajewski; Emrani *J. Am. Chem. Soc.* **1984**, *106*, 5733; Cameron; Knight *J. Chem. Soc., Perkin Trans. 1* **1986**, 161. See also Wilcox; Babston *J. Am. Chem. Soc.* **1986**, *108*, 6636.

Alternatively, the silylketene acetal $R^3R^2C=C(OSiR_3)OCH_2CH=CHR^1$ is often used instead of **126**.[509] This rearrangement also proceeds at room temperature. By either procedure, the reaction is called the *Ireland–Claisen rearrangement*. Note the presence of the negative charge in **126**. As with the oxy-Cope rearrangement (in **8-34**), negative charges generally accelerate the Claisen reaction,[510] though the extent of the acceleration can depend on the identity of the positive counterion.[511] The Ireland–Claisen rearrangement has been made enantioselective by converting **126** to an enol borinate in which the boron is attached to a chiral group.[512]

A number of expected analogs of the Claisen rearrangement are known, e.g., rearrangement of $ArNHCH_2CH=CH_2$,[513] of N-allylic enamines $R_2C=CRNRCR_2CR=CR_2$,[514] of allylic imino esters $RC(OCH_2CH=CH_2)=NR$[515] (these have often been rearranged with transition metal catalysts[516]), and of $RCH=NRCHRCH_2CH=CH_2$. These rearrangements of nitrogen-containing compounds are often called *aza-Cope rearrangements*.[517] An *azo-Cope* rearrangement: $CH_2=CHCR_2^1CR_2^2N=NAr \rightarrow R_2^1CH=CHCH_2NArN=CR_2^2$ has been reported.[518] Propargylic vinylic compounds give allenic aldehydes, ketones, esters, or amides:[519]

$$A = H, R, OR, NR_2$$

The conversion of allylic aryl thioethers $ArSCH_2CH=CH_2$ to *o*-allylic thiophenols (the *thio-Claisen rearrangement*) is not feasible, because the latter are not stable[520] but react to give bicyclic compounds.[521] However, many allylic vinylic sulfides do give the rearrangement.[522] Allylic vinylic sulfones, e.g., $H_2C=CRCH_2-SO_2-CH=CH_2$, rearrange, when

[509]Ref. 508; Ireland; Wipf; Armstrong *J. Org. Chem.* **1991**, *56*, 650.

[510]See, for example, Denmark; Harmata *Tetrahedron Lett.* **1984**, *25*, 1543; Denmark; Harmata; White *J. Am. Chem. Soc.* **1989**, *111*, 8878.

[511]Koreeda; Luengo *J. Am. Chem. Soc.* **1985**, *107*, 5572; Kirchner; Pratt; Hopkins *Tetrahedron Lett.* **1988**, *29*, 4229.

[512]Corey; Lee *J. Am. Chem. Soc.* **1991**, *113*, 4026.

[513]Marcinkiewicz; Green; Mamalis *Tetrahedron* **1961**, *14*, 208; Inada; Ikado; Okazaki *Chem. Lett.* **1973**, 1213; Schmid; Hansen; Schmid *Helv. Chim. Acta* **1973**, *56*, 105; Jolidon; Hansen *Helv. Chim. Acta* **1977**, *60*, 978.

[514]Ficini; Barbara *Tetrahedron Lett.* **1966**, 6425; Hill; Gilman *Tetrahedron Lett.* **1967**, 1421; Ireland; Willard *J. Org. Chem.* **1974**, *39*, 421; Hill; Khatri *Tetrahedron Lett.* **1978**, 4337. For the reverse of this rearrangement, see Wu; Fowler *J. Org. Chem.* **1988**, *53*, 5998.

[515]For examples, see Synerholm; Gilman; Morgan; Hill *J. Org. Chem.* **1968**, *33*, 1111; Black; Eastwood; Okraglik; Poynton; Wade; Welker *Aust. J. Chem.* **1972**, *25*, 1483; Overman *J. Am. Chem. Soc.* **1974**, *96*, 597; Metz; Mues *Tetrahedron* **1988**, *44*, 6841.

[516]See Schenck; Bosnich *J. Am. Chem. Soc.* **1985**, *107*, 2058, and references cited therein.

[517]For a review, see Przheval'skii; Grandberg *Russ. Chem. Rev.* **1987**, *56*, 477-491. For reviews of [3,3] sigmatropic rearrangements with hetero atoms present, see Blechert *Synthesis* **1989**, 71-82; Winterfeldt *Fortschr. Chem. Forsch.* **1970**, *16*, 75-102. For a review of [3,3] rearrangements of iminium salts, see Heimgartner; Hansen; Schmid *Adv. Org. Chem.* **1979**, *9*, pt. 2, 655-731.

[518]Mitsuhashi *J. Am. Chem. Soc.* **1986**, *108*, 2400.

[519]For reviews of Claisen rearrangements involving triple bonds, see Schuster; Coppola, Ref. 124, pp. 337-343; Viola et al., Ref. 460; Théron et al., Ref. 460, pp. 421-428. See also Henderson; Heathcock *J. Org. Chem.* **1988**, *53*, 4736.

[520]They have been trapped: See, for example, Mortensen; Hedegaard; Lawesson *Tetrahedron* **1971**, *27*, 3831; Kwart; Schwartz *J. Org. Chem.* **1974**, *39*, 1575.

[521]Kwart; Hackett *J. Am. Chem. Soc.* **1962**, *84*, 1754; Meyers; Rinaldi; Banoli *J. Org. Chem.* **1963**, *28*, 2440; Makisumi *Tetrahedron Lett.* **1966**, 6399; Kwart; Cohen *J. Org. Chem.* **1967**, *32*, 3135, *Chem. Commun.* **1968**, 319; Makisumi; Murabayashi *Tetrahedron Lett.* **1969**, 1971, 2449.

[522]See, for example, Schuijl; Brandsma *Recl. Trav. Chim. Pays-Bas* **1968**, *87*, 929, **1969**, *88*, 1201; Corey; Shulman *J. Am. Chem. Soc.* **1970**, *92*, 5522; Kondo; Ojima *Chem. Commun.* **1972**, 62; Meijer; Vermeer; Bos; Brandsma *Recl. Trav. Chim. Pays-Bas* **1974**, *93*, 26; Morin; Paquer; Smadja *Recl. Trav. Chim. Pays-Bas* **1976**, *95*, 179; Schaumann; Grabley *Liebigs Ann. Chem.* **1979**, 1746; Metzner; Pham; Vialle *Tetrahedron* **1986**, *42*, 2025; Beslin; Perrio *Tetrahedron* **1991**, *47*, 6275.

heated in the presence of ethanol and pyridine, to unsaturated sulfonate salts $CH_2=CRCH_2CH_2CH_2SO_3^-$, produced by reaction of the reagents with the unstable sulfene intermediates $CH_2=CRCH_2CH_2CH=SO_2$.[523] Allylic vinylic sulfoxides rapidly rearrange at room temperature or below.[524]

OS **III**, 418; **V**, 25; **VI**, 298, 491, 507, 584, 606; **VII**, 177; **66**, 22, 29.

8-36　The **Fischer Indole Synthesis**

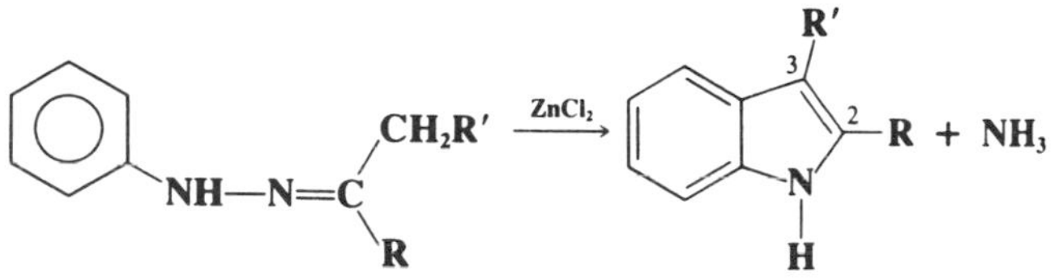

When arylhydrazones of aldehydes or ketones are treated with a catalyst, elimination of ammonia takes place and an indole is formed, in the *Fischer indole synthesis*.[525] Zinc chloride is the catalyst most frequently employed, but dozens of others, including other metal halides, proton and Lewis acids, and certain transition-metals have also been used. Arylhydrazones are easily prepared by the treatment of aldehydes or ketones with phenylhydrazine (**6-2**) or by aliphatic diazonium coupling (**2-7**). However, it is not necessary to isolate the arylhydrazone. The aldehyde or ketone can be treated with a mixture of phenylhydrazine and the catalyst; this is now common practice. In order to obtain an indole, the aldehyde or ketone must be of the form $RCOCH_2R'$ (R = alkyl, aryl, or hydrogen).

At first glance the reaction does not seem to be a rearrangement. However, the key step of the mechanism is a [3,3] sigmatropic rearrangement:[526]

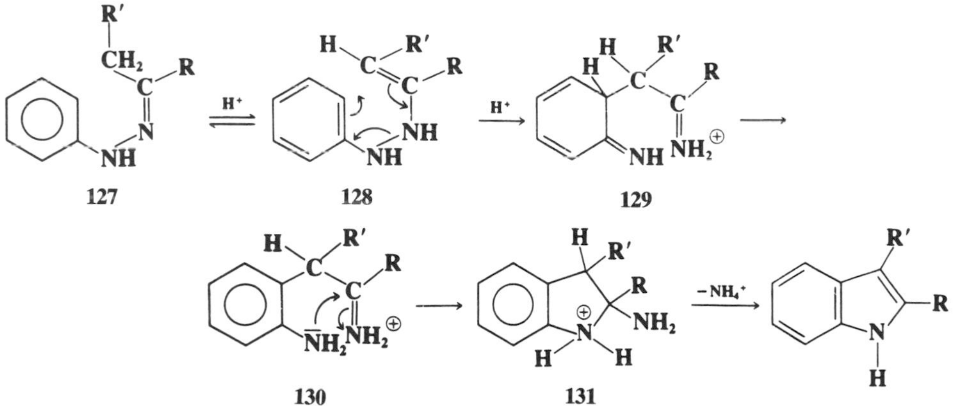

[523]King; Harding *J. Am. Chem. Soc.* **1976**, *98*, 3312.
[524]Block; Ahmad *J. Am. Chem. Soc.* **1985**, *107*, 6731.
[525]For a monograph, see Robinson *The Fischer Indole Synthesis*; Wiley: New York, 1983. For reviews, see Grandberg; Sorokin *Russ. Chem. Rev.* **1974**, *43*, 115-128; Shine *Aromatic Rearrangements*, Ref. 489, pp. 190-207; Sundberg *The Chemistry of Indoles*; Academic Press: New York, 1970, pp. 142-163; Robinson *Chem. Rev.* **1969**, *69*, 227-250. For reviews of some abnormal Fischer indole syntheses, see Ishii *Acc. Chem. Res.* **1981**, *14*, 275-283; Fusco; Sannicolo *Tetrahedron* **1980**, *36*, 161-170.
[526]This mechanism was proposed by Robinson; Robinson *J. Chem. Soc.* **1918**, *113*, 639.

There is much evidence for this mechanism, e.g., (1) the isolation of **131**,[527] (2) the detection of **130** by ^{13}C and ^{15}N nmr,[528] (3) the isolation of side products that could only have come from **129**,[529] and (4) ^{15}N labeling experiments that showed that it was the nitrogen farther from the ring that is eliminated as ammonia.[530] The main function of the catalyst seems to be to speed the conversion of **127** to **128**. The reaction can be performed without a catalyst.

OS **III**, 725; **IV**, 884. Also see OS **IV**, 657.

8-37   [2,3] Sigmatropic Rearrangements
         (2/S-3/)→(1/5/)-*sigma*-Migration

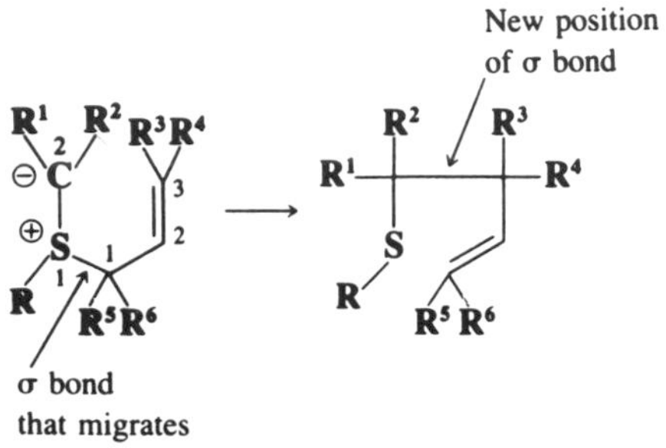

New position
of σ bond

σ bond
that migrates

Sulfur ylides bearing an allylic group are converted on heating to unsaturated sulfides.[531] This is a concerted [2,3] sigmatropic rearrangement[532] and has also been demonstrated for the analogous cases of nitrogen ylides[533] and the conjugate bases of allylic ethers (in the last

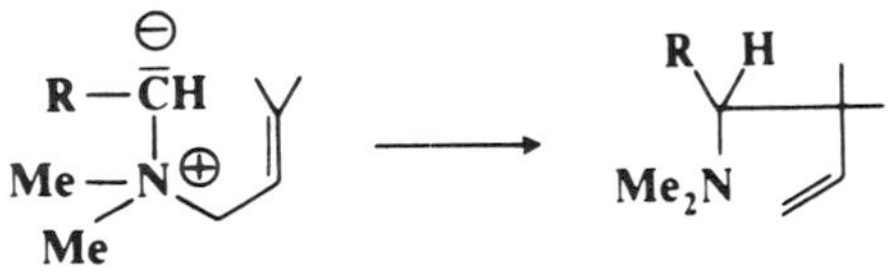

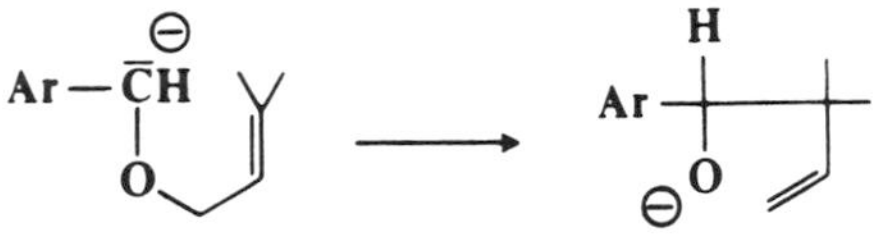

[527]Southwick; McGrew; Engel; Milliman; Owellen *J. Org. Chem.* **1963**, *28*, 3058; Southwick; Vida; Fitzgerald; Lee *J. Org. Chem.* **1968**, *33*, 2051; Forrest; Chen *J. Chem. Soc., Chem. Commun.* **1972**, 1067.
[528]Douglas *J. Am. Chem. Soc.* **1978**, *100*, 6463, **1979**, *101*, 5676.
[529]Robinson; Brown *Can. J. Chem.* **1964**, *42*, 1940; Bajwa; Brown *Can J. Chem.* **1968**, *46*, 1927, 3105, **1969**, *47*, 785, **1970**, *48*, 2293.
[530]Clausius; Weisser *Helv. Chim. Acta* **1952**, *35*, 400.
[531]For example, see Blackburn; Ollis; Plackett; Smith; Sutherland; *Chem. Commun.* **1968**, 186; Trost; LaRochelle *Tetrahedron Lett.* **1968**, 3327; Baldwin; Hackler; Kelly *Chem. Commun.* **1968**, 537, 538, 1083; Bates; Feld *Tetrahedron Lett.* **1968**, 417; Kirmse; Kapps *Chem. Ber.* **1968**, *101*, 994, 1004; Biellmann; Ducep *Tetrahedron Lett.* **1971**, 33; Ceré; Paolucci; Pollicino; Sandri; Fava *J. Org. Chem.* **1981**, *46*, 3315; Kido; Sinha; Abiko; Yoshikoshi *Tetrahedron Lett.* **1989**, *30*, 1575. For a review as applied to ring expansions, see Vedejs *Acc. Chem. Res.* **1984**, *17*, 358-364.
[532]For a review of the stereochemistry of these reactions, see Hoffmann *Angew. Chem. Int. Ed. Engl.* **1979**, *18*, 563-572 [*Angew. Chem. 91*, 625-634].
[533]For example, see Jemison; Ollis *Chem. Commun.* **1969**, 294; Rautenstrauch *Helv. Chim. Acta* **1972**, *55*, 2233; Mageswaran; Ollis; Sutherland; Thebtaranonth *J. Chem. Soc., Chem. Commun.* **1973**, 651; Ollis; Sutherland; Thebtaranonth *J. Chem. Soc., Chem. Commun.* **1973**, 657; Mander; Turner *J. Org. Chem.* **1973**, *38*, 2915; Stévenart-De Mesmaeker; Merényi; Viehe *Tetrahedron Lett.* **1987**, *28*, 2591; Honda; Inoue; Sato *J. Am. Chem. Soc.* **1990**, *112*, 1999.

case it is called the [*2,3*] *Wittig rearrangement*).[534] The reaction has been extended to certain other systems,[535] even to an all-carbon system.[536]

Since the reactions involve migration of an allylic group from a sulfur, nitrogen, or oxygen atom to an adjacent negatively charged carbon atom, they are special cases of the Stevens or Wittig rearrangements (**8-22, 8-23**). However, in this case the migrating group *must* be allylic (in **8-22** and **8-23** other groups can also migrate). Thus, when the migrating group is allylic, there are two possible pathways: (1) the radical-ion or ion-pair mechanisms (**8-22, 8-23**) and (2) the concerted pericyclic [2,3] sigmatropic rearrangement. These can easily be told apart, since the latter always involves an allylic shift (as in the Claisen rearrangement), while the former pathway does not.

Of these reactions, the [2,3] Wittig rearrangement in particular has often been used as a means of transferring chirality. The product of this reaction has potential chiral centers at C-3 and C-4 (if $R^5 \neq R^6$), and if the starting ether is optically active because of a chiral

center at C-1, the product may be optically active as well. Many examples are known in which an optically active ether was converted to a product that was optically active because of chirality at C-3, C-4, or both.[537] If a suitable chiral center is present in R^1 (or if a functional group in R^1 can be so converted), then stereocontrol over three contiguous chiral centers can be achieved. Stereocontrol of the new double bond (*E* or *Z*) has also been accomplished.

If an OR or SR group is attached to the negative carbon, the reaction becomes a method for the preparation of β,γ-unsaturated aldehydes, because the product is easily hydrolyzed.[538]

Another [2,3] sigmatropic rearrangement converts allylic sulfoxides to allylically rearranged alcohols by treatment with a thiophilic reagent such as trimethyl phosphite.[539] In this

[534]See, for example, Makisumi; Notzumoto *Tetrahedron Lett.* **1966**, 6393; Schöllkopf; Fellenberger; Rizk *Liebigs Ann. Chem.* **1970**, *734*, 106; Rautenstrauch *Chem. Commun.* **1970**, 4. For a review, see Nakai; Mikami *Chem. Rev.* **1986**, *86*, 885-902. For a list of references, see Ref. 106, pp. 521-522.

[535]See, for example, Baldwin; Brown; Höfle *J. Am. Chem. Soc.* **1971**, *93*, 788; Yamamoto; Oda; Inouye *J. Chem. Soc., Chem. Commun.* **1973**, 848; Ranganathan; Ranganathan; Sidhu; Mehrotra *Tetrahedron Lett.* **1973**, 3577; Murata; Nakai *Chem. Lett.* **1990**, 2069. For reviews with respect to selenium compounds, see Reich, in Liotta *Organoselenium Chemistry*; Wiley: New York, 1987, pp. 365-393; Reich, in Trahanovsky *Oxidation in Organic Chemistry*, pt. C; Academic Press: New York, 1978, pp. 102-111.

[536]Baldwin; Urban *Chem. Commun.* **1970**, 165.

[537]For reviews of stereochemistry in this reaction, see Mikami; Nakai *Synthesis* **1991**, 594-604; Nakai; Mikami, Ref. 534, pp. 888-895. See also Nakai; Nakai *Tetrahedron Lett.* **1988**, *29*, 4587; Balestra; Kallmerten *Tetrahedron Lett.* **1988**, *29*, 6901; Brückner *Chem. Ber.* **1989**, *122*, 193, 703; Scheuplein; Kusche; Brückner; Harms *Chem. Ber.* **1990**, *123*, 917; Wu; Houk; Marshall *J. Org. Chem.* **1990**, *55*, 1421; Marshall; Wang *J. Org. Chem.* **1990**, *55*, 2995.

[538]Huynh; Julia; Lorne; Michelot *Bull. Soc. Chim. Fr.* **1972**, 4057.

[539]Bickart; Carson; Jacobus; Miller; Mislow *J. Am. Chem. Soc.* **1968**, *90*, 4869; Tang; Mislow *J. Am. Chem. Soc.* **1970**, *92*, 2100; Grieco *J. Chem. Soc., Chem. Commun.* **1972**, 702; Evans; Andrews *Acc. Chem. Res.* **1974**, *7*, 147-155; Isobe; Iio; Kitamura; Goto *Chem. Lett.* **1978**, 541; Hoffmann; Goldmann; Maak; Gerlach; Frickel; Steinbach *Chem. Ber.* **1980**, *113*, 819; Sato; Otera; Nozaki *J. Org. Chem.* **1989**, *54*, 2779.

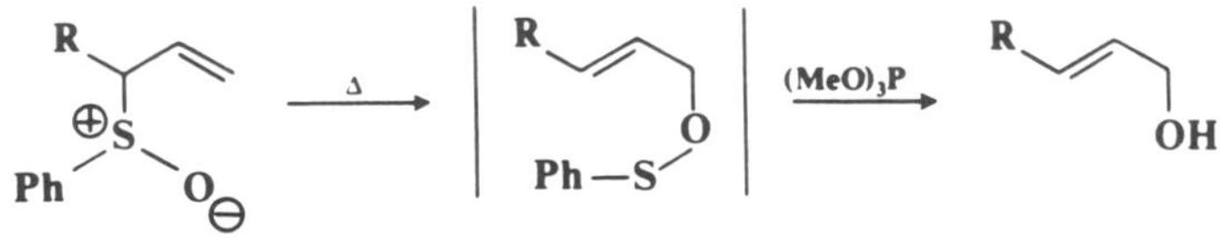

case the migration is from sulfur to oxygen. [2,3] oxygen-to-sulfur migrations are also known.[540] The Sommelet–Hauser rearrangement (**3-26**) is also a [2,3] sigmatropic arrangement.

OS **65**, 159.

8-38 The **Benzidine Rearrangement**

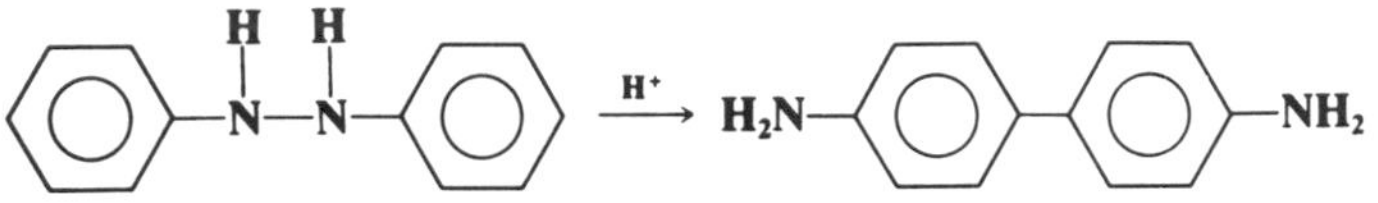

When hydrazobenzene is treated with acids, it rearranges to give about 70% 4,4′-diaminobiphenyl (**132**, benzidine) and about 30% 2,4′-diaminobiphenyl (**133**). This reaction is called the *benzidine rearrangement* and is general for N,N′-diarylhydrazines.[541] Usually, the major product is the 4,4′-diaminobiaryl, but four other products may also be produced. These are the 2,4′-diaminobiaryl (**133**), already referred to, the 2,2′-diaminobiaryl (**134**), and the *o*- and *p*-arylaminoanilines (**135** and **136**), called *semidines*. The **134** and **136** com-

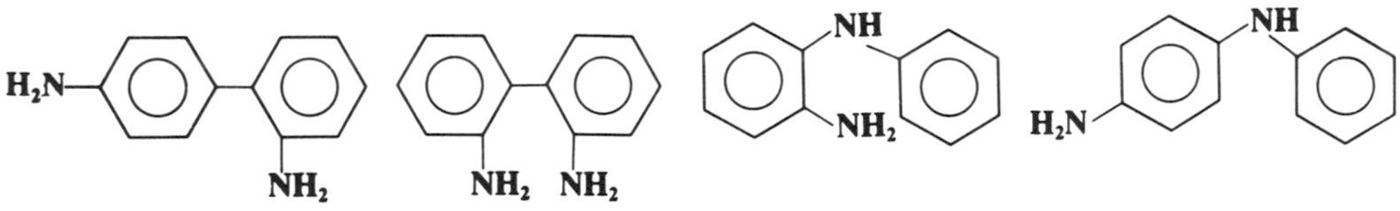

pounds are formed less often and in smaller amounts than the other two side products. Usually, the 4,4′-diaminobiaryl predominates, except when one or both para positions of the diarylhydrazine are occupied. However, the 4,4′-diamine may still be produced even if the para positions are occupied. If SO₃H, COOH, or Cl (but not R, Ar, or NR₂) is present in the para position, it may be ejected. With dinaphthylhydrazines, the major products are not the 4,4′-diaminobinaphthyls, but the 2,2′ isomers. Another side reaction is disproportionation to ArNH₂ and ArN=NAr. For example, *p,p′*-PhC₆H₄NHNHC₆H₄Ph gives 88% disproportionation products at 25°C.[542]

[540]Braverman; Mechoulam *Isr. J. Chem.* **1967**, *5*, 71, Braverman; Stabinsky *Chem. Commun.* **1967**, 270; Rautenstrauch *Chem. Commun.* **1970**, 526; Smith; Stirling *J. Chem. Soc. C* **1971**, 1530; Tamaru; Nagao; Bando; Yoshida *J. Org. Chem.* **1990**, *55*, 1823.
[541]For reviews, see, in Patai *The Chemistry of the Hydrazo, Azo, and Azoxy Groups*, pt. 2; Wiley: New York, 1975, the reviews by Cox; Buncel, pp. 775-807; Koga; Koga; Anselme, pp. 914-921; Williams, in Bamford; Tipper, Ref. 365, vol. 13, 1972, pp. 437-448; Shine *Mech. Mol. Migr.* **1969**, *2*, 191-247, *Aromatic Rearrangements*, Ref. 489, pp. 126-179; Banthorpe *Top. Carbocyclic Chem.* **1969**, *1*, 1-62; Lukashevich *Russ. Chem. Rev.* **1967**, *36*, 895-902.
[542]Shine; Stanley *J. Org. Chem.* **1967**, *32*, 905. For investigations of the mechanism of the disproportionation reactions, see Shine; Habdas; Kwart; Brechbiel; Horgan; San Filippo *J. Am. Chem. Soc.* **1983**, *105*, 2823; Rhee; Shine *J. Am. Chem. Soc.* **1986**, *108*, 1000, **1987**, *109*, 5052.

The mechanism has been exhaustively studied and several mechanisms have been proposed.[543] At one time it was believed that NHAr broke away from ArNHNHAr and became attached to the para position to give the semidine (**136**), which then went on to product. The fact that semidines could be isolated lent this argument support, as did the fact that this would be analogous to the rearrangements considered in Chapter 11 (**1-32** to **1-36**). However, this theory was killed when it was discovered that semidines could not be converted to benzidines under the reaction conditions. Cleavage into two independent pieces (either ions or radicals) has been ruled out by many types of crossover experiments, which always showed that the two rings of the starting material are in the product; that is, ArNHNHAr' gives no molecules (of any of the five products) containing two Ar groups or two Ar' groups, and mixtures of ArNHNHAr and Ar'NHNHAr' give no molecules containing both Ar and Ar'. An important discovery was the fact that, although the reaction is always first order in substrate, it can be either first[544] or second[545] order in [H⁺]. With some substrates the reaction is entirely first order in [H⁺], while with others it is entirely second order in [H⁺], regardless of the acidity. With still other substrates, the reaction is first order in [H⁺] at low acidities and second order at higher acidities. With the latter substrates fractional orders can often be observed,[546] because at intermediate acidities, both processes take place simultaneously. These kinetic results seem to indicate that the actual reacting species can be either the monoprotonated substrate ArNH$\overset{\oplus}{N}$H₂Ar or the diprotonated Ar$\overset{\oplus}{N}$H₂$\overset{\oplus}{N}$H₂Ar.

Most of the proposed mechanisms[547] attempted to show how all five products could be produced by variations of a single process. An important breakthrough was the discovery that the two main products, **132** and **133,** are formed in entirely different ways, as shown by isotope-effect studies.[548] When the reaction was run with hydrazobenzene labeled with ¹⁵N at both nitrogen atoms, the isotope effect was 1.022 for formation of **132,** but 1.063 for formation of **133.** This showed that the N—N bond is broken in the rate-determining step in both cases, but the steps themselves are obviously different. When the reaction was run with hydrazobenzene labeled with ¹⁴C at a para position, there was an isotope effect of 1.028 for formation of **132,** but essentially no isotope effect (1.001) for formation of **133.** This can only mean that for **132** formation of the new C—C bond *and* breaking of the N—N bond both take place in the rate-determining step; in other words, the mechanism is concerted. The following [5.5] sigmatropic rearrangement accounts for this:[549]

137

[543]For a history of the mechanistic investigations and controversies, see Shine *J. Phys. Org. Chem.* **1989**, *2*, 491.

[544]Banthorpe; Hughes; Ingold *J. Chem. Soc.* **1962**, 2386, 2402, 2407, 2413, 2418, 2429; Shine; Chamness *J. Org. Chem.* **1963**, *28*, 1232; Banthorpe; O'Sullivan *J. Chem. Soc. B* **1968**, 627.

[545]Hammond; Shine *J. Am. Chem. Soc.* **1950**, *72*, 220; Banthorpe; Cooper *J. Chem. Soc. B* **1968**, 618; Banthorpe; Cooper; O'Sullivan *J. Chem. Soc. B* **1971**, 2054.

[546]Carlin; Odioso *J. Am. Chem. Soc.* **1954**, *76*, 100; Banthorpe; Ingold; Roy *J. Chem. Soc. B* **1968**, 64; Banthorpe; Ingold; O'Sullivan *J. Chem. Soc. B* **1968**, 624.

[547]For example, see the "polar-transition-state mechanism:" Banthorpe; Hughes; Ingold *J. Chem. Soc.* **1964**, 2864, and the "π-complex mechanism:" Dewar, in Mayo, Ref. 114, vol. 1, pp. 323-344.

[548]Shine; Zmuda; Park; Kwart; Horgan; Collins; Maxwell *J. Am. Chem. Soc.* **1981**, *103*, 955; Shine; Zmuda; Park; Kwart; Horgan; Brechbiel *J. Am. Chem. Soc.* **1982**, *104*, 2501.

[549]This step was also part of the "polar-transition-state mechanism"; see Ref. 547.

The diion **137** was obtained as a stable species in super-acid solution at $-78°C$ by treatment of hydrazobenzene with $FSO_3H–SO_2$ (SO_2ClF).[550] Though the results just given were obtained with hydrazobenzene, which reacts by the diprotonated pathway, monoprotonated substrates have been found to react by the same [5,5] sigmatropic mechanism.[551] Some of the other rearrangements in this section are also sigmatropic. Thus, formation of the *p*-semidine **136** takes place by a [1,5] sigmatropic rearrangement,[552] and the conversion of 2,2'-hydrazonaphthalene to 2,2'-diamino-1,1'-binaphthyl by a [3,3] sigmatropic rearrangement.[553]

133 is formed by a completely different mechanism, though the details are not known. There is rate-determining breaking of the N—N bond, but the C—C bond is not formed during this step.[554] The formation of the *o*-semidine **135** also takes place by a nonconcerted pathway.[555] Under certain conditions, benzidine rearrangements have been found to go through radical cations.[556]

C. Other Cyclic Rearrangements

8-39 Metathesis of Olefins
Alkene metathesis

$$CH_3CH{=}CHCH_2CH_3 \xrightarrow[\text{WCl}_6-\text{EtOH}]{\text{EtAlCl}_2} CH_3CH{=}CHCH_3 + CH_3CH_2CH{=}CHCH_2CH_3$$

When olefins are treated with certain catalysts (most often tungsten, molybdenum, or rhenium complexes), they are converted to other olefins in a reaction in which the alkylidene groups ($R^1R^2C{=}$) have become interchanged by a process schematically illustrated by the equation:

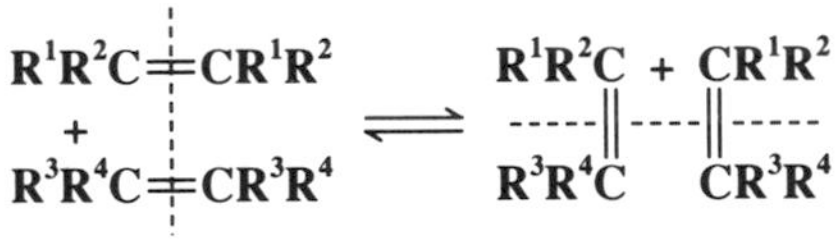

$$\begin{array}{c} R^1R^2C \dotplus CR^1R^2 \\ + \\ R^3R^4C \dotplus CR^3R^4 \end{array} \rightleftharpoons \begin{array}{c} R^1R^2C + CR^1R^2 \\ \text{----}\|\text{----}\|\text{----} \\ R^3R^4C \quad CR^3R^4 \end{array}$$

The reaction is called *metathesis* of olefins.[557] In the example shown above, 2-pentene (either cis, trans, or a cis–trans mixture) is converted to a mixture of about 50% 2-pentene, 25% 2-butene, and 25% 3-hexene. The reaction is an equilibrium and the same mixture can be obtained by starting with equimolar quantities of 2-butene and 3-hexene.[558] In general, the

[550]Olah; Dunne; Kelly; Mo *J. Am. Chem. Soc.* **1972**, *94*, 7438.

[551]Shine; Park; Brownawell; San Filippo *J. Am. Chem. Soc.* **1984**, *106*, 7077.

[552]Heesing; Schinke *Chem. Ber.* **1977**, *110*, 3319; Shine; Zmuda; Kwart; Horgan; Brechbiel *J. Am. Chem. Soc.* **1982**, *104*, 5181.

[553]Shine; Gruszecka; Subotkowski; Brownawell; San Filippo *J. Am. Chem. Soc.* **1985**, *107*, 3218.

[554]See Rhee; Shine, Ref. 542.

[555]Rhee; Shine *J. Org. Chem.* **1987**, *52*, 5633.

[556]See, for example, Nojima; Ando; Tokura *J. Chem. Soc., Perkin Trans. 1* **1976**, 1504.

[557]For monographs, see Drǎguţan; Balaban; Dimonie *Olefin Metathesis and Ring-Opening Polymerization of Cyclo-Olefins*; Wiley: New York, 1985; Ivin *Olefin Metathesis*; Academic Press: New York, 1983. For reviews, see Feast; Gibson, in Hartley, Ref. 313, vol. 5, 1989, pp. 199-228; Streck *CHEMTECH* **1989**, 498-503; Schrock *J. Organomet. Chem.* **1986**, *300*, 249-262; Grubbs, in Wilkinson *Comprehensive Organometallic Chemistry*, vol. 8; Pergamon: Elmsford, NY, 1982, pp. 499-551; Basset; Leconte, *CHEMTECH* **1980**, 762-767; Banks, *CHEMTECH* **1979**, 494-500, *Fortschr. Chem. Forsch.* **1972**, *25*, 39-69; Calderon; Lawrence; Ofstead *Adv. Organomet. Chem.* **1979**, *17*, 449-492; Grubbs *Prog. Inorg. Chem.* **1978**, *24*, 1-50; Calderon, in Patai *The Chemistry of Functional Groups: Supplement A*, pt. 2; Wiley: New York, 1977, pp. 913-964, *Acc. Chem. Res.* **1972**, *5*, 127-132; Katz *Adv. Organomet. Chem.* **1977**, *16*, 283-317; Haines; Leigh *Chem. Soc. Rev.* **1975**, *4*, 155-188; Hocks *Bull. Soc. Chim. Fr.* **1975**, 1893-1903; Mol; Moulijn *Adv. Catal.* **1974**, *24*, 131-171; Hughes *Organomet. Chem. Synth.* **1972**, *1*, 341-374; Khidekel', Shebaldova; Kalechits *Russ. Chem. Rev.* **1971**, *40*, 669-678; Bailey, *Catal. Rev.* **1969**, *3*, 37-60.

[558]Calderon; Chen; Scott *Tetrahedron Lett.* **1967**, 3327; Wang; Menapace *J. Org. Chem.* **1968**, *33*, 3794; Hughes *J. Am. Chem. Soc.* **1970**, *92*, 532.

reaction can be applied to a single unsymmetrical olefin, giving a mixture of itself and two other olefins, or to a mixture of two olefins, in which case the number of different molecules in the product depends on the symmetry of the reactants. As in the case above, a mixture of $R^1R^2C{=}CR^1R^2$ and $R^3R^4C{=}CR^3R^4$ gives rise to only one new olefin ($R^1R^2C{=}CR^3R^4$), while in the most general case, a mixture of $R^1R^2C{=}CR^3R^4$ and $R^5R^6C{=}CR^7R^8$ gives a mixture of ten olefins: the original two plus eight new ones. With simple alkenes the proportions of products are generally statistical,[559] which limits the synthetic utility of the reaction since the yield of any one product is low. However, in some cases one alkene may be more or less thermodynamically stable than the rest, so that the proportions are not statistical. Furthermore, it may be possible to shift the equilibrium. For example, 2-methyl-1-butene gives rise to ethylene and 3,4-dimethyl-3-hexene. By allowing the gaseous ethylene to escape, the yield of 3,4-dimethyl-3-hexene can be raised to 95%.[560]

Many catalysts, both homogeneous[561] and heterogeneous,[562] have been used for this reaction. Some of the former[563] are WCl_6–EtOH–$EtAlCl_2$,[559] $MoCl_2(NO)_2(Ph_3P)_2$–Et-$AlCl_2$,[564] WCl_6–BuLi,[565] and WCl_6–$LiAlH_4$,[566] while among the latter are oxides of Mo, W, and Re deposited on alumina or silica gel.[567] In general, the former group are more useful for synthetic purposes. By choice of the proper catalyst, the reaction has been applied to terminal and internal alkenes, straight chain or branched. The effect of substitution on the ease of reaction is $CH_2{=}>RCH_2CH{=}>R_2CHCH{=}>R_2C{=}$.[568] Dienes can react intermolecularly or intramolecularly,[569] e.g.,

Cyclic olefins give dimeric dienes,[570] e.g.,

However, the products can then react with additional monomers and with each other, so that polymers are generally produced, and the cyclic dienes are obtained only in low yield.

[559]Calderon; Ofstead; Ward; Judy; Scott *J. Am. Chem. Soc.* **1968**, *90*, 4133.

[560]Knoche, Ger. Pat.(Offen.) 2024835, 1970 [*Chem. Abstr.* **1971**, *74*, 44118b]. See also Chevalier; Sinou; Descotes *Bull. Soc. Chim. Fr.* **1976**, 2254; Bespalova; Babich; Vdovin; Nametkin *Doklad. Chem.* **1975**, *225*, 668; Ichikawa; Fukuzumi *J. Org. Chem.* **1976**, *41*, 2633; Baker; Crimmin *Tetrahedron Lett* **1977**, 441.

[561]First reported by Calderon; Chen; Scott, Ref. 558.

[562]First reported by Banks; Bailey *Ind. Eng. Chem., Prod. Res. Dev.* **1964**, *3*, 170. See also Banks *CHEMTECH* **1986**, 112-117.

[563]For a lengthy list, see Hughes *Organomet. Chem. Synth.*, Ref. 557, pp. 362-368. For a homogeneous rhenium catalyst, see Toreki; Schrock *J. Am. Chem. Soc.* **1990**, *112*, 2448.

[564]Zuech; Hughes; Kubicek; Kittleman *J. Am. Chem. Soc.* **1970**, *92*, 528; Hughes, Ref. 558.

[565]Wang; Menapace, Ref. 558.

[566]Chatt; Haines; Leigh *J. Chem. Soc., Chem. Commun.* **1972**, 1202; Matlin; Sammes *J. Chem. Soc., Perkin Trans. 1* **1978**, 624.

[567]For a list of heterogeneous catalysts, see Banks, *Fortschr. Chem. Forsch.*, Ref. 557, pp. 41-46.

[568]For an explanation for this order, see McGinnis; Katz; Hurwitz *J. Am. Chem. Soc.* **1976**, *98*, 605; Casey; Tuinstra; Saeman *J. Am. Chem. Soc.* **1976**, *98*, 608.

[569]Kroll; Doyle *Chem. Commun.* **1971**, 839; Zuech et al., Ref. 564.

[570]Calderon; Ofstead; Judy *J. Polym. Sci., Part A-1* **1967**, *5*, 2209; Wasserman; Ben-Efraim; Wolovsky *J. Am. Chem. Soc.* **1968**, *90*, 3286; Wolovsky; Nir *Synthesis* **1972**, 134.

The reaction between a cyclic and a linear olefin can give an ring-opened diene:[571]

Olefins containing functional groups[572] do not give the reaction with most of the common catalysts, but some success has been reported with WCl_6–$SnMe_4$[573] and with certain other catalysts.

The reaction has also been applied to internal triple bonds:[574]

$$2RC{\equiv}CR' \;\rightleftharpoons\; RC{\equiv}CR + R'C{\equiv}CR'$$

but it has not been successful for terminal triple bonds.[575] An intramolecular reaction of a double bond with a triple bond has been reported.[576]

The generally accepted mechanism is a chain mechanism, involving the intervention of a metal-carbene complex (**138**)[577] and a four-membered ring containing a metal[578] (**139**).[579]

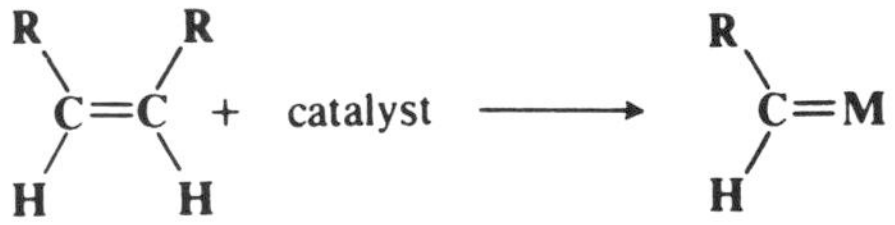

[571]Wasserman; Ben-Efraim; Wolovsky, Ref. 570; Ray; Crain, Fr. Pat. 1511381, 1968 [*Chem. Abstr.* **1969,** *70,* 114580q]; Mango, U.S. Pat. 3424811, 1969 [*Chem. Abstr.* **1969,** *70,* 106042a]; Rossi; Diversi; Lucherini; Porri *Tetrahedron Lett.* **1974,** 879; Lal; Smith *J. Org. Chem.* **1975,** *40,* 775.

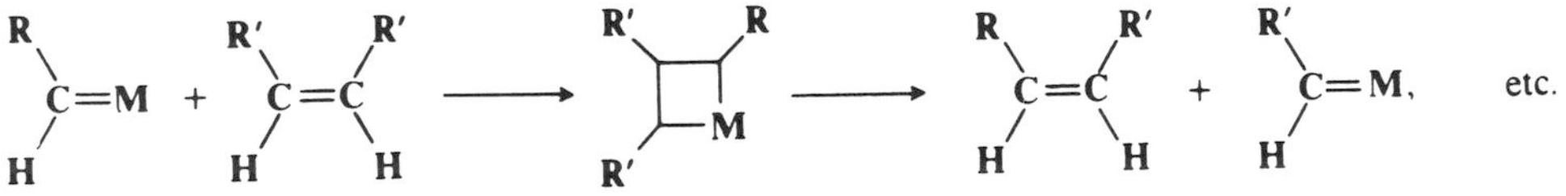

[572]For a review, see Mol *CHEMTECH* **1983,** 250-255. See also Bosma; van den Aardweg; Mol *J. Organomet. Chem.* **1983,** *255,* 159, **1985,** *280,* 115; Xiaoding; Mol *J. Chem. Soc., Chem. Commun.* **1985,** 631; Crisp; Collis *Aust. J. Chem.* **1988,** *41,* 935.

[573]First shown by van Dam; Mittelmeijer; Boelhouwer *J. Chem. Soc., Chem. Commun.* **1972,** 1221.

[574]Pennella; Banks; Bailey *Chem. Commun.* **1968,** 1548; Mortreux; Petit; Blanchard *Tetrahedron Lett.* **1978,** 4967; Devarajan; Walton; Leigh *J. Organomet. Chem.* **1979,** *181,* 99; Wengrovius; Sancho; Schrock *J. Am. Chem. Soc.* **1981,** *103,* 3932; Villemin; Cadiot *Tetrahedorn Lett.* **1982,** *23,* 5139; McCullough; Schrock *J. Am. Chem. Soc.* **1984,** *106,* 4067.

[575]McCullough; Listemann; Schrock; Churchill; Ziller *J. Am. Chem. Soc.* **1983,** *105,* 6729.

[576]Trost; Trost *J. Am. Chem. Soc.* **1991,** *113,* 1850.

[577]For a review of these complexes and their role in this reaction, see Crabtree *The Organometallic Chemistry of the Transition Metals*; Wiley: New York, 1988, pp. 244-267.

[578]For reviews of metallocycles, see Collman; Hegedus; Norton; Finke *Principles and Applications of Organotransition Metal Chemistry*, 2nd ed.; University Science Books: Mill Valley, CA; 1987, pp. 459-520; Lindner *Adv. Heterocycl. Chem.* **1986,** *39,* 237-279.

[579]For reviews of the mechanism, see Grubbs, *Prog. Inorg. Chem.*, Ref. 557; Katz, Ref. 557; Calderon; Ofstead; Judy *Angew. Chem. Int. Ed. Engl.* **1976,** *15,* 401-409 [*Angew. Chem.* **88,** 443-442]. See also McLain; Wood; Schrock *J. Am. Chem. Soc.* **1977,** *99,* 3519; Casey; Polichnowski *J. Am.Chem. Soc.* **1977,** *99,* 6097; Mango *J. Am. Chem. Soc.* **1977,** *99,* 6117; Stevens; Beauchamp *J. Am. Chem. Soc.* **1979,** *101,* 6449; Lee; Ott; Grubbs *J. Am. Chem. Soc.* **1982,** *104,* 7491; Levisalles; Rudler; Villemin *J. Organomet. Chem.* **1980,** *193,* 235; Iwasawa; Hamamura *J. Chem. Soc., Chem. Commun.* **1983,** 130; Rappé; Upton *Organometallics* **1984,** *3,* 1440; Kress; Osborn; Greene; Ivin; Rooney *J. Am. Chem. Soc.* **1987,** *109,* 899; Feldman; Davis; Schrock *Organometallics* **1989,** *8,* 2266.

8-40   Metal-Ion-Catalyzed σ-Bond Rearrangements

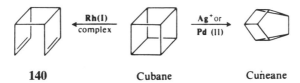

140   Cubane   Cuneane

Many highly strained cage molecules undergo rearrangement when treated with metallic ions such as Ag[+], Rh(I), or Pd(II).[580] The bond rearrangements observed can be formally classified into two main types: (1) 2 + 2 ring openings of cyclobutanes and (2) conversion

Type 1   Type 2

of a bicyclo[2.2.0] system to a bicyclopropyl system. The molecule cubane supplies an example of each type (see above). Treatment with Rh(I) complexes converts cubane to tricyclo[4.2.0.0²·⁵]octa-3,7-diene (**140**),[581] an example of type 1, while Ag[+] or Pd(II) causes the second type of reaction, producing cuneane.[582] Other examples are:

Type 1

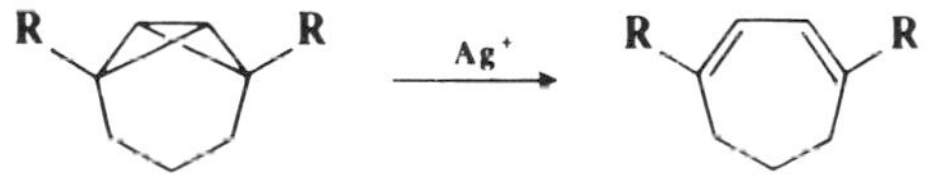

R ⟶ R   Ag[+]   R ⟶ R   Ref. 583

141   **142**

Type 2[584]

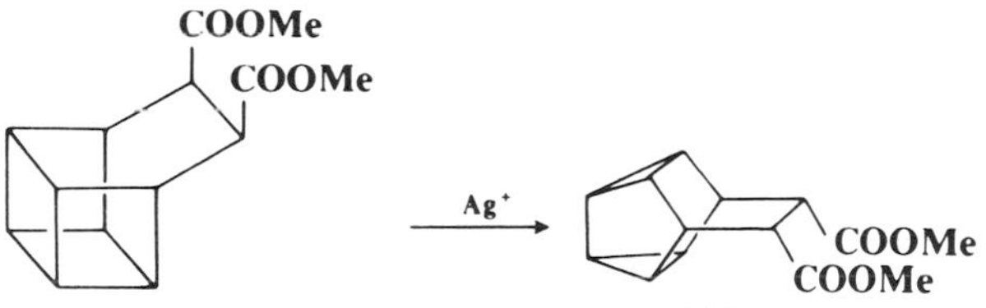

COOMe / COOMe   Ag[+]   COOMe / COOMe   Ref. 585

143

Type 2

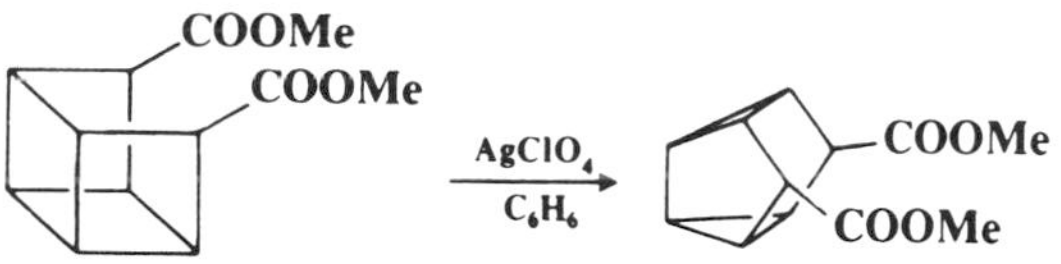

COOMe / COOMe   AgClO₄ / C₆H₆   COOMe / COOMe   Ref. 586

[580]For reviews, see Halpern, in Wender; Pino *Organic Syntheses via Metal Carbonyls*, vol. 2; Wiley: New York, 1977, pp. 705-721; Bishop *Chem. Rev.* **1976**, *76*, 461-486; Cardin; Cetinkaya; Doyle; Lappert *Chem. Soc. Rev.* **1973**, *2*, 99-144, pp. 132-139; Paquette *Synthesis* **1975**, 347-357, *Acc. Chem. Res.* **1971**, *4*, 280-287.
[581]Cassar; Eaton; Halpern *J. Am. Chem. Soc.* **1970**, *92*, 3515; Eaton; Chakraborty *J. Am. Chem. Soc.* **1978**, *100*, 3634.
[582]Cassar; Eaton; Halpern *J. Am. Chem. Soc.* **1970**, *92*, 6336.
[583]Paquette; Allen; Henzel *J. Am. Chem. Soc.* **1970**, *92*, 7002; Gassman; Atkins *J. Am. Chem. Soc.* **1971**, *93*, 4579, **1972**, *94*, 7748; Sakai; Westberg; Yamaguchi; Masamune *J. Am. Chem. Soc.* **1972**, *93*, 4611; Paquette; Wilson; Henzel *J. Am. Chem. Soc.* **1972**, *94*, 7771.
[584]The starting compound here is a derivative of basketane, or 1,8-bishomocubane. For a review of homo-, bis-homo-, and trishomocubanes, see Marchand *Chem. Rev.* **1989**, *89*, 1011-1033.
[585]See, for example, Furstoss; Lehn *Bull. Soc. Chim. Fr.* **1966**, 2497; Paquette; Stowell *J. Am. Chem. Soc.* **1970**, *92*, 2584, **1971**, *93*, 2459; Dauben; Kielbania *J. Am. Chem. Soc.* **1971**, *93*, 7345; Paquette; Beckley; Farnham *J. Am. Chem. Soc.* **1975**, *97*, 1089.
[586]Paquette; Beckley; McCreadie *Tetrahedron Lett.* **1971**, 775; Dauben; Schallhorn; Whalen *J. Am. Chem. Soc.* **1971**, *93*, 1446.

143 is the 9,10-dicarbomethyoxy derivative of *snoutane* (pentacyclo[3.3.2.0^{2,4}.0^{3,7}.0^{6,8}]-decane).

The mechanisms of these reactions are not completely understood, although relief of strain undoubtedly supplies the driving force. The reactions are thermally forbidden by the orbital-symmetry rules, and the role of the catalyst is to provide low-energy pathways so that the reactions can take place. The type 1 reactions are the reverse of the catalyzed 2 + 2 ring closures discussed at **5-49**. The following mechanism, in which Ag$^+$ attacks one of the edge bonds, has been suggested for the conversion of **141** to **142**.[587]

Simpler bicyclobutanes can also be converted to dienes, but in this case the products usually result from cleavage of the central bond and one of the edge bonds.[588] For example, treatment of **144** with AgBF$_4$,[589] (C$_6$F$_5$Cu)$_4$,[590] or [(π-allyl)PdCl]$_2$[591] gives a mixture of the

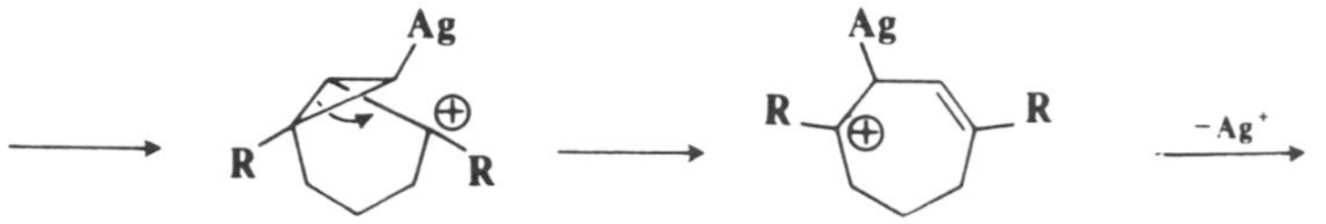

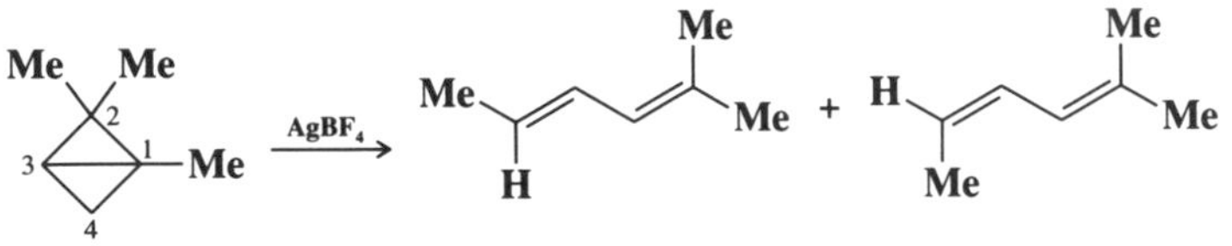

144

two dienes shown, resulting from a formal cleavage of the C$_1$—C$_3$ and C$_1$—C$_2$ bonds (note that a hydride shift has taken place).

8-41   The Di-π-methane and Related Rearrangements
Di-π-methane rearrangement

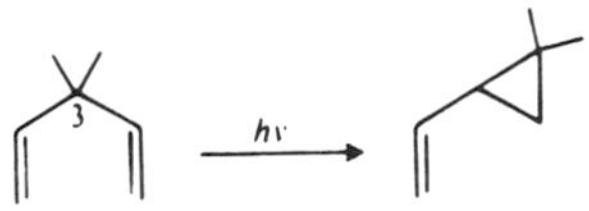

1,4-Dienes carrying alkyl or aryl substituents on C-3[592] can be photochemically rearranged to vinylcyclopropanes in a reaction called the *di-π-methane rearrangement*.[593] An example is conversion of **145** to **146**.[594] For most 1,4-dienes it is only the singlet excited states that

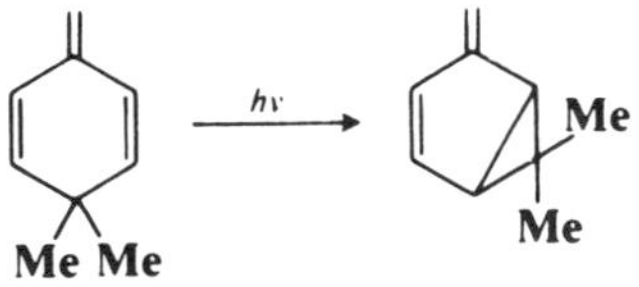

145   **146**

[587]Gassman; Atkins, Ref. 583; Sakai et al., Ref. 583.
[588]**141** can also be cleaved in this manner, giving a 3-methylenecyclohexene. See, for example, Gassman; Atkins *J. Am. Chem. Soc.* **1971**, *93*, 1042; Dauben; Kielbania *J. Am. Chem. Soc.* **1972**, *94*, 3669; Gassman; Reitz *J. Am. Chem. Soc.* **1973**, *95*, 3057; Paquette; Zon *J. Am. Chem. Soc.* **1974**, *96*, 203, 224.
[589]Paquette; Henzel; Wilson *J. Am. Chem. Soc.* **1971**, *93*, 2335.
[590]Gassman; Williams *Tetrahedron Lett.* **1971**, 1409.
[591]Gassman; Meyer; Williams *Chem. Commun.* **1971**, 842.
[592]Zimmerman; Pincock *J. Am. Chem. Soc.* **1973**, *95*, 2957.
[593]For reviews, see Zimmerman *Org. Photochem.* **1991**, *11*, 1-36; Zimmerman, in Mayo, Ref. 1, vol. 3, pp. 131-166; Hixson; Mariano; Zimmerman *Chem. Rev.* **1973**, *73*, 531-551.
[594]Zimmerman; Hackett; Juers; McCall; Schröder *J. Am. Chem. Soc.* **1971**, *93*, 3653.

give the reaction; triplet states generally take other pathways.[595] For unsymmetrical dienes, the reaction is regioselective. For example, **147** gave **148**, not **149**:[596]

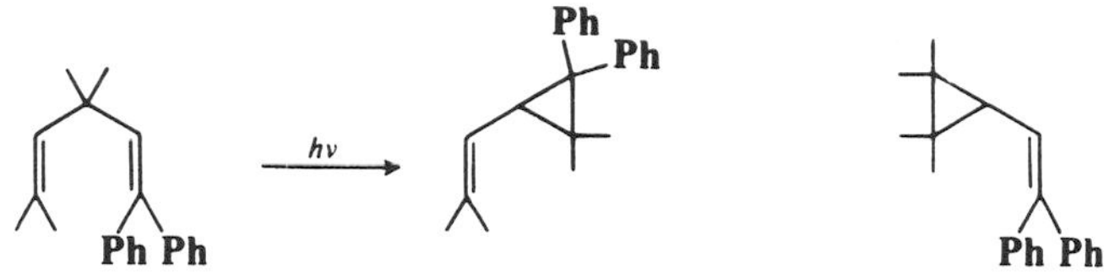

147   **148**   **149**

Not formed

The mechanism can be described by the diradical pathway given[597] (the C-3 substituents act to stabilize the radical), though the species shown are not necessarily intermediates, but

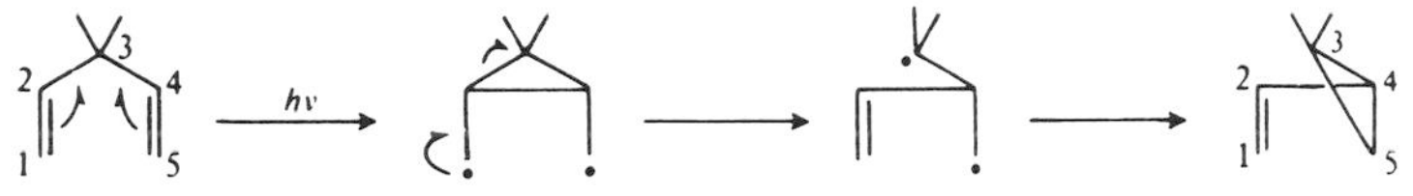

150

may be transition states. It has been shown, for the case of certain substituted substrates, that configuration is retained at C-1 and C-5 and inverted at C-3.[598]

The reaction has been extended to allylic benzenes[599] (in this case C-3 substituents are not required), to β,γ unsaturated ketones[600] (the latter reaction, which is called the *oxa di*

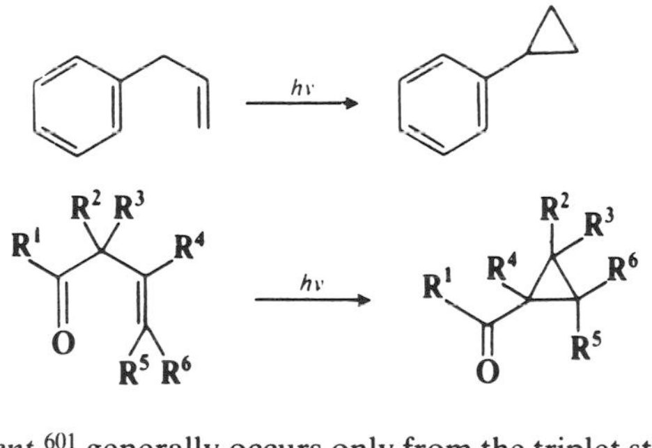

π-methane rearrangement,[601] generally occurs only from the triplet state), to β,γ-unsaturated imines,[602] and to triple-bond systems.[603]

---

[595]However, some substrates, generally rigid bicyclic molecules, (e.g., barrelene, p. 1136, which is converted to semi-bullvalene) give the di-π-methane rearrangement only from triplet states.

[596]Zimmerman; Pratt *J. Am. Chem. Soc.* **1970**, *92*, 6259, 6267; Zimmerman; Baum *J. Am. Chem. Soc.* **1971**, *93*, 3646. See also Zimmerman; Welter *J. Am. Chem. Soc.* **1978**, *100*, 4131; Alexander; Pratt; Rowley; Tipping *J. Chem. Soc., Chem. Commun.* **1978**, 101; Paquette; Bay; Ku; Rondan; Houk *J. Org. Chem.* **1982**, *47*, 422.

[597]See Zimmerman; Werthemann; Kamm *J. Am. Chem. Soc.* **1974**, *96*, 439; Zimmerman; Little *J. Am. Chem. Soc.* **1974**, *96*, 5143; Zimmerman; Boettcher; Buehler; Keck *J. Am. Chem. Soc.* **1975**, *97*, 5635. For an argument against the intermediacy of **150**, see Adam; De Lucchi; Dörr *J. Am. Chem. Soc.* **1989**, *111*, 5209.

[598]Zimmerman; Robbins; McKelvey; Samuel; Sousa *J. Am. Chem. Soc.* **1974**, *96*, 4630.

[599]For example, see Griffin; Covell; Petterson; Dodson; Klose *J. Am. Chem. Soc.* **1965**, *87*, 1410; Hixson *J. Am. Chem. Soc.* **1972**, *94*, 2507; Cookson; Ferreira; Salisbury *J. Chem. Soc., Chem. Commun.* **1974**, 665; Fasel; Hansen *Chimia* **1982**, *36*, 193; Paquette; Bay *J. Am. Chem. Soc.* **1984**, *106*, 6693; Zimmerman; Swafford *J. Org. Chem.* **1984**, *49*, 3069.

[600]For reviews of photochemical rearrangements of unsaturated ketones, see Schuster, in Mayo, Ref. 1, vol. 3, pp. 167-279; Houk *Chem. Rev.* **1976**, *76*, 1-74; Schaffner *Tetrahedron* **1976**, *32*, 641-653; Dauben; Lodder; Ipaktschi *Top. Curr. Chem.* **1975**, *54*, 73-114.

[601]For a review, see Demuth *Org. Photochem.* **1991**, *11*, 37-109.

[602]See Armesto; Horspool; Langa; Ramos *J. Chem. Soc., Perkin Trans. 1* **1991**, 223.

[603]See Griffin; Chihal; Perreten; Bhacca *J. Org. Chem.* **1976**, *41*, 3931.

When photolyzed, 2,5-cyclohexadienones can undergo a number of different reactions, one of which is formally the same as the di-π-methane rearrangement.[604] In this reaction, photolysis of the substrate **151** gives the bicyclo[3.1.0]hexenone **156**. Though the reaction is formally the same (note the conversion of **145** to **146** above), the mechanism is different

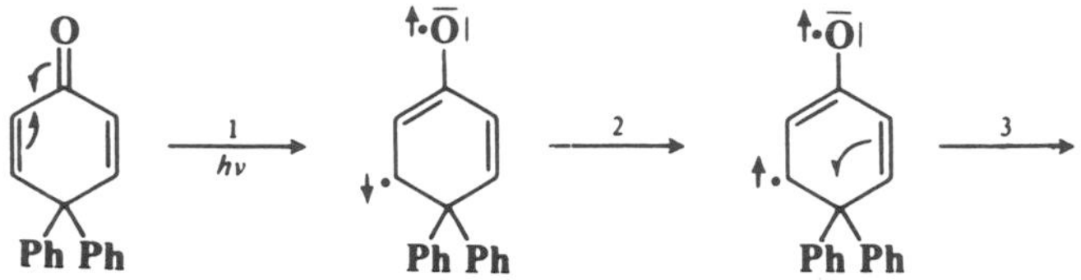

from that of the di-π-methane rearrangement, because irradiation of a ketone can cause an $n \to \pi^*$ transition, which is of course not possible for a diene lacking a carbonyl group. The mechanism[605] in this case has been formulated as proceeding through the excited triplet states **153** and **154**. In step 1, the molecule undergoes an $n \to \pi^*$ excitation to the singlet species **152,** which cross to the triplet **153**. Step 3 is a rearrangement from one excited state to another. Step 4 is a $\pi^* \to n$ electron demotion (an intersystem crossing from $T_1 \to S_0$, see p. 239). The conversion of **155** to **156** consists of two 1,2 alkyl migrations (a one-step process would be a 1,3 migration of alkyl to a carbocation center, see p. 1062): The old C_6—C_5 bond becomes the new C_6—C_4 bond and the old C_6—C_1 bond becomes the new C_6—C_5 bond.[606]

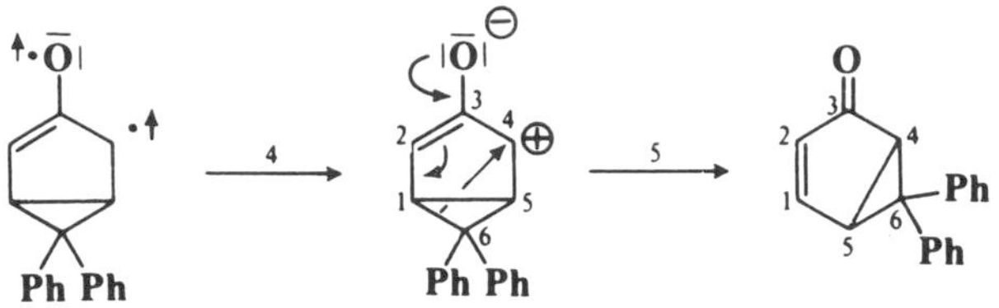

2,4-Cyclohexadienones also undergo photochemical rearrangements, but the products are different, generally involving ring opening.[607]

8-42 The Hofmann–Löffler and Related Reactions

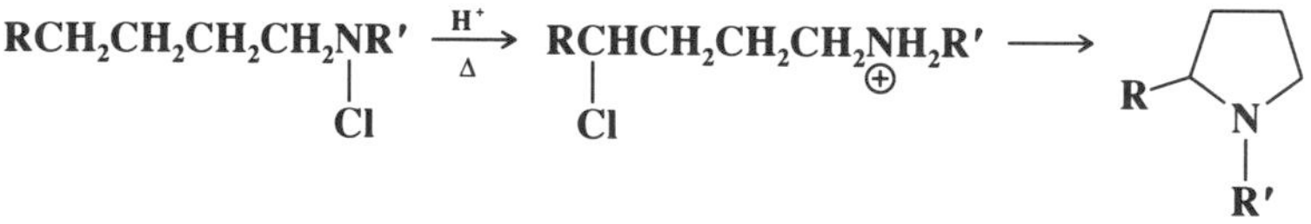

[604]For reviews of the photochemistry of 2,5-cyclohexadienones and related compounds, see Schaffner; Demuth, in Mayo, Ref. 1, vol. 3, pp. 281-348; Zimmerman *Angew. Chem. Int. Ed. Engl.* **1969**, *8*, 1-11 [*Angew. Chem. 81*, 45-55]; Kropp *Org. Photochem.* **1967**, *1*, 1-90; Schaffner *Adv. Photochem.* **1966**, *4*, 81-112. For synthetic use, see Schultz; Lavieri; Macielag; Plummer *J. Am. Chem. Soc.* **1987**, *109*, 3991, and references cited therein.

[605]Zimmerman; Schuster *J. Am. Chem. Soc.* **1961**, *83*, 4486; Schuster; Patel *J. Am. Chem. Soc.* **1968**, *90*, 5145; Schuster *Acc. Chem. Res.* **1978**, *11*, 65-73; Zimmerman; Pasteris *J. Org. Chem.* **1980**, *45*, 4864, 4876; Schuster; Liu *Tetrahedron* **1981**, *37*, 3329.

[606]Zimmerman; Crumine; Döpp; Huyffer *J. Am. Chem. Soc.* **1969**, *91*, 434.

[607]For reviews, see Schaffner; Demuth, Ref. 604; Quinkert *Angew. Chem. Int. Ed. Engl.* **1972**, *11*, 1072-1087 [*Angew. Chem. 84*, 1157-1173]; Kropp, Ref. 604.

A common feature of the reactions in this section[608] is that they serve to introduce functionality at a position remote from functional groups already present. As such, they have proved very useful in synthesizing many compounds, especially in the steroid field (see also **9-2** and **9-16**). When N-haloamines in which one alkyl group has a hydrogen in the 4 or 5 position are heated with sulfuric acid, pyrrolidines or piperidines are formed, in a reaction known as the *Hofmann–Löffler reaction* (also called the *Hofmann–Löffler–Freytag reaction*).[609] R' is normally alkyl, but the reaction has been extended to R' = H by the use of concentrated sulfuric acid solution and ferrous salts.[610] The first step of the reaction is a rearrangement, with the halogen migrating from the nitrogen to the 4 or 5 position of the alkyl group. It is possible to isolate the resulting haloamine salt, but usually this is not done, and the second step, the ring closure (**0-43**), takes place.Though the reaction is most often induced by heat, this is not necessary, and irradiation and chemical initiators (e.g., peroxides) have been used instead. The mechanism is of a free-radical type, with the main step involving an internal hydrogen abstraction.[611]

Initiation

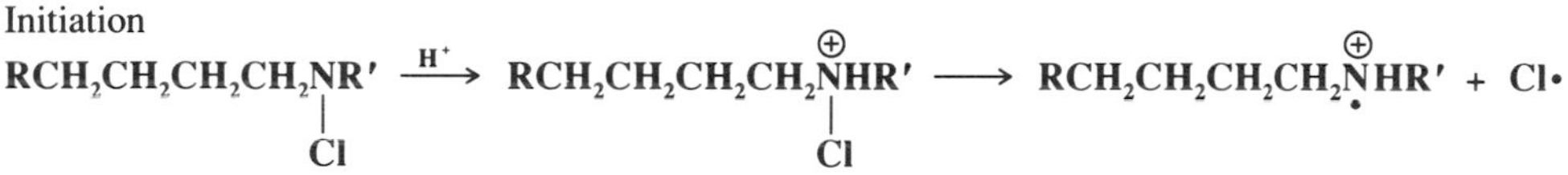

$$RCH_2CH_2CH_2CH_2NR' \xrightarrow{H^+} RCH_2CH_2CH_2CH_2\overset{\oplus}{N}HR' \longrightarrow RCH_2CH_2CH_2CH_2\overset{\oplus}{\underset{\bullet}{N}}HR' + Cl\cdot$$
$$\qquad\qquad | \qquad\qquad\qquad\qquad\qquad | $$
$$\qquad\qquad Cl \qquad\qquad\qquad\qquad\qquad Cl$$

Propagation

$$\underset{\bullet}{R\overset{\oplus}{C}}HCH_2CH_2CH_2\overset{\oplus}{N}H_2R' + RCH_2CH_2CH_2CH_2\overset{\oplus}{N}HR' \longrightarrow$$
$$\qquad\qquad\qquad\qquad\qquad\qquad\qquad\qquad\qquad |$$
$$\qquad\qquad\qquad\qquad\qquad\qquad\qquad\qquad\qquad Cl$$

$$RCHCH_2CH_2CH_2\overset{\oplus}{N}H_2R' + RCH_2CH_2CH_2CH_2\overset{\oplus}{\underset{\bullet}{N}}HR'$$
$$\,|$$
$$Cl$$

etc.

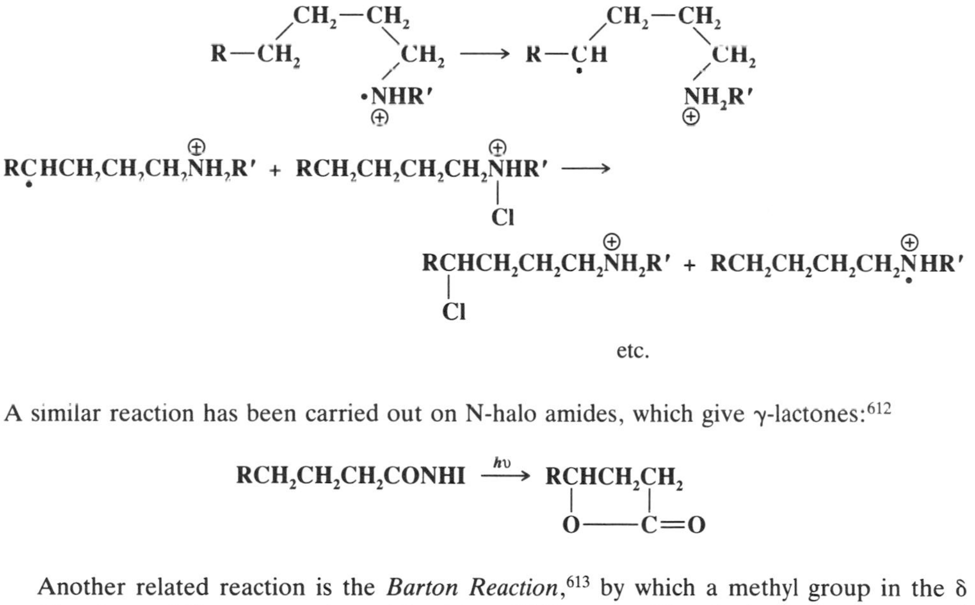

A similar reaction has been carried out on N-halo amides, which give γ-lactones:[612]

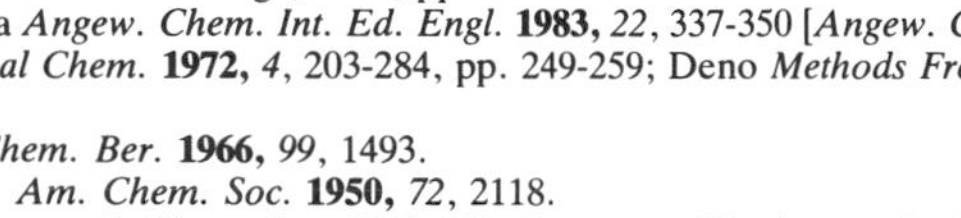

$$RCH_2CH_2CH_2CONHI \xrightarrow{h\upsilon} RCHCH_2CH_2$$
$$\qquad\qquad\qquad\qquad\qquad\qquad | \qquad\quad |$$
$$\qquad\qquad\qquad\qquad\qquad\qquad O\!-\!\!-\!\!-C\!=\!O$$

Another related reaction is the *Barton Reaction*,[613] by which a methyl group in the δ position to an OH group can be oxidized to a CHO group. The alcohol is first converted

[608]For a review of the reactions in this section, see Carruthers *Some Modern Methods of Organic Synthesis*, 3rd ed.; Cambridge University Press: Cambridge, 1986, pp. 263-279.

[609]For reviews, see Stella *Angew. Chem. Int. Ed. Engl.* **1983**, *22*, 337-350 [*Angew. Chem. 95*, 368-380]; Sosnovsky; Rawlinson *Adv. Free-Radical Chem.* **1972**, *4*, 203-284, pp. 249-259; Deno *Methods Free-Radical Chem.* **1972**, *3*, 135-154, pp. 136-143.

[610]Schmitz; Murawski *Chem. Ber.* **1966**, *99*, 1493.

[611]Wawzonek; Thelan *J. Am. Chem. Soc.* **1950**, *72*, 2118.

[612]Barton; Beckwith; Goosen *J. Chem. Soc.* **1965**, 181; Petterson; Wambsgans *J. Am. Chem. Soc.* **1964**, *86*, 1648; Neale; Marcus; Schepers *J. Am. Chem. Soc.* **1966**, *88*, 3051. For a review of N-halo amide rearrangements, see Neale *Synthesis* **1971**, 1-15.

[613]For reviews, see Hesse *Adv. Free-Radical Chem.* **1969**, *3*, 83-137; Barton *Pure Appl. Chem.* **1968**, *16*, 1-15.

to the nitrite ester. Photolysis of the nitrite results in conversion of the nitrite group to the OH group and nitrosation of the methyl group. Hydrolysis of the oxime tautomer gives the aldehyde, e.g.,[614]

This reaction takes place only when the methyl group is in a favorable steric position.[615] The mechanism is similar to that of the Hofmann–Löffler reaction.[616]

This is one of the few known methods for effecting substitution at an angular methyl group. Not only CH_3 groups but also alkyl groups of the form RCH_2 and R_2CH can give the Barton reaction if the geometry of the system is favorable. An RCH_2 group is converted to the oxime $R(C=NOH)$ (which is hydrolyzable to a ketone) or to a nitroso dimer, while an R_2CH group gives a nitroso compound $R_2C(NO)$. With very few exceptions, the only carbons that become nitrosated are those in the position δ to the original OH group, indicating that a six-membered transition state is necessary for the hydrogen abstraction.[617]

OS **III**, 159.

D. Noncyclic Rearrangements

8-43   Hydride Shifts

[614]Barton; Beaton *J. Am. Chem. Soc.* **1961**, *83*, 4083. Also see Barton; Beaton; Geller; Pecket *J. Am. Chem. Soc.* **1960**, *82*, 2640.

[615]For a discussion of which positions are favorable, see Burke; Silks; Strickland *Tetrahedron Lett.* **1988**, *29*, 2761.

[616]Kabasakalian; Townley *J. Am. Chem. Soc.* **1962**, *84*, 2711; Akhtar; Barton; Sammes *J. Am. Chem. Soc.* **1965**, *87*, 4601. See also Nickon; Ferguson; Bosch; Iwadare *J. Am. Chem. Soc.* **1977**, *99*, 4518; Barton; Hesse; Pechet; Smith *J. Chem. Soc., Perkin Trans. 1* **1979**, 1159; Green; Boyle; Vairamani; Mukhopadhyay; Saunders; Bowen; Allinger *J. Am. Chem. Soc.* **1986**, *108*, 2381.

[617]For a discussion, see Nickon et al., Ref. 616.

The above is a typical example of a transannular hydride shift. The 1,2-diol is formed by a normal epoxide hydrolysis reaction (**0-7**). For a discussion of 1,3 and longer hydride shifts, see p. 1062.

8-44 The Chapman Rearrangement
         1/O→3/N-Aryl-migration

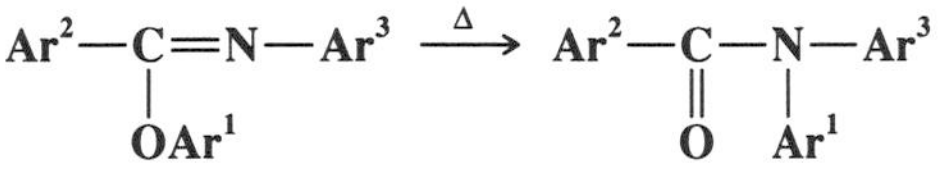

$$Ar^2-C=N-Ar^3 \xrightarrow{\Delta} Ar^2-C-N-Ar^3$$
$$\quad\quad | \quad\quad\quad\quad\quad\quad || \quad |$$
$$\quad OAr^1 \quad\quad\quad\quad\quad O \quad Ar^1$$

In the *Chapman rearrangement*, N,N-diaryl amides are formed when aryl imino esters are heated.[618] Best yields are obtained in refluxing tetraethylene glycol dimethyl ether (tetraglyme),[619] though the reaction can also be carried out without any solvent at all. Many groups may be present in the rings, e.g., alkyl, halo, OR, CN, COOR, etc. Aryl migrates best when it contains electron-withdrawing groups. On the other hand, electron-withdrawing groups in Ar^2 or Ar^3 decrease the reactivity. The products can be hydrolyzed to diarylamines, and this is a method for preparing these compounds. The mechanism probably involves an intramolecular[620] aromatic nucleophilic substitution, resulting in a 1,3 oxygen-to-nitrogen

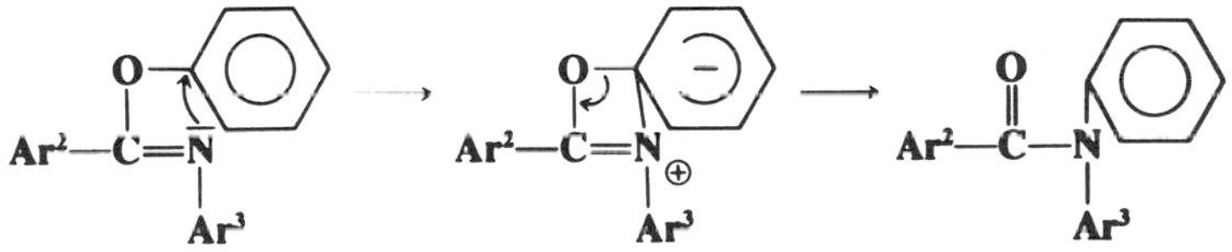

shift. Aryl imino esters can be prepared from N-aryl amides by reaction with PCl_5, followed by treatment of the resulting imino chloride with an aroxide ion.[621] Imino esters with any

$$Ar^2-C-NH-Ar^3 + PCl_5 \longrightarrow Ar^2-C=N-Ar^3 \xrightarrow{Ar^1O^-} Ar^2-C=N-Ar^3$$
$$\quad\quad ||\quad\quad\quad\quad\quad\quad\quad\quad\quad\quad | \quad\quad\quad\quad\quad\quad\quad\quad\quad |$$
$$\quad\quad O \quad\quad\quad\quad\quad\quad\quad\quad\quad\quad Cl \quad\quad\quad\quad\quad\quad\quad\quad OAr^1$$

or all of the three groups being alkyl also rearrange, but they require catalysis by H_2SO_4 or a trace of methyl iodide or methyl sulfate.[622] The mechanism is different, involving an intermolecular process.[623] This is also true for derivatives for formamide (Ar^2 = H).

8-45 The Wallach Rearrangement

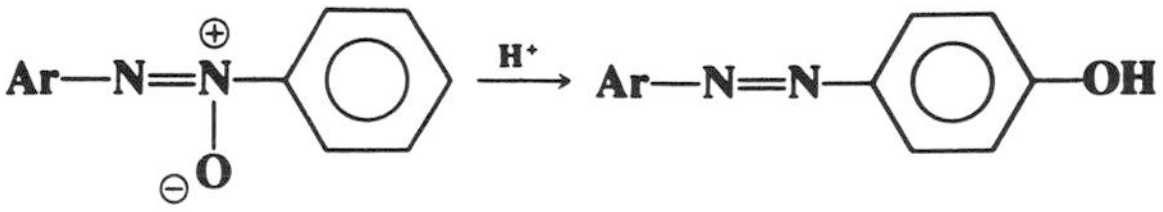

$$Ar-N=\overset{\oplus}{N}-\bigcirc \xrightarrow{H^+} Ar-N=N-\bigcirc-OH$$
$$\quad\quad\quad\quad |$$
$$\quad\quad\quad\ominus O$$

         157   **158**

[618]For reviews, see Schulenberg; Archer *Org. React.* **1965**, *14*, 1-51; McCarty, in Patai, Ref. 237, pp. 439-447; McCarty; Garner, in Patai *The Chemistry of Amidines and Imidates*; Wiley: New York, 1975, pp. 189-240. For a review of 1,3 migrations of R in general, see Landis *Mech. Mol. Migr.* **1969**, *2*, 43-63.
[619]Wheeler; Roman; Santiago; Quiles *Can. J. Chem.* **1969**, *47*, 503.
[620]For evidence for the intramolecular character of the reaction, see Wiberg; Rowland *J. Am. Chem. Soc.* **1955**, 77, 2205; Wheeler; Roman; Rosado *J. Org. Chem.* **1969**, *34*, 966; Kimura *J. Chem. Soc., Perkin Trans. 2* **1987**, 205.
[621]For a review of the formation and reactions of imino chlorides, see Bonnett, in Patai, Ref. 237, pp. 597-662.
[622]Landis, Ref. 618.
[623]See Challis; Frenkel *J. Chem. Soc., Perkin Trans. 2* **1978**, 192.

The conversion of azoxy compounds, on acid treatment, to *p*-hydroxy azo compounds (or sometimes the *o*-hydroxy isomers[624]) is called the *Wallach rearrangement*.[625] When both para positions are occupied, the *o*-hydroxy product may be obtained, but ipso substitution at one of the para positions is also possible.[626] Although the mechanism[627] is not completely settled, the following facts are known: (1) The para rearrangement is intermolecular.[628] (2) When the reaction was carried out with an azoxy compound in which the N—O nitrogen was labeled with ^{15}N, *both* nitrogens of the product carried the label equally,[629] demonstrating that the oxygen did not have a preference for migration to either the near or the far ring. This shows that there is a symmetrical intermediate. (3) Kinetic studies show that two protons are normally required for the reaction.[630] The following mechanism,[631] involving the symmetrical intermediate **160**, has been proposed to explain the facts.[632]

$$157 \; \underset{}{\overset{H^+}{\rightleftharpoons}} \; \underset{\underset{\underset{\underset{159}{A-H}}{OH}}{|}}{Ar-N=\overset{\oplus}{N}-Ar} \; \longrightarrow \; Ar-\overset{\oplus}{N}\equiv\overset{\oplus}{N}-Ar \; \underset{\text{or } H_2SO_4}{\overset{H_2O}{\longrightarrow}} \; 158$$

$$160$$

It has proved possible to obtain **159** and **160** as stable species in super-acid solutions.[550] Another mechanism, involving an intermediate with only one positive charge, has been proposed for certain substrates at low acidities.[633]

A photochemical Wallach rearrangement[634] is also known: The product is the *o*-hydroxy azo compound, the OH group is found in the farther ring, and the rearrangement is intramolecular.[635]

8-46 Dyotropic Rearrangements
   1/*C*-Trialkylsilyl,2/*O*-trialkylsilyl-interchange

$$R^1-\underset{\underset{O-SiR_3^4}{|}}{\overset{\overset{R^2}{|}}{C}}-SiR_3^3 \; \rightleftharpoons \; R^1-\underset{\underset{O-SiR_3^3}{|}}{\overset{\overset{R^2}{|}}{C}}-SiR_3^4$$

[624]For example, see Dolenko; Buncel *Can. J. Chem.* **1974**, *52*, 623; Yamamoto; Nishigaki; Umezu; Matsuura *Tetrahedron* **1980**, *36*, 3177.
[625]For reviews, see Buncel *Mech. Mol. Migr.* **1968**, 1, 61-119; Shine *Aromatic Rearrangements*, Ref. 489, pp. 272-284, 357-359; Cox; Buncel, Ref. 541, pp. 808-837.
[626]See, for example, Shimao; Oae *Bull. Chem. Soc. Jpn.* **1983**, *56*, 643.
[627]For reviews, see Furin *Russ. Chem. Rev.* **1987**, *56*, 532-545; Williams; Buncel *Isot. Org. Chem.* **1980**, *5*, 184-197; Buncel *Acc. Chem. Res.* **1975**, *8*, 132-139.
[628]See, for example, Oae; Fukumoto; Yamagami *Bull. Chem. Soc. Jpn.* **1963**, *36*, 601.
[629]Shemyakin; Maimind; Vaichunaite *Chem. Ind.* (*London*) **1958**, 755, *Bull. Acad. Sci. USSR, Div. Chem. Sci.* **1960**, 808. Also see Behr; Hendley *J. Org. Chem.* **1966**, *31*, 2715.
[630]Buncel; Lawton *Chem. Ind.* (*London*) **1963**, 1835; Hahn; Lee; Jaffé *J. Am. Chem. Soc.* **1967**, *89*, 4975; Cox *J. Am. Chem. Soc.* **1974**, *96*, 1059.
[631]Buncel; Lawton *Can. J. Chem.* **1965**, *43*, 862; Buncel; Strachan *Can. J. Chem.* **1970**, *48*, 377; Cox, Ref. 630; Buncel; Keum *J. Chem. Soc., Chem. Commun.* **1983**, 578.
[632]For other proposed mechanisms, see Shemyakin; Maimind; Agadzhanyan *Chem. Ind.* (*London*) **1961**, 1223; Shemyakin; Agadzhanyan; Maimind; Kudryavtsev *Bull. Acad. Sci. USSR, Div. Chem. Sci.* **1963**, 1216; Hahn; Lee; Jaffé, Ref. 630; Duffey; Hendley *J. Org. Chem.* **1968**, *33*, 1918; Hendley; Duffey *J. Org. Chem.* **1970**, *35*, 3579.
[633]Cox; Dolenko; Buncel *J. Chem. Soc., Perkin Trans. 2* **1975**, 471; Cox; Buncel *J. Am. Chem. Soc.* **1975**, *97*, 1871.
[634]For a thermal rearrangement (no catalyst), see Shimao; Hashidzume *Bull. Chem. Soc. Jpn.* **1976**, *49*, 754.
[635]For discussions of the mechanism of the photochemical reaction, see Goon; Murray; Schoch; Bunce *Can. J. Chem.* **1973**, *51*, 3827; Squire; Jaffé *J. Am. Chem. Soc.* **1973**, *95*, 8188; Shine; Subotkowski; Gruszecka *Can. J. Chem.* **1986**, *64*, 1108.

A *dyotropic rearrangement*[636] is an uncatalyzed process in which two σ bonds simultaneously migrate intramolecularly.[637] There are two types. The above is an example of Type 1, which consists of reactions in which the two σ bonds interchange positions. In Type 2, the two σ bonds do not interchange positions. An example is

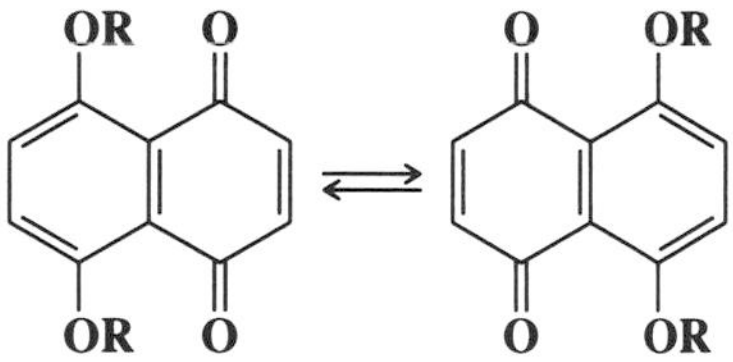

Some other examples are

Type 1

Nitrile oxide   Isocyanate

Ref. 638

Type 1

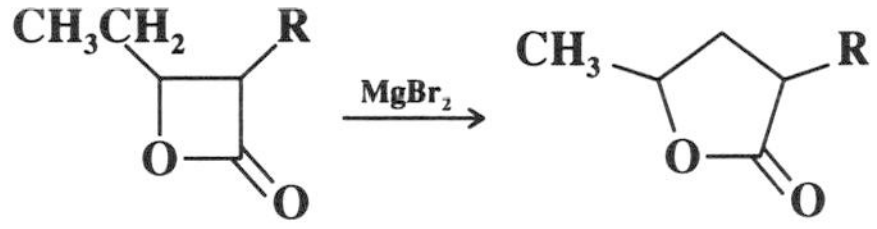

Ref. 639

Type 2

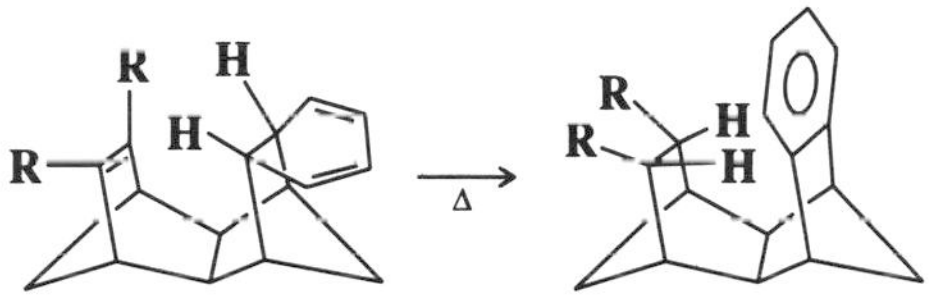

Ref. 640

[636]Reetz *Angew. Chem. Int. Ed. Engl.* **1972,** *11,* 129, 130 [*Angew. Chem. 84,* 161, 163].

[637]For reviews, see Minkin; Olekhnovich; Zhdanov *Molecular Design of Tautomeric Compounds*; D. Reidel Publishing Co.: Dordrecht, 1988, pp. 221-246; Minkin *Sov. Sci. Rev., Sect. B* **1985,** *7,* 51-98; Reetz *Adv. Organomet. Chem.* **1977,** *16,* 33-65.

[638]See, for example, Taylor *J. Chem. Soc., Perkin Trans. 1* **1985,** 1181.

[639]See Black; Hall; Sheu *J. Org. Chem.* **1988,** *53,* 2371; Black; Fields *Synth. Commun.* **1988,** *18,* 125.

[640]See Mackenzie; Proctor; Woodnutt *Tetrahedron* **1987,** *43,* 5981, and references cited therein.

19

OXIDATIONS AND REDUCTIONS

First we must examine what we mean when we speak of oxidation and reduction. Inorganic chemists define oxidation in two ways: loss of electrons and increase in oxidation number. In organic chemistry, these definitions, while still technically correct, are not easy to apply. While electrons are directly transferred in some organic oxidations and reductions, the mechanisms of most of these reactions do not involve a direct electron transfer. As for oxidation number, while this is easy to apply in some cases, e.g., the oxidation number of carbon in CH_4 is -4, in most cases attempts to apply the concept lead to fractional values or to apparent absurdities. Thus carbon in propane has an oxidation number of -2.67 and in butane of -2.5, though organic chemists seldom think of these two compounds as being in different oxidation states. An improvement could be made by assigning different oxidation states to different carbon atoms in a molecule, depending on what is bonded to them (e.g., the two carbons in acetic acid are obviously in different oxidation states), but for this a whole set of arbitrary assumptions would be required, since the oxidation number of an atom in a molecule is assigned on the basis of the oxidation numbers of the atoms attached to it. There would seem little to be gained by such a procedure. The practice in organic chemistry has been to set up a series of functional groups, in a qualitative way, arranged in order of increasing oxidation state, and then to define oxidation as *the conversion of a functional group in a molecule from one category to a higher one*. Reduction is the opposite. For the simple functional groups this series is shown in Table 19.1.[1] It should be noted that this classification applies only to a single carbon atom or to two adjacent carbon atoms. Thus 1,3-dichloropropane is in the same oxidation state as chloromethane, but 1,2-dichloropropane is in a higher one. Obviously, such distinctions are somewhat arbitrary, and if we attempt to carry them too far, we shall find ourselves painted into a corner. Nevertheless, the basic idea has served organic chemistry well. It should be noted that conversion of any compound to another in the same category is not an oxidation or a reduction. Most oxidations in organic chemistry involve a gain of oxygen and/or a loss of hydrogen (Lavoisier's original definition of oxidation). The reverse is true for reductions.

Of course, there is no oxidation without a concurrent reduction. However, we classify reactions as oxidations or reductions depending on whether the *organic compound* is oxidized or reduced. In some cases both the oxidant and reductant are organic; those reactions are treated separately at the end of the chapter.

MECHANISMS

It must be noted that our definition of oxidation has nothing to do with mechanism. Thus the conversion of bromomethane to methanol with KOH (**0-1**) and to methane with $LiAlH_4$ (**0-76**) have the same S_N2 mechanisms, but one is a reduction (according to our definition)

---

[1]For more extensive tables, with subclassifications, see Soloveichik; Krakauer *J. Chem. Educ.* **1966**, *43*, 532-535.

TABLE 19.1   Categories ot simple functional groups arranged according to oxidation state

Oxidation is the conversion of a functional group in a molecule to a higher category; reduction is conversion to a lower one. Conversions within a category are neither oxidations nor reductions. The numbers given at the bottom are only approximations

RH	$-\overset{|}{C}=\overset{|}{C}-$	$-C\equiv C-$	$R-\underset{\overset{\|}{O}}{C}-OH$	CO_2
	ROH	$R-\underset{\overset{\|}{O}}{C}-R$		CCl_4
	RCl		$R-\underset{\overset{\|}{O}}{C}-NH_2$	
	RNH_2			
	etc.	$-\overset{|}{\underset{Cl}{C}}-Cl$	$-\overset{Cl}{\underset{Cl}{C}}-Cl$	
		$-\overset{|}{\underset{Cl}{C}}-\overset{|}{\underset{Cl}{C}}-$	etc.	
		$-\overset{|}{\underset{OH}{C}}-\overset{|}{\underset{OH}{C}}-$		
		etc.		

Approximate oxidation number

-4	-2	0	$+2$	$+4$

and the other is not. It is impractical to consider the mechanisms of oxidation and reduction reactions in broad categories in this chapter as we have done for the reactions considered in Chapters 10 to 18.[2] The main reason is that the mechanisms are too diverse, and this in turn is because the bond changes are too different. For example, in Chapter 15, all the reactions involved the bond change C=C → W—C—C—Y and a relatively few mechanisms covered all the reactions. But for oxidations and reductions the bond changes are far more diverse. Another reason is that the mechanism of a given oxidation or reduction reaction can vary greatly with the oxidizing or reducing agent employed. Very often the mechanism has been studied intensively for only one or a few of many possible agents.

Though we therefore do not cover oxidation and reduction mechanisms in the same way as we have covered other mechanisms, it is still possible to list a few broad mechanistic categories. In doing this, we follow the scheme of Wiberg.[3]

1. *Direct electron transfer.*[4]   We have already met some reactions in which the reduction is a direct gain of electrons or the oxidation a direct loss of them. An example is the Birch reduction (**5-10**), where sodium directly transfers an electron to an aromatic ring. An example from this chapter is found in the bimolecular reduction of ketones (**9-62**), where again it is

---

[2]For monographs on oxidation mechanisms, see Bamford; Tipper *Comprehensive Chemical Kinetics*, vol. 16; Elsevier: New York, 1980; *Oxidation in Organic Chemistry*; Academic Press: New York, pt. A [Wiberg], 1965, pts. B. C. and D [Trahanovsky], 1973, 1978, 1982; Waters *Mechanisms of Oxidation of Organic Compounds*; Wiley: New York, 1964; Stewart *Oxidation Mechanisms*; W. A. Benjamin: New York, 1964. For a review, see Stewart *Isot. Org. Chem.* **1976**, *2*, 271-310.

[3]Wiberg *Surv. Prog. Chem.* **1963**, *1*, 211-248.

[4]For a monograph on direct electron-transfer mechanisms, see Eberson *Electron Transfer Reactions in Organic Chemistry*; Springer: New York, 1987. For a review, see Eberson *Adv. Phys. Org. Chem.* **1982**, *18*, 79-185. For a review of multistage electron-transfer mechanisms, see Deuchert; Hünig *Angew. Chem. Int. Ed. Engl.* **1978**, *17*, 875-886 [*Angew. Chem. 90*, 927-938].

a metal that supplies the electrons. This kind of mechanism is found largely in three types of reaction:[5] (a) the oxidation or reduction of a free radical (oxidation to a positive or reduction to a negative ion), (b) the oxidation of a negative ion or the reduction of a positive ion to a comparatively stable free radical, and (c) electrolytic oxidations or reductions (an example is the Kolbe reaction, **4-38**). An important example of (b) is oxidation of amines and phenolate ions:

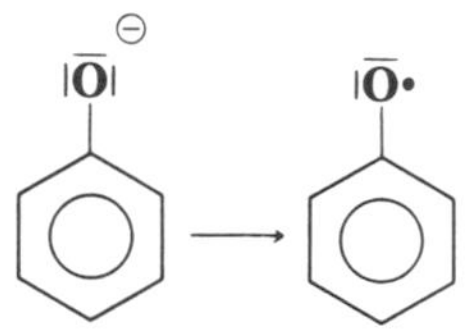

$$
\begin{array}{ccc}
|\overset{\ominus}{\underline{\underline{O}}}| & & |\overline{\underline{O}}\cdot \\
& \longrightarrow &
\end{array}
$$

These reactions occur easily because of the relative stability of the radicals involved.[6] The single electron transfer mechanism (SET), which we have met several times (e.g., p. 307) is an important case.

2. *Hydride transfer.*[7] In some reactions a hydride ion is transferred to or from the substrate. The reduction of epoxides with $LiAlH_4$ is an example (**0-80**). Another is the Cannizzaro reaction (**9-69**). Reactions in which a carbocation abstracts a hydride ion belong in this category:[8]

$$R^+ + R'H \longrightarrow RH + R'^+$$

3. *Hydrogen-atom transfer.* Many oxidation and reduction reactions are free-radical substitutions and involve the transfer of a hydrogen atom. For example, one of the two main propagation steps of **4-1** involves abstraction of hydrogen:

$$RH + Cl\cdot \longrightarrow R\cdot + HCl$$

This is the case for many of the reactions of Chapter 14.

4. *Formation of ester intermediates.* A number of oxidations involve the formation of an ester intermediate (usually of an inorganic acid), and then the cleavage of this intermediate:

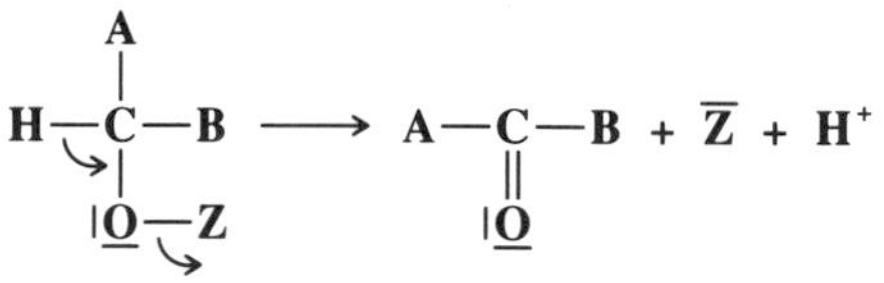

$$
\begin{array}{c}
\overset{\displaystyle A}{\underset{\displaystyle |\underline{O}-Z}{H-C-B}} \longrightarrow \overset{\displaystyle A}{\underset{\displaystyle |\underline{O}}{A-C-B}} + \overline{Z} + H^+
\end{array}
$$

Z is usually CrO_3H, MnO_3, or a similar inorganic acid moiety. One example of this mechanism was seen in **4-6**, where A was an alkyl or aryl group, B was OH, and Z was CrO_3H.

[5]Littler; Sayce *J. Chem. Soc.* **1964**, 2545.

[6]For a review of the oxidation of phenols, see Mihailović; Čeković, in Patai *The Chemistry of the Hydroxyl Group*, pt. 1; Wiley: New York, 1971, pp. 505-592.

[7]For a review, see Watt *Adv. Phys. Org. Chem.* **1988**, *24*, 57-112.

[8]For a review of these reactions, see Nenitzescu, in Olah; Schleyer *Carbonium Ions*, vol. 2; Wiley: New York, 1970, pp. 463-520.

Another is the oxidation of a secondary alcohol to a ketone (**9-3**), where A and B are alkyl or aryl groups and Z is also CrO_3H. In the lead tetraacetate oxidation of glycols (**9-7**) the mechanism also follows this pattern, but the positive leaving group is carbon instead of hydrogen. It should be noted that the cleavage shown is an example of an E2 elimination.

5. *Displacement mechanisms.*   In these reactions the organic substrate uses its electrons to cause displacement on an electrophilic oxidizing agent. One example is the addition of bromine to an olefin (**5-26**).

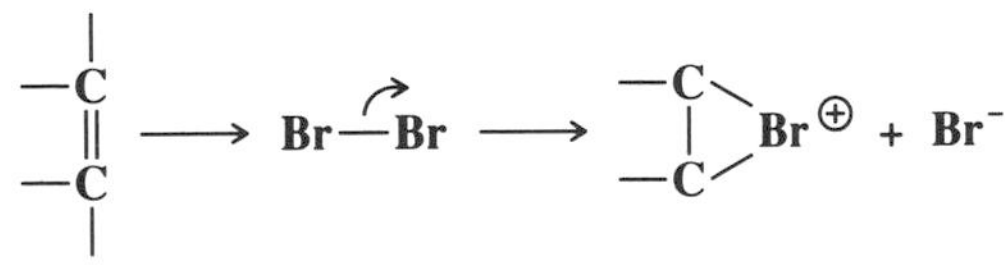

An example from this chapter is found in **9-28**:

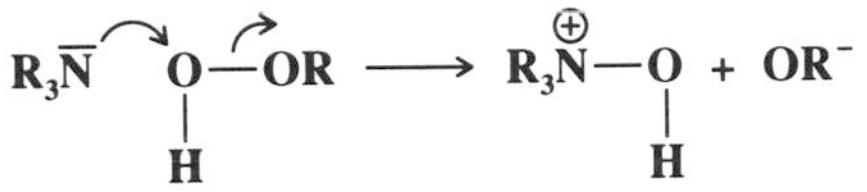

6. *Addition–elimination mechanisms.*   In the reaction between α,β-unsaturated ketones and alkaline peroxide (**5-36**), the oxidizing agent adds to the substrate and then part of it is lost:

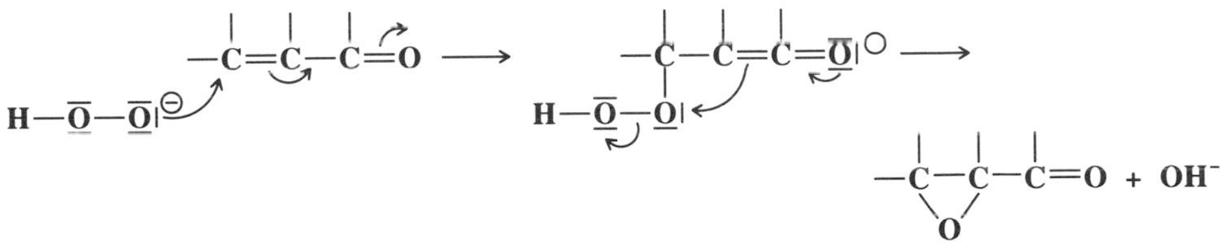

In this case the oxygen of the oxidizing agent is in oxidation state -1 and the OH^- departs with its oxygen in the -2 state, so it is reduced and the substrate oxidized. There are several reactions that follow this pattern of addition of an oxidizing agent and the loss of part of the agent, usually in a different oxidation state. Another example is the oxidation of ketones with SeO_2 (**9-16**). This reaction is also an example of category 4, since it involves formation and E2 cleavage of an ester. This example shows that these six categories are not mutually exclusive.

REACTIONS

In this chapter, the reactions are classified by the type of bond change occurring to the organic substrate, in conformity with our practice in the other chapters.[9] This means that there is no discussion in any one place of the use of a particular oxidizing or reducing agent, e.g., acid dichromate or $LiAlH_4$ (except for a discussion of selectivity of reducing agents, p. 1206). Some oxidizing or reducing agents are fairly specific in their action, attacking only

---

[9]For a table of oxidation and reduction reactions, and the oxidizing and reducing agents for each, see Hudlicky *J. Chem. Educ.* **1977**, *54*, 100-106.

one or a few types of substrate. Others, like acid dichromate, permanganate, $LiAlH_4$, and catalytic hydrogenation, are much more versatile.[10]

When an oxidation or a reduction could be considered in a previous chapter, this was done. For example, the catalytic hydrogenation of olefins is a reduction, but it is also an addition to the C=C bond and was treated in Chapter 15. In this chapter are discussed only those reactions that do not fit into the nine categories of Chapters 10 to 18. An exception to this rule was made for reactions that involve elimination of hydrogen (**9-1** to **9-6**) which were not treated in Chapter 17 because the mechanisms generally differ from those in that chapter.

Oxidations[11]

The reactions in this section are classified into groups depending on the type of bond change involved. These groups are: (A) eliminations of hydrogen, (B) reactions involving cleavage of carbon–carbon bonds, (C) reactions involving replacement of hydrogen by oxygen, (D) reactions in which oxygen is added to the substrate, and (E) oxidative coupling.

A. Eliminations of Hydrogen

9-1 Aromatization of Six-Membered Rings
Hexahydro-terelimination

$$\text{(cyclohexane)} \xrightarrow{\text{Pt}} \text{(benzene)} + 3H_2$$

[10]For books on certain oxidizing agents, see Mijs; de Jonge *Organic Synthesis by Oxidation with Metal Compounds*; Plenum: New York, 1986; Cainelli; Cardillo *Chromium Oxidations in Organic Chemistry*; Springer: New York, 1984; Arndt *Manganese Compounds as Oxidizing Agents in Organic Chemistry*; Open Court Publishing Company: La Salle, IL, 1981; Lee *The Oxidation of Organic Compounds by Permanganate Ion and Hexavalent Chromium*; Open Court Publishing Company: La Salle, IL, 1980. For some reviews, see Curci *Adv. Oxygenated Processes* **1990**, *2*, 1-59 (dioxiranes); Adam; Curci; Edwards *Acc. Chem. Res.* **1989**, *22*, 205-211 (dioxiranes); Murray *Chem. Rev.* **1989**, *89*, 1187-1201, *Mol. Struc. Energ.* **1988**, *5*, 311-351 (dioxiranes); Kafafi; Martinez; Herron *Mol. Struc. Energ.* **1988**, *5*, 283-310 (dioxiranes); Krief; Hevesi *Organoselenium Chemistry I*; Springer: New York, 1988, pp. 76-103 (seleninic anhydrides and acids); Ley, in Liotta *Organoselenium Chemistry*; Wiley: New York, 1987, pp. 163-206 (seleninic anhydrides and acids); Barton; Finet *Pure Appl. Chem.* **1987**, *59*, 937-946 [bismuth(V)]; Fatiadi *Synthesis* **1987**, 85-127 (KMnO₄); Rubottom, in Trahanovsky, Ref. 2, pt. D, 1982, pp. 1-145 (lead tetraacetate); Fatiadi, in Pizey *Synthetic Reagents*, vol. 4; Wiley: New York, 1981, pp. 147-335, *Synthesis* **1974**, 229-272 (HIO₄); Fatiadi *Synthesis* **1976**, 65-104, 133-167 (MnO₂); Ogata; in Trahanovsky, Ref. 2, pt. C, pp. 295-342, 1978 (nitric acid and nitrogen oxides); McKillop, *Pure Appl. Chem.* **1975**, *43*, 463-479 (thallium nitrate); Pizey *Synthetic Reagents*, vol. 2; Wiley: New York, 1974, pp. 143-174 (MnO₂); George; Balachandran *Chem. Rev.* **1975**, *75*, 491-519 (nickel peroxide); Courtney; Swansborough *Rev. Pure Appl. Chem.* **1972**, *22*, 47-54 (ruthenium tetroxide); Ho *Synthesis* **1973**, 347-354 (ceric ion); Aylward *Q. Rev., Chem. Soc.* **1971**, *25*, 407-429 (lead tetraacetate); Meth-Cohn; Suschitzky *Chem. Ind.* (*London*) **1969**, 443-450 (MnO₂); Sklarz *Q. Rev. Chem. Soc.* **1967**, *21*, 3-28 (HIO₄); Korshunov; Vereshchagin *Russ. Chem. Rev.* **1966**, *35*, 942-957(MnO₂); Weinberg; Weinberg *Chem. Rev.* **1968**, *68*, 449-523 (electrochemical oxidation). For reviews of the behavior of certain reducing agents, see Keefer; Lunn *Chem. Rev.* **1989**, *89*, 459-502 (Ni–Al alloy); Málek *Org. React.* **1988**, *36*, 249-590, **1985**, *34*, 1-317 (metal alkoxyaluminum hydrides); Alpatova; Zabusova; Tomilov *Russ. Chem. Rev.* **1986**, *55*, 99-112 (solvated electrons generated electrochemically); Caubère *Angew. Chem. Int. Ed. Engl.* **1983**, *22*, 599-613 [*Angew. Chem. 95*, 597-611] (modified sodium hydride); Nagai *Org. Prep. Proced. Int.* **1980**, *12*, 13-48 (hydrosilanes); Pizey *Synthetic Reagents*, vol. 1; Wiley: New York, 1974, pp. 101-294 (LiAlH₄); Winterfeldt *Synthesis* **1975**, 617-630 (diisobutylaluminum hydride and triisobutylaluminum); Hückel *Fortschr. Chem. Forsch.* **1966**, *6*, 197-250 (metals in ammonia or amines). For books on reductions with metal hydrides, see Seyden-Penne *Reductions by the Alumino- and Borohydrides*; VCH: New York, 1991; Štrouf; Čásenský; Kubánek *Sodium Dihydrido-bis(2-methoxyethoxo)aluminate (SDMA)*; Elsevier: New York, 1985; Hajós *Complex Hydrides*; Elsevier: New York, 1979. Also see House *Modern Synthetic Reactions*, 2nd ed.; W. A. Benjamin: New York, 1972; Refs. 9 and 11.

[11]For books on oxidation reactions, see Hudlický *Oxidations in Organic Chemistry*; American Chemical Society: Washington, 1990; Haines *Methods for the Oxidation of Organic Compounds*, 2 vols.; Academic Press: New York, 1985, 1988 [The first volume (we refer to this as Haines-1985) pertains to hydrocarbon substrates; the second (Haines-1988) mostly to oxygen- and nitrogen-containing substrates]; Chinn *Selection of Oxidants in Synthesis*; Marcel Dekker: New York, 1971; Augustine; Trecker *Oxidation*, 2 vols.; Marcel Dekker: New York, 1969, 1971; Ref. 2.

Six-membered alicyclic rings can be aromatized in a number of ways.[12] Aromatization is accomplished most easily if there are already one or two double bonds in the ring or if the ring is fused to an aromatic ring. The reaction can also be applied to heterocyclic five- and six-membered rings. Many groups may be present on the ring without interference, and even *gem*-dialkyl substitution does not always prevent the reaction: In such cases one alkyl group often migrates or is eliminated. However, more drastic conditions are usually required for this. In some cases OH and COOH groups are lost from the ring. Cyclic ketones are converted to phenols. Seven-membered and larger rings are often isomerized to six-membered aromatic rings, though this is not the case for partially hydrogenated azulene systems (which are frequently found in nature); these are converted to azulenes.

There are three types of reagents most frequently used to effect aromatization.

1. Hydrogenation catalysts,[13] such as platinum, palladium, nickel, etc. In this case the reaction is the reverse of double-bond hydrogenation (**5-9** and **5-11**), and presumably the mechanism is also the reverse, though not much is known.[14] Cyclohexene has been detected as an intermediate in the conversion of cyclohexane to benzene, using Pt.[15] The substrate is heated with the catalyst at about 300 to 350°C. The reactions can often be carried out under milder conditions if a hydrogen acceptor, such as maleic acid, cyclohexene, or benzene, is present to remove hydrogen as it is formed. The acceptor is reduced to the saturated compound. It has been reported that dehydrogenation of 1-methylcyclohexene-1-^{13}C over an alumina catalyst gave toluene with the label partially scrambled throughout the aromatic ring.[16]

2. The elements sulfur and selenium, which combine with the hydrogen evolved to give, respectively, H_2S and H_2Se. Little is known about this mechanism either.[17]

3. Quinones,[18] which become reduced to the corresponding hydroquinones. Two important quinones often used for aromatizations are chloranil (2,3,5,6-tetrachloro-1,4-benzoquinone) and DDQ (2,3-dichloro-5,6-dicyano-1,4-benzoquinone).[19] The latter is more reactive and can be used in cases where the substrate is difficult to dehydrogenate. It is likely that the mechanism involves a transfer of hydride to the quinone oxygen, followed by the transfer of a proton to the phenolate ion:[20]

[12]For reviews, see Haines-1985, Ref. 11, pp. 16-22, 217-222; Fu; Harvey *Chem. Rev.* **1978**, *78*, 317-361; Valenta, in Bentley; Kirby *Elucidation of Chemical Structures by Physical and Chemical Methods* (vol. 4 of Weissberger *Techniques of Chemistry*), 2nd ed., pt. 2; Wiley: New York, 1973, pp. 1-76; House, Ref. 10, pp. 34-44.

[13]For a review, see Rylander *Organic Synthesis with Noble Metal Catalysts*; Academic Press: New York, 1973, pp. 1-59.

[14]For a discussion, see Tsai; Friend; Muetterties *J. Am. Chem. Soc.* **1982**, *104*, 2539. See also Augustine; Thompson *J. Org. Chem.* **1987**, *52*, 1911.

[15]Land; Pettiette-Hall; McIver; Hemminger *J. Am. Chem. Soc.* **1989**, *111*, 5970.

[16]Marshall; Miiller; Ihrig *Tetrahedron Lett.* **1973**, 3491.

[17]House; Orchin *J. Am. Chem. Soc.* **1960**, *82*, 639; Silverwood; Orchin *J. Org. Chem.* **1962**, *27*, 3401.

[18]For reviews, see Becker; Turner, in Patai; Rappoport *The Chemistry of the Quinonoid Compounds*, vol. 2, pt. 2; Wiley: New York, 1988, pp. 1351-1384; Becker, in Patai *The Chemistry of the Quinonoid Compounds*, vol. 1, pt. 1, Wiley: New York, 1974, pp. 335-423.

[19]For reviews of DDQ, see Turner, in Pizey, Ref. 10, vol. 3, 1977, pp. 193-225; Walker; Hiebert *Chem. Rev.* **1967**, *67*, 153-195.

[20]Braude; Jackman; Linstead; Lowe *J. Chem. Soc.* **1960**, 3123, 3133; Trost *J. Am. Chem. Soc.* **1967**, *89*, 1847; Ref. 18. See also Stoos; Roček *J. Am. Chem. Soc.* **1972**, *94*, 2719; Hashish; Hoodless *Can. J. Chem.* **1976**, *54*, 2261; Müller; Joly; Mermoud *Helv. Chim. Acta* **1984**, *67*, 105; Radtke; Hintze; Rösler; Heesing *Chem. Ber.* **1990**, *123*, 627.

Among other reagents[21] that have been used are atmospheric oxygen, MnO_2,[22] SeO_2, various strong bases,[23] chromic acid,[24] and activated charcoal.[25] The last-mentioned reagent also dehydrogenates cyclopentanes to cyclopentadienes. In some instances the hydrogen is not released as H_2 or transferred to an external oxidizing agent but instead serves to reduce another molecule of substrate. This is a disproportionation reaction and can be illustrated by the conversion of cyclohexene to cyclohexane and benzene.

OS **II**, 214, 423; **III**, 310, 358, 729, 807; **IV**, 536; **VI**, 731. Also see OS **III**, 329.

9-2 Dehydrogenations Yielding Carbon–Carbon Double Bonds
Dihydro-elimination

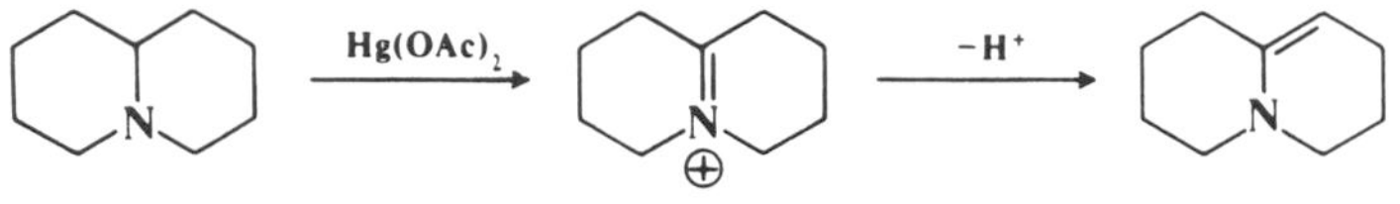

1

Dehydrogenation of an aliphatic compound to give a double bond in a specific location is not usually a feasible process, though industrially mixtures of olefins are obtained in this way from mixtures of alkanes (generally by heating with chromia–alumina catalysts). There are, however, some notable exceptions, and it is not surprising that these generally involve cases where the new double bond can be in conjugation with a double bond or with an unshared pair of electrons already present.[26] One example is the synthesis developed by Leonard and co-workers,[27] in which tertiary amines give enamines when treated with mercuric acetate[28] (see the example above). In this case the initial product is the iminium ion **1** which loses a proton to give the enamine. In another example, the oxidizing agent SeO_2 can in certain cases convert a carbonyl compound to an α,β-unsaturated carbonyl compound by removing H_2[30] (though this reagent more often gives **9-16**). This reaction has been most often applied in the steroid series, an example being[31]

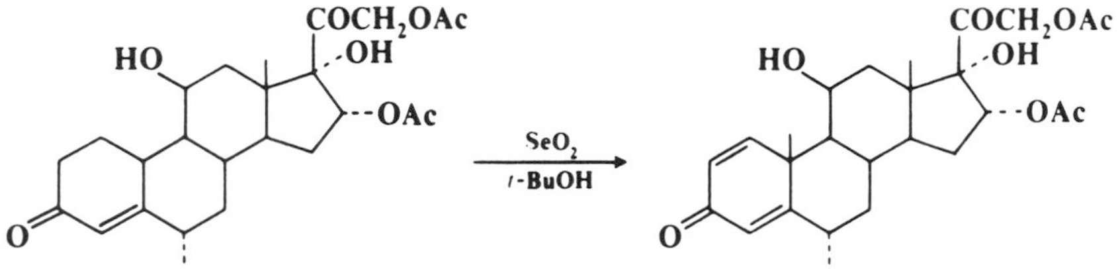

[21]For a list of reagents, with references, see Larock *Comprehensive Organic Transformations*; VCH: New York, 1989, pp. 93-97.
[22]See, for example, Leffingwell; Bluhm *Chem. Commun.* **1969**, 1151.
[23]For a review, see Pines; Stalick *Base-Catalyzed Reactions of Hydrocarbons and Related Compounds*; Academic Press: New York, 1977, pp. 483-503. See also Reetz; Eibach *Liebigs Ann. Chem.* **1978**, 1598; Trost; Rigby *Tetrahedron Lett.* **1978**, 1667.
[24]Müller; Pautex; Hagemann *Chimia* **1988**, *42*, 414.
[25]Shuikin; Naryschkina *J. Prakt. Chem.* **1961**, [4] *13*, 183.
[26]For a review, see Haines-1985, Ref. 11, pp. 6-16, 206-216. For lists of examples, with references, see Ref. 21, pp. 129-131.
[27]For example, see Leonard; Hay; Fulmer; Gash *J. Am. Chem. Soc.* **1955**, *77*, 439; Leonard; Musker *J. Am. Chem. Soc.* **1959**, *81*, 5631, **1960**, *82*, 5148.
[28]For reviews, see Haynes; Cook, in Cook *Enamines*, 2nd ed. Marcel Dekker: New York, 1988, pp. 103-163; Lee, in Augustine, Ref. 11, vol. 1, pp. 102-107.
[29]This reaction can also be accomplished with I_2: Wadsworth; Detty; Murray; Weidner; Haley *J. Org. Chem.* **1984**, *49*, 2676.
[30]For reviews, see Back, in Patai *The Chemistry of Organic Selenium and Tellurium Compounds*, pt. 2; Wiley: New York, 1987, pp. 91-213, pp. 110-114; Jerussi *Sel. Org. Transform.* **1970**, *1*, 301-326, pp. 315-321; Trachtenberg, in Augustine, Ref. 11, pp. 166-174.
[31]Bernstein; Littell *J. Am. Chem. Soc.* **1960**, *82*, 1235.

Similarly, SeO_2 has been used to dehydrogenate 1,4-diketones[32] ($RCOCH_2CH_2COR \rightarrow$ $RCOCH=CHCOR$) and 1,2-diarylalkanes ($ArCH_2CH_2Ar \rightarrow ArCH=CHAr$). These conversions can also be carried out by certain quinones, most notably DDQ (see **9-1**).[19] Simple aldehydes and ketones have been dehydrogenated (e.g., cyclopentanone $\rightarrow$ cyclopentenone) by $PdCl_2$,[33] by $FeCl_3$,[34] and by benzeneseleninic anhydride[35] (this reagent also dehydrogenates lactones in a similar manner), among other reagents.

In an indirect method of achieving this conversion, the silyl enol ether of a simple ketone is treated with DDQ[36] or with triphenylmethyl cation[37] (for another indirect method, see **7-12**). Simple linear alkanes have been converted to alkenes by treatment with certain transition metal compounds.[38]

An entirely different approach to specific dehydrogenation has been reported by R. Breslow[39] and by J. E. Baldwin.[40] By means of this approach it was possible, for example, to convert 3α-cholestanol (**2**) to 5α-cholest-14-en-3α-ol (**3**), thus introducing a double bond

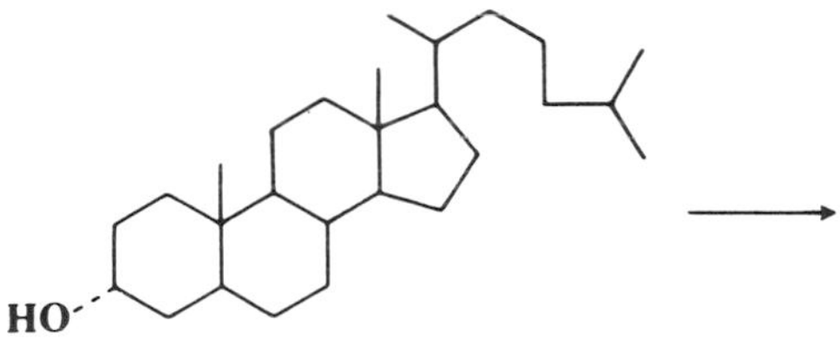

2

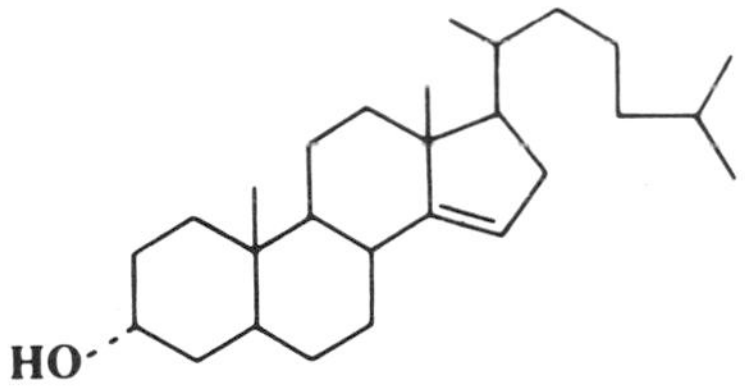

3

at a specific site remote from any functional group.[41] This was accomplished by conversion of **2** to the ester **4**, followed by irradiation of **4**, which gave 55% **6**, which was then hydrolyzed

[32]For example, see Barnes; Barton *J. Chem. Soc.* **1953**, 1419.

[33]Bierling; Kirschke; Oberender; Schultz *J. Prakt. Chem.* **1972**, *314*, 170; Kirschke; Müller; Timm *J. Prakt. Chem.* **1975**, *317*, 807; Mincione; Ortaggi; Sirna *Synthesis* **1977**, 773; Mukaiyama; Ohshima; Nakatsuka *Chem. Lett.* **1983**, 1207. See also Heck *Palladium Reagents in Organic Synthesis*; Academic Press: New York, 1985, pp. 103-110.

[34]Cardinale; Laan; Russell; Ward *Recl. Trav. Chim. Pays-Bas* **1982**, *101*, 199.

[35]Barton; Hui; Ley; Williams *J. Chem. Soc., Perkin Trans. 1* **1982**, 1919; Barton; Godfrey; Morzycki; Motherwell; Ley *J. Chem. Soc., Perkin Trans. 1* **1982**, 1947.

[36]Jung; Pan; Rathke; Sullivan; Woodbury *J. Org. Chem.* **1977**, *42*, 3961.

[37]Ryu; Murai; Hatayama; Sonoda *Tetrahedron Lett.* **1978**, 3455. For another method, which can also be applied to enol acetates, see Tsuji; Minami; Shimizu *Tetrahedron Lett.* **1983**, *24*, 5635, 5639.

[38]See Burchard; Felkin *Nouv. J. Chim.* **1986**, *10*, 673; Burk; Crabtree *J. Am. Chem. Soc.* **1987**, *109*, 8025; Renneke; Hill *New J. Chem.* **1987**, *11*, 763, *Angew. Chem. Int. Ed. Engl.* **1988**, *27*, 1526 [*Angew. Chem. 100*, 1583], *J. Am. Chem. Soc.* **1988**, *110*, 5461; Maguire; Boese; Goldman *J. Am. Chem. Soc.* **1989**, *111*, 7088; Sakakura; Ishida; Tanaka *Chem. Lett.* **1990**, 585; and references cited in these papers.

[39]Breslow; Baldwin *J. Am. Chem. Soc.* **1970**, *92*, 732. For reviews, see Breslow *Chemtracts: Org. Chem.* **1988**, *1*, 333-348, *Acc. Chem. Res.* **1980**, *13*, 170-177, *Isr. J. Chem.* **1979**, *18*, 187-191, *Chem. Soc. Rev.* **1972**, *1*, 553-580.

[40]Baldwin; Bhatnagar; Harper *Chem. Commun.* **1970**, 659.

[41]For other methods of introducing a remote double bond, see Čeković; Cvetković *Tetrahedron Lett.* **1982**, *23*, 3791; Czekay; Drewello; Schwarz *J. Am. Chem. Soc.* **1989**, *111*, 4561. See also Bégué *J. Org. Chem.* **1982**, *47*, 4268; Nagata; Saito *Synlett* **1990**, 291-300.

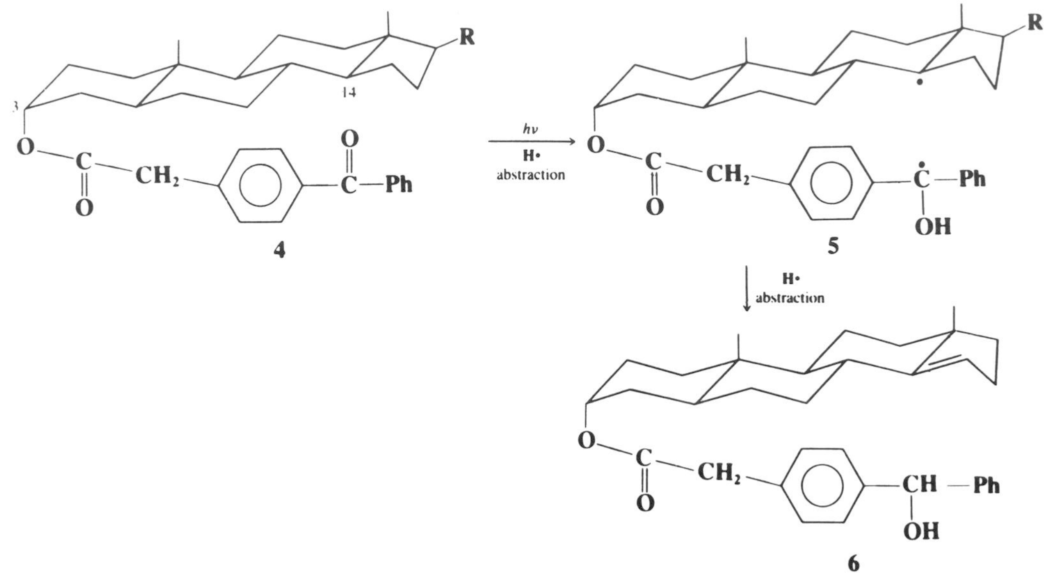

to **3**. The radiation excites the benzophenone portion of **4** (p. 246), which then abstracts hydrogen from the 14 position to give the diradical **5** which undergoes another internal abstraction to give **6**. In other cases, diradicals like **5** can close to a macrocyclic lactone (**9-16**). In an alternate approach,[42] a 9(11) double bond was introduced into a steroid nucleus by reaction of the *m*-iodo ester **7** with PhICl$_2$ and uv light, which results in hydrogen being

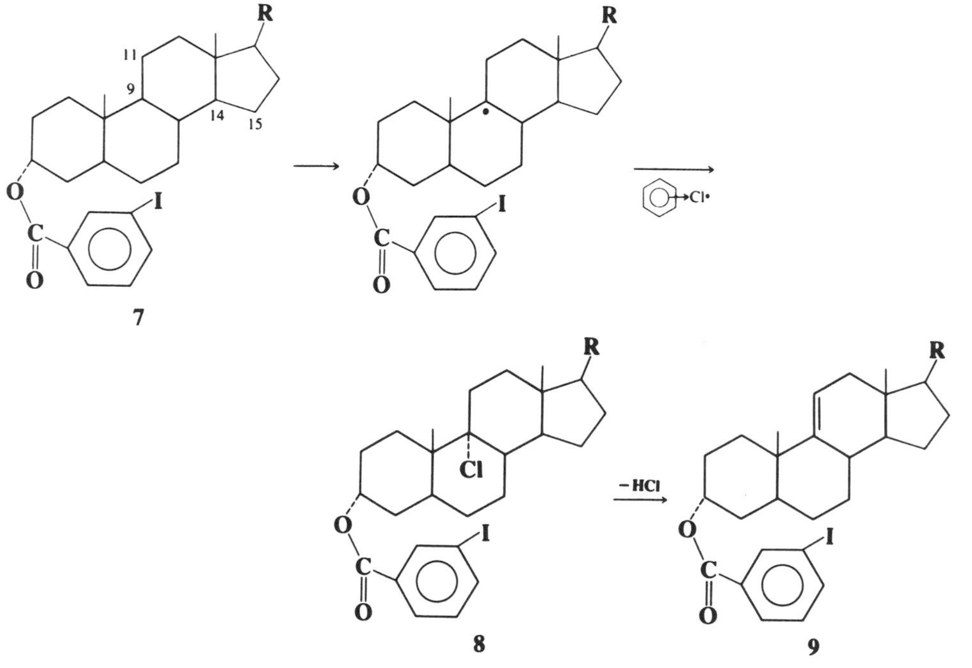

[42]Breslow; Corcoran; Snider; Doll; Khanna; Kaleya *J. Am. Chem. Soc.* **1977**, *99*, 905. For related approaches, see Wolner *Tetrahedron Lett.* **1979**, 4613; Breslow; Heyer *J. Am. Chem. Soc.* **1982**, *104*, 2045; Breslow; Guo *Tetrahedron Lett.* **1987**, *28*, 3187; Breslow; Brandl; Hunger; Adams *J. Am. Chem. Soc.* **1987**, *109*, 3799; Batra; Breslow *Tetrahedron Lett.* **1989**, *30*, 535; Orito; Ohto; Suginome *J. Chem. Soc., Chem. Commun.* **1990**, 1076.

abstracted regioselectively from the 9 position, resulting in chlorination at that position. Dehydrohalogenation of **8** gives the 9(11)-unsaturated steroid **9**. In contrast, use of the para isomer of **7** results in chlorination at the 14 position and loss of HCl gives the 14-unsaturated steroid. These reactions are among the very few ways to introduce functionality at a specific site remote from any functional group (see also **9-16**).

Certain 1,2-diarylalkenes ArCH=CHAr' have been converted to the corresponding alkynes ArC≡CAr' by treatment with *t*-BuOK in DMF.[42a]

A different kind of dehydrogenation was used in the final step of Paquette's synthesis of dodecahedrane:[43]

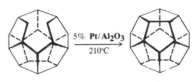

dodecahedrane

OS **V**, 428, **VII**, 4, 473.

9-3 Oxidation or Dehydrogenation of Alcohols to Aldehydes and Ketones
C,O-**Dihydro-elimination**

$$RCH_2OH \xrightarrow[\text{chromite}]{\text{copper}} RCHO$$

$$RCHOHR' \xrightarrow[\text{H}_2\text{SO}_4]{\text{K}_2\text{Cr}_2\text{O}_7} RCOR'$$

Primary alcohols can be converted to aldehydes and secondary alcohols to ketones in four main ways:[44]

1. *With strong oxidizing agents.*[45] Secondary alcohols are easily oxidized to ketones by acid dichromate[46] at room temperature or slightly above. Though this is the most common reagent, many other strong oxidizing agents (e.g., KMnO₄, Br₂, MnO₂, ruthenium tetroxide,[47] etc.) have also been employed. A solution of chromic acid and sulfuric acid in water is known as the *Jones reagent*.[48] When secondary alcohols are dissolved in acetone, titration with the Jones reagent oxidizes them to ketones rapidly and in high yield without disturbing any double or triple bonds that may be present (see **9-10**) and without epimerizing an adjacent

[42a]Akiyama; Nakatsuji; Nomura; Matsuda; Nakashima *J. Chem. Soc., Chem. Commun.* **1991**, 948.
[43]Paquette; Ternansky; Balogh; Kentgen *J. Am. Chem. Soc.* **1983**, *105*, 5446; Paquette; Weber; Kobayashi; Miyahara *J. Am. Chem. Soc.* **1988**, *110*, 8591. For a monograph on dodecahedrane and related compounds, see Paquette; Doherty *Polyquinane Chemistry*; Springer: New York, 1987. For reviews, see, in Olah *Cage Hydrocarbons*; Wiley: New York, 1990, the reviews by Paquette, pp. 313-352, and by Fessner; Prinzbach, pp. 353-405; Paquette *Chem. Rev.* **1989**, *89*, 1051-1065, *Top. Curr. Chem.* **1984**, *119*, 1-158, in Lindberg *Strategies and Tactics in Organic Synthesis*; Academic Press: New York, 1984, pp. 175-200.
[44]For reviews, see Hudlický, Ref. 11, pp. 114-126, 132-149; Haines-1988, Ref. 11, pp. 5-148, 326-390; Müller, in Patai *The Chemistry of Functional Groups, Supplement E*; Wiley: New York, 1980, pp. 469-538; Cullis; Fish, in Patai *The Chemistry of the Carbonyl Group*, vol. 1; Wiley: New York, 1966, pp. 129-157. For a lengthy list of reagents, with references, see Ref. 21, pp. 604-615.
[45]For thorough discussions, see Lee, Ref. 28, pp. 56-81; and (with respect to chromium and managanese reagents) House, Ref. 10, pp. 257-273.
[46]Various forms of H₂CrO₄ and of CrO₃ are used for this reaction. For a review, see Cainelli; Cardillo, Ref. 10, pp. 118-216. For discussions, see Fieser; Fieser *Reagents for Organic Synthesis*, vol. 1; Wiley: New York, 1967, pp. 142-147, 1059-1064, and subsequent volumes in this series.
[47]For a review, see Lee; van den Engh, in Trahanovsky, Ref. 2, pt. B, pp. 197-222.
[48]Bowden; Heilbron; Jones; Weedon *J. Chem. Soc.* **1946**, 39; Bowers; Halsall; Jones; Lemin *J. Chem. Soc.* **1953**, 2548.

chiral center.[49] The Jones reagent also oxidizes primary allylic alcohols to the corresponding aldehydes.[50] Three other Cr(VI) reagents commonly used[51] are dipyridine Cr(VI) oxide (Collins's reagent),[52] pyridinium chlorochromate (Corey's reagent),[53] and pyridinium dichromate.[54] MnO_2 is also a fairly specific reagent for OH groups and is often used to oxidize allylic alcohols to α,β-unsaturated aldehydes or ketones. For acid-sensitive compounds CrO_3 in HMPA,[55] a CrO_3–pyridine complex,[56] or trimethylsilyl chromates[57] can be used. Sodium hypochlorite in acetic acid is useful for oxidizing larger amounts of secondary alcohols.[58] The oxidizing agent can be supported on a polymer.[59] Both chromic acid[60] and permanganate[61] have been used in this way (see p. 421). Phase transfer catalysis has also been used with permanganate,[62] chromic acid,[63] and ruthenium tetroxide.[64] Phase transfer catalysis is particularly useful because the oxidizing agents are insoluble in most organic solvents, while the substrates are generally insoluble in water (see p. 362). Ultrasound has been used for $KMnO_4$ oxidations.[65]

Most of these oxidizing agents have also been used to convert primary alcohols to aldehydes, but precautions must be taken that the aldehyde is not further oxidized to the carboxylic acid (**9-22**).[66] One way to halt oxidation is by distillation of the aldehyde as it is formed. The following are among the oxidizing agents that have been used to convert at least some primary alcohols to aldehydes:[67] dimethyl sulfoxide (see **9-20**), Collins's reagent, Corey's reagent, pyridinium dichromate, tetrapropylammonium perruthenate Pr_4N^+ RuO_4^-,[68] ceric ammonium nitrate (CAN),[69] $Na_2Cr_2O_7$ in water,[70] Ag_2CO_3-on-celite,[71] hot

[49]For example, see Djerassi; Hart; Warawa *J. Am. Chem. Soc.* **1964**, *86*, 78.
[50]Harding; May; Dick *J. Org. Chem.* **1975**, *40*, 1664.
[51]For a comparative study of Jones's, Collins's, and Corey's reagents, see Warrener; Lee; Russell; Paddon-Row *Aust. J. Chem.* **1978**, *31*, 1113.
[52]Collins; Hess; Frank *Tetrahedron Lett.* **1968**, 3363; Ratcliffe; Rodehorst *J. Org. Chem.* **1970**, *35*, 4000; Stensiö *Acta Chem. Scand.* **1971**, *25*, 1125; Collins; Hess *Org. Synth. VI*, 644; Sharpless; Akashi *J. Am. Chem. Soc.* **1975**, *97*, 5927.
[53]Corey; Suggs *Tetrahedron Lett.* **1975**, 2647. For reviews of this and related reagents, see Luzzio; Guziec *Org. Prep. Proceed. Int.* **1988**, *20*, 533-584; Piancatelli; Scettri; D'Auria *Synthesis* **1982**, 245-258. For an improved method of preparing this reagent, see Agarwal; Tiwari; Sharma *Tetrahedron* **1990**, *46*, 4417.
[54]Coates; Corrigan *Chem. Ind. (London)* **1969**, 1594; Corey; Schmidt *Tetrahedron Lett.* **1979**, 399; Czernecki; Georgoulis; Stevens; Vijayakumaran *Tetrahedron Lett.* **1985**, *26*, 1699.
[55]Cardillo; Orena; Sandri *Synthesis* **1976**, 394.
[56]Poos; Arth; Beyler; Sarett *J. Am. Chem. Soc.* **1953**, *75*, 422.
[57]Moiseenkov; Cheskis; Veselovskii; Veselovskii; Romanovich; Chizhov *J. Org. Chem. USSR* **1987**, *23*, 1646.
[58]Stevens; Chapman; Weller *J. Org. Chem.* **1980**, *45*, 2030. See also Schneider; Weber; Faller *J. Org. Chem.* **1982**, *47*, 364; Mohrig; Nienhuis; Linck; van Zoeren; Fox; Mahaffy *J. Chem. Educ.* **1985**, *62*, 519.
[59]For a review of oxidations and other reactions with supported reagents, see McKillop; Young *Synthesis* **1979**, 401-422.
[60]Cainelli; Cardillo; Orena; Sandri *J. Am. Chem. Soc.* **1976**, *98*, 6737; Santaniello; Ponti; Manzocchi *Synthesis* **1978**, 534. See also San Filippo; Chern *J. Org. Chem.* **1977**, *42*, 2182.
[61]Regen; Koteel *J. Am. Chem. Soc.* **1977**, *99*, 3837; Noureldin; Lee *Tetrahedron Lett.* **1981**, *22*, 4889. See also Menger; Lee *J. Org. Chem.* **1979**, *44*, 3446.
[62]For a review of phase transfer assisted permanganate oxidations, see Lee, in Trahanovsky, Ref. 2, pt. D, pp. 147-206.
[63]See for example, Hutchins; Natale; Cook *Tetrahedron Lett.* **1977**, 4167; Landini; Montanari; Rolla *Synthesis* **1979**, 134; Pletcher; Tait *J. Chem. Soc., Perkin Trans. 2* **1979**, 788.
[64]Morris; Kiely *J. Org. Chem.* **1987**, *52*, 1149.
[65]Yamawaki; Sumi; Ando; Hanfusa *Chem. Lett.* **1983**, 379.
[66]Though ketones are much less susceptible to further oxidation than aldehydes, such oxidation is possible (**9-8**), and care must be taken to avoid it, usually by controlling the temperature and/or the oxidizing agent.
[67]For some other reagents, not mentioned here, see Kaneda; Kawanishi; Teranishi *Chem. Lett.* **1984**, 1481; Semmelhack; Schmid; Cortés; Chou *J. Am. Chem. Soc.* **1984**, *106*, 3374; Cameron; Bocarsly *J. Am. Chem. Soc.* **1985**, *107*, 6116; Anelli; Biffi; Montanari; Quici *J. Org. Chem.* **1987**, *52*, 2559; Bilgrien; Davis; Drago *J. Am. Chem. Soc.* **1987**, *109*, 3786; Nishiguchi; Asano *J. Org. Chem.* **1989**, *54*, 1531; Dess; Martin *J. Am. Chem. Soc.* **1991**, *113*, 7277. See also Ref. 21, pp. 604-615.
[68]Griffith; Ley; Whitcombe; White *J. Chem. Soc., Chem. Commun.* **1987**, 1625; Griffith; Ley *Aldrichimica Acta* **1990**, *23*, 13-19.
[69]Trahanovsky; Young *J. Chem. Soc.* **1965**, 5777; Trahanovsky; Young; Brown *J. Org. Chem.* **1967**, *32*, 3865.
[70]Lee; Spitzer *J. Org. Chem.* **1970**, *35*, 3589. See also Rao; Filler *J. Org. Chem.* **1974**, *39*, 3304; Lou *Synth. Commun.* **1989**, *19*, 1841, *Chem. Ind. (London)* **1989**, 312.
[71]Fetizon; Golfier *C. R. Acad. Sci., Ser. C* **1968**, *267*, 900; Kakis; Fetizon; Douchkine; Golfier; Mourgues; Prange *J. Org. Chem.* **1974**, *39*, 523.

HNO$_3$ in aqueous glyme,[72] O$_2$–pyridine–CuCl,[73] Pb(OAc)$_4$–pyridine,[74] and benzoyl peroxide–NiBr$_2$.[75] Most of these reagents also oxidize secondary alcohols to ketones. Reagents that can be used specifically to oxidize a secondary OH group in the presence of a primary OH group[76] are Cl$_2$–pyridine,[77] H$_2$O$_2$–ammonium molybdate,[78] NaBrO$_3$–CAN,[79] and NaOCl in HOAc,[80] while RuCl$_2$(PPh$_3$)$_3$–benzene,[81] osmium tetroxide,[82] 2,2′-bipyridylchromium peroxide,[83] and Br$_2$–Ni(OBz)$_2$[84] oxidize primary OH groups in the presence of a secondary OH group.[85] Benzylic and allylic alcohols have been selectively oxidized to the aldehydes in the presence of saturated alcohols by the use of potassium manganate K$_2$MnO$_4$ under phase transfer conditions.[86] On the other hand, Fremy's salt (see **9-4**) selectively oxidizes benzylic alcohols and not allylic or saturated ones.[87] Benzylic alcohols can also be oxidized to aldehydes by NH$_4$NO$_3$ or NaNO$_2$ in aqueous F$_3$CCOOH,[88] by H$_2$O$_2$–HBr,[89] and by *m*-chloroperbenzoic acid–HCl–DMF,[90] among other reagents. Certain zirconocene complexes can selectively oxidize only one OH group of a diol, even if both are primary.[91]

   2. *By catalytic dehydrogenation.* For the conversion of primary alcohols to aldehydes, dehydrogenation catalysts have the advantage over strong oxidizing agents that further oxidation to the carboxylic acid is prevented. Copper chromite is the agent most often used, but other catalysts, e.g., silver and copper, have also been employed. Many ketones have also been prepared in this manner. Catalytic dehydrogenation is more often used industrially than as a laboratory method. However, convenient laboratory procedures using copper oxide,[92] Raney nickel,[93] and palladium acetate (under phase transfer conditions)[94] have been reported.

   3. *The Oppenauer oxidation.* When a ketone in the presence of base is used as the oxidizing agent (it is reduced to a secondary alcohol), the reaction is known as the *Oppenauer oxidation.*[95] This is the reverse of the Meerwein–Pondorf–Verley reaction (**6-25**), and the mechanism is also the reverse. The ketones most commonly used are acetone, butanone, and cyclohexanone. The most common base is aluminum *t*-butoxide. The chief advantage of the method is its high selectivity. Although the method is most often used for the preparation of ketones, it has also been used for aldehydes.

   4. *With N-bromosuccinimide or related compounds.* These compounds are chemose-

[72]McKillop; Ford *Synth. Commun.* **1972,** *2*, 307.
[73]Jallabert; Riviere *Tetrahedron Lett.* **1977,** 1215.
[74]Partch *Tetrahedron Lett.* **1964,** 3071; Partch; Monthony *Tetrahedron Lett.* **1967,** 4427. See also Brocksom; Ferreira *J. Chem. Res. (S)* **1980,** 412; Mihailović; Konstantinović; Vukićević *Tetrahedron Lett.* **1986,** *27*, 2287.
[75]Doyle; Patrie; Williams *J. Org. Chem.* **1979,** *44*, 2955.
[76]For other methods, see Jung; Speltz *J. Am. Chem. Soc.* **1976,** *98*, 7882; Jung; Brown *Tetrahedron Lett.* **1978,** 2771; Kaneda; Kawanishi; Jitsukawa; Teranishi *Tetrahedron Lett.* **1983,** *24*, 5009; Siedlecka; Skarżewski; Młochowski *Tetrahedron Lett.* **1990,** *31*, 2177.
[77]Wicha; Zarecki *Tetrahedron Lett.* **1974,** 3059.
[78]Trost; Masuyama *Isr. J. Chem.* **1984,** *24*, 134. For a method involving H$_2$O$_2$ and another catalyst, see Sakata; Ishii *J. Org. Chem.* **1991,** *56*, 6233.
[79]Tomioka; Oshima; Nozaki *Tetrahedron Lett.* **1982,** *23*, 539.
[80]Stevens; Chapman; Stubbs; Tam; Albizati *Tetrahedron Lett.* **1982,** *23*, 4647.
[81]Tomioka; Takai; Oshima; Nozaki *Tetrahedron Lett.* **1981,** *22*, 1605.
[82]Maione; Romeo *Synthesis* **1984,** 955.
[83]Firouzabadi; Iranpoor; Kiaeezadeh; Toofan *Tetrahedron* **1986,** *42*, 719
[84]Doyle; Bagheri *J. Org. Chem.* **1981,** *46*, 4806; Doyle; Dow; Bagheri; Patrie *J. Org. Chem.* **1983,** *48*, 476.
[85]For a list of references to the selective oxidation of various types of alcohol, see Kulkarni; Mathew T *Tetrahedron* **1990,** *31*, 4497.
[86]Kim; Chung; Cho; Hahn *Tetrahedron Lett.* **1989,** *30*, 2559. See also Kim; Song; Lee; Hahn *Tetrahedron Lett.* **1986,** *27*, 2875.
[87]Morey; Dzielenziak; Saa *Chem. Lett.* **1985,** 263.
[88]Rodkin; Shtern; Cheprakov; Makhon'kov; Mardaleishvili; Beletskaya *J. Org. Chem. USSR* **1988,** *24*, 434.
[89]Dakka; Sasson *Bull. Soc. Chim. Fr.* **1988,** 756.
[90]Kim; Jung; Kim; Ryu *Synth. Commun.* **1990,** *20*, 637.
[91]Nakano; Terada; Ishii; Ogawa *Synthesis* **1986,** 774.
[92]Sheikh; Eadon *Tetrahedron Lett.* **1972,** 257.
[93]Krafft; Zorc *J. Org. Chem.* **1986,** *51*, 5482.
[94]Choudary; Reddy; Kantam; Jamil *Tetrahedron Lett.* **1985,** *26*, 6257.
[95]For a review, see Djerassi *Org. React.* **1951,** *6*, 207-272.

lective oxidizing agents and often oxidize OH groups without disturbing other oxidizable groups.[96] N-Bromosuccinimide does not oxidize aliphatic primary alcohols, but N-chlorosuccinimide does. With these reagents it is often possible to oxidize only one of several OH groups that may be present in a molecule. The combination of N-iodosuccinimide and Bu_4N^+ I^- oxidizes primary (to aldehydes) and secondary alcohols in high yields.[97]

Primary and secondary alcohols can also be oxidized, indirectly, through their esters (see **9-20**). In some cases, isolation of the ester is not required and the alcohol can then be oxidized to the aldehyde or ketone in one step.

The mechanism of oxidation with acid dichromate has been intensely studied.[98] The currently accepted mechanism is essentially that proposed by Westheimer.[99] The first two steps constitute an example of category 4 (p. 1160).

$$R_2C-H + HCrO_4^- + H^+ \rightleftharpoons R_2C-H$$
$$\underset{OH}{|} \qquad\qquad\qquad\qquad \underset{OCrO_3H}{|}$$

$$R_2C{\underset{O-CrO_3H}{|}}H \xrightarrow{\text{base}} R_2CO + HCrO_3^-[Cr(IV)] + base-H^+$$

$$R_2CHOH + Cr(IV) \longrightarrow R_2\overset{\bullet}{C}OH + Cr(III)$$

$$R_2\overset{\bullet}{C}OH + Cr(VI) \longrightarrow R_2CO + Cr(V)$$

$$R_2CHOH + Cr(V) \longrightarrow R_2CO + Cr(III)$$

The base in the second step may be water, though it is also possible[100] that in some cases no external base is involved and that the proton is transferred directly to one of the CrO_3H

$$\begin{array}{c} R_2C-H \\ O \diagdown \diagup O \\ \diagup Cr \diagdown \\ O \qquad OH \end{array} \longrightarrow R_2C=O + H_2CrO_3$$

oxygens in which case the Cr(IV) species produced would be H_2CrO_3. Part of the evidence for this mechanism was the isotope effect of about 6 found on use of MeCDOHMe, showing that the α hydrogen is removed in the rate-determining step.[101] Note that, as in **4-6**, the substrate is oxidized by three different oxidation states of chromium.[102]

[96]For a review, see Filler *Chem. Rev.* **1963**, *63*, 21-43, pp. 22-28.

[97]Hanessian; Wong; Therien *Synthesis* **1981**, 394.

[98]See Müller *Chimia* **1977**, *31*, 209-218; Wiberg, in Wiberg, Ref. 2, pp. 142-170; Venkatasubramanian *J. Sci. Ind. Res.* **1963**, *22*, 397-400; Waters, Ref. 2, pp. 49-71; Stewart, Ref. 2, pp. 37-48; Durand; Geneste; Lamaty; Moreau; Pomarès; Roque *Recl. Trav. Chim. Pays-Bas* **1978**, *97*, 42; Sengupta; Samanta; Basu *Tetrahedron* **1985**, *41*, 205.

[99]Westheimer *Chem. Rev.* **1949**, *45*, 419-451, p. 434; Holloway; Cohen; Westheimer *J. Am. Chem. Soc.* **1951**, *73*, 65.

[100]Kwart; Francis *J. Am. Chem. Soc.* **1959**, *81*, 2116; Stewart; Lee *Can. J. Chem.* **1964**, *42*, 439; Awasthy; Roček; Moriarty *J. Am. Chem. Soc.* **1967**, *89*, 5400; Kwart; Nickle *J. Am. Chem. Soc.* **1973**, *95*, 3394, **1974**, *96*, 7572, **1979**, *98*, 2881; Sengupta; Samanta; Basu *Tetrahedron* **1986**, *42*, 681. See also Müller; Perlberger *Helv. Chim. Acta* **1974**, *57*, 1943; Agarwal; Tiwari; Sharma *Tetrahedron* **1990**, *46*, 1963.

[101]Westheimer; Nicolaides *J. Am. Chem. Soc.* **1949**, *71*, 25. For other evidence, see Brownell; Leo; Chang; Westheimer *J. Am. Chem. Soc.* **1960**, *82*, 406; Roček; Westheimer; Eschenmoser; Moldoványi; Schreiber *Helv. Chim. Acta* **1962**, *45*, 2554; Lee; Stewart *J. Org. Chem.* **1967**, *32*, 2868; Wiberg; Schäfer *J. Am. Chem. Soc.* **1967**, *89*, 455; **1969**, *91*, 927, 933; Müller *Helv. Chim. Acta* **1970**, *53*, 1869, **1971**, *54*, 2000, Lee; Raptis *Tetrahedron* **1973**, *29*, 1481.

[102]Rahman; Roček *J. Am. Chem. Soc.* **1971**, *93*, 5455, 5462; Doyle; Swedo; Roček *J. Am. Chem. Soc.* **1973**, *95*, 8352; Wiberg; Mukherjee *J. Am. Chem. Soc.* **1974**, *96*, 1884, 6647.

With other oxidizing agents, mechanisms are less clear.[103] It seems certain that some oxidizing agents operate by a hydride-shift mechanism,[104] e.g., dehydrogenation with triphenylmethyl cation[105] and the Oppenauer oxidation, and some by a free-radical mechanism, e.g., oxidation with $S_2O_8^{2-}$ [106] and with VO_2^+.[107] A summary of many proposed mechanisms is given by Littler.[108]

Secondary alkyl ethers can be oxidized to ketones by bromine (e.g., $Me_2CHOCHMe_2$ + $Br_2 \rightarrow Me_2CO$).[109] Primary alkyl ethers give carboxylic acids (**9-22**) with bromine, but can be cleaved to aldehydes with 1-chlorobenzotriazole.[110]

OS **I**, 87, 211, 241, 340; **II**, 139, 541; **III**, 37, 207; **IV**, 189, 192, 195, 467, 813, 838; **V**, 242, 310, 324, 692, 852, 866; **VI**, 218, 220, 373, 644, 1033; **VII**, 102, 112, 114, 177, 258, 297; **65**, 81; **68**, 175; **69**, 212. Also see OS **IV**, 283; **65**, 243; **66**, 14.

9-4 Oxidation of Phenols and Aromatic Amines to Quinones
1/*O*,6/*O*-Dihydro-elimination

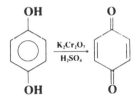

Ortho and para diols are easily oxidized to *ortho-* and *para*-quinones, respectively.[111] Either or both OH groups can be replaced by NH_2 groups to give the same products, though for the preparation of *ortho*-quinones only OH groups are normally satisfactory. The reaction has been successfully carried out with other groups para to OH or NH_2; halogen, OR, Me, *t*-Bu, and even H, though with the last yields are poor. Many oxidizing agents have been used: acid dichromate,[112] silver oxide, silver carbonate, lead tetraacetate, HIO_4, and atmospheric oxygen, to name a few. A particularly effective reagent for rings with only one OH or NH_2 group is $(KSO_3)_2N$—O• (dipotassium nitrosodisulfonate; Fremy's salt), which is a stable free radical.[113] Phenols, even some whose para positions are unoccupied, can be oxidized to *ortho*-quinones with diphenylseleninic anhydride.[114]

---

[103]For a review, see Cockerill; Harrison, in Patai *The Chemistry of Functional Groups, Supplement A*, pt. 1; Wiley: New York, 1977, pp. 264-277.

[104]See Barter; Littler *J. Chem. Soc. B.* **1967**, 205. For evidence that oxidation by HNO_2 involves a hydride shift, see Moodie; Richards *J. Chem. Soc., Perkin Trans. 2* **1986**, 1833; Ross; Gu; Hum; Malhotra *Int. J. Chem. Kinet.* **1986**, *18*, 1277.

[105]Bonthrone; Reid *J. Chem. Soc.* **1959**, 2773.

[106]Ball; Crutchfield; Edwards *J. Org. Chem.* **1960**, *25*, 1599; Bida; Curci; Edwards *Int. J. Chem. Kinet.* **1973**, *5*, 859; Snook; Hamilton *J. Am. Chem. Soc.* **1974**, *96*, 860; Walling; Camaioni *J. Org. Chem.* **1978**, *43*, 3266; Clerici; Minisci; Ogawa; Surzur *Tetrahedron Lett.* **1978**, 1149; Beylerian; Khachatrian *J. Chem. Soc., Perkin Trans. 2* **1984**, 1937.

[107]Littler; Waters *J. Chem. Soc.* **1959**, 4046.

[108]Littler *J. Chem. Soc.* **1962**, 2190.

[109]Deno; Potter *J. Am. Chem. Soc.* **1967**, *89*, 3550, 3555. See also Miller; Wolf; Mayeda *J. Am. Chem. Soc.* **1971**, *93*, 3306; Saigo; Morikawa; Mukaiyama *Chem. Lett.* **1975**, 145; Olah; Gupta; Fung *Synthesis* **1980**, 897.

[110]Pojer *Aust. J. Chem.* **1980**, *32*, 2787.

[111]For reviews, see Haines-1988, Ref. 11, pp. 305-323, 438-447; Naruta; Maruyama in Patai, Rappoport, Ref. 18, pt. 1, pp. 247-276; Thomson, in Patai, Ref. 18, pt. 1, pp. 112-132.

[112]For a review of this oxidation with chromium reagents, see Cainelli; Cardillo, Ref. 10, pp. 92-117.

[113]For a review of oxidation with this salt, see Zimmer; Lankin; Horgan *Chem. Rev.* **1971**, *71*, 229-246.

[114]Barton; Brewster; Ley; Rosenfeld *J. Chem. Soc., Chem. Commun.* **1976**, 985; Barton; Ley, in *Further Perspectives in Organic Chemistry*; North Holland Publishing Co.: Amsterdam, 1979, pp. 53-66. For another way of accomplishing this, see Krohn; Rieger; Khanbabaee *Chem. Ber.* **1989**, *122*, 2323.

Less is known about the mechanism than is the case for **9-3,** but, as in that case, it seems to vary with the oxidizing agent. For oxidation of catechol with $NaIO_4$, it was found that the reaction conducted in $H_2^{18}O$ gave unlabeled quinone,[115] so the following mechanism[116] was proposed:

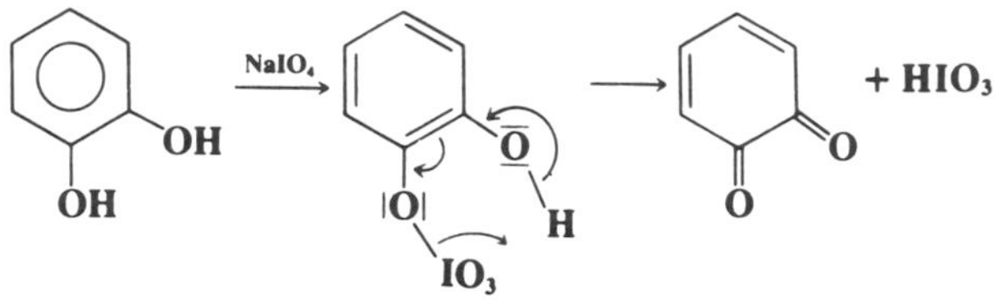

When catechol was oxidized with MnO_4^- under aprotic conditions, a semiquinone radical ion intermediate was involved.[117] For autoxidations (i.e., with atmospheric oxygen) a free-radical mechanism is known to operate.[118]

OS **I,** 383, 482, 511; **II,** 175, 254, 430, 553; **III,** 663, 753; **IV,** 148; **VI,** 412, 480, 1010.

9-5 Dehydrogenation of Amines

1/1/N,2/2/C-Tetrahydro-bielimination

$$RCH_2NH_2 \xrightarrow[2.\,H_2O]{1.\,IF_5} RCN$$

Primary amines at a primary carbon can be dehydrogenated to nitriles. The reaction has been carried out with a variety of reagents, among others, IF_5,[119] lead tetraacetate,[120] nickel peroxide,[121] NaOCl in micelles,[122] $K_2S_2O_8$–$NiSO_4$,[123] and $CuCl$–O_2–pyridine.[124] Several methods have been reported for the dehydrogenation of secondary amines to imines.[125] Among them[126] are treatment with(1) iodosylbenzene PhIO alone or in the presence of a ruthenium complex,[127] (2) Me_2SO and oxalyl chloride,[128] and (3) t-BuOOH and a rhenium catalyst.[129]

A reaction that involves dehydrogenation to an imine which then reacts further is the reaction of primary or secondary amines with palladium black.[130] The imine initially formed by the dehydrogenation reacts with another molecule of the same or a different amine to give an aminal, which loses NH_3 or RNH_2 to give a secondary or tertiary amine. An example

[115]Adler; Falkehag; Smith *Acta Chem. Scand.* **1962,** *16,* 529.
[116]This mechanism is an example of category 4 (p. 1160).
[117]Bock; Jaculi *Angew. Chem. Int. Ed. Engl.* **1984,** *23,* 305 [*Angew. Chem. 96,* 298].
[118]Sheldon; Kochi *Metal-Catalyzed Oxidations of Organic Compounds*; Academic Press: New York, 1981, pp. 368-381; Walling *Free Radicals in Solution*; Wiley: New York, 1957, pp. 457-461.
[119]Stevens *J. Org. Chem.* **1961,** *26,* 2531.
[120]Stojiljković; Andrejević; Mihailović *Tetrahedron* **1967,** *23,* 721.
[121]Nakagawa; Tsuji *Chem. Pharm. Bull.* **1963,** *11,* 296. See also Xu; Yamaguchi; Tanabe *Chem. Lett.* **1988,** 281.
[122]Juršić *J. Chem. Res. (S)* **1988,** 168.
[123]Yamazaki; Yamazaki *Bull. Chem. Soc. Jpn.* **1990,** *63,* 301.
[124]Kametani; Takahashi; Ohsawa; Ihara *Synthesis* **1977,** 245; Capdevielle; Lavigne; Maumy *Synthesis* **1989,** 453, *Tetrahedron* **1990,** 2835; Capdevielle; Lavigne; Sparfel; Baranne-Lafont; Cuong; Maumy *Tetrahedron Lett.* **1990,** *31,* 3305.
[125]For a review, see Dayagi; Degani, in Patai *The Chemistry of the Carbon–Nitrogen Double Bond*; Wiley: New York, 1970, pp. 117-124.
[126]For other methods, see Cornejo; Larson; Mendenhall *J. Org. Chem.* **1985,** *50,* 5382; Nishinaga; Yamazaki; Matsuura *Tetrahedron Lett.* **1988,** *29,* 4115.
[127]Müller; Gilabert *Tetrahedron* **1988,** *44,* 7171.
[128]Keirs; Overton *J. Chem. Soc., Chem. Commun.* **1987,** 1660.
[129]Murahashi; Naota; Taki *J. Chem. Soc., Chem. Commun.* **1985,** 613.
[130]Murahashi; Yoshimura; Tsumiyama; Kojima *J. Am. Chem. Soc.* **1983,** *105,* 5002. See also Wilson; Laine *J. Am. Chem. Soc.* **1985,** *107,* 361.

is the reaction between N-methylbenzylamine and butylmethylamine, which produces 95%
N-methyl-N-butylbenzylamine.

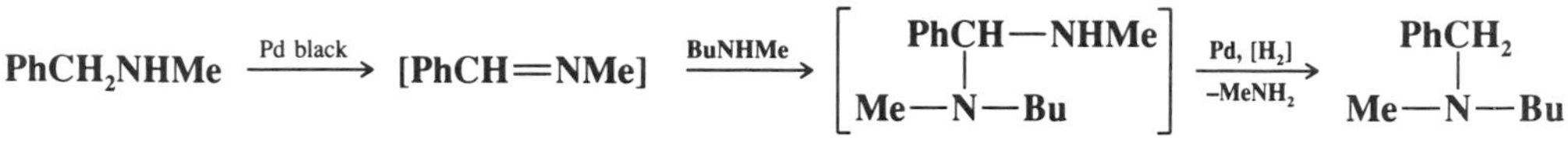

$$PhCH_2NHMe \xrightarrow{Pd\ black} [PhCH{=}NMe] \xrightarrow{BuNHMe} \left[\begin{array}{c} PhCH{-}NHMe \\ | \\ Me{-}N{-}Bu \end{array} \right] \xrightarrow[-MeNH_2]{Pd,\ [H_2]} \begin{array}{c} PhCH_2 \\ | \\ Me{-}N{-}Bu \end{array}$$

Another method for the conversion of primary to secondary amines ($2RNH_2 \rightarrow R_2NH$)
involves treatment with a catalytic amount of sodium hydride.[131] This reaction also involves
an imine intermediate.

9-6 Oxidation of Hydrazines, Hydrazones, and Hydroxylamines
   1/*N*,2/*N*-Dihydro-elimination

$$Ar{-}NH{-}NH{-}Ar \xrightarrow{NaOBr} Ar{-}N{=}N{-}Ar$$

N,N'-Diarylhydrazines (hydrazo compounds) are oxidized to azo compounds by several
oxidizing agents, including NaOBr, HgO,[132] $K_3Fe(CN)_6$ under phase transfer conditions,[133]
benzeneseleninic anhydride,[134] MnO_2 (this reagent yields *cis* azobenzenes),[135] $CuCl_2$, and
air and NaOH.[136] The reaction is also applicable to N,N'-dialkyl- and N,N'-diacylhydrazines.
Hydrazines (both alkyl and aryl) substituted on only one side also give azo compounds,[137]
but these are unstable and decompose to nitrogen and the hydrocarbon:

$$Ar{-}NH{-}NH_2 \longrightarrow [Ar{-}N{=}NH] \longrightarrow ArH + N_2$$

When hydrazones are oxidized with HgO, Ag_2O, MnO_2, lead tetraacetate, or certain other
oxidizing agents, diazo compounds are obtained[138] (see also **7-28**):

$$R_2C{=}N{-}NH_2 \xrightarrow{HgO} R_2C{=}\overset{\oplus}{N}{=}\overset{\ominus}{N}$$

Hydrazones of the form ArCH=NNH$_2$ react with HgO in solvents such as diglyme or ethanol
to give nitriles ArCN.[139] Aromatic hydroxylamines are easily oxidized to nitroso compounds,
most commonly by acid dichromate.[140]

$$Ar{-}NH{-}OH \xrightarrow{CrO_3} Ar{-}N{=}O$$

OS **II**, 496; **III**, 351, 356, 375, 668; **IV**, 66, 411; **V**, 96, 160, 897; **VI**, 78, 161, 334, 392,
803, 936; **VII**, 56. Also see OS **V**, 258.

[131]Richey; Erickson *Tetrahedron Lett.* **1972**, 2807; Erickson; Richey *Tetrahedron Lett.* **1972**, 2811.
[132]For a review of HgO, see Pizey, Ref. 10, vol. 1, 1974, pp. 295-319.
[133]Dimroth; Tüncher *Synthesis* **1977**, 339.
[134]Barton; Lester; Ley *J. Chem. Soc., Chem. Commun.* **1978**, 276; Back *J. Chem. Soc., Chem. Commun.* **1978**, 278.
[135]Hyatt *Tetrahedron Lett.* **1977**, 141.
[136]For a review, see Newbold, in Patai *The Chemistry of the Hydrazo, Azo, and Azoxy Groups*, pt. 1; Wiley: New York, 1975, pp. 543-557, 564-573.
[137]See Mannen; Itano *Tetrahedron* **1973**, 29, 3497.
[138]For a review, see Regitz; Maas *Diazo Compounds*; Academic Press: New York, 1986, pp. 233-256.
[139]Mobbs; Suschitzky *Tetrahedron Lett.* **1971**, 361.
[140]For a review, see Hudlický, Ref. 11, pp. 231-232.

B.   Oxidations Involving Cleavage of Carbon–Carbon Bonds[141]

9-7   Oxidative Cleavage of Glycols and Related Compounds
2/ *O*-De-hydrogen-uncoupling

$$\begin{array}{c} | \quad | \\ -\text{C}-\text{C}- \\ | \quad | \\ \text{OH} \ \ \text{OH} \end{array} \xrightarrow[\text{Pb(OAc)}_4]{\text{HIO}_4 \text{ or}} \begin{array}{c} | \quad | \\ -\text{C} + \text{C}- \\ \| \quad \| \\ \text{O} \quad \text{O} \end{array}$$

1,2-Glycols are easily cleaved under mild conditions and in good yield with periodic acid or lead tetraacetate.[142] The products are 2 moles of aldehyde, or 2 moles of ketone, or 1 mole of each, depending on the groups attached to the two carbons. The yields are so good that olefins are often converted to glycols (**5-35**) and then cleaved with HIO_4 or $Pb(OAc)_4$ rather than being cleaved directly with ozone (**9-9**) or dichromate or permanganate (**9-10**). A number of other oxidizing agents also give the same products, among them[143] activated MnO_2,[144] thallium(III) salts,[145] pyridinium chlorochromate,[146] and O_2 catalyzed by Co(III) salts.[147] Permanganate, dichromate, and several other oxidizing agents[148] also cleave glycols, giving carboxylic acids rather than aldehydes, but these reagents are seldom used synthetically. Electrochemical oxidation is an efficient method, and is useful not only for diols, but also for their mono- and dimethoxy derivatives.[149]

The two reagents (periodic acid and lead tetraacetate) are complementary, since periodic acid is best used in water and lead tetraacetate in organic solvents. When three or more OH groups are located on adjacent carbons, the middle one (or ones) is converted to formic acid.

Similar cleavage is undergone by other compounds that contain oxygens or nitrogens on adjacent carbons:

$$\begin{array}{c} \text{C}-\text{C} \\ | \quad | \\ \text{OH} \ \ \text{NH}_2 \end{array} \qquad \begin{array}{c} \text{C}-\text{C} \\ | \quad | \\ \text{OH} \ \ \text{NHR} \end{array} \qquad \begin{array}{c} \text{C}-\text{C} \\ | \quad \| \\ \text{OH} \quad \text{O} \end{array} \qquad \begin{array}{c} \text{C}-\text{C} \\ \| \quad \| \\ \text{O} \quad \text{O} \end{array}$$

| β-Amino alcohols | | α-Hydroxy aldehydes and ketones | α-Diketones α-Keto aldehydes Glyoxal |

α-Diketones and α-hydroxy ketones are also cleaved by alkaline H_2O_2.[150] HIO_4 has been used to cleave epoxides to aldehydes,[151] e.g.,

$$\text{(cyclohexene oxide)} \xrightarrow{\text{HIO}_4} \text{HCO}-(\text{CH}_2)_4-\text{CHO}$$

[141]For a review, see Bentley, in Bentley; Kirby, Ref. 12, pp. 137-254.
[142]For reviews covering both reagents, see Haines-1988, Ref. 11, pp. 277-301, 432-437; House, Ref. 10, pp. 3353-363; Perlin, in Augustine *Oxidation*, vol. 1; Marcel Dekker: New York, 1969, pp. 189-212; Bunton, in Wiberg, Ref. 2, pp. 367-407. For reviews of lead tetraacetate, see Rubottom, Ref. 10; Aylward, Ref. 10. For reviews of HIO_4, see Fatiadi, Ref. 10; Sklarz, Ref. 10.
[143]For a list of reagents, with references, see Ref. 21, pp. 615-616.
[144]Adler; Becker *Acta Chem. Scand.* **1961**, *15*, 849; Ohloff; Giersch *Angew. Chem. Int. Ed. Engl.* **1973**, *12*, 401 [*Angew. Chem. 85*, 401].
[145]McKillop; Raphael; Taylor *J. Org. Chem.* **1972**, *37*, 4204.
[146]Cisneros; Fernández; Hernández *Synth. Comm.* **1982**, *12*, 833.
[147]de Vries; Schors *Tetrahedron Lett.* **1968**, 5689.
[148]For a list of reagents, with references, see Ref. 21, pp. 836-837.
[149]For a review, see Shono *Electroorganic Chemistry as a New Tool in Organic Synthesis*; Springer: New York, 1984, pp. 31-37. See also Ruholl; Schäfer *Synthesis* **1988**, 54.
[150]See, for example, Ogata; Sawaki; Shiroyama *J. Org. Chem.* **1977**, *42*, 4061.
[151]Nagarkatti; Ashley *Tetrahedron Lett.* **1973**, 4599.

α-Hydroxy acids and α-keto acids are not cleaved by HIO_4 but are cleaved by $Pb(OAc)_4$, alkaline H_2O_2, and other reagents. These are oxidative decarboxylations. α-Hydroxy acids give aldehydes or ketones, and α-keto acids give carboxylic acids:

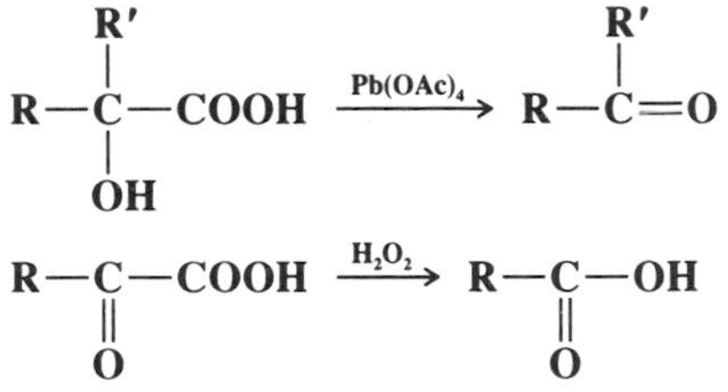

$$\underset{\overset{|}{OH}}{\overset{\overset{R'}{|}}{R-C-COOH}} \xrightarrow{Pb(OAc)_4} \overset{\overset{R'}{|}}{R-C=O}$$

$$\underset{\overset{||}{O}}{R-C-COOH} \xrightarrow{H_2O_2} \underset{\overset{||}{O}}{R-C-OH}$$

Also see **9-13** and **9-14**.

The mechanism of glycol oxidation with $Pb(OAc)_4$ was proposed by Criegee:[152]

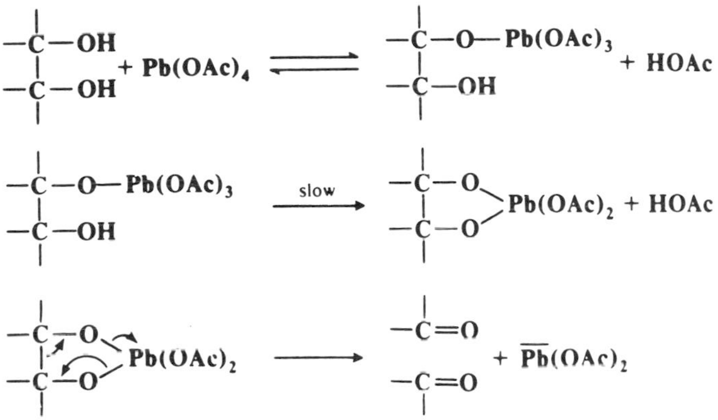

$$\underset{\overset{|}{-C-OH}}{\overset{\overset{|}{-C-OH}}{}} + Pb(OAc)_4 \; \rightleftharpoons \; \underset{\overset{|}{-C-OH}}{\overset{\overset{|}{-C-O-Pb(OAc)_3}}{}} + HOAc$$

$$\underset{\overset{|}{-C-OH}}{\overset{\overset{|}{-C-O-Pb(OAc)_3}}{}} \xrightarrow{\text{slow}} \underset{\overset{|}{-C-O}}{\overset{\overset{|}{-C-O}}{}} Pb(OAc)_2 + HOAc$$

$$\underset{\overset{|}{-C-O}}{\overset{\overset{|}{-C-O}}{}} Pb(OAc)_2 \longrightarrow \underset{\overset{|}{-C=O}}{\overset{\overset{|}{-C=O}}{}} + \overline{Pb}(OAc)_2$$

This mechanism is supported by these facts: (1) the kinetics are second order (first order in each reactant); (2) added acetic acid retards the reaction (drives the equilibrium to the left); and (3) cis glycols react much more rapidly than trans glycols.[153] For periodic acid the mechanism is similar, with the intermediate[154]

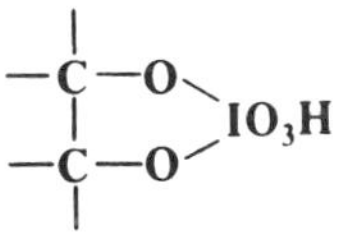

$$\underset{\overset{|}{-C-O}}{\overset{\overset{|}{-C-O}}{}} IO_3H$$

However, the cyclic-intermediate mechanism cannot account for all glycol oxidations, since some glycols that cannot form such an ester (e.g., **10**) are nevertheless cleaved by lead

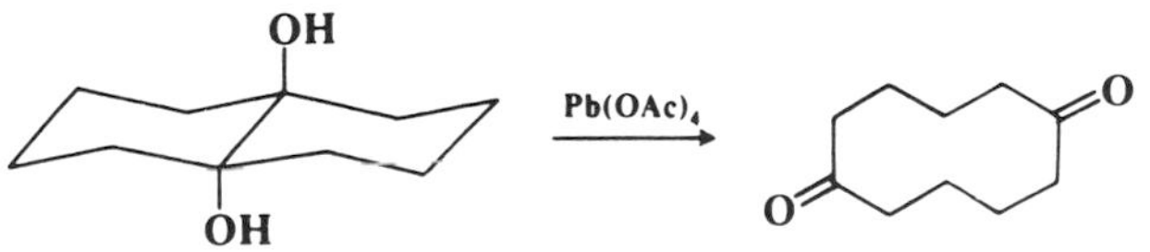

10

---

[152]Criegee; Kraft; Rank *Liebigs Ann. Chem.* **1933,** *507,* 159. For reviews, see Waters, Ref. 2, pp. 72-81; Stewart, Ref. 2, pp. 97-106.

[153]For example, see Criegee; Höger; Huber; Kruck; Marktscheffel; Schellenberger *Liebigs Ann. Chem.* **1956,** *599,* 81.

[154]Buist; Bunton; Miles *J. Chem. Soc.* **1959,** 743; Buist; Bunton; Hipperson *J. Chem. Soc. B* **1971,** 2128.

tetraacetate (though other glycols that cannot form cyclic esters are *not* cleaved, by either reagent[155]). To account for cases like **10**, a cyclic transition state has been proposed:[153]

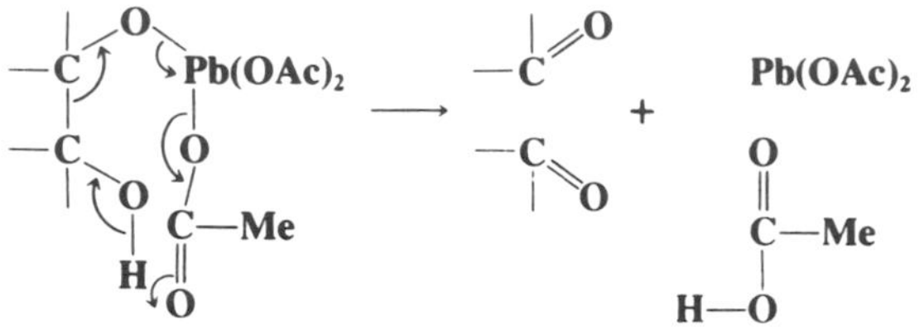

OS **IV**, 124; **VII**, 185; **68**, 162.

9-8 Oxidative Cleavage of Ketones, Aldehydes, and Alcohols
Cycloalkanone oxidative ring opening

$$\text{HOOC}-(CH_2)_4-\text{COOH}$$

Oxidative cleavage of open-chain ketones or alcohols[156] is seldom a useful preparative procedure, not because these compounds do not undergo oxidation (they do, except for diaryl ketones) but because the result is generally a hopeless mixture. However, the reaction is quite useful for cyclic ketones and the corresponding secondary alcohols, the dicarboxylic acid being prepared in good yield. The formation of adipic acid from cyclohexanone (shown above) is an important industrial procedure. Acid dichromate and permanganate are the most common oxidizing agents, though autoxidation (oxidation with atmospheric oxygen) in alkaline solution[157] and potassium superoxide under phase transfer conditions[158] have also been used. The last-mentioned reagent has also been used to cleave open-chain ketones to give carboxylic acid products in good yield.[158]

Cyclic ketones can also be cleaved by treatment with NOCl and an alcohol in liquid SO_2 to give an ω-oximinocarboxylic ester, e.g.,[159]

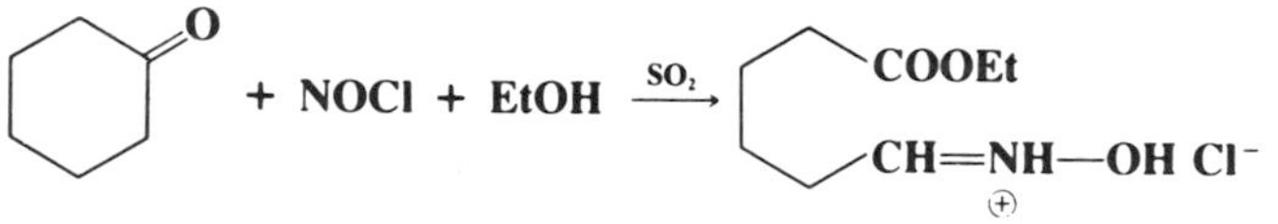

Cyclic 1,3-diketones, which exist mainly in the monoenolic form, can be cleaved with sodium periodate with loss of one carbon, e.g.,[160]

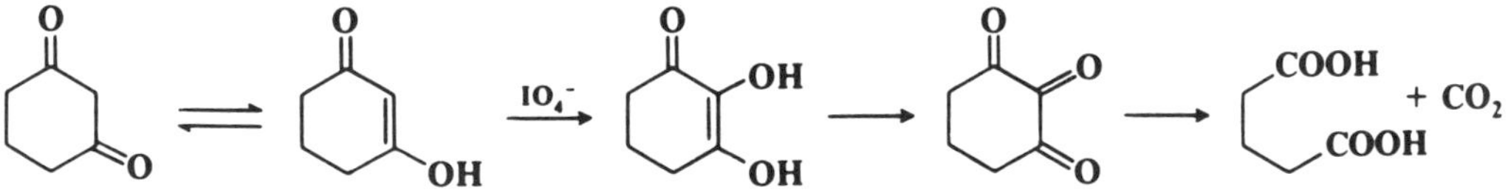

[155]Angyal; Young *J. Am. Chem. Soc.* **1959**, *81*, 5251.
[156]For a review of metal ion-catalyzed oxidative cleavage of alcohols, see Trahanovsky *Methods Free-Radical Chem.* **1973**, *4*, 133-169. For a review of the oxidation of aldehydes and ketones, see Verter, in Zabicky *The Chemistry of the Carbonyl Group*, pt. 2; Wiley: New York, 1970, pp. 71-156.
[157]Wallace; Pobiner; Schriesheim *J. Org. Chem.* **1965**, *30*, 3768. See also Osowska-Pacewicka; Alper *J. Org. Chem.* **1988**, *53*, 808.
[158]Lissel; Dehmlow *Tetrahedron Lett.* **1978**, 3689; Sotiriou; Lee; Giese *J. Org. Chem.* **1990**, *55*, 2159.
[159]Rogić; Vitrone; Swerdloff *J. Am. Chem. Soc.* **1977**, *99*, 1156; Moorhoff; Paquette *J. Org. Chem.* **1991**, *56*, 703.
[160]Wolfrom; Bobbitt *J. Am. Chem. Soc.* **1956**, *78*, 2489.

The species actually undergoing the cleavage is the triketone, so this is an example of **9-7**. OS **I**, 18; **IV**, 19; **VI**, 690. See also OS **VI**, 1024.

9-9 Ozonolysis
Oxo-uncoupling

11

When compounds containing double bonds are treated with ozone, usually at low temperatures, they are converted to compounds called *ozonides* (**11**) that can be isolated but, because some of them are explosive, are more often decomposed with zinc and acetic acid or catalytic hydrogenation to give 2 moles of aldehyde, or 2 moles of ketone, or 1 mole of each, depending on the groups attached to the olefin.[161] The decomposition of **11** has also been carried out with many other reducing agents, among them trimethyl phosphite,[162] thiourea,[163] and dimethyl sulfide.[164] However, ozonides can also be *oxidized* with oxygen, peracids, or H_2O_2 to give ketones and/or carboxylic acids or *reduced* with $LiAlH_4$, $NaBH_4$, BH_3, or catalytic hydrogenation with excess H_2 to give 2 moles of alcohol.[165] Ozonides can also be treated with ammonia, hydrogen, and a catalyst to give the corresponding amines,[166] or with an alcohol and anhydrous HCl to give the corresponding carboxylic esters.[167] Ozonolysis is therefore an important synthetic reaction.

A wide variety of olefins undergo ozonolysis, including cyclic ones, where cleavage gives rise to one bifunctional product. Olefins in which the double bond is connected to electron-donating groups react many times faster than those in which it is connected to electron-withdrawing groups.[168] The reaction has often been carried out on compounds containing more than one double bond; generally all the bonds are cleaved. In some cases, especially when bulky groups are present, conversion of the substrate to an epoxide (**5-36**) becomes an important side reaction and can be the main reaction.[169] Ozonolysis of triple bonds[170] is less common and the reaction proceeds less easily, since ozone is an electrophilic agent[171]

[161]For monographs, see Razumovskii; Zaikov *Ozone and its Reactions with Organic Compounds*; Elsevier: New York, 1984; Bailey *Ozonation in Organic Chemistry*, 2 vols.; Academic Press: New York, 1978, 1982. For reviews, see Odinokov; Tolstikov *Russ. Chem. Rev.* **1981**, *50*, 636-657; Belew, in Augustine; Trecker, Ref. 11, vol. 1, pp. 259-335; Menyailo; Pospelov *Russ. Chem. Rev.* **1967**, *36*, 284-294. For a review with respect to vinylic ethers, see Kuczkowski *Adv. Oxygenated Processes* **1991**, *3*, 1-42. For a review with respect to haloalkenes, see Gillies; Kuczkowski *Isr. J. Chem.* **1983**, *23*, 446-450.
[162]Knowles; Thompson *J. Org. Chem.* **1960**, *25*, 1031.
[163]Gupta; Soman; Dev *Tetrahedron* **1982**, *38*, 3013.
[164]Pappas; Keaveney; Gancher; Berger *Tetrahedron Lett.* **1966**, 4273.
[165]Sousa; Bluhm *J. Org. Chem.* **1960**, *25*, 108; Diaper; Mitchell *Can. J. Chem.* **1960**, *38*, 1976; Diaper; Strachan *Can. J. Chem.* **1967**, *45*, 33; White; King; O'Brien *Tetrahedron Lett.* **1971**, 3587; Flippin; Gallagher; Jalali-Araghi *J. Org. Chem.* **1989**, *54*, 1430.
[166]Diaper; Mitchell *Can. J. Chem.* **1962**, *40*, 1189; Benton; Kiess *J. Org. Chem.* **1960**, *25*, 470; Pollart; Miller *J. Org. Chem.* **1962**, *27*, 2392; White; King; O'Brien *Tetrahedron Lett.* **1971**, 3591.
[167]Neumeister; Keul; Saxena; Griesbaum *Angew. Chem. Int. Ed. Engl.* **1978**, *17*, 939 [*Angew. Chem. 90*, 999]. See also Schreiber; Claus; Reagan *Tetrahedron Lett.* **1982**, *23*, 3867; Cardinale; Grimmelikhuysen; Laan; Ward *Tetrahedron* **1984**, *40*, 1881.
[168]Pryor; Giamalva; Church *J. Am. Chem. Soc.* **1983**, *105*, 6858, **1985**, *107*, 2793.
[169]See, for example, Bailey; Lane *J. Am. Chem. Soc.* **1967**, *89*, 4473; Gillies *J. Am. Chem. Soc.* **1975**, *97*, 1276; Bailey; Hwang; Chiang *J. Org. Chem.* **1985**, *50*, 231.
[170]For a discussion of the mechanism of ozonolysis of triple bonds, see Pryor; Govindan; Church *J. Am. Chem. Soc.* **1982**, *104*, 7563.
[171]See, for example, Wibaut; Sixma *Recl. Trav. Chim. Pays-Bas* **1952**, *71*, 761; Williamson; Cvetanović *J. Am. Chem. Soc.* **1968**, *90*, 4248; Razumovskii; Zaikov *J. Org. Chem. USSR* **1972**, *8*, 468, 473; Klutsch; Fliszár *Can. J. Chem.* **1972**, *50*, 2841.

and prefers double to triple bonds (p. 748). Compounds that contain triple bonds generally give carboxylic acids, though sometimes ozone oxidizes them to α-diketones (**9-27**). Aromatic compounds are also attacked less readily than olefins, but have often been cleaved. Aromatic compounds behave as if the double bonds in the Kekulé structures were really there. Thus benzene gives 3 moles of glyoxal (HCOCHO), and o-xylene gives a glyoxal/MeCOCHO/MeCOCOMe ratio of 3:2:1, which shows that in this case cleavage is statistical. With polycyclic aromatic compounds the site of attack depends on the structure of the molecule and on the solvent.[172]

Although a large amount of work has been done on the mechanism of ozonization (formation of **11**), not all the details are known. The basic mechanism was formulated by Criegee.[173] The first step of the Criegee mechanism is a 1,3 dipolar addition (**5-46**) of ozone to the substrate to give the "initial" or "primary" ozonide, the structure of which has been shown to be the 1,2,3-trioxolane **12** by microwave and other spectral methods.[174] However,

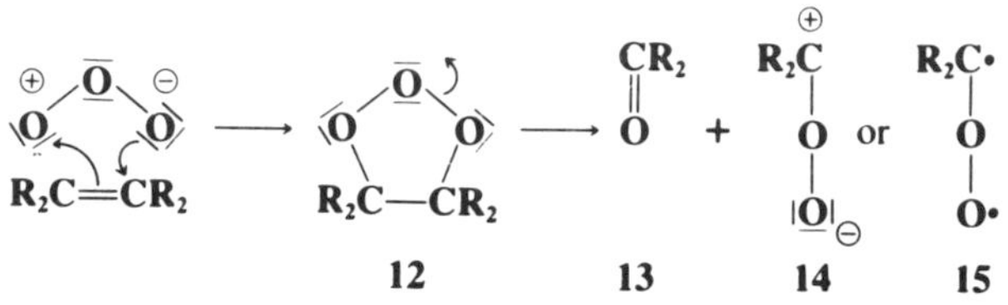

12 is highly unstable and cleaves to an aldehyde or ketone (**13**) and an intermediate which Criegee showed as a zwitterion (**14**) but which may be a diradical (**15**). This compound is usually referred to as a carbonyl oxide.[175] The carbonyl oxide (which we will represent as **14**) can then undergo various reactions, three of which lead to normal products. One is a recombination with **13**, the second a dimerization to the bisperoxide **16**, and the third a

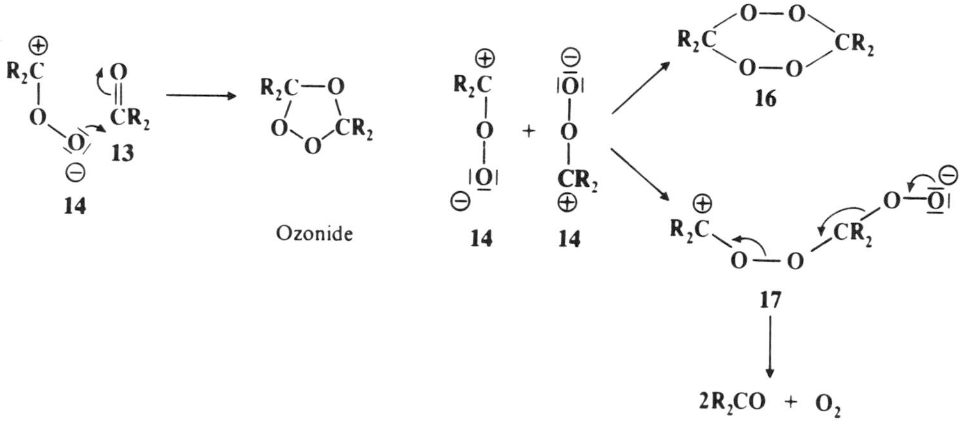

[172]Dobinson; Bailey *Tetrahedron Lett.* **1960** (No. 13) 14; O'Murchu *Synthesis* **1989**, 880.
[173]For reviews, see Kuczkowski *Acc. Chem. Res.* **1983**, *16*, 42-47; Razumovskii; Zaikov *Russ. Chem. Rev.* **1980**, *49*, 1163-1180; Criegee *Angew. Chem. Int. Ed. Engl.* **1975**, *14*, 745-752 [*Angew. Chem. 87*, 765-771]; Murray *Acc. Chem. Res.* **1968**, *1*, 313-320.
[174]Gillies; Gillies; Suenram; Lovas *J. Am. Chem. Soc.* **1988**, *110*, 7991. See also Criegee; Schröder *Chem. Ber.* **1960**, *93*, 689; Durham; Greenwood *J. Org. Chem.* **1968**, *33*, 1629; Bailey; Carter; Fischer; Thompson *Can. J. Chem.* **1973**, *51*, 1278; Hisatsune; Shinoda; Heicklen *J. Am. Chem. Soc.* **1979**, *101*, 2524; Mile; Morris; Alcock *J. Chem. Soc., Perkin Trans. 2* **1979**, 1644; Kohlmiller; Andrews *J. Am. Chem. Soc.* **1981**, *103*, 2578; McGarrity; Prodolliet *J. Org. Chem.* **1984**, *49*, 4465.
[175]For reviews of carbonyl oxides, see Sander *Angew. Chem. Int. Ed. Engl.* **1990**, *29*, 344-354 [*Angew. Chem. 102*, 362-372]; Brunelle *Chem. Rev.* **1991**, *91*, 335-362.

kind of dimerization to **17**.[176] If the first path is taken (this is normally possible only if **13** is an aldehyde; most ketones do not do this[177]), hydrolysis of the ozonide gives the normal products. If **16** is formed, hydrolysis of it gives one of the products, and of course **13**, which then does not undergo further reaction, is the other. **17**, if formed, can decompose directly, as shown, to give the normal products and oxygen. In protic solvents, **14** is converted to a hydroperoxide, and these have been isolated, for example, Me_2C—OMe from Me_2C=CMe_2

$$\overset{|}{O}OH$$

in methanol. Further evidence for the mechanism is that **16** can be isolated in some cases, e.g., from Me_2C=CMe_2. But perhaps the most impressive evidence comes from the detection of cross products. In the Criegee mechanism, the two parts of the original olefin break apart and then recombine to form the ozonide. In the case of an unsymmetrical olefin RCH=CHR' there should be three ozonides:

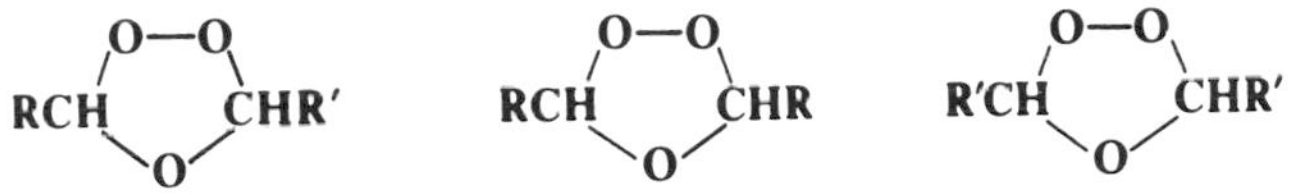

since there are two different aldehydes **13** and two different species **14**, and these can recombine in the three ways shown. Actually *six* ozonides, corresponding to the cis and trans forms of these three, were isolated and characterized for methyl oleate.[178] Similar results have been reported for smaller olefins, e.g., 2-pentene, 4-nonene, and even 2-methyl-2-pentene.[179] The last-mentioned case is especially interesting, since it is quite plausible that this compound would cleave in only one way, so that only one ozonide (in cis and trans versions) would be found; but this is not so, and three were found for this case too. However, terminal olefins give little or no cross ozonide formation.[180] In general, the less alkylated end of the olefin tends to go to **13** and the other to **14**. Still other evidence[181] for the Criegee mechanism is: (1) When Me_2C=CMe_2 was ozonized in the presence of HCHO, the ozonide **18** could be isolated;[182] (2) **14** prepared in an entirely different manner (photooxidation of

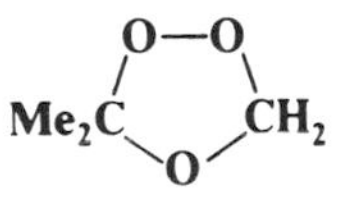

18

diazo compounds), reacted with aldehydes to give ozonides;[183] and (3) cis and trans olefins generally give the same ozonide, which would be expected if they cleave first.[184] However,

[176]Fliszár; Gravel; Cavalieri *Can. J. Chem.* **1966**, *44*, 67, 1013; Fliszár; Chylińska *Can. J. Chem.* **1967**, *45*, 29, **1968**, *46*, 783.

[177]It follows that tetrasubstituted alkenes do not normally give ozonides. However, they do give the normal cleavage products (ketones) by the other pathways. For the preparation of ozonides from tetrasubstituted alkenes by ozonolysis on polyethylene, see Griesbaum; Volpp; Greinert; Greunig; Schmid; Henke *J. Org. Chem.* **1989**, *54*, 383.

[178]Riezebos; Grimmelikhuysen; van Dorp *Recl. Trav. Chim. Pays-Bas* **1963**, *82*, 1234; Privett; Nickell *J. Am. Oil Chem. Soc.* **1964**, *41*, 72.

[179]Loan; Murray; Story *J. Am. Chem. Soc.* **1965**, *87*, 737; Lorenz; Parks *J. Org. Chem.* **1965**, *30*, 1976.

[180]Murray; Williams *J. Org. Chem.* **1969**, *34*, 1891.

[181]For further evidence, see Keul; Choi; Kuczkowski *J. Org. Chem.* **1985**, *50*, 3365; Mori; Nojima; Kusabayashi *J. Am. Chem. Soc.* **1987**, *109*, 4407; Pierrot; El Idrissi; Santelli *Tetrahedron Lett.* **1989**, *30*, 461; Wojciechowski; Chiang; Kuczkowski *J. Org. Chem.* **1990**, *55*, 1120; Paryzek; Martynow; Swoboda *J. Chem. Soc., Perkin Trans. 1* **1990**, 1220; Murray; Morgan *J. Org. Chem.* **1991**, *56*, 684, 6123.

[182]Even ketones can react with **14** to form ozonides, provided they are present in large excess: Criegee; Korber *Chem. Ber.* **1971**, *104*, 1812.

[183]Murray; Suzui *J. Am. Chem. Soc.* **1973**, *95*, 3343; Higley; Murray *J. Am. Chem. Soc.* **1974**, *96*, 3330.

[184]See, for example, Murray; Williams *J. Org. Chem.* **1969**, *34*, 1896.

this was not true for $Me_3CCH=CHCMe_3$, where the cis olefin gave the cis ozonide (chiefly), and the trans gave the trans.[185] The latter result is not compatible with the Criegee mechanism. Also incompatible with the Criegee mechanism was the finding that the cis/trans ratios of symmetrical (cross) ozonides obtained from *cis*- and *trans*-4-methyl-2-pentene were not the same.[186]

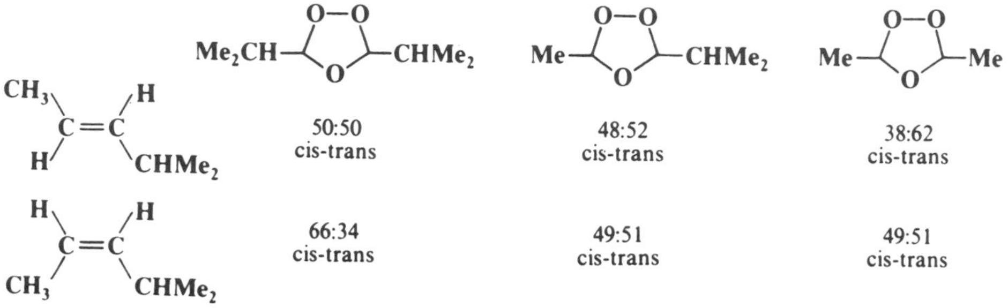

If the Criegee mechanism operated as shown above, the cis/trans ratio for each of the two cross ozonides would have to be identical for the cis and trans olefins, since in this mechanism they are completely cleaved.

The above stereochemical results have been explained[187] on the basis of the Criegee mechanism with the following refinements: (1) The formation of **12** is stereospecific, as expected from a 1,3 dipolar cycloaddition. (2) Once they are formed, **14** and **13** remain attracted to each other, much like an ion pair. (3) **14** exists in syn and anti forms, which are produced in different amounts and can hold their shapes, at least for a time. This is

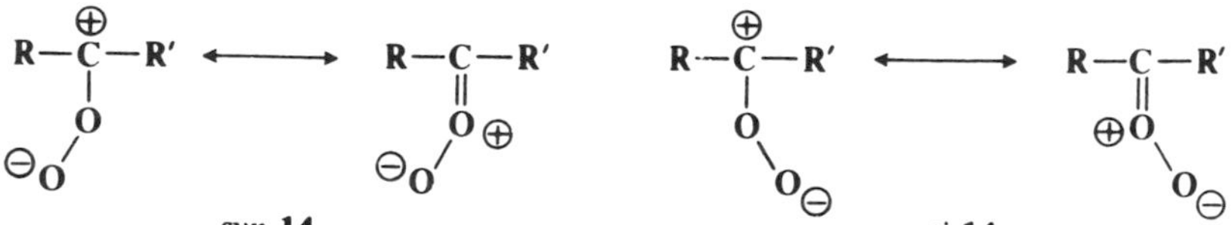

plausible if we remember that a $C=O$ canonical form contributes to the structure of **14**. (4) The combination of **14** and **13** is also a 1,3 dipolar cycloaddition, so configuration is retained in this step too.[188]

Evidence that the basic Criegee mechanism operates even in these cases comes from ^{18}O labeling experiments, making use of the fact, mentioned above, that mixed ozonides (e.g., **18**) can be isolated when an external aldehyde is added. Both the normal and modified Criegee mechanisms predict that if ^{18}O-labeled aldehyde is added to the ozonolysis mixture, the label will appear in the ether oxygen (see the reaction between **14** and **13**), and this is what is found.[189] There is evidence that the anti-**14** couples much more readily than the syn-**14**.[190]

[185]Schröder *Chem. Ber.* **1962**, *95*, 733; Kolsaker *Acta Chem. Scand., Ser. B* **1978**, *32*, 557.
[186]Murray; Youssefyeh; Story *J. Am. Chem. Soc.* **1966**, *88*, 3143, 3655; Story; Murray; Youssefyeh *J. Am. Chem. Soc.* **1966**, *88*, 3144. Also see Greenwood *J. Am. Chem. Soc.* **1966**, *88*, 3146; Choe; Srinivasan; Kuczkowski *J. Am. Chem. Soc.* **1983**, *105*, 4703.
[187]Bauld; Thompson; Hudson; Bailey *J. Am. Chem. Soc.* **1968**, *90*, 1822; Bailey; Ferrell *J. Am. Chem. Soc.* **1978**, *100*, 899; Keul; Kuczkowski *J. Am. Chem. Soc.* **1985**, *50*, 3371.
[188]For isotope-effect evidence that this step is concerted in some cases, see Choe; Painter; Kuczkowski *J. Am. Chem. Soc.* **1984**, *106*, 2891. However, there is evidence that it may not always be concerted: See, for example, Murray; Su *J. Org. Chem.* **1983**, *48*, 817.
[189]Bishop; Denson; Story *Tetrahedron Lett.* **1968**, 5739; Fliszár; Carles; Renard *J. Am. Chem. Soc.* **1968**, *90*, 1364; Fliszár; Carles *J. Am. Chem. Soc.* **1969**, *91*, 2637; Gillies; Kuczkowski *J. Am. Chem. Soc.* **1972**, *94*, 7609; Higley; Murray *J. Am. Chem. Soc.* **1976**, *98*, 4526; Mazur; Kuczkowski *J. Org. Chem.* **1979**, *44*, 3185.
[190]Mile; Morris *J. Chem. Soc., Chem. Commun.* **1978**, 263.

The ozonolysis of ethylene in the liquid phase (without a solvent) was shown to take place by the Criegee mechanism.[191] This reaction has been used to study the structure of the intermediate **14** or **15**. The compound dioxirane (**19**) was identified in the reac-

$$H_2C \underset{O}{\overset{O}{\diagdown}} \rvert \;\; \rightleftharpoons \;\; \overset{O}{C} H_2 \overset{O}{\diagdown} \rvert$$

19   **15(R = H)**

tion mixture[192] at low temperatures and is probably in equilibrium with the biradical **15** (R = H).

Ozonolysis in the gas phase is not generally carried out in the laboratory. However, the reaction is important because it takes place in the atmosphere and contributes to air pollution.[193] There is much evidence that the Criegee mechanism operates in the gas phase too, though the products are more complex because of other reactions that also take place.[194]

OS **V**, 489, 493; **VI**, 976; **VII**, 168. Also see OS **IV**, 554. For the preparation of ozone, see OS **III**, 673.

9-10 Oxidative Cleavage of Double Bonds and Aromatic Rings
Oxo-de-alkylidene-bisubstitution, etc.

$$R_2C{=}CHR \xrightarrow{\text{CrO}_3} R_2C{-}O + RCOOH$$

Double bonds can be cleaved by many oxidizing agents,[195] the most common of which are neutral or acid permanganate and acid dichromate. The products are generally 2 moles of ketone, 2 moles of carboxylic acid, or 1 mole of each, depending on what groups are attached to the olefin. With ordinary solutions of permanganate or dichromate yields are generally low, and the reaction is seldom a useful synthetic method; but high yields can be obtained by oxidizing with $KMnO_4$ dissolved in benzene containing the crown ether dicyclohexano-18-crown-6 (see p. 82).[196] The crown ether coordinates with K^+, permitting the $KMnO_4$ to dissolve in benzene. Another reagent frequently used for synthetic purposes is the *Lemieux–von Rudloff reagent*: HIO_4 containing a trace of MnO_4^-.[197] The MnO_4^- is the actual oxidizing agent, being reduced to the manganate stage, and the purpose of the $IIIO_4$ is to reoxidize the manganate back to MnO_4^-. Another reagent that behaves similarly is $NaIO_4$–ruthenium tetroxide.[198]

[191]Fong; Kuczkowski *J. Am. Chem. Soc.* **1980**, *102*, 4763.

[192]Suenram; Lovas *J. Am. Chem. Soc.* **1978**, *100*, 5117. See, however, Ishiguro; Hirano; Sawaki *J. Org. Chem.* **1988**, *53*, 5397.

[193]For a review of the mechanisms of reactions of organic compounds with ozone in the gas phase, see Atkinson; Carter *Chem. Rev.* **1984**, *84*, 437-470.

[194]See Ref. 193, pp. 452-454; Kühne; Forster; Hulliger; Ruprecht; Bauder; Günthard *Helv. Chim. Acta* **1980**, *63*, 1971; Martinez; Herron *J. Phys. Chem.* **1988**, *92*, 4644.

[195]For a review of the oxidation of C=C and C=N bonds, see Henry; Lange, in Patai, Ref. 103, pp. 965-1098. For a review of oxidative cleavages of C=C double bonds and aromatic rings, see Hudlický, Ref. 11, pp. 77-84, 96-98. For reviews with respect to chromium reagents, see Badanyan; Minasyan; Vardapetyan *Russ. Chem. Rev.* **1987**, *56*, 740-755; Cainelli; Cardillo, Ref. 10, pp. 59-92. For a list of reagents, with references, see Ref. 21, p. 828.

[196]Sam; Simmons *J. Am. Chem. Soc.* **1972**, *94*, 4024. See also Lee; Chang *J. Org. Chem.* **1978**, *43*, 1532.

[197]Lemieux; Rudloff *Can. J. Chem.* **1955**, *33*, 1701, 1710; Rudloff *Can. J. Chem.* **1955**, *33*, 1714, **1956**, *34*, 1413, **1965**, *43*, 1784.

[198]For a review, see Lee; van den Engh, in Trahanovsky, Ref. 2, pt. B, pp. 186-192. For the use of $NaIO_4$–OsO_4, see Cainelli; Contento; Manescalchi; Plessi *Synthesis* **1989**, 47.

The *Barbier–Wieland procedure* for decreasing the length of a chain by one carbon involves oxidative cleavage by acid dichromate ($NaIO_4$–ruthenium tetroxide has also been used), but this is cleavage of a 1,1-diphenyl olefin, which generally gives good yields:

$$RCH_2COOH \xrightarrow[\substack{H^+ \\ 0\text{-}23}]{EtOH} RCH_2COOEt \xrightarrow[6\text{-}31]{PhMgBr}$$

$$\underset{\underset{OH}{|}}{RCH_2CPh_2} \xrightarrow[7\text{-}1]{\Delta} RCH{=}CPh_2 \xrightarrow[\substack{or \\ NaIO_4\text{-}RuO_4}]{CrO_3} RCOOH + Ph_2CO$$

With certain reagents, the oxidation of double bonds can be stopped at the aldehyde stage, and in these cases the products are the same as in the ozonolysis procedure. Among these reagents are chromyl trichloroacetate,[199] *t*-butyl iodoxybenzene,[200] $KMnO_4$ in $THF–H_2O$,[201] and $NaIO_4–OsO_4$.[202] Enol ethers $RC(OR'){=}CH_2$ have been cleaved to carboxylic esters $RC(OR'){=}O$ by atmospheric oxygen.[203]

The mechanism of oxidation probably involves in most cases the initial formation of a glycol (**5-35**) or cyclic ester,[204] and then further oxidation as in **9-7**.[205] In line with the electrophilic attack on the olefin, triple bonds are more resistant to oxidation than double bonds. Terminal triple-bond compounds can be cleaved to carboxylic acids ($RC{\equiv}CH \rightarrow RCOOH$) with thallium(III) nitrate[206] or with [bis(trifluoroacetoxy)iodo]pentafluorobenzene $C_6F_5I(OCOCF_3)_2$,[207] among other reagents.

Aromatic rings can be cleaved with strong enough oxidizing agents. An important laboratory reagent for this purpose is ruthenium tetroxide along with a cooxidant such as $NaIO_4$ or $NaOCl$ (household bleach can be used).[208] Examples[209] are the oxidation of naphthalene to phthalic acid[210] and, even more remarkably, of cyclohexylbenzene to cyclohexanecar-

boxylic acid[211] (note the contrast with **9-11**). The latter conversion was also accomplished with ozone.[212] Another reagent that oxidizes aromatic rings is air catalyzed by V_2O_5. The

[199]Schildknecht; Föttinger *Liebigs Ann. Chem.* **1962**, *659*, 20.
[200]Ranganathan; Ranganathan; Singh *Tetrahedron Lett.* **1985**, *26*, 4955.
[201]Viski; Szeverényi; Simándi *J. Org. Chem.* **1986**, *51*, 3213.
[202]Pappo; Allen; Lemieux; Johnson *J. Org. Chem.* **1956**, *21*, 478.
[203]Taylor *J. Chem. Res. (S)* **1987**, 178. For a similar oxidation with RuO_4, see Torii; Inokuchi; Kondo *J. Org. Chem.* **1985**, *50*, 4980.
[204]See, for example, Lee; Spitzer *J. Org. Chem.* **1976**, *41*, 3644; Lee; Chang; Helliwell *J. Org. Chem.* **1976**, *41*, 3644, 3646.
[205]There is evidence that oxidation with Cr(VI) in aqueous acetic acid involves an epoxide intermediate: Awasthy; Roček *J. Am. Chem. Soc.* **1969**, *91*, 991; Roček; Drozd *J. Am. Chem. Soc.* **1970**, *92*, 6668.
[206]McKillop; Oldenziel; Swann; Taylor; Robey *J. Am. Chem. Soc.* **1973**, *95*, 1296.
[207]Moriarty; Penmasta; Awasthi; Prakash *J. Org. Chem.* **1988**, *53*, 6124.
[208]Ruthenium tetroxide is an expensive reagent, but the cost can be greatly reduced by the use of an inexpensive cooxidant such as $NaOCl$, the function of which is to oxidize RuO_2 back to ruthenium tetroxide.
[209]For other examples, see Piatak; Herbst; Wicha Caspi *J. Org. Chem.* **1969**, *34*, 116; Wolfe; Hasan; Campbell *Chem. Commun.* **1970**, 1420; Ayres; Hossain *Chem. Commun.* **1972**, 428; Nuñez; Martín *J. Org. Chem.* **1990**, *55*, 1928.
[210]Spitzer; Lee *J. Org. Chem.* **1974**, *39*, 2468.
[211]Caputo; Fuchs *Tetrahedron Lett.* **1967**, 4729.
[212]Klein; Steinmetz *Tetrahedron Lett.* **1975**, 4249. For other reagents that convert an aromatic ring to COOH and leave alkyl groups untouched, see Deno; Greigger; Messer; Meyer; Stroud *Tetrahedron Lett.* **1977**, 1703; Liotta; Hoff *J. Org. Chem.* **1980**, *45*, 2887; Chakraborti; Ghatak *J. Chem. Soc., Perkin Trans. 1* **1985**, 2605.

oxidations of naphthalene to phthalic anhydride and of benzene to maleic anhydride (p. 794) by this reagent are important industrial procedures.[213] *o*-Diamines have been oxidized with nickel peroxide, with lead tetraacetate,[214] and with O_2 catalyzed by CuCl:[215]

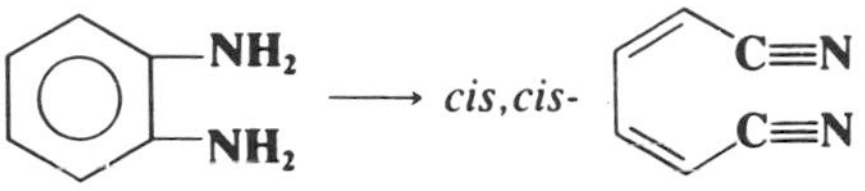

The last-named reagent also cleaves *o*-dihydroxybenzenes (catechols) to give, in the presence of MeOH, the monomethylated dicarboxylic acids HOOC—C=C—C=C—COOMe.[216]

OS **II**, 53, 523; **III**, 39, 234, 449; **IV**, 136, 484, 824; **V**, 393; **VI**, 662, 690; **VII**, 397; **66**, 180; **68**, 41. Also see OS **II**, 551.

9-11   Oxidation of Aromatic Side Chains
   Oxo,hydroxy-de-dihydro,methyl-tersubstitution

$$\text{ArR} \xrightarrow{\text{KMnO}_4} \text{ArCOOH}$$

Alkyl chains on aromatic rings can be oxidized to COOH groups by many oxidizing agents, including permanganate, nitric acid, and acid dichromate.[218] The method is most often applied to the methyl group, though longer side chains can also be cleaved. However, tertiary alkyl groups are resistant to oxidation, and when they *are* oxidized, ring cleavage usually occurs too.[219] It is usually difficult to oxidize an R group on a fused aromatic system without cleaving the ring or oxidizing it to a quinone (**9-19**). However, this has been done (e.g., 2-methylnaphthalene was converted to 2-naphthoic acid) with aqueous $Na_2Cr_2O_7$.[220] Functional groups can be present anywhere on the side chain and, if in the α position, greatly increase the ease of oxidation. An exception is an α phenyl group. In such cases the reaction stops at the diaryl ketone stage. Molecules containing aryl groups on different carbons cleave so that each ring gets one carbon atom, e.g.,

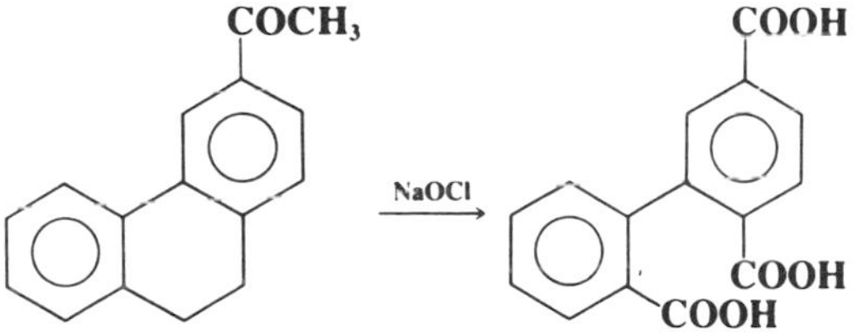

It is possible to oxidize only one alkyl group of a ring that contains more than one. The order of reactivity[221] toward most reagents is $CH_2Ar > CHR_2 > CH_2R > CH_3$.[222] Groups

[213]For a review, see Pyatnitskii *Russ. Chem. Rev.* **1976**, *45*, 762-776.
[214]Nakagawa; Onoue *Tetrahedron Lett.* **1965**, 1433, *Chem. Commun.* **1966**, 396.
[215]Kajimoto; Takahashi; Tsuji *J. Org. Chem.* **1976**, *41*, 1389.
[216]Tsuji; Takayanagi *Tetrahedron* **1978**, *34* 641; Bankston *Org. Synth.* 66, 180.
[217]This is the name if R = ethyl. The IUPAC names will obviously differ, depending on the R group.
[218]For many examples, see Hudlický, Ref. 11, pp. 105-109; Lee, Ref. 10, pp. 43-64. For a review with chromium oxidizing agents, see Cainelli; Cardillo, Ref. 10, pp. 23-33.
[219]Brandenberger; Maas; Dvoretzky *J. Am. Chem. Soc.* **1961**, *83*, 2146.
[220]Friedman; Fishel; Shechter *J. Org. Chem.* **1965**, *30*, 1453.
[221]Oxidation with Co(III) is an exception. The methyl group is oxidized in preference to the other alkyl groups: Onopchenko; Schulz; Seekircher *J. Org. Chem.* **1972**, *37*, 1414
[222]For example, see Foster; Hickinbottom *J. Chem. Soc.* **1960**, 680; Ferguson; Wims *J. Org. Chem.* **1960**, *25*, 668.

on the ring susceptible to oxidation (OH, NHR, NH_2, etc.) must be protected. The oxidation can be performed with oxygen, in which case it is autoxidation, and the mechanism is like that in **4-9**, with a hydroperoxide intermediate. With this procedure it is possible to isolate ketones from $ArCH_2R$, and this is often done.[223]

The mechanism has been studied for the closely related reaction: $Ar_2CH_2 + CrO_3 \rightarrow Ar_2C{=}O$.[224] A deuterium isotope effect of 6.4 was found, indicating that the rate-determining step is either $Ar_2CH_2 \rightarrow Ar_2CH\bullet$ or $Ar_2CH_2 \rightarrow Ar_2CH^+$. Either way this explains why tertiary groups are not converted to COOH and why the reactivity order is $CHR_2 > CH_2R > CH_3$, as mentioned above. Both free radicals and carbocations exhibit this order of stability (Chapter 5). The two possibilities are examples of categories 2 and 3 (p. 1160). Just how the radical or the cation goes on to the product is not known.

OS **I**, 159, 385, 392, 543; **II**, 135, 428; **III**, 334, 420, 740, 791, 820, 822; **V**, 617, 810.

9-12 Oxidative Cleavage of Alkyl Groups from Rings
Hydroxy-de-alkyl-*cine*-substitution

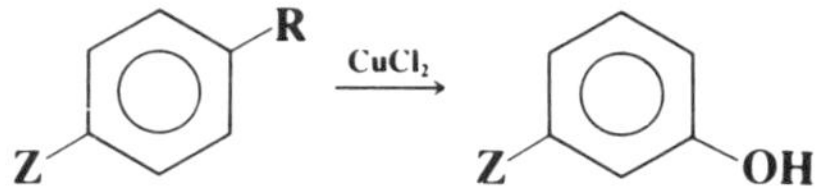

It is possible to replace an alkyl group on a ring by an OH group. When the alkyl group is one oxidizable to COOH (**9-11**), cupric salts are oxidizing agents, and the OH group is found in a position ortho to that occupied by the alkyl group.[225] This reaction is used industrially to convert toluene to phenol.

In another kind of reaction, an aromatic aldehyde ArCHO or ketone ArCOR' is converted to a phenol ArOH on treatment with alkaline H_2O_2,[226] but there must be an OH or NH_2 group in the ortho or para position. This is called the *Dakin reaction*.[227] The mechanism may be similar to that of the Baeyer–Villiger reaction (**8-20**):[228]

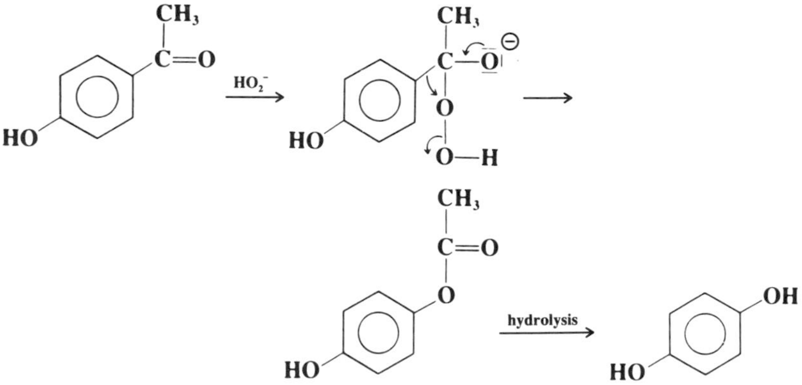

[223]For a review, see Pines; Stalick, Ref. 23, pp. 508-543.
[224]Wiberg; Evans *Tetrahedron* **1960**, *8*, 313.
[225]Kaeding *J. Org. Chem.* **1961**, *26*, 3144. For a discussion, see Lee; van den Engh, in Trahanovsky, Ref. 2, pt B. pp. 91-94.
[226]For a convenient procedure, see Hocking *Can. J. Chem.* **1973**, *51*, 2384.
[227]See Schubert; Kintner, in Patai *The Chemistry of the Carbonyl Group*, Ref. 44, pp. 749-752.
[228]For a discussion, see Hocking; Bhandari; Shell; Smyth *J. Org. Chem.* **1982**, *47*, 4208.

The intermediate **20** has been isolated.[229] The reaction has been performed on aromatic aldehydes with an alkoxy group in the ring, and no OH or NH_2. In this case acidic H_2O_2 was used.[230]

OS **I**, 149; **III**, 759.

9-13   Oxidative Decarboxylation

$$RCOOH \xrightarrow{Pb(OAc)_4} ROAc$$

Acetoxy-de-carboxy-substitution

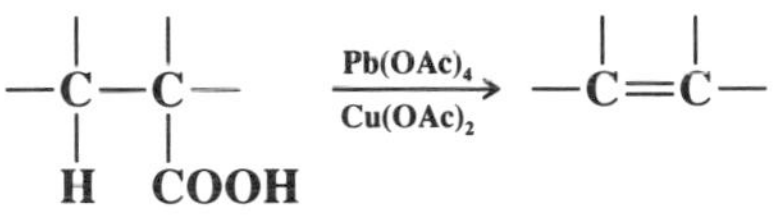

$$-\overset{|}{\underset{H}{C}}-\overset{|}{\underset{COOH}{C}}- \xrightarrow[Cu(OAc)_2]{Pb(OAc)_4} -\overset{|}{C}=\overset{|}{C}-$$

Hydro-carboxy-elimination

Carboxylic acids can be decarboxylated[231] with lead tetraacetate to give a variety of products, among them the ester ROAc (formed by replacement of COOH by an acetoxy group), the alkane RH (see **2-40**), and, if a β hydrogen is present, the alkene formed by elimination of H and COOH, as well as numerous other products arising from rearrangements, internal cyclizations,[232] and reactions with solvent molecules. When R is tertiary, the chief product is usually the alkene, which is often obtained in good yield. High yields of alkenes can also be obtained when R is primary or secondary, in this case by the use of $Cu(OAc)_2$ along with the $Pb(OAc)_4$.[233] In the absence of $Cu(OAc)_2$, primary acids give mostly alkanes (though yields are generally low) and secondary acids may give carboxylic esters or alkenes. Carboxylic esters have been obtained in good yields from some secondary acids, from β,γ-unsaturated acids, and from acids in which R is a benzylic group. Other oxidizing agents,[234] including Co(III), Ag(II), Mn(III), and Ce(IV), have also been used to effect oxidative decarboxylation.[235]

The mechanism with lead tetraacetate is generally accepted to be of the free-radical type.[236] First there is an interchange of ester groups:

$$Pb(OAc)_4 + RCOOH \longrightarrow Pb(OAc)_3OCOR \qquad or \qquad Pb(OAc)_2(OCOR)_2$$

$$\qquad\qquad\qquad\qquad\qquad\qquad\quad \textbf{21} \qquad\qquad\qquad\qquad\qquad\qquad \textbf{22}$$

[229]Hocking; Ko; Smyth *Can. J. Chem.* **1978**, *56*, 2646.

[230]Matsumoto; Kobayashi; Hotta *J. Org. Chem.* **1984**, *49*, 4740.

[231]For reviews, see Serguchev; Beletskaya *Russ. Chem. Rev.* **1980**, *49*, 1119-1134; Sheldon; Kochi *Org. React.* **1972**, *19*, 279-421.

[232]For examples, see Moriarty; Walsh; Gopal *Tetrahedron Lett.* **1966**, 4363; Davies; Waring *J. Chem. Soc. C.* **1968**, 1865, 2337.

[233]Bacha; Kochi *Tetrahedron* **1968**, *24*, 2215; Ogibin; Katzin; Nikishin *Synthesis* **1974**, 889.

[234]For references, see Trahanovsky; Cramer; Brixius *J. Am. Chem. Soc.* **1974**, *96*, 1077; Kochi *Organometallic Mechanisms and Catalysis*; Academic Press: New York, 1978, pp. 99-106. See also Dessau; Heiba *J. Org. Chem.* **1975**, *40*, 3647; Fristad; Fry; Klang *J. Org. Chem.* **1983**, *48*, 3575; Barton; Crich; Motherwell *J. Chem. Soc., Chem. Commun.* **1984**, 242; Toussaint; Capdevielle; Maumy *Tetrahedron Lett.* **1984**, *25*, 3819.

[235]For another method, see Barton; Bridon; Zard *Tetrahedron* **1989**, *45*, 2615.

[236]Kochi *J. Am. Chem. Soc.* **1965**, *87*, 1811, 3609; Starnes *J. Am. Chem. Soc.* **1964**, *86*, 5603; Davies; Waring *Chem. Commun.* **1965**, 263; Kochi; Bacha; Bethea *J. Am. Chem. Soc.* **1967**, *89*, 6538; Cantello; Mellor; Scholes *J. Chem. Soc., Perkin Trans. 2* **1974**, 348; Beckwith; Cross; Gream *Aust. J. Chem.* **1974**, *27*, 1673, 1693.

There follows a free-radical chain mechanism (shown for **21** though **22** and other lead esters can behave similarly)

Initiation $$Pb(OAc)_3OCOR \longrightarrow \overset{\bullet}{Pb}(OAc)_3 + R\cdot + CO_2$$

21

Propagation $$R\cdot + Pb(OAc)_3OCOR \longrightarrow R^+ + \overset{\bullet}{Pb}(OAc)_2OCOR + OAc^-$$

$$\overset{\bullet}{Pb}(OAc)_2OCOR \longrightarrow Pb(OAc)_2 + R\cdot + CO_2$$

Products can then be formed either from $R\cdot$ or R^+. Primary $R\cdot$ abstract H from solvent molecules to give RH. R^+ can lose H^+ to give an alkene, react with HOAc to give the carboxylic ester, react with solvent molecules or with another functional group in the same molecule, or rearrange, thus accounting for the large number of possible products. $R\cdot$ can also dimerize to give RR. The effect of Cu^{2+} ions[237] is to oxidize the radicals to alkenes,

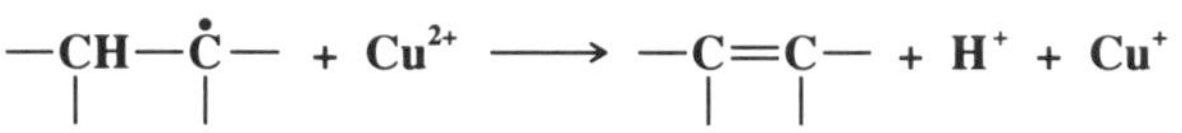

$$-CH-\overset{\bullet}{C}- + Cu^{2+} \longrightarrow -C=C- + H^+ + Cu^+$$

thus producing good yields of alkenes from primary and secondary substrates. Cu^{2+} has no effect on tertiary radicals, because these are efficiently oxidized to alkenes by lead tetraacetate.

In another type of oxidative decarboxylation, arylacetic acids can be oxidized to aldehydes with one less carbon ($ArCH_2COOH \rightarrow ArCHO$) by tetrabutylammonium periodate.[238] Simple aliphatic carboxylic acids were converted to nitriles with one less carbon ($RCH_2COOH \rightarrow RC\equiv N$) by treatment with trifluoroacetic anhydride and $NaNO_2$ in F_3CCOOH.[239]

See also **4-39**.

9-14 Bisdecarboxylation
Dicarboxy-elimination

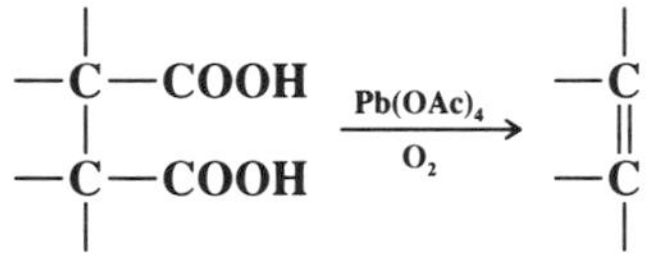

$$\begin{matrix} -\overset{|}{C}-COOH \\ -\underset{|}{C}-COOH \end{matrix} \xrightarrow[O_2]{Pb(OAc)_4} \begin{matrix} -\overset{|}{C} \\ \parallel \\ -\underset{|}{C} \end{matrix}$$

Compounds containing carboxyl groups on adjacent carbons (succinic acid derivatives) can be bisdecarboxylated with lead tetraacetate in the presence of O_2.[231] The reaction is of wide scope. The elimination is stereoselective, but not stereospecific (both *meso*- and *dl*-2,3-diphenylsuccinic acid gave *trans*-stilbene);[240] a concerted mechanism is thus unlikely. The

[237]Bacha; Kochi *J. Org. Chem.* **1968**, *33*, 83; Kochi; Bacha *J. Org. Chem.* **1968**, *33*, 2746; Torssell *Ark. Kemi* **1970**, *31*, 401.

[238]Santaniello; Ponti; Manzocchi *Tetrahedron Lett.* **1980**, *21*, 2655. For other methods of accomplishing this and similar conversions, see Cohen; Song; Fager; Deets *J. Am. Chem. Soc.* **1967**, *89*, 4968; Wasserman; Lipshutz *Tetrahedron Lett.* **1975**, 4611; Kaberia; Vickery *J. Chem. Soc., Chem. Commun.* **1978**, 459; Doleschall; Tóth *Tetrahedron* **1980**, *36*, 1649.

[239]Smushkevich; Usorov; Suvorov *J. Org. Chem. USSR* **1975**, *11*, 653.

[240]Corey; Casanova *J. Am. Chem. Soc.* **1963**, *85*, 165.

following mechanism is not inconsistent with the data:

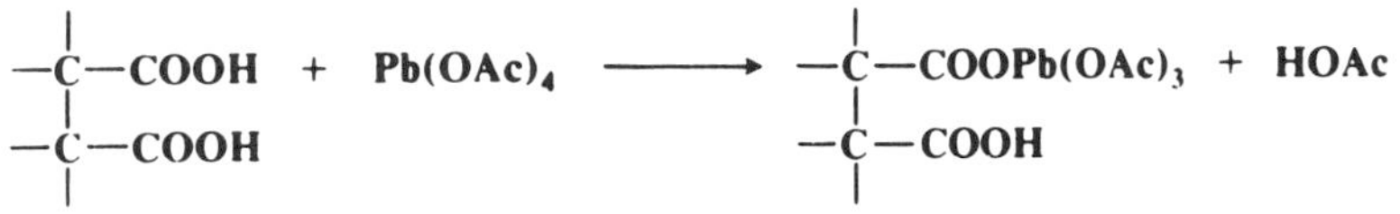

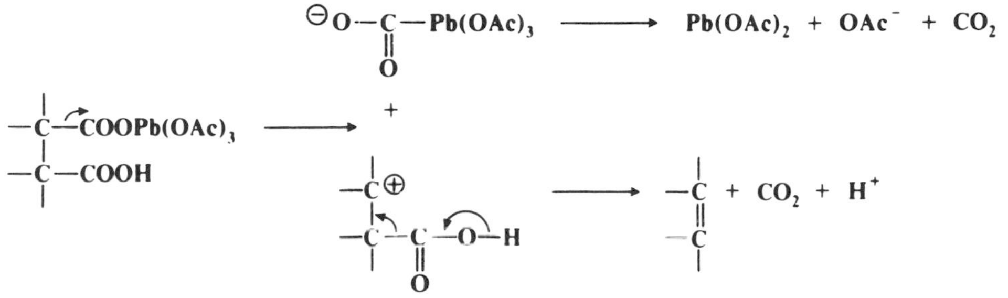

$$-\overset{|}{\underset{|}{C}}-COOH \ + \ Pb(OAc)_4 \ \longrightarrow \ -\overset{|}{\underset{|}{C}}-COOPb(OAc)_3 \ + \ HOAc$$
$$-\overset{|}{\underset{|}{C}}-COOH \qquad\qquad\qquad -\overset{|}{\underset{|}{C}}-COOH$$

$$\overset{\ominus}{O}-\underset{\underset{O}{\parallel}}{C}-Pb(OAc)_3 \ \longrightarrow \ Pb(OAc)_2 \ + \ OAc^- \ + \ CO_2$$

$$-\overset{|}{\underset{|}{C}}\overset{\frown}{-}COOPb(OAc)_3 \ \longrightarrow \ +$$
$$-\overset{|}{\underset{|}{C}}-COOH$$

$$-\overset{|}{\underset{|}{C}}{}^{\oplus} \qquad\qquad \longrightarrow \quad -\overset{|}{\underset{\parallel}{C}} \ + \ CO_2 \ + \ H^+$$
$$-\overset{|}{\underset{\underset{O}{\parallel}}{C}}-\overset{}{C}\overset{\frown}{-O}-H \qquad\qquad -C$$

though a free-radical mechanism seems to hold in some cases. Bisdecarboxylation of succinic acid derivatives to give alkenes[241] has also been carried out by other methods, including treatment of the corresponding anhydrides with nickel, iron, or rhodium complexes,[242] by decomposition of the corresponding bis peresters,[243] and electrolytically.[244]

Compounds containing geminal carboxyl groups (disubstituted malonic acid derivatives) can also be bisdecarboxylated with lead tetraacetate,[245] *gem*-diacetates (acylals) being produced, which are easily hydrolyzable to ketones:[246]

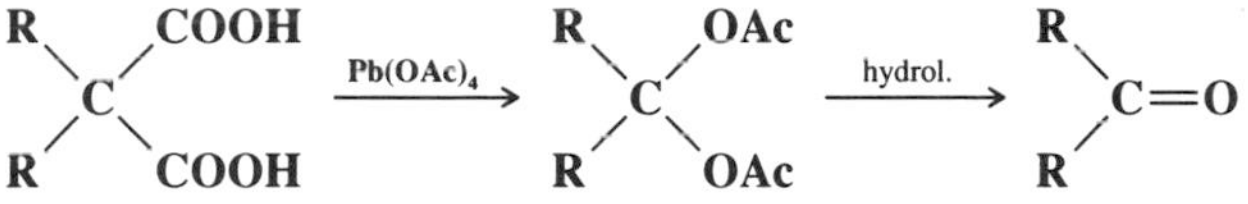

$$\underset{R}{\overset{R}{\diagdown}}C\underset{COOH}{\overset{COOH}{\diagup}} \ \xrightarrow{Pb(OAc)_4} \ \underset{R}{\overset{R}{\diagdown}}C\underset{OAc}{\overset{OAc}{\diagup}} \ \xrightarrow{hydrol.} \ \underset{R}{\overset{R}{\diagdown}}C=O$$

9-15 Oxidative Decyanation
Oxo-de-hydro,cyano-bisubstitution

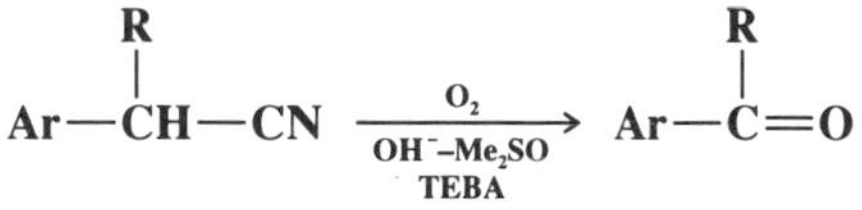

$$\underset{Ar-\overset{\overset{R}{|}}{C}H-CN}{} \ \xrightarrow[\substack{OH^- - Me_2SO \\ TEBA}]{O_2} \ Ar-\overset{\overset{R}{|}}{C}=O$$

α-Substituted aryl nitriles having a sufficiently acidic α hydrogen can be converted to ketones by oxidation with air under phase transfer conditions.[247] The nitrile is added to NaOH in benzene or Me_2SO containing a catalytic amount of triethylbenzylammonium chloride

[241]For a review, see De Lucchi; Modena *Tetrahedron* **1984**, *40*, 2585-2632, pp. 2591-2608.
[242]Trost; Chen *Tetrahedron Lett.* **1971**, 2603.
[243]Cain; Vukov; Masamune *Chem. Commun.* **1969**, 98.
[244]Plieninger; Lehnert *Chem. Ber.* **1967**, *100*, 2427; Radlick; Klem; Spurlock; Sims; van Tamelen; Whitesides *Tetrahedron Lett.* **1968**, 5117; Westberg; Dauben *Tetrahedron Lett.* **1968**, 5123. For additional references, see Fry *Synthetic Organic Electrochemistry*, 2nd ed.; Wiley: New York, 1989, pp. 253-254.
[245]For a similar reaction with ceric ammonium nitrate, see Salomon; Roy; Salomon *Tetrahedron Lett.* **1988**, *29*, 769.
[246]Tufariello; Kissel *Tetrahedron Lett.* **1966**, 6145.
[247]For other methods of achieving this conversion, with references, see Ref. 21, p. 618.

(TEBA).[248] This reaction could not be applied to aliphatic nitriles, but an indirect method for achieving this conversion is given in **9-57**. α-Dialkylamino nitriles can be converted to ketones [$R_2C(NMe_2)CN \rightarrow R_2C{=}O$] by hydrolysis with $CuSO_4$ in aqueous methanol[249] or by autoxidation in the presence of t-BuOK.[250]

C. Reactions Involving Replacement of Hydrogen by Oxygen

9-16 Oxidation of Methylene to Carbonyl
Oxo-de-dihydro-bisubstitution

$$\underset{O}{R-\overset{\|}{C}-CH_2-R'} \xrightarrow{\text{SeO}_2} \underset{O\ \ O}{R-\overset{\|}{C}-\overset{\|}{C}-R'}$$

Methyl or methylene groups α to a carbonyl can be oxidized with selenium dioxide to give, respectively, α-keto aldehydes and α-diketones.[251] The reaction can also be carried out α to an aromatic ring or to a double bond, though in the latter case, hydroxylation (see **4-4**) is the more common result. Although SeO_2 is the reagent most often used, the reaction has also been carried out with N_2O_3 and other oxidizing agents.[252] Substrates most easily oxidized contain two aryl groups on CH_2, and these substrates can be oxidized with many oxidizing agents (see **9-11**). Monoaryl alkanes have been oxidized to alkyl aryl ketones with several oxidizing agents, including CrO_3–acetic acid,[253] the Jones reagent,[254] pyridinium chlorochromate,[255] ceric ammonium nitrate,[256] benzeneseleninic anhydride PhSe(O)OSe(O)Ph,[257] a silver ion–persulfate couple,[258] and DDQ,[259] as well as with SeO_2. Alkenes of the form $C{=}C{-}CH_2$ have been oxidized to α,β-unsaturated ketones[260] by sodium dichromate in HOAc–Ac_2O, by aqueous Na_2O_2 (α,β-unsaturated alkenes),[261] by t-BuOOH and chromium compounds,[262] by 2-pyridineseleninic anhydride,[263] by CrO_3–pyridine complex,[264] and by mercuric salts,[265] among other reagents, as well as electrolytically.[266] CrO_3–pyridine[267] and t-BuOOH–chromium compounds[268] have also been used to convert alkynes of the form $C{\equiv}C{-}CH_2$ to α-keto acetylenes. Methyl ketones RCOMe react with ammonium peroxy-

[248]Masuyama; Ueno; Okawara *Chem. Lett.* **1977**, 1439; Donetti; Boniardi; Ezhaya *Synthesis* **1980**, 1009; Kulp; McGee *J. Org. Chem.* **1983**, *48*, 4097.

[249]Büchi; Liang; Wüest *Tetrahedron Lett.* **1978**, 2763.

[250]Chuang; Yang; Chang; Fang *Synlett* **1990**, 733.

[251]For reviews of oxidation by SeO_2, see Krief; Hevesi, Ref. 10, pp. 115-180; Krongauz *Russ. Chem. Rev.* **1977**, *46*, 59-75; Rabjohn *Org. React.* **1976**, *24*, 261-415; Trachtenberg, in Augustine; Trecker, Ref. 11, pp. 119-187.

[252]For other methods, see Wasserman; Ives *J. Org. Chem.* **1978**, *43*, 3238, **1985**, *50*, 3573; Rao; Stuber; Ulrich *J. Org. Chem.* **1979**, *44*, 456.

[253]For example, see Harms; Eisenbraun *Org. Prep. Proced. Int.* **1972**, *4*, 67.

[254]Rangarajan; Eisenbraun *J. Org. Chem.* **1985**, *50*, 2435.

[255]Rathore; Saxena; Chandrasekaran *Synth. Commun.* **1986**, *16*, 1493.

[256]Syper *Tetrahedron Lett.* **1966**, 4493.

[257]Barton; Hui; Ley *J. Chem. Soc., Perkin Trans. 1* **1982**, 2179.

[258]Daniher *Org. Prep. Proced.* **1970**, *2*, 207; Bhatt; Perumal *Tetrahedron Lett.* **1981**, *22*, 2605.

[259]Lee; Harvey *J. Org. Chem.* **1988**, *53*, 4587.

[260]For a review, see Muzart *Bull. Soc. Chim. Fr.* **1986**, 65-77. For a list of reagents, with references, see Ref. 21, pp. 592-593.

[261]Holland; Daum; Riemland *Tetrahedron Lett.* **1981**, *22*, 5127.

[262]Pearson; Chen; Han; Hsu; Ray *J. Chem. Soc., Perkin Trans. 1* **1985**, 267; Muzart *Tetrahedron Lett.* **1987**, *28*, 2131; Chidambaram; Chandrasekaran *J. Org. Chem.* **1987**, *52*, 5048.

[263]Barton; Crich *Tetrahedron* **1985**, *41*, 4359.

[264]Dauben; Lorber; Fullerton *J. Org. Chem.* **1969**, *34*, 3587; Fullerton; Chen *Synth. Commun.* **1976**, *6*, 217.

[265]Arzoumanian; Metzger *Synthesis* **1971**, 527-536; Charavel; Metzger *Bull. Soc. Chim. Fr.* **1968**, 4102.

[266]Madurro; Chiericato; De Giovani; Romero *Tetrahedron Lett.* **1988**, *29*, 765.

[267]Shaw; Sherry *Tetrahedron Lett.* **1971**, 4379; Sheats; Olli; Stout; Lundeen; Justus; Nigh *J. Org. Chem.* **1979**, *44*, 4075.

[268]Muzart; Piva *Tetrahedron Lett.* **1988**, *29*, 2321.

disulfate $(NH_4)_2S_2O_8$ and a catalytic amount of diphenyl diselenide in MeOH to give α-keto acetals $RCOCH(OMe_2)$.[269]

Two mechanisms have been suggested for the reaction with SeO_2. One of these involves a selenate ester of the enol:[270]

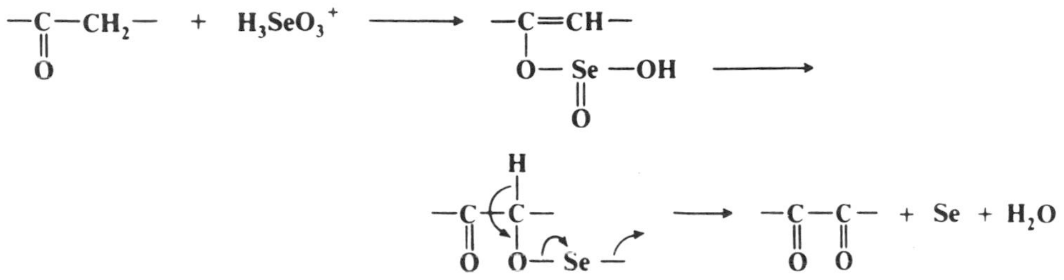

In the other proposed mechanism,[271] the principal intermediate is a β-ketoseleninic acid

$$-\overset{O}{\underset{\|}{C}}\overset{O}{\underset{\|}{C}}HSeOH,$$ and a selenate ester is not involved.

It has proved possible to convert CH_2 to $C=O$ groups, even if they are not near any functional groups, indirectly, by the remote oxidation method of Breslow[39] (see **9-2**). In a typical example, the keto ester **23** was irradiated to give the hydroxy lactone **24,** which was

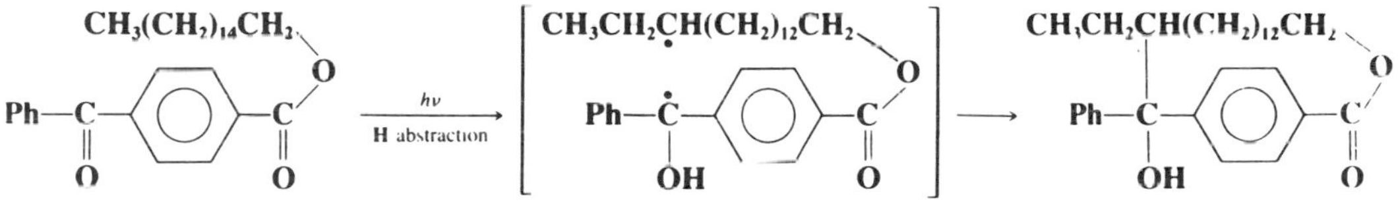

dehydrated to **25.** Ozonolysis of **25** gave the diketo ester **26,** in which the C-14 CH_2 group of **23** has been oxidized to a $C=O$ group.[272] The reaction was not completely regioselective: **26** comprised about 60% of the product, with the remainder consisting of other compounds in which the keto group was located at C-12, C-15, and other positions along the carbon chain. Greater regioselectivity was achieved when the aromatic portion was connected to the chain at two positions.[273] In the method so far described, the reaction takes place because one portion of a molecule (the benzophenone moiety) abstracts hydrogen from another

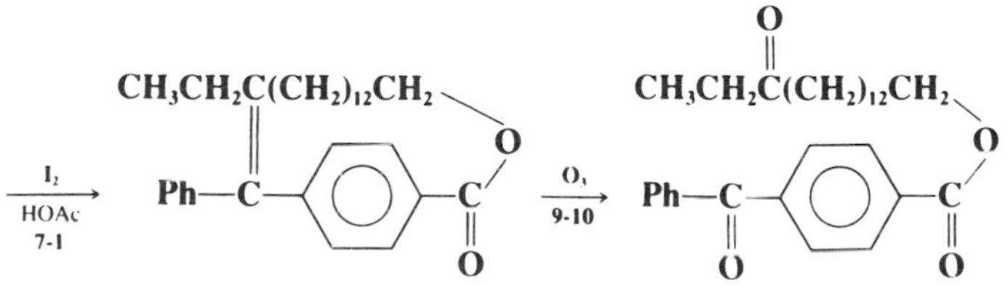

[269]Tiecco; Testaferri; Tingoli; Bartoli *J. Org. Chem.* **1990**, *55*, 4523.
[270]Corey; Schaefer *J. Am. Chem. Soc.* **1960**, *82*, 918.
[271]Sharpless; Gordon *J. Am. Chem. Soc.* **1976**, *98*, 300.
[272]Breslow; Winnik *J. Am. Chem. Soc.* **1969**, *91*, 3083; Breslow; Rothbard; Herman; Rodriguez *J. Am. Chem. Soc.* **1978**, *100*, 1213.
[273]Breslow; Rajagopalan; Schwarz *J. Am. Chem. Soc.* **1981**, *103*, 2905.

portion of the same molecule, i.e., the two portions are connected by a series of covalent bonds. However, the reaction can also be carried out where the two reacting centers are actually in different molecules, providing the two molecules are held together by hydrogen bonding. For example, one of the CH_2 groups of *n*-hexadecanol monosuccinate $CH_3(CH_2)_{14}CH_2OCOCH_2CH_2COOH$ was oxidized to a $C=O$ group by applying the above procedure to a mixture of it and benzophenone-4-carboxylic acid *p*-$PhCOC_6H_4COOH$ in CCl_4.[274]

Other remote oxidations[275] have also been reported. Among these are conversion of aryl ketones $ArCO(CH_2)_3R$ to 1,4-diketones $ArCO(CH_2)_2COR$ by photoirradiation in the presence of such oxidizing agents as $K_2Cr_2O_7$ or $KMnO_4$,[276] and conversion of alkyl ketones $RCO(CH_2)_3R'$ to 1,3- and 1,4-diketones with $Na_2S_2O_8$ and $FeSO_4$.[277]

It is possible to perform the conversion $CH_2 \rightarrow C=O$ on an alkane, with no functional groups at all, though the most success has been achieved with substrates in which all CH_2 groups are equivalent, such as unsubstituted cycloalkanes. One method uses H_2O_2 and bis(picolinato)iron(II). With this method, cyclohexane was converted with 72% efficiency to give 95% cyclohexanone and 5% cyclohexanol.[278] The same type of conversion, with lower yields (20-30%), has been achieved with the *Gif system*.[279] There are several variations. One consists of pyridine–acetic acid, with H_2O_2 as oxidizing agent and tris(picolinato)iron(III) as catalyst.[280] Other Gif systems use O_2 as oxidizing agent and zinc as a reductant.[281] The selectivity of the Gif systems towards alkyl carbons is $CH_2 > CH \geq CH_3$, which is unusual, and shows that a simple free-radical mechanism (see p. 683) is not involved.[282] Another reagent that can oxidize the CH_2 of an alkane is methyl(trifluoromethyl)dioxirane, but this produces CH—OH more often than C=O (see **4-4**).[283]

OS **I**, 266; **II**, 509; **III**, 1, 420, 438; **IV**, 189, 229, 579; **VI**, 48. Also see OS **IV**, 23.

9-17 Oxidation of Arylmethanes
Oxo-de-dihydro-bisubstitution

$$ArCH_3 \xrightarrow{CrO_2Cl_2} ArCHO$$

Methyl groups on an aromatic ring can be oxidized to the aldehyde stage by several oxidizing agents. The reaction is a special case of **9-16**. When the reagent is chromyl chloride (CrO_2Cl_2), the reaction is called the *Étard reaction*[284] and the yields are high.[285] Another oxidizing agent is a mixture of CrO_3 and Ac_2O. In this case the reaction stops at the aldehyde stage because

[274]Breslow; Scholl *J. Am. Chem. Soc.* **1971**, *93*, 2331. See also Breslow; Heyer *Tetrahedron Lett.* **1983**, *24*, 5039.
[275]See also Beckwith; Duong *J. Chem. Soc., Chem. Commun.* **1978**, 413.
[276]Mitani; Tamada; Uehara; Koyama *Tetrahedron Lett.* **1984**, *25*, 2805.
[277]Nikishin; Troyansky; Lazareva *Tetrahedron Lett.* **1984**, *25*, 4987.
[278]Sheu; Richert; Cofré; Ross; Sobkowiak; Sawyer; Kanofsky *J. Am. Chem. Soc.* **1990**, *112*, 1936. See also Sheu; Sobkowiak; Jeon; Sawyer *J. Am. Chem. Soc.* **1990**, *112*, 879; Tung; Sawyer *J. Am. Chem. Soc.* **1990**, *112*, 8214.
[279]Named for Gif-sur-Yvette, France, where it was discovered.
[280]About-Jaudet; Barton; Csuhai; Ozbalik *Tetrahedron Lett.* **1990**, *31*, 1657.
[281]See Barton; Boivin; Gastiger; Morzycki; Hay-Motherwell; Motherwell; Ozbalik; Schwartzentruber *J. Chem. Soc., Perkin Trans. 1* **1986**, 947; Barton; Csuhai; Ozbalik *Tetrahedron* **1990**, *46*, 3743.
[282]Barton; Csuhai; Doller; Ozbalik; Senglet *Tetrahedron Lett.* **1990**, *31*, 3097. For mechanistic studies, see Barton; Csuhai; Ozbalik *Tetrahedron Lett.* **1990**, *31*, 2817; Barton; Csuhai; Doller; Balavoine *J. Chem. Soc., Chem. Commun.* **1990**, 1787; Barton; Doller; Geletii *Tetrahedron Lett.* **1991**, *32*, 3811; Knight; Perkins *J. Chem. Soc., Chem. Commun.* **1991**, 925.
[283]Mello; Fiorentino; Fusco; Curci *J. Am. Chem. Soc.* **1989**, *111*, 6749.
[284]The name Étard reaction is often applied to any oxidation with chromyl chloride, e.g., oxidation of glycols (**9-7**), olefins (**9-10**), etc.
[285]For a review, see Hartford; Darrin *Chem. Rev.* **1958**, *58*, 1-61, pp. 25-53.

the initial product is ArCH(OAc)$_2$ (an acylal), which is resistant to further oxidation. Hydrolysis of the acylal gives the aldehyde.

Among other oxidizing agents[286] that have been used to accomplish the conversion of ArCH$_3$ to ArCHO are ceric ammonium nitrate,[287] ceric trifluoroacetate,[288] benzeneseleninic anhydride,[257] KMnO$_4$–Et$_3$N,[289] and silver(II) oxide.[290] Oxidation of ArCH$_3$ to carboxylic acids is considered at **9-11**.

Conversion of ArCH$_3$ to ArCHO can also be achieved indirectly by bromination to give ArCHBr$_2$ (**4-1**), followed by hydrolysis (**0-2**).

The mechanism of the Étard reaction is not completely known.[291] An insoluble complex is formed on addition of the reagents, which is hydrolyzed to the aldehyde. The complex is probably a kind of acylal, but what the structure is is not fully settled, though many proposals have been made as to its structure and as to how it is hydrolyzed. It is known that ArCH$_2$Cl is not an intermediate (see **9-20**), since it reacts only very slowly with chromyl chloride. Magnetic susceptibility measurements[292] indicate that the complex from toluene is **27**, a structure first proposed by Étard. According to this proposal the reaction stops after

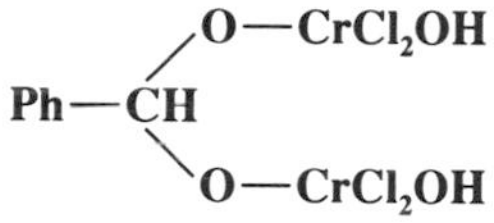

$$Ph-CH \Big\langle \begin{array}{l} O-CrCl_2OH \\ O-CrCl_2OH \end{array}$$

27

only two hydrogens have been replaced because of the insolubility of **27**. There is a disagreement on how **27** is formed, assuming that the complex has this structure. Both an ionic[293] and a free-radical[294] process have been proposed. An entirely different structure for the complex was proposed by Nenitzescu and co-workers.[295] On the basis of esr studies they proposed that the complex is PhCH$_2$OCrCl$_2$OCrOCl$_2$OH, which is isomeric with **27**. However, this view has been challenged by Wiberg and Eisenthal,[294] who interpret the esr result as being in accord with **27**. Still another proposal is that the complex is composed of benzaldehyde coordinated with reduced chromyl chloride.[296]

OS **II**, 441; **III**, 641; **IV**, 31, 713.

9-18 Oxidation of Ethers to Carboxylic Esters and Related Reactions
       Oxo-de-dihydro-bisubstitution

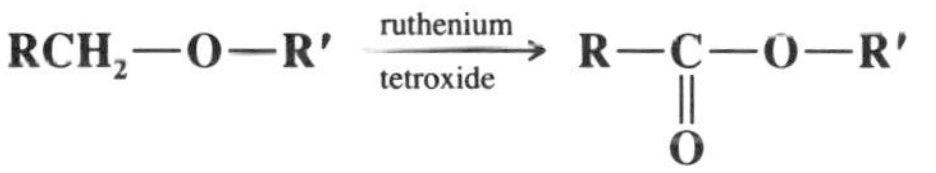

$$RCH_2-O-R' \xrightarrow[\text{tetroxide}]{\text{ruthenium}} R-\underset{\underset{O}{\|}}{C}-O-R'$$

[286]For a review of the use of oxidizing agents that are regenerated electrochemically, see Steckhan *Top. Curr. Chem.* **1987**, *142*, 1-69; pp. 12-17.

[287]Trahanovsky; Young *J. Org. Chem.* **1966**, *31*, 2033; Radhakrishna Murti; Pati *Chem. Ind.* (*London*) **1967**, 702; Ref. 256.

[288]Marrocco; Brilmyer *J. Org. Chem.* **1983**, *48*, 1487. See also Kreh; Spotnitz; Lundquist *J. Org. Chem.* **1989**, *54*, 1526.

[289]Li; Liu *Synthesis* **1989**, 293.

[290]Syper *Tetrahedron Lett.* **1967**, 4193.

[291]For a review, see Nenitzescu *Bull. Soc. Chim. Fr.* **1968**, 1349-1357.

[292]Wheeler *Can. J. Chem.* **1960**, *38*, 2137. See also Makhija; Stairs *Can. J. Chem.* **1968**, *46*, 1255.

[293]Stairs *Can. J. Chem.* **1964**, *42*, 550.

[294]Wiberg; Eisenthal *Tetrahedron* **1964**, *20*, 1151. See also Gragerov; Ponomarchuk *J. Org. Chem. USSR* **1969**, *6*, 1125.

[295]Necşoiu; Balaban; Pascaru; Sliam; Elian; Nenitzescu *Tetrahedron* **1963**, *19*, 1133; Necşoiu; Przemetchi; Ghenciulescu; Rentea; Nenitzescu *Tetrahedron* **1966**, *22*, 3037.

[296]Duffin; Tucker *Chem. Ind.* (*London*) **1966**, 1262, *Tetrahedron* **1968**, *24*, 6999.

Ethers in which at least one group is primary alkyl can be oxidized to the corresponding carboxylic esters in high yields with ruthenium tetroxide.[297] Cyclic ethers give lactones. The reaction, a special case of **9-16**, has also been accomplished with CrO_3 in sulfuric acid,[298] with benzyltriethylammonium permanganate,[299] and with trichloroisocyanuric acid in the presence of an excess of water.[300] In a similar reaction, cyclic tertiary amines (e.g., **28**) can

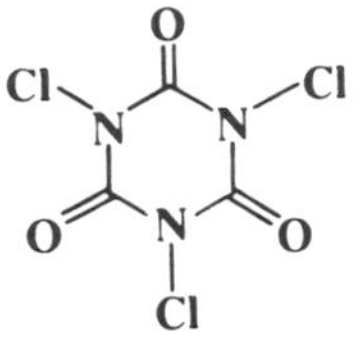

Trichloroisocyanuric acid

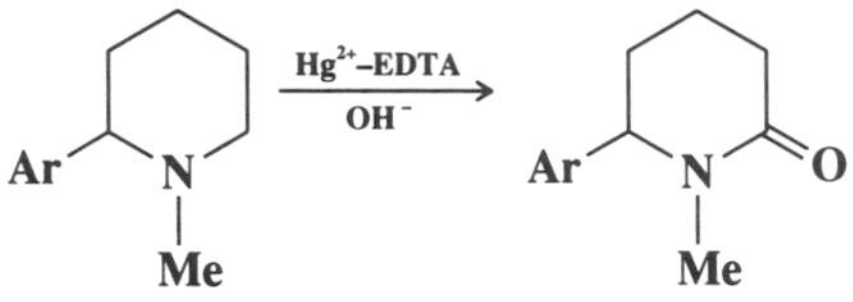

28

be converted to lactams by oxidation with Hg(II)–EDTA complex in basic solution.[301] Lactams, which need not be N-substituted (e.g., **29**), can be converted to cyclic imides by

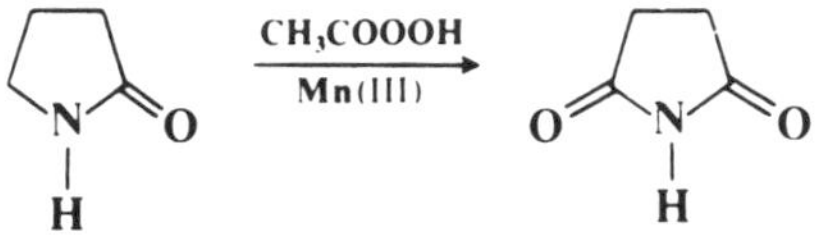

29

oxidation with a hydroperoxide or peracid and an Mn(II) or Mn(III) salt.[302] Certain tertiary amines containing a methyl group can be oxidized[303] to formamides ($R_2NCH_3 \rightarrow R_2NCHO$) by MnO_2,[304] CrO_3–pyridine,[305] O_2 and platinum,[306] or other oxidizing agents, but the reaction is not general.

9-19 Oxidation of Aromatic Hydrocarbons to Quinones
Arene–quinone transformation

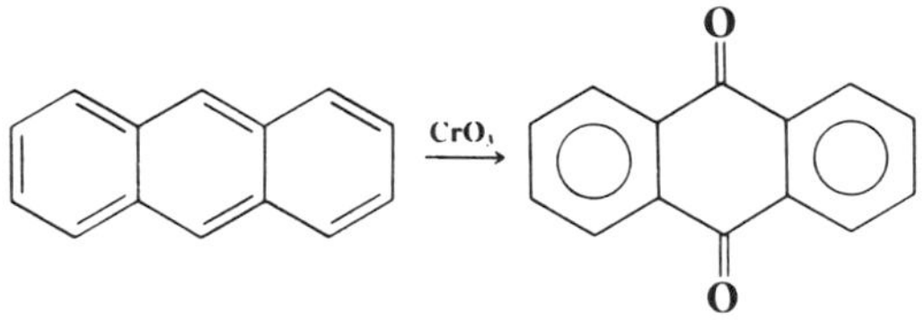

[297]Berkowitz; Rylander *J. Am. Chem. Soc.* **1958**, *80*, 6682; Lee; van den Engh, in Trahanovsky, Ref. 2, pt. B. pp. 222-225; Smith; Scarborough *Synth. Commun.* **1980**, *10*, 205; Carlsen; Katsuki; Martin; Sharpless *J. Org. Chem.* **1981**, *46*, 3936.

[298]Henbest; Nicholls *J. Chem. Soc.* **1959**, 221, 227; Harrison; Harrison *Chem. Commun* **1966**, 752.

[299]Schmidt; Schäfer *Angew. Chem. Int. Ed. Engl.* **1979**, *18*, 69 [*Angew. Chem. 91*, 78].

[300]Juenge; Beal *Tetrahedron Lett.* **1968**, 5819; Juenge; Corey; Beal *Tetrahedron* **1971**, *27*, 2671.

[301]Wenkert; Angell *Synth. Commun.* **1988**, *18*, 1331.

[302]Doumaux; McKeon; Trecker *J. Am. Chem. Soc.* **1969**, *91*, 3992; Doumaux; Trecker *J. Org. Chem.* **1970**, *35*, 2121.

[303]See also Bettoni; Carbonara; Franchini; Tortorella *Tetrahedron* **1981**, *37*, 4159; Schmidt; Schäfer *Angew. Chem. Int. Ed. Engl.* **1981**, *20*, 109 [*Angew. Chem. 93*, 124].

[304]See, for example, Henbest; Thomas *J. Chem. Soc.* **1957**, 3032; Henbest; Stratford *J. Chem. Soc. C* **1966**, 995.

[305]Cavé; Kan-Fan; Potier; Le Men; Janot *Tetrahedron* **1967**, *23*, 4691.

[306]Davis; Rosenblatt *Tetrahedron Lett.* **1968**, 4085.

Condensed aromatic systems (including naphthalenes) can be directly oxidized to quinones by various oxidizing agents.[307] Yields are generally not high, though good yields have been reported with ceric ammonium sulfate.[308] Benzene cannot be so oxidized by strong oxidizing agents but can be electrolytically oxidized to benzoquinone.[309]

OS **IV**, 698, 757. Also see OS **II**, 554.

9-20 Oxidation of Primary Halides and Esters of Primary Alcohols to Aldehydes[310]
Oxo-de-hydro,halo-bisubstitution

$$RCH_2Cl \xrightarrow{(CH_3)_2SO} RCHO$$

Primary alkyl halides (chlorides, bromides, and iodides) can be oxidized to aldehydes easily and in good yields with dimethyl sulfoxide.[311] Tosyl esters of primary alcohols can be similarly converted to aldehydes,[312] and epoxides[313] give α-hydroxy ketones or aldehydes.[314] The reaction with tosyl esters is an indirect way of oxidizing primary alcohols to aldehydes (**9-3**). This type of oxidation can also be carried out without isolation of an intermediate ester: The alcohol is treated with dimethyl sulfoxide, dicyclohexylcarbodiimide (DCC),[315] and anhydrous phosphoric acid.[316] In this way a primary alcohol can be converted to the aldehyde with no carboxylic acid being produced.

Similar oxidation of alcohols has been carried out with dimethyl sulfoxide and other reagents[317] in place of DCC: acetic anhydride,[318] SO$_3$–pyridine–triethylamine,[319] trifluoroacetic anhydride,[320] oxalyl chloride,[321] tosyl chloride,[322] chlorine,[323] bromine,[324] AgBF$_4$–Et$_3$N,[325] P$_2$O$_5$, Et$_3$N,[326] phenyl dichlorophosphate,[327] trichloromethyl chloroformate,[328] tri-

[307]For reviews, see Naruta; Maruyama, in Patai; Rappoport, Ref. 18, vol. 2, pt. 1, 1988, pp 242-247; Hudlický, Ref. 11, pp. 94-96; Haines-1985, Ref. 11, pp. 182-185, 358-360; Thomson, in Patai, Ref. 18, 1974, pp. 132-134. See also Šket; Zupan *Synth. Commun* **1990**, *20*, 933; Ref. 112.

[308]Periasamy; Bhatt *Synthesis* **1977**, 330; Balanikas; Hussain; Amin; Hecht *J. Org. Chem.* **1988**, *53*, 1007.

[309]See, for example, Ito; Katayama; Kunai; Sasaki *Tetrahedron Lett.* **1989**, *30*, 205.

[310]For reviews of the reactions in this section, see Tidwell *Org. React.* **1990**, *39*, 297-572, *Synthesis* **1990**, 857-870; Haines-1988, Ref. 11, pp. 171-181, 402-406; Durst *Adv. Org. Chem.* **1969**, *6*, 285-388, pp. 343-356; Epstein; Sweat *Chem. Rev.* **1967**, *67*, 247-260; Moffatt, in Augustine; Trecker, Ref. 11, vol. 2, pp. 1-64. For a list of reagents, with references, see Ref. 21, pp. 599-600.

[311]Nace; Monagle *J. Org. Chem.* **1959**, *24*, 1792; Kornblum; Jones; Anderson *J. Am. Chem. Soc.* **1959**, *81*, 4113.

[312]Kornblum; Jones; Anderson, Ref. 311.

[313]Epoxides can be converted to α-halo ketones by treatment with bromodimethylsulfonium bromide: Olah; Vankar; Arvanaghi *Tetrahedron Lett.* **1979**, 3653.

[314]Cohen; Tsuji *J. Org. Chem.* **1961**, *26*, 1681; Tsuji *Tetrahedron Lett.* **1966**, 2413; Santosusso; Swern *Tetrahedron Lett.* **1968**, 4261, *J. Org. Chem.* **1975**, *40*, 2764.

[315]The DCC is converted to dicyclohexylurea, which in some cases is difficult to separate from the product. One way to avoid this problem is to use a carbodiimide linked to an insoluble polymer: Weinshenker; Shen *Tetrahedron Lett.* **1972**, 3285.

[316]Pfitzner; Moffatt *J. Am. Chem. Soc.* **1965**, *87*, 5661, 5670; Fenselau; Moffatt *J. Am. Chem. Soc.* **1966**, *88*, 1762; Albright; Goldman *J. Org. Chem.* **1965**, *30*, 1107.

[317]For a review of activated Me$_2$SO reagents and their use in this reaction, see Mancuso; Swern *Synthesis* **1981**, 165-185.

[318]Albright; Goldman *J. Am. Chem. Soc.* **1967**, *89*, 2416.

[319]Parikh; Doering *J. Am. Chem. Soc.* **1967**, *89*, 5507.

[320]Huang; Omura; Swern *Synthesis* **1978**, 297.

[321]Omura; Swern *Tetrahedron* **1978**, *34*, 1651. See also Marx; Tidwell *J. Org. Chem.* **1984**, *49*, 788.

[322]Albright *J. Org. Chem.* **1974**, *39*, 1977.

[323]Corey; Kim *Tetrahedron Lett.* **1973**, 919.

[324]Munavu *J. Org. Chem.* **1980**, *45* 3341.

[325]Ganem; Boeckman *Tetrahedron Lett.* **1974**, 917.

[326]Taber; Amedio; Jung *J. Org. Chem.* **1987**, *52*, 5621.

[327]Liu; Nyangulu *Tetrahedron Lett.* **1988**, *29*, 3167.

[328]Takano; Inomata; Tomita; Yanase; Samizu; Ogasawara *Tetrahedron Lett.* **1988**, *29*, 6619.

methylamine oxide,[329] KI and NaHCO$_3$,[330] and methanesulfonic anhydride,[322] among others. When oxalyl chloride is used, the method is called *Swern oxidation*.

The mechanism of these dimethyl sulfoxide oxidations is probably as follows:[331]

$$Me_2SO + RCH_2X \xrightarrow[-X^-]{S_N2} Me-\overset{\oplus}{\underset{\underset{Me}{|}}{S}}-O-\underset{\underset{H}{|}}{C}HR \xrightarrow[-H^+]{base}$$

30

$$Me-\overset{\oplus}{\underset{\underset{\ominus CH_2}{|}}{S}}-O-CHR \longrightarrow Me_2S + RCHO$$

31

though in some cases the base abstracts a proton directly from the carbon being oxidized, in which case the ylide **31** is not an intermediate. Alkoxysulfonium salts **30** have been isolated.[332] This mechanism predicts that secondary compounds should be oxidizable to ketones, and this is the case. In a related procedure for the oxidation of alcohols, the intermediate **30**[333] is formed without the use of dimethyl sulfoxide by treating the substrate with a complex generated from chlorine or N-chlorosuccinimide and dimethyl sulfide.[334]

Another way to oxidize primary alkyl halides to aldehydes is by the use of hexamethylenetetramine followed by water. However, this reaction, called the *Sommelet reaction*,[335] is limited to benzylic halides. The reaction is seldom useful when the R in RCH$_2$Cl is alkyl. The first part of the reaction is conversion to the amine ArCH$_2$NH$_2$ (**0-44**), which can be isolated. Reaction of the amine with excess hexamethylenetetramine gives the aldehyde. It is this last step that is the actual Sommelet reaction, though the entire process can be conducted without isolation of intermediates. Once the amine is formed, it is converted to an imine (ArCH$_2$N=CH$_2$) with formaldehyde liberated from the reagent. The key step then follows: transfer of hydrogen from another mole of the arylamine to the imine:

$$ArCH_2N=CH_2 + ArCH_2NH_2 \longrightarrow ArCH_2NHCH_3 + ArCH=NH$$

This last imine is then hydrolyzed by water to the aldehyde. Alternatively, the benzylamine may transfer hydrogen directly to hexamethylenetetramine.

Other reagents that convert benzylic halides to aldehydes are 2-nitropropane–NaOEt in EtOH,[336] mercury(I) nitrate followed by ethanolic alkali,[337] and pyridine followed by *p*-nitrosodimethylaniline and then water. The last procedure is called the *Kröhnke reaction*. Primary halides in general have been oxidized to aldehydes by trimethylamine oxide,[338] by

[329]Godfrey; Ganem *Tetrahedron Lett.* **1990**, *31*, 4825.
[330]Bauer; Macomber *J. Org. Chem.* **1975**, *40*, 1990.
[331]Pfitzner; Moffatt *J. Am. Chem. Soc.* **1965**, *87*, 5661; Johnson; Phillips *J. Org. Chem.* **1967**, *32*, 1926; Torssell *Acta Chem. Scand.* **1967**, *21*, 1.
[332]Torssell *Tetrahedron Lett.* **1966**, 4445; Johnson; Phillips, Ref. 331; Khuddus; Swern *J. Am. Chem. Soc.* **1973**, *95*, 8393.
[333]It has been suggested that in the DCC reaction, **30** is not involved, but the ylide **31** is formed directly from a precursor containing DCC and dimethyl sulfoxide: Torssell, Ref. 332; Moffatt *J. Org. Chem.* **1971**, *36*, 1909.
[334]Vilsmaier; Sprügel *Liebigs Ann. Chem.* **1971**, *747*, 151; Corey; Kim *J. Am. Chem. Soc.* **1972**, *94*, 7586, *J. Org. Chem.* **1973**, *38*, 1233; McCormick *Tetrahedron Lett.* **1974**, 1701; Katayama; Fukuda; Watanabe; Yamauchi *Synthesis* **1988**, 178.
[335]For a review, see Angyal *Org. React.* **1954**, *8*, 197-217.
[336]Hass; Bender *J. Am. Chem. Soc.* **1949**, *71*, 1767.
[337]McKillop; Ford *Synth. Commun.* **1974**, *4*, 45.
[338]Franzen; Otto *Chem. Ber.* **1961**, *94*, 1360.

4-dimethylaminopyridine-N-oxide,[339] by other amine oxides (for allylic chlorides)[340] and by K_2CrO_4 in HMPA in the presence of a crown ether.[341] The first of these procedures has also been applied to primary tosylates.[338]

OS **II**, 336: **III**, 811; **IV**, 690, 918, 932; **V**, 242, 668, 825, 852, 872. Also see OS **V**, 689; **VI**, 218.

9-21   Oxidation of Amines or Nitro Compounds to Aldehydes, Ketones, or Dihalides
Oxo-de-hydro,amino-bisubstitution (overall transformation)

$$RR'CH-NH_2 \xrightarrow[\text{aq. NaOH}]{\text{AgNO}_3\cdot\text{Na}_2\text{S}_2\text{O}_8} \left[\begin{array}{c} R-\underset{\underset{NH}{\|}}{C}-R' \end{array} \right] \longrightarrow R-\underset{\underset{O}{\|}}{C}-R'$$

Primary aliphatic amines can be oxidized to aldehydes or ketones[342] by reaction with Ag(II) prepared in situ by treatment of silver nitrate with sodium persulfate.[343] The reaction consists of dehydrogenation to the imine (**9-5**) followed by hydrolysis. Other reagents used[344] have been nitrosobenzene[345] or N-bromoacetamide[346] (for benzylic amines), 3,5-di-t-butyl-1,2-benzoquinone,[347] m-trifluoromethylbenzenesulfonyl peroxide,[348] diphenylseleninic anhydride,[349] PdCl₂ or AuCl₃,[350] and aqueous NaOCl with phase-transfer catalysts.[351] Benzylic amine salts PhCHRNR₂H⁺ Cl⁻ (R,R' = H or alkyl) give benzaldehydes or aryl ketones when heated in Me₂SO.[352] Several indirect methods for achieving the conversion RR'CHNH₂ → RR'C=O (R' = alkyl, aryl, or H) have been reported.[353]

Primary, secondary, and tertiary aliphatic amines have been cleaved to give aldehydes, ketones, or carboxylic acids with aqueous bromine[354] and with neutral permanganate.[355] The other product of this reaction is the amine with one less alkyl group.

In a different type of procedure, primary alkyl primary amines can be converted to *gem*-dihalides [RCH₂NH₂ → RCHX₂ (X = Br or Cl)] by treatment with an alkyl nitrite and the anhydrous copper(I) halide.[356]

Primary and secondary aliphatic nitro compounds have been oxidized to aldehydes and ketones, respectively (RR'CHNO₂ → RR'C=O) with sodium chlorite under phase transfer conditions,[357] as well as with other reagents.[358]

[339]Mukaiyama; Inanaga; Yamaguchi *Bull. Chem. Soc. Jpn.* **1981,** *54,* 2221.
[340]Suzuki; Onishi; Fujita; Misawa; Otera *Bull. Chem. Soc. Jpn.* **1986,** *59,* 3287.
[341]Cardillo; Orena; Sandri *J. Chem. Soc. Chem. Commun.* **1976,** 190, *Tetrahedron Lett.* **1976,** 3985. For related procedures, see Landini; Rolla *Chem. Ind. (London)* **1979,** 213; Thuy; Maitte *Bull. Soc. Chim. Belg.* **1989,** *98,* 221.
[342]For a review, see Haines-1988, Ref. 11, pp. 200-220, 411-415.
[343]Bacon; Stewart *J. Chem. Soc. C* **1966,** 1384. See also Lee; Clarke *Tetrahedron Lett.* **1967,** 415.
[344]For lists of reagents, with references, see Ref. 21, pp. 601-602; Hudlický, Ref. 11, p. 240.
[345]Suzuki; Weisburger *Tetrahedron Lett.* **1966,** 5409, *J. Chem. Soc. C* **1968,** 199.
[346]Banerji *Bull. Chem. Soc. Jpn.* **1988,** *61,* 3717.
[347]Corey; Achiwa *J. Am. Chem. Soc.* **1969,** *91,* 1429. For a study of the mechanism, see Klein; Bargas; Horak *J. Org. Chem.* **1988,** *53,* 5994.
[348]Hoffman; Kumar *J. Org. Chem.* **1984,** *49,* 4011.
[349]Czarny *J. Chem. Soc., Chem. Commun.* **1976,** 81. See also Czarny *Synth. Commun.* **1976,** *6,* 285.
[350]Kuehne; Hall *J. Org. Chem.* **1976,** *41,* 2742.
[351]Lee; Freedman *Tetrahedron Lett.* **1976,** 1641.
[352]Traynelis; Ode *J. Org. Chem.* **1970,** *35,* 2207. For other methods, see Takabe; Yamada *Chem. Ind. (London)* **1982,** 959; Azran; Buchman; Pri-Bar *Bull. Soc. Chim. Belg.* **1990,** *99,* 345.
[353]See, for example, Dinizio; Watt *J. Am. Chem. Soc.* **1975,** *97,* 6900; Black; Blackman *Aust. J. Chem.* **1975,** *28,* 2547; Scully; Davis *J. Org. Chem.* **1978,** *43,* 1467; Doleschall *Tetrahedron Lett.* **1978,** 2131; Babler; Invergo *J. Org. Chem.* **1981,** *46,* 1937.
[354]Deno; Fruit *J. Am. Chem. Soc.* **1968,** *90,* 3502.
[355]Rawalay; Shechter *J. Org. Chem.* **1967,** *32,* 3129. For another procedure, see Monković; Wong; Bachand *Synthesis* **1985,** 770.
[356]Doyle; Siegfried *J. Chem. Soc., Chem. Commun.* **1976,** 433.
[357]Ballini; Petrini *Tetrahedron Lett.* **1989,** *30,* 5329.
[358]For a list of reagents, with references, see Ref. 21, p. 603.

9-22   Oxidation of Primary Alcohols to Carboxylic Acids or Carboxylic Esters
Oxo-de-dihydro-bisubstitution

$$RCH_2OH \xrightarrow{CrO_3} RCOOH$$

Primary alcohols can be oxidized to carboxylic acids by many strong oxidizing agents including chromic acid, permanganate, and nitric acid.[359] The reaction can be looked on as a combination of **9-3** and **4-6**. When acidic conditions are used, a considerable amount of carboxylic ester $RCOOCH_2R$ is often isolated, though this is probably not formed by a combination of the acid with unreacted alcohol, but by a combination of intermediate aldehyde with unreacted alcohol to give an acetal or hemiacetal, which is oxidized to the ester.[360] $RCOOCH_2R$ can be made the main product by treating the alcohol with (1) $Ru_3(CO)_{12}$ and diphenylacetylene, or with a complex formed from these two reagents;[361] (2) Pd salts and CCl_4 in the presence of K_2CO_3;[362] or (3) $RuH_2(PPh_3)_4$.[363] Primary alcohols RCH_2OH can be directly oxidized to acyl fluorides RCOF with cesium fluoroxysulfate.[364] Lactones can be prepared by oxidizing diols in which at least one OH is primary.[365]

Primary alkyl ethers can be selectively cleaved to carboxylic acids by aqueous Br_2 ($RCH_2OR' \rightarrow RCOOH$).[109] Aldehydes RCHO can be directly converted to carboxylic esters RCOOR' by treatment with Br_2 in the presence of an alcohol.[366]

OS **I**, 138, 168; **IV**, 499, 677; **V**, 580; **VII**, 406. Also see OS **III**, 745.

9-23   Oxidation of Olefins to Aldehydes and Ketones
1/Oxo-(1/→2/hydro)-*migro*-attachment

Monosubstituted and 1,2-disubstituted olefins can be oxidized to aldehydes and ketones by palladium chloride and similar salts of noble metals.[367] 1,1-Disubstituted olefins generally give poor results. The reaction is used industrially to prepare acetaldehyde from ethylene

[359]For reviews, see Hudlický, Ref. 11, pp. 127-132; Haines-1988, Ref. 11, 148-165, 391-401. For a list of reagents, with references, see Ref. 21, pp. 834-835.
[360]Craig; Horning *J. Org. Chem.* **1960**, *25*, 2098. See also Berthon; Forestiere; Leleu; Sillion *Tetrahedron Lett.* **1981**, *22*, 4073; Nwaukwa; Keehn *Tetrahedron Lett,* **1982**, *23*, 35.
[361]Blum; Shvo *J. Organomet. Chem.* **1984**, *263*, 93, *Isr. J. Chem.* **1984**, *24*, 144.
[362]Nagashima; Sato; Tsuji *Tetrahedron* **1985**, *41*, 5645.
[363]Murahashi; Naota; Ito; Maeda; Taki *J. Org. Chem.* **1987**, *52*, 4319. For another method, see Markó; Mekhalfia; Ollis *Synlett* **1990**, 347.
[364]Stavber; Planinšek; Zupan *Tetrahedron Lett.* **1989**, *30*, 6095.
[365]For examples of the preparation of lactones by oxidation of diols, see Doyle; Bagheri *J. Org. Chem.* **1981**, *46*, 4806; Ishii; Suzuki; Ikariya; Saburi; Yoshikawa *J. Org. Chem.* **1986**, *51*, 2822; Jefford; Wang *J. Chem. Soc., Chem. Commun.* **1988**, 634; Jones; Jakovac *Org. Synth.* **VII**, 406. For a list of reagents used to effect this conversion, with references, see Ref. 21, pp. 837-838.
[366]Williams; Klingler; Allen; Lichtenthaler *Tetrahedron Lett.* **1988**, *29*, 5087; Al Neirabeyeh; Pujol *Tetrahedron Lett.* **1990**, *31*, 2273. For other methods, see Sundararaman; Walker; Djerassi *Tetrahedron Lett.* **1978**, 1627; Grigg; Mitchell; Sutthivaiyakit *Tetrahedron* **1981**, *37*, 4313; Massoui; Beaupère; Nadjo; Uzan *J. Organomet. Chem.* **1983**, *259*, 345; O'Connor; Just *Tetrahedron Lett.* **1987**, *28*, 3235; McDonald; Holcomb; Kennedy; Kirkpatrick; Leathers; Vanemon *J. Org. Chem.* **1989**, *54*, 1212. For a list of reagents, with references, see Ref. 21, pp. 840-841.
[367]For a monograph, see Henry *Palladium Catalyzed Oxidation of Hydrocarbons*; D. Reidel Publishing Co.: Dordrecht, 1980. For reviews, see Tsuji *Organic Synthesis with Palladium Compounds*; Springer: New York, 1980, pp. 6-12, *Synthesis* **1990**, 739-749, **1984**, 369-384, *Adv. Org. Chem.* **1969**, *6*, 109-255, pp. 119-131; Heck *Palladium Reagents in Organic Syntheses*; Academic Press: New York, 1985, pp. 59-80; Sheldon; Kochi, Ref. 118, pp. 189-193, 299-303; Henry *Adv. Organomet. Chem.* **1975**, *13*, 363-452, pp. 378-388; Jira; Freiesleben *Organomet. React.* **1972**, *3*, 1-190, pp. 1-44; Khan; Martell *Homogeneous Catalysis by Metal Complexes*, vol. 2; Academic Press: New York, 1974, pp. 77-91; Hüttel *Synthesis* 225-255, **1970**, pp. 225-236; Aguiló *Adv. Organomet. Chem.* **1967**, *5*, 321-352; Bird *Transition Metal Intermediates in Organic Synthesis*; Academic Press: New York, 1967, pp. 88-111.

(the *Wacker process*), but it is also suitable for laboratory preparations. The palladium chloride is reduced to palladium. Because the reagent is expensive, the reaction is usually carried out with a cooxidant, most often $CuCl_2$, whose function is to reoxidize the Pd to Pd(II). The $CuCl_2$ is reduced to Cu(I), which itself is reoxidized to Cu(II) by air, so that atmospheric oxygen is the only oxidizing agent actually used up. Many other cooxidants have been tried, among them O_3, Fe^{3+}, and PbO_2. The principal product is an aldehyde only from ethylene: With other olefins Markovnikov's rule is followed, and ketones are formed predominantly. The generally accepted mechanism involves π complexes of palladium.[368]

$$C_2H_4 + PdCl_4{}^{2-} \xrightarrow{-Cl^-} \left[\begin{array}{c} CH_2 \\ \| \\ CH_2 \end{array} \xrightarrow{} \begin{array}{c} Cl \\ | \\ Pd-Cl \\ | \\ Cl \end{array} \right]^- \underset{H_2O, -HCl}{\rightleftharpoons} \left[\begin{array}{c} CH_2 \\ \| \\ CH_2 \end{array} \xrightarrow{} \begin{array}{c} Cl \\ | \\ Pd-Cl \\ | \\ OH \end{array} \right] \xrightarrow{H_2O} $$

$$\left[\begin{array}{c} Cl_2Pd-CH_2-CH_2-OH \\ | \\ OH_2 \end{array} \right]^- \rightleftharpoons \left[\begin{array}{c} CH_2\!\!=\!\!CH-OH \\ \downarrow \\ HPdCl_2 \end{array} \right]^- \rightleftharpoons \left[\begin{array}{c} CH_3-CH-PdCl_2 \\ | \\ OH \end{array} \right] \xrightarrow{H_2O}$$

$$CH_3-\underset{\underset{O}{\|}}{C}-H + H_3O^+ + Pd + 2Cl^-$$

This mechanism accounts for the fact, established by deuterium labeling, that the four hydrogens of the acetaldehyde all come from the original ethylene and none from the solvent.

Similar reactions have been carried out with other oxidizing agents. An example involving migration of an alkyl group instead of hydrogen is oxidation of $Me_2C\!\!=\!\!CMe_2$ with peroxytrifluoroacetic acid–boron trifluoride to give Me_3COMe (pinacolone).[369] This reaction consists of epoxidation (**5-36**) followed by pinacol rearrangement of the epoxide (**8-2**). A migration is also involved in the conversion of $ArCH\!\!=\!\!CHCH_3$ to $ArCH(CH_3)CHO$ by treatment with I_2–Ag_2O in aqueous dioxane.[370]

Other reagents used have been chromyl chloride[371] (e.g., $Me_3CCH_2CMe\!\!=\!\!CH_2 \rightarrow Me_3CCH_2CHMeCHO$), $Pb(OAc)_4$–F_3CCOOH[372] (e.g., $PhCH\!\!=\!\!CH_2 \rightarrow PhCH_2CHO$), thallium(III) nitrate–methanol[373] (e.g, cyclohexene $\rightarrow$ cyclopentanecarboxaldehyde), Cl_2 or Br_2 and $AgNO_3$,[374] disiamylborane followed by pyridinium chlorochromate,[375] H_2O_2 and a Pd catalyst,[376] H_2O–$PdCl_2$–polyethylene glycol,[377] O_2 and a catalyst,[378] CrO_3–H_2SO_4–Hg(II)

[368]Henry *J. Am. Chem. Soc.* **1966**, *88*, 1595, **1972**, *94*, 4437; Jira; Sedlmeier; Smidt *Liebigs Ann. Chem.* **1966**, *693*, 99; Hosokawa; Maitlis *J. Am. Chem. Soc.* **1973**, *95*, 4924; Moiseev; Levanda; Vargaftik *J. Am. Chem. Soc.* **1974**, *96*, 1003; Bäckvall; Åkermark; Ljunggren *J. Chem. Soc., Chem. Commun.* **1977**, 264, *J. Am. Chem. Soc.* **1979**, *101*, 2411; Zaw; Henry *J. Org. Chem.* **1990**, *55*, 1842.
[369]Hart; Lerner *J. Org. Chem.* **1967**, *32*, 2669.
[370]Kikuchi; Kogure; Toyoda *Chem. Lett.* **1984**, 341.
[371]Freeman; Cameron; DuBois *J. Org. Chem.* **1968**, *33*, 3970; Freeman; Arledge *J. Org. Chem.* **1972**, *37*, 2656. See also Sharpless; Teranishi; Bäckvall *J. Am. Chem. Soc.* **1977**, *99*, 3120.
[372]Lethbridge; Norman; Thomas *J. Chem. Soc., Perkin Trans. 1* **1973**, 35.
[373]McKillop; Hunt; Kienzle; Bigham; Taylor *J. Am. Chem. Soc.* **1973**, *95*, 3635. See also Grant; Liau; Low *Aust. J. Chem.* **1975**, *28*, 903.
[374]Kakis; Brase; Oshima *J. Org. Chem.* **1971**, *36*, 4117.
[375]Brown; Kulkarni; Rao *Synthesis* **1980**, 151.
[376]Roussel; Mimoun *J. Org. Chem.* **1980**, *45*, 5387.
[377]Alper; Januszkiewicz; Smith *Tetrahedron Lett.* **1985**, *26*, 2263.
[378]See, for example, Zombeck; Hamilton; Drago *J. Am. Chem. Soc.* **1982**, *104*, 6782; Januszkiewicz; Alper *Tetrahedron Lett.* **1983**, *24*, 5159, 5163; Bäckvall; Hopkins *Tetrahedron Lett.* **1988**, *29*, 2885; Chipperfield; Shana'a; Webster *J. Organomet. Chem.* **1988**, *341*, 511; Sage; Gore; Guilmet *Tetrahedron Lett.* **1989**, *30*, 6319.

salts,[379] $HgSO_4$–H_2O,[380] and $Hg(OAc)_2$ followed by $PdCl_2$.[381] The reaction has also been accomplished electrochemically.[382]

Alkenes have also been converted to more-highly-oxidized products. Examples are: (1) Treatment with $KMnO_4$ in aqueous acetone containing acetic acid gives α-hydroxy ketones.[383] (2) 1,2-Disubstituted and trisubstituted alkenes give α-chloro ketones when oxidized with chromyl chloride in acetone: $RCH=CR'R'' \rightarrow RCOCClR'R''$.[384] (3) α-Iodo ketones can be prepared by treating alkenes with bis(*sym*-collidine)iodine(I) tetrafluoroborate.[385] (4) $KMnO_4$ in acetic anhydride oxidizes large-ring cycloalkenes to 1,2-diketones.[386]

Enol ethers are oxidized to carboxylic esters ($RCH=CHOR' \rightarrow RCH_2COOR'$) with pyridinium chlorochromate[387] and enamines to α-amino ketones ($R^1CH=CR^2NR_2^3 \rightarrow R^1COCR^2NR_2^3$) with N-sulfonyloxaziridines.[388] Enamines $R^1R^4C=CR^2NR_2^3$ ($R^4 \neq H$) do not give these products, but lose the amino group to give α-hydroxy ketones $R^1R^4C(OH)COR^2$.[388] Carboxylic acids can be prepared from terminal alkynes ($RC\equiv CH \rightarrow RCH_2COOH$) by conversion of the alkyne to its thiophenyl ether ($RC\equiv CSPh$) and treatment of this with $HgSO_4$ in $HOAc$–H_2SO_4.[389]

OS **VI**, 1028; **VII**, 137; **67**, 121.

9-24   Oxidation of Amines to Nitroso Compounds and Hydroxylamines
N-Oxo-de-dihydro-bisubstitution

$$ArNH_2 \xrightarrow{H_2SO_5} Ar—N{=}O$$

Primary aromatic amines can be oxidized[390] to nitroso compounds. Most often the conversion is accomplished by Caro's acid (H_2SO_5) or with H_2O_2 in $HOAc$.[391] Hydroxylamines, which are probably intermediates in most cases, can sometimes be isolated, but under the reaction conditions are generally oxidized to the nitroso compounds. Primary aliphatic amines can be oxidized in this manner, but the nitroso compound is stable only if there is no α hydrogen. If there is an α hydrogen, the compound tautomerizes to the oxime.[392] Among the reagents used for this oxidation are sodium perborate[393] and Na_2WO_4–H_2O_2.[394] The mechanism with H_2SO_5 has been postulated to be an example of category 5 (p. 1161).[395]

$$R{-}\overset{\displaystyle H}{\underset{\displaystyle H}{N}}{\overset{\frown}{I}}\;\;O{-}\overset{\frown}{O}{-}SO_3^- \longrightarrow H^+ + SO_4^{2-} + R{-}\underset{\displaystyle H}{\overset{\displaystyle H}{N}}{-}OH \longrightarrow \text{further oxidation}$$

[379]Rogers; McDermott; Whitesides *J. Org. Chem.* **1975**, *40*, 3577.
[380]Arzoumanian; Aune; Guitard; Metzger *J. Org. Chem.* **1974**, *39*, 3445.
[381]Rodeheaver; Hunt *Chem. Commun.* **1971**, 818. See also Hunt; Rodeheaver *Tetrahedron Lett.* **1972**, 3595.
[382]See Tsuji; Minato *Tetrahedron Lett.* **1987**, *28*, 3683.
[383]Srinivasan; Lee *Synthesis* **1979**, 520. See also Baskaran; Das; Chandrasekaran *J. Org. Chem.* **1989**, *54*, 5182.
[384]Sharpless; Teranishi *J. Org. Chem.* **1973**, *38*, 185. See also Cardillo; Shimizu *J. Org. Chem.* **1978**, *42*, 4268; D'Ascoli; D'Auria; Nucciarelli; Piancatelli; Scettri *Tetrahedron Lett.* **1980**, *21*, 4521; Kageyama; Tobito; Katoh; Ueno; Okawara *Chem. Lett.* **1983**, 1481; Lee; Ha *Tetrahedron Lett.* **1989**, *30*, 193.
[385]Evans; Schauble *Synthesis* **1986**, 727.
[386]Sharpless; Lauer; Repič; Teranishi; Williams, *J. Am. Chem. Soc.* **1971**, *93*, 3303; Jensen; Sharpless *J. Org. Chem.* **1974**, *39*, 2314.
[387]Piancatelli; Scettri; D'Auria *Tetrahedron Lett.* **1977**, 3483. When $R^1CR^2C=CR^3OR^4$ are used, cleavage of the double bond takes place instead: Baskaran; Islam; Raghavan; Chandrasekaran *Chem. Lett.* **1987**, 1175.
[388]Davis; Sheppard *Tetrahedron Lett.* **1988**, *29*, 4365.
[389]Abrams *Can. J. Chem.* **1983**, *61*, 2423.
[390]For reviews on the oxidation of amines, see Rosenblatt; Burrows, in Patai The *Chemistry of Functional Groups, Supplement F*, pt. 2; Wiley: New York, 1982, pp. 1085-1149; Challis; Butler, in Patai The *Chemistry of the Amino Group*; Wiley: New York, 1968, pp. 320-338. For reviews confined to primary aromatic amines, see Hedayatullah *Bull. Soc. Chim. Fr.* **1972**, 2957; Surville; Jozefowicz; Buvet *Ann. Chem. (Paris)* **1967**, [14] *2*, 149-157.
[391]Holmes; Bayer *J. Am. Chem. Soc.* **1960**, *82*, 3454.
[392]For example, see Kahr; Berther *Chem. Ber.* **1960**, *93*, 132.
[393]Zajac; Darcy; Subong; Buzby *Tetrahedron Lett.* **1989**, *30*, 6495.
[394]Corey; Gross *Org. Synth.* 65, 166.
[395]Gragerov; Levit *J. Gen Chem. USSR* **1960**, *30*, 3690.

Secondary amines R_2NH are oxidized to hydroxylamines R_2NHOH (which are resistant to further oxidation) by dimethyldioxirane[396] and by benzoyl peroxide and Na_2HPO_4.[397]

OS **III**, 334; **65**, 166.

9-25   Oxidation of Primary Amines, Oximes, Azides, Isocyanates, or Nitroso Compounds to Nitro Compounds

$$R_3CNH_2 \xrightarrow{\text{KMnO}_4} R_3CNO_2$$

$$R_2C\text{=}NOH \xrightarrow{\text{F}_3\text{CCOOOH}} R_2CHNO_2$$

Tertiary alkyl primary amines can be oxidized to nitro compounds in excellent yields with $KMnO_4$.[398] This type of nitro compound is not easily prepared in other ways. All classes of primary amine (including primary, secondary, and tertiary alkyl as well as aryl) are oxidized to nitro compounds in high yields with dimethyldioxirane.[399] Other reagents that oxidize various types of primary amines to nitro compounds are dry ozone,[400] various peracids,[401] including peracetic and peroxytrifluoroacetic acids, t-butyl hydroperoxide in the presence of certain molybdenum and vanadium compounds,[402] F_2–H_2O McCN,[402a] and sodium perborate.[403]

Dimethyldioxirane in wet acetone oxidizes isocyanates to nitro compounds ($RNCO \rightarrow RNO_2$).[404] Oximes can be oxidized to nitro compounds with peroxytrifluoroacetic acid, among other ways.[398] Primary and secondary alkyl azides have been converted to nitro compounds by treatment with Ph_3P followed by ozone.[405] Aromatic nitroso compounds are easily oxidized to nitro compounds by many oxidizing agents.[406]

OS **III**, 334; **V**, 367, 845; **VI**, 803.

9-26   Oxidation of Thiols and Other Sulfur Compounds to Sulfonic Acids
Thiol-sulfonic acid oxidation

$$RSH \xrightarrow{\text{HNO}_3} RSO_3H$$

Thiols, sulfoxides, sulfones, disulfides,[407] and other sulfur compounds can be oxidized to sulfonic acids with many oxidizing agents, though for synthetic purposes the reaction is most important for thiols.[408] Among oxidizing agents used are boiling nitric acid and barium

[396]Murray; Singh *Synth. Commun.* **1989**, *19*, 3509. This reagent also oxidizes primary amines to hydroxylamines: Wittman; Halcomb; Danishefsky *J. Org. Chem.* **1990**, *55*, 1981.

[397]Biloski; Ganem *Synthesis* **1983**, 537.

[398]Larson, in Feuer *The Chemistry of the Nitro and Nitroso Groups*, vol. 1; Wiley: New York, 1969, pp. 306-310. See also Barnes; Patterson *J. Org. Chem.* **1976**, *41*, 733. For reviews of oxidations of nitrogen compounds, see Butler *Chem. Rev.* **1984**, *84*, 249-276; Boyer *Chem. Rev.* **1980**, *80*, 495-561.

[399]Murray; Rajadhyaksha; Mohan *J. Org. Chem.* **1989**, *54*, 5783. See also Zabrowski; Moorman; Beck *Tetrahedron Lett.* **1988**, *29*, 4501.

[400]Keinan; Mazur *J. Org. Chem.* **1977**, *42* 844; Bachman; Strawn *J. Org. Chem.* **1968**, *33*, 313.

[401]Emmons *J. Am. Chem. Soc.* **1957**, *79*, 5528; Gilbert; Borden *J. Org. Chem.* **1979**, *44*, 659.

[402]Howe; Hiatt *J. Org. Chem.* **1970**, *35*, 4007. See also Nielsen; Atkins; Norris; Coon; Sitzmann *J. Org. Chem.* **1980**, *45*, 2341.

[402a]Kol; Rozen *J. Chem. Soc., Chem. Commun.* **1991**, 567.

[403]McKillop; Tarbin *Tetrahedron* **1987**, *43*, 1753.

[404]Eaton; Wicks *J. Org. Chem.* **1988**, *53*, 5353.

[405]Corey; Samuelsson; Luzzio *J. Am. Chem. Soc.* **1984**, *106*, 3682.

[406]See Boyer, in Feuer, Ref. 398, pp. 264-265.

[407]For a review of the oxidation of disulfides, see Savige; Maclaren, in Kharasch; Meyers *Organic Sulfur Compounds*, vol. 2; pp. 367-402, Pergamon, New York, 1966.

[408]For a general review of the oxidation of thiols, see Capozzi; Modena, in Patai *The Chemistry of the Thiol Group*, pt. 2; Wiley: New York, 1974, pp. 785-839. For a review specifically on the oxidation to sulfonic acids, see Gilbert *Sulfonation and Related Reactions*; Wiley: New York, 1965, pp. 217-239.

permanganate. Autoxidation (oxidation by atmospheric oxygen) can be accomplished in basic solution.[409] Oxidation of thiols with chlorine and water gives sulfonyl chlorides directly.[410] Thiols can also be oxidized to disulfides (**9-35**).

OS **II**, 471; **III**, 226. Also see OS **V**, 1070.

D. Reactions in Which Oxygen is Added to the Substrate

9-27 The Oxidation of Alkynes to α-Diketones
Dioxo-biaddition

$$R-C\equiv C-R' \xrightarrow[\text{tetroxide}]{\text{ruthenium}} R-\underset{\underset{O}{\|}}{C}-\underset{\underset{O}{\|}}{C}-R'$$

Internal alkynes have been oxidized[411] to α-diketones by several oxidizing agents,[412] including ruthenium tetroxide,[413] neutral $KMnO_4$,[414] SeO_2 with a small amount of H_2SO_4,[415] bis(trifluoroacetoxy)iodobenzene,[416] $NaIO_4$–RuO_2,[417] I_2–Me_2SO,[418] and thallium(III) nitrate,[206] as well as by electrooxidation.[419] Ozone generally oxidizes triple-bond compounds to carboxylic acids (**9-9**), but α-diketones are sometimes obtained instead. SeO_2 with a small amount of H_2SO_4 oxidizes arylacetylenes to α-keto acids ($ArC\equiv CH \rightarrow ArCOCOOH$),[415] while H_2O_2–$Hg(OAc)_2$ together with a molybdenate salt oxidizes them to α-keto aldehydes, though yields are not high.[420]

9-28 Oxidation of Tertiary Amines to Amine Oxides
N-Oxygen-attachment

$$R_3N \xrightarrow{H_2O_2} R_3\overset{\oplus}{N}-\overset{\ominus}{O}$$

Tertiary amines can be converted to amine oxides by oxidation. Hydrogen peroxide is often used, but peracids are also important reagents for this purpose. Pyridine and its derivatives are oxidized only by peracids.[421] In the attack by hydrogen peroxide there is first formed a trialkylammonium peroxide, a hydrogen-bonded complex represented as $R_3N \cdot H_2O_2$, which can be isolated.[422] The decomposition of this complex probably involves an attack by the

[409]Wallace; Schriesheim *Tetrahedron* **1965**, *21*, 2271.
[410]For a review, see Gilbert, Ref. 408, pp. 202-214.
[411]For a review of this reaction, see Haines-1985, Ref. 11, pp. 153-162, 332-338. For a review of oxidations of triple bonds in general, see Simándi, in Patai; Rappoport *The Chemistry of Functional Groups, Supplement C*, pt. 1; Wiley: New York, 1983, pp. 513-570.
[412]For a list of reagents, with references, see Hudlický, Ref. 11, p. 92.
[413]Gopal; Gordon *Tetrahedron Lett.* **1971**, 2941.
[414]Khan; Newman *J. Org. Chem.* **1952**, *17*, 1063; Srinivasan; Lee *J. Org. Chem.* **1979**, *44*, 1574; Lee; Lee; Chandler *J. Org. Chem.* **1985**, *50*, 4306.
[415]Sonoda; Yamamoto; Murai; Tsutsumi *Chem. Lett.* **1972**, 229.
[416]Vasil'eva; Khalfina; Karpitskaya; Merkushev *J. Org. Chem. USSR* **1987**, *23*, 1967.
[417]Zibuck; Seebach *Helv. Chim. Acta* **1988**, *71*, 237.
[418]Yusybov; Filimonov *Synthesis* **1991**, 131.
[419]Torii; Inokuchi; Hirata *Synthesis* **1987**, 377.
[420]Ballistreri; Failla; Tomaselli *J. Org. Chem.* **1988**, *53*, 830.
[421]For reviews, see Albini; Pietra *Heterocyclic N-Oxides*; CRC Press: Boca Raton, FL, 1991, pp. 31-41; Katritzky; Lagowski *Chemistry of the Heterocyclic N-Oxides*; Academic Press: New York, 1971, pp. 21-72, 539-542.
[422]Oswald; Guertin *J. Org. Chem.* **1963**, *28*, 651.

OH$^+$ moiety of the H$_2$O$_2$. Oxidation with Caro's acid has been shown to proceed in this manner:[423]

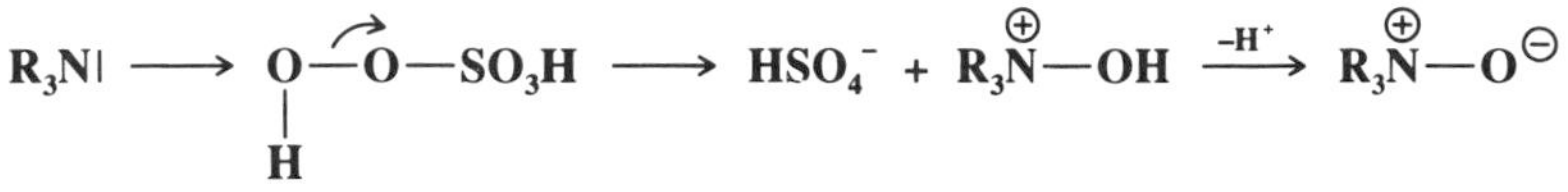

$$R_3NI \longrightarrow \underset{H}{O}{\overset{\frown}{-}}O-SO_3H \longrightarrow HSO_4^- + R_3\overset{\oplus}{N}-OH \xrightarrow{-H^+} R_3\overset{\oplus}{N}-O^{\ominus}$$

This mechanism is the same as that of **9-24**; the products differ only because tertiary amine oxides cannot be further oxidized. The mechanism with other peracids is probably the same. Racemic β-hydroxy tertiary amines have been resolved by oxidizing them with *t*-BuOOH and a chiral catalyst—one enantiomer reacts faster than the other.[424] This kinetic resolution gives products with enantiomeric excesses of >90%.

OS **IV**, 612, 704, 828; **VI**, 342, 501; **69**, 226.

9-29 Oxidation of Azobenzenes to Azoxybenzenes
N-Oxygen-attachment

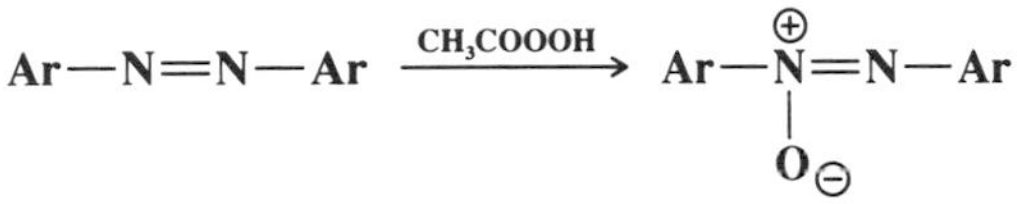

$$Ar-N{=}N-Ar \xrightarrow{CH_3COOOH} Ar-\underset{\underset{O_{\ominus}}{|}}{\overset{\oplus}{N}}{=}N-Ar$$

Azo compounds can be oxidized to azoxy compounds by peracids[425] or by hydroperoxides and molybdenum complexes.[426] The mechanism is probably the same as that of **9-28**.[427]

9-30 Oxidation of Isocyanides to Isocyanates
Oxygen-attachment

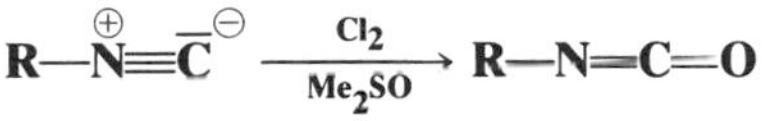

$$R-\overset{\oplus}{N}{\equiv}\overset{\ominus}{C} \xrightarrow[Me_2SO]{Cl_2} R-N{=}C{=}O$$

Isocyanides have been oxidized to isocyanates with HgO and with O$_3$, as well as with a halogen and dimethyl sulfoxide (or pyridine N-oxide).[428] In the latter case the oxidizing agent is the halogen, which converts the isocyanide to R—N=CCl$_2$ which is hydrolyzed to the isocyanate.[429] Cyanide ion has been oxidized to cyanate ion with many oxidizing agents.

Isocyanides can be converted to isothiocyanates (RNC → RNCS) by treatment with a disulfide such as PhCOSSCOPh and thallium(I) acetate or lead(II) acetate.[430]

9-31 Oxidation of Thioethers to Sulfoxides and Sulfones
S-Oxygen-attachment

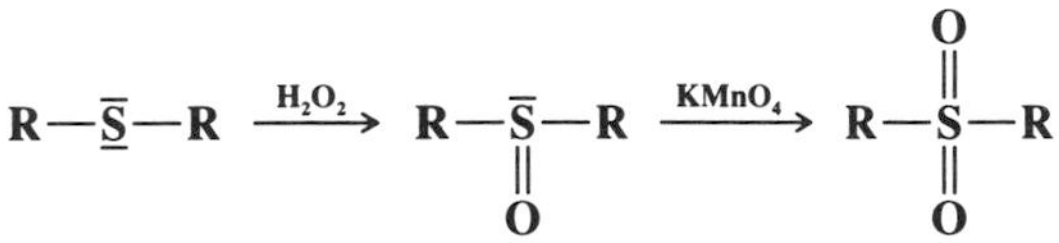

$$R-\bar{S}-R \xrightarrow{H_2O_2} R-\underset{\underset{O}{\|}}{\bar{S}}-R \xrightarrow{KMnO_4} R-\underset{\underset{O}{\|}}{\overset{\overset{O}{\|}}{S}}-R$$

[423]Ogata; Tabushi *Bull. Chem. Soc. Jpn.* **1958**, *31*, 969.
[424]Miyano; Lu; Viti; Sharpless *J. Org. Chem.* **1985**, *50*, 4350.
[425]For reviews, see Yandovskii; Gidaspov; Tselinskii *Russ. Chem. Rev.* **1981**, *50*, 164-179; Newbold, Ref. 136, pp. 557-563, 573-593.
[426]Johnson; Gould *J. Org. Chem.* **1974**, *39*, 407.
[427]Mitsuhashi; Simamura; Tezuka *Chem. Commun.* **1970**, 1300.
[428]For a review, see Simándi, Ref. 411, pp. 559-562.
[429]Johnson; Daughhetee *J. Org. Chem.* **1964**, *29*, 246; Johnson; Krutzsch *J. Org. Chem.* **1967**, *32*, 1939.
[430]Tanaka; Uemura; Okano *Bull. Chem. Soc. Jpn.* **1977**, *50*, 2785.

Thioethers can be oxidized to sulfoxides by 1 mole of 30% H_2O_2 or by many other oxidizing agents,[431] including $NaIO_4$,[432] t-BuOCl,[433] calcium hypochlorite $Ca(OCl)_2$,[434] sodium chlorite $NaClO_2$,[434] sodium hypochlorite $NaOCl$,[435] dioxiranes,[436] HNO_3 and an $AuCl_4^-$ catalyst,[437] O_2 and a ceric ammonium nitrate catalyst,[438] acyl nitrites,[439] sodium perborate,[403] and peracids.[440] Sulfoxides can be further oxidized to sulfones by another mole of H_2O_2, $KMnO_4$, sodium perborate, potassium hydrogen persulfate $KHSO_5$,[441] or a number of other agents. If enough oxidizing agent is present, thioethers can be directly converted to sulfones without isolation of the sulfoxides.[442] These reactions give high yields, and many functional groups do not interfere.[443] As with tertiary amines (**9-28**), racemic thioethers can be kinetically resolved by oxidation to sulfoxides with an optically active reagent, and this has often been done.[444] Selenides R_2Se can be oxidized to selenoxides and selenones.[445]

When the oxidizing agent is a peroxide, the mechanism[446] of oxidation to the sulfoxide is similar to that of **9-28**.[447]

$$R-\overset{|}{\underset{R}{\bar{S}}}\vert \to O\underset{H}{-}O-R' \longrightarrow R-\overset{\oplus}{\underset{R}{\bar{S}}}-O-H + R'O^- \xrightarrow[\text{transfer}]{\text{rapid proton}} R-\overset{|}{\underset{R}{\bar{S}}}=O + R'OH$$

The second oxidation, which is normally slower than the first[448] (which is why sulfoxides are so easily isolable), has the same mechanism in neutral or acid solution, but in basic solution it has been shown that the conjugate base of the peroxy compound ($R'OO^-$) also attacks the SO group as a nucleophile:[449]

$$R-\overset{|}{\underset{R}{\bar{S}}}=\bar{O}\vert + {}^{\ominus}OOR' \longrightarrow R-\overset{|}{\underset{R}{S}}\overset{\vert\bar{O}\vert^{\ominus}}{\underset{}{\bar{O}}}-O-R' \longrightarrow R-\overset{\vert\bar{O}}{\underset{R}{\overset{\vert\vert}{S}}}=\bar{O}\vert + R'O^-$$

[431]For reviews, see Hudlický, Ref. 11, pp. 252-263; Drabowicz; Kiełbasinski; Mikołajczyk, in Patai; Rappoport; Stirling *The Chemistry of Sulphones and Sulphoxides;* Wiley: New York, 1988, pp. 233-378, pp. 235-255; Madesclaire *Tetrahedron* **1986**, *42*, 5459-5495; Block, in Patai *Supplement E*, Ref. 44, pt. 1, pp. 539-608. For reviews on methods of synthesis of sulfoxides, see Drabowicz; Mikołajczyk *Org. Prep. Proced. Int.* **1982**, *14*, 45-89; Oae, in Oae *The Organic Chemistry of Sulfur;* Plenum: New York, 1977, pp. 385-390. For a review with respect to enzymic oxidation, see Holland *Chem. Rev.* **1988**, *88*, 473-485.

[432]Leonard; Johnson *J. Org. Chem.* **1962**, *27*, 282; Hiskey; Harpold *J. Org. Chem.* **1967**, *32*, 3191.

[433]Walling; Mintz *J. Org. Chem.* **1967**, *32*, 1286; Skattebøl; Boulette; Solomon *J. Org. Chem.* **1967**, *32*, 3111.

[434]Weber; Scheider; Salami; Paquer *Recl. Trav. Chim. Pays-Bas* **1986**, *105*, 99.

[435]Ramsden; Drago; Riley *J. Am. Chem. Soc.* **1989**, *111*, 3958.

[436]Colonna; Gaggero *Tetrahedron Lett.* **1989**, *30*, 6233.

[437]Gasparrini; Giovannoli; Misiti; Natile; Palmieri *J. Org. Chem.* **1990**, *55*, 1323.

[438]Riley; Smith; Correa *J. Am. Chem. Soc.* **1988**, *110*, 177.

[439]Louw; Vermeeren; van Asten; Ultée *J. Chem. Soc., Chem. Commun.* **1976**, 496.

[440]For lists of some of the many oxidizing agents used in this reaction, see Ref. 431 and Block *Reactions of Organosulfur Compounds*; Academic Press: New York, 1978, p. 16.

[441]Trost; Curran *Tetrahedron Lett.* **1981**, *22*, 1287.

[442]For a review, see Schank, in Patai; Rappoport; Stirling, Ref. 431, pp. 165-231, pp. 205-213.

[443]For a review of the oxidation of α-halo sulfides, see Venier; Barager *Org. Prep. Proced. Int.* **1974**, *6*, 77-102, pp. 85-86.

[444]For reviews, see Kagan; Rebiere *Synlett* **1990**, 643-650; Drabowicz; Kiełbasinski; Mikołajczyk, Ref. 431, pp. 288-297; Madesclaire, Ref. 431, pp. 5481-5488. See also Zhao; Samuel; Kagan *Tetrahedron* **1987**, *43*, 5135; Glahsl; Herrmann *J. Chem. Soc., Perkin Trans. 1* **1988**, 1753; Davis; ThimmaReddy; Weismiller *J. Am. Chem. Soc.* **1989**, *111*, 5964; Di Furia; Licini; Modena; Valle *Bull. Soc. Chim. Fr.* **1990**, 734; Ref. 436.

[445]See Reich; in Trahanovsky, Ref. 5, pt. C, pp. 7-13; Davis; Stringer; Billmers *Tetrahedron Lett.* **1983**, *24*, 1213; Kobayashi; Ohkubo; Shimizu *Bull. Chem. Soc. Jpn.* **1986**, *59*, 503.

[446]For discussions of the mechanism with various other agents, see Rajasekaran; Baskaran; Gnanasekaran *J. Chem. Soc., Perkin Trans. 2* **1984**, 1183; Srinivasan; Chellamani; Rajagopal *J. Org. Chem.* **1985**, *50*, 1201; Agarwal; Bhatt; Banerji *J. Phys. Org. Chem.* **1990**, *3*, 174; Lee ; Chen *J. Org. Chem.* **1991**, *56*, 5346.

[447]Modena; Todesco *J. Chem. Soc.* 4920, **1962**, and references cited therein.

[448]There are some reagents that oxidize sulfoxides in preference to sulfides, e.g., $NaMnO_4$: see Henbest; Khan *Chem. Commun.* **1968**, 1036.

[449]Curci; Modena *Tetrahedron Lett.* **1963**, 1749, *Tetrahedron Lett.* **1966**, *22*, 1227; Curci; Di Furia; Modena *J. Chem. Soc., Perkin Trans. 2* **1978**, 603. See also Oae; Takata *Tetrahedron Lett.* **1980**, *21*, 3213; Akasaka; Ando *J. Chem. Soc., Chem. Commun.* **1983**, 1203.

OS **V**, 791; **VI**, 403, 404, 482; **VII**, 453, 491; **67**, 157; **68**, 49. Also see OS **V**, 723; **VI**, 23.

9-32   Oxidation of Carboxylic Acids to Peroxy Acids
Peroxy-de-hydroxy-substitution

$$\underset{\underset{O}{\parallel}}{R-C}-OH + H_2O_2 \overset{H^+}{\rightleftharpoons} \underset{\underset{O}{\parallel}}{R-C}-O-O-H + H_2O$$

The oxidation of carboxylic acids with H_2O_2 and an acid catalyst is the best general method for the preparation of peroxy acids.[450] The most common catalyst for aliphatic R is concentrated sulfuric acid. The reaction is an equilibrium and is driven to the right by removal of water or by the use of excess reagents. For aromatic R the best catalyst is methanesulfonic acid, which is also used as the solvent.

E. Oxidative Coupling

9-33   Coupling Involving Carbanions
De-hydro,chloro-coupling

$$\underset{\underset{Cl}{\mid}}{2R-CH}-Z + KOH \longrightarrow \underset{\underset{Z}{\mid}\,\,\underset{Z}{\mid}}{R-C=C-R}$$

Alkyl halides with an electron-withdrawing group on the halogen-bearing carbon can be dimerized to olefins by treatment with bases. Z may be nitro, aryl, etc. It is likely that in most cases the mechanism[451] involves nucleophilic substitution followed by elimination[452] (illustrated for benzyl chloride):

$$PhCH_2Cl \xrightarrow{base} Ph\overset{\ominus}{C}HCl \xrightarrow[S_N]{PhCH_2Cl} PhCHClCH_2Ph \xrightarrow{-HCl} PhCH=CHPh$$

α,α-Dibromotoluenes $ArCHBr_2$ give tolanes $ArC\equiv CAr$, by debromination of the intermediates $ArCBr=CBrAr$.[453] In a related reaction, diarylmethane dihalides Ar_2CX_2 have been dimerized to tetraaryl alkenes $Ar_2C=CAr_2$ with sodium selenide,[454] with copper,[455] with iron(II) oxalate dihydrate,[456] and with iron pentacarbonyl.[457]

A somewhat different type of coupling is observed when salts of β-keto esters, arylacetonitriles $ArCH_2CN$, and other compounds of the form ZCH_2Z' are treated with an

[450]For a review of the preparation of peroxy acids, see Swern, in Swern *Organic Peroxides*, vol. 1; Wiley: New York, 1970, pp. 313-516.

[451]For discussion, see Saunders; Cockerill *Mechanisms of Elimination Reactions*; Wiley: New York, 1973, pp. 548-554.

[452]For example, see Hauser; Brasen; Skell; Kantor; Brodhag *J. Am. Chem. Soc.* **1956**, *78*, 1653; Hoeg; Lusk *J. Organomet. Chem.* **1966**, *5*, 1; Reisdorf; Normant *Organomet. Chem. Synth.* **1972**, *1*, 375; Hanna; Wideman *Chem. Ind. (London)* **1968**, 486. In some cases a radical anion chain mechanism can take place: Bethell; Bird *J. Chem. Soc., Perkin Trans. 2* **1977**, 1856.

[453]Vernigor; Shalaev; Luk'yanets *J. Org. Chem. USSR* **1981**, *17*, 317.

[454]Okamoto; Yano *J. Org. Chem.* **1969**, *34*, 1492.

[455]Buckles; Matlack *Org. Synth. IV*, 914.

[456]Khurana; Maikap; Mehta *Synthesis* **1990**, 731.

[457]Coffey *J. Am. Chem. Soc.* **1961**, *83*, 1623.

oxidizing agent such as iodine,[458] PbO_2,[459] Ag_2O,[460] Cu(II) salts,[461] or a Cu–amine–O_2 system,[462] e.g.,

$$Ar_2CH_2 \xrightarrow[\text{liq. } NH_3]{NaNH_2} Ar\overset{\ominus}{CH} + I_2 \longrightarrow Ar_2CH\!-\!CHAr_2$$

In this case the product is a substituted alkane rather than an alkene. This reaction has been used to close rings.[463] Arylmethanesulfonyl chlorides $ArCH_2SO_2Cl$ couple to give $ArCH\!=\!CHAr$ when treated with Et_3N.[464]

OS **II**, 273; **IV**, 372, 869, 914; **68**, 198. Also see OS **I**, 46; **IV**, 877.

9-34 Dimerization of Silyl Enol Ethers or of Lithium Enolates
3/*O*-De-trimethylsilyl-1/*C*-coupling

$$R^1\!-\!\underset{\underset{OSiMe_3}{|}}{\overset{\overset{R^2}{|}}{C}}\!=\!\underset{}{\overset{}{C}}\!-\!R^3 \xrightarrow[Me_2SO]{Ag_2O} R^1\!-\!\underset{\overset{|}{O}}{\overset{\overset{R^2}{|}}{C}}\!-\!\underset{\overset{|}{R^3}}{\overset{\overset{R^2}{|}}{C}}\!-\!\underset{\overset{|}{R^3}}{\overset{|}{C}}\!-\!\underset{\overset{||}{O}}{\overset{}{C}}\!-\!R^1$$

32

Silyl enol ethers can be dimerized to symmetrical 1,4-diketones by treatment with Ag_2O in dimethyl sulfoxide or certain other polar aprotic solvents.[465] The reaction has been performed with R^2, R^3 = hydrogen or alkyl, though best yields are obtained when $R^2 = R^3 = H$. In certain cases, unsymmetrical 1,4-diketones have been prepared by using a mixture of two silyl enol ethers. Other reagents that have been used to achieve either symmetrical or cross-coupled products are iodosobenzene–BF_3–Et_2O,[466] ceric ammonium nitrate,[467] and lead tetraacetate.[468] If R^1 = OR (in which case the substrate is a ketene silyl acetal), dimerization with $TiCl_4$ leads to a dialkyl succinate (**32**, R^1 = OR).[469]

In a similar reaction, lithium enolates $RC(Li)\!=\!CH_2$ were dimerized to 1,4-diketones $RCOCH_2CH_2COR$ with $CuCl_2$, $FeCl_3$, or copper(II) triflate, in a nonprotic solvent.[470]

OS **69**, 173.

9-35 Oxidation of Thiols to Disulfides
S-De-hydrogen-coupling

$$2RSH \xrightarrow{H_2O_2} RSSR$$

[458]See, for example, Kaiser *J. Am. Chem. Soc.* **1967**, *89*, 3659; Belletire; Spletzer; Pinhas *Tetrahedron Lett.* **1984**, *25*, 5969; Mignani; Lahousse; Merényi; Janousek; Viehe *Tetrahedron Lett.* **1985**, *26*, 4607; Aurell; Gil; Tortajada; Mestres *Synthesis* **1990**, 317.

[459]Brettle; Seddon *J. Chem. Soc., C.* **1970**, 1320.

[460]Ito; Fujii; Konoike; Saegusa *Synth. Commun.* **1976**, *6*, 429.

[461]Rathke; Lindert *J. Am. Chem. Soc.* **1971**, *93*, 4605; Baudin; Julia; Rolando; Verpeaux *Bull. Soc. Chim. Fr.* **1987**, 493.

[462]de Jongh; de Jonge; Mijs *J. Org. Chem.* **1971**, *36*, 3160.

[463]Chung; Dunn *J. Org. Chem.* **1983**, *48*, 1125.

[464]King; Durst *Tetrahedron Lett.* **1963**, 585; King; Harding *Can. J. Chem.* **1976**, *54*, 2652; Nakayama; Tanuma; Honda; Hoshino *Tetrahedron Lett.* **1984**, *25*, 4553.

[465]Ito; Konoike; Saegusa *J. Am. Chem. Soc.* **1975**, *97*, 649.

[466]Moriarty; Prakash; Duncan *J. Chem. Soc., Perkin Trans. 1* **1987**, 559.

[467]Baciocchi; Casu; Ruzziconi *Tetrahedron Lett.* **1989**, *30*, 3707.

[468]Moriarty; Penmasta; Prakash *Tetrahedron Lett.* **1987**, *28*, 873.

[469]Inaba; Ojima *Tetrahedron Lett.* **1977**, 2009. See also Totten; Wenke; Rhodes *Synth. Commun.* **1985**, *15*, 291, 301.

[470]Ito; Konoike; Harada; Saegusa *J. Am. Chem. Soc.* **1977**, *99*, 1487; Kobayashi; Taguchi; Tokuno *Tetrahedron Lett.* **1977**, 3741; Frazier; Harlow *J. Org. Chem.* **1980**, *45*, 5408.

Thiols are easily oxidized to disulfides.[471] Hydrogen peroxide is the most common reagent,[472] but many oxidizing agents give the reaction, among them thallium(III) acetate,[473] Me_2SO–I_2,[474] Br_2 under phase transfer conditions,[475] methoxytributyltin–$FeCl_3$,[476] sodium perborate,[477] NO,[478] and NO_2.[478] It can also be done electrochemically.[479] However, strong oxidizing agents may give **9-26**. Even the oxygen in the air oxidizes thiols on standing, if a small amount of base is present. The reaction is reversible (see **9-61**), and the interconversion between cysteine and cystine is an important one in biochemistry.

The mechanism has been studied for several oxidizing agents and varies with the agent.[480] For oxygen it is [481]

$$RSH + B^- \rightleftharpoons RS^- + BH$$

$$RS^- + O_2 \longrightarrow RS\cdot + \overset{\cdot}{O}_2{}^-$$

$$RS^- + \overset{\cdot}{O}_2{}^- \longrightarrow RS\cdot + O_2{}^{2-}$$

$$2RS\cdot \longrightarrow RSSR$$

$$O_2{}^{2-} + BH \longrightarrow OH^- + B^- + O_2$$

With respect to the sulfur, this mechanism is similar to that of **4-17**, involving as it does loss of a proton, oxidation to a free radical, and radical coupling.

Unsymmetrical disulfides can be prepared[482] by treatment of a thiol RSH with diethyl azodicarboxylate EtOOCN=NCOOEt to give an adduct, to which another thiol R'SH is then added, producing the disulfide RSSR'.[483]

OS **III**, 86, 116.

9-36   Oxidation of Amines to Azo or Azoxy Compounds
N-De-bishydrogen-coupling

$$ArNH_2 \xrightarrow{MnO_2} Ar—N{=}N—Ar$$

Primary aromatic amines have been oxidized to azo compounds by a variety of oxidizing agents, among them MnO_2, lead tetraacetate, O_2 and a base, barium permanganate,[484] and sodium perborate in acetic acid. *t*-Butyl hydroperoxide has been used to oxidize certain primary amines to azoxy compounds.[485]

OS **V**, 341.

[471]For a review, see Capozzi; Modena, Ref. 408, pp. 785-839. For a list of reagents, with references, see Block, Ref. 440.

[472]It has been pointed out that, nevertheless, H_2O_2 is not a very good reagent for this reaction, since it gives sulfonic acids (**9-26**) as well as disulfides: Evans; Doi; Musker *J. Org. Chem.* **1990**, *55*, 2337.

[473]Uemura; Tanaka; Okano *Bull. Chem. Soc. Jpn.* **1977**, *50*, 220.

[474]Aida; Akasaka; Furukawa; Oae *Bull. Chem. Soc. Jpn.* **1976**, *49*, 1441. See also Fristad; Peterson *Synth. Commun.* **1985**, *15*, 1.

[475]Drabowicz; Mikołajczyk *Synthesis* **1980**, 32.

[476]Sato; Otera; Nozaki *Tetrahedron Lett.* **1990**, *31*, 3591.

[477]McKillop; Koyuncu *Tetrahedron Lett.* **1990**, *31*, 5007.

[478]Pryor; Church; Govindan; Crank *J. Org. Chem.* **1982**, *47*, 156.

[479]See, for example, Leite; Pardini; Viertler *Synth. Commun.* **1990**, *20*, 393. For a review, see Shono, Ref. 149, pp. 38-43.

[480]See Tarbell, in Kharasch, *Organic Sulfur Compounds*; Pergamon: Elmsford, NY, 1961, pp. 97-102.

[481]Wallace; Schriesheim; Bartok *J. Org. Chem.* **1963**, *28*, 1311.

[482]Mukaiyama; Takahashi *Tetrahedron Lett.* **1968**, 5907.

[483]For other methods, see Boustany; Sullivan *Tetrahedron Lett.* **1970**, 3547; Harpp; Ash; Back; Gleason; Orwig; VanHorn; Snyder *Tetrahedron Lett.* **1970**, 3551; Oae; Fukushima; Kim *J. Chem. Soc., Chem. Commun.* **1977**, 407.

[484]Firouzabadi; Mostafavipoor *Bull. Chem. Soc. Jpn.* **1983**, *56*, 914.

[485]Kosswig *Liebigs Ann. Chem.* **1971**, *749*, 206.

Reductions: Selectivity[486]

It is often necessary to reduce one group in a molecule without affecting another reducible group. It is usually possible to find a reducing agent that will do this. The most common broad-spectrum reducing agents are the metal hydrides[487] and hydrogen (with a catalyst).[488] Many different metal-hydride systems and hydrogenation catalysts have been investigated in order to find conditions under which a given group will be reduced chemoselectively. Tables 19.2, 19.3, and 19.4 list the reactivity of various functional groups toward catalytic hydrogenation, LiAlH$_4$, and BH$_3$, respectively.[489] Table 19.5 shows which groups can be reduced by catalytic hydrogenation and various metal hydrides.[490] Of course, the tables cannot be exact, because the nature of R and the reaction conditions obviously affect reactivity. Nevertheless, the tables do give a fairly good indication of which reagents reduce

TABLE 19.2 The ease of reduction of various functional groups toward catalytic hydrogenation[489]

The groups are listed in approximate order of ease of reduction

Reaction	Substrate	Product	
0-83	RCOCl	RCHO	Easiest
9-47	RNO$_2$	RNH$_2$	
5-9	RC≡CR	RCH=CHR	
6-25	RCHO	RCH$_2$OH	
5-9	RCH=CHR	RCH$_2$CH$_2$R	
6-25	RCOR	RCHOHR	
0-79	ArCH$_2$OR	ArCH$_3$ + ROH	
6-27	RC≡N	RCH$_2$NH$_2$	
5-10	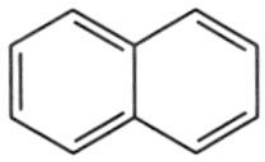	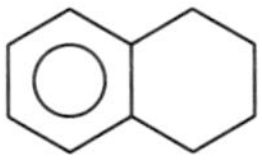	
9-42	RCOOR'	RCH$_2$OH + R'OH	
9-39	RCONHR'	RCH$_2$NHR	
5-10			Most difficult
9-38	RCOO⁻		Inert

[486]For monographs on reductions in general, see Hudlický *Reductions in Organic Chemisty*; Wiley: New York, 1984; Augustine *Reduction*; Marcel Dekker: New York, 1968. For a review, see Candlin; Rennie, in Bentley; Kirby, Ref. 12, pp. 77-135.

[487]For discussions of selectivity with metal hydride reducing agents, see Brown; Krishnamurthy *Tetrahedron*, **1979**, *35*, 567-607; Walker *Chem. Soc. Rev.* **1976**, *5*, 23-50; Brown *Boranes in Organic Chemistry*; Cornell University Press: Ithaca, NY, 1972, pp. 209-251, Rerick, in Augustine, Ref. 486. For books, see, in Ref. 10, the works by Seyden-Penne, Štrouf et al., and Hajós.

[488]For a discussion of catalyst selectivity for hydrogenations, see Rylander *Aldrichimica Acta* **1979**, *12*, 53-57. See also Rylander *Hydrogenation Methods*; Academic Press: New York, 1985.

[489]Table 19.2 is from House, Ref. 10, p. 9. Tables 19.3 and 19.4 are from Brown, Ref. 487, pp. 213 and 232, respectively.

[490]The first ten columns are from Brown; Krishnamurthy, Ref. 487, p. 604. The column on (i-Bu)$_2$AlH is from Yoon; Gyoung *J. Org. Chem.* **1985**, *50*, 2443; the one on NaAlEt$_2$H$_2$ from Stinson, *Chem. Eng. News* Nov. 3, **1980**, *58*, No. 44, 19; and the one on LiBEt$_3$H from Brown; Kim; Krishnamurthy *J. Org. Chem.* **1980**, *45*, 1. For similar tables that show additional reducing agents, see Pelter; Smith; Brown, Ref. 494, p. 129; Hajós, Ref. 10, pp. 16-17. For tables showing which agents reduce a wide variety of functional groups, see Hudlický, Ref. 486, pp. 177-200.

TABLE 19.3 The ease of reduction of various functional groups with LiAlH$_4$ in ether[489]

However, LiAlH$_4$ is a very powerful reagent, and much less chemoselectivity is possible here than with most of the other metal hydrides

Reaction	Substrate	Product	
6-25	RCHO	RCH$_2$OH	Easiest
6-25	RCOR	RCHOHR	
9-45	RCOCl	RCH$_2$OH	
9-42	Lactone	Diol	
0-80	RCH—CHR \O/	RCH$_2$CHOHR	
9-42	RCOOR′	RCH$_2$OH + R′OH	
9-38	RCOOH	RCH$_2$OH	
9-38	RCOO⁻	RCH$_2$OH	
9-39	RCONR′$_2$	RCH$_2$NR′$_2$	
6-27	RC≡N	RCH$_2$NH$_2$	
9-47	RNO$_2$	RNH$_2$	
9-67	ArNO$_2$	ArN=NAr	Most difficult
5-9	RCH=CHR		Inert

which groups.[491] LiAlH$_4$ is very powerful and unselective reagent.[492] Consequently, other metal hydrides are generally used when chemoselectivity is required. As mentioned on p. 917, a number of less reactive (and more selective) reagents have been prepared by replacing some of the hydrogens of LiAlH$_4$ with alkoxy groups (by treatment of LiAlH$_4$ with ROH).[493] Most of the metal hydrides are nucleophilic reagents and attack the carbon atom of a carbon-hetero single or multiple bond. However, BH$_3$[494] and AlH$_3$[495] are electrophiles (Lewis acids)

TABLE 19.4 The ease of reduction of various functional groups with borane[489]

It is evident that this reagent and LiAlH$_4$ (Table 19.3) complement each other

Reaction	Substrate	Product	
9-38	RCOOH	RCH$_2$OH	Easiest
5-12	RCH=CHR	(RCH$_2$CHR)$_3$B	
6-25	RCOR	RCHOHR	
6-27	RCN	RCH$_2$NH$_2$	
0-80	RCH—CHR \O/	RCH$_2$CHOHR	
9-42	RCOOR′	RCH$_2$OH + R′OH	Most difficult
0-83, 9-45	RCOCl		Inert

[491]See also the table in Ref. 9.

[492]For a review of LiAlH$_4$, see Pizey, Ref. 10, vol. 1 pp. 101-194.

[493]For reviews of reductions by these reagents, see Málek Ref. 10; Málek; Černy *Synthesis* **1972**, 217-234.

[494]See Brown; Heim; Yoon *J. Am. Chem. Soc.* **1970**, *92*, 1637; Cragg *Organoboranes in Organic Synthesis*; Marcel Dekker: New York, 1973, pp. 319-371. For reviews of reductions with BH$_3$, see Wade *J. Mol. Catal.* **1983**, *18*, 273-297 (BH$_3$ and a catalyst); Lane *Chem. Rev.* **1976**, *76*, 773-799, *Aldrichimica Acta* **1977**, *10*, 41-51; Brown; Krishnamurthy *Aldrichimica Acta* **1979**, *12*, 3-11. For reviews of reduction with borane derivatives, see Pelter; Smith; Brown *Borane Reagents*; Academic Press: New York, 1988, pp. 125-164; Pelter *Chem. Ind.* (*London*) **1976**, 888-896.

[495]See Brown; Yoon *J. Am. Chem. Soc.* **1966**, *88*, 1464; Yoon; Brown *J. Am. Chem. Soc.* **1968**, *90*, 2927.

TABLE 19.5 Reactivity of various functional groups with some metal hydrides and toward catalytic hydrogenation.[490] ± indicates a borderline case.

Reaction		NaBH₄ in EtOH	NaBH₄ + LiCl in diglyme	NaBH₄ + AlCl₃ in diglyme	BH₃–THF[487]	Bis-3-methyl-2-butyl-borane (disiamylborane) in THF[498]	9-BBN[499]	LiAlH(O-t-Bu)₃ in THF	LiAlH(OMe)₃ in THF	LiAlH₄ in ether	AlH₃ in THF[495]	LiBEt₃H[501]	(i-Bu)₂AlH (DIBALH)	NaAlEt₂H₂	Catalytic hydrogenation
6-25	RCHO → RCH₂OH	+	+	+	+	+	+	+	+	+	+	+	+	+	+
6-25	RCOR → RCHOHR	+	+	+	+	+	+	+	+	+	+	+	+	+	+
0-83, 9-45	RCOCl → RCHO / RCH₂OH	+[496]	+	+	−	−	+	+	+	+	+	+	+	+	+
9-42	Lactone → diol	−	+	+	+	+	+	±	+	+	+	+	+	+	+
0-80	Epoxide → alcohol	−	+	+	+	±	±	±	+	+	+	+	+	+	+
9-42	RCOOR' → RCH₂OH + R'OH	−	+	+	±	−	±	±	+	+	+	+	+	+	+
9-38	RCOOH → RCH₂OH	−	+	+	+	−	−	−	+	+	+	−	+	+	−
9-38	RCOO⁻ → RCH₂OH	−	+	+	−	−	−	−	+	+	+	−	−	+	−
9-39, 0-85	RCONR'₂ → RCH₂NR'₂ / RCHO	−	−	−	+	+	+	−	+	+	+	+	+	+	+
6-27	RC≡N → RCH₂NH₂	−	−	−	+	−	±	−	+	+	+	±	+[497]	+	+
9-47	RNO₂ → RNH₂	−	−	−	−	−	−	−	+	+	−	−	+[500]	+	+
9-67	RN=NR	−	−	−	−	−	−	−	+	+	−	−	−	+	+
5-9	RCH=CHR → RCH₂CH₂R	−	−	−	+	+	+	−	−	−	−	−	−	−	+

and attack the hetero atom. This accounts for the different patterns of selectivity shown in the tables.

The reactions in this section are grouped into classifications based on bond changes, similar to those used for the oxidation reactions. These sections are: (A) reactions involving replacement of oxygen by hydrogen, (B) reactions in which oxygen is removed from the substrate, (C) reduction with cleavage, and (D) reductive coupling.

A. Reactions Involving Replacement of Oxygen by Hydrogen. In reactions **9-37** to **9-41**, a C=O is reduced to a CH$_2$ group.

9-37 Reduction of Carbonyl to Methylene in Aldehydes and Ketones
Dihydro-de-oxo-bisubstitution

$$ R-\underset{\underset{O}{\|}}{C}-R' \xrightarrow[\text{HCl}]{\text{Zn-Hg}} R-CH_2-R' $$

There are various ways of reducing the C=O group of aldehydes and ketones to CH$_2$.[502] The two oldest, but still very popular, methods are the *Clemmensen reduction* and the *Wolff–Kishner reduction*. The Clemmensen reduction consists of heating the aldehyde or ketone with zinc amalgam and aqueous HCl.[503] Ketones are reduced more often than aldehydes. In the Wolff–Kishner reduction,[504] the aldehyde or ketone is heated with hydrazine hydrate and a base (usually NaOH or KOH). The *Huang–Minlon modification*[505] of the Wolff–Kishner reaction, in which the reaction is carried out in refluxing diethylene glycol, has completely replaced the original procedure. The reaction can also be carried out under more moderate conditions (room temperature) in dimethyl sulfoxide with potassium *t*-butoxide as base.[506] The Wolff–Kishner reaction can also be applied to the semicarbazones of aldehydes or ketones. The Clemmensen reduction is usually easier to perform, but it fails for acid-sensitive and high-molecular-weight substrates. For these cases the Wolff–Kishner reduction is quite useful. For high-molecular-weight substrates, a modified Clemmensen reduction, using activated zinc and gaseous HCl in an organic solvent such as ether or acetic anhydride, has proved successful.[507] The Clemmensen and Wolff–Kishner reactions are complementary, since the former uses acidic and the latter basic conditions.

Both methods are fairly specific for aldehydes and ketones and can be carried out with many other functional groups present. However, certain types of aldehydes and ketones do not give normal reduction products. Under Clemmensen conditions,[508] α-hydroxy ketones give either ketones (hydrogenolysis of the OH, **0-78**) or olefins, and 1,3-diones usually

[496]Reacts with solvent, reduced in aprotic solvents.
[497]Reduced to aldehyde (**6-28**).
[498]Brown; Bigley; Arora; Yoon *J. Am. Chem. Soc.* **1970,** *92*, 7161. For reductions with thexylborane, see Brown; Heim; Yoon *J. Org. Chem.* **1972,** *37*, 2942.
[499]Brown; Krishnamurthy; Yoon *J. Org. Chem.* **1976,** *41*, 1778.
[500]Reduced to hydroxylamine (**9-49**).
[501]Brown; Kim; Krishnamurthy, Ref. 490. For a review of the synthesis of alkyl-substituted borohydrides, see Brown; Singaram; Singaram *J. Organomet. Chem.* **1982,** *239*, 43-64.
[502]For a review, see Reusch, in Augustine, Ref. 486, pp. 171-211.
[503]For a review, see Vedejs *Org. React.* **1975,** *22*, 401-422. For a discussion of experimental conditions, see Fieser; Fieser, Ref. 46, vol. 1, pp. 1287-1289.
[504]For a review, see Todd *Org. React.* **1948,** *4*, 378-422.
[505]Huang-Minlon *J. Am. Chem. Soc.* **1946,** *68*, 2487, **1949,** *71*, 3301.
[506]Cram; Sahyun; Knox *J. Am. Chem. Soc.* **1962,** *84*, 1734.
[507]Yamamura; Ueda; Hirata *Chem. Commun.* **1967,** 1049; Toda; Hayashi; Hirata; Yamamura *Bull. Chem. Soc. Jpn.* **1972,** *45*, 264.
[508]For a review of Clemmensen reduction of diketones and unsaturated ketones, see Buchanan; Woodgate *Q. Rev. Chem. Soc.* **1969,** *23*, 522-536.

undergo rearrangement, e.g., $MeCOCH_2COMe \rightarrow MeCOCHMe_2$.[509] Neither method is suitable for α,β-unsaturated ketones. These give pyrazolines[510] under Wolff-Kishner conditions, while under Clemmensen conditions both groups of these molecules may be reduced or if only one group is reduced, it is the C=C bond.[511] Sterically hindered ketones are resistant to both the Clemmensen and Huang–Minlon procedures but can be reduced by vigorous treatment with anhydrous hydrazine.[512] In the Clemmensen reduction, pinacols (**9-62**) are often side products.

Other reagents have also been used to reduce the C=O of aldehydes and ketones to CH_2.[513] Among these are H_2 and a catalyst at 180 to 250°C,[514] triisopropyl phosphite $P(O-i\text{-}Pr)_3$,[515] and, for aryl ketones (ArCOR and ArCOAr), $LiAlH_4$–$AlCl_3$,[516] $LiAlH_4$–P_2I_4,[517] Li–NH_3,[518] $NaBH_4$–F_3CCOOH,[519] $NaBH_4$–$AlCl_3$,[520] BH_3–t-$BuNH_2$–$AlCl_3$,[521] CO–Se–H_2O,[522] $HCOONH_4$–Pd–C,[523] or trialkylsilanes in F_3CCOOH.[524] Most of these reagents also reduce aryl aldehydes ArCHO to methylbenzenes $ArCH_3$.[525] Aliphatic aldehydes RCHO can be reduced to RCH_3 with titanocene dichloride$(C_5H_5)_2TiCl_2$.[526] One carbonyl group of 1,2-diketones can be selectively reduced by H_2S with an amine catalyst[527] or by HI in refluxing acetic acid.[528] One carbonyl group of quinones can be reduced with copper and sulfuric acid or with tin and HCl:[529]

One carbonyl group of 1,3-diketones was selectively reduced by catalytic hydrogenolysis.[530]

An indirect method of accomplishing the reaction is reduction of tosylhydrazones $(R_2C=N-NHTs)$ to R_2CH_2 with $NaBH_4$, BH_3, catecholborane, bis(benzyloxy)borane,

[509]Cusack; Davis *J. Org. Chem.* **1965,** *30,* 2062; Wenkert; Kariv *Chem. Commun.* **1965,** 570; Galton; Kalafer; Beringer *J. Org. Chem.* **1970,** *35,* 1.

[510]Pyrazolines can be converted to cyclopropanes; see **7-46.**

[511]See, however, Banerjee; Alvárez; Santana; Carrasco *Tetrahedron* **1986,** *42,* 6615.

[512]Barton; Ives; Thomas *J. Chem. Soc.* **1955,** 2056.

[513]For a list, with references, see Ref. 21, pp. 35-38.

[514]See for example, Maier; Bergmann; Bleicher; Schleyer *Tetrahedron Lett.* **1981,** *22,* 4227. For a review of the mechanism, see Pavlenko *Russ. Chem. Rev.* **1989,** *58,* 453-469.

[515]Olah; Wu *Synlett* **1990,** 54.

[516]Nystrom; Berger *J. Am. Chem. Soc.* **1958,** *80,* 2896. See also Volod'kin; Ershov; Portnykh *Bull. Acad. Sci. USSR, Div. Chem. Sci.* **1967,** 384.

[517]Suzuki; Masuda; Kubota; Osuka *Chem. Lett.* **1983,** 909.

[518]Hall; Lipsky; McEnroe; Bartels *J. Org. Chem.* **1971,** *36,* 2588.

[519]Gribble; Nutaitis *Org. Prep. Proced. Int.* **1985,** *17,* 317-384.

[520]Ono; Suzuki; Kamimura *Synthesis* **1987,** 736.

[521]Lau; Tardif; Dufresne; Scheigetz *J. Org. Chem.* **1989,** *54,* 491.

[522]Nishiyama; Hamanaka; Ogawa; Kambe; Sonoda *J. Org. Chem.* **1988,** *53,* 1326.

[523]Ram; Spicer *Tetrahedron Lett.* **1988,** *29,* 3741.

[524]Kursanov; Parnes; Loim *Bull. Acad. Sci. USSR, Div. Chem. Sci.* **1966,** 1245; West; Donnelly; Kooistra; Doyle *J. Org. Chem.* **1973,** *38,* 2675. See also Fry; Orfanopoulos; Adlington; Dittman; Silverman *J. Org. Chem.* **1978,** *43,* 374; Olah; Arvanaghi *Synthesis* **1973,** 770.

[525]See, for example, Hall; Bartels; Engman *J. Org. Chem.* **1972,** *37,* 760; Kursanov; Parnes; Loim; Bakalova *Doklad. Chem.* **1968,** *179,* 328; Zahalka; Alper *Organometallics* **1986,** *5,* 1909.

[526]van Tamelen; Gladys *J. Am. Chem. Soc.* **1974,** *96,* 5290.

[527]Mayer; Hiller; Nitzschke; Jentzsch *Angew. Chem. Int. Ed. Engl.* **1963,** *2,* 370-373 [*Angew. Chem.* **75,** 1011-1014].

[528]Reusch; LeMahieu *J. Am. Chem. Soc.* **1964,** *86,* 3068.

[529]Meyer *Org. Synth. I,* 60; Macleod; Allen *Org. Synth. II,* 62.

[530]Cormier; McCauley *Synth. Commun.* **1988,** *18,* 675.

NaBH$_3$CN, or bis(triphenylphosphine)copper(I) tetrahydroborate.[531] The reduction of α,β-unsaturated tosylhydrazones with NaBH$_3$CN, with NaBH$_4$–HOAc, or with catecholborane proceeds with migration of the double bond to the position formerly occupied by the carbonyl carbon, even if this removes the double bond from conjugation with an aromatic ring,[532] e.g.,

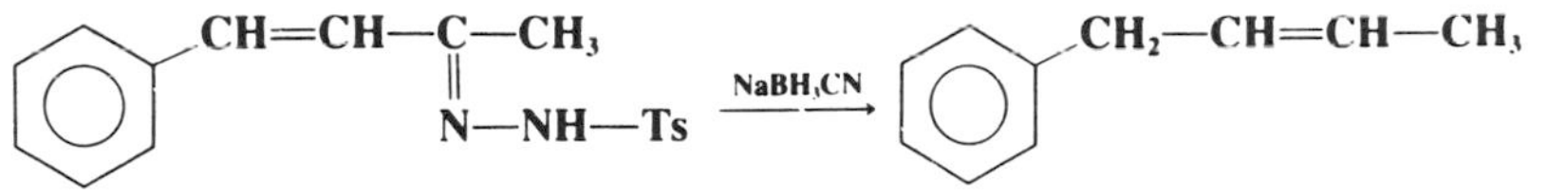

A cyclic mechanism is apparently involved:

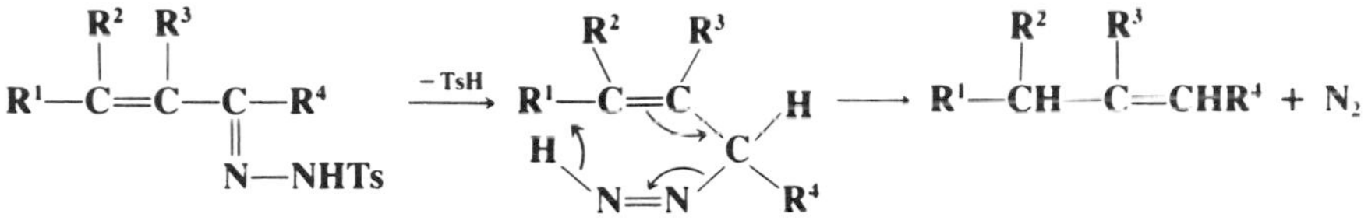

Another indirect method is conversion of the aldehyde or ketone to a dithioacetal or ketal, and desulfurization of this (**4-36**).

The first step in the mechanism[533] of the Wolff–Kishner reaction consists of formation of the hydrazone (**6-20**).

$$R_2C{=}O \longrightarrow R_2C{=}N{-}NH_2$$

It is this species that undergoes reduction in the presence of base, most likely in the following manner:

$$R_2C{=}N{-}NH_2 \xrightarrow{OH^-} R_2CH{-}N{=}NH + OH^- \longrightarrow R_2CH{-}N{=}N^{\ominus} + H_2O$$

$$R_2CH{-}N{=}N^{\ominus} \xrightarrow{-N_2} R_2\bar{C}H^{\ominus} \xrightarrow{H_2O} R_2CH_2 + OH^-$$

Not much is known about the mechanism of the Clemmensen reduction. Several mechanisms have been proposed,[534] including one going through a zinc–carbene intermediate.[535] One thing reasonably certain is that the corresponding alcohol is not an intermediate, since alcohols prepared in other ways fail to give the reaction. Note that the alcohol is not an intermediate in the Wolff-Kishner reduction either.

OS **I**, 60; **II**, 62, 499, **III**, 410, 444, 513, 786; **IV**, 203, 510; **V**, 533, 747; **VI**, 62, 293, 919; **VII**, 393. Also see OS **IV**, 218; **VII**, 18.

[531]Caglioti; Magi *Tetrahedron* **1963**, *19*, 1127; Fischer; Pelah; Williams; Djerassi *Chem. Ber.* **1965**, *98*, 3236; Elphimoff-Felkin; Verrier *Tetrahedron Lett.* **1968**, 1515; Hutchins; Milewski; Maryanoff *J. Am. Chem. Soc.* **1973**, *95*, 3662; Cacchi; Caglioti; Paolucci *Bull. Chem. Soc. Jpn.* **1974**, *47*, 2323; Lane *Synthesis* **1975**, 135-146, pp. 145-146; Kabalka; Yang; Chandler; Baker *Synthesis* **1977**, 124; Kabalka; Summers *J. Org. Chem.* **1981**, *46*, 1217; Fleet; Harding; Whitcombe *Tetrahedron Lett.* **1980**, *21*, 4031; Miller; Yang; Weigel; Han; Liu *J. Org. Chem.* **1989**, *54*, 4175.

[532]Hutchins; Kacher; Rua *J. Org. Chem.* **1975**, *40*, 923; Kabalka; Yang; Baker *J. Org. Chem.* **1976**, *41*, 574; Taylor; Djerassi *J. Am. Chem. Soc.* **1976**, *98*, 2275; Hutchins; Natale *J. Org. Chem.* **1978**, *43*, 2299; Greene *Tetrahedron Lett.* **1979**, 63.

[533]For a review of the mechanism, see Szmant *Angew. Chem. Int. Ed. Engl.* **1968**, *7*, 120-128 [*Angew. Chem. 80*, 141-149].

[534]See, for example, Horner; Schmitt *Liebigs Ann. Chem.* **1978**, 1617; Poutsma; Wolthius *J. Org. Chem.* **1959**, *24*, 875; Nakabayashi *J. Am. Chem. Soc.* **1960**, *82*, 3900, 3906; Di Vona; Rosnati *J. Org. Chem.* **1991**, *56*, 4269.

[535]Burdon; Price *J. Chem. Soc., Chem. Commun.* **1986**, 893.

9-38 Reduction of Carboxylic Acids to Alcohols
Dihydro-de-oxo-bisubstitution

$$RCOOH \xrightarrow{\text{LiAlH}_4} RCH_2OH$$

Carboxylic acids are easily reduced to primary alcohols by LiAlH$_4$.[536] The reaction does not stop at the aldehyde stage (but see **0-84**). The conditions are particularly mild, the reduction proceeding quite well at room temperature. Other hydrides have also been used,[537] but not NaBH$_4$ (see Table 19.5).[538] Catalytic hydrogenation is also generally ineffective.[539] Borane is particularly good for carboxyl groups (Table 19.4) and permits selective reduction of them in the presence of many other groups (though the reaction with double bonds takes place at about the same rate).[540] Borane also reduces carboxylic acid salts.[541] Aluminum hydride reduces COOH groups without affecting carbon–halogen bonds in the same molecule. The reduction has also been carried out with SmI$_2$ in basic media.[541a]

OS **III**, 60; **VII**, 221; 530; **65**, 173; **66**, 160; **68**, 77.

9-39 Reduction of Amides to Amines
Dihydro-deoxo-bisubstitution

$$RCONH_2 \xrightarrow{\text{LiAlH}_4} RCH_2NH_2$$

Amides can be reduced[542] to amines with LiAlH$_4$ or by catalytic hydrogenation, though high temperatures and pressures are usually required for the latter. Even with LiAlH$_4$ the reaction is more difficult than the reduction of most other functional groups, and other groups often can be reduced without disturbing an amide function. NaBH$_4$ by itself does not reduce amides, though it does so in the presence of certain other reagents.[543] Substituted amides can be similarly reduced:

$$RCONHR' \longrightarrow RCH_2NHR'$$

$$RCONR_2' \longrightarrow RCH_2NR_2'$$

Borane[544] and sodium in 1-propanol[545] are good reducing agents for all three types of amides. Another reagent that reduces disubstituted amides to amines is trichlorosilane.[546] Sodium

---

[536]For a review, see Gaylord *Reduction with Complex Metal Hydrides*; Wiley: New York, 1956, pp. 322-373.
[537]For a list of reagents, with references, see Ref. 21, pp. 548-549.
[538]NaBH$_4$ in the presence of Me$_2$N=CHCl$^+$ Cl$^-$ reduces carboxylic acids to primary alcohols chemoselectively in the presence of halide, ester, and nitrile groups: Fujisawa; Mori; Sato *Chem. Lett.* **1983**, 835.
[539]See Rylander *Hydrogenation Methods*, Ref. 488, pp. 78-79.
[540]Brown; Korytnyk *J. Am. Chem. Soc.* **1960**, *82*, 3866; Batrakov; Bergel'son *Bull. Acad. Sci. USSR, Div. Chem. Sci.* **1965**, 348; Pelter; Hutchings; Levitt; Smith *Chem. Commun.* **1970**, 347; Brown; Stocky *J. Am. Chem. Soc.* **1977**, *99*, 8218.
[541]Yoon; Cho *Tetrahedron Lett.* **1982**, *23*, 2475.
[541a]Kamochi; Kudo *Chem. Lett.* **1991**, 893.
[542]For a review, see Challis; Challis, in Zabicky *The Chemistry of Amides*; Wiley: New York, 1970, pp. 795-801. For a review of the reduction of amides, lactams, and imides with metallic hydrides, see Gaylord, Ref. 536, pp. 544-636. For a list of reagents, with references, see Ref. 21, pp. 432-433.
[543]See, for example, Satoh; Suzuki; Suzuki; Miyaji; Imai *Tetrahedron Lett.* **1969**, 4555; Rahman; Basha; Waheed; Ahmed *Tetrahedron Lett.* **1976**, 219; Kuehne; Shannon *J. Org. Chem.* **1977**, *42*, 2082; Wann; Thorsen; Kreevoy *J. Org. Chem.* **1981**, *46*, 2579; Mandal; Giri; Pakrashi *Synthesis* **1987**, 1128; Akaboro; Takanohashi *Chem. Lett.* **1990**, 251.
[544]Brown; Heim *J. Org. Chem.* **1973**, *38*, 912; Brown; Narasimhan; Choi *Synthesis* **1981**, 441, 996; Krishnamurthy *Tetrahedron Lett.* **1982**, *23*, 3315; Bonnat; Hercouet; Le Corre *Synth. Commun.* **1991**, *21*, 1579.
[545]Bhandari; Sharma; Chatterjee *Chem. Ind. (London)* **1990**, 547.
[546]Nagata; Dohmaru; Tsurugi *Chem. Lett.* **1972**, 989. See also Benkeser; Li; Mozdzen *J. Organomet. Chem.* **1979**, *178*, 21.

(dimethylamino)borohydride reduces unsubstituted and disubstituted, but not monosubstituted amides.[547] With some RCONR$_2'$, LiAlH$_4$ causes cleavage, and the aldehyde (**0-85**) or alcohol is obtained. Lithium triethylborohydride produces the alcohol with most N,N-disubstituted amides, though not with unsubstituted or N-substituted amides.[548] Lactams are reduced to cyclic amines in high yields with LiAlH$_4$, though cleavage sometimes occurs here too. Imides are generally reduced on both sides, though it is sometimes possible to stop with just one. Both cyclic and acyclic imides have been reduced in this manner, though with acyclic imides cleavage is often obtained, e.g.,[549]

$$PhN(COMe)_2 \longrightarrow PhNHEt$$

Acyl sulfonamides have been reduced (RCONHSO$_2$Ph → RCH$_2$NHSO$_2$Ph) with BH$_3$–SMe$_2$.[550]

OS **IV**, 339, 354, 564; **VI**, 382; **VII**, 41.

9-40 Reduction of Carboxylic Esters to Ethers
Dihydro-de-oxo-bisubstitution

$$RCOOR' \xrightarrow[\text{LiAlH}_4]{\text{BF}_3\text{-etherate}} RCH_2OR'$$

Carboxylic esters and lactones have been reduced to ethers, though the more usual course is the obtention of 2 moles of alcohol (**9-42**). Reduction to ethers has been accomplished with a reagent prepared from BF$_3$–etherate and either LiAlH$_4$, LiBH$_4$, or NaBH$_4$,[551] with trichlorosilane and uv light,[552] and with catalytic hydrogenation. The reaction with the BF$_3$ reagent apparently succeeds with secondary R', but not with primary R', which give **9-42**. Lactones give cyclic ethers.[553] Thiono esters RCSOR' can be reduced to ethers RCH$_2$OR' with Raney nickel (**4-36**).[554] Since the thiono esters can be prepared from carboxylic esters (**6-11**), this provides an indirect method for the conversion of carboxylic esters to ethers. Thiol esters RCOSR' have been reduced to thioethers RCH$_2$SR'.[555]

See also **9-43, 0-81**.

9-41 Reduction of Cyclic Anhydrides to Lactones
Dihydro-de-oxo-bisubstitution

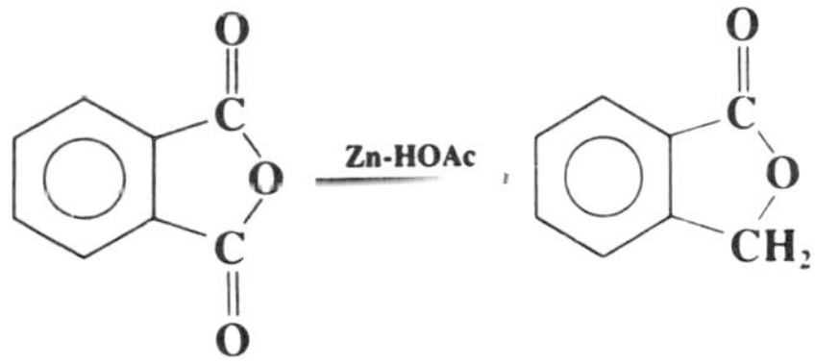

[547]Hutchins; Learn; El-Telbany; Stercho *J. Org. Chem.* **1984**, *49*, 2438.
[548]Brown; Kim *Synthesis* **1977**, 635.
[549]Witkop; Patrick *J. Am. Chem. Soc.* **1952**, *74*, 3861.
[550]Belletire; Fry *Synth. Commun.* **1988**, *18*, 29.
[551]Pettit; Ghatak; Green; Kasturi; Piatak *J. Org. Chem.* **1961**, *26*, 1685; Pettit; Green; Kasturi; Ghatak *Tetrahedron* **1962**, *18*. 953; Ager; Sutherland *J. Chem. Soc., Chem. Commun* **1982**, 248. See also Dias; Pettit *J. Org. Chem.* **1971**, *36*, 3485.
[552]Tsurugi; Nakao; Fukumoto *J. Am. Chem. Soc.* **1969**, *91*, 4587; Nagata; Dohmaru; Tsurugi *J. Org. Chem.* **1973**, *38*, 795; Baldwin; Doll; Haut *J. Org. Chem.* **1974**, *39*, 2470; Baldwin; Haut *J. Org. Chem.* **1975**, *40*, 3885. See also Kraus; Frazier; Roth; Taschner; Neuenschwander *J. Org. Chem.* **1981**, *46*, 2417.
[553]See, for example, Pettit; Kasturi; Green; Knight *J. Org. Chem.* **1961**, *26*, 4773; Edward; Ferland *Chem. Ind.* (*London*) **1964**, 975.
[554]Baxter; Bradshaw *J. Org. Chem.* **1981**, *46*, 831.
[555]Eliel; Daignault *J. Org. Chem.* **1964**, *29*, 1630; Bublitz *J. Org. Chem.* **1967**, *32*, 1630.

Cyclic anhydrides can give lactones if reduced with Zn–HOAc, with hydrogen and platinum or $RuCl_2(Ph_3P)_3$,[556] with $NaBH_4$,[557] or even with $LiAlH_4$, though with the last-mentioned reagent diols are the more usual product (**9-44**). With some reagents the reaction can be accomplished regioselectively, i.e., only a specific one of the two C=O groups of an unsymmetrical anhydride is reduced.[558] Open-chain anhydrides either are not reduced at all (e.g., with $NaBH_4$) or give 2 moles of alcohol.

There are no *Organic Syntheses* references, but see OS **II**, 526, for a related reaction.

9-42 Reduction of Carboxylic Esters to Alcohols
Dihydro,hydroxy-de-oxo,alkoxy-tersubstitution

$$RCOOR' \xrightarrow{\text{LiAlH}_4} RCH_2OH + R'OH$$

$LiAlH_4$ reduces carboxylic esters to give 2 moles of alcohol.[559] The reaction is of wide scope and has been used to reduce many esters. Where the interest is in obtaining R'OH, this is a method of "hydrolyzing" esters. Lactones yield diols. Among the reagents that give the same products[560] are DIBALH, lithium triethylborohydride, and BH_3–SMe_2 in refluxing THF.[561] $NaBH_4$ reduces phenolic esters, especially those containing electron-withdrawing groups,[562] but its reaction with other esters is usually so slow that such reactions are seldom feasible (though exceptions are known[563]), and it is generally possible to reduce an aldehyde or ketone without reducing an ester function in the same molecule. However, $NaBH_4$ reduces esters in the presence of certain compounds (see Table 19.5).[564] Carboxylic esters can also be reduced to alcohols by hydrogenation over copper chromite catalysts,[565] though high pressures and temperatures are required. Ester functions generally survive low-pressure catalytic hydrogenations. Before the discovery of $LiAlH_4$, the most common way of carrying out the reaction was with sodium in ethanol, a method known as the *Bouveault–Blanc procedure*. This procedure is still sometimes used where selectivity is necessary. See also **9-40, 9-43**, and **0-81**.

OS **II**, 154, 325, 372, 468; **III**, 671; **IV**, 834; **VI**, 781; **VII**, 356; **68**, 92.

9-43 Reduction of Carboxylic Acids and Esters to Alkanes
Trihydro-de-alkoxy,oxo-tersubstitution, etc.

$$RCOOR' \xrightarrow{(C_5H_5)_2TiCl_2} RCH_3 + R'OH$$

The reagent titanocene dichloride reduces carboxylic esters in a different manner from that of **0-81, 9-40**, or **9-42**. The products are the alkane RCH_3 and the alcohol R'OH.[526] The mechanism probably involves an alkene intermediate. Aromatic acids can be reduced to methylbenzenes by a procedure involving refluxing first with trichlorosilane in MeCN, then

[556]Lyons *J. Chem. Soc., Chem. Commun.* **1975**, 412; Morand; Kayser *J. Chem. Soc., Chem. Commun.* **1976**, 314. See also Hara; Wada *Chem. Lett.* **1991**, 553.

[557]Bailey; Johnson *J. Org. Chem.* **1970**, *35*, 3574.

[558]See, for example, Kayser; Salvador; Morand *Can. J. Chem.* **1983**, *61*, 439; Ikariya; Osakada; Ishii; Osawa; Saburi; Yoshikawa *Bull. Chem. Soc. Jpn.* **1984**, *57*, 897; Soucy; Favreau; Kayser *J. Org. Chem.* **1987**, *52*, 129.

[559]For a review, see Gaylord, Ref. 536, pp. 391-531.

[560]For a list of reagents, with references, see Ref. 21, pp. 549-551.

[561]Brown; Choi *Synthesis* **1981**, 439; Brown; Choi; Narasimhan *J. Org. Chem.* **1982**, *47*, 3153.

[562]Takahashi; Cohen *J. Org. Chem.* **1970**, *35*, 1505.

[563]For example, see Brown; Rapoport *J. Org. Chem.* **1963**, *28*, 3261; Bianco; Passacantilli; Righi *Synth. Commun.* **1988**, *18*, 1765.

[564]See also Kikugawa *Chem. Lett.* **1975**, 1029; Santaniello; Ferraboschi; Sozzani *J. Org. Chem.* **1981**, *46*, 4584; Brown; Narasimhan; Choi *J. Org. Chem.* **1982**, *47*, 4702; Soai; Oyamada; Takase; Ookawa *Bull. Chem. Soc. Jpn.* **1984**, *57*, 1948; Guida; Entreken; Guida *J. Org. Chem.* **1984**, *49*, 3024.

[565]For a review, see Adkins, *Org. React.* **1954**, *8*, 1-27.

with tripropylamine added, and finally with KOH and MeOH (after removal of the MeCN).[566] The following sequence has been suggested:[566]

$$\text{ArCOOH} \xrightarrow{\text{SiHCl}_3} \text{(ArCO)}_2\text{O} \xrightarrow[\text{R}_3\text{N}]{\text{SiHCl}_3} \text{ArCH}_2\text{SiCl}_3 \xrightarrow[\text{MeOH}]{\text{KOH}} \text{ArCH}_3$$

Esters of aromatic acids are not reduced by this procedure, so an aromatic COOH group can be reduced in the presence of a COOR' group.[567] However, it is also possible to reduce aromatic ester groups, by a variation of the trichlorosilane procedure.[568] *o*- and *p*-hydroxybenzoic acids and their esters have been reduced to cresols $\text{HOC}_6\text{H}_4\text{CH}_3$ with sodium bis(2-methoxyethoxy)aluminum hydride $\text{NaAlH}_2(\text{OC}_2\text{H}_4\text{OMe})_2$ (*Red-Al*).[569]

Carboxylic acids can also be converted to alkanes, indirectly,[570] by reduction of the corresponding tosylhydrazides RCONHNH_2 with LiAlH_4 or borane.[571]

OS **VI**, 747.

9-44 Reduction of Anhydrides to Alcohols

$$\text{RCOOCOR} \xrightarrow{\text{LiAlH}_4} 2\text{RCH}_2\text{OH}$$

LiAlH_4 usually reduces open-chain anhydrides to give 2 moles of alcohol. With cyclic anhydrides the reaction with LiAlH_4 can be controlled to give either diols or lactones[572] (see **9-41**). NaBH_4 in THF, with dropwise addition of methanol, reduces open-chain anhydrides to one mole of primary alcohol and one mole of carboxylic acid.[573]

OS **VI**, 482.

9-45 Reduction of Acyl Halides to Alcohols

Dihydro,hydroxy-de-halo,oxo-tersubstitution

$$\text{RCOCl} \xrightarrow{\text{LiAlH}_4} \text{RCH}_2\text{OH}$$

Acyl halides are reduced[574] to alcohols by LiAlH_4 or NaBH_4, as well as by other metal hydrides (Table 19.5), but not by borane. The reaction may be regarded as a combination of **9-37** and **0-76**.

OS **IV**, 271.

9-46 Complete Reduction of Epoxides

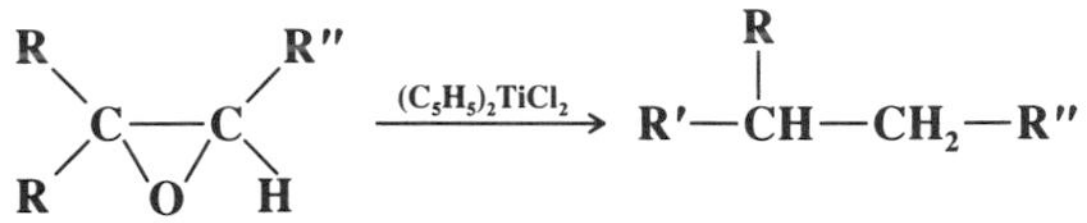

---

[566]Benkeser; Foley; Gaul; Li *J. Am. Chem. Soc.* **1970**, *92*, 3232.

[567]Benkeser; Ehler *J. Org. Chem.* **1973**, *38*, 3660.

[568]Benkeser; Mozdzen; Muth *J. Org. Chem.* **1979**, *44*, 2185.

[569]Černý; Málek *Tetrahedron Lett.* **1969**, 1739, *Collect. Czech. Chem. Commun.* **1970**, *35*, 2030.

[570]For another indirect method, which can also be applied to acid derivatives, see Degani; Fochi *J. Chem. Soc., Perkin Trans. 1* **1978**, 1133. For a direct method, see Le Deit; Cron; Le Corre *Tetrahedron Lett.* **1991**, *32*, 2759.

[571]Attanasi; Caglioti; Gasparrini; Misiti *Tetrahedron* **1975**, *31*, 341, and references cited therein.

[572]Bloomfield; Lee *J. Org. Chem.* **1967**, *32*, 3919.

[573]Soai; Yokoyama; Mochida *Synthesis* **1987**, 647.

[574]For a review of the reduction of acyl halides, see Wheeler, in Patai *The Chemistry of Acyl Halides*; Wiley: New York, 1972, pp. 231-251. For a list of reagents, with references, see Ref. 21, p. 549.

Though the usual product of epoxide reductions is the alcohol (**0-80**), epoxides are reduced all the way to the alkane by titanocene dichloride[526] and by $Et_3SiH-BH_3$.[575]

9-47 Reduction of Nitro Compounds to Amines

$$RNO_2 \xrightarrow[HCl]{Zn} RNH_2$$

Both aliphatic[576] and aromatic nitro compounds can be reduced to amines, though the reaction has been applied much more often to aromatic nitro compounds, owing to their greater availability. Many reducing agents have been used to reduce aromatic nitro compounds, the most common being Zn, Sn, or Fe (or sometimes other metals) and acid, and catalytic hydrogenation.[577] Among other reagents used[578] have been AlH_3-AlCl_3, hydrazine and a catalyst,[579] $TiCl_3$,[580] $Al-NiCl_2-THF$,[581] formic acid and Pd-C,[582] and sulfides such as NaHS, $(NH_4)_2S$, or polysulfides. The reaction with sulfides or polysulfides is called the *Zinin reduction*.[583] The reagent sodium dihydro(trithio)borate $NaBH_2S_3$ reduces aromatic nitro compounds to amines,[584] but aliphatic nitro compounds give other products (see **9-58**). In contrast, $LiAlH_4$ reduces aliphatic nitro compounds to amines, but with aromatic nitro compounds the products with this reagent are azo compounds (**9-67**). Most metal hydrides, including $NaBH_4$ and BH_3, do not reduce nitro groups at all, though both aliphatic and aromatic nitro compounds have been reduced to amines with $NaBH_4$ and various catalysts, such as $NiCl_2$ or $CoCl_2$.[585] Treatment of aromatic nitro compounds with $NaBH_4$ alone has resulted in reduction of the *ring* to a cyclohexane ring with the nitro group still intact[586] or in cleavage of the nitro group from the ring.[587] With $(NH_4)_2S$ or other sulfides or polysulfides it is often possible to reduce just one of two or three nitro groups on an aromatic ring or on two different rings in one molecule.[588] The nitro groups of N-nitro compounds can also be reduced to amino groups, e.g., nitrourea $NH_2CONHNO_2$ gives semicarbazide $NH_2CONHNH_2$.

With some reducing agents, especially with aromatic nitro compounds, the reduction can be stopped at an intermediate stage, and hydroxylamines (**9-49**), hydrazobenzenes (**9-68**),

[575]Fry; Mraz *Tetrahedron Lett.* **1979**, 849.

[576]For a review of selective reduction of aliphatic nitro compounds without disturbance of other functional groups, see Ioffe; Tartakovskii; Novikov *Russ. Chem. Rev.* **1966**, *35*, 19-32.

[577]For reviews, see Rylander *Hydrogenation Methods*, Ref. 488, pp. 104-116, *Catalytic Hydrogenation over Platinum Metals*; Academic Press: New York, 1967, pp. 168-202.

[578]For a list of reagents, with references, see Ref. 21, pp. 411-415.

[579]An explosion has been reported with *o*-chloronitro compounds: Rondestvedt; Johnson *Synthesis* **1977**, 851. For a review of the use of hydrazine, see Furst; Berlo; Hooton *Chem. Rev.* **1965**, *65*, 51-68, pp. 52-60. See also Yuste; Saldaña; Walls *Tetrahedron Lett.* **1982**, *23*, 147; Adger; Young *Tetrahedron Lett.* **1984**, *25*, 5219.

[580]Ho; Wong *Synthesis* **1974**, 45. See also George; Chandrasekaran *Synth. Commun.* **1983**, *13*, 495.

[581]Sarmah; Barua *Tetrahedron Lett.* **1990**, *31*, 4065.

[582]Entwistle; Jackson; Johnstone; Telford *J. Chem. Soc., Perkin Trans. 1* **1977**, 443. See also Terpko; Heck *J. Org. Chem.* **1980**, *45*, 4992; Babler; Sarussi *Synth. Commun.* **1981**, *11*, 925.

[583]For a review of the Zinin reduction, see Porter *Org. React.* **1973**, *20*, 455-481.

[584]Lalancette; Brindle *Can. J. Chem.* **1971**, *49*, 2990. See also Maki; Sugiyama; Kikuchi; Seto *Chem. Lett.* **1975**, 1093.

[585]See, for example, Jardine; McQuillin *Chem. Commun.* **1970**, 626; Hanaya; Muramatsu; Kudo; Chow *J. Chem. Soc., Perkin Trans. 1* **1979**, 2409; Ono; Sasaki; Yaginuma *Chem. Ind. (London)* **1983**, 480; Osby; Ganem *Tetrahedron Lett.* **1985**, *26*, 6413; Petrini; Ballini; Rosini *Synthesis* **1987**, 713; He; Zhao; Pan; Wang *Synth. Commun.* **1989**, *19*, 3047.

[586]Severin; Schmitz *Chem. Ber.* **1962**, *95*, 1417; Severin; Adam *Chem. Ber.* **1963**, *96*, 448.

[587]Kaplan *J. Am. Chem. Soc.* **1964**, *86*, 740. See also Swanwick; Waters *Chem. Commun.* **1970**, 63.

[588]This result has also been achieved by hydrogenation with certain catalysts [Lyle; LaMattina, *Synthesis* **1974**, 726; Knifton *J. Org. Chem.* **1976**, *41*, 1200; Ono; Terasaki; Tsuruoka *Chem. Ind. (London)* **1983**, 477], and with hydrazine hydrate and Raney nickel: Ayyangar; Kalkote; Lugade; Nikrad; Sharma *Bull. Chem. Soc. Jpn.* **1983**, *56*, 3159.

azobenzenes (**9-67**), and azoxybenzenes (**9-66**) can be obtained in this manner. However, nitroso compounds, which are often postulated as intermediates, are too reactive to be isolated, if indeed they are intermediates (see however, **9-48**). Reduction by metals in mineral acids cannot be stopped, but always produces the amine. Amines are also the products when nitro compounds, both alkyl and aryl, are reduced with $HCOONH_4$–Pd–C.[589] Many other functional groups (e.g., COOH, COOR, CN, amide) are not affected by this reagent (though ketones are reduced—see **9-37**). With optically active alkyl substrates this method gives retention of configuration. [590]

The mechanisms of these reductions have been very little studied, though it is usually presumed that, at least with some reducing agents, nitroso compounds and hydroxylamines are intermediates. Both of these types of compounds give amines when exposed to most of these reducing agents (**9-50**), and hydroxylamines can be isolated (**9-49**). With metals and acid the following path has been suggested:[591]

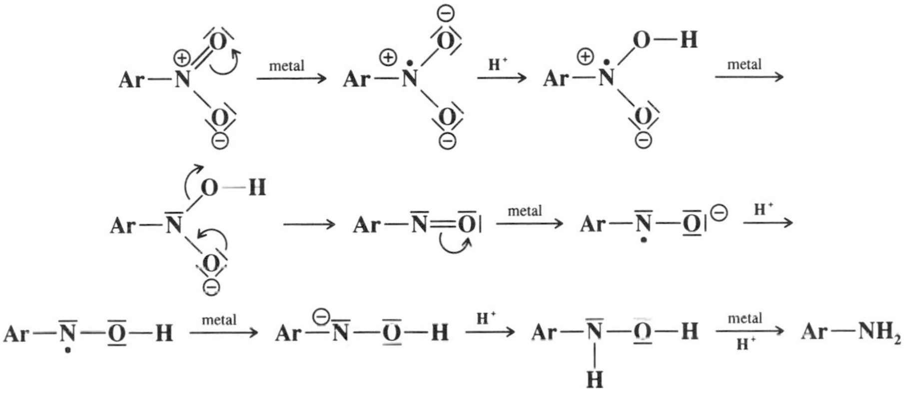

OS **I**, 52, 240, 455, 485; **II**, 130, 160, 175, 254, 447, 471, 501, 617; **III**, 56, 59, 63, 69, 73, 82, 86, 239, 242, 453; **IV**, 31, 357; **V**, 30, 346, 552, 567, 829, 1067, 1130.

9-48　Reduction of Nitro Compounds to Nitroso Compounds
N-Oxygen-detachment

$$ArNO_2 \xrightarrow[H_2O]{h\nu,\ CN^-} ArNO$$

Certain aromatic nitroso compounds can be obtained in good yields by irradiation of the corresponding nitro compounds in 0.1 M aqueous KCN with uv light.[592] The reaction has also been performed electrochemically.[593] When nitro compounds are treated with most reducing agents, nitroso compounds are either not formed or react further under the reaction conditions and cannot be isolated. N-Nitroamines have been reduced to N-nitrosoamines with Bu_3SnH–azobisisobutyrylnitrile.[594]

[589]Ram; Ehrenkaufer *Tetrahedron Lett.* **1984**, *25*, 3415.
[590]Barrett; Spilling *Tetrahedron Lett.* **1988**, *29*, 5733.
[591]House, Ref. 10, p. 211.
[592]Petersen; Letsinger *Tetrahedron Lett.* **1971**, 2197; Vink; Cornelisse; Havinga *Recl. Trav. Chim. Pays-Bas* **1971**, *90*, 1333.
[593]Lamoureux; Moinet *Bull. Soc. Chim. Fr.* **1988**, 59.
[594]de Armas; Francisco; Hernández; Suárez *Tetrahedron Lett.* **1986**, *27*, 3195.

9-49 Reduction of Nitro Compounds to Hydroxylamines

$$ArNO_2 \xrightarrow[H_2O]{Zn} ArNHOH$$

When aromatic nitro compounds are reduced with zinc and water under neutral conditions,[595] hydroxylamines are formed. Among other reagents used for this purpose have been SmI_2,[596] N_2H_4–Rh–C,[597] and $NaBH_4$–Se.[598] Borane in THF reduces aliphatic nitro compounds (in the form of their salts) to hydroxylamines:[599]

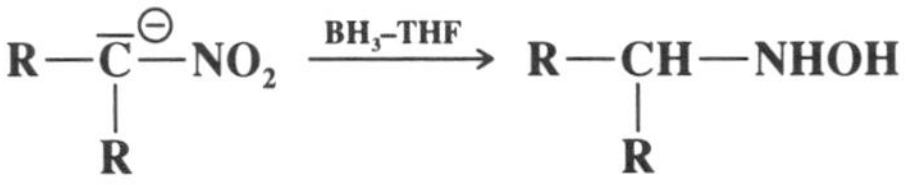

$$R-\overset{\ominus}{\underset{\underset{R}{|}}{C}}-NO_2 \xrightarrow{BH_3-THF} R-\underset{\underset{R}{|}}{CH}-NHOH$$

Nitro compounds have been reduced electrochemically, to hydroxylamines as well as to other products.[600]

OS **I**, 445; **III**, 668; **IV**, 148; **VI**, 803; **67**, 187.

9-50 Reduction of Nitroso Compounds and Hydroxylamines to Amines

$$RNO \xrightarrow[HCl]{Zn} RNH_2$$

N-Dihydro-de-oxo-bisubstitution

$$RNHOH \xrightarrow[HCl]{Zn} RNH_2$$

N-Hydro-de-hydroxylation or **N-Dehydroxylation**

Nitroso compounds and hydroxylamines can be reduced to amines by the same reagents that reduce nitro compounds (**9-47**). N-Nitroso compounds are similarly reduced to hydrazines:[601]

$$R_2N-NO \longrightarrow R_2N-NH_2$$

OS **I**, 511; **II**, 33, 202, 211, 418; **III**, 91; **IV**, 247. See also OS **65**, 166.

9-51 Reduction of Oximes to Primary Amines or Aziridines

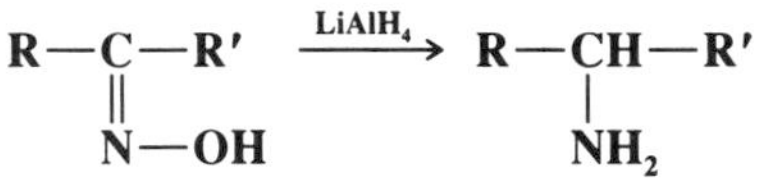

$$R-\underset{\underset{N-OH}{||}}{C}-R' \xrightarrow{LiAlH_4} R-\underset{\underset{NH_2}{|}}{CH}-R'$$

Both aldoximes and ketoximes can be reduced to primary amines with $LiAlH_4$. The reaction is slower than with ketones, so that, for example, PhCOCH=NOH gave 34% Ph-

---

[595]For some other methods of accomplishing this conversion, see Rondestvedt; Johnson *Synthesis* **1977**, 850; Entwistle; Gilkerson; Johnstone; Telford *Tetrahedron* **1978**, *34*, 213.

[596]Kende; Mendoza *Tetrahedron Lett.* **1991**, *32*, 1699.

[597]Oxley; Adger; Sasse; Forth *Org. Synth.* 67, 187.

[598]Yanada; Yamaguchi; Meguri; Uchida *J. Chem. Soc., Chem. Commun.* **1986**, 1655.

[599]Feuer; Bartlett; Vincent; Anderson *J. Org. Chem.* **1965**, *31*, 2880.

[600]For reviews of the electroreduction of nitro compounds, see Fry, Ref. 244, pp. 188-198; Lund, in Baizer; Lund *Organic Electrochemistry*; Marcel Dekker: New York, 1983, pp. 285-313.

[601]For examples of this reduction, accomplished with titanium reagents, see Entwistle; Johnstone; Wilby *Tetrahedron* **1982**, *38*, 419; Lunn; Sansone; Keefer *J. Org. Chem.* **1984**, *49*, 3470.

CHOHCH=NOH.[602] Among other reducing agents that give this reduction[603] are zinc and acetic acid, sodium ethoxide, BH_3,[604] $NaBH_3CN$–$TiCl_3$,[605] and sodium and an alcohol.[606] Catalytic hydrogenation is also effective.[607] The reduction has been performed enantioselectively with baker's yeast[608] and with Ph_2SiH_2 and an optically active rhodium complex catalyst.[609]

When the reducing agent is DIBALH, the product is a secondary amine, arising from a rearrangement:[610]

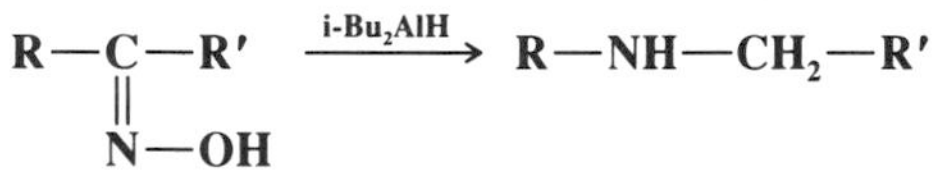

$$R-\underset{\underset{N-OH}{\|}}{C}-R' \xrightarrow{\text{i-Bu}_2\text{AlH}} R-NH-CH_2-R'$$

With certain oximes (e.g., those of the type $ArCH_2CR{=}NOH$), treatment with $LiAlH_4$ gives aziridines,[611] e.g.,

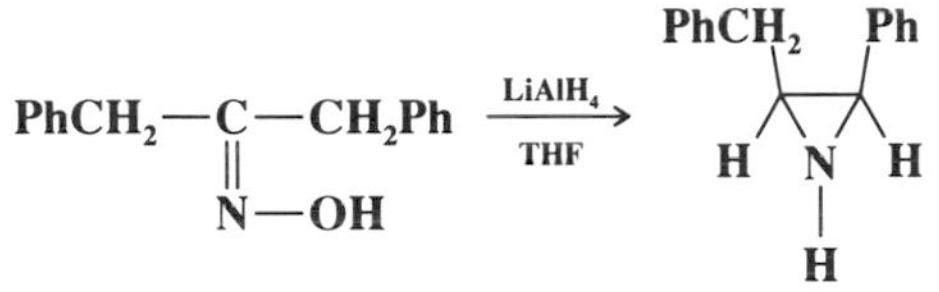

$$PhCH_2-\underset{\underset{N-OH}{\|}}{C}-CH_2Ph \xrightarrow[\text{THF}]{\text{LiAlH}_4} \begin{array}{c} PhCH_2 \quad Ph \\ \diagup\!\!\!\triangle\!\!\!\diagdown \\ H \quad N \quad H \\ | \\ H \end{array}$$

Hydrazones, arylhydrazones, and semicarbazones can also be reduced to amines with various reducing agents, including Zn–HCl and H_2 and Raney nickel.

Oximes have been reduced in a different way, to give imines ($RR'C{=}NOH \rightarrow RR'C{=}NH$), which are generally unstable but which can be trapped to give useful products. Among reagents used for this purpose have been Bu_3P–SPh_2[612] and $Ru_3(CO)_{12}$.[613]

Oximes can also be reduced to hydroxylamines (**6-26**).

OS **II**, 318; **III**, 513; **V**, 32, 83, 373, 376.

9-52   Reduction of Azides to Primary Amines
N-Dihydro-de-diazo-bisubstitution

$$RN_3 \xrightarrow{\text{LiAlH}_4} RNH_2$$

Azides are easily reduced to primary amines by $LiAlH_4$, as well as by a number of other reducing agents,[614] including $NaBH_4$, PPh_3 (with this reagent, the process is called the

[602]Felkin *C. R. Acad. Sci.* **1950**, *230*, 304.
[603]For a list of reagents, with references, see Ref. 21, p. 424.
[604]Feuer; Braunstein *J. Org. Chem.* **1969**, *34*, 1817.
[605]Leeds; Kirst *Synth. Commun.* **1988**, *18*, 777.
[606]For example, see Sugden; Patel *Chem. Ind. (London)* **1972**, 683.
[607]For a review, see Rylander *Catalytic Hydrogenation over Platinum Metals*, Ref. 577, pp. 139-159.
[608]Gibbs; Barnes *Tetrahedron Lett.* **1990**, *31*, 5555.
[609]Brunner; Becker; Gauder *Organometallics* **1986**, *5*, 739.
[610]Sasatani; Miyazaki; Maruoka; Yamamoto *Tetrahedron Lett.* **1983**, *24*, 4711. See also Rerick; Trottier; Daignault; DeFoe *Tetrahedron Lett.* **1963**, 629; Petrarca; Emery *Tetrahedron Lett.* **1963**, 635; Graham; Williams *Tetrahedron* **1965**, *21*, 3263.
[611]For a review, see Kotera; Kitahonoki *Org. Prep. Proced.* **1969**, *1*, 305-324. For examples, see Shandala; Solomon; Waight *J. Chem. Soc.* **1965**, 892; Kitahonoki; Takano; Matsuura; Kotera *Tetrahedron* **1969**, *25*, 335; Landor; Sonola; Tatchell *J. Chem. Soc., Perkin Trans. 1* **1974**, 1294; Ferrero; Rouillard; Decouzon; Azzaro *Tetrahedron Lett.* **1974**, 131; Diab; Laurent; Mison *Tetrahedron Lett.* **1974**, 1605.
[612]Barton; Motherwell; Simon; Zard *J. Chem. Soc., Chem. Commun.* **1984**, 337.
[613]Akazome; Tsuji; Watanabe *Chem. Lett.* **1990**, 635.
[614]For a review, see Scriven; Turnbull *Chem. Rev.* **1988**, *88*, 297-368, pp. 321-327. For lists of reagents, with references, see Ref. 21, pp. 409-410; Rolla *J. Org. Chem.* **1982**, *47*, 4327.

Staudinger reaction),[615] H$_2$ and a catalyst, Mg or Ca in MeOH,[616] N$_2$H$_4$–Pd,[617] and tin complexes prepared from SnCl$_2$ or Sn(SR)$_2$.[618] This reaction, combined with RX → RN$_3$ (**0-61**), is an important way of converting alkyl halides RX to primary amines RNH$_2$; in some cases the two procedures have been combined into one laboratory step.[619] Sulfonyl azides RSO$_2$N$_3$ have been reduced to sulfonamides RSO$_2$NH$_2$ by irradiation in isopropyl alcohol[620] and with NaH.[621]

OS **V**, 586; **VII**, 433.

9-53   Reduction of Miscellaneous Nitrogen Compounds

$$R-N{=}C{=}O \xrightarrow{\text{LiAlH}_4} R-NH-CH_3$$ **Isocyanate-methylamine transformation**

$$R-N{=}C{=}S \xrightarrow{\text{LiAlH}_4} R-NH-CH_3$$ **Isothiocyanate-methylamine transformation**

$$Ar-N{=}N-Ar \xrightarrow[\text{catalyst}]{\text{H}_2} Ar-NH-NH-Ar$$ **N,N-Dihydro-addition**

$$ArN_2^+ \; Cl^- \xrightarrow{\text{Na}_2\text{SO}_3} ArNHNH_2$$ **Diazonium-arylhydrazone reduction**

$$R_2N-NO \xrightarrow[\text{Ni}]{\text{H}_2} R_2NH$$ **N-Hydro-de-nitroso-substitution**

Isocyanates and isothiocyanates are reduced to methylamines on treatment with LiAlH$_4$. LiAlH$_4$ does not usually reduce azo compounds[622] (indeed these are the products from LiAlH$_4$ reduction of nitro compounds, **9-67**), but these can be reduced to hydrazo compounds by catalytic hydrogenation or with diimide[623] (see **5-9**). Diazonium salts are reduced to hydrazines by sodium sulfite. This reaction probably has a nucleophilic mechanism.[624]

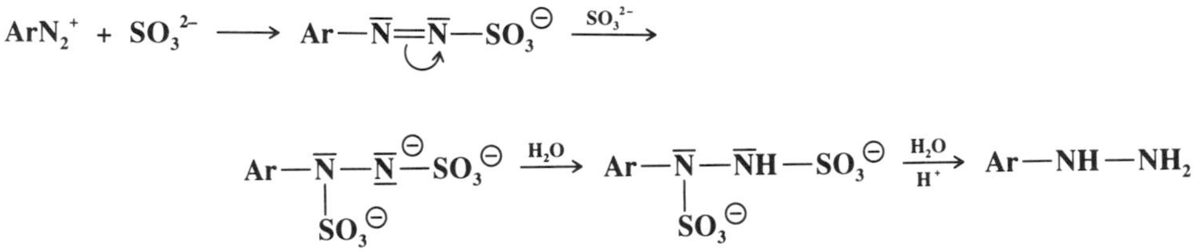

$$ArN_2^+ + SO_3^{2-} \longrightarrow Ar-\overline{N}{=}\overline{N}-SO_3^{\ominus} \xrightarrow{\text{SO}_3^{2-}}$$

$$Ar-\overline{N}-\underset{\underset{SO_3^{\ominus}}{|}}{\overline{N}}{}^{\ominus}-SO_3^{\ominus} \xrightarrow{\text{H}_2\text{O}} Ar-\overline{N}-\underset{\underset{SO_3^{\ominus}}{|}}{\overline{N}}H-SO_3^{\ominus} \xrightarrow[\text{H}^+]{\text{H}_2\text{O}} Ar-NH-NH_2$$

The initial product is a salt of hydrazinesulfonic acid, which is converted to the hydrazine by acid treatment. Diazonium salts can also be reduced to arenes (**4-24**). N-Nitrosoamines

[615]First reported by Staudinger; Meyer *Helv. Chim. Acta* **1919**, *2*, 635.
[616]Maiti; Spevak; Narender Reddy *Synth. Commun.* **1988**, *18*, 1201.
[617]Malik; Preston; Archibald; Cohen; Baum *Synthesis* **1989**, 450.
[618]Bartra; Romea; Urpí; Vilarrasa *Tetrahedron* **1990**, *46*, 587.
[619]See, for example, Koziara; Osowska-Pacewicka; Zawadzki; Zwierzak *Synthesis* **1985**, 202, **1987**, 487. The reactions **0-67**, **0-61**, and **9-52** have also been accomplished in one laboratory step: Koziara *J. Chem. Res. (S)* **1989**, 296.
[620]Reagen; Nickon *J. Am. Chem. Soc.* **1968**, *90*, 4096.
[621]Lee; Closson *Tetrahedron Lett.* **1974**, 381.
[622]For a review see Newbold, in Patai, Ref. 136, pt. 2, pp. 601, 604-614.
[623]For example, see Ioffe; Sergeeva; Dumpis *J. Org. Chem. USSR* **1969**, *5*, 1683.
[624]Huisgen; Lux *Chem. Ber.* **1960**, *93*, 540.

can be denitrosated to secondary amines by a number of reducing agents, including H_2 and a catalyst,[625] BF_3–THF–$NaHCO_3$,[626] and $NaBH_4$–$TiCl_4$,[627] as well as by hydrolysis.[628]

A cyano group can be reduced to a methyl group by treatment with a terpene such as limonene (which acts as reducing agent) in the presence of palladium–charcoal.[629] H_2 is also

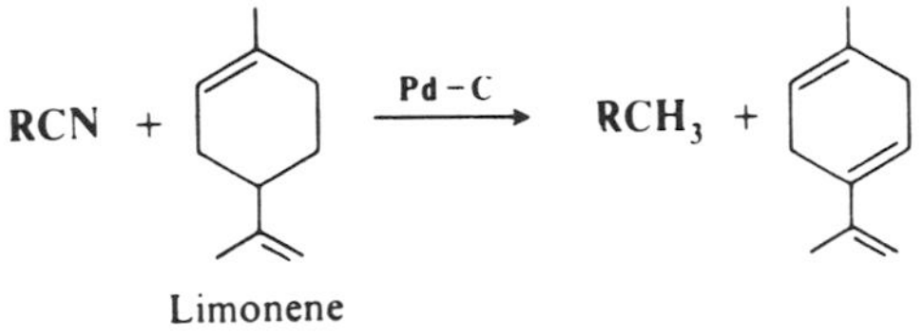

$$RCN \; + \quad \overset{Pd-C}{\longrightarrow} \quad RCH_3 \; +$$

Limonene

effective,[630] though higher temperatures are required. R may be alkyl or aryl. Cyano groups CN have also been reduced to CH_2OH, in the vapor phase, with 2-propanol and zirconium oxide.[631]

OS **I**, 442; **III**, 475. Also see OS **V**, 43.

9-54   Reduction of Sulfonyl Halides and Sulfonic Acids to Thiols

$$RSO_2Cl \xrightarrow{LiAlH_4} RSH$$

Thiols can be prepared by the reduction of sulfonyl halides[632] with $LiAlH_4$. Usually, the reaction is carried out on aromatic sulfonyl chlorides. Zinc and acetic acid, and HI, also give the reduction. Sulfonic acids have been reduced to thiols with a mixture of triphenylphosphine and either I_2 or a diaryl disulfide.[633] Disulfides RSSR can also be produced.[634] For the reduction of sulfonyl chlorides to sulfinic acids, see **0-118**.

OS **I**, 504; **IV**, 695; **V**, 843.

B.   Reactions in Which an Oxygen Is Removed from the Substrate

9-55   Reduction of Amine Oxides and Azoxy Compounds
N-Oxygen-detachment

$$R_3\overset{\oplus}{N}{-}\overset{\ominus}{O} \xrightarrow{PPh_3} R_3N$$

$$Ar{-}N{=}\overset{\oplus}{N}{-}Ar \xrightarrow{PPh_3} Ar{-}N{=}N{-}Ar$$
$$\quad\quad\quad \underset{O \ominus}{|}$$

[625]Enders; Hassel; Pieter; Renger; Seebach *Synthesis* **1976**, 548.
[626]Jeyaraman; Ravindran *Tetrahedron Lett.* **1990**, *31*, 2787.
[627]Kano; Tanaka; Sugino; Shibuya; Hibino *Synthesis* **1980**, 741.
[628]Fridman; Mukhametshin; Novikov *Russ. Chem. Rev.* **1971**, *40*, 34-50, pp. 41-42.
[629]Kindler; Lührs *Chem. Ber.* **1966**, *99*, 227, *Liebigs Ann. Chem.* **1967**, *707*, 26.
[630]See also Andrade; Maier; Zapf; Schleyer *Synthesis* **1980**, 802; Brown; Foubister *Synthesis* **1982**, 1036.
[631]Takahashi; Shibagaki; Matsushita *Chem. Lett.* **1990**, 311.
[632]For a review, see Wardell, in Patai, Ref. 408, pp. 216-220.
[633]Oae; Togo *Bull. Chem. Soc. Jpn.* **1983**, *56*, 3802, **1984**, *57*, 232.
[634]For example, see Alper *Angew. Chem. Int. Ed. Engl.* **1969**, *8*, 677 [*Angew. Chem. 81*, 706]; Chan; Montillier; Van Horn; Harpp *J. Am. Chem. Soc.* **1970**, *92*, 7224. See also Olah; Narang; Field; Karpeles *J. Org. Chem.* **1981**, *46*, 2408; Oae; Togo *Synthesis* **1982**, 152, *Bull. Chem. Soc. Jpn.* **1983**, *56*, 3813; Suzuki; Tani; Osuka *Chem. Lett.* **1984**, 139; Babu; Bhatt *Tetrahedron Lett.* **1986**, *27*, 1073; Narayana; Padmanabhan; Kabalka *Synlett* **1991**, 125.

Amine oxides[635] and azoxy compounds (both alkyl and aryl)[636] can be reduced practically quantitatively with triphenylphosphine.[637] Other reducing agents, e.g., LiAlH$_4$, H$_2$–Ni, PCl$_3$, CS$_2$,[638] NaHTe,[639] TiCl$_3$,[640] TiCl$_4$ with LiAlH$_4$, SbCl$_2$, or NaI,[641] and sulfur have also been used. Nitrile oxides[642] R—C$\equiv$$\overset{\oplus}{N}$—$\overset{\ominus}{O}$ can be reduced to nitriles with trialkylphosphines,[643] and isocyanates RNCO to isocyanides RNC with Cl$_3$SiH–Et$_3$N.[644]

OS **IV**, 166. See also OS **67**, 20.

9-56 Reduction of Sulfoxides and Sulfones

S-Oxygen-detachment

$$R-\underset{\underset{O}{\|}}{S}-R \xrightarrow{\text{LiAlH}_4} R-S-R$$

Sulfoxides can be reduced to sulfides by many reagents,[645] among them LiAlH$_4$, HI, Bu$_3$SnH,[646] TiCl$_2$,[647] MeSiCl$_3$–NaI,[648] H$_2$–Pd–C,[649] NaBH$_4$–FeCl$_3$,[650] NaBr,[651] TiCl$_4$–NaI,[652] Ph$_3$P,[653] and *t*-BuBr.[654] Sulfones, however, are usually stable to reducing agents, though they have been reduced to sulfides with DIBALH (i-Bu)$_2$AlH.[655] A less general reagent is LiAlH$_4$, which reduces some sulfones to sulfides, but not others.[656] Both sulfoxides and sulfones can be reduced by heating with sulfur (which is oxidized to SO$_2$), though the reaction with sulfoxides proceeds at a lower temperature. It has been shown by using substrate labeled with ^{35}S that sulfoxides simply give up the oxygen to the sulfur, but that the reaction with sulfones is more complex, since about 75% of the original radioactivity of the sulfone is lost.[657] This indicates that most of the sulfur in the sulfide product comes in this case from the *reagent*. There is no direct general method for the reduction of sulfones to sulfoxides,

[635]For reviews of the reduction of heterocyclic amine oxides, see Albini; Pietra, Ref. 421, pp. 120-134; Katritzky; Lagowski, Ref. 421, pp. 166-231.

[636]For a review, see Newbold, in Patai, Ref. 136, pt. 2, pp. 602-603, 614-624.

[637]For a review, see Rowley, in Cadogan *Organophosphorus Reagents in Organic Synthesis*; Academic Press: New York, 1979, pp. 295-350.

[638]Yoshimura; Asada; Oae *Bull. Chem. Soc. Jpn.* **1982**, *55*, 3000.

[639]Barton; Fekih; Lusinchi *Tetrahedron Lett.* **1985**, *26*, 4603.

[640]Kuz'min; Mizhiritskii; Kogan *J. Org. Chem. USSR* **1989**, *25*, 596.

[641]Malinowski; Kaczmarek *Synthesis* **1987**, 1013; Kaczmarek; Malinowski; Balicki *Bull. Soc. Chim. Belg.* **1988**, *97*, 787.

[642]For reviews of the chemistry of nitrile oxides, see Torssell *Nitrile Oxides, Nitrones, and Nitronates in Organic Synthesis*; VCH: New York, 1988, pp. 55-74; Grundmann *Fortschr. Chem. Forsch.* **1966**, *7*, 62-127.

[643]Grundmann; Frommeld *J. Org. Chem.* **1965**, *30*, 2077.

[644]Baldwin; Derome; Riordan *Tetrahedron* **1983**, *39*, 2989.

[645]For reviews, see Kukushkin *Russ. Chem. Rev.* **1990**, *59*, 844-852; Madesclaire *Tetrahedron* **1988**, *44*, 6537-6580; Drabowicz; Togo; Mikołajczyk; Oae *Org. Prep. Proced. Int.* **1984**, *16*, 171-198; Drabowicz; Numata; Oae *Org. Prep. Proced. Int.* **1977**, *9*, 63-83. For a list of reagents, with references, see Block, Ref. 440.

[646]Kozuka; Furumai; Akasaka; Oae *Chem. Ind. (London)* **1974**, 496.

[647]Drabowicz; Mikołajczyk *Synthesis* **1978**, 138. For the use of TiCl$_3$, see Ho; Wong *Synth. Commun.* **1973**, *3*, 37.

[648]Olah; Husain; Singh; Mehrotra *J. Org. Chem.* **1983**, *48*, 3667. See also Schmidt; Russ *Chem. Ber.* **1981**, *114*, 822.

[649]Ogura; Yamashita; Tsuchihashi *Synthesis* **1975**, 385.

[650]Lin; Zhang *Synth. Commun.* **1987**, *17*, 1403.

[651]Bernard; Caredda; Piras; Serra *Synthesis* **1990**, 329.

[652]Balicki *Synthesis* **1991**, 155.

[653]For a review, see Ref. 637, pp. 301-304.

[654]Tenca; Dossena; Marchelli; Casnati *Synthesis* **1981**, 141.

[655]Gardner; Kaiser; Krubiner; Lucas *Can. J. Chem.* **1973**, *51*, 1419.

[656]Bordwell *J. Am. Chem. Soc.* **1951**, *73*, 2251; Whitney; Cram *J. Org. Chem.* **1970**, *35*, 3964; Weber; Stromquist; Ito *Tetrahedron Lett.* **1974**, 2595.

[657]Oae; Kawamura *Bull. Chem. Soc. Jpn.* **1963**, *36*, 163; Kiso; Oae *Bull. Chem. Soc. Jpn.* **1967**, *40*, 1722. See also Oae; Nakai; Tsuchida; Furukawa *Bull. Chem. Soc. Jpn.* **1971**, *44*, 445.

but an indirect method has been reported.[658] Selenoxides can be reduced to selenides with a number of reagents.[659]

9-57 Reduction of Hydroperoxides

$$R\!-\!O\!-\!O\!-\!H \xrightarrow{\text{LiAlH}_4} ROH$$

Hydroperoxides can be reduced to alcohols with $LiAlH_4$ or Ph_3P[660] or by catalytic hydrogenation. This functional group is very susceptible to catalytic hydrogenation, as shown by the fact that a double bond may be present in the same molecule without being reduced.[661]

The reaction is an important step in a method for the oxidative decyanation of nitriles containing an α hydrogen.[662] The nitrile is first converted to the α-hydroperoxy nitrile by treatment with base at $-78°C$ followed by O_2. The hydroperoxy nitrile is then reduced to

$$RR'CHCN \xrightarrow[-78°C]{\text{LiN(i-Pr)}_2} \left[RR'\overset{\ominus}{C}CN\right] \xrightarrow[-78°C]{O_2} RR'\overset{\displaystyle OO^-}{\underset{|}{C}}CN \xrightarrow[\text{2. Sn}^{2+}]{\text{1. H}^+} RR'\overset{\displaystyle OH}{\underset{|}{C}}\!-\!CN \xrightarrow{OH^-} RR'CO$$

the cyanohydrin, which is cleaved (the reverse of **6-49**) to the corresponding ketone. The method is not successful for the preparation of aldehydes (R' = H).

9-58 Reduction of Aliphatic Nitro Compounds to Oximes or Nitriles

$$RCH_2NO_2 \xrightarrow[\text{HOAc}]{\text{Zn}} RCH\!\!=\!\!NOH$$

Nitro compounds that contain an α hydrogen can be reduced to oximes with zinc dust in HOAc[663] or with other reagents, among them Co–Cu(II) salts in alkanediamines,[664] CS_2–Et_3N,[665] $CrCl_2$,[666] and (for α-nitro sulfones) $NaNO_2$.[667] α-Nitro alkenes have been converted to oximes ($-\overset{|}{C}\!\!=\!\!\overset{|}{C}\!\!-\!\!NO_2 \rightarrow -\overset{|}{C}H\!\!-\!\!\overset{|}{C}\!\!=\!\!NOH$) with sodium hypophosphite and with Pb–HOAc–DMF, as well as with certain other reagents.[668]

Primary aliphatic nitro compounds can be reduced to nitriles with sodium dihydro(trithio)borate.[584] Secondary compounds give mostly ketones (e.g., nitrocyclohexane

$$RCH_2NO_2 \xrightarrow{\text{NaBH}_2\text{S}_3} RC\!\!\equiv\!\!N$$

[658]Still; Ablenas *J. Org. Chem.* **1983**, *48*, 1617.

[659]See for example, Sakaki; Oae *Chem. Lett.* **1977**, 1003; Still; Hasan; Turnbull *Can. J. Chem.* **1978**, *56*, 1423; Denis; Krief *J. Chem. Soc., Chem. Commun.* **1980**, 544.

[660]For a review, see Ref. 637, pp. 318-320.

[661]Rebeller; Clément *Bull. Soc. Chim. Fr.* **1964**, 1302.

[662]Freerksen; Selikson; Wroble; Kyler; Watt *J. Org. Chem.* **1983**, *48*, 4087. This paper also reports several other methods for achieving this conversion.

[663]Johnson; Degering *J. Am. Chem. Soc.* **1939**, *61*, 3194.

[664]Knifton *J. Org. Chem.* **1973**, *38*, 3296.

[665]Barton; Fernandez; Richard; Zard *Tetrahedron* **1987**, *43*, 551; Albanese; Landini; Penso *Synthesis* **1990**, 333.

[666]Hanson; Organ *J. Chem. Soc. C* **1970**, 1182; Hanson *Synthesis* **1974**, 1-8, pp. 7-8.

[667]Zeilstra; Engberts *Synthesis* **1974**, 49.

[668]See Varma; Varma; Kabalka *Synth. Commun.* **1986**, *16*, 91; Kabalka; Pace; Wadgaonkar *Synth. Commun.* **1990**, *20*, 2453; Sera; Yamauchi; Yamada; Itoh *Synlett* **1990**, 477.

gave 45% cyclohexanone, 30% cyclohexanone oxime, and 19% N-cyclohexylhydroxylamine). Tertiary aliphatic nitro compounds do not react with this reagent. See also **9-47**.
OS **IV**, 932.

C. Reduction with Cleavage

9-59   Reduction of Azo, Azoxy, and Hydrazo Compounds to Amines

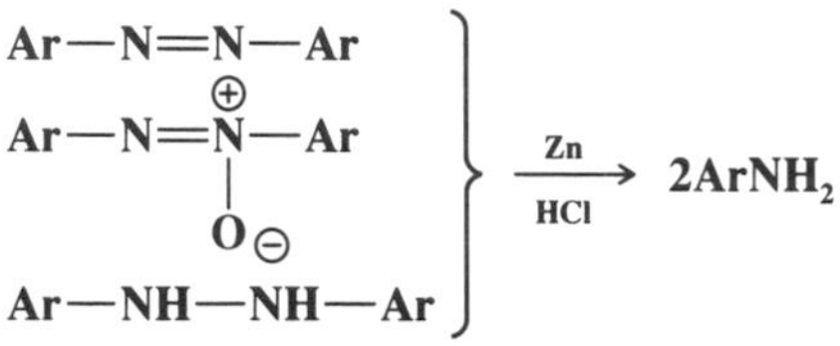

$$
\left.
\begin{array}{c}
Ar-N{=}N-Ar \\[2pt]
Ar-\overset{\oplus}{N}{=}N-Ar \\[-2pt]
| \\[-2pt]
\underset{\ominus}{O} \\[2pt]
Ar-NH-NH-Ar
\end{array}
\right\}
\xrightarrow[\text{HCl}]{\text{Zn}} 2ArNH_2
$$

Azo, azoxy, and hydrazo compounds can all be reduced to amines.[669] Metals (notably zinc) and acids, and $Na_2S_2O_4$, are frequently used as reducing agents. Borane reduces azo compounds to amines, though it does not reduce nitro compounds.[670] LiAlH$_4$ does not reduce hydrazo compounds or azo compounds, though with the latter, hydrazo compounds are sometimes isolated. With azoxy compounds, LiAlH$_4$ gives only azo compounds (**9-55**).
OS **I**, 49; **II**, 35, 39; **III**, 360. Also see OS **II**, 290.

9-60   Reduction of Peroxides
O-Hydrogen-uncoupling

$$\textbf{R{-}O{-}O{-}R} \xrightarrow{\text{LiAlH}_4} \textbf{2ROH}$$

Peroxides are cleaved to 2 moles of alcohols by LiAlH$_4$ or by catalytic hydrogenation. Peroxides can be reduced to ethers with $P(OEt)_3$.[671]

$$\textbf{ROOR} + \textbf{P(OEt)}_3 \longrightarrow \textbf{ROR} + \textbf{OP(OEt)}_3$$

In a similar reaction, disulfides RSSR′ can be converted to sulfides RSR′ by treatment with tris(diethylamino)phosphine $(Et_2N)_3P$.[672]
OS **VI**, 130.

9-61   Reduction of Disulfides to Thiols
S-Hydrogen-uncoupling

$$\textbf{RSSR} \xrightarrow{\text{Zn}}_{\text{H}^+} \textbf{2RSH}$$

Disulfides can be reduced to thiols by mild reducing agents,[673] such as zinc and dilute acid or Ph$_3$P and H$_2$O.[674] The reaction can also be accomplished simply by heating with alkali.[675]

[669]For a review, see Newbold, in Patai, Ref. 136, pt. 2, pp. 629-637.
[670]Brown; Subba Rao *J. Am. Chem. Soc.* **1960**, *82*, 681.
[671]Horner; Jurgeleit *Liebigs Ann. Chem.* **1955**, *591*, 138. See also Ref. 637, pp. 320-322.
[672]Harpp; Gleason; Snyder *J. Am. Chem. Soc.* **1968**, *90*, 4181; Harpp; Gleason *J. Am. Chem. Soc.* **1971**, *93*, 2437.
For another method, see Comasseto; Lang; Ferreira; Simonelli; Correia *J. Organomet. Chem.* **1987**, *334*, 329.
[673]For a review, see Wardell, in Patai, Ref. 408, pp. 220-229.
[674]Overman; Smoot; Overman *Synthesis* **1974**, 59.
[675]For discussions, see Danehy; Hunter *J. Org. Chem.* **1967**, *32*, 2047; Danehy, in Ref. 407, pp. 337-349.

Among other reagents used have been LiAlH$_4$, KBH(O-i-Pr)$_3$,[676] and hydrazine or substituted hydrazines.[678]

OS **II**, 580. Also see OS **IV**, 295.

D. Reductive Coupling

9-62 Bimolecular Reduction of Aldehydes and Ketones to 1,2-Diols
   2/O-Hydrogen-coupling

$$2RCOR \xrightarrow{\text{Na-Hg}} R_2C \underset{\underset{OH}{|}}{\quad} CR_2 \atop \underset{OH}{|}$$

1,2-Diols (pinacols) can be synthesized by reduction of aldehydes and ketones with active metals such as sodium, magnesium, or aluminum.[679] Aromatic ketones give better yields than aliphatic ones. The use of a Mg–MgI$_2$ mixture has been called the *Gomberg–Bachmann pinacol synthesis*. As with a number of other reactions involving sodium, there is a direct electron transfer here, converting the ketone or aldehyde to a ketyl, which dimerizes.

$$R-\underset{\underset{|O|}{\|}}{C}-R + Na \longrightarrow R-\overset{\cdot}{\underset{\underset{|O|_\ominus}{|}}{C}}-R \longrightarrow R-\overset{R}{\underset{\underset{\ominus|O|}{|}}{C}}-\overset{R}{\underset{\underset{|O|_\ominus}{|}}{C}}-R$$

Other reagents have been used,[680] including SmI$_2$,[681] Ce–I$_2$,[682] Yb,[683] and a reagent prepared from TiCl$_4$ and Mg amalgam[684] (a low-valent titanium reagent; see **9-64**). Dialdehydes have been cyclized by this reaction (with TiCl$_3$) to give cyclic 1,2-diols in good yield.[685] Unsymmetrical coupling between two different aldehydes has been achieved by the use of a vanadium complex,[686] while TiCl$_3$ in aqueous solution has been used to couple two different ketones.[687]

The dimerization of ketones to 1,2-diols can also be accomplished photochemically; indeed, this is one of the most common photochemical reactions.[688] The substrate, which

[676]Brown; Nazer; Cha *Synthesis* **1984**, 498.

[677]Krishnamurthy; Aimino *J. Org. Chem.* **1989**, *54*, 4458.

[678]Maiti; Spevak; Singh; Micetich; Narender Reddy *Synth. Commun.* **1988**, *18*, 575.

[679]For efficient methods, see Schreibmann *Tetrahedron Lett.* **1970**, 4271; Fürstner; Csuk; Rohrer; Weidmann *J. Chem. Soc., Perkin Trans. 1* **1988**, 1729.

[680]For a list of reagents, with references, see Ref. 21, pp. 547-548.

[681]Namy; Souppe; Kagan *Tetrahedron Lett.* **1983**, *24*, 765.

[682]Imamoto; Kusumoto; Hatanaka; Yokoyama *Tetrahedron Lett.* **1982**, *23*, 1353.

[683]Hou; Takamine; Fujiwara; Taniguchi *Chem. Lett.* **1987**, 2061.

[684]Corey; Danheiser; Chandrasekaran *J. Org. Chem.* **1976**, *41*, 260; Pons; Zahra; Santelli *Tetrahedron Lett.* **1981**, *22*, 3965. For some other titanium-containing reagents, see Clerici; Porta *J. Org. Chem.* **1985**, *50*, 76; Handa; Inanaga *Tetrahedron Lett.* **1987**, *28*, 5717. For a review of such coupling with Ti and V halides, see Lai *Org. Prep. Proced. Int.* **1980**, *12*, 363-391.

[685]McMurry; Rico *Tetrahedron Lett.* **1989**, *30*, 1169. For other cyclization reactions of dialdehydes and ketoaldehydes, see Molander; Kenny *J. Am. Chem. Soc.* **1989**, *111*, 8236; Raw; Pedersen *J. Org. Chem.* **1991**, *56*, 830; Chiara; Cabri; Hanessian *Tetrahedron Lett.* **1991**, *32*, 1125.

[686]Freudenberger; Konradi; Pedersen *J. Am. Chem. Soc.* **1989**, *111*, 8014.

[687]Clerici; Porta *J. Org. Chem.* **1982**, *47*, 2852, *Tetrahedron* **1983**, *39*, 1239. For some other unsymmetrical couplings, see Hou; Takamine; Aoki; Shiraishi; Fujiwara; Taniguchi *J. Chem. Soc., Chem. Commun.* **1988**, 668; Delair; Luche *J. Chem. Soc., Chem. Commun.* **1989**, 398; Takahara; Freudenberger; Konradi; Pedersen *Tetrahedron Lett.* **1989**, *30*, 7177.

[688]For reviews, see Schönberg *Preparative Organic Photochemistry*; Springer: New York, 1968, pp. 203-217; Neckers *Mechanistic Organic Photochemistry*; Reinhold: New York, 1967, pp. 163-177; Calvert; Pitts *Photochemistry*; Wiley: New York, 1966, pp. 532-536; Turro *Modern Molecular Photochemistry*; W.A. Benjamin: New York, 1978, pp. 363-385; Kan *Organic Photochemisty*; McGraw-Hill: New York, 1966, pp. 222-229.

is usually a diaryl or aryl alkyl ketone (though a few aromatic aldehydes and dialkyl ketones have been dimerized), is irradiated with uv light in the presence of a hydrogen donor such as isopropyl alcohol, toluene, or an amine.[689] In the case of benzophenone, irradiated in the presence of 2-propanol, the ketone molecule initially undergoes $n \rightarrow \pi^*$ excitation, and the singlet species thus formed crosses to the T_1 state with a very high efficiency. The T_1

$$Ph_2CO \xrightarrow{h\upsilon} [Ph_2CO] \longrightarrow [Ph_2CO] \xrightarrow{\text{iso-PrOH}} Ph-\overset{\cdot}{\underset{OH}{C}}-Ph \longrightarrow Ph_2\overset{}{\underset{OH}{C}}\text{---}\overset{}{\underset{OH}{C}}Ph_2$$

$$\text{singlet} \qquad \text{triplet} \qquad\qquad\qquad \mathbf{33}$$

species abstracts hydrogen from the alcohol (p. 246) and then dimerizes. The iso-PrO• radical, which is formed by this process, donates H• to another molecule of ground-state benzophenone, producing acetone and another molecule of **33**. This mechanism[690] predicts that the quantum yield for the disappearance of benzophenone should be 2, since each quantum

$$\text{iso-PrO•} + Ph_2CO \longrightarrow Me_2CO + Ph_2\overset{\cdot}{C}OH$$
$$\mathbf{33}$$

of light results in the conversion of 2 moles of benzophenone to **33**. Under favorable experimental conditions the observed quantum yield does approach 2. Benzophenone abstracts hydrogen with very high efficiency. Other aromatic ketones are dimerized with lower quantum yields, and some (e.g., *p*-aminobenzophenone, *o*-methylacetophenone) cannot be dimerized at all in 2-propanol (though *p*-aminobenzophenone, for example, can be dimerized in cyclohexane[691]). The reaction has also been carried out electrochemically.[692]

In a similar type of process, imines have been dimerized to give 1,2-diamines, by a number of procedures, including treatment with Al–PbBr$_2$,[693] with TiCl$_4$–Mg,[694] with SmI$_2$,[695]

$$2R-\underset{\underset{NR'}{\|}}{C}-H \xrightarrow[\text{PbBr}_2]{\text{Al}} R\underset{\underset{NHR'}{|}}{CH}\text{---}\underset{\underset{NHR'}{|}}{CHR}$$

and (for silylated imines) NbCl$_4$(THF)$_2$.[696] When electroreduction was used, it was even possible to obtain cross products, by coupling a ketone to an O-methyl oxime:[697]

$$R^1-\underset{\underset{O}{\|}}{C}-R^2 + R^3-\underset{\underset{NOMe}{\|}}{C}-R^4 \xrightarrow{\text{electroreduction}} R^1-\underset{\underset{OH}{|}}{\overset{\overset{R^2}{|}}{C}}\text{---}\underset{\underset{NHOMe}{|}}{\overset{\overset{R^3}{|}}{C}}-R^4 \xrightarrow[\text{cat.}]{H_2} R^1-\underset{\underset{OH}{|}}{\overset{\overset{R^2}{|}}{C}}\text{---}\underset{\underset{NH_2}{|}}{\overset{\overset{R^3}{|}}{C}}-R^4$$

[689]For a review of amines as hydrogen donors in this reaction, see Cohen; Parola; Parsons *Chem. Rev.* **1973**, *73*, 141-161.

[690]For some of the evidence for this mechanism, see Pitts; Letsinger; Taylor; Patterson; Recktenwald; Martin *J. Am. Chem. Soc.* **1959**, *81*, 1068; Hammond; Moore *J. Am. Chem. Soc.* **1959**, *81*, 6334; Moore; Hammond; Foss *J. Am. Chem. Soc.* **1961**, *83*, 2789; Huyser; Neckers *J. Am. Chem. Soc.* **1963**, *85*, 3641.

[691]Porter; Suppan *Proc. Chem. Soc.* **1964**, 191.

[692]For reviews, see Fry, Ref. 244, pp. 174-180; Shono, Ref. 149, pp. 137-140; Baizer; Petrovich *Prog. Phys. Org. Chem.* **1970**, *7*, 189-227. For a review of electrolytic reductive coupling, see Baizer, in Baizer, Lund, Ref. 600, pp. 639-689.

[693]Tanaka; Dhimane; Fujita; Ikemoto; Torii *Tetrahedron Lett.* **1988**, *29*, 3811.

[694]Betschart; Seebach *Helv. Chim. Acta* **1987**, *70*, 2215; Betschart; Schmidt; Seebach *Helv. Chim. Acta* **1988**, *71*, 1999; Mangeney; Tejero; Alexakis; Grosjean; Normant *Synthesis* **1988**, 255.

[695]Enholm; Forbes; Holub *Synth. Commun.* **1990**, *20*, 981; Imamoto; Nishimura *Chem. Lett.* **1990**, 1141.

[696]Roskamp; Pedersen *J. Am. Chem. Soc.* **1987**, *109*, 3152.

[697]Shono; Kise; Fujimoto *Tetrahedron Lett.* **1991**, *32*, 525.

The N-methoxyamino alcohol could then be reduced to the amino alcohol.[697]
OS **I**, 459; **II**, 71.

9-63   Bimolecular Reduction of Aldehydes and Ketones to Epoxides
Aldehyde-oxirane transformation

$$2ArCHO \xrightarrow{(Me_2N)_3P} ArCH\!\!-\!\!CHAr + (Me_2N)_3PO$$

ArCH—CHAr with O bridge;   **HMPA**

Aromatic aldehydes can be dimerized to epoxides by treatment with hexamethylphosphorus triamide.[698] The reagent[699] is converted to hexamethylphosphoric triamide (HMPA). The reaction can be used for the preparation of mixed epoxides by the use of a mixture of two aldehydes in which the less reactive aldehyde predominates. Epoxides have also been pre-pared by treatment of aromatic aldehydes or ketones with the anions $(Me_2N)_2\overset{\ominus}{P}\!\!=\!\!O$ and $(EtO)_2\overset{\ominus}{P}\!\!=\!\!O$ (derived, respectively, by treatment with an alkali metal of HMPA or triethyl phosphite).[700]
OS **V**, 358.

9-64   Bimolecular Reduction of Aldehydes or Ketones to Alkenes
De-oxygen-coupling

$$R\!-\!\underset{\underset{O}{\|}}{C}\!-\!R' \xrightarrow[Zn-Cu]{TiCl_3} R\!-\!\underset{\underset{R'}{|}}{C}\!=\!\underset{\underset{R'}{|}}{C}\!-\!R$$

Aldehydes and ketones, both aromatic and aliphatic (including cyclic ketones), can be converted in high yields to dimeric alkenes by treatment with $TiCl_3$ and a zinc–copper couple.[701] This is called the *McMurry reaction.*[702] The reagent produced in this way is called a *low-valent titanium reagent*, and the reaction has also been accomplished[703] with low-valent titanium reagents prepared in other ways, e.g., from Mg and a $TiCl_3$–THF complex,[704] from $TiCl_4$ and Zn or Mg,[705] from $TiCl_3$ and $LiAlH_4$,[706] from $TiCl_3$ and lamellar potassium graphite,[707] from $TiCl_3$ and K or Li,[708] as well as with Zn–Me_3SiCl[709] and with certain compounds prepared from WCl_6 and either lithium, lithium iodide, $LiAlH_4$, or an

[698]Mark *J. Am. Chem. Soc.* **1963**, 85, 1884; Newman; Blum *J. Am. Chem. Soc.* **1964**, *86*, 5598.
[699]For the preparation of the reagent, see Mark *Org. Synth. V*, 602.
[700]Normant *Bull. Soc. Chim. Fr.* **1966**, 3601.
[701]McMurry; Fleming; Kees; Krepski *J. Org. Chem.* **1978**, *43*, 3255. For an optimized procedure, see McMurry; Lectka; Rico *J. Org. Chem.* **1989**, *54*, 3748.
[702]For reviews, see McMurry *Chem. Rev.* **1989**, *89*, 1513-1524, *Acc. Chem. Res.* **1983**, *16*, 405-511; Lenoir *Synthesis* **1989**, 883-897; Betschart; Seebach *Chimia* **1989**, *43*, 39-49; Lai, Ref. 684. For related reviews, see Kahn; Rieke *Chem. Rev.* **1988**, *88*, 733-745; Pons; Santelli *Tetrahedron* **1988**, *44*, 4295-4312.
[703]For a list of reagents, with references, see Ref. 21, pp. 160-161.
[704]Tyrlik; Wolochowicz *Bull. Soc. Chim. Fr.* **1973**, 2147.
[705]Mukaiyama; Sato; Hanna *Chem. Lett.* **1973**, 1041; Lenoir *Synthesis* **1977**, 553; Lenoir; Burghard *J. Chem. Res. (S)* **1980**, 396; Carroll; Taylor *Aust. J. Chem.* **1990**, *43*, 1439.
[706]McMurry; Fleming *J. Am. Chem. Soc.* **1974**, *96*, 4708; Dams; Malinowski; Geise *Bull. Soc. Chim. Belg.* **1982**, *91*, 149, 311; Bottino; Finocchiaro; Libertini; Reale; Recca *J. Chem. Soc., Perkin Trans. 2* **1982**, 77. This reagent has been reported to give capricious results; see McMurry; Fleming, Ref. 708.
[707]Fürstner; Weidmann *Synthesis* **1987**, 1071.
[708]McMurry; Fleming *J. Org. Chem.* **1976**, *41*, 896; Richardson *Synth. Commun.* **1981**, *11*, 895.
[709]Banerjee; Sulbaran de Carrasco; Frydrych-Houge; Motherwell *J. Chem. Soc., Chem. Commun.* **1986**, 1803.

alkyllithium[710] (see **7-21**). The reaction has been used to convert dialdehydes and diketones to cycloalkenes.[711] Rings of 3 to 16 and 22 members have been closed in this way, e.g.,[712]

$$Ph-\overset{\underset{\displaystyle Me}{|}}{\underset{\displaystyle O}{\overset{\displaystyle \|}{C}}}-\overset{\underset{\displaystyle Me}{|}}{\underset{\displaystyle Me}{C}}-\overset{\underset{\displaystyle O}{\overset{\displaystyle \|}{C}}}{C}-Ph \xrightarrow[\text{LiAlH}_4]{\text{TiCl}_3} 46\%$$

The same reaction on a keto ester gives a cycloalkanone.[713]

$$R-\overset{\underset{\displaystyle O}{\overset{\displaystyle \|}{C}}}{C}-(CH_2)_n-COOEt \xrightarrow[\text{2. H}^+]{\text{1. TiCl}_3\text{-LiAlH}_4} R-C\underset{(CH_2)_n}{\overset{O}{\diagup}}C\overset{\displaystyle O}{\diagdown}$$

Unsymmetrical alkenes can be prepared from a mixture of two ketones, if one is in excess.[714] The mechanism consists of initial coupling of two radical species to give a 1,2-dioxygen compound (a titanium pinacolate), which is then deoxygenated.[715]
OS **VII**, 1.

9-65 Acyloin Ester Condensation

$$2RCOOR' \xrightarrow[\text{xylene}]{\text{Na}} R-\overset{\underset{\displaystyle ONa}{|}}{C}=\overset{\underset{\displaystyle ONa}{|}}{C}-R \xrightarrow{\text{H}_2\text{O}} R-\overset{\underset{\displaystyle OH}{|}}{C}H-\overset{\underset{\displaystyle O}{\overset{\displaystyle \|}{C}}}{C}-R$$
$$\textbf{34}$$

When carboxylic esters are heated with sodium in refluxing ether or benzene, a bimolecular reduction takes place, and the product is an α-hydroxy ketone (called an acyloin).[716] The reaction, called the *acyloin ester condensation*, is quite successful when R is alkyl. Acyloins with long chains have been prepared in this way, for example, R = $C_{17}H_{35}$, but for high-molecular-weight esters, toluene or xylene is used as the solvent. The acyloin condensation has been used with great success, in boiling xylene, to prepare cyclic acyloins from diesters.[717] The yields are 50 to 60% for the preparation of 6- and 7-membered rings, 30 to 40% for 8- and 9-membered, and 60 to 95% for rings of 10 to 20 members. Even larger rings have been closed in this manner. This is one of the best ways of closing rings of 10 members or more. The reaction has been used to close 4-membered rings,[718] though this is generally not

[710]Sharpless; Umbreit; Nieh; Flood *J. Am. Chem. Soc.* **1972**, *94*, 6538; Fujiwara; Ishikawa; Akiyama; Teranishi *J. Org. Chem.* **1978**, *43*, 2477; Dams; Malinowski; Geise, Ref. 706. See also Petit; Mortreux; Petit *J. Chem. Soc., Chem. Commun.* **1984**, 341; Chisholm; Klang *J. Am. Chem. Soc.* **1989**, *111*, 2324.

[711]Baumstark; Bechara; Semigran *Tetrahedron Lett.* **1976**, 3265; McMurry; Fleming; Kees; Krepski, Ref. 701.

[712]Baumstark; McCloskey; Witt *J. Org. Chem.* **1978**, *43*, 3609.

[713]McMurry; Miller *J. Am. Chem. Soc.* **1983**, *105*, 1660.

[714]McMurry; Fleming; Kees; Krepski, Ref. 701; Nishida; Kataoka *J. Org. Chem.* **1978**, *43*, 1612; Coe; Scriven *J. Chem. Soc., Perkin Trans. 1* **1986**, 475; Chisholm; Klang, Ref. 710.

[715]McMurry; Fleming; Kees; Krepski, Ref. 701; Dams; Malinowski; Westdorp; Geise *J. Org. Chem.* **1982**, *47*, 248.

[716]For a review, see Bloomfield; Owsley; Nelke *Org. React.* **1976**, *23*, 259-403. For a list of reactions, with references, see Ref. 21, pp. 645-646.

[717]For a review of cyclizations by means of the acyloin condensation, see Finley, *Chem. Rev.* **1964**, *64*, 573-589.

[718]Cope; Herrick *J. Am. Chem. Soc.* **1950**, *72*, 983; Bloomfield; Irelan *J. Org. Chem.* **1966**, *31*, 2017.

successful. The presence of double or triple bonds does not interfere.[719] Even a benzene ring can be present, and many paracyclophane derivatives (**35**) with n = 9 or more have

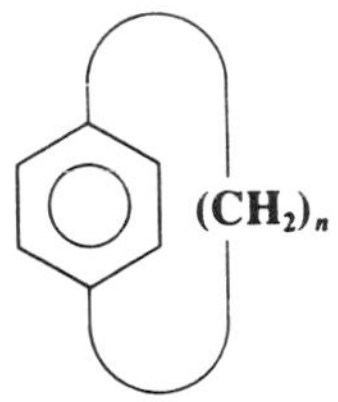

been synthesized in this manner.[720]

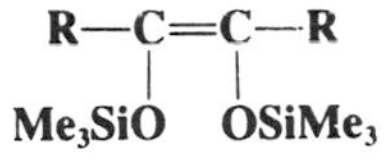

Yields in the acyloin condensation can be improved by running the reaction in the presence of chlorotrimethylsilane Me$_3$SiCl, in which case the dianion **34** is converted to the bis silyl enol ether **36,** which can be isolated and subsequently hydrolyzed to the acyloin with aqueous acid.[721] This is now the standard way to conduct the acyloin condensation. Among other things, this method inhibits the Dieckmann condensation[722] (**0-108**), which otherwise competes with the acyloin condensation when a 5-, 6-, or 7-membered ring can be closed (note that the ring formed by a Dieckmann condensation is always one carbon atom smaller than that formed by an acyloin condensation of the same substrate). The Me$_3$SiCl method is especially good for the closing of four-membered rings.[723] Yields of 4-, 5-, and 6-membered rings are improved by the use of ultrasound.[724]

The mechanism is not known with certainty, but it is usually presumed that the diketone RCOCOR is an intermediate,[725] since small amounts of it are usually isolated as side products, and when it is resistant to reduction (e.g., *t*-Bu—COCO—*t*-Bu), it is the major product. A possible sequence (analogous to that of **9-62**) is

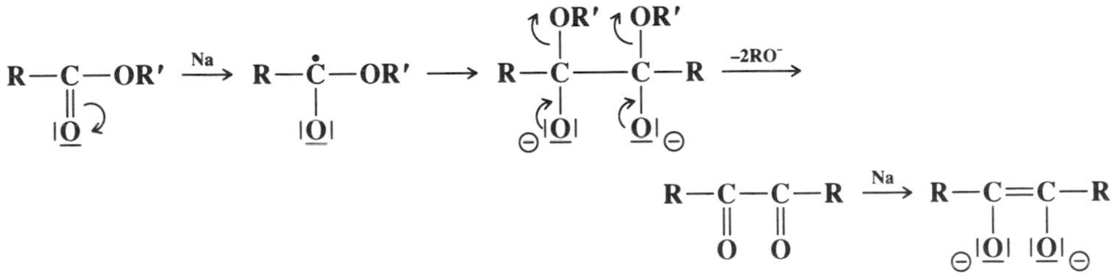

In order to account for the ready formation of large rings, which means that the two ends of the chain must approach each other even though this is conformationally unfavorable for

[719]Cram; Gaston *J. Am. Chem. Soc.* **1960,** *82,* 6386.

[720]For a review, see Cram *Rec. Chem. Prog.* **1959,** *20,* 71.

[721]Schräpler; Rühlmann *Chem. Ber.* **1964,** *97,* 1383. For a review of the Me$_3$SiCl method, see Rühlmann *Synthesis* **1971,** 236-253.

[722]Bloomfield *Tetrahedron Lett.* **1968,** 591.

[723]Bloomfield *Tetrahedron Lett.* **1968,** 587; Gream; Worthley *Tetrahedron Lett.* **1968,** 3319; Wynberg; Reiffers; Strating *Recl. Trav. Chim. Pays-Bas* **1970,** *89,* 982; Bloomfield; Martin; Nelke *J. Chem. Soc., Chem. Commun.* **1972,** 96.

[724]Fadel; Canet; Salaün *Synlett* **1990,** 89.

[725]Another mechanism, involving addition of the ketyl to another molecule of ester (rather than a dimerization of two ketyl radicals), in which a diketone is not an intermediate, has been proposed: Bloomfield; Owsley; Ainsworth; Robertson *J. Org. Chem.* **1975,** *40,* 393.

long chains, it may be postulated that the two ends become attached to nearby sites on the surface of the sodium.

In a related reaction, aromatic carboxylic acids were condensed to α-diketones (2ArCOOH → ArCOCOAr) on treatment with excess Li in dry THF in the presence of ultrasound.[726]

The acyloin condensation was used in an ingenious manner to prepare the first reported catenane (see p. 91).[727] The catenane (**39**) was prepared by a statistical synthesis (p. 91) in the following manner: An acyloin condensation was performed on the diethyl ester of the C_{34} dicarboxylic acid (tetratriacontandioic acid) to give the cyclic acyloin **37**. This was reduced by a Clemmensen reduction with DCl in D_2O instead of HCl in H_2O, thus producing a C_{34} cycloalkane containing deuterium (**38**):[728]

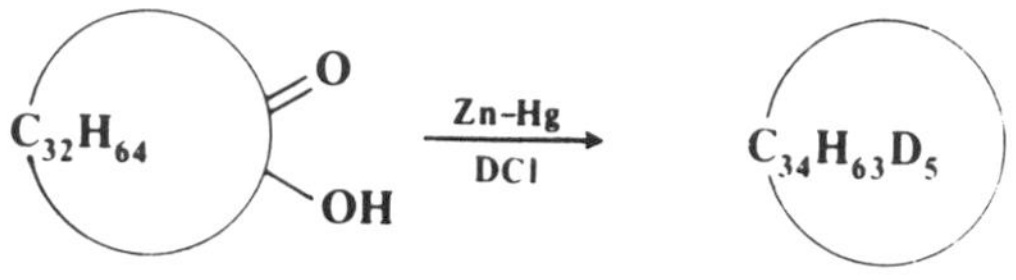

38 contained about five atoms of deuterium per molecule. The reaction was then repeated, this time *in a 1:1 mixture of xylene and* **38** *as solvent*. It was hoped that some of the molecules of ester would be threaded through **38** before they closed:

The first thing that was done with the product was to remove by chromatography the **38** that had been used as the solvent. The remaining material still contained deuterium, as determined by ir spectra, even with all the **38** gone. This was strong evidence that the material consisted not only of **37**, but also of **39**. As further evidence, the mixture was oxidized to open up the acyloin rings (**9-7**). From the oxidation product was isolated the C_{34} diacid (as expected) *containing no deuterium*, and **38**, containing deuterium. The total yield of **39** and **37** was 5 to 20%, but the percentage of **39** in this mixture was only about 1 to 2%.[728] This synthesis of a catenane produced only a small yield and relied on chance, on the probability that a diester molecule would be threaded through **38** before it closed.

Several *directed* syntheses of catenanes have also been reported. One of these relies upon coordination of ligands to a metallic ion to achieve the proper geometry. An example is shown in Figure 19.1.[729] 2,9-Bis(p-hydroxyphenyl)-1,10-phenanthroline **40** is converted to the macrocycle **42** by treatment with the polyether **41**, in a Williamson reaction, under

[726]Karaman; Fry *Tetrahedron Lett.* **1989**, *30*, 6267.
[727]For reviews of the synthesis of catenanes, see Sauvage *Acc. Chem. Res.* **1990**, *23*, 319-327, *Nouv. J. Chim.* **1985**, *9*, 299-310; Dietrich-Buchecker; Sauvage *Chem. Rev.* **1987**, *87*, 795-810.
[728]This work was done by Wasserman *J. Am. Chem. Soc.* **1960**, *82*, 4433. For other statistical syntheses, see Wolovsky *J. Am. Chem. Soc.* **1970**, *92*, 2132; Ben-Efraim; Batich; Wasserman *J. Am. Chem. Soc.* **1970**, *92*, 2133; Agam; Zilkha *J. Am. Chem. Soc.* **1976**, *98*, 5214; Schill; Schweickert; Fritz; Vetter *Angew. Chem. Int. Ed. Engl.* **1983**, *22*, 889 [*Angew. Chem. 95*, 909].
[729]Dietrich-Buchecker; Sauvage; Kintzinger *Tetrahedron Lett.* **1983**, *24*, 5095; Dietrich-Buchecker; Sauvage; Kern *J. Am. Chem. Soc.* **1984**, *106*, 3043; Dietrich-Buchecker; Sauvage *Tetrahedron* **1990**, *46*, 503.

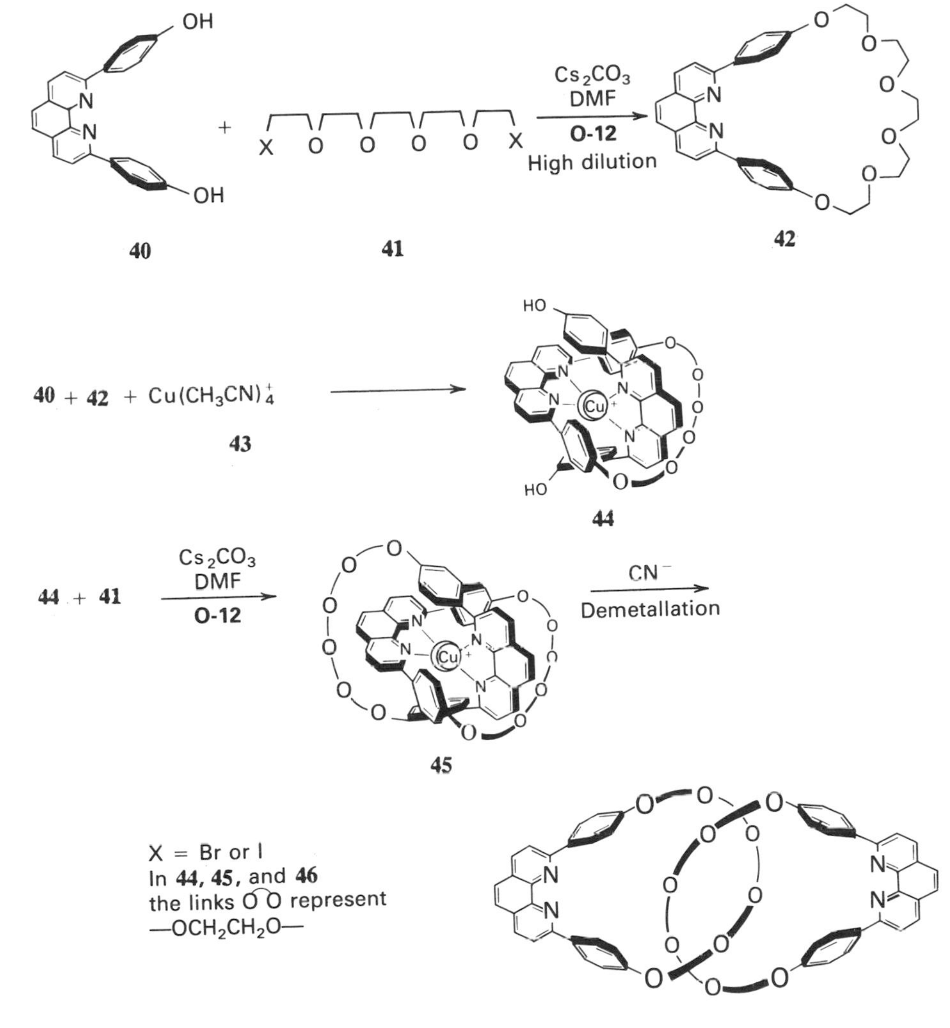

FIGURE 19.1 Synthesis of the catenane **46**.[729]

high dilution conditions. **42**, combined with another molecule of **40**, forms a coordination compound **44** when a mixture of **40** and **42** is treated with a copper complex **43**. The two phenolic OH groups of **44** are in proper positions so that when combined with additional **41**, the interlocking copper complex **45** (called a *catenate*) is formed. In the final step, the copper is removed by treatment with CN⁻ (which preferentially coordinates with Cu⁺) to give the catenane **46**. **45** was obtained in 42% yield from **42**. **45** was also obtained more directly, by treatment of **40** with the complex **43**, which forms another complex in which the two molecules of **40** coordinate with the copper. This, treated with two moles of **41**, generates **45**. A similar strategy was used to prepare [3]catenanes.[730]

[730]Sauvage; Weiss *J. Am. Chem. Soc.* **1985**, *107*, 6108. For other preparations of [3]catenanes, see Dietrich-Buchecker; Hemmert; Khémiss; Sauvage *J. Am. Chem. Soc.* **1990**, *112*, 8002; Ashton et al. *Angew. Chem. Int. Ed. Engl.* **1991**, *30*, 1039 [*Angew. Chem. 103*, 1055]. See also Guilhem; Pascard; Sauvage; Weiss *J. Am. Chem. Soc.* **1988**, *110*, 8711.

Another directed synthesis of catenanes[731] does not use a metallic ion. The key step in this approach[732] was formation of a tertiary amine by **0-43**. Sterically, one of the halide groups of **47** is above the plane, and the other below it, so that ring closure must occur

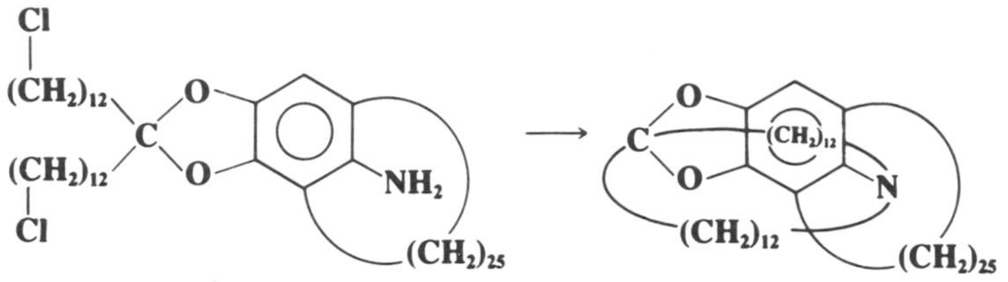

47   **48**

through the 28-membered ring. After **48** was formed, the acetal was cleaved (**0-6**). It was then necessary to cleave the remaining bond holding the two rings together, i.e., the C—N bond. This was done by oxidation to the *ortho*-quinone (**9-4**), which converted the amine function to an enamine, which was hydrolyzable (**6-2**) with acid to give the catenane (**49**):

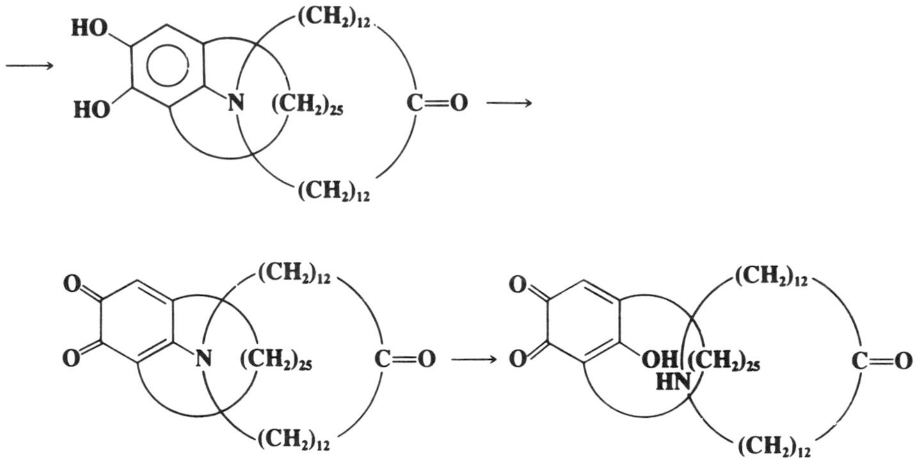

49

OS **II**, 114; **IV**, 840; **VI**, 167.

9-66 Reduction of Nitro to Azoxy Compounds
Nitro-azoxy reductive transformation

$$2ArNO_2 \xrightarrow{Na_3AsO_3} Ar-\overset{\oplus}{N}=N-Ar$$
$$\underset{O_\ominus}{\overset{|}{}}$$

[731]For still others, see Schill; Schweickert; Fritz; Vetter *Chem. Ber.* **1988**, *121*, 961; Ashton; Goodnow; Kaifer; Reddington; Slawin; Spencer; Stoddart; Vicent; Williams *Angew. Chem. Int. Ed. Engl.* **1989**, *28*, 1396 [*Angew. Chem.* *101*, 1404]; Brown; Philp; Stoddart *Synlett* **1991**, 459.
[732]Schill; Lüttringhaus *Angew. Chem. Int. Ed. Engl.* **1964**, *3*, 546 [*Angew. Chem.* *76*, 567]; Schill *Chem. Ber.* **1965**, *98*, 2906, **1966**, *99*, 2689, **1967**, *100*, 2021; Logemann; Rissler; Schill; Fritz *Chem. Ber.* **1981**, *114*, 2245. For the preparation of [3]catenanes by a similar approach, see Schill; Zürcher *Chem. Ber.* **1977**, *110*, 2046; Rissler; Schill; Fritz; Vetter *Chem. Ber.* **1986**, *119*, 1374.

Azoxy compounds can be obtained from nitro compounds with certain reducing agents, notably sodium arsenite, sodium ethoxide, NaTeH,[733] lead,[734] $NaBH_4$–PhTeTePh,[735] and glucose. The most probable mechanism with most reagents is that one molecule of nitro compound is reduced to a nitroso compound and another to a hydroxylamine (**9-49**), and these combine (**2-53**). The combination step is rapid compared to the reduction process.[736] Nitroso compounds can be reduced to azoxy compounds with triethyl phosphite or triphenylphosphine[737] or with an alkaline aqueous solution of an alcohol.[738]

OS **II**, 57.

9-67   Reduction of Nitro to Azo Compounds
N-De-bisoxygen-coupling

$$2ArNO_2 \xrightarrow{\text{LiAlH}_4} Ar—N{=}N—Ar$$

Nitro compounds can be reduced to azo compounds with various reducing agents, of which $LiAlH_4$ and zinc and alkali are the most common. With many of these reagents, slight differences in conditions can lead either to the azo or azoxy (**9-66**) compound. Analogously to **9-66,** this reaction may be looked on as a combination of ArN=O and ArNH$_2$ (**2-52**). However, when the reducing agent was HOCH$_2$CH$_2$ONa[739] or NaBH$_4$,[740] it was shown that azoxy compounds were intermediates. Nitroso compounds can be reduced to azo compounds with $LiAlH_4$.

OS **III**, 103.

9-68   Reduction of Nitro to Hydrazo Compounds
N-Hydrogen-de-bisoxygen-coupling

$$2ArNO_2 \xrightarrow[\text{NaOH}]{\text{Zn}} Ar—NH—NH—Ar$$

Nitro compounds can be reduced to hydrazo compounds with zinc and sodium hydroxide, with hydrazine hydrate and Raney nickel,[741] or with $LiAlH_4$ mixed with a metal chloride such as $TiCl_4$ or VCl_3.[742] The reduction has also been accomplished electrochemically.

Reactions in Which an Organic Substrate is Both Oxidized and Reduced

Some reactions that belong in this category have been considered in earlier chapters. Among these are the Tollens' condensation (**6-46**), the benzil–benzilic acid rearrangement (**8-6**), and the Wallach rearrangement (**8-45**).

9-69   The Cannizzaro Reaction
Cannizzaro Aldehyde Disproportionation

$$2ArCHO \xrightarrow{\text{NaOH}} ArCH_2OH + ArCOO^-$$

[733]Osuka; Shimizu; Suzuki *Chem. Lett.* **1983**, 1373.
[734]Azoo; Grimshaw *J. Chem. Soc. C* **1968**, 2403.
[735]Ohe; Uemura; Sugita; Masuda; Taga *J. Org. Chem.* **1989**, *54*, 4169.
[736]Ogata; Mibae *J. Org. Chem.* **1962**, *27*, 2048.
[737]Bunyan; Cadogan *J. Chem. Soc.* **1963**, 42.
[738]See, for example, Hutton; Waters *J. Chem. Soc. B* **1968**, 191. See also Porta; Pizzotti; Cenini *J. Organomet. Chem.* **1981**, *222*, 279.
[739]Tadros; Ishak; Bassili *J. Chem. Soc.* **1959**, 627.
[740]Hutchins; Lamson; Rufa; Milewski; Maryanoff *J. Org. Chem.* **1971**, *36*, 803.
[741]Furst; Moore *J. Am. Chem. Soc.* **1957**, *79*, 5492.
[742]Olah *J. Am. Chem. Soc.* **1959**, *81*, 3165.

Aromatic aldehydes, and aliphatic ones with no α hydrogen, give the *Cannizzaro reaction* when treated with NaOH or other strong bases.[743] In this reaction one molecule of aldehyde oxidizes another to the acid and is itself reduced to the primary alcohol. Aldehydes with an α hydrogen do not give the reaction, because when these compounds are treated with base the aldol reaction (**6-39**) is much faster.[744] Normally, the best yield of acid or alcohol is 50% each, but this can be altered in certain cases. When the aldehyde contains a hydroxide group in the ring, excess base oxidizes the alcohol formed and the acid can be prepared in high yield (the OH⁻ is reduced to H₂). On the other hand, high yields of alcohol can be obtained from almost any aldehyde by running the reaction in the presence of formaldehyde. In this case the formaldehyde reduces the aldehyde to alcohol and is itself oxidized to formic acid. In such a case, where the oxidant aldehyde differs from the reductant aldehyde, the reaction is called the *crossed Cannizzaro reaction*. The Tollens' condensation (**6-46**) includes a crossed Cannizzaro reaction as its last step. A Cannizzaro reaction run on 1,4-dialdehydes (note that α hydrogens are present here) with a rhodium phosphine complex catalyst gives ring closure, e.g.,[745]

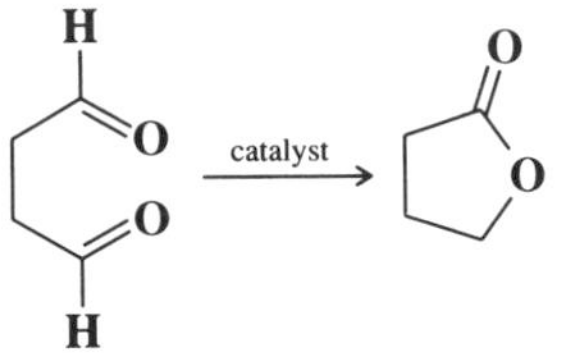

The product is the lactone derived from the hydroxy acid that would result from a normal Cannizzaro reaction.

α-Keto aldehydes give internal Cannizzaro reactions:

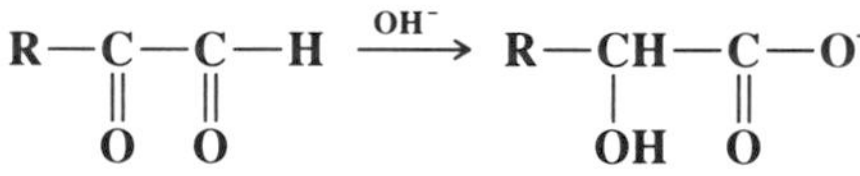

This product is also obtained on alkaline hydrolysis of compounds of the formula RCOCHX₂. Similar reactions have been performed on α-keto acetals[746] and γ-keto aldehydes.

The mechanism[747] of the Cannizzaro reaction[748] involves a hydride shift (an example of mechanism type 2, p. 1160). First OH⁻ adds to the C=O to give **50**, which may lose a proton in the basic solution to give the diion **51**.

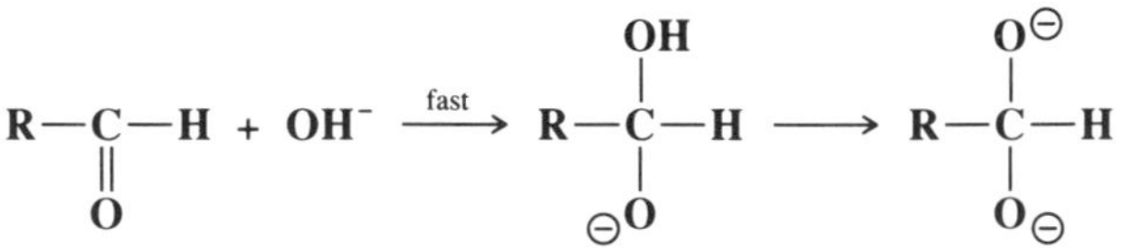

[743]For a review, see Geissman *Org. React.* **1944**, *2*, 94-113.
[744]An exception is cyclopropanecarboxaldehyde: van der Maeden; Steinberg; de Boer *Recl. Trav. Chim. Pays-Bas* **1972**, *91*, 221.
[745]Bergens; Fairlie; Bosnich *Organometallics* **1990**, *9*, 566.
[746]Thompson *J. Org. Chem.* **1967**, *32*, 3947.
[747]For evidence that an SET pathway may intervene, see Ashby; Coleman; Gamasa *J. Org. Chem.* **1987**, *52*, 4079; Fuentes; Marinas; Sinisterra *Tetrahedron Lett.* **1987**, *28*, 2947.
[748]See for example, Swain; Powell; Sheppard; Morgan *J. Am. Chem. Soc.* **1979**, *101*, 3576; Watt *Adv. Phys. Org. Chem.* **1988**, *24*, 57-112, pp. 81-86.

The strong electron-donating character of O^- greatly facilitates the ability of the aldehydic hydrogen to leave with its electron pair. Of course, this effect is even stronger in **51**. When the hydride does leave, it attacks another molecule of aldehyde. The hydride can come from **50** or **51**:

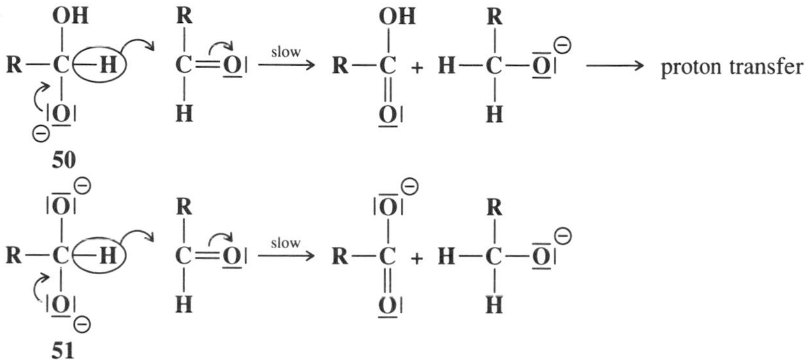

If the hydride ion comes from **50**, the final step is a rapid proton transfer. In the other case, the acid salt is formed directly, and the alkoxide ion acquires a proton from the solvent. Evidence for this mechanism is: (1) The reaction can be first order in base and second order in substrate (thus going through **50**) or, at higher base concentrations, second order in each (going through **51**); and (2) when the reaction was run in D_2O, the recovered alcohol contained no α deuterium,[749] indicating that the hydrogen comes from another mole of aldehyde and not from the medium.[750]

OS **I**, 276; **II**, 590; **III**, 538; **IV**, 110.

9-70 The Tishchenko Reaction
Tishchenko aldehyde-ester disproportionation

$$2RCHO \xrightarrow{\text{Al(OEt)}_3} RCOOCH_2R$$

When aldehydes, with or without α hydrogen, are treated with aluminum ethoxide, one molecule is oxidized and another reduced, as in **9-69**, but here they are found as the ester. The process is called the *Tishchenko reaction*. Crossed Tishchenko reactions are also possible. With more strongly basic alkoxides, such as magnesium or sodium alkoxides, aldehydes with an α hydrogen give the aldol reaction. Like **9-69**, this reaction has a mechanism that involves hydride transfer.[751] The Tishchenko reaction can also be catalyzed[752] by ruthenium complexes,[753] by boric acid,[754] and, for aromatic aldehydes, by disodium tetracarbonylferrate $Na_2Fe(CO)_4$.[755]

OS **I**, 104.

[749]Fredenhagen; Bonhoeffer *Z. Phys. Chem., Abt. A* **1938**, *181*, 379; Hauser; Hamrick; Stewart *J. Org. Chem.* **1956**, *21*, 260.

[750]When the reaction was run at 100°C in MeOH–H_2O, isotopic exchange was observed (the product from PhCDO had lost some of its deuterium): Swain; Powell; Lynch; Alpha; Dunlap *J. Am. Chem. Soc.* **1979**, *101*, 3584. Side reactions were postulated to account for the loss of deuterium. See, however, Chung *J. Chem. Soc., Chem. Commun.* **1982**, 480.

[751]See, for example, Zakharkin; Sorokina *J. Gen. Chem. USSR* **1967**, *37*, 525; Saegusa; Ueshima; Kitagawa *Bull. Chem. Soc. Jpn.* **1969**, *42*, 248; Ogata; Kishi *Tetrahedron* **1969**, *25*, 929.

[752]For a list of reagents, with references, see Ref. 21, p. 840.

[753]Ito; Horino; Koshiro; Yamamoto *Bull. Chem. Soc. Jpn.* **1982**, *55*, 504.

[754]Stapp *J. Org. Chem.* **1973**, *38*, 1433.

[755]Yamashita; Watanabe; Mitsudo; Takegami *Bull. Chem. Soc. Jpn.* **1976**, *49*, 3597.

9-71 The Pummerer Rearrangement

Pummerer methyl sulfoxide rearrangement

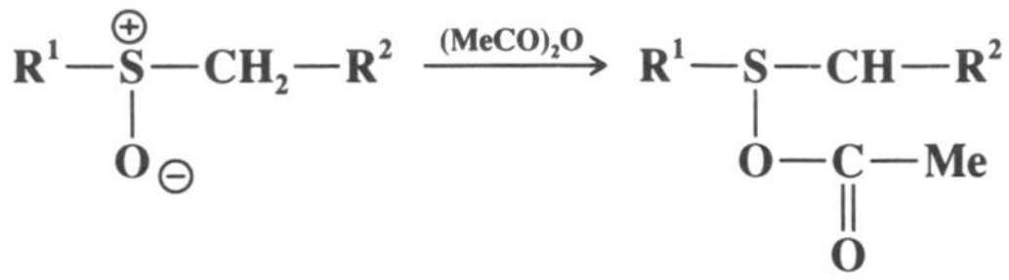

$$R^1-\overset{\oplus}{\underset{O_\ominus}{S}}-CH_2-R^2 \xrightarrow{(MeCO)_2O} R^1-S-\underset{\underset{O}{\overset{|}{C}-Me}}{\overset{|}{C}H}-R^2$$

When sulfoxides bearing an α hydrogen are treated with acetic anhydride, the product is an α-acetoxy sulfide. This is one example of *the Pummerer rearrangement*, in which the sulfur is reduced while an adjacent carbon is oxidized.[756] The product is readily hydrolyzed (**0-6**) to the aldehyde R^2CHO.[757] Besides acetic anhydride, other anhydrides and acyl halides give similar products. Inorganic acids such as HCl also give the reaction, and $RSOCH_2R'$ can be converted to $RSCHClR'$ in this way. Sulfoxides can also be converted to α-halo sulfides[758] by other reagents, including sulfuryl chloride, N-bromosuccinimide, and N-chlorosuccinimide.

The following 4-step mechanism has been proposed for the reaction between acetic anhydride and dimethyl sulfoxide:[759]

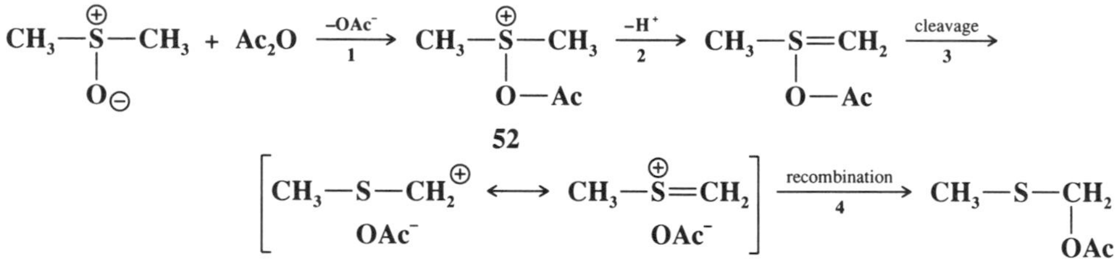

$$CH_3-\overset{\oplus}{\underset{O_\ominus}{S}}-CH_3 + Ac_2O \xrightarrow[1]{-OAc^-} CH_3-\overset{\oplus}{\underset{O-Ac}{S}}-CH_3 \xrightarrow[2]{-H^+} CH_3-\underset{O-Ac}{\overset{||}{S}=CH_2} \xrightarrow[3]{cleavage}$$

$$\underset{52}{}$$

$$\left[CH_3-S-\overset{\oplus}{CH_2} \underset{OAc^-}{\longleftrightarrow} CH_3-\overset{\oplus}{\underset{OAc^-}{S}}=CH_2\right] \xrightarrow[4]{recombination} CH_3-S-\underset{OAc}{\overset{|}{C}H_2}$$

For dimethyl sulfoxide and acetic anhydride, step 4 is intermolecular, as shown by ^{18}O isotopic labeling studies.[760] With other substrates, however, step 4 can be inter- or intramolecular, depending on the structure of the sulfoxide.[761] Depending on the substrate and reagent, any of the first three steps can be rate-determining. In the case of Me_2SO treated with $(F_3CCO)_2O$ the intermediate corresponding to **52**[762] could be isolated at low temperature, and on warming gave the expected product.[763] There is much other evidence for this mechanism.[764]

9-72 The Willgerodt Reaction

Willgerodt carbonyl transformation

$$ArCOCH_3 \xrightarrow{(NH_4)_2S_x} ArCH_2CONH_2 + ArCH_2COO^- NH_4^+$$

[756]For reviews, see De Lucchi; Miotti; Modena *Org. React.* **1991**, *40*, 157-405; Warren *Chem. Ind. (London)* **1980**, 824-828; Oae; Numata *Isot. Org. Chem.* **1980**, *5*, 45-102; Block, Ref. 440, pp. 154-162.

[757]See, for example, Sugihara; Tanikaga; Kaji *Synthesis* **1978**, 881.

[758]For a review of α-chloro sulfides, see Dilworth; McKervey *Tetrahedron* **1986**, *42*, 3731-3752.

[759]See, for example, Numata; Itoh; Yoshimura; Oae *Bull. Chem. Soc. Jpn.* **1983**, *56*, 257.

[760]Oae; Kitao; Kawamura; Kitaoka *Tetrahedron* **1963**, *19*, 817.

[761]See, for example, Itoh; Numata; Yoshimura; Oae *Bull. Chem. Soc. Jpn.* **1983**, *56*, 266; Oae; Itoh; Numata; Yoshimura *Bull. Chem. Soc. Jpn.* **1983**, *56*, 270.

[762]For a review of sulfur-containing cations, see Marino *Top. Sulfur Chem.* **1976**, *1*, 1-110.

[763]Sharma; Swern *Tetrahedron Lett.* **1974**, 1503.

[764]See Block, Ref. 440, pp. 154-156; Oae; Numata, Ref. 756, pp. 48-71; Wolfe; Kazmaier *Can. J. Chem.* **1979**, *57*, 2388, 2397; Russell; Mikol *Mech. Mol. Migr.* **1968**, *1*, 157-207.

In the *Willgerodt reaction* a straight- or branched-chain aryl alkyl ketone is converted to the amide and/or the ammonium salt of the acid by heating with ammonium polysulfide.[765] The carbonyl group of the product is always at the end of the chain. Thus $ArCOCH_2CH_3$ gives the amide and the salt of $ArCH_2CH_2COOH$, and $ArCOCH_2CH_2CH_3$ gives derivatives of $ArCH_2CH_2CH_2COOH$. However, yields sharply decrease with increasing length of chain. The reaction has also been carried out on vinylic and ethynyl aromatic compounds and on aliphatic ketones, but yields are usually lower in these cases. The use of sulfur and a dry primary or secondary amine (or ammonia) as the reagent is called the *Kindler modification* of the Willgerodt reaction.[766] The product in this case is $Ar(CH_2)_nCSNR_2$,[767] which can be hydrolyzed to the acid. Particularly good results are obtained with morpholine as the amine. For volatile amines the HCl salts can be used instead, with NaOAc in DMF at 100°C.[768] Dimethylamine has also been used in the form of dimethylammonium dimethylcarbamate Me_2NCOO $Me_2NH_2^+$.[769] The Kindler modification has also been applied to aliphatic ketones.[770]

Alkyl aryl ketones can be converted to arylacetic acid derivatives in an entirely different manner. The reaction consists of treatment of the substrate with silver nitrate and I_2 or Br_2,[771] or with thallium nitrate, MeOH, and trimethyl orthoformate adsorbed on K-10, an acidic clay.[772]

$$ArCOCH_2R \xrightarrow[I_2 \text{ or } Br_2]{AgNO_3} Ar\overset{\overset{\displaystyle R}{|}}{C}H\!-\!COOCH_3 \qquad R = H, Me, Et$$

The mechanism of the Willgerodt reaction is not completely known, but some conceivable mechanisms can be excluded. Thus, one might suppose that the alkyl group becomes completely detached from the ring and then attacks it with its other end. However, this possibility is ruled out by experiments such as the following: When isobutyl phenyl ketone (**53**) is subjected to the Willgerodt reaction, the product is **54**, not **55**, which would arise if the end carbon of the ketone became bonded to the ring in the product:[773]

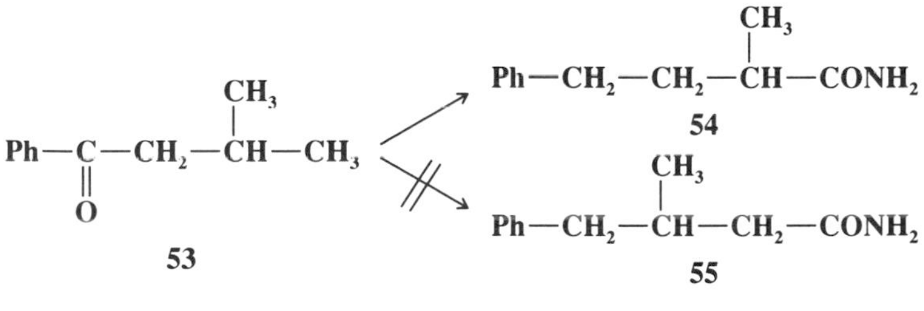

[765]For a review, see Brown *Synthesis* **1975**, 358-375.

[766]For a review, see Mayer, in Oae, Ref. 431, pp. 58-63. For a study of the optimum conditions for this reaction, see Lundstedt; Carlson; Shabana *Acta Chem. Scand., Ser. B* **1987**, *41*, 157, and other papers in this series. See also Carlson; Lundstedt *Acta Chem. Scand., Ser. B* **1987**, *41*, 164.

[767]The reaction between ketones, sulfur, and ammonia can also lead to heterocyclic compounds. For a review, see Asinger; Offermanns *Angew. Chem. Int. Ed. Engl.* **1967**, *6*, 907-919 [*Angew. Chem. 79*, 953-965].

[768]Amupitan *Synthesis* **1983**, 730.

[769]Schroth; Andersch *Synthesis* **1989**, 202.

[770]See Dutron-Woitrin; Merényi; Viehe *Synthesis* **1985**, 77.

[771]Higgins; Thomas *J. Chem. Soc., Perkin Trans. 1* **1982**, 235. See also Higgins; Thomas *J. Chem. Soc., Perkin Trans. 1* **1983**, 1483.

[772]Taylor; Chiang; McKillop; White *J. Am. Chem. Soc.* **1976**, *98*, 6750; Taylor; Conley; Katz; McKillop *J. Org. Chem.* **1984**, *49*, 3840.

[773]King; McMillan *J. Am. Chem. Soc.* **1946**, *68*, 632.

This also excludes a cyclic-intermediate mechanism similar to that of the Claisen rearrangement (**8-35**). Another important fact is that the reaction is successful for singly branched side chains, such as **53,** but not for doubly branched side chains, as in $PhCOCMe_3$.[773] Still another piece of evidence is that compounds oxygenated along the chain give the same products; thus $PhCOCH_2CH_3$, $PhCH_2COMe$, and $PhCH_2CH_2CHO$ all give $Ph-CH_2CH_2CONH_2$.[774] All these facts point to a mechanism consisting of consecutive oxidations and reductions along the chain, though just what form these take is not certain. Initial reduction to the hydrocarbon can be ruled out, since alkylbenzenes do not give the reaction. In certain cases imines[775] or enamines[776] have been isolated from primary and secondary amines, respectively, and these have been shown to give the normal products, leading to the suggestion that they may be reaction intermediates.

[774]For an example of this type of behavior, see Asinger; Saus; Mayer *Monatsh. Chem.* **1967,** *98,* 825.
[775]Asinger; Halcour *Monatsh. Chem.* **1964,** *95,* 24. See also Nakova; Tolkachev; Evstigneeva *J. Org. Chem. USSR* **1975,** *11,* 2660.
[776]Mayer, in Janssen *Organosulfur Chemistry*; Wiley: New York, 1967, pp. 229-232.

Appendix A
THE LITERATURE OF ORGANIC CHEMISTRY

All discoveries in the laboratory must be published somewhere if the information is to be made generally available. A new experimental result that is not published might as well not have been obtained, insofar as it benefits the entire chemical world. The total body of chemical knowledge (called *the literature*) is located on the combined shelves of all the chemical libraries in the world. Anyone who wishes to learn whether the answer to any chemical question is known, and, if so, what the answer is, has only to turn to the contents of these shelves. Indeed the very expressions "is known," "has been done," etc., really mean "has been published." To the uninitiated, the contents of the shelves may appear formidably large, but fortunately the process of extracting information from the literature of organic chemistry is usually not difficult. In this appendix we shall examine the literature of organic chemistry, confining our attention chiefly to the results of laboratory work, rather than those of industrial organic chemistry.[1] The literature can be divided into two broad categories: primary sources and secondary sources. A *primary source* publishes the original results of laboratory investigations. Books, indexes, and other publications that cover material that has previously been published in primary sources are called *secondary sources*. It is because of the excellence of the secondary sources in organic chemistry (especially *Chemical Abstracts* and Beilstein) that literature searching is comparatively not difficult. The two chief kinds of primary source are journals and patents. There are several types of secondary source.

PRIMARY SOURCES

Journals

For well over a hundred years, nearly all new work in organic chemistry (except for that disclosed in patents) has been published in journals. There are thousands of journals that publish chemical papers, in many countries and in many languages. Some print papers covering all fields of science; some are restricted to chemistry; some to organic chemistry; and some are still more specialized. Fortunately for the sanity of organic chemists, the vast majority of important papers in "pure" organic chemistry (as opposed to "applied") are published in relatively few journals, perhaps 50 or fewer. Of course, this is still a large number, especially since some are published weekly and some semimonthly, but it is considerably smaller than the total number of journals (perhaps as high as 10,000) that publish chemical articles.

---

[1]For books on the chemical literature, see Wolman *Chemical Information*, 2nd ed.; Wiley: New York, 1988; Maizell *How to Find Chemical Information*, 2nd ed.; Wiley: New York, 1987; Mellon *Chemical Publications*, 5th ed.; McGraw-Hill: New York, 1982; Skolnik *The Literature Matrix of Chemistry*; Wiley: New York, 1982; Antony *Guide to Basic Information Sources in Chemistry*; Jeffrey Norton Publishers: New York, 1979; Bottle *Use of the Chemical Literature*; Butterworth: London, 1979; Woodburn *Using the Chemical Literature*; Marcel Dekker: New York, 1974. For a three-part article on the literature of organic chemistry, see Hancock *J. Chem. Educ.* **1968**, *45*, 193-199, 260-266, 336-339.

In addition to ordinary papers, there are two other types of publications in which original work is reported: *notes* and *communications*. A note is a brief paper, often without a summary (nearly all papers are published with summaries or abstracts prepared by the author). Otherwise, a note is similar to a paper.[2] Communications (also called *letters*) are also brief and usually without summaries (though some journals now publish summaries along with their communications, a welcome trend). However, communications differ from notes and papers in three respects:

1. They are brief, not because the work is of small scope, but because they are condensed. Usually they include only the most important experimental details or none at all.

2. They are of immediate significance. Journals that publish communications make every effort to have them appear as soon as possible after they are received. Some papers and notes are of great importance, and some are of lesser importance, but all communications are supposed to be of high importance.

3. Communications are preliminary reports, and the material in them may be republished as papers at a later date, in contrast to the material in papers and notes, which cannot be republished.

Although papers (we use the term in its general sense, to cover notes and communications also) are published in many languages, the English-speaking chemist is in a fairly fortunate position. At present well over half of the important papers in organic chemistry are published in English. Not only are American, British, and British Commonwealth journals published almost entirely in English, but so are many others around the world. There are predominantly English-language journals published in Japan, Italy, Czechoslovakia, Sweden, the Netherlands, Israel, and other countries, and even such traditionally German or French journals as *Chemische Berichte, Liebigs Annalen der Chemie*, and *Bulletin de la Société Chimique de France* now publish some papers in English. Most of the articles published in other languages have summaries printed in English also. Furthermore, the second most important language (in terms of the number of organic chemical papers published) is Russian, and most of these papers are available in English translation, though in most cases, six months to a year later. A considerable number of important papers are published in German and French; these are generally not available in translation, so that the organic chemist should have at least a reading knowledge of these languages. An exception is the journal *Angewandte Chemie*, which in 1962 became available in English under the title *Angewandte Chemie International Edition in English*. Of course, a reading knowledge of French and German (especially German) is even more important for the older literature. Before about 1920, more than half of the important chemical papers were in these languages. It must be realized that the original literature is never obsolete. Secondary sources become superseded or outdated, but nineteenth century journals are found in most chemical libraries and are still consulted. Table A.1 presents a list of the more important current journals that publish original papers[3] and communications in organic chemistry. Some of them also publish review articles, book reviews, and other material. Changes in journal title have not been infrequent; footnotes to the table indicate some of the more important, but some of the other journals listed have also undergone title changes.

The primary literature has grown so much in recent years that attempts have been made to reduce the volume. One such attempt is the *Journal of Chemical Research*, begun in 1977. The main section of this journal, called the "Synopses," publishes synopses, which are essentially long abstracts, with references. The full texts of most of the papers are published only in microfiche and miniprint versions. For some years, the American Chemical

---

[2]In some journals notes are called "short communications," an unfortunate practice, because they are not communications as that term is defined in the text.

[3]In Table A.1 notes are counted as papers.

TABLE A.1 A list of the more important current journals that publish original papers in organic chemistry, listed in alphabetical order of *Chemical Abstracts* abbreviations, which are indicated in boldface. Also given are the year of founding, number of issues per year as of 1991, and whether the journal primarily publishes papers (P), communications (C), or both

No.	Name	Papers or communications	Issues per year
1	**Acta Chem**ica **Scand**inavica (1947)	P	10
2	**Angew**andte **Chem**ie (1888)[4]	C[5]	12
3	**Aust**ralian **J**ournal of **Chem**istry (1948)	P	12
4	**Bioorg**anic **Chem**istry (1971)	P[5]	4
5	**Bioorg**anic & **Med**icinal **Chem**istry **Lett**ers (1991)	C	12
6	**Bull**etin of the **Chem**ical **Soc**iety of **Japan** (1926)	P	12
7	**Bull**etin des **Soc**iétés **Chim**ique **Belges** (1887)	P	12
8	**Bull**etin de la **Soc**iété **Chim**ique de **France** (1858)	P[5]	6
9	**Can**adian **J**ournal of **Chem**istry (1929)	PC	12
10	**Carbohydr**ate **Res**earch (1965)	PC	22
11	**Chem**ische **Ber**ichte (1868)[6]	P	12
12	**Chem**istry and **Ind**ustry **(London)** (1923)	C	24
13	**Chem**istry **Lett**ers (1972)	C	12
14	**Chim**ia (1947)	C[5]	12
15	**Coll**ection of **Czech**oslovak **Chem**ical **Commun**ications (1929)	P	12
16	**Dokl**ady **Akad**emii **Nauk SSSR** (1922)[4]	C	36
17	**Gazz**etta **Chim**ica **Ital**iana (1871)	P	12
18	**Helv**etica **Chim**ica **Acta** (1918)	P	8
19	**Heteroat**om **Chem**istry (1990)	P	6
20	**Heterocycles** (1973)	C[5]	12
21	**Int**ernational **J**ournal of **Chem**ical **Kinet**ics (1969)	P	12
22	**Israel J**ournal of **Chem**istry (1963)	P[7]	4
23	**Izv**estiya **Akad**emii **Nauk SSSR, Ser**iya **Khim**icheskaya (1936)[4]	PC	12
24	**J**ournal of the **Am**erican **Chem**ical **Soc**iety (1879)	PC	26
25	**J**ournal of **Chem**ical **Res**earch, **Synop**ses (1977)	P	12
26	**J**ournal of the **Chem**ical **Soc**iety, **Chem**ical **Commun**ications (1965)	C	24
27	**J**ournal of the **Chem**ical **Soc**iety, **Perkin Trans**actions **1**: Organic and Bio-Organic Chemistry (1841)[8]	PC	12
28	**J**ournal of the **Chem**ical **Soc**iety, **Perkin Trans**actions 2: Physical Organic Chemistry (1841)[8]	P	12

[4]These journals are available in English translation; see Table A.2.
[5]These journals also publish review articles regularly.
[6]Former title. **Ber**ichte der deutschen chemischen Gesellschaft.
[7]Each issue of this journal is devoted to a specific topic.
[8]Beginning with 1966 and until 1971, *J. Chem. Soc.* was divided into three sections: *A*, *B*, and *C*. Starting with 1972, Section *B* became *Perkin Trans. 2* and Section *C* became *Perkin Trans. 1*. Section *A* (Physical and Inorganic Chemistry) was further divided into *Faraday* and *Dalton Transactions*.

TABLE A.1 *(Continued)*

No.	Name	Papers or communications	Issues per year
29	Journal of **Fluorine Chem**istry (1971)	PC	12
30	Journal of **Heterocyclic Chem**istry (1964)	PC	12
31	Journal of the **Indian Chem**ical Society (1924)	P	12
32	Journal of **Med**icinal **Chem**istry (1958)	PC	12
33	Journal of **Mol**ecular **Struct**ure (1967)	PC	16
34	Journal of **Organomet**allic **Chem**istry (1963)	PC	48
35	Journal of **Org**anic **Chem**istry (1936)	PC	26
36	Journal of **Photochem**istry and **Photobiol**ogy, **A**: Chemistry (1972)	P	12
37	Journal of **Phys**ical **Org**anic **Chem**istry (1988)	P	12
38	Journal für **Prakt**ische **Chem**ie (1834)	P	6
39	**Khim**iya **Geterotsikl**icheskikh **Soedin**enii (1965)[4]	P	12
40	**Liebigs Ann**alen der **Chem**ie (1832)	P	12
41	**Mendeleev Commun**ications (1991)	C	8
42	**Metalloorg**anicheskaya **Khim**iya (1988)[4]	PC	6
43	**Monatsh**efte für **Chem**ie (1870)	P	12
44	**New J**ournal of **Chem**istry (1977)[9]	P	11
45	**Organometallics** (1982)	PC	12
46	**Org**anic **Mass Spectrom**etry (1968)	PC	12
47	**Org**anic **Prep**arations and **Proced**ures **Int**ernational (1969)	P[5]	6
48	**Photochem**istry and **Photobiol**ogy (1962)	P[5]	12
49	**Pol**ish **J**ournal of **Chem**istry (1921)[10]	PC	12
50	**Pure** and **Appl**ied **Chem**istry (1960)	[11]	12
51	**Rec**ueil des **Trav**aux **Chim**iques des **Pays-Bas** (1882)	PC	12
52	**Res**earch on **Chem**ical **Intermed**iates (1973)[12]	P[5]	6
53	**Sulf**ur **Lett**ers (1982)	C	6
54	**Synlett** (1989)	C[5]	12
55	**Synth**etic **Commun**ications (1971)	C	22
56	**Synthesis** (1969)	P[5]	12
57	**Tetrahedron** (1958)	P[5]	48
58	**Tetrahedron: Asymmetry** (1990)	PC	12
59	**Tetrahedron Lett**ers (1959)	C	52
60	**Zhurnal Obshchei Khim**ii (1869)[4]	PC	12
61	**Zhurnal Organ**icheskoi **Khim**ii (1965)[4]	PC	12

Society journals, including *J. Am. Chem. Soc.* and *J. Org. Chem.*, have provided supplementary material for some of their papers. This material is available from the Microforms and Back Issues Office at the ACS Washington office, either on microfiche or as a photocopy. These practices have not yet succeeded in substantially reducing the total volume of the world's primary chemical literature.

[9]Before 1987 this journal was called **Nouveau J**ournal de **Chim**ie.
[10]Before 1978 this journal was called **Roczniki Chem**ii.
[11]*Pure Appl. Chem.* publishes IUPAC reports and lectures given at IUPAC meetings.
[12]Before 1989 this journal was called **Rev**iews of **Chem**ical **Intermed**iates.

TABLE A.2   Journals from Table A.1 available in English translation. The numbers are keyed to those of Table A.1. The year of first translation is given

2	**Angew**andte **Chem**ie, **International Edition in Engl**ish (1962)
16	**Dokl**ady **Chem**istry **(English Transl**ation) (1956)
23	**Bull**etin of the **Acad**emy of **Sci**ences of the **USSR, Di**vision of **Chem**ical **Sci**ence (1952)
39	**Chem**istry of **Heterocyc**lic **Comp**ounds **(English Transl**ation) (1965)
42	**Organomet**allic **Chem**istry in the **USSR** (1988)
60	Journal of **Gen**eral **Chem**istry of the **USSR** (1949)
61	Journal of **Org**anic **Chem**istry of the **USSR** (1949)

Patents

In many countries, including the United States, it is possible to patent a new compound or a new method for making a known compound (either laboratory or industrial procedures), as long as the compounds are useful. It comes as a surprise to many to learn that a substantial proportion of the patents granted (perhaps 20 to 30%) are chemical patents. Chemical patents are part of the chemical literature, and both U.S. and foreign patents are regularly abstracted by *Chemical Abstracts*. In addition to learning about the contents of patents from this source, chemists may consult the *Official Gazette* of the U.S. Patent Office, which, published weekly and available in many libraries, lists titles of all patents issued that week. Bound volumes of all U.S. patents are kept in a number of large libraries, including the New York Public Library, which also has an extensive collection of foreign patents. Photocopies of any U.S. patent and most foreign patents can be obtained at low cost from the U.S. Patent and Trademark Office, Washington, D.C., 20231. In addition, *Chemical Abstracts* lists, in the introduction to the first issue of each volume, instructions for obtaining patents from 26 countries.

Although patents are often very useful to the laboratory chemist, and no literature search is complete that neglects relevant patents, as a rule they are not as reliable as papers. There are two reasons for this:

1. It is in the interest of the inventor to claim as much as possible. Therefore he or she may, for example, actually have carried out a reaction with ethanol and with 1-propanol, but will claim all primary alcohols, and perhaps even secondary and tertiary alcohols, glycols, and phenols. An investigator repeating the reaction on an alcohol that the inventor did not use may find that the reaction gives no yield at all. In general, it is safest to duplicate the actual examples given, of which most chemical patents contain one or more.

2. Although legally a patent gives an inventor a monopoly, any alleged infringements must be protected in court, and this may cost a good deal of money. Therefore some patents are written so that certain essential details are concealed or entirely omitted. This practice is not exactly cricket, because a patent is supposed to be a full disclosure, but patent attorneys are generally skilled in the art of writing patents, and procedures given are not always sufficient to duplicate the results.

Fortunately, the above statements do not apply to all chemical patents: many make full disclosures and claim only what was actually done. It must also be pointed out that it is not always possible to duplicate the work reported in every paper in a journal. In general, however, the laboratory chemist must be more wary of patents than of papers.

SECONDARY SOURCES

Journal articles and patents contain virtually all of the original work in organic chemistry. However, if this were all—if there were no indexes, abstracts, review articles, and other secondary sources—the literature would be unusable, because it is so vast that no one could hope to find anything in particular. Fortunately, the secondary sources are excellent. There are various kinds and the categories tend to merge. Our classification is somewhat arbitrary.

Listings of Titles

The profusion of original papers is so great that publications that merely list the titles of current papers find much use. Such lists are primarily methods of alerting the chemist to useful papers published in journals that he or she does not normally read. There are two "title" publications covering the whole of chemistry. One of these, *Current Contents Physical, Chemical & Earth Sciences*,[13] which began in 1967 and appears weekly, contains the contents pages of all issues of about 800 journals in chemistry, physics, earth sciences, mathematics, and allied sciences. Each issue contains an index of important words taken from the titles of the papers listed in that issue, and an author index, which, however, lists only the first-named author of each paper. The author's address is also given, so that one may write for reprints. *Current Contents* is also available on computer discs, with "keywords"—words taken from the title and the interior of the paper. The discs can be searched for the keywords, allowing the user to find papers containing specific topics of interest.

The other "title" publication is *Chemical Titles*, published by Chemical Abstracts Service. This biweekly publication, begun in 1961, lists, in English, all titles from more than 700 journals, all in the field of chemistry. The most useful aspect of this publication is the way the titles are given. They are listed in alphabetical order of *every word in the title*, except for such words as "the," "of," "investigation," "synthesis," etc. (each issue contains a list of words prevented from indexing). This means that a title containing seven significant words is listed seven times. These words are also called "keywords". Furthermore, at each listing are given the words that immediately precede and follow the keyword. In the second section of each issue (called the Bibliography) the complete titles and the authors are given. Incidentally, this Bibliography duplicates, for the journals they both cover, the listings in *Current Contents Physical, Chemical, & Earth Sciences*, since the complete contents of journals are given in order of page number. Each issue of *Chemical Titles* has an author index, covering all authors, not just the first author. Addresses are not given.

Abstracts

Listings of titles are valuable, as far as they go, but they do not tell what is in the paper, beyond the implications carried by the titles. From the earliest days of organic chemistry, abstracts of papers have been widely available, often as sections of journals whose principal interests lay elsewhere.[14] At the present time there are only two publications entirely devoted to abstracts covering the whole field of chemistry. One of these, *Referativnyi Zhurnal, Khimiya*, which began in 1953, is published in Russian and is chiefly of interest to Russian-

[13]Title pages of organic chemistry journals are also carried by *Current Contents Life Sciences*, which is a similar publication covering biochemistry and medicine.

[14]For example, *Chem. Ind. (London)* publishes abstracts of papers that appear in other journals. In the past, journals such as *J. Am. Chem. Soc., J. Chem Soc.,* and *Ber.* also did so.

speaking chemists. The other is *Chemical Abstracts*. This publication, which appears weekly, prints abstracts in English of virtually every paper containing original work in pure or applied chemistry published anywhere in the world.[15] Approximately 18,000 journals are covered, in many languages. In addition, *CA* publishes abstracts of every patent of chemical interest from 18 countries, including the United States, United Kingdom, Germany, and Japan, as well as many patents from eight additional countries. *CA* lists and indexes but does not abstract review articles and books. The abstracts currently appear in 80 sections, of which sections 21 to 34 are devoted to organic chemistry, under such headings as Alicyclic Compounds, Alkaloids, Physical Organic Chemistry, Heterocyclic Compounds (One Hetero Atom), etc. Each abstract of a paper begins with a heading that gives (1) the abstract number;[16] (2) the title of the paper; (3) the authors' names as fully as given in the paper; (4) the authors' address; (5) the abbreviated name of the journal (see Table A.1);[17] (6) the year, volume, issue, and page numbers; and (7) the language of the paper. In earlier years *CA* gave the language only if it differed from the language of the journal title. Abstracts of patents begin with the abstract number, title, inventor and company (if any), patent number, patent class number, date patent issued, country of priority, patent application number, date patent applied for, and number of pages in the patent. The body of the abstract is a concise summary of the information in the paper. For many common journals the author's summary (if there is one) is used in *CA* as it appears in the original paper, with perhaps some editing and additional information. Each issue of *CA* contains an author index, a patent index, and an index of keywords taken from the titles and the texts or contexts of the abstracts. The patent index lists all patents in order of number. The same compound or method is often patented in several countries. *CA* abstracts only the first patent, but does list the patent numbers of the duplicated patents in the patent index along with all previous patent numbers that correspond to it. Before 1981 there were separate Patent Number Indexes and Patent Concordances (the latter began in 1963).

At the end of each section of *CA* there is a list of cross-references to related papers in other sections.

Chemical Abstracts is, of course, highly used for "current awareness"; it allows one to read, in one place, abstracts of virtually all new work in chemistry, though its large size puts a limit on the extent of this type of usefulness.[18] *CA* is even more useful as a repository of chemical information, a place for finding out what was done in the past. This value stems from the excellent indexes, which enable the chemist in most cases to ascertain quickly where information is located. From the time of its founding in 1907 until 1961, *CA* published annual indexes. Since 1962 there are two volumes published each year, and a separate index is issued for each volume. For each volume there is an index of subjects, authors, formulas, and patent numbers. Beginning in 1972 the subject index has been issued in two parts, a chemical substance index and a general subject index, which includes all entries that are not the names of single chemical substances. However, the indexes to each volume become essentially superseded as collective indexes are issued. The first collective indexes are ten-year (decennial) indexes, but the volume of information has made five-year indexes necessary since 1956. Collective indexes so far published are shown in Table A.3. Thus a user of the indexes at the time of this writing would consult the collective indexes through 1986 and the semiannual indexes thereafter. The 12th collective index (covering 1987 through 1991) is scheduled to appear in 1992.

---

[15]For a guide to the use of *CA*, see Schulz *From CA to CAS ONLINE*; VCH: New York, 1988.
[16]Beginning in 1967. See p. 1247.
[17]These abbreviations are changed from time to time. Therefore the reader may notice inconsistencies.
[18]It is possible to subscribe to *CA Selects*, which provides copies of all abstracts within various narrow fields, such as organofluorine chemistry, organic reaction mechanisms, organic stereochemistry, etc.

TABLE A.3   *CA* collective indexes so far published

Coll. index	Subject General subject	Chemical substance	Author	Formula	Patents
1	1907-1916		1907-1916		
2	1917-1926		1917-1926		1907-1936
3	1927-1936		1927-1936	1920-1946	
4	1937-1946		1937-1946		1937-1946
5	1947-1956		1947-1956	1947-1956	1947-1956
6	1957-1961		1957-1961	1957-1961	1957-1961
7	1962-1966		1962-1966	1962-1966	1962-1966
8	1967-1971		1967-1971	1967-1971	1967-1971
9	1972-1976	1972-1976	1972-1976	1972-1976	1972-1976
10	1977-1981	1977-1981	1977-1981	1977-1981	1977-1981
11	1982-1986	1982-1986	1982-1986	1982-1986	1982-1986

Beginning with the eighth collective index period, *CA* has published an *Index Guide*. This publication gives structural formulas and/or alternate names for thousands of compounds, as well as many other cross-references. It is designed to help the user efficiently and rapidly to find *CA* references to subjects of interest in the general subject, formula, and chemical substance indexes. Each collective index contains its own *Index Guide*. A new *Index Guide* is issued every 18 months. The *Index Guide* is necessary because the *CA* general subject index is a "controlled index", meaning it restricts its entries only to certain terms. For example, anyone who looks for the term "refraction" in the general subject index will not find it. The *Index Guide* includes this term, and directs the reader to "Electromagnetic wave, refraction of", "Sound and ultrasound, refraction of", and other terms, all of which will be found in the general subject index. Similarly, the chemical substance index usually lists a compound only under one name—the approved *CA* name. Trivial and other names will be found in the *Index Guide*. For example, the term *"methyl carbonate"* is not in the chemical substance index, but the *Index Guide* does have this term, and tells us to look for it in the chemical substance index under the headings "Carbonic acid, esters, dimethyl ester" (for Me_2CO_3) and "Carbonic acid, esters, monomethyl ester" (for $MeHCO_3$). Furthermore, the *Index Guide* gives terms related to the chosen term, helping users to broaden a search. For example, one who looks for "Atomic orbital" in the *Index Guide* will find the terms "Energy level", "Molecular orbital", "Atomic integral", and "Exchange, quantum mechanical, integrals for", all of which are controlled index terms.

Along with each index (annual, semiannual, or collective) appears an index of ring systems. This valuable index enables the user to ascertain immediately if any ring system appears in the corresponding subject or chemical substance index and under what names. For example, someone wishing to determine whether any compounds containing this ring system

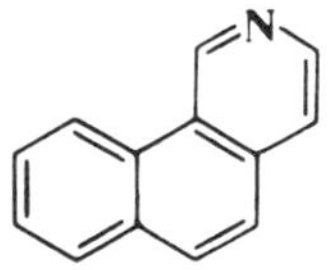

Benz(*h*)isoquinoline

are reported in the 1982-1986 collective index (even if he or she did not know the name) would locate, under the heading "3-ring systems," the listing **6, 6, 6** (since the compound has three rings of six members each), under which he or she would find the sublisting C_5N—C_6—C_6 (since one ring contains five carbons and a nitrogen while the others are all-carbon), under which is listed the name benz(h)isoquinoline, as well as the names of 30 other systems C_5N—C_6—C_6. A search of the chemical substance index under these names will give all references to these ring systems that have appeared in *CA* from 1982 to 1986.

Before 1967, *CA* used a two-column page, with each column separately numbered. A row of letters from *a* to *h* appeared down the center of the page. These letters are for the guidance of the user. Thus an entry 7337*b* refers to the *b* section of column 7337. In early years superscript numbers, e.g., 4327^5, were used in a similar manner. In very early years these numbers were not printed on the page at all, though they are given in the decennial indexes, so that the user must mentally divide the page into nine parts. Beginning with 1967, abstracts are individually numbered and column numbers are discarded. Therefore, beginning with 1967, index entries give abstract number rather than column number. The abstract numbers are followed by a letter that serves as a check character to prevent miscopying errors in computer handling. To use the *CA* general subject, chemical substance, and formula indexes intelligently requires practice, and the student should become familiar with representative volumes of these indexes and with the introductory sections to them, as well as with the *Index Guides*.

In the *CA* formula indexes formulas are listed in order of (1) number of carbon atoms; (2) number of hydrogen atoms; (3) other elements in alphabetic order. Thus, all C_3 compounds are listed before any C_4 compound; all C_5H_7 compounds before any C_5H_8 compound; $C_7H_{11}Br$ before $C_7H_{11}N$; $C_9H_6N_4S$ before C_9H_6O, etc. Deuterium and tritium are represented by D and T and treated alphabetically, e.g., C_2H_5DO after C_2H_5Cl and before C_2H_5F or C_2H_6.

Since 1965, *CA* has assigned a Registry Number to each unique chemical substance. This is a number of the form [766-51-8] that remains invariant, no matter what names are used in the literature. More than 10 million numbers have already been assigned and thousands are added each week. Registry Numbers are primarily for computer use. All numbers so far have been published with the *CA* preferred names in a multivolume "Registry Handbook."

For abstracts printed since 1967 (the eighth collective period and later) *CA* can be searched by computer online. For a discussion of online searching see pp. 1260-1266.

Although *CA* and *Referativnyi Zhurnal, Khimya* are currently the only chemical abstracting publications that cover the entire field of chemistry, there were a number of earlier abstracting publications now defunct. The most important are *Chemisches Zentralblatt* and *British Abstracts*. These publications are still valuable because they began before *CA* and can therefore supply abstracts for papers that appeared before 1907. Furthermore, even for papers published after 1907, *Zentralblatt* and *British Abstracts* are often more detailed. *Zentralblatt* was published, under various names, from 1830 to 1969.[19] *British Abstracts* was a separate publication from 1926 to 1953, but earlier abstracts from this source are available in the *Journal of the Chemical Society* from 1871 to 1925.

Beilstein

This publication is so important to organic chemistry that it deserves a section by itself. Beilstein's "Handbuch der organischen Chemie," usually referred to as *Beilstein*, lists all the known organic compounds reported in the literature during its period of coverage. For

---

[19]An "obituary" of *Zentralblatt* by Weiske, which gives its history and statistical data about its abstracts and indexes, was published in the April 1973 issue of *Chem. Ber.* (pp. I-XVI).

each compound are given: all names; the molecular formula; the structural formula; all methods of preparation (briefly, e.g., "by refluxing 1-butanol with NaBr and sulfuric acid"); physical constants such as melting point, refractive index, etc.; other physical properties; chemical properties including reactions; occurrence in nature (i.e., which species it was isolated from); biological properties, if any; derivatives with melting points; analytical data, and any other information that has been reported in the literature.[20] Equally important, for every piece of information, a reference is given to the original literature. Furthermore, the data in Beilstein have been critically evaluated. That is, all information is carefully researched and documented, and duplicate and erroneous results are eliminated. Some compounds are discussed in two or three lines and others require several pages. The value of such a work should be obvious.

The first three editions of Beilstein are obsolete. The fourth edition (*vierte Auflage*) covers the literature from its beginnings through 1909. This edition, called *das Hauptwerk*, consists of 27 volumes. The compounds are arranged in order of a system too elaborate to discuss fully here.[21] The compounds are divided into three divisions which are further subdivided into "systems":

Division	Volumes	System numbers
I. Acyclic compounds	1–4	1–449
II. Carbocyclic compounds	5–16	450–2359
III. Heterocyclic compounds	17–27	2360–4720

Das Hauptwerk is still the basis of Beilstein and has not been superseded. The later literature is covered by supplements that have been arranged to parallel *das Hauptwerk*. The same system is used, so that the compounds are treated in the same order. The first supplement (*erstes Ergänzungswerk*) covers 1910-1919; the second supplement (*zweites Ergänzungwerk*) covers 1920-1929; the third supplement (*drittes Ergänzungswerk*) covers 1930-1949; the fourth supplement (*viertes Ergänzungswerk*) covers 1950-1959, and the fifth supplement covers 1960-1979. Like *das Hauptwerk*, each supplement contains 27 volumes,[22] except that supplements 3 and 4 are combined for vols. 17 to 27, so that for these volumes the combined third and fourth supplement covers the years 1930-1959. Each supplement has been divided into volumes in the same way as *das Hauptwerk*, and, for example, compounds found in vol. 3, system number 199 of *das Hauptwerk* will also be found in vol. 3, system number 199 of each supplement. To make cross-referencing even easier, each supplement gives, for each compound, the page numbers at which the same compound can be found in the earlier books. Thus, on page 554 of vol. 6 of the fourth supplement, under the listing phenetole are found the symbols (H 140; E I 80; E II 142; E III 545) indicating that earlier information on phenetole is given on page 140 of vol. 6 of *das Hauptwerk*, on page 80 of the first, page 142 of the second, and page 545 of the third supplement. Furthermore, each page of the

[20]For a discussion of how data are processed for inclusion in Beilstein, see Luckenbach; Ecker; Sunkel *Angew. Chem. Int. Ed. Engl.* **1981**, *20*, 841-849 [*Angew. Chem. 93*, 876-885].

[21]For descriptions of the Beilstein system and directions for using it, see Sunkel; Hoffmann; Luckenbach *J. Chem. Educ.* **1981**, *58*, 982; Luckenbach *CHEMTECH* **1979**, 612-621. The Beilstein Institute has also published two English-language guides to the system. One, available free, is *How to Use Beilstein*; Beilstein Institute: Frankfurt/Main, 1979. The other is by Weissbach *A Manual for the Use of Beilstein's Handbuch der Organischen Chemie*; Springer: New York, 1976. An older work, which many students will find easier to follow, is by Huntress *A Brief Introduction to the Use of Beilstein's Handbuch der Organischen Chemie*, 2nd ed.; Wiley: New York, 1938.

[22]In some cases, to keep the system parallel and to avoid books that are too big or too small, volumes are issued in two or more parts, and, in other cases, two volumes are bound as one.

supplements contains, at the top center, the corresponding page numbers of *das Hauptwerk*. Since the same systematic order is followed in all six series, location of a compound in any one series gives its location in the other five. If a compound is found, for example, in vol. 5 of *das Hauptwerk*, one has but to note the page number and scan vol. 5 of each supplement until that number appears in the top center of the page (the same number often covers several pages). Of course, many compounds are found in only one, two, three, four, or five of the series, since no work may have been published on that compound during a particular period covered.

From *das Hauptwerk* to the fourth supplement, Beilstein is in German, though it is not difficult to read since most of the words are the names of compounds (a Beilstein German–English Dictionary, available free from the publisher, is in many libraries). For the fifth supplement (covering 1960-1979), which is in English, publication of Division III began before the earlier divisions. At the time of this writing, vols. 17 to 22 (totaling 70 separate parts exclusive of index volumes) of this supplement have been published, as well as a combined index for volumes 17-19. This index covers only the fifth supplement. The subject portion of this index, which lists compound names only, gives these names in English.

Volumes 28 and 29 of Beilstein are subject and formula indexes, respectively. The most recent complete edition of these volumes is part of the second supplement and covers only *das Hauptwerk* and the first two supplements (though complete indexes covering *das Hauptwerk* and the first four supplements have been announced to appear in the next few years). For vol. 1 there is a cumulative subject and a cumulative formula index, which combine *das Hauptwerk* and the first four supplements.[23] Similar index volumes, covering all four supplements, have been issued for the other volumes, 2 to 27. Some of these are combined, e.g., 2-3, 12-14, and 23-25. For English-speaking chemists (and probably for many German-speaking chemists) the formula indexes are more convenient. Of course (except for the fifth supplement indexes), one must still know some German, because most formula listings contain the names of many isomers. If a compound is found only in *das Hauptwerk*, the index listing is merely the volume and page numbers, e.g., **1**, 501. Roman numbers are used to indicate the supplements, for example, **26**, 15, I 5, II 7. Thus the subject and formula indexes lead at once to locations in *das Hauptwerk* and the first four supplements. The Beilstein formula indexes are constructed the same way as the *CA* indexes (p. 1247).

There is also a fourth division of Beilstein (systems 4721 to 4877) that covers natural products of uncertain structure: rubbers, sugars, etc. These are treated in vols. 30 and 31, which do not go beyond 1935 and which are covered in the collective indexes. These volumes will not be updated. All such compounds are now included in the regular Beilstein volumes.

Like *CA*, Beilstein is available online.

Compendia and Tables of Information

In addition to Beilstein, there are many other reference works in organic chemistry that are essentially compilations of data. These books are very useful and often save the research worker a great deal of time. In this section we discuss some of the more important of such works.

1. The fifth edition of "Heilbron's Dictionary of Organic Compounds," J. Buckingham, Ed., 7 vols., Chapman and Hall, London, 1982, contains brief listings of more than 150,000

---

[23]Most page number entries in the combined indexes contain a letter, e.g., $CHBr_2Cl$ 67f, II 33a, III 87d, IV, 81. These letters tell where on the page to find the compound and are useful because the names given in the index are not necessarily those used in the earlier series. The letter "a" means the compound is the first on its page, "b" is the second, etc. No letters are given for the fourth supplement.

organic compounds, giving names, structural formulas, physical properties, and derivatives, with references. For many entries additional data concerning occurrence, biological activity, and toxicity hazard information are also given. The arrangement is alphabetical. The dictionary contains indexes of names, formulas, hetero atoms, and *CA* Registry Numbers. Annual supplements, with cumulative indexes, have appeared since 1983. A similar work, devoted to organometallic compounds, is "Dictionary of Organometallic Compounds," 3 vols. with supplements, published by Chapman and Hall beginning in 1984. Another, "Dictionary of Steroids," 2 vols., 1991, is also published by Chapman and Hall.

2. A multivolume compendium of physical data is Landolt-Börnsteins's "Zahlenwerte und Funktionen aus Physik, Chemie, Astronomie, Geophysik, und Technik," 6th ed., Springer, Berlin, 1950-. There is also a "New Series," for which the volumes are given the English title "Numerical Data and Functional Relationships in Science and Technology," as well as the German title. This compendium, which is not yet complete, lists a great deal of data, some of which are of interest to organic chemists, e.g., indexes of refraction, heats of combustion, optical rotations, and spectral data. Literature references are given for all data.

3. "The Handbook of Chemistry and Physics," CRC Press, Boca Raton, FL (called the "rubber handbook"), which is revised annually (71st ed., 1990-91), is a valuable repository of data quickly found. For organic chemists the most important table is "Physical Constants of Organic Compounds," which lists names, formulas, color, solubilities, and physical properties of thousands of compounds. However, there are many other useful tables. A similar work is Lange's "Handbook of Chemistry," 13th ed., McGraw-Hill, New York, 1985. Another such handbook, but restricted to data of interest to organic chemists, is Dean, "Handbook of Organic Chemistry," McGraw-Hill, New York, 1987. This book also contains a long table of "Physical Constants of Organic Compounds," and has much other information including tables of thermodynamic properties, spectral peaks, pK_a values, bond distances, and dipole moments.

4. A list of most of the known natural compounds, e.g., terpenes, alkaloids, carbohydrates, to which structures have been assigned, along with structural formulas, melting points, optical rotations, and references, is provided in Devon and Scott, "Handbook of Naturally Occurring Compounds," 3 vols., Academic Press, New York, 1972.

5. Dreisbach, "Physical Properties of Chemical Compounds," Advances in Chemistry Series nos. 15, 22, 29, American Chemical Society, Washington, 1955-1961 lists many physical properties of more than 1000 organic compounds.

6. Physical properties of thousands of organometallic compounds, with references, are collected in four large compendia: the "Dictionary of Organometallic Compounds," mentioned under item 1, above; Dub, "Organometallic Compounds," 2nd ed., 3 vols. with supplements and index, Springer, New York, 1966-1975; Hagihara, Kumada, and Okawara, "Handbook of Organometallic Compounds," W. A. Benjamin, New York, 1968; and Kaufman, "Handbook of Organometallic Compounds," Van Nostrand, Princeton, NJ, 1961.

7. The "Merck Index," 11th ed., Merck and Company, Rahway, NJ, 1989, is a good source of information about chemicals of medicinal importance. Many drugs are given three types of name: *chemical name* (which is the name an organic chemist would give it; of course, there may well be more than one); *generic name*, which must be placed on all containers of the drug; and *trade names*, which are different for each company that markets the drug. For example, the generic name for 1-(4-chlorobenzhydryl)-4-methylpiperazine is chlorcyclazine. Among the trade names for this drug, which is an antihistamine, are Trihistan, Perazyl, and Alergicide. The "Merck Index" is especially valuable because it gives all known names of all three types for each compound and the names are cross-indexed. Also given, for each compound, are the structural formula, *CA* preferred name and Registry Number, physical properties, medicinal and other uses, toxicity indications, and references to methods of synthesis. There are indexes of formulas and Registry Numbers, and miscellaneous tables. The 10th edition of the "Merck Index" (1983) also includes a lengthy list of organic name reactions, with references, but the 11th edition omits this list.

8. There are two publications that list properties of azeotropic mixtures. Timmermans, "The Physico-Chemical Constants of Binary Systems in Concentrated Solutions," 4 vols., Interscience, New York, 1959-1960, is by far the more comprehensive. The other is "Azeotropic Data," 2 vols., Advances in Chemistry Series no. 6 and no. 35, American Chemical Society, Washington, 1952, 1962.

9. Thousands of dipole moments, with references, are collected in McClellan, "Tables of Experimental Dipole Moments," vol. 1, W.H. Freeman, San Francisco, CA, 1963; vol. 2, Rahara Enterprises, El Cerrita, CA, 1974.

10. "Tables of Interatomic Distances and Configurations in Molecules and Ions," London Chemical Society Special Publication no. 11, 1958, and its supplement, Special Publication no. 18, 1965, include bond distances and angles for hundreds of compounds, along with references.

11. The "Ring Systems Handbook," published in 1988 by the Chemical Abstracts Service, provides the names and formulas of ring and cage systems that have been published in *CA*. The ring systems are listed under a system essentially the same as that used for the *CA* index of ring systems (p. 1246). Each entry gives the *CA* index name and Registry Number for that ring system. In many cases a *CA* reference is also given. There is a separate Formula Index (for the parent ring systems) and a Ring Name Index. Cumulative supplements are issued twice a year. The "Ring Systems Handbook" supersedes earlier publications called "The Parent Compound Handbook" and "The Ring Index".

12. The Sadtler Research Laboratories publish large collections of ir, uv, nmr, and other spectra, in loose-leaf form. Indexes are available.

13. Infrared, uv, nmr, Raman, and mass spectral data, as well as melting-point, boiling-point, solubility, density, and other data for more than 30,000 organic compounds are collected in the "CRC Handbook of Data on Organic Compounds," 2nd ed., 9 vols., CRC Press, Boca Raton, FL, 1988, edited by Weast and Grasselli. It differs from the Sadtler collection in that the data are given in tabular form (lists of peaks) rather than reproduction of the actual spectra, but this book has the advantage that all the spectral and physical data for a given compound appear at one place. References are given to the Sadtler and other collections of spectra. Volumes 7 to 9 contain indexes of spectral peaks for ir, uv, nmr, ^{13}C nmr, mass, and Raman spectra, as well as indexes of other names, molecular formulas, molecular weights, and physical constants. Annual updates began appearing in 1990 (the first one is called volume 10).

14. The "Aldrich Library of Infrared Spectra," 3rd ed., Aldrich Chemical Company, Milwaukee, WI, 1981, by Pouchert contains more than 12,000 ir spectra so arranged that the user can readily see the change that takes place in a given spectrum when a slight change is made in the structure of a molecule. The same company also publishes the "Aldrich Library of FT-IR Spectra" and the "Aldrich Library of NMR Spectra", both also by Pouchert. A similar volume, which has ir and Raman spectra of about 1000 compounds, is "Raman/Infrared Atlas of Organic Compounds," 2nd ed., VCH, New York, 1989, by Schrader.

15. An extensive list of visible and uv peaks is given in "Organic Electronic Spectral Data," Wiley, New York. Twenty-six volumes have appeared so far, covering the literature through 1984.

16. A collection of 500 ^{13}C nmr spectra is found in Johnson and Jankowski, "Carbon-13 NMR Spectra," Wiley, New York, 1972.

Reviews

A review article is an intensive survey of a rather narrow field; e.g., the titles of some recent reviews are "Preparation, Properties, and Reactions of Carbonyl Oxides,"[24] "Enantiose-lective Addition of Organometallic Reagents to Carbonyl Compounds: Chirality Transfer, Multiplication, and Amplification,"[25] 1,3-Dipolar Cycloadditions of Diazoalkanes to some Nitrogen Containing Heteroaromatic Systems,"[26] and "Alkyl and Aryl-Substituted Main-Group Metal Amides."[27] A good review article is of enormous value, because it is a thorough survey of all the work done in the field under discussion. Review articles are printed in review journals and in certain books. The most important review journals in organic chem-istry (though most are not exclusively devoted to organic chemistry) are shown in Table A.4. Some of the journals listed in Table A.1, for example, the *Bull Soc. Chim. Fr.* and *J. Organomet. Chem.* also publish occasional review articles.

There are several open-ended serial publications that are similar in content to the review journals but are published irregularly (seldom more often than once a year) and are hard-bound. Some of these publish reviews in all fields of chemistry; some cover only organic chemistry; some specialize further. The coverage is indicated by the titles. Table A.5 shows some of the more important such publications, with *CA* abbreviations.

There are several publications that provide listings of review articles in organic chemistry. The most important is the *J. Org. Chem.*, which began to list review articles in 1978 (the first list is at *J. Org. Chem. 43*, 3085), suspended the listings in 1985, and resumed them in 1990 (at *J. Org. Chem. 55*, 398). These lists, which appear about four times a year, give the titles and reference sources of virtually all review articles in the field of organic chemistry that have appeared in the preceding three months, including those in the review journals and serials mentioned above, as well as those in monographs and treatises. There is also a listing of new monographs on a single subject. Each list includes a subject index.

TABLE A.4　Review journals, with year of founding and issues per year as of 1991

Accounts of Chemical Research (1968)	12
Aldrichimica Acta (1968)	4
Angewandte Chemie (1888)	12
and its English Translation:	
Angewandte Chemie, International Edition in English (1962)	12
Chemical Reviews (1924)	8
Chemical Society Reviews (1947)[28]	4
Heterocycles (1973)	12
Natural Product Reports (1984)	6
Soviet Scientific Reviews, Section B, Chemistry Reviews (1979)	Irreg.
Sulfur Reports (1980)	6
Synthesis (1969)	12
Tetrahedron (1958)	48
Topics in Current Chemistry (1949)[29]	Irreg.
Uspekhi Khimii (1932)	12
and its English translation: Russian Chemical Reviews (1960)	12

[24]Bunnelle *Chem. Rev.* **1991**, *91*, 335-362.
[25]Noyori; Kitamura *Angew. Chem. Int. Ed. Engl.* **1991**, *30*, 49-69 [*Angew. Chem. 103* 34-55].
[26]Stanovnik *Tetrahedron* **1991**, *47*, 2925-2945.
[27]Veith *Adv. Organomet. Chem.* **1990**, *31*, 269-300.
[28]Successor to *Quarterly Reviews* (abbreviated as *Q. Rev., Chem. Soc.*).
[29]Formerly called **Fortschritte der Chemischen Forschung**.

TABLE A.5   Irregularly Published Serial Publications

Advances in **Carbocation Chem**istry	**Fortshritte** der **Chem**ie **Org**anischer
Advances in **Carbohydr**ate **Chem**istry and	**Naturst**offe
Biochemistry	**Iso**topes in **Org**anic **Chem**istry
Advances in **Catal**ysis	**Mol**ecular **Structure** and **Energ**etics
Advances in **Cycload**dition	**Organic Photochem**istry
Advances in **Free Radical Chem**istry	**Organomet**allic **Reac**tions
Advances in **Heterocyc**lic **Chem**istry	**Organic Reac**tions
Advances in **Metal-Org**anic **Chem**istry	**Organic Synth**esis: **Theory** and **App**lications
Advances in **Mol**ecular **Model**ing	**Prog**ress in **Heterocyc**lic **Chem**istry
Advances in **Organomet**allic **Chem**istry	**Prog**ress in **Macrocyc**lic **Chem**istry
Advances in **Oxygen**ated **Processes**	**Prog**ress in **Physical Org**anic **Chem**istry
Advances in **Photochem**istry	**Reac**tive **Intermediates** (Plenum)
Advances in **Physical Org**anic **Chem**istry	**Reac**tive **Intermediates** (Wiley)
Advances in **Protein Chem**istry	**Surv**ey of **Prog**ress in **Chem**istry
Advances in **Theor**etically **Interesting**	**Top**ics in **Physical Organomet**allic **Chem**istry
Molecules	**Top**ics in **Stereochem**istry
Fluorine Chemistry **Rev**iews	

Another publication is the "Index of Reviews in Organic Chemistry," complied by Lewis, Chemical Society, London, a classified listing of review articles. The first volume, published in 1971, lists reviews from about 1960 (in some cases much earlier) to about 1970 in alphabetical order of topic. Thus four reviews are listed under "Knoevenagel condensation," five under "Inclusion compounds," and one under "Vinyl ketones." There is no index. A second volume (1977) covers the literature to 1976. Annual or biannual supplements appeared from 1979 until the publication was terminated in 1985. Classified lists of review articles on organometallic chemistry are found in articles by Smith and Walton[30] and by Bruce.[31] A similar list for heterocyclic chemistry is found in articles by Katritzky and others.[32] See also the discussion of the *Index of Scientific Reviews*, p. 1267.

Annual Reviews

The review articles discussed in the previous section are each devoted to a narrow topic covering the work done in that area over a period of years. An annual review is a publication that covers a broad area but limits the period covered, usually to 1 or 2 years.

1. The oldest annual review publication still publishing is *Annual Reports on the Progress of Chemistry*, published by the Royal Society of Chemistry (formerly the Chemical Society), which began in 1905 and which covers the whole field of chemistry. Since 1967 it has been divided into sections. Organic chemistry is found in Section B.

2. Because the number of papers in chemistry has become so large, the Royal Society of Chemistry publishes annual-review-type volumes of smaller scope, called *Specialist Periodical Reports*. Among those of interest to organic chemists are "Carbohydrate Chemistry" (vol. 22 covers 1988); "Photochemistry" (vol. 21 covers 1988-1989); and "General and Synthetic Methods," (vol. 12 covers 1987).

[30]Smith; Walton *Adv. Organomet. Chem.* **1975**, *13*, 453-558.
[31]Bruce *Adv. Organomet. Chem.* **1972**, *10*, 273-346, **1973**, *11*, 447-471, **1974**, *12*, 380-407.
[32]Belen'kii *Adv. Heterocycl. Chem.* **1988**, *44*, 269-396; Katritzky; Jones *Adv. Heterocycl. Chem.* **1979**, *25*, 303-391; Katritzky; Weeds *Adv. Heterocycl. Chem.* **1966**, 7, 225-299.

3. "Organic Reaction Mechanisms," published by Wiley, New York, is an annual survey that covers the latest developments in the field of mechanisms. The first volume, covering 1965, appeared in 1966.

4. There are two annual reviews devoted to progress in organic synthesis. Theilheimer, "Synthetic Methods of Organic Chemistry," S. Karger Verlag, Basel, is an annual compilation, beginning in 1946, of new methods for the synthesis of organic compounds, arranged according to a system based on bond closings and bond breakings. Equations, brief procedures, yields, and literature references are given. Volume 44 was issued in 1990. Volumes 3 and 4 are available only in German, but all the rest are in English. There is an index to each volume. Cumulative indexes appear in every fifth volume. Beginning with vol. 8, each volume includes a short summary of trends in synthetic organic chemistry. A more recent series is "Annual Reports in Organic Synthesis," Academic Press, New York, which has covered the literature of each year since 1970. Equations are listed with yields and references according to a fairly simple system.

5. The *Journal Of Organometallic Chemistry* several times a year publishes annual surveys arranged according to metallic element. For example, vol. 404, published in February 1991, contains annual surveys for 1989 of organic compounds containing Sb, Bi, and Fe, and the use of transition metals in organic synthesis, and surveys for 1988 covering B, Ru, and Os.

Awareness Services

Besides the annual reviews and the title and abstract services previously mentioned, there exist a number of publications designed to keep readers aware of new developments in organic chemistry or in specific areas of it.

1. *Chemtracts: Organic Chemistry* is a bimonthly periodical, begun in 1988, that prints abstracts of certain recently published papers (those that the editors consider most important), with commentaries on these papers by distinguished organic chemists. Each issue deals with about 20 papers, and also includes a review article.

2. The Institute for Scientific Information (ISI), besides publishing *Current Contents* (p. 1244) and the *Science Citation Index* (p. 1266), also publishes *Index Chemicus* (formerly called *Current Abstracts of Chemistry and Index Chemicus*). This publication, begun in 1960 and appearing weekly, is devoted to printing structural formulas of all new compounds appearing in more than 100 journals, along with equations to show how they were synthesized and an author's summary of the work. Each issue contains five indexes: author, journal, biological activity, labeled compounds, and unisolated intermediates. These indexes are cumulated annually.

3. Theilheimer and the "Annual Reports on Organic Synthesis," mentioned in the previous section, list new synthetic methods once a year. There are several publications that do this monthly. Among these are *Current Chemical Reactions* (begun in 1979 and published by ISI), *Journal of Synthetic Methods* (begun in 1975 and published by Derwent Publications), and *Methods in Organic Synthesis*, begun in 1984 and published by the Royal Society of Chemistry. *Methods in Organic Synthesis* also lists books and review articles pertaining to organic synthesis.

4. *Natural Product Updates*, a monthly publication begun in 1987 and published by the Royal Society of Chemistry, lists recent results in the chemistry of natural products, along with structural formulas. It covers new compounds, structure determinations, new properties and total syntheses, among other topics.

General Treatises

There are a number of large-scale multivolume treatises that cover the whole field of organic chemistry or large areas of it.

1. "Rodd's Chemistry of Carbon Compounds," edited by Coffey, Elsevier, Amsterdam, is a treatise consisting of five main volumes, each of which contains several parts. Publication began in 1964 and is not yet complete. The organization is not greatly different from most textbooks, but the coverage is much broader and deeper. Supplements to many of the volumes have appeared. An earlier edition, called "Chemistry of Carbon Compounds," edited by Rodd, was published in 10 parts from 1951 to 1962.

2. Houben-Weyl's "Methoden der organischen Chemie," Georg Thieme Verlag, Stuttgart, is a major treatise in German devoted to laboratory methods. The fourth edition, which was begun in 1952 and consists of 20 volumes, most of them in several parts, is edited by E. Muller. The series includes supplementary volumes. The first four volumes contain general laboratory methods, analytical methods, physical methods, and general chemical methods. The later volumes are devoted to the synthesis of specific types of compounds, e.g., hydrocarbons, oxygen compounds, nitrogen compounds, etc. Beginning in 1990 parts of the series have appeared in English.

3. "Comprehensive Organic Chemistry," Pergamon, Elmsford, NY, 1979, is a six-volume treatise on the synthesis and reactions of organic compounds. The first three volumes cover the various functional groups, vol. 4, heterocyclic compounds, and vol. 5, biological compounds such as proteins, carbohydrates, and lipids. Probably the most useful volume is vol. 6, which contains formula, subject, and author indexes, as well as indexes of reactions and reagents. The last two of these not only refer to pages within the treatise, but directly give references to review articles and original papers. For example, on p. 1129, under "Chromic acid–sulphuric acid (Jones reagent), oxidation, alcohols," are listed 13 references to original papers. Several similar treatises, including the nine-volume "Comprehensive Organometallic Chemistry" (1982), the eight-volume "Comprehensive Heterocyclic Chemistry" (1984), and the six-volume "Comprehensive Medicinal Chemistry" (1989) are also published by Pergamon. The indexes to these works also include references.

4. A major treatise devoted to experimental methods of chemistry is "Techniques of Chemistry," edited first by Weissberger and then by Saunders, Wiley, New York. This publication, which began in 1970, so far consists of 21 volumes, most of them in several parts, covering such topics as electrochemical and spectral methods, kinetic methods, photochromism, and organic solvents. "Techniques of Chemistry" is a successor to an earlier series, called "Techniques of Organic Chemistry," which appeared in 14 volumes, some of them in more than one edition, from 1945 to 1969.

5. "Comprehensive Chemical Kinetics," edited by Bamford and Tipper, 1969–, Elsevier, Amsterdam, is a multivolume treatise covering the area of reaction kinetics. Six of these volumes (not all published at the time of writing) deal with the kinetics and mechanisms of organic reactions in a thorough and comprehensive manner.

6. Three multivolume treatises that cover specific areas are Elderfield, "Heterocyclic Compounds," Wiley, New York, 1950–; Manske and Holmes, "The Alkaloids," Academic Press, New York, 1950–; and Simonson, Owen, Barton, and Ross, "The Terpenes," Cambridge University Press, London, 1947–1957.

Monographs and Treatises on Specific Areas

Organic chemistry is blessed with a large number of books devoted to a thorough coverage of a specific area. Many of these are essentially very long review articles, differing from ordinary review articles only in size and scope. Some of the books are by a single author, and others have chapters by different authors but all are carefully planned to cover a specific area. Many of these books have been referred to in footnotes in appropriate places in this book. There have been several series of monographs, one of which is worth special mention: "The Chemistry of Functional Groups," under the general editorship of Patai, published by Wiley, New York. Each volume deals with the preparation, reactions, and physical and chemical properties of compounds containing a given functional group. Volumes covering more than 20 functional groups have appeared so far, including books on alkenes, cyano compounds, amines, carboxylic acids and esters, quinones, etc.

Textbooks

There are many excellent textbooks in the field of organic chemistry. We restrict ourselves to listing only a few of those published, mostly since 1985. Some of these are first-year texts and some are advanced (advanced texts generally give references; first-year texts do not, though they may give general bibliographies, suggestions for further reading, etc.); some cover the whole field, and others cover reactions, structure, and/or mechanism only. All the books listed here are not only good textbooks but valuable reference books for graduate students and practicing chemists.

Baker and Engel, "Organic Chemistry," West Publishing Co., St. Paul, MN, 1992.

Carey, "Organic Chemistry," 2nd ed., McGraw-Hill, New York, 1992.

Carey and Sundberg, "Advanced Organic Chemistry," 2 vols., Plenum, New York, 3rd. ed., 1990.

Carruthers, "Some Modern Methods of Organic Synthesis," 3rd ed., Cambridge University Press, Cambridge, 1986.

Ege, "Organic Chemistry," 2nd ed., D.C. Heath, New York, 1989.

Fessenden and Fessenden, "Organic Chemistry," 4th ed., Brooks/Cole, Monterey, CA, 1990.

House, "Modern Synthetic Reactions," 2nd ed., W. A. Benjamin, New York, 1972.

Ingold, "Structure and Mechanism in Organic Chemistry," 2nd ed., Cornell University Press, Ithaca, NY, 1969.

Isaacs, "Physical Organic Chemistry," Wiley, New York, 1987.

Jones, "Physical and Mechanistic Organic Chemistry," 2nd ed., Cambridge University Press, Cambridge, 1984.

Loudon, "Organic Chemistry," 2nd ed., Benjamin/Cummings, Menlo Park, CA, 1988.

Lowry and Richardson, "Mechanism and Theory in Organic Chemistry," 3rd ed., Harper and Row, New York, 1987.

McMurry, "Organic Chemistry," 2nd ed., Brooks/Cole, Monterey, CA, 1988.

Maskill, "The Physical Basis of Organic Chemistry," Oxford University Press, Oxford, 1985.

Morrison and Boyd, "Organic Chemistry," 6th ed., Prentice-Hall, Englewood Cliffs, NJ, 1992.

Pine, "Organic Chemistry," 5th ed., McGraw-Hill, New York, 1987.

Ritchie, "Physical Organic Chemistry," 2nd ed., Marcel Dekker, New York, 1989.

Solomons, "Organic Chemistry," 5th ed., Wiley, New York, 1992.

Streitwieser, Heathcock, and Kosower,"Introduction to Organic Chemistry," 4th ed., Macmillan, New York, 1992.

Sykes, "A Guidebook to Mechanism in Organic Chemistry," 6th ed., Longmans Scientific and Technical, Essex, 1986.

Vollhardt, "Organic Chemistry," W.H. Freeman, San Francisco, 1987.
Wade, "Organic Chemistry," 2nd ed., Prentice-Hall, Englewood Cliffs, NJ, 1991.

Other Books

In this section we mention several books that do not fit conveniently into the previous categories. All but the last have to do with laboratory synthesis.

1. *Organic Syntheses*, published by Wiley, New York is a collection of procedures for the preparation of specific compounds. The thin annual volumes have appeared each year since 1921. For the first 59 volumes, the procedures for each 10- (or 9-) year period are collected in cumulative volumes. Beginning with vol. 60, the cumulative volumes cover five-year periods. The cumulative volumes published so far are:

Annual volumes	Collective volumes
1–9	I
10–19	II
20–29	III
30–39	IV
40–49	V
50–59	VI
60–64	VII

The advantage of the procedures in *Organic Syntheses*, compared with those found in original journals, is that these procedures are *tested*. Each preparation is carried out first by its author and then by a member of the *Organic Syntheses* editorial board, and only if the yield is essentially duplicated is the procedure published. While it is possible to repeat most procedures given in journals, this is not always the case. All *Organic Syntheses* preparations are noted in Beilstein and in *CA*. In order to locate a given reaction in *Organic Syntheses*, the reader may use the OS references given in the present volume (through OS **69**); the indexes in *Organic Syntheses* itself; Shriner and Shriner, "Organic Syntheses Collective Volumes I, II, III, IV, V Cumulative Indices," Wiley, New York, 1976, or Sugasawa and Nakai; "Reaction Index of Organic Syntheses," Wiley, New York, 1967 (through OS **45**). Another book classifies virtually all the reactions in *Organic Syntheses* (collective vols. I to VII and annual vols. 65 to 68) into eleven categories: annulation, rearrangement, oxidation, reduction, addition, elimination, substitution, C—C bond formation, cleavage, protection/deprotection, and miscellaneous. This is "Organic Syntheses: Reaction Guide," by Liotta and Volmer, published by Wiley, New York, in 1991. Some of the categories are subdivided further, and some reactions are listed in more than one category. What is given under each entry are the equation and the volume and page reference to *Organic Syntheses*.

2. Volume 1 of "Reagents for Organic Synthesis," by Fieser and Fieser, Wiley, New York, 1967, is a 1457-page volume which discusses, in separate sections, some 1120 reagents and catalysts. It tells how each reagent is used in organic synthesis (with references) and, for each, tells which companies sell it, or how to prepare it, or both. The listing is alphabetical. Fourteen additional volumes have so far been published, which continue the format of vol. 1 and add more recent material. A cumulative index for vols. 1 to 12, by Smith and Fieser, was published in 1990.

3. "Comprehensive Organic Transformations," by Larock, VCH, New York, 1989, has been frequently referred to in footnotes in Part 2 of this book. This compendium is devoted to listings of methods for the conversion of one functional group into another, and covers the literature through 1987. It is divided into nine sections covering the preparation of alkanes and arenes, alkenes, alkynes, halides, amines, ethers, alcohols and phenols, aldehydes and ketones, and nitriles, carboxylic acids and derivatives. Within each section are given many methods for synthesizing the given type of compound, arranged in a logical system. A schematic equation is given for each method, and then a list of references (without author names, to save space) for locating examples of the use of that method. When different reagents are used for the same functional group transformation, the particular reagent is shown for each reference. There is a 164-page index of group transformations.

4. "Survey of Organic Synthesis," by Buehler and Pearson, Wiley, New York, 2 vols., 1970, 1977, discusses hundreds of reactions used to prepare the principal types of organic compounds. The arrangement is by chapters, each covering a functional group, e.g., ketones, acyl halides, amines, etc. Each reaction is thoroughly discussed and brief synthetic procedures are given. There are many references.

5. A similar publication is Sandler and Karo, "Organic Functional Group Preparations," 2nd ed., 3 vols., Academic Press, New York, 1983-1989. This publication covers more functional groups than Buehler and Pearson.

6. "Compendium of Organic Synthetic Methods," Wiley, New York, contains equations describing the preparation of thousands of monofunctional and difunctional compounds with references. Seven volumes have been published so far (1971 and 1974, edited by Harrison and Harrison; 1977, edited by Hegedus and Wade; 1980 and 1984, edited by Wade; 1988 and 1992, edited by Smith).

7. "The Vocabulary of Organic Chemistry," by Orchin, Kaplan, Macomber, Wilson, and Zimmer, Wiley, New York, 1980, presents definitions of more than 1000 terms used in many branches of organic chemistry, including stereochemistry, thermodynamics, wave mechanics, natural products, and fossil fuels. There are also lists of classes of organic compounds, types of mechanism, and name reactions (with mechanisms). The arrangement is topical rather than alphabetical, but there is a good index. "Compendium of Chemical Terminology," by Gold, Loening, McNaught, and Sehmi (the "Gold book"), published by Blackwell Scientific Publications, Oxford, in 1987, is an official IUPAC list of definitions of terms in several areas of chemistry, including organic.

LITERATURE SEARCHING

Until recently searching the chemical literature meant looking only at printed materials (some of which might be on microfilm or microfiche). Now, however, much of the literature can be searched online, including some of the most important. Whether the search is online or uses only the printed material, there are two basic types of search, (1) searches for information about one or more specific compounds or classes of compounds, and (2) other types of searches. First we will discuss searches using only printed materials, and then online searching.[32a]

Literature Searching Using Printed Materials

Searching for specific compounds. Organic chemists often need to know if a compound has ever been prepared and if so, how, and/or they may be seeking a melting point, an ir

[32a]For a monograph that covers both online searching and searches using printed materials, see Wiggins *Chemical Information Sources*; McGraw-Hill: New York, 1991.

spectrum, or some other property. Someone who wants all the information that has ever been published on any compound begins by consulting the formula indexes in Beilstein (p. 1249). At this time there are two ways to do this. (1) The formula index to the second supplement (Vol. 29, see p. 1249) will quickly show whether the compound is mentioned in the literature through 1929. If it is there, the searcher turns to the pages indicated, where all methods used to prepare the compound are given, as well as all physical properties, with references. Use of the page heading method described on p. 1249 will then show the locations, if any, in the third and later supplements. (2) If one has an idea which volume of Beilstein the compound is in (and the tables of contents at the front of the volumes may help), one may search the cumulative index for that volume. If not sure, one may consult several indexes. One of these two procedures will locate all compounds mentioned in the literature through 1959. If the compound is heterocyclic, it may be in the fifth supplement. If it is in vols. 17-19 (or in a later volume whose index has been published), the corresponding indexes may be consulted. If not, the page heading method will find it, if it was reported before 1960.[33] There is a way by which all of the above can be avoided. A computer program, called SANDRA (available from the Beilstein publisher), allows the user to find the Beilstein location by using a mouse to draw the structural formula of the compound sought. At this point the investigator will know (1) all information published through 1959 or 1979,[34] or (2) that the compound is not mentioned in the literature through 1959 or 1979.[34] In some cases, scrutiny of Beilstein will be sufficient, perhaps if only a boiling point or a refractive index is required. In other cases, especially where specific laboratory directions are needed, the investigator will have to turn to the original papers.

To carry the search past 1959 (or 1979), the chemist next turns to the collective formula indexes of *Chemical Abstracts:* 1957-1961; 1962-1966; 1967-1971; 1972-1976; 1977-1981; 1982-1986; such later collective indexes as have appeared; and the semiannual indexes thereafter. If a given formula index contains only a few references to the compound in question, the pages or abstract numbers will be given directly in the formula index. However, if there are many references, the reader will be directed to see the chemical substance index or (before 1972) the subject index for the same period; and here the number of page or abstract numbers may be very large indeed. Fortunately, numerous subheadings are given, and these often help the user to narrow the search to the more promising entries. Nevertheless, one will undoubtedly turn to many abstracts that do not prove to be helpful. In many cases, the information in the abstracts will be sufficient. If not, the original references must be consulted. In some cases (the index entry is marked by an asterisk or a double asterisk) the compound is not mentioned in the abstract, though it is in the original paper or patent. Incidentally, all entries in the *CA* indexes that refer to patents are prefixed by the letter P. Since 1967, the prefixes B and R have also been used, to signify books and reviews, respectively.

By the procedure outlined above, all information regarding a specific compound that has been published up to about a year before the search can be found by a procedure that is always straightforward and that in many cases is rapid (if the compound has been reported only a few times). Equally important, if the compound has not been reported, the investigator will know that, too. It should be pointed out that for common compounds, such as benzene, ether, acetone, etc., trivial mentions in the literature are not indexed (so they will not be found by this procedure), only significant ones. Thus, if acetone is converted to another compound, an index entry will be found, but not if it is used as a solvent or an eluent in a common procedure.

---

[33]Compounds newly reported in the fifth supplement that are in a volume whose index has not yet been published will not be found by this procedure. To find them in Beilstein it is necessary to know something about the system (see Ref. 21), but they may also be found by consulting *CA* indexes beginning with the sixth collective index, or by using Beilstein online.

[34]For those heterocyclic compounds that would naturally belong to a volume for which the fifth supplement has been published.

The best way to learn if a compound is mentioned in the literature after the period covered by the latest semiannual formula index of *CA* is to use the online services (p. 1261). However, if one lacks access to these, one may consult *Chemical Titles* and the keyword index (p. 1244) at the end of each issue of *CA*. In these cases, of course, it is necessary to know what name might be used for the compound. The name is not necessary for *Index Chemicus* (p. 1254); one consults the formula indexes. However, these methods are far from complete. *Index Chemicus* lists primarily new compounds, those which would not have been found in the earlier search. As for *Chemical Titles*, the compound can be found only if it is mentioned in the title. The keyword indexes in *CA* are more complete, being based on internal subject matter as well as title, but they are by no means exhaustive. Furthermore, all three of these publications lag some distance behind the original journals. To locate all references to a compound after the period covered by the latest semiannual formula index of *CA*, it is necessary to use *CA* online.

The complete procedure described above may not be necessary in all cases. Often all the information one needs about a compound will be found in one of the handbooks (p. 1250), in the "Dictionary of Organic Compounds" (p. 1249), or in one of the other compendia listed in this chapter, most of which give references to the original literature.

Other Searches[35]

There is no definite procedure for making other literature searches using only printed materials. Any chemist who wishes to learn all that is known about the mechanism of the reaction between aldehydes and HCN, or which compounds of the general formula Ar_3CR have been prepared, or which are the best catalysts for Friedel–Crafts acylation of naphthalene derivatives with anhydrides, or where the group $—C(NH_2)=N—$ absorbs in the ir, is dependent on his or her ingenuity and knowledge of the literature. If a specific piece of information is needed, it may be possible to find it in one of the compendia mentioned previously. If the topic is more general, the best procedure is often to begin by consulting one or more monographs, treatises, or textbooks that will give general background information and often provide references to review articles and original papers. In many cases this is sufficient, but when a complete search is required, it is necessary to consult the *CA* subject and/or chemical substance indexes, where the ingenuity of the investigator is most required, for now it must be decided which words to look under. If one is interested in the mechanism of the reaction between aldehydes and HCN, one might look under "aldehydes," or "hydrogen cyanide," or even under "acetaldehyde" or "benzaldehyde," etc., but then the search is likely to prove long. A better choice in this case would be "cyanohydrin," since these are the normal products and references there would be fewer. It would be a waste of time to look under "mechanism." In any case, many of the abstracts would not prove helpful. Literature searching of this kind is necessarily a wasteful process. Of course, the searcher would not consult the *CA* annual indexes but only the collective indexes as far as they go and the semiannual indexes thereafter. If it is necessary to search before 1907 (and even before 1920, since *CA* was not very complete from 1907 to about 1920), recourse may be made to *Chemisches Zentralblatt* (p. 1247) and the abstracts in the *Journal of the Chemical Society* (p. 1247).

Literature Searching Online[32a]

Online searching means using a computer terminal to search a *database*. Although databases in chemistry are available from several organizations, by far the most important such or-

---

[35]This discussion is necessarily short. For much more extensive discussions, consult the books in Refs. 1 and 15.

ganization is STN International (The Scientific & Technical Information Network), which is available in many countries. STN has dozens of databases, including many that cover chemistry and chemical engineering. To access these databases a chemistry department, a library, or an individual subscribes to STN (for a nominal fee), and receives code numbers that will permit access to the system. Then all one needs is a computer and a modem. STN charges for each use, depending on which databases are used, for how long, and what kind of information is requested. One of the nice features of STN is that the same command language is used for all databases, so when one has mastered the language for one database, one can use it for all the others. In this section we will discuss literature searching using *CA* online, which is one of the databases available from STN. One thing that must be remembered is that *CA* online is complete only from 1967 to the present,[36] so that searches for earlier abstracts must use the printed volumes. However, for the period since 1967, not only is online searching a great deal faster than searching the printed *CA*, but, as we shall see, one can do kinds of searches online that are simply not possible using only the printed volumes. Furthermore, the online files are updated every two weeks, so that one will find all the abstracts online well past the appearance of the latest semiannual indexes, often even before the library has received the latest weekly printed issue of *CA*. *CA* online is extremely flexible; one can search in a great many ways. It is beyond the scope of this book to discuss the system in detail[37] (*CA* conducts workshops on its use), but even with the few commands we will give here, a user can often find all that he or she is looking for. *CA* online has two major files, *the CA File* and *the Registry File*.[38] These are so different that we discuss them separately.

The CA File
This file is accessed with the command FILE CA. Once in the file, the user uses the command SEARCH (or SEA or S)[39] to look for references to specific terms. For example, one may type SEA SEMIPINACOL. On the screen will appear something like

L1   4 SEMIPINACOL

The L1 means that this is line one. Future answers from the system will number the lines in consecutive order. The 4 means that the system has four abstracts that contain the word semipinacol. The word may be in the title, an index entry, or a keyword. The search term may be the name of a compound, which means that individual compounds can be searched for in this way. If the name used is the *CA* indexing name, all the abstracts mentioning that compound will be retrieved. However, common names or other names can also be searched (e.g., catechol), and if they are mentioned in the title of the paper, or, for example as a keyword, those abstracts will be retrieved.

Compounds can also be searched for by using the Registry Number, e.g.,

SEARCH 126786-44-3

Let us return to the example of semipinacol. The system told us there were four abstracts for this term. We may now see these abstracts by using the display command (DISPLAY or DIS or D), e.g.,

D L1 1 BIB ABS

---

[36]There is also a file called CAOLD that has some papers earlier than 1967.
[37]For a discussion of *CA* online, see Ref. 15.
[38]There is also a file, LCA, which is used for learning the system. It includes only a small fraction of the papers in the CA File, and is not updated. There is no charge for using the LCA File, except for a small hourly fee.
[39]Most commands can be used in three ways, as shown here. When the full term is spelled out (SEARCH), the system assumes an unsophisticated user, and gives more help. If the command is S, the system assumes the user is knowledgable about the system.

L1 means we are asking the system to display material pertaining to semipinacol, which is on line L1. If we fail to insert this information, the system will display items pertaining to the last L number shown.

1 means we are asking for information on the first of the four papers. The papers are listed in reverse chronological order, meaning paper 1 will be the latest of the four. Similarly, we can ask to see the information on any of the others.

BIB ABS means we are asking for bibliographic data (abstract number, title of paper, authors' names, journal, year, etc.) and for a display of the full abstract.[40] There are other choices. Instead of BIB ABS we could have typed CAN which would give us the abstract number only (we might then choose to find the other information in the printed *CA*). Or, we could have typed IND, which would give us the abstract number and the index terms for this paper, or ALL which would give everything we get from BIB ABS plus the index terms. In all, there are nine or ten ways to ask for display material. Our choice will depend on how much we need to know, and on the cost, since the more information requested, the higher the cost.

As so far described, online searching is faster then searching the printed *CA*, but gives us essentially the same information. The scope of the online method is much greater than that, for it allows us to combine words, in a number of ways. One such way is by the use of the terms AND, NOT, and OR. If we search AMBIDENT AND NUCLEOPHILE, we will get something like this:

$$332 \text{ AMBIDENT}$$

$$3275 \text{ NUCLEOPHILE}$$

L2   42 AMBIDENT AND NUCLEOPHILE

This means there are 42 entries that have the words AMBIDENT and NUCLEOPHILE somewhere in them; in the titles, keywords, or index entries. We can now, if we wish, display any or all of them. But a particular entry might have these two words in unrelated contexts, e.g., it might be a paper about ambident electrophiles, but which also has NUCLEOPHILE as an index term. We would presumably get fewer papers, but with a higher percentage of relevent ones, if we could ask for AMBIDENT NUCLEOPHILE, and in fact, the system does allow this. If we type S AMBIDENT(W)NUCLEOPHILE, we will get only those papers in which the term NUCLEOPHILE directly follows AMBIDENT, with no words in between.[41] This is called proximity searching, and there are other, similar commands. For example, the use of (4A) instead of (W) will give all instances in which the two words appear 4 or fewer words apart, in either order.

Another important option is a truncation symbol. If we ask for NUCLEOPHILE we will find all entries that contain the term nucleophile, but not those that contain a different form of this term, e.g., nucleophilic. We can take care of this by using NUCLEOPHIL? as a search term instead of NUCLEOPHILE. This will retrieve all terms that start with the letters NUCLEOPHIL, no matter what other letters follow, thus retrieving nucleophilic, nucleophilicity, nucleophiles, etc., as well as nucleophile. The question mark is one of several truncation symbols, each of which serves a different function.

The words AND, NOT, and OR are called *Boolean operators*. They may be combined in many ways, e.g.,

S ORTHO AND EFFECT AND HAMMETT

S (CARBON(W)DIOXIDE OR CARBON(W)DISULFIDE) AND CATALY? NOT ACID

S HYDANTOIN AND (METHYL OR ETHYL) NOT (VINYL? OR PHENYL)

---

[40]For some papers in the late 1960s only the bibliographic data, and not the abstracts, are available online.
[41]If we ask for AMBIDENT NUCLEOPHILE without the (W), the system treats it as if we asked for AMBIDENT(W)NUCLEOPHILE, and gives the same answers.

A particular search command can contain dozens of such terms. Obviously, if one is careful about choosing the proper search terms, one can focus in on just the relevent papers, and leave out those that will not be useful. However, there will often be far more papers than can conveniently be handled, and there are other ways to limit searches. One such way is by using a narrow field. For example, a synthetic chemist may wish to find references in which a given compound is synthesized, but find, when he or she searches for that compound, that most of the references concern biological or medicinal uses of the compound. By using the command

SEA 3489-26-7/ORG

the search will retrieve only those papers for the compound of Registry Number 3489-26-7 that have been abstracted in the organic (ORG) sections of *CA* (Sections 21 to 34), and will not retrieve papers from the biochemical sections, which are more likely to stress biological or medicinal uses. There are many other ways to limit searches. It is possible to search only for papers in a single section of *CA*, only those that appeared in a given year or range of years, only those in which the search term appears in the title of the paper, only those by a given author, etc.

Besides subject terms, the CA File also contains bibliographical information, such as author names, location of the laboratory in which the work was done, language of the paper, etc., and these can be searched. For example, S ROBERTS, J?/AU will find all papers published by any authors named Roberts whose first name begins with J. These terms can be combined with subject terms in Boolean searches.

The Registry File
The Registry File is entered with the command FILE REGISTRY. This can be done at any time, and it is possible to go back and forth between the CA and Registry Files at will. The Registry File uses the same commands (including Boolean) as the CA File, but instead of displaying abstracts and bibliographical information, it displays information about compounds. Its most useful feature is that it allows the user to build a structure, and then gives information about compounds that possess that structure, even if the structure is only part of a larger structure (see below).

The procedure for building a structure can be long and complex, if the structure is large and complex, but the commands are simple. We will illustrate by building the structure for 3-ethylpipcridine, which uses the most important commands. We begin with the command

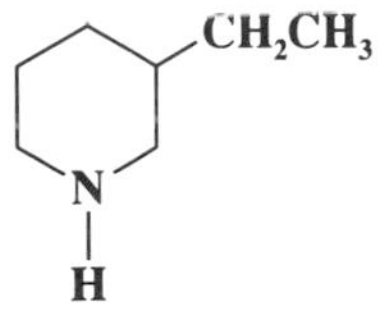

3-Ethylpiperidine

STRUCTURE. The system will ask if we wish to build on a structure previously used. If we say no, we will then get the prompt

ENTER (DIS), GRA, NOD, BON OR ?:

GRA is used for putting in chains or rings. We enter GRA R6, DIS (the DIS must be typed if the structure is to be displayed), and get the structure **1** shown in Figure A.1.[42] R6 specifies

---

[42]The structures shown in Figure A.1 are those received by ordinary computer terminals. Better-looking structures, more like those printed in books, are obtained with certain types of terminals. The system always asks the user to specify which type of terminal is being used.

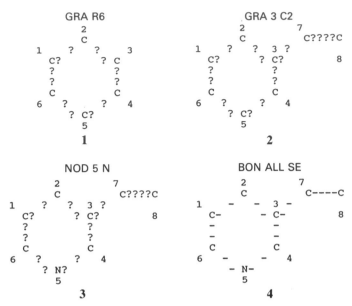

FIGURE A.1 Steps in building the structure of 3-ethylpiperidine in the Registry File. Above each structure is the command that gave that structure.

a 6-membered ring. If we had simply entered 6 we would have created a six-atom chain. The six numbers shown are purely arbitrary and have no connection with the way the positions are actually numbered in any nomenclature scheme. Immediately after this structure is displayed, the same prompt reappears, as it does after every structure. We wish to introduce a two-atom side chain (the ethyl group), so we enter GRA 3 C2, DIS, and get structure **2** in Figure A.1. The C2 indicates a two-atom chain, and the 3 means that we want it attached to atom 3 (in this case the atom number is completely arbitrary, since attachment to any atom would give an equivalent result). Note that the system has numbered the new atoms 7 and 8. We will not be introducing any more atoms into our structure, but if we were they would be numbered consecutively, in the order in which they were introduced. We have all the atoms we need (we do not indicate the hydrogens because the system assumes that all nonspecified valences are connected to hydrogen, unless we tell it otherwise), but we still have not told the system about the nitrogen. Although the system uses C to specify all atoms, they will only remain carbon atoms until we instruct the system differently. To get 3-ethylpiperidine the atom in the 5 position must be nitrogen. A C can be changed to another element by using NOD (for node), so we now type NOD 5 N, DIS. This changes C-5 to N-5, giving all the atoms in their final positions (**3** in Figure A.1). However, the structure is still not complete because the bonds have not been specified. By using the BON command we can make any bond single, double, or triple, and can even indicate aromaticity or other resonance. In this case we want all the bonds to be single bonds, so we type BON ALL SE, DIS (SE is used for single bonds), and get our final structure **4** in Figure A.1.

At this point we type END, and get

L1   STRUCTURE CREATED

The structure is now ready to be searched. At the command SEARCH L1, the system will give the prompts

ENTER TYPE OF SEARCH (SSS), CSS, FAMILY, OR EXACT:

ENTER SCOPE OF SEARCH (SAMPLE), FULL, RANGE, OR SUBSET:

The system asks for the desired scope, because a full search (of the entire Registry File of more than 10 million substances) may cost a lot of money and may not be worth it if the desired answers can be obtained from a more limited search.

As shown by the first prompt there are four types of search, of which we will discuss two: exact and substructure (SSS). In an exact search, only information regarding exactly the structure given will be retrieved, but even so there may well be several answers, because *CA* treats stereoisomers and isotopically-substituted compounds as separate answers. At the conclusion of the search the system gives the number of answers, e.g., 4. We may now look at the four answers by using the display command. As in the CA File, there is a choice of display formats, but if we choose SUB we will get (1) the Registry Number, (2) the approved *CA* index name, (3) other names that have appeared in *CA* for that compound, (4) a structural formula, and (5) the number of *CA* references since 1967, along with a notation as to whether the compound is found in the CAOLD File. By using other display formats, we can also obtain the bibliographic information and abstracts for the latest 10 references. If there are more than 10 *CA* references, we can of course switch to the CA File, and use the Registry Number to search that file completely. If the exact structure search yields no answers, we know that no references to it have appeared in *CA* since 1967.

Although an exact search can be useful, in most cases it does not give any more information than can be obtained from the printed *CA*. Substructure searches (SSS) are far more important, because there is no other way to get this information. If we do a substructure search on **4** in Figure A.1, we not only get all the answers we would get in an an exact search, but all substances that contain, anywhere within their structure, the arrangement of atoms and bonds shown in **4**. For example, **5, 6, 7,** and **8** would all be retrieved in this

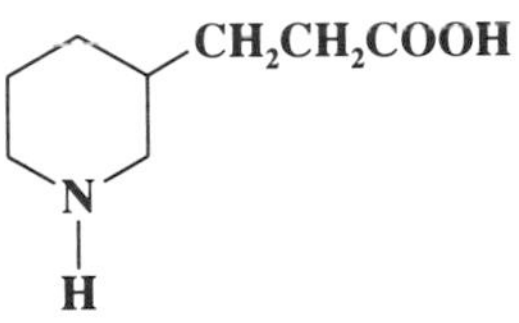

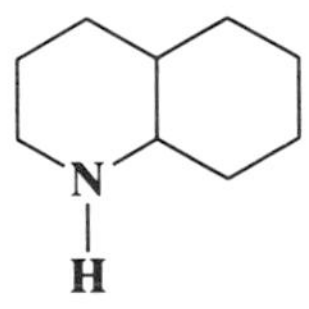

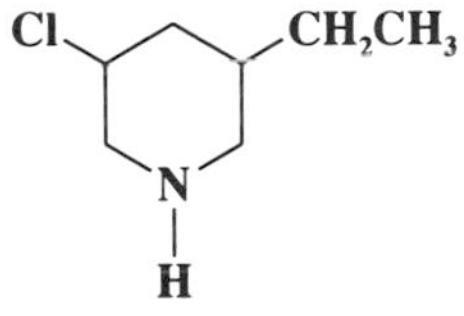

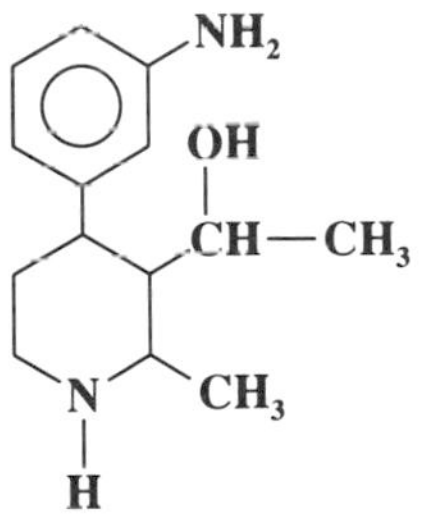

search, but **9** would not be. SSS searches typically retrieve from tens to hundreds of times as many answers as exact searches of the same structure. Furthermore, the scope can be widened by the use of variable nodes. For example, the symbol X means any halogen, the symbol M any metal, and the symbol G allows the user to specify his or her own variable at that point (e.g., $G = Cl$ or NO_2 or Ph). As with an exact search, each answer can be displayed as described above.

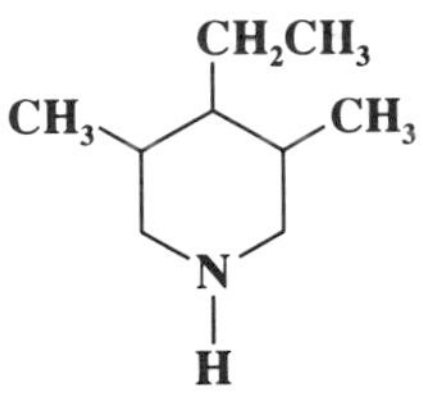

As mentioned above, building structures can be very complicated, and, because there is great flexibility in the system, there are a great many ways to use the commands, but the rewards are the retrieval of information that cannot be obtained in any other way. We have given here only a hint of the possibilities in using this system.

It is not necessary to build structures to use the Registry File. Compounds can also be searched for by using names, combinations of name fragments, Registry Numbers, molecular formulas, and in other ways. The display methods are the same.

Other Databases

Several of the other databases carried by STN are of interest to organic chemists. Among these are BEILSTEIN, which allows Beilstein to be searched online (SSS and EXACT searches can also be done in this database); CASREACTS, in which the user can specify a starting compound and a product, usually by giving Registry Numbers, and the system tells whether that transformation has been reported in the literature (beginning in 1985), and if so gives reagents and references; and CJACS, which gives the complete texts (but not the display material, such as tables and displayed equations) of all papers published in about 20 journals published by the American Chemical Society (including *J. Am. Chem. Soc.*, *J. Org. Chem.*, and *Chem. Rev.*) since 1982. Chemical journals of several other publishers, including Elsevier (*J. Organomet. Chem.* etc.), VCH (*Angew. Chem. Int. Ed. Engl.*), and the Royal Society of Chemistry (*J. Chem. Soc., Perkin Trans. 1* etc.), are also available online in a similar manner. Having these journals online is particularly useful because their texts can be seached for keywords, author's names, Registry Numbers, and other types of information.

Science Citation Index

A publication that can greatly facilitate literature searching is *Science Citation Index (SCI)*, begun in 1961. This publication, which is quite different from any other mentioned in this chapter, gives a list of all papers in a given year that have cited a given paper, patent, or book. Its utility lies in the fact that it enables the user to search *forward* from a given paper or patent, rather than backward, as is usually the case. For example, suppose a chemist is familiar with a paper by Jencks and Gilchrist (*J. Am. Chem. Soc.* **1968,** 90, 2622) entitled "Nonlinear Structure–Reactivity Correlations. The Reactivity of Nucleophilic Reagents toward Esters." The chemist is easily able to begin a search for earlier papers by using references supplied in this paper and can then go further backward with the aid of references in those papers, etc. But for obvious reasons the paper itself supplies no way to locate *later* papers. *SCI* is designed to make up for this gap. The citation index of *SCI* lists all papers, patents, or books cited in a given year or 2-month period (by first author only) and then gives a list of papers that have done the citing. The index is published bimonthly and cumulated annually. For example, column 43901 of the 1989 citation index shows that the Jencks paper mentioned above was cited as a footnote in 16 papers published in 1989. It is reasonable to assume that most of the papers that cited the Jencks paper were on closely related subjects. For each of the 16 papers are listed the first author, journal abbreviation, volume and page numbers, and year. In a similar manner, if one consulted *SCI* for all the years from 1968 on, one would have a complete list of papers that cited that paper. One could obviously broaden the search by then consulting *SCI* (from 1989 on) for papers that cited these 16 papers and so on. Papers, patents, or books listed, for example, in the 1989 *SCI* may go back many years, e.g., papers published by Einstein in 1905 and 1906 are included. The only requirement is that a paper published in 1989 (or late 1988) has mentioned the earlier paper in a footnote. The arrangement of cited papers or books is alphabetical by cited first author and then by cited year. Cited patents are listed in a separate table, in order of patent number, though the inventor and country are also given.

SCI covers about 3200 journals in the physical and biological sciences, as well as in medicine, agriculture, and technology. In addition to the citation index, each bimonthly and annual *SCI* also includes three other indexes. One of these, called *Source Index*, is similar to the *CA* author index. It lists the titles, journal abbreviations, volume, issue, page numbers, and year of all papers published by a given author during that two-month period or year. All authors are listed; not just first authors. The second, called the *Corporate Index*, lists all publications that have been published from a given institution during that period, by first author. Thus, the corporate index for 1989 lists 63 papers by 45 different first authors emanating from the Department of Chemistry of Rutgers University, New Brunswick, NJ. The main section of the corporate index (the Geographic Section) lists institutions by country or (for the U.S.) by state. There is also an Organization Section, which lists the names of institutions alphabetically, and for each gives the location, so it can be found in the geographic section. The third index included in *SCI* is the *Permuterm*[43] *Subject Index*. This index alphabetically lists every significant word in the titles of all papers published in that year or bimonthly period, paired with all other significant words in the same title. Thus, for example, a title with seven significant words appears at 42 separate places in the index. Each of the seven words appears six times as the main word, each time paired with a different word as the co-word. The user is then led to the *Source Index*, where the full reference is given. *SCI* is also available online (though not through STN) and on CD-ROM discs. A version of *SCI* that is restricted to chemistry but also includes searchable abstracts, is available only in the CD-ROM format.

The publishers of *SCI* also produce another publication, called *Index to Scientific Reviews*, that appears semiannually. This publication, which began in 1974, is very similar to *SCI*, but confines itself to listing citations to review articles. The citations come from about 2500 journals in the same general areas as are covered by *SCI*. The review articles cited appeared in about 215 review journals and books, as well as in those journals that publish occasional review articles. Like *SCI*, the *Index to Scientific Reviews* contains citation, source, corporate, and Permuterm indexes. It also contains a "Research Front Specialty Index," which classifies reviews by subject.

How to Locate Journal Articles

Having obtained a reference from Beilstein, *SCI, CA*, a treatise, or some other source, one often needs to consult the original journal (the location of patents is discussed on p. 1243). The first step is to ascertain the full name of the journal, since it is the abbreviation that is generally given. Of course, everyone should be familiar with the abbreviations of the very important journals, such as *J. Org. Chem., Chem. Ber.*, etc., but references are often found to journals whose titles are not at all familiar (e.g., *K. Skogs Lantbruksakad. Tidskr.* or *Nauchn. Tr. Mosk. Lesotekh. Inst.*). In such cases, one consults the *Chemical Abstracts Service Source Index (CASSI)*, 1989 edition, which contains the names of all the journals covered by *CA* from 1907 to 1989 (even those no longer published), with the most recent abbreviations in bold print. CASSI also lists journals covered by *Chemisches Zentralblatt* and its predecessors from 1830 to 1969, and journals cited in Beilstein before 1907. The journals are listed in alphabetical order of the *abbreviations*, not of the titles. Journal title changes have not been infrequent, and *CASSI* also contains all former names, with cross-references to the current names. Quarterly supplements, cumulated annually, to *CASSI* have appeared since 1990 listing new journals and recent changes in journal titles. It should be pointed out that, while many publications use the *CA* abbreviations, not all do. The

---

[43]Registered trade name.

student will find that usages vary from country to country, and even from journal to journal within a country. Futhermore, the *CA* abbreviations have changed from time to time.

Once the complete title is known, the journal can easily be obtained if it is in the library customarily used by the chemist. If not, one must use another library, and the next step is to find out which libraries carry the journal. *CASSI* answers this question too, since it carries a list of some 360 libraries in the United States and other countries, and *for each journal it tells which of these libraries carries it*, and furthermore, if the holdings are incomplete, which volumes of that journal are carried by each library. It may be possible to visit the closest library personally. If not, a copy of the article can usually be obtained through interlibrary loan. *CASSI* also includes lists of journal publishers, sales agents, and document depositories. Photocopies of most documents cited in *CA* can be obtained from Chemical Abstracts Document Delivery Service, Customer Services, 2540 Olentangy River Road, Columbus OH, 43210, U.S.A. Orders for documents can be placed by mail, telephone, Telex, fax, or online through STN or other services.

Appendix B
CLASSIFICATION OF REACTIONS BY TYPE OF COMPOUND SYNTHESIZED

Acetals and Ketals

0-12 Reaction between alkoxides and *gem*-dihalides (Williamson) or α-halo ethers

0-15 Reaction between diazoalkanes and alcohols

0-17 Transetherification

0-79 Reduction of ortho esters

0-92 Reaction between Grignard reagents and ortho esters

4-7 Electrolytic alkoxylation of ethers

4-8 Cyclization of β-hydroxy ethers

5-4 Addition of alcohols or phenols to triple bonds

6-6 Addition of alcohols to aldehydes or ketones

6-53 Addition of aldehydes to olefins (Prins)

6-57 Trimerization and polymerization of aldehydes

Acetoxy Sulfides

9-71 Pummerer rearrangement

Acetylenes (*see* Alkynes)

Acids (*see* Carboxylic Acids, Sulfonic Acids)

Acylals

5-5 Addition of carboxylic acids to alkynes

6-56 Acylation of aldehydes or ketones

9-14 Bisdecarboxylation of malonic acids

9-17 Oxidation of arylmethanes with CrO$_3$ and Ac$_2$O

Acyl Halides

0-3 Reaction between 1,1,1-trihalides and SO$_3$

0-74 From carboxylic acids

0-75 Conversion of acid derivatives to acyl halides

4-3 Halogenation of aldehydes

5-1 Addition of hydrogen halides to ketenes

5-22 Free-radical addition of acyl halides to olefins

9-22 Oxidation of alcohols

Acyloxy Ketones

5-44 Addition of an acyl and an acyloxy group to a double bond

Acyloins (*see* Hydroxy Aldehydes and Ketones)

Alcohols (*see also* Diols, Hydroxy Esters, etc.)

0-1 Hydrolysis of alkyl halides

0-4 Hydrolysis of inorganic esters

0-6 Hydrolysis of enol ethers, acetals, or ortho esters

0-10 Hydrolysis of carboxylic esters

0-17 Transetherification

0-18 Payne rearrangement

0-23 Transesterification

0-55 Ammonolysis of carboxylic esters

0-68 Cleavage of ethers with concentrated acids

0-79 Reduction of acetals or ortho esters

0-80 Reduction of epoxides

0-92 Cleavage of acetals or ortho esters with Grignard reagents

Aldehydes (continued)

2-40 Decarboxylation of glycidic acids

3-15 Carbonylation of aryl iodides

3-17 Vicarious substitution of aryl nitro compounds

4-16 Cross-coupling of alkanes with trioxane

4-20 Arylation of allylic alcohols

4-31 Reaction of diazonium salts with oximes, followed by hydrolysis

5-2 Cleavage of activated olefins with water

5-3 Hydration of acetylene

5-9 Selective reduction of unsaturated aldehydes

5-12 Oxidation of boranes; hydrolysis of unsaturated boranes

5-18 Addition of organometallic compounds to unsaturated aldehydes

5-19 Addition of boranes to unsaturated aldehydes

5-24 Hydroformylation of olefins (oxo process)

6-2 Hydrolysis of imines, oximes, hydrazones, or other C=N compounds

6-4 Hydrolysis of primary nitro compounds (Nef)

6-28 Reduction of nitriles

6-32 Addition of Grignard reagents to formamides

6-41 Reaction of aldehydes or ketones with boron methides

6-69 Hydrolysis of metalated aldimines

7-1 Dehydration of 1,2-diols

7-2 Pyrolysis of vinylic ethers

7-32 Fragmentation of γ-amino or γ-hydroxy halides

7-33 Fragmentation of 1,3-diols

7-38 Fragmentation of certain ketoximes

7-43 Pyrolysis of β-hydroxy olefins

7-44 Pyrolysis of allylic ethers

8-2 Rearrangements of diols (pinacol)

8-9 Homologation of aldehydes

8-14 Reaction between α-hydroxy or α-halo amides and NaOBr (Hofmann)

8-21 Cleavage of hydroperoxides

8-23 Rearrangement of allylic ethers

8-26 Treatment of boranes with CO and LiAl(OMe)$_3$

8-32 [1,3] Sigmatropic rearrangements of allylic vinylic ethers

8-42 Photolysis of nitrites, followed by hydrolysis (Barton)

9-3 Oxidation of primary alcohols

9-7 Oxidative cleavage of glycols or related compounds

9-9 Ozonolysis of olefins

9-13 Oxidation of arylacetic acids

9-16 Oxidation of activated methyl groups

9-17 Oxidation of arylmethanes (Étard)

9-20 Oxidation of primary halides or esters of primary alcohols

9-21 Oxidation of amines or nitro compounds

9-23 Oxidation of olefins with noble-metal salts

9-71 Hydrolysis of α-acetoxy sulfides

Alicyclic Compounds

0-86 Internal coupling (Wurtz)

0-88 Cyclization of diallylic halides

0-90 Cyclization of 1,3-diols

0-94 Internal malonic ester synthesis

0-102 Carbonylation of 1,4-dihalides

0-108 Internal condensation of diesters (Dieckmann)

0-113 Ketonic decarboxylation of dicarboxylic acids

1-12 Intramolecular Friedel–Crafts alkylation

1-13 Scholl ring closure

1-14 Intramolecular Friedel–Crafts acylation

1-23 Cyclodehydration of aldehydes and ketones

2-16 Intramolecular insertion of carbocations

2-20 Intramolecular insertion of carbenes

3-16 Cyclization of dihalobiphenyls

4-17 Coupling of terminal diynes (cycloalkynes)

4-18 Intramolecular arylation (Pschorr)

4-33 Cyclization of dimagnesio compounds

5-10 Reduction of aromatic rings

5-15 Cyclization of dienes or diynes

5-18 Cyclization of unsaturated Grignard reagents

5-20 Free radical cyclization of alkenes with tin or mercury halides

Alkyl Halides (continued)

5-26 Addition of halogens to olefins or alkynes

5-33 Addition of alkyl or aryl halides to olefins

6-24 Reductive halogenation of aldehydes

6-29 Addition of methylniobium reagents to ketones

7-39 Reaction of N-substituted amides with PCl_5 (von Braun)

Alkynes (*see also* Alkynyl Halides, Alkynyl Ethers)

0-78 Reduction of acetylenic alcohols

0-87 From allenic substrates, with organocopper reagents

0-88 Propargylation of alkyl halides

0-100 Alkylation at an alkynyl carbon

2-2 Triple-bond migration

2-40 Decarboxylation of acetylenic acids

3-13 Reaction between aryl iodides and copper acetylides

4-17 Coupling of alkynes (Eglinton)

4-20 Arylation of alkynes

4-33 Dimerization of alkynyl organometallic compounds

4-34 Coupling of alkynyl borates

7-6 Pyrolysis of bisquaternary ammonium hydroxides

7-12 Cleavage of selenoxides

7-13 Dehydrohalogenation of dihalides or vinylic halides

7-17 Elimination of the elements of CH_4 from certain alkenes

7-25 Decomposition of thiiren-1,1-dioxides

7-28 Reaction of bistosylhydrazones with metallic oxides

7-29 Dehalogenation of tetrahalides

8-28 From boranes and lithium acetylides

8-39 Metathesis of alkynes

9-2 Dehydrogenation of certain diaryl alkenes

9-33 Oxidation of dihalotoluenes

Alkynyl Ethers

7-13 Reaction between vinylidine dihalides and amide ion

Alkynyl Halides

2-30 Reaction of acetylide ions with halogens

Allenes

0-76 Reduction of propargyl halides

0-81 Reduction of propargyl acetates

0-88 Alkylation of propargyl halides

0-89 Alkylation of propargyl tosylates

0-91 Reaction between propargyl esters and organometallic reagents

0-92 Cleavage of propargyl ethers by Grignard reagents

2-2 Rearrangement of alkynes

6-47 Reaction of phosphoranes with ketenes or CO_2

7-13 Dehydrohalogenation of dihalides

7-29 Dehalogenation of tetrahalides or dihaloalkenes

7-43 Pyrolysis of β-hydroxy alkynes

8-3 Contraction of three-membered rings

8-35 Rearrangement of propargylic vinyl compounds

Amidals (*see* Bisamides)

Amides (*see also* Bisamides)

0-11 Cleavage of an alkyl group from N-*t*-butyl amides

0-51 Reaction between secondary amines and chloroform

0-52 Amination of acyl halides

0-53 Amination of anhydrides

0-54 Amination of carboxylic acids

0-55 Amination of carboxylic esters

0-56 Amination of amides

0-57 Amination of other acid derivatives

0-58 N-Alkylation of amides

0-103 Carbonylation of alkyl halides

1-6 Amidation of aromatic rings with hydroxamic acids

1-19 Carbamoylation of aromatic rings (Gatterman)

1-21 Amidation of aromatic rings with isocyanates

1-25 Amidoalkylation of aromatic rings

1-35 Rearrangement of N-halo-N-acyl aromatic amines (Orton)

2-12 Insertion by nitrenes

2-31 Indirectly from aldehydes

2-32 From imines, CO, and a borane

2-42 Reaction between amino acids and anhydrides (Dakin–West)

2-46 Cleavage of ketones with amide ion (Haller–Bauer)

Amines (continued)

2-31 Conversion of organometallic compounds to amines

2-40 Decarboxylation of amino acids

2-48 Decyanation of cyanoamines

3-6 Arylation of ammonia or amines

3-7 Reaction between naphthols, bisulfite ion, and ammonia or amines (Bucherer)

3-18 Amination of heterocyclic nitrogen compounds (Chichibabin)

3-19 Direct amination of activated aromatic rings

3-26 Rearrangement of benzylic quaternary ammonium salts (Sommelet–Hauser)

3-27 Rearrangement of aryl hydroxylamines

4-10 Demethylation of tertiary amines

4-36 Desulfurization of thioamides

5-7 Addition of ammonia or amines to olefins

5-18 Addition of organometallic compounds to allylic amines

5-22 Free-radical addition of amines to olefins

5-41 Diamination of alkenes

5-43 Addition of R_2N and SR to double bonds

6-2 Hydrolysis of imines, enamines, and iminium ions

6-3 Hydrolysis of isocyanates or isothiocyanates

6-5 Hydrolysis of cyanamides

6-13 Addition of ammonia to aldehydes

6-15 Reductive alkylation of ammonia or amines

6-16 Reaction between aldehydes, ammonia or amines, and an active hydrogen compound (Mannich)

6-26 Reduction of imines, hydrazones, or other compounds containing the C=N bond

6-27 Reduction of nitriles or nitrilium ions

6-29 Addition of organometallic compounds to amides

6-32 Addition of Grignard reagents to formamides

6-35 Addition of Grignard reagents to imines

6-66 Reduction of isocyanides

7-6 Cleavage of quaternary ammonium hydroxides (Hofmann)

7-7 Cleavage of quaternary ammonium salts

7-38 Fragmentation of certain ketoximes

8-14 Reaction between amides and NaOBr (Hofmann)

8-15 Rearrangement of acyl azides in the presence of water (Curtius)

8-16 Rearrangement of hydroxamic acids and acyl halides (Lossen)

8-17 Addition of hydrazoic acid to carboxylic acids (Schmidt)

8-19 Rearrangement of N-haloamines

8-22 Rearrangement of quaternary ammonium salts and tertiary benzylic amines (Stevens)

8-37 [2,3] Sigmatropic rearrangements of quaternary ammonium salts

8-38 Rearrangement of benzidines

8-42 Hofmann–Löffler and related reactions

9-5 Conversion of primary to secondary amines by dehydrogenation

9-9 Reaction between ozonides, ammonia, and hydrogen

9-21 Oxidative cleavage of amines

9-39 Reduction of amides

9-47 Reduction of nitro compounds

9-50 Reduction of nitroso compounds or hydroxylamines

9-51 Reduction of oximes

9-52 Reduction of azides

9-53 Reduction of isocyanates, isothiocyanates, or N-nitroso compounds

9-55 Reduction of amine oxides

9-59 Reduction of azo, azoxy, or hydrazo compounds

9-62 Bimolecular reduction of imines (1,2-diamines)

Amino Acids and Esters

0-11 Hydrolysis of lactams

0-43 Amination of halo acids

0-55 Ammonolysis of β-lactones

0-94 Alkylation of N-acetylaminomalonic esters (Sorensen)

2-8 Nitrosation at a carbon bearing an active hydrogen and reduction of the

Azo Compounds

1-4   Coupling of diazonium salts with aromatic rings
1-34   Rearrangement of aryl triazenes
2-7   Aliphatic diazonium coupling
2-52   Reaction of amines with nitroso compounds (Mills)
2-53   From aromatic nitro compounds
4-29   Coupling of aryl diazonium salts
8-45   Rearrangement of azoxy compounds (Wallach)
9-6   Oxidation of hydrazines
9-36   Oxidation of amines
9-55   Reduction of azoxy compounds
9-67   Reduction of nitro compounds

Azoxy Compounds

0-64   Reaction between alkyl halides and alkanediazotates
2-53   Reaction of nitroso compounds with hydroxylamines
9-29   Oxidation of azo compounds
9-36   Oxidation of amines
9-66   Reduction of nitro or nitroso compounds; reaction between nitroso compounds and hydroxylamines

Benzoins (*see* Hydroxy Aldehydes and Ketones)

Bisamides

4-16   Coupling of amides
6-14   Addition of amides to aldehydes or ketones
6-67   Reaction between isocyanides, acids, amines, and aldehydes or ketones (Ugi)

Bis(trimethylsilyl)amines

0-63   Reaction between halides or tosylates and $(Me_3Si)_2NNa$

Bisulfite Addition Compounds (*see* Hydroxy Sulfonic Acids)

Boranes

2-35   Reaction between boron halides and Grignard reagents
5-12   Hydroboration of olefins or alkynes
5-19   Reaction of borinates with organometallic compounds

7-15   Exchange reaction between boranes and olefins
8-11   Migration of boron

Bunte Salts

0-39   Reaction between alkyl halides and thiosulfate ion

Carbamates

0-24   Reaction between K_2CO_3, amines, and halides
0-52   Reaction between chloroformates and primary amines
0-62   Reaction between alkyl halides, ethanol, and thiocyanate ion
0-72   Cleavage of tertiary amines with ClCOOPh
2-12   Insertion by nitrenes
2-55   Carbonylation of amines or nitro or nitroso compounds
6-8   Addition of alcohols to isocyanates
6-9   Reaction of alcohols with ClCN
6-68   Addition of alkyl hypochlorites to isocyanides
8-14   Reaction between amides, bromine, and alkoxides (Hofmann), and similar rearrangement reactions
8-15   Rearrangement of acyl azides in the presence of alcohols (Curtius)

Carbodiimides

6-58   Addition of isocyanates to isocyanates
7-42   Dehydration of ureas and thioureas

Carbonates

0-20   Alcoholysis of phosgene
0-24   Reaction between alkyl halides and carbonate salts

Carboxylic Acids

0-3   Hydrolysis of 1,1,1-trihalides
0-6   Hydrolysis of ortho esters
0-8   Hydrolysis of acyl halides
0-9   Hydrolysis of anhydrides
0-10   Hydrolysis of carboxylic esters
0-11   Hydrolysis of amides
0-70   Cleavage of carboxylic esters with LiI
0-81   Reductive cleavage of carboxylic esters

Carboxylic Esters (continued)

3-17 Vicarious substitution of aryl nitro compounds

4-11 Free-radical acyloxylation

4-23 Carbalkoxylation of nitrogen heterocycles

4-39 Reaction between silver salts and iodine (Simonini)

5-3 Hydration of acetylenic ethers

5-4 Addition of alcohols or phenols to ketenes

5-5 Addition of carboxylic acids or acyl peroxides to olefins

5-17 Addition of carboxylic esters to activated olefins (Michael)

5-18 Addition of organometallic compounds to unsaturated esters

5-20 Addition of tin and mercury hydrides to unsaturated ketones

5-22 Free-radical addition of carboxylic esters to olefins

5-23 Hydrocarboxylation of olefins in the presence of alcohols

5-35 Addition of carboxylic acid salts to olefins

5-43 Addition of OAc and SR to double bonds

5-54 Dicarbalkoxylation of olefins and acetylenes

6-7 Reductive acylation of ketones

6-9 Alcoholysis of nitriles

8-7 Rearrangement of α-halo ketones (Favorskii)

8-8 Rearrangement of diazo ketones in the presence of alcohols (Arndt–Eistert)

8-20 Reaction between ketones and peroxy compounds (Baeyer-Villiger)

9-8 Cleavage of cyclic ketones with NOCl and an alcohol

9-9 From ozonides

9-10 Oxidative cleavage of enol ethers

9-13 Reaction between carboxylic acids and lead tetraacetate

9-18 Oxidation of ethers

9-22 Oxidation of primary alcohols or aldehydes

9-23 Oxidation of enol ethers

9-70 Reaction between aldehydes and aluminum ethoxide (Tishchenko)

9-72 Reaction of acetophenones with $AgNO_3$–I_2 or other reagents

Catenanes

9-65 Acyloin condensation or other methods

Cyanamides

0-45 Reaction between alkyl halides and cyanamide

0-73 Cleavage of tertiary amines with cyanogen bromide (von Braun)

7-39 Dehydration of disubstituted ureas

Cyanates

0-12 Reaction of aroxides and cyanogen halides

Cyanoamines

0-46 Amination of cyanohydrins

1-28 Cyanation of aromatic amines

2-17 Cyanation of secondary amines

6-16 Reaction between aldehydes, ammonia, and nitriles (Mannich)

6-50 Addition of cyanide and ammonium ions to aldehydes or ketones (Strecker)

6-51 Addition of HCN to C=N or C≡N bonds

Cyano Carbonyl Compounds

0-94 Akylation of cyano carbonyl compounds

0-107 Acylation of nitriles by acyl halides

0-109 Acylation of nitriles by carboxylic esters

0-111 Reaction between acyl halides and CuCN

2-17 Cyanation of ketones

2-19 Cyanoethylation of enamines; reaction of enamines with cyanogen chloride

3-14 Arylation of cyano carbonyl compounds

5-17 Addition of olefins (Michael)

5-21 Acylation of unsaturated nitriles

5-25 Addition of HCN to unsaturated aldehydes, ketones, or carboxylic esters

6-41 Addition of cyano carbonyl compounds to aldehydes or ketones (Knoevenagel)

6-48 Condensation of nitriles (Thorpe)

9-33 Dimerization of cyano carbonyl compounds

Cyanohydrins (*see* Hydroxy Nitriles)

Cycloalkanes and Alkenes (*see* Alicyclic Compounds)

Dialdehydes (*see* Dicarbonyl Compounds)

Diazo Compounds

0-112 Reaction between acyl halides and diazomethane

2-9   Reaction of active hydrogen compounds with tosyl azide

2-49   Diazotization of α-amino esters and similar compounds

6-41   Addition of diazo esters to aldehydes

7-45   Elimination from N-nitroso-N-alkyl compounds

9-6   Oxidation of hydrazones

Diazonium Salts

1-5   Direct diazotization of aromatic rings

2-49   Diazotization of primary amines

1,2-Dicarbonyl Compounds

0-103 Dicarbonylation of halides

0-106 Dimerization of acyl halides

0-109 Acylation of 1,3-dithianes, followed by hydrolysis

6-29   Addition of RLi and CO to carboxylic esters

6-69   Reaction of metalated aldimines with CO_2

9-9   Ozonization of alkynes or aromatic rings

9-16   Oxidation of ketones with selenium dioxide

9-21   Oxidative cleavage of α-amino ketones

9-23   Oxidation of olefins

9-27   Oxidation of alkynes

9-65   Reductive condensation of aromatic carboxylic acids

1,3-Dicarbonyl Compounds

0-94   Alkylation at a carbon bearing an active hydrogen

0-107 Acylation at a carbon bearing an active hydrogen

0-108 Acylation of carboxylic esters by carboxylic esters (Claisen; Dieckmann)

0-109 Acylation of ketones by carboxylic esters

0-110 Acylation of carboxylic acid salts

1-22   Reaction between aromatic compounds and diethyl oxomalonate

2-15   Acylation of acetals or ketals followed by hydrolysis

2-16   Alkoxycarbonylalkylation of aldehydes

2-19   Acylation of enamines followed by hydrolysis (Stork)

3-14   Arylation at a carbon bearing an active hydrogen

5-2   Cleavage of activated olefins with water

5-17   Addition of active hydrogen compounds to olefins (Michael)

5-22   Free-radical addition of 1,3-dicarbonyl compounds to olefins

6-30   Reaction between nitriles, zinc, and α-halo esters (Blaise)

6-41   Addition of 1,3-dicarbonyl compounds to aldehydes or ketones (Knoevenagel)

6-43   Carboxylation of ketones and carboxylic esters

7-20   Cleavage of Michael adducts

7-50   Extrusion of sulfur from β-keto thiol esters

8-2   Rearrangement of epoxy ketones

8-9   Reaction of ketones with ethyl diazoacetate

9-16   Remote oxidation of ketones

9-33   Dimerization of β-keto esters or similar compounds

1,4-Dicarbonyl Compounds

0-6   Cleavage of furans

1-14   Acylation of aromatic rings by succinic anhydride

4-16   Coupling of ketones, carboxylic acids, and esters

5-21   Acylation of unsaturated ketones or alkynes

5-54   Dicarbalkoxylation of olefins and acetylenes

9-16   Remote oxidation of ketones

9-34   Dimerization of silyl enol ethers or of lithium enolates

1,5-Dicarbonyl Compounds

5-17   Addition of silyl enol ethers or silyl ketene acetals to unsaturated ketones or esters

Dicarboxylic Acids (*see* Dicarbonyl Compounds, Carboxylic Acids)

Dicyano Compounds

0-94 Alkylation of malononitriles
3-14 Arylation of malononitriles
5-17 Addition of nitriles to unsaturated nitriles (Michael)
5-25 Addition of HCN to triple bonds
6-41 Addition of malononitriles to aldehydes or ketones (Knoevenagel)
6-51 Addition of HCN to nitriles
9-10 Oxidation of *o*-diamines

Diesters (*see* Dicarbonyl Compounds)

Dihalides and Polyhalides

0-69 Treatment of epoxides with $SOCl_2$, Ph_3P and CCl_4 or Ph_3PCl_2
0-76 Reduction of trihalides
0-87 Coupling of halides with trihalides
2-40 Decarboxylation of trihalo acids
2-44 The haloform reaction
3-17 Vicarious substitution of aryl nitro compounds
4-1 Free-radical halogenation
5-1 Addition of hydrogen halides to alkynes
5-26 Addition of halogens to olefins or alkynes
5-33 Free-radical addition of polyhalides to olefins
6-24 Reaction of PCl_5, SF_4, or other reagents with aldehydes, ketones, or other C=O compounds
9-21 Treatment of amines with CuX and alkyl nitrites

Diketones (*see* Dicarbonyl Compounds)

Dinitro Compounds

4-13 Nitration of alkanes or nitro compounds
5-40 Addition of N_2O_4 to olefins

***gem*-Diols (Hydrates)**

6-1 Hydration of aldehydes

1,2-Diols

0-7 Hydrolysis of epoxides
4-16 Coupling of alcohols
5-35 Hydroxylation of olefins

6-29 Addition of a masked Grignard reagent to an aldehyde or ketone
6-41 From aromatic aldehydes and carbanions
9-62 Bimolecular reduction of aldehydes or ketones

1,3-Diols

6-46 Condensation between formaldehyde and aldehydes or ketones (Tollens)
6-53 Addition of aldehydes to olefins (Prins)

Disulfides

0-38 Reaction between alkyl halides and disulfide ion
3-5 Reaction between aryl halides and disulfide ion
3-28 The Smiles rearrangement
5-28 Addition of ArSSCl to alkenes
9-35 Oxidation of thiols
9-54 Reduction of sulfonyl halides

Dithioacetals

0-36 From *gem*-dihalides or acetals and thiolate ions
5-6 Addition of thiols to alkynes
6-11 Addition of thiols to aldehydes or ketones

Dithiols

5-38 Reaction of alkenes with a disulfide and BF_3 etherate
6-11 Addition of H_2S to carbonyl compounds or imines

Enamines

0-97 Alkylation of enamines
5-7 Addition of amines to triple-bond compounds
6-14 Addition of amines to aldehydes or ketones
6-32 Reaction between Grignard reagents and formamides
6-47 Reaction of phosphonates with aldehydes or ketones
7-18 Dehydrocyanation of cyano amines
9-2 Dehydrogenation of tertiary amines

Enolate Ions

0-95 From enol acetates
2-3 Treatment of aldehydes or ketones with base
2-22 Treatment of active hydrogen compounds with base

Enol Carbamates

5-5 Reaction between alkynes, CO, and an amine

Enol Ethers and Esters

0-15 O-Alkylation of carbonyl compounds with diazo alkanes
0-17 Transetherification
0-20 Reaction between acyl halides and active hydrogen compounds
0-23 Transesterification
0-24 Acylation of vinylic halides
0-94 Alkylation with ortho esters
0-107 O-Acylation of 1,3-dicarbonyl compounds
5-4 Addition of alcohols or phenols to alkynes; addition of aldehydes or ketones to ketene
5-5 Addition of carboxylic acids to alkynes
6-6 Addition of alcohols or anhydrides to aldehydes or ketones
6-33 Reaction between carboxylic esters and Tebbe's reagent or metal carbene complexes
6-47 Reaction of α-alkoxy phosphoranes with aldehydes or ketones
7-2 Cleavage of acetals
7-31 Elimination from β-halo acetals

Enols (*see* Unsaturated Alcohols and Phenols)

Enol Thioethers

5-6 Addition of thiols to alkynes
6-11 Reaction of aldehydes or ketones with thiols
9-2 Dehydrogenation and reduction of sulfoxides

Enynes

5-15 Dimerization of alkynes

Episulfides

0-36 Reaction between epoxides and phosphine sulfides
5-28 Cyclization of β-halo disulfides
6-62 Reaction of diazoalkanes with sulfur or thioketones

Epoxides

0-13 Cyclization of halohydrins
0-16 Cyclization of 1,2-diols
0-18 Payne rearrangement of 2,3-epoxy alcohols
4-8 Epoxidation of a secododecahedrane
5-36 Epoxidation of olefins
6-29 Reaction of carbonyl compounds with *gem*-dihalides and Li or BuLi
6-45 Condensation between aldehydes and α-halo esters, ketones, or amides (Darzens)
6-61 Addition of sulfur ylides or diazomethane to aldehydes or ketones
9-63 Bimolecular reduction of aldehydes or ketones

Esters (*see* Carboxylic Esters, Inorganic Esters)

Ethers (*see also* Hydroxy Ethers, etc.)

0-6 Cleavage of oxonium ions
0-10 Reaction between carboxylic esters and alkoxide ion
0-12 Reaction between alkoxides or aroxides and alkyl halides (Williamson)
0-14 Reaction between alkoxides or aroxides and inorganic esters
0-15 Alkylation of alcohols or phenols with diazo compounds
0-16 Dehydration of alcohols
0-17 Transetherification
0-19 Alkylation of alcohols with onium salts
0-29 Exchange of ethers and oxonium salts
0-30 Reaction of halides with oxide ion
0-68 Cleavage of oxonium salts
0-79 Reduction of acetals or ketals
0-92 Reaction between Grignard reagents and acetals or ketals; dimerization of acetals

Ethers (continued)

2-23   Reaction between Grignard reagents and *t*-butyl peresters

3-4   Reaction between aryl halides and alkoxides or aroxides

4-8   Cyclization of alcohols with lead tetraacetate

4-36   Desulfurization of thiono esters

5-4   Addition of alcohols or phenols to olefins

5-22   Free-radical addition of ethers to olefins

6-7   Reductive alkylation of alcohols

9-40   Reduction of carboxylic esters

9-60   Reduction of peroxides

Glycidic Esters

5-36   Epoxidation of α,β-unsaturated esters

6-45   Condensation between aldehydes or ketones and α-halo esters (Darzens)

Grignard Reagents (*see* Organometallic Compounds)

Halo Acids, Esters, Aldehydes, Ketones (*see* Halo Carbonyl Compounds)

Haloamines

5-29   Addition of N-haloamines to unsaturated compounds

N-Haloamines and Amides

2-54   Halogenation of amines or amides

Halo Carbonyl Compounds

0-69   Reaction of acyl chlorides with ethylene oxide and NaI

0-71   Reaction of diazo ketones with hydrohalic acids

2-4   Halogenation of aldehydes or ketones

2-5   Halogenation of carboxylic acids (Hell–Volhard–Zelinskii) and acid derivatives

5-26   Addition of halogens to ketenes

5-27   Addition of HOBr or HOCl to triple bonds; addition of chlorine acetate or other reagents to olefins

5-34   Addition of acyl halides to olefins

8-10   Rearrangement of halo epoxides

9-23   Oxidation of certain alkenes

Halo Ethers and Acetals

5-27   Addition of hypohalites to double bonds

6-23   Addition of alcohols and hydrogen halides to aldehydes or ketones

6-24   Reaction of carboxylic esters with ClF or other reagents

Haloformic Esters

0-20   Alcoholysis of phosgene

Halohydrins

0-69   Cleavage of epoxides with hydrogen halides

5-27   Addition of hypohalous acids to olefins

Halo Sulfides, Sulfoxides, and Sulfones

2-6   Halogenation of sulfoxides and sulfones

5-29   Addition of sulfonyl halides to olefins

9-71   Pummerer rearrangements

Hemiacetals

4-4   Electrolytic oxidation of tetrahydrofuran

6-6   Addition of alcohols to aldehydes or ketones

Hemiaminals

6-13   Reaction between aldehydes or ketones and ammonia

6-14   Reaction between aldehydes or ketones and amines

Hemimercaptals

6-11   Addition of thiols to aldehydes or ketones

Heterocyclic Compounds (*see also* Anhydrides, Aziridines, Epoxides, Episulfides, Imides, Lactams, Lactones)

0-13   Cyclization of halohydrins (cyclic ethers)

0-16   Cyclization of glycols (cyclic ethers; furans)

0-17   Reaction of diols with acetals (cyclic acetals)

0-36   Reaction of dihalides with sulfide ion (cyclic sulfides)

Heterocyclic Compounds (continued)

0-43 Cyclization of haloamines (cyclic amines); dealkylation of quaternary salts of nitrogen heterocycles

0-45 Reaction between dihalides and cyanamide (cyclic amines)

0-59 Reaction between ureas and malonic esters (cyclic ureides)

1-9 Sulfurization of aromatic rings (cyclic sulfides)

1-14 Intramolecular acylation

1-21 Intramolecular amidation of aromatic rings

1-23 Cyclization of amides with $POCl_3$ (isoquinolines)

2-12 Intramolecular nitrene insertion

3-6 Intramolecular arylation of amines (cyclic amines)

3-14 Intramolecular arylation of active hydrogen compounds

3-17 Arylation of heterocyclic nitrogen compounds

3-18 Amination of heterocyclic nitrogen compounds

4-8 Cyclization of alcohols with lead tetraacetate (tetrahydrofurans)

4-15 Cyclization of N-tosyl malonic esters

4-18 Intramolecular arylation (Pschorr)

4-23 Alkylation, arylation, and carbalkoxylation of nitrogen heterocycles

5-7 Addition of ammonia or primary amines to conjugated diynes (pyrroles)

5-10 Hydrogenation of heterocyclic aromatic rings

5-12 Addition of boranes to dienes (cyclic boranes)

5-37 Photooxidation of dienes (cyclic peroxides)

5-38 Cyclization of unsaturated alcohols with sulfenyl chlorides (tetrahydrofurans)

5-42 Addition of aminonitrenes to triple bonds (1-azirines)

5-46 1,3-Dipolar addition to double or triple bonds

5-47 Diels–Alder addition involving hetero atoms

5-50 Expansion of heterocyclic rings upon treatment with carbenes

5-52 Other cycloaddition reactions

6-6 Formation of cyclic acetals; reaction between diketones and acids (furans, pyrans)

6-11 Addition of H_2S to aldehydes or ketones (cyclic thioacetals)

6-13 Reaction between aldehydes and ammonia (cyclic amines)

6-14 Intramolecular addition of amines to carbonyl groups (cyclic imines)

6-18 Reaction of dinitriles with ammonia (cyclic imidines)

6-20 Reaction between hydrazines and β-diketones or β-keto esters (pyrazoles; pyrazolones)

6-38 Ring expansion of thiono lactones (cyclic ethers)

6-41 Reaction of ketones with tosylmethylisocyanide (oxazolines)

6-53 Reaction between alcohols and aldehydes (dioxanes)

6-57 Trimerization of aldehydes (trioxanes)

6-60 Trimerization of nitriles (triazines)

6-63 Addition of olefins to aldehydes or ketones (oxetanes)

7-25 Reaction of dichlorobenzyl sulfones with base (thiiren-1,1-dioxides)

7-47 Extrusion of CO_2 from benzoxadiazepinones (indazoles)

7-51 Condensation of thiobenzilic acid with aldehydes or ketones (oxathiolan-5-ones)

8-15 Curtius rearrangement of cycloalkyl or aryl azides

8-19 Rearrangement of N-haloamines (cyclic amines)

8-22 Ring enlargement of cyclic quaternary ammonium salts (cyclic amines)

8-33 Ring expansion of N-acylaziridines (oxazoles)

8-36 Cyclization of arylhydrazones (Fischer indole synthesis)

8-42 Acid-catalyzed rearrangement of N-haloamines (pyrrolidines; piperidines—Hofmann–Löffler)

9-1 Aromatization of heterocyclic rings

9-37 Reduction of α,β-unsaturated ketones (pyrazolones)

9-39 Reduction of lactams (cyclic amines)

9-40 Reduction of lactones (cyclic ethers)

Hydroxyamines and Amides (continued)

1-29 Hydroxylation of amines
3-27 Rearrangement of aryl hydroxylamines (Bamberger)
4-4 Hydroxylation of amides
5-39 Oxyamination of double bonds; aminomercuration of alkenes, followed by hydrolysis
6-13 Addition of ammonia to aldehydes or ketones
6-14 Addition of amines or amides to aldehydes or ketones
6-30 Reaction between aldehydes or ketones, zinc, and halo amides
6-41 Reaction of aldehydes with the conjugate base of formamide; reaction of ketones with imines
6-67 Reaction between isocyanides, $TiCl_4$ and aldehydes or ketones, followed by hydrolysis
9-62 Coupling of ketones and O-methyl oximes

Hydroxy Esters

0-23 Transesterification of lactones
0-25 Acylation of epoxides
4-4 Hydroxylation of carboxylic esters
6-30 Reaction between aldehydes or ketones, zinc, and α-halo esters (Reformatsky)
6-40 Condensation between carboxylic esters and aldehydes or ketones
6-41 Addition of α-metalated esters to ketones

Hydroxy Ethers

0-18 Alcoholysis of epoxides

Hydroxylamines

5-7 Addition of hydroxylamine to olefins
6-26 Reduction of oximes
6-35 Addition of alkyllithium compounds to oximes
7-8 Cleavage of amine oxides (Cope)
8-22 Rearrangement of N-oxides (Meisenheimer)
9-24 Oxidation of amines
9-49 Reduction of nitro compounds

Hydroxy Nitriles

0-101 Reaction between epoxides and cyanide ion
4-4 Hydroxylation of nitriles
6-30 Reaction between aldehydes and ketones, zinc, and halo nitriles
6-41 Addition of nitriles to ketones
6-49 Addition of HCN to aldehydes or ketones

Hydroxy Sulfonic Acids

0-41 Reaction between epoxides and bisulfite ion
6-12 Addition of bisulfite ion to aldehydes or ketones

Hydroxy Thiols and Thioethers

0-35 Reaction between epoxides and NaSH
0-36 Reaction between epoxides and thiolate ions
1-26 Thioalkylation of phenols
5-38 Hydroxysulfenylation af alkenes
6-11 Addition of H_2S to aldehydes or ketones

Imides (including Ureides)

0-52 Reaction between acyl halides and lithium nitride
0-53 Amination of anhydrides
0-58 N-Alkylation of imides
0-59 N-Acylation of amides or imides
5-7 Addition of imides to olefins
5-23 Hydrocarboxylation of unsaturated amides
6-68 Addition of N-halo amides to isocyanides
8-14 Reaction between amides and NaOBr (Hofmann)
8-15 Rearrangement of acyl azides in the presence of water (Curtius)
9-18 Oxidation of lactams

Imines

2-8 Reaction between active hydrogen compounds and nitroso compounds
2-19 Treatment of enamines with nitrilium salts
5-7 Addition of amines to triple-bond compounds

Ketenimines

6-47 Reaction between phosphoranes and isocyanates

7-1 Dehydration of amides

Keto Acids, Aldehydes, and Esters (*see* Dicarbonyl Compounds)

Ketones (*see also* Dicarbonyl Compounds, Unsaturated Carbonyl Compounds, etc.)

0-1 Hydrolysis of vinylic halides
0-2 Hydrolysis of *gem*-dihalides
0-4 Hydrolysis of enol esters of inorganic acids
0-6 Hydrolysis of enol ethers, ketals, thioketals, etc.
0-10 Hydrolysis of enol esters
0-76 Reduction of halo ketones
0-78 Reduction of hydroxy ketones
0-82 Reduction of diazo ketones or nitro ketones
0-87 Coupling of halo ketones with lithium alkylcopper reagents
0-94 Acetoacetic ester synthesis and similar reactions
0-95 Alkylation of ketones
0-97 Alkylation and hydrolysis of dithianes and similar compounds
0-98 Alkylation and hydrolysis of oxazines
0-99 Reaction of halo ketones or diazo ketones with boranes
0-102 Carbonylation of alkyl halides
0-104 Reaction between acyl halides and organometallic compounds
0-105 Reaction between other acid derivatives and organometallic compounds
0-107 Acylation of active hydrogen compounds followed by cleavage
0-109 Reduction of β-keto sulfoxides
0-110 Acylation of carboxylic acid salts followed by cleavage
0-113 Ketonic decarboxylation
1-14 Acylation of aromatic rings (Friedel–Crafts)
1-19 Reaction between aromatic rings and phosgene
1-27 Acylation of aromatic rings with nitriles (Hoesch)

1-30 Rearrangement of phenolic ethers (Fries)
1-36 Photolysis of acylated arylamines
2-2 Rearrangement of hydroxy olefins
2-16 Reaction between aldehydes and boron-stabilized carbanions
2-19 Alkylation of enamines followed by hydrolysis (Stork)
2-25 Oxidation of *gem*-dimetallic compounds
2-32 Carbonylation of organometallic compounds
2-40 Decarboxylation of β-keto acids or esters
2-41 Cleavage of tertiary alkoxides
2-42 Reaction between amino acids and anhydrides (Dakin–West)
2-43 Basic cleavage of β-diketones
3-14 Arylation of ketones
3-15 Acylation of aryl iodides
4-20 Arylation of allylic alcohols
4-23 Acylation of nitrogen heterocycles
4-31 Reaction of diazonium salts with oximes, followed by hydrolysis; or with R₄Sn and CO; or with silyl enol ethers
5-3 Hydration of alkynes or allenes
5-9 Selective reduction of unsaturated ketones
5-10 Reduction of phenols
5-12 Oxidation of boranes; hydrolysis of unsaturated boranes
5-17 Addition of ketones to activated olefins (Michael)
5-18 Addition of organometallic compounds to unsaturated ketones
5-19 Addition of boranes to unsaturated ketones
5-20 Addition of tin and mercury hydrides to unsaturated ketones
5-22 Free-radical addition of aldehydes or ketones to olefins
5-24 Hydroacylation of alkenes
5-50 Hydrolysis of bicyclo[4.1.0]heptanes
6-2 Hydrolysis of imines, oximes, hydrazones, and other C=N compounds
6-4 Hydrolysis of secondary aliphatic nitro compounds (Nef)
6-31 Reaction between lithium carboxylates and alkyllithium compounds

Ketones (continued)

6-33 Indirectly, from carboxylic esters

6-37 Addition of Grignard reagents to nitriles

6-42 Hydrolysis of epoxy silanes

6-69 Reaction of alkyl halides with metalated aldimines

7-1 Dehydration of 1,2-diols

7-32 Fragmentation of γ-amino or γ-hydroxy halides

7-33 Fragmentation of 1,3-diols

7-38 Fragmentation of certain ketoximes

7-43 Pyrolysis of β-hydroxy olefins

7-44 Pyrolysis of allylic ethers

8-2 Rearrangement of glycols and related compounds (pinacol)

8-3 Ring expansion of certain hydroxyamines (Tiffeneu–Demyanov)

8-4 Acid-catalyzed ketone rearrangements

8-9 Homologation of aldehydes or ketones

8-14 Reaction between α-hydroxy or α-halo amides and NaOBr (Hofmann)

8-21 Cleavage of hydroperoxides

8-25 Treatment of boranes with CO and H_2O, followed by NaOH and H_2O_2; or with CN^- followed by trifluoroacetic anhydride; from dialkylchloroboranes

8-28 Treatment of lithium alkynyltrialkylborates with electrophiles

8-32 [1,3] Sigmatropic rearrangements of allylic vinylic ethers

9-3 Oxidation of secondary alcohols

9-7 Oxidative cleavage of glycols and related compounds

9-9 Ozonolysis of olefins

9-10 Oxidative cleavage of olefins

9-11 Oxidation of diarylmethanes

9-14 Bisdecarboxylation of malonic acids

9-15 Oxidative decyanation of nitriles

9-16 Oxidation of activated or unactivated methylene groups

9-20 Oxidation of secondary alkyl halides and tosylates

9-21 Oxidation of amines or nitro compounds

9-23 Oxidation of olefins with noble-metal salts

9-37 Reduction of diketones or quinones

9-57 Indirect oxidative decyanation of nitriles

Lactams

0-54 Cyclization of amino acids

0-55 Reaction between lactones and ammonia or amines; ring expansion of lactams

0-58 Cyclization of halo amides

5-7 Addition of lactams to olefins

5-23 Hydrocarboxylation of unsaturated amines

6-31 Reaction between imines, zinc, and halo esters

6-47 Reaction between imides and phosphoranes

6-64 Addition of ketenes to imines; addition of enamines to isocyanates

8-17 Reaction between cyclic ketones and hydrazoic acid (Schmidt)

8-18 Rearrangement of oximes of cyclic ketones (Beckmann)

8-19 Expansion of aminocyclopropanols

9-18 Oxidation of cyclic tertiary amines

Lactones

0-22 Cyclization of hydroxy acids

0-24 Cyclization of halo acids

0-89 Intramolecular coupling

2-43 Cleavage of cyclic α-cyano ketones

5-4 Internal addition of alcohols to a ketene function

5-5 Cyclization of olefinic acids

5-23 Hydrocarboxylation of unsaturated alcohols

5-27 Halolactonization

5-45 Reaction of alkenes with manganese(III) acetate

6-47 Reaction of anhydrides with phosphoranes

6-63 Addition of ketenes to aldehydes or ketones

7-47 Extrusion of CO_2 from 1,2-dioxolane-3,5-diones

7-49 Decarboxylation of cyclic peroxides (Story)

8-20 Reaction between cyclic ketones and peroxy compounds (Baeyer–Villiger)

8-42 Rearrangement of N-halo amides

9-18 Oxidation of cyclic ethers

Lactones (continued)
9-22   Oxidation of diols
9-41   Reduction of cyclic anhydrides
9-69   Oxidative–reductive ring closure of dialdehydes

Mercaptals (*see* Thioacetals)

Mercaptans (*see* Thiols)

Metallocenes

2-35   Reaction between sodium cyclopentadienylide and metal halides

Monoesters of Dicarboxylic Acids

0-21   Alcoholysis of cyclic anhydrides
0-23   Equilibration of dicarboxylic acids and esters
6-9   Alcoholysis of cyano acids
9-10   Oxidative cleavage of catechols

Nitrile Oxides

7-40   Oxidation of nitro compounds

Nitriles (*see also* Dicyano Compounds, Cyano Carbonyl Compounds, etc.)

0-95   Alkylation of nitriles
0-99   Reaction of halo nitriles or diazo nitriles with boranes
0-101 Reaction between alkyl halides and cyanide ion
1-28   Cyanation of aromatic rings
2-17   Cyanation of ketones, nitro compounds, or benzylic compounds
2-33   Cyanation of organometallic compounds
2-40   Decarboxylation of α-cyano acids
3-11   Reaction between aryl halides and CuCN (Rosenmund–von Braun)
3-12   Cyanide fusion of sulfonic acid salts
3-14   Arylation of nitriles
3-17   Vicarious substitution of aryl nitro compounds
4-28   Reaction between diazonium salts and CuCN (Sandmeyer)
4-39   Reaction of acyl peroxides with copper cyanide
4-41   Decarbonylation of aromatic acyl cyanides
5-17   Addition to activated olefins (Michael)
5-19   Addition of boranes to acrylonitrile

5-20   Addition of tin and mercury hydrides to unsaturated nitriles
5-22   Free-radical addition of nitriles to olefins
5-25   Addition of HCN to olefins
5-43   Addition of CN and SR to double bonds
6-22   From aldehydes or carboxylic esters
6-41   Reaction of ketones with tosylmethylisocyanide
6-51   Addition of KCN to sulfonyl hydrazones
6-59   Reaction between acid salts and BrCN
7-37   Dehydration of aldoximes and similar compounds
7-38   Fragmentation of ketoximes
7-39   Dehydration of amides
7-40   From primary nitro compounds or azides
8-22   Rearrangement of isocyanides
9-5   Dehydrogenation of amines
9-6   Oxidation of hydrazones
9-13   Treatment of carboxylic acids with trifluoroacetic anhydride and $NaNO_2$
9-55   Reduction of nitrile oxides
9-58   Reduction of nitro compounds with $NaBH_2S_3$

Nitro Compounds

0-60   Reaction between alkyl halides and nitrite ion
0-94   Alkylation of nitro compounds
1-2   Nitration of aromatic rings
1-32   Rearrangement of N-nitro aromatic amines
2-40   Decarboxylation of α-nitro acids
2-51   N-Nitration of amines or amides
3-17   Alkylation of aromatic nitro compounds
4-13   Nitration of alkanes
4-26   Reaction between diazonium salts and sodium nitrite
5-7   Nitromercuration–reduction of alkenes
5-9   Reduction of unsaturated nitro compounds
5-17   Addition to activated olefins (Michael)

Oximes (continued)

6-21 Addition of hydroxylamine to aldehydes or ketones

6-35 Addition of Grignard reagents to the conjugate bases of nitro compounds

8-42 Photolysis of nitrites (Barton)

9-8 Cleavage of cyclic ketones with NOCl and an alcohol

9-24 Oxidation of aliphatic primary amines

9-58 Reduction of nitro compounds

Oxiranes (*see* Epoxides)

Oxonium Salts

0-29 Reaction between alkyl halides and ethers or ketones

Ozonides

9-9 Ozonolysis of olefins

Peptides

0-54 Coupling of amino acids

Peroxides (*see also* Hydroperoxides, Peroxy acids)

0-31 Reaction of alkyl and acyl halides with peroxide ion

4-10 Reaction between hydroperoxides and susceptible hydrocarbons

5-4 Oxymercuration–reduction of alkenes in the presence of a hydroperoxide

5-37 Photooxidation of dienes

7-49 Reaction of ketones with H_2O_2

Peroxy Acids

9-32 Oxidation of carboxylic acids

Phenols

0-10 Hydrolysis of phenolic esters

0-32 Cleavage of phenolic ethers with sulfonic acids

0-36 Cleavage of phenolic ethers

0-46 Cleavage of phenolic ethers

0-68 Cleavage of phenolic ethers with HI or HBr

1-29 Electrophilic hydroxylation of aromatic rings

1-30 Rearrangement of phenolic esters (Fries)

1-31 Rearrangement of phenolic ethers

2-25 Oxidation of aryl organometallic compounds

2-26 Oxidation of arylthallium compounds

3-1 Hydrolysis of aryl halides and other compounds

3-2 Reaction between naphthylamines and bisulfite ion (Bucherer)

3-3 Alkali fusion of sulfonate ions

3-20 Hydrolysis of diazonium salts

3-27 Rearrangement of N-hydroxylamines

4-5 Free-radical hydroxylation of aromatic rings

4-21 Phenylation of phenols

5-50 Ortho methylation of phenols

6-25 Reduction of quinones

8-5 The dienone–phenol rearrangement

8-20 Cleavage of aryl ketones with peracids (Baeyer–Villiger)

8-21 Rearrangement of aralkyl peroxides

8-35 Rearrangement of allylic aryl ethers (Claisen)

8-45 Rearrangement of azoxy compounds (Wallach)

9-1 Aromatization of cyclic ketones

9-12 Oxidative cleavage of alkylbenzenes or aromatic aldehydes

9-42 Reduction of phenolic esters

9-43 Reduction of certain acids and esters

Phosphines

0-43 Reaction between alkyl halides and phosphine

0-82 Reduction of quaternary phosphonium salts

2-35 Reaction between phosphorus halides and Grignard reagents

Phosphonates

6-47 Reaction between alkyl halides and phosphites (Arbuzov)

Phosphoranes

6-47 Treatment of phosphonium ions with alkyllithiums

Quaternary Ammonium and Phosphonium Salts

0-43 Alkylation of amines (Menschutkin) or phosphines

Quaternary Ammonium and Phosphonium Salts (continued)

5-7   Addition of tertiary amines to alkenes

6-47   Reaction of phosphines with Michael olefins or with alkyl halides

Quinones

1-14   Intramolecular Friedel–Crafts acylation of diaryl ketones

9-4   Oxidation of phenols or aromatic amines

9-19   Oxidation of aromatic hydrocarbons

Schiff Bases (see Imines)

Selenides

0-36   Selenylation of alkyl halides

2-13   Selenylation of aldehydes, ketones, and carboxylic esters

2-29   Selenylation of organometallic compounds

9-56   Reduction of selenoxides

Semicarbazones

6-20   Addition of semicarbizide to aldehydes or ketones

Silyl Enol Ethers

2-23   Trialkylsilylation of ketones or aldehydes

2-27   Reaction betweeen vinylic lithium compounds and silyl peroxides

5-18   Michael-type reaction in the presence of Me_3SiCl

Sulfenyl Chlorides

4-12   Chlorosulfenation

Sulfides (see Thioethers)

Sulfinic Acids and Esters

0-118   Reduction of sulfonyl chlorides

2-29   Reaction of Grignard reagents with SO_2

3-28   The Smiles rearrangement

4-27   Reaction of diazonium salts with $FeSO_4$ and Cu

7-12   Cleavage of sulfones

Sulfonamides

0-58   N-Alkylation of sulfonamides

0-94   Alkylation of sulfonamides

0-99   Reaction of halo sulfonamides with boranes

0-116   Reaction between sulfonyl halides and ammonia or amines

3-17   Vicarious substitution of aryl nitro compounds

5-7   Addition of sulfonamides to olefins

9-39   Reduction of acyl sulfonamides

9-53   Reduction of sulfonyl azides

Sulfones

0-40   Reaction between alkyl halides and sulfinates

0-94   Alkylation of sulfones

0-95   Alkylation of sulfones

0-99   Reaction of halo sulfones with boranes

0-109   Reaction between carboxylic esters and methylsulfonyl carbanion

0-119   Reaction between sulfonic acid derivatives and organometallic compounds

1-10   Sulfonylation of aromatic rings

3-5   Reaction between aryl halides and sulfinate ions

3-17   Vicarious substitution of aryl nitro compounds

5-17   Addition of sulfones to activated olefins (Michael)

5-18   Addition of organometallic compounds to unsaturated sulfones

5-28   Addition of sulfonyl halides to olefins

6-41   Addition of sulfones to aldehydes or ketones (Knoevenagel)

9-31   Oxidation of thioethers or sulfoxides

Sulfonic Acid Esters

0-32   Reaction between alcohols or ethers and sulfonic acids

0-94   Alkylation of sulfonic acid esters

0-95   Alkylation of sulfonic acid esters

0-99   Reaction of halo sulfonic acid esters with boranes

0-115   Alcoholysis of sulfonic acid derivatives

3-17   Vicarious substitution of aryl nitro compounds

Sulfonic Acid Esters (continued)

6-41 Addition of sulfonic acid esters to aldehydes or ketones (Knovevenagel)

Sulfonic Acids

0-41 Reaction between alkyl halides and sulfite ion

0-114 Hydrolysis of sulfonic acid derivatives

1-7 Sulfonation of aromatic rings

1-40 Sulfonation with rearrangement (Jacobsen)

2-14 Sulfonylation of aldehydes, ketones, or carboxylic acids

3-5 Reaction between aryl halides and sulfite ion

9-26 Oxidation of thiols or other sulfur compounds

Sulfonium Salts

0-36 Reactions between alkyl halides and thioethers

Sulfonyl Azides

0-116 Reaction between sulfonyl halides and azide ion

Sulfonyl Halides

0-117 From sulfonic acids and derivatives

1-8 Halosulfonation of aromatic rings

2-29 Reaction of Grignard reagents with sulfuryl chloride or with SO_2 followed by X_2

4-12 Free-radical halosulfonation (Reed)

4-27 Reaction of diazonium salts with SO_2 and $CuCl_2$

Sulfoxides

0-94 Alkylation of sulfoxides

0-109 Reaction between carboxylic esters and methysulfinyl anion

1-9 Sulfurization of aromatic rings with thionyl chloride

2-29 Reaction of Grignard reagents with sulfinic esters

5-18 Addition of organometallic compounds to unsaturated sulfoxides

5-38 Treatment of alkenes with O_2 and RSH

6-41 Addition of sulfoxides to aldehydes or ketones (Knoevenagel)

9-31 Oxidation of thioethers

9-56 Indirectly, from sulfones

Thioamides

1-21 Amidation of aromatic rings with isothiocyanates

4-14 From thioaldehydes generated in situ

6-36 Addition of Grignard reagents to isothiocyanates

9-72 Reaction of ketones with sulfur and ammonia or amines

Thiocarbamates

6-5 Hydrolysis of thiocyanates

6-8 Addition of alcohols to isothiocyanates

Thiocyanates

0-42 Reaction between alkyl halides and thiocyanate ion

3-5 Reaction between aryl halides and thiocyanate ion

3-21 Reaction between diazonium salts and thiocyanate ion

4-39 Reaction between acyl peroxides and copper thiocyanate

5-28 Addition of halogen and SCN to alkenes

Thioethers

0-36 Reaction between alkyl halides and thiolate ions or Na_2S

0-97 Alkylation of thioethers

1-9 Sulfurization of aromatic rings

1-26 Thioalkylation of aromatic rings

2-13 Sulfenylation of ketones, carboxylic esters, and amides

2-29 Reaction between Grignard reagents and sulfur or disulfides

3-5 Reaction between aryl halides and thiolate ions

3-21 Reaction bewteen diazonium salts and thiolate ions or Na_2S

4-36 Reduction of dithioacetals

5-6 Addition of thiols to olefins

5-28 Addition of sulfenyl chlorides to olefins

Thioethers (continued)

5-43 Diarylamino-arylthio-addition to double bonds

6-11 Reductive alkylation of thiols

7-11 Cleavage of sulfonium compounds

8-22 Rearrangement of sulfonium salts (Stevens)

8-37 [2,3] Sigmatropic rearrangements of sulfur ylides

9-40 Reduction of thiol esters

9-56 Reduction of sulfoxides or sulfones

9-60 Reduction of disulfides

Thiol Acids and Esters

0-36 Reaction between alcohols and thiol acids

0-37 Reaction between acid derivatives and thiols or H_2S

1-27 Reaction between aromatic rings and thiocyanates

5-3 Hydration of acetylenic thioethers

5-6 Addition of thiol acids to olefins; addition of thiols to ketenes

5-23 Hydrocarboxylation of olefins in the presence of thiols

6-11 From carboxylic acids, alcohols, and P_4S_{10}

6-38 Addition of Grignard reagents to carbon disulfide

7-50 From thiol acids and α-halo ketones

Thiols

0-10 Hydrolysis of thiol esters

0-35 Reaction of alkyl halides with NaSH; cleavage of isothiuronium salts

1-9 Sulfurization of aromatic compounds (Herz)

2-29 Reaction between Grignard reagents and sulfur

3-5 Reaction between aryl halides and NaSH

3-21 Reaction between diazonium salts and NaSH

5-6 Addition of H_2S to olefins

6-38 Addition of lithium dialkylcopper reagents to dithiocarboxylic esters

9-54 Reduction of sulfonic acids or sulfonyl halides

9-61 Reduction of disulfides

Thioketones

6-11 From ketones

Thiono Esters and Thioamides

6-11 From carboxylic esters or amides

6-64 Addition of imines to thioketenes (β-thiolactams)

Thioureas (*see* Ureas)

Triazenes

1-4 Reaction between aromatic amines and diazonium salts

2-51 Reaction between amines and diazonium salts

Unsaturated Acids, Esters, Aldehydes, Ketones (*see* Unsaturated Carbonyl Compounds)

Unsaturated Alcohols and Phenols

2-2 Isomerization of allylic alcohols (formation of enols)

4-4 Allylic hydroxylation

5-10 Selective reduction of α,β-unsaturated aldehydes or ketones

5-18 Addition of organometallic compounds to propargylic alcohols

6-25 Selective reduction of α,β-unsaturated aldehydes or ketones

6-29 Addition of vinylic or alkynyl organometallic compounds to aldehydes or ketones

6-41 Condensation of alkyne salts with aldehydes or ketones

6-47 Reaction of certain ylides with aldehydes (scoopy reactions)

6-53 Addition of aldehydes to olefins (Prins)

7-2 Reaction of epoxides with strong bases

7-12 From epoxides or alkenes via selenoxide cleavage

8-3 Ring opening of cycloalkyl carbocations

8-33 Rearrangement of Li salts of 2-vinylcyclopropanols

8-35 Rearrangement of allylic aryl ethers (Claisen)

8-37 [2,3] Sigmatropic rearrangements

Unsaturated Carbonyl Compounds

0-95 Vinylation of ketones or carboxylic esters

0-97 Hydrolysis of bis(methylthio)-alkenes

2-2 Isomerization of α-hydroxy alkynes and alkynones

2-15 Acylation of olefins

2-27 From lithium acetylides

2-32 From vinylic organometallic compounds

2-55 From allylic amines and CO

4-6 Oxidation of unsaturated aldehydes

4-40 Decarboxylative allylation of keto acids

5-17 Addition to activated alkynes (Michael)

5-18 Addition of vinylic organometallic compounds to unsaturated carbonyl compounds; addition of organometallic compounds to acetylenic carbonyl compounds

5-19 Addition of unsaturated boranes to methyl vinyl ketones

5-23 Hydrocarboxylation of triple bonds

5-34 Addition of acyl halides to triple bonds

5-35 1,4-Addition of acetals to dienes

6-16 Reaction between aldehydes, ammonia, and aldehydes, ketones, or carboxylic esters (Mannich)

6-30 Reaction between aldehydes or ketones, zinc, and α-halo esters (Reformatsky)

6-39 Condensation of aldehydes and/or ketones (aldol)

6-40 Condensation between carboxylic esters and aldehydes or ketones

6-41 Condensation between active-hydrogen compounds and aldehydes or ketones (Knoevenagel)

6-44 Condensation between anhydrides and aldehydes (Perkin)

6-47 Condensation between β-carboxy phosphoranes and aldehydes or ketones

7-3 Pyrolysis of lactones

7-12 Cleavage of carbonyl-containing selenoxides and sulfones

7-35 Fragmentation of epoxy hydrazones

8-31 Rearrangement of vinylic hydroxy-cyclopropanes

8-34 Rearrangement of 3-hydroxy-1,5-dienes (oxy-Cope)

8-35 Rearrangement of allylic vinylic ethers (Claisen)

8-37 [2,3] Sigmatropic rearrangements

9-2 Dehydrogenation of aldehydes or ketones

9-16 Oxidation of a methylene group α to a double or triple bond

Unsaturated Ethers and Thioethers

0-97 Alkylation of allylic ethers

7-31 Elimination of X and OR from β-halo acetals

8-37 [2,3] Sigmatropic rearrangement of allylic sulfur ylides

Unsaturated Nitriles, Nitro Compounds, and Sulfonic Acids and Esters

2-33 Cyanation of vinylic organometallic compounds

5-17 Addition to activated alkynes (Michael)

5-18 Addition of organometallic compounds to activated alkynes

5-25 Addition of HCN to alkynes

5-33 Addition of nitryl chloride to triple bonds

6-41 Condensation between active hydrogen compounds and aldehydes or ketones (Knoevenagel)

7-18 Cleavage of H and HgCl from β-nitro mercuric halides

8-35 Rearrangement of allylic vinylic sulfones and sulfoxides

Ureas and Thioureas

0-56 Exchange of ureas

2-55 Carbonylation of amines

6-17 Addition of amines to isocyanates or isothiocyanates

6-19 Addition of amines to CO_2 or CS_2

6-55 Addition of alcohols or other carbocation sources to cyanamides (Ritter)

8-14 Reaction between amides and lead tetraacetate

Ureides (see Imides)

AUTHOR INDEX

Entries in this index refer to chapter number (**boldface**) and footnote number, except for citations that do not occur in a numbered chapter, for which page numbers are given instead.

Müller, P. **4** 325, **10** 10, 292, 299, 347, 539, 1128, 1152, 1587, **15** 868, **19** 20, 24, 44, 98, 100, 101, 127
Müller, R.K. **17** 373
Müller, S. **4** 359
Müller, W. **10** 957, **14** 143
Müller, W.M. **3** 69, 87, **4** 174
Müllhofer, G. **17** 54, 57
Mullholland, D.L. **2** 247
Mulligan, P.J. **18** 145
Mulliken, R.S. **1** 23, **2** 262, **3** 40, 59, **7** 13
Mullin, A. **10** 246
Mullineaux, R.D. **18** 68
Mullins, S.T. **12** 347
Mulvey, R.E. **13** 202
Mulzer, J. **16** 520, **17** 365, 367, 369
Munavilli, S. **11** 382
Munavu, R.M. **19** 324
Munchausen, L.L. **15** 832
Mundhenke, R.F. **11** 224
Mundy, B.P. **16** 550
Munekata, T. **17** 252
Muneyuki, R. **10** 137
Mungall, W.S. **4** 135, **15** 732
Munger, P. **12** 356
Munjal, R,C. **2** 128
Muñoz, L. **12** 164
Munro, M.H.G. **15** 272
Munson, M.S.B. **8** 144, 152
Münsterer, H. **12** 173
Munyemana, F. **10** 998
Münzel, N. **2** 141
Mura, A.J. Jr. **10** 1518
Mura, L.A. **10** 1551
Murabayashi, A. **18** 521
Muraglia, V. **10** 135
Murahashi, S. **2** 154, **10** 819, 824, 942, 945, 1108, 1167, 1299, 1351, 1371, 1579, **12** 583, **14** 233, 234, 313, 399, 463, **16** 198, **19** 129, 130, 363
Murai, A. **10** 1355
Murai, S. **10** 1408, **12** 190, 374, **13** 185, **15** 249, 693, **19** 37, 415
Murai, T. **10** 1408
Muraka, K. **10** 1112
Murakami, H. **16** 808
Murakami, M. **10** 1385, **16** 512, 807
Murakata, M. **10** 1469
Murakawa, K. **12** 566
Muraki, M. **10** 1224, 1235
Muralidharan, V.P. **16** 782
Muralimohan, K. **13** 165
Muramatsu, H. **15** 233
Muramatsu, I. **10** 631
Muramatsu, T. **19** 585

Muraoka, K. **10** 1320
Muraoka, M. **16** 112, 540
Murari, M.P. **10** 1426
Murata, I. **2** 78, 231, **7** 19
Murata, S. **16** 512, **17** 165
Murata, Y. **18** 535
Murati, M.P. **10** 1367
Muratov, N.N. **16** 241
Murawski, D. **18** 610
Murayama, D.R. **2** 154
Murchie, M.P. **3** 70
Murcko, M.A. **4** 197
Murdoch, J.R. **12** 352
Murov, S. **16** 783
Murphy, D.K. **10** 440
Murphy, P.J. **16** 534, 689
Murphy, R.B. **11** 438
Murphy, T.J. **15** 999
Murphy, W.S. **10** 1416, **12** 359, **18** 96
Murr, B.L. **6** 65, 73, **15** 790
Murray, B.J. **19** 29
Murray, C.D. **14** 304
Murray, C.J. **8** 103, **15** 201
Murray, K.J. **12** 293, **15** 287
Murray, K.K. **5** 204
Murray, M.A. **13** 128
Murray, N.G. **14** 452, **18** 635
Murray, R.E. **12** 395
Murray, R.W. **5** 201, **10** 840, **14** 139, 216, 217, 220, **18** 416, **19** 10, 173, 179-181, 183, 184, 186, 188, 189, 396, 399
Murrell, J.N. **1** 1
Mursakulov, I.G. **4** 263
Murthy, K.S.K. **10** 660
Muscio, O.J. Jr. **13** 16, **15** 950
Musco, A. **4** 61
Musgrave, O.C. **10** 1015
Musgrave, W.K.R. **11** 368
Musker, W.K. **2** 52, **17** 189, **19** 27, 472
Musliner, W.J. **13** 130
Musso, H. **4** 118, 119, 262, 299, 301, **14** 365, 367, **18** 476
Musumarra, G. **2** 57, **10** 58, 63, 348
Muszkat, K.A. **15** 901
Mutai, K. **6** 10
Muth, C.L. **19** 568
Muthard, J.L. **2** 251
Mutter, L. **17** 139
Mutter, M. **10** 884, **16** 671, 675
Muzart, J. **16** 506, **19** 260, 262, 268
Myers, J.D. **17** 327
Myers, M. **15** 488, 1078
Myers, M.M. **15** 501
Myers, R.J. **1** 73

Myhre, P.C. **5** 14, 59, **10** 149, 166, 173, 454, **11** 10, 51, 53, 87, **15** 88
Myles, D.C. **10** 702
Myshenkova, T.N. **10** 1604

Naab, P. **15** 4, 29, 33
Naae, D.G. **15** 16
Naan, M.P. **17** 51
Nababsing, P. **14** 14
Nabeya, A. **10** 1536, **16** 575
Nace, H.R. **17** 118, 179, 184, **19** 311
Nachtigall, G.W. **10** 45, 101
Nadelson, J. **10** 1635
Nadjo, L. **19** 366
Nadler, E.B. **2** 274, 276
Naef, R. **15** 519
Naegele, W. **15** 99
Naemura, K. **4** 51, 326, 349, **10** 1373
Näf, F. **10** 1492, **15** 528
Nafti, A. **12** 173
Nagahara, S. **18** 109
Nagai, K. **10** 1336
Nagai, M. **5** 84
Nagai, T. **5** 155, **10** 1729, **12** 179, **14** 68
Nagai, Y. **10** 450, **15** 692, 826, **19** 10
Nagao, K. **18** 540
Nagao, S. **14** 130
Nagao, Y. **4** 88, 239
Nagarajan, K. **12** 76
Nagareda, K. **16** 434, 665
Nagarkatti, J.P. **19** 151
Nagasawa, K. **10** 623
Nagasawa, N. **15** 506
Nagase, H. **10** 595, **17** 301
Nagase, S. **10** 168, 1320, **16** 665
Nagashima, H. **14** 313, **15** 466, 692, **19** 362
Nagata, J. **15** 937
Nagata, R. **19** 41
Nagata, T. **14** 198
Nagata, W. **11** 330, **15** 588, 591
Nagata, Y. **19** 546, 552
Nagira, K. **14** 383, 385
Nagraba, K. **15** 784
Nagubandi, S. **11** 334
Naguib, Y.M.A. **14** 127
Nagumo, M. **4** 63
Nagy, J.B. **10** 901
Nagy, O.B. **10** 901
Nahabedian, K.V. **11** 439
Nahm, S. **10** 1673, **15** 441
Naidenov, S.V. **11** 26
Naik, R.G. **15** 384
Naik, S.N. **12** 555
Nair, V. **16** 587

SUBJECT INDEX

PLEASE SIGN AND DATE   CARD